Superconductivity

K.H. Bennemann, J.B. Ketterson: Superconductivity

Volume 1: Conventional and Unconventional Superconductors
Volume 2: Novel Superconductors

K.H. Bennemann
J.B. Ketterson
Editors

Superconductivity

Novel Superconductors

Volume 2

Springer

Professor Dr. K.H. Bennemann
FU Berlin
Department of Physics
Arnimallee 14
14195 Berlin, Germany
e-mail: khb@physik.fu-berlin.de

Professor Dr. John B. Ketterson
Northwestern University
Department of Physics & Astronomy
Sheridan Road 2145
60208-3112 Evanston, USA
e-mail: j-ketterson@nwu.edu

ISBN 978-3-540-73252-5 e-ISBN 978-3-540-73253-2

DOI 10.1007/978-3-540-73253-2

Library of Congress Control Number: 2008926416

Typesetting: PTP-Berlin Protago-TEX-Production GmbH, Germany
Production: le-tex publishing services oHG, Leipzig, Germany
Cover design: WMXDesign, Heidelberg, Germany

Printed on acid-free paper

9 8 7 6 5 4 3 2 1

springer.com

Preface

Superconductivity is a classical problem of physics. Since the discovery by Kamerlingh-Onnes in 1911 it has undergone tremendous advances in recent years. Complimentary theoretical and experimental studies have advanced our understanding of systems with strong electronic correlations and the associated superconducting phase transitions. New experimental techniques and field-theoretical methods have emerged. The symmetry of the superconducting order-parameter in certain material classes seems different from the usual singlet, *s*-wave form found in conventional superconductors. There is strong evidence of a Cooper-pairing mechanism differing from the conventional one arising from the electron–phonon interaction. In view of the simultaneous occurrence of an enhanced magnetic response in many novel superconductors (cuprates, ruthenates and heavy-fermion metals), pairing due to exchange of spin-excitations has been proposed and studied. With this mechanism, triplet Cooper-pairing can also occur, in addition to singlet Cooper-pairing. This may be the case in Sr_2RuO_4 and heavy-fermion metals like UPt_3.

Most notable of course was the discovery of high temperature superconductivity in the cuprates. This discovery resulted in an enormous activity. Combining this with the discovery of superconductivity in other materials such as heavy-Fermion metals, MgB_2, various organics, fullerenes, magnetic metals under pressure, and presumably non *s*-symmetry triplet Cooper-pairing in ruthenates (Sr_2RuO_4) we have an enrichment of the phenomenon superconductivity.

Many of the new superconductors are being called unconventional, either with respect to the pairing mechanism or the symmetry of the underlying order parameter. A central problem is the definite identification of the pairing mechanism. Yet in spite of the new superconductors continuing studies of conventional electron–phonon based superconductivity remain interesting.

With the aid of historical hindsight, we expect the field to continue to advance in unexpected directions. Nonetheless it is our view that a state-of-the-art treatise on superconductivity is justified at this time, not only to summarize the present understanding, but also to introduce newcomers to the field, as was the case with the two justly-famous 1968 "bibles of superconductivity" by our former colleague R.D. Parks for an earlier generation. It is hoped that the present book will help to point the way for future work, research and discoveries.

The treatise starts with a historical introduction and with the basic phenomenological and microscopic theory of conventional superconductors. This should be particularly useful for students and beginning researchers. This and the following chapters reflect the somewhat refined present state of the understanding of phonon-mediated superconductivity. However, many of the results, in particular the Cooper-pairing theory itself, have much wider validity. The phenomenological models and many-body theories are also useful for the study of other pairing mechanisms. As an example see the treatise on electron–phonon superconductivity, studies using Ginzburg–Landau theory and fluctuation phenomena. Note that appropriately generalized Ginzburg–Landau theories are useful for analyzing multi-component and competing order parameters in unconventional superconductors and for non-equilibrium behavior in general.

Several chapters discuss recent developments of superconductors containing impurities. Of particular interest are nanostructured superconductors. The important interplay of magnetism and superconductivity is also discussed. Later chapters (see Volume II) focus on novel superconductors, in particular on cuprate superconductivity with its almost universally accepted d-wave symmetry order parameter. A wide variety of important experimental results on high superconductivity are described. In particular, experiments determining the symmetry of the superconducting order-parameter are discussed. The theory analyses the pairing mechanism, important properties and in particular vortices in high-T_c superconductors. Note that the study of vortex matter is also of interest for applications. Several chapters present a theory for cuprate superconductivity assuming Cooper-pairing due to the exchange of spin-excitations. A critical comparison with experiments is presented. Moreover, superconductivity in various novel superconductors like ruthenates, heavy-Fermions and organics is discussed. Many of the properties involve an interesting interplay of magnetic and superconducting excitations and interactions. The use of neutrons is very important for the analysis of the elementary excitations and of magnetic activity, in particular regarding novel superconductors. Still important open problems are discussed. This should stimulate in particular current and future research.

The theoretical studies illustrate our present understanding of novel superconductors and should be helpful for further theoretical development towards a general theory that plays the role of the BCS theory in the past. The connections with superfluid ^{3}He, the best understood unconventional Fermion superfluid, and with Bose–Einstein condensation are also discussed.

This treatise on *superconductivity* is a follow up of the previous books on the physics of superconductors published a few years ago. Several important new chapters are presented and most of the other chapters have been largely rewritten, add new results and are generally updated. Thus it is hoped that this book, like Park's famous books on conventional superconductors, describe our present more or less definite knowledge on superconductivity one of the most interesting problem in physics.

It is hoped that the book will remain useful for a long time, as was the case with the Parks treatise. A parallel goal is that the book will be helpful for both students and researchers.

We wish to thank the authors for their careful work and great patience. Special thanks go to Christof Bennemann for his essential help in preparing the book. Finally we thank Dr. Ascheron from Springer for his patience and help during editing.

Berlin, February 2008

K. H. Bennemann and J. B. Ketterson

K. H. Bennemann is Professor of Physics at the Physics Department FUB (Berlin,Germany), Sloan Fellow, and internationally known for his research on magnetism, superconductivity, and non-equilibrium phenomena.
J. B. Ketterson is Professor of Physics at Northwestern University (Evanston, Ill. USA), internationally known for his research in the area of Solid State Physics, in particular of Low Temperature Physics, on Helium. Both are Editors of the well-known books The Physics of Solid and Liquid Helium.

Contents

Volume I. Superconductivity: Conventional and Unconventional Superconductors

Volume II. Superconductivity: Novel Superconductors

List of Contributors

Bennemann, K.H.
Institut für Theoretische Physik,
Freie Universität Berlin,
Berlin, Germany
khb@physik.fu-berlin.de

Campuzano, J.C.
Dept. of Physics,
University of Illinois at Chicago, USA and
Materials Science Division, Argonne
National Lab., Argonne, USA
jcc@uic.edu

Carlson, E.W.
Department of Physics, Purdue University,
525 Northwestern Avenue
West Lafayette, IN 47907-2036, USA
ewcarlson@physics.purdue.edu

Chubukov, A.V.
University of Wisconsin Dept. of Physics,
1150 University Ave Madison, WI 53706, USA
chubukov@skazka.physics.wisc.edu

Emery, V.J.
Brookhaven National Laboratory,
Department of Physics,
New York, USA
emery@cmth.phy.dnl.gov

Eremin, I.
Institut für Mathematische Physik,
TU-Braunschweig , Germany
i.eremin@tu-braunschweig.de

Hayden, S.M.
H.H. Wills Physics Laboratory,
University of Bristol, United Kingdom
S.Hayden@bristol.ac.uk

Kirtley, J.B.
IBM Thomas J. Watson Research Center,
P.O. Box 218 Yorktown Heights,
NewYork 10598, USA
kirtley@us.ibm.com.

Kivelson, S.A.
Stanford University, Dept. of Physics,
GLAM McCullough Bldg 476
Lomita Mall Stanford University,
Stanford, CA 94305-4045, California, USA
stevek@physics.ucla.edu

Lang, M.
Physik-Institut der
Johann Wolfgang Goethe Universität,
Frankfurt/Main, Germany
michael.lang@physik.uni-frankfurt.de

Leggett, A.J.
Dept. of Physics, University of Illinois,
Illinois, USA
aleggett@uiuc.edu

Manske, D.
Max-Plank Institute for Solid State Research,
Stuttgart, Germany
D.Manske@fkf.mpg.de

Müller, J.
Max-Planck Institute for Chemical Physics of Solids,
Dresden, Germany
mueller@cpfs.mpg.de

Norman, M. R.
Materials Science Division,
Argonne National Laboratory,
Argonne, USA
norman@anl.gov

Orgad, D.
The Hebrew Univ. of Jerusalem,
Racah Institute of Physics, Israel
orgad@phys.huji.ac.il

Ott, H. R.
ETH Zürich, Laboratorium für Festkörperphysik,
ETH Hönggerberg CH-8093, Switzerland
ott@solid.phys.ethz.ch

Pines, D.
Los Alamos National Laboratory,
CNLS Mail Stop B258 Los Alamos, NM 87545, USA
pines@santafe.edu.

Randeria, M.
Tata Institute of Fundamental Research,
Mumbai, India
randeria@theory.tifr.res.in

Riseborough, P. S.
Department of Physics,
Polytech University,
New York, USA
prise@poly.edu

Schmalian, J.
Ames Laboratory, Dept. of Physics and Astronomy,
1 Osborn Drive Ames, IA 50011, USA
schmalian@ameslab.gov.

Schmiedeshoff, G.M.
Occidental College, Department of Physics,
California, USA
gms@oxy.edu

Smith, J.L.
National Laboratory Los Alamos,
Nevada, USA
jlsmith@lanl.gov

Tsuei, C.C.
IBM Thomas J. Watson Research Center,
P.O. Box 218 Yorktown Heights,
New York 10598, USA
tsuei@watson.ibm.com.

Zasadzinski, J.F.
Illinois Institute of Technology,
Physics Department,
3301 Dearborn, Chicago,
IL 60616, USA
john_zasadzinski@anl.gov

Volume II

Superconductivity:

Novel Superconductors

14 High-T_c Superconductivity

H.R. Ott ETH Zürich, Laboratorium für Festkörperphysik, Switzerland

14.1 Introduction

The unexpected and really amazing discovery of J.G. Bednorz and K.A. Müller [1], suggesting that certain copper-oxide compounds enter a superconducting state at temperatures exceeding 30 K, provoked an avalanche of research activities worldwide. Early confirmations of the conjecture [2–4], as well as rapid progress in further enhancing the critical temperatures T_c in related materials, i.e., other types of copper oxides, to temperatures above the boiling point of nitrogen [5,6], pushed research in superconductivity into the limelight of science and public interest. This unprecedented development in the field of superconductivity is best illustrated by plotting the critical temperature T_c as a function of time (see Fig. 14.1), exemplifying the role of the cuprate materials.

It is therefore not surprising that referring to high-T_c superconductivity, is very often understood as considering superconductivity of copper oxides only. This is, however, not quite fair, because other materials, such as Pb/Bi-type oxides and, subsequent discoveries, such as the doped fullerites or fullerides and, more recently, MgB_2, also exhibit superconduc-

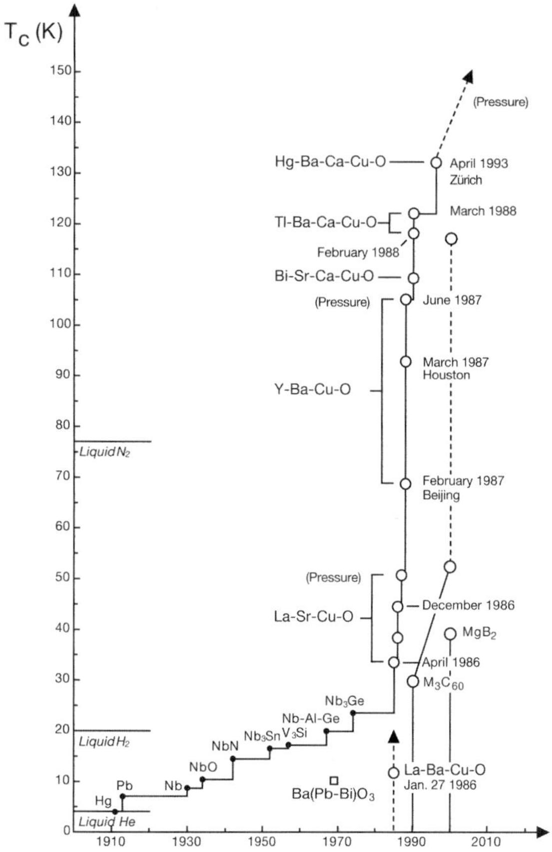

Fig. 14.1. Enhancement of critical temperatures of superconductors as function of time.

tivity at amazingly high temperatures. It is for this reason that in this review, although for the most part dedicated to the outstanding properties of a variety of cuprate materials, short overviews of physical properties related with superconductivity of Ba-based Bi-oxide perovskites, doped C_{60} materials and the amazing binary compound MgB_2, are also included. The class of quaternary borocarbide compounds for which also maximum critical temperatures for superconductivity exceeding 20 K have been reported, will be treated in a special chapter of this treatise and is therefore not included in this overview.

The notion of high temperature superconductivity was introduced by V.L. Ginzburg quite some time before the essential breakthrough in 1986. Research led by him, concentrating on identifying alternative electron-phonon interactions in solids that might lead to superconductivity at elevated temperatures, started in 1964 and resulted in a compendium with the title "High Temperature Superconductivity", published 1977 in Russian in Moscow. An English translation of this work appeared 1982 in London [7]. Since at present even the highest critical temperatures that have been achieved, by common standards still have to be regarded as low temperatures ($\sim$−140 °C), we rather use the terminology of high-T_c superconductivity in our context.

In spite of substantial efforts in both experimental and theoretical research, the mysteries behind the occurrence of superconductivity in cuprate materials up to temperatures exceeding 100 K are still to be regarded as conspiring to pose one of the major unsolved problems of contemporary physics. Most of the physical properties of the Cu oxides have experimentally been established with a high degree of reliability and advances in preparing the materials are such that excuses for the lack of understanding as being related to spurious effects and uncertainties in materials compositions, homogeneities and impurity content can no longer be accepted. However, it is still not possible to control the critical temperature T_c for superconductivity of these materials by suitably tailoring the chemical composition of the compounds and alloys to the extent that T_c can be enhanced at will. This unsatisfactory situation is, at least partly, due to the fact that no real basic understanding of the essential processes that lead to superconductivity in these cuprates at temperatures far above those for common metals and alloys exists and therefore no unfailing guidance for the design and subsequent synthesis of new and promising materials with even higher critical temperatures is available.

Simple intuitive approaches for improving the situation are made difficult by the sheer chemical complexity, of the known materials, at least when viewed at the level of conventional standards of condensed matter physics. The simplest, i.e., ternary compounds have to be doped with charge carriers in order to exhibit metallic conductivity, giving way to superconductivity at low temperatures. Most of these superconductors contain four or more constituents in their chemical composition and their crystallographic structures, although being rather similar for all of them, exhibit some intricate details whose importance is still debated. It is for this reason that, before presenting and discussing a selection of physical properties of high-T_c superconductors, a section on materials aspects including chemical compositions and crystal structures is inserted at the beginning of this chapter. An additional justification for this section is the fact that more recently, solids with similar structural properties have attracted a lot of attention in other areas of condensed matter research, involving spin charge and orbital ordering phenomena [8].

With respect to physical properties in general and to features of the superconducting state in particular, we shall concentrate on some typical aspects rather than list and present many details. For instance, the entire field of vortex physics that has emerged and has attracted a lot of attention in connection with high-T_c superconductors, is discussed in a special chapter of this treatise.

It was recognized very early [9] that the cuprate materials, which exhibit the highest critical temperatures for superconductivity at present, cannot simply be regarded as common metals, because even the normal state of these materials exhibits anomalous features that are difficult to understand. Therefore, some attention is also given to properties of the normal state above T_c.

14.2 Typical Structural Characteristics

Although, as pointed out above, no consensus about the real causes for superconductivity at elevated temperatures has yet been achieved, all relevant materials to be discussed here, with the exception of the fullerenes and MgB_2, in some way share a common structural feature and this is the unit cell of the perovskite structure shown in Fig. 14.2. This structure is adopted by ABO_3 compounds in which A is a fairly large cation and B, a metal element, helps to form a three-dimensional array of corner-sharing BO_6 octahedra. The undistorted version is cubic, as indicated in Fig. 14.2, and only very few compounds for which the ionic radii of the A atoms are of sufficient size, are

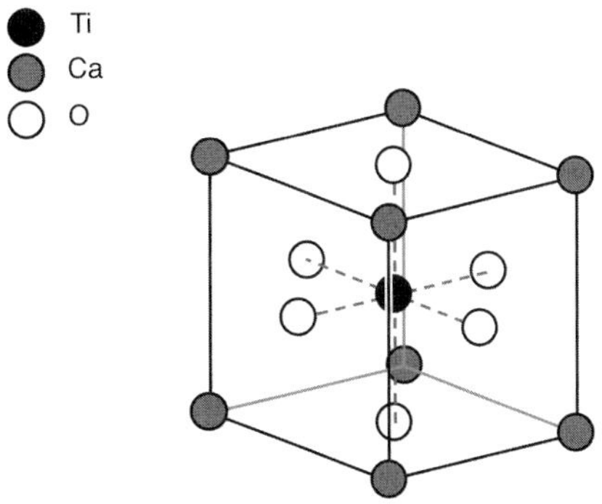

Fig. 14.2. Schematic representation of the crystallographic unit cell of $CaTiO_3$ (perovskite)

known to adopt the truly cubic perovskite structure. Most of the so called perovskite materials crystallize in a distorted version, the most common being an orthorhombic distortion. Most of the fullerenes also crystallize in cubic structures (fcc) but the occupation per lattice site is given by a rather large basis, formed by the C_{60}.

During the time, when investigations of superconducting materials were focused on either chemical elements or, at most binary compounds and alloys, it was argued [10] that a high symmetry of the crystal lattice was favorable for achieving high critical temperatures. Superconductivity of the cubic A15 compounds with critical temperatures between 15 and 20 K was taken as the show case for this conjecture. The most recent and also rather surprising exception from this trend is, no doubt, MgB_2 which adopts a structure with hexagonal symmetry [11]. Its critical temperature, of the order of 40 K, is by now no longer a top value in general, but for simple binary compounds, it most certainly is.

14.2.1 $BaPb_{1-x}Bi_xO_3$, $BaPb_{1-x}Sb_xO_3$, $BaBi_{1-x}K_xO_3$

The structure of these non cuprate materials of the type $BaPb_{1-x}Bi_xO_3$ and $Ba_{1-x}K_xBiO_3$ varies with the parameter x and different varieties of distorted perovskite type arrangements of the atoms are observed. The Pb/Bi alloy series exhibits a number of structural phase transitions, starting with an orthorhombic lattice for metallic $BaPbO_3$ [12], changing to tetragonal at x $\sim$ 0.1, back to orthorhombic for x $\sim$ 0.35 [13]

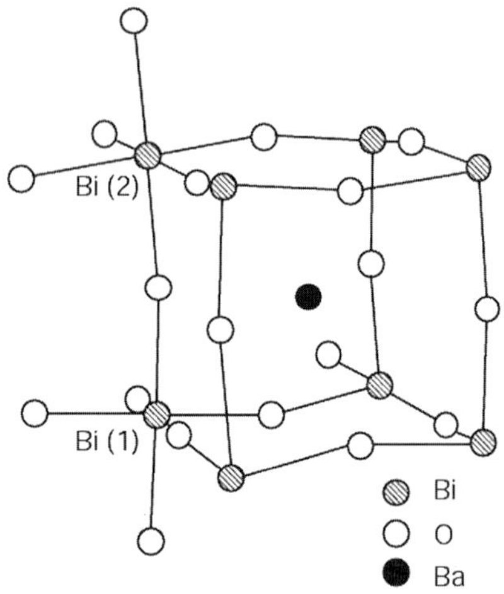

Fig. 14.3. Schematic representation of the crystallographic unit cell of $BaBiO_3$, as reported in [14]

and finally distorting to a monoclinic lattice structure at x ~ 0.9 [14]. In the range of $0.1 < x < 0.3$, also the Pb/Sb alloys adopt a tetragonally distorted perovskite-type structure. $BaBiO_3$ is electrically insulating. Its crystallographic unit cell is shown in Fig. 14.3, revealing the distorted perovskite-type arrangement of the atoms by the quasi-octahedral coordination of oxygen atoms around the Bi atoms on inequivalent sites. On site B(1), one of the Bi-O distances is much shorter than on site B(2) and this freezing of a breathing mode is thought to be the reason for the insulating ground state. The substitution of Ba with K atoms in $BaBiO_3$ to form $Ba_{1-x}K_xBiO_3$, eventually triggers a transition to a cubic crystal lattice and a metallic ground state [15,16]. It is obvious that for these oxides, structural and electronic properties are intimately coupled, a feature that is also observed for the cuprate materials to be discussed below.

14.2.2 Copper Oxide Superconductors

The crystal structures of all superconducting Cu oxides are more or less evidently related to the perovskite structure. These Cu compounds belong to a large class of mixed-valency Cu oxides where the Cu cations may adopt different ionic configurations (2+ or 3+). In some cases rather the concept of intermediate valence (between 2+ and 3+) seems appropriate. The three-dimensional character of the original perovskite structure, which may also be viewed as being built by a stacking of AO and BO_2 planes, is lost because these planes are now stacked in different sequences, leading to a more or less sizeable anisotropy between the directions parallel and perpendicular to these planes. A fairly transparent case in this respect is the parent compound of the material where superconductivity of the cuprates was discovered by Bednorz and Müller [1].

$La_{2-x}A_xCuO_{4+\delta}$ (A = Sr, Ba)

An early study [17] of substances of this type with A = Ca, Sr, Ba and Pb revealed the metallic conductivity for these materials in the form of a decreasing electrical resistivity with decreasing temperature, i.e., $\partial\rho/\partial T > 0$. For unknown reasons, this study was limited to temperatures above 200 K, but it contained also important information concerning the crystal structure of this series of compounds. Later a more detailed investigation [18] confirmed and extended these findings. As is already legend by now, it was this type of materials where Bednorz and Müller [1] found the first evidence for the onset of superconductivity between 30 and 40 K.

Ternary La_2CuO_4, an insulating antiferromagnet, crystallizes in a tetragonal K_2NiF_4-type structure but at lower temperatures adopts an orthorhombically distorted version of this structure, induced by a cooperative alternating tilting of the CuO_6 octahedra about the [110] tetragonal axis, as shown in Fig. 14.4 [19,20]. For x = 0 the tetragonal to orthorhombic transition occurs at approximately 530 K. The onset temperature of the orthorhombic distortion T_d can be reduced by partly replacing La by an alkaline earth element A and above a critical concentration of x ~ 0.2, the tetragonal K_2NiF_4-type structure is stable down to very low temperatures [21,22]. The tetragonal arrangement, also denoted as T-structure and shown in Fig. 14.5, may be regarded as a stacking of different planes that are also contained in the original perovskite structure. Between two $CuO_2(BO_2)$ planes, two LaO (AO) planes instead of only one are inserted, hence weakening the three-dimensional

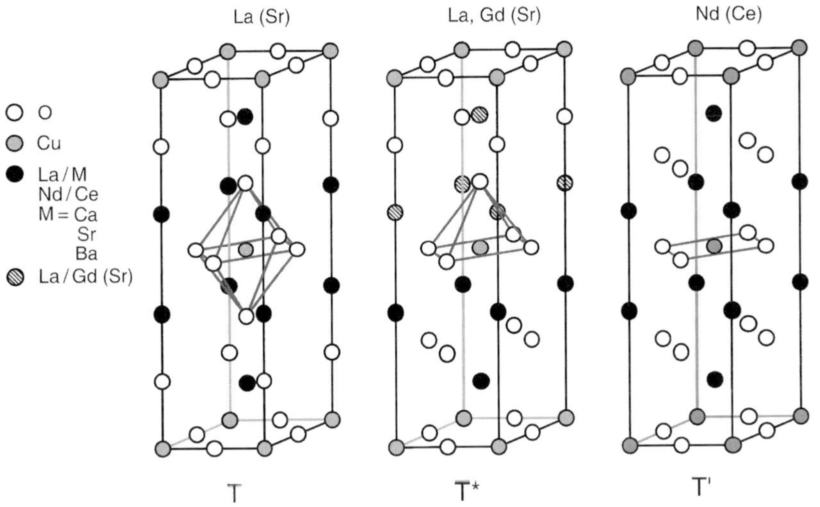

Fig. 14.5. Schematic representation of the tilting of the oxygen octa-hedra in $La_{2-x}Sr_xCuO_4$ (see [36])

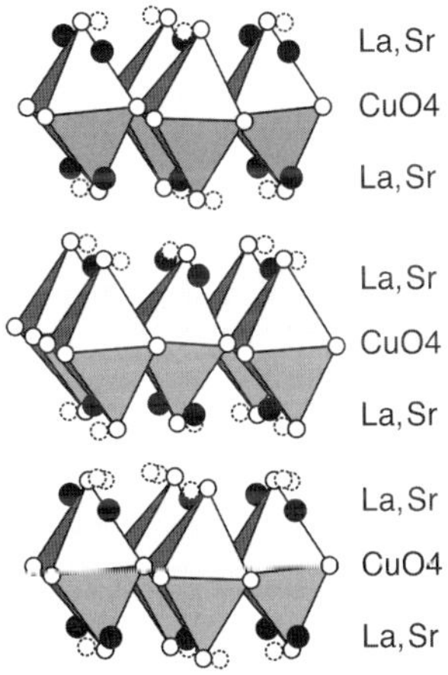

Fig. 14.4. Schematic representation of the tilting of the oxygen octahedra in $La_{2-x}Sr_xCuO_4$ (see [36])

character of the structure. The essential subunits are clearly the CuO_2 planes, a major characteristics of all known cuprate superconductors. For A = Ba, the appearance of a low-temperature tetragonal phase has been reported for x values around 0.125 [23]. As is well known, enhancing x eventually leads to a metallic behavior and superconductivity. Optimal conditions for superconductivity are reached for materials exhibiting the orthorhombic crystal structure at $T \sim 4T_c$. Structural effects are particularly pronounced in $La_2CuO_{4+\delta}$, containing excess oxygen on interstitial sites. Also oxygen rich material undergoes a transition from a tetragonal high-temperature to an orthorhombic low-temperature phase [24]. The transition temperature T_d is only weakly reduced upon increasing δ. For low oxygen surplus, a macroscopic phase separation phenomenon is observed below room temperature. The material separates into an oxygen-rich metallic phase and an oxygen-poor insulating phase with an antiferromagnetically ordered ground state [24].

$Ln_{2-x}Ce_xCuO_{4-\delta}$ (Ln = Pr, Nd, Sm)

Cu-oxides with a 214-type composition as mentioned above also form if La is replaced by rare-earth or lanthanide (Ln) elements. These compounds, however, adopt a somewhat different crystal structure [25], the so called T' structure and they keep the tetragonal structure down to low temperatures. It is depicted in Fig. 14.5. Here, the oxygen environment of each Cu atom in the form of a planar square is distinctly different from that of an octahedron in the T structure. As will be discussed later, the ternary compounds are again insulating antiferromagnets. Both the Cu spins and the localized Ln moments are involved in magnetic-ordering phenomena, at different temperatures, however. Metallic behavior and superconductivity is obtained by a partial replacement of the trivalent Ln element by tetravalent Ce and, in addition, a slight reduction of the oxygen content.

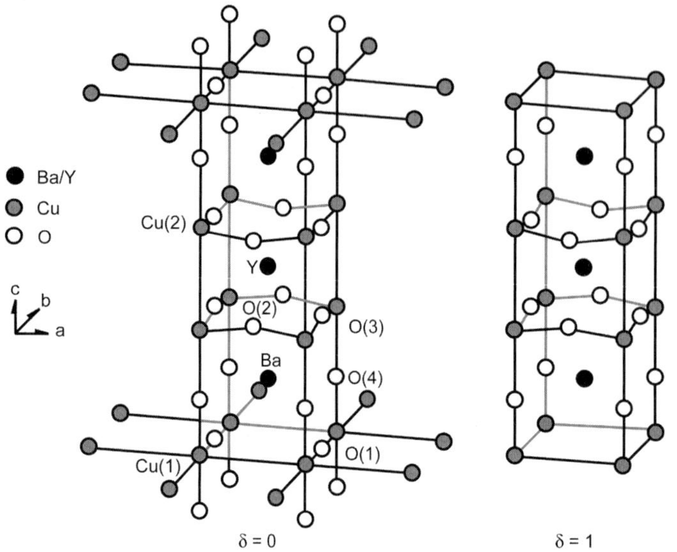

Fig. 14.6. Schematic representation of the unit cells of the crystal structure of $YBa_2CuO_{7-\delta}$ for **(a)** $\delta = 0$, and **(b)** $\delta = 1$

$Ln_{2-x-y}Ce_xSr_yCuO_{4-\delta}$

Yet another type of Cu-O coordination is obtained in this type of compounds where part of the Ln sites are occupied by Ce and Sr [26]. The resulting unit cell of the structure is shown in Fig. 14.5. A remarkable feature of this T*-structure is the lack of an inversion center, obviously implying polarity but, nevertheless, with the proper values of x,y, and δ, superconductivity may still be achieved. The Sr content must be large enough in order to allow for a sequential ordering (Sr,Ce) and (Nd,Ce) planes as indicated in Fig. 14.5.

$MBa_2Cu_3O_{7-\delta}$, $MBa_2Cu_4O_8$

Structurally, these two types of compounds, where M may be Yttrium or any element of the rare-earth series except Ce and Tb, are related but chemically, the second type of materials is much more stable than the first. Again the structures may be viewed as stackings of different layers with different chemical compositions. For $YBa_2Cu_3O_{7-\delta}$ (YBCO-123), the stacking sequence in the crystallographic unit cell is Y-CuO_2-BaO-CuO_x-BaO-CuO_2-Y, reminiscent of the perovskite structure [27]. The resulting orthorhombic unit cell is shown in Fig. 14.6a for the ideal case where $\delta = 0$, i.e., x = 1. We note two inequivalent Cu sites which can most easily be distinguished by their different oxygen environments. In detail the M atoms separate two identical blocks which contain two CuO_2 planes with the Cu(2) sites, two BaO planes and a plane formed by Cu-O chains along the b direction of the orthorhombic structure involving the Cu(1) sites. As may be seen from Fig. 14.6a, the oxygen coordination of the two Cu sites is pyramidally for the Cu(2) sites and linear for the Cu(1) sites. It is the missing oxygen between the chains which gives rise to the orthorhombic distortion of the lattice. This low temperature orthorhombic structure develops out of a tetragonal high temperature structure at about 750 °C by a rearrangement of atoms within the oxygen sublattice. This rearrangement can be reversed at low temperatures by reducing the oxygen content of the material. For $\delta = 0.6$, the structure is again tetragonal also at low temperatures. Upon depletion, oxygen vacancies are formed within the Cu-O chains and they redistribute in a fashion such that on the average, oxygen atoms occupy sites along the *a* and *b* directions with equal probability [28]. For $\delta = 1$, i.e., $x = 0$, the former Cu-O chains are fully depleted of oxygen. The resulting structure is shown in Fig. 14.6b. This variation of the oxygen content not only influences the structure of the crystal lattice but also many other properties of different ground states, as we shall see below. A particular aspect of

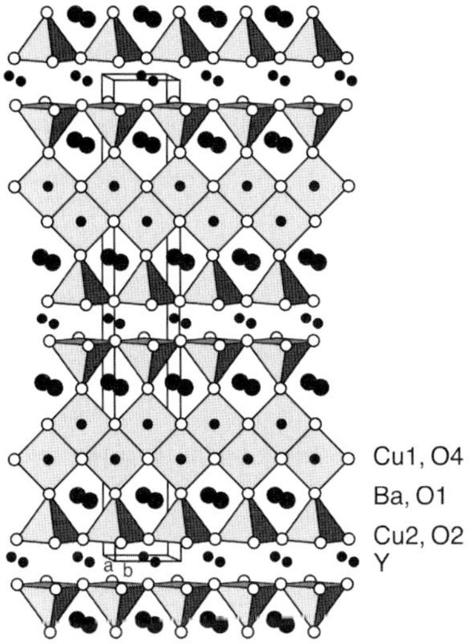

Fig. 14.7. Schematic representation of the unit cell of the crystal structure of $YBa_2Cu_4O_8$ (see [36])

the crystal structure in this YBCO-123 series is the formation of superstructures in different regions of oxygen content, leading to inhomogeneities and different orthorhombic phases at different values of δ. Since a detailed discussion of these aspects is beyond the scope of this review, we refer to relevant work in the literature [29]. The unit cell of the $MBa_2Cu_4O_8$ (YBCO-124) compounds [30,31] is shown in Fig. 14.7. As mentioned above this type of compound is chemically less fragile than the YBCO-123 variety and crystals may be grown without twin boundaries, a major problem in the growth of YBCO-123 compounds. The insertion of a second Cu-O chain element, forming Cu-O ribbons rather than chains, enhances the anisotropy, especially within the planes perpendicular to the c axis. In this Y–Ba–Cu–O quaternary system, an additional stable compound with yet another although related structure and composition, namely $Y_2Ba_4Cu_7O_{14+\delta}$, in short YBCO-123.5, has been identified [32,33]. The corresponding unit cell is obtained by stacking the unit cells of YBCO-123 and YBCO-124 on top of each other. Concerning the Cu-O structural elements, the unit cell of the atomic arrangement thus contains, apart from the notorious Cu-O planes, both Cu-O ribbons and Cu-O chains.

$A_mM_2R_{n-1}Cu_nO_{3n+m+1}$

A fairly large number of high-T_c cuprate superconductors may be classified according to the schematic chemical composition given in the subtitle. This type of compounds form with A = Hg, Tl or Bi, M = Sr or Ba and R = Ca or a heavy rare earth. These materials are of particular importance because some of them exhibit the highest critical temperatures that have been achieved until today. For A = Hg, m may be 1, for Bi only 2 and for Tl 1 or 2. Although Bi-

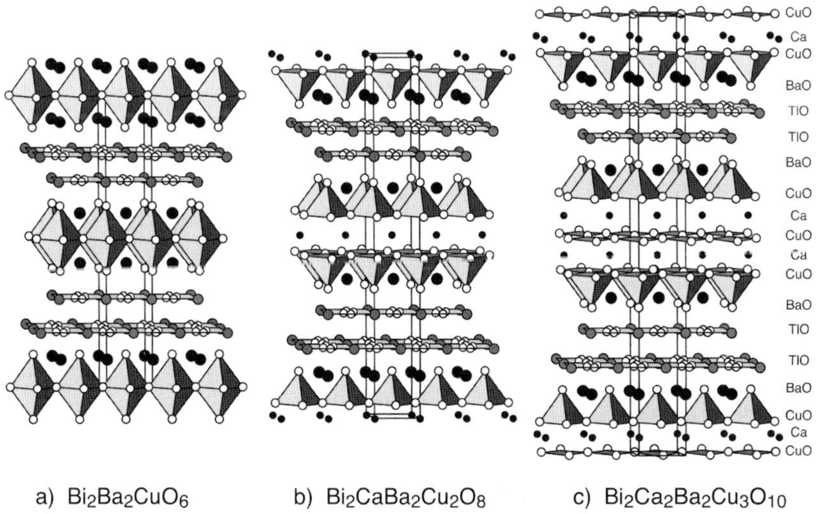

Fig. 14.8. Schematic representation of the crystal structure of some Bi-based Cooper oxide superconductors: Bi-2201, Bi-2212 and Bi-2223 (see [36])

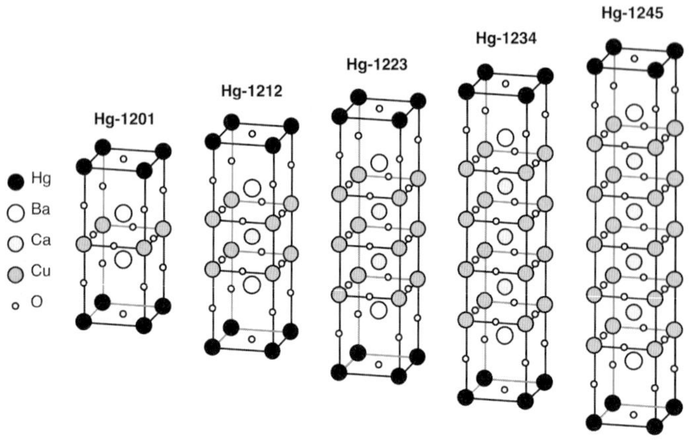

Fig. 14.9. Schematic representation of the crystallographic unit cells of Hg-12(n-1)n copper oxides

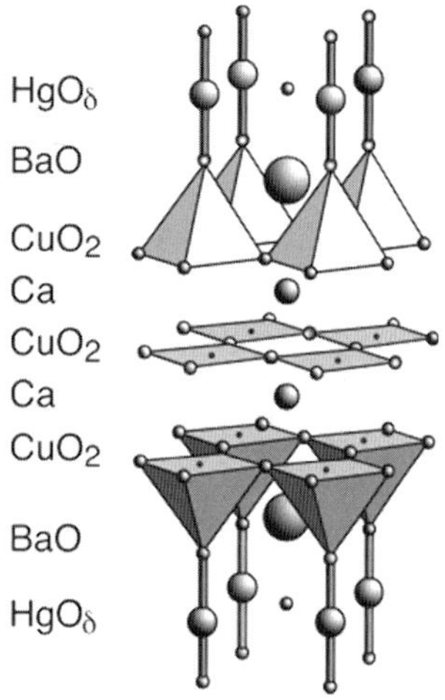

Fig. 14.10. Schematic representation of the crystallographic unit cell of $HgBa_2Ca_2Cu_3O_{8+\delta}$

based cuprates have been synthesized with the 1222-type structure, all of them turned out to be non superconducting above 1.5 K [34]. In most cases the parameter n varies from 1 to 4 but may also adopt higher values. The parameter n indicates the number of CuO_2 planes within one unit cell, separated from each other by n-1 Ca layers, and as such it is related with some physical properties to be discussed below. The structures of these compounds may be viewed as sequential stackings of elements of the perovskite as well as the NaCl structure [35]. Very often, crystallographic shears, i.e., shifts between adjacent layers are introduced in the stacking sequences. A nice compilation of possible atomic arrangements in the Tl-based and Bi-based compounds is given, e.g., in [36] and therefore we present only three examples of the physically most important Bi compounds in Fig. 14.8 (on page 771). It is also possible to partially replace some of the above mentioned constituents by other elements. One such example is Pb substituting for Bi, which in some cases provides a way to stabilize a favorable phase, such as the Bi-2223 phase [37]. The latest progress in raising T_c has been made in the series $HgBa_2Ca_{n-1}Cu_nO_{3n+2}$ [38] and it also has been possible to synthesize thin layers of material where $n \geq 4$ [39]. A schematic representation of the stacking with increasing n is shown in Fig. 14.9 and a somewhat more detailed drawing of the unit cell of $HgBa_2Ca_2Cu_3O_{8+\delta}$, the material for which the highest values of T_c have as yet been observed, is presented in Fig. 14.10.

$Ca_{1-x}Sr_xCuO_2$

The structure of this type of compound is of interest here, because it may be viewed as the result of enhancing the number n of CuO_2 layers per unit cell in the compounds discussed in the previous section towards infinity. Taking into account the $n-1$ Ca interlayers, for large values of n, the stoichiometry will approach that of $CaCuO_2$. With respect to the perovskite structure, the (Ca/Sr) layers may be regarded as AO layers from which all the oxygen atoms have been removed. $CaCuO_2$, a hypothetical tetrago-

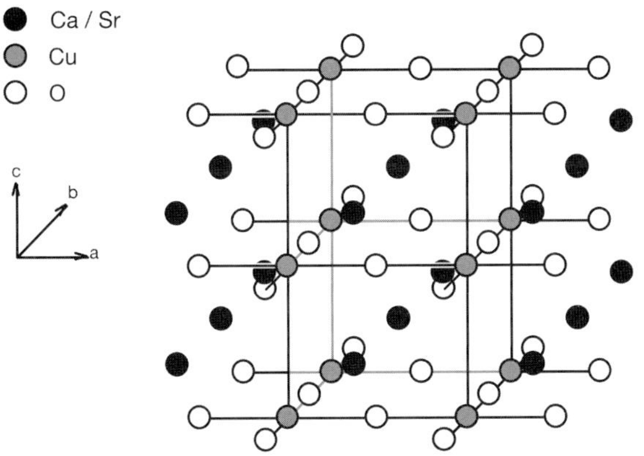

Fig. 14.11. Schematic representation of the crystal structure of $Ca_{1-x}Sr_xCuO_2$ (see [40])

nal compound does not form but it has been found that small amounts of Sr on Ca sites can stabilize the structure, which is shown in Fig. 14.11, and single crystals may be grown [40]. Subsequently it has been found that a high pressure synthesis helps to extend the compositional range of stability considerably [41].

General Remarks

This brief overview of some crystal structures of relevant cuprate materials demonstrates quite clearly that one ingredient, namely the CuO_2 planes, are a common building element for all the structures presented in the previous subsections. Even before discussing the physical properties of these materials we note that the electronic features are, to a large extent, dominated by these CuO_2 planes. They are separated from each other by additional building blocks that serve both to stabilize the structure and as charge reservoirs, which control the number of itinerant charge carriers in the planes containing Cu atoms. In this sense, the structures of the compounds mentioned in the first three subsections contain one CuO_2 plane per unit cell. Two CuO_2 planes per unit cell are the hallmark of the compounds mentioned in the fourth subsection. Structurally the most simple member of this type of double-layer compounds, but not mentioned in detail here, is $La_{2-x}Sr_xCaCu_2O_6$ [42]. The compounds with the possibility of accommodating more than two CuO_2 planes per unit cell are mentioned in Sect. 14.5. Another important compound, with all the features of the $m = 2$ and $n = 2$ Tl-based or Bi-based compounds, except for an inserted oxygen depleted Cu layer, is $Pb_2Sr_2YCu_3O_8$ [43].

14.2.3 Fullerites, Fullerides

Fullerites are solids composed by a regular arrangement of C_{60} molecules, which belong to a large class of fullerenes, i.e., stable molecules formed by a closed carbon network (C_{60}, C_{70}, C_{540}, etc.) [44]. For our purposes, only C_{60} molecules are of interest, because up to now, bulk superconductivity has only been observed in solids formed by them. A pure C_{60} solid is, under normal circumstances, an electrical insulator [45]. It adopts an fcc structure which, with decreasing temperature, transforms to a simple cubic structure, at 260 K, accompanied by an orientational order of the individual C_{60} molecules [46]. Metallic C_{60}-based compounds, the fullerides, may be obtained by doping [47] and the most relevant materials in our context are A_3C_{60}, where A is K, Rb, Cs or some combination of these three alkali elements and Na_2AC_{60}, where A may again be one of the other alkali elements or a combination of them [45]. The structure of these materials is of fcc type with

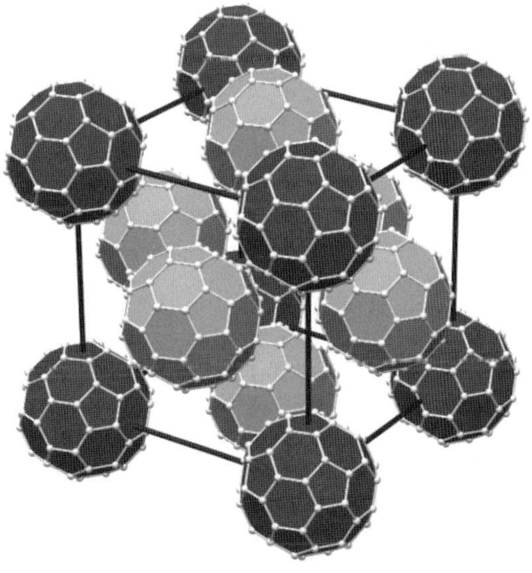

Fig. 14.12. Schematic representation of the unit cell of fcc C_{60}

varying space group symmetries, however. Non cubic structures of superconducting fullerides have been reported for the cases of $Cs_3\ C_{60}$ and $NH_3\ K_3\ C_{60}$ under pressure. The compounds with chemical composition of the type A_4C_{60} usually adopt a bct cubic structure and are insulators. A schematic drawing of the cubic structure of the A_3C_{60} type materials is shown in Fig. 14.12. More on structural aspects of C_{60} material may be found in [48].

14.2.4 MgB_2

Although this compound has been known for a long time, it obviously had never been tested for superconductivity until late in 2000. The discovery of its unexpectedly high critical temperature of the order of 40 K by Akimitsu and co-workers [11] came as a real surprise. MgB_2 crystallizes in the hexagonal AlB_2-type structure, shown in Fig. 14.13, which exhibits clear 2D-type features. The B atoms form layers that are equivalent to the layered structure of graphite. It has been suggested [49] that this apparent two-dimensionality might be the key to the high value of T_c of this material, reminiscent of the case of the cuprates.

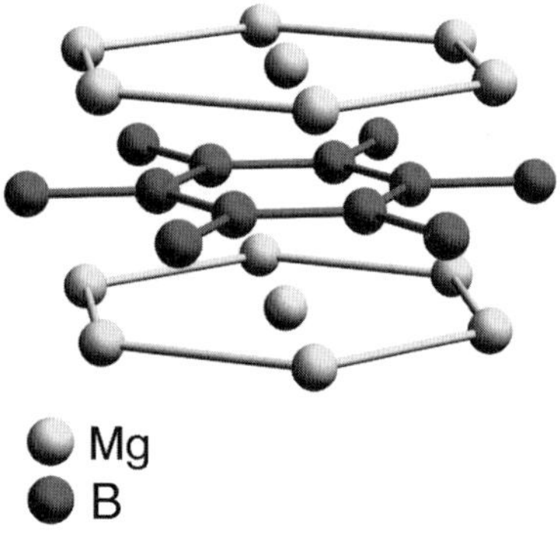

Fig. 14.13. Schematic representation of the crystallographic unit cell of MgB_2

14.3 Occurrence of Superconductivity

For all the materials that in this review are considered in relation with high-T_c superconductivity, a superconducting ground state is achieved only in limited regions of chemical composition. In this section, we briefly discuss the relation between chemical composition and the observation of superconductivity, including the established critical temperatures T_c, for a few instructive cases.

14.3.1 $BaPb_{1-x}Bi_xO_3$, $Ba_{1-x}K_xBiO_3$

Superconductivity in the Pb/Bi compound series [50] is restricted to values of x that are close to 0.25 where a maximum critical temperature of approximately 12 K is observed [51,52]. This value of T_c is unexpectedly high and remarkable because of the absence of any element of the d-transition range in the chemical composition. Figure 14.14 demonstrates that, although partial superconducting transitions, measured magnetically and resistively, have been observed around x = 0.1 and x = 0.3, narrow and complete transitions are monitored only for x = 0.25 [52]. It is therefore quite obvious that the optimal conditions for superconductivity are reached for $BaPb_{0.75}\ Bi_{0.25}\ O_3$ and any deviation from this composition leads to a rapid deterioration of the superconducting properties and, in addition, favors inhomogeneities in the chemical composition. It is conceivable that the occurrence of superconductivity is extremely dependent on the doping level and that compositional inhomogeneities lead to the spread in T_c displayed in Fig. 14.14.

Replacing Bi with Sb leads to a less favorable situation for superconductivity [53]. Maximum critical temperatures are again reached for $x \sim 0.25$ but they barely exceed 3 K. It is not quite clear why T_c is so low in this case.

More successful in this respect were experiments with material that was prepared by partly replacing the divalent Ba by the monovalent alkaline metal K in $BaBiO_3$. This type of substitution and the concomitant reduction of the electron concentration induces the already mentioned structural phase transition and the symmetry of the crystal lattice changes from monoclinic to cubic. Likewise, the chemical substitution leads from an insulating to a metallic state. K-doped $BaBiO_3$ is electronically very similar to Bi-doped $BaPbO_3$. Optimal conditions are reached at the composition $Ba_{0.6}K_{0.4}BiO_3$ for which T_c is approximately 30 K [54,55].

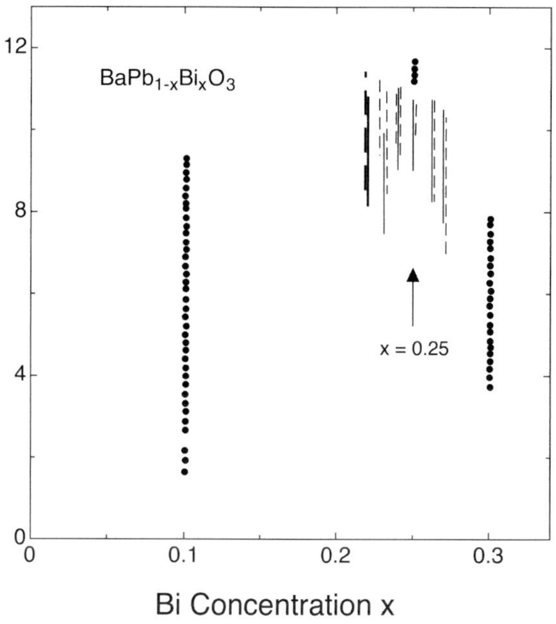

Fig. 14.14. Onset and widths of the superconducting transition of $BaPb_{1-x}Bi_xO_3$ (see [52])

14.3.2 Copper Oxides

Very shortly after the discovery of superconductivity in Cu oxides it was realized that in these materials, the occurrence of superconductivity is intimately related with the number of itinerant charge carriers which, in most cases, can be varied rather easily by simply altering the chemical compositions of the materials in specific ways. For some types of compounds, the result of this procedure in the form of a transition from an insulating to a metallic behavior and superconductivity, or vice versa, may be illustrated very well, for others it proves to be less obvious. In what follows, four rather instructive cases, representative for the cuprate superconductors, are presented.

$La_{2-x}Sr_xCuO_{4+\delta}$

The ternary compound La_2CuO_4 has long been known to be an electrical insulator. Early measurements revealed peak features in the temperature dependence of the magnetic susceptibility which indicated some type of magnetic order below approximately 200 K [56]. It was only after the discovery of superconductivity in exactly this class of cuprates, when the physical properties of these materials were studied in great detail. The results of these investigations are being reviewed in some of the following sections but we summarize part of them in the form of low temperature phase diagrams, displaying the variation of the physical behavior as a function of the concentration of itinerant charge carriers. The charge carrier doping may be accomplished in two ways, either by enhancing the oxygen content to above four atoms per formula unit or by replacing part of the trivalent La ions by divalent ions of the alkaline earth elements Sr and Ba. No successful attempts to replace La by Ca have been reported.

Figure 14.15 captures the low temperature phases that are identified if La is replaced by Sr [57]. The diagram is somewhat more complicated if Ba is chosen as the substitute. In the region of $x \sim 0.12$ where a slight indentation in $T_c(x)$ is apparent for the Sr doping (see Fig. 14.15), T_c is suppressed completely if the dopant is Ba (see also below). As mentioned above, the ternary compound La_2CuO_4 is an insulator with an antiferromagnetically ordered ground state. Upon doping with Sr^{2+}, the Néel temperature T_N of 325 K for $x = 0$ is efficiently reduced, giving way to a spin-glass type ground state for $0.02 < x < 0.05$. Once

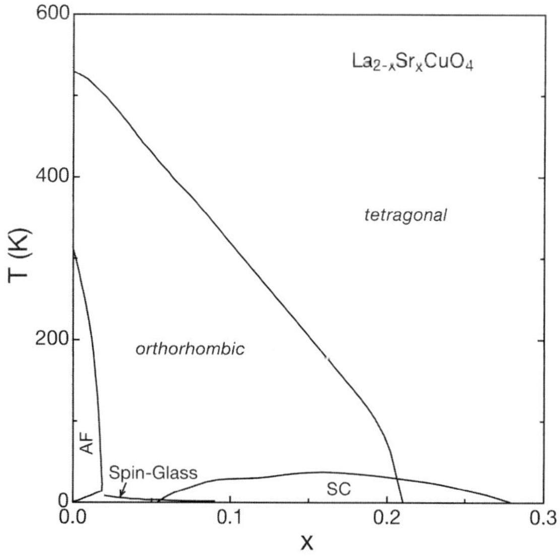

Fig. 14.15. Low temperature phase diagram of $La_{2-x}Sr_xCuO_4$ (see [57])

the Sr concentration exceeds 5%, superconductivity sets in at low temperatures. The optimal doping, i.e., the highest transition temperature T_c of about 40 K is reached at $x \sim 0.15$. Further increasing x reduces T_c and superconductivity is lost for $x > 0.25$. Since, as mentioned above, the structural transition from a tetragonal to an orthorhombic structure has been reported to disappear at $x \sim 0.21$, bulk superconductivity seems to persist into the tetragonal structure regime. This observation is important in view of discussing relations between the crystal structure symmetry and the occurrence of superconductivity, as illustrated in the following case. A tiny dip in the $T_c(x)$ curve at $x \sim 0.125$, displayed in Fig. 14.15, signals an effect that is much more pronounced in $La_{1-x}Ba_xCuO_4$. Again around a concentration of $x \sim 0.125$, T_c of the Ba-doped series is suppressed to zero and static magnetic order is observed instead [58,59]. Detailed studies have traced this back to the occurrence of a low temperature tetragonal structure [23,59,60].

As already mentioned in a previous section enhancing the oxygen content leads to phase separation phenomena in $La_2CuO_{4+\delta}$. The oxygen rich metallic

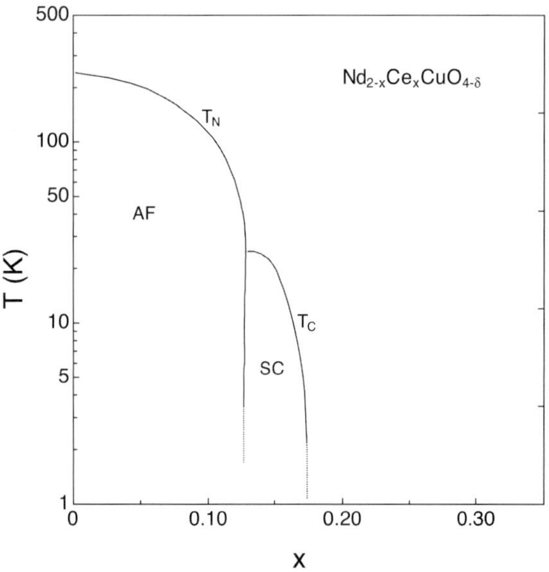

Fig. 14.16. Low temperature phase diagram of $Nd_{2-x}Ce_xCuO_{4-\delta}$ -type compounds (see [63])

phase is superconducting at low temperatures and again, T_c values exceeding 30 K are achieved. An instructive discussion of the ground state properties of $La_2CuO_{4+\delta}$ is offered in [57].

$Ln_{2-x}Ce_xCuO_{4-\delta}$ (Ln = Pr, Nd, Sm)

This variety of compounds is, with respect to electronic properties, distinctly different from all the other Cu-oxide compounds. Here, the doping is not with holes but with electrons [61]. This difference is also reflected in the phase diagram capturing the occurrence of superconductivity [62,63]. The antiferromagnetic phase is again rapidly suppressed upon doping and gives way to a superconducting ground state rather abruptly, as may be seen in Fig. 14.16. The maximum value of T_c is found next to the phase boundary and a further enhancement of doping leads to a reduction of the critical temperature and, again, to a rather abrupt disappearance of superconductivity. The range of doping where superconductivity has been observed is much more narrow than in the hole doped cuprates. The discovery of electron doped superconductors was initiated by the preceding observation of a superconducting ground state in the chemically similar but somewhat more complicated series adopting the T^* structure [64] (see also Sect. 14.2.2).

$YBa_2Cu_3O_{6+\delta}$

In this series of compounds, tetragonal $YBa_2Cu_3O_6$ represents the electrically insulating and antiferromagnetically ordering parent compound [65]. The doping of this material is accomplished by enhancing the oxygen content. Superconductivity sets in at oxygen concentrations close to those where the structural instability leads to an orthorhombic crystal lattice, i.e., $\delta \sim 0.45$. The critical temperature increases roughly in two steps, with two plateaus at approximately 60 and 90 K [66]. Optimal values for T_c, of the order 91 to 92 K are obtained for δ between 0.9 and 0.95. Further enhancing the oxygen content leads to a slight decrease of T_c, by 1 or 2 K. The schematic $T_c(x)$ variation is displayed in Fig. 14.17.

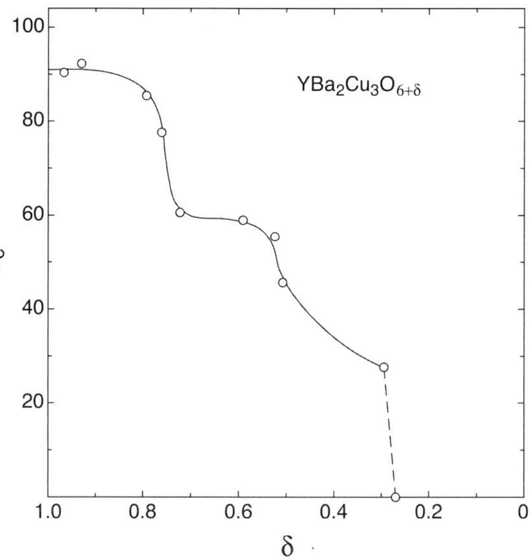

Fig. 14.17. Variation of the critical temperature T_c as a function of δ in $YBa_2Cu_3O_{6+\delta}$ (see [66])

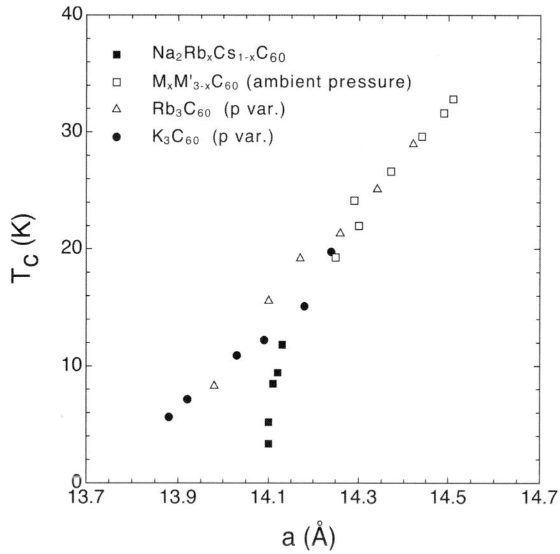

Fig. 14.18. Variation of the critical temperature $T_c$of fullerides as a function of the lattice parameter **a**, achieved either by doping or by application of external pressure (p) (see [45])

Bi-Based and Tl-Based Copper Oxides

Most of what is known about superconductivity of the Bi-based compounds [67–69] is very similar to what we have discussed in Sect. 14.3.2 concerning the hole doped La-cuprates. Also here, underdoped and overdoped regimes embrace the situation of optimal doping, more or less independent of the maximum value of T_c which depends on the chemical composition of the compound, i.e., on the number of Cu-O planes per unit cell. It is rather amazing that the maximum T_c's of the individual series of alloys ranges from 15 to 125 K. Again it is observed that strong overdoping eventually leads to the disappearance of superconductivity. The corresponding data base for superconducting Tl-based compounds [70–72] is much less extensive but there is no evidence for a significantly different behavior.

14.3.3 Fullerides

The secret of inducing superconductivity in these types of materials is again doping, with electrons in this case [73,74]. This is, e.g., achieved with the introduction of various numbers, between 1 and 3, of atoms of alkaline metal elements per unit cell. The main result of numerous investigations is the distinct variation of the critical temperature with the lattice parameter, seemingly a universal feature and, with a few exceptions, independent of the chosen material. An increase of the separation **a** of the atoms leads to an enhancement of T_c [75] and, vice versa, a reduction of T_c upon applying external pressure is observed [76]. This has been verified by changing the lattice parameter both via variation of the chemical composition and by applying external pressure. A schematic representation of T(**a**) for a selection of these materials is shown in Fig. 14.18. This plot suggests that an increase of T_c may be achieved by simply enhancing the lattice parameter **a**. Slight deviations from a universal T(**a**) behavior under pressure have, however, been reported [77,78]. Considering an extended set of data, and plotting T_c /T_c^{max} versus the concentration n of valence electrons, whereby T_c^{max} is the maximum critical temperature in a single alloy series, it has been claimed that the maxima of all the considered T_c 's is reached if $n = 3$ and that no superconductivity is observed if either $n < 2.5$ or $n > 4.0$ [79]. The maximum value of the critical temperature reported for bulk material is that of Cs_3C_{60}, apparently a non cubic compound with a T_c of about 40 K [80].

14.3.4 MgB_2

This compound does not seem to allow for large variations of the chemical composition without loosing superconductivity [81]. So far it appears that the binary compound with the proper stoichiometry reveals the highest T_c. Most attempts of alloying have led to substantial reductions of T_c and all other diborides with different cation elements have, so far, resisted to exhibit superconductivity at similarly high critical temperatures. An exception is Zn for which a tiny enhancement of T_c at low doping concentration may have been observed [82,83].

14.4 Physical Properties of Copper Oxides

As we have indicated in the preceding section, the general properties of the relevant cuprate compounds and alloys depend crucially on the chemical composition of the investigated materials. The same is true for those materials that actually do exhibit high-T_c superconductivity. Already early results of measurements probing the physical properties of these materials indicated anomalous features also in the normal state preceding the superconducting ground state and it soon became clear that the occurrence of this superconductivity at unusually high temperatures would not be explained without a sound understanding of the character of the normal state [84]. It is for this reason that we also briefly discuss those normal state properties that are difficult to understand on the basis of the very successful Fermi liquid concept that has, in the past, been used for describing the characteristic electronic properties of common metals. To keep this review tractable, only few examples that are believed to capture the essential issues of both the normal and the superconducting state of this type of materials are presented and discussed. Most of these examples, if they are not intended to emphasize the influence of the crystal structure or the chemical composition, are chosen to represent those materials with so called optimal doping, i.e., with the composition for which a maximum T_c has been observed. In these cases, some evidence for the existence of a boundary in the quasiparticle excitation spectrum that may be identified as a Fermi surface has been obtained from both conventional [85] and angle resolved photoemission spectroscopy (ARPES) [86,87] with high energy resolution. A schematic view of the cross sections of the 3D Fermi surfaces of Bi cuprates, deduced from ARPES measurements, is shown in Fig. 14.19. This evidence is more obvious for overdoped materials than it is for optimally doped or underdoped compounds although more recently it has been claimed that also for underdoped material, resolution limited signals, indicating well defined quasiparticle excitations, have been obtained [88]. Measurements with further improved energy resolution, however, seem to reveal that quasiparticle-type signals are only well identified if the investigated material is in the superconducting state [89]. It thus appears that it is still unclear whether well defined quasiparticles do exist in the normal state. At any rate, their manifestation certainly depends on the doping level of the material. As we shall see below, the difficulties encountered with the spectra of underdoped cuprates may arise because of the formation of a pseudogap

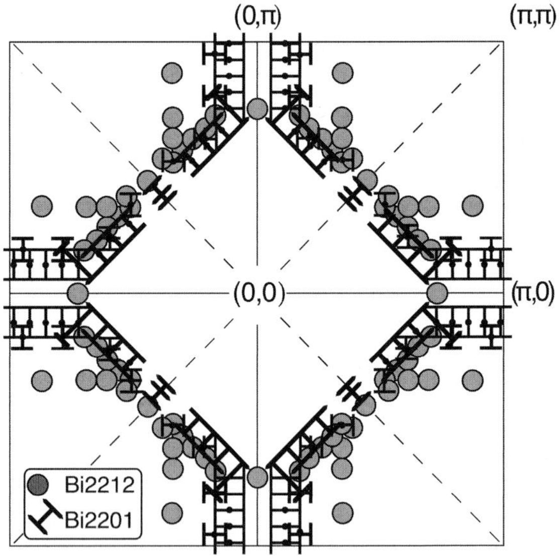

Fig. 14.19. Mapping of the Fermi surface of Bi-based superconducting cuprates by ARPES (see [86])

in the quasiparticle excitation spectrum already in the normal state of these materials.

In this context it is important to note that, because of the particular crystal structures, featuring the Cu-O planes as the essential structural element, most of the physical properties of these cuprates exhibit distinct anisotropies.

14.4.1 Normal State Properties

The first indications of anomalous features of the normal state of high-T_c cuprates were provided by measurements of the temperature dependence of the electrical resistivity $\rho(T)$ [90] of some of these materials, both parallel and perpendicular to the Cu-O planes [91]. These early transport results were confirmed in many cases and with improving sample quality, and complemented by additional data, mainly concerning the thermal conductivity and the Hall effect, both in the normal and the superconducting state.

Electrical Resistivity

First it seems instructive to consider the variation of ρ(T), measured for currents parallel to the Cu-O planes, upon doping. A well studied case is $La_{2-x}Sr_xCuO_4$ and some selected data are shown in Fig. 14.20 [92]. What is amazing, is the strong temperature dependence of the resistivity even at large absolute values. The increasing concentration of hole doping with increasing x is reflected in both the magnitude of ρ at room temperature and the slope $\partial\rho/\partial T$. Another typical feature is the extended region of the linear variation of ρ with temperature for $x \leq 0.18$. For common metals, the linear in T variation of the resistivity is restricted to temperatures around 200 K or above, but usually is not observed much below 100 K. It may also be seen that for higher doping levels, this particular feature is lost and the change of $\rho(T)$ with increasing x is much less pronounced than for lower values of x. Considering that the maximum value of T_c for this compound series is reached at optimal doping with $x \approx 0.15$, $\rho(T)$ is clearly anomalous in what has been termed the underdoped regime and progressively adopts a more common behavior in the so called overdoped region. This linear proportionality of the resistivity with respect to the temperature is one of the common features of the normal state of cuprate superconductors, most clearly borne out in the underdoped regime. An important case in this respect is the compound $Bi_2Sr_2CuO_6$ (Bi2201) where the linear in T variation of ρ may be observed from low temperatures, because T_c is only of the order of 10 K, up to 700 K [93]. The persistence of this temperature dependence of ρ or, equivalently, of the scattering rate τ^{-1} of the electronic quasiparticles down to T_c, is taken as evidence for an anomalous metallic state, not compatible with expectations of the Fermi liquid model, where a T^2 dependence of τ^{-1} is expected. In particular, the absence of any onset of residual resistance, definitely expected to be visible if the absolute magnitude of the measured values of ρ are considered, is difficult to understand.

Another unusual feature of the electrical resistivity is the generally observed large anisotropies and

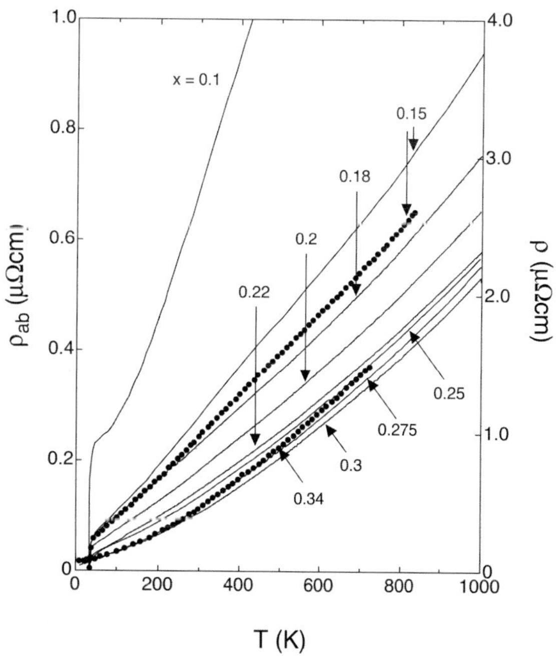

Fig. 14.20. Temperature dependence of the electrical resistivity of $La_{2-x}Sr_xCuO_4$ in polycrystalline form (*solid lines*) and for currents in the basal (*ab*) plane of single crystalline films (see [92])

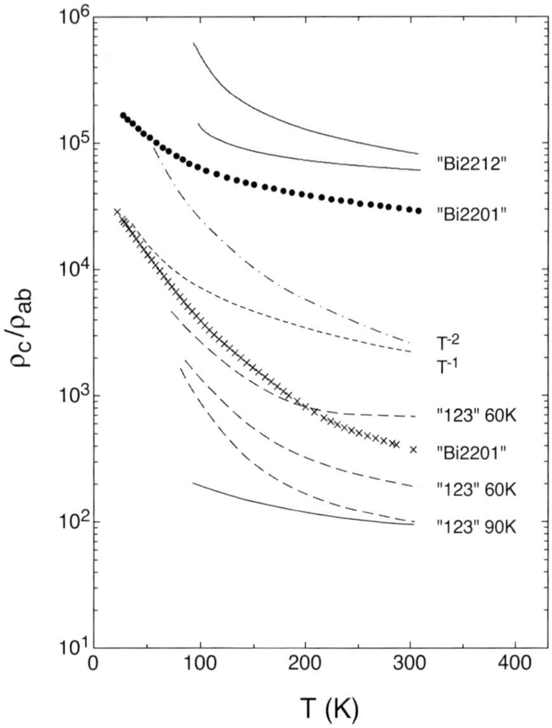

Fig. 14.21. Temperature variation of the ratio of ρ_c and ρ_{ab} for various cuprate superconductors. The general trend for T^{-1} and T^{-2} behavior is also indicated (see, e.g., [94])

their temperature dependences if the resistivities are measured either parallel or perpendicular to the Cu-O planes [94]. As may be expected, these anisotropies are strongly influenced by details of the crystal structure and examples are shown in Fig. 14.21. These large anisotropies imply that the motion of the quasiparticles is confined to the Cu-O planes and that they tend to localize in the motion perpendicular to it. The localization of quasiparticles in only one direction cannot occur in a Fermi liquid and therefore the observed anisotropies indicate the failure of the Fermi liquid model for describing the metallic state of cuprates above T_c, at least in the underdoped and optimally doped regime. Large anisotropies, as those of the electrical conductivities, are also observed in many other, physically relevant parameters. This has led to the notion that at least electronically, these cuprates should be regarded as quasi-two dimensional systems. This view has guided many theoretical attempts to understand the physical properties of these materials.

The Hall effect

Already early experiments [95] indicated anomalous features of the Hall response in the normal state of cuprate superconductors, exhibiting a strong increase of the Hall coefficient R_H with decreasing temperature above T_c. The Hall resistivity ρ_{xy} and the Hall constant R_H are related via

$$\rho_{xy} = R_H \cdot H \,, \tag{14.1}$$

where H is the externally applied magnetic field. Various later experiments confirmed this trend and experiments made with single-crystalline samples of $YBa_2Cu_3O_{7-x}$ (YBCO) revealed that the Hall constant, for currents parallel to the Cu-O planes and external magnetic fields oriented perpendicularly to these planes, is positive and varies as T^{-1} [96]. The out of plane Hall constant, however, obtained for the arrangement that both the current and the field direction are in the plane but perpendicular to each other, providing a Hall field E_H along the c direction of the crystal lattice, remained constant and negative between 400 K and T_c [97]. As an example for the distinct temperature dependence of the in plane Hall constant we show, in Fig. 14.22, R_H^{-1} versus T, measured on single crystalline samples of YBCO-123 and $Bi_2Sr_2CaCu_2O_8$ (Bi2212) [98]. Equally intriguing is the temperature dependence of $\cot\Theta_H = \rho_{xx}/\rho_{xy}$ which, as demonstrated in Fig. 14.23, varies almost as T^2 over a wide temperature range [97,98]. The in-plane Hall angle is defined as

$$\Theta_H = arctg\left(\frac{\rho_{xy}}{\rho_{xx}}\right) . \tag{14.2}$$

These two phenomena, apparently quite universal for high-T_c copper oxides, were interpreted by Anderson [99] to provide evidence for the anomalous features of the electronic ensemble in these cuprates. He argued that the electronic transport properties appear to be those of a two-dimensional Luttinger liquid, rather than those of a Fermi liquid. In essence there are two different scattering rates, τ_{tr}^{-1} and τ_H^{-1}

that have to be considered. The longitudinal transport scattering rate is $\tau_{tr}^{-1} \sim T$, as implied by $\rho(T)$ discussed above and for the transverse scattering rate τ_H^{-1}, in this model representing the intrinsic scattering rate for quasiparticle-like excitations of the Luttinger liquid, so called spinons, a T^2 dependence is predicted. If, in addition, impurities are considered, the model predicts that $\cot\Theta_H = aT^2 + C$, where a is a constant and C increases with increasing impurity content. This predicted behavior has actually also been verified experimentally, supporting again the view that the normal state of these cuprates is quite different from that of a common metal.

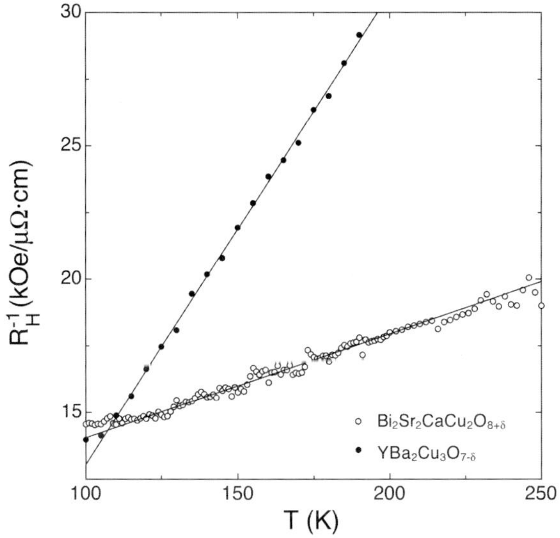

Fig. 14.22. Temperature dependence of the in plane Hall constant of single crystalline $YBa_2Cu_3O_7$ and $Bi_2Sr_2CaCu_2O_8$in the form of R_H^{-1} vs T

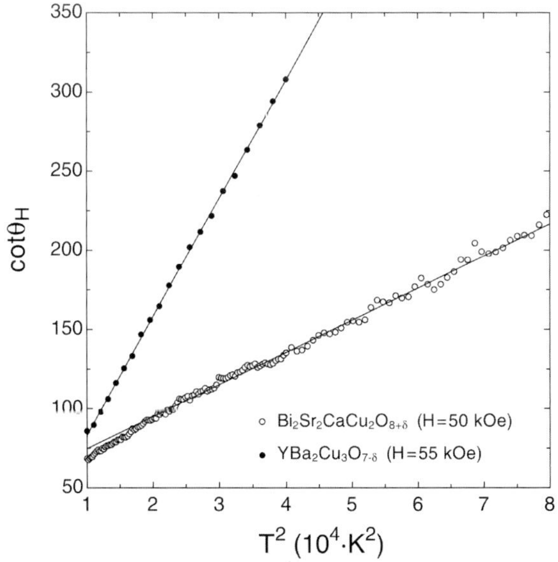

Fig. 14.23. Temperature dependence of the in plane Hall angle Θ of single crystalline $YBa_2Cu_3O_7$ and $Bi_2Sr_2CaCu_2O_8$ in the form of $\cot\Theta$ versus T^2

Magnetic Susceptibility

The temperature dependence of the bulk magnetic susceptibility $\chi(T)$ is very much dependent on the doping concentration. This may well be seen in Fig. 14.24, where $\chi(T)$ of a series of $La_{2-x}Sr_xCuO_4$ compounds with varying x is displayed. While for low doping, $\partial\chi/\partial T$ is positive across the whole covered temperature range, we note the appearance of a maximum in $\chi(T)$ with increasing Sr content x, the maximum shifting to lower temperatures with increasing x. A similar behavior of $\chi(T)$ with varying doping concentration was also observed for the YBCO series, where also in the underdoped regime, a strong decrease of χ with decreasing temperature

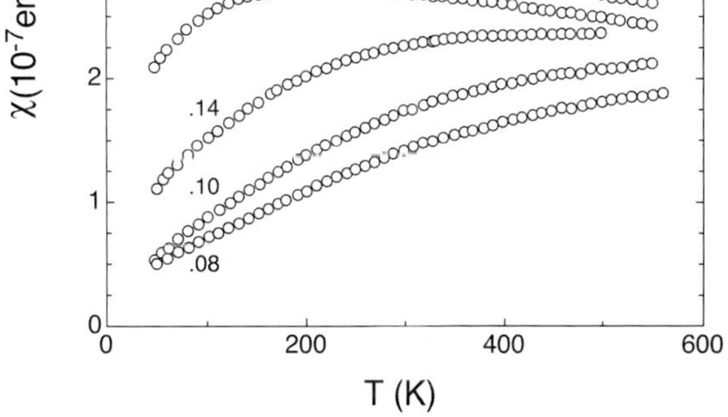

Fig. 14.24. Temperature dependence of the normal state magnetic susceptibility of $La_{2-x}Sr_xCuO_4$ (see [8])

was observed. Naturally there are many ways to compare these experimental data with model calculations but it seems clear that a simple description invoking only a Pauli-type electronic magnetic susceptibility, which is a measure of the electronic density of states at the Fermi energy $D(E_F)$, usually independent of temperature, is not appropriate. The low temperature negative slope $\partial\chi/\partial T$ indicates a loss of magnetic degrees of freedom with decreasing temperature and in retrospect, these measurements of $\chi(T)$ gave the first hint of a phenomenon that is still of high current interest: the opening of a pseudogap in the quasiparticle excitation spectrum, possibly due to the formation of pairs in some incoherent fashion, at temperatures distinctly above T_c. As we shall discuss in more detail below, a lot of additional experimental evidence for the formation of a pseudogap or, in other words, a loss of degrees of freedom in the quasiparticle excitation spectrum, has since been accumulated.

NMR: Knight Shift, Spin-Lattice Relaxation

In common metals the NMR response in the normal state is, to a large extent, dictated by the presence of conduction electrons. First, they provoke a shift of the resonance signal, the well known Knight shift $K = \Delta H/H_0$, where H_0 is the resonance field for the nuclear spins in a non metallic environment, and ΔH is the shift in magnetic field if the resonance is probed at constant frequency. In simple metals, K is given by [100]

$$K = \left(\frac{8\pi}{3}\right) \cdot < |u_k(0)|^2 > \cdot \chi_P \,, \qquad (14.3)$$

hence is related to the spin susceptibility of the conduction electrons

$$\chi_P = 2\mu_B^2 D(E_F) \,, \qquad (14.4)$$

and therefore expected to be essentially temperature independent. Second, the conduction electrons also dominate the spin-lattice relaxation process, leading to a Korringa-type behavior of the relaxation rate T_1^{-1}, such that $(T_1T)^{-1}$ is approximately constant. This so called Korringa constant is related to the Knight shift via

$$(T_1T)^{-1} \sim K^2 \,. \qquad (14.5)$$

It is not at all surprising that distinct deviations from these types of behavior have been identified in NMR studies of the cuprate materials. Complications arise, particularly in the evaluation of the Knight shift, because in structurally and electronically complicated materials, such as the cuprates, this shift is the result of various different contributions, also involving orbital degrees of freedom. Since NMR is a local probe, the response of different nuclei occupying different lattice sites may be probed and all the results have to be interpreted by taking into account the local environment of chosen species of nuclei. For the cuprates,

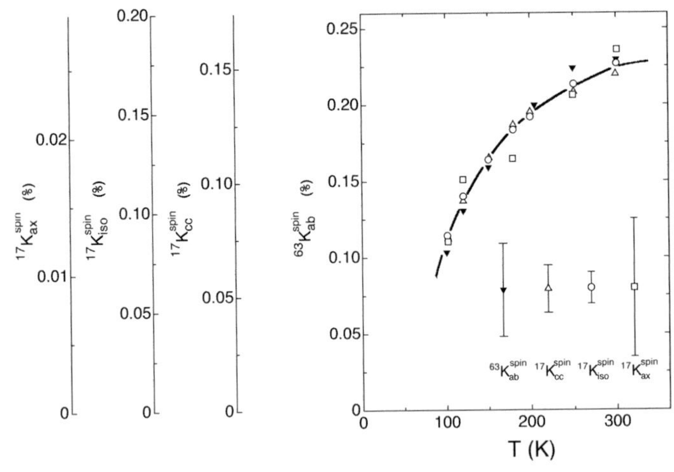

Fig. 14.25. Temperature dependence of various components of the magnetic Knight shift of Cu and O NMR signals in $YBa_2Cu_4O_8$ (see [101])

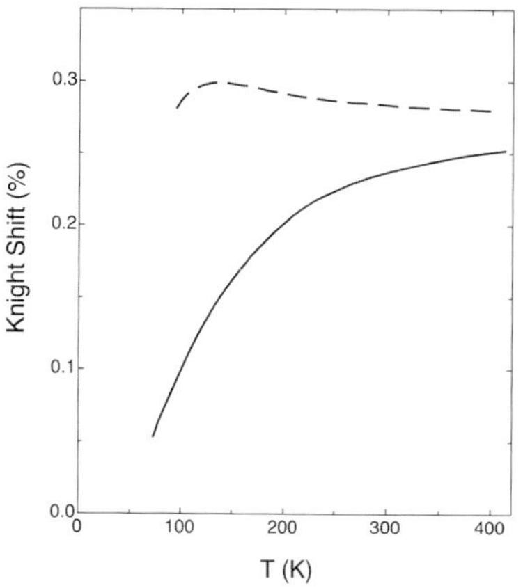

Fig. 14.26. Average temperature dependence of the magnetic shift of the NMR signals probing the nuclei on planar Cu sites of $YBa_2Cu_3O_7$ (*broken line*) and $YBa_2Cu_4O_8$ (solid line) (see [102])

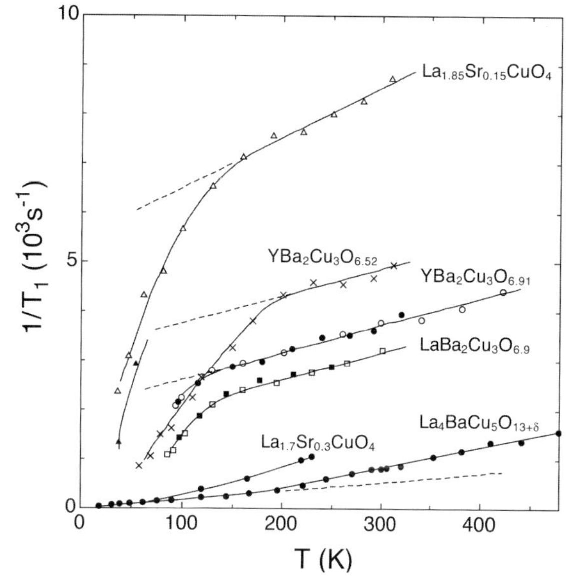

Fig. 14.27. Temperature dependence of ^{63}Cu NMR spin lattice relaxation rates for nuclei on planar Cu sites in different copper oxide compounds. The *solid lines* are to guide the eye and the *broken lines* indicate the Korringa-type variation, depending linearly on T (see [103])

the bulk of data has been obtained from nuclei of Cu and O isotopes.

First of all, there are clear cases where the Knight shift exhibits a strong temperature dependence, not really surprising if we recall the behavior of $\chi(T)$ mentioned above. Examples of different components of $K(T)$ are shown in Fig. 14.25. Here, these K components have been evaluated for Cu and O nuclei in the compound $YBa_2Cu_4O_8$ (YBCO-124) [101]. It is obvious that these different shift components are substantially reduced with decreasing temperature below 300 K, i.e., at temperatures far above T_c. Similar observations were made in NMR experiments probing underdoped YBCO and Bi-2212. The situation is quite different, if K is measured in optimally or overdoped material. In Fig. 14.26 we show this difference schematically for the planar Cu nuclei in YBCO-123 and YBCO-124 [102]. It is now accepted that the strong temperature dependence of K is related with the low doping of the corresponding materials and most likely also reflects the quasi-gap in the excitation spectrum mentioned above.

It turns out that also the different experimental data concerning the temperature dependence of the spin-lattice relaxation rate $T_1^{-1}(T)$ are not compatible with the simplest expectations for a metal. Examples of experimental results [103] for the relaxation rate, measured for ^{63} Cu nuclei on planar Cu sites in various cuprates at temperatures above the respective critical temperatures, are shown in Fig. 14.27. For comparison, the data for superconducting and non superconducting Cu oxides are shown. First we note that the relaxation is much faster in those materials which exhibit high-T_c superconductivity. It may also be seen that the relaxation in the non superconducting compounds is indeed of Korringa-type, at least at low temperatures. This is clearly not the case for the superconducting compounds. At elevated temperatures, the relaxation rate is rather well approximated by $T_1^{-1} = D + cT$, indicating at least two relaxation channels. While the second contribution may be interpreted as representing some sort of Korringa relaxation, the first and temperature independent term has to be associated with another contribution, most likely due to the dynamics of localized d-electron spins on the Cu atoms. If the Korringa-type term is subtracted from the experimental T_1^{-1} values, the resulting d-electron contributions, shown in Fig. 14.28, are obtained [103]. A temperature independent relaxation rate may be related with fluctuations of local-

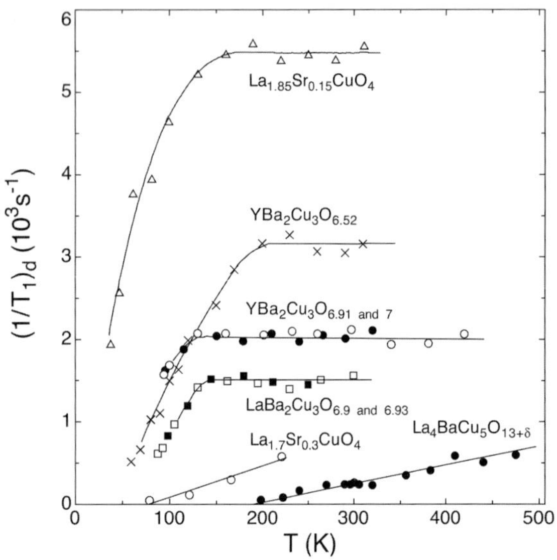

Fig. 14.28. Same data as in Fig. 14.27, after subtraction of the Korringa type contribution varying linearly with temperature, as indicated by the *broken lines* in Fig. 14.27. These data are interpreted as representing the relaxation due to Cu d electrons. The *solid lines* are to guide the eye (see [103])

ized spins but may also be accounted for by spin-density fluctuations in relation with a 2D Fermi surface. Fig. 14.28, however, reveals that also this contribution suffers from a temperature dependent reduction, starting well above the corresponding critical temperatures, at least in some cases. It seems quite reasonable to ascribe this feature again to the presence of the pseudo gap discussed above. By now it is clear that the onset temperature of this gap formation tends to grow with decreasing doping.

Specific Heat

The outstanding problem in connection with the specific heat of copper oxide superconductors in the normal state is the evaluation of the electronic contribution to the total specific heat $C_{\text{tot}}(T)$, which in the range of T_c is completely dominated by the specific heat due to lattice excitations. The lattice specific heat C^{ph} (T) may be estimated by measurements of C_{tot} (T) of one particular material in external magnetic fields or, in an attempt to only change the electronic contribution, by slightly changing the chemical composition, such as to vary the doping level. In both cases it has to be assumed, that C^{ph} (T) is unaffected by these measures, not easy to verify in a rigorous way. The electronic part C^{el} (T), i.e., $(C_{\text{tot}} - C^{ph})(T)$ is usually assumed to be proportional to T or, in other words, it is assumed that the ratio C^{el}/T is a good measure for the electronic specific heat parameter γ. For the formal representation of C^{ph} (T), various approximation procedures have been suggested and a number of relevant articles discuss this issue [104, 105]. For optimally doped YBCO-123 an average and reasonable value for the electronic parameter in the normal state is $\gamma = 25$ mJ/moleK^2. For Bi-based copper oxides, the corresponding molar values for γ are somewhat smaller but still of the order of 10 mJ/moleK^2. As we shall see below, the situation is more complicated in underdoped material where C^{el}/T is not a constant in the normal state. In comparison with simple metals, the γ value quoted above seems fairly large but still, the corresponding density of electronic states $D(E_F)$ is not exceptionally high in order to justify the high values of T_c in a straightforward manner.

A common way to estimate the γ value is based on the weak coupling BCS relation [106]

$$\frac{(C_s^{\text{el}} - C_n^{\text{el}})(T_c)}{C_n^{\text{el}}(T_c)} = 1.43\,, \tag{14.6}$$

where $C_n^{\text{el}}(T_c) = \gamma \cdot T_c$, and experimental data concerning the specific heat anomaly at T_c . Since this approach is valid only for mean-field type transitions and the weak coupling limit, most of these calculations may serve as order of magnitude estimates at most. As we shall see below, none of the transitions in high-T_c cuprates is really of mean-field type and in most cases the characteristics of the transition are really quite different, mainly because of rather strong fluctuation effects. In addition and complicating the situation, strong coupling effects are indicated by experimentally determined ratios of the Δ and T_c and therefore the use of Eq. (14.6) for estimating the γ parameter is more than questionable.

Characteristics of the Normal State

From this brief compilation of some anomalies that are encountered in the results of measurements of some selected physical quantities, it may safely be concluded, that very simple approaches, usually employed for the description of common metals, are inadequate for treating the features of these cuprate materials. Various theoretical attempts to cope with these problems are the subject of other chapters in this treatise and are discussed in more detail there. Some of them start with the notion that the conventional Fermi liquid model for describing the metallic normal state has to be abandoned here and that other quasiparticle-like excitations have to be introduced. Many of the experimental facts mentioned above and others, not considered here, seem to justify this view. What is particularly hard to grasp, however, is the fact that the character of these excitations seems to vary with the doping level. There is not much doubt that a common quasiparticle picture is adequate for describing the properties of overdoped materials but it is still much debated how the underdoped regime should be treated. Here, the appearance of the often quoted pseudogap, to be discussed in some more detail in the next subsection, adds another complication to the state of affairs. If finally it should turn out that the electronic properties across the entire doping range where superconductivity occurs may still be understood on the basis of the Fermi liquid model, a lot of additional input concerning details of that model is probably needed.

Pseudogap, Additional Evidence

In this subsection we are briefly discussing additional experimental evidence for the pseudogap in the normal state of cuprate superconductors. We have chosen results of experiments whose interpretation leads to conclusions that are somewhat controversial. This should serve the purpose to indicate that these issues have not yet been settled completely. To set the stage we display a schematic phase diagram for the cuprate superconductors, where also the doping dependence of T^*, the temperature below which experimental evidence for the presence of a pseudogap exists, in Fig. 14.29.

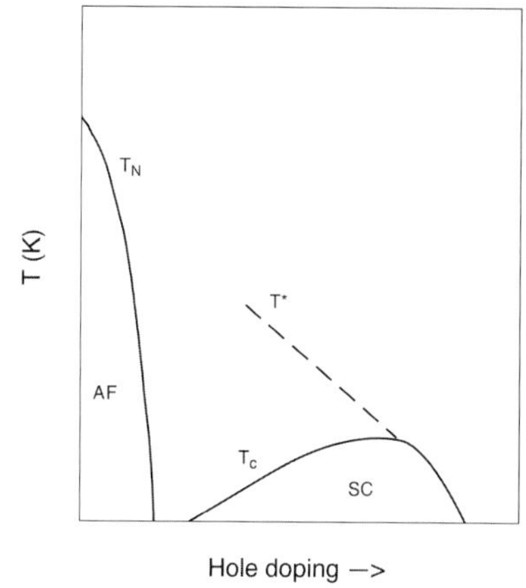

Fig. 14.29. Schematic low temperature phase diagram of cuprate superconductors. Emphasized are the antiferromagnetically ordered phase at low doping, the superconducting phase and the approximate location of $T^*(x)$, indicating the pseudogap formation

First we mention the results of specific heat measurements [107], covering the temperature range above and below T_c of a series of samples of $Y_{0.8}Ca_{0.2}Ba_2Cu_3O_{7-\delta}$ with different oxygen contents, i.e., varying parameter δ. This choice of material allows to extend the overdoped regime for YBCO-123 type material; optimal doping conditions are reached for $\delta \sim 0.32$. In Fig. 14.30 we show the temperature dependence of the electronic specific heat of different samples in the form of C^{el} /T. The electronic contribution has been obtained by subtracting the lattice contribution from the raw experimental data in an apparently controlled way [108]. In Fig. 14.30a it may be seen that in the normal state of overdoped material, C^{el}/T is temperature independent above the respective critical temperatures and also does not depend on δ. The upturns, appearing somewhat above T_c, most likely reflect fluctuation contributions to C^{el}. At this point of our discussion we do not address the temperature dependence of C^{el} at very low temperatures, leaving this point to be addressed in another section below. The situation is completely different for underdoped material. Here, as may be seen in Fig. 14.30b, the different C^{el}/T versus T plots start to diverge above T_c in a manner that C^{el}/T decreases with decreasing T if $\delta > 0.5$, thus indicating a loss of entropy already above the onset of superconductivity, distinctly different from the behavior displayed in Fig. 14.30a. An additional clear difference in the (b) panel of the figure is the sudden drop in the peak height of C^{el}/T for $\delta > 0.32$, although no corresponding large change in T_c is observed. In [109], a detailed discussion of the temperature dependence of the related entropies and of similar data, obtained for a series of $La_{2-x}Sr_xCuO_4$ compounds, leads to the conclusion that the pseudogap phenomenon should not be regarded as being due to a precursor formation of Cooper pairs and therefore is not related with the gap opening in the superconducting state. In particular it is argued that the pseudogap related loss of low-energy normal state spectral weight actually weakens the superconducting condensate. To some extent this is compatible with the fact that contrary to the decrease of T_c with decreasing doping, the onset temperature T^* increases.

A quite different view has been obtained from ARPES experiments on single crystals of underdoped Bi-2212. With this type of experiments it has been possible to map the Fermi surface of this type of materials and, because of much improved energy resolutions, to verify the onset of the gap related with the superconducting ground state. A detailed presentation and discussion of ARPES experiments on high-T_c superconductors is given elsewhere in this compendium. Nevertheless we note here that measurements of this type of material mentioned above and as a function of temperature reveal that a gap signature at E_F in different directions of k space appears at different temperatures. In this sense, the data indicate a gradual loss of Fermi surface with decreasing temperature, starting well above T_c [110]. The evolution, ending in the gap features of the superconducting state below T_c appears to be smooth and therefore, it is concluded that the gap formation in the normal state is most likely due to a precursor pairing of quasiparticles, i.e., definitely related with the appearance of superconductivity below T_c [111]. A schematic view of the temperature dependent shrinking of the Fermi surface in k-space is displayed in Fig. 14.31, indicating that the pseudogap in the normal state and the finally resulting gap of the superconducting state reveal the same anisotropy in k-space. A k-dependent incoherent pairing is, of course, not compatible with conventional superconductivity, where the instability of the whole Fermi surface occurs at once at T_c. Results of the same kind of experiments on some overdoped material were interpreted as to confirm the previously established view of the absence of a pseudogap above T_c [88].

Traditionally, various kinds of tunneling experiments have been instrumental for qualitative and quantitative measurements of gaps in electronic

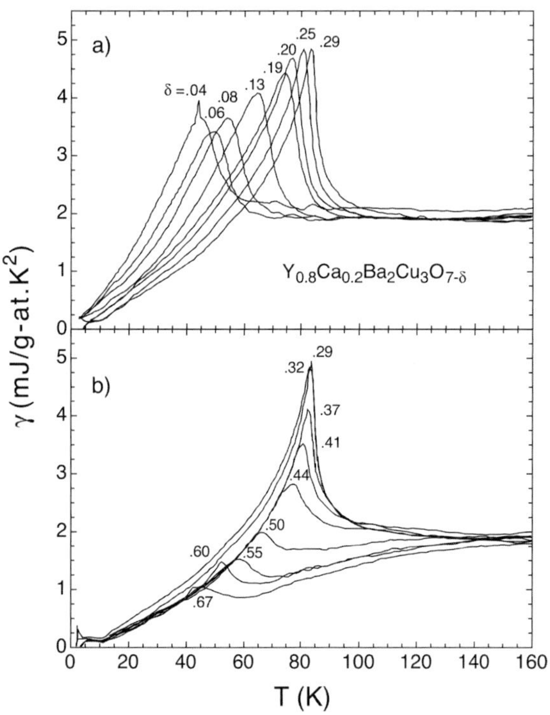

Fig. 14.30. Temperature dependence of the electronic specific heat of $Y_{0.8}Ca_{0.2}Ba_2Cu_3O_{7-\delta}$ in the overdoped (**a**) and underdoped (**b**) regime (see [107])

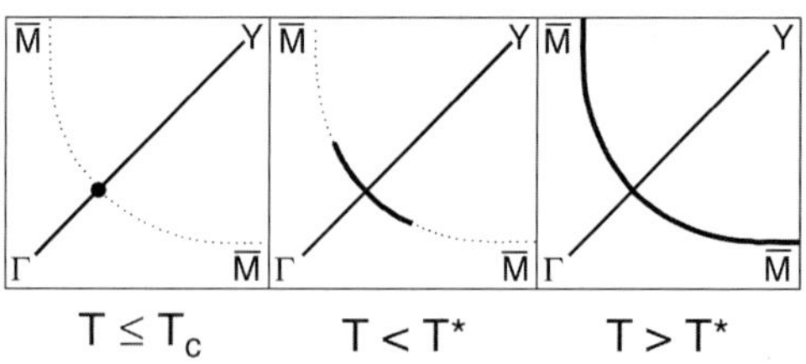

Fig. 14.31. Schematic representation of the growth of the Fermi surface of Bi-2212 with increasing temperature, as obtained from ARPES experiments (see [110])

quasiparticle excitation spectra, especially in connection with superconductivity [112]. With respect to the present problem we quote the results of some recent measurements employing the technique of scanning tunneling microscopy (STM), providing the possibility of local tunneling experiments. Information about gap features is obtained from measurements of the tunneling current I as a function of the applied voltage V and by plotting the tunneling conductance $\partial I/\partial V$ versus V. The voltage dependence of the conductance is, to some extent, representing the energy dependence of $D(E_F)$, the density of states for quasiparticle excitations at the Fermi energy. Early STM experiments [113], again on single crystals of Bi-2212 compounds, confirmed the presence of a pseudogap in the normal state, but now it was claimed that this feature was present for *underdoped and overdoped material* [114] and that the temperature dependence of the experimental $\partial I/\partial V$ versus V curves was such that the pseudogap had to be a precursor of the genuine superconducting gap. More recent experiments on the single Cu-O layer material Bi-2201 confirmed this view [115]. In Fig. 14.32 we show the reported results of the temperature evolution of conductance versus voltage curves for overdoped $Bi_{2.1}Sr_{1.9}CuO_{6+d}$. It may well be seen that the reduction of the conductance at low energies that is believed to indicate a gap feature persists to temperatures well above the critical temperature of the bulk. An intimate relation between the two gaps in either the normal or the superconducting state is also deduced from the fact that the gap width is more or less temperature independent [113]. In addition it has been pointed out that the gap-width variations from one material to the other is the same for the superconducting and the normal state, again suggesting a common origin of the two features. The data imply a filling of the gap with increasing temperature, rather than the commonly expected narrowing.

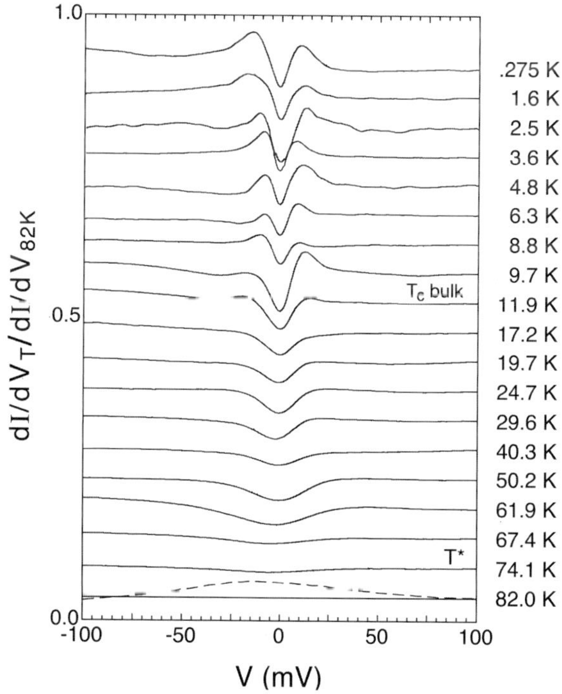

Fig. 14.32. Temperature dependence of the normalized conductance vs. voltage of overdoped Bi-2201, as measured by STM. The *broken line* represents the background at 82 K to which the curves are normalized. The gap feature persists to temperatures substantially exceeding T_c (see [115])

As pointed out above, the STM results seem to confirm some of the conclusions that are drawn from the ARPES data but in addition and unlike a number of other experiments, they claim the persistence of the normal state pseudogap well into the overdoped regime.

14.4.2 Superconductivity

General Remarks

The unexpected occurrence of superconductivity in these cuprate materials at unprecedented high temperatures almost immediately raised questions with regard to possible mechanisms that might trigger these transitions. Most researchers would agree that this issue is not yet definitely settled, as also the theoretical chapters of this treatise may demonstrate. The main controversy addresses the question whether lattice degrees of freedom are in any way involved in the occurrence of superconductivity in copper oxides or whether this phenomenon is entirely based on interactions within the, however anomalous, electronic subsystem. Early flux-trapping experiments [116] made it clear, at least, that the objects responsible for the superconducting state carry a charge of twice the charge of a single electron, i.e., $2|e|$, just as the Cooper pairs in conventional superconductors. As has long been known, of course, pairing states need not be of the simplest variety with zero orbital momentum and a spin singlet configuration. Although of little interest at first, the pairing configuration became an issue in the wake of some theoretical debates, and was finally addressed experimentally by a number of different types of measurements. Another phenomenon that is intimately related with superconductivity and is not observed for normal conductors, is the screening of static external magnetic fields H from the interior of a singly connected superconducting specimen, i.e., the decay of H over a characteristic length, the London penetration depth λ_L. Early experiments using muon-spin rotation (μSR) techniques established beyond any doubts that also for the cuprate superconductors this field expulsion, the Meissner–Ochsenfeld effect, is present [117].

The major difficulty, both in experiment and in theory, however, is still the unequivocal identification of the objects that are to be paired. At least part of our discussion concerning the normal state properties have made it clear that these objects are not yet really well defined. Because of the structural characteristics of the cuprates and the corresponding anisotropies, a large part of the discussion of their superconductivity concentrated on the properties of the Cu-O planes and the possible influence of the two-dimensionality of these systems. In what follows we try to achieve a dense presentation of some of the most significant aspects concerning superconductivity in Cu oxide materials, now concentrating on the characterization of the superconducting state. A lot of experimental evidence has been accumulated that the superconductivity of cuprates is of unconventional nature. Above we have already emphasized the importance of the concentration n_c of itinerant charge carriers, dictated by the degree of doping the electrically insulating parent compounds and its influence on the magnitude of the critical temperature. That the behavior of T_c (n_c) is, to some extent, quite universal, as may be seen in Fig. 14.33, has been pointed out by Uemura [118]. The magnitude of T_c is, to a large extent, also dictated by the number n of Cu-O planes per unit cell. It turns out that for one class of materials, T_c may be enhanced by enhancing n, but

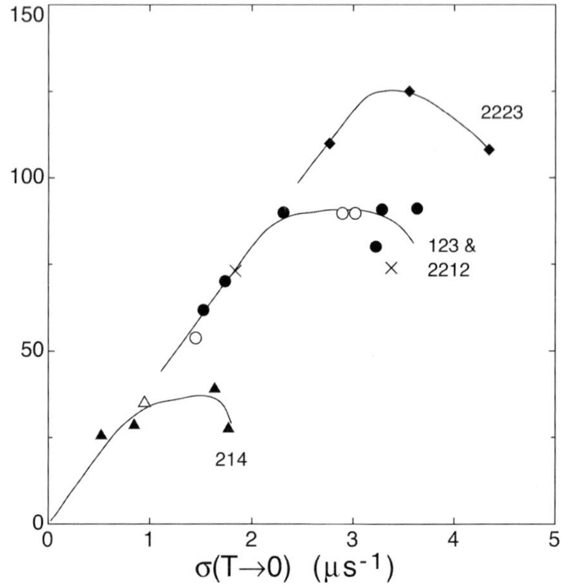

Fig. 14.33. The variation of the critical temperature T_c of cuprate superconductors vs the low temperature μSR relaxation rate σ. The *closed and open triangles* are for La-214 compounds, the *closed and open circles* are for Y-123 material, the *crosses* represent data of Bi-2212 material and the *closed diamonds* are for Bi-2223 compounds. The *solid lines* are to guide the eye (see [118])

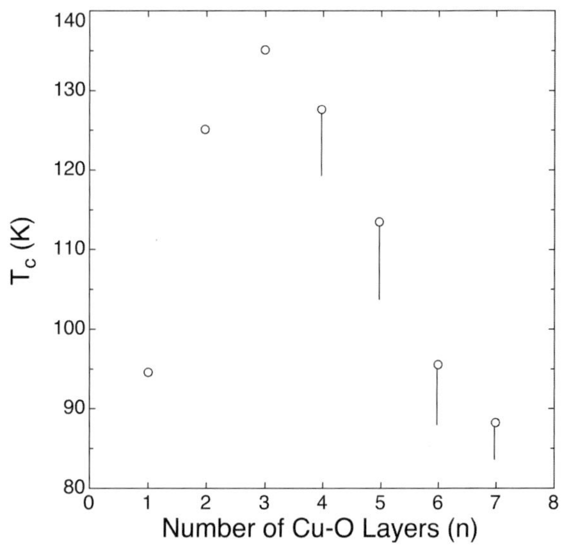

Fig. 14.34. Variation of the critical temperature T_c of mercury based superconductors $HgBa_2Ca_{n-1}Cu_nO_{2n+2+\delta}$ as a function of n, the number of Cu-O planes (see [39])

this trend is usually only obeyed up to $n = 3$. This latter trend has very nicely been demonstrated for Hg-based cuprates, as may be seen in Fig. 14.34 [39].

Inspecting the behavior of electron or hole doped bulk cuprate superconductors, it seems obvious that hole doping of insulating cuprates results in significantly higher values of T_c than electron doping. This naturally forces the question in what way the degree of the occupation of electronic states contributing to electronic conduction influences the magnitude of the critical temperature for the onset of superconductivity in general. In this context it is at least amusing to note that even in the case of superconductivity of common metal elements most of them have a normal-state Hall constant with a positive sign, which is typical for hole dominated conductivity. A theoretical model, emphasizing "hole superconductivity", has been promoted by Hirsch and coworkers [119].

As most non-elemental superconductors, the cuprates are type II superconductors with extremely large ratios H_{c2}/H_{c1} between the upper and the lower critical fields or, in other words, resulting in a very large value of the Ginzburg–Landau parameter $\kappa = \lambda/\xi$. This may qualitatively be recognized from magnetization curves $M(H)$, such as the one displayed in Fig. 14.35 for YBCO-123 [120]. Taking into account that by definition

$$H_{c1} = \left(\Phi_0 4\pi \lambda_L{}^2\right) \ln \kappa \tag{14.7}$$

and

$$H_{c2} = \frac{\Phi_0}{2\pi\xi^2}, \tag{14.8}$$

the shape of the curve readily implies that the penetration depth λ_L for magnetic fields is much larger than the superconducting coherence length ξ. For various reasons, the experimental evaluations of these critical fields of cuprate superconductors are not very precise or even impossible. In particular, H_{c2} is often regarded as a crossover field only, without manifesting a true phase transition. Because of these uncertainties we refrain from presentations and discussions of critical field data and their temperature dependence. However, both the magnetic-field penetration depth λ_L and the coherence length ξ may be evaluated with an adequate certainty from $M(T, H)$ data [121,122]. It turns out that these two characteristic lengths are very anisotropic, of no surprise when considering the structural properties of these materials. This also means that the magnitudes of the critical fields for magnetic-field orientations parallel or perpendicular to the Cu-O planes are very different. In most cases, the in-plane components of the two lengths that are of interest. Since the values for the in-plane penetration depths are between 1000 and 3000 Å for most of these cuprates [123], it is clear that the in-plane coherence lengths are of the order of a few in-plane interatomic distances, i.e., extremely short. The situation is even more extreme for the direction perpendicular to the Cu-O plane. Here, λ_L may of the order of a few 10^4 Å and ξ is barely as long as an out of plane interatomic distance or even less. These anisotropies, often expressed by the ratio $\Gamma = (m_c/m_{ab})^{1/2}$ between the effective masses entering the penetration depth equation, and the short coherence lengths create a situation that is very different from that encountered in conventional superconductors and, as we shall see, provoke special features in some

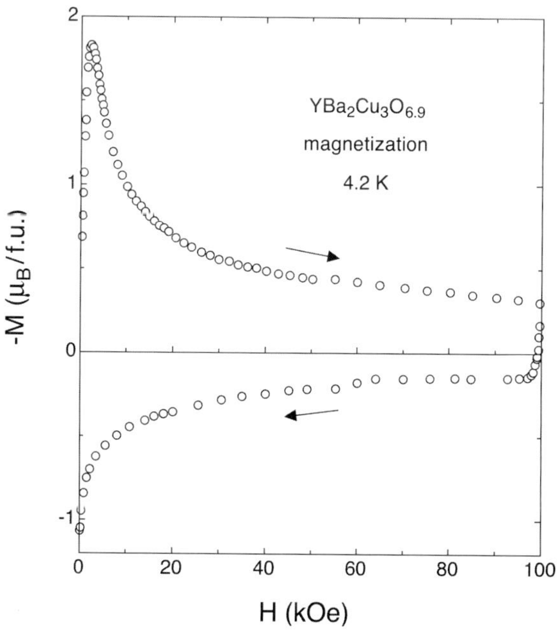

Fig. 14.35. Negative magnetization of a ceramic sample of superconducting $YBa_2Cu_3O_{7-\delta}$ as a function of applied magnetic field up to 100 kOe. The *arrows* indicate the direction of the field variation

of the observations to be discussed below. Because of the short coherence lengths, the coherence volume is rather small and may host only a low number of Cooper pairs. This situation favors large fluctuation effects; their importance is discussed in a special chapter of this book.

Pressure Effect, Isotope Effect

So far we have only considered T_c variations induced by changes of the chemical composition of the materials. Other external influences, apart from applying external magnetic fields, that may affect the critical temperature, are the application of external pressure or replacing part of the atoms with other isotopes of the same element. Historically, the strong enhancement of T_c of the first La-based copper oxide superconductors of the 214 family upon applying external pressure [124] led eventually to the discovery of the YBCO-123 superconductor, by following up the idea that the smaller ionic radius of the Y^{3+} ions might enhance the internal pressure of La^{3+}-based compounds and thus lead to higher critical temperatures [5]. Since the reduction of the volume by exerting external pressure may affect the spectra of both the lattice and the electronic excitations, direct and chemical pressure may not always induce equivalent variations of physical properties. In most common superconductors, external pressure usually reduces T_c and this trend is interpreted as being due to a pressure induced weakening of the electron phonon coupling constant λ_{ph} which in essence determines the magnitude of T_c of materials where superconductivity is due to the electron phonon interaction. It has been shown [125] that to a good approximation

$$\lambda_{ph} = \frac{A}{(M\Theta_D^2)}, \qquad (14.9)$$

where A is a constant for a given crystal structure, M is the ionic mass and Θ_D is the Debye temperature. The pressure induced stiffening of the lattice and the corresponding enhancement of Θ_D thus account for the observed decrease of T_c with increasing pressure. Exceptions from this trend of $\partial T_c/\partial p < 0$ are only a few, the most prominent being that of La metal where T_c increases considerably with increasing pressure [126], and always have been considered as indicating some anomalous pressure response of the electronic subsystem. Thus, in relation with common superconductors, the generally negative influence of external pressure on T_c is usually interpreted as confirming the importance of the electron phonon interaction for triggering the phase transition.

For optimally doped Cu-O superconductors, the pressure effect on T_c is, in most cases, negligible. Prominent exceptions are YBCO-124 and Hg-1223, the compound with the highest yet achieved critical temperature under ambient pressure conditions. An unusually strong enhancement of T_c from 80 K at ambient pressure to 108 K at 10 GPa has been reported for the YBCO-124 compound [127] and was interpreted as being due to a sizeable shortening of the Cu2-O1 bond, leading to a charge transfer into the Cu-O planes. Also for Hg-1223, the application of external pressure raises T_c considerably. Although different claims for the T_c enhancement, up to 160 K at pressures between 150 and 300 kbar, have been made [128–130], it is undisputed that the pressure induced enhancement is significant and can apparently not be achieved by means of variations of the chemical composition. It has been argued that the three Cu-O planes per unit cell of this compound are not all optimally doped at ambient pressure and that the situation can only be improved by a pressure induced transfer of charge carriers between the planes, not achievable by chemical means. This automatically means that the large positive value of $\partial T_c/\partial p$ is most likely not a lattice effect and certainly cannot be used to make a case for a significant influence of an electron phonon interaction in the creation of a superconducting state.

Much more clean in regard of identifying the significance of lattice effects being involved in the superconductivity of any material is the controlled exchange of isotopes of some or all the atoms involved in the chemical composition of a superconductor, because the different ionic masses will only affect the lattice properties of the material. Indeed, measurements of the isotope effect on T_c gave the first unequivocal experimental evidence for the importance of the lattice dynamics on superconductivity in simple metals [131, 132]. For elements the situation is particularly simple because

$$T_c \sim M^{-\alpha} \quad (14.10)$$

and in first approximation [133] it is expected that $\alpha = 1/2$. This result has been verified in a number of cases [134], but deviations from this most simple behavior, the most drastic for α-U, implying negative values of α [135], have been identified even for elemental superconductors. The interpretation of isotope effect experiments is increasingly more complicated, if the chemical composition of a material involves four or more constituents, as is the case for many high-T_c superconductors. Nevertheless, the presence of an isotope effect on T_c almost inevitably indicates the influence, even if of limited significance, of the crystal lattice on the mechanism triggering the superconductivity.

Reliable results on the isotope effect on T_c of cuprates are difficult to achieve. This is because apart from exchanging the chosen isotope, any other variations in the composition of the material and its structural properties must be avoided. In copper oxide compounds, the element of choice for a controlled isotope exchange is, of course, oxygen. Careful work in exchanging ^{16}O with ^{18}O and back has been made on $La_{2-x}Sr_xCuO_4$ with $0.1 \leq x \leq 0.15$ [136]. From these experiments it is concluded that the compound with ^{18}O exhibits a distinctly lower T_c than if the oxygenation is accomplished with ^{16}O. While the analysis of the measurements indicates that there is no isotope effect on the charge carrier concentration, as one might expect, a significant shift of the effective mass of the carriers, which are identified as being of polaronic character, is deduced. Since similar results have been obtained in a number of other analogous investigations [137, 138], it has to be accepted that at least in some of the high-T_c superconductors, the isotope effect is not negligible.

More recently it has been claimed that isotope variations are also affecting the onset temperature T^* of the pseudogap formation discussed above. The corresponding experiments involved the measurements of linewidths of crystal field excitations probed by inelastic neutron scattering on $HoBa_2Cu_4O_8$ [139]. While the isotope effect on the critical temperature is, in accordance with other work, found to be rather small with $\alpha \approx 0.05$, the variation of T^* with the exchange of ^{16}O and ^{18}O is considerable and of opposite sign, i.e., T^* of the ^{18}O variety of the compound is distinctly higher than T* of the ^{16}O material. The corresponding α parameter in $T^* \sim M^{-\alpha}$ is therefore negative. This again seems to indicate that the lattice dynamics are, to a certain extent, involved in the formation of the pseudogap of the normal state. It may well be that understanding the significance of the sign difference in the α parameters for T_c and T^* will also clarify the relation between the pseudogap of the normal state and superconductivity below T_c. In view of these experimental results it seems difficult to deny that superconductivity in copper oxides involves, to some extent, the lattice dynamics of these materials. In this regard it has been argued [140] that these lattice dynamics lead to structural inhomogeneities in these materials, better known under the names of phase separation and stripe formation. Microscopic evidence for stripe formation in some of these materials has definitely been obtained [141, 142], but it is not yet clear to what extent this phenomenon is essential for the occurrence of superconductivity at elevated temperatures. In interpreting experimental data obtained with different techniques it should be kept in mind that the process of stripe formation may also be highly dynamical in the sense that the structural inhomogeneities are not static. This might explain some controversies that arise in comparing experimental data from measurements that probe in different energy or time windows. It seems clear that structural inhomogeneities will also generate electronic inhomogeneities and it has not yet been worked out how these have to be taken into account in the interpretation of some of the anomalous features of the normal state properties discussed above.

Phase Transition Aspects

The short coherence lengths, even within the Cu-O planes, of the cuprate superconductors, play an essential role in all experimental verifications of the superconducting phase transition. This is mainly because even in zero magnetic field, fluctuation effects are non negligible to an extent that distinct deviations from the mean field behavior, usually observed for clean common superconductors, are observed. The fundamental parameter governing the strength of these fluctuations is the Ginzburg number [143]

$$Gi = \frac{\left[\frac{T_c}{H_c^2(0)\xi_{ab}^2\xi_c}\right]^2}{2}, \qquad (14.11)$$

where ξ_{ab} and ξ_c are the zero temperature coherence lengths parallel and perpendicular to the Cu-O planes, respectively. This number, which is obviously large if the ξ values are small, fixes the temperature range around the critical temperature where fluctuation effects are significant, i.e., critical, via $|T - T_c| < T_c \cdot Gi$. For common superconductors, Gi is of the order of 10^{-6} or less and therefore, deviations from mean field behavior are usually not important. The situation is completely different in cuprates and further complications arise, because of quasi two-dimensional fluctuation effects in particularly anisotropic cuprate systems. In that case the relevant Ginzburg number is [143]

$$Gi^{2D} = \frac{T_c}{\sqrt{2} \cdot \varepsilon d}, \qquad (14.12)$$

where $\varepsilon = (\Phi_0 / 4\pi\lambda)^2$ and d is the thickness of one layer. Fluctuation effects in the temperature dependences of the electrical resistivity $\rho(T)$ and the magnetization $M(T)$ are somewhat annoying because they mainly hamper an exact evaluation of the zero field critical temperature. In Fig. 14.36 we show an example of $\rho(T)$ of Bi-2212 which demonstrates the gradual decrease of ρ above T_c. A thorough discussion of fluctuation effects on the electrical transport is beyond the scope of this more general review. Instead we briefly consider the temperature dependence of the specific heat $C(T)$ in zero magnetic field and in the vicinity of T_c. As an example we first present the result of a relevant experiment on single crystalline YBCO-123 in Fig. 14.37, revealing the anomaly in $C(T)$ at the transition [144]. As may be seen from the scale on the left vertical border of the figure, this anomaly rides on a large background which at temperatures above the boiling point of liquid nitrogen, is due to the specific heat of the lattice. In order to achieve the experimental resolution that is demonstrated here and which is necessary to obtain any reliable comparison with theoretical models, special methods of measurement have to be involved. One particular approach, a differential thermal analysis (DTA) adapted for measurements on small single crystals, is described in [145]. From the data close to T_c both in the normal and the superconducting state, clear deviations from ideal mean field behavior may be recognized. The solid line in Fig. 14.37 is a fit on the basis of the 3D XY model for phase transitions [146]. A detailed discussion of the relevance of this model with respect to the critical regime of the zero field phase transition of cuprate supercon-

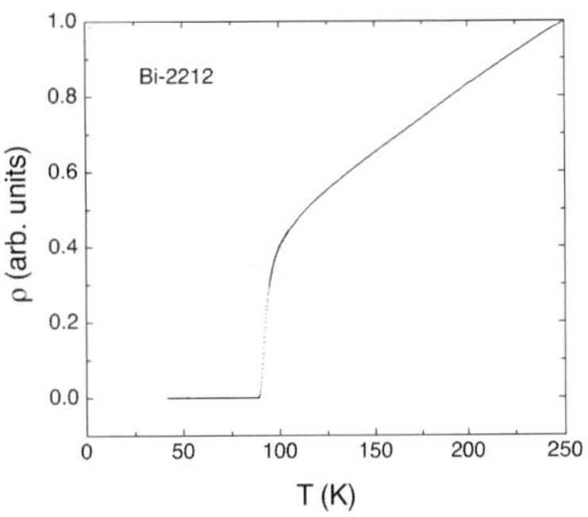

Fig. 14.36. Temperature dependence of the electrical resistivity of Bi-2212. Note the growing slope of $\partial\rho/\partial T$ far above T_c

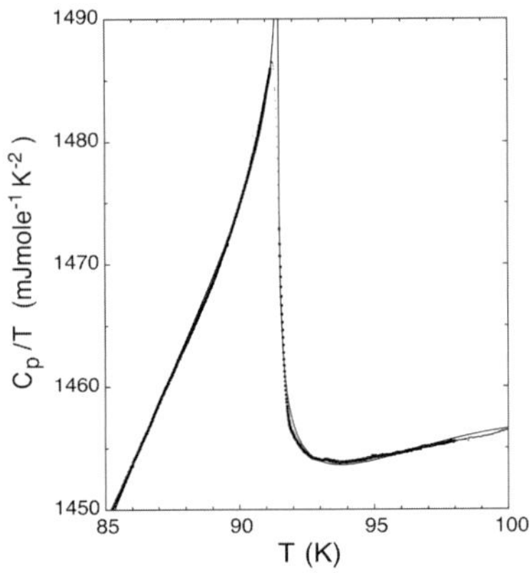

Fig. 14.37. Specific heat anomaly at the superconducting transition of single crystalline $YBa_2Cu_3O_{7-\delta}$ in zero magnetic field. The *solid line* is a fit based on the 3D XY model, as explained in the text

ductors has been given in [147]. Here we only note that the experimental data can reasonably well be described on the basis of this model. The analysis is by no means straightforward. The universal temperature dependence $C(T)$ is given by

$$C(T) = C_{\text{lattice}}(T) + \Delta C + A^{-} \ln \left| \left(\frac{T}{T_c}\right) - 1 \right|, \quad \text{for } T < T_c$$
$$= C_{\text{lattice}}(T) + A^{+} \ln \left| \left(\frac{T}{T_c}\right) - 1 \right|, \quad \text{for } T > T_c . \tag{14.13}$$

The parameters A^+ and A^- are related via the universal ratio $A^+/A^- \cong 1.05$. First of all, the lattice part of the specific heat has to be evaluated with some degree of confidence. This can be achieved by additional measurements in external magnetic fields. Second it has to be tested to what extent the values of the fit parameters depend on the temperature window chosen for the fits. In the case presented here, the fit quality suggests that the critical regime appears to be extended over a range of ± 8 K around T_c. Without going into further details, we note that attempts with fits on the basis of other approaches for dealing with fluctuations in the critical region around T_c gave less convincing results.

As may be expected, the specific heat anomaly is considerably reduced upon the application of an external magnetic field. This is demonstrated in Fig. 14.38, where $C(T, H)$ data, measured on the same YBCO-123 sample in various external magnetic fields, oriented perpendicularly to the Cu-O planes, are displayed. In view of the success in describing the critical fluctuations on the basis of the 3D XY model in zero field, it seems justified to test a related scaling law that is inherent in this model for non zero magnetic fields. The relevant scaling relation is

$$[C(T, H) - C(T, 0)]\, H^{\frac{\alpha}{2\nu}} = \mathcal{C}\left(\left(\frac{T}{T_c} - 1\right) H^{\frac{-1}{2\nu}}\right), \tag{14.14}$$

where $\mathcal{C}(x)$ is the unknown scaling function [148]. The parameters $\alpha \sim 0.007$ and $\nu \sim 0.669$ are the 3D XY critical exponents of the specific heat and the coherence length, respectively. In Fig. 14.39, this scaling is tested by plotting the appropriately scaled quanti-

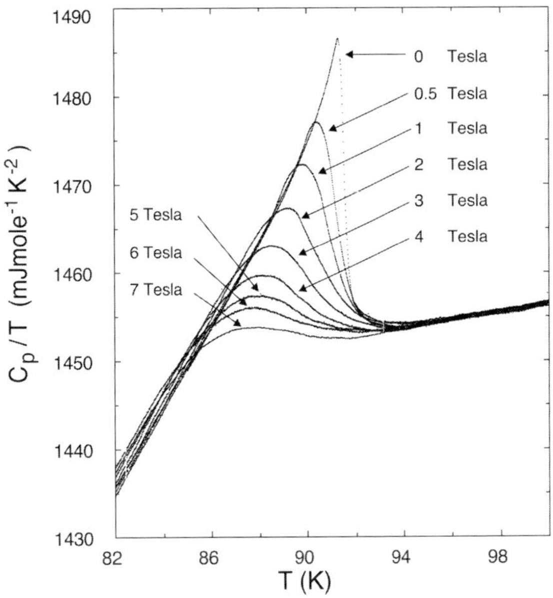

Fig. 14.38. Influence of external magnetic fields on the specific heat anomaly at the superconducting phase transition of single crystalline $YBa_2Cu_3O_{7-\delta}$

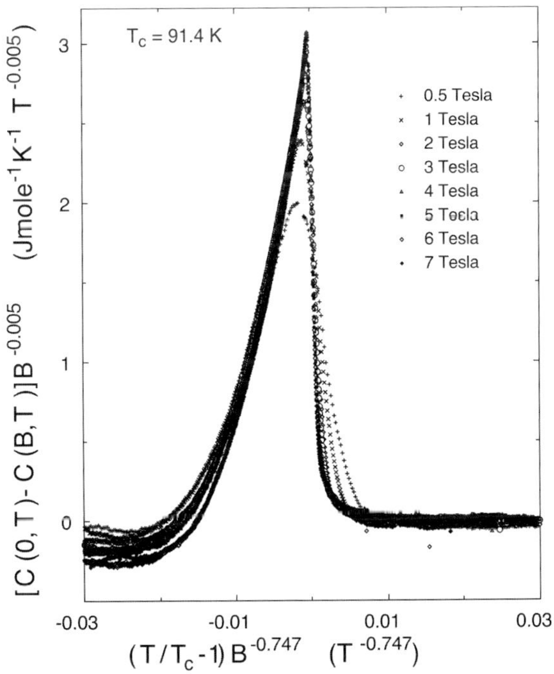

Fig. 14.39. Specific heat anomalies of $YBa_2Cu_3O_{7-\delta}$ plotted in variables that are consistent with 3D XY-scaling (see Eq. (14.14)). $B = \mu_0 H$

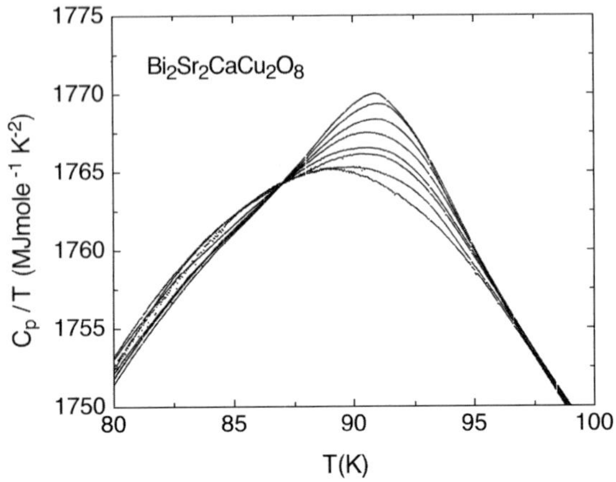

Fig. 14.40. Specific heat of textured $Bi_2Sr_2CaCu_2O_8$ in the temperature range around $T_c = 91$ K in different magnetic fields. The plots $C_p(T)/T$ are for $\mu_0 H = 0, 0.1, 0.5, 1, 2, 3, 4$ and $7T$ from top to bottom. Note the crossing point of all the curves at $T \approx 87$ K

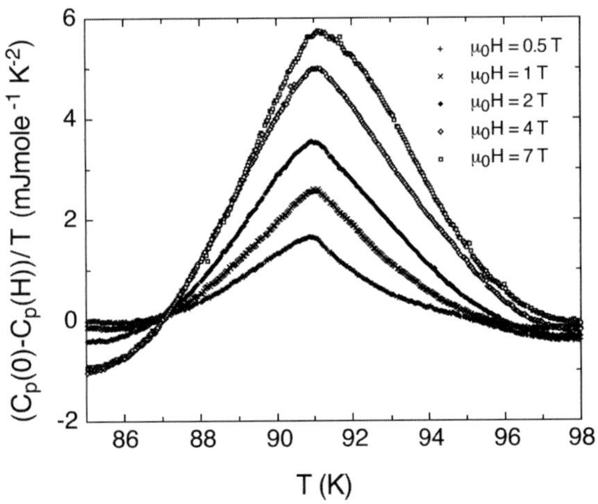

Fig. 14.41. Magnetic field dependence of the specific heat anomaly of superconducting Bi-2212 if plotted in the form $[C_p(0) - C_p(H)]/T$

ties. It may be seen, that the scaling is not well obeyed, indicating that the use of the 3D XY model for describing the critical fluctuations in external magnetic fields exceeding 1 T, is inappropriate. Results of other measurements on other single crystals of YBCO-123 are in conflict with this conclusion, claiming a 3D XY type scaling to be valid in fields up to 14 T [149,150].

Our second example is C(T) of a Bi-based copper oxide material whose anisotropy is far more excessive than that of YBCO-123. For the latter, the anisotropy parameter Γ is estimated to be less than 10; for superconducting Bi-2212, Γ ~ 350. In that case it may be expected that fluctuation effects are drastically influencing the temperature dependence of the specific heat in the vicinity of the transition. That this is indeed the case may be seen in Fig. 14.40, where $C(T, H)$ of Bi-2212 is shown for a set of different fields H. Even in zero external field no trace of a discontinuity at T_c, still discernible in the zero field data of YBCO-123, may be recognized. More instructive is the plot of $[C(0) - C(H)]/T$ versus T, shown in Fig. 14.41. It may be seen that the anomaly adopts the shape of a triangle whose apex is, amazingly, not shifted in temperature with increasing magnetic field. These data cannot be brought into agreement with predictions of the 3D XY model [150,151].

From our brief description of specific heat data of two typical cuprate superconductors with very similar critical temperatures but differing a lot with respect to their anisotropy, it is clear that the character of the phase transition depends very strongly on the anisotropy parameter Γ. For small values of Γ, i.e., less than 10, the phase transition seems, at least in zero magnetic field, to belong to the 3D XY universality class. This is certainly not the case for strongly anisotropic, quasi-2D superconductors. Unfortunately, YBCO-123 is probably the only cuprate superconductor with a low value of Γ; most of the others are far more anisotropic. It has been argued that this trend might indicate that in the quasi-2D systems, the bosonic pairs are preformed and simply condense at T_c, thus implying an altogether different universality class of the phase transition.

The characteristics of the phase transition may also be discussed on the basis of magnetization $M(T, H)$ data. A particularly intriguing phenomenon has been observed in $M(T, H)$ of cuprate superconductors with a large anisotropy parameter Γ. A data set of the temperature dependence of the magnetization of a single crystal of Bi-2212, measured in different external magnetic fields [152], is shown in Fig. 14.42. It may be seen that the $M(T, H)$ curves all cross in a single point at a temperature $T_{cr} < T_c$, implying that the magnetization is field independent at this single temperature T_{cr}. In practice it is found that T_{cr} is always a few degrees below the zero field T_c. The same phenomenon is also observed for Bi-2201 [153], another extremely anisotropic superconductor, but not for YBCO-123. This crossing feature indicates that the field dependence of the magnetization $\partial M/\partial H$ changes sign at T_{cr}, certainly not expected for the usual magnetization of a superconductor in the mixed state. The sign change may be understood, however, by considering two contributions to the magnetization, such that

$$M = M_0 + M_{fl}\,, \tag{14.15}$$

where M_0 is the usual field dependent magnetization of the mixed state, calculated using a modified London equation with the appropriate approximations, and M_{fl} is due to thermal fluctuations of the vortices below T_c [154]. Using a modified London model, also M_{fl} (T) has been calculated with different approximations in high and low fields. Inserting the appropriate expressions for M_0 and M_{fl} into Eq. (14.15) and setting $\partial M/\partial lnH \equiv 0$, one obtains

$$M^{cr} \equiv M(T_{cr}) = -\left[\frac{k_B T_{cr}}{\Phi_0 s}\right] \ln\left(\frac{\eta\alpha}{\sqrt{e}}\right) \tag{14.16}$$

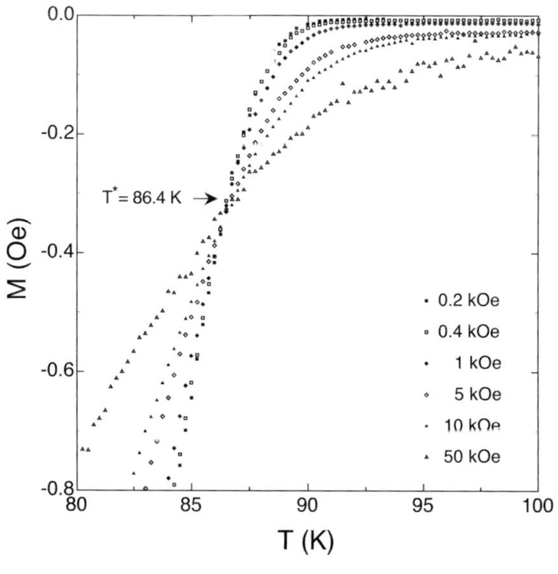

Fig. 14.42. Temperature dependence of the magnetization of single crystalline Bi-2212 in the range of the superconducting transition temperature in various external magnetic fields. Note the crossing of all curves at $T^* = 86.4$ K

and

$$T_{cr} = \left[\frac{\Phi_0^2 s}{32 k_B \pi^2 \lambda_{ab}^2(T_{cr})}\right], \tag{14.17}$$

where k_B is the Boltzmann constant and s is taken as the interlayer distance, which is of the order of 1 nm. The numerical values of η and α are not easy to evaluate and often it is assumed that $\ln(\eta\alpha/\sqrt{e}) = 1$ [155]. For this reason it is difficult to check the quantitative agreement between the experimental and calculated values of M_{cr} and T_{cr}.

The crossing of the $M(T,H)$ curves at a single temperature, experimentally observed to occur in an unexpectedly wide range of magnetic fields [152], indicates that some scaling should be obeyed. This is indeed the case, as may be seen in Fig. 14.43. The analysis is based on scaling functions that have been deduced from a Ginzburg–Landau free energy functional for quasi-2D type II superconduc-

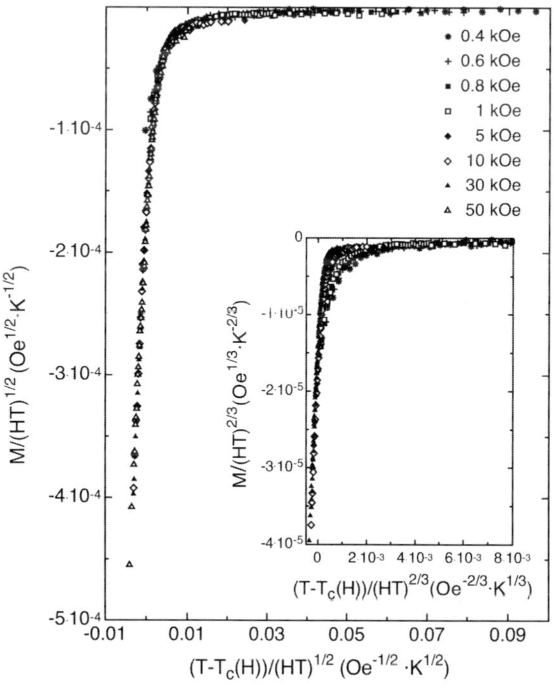

Fig. 14.43. Scaling of the magnetization of Bi-2212 at different temperatures and in different magnetic fields. The scaling variables of the main frame are from a 2D Ginzburg–Landau, lowest Landau level (GL–LLL) type approximation. In the inset, the same data are plotted assuming a 3D version of the GL–LLL scaling (see text)

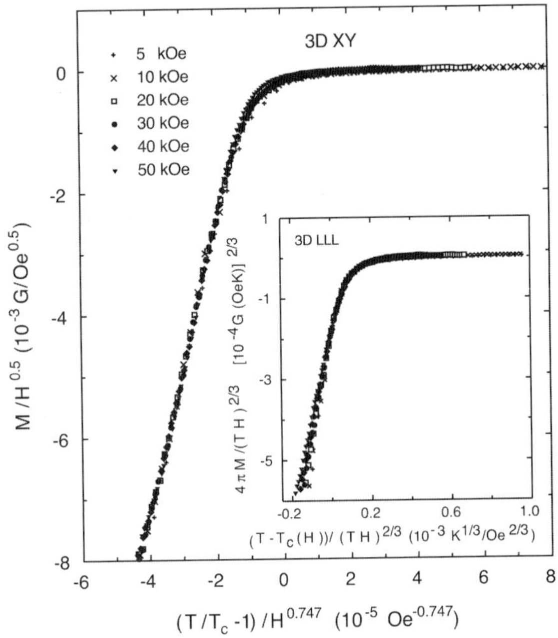

Fig. 14.44. Scaling of the magnetization of Y-123. The scaling variables in the main frame are compatible with the 3D XY model while those in the inset are consistent with 3D GL-LLL scaling

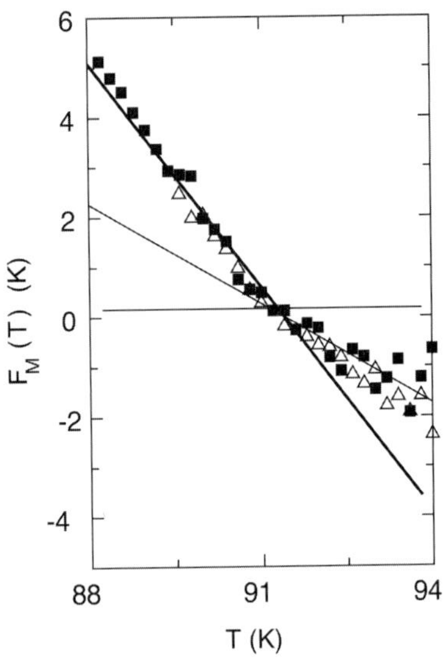

Fig. 14.45. F_M(T), as defined in Eq. (14.18), vs T. *Open triangles:* $\mu_0 H = 0.75$ T. *Full squares,* $\rho_0 H = 15$ T. The plot indicates that the critical exponent ν is not the same above and below T_c (see text)

tors [155]. In this case, the scaling function variable is $[T - T_c(H)]/(TH)^{1/2}$. The corresponding plot of the renormalized magnetization in Fig. 14.43 reveals that the scaling, based on fluctuations in two dimensions, is rather well obeyed. In addition, the same plot for 3D scaling, shown in the inset of Fig. 14.43, implies that the corresponding approach is not valid in this case. A 3D scaling of this same type has, however, been verified [156] for analogous data of YBCO-123 where, as mentioned above, the crossing phenomenon is absent. A scaling plot, based on the scaling variable valid for the 3D situation is shown in Fig. 14.44.

Another possibility for checking the scaling of $M(T, H)$ in the critical regime is offered in the following way. The unknown scaling function may be eliminated by evaluating

$$F_M = \frac{2H\left[\frac{\partial M(T,H)}{\partial H} - \frac{M(T,H)}{2H}\right]}{\left[\frac{\partial M(T,H)}{\partial T}\right]} \equiv (T_c - T)/\nu\,, \quad (14.18)$$

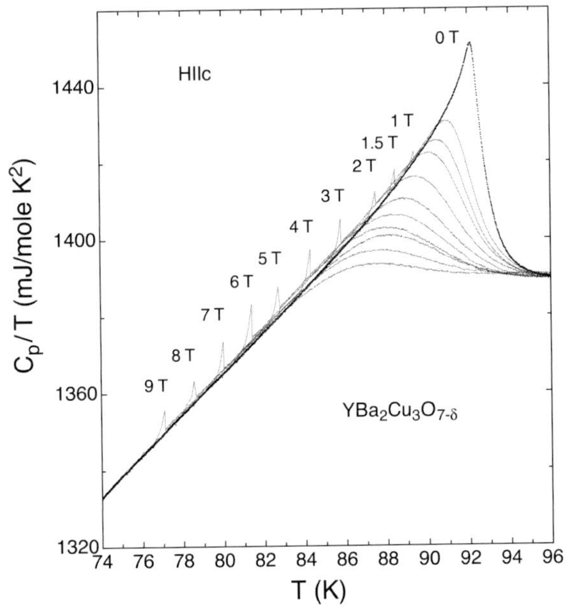

Fig. 14.46. Temperature dependence of the specific heat of single crystalline Y-123 in various external magnetic fields oriented along the orthorhombic c-axis. Note the sharp anomalies at decreasing transition temperatures upon increasing magnetic field (see [162])

i.e., the temperature dependence of F_M is linear in T and the slope relates to the critical exponent ν. A plot of $F_M(T)$, shown in Fig. 14.45, reveals that a temperature independent scaling exponent ν cannot account for the data and therefore confirms the failure of the 3D XY model in external fields of 1 T and above. In relation with the superconducting transition in non zero external magnetic fields it has been realized, also on the basis of the $C(T, H)$ data presented above, that the reduced anomalies are due to some crossover phenomenon rather than a true phase transition. Guided by theoretical insight [157] a distinct phase transition of first order has then be identified by high resolution measurements of thermodynamical properties, such as the magnetization [158, 159] and the specific heat [160, 161], both as a function of temperature in static magnetic fields. Relevant $C(T, H)$ data [162] is shown in Fig. 14.46. The first order nature of the transition is confirmed by the observation of discontinuities in the magnetization at the same temperatures where the $C(T)$ anomalies occur, requesting the release of some latent heat on general grounds. The transition is believed to be associated with the ensemble of the vortices of the mixed state.

Transport Below T_c(H)

(a) Resistivity

As has already been pointed out above, the zero field resistive transition is significantly influenced by fluctuation effects, particularly in those cases where the anisotropy parameter Γ is large. It is also interesting to note that in the temperature range where the in-plane resistivity ρ_{ab} exhibits the fluctuation induced paraconductivity, often a substantial increase of the out of plane resistivity ρ_c is observed [163], i.e., at T_c, $\partial\rho_c/\partial T$ is negative. An example of this feature, observed for Bi-2212, is shown in Fig. 14.47. Nevertheless, even in these cases, ρ_c vanishes in a narrow range of temperature, confirming that even for large values of Γ, superconductivity is still a 3D phenomenon.

Quite unusual, at least in comparison with common superconductors, is the influence of external magnetic fields on the resistive transition. First we consider the influence of external magnetic fields, oriented along the c-axis, on the in plane charge transport, i.e., ρ_{ab} (T,H). It may be seen from Fig. 14.48, where results of measurements [164] of the in plane resistivity of Bi-2212 are presented, that an increasing field H does not simply shift the transition to lower temperatures. Instead, the decrease of ρ_{ab} occurs in a temperature range of increasing width and from these data it is not a priori clear, where the resistivity actually vanishes. The extension of the temperature range of the resistivity drop is particularly large, if the anisotropy parameter is large, as in this case. This diagram confirms the above mentioned view that a rigorous definition of the upper critical field H_{c2} from resistivity data is difficult. It turns out that again these features are intimately related to the behavior of the vortices in the mixed state. Naturally also $\rho_c(T)$ is strongly affected by external magnetic fields. In Fig. 14.49 it may be seen that, because of the magnetic field shifted transition, ρ_c of Bi-2212 continues to increase with decreasing temperature, confirming the characteristically different charge transport in the two different crystallographic directions [163]. Also here, the subsequent drop of the resistivity is rather gradual, indicating a cross over behavior rather than a distinct phase transition.

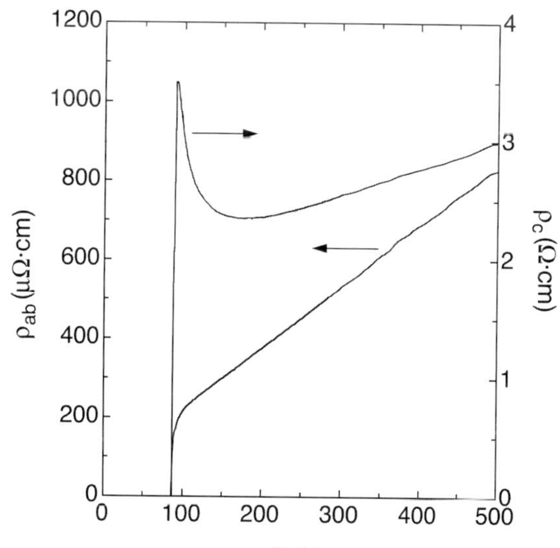

Fig. 14.47. Anisotropy of the temperature dependence of the electrical resistivities of single crystalline Bi-2212, measured in the ab plane (ρ_{ab}) and along the c-axis (ρ_c), respectively. Note the different scales for ρ_{ab} and ρ_c (see [163])

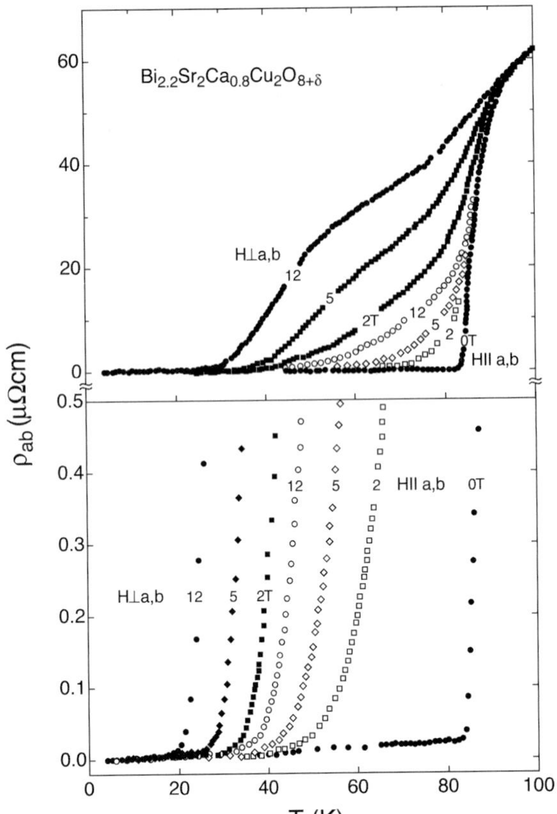

Fig. 14.48. Magnetic field dependence of the in plane electrical resistivity $\rho_{ab}(T)$ at the superconducting phase transition of single crystalline Bi-2212. Note the difference in the features, depending on the orientation of the external field. The *lower frame* emphasizes the low resistance regime (see [164])

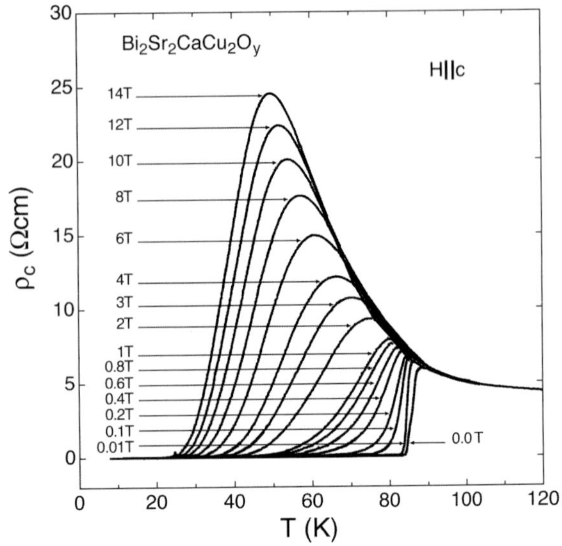

Fig. 14.49. Magnetic field dependence of the out of plane electrical resistivity $\rho_c(T)$ of single crystalline Bi-2212 (see [163])

(b) Thermal conductivity

Because of the gap formation in the excitation spectrum of electronic quasiparticles of superconductors, the transport of energy or heat is naturally affected by the phase transition. In common superconductors, because of the gap formation, the electronic transport of energy, i.e., the corresponding thermal conductivity λ_{el}, vanishes gradually below T_c [165]. Because of the reduced rate of scattering of phonons with electronic quasiparticles, the lattice contribution to the thermal conduction is expected [165] and observed [166] to increase with decreasing temperature below T_c, at least as long as no other scattering mechanisms for phonons dictate their mean free path. The situation seems to be somewhat different for the cuprate superconductors. In Fig. 14.50 [167] we show experimental data for the temperature dependence of the thermal conductivity $\lambda(T)$ for the transport of energy parallel to the Cu-O planes of YBCO-123. Apart from some intrinsic anisotropy between the a and b direction, the most obvious features are the discontinuous change of the slope $\partial\lambda/\partial T$ at T_c and the distinct maxima of λ at approximately $T_c/2$, for both crystallographic directions. The anisotropy ratio of λ_a/λ_b is constant above T_c and the enhanced thermal conductivity along the b-direction is attributed to the additional electronic contribution due to quasiparticle transport along the Cu-O chains that are, according to Fig. 14.6a, extended only along the b direction of the crystal lattice. The slight increase of λ with decreasing temperature in the normal state is interpreted as being due to itinerant lattice excitations. Although at first sight it is tempting to ascribe the anomalous increase of λ below T_c to an enhanced phonon conductivity, it turns out [168] that the observed behavior is rather due to an enhanced electronic heat conduction, obviously again a rather unusual phenomenon but, to a certain ex-

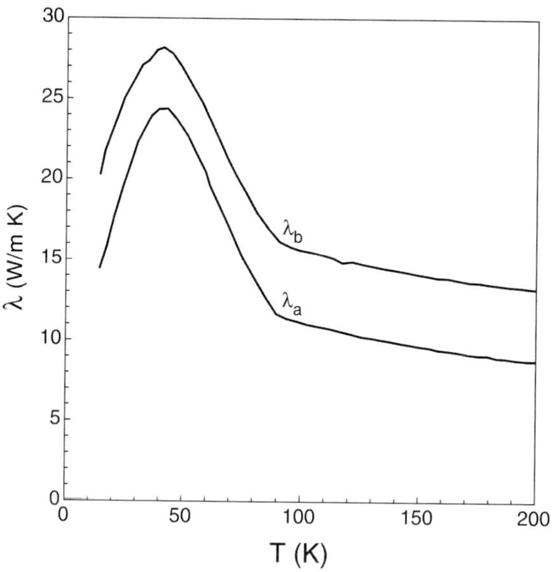

Fig. 14.50. Temperature dependences and anisotropy of the thermal conductivities of single crystalline Y-123, measured along the *a*-axis and the *b*-axis, respectively (see [167])

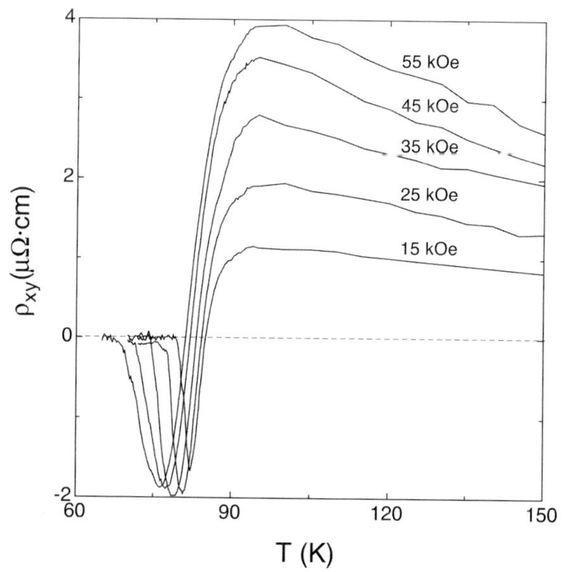

Fig. 14.51. Temperature dependence of the in plane Hall resistivity ρ_{xy} of single crystalline Y-123 in various external magnetic fields oriented along the *c*-axis between 60 K and 150 K

tent, also observed in strongly coupled conventional superconductors [169, 170]. For the cuprates it is ascribed [168, 171] to an anomalous decrease of the quasiparticle scattering rate in the superconducting state which at first was revealed in experiments probing the surface impedance of cuprate superconductors below T_c [172, 173]. Assuming that the temperature dependence of the phonon conductivity is not significantly altered in the superconducting state, the maxima in λ are then a consequence of the compensation of the scattering rate reduction by the reduced number of quasiparticle excitations because of the gap formation in the superconducting state.

(c) The Hall Effect

It turns out that magnetoresistive effects in these cuprate materials, at least in experimentally achievable magnetic fields, are not really extraordinary. This is not quite true for the results of measurements of the Hall resistivity ρ_{xy} . Even for common type II superconductors, the Hall response in the mixed state is somewhat complicated and different model calculations of what should be expected, lead to different results [174, 175]. This is even more so for the cuprate superconductors where again, some anomalous features are particularly well developed. Since the observed anomalous features seem again to be strongly linked with the properties of the vortex state, we restrict ourselves to a brief presentation of the most important observations. For this purpose we show, in Fig. 14.51, the temperature dependence of the Hall resistivity $\rho_{xy}(T)$ of a single crystal of slightly underdoped YBCO-123 in various external magnetic fields [176]. Most obvious is the sign change of ρ_{xy} at field dependent temperatures that are close to but always less than the zero field critical temperature $T_c(0)$ [177]. There is a clear negative shift in temperature where the minimum of ρ_{xy} is observed, with increasing external magnetic field. At this point it should be noted that sign changes of ρ_{xy}, although somewhat less pronounced, have also been observed in conventional type II superconductors [178]. It is therefore most likely, that this phenomenon is tied to the vortex state and its dynamics as such and not to special intrinsic properties of the cuprate superconductors. Nevertheless, more Although various

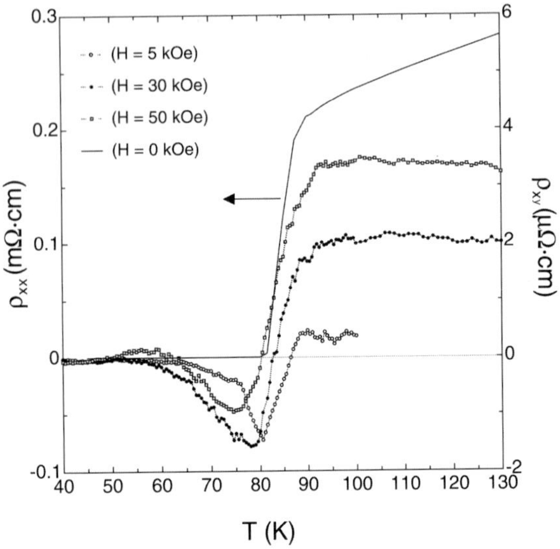

Fig. 14.52. Temperature dependence of the in plane Hall resistivity ρ_{xy} of single crystalline Bi-2212 in various magnetic fields oriented along the c axis between 40 and 130 K. Also shown, as the *solid line*, is the zero field in plane electrical resistivity ρ_{xx}

explanations for these phenomena have been suggested, it is most likely that the Hall conductivity detailed investigations indicate that again the degree of anisotropy and, in addition, the concentration of holes introduced by doping, play an important role. Another set of ρ_{xy} data, obtained from measurements on a single crystal of Bi-2212 material [179], is shown in Fig. 14.52. The difference to the results shown in the previous figure is yet another sign change of ρ_{xy} at temperatures much below T_c [180]. Apart from this observation we note that the deviations from the T^{-1} behavior of R_H start at temperatures well above the zero field T_c. This seems to indicate that again some fluctuation effects are significant.

$$\sigma_{xy} = \frac{\rho_{xy}}{(\rho_{xx}^2 + \rho_{xy}^2)} \qquad (14.19)$$

is given by the sum of at least two different contributions [181]. One of them is related to the superconducting matrix of the mixed state and the second would represent the contribution from excited quasiparticles in the cores of the vortices. There are obviously different ways for theoretical treatments of this problem. A number of publications, considering either the pinning force, fluctuation effects, a Magnus-type force, two band effects, or charge imbalances between the vortex cores and the superconducting matrix are available in the literature [182–187]. A thorough discussion of all of them is beyond the scope of this review.

Characteristics of the Superconducting State (Gap Function)

The stability of the superconducting state is, in general terms, guaranteed by the formation of a gap in the quasiparticle excitation spectrum below T_c. At present we do not consider the very special case of gapless superconductivity [188], sometimes observed in cases where the quasiparticle mean free path is much shorter than the coherence length. This condition is not fulfilled in typical cuprate superconductors. In the most simple approximation, this gap is assumed to be isotropic in k space and its magnitude at zero temperature, $\Delta(0)$, is a measure for the free energy difference between the normal and the superconducting state, the condensation energy E_{cond}, via [106]

$$E_{\text{cond}} = \frac{D(E_F) \cdot \Delta^2(0)}{2} . \qquad (14.20)$$

Since this condensation energy is another manifestation of the Meissner–Ochsenfeld effect, its experimental observation, to be discussed in more detail below, implies the existence of this gap also in cuprate superconductors. The opening of this gap at T_c is also reflected in the almost discontinuous enhancement of the specific heat at the critical temperature of YBCO-123, measured in zero magnetic field.

All these results from measurements of macroscopic properties give only indirect evidence for the presence of a gap and the real way to go are experiments that probe the excitation spectrum of the electronic quasiparticles directly. Many early tunneling data revealed the expected loss of excited states in a narrow energy range around zero energy [189]. A new way of exploring the occupation of quasiparticle states near the Fermi energy of superconductors was then offered by a significant progress in improving

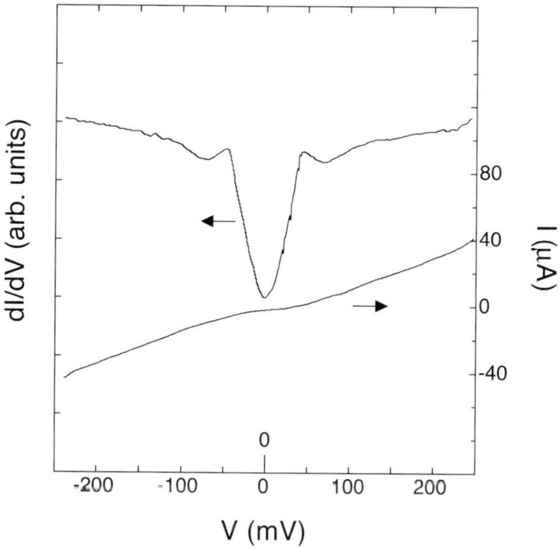

Fig. 14.53. Current and conductance vs. voltage characteristics of a break junction of a Bi-based copper oxide superconductor, reflecting the loss of available states close to E_F(see [190])

the energy resolution in measurements of photoemission spectra near E_F. In Fig. 14.53 we show an early result of the voltage dependence of the differential conductivity $\partial I/\partial V$ of a break junction of Bi-based copper oxide, reflecting the loss of achievable states close to E_F [190]. Before we address the issue of a possible k dependence of the gap function, we consider the information that essentially all these experiments provide, the amplitude Δ_0 of the gap function and its temperature dependence. The zero temperature value $\Delta_0(0)$ is of interest because, together with the magnitude of T_c, it reveals the strength of the interaction that provides the superconducting ground state. In the weak coupling limit, the relevant ratio, following from the original BCS theory [106] is

$$\frac{2\Delta_0(0)}{k_B T_c} = 3.56 \,. \tag{14.21}$$

Larger values, following from direct experimental evaluations of T_c and $\Delta_0(\mathrm{T})$ are usually interpreted as a manifestation of strong coupling effects. Without going into details, most of the results of experiments that are briefly discussed below, indicate ratios of $2\Delta_0\ /k_B\ T_c$ that are factors of 2 to 3 larger than the value quoted in Eq. (14.21), at least when probing the superconducting state in the Cu-O planes, i.e., we are dealing with superconductivity invoking a very strong coupling.

Although there was, even before these microscopic experiments were done, never any serious doubt that the superconducting gap is well established in cuprate superconductors, there was, at first, only limited interest and correspondingly little information concerning the k dependence $\Delta(k)$ of the gap in momentum space. The shape or symmetry of $\Delta(k)$ is, to a certain extent at least, tied to the symmetry of the pairing configuration and the corresponding order parameter of the superconducting state. Therefore it is clear that experimental verifications of the shape of $\Delta(k)$ are important and useful for the interpretation of a variety of experimental data, but especially also for testing predictions of theoretical models. Considering the overwhelming evidence that superconductivity in these copper oxides is mainly due to the special properties of the Cu-O planes and also taking into account the atomic arrangement of Cu and O in these planes, it seems suggestive that the most likely configuration of the gap function $\Delta(k)$ has a $d_{x^2-y^2}$ symmetry, also compatible with predictions assuming non electron–phonon pairing mechanisms [191]. In the following, a short presentation of various different experimental attempts to evaluate $\Delta(k)$ and the corresponding main results that have been achieved, is given.

Experiments Probing the Gap Anisotropy

The k-dependence of the gap may have trivial causes. Its anisotropy may be tied to the anisotropy or shape of the Fermi surface. It is quite conceivable that an isotropic gap, which essentially relies on the choice of a k-independent pairing interaction, is not strictly realized even in conventional superconductors. This type of anisotropic gaps, based on a pairing interaction whose strength is not the same on all parts of the Fermi surface, may also be present in conventional superconductors and many indirect evidences for this have been accumulated a long time ago [192]. Of much more interest are cases where the anisotropy of the Fermi surface and of $\Delta(k)$ do not match. This may best be discussed by invoking the concept of symmetry breaking at phase transitions, introduced by Landau [193] and, in more general terms, by Anderson [194]. For conventional superconductors, the formation of a phase-coherent pair condensate invokes the breaking of the one dimensional global gauge symmetry U(1). A superconductor is termed *unconventional* if the corresponding pair state breaks yet another symmetry, which may, e.g., be the time reversal symmetry T, the spin rotation symmetry R or the translational symmetry X of the crystal lattice. The resulting symmetry of the pair state is reflected in the symmetry of the order parameter and also of the gap function. Even if the symmetry of the gap function or the order parameter is known, it is still difficult or even impossible to conclude rigourously, which kind of interaction has produced the observed type of symmetry breaking. Aspects of unconventional superconductivity are also addressed in the chapter on superconductivity of heavy electron compounds and have extensively been discussed in [195].

Additional symmetry breaking and corresponding unconventional pairing configurations often lead to gap functions $\Delta(k)$ with nodes, i.e., zero gap amplitudes at points or on lines on the Fermi surface. Such nodes have a significant influence on the temperature dependence of physical properties that are dominated by electronic excitations in the superconducting state well below T_c. For superconductors with conventional pairing, gap nodes may form only accidentally and usually the gap assumes non zero values across the entire Fermi surface. In these cases, all properties of the superconducting state which are related with the excitation of electronic quasiparticles, exhibit an exponential temperature dependence at $T/T_c << 1$. This is, of course, not the case if gap nodes are present. Contrary to conventional superconductors for which at zero temperature the quasiparticle density of states $D(E)$ is zero between $E = E_F = 0$ and $E = \pm\Delta$, this is not so for many configurations of unconventional pairing. A well known exception is the superfluid B phase of liquid ^{3}He, where an unconventional pair wavefunction provides a complete non-zero gap [196]. Possible forms of D(E) close to E = 0 for unconventional

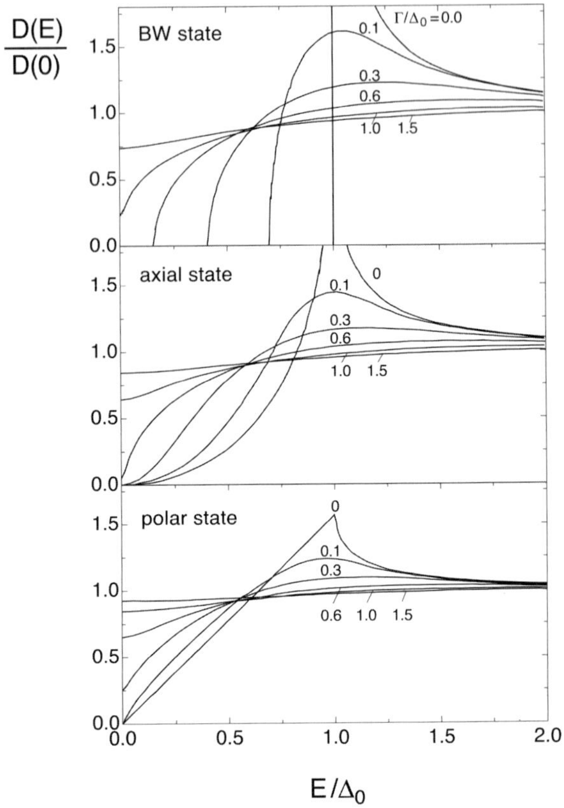

Fig. 14.54. Schematic representation of the energy dependence of the electronic density of states D(E) in the gap region of unconventional superconductors. The Balian–Werthamer (BW) state is a nodeless *p*-wave state. The axial and the polar state imply point and lines of nodes of the gap function, respectively. The parameter Γ/Δ_0 is a measure for impurity induced pair breaking (see [195])

pairing states are shown in Fig. 14.54. As may be seen, D(E) vanishes in power laws of E as $E \to 0$. The nodes are not always strictly on points or lines on the Fermi surface. Impurity scattering may widen these regions where the gap function vanishes. In some cases this happens even for tiny amounts of impurities or defects, in other cases, a critical value of impurity concentration has to be exceeded in order to achieve a gapless situation.

As an immediate consequence, the above mentioned exponential temperature dependence of physical properties is replaced by power laws in T. As an example we may quote the electronic specific heat.

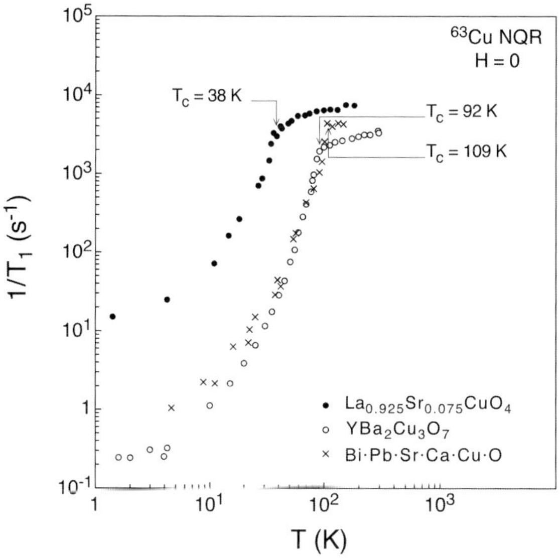

Fig. 14.55. Temperature dependence of the ^{63}Cu NQR spin lattice relaxation rate $T_1^{-1}(T)$ for three different cuprate superconductors

While for conventional superconductors it is varying as $C(T) \sim \exp(-\Delta/T)$, a power law $C(T) \sim T^n$ is expected for unconventional superconductors. The value of the exponent n depends on the configuration of the gap nodes. It is therefore possible to identify unconventional superconductors by measurements of the temperature variation of selected physical properties well below T_c. Relevant examples are presented below.

Another feature that is typical for conventional superconductors but is absent in those of unconventional varieties, is the observation of coherence effects that are a consequence of the special configuration of conventional Cooper pairs that are formed between states with opposite momenta and spins, i.e., $(k\uparrow, -k\downarrow)$, a major ingredient of the original BCS theory [106]. The observations of these coherence effects, which depend on the interaction exerted by the measuring technique were, at the time, considered as convincing confirmations for the validity of the BCS theory [197, 198]. Some of the first indications for unconventional superconductivity in the cuprates were related with the absence of the coherence effect that is expected to be observed in measurements of the temperature dependence of NMR relaxation rate just below T_c [197]. An example for this kind of data is shown in Fig. 14.55 [199]. The drop of the relaxation rate T_1^{-1}, measured for Cu nuclear spins, is abrupt and quite in contrast to the expected rise of T_1^{-1} with an onset at T_c, passing over a distinct maximum somewhat below the critical temperature and finally decreasing exponentially at $T \ll T_c$. This latter behavior, clearly identified for simple superconductors [200], is a typical manifestation of coherence effects. The same set of data, shown in Fig. 14.55, has also been claimed to provide evidence for unconventional superconductivity via the power-law type temperature dependence of the relaxation rate below T_c.

(a) Penetration Depth

With respect to cuprate superconductors, an early indication for unconventional superconductivity was obtained from measurements of the temperature dependence of the penetration depth $\lambda_L(T)$ for external magnetic fields well below T_c [201]. In simplest terms this characteristic length is given by

$$\lambda_L^2 = \frac{mc^2}{4\pi n_s e^2}\,, \tag{14.22}$$

i.e., it depends on the ratio between the mass m of the quasiparticles and the density n_s of quasiparticles condensed into pairs. The latter is temperature dependent and reaches its maximum value at T = 0K. As mentioned above, the penetration depth may be measured using the technique of μSR. For high resolution measurements aiming at establishing the temperature dependence of the penetration depth at low temperatures, other experiments seem more appropriate, however. Based on microwave absorption experiments, which are well suited to measure the temperature induced variation of λ_L with respect to a given value, arbitrarily set to zero, at some chosen temperature T_0, quite reliable values of $\Delta\lambda_L = \lambda_L(T) - \lambda_L(T_0)$ may be obtained. Corresponding results, obtained for high quality single crystals of YBCO-123 at temperatures above $T_0 = 1.3$ K [201] (see Fig. 14.56), have demonstrated that $\Delta\lambda_L$ (λ_L^{ab} in this case) increases linearly with T, at least between 1.3 and 20 K, i.e., well below T_c of about 90 K. This temperature dependence of $\Delta\lambda_L$, is a clear indication that the

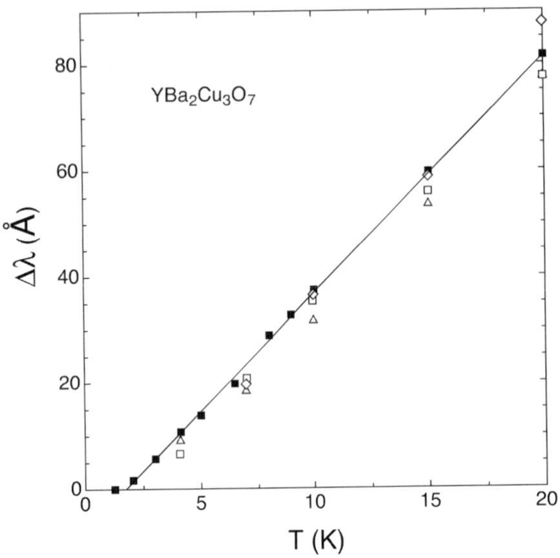

Fig. 14.56. Temperature dependence of the incremental increase of the in-plane penetration depth of single crystalline Y-123. The different symbols represent the data for different samples (see [201])

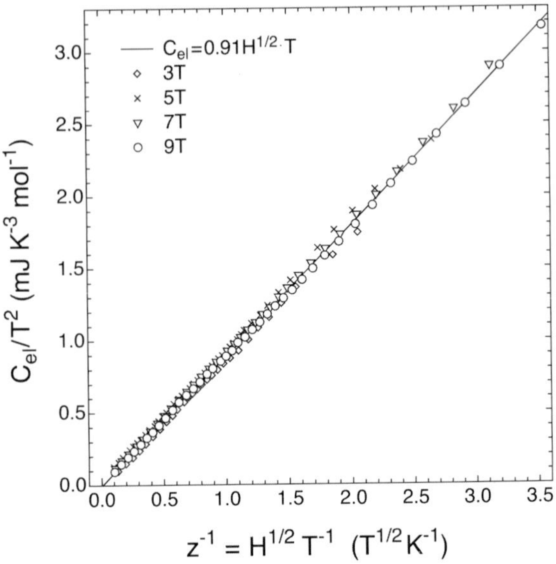

Fig. 14.57. Scaling of the low temperature specific heat of $YBa_2Cu_3O_{7-\delta}$ at different temperatures and in varying magnetic fields (see [206])

gap function of YBCO-123 has nodes, most likely on lines on the Fermi surface. The claim is quite convincing because a control experiment of the same type, using a sample of a PbSn alloy, a conventional type II superconductor, resulted in an exponential increase of $\Delta\lambda_L$ at temperatures well below $T_c = 7.2$ K of this material. Later experiments [202] revealed that shortening the quasiparticles' mean free path by introducing impurities results in $\Delta\lambda_L \sim T^2$.

(b) Specific Heat

Since the specific heat $C(T)$ is solely dependent on the density of states of possible thermally induced excitations, the temperature dependence of the contribution to the specific heat due to excited electronic quasiparticles, $C_e(T)$, is bound to reflect the presence of gap nodes. In general, point-type or line-type gap nodes are expected to result in $C_e(T)$ being proportional to T^3 and T^2, respectively, at very low temperatures at least. Although this verification appears as being rather trivial in principle, the practical execution is hampered by the fact, that the subtraction of not well known background contributions to $C(T)$ is less simple as it may seem. Nevertheless, attempts of this sort have been made, mainly by invoking measurements of the specific heat of close to optimally doped YBCO-123, both in zero and non-zero magnetic fields H. In some work, a T^2 variation of C_e at $T \ll T_c$ and $H = 0$, indicating line nodes of the gap [203], was claimed to have been verified [204]. Even more involved is the analysis of specific heat data obtained in non zero external magnetic fields, because the structure of the vortices and their normal cores add to the already present complications. However, the measured H and T dependencies may be compared with rigorous theoretical predictions for the consequences of unconventional gap functions. In particular, line nodes are predicted [205] to provoke the appearance of a new term representing the contribution due to the quasiparticle density of states C_e, varying as $H^{1/2}T^{-1}$ and replacing the T^2 contribution at low temperatures. The predictions for line nodes include a crossover behavior at a critical value $z_c = H_c 2^{-1/2}\, T_c$ to a high-T, low-H regime with two separate contributions, an H independent T^2 term and a temperature independent contribution varying linearly with H. Both predictions have been verified experimentally [204, 206]. One of the verifications of the predicted scaling of C_e in the low temperature regime is shown in

Fig. 14.57. Another prediction [207,208] for the line node configuration, the scaling of the anisotropic component of the field induced specific heat, given by $C_e\ (T, H||c) - C_e(T, H\perp c)$ has also been investigated and confirmed experimentally [209]. Analogous experiments with similar results and conclusions, namely the presence of line nodes in the gap function, have also been made on superconducting $La_{1.85}Sr_{0.15}CuO_4$ [210].

(c) Thermal Conductivity

As we have seen in one of the previous subsections, the transport of energy or heat in the superconducting state of cuprate materials is mainly carried by excited electronic quasiparticles. It may therefore be expected that at very low temperatures, measurements of the temperature dependence of the thermal conductivity at temperatures much below T_c again provide results that are dictated by the gap configuration $\Delta(k)$. In particular, again a rather general theoretical prediction is available for the case of d-wave type gap function with line nodes in two dimensions [211]. It addresses the quasiparticle transport in such a system and states that this transport should become independent of the scattering rate as T approaches 0 K. For a d-wave type gap configuration it is known that impurities widen the zero gap regions around the line nodes, thereby enhancing the amount of remnant normal quasiparticles even at $T = 0$ K. This enhancement is compensated by the impurity imposed reduction of the mean free path, thus resulting in a universal limit of the heat transport. Relevant experimental data for testing this prediction are available for heat transport parallel to the Cu-O planes of YBCO-123 and Bi-2212 single crystals [212,213]. In Fig. 14.58 we show the measured T dependence of the thermal conductivity along the a axis of YBCO-123 single crystals, doped with different amounts of Zn impurities. It may be seen that at the lowest temperatures, the ratio λ/T reaches a value that is independent of the amount of impurities and hence the scattering rate. From these results it is claimed that a universal value of λ/T does exist at very low concentrations of impurities [212]. In Fig. 14.59, a comparison of data obtained for the in plane thermal conductivity of YBCO-123 and Bi-2212 single crystals confirms that a remnant normal component λ_0, varying linearly with T, is observed in cuprate superconductors at T close to 0 K [213]. For

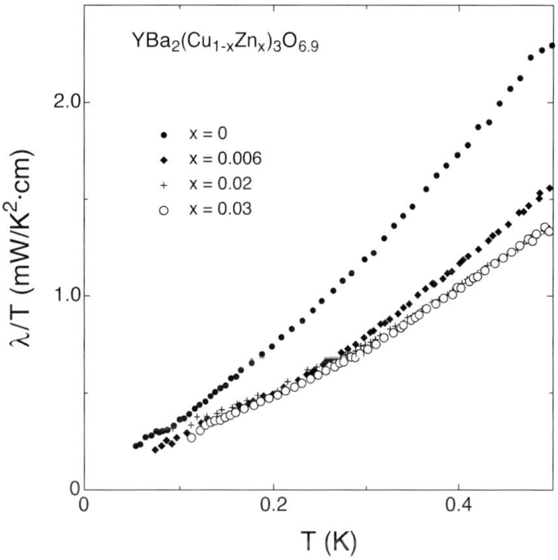

Fig. 14.58. Variation of the low temperature thermal conductivity along the a-axis of single crystalline Y-123 upon replacing Cu by Zn (see [212])

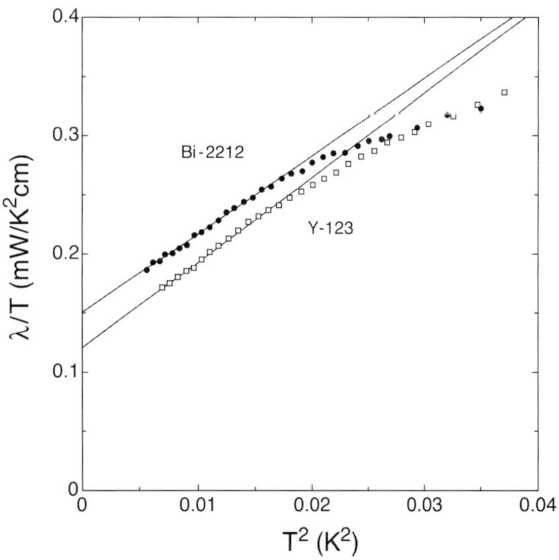

Fig. 14.59. Low temperature thermal conductivity of single crystalline Y-123 and Bi-2212. The solid lines emphasize the two contributions varying as T and T^3, respectively (see [213])

a d-wave gap configuration, the ratio λ_0/T may approximately be expressed as [214]

$$\frac{\lambda_0}{T} = \left(\frac{k_B^2 \cdot n}{3h \cdot d}\right)\left(\frac{v_F}{v_2}\right), \qquad (14.23)$$

where n/d is the stacking density of Cu-O planes, v_F and v_2 are the Fermi velocity, directed normal to the constant-energy contour, and the quasiparticle velocity tangential to the Fermi surface at each node, respectively. It turns out that the ratio v_F/v_2 can be deduced from direct measurements of the k-dependence of the gap function (see below) with ARPES [215]. The values of v_F/v_2, calculated either from λ_0/T or from the ARPES data differ only insignificantly, i.e., safely within experimental uncertainties [213].

While these experimental findings strongly support that the pairing state of cuprate superconductors implies a gap function with line nodes in general and a d-wave configuration in particular, the good agreement of the values for v_F/v_2 from transport and spectroscopic measurements tends to imply that the quasiparticles act like those of a Fermi liquid, at least at low energies. The pros and cons related with the Fermi liquid description of the normal state have been addressed in Sect. 14.4.1 above.

(d) Nuclear Magnetic Resonance

The first indications for unconventional superconductivity from NMR measurements probing the spin lattice relaxation of Cu and O nuclear spins have briefly been mentioned in the introductory to section of this subchapter. Here we first simply recall that also the relaxation of nuclear spins via interaction with electronic quasiparticles is expected to reveal the presence of a vanishing gap. This relaxation rate provided by this channel is expected to vanish following a power law in T with decreasing temperature and not exponentially as in the case of a nodeless gap. That this is indeed observed is shown in Fig. 14.55, where it may be seen that the reduction of T_1^{-1} below T_c is definitely not exponential in $1/T$.

Additional relevant information concerning the gap and even the spin configuration of the pair state can be obtained from the temperature dependence of the Knight shift in the superconducting state. As

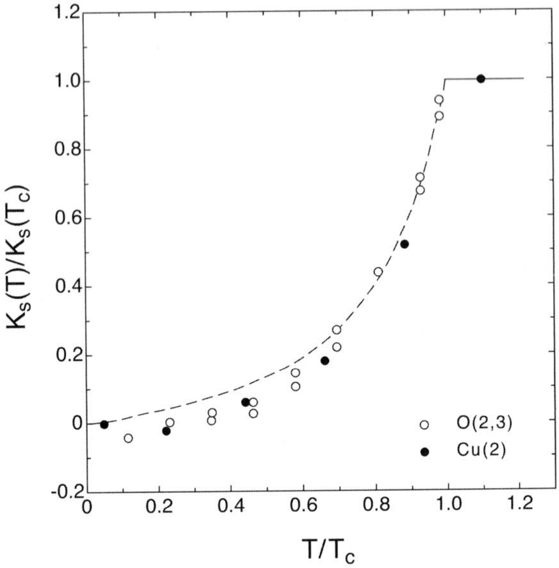

Fig. 14.60. Temperature dependence of the normalized Knight shifts of Cu and O NMR signals in the superconducting state of Y-123. The *broken line* represents a calculation based on gap function with d-wave symmetry (see [191])

we have mentioned in Sect. 14.4.1 the Knight shift K is, via its proportionality to the spin susceptibility χ_P of the quasiparticles, proportional to the density of quasiparticle states at the Fermi energy. For more complex materials, such as the cuprate superconductors, other contributions to the magnetic susceptibility have to be taken into account.

For conventional superconductors, where K may simply be assumed to be proportional to the magnetic susceptibility and, below T_c, the temperature dependence $K(T)$ is given by the so called Yosida function [216]. It results from the calculation of the Pauli-type magnetic susceptibility in the superconducting state $\chi_s(T)$ which is given by

$$\chi_s = -4\mu_B^2 \int D_{\mathrm{BCS}}(E)\left(\frac{\partial f}{\partial E}\right) \mathrm{d}E\,, \qquad (14.24)$$

where $D_{\mathrm{BCS}} = D(E_F)\cdot|E|/(E^2-\Delta^2)$, and f is the Fermi–Dirac distribution function. If the gap function Δ is nodeless and the pairing is of spin singlet type, χ_s vanishes exponentially with decreasing temperature for $T \to 0$ K. For spin triplet pairing the Knight shift is unaffected by the superconducting transition, i.e., is not significantly reduced below T_c [217]. The main influence of a gap function with nodes is again visible in a non exponential decay of $K(T)$ at low temperatures.

In Fig. 14.60 a selection of results of Knight shift measurements involving nuclei on different lattice sites of YBCO-123 is presented [218,219]. The strong reduction of K in the superconducting state clearly reflects spin singlet pairing. Together with the information suggesting the existence of line nodes in the gap function, there is strong evidence for a d-wave type pairing being established in cuprate superconductors. Theoretical model calculations to account for the experimental data gave better agreement, if d-wave rather than s-wave pairing was assumed [191]. One such attempt is shown as the broken line in Fig. 14.60.

(e) Angle Resolved Photoemission

As was mentioned above, significant progresses in measurements employing photoemission spectroscopy techniques, in particular ARPES with high energy and momentum resolution, were responsible for the evaluation of the excitation spectrum of quasiparticles in cuprate superconductors. While integrative photoemission has been able to demonstrate the formation of a gap below T_c [85], ARPES also provides information concerning the k-dependence of the gap. Since a whole chapter of this compendium is dedicated to this type of experiments, it suffices for our purposes to present the results of ARPES measurements probing $\Delta(k)$ (see Fig. 14.61). The fourfold symmetry of the node configuration expected for a gap function with $d_{x^2-y^2}$ symmetry is very well borne out [87].

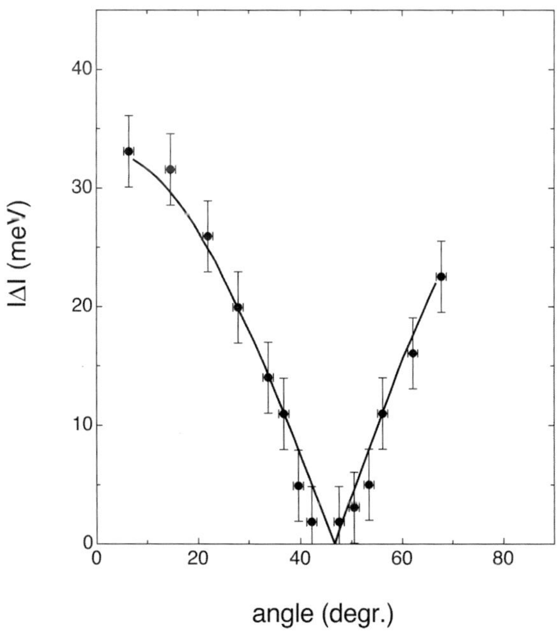

Fig. 14.61. Angle dependence of the leading-edge shift in ARPES spectra of Bi-2212, relative to the position of the reference Fermi edge of Pt. The *solid line* represents a calculation assuming a d-wave type gap function (see [87])

(f) Phase Sensitive Measurements

The experiments cited and discussed above all confirm that the gap functions in the investigated cuprate superconductors exhibit nodes, most likely on lines, and that the pairing state of the quasiparticles is of a spin-singlet variety. All this taken together, also taking into account the essential electronic wave functions in the Cu-O planes and considering various theoretical arguments based on phenomenological and computational approaches [191] leads to the conjecture that the pair state and hence the order parameter as well as the gap function exhibit a $d_{x^2-y^2}$ symmetry. This implies that the amplitude of the order parameter changes sign at the nodes or, in other words, the phase of the order parameter changes discontinuously by π. None of the above experiments is sensitive to this phase change and other types of measurements have to be invoked to settle this issue. These measurements have to be devised such that they probe the relative phase of the gap function on different parts of the Fermi surface.

Since phase information is required, this type of experiments is based on the macroscopic quantum coherence in superconductors, most distinctly manifested in the Josephson effects and the flux quantization. The concept for this kind of tests has first been put forward by Geshkenbein and Larkin [220], in connection with probing the unconventional nature of superconductivity of heavy electron metals. They suggested to investigate the phase coherence of specially tailored superconducting loops containing a combination of the superconductor under investigation and conventional superconducting materials. This idea was later adapted by Sigrist and Rice [221] for a proposal to test the $d_{x^2-y^2}$ character of the gap function of cuprate superconductors and the schematic layout of the required superconducting loop is shown in Fig. 14.62. The idea rests on the fact that at a Josephson contact between two superconductors with opposite signs of their order parameter amplitudes, a so called π-junction [222], where the phase of the order parameter changes discontinuously by π, is generated. It is well known that the magnetic flux through the cavity of a closed loop of a single conventional superconductor is quantized [223]; the flux quantum is $\Phi_0 = hc/2e$. It turns out that the response to external magnetic fields of loops containing either an even (including zero) or odd number of π-junctions is significantly different. In the first case the highest stability is achieved if an integer number of flux quanta $\Phi = \nu \cdot \Phi_0$ is trapped in the loop. In the second case, however, the stability criterion requires that $\Phi = (\nu + 1/2) \cdot \Phi_0$. Essentially two different types of experiments have been designed to obtain the requested information. In this more general review, none of the intricacies of these experiments and their analyses will be addressed in detail. This has, to a large extent, been done in more specialized reviews on this topic [224,225] (see below).

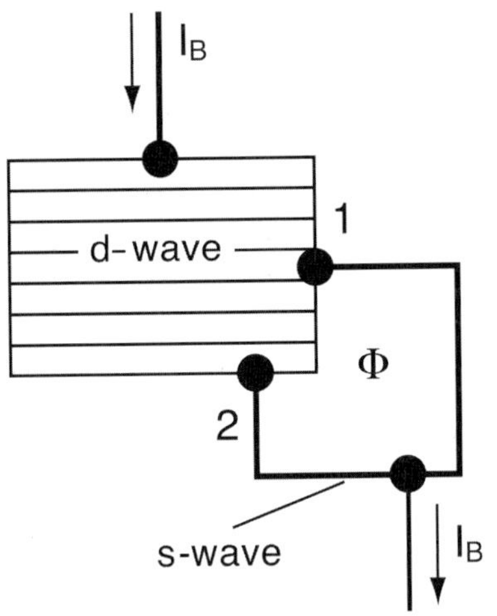

Fig. 14.62. Schematic layout of a superconducting loop for phase sensitive tests of the order parameter symmetry of cuprate superconductors, employing a SQUID type configuration (see [221])

(f1) SQUID-Type Experiments

Superconducting quantum interference devices (SQUIDs) are based on the Josephson effect, i.e., on the tunneling of pairs as a whole and usually contain two Josephson contacts in parallel. As the name indicates, SQUIDs offer the possibility of experimental control of the phase of the macroscopic wave function describing the superconducting state. For our purposes the chosen loop may be manufactured in a SQUID configuration, as is schematically outlined in Fig. 14.62. The design is such that one, two or none of the contacts may form as a π-junction. The critical current I_c, i.e., the maximum supercurrent, varies periodically as a function of the total magnetic flux inside the loop. The location of the maxima or minima of I_c depends critically on the phase coherence around the loop. If the SQUID loop contains none or two π-junctions, an $I_c(\Phi)$ pattern as shown in Fig. 14.63a is expected. The $I_c(\Phi)$ diagram for a SQUID with only one π-contact, however, is shifted by π along the horizontal Φ-axis, as shown in Fig. 14.63(b). The first realization of this type of SQUID experiments was accomplished by Wollman and co-workers [226] by studying the critical current behavior of SQUIDs that were fabricated from single crystalline YBCO-123 and Pb. The Pb part was in the form of a thin film, deposited in two different configurations. In the first, the film made two contacts around the corner of a single crystal, such that one of them is along the a direction and the other along the b direction of the crystal lattice. The second configuration had both contacts onto the same edge of the crystal, facing the same crystallographic direction. The schematic sample configuration for this pioneering work is shown in Fig. 14.64. Considering the crystal structure of YBCO-123 and the expected $d_{x^2-y^2}$ symmetry of the order parameter, the first configuration is expected to contain one contact acting as a π-junction. For the second case either none or two π-junctions are likely to be established. Measurements of the dynamical resistances of these SQUIDs as a function of applied magnetic field were used to monitor the modulation of the $I_c(\Phi)$ curves. The analysis of the data indicated a tendency of a π-shift of $I_c(\Phi)$ for loops around the corner of the crystals with respect to $I_c(\Phi)$ obtained for edge-type SQUIDs.

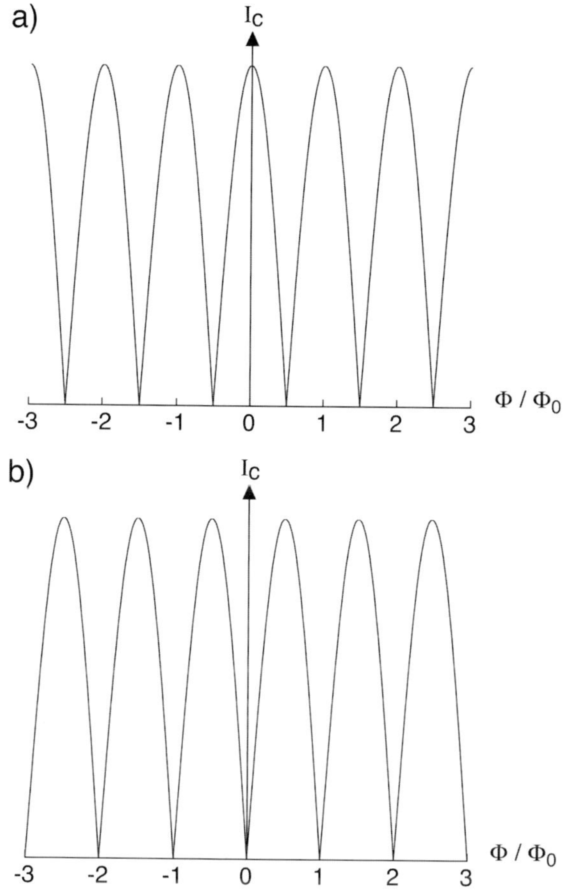

Fig. 14.63. Interference patterns of the critical current I_c of a superconducting loop as shown schematically in Fig. 14.62, upon varying the magnetic flux through the loop **(a)** with an even or **(b)** an odd number of π-junctions (see text)

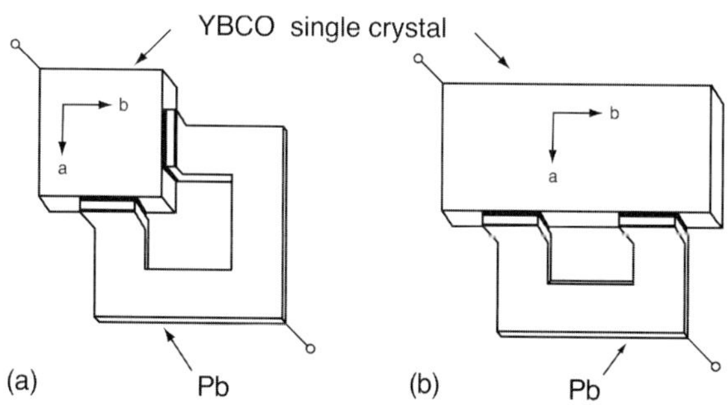

Fig. 14.64. Schematic view of the experimental realization of SQUID type configurations forming **(a)** a corner SQUID and **(b)** an edge SQUID for phase aensitive tests of the order parameter symmetry of Y-123 single crystals see [224]

The result was thus regarded as a confirmation of the $d_{x^2-y^2}$ character of the order parameter of this material. In another approach [227], a direct comparison of the field induced critical current variation of a hybrid SQUID fabricated from single crystalline YBCO-123 and Nb, expected to contain one π-junction, with $I_c(\Phi)$ of two control SQUIDs made of Nb alone was made. This comparison of $I_c(\Phi)$ of the two configurations, accomplished by simultaneous measurements of the dynamical resistances, again gave evidence for a π-shift between the $I_c(\Phi)$ curves of the two types of SQUIDs. Additional experiments [228] of the same type confirmed the main result of the quoted first two investigations.

(f2) Single Junction Experiments

The magnetic field induced modulation of the critical current I_c of a single Josephson junction between a conventional s-wave superconductor and a cuprate superconductor may also give information concerning the order parameter symmetry of the latter. This possibility has first been discussed and finally realized again by Wollman and coworkers [229]. They argued that the $I_c(\Phi)$ curves for single Josephson contacts, either straddling the corner separating the a and the b direction of a single crystal of YBCO-123 or placed along the edge of a crystal, would again differ significantly. They also pointed to some advantages using this technique, avoiding some problems that are related with the use of the SQUID technique mentioned in (f1) above.

For extended Josephson junctions between two conventional superconductors, the critical current is strongly influenced by external magnetic fields and it has since long been established that $I_c(\Phi)$, where Φ is the magnetic flux threading the barrier region of the contact, exhibits a Fraunhofer type pattern (see Eq. (14.25)), as shown in Fig. 14.65. For junctions with extensions that are still smaller than a characteristic length, such that self-field effects can be neglected, the resulting pattern has indeed the shape described by

$$I_c(\Phi) \sim \left| \frac{\sin\left(\frac{\pi\Phi}{\Phi_0}\right)}{\frac{\pi\Phi}{\Phi_0}} \right| \cdot \left| \frac{\sin\left(\frac{\pi\Phi}{\Phi_0}\right)}{\frac{\pi\Phi}{\Phi_0}} \right| . \qquad (14.25)$$

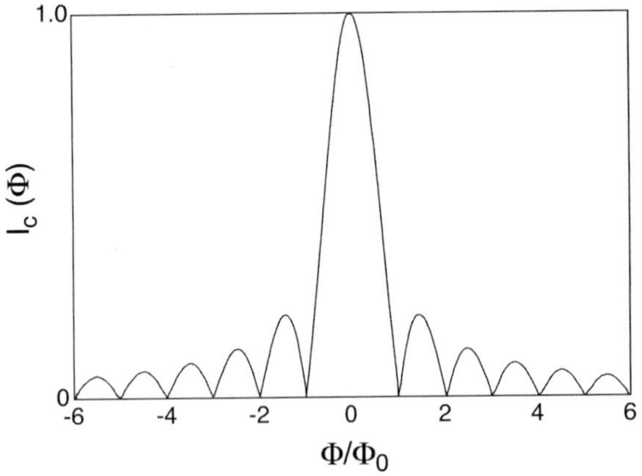

Fig. 14.65. Fraunhofer type interference pattern of an extended Josephson junction as a function of magnetic flux through the junction (see Eq. (14.25))

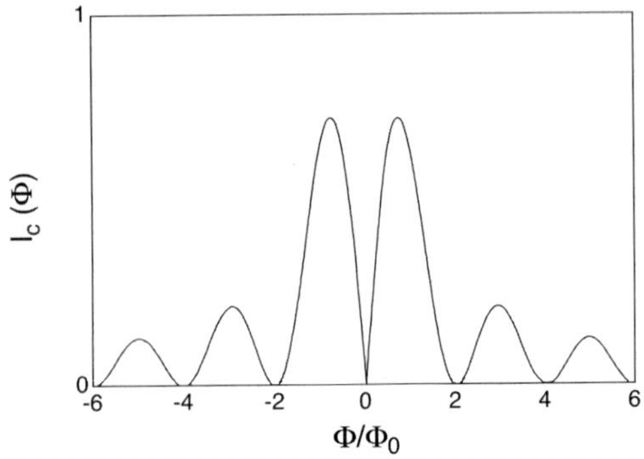

Fig. 14.66. Expected schematic interference pattern for a single Josephson junction straddling the corner of a Y-123 single crystal, assuming a d-wave symmetry of the superconducting order parameter (see [229])

For individual junctions based on contacts between a conventional superconductor and a cuprate material, essentially the same I_c behavior is anticipated for the above mentioned single edge contacts. In case of an anisotropic order parameter with $d_{x^2-y^2}$ symmetry, in contacts straddling a crystal corner in a more or less symmetrical way, the compensation of the positive and the negative currents is expected to result in a zero critical current at $\Phi = 0$ (mod.π) and hence [229]

$$I_c(\Phi) \sim \left| \frac{\sin^2\left(\frac{\pi\Phi}{2\Phi_0}\right)}{\left(\frac{\pi\Phi}{2\Phi_0}\right)} \right| \cdot \left| \frac{\sin^2\left(\frac{\pi\Phi}{2\Phi_0}\right)}{\left(\frac{\pi\Phi}{2\Phi_0}\right)} \right| . \quad (14.26)$$

This behavior is displayed in Fig. 14.66. Again, contacts along a single edge ought to show a flux dependence according to Eq. (14.25). Since in practice these corner contacts can never be made exactly symmetrical, some asymmetry in the patterns must be anticipated. The key feature, however, remains whether a dip or a peak in $I_c(\Phi)$ is observed for single corner junctions at $\Phi = 0$. In Fig. 14.67 the result of one of this type of experiments between Pb and an underdoped single crystal of YBCO-123 is shown [230], confirming the earlier results of [229], obtained on optimally doped YBCO-123. The experimental attempts based on the techniques briefly addressed in (f1) and (f2) have been reviewed in [224].

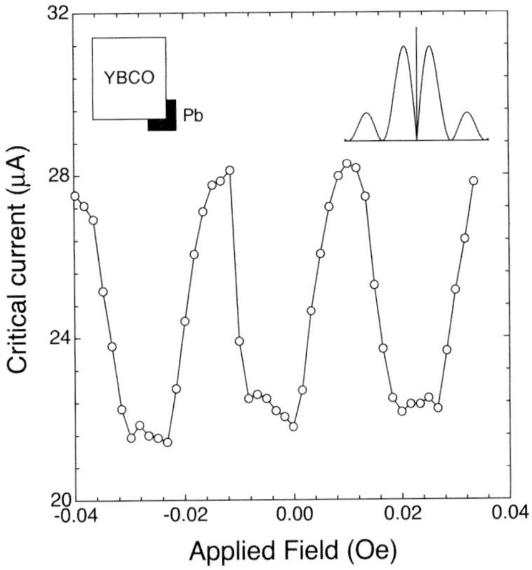

Fig. 14.67. Magnetic field dependence of the critical current I_c of a corner junction between an underdoped Y-123 single crystal and Pb. The insets show the schematic experimental configuration and the ideal expected pattern. Note that I_c adopts a minimum at zero external magnetic field

(f3) Scanning SQUID Magnetometry

For a simple closed superconducting loop, the maxima of I_c at $\Phi = \nu \cdot \Phi_0$ imply that the loop may trap flux quantities given by an integer number of flux quanta, a situation with zero trapped flux is also possible. A loop with an odd number of π-junctions, however, will trap fluxes $\Phi = (\nu+1/2) \cdot \Phi_0$. This means that in order to stabilize, the loop will always generate a spontaneous current, thus inducing a magnetic moment corresponding to at least one half of a flux quantum if the external field is set to zero.

In order to test this consequence of an odd number of π-junctions in a loop arrangement, Tsuei, Kirtley and coworkers designed a really ingenious experiment [231]. In their first attempts, they deposited small ring-shaped films of YBCO-123 on suitably tailored $SrTiO_3$ substrates. These substrates were composed of differently oriented single crystals, mended in such a way that would allow the preparation of different rings, epitaxially deposited on the substrates, with either an even, including zero, or an odd number of π-junctions. The amount of trapped flux, induced in different ways, was monitored with a scanning micro SQUID probe. It was convincingly shown that loops with an odd number of π-junctions behaved differently from those with an even number or none at all. The original layout of the samples is schematically shown in Fig. 14.68. With this relative orientation of the crystals, the relative orientations of the $d_{x^2-y^2}$ lobes of the order parameter in the corresponding segments of the YBCO-123 films as shown in Fig. 14.69, are anticipated. With these orientations, the prediction is that only the loop with its center at the location where all three crystals meet should exhibit the trapping of half integer fluxes. This is exactly what the outcome of the experiment confirmed. A number of additional experiments with either other choices for the relative orientations of the substrate crystals [232] or with films of other cuprate superconductors, such as Tl-2201 [233] and Bi-2212 [234], all confirmed the above result and helped to rule out extrinsic causes for the observation of half integer fluxes where expected. The same type of experiments also settled an earlier debate concerning the order parameter symmetry of electron doped cuprate superconductors. For a while, based on penetration depth experiments, such as mentioned in a previous subsection, it was claimed that the superconducting state of this variety of copper oxides has to be characterized by an s-wave type pairing. Phase sensitive measurements using the scanning SQUID method [235] have shown that also electron doped cuprates adopt a superconducting state with a gap of $d_{x^2-y^2}$ symmetry. An extensive review covering

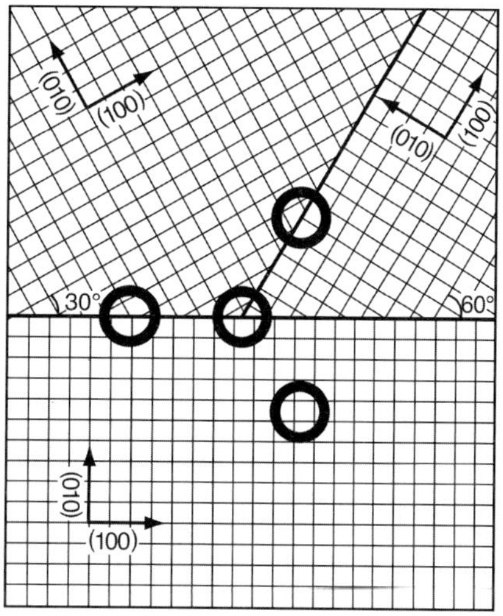

Fig. 14.68. Schematic view of superconducting loops of ring shaped films of Y-123, deposited epitaxially on a tricrystal substrate with specially chosen orientations. The central ring is expected to have one π-junction (see [231])

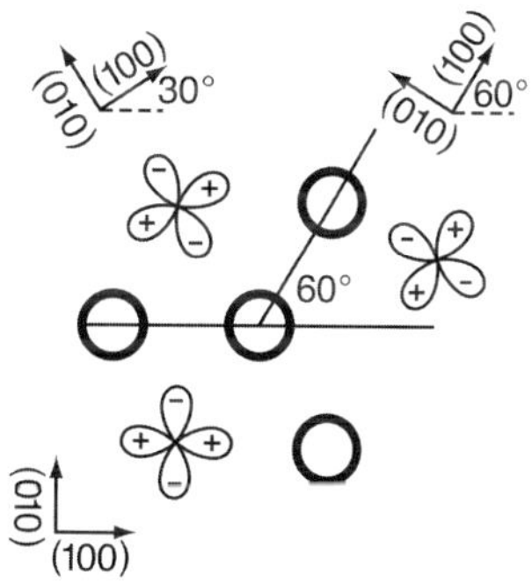

Fig. 14.69. Schematic relative orientation of the d-wave type order parameter contour in the ring shaped films (see [225])

the experiments based on the scanning SQUID technique is given in [225].

(f4) Related Experiments and Problems

In connection with these tunneling experiments is should also be mentioned that not all of them give straightforward evidences for the pairing and hence gap anisotropies discussed above. Results of early experiments probing the Josephson tunneling through grain boundaries [236] were incompatible with d-wave pairing in the sense that no complete suppression of the critical current in the relevant geometry of two order parameter lobes with different signs opposing each other on either side of the grain boundary could be achieved. It has been suggested [237] that in the long junction limit, the junction roughness may play an important role and that the appearance of local vortices at the grain boundary may weaken the expected frustration and therefore allow for non zero Josephson currents, i.e., currents at zero voltage, in this case.

Somewhat more complicated is the interpretation of tunneling experiments where the tunneling current is directed along the **c** -direction of the crystal lattice. Various experiments investigating this type of arrangement with contacts between Pb and YBCO-123 single crystals as well as epitaxial films have been made [238–240], and non zero Josephson currents through these junctions were observed. In case of a strict d-wave anisotropy of the pairing in the plane, such non zero currents should, of course, be absent. This apparent inconsistency has been ascribed to the orthorhombicity of the crystal lattice of YBCO-123 and the corresponding loss of the fourfold rotation symmetry in the plane [241]. This situation allows for a mixing of d and s symmetry in the pairing, a scenario that was early suggested in [242]. Therefore, the observation of non zero Josephson currents may be explained in this way. More detailed discussions of these issues are available in [224] and [225].

Consequences of *d*-Wave Pairing

The now fairly well established dominant d-wave pairing of quasiparticles in cuprate superconductors does not automatically provide a clear cut conclusion concerning the nature of the pairing interaction. An unequivocal identification of the pairing mechanism in cuprates is a formidable task, also in view of the non Fermi liquid behavior. It was believed that, because of the appearance of gap nodes, at least an electron phonon mediated pairing could be ruled out. Recent Eliashberg-type calculations may indicate that this may not necessarily be true [243].

Nevertheless, the earliest suggestions of a $d_{x^2-y^2}$-type pairing symmetry were based on the assumption that the pairing was achieved by exchange of antiferromagnetic spin excitations. This work has been reviewed in [191] and [244], see also chapters by Chubukov et al. and Manske et al. in this volume. Since in this case in the gap equation

$$\Delta(k) = -\Sigma \frac{V(k-k')\Delta(k')}{2E(k')} \qquad (14.27)$$

the interaction $V(k-k')$ is always positive and has its maxima at the (π,π) points of the Brillouin zone, the resulting gap function adopts a $d_{x^2-y^2}$-type symmetry of the form

$$\Delta(k) = \Delta_0(\cos k_x - \cos k_y)\,. \qquad (14.28)$$

These simple considerations have been complemented by extensive numerical studies cited in [191] which, for this type of interaction, confirm the stability of a superconducting state with $d_{x^2-y^2}$ pairing.

Among other consequences, the d-wave pairing also has some direct influence on the configuration of the vortices in the mixed state [245, 246]. First of all, singular vortices are no longer expected to exhibit a circular cross section. Second, the vortex lattice, exhibiting a hexagonal symmetry in conventional superconductors, has been shown experimentally [247, 248] to adopt a rhombohedral symmetry where the two unit cell vectors are of approximately equal length, forming an angle of 77 degrees. One may argue that this again is a consequence of the d-wave pairing, considering both the distorted cross section of the single vortices and possible resulting changes in the interaction between the vortices.

In view of some isotope effect experiments, it seems that interactions between itinerant quasiparticles and the lattice are of some importance for the superconductivity in cuprates as well. In general it is assumed that electron phonon interactions are not favorable for d-wave pairing. Therefore, it has been proposed that the pairing or gap symmetry is a combination mainly of d-wave and s-wave character [242]. Even to date the situation is far from being clear.

14.4.3 Coexistence of Superconductivity and Magnetic Order

The phenomenon of coexistence of magnetic order and superconductivity was long thought to be impossible, especially in view of the pair breaking influence of magnetic fields and magnetic impurities established for conventional superconductors [249]. Nevertheless, the first examples of the coexistence of antiferromagnetic order and superconductivity were reported in the mid seventies in two classes of compounds, the Chevrel phases (molybdenum sulfides and selenides) [250] and the Rhodium–Boron compounds [251]. In these series, compounds with rare-earth ions occupying regular lattice sites may be prepared. Because of the incompletely filled 4f electron shells, most of these ions carry respectable ionic moments resulting from spin and orbital degrees of freedom. Since these materials are metallic, it is no surprise that these moments order spontaneously at low temperatures via the Ruderman–Kittel–Kasuya–Yoshida (RKKY) interaction mediated by the conduction electrons [252]. What is surprising here is that both the Néel temperature, fixing the onset of antiferromagnetic order, and T_c for superconductivity are of similar magnitude. This situation has been studied in great detail for a number of different materials and reviews of the main results are available in [253–255], see als chapters by Kulić and Buzdin in volume one of this book.

As we mentioned in Sect. 14.2.2, some cuprates with rare-earth ions on regular lattice sites may also be synthesized in the classes of YBCO-123 and YBCO-124 compounds. Most of the rare-earth elements may be used to replace Y in these materials. At optimal doping conditions, i.e., close to seven and eight oxygen atoms per unit formula, respectively, the values of the critical temperatures of these rare-earth cuprates are very close to those of the corresponding Y compounds. In this sense there are no significant pair breaking effects to speak of. The observation of antiferromagnetic-type of ordering has been reported for some of them. The ordering temperatures are rather low and reach a maximum of approximately 2.2 K for fully oxygenated GdBCO-123 [256]. In most of the other of these compounds the ordering sets in below 1 K [257].

It is most likely that both the coexistence phenomenon and the low magnetic transition temperatures are due to the rare-earth site in the YBCO-123 crystal lattice which is situated between the two essential Cu-O planes (see Fig. 14.6(a)). These sites are electronically quite well separated from the charge carrier system in the planes and therefore have little influence on superconductivity. Likewise the RKKY interaction between these ions is, for the same reason, rather weak. Together with the reduced number of available excited quasiparticles this implies a priori a low ordering temperature. It turns out that the concentration of itinerant charge carriers at the rare-earth sites is so low that it is doubtful whether the RKKY interaction is of any significance at all. The Néel temperature for antiferromagnetic order in $HoBa_2Cu_3O_7$ is of the order of 0.1 K [258], most likely induced by the hyperfine field of the Ho nuclei. This is no surprise since the CEF splitting of the Ho^{3+} Hund's rule $J = 8$ multiplet leaves a singlet ground state and the ordered moments of approximately $2\mu_B$/Ho ion turn out to be much reduced from the free ion value.

14.5 Physical Properties of Non-Cuprate High-T_c Superconductors

In this section, some selected physical properties of superconductors with unusually high critical temperatures are presented and discussed. None of them, reaches values that exceed the boiling point of liquid nitrogen. Nevertheless, the discovery of superconductivity of these materials and the magnitude of T_c were and in some cases still are rather surprising, and therefore of interest.

14.5.1 Ba-Based Oxides

$BaPbO_3$, in spite of in principle compensating valences of its chemical constituents, is metallic. The electrical resistivity is almost temperature independent and of the order of a few hundred $\mu\Omega$cm. Gradually replacing Pb by Bi leads to a structural transition if about 10% of the Pb content is replaced by Bi. For $BaPb_{1-x}Bi_x$, the electrical resistivity ρ increases continuously with increasing x but still stays more or less temperature independent [51]. According to measurements of the Hall constant R_H [51], the charge carrier concentration increases with x, passing through a maximum at $x \approx 0.25$, which appears to be the optimal doping situation for the occurrence of superconductivity. At this x value, the density of itinerant charge carriers and the sign of R_H are compatible with one extra itinerant electron per Bi atom. Conventionally viewed, this concentration of conduction electrons is rather low for a superconductor with a T_c that exceeds 10 K. In contrast to other oxide superconductors, in particular also the cuprate superconductors, the bulk magnetic susceptibility χ of the normal metallic phase of $BaPb_{1-x}Bi_xO_3$ is negative and largest in magnitude at $x \approx 0.25$. This suggests that the paramagnetic contribution to χ is weakest at this composition.

As shown in Fig. 14.70, clear resistive transitions to superconductivity are observed for $0.15 < x < 0.35$, but the sharpest transitions at maximum values for T_c are observed for $x = 0.25$ [51, 259]. A further increase of x provokes a reduction of T_c and eventually leads, as indicated by both ρ (see Fig. 14.70) and R_H data, to a metal insulator transition which is accompanied by the already mentioned structural transition back to an orthorhombic symmetry of the crystal structure.

Investigations of the superconducting state [52] revealed that (BPBO) $BaPb_{0.75}Bi_{0.25}O_3$ is an extreme type II superconductor with a Ginzburg Landau parameter κ of the order of 70÷80. It appears that the electronic specific parameter γ, estimated from a barely resolvable specific heat anomaly at T_c and using Eq. (14.6), is rather small, i.e., $\gamma \approx 1.5$ mJ/molK^2, indicating a low density of electronic states $D(E_F)$ at the Fermi energy, usually unfavorable for high critical temperatures for the onset of superconductivity. Tunneling data seem to indicate strong electron–phonon interactions extending in energy into the range of the superconducting gap in the excitation spectrum [52, 259]. If strong coupling effects are important, the γ value quoted above is significantly reduced and $D(E_F)$ turns out to be even smaller. From the theoretical point of view [260] there is no consensus, how the relatively high critical temperature is to be explained, because the strong-coupling concept is not compatible with band structure calculations [261] and their connection with the metal to insulator transition mentioned above. In addition, the ratio $2\Delta_0(0)/k_BT_c$ has been estimated to be 3.5, practically identical with the weak coupling BCS value.

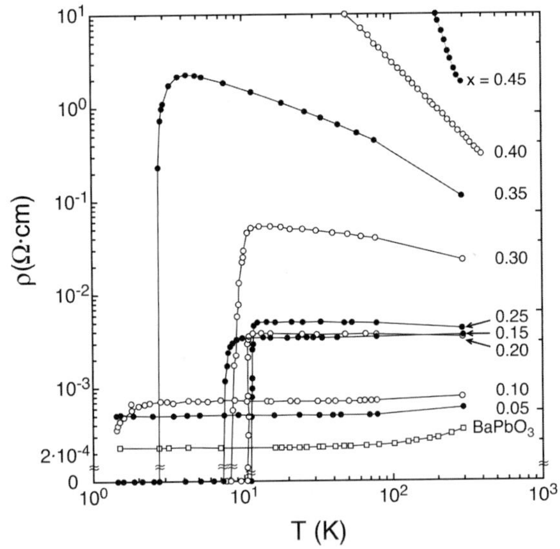

Fig. 14.70. Variation of the temperature dependence of the electrical resistivity of $BaPb_{1-x}Bi_xO_3$ (see [51])

Some problems are also encountered in the interpretation of measurements of the upper critical field H_{c2}. In the dirty limit, a reasonable choice considering the resistivity data displayed in Fig. 14.70, the limiting orbital critical field is given by [262]

$$H_{c2}(0) = -0.691 \left(\frac{\partial H_{c2}}{\partial T} \right)_{T_c} \cdot T_c \,. \qquad (14.29)$$

It appears that the extrapolated experimental limit of $H_{c2}(0)$, based on measurements of H_{c2} (T) [52], shown in Fig. 14.71, by far exceeds the one that is calculated from the measured initial temperature dependence of H_{c2} just below T_c, indicated by the broken line in Fig. 14.71. A clear result is, however, obtained from measurements of the isotope effect on T_c by exchanging ^{16}O and ^{18}O on the oxygen sites [263]. The heavier oxygen isotope leads to a significant reduction of the critical temperature and the value of the parameter α entering Eq. (14.10) has been calculated to be $\alpha = 0.22$. Although this result seems to confirm that most likely the occurrence of superconductivity in the Pb/Bi oxides is due to an electron phonon interaction, it has been argued that electronic excitations involving atomic displacements due to charge redistributions are mediating the pairing interaction [263]. This conclusion is based on considering the relatively high value of T_c and indications that pairing in BPBO is a weak coupling effect.

The observation of critical temperatures that are more than twice as large as those of the Pb/Bi alloys and comparable to some of the cuprate superconductors, i.e., of the order of 30 K, for $Ba_{0.6}K_{0.4}BiO_3$ (BKBO) [54,55], naturally led to additional experiments probing the physical properties of this compound [263,264]. In particular, a comparison with some of the properties of the cuprate superconductors was attempted. The total normal state magnetic susceptibility again reveals a strong core diamagnetism which, overall, masks a very weak paramagnetic contribution [263,264]. Normal state Hall effect measurements [264] revealed that R_H is weakly temperature dependent and negative at all temperatures below 300 K. This confirms that the Bi-based oxides, without Cu, however, are electronically and magnetically quite different species than the hole doped copper oxides.

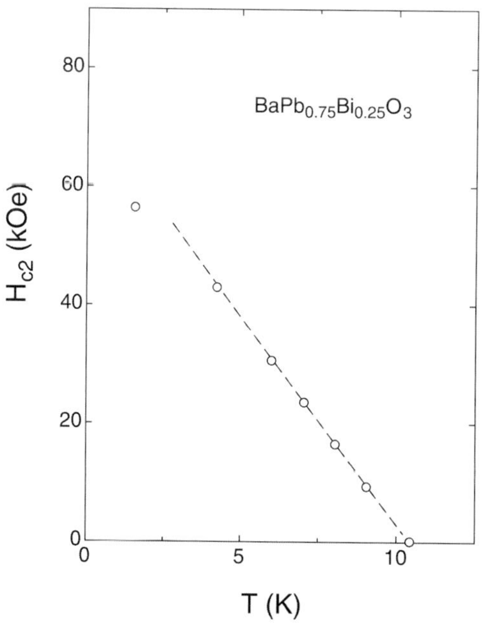

Fig. 14.71. Temperature dependence of the upper critical field H_{c2} of $BaPb_{0.75}Bi_{0.25}O_3$. The *broken line* is to guide the eye (see [52])

The response of the superconducting state to external magnetic fields is rather similar to that of the Pb/Bi alloys, leading to almost the same values of $(\partial H_{c2}/\partial T)_{T_c}$, of the order of −5 kOe/K. Estimates of the Ginzburg–Landau parameter give $\kappa \approx 45$, again a large value. As for the Pb/Bi alloys this is compatible with a low density of itinerant carriers and consequently a large penetration depth λ. From available experimental results covering the upper and lower critical fields, $H_{c2}(T)$ and $H_{c1}(T)$, estimates of the initial temperature dependence of the thermodynamical critical field $(\partial H_c/\partial T)_{T_c}$ can be made. Finally, using the law of corresponding states allows to calculate the electronic specific heat parameter γ or effective density of states $D(E_F)$. Although somewhat larger than the corresponding value calculated for the Pb/Bi alloy, also $D(E_F)$ for the Ba/K alloy is much too small to account for a T_c of the order of 30 K in any conventional way. Nevertheless, also for this material, a significant oxygen isotope effect confirms that some type of electron phonon interaction must be important for the onset of superconductivity [263]. The values for the α parameter differ somewhat in different reports; α-values between 0.21 and 0.35 have been calculated [263,264]. Again, conventional strong coupling effects have been ruled out.

For $(Ba/K)BiO_3$, measurements aiming at probing the symmetry of the superconducting state have been made [265]. Corner-junction type experiments, exploring the field dependence of the critical current $I_c(\Phi)$(see Eq. (14.26)), indicate that the superconducting state of this alloy is characterized by a nodeless gap function or order parameter.

14.5.2 C_{60}-Based Materials

As we have briefly mentioned in Sect. 14.3.3 the simplest way to achieve a metallic state in bulk C_{60}-based material is via electron doping with atoms of the alkaline metal elements. Most of the experimental results that are briefly discussed below have been obtained for compounds of the type A_3C_{60}. Some combinations of this type turn out to be electrical insulators. Based on calculations of the widths W of the essential electronic energy bands and the onsite Coulomb correlation energies U, indicating that

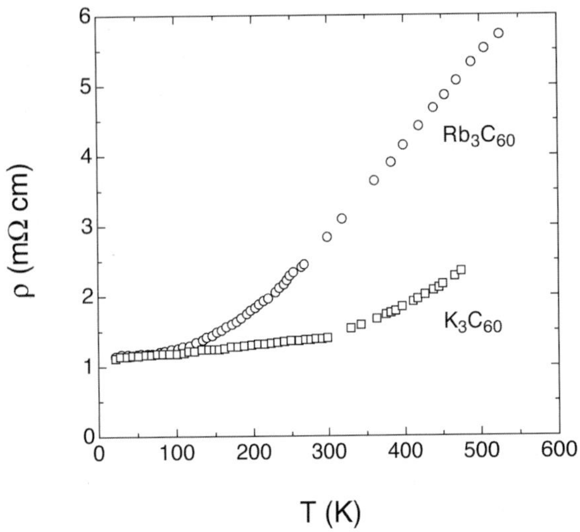

Fig. 14.72. Temperature dependence of the electrical resistivity of thin films of K_3C_{60} and Rb_3C_{60}(see [269])

the ration U/W significantly exceeds the value of 1, it is argued that the metallic species are on the metallic side of a metal insulator transition of Mott–Hubbard type and therefore, correlation effects have to be taken into account [266,267]. Another difficulty arises, because the energies of the relevant phonon branches are not very different from W and hence, invoking Migdal's theorem [268] for justifying the electron pairing via a retarded electron–phonon interaction is questionable.

The absolute values of the electrical resistivities ρ are rather high and correspond to electronic mean free paths of the order or even less than the separation between single molecules. Examples of $\rho(T)$ of superconducting A_3 C_{60} films are shown in Fig. 14.72 [269]. Even at very low temperatures and for single crystalline samples, the absolute values of ρ are of the order of 1 mΩ cm. In most cases the low temperature electrical resistivities vary as T^2 [270] but also linear-in-T or T-independent variations down to low temperatures, depending on the sample configuration (film, crystalline, granular) have been reported. It is therefore clear that a theoretical treatment of this transport property is difficult, even more so, because it seems to matter whether $\rho(T)$ is measured at constant volume or at constant pressure. Nevertheless, calculations invoking the interactions between conduction electrons and various lattice modes may be brought into reasonable agreement between experiments, carried out under constant pressure conditions, revealing $\rho \sim T^2$, and theory. A renormalization of the same data to constant volume conditions reveals that ρ varies approximately linearly with T above 100 K [271]. This result has apparently not yet found a theoretical explanation. Although the very short mean free path at temperatures of the order of 300 K might suggest a saturation behavior in $\rho(T)$ at even more elevated temperatures, this phenomenon is, very much like in the case of the cuprates, not observed. In spite of the similarity of these features in $\rho(T)$ of the two types of materials, it seems quite unlikely that the same scattering processes are the cause for this observation.

Quite generally, the magnetic susceptibility χ in the normal state is positive and almost temperature independent [272], as is also found for the resonance shift in NMR measurements [273]. Comparing the magnitude of experimental data with χ values calculated from band structure data suggests that a rather substantial enhancement of χ must be considered [272]. The corresponding density of states $D(E_F)$ (see Eq. (14.4)) is larger by a factor of 3 than the bare density of states D_b (E_F) that follows from the band structure, but in fair agreement with estimates of $D(E_F)$ based on NMR spin-lattice relaxation data [273]. As expected for metals, the spin-lattice relaxation rate T_1^{-1} is approximately linear in T in the normal state, however, the hyperfine coupling constant is not known very well.

Apart from the clear manifestations of the superconducting transition in $\rho(T)$ and $\chi(T)$ the transition is, as usual, also manifest in experimental data of a number of other properties. This is certainly the case for measurements probing the specific heat. Because of the fragility of the materials, especially with respect to air exposure, these measurements [272] are technically not trivial. As demonstrated in Fig. 14.73, the specific heat anomaly at T_c is masked by a large background of non electronic origin, but estimates concerning its magnitude can still be made. For K_3C_{60}, the ratio of $\Delta C(T_c)/T_c$ is of the order of 70 mJ/mole K^2, a rather small value but confirming the rather low effective electronic density of states at E_F.

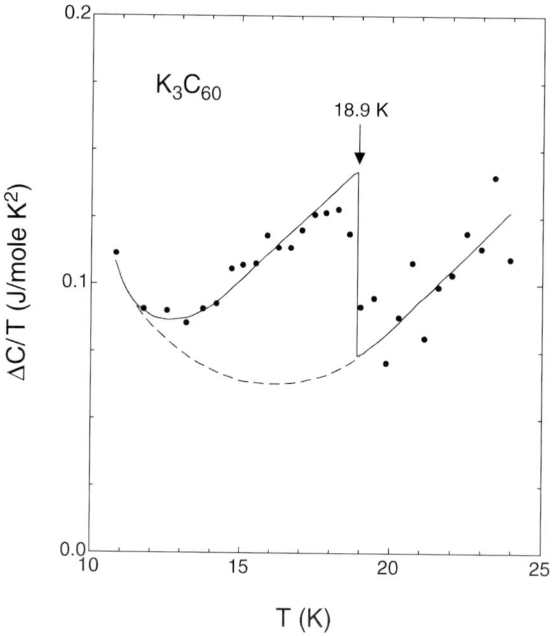

Fig. 14.73. Anomaly of the specific heat at the superconducting transition of K_3C_{60}. The *solid line* is to guide the eye and the *broken line* indicates the estimated background (see [272])

Considering the electronic specific heat parameter γ that is related to $\Delta C(T_c)/T_c$ and the enhancement of the magnetic susceptibility, estimates of the electron mass enhancement $(1+\lambda_{ph})$ due to the electron phonon interaction can be made. Depending on details of the approximations, λ_{ph} values between 0,5 and 0.7 have been calculated and it was concluded that the pairing via the electron–phonon interaction could not be achieved by invoking low energy phonon excitations. Instead a coupling to high energy phonons and an upper bound of $\lambda_{ph} \approx 1.0$ have been postulated [272].

The participation of phonons in the pairing of electrons is put in evidence by isotope-effect experiments where ^{12}C is replaced by ^{13}C. Although there is some scatter in the data, clear variations of T_c upon isotope exchange have been identified [274,275] and α-values between 0.3 and 0.4 seem to be the most reasonable. The importance of the phonons related with excitation modes of the C_{60} molecules is emphasized by the absence of an isotope effect if the isotope exchange involves the alkaline metal elements. This indicates that apart from the direct influence on the density of itinerant electrons via their ionic radii, the alkaline-metal atoms have only a weak influence on the magnitude of T_c [276,277].

The onset of superconductivity is clearly reflected in a distinct reduction of the NMR spin lattice relaxation rate T_1^{-1} at low temperatures [273]. Results of early measurements gave no evidence for a Hebel–Slichter type enhancement of the relaxation rate at temperatures close to but below T_c, expected if the above mentioned coherence effects are effective. Even for well known conventional superconductors, the absence of this enhancement may be caused by strong coupling and other trivial effects. However, a later investigation clearly revealed an increase of T_1^{-1} with decreasing temperature below T_c, as displayed in Fig. 14.74, providing convincing evidence

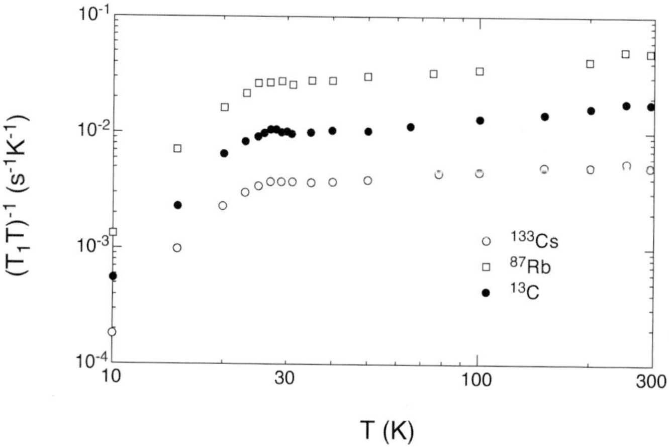

Fig. 14.74. Temperature dependence of NMR spin lattice relaxation rates, plotted as $(T_1T)^{-1}$ vs T, in the normal and superconducting states of Rb_2CsC_60, providing some evidence for coherence effects just below $T_c \approx 31$ K (see [278])

for coherence effects in a at most moderately coupled superconductor [278]. The same type of feature has been observed in μSR experiments [279], and therefore it seems reasonable to qualify these alkaline-metal doped C_{60} materials as s-wave type superconductors with medium strength electron phonon coupling. Another manifestation of the superconducting state is apparent in the reduction of the spin susceptibility χ_S, which in principle can be deduced from the temperature dependence of the spin part K_S of the resonance shift [273,278]. Measurements of total resonance shifts of nuclei occupying inequivalent lattice sites made it possible to extract the temperature dependence of the Knight shift $K_S(T)$ and the results are compatible with expectations that are valid for weak coupling superconductors and s-wave pairing, in particular via the evaluation of the ratio $2\Delta_0(0)/k_B T_c$ [278]. Different values of this ratio have been extracted from different NMR data sets, ranging from 3.0 to 4.3. Estimates of this ratio based on optical experiments [280,281] gave values between 3.4 and 4.2. Somewhat more enhanced values of 6.3 and 6.4 have been extracted from results of tunneling experiments [282,283]. These values would imply a substantial enhancement of the effective density of states and $\lambda_{ph} \approx 3.0$. This value is far beyond any of the other estimates of this value, such that the results of the tunnelling experiments seem more than questionable.

Values for the characteristic lengths of typical C_{60}-based superconductors have been evaluated from μSR [284,285], NMR [273] and optical data [286], as well as from results of measurements of the upper critical field H_{c2} [287]. Evaluations of H_{c2} (T) from observing resistive transitions are hampered by a sizeable increase of the transition width in external magnetic fields [269]. The spread of the results concerning λ_L and ξ is sizeable. For the penetration depth $\lambda_L(0)$, values between 4000 and 8000 Å have been obtained. The effective coherence length ξ_0, taking into account the very short electronic mean free path, turns out to be of the order of 30 Å. The ratio of these two quantities, defining the GL parameter κ, indicates that these C_{60} superconductors are also of the extreme type II variety. The large values of the penetration depths are compatible with the previous indications for a low concentration of itinerant charge carriers and a substantial enhancement of their effective mass.

If all these experimental results providing information on the superconducting state of these C_{60}-based bulk materials are taken together [45, 288], it seems inevitable to conclude that one is dealing with conventional superconductors of the weak, or at best medium coupled variety invoking some sort of electron–phonon interaction.

14.5.3 MgB_2

In the class of seemingly simple binary intermetallic compounds, MgB_2 exhibits by far the highest onset temperature for superconductivity. For comparison, the cubic binary transition metal compounds of the type A_3B, crystallizing with the β tungsten or A15 crystal structure, exhibit T_c's up to 18 K [289] and only if they are prepared under special conditions or in special forms, such as quench condensed films, they exhibit superconductivity at temperatures exceeding the temperature of the boiling point of liquid hydrogen, i.e., 20.4 K, by about 3 K [290] (see also Fig. 14.1). The again completely unexpected discovery of a critical temperature close to 40 K for a material that simple, naturally raises the question whether this feature may already be recognized in the normal state properties of the compound. The magnitude of T_c also provokes experiments aiming at clarifying the characteristics of the superconducting state and, in particular, to identify the mechanism leading to the pairing of electrons at this relatively high temperature. In this context also rather challenging questions need to be answered by theorists. In addition, for MgB_2, some properties that are crucial for potential applications seem to show favorable behavior and therefore this material is also of high current interest for more technically oriented developments.

The first report on superconductivity of MgB_2 dates from mid-January of 2001 [11] and, although a lot of subsequent work has appeared either in print or in e-print archives, not all the open questions that are indicated above have been answered up to now. This brief review, by far not complete and aiming at presenting typical features, considers work that has become available until the end of July 2001. Samples for the many investigations were prepared in many different forms, including single crystals, mostly of rather small size, polycrystals, powder, films, wires and tapes. A first brief review of various properties of superconducting MgB_2 has been presented in [81].

The Normal State

The electrical resistivity below room temperature is that of a typical metal with ρ of the order of 10 $\mu\Omega$cm at room temperature and $\partial\rho/\partial T > 0$ at all temperatures between T_c and 300 K. Resistance ratios RR $= \rho(300)/\rho(T_c)$ of the order of 6 to 7 have been reported [291] for single crystals for currents along the basal plane of the hexagonal crystal lattice. An example of $\rho(T)$, measured for a current in the basal plane of a single crystal between 2 and 300 K is shown in Fig. 14.75 [292]. Amazingly large ratios of up to 25 have been obtained for polycrystalline material and even wires [293,294]. Since there are no reliable data yet available for the electrical conductivity along the hexagonal c-axis, it is not clear whether this difference might be due to a strongly anisotropic conductivity or, in other words, a much lower resistivity for currents along the c-axis. In all cases, however, the regime of residual resistivity, of the order of 1 $\mu\Omega$ cm or less, seems to be reached at T_c. While practically absent at temperatures above 200 K, a fairly large positive magnetoresistance at $T \approx T_c$, with almost a doubling of the zero field resistance in an external field of 10 T has been reported for polycrystalline material [293]. This effect is less pronounced for single crystals, at least for ρ measured in the hexagonal basal plane [291,292]. Hall effect data are scarce and they have not been obtained for well defined geometrical configurations. Nevertheless, the available set of results [81,295] is entirely consistent with holes being the itinerant charge carriers in MgB_2. No agreement has yet been established concerning the question whether the positive Hall constant R_H varies with decreasing temperature between room temperature and T_c. It is interesting to note [296] that for other, isostructural metallic diboride compounds, for which no occurrence of superconductivity has yet been reported, R_H is negative [297,298].

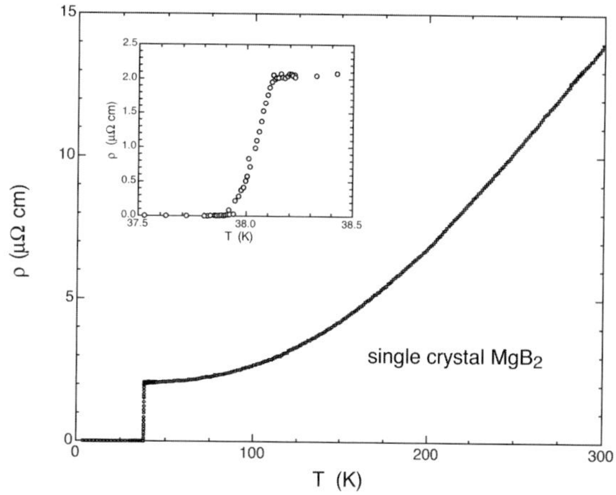

Fig. 14.75. Temperature dependence of the in plane electrical resistivity of single crystalline MgB_2. The *inset* emphasizes the superconducting transition region

Significant anisotropies have been revealed in measurements of the lattice parameters, both as a function of external pressure at room temperature [299] and as a function of temperature at ambient pressure [300]. In both cases the response along the c axis is much more pronounced than along the a axis, confirming that the out-of-plane Mg-B bonds are significantly weaker than the in-plane Mg-Mg bonds. As a consequence the c/a ratio decreases with increasing pressure and with decreasing temperature. Band structure calculations [49,301] indicate that the in-plane B-B bonds are dominantly covalent.

As usual in cases where superconductivity sets in at relatively high temperatures, it is not straightforward to obtain information on the normal state electronic properties, such as $D(E_F)$, from measurements of the specific heat. This problem is also met in the case of MgB_2 and several ways of solving it have been followed. An analysis [302] of measurements of $C_p(T)$ in external fields between 0 and 9 T provides the following important parameters: (i) the electronic specific heat parameter $\gamma = 2.6$ mJ/mole K^2 and (ii) a Debye temperature $\Theta_D = 1050$ K. Another set of similar measurements, including results obtained in magnetic fields up to 16 T give $\gamma = 2.7$ mJ/mole K^2 [303]. Since in this work, the lattice contribution is ascribed to a continuum of modes, no

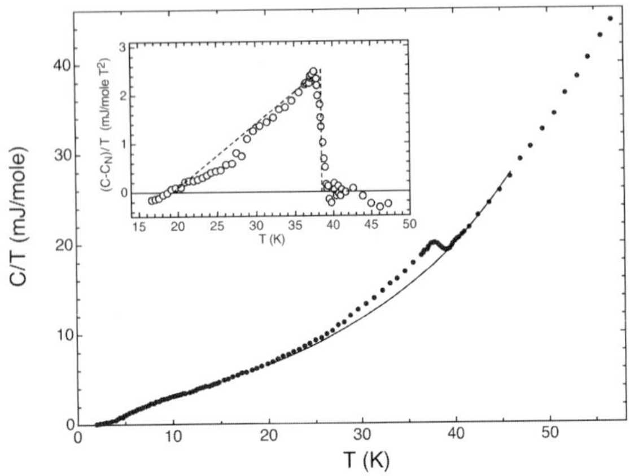

Fig. 14.76. Temperature dependence of the specific heat of a pressed-powder sample of MgB_2, plotted as C/T vs T. The *solid line* represents an estimated normal state background C_N. The *inset* emphasizes the anomaly at and below the transition temperature by plotting $(C - C_N)/T$ vs T

value for Θ_D is provided. In yet another approach, concentrating on a reliable estimate of the non-electronic background contributions in zero field, the same parameters turn out to be $\gamma = 3.8$ mJ/mole K^2 and $\Theta_D = 750$ K [304]. An example of C_p (T) of pressed non-commercial powder of MgB_2 below 60 K is shown in Fig. 14.76 [305]. The normal state NMR response of MgB_2, monitored on ^{11}B nuclei is compatible with the overall metallic character of the compound [306, 307]. The Knight shift K is very small and positive at all temperatures. Different reports do not agree with respect to the magnitude of the shift. $K(T)$ is approximately constant between 150 and 300 K. According to [307], K decreases steadily with decreasing temperature below 150 K and the decrease continues smoothly through the transition at $T_c(H)$. A distinct change in the slope of $K(T)$ at T_c has been reported, however, in [308]. The spin-lattice relaxation is well described by assuming a Korringa type relaxation via conduction electrons between room temperature and T_c with $T_1T \approx 170$ s, an average value resulting from a number of different investigations [306–308]. The Korringa constant, incorporating the Knight shift, is substantially smaller than the expected value for a relaxation via s-electrons and indicates that orbital degrees of freedom are dominant in the relaxation process [309].

Superconductivity

Rather narrow transition widths ΔT have been observed in resistivity measurements. For single crystals, values of ΔT of the order of 0.2 K have been achieved [291, 292]. If RR, defined above, exceeds the value of 5, the reported critical temperatures only vary within a range of experimental uncertainty. However, considerable variations of T_c may be induced by applying external pressure or by exchange of isotopes, such as ^{10}B for ^{11}B or ^{24}Mg for ^{26}Mg. At this point we may also mention that many different types of chemical substitutions have been tested. None of them, except one, have induced an enhancement of T_c. So far, the exceptional case is Zn for which a small enhancement of T_c at low concentrations of Zn atoms has been reported [310, 311].

As in most common superconductors, the application of quasi-hydrostatic pressure causes a reduction of T_c. We note a considerable scatter in the available data set [81] which may be due to non hydrostatic pressure conditions in some of the experiments, combined with an almost certain anisotropic response of T_c versus uniaxial stress. On the average the pressure derivative of the critical temperature, $\partial T_c/\partial p$ is of the order of 0.8 K/GPa. Clear evidence for isotope effects have been obtained for isotope exchanges on both the B [312, 313] and the Mg [313] sites. As may be expected, the Boron isotope effect is particularly prominent and from corresponding experimental results, the measured shift of the magnetic transition is displayed in Fig. 14.77, the mass exponent α_B in Eq. (14.10) related with the isotopes ^{10}B and ^{11}B, is deduced to lie between 0.26 and 0.3. This seems to confirm that lattice modes involving B atoms are involved in interactions providing the pairing of electrons. It is of no surprise that the exchange of the two Mg isotopes mentioned above leads to an only marginal shift of T_c and hence the Mg mass exponent $\alpha_{Mg} \approx 0.02$. While the pressure dependence of T_c may partly be due to a change in the electronic excitation spectrum, the isotope effect, although reduced to below its classical value [313], provides convincing evidence that the lattice and hence the electron–phonon interaction are essential for this superconductivity at moderately high temperatures.

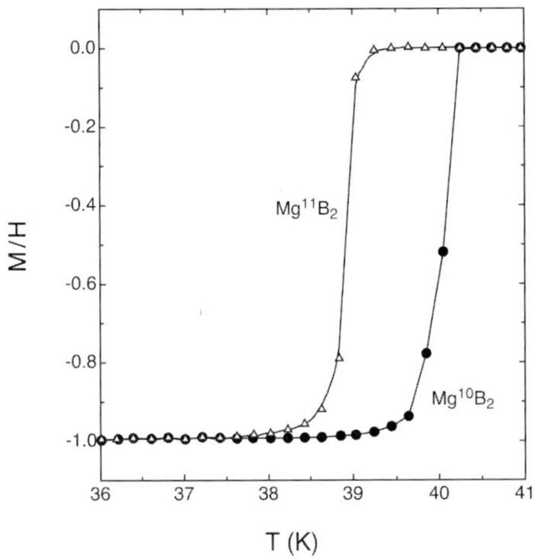

Fig. 14.77. Boron isotope effect on the magnetically probed superconducting transition of MgB_2 (see [312])

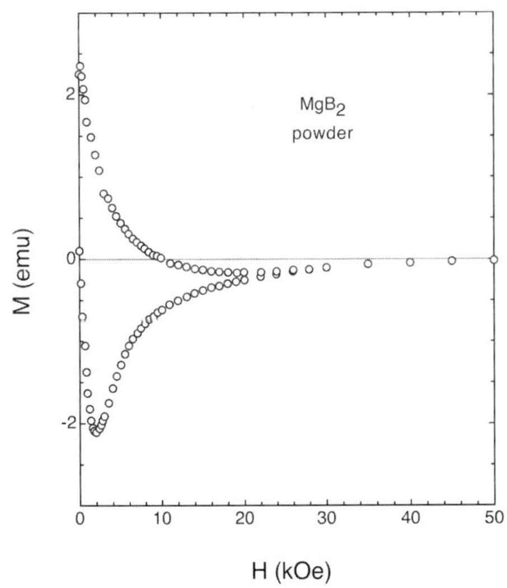

Fig. 14.78. Magnetization M(H) curve of a powder sample of superconducting MgB_2

Simple magnetization M(H) curves (see Fig. 14.78) [314] reveal at once that MgB_2 is a type II superconductor, adopting a fairly large GL parameter κ. Many attempts to measure the temperature dependence of the upper critical field H_{c2} (T) have been made. The available data sets on H_{c2} (0) and $(\partial H_{c2}/\partial T)_{Tc}$ cover a wide range of values for both quantities [81]. It appears that the claimed H_{c2} values are systematically higher if measured on thin film samples rather than on bulk crystalline specimens and reported values for $H_{c2}(0)$ vary between 15 and 40 T. Because of these uncertainties, which partly may be caused by anisotropy effects, we concentrate on the few experimental results that have been obtained using single crystals [291,292]. Unfortunately these experiments, measuring the resistive transition for currents in the *ab* plane and external fields oriented either parallel or perpendicular to the c-axis, have not been carried to very high fields and therefore any extrapolation to low temperatures is questionable. Nevertheless, a clear anisotropy is discernible in these data. The destruction of the superconducting state is more easily achieved, if the external magnetic field is applied along the direction of the c-axis of the crystal lattice. Overall, an anisotropy ratio $H_{c2}(H||ab)/H_{c2}(H||c) \approx 2$ is observed. Near T_c, $H_{c2}(T)$ derived from monitoring resistive transitions, definitely exhibits a positive curvature for both field orientations. From these experimental data, the corresponding coherence lengths have been estimated using Eq. (14.8). The respective values are $\xi_{ab}(0) = 65$ Å and $\xi_c(0) = 25$ Å. A comparison with the electronic mean free path of the order of 500 to 700 Å, that may be estimated from the normal state transport properties, indicates that MgB_2 may be treated as a clean-limit superconductor. Various estimates [81,303] of average penetration depths λ_L (0) at zero temperature range between 1200 and 1850 Å.

Estimates of the anisotropy of superconducting properties using electron spin resonance (ESR) and magnetization data from measurements on powder samples indicate an anisotropy parameter Γ between 6 and 9, i.e., much larger than what is quoted above from measurements of the upper critical fields on single crystals [315].

The bulk transition to the superconducting state has also been verified by measurements of the specific heat. Distinct anomalies indicating the discontinuity $\Delta C = C_s^{el} - C_n^{el}$ at T_c have been reported [302–304]. In magnitude these discontinuities are of the order $\Delta C/T_c \approx 3$ mJ/mole K^2, a value that is distinctly smaller that what is expected if the weak coupling BCS limit is considered. Electron phonon coupling parameters λ between 0.6 and 1.8 have been derived. These values are of the same order of magnitude as results from band structure calculations [316]. Most estimates of the ratio $\Delta_0(0)/T_c$ imply that MgB_2 should be regarded as a weakly coupled superconductor, rather surprising in view of the relatively high critical temperature. Specific heat data at temperatures well below T_c have been used to reach conclusions concerning details of the gap function. While some reports claim that the data are consistent with a common s-wave type pairing and a gap without nodes [304], a more detailed analysis suggests that at least two gap values, differing by a factor of 4, have to be considered [302,303]. In [303], the C_s^{el} data obtained at very low temperatures are claimed to suggest line nodes of the gap and also the field dependence of C_s^{el} indicates a d-wave pairing. Similarly inconclusive are available tunneling data [317,318]. In this context, however, a first principles calculation of the electron phonon coupling in MgB_2 leads to the conclusion that two gaps, related with different coupling strengths on different parts of the Fermi surface, should develop [319].

The NMR response in the superconducting state has also been investigated by a number of groups. The Knight shift data [307,308] exhibit a reduction of K below T_c but a rigorous analysis is very much hampered by the small size of the shift in the first place [309]. In one of the studies the observation of a coherence induced enhancement of the spin lattice relaxation rate T_1^{-1} below T_c and a subsequent drastic reduction of T_1^{-1} has been reported [307]. The result was interpreted as reflecting a strong coupling s-wave type superconducting state of MgB_2. In other work [308,309], no evidence for a coherence peak was identified and only minor reductions of T_1^{-1} below T_c were observed in rather low fields. It cannot be ruled out that problems with sample purities and inhomogeneities are responsible for these discrepancies.

First reports on probing the temperature dependence of the occupied electronic states close to E_F above and below T_c using photoelectron spectroscopy on polycrystalline material [320] indicate that the gap formation is close to isotropic in k-space, consistent with s-wave pairing. The temperature dependence of the gap amplitude is well reproduced with the original BCS expression of $\Delta(T)$ and $2\Delta(0)/k_B T_c = 3.0$, a value indicating weak coupling.

At this point is seems fair to say that the situation concerning the basic aspects of superconductivity of MgB_2 has not yet been settled. Although there is some consensus that the electron pairing in the superconducting state of MgB_2 is, in view of the high critical temperature, caused by an amazingly weak electron phonon interaction, many details need to be clarified. To some extent this situation is due to the present lack of experimental data obtained from single crystalline specimens. In particular, any aspects concerning the anisotropy of normal state and superconducting properties of this material must be treated by using single crystals.

Polycrystalline or powder samples of high purity, however, may be used for exploring the potential of MgB_2 for applications [293]. In this context some very encouraging results have been obtained [81]. Measurements of critical current densities j_c as a function of temperature and magnetic field in samples of different configurations have been made. Contrary to the Cu-oxide materials, the critical currents in MgB_2 seem to be controlled by the pinning strength for trapping vortices and not by effects of weak links at grain boundaries [321]. In zero magnetic field, j_c values of the order of 10^7 A/cm^2 have been achieved in films [322] and about an order of magnitude less in bulk samples, wires and tapes [81]. In wires and tapes, the critical current values are reduced rather rapidly in external magnetic fields, even at 4.2 K, much more than for the technically important superconductors of Nb-Ti alloy and Nb_3 Sn, which exhibit much lower critical temperatures. It has been established, however, that at the same fields, MgB_2 films may sustain much higher current densities and, in the range of liquid Helium temperatures, perform equally well as commercial Nb-Ti and Nb_3Sn wires [322]. This nurtures hopes that also the performance of wires and tapes may considerably be improved. Last but not least, the fabrication of superconducting MgB_2-based conductors appears to be a rather low cost affair.

14.6 Final Remarks

The field of superconductivity has experienced a number of unexpected discoveries during the last 20 years. Only some of them are discussed in this chapter, others are discussed in different sections of this book.

The appearance of superconductivity at unprecedented high temperatures has not only triggered research activities worldwide, but has also influenced the thinking with respect to the common understanding of the metallic state in solids in general. While there is consensus that the cuprate materials exhibit some unconventional superconductivity, it is yet not clear what is the pairing mechanisms. Note, recent ARPES experiments [321,323] observe interesting changes in the dispersions of some high-T_c cuprate superconductors.

There is little doubt that electron phonon interactions are essential for superconductivity in the considered non-cuprate oxides, the fullerides and MgB_2. Nevertheless, there are still open questions with regard to the characterization of the superconducting state in these materials.

Acknowledgement

I thank H. Thomas and S. Sigrist for their help in preparing the manuscript.

References

1. J.G. Bednorz and K.A. Müller, Z. Phys. B **64**, 189 (1986)
2. H. Takagi, S. Uchida, K. Kitazawa, and S. Tanaka, Jpn. J. Appl. Phys. **26**, L123 (1987)
3. C.W. Chu, P.H. Hor, R.L. Meng, L. Gao, Z.J. Huang, and Y.Q. Wang, Phys. Rev. Lett. **58**, 405 (1987)
4. R.J. Cava, R.B. van Dover, B. Batlogg, and E.A. Rietman, Phys. Rev. Lett. **58**, 408 (1987)
5. M.K. Wu, J.R. Ashburn, C.J. Torng, P.H. Hor, R.L. Meng, L. Gao, Z.J. Huang, Y.Q. Wang, and C.W. Chu, Phys. Rev. Lett. **58**, 908 (1987)
6. R.J. Cava, B. Batlogg, R.B. van Dover, D.W. Murphy, S. Sunshine, T. Siegrist, J.P. Remeika, E.A. Rietman, S. Zahurak, and G.P. Espinosa, Phys. Rev. Lett. **58**, 1676 (1987)
7. In: *High-Temperature Superconductivity*, ed. by V.L. Ginzburg and D.A. Kirzhnits (Consult. Bureau, London 1982)
8. M. Imada, A. Fujimori, and Y. Tokura, Rev. Mod. Phys. **70**, 1039 (1998)
9. see, e.g., P.W. Anderson and Y. Ren, In: *High Temperature Superconductivity*, The Los Alamos Symposium 1989, ed. by, K.S. Bedell, D. Coffey, D.E. Meltzer, D. Pines, and J.R. Schrieffer (Addison-Wesley, Redwood City, 1990) p.3.
10. B.T. Matthias, In: *Ternary Superconductors*, ed. by G.K. Shenoy, B.D. Dunlap, and F.Y. Fradin (North-Holland, Amsterdam, 1981) p.3.
11. J. Nagamatsu, N. Nakagawa, T. Muranaka, Y. Zenitani, and J. Akimitsu, Nature (London) **410**, 63 (2001)
12. R.D. Shannon and P.E. Bierstedt, J. Am. Ceram. Soc. **53**, 635 (1970)
13. D.E. Cox and A.W. Sleight, In: *Proc. of the Conf. of Neutron Scattering*, Gatlinburg, Tenn., 1976, ed. by R.M. Moon (Nat. Tech. Inf. Serv., Springfield, VA, 1976) p. 45.
14. D.E. Cox and A.W. Sleight, Acta Crystall. B **35**, 1 (1979)
15. L.F. Mattheiss, E.M. Giorgi, and D.W. Johnson, Phys. Rev. B **37**, 3745 (1988)
16. R.J. Cava, B. Batlogg, J.J. Krajewski, R.C. Farrow, L.W. Rupp, A.E. White, K.T. Short, W.F. Peck, and T.Y. Kometani, Nature (London) **332**, 814 (1988)
17. I.S. Shaplygin, B.G. Kakhan, and V.B. Lazarev, Zh. Neorg. Khimii **24**, 1478 (1979)
18. C. Michel and B. Raveau, Rev. Chim. Miner. **21**, 407 (1984)
19. B. Grande, H.R. Müller-Buschbaum, and M. Schweizer, Z. Anorg.Allg. Chem. **428**, 120 (1971)
20. J.M. Longo and P.M. Racah, J. Solid State Chem. **6**, 526 (1973)
21. R.M. Fleming, B. Batlogg, R.J. Cava, and E.A. Rietman, Phys. Rev. B **35**, 7191 (1987)
22. D.C. Johnston, In: *Magnetic Materials*, ed. by K.H.J. Buschow (North-Holland, Amsterdam)
23. J.D. Axe, A.H. Mouddon, D. Hohlwein, D.E. Cox, K.M. Mohanty, A.R. Moodenbaugh, and Youwen Xu, Phys. Rev. Lett. **62**, 2751 (1989)

24. J.D. Jorgensen, B. Dabrowski, S. Pei, D.G. Hinks, L. Soderholm, B. Morosin, J.E. Schirber, E.L. Venturini, and D.S. Ginley, Phys. Rev. B **38**, 11337 (1988)
25. H.R. Müller-Buschbaum and W. Wollschläger, Z. Anorg. Allg.Chem. **44**, 76 (1975)
26. H. Sawa, S. Suzuki, M. Watanabe, J. Akimitsu, H. Matsubara, H. Watabe, S. Uchida, K. Kokusho, H. Asano, F. Izumi, and E. Takayama-Muromachi, Nature (London) **337**, 347 (1989)
27. T. Siegrist, S. Sunshine, D.W. Murphy, R.J. Cava, and S.M. Zahurak, Phys. Rev. B **35**, 7137 (1987)
28. J.D. Jorgensen, M.A. Beno, D.G. Hinks, L. Soderholm, K.J. Volin, R.L. Hitterman, J.D. Grace, I.K. Schuller, C.U. Segrè, K. Zhang and M.S. Kleefisch, Phys. Rev. B **36**, 3608 (1987)
29. see, e.g., D. Hohlwein, In: *Materials and Crystallographic Aspects of High-T_c Superconductivity*, ed. by E. Kaldis, NATO ASI Series, Series E: Applied Sciences, Vol. 263 (Kluwer, Dordrecht, 1994) p. 65 and references therein
30. K. Char, M.Lee, R.W. Barton, A.F. Marshall, R.H. Bozovic, R.H. Hammond, M.R. Beasley, T.H. Geballe, A. Kapitulnik, and S.S. Laderman, Phys. Rev. B **38**, 834 (1988)
31. P. Marsh, R.M. Fleming, M.L. Mandich, A.M. DeSantolo, J. Kwo, M. Hong, and L.J. Martinez-Miranda, Nature (London) **334**, 141 (1988)
32. J. Karpinski, C. Beeli, E. Kaldis, A. Wisard, and E. Jilek, Physica C **153-155**, 830 (1988)
33. P. Bordet, C. Chaillout, J. Chenavas, J.L. Hodeau, M. Marezio, J. Karpinski, and E. Kaldis, Nature (London) **334**, 596 (1988)
34. A. Schilling, J.D. Guo, M. Cantoni, F. Hulliger, B. Xue, and H.R. Ott, Materials Lett. **15**, 141 (1992)
35. see, e.g., M. Marezio and C. Chaillout, Ref. [29] p.3
36. A.W. Hewat, Ref. [29] p. 17.
37. M. Takano, J. Takada. K. Oda, H. Kitaguchi, Y. Miura, Y. Ikeda, Y. Tomii, and H. Mazaki, Jpn. J. Appl. Phys. **27**, L1041 (1988)
38. A. Schilling, M. Cantoni, J.D. Guo, and H.R. Ott, Nature (London) **363**, 56 (1993)
39. B.A. Scott, E.Y. Suard, C.C. Tsuei, D.B. Mitzi, T.R. McGuire, B.-H. Chen, and D. Walker, Physica C **230**, 239 (1994)
40. T. Siegrist, S.M. Zahurak, D.W. Murphy, and R.S. Roth, Nature (London) **334**, 231 (1988)
41. M. Takano, Y. Takeda, H. Okada, M. Miyamoto, and T. Kusaka, Physica C **159**, 375 (1989)
42. R.J. Cava, B. Batlogg, R.B. van Dover, J.J. Krajewski, J.V. Waszczak, R.M. Fleming, W.F. Peck, Jr., L.W. Rupp, Jr., P. Marsh, A.C.W.P. James, and L.F. Schneemeyer, Nature (London) **345**, 602 (1990)
43. R.J. Cava, B. Batlogg, J.J. Krajewski, L.W. Rupp, Jr., L.F. Schneemeyer, T, Siegrist, R.B. van Dover, P. Marsh, W.F. Peck, Jr., P.K. Gallagher, S.H. Glarum, J.H. Marshall, R.C. Farrow, J.V. Waszczak, R. Hull, and P. Trevor, Nature (London) **336**, 211 (1988)
44. H.W. Kroto, J.R. Heath, S.C. O'Brien, R.F. Curl, and R.E. Smalley, Nature (London) **318**, 162 (1985)
45. see, e.g., O. Gunnarsson, Rev. Mod. Phys. **69**, 575 (1997) and references therein
46. P.A. Heiney, J.E. Fischer, A.R. McGhie, W.J. Romanow, A.M. Denenstein, J.P. McCauley, Jr., A.B. Smith, and D.E. Cox, Phys. Rev. Lett. **66**, 2911 (1991)
47. R.C. Haddon, A.F. Hebard, M.J. Rosseinsky, D.W. Murphy, S.J. Duclos, K.B. Lyons, B. Miller, J.M. Rosamilia, R.M. Fleming, A.R. Kortan, S.H. Glarum, A.V. Makhija, A.J. Muller, R.H. Eick, S.M. Zahurak, R. Tycko, G. Dabbagh, and R.F. Thiel, Nature (London) **350**, 320 (1991)
48. J.D. Axe, S.C. Moss, and D.A. Neumann, In: *Solid State Physics,* Vol. 48, ed. by H. Ehrenreich and F. Spaepen, (Academic Press, Boston 1994) p. 149.
49. J.M. An and W.E. Pickett, Phys. Rev. Lett. **86**, 4366 (2001)
50. A.W. Sleight, J.L. Gillson and P.E. Bierstedt, Solid State Commun. **17**, 27 (1975)
51. T.D. Tranh, A. Koma, and S. Tanaka, Appl. Phys. **22**, 205 (1980)
52. B. Batlogg, Physica B **126**, 275 (1984)
53. R.J. Cava, B. Batlogg, G.P. Espinosa, A.P. Ramirez, J.J. Krajewski, W.F: Peck Jr., W. Rupp Jr., and A.S. Cooper, Nature (London) **339**, 291 (1989)
54. L.F. Mattheiss, E.M. Gyorgy, and D.W. Johnson, Phys. Rev. B **37**,3745 (1988)
55. R.J. Cava, B. Batlogg, J.J. Krajewski, R.C. Farrow, L.W. Rupp Jr, A.E. White, K.T. Short, W.F. Peck Jr., and T.Y Kometani, Nature (London) **332**, 814 (1988)
56. P. Ganguly and C.N.R. Rao, Mater. Res. Bull. **8**, 405 (1973)

57. see, e.g., D.C. Johnston, F. Borsa, P. Carretta, J.H. Cho, F.C. Chou, M. Corti, R.J. Gooding, E. Lai, A. Lascialfari, L.L.Miller, N.M. Salem, B.J. Suh, D.R. Torgeson, D. Vaknin, K.J.E.Vos, and J.L. Zarestky, In: *NATO ASI Series E*, vol.343, ed. by E.Kaldis, E. Liarokapis, and K.A. Müller (Kluwer, Dordrecht, 1997) p. 311
58. A.R. Moodenbaugh, Y. Xu, M. Suenaga, T.J. Folkerts, and R.N. Shelton, Phys. Rev. B **38**, 4596 (1988)
59. G.M. Luke, L.P. Le, B.J. Sternlieb, W.D. Wu, Y.J. Uemura, J.H. Brewer, T.M. Riseman, S. Ishibashi, and S. Uchida, Physica C **185-189**, 1175 (1991)
60. Y. Maeno, Ref. [29] p. 203.
61. Y. Tokura, H. Takagi, and S. Uchida, Nature (London) **337**, 345 (1989)
62. H. Takagi, S. Uchida, and Y. Tokura, Phys. Rev. Lett. **62**, 1197 (1989)
63. S. Uchida, H. Takagi, Y. Tokura, N. Koshihara, and T. Arima, In: *Strong Correlation and Superconductivity*, Springer series in Solid-State Sciences 89, ed. by H. Fukuyama, S. Maekawa and A.P. Malozemoff, (Springer, Berlin, 1989) p.194
64. J. Akimitsu, S. Suzuki, M. Watanabe, and H. Sawa, Jap. J. Appl. Phys. **27**, L1859 (1988)
65. P. Burlet, C. Vettier, M.J.G.M. Jurgens, J.Y. Henry, J. Rossat-Mignod, H. Noel, M. Potel, P. Gougeon, and J.C. Levet, Physica C **153-155**, 1115 (1988)
66. R.J. Cava, A.W. Hewat, E.A. Hewat, B. Batlogg, M. Marezio, K.M. Rabe, J.J. Krajewski, W.F. Peck Jr., and L.W. Rupp Jr., Physica C **165**, 419 (1990)
67. C. Michel, M. Hervieu, M.M. Borel, A. Grandin, F. Deslandes, J. Provost, and B. Raveau, Z. Phys. B **68**, 421 (1987)
68. H. Maeda, Y. Tanaka, M. Fukutomi, and T. Asano, Jpn. J. Appl. Phys. **27**, L209 (1988)
69. see, e.g., J. Tarascon, P. Barboux, L.H. Greene, B.G. Bagley, and G.W. Hull, Physica C **153-155**, 566 (1988)
70. Z.Z. Sheng and A.M. Hermann, Nature (London) **332**, 55 (1988)
71. R.M. Hazen, L.W. Finger, R.J. Angel, C.T. Prewitt, N.L. Ross, C.G. Hadidiacos, P.J. Heaney, D.R. Veblen, Z.Z. Sheng, A. ElAli, and A.M. Hermann, Phys. Rev. Lett. **60**, 1657 (1988)
72. S.S.P. Parkin, V.Y. Lee, E.M. Engler, A.I. Nazzal, T.C. Huang, G. Gorman, R. Savoy, and R. Beyers, Phys. Rev. Lett. **60**, 2539 (1988)
73. A.F. Hebard, M.J. Rosseinsky, R.C. Haddon, D.W. Murphy, S.H. Glarum, T.T.M. Palstra, A.P. Ramirez, and A.R. Kortan, Nature (London) **350**, 600 (1991)
74. R.M. Fleming, A.P. Ramirez, M.J. Rosseinsky, D.W. Murphy, R.C. Haddon, S.M. Zahurak, and A.V. Makhija, Nature (London) **352**, 787 (1991)
75. T. Yildirim, J.E. Fischer, R. Dinnebier, P.W. Stephens, and C.L.Lin, Solid State Commun. **93**, 269 (1995)
76. O. Zhou, G.B.M. Vaughn, Q. Zhu, J.E. Fischer, P.A. Heiney, N. Coustel, J.P. McCauley Jr., and A.B. Smith III, Science **255**, 833 (1992)
77. J.E. Schirber, D.L. Overmeyer, W.R. Bayless, M.J. Rosseinsky, D.W. Murphy, Q. Zhu, O. Zhou, K. Kniaz, and J.E. Fischer, J. Phys. Chem. Solids **54**, 1427 (1993)
78. J. Diederichs, J.S. Schilling, K.W. Herwig, and W.B. Yelon, J. Phys. Chem. Solids **58**, 123 (1997)
79. T. Yildirim, L. Barbedette, J.E. Fischer, C.L. Lin, J. Robert, P. Petit, and T.T.M. Palstra, Phys. Rev. Lett. **77**, 167 (1996)
80. T.T.M. Palstra, O. Zhou, Y. Iwasa, P.E. Sulewski, R.M. Fleming, and B.R. Zegarski, Solid State Commun. **93**, 327 (1995)
81. C. Buzea and T. Yamashita, Supercond. Sci. Technol. **14**, R115 (2001)
82. S.M. Kazakov, M. Angst, and J. Karpinski, I.M. Fita, and R. Puzniak, Solid State Commun. **119**, 1 (2001)
83. Y. Moritomo and S. Xu, cond-mat/0104568 (unpublished)
84. P.W. Anderson, Ref. [63] p.2.
85. J.M. Imer, F. Patthey, B. Dardel, W.D. Schneider, Y. Baer, Y. Petroff, and A. Zettl, Phys. Rev. Lett. **62**, 336 (1989)
86. Z.X. Shen, W.E. Spicer, D.M. King, D.S. Dessau, and B.O. Wells, Science **267**, 343 (1995)
87. J.C. Campuzano, M. Randeria, M. Norman, and H. Ding, In: *The Gap Symmetry and Fluctuations in high-T_c Superconductors*, NATO-ASI, Series B, Vol. 371, ed. by J. Bok, G. Deutscher, D. Pavuna and S.A. Wolf, (Plenum, New York 1998) p.229.
88. M.R. Norman, H. Ding, J.C. Campuzano, T. Takeuchi, M. Randeria, T. Yokoya, T. Takahashi, T. Mochiku, and K. Kadowaki, Phys. Rev. Lett. **79**, 3506 (1997)
89. A. Kaminski, J. Mesot, H. Fretwell, J.C. Campuzano, M.R. Norman, M. Randeria, H. Ding, T. Sato, T. Takahashi, T. Mochiku, K. Kadowaki, and H. Hoechst, Phys. Rev. Lett. **84**, 1788 (2000)
90. R.J. Cava, B. Batlogg, R.B. van Dover, D.W. Murphy, S. Sunshine, T. Siegrist, J.P. Remeika, E.A. Rietman, S. Zahurak, and G.P. Espinosa, Phys. Rev. Lett. **58**, 1676 (1987)

91. S.J. Hagen, T.W. Jing, Z.Z. Wang, J. Horvath, and N.P. Ong, Phys. Rev. B **37**, 7928 (1988)
92. H. Takagi, B. Batlogg, H.L. Kao, J. Kwo, R.J. Cava, J.J. Krajewski, and W.F. Peck Jr., Phys. Rev. Lett. **69**, 2975 (1992)
93. A. Fiory, S. Martin, R. Fleming, L. Schneemeyer, J. Waszczak, A. Hebard, and S. Sunshine, Physica C **162-164**, 1195 (1989)
94. see, e.g., S. Martin, A.T. Fiory, R.M. Fleming, L.F. Schneemeyer, and J.V. Waszczak, Phys. Rev. Lett. **60**, 2194 (1988)
95. N.P. Ong, Z.Z. Wang, J. Clayhold, J. Tarascon, L.H. Greene, W.R. McKinnon, Phys. Rev. B **35**, 8807 (1987)
96. Z.Z. Wang, J. Clayhold, N.P. Ong, J.M. Tarascon, L.H. Greene, W.R. McKinnon, and G.W. Hull, Phys. Rev. B **36**, 7222 (1987)
97. J.M. Harris, Y.F. Yan, and N.P. Ong, Phys. Rev. B **46**, 14293 (1992)
98. R. Jin, Doctoral Thesis, ETH nr.11738, 1996, unpublished.
99. P.W. Anderson, Phys. Rev Lett. **67**, 2092 (1991)
100. see, e.g., C.P. Slichter, *Principles of Magnetic Resonance*, Springer Series in Solid State Sciences 1, (Springer, Berlin, 1990)
101. I. Mangelschots, M. Mali, J. Roos, D. Brinkmann, S. Rusiecki, J. Karpinski, and E. Kaldis, Physica C **194**, 277 (1992)
102. D. Brinkmann, Ref. [29] p. 225
103. H. Yasuoka, T. Imai, and T. Shimizu, Ref. [63] p. 254
104. A. Junod, In: *Physical Properties of High Temperature Superconductors*, ed. by D.M. Ginsberg, (World Scientific, Singapore 1990) p.13.
105. N.E. Phillips, R.A. Fisher, and J.E. Gordon, *Progr. Low Temp.Phys.*13, 267 (1992)
106. J. Bardeen, L.N. Cooper, and J.R. Schrieffer, Phys. Rev. **108**, 1175 (1957)
107. J.W. Loram, K.A. Mirza, J.R. Cooper, and J.L. Tallon, Physica C **282-287**, 1405 (1997)
108. J.W. Loram, K.A. Mirza, J.R. Cooper, and W.Y. Liang, Phys. Rev. Lett. **71**, 1740 (1993)
109. J.W. Loram, K.A. Mirza, J.R. Cooper, and J.L. Tallon, J. Phys. Chem. Solids **59**, 2091 (1998)
110. H. Ding, T. Yokoya, J.C. Campuzano, T. Takahashi, M. Randeria, M.R. Norman, T. Mochiku, K. Kadowaki, and J. Giapintzakis, Nature (London) **382**, 51 (1996)
111. M.R. Norman, H. Ding, M. Randeria, J.C. Campuzano, T. Yokoya, T. Takeuchi, T. Takahashi, T. Mochiku, K. Kadowaki, P. Gupta-Sarma, and D.G. Hinks, Nature (London) **392**, 157 (1998)
112. I. Giaever, Phys. Rev. Lett. **5**, 147, 464 (1960)
113. Ch. Renner, B. Revaz, J.-Y. Genoud, K. Kadowaki, and Ø. Fischer, Phys. Rev. Lett. **80**, 149 (1998)
114. Y. DeWilde, N. Miyakawa, P. Guptasarma, M. Iavarone, L. Ozyuzer, J.F. Zasadzikski, P. Romano, D.G. Hinks, C. Kendziora, G.W. Crabtree, and K.E. Gray, Phys. Rev. Lett. **80**, 153 (1998)
115. M. Kugler, Ø. Fischer, Ch. Renner, S. Ono, and Y. Ando, Phys.Rev. Lett. **86**, 4911 (2001)
116. C.E. Gough, M.S. Colclough, E.M. Forgan, R.G. Jordan, M. Keene, C.M. Muirhead, A.I.M. Rae, N. Thomas, J.S. Abbell, and S. Sutton, Nature (London) **326**, 855 (1987)
117. see, e.g., A. Schenck, Physica C **153-155**, 1127 (1988)
118. Y.J. Uemura, G.M. Luke, B.J. Sternlieb, J.H. Brewer, J.F. Carolan, W.N. Hardy, R. Kadono, J.R. Kempton, R.F. Kiefl, S.R. Kreitzman, P. Mulhern, T.M. Riseman, D. Ll. Williams, B.X. Yang, S. Uchida, H. Takagi, J. Gopalakrishnan, A.W. Sleight, M.A. Subramanian, C.L. Chien, M.Z. Cieplak, G. Xiao, V.Y. Lee, B.W. Statt, C.E. Stronach, W.J. Kossler, and X.H. Yu, Phys. Rev. Lett. **62**, 2317 (1989)
119. J.E. Hirsch and F. Marsiglio, Phys. Rev. B **43**, 424 (1991)
120. F. Hulliger and H.R. Ott, Z. Phys. B **67**, 291 (1987)
121. Z. Hao and J.R. Clem, Phys. Rev. Lett. **67**, 2371 (1991)
122. R. Jin, H.R. Ott, and A. Schilling, Physica C **235-240**, 1923 (1994)
123. A. Schilling, Doctoral thesis ETH nr.9645, 1992, unpublished.
124. C.W. Chu, P.H. Hor, R.L. Meng, L. Gao, and Z.J. Huang, Science **235**, 567 (1987)
125. W.L. McMillan, Phys. Rev. **167**, 331 (1968)
126. T.F. Smith and W.E. Gardner, Phys. Rev. **146**, 291 (1966)
127. E.N. Van Eenige, R. Griessen, R.J. Wijngaarden, J. Karpinski, E. Kaldis, S. Rusiecki, and E. Jilek, Physica C **168**, 482 (1990)
128. M. Nuñez-Regueiro, J.L. Tholence, E.V. Antipov, J.J. Capponi, and M. Marezio, Science **262**, 97 (1993)

129. L. Gao, Y.Y. Xue, F. Chen, Q. Xiong, R.L. Meng, D. Ramirez, C.W. Chu, J.H. Eggert, and H.K. Mao, Phys. Rev. B **50**, 4260 (1994)
130. D. Tristan Jover, R.J. Wijngarden, H. Wilhelm, R. Griessen, S.M. Loureiro, J.J. Capponi, A. Schilling, and H.R. Ott, Phys. Rev. B **54**, 4265 (1996)
131. E. Maxwell, Phys. Rev. B **78**, 477 (1950)
132. C.A. Reynolds, B. Serin, W.H. Wright, and L.B. Nesbitt, Phys. Rev. **78**, 487 (1950)
133. H. Fröhlich, Phys. Rev. **79**, 845 (1950)
134. R.A. Hein and J.W. Gibson, Phys. Rev. **131**, 1105 (1963)
135. R.D. Fowler, J.D.G. Lindsay, R.W. White, H.H. Hill, and B.T. Matthias, Phys. Rev. Lett. **19**, 892 (1967)
136. Guo-meng Zhao, M.B. Hunt, H. Keller, and K.A. Müller, Nature (London) **385**, 236 (1997)
137. M.K. Crawford, W.E. Farneth, E.M. McCarron, R.L. Harlow, and A.H. Mouddon, Science **250**, 1390 (1990)
138. H.J. Bornemann and D.E. Morris, Phys. Rev. B **44**, 5322 (1991)
139. D. Rubio Temprano, J. Mesot, S. Janssen, K. Conder, A. Furrer, H. Mutka, and K.A. Müller, Phys. Rev. Lett. **84**, 1990 (2000)
140. V.J. Emery and S.A. Kivelson, Phys. Rev. Lett. **64**, 475 (1900)
141. A. Bianconi, N.L. Sini, A. Lanzara, M. Missori, R.T. Rosetti, H. Oyanagi, H. Yamaguchi, K.Ota, and T. Ito, Phys. Rev. Lett. **76**, 3412 (1996)
142. see, e.g., T. Egami, Ref. [29] p.45 and references therein.
143. G. Blatter, M.V. Feigel'man, V.B. Geshkenbein, and V.M. Vinokur, Rev. Mod. Phys. **66**, 1125 (1994)
144. O. Jeandupeux, A. Schilling, H.R. Ott, and A. van Otterlo, Phys. Rev. B **53**, 12475 (1996)
145. A. Schilling and O. Jeandupeux, Phys. Rev. B **52**, 9714 (1995)
146. D.S. Fisher, M.P.A. Fisher and D.A. Huse, Phys. Rev. B **43**, 130 (1991)
147. T. Schneider and H. Keller, Physica C **207**, 366 (1993) and references therein.
148. M.B. Salamon, J. Shi, N. Overend, and M.A. Howson, Phys. Rev. B **47**, 5520 (1993)
149. M. Roulin, A. Junod, and E. Walker, Physica C **260**, 257 (1996)
150. A. Junod, Ref. [90] p.403
151. O. Jeandupeux, Ch. Wälti, H.R. Ott, and A. van Otterlo, Czech. J. of Physics **46** - Suppl., 1211 (1996)
152. R. Jin, A. Schilling, and H.R. Ott, Phys. Rev. B **49**, 9218 (1994)
153. R. Jin, H.R. Ott, and A. Schilling, Physica C **228**, 401 (1994)
154. L.N. Bulaevskii. M. Ledvij, and V.G. Kogan, Phys. Rev. Lett. **68**, 3773 (1992)
155. Z. Tesanovic, L. Xing, L. Bulaevskii, Q. Li, and M. Suenaga, Phys. Rev. Lett. **69**, 3563 (1992)
156. U. Welp, S. Fleshler, W.K. Kwok, R.A. Klemm, V.M. Vinokur, J. Downey, and B. Veal, Phys. Rev. Lett. **67**, 3180 (1991)
157. R.E. Hetzel, A. Sudbo, and D.A. Huse, Phys. Rev. Lett. **69**, 518 (1992)
158. H. Pastoriza, M.F. Goffman, A. Arribére, and F. de la Cruz, Phys. Rev. Lett. **72**, 2951 (1994)
159. E. Zeldov, D. Majer, M. Konczykowski, V.B. Geshkenbein, V.M. Vinokur, and H. Shtrikman, Nature (London) **375**, 373 (1995)
160. A. Schilling, R.A. Fisher, N.E. Phillips, U. Welp, D. Dasgupta, W.K. Kwok, and G.W. Crabtree, Nature (London) **382**, 791 (1996)
161. M. Roulin, A. Junod, and E. Walker, J. Low Temp. Phys. **105**, 1099 (1996)
162. A. Schilling, R.A. Fisher, N.E. Phillips, U. Welp, W.K. Kwok, and G.W. Crabtree, Phys. Rev. Lett. **78**, 4833 (1997)
163. N.P. Ong, Y.F. Yan, and J.M. Harris, In: *Abstracts: CCAST Symposium on High-T_c Superconductivity and the C_{60} Family* (Gordon and Breach, Beijing, 1994)
164. T.T.M. Palstra, B. Batlogg, R.B. Van Dover, L.F. Schneemeyer, and J.V. Waszczak, Phys. Rev. Lett. **61**, 1662 (1988)
165. J. Bardeen, G. Rickayzen, and L. Tewordt, Phys. Rev. **113**, 982 (1958)
166. see, e.g., W. Odoni, H.R. Ott, and M.B. Maple, J. Low Temp. Phys. **51**, 505 (1983)
167. R.C. Yu, M.B. Salamon, J.P. Lu, and W.C. Lee, Phys. Rev. Lett. **69**, 1431 (1992)
168. K. Krishana, J.M. Harris, and N.P. Ong, Phys. Rev. Lett. **75**, 3529 (1995)
169. M.H. Jericho, W. Odoni, and H.R. Ott, Phys. Rev. B **31**, 3124 (1985)
170. J. Beyer Nielsen and H. Smith, Phys. Rev. Lett. **49**, 689 (1982)
171. R. Gagnon, S. Pu, B. Ellman, and L. Taillefer, Phys. Rev. Lett. **78**, 1976 (1997)
172. D.A. Bonn, P. Dosanjh, R. Liang, and W.N. Hardy, Phys. Rev. Lett. **68**, 2390 (1992)

173. D.A. Bonn, S. Kamal, K. Zhang, R. Liang, D.J. Baar, E. Klein, and W.N. Hardy, Phys. Rev. B **50**, 4051 (1994)
174. J. Bardeen and M.J. Stephen, Phys. Rev. **140**, A1197 (1965)
175. P. Nozières and W.F. Vinen, Phil. Mag. **14**, 667 (1966)
176. R. Jin and H.R. Ott, Phys. Rev. B **57**, 13872 (1998)
177. J. Luo, T.P. Orlando, J.M. Graybeal, X.D. Wu, and R. Münchhausen, Phys. Rev. Lett. **68**, 690 (1992)
178. K. Noto, S. Shinazawa and Y. Muto, Solid State Commun. **18**, 1081 (1976)
179. R. Jin and H.R. Ott, Phys. Rev. B **53**, 9406 (1996)
180. A.V. Samoilov, Z.G. Ivanov, and L.G. Johansson, Phys. Rev. B **49**, 3667 (1994)
181. R.J. Troy and A. Dorsey, Phys. Rev. B **47**, 2715 (1993)
182. Z.D. Wang, J. Dong, and C.S. Ting, Phys. Rev. Lett. **72**, 3875 (1994)
183. J.M. Harris, N.P. Ong, and Y.F. Yan, Phys. Rev. Lett. **71**, 1455 (1993)
184. P. Ao and D. Thouless, Phys. Rev. Lett. **70**, 2158 (1993)
185. J.E. Hirsch and F. Marsiglio, Phys. Rev. B **43**, 424 (1991)
186. M.V. Feigel'man, V.B. Geshkenbein, A.I. Larkin, and V.M. Vinokur, Pis'ma Zh. Eksp. Teor. Fiz. **62**, 811 (1995) [JETP Lett. **62**, 834 (1995)]
187. D.I. Khomskii and A. Freimuth, Phys. Rev. Lett. **75**, 1384 (1995)
188. see, e.g., P.G. De Gennes, In: *Superconductivity of Metals and Alloys*, (Addison Wesley, Redwood City 1989) p. 255.
189. J.S. Tsai, I. Takeuchi, J. Fujita, T. Yoshitake, S. Miura, S. Tanaka, T. Terashima, Y. Bando, K. Iijima, and K. Yamamoto, Physica C **153-155**, 1385 (1988)
190. J.S. Tsai, I. Takeuchi, J. Fujita, T. Yoshitake, S. Miura, S. Tanaka, T. Terashima, Y. Bando, K. Iijima, and K. Yamamoto, In: *Mechanisms of High Temperature Superconductivity*, Springer Series in Materials Science 11, ed. by H. Kamimura and Oshiyama (Springer, Berlin, 1989) p.229.
191. D.J. Scalapino, Phys. Rep. **250**, 330 (1995)
192. see, e.g., In: *Anisotropy Effects in Superconductors*, ed. H.W. Weber, (Plenum Press, New York 1977)
193. see, e.g., L.D. Landau and E.M. Lifshitz, In: *Statistical Physics* (Pergamon, Elmsford, 1969)
194. P.W. Anderson, In: *Basic Notions of Condensed Matter Physics,* (Benjamin, London 1984)
195. M. Sigrist and K. Ueda, Rev. Mod. Phys. **63**, 239 (1991)
196. R. Balian and N.R. Werthamer, Phys. Rev. **131**, 1553 (1963)
197. L.C. Hebel and C.P. Slichter, Phys. Rev. **107**, 901 (1957)
198. R.W. Morse, In: *Progress in Cryogenics*, vol.I, ed. by K. Mendelssohn, (Heywood, London 1959) p.219.
199. Y. Kitaoka, K. Ishida, F. Fujiwara, T. Kondo, K. Asayama, M. Horvatic, Y. Berthier, P. Butaud, P. Segransan, C. Berthier, H. Katayama-Yoshida, Y. Okabe, and T. Takahashi, Ref. [63] p.262.
200. see, e.g., D.E. MacLaughlin, In: *Solid State Physics*, vol.31, ed. by H. Ehrenreich, F. Seitz, and D. Turnbull (Academic Press, New York 1976) p.1.
201. W.N. Hardy, D.A. Bonn, D.C. Morgan, R. Liang, and K. Zhang, Phys. Rev. Lett. **70**, 3999 (1993)
202. see, e.g., D.A. Bonn and W.N. Hardy, In: *Physical Properties of High Temperature Superconductors*, Vol. V, ed. by D.M. Ginsberg (World Scientific, Singapore 1996) Chap.2, p.7, and references therein
203. N.B. Kopnin and G.E. Volovik, JETP Lett. **64**, 690 (1996)
204. K.A. Moler, D.J. Baar, J.S. Urbach, R. Liang, W.N. Hardy, and Kapitulnik, Phys. Rev. Lett. **73**, 2744 (1994)
205. G.E. Volovik, JETP Lett. **58**, 469 (1993)
206. D.A. Wright, J.P. Emerson, B.F. Woodfield, J.E. Gordon, R.A. Fisher, and N.E. Phillips, Phys. Rev. Lett. **82**, 1550 (1999)
207. S.H. Simon and P.A. Lee, Phys. Rev. Lett. **78**,1548 (1997)
208. G.E. Volovik, JETP Lett. **65**, 491 (1997)
209. B. Revaz, J.Y. Genoud, A. Junod, K. Neumaier, A. Erb, and E. Walker, Phys. Rev. Lett. **80**, 3364 (1998)
210. R.A. Fisher, N.E. Phillips, A. Schilling, B. Buffeteau, R. Calemczuk, K.W. Dennis, R.W. McCallum, and A.S. O'Conner, Phys. Rev. B **61**, 1473 (2000)
211. P.A. Lee, Phys. Rev. Lett. **71**, 1887 (1993)
212. L. Taillefer, B. Lussier, R. Gagnon, K. Behnia, and H. Aubin, Phys. Rev. Lett. **79**, 483 (1997)
213. M. Chiao, R.W. Hill, Ch. Lupien, L. Taillefer, P. Lambert, R. Gagnon, and R. Fournier, Phys. Rev. B **62**, 3554 (2000)
214. M.J. Graf, S.K. Kip, J.A. Sauls, and D. Rainer, Phys. Rev. B **53**, 15147 (1996)

215. J. Mesot, M.R. Norman, H. Ding, M. Randeria, J.C. Campuzano, Paramekanti, H.M. Fretwell, A. Kaminski, T. Takeuchi, T. Yokoya, T. Sato, T. Takahashi, T. Mochiku, and K. Kadowaki, Phys. Rev. Lett. **83**, 840 (1999)
216. K. Yosida, Phys. Rev. **110**, 769 (1958)
217. see, e.g., A.J. Leggett, Rev. Mod. Phys. **47**, 331 (1975)
218. S.E. Barrett, D.J. Durand. C.H. Pennington, C.P. Slichter, T.A. Friedmann, J.P. Rice, and D.M. Ginsberg, Phys. Rev. B **41**, 6283 (1990)
219. M. Takigawa, P.C. Hammel, R.H. Heffner, and Z. Fisk, Phys. Rev. B **39**, 7371 (1989)
220. V.B. Geshkenbein and A.I. Larkin, JETP Lett. **43**, 395 (1986)
221. M. Sigrist and T.M. Rice, J. Phys. Soc. Jpn. **61**, 4283 (1992)
222. L.N. Bulaevskii, V.V. Kuzii, and A.A. Sobyanin, Pis'ma Eksp. Teor. Fiz. **25**, 314 (1977); JETP Lett. **25**, 290 (1977)
223. see, e.g., M. Tinkham, *Introduction to Superconductivity*, 2nd Edition, (McGraw-Hill, New York, 1996)
224. D.J. Van Harlingen, Rev. Mod. Phys. **67**, 515 (1995)
225. C.C. Tsuei and J.R. Kirtley, Rev. Mod. Phys. **72**, 969 (2000)
226. D.A. Wollman, D. Van Harlingen, W.C. Lee, D.M. Ginsberg, and A.J. Leggett, Phys. Rev. Lett. **71**, 2134 (1993)
227. D.A. Brawner and H.R. Ott, Phys. Rev. B **50**, 6530 (1994)
228. A. Mathai, Y. Gim, R.C. Black, A. Amar, and F.C. Wellstood, Phys. Rev. Lett. **74**, 4523 (1995)
229. D.A. Wollman, D.J. Van Harlingen, J. Giapintzakis, and D.M. Ginsberg, Phys. Rev. Lett. **74**, 797 (1995)
230. D.A. Brawner, and H.R. Ott, Phys. Rev. B **53**, 8249 (1996)
231. C.C. Tsuei, J.R. Kirtley, C.C. Chi, L.S. Yu-Jahnes, A. Gupta, T. Shaw, J.Z. Sun, and M.B. Ketchen, Phys. Rev. Lett. **73**, 593 (1994)
232. J.R. Kirtley, C.C. Tsuei, J.Z. Sun, C.C. Chi, L.S. Yu-Jahnes, Gupta, M. Rupp, and M.B. Ketchen, Nature (London) **373**, 225 (1995)
233. C.C. Tsuei, J.R. Kirtley, M. Rupp, J.Z. Sun, A. Gupta, and M.B. Ketchen, Science **272**, 329 (1996)
234. J.R. Kirtley, C.C. Tsuei, H. Raffy, Z.Z. Li, A. Gupta, J.Z. Sun, and S. Megert, Europhys. Lett. **36**, 707 (1996)
235. C.C. Tsuei and J.R. Kirtley, Phys. Rev. Lett. **85**, 182 (2000)
236. P. Chaudhari, and S.Y. Lin, Phys. Rev. Lett. **72**, 1084 (1994)
237. A.J. Millis, Phys. Rev. B **49**, 15408 (1994)
238. A.G. Sun, D.A. Gajewski, M.B. Maple, and R.C. Dynes, Phys. Rev. Lett. **72**, 2267 (1994)
239. A.S. Katz, A.G. Sun, R.C. Dynes, and K. Char, Appl. Phys. Lett. **66**, 105 (1995)
240. J. Lesueur, M. Aprili, A. Goulon, T.J. Horton, and L. Dumoulin, Phys. Rev. B **55**, R3398 (1997)
241. see the extensive discussion on this aspect in Ref. [225].
242. K.A. Müller. Nature (London) **377**, 133 (1995)
243. M. Weger, M. Peter, and L.P. Pitaevski, Z. Phys. B **101**, 573 (1996)
244. D. Pines, Ref. [90] p. 111
245. J.H. Xu, J.H. Miller, Jr., and C.S. Ting, Phys. Rev. B **53**, R2991 (1996)
246. K. Maki, H. Won, M. Kohmoto, J. Shiraishi, Y. Morita, and G.F. Wang, Physica C **317-318**, 353 (1999) and references therein.
247. B. Keimer, W.Y. Shih, R.W. Erwin, J.W. Lynn, F. Dogan, and I.A. Aksay, Phys. Rev. Lett. **73**, 3459 (1994)
248. I. Maggio-Aprile, Ch. Renner, A. Erb, E. Walker, and Ø. Fischer, Phys. Rev. Lett. **75**, 2754 (1995)
249. A.A. Abrikosov and L.P. Gor'kov, Zh. Eksperim. i Teor. Fiz. **39**, 1781 (1960); Sov. Phys. JETP **12**, 1243 (1961)
250. M. Ishikawa, Ø. Fischer, and J. Muller, J. de Physique C **6**, 1379 (1978)
251. M.B. Maple, J. de Physique C **6**, 1374 (1978)
252. T. Kasuya, Progr. Theoret. Phys. (Kyoto) **16**, 58 (1956); K. Yosida, Phys. Rev. **107**, 396 (1957)
253. In: *Superconductivity in Ternary Compounds I*, Topics in Current Physics, Vol. **32**, ed. by Ø. Fischer and M.B. Maple (Springer, Berlin, 1982)
254. In: *Superconductivity in Ternary Compounds II, Topics in Current Physics*, Vol. **34**, ed. by M.B. Maple and Ø. Fischer (Spinger, Berlin, 1982)
255. P. Fulde and J. Keller, Ref. [253]
256. D.M. Paul, H.A. Mook, A.W. Hewat, B.C. Sales, L.A. Boatner, J.R. Thomson, and M. Mostoller, Phys. Rev. B **37**, 2341 (1988)

257. see, e.g., B. Roessli, P. Allenspach, P. Fischer, J. Mesot, U. Staub, H. Maletta, P. Brüesch, C. Ritter, and A.W. Hewat. Physica B **180** and **181**, 396 (1992)
258. B. Roessli, P. Fischer, U. Staub, M. Zolliker, and A. Furrer, J. Appl. Phys. **75**, 6337 (1994)
259. B.Batlogg, J.P. Remeika, R.C. Dynes, H. Barz, A.S. Cooper, and J.P. Garno, Proc. IV. Conference on *Superconductivity in d-and f-Band Metals* , ed. by W. Buckel and W. Weber (KFZ, Karlsruhe, 1982) p. 401
260. T.M. Rice, In: *Superconductivity in Magnetic and Exotic Materials* , ed. by T. Matsubara and A. Kotani (Springer, Berlin 1984) p. 178
261. L.F. Mattheiss and D.R. Hamann, Phys. Rev. B **26**, 2686 (1982); *ibid.* B **28**, 4227 (1983)
262. see, e.g., R.R. Hake, Appl. Phys. Lett. **10**, 186 (1967)
263. B. Batlogg, R.J. Cava, L.W. Rupp, Jr., A.M. Mujsce, J.J. Krajewski, J.P. Remeika, W.F. Peck, Jr., A.S. Cooper, and G.P. Espinosa, Phys. Rev. Lett. **61**, 1670 (1988)
264. M. Sato, S. Kondoh, and M. Sera, Ref. [63] p. 341
265. D.A. Brawner, C. Manser, and H.R. Ott, Phys. Rev. B **55**, 2788 (1997)
266. A. Georges, G. Kotliar, W. Krauth, and M.J. Rozenberg, Rev. Mod. Phys. **68**, 13 (1996)
267. R.W. Lof, M.A. van Veenendaal, B. Koopmans, H.T. Jonkman, and G.A. Sawatzky, Phys. Rev. Lett. **68**, 3924 (1992)
268. A.B. Migdal, Zh. Eksp. i Teor. Fiz. **34**, 1438 (19589, Sov. Phys. JETP **7**, 996 (1958)
269. T.T.M. Palstra, A.F. Hebard, R.C. Haddon, and P.B. Littlewood, Phys. Rev. B **50**, 3462 (1994)
270. X.D. Xiang, J.G. Hou, G. Briceno, W.A. Vareka, R. Mostovoy, Zettl, V.H. Crespi, and M.L. Cohen, Science **256**, 1190 (1992)
271. W.A. Vareka, and A. Zettl, Phys. Rev. Lett. **72**, 4121 (1994)
272. A.P. Ramirez, M.J. Rosseinsky, D.W. Murphy, and R.C. Heddon, Phys. Rev. Lett. **69**, 1687 (1992)
273. R. Tycko, G. Dabbagh, M.J. Rosseinsky, D.W. Murphy, A.P. Ramirez, and R.M. Fleming, Phys. Rev. Lett. **68**, 1912 (1992)
274. A.P. Ramirez, A.R. Kortan, M.J. Rosseinsky, S.J. Duclos, A.M. Mujsce, R.C. Haddon, D.W. Murphy, A.V. Makhija, S.M. Zahurak, and K.B. Lyons, Phys. Rev. Lett. **68**, 1058 (1992)
275. C.C. Chen and C.M. Lieber, J. Am. Chem. Soc. **114**, 3141 (1992)
276. T.W. Ebbesen, J.S. Tsai, K. Tanigaki, H. Hiura, Y. Shimakawa, Y. Kubo, I. Hirosawa, and J. Mizuki, Physica C **203**, 163 (1992)
277. B. Burk, V.H. Crespi, A. Zettl, and M.L. Cohen, Phys. Rev. Lett. **72**, 3706 (1994)
278. V.A. Stenger, C.H. Pennington, D.R. Buffinger, and R.P. Ziebarth, Phys. Rev. Lett. **74**, 1649 (1995)
279. R.F. Kiefl, W.A. MacFarlane, K.H. Chow, S. Dunsinger, T.L. Duty, T.M.S. Johnston, J.W. Schneider, J. Sonier, L. Brard, R.M. Strongin, J.E. Fischer, and A.B. Smith III, Phys. Rev. Lett. **70**, 3987 (1993)
280. L. Degiorgi, G. Briceno, M.S. Fuhrer, A. Zettl, and P. Wachter, Nature (London) **369**, 541 (1994)
281. D. Koller, M.C. Martin, and L. Mihaly, Phys. Rev. Lett. **77**, 4082 (1996)
282. Z. Zhang, C.C. Chen, S.P. Kelty, H. Dai, and C.M. Lieber, Nature (London) **353** , 333 (1991)
283. P. Jess, S. Behler, M. Bernasconi, V. Thommen-Geiser, H.P. Lang, M. Baenitz, K. Lüders, and H.J. Güntherodt, Physica C **235-240**, 2499 (1994)
284. Y.J. Uemura, A. Keren, L.P. Le, G.M. Luke, B.J. Sternlieb, W.D. Wu, J.H. Brewer, R.L. Whetten, S.M. Huang, S. Lin, R.B. Kaner, F. Diederich, S. Donovan, G. Grüner, and K. Holczer, Nature (London) **352**, 605 (1991)
285. Y.J. Uemura, L.P. Le, and G.M. Luke, Synth. Met. **56**, 2845 (1993)
286. L. Degiorgi, P. Wachter, G. Grüner, S.M. Huang, J. Wiley, and R.B. Kaner, Phys. Rev. Lett. **69**, 2987 (1992)
287. see, e.g., V. Buntar and H.W. Weber, Supercond. Sci. Technol. **9**, 599 (1996)
288. A.P. Ramirez, Supercond. Rev. **1**, 1 (1994)
289. B.W. Roberts, J. Phys. Chem. Ref. Data **84**, 581 (1976)
290. J.R. Gavaler, M.A. Janocko, and C.K. Jones, J. Appl. Phys. **47**, 3009 (1974)
291. M. Xu, H. Kitazawa, Y. Takano, J. Ye, K. Nishida, H. Abe, A.Matsushita, N. Tsujii, and G. Kido. Appl. Phys. Lett. **79**, 2779 (2001)
292. A.V. Sologubenko, J. Jun, S. M. Kazakov, J. Karpinski, and H.R. Ott, Phys. Rev. B **65**, 180505 (2002)
293. D.K. Finnemore, J.E. Ostenson, S.L. Bud'ko, G. Lapertot, and P.C. Canfield, Phys. Rev. Lett. **86**, 2420 (2001)
294. P.C. Canfield, D.K. Finnemore, S.L. Bud'ko, J. Ostenson, G. Lapertot, C.E. Cunningham, and C. Petrovic, Phys. Rev. Lett. **86**, 2423 (2001)
295. R. Jin, M. Paranthaman, H.Y. Zhai, H.M. Christen, D.K. Christen, and D. Mandrus, cond-mat/0104411

296. J.E. Hirsch, cond-mat/0102115
297. R.W. Johnson and A.H. Daane, J. Chem. Phys. **38**, 425 (1963)
298. H.J. Juretschke and R. Steinitz, J. Phys. Chem. Solids **4**, 118 (1958)
299. P. Bordet, M. Mezouar, M. Nunez-Regueiro, M. Monteverde, M.D. Nunez.Regueiro, N. Rogado, K.A. Regan, M.A. Hayward, T. He, S.M. Loureiro, and R.J. Cava, cond-mat/0106585
300. J.D. Jorgensen, D.G. Hinks, and S. Short, Phys. Rev. B **63**, 224522 (2001)
301. J. Kortus, I.I. Mazin, K.D. Belashchenko, V.P. Antropov, and L.L. Boyer, Phys. Rev. Lett. **86**, 4656 (2001)
302. F. Bouquet, R.A. Fisher, N.E. Phillips, D.G. Hinks, and J.D. Jorgensen, Phys. Rev. Lett. **87**, 047001-1 (2001)
303. Y. Wang, T. Plackowski, and A. Junod, Physica C **355**, 179 (2001)
304. Ch. Wälti, E. Felder, C. Degen, G. Wigger, R. Monnier, Delley, and H.R. Ott, Phys. Rev. B **64**, 172515 (2001)
305. E. Felder, S.M. Kazakov, J. Karpinski, and H.R. Ott, unpublished
306. H. Kotegawa, K. Ishida, Y. Kitaoka, T. Muranaka, and J. Akimitsu, cond-mat/0102334
307. J.K. Jung, S.H. Baek, F. Borsa, S.L. Bud'ko, G. Lapertot, and P.C. Canfield, cond-mat/0103040
308. A. Gerashenko, K. Mikhalev, S. Verkhovskii, T. D'yachkova, Tyutyunnik, and V. Zubkov, cond-mat/0102421
309. E. Pavarini and I.I. Mazin, cond-mat/0105350
310. Y. Moritomo, and S. Xu, cond-mat/0104568
311. S.M. Kazakov, M. Angst, J. Karpinski, I.M. Fita, and R. Puzniak, Solid State Commun. **119**, 1 (2001)
312. S.L. Bud'ko, G. Lapertot, C. Petrovic, C.E. Cunningham, N. Anderson, and P.C. Canfield, Phys. Rev. Lett. **86**, 1877 (2001)
313. D.G. Hinks, H. Claus, and J.D. Jorgensen, Nature (London) **411**,457 (2001)
314. G. Wigger, Ch. Wälti, and H.R. Ott, unpublished
315. F. Simon, A. Janossy, T. Feher, F. Muranyi, S. Garaj, L. Forro, Petrovic, S.L. Bud'ko, G. Lapertot, V.G. Kogan, and P.C. Canfield, Phys. Rev. Lett. **87**, 04002-1 (2001)
316. Y. Kong, O.V. Dolgov, O. Jepsen, and O.K. Andersen, Phys. Rev. B **64**, R020501 (2001)
317. G. Rubio-Bollinger, H. Suderow, and S. Vieira, Phys. Rev. Lett. **86**, 5582 (2001)
318. G. Karapetrov, M. Iavarone, W.K. Kwok, G.W. Crabtree, and D.G. Hinks, Phys. Rev. Lett. **86**, 4374 (2001)
319. A.Y. Liu, I.L. Mazin, and J. Kortus, cond-mat/0103570
320. T. Takahashi, T. Sato, S. Souma, T. Muranaka, and J. Akimitsu, Phys. Rev. Lett. **86**, 4915 (2001)
321. D.C. Larbalestier, L.D. Cooley, M.O. Rikel, A.A. Polyanskii, J.Jiang, S. Patnaik, X.Y. Cai, D.M. Feldman, A. Gurevich, A.A. Squitieri, M.T. Naus, C.B. Eom, E.E. Hellstrom, R.J. Cava, K.A. Regan, N. Rogado, M.A. Hayward, T. He, J.S. Slusky, P. Khalifah, K. Inumaru, and M. Haas, Nature (London) **410**, 186 (2001)
322. H.J. Kim, W.N. Kang, E.M. Choi, M.S. Kim, K.H.P. Kim, and S.I. Lee, cond-mat/0105363
323. A. Lanzara, P.V. Bogdanov, X.J. Zhou, S.A. Kellar, D.L. Feng, E.D. Lu, T. Yoshida, H. Eisaki, A. Fujimori, K. Kishio, J.I. Shimoyama, T. Noda, S. Uchida, Z. Hussain, and Z.X. Shen, Nature (London) **412**, 510 (2001)

15 Tunneling Spectroscopy of Conventional and Unconventional Superconductors

J. Zasadzinski Illinois Institute of Technology, Physics Dept. Chicago, USA

15.1 Introduction

Tunneling spectroscopy of conventional superconductors [1] such as Pb [2] leads to a complete description of the superconducting state. From the tunneling conductance, $\frac{dI}{dV}$ vs. V (appropriately normalized), one can obtain the quasiparticle density of states, $N(E)$. This gives an implicit measure of the complex, superconducting gap parameter, $\Delta(E)$. Using Migdal–Eliashberg theory [3] the gap parameter can be "inverted" by the iterative McMillan–Rowell procedure (MR) [2] to obtain the microscopic interactions responsible for superconductivity, namely, the electron–phonon spectral function, $\alpha^2F(\omega)$, and the renormalized coulomb pseudopotential, μ^*. These quantities can then be used to determine the transition temperature, T_c, as well as the electron mass enhancement, $1+\lambda$, which enters normal state thermodynamic properties such as the specific heat [4]. In some cases the $\alpha^2F(\omega)$ obtained from tunneling has been used to provide a very good fit of the temperature dependent electrical resistivity far above

T_c [5]. The fact that both superconducting and normal state quantities can be determined demonstrates both the accuracy of strong-coupling superconductivity theory and the power of the tunneling method. The various manifestations of the electron–phonon interaction can be found in the review articles of Carbotte [3] and Allen [4].

Using a straightforward extension of the basic tunneling measurement to normal metal superconductor (N/S) bi-layers, the quantities $\alpha^2F(\omega)$ and μ^* have been obtained for some transition metal elements (Nb, V, Ta), alloys (NbZr) and A-15 compounds such as Nb_3Sn as well as nonsuperconducting materials such as Mg [1]. Furthermore, such proximity effect junctions of the form Nb/Al/Al-oxide/Nb are useful for a variety of devices from sensitive photon detectors to fast switches and are the mainstay of a successful superconducting electronics industry.

Considering the potential of tunneling spectroscopy for basic research and applications, it is not surprising that many groups around the world have attempted to fabricate junctions on exotic superconductors including the high-T_c cuprates and bismuthates, organics, heavy fermion materials and the recently discovered MgB_2. Successful inversion of tunneling spectra using a variation of the traditional MR procedure has been accomplished for the bismuthate, $Ba_{1-x}K_xBiO_3$, and the relatively lower T_c, electron-doped cuprate $Nd_{1.85}Ce_{0.15}CuO_4$ [6]. The resulting $\alpha^2F(\omega)$ spectra strongly suggest phonon mediated pairing, however there remain unresolved questions such as whether this interaction is compatible with d-wave symmetry [7]. For all of the other classes of exotic materials the current state-of-the-art tunneling spectroscopy does not include a quantitative determination of the pairing mechanism. This is mostly a consequence of their remarkable complexity. Tunneling in these materials has, nevertheless, provided a wealth of important information on the superconducting gap magnitude, pseudogap phenomena, pairing symmetry as well as atomic scale imaging of crystal structure and defects. For example, a highly unusual doping dependence of the measured gap in the cuprate $Bi_2Sr_2CaC_2O_{8+\delta}$ (Bi2212) has been revealed by tunneling and appears to be related to pseudogap phenomena observed above T_c. Higher energy spectral features, analogous to phonon structures, are reproducibly observed in some cuprates as well as heavy fermion materials which have been linked to spin excitations. The large ratios of gap magnitude to T_c in organic compounds are similar to those found on underdoped cuprates suggesting the possibility of common origins of pairing in hole doped cuprate, heavy fermion and organic superconductors.

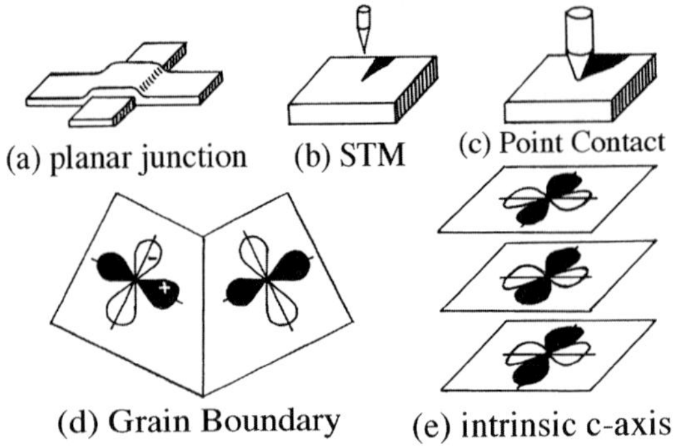

Fig. 15.1. Schematic representation of junction types

It is impossible to adequately review all of the tunneling literature, even on cuprate superconductors alone. The inclusion of conventional superconductors as well as other interesting materials makes the task even more difficult. To narrow the scope, the focus of this review will be on those quasiparticle tunneling experiments which directly address the issue of unconventionality, i.e., gap size vs. T_c, gap symmetry, pseudogap and pairing mechanism. An attempt has been made to select those results which are representative and exhibit features reproduced by other tunneling groups or are consistent with other (non-tunneling) experiments. The brief review of conventional superconductors is structured to serve as a baseline for understanding the more complicated tunneling spectra found in unconventional superconductors. Consequently a great deal of important work on other aspects of the field (e.g. high-T_c junction devices) has been left out. Furthermore, the large volume of publications necessitates the use of monographs and review articles as references whenever possible.

A schematic of the junction types considered here is shown in Fig. 15.1. Figures 15.1(a)–(c) are the standard junction geometries for all superconductors, however almost all of the $\alpha^2F(\omega)$ spectra have been obtained using planar junctions. Figures 15.1(d) and (e) are junction types that are associated with cuprate superconductors.

15.2 Basic Tunneling Phenomenology

In this section an overview is given of the general phenomena found for tunneling into s-wave and d-wave superconductors. In addition, the theoretical frameworks for understanding the various spectra are discussed.

15.2.1 Elastic Tunneling Current

Analysis of experimental tunneling data often begins with the standard expression for the junction current between two metals separated by a thin barrier (typically 1–2 nm) either due to a vacuum as in STM or a surface insulating layer as in planar or point contact junctions [1]. Assuming elastic tunneling,one has

$$I = \frac{2\pi e}{\hbar} \int_{-\infty}^{+\infty} |T|^2 N_1(E+eV)N_2(E)[f(E) - f(E+eV)]\mathrm{d}E\,.$$

Here, $N(E)$ is the electronic density of states (DOS) and $|T|^2$ is the tunneling matrix element. All information has been lost about the various approximations and assumptions that take the three-dimensional sum over electron momenta, $\boldsymbol{k}, \boldsymbol{p}$ in each electrode and collapse it into a single integral over the quasiparticle energy, E. Thus what is not clear is whether $N(E)$ represents the full quasiparticle DOS, including band structure effects, or is just that part that is due to correlation effects which arise from e.g., superconductivity, localization, charge density waves etc. This is discussed in more detail later. For now we make a standard assumption that the energy and voltage dependence of $|T|^2$ is weak and can be taken out of the integral to obtain

$$I = \frac{2\pi e}{\hbar}|T|^2 \int_{-\infty}^{+\infty} N_1(E+eV)N_2(E)[f(E) - f(E+eV)]\mathrm{d}E\,. \tag{15.1}$$

This equation is used for both superconductor-insulator-normal metal (SIN) junctions as well as SIS junctions. Considering for the moment that N_1 is a normal metal with a constant DOS near the Fermi energy, one obtains,

$$I \propto N_{1n}(0) \int_{-\infty}^{+\infty} N_{2s}(E)[f(E) - f(E+eV)]\mathrm{d}E\,.$$

Assuming $T = 0$ and taking the derivative, $\mathrm{d}I/\mathrm{d}V$, in both the superconducting and normal states one obtains,

$$\sigma(V) = \frac{\frac{\mathrm{d}I_{sn}}{\mathrm{d}V}}{\frac{\mathrm{d}I_{nn}}{\mathrm{d}V}} = \frac{N_{1n}(0)N_{2s}(eV)}{N_{1n}(0)N_{2n}(0)} = \frac{N_{2s}(eV)}{N_{2n}(0)}\,.$$

For an s-wave, BCS superconductor, the DOS can be written as a simple product

$$N_{2s}(E) = N_{2n}(0)\sigma_{BCS}(E) = N_{2n}(0)\frac{|E|}{\sqrt{E^2-\Delta^2}}\,,$$

where $N_{2n}(0)$ is the DOS of the superconductor in the normal state, assumed to be a constant given by its value at the Fermi energy, $E = 0$. Under these assumptions the normalized conductance, $\sigma(V)$ is therefore a direct measure of the superconducting part of the DOS

$$\sigma_{BCS} = \frac{|E|}{\sqrt{E^2-\Delta^2}}\,. \tag{15.2}$$

Note that the symbol σ has been used to denote both the normalized conductance $\sigma(V)$ and the superconducting DOS, σ_{BCS}, since they are equivalent at zero temperature. For non-zero T it is

$$\sigma(V) = \int_{-\infty}^{+\infty} \sigma_{BCS}\frac{\mathrm{d}}{\mathrm{d}V}[f(E-eV)-f(E)]\mathrm{d}E\,. \tag{15.3}$$

The superconducting DOS is convoluted with the derivative of the fermi function, a gaussian shaped curve with a half-width of approximately 3.5 kT. Thus the measured normalized conductance gives a thermally smeared superconducting DOS.

15.2.2 Superconducting Gap Structure

The justification for all assumptions about the tunnel current lies in the degree to which the measured normalized conductance agrees with equations of the previous section. For example, the SIN gap region conductance for the bismuthate, $Ba_{1-x}K_xBiO_3$ (BKBO) is shown in Fig. 15.2 where it is clear that an ideal fit is obtained using Eqs. (15.2) and (15.3) [6]. Thus, BKBO with a T_c near 28 K appears to be an isotropic, s-wave superconductor. The strong coupling ratio, $\frac{2\Delta}{kT_c}$, is in the range 3.7–3.8, which is slightly more than the BCS weak coupling limit, and indicates moderately strong coupling.

While the issue of pairing symmetry in high-T_c cuprates is not entirely settled, there is certainly considerable evidence to suggest that the order parameter has $d_{x^2-y^2}$ (d-wave) symmetry. Thus it is useful to consider the d-wave analogue of the BCS density of states. For conventional superconductors, the energy gap, Δ, is a constant in $\mathbf{k}$ space (isotropic s-wave pairing symmetry). However, for superconductors with d-wave pairing symmetry, the energy gap is anisotropic on the Fermi surface. In a two dimensional superconductor with a BCS-type interaction, a simple d-wave DOS suggested by Won and Maki [8] is given by

$$\sigma(E, \mathbf{k}) = \mathrm{Re}\left\{\frac{E - i\Gamma}{\sqrt{(E - i\Gamma)^2 - \Delta(\mathbf{k})^2}}\right\}, \qquad (15.4)$$

where $\Delta(\mathbf{k}) = \Delta\cos(2\phi)$ is the $\mathbf{k}$ dependent energy gap and ϕ is the polar angle in $\mathbf{k}$-space. Note that when d-wave symmetry is assumed, the parameter Δ refers to the maximum value of the gap. The DOS, $\sigma(E)$, is found by an integral over the polar angle ϕ. Here, E is replaced by $E - i\Gamma$, where Γ is a smearing parameter to account for quasiparticle lifetime. This parameter was originally introduced into the BCS DOS by Kaplan et al. [9] as a convenient way to approximately account for scattering and recombination effects on the quasiparticle lifetime within conventional Eliashberg theory. Experimental verification of this modified DOS was made by Dynes et al. [10], where Γ was shown to be significant only near T_c. However, this Γ term has taken on an expanded role, serving as a phenomenological smearing term for the superconducting DOS, irrespective of origin.

In Fig. 15.3 is shown the density of states DOS for a d-wave superconductor with a maximum gap mag-

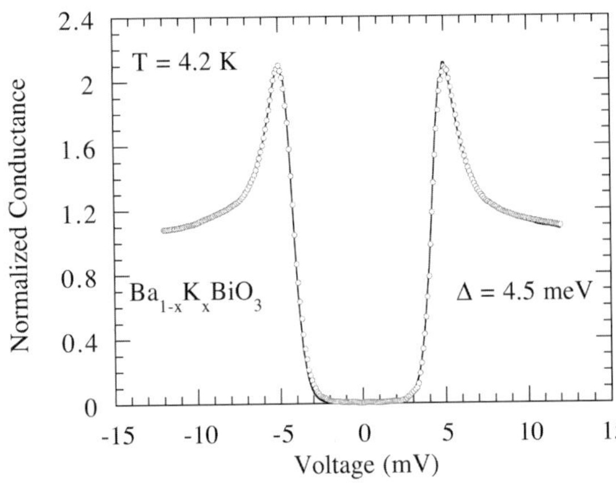

Fig. 15.2. Normalized tunneling conductance and BCS fit for $Ba_{1-x}K_xBiO_3$ from Huang et al. [6]

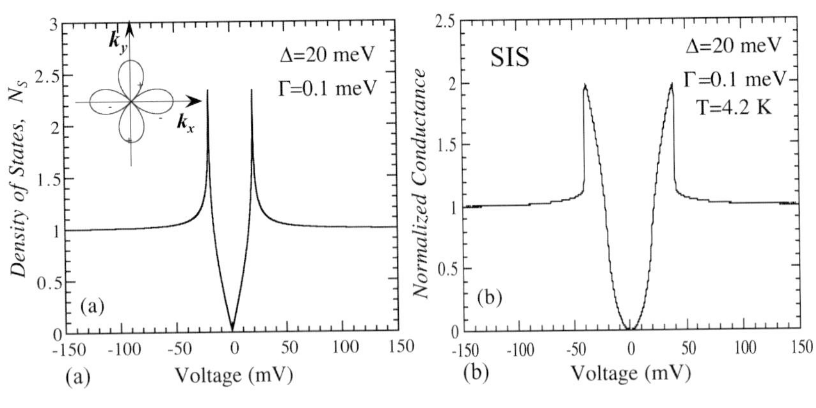

Fig. 15.3. Characteristic features of (**a**) BCS density of states and (**b**) SIS spectrum for d-wave superconductors

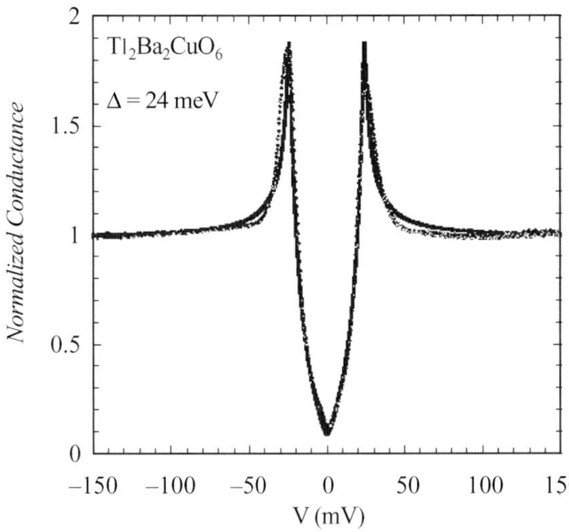

Fig. 15.4. Normalized conductance (*dots*) and *d*-wave fit (*solid line*) for Tl2201 from Ozyuzer et al. [12]

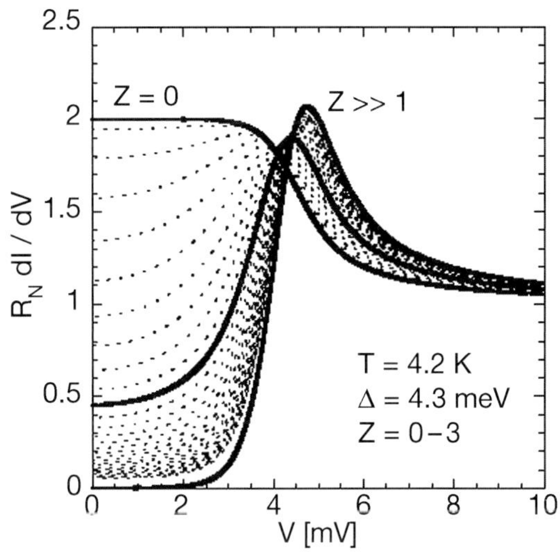

Fig. 15.5. Calculated conductances for a point contact between a superconducting and normal metal electrode in the BTK model for various Z parameters. *Dark solid lines* are for the special cases of $Z = 0, 1, \gg 1$. *Dashed lines* show smooth evolution of the curves from $Z = 0$ to 3

nitude of 20 meV and the expected SIS spectrum for two identical superconductors. The geometry of the SIS junction is very important as will be discussed later. Here it is assumed the barrier normal is along the c-axis, similar to the intrinsic junction shown in Fig. 15.1(e). The characteristic feature of the DOS is a cusp at the Fermi energy (zero bias) and a relatively small ratio of the peak height to background (PHB) which is about 2. The SIS curve is characterized by a similar PHB ratio and a sharp drop of the conductance at the voltage, $V = \frac{2}{e}\Delta$.

These characteristics are certainly distinctive and should be observed in tunnel junctions on d-wave superconductors. However, cuprates superconductors generally exhibit much more complicated spectra as will be shown. The one high-T_c cuprate system that most closely and reproducibly displays the d-wave characteristics in the tunneling conductance is $Tl_2Ba_2CuO_6$ [11], an example of which is shown in Fig. 15.4 [12].

15.2.3 BTK Model

Many junctions on novel superconductors utilize the point contact geometry (see Fig. 15.1(c)) where the force from a sharp, normal metal tip on a superconductor can be varied. The contact between tip and sample is a nanoconstriction and can be of the tunneling type if the surface oxides of the electrodes provide a suitable barrier, but can also be metallic. The latter occurs if the surface barrier is perforated or if the electrodes are noble metals without a native insulating surface. The model of Blonder–Tinkham–Klapwijk (BTK) [13] provides a smooth interpolation between the SIN tunneling and SN metallic regimes. A delta function tunnel barrier is assumed and the strength is given by a dimensionless parameter, Z. The transmission probability is assumed to be the same for all electrons impinging on the interface between the electrodes.

The calculated conductance spectra for various Z parameters is shown in Fig. 15.5 from the work of Schmidt [14]. For the pure metallic contact ($Z = 0$) there is an increase in conductance near zero bias by a factor of 2 over the high bias value. This is a consequence of Andreev reflection. An electron in the normal electrode incident on the superconductor cannot propagate and gets reflected back as a hole, producing a dissipationless pair current in the superconductor. In steady state, electrons and holes are moving in opposite directions, adding to the conductance of the normal electrode. Physically, it is as though only half of the series resistance for quasiparticle excitations is present, and thus the conductance is doubled. This occurs only for voltages such that $eV < \Delta$. For voltages beyond this value quasiparticles are transmitted directly into the superconductor and there is an abrupt decrease of conductance back to the normal state conductance of the two electrodes. Thus the pure metallic SN contact provides a measure of the "coherent" gap associated with superconductivity.

For $Z > 1$ the tunneling regime develops and the standard SIN quasiparticle gap feature is observed. It is therefore useful to explore both regimes with point contacts. Quasiparticle excitation gaps observed in tunneling can arise from a variety of sources including charge or spin density waves (CDW or SDW) as well as from superconductivity. But the Andreev process described here only occurs for the phase coherent superconducting state.

15.2.4 Novel *d*-Wave Effects

The BTK model has been extended to *d*-wave superconductors. In particular, novel effects are found for junctions where the barrier plane is perpendicular to the *ab* plane of cuprates. An example of this type of junction is shown schematically in Fig. 15.1(d). This is due to the fact that the quasiparticles in the *ab* plane which undergo a specular reflection from the superconductor/insulator interface may experience a sign change of the order parameter. For example, the scattering is from $-\Delta$ to $+\Delta$ and therefore the formalism which describes such a reflection is similar to that of the proximity effect interface in S/N bilayers, described later. The general consequence of

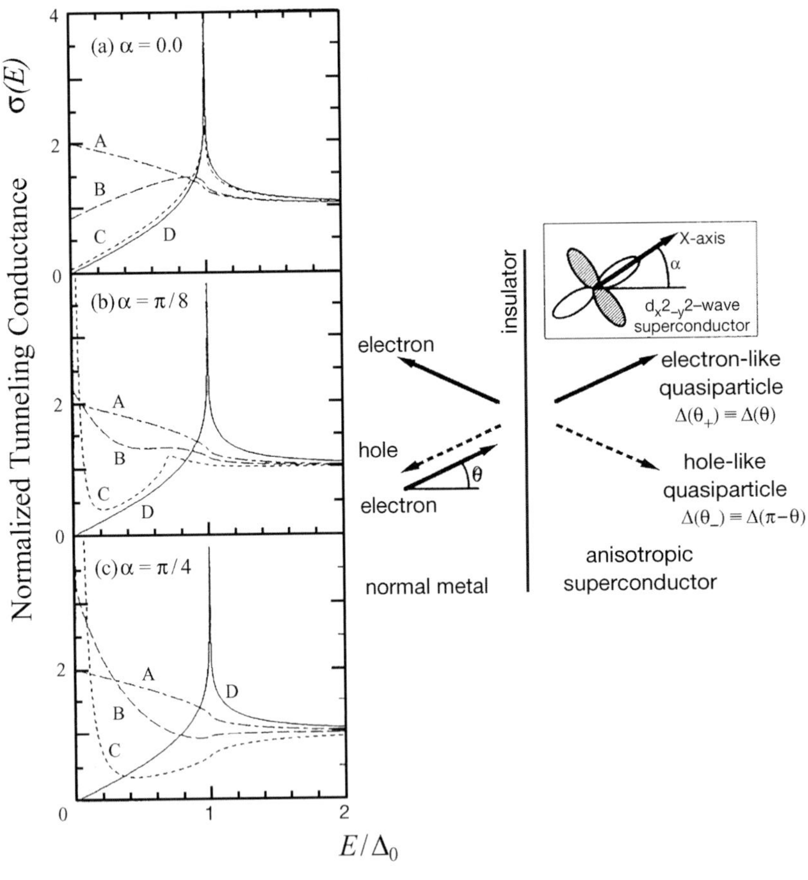

Fig. 15.6. *Right*: schematic of the tunneling geometry for *ab* plane junctions. *Left*: calculations of the junction conductances for various angles α and barrier strengths Z, from Tanaka and Kashiwaya [16]

such scattering events is the development of Andreev bound states in the quasiparticle excitation spectrum at the Fermi energy. These bound states exist only at the interface but nevertheless can have important effects on the measured tunneling spectrum, leading to a zero bias conductance peak (ZBCP) in certain geometries. These effects for d-wave superconductors were first discussed by C.R. Hu [15] and later developed within a generalized BTK framework by Tanaka and Kashiwaya [16].

The d-wave BTK model is shown in Fig. 15.6 where the right panel indicates the geometry considered. The angle α is between the barrier normal and the lobe direction of the d-wave order parameter. Consider first the curves labeled A in the left panel of Fig. 15.6 which correspond to $Z = 0$. These curves represent the d-wave analogue of the Andreev reflection curve in Fig. 15.5 and display a less abrupt decrease of the conductance near $eV = \Delta$ compared to the s-wave case which is due to the gap nodes. There is no α dependence to the $Z = 0$ curves because all incident quasiparticles are assumed to have equivalent transmission probabilities. The effect of a tunnel barrier is seen in the curves labeled C. For non-zero values of α a sharp peak at zero bias is observed and the coherence peaks at the gap voltage are diminished considerably. Spectral weight in the DOS has been shifted from the coherence peaks to the Andreev bound states which produce the ZBCP. There is no ZBCP for c-axis tunnel junctions (Fig. 15.1(e)) because the barrier is parallel to the ab plane and specular reflections off the interface have no effect on in-plane wavevector and thus no sign change occurs. It thus turns out that c-axis tunnel junctions are the only ones capable of revealing the intrinsic quasiparticle DOS.

15.2.5 Strong-Coupling Effects: Phonon Structure

For s-wave superconductors the extension of Eq. (15.2) to the strong-coupling limit has been worked out in Migdal–Eliashberg theory. [1] The superconducting part of the DOS is given by

$$\sigma(E) = Re\frac{|E|}{\sqrt{E^2 - \Delta^2(E)}} \quad , \qquad (15.5)$$

where now it is understood that $\Delta(E)$ is a complex function of the quasiparticle energy, E, in the superconducting state. The imaginary part is related to the finite lifetime of the quasiparticle due to emission of phonons and thus becomes largest near peaks in the phonon density of states. This gives rise to structures in $\sigma(V)$ as shown in Fig. 15.7 for PCT junction on single crystal Nb. [17] There are several noteworthy points to be made here. First, the normal state conductance, often referred to as the background conductance, is rather featureless, showing a weak,

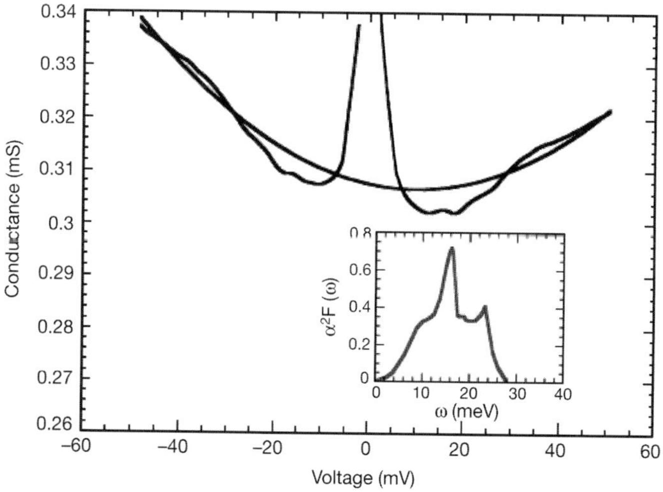

Fig. 15.7. Tunneling conductance of Nb in the superconducting and normal states obtained by PCT in the region of phonon structures, from Huang et al. [17]. Peaks in $\alpha^2F(\omega)$ (*inset*) give rise to dip structures in the superconducting state dynamic conductance

parabolic increase which is due to the barrier. Second, the superconducting conductance near the phonon structures is less than the normal state by about 1.5 percent. This leads to a dip in the normalized conductance or superconducting DOS. Physically this means that states have been lost in this region due to phonon emission and have been transferred primarily to lower energies leading to a pile-up at the gap edge. This effect provides a straightforward means of identifying strong coupling effects. Inversion of the normalized conductance using the MR procedure [1,2], leads to $\alpha^2F(\omega)$ shown in the inset of Fig. 15.7. The dip minima in $\sigma(V)$ are close to the peaks in $\alpha^2F(\omega)$, but more precisely it is the maxima of the second derivative, $\frac{\mathrm{d}\sigma(V)}{\mathrm{d}V}$, that correspond to the peaks.

15.2.6 Inelastic Tunneling

Until now the only conductance channel assumed for the junction was elastic tunneling where initial and final state energies are the same. However, transport across a junction can occur by other means such as inelastic tunneling [1,18], or resonant tunneling [19]. Resonant tunneling is a multistep process whereby the electron first tunnels to a bound orbital state of an impurity atom or ion in the tunnel barrier and then tunnels again into the other electrode. This can be elastic or inelastic and may involve more than one intermediate steps. Resonant tunneling will not be discussed since it tends to wash out the quasiparticle DOS and the resulting spectra are indicative of the barrier states. On the other hand, the spectra presented here are clearly associated with the gap feature in the DOS of the superconducting electrode since they disappear above T_c. However, the resonant tunneling process is believed to be playing an important role in grain-boundary junctions on cuprates [20].

Of more relevance to the issues of concern here is inelastic tunneling. With this channel it is assumed that the electron undergoes either absorption or emission of an excitation of energy $\hbar\omega$ simultaneously with tunneling through the barrier. At liquid helium temperatures, the emission process dominates, and this occurs at a threshold voltage $eV = \hbar\omega$ when the tunneling electron has sufficient energy to create the excitation. At this voltage the inelastic conductance channel opens up and there is a step increase in the conductance. For a flat continuum of excitations the sum of such steps leads to a linear increase in conductance [21].

Strong linear increases in background conductance were often seen in the early tunneling literature on cuprate and bismuthate superconductors [22]. An example of this effect is seen in Fig. 15.8 for $Pr_{1.85}Th_{0.85}CuO_4$ which is similar to the linear backgrounds observed in other electron doped cuprates [23]. The V-shaped background exhibits an increase in conductance of 100% or more at 60 mV compared to the extrapolated value at zero bias. This should be compared to the ordinary parabolic shaped background for elastic tunneling in Nb shown in Fig. 15.7. In that case the increase at 60 mV is only about 10% and is due to the barrier effect from the Nb-oxide layer. For Al-Al oxide-Pb junctions the increase would be about 2%. Thus it is clear the strong V-shaped background is not a consequence of ordinary barrier effects.

There was early speculation that the linear tunneling background might be intrinsic in oxide superconductors, owing possibly to the Marginal Fermi Liquid

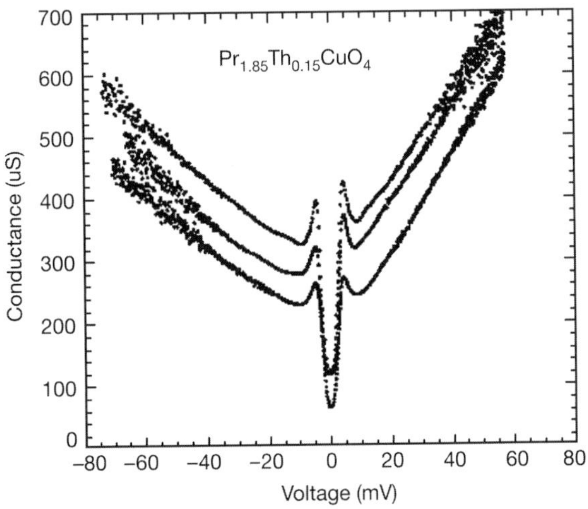

Fig. 15.8. Measured conductances of three different point contact tunnel junctions on a polycrystal of $Pr_{1.85}Th_{0.85}CuO_4$ from the Argonne group (unpublished). Similar strong linear increases are observed in other electron-doped cuprates as seen in [23]

behavior of the quasiparticle self-energy evident in many experiments on optimal doped cuprates. However, such self-energy effects do not lead directly to linear increases in the total electronic DOS, $N(E)$, about the Fermi energy. Furthermore, Kirtley showed that such an inelastic channel was easily induced in ordinary tunnel barriers by introducing magnetic impurities into the oxide [24]. The incoherent spin fluctuations of the weakly coupled spins gave the flat continuum of excitations. Further evidence that the linear background is due to an additional channel can be found by considering the ratio of the conductance peak height to background (PHB) which is typically a value of 2 or more for high quality SIN junctions (e.g. see Figs. 15.2 and 15.4). When strong V-shaped backgrounds are observed the PHB ratio is significantly smaller than 2 as can be seen in Fig. 15.8. This seems to be a general result.

15.3 Tunneling and Strong-Coupling Effects: Microscopic Picture

Here we provide a more detailed theoretical treatment of the tunneling process and examine the various approximations that go into the expression for the tunnel current. We follow the transfer Hamiltonian approach of Bardeen [25]. The total Hamiltonian for the system is

$$H = H_l + H_r + H_T\,, \qquad (15.6)$$

where H_l and H_r are the Hamiltonians for the left and right metals in the absence of tunneling. The perturbing Hamiltonian H_T for the weak virtual coupling is given by

$$H_T = \sum_{k,q} T_{bfk,q}\{c_q^\dagger c_k + c_k^\dagger c_q\}\,, \qquad (15.7)$$

where $c_k^\dagger$, c_k, $c_q^\dagger$ and c_q are creation and annihilation operators for quasiparticles in the eigen states φ_k and φ_q of the left and right metals, respectively. The tunneling matrix element $\left|T_{k,q}\right|^2$ was obtained by Harrison [26] as

$$\left|T_{\mathbf{k},\mathbf{q}}\right|^2 = \frac{L}{4\pi^2}\frac{\delta_{k\parallel}\delta_{q\parallel}}{\rho_{k\perp}\rho_{q\perp}}\exp[-2\int_{z_l}^{z_r} K_\perp(z)\mathrm{d}z]\,, \qquad (15.8)$$

where $\rho_{k_\perp} = (L/\pi)dk_\perp/d\xi_\perp$ is the one-dimensional band structure density of states (proportional to the inverse of the group velocity normal to the barrier) and L is the length of the metal in the normal direction to the oxide surface. The group velocity normal to the barrier enters because the tunnel current is a rate of charge transfer which necessarily involves an attempt frequency of electrons on the tunnel barrier. Here $K_\perp(z) = \sqrt{2m[\varphi(z) - E_z]/\hbar^2}$, $E_z = E - \hbar^2(k_x^2 + k_y^2)/2m$ is the kinetic energy of the electron in the z-direction (normal to interface), and $\varphi(z)$ is the barrier potential. Here E is measured from the bottom of the conduction band.

Due to the exponential dependence on the normal component of energy, the tunneling matrix element heavily weights the contributions of electrons normal to the barrier. There is also a voltage dependence to the matrix element as the shape of the barrier changes with V, e.g. an initially rectangular barrier becomes trapezoidal with applied voltage. For values of $eV \ll \phi(z)$ this barrier effect is negligible and the tunneling conductance for two normal metal electrodes in nearly constant. However, as the applied voltage becomes a reasonable fraction of the barrier height, a parabolic term is added to the background conductance, [27] an effect which can be observed in Fig. 15.7 since the barrier height of Nb oxide is about 250 meV [17].

The standard Fermi's Golden Rule for the transition rate $\omega_{r\leftarrow l}$ from an initial state i on the left metal to a final state f of equal energy on right metal can be obtained by substituting Eq. (15.7) into the time-dependent Schrödinger equation. Subtracting the rate from right to left leads to the following form for the tunnel current

$$I = 2e\sum_{\mathbf{kq}} |T_{\mathbf{kq}}|^2 \int_{-\infty}^{\infty} \frac{\mathrm{d}\omega}{2\pi} A_R(\mathbf{k},\omega)A_L(\mathbf{q},\omega + eV) \times [f(\omega) - f(\omega + eV)]\,.$$

This form for the tunnel current is particularly useful for it incorporates the quasiparticle spectral weight function

$$A(\mathbf{k},\omega) = \frac{1}{\pi}\mathrm{Im}\left(\frac{Z\omega + \varepsilon_\mathbf{k}}{Z^2[\omega^2 - \Delta(\mathbf{k})^2] - \varepsilon_\mathbf{k}^2}\right)\,, \qquad (15.9)$$

a quantity that is measured directly in angle-resolved photoemission spectroscopy (ARPES). The spectral weight contains all information about electron correlation effects via the renormalization parameter, $Z(\mathbf{k}, E)$ and gap parameter, $\Delta(\mathbf{k}, E)$. The general problem of a momentum dependent pairing interaction leading to a strong-coupled d-wave superconductor has not been solved self-consistently as with Migdal–Eliashberg theory for s-wave superconductors. Consequently, calculations of $A(\mathbf{k}, \omega)$ usually involve some type of approximation scheme. The connection to the quasiparticle DOS is given by

$$N(\omega) = \sum_{\mathbf{k}} A(\mathbf{k}, \omega) ,$$

which can be measured in tunneling. The simplest case scenario is to assume the tunneling matrix element is a constant, independent of energy, voltage and momentum, and can be taken out of the summation. Performing the sum over momenta leads directly to Eq. (15.1), with $N(\omega)$ corresponding to the full, three dimensional quasiparticle DOS including band structure and correlation effects. However,this scenario is unrealistic. The matrix element of Eq. (15.8) is clearly not constant and heavily weights quasiparticle momenta normal to the barrier. If both electrodes are assumed to be nearly free electron metals with ellipsoidal Fermi surfaces, then the dominance of contributions to the tunnel current from near-normal incident quasiparticles leads to a collapse of the transverse momentum sum in Eq. (15.9). The resulting momentum sum is over $k_\perp$ and $q_\perp$ which gives the one-dimensional band structure DOS and this is canceled exactly by the group velocity term in the matrix element. For superconducting electrodes, the resulting expression is identical to Eq. (15.1), but with the superconducting DOS, $\sigma(E)$ used instead of N(E). This cancelation by the group velocity, first pointed out by Harrison, leads to the commonly held view that tunneling spectroscopy does not probe band structure effects. However, this is a consequence of several contributing factors, and is not necessarily true in general.

Cuprate superconductors are composed of incoherently coupled Cu-O layers, giving rise to a quasi two-dimensional system and Fermi surface. Even assuming some weak coherent coupling which would give rise to dispersion along the c-direction, the resulting Fermi surface would have open faces perpendicular to the $\mathbf{k}_z$-direction. For STM or point contact configurations where the tip is pointing along the c-axis of cuprate superconductors there can be no tunneling of electrons with momenta normal to the barrier as there are no such states on the Fermi surface. Thus it is expected that the tunneling current from tip to sample in an STM junction or from Cu-O plane to Cu-O plane in an intrinsic junction is made up of electrons with momenta transverse to the barrier. This is quite different from conventional superconducting metals discussed earlier and the key result is that there is no cancelation of band structure effects from the group velocity normal to the barrier. The resulting $N(E)$ should be the 2D density of states in the Cu-O plane including band structure effects.

One of the dominant features of the band structure is the extended saddle point near the $(\pi, 0)$ direction which gives rise to a peak, or van Hove singularity (VHS), in the DOS [28]. Tunneling data out to very high bias voltages has provided some evidence for this feature as shown in Fig. 15.9 which is an STM tunneling conductance on a cleaved surface of Bi2212 [29]. In addition to an asymmetric background there appears to be a broad peak in the conductance centered near zero bias. It is difficult to locate the VHS peak because the superconducting gap feature (including dip structures) is emerging out of this background. This broad peak has a width of at least 800 meV which is consistent with the expected bandwidth.

A similar broad peak is found in PCT conductances of Bi2212 measured by the Argonne tunneling group (including some unpublished spectra, see, for example, [30]). These are shown in Fig. 15.10 where spectra from three different crystals with slightly different oxygen dopings are presented. Again a broad peak with a characteristic width of about 800 meV can be inferred from the data. Yusof et al. have used the fits of ARPES data to generate the band structure DOS for Bi2212 and this is shown in Fig. 15.11 for three different values of the quasiparticle scattering rate [31]. It is not clear what value of Γ should be used. Above T_c the ARPES spectral weight peak

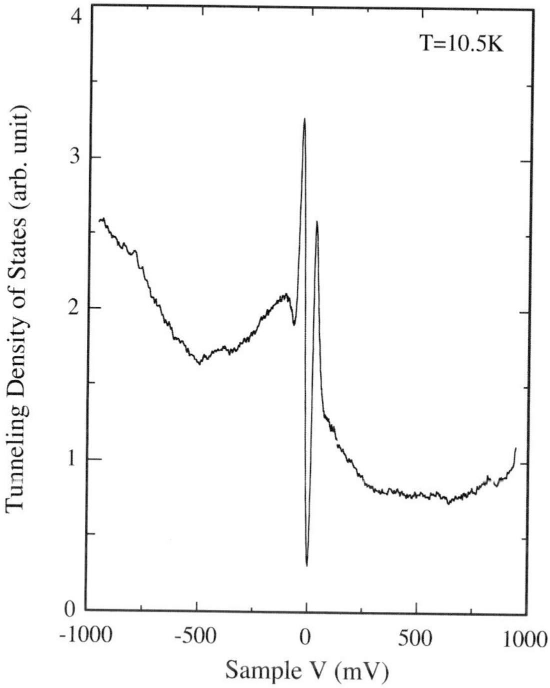

Fig. 15.9. STM spectra of superconducting Bi2212 out to high bias voltage. From Sugita et al. [29]

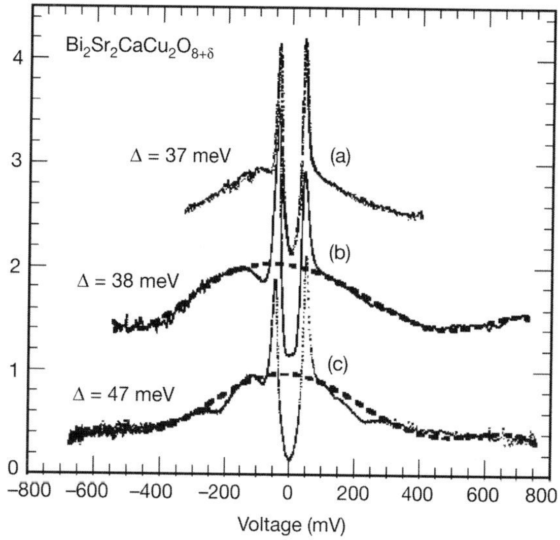

Fig. 15.10. High bias PCT spectra of superconducting Bi2212, obtained from three different crystals. **(a)** and **(b)** correspond to optimally doped and **(c)** is from an underdoped crystal. *Dashed lines* through **(b)** and **(c)** are sixth order polynomial fits through the data

is considerably broadened, but is resolution limited in the superconducting state. From observation of Fig. 15.9 and Fig. 15.10 it appears that a Γ value near the superconducting gap energy would provide a reasonable fit to the measured high bias spectra.

Thus it appears that the decreasing background conductance that is commonly observed in the tunneling conductances of hole-doped cuprates is due to the underlying band structure DOS. However, the VHS peak itself is never directly observed and this is probably due to the fact that it is pinned somewhere close to the Fermi energy and is thus inside the superconducting gap. From the fits to the ARPES data on near optimal doped Bi2212, the peak in the band structure (b.s.) DOS is found about 30 meV below the Fermi energy so this is consistent with such a feature getting masked when a superconducting gap of $\Delta = 35$ meV develops in the superconducting state. Above T_c the scattering rate is increased considerably and there might exist a pseudogap, both effects tending to obscure the observation of the VHS.

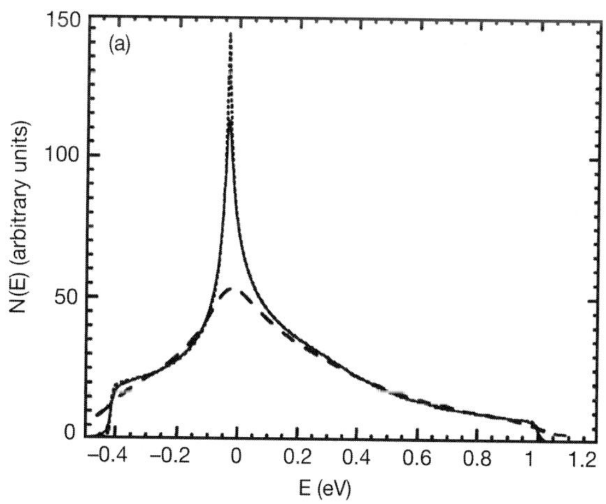

Fig. 15.11. Band structure density of states of Bi2212 as obtained from fitting ARPES spectra, from [31]. The different curves correspond to different values of the quasiparticle scattering rate, Γ. The *dotted, solid* and *dashed lines* correspond to Γ=3, 8, 80 meV, respectively

15.4 Tunneling Spectroscopy of Conventional Superconductors

In the previous section the basic formalism for elastic tunneling was presented. For conventional superconductors with closed Fermi surfaces the tunnel current in SIN junctions is dominated by electrons with momenta normal to the barrier. Consequently the band structure DOS is canceled by the group velocity and only the superconducting part, $\sigma(E)$, rescaled by a constant enters Eq. (15.2). This technical point is generally ignored for conventional superconductors, but as was shown in the previous section the role of band structure effects can be quite important for understanding the tunneling spectra of cuprate superconductors.

Since most conventional superconductors, especially s-p metals and alloys, have a weak energy dependence of the b.s. DOS near the Fermi energy, then even if the $N(E)$ was the full DOS, the band structure part would be expected to be nearly constant over the voltage range of interest (approximately 50 meV). Normalization of the tunneling data would be expected to give $\sigma(E)$ as before, giving an implicit measure of $\Delta(E)$ via Eq. (15.5). An example of such data for ZrN, and the corresponding analysis, is given below.

Before proceeding it is reasonable to ask whether there is any evidence that the tunnel current in conventional superconductors is dominated by electrons normal to the barrier. Noteworthy here is the fact that planar junctions on single crystals of Nb exhibit reproducibly different $\alpha^2F(\omega)$ for junctions fabricated on (100),(110) and (111) crystallographic planes. This is perhaps the most direct evidence for tunneling directionality effects [1]. Also, junctions on (N/S) bilayers are consistent with proximity effect models which assume specular tunneling normal to the barrier. This is discussed later.

15.4.1 McMillan–Rowell Analysis

There is an exhaustive literature on conventional, lower T_c superconductors, detailing how the complex gap parameter, $\Delta(E)$, can be inverted using Migdal–Eliashberg theory [1] to obtain the microscopic interactions responsible for superconductivity. The procedure of McMillan and Rowell (MR) [2] is iterative, beginning with an educated guess for the electron–phonon spectral function, $\alpha^2F(\omega)$ and the renormalized coulomb pseudopotential, μ^*. These parameters are then refined by comparison of the calculated $\sigma(E)$ with the measured normalized conductance. We will not review the early works but rather show some relatively recent data on the binary compound, ZrN, which has a T_c of 9.4 K and is exemplary of the MR analysis [32]. This material is also of interest because it exhibits relatively high frequency optical phonon modes and the issues concerning whether and how electrons couple to such modes is of great interest. ZrN is one member of a class of transition metal compounds with the NaCl structure such as NbN, NbC, and HfN. These materials have excellent mechanical properties as well as superconducting transition temperatures as high as 18 K.

Thin films of ZrN were prepared by reactive sputtering and then oxidized in laboratory air for 50 h to form the tunnel barrier. The planar junction was completed by evaporation of an In counterelectrode. In Fig. 15.12 is shown the dynamic conductance $\frac{dI}{dV}$ vs. V in the superconducting and normal state. Acoustic mode phonon structures are observed in the range 20-30 mV and a distinct dip feature is found near 60 mV which is in the range of optical phonons. Note that these phonon structures in the superconducting state conductance lie below the normal state curve and this is a critical feature of strong coupling effects, arising from the imaginary part of the gap function $\Delta(E)$ and reflecting a loss of states in this region. Had this feature been due to an inelastic channel opening up there would be an increase of conductance in both the superconducting and normal state spectrum.

In Fig. 15.13 is shown the reduced conductance $(\frac{\sigma(V)}{\sigma_{BCS}}$-1) and the fit obtained from the MR inversion analysis. Also shown is the resulting $\alpha^2F(\omega)$ spectrum which is found to be in very good agreement with the phonon density of states $F(\omega)$ measured by inelastic neutron scattering. The coulomb pseudopotential $\mu^* = 0.1$ is in the expected range and the calculated $T_c = 8.4$ K is reasonably close to the experimental value. This experiment provides clear-

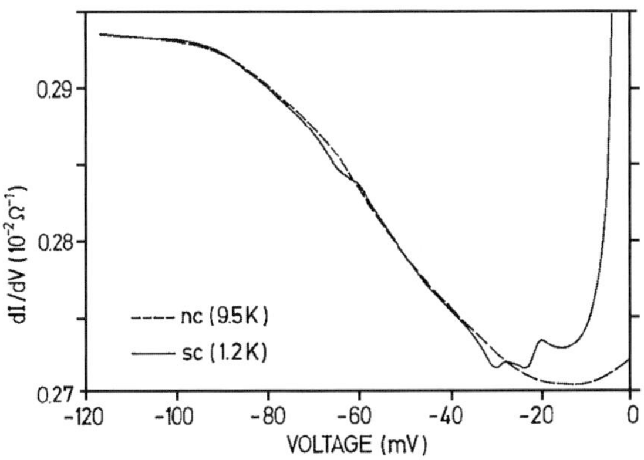

Fig. 15.12. Superconducting (*solid line*) and normal state (*dashed line*) conductance for planar junction, ZrN/ZrN-oxide/In. The superconductivity of In has been quenched with a small magnetic field

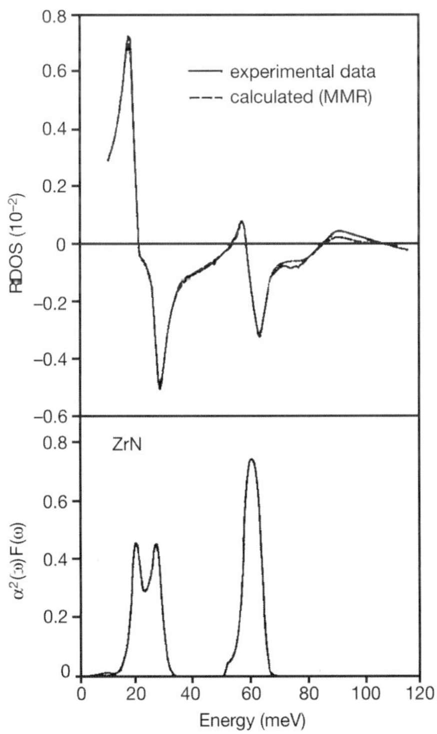

Fig. 15.13. *Top panel*: measured and calculated reduced conductance. *Bottom panel*: the resulting $\alpha^2 F(\omega)$ from MR inversion

cut evidence for electron coupling to high frequency optical modes. It is for this reason that such a relatively high-T_c could be obtained with a rather moderate coupling strength λ=0.62. It is thus clear that for oxide superconductors, which have even higher frequency optical phonon modes, the possible role of electron–phonon coupling must be considered.

15.4.2 Proximity Effect Tunneling in Conventional Superconductors

Tunneling investigations of (N/S) bilayers has had a tremendous impact on the field of conventional superconductivity from both a fundamental and applied perspective. Such studies have provided a more complete understanding of the electron–phonon interaction (even in non-superconducting elements such as Mg) and have generated numerous electronic devices, such as high-speed A/D converters and sensitive photon detectors. A brief discussion of this area is therefore warranted. Furthermore, the field is mature and well understood so that speculations about proximity effects in unconventional superconductors such as high-T_c cuprates can be critically examined using this information base.

We imagine here junctions of the form C-I-NS where C is a convenient counterelectrode, either a normal metal for spectroscopy studies of the N and S layer, or a superconductor for studies of the Josephson effect and related device phenomena. Thus tunneling is into (or out of) the N side of the normal metal/superconductor sandwich as shown schematically in the inset of Fig. 15.14. By using N layers of known thickness and properties, such structures have extended the standard tunneling spectroscopy and MR inversion procedure to include superconducting elements, alloys and compounds which themselves do not form suitable surface oxides for tunnel barriers or which are strongly affected by proximity to non-native oxides, such as from the counterelectrode [1,33].

Simple examples are Nb and V whose native surface oxides are complex, consisting of various oxide phases (including e.g. NbO which is metallic). These complex oxides also have defects, e.g. voids or ionized atoms, which increase the likelihood of mul-

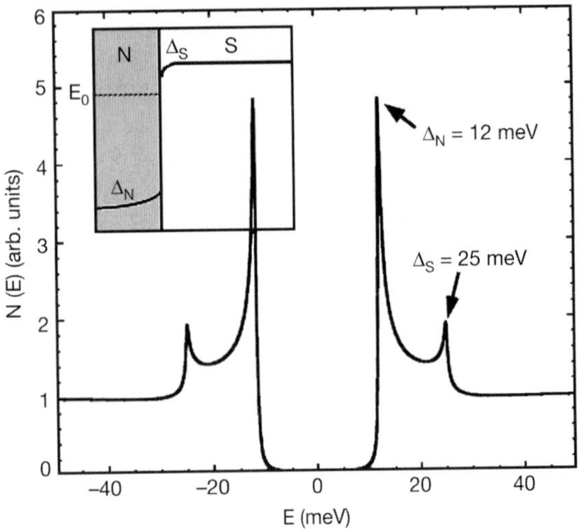

Fig. 15.14. McMillan model for the density of states $N_N(E)$ obtained for tunneling into the N side of an NS bilayer. *Inset*: schematic representation of the superconducting gap parameters in the proximity sandwich

tistep, resonant tunneling processes which compete with the desired direct elastic tunneling between the two metal electrodes. [34] In addition, even small amounts of diffused oxygen are known to rapidly decrease the superconducting T_c of Nb and V (approximately 1 K per atomic percent of oxygen) which means the standard fabrication method of evaporating or sputtering Nb or V onto oxidized films of Al does not work. The resulting gap region characteristics are extremely broadened and all phonon structures are smeared out, most likely due to the fact that the resulting junction is of the form Al/Al-oxide/S'S where S' is an inhomogeneous, disordered layer of reduced superconductivity on the surface of the bulk superconductor of interest.

To circumvent such problems, a new method was developed during the mid 1970s known as proximity electron tunneling spectroscopy (PETS) [35], which allowed a quantitative fit to the tunneling spectra of N/S bilayers using the proximity effect theory of Arnold. The two relevant models for proximity effect tunneling are those of McMillan and Arnold which we discuss briefly. The McMillan model assumes the N/S interface has a weak, delta function barrier for diffuse tunneling to occur, allowing a perturbation treatment of the problem. The calculation generates a set of coupled non-linear equations which determine the gap functions, $\Delta_S(E)$ in the S layer (reduced from its bulk value) and $\Delta_N(E)$ induced in the normal layer. The quasiparticle density of states in each layer is that given by Eq. (15.5) but with the appropriate gap parameter. Assuming the N layer has no intrinsic pairing interaction and is much thinner than S, then the McMillan DOS in the N layer is determined by two parameters, the bulk superconducting gap and a scattering rate, Γ_N, which is inversely proportional to the N layer thickness.

An example of the expected tunneling density of states in the normal metal from the McMillan model is shown in Fig. 15.14 assuming a superconductor with a bulk gap of 25 meV, a value typical for high-T_c cuprates. While the McMillan model has intuitive appeal for understanding the proximity effect attempts to extract $\alpha^2F(\omega)$ in either the N or S layer were unsuccessful [33]. The PETS method utilizes a clean N/S interface and this is described by the Arnold model, which, in the limit of small N-layer thickness, allows a direct measure of the underlying S-layer gap function [33]. The specular N/S interface also results in resonance effects due to Andreev reflection, and this leads to quasiparticle bound states in the N layer as well as higher energy oscillations of the DOS. The relevant parameter which determines these effects is $R = \left(\frac{1+r}{1-r}\right)\frac{2d}{h v_f}$. Here r is the reflection coefficient between N and S layers due to differences in Fermi momenta, v_f is the fermi velocity in the N layer and d is the N-layer thickness. The first bound state energy is given by, $E_0 = \Delta_S(1 - \frac{1}{2}(R\Delta_S^2)$, and is shown schematically in Fig. 15.14. An example of the DOS for very thick N layers is shown in Fig. 15.15 where the bound states and oscillations are clearly visible.

Of more relevance to the issue of pairing interactions is the Arnold model in the ultra thin N layer limit, i.e. $R \approx 0$. In this case it can be shown that phonon structures from the underlying S layer as embedded in $\Delta_S(E)$ can be observed as if tunneling directly into S, but with a reduced amplitude scaled by the factor $e^{-\frac{d}{l}}$. Thus the standard MR inversion procedure can be used but modified to include an additional free parameter, d/l, where l is a quasiparticle mean-free-path in the N layer. This is referred

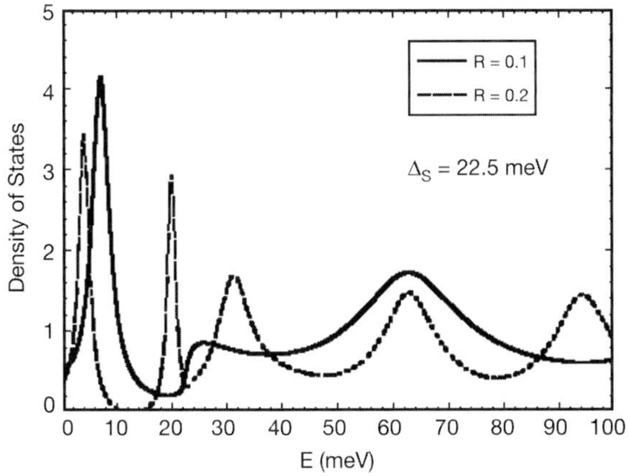

Fig. 15.15. Arnold density of states for relatively large R values. Note the presence of two bound states for $R = 0.2$

to as the modified McMillan–Rowell or MMR procedure.

From the experimental side the idea was to form a nearly perfect N/S interface by depositing a thin layer of Al onto a clean single crystal superconductor, such as Nb, in an ultra-high vacuum environment. The Al layer was known to provide an ideal, self-limiting oxide tunnel barrier and at the same time the unoxidized portion would provide a protective layer for the superconductor. The first successful junctions of this type on Nb and V showing clear phonon structures were published in 1978 [35] and a full quantitative analysis using the Arnold proximity effect model appeared a few years later [36,37]. These results showed that the standard MR inversion procedure could be extended to (N/S) bilayers allowing a determination of $\alpha^2F(\omega)$ in transition metals such as Nb and V. Examples of these spectra using the PETS method are shown in Fig. 15.16. These and other $\alpha^2F(\omega)$ spectra have been used successfully to determine the phonon contribution to the temperature dependent electrical resistivity above T_c in several superconducting elements and compounds. An example of the fit of Nb data is shown in the right panel of Fig. 15.16 taken from the work of Tralshawala et al. [5]. The PETS method has been used to examine alloys such as NbZr as well as A-15 compounds such as Nb_3Sn [1].

After the discovery that a thin Al overlayer could provide a pinhole free oxide barrier on Nb, the proximity effect method was extended to SIS junctions of the type Nb/Al/Al-oxide/Nb where it was shown that excellent Josephson tunnel junctions could be obtained. Further development of this method included the patterning of arrays of such junctions which are now the basis for all superconducting electronic devices [38].

15.5 Tunneling in High-Temperature Superconductors

Since the discovery of high temperature superconductivity in layered copper oxides by Bednorz and Müller [39] in 1986 there has been a worldwide effort to obtain tunnel junctions on these materials. Many of the early efforts suffered from complications owing to the ceramic nature of cuprates and the fact that the superconductivity exists only over a narrow

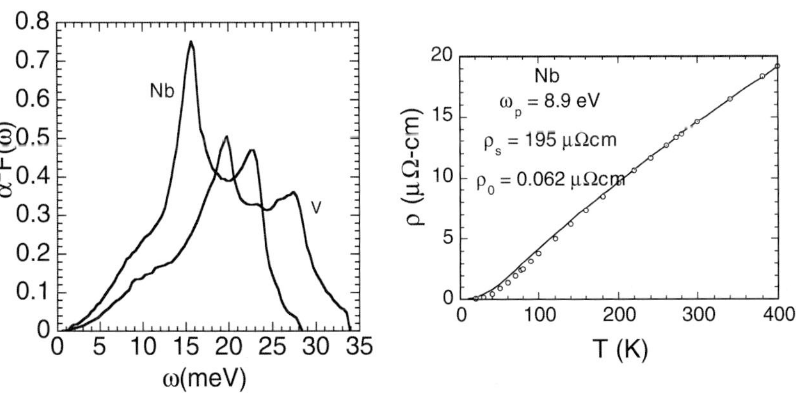

Fig. 15.16. *Left*: $\alpha^2F(\omega)$ for Nb and V obtained from N/S proximity bilayers using Al overlayers. *Right*: the fit to Nb resistivity using the electron–phonon spectral function, from Tralshawala et al. [5]

range of the composition phase diagram. All nearby chemical phases are insulating and the parent phase is an antiferromagnetic insulator. Thus the early ceramic materials had thick, insulating dead layers on the surfaces (thwarting attempts at STM measurements) along with nanoscale sized grains which produced charging effects in tunnel junctions. One such charging effect, the Coulomb's staircase, exhibited current-voltage (I-V) characteristics quite similar to that expected from an s-wave superconducting gap, leading to much confusion over the magnitude and pairing state symmetry of the gap. A review of the early tunneling efforts was given by Gray et al. [40] and a later review by Kirtley [41] showed improvements in junction quality resulting from the growth of HTS crystals, such as $La_{2-x}Sr_xCuO_4$.

By 1992 most of the interesting oxide superconductors had been studied by some form of tunneling spectroscopy using one or more of the junction configurations of Fig.řefeps0. The review by Hasegawa and Kitazawa [22] published at that time showed that there was mixed success. The bismuthate, $Ba_{1-x}K_xBiO_3$ (BKBO), reproducibly exhibited a clean, BCS-like gap structure (see Fig. 15.2) indicating s-wave superconductivity. Higher energy spectral features in the range 20 meV – 70 meV resembled strong-coupling phonon structures which were analyzed using the MMR procedure to obtain $\alpha^2F(\omega)$ [6]. But this was controversial because intergrain Josephson switching produced structures in this energy range as well [42]. The issue was resolved by Samuely et al. [43] who applied a large magnetic field to quench any Josephson switching and the resultant $\alpha^2F(\omega)$ spectra were in agreement with Huang et al. [6]. Further support for electron–phonon coupling in $Ba_{1-x}K_xBiO_3$ was found when the tunneling spectral function was successfully used to fit the d.c. electrical resistivity in overdoped compounds with T_c near 20 K [5]. However, BKBO is not completely understood. The optical response is inconsistent with standard electron–phonon coupling and can only be explained by assuming weak coupling to an unspecified high energy excitation near 0.4 eV [44]. Such an excitation might be related to the nearby CDW phase that develops with even slight underdoping from the optimal T_c value of 30 K.

Comparing the state of the field in 1992 to the present, some interesting developments and changes in perspective have occurred. Early measurements on the electron-doped cuprate, $Nd_{2-x}Ce_xCuO_4$, including tunneling, penetration depth, Raman and photoemission were all consistent with s-wave pairing [6, 45], however more recent results, including tri-crystal ring experiments, indicate d-wave symmetry [7]. Thus there is a resurgence of interest in this cuprate. Although high quality single crystals and epitaxial thin films of $YBa_2Cu_3O_7$ had been developed early, the tunneling spectra were confusing. The energy gap structure appeared very much broadened and in many cases a large ZBCP anomaly was found. It was later discovered that this ZBCP was due to the unusual effects that a d-wave gap has on quasiparticle scattering processes (see the section on novel d-wave effects). Now studies of the ZBCP are being used to search for more exotic pairing states, including broken time-reversed symmetry states.

Much clearer tunneling gap features were generally found on $Bi_2Sr_2CaCu_2O_8$ (Bi2212) which was attributed to the ability to cleave such crystals, leaving an atomically smooth surface for vacuum or point contact tunneling. Using STM or PCT methods resulted in c-axis tunneling which eliminated any ZBCP anomalies as well. Nevertheless, there was much scatter in the magnitude of the energy gap reported by Hasegawa et al. [22]. It is now understood that this spread of gaps is due to a strong dependence of the intrinsic gap on doping concentration and it has taken a new generation of high quality Bi2212 crystals to unravel this mystery. At the present time there exists a great deal of reproducibility among tunneling spectra in Bi2212, not only in the size of the energy gap but in the overall spectral shape. Therefore this material will be discussed in detail. The other cuprates mentioned will also be given further attention as they are exemplary, providing a focus for the specific issues of pairing mechanism and gap symmetry.

15.5.1 $Nd_{2-x}Ce_xCuO_4$

High quality tunneling spectroscopy measurements of $Nd_{2-x}Ce_xCuO_4$ (NCCO) were reported by Huang et al. in 1990 [6]. At that time there was little dis-

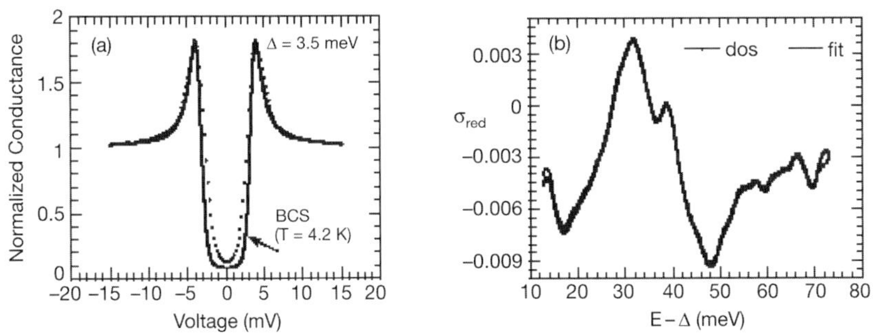

Fig. 15.17. (**a**) Normalized tunneling conductance of NCCO in the gap region (*dots*) and *s*-wave BCS fit (*solid line*). (**b**) Phonon fine structure observed at higher bias voltages in the reduced conductance and corresponding fit from the MMR inversion

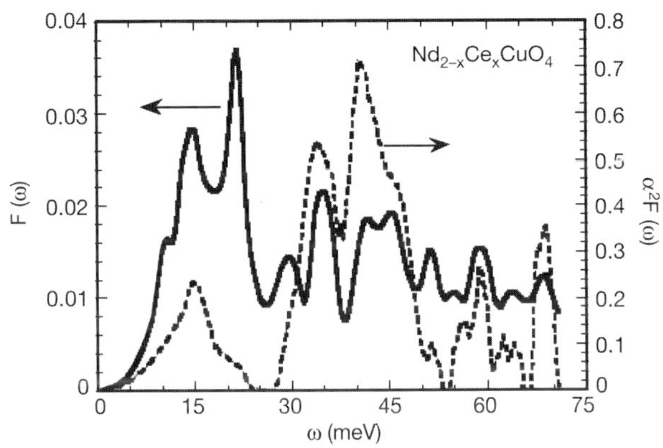

Fig. 15.18. Comparison of phonon density of states obtained from neutron scattering with $\alpha^2F(\omega)$, from Tralshawala et al. [5]

cussion of *d*-wave superconductivity. The gap region tunneling conductance shown in Fig. 15.17 is reasonably described by a conventional, *s*-wave BCS fit with thermal smearing and a small leakage conductance of 10%. However, the fit to BCS is not ideal as is observed with BKBO. The gap parameter, $\Delta =$ 3.5 meV–3.7 meV, was found on 20 junctions measured on three different crystals. Using 3.6 meV as the average leads to $\frac{2\Delta}{kT_c} = 3.8$ indicating moderate coupling strength within an *s*-wave picture.

Inversion of the fine structure in the high bias conductance was done using a modified version of the MR procedure to include the small damping effect of a proximity layer on the surface. This MMR procedure converged and led to the $\alpha^2F(\omega)$ spectrum in Fig. 15.18. This is compared to the measured neutron $F(\omega)$ on the parent Nd_2CuO_4 where good agreement is found. It is important to stress that the MMR procedure is iterative and if the high bias fine structures are due to extrinsic sources then this procedure will not converge. Convergence is thus the first sign that there is physical significance to the fine structure. More importantly the $\alpha^2F(\omega)$ spectrum leads to a calculated T_c that is very close to the measured value of 22 K [6]. The resulting coupling constant, $\lambda = 1.0$, is close to that expected based on the measured strong-coupling ratio. Thus there are no inconsistencies in the tunneling data or the *s*-wave analysis, and the conclusion of predominantly phonon-mediated pairing in NCCO seems justified.

The observation of half-integral flux quanta in tri-crystal ring experiments on NCCO has led to the conclusion that this a *d*-wave superconductor [46]. While *d*-wave symmetry might explain the persistent quasi-particle sub-gap tunneling conductance in NCCO, it appears that electron–phonon coupling still plays a role in the superconductivity of this material. The present state of affairs thus has contradictory aspects, since it is not obvious how to make *d*-wave superconductivity and the observed electron–phonon coupling compatible. For example, the measured strong-coupling ratio in NCCO makes sense in an *s*-wave picture but it is significantly smaller than the weak coupling limit value of 4.26 for d-wave pairing [8]. This is to be contrasted with higher T_c hole-doped cuprates where the measured strong-coupling ratios are larger than 4.26, as expected, and where such fine structure from phonons is not observed in any reproducible manner. Instead, the measured tunneling spectra are consistent with pairing interactions that are dominated by electron–electron interactions that naturally give rise to d-wave superconductivity.

Perhaps in NCCO electron–electron interactions are present, but necessarily weaker, since T_c is quite small compared to the hole-doped counterparts. Nevertheless, it might be energetically favorable for the competing interactions to settle on a d-wave pairing state. However, it is unlikely that the resultant momentum dependent gap would have the simple BCS form considered in Sect. 15.2.1. Furthermore, since d-wave superconductivity is sensitive to disorder, there might be local regions where a competing s-wave state wins out. It is safe to say at present that there are new questions to be resolved for NCCO.

15.5.2 $Bi_2Sr_2CaCu_2O_8$

The generic spectral features of Bi2212 have already been discussed for STM and PCT respectively, see Fig. 15.20. These features can be separated according to energy scale. When viewed at the highest energies (out to 1 eV) the conductances are consistent with a broad peak centered near the Fermi energy which is most likely due to the band structure DOS [31]. While the peak itself is obscured by the superconducting gap, its effect is to produce a decreasing background out to 300 meV. The sharp conductance peaks indicate the gap for quasiparticle excitations. At energies of the order two and three times the gap energy there are dip and hump features with a strength that is generally asymmetric in bias voltage. However, the degree of asymmetry of the dip feature seems to depend on the type of junction. As will be shown, some STM measurements show nearly symmetric dip features in the normalized data, whereas PCT junctions, which average over a much larger contact area, always display a large asymmetry.

These basic characteristics are evident for all doping levels, changing only in energy scale. An example of PCT conductance data for a slightly overdoped Bi2212 crystal is shown in Fig. 15.19 from the work of Ozyuzer et al. [47]. Note the strong asymmetry of the dip feature, being most pronounced for bias voltages which remove electrons from the superconductor. The raw data have been compared to a BCS d-wave fit which provides an excellent fit to the subgap conductance. This indicates that the normal state (or background) conductance is approximately constant in the subgap region. The cusp at zero bias is a characteristic of a d-wave gap and this feature has been seen in the PCT data of Tl2201 (see Fig. 15.4) as well as in many spectra on Bi2212. However, for bias voltages $eV > \Delta$ there is no agreement. The simple BCS d-wave fit cannot explain the asymmetric dip

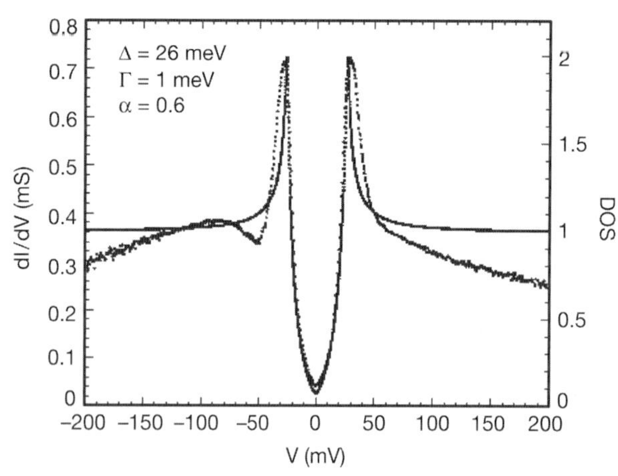

Fig. 15.19. PCT tunneling conductance of overdoped Bi2212 (*dots*) and d-wave fit (*solid line*), from Ozyuzer et al. [47]

feature or the decreasing background. In addition the conductance peaks are wider than the fit curve and this is generally true for Bi2212 data measured by STM, PCT, break junctions etc. A closer examination of Tl2201 in Fig. 15.4 reveals an interesting similarity in the deviation of the conductance data with the d-wave fit.

The background conductance in Bi2212 cannot be measured to normalize away those features not associated with strong coupling effects as done with Nb or ZrN shown earlier. This is due to the fact that the temperatures required to reach the normal state are >90 K and the background can change when compared to low temperatures. More importantly, the superconducting gap might be evolving into a pseudogap above T_c in which case one needs to exceed T^* to reach the "true" normal state. For the analysis of high-T_c cuprates it becomes clear that the all effects which might be contributing to the low temperature conductance, e.g. d-wave superconductivity, strong coupling effects, band structure DOS, pseudogaps, etc., will have to be built in to any model of the tunneling conductance. Such a model must also explain all asymmetries. This situation is obviously much more complex than found in conventional superconductors and leads to ambiguities in interpretation of tunneling spectra.

In order to gain insight into the spectral features the Argonne group has performed tunneling measurements of Bi2212 over a wide doping range from underdoped (T_c = 70 K) to optimal doped (T_c = 95 K) to overdoped (T_c = 48 K) [48,49]. Doping de-

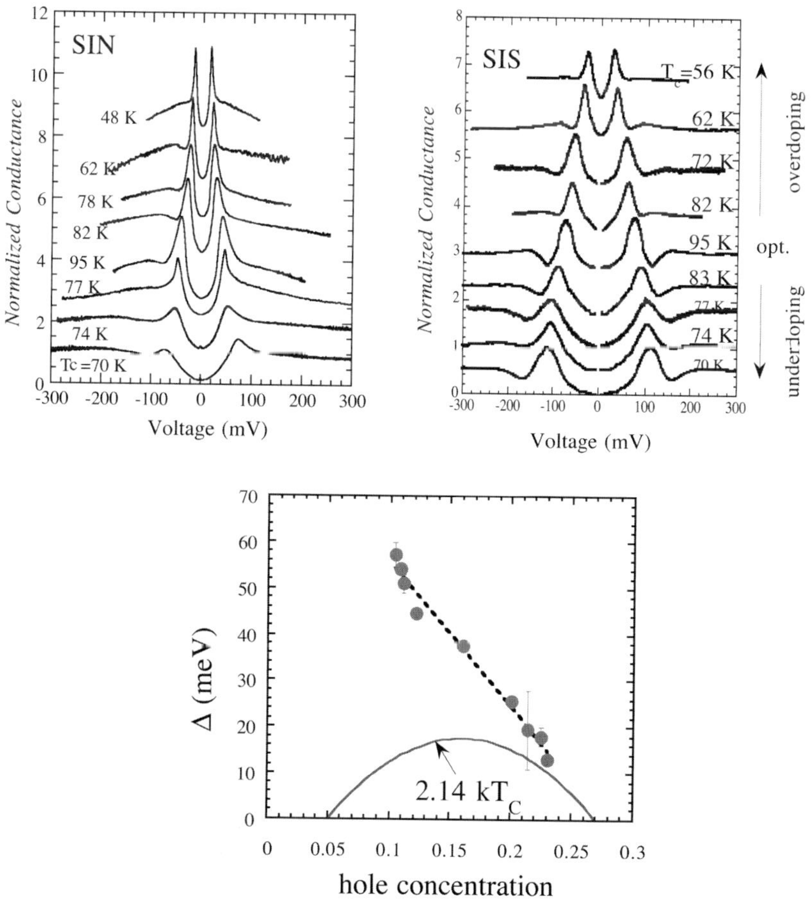

Fig. 15.20. *Top*: SIN and SIS tunneling conductances vs. doping obtained from PCT junctions on Bi2212. *Bottom*: Gap parameter vs. doping obtained from all junction types. From Miyakawa et al. [48,56]

pendent trends have been observed in the energy gap and dip/hump features. The hole concentration has been varied by oxygen content. Starting with optimal doped crystals, overdoping (underdoping) is achieved by annealing in high (low) pressure oxygen. Representative SIN conductances and SIS break junctions are shown in Fig. 15.20. What is evident is that the energy gap increases monotonically with decreased doping, even in the underdoped region where T_c decreases. This is shown more clearly in the bottom panel of Fig. 15.20 where the data points represent averages of a large number of SIN and SIS junctions. The error bars are small and this indicates a high degree of homogeneity within a given crystal. These gap values have also been reproduced in the STM studies of Bi2212 by DeWilde et al. [30] and Renner et al. [50,51] and therefore Fig. 15.20 can be considered intrinsic from the perspective of tunneling. There is always a question that tunneling might be sensitive to surface effects, but it should be noted that the same magnitudes and trend of gap vs. doping has also been seen (over a more limited doping range) in Raman spectroscopy which is a bulk probe [52].

Such an unusual doping dependence to the measured quasiparticle gap has stirred up considerable debate, especially since it is directly related to the issue of a pseudogap observed above T_c in various experimental probes of cuprate superconductors. The various interpretations of the pseudogap can be grouped into two broad classes [53] as shown in the schematic phase diagrams in Fig. 15.21. In the top picture, the pseudogap is depicted as a separate energy scale, distinct from superconductivity. Possible origins of such a pseudogap are charge density waves (CDW), spin density waves (SDW) or more exotic orbital antiferromagnetism or d-wave density waves (dDW) [54,55]. In such a view the superconducting gap more closely follows the doping dependence of T_c and therefore the measured quasiparticle excitation gap in Fig. 15.20 must represent a smooth change from purely superconducting character in the overdoped region to some other type of gap in the underdoped region. Tallon and Loram have suggested that various bulk experimental probes such as specific heat are consistent with this picture [53]. However, it was argued by Miyakawa et al. [48,56] that such

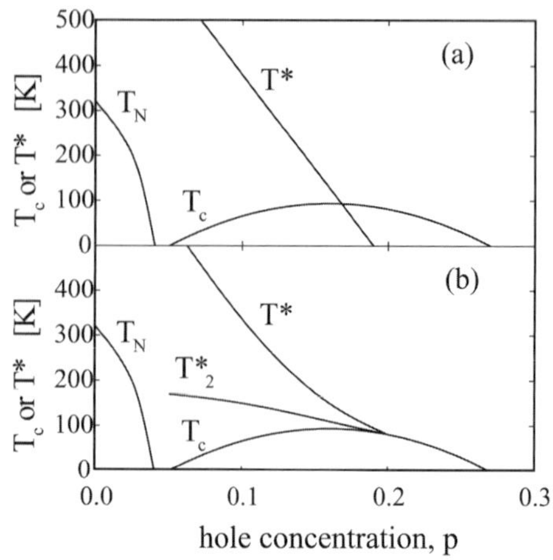

Fig. 15.21. Schematic phase diagrams for cuprate superconductors. In (**a**) the pseudogap is a separate energy scale, distinct from superconductivity. In (**b**) the pseudogap temperature represents some form of precursor superconductivity

a view is not supported when the tunneling spectra are examined in detail. In particular, two-gap models should exhibit two distinct features in the DOS and therefore in the tunneling conductance, whereas only a single, d-wave type gap feature is observed. Although the dDW state allows coherent mixing of the two gaps, this is true only when mean-field theory is applicable. The same disorder (fluctuations) that causes the mean-field dDW gap to become a pseudogap would most likely affect the coherent mixing in the superconducting state, leading again to distinct spectral features in the DOS. Furthermore, robust Josephson currents are observed in heavily underdoped Bi2212 crystals, indicating that the excitation gap is linked to a superconducting energy scale [57].

The second viewpoint is indicated in the bottom graph of Fig. 15.21. Here the crossover temperatures T^* and T^M smoothly connect to the T_c curve suggesting there is break between temperature scales associated with electron pairing and long-range phase coherence, as is likely in cuprates due to the low phase stiffness [58]. In such pictures, pseudogap phenomena are associated with some form of precursor superconductivity and therefore the measured quasiparticle gap of Fig. 15.20 is the true pairing gap associated with superconductivity [59]. This viewpoint is

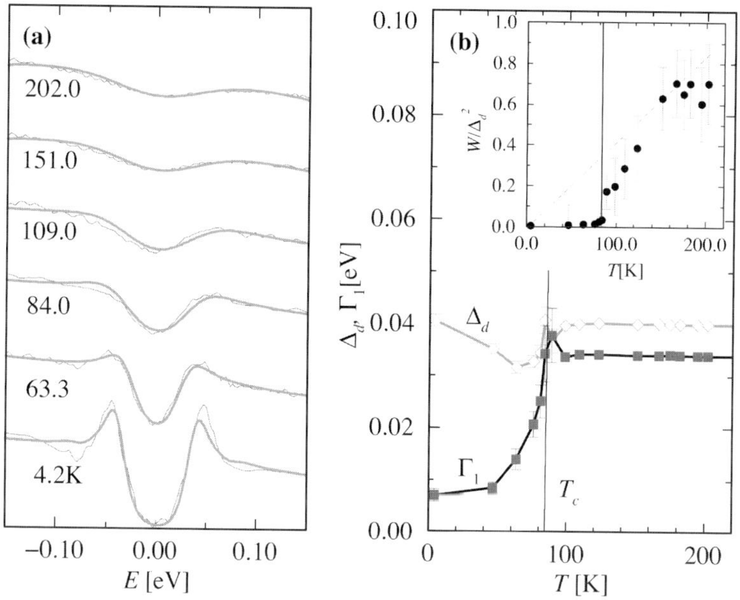

Fig. 15.22. (a) STM conductances of Bi2212 at various temperatures from Renner et al. [62] and the KT fits of Franz and Millis, [63]. (b) Resultant energy gap and scattering rates as a function of temperature

consistent with the fact that $\frac{2\Delta}{kT^*} = 4.5\text{-}5.5$, a reasonable strong-coupling value [48, 60] whereas putting the measured T_c in the denominator leads to unphysically large values, as high as 20. It should be mentioned here that the notation for the crossover temperatures differs throughout the literature. The notation used by Tallon et al. [53] is such that the upper crossover is labeled as T^*. This crossover corresponds to the temperature where the uniform susceptibility flattens out and is associated with the development of short range magnetic order. More often it is referred to as T^M (as it is here) or T^0 and it is more common to see the lower crossover listed as T^*. It is also possible that the upper crossover corresponds to the development of some type of pseudogap associated with magnetic order, such as an incipient SDW [61]. This would then suggest that there might be two pseudogaps, with the lower crossover temperature T^* generally associated with the development of a pairing gap from superconductivity.

It might seem that a measure of the tunneling gap vs. temperature would resolve the debate over the nature of the pseudogap, however such measurements are not easily interpreted. In Fig. 15.22 is shown the temperature dependent tunneling conductances on slightly underdoped Bi2212 by Renner et al. using STM [62]. Qualitatively, the data appear to show a superconducting gap smoothly merging into a pseudogap of similar energy scale, indicating some form of precursor superconducting state. The data have been fit by Franz and Millis [63] using a Kosterlitz–Thouless (KT) model to describe the phase fluctuations in the pseudogap state. In this model, long range phase coherence is lost above T_c by the presence of vortex/antivortex pairs which also serve to scatter quasiparticles and turn the pairing gap into a pseudogap. Support for this type of phase fluctuation model for the pseudogap comes from a.c. conductivity measurements of Bi2212 up to 600 GH_z [64]. The temperature dependence of the tunneling gap that results from this analysis is of particular interest. $\Delta(T)$ decreases by more than 20 percent as T_c is approached from below but does not go to zero. This behavior is also seen by SIS break junctions (Miyakawa et al., Ref. [48]) and some intrinsic c-axis junctions on Bi2212 (as will be shown in a later section). The break junction data are shown in Fig. 15.23 for an underdoped Bi2212 crystal with a $T_c = 77$ K. In the Miyakawa et al. work there was no attempt to fit the data just above T_c which was a well-defined temperature where the Josephson current went to zero. Thus in several tunneling studies of Bi2212, there appears

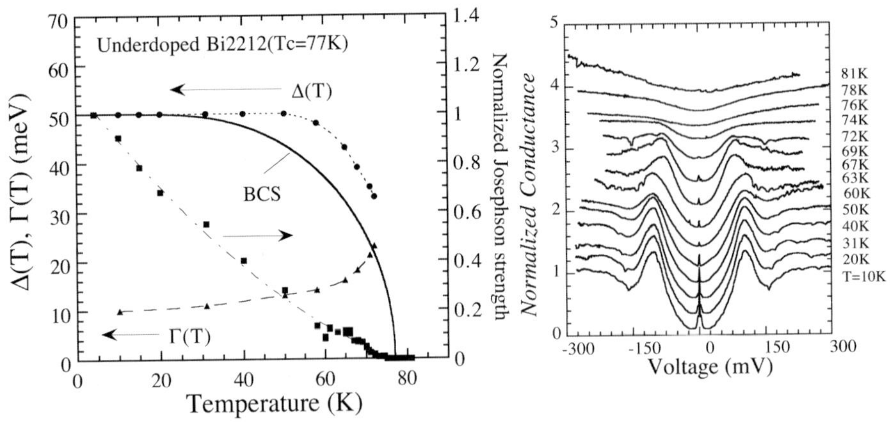

Fig. 15.23. Measured superconducting gap vs. temperature from SIS break junctions in underdoped Bi2212, from Miyakawa et al. [48]

to be evidence that the pairing gap magnitude persists into the pseudogap state, supporting the view of precursor superconductivity.

However, an interesting effect is found in Fig. 15.22 just above T_c where the gap parameter which best fits the data actually increases. The authors considered this to be an artifact of fitting the broad tunneling conductance and fixed the gap magnitude to that of the low temperature value. Their interest was mainly in the quasiparticle scattering rate [63]. In this KT model, the broadening of spectral features above T_c is due to the fact that the scattering of quasiparticles has the effect of smearing out the spectral weight function. More striking behavior is found in other analyses of temperature dependent gap region tunneling conductances which use the ordinary quasiparticle scattering rate, Γ, along with a d-wave gap parameter to fit the data [65]. In those studies the gap magnitude decreases as T_c is approached from below but then increases substantially above T_c, often reaching values that exceed the gap magnitude at low temperatures. It is difficult to understand this in the context of any superconducting fluctuation model, especially since the superconducting gap magnitude should continue to decrease above T_c. Rather, these temperature dependent studies are suggestive of a second gap in the quasiparticle spectrum, perhaps associated with the upper crossover temperature, T^M.

In general it can be said that tunneling data from STM, PCT, break junctions and some intrinsic junctions (as will be shown) on Bi2212 are consistent and support the phase diagram shown in Fig. 15.21(b). Thus the pairing gap persists into the pseudogap state but there is no long range phase coherence. The inability of tunneling to clearly and reproducibly map out the temperature evolution of the superconducting gap is probably due to the upper crossover temperature, T^M, which also gives rise to a pseudogap. This second pseudogap has an energy scale which is several times larger than the low-T superconducting gap, but is a much weaker effect on the DOS and is therefore more difficult to detect experimentally. Support for this view is also found in photoemission studies of Bi2212 [66] and in theoretical spin-fermion models [61].

The dip/hump feature shown in Fig. 15.19 is also seen throughout the doping range in Fig. 15.20. In PCT junctions the dip is highly asymmetric with bias voltage, always being very strong for bias voltages which correspond to removal of quasiparticles from the superconductor. This dip feature was observed in some of the earliest tunneling studies of Bi2212 and at present there is still no consensus on its origin [67]. Because of the asymmetry, the positive bias feature was ignored and the dip was thought to be part of the background tunneling conductance, perhaps a consequence of the van Hove singularity in the DOS. However, later STM and PCT studies clearly showed that the dip occurred for both bias polarities which was more consistent with a strong-coupling effect similar to that found with phonon structures (e.g., see Fig. 15.7). For example we show the vacuum

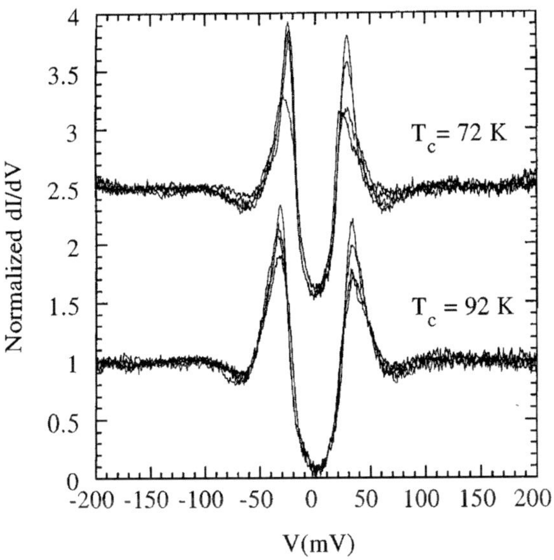

Fig. 15.24. Normalized STM conductance of Bi2212 for optimal doped and overdoped samples from DeWilde et al. [30]

tunneling results of DeWilde et al. [30] in Fig. 15.24 for two crystals, one near optimal doping (T_c = 93 K) and another that is overdoped (T_c = 77 K). The data from many junctions at various locations on the surface have been normalized by a smooth background and plotted together. There is generally good agreement among the different junctions. The normalization has removed some of the asymmetry but not all of it. There remains a difference in the peak heights and the strength of the dip but it is clearly seen for both voltage polarities and is approximately at 2Δ. Also, the normalized data indicates that the dip drops below unity by about 15%, again consistent with strong coupling effects. This magnitude is about ten times larger than observed in Nb, but this is consistent with the much larger gap size.

One of the earliest notions about the dip feature was that it might be a proximity effect. As Figs. 15.14 and 15.15 show, dip features can arise from the proximity effect however in those cases the main conductance peak is either an induced gap (McMillan model) or a bound state (Arnold model) in the N layer and therefore would always be at a lower energy than the true gap. Also the conductance peak voltage would depend on details of the surface N layer and thus would be material dependent, an effect which is not observed. Most importantly, the dip/hump feature in proximity effect models would be a closer measure of the underlying superconducting gap and for tunneling data on Bi2212 this would give an overestimate of the gap value by a factor of two or three. As noted before, Raman scattering which is bulk probe, matches the gap vs. doping curve in Fig. 15.20 and therefore the tunneling conductance peak gives the correct value of the superconducting gap. Thus the

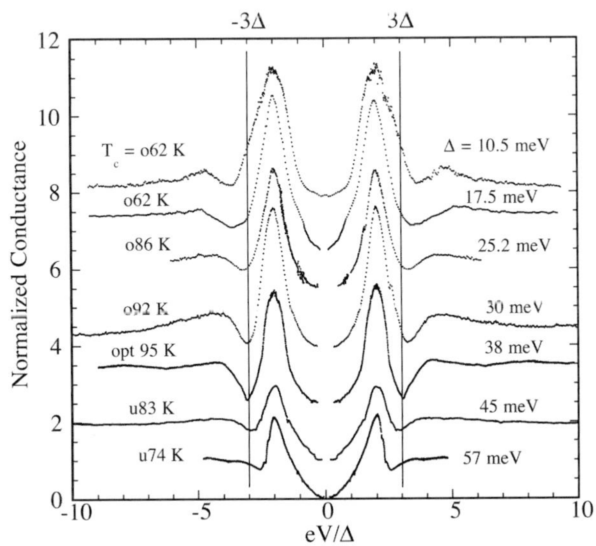

Fig. 15.25. SIS tunneling conductances at various doping levels showing the shift of the dip feature with T_c. Voltage axis is rescaled in units of Δ/e, from [68]

dip/hump features are not consistent with any proximity effect interpretation.

Further insight into the dip is found in SIS break junctions where the feature is much more pronounced. More recent data from the Argonne group [68] is shown in Fig. 15.25. Here the voltage axis has been rescaled in units of Δ/e. If the dip in the SIN spectra was exactly at $E = 2\Delta$, then it should appear at 3Δ in the SIS plots but Fig. 15.25 shows that there is a systematic shift of the dip feature with doping. This dip in the SIS junctions is found to be at $2\Delta + \Omega$ where $\Omega = 4.9\ kT_c$. This scaling is nearly the same as found in neutron scattering for the resonance spin excitation [69,70]. Thus it appears that the dip feature can be considered a strong coupling effect analogous to phonon structures, but the bosonic excitation that is responsible for this feature, the resonance spin excitation, is itself well defined only in the superconducting state. As will be shown later, a similar effect is seen in a heavy fermion superconductor.

15.5.3 $Bi_2Sr_2CaCu_2O_8$ S™ Mapping

There has been a systematic improvement in the quality of STM data on HTS cuprates in general and on Bi2212 in particular. These are a consequence of improved quality of crystals as well as vacuum tunneling methods. In general, the Bi2212 crystals can be cleaved in-situ in a UHV environment and then measured at low temperatures. These latest generation STM measurements are providing unprecedented atomic resolution images of the cleaved surfaces which have an exposed Bi-O layer. An example is shown in Fig. 15.26 where the Bi and O atoms can be resolved [71]. The modulation of the Bi-O structure along the b-axis is indicated by the light and dark fringe pattern and along the light fringes there is a sinusoidal pattern which is not understood. Thus such atomically resolved images provide a unique opportunity to examine the local superconducting DOS and perhaps link the spectra to particular kinds of defects. By examining the superconducting conductance spectra at different locations, a defect structure is found that appears to be intrinsic to the Bi2212 material.

Two such spectra are shown in Fig. 15.27 for a slightly overdoped Bi2212 crystal from Hudson et al. [71] The dashed line conductance looks very similar to superconducting spectra obtained by PCT and previous STM studies. Note the dip features found at about 2Δ with the same strength as observed earlier by DeWilde in Fig. 15.24. However, there is an additional feature, a small peak, observed just before the dip. The origin of this feature is not understood, but it bears a resemblance to general strong coupling features, and it might be that the improved resolution of this STM spectrum is providing more detailed strong coupling features. This gives further credence to the notion that the spectral region near the dip is due to some type of quasiparticle emission process and Fig. 15.25 is suggesting that the boson emitted seems to be the resonance spin excitation.

The solid line curve of Fig. 15.26 has a slightly different spectral shape, generally weaker superconducting features and a small peak near zero bias. These types of ZBCP are common for junctions where the tunneling direction is into the *ab* plane as was discussed in the section on novel *d*-wave effects. They correspond to surface induced Andreev bound states which result when a quasiparticle scattering off a vacuum interface changes the sign of its

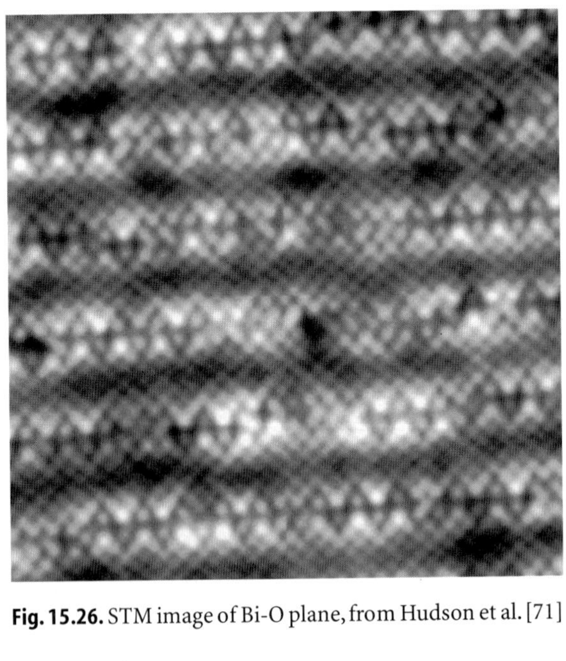

Fig. 15.26. STM image of Bi-O plane, from Hudson et al. [71]

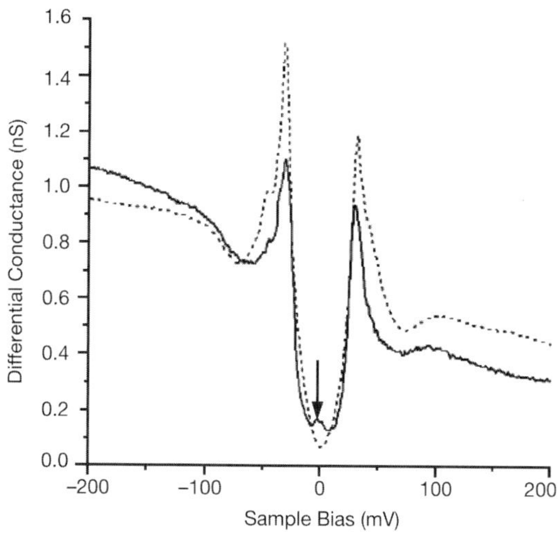

Fig. 15.27. STM conductance spectra for the tip over a superconducting region (*dashed line*) and a defect region (*solid line*). Note the zero bias conductance peak for the spectrum at a defect, from [71]

order parameter as can happen with d-wave symmetry. Similar effects are found for scattering off defects either adjacent to the Cu-O plane or within it. Even for elastic scattering, the quasiparticle can change its wavevector such that the sign of the order parameter is changed. The intrinsic defect regions are probably associated with defects that originate within the Bi-O layer where the excess oxygen has been placed to dope the material. Such a defect can still cause potential scattering within the Cu-O plane.

A more striking result is obtained when a Ni or Zn impurity is introduced directly into the Cu-O plane [72]. Scattering off such defects leads to large a ZBCP in the local density of states due to Andreev bound states and a suppression of superconductivity. An example of the spatial profile over a Zn impurity is shown in Fig. 15.28. It is important to understand that this is not a standard atomic image but is a spatial plot of the zero bias conductance and therefore the insulating Bi-O/Sr-O layer is acting as part of the tunnel barrier and superconductivity of the underlying Cu-O plane is being probed. The superconducting regions far from the Zn impurity have a low conductance since this is inside the superconducting gap and appear as dark regions. The brightest regions correspond to the largest ZBCP where superconductivity is weakest and the bound states exist. The dominant peak corresponds to the tip being just above the Zn atom. A striking feature of this image is the fourfold symmetric pattern which is a consequence of d-wave symmetry.

15.5.4 Intrinsic c-Axis Junctions on Bi2212

The cuprates are layered materials, with Cu-O planes separated by charge reservoir layers. Thus the doped cuprates can be viewed as stacks of metallic layers separated by thin insulating regions. Figure 15.1(e) is a schematic representation of such a stack for a single Cu-O plane per unit cell as in Tl2201. Whether intrinsic tunneling phenomena can be observed in experiments depends on the nature and magnitude of the tunneling matrix element. For example, coherent tunneling occurs when transverse wavevector is conserved and this leads to the standard Bloch waves and band dispersion along the c-direction. Nothing unconventional would be expected in this case. Transport along the c-direction would be of the Drude type. In Bi2212, the c-axis resistivity is larger than the maximum metallic resistivity and therefore adjacent Cu-O bi-layers are coupled incoherently. A single crystal along the c-axis can be viewed as a set of incoherently coupled tunnel junctions [73]. Lithographic methods have been developed to isolate stacks of junctions in a mesa configuration which contain a relatively small number ($\approx$ 5–20) of such SIS junctions in series. For example, three Cu-O bilayers in Bi2212 would act as two SIS junctions in series. These stacks exhibit Josephson effects and when the Josephson critical current for an individual junction is exceeded, that junction switches over to the single-electron (or quasiparticle) branch. Studies of such intrinsic junctions were pioneered by Kleiner et al. [74,75] and have led to an active field of study. A thorough review of the extensive literature has been given by Yurgens [76].

It might be assumed that such a stack of junctions would provide ideal quasiparticle characteristics since the Cu-O layers are perfectly aligned and there are no exposed surfaces which potentially could effect standard tunneling geometries. However, these intrinsic junctions have highly transmissive tunnel

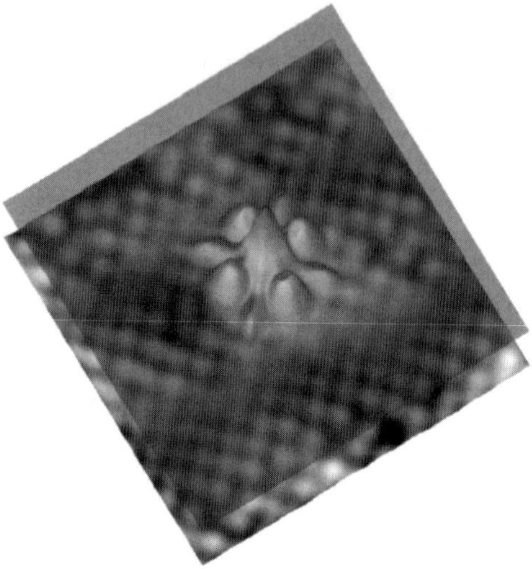

Fig. 15.28. STM mapping of the zero bias conductance near a Zn impurity. The large central peak corresponds to the tip right above the Zn atom. From [72]

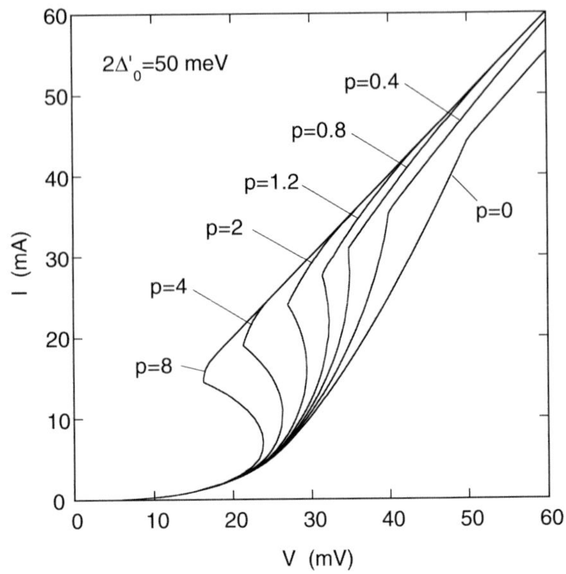

Fig. 15.29. Effects of high injection rates on the I-V characteristics of an SIS tunnel junction between two d-wave superconductors. For each curve the gap suppression is proportional to the square of the junction current and p is the proportionality constant. From Suzuki et al. [77]

barriers with a conductance per unit area greater than $10^5(\frac{S}{\text{cm}^2})$. This value is at least two orders of magnitude larger than found in planar junctions on conventional superconductors. At the typical bias voltages of interest, i.e., eV=2Δ=80 meV per SIS junction, it is easy to realize currents of $10^4(\frac{A}{\text{cm}^2})$ which is the typical value for various nonequilibrium phenomena to occur [77]. These include heating effects where the electronic and bosonic (e.g. phonons) distribution functions have an effective temperature, T^*, higher than the bath temperature (this T^* should not be confused with the pseudogap temperature). Another way to view this is to consider the electrical power developed across a junction. The voltage scale of interest is set by the gap value and thus the power is $\frac{V^2}{R}$ which rises as the junction resistance is lowered. The more transmissive a barrier, the higher the heating effects become.

Another consequence of high injection rates is that the electron distribution function is described by a chemical potential greater than the equilibrium value. These non-equilibrium distributions of electrons and phonons both tend to reduce the superconducting gap value. Suzuki et al. [77] have modeled these non-equilibrium effects using three different phenomenological expressions for the reduction of the energy gap with current across the junction. The effects for the different models are qualitatively similar and lead to backbending of the I-V curve in the sub-gap region, similar to what is often observed. The effect of quasiparticle injection is shown in Fig. 15.29 assuming a quadratic reduction of the gap ($\propto pI^2$) with junction current. Note the reduced size of the measured gap value compared to the bulk value and the backbending of the I-V curves for larger values of p. In Fig. 15.30 is shown a comparison of two different nonequilibrium models for the reduction of the gap value with measured data on a mesa stack. The two models are a linear and quadratic gap suppression (LGS and QGS) with junction current. Another important result is that I-V curve rises sharply at the gap voltage just at the threshold of backbending. This can give rise to extremely sharp conductance peaks.

Various attempts have been made to reduce the heating effects in mesa stacks. Considering the poor thermal conductivity of Bi2212, one way to reduce heating effects is to shrink the lateral area of the mesa. This improves heat conduction away from the

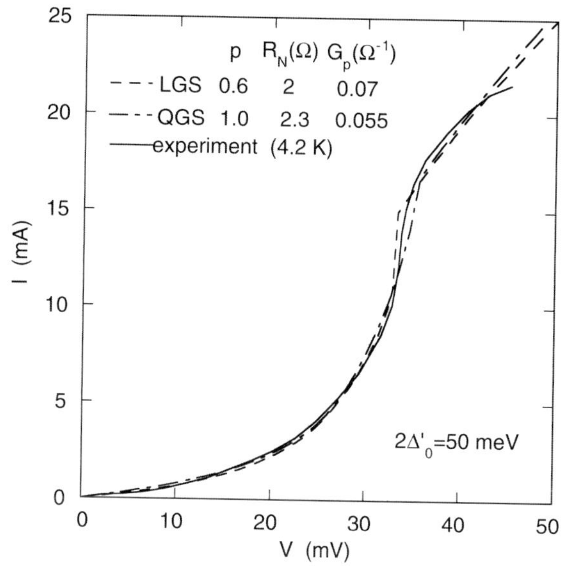

Fig. 15.30. Comparison of measured I-V characteristics of a mesa stack of intrinsic junction on Bi2212 with two nonequilibrium models for the suppression of the superconducting gap with temperature. From Suzuki et al. [77]

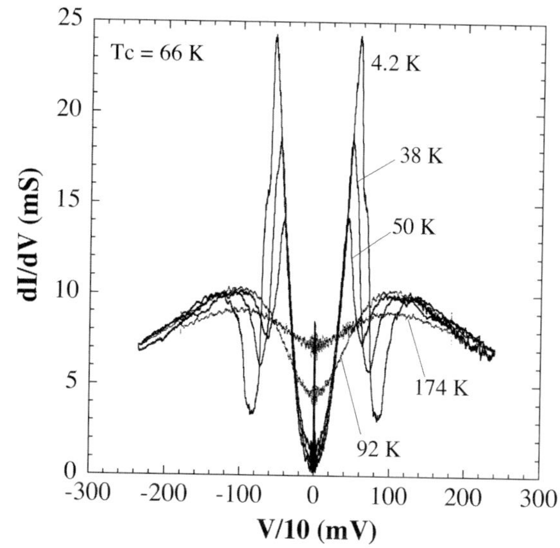

Fig. 15.31. Dynamic conductances at different temperatures for a stack of intrinsic junctions on a Bi2212 mesa intercalated with HgB_2. From Yurgens et al. [78]

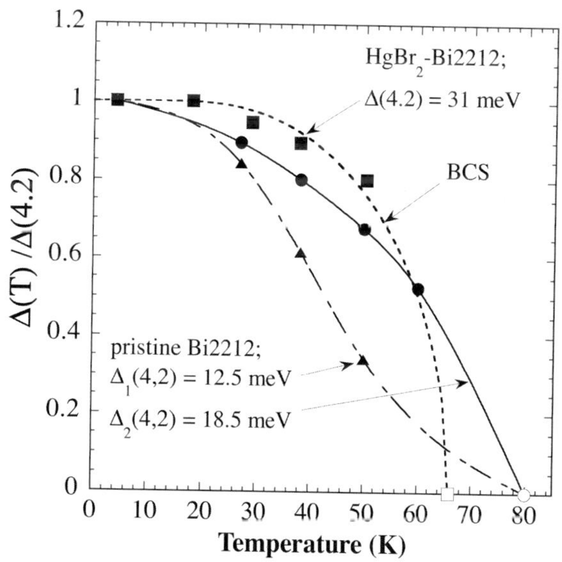

Fig. 15.32. Extracted gap vs. temperature in intercalated and pristine intrinsic junctions. From [78]

mesa. However, the junctions still have the same current per unit area and thus direct injection can still lead to nonequilibrium electron distributions. Pulsed current methods also help. A more suitable way to reduce nonequilibrium effects is to increase the junction resistance and thereby reduce the source of nonequilibrium effects. Yurgens et al. [78] have shown that this can be accomplished by intercalating HgB_2 between the double Bi-O layers in Bi2212 resulting in up to a 100 fold increase in resistance. The resulting quasiparticle spectra and T-dependent gap values, $\Delta(T)$, are very similar to break junctions shown earlier. Note the strong dip features, similar to break junctions shown in Fig. 15.25 which abruptly disappear near T_c. The dynamic conductance at various temperatures is shown in Fig. 15.31 and the extracted $\Delta(T)$ is shown in Fig. 15.32. These intercalated mesas do not exhibit any obvious closing of the gap at T_c a result similar to STM and break junctions shown earlier. Pristine junctions on the other hand show a much more rapid decrease of the gap with temperature than expected from BCS theory.

Thus it can be said that intrinsic junctions on fully intercalated Bi2212 mesas (with 100-fold decreases in conductance per unit area) display conductance spectra (including T-dependence) that are very similar to those obtained from single junction methods such as STM and break junctions. Such agreement and reproducibility among the very different junction types on Bi2212 points toward these spectra as revealing intrinsic properties. However, *c*-axis mesa junctions on pristine crystals can exhibit markedly different behavior, including very sharp conductance peaks, no dip features and a rapid decrease of the gap with temperature. [79] Not surprisingly, these [79] and other studies [80] have led to very different interpretations of the evolution of gap features with temperature, concluding that the superconducting gap closes at T_c and that there is a single pseudogap which has a different origin than the superconducting gap. Given the propensity of pristine junctions to exhibit non-equilibrium effects, such conclusions must be treated with caution at present. Further studies which compare single-junction methods and *c*-axis, intrinsic junctions on the same Bi2212 crystals are needed.

15.5.5 $YBa_2Cu_3O_7$

Much of the early tunneling work on high-T_c cuprates involved $YBa_2Cu_3O_7$ (YBCO) since this compound was easily grown as single crystals or thin films [22]. However, interpretation of the results has been more challenging than for Bi2212 most likely due to the presence of another metallic layer in the crystal structure arising from the Cu-O chains. As discussed earlier, such normal metal layers can be induced into the superconducting state via the proximity effect. But of course this effect must now be treated with a *d*-wave order parameter, where little is known either from the experimental or theoretical side. From planar junctions on etched single crystals, a reproducible gap-like feature near 5 meV in YBCO has been observed in addition to a higher energy gap feature near 20 meV attributed to the Cu-O plane [81]. These structures have been verified by STM and it was conjectured that the small gap was due to the chain layer, either an induced superconducting gap or a CDW gap [82]. Strong linear increases in background conductance were also observed but as discussed earlier, these are probably due to inelastic tunneling channels.

Tunneling studies on YBCO films were among the first to display a large ZBCP anomaly, an example of which is shown in Fig. 15.33. This observation further complicated the interpretation of tunneling spectra. Several reviews of the various types of ZBCP phenomena in different junction types and in different cuprates can be found [83–85]. A key observation was that the ZBCP was found only on films of YBCO oriented such that the *ab* plane was not parallel to the substrate plane. Thus the ZBCP was associated with tunneling into the *ab* plane not along the *c*-axis [86]. As discussed in the section on novel *d*-wave effects, the ZBCP could thus be associated with Andreev bound states at the interface arising from quasiparticle scattering events which change the sign of the order parameter. The tunnel junction probes these surface bound states. The ZBCP in Fig. 15.33 could be split by a magnetic field and this was explained by a doppler shift of the quasiparticle energy due to the superfluid currents [87].

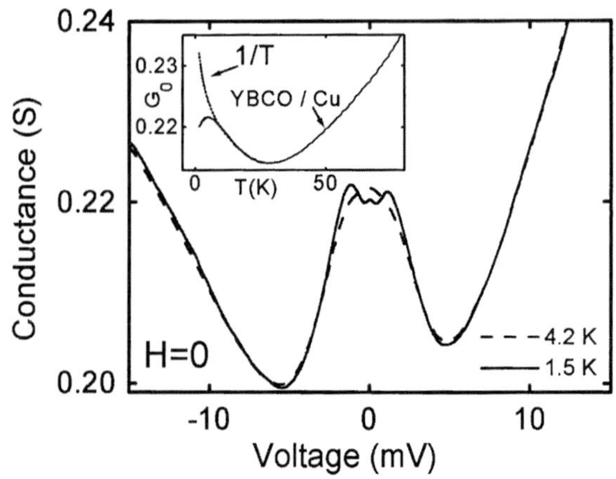

Fig. 15.33. Splitting of ZBCP in zero field for YBCO junctions as reported, Greene et al. [88]

Interestingly there appears to be a splitting of the ZBCP in zero field. Since magnetic fields produce currents with a definite rotational direction, this inherently breaks time reversal symmetry. The observation of splitting in zero field thus suggests the possibility of spontaneously broken time reversal symmetry (BTRS). This can occur if the pure *d*-wave state changes to a form $d+id'$ or $d+is$. In these cases there is a continuous change of phase with rotation of the wavevector, but without any nodes, and this must be accompanied by supercurrents. A review of BTRS

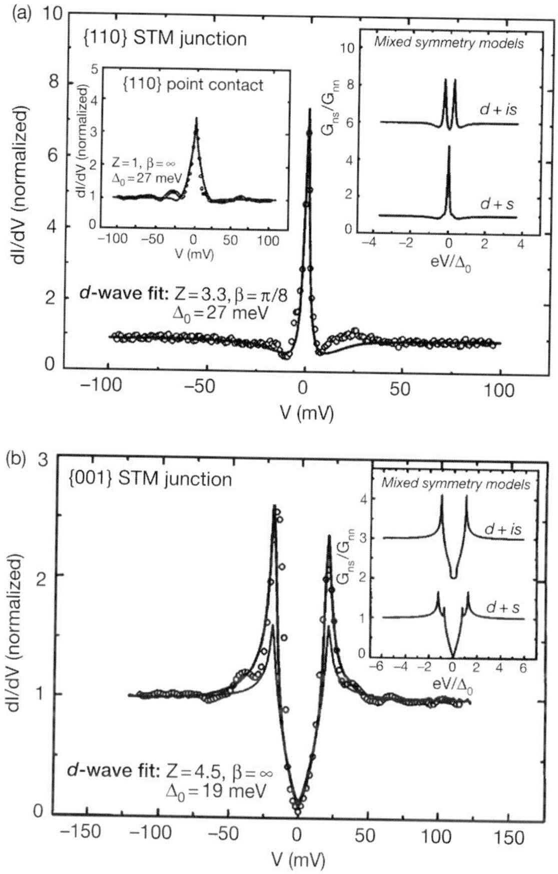

Fig. 15.34. (**a**) Vacuum and point contact junctions along the (110) direction of YBCO. (**b**) Vacuum junction along the *c*-axis. Data from [90]

and the use of Andreev bound state spectroscopy to probe such exotic mixed pairing symmetries has been given by Greene et al. [88]. In the case of YBCO the BTRS state is probably due to the interface and is not a bulk effect. Interface scattering which is accompanied by sign changes of the gap leads to depairing effects and therefore suppression of the *d*-wave gap near the interface. This effect is similar to the proximity effect in N/S bilayers discussed earlier. If the suppression of the *d*-wave gap is strong enough, a sub-dominant pairing channel can develop, leading to the mixed symmetry states and BTRS. However, mixed symmetry can also be a property of the bulk and whether this is occurring in cuprates is an area of current research [89].

Another STM study of YBCO crystals attempted to sort out the differences between *c*-axis and *ab* tunneling as well as impedance effects due to the *Z* parameter of the *d*-wave BTK model (see the section on novel *d*-wave effects) [90]. In Fig. 15.34(a) is shown a vacuum tunnel contact and point contact junction along the (110) direction. The large ZBCP is due to Andreev bound states but there is no splitting as was found in the planar junctions on YBCO thin films. This might be another indication that the BTRS state is a surface effect which depends on details such as surface roughness. In Fig. 15.34(b) is shown a tunnel junction along the *c*-axis. As discussed earlier, this orientation should probe the full *d*-wave DOS in the Cu-O plane without any ZBCP. The experimental data show a cusp like feature consistent with a *d*-wave DOS and reasonable overall agreement with a gap parameter, $\Delta = 19$ meV.

The gap value measured by STM is in agreement with other tunneling studies of YBCO and is considerably smaller than that found in optimal doped Bi2212 where $\Delta = 37$ meV. The origin of this rather large discrepancy in two cuprates with similar T_c is not completely understood. One possibility is that the metallic chain layers in YBCO are reducing the gap size in the Cu-O layers due to the proximity effect. Another possibility is that there is no universal curve relating Δ to T_c. The latter is likely especially since the two quantities have such different doping dependencies in Bi2212. The dip/hump feature that is common to Bi2212 is also seen reproducibly in STM measurements of YBCO films [91]. In that work the gap parameter, Δ=20-25 meV, and the dip is approximately at twice the gap as is found in Bi2212.

15.6 Heavy Fermion Superconductors

The same electron correlation effects which lead to the large effective masses in heavy fermion superconductors (HFS) are expected to give rise to unconventional superconducting order parameters. However, experimental verification, especially from tunneling, is lacking. Furthermore, HFS exhibit distinctly non-universal behavior [92]. For the cuprates, tunneling evidence for *d*-wave superconductivity is substantial [7]. It includes the cusp feature and subgap shape

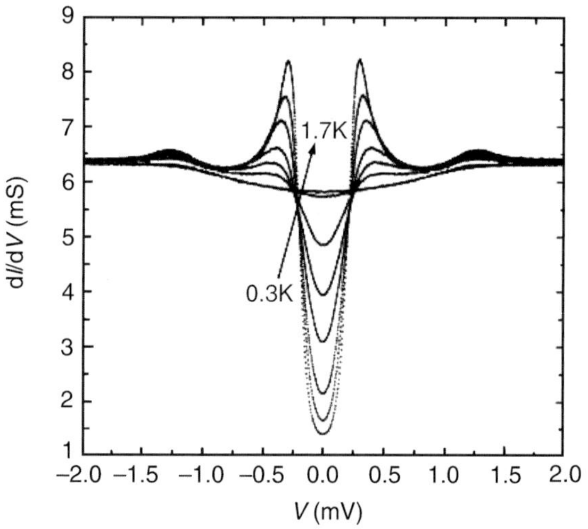

Fig. 15.35. Tunneling conductance from SIN junctions on UPd_2Al_3 for various temperatures. From Jourdan et al. [97]

of the quasiparticle DOS, a reproducible ZBCP found on *ab* plane junctions, and the fourfold symmetry of the zero bias STM mapping near Zn impurities in Bi2212. In addition there are phase sensitive, corner SQUID measurements. No such analogue exists for HFS. All of the point contact measurements on HFS are in the metallic regime. The differences between the shape of the Andreev reflection peaks for *s*-wave and *d*-wave are subtle. In a comparative Andreev-reflection study DeWilde et al. [93] found no evidence for nodes in $CeCu_2Si_2$ or URu_2Si_2. However, deviations from *s*-wave BTK theory in UPt_3 were attributed to nodes in the order parameter. The observation of Josephson currents between an Al tip and $CeCu_2Si_2$ by Poppe [94] is consistent with the absence of nodes, while the absence of such currents in UPt_3 supports the conclusion of DeWilde et al..

Point contact (N/S) studies of UBe_{13} are not easily reconciled. Early work using Pt and W tips [95] indicated $\Delta = 0.13$ meV, near the BCS value, but with a more rapid suppression with temperature than expected for $T_c = 0.89$ K. The authors expressed concern over possible heating effects due to high current injection. The conductance curves did not show an abrupt drop at the gap voltage and therefore this could be a sign of nodes in the order parameter. More recent studies using a Au tip revealed considerably different behavior [96]. A much larger energy gap was found, giving $\frac{2\Delta}{kT_c} = 6.7$, far in excess of the BCS value and a giant ZBCP was observed, similar to that expected from Andreev bound states. Both of these observations are intriguingly similar with that found on high-T_c cuprates.

A successful tunnel junction study of UPd_2Al_3 was realized by Jourdan et al. [97] using a more conventional planar junction geometry. This was possible due to the ability to sputter deposit high quality films of UPd_2Al_3 onto $LaAlO_3$ substrates with a $T_c = 1.6$ K, slightly reduced from the bulk value. A thin Al overlayer was then used to provide the tunnel barrier along with Pb counterelectrode, its superconductivity quenched by a small magnetic field. The tunneling conductances are shown in Fig. 15.35 for temperatures up to T_c.

In addition to the conductance peaks, which give a gap value $\Delta = 0.235$ meV, there are higher bias dip and hump features at characteristic energy scales of 2 to 4 times the gap value. These effects are strikingly similar to that found on Bi2212 although scaled down in energy by a factor of more than 100. More importantly, the dip feature appears to be connected to a new spin excitation, a magnetic exciton, that develops below T_c. This is similar to the evidence for a resonance mode in Bi2212 seen in neutron scattering and correlated with the tunneling dip feature. Also, an effect which was not commented on by Jourdan et al., is an apparent pseudogap type feature observed at T_c. All of these phenomena point toward a possible common mechanism for pairing between high-T_c cuprates and at least one HFS. Also, the tunneling data are compatible with a *d*-wave gap symmetry which is allowed from group symmetry considerations as well as from the magnetic interactions suggested as the pairing mechanism.

15.7 Organic Superconductors

When superconductivity was discovered in organic solids based on the bis(ethylenedithio)-tetrathiafulvalene (BEDT-TTF) molecule, it was not clear at all that these materials might be related to high temperature superconducting cuprates. However, the accumulated evidence to date shows that they bear intriguing similarities, including a similar

Table 15.1. Energy gap parameters obtained from tunneling for several organic superconductors compared with measured T_c

Compound	T_c (K)	$\frac{2\Delta}{kT_c}$	Reference
$\kappa-(\mathrm{BEDT-TTF})_2\mathrm{Cu(NCS)}_2$	11	5.1	Bando et al. (1990)
$\beta-(\mathrm{BEDT-TTF})_2\mathrm{I}_3$	1.35	8	Nowack et al. (1987)
$\beta-(\mathrm{BEDT-TTF})_2\mathrm{IAuI}$	4.1	4	Nowack et al. (1987)
$\beta-(\mathrm{BEDT-TTF})_2\mathrm{AuI}_2$	3.8	16	Hawley et al. (1986)

phase diagram [98]. A schematic of the "temperature vs. hole doping" phase diagram for cuprates has been shown in Fig. 15.21. Changing the hole concentration is generally straightforward, e.g. changing oxygen concentration in Bi2212. For the BEDT-TTF family, pressure plays the role of doping. The family $(\mathrm{BEDT-TTF})_2\mathrm{X}$ consists of conducting layers of BEDT-TTF molecules sandwiched between insulating layers of anions, for example, $\mathrm{X} = \mathrm{Cu[N(CN)_2]Br}$. Similar to cuprates, the insulating layers dope holes onto the organic layer making them metallic and the entire system highly anisotropic. This hole concentration can be changed with X as well as with increasing pressure. A Greek prefactor symbol denotes the packing pattern of the BEDT-TTF molecules. The parent insulating phase is antiferromagnetic and for some compounds the state above T_c can be an unconventional metal with evidence of a pseudogap. Tunneling measurements on organic superconductors have not displayed the requisite sharpness to discriminate among various gap symmetries. However, good estimates of gap magnitudes can be obtained and these have been used to generate the strong coupling ratios shown in Table 15.1 [99–101].

What is immediately clear is that in three of the organic materials, $\frac{2\Delta}{kT_c}$ is significantly larger than the BCS value of 3.53. The extraordinarily large value found in $\beta-(\mathrm{BEDT-TTF})_2\mathrm{AuI}_2$ was considered by some to be unphysical. However, these large values are reminiscent of that found in underdoped Bi2212 where values as high as 20 are obtained. This suggests another important similarity to cuprates which might help explain the anomalies. In cuprates the anomalously large strong coupling ratios are generally understood to be due to fluctuation effects which are stronger in two dimensions. In addition, d-wave symmetry is more susceptible to scattering of any type and such fluctuations therefore have a more severe effect on reducing T_c. Thus it is not that the gap is too large but rather that the measured T_c is too small. Such large ratios found in organics thus are consistent with pseudogap phenomena and possibly with unconventional order parameters.

15.8 Other Materials

While the high-T_c cuprates and bismuthates have captured most of the attention over the past 15 years, along with organic and heavy fermion superconductors, a number of other interesting superconductors have been discovered and are worthy of some comment. The fullerenes certainly represent an exciting class of materials. A comprehensive review of the chemical, structural, electronic and optical properties of these systems is available [102]. Another review which focuses on the electronic properties indicates that while electron–phonon coupling is viable as a mechanism for superconductivity, a quantitative tunneling spectroscopy measurement does not exist [103]. Inelastic tunneling measurements which utilize C_{60} and C_{70} as part of a multilayer tunnel junction barrier have revealed the intramolecular vibrational modes which extend up to nearly 200 meV [104]. If these extremely high frequency modes were participating in the pairing, they would be very difficult to detect in superconducting tunneling spectroscopy since the strength of the phonon structure is proportional to $\frac{\Delta^2}{E^2}$ [1].

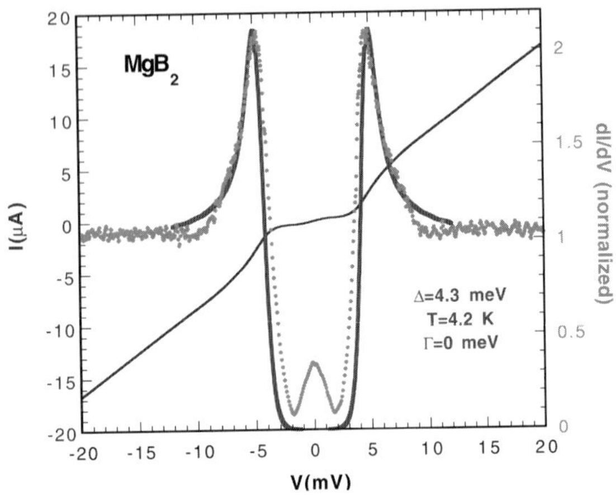

Fig. 15.36. Tunneling conductance for MgB_2 from a point contact junction. Data from [109]

Other compounds of interest are the intermetallic borocarbides RNi_2B_2C (R = rare earth). These layered systems display rich phase diagrams showing coexistence or competition between superconductivity and magnetic order of the rare earth spins. At present there seem to be no surprises in the tunneling spectra, with most measurements of the gap consistent with weak coupling, BCS theory [105,106]. Recent tunneling measurements of $TmNi_2B_2C$ exhibit very clean gap region spectra, consistent with BCS, and which show no particular change below 1.5 K where antiferromagnetic order sets in [107].

The discovery of superconductivity at 39 K in the simple intermetallic compound MgB_3 has led to a flurry of activity [108]. At the time of this writing there exist several reports of the energy gap from tunneling. Values of the gap parameter Δ vary from 2 to 8 meV which could be due to a number of reasons including compositional inhomogeneities. An example of a relatively clean gap feature is shown in Fig. 15.36 from PCT measurements [109].

Although the authors assumed this was SIN tunneling, the presence of a zero bias feature in this and other junctions suggests an incipient Josephson current and possibly SIS tunneling. In the latter case the gap value would be $\Delta = 2.2$ meV far below the BCS value of 5.9 meV but consistent with STM measurements which display a clean gap [110]. Unpublished reports indicate that this small gap nonetheless closes at the bulk T_c which is a strong indication for multiband superconductivity in this material [111].

15.9 Conclusions

Conventional superconductors such as Pb are well understood with- in Migdal–Eliashberg theory, which is essentially BCS theory in the strong electron–phonon coupling limit. The use of proximity effect tunneling (PETS) methods has allowed the determination of the electron–phonon spectral function, $\alpha^2F(\omega)$, in a number of transition metal elements (e.g. Nb, V, Ta), alloys and intermetallic compounds. Tunneling spectroscopy measurements on binary compounds with light elements such as ZrN have shown that the strong coupling theory also provides a good description when high frequency optical phonons ($\approx$ 60 meV) play an important role in the pairing. The coupling of electrons to optical phonons near 40 meV appears to explain T_c values up to 30 K in the bismuthate, BKBO, and also describes well the normal state d.c. resistivity in overdoped samples. The optical properties of optimal doped BKBO are not easily understood which might be due to the nearby CDW phase which develops with even slight underdoping.

Tunneling measurements on hole-doped, higher T_c cuprates such as Bi2212, Tl2201 and YBCO have revealed common, reproducible features including strong evidence for d-wave gap symmetry and spectral dip/hump features. Since Bi2212 can be doped over a wide range, tunneling data on this material have revealed important doping dependent trends. The energy gap increases monotonically with decreasing doping, scaling with the pseudogap temperature, T^*. Although not a proof, this trend indicates that some sort of precursor superconducting state exists well above T_c. Such a state leads to a pseudogap and can explain the non-Fermi-liquid behavior of the quasiparticles in the normal state. However, tunneling is also providing some evidence for a weaker, larger scale pseudogap that is about three times the superconducting gap. This effect may be causing some of the confusion over the nature and energy scale of the pseudogap.

The doping dependence of the dip feature in Bi2212 follows that of the resonance spin excitation, suggesting spin fluctuations play an important role in the pairing mechanism. The size and location of the dip/hump features in Bi2212 are highly reproducible among various tunneling groups and junction methods. This is in contrast to phonon fine structures where evidence is sporadic. A similar dip feature is found in the heavy fermion superconductor UPd_2Al_3 which is also linked to a feature that develops in the spin excitation spectrum below T_c. This commonality is striking and suggests that in both cases, superconductivity is generating a magnetic exciton, a purely electronic phenomenon. This boson is potentially mediating the pairing as well.

The electron-doped cuprate NCCO presents new mysteries. A successful inversion of tunneling data in this material showed that electron phonon coupling adequately describes the superconductivity. However, more recent measurements indicate d-wave pairing symmetry, similar to that found on hole-doped cuprates with much higher T_c. In contrast to hole-doped cuprates, both electron–phonon and electron–electron interactions may be in competition in this cuprate.

The large ratios of energy gap to T_c in the organic superconductors is a mystery that might have been resolved by advancements in the field of cuprate superconductivity. This large ratio is reminiscent of that found in Bi2212, especially in the underdoped region. The fact that the organics display many other similarities to cuprates, including a layered structure and similar phase diagram (including antiferromagnetism), suggests there is an important linkage. It is possible that high-T_c cuprates, organics and some heavy fermion superconductors might all share a common pairing interaction that is magnetic in origin.

Acknowledgements

The author would like to thank K.E. Gray and the Materials Science Division of Argonne National Laboratory for supporting a sabbatical leave appointment during which much of the manuscript was prepared and written, and also H. Schmidt and L. Ozyuzer for help in preparation. The author has benefitted from the experimental work of former and current IIT graduate students: Q. Huang, N. Tralshawala, S. Okur, Z. Yusof, L. Ozyuzer, and H. Schmidt as well as visiting scientists: June Chen, P. Romano and N. Miyakawa. Dave Hinks was instrumental in providing high quality materials for the Argonne tunneling program and C. Kendziora assisted in the overdoping of Bi2212. Important help in understanding the theoretical background was received from L. Coffey, M. Norman, A. Chubukov and B. Janko. Research support has been received from NSF Science and Technology Center for Superconductivity and DOE Basic Energy Sciences.

References

1. E.L. Wolf, In: *Principles of Electron Tunneling Spectroscopy* (Oxford University Press, Oxford, 1985)
2. W.L. McMillan and J.M. Rowell, In: *Superconductivity*, ed. by R.D. Parks (Dekker, New York, 1969) pp. 561–331
3. J.P. Carbotte, Rev. Mod. Phys. **62**, 1027 (1990)
4. P. Allen, Electron–phonon coupling constants'. In: *Handbook of Superconductivity*, ed. by C.P. Poole (Academic Press, San Diego, 2000) pp. 478–489
5. N. Tralshawala, J.F. Zasadzinski, L. Coffey, W. Gai, M. Romalis, Q. Huang, R. Vaglio, and K.E. Gray, Phys. Rev. B **51**, 3812 (1995)
6. Q. Huang, J.F. Zasadzinski, N. Tralshawala, K.E. Gray, D.G. Hinks, J.L. Peng and R.L. Greene, Nature (London) **347**, 369 (1990)
7. C.C. Tsuei and J.R. Kirtley, Rev. Mod. Phys. **72**, 969 (2000)
8. H. Won, K. Maki, Phys. Rev. B **49**, 1397 (1994)
9. S.B. Kaplan, C.C. Chi, D.N. Langenberg, J.J. Chang, S. Jafarey and D.J. Scalapino , Phys. Rev. B **14**, 4854 (1976)
10. R.C. Dynes, V. Naraynamurti and J.P. Garno, Phys. Rev. Lett. **41**, 1509 (1978)

11. L. Ozyuzer, Z. Yusof, J.F. Zasadzinski, R. Mogilevsky, D.G. Hinks, K.E. Gray, Phys. Rev. B **57**, 3245 (1998)
12. Lutfi Ozyuzer, Zikri Yusof, John F. Zasadzinski, Ting-Wei Li, Dave G.Hinks, K.E. Gray, Physica C **320**, 9 (1999)
13. G.E. Blonder, M. Tinkham, and T.M. Klapwijk, Phys. Rev. B **25**, 4515 (1982)
14. Herbert Schmidt et al., unpublished
15. C.-R. Hu, Phys. Rev. Lett. **72**, 1526 (1994)
16. Y. Tanaka and S. Kashiwaya, Phys. Rev. Lett. **74**, 3451 (1995)
17. Q. Huang, J.F. Zasadzinski and K.E. Gray, Phys. Rev. B **42**, 7953 (1990)
18. R.C. Jaklevi and J. Lambe, Phys. Rev. **165**, 821 (1968)
19. J. Halbritter, Surface Science **122**, 80 (1982)
20. R. Gross, In: *Interfaces in Superconducting Systems*, ed. by S.L. Shinde and D. Rudman (Springer, New York, 1994) p. 176
21. J.R. Kirtley and D.J. Scalapino, Phys. Rev. Lett. **65**, 798 (1990)
22. Tetsuya Hasegawa, Hiroshi Ikuta, and Koichi Kitazawa, In: *Physical Properties of High Temperature Superconductors III*, ed. by D.M. Ginsberg (World Scientific, Singapore, 1992)
23. John Zasadzinski, L. Ozyuzer, Z. Yusof, June Chen, K.E. Gray, R. Mogilevsky, D.G. Hinks, J.L. Cobb and J.T. Markert, In: *Spectroscopic Studies of Superconductors*, ed. by Ivan Bozovic and Dirk van der Marel (SPIE, Bellingham, 1996)
24. J.R. Kirtley, S. Washburn, D.J. Scalapino, Phys. Rev. B **45**, 336 (1992)
25. J.R. Schrieffer, In: *Theory of Superconductivity* (Addison-Wesley, Reading, MA, 1964)
26. Walter A. Harrison, Phys. Rev. **123**, 85 (1961)
27. W.F. Brinkman, R.C. Dynes and J.M. Rowell, J. Appl. Phys. **41**, 1915 (1970)
28. M.R. Norman, M. Randeria, H. Ding, J.C. Campuzano, Phys. Rev. B **52**, 615 (1995)
29. Satoshi Sugita, Takao Watanabe and Azusa Matsuda, Phys. Rev. B **62**, 8715 (2000)
30. Y. DeWilde, N. Miyakawa, P. Guptasarma, M. Iavarone, L. Ozyuzer, J.F. Zasadzinski, P. Romano, D.G. Hinks, C. Kendziora, G.W. Crabtree, K.E. Gray, Phys. Rev. Lett. **80**, 153 (1998)
31. Z. Yusof, J.F. Zasadzinski, L. Coffey, N. Miyakawa, Phys. Rev. B **58**, 514 (1998)
32. J. Geerk, G. Linker and R. Smithey, Phys. Rev. Lett. **57**, 3284 (1986)
33. E.L. Wolf and G.B. Arnold, Phys. Reports **91**, 31 (1982)
34. J. Halbritter, J. Appl. Phys. **58**, 1320 (1985)
35. E.L. Wolf and J.F. Zasadzinski, Phys. Lett. **62A**, 165 (1978)
36. E.L. Wolf, J. Zasadzinski, J.W. Osmun and G.B. Arnold, J. Low Temp. Phys. **40**, 19 (1980)
37. J. Zasadzinski, D.M. Burnell, E.L. Wolf, and G.B. Arnold, Phys. Rev. B **25**, 1622 (1982)
38. see for example In: *Superconducting Devices*, ed. by S.T. Ruggiero and D.A. Rudman (Academic Press, San Diego, CA, 1990)
39. J.G. Bednorz and K.A. Müller, Z. Phys. B **64**, 189 (1986)
40. K.E. Gray, M.E. Hawley and E.R. Moog, In: *Novel Superconductivity*, ed. by S.A. Wolf and V.Z. Krezin (Plenum, New York, 1987) p. 611
41. J.R. Kirtley, Int. J. Mod. Phys. **B4**, 201 (1990)
42. F. Shafiri, A.N. Pargellis and R.C. Dynes, Phys. Rev. Lett. **67**, 509 (1991)
43. P. Samuely, N.L. Bobrov, A.G.M. Jansen, P. Wyder, S.N. Barilo and S.V. Shiryaev, Phys. Rev. B **48**, 13,904 (1993)
44. H.J. Kaufmann, Oleg V. Dolgov, and E.K.H. Salje, Phys. Rev. B **58**, 9479 (1998)
45. B. Stadlober, G. Krug, R. Nemetschek, R. Hackl, J.L. Cobb, J.T. Markert, Phys. Rev. Lett. **74**, 4911 (1995)
46. C.C. Tsuei and J.R. Kirtley, Phys. Rev. Lett. **85**, 182 (2000)
47. Lutfi Ozyuzer, John F. Zasadzinski, Chris Kendziora, K.E. Gray, Phys. Rev. B **61**, 3629 (2000)
48. N. Miyakawa, J.F. Zasadzinski, L. Ozyuzer, P. Guptasarma, D.G. Hinks, C. Kendziora and K.E. Gray, Phys. Rev. Lett. **83**, 1018 (1999)
49. L. Ozyuzer, J.F. Zasadzinski and N. Miyakawa, Int. J. Mod. Phys. B **29-31**, 3721 (1999)
50. C. Renner and Ø. Fischer, Phys. Rev. B **51**, 9208 (1995)
51. C. Renner et al., J. Low Temp. Phys. **105**, 1083 (1996)
52. H.L. Liu, G. Blumberg, M.V. Klein, P. Guptasarma and D.G. Hinks, Phys. Rev. Lett. **82**, 9208 (1999)
53. J.L. Tallon and J.W. Loram, cond-mat/0005063
54. R.S. Markiewicz and C. Kusko, Phys. Rev. Lett. **84**, 5674 (2000)

55. Sudip Chakravarty, R.B. Laughlin, Dirk K. Morr and Chetan Nayak, cond-mat/0005443
56. John Zasadzinski and Nobuaki Miyakawa, Phys. Rev. Let. **84**, 5675 (2000)
57. N. Miyakawa, J.F. Zasadzinski, L. Ozyuzer, P. Guptasarma, C. Kendziora, D.G. Hinks, T. Kaneko and K.E. Gray, Physica C **341-348**, 835 (2000)
58. V.J. Emery and S.A. Kivelson, Nature (London) **374**, 434 (1998)
59. Qijin Chen, Ioan Kosztin, Boldizsar Janko and K. Levin, Phys. Rev. Lett. **81**, 4708 (1998)
60. T. Nakano, N. Momono, M. Oda and M. Ido, J. Phys. Soc. Jpn. **67**, 2622 (1998)
61. J. Schmalian, D. Pines, and B. Stojkovic, Phys. Rev. Lett. **80**, 3839 (1998)
62. Ch. Renner, B. Revaz, J.-Y. Genoud, K. Kadowaki and O. Fischer, Phys. Rev. Lett. **80**, 149 (1998)
63. M. Franz and A.J. Millis, Phys. Rev. B **58**, 14,572 (1998)
64. J. Corson, R. Mallozzi, J. Orenstein, J.N. Eckstein, and I. Bozovic, Nature (London) **398**, 221 (1999)
65. Toshikazu Ekino, Yoshie Sezaki and Hironobu Fujii, Phys. Rev. B **60**, 6916 (1999)
66. T. Takahashi, T. Sato, T. Yokoya, T. Kamiyama, T. Naitoh, T. Mochiku, K. Yamada, Y. Endoh and K. Kadowaki, J. Phys. Chem. Solids **62**, 41 (2001)
67. Q. Huang, J.F. Zasadzinski, K.E. Gray, J.Z. Liu and H. Claus, Phys. Rev. B **40**, 9366 (1989)
68. J.F. Zasadzinski, L. Qzyuzer, N. Miyakawa, K.E. Gray, D.G. Hinks and C. Kendziora, Phys. Rev. Lett. **87**, 067005 (2001)
69. H.F. Fong, B. Keimer, D.L. Milius and I.A. Aksay, Phys. Rev. Lett. **78**, 713 (1997)
70. Pengcheng Dai, H.A. Mook, S.M. Hayden, G. Aeppli, T.G. Terring, R.D. Hunt and F. Dogan, Science **284**, 1344 (1999)
71. E.W. Hudson, S.H. Pan, A.K. Gupta, K.-W. Ng, J.C. Davis, Science **5285**, 88 (1999)
72. S.H. Pan et al., Nature (London) **403**, 746 (2000)
73. S.L. Cooper and K.E. Gray, In: *Physical Properties of High Temperature Superconductors*, ed. by D.M. Ginsberg (World Scientific, Singapore, 1994) p. 61.
74. R. Kleiner, F. Steinmeyer, G. Kunkel and P. Müller, Phys. Rev. Lett. **68**, 2394 (1992)
75. R. Kleiner and P.Müller, Phys. Rev. B **49**, 1327 (1994)
76. A.A. Yurgens, Sup. Sci. Tech. **13**, R85 (2000)
77. Minoru Suzuki and Keiichi Tanabe, Jpn. J. Appl. Phys. **35**, L482(1996)
78. A. Yurgens, D. Winkler, T. Claeson, S-J Hwang and J-H Choy, Int. J. Mod. Phys. B **13**, 3758 (1999)
79. V.M. Krasnov, A. Yurgens, D. Winkler, P. Delsing and T. Claeson, Phys. Rev. Lett. **84**, 5860 (2000)
80. Minoru Suzuki and Takao Watanabe, Phys. Rev. Lett. **85**, 4787 (2000)
81. M. Gurvitch, J.M. Valles Jr., A.M. Cuculo, R.C. Dynes, J.P. Garno, L.F. Schneemeyer and J.V. Waszczak, Phys. Rev. Lett. **63**, 1008 (1989)
82. H.L. Edwards, D.J. Derro, A.L. Barr, J.T. Markert and A.L. de Lozanne, Phys. Rev. Lett. **75**, 1387 (1995)
83. T. Walsh, Int. J. Mod. Phys. B **6**, 126 (1992)
84. L. Alff, S. Kleefisch, U. Schoop, M. Zittartz, T. Kemen, T. Bauch, A. Marx and R. Gross, Eur. Phys. J. B **5**, 423 (1998)
85. Saion Sinha and K.W. Ng, Superlatt. Microstruct. **25**, 1055 (1999)
86. M. Covington, M. Aprili, E. Paraoanu, L.H. Greene, F. Xu, J. Zhu, and C.A. Mirkin, Phys. Rev. Lett. **79**, 277 (1997)
87. M. Fogelström, D. Rainer and J.A. Sauls, Phys. Rev. Lett. **79**, 281 (1997)
88. L.H. Greene, M. Covington, M. Aprili, E. Badica, D.E. Pugel, Physica B **280**, (2000)
89. M. Sigrist, Physica C **341-348**, 695 (2000)
90. J.Y.T. Wei, N.-C. Yeh, D.F. Garrigus, and M. Strasik, Phys. Rev. Lett. **81**, 2542 (1998)
91. T. Cren, D. Roditchev, W. Sacks, J. Klein, Europhys. Lett. **52**, 203 (2000)
92. F. Steglich, N. Sato, T. Tayama, T. Lühmann, C. Langhammer, P. Gegenwart, P. Hinze, C. Geibel, M. Lang, G. Spain and W. Assmus, Physica C **341-348**, 691 (2000)
93. Y. DeWilde, J. Heil, A.G.M. Jansen, P. Wyder, R. Deltour, W. Assmus, A. Menorsky, W. Sun and L. Taillefer, Phys. Rev. Lett. **72**, 2278 (1994)
94. U. Poppe, J. Mag. Magn. Mat. **52**, 157 (1985)
95. A. Nowack, A. Heinz, F. Oster, D. Wohlleben, G. Güntherodt, Z. Fisk and A. Menorsky, Phys. Rev. B **36**, 2436 (1987)
96. Ch. Wälti, H.R. Ott, Z. Fisk, J.L. Smith, Phys. Rev. Lett. **84**, 5616 (2000)
97. M. Jourdan, M. Huth and H. Adrian, Nature **398**, 47 (1999)
98. Ross H. McKenzie, Science **278**, 820 (1997)
99. H. Bando, S. Kashiwaya, H. Tokumoto, H. Anzai, N. Kinoshita and K. Kajimura, J. Vac. Sci. Technol. A **8**, 479 (1990)

100. A. Nowack, U. Poppe, M. Weger, D. Schweitzer and H. Schwenk, Z. Phys. B - Condensed Matter **68**, 41 (1987)
101. M.E. Hawley, K.E. Gray, B.D. Terris, H.H. Wang, K.D. Carlson and Jack M. Williams, Phys. Rev. Lett. **57**, 629 (1986)
102. M. Dresselhaus, G. Dresselhaus, P. Eklund, *Science of Fullerenes and Carbon Nanotubes* (Academic Press, San Diego, CA, 1996)
103. Laszlo Forro and Laszlo Mihaly, Rep .Prog. Phys. **64**, 649 (2001)
104. S. Nolen, S.T. Ruggiero, Chem. Phys. Lett. **300**, 656 (1999)
105. Y. DeWilde, M. Iavarone, U. Welp, V. Metlushko, A.E. Koshelev, I. Aranson, G.W. Crabtree and P.C. Canfield, Phys. Rev. Lett. **78**, 4273 (1997)
106. Toshikazu Ekino, H. Fujü, M. Kosugi, Y. Zenitani and J. Akimitsu,Phys. Rev. B **53**, 5640 (1996)
107. H. Suderow, P. Martinez-Samper, N. Luchier, J.P. Brison, S. Vieira and P.C. Canfield, Phys. Rev. B **64**, 020503 (2001)
108. J. Nagamatsu, N. Nakagawa, T. Muranaka, Y. Zenitani, J. Akimitsu, Nature **410**, 63 (2001)
109. H. Schmidt, J.F. Zasadzinski, K.E. Gray and D.G. Hinks, Phys. Rev. B **63**, 220504 (2001)
110. G. Rubio-Bollinger, H. Suderow and S. Vieira, Phys. Rev. Lett. **86**, 5582 (2001)
111. A.Y. Liu, I.I. Mazin and J. Kortus, cond-mat/0105146

16 Phase-Sensitive Tests of Pairing Symmetry in Cuprate Superconductors

C. C. Tsuei IBM Thomas J. Watson Research Center, New York, USA
J. R. Kirtley IBM Thomas J. Watson Research Center, New York, USA

16.1 Introduction

The discovery of high-temperature superconductivity (HTS) in the cuprates [1] has stimulated a great deal of re-thinking of some of the most basic concepts in condensed matter physics. Despite the progress that has been made in the studies of the superconducting and normal-state properties of cuprate superconductors, there is still no consensus on the extent to which such basic conventional paradigms as the concept of the Fermi liquid or the Bardeen–Cooper–Schrieffer (BCS) [2] theory can be used to understand the mechanism of high-temperature superconductivity [3]. An important part of the HTS problem, the symmetry of the pair wavefunction in cuprates, was a controversial topic for more than a decade [4,5]. Recently, a new class of experiments have emerged that provide phase-sensitive tests of this symmetry, producing compelling evidence for a $d_{x^2-y^2}$-wave pair state in several optimally hole- and electron-doped cuprate superconductors [6,7]. In this chapter, we will describe the fundamental aspects and present the main results of these experiments. The implications of the universally observed d-wave pairing for understanding various anomalous properties of the high-T_c superconductors will be discussed. We will conclude with potential applications of d-wave superconductivity in areas such as SQUIDs and quantum computers.

16.1.1 Superconducting Order Parameter and Symmetry Breaking

The order parameter is a fundamental concept for characterizing the ordered state of various phase transitions [8,9]. Ginzburg and Landau [10] introduced the idea of a superconducting order parameter, based on the Landau theory of second-order phase transitions, to represent the extent of macroscopic phase coherence in a superfluid condensate. In their phenomenological description of the superconducting state, the thermodynamic and magnetic properties of a superconductor are described by a complex position-dependent order parameter $\Psi(\mathbf{r}) = |\Psi(\mathbf{r})|e^{i\varphi(\mathbf{r})}$, characterized by a phase $\varphi(\mathbf{r})$ and a modulus $|\Psi(\mathbf{r})|$. The local superfluid density $n_s(\mathbf{r})$ is equal to $|\Psi(\mathbf{r})|^2$, suggesting that $\Psi(\mathbf{r})$ is a wavefunction of some sort. In the Ginzburg–Landau formalism, the total free energy of a superconductor is expressed in terms of the order parameter $\Psi(r)$ and the vector potential $A(r)$. Minimization of the free energy with respect to variations in $\Psi(r)$ and $A(r)$ leads to the two celebrated Ginzburg–Landau differential equations. These equations can be solved, with appropriate boundary conditions, to determine the order parameter for describing the macroscopic properties such as critical field, critical current, and flux lattice dynamics. Thus the superconducting order parameter is, in principle, defined phenomenologically in the context of the Ginzburg–Landau theory. There is a huge literature on the applications of Ginzburg–Landau theory and its extensions to a variety of superconducting systems [11,12].

The microscopic significance of the phenomenological Ginzburg–Landau order parameter was established by Gor'kov [13] shortly after the publication of the BCS theory of superconductivity. He showed that, near T_c, the Ginzburg–Landau equations can be derived from the BCS theory. The Ginzburg–Landau order parameter $\Psi(\mathbf{r})$ is then identified with the pair wavefunction, and is proportional to the energy gap $\Delta(\mathbf{r})$. In principle, Ginzburg–Landau theory is only valid in a temperature range near T_c, due to the inherent assumption that the order parameter is small and slowly varying close to the phase transition. In practice, the Ginzburg–Landau theory is often successfully applied well beyond its range of validity. The equivalence between the energy gap $\Delta(\mathbf{r})$ and the Ginzburg–Landau order parameter is expected to be valid for all temperatures below T_c, since the superconducting order parameter represents the degree of long-range phase coherence in the pair state, regardless of whether it is defined at a phenomenological or a microscopic level.

The superconducting transition temperature, T_c, signals the onset of a macroscopic phase-coherent pair state [2]. An essential condition of macroscopic quantum phenomena such as superfluidity and superconductivity is the occurrence of off-diagonal long-range order (ODLRO) [14]. In BCS superconductors, the ODLRO stems from a non-vanishing anomalous expectation value of the local pair amplitude $< \Psi_\downarrow^\dagger(r)\Psi_\uparrow^\dagger(r) >$, which is basically the Ginzburg–Landau order parameter. In the momentum space representation, ODLRO corresponds to a non-zero expectation value of $< c_{k\uparrow}c_{-k\downarrow} >$, which is proportional to the gap potential $\Delta(\boldsymbol{k})$, the microscopic order parameter. The normal-to-superconducting phase transition, since it involves the onset of long-range order, is accompanied by a lowering in symmetry. As in any second-order phase transition, the symmetries above and below the normal-to-superconducting state phase transition are related, since the symmetry breaking across the transition is continuous. In this context, the order parameter is just a measure of the amount of symmetry breaking in the ordered state. The symmetry group H describing the superconducting state must be a subgroup of the full symmetry group G describing the normal state:

$$G = X \times R \times U(1) \times T \quad \text{for } T > T_c \tag{16.1}$$

and

$$H \subset G \quad \text{for } T \leq T_c \,, \tag{16.2}$$

where X is the symmetry group of the crystal lattice, R the symmetry group of spin rotation, $U(1)$ the one-dimensional global gauge symmetry, and T the time-reversal symmetry operation.

The establishment of ODLRO in superconductors below T_c leads to a phase-coherent pair condensate which is always characterized by a spontaneously broken global gauge symmetry, $U(1)$. Macroscopic phase-coherent quantum phenomena, such as the Meissner effect, flux quantization, and the Josephson effects, are all manifestations of the global gauge symmetry violation in the superconducting state. Only $U(1)$ is broken in a conventional superconductor as a result of the normal-to-superconducting phase transition. In an unconventional superconductor, one or more symmetries in addition to $U(1)$ can be spontaneously broken at and below T_c. The degree of symmetry breaking occurring in the pair state is reflected in the symmetry properties of the order parameter.

16.1.2 Symmetry Classifications of the Pair States

The pair wavefunction is antisymmetric under particle exchange as a consequence of the Pauli exclusion principle. Therefore, spin singlet pairs, with total spin $S = 0$, can only have angular momentum $L = 0, 2, 4, \ldots$ quantum numbers, while spin triplet pair states can only have odd-number angular momentum $L = 1, 3, \ldots$ quantum numbers. All known bulk superconductors, including the heavy-fermion and cuprate systems, have crystal structures with a center of inversion. Therefore superconductors can be classified by the parity of the pair state: The spin triplet state (total spin $S = 1$) has a superconducting order parameter (gap function) with odd parity; the spin-singlet pair state ($S = 0$) has an orbital pair wavefunction $\Psi(\mathbf{k}) \propto \Delta(\mathbf{k})$ with even parity: i.e. $\Delta(\mathbf{k}) = \Delta(-\mathbf{k})$. The mixing of singlet and triplet pair states induced by spin-orbit interaction, which may be important in heavy-fermion superconductors, can result in an observable non-linear magneto-optical effect due to the absence of inversion symmetry close to a surface [15]. In cuprate superconductors, the spin-orbit coupling is expected to be relatively small. Thus, the spin singlet and triplet pair states are well-defined. It is now well-established that spin-singlet Cooper pairing prevails in cuprate superconductors, based on evidence from Andreev reflection and spin susceptibility measurements [5,7].

Further classification of superconductors requires a knowledge of possible lattice symmetries, X in (16.1), that might be broken in the superconducting state. This assumes that the symmetry of the superconducting pair wavefunction reflects that of the underlying crystal lattice.

The Landau theory of second order phase transitions states that the order parameter describing the transition must transform according to one of the irreducible representations of the symmetry group of the high-temperature phase [16]. Therefore the possible forms of the order parameter can be categorized by decomposing the representation of the normal-state symmetry group into irreducible representations [17].

Point-group symmetry classification of pair states has been extensively studied in superfluid He [18], heavy-fermion superconductors [19–27], and also in cuprate superconductors [5,17,28–32]. Nearly all group theoretic classifications of superconducting states are based on point group symmetry. There are exceptions when translational invariance is broken. For example, the presence of spin density waves in certain heavy-fermion superconductors reduces the number of possible pair states [17,33].

The gap function can be expressed as a linear combination of the basis functions (χ_μ^j) of the irreducible representation (Γ^j):

$$\Delta(\mathbf{k}) = \sum_{\mu=1}^{l_j} \eta_\mu \chi_\mu^j(\mathbf{k}) \,, \tag{16.3}$$

where l_j is the dimensionality of Γ^j, and the complex number η_μ is invariant under all symmetry operations of the normal-state group G in (16.1).

Table 16.1. Spin-singlet even-parity pair states in a tetragonal crystal with point group D_{4h}

Wavefunction Name	Group Theoretic Notation, T_j	Residual Symmetry	Basis Function	Nodes
s–wave	A_{1g}	$D_{4h} \times T$	$1,(x^2+y^2),z^2$	none
g	A_{2g}	$D_4[C_4] \times C_i \times T$	$xy(x^2-y^2)$	line
$d_{x^2-y^2}$	B_{1g}	$D_4[D_2] \times C_i \times T$	x^2-y^2	line
d_{xy}	B_{2g}	$D_4[D_2'] \times C_i \times T$	xy	line
$e_{(1,0)}$	$E_g(1,0)$	$D_4[C_2'] \times C_i \times T$	xz	line
$e_{(1,1)}$	$E_g(1,1)$	$D_2[C_2''] \times C_i \times T$	(x+y)z	line
$e_{(1,i)}$	$E_g(1,i)$	$D_4[E] \times C_i$	(x+iy)z	line

Table 16.2. Spin-singlet even-parity pair states in an orthorhombic crystal (point group D_{2h})

Group Theoretic Notations	Residual Symmetry	Basis Function	Nodes
A_{1g}	$D_{2h} \times T$	1	—
B_{1g}	$D_2[C_2^z] \times C_i \times T$	xy	line
B_{2g}	$D_2[C_2^y] \times C_i \times T$	xz	line
B_{3g}	$D_2[C_2^x] \times C_i \times T$	yz	line

The basis functions χ_μ^j can always be selected to be real, so that $\eta_\mu^* \eta_\mu$ will contain information about the time-reversal symmetry [26]. Furthermore, as shown by Yip and Garg [26], time-reversal symmetry (T in (16.1)) can be broken only when the representation is multidimensional. The basis functions of the irreducible representations for various symmetry groups are well-tabulated in the literature [17,22,25] for various crystal structures.

The crystal structures of cuprates have been intensively studied and well-documented [34, 35]. All cuprate superconductors have a generic layered structure with a Cu-O square lattice (the CuO_2 planes) sandwiched between charge reservoir blocks. The crystal structure of the cuprates can be generally divided into two categories based on their atomic arrangements: a tetragonal lattice with point group symmetry D_{4h}, and an orthorhombic lattice with D_{2h}. Examples of the tetragonal superconductors include: $La_{2-x}Sr_xCuO_4$ (LSCO), $Tl_2Ba_2CaCu_2O_8$ (Tl-2212), $HgBa_2CaCu_2O_6$ (Hg-1212), and $Nd_{1.85}Ce_{0.15}CuO_{4-\delta}$ (NCCO). Structural distortions arising from Cu-O chains, as in $YBa_2Cu_3O_7$ (YBCO), or the incommensurate superlattice modulation in the BiO layers, as in $Bi_2Sr_2CaCu_2O_8$ (Bi-2212), results in an orthorhombic variant of the basic tetragonal crystal structure.

Allowed pair states for a tetragonal superconductor are listed in Table 16.1, for spin singlet even-parity pairing under the standard group theoretic constraints [29]. The notation in Tables 16.1 and 16.2 is adopted from [12]. The superconducting order parameter should transform like the basis function of an irreducible representation of the relevant point group. However, the basis function is not necessarily unique [26]. An example is the case of the A_{1g} (s-wave) pair state in Table 16.1. Each of the four one-dimensional irreducible representations corresponds to a single, scalar gap function of complex numbers. Therefore, these pair states should exhibit only one superconducting transition. For the two-dimensional representation (E_g), there are three possible states characterized by different residual symmetries (see Table 16.1). Of the three states of E_g, only $E_g(1,i)$ has broken time reversal symmetry. The gap function $\Delta(\mathbf{k})$ for each pair state can be expanded as a function of k_x, k_y, and k_z, the wavevector components along the principal axes in the Brillouin zone [30] using the basis functions listed in Table 16.1. With the exception of the s-wave pair state, the order parameters have basis functions with node lines. However, the number and the location of the nodes at the Fermi surface depends on the Fermi surface topology, as well as the band filling of a given band structure [36].

In addition to the pure states listed in Table 16.1, the order parameter of various mixed pair states can be formed by combining a real subcomponent from one 1D-representation with an imaginary subcomponent from another 1D-representation [30]. Time-reversal symmetry is broken in all these mixed states. We reiterate that such states can only occur if the superconductivity is first order, or as a result of two successive phase transitions. This follows from the ideas of the Landau theory of second order phase transitions stated earlier.

The symmetry properties of the allowed spin-singlet even-parity pair states for a standard orthorhombic superconductor are tabulated in Table 16.2 [29]. Note that both the s and $d_{x^2-y^2}$-wave pair states in the orthorhombic case belong to the same irreducible representation (A_{1g}). Hence, an admixture of these two states is allowed and only one single superconducting transition should be observed. This is apparently the case for YBCO.

Regardless of the crystal structure, tetragonal or orthorhombic, all cuprates share one common structural feature: the CuO_2 planes. These planar Cu-O lattices give rise to the strong anisotropy and electron correlation effects which dictate many of the normal-state and the superconductivity properties of this class of materials. In particular, there is a general consensus, based on over a decade of extensive research on HTS, that superconductivity in all cuprates basically originates from the CuO_2 layers. Recent studies of interplane dc and ac intrinsic Josephson effects have convincingly demonstrated that highly-anisotropic high-T_c superconductors such as Bi-2212 can be modeled as stacks of two-dimensional superconducting CuO_2-based layers coupled by Josephson interactions [37,38]. Also, the vortex state can be understood in terms of stacks of two-dimensional pancake vortices with cores localized in the CuO_2 layers, which are connected by Josephson vortices with cores confined in the non-superconducting charge reservoir layers [39,40]. Investigations of c-axis charge dynamics and transport have also produced strong evidence for charge confinement in the CuO_2 layers [41–44]. Similar supporting evidence is abundant in the literature.

In view of the importance of CuO_2 planes in the HTS problem, one expects that pairing symmetry reflects the point-group symmetry of the underlying Cu-O lattices. In the tetragonal cuprates such as LSCO, NCCO, Hg-1201, Hg-1212, Tl-2201, ..., etc., the Cu and O atoms arrange themselves in a simple square lattice with point group symmetry C_{4v}. The point group C_{4v} consists of the following symmetry operations: mirror reflections with respect to the principle crystallographic axes ($x = 0, y = 0$) and the diagonals ($x = \pm y$); and a fourfold (C_4) and a twofold (C_2) rotation about the c-axis. In orthorhombic cuprates such as YBCO, the CuO_2 plane takes the form of a CuO rectangular lattice with point group symmetry C_{2v}. The symmetry elements of C_{2v} differ from those of C_{4v} by the absence of C_4 and reflections with respect to the diagonals of the rectangular unit cell. A schematic representation in $\boldsymbol{k}$-space for possible spin-singlet pair states in a simple square lattice with point group symmetry C_{4v} is shown in Fig. 16.1. Each of these candidates corresponds to a distinctly different irreducible representation (see (16.3)). Thus pure d-wave and s-wave pair states are not allowed to mix in the tetragonal cuprates. In the case of the Cu-O rectangular lattice both the s and d-wave pair states belong to the same irreducible representation (A_{1g}) because of the lower crystal symmetry. Therefore an $s + d$ mixed pair state is allowed.

In short, we have classified and enumerated possible spin-singlet pair states in the cuprates based on crystal symmetry. Thus the stage is set for experimentally probing pairing symmetry in various cuprate superconductors.

Group-theoretic notation	A_{1g}	A_{2g}	B_{1g}	B_{2g}
Order parameter basis function	constant	$xy(x^2 - y^2)$	$x^2 - y^2$	xy
Wavefunction name	s-wave	g	$d_{x^2-y^2}$	d_{xy}
Schematic representation of Δ(k) in B.Z.				

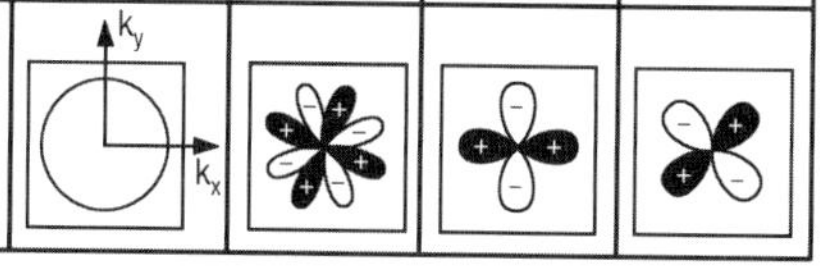

Fig. 16.1. k-space representation of allowed symmetry basis functions for the C_{4v} symmetry appropriate for the CuO_2 planes in the high-T_c superconductors

16.2 Phase Sensitive Tests: Theoretical Background

Symmetry tests using a variety of non-phase sensitive techniques, including nuclear magnetic resonance [4,45–49], penetration depth [50–52], angle-resolved photoemission [53–56], Josephson tunneling spectroscopy [57,58], specific heat [59–62], Raman scattering [63–65], and thermal conductivity [66–68], provide strong evidence that the gap in cuprates such as YBCO and Bi-2212 is highly anisotropic with nodes, consistent with d-wave pairing symmetry. However, these tests are not sensitive to the phase of the order parameter, and often depend on modeling details, making for much controversy. Nevertheless, these studies have produced a large body of indirect evidence for d-wave pairing.

In recent years, there has arisen a new class of phase-sensitive pairing symmetry tests. These symmetry probing techniques are based on two macroscopic quantum coherence phenomena: Josephson tunneling and flux quantization.

16.2.1 Josephson Supercurrents Between Unconventional Superconductors

The symmetry of a pair wavefunction can best be probed at the junction interface as the Cooper pairs tunnel across a Josephson junction or weak link. A schematic drawing of a Josephson tunnel junction is depicted in Fig. 16.2, showing the tunnel barrier sandwiched between two junction electrodes, with order parameters $\Delta_i(\mathbf{k}_i) = |\Delta_i| e^{i\varphi_i}$, where the subscript $i = L, R$. The concept of Josephson tunneling is not just limited to superconductor–insulator–superconductor (SIS) junctions, but has been generalized to all weak-link structures consisting of two superconductors (not necessarily identical) coupled by a small region of depressed order parameter [12,69,70]. The supercurrent I_s, proportional to the tunneling rate of Cooper pairs through the barrier, as given by Josephson, can be expressed by:

$$I_s = I_c \sin\phi\,, \tag{16.4}$$

where I_c is the Josephson critical current, and ϕ is the gauge-invariant phase difference [12] at the junction:

$$\phi = \varphi_L - \varphi_R + \frac{2\pi}{\Phi_0}\int_L^R \mathbf{A}\cdot d\boldsymbol{\ell}\,, \tag{16.5}$$

where $\mathbf{A}$ is the vector potential, and $d\boldsymbol{\ell}$ is the element of line integration from the left electrode (L) to the right electrode (R) across the tunneling barrier. There have been numerous studies on the Josephson effects and their applications [70]. We wish here only to concentrate on the fact that pair tunneling, as measured by I_s, is sensitive to the gap symmetry and the relative orientation of the junction electrodes. To illustrate this point, we use the tunneling Hamiltonian approach first introduced by Cohen et al. [71]. The supercurrent I_s at zero temperature is given by:

$$I_s \propto \sum_{\mathbf{k},\mathbf{l}} |T_{\mathbf{k},\mathbf{l}}|^2 \frac{\Delta_L(\mathbf{k})\Delta_R(\mathbf{l})}{E_L(\mathbf{k})E_R(\mathbf{l})}\frac{1}{(E_L(\mathbf{k})+E_R(\mathbf{l}))} \times \sin(\phi_L - \phi_R)\,, \tag{16.6}$$

where $T_{\mathbf{k},\mathbf{l}}$ is the time-reversal symmetry invariant tunneling matrix element, $E_i(\mathbf{k}) = \sqrt{\epsilon_i^2(\mathbf{k}) + \Delta_i^2(\mathbf{k})}$, and $\epsilon(\mathbf{k})$ is the one-electron energy.

The tunneling matrix element $T_{\mathbf{k},\mathbf{l}}$ is in general dependent on the characteristics of the junction electrodes *and* the tunnel barrier in between them. Therefore, it is not always straightforward to decipher information about pairing symmetry from measurements of supercurrent alone, but it is possible to gain some insights into the nature of the pair state. For tunnel junctions between spin-singlet (even parity) and spin-triplet (odd-parity) superconductors, Pals et al. [72] proved that the supercurrent I_s always vanishes, up to second order in $T_{\mathbf{k},\mathbf{l}}$,

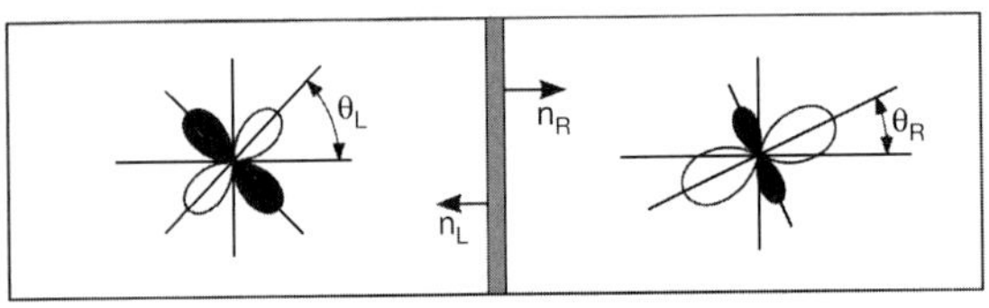

Fig. 16.2. Schematic diagram of a Josephson junction between a pure $d_{x^2-y^2}$ superconductor on the *left*, and a superconductor with some admixture of s in a predominantly $d_{x^2-y^2}$ state on the *right*. The gap states are assumed to align with the crystalline axes, which are rotated by angles θ_L and θ_R with respect to the junction normals $\mathbf{n}_L$ and $\mathbf{n}_R$ on the *left* and *right* hand sides, respectively

Table 16.3. Basis functions $\chi(\mathbf{n})$ for a Josephson junction electrode with tetragonal crystal symmetry

Irreducible Representation	A_{1g} s	A_{2g} g	B_{1g} $d_{x^2-y^2}$	B_{2g} d_{xy}
Basis Function $\chi(\mathbf{n})$	1	$n_x n_y(n_x^2 - n_y^2)$	$n_x^2 - n_y^2$	$n_x n_y$

if the spin-orbit coupling is not large (which is apparently the case for the cuprate superconductors). The fact that pair tunneling has been observed in many Josephson junctions made between a cuprate superconductor and a low-T_c conventional superconductor such as Pb or Nb constitutes an experimental proof that the cuprates have even parity in the superconducting state. As can be concluded from (16.6), a $\boldsymbol{k}$-independent tunneling matrix element would imply $I_s \equiv 0$ for Josephson junctions made with d-wave superconductors. There have been several theoretical studies on the Josephson effect in d-wave junctions with emphasis on the consequences of direction-dependent tunneling matrix elements [73–75]. In a systematic experimental study of the Josephson effect in YBCO bicrystal grain boundary junctions, the observed highly anisotropic supercurrent transport was indeed found to be consistent with a combined effect of unconventional pairing symmetry such as that of the $d_{x^2-y^2}$-wave pair state and anisotropy of the tunneling matrix elements [76].

Although various theoretical studies [73–75] on Josephson junctions of unconventional superconductors have predicted well-defined Cooper pair and quasiparticle tunneling characteristics that, in principle, can be used for discriminating between d-wave and s-wave pairing symmetries, it is difficult in practice to make a quantitative comparison between theory and experiment because nearly all theoretical treatments on this topic are based on the implicit assumption that the junction interface is smooth and uniform. However, there are suggestions that supercurrent transport across a junction between unconventional superconductors can be significantly altered by the presence of roughness, pair breaking states or other microstructural defects at the junction interface [57,77]. This problem is also present even for isotropic s-wave Josephson junctions, but is more acute for the d-wave case due to the directional dependence of the tunneling. Therefore, the use of the magnitude of I_s as a probe of pairing symmetry is sample-dependent and is not reliable for deciphering information about pairing symmetry.

To circumvent this difficulty, phase-sensitive symmetry tests probe instead the sign change in I_s as a consequence of directional tunneling processes in junctions made of at least one unconventional superconductor. I_s can be related to the group symmetry properties of the gap functions $\Delta_{R,L}(\boldsymbol{k})$ in terms of the irreducible representations $\chi_{R,L}$ [17,25,78–81]. By minimizing the total free energy with respect to ψ_L and ψ_R, the Josephson current density J_s, flowing perpendicular to the junction interface from superconductor L to R, is given by [81]:

$$J_s = t_{L,R}\chi_L(\mathbf{n})\chi_R(\mathbf{n}) \mid \eta_L \mid\mid \eta_R \mid \sin\phi = J_c \sin\phi\,, \tag{16.7}$$

where J_c is the critical current density and $\chi_{L,R}$, the basis function, is related to the gap function $\Delta(\mathbf{k})$ through (16.3), $\Delta_{L,R}(\mathbf{n}) = \eta_{L,R}(\mathbf{n})\chi_{L,R}(\mathbf{n})$, $\mathbf{n}$ is the unit vector normal to the junction interface, $\eta_{L,R}(\mathbf{n}) = \mid \eta_{L,R} \mid e^{i\varphi_{L,R}}$, and ϕ is the gauge-invariant phase difference as defined in (16.5). The quantity $t_{L,R}$ is a constant characteristic of a given junction configuration and is closely related to the tunneling matrix element in (16.6). The basis functions $\chi(\mathbf{n})$ for a Josephson junction electrode with tetragonal crystal symmetry (point group C_{4v}), are listed in Table 16.3, where n_x, n_y are the projections of the unit vector $\mathbf{n}$ onto the crystallographic axes $\mathbf{x}$ and $\mathbf{y}$ respectively.

For Josephson junctions between two d-wave superconductors $\chi(\hat{\mathbf{n}}) = n_x^2 - n_y^2$, (16.7) reduces to the well-known Sigrist and Rice [82] formula:

$$J_s = A_s(n_x^2 - n_y^2)_L(n_x^2 - n_y^2)_R \sin\phi\,, \tag{16.8}$$

or in terms of θ_L, θ_R, the angles of the crystallographic axes with respect to the interface:

$$J_s = A_s \cos(2\theta_L)\cos(2\theta_R)\sin\phi, \quad \text{(clean)}\,, \tag{16.9}$$

where A_s is a constant characteristic of the junction. It is implicitly assumed in the derivation of the Sigrist–Rice formula that the junction interface is uniform and smooth, and that only the pair tunneling process perpendicular to the junction interface needs to be considered. In real Josephson junctions, especially those made with cuprates, the electron wave vector normal to the junction interface can be significantly distorted by interface roughness, oxygen deficiency, faceting, strain, etc. In particular, cuprate grain boundary junctions have inhomogeneous and meandering junction interfaces which depend on the fabrication conditions [77,83–86]. A maximum disorder formula for the Josephson current can be derived [87], by allowing a broad distribution of angular deviations at the tunnel barrier, and by taking into account the fact that, due to the fourfold symmetry of CuO_2 planes, the maximum angle of deviation is $\pi/4$:

$$J_s = A_s \cos 2(\theta_L + \theta_R) \sin\phi \qquad \text{(dirty)}. \tag{16.10}$$

The difference between (16.10) and (16.9) shows that it is important to consider the effects of disorder at the junction interface in using Josephson tunneling to determine pairing symmetry. A series expansion of trigonometric functions of $\theta_{L,R}$ is needed for a general description of the angular dependence of the Josephson current. Such an expression was obtained by Walker and Luettmer–Strathmann [80] by writing the Ginzburg–Landau free energy of Josephson coupling in the form:

$$F_J = C(\theta_L, \theta_R) \cos\phi , \tag{16.11}$$

and by imposing the symmetry requirements for the tetragonal lattice: $C(\theta_L, \theta_R) = C(\theta_R, \theta_L) = C(-\theta_L, -\theta_R) = C(\theta_L + \pi, \theta_R)$; and $C(\theta_L + \frac{\pi}{2}, \theta_R) = C(\theta_L, \theta_R)$ for s-wave pairing symmetry, $C(\theta_L + \frac{\pi}{2}, \theta_R) = -C(\theta_L, \theta_R)$ for d-wave superconductivity. With these symmetry constraints, the supercurrent for a junction between two generalized s-wave superconductors can be given as:

$$J_s^s = \sum_{n,n'} [C_{4n,4n'} \cos(4n\theta_L) \cos(4n'\theta_R) + S_{4n,4n'} \sin(4n\theta_L) \sin(4n'\theta_R)]\, sin\,\phi , \tag{16.12}$$

where n, n' are positive integers including zero. Alternately, (16.13) can be re-written as:

$$J_s^s = \{C_{0,0} + C_{4,0}[\cos(4\theta_L) + \cos(4\theta_R)] + \cdots\} \sin\phi . \tag{16.13}$$

For a Josephson junction between d-wave superconductors,

$$\begin{aligned} J_s^d = \{&C_{4n+2,4n'+2} \\ &\times \cos[(4n+2)\theta_L] \cos[(4n'+2)\theta_R] \\ &+ S_{4n+2,4n'+2} \\ &\times \sin[(4n+2)\theta_L] \sin[(4n'+2)\theta_R]\} \sin\phi , \end{aligned} \tag{16.14}$$

which can be re-written as:

$$\begin{aligned} J_s^d = \{&C_{2,2} \cos(2\theta_L) \cos(2\theta_R) \\ &+ S_{2,2} \sin(2\theta_L) \sin(2\theta_R) + \cdots\} \sin\phi . \end{aligned} \tag{16.15}$$

It is interesting to note that, in (16.16), the first term is just the Sigrist–Rice clean limit formula (16.9). If $S_{2,2} = -C_{2,2}$, the sum of the first two terms leads to the dirty limit formula (16.10).

16.2.2 Physics of Frustrated Superconducting Loops

For the purposes of this chapter, a frustrated superconducting loop is one that, in the absence of an externally applied field, has a local maximum in its free energy with zero circulating supercurrent (see, e.g. (16.17)). Sign changes in I_c resulting from the direction dependence of the pair wavefunction can be systematically examined by looking for the experimental signatures of such frustrated loops, providing a powerful tool for studying the internal symmetry of Cooper pairs. Sign changes in I_c due to pairing symmetry are arbitrary for a particular junction, since an arbitrary phase can always be added to both sides of the junction. However, the signs of the critical currents in a closed ring of superconductors interrupted by Josephson weak links can always be assigned self-consistently. Counting these sign changes provides a convenient way to determine if a particular geometry is frustrated. A geometry with an odd number of sign changes in the normal component of the order parameter across the Josephson weak links around a closed loop is a frustrated or π geometry. A geometry with an even number of such sign changes is an unfrustrated or 0 geometry. A negative pair tunneling critical current I_c can be thought of (for counting purposes) as a phase shift of π at the junction interface ($I_s = -\mid I_c \mid \sin\phi = \mid I_c \mid \sin(\phi + \pi)$). Such π phase-shifts were theoretically predicted between Josephson junctions involving unconventional superconductors such as heavy fermion superconducting systems [78,88] and cuprate superconductors [82]. In addition, it has been suggested that a π-phase shift can be realized within a junction by mechanisms unrelated to pairing symmetry such as spin–flip scattering by magnetic impurities [89] or indirect electron tunneling [90]. There is now experimental evidence for this second mechanism for phase shifts [91–93]. However, as we will describe below, there is ample evidence that the experiments discussed in this chapter depend on pairing symmetry dependent phase shifts. Historically, the term π-junction has been used to describe both the situations in which phase shifts within the superconductor are caused by pairing symmetry, and in which phase shifts within a single Josephson junction are caused by the tunneling mechanism. We will restrict our use of the term π-junction to the latter. However, a superconducting ring with an odd number of π-shifts is frustrated [87], and will show the special effects described below, independent of whether the π-shifts result from symmetry or tunneling mechanism effects. Such frustrated rings will be referred to as π-rings, independent of the mechanism for the π-phase shifts.

It is useful to distinguish two classes of geometry for phase-sensitive tests of pairing symmetry in the cuprate superconductors: "rings" and "junctions". For the present purposes a "ring" is a closed superconducting loop interrupted by one or more Josephson weak links; a "junction" is two slabs of superconductor separated by a Josephson weak link.

The most striking property of frustrated superconducting rings and junctions is that they will spontaneously generate circulating supercurrents and magnetic flux when cooled through the superconducting transition temperature in zero applied external field. Consider first a ring with inductance L interrupted by a single Josephson junction with critical current I_c, in an externally applied field sufficient to produce Φ_a of total flux through the ring (in the absence of screening from a supercurrent I_s circulating around the ring). Writing the total flux Φ through the ring as $\Phi = \Phi_a + I_s L$, the free energy of the ring has two contributions, from the inductive energy in the supercurrents, and the Josephson coupling energy:

$$U(\Phi,\Phi_a) = \frac{\Phi_0^2}{2L}\left\{\left(\frac{\Phi-\Phi_a}{\Phi_0}\right)^2 - \left(\frac{L\mid I_c\mid}{\pi\Phi_0}\right)\cos[\frac{2\pi\Phi}{\Phi_0}+\Theta]\right\}, \tag{16.16}$$

where $\Phi_0 = hc/2e$ is the superconducting flux quantum, and Θ=0, π for a 0 and a π-ring, respectively.

The ground state of the single-junction ring can be obtained by minimizing $U(\Phi,\Phi_a)$ to obtain Φ as a function of Φ_a [82] (Fig. 16.3). For a small applied flux Φ_a, a 0-ring has $\Phi < \Phi_a$ (diamagnetic shielding, Fig. 16.3(a)), while a π-ring has $\Phi > \Phi_a$ (paramagnetic shielding, Fig. 16.3(b)). Paramagnetic shielding in granular samples [94–99] was one of the early hints that the cuprates had unconventional pairing symmetry [81,82].

Plots of the free-energy vs Φ at $\Phi_a = 0$ (Fig. 16.4) shows that the 0-ring has a ladder of metastable states centered at the ground state $\Phi = 0$, while the π-ring has a ladder of states shifted by $\Phi_0/2$ from those of the 0-state, centered on a doubly degenerate ground state. In the limit $\beta_e = 2\pi LI_c/\Phi_0 \gg 1$, the 0-ring has allowed flux states that are integer multiples of the superconducting flux quantum

$$\Phi = n\Phi_0 \qquad n = 0, \pm1, \pm2, \ldots, \tag{16.17}$$

while the π-ring has allowed flux states that are half-integer multiples

$$\Phi = (n+1/2)\Phi_0 \qquad n = 0, \pm1, \pm2, \ldots \tag{16.18}$$

The total ground state spontaneous flux Φ at zero applied flux Φ_a is plotted as a function of $2\pi LI_c/\Phi_0$ in Fig. 16.5.

A frustrated junction can be modeled as having two sections, one with no intrinsic phase shift (a 0-section), and one with an intrinsic π-phase shift (a π-section). The phase shift $\phi(r_i)$ across the ith section of the junction is a solution of the sine-Gordon equation [100,101]

$$\nabla^2\phi(r_i) = \sin(\phi(r_i) + \Theta(r_i))/\lambda_{Ji}^2, \tag{16.19}$$

where r_i is the position along the i-th section of the junction, $\lambda_{Ji} = (\hbar c^2/8\pi e dJ_1)^{1/2}$ is the Josephson penetration depth of the i-th section of the junction,

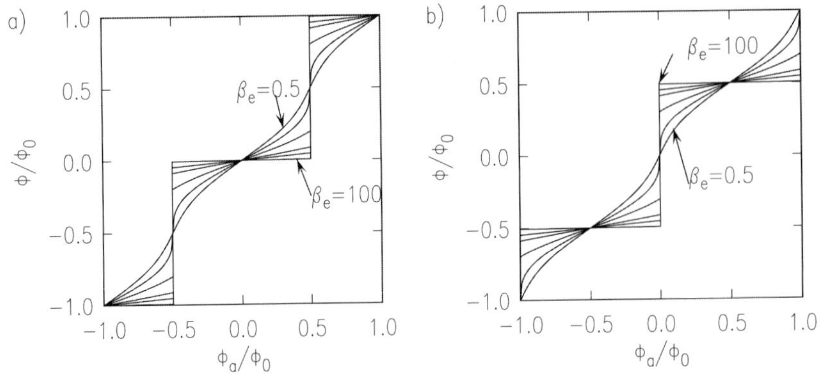

Fig. 16.3. Minimum energy solution of (16.17) for the total flux Φ/Φ_0 as a function of the externally applied flux Φ_a/Φ_0 for a 0-ring (**a**) and π-ring (**b**), for $\beta_e = 2\pi \mathrm{LI}_c/\Phi_0 = 0.5,1,2,5,10,$ and 100. For small applied fields and $\beta_e < 1$ the shielding currents oppose the applied flux for the 0-ring (diamagnetic shielding) and are aligned with the applied flux for the π-ring (paramagnetic shielding). For $\beta_e > 1$ the π-ring has spontaneous magnetization with the same sign as small externally applied fields, approaching total flux $\Phi = \Phi_0/2$ for $\Phi_a = 0$ as β_e becomes large

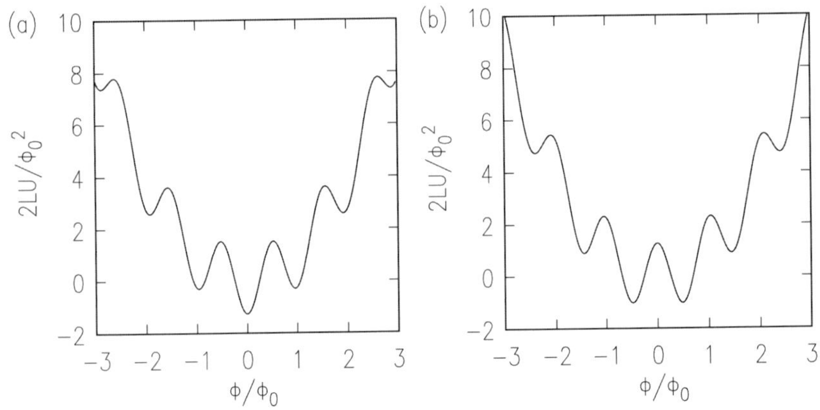

Fig. 16.4. Free energy U of a superconducting ring with a single junction in (**a**) a 0-ring, or (**b**) a π-ring configuration, with zero external applied field (16.17), as a function of the total flux Φ in the ring. Here β_e=$2\pi \mathrm{LI}_c/\Phi_0 = 5$

where d is the spacing between superconductors, and J_1 is the supercurrent density in the junction, and Θ is 0 for the 0-section, and π for the π section. As discussed above, the assignment of 0 or π for the intrinsic phase shift of a particular section of the junction is arbitrary: identical results are obtained if the assignments are reversed. $\phi(r_i)$ is related to the current flowing through the junction by the Josephson relation $j_s = j_{ci}\sin(\phi(r_i) + \Theta(r_i))$.

The lowest energy solution of (16.19) for a 0–0-junction is $\phi(r_i) = 0$: there is no spontaneous magnetization. For a $0-\pi$-junction with width W much wider than the λ_{Ji}s, the solution of (16.19) is [102]

$$\phi(x) = \begin{cases} 4\tan^{-1}[(\sqrt{2}-1)e^{r_i/\lambda_{Ji}}] & r_i < 0\,, \\ 4\tan^{-1}[(\sqrt{2}+1)e^{r_i/\lambda_{Ji}}] - \pi & r_i > 0\,. \end{cases} \tag{16.20}$$

This is the equation for a Josephson vortex with $\Phi_0/2$ total flux. If the width of the junction is comparable to the Josephson penetration depth, the solution of (16.19) is more difficult [100]. Numerical results [103] for the total spontaneous flux for a symmetric $0-\pi$ junction (equal widths of intrinsic 0 and π phase shifts) as a function of the reduced width $W/2\lambda_j$ are shown as the open symbols in Fig. 16.5. As discussed below, the spontaneous magnetization in frustrated rings and junctions has been directly observed by magnetic imaging using scanning SQUID microscopes.

One can also test for frustration in a superconducting loop by connecting leads to it, and measuring the superconducting critical current through the

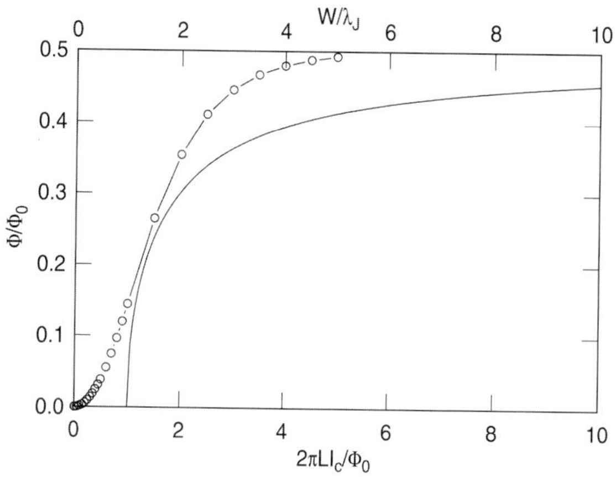

Fig. 16.5. Total spontaneous flux Φ/Φ_0, for a frustrated superconducting ring in zero applied flux Φ_a, as a function of $\beta_e = 2\pi LI_c/\Phi_0$ *(solid line, lower scale)*, and for a symmetric $0-\pi$ junction as a function of $W/2\lambda_J$ *(open circles, upper scale)*, where W is the total width of the junction, and λ_J is the Josephson penetration depth

loop as a function of the applied flux Φ_a. Consider a closed ring of superconductor interrupted by two Josephson weak links labeled a and b. The measured critical current of the SQUID is the maximum of

$$I_s = I_a \sin\phi_a + I_b \sin\phi_b\,, \tag{16.21}$$

where I_a (I_b) and ϕ_a (ϕ_b) are the critical currents and phases of the a (b)-junctions, respectively. This maximum must be calculated subject to the constraint that the phase be single valued:

$$2\pi n = \phi_a - \phi_b + \Theta + 2\pi\left(\frac{I_a L_a}{\Phi_0} - \frac{I_b L_b}{\Phi_0} + \frac{\Phi_a}{\Phi_0}\right), \tag{16.22}$$

where L_a and L_b are the effective self-inductances of the two arms of the ring, and $\Theta = 0$ or π for a 0-ring or a π-ring, respectively. The dependence of the SQUID critical current I_{SQUID} vs. applied flux Φ_a is plotted in Fig. 16.6 for a number of different $\beta_e = 2\pi LI_c/\Phi_0$ values. For these plots we have assumed $I_0 = I_a = I_b$, and $L_a = L_b$. Asymmetries in the junction critical current or arm inductances produce phase shifts in the interference patterns, and reductions in the modulation depths. The modulation depth decreases as the β_e increases.

For junctions with uniform Josephson current density, and in the "short" junction limit where the junction size is much smaller than the Josephson penetration depth, the critical current should follow the standard Fraunhofer pattern $I_c(\Phi) = I_0|\sin(\pi\Phi/\Phi_0)/(\pi\Phi/\Phi_0)|$ as a function of the flux Φ threading the junction, for an unfrustrated (0–0) junction. This pattern has a maximum in the critical current at zero applied field (see Fig. 16.7). In contrast, a "short" frustrated $(0-\pi)$ junction should show the interference pattern $I_c(\Phi) = I_0|\sin^2(\pi\Phi/2\Phi_0)/(\pi\Phi/2\Phi_0)|$, with a minimum at zero applied flux. As can be seen from Fig. 16.7, as the ratio of the junction width to the Josephson penetration depth gets larger, the modulation depth of these interference patterns decrease. Further, inhomogeneities in the junction current densities will

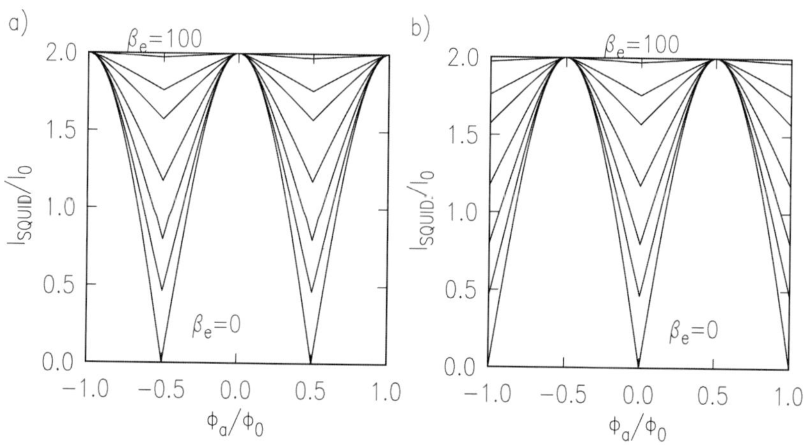

Fig. 16.6. Dependence of the total critical current I_{SQUID} through a symmetric conventional (**a**) and frustrated (**b**) SQUID, normalized by the single junction critical current I_0, as a function of the normalized applied flux Φ_a/Φ_0 threading through the SQUID ring, for values of $\beta_e = 2\pi LI_c/\Phi_0 = 0, 0.5, 1, 2, 5, 10$, and 100

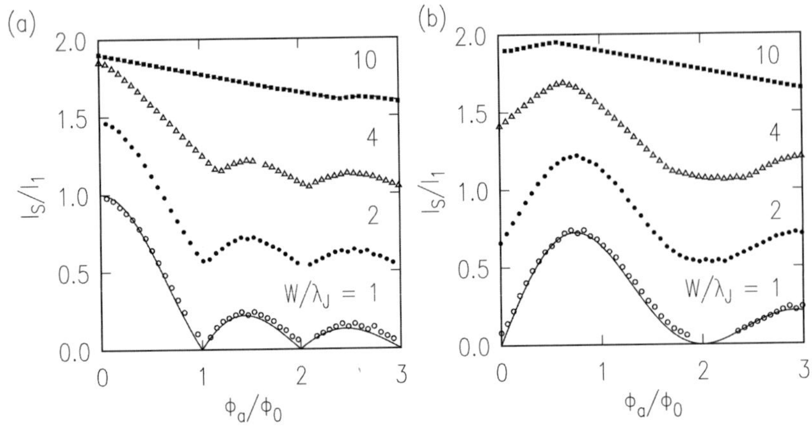

Fig. 16.7. Dependence of the total critical current I_s through a single symmetric conventional (0–0, (**a**) and a frustrated $(0-\pi$, (**b**) junction, normalized by the single (0–0) junction critical current I_0, as a function of the normalized applied flux Φ_a/Φ_0 threading through the junction, for values $W/\lambda_J =$ of 0 (*solid lines*), 1,2,4, and 10 (*symbols*), where W is the junction width, and λ_J is the Josephson penetration depth. Each successive curve has been offset by 0.5 units for clarity

modify the interference patterns, in general further reducing the modulation depths.

To conclude this section, we note that there are two ways to test for frustration in superconducting loops: by measuring the spontaneous magnetization at zero applied magnetic flux, or by measuring the critical current through the loop (or junction) as a function of applied magnetic field. These two types of test are complementary, in the sense that the spontaneous magnetization is largest when the ring inductance or junction width is big, but the magnetic interference patterns are the most clear when the ring inductance or junction width is small.

16.3 Phase-Sensitive Tests: Experiments

16.3.1 SQUID Interferometry

Wollman et al. reported the results of the first phase-sensitive test of pairing symmetry in a controlled configuration [104] by measuring the quantum interference effects in a YBCO-Pb dc-SQUID, in the "corner SQUID" geometry, proposed independently by Sigrist and Rice [82]. (Fig. 16.8(a)). Josephson weak links were made between Pb thin films and two orthogonally oriented *ac* (or *bc*) plane faces of single crystals of YBCO. If YBCO is a *d*-wave superconductor, there should be a π-phase shift between weak links on adjacent faces of the crystal. Annett et al. [5] have argued that it is plausible that the pair transfer matrix $T_{\mathrm{k,l}}$ (16.6) is strongly peaked in the forward direction for both *SNS* and tunneling weak

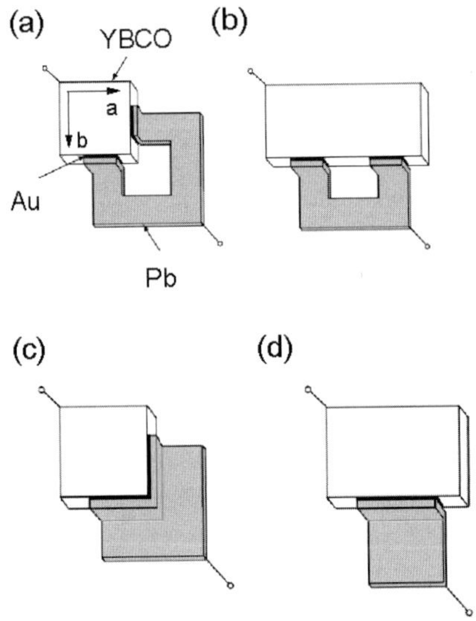

Fig. 16.8. Experimental geometry used for the experiments of Wollman et al. [104, 105]. (**a**) Corner SQUID configuration, (**b**) edge SQUID configuration, (**c**) corner junction, and (**d**) edge junction

links, making the Sigrist–Rice clean relation (16.9) applicable. Wollman et al. [104] tested for this phase shift by measuring the SQUID critical current as a function of Φ_a, the externally applied magnetic flux through the SQUID, as described in the discussion of Fig. 16.6. In the experiments of Wollman et al. [104], the junction I_c's and L's were not necessarily balanced, leading to shifts in the I_c vs Φ_a characteristics. Noise rounding of the current-voltage characteristics of these SQUIDs made it possible for Wollman et al. to do these measurements at several values of the d.c. applied current through the SQUID, and correct for these self-field effects. This was done by plotting the phase shift as a function of d.c. current through the SQUID, and then extrapolating to zero current to infer the intrinsic phase shift. The intercepts for the "corner SQUIDs" varied between 0.3 and 0.6 Φ_0. When these experiments were repeated with "edge SQUIDs", Fig. 16.8(b), with two junctions on the same *ac* (or *bc*) face of the crystal. the intercepts centered around zero (Fig. 16.9(a)).

There were a number of complicating factors in the interpretation of the SQUID experiments of Wollman et al. [104]. The issue of twinning will be discussed in detail in Sect. 16.3.7. Briefly, if the pair order parameter phase of a *d*-wave superconductor is locked to the crystal axes (for example, with the positive lobe always pointing along the ***a*** direction), the sign in a particular *ab* plane direction would be expected to alternate from twin to twin, washing out the phase-sensitive effects described here. This does not happen, which indicates that the *d*-component of the order parameter locks across twin boundaries. Other issues in the interpretation of these experiments were the linear extrapolation in d.c. current used [106, 107], the comparison of sample geometries with corners with those without corners [108], and the effects of flux trapping. However, subsequent phase-sensitive experiments without these problems [87, 109] have shown that these issues were not serious enough to affect the qualitative conclusion of predominantly *d*-wave orbital pairing symmetry in YBCO made by Wollman et al..

Brawner and Ott [110] performed a second SQUID interference experiment on single crystals of YBCO. In contrast to the Wollman et al. experiments [104],

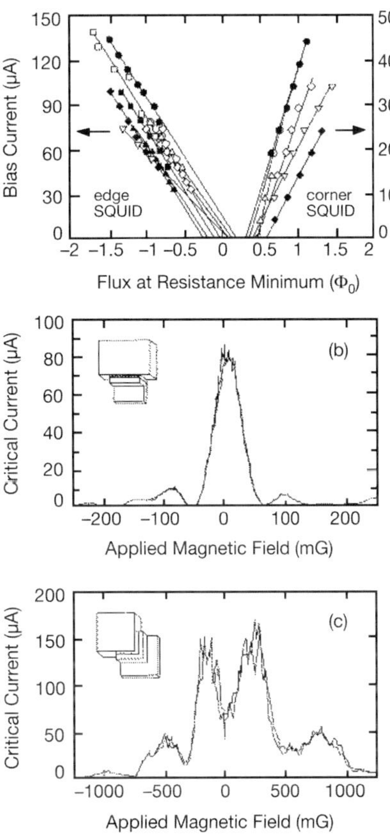

Fig. 16.9. Summary of the experimental results of Wollman et al. [104, 105]. (**a**) Extrapolation of the measured SQUID resistance minimum vs flux to zero bias current for a corner SQUID and an edge SQUID on the same crystal. Each curve represents a different cooldown of the sample. (**b**) Measured critical current vs applied magnetic field for an "edge" junction and (**c**) a "corner" junction

instead of using evaporated thin films for the second electrode, they made a SQUID with two bulk point contact junctions of niobium, producing Josephson weak links on adjoining orthogonal faces of an un twinned YBCO single crystal. Two all niobium control SQUIDs on either side of the sample SQUID were used to measure the relative phase shifts. The junction critical currents and the inductances of the two arms of the SQUIDs were asymmetric, so that as with the Wollman et al. [104] experiments, corrections for self-field effects had to be made. To do this Brawner and Ott [110] measured the dynamic resistance of

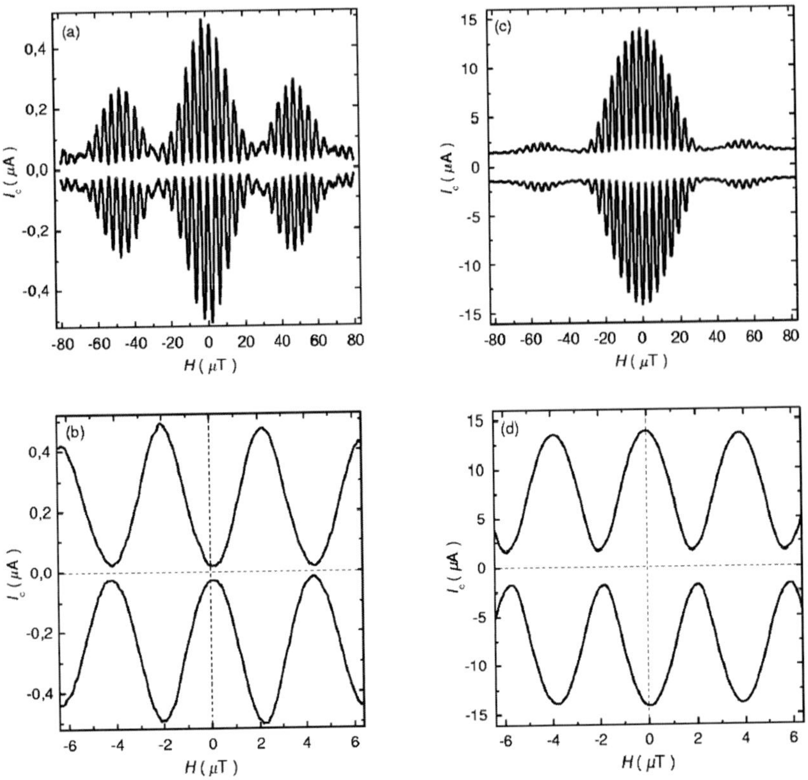

Fig. 16.10. Magnetic interference patterns for a thin film, spatially distributed junctions, low inductance π-SQUID (**a**, **b**), and a 0-SQUID (**c**, **d**). The characteristics show a nearly ideal interference pattern over a small field range (**b**, **d**), with a shift of $\Phi_0/2$, as expected. An interference envelope due to the finite size of the junctions is apparent over a larger field range (**a**, **c**). From [109]

their SQUIDs as a function of d.c. applied current and extrapolated their results to zero current. They found consistent phase shifts between their control and sample SQUIDs of $160 \pm 20°$, consistent with the $180°$ degree phase shift expected between the normal components to the order parameter on the two faces of the YBCO crystal for a $d_{x^2-y^2}$ superconductor.

Schulz et al. [109] produced all high-T_c 0 and π-ring SQUIDs based on the spatially distributed junctions, low inductance SQUID design of Chesca [111], using YBCO thin films epitaxially grown on bicrystal and tetracrystal (e.g. Fig. 16.22(a)) $SrTiO_3$ substrates. These devices, which had the advantages of very small sample volumes (so that flux trapping was not an issue) and small I_cL products (so that self–field effects were negligible), showed nearly ideal dependences of the critical current on applied field, with a minimum at zero applied field for the π-ring SQUID, as expected for a d-wave superconductor in the tetracrystal geometry used (see Fig. 16.10).

16.3.2 Single Josephson Junction Modulation

In addition to measurements in a SQUID geometry, Wollman et al. also performed phase-sensitive measurements in a single junction geometry [104, 105] (Figs. 16.8(c) and (d)). This geometry had the advantage of being less sensitive to the effects of flux trapping and sample asymmetry. It is expected that the critical current in the single junction geometry should have the interference patterns shown in Fig. 16.7 for junctions with uniform Josephson current density. Wollman et al. [104, 105] reported consistent results of a maximum in the interference pattern for "edge" junctions(Fig. 16.9(b)), and a minimum for "corner" junctions(Fig. 16.9(c)), consistent with $d_{x^2-y^2}$ symmetry. Similar evidence for d-wave pairing was also reported by Iguchi and Wen [112].

As can be seen from a comparison of Figs. 16.9 and 16.7, the interference patterns reported by Wollman et al. [104, 104] did not agree closely with the ideal expressions. They attributed this discrepancy to asymmetries in the current densities in the junctions and flux trapping, and reported good agreement with modeling including these effects [6]. However, qualitatively similar interference patterns have been reported for square [113] and annular [114–116] Josephson junctions with a single vortex trapped in them. The Illinois group reported that symmetric interference patterns, with minima at zero applied field, such as they observe, cannot be reproduced in modeling with an s-wave superconductor in their geometry. In addition, it seems unlikely that magnetic flux consistently trapped in the Illinois "corner" junctions in such a way as to mimic d-wave superconductivity.

In an experiment analogous to the Illinois "corner" single junction experiments, Miller et al. [117] used frustrated thin-film tricrystal samples to probe the pairing symmetry in YBCO. The tricrystal geometry will be discussed in detail in the next section. Miller et al. measured the dependence on magnetic field of the critical current of a 3μm wide micro bridge spanning the tricrystal point. They found a minimum in the critical current at zero applied field, as expected for a d-wave superconductor in this geometry, in the "short junction" limit (where the width W of the bridge is much shorter than the Josephson penetration depth λ_J [100] (see Fig. 16.7). They found that junctions in an unfrustrated geometry, or wide junctions ($L \gg \lambda_J$) in a frustrated geometry, showed maxima in their critical currents at zero magnetic fields [102], as expected.

16.3.3 Tricrystal Magnetometry

Controlled-Orientation Multi-Crystals

In this class of experiments, a multiple-junction ring is made, consisting of deliberately oriented cuprate crystals that define the direction of the pair wavefunction. The presence or absence of the half-integer flux quantum effect in such samples as a function of the tricrystal geometry is used as an unambiguous signature for a certain pairing symmetry. In the first such phase-sensitive experiment of this type for testing the d-wave pairing symmetry, tricrystal (100) $SrTiO_3$ substrates with controlled orientations were designed and fabricated (Fig. 16.11). The c-axis oriented epitaxial cuprate films were deposited and photolithographically patterned into rings (Fig. 16.12). The ring centered at the tricrystal meeting point is interrupted by three grain boundary Josephson junctions. In accordance with an assumed pair state, the three-junction ring is expected to exhibit conventional integer flux quantization (0-ring) or the half-integer flux quantum effect (π-ring) depending on the misorientation angles α_{12}, α_{31}, and the angle β between the grain boundaries 23 and 31 as defined in Fig. 16.11(a). For testing of d-wave pairing symmetry, based on the Sigrist–Rice (clean) formula (16.9), the three-junction ring is a 0-ring if $\cos 2(\alpha_{12}+\beta)\cos 2(\alpha_{12}-\beta)$ is positive (with the condition $\alpha_{12} = \pi/2 - \alpha_{31}$ assumed for simplicity), and a π-ring if $\cos 2(\alpha_{12}+\beta)\cos 2(\alpha_{12}-\beta)$ is negative. In Fig. 16.11(b), the tricrystal design parameters range (α_{12}, β) for π-rings (which can be thought of as having an odd number of sign changes in the I_c's) are plotted as open area; and that for 0-rings (an even number of sign changes) are shown in the shaded areas. As emphasized in Sect. 16.2.1, the effect of disorder at the junction interface must be considered in the design of any viable phase-sensitive experiment. It can be shown (16.10) [87] that in the maximum disorder limit the three-junction ring is a π-ring if $\cos(2\alpha_{12})\cos(2\alpha_{31})\cos(\alpha_{12}-\alpha_{31})$ is negative. The design parameter space for the d wave 0 and π-ring configuration with maximum disorder taken into account is shown in Fig. 16.11(c).

The design parameters selected for the original tricrystal phase-sensitive experiment [87] were $\alpha_{12} = 30°$, $\alpha_{31} = 60°$, and $\beta = 60°$ (Fig. 16.12), corresponding to the solid dot in Figs. 16.11(b) (the clean limit) and 16.11(c) (the dirty limit), well within the bounds of the d-wave π-ring regime. If a cuprate under the symmetry test is indeed a d-wave superconductor, such tricrystal rings should show half-integer flux quantization, regardless of whether the junction interface is in the clean or dirty limit. The situation in an actual grain boundary junction of cuprate superconductors is expected to fall somewhere in between. Tricrystal rings with design parameters located within the 0-ring regime (shaded areas in Figs. 16.11(b) and 16.11(c)) should display

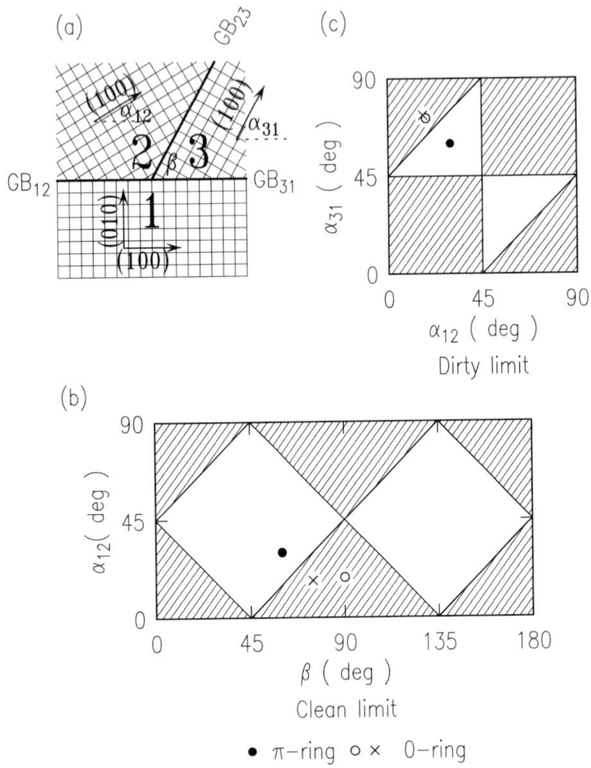

Fig. 16.11. Tricrystal geometry (**a**), regions of the design parameters which give 0 (*shaded areas*) and π (*open areas*) rings in the clean (**b**) and dirty (**c**) limits. The • is the design point for the frustrated three-junction ring samples (Fig. 16.16a). The ○ (Fig. 16.16b) and × (Fig. 16.16c) are the design points for the unfrustrated three-junction ring samples

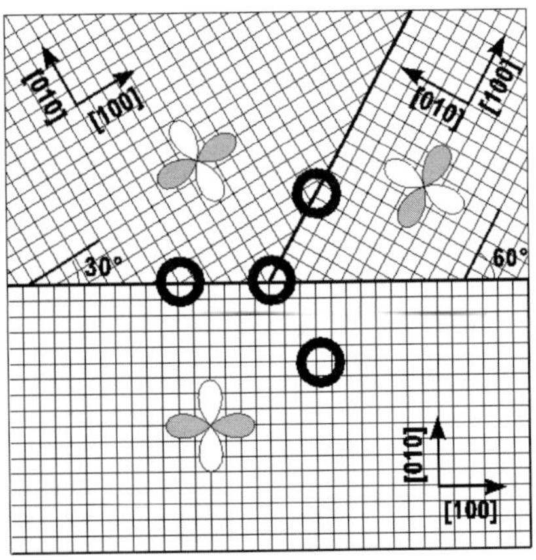

Fig. 16.12. Experimental configuration for the π-ring tricrystal experiment of Tsuei et al. [87]. The central, three-junction ring is a π-ring, which should show half-integer flux quantization for a $d_{x^2-y^2}$ superconductor, and the two-junction rings and 0-junction ring are 0-rings, which should show integer flux quantization, independent of the pairing symmetry

only the regular integer flux quantization, for rings made of a d-wave superconductor. The absence of the half-integer flux quantum effect in these three-junction rings designed to be 0-rings represent evidence for d-wave order parameter symmetry in addition to that provided by the presence of the half-integer flux quantum effect in three-junction rings designed to be π-rings.

In the tricrystal experiment of Tsuei et al. [87], an epitaxial YBCO film (1200 Å thick) was deposited using laser ablation on a tricrystal (100) $SrTiO_3$ substrate, and rings (48μm inner diameter, 10μm in width) were patterned by a standard ion-milling photolithographic technique. In addition to the three-junction ring located at the tricrystal meeting point, two two-junction rings and one ring with no junction were also made as controls (see Fig. 16.12). The control rings are in the 0-ring configuration and should exhibit the standard integer flux quantization. Measurements of the critical current densities of these grain boundaries, combined with the estimated self-inductance of the rings $L = 100$ pH, indicated that the I_cL product was about $100\Phi_0$, easily satisfying the condition $I_cL >> \Phi_0$ for observing $\Phi = \Phi_0/2$ flux quantization (see the discussion in Sect. 16.2.2).

Magnetic Flux Imaging

The magnetic flux threading through the superconducting cuprate rings in the tricrystal experiments was measured using a scanning SQUID microscope (SSM) [118]. The magnetic field sensor used is an integrated miniature SQUID magnetometer with system noise about $2\times10^{-6}\Phi_0/\mathrm{Hz}^{1/2}$. Shown in Fig. 16.13 is an SSM image of four rings of c-axis oriented epitaxial YBCO film deposited on a tricrystal (100) $SrTiO_3$ substrate three-junction YBCO ring with the configuration shown in Fig. 16.12. As presented in the following discussion, four different ways of analyzing the SSM data of the type shown in Fig. 16.13 have lead to the conclusion that the magnetic flux through the three-junction ring is indeed $\Phi_0/2$, while the other rings contain no flux, providing strong evidence for d-wave order parameter symmetry in YBCO.

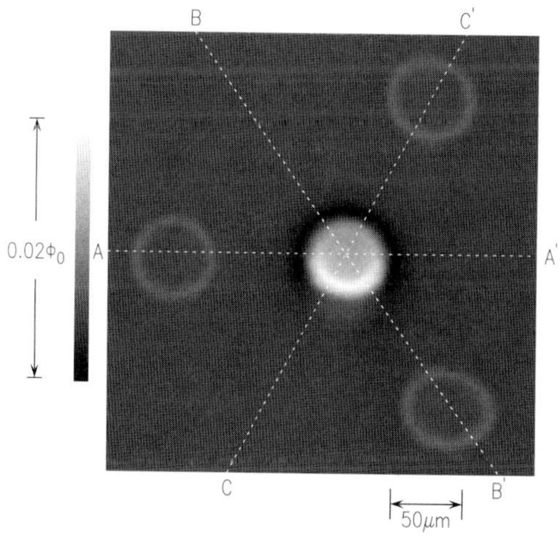

Fig. 16.13. Top view of a scanning SQUID microscope image of a thin-film YBCO tricrystal ring sample, cooled and imaged in nominally zero magnetic field. The outer control rings have no flux in them; the central three-junction ring has half of a superconducting quantum of flux spontaneously generated in it

The four techniques for calibrating the amount of flux through the superconducting rings can be briefly summarized as follows. The details can be found in [7].

(a) Direct calculation: A given flux threading a ring with self-inductance L induces a circulating current $I_r = \Phi_0/L$ around the ring, which in turn produces a flux $\Phi_s(\rho) = M(\rho)\Phi/L$ in the SQUID pickup loop. The calculated inductance of the rings of Fig. 16.13 is $L = 99 \pm 5$ pH.

The mutual inductance $M(\rho)$ between a pickup loop tilted at an angle θ from the sample $(x-y)$ plane in the $x-z$ plane and a circular wire of radius R at the origin can be written as:

$$M(\rho) = \frac{\mu_0 R}{4\pi} \int \mathrm{d}x\mathrm{d}y \int_0^{2\pi} d\phi \times \frac{\cos\theta(R - y\sin\phi - x\cos\phi) - sin\theta(z\cos\phi)}{(x^2+y^2+z^2-2xR\cos\phi-2yR\sin\phi)^{3/2}}, \tag{16.23}$$

where the integral $\mathrm{d}x\mathrm{d}y$ is over the plane of the pickup loop, and the vector ρ specifies the displacement of the pickup loop with respect to the ring in the x–y plane.

The solid lines in Fig. 16.14a are model calculations, based on (16.23) and the pickup loop geometry used in the experiment, for three cross-section through the SSM image, assuming $\Phi = \Phi_0/2$. The asymmetry in the image can be attributed to the tilt of the pickup loop, as well as the asymmetric pickup area from the unshielded section of the leads. A best fit of the SSM data yields $\Phi = 0.57 \pm 0.1\Phi_0$, using a doubling of the χ^2 as a criterion for assigning statistical errors.

(b) Magnetic "oil drop": The changes in the SQUID signal as a function of applied magnetic field are monitored as individual single flux quanta enter the superconducting rings. A staircase pattern in the SQUID output vs. field thus observed (see Fig. 16.14 (b)) can be used for calibration. Using the heights of the single flux-quantum steps in Fig. 16.14(b) as a calibration leads to values for the three-junction ring flux of $0.46 \pm 0.09\Phi_0$ in its lowest allowed flux state.

(c) Measurements of the absolute values of the difference between the SQUID signal when the pickup loop

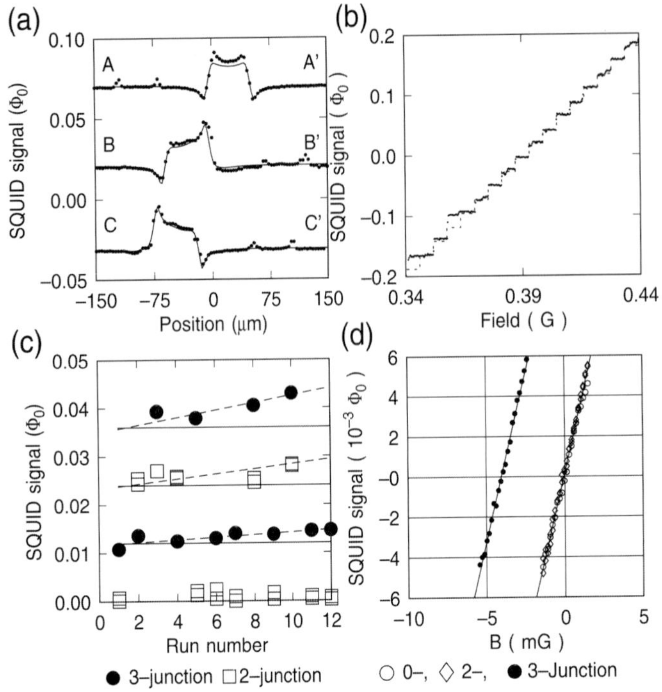

Fig. 16.14. Four techniques for demonstrating the half-flux quantum effect in the tricrystal ring samples: (**a**) Direct calculation, assuming the central ring has $\Phi_0/2 = h/4e$ flux in it. (**b**) Observation of the change in the SQUID signal as individual vortices enter the three-junction ring, with the pickup loop centered on the ring. (**c**) Measurements of the absolute values of the pickup loop flux when it is directly above the 0-junction ring minus that above the two-junction (*open squares*) and three-junction (*dots*) rings, for a number of cool downs. (**d**) Measurements of the SQUID signal directly above the rings, as a function of externally applied field

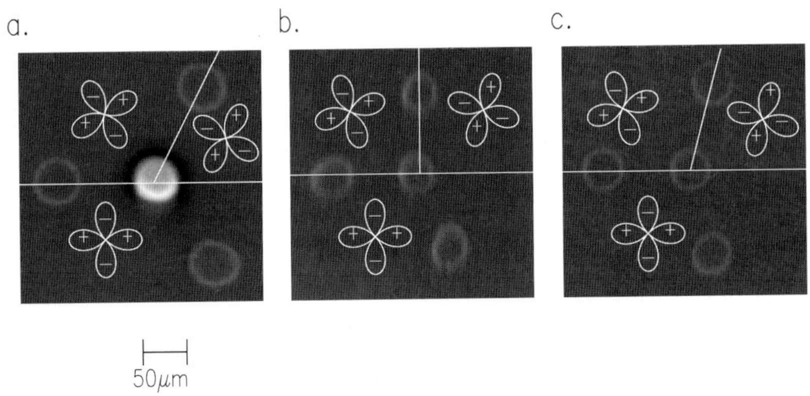

Fig. 16.15. (**a**)–(**c**) Three different $SrTiO_3$ tricrystal geometries and scanning SQUID images of YBCO ring samples fabricated on these substrates to elucidate the origin of the observed half-flux quantum effect in YBCO. Both the presence *and* the absence of this effect represent strong evidence for a *d*-wave pair state in YBCO

is directly above the 0-junction ring minus that above the two-junction (open squares in Fig. 16.14(c)) and three-junction (dots) rings, for a number of cooldowns of the tricrystal sample. The solid lines are the expected values for the flux difference, calculated from (16.23) as described above. The values of the SQUID signal always fall close to $(N + 1/2)\Phi_0$ for the three-junction rings, and $N\Phi_0$ for the two-junction rings (N an integer). The upward drift in the data as a function of increasing run number was due to tip wear of the SQUID assembly. The dashed lines in Fig. 16.14(c), including this correction agree with the data.

(d) "Magnetic field titration": The SQUID signal at the center of a superconducting ring, relative to the background outside the ring, for all rings in the experiments, can be determined by using the SSM

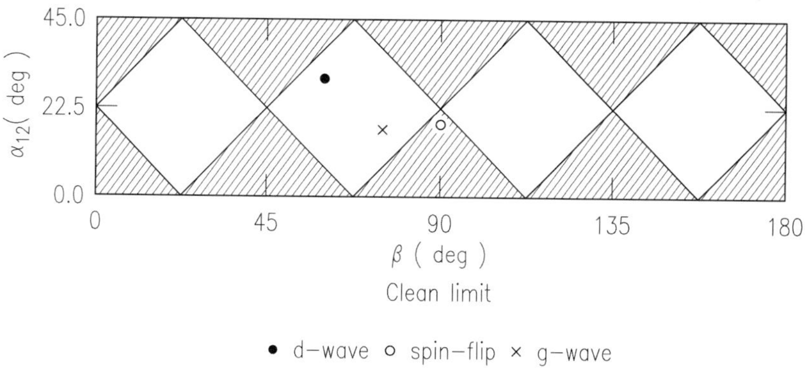

Fig. 16.16. Parameter space for the presence (*open*) or absence (*shaded*) of the half-integer flux quantum effect for tricrystal geometries, for an assumed symmetric g-wave pairing symmetry $\Delta \sim \cos k_x + \cos k_y$. The *solid symbol* represents the design of Fig. 16.15(a), the *open symbol* is that of Fig. 16.15(b), and the cross is that of Fig. 16.15(c)

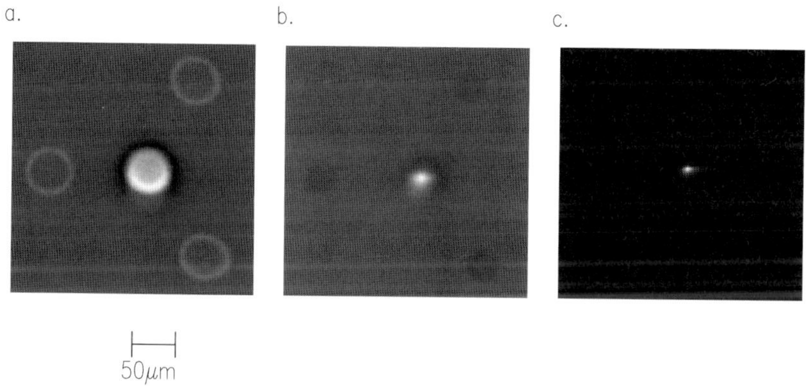

Fig. 16.17. (a)–(c) Scanning SQUID microscope images of three thin film samples of YBCO epitaxially grown on identical d-wave testing tricrystal substrates to show the half-flux quantum effect is independent of macroscopic sample geometry. In all cases there is a half-flux quantum of total flux spontaneously generated at the tricrystal point, but the spatial distributions of supercurrent are different

while varying the externally applied magnetic field until the SQUID signal outside the ring is exactly nullified by the screening current induced field (see Fig. 16.14(d)). The result of such a procedure leads to a value of $0.49 \pm 0.015\Phi_0$ more flux threading through the three-junction ring that the 0-junction or two-junction rings.

On a separate note, it should be mentioned that the reason that the zero-flux state of the control rings (i.e. the two-junction and 0-junction rings) are visible in the SSM (see Fig. 16.13) is that the high sensitivity of SSM allows it to detect very small changes in the inductance of the SQUID when the pickup loop passes over the ring, even when it contains no flux.

Shown in Fig. 16.15 are the results of a series of tricrystal experiments with various geometrical configurations to clarify further the nature of the observed half-flux quantum effect. As expected, the sample in panel (a), which had a frustrated geometry for a $d_{x^2-y^2}$ superconductor ($\alpha_{12} = 30°, \alpha_{31} = 60°, \beta = 60°$, see Fig. 16.11) showed half-integer flux quantization, while the samples in panels (b) ($\alpha_{12} = 108.4°$, $\alpha_{13} = 71.6°, \beta = 90°$) and (c) ($\alpha_{12} = 107°, \alpha_{13} = 73°$, $\beta = 75°$) [119,120] which had 0-ring configurations (i.e. shaded areas in Figs. 16.11(b) and 16.11(c)), did not. The absence of the half-integer flux quantum effect in these two tricrystal experiments has ruled out any symmetry-independent mechanisms [89,90] as the cause of the π-phase shift. In addition, the tricrystal shown in panel (c) rules out even-parity g-wave pairing symmetry with order parameter varying as $(\cos k_x + \cos k_y)$ (see Fig. 16.16).

As concluded from the discussion in Sect. 16.2.2, half-integer flux quantization is a manifestation of the ground state in any singly connected superconducted loop with an intrinsic phase shift of π. The observation of such an effect should be independent of the macroscopic form that the sample takes. In-

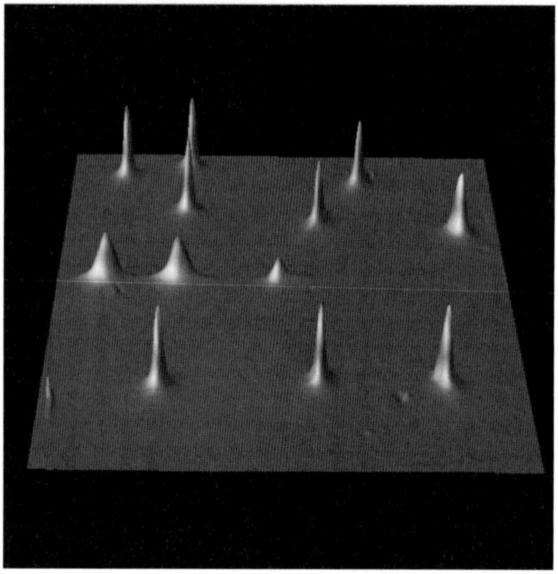

Fig. 16.18. Three-dimensional rendering of SQUID microscopy image of several types of vortex trapped in a YBCO film on a tricrystal $SrTiO_3$ substrate designed to generate the half-flux quantum effect in a $d_{x^2-y^2}$ superconductor

deed, the half-flux quantum effect has been observed directly with the SSM in tricrystal YBCO disks and blanket films (Figs. 16.17(b) and (c) and Fig. 16.18), in addition to rings (Fig. 16.17(a)).

Integer and Half-Integer Josephson Vortices

The unpatterned (blanket) tricrystal sample geometry has the advantage that no photolithographic processing has to be done after the cuprate film is deposited. It has the disadvantage that the total flux generated at the tricrystal point is more difficult to calibrate, compared with the ring geometry. As noted in Sect. 16.2.2, superconducting rings interrupted by Josephson weak links with large values of $\beta_e = 2\pi LI_c/\Phi_0$ have total magnetic fluxes close to $N\Phi_0$ or $(N + 1/2)\Phi_0$ (N an integer), depending on whether there is an intrinsic phase shift of 0 or π in the ring respectively. In contrast, an unpatterned tricrystal sample has only two allowed flux states $\Phi = \pm\Phi_0/2$ at the tricrystal point in the frustrated geometry: It is energetically favorable for higher quantized flux states (e.g. $\Phi = \pm 3\Phi_0/2, \pm 5\Phi_0/2$, etc.) to split into a half-flux quantum at the tricrystal point, and some number of integer Josephson vortices at a distance from the tricrystal point [121]. In fact, Josephson vortices with flux different from $\pm\Phi_0/2$ have never been observed at the tricrystal point in a frustrated tricrystal geometry.

Figure 16.19 compares images of a bulk vortex (a), Φ_0 Josephson vortices trapped along the diagonal (b) and horizontal (c) grain boundaries, and a $\Phi_0/2$ half-vortex trapped at the center of a π-ring thin-film tricrystal sample of YBCO [122]. Contour lines have been placed on the data at 0.1, 0.3, 0.5, 0.7, and 0.9 of the full scale amplitudes. Superposed on panel (a) is a schematic of the 4 μm diameter octagonal pickup loop used for these measurements. The dots in Fig. 16.20 are cross-sections through the images of Fig. 16.19 as indicated by the contrasting lines and letters in this figure.

To date all experiments on tricrystal samples have been done with samples with film thicknesses comparable to the London penetration depth. It is possible to derive expressions for the fields from a Josephson integer or half-integer vortex in samples with thin films [123], but these expressions are cumbersome, and all detailed analyses of experimental data to date have used a simpler slab model. In this picture the (thin-film) crystals making up the tricrystal point are modeled as semi-infinite wedges meeting at the tricrystal line $x = 0, y = 0, z$. The phase shift $\phi(r_i)$ across the i-th grain boundary is described by the sine-Gordon equation (16.19) [100,101]. The solution to (16.19) for a $h/4e$ half-vortex with $\theta(r_i) = 0, r_i < 0$ and $\theta(r_i) = \pi, r_i > 0$ was given in (16.20). The solution to (16.19) which describes a $h/2e$ Josephson vortex ($\theta(r_i) = 0$ everywhere), centered about $r_i = 0$ along a single grain boundary is

$$\phi(r_i) = 4\tan^{-1}(e^{r_i/\lambda_{Ji}}) . \qquad (16.24)$$

The general expression for the phase ϕ at the tricrystal point is complicated, if the grain boundaries have different supercurrent densities, but the magnetic flux/unit length in the i-th branch of the vortex can be written as:

$$\frac{d\Phi(r_i)}{dr_i} = \frac{\Phi_0}{2\pi}\frac{d\phi}{dr_i} = \frac{\Phi_0}{2\pi}\frac{-4a_i}{\lambda_{Ji}}\frac{e^{-r_i/\lambda_{Ji}}}{1 + a_i^2 e^{-2r_i/\lambda_{Ji}}} , \qquad (16.25)$$

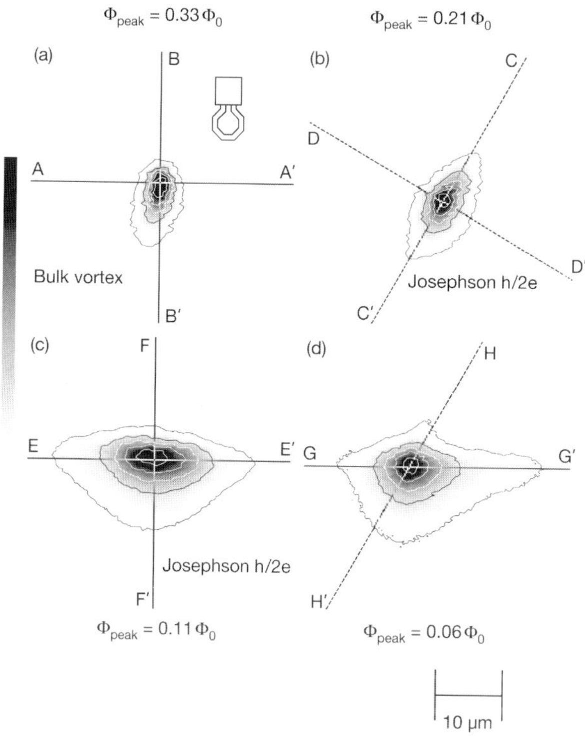

Fig. 16.19. (**a**)–(**d**) High resolution images of several types of vortex trapped in a YBCO film on a tricrystal $SrTiO_3$ substrate designed to generate the half-flux quantum effect in a $d_{x^2-y^2}$ superconductor

where in this case $r_i = 0$ for all grain boundaries at the tricrystal point, and r_i is restricted to values greater than 0. Inside the superconductors, the London theory gives $\nabla^2 \boldsymbol{B} = \boldsymbol{B}/\lambda_L^2$. Using London theory to describe the field inside a superconductor is strictly valid only if no current flows through the sample boundary. This is a reasonable approximation if $\lambda_L << \lambda_J$. Neglecting the derivative parallel to the grain boundaries in the Laplacian leads to

$$B_z(r_i, r_\perp) = \frac{\Phi_0 a_i}{\pi \lambda_L \lambda_{Ji}} \frac{e^{-r_i/\lambda_{Ji}}}{1 + a_i^2 e^{-2r_i/\lambda_{Ji}}} e^{-|r_{i\perp}|/\lambda_L} , \quad (16.26)$$

where $r_{i\perp}$ is the perpendicular distance from the i-th grain boundary. A Josephson vortex with $h/2e$ total flux in it, and with a single penetration depth, has

$$B_z(r_i, r_{i\perp}) = (\Phi_0/2\pi\lambda_L\lambda_J)\mathrm{sech}(r_i/\lambda_J)e^{-|r_{i\perp}|/\lambda_L} . \quad (16.27)$$

For the tricrystal, the a_i's are normalization constants determined by two conditions: First, the condition that the magnetic induction at the tricrystal point varies smoothly implies that $a_i/(\lambda_{Ji}(1 + a_i^2))$ is the same for each grain boundary. Second, the total flux at the tricrystal point is given by $\Phi = \sum_i (\Phi_0/2\pi) 4\tan^{-1}(a_i) = \Phi_0/2$.

The magnetic inductions at a height z above the surface are determined from those at the surface by noting that for magnetic inductions derived from a two-dimensional current distribution,

$$b_z(k_x, k_y, z) = \exp(-\sqrt{k_x^2 + k_y^2} z) b_z(k_x, k_y, 0) , \quad (16.28)$$

where b_z is the two-dimensional Fourier transform of the field $B_z(r)$, and k_x and k_y are the wavevectors in the x and y directions respectively [124]. The flux is derived by integrating the induction over the area of the pickup loop. The effective height of the pickup loop above the sample surface is obtained by fitting data from $h/2e$ bulk vortices, which are resolution limited since, except very close to T_c, the London penetration depth λ_L is much smaller than the size of the pickup loop. In this limit the magnetic induc-

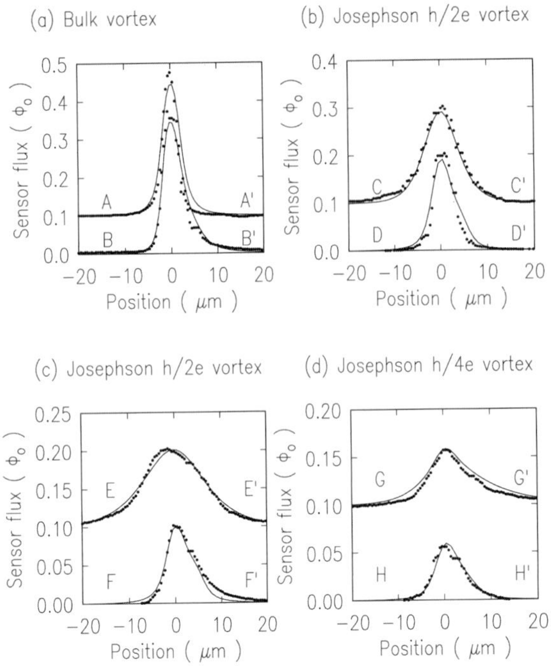

Fig. 16.20. (**a**)–(**d**) The *dots* are cross-sections through the vortex images of Fig. 16.19 as indicated by the contrasting lines and letters. The *lines* are fits to this data as described in the text, assuming that the bulk vortex and Josephson vortices along the grain boundaries have $h/2e$ of total flux in them, and the vortex at the tricrystal point has $h/4e$ total flux

tion in vacuum at a distance r away from the center of a superconducting bulk vortex is given by [125–127]

$$B(r) = (\Phi_0/2\pi)r/\,|\,r\,|^3\;. \tag{16.29}$$

The solid lines AA' and BB' in Fig. 16.20(a) are fits to the data for a bulk vortex using the fields of (16.29). The effective height determined from this fit fixed the height for the rest of the modeling. The solid lines $CC' - FF'$ in Figs. 16.20(b) and (c) are fits of the model above to the data, with $\lambda_L = 150$ nm, and using the Josephson penetration depth λ_J as the only fitting parameter. The lines GG' and HH' in Fig. 16.20(d) are fits to the half-vortex at the tricrystal point, using the three Josephson penetration depths as fitting parameters. This agreement between experiment and modeling shows that there is approximately $\Phi_0/2$ total flux at the tricrystal point. For a tricrystal sample in a geometry that is frustrated for a d-wave superconductor, if the sample is cooled repeatedly in fields close to zero, the half-flux quantum Josephson vortex is always present, either oriented with fields pointing out of or into the sample.

A recent SSM flux imaging tricrystal experiment by Iguchi's group [128] also observed the half-integer flux quantum effect (see Fig. 16.21), providing independent phase-sensitive evidence for d-wave pairing symmetry in YBCO.

The tricrystal magnetometry experiments directly image the half-integer flux quantum effect, in the ground state, with a tunable geometry to ensure that the pairing symmetry is being probed.

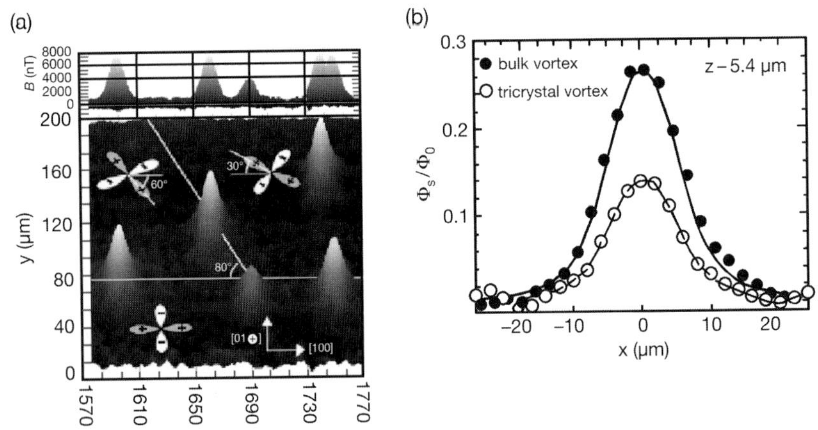

Fig. 16.21. (**a**) Three-dimensional rendering of the magnetic fields above an unpatterned YBCO film on a tricrystal substrate with geometry designed to be frustrated for a d-wave superconductor. (**b**) Modeling (*solid lines*) and cross-section data (*symbols*) for a h/2e vortex away from the grain boundaries, and a $h/4e$ Josephson vortex at the tricrystal point. (After Sugimoto et al. [128])

There are no complications associated with self-field effects, since no currents are applied externally. The tricrystal ring samples have very small sample volumes, making flux trapping highly unlikely. Any flux trapping to occur is imaged with high sensitivity. These experiments do use grain boundary junctions, which have moderately rough interfaces. However, care was taken to account for this interface roughness. Further, these experiments are extremely reproducible: dozens of samples have been made and measured in hundreds of cool downs. All tricrystal or tetracrystal samples that had sufficiently high supercurrent densities across the grain boundaries ($\lambda_J < 50\mu$m) gave results (producing either 0 or π-rings) that were as expected for predominantly d-wave pairing symmetry.

Evidence for Pure d-Wave Pairing Symmetry

The phase-sensitive pairing symmetry tests described above have yielded convincing evidence for a predominantly d-wave order parameter in YBCO. However, none of these experiments were able to unambiguously eliminate the possibility of a $d+s$ mixed pair state. An admixture of s-wave and d-wave pairing symmetries is allowed in orthorhombic YBCO. The orthorhombic crystal structure of YBCO is derived from the inequivalent lattice distortions along the a, b direction in the CuO_2 planes. As a consequence, the relevant group symmetry for the Cu-O rectangular lattice is C_{2v}, instead of C_{4v} as appropriate for the Cu-O square lattice planes in cuprates such as LSCO, Hg-1201, Tl-2201, etc.. As discussed in Sect. 16.1.2, the s-wave and d-wave spin-singlet order parameter symmetries belong to the same irreducible representation (A_{1g}) of the point group C_{2v}. Hence, the $d+s$ mixed pairing is an unavoidable consequence of orthorhombicity [80,129]. The question is: What is the amount of the s-wave pairing in the mixed pair state? In a $(d+s)$-wave superconductor the order parameter $\Delta(\boldsymbol{k})$ varies as $d(k_x^2-k_y^2)+s(k_x^2+k_y^2)$ to reflect the underlying crystal lattice symmetry, where s and d represent the size of the s-wave and d-wave subcomponents in the admixture respectively. The node lines ($\Delta = 0$), given by $k_y = \pm(\frac{d+s}{d-x})^{1/2}k_x$, exist only under the condition that the d-wave order parameter is dominant (i.e. $d/s > 1$). The slope of the node lines deviates from that of pure d-wave $k_x = \pm k_y$, the diagonals of the Brillouin zone, depending on the degree of the $s-d$ admixture.

The Pb-YBCO corner SQUID (or single Josephson junction) interference experiments rely only on a sign change of the order parameter between the a and b faces of the YBCO single crystal, and therefore can not discriminate pure d-wave from $d+s$ pair states as long as $d/s \geq 1$. In principle, the tricrystal experiments with YBCO and Tl-2201 are capable of locating the nodes on the Fermi surface of a $(d+s)$-wave superconductor. However, this would require a systematic series of tricrystal experiments based on a knowledge of the bandstructure and a detailed model describing the complex process of pair tunneling across a realistic grain boundary junction. In the face of these difficulties in determining the degree of $s+d$ mixing in pairing, it is important to demonstrate unambiguously the existence of a pure $d_{x^2-y^2}$-wave cuprate superconductor. This is significant in view of the substantial amount of theoretical work [80,129] on the effect of $s+d$ wave pairing on the properties of cuprate superconductors, including the origin of high temperature superconductivity. The demonstration of a pure d-wave pair state can help to understand whether the coexistence of d and s pairing channels is essential or just accidental to high-temperature superconductivity. In the following, phase-sensitive evidence for pure d-wave pairing symmetry will be presented for two high-T_c cuprate systems:

(a) Tetragonal $Tl_2Ba_2CuO_{6+\delta}$

Stoichiometric $Tl_2Ba_2CuO_{6+\delta}$ (Tl-2201) can be prepared in the form of c-axis epitaxial films with a tetragonal crystal structure. Unlike YBCO, the tetragonal Tl-2201 is intrinsically twin-free and has only one CuO_2 layer per unit cell. Such a relatively simple lattice structure makes Tl-2201 a model system for studying many normal-state and superconducting properties of the high-T_c cuprates.

To demonstrate the existence of a pure d-wave cuprate superconductor, a tetracrystal phase-sensitive experiment, with tetragonal Tl-2201 [130], adopted from a suggestion by Walker and Luettmer-

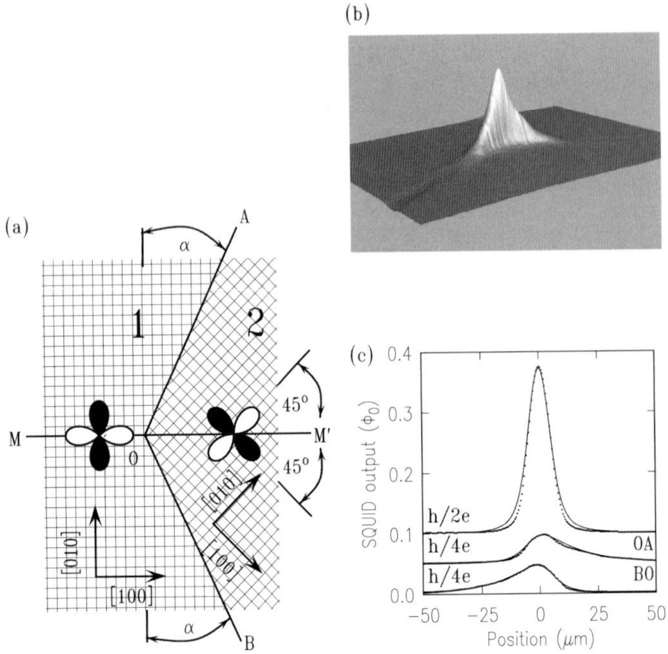

Fig. 16.22. (**a**) Tetracrystal geometry. (**b**) SSM image of the crystal meeting point of a film of Tl2201 epitaxially grown on a $SrTiO_3$ substrate with the geometry of (**a**), cooled in nominally zero field, and imaged at 4.2 K with an 8.2μm square pick-up loop. (**c**) Cross-sections through a bulk Abrikosov vortex, and through the half-vortex of (**b**), along the directions indicated in (**a**). The *dots* are the experimental data, and the *solid lines* are modeling, assuming the Abrikosov vortex has $h/2e$ flux trapped in it, and the vortex at the tetracrystal point has $h/4e$ flux

Strathmann [80], was carried out. Shown in Fig. 16.22 is the geometrical configuration of a four-crystal (100) $SiTiO_3$ substrate used in the experiment. The zero misorientation angle at the grain boundary MOM' makes the tetracrystal substrate effectively a bicrystal made of two crystals rotated about the normal (c-axis) to the substrate plane by $\pi/4$ with respect to each other. The built-in reflection symmetry with respect to the equal-partition line MOM' assures that, for a pair state with $d_{x^2-y^2}$ or d_{xy} symmetry, the pair tunneling currents across the two grain boundaries OA and OB are equal in magnitude but opposite in sign.

$$(I_s)_{OA} = -(I_s)_{OB} \,. \qquad (16.30)$$

This conclusion can also be reached using the general formula for the supercurrent of a Josephson junction between two d-wave superconductors, (16.16), [80]:

$$\begin{aligned} I_s \propto\, & C_{2,2}\cos(2\theta_1)\cos(2\theta_2) \\ & + S_{2,2}\sin(2\theta_1)\sin(2\theta_2) + \cdots \end{aligned} \qquad (16.31)$$

By reflection symmetry, the angles θ_1 and θ_2 are given by:

$$\theta_1 = \binom{+\alpha}{-\alpha}\,, \; \theta_2 = \binom{\pi/4-\alpha}{\pi/4+\alpha}\,, \qquad (16.32)$$

for grain boundary junctions

$$\binom{OA}{OB}\,,$$

respectively.

When this reflection symmetry operation is applied to (16.31), every term changes its sign, reducing to (16.30). Therefore, any superconducting loop around the tetracrystal meeting point O will be in a π-ring configuration, giving rise to the half-flux quantum effect. On the other hand, for s-wave and g-wave pairing symmetries, corresponding to A_{1g} and A_{2g} irreducible representations respectively, there is no sign reversal in I_s. Standard integer flux quantization is expected in any superconducting loop enclosing the wedge tip O. Shown in Fig. 16.22(b) is the scanning SQUID image of an epitaxial Tl-2201 film (T_c = 83 K) on the tetracrystal STO substrate (Fig. 16.22(a)). The tetragonal crystal structure of the single phase Tl-2201 epitaxial films used in the experiment was confirmed by micro-Raman spectroscopy, X-ray diffraction, bright-field imaging transmission electron microscopy, selected-area electron diffraction, and convergent-beam electron diffraction [131,132]. A detailed modeling [122,130] of the image data (Fig. 16.22(c)) indicates the total amount of flux threading the wedge tip is indeed $\Phi_0/2$. The fact that the half-flux quantum is the only vortex in a relatively large area suggests that it is spontaneously generated, as expected. Since the conclusion about the sign change in the supercurrent is based on a rigorous model-independent symmetry argument, the observation of the half-flux quantum effect represents strong phase-sensitive evidence for d-wave pairing symmetry in Tl-2201.

(b) Orthorhombic $Bi_2Sr_2CaCu_2O_{\delta-x}$

The high-T_c superconducting systems YBCO and Bi-2212 are probably the most investigated members of the cuprate family, owing to their importance in fundamental studies and practical applications. Both are characterized by an orthorhombic layered crystal structure. The YBCO system derives its orthorhombicity from the presence of Cu-O linear chains which distort the CuO_2 square lattice, resulting in lattice constants $a \neq b$. In the case of Bi-2212, an incommensurate superlattice modulation in the BiO layers, along the b direction, gives rise to unequal lattice constants a and b in the CuO_2 planes [34,133]. An important difference between the two cuprate systems is that the in-plane Cu-O bonds coincide with the inequivalent a- and b axes in YBCO, but not in Bi-2212. As for the case of YBCO, early pairing symmetry results on Bi-2212 using quasiparticle tunneling spectroscopy and ARPES produced controversial results, supporting s-wave as much as d-wave symmetry. However, more recent ARPES studies on gap anisotropy have yielded important evidence suggesting d-wave pairing symmetry in Bi-2212 [53,54,134]. A phase-sensitive experiment is needed to determine the pair order parameter symmetry unambiguously. The results of a tricrystal flux imaging experiment are presented in Figs. 16.23(b) and (c). Shown in Fig. 16.23b is a scanning SQUID microscopy image of a c-axis oriented Bi-2212 film deposited on a tricrystal STO substrate designed for testing d-wave pairing symmetry (the same as in Fig. 16.12). With the tricrystal sample cooled in a field of 0.4μT and imaged at 4.2 K, many Abrikosov vortices are found randomly distributed in the grains, while Josephson vortices decorated the grain boundaries. At the tricrystal point, a relatively weak vortex is always observed. Detailed modeling of the SSM imaging data shows that the vortex at the tricrystal point has a total flux of $\Phi_0/2$, while all others are single vortices with the full flux quantum Φ_0 (Fig. 16.23(d)). To prove that the half-flux quantum at the tricrystal point is indeed generated spontaneously, the sample was cooled to 4.2 K in a nominally zero ambient field and imaged. The fact that only the half-flux quantum at the tricrystal point can be found (see Figs. 16.23(c) and (d)) provides the proof.

As shown in Fig. 16.24, the Brillouin zone for Bi-2212 is rotated by exactly $\pi/4$ with respect to that of the orthorhombic YBCO. As a result, the in equivalent axes $\boldsymbol{a}$ and $\boldsymbol{b}$ coincide with the node lines of the anisotropic gap established by the ARPES measurements [53,54,134]. From the viewpoint of the point group symmetry operations, s-wave and $d_{x^2-y^2}$-wave pair states correspond to two distinct irreducible representations A_{1g} and B_{1g}, respectively. Therefore, Bi-2212 can have either a pure s-wave or a pure d-wave pairing symmetry. The results of the tricrystal experiment and the ARPES measurements together argue strongly in favor of a pure d-wave order parameter symmetry for Bi-2212. This conclusion is also supported by the observation of zero [135,136] or nearly zero [137] c-axis pair tunneling in Bi 2212.

16.3.4 Electron-Doped Cuprate Superconductors

The chemical composition of the electron-doped cuprates [138] is given by $Ln_{2-x}Ce_xCuO_{4-y}$, Ln=Nd,Pr,Eu, or Sm; y$\simeq$0.04. Electron doping is accomplished by substituting the lanthanides Ln^{3+} by Ce^{4+} and by removing oxygen. Of these compounds, $Nd_{1.85}Ce_{0.15}CuO_{4-y}$ (NCCO) and $Pr_{1.85}Ce_{0.15}CuO_{4-y}$ (PCCO) have been the subjects of most studies. All the cuprates are characterized by the Cu-O square lattice layers in their crystal structures. However, the electron doped ones are noted for the absence of the apical oxygen atoms which form two-dimensional arrays of Cu-O pyramids or octahedra with the CuO_2 planes in the hole-doped cuprates such as LCSO, YBCO or Bi-2212. Superconductivity in electron-

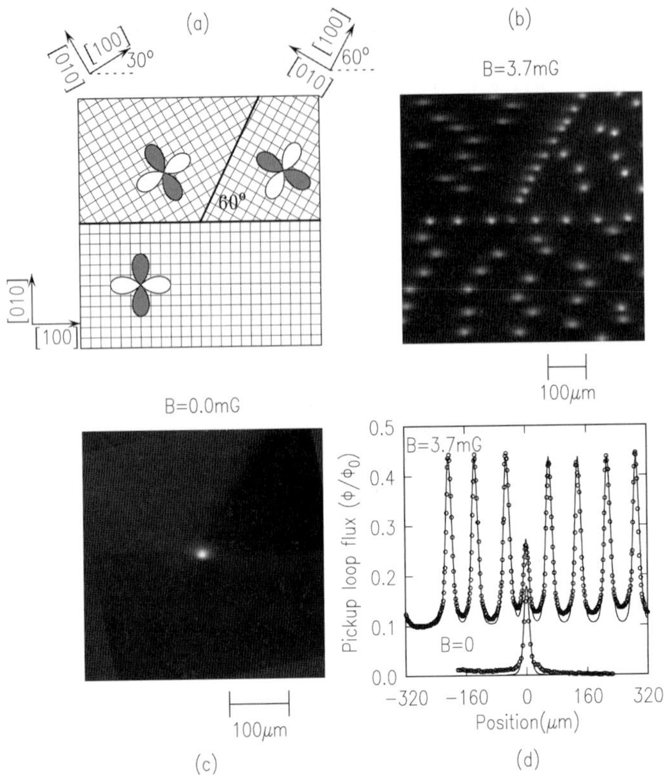

Fig. 16.23. SSM data on Bi2212. (**a**) Tricrystal geometry, including polar plots of assumed $d_{x^2-y^2}$ order parameters aligned along the crystalline axes. (**b**) SSM image of 640 × 640μm^2 area of an epitaxial film of Bi2212 on a $SrTiO_3$ tricrystal substrate with the geometry of (**a**), cooled in a field of 0.4μT, and imaged at 4.2 K with an 8.2μm square pick-up loop. (**c**) 400 × 400μm^2 image of the same sample, cooled in nominal zero field. (**d**) Comparison of horizontal cross-sections through the images (**b**) and (**c**) with modeling assuming that all vortices have Φ_0 of flux, except for at the tricrystal point, which has $\Phi_0/2$ flux spontaneously generated

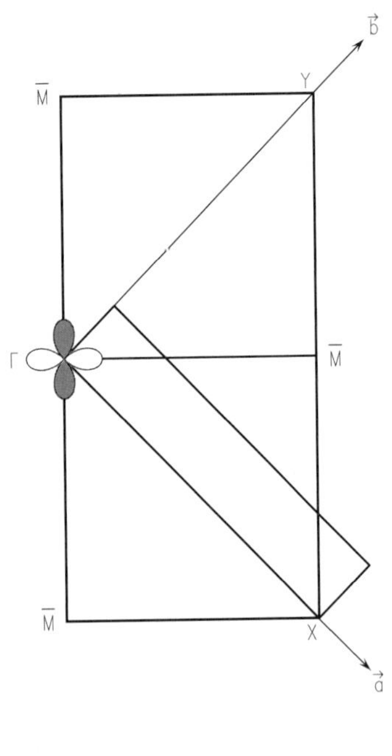

Fig. 16.24. Brillouin zone for Bi2212. The reduced zone as a result of the incommensurate super-lattice modulation with a period of 4.7b (b=5.41/AA) along the b direction is outlined schematically. The orientation of the $d_{x^2-y^2}$-wave polar plots is established by the ARPES data

doped cuprate systems occurs in a very narrow doping range (0.14 ≤ x ≤ 0.17 for NCCO [138], and 0.13 ≤ x ≤ 0.2 for PCCO [139,140]); in the hole-doped counterpart LSCO the range is broader (0.05 < x < 0.3). The highest T_c values in the hole-doped are over five times those in the electron-doped cuprates. In optimally doped YBCO and LSCO the in-plane resistivity exhibits a linear temperature dependence over a wide range of temperature, with small or nearly zero extrapolated values at $T = 0$. In NCCO and PCCO ($x = 0.15$), the in-plane resistivity is quadratic in temperature with a large residual resistivity [141]. The results of ARPES have revealed CuO_2-plane derived flat energy bands near the Fermi surface of the high-T_c hole-doped cuprates such as YBCO [142] and Bi-2212 [143]; but not within 300 meV of the Fermi surface of NCCO [144]. Other physical properties such as the Hall coefficient, thermopower, and the pressure dependence of T_c [145], are also different. For more details, reviews are available in the literature [145, 146].

In view of these remarkable differences between the hole- and electron-doped cuprates, it is only logical to expect that the pairing symmetries of these two classes of superconductors are also different. In fact, based on the measurements of in-plane penetration depth λ_{ab}, tunneling characteristics, including the absence of zero-bias conductance peaks (ZBCP), ..., it was indeed widely believed that NCCO and PCCO were s-wave superconductors [147]. In recent years, a number of new experiments and new insight into the data analysis have cast doubts about the s-wave conclusion for the electron-doped superconductors. After taking into account the effect of paramagnetic Nd^{3+} ions [148], for example, a linear temperature dependence of λ_{ab} was found for NCCO [148–150], expected for unconventional superconductors with a line of nodes in the gap function. However, this point remains controversial [151, 152]. The absence of a ZBCP in NCCO [151, 153, 154] was taken as evidence for s-wave pairing symmetry, despite the strong resemblance between the tunneling characteristics of both types of cuprate superconductors. More evidence against s-wave symmetry comes from the extremely small I_cR_n product (~6μV) observed in c-axis pair tunneling in Pb/NCCO single crystal junctions [155], almost three orders of magnitude smaller than the 3 mV Ambegaokar–Baratoff limit expected for an s-wave superconductor.

As in the case of the hole-doped cuprates, the conflicting results from indirect symmetry studies with NCCO and PCCO emphasize the need for a phase-sensitive test. Tricrystal experiments are made difficult in these materials by the extremely small supercurrent density across the grain boundaries. For example, the J_c for a [100] tilt grain boundary junction with a 30° misorientation in NCCO is about 1A/cm^2 at 4.2 K, five orders of magnitude smaller than for YBCO [156, 157]. This low critical current density leads to a very long Josephson penetration depth, making a half-flux quantum Josephson vortex at the tricrystal point difficult to observe. This difficulty can be circumvented by making thick-film tricrystal samples, provided that sample homogeneity and c-axis epitaxy can be maintained. Pulsed laser-ablated c-axis oriented epitaxial films (~1μm thick) of NCCO (T_c ~22–25 K) and PCCO (T~22–23 K) have been developed for various tricrystal experiments [158]. A J_c of about 10A/cm^2 at 4.2 was achieved with these thick films. Shown in Fig. 16.25a is a SSM image of such a NCCO film deposited on a tricrystal (100) STO substrate with the geometry shown in Fig. 16.12 for testing $d_{x^2-y^2}$ pairing symmetry regardless of whether the junction interface is in the clean or dirty limit [87]. The Josephson vortex centered at the tricrystal point spreads along the grain boundaries for about 50μm, a measure of the Josephson penetration depth λ_J. Such a long λ_J results from the small J_c ($\lambda_J \propto J_c^{-1/2}$). This in turn results in small energy barriers $I_c\Phi_0/\pi$ between the two degenerate half-flux quantum states with fields pointing in to, and out of, the sample. Therefore small disturbances, such as an external field as small as 0.2 mG, or the inductive coupling between the sample and the SQUID pick-up coil, can cause the Josephson vortex at the tricrystal point to flip to the opposite field direction. Under any conditions of cooldown and externally applied field there is always a vortex at the tricrystal point. Figure 16.25a shows a three-dimensional rendering of the SSM images obtained by subtracting the two vortices with field directions pointing into or out of the sample plane and dividing by two. This procedure significantly reduces spurious signals in the SSM data. A detailed analysis of the SSM data indicates that the Josephson vortex at the tricrystal point has total flux $\Phi_0/2$ (Figs. 16.25(b) and (c)) [158]. The value of λ_J from this analysis was 48 μm, corresponding to $J_c = \hbar c^2/8\pi ed\lambda_J^2$ ~ 20A/cm^2 (d=2λ_{ab}=0.5μm), in agreement with a 4-point probe measurement. Tricrystal experiments with NCCO in two other configurations, designed to be unfrustrated for a d-wave superconductor, did not show the half-flux quantum effect. Identical results were obtained in a series of tricrystal experiments with PCCO [158].

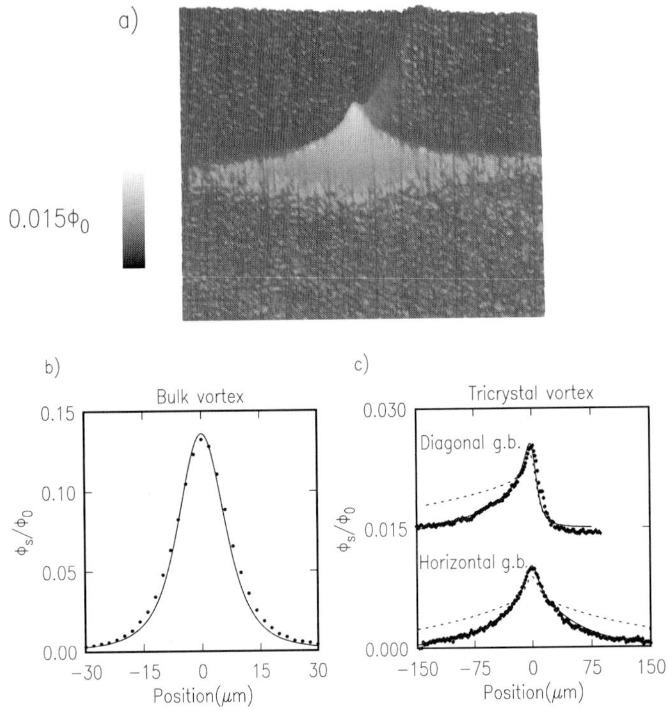

Fig. 16.25. (**a**) Scanning SQUID microscope image of the central area of a tricrystal sample of NCCO in a π-ring geometry for a $d_{x^2-y^2}$ superconductor, cooled and imaged in zero applied field. (**b**) and (**c**) are modeling of a bulk vortex (**b**) and two cross sections through the half-vortex (**c**). The dashed curve in (**c**) is the best fit assuming a total flux of Φ_0 in the Josephson vortex at the tricrystal point. The *solid line*, assuming $\Phi_0/2$ of flux, is a much better fit

In short, the tricrystal experiments described above have provided strong phase-sensitive evidence for d-wave pairing symmetry in the electron-doped superconductors NCCO and PCCO. Recently, there has been supporting evidence for a d-wave pairing state in various electron-doped cuprate superconductors from ARPES measurements [159, 160], temperature dependent penetration depth measurements [149,150,161], measurements of complex conductivity [162], the observation of zero bias conductance peaks by point contact spectroscopy [163], SQUID characteristics [164], and low-temperature specific heat measurements [165].

16.3.5 Thin-Film SQUID Magnetometry

Mathai et al. [106] performed pairing symmetry tests in YBCO by imaging the flux in thin-film YBCO-Pb (sample) SQUIDs using a scanning SQUID microscope with a low-T_c Nb-Pb (sensor) SQUID. In analogy with the experiments of Wollman et al. [104], the sample SQUIDs had either a "corner" or "edge" configuration (see, e.g. Figs. 16.8(a) and (b)): the "corner" SQUIDs should be π-rings if YBCO is a d-wave superconductor, while the "edge" SQUIDs should be 0-rings, independent of the superconducting pairing symmetry. The sample SQUIDs had a self-inductance of about 75pH, and a $\beta_e = 2\pi \mathrm{LI}_c/\Phi_0 \sim 1$. This relatively small β_e value meant that the spontaneous flux and circulating supercurrents at zero external field (see Fig. (16.5)) generated by the π-ring SQUIDs were quite small. Further, there was a large mutual inductance between the sample and sensor SQUIDs. Therefore the influence of the Nb-Pb sensor SQUID on the YBCO-Pb sample SQUID screening currents had to be corrected for in these measurements.

Rather than directly measuring the amplitudes of the spontaneous flux at zero applied field, they measured the superconducting phase shift in their rings by measuring the dependence of the screening currents in the YBCO-Pb sample SQUIDs as a function of externally applied field, tracing out a number of branches, each branch corresponding to a different number of vortices in the YBCO-Pb sample SQUID.

The measurements were then repeated with the leads to the Nb-Pb sensor SQUID and the external magnet reversed, equivalent to a time-reversal operation. If time-reversal symmetry invariance is satisfied, a 0-ring YBCO-Pb sample SQUID in branch n will map under time-reversal onto a branch n' with $\Delta n = n-n'$ an even integer, while a π-ring will have Δn odd. Mathai et al. [106] found that the "corner" SQUIDs had Δn odd while the "edge" SQUIDs had Δn even, as expected for d-wave superconductivity. Further, Δn was close to an integer for both types of sample SQUIDs, implying that time-reversal invariance was closely satisfied. These experiments had the advantages that they directly imaging trapped flux, so that its influence could be avoided; and they allowed a phase shift of π to be measured self-consistently using just the "corner" SQUID, without reference to the "edge" SQUID.

Later work by Gim et al. [166] tested geometries intermediate between the "corner" SQUID and "edge" SQUID in an attempt to determine the momentum dependence of the phase shift in the order parameter. In these SQUIDs one of the junction normals was parallel to the YBCO b axis, while the other was at an angle θ relative to the a axis. If YBCO were a pure d-wave superconductor, these SQUIDs would be expected to have a 0-phase shift for $0< \theta <45^o$, and a π-phase shift for $45^\circ < \theta <90^\circ$. The results of these experiments were complicated by interface roughness introduced by the etching process used. However, the authors concluded that the angular dependence of the phase of the order parameter follows $d_{x^2-y^2}$ symmetry, although it was necessary to understand the details of the junction fabrication process to reach this conclusion.

16.3.6 Spontaneous Magnetization in Facetted Grain Boundaries

Although grain boundaries in cuprate superconductors often act like conventional Josephson weak links, an exception is seen in [001] tilt boundaries in YBCO with very large misorientation angles, in particular in samples with the [100] or [010] axis oriented normal to the interface on one side, with the [110] axis normal to the interface on the other side. Anomalous dependences of the Josephson critical current on applied magnetic field have been observed in such junctions [167, 168]. These anomalies can be understood using a model which combines d-wave superconductivity and facetting of the grain boundary. In this case the sign of the component of the gap function normal to the local boundary line can alternate from facet to facet, producing local π-rings. In such a model the Josephson currents from adjacent facets tend to cancel, reducing or eliminating the normally seen maximum in the junction critical current at zero applied magnetic field, in agreement with experiment. An additional implication of this model is that magnetic fields should be spontaneously generated in the grain boundaries, which act as a linear array of π-rings. This is in fact what is observed [169]. An example of this effect is shown in Fig. 16.26. The upper panel shows a SQUID microscopy image of a 0-45° [001] tilt grain boundary in YBCO. There are 4 vortices in the grains to the right of the image, but for the most part the image is dominated by spontaneous magnetization, of both signs, in the grain boundary. The lower panel shows a cross-section through the image along the grain boundary. The presence of this spontaneous magnetization in these grain boundaries, and not in samples with lower misorientation angles, provides additional strong evidence for d-wave superconductivity.

Mints and co-workers [170–173] have proposed that a linear array of π-rings with spacing smaller than a local Josephson penetration depth (given by $\lambda_{\text{loc}} = (c\Phi_0/16\pi^2\lambda_L <| j_c(x) |>)^{1/2}$, where λ_L is the London penetration depth of the grains, and $<| j_c(x) |>$ is an average of the absolute values of the critical current densities along the facet) will exhibit pairs of Josephson vortices, each with flux different from Φ_0, but with their fluxes summing to Φ_0. The technology for producing junctions with controlled 0 and π-facets, and SQUIDs with controlled 0 and π-junctions is now being developed [174–176].

16.3.7 c-Axis Josephson Tunneling

The phase-sensitive pairing symmetry tests described above have all made use of pair tunneling

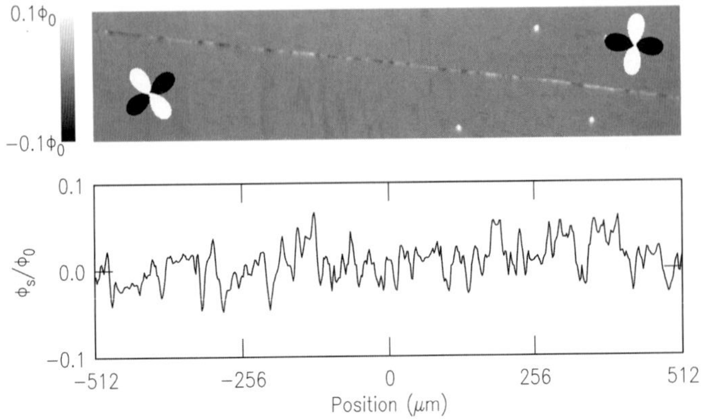

Fig. 16.26. SQUID microscope image of an asymmetric 0–45deg YBCO [001] tilt grain boundary, cooled and imaged in nominal zero field with a 4 μm diameter SQUID pickup loop. The polar plots indicate the orientation of an assumed $d_{x^2-y^2}$ gap function on the two sides of the grain boundary

parallel to the CuO_2 planes. Several interesting experiments have exploited Josephson tunneling along the c-axis, perpendicular to the CuO_2 planes. As in basal-plane Josephson tunneling, the magnitude of the c-axis supercurrent can be estimated from (16.6). There should be no c-axis pair tunneling between an s-wave conventional superconductor and a pure d-wave cuprate superconductor due to an exact cancelation of the contributions from the positive and negative lobes of the d-wave gap function. A finite critical current (I_c) is expected if the cuprate junction electrode is a $d+s$-mixed wave superconductor. As a reference point, the value of I_c at $T = 0$ K for a tunnel junction between two identical s-wave superconductors is given by [177]:

$$I_c = \frac{\pi \Delta_1 \Delta_2}{eR_n(\Delta_1 + \Delta_2)} K\left(\frac{\Delta_1 - \Delta_2}{\Delta_1 + \Delta_2}\right), \qquad (16.33)$$

where R_n is the normal-state junction resistance, Δ_1 and Δ_2 are the BCS energy gaps of the junction electrodes, and K is the complete elliptic integral.

The observation of a non-zero c-axis supercurrent in a Pb/YBCO tunnel junction suggests the existence of an s-wave component of the order parameter in YBCO. Since second order tunneling [178] has been ruled out by the absence of half-integer microwave induced Shapiro steps in such c-axis tunneling experiments [179], the ratio of the experimentally determined value of I_cR_n to that predicted by (16.33) can serve as a measure of the degree of the $d+s$ admixture in the pair state.

Effects of Twinning

Most of the c-axis tunneling experiments were done with orthorhombic cuprates such as YBCO and Bi-2212, which are usually twinned due to the in-plane anisotropy in the CuO_2 layers, unless they are specially treated. For example, the twin boundaries in YBCO run at 45° with respect to the major crystallographic axes a and b. The two crystal lattice domains on each side of a twin boundary are related via a mirror reflection symmetry (see Fig. 16.27). In polycrystalline YBCO, the spacing between parallel twins can be as small as 10 nm [34,180]. Most of the YBCO samples studied in phase-sensitive pairing symmetry experiments were also twinned. However, the fact that strong evidence for d-wave order parameter symmetry has been invariably observed in phase-sensitive experiments with a tetragonal cuprate superconductor such as Tl-2201 [130,181] and untwinned YBCO samples [104,110] make the complications arising from twinning unimportant to the basic conclusion of predominantly d-wave pairing symmetry in the cuprate superconductors. On the other hand, twinning is expected to play a crucial role in c-axis Josephson tunneling, especially in determining the value of I_cR_n. To understand the effect of twinning on these experiments in such heavily twinned samples, it is extremely important to know how the order parameter varies across a single twin boundary in a YBCO single crystal. There are two alternatives in YBCO [5,182,183]: (1) The d-wave component of the order parameter changes sign across the twin boundary,

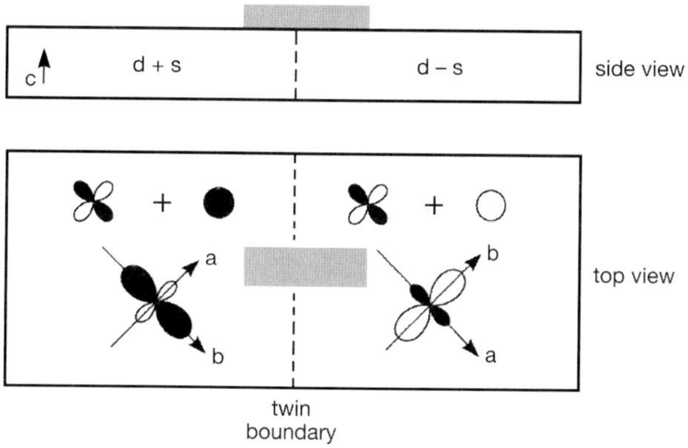

Fig. 16.27. Experimental geometry for c-axis tunneling into a conventional superconductor (Pb) from single crystal YBCO, with the junction spanning a single twin boundary. YBCO is an orthorhombic superconductor with an admixture of d-wave and s-wave pairing symmetries. The diagram shows the $d_{x^2-y^2}$ components with the same phase across the twin boundary, but the s-wave component with opposite signs across the twin boundary. There is therefore a change in sign of the net c-axis coupling across the twin boundary. Adapted from Kouznetsov et al. [184]

whereas the sign of the s-wave component remains unchanged; or (2) the converse. The interpretation of the symmetry experiments with twinned YBCO single crystals or films depends on which alternative exists. The fact that a d-wave signature has been consistently and reproducibly observed in both orthorhombic and tetragonal cuprates argues strongly that the twin boundaries in YBCO have odd reflection symmetry [182]. The d-wave component of the order parameter continues across the twin boundary whereas the s-wave counterpart changes its sign. An even-reflection twin boundary corresponds to a predominantly s-wave pair state, which is not observed in any of the YBCO experiments.

Kouznetsov et al. [184], demonstrated the sign reversal of the s-wave component across the twin boundary in a study of a c-axis Josephson tunnel junction that straddled a single YBCO twin boundary. In Fig. 16.27 the polar plots of the gap function in the twins (top view) represent the mirror reflection symmetry across the twin boundary (dotted line). Only the s-wave component changes its sign across the twin boundary. As a result, the signs of the c-axis supercurrent switch across the twin boundary. This is analogous to the Pb/YBCO corner junction (Wollman et al. [104, 105]. When a magnetic field B is applied parallel to the twin boundary, the field variation of the critical current I_c(B) should show a Fraunhofer pattern with the maximum critical current peak shifted to the field value corresponding to a half-flux quantum. Furthermore, as the angle ϕ between the direction of B and the twin boundary varies from 0 to 90°, the Fraunhofer pattern I_c(B) should gradually recover its standard form. This is as observed (Fig. 16.28). These results provide strong evidence for a d-wave dominant (d+s) mixed pairing state in YBCO, and for a sign reversal of the s-wave component across the twin boundary. This finding is consistent with the conclusions drawn from all the phase-sensitive pairing symmetry experiments presented in the previous sections.

I_cR_n-Products

As discussed earlier, the I_cR_n-product of a c-axis Josephson junction can in principle be used as a measure of the strength of the s-wave component of the $d + s$ mixed order parameter. In practice, such a determination is unreliable since I_cR_n products of Pb/YBCO c-axis tunnel junctions vary from ~1 mV to a few μ V. For example, the I_cR_n products are 1-1.3 mV for untwinned YBCO crystals [179]; 0.3–0.9 mV [185] or 0.1–0.3mV [179] for twinned crystals; and ~10μV for heavily twinned thin films [179, 186, 187]. Microstructural defects at junction interfaces such as atomic steps and etch pits could give rise to ab plane tunneling and contribute a random quantity to the apparent c-axis tunneling current. A systematic study of the direction of tunneling in various Pb/YBCO tunnel junctions seems to have ruled out such a possibility. The observed large variation can be due to twinning. The c-axis tunneling experiments of Kouznetsov et al. [184] has clearly demonstrated that the d-wave component is dominant and contin-

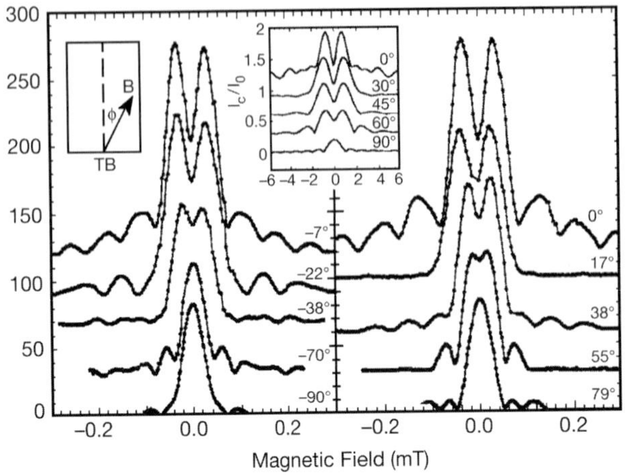

Fig. 16.28. Josephson interference patterns from the sample of Fig. 16.27, with the magnetic field in the plane of the junction, as a function of field angle relative to the twin boundary. These characteristics are consistent with the pairing phases as diagrammed in Fig. 16.27. The *inset* is modeling. Adapted from Kouznetsov et al. [184]

uous across the twin boundary in YBCO, whereas the *s*-wave component changes its sign. Thus the magnitude of the I_cR_n products in twinned YBCO depends sensitively on the size and relative distribution of the $d + s$ and $d - s$ domains, as suggested by several theoretical studies [80, 182, 188]. Indeed, this is consistent with the observation that I_cR_n products in untwinned samples are invariably larger than those of their untwinned counterparts. The nearly-zero I_cR_n products observed in heavily twinned thin film samples apparently can be attributed to the cancelation arising from the nearly equal abundance of the two types of twin domains. Obviously, one needs precise knowledge of the relative population of the twin domains to quantitatively understand the large scatter in the I_cR_n data. Only data from untwinned junctions is reliable for an estimate of the strength of the *s*-wave pairing symmetry in YBCO.

A comparison of I_cR_n values from untwinned junctions with the Ambegaokar-Baratoff limit (16.33) provides an estimate of as much as 20% *s*-wave component in the $d + s$ admixture for YBCO. However, estimates from other measurements are significantly smaller. For example, an upper bound limit of 10% *s*-wave was concluded from a study of the angular magnetic dependence of thermal conductivity in a detwinned YBCO single crystal [68]. A quantitative data analysis of directional tunneling and Andreev-reflection measurements on YBCO single crystals indicated a $d_{x^2-y^2}$-wave dominant gap parameter with less than 5% *s*-wave component in either the $d + s$ or $d + is$ scenarios [189, 190].

Vanishingly small I_cR_n products ($\sim$ a few μV) have been observed in *c*-axis tunneling between Pb and Bi-2212 [137] or the electron-doped NCCO [155] single crystals. A $d + s$ mixed pair state is not allowed in tetragonal NCCO or orthorhombic Bi2212, based on group theory symmetry arguments (see Sects. 16.1.2 and 16.3.3). Phase-sensitive evidence from tricrystal experiments (see Sect. 16.3.3) with Bi-2212 [191] and NCCO [158] has shown that both cuprates are *d*-wave superconductors. Thus it is logical to conclude that the small observed I_cR_n product is derived from a small *s*-wave component induced at twin-boundary or junction interfaces, as suggested by several theoretical studies [192, 193].

c-Axis Twist Junctions

In principle, one can test the order parameter symmetry by measuring the angular dependence of the critical current $I_c(\phi)$ of (001) basal-plane grain boundary junctions between two identical cuprate superconductors with a twist angle ϕ about the *c*-axis direction (see the inset in Fig. 16.29). If the in-plane momentum is conserved in the *c*-axis tunneling, (16.6) leads, for the case of a pure *d*-wave superconductor, to

$$I_c(\phi) = K(\phi)\Delta_0^2 cos2\phi \,, \tag{16.34}$$

where $K(\phi)$ is proportional to the ϕ-dependent tunneling matrix element squared, and Δ_0 is the value of the d-wave gap in the a, b directions. The parameter $K(\phi)$ is not readily determined, but (16.34) predicts that $I_c \equiv 0$ for $\phi = \pi/4$ for $d_{x^2-y^2}$-wave pairing symmetry. There have been several theoretical studies [194–203] and experiments [204,205] which emphasized the importance of including incoherent (momentum non-conserving) tunneling in the calculation of $I_c(\phi)$. It has been argued that the condition $I_c(\phi = \pi/4) = 0$ should always be satisfied for a d-wave pair state, regardless of the amount of coherence in the tunneling processes [203].

The critical currents of c-axis twist junctions have been measured in the Bi-2212 system. Although early measurements reported a strong dependence of I_c on ϕ [207,208], the results of one group remarkably showed no angular dependence [209–211]. This has been taken as evidence against d-wave pairing symmetry in Bi-2212. However, these experiments were done with macroscopic samples with junction areas $\sim 10^5\,\mu m^2$, and critical currents of 50-150mA [210]. In such experiments the c-axis tunneling current can be limited by local heating, self-field, and bulk critical current effects, masking the intrinsic ϕ-dependence of I_c. To clarify these issues, a recent experiment by Tachiki et al. [206] has repeated the previous experiment [210], with Bi-2212 single crystal whiskers. The Bi-2212 whisker twist junctions have a junction area of about $10^3\,\mu m^2$ and critical currents one to two orders of magnitude smaller than in the experiments of Li et al. [210]. As shown in Fig. 16.29, $I_c(\phi)$ is clearly a strong function of the twist angle ϕ, approaching zero at $\phi = \pi/4$. In sharp contrast with the results obtained in a previous experiment [210], the vanishingly small I_cR_n value observed at $\phi = \pi/4$ is particularly interesting in view of the theoretical prediction that I_cR_c should be zero for d-wave and finite with no significant angular dependence near $\phi = \pi/4$ is expected for the case of an anisotropic s-wave pair state [203]. Shown as a dotted line in Fig. 16.29 is the $|\, cos2\phi\, |$ dependence expected for a pure d-wave superconductor with no dependence of the tunneling probabilities on ϕ. In-plane tilt grain boundary junctions in the cuprates show an exponential dependence of the critical current density on misorientation angle $J_c(\Theta) = C_i e^{-\Theta/\Theta_i}$ with $\Theta_i \sim 5°$ for the hole-doped cuprates, and $\Theta_i \sim 2°$ for the electron-doped cuprates [156,157,212]. It is believed that this exponential dependence is dominated by the strong dependence of the tunneling matrix elements on misorientation angle, due to microstructural disorder effects [77,213]. The dashed line in Fig. 16.29 shows for comparison a fit of the data to an exponential dependence of the critical current on twist angle $I_c \sim e^{(\phi-90)/\delta\phi}$, with $\delta\phi = 9.3°$. The good agreement between this exponential dependence and experiment makes it appear likely that the observed angular dependence of I_c can be explained in terms of Fermi surface or ϕ-dependent tunneling matrix element effects [203], in a d-wave superconductor, as opposed to an anisotropic s-wave superconductor.

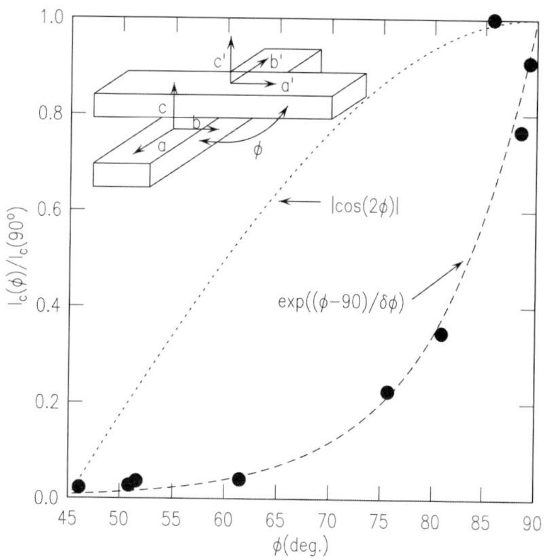

Fig. 16.29. Measured dependence of the critical current of c-axis twist grain boundary junctions between single crystal whiskers of Bi-2212 (*solid symbols*) [206]. The functional forms $|\, cos2\phi\, |$ (*dotted line*) and $e^{\phi-90/\delta\phi}$ (*dashed line*) are shown for comparison

16.4 Angle-Resolved Determination of Gap Anisotropy in YBCO

It is clear from the work described so far that an admixture of $d_{x^2-y^2}$ and s-wave order parameters is well established in orthorhombic YBCO. Such an anisotropic $d+s$ pair state is required by the group

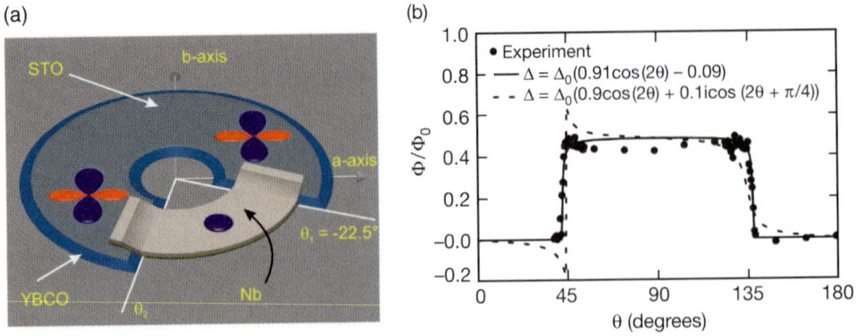

Fig. 16.30. (**a**) Schematic of two-junction YBCO/Nb rings used for angle-resolved determination of the gap symmetry in optimally doped YBCO [214]. One ramp edge junction angle was held fixed from ring to ring relative to the YBCO crystalline *a* or *b* axis while the other was varied. (**b**) Integrated spontaneous flux generated by these rings, as determined by scanning SQUID microscopy, for a sample with the first junction normal parallel to the majority twin *a*-axis direction, as a function of the second junction normal angle (θ) relative to this direction. The *solid points* are the data; the *solid line* is a best fit using a predominant $d_{x^2-y^2}$ component to the gap with a small *s* component. The *dashed line* shows much worse fit results when a small imaginary component is added to the gap functional form

theoretic argument discussed in Sects. 16.1.2 and 16.3.3. The size of this gap anisotropy was measured recently by locating the nodes in the in-plane gap function for optimally doped YBCO [214] by measuring the spontaneously generated magnetic flux in a set of 0 or π-rings with varying junction angles. This work was based on an experimental configuration originally suggested for testing for time-reversal symmetry breaking (TRSB) in cuprate superconductors [215,216]. A set of two-junction YBCO/Nb rings were fabricated with one junction normal angle fixed relative to a YBCO *a* or *b* crystallographic axis while the second junction normal angle was rotated with respect to the first in 5 degree or 0.5 degree increments (see Fig. 16.30) from ring to ring. Arrays of such thin film YBCO/Nb two-junction rings were fabricated using a ramp-type Josephson junction technology [217]. The magnetic flux state of these rings, cooled in various magnetic fields, were imaged at 4.2 K with a scanning SQUID microscope. The spontaneously generated flux Φ in the rings, when cooled in zero field, alternated systematically as a function of the second junction angle between nearly zero flux ($N = 0$) and the half-flux quantum state $\Phi = \Phi_0/2$ ($N = 1/2$) in a manner consistent with a predominantly $d_{x^2-y^2}$ pairing symmetry. The transitions between the $N = 0$ and $N = 1/2$ states occur at angles slightly different from the (2m+1)45° expected for a pure *d*-wave superconductor, and is in fact consistent with an s/d ratio of 0.1 ± 0.02 (Fig. 16.30). A three-dimensional rendering of the SSM images of the rings is presented in Fig. 16.31, along with a polar plot of the corresponding anisotropic $d + s$ pairing function. A s/d ratio of 0.1 corresponds to a gap magnitude along the *b*-axis (along the Cu-O chains) direction about 20% larger than in the *a*-axis direction. This is somewhat smaller than obtained from measurements of the Josephson critical current density as a function of junction angle in single junctions produced using the same technology, [218] perhaps because of incomplete detwinning of the YBCO films used in the ring samples.

16.5 Universality of the *d*-Wave Pair State

In two-dimensional strongly correlated electron systems, especially near the quantum critical point, phases and phase-transitions abound due to the competition between various ordered states [219]. The pairing symmetry tests presented so far have firmly established that the ground state of several optimally-doped cuprates superconductors has an order parameter with $d_{x^2-y^2}$ pairing symmetry. However, there are many theoretical studies that suggest

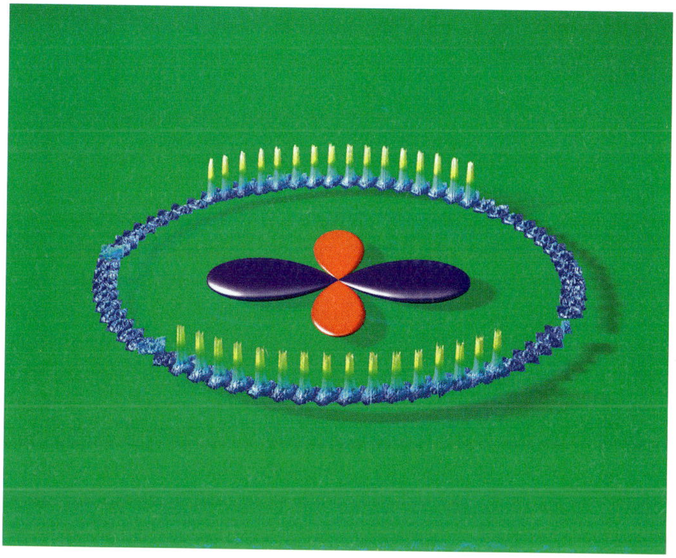

Fig. 16.31. Three-dimensional rendering of scanning SQUID microscope images of 72 YBCO/Nb two-junction rings with one junction normal angle fixed at -22.5° relative to the majority twin b-axis and the second junction normal angle varied in 5° intervals [214]. The rings were cooled and imaged in zero field. This resulted in rings in either the $N = 0$ or $N = 1/2$ flux quantum state, with the transitions at angles slightly different from $(2m+1)\,45^\circ$, m an integer, because of a small s component added to the predominantly $d_{x^2-y^2}$ gap functional form in optimally doped YBCO

that the d-wave pair state is not stable against various competing pairing symmetries, depending on variations in doping, temperature, and time-reversal symmetry breaking (TRSB) at surfaces and interfaces [220–229]. It is therefore important to experimentally determine whether the d-wave state is robust to a change in these variables. These universality issues can be investigate by doing the phase-sensitive symmetry tests with various cuprates systems at optimal doping, and for a given cuprate system as functions of composition and temperature. Possible evidence for TRSB at the surfaces and interfaces of d-wave superconductors should also be examined.

16.5.1 Optimally-Doped Cuprates

The bulk of pairing symmetry tests using phase-sensitive methods was done in various cuprate systems at optimal doping. A complete list of the cuprate superconductors tested (with references) can be found in a recent review [7].

Hole-doped Cuprates

Evidence for predominantly d-wave pairing has been found in all the hole-doped cuprate superconductors successfully tested using phase-sensitive methods, including YBCO [87, 104–106, 110, 117], $GdBa_2Cu_3O_{7-\delta}$ [230], Tl-2201 [130, 181], and Bi-2212 [191]. Given a choice between the s-wave and d-wave pairing channels, a number of numerical studies have concluded that gap symmetry depends on band filling and certain band parameters. Based on the t-t'-J model, with $t'/t \geq -0.4$, the d-wave pair state prevails in a doping range of the band centered around half-filling, and s-wave pairing symmetry is more stable in the under- and over-doped regimes [220, 222, 225]. The phase-sensitive evidence obtained so far is generally consistent with these conclusions: the d-wave channel is favored near half-filling where T_c is optimized.

Electron-doped Cuprates

As described in Sect. 16.3.4, strong evidence has recently been found for $d_{x^2-y^2}$ symmetry in the T_c optimized electron-doped cuprate superconductors NCCO and PCCO from both tricrystal experiments and conventional symmetry probing measurements. It is quite remarkable that the d-wave pair state predominates in both the electron-doped and hole-doped cuprate superconductors in spite of their significant differences in many of the superconducting and normal state properties. On the other hands, the observed universal d-wave pairing in cuprate superconductors is not unexpected, because the basic theoretical models for describing high-temperature superconductivity, such as the one-band Hubbard model and its large-U limit, the t-J model, are all intrinsically electron-hole symmetrical [231]. Electron-hole asymmetry can be incorporated into various $t - t' - J$ models [232,233] and three-band Hubbard models [234]. For example, the differences in doping characteristics between the electron- and hole-doped cuprates can be explained in terms of band structure effects based on a $t - t' - J$ model [233] or a three-band Hubbard model [235]. However, this feat and more can be accomplished by taking into account the different dispersions for the two types of charge carriers, and by using an effective one-band Hubbard model [236]. Therefore, it appears that the apparent electron-hole asymmetry, as manifested in many anomalous properties, stems largely from band structure effects and is not relevant in determining pairing symmetry. This finding and the small in-plane coherence length (∼20Å) characteristic to all cuprate superconductors, suggests that the universal d-wave pairing occurs on a relatively local scale.

16.5.2 Extended *s*-Wave Pair State

As mentioned in Sect. 16.2, there has been an enormous amount of work on tests of the pairing symmetry in the cuprate superconductors using various phase insensitive techniques [4,5,7]. Although this work has resulted in a general consensus that the pairing wavefunction has lines of nodes, these techniques, with the exception of angle resolved photoemission, are relatively insensitive to the momentum dependence of the nodal structure. This makes it difficult for them to distinguish, for example, between $d_{x^2-y^2}$ symmetry (with four lines of nodes), and extended s-wave symmetry (with eight lines of nodes). There has recently been support for the latter symmetry from analysis of various phase insensitive tests of pairing symmetry [237,238]. These authors attempt to reconcile this conclusion with the phase-sensitive experiments, which definitively favor $d_{x^y-y^2}$ symmetry, by arguing that the cuprates can have different bulk and surface symmetries [192]. In this view the phase-insensitive techniques probe the bulk extended s-wave symmetry, while the phase-sensitive techniques, which depend on surface sensitive Josephson tunneling, probe d-wave pairing symmetry.

This view seems untenable for several reasons. First, the consistency of the tricrystal experiments [7], both in that samples with a π-ring geometry for a d-wave superconductor always show the half-flux quantum effect, and that the half-flux quantum effect can be turned on and off by changing the tricrystal geometry in a way that is consistent with $d_{x^2-y^2}$ pairing symmetry, make it appear extremely unlikely that these experiments are controlled by surface effects. Second, the tricrystal experiments have shown that the half-flux quantum effect persists, with no change in the total flux within experimental accuracy, to within a degree of T_c [239]. At these temperatures the *ab* plane coherence length can be ten times the unit cell dimension, meaning that these experiments are not surface sensitive but rather probe the order parameter deeply into the bulk: nevertheless the results strongly favor d-wave pairing symmetry. Further, although the $I_c R_n$ product for the grain boundary junctions used in the tricrystal experiments are relatively small, experiments with π-SQUIDs and facetted π-junctions in YBCO-Nb ramp junctions produce convincing evidence for d-wave symmetry with junction $I_c R_n$ products within a factor of four of the Ambegaokar–Baratoff limit [174–176]. This means that the superconducting gap is not significantly reduced by surface effects in these samples.

The tetracrystal experiments [130] have provided strong evidence for pure d-wave pairing symmetry in tetragonal Tl-2201. It also rules out anisotropic s-wave pairing symmetry. The tetragonal crystal structure of the Tl-2201 epitaxial films used in this experiment were checked and confirmed with three techniques for microstructural analysis [130]. It is well established that Tl-2201 can be fabricated with either orthorhombic or tetragonal crystal structure with identical superconducting properties (including T_c ∼80 K) [240–242]. Furthermore, it has been shown that there is no temperature-dependent structural phase transition in the temperature range

50(well below T_c)–300 K in either form of Tl-2201. This observation, as well as the results of neutron scattering and x-ray diffraction, leads to the conclusion that the orthorhombic to tetragonal structure change is due to static compositional effects, and is not driven by any thermal structural instability.

The combination of the observations above lead us to conclude with confidence that the pairing symmetry in the cuprate superconductors has $d_{x^2-y^2}$ symmetry, and not the extended s-wave symmetry.

16.5.3 The Effect of Doping

The competition between pairing symmetries in various superconducting cuprate systems as a function of doping has been emphasized repeatedly [219,220,222,223,225,243–246]. In particular, a transition from a d to an s-wave pair state induced by a change in doping has been suggested. For references, see Sect. 16.4. The tricrystal pairing symmetry test is especially well adapted for a systematic study of the effect of doping on pairing symmetry.

Recently, we have carried out a series of tricrystal experiments with a c-axis oriented Bi-2212 tricrystal thin-film sample as a function of doping [247]. The systematic pairing symmetry study started with an over-doped Bi-2212 tricrystal sample with a T_c of 69 K. The doping level was then reduced by annealing the sample in flowing Argon at temperatures between 350 C and 420 C for various lengths of time to achieve an optimum T_c of 82 K and a T_c of 40 K in the under-doped regime. The half-flux quantum effect at the tricrystal point was monitored as a function of such annealing. The d-wave pairing symmetry was found to persist throughout the entire doping range studied so far. There are also similar phase-sensitive evidences for d-wave pairing symmetry in both the under-doped and over-doped Tl-2201 samples, overdoped $Y_{0.7}Ca_{0.3}Ba_2Cu_3O_{7-\delta}$, and underdoped $La_{1.85}Sr_{0.15}CuO_y$ [247]. A possible $d_{x^2-y^2}$ to $d_{x^2-y^2}+s$ symmetry transition in Ca-doped YBCO has been reported [248].

In addition to the tricrystal experiments, measurements of low-temperature thermal conductivity [249,250] have provided strong evidence in support of a d-wave pair state that prevails throughout the entire phase diagram from underdoped to overdoped regimes for several cuprate systems.

16.5.4 Temperature Dependence

In view of the possible changes in pairing symmetry as a function of temperature [222,223,225,251–253], it is important to experimentally determine the temperature dependence of the pairing symmetry in a d-wave superconductor such as YBCO. It is difficult to study the temperature variation of the pairing symmetry in an unconventional superconductor using conventional techniques, since they become obscured by thermal excitations at higher temperatures. Phase-sensitive pairing symmetry probes that employ Josephson pair tunneling are in principle immune to thermal broadening. In practice most of these techniques also fail at higher temperatures because they depend on tunneling into low-T_c superconductors [104–106, 110]. Only phase-sensitive symmetry tests, such as the tricrystal experiments, which use all-cuprate grain boundary Josephson junctions can be used for determining pairing symmetry at all temperatures below T_c. Kirtley et al. [239] used a variable sample temperature scanning SQUID microscope [254] to measure the total spontaneous flux at the tricrystal point of a YBCO thin film in a frustrated tricrystal geometry as a function of temperature. With this microscope the tricrystal sample was heated to different temperatures while the SQUID sensor was kept below its superconducting temperature of ~9 K. Shown in Fig. 16.32 are scanning SQUID microscopy images of the d-wave signature, the half-flux quantum, at the tricrystal point at selected temperatures.

These experiments showed that the half-integer flux quantum effect at the tricrystal meeting point persists from $T = 0.5$ K through T_c (~ 90 K). Careful modeling (Fig. 16.33) showed that the total flux in the half-vortex was equal to $\Phi_0/2$ to within experimental errors at all temperatures. This implies that d-wave pairing symmetry dominates in this cuprate, with no observable time-reversal symmetry breaking, over the entire temperature range.

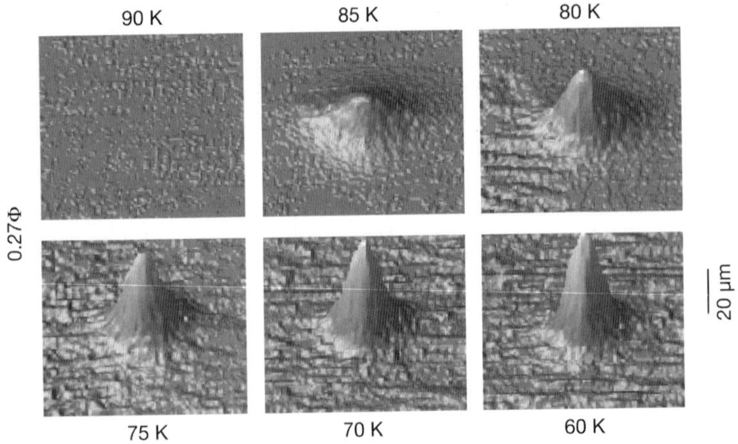

Fig. 16.32. Scanning SQUID microscope images of the development with temperature of the half-flux quantum vortex at the tricrystal point for a thin film sample of YBCO in a geometry designed to be a π-ring for a d-wave superconductor

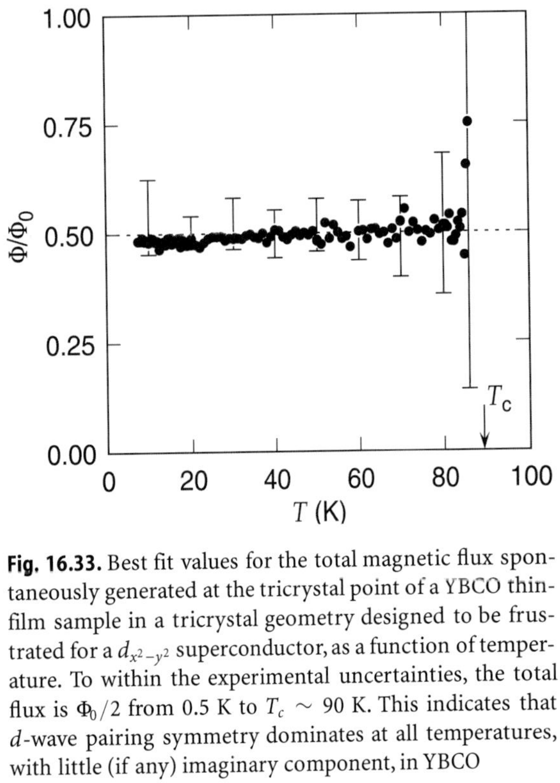

Fig. 16.33. Best fit values for the total magnetic flux spontaneously generated at the tricrystal point of a YBCO thin-film sample in a tricrystal geometry designed to be frustrated for a $d_{x^2-y^2}$ superconductor, as a function of temperature. To within the experimental uncertainties, the total flux is $\Phi_0/2$ from 0.5 K to $T_c \sim 90$ K. This indicates that d-wave pairing symmetry dominates at all temperatures, with little (if any) imaginary component, in YBCO

16.5.5 Time-Reversal Symmetry Breaking

It has been suggested that as temperature decreases below a certain characteristic value, a pure d-wave pair state is not stable against the formation of mixed pair states such as $d_{x^2-y^2} + is$ or $d_{x^2-y^2} + id_{xy}$ [129, 251, 255–261]. Due to the coupling between the two subcomponents of the order parameter, the mixed pair state is fully-gapped, much like that in an s-wave superconductor [251, 257]. The power-law temperature dependence of observables such as specific heat, thermal conductivity, and penetration depth characteristic of d-wave superconductors reverts to the exponential behavior typical for s-wave superconductors. Measurements of the penetration depth of Zn substituted LSCO [262] shows that it retains a T^2 dependence, indicative of d-wave pairing in the presence of scattering, until T_c is suppressed to zero. This argues against an admixture of s-wave pairing at any T_c. There has been a report of a temperature-dependent time reversal symmetry broken pair state (i.e. $d_{x^2-y^2} + is$ state) in YBCO near T_c [263]. Furthermore, there is thermal transport evidence suggesting that a phase transition from a pure $d_{x^2-y^2}$-wave state into time-reversal symmetry broken states such as $d + is$ or $d_{x^2-y^2} + id_{xy}$ can be induced in Bi-2212 at low temperatures by magnetic field [264] or by magnetic impurities [265]. However, there are new experiments [266–268] and calculations [269] indicating the interpretation of the original experiments are much more involved.

Further, it has been shown theoretically that broken time-reversal symmetry could occur locally in a $d_{x^2-y^2}$ superconductor at certain surfaces and interfaces, or in the presence of magnetic impurities [270–277]. Evidence for broken time reversal symmetry in YBCO has been observed from the splitting of the zero bias conductance peak (ZBCP) induced by the Andreev surface bound states [154,278,279], and by a small spontaneous magnetization at the superconducting transition [280].

A ZBCP in the quasiparticle tunneling spectrum, due to the formation of zero-energy Andreev-bound states associated with sign changes across the nodes of the $\boldsymbol{k}$ dependent order parameter, can be used as a signature for d-wave pairing symmetry [57,58,281,282]. A split in the ZBCP is expected in the presence of TRSB in mixed pair states such as $d_{x^2-y^2} + is$ or $d_{x^2-y^2} + id_{xy}$ [283]. Recent theoretical treatments have pointed out that there are possibilities other than TRSB that can lead to ZBCP splitting in pure d-wave superconductors [284,285].

In the tricrystal experiments with YBCO [119] and Tl-2201 [181], the applied flux needed to nullify the SQUID difference signal in the three-junction ring was always found to be within 3% of $\Phi_0/2$. Tricrystal experiments with electron-doped NCCO and PCCO at 4.2 [158] and YBCO at temperatures close to T_c [239] have all demonstrated that the total amount of flux spontaneously generated at the tricrystal point remains the same (within the experimental accuracy of $\sim$ 3%), regardless of which of the two time-reversed half-flux quantum states the tricrystal sample is in. All these observations imply that the gauge-invariant phase shift as defined in (16.5) is very close to π, in turn suggesting that any out-of-phase component of the order parameter must have an amplitude less than a few percent of the primary component. Such a mixed pair state (e.g. $d + is$) would lead to a phase shift $0 < \phi_J < \pi$, and also a fractional flux quantum $x\Phi_0$ with $0 < x < 1/2$ [80,193].

The observation of total flux different from $\Phi_0/2$ in the multi-crystal experiments in a ring geometry would not necessarily imply broken time reversal symmetry. As seen in Fig. 16.5, if the LI_c product is not sufficiently large compared with Φ_0, then the total flux in a ring geometry will be less than $\Phi_0/2$. For example, in a tricrystal experiment with Tl-2201 [181], in which the I_cL product was low, and the amount of the spontaneously generated flux through the ring was indeed noticeably smaller than $\frac{1}{2}\Phi_0$ ($x \sim 0.3$). However, multi-crystal experiments in a blanket film geometry have always been done with the sample size much larger than the Josephson penetration depth λ_J. In this case, the total flux in the Josephson vortex at the multi-crystal point in a frustrated geometry should be exactly $\Phi_0/2$. Mathai et al. [106] estimated that the time-reversal symmetry breaking component of the pairing order parameter was less than 5%, in agreement with the conclusions by Kirtley et al. [119] from "magnetic titration" of a tricrystal ring. Recent studies on the magnetic field modulation of critical currents in grain boundary junctions [286] and π-SQUIDs [176] have produced similar evidence against TRSB at the junction interface.

Tafuri and Kirtley [287] have found evidence for spontaneous magnetization, associated with the superconducting transition, that appears to be correlated with defects in some c-axis oriented YBCO thin films. The association of spontaneous magnetization with defects could help to explain why evidence for time-reversal symmetry breaking is seen in some samples, using some techniques, but not in others.

16.6 Implications of *d*-Wave Pairing Symmetry

The implications of $d_{x^2-y^2}$-wave pairing symmetry for understanding HTS and for its applications are numerous. Due to space limitations, we shall briefly discuss only some of the topics here.

16.6.1 Universal *d*-Wave Pair State in Cuprates

The phase-sensitive experiments, as presented in Sects. 16.3 and 16.4 above, augmented with several conventional order parameter symmetry probing techniques, have established the predominance of $d_{x^2-y^2}$-wave pairing symmetry in both the hole- and electron-doped cuprate superconductors. The tricrystal and tetracrystal experiments have also provided phase-sensitive evidence, based on group-theoretic symmetry arguments, for the existence of a pure d-wave state in Bi-2212 and Tl-2201 (Sect. 16.3.3). Furthermore, a series of tricrystal experiments have also demonstrated that the d-wave pair state in several high-T_c cuprate superconductors persists in spite of deviations from optimal doping (Sect. 16.4), and variation in temperature over a wide range in YBCO (Sect. 16.4). The results of several phase-sensitive experiments (Sect. 16.4) also consistently suggest that a subcomponent of the d-wave order parameter in the form of $d_{x^2-y^2} + is$ or $d_{x^2-y^2} + id_{xy}$, if it exists, is very small (less than a few percent at most). The robustness of d-wave pairing symmetry has been demonstrated also on the surface of a Bi-2212 single crystal. In a recent c-axis tunneling study of stacks of intrinsic Josephson junctions formed on Bi-2212 single crystals, the quasiparticle branch of the I-V characteristics indicate that the surface layer has a suppressed T_c, but its d-wave parameter symmetry remains intact [288]. In short, the experimental evidence is that the d-wave pair state is remarkably ubiquitous and robust, despite the fragility suggested in several theoretical studies. Such universal characteristics foretell that the d-wave pairing in cuprates has an intrinsic common origin.

It is well established by theory [221, 289] and experiment [290, 291] that two-dimensional strong Coulomb correlations in the CuO_2 layers play a fundamental role in the physics governing the superconducting and normal-state properties of the cuprates. All cuprates are characterized by a strong on-site Coulomb repulsion which results in the universally observed Mott metal-insulator transition at near-zero doping. The strong on-site Coulomb repulsion rules out simple s-wave pairing symmetry. For pairing interactions of range up to next nearest neighbors, a strong Coulomb repulsion also rules out all pair state candidates (see Sect. 16.1.2) other than $d_{x^2-y^2}$ pairing in models based on a van Hove singularity in the 2D band structure of the CuO_2 planes [292]. For related work, see also [293–295]. Robust pairing correlations in the d-wave channel are found in a variety of theoretical models treating pairing between holes moving in a background of antiferromagnetic fluctuations [296–299]. Of particular interest, long-range Coulomb repulsion interaction is found to enhance d-wave pairing [300] and to produce d-wave pairing even in the presence of charge stripes [301]. These works have provided a basis for understanding why d-wave pairing persists in the under-doped regime. In short, the pervasive nature of the d-wave superconductivity in cuprates is a natural consequence of strong Coulomb correlations in the CuO_2 planes.

16.6.2 Constraints on Pairing Interaction

Although the establishment of d-wave pairing symmetry in high-T_c cuprate superconductors does not necessarily specify a particular high-T_c mechanism, it does impose strong constraints on possible models for the high-T_c mechanism. The pairing interaction responsible for superconductivity can be defined with the aid of a generalized BCS Hamiltonian:

$$H = \sum_{\mathbf{k},\sigma} \epsilon(\mathbf{k}) c^{\dagger}_{\mathbf{k},\sigma} c_{\mathbf{k},\sigma} + H_{\text{pairing}} , \qquad (16.35)$$

where $\epsilon(\mathbf{k})$ is the one-electron dispersion of the quasiparticles in the CuO_2 band. The pairing interaction $H_{pairing}$ must be compatible with the underlying point group symmetry and can be expressed in terms of the momentum transfer $\mathbf{k}\text{-}\mathbf{k}'=\mathbf{q}$ (see Fig. 16.34). For scalar pairing interaction,

$$H_{\text{pairing}} = \sum V(\mathbf{q}) c^{\dagger}_{\mathbf{k}+\mathbf{q},\uparrow} c^{\dagger}_{-\mathbf{k}-\mathbf{q},\downarrow} c_{-\mathbf{k},\downarrow} c_{\mathbf{k},\uparrow} . \qquad (16.36)$$

The BCS gap equation can be expressed as:

$$\Delta(\mathbf{k}) = -\sum V(\mathbf{k}-\mathbf{k}') \frac{\Delta(\mathbf{k}')}{2E(\mathbf{k}')} \tanh\left[\frac{E(\mathbf{k}')}{2k_B T}\right] , \qquad (16.37)$$

where the quasiparticle energy is defined as $E(\mathbf{k}) = \sqrt{\epsilon^2(\mathbf{k}) + \Delta^2(\mathbf{k})}$.

For the case of d-wave pairing, the $\mathbf{k}$-dependent gap potential can be expressed as:

$$\Delta(\mathbf{k}) = \Delta_0(\cos k_x - \cos k_y) . \qquad (16.38)$$

To satisfy the conditions imposed by the d-wave order parameter, (16.38), and the BCS gap equation, (16.37), it is required that the pairing interaction $V(q)$ is attractive (or less repulsive) for small momentum transfer ($\mathbf{q}$) and more repulsive for large $\mathbf{q}$. For the case of d-wave pairing, a constant can be added to $V(q)$ without affecting the BCS gap equation [2]. Hence, an all-repulsive $V(q)$ can pair. A general form of the pairing interaction of a d-wave superconductor is shown in Fig. 16.34.

d-wave pairing is consistent with a high-T_c mechanism based on magnetic fluctuation models. In fact,

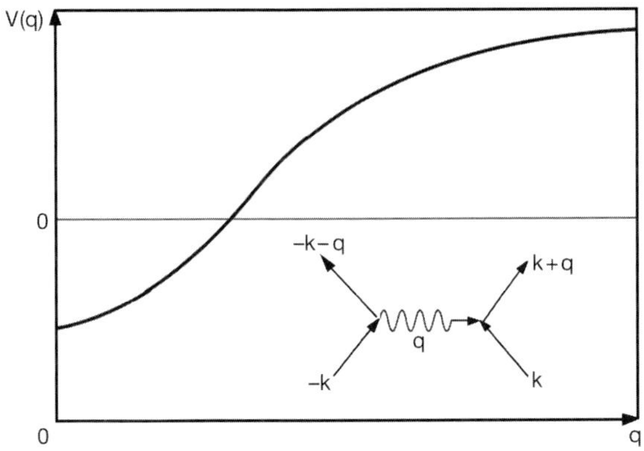

Fig. 16.34. Schematic representation of the BCS pairing interaction $V(\mathbf{q})$, constrained by a $d_{x^2-y^2}$ order parameter, as a function of the momentum transfer $\mathbf{q} = \mathbf{k} - \mathbf{k}'$

the concept of d-wave pairing mediated by magnetic fluctuations was proposed prior to the discovery of high temperature superconductivity [302,303]. This work has stimulated much of the ensuing theoretical work on high-T_c mechanisms based on spin fluctuations in strongly correlated cuprate superconductors [296,297,304–309]. The tremendous amount of work developed in this field in conjunction with d-wave pairing symmetry and HTS has been reviewed by Scalapino [4] and by Chubokov, Pines, and Schmalian in Chap. 22 of the present book.

The pairing interaction between charge carriers and antiferromagnetic spin fluctuations can be expressed in terms of a q-dependent spin susceptibility $\chi(\mathbf{q})$:

$$V(\mathbf{k}-\mathbf{k}') = V(\mathbf{q}) = g^2\chi(\mathbf{q}), \qquad (16.39)$$

where g is a measure of the coupling strength between charge carriers and spin fluctuations. With the assumption that the cuprates can be considered as a nearly antiferromagnetic Fermi-liquid system, the spin susceptibility function $\chi(q)$ is sharply peaked at or near $q = (\pi, \pi)$. Such a repulsive pairing interaction, when entered in the BCS gap equation (16.37), always results in an order parameter with $d_{x^2-y^2}$ symmetry, i.e., a gap with sign changes along the diagonals of the Brillouin zone.

One can calculate $\chi(\mathbf{q})$ from phenomenological models to estimate the magnitude of T_c. A $\chi(\mathbf{q})$ based on NMR data on $YBa_2Cu_3O_{7-\delta}$ predicts a T_c of 100, while a $\chi(\mathbf{q})$ determined from neutron scattering data yields a $T_c \leq 20$ K [310].

The electron-phonon pairing interaction, which gives rise to conventional s-wave superconductivity, does not naturally lead to a $d_{x^2-y^2}$ gap symmetry. The early observation of a vanishingly small isotope effect in optimally doped high-T_c cuprate superconductors [311] supports a non-phonon high-T_c mechanism. However, recent systematic studies of the isotope effects in several cuprate systems suggest an important role for electron-phonon interaction in HTS [312,313]. The electron-phonon coupling strength deduced from dynamical conductivity and dc resistivity data is too small to support the HTS observed in cuprates. However, these conclusions are modified when one considers strong Coulomb correlation effects. For example, it is suggested that a strong electron-phonon interaction renormalized by electronic correlations is responsible for the apparently small values of electron-phonon coupling strength concluded from the transport measurements [314–316]. The doping dependence of the isotope effects can be understood on the basis of electron-phonon interaction in the presence of strong Coulomb correlations [317–320].

All high-T_c cuprate superconductors are characterized by a relatively low carrier density. As a result, the ratio of $k_B T_c$ to E_F is in the range of 10^{-1}–10^{-2}, which is 10^2–10^3 times larger than that in unconventional superconductors. The Migdal theorem therefore holds for neither a spin-fluctuation nor a phonon mediated high-T_c mechanism [310, 319,321–323], and vertex corrections have to be included in any HTS models. With the breakdown of the Migdal approximation taken into account, d-wave pairing symmetry has been found in the theory of electron-phonon pairing interaction renormalized by strong Coulomb correlations [316]. In particular, the pairing interaction $V(q)$ has a momentum dependence shown in Fig. 16.34, i.e. a dominant attractive electron-phonon forward scattering in the small q regime, and a repulsive pairing interaction for large q's dominated by strong correlations [314,316,324–327]. In other words, the on-site Coulomb repulsion prevents pairing in the s-wave channel while the electron-phonon coupling is enhanced in the small q regime, resulting in the next available pair state: $d_{x^2-y^2}$ superconductivity. The role of the electron-phonon interaction in HTS is also emphasized in recent studies of angle-resolved photoemission spectroscopy [328].

16.6.3 Nodal Excitations

In a conventional BCS superconductor, a minimum energy of 2Δ, the energy gap, is required to break a pair to produce quasiparticle (qp) above the ground state of the superfluid condensate. At low temperatures ($T \ll T_c$), the number of thermally activated quasiparticles is exponentially small, leading to exponentially reduced scattering processes. In an unconventional superconductor, however, low energy quasiparticles can exist in significant numbers due to the presence of nodes ($\Delta = 0$) in the $\boldsymbol{k}$-dependent gap function $\Delta(\boldsymbol{k})$. These nodal quasiparticles are responsible for the well-established power-law low-temperature properties of the cuprate superconductors [51].

Recently, tremendous effort has been devoted to the study of low energy nodal quasiparticles and qp dynamics. With significant progress in both the theoretical and experimental fronts, this field has developed into an important area of d-wave superconductivity. In the following, some of the important advances will be reviewed.

Superconductivity is assumed to prevail mostly in the CuO_2 planes in the cuprate superconductors, with four nodes on the Fermi surface, located at $\boldsymbol{k}_x = \pm\boldsymbol{k}_y$ in the Brillouin zone (see Fig. 16.35) of the two-dimensional Cu-O square lattice. The d-wave order parameter can be written as $\Delta(\boldsymbol{k}) = \Delta_0(k_x^2 - k_y^2)/2$. In the vicinity of these nodes, the qp excitations spectrum can be approximated by a cone-like Dirac dispersion [329,330]:

$$E(k) = (v_F^2 \boldsymbol{k}_1^2 + v_\Delta^2 \boldsymbol{k}_2)^{1/2}, \qquad (16.40)$$

where $\boldsymbol{k}_1, \boldsymbol{k}_2$ are local momentum variables at each node; the Fermi velocity v_F and the gap velocity v_Δ are the components of the qp group velocity normal (parallel to $\boldsymbol{k}_1$) and perpendicular (parallel to $\boldsymbol{k}_2$) to the Fermi surface respectively. The gap velocity v_Δ owes its origin to the $\boldsymbol{k}$ dependence of the gap, $\Delta(\boldsymbol{k})$. The ratio v_F/v_Δ is an important measure of the anisotropy of the low temperature qp state. There are many important theoretical and experimental studies of the nodal qp-dominated low energy physics in 2D d-wave superconductors. Examples include: qp transport [66,67,329–331], specific heat [59,61,62,332], qp localization [333–340], qp extended states [341,342], qp-qp interaction [343], quantum critical phenomena [243,344], and vortex state dynamics [345–349]. In particular, the dominance of low energy nodal quasiparticles at low temperatures has led to the prediction of novel thermodynamics and transport effects [329,332], which were recently confirmed experimentally [59,66].

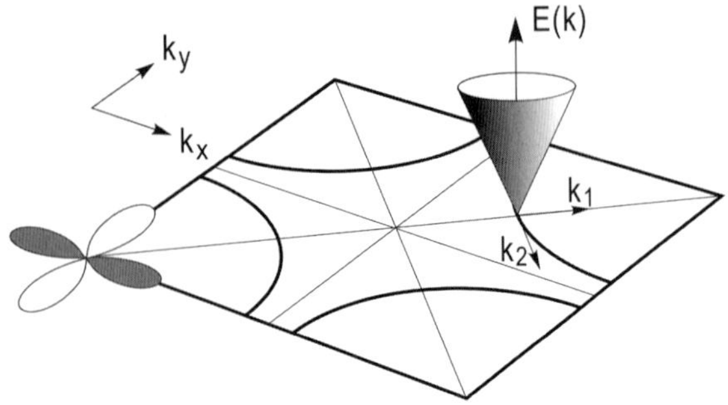

Fig. 16.35. A schematic representation of the quasiparticle dispersion E($\boldsymbol{k}$) at a nodal point in the Brillouin zone of a 2D d-wave superconductor. Three other equivalent nodes are not shown for clarity

Low Temperature Specific Heat

In the presence of a magnetic field H, the Dirac dispersion (16.40) is modified, resulting in a Doppler shift of the qp excitations outside the vortex core. In the low temperature limit, $T/T_c \ll \sqrt{H/H_{c2}}$, Volovik predicts that the electronic specific heat is dominated by a term proportional to $T\sqrt{H/H_{c2}}$. This effect was observed in YBCO by Moler et al. [59], B. Revaz et al. [62], and Wright et al. [61].

These results represent important evidence in support of d-wave pairing symmetry in YBCO. Further, they are an important validation of the concept of quasiparticles in an unconventional superconductor.

Universal Transport Limit

In a d-wave superconductor, the presence of impurities generally induces a finite zero-energy density of quasiparticle states, concentrated at the nodes on the Fermi surface [329,350]. It was predicted by Lee [329] that the residual conductivity of a 2D d-wave superconductor should be independent of impurity concentration. This universal transport limit is a consequence of cancelation between the increase in qp density of states and the decrease in qp mean free path as impurity concentration is increased. This novel effect can be generalized to include thermal and spin conductivities [330,351]. Durst and Lee [330] concluded that vertex corrections modify electrical conductivity, and Fermi-liquid corrections renormalize both electrical and spin conductivities. Only thermal conductivity maintains its universal value obtained in the "bare bubble" calculations, and is not affected by any of these corrections. Thermal conductivity measurement is a well-established probe of the bulk properties, and is free of surface effects. The phonon contribution to thermal conductivity can easily be separated from its electronic counterpart. All these attributes make it an important tool for studying the quasiparticle state at low temperatures.

The residual thermal conductivity for a d-wave cuprate superconductor is given by [330,351]:

$$\frac{\kappa}{T} = \frac{k_B^2}{3\hbar}\frac{n}{d}\left(\frac{v_F}{v_\Delta} + \frac{v_\Delta}{v_F}\right), \qquad (16.41)$$

where n/d is the stacking density of the CuO_2 planes.

Taillefer et al. [66] reported the observation of residual thermal conductivity in Zn-doped $YBa_2Cu_3O_{6.9}$. They found that the value of κ/T in the zero temperature limit is independent of the Zn concentration and in quantitative agreement with the quasiclassical calculation for unconventional superconductors [351]. Similar results were obtained with Bi-2212 at optimal doping [352]. In the case of Bi-2212, the value of the anisotropy parameter v_F/v_Δ deduced from the thermal conductivity measurement can be checked with the results of ARPES measurements [53,353]. An excellent agreement was found [67]: For Bi-2212,

$$\begin{aligned} v_F/v_\Delta &= 19 \quad \text{(thermal conductivity results)},\\ &= 20 \quad \text{(ARPES data)}. \end{aligned}$$

Similar measurements on over-doped Tl-2201 single crystals led to the conclusion that a large residual thermal conductivity is observed and the Weidemann-Franz law is verified [354].

The overall consistency between thermal transport and ARPES data, and the agreement with the d-wave theory, is quite remarkable. One can therefore conclude that the d-wave pair state in high-T_c cuprate superconductors is well-described by the Fermi-liquid theory at low energy in the optimal and over-doped regimes. However, exceptions to the above have appeared in the literature. For example, the universal thermal conduction was not observed at low temperature in the electron-doped PCCO at optimal doping [355], and the intrinsically under-doped $YBa_2Cu_4O_8$ ($T_c = 80$) [356]. The residual thermal conductivity was recently observed in LSCO cuprates over the entire range of superconducting doping phase, but there are indications that the Wiedemann-Franz law is violated in the under-doped regime in the LSCO system [357].

A clear deviation from the Wiedemann-Franz law was found in the normal state of PCCO [355]. While these results may indicate the breakdown of Fermi-liquid theory in cuprate superconductors, equally interesting alternatives are possible. For example, qp localization can lead to a thermal insulator phase within the d-wave condensate [337]. Another possibility is via time reversal symmetry breaking in the form of pair states such as $d_{x^2-y^2} + id_{xy}$ or $d_{x^2-y^2} + is$, which are gapped [343,344].

16.6.4 Applications of d-Wave Superconductivity

Several potential applications of d-wave superconductivity arise from the momentum dependence of the sign of the pairing wavefunction. First, this enables the design of π-rings, with two degenerate states that are identical, aside from reversing the sign of time, at zero applied magnetic field. Although it is possible to create degenerate flux states in a conventional 0-ring by applying an external field equivalent to $\Phi_0/2$ of magnetic flux, the two states in this case are different because of the effects of Meissner screening. These differences can have profound effects in, for example, attempts to observe long-range antiferromagnetic coupling in arrays of conventional superconducting rings [358].

The possibility of degenerate, time-reversed states has been used as a basis for a suggestion of complementary Josephson logic (memory) circuits by Terzioglu and Beasley [359], and in several suggestions for qubits, the fundamental cell for quantum computation [360, 361]. A qubit operates by superposing two degenerate quantum states [362]. In most proposals for superconducting qubits these states can have either distinct charge- or flux- states. The proposal of Ioffe et al. [363] makes use of the possible deviations from the Josephson $I_c = I_0 \sin\phi$ current–phase relationship in junctions with d-wave superconductors to produce a "quiet" qubit, in which the information is expressed through the phase degree of freedom. In the "quiet" qubit the phase-state itself carries no charge and no current, and therefore has minimal interaction with states external to the qubit, reducing dissipation [363]. Several designs take advantage of the degeneracy of the ground state of a frustrated ring with d-wave superconductors and the spontaneous currents and fluxes in equilibrium [364, 365]. Alternative concepts are based on the occurrence of Andreev bound-states in different types of Josephson junctions [366].

Macroscopic quantum tunneling and energy level quantization have been observed in bi-epitaxial grain boundary Josephson junctions [367, 368]. Evidence of macroscopic quantum tunneling in intrinsic junctions of $Bi_2Sr_2CaCu_2O_{8+\delta}$ have also been reported [369, 370]. The demonstration of macroscopic quantum effects in d-wave Josephson junctions has shown that dissipation arising from low-lying excitations of nodal quasiparticles does not necessarily have to be a significant source of decoherence. Several theoretical studies on this topic exist [371–373]. Using a novel YBCO/Nb ramp-type Josephson junction technology [217], fabrication of 25,000 π-loop arrays on a single chip, and magnetic coupling between the half-fluxons in the array have been demonstrated experimentally [374]. Furthermore, flip-flopping fractional quantum devices have been demonstrated using the d-wave superconductor YBCO [375]. All of these advances represent important steps towards the realization of d-wave superconducting qubits.

16.7 Conclusions

A predominantly $d_{x^2-y^2}$-wave order parameter has been established in the hole- and electron-doped cuprate superconductors through a combination of phase-sensitive and conventional techniques for probing pairing symmetry. In particular, the phase-sensitive techniques have demonstrated an amazing robustness of the d-wave pairing symmetry, which survives in the face of significant deviations from optimal doping, through many classes of cuprate compounds, from hole- to electron-doping, and over a wide variation in temperature. Furthermore, the phase-sensitive pairing symmetry tests have shown that there cannot be more than a few percent of the subdominant order parameter in mixed pair states such as $d_{x^2-y^2} + id_{xy}$ or $d_{x^2-y^2} + is$. The resilience of the d-wave pair state is remarkable in view of the fact that cuprate superconductors are known to exhibit a great variety in many of their normal-state and superconducting properties. The basic origin of the observed universal d-wave pairing symmetry may stem from the strong 2D Coulomb correlations, characteristic of all cuprates. Furthermore, a HTS model has recently been proposed that is decoupled from the on-site Coulomb repulsion and based on a strong, local, nonlinear interaction between electrons within the Cu–O–Cu bond in the CuO_2 plane, and the oxygen vibrational degree of freedom, termed the Fluctuating Bond Model [376], which captures the essence of both the pseudogap and the superconducting states of HTS.

Although the establishment of d-wave pairing symmetry in the cuprates does not pin down a specific mechanism for high temperature superconductivity, it does impose well-defined constraints on possible HTS models. A pair state with $d_{x^2-y^2}$ symmetry occurs naturally in spin-fluctuation HTS models. However, one can not rule out a HTS mechanism based on electron-phonon interaction renormalized with strong correlation effects.

Research on the low-lying quasiparticle excitations near the nodes of the d-wave gap function has led to new effects in the low temperature, low energy quasiparticle state, in support of d-wave pairing symmetry. Further, these studies have produced strong evidence for the validity of the concept of the quasiparticle in the context of BCS Fermi liquid theory in some optimally and over-doped cuprates. However, recent examples of exceptions to this existence proof for the Fermi-liquid theory remind us vividly of the richness and complexity of the underlying physics in the cuprates.

Potential applications of d-wave superconductivity in areas of π-SQUIDs and quantum computation, while rich and exciting, are in the infancy of their development, and are only mentioned briefly here.

Acknowledgments

The authors wish to thank A. Barone, B. Chesca, H. Hilgenkamp, I. Iguchi, R. Koch, J. Mannhart, K.A. Moler, D.M. Newns, H. Raffy, J.Z. Sun, M. Tachiki, F. Tafuri, L. Taillefer, M.B. Walker, and J.Y.T Wei for useful discussions during the preparation of this manuscript.

References

1. J.G. Bednorz, K.A. Müller, Z. Phys. B **61**, 189 (1986)
2. J. Bardeen, L.N. Cooper, J.R. Schrieffer, Phys. Rev. **108**, 1175 (1959)
3. P.W. Anderson, *The Theory of Superconductivity in the High-T_c Cuprates* (Princeton University Press, Princeton, New Jersey, 1997)
4. D.J. Scalapino, Phys. Rep. **250**, 329 (1995)
5. J.F. Annett, N.D. Goldenfeld, A. Leggett, In: *Physical Properties of High Temperature Superconductors*, ed. by D.M. Ginsberg, Vol. 5 (World Scientific, Singapore, 1996)
6. D.J. van Harlingen, Rev. Mod. Phys. **67**, 515 (1995)
7. C.C. Tsuei, J.R. Kirtley, Rev. Mod. Phys. **72**, 969 (2000)
8. R.M. White, T.H. Geballe, *Long Range Order in Solids* (Academic Press, New York, 1979)
9. P.W. Anderson, *Basic Notions of Condensed Matter Physics* (Benjamin-Cummings, San Francisco, CA, 1984)
10. V.L. Ginzburg, L.G. Landau, Zh. Eksp. Teor. Fiz. **20**, 1064 (1950)
11. M. Cyrot, Rep. Prog. Phys. **36**, 103 (1973)
12. M. Tinkham, *Introduction to Superconductivity*, 2nd Edition (McGraw-Hill, New York, 1996)
13. L.P. Gor'kov, Zh. Eksperim. i. Teor. Fiz **36**, 1918 (1959)
14. C.N. Yang, Rev. Mod. Phys **34**, 694 (1962)
15. J. Schmalian, W. Hübner, Phys. Rev. B **53**, 11 860 (1996)
16. L.D. Landau, E.M. Lifshitz, *Statistical Physics* (Pergamon, New York, 1979)
17. J.F. Annett, Adv. Phys. **39**, 83 (1991)
18. E.I. Blount, Phys. Rev. B **32**, 2935 (1985)
19. M. Ozaki, K. Machida, T. Ohmi, Prog. Theor. Phys. **74**, 221 (1985)
20. M. Ozaki, K. Machida, T. Ohmi, Prog. Theor. Phys. **75**, 442 (1986)
21. K. Ueda, T.M. Rice, Phys. Rev. B **31**, 7114 (1985)
22. G.E. Volovik, L.P. Gor'kov, Sov. Phys. JETP **61**, 843 (1985)
23. L.P. Gor'kov, Sov. Sci. Rev. (Sect. A), Phys. Rev. **9**, 1 (1987)
24. P. Fulde, J. Keller, G. Zwicknagl, Solid State Physics **41**, 1 (1988)
25. M. Sigrist, K. Ueda, Rev. Mod. Phys. **63**, 239 (1991)
26. S. Yip, A. Garg, Phys. Rev. B **48**, 3304 (1993)
27. J.A. Sauls, Adv. Phys. **43**, 113 (1994)
28. M. Sigrist, T.M. Rice, Z. Phys. B **68**, 9 (1987) .
29. J.F. Annett, N.D. Goldenfeld, S.R. Renn, In: *Physical Properties of High Temperature Superconductors*, ed. by D. Ginsberg, Vol. 2 (World Scientific, Singapore, 1990) p. 571
30. F. Wenger, S. Östlund, Phys. Rev. B **47**, 5977 (1993)
31. Q.P. Li, B.E. Koltenbah, R. Joynt, Phys. Rev. B **48**, 437 (1993)
32. S.S. Jha, A.K. Rajagopal, Phys. Rev. B **55**, 15 248 (1997)

33. M. Ozaki, K. Machida, Phys. Rev. B **39**, 4145 (1989)
34. R. Beyers, T.M. Shaw, In: *Solid State Physics*, ed. by H. Ehrenreich, D. Turnbull, Vol. 42 (Academic, San Diego, CA, 1988) p. 135
35. H. Shaked, P.M. Keane, J.C. Rodriguez, F.F. Owen, R.L. Hitterman, J.D. Jorgensen, *Crystal Structure of the High-T_c Superconducting Cooper-Oxides* (Elsevier, Amsterdam, 1994)
36. L. Chen, C.J. Wei, A.M.S. Tremblay, Phys. Rev. B **47**, 15 316 (1993)
37. R. Kleiner, P. Müller, Phys. Rev. B **49**, 1327 (1994)
38. N. Mros, V.M. Krasnov, A. Yurgens, D. Winkler, T. Claeson, Phys. Rev. B **57**, R8135 (1998)
39. J.R. Clem, Supercond. Sci. Technol. **11**, 909 (1998)
40. R.J. Drost, C.J. can der Beek, H.W. Zandbergen, M. Konczykowski, A.A. Menovsky, P.H. Kes, Phys. Rev. B **59**, 13 612 (1999)
41. K.E. Gray, D. Kim, Phys. Rev. Lett. **70**, 1693 (1993)
42. D.G. Clarke, S.P. Strong, P.W. Anderson, Phys. Rev. Lett. **74**, 4499 (1995)
43. A. Yurgens, D. Winkler, N.V. Zavaritsky, T. Claeson, Phys. Rev. Lett. **79**, 5122 (1997)
44. A. Yurgens, D. Winkler, T. Claeson, G. Yang, I.F.G. Parker, C.E. Gough, Phys. Rev. B **59**, 7196 (1999)
45. R.E. Walstedt, W.W.W. Jr, Science **248**, 1082 (1990)
46. C.H. Pennington, C.P. Slichter, In: *Physical Properties of the High Temperature Superconductors*, ed. by D.M. Ginsberg, Vol. 2 (World Scientific, Singapore, 1990)
47. N. Bulut, D.J. Scalapino, Phys. Rev. Lett. **68**, 706 (1992)
48. N. Bulut, , D.J. Scalapino, Phys. Rev. B **45**, 2371 (1992)
49. N. Bulut, D.J. Scalapino, Phys. Rev. Lett. **67**, 2898 (1991)
50. W.N. Hardy, D.A. Bonn, D.C. Morgan, R. Liang, K. Zhang, Phys. Rev. Lett. **70**, 3999 (1993)
51. D.A. Bonn, W.N. Hardy, In: *Physical Properties of High Temperature Superconductors*, ed. by D.M. Ginsberg, Vol. 5 (World Scientific, Singapore, 1996)
52. S. Kamal, R. Liang, A. Hosseini, D.A. Bonn, W.N. Hardy, Phys. Rev. B **58**, R8933 (1998)
53. Z.X. Shen, D.S. Dessau, B.O. Wells, D.M. King, W.E. Spicer, A.J. Arko, D. Marshall, L.W. Lombardo, A. Kapitulnik, P. Diskinson, S. Doniach, J. DiCarlo, A.G. Loeser, C.H. Park, Phys. Rev. Lett. **70**, 1553 (1993)
54. H. Ding, M.R. Norman, J.C. Campuzano, M. Randeria, A.F. Bellman, T. Yokoya, T. Takahashi, T. Mochiku, K. Kadowaki, Phys. Rev. B **54**, R9678 (1996)
55. J. Ma, C. Quitmann, R.J. Kelley, H. Berger, G. Margaritondo, M. Onellion, Science **267**, 862 (1995)
56. Z.X. Shen, D.S. Dessau, Phys. Reports **253**, 1 (1995)
57. S. Kashiwaya, Y. Tanaka, Rep. Prog. Phys. **63**, 1641 (2001)
58. T. Löfwander, V.S. Shumeiko, G. Wendin, Supercond. Sci. Technol. **14**, 53 (2001)
59. K.A. Moler, D.J. Baar, J.S. Urbach, R. Liang, W.N. Hardy, A. Kapitulnik, Phys. Rev. Lett. **73**, 2744 (1994)
60. K.A. Moler, D.L. Sisson, J.S. Urbach, M.R. Beasley, A. Kapitulnik, D.J. Baar, R. Liang, W.N. Hardy, Phys. Rev. B **55**, 3954 (1997)
61. D.A. Wright, J.P. Emerson, B.F. Woodfield, J.E. Gordon, R.A. Fisher, N.E. Phillips, Phys. Rev. Lett. **82**, 1550 (1999)
62. B. Revaz, J.Y. Genoud, A. Junod, K. Neumaier, A. Erb, E. Walker, Phys. Rev. Lett. **80**, 3364 (1998)
63. T.P. Devereaux, D. Einzel, B. Stadlober, R. Hackl, D.H. Leach, J.J. Neumeier, Phys. Rev. Lett. **72**, 396 (1994)
64. T.P. Devereaux, D. Einzel, Phys. Rev. B **51**, 16 336 (1995)
65. T. Strohm, M. Cardona, Solid State Commun. **104**, 233 (1997)
66. L. Taillefer, B. Lussier, R. Gagnon, K. Behnia, H. Aubin, Phys. Rev. Lett. **79**, 483 (1997)
67. M. Chiao, R.W. Hill, C. Lupien, L. Taillefer, P. Lambert, R. Gagnon, P. Fournier, Phys. Rev. B **62**, 3554 (2000)
68. H. Aubin, K. Behnia, M. Ribault, R. Gagnon, L. Taillefer, Phys. Rev. Lett. **78**, 2624 (1997)
69. K.K. Likharev, Rev. Mod. Phys. **51**, 101 (1979)
70. A. Barone, G. Paterno, *Physics and Applications of the Josephson Effect* (Wiley, New York, 1982)
71. L.M.F. M H Cohen, J.C. Phillips, Phys. Rev. Lett. **8**, 316 (1962)
72. J.A. Pals, van Haeringen, M.H. van Marren, Phys. Rev. B **15**, 2592 (1977)
73. C. Bruder, G.Z. A van Otterlo, Phys. Rev. B **51**, 12 904 (1995)
74. Y.S. Barash, A.V. Galaktionov, A. Zaikin, Phys. Rev. B **52**, 665 (1995)
75. A. Millis, D. Rainer, J.A. Sauls, Phys. Rev. B **38**, 4504 (1988)

76. Z.G. Ivanov, E.A. Stepantsov, T. Claeson, F. Wenger, S.Y. Lin, N. Khare, P. Chaudhari, Phys. Rev. B **57**, 602 (1998)
77. H. Hilgenkamp, J. Mannhart, Rev. Mod. Phys. **74**, 485 (2002)
78. V.B. Geshkenbein, A.I. Larkin, JETP Lett. **43**, 395 (1986)
79. S.K. Yip, O.F.D.A. Bonfim, P. Kumar, Phys. Rev. B **41**, 11 214 (1990)
80. M.B. Walker, J. Luettmer-Strathmann, Phys. Rev. B **54**, 588 (1996)
81. M. Sigrist, T.M. Rice, Rev. Mod. Phys. **67**, 503 (1995)
82. M. Sigrist, T.M. Rice, J. Phys. Soc. Japan **61**, 4283 (1992)
83. D.J. Miller, T.A. Roberts, J.H. Kana, J. Talvacchio, D.B. Bucholz, R.P.H. Chang, Appl. Phys. Lett. **66**, 2561 (1995)
84. B.H. Moeckly, D.K. Lathrop, R.A. Buhrman, Phys. Rev. B **47**, 400 (1993)
85. L. Antognazza, B.H. Moeckly, T.H. Geballe, K. Char, Phys. Rev. B **52**, 4559 (1995)
86. F. Tafuri, F.M. Granozio, F. Carillo, A.D. Chiara, K. Verbist, G. van Tendeloo, Phys. Rev. B **59**, 11 523 (1999)
87. C.C. Tsuei, J.R. Kirtley, C.C. Chi, L.S. Yu-Jahnes, A. Gupta, T. Shaw, J.Z. Sun, M.B. Ketchen, Phys. Rev. Lett. **73**, 593 (1994)
88. V.B. Geshkenbein, A.I. Larkin, A. Barone, Phys. Rev. B **36**, 235 (1987)
89. L.N. Bulaevskii, V.V. Kuzii, A.A. Sobyanin, JETP Lett. **25**, 290 (1977)
90. B.I. Spivak, S. Kivelson, Phys. Rev. B **43**, 3740 (1991)
91. V.V. Ryanazov, V.A. Obuznov, A.V. Veretennikov, A.A. Golubov, J. Aarts, Phys. Rev. Lett. **86**, 2427 (2001)
92. E. Il'ichev, M. Grajcar, R. Hlubina, R.P.J. IJsselsteijn, H.E. Hoenig, H.G. Meyer, A. Golubov, M.H.S. Amin, A.M. Zagoskin, A.N. Omelyanchouk, M.Y. Kupriyanov, Phys. Rev. Lett. **86**, 5359 (2001)
93. V.V. Ryazanov, V.A. Obuznov, A.V. Veretennikov, A.Y. Rusanov, Phys. Rev. B **65**, 020 501 (2002)
94. P. Svendlindh, K. Niskanen, P. Norling, P. Nordblad, L. Lundgren, B. Lönnberg, T. Lundström, Physica C **162–164**, 1365 (1989)
95. W. Braunisch, N. Knauf, V. Kateev, S. Neuhausen, A. Grütz, A. Koch, B. Roden, D. Khomskii, D. Wohlleben, Phys. Rev. Lett. **68**, 1908 (1992)
96. W. Braunisch, N. Knauf, G. Bauer, A. Kock, A. Becker, B. Freitag, A. Grütz, V. Kataev, S. Neuhausen, B. Roden, D. Khomskii, D. Wohlleben, Phys. Rev. B **48**, 4030 (1993)
97. C. Heinzel, T. Theilig, P. Ziemann, Phys. Rev. B **48**, 3445 (1993)
98. K. Niskanen, J. Magnusson, P. Nordblad, P. Svedlindh, A.S. Ullström, T. Lundström, Physica B **194-196**, 1549 (1994)
99. K.N. Shrivastava, Solid State Commun. **90**, 589 (1994)
100. C.S. Owen, D.J. Scalapino, Phys. Rev. **164**, 538 (1967)
101. J.R. Clem, M.W. Coffey, Phys. Rev. B **42**, 6209 (1990)
102. J.H. Xu, J.H.M. Jr, C.S. Ting, Phys. Rev. B **51**, 11 958 (1995)
103. J.R. Kirtley, K.A. Moler, D. Scalapino, Phys. Rev. B **56**, 886 (1997)
104. D.A. Wollman, D.J. van Harlingen, W.C. Lee, D.M. Ginsberg, A.J. Leggett, Phys. Rev. Lett. **71**, 2134 (1993)
105. D.A. Wollman, D.J. van Harlingen, J. Giapintzakis, D.M. Ginsberg, Phys. Rev. Lett. **74**, 797 (1995)
106. A. Mathai, Y. Gim, R.C. Black, A. Amar, F.C. Wellstood, Phys. Rev. Lett. **74**, 4523 (1995)
107. B.M. Hinaus, M.S. Rzchowski, N. Heinig, X.Y. Cai, D.L. Kaiser, Phys. Rev. B **54**, 6770 (1996)
108. R.A. Klemm, Phys. Rev. Lett. **73**, 1871 (1994)
109. R.R. Schulz, B. Chesca, B. Goetz, C.W. Schneider, A. Schmehl, H. Bielefeldt, H. Hilgenkamp, J. Mannhart, C. Tsuei, Appl. Phys. Lett. **76**, 912 (2000)
110. D.A. Brawner, H.R. Ott, Phys. Rev. B **50**, 6530 (1994)
111. B. Chesca, Ann. Phys. **8**, 511 (1999)
112. I. Iguchi, Z. Wen, Phys. Rev. B **49**, 12 388 (1994)
113. O.B. Hyun, J.R. Clem, D. Finnemore, Phys. Rev. B **40**, 175 (1989)
114. I.V. Vernik, S. Keil, N. Thyssen, T. Doderer, A.V. Ustinov, H. Kohlstedt, R.P. Huebener, J. Appl. Phys. **81**, 1335 (1997)
115. C. Nappi, R. Cristiano, Appl. Phys. Lett. **70**, 1320 (1997)
116. C. Nappi, R. Cristiano, M.P. Lisitskii, Phys. Rev. B **58**, 11 685 (1998)
117. J.H. Miller Jr., Q.Y. Ying, Z.G. Zou, N.Q. Fan, J.H. Xu, M.F. Davis, J.C. Wolfe, Phys. Rev. Lett. **74**, 2347 (1995)
118. J.R. Kirtley, M.B. Ketchen, K.G. Stawiasz, J.Z. Sun, W.J. Gallagher, S.H. Blanton, S.J. Wind, Appl. Phys. Lett. **66**, 1138 (1995)
119. J.R. Kirtley, C.C. Tsuei, J.Z. Sun, C.C. Chi, L.S. Yu-Jahnes, A. Gupta, M. Rupp, M. Ketchen, Nature **373**, 225 (1995)

120. C.C. Tsuei, J.R. Kirtley, M. Rupp, J.Z. Sun, L.S. Yu-Jahnes, C.C. Chi, A. Gupta, M. Ketchen, J. Phys. Chem Solids **56**, 1787 (1995)
121. V.G. Kogan, J.R. Clem, J.R. Kirtley, Phys. Rev. B **61**, 9122 (2000)
122. J.R. Kirtley, C.C. Tsuei, M. Rupp, J.Z. Sun, L.S. Yu-Jahnes, A. Gupta, M.B. Ketchen, K.A. Moler, M. Bhushan, Phys. Rev. Lett. **76**, 1336 (1996)
123. V.G. Kogan, V.V. Dobrovitski, J.R. Clem, Y. Mawatari, R.G. Mints, Phys. Rev. B **63**, 144 501 (2001)
124. B.J. Roth, N.G. Sepulveda, J.P. Wikswo, J. Appl. Phys. **65**, 361 (1988)
125. J. Pearl, J. Appl. Phys. **37**, 4139 (1966)
126. V.G. Kogan, A.Y. Siminov, M. Ledvij, Phys. Rev. B **48**, 392 (1993)
127. A.M. Chang, H.D. Hallen, H.F. Hess, H.L. Kao, J. Kwo, A. Sudbo, T.Y. Chang, Europhys. Lett. **20**, 645 (1992)
128. A. Sugimoto, T. Yamaguchi, I. Iguchi, S. Kashigawa, Physica C **367**, 28 (2002)
129. M.T. Béal-Monod, Phys. Rev. B **58**, 8830 (1998)
130. C.C. Tsuei, J.R. Kirtley, Z.F. Ren, J.H. Wang, H. Raffy, Z.Z. Li, Nature **387**, 481 (1997)
131. C.A. Wang, Z.F. Ren, J.H. Wang, D.K. Petrov, M.J. Naughton, W.Y. Yu, A. Petrou, Physica C **262**, 98 (1996)
132. Z.F. Ren, J.H. Wang, D.J. Miller, Appl. Phys. Lett. **69**, 1798 (1996)
133. C.H. Chen, In: *Physical Properties of High Temperature Superconductors*, ed. by D.M. Ginsberg, Vol. 3 (World Scientific, Singapore, 1992) p. 200
134. H. Ding, M.R. Norman, T. Mochiku, K. Kadowaki, J. Giapinzakis, Nature **382**, 51 (1996)
135. H.Z. Durusoy, D. Lew, L. Lombardo, A. Kapitulnik, T.H. Geballe, M.R. Beasley, Physica C **266**, 253 (1996)
136. S. Sinha, K.W. Ng, Europhys. Lett. **44**, 648 (1998)
137. M. Mössle, R. Kleiner, Phys. Rev. B **59**, 4486 (1999)
138. Y. Tokura, H. Takagi, S. Uchida, Nature **337**, 345 (1989)
139. H. Takagi, S. Uchida, Y. Tokura, Phys. Rev. Lett. **62**, 1197 (1989)
140. J.L. Peng, E. Maiser, T. Venkatesan, R.L. Greene, G. Czysek, Phys. Rev. B **55**, R6145 (1997)
141. C.C. Tsuei, A. Gupta, G. Koren, Physica C **161**, 415 (1989)
142. A.A. Abrikosov, J.C. Campuzano, K. Gofron, Physica C **214**, 73 (1993)
143. D.S. Dessau, Z.X. Shen, D.M. King, D.S. Marshall, L.W. Lombardo, P.H. Dickinson, A.G. Loeser, J. DiCarlo, C.H. Park, A. Kapitulnik, W.E. Spicer, Phys. Rev. Lett. **71**, 2781 (1993)
144. D.M. King, Z.X. Shen, D.S. Dessau, B.O. Wells, W.E. Spicer, A.J. Arko, D.S. Marshall, J. DiCarlo, A.G. Loeser, C.H. Park, E.R. Ratner, J.L. Peng, Z.Y. Li, R.L. Greene, Phys. Rev. Lett. **70**, 3159 (1993)
145. J. Fontcuberta, L. Fàbrega, In: *Studies of High Temperature Superconductors*, ed. by A.V. Narlikar, Vol. 16 (Nova Science, New York, 1995) pp. 185–241
146. C.C. Almasan, M.B. Maple, In: *Chemistry of High Temperature Superconductors*, ed. by C.N.R. Rao (World Scientific, New York, 1992) p. 205.
147. P. Fournier, E. Maiser, R.L. Greene, In: *The Gap Symmetry and Fluctuations in High-T_c Superconductors*, ed. by J. Bok, G. Deutscher, D. Pavuna, S.A. Wolf (Plenum Press, New York, 1998) pp. 145–158
148. J.R. Cooper, Phys. Rev. B. **54**, R3753 (1996)
149. J.D. Kokales, P. Fournier, L.V. Mercaldo, V.V. Talanov, R.L. Greene, S.M. Anlage, Phys. Rev. Lett. **85**, 3696 (2000)
150. R. Prozorov, R.W. Giannetta, P. Fournier, R.L. Greene, Phys. Rev. Lett. **85**, 3700 (2000)
151. L. Alff, A. Beck, R. Gross, A. Marx, S. Kleefisch, T. Bauch, H. Sato, M. Naito, G. Koren, Phys. Rev. B **58**, 11 197 (1998)
152. J.A. Skinta, T.R. Lemberger, T. Greibe, M. Naito, cond-mat/0108545 (2001)
153. S. Kashiwaya, T. Ito, K. Oka, S. Ueno, H. Takashima, M. Koyanagi, Y. Tanaka, K. Kajimura, Phys. Rev. B **57**, 8680 (1998)
154. M. Covington, M. Aprili, E. Paraoanu, L.H. Greene, F. Xu, J. Zhu, C.A. Mirkin, Phys. Rev. Lett. **79**, 277 (1997)
155. S.I. Woods, A.S. Katz, T.L. Kirk, M.C. de Andrade, M.B. Maple, R.C. Dynes, IEEE Trans. on Appl. Supercond. (ASC 98) **9**, 3917 (1999)
156. J. Hilgenkamp, J. Mannhart, IEEE Trans. Appl. Supercond. **9**, 3405 (1999)
157. U. Schoop, S. Kleefisch, S. Meyer, A. Marx, A. Alff, R. Gross, M. Naito, H. Sato, IEEE Trans. Appl. Supercond. **9**, 3409 (1999)
158. C.C. Tsuei, J.R. Kirtley, Phys. Rev. Lett. **85**, 182 (2000)
159. N.P. Armitage, D.H. Lu, D.L. Feng, C. Kim, A. Damascelli, K.M. Shen, F. Ronning, Z.X. Shen, Y. Onose, Y. Taguchi, Y. Tokura, Phys. Rev. Lett. **86**, 1126 (2001)

160. T. Sato, T. Kamiyama, T. Takahashi, K. Kurakashi, K. Yamada, Science **291**, 1517 (2001)
161. H.C. Chiang, Y.Y. Hsu, B.C. Chang, B.N. Lin, T.I. Hsu, H.C. Ku, J. App. Phys. **89**, 7660 (2001)
162. A. Pimenov, A.V. Pronin, A. Loidl, U. Michelucci, A.P. Kampf, S.I. Krasnosvobodtsev, V.S. Nozdrin, D. Rainer, Phys. Rev. B **62**, 9822 (2000)
163. A. Biswas, P. Fournier, M.M. Qazilbash, V.N. Smolyaninovna, H. Balci, R.L. Greene, cond-mat/0111544 (2001)
164. B. Chesca, K. Ehrhardt, M. Mö"ßle, R. Straub, D. Koelle, R. Kleiner, A. Tsukada, Phys. Rev. Lett. **90**, 057004 (2003)
165. W. Yu, B. Liang, R.L. Greene, Phys. Rev. B **72**, 212512 (2005)
166. Y. Gim, A. Mathai, R. Black, A. Amar, F.C. Wellstood, IEEE Trans. Appl. Supercond. **7**, 2331 (1997)
167. C.A. Copetti, F. Rueders, B. Oleze, C. Buchal, B. Kobius, Physica C **253**, 63 (1995)
168. J. Mannhart, B. Mayer, H. Hilgenkamp, Z. Phys. B **101**, 175 (1996)
169. J. Mannhart, H. Hilgenkamp, B. Mayer, C. Gerber, J.R. Kirtley, K.A. Moler, M. Sigrist, Phys. Rev. Lett. **77**, 2782 (1996)
170. R.G. Mints, V.G. Kogan, Phys. Rev. B **55**, R8682 (1997)
171. R.G. Mints, Phys. Rev. B **57**, 3221 (1998)
172. R.G. Mints, I. Papiashvili, Phys. Rev. B **64**, 134 501 (2001)
173. R.G. Mints, I. Papiashvili, J.R. Kirtley, H. Hilgenkamp, G. Hammerl, and J. Mannhart, Phys. Rev. Lett. **89**, 067004 (2002)
174. H.J.H. Smilde, H. Hilgenkamp, G.J. Gerritsma, D.H.A. Blank, H. Rogalla, IEEE Trans. Appl. Supercond. **11**, 501 (2001)
175. H.J.H. Smilde, H. Hilgenkamp, G.J. Gerritsma, D.H.A. Blank, H. Rogalla, Physica (Amsterdam) **350C**, 269 (2001)
176. H.J.H. Smilde, Ariando, D.H.A. Blank, G.J. Gerritsma, H. Hilgenkamp, H. Rogalla, Phys. Rev. Lett. **88**, 057 004 (2002)
177. P.W. Anderson, In: *Lectures on the Many-Body Problem*, ed. by E.R. Caianiello, Vol. 2 (Academic Press, New York, 1964) p. 113
178. Y. Tanaka, Phys. Rev. Lett. **72**, 3871 (1994)
179. R. Kleiner, A.S. Katz, A.G. Sun, R. Summer, D.A. Gajewski, S.H. Han, S.I. Woods, E. Dantsker, B. Chen, K. Char, M.B. Maple, R.C. Dynes, J. Clarke, Phys. Rev. Lett. **76**, 2161 (1996)
180. T.M. Shaw, S.L. Shinde, D. Dimos, R.F. Cook, P.R. Duncombe, R. Kroll, J. Mater. Res. **4**, 248 (1989)
181. C.C. Tsuei, J.R. Kirtley, M. Rupp, J.Z. Sun, A. Gupta, M.B. Ketchen, C.A. Wang, Z.F. Ren, J.H. Wang, M. Bhushan, Science **272**, 329 (1996)
182. M.B. Walker, Phys. Rev. B **53**, 5835 (1996)
183. M.B. Walker, J. Luettmer-Strathmann, J. Low. Temp. Phys. **105**, 483 (1996)
184. K.A. Kouznetsov, A.G. Sun, B. Chen, A.S. Katz, S.R. Bahcall, J. Clarke, R.C. Dynes, D.A. Gajewski, S.H. Han, M.B. Maple, J. Giapintzakis, J.T. Kim, D.M. Ginsberg, Phys. Rev. Lett. **79**, 3050 (1997)
185. A.G. Sun, D.A. Gajewski, M.B. Maple, R.C. Dynes, Phys. Rev. Lett. **72**, 2267 (1994)
186. A.S. Katz, A.G. Sun, R.C. Dynes, K. Char, Appl. Phys. Lett. **66**, 105 (1995)
187. J.M. Lesueur, J.M. Aprili, A. Goulon, T.J. Horton, L. Dumoulin, Phys. Rev. B **55**, R3398 (1997)
188. C. O'Donovan, M.D. Lumsden, B.D. Gaulin, J.P. Carbotte, Phys. Rev. B **55**, 9088 (1997)
189. J.Y.T. Wei, C.C. Tsuei, P.J.M. van Bentum, Q. Xiong, C.W. Chu, M.K. Wu, Phys. Rev. B **57**, 3650 (1998)
190. J.Y.T. Wei, N.C. Yeh, D.F. Garrigus, M. Strasik, Phys. Rev. Lett. **81**, 2542 (1998)
191. J.R. Kirtley, C.C. Tsuei, H. Raffy, Z.Z. Li, A. Gupta, J.Z. Sun, S. Megert, Europhys. Lett. **36**, 707 (1996)
192. S.R. Bahcall, Phys. Rev. Lett. **76**, 3634 (1996)
193. M. Sigrist, Prog. Theor. Phys. **99**, 899 (1998)
194. M.J. Graf, D. Rainer, J. Sauls, Phys. Rev. B **47**, 12 089 (1993)
195. M.J. Graf, M. Palumbo, D. Rainer, J. Sauls, Phys. Rev. B **52**, 10 588 (1995)
196. A.G. Rojo, K. Levin, Phys. Rev. B **48**, 16 861 (1993)
197. P.J. Hirschfeld, W.O. Puttika, D.J. Scalapino, Phys. Rev. B **50**, 10 250 (1994)
198. P.J. Hirschfeld, S.M. Quinlan, D.J. Scalapino, Phys. Rev. B **55**, 12 742 (1997)
199. A.A. Abrikosov, Physica C **258**, 53 (1996)
200. R.J. Radtke, V.N. Kostur, K. Levin, Phys. Rev. B **53**, R522 (1996)
201. S.D. Sarma, E.H. Hwang, Phys. Rev. Lett. **80**, 4753 (1998)
202. R.A. Klemm, G. Arnold, C.T. Rieck, K. Scharnberg, Phys. Rev. B **58**, 14 203 (1998)
203. A. Bille, R.A. Klemm, K. Scharnberg, Phys. Rev. B **64**, 174 507 (2001)
204. S.L. Cooper, K.E. Gray, In: *Physical Properties of High Temperature Superconductors*, ed. by D. Ginsberg, Vol. 4 (World Scientific, Singapore, 1994) p. 61

205. A. Hosseini, S. Kamal, D.A. Bonn, R. Liang, W.N. Hardy, Phys. Rev. Lett. **81**, 1298 (1998)
206. M. Tachiki, Y. Takano, T. Hatano, Physica C **367**, 343 (2002)
207. N. Tomita, Y. Takahashi, M. Mori, Y. Uchida, Jpn. J. Appl. Phys. **31**, L942 (1992)
208. J.L. Wang, X.Y. Cai, R.J. Kelley, M.D. Vaudin, S.E. Babcock, D.C. Larbalestier, Physica C **230**, 189 (1994)
209. Q. Li, Y.N. Tsay, M. Suenaga, G.D. Gu, N. Koshizuka, Physica C **282-287**, 1495 (1997)
210. Q. Li, Y.N. Tsay, M. Suenaga, R.A. Klemm, G.D. Gu, N. Koshisuka, Phys. Rev. Lett. **83**, 4160 (1999)
211. Y. Zhu, Q. Li, Y.N. Tsay, M. Suenaga, G.D. Gu, N. Koshizuka, Phys. Rev. B **57**, 8601 (1998)
212. D. Dimos, P. Chaudhari, J. Mannhart, Phys. Rev. B **41**, 4038 (1990)
213. A. Gurevich, E. Pashitskii, Phys. Rev. B **57**, 13 878 (1998)
214. J.R. Kirtley, C.C. Tsuei, A. Ariando, C.J.M. Verwijz and S. Harkema and H. Hilgenkamp, Nature Physics **2**, 190 (2006)
215. M.R. Beasley, D. Lew, R.B. Laughlin, Phys. Rev. B **49**, 12330 (1994)
216. T.-K. Ng, CM Varma, Phys. Rev. B **70**, 054514 (2004)
217. H.J.H. Smilde, H. Hilgenkamp, G. Rijnders, H. Rogalla, D.H.A. Blank, Appl. Phys. Lett. **80**, 4579 (2002)
218. H.J.H. Smilde, A.A. Golubov, Ariando, G. Rijnders, J.M. Dekkers, S. Harkema, D.H.A. Blank, H. Rogalla, H. Hilgenkamp, Phys. Rev. Lett. **95**, 257001 (2005)
219. S. Sachdev, Science **475**, 475 (2000)
220. J. Wheatley, T. Xiang, Solid State Commun. **88**, 593 (1993)
221. E. Dagotto, Rev. Mod. Phys. **6**, 763 (1994)
222. C. O'Donovan, J.P. Carbotte, Physica C **252**, 87 (1995)
223. C. O'Donovan, J.P. Carbotte, Phys. Rev. B **52**, 16 208 (1995)
224. K. Sakai, Y. Yokoya, Y.O. Nakamura, Physica C **279**, 127 (1997)
225. B.E.C. Koltenbah, R. Joynt, Rep. Prog. Phys. **60**, 23 (1997)
226. K.I. Wysokinski, Physica C **282-287**, 1701 (1997)
227. G. Varelogiannis, Phys. Rev. B **57**, 13 743 (1998)
228. G. Varelogiannis, Solid State Commun. **107**, 427 (1998)
229. E.A. Pashitskii, V.I. Pentegov, A.V. Semenov, Low Temp. Phys. **23**, 101 (1997)
230. C.C. Tsuei, J.R. Kirtley, J. Alloys and Compounds **250**, 615 (1997)
231. C.P. Poole, H.A. Farach, R.J. Creswick, *Superconductivity* (Academic Press, New York, 1995)
232. R.J. Gooding, K.E. Vos, P.W. Leung, Phys. Rev. B **50**, 12 866 (1994)
233. T. Tohyama, S. Maekawa, Phys. Rev. B **49**, 3596 (1994)
234. V.J. Emery, Phys. Rev. Lett. **58**, 2794 (1987)
235. M. Sugihara, M.A. Ikeda, P. Entel, Phys. Rev. B **57**, 11 760 (1998)
236. D. Manske, I. Eremin, K.H. Bennemann, Phys. Rev. B **62**, 13 922 (2000)
237. G.M. Zhao, Phys. Rev. B **64**, –24 503 (2001)
238. B.H. Brandow, Phys. Rev. B **65**, 054 503 (2002)
239. J.R. Kirtley, C.C. Tsuei, K.A. Moler, Science **285**, 1373 (1999)
240. Y. Shimakawa, et al., NEC Res. and Develop. **34**, 35 (1993)
241. Y. Shimakawa, Physica C **204**, 247 (1993)
242. M.A.C. Aranda, D.C. Sinclair, J.P. Attfield, Phys. Rev. B **51**, 12 747 (1995)
243. M. Vojta, Y. Zhang, S. Sachdev, Phys. Rev. Lett. **85**, 4940 (2000)
244. J.P. Wallington, J.F. Annett, Phys. Rev. B **61**, 1433 (2000)
245. A.A. Ovchinnikov, M.Y. Ovchinnikova, Phys. Lett. A **249**, 531 (1998)
246. S. Sachdev, cond-mat/0109419 (2001)
247. C.C. Tsuei, J.R. Kirtley, H. Raffy, Z.Z. Li, to be published.
248. N.C. Yeh, C.T. Chen, G. Hammert, J. Mannhart, A. Schmehl, C.W. Schneider, P.R. Schulz, S. Tajima, K. Yoshid, D. Garrigus, M. Strasik, Phys. Rev. Lett. **87**, 087 003 (2001)
249. M. Sutherland, D.G. Hawthorne, R.W. Hill, F. Ronning, S. Wakimoto, H. Zhang, C. Proust, E. Boaknin, C. Lupien, L. Taillefer, R. Liang, D.A. Bonn, W.N Hardy, R. Gagnon, N.E. Hussey, T. Kimura, N. Nohara and H. Takagi, Phys. Rev. B **67**, 174520 (2003)
250. D.G. Hawthorne, S.Y. Li, M. Sutherland, E. Boaknin, R.W. Hill, C. Proust, F. Ronning, M.A. Tanatar, J. Paglione, D. Peets, R. Liang, D.A. Bonn, W.N. Hardy, N.N. Kolesnikov, Louis Taillefer, cond-mat/0502273 (2005)

251. A. Ghosh, K. Adhikari, Physica C **309**, 251 (1998)
252. Y.G. Wang, M. Liu, Y.H. Yang, D.Y. Xing, Phys. Lett. A **291**, 433 (2001)
253. E.A. Shapoval, Physica C **341-348**, 1697 (2000)
254. J.R. Kirtley, C.C. Tsuei, K.A. Moler, V.G. Kogan, J.R. Clem, A.J. Turberfield, Appl. Phys. Lett. **74**, 4011 (1999)
255. G. Kotliar, Phys. Rev. B **37**, 3664 (1988)
256. M. Liu, D.Y. Xing, Z.D. Wang, Phys. Rev. B **55**, 3181 (1997)
257. A. Ghosh, K. Adhikari, J. Phys. Condens. Matter **10**, L319 (1998)
258. R.B. Laughlin, Physica C **234**, 280 (1994)
259. R.B. Laughlin, Phys. Rev. Lett. **80**, 5188 (1998)
260. J.X. Zhu, C.S. Ting, Phys. Rev. B **57**, 3038 (1998)
261. J.X. Zhu, W. Kim, C.S. Ting, Phys. Rev. B **57**, 13 410 (1998)
262. K. Karpińska, M.Z. Cieplak, S. Guha, A. Malinowski, T. Skośkiewicz, W. Plesiewicz, M. Berkowski, B. Boyce, T.R. Lemberger, P. Lindenfeld, Phys. Rev. Lett. **84**, 155 (2000)
263. G. Koren, E. Polturak, N. Levy, G. Deutscher, N. Zakaharov, Appl. Phys. Lett. **73**, 3763 (1998)
264. K. Krishana, N.P. Ong, Q. Li, G.D. Gu, N. Koshizuka, Science **277**, 83 (1997)
265. R. Movshovich, M.A. Hubbard, M.B. Salamon, A.V. Balatsky, R. Yoshizaki, J.L. Sarrao, M. Jaime, Phys. Rev. Lett. **80**, 1968 (1998)
266. H. Aubin, K. Behnia, S. Ooi, T. Tamegai, K. Krishna, N.P. Ong, Q. Li, G. Gu, N. Koshizuka, Science **280**, 11 (1998)
267. H. Aubin, K. Behnia, S. Ooi, T. Tamegai, Phys. Rev. Lett. **82**, 624 (1999)
268. K. Krishana, N.P. Ong, Q. Li, G.D. Gu, N. Koshizu, J. Phys. Chem. Solids **59**, 2088 (1998)
269. P.W. Anderson, cond-mat/9812063 (1998)
270. M. Matsumoto, H. Shiba, J. Phys. Soc. Jpn. **64**, 1703 (1995)
271. M. Matsumoto, H. Shiba, J. Phys. Soc. Jpn. **64**, 3384 (1995)
272. M. Matsumoto, H. Shiba, J. Phys. Soc. Jpn. **64**, 4867 (1995)
273. A.B. Kuklov, M. Sigrist, Int. J. Mod. Phys. B **11**, 1113 (1997)
274. A.B. Kuklov, Phys. Rev. B **52**, R7002 (1995)
275. M.E. Simon, C.M. Varma, Phys. Rev. B **60**, 9744 (1999)
276. A.V. Balatsky, Phys. Rev. Lett. **80**, 1972 (1998)
277. J.X. Zhu, C.S. Ting, Phys. Rev. B **60**, R 3739 (1999)
278. Y. Dagan, G. Deutscher, Phys. Rev. Lett. **87**, 177 004 (2001)
279. Y. Dagan, G. Deutscher, Phys. Rev. B **64**, 092 509 (2001)
280. R. Carmi, E. Polturak, E. Koren, A. Auerbach, Nature **404**, 853 (2000)
281. C.R. Hu, Phys. Rev. Lett. **72**, 1526 (1994)
282. Y. Tanaka, S. Kashiwaya, Phys. Rev. Lett. **74**, 3451 (1995)
283. M. Fogelström, D. Rainer, J. Sauls, Phys. Rev. Lett. **79**, 281 (1997)
284. A. Golubov, F. Tafuri, Phys. Rev. B **62**, 15 200 (2000)
285. D.K. Morr, E. Delmer, cond-mat/0010460 (2000)
286. W.K. Neils, D.J. van Harlingen, Phys. Rev. Lett. **88**, 47 001 (2002)
287. F. Tafuri, J.R. Kirtley, Phys. Rev. B **63**, 13 924 (2000)
288. N. Kim, Y.J. Doh, H.S. Chang, H.J. Lee, Phys. Rev. B **59**, 14 639 (1999)
289. P.W. Anderson, Science **235**, 1196 (1987)
290. M. Imada, A. Fujimori, Y. Tokura, Rev. Mod. Phys. **70**, 1039 (1998)
291. J. Orenstein, A.J. Millis, Science **288**, 468 (2000)
292. C.C. Tsuei, D.M. Newns, J.R. Kirtley, P.C. Pattnaik, In: *High-T_c Superconductivity and the C_{60} Family*, ed. by S. Feng, H.C. Ren (Gordon and Breach, New York, 1995) p. 293
293. E. Dagotto, A. Nazarenko, M. Boninsegni, Phys. Rev. Lett. **73**, 728 (1994)
294. K. Levin, J.H. Kim, J.P. Lu, Q. Si, Physica C **175**, 449 (1991)
295. D.M. Newns, C.C. Tsuei, P.C. Pattnaik, Phys. Rev. B **52**, 13 611 (1995)
296. N.E. Bickers, D.J. Scalapino, R.T. Scalettar, Int. J. Mod. Phys. B **1**, 687 (1987)
297. N.E. Bickers, D.J. Scalapino, S.R. White, Phys. Rev. Lett. **62**, 961 (1989)
298. D. Poiblanc, Phys. Rev. B **49**, 1477 (1994)

299. M. Moraghebi, S. Yunoki, A. Moreo, cond-mat/0110382 (2001)
300. D.Z. Lin, K. Levin, Physica C **275**, 81 (1997)
301. E. Arrigoni, A.P. Harji, W. Hanke, B. Brendel, S.A. Kivelson, cond-mat/0012293 (2000)
302. D.J. Scalapino, E.L. Jr, J.E. Hirsch, Phys. Rev. B **34**, 8190 (1986)
303. K. Miyake, S. Schmitt-Rink, C.M. Varma, Phys. Rev. B **34**, 6554 (1986)
304. D.J. Scalapino, E.L. Jr, J.E. Hirsch, Phys. Rev. B **35**, 6694 (1987)
305. A.J. Millis, H. Monien, D. Pines, Phys. Rev. B **42**, 167 (1990)
306. T. Moriya, Y. Takahasi, K. Udea, J. Phys. Soc. Jpn. **59**, 2905 (1990)
307. P. Monthoux, A.V. Balatsky, D. Pines, Phys. Rev. Lett. **67**, 3448 (1991)
308. P. Monthoux, D. Pines, Phys. Rev. Lett. **69**, 961 (1992)
309. P. Monthoux, D. Pines, Phys. Rev. B **47**, 6069 (1993)
310. H.B. Schüttler, M.R. Norman, Phys. Rev. B **54**, 13 295 (1996)
311. J.P. Franck, In: *Physical Properties of High Temperature Superconductors*, ed. by D.M. Ginsberg, Vol. 4 (World Scientific, Singapore, 1994) p. 189
312. G.M. Zhao, H. Keller, K. Conder, J. Phys. Condens. Matter **13**, R569 (2001)
313. A. Bussmann-Holder, K.A. Müller, R. Micans, H. Buttner, A. Simon, A.R. Bishop, T. Egami, J. Phys. Cond. Matt. **13**, L169 (2001)
314. R. Zeyher, M.L. Kulić, Phys. Rev. B **53**, 2850 (1996)
315. M.L. Kulić, O.V. Dolgov, cond-mat/9902232 (1999)
316. M.L. Kulić, Phys. Rept. **338**, 1 (2000)
317. J.H. Kim, Z. Tešanović, Phys. Rev. Lett. **71**, 4218 (1993)
318. A. Nazarenko, E. Dagatto, Phys. Rev. B **53**, R2987 (1996)
319. G.A. Ummarino, R.S. Gonnelli, Phys. Rev. B **56**, R14 279 (1997)
320. T.S. Nunner, J. Schmalian, K.H. Benneman, Phys. Rev. B **59**, 8859 (1999)
321. J.R. Schrieffer, Solid State Commun. **92**, 129 (1994)
322. J.R. Schrieffer, J. Low. Temp. Phys. **99**, 397 (1995)
323. E. Cappelluti, L. Pietronero, Europhys. Lett. **36**, 619 (1996)
324. M. Grilli, C. Castellani, Phys. Rev. B **50**, 16 880 (1994)
325. O.V. Danylenko, O.V. Dolgov, M.L. Kulić, V. Oudovenko, Eur. Phys. J. B **9**, 201 (1999)
326. G. Varelogiannis, Physica C **317-318**, 238 (1999)
327. P. Paci, C. Grimaldi, L. Pietronero, E. Cappelluti, Int. J. Mod. Phys. B **14**, 2982 (2000)
328. A. Lanzara, P.V. Bogdanov, X.J. Zhou, S.A. Kellar, D.L. Feng, E.D. Lu, T. Yoshida, H. Elsaki, H. Fujimori, K. Kishio, J.I. Shimoyama, T. Noda, S. Uchida, Z. Hussain, Z.X. Shen, Nature **412**, 510 (2001)
329. P.A. Lee, Phys. Rev. Lett. **71**, 1887 (1993)
330. A.C. Durst, P. Lee, Phys. Rev. B **62**, 1270 (2000)
331. Y. Sun, K. Maki, Europhys. Lett. **32**, 355 (1995)
332. G.E. Volovik, JETP Lett. **58**, 469 (1993)
333. T. Senthil, M.P.A. Fisher, L. Balents, C. Nayak, Phys. Rev. Lett. **81**, 4704 (1998)
334. T. Senthil, M.P.A. Fisher, Phys. Rev. B **60**, 6893 (1999)
335. A.A. Nersesyan, A.M. Tsvelik, F. Wenger, Phys. Rev. Lett. **72**, 2628 (1994)
336. W.A. Atkinson, P.J. Hirschfeld, A.H. MacDonald, Phys. Rev. Lett. **85**, 3922 (2000)
337. S. Vishveshwara, T. Senthil, M.P.A. Fisher, Phys. Rev. B **61**, 6966 (2000)
338. J.X. Zhu, D.N. Zheng, C.S. Ting, Phys. Rev. Lett. **85**, 4944 (2000)
339. T. Fukui, Phys. Rev. B **62**, R9279 (2000)
340. S. Haas, K. Maki, Phys. Rev. Lett. **85**, 2172 (2000)
341. A.V. Balatsky, M.I. Salkola, A. Rosengren, Phys. Rev. B **51**, 15 547 (1995)
342. A.V. Balatsky, M.I. Salkola, Phys. Rev. Lett. **76**, 2386 (1996)
343. D.V. Khveshchenko, A.G. Yashenkin, I.V. Gornyi, Phys. Rev. Lett. **86**, 4668 (2001)
344. D.V. Khveshchenko, J. Paaske, Phys. Rev. Lett. **86**, 4672 (2001)
345. L. Marinelli, B.I. Halperin, S.H. Simon, Phys. Rev. B **62**, 3488 (2000)
346. M. Franz, Z. Tešanović, Phys. Rev. Lett. **84**, 554 (2000)

347. O. Vafek, A. Melikyan, Z. Tešanovič, Phys. Rev. B **64**, 224 508 (2001)
348. M. Franz, O. Vafek, cond-mat/0107521 (2001)
349. M. Kao, K. Maki, J. Magn. Magn. Mater. **226-230**, 280 (2001)
350. L.P. Gorkov, P.A. Kalugin, JETP Lett. **41**, 253 (1985)
351. M.J. Graf, S.K. Yip, J.A. Sauls, D. Rainer, Phys. Rev. B **53**, 15 147 (1996)
352. M. Chiao, K. Behnia, S. Belin, H. Aubin, F. Rullier-Albenque, S. Ooi, T. Tameqai, A. Deluzer, P. Batail, J. Low. Temp. Phys. **117**, 1089 (1999)
353. J. Mesot, M.R. Norman, H. Ding, M. Randeria, J.C. Campuzano, A. Paramekanti, H.M. Fretwell, A. Kaminski, T. Takeuchi, T. Yokoya, T. Sato, T. Takahashi, T. Mochiku, K. Kadowaki, Phys. Rev. Lett. **83**, 840 (1999)
354. C. Proust, E. Boaknin, R.W. Hill, L. Taillefer, A.P. Mackenzie, cond-mat/0202101 (2001)
355. R.W. Hill, C. Proust, L. Taillefer, P. Fournier, R.L. Greene, NATURE **414**, 711 (2001)
356. N.E. Hussey, S. Nakamae, K. Behnia, H. Takagi, C. Urano, S. Adachi, S. Tajima, Phys. Rev. Lett. **85**, 4140 (2000)
357. J. Takeya, Y. Ando, S. Komiya, X.F. Sun, cond-mat/0108055 v2 (2001)
358. D. Davidović, S. Kumar, D.H. Reich, J. Siegel, S.B. Field, R.C. Tiberio, R. Hey, K. Ploog, Phys. Rev. B **55**, 6518 (1997)
359. E. Terzioglu, M.R. Beasley, IEEE Trans. Appl. Supercond. **8**, 48 (1998)
360. C.H. Bennett, Phys. Today **48(10)**, 24 (1995)
361. C.H. Bennett, D.P. DiVincenzo, Nature **404**, 247 (2000)
362. M.A. Nelson, I.L. Chuang, *Quantum Computation and Quantum Information* (Cambridge University Press, Cambridge, 2000)
363. L.B. Ioffe, V.B. Geshkenbein, M.V. Feigel'man, A.L. Fauchère, G. Blatter, Nature **398**, 679 (1999)
364. M.H.S. Amin, A.Yu Smirnov, A.M Zagoskin, T. Lindström, S.A. Charlebois, T. Claeson, A.Ya Tzalenchuk, Phys. Rev. B **71**, 064516 (2005)
365. C.C. Tsuei, J.R. Kirtley, Physica C **367**, 1 (2002)
366. T. Löfwander, G. Johansson, V. Shumeiko, G. Wendin, M. Hurd, Superlatt. and Microstructures **25**, 1115 (1999)
367. T. Bauch, F. Lombardi, F. Tafuri, A. Barone, G. Rotoli, P. Delsing, T. Claeson, Phys. Rev. Lett. **94**, 087003 (2005)
368. T. Bauch, T. Lindström, F. Tafuri, G. Rotoli, P. Delsin, T. Claeson, F. Lombardi, Science **311**, 57 (2006)
369. K. Inomata, S. Sato, K. Nakajima, A. Tanaka, Y. Takano, H.B. Wang, M. Nagao, H. Hatano, S. Kawabata, Phys. Rev. Lett. **95**, 107005 (2005)
370. X.Y. Jin, J. Lisenfeld, Y. Koval, A. Lukashenko, A.V. Ustinov, P. Müller, Phys. Rev. Lett. **96**, 177003 (2006)
371. Y.V. Fominov, A.A. Golubov, M.Y. Kupriyanov, JETP Lett. **77**, 587 (2003)
372. M.H.S. Amin, A.Yu Smirnov, Phys. Rev. Lett. **92**, 017001(2004)
373. S. Kawabata, S. Kashiwaya, Y. Asano, Y. Tanaka, Phys. Rev. B **72**, 052506 (2005)
374. H. Hilgenkamp, Ariando, H.-J Smilde, D.H.A. Blank, H. Rogalla, J.R. Kirtley, C.C. Tsuei, Nature **422**, 50 (2003)
375. T. Ortlepp, Ariando, O. Mielke, C.J.M. Verwijs, K.F.K. Foo, H. Rogalla, H.F. Uhlmann, H. Hilgenkamp, Science **312**, 1495 (2006)
376. D.M. Newns and C.C. Tsuei, Nature Physics **3**, 184 (2007)

17 Photoemission in the High-T_c Superconductors

J. C. Campuzano Department of Physics, University of Illinois at Chicago, Chicago, USA
M. R. Norman Materials Science Division, Argonne National Laboratory, Argonne, USA
M. Randeria Tata Institute of Fundamental Research, Mumbai, India

17.1 Introduction

Angle resolved photoemission spectroscopy (ARPES) has played a major role in the elucidation of the electronic excitations in the high temperature cuprate superconductors. Several reasons have contributed to this development. First, the great improvement in experimental resolution, both in energy and momentum, aided by the large energy scales present in the cuprates, allows one to see features on the scale of the superconducting gap. More recently the resolution has improved to such an extent, that now features in traditional superconductors like Nb and Pb, with energy scales of a meV, can be observed by ARPES [1]. Second, most studies have focused on $Bi_2Sr_2CaCu_2O_8$ (Bi2212) and its single layer counterpart, $Bi_2Sr_2CuO_6$ (Bi2201). These materials are characterized by weakly coupled BiO layers, with the longest interplanar separation in the cuprates. This results in a natural cleavage plane, with minimal charge transfer. This is crucial for ARPES, since it is a surface sensitive technique; for the photon energies typically used, the escape depth of the outgoing electrons is only of the order of $\sim$ 10 Å. For this reason, we elect to concentrate on these materials in the current article, since the data are known to be reproducible among the various groups.

The third reason is the quasi-two-dimensionality of the electronic structure of the cuprates, which permits one to unambiguously determine the momentum of the initial state from the measured final state momentum, since the component parallel to the surface is conserved in photoemission. Moreover, in two dimensions, ARPES directly probes the single particle spectral function, and therefore offers a complete picture of the many body interactions inherent in these strongly correlated systems.

There are other reasons to point out as well. For example, the very incoherent nature of the excitations near the Fermi energy in the normal state does not allow the application of traditional techniques, such as the de Haas van Alphen effect. The electrons simply do not live long enough to complete a cyclotron orbit. Another important reason why ARPES has played a major role is the highly anisotropic nature of the electronic excitations in the cuprates, which means that momentum resolved probes are desirable when attempting to understand these materials.

It is satisfying to compare the earlier ARPES papers to more recent ones, and observe the remarkable progress the field has undergone. But this satisfaction must be tempered by the knowledge that the current literature is still full of unresolved issues. Undoubtedly, some of what is described here will be viewed in a new light in the future, and we will try to point out the contentious points, but the reader must keep in mind that the writers, by necessity, approach this task with their own set of biases. For a complementary view, the reader is referred to the recent review by Damascelli, Shen and Hussain [2]. An earlier review of ARPES studies of high T_c superconductors can be found in the book by Lynch and Olson [3].

17.2 Basics of Angle-Resolved Photoemission

The basics of angle-resolved photoemission have been described in detail in the literature [4]. We will limit ourselves to a brief review of some salient points, emphasizing those aspects of the technique which will be useful in understanding ARPES studies of high T_c superconductors. We start by looking at a simple independent particle picture, and subsequently include the effects of strong interactions.

The simplest model of ARPES is the three step model [5], which separates the process into photon absorption, electron transport through the sample, and emission through the surface. In the first step, the incident photon with energy $h\nu$ is absorbed by an electron in an occupied initial state, causing it to be promoted to an unoccupied final state, as shown in Fig. 17.1(a). There is conservation of energy, such that

$$h\nu = BE + \Phi + E_{\mathrm{kin}} , \tag{17.1}$$

where Φ is the work function, and BE and E_{kin} are the binding and kinetic energies of the electron, respectively. We will not discuss the subsequent two steps of the three step model, photoelectron transport to the surface and transmission through the surface into the vacuum, as they only affect the number of emitted electrons, and thus the absolute intensity [4].

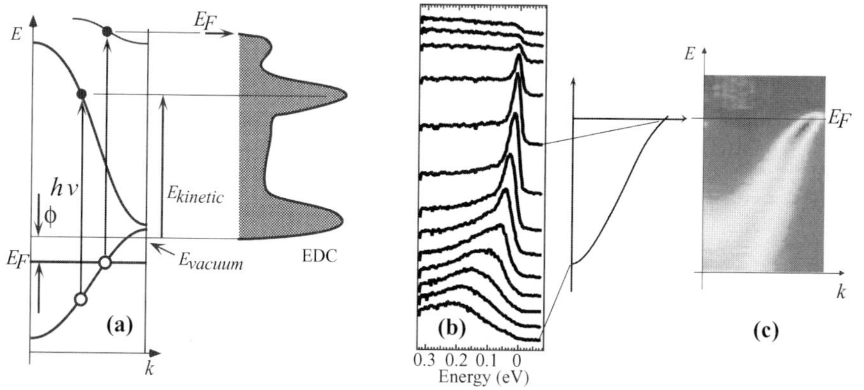

Fig. 17.1. (**a**) Independent particle approximation view of ARPES; to the right is the resulting energy distribution curve (EDC). (**b**) Experimental EDCs along a path in momentum space in Bi2212. (**c**) Intensity map from these data

The kinetic energy of the electrons is measured by an electron energy analyzer. If the number of emitted electrons is then plotted as a function of their kinetic energy, as shown in Fig. 17.1, peaks are found whenever an allowed transition takes place. Eq. (17.1) then yields the kinetic energy of the electron if the work function is known. Measuring Φ accurately is not a trivial task, but fortunately, in the case of a metallic sample one does not need to know Φ. By placing the sample in electrical contact with a good metal (e.g., polycrystalline gold) one can measure the binding energy in the sample with respect to the chemical potential (Fermi level E_f) of Au. The photoemission signal from Au will simply be a Fermi function convolved with the experimental energy resolution, and from the mid point of its leading edge one estimates E_f.

The existence of the sample surface breaks (discrete) translational invariance in the direction normal to the surface. However, two-dimensional translational invariance in the directions parallel to the surface is still preserved, and thus $\mathbf{k}_\parallel$, the component of the electron momentum parallel to the surface, is conserved in the emission process. This allows us to obtain the in-plane momentum of the initial state by identifying it with the parallel momentum of the emitted electron

$$k_\parallel = \sqrt{\frac{2mE_{\text{kin}}}{\hbar}} \sin\theta \cos\phi \qquad (17.2)$$

of the outgoing electron emitted along the (θ, ϕ) direction with a kinetic energy E_{kin}, as shown in Fig. 17.2(a).

The momentum perpendicular to the sample surface k_z is not conserved, and thus knowledge of the final state k_z^f does not permit one to say anything useful about the initial state k_z^i, except in the case where $\mathbf{k}_\parallel = 0$, in which case $k_z^i = k_z^f$. We should emphasize that a given outgoing electron corresponds to a fixed, but a priori unknown, k_z^i. In materials with three-dimensional electronic dispersion it is, of course, essential to fully characterize the initial state, and many techniques have been developed to estimate k_z^i; see [4]. Aside from making $\mathbf{k}_\parallel = 0$ as described above, one can vary the incident photon energy $h\nu$ thereby changing the kinetic energy of the outgoing electron. For fixed $\mathbf{k}_\parallel$ this amounts to changing the initial state k_z. (One also has to take into account the changes in intensity brought about by the photon energy dependence of the matrix elements).

We will not be much concerned about methods of k_z determination here, since the high T_c cuprates are quasi-two-dimensional (2D) materials with, in many cases (e.g., Bi2212), no observable k_z-dispersion. In fact, this makes ARPES data from 2D materials much easier to interpret, and is one of the reasons for the great success of the ARPES technique for the cuprates. In the remainder of this article, we will use the symbol $\mathbf{k}$ to simply denote the two-dimensional momentum parallel to the sample surface, unless explicitly stated otherwise.

In the independent particle approximation, the ARPES intensity as a function of momentum $\mathbf{k}$ (in the 2D Brillouin zone) and energy ω (measured with respect to the chemical potential) is given by Fermi's Golden rule as

$$I(k, \omega) \propto \left|\langle \psi_f | \mathbf{A} \cdot \mathbf{p} | \psi_i \rangle\right|^2 f(\omega)\delta(\omega - \varepsilon_k) , \quad (17.3)$$

where ψ_i and ψ_f are the initial and final states, $\mathbf{p}$ is the momentum operator, and $\mathbf{A}$ the vector potential of the incident photon. (In general, the matrix element also contains a $\mathbf{p} \cdot \mathbf{A}$ term. Above the bulk plasma frequency, the light penetrates deeply into the solid, making this term small. This "surface photoemission" has not been prominent in the many studies of the cuprates, and we will henceforth ignore it.) Here $f(\omega) = 1/[\exp(\omega/T) + 1]$ is the Fermi function at a temperature T in units where $\hbar = k_B = 1$. It ensures the physically obvious constraint that photoemission only probes *occupied* electronic states.

For noninteracting electrons, the emission at a given $\mathbf{k}$ is at a sharp energy $\epsilon_\mathbf{k}$ corresponding to the initial state dispersion. As we will discuss below, the effect of interactions (either electron–electron or electron–phonon) is to replace the delta function by the one-particle spectral function, which is a non-trivial function of both energy and momentum. Although now the peaks will in general shift in energy and acquire a width, the overall intensity is still governed by the same matrix element as in the non-interacting case. Thus the lessons learned from studying general properties of the matrix elements are equally applicable in the fully interacting case.

17.2.1 Matrix Elements and Selection Rules

The dipole matrix element $M_{fi} = \langle \psi_f | \mathbf{A} \cdot \mathbf{p} | \psi_i \rangle$ is in general a function of the momentum $\mathbf{k}$, of the incident photon energy $h\nu$, and of the polarization $\mathbf{A}$ of the incident light. Here we discuss dipole selection rules which arise from very general symmetry constraints imposed on M_{fi}, and which are very useful in interpreting ARPES data. Our approach is a simplified version of Hermanson's analysis [6].

Let the photon beam be incident along a plane of mirror symmetry of the sample ($\mathcal{M}$), with detector placed in the same mirror plane; see Fig. 17.2(a). The final state ψ_f must then be even with respect to reflection in $\mathcal{M}$, because if it were odd the wave function would vanish at the detector. (For this discussion it is simpler to imagine changing the photon polarization, keeping the detector fixed. In the actual experiment, however, it is the polarization which is fixed and the detector is moved when checking dipole selection rules.)

The dipole transition is allowed if the entire matrix element has an overall even symmetry. Thus two possibilities arise. (1) For an initial state ψ_i which is even with respect to $\mathcal{M}$, the light polarization $\mathbf{A}$ must also be even, i.e., parallel to $\mathcal{M}$. (2) For an initial state odd with respect to $\mathcal{M}$, $\mathbf{A}$ must also be odd, i.e., perpendicular to $\mathcal{M}$. This can be summarized as:

$$\langle \psi_f | \mathbf{A} \cdot \hat{\mathbf{p}} | \psi_i \rangle \begin{cases} \psi_i \text{ even } \langle + | + | + \rangle \Rightarrow \mathbf{A} \text{ even} , \\ \psi_i \text{ odd } \langle + | - | - \rangle \Rightarrow \mathbf{A} \text{ odd} . \end{cases} \quad (17.4)$$

Consider hybridized Cu3d-O2p initial states, as shown in Fig. 17.2(b), which have a $d_{x^2-y^2}$ symmetry about a Cu site. These states are even under reflection in $(0, 0) - (\pi, 0)$ (i.e., the plane defined by this symmetry axis and the z-axis) and odd with respect

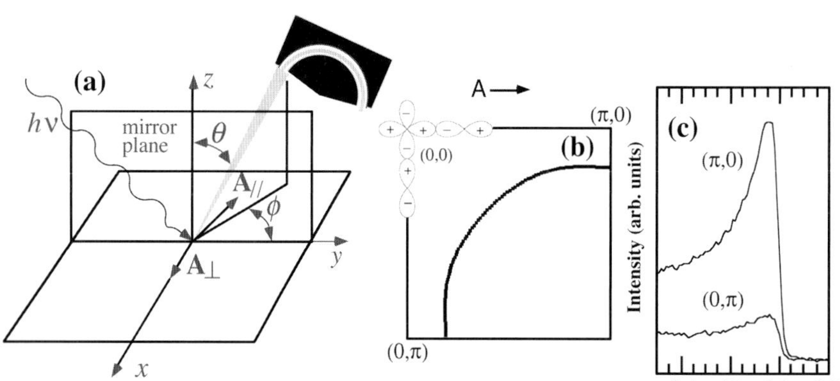

Fig. 17.2. (a) Arrangement of the photon beam and detector in order to make use of the photoemission selection rules. (b) Symmetry of the Cu3$d_{x^2-y^2}$ orbitals hybridized with the O2p orbitals. (c) EDCs showing the symmetry of the orbitals in (b) obtained at $\hbar\nu$ =22 eV

to $(0, 0) - (\pi, \pi)$. Therefore, measurement along the $(0, 0) - (\pi, 0)$ direction will be dipole-allowed (forbidden) if the polarization vector **A** is parallel (perpendicular) to this axis. Figure 17.2(c) shows that, consistent with an initial state which is even about $(0, 0) - (\pi, 0)$, the signal is maximized when **A** lies in the mirror plane and minimized when **A** is perpendicular to this plane. (The reasons for small nonzero intensity in the dipole forbidden geometry are the small, but finite, **k**-window of the experiment and the possibility of a small misalignment of the sample). Similarly, we have checked experimentally (for Bi2212 in the Y-quadrant where there are no superlattice complications) that the initial state is consistent with odd symmetry about $(0, 0) - (\pi, \pi)$ (see also [7]).

While the dipole matrix elements are strongly photon energy dependent, the selection rules are, of course, independent of photon energy. This has been checked by measurements at 22 eV and 34 eV. [8] All of these results are consistent with the fact that we are probing Cu3*d*-O2*p* initial one-electron states with $d_{x^2-y^2}$ symmetry. We will see below that these selection rules are extremely useful in disentangling the main CuO_2 "band" from its umklapp images due to the superlattice in Bi2212.

17.2.2 The One-Step Model

The three-step model of photoemission gives a very useful zeroth order description of the photoemission process, but it needs to be put on a firmer footing, both as regards the calculation of photoemission intensities from ab initio electronic structure calculations, and also for developing a deeper understanding of how many-body effects influence ARPES spectra. We now briefly review the so called "one-step" model of photoemission with these goals in mind.

Ever since Hertz and Hallwachs discovered photoemission [9], it is known that the photo-electron current at the detector is proportional to the incident photon flux, i.e., to the *square* of the vector potential. Thus photoemission measures a *nonlinear* response function, and the relevant correlation function is a three current correlator, as first emphasized by Schaich and Ashcroft [10]. It is instructive to briefly review their argument. As in standard response function calculations, let's look at an expansion of the current at the detector (the response) in powers of the vector potential of incident photons (the perturbation). Let **R** be the location of the detector in vacuum and **r** denote points inside the sample. The zeroth order contribution $\langle 0|j_\alpha(\mathbf{R}, t)|0\rangle$ vanishes as there are no currents flowing in the ground state $|0\rangle$ of the unperturbed system. The linear response also vanishes, with $\langle 0|j_\alpha(\mathbf{R}, t)j_\beta(\mathbf{r}, t')|0\rangle = 0$ and $\langle 0|j_\alpha(\mathbf{r}, t')j_\beta(\mathbf{R}, t)|0\rangle = 0$, since there are no particles at the detector, in absence of the electromagnetic field, and $j_\beta(\mathbf{R}, t)|0\rangle = 0$. Thus the leading term which survives is

$$\langle j_Y(\mathbf{R}, t)\rangle \propto \int \mathrm{d}\mathbf{r}'\mathrm{d}t'\mathrm{d}\mathbf{r}''\mathrm{d}t'' A_\alpha(\mathbf{r}', t')A_\beta(\mathbf{r}'', t'') \times \langle 0|j_\alpha(\mathbf{r}', t')j_Y(\mathbf{R}, t)j_\beta(\mathbf{r}'', t'')|0\rangle \,, \tag{17.5}$$

where only current operators *inside* the sample act on the unperturbed ground state on either side and the current at the detector is sandwiched in between.

The three current correlation function can be represented by the triangle diagram shown in Fig. 17.3(A) [11], where the line between the two external photon vertices is the Greens function of the "initial state" or "photo-hole" and the two lines connecting the photon vertex to the current at the detector represent the "photo-electron" which is emitted from the solid. There is a large literature on the evaluation of the (bare) triangle diagram incorporating the results of ab initio electronic structure calculations of a semi-infinite solid, including the effects of realistic surface termination and multiple scattering effects in the photo-electron final states. Detailed calculations using this formalism were first carried out by Pendry [12] for a system with one atom per unit cell, and later generalized to more complex crystals [13–15].

This approach is very reliable for calculating photoemission intensities, and gives important information about "matrix element effects", i.e., the dependence of the ARPES intensity on **k**, and on the incident photon energy and polarization. (It does not, however, shed light on the many-body aspects of the lineshape). Such ab initio methods have been exten-

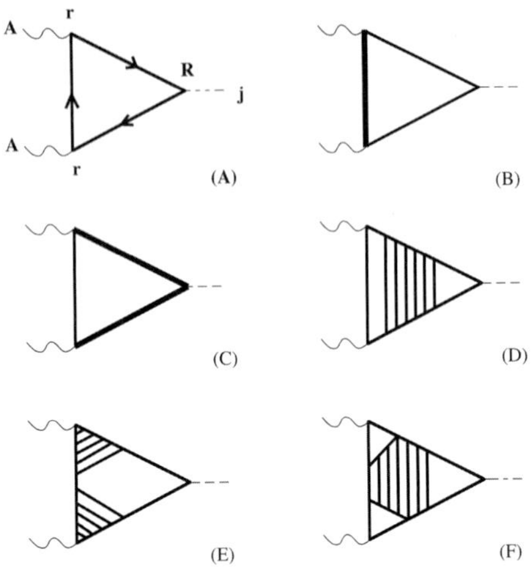

Fig. 17.3. Diagrams contributing to the three-current correlator of ARPES. The physical processes that each diagram represents is discussed in the text

sively used by Bansil, Lindroos and coworkers [16] for the high T_c cuprates.

As an example of the usefulness of ab initio methods, we show the comparison between the observed and calculated ARPES intensities for YBCO in Fig. 17.4. Since the calculated intensity is a sensitive function of the termination plane, such a comparison suggests that the crystal breaks at the chains, a conclusion later confirmed by STM measurements [17]. In fact, it was the complexity of the termination in YBCO, plus the fact that none of the possible cleavage planes of YBCO are charge neutral, and thus involve significant charge transfer, that convinced us to focus primarily on the Bi-based cuprates (Bi2212 and Bi2201). These materials have two adjacent BiO planes which are van der Waals bonded, and thus cleavage leads to neutral surfaces which are not plagued by uncontrolled surface effects. The highly two-dimensional Bi-based cuprates are thus ideal from the point of view of ARPES studies.

17.2.3 Single-Particle Spectral Function

Although the one step model gives a reasonable interpretation of the overall *intensity* in the photoemission process, much of what we really want to learn from the experiment relates to the spectral *lineshape*, which, as we will show, is strongly influenced by correlations. We must then ask ourselves the important question of how the initial state lineshape enters the ARPES intensity and to what extent this is revealed in the observed lineshape. Although a fully rigorous justification of a simple interpretation of ARPES spectra is not available at the present time, a reasonable case can be made for analyzing the data in terms of the single particle spectral function of the initial state.

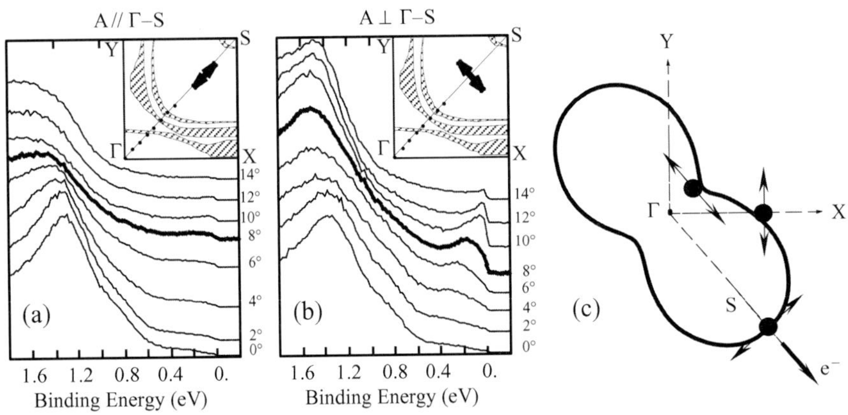

Fig. 17.4. Comparison of the photoemission intensity of states along the zone diagonal in YBCO with (**a**) even polarization and (**b**) odd polarization, showing that this state is odd with respect to reflection about the diagonal to the Cu-O bond direction. (**c**) Comparison of the photoemission intensity to the observed one as a function of polarization. The detailed polarization comparison gives confidence to the interpretation of the observed peak as arising from the Cu-O plane states

We begin by considering the various many-body renormalizations of the bare triangle diagram which are shown in Fig. 17.3. These renormalizations can arise, in principle, from either electron–phonon or electron–electron interactions. These self-energy effects and vertex corrections are easy to draw, but impossible to evaluate in any controlled calculation. Nevertheless, they are useful in obtaining a qualitative understanding of the various processes and estimating their relative importance. Diagram (B) represents the self-energy corrections to the occupied initial state that we are actually interested in studying; (C) and (D) represent final state line-width broadening and inelastic scattering; (E) is a vertex correction that describes the interaction of the escaping photo-electron with the photo-hole in the solid; (F) is a vertex correction which combines features of (D) and (E). (An additional issue in a quantitative theory of photoemission is related to the modification of the external vector potential inside the medium, i.e., renormalizations of the photon line. These are considered in detail in [16]).

If the sudden approximation is valid, we can neglect the vertex corrections: the outgoing photo-electron is moving so fast that it has no time to interact with the photo-hole. Let us make simple time scale estimates for the cuprates with 15–30 eV (ultraviolet) incident photons. The time t spent by the escaping photo-electron in the vicinity of the photo-hole is the time available for their interaction. A photoelectron with a kinetic energy of (say) 20 eV has a velocity $v = 3 \times 10^8$ cm/s. The relevant length scale, which is the smaller of the screening radius (of the photo-hole) and the escape depth, is $\sim$ 10 Å. Thus $t = 3 \times 10^{-16}$ s, which should be compared with the time scale for electron–electron interactions (which are the dominant source of interactions at the high energies of interest): $t_{ee} = (2\pi)/\omega_p = 4 \times 10^{-15}$ s, using a plasma frequency $\omega_p = 1$ eV for the cuprates (this would be even slower if c-axis plasmons are involved). *If* $t \ll t_{ee}$, then we can safely ignore vertex corrections. From our very crude estimate $t/t_{ee} = 0.1$, so that the situation with regard to the validity of the impulse approximation is not hopeless, but clearly, experimental checks are needed, and we present these in the next subsection.

Very similar estimates can be made for renormalizations of the outgoing photo-electron due to its interaction with the medium; again electron–electron interactions dominate at the energies of interest. The relevant length scale here is the escape depth, which leads to a process of self-selection: those electrons that actually make it to the detector with an appreciable kinetic energy have suffered no collisions in the medium. Such estimates indicate that the "inelastic background" must be small and we will show how to experimentally obtain its precise dependence on $\mathbf{k}$ and ω later.

Finally, one expects that final state linewidth corrections are small for quasi-2D materials based on the estimates made by Smith et al. [18] In fact, this is yet another reason for the ease of interpreting ARPES data in quasi-2D materials. A clear experimental proof that these effects are negligible for Bi2212 will be presented later, where it will be seen that deep in the superconducting state, a resolution limited leading edge is obtained for the quasiparticle peak.

17.2.4 Spectral Functions and Sum Rules

Based on the arguments presented in the preceding subsection, we assume the validity of the sudden approximation and ignore both the final-state linewidth broadening and the additive extrinsic background. Then (B) is the only term that survives from all the terms described in Fig. 17.3 and the ARPES intensity is given by [19,20]

$$I(\mathbf{k}, \omega) = I_0(\mathbf{k}) f(\omega) A(\mathbf{k}, \omega) , \tag{17.6}$$

where $\mathbf{k}$ is the initial state momentum in the 2D Brillouin zone and ω the energy relative to the chemical potential. The prefactor $I_0(\mathbf{k})$ includes all the kinematical factors and the square of the dipole matrix element (shown in Eq. (17.3)), $f(\omega)$ is the Fermi function, and $A(\mathbf{k}, \omega)$ is the one-particle spectral function which will be described in detail below.

We first describe some general consequences of Eq. (17.6) based on sum rules and their experimental checks. The success of this strategy employed by Randeria et al. [20] greatly strengthens the case for a simple $A(\mathbf{k}, \omega)$ interpretation of ARPES data.

The one-particle spectral function represents the probability of adding or removing a particle from the interacting many-body system, and is defined as $A(\mathbf{k},\omega) = -(1/\pi)\mathrm{Im}G(\mathbf{k},\omega + i0^+)$ in terms of the Green's function. It can be written as the sum of two pieces $A(\mathbf{k},\omega) = A_-(\mathbf{k},\omega) + A_+(\mathbf{k},\omega)$, where the spectral weight to add an electron to the system is given by $A_+(\mathbf{k},\omega) = Z^{-1}\sum_{m,n} e^{-\beta\mathcal{E}_m}|\langle n|c^\dagger_\mathbf{k}|m\rangle|^2 \times \delta(\omega + \mathcal{E}_m - \mathcal{E}_n)$, and that to extract an electron is $A_-(\mathbf{k},\omega) = Z^{-1}\sum_{m,n} e^{-\beta\mathcal{E}_m}|\langle n|c_\mathbf{k}|m\rangle|^2\delta(\omega+\mathcal{E}_n-\mathcal{E}_m)$. Here $|m\rangle$ is an exact eigenstate of the many-body system with energy $\mathcal{E}_m$, Z is the partition function and $\beta = 1/T$. It follows from these definitions that $A_-(\mathbf{k},\omega) = f(\omega)A(\mathbf{k},\omega)$ and $A_+(\mathbf{k},\omega) = \left[1 - f(\omega)\right]A(\mathbf{k},\omega)$, where $f(\omega) = 1/[\exp(\beta\omega) + 1]$ is the Fermi function. Since an ARPES experiment involves removing an electron from the system, the simple golden rule Eq. (17.3) can be generalized to yield an intensity proportional to $A_-(\mathbf{k},\omega)$

We now discuss various sum rules for $A(\mathbf{k},\omega)$ and their possible relevance to ARPES intensity $I(\mathbf{k},\omega)$. While the prefactor I_0 depends on $\mathbf{k}$ and also on the incident photon energy and polarization, it does *not* have any significant ω or T dependence. Thus the energy dependence of the ARPES lineshape and its T dependence are completely characterized by the spectral function and the Fermi factor. The simplest sum rule $\int_{-\infty}^{+\infty} d\omega A(\mathbf{k},\omega) = 1$ is not useful for ARPES since it involves both occupied (through A_-) *and unoccupied* states (A_+). Next, the density of states (DOS) sum rule $\sum_\mathbf{k} A(\mathbf{k},\omega) = N(\omega)$ is also not directly useful since the prefactor I_0 has very strong $\mathbf{k}$-dependence. However it may be useful to $\mathbf{k}$-sum ARPES data $N_p(\omega) = \sum_\mathbf{k} I_0(\mathbf{k})A(\mathbf{k},\omega)$ in an attempt to simulate the angle-integrated photoemission intensity.

The important sum rule for ARPES is

$$\int_{-\infty}^{+\infty} d\omega f(\omega)A(\mathbf{k},\omega) = n(\mathbf{k}) \,, \qquad (17.7)$$

which directly relates the energy-integrated ARPES intensity to the momentum distribution $n(\mathbf{k}) = \langle c^\dagger_\mathbf{k} c_\mathbf{k}\rangle$. (The sum over spins is omitted for simplicity). Somewhat surprisingly, the usefulness of this sum rule has been overlooked in the ARPES literature prior to [20].

We first focus on the Fermi surface $\mathbf{k} = \mathbf{k}_F$. One of the major issues, that we will return to several times in the remainder of this article, will be the question of how to define "$\mathbf{k}_F$" at finite temperatures in a strongly correlated system which may not even have well-defined quasiparticle excitations, and how to determine it experimentally. For now, we simply define the Fermi surface to be the locus of gapless excitations in $\mathbf{k}$-space in the normal state, so that $A(\mathbf{k}_F,\omega)$ has a peak at $\omega = 0$.

To make further progress with Eq. (17.7), we need to make a weak particle–hole symmetry assumption: $A(\mathbf{k}_F,-\omega) = A(\mathbf{k}_F,\omega)$ for "small" ω, where "small" means those frequencies for which there is significant T-dependence in the spectral function. It then follows that [20] $\partial n(\mathbf{k}_F)/\partial T = 0$, i.e., *the integrated area under the EDC at* $\mathbf{k}_F$ *is independent of temperature*. To see this, rewrite Eq. (17.7) as $n(\mathbf{k}_F) = \frac{1}{2} - \frac{1}{2}\int_0^\infty d\omega \tanh(\omega/2T)\,[A(\mathbf{k}_F,\omega) - A(\mathbf{k}_F,-\omega)]$, and take its T-derivative. It should be emphasized that we cannot say anything about the *value* of $n(\mathbf{k}_F)$, only that it is T-independent. (A much stronger assumption, $A(\mathbf{k}_F,-\omega) = A(\mathbf{k}_F,\omega)$ for *all* ω, is sufficient to give $n(\mathbf{k}_F) = 1/2$ independent of T). We emphasize the approximate nature of the $\mathbf{k}_F$-sum-rule since there is no exact symmetry that enforces it.

We note that we did not make use of any properties of the spectral function other than the weak particle–hole symmetry assumption, and to the extent that this is also valid in the superconducting state, our conclusion $\partial n(\mathbf{k}_F)/\partial T = 0$ holds equally well below T_c. There is the subtle issue of the meaning of "$\mathbf{k}_F$" in the superconducting state. In analogy with the Fermi surface as the "locus of gapless excitations" above T_c, we can define the "minimum gap locus" below T_c. We will describe this in great detail in Sect. 17.2.5 below; it suffices to note here that "$\mathbf{k}_F$" is independent of temperature, within experimental errors, in both the normal and superconducting states of the systems studied thus far [21].

In Fig. 17.5(a) we show ARPES spectra for near-optimal Bi2212 ($T_c = 87$ K) at $\mathbf{k}_F$ along $(\pi,0)$ to (π,π) at two temperatures: $T = 13$ K, which is well below T_c, and $T = 95$ K, which is in the normal state. The two data sets were normalized in the positive energy region, which after normalization was cho-

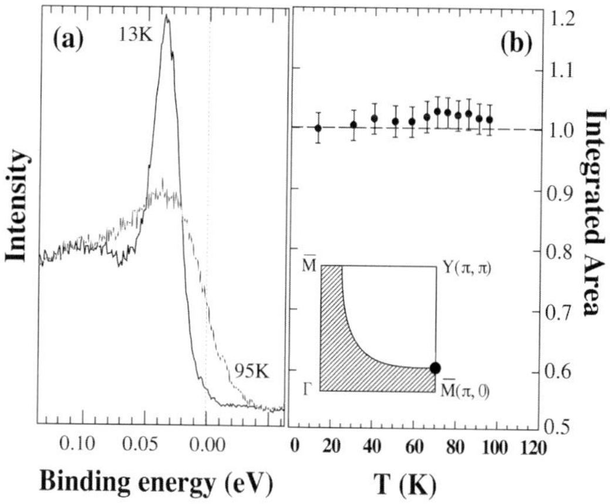

Fig. 17.5. (**a**) ARPES spectra for Bi2212 at the $\mathbf{k}_F$ point shown in the zone inset in the normal and superconducting states. (**b**) Integrated intensity as a function of temperature at the same k point

sen to be the common zero baseline. (The essentially ω-independent emission at positive energies, which is not cut off by the Fermi function, is due to higher harmonics of the incident photon beam, called "second order light".) For details, see [20].

In Fig. 17.5(b) we plot the integrated intensity as a function of T and find that, in spite of the remarkable changes in the lineshape from 95 K to 13 K, the integrated intensity at $\mathbf{k}_F$ is very weakly T-dependent, verifying the sum rule $\partial n(\mathbf{k}_F)/\partial T = 0$. The error bars come from the normalization due to the low count rate in the $\omega > 0$ background.

Let us discuss several other potential complications in testing the T-independence of the integrated intensity at $\mathbf{k}_F$. Note that matrix elements have no effect on this result, since they are T-independent. The same is true for a T-independent additive extrinsic background. In an actual experiment the observed intensity will involve convolution of Eq. (17.6) with the energy resolution and a sum over the $\mathbf{k}$-values within the momentum window. While energy resolution is irrelevant to an integrated quantity, sharp $\mathbf{k}$-resolution is of the essence.

The T-independence of the integrated intensity is insensitive to the choice of the integration cutoff at negative ω, provided it is chosen beyond the dip feature. It has been observed experimentally that the normalized EDCs at $\mathbf{k}_F$ have identical intensities for all $\omega < -100$ meV. This is quite reasonable, since we expect the spectral functions to be the same for energies much larger than the scale associated with superconductivity. The fact that one has to go to 100meV (much larger than $T_c \sim 10$ meV) in order to satisfy the sum rule suggests that electron–electron interactions are involved in superconductivity.

17.2.5 Analysis of ARPES Spectra: EDCs and MDCs

In the preceding subsections we have presented evidence in favor of a simple spectral function interpretation of ARPES data on the quasi 2D high T_c cuprates. In the process, we saw that the ARPES intensity in Fig. 17.5 has a nontrivial lineshape which has a significant temperature dependence. We now introduce some of the basic ideas which will be used throughout the rest of the article to analyze and understand the ARPES lineshape.

The one-electron Green's function can be generally written as $G^{-1}(\mathbf{k},\omega) = G_0^{-1}(\mathbf{k},\omega) - \Sigma(\mathbf{k},\omega)$ where $G_0(\mathbf{k},\omega) = 1/[\omega - \varepsilon_\mathbf{k}]$ is the Green's function of the noninteracting system, $\varepsilon_\mathbf{k}$ is the "bare" dispersion, and the (complex) self-energy $\Sigma(\mathbf{k},\omega) = \Sigma'(\mathbf{k},\omega) + i\Sigma''(\mathbf{k},\omega)$ encapsulates the effects of all the many-body interactions. Then using its definition in terms of ImG, we obtain the general result

$$A(\mathbf{k},\omega) = \frac{1}{\pi}\frac{\Sigma''(\mathbf{k},\omega)}{[\omega - \varepsilon_\mathbf{k} - \Sigma'(\mathbf{k},\omega)]^2 + [\Sigma''(\mathbf{k},\omega)]^2}\,. \tag{17.8}$$

We emphasize that this expression is entirely general, and does not make any assumptions about the validity of perturbation theory or of Fermi liquid theory.

New electron energy analyzers, which measure the photoemitted intensity as a function of energy and momentum simultaneously, allow the direct visualization of the spectral function, as shown in Fig. 17.6, and have also suggested new ways of plotting and analyzing ARPES data.

(a) In the traditional energy distribution curves (EDCs), the measured intensity $I(\mathbf{k},\omega)$ is plotted as a function of ω (binding energy) for a fixed value of $\mathbf{k}$; and

(b) In the new [22] momentum distribution curves (MDCs), $I(\mathbf{k},\omega)$ is plotted at fixed ω as a function of $\mathbf{k}$.

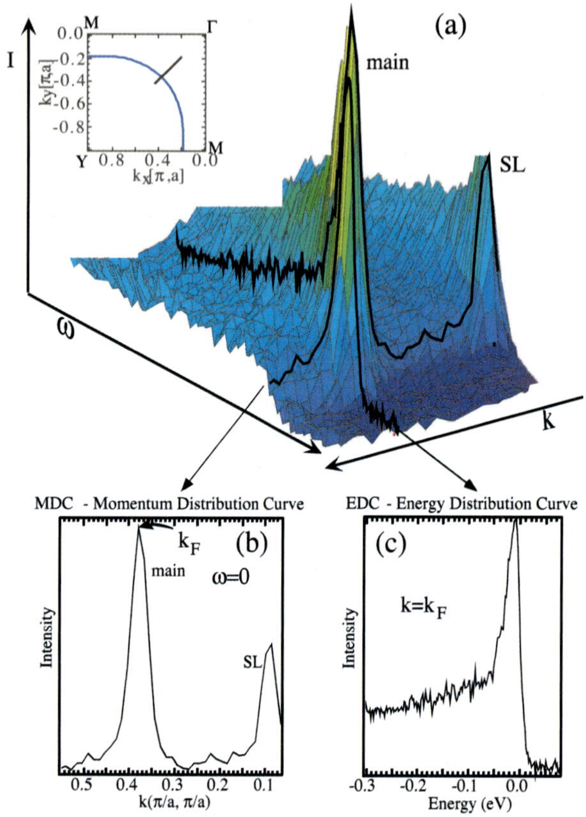

Fig. 17.6. (a) The ARPES intensity as a function of $\mathbf{k}$ and ω at $h\nu = 22$ eV and $T = 40$ K. main is the main band, and SL a superlattice image. (**b**) A constant ω cut (MDC) from (**a**). (**c**) A constant $\mathbf{k}$ cut (EDC) from (**a**). The *diagonal line* in the *zone inset* shows the location of the $\mathbf{k}$ cut; the *curved line* is the Fermi surface

Until a few years ago, the only data available were in the form of EDCs, and even today this is the most useful way to analyze data corresponding to gapped states (superconducting and pseudogap phases). These analyses will be discussed in great detail in subsequent sections. We note here some of the issues in analyzing EDCs and then contrast them with the MDCs. First, note that the EDC lineshape is non-Lorentzian as a function of ω for two reasons. The trivial reason is the asymmetry introduced by the Fermi function $f(\omega)$ which chops off the positive ω part of the spectral function. (We will discuss later on ways of eliminating the effect of the Fermi function). The more significant reason is that the self energy has non-trivial ω dependence and this makes even the full $A(\mathbf{k}, \omega)$ non-Lorentzian in ω as seen from Eq. (17.8). Thus one is usually forced to model the self energy and make fits to the EDCs. At this point one is further hampered by the lack of detailed knowledge of the additive extrinsic background which itself has ω-dependence. (Although, as we shall see, the MDC analysis give a new way of determining this background).

The MDCs obtained from the new analyzers have certain advantages in studying gapless excitations near the Fermi surface [22–24]. In an MDC the intensity is plotted as a function of k varying normal to the Fermi surface in the vicinity of a fixed $\mathbf{k}_F(\theta)$, where θ is the angle parameterizing the Fermi surface. For k near k_F we may linearize the bare dispersion $\varepsilon_\mathbf{k} \simeq v_F^0(k - k_F)$, where $v_F^0(\theta)$ is the bare Fermi velocity. We will not explicitly show the θ dependences of k_F, v_F^0, or other quantities considered below, in order to simplify the notation.

Next we make certain simplifying assumptions about the remaining $\mathbf{k}$-dependences in the intensity $I(\mathbf{k}, \omega)$. We assume that: (i) the self-energy Σ is essentially independent of k normal to the Fermi surface, but can have arbitrary dependence on θ along the Fermi surface; and (ii) the prefactor $I_0(\mathbf{k})$ does not have significant k dependence over the range of interest. It is then easy to see from Eqs. (17.6) and (17.8) that $I(\mathbf{k}, \omega)$ plotted as function k (with fixed ω and θ) has the following lineshape. The MDC is a Lorentzian:

(a) centered at $k = k_F + [\omega - \Sigma'(\omega)]/v_F^0$; with

(b) width (HWHM) $W_M = |\Sigma''(\omega)|/v_F^0$.

Thus the MDC has a very simple lineshape, and its peak position gives the renormalized dispersion, while its width is proportional to the imaginary self energy. The consistency of the assumptions made in reaching this conclusion may be tested by simply checking whether the MDC lineshape is fit by a Lorentzian or not. Experimentally, excellent Lorentzian fits are invariably obtained (except when one is very near the bottom of the "band" or in a gapped state [25]).

Finally, note that the external background in the case of MDCs is also very simple. One can fit the MDC (at each ω) to a Lorentzian plus a constant (at worst Lorentzian plus linear in k) background. From this one obtains the value of the external background including its ω dependence. Now this ω-dependent background can be subtracted off from the EDC also, if one wishes to. Note that estimating this background was not possible from an analysis of the EDCs alone.

17.3 The Valence Band

The basic unit common to all cuprates is the copper-oxide plane, CuO_2. Some compounds have a tetragonal cell, $a = b$, such as the *Tl* compounds, but most have an orthorhombic cell, with a and b differing by as much as 3% in YBCO. There are two notations used in the literature for the reciprocal cell. The one used here, appropriate for Bi2212 and Bi2201, has $\Gamma - M$ along the $Cu{-}O$ bond direction, with $M \equiv (\pi, 0)$, and $\Gamma - X(Y)$ along the diagonal, with $Y \equiv (\pi, \pi)$. The other notation, appropriate to YBCO, has $\Gamma - X(Y)$ along the $Cu - O$ bond direction and $\Gamma - S$ along the diagonal. This difference occurs because the orthorhombic distortion in one compound is rotated 45° with respect to the other. The main effect of the orthorhombicity in Bi2212 and Bi2201 is the superlattice modulation along the b axis, with Q_{SL} parallel to $\Gamma - Y$. Except when referring to this modulation, we will assume tetragonal symmetry in our discussions. For a complete review of the electronic structure of the cuprates, see [26].

The Cu ions are four fold coordinated to planar oxygens. Apical (out of plane oxygens) exist in some structures (LSCO), but not in others. Either way, the apical bond distance is considerably longer than the planar one, so in all cases, the cubic point group symmetry of the Cu ions is lowered, leading to the highest energy Cu state having $d_{x^2-y^2}$ symmetry. As the atomic $3d$ and $2p$ states are nearly degenerate, a characteristic which distinguishes cuprates from other 3d transition metal oxides, the net result is a strong bonding–antibonding splitting of the Cu $3d_{x^2-y^2}$ and O $2p$ σ states, with all other states lying in between. In the stochiometric (undoped) material, Cu is in a d^9 configuration, leading to the upper (antibonding) state being half filled. According to band theory, the system should be a metal. But in the undoped case, integer occupation of atomic orbitals is possible, and correlations due to the strong on-site Coulomb repulsion on the Cu sites leads to an insulating state. That is, the antibonding band "Mott–Hubbardizes" and splits into two, one completely filled (lower Hubbard band), the other completely empty (upper Hubbard band) [27].

On the other hand, for dopings characteristic of the superconducting state, a large Fermi surface is observed by ARPES (as discussed in the next section). Thus, to a first approximation, the basic electronic structure in this doping range can be understood from simple band theory considerations. The simplest approximation is to consider the single Cu $3d_{x^2-y^2}$ and two O $2p$ (x,y) orbitals. The resulting secular equation is [28]

$$\begin{pmatrix} \epsilon_d & 2t\sin\left(\frac{k_xa}{2}\right) & -2t\sin\left(\frac{k_xa}{2}\right) \\ 2t\sin\left(\frac{k_xa}{2}\right) & \epsilon_p & 0 \\ -2t\sin\left(\frac{k_xa}{2}\right) & 0 & \epsilon_p \end{pmatrix}, \quad (17.9)$$

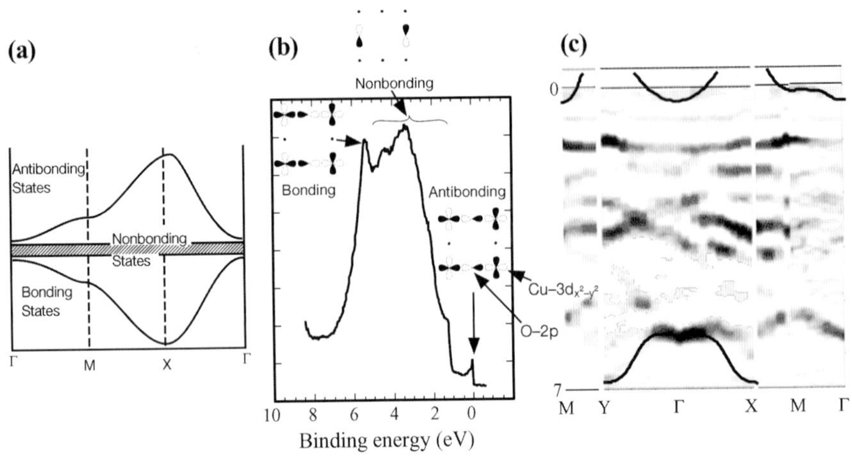

Fig. 17.7. (**a**) Simple three-band estimate of the electronic structure of the Cu-O plane states. (**b**) EDC showing the whole valence band at the $(\pi, 0)$ point. (**c**) Intensity map of the whole valence band obtained by taking the second derivative of spectra such as the one in (**b**). The orbitals in (**b**) are based on the three-band model, where black and white lobes corresponding to positive and negative wavefunctions

where ϵ_d is the atomic $3d$ orbital energy, ϵ_p is the atomic $2p$ orbital energy, and t is the Cu-O hopping integral. Diagonalization of this equation leads to a non-bonding eigenvalue $E_{nb} = \epsilon_p$, and

$$E_{\pm} = \frac{\epsilon_p + \epsilon_d}{2} \pm \sqrt{\left(\frac{\epsilon_p - \epsilon_d}{2}\right)^2 + 4t^2\left(\sin^2\left(\frac{k_x a}{2}\right) + \sin^2\left(\frac{k_x a}{2}\right)\right)}, \qquad (17.10)$$

where + refers to antibonding to bonding. This dispersion is shown in Fig. 17.7(a)

In Fig. 17.7(b), we show an ARPES spectrum obtained at the $(\pi, 0)$ point of the Brillouin zone for Bi2212. Three distinct features can be observed: the bonding state at roughly -6eV, the antibonding state near the Fermi energy, and the rest of the states in between. This rest consists of the non-bonding state mentioned above, as well as the remainder of the Cu $3d$ and O $2p$ orbitals, plus states originating from the other (non Cu-O) planes. It is difficult to identify all of these "non-bonding" states, as their close proximity and broadness causes them to overlap in energy. Perhaps surprisingly, the overall picture of the electronic structure of the valence band has the structure that one would predict by the simple chemical arguments given above, as shown in Fig. 17.7(c). The most important conclusion that one can derive from Fig 17.7 is the early prediction by Anderson [29], namely that there is a single state relevant to transport and superconducting properties. This state, the antibonding state in Fig 17.7, is well separated from the rest of the states, and therefore any reasonable theoretical description of the physical properties of these novel materials should arise from this single state.

Despite these simple considerations, correlation effects do play a major role, even in the doped state. The observed antibonding band width is about a factor of 2–3 narrower than that predicted by band theory [30]. In fact, the correlation effects are the ones we are most interested in, as they give rise to the many unusual properties of the cuprates, including the superconducting state.

As mentioned above, in the magnetic insulating regime, the picture is quite different. The antibonding band splits into two, with the chemical potential lying in the gap. (The electronic structure of the insulator has been extensively studied by the Stanford group, confirming beautifully the predicted valence band maximum at $(\pi/2, \pi/2)$, see [31].) Upon doping with electrons or holes, the chemical potential would move up or down. One would then expect to observe small Fermi surfaces, hole pockets centered at $(\pi/2, \pi/2)$ with a volume equal to the hole doping, x. These can be generated at the mean field level by folding Fig. 17.7(a) back into the magnetic zone with $Q=(\pi, \pi)$, and turning on an interaction, U_{eff}, between the two folded bands. This can be contrasted with the large Fermi surface of the unfolded case, typically centered at (π, π) with a volume $1 + x$. It is still an open question whether there is a continuous evolution between these two limits, or whether there is a discontinuous change at a metal–insulator transition point. In any case, a proper description of the electronic structure must take the strong electron–electron correlations into account, even in the superconducting regime.

17.4 Normal State Dispersion and the Fermi Surface

The Fermi surface is one of the central concepts in the theory of metals, with electronic excitations near the Fermi surface dominating all the low energy properties of the system. In this Section we describe the use of ARPES to elucidate the electronic structure and the Fermi surface of the high T_c superconductors.

It is important to discuss these results in detail because ARPES is the only experimental probe which has yielded useful information about the electronic structure and the Fermi surface of the planar Cu-O states which are important for high T_c superconductivity. Traditional tools for studying the Fermi surface such as the de Haas–van Alphen effect have not yielded useful information about the cuprates, because of the need for very high magnetic fields, and possibly because of the lack of well defined quasiparticles. Other Fermi surface probes like positron annihilation are hampered by the fact that the positrons appear to preferentially probe spatial regions other than the Cu-O planes.

The first issue facing us is: what do we mean by a Fermi surface in a system at high temperatures where there are no well-defined quasiparticles? (Recall that quasiparticles, if they exist, manifest themselves as sharp peaks in the one-electron spectral function whose width is less than their energy, and lead to a jump discontinuity in the momentum distribution at $T = 0$.) Clearly, the traditional $T = 0$ definition of a Fermi surface defined by the jump discontinuity in $n(\mathbf{k})$ is not useful for the cuprates. First, the systems of interest are superconducting at low temperatures. But even samples which have low T_c's have normal state peak widths at E_f which are an order of magnitude broader than the temperature [24,32]. If, as indicated both by ARPES and transport, sharp quasiparticle excitations do not exist above T_c, there is no possibility of observing a thermally-smeared, resolution-broadened, discontinuity in $n(\mathbf{k})$.

It is an experimental fact that in the cuprates ARPES sees broad peaks which disperse as a function of momentum and go through the chemical potential at a reasonably well-defined momentum. We can thus adopt a practical definition of the "Fermi surface" in these materials as "the locus of gapless excitations".

Historically, the first attempts to determine the Fermi surface in cuprates were made on YBCO [33], however, surface effects as well as the presence of chains appear to complicate the picture, so we will focus principally on Bi2212 and Bi2201, which have been studied the most intensively. Other cuprates which have also been studied by ARPES, include the electron-doped material NCCO [34] and, more recently, LSCO as a function of hole doping [35].

We discuss below various methods used for the determination of the spectral function peaks in the vicinity of E_f. In addition, we supplement these methods with momentum distribution studies, taking due care of matrix element complications. We will then discuss three topics: the extended saddle-point in the dispersion, the search for bilayer splitting in Bi2212, and (in Sect. 17.6.4) the doping dependence of the Fermi surface.

17.4.1 Normal State Dispersion in Bi2212: A First Look

We begin with the results obtained by using the traditional method of deducing the dispersion and Fermi surface by studying the EDC peaks as a function of momentum. This method was used for the cuprates by Campuzano et al. [33], Olson et al. [30], and Shen and Dessau [7], culminating in the very detailed study of Ding et al. [36]. The use of EDC peak dispersion has some limitations which we discuss below. Nevertheless, it has led to very considerable understanding of the overall electronic structure, Fermi surface, and of superlattice effects in Bi2212, and therefore it is worthwhile to review its results first, before turning to more refined methods.

The main results of Ding et al. [36] on the electronic dispersion and the Fermi surface in the nor-

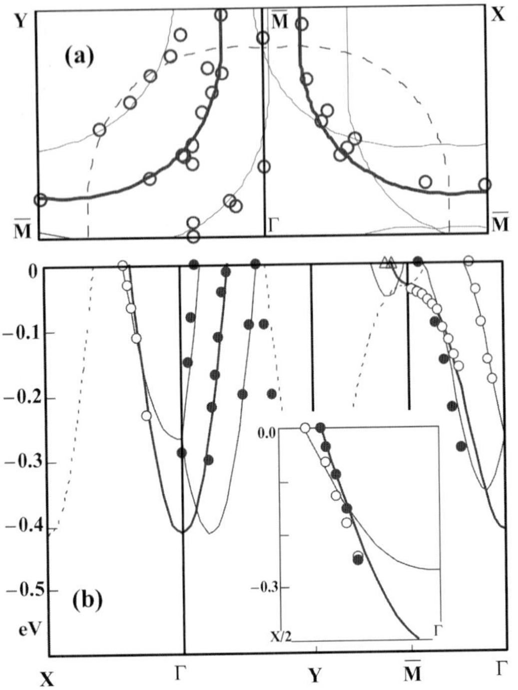

Fig. 17.8. Fermi surface (**a**) and dispersion (**b**) obtained from normal state measurements. The *thick lines* are obtained by a tight binding fit to the dispersion data of the main band with the *thin lines* $(0.21\pi, 0.21\pi)$ umklapps and the *dashed lines* (π, π) umklapps of the main band. *Open circles* in (**a**) are the data. In (**b**), *filled circles* are for odd initial states (relative to the corresponding mirror plane), *open circles* for even initial states, and *triangles* for data taken in a mixed geometry. The *inset* of (**b**) is a blowup of ΓX

mal state ($T = 95$ K) of near-optimal OD Bi2212 ($T_c = 87$ K) using incident photon energies of 19 and 22 eV are summarized in Fig. 17.8. The peak positions of the EDC's as a function of **k** are marked with various symbols in Fig. 17.8(b). The filled circles are for odd initial states (relative to the corresponding mirror plane), open circles for even initial states, and triangles for data taken in a mixed geometry (i.e., the photon polarization was at 45° to the mirror plane). The Fermi surface crossings corresponding to these dispersing states are estimated from the **k**-point at which the EDC peak positions go through the chemical potential when extrapolated from the occupied side. The k_F estimates are plotted as open symbols in Fig. 17.8(a).

We use the following square lattice notation for the 2D Brillouin zone of Bi2212: $\Gamma\bar{M}$ is along the CuO bond direction, with $\Gamma = (0, 0)$, $\bar{M} = (\pi, 0)$, $X = (\pi, -\pi)$ and $Y = (\pi, \pi)$ in units of $1/a^*$, where $a^* = 3.83$ Å is the separation between near-neighbor planar Cu ions. (The orthorhombic a axis is along X and b axis along Y).

In addition to the symbols obtained from data in Fig. 17.8, there are also several curves which clarify the significance of all of the observed features. The thick curve in Fig. 17.8(b) is $\epsilon(\mathbf{k})$, a six-parameter tight-binding fit [37] to the dispersion data in the Y-quadrant; this represents the main CuO_2 "band". It cannot be overemphasized that, although this dispersion looks very much like band theory (except for an overall renormalization of the bandwidth by a factor of 2 to 3), the actual normal state lineshape is highly anomalous. As discussed in Sect. 17.7 below, there are no well-defined quasiparticles in the normal state.

The thin curves in Fig. 17.8(b) are $\epsilon(\mathbf{k} \pm \mathbf{Q})$, obtained by shifting the main band fit by $\pm\mathbf{Q}$ respectively, where $\mathbf{Q} = (0.21\pi, 0.21\pi)$ is the superlattice (SL) vector known from structural studies [38]. We also have a few data points lying on a dashed curve $\epsilon(\mathbf{k}+\mathbf{K}_\pi)$ with $\mathbf{K}_\pi = (\pi, \pi)$; this is the "shadow band" discussed below.

The thick curve in Fig. 17.8(a) is the Fermi surface contour obtained from the main band fit, while the Fermi surfaces corresponding to the SL bands are the thin lines and that for the shadow band is dashed. It is very important to note that the shifted dispersion curves and Fermi surfaces provide an excellent description of the data points that do not lie on main band. We note that the main Fermi surface is a large hole-like barrel centered about the (π, π) point whose enclosed area corresponds to approximately 0.17 holes per planar Cu. One of the key questions is why *only one* CuO main band is found in Bi2212 which is a bilayer material with two CuO planes per unit cell. We postpone discussion of this important issue to end of this section.

The "shadow bands" seen above, were first observed by Aebi et al. [39] in ARPES experiments done in a mode similar to the MDCs by measuring as a function of **k** the intensity $\int_{\delta\omega} d\omega f(\omega) A(\mathbf{k}, \omega)$ integrated over a small range $\delta\omega$ near $\omega = 0$. The physical origin of these "shadow bands" is not certain at the present time. They were predicted early on to arise from short ranged antiferromagnetic correlations [40]. In this case the effect should become stronger with underdoping toward the AFM insulator, for which there is little experimental evidence [41]. An alternative explanation is that the shadow bands are of structural origin: Bi2212 has a face-centered orthorhombic cell with two inequivalent Cu sites per plane, which by itself could generate a (π, π) foldback. Interestingly, it has been recently observed that the shadow band intensity is maximal at optimal doping [42].

We now turn to the effect of the superlattice (SL) on the ARPES spectra. This is very important, since a lack of understanding of these effects has led to much confusion regarding such basic issues as the Fermi surface topology (see below), and the anisotropy of the SC gap (see Sect. 17.5). The data strongly suggest that these "SL bands" arise due to diffraction of the outgoing photoelectron off the structural superlattice distortion (which lives primarily) on the Bi–O layer, thus leading to "ghost" images of the electronic structure at $\epsilon_{\mathbf{k}\pm\mathbf{Q}}$. (Note, this would allow us to understand (1) why polarization selection rules are obeyed for the $\Gamma\bar{M}$ mirror plane, and (2) qualitatively, why the intensities for odd and even polarizations along ΓX are comparable.)

We emphasize the use of polarization selection rules (discussed in Sect. 17.2.1) for ascertaining various important points. First, we use them to carefully check the absence of a Fermi crossing for the main band along $(0, 0) - (\pi, 0)$, i.e., $\Gamma\bar{M}$. Note that the Fermi crossing that we do see near $(\pi, 0)$ along $\Gamma\bar{M}$ in Fig. 17.8(a) is clearly associated with a superlattice umklapp band, as seen both from the dispersion data in Fig. 17.8(b) and its polarization analysis. This Fermi crossing is only seen in the $\Gamma\bar{M} \perp$ (odd) geometry both in our data and in earlier work [43]. Emission from the main $d_{x^2-y^2}$ band, which is even about $\Gamma\bar{M}$, is dipole forbidden, and one only observes a weak superlattice signal crossing E_f. (We will return below to newer data at different incident photon energies where the possibility of a Fermi crossing along $\Gamma\bar{M}$ is raised again).

Second, we use polarization selection rules to disentangle the main and SL bands in the X-quadrant where the main and umklapp Fermi surfaces are very close together; see Fig. 17.8(a). The point is that ΓX (together with the z-axis) and, similarly ΓY, are mirror planes, and an initial state arising from an orbital which has $d_{x^2-y^2}$ symmetry about a planar Cu-site is odd under reflection in these mirror planes. With the detector placed in the mirror plane the final state is even, and one expects a dipole-allowed transition when the photon polarization $\mathbf{A}$ is perpendicular to (odd about) the mirror plane, but no emission when the polarization is parallel to (even about) the mirror plane. While this selection rule is obeyed along ΓY it is violated along ΓX. In fact this apparent violation of selection rules in the X quadrant was a puzzling feature of all previous studies [7] of Bi2212. It was first pointed out in [44], and then experimentally verified in [36], that this "forbidden" $\Gamma X||$ emission originates from the SL umklapps. We will come back to the $\Gamma X||$ emission in the superconducting state below.

17.4.2 Improved Methods for Fermi Surface Determination

We now discuss more recently developed methodologies for Fermi surface determination. The need for these improvements arises in part from the practical difficulty of determining precisely where in $\mathbf{k}$-space a state goes through E_f. This problem is particularly severe in the vicinity of the $(\pi, 0)$ point of the zone where one is beset by the following complications in both Bi2201 and 2212:

(1) The normal state ARPES peaks are very broad. This has important implications about the (non-Fermi-liquid) nature of the normal state, as discussed later (Sect. 17.7).

(2) The electronic dispersion near $(\pi, 0)$ is anomalously flat ("extended saddle-point").

(3) In Bi2212 the superlattice structure complicates the interpretation of the data as described above. Fortunately this complication is greatly reduced in Pb-doped Bi2201 and 2212.

(4) The final complication comes from the strong variation of the $\mathbf{k}$-dependent ARPES matrix elements

with incident photon energy. This makes the use of changes in absolute intensity as a function of **k** to estimate Fermi surface crossings highly questionable.

Note that the first two points are intrinsic problems intimately related to the physics of high T_c superconductivity, the third is a material-specific problem, while the last is specific to the technique of ARPES. Nevertheless, all of these issues must be dealt with adequately before ARPES data on Bi2212 can be used to yield definitive results on the Fermi surface.

Eliminating the Fermi Function

Recall that the peak of the EDC as a function of ω corresponds to that of $f(\omega)A(\mathbf{k}, \omega)$, which in general does *not* coincide with the peak of the spectral function $A(\mathbf{k}, \omega)$. If one has a broad spectral function, which at k_F is centered about $\omega = 0$, then the peak of the EDC will be at $\omega < 0$ (positive binding energy), produced by the Fermi function $f(\omega)$ chopping off the peak of A, in addition to resolution effects.

Since the goal is to study the dispersive peaks in $A(\mathbf{k}, \omega)$, rather than in the EDC, one must effectively eliminate the Fermi function from the observed intensity. We present two ways of achieving this goal, and illustrate it with data on Bi2201 where it permits us to study the broad and weakly dispersive spectral peaks (points (1) and (2) above) near $(\pi, 0)$ without the additional complication of the superlattice.

One approach is simply to divide the data by the Fermi function; more accurately one divides the measured intensity $I(\mathbf{k}_F, \omega)$ by $f(\omega) * R(\omega)$, the convolution of the Fermi function with the energy resolution. Although this does not rigorously give the spectral function (because of the convolution), this is a good approximation in situations where the energy resolution is very sharp. An excellent example of such an approach can be seen in the work of Sato et al. [45] on Pb-doped Bi2201 ($Bi_{1.80}Pb_{0.38}Sr_{2.01}CuO_{6-\delta}$) which is overdoped with a $T_c < 4K$. In Fig. 17.9 we show the raw data in the vicinity of M at various temperatures in the left panel and the corresponding data "divided by the Fermi function" in the right panel.

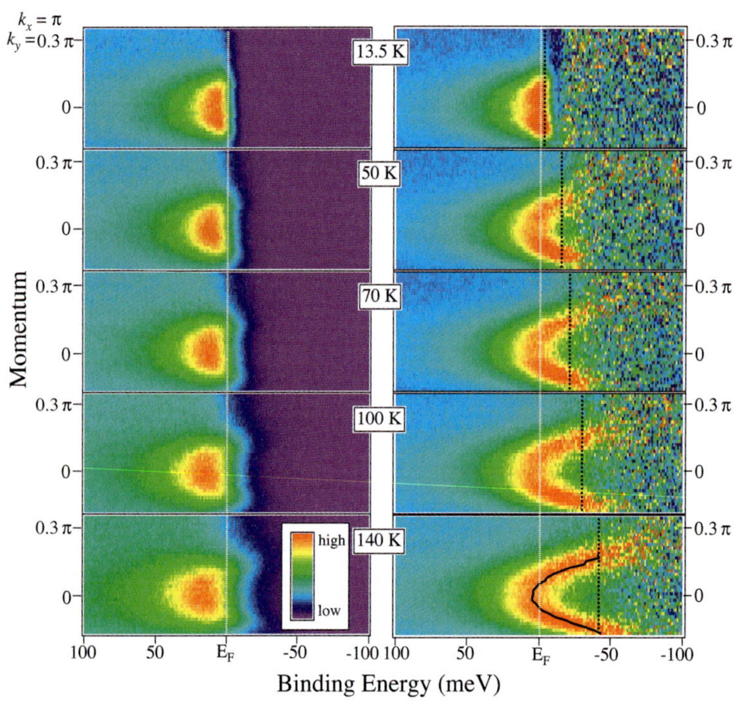

Fig. 17.9. *Left*: temperature dependence of ARPES intensity of OD Bi2201 ($T_c <$ 4 K) along the $(\pi, 0)-(\pi, \pi)$ cut. *Right*: same divided by the Fermi function at each temperature convoluted with a Gaussian of width 111meV (energy resolution). *Dotted lines* show the energy above E_f at which the Fermi function takes a value of 0.03. The *solid line* in the 140 K data indicates the peak positions obtained by fitting with a Lorentzian

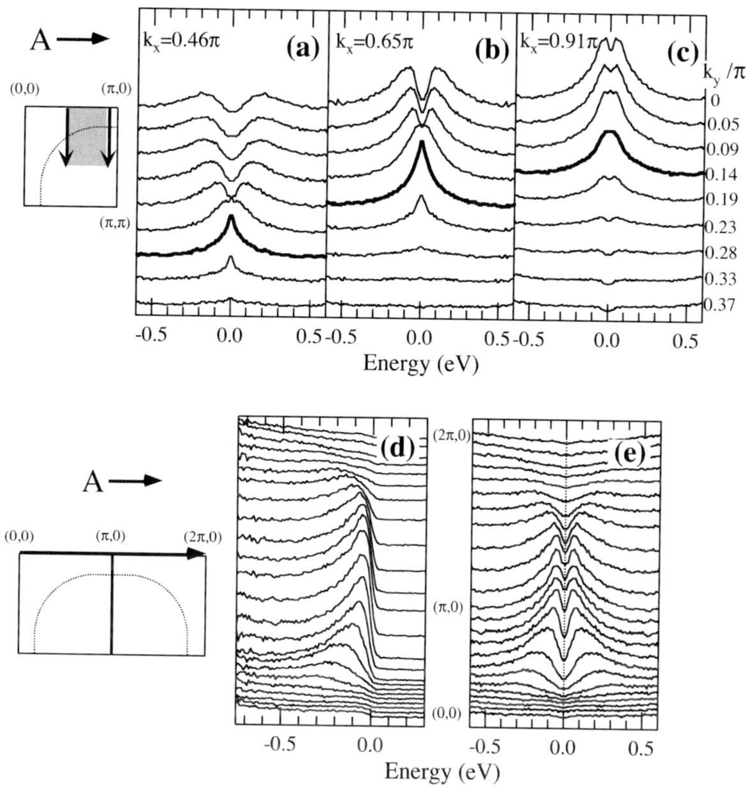

Fig. 17.10. (**a**), (**b**), **c**) Symmetrized EDCs along cuts parallel to $(\pi, 0)$ to (π, π) for OD 23K Bi2201 at $T = 25K$ using $h\nu = 22eV$ photons. The *curve* corresponding to $\mathbf{k}_F$ along each cut is determined as explained in the text and is shown by a *thick line*. (**d**) EDCs along the $(0, 0) \rightarrow (2\pi, 0)$ direction. (**e**) Symmetrized curves corresponding to the data in (**d**), showing the absence of a Fermi surface crossing along $(0, 0)$ to $(\pi, 0)$

The "divided" data permits one to clearly observe the dispersion well above E_f particularly at high temperatures, and thereby identify the Fermi crossings with a great degree of confidence.

An alternative approach [8] to eliminating the Fermi function is to symmetrize the data. For each $\mathbf{k}$ define the symmetrized intensity by $I_{\text{sym}}(\mathbf{k}_F, \omega) = I(\mathbf{k}_F, \omega) + I(\mathbf{k}_F, -\omega) = I_0\, A_{sym}(\mathbf{k}_F, \omega)$. It is easy to show that $I_{\text{sym}}(\omega)$ will exhibit a local minimum, or dip, at $\omega = 0$ for an occupied $\mathbf{k}$ state, while it will show a local maximum at $\omega = 0$ for an unoccupied $\mathbf{k}$ state. In practice, then, the Fermi crossing $\mathbf{k}_F$ is determined as follows: All EDCs along a cut are symmetrized and $\mathbf{k}_F$ is identified as the boundary in $\mathbf{k}$-space between points where I_{sym} has a local maximum versus a local minimum at $\omega = 0$. As shown in [8], these arguments work even in the presence of finite resolution effects. We note that this method, and the one presented above, for eliminating the Fermi function require a very accurate determination of the chemical potential (zero of binding energy).

In Fig. 17.10 we show the results [8] of a symmetrization analysis for an OD 23K Bi2201 ($Bi_{1.6}Pb_{0.4}Sr_2CuO_{6-\delta}$) sample. From the raw data along cuts parallel to $(\pi, 0)$ to (π, π) (see Fig. 7 of [8]) and along $(0, 0) \rightarrow (2\pi, 0)$ (shown in Fig. 17.10(d)) one sees broad peaks whose dispersion is very flat near $(\pi, 0)$, thus making it hard to determine $\mathbf{k}_F$ from EDC dispersion alone. Nevertheless, the symmetrized data provide completely unambiguous results: in the top panels (a, b, and c) of Fig. 17.10 we illustrate the use of symmetrized data to determine $\mathbf{k}_F$ using the criterion described above. Two other features about this analysis are worth noting. First, on approaching $\mathbf{k}_F$ from the occupied side, resolution effects are expected to lead to a flat topped symmetrized spectrum. Second, one expects an intensity drop in the symmetrized spectrum upon crossing

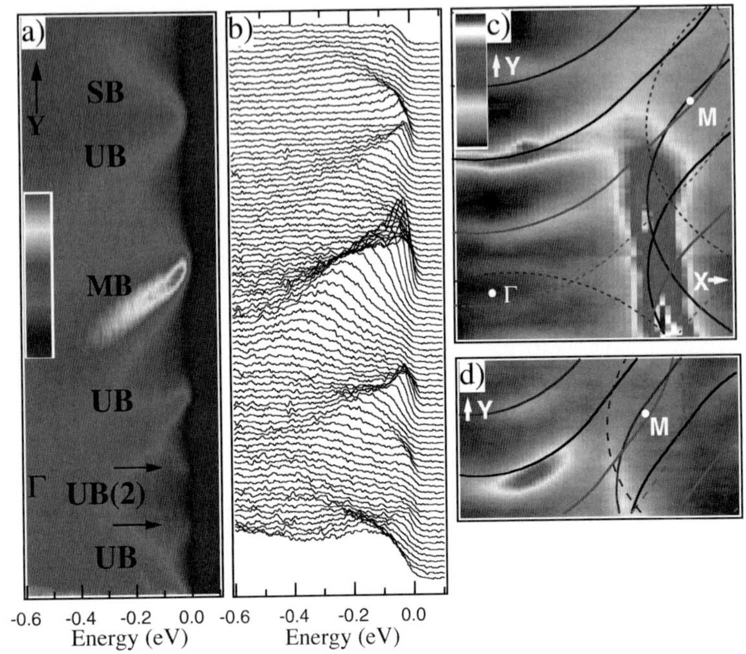

Fig. 17.11. (**a**) Intensity $I(\mathbf{k}, \omega)$ and (**b**) EDCs along ΓY measured on an optimally doped sample ($T_c = 90$ K) at $T = 40$ K with 33 eV photons polarized along ΓX. (**c**) Integrated intensity (−100 to +100 meV) covering the X and Y quadrants of the Brillouin zone. Data were collected on a regular lattice of **k** points (spacing 1° along ΓX and 0.26° along ΓY). (**d**) Integrated intensity (± 40 meV) as in (**c**), but in the normal state ($T = 150$ K). Overlaid on (**c**) is the main band (*black*), ± umklapps (*blue/red*), and ± 2nd order umklapps (*dashed blue/red lines*) Fermi surfaces from a tight binding fit [37]

$\mathbf{k}_F$, assuming that matrix elements are not strongly $\mathbf{k}$-dependent. Both of these effects are indeed seen in the data and further help in deducing $\mathbf{k}_F$.

It is equally important to be able to ascertain the *absence* of a Fermi crossing along a cut. In this respect, the raw data along $(0,0)-(\pi,0)-(2\pi,0)$ in Fig. 17.10(d) is difficult to interpret: the "flat band" remains extremely close to E_f but does it cross E_f? It is simple to see from the symmetrized data in Fig. 17.10(e) the answer is "no". The symmetrized data do not show a peak centered at $\omega = 0$ for any $\mathbf{k}$, thus establishing the absence of a Fermi crossing along this cut.

Intensity Plots

A different method for determining the Fermi surface is to make a 2D plot as a function of $\mathbf{k}$ of $\int_{\delta\omega} d\omega I(\mathbf{k},\omega)$, the observed intensity integrated over a suitably chosen narrow energy interval $\delta\omega$ about the Fermi energy. At first sight this seems to be a very direct way to find out the $\mathbf{k}$-space locus on which the low energy excitations live. However, as we discuss below, one has to be very careful in interpreting such plots since one is now focusing on the *absolute* intensity of the ARPES signal, which can be strongly affected by the $\mathbf{k}$-dependence of the photoemission matrix elements.

This approach was pioneered in the cuprates by Aebi and coworkers [39], and in recent times with the availability of Scienta detectors with dense $\mathbf{k}$-sampling it has been used by several groups [46–50]. We show as an example in Fig. 17.11 results from our group [49] on an optimally doped $T_c = 90$ K Bi2212 sample using 33eV incident photons. (Note, the data in [49] is taken in the superconducting state ($T = 40$ K) of the $T_c = 90$ K sample. We note that the integration range (±100 meV) employed for the intensity patterns is much larger than the gap energy scale, and the minimum gap locus (to be discussed below in Sect. 17.6) in the superconducting state is essentially identical to the normal state Fermi surface.)

Before commenting on the controversies about the Fermi surface crossing near the $(\pi, 0)$ point, we first examine EDCs along the ΓY direction (middle panel of Fig. 17.11, where the left panel shows a two dimensional plot of the energy and momentum dependent

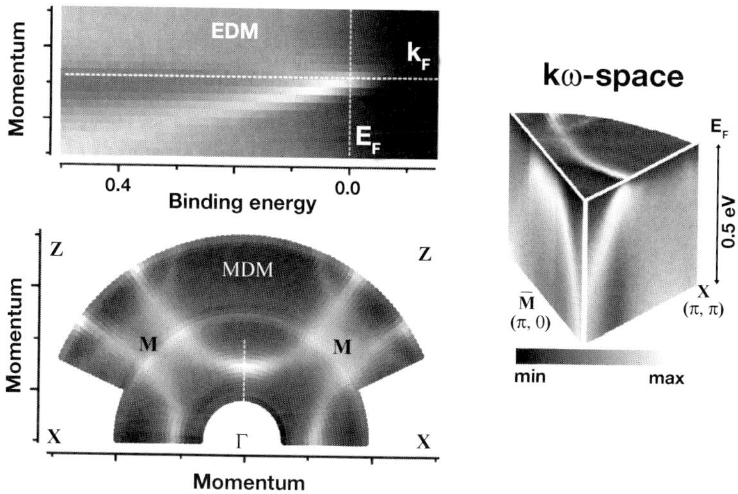

Fig. 17.12. *Upper panel*: Energy distribution map from the $\Gamma - (\pi, \pi)$ direction in the Brillouin zone of Pb-doped BSCCO recorded at room temperature. *Lower panel*: momentum distribution map of Pb-doped BSCCO, recorded at room temperature (raw data). The *White horizontal dashed line* represents a $\mathbf{k}_F$-EDC, *vertical lines* correspond to an E_F-MDC. In both cases the *gray scale* represents the photoemission intensity as indicated. The *inset* shows the three-dimensional (k_x, k_y, ω) space, which is probed in ARPES of quasi-2D systems (from [50])

intensity of photoelectrons along the ΓY cut). All of the features — main band (MB), superlattice umklapp bands (UB) and the shadow band (SB) — seen before [36] and discussed above are confirmed. We also see weaker, second order umklapps from the superlattice (corresponding to $\pm(0.42, 0.42)$, twice the superlattice wavevector), which confirms the diffraction origin of the superlattice bands.

We now turn to panels (c,d) of Fig. 17.11 where we plot the integrated intensity within a ± 100 meV window about the chemical potential. We note the very rapid suppression of intensity beyond $\sim 0.8\Gamma\mathrm{M}$ [46], which does not occur in data taken with 22eV incident photons. This has led some authors [48,51] to suggest the existence of an electron-like Fermi surface with a crossing at this point. However, Fretwell et al. [49] and independently Borisenko et al. [50] have argued that this Fermi crossing along $(0, 0)$ to $(\pi, 0)$ is actually due to one of the umklapp bands, and the near optimally doped Fermi surface is indeed hole like as earlier shown by Ding et al. [36].

This can be seen most clearly in Pb-doped samples, where the umklapp bands are not visible. In Fig. 17.12, we show the Fermi energy intensity map of Borisenko et al. [50], where the hole surface centered around (π, π) and its shadow band partner are quite apparent.

One of the main reasons for the controversy surrounding the topology of the optimal doped Fermi surface is the fact that data taken at different incident photon energies $h\nu$ lead to different intensity patterns. Our assertion, based on [8,49], is that the superlattice umklapp band is more noticeable at $h\nu = 33$ eV compared with 22 eV since matrix element effects suppress the main band intensity at 33 eV. For a detailed discussion about how to discriminate between a main band and superlattice Fermi crossing, we refer the reader to the cited papers, and also to [8,36] for the use of polarization selection rules for this purpose.

Matrix Element Effects

There is an important general lesson to be learned from the above discussion which is equally relevant for the $n(\mathbf{k})$ methods to be discussed below. Changes in the ARPES intensities (either integrated over a small energy window or over a large energy range) as a function of $\mathbf{k}$ can be strongly affected by matrix element effects. This is, of course, obvious from the expression for the ARPES intensity: $I(\mathbf{k}, \omega) = I_0(\mathbf{k}; \nu; \hat{\mathbf{A}}) f(\omega) A(\mathbf{k}, \omega)$. The key question is: after integration over the appropriate range in ω, how do we differentiate between the $\mathbf{k}$ dependence coming from the matrix elements I_0 (which we are not interested in per se) from the $\mathbf{k}$ dependence coming from the spectral function?

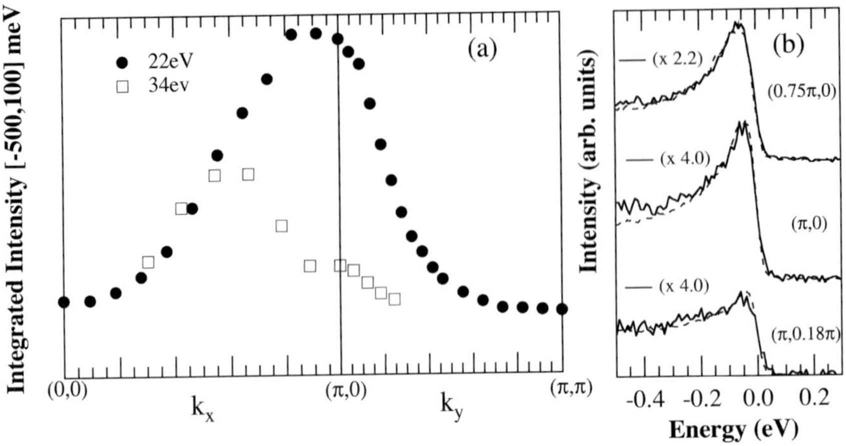

Fig. 17.13. Bi2201–OD23K. (**a**) Integrated intensity (over the range is −500 to 100 meV) along the $(0,0) \rightarrow (\pi,9) \rightarrow (\pi,\pi)$ directions for two incident photon energies $h\nu = 22$ eV and 34 eV. (**b**) Comparison of the ARPES lineshape measured at 22 eV (*dashed lines*) and 34 eV (*solid lines*) at three different **k** points. One *curve* has been scaled by the multiplicative constant indicated to make it lie on top of the other

One possibility is to have a priori information about the matrix elements from electronic structure calculations [16]. But as we now show, even in the absence of such information, one can experimentally separate the effects of a strong **k**-variation of the matrix element from a true Fermi surface crossing. The basic idea is to exploit the fact that changing the incident photon energy one only changes the ARPES matrix elements and not the spectral function (or the resulting momentum distribution) of the initial states.

We will use Bi2201 to illustrate our point since it has all the complications (points (1),(2) and (4)) listed above without the superlattice (point (3)). Figure 17.13(a), from [8], shows the intensity (integrated over a large energy range) as a function of **k** for OD 23 K Bi2201, and highlights the differences between data obtained at 221 eV and 34 eV incident photon energies. At 22 eV the maximum intensity occurs close to $(\pi, 0)$ and decreases both toward $(0, 0)$ and (π, π), while at 34 eV there is a strong depression of intensity on approaching $(\pi, 0)$, resulting in a shift of the intensity maximum away from $(\pi, 0)$.

From the discussion of Sect. 17.2.4, we can write the integrated intensity as: $I(\mathbf{k}) = \int_{-\infty}^{+\infty} d\omega I(\mathbf{k}, \omega) = I_0(\mathbf{k}; \nu; \mathbf{A}) n(\mathbf{k})$. We attribute this loss in intensity around $(\pi, 0)$ at 34 eV seen in Fig. 17.13(a) to strong **k**-dependence of I_0 rather than $n(\mathbf{k})$. Experimentally we prove this by showing that the EDCs at the same point in the Brillouin zone obtained at the two different photon energies show exactly the same lineshape, i.e. one can be rescaled onto the other as shown in Fig. 17.13(b). As an independent check of this, we have also shown that the symmetrization analysis leads to the same conclusion that there is no Fermi crossing along $(0, 0) \rightarrow (\pi, 0)$; see Fig. 9 of [8].

For completeness, we note that another possible source of incident $h\nu$-dependence in ARPES is k_z-dispersion. If there was c-axis dispersion, different photon energies would probe initial states with different k_z values consistent with energy conservation. However the scaling shown in Fig. 17.13(b) proves that it is the *same* two-dimensional (k_z-independent) initial state which is being probed in the data shown here, and the $h\nu$-dependence arises entirely from the different final states that the matrix element couples to.

Methods Based on the Momentum Distribution

Finally we turn to the use of the momentum distribution sum rule [20] (discussed in Sect. 17.2.4) in determining the Fermi surface. In principle, locating the rapid variation of $n(\mathbf{k})$ offers a very direct probe of the Fermi surface which we emphasize is not restricted to Fermi liquids. (The $T = 0$ momentum distribution for known non-Fermi liquid systems, such as Luttinger liquids in one dimension, do show a inflection point singularity at $\mathbf{k}_F$.) However in practice, one needs to be very careful about the **k**-dependence

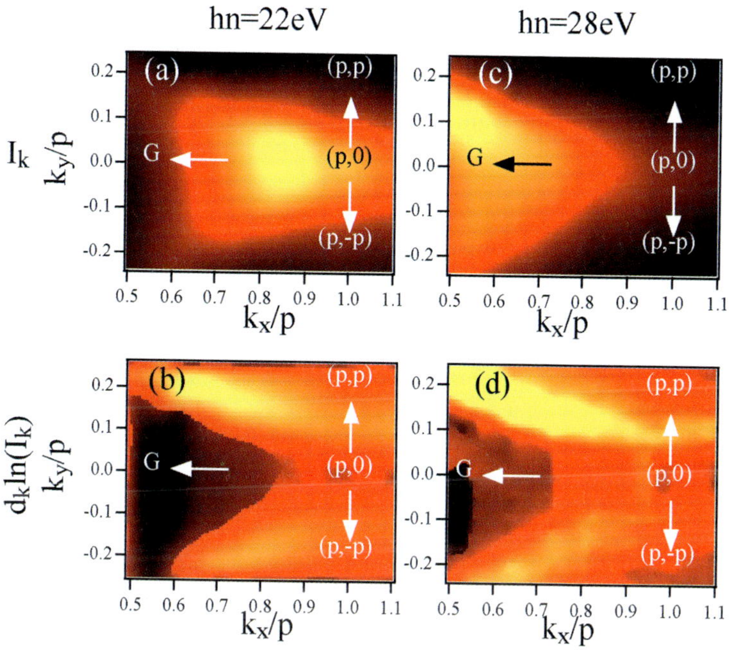

Fig. 17.14. Bi2201–OD0K. (**a** and **c**) Integrated intensity (over −350 meV to +50 meV) $I(\mathbf{k})$ measured at $\hbar\nu = 22$ eV and 28 eV around the $(\pi, 0)$ point. Notice that the intensity maximum depends strongly upon the photon energy $\hbar\nu$. (**b** and **d**) Corresponding gradient of the logarithm, $|\nabla_\mathbf{k} I(\mathbf{k})|/I(\mathbf{k})$, the maxima which correspond to Fermi crossings and clearly show that, *independent of the photon energy*, the Fermi surface consists of a hole barrel centered around (π, π)

of the matrix elements, as clearly recognized in the original proposal [20].

Here we discuss two approaches using the $n(\mathbf{k})$ method to obtain information about the Fermi surface. The first method is to study the $\mathbf{k}$-space gradient of the logarithm of the integrated intensity. The second method is to study the temperature-dependence of the integrated intensity and use the approximate sum rule [20] $\partial n(\mathbf{k}_F)/\partial T = 0$ discussed in Sect. 17.2.4.

The gradient method was used in our early work [36,52] where $\mathbf{k}_F$ was estimated from the location of max $|\nabla_\mathbf{k} n(\mathbf{k})|$. The same method has also been successfully used later by other authors [53,54]. In the presence of strong matrix element effects, it is even more useful to plot the magnitude of the logarithmic gradient: $|\nabla_\mathbf{k} I(\mathbf{k})|/I(\mathbf{k})$ which emphasizes the rapid changes in the integrated intensity. The logarithmic gradient filters out the less abrupt changes in the matrix elements and helps to focus on the intrinsic variations in $n(\mathbf{k})$.

As an example we show in Fig. 17.14 the results [8] of such an analysis for an OD 0K Bi2201 sample. In the top panels (a) and (c) we show the integrated intensity $I(\mathbf{k})$ around the $(\pi, 0)$ point obtained at two different photon energies: 22 eV and 28 eV respectively. In the lower panels 11(b) and (d), we plot $|\nabla_\mathbf{k} I(\mathbf{k})|/I(\mathbf{k})$. Note that there are large differences between the two top panels, due to different matrix elements at 22 eV and 28 eV. However, as explained above, the logarithmic gradients in the bottom panels, which are more influenced by the intrinsic $n(\mathbf{k})$, are much more similar. The Fermi surface can be clearly seen as two high intensity arcs curving away from the $(\pi, 0)$ point. Once matrix element effects are taken care of, the Fermi surface results obtained at the two different photon energies are quite similar, and in good agreement with the results obtained from independent methods like symmetrization on the same data set [8].

Our final method for the determination of a Fermi crossing goes back to the sum rule [20] that we had introduced earlier, $\partial n(\mathbf{k}_F)/\partial T = 0$. Assuming that the matrix elements are T-independent on the temperature scales of interest, this immediately implies the integrated intensity at (and only at) $\mathbf{k}_F$ is T-independent. In Sect. 17.2.4, we had used this sum rule to get confidence in the validity of the single-

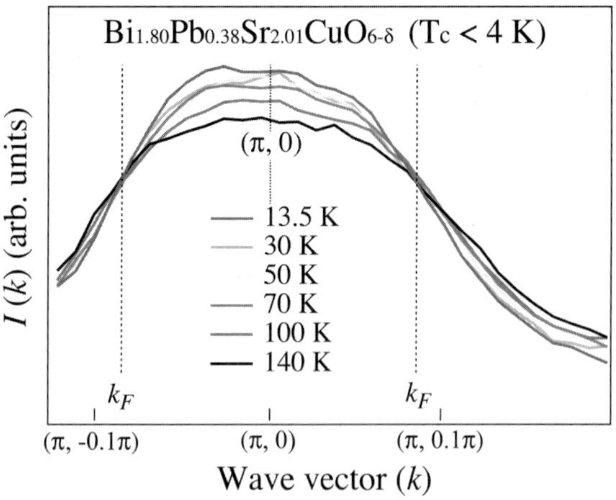

Fig. 17.15. Temperature dependence of the integrated ARPES intensity $I(\mathbf{k})$ along the $(\pi,0)-(\pi,\pi)$ cut, obtained by integrating the ARPES intensity from -100 meV to 100 meV. (An angle-integrated temperature-independent background is subtracted before integration)

particle spectral function interpretation of ARPES by verifying at $\mathbf{k}_F$ assuming that $\mathbf{k}_F$ was known (by some other means). Now we can invert the logic: we can look at the T dependence of the integrated intensity, and identify $\mathbf{k}_F$ as that point in $\mathbf{k}$-space where the integrated intensity is T-independent. This is illustrated in Fig. 17.15 from the work of Sato et al. [45], who determine $\mathbf{k}_F$ along the $(\pi,0)-(\pi,\pi)$ cut for a highly OD Bi2201 sample. The result is completely consistent with that obtained by other methods, such as "division by the Fermi function" on the same sample (see Fig. 17.9).

EDCs vs MDCs

To conclude this discussion, let us note a very recent development for determining k_F and the near-E_f dispersion based on the MDCs, which are plots of the ARPES intensity as a function of k (in this case normal to the expected Fermi surface), at various fixed values of ω. As shown in Sect. 17.2.5, the MDC peak position in the vicinity of the Fermi surface, i.e, near $(\mathbf{k}=\mathbf{k}_F, \omega=0)$ is given by: $k = k_F + [\omega - \Sigma'(\omega)]/v_F^0$.

Thus k_F is determined by the peak location of the MDC at $\omega = 0$. The fully renormalized Fermi velocity $v_F = v_F^0/[1-\partial\Sigma'/\partial\omega]$ is given by the slope of the MDC peak dispersion. We note that the factor arising from the k-dependence of the self-energy is already included in v_f^0, so that $v_f^0 = v_f^{\text{bare}}[1+\partial\Sigma'/\partial\varepsilon_{\mathbf{k}}]$. (To see this, note that the analysis of Sect. 17.2.5 can be easily generalized to retain the first order term $\partial\Sigma'/\partial\varepsilon_{\mathbf{k}}$ without spoiling the Lorentzian lineshape of the MDC provided this k-dependence does not enter Σ'').

As discussed earlier, the above results derive from the Lorentzian lineshape of the MDC which arises when three conditions are satisfied: the matrix elements do not depend on k, the self energy does not depend on k (except for the $(k-k_F)$ variation of Σ' noted above) and dispersion can be linearized near the Fermi surface. The validity of these assumptions can be checked self-consistently by the Lorentzian MDC lineshape and the dispersion deduced from the data.

The significance of this approach is that, as emphasized by Kaminski et al. [24], the dispersions of the EDC and MDC peak positions are actually different in the cuprates; see Fig. 17.54(a) in Sect. 17.7.7. This difference arises due to the non-Fermi liquid nature of the normal state, so that the EDC peak dispersion is *not* given by the condition $\omega - v_F^0(k-k_F) - \Sigma' = 0$ but also involves in general Σ''. In contrast the MDC peak dispersion is rigorously described by the expression described above, and is much simpler to interpret.

We expect that the MDC method for determining the dispersion and k_F which has thus far been used mainly along the zone diagonal, will eventually be the method of choice, except when one is very close to the bottom of the band where linearization fails.

17.4.3 Summary of Results on the Optimally Doped Fermi Surface

We have discussed a large number of methods for determination of the Fermi surface in Bi2201 and Bi2212 in the previous subsection. These include: (a) dispersion of EDC peaks through E_f, (b) dispersion of peaks after division by the Fermi function, (c) symmetrization, (d) maps of intensities at E_f, (e) gradient of $n(\mathbf{k})$, (f) T-dependence of $n(\mathbf{k})$, and (g) MDC dispersion. In addition we also discussed using the $h\nu$-dependence of the data and polarization selection rules to eliminate matrix element effects and to identify superlattice Fermi crossings.

The reader might well ask: why so many different methods? The reason is that the development of all of these methods has taken place to deal with the complications of accurately identifying the Fermi surface in the presence of the four problems listed at the beginning of the preceding subsection. Each method has its pros and cons, so that some, like (b) and (c) require very accurate E_f determination, which is not the case in (e) and (f) which use energy-integrated intensities. Most methods require dense sampling in **k** space, while method (f) requires in addition data at several temperatures.

Given the complications of the problem at hand it is important to look for crosschecks and consistency between various ways of determining the Fermi surface. We believe that for optimally doped Bi2201 and 2212 there is unambiguous evidence for a single hole barrel centered about the (π, π) point enclosing a Luttinger volume of $(1 + x)$ holes where x is the hole doping. We discuss further below the issues of the doping-dependence of the Fermi surface and of bilayer splitting in Bi2212.

17.4.4 Extended Saddle Point Singularity

The very flat dispersion near the $(\pi, 0)$ point observed in all of the data is striking. Specifically, along $(0, 0)$ to $(\pi, 0)$ there is an intense spectral peak corresponding to the main band, which disperses toward E_F but stays just below it. This is often called the "flat band" or "extended saddle point", and appears to exist in all cuprates, though at different binding energies in different materials [7,43,55].

In our opinion this flat band is not a consequence of the bare electronic structure, but rather a many-body effect, because a tight-binding description of such a dispersion requires fine-tuning (of the ratio of the next-near-neighbor hopping to the near-neighbor hopping) which would be unnatural even in one material, let alone many.

An important related issue is whether this flat band leads to a singular density of states. It is very important to recognize that, while Fig. 17.8(b) *looks like* a conventional band structure, the dispersing states whose "peak positions" are plotted are extremely broad, with a width comparable to binding energy, and these simply cannot be thought of as quasiparticles. This general point is true at all **k**'s, but specifically for the flat band region it has the effect of spreading out the spectral weight over such a broad energy range that any singularity in the DOS would be washed out. This is entirely consistent with the fact that other probes (tunneling, optics, etc.) do not find any evidence for a singular density of states either.

17.4.5 Bilayer Splitting?

On very general grounds, one expects that the two CuO_2 layers in a unit cell of Bi2212 should hybridize to produce two electronic states which are even and odd under reflection in a mirror plane mid-way between the layers. Where are these two states? Why then did we find only one main "band" and only one Fermi surface in Bi2212?

Let us first recall the predictions of electronic structure calculations [56]. In systems like Bi2212, the intra-bilayer hopping as a function of the in-plane momentum **k** is of the form [28,57] $t_\perp(\mathbf{k}) = -t_z(\cos k_x - \cos k_y)^2$. Thus the two bilayer states are necessarily degenerate along the zone diagonal. However they should have a maximum splitting at $\bar{M} = (\pi, 0)$ of order 0.25 eV, which may be somewhat reduced by many-body interactions.

Depending on the exact doping levels and on the presence of Bi–O Fermi surface pockets, which are neither treated accurately in the theory nor observed in the ARPES data, we must obtain one of the two following situations: (1) the bilayer antibonding (A) state is unoccupied while the bonding (B) state is occupied at $(\pi, 0)$. This would lead to an A Fermi crossing along $(0, 0)-(\pi, 0)$ and a B Fermi crossing along $(\pi, 0)-(\pi, \pi)$. As described at great length above we did not find evidence for a main band Fermi crossing along $(0, 0) - (\pi, 0)$ at least for the near optimal doped sample, therefore this possibility is ruled out.

(2) The second possibility is that both the A and B bilayer states are occupied at the $(\pi, 0)$. In this case, there should be two (in principle, distinct) Fermi crossings along $(\pi, 0) - (\pi, \pi)$, although they might

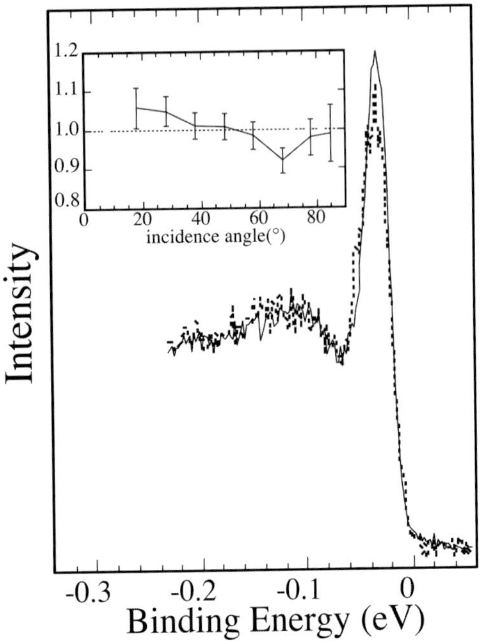

Fig. 17.16. Low temperature ($T = 13$ K) EDC's of near optimal $T_c = 87$ K Bi2212 at $(\pi, 0)$ for various incident photon angles. The *solid (dashed) line* is 18° (85°) from the normal. The *inset* shows the height of the sharp peak for data normalized to the broad bump, at different incident angles

be difficult to resolve in practice. Nevertheless, one would definitely expect to see two distinct occupied states at the $(\pi, 0)$ point. Unfortunately, the normal state spectrum at $(\pi, 0)$ is so broad in the optimally doped and underdoped materials that it is hard to make a clear case for bilayer splitting. Thus an effort was made to search for this effect in the superconducting state at $T \ll T_c$, when a sharp feature (quasiparticle peak) is seen (see Fig. 17.16) and one might hope that the bilayer splitting should be readily observable.

The issue then is how to interpret the peak/dip/hump structure seen in the ARPES lineshape at $(\pi, 0)$ in Fig. 17.16. The peak/dip/hump structure will be discussed at length in Sect. 17.7 below. Nevertheless, here we will briefly address the question of whether: (I) the peak and the hump are the two bilayer split states, which are resolved below T_c once the peak becomes sharp? Or (II) is the non-trivial line shape due to many-body effects in a single spectral function $A(\mathbf{k}, \omega)$?

Three pieces of evidence will be offered in favor of hypothesis (II) as opposed to (I), so that no bilayer splitting is observable even in the superconducting state of near optimal doped Bi2212. The first evidence comes from studying the polarization dependence of the ARPES matrix elements. For case (I) there are two independent matrix elements which, in general, should vary differently with photon polarization $\mathbf{A}$, and thus the intensities of the two features should vary independently as $\mathbf{A}$ is varied. On the other hand, for case (II), the intensities of the two features are governed by a single matrix element. As shown in Fig. 17.16 it was found in [36] that by varying the z-component of , the peak and hump intensities scale together, and thus the peak/dip/hump are all part of a single spectral function for Bi2212.

A second piece of evidence comes from a comparison [58] of the normal and superconducting state dispersions near the $(\pi, 0)$ point, which will be discussed in detail in connection with Figs. 17.43 and 17.44 of Sect. 17.7.4. From these data, we argue that there is no evidence for a feature above T_c, which would correspond to the dispersionless quasiparticle peak below T_c. Thus the dispersionless peak must be of many-body origin.

The third and final piece of evidence comes from both ARPES and SIS tunneling. In the ARPES data [59] shown in Fig. 17.50 of Sect. 17.7.6 one sees the striking fact that while the energy scales of both the quasiparticle peak and the $(\pi, 0)$ hump increase with underdoping, their ratio is essentially doping independent. Since the location of the peak is the (maximum) superconducting gap — as discussed in detail in Sect. 17.5 — the $(\pi, 0)$ hump energy scales with the gap. SIS tunneling data also finds the same correlation on a very wide range of materials (including some which have a single CuO plane) whose gap energies vary by a factor of 30 [60]. This provides very strong evidence that both the peak and hump are related to many-body effects and not manifestations of bilayer splitting. We note that Anderson [61] had predicted that many-body effects within a single layer could destroy both the quasiparticles and the coherent bilayer splitting *in the normal state*. But why the

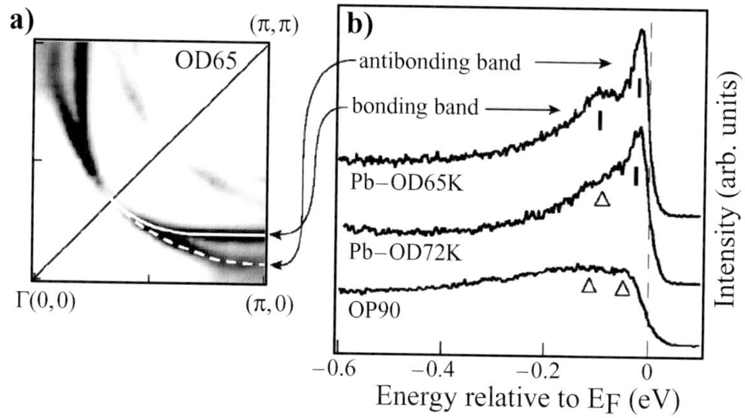

Fig. 17.17. (**a**) Bilayer-split Fermi surfaces of heavily overdoped OD65; the two weaker features are their superstructure counter parts. *Solid* and *dashed lines* represent the bonding and antibonding Fermi surfaces, respectively. (**b**) Normal state photoemission spectra of Bi2212 taken at $(\pi, 0)$ for three different doping levels. Data were taken with $h\nu = 22.7$ eV photon. *Bars* indicate identified feature positions, and *triangles* indicate possible feature positions (from [62])

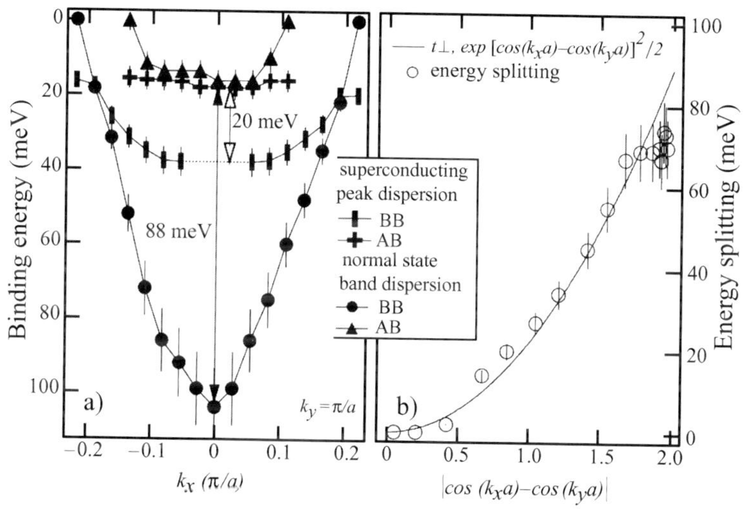

Fig. 17.18. (**a**) Dispersion extracted from heavily overdoped OD65. (**b**) Energy splitting along the antibonding Fermi surface, which is obtained from data shown in Fig. 17.17. It is simply the binding energy of the bonding band, since the binding energy of the antibonding band is zero at its Fermi surface. The curve is $\frac{1}{2}t_{\perp,\text{exp}}[\cos(k_x a) - \cos(k_y a)]^2$, where $t_{\perp,\text{exp}} = 44 \pm 5$ meV. Error bars are due to the uncertainties in determining the energy position (from [62])

splitting should not be visible in the superconducting state, where sharp quasiparticles do exist, is not clear from a theoretical point of view.

Recently, the above picture has been challenged by a number of authors [62–64]. What has become clear is that bilayer splitting is indeed present for heavily overdoped Bi2212 samples, and has been seen now by several groups, including our own. In Fig. 17.17, we show (a) the bilayer split Fermi surfaces and (b) the bilayer split EDCs observed by the Stanford group for a heavily overdoped ($T_c = 65$ K) Bi2212 sample. Note that the bilayer splitting can even be seen in the umklapp bands. The resulting dispersion is reproduced in Fig. 17.18, where one sees that the momentum dependence of the splitting follows that expected from electronic structure considerations [28,57].

How this effect evolves as a function of doping, though, is still controversial. In particular, if it is present for optimal doping, it is difficult to resolve. Moreover, some authors who advocate bilayer splitting still argue that the peak/dip/hump structure in the superconducting state is largely a many-body effect [62] as we advocate here. This is supported by the fact that the same lineshape is seen in trilayer Bi2223 [65], where one would expect three features if layer splitting were causing these effects. And similar lineshapes have been seen by tunneling in single layer systems, in support of a many-body interpretation. The reader is referred to [2], where the issue of bilayer splitting is discussed in greater detail than here.

17.5 Superconducting Energy Gap

In this Section, we will first establish how the superconducting (SC) gap manifests itself in ARPES spectra, and then directly map out its variation with **k** along the Fermi surface. This is the only available technique for measuring the momentum dependence of the energy gap, and complements phase-sensitive tests of the order parameter symmetry [66]. Thus ARPES has played an important role [67], [68] in establishing the d-wave order parameter in the high T_c superconductors [66]. At the end of the section, we will discuss the doping dependence of the SC gap and its anisotropy, and the implications of this study for various low temperature experiments like thermal conductivity and penetration depth.

17.5.1 Particle–Hole Mixing

To set the stage for the experimental results it may be useful to recall particle–hole (p–h) mixing in the BCS framework (even though, as we shall see in Sect. 17.7, there are aspects of the data which are dominated by many body effects beyond weak coupling BCS theory). The *BCS* spectral function is given by

$$A(\mathbf{k},\omega) = u_\mathbf{k}^2\Gamma/\pi((\omega - E_\mathbf{k})^2 + \Gamma^2) + v_\mathbf{k}^2\Gamma/\pi((\omega + E_\mathbf{k})^2 + \Gamma^2)\,, \quad (17.11)$$

where the coherence factors are $v_\mathbf{k}^2 = 1 - u_\mathbf{k}^2 = \frac{1}{2}(1 - \epsilon_\mathbf{k}/E_\mathbf{k})$ and Γ is a phenomenological linewidth. The normal state energy $\epsilon_\mathbf{k}$ is measured from E_f and the Bogoliubov quasiparticle energy is $E_\mathbf{k} = \sqrt{\epsilon_\mathbf{k}^2 + |\Delta(\mathbf{k})|^2}$, where $\Delta(\mathbf{k})$ is the gap function. Note that only the second term in Eq. (17.11), with the $v_\mathbf{k}$-coefficient, would be expected to make a significant contribution to the EDCs at low temperatures.

In the normal state above T_c, the peak of $A(\mathbf{k},\omega)$ is at $\omega = \epsilon_\mathbf{k}$ as can be seen by setting $\Delta = 0$ in Eq. (17.11). We would thus expect to see in ARPES a spectral peak which disperses through zero binding energy as **k** goes through $\mathbf{k}_F$ (the Fermi surface).

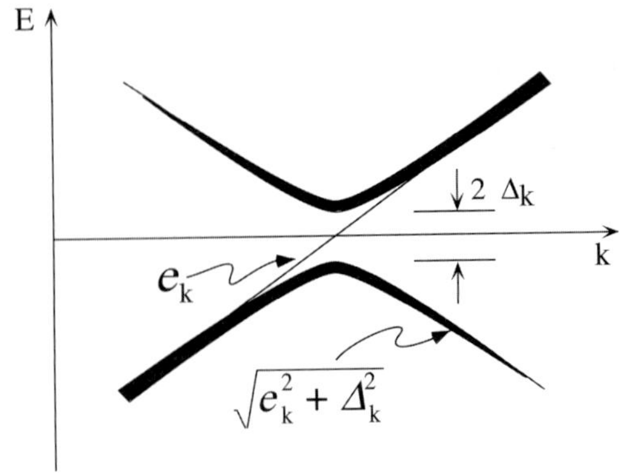

Fig. 17.19. Schematic dispersion in the normal (*thin line*) and superconducting (*thick lines*) states following BCS theory. The thickness of the superconducting state lines indicate the spectral weight given by the BCS coherence factors u and v

In the superconducting state, the spectrum changes from $\epsilon_\mathbf{k}$ to $E_\mathbf{k}$; see Fig. 17.19. As **k** approaches the Fermi surface the spectral peak shifts towards lower binding energy, but no longer crosses E_f. Precisely at $\mathbf{k}_F$ the peak is at $\omega = |\Delta(\mathbf{k}_F)|$, which is the closest it gets to E_f. This is the manifestation of the gap in ARPES. Further, as **k** goes beyond $\mathbf{k}_F$, in the region of states which were unoccupied above T_c, the spectral peak *disperses back*, receding away from E_f, although with a decreasing intensity (see Eq. (17.11)). This is the signature of p–h mixing.

Experimental evidence for particle–hole mixing in the SC state was first given in Ref. [52]. In Fig. 17.20 we show normal and SC state spectra for Bi2212 for **k**'s along the cut shown in the inset. In the normal state data in panel (b) we see the electronic state dispersing through E_f: the **k**'s go from occupied (top of panel) to unoccupied states (bottom of panel). The normal state dispersion is plotted as black dots in Fig. 17.21 (b). The $\mathbf{k}_F$ obtained from this dispersion is in agreement with that estimated from the $|\nabla_\mathbf{k} n(\mathbf{k})|$ analysis of the normal state data shown in Fig. 17.21(a).

We see from Fig. 17.20 (a) that the SC state spectral peaks do not disperse through the chemical potential, rather they first approach $\omega = 0$ and then recede away from it. The difference between the normal and SC state dispersions is clearly shown in Fig. 17.21 (b).

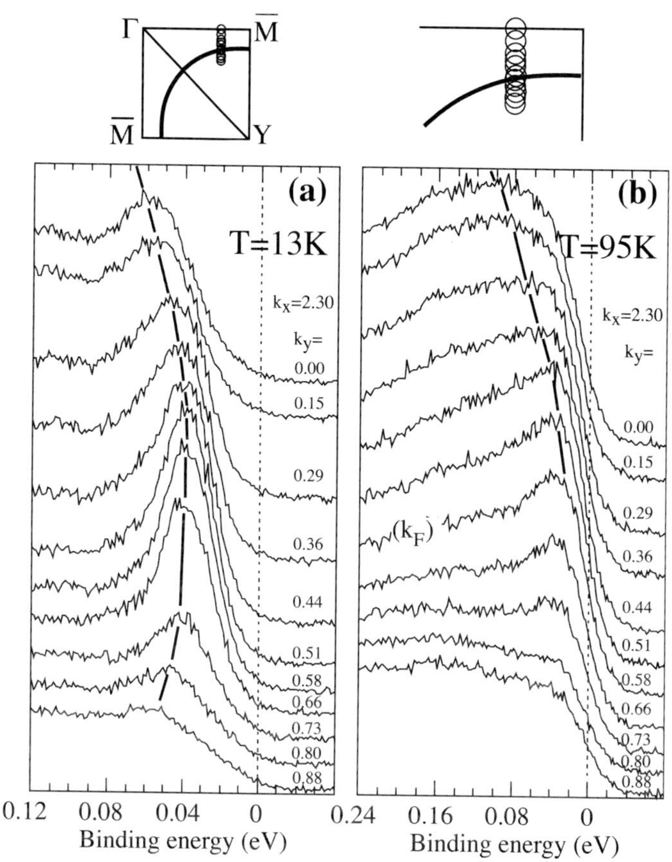

Fig. 17.20. (**a**) Superconducting state and (**b**) normal state EDC's for a near optimal $T_c = 87$ K Bi2212 sample for a set of **k** values (in units of $1/a$) shown in the Brillouin zone at the *top*. Note the different binding energy scales in (**a**) and (**b**)

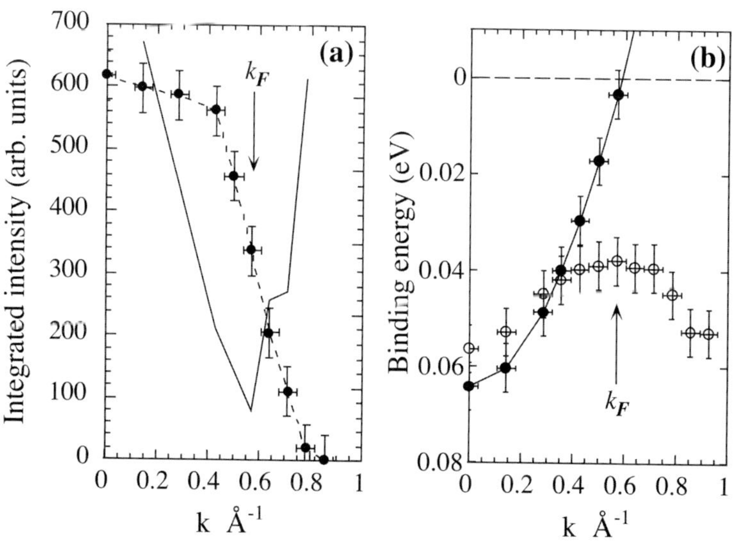

Fig. 17.21. (**a**) Integrated intensity versus **k** from the normal state data of Fig. 17.20 (**b**) shown by *black dots* gives information about the momentum distribution $n(\mathbf{k})$. The derivative of the integrated intensity is shown by the *black curve* (arbitrary scale). The Fermi surface crossing $\mathbf{k} = \mathbf{k}_F$ is identified by the minimum in the derivative. (**b**) Normal state dispersion (*closed circles*) and SC state dispersion (*open circles*) obtained from EDC's of Fig. 17.20. Note the back-bending of the SC state dispersion for **k** beyond $\mathbf{k}_F$, which is a clear indication of particle–hole mixing. The SC state EDC peak position at $\mathbf{k}_F$ is an estimate of the SC gap at that point on the Fermi surface

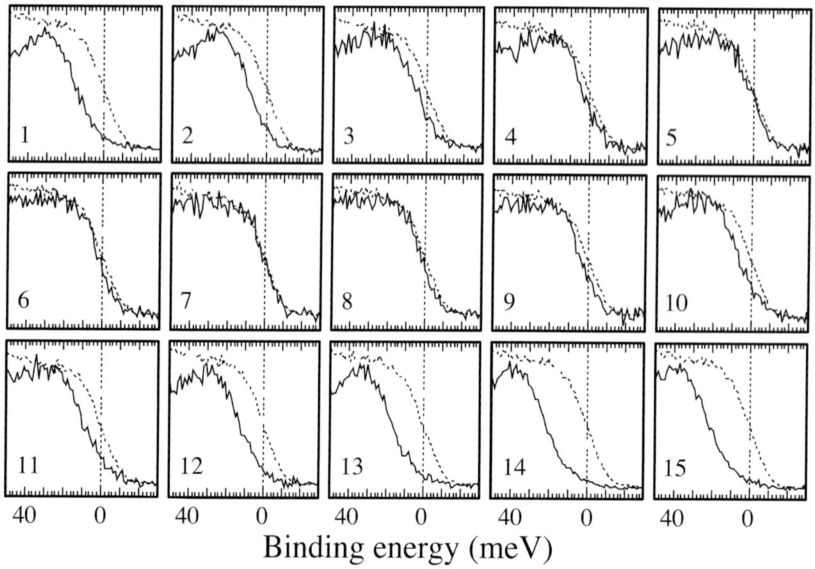

Fig. 17.22. Bi2212 spectra (*solid lines*) for an 87 K T_c sample at 13 K and Pt spectra (*dashed lines*) versus binding energy (meV) along the Fermi surface in the Y quadrant. The photon polarization and BZ locations of the data points are shown in the *inset* to Fig. 17.23

There are three important conclusions to be drawn from Fig. 17.21 (b). First, the bending back of the SC state spectrum for $\mathbf{k}$ beyond $\mathbf{k}_F$ is direct evidence for p–h mixing in the SC state. Second, the energy of closest approach to $\omega = 0$ is related to the SC gap that has opened up at the FS, and a quantitative estimate of this gap will be described below. Third, the location of closest approach to $\omega = 0$ ("minimum gap") coincides, within experimental uncertainties, with the $\mathbf{k}_F$ obtained from analysis of normal state data.

In fact by taking cuts in $\mathbf{k}$-space which are perpendicular to the normal state Fermi surface one can map out the "minimum gap locus" in the SC state, or for that matter in any gapped state (e.g., the pseudogap regime to be discussed in the following Section). We emphasize that particle–hole mixing leads to the appearance of the "minimum gap locus" and this locus in the gapped state gives information about the underlying Fermi surface. (By this we mean the Fermi surface on which the SC state gap appears below T_c). In fact, the observation of p–h mixing in the ARPES spectra is a clear way of asserting that the gap seen by ARPES is due to superconductivity rather than of some other origin, e.g., charge or spin-density wave formation.

17.5.2 Quantitative Gap Estimates

The first photoemission measurements of the SC gap in the cuprates was by Imer et al. [69] using angle-integrated photoemission, and by Olson and coworkers [70] using angle-resolved photoemission. The first identification of a large gap anisotropy consistent with d-wave pairing was made by Shen and coworkers [67]. Ding et al. [21,68] subsequently made quantitative fits to the SC state spectral function to study the gap anisotropy in detail.

We now discuss the quantitative extraction of the gap at low temperatures ($T \ll T_c$) following Ding et al. [68]. In Fig. 17.23, we show the $T = 13$K EDCs for an 87 K T_c sample for various points on the main band FS in the Y-quadrant. Each spectrum shown corresponds to the minimum observable gap along a set of $\mathbf{k}$ points normal to the FS, obtained from a dense sampling of $\mathbf{k}$-space. (The large dispersion along ΓY, of about 60 meV within our momentum window $\delta\mathbf{k} = 0.045\pi/a^*$, makes it hard to locate $\mathbf{k}_F$ accurately and to map out the nodal region. To this end the we use a step size of $\delta\mathbf{k}/2$ normal to the Fermi surface and $\delta\mathbf{k}$ along it.) We used 22 eV photons in a $\Gamma Y \perp$ polarization, with a 17 meV (FWHM) energy resolution, and a $\mathbf{k}$-window of radius $0.045\pi/a^*$.

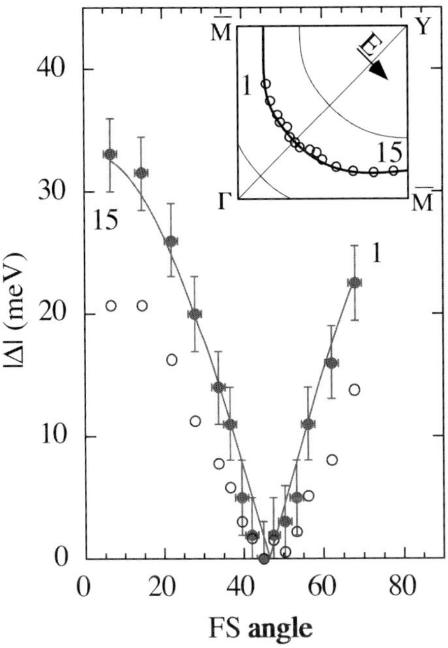

Fig. 17.23. *Y* quadrant gap in meV versus angle on the Fermi surface (*filled circles*) from fits to the data of Fig. 17.22. *Open symbols* show leading edge shift with respect to Pt reference. The *solid curve* is a *d*-wave fit to the filled symbols

The simplest gap estimate is obtained from the mid-point shift of the leading edge of Bi2212 relative to a good metal (here Pt) in electrical contact with the sample. This has no obvious quantitative validity, since the Bi2212 EDC is a spectral function while the polycrystalline Pt spectrum (dashed curve in Fig. 17.22) is a weighted density of states whose leading edge is an energy-resolution limited Fermi function. We see that the shifts (open circles in Fig. 17.23) indicate a highly anisotropic gap which vanishes in the nodal directions, confirming earlier results by Shen et al. [67]. These results are qualitatively similar to one obtained from the fits described below.

Next we turn to modeling [21,68] the SC state data in terms of spectral functions. It is important to ask how can we model the non-trivial line shape (with the dip-hump structure at high ω) in the absence of a detailed theory, and, second, how do we deal with the extrinsic background? We argue as follows: in the large gap region near $(\pi, 0)$, we see a linewidth collapse for frequencies smaller than $\sim 3\Delta$ upon cooling well below T_c. Thus for estimating the SC gap at the low temperature, it is sufficient to look at small frequencies, and to focus on the coherent piece of the spectral function with a resolution-limited leading edge. (Note this argument fails at higher temperatures, e.g., just below T_c). This coherent piece is modeled by the BCS spectral function, Eq. (17.11).

The effects of experimental resolution are taken into account via

$$\tilde{I}(\mathbf{k}, \omega) = I_0 \int_{\delta\mathbf{k}} \mathrm{d}\mathbf{k}' \int_{-\infty}^{+\infty} \mathrm{d}\omega' R(\omega - \omega') f(\omega') A(\mathbf{k}', \omega') \,, \tag{17.12}$$

where $R(\omega)$, the energy resolution, is a normalized Gaussian and $\delta\mathbf{k}$ is the $\mathbf{k}$-window of the analyzer. In so far as the fitting procedure is concerned, all of the incoherent part of the spectral function is lumped together with the experimental background into one function which is added to the $\tilde{I}$ above. Since the gap is determined by fitting the resolution-limited leading edge of the EDC, its value is insensitive to this drastic simplification. To check this, we have made an independent set of fits to the small gap data where we do not use any background fitting function, and only try to match the leading edges, not the full spectrum. The two gap estimates are consistent within a meV. Once the insensitivity of the gap to the assumed background is established, there are only two free parameters in the fit at each $\mathbf{k}$: the overall intensity I_0 and the gap $|\Delta|$; the dispersion $\epsilon_{\mathbf{k}}$ is known from the normal state study, the small linewidth Γ is dominated by the resolution.

The other important question is the justification for using a coherent spectral function to model the rather broad EDC along and near the zone diagonal. As far as the early data being discussed here is concerned, such a description is self-consistent [21,68], though perhaps not unique, with the entire width of the EDC accounted for by the large dispersion (of about 60 meV within our $\mathbf{k}$-window) along the zone diagonal. More recent data taken along $(0, 0)$ to (π, π) with a momentum resolution of $\delta\mathbf{k} \simeq 0.01\pi/a^*$ fully justifies this assumption by resolving coherent nodal quasiparticles in the SC state; see Sect. 17.7.7.

The gaps extracted from fits to the spectra of Fig. 17.22 are shown as filled symbols in Fig. 17.23. For a detailed discussion of the error bars (both on the gap value and on the Fermi surface angle), and also of sample-to-sample variations in the gap estimates, we refer the reader to [68]. The angular variation of the gap obtained from the fits is in excellent agreement with $|\cos(k_x) - \cos(k_y)|$ form. The ARPES experiment cannot of course measure the phase of the order parameter, but this result is in strong support of $d_{x^2-y^2}$ pairing [66]. Such an order parameter arises naturally in theories with strong correlations and/or antiferromagnetic spin fluctuations [71]. Moreover, the functional form of the anisotropy we find is consistent with electrons in the Cooper pair residing on neighboring Cu sites. That is, ARPES gives information on the spatial range of the pair interaction which is difficult to obtain from other techniques.

Finally we comment on the temperature dependence of the gap. Unfortunately with increasing temperature the linewidth grows and a simple BCS-type modeling (valid for $T \ll T_c$) of just the coherent part of the spectral function is not possible. While T-dependence of the leading edge can certainly be used as a rough guide, this estimate is affected by both the gap and the linewidth (the diagonal self-energy). We will discuss methods for modeling the SC state data in Sect. 17.7.

For completeness, we add a remark clarifying the earlier observation of two nodes in the X-quadrant [21], and the related non-zero gap along ΓX in the $\Gamma X||$ geometry [21,72]. It was realized soon afterwards that these observations were related to gaps on the superlattice bands [44], and not on the main band. To prove this experimentally, the X-quadrant gap has been studied in the $\Gamma X \perp$ geometry [68] and found to be consistent with Y-quadrant $d_{x^2-y^2}$ result described above.

17.5.3 Doping Dependence of the SC Gap

There are two important issues to be addressed about the doping dependence: How does the maximum gap change with doping? and how does the gap anisotropy evolve with doping?

ARPES results on underdoped samples show that the maximum gap *increases* [73,74] with decreasing hole concentration. For a recent compilation of gap versus doping results [59] in Bi2212, see Fig. 17.50 of Sect. 17.7.6. Identical results have also been obtained by tunneling spectroscopy [75]. This was at first quite unexpected since T_c decreases as one underdopes from optimality. The fact that $2\Delta/k_B T_c$ is not constant with doping and can become an order of magnitude larger than its BCS weak-coupling value is very clear evidence that the SC phase transition on the underdoping side is qualitatively different from the BCS transition, a point that we will return to in the next Section on pseudogaps.

We next turn to the question of the SC gap anisotropy as a function of doping. In this case it is the behavior of the gap in the vicinity of the node which is the most important since all low temperature $T \ll T_c$ thermodynamic and transport properties are controlled by thermal excitation of quasiparticles in the vicinity of the nodes (where the SC gap vanishes on the Fermi surface). These low energy excitations have a Dirac-like spectrum $E(\mathbf{k}) = \sqrt{v_F^2 k_\perp^2 + v_2^2 k_\parallel^2}$ where $k_\perp$ ($k_\parallel$) are the components of the $\mathbf{k}$ perpendicular (tangential) to the Fermi surface, and measured from the nodal point. Here v_F is the nodal Fermi velocity controlling the dispersion perpendicular to the Fermi surface, while $v_2 = \partial\Delta_\mathbf{k}/\partial k_\parallel$ is the gap slope at the nodal point.

The density of states for low energy excitations is then given by $N(\omega) = 2\omega/(\pi v_F v_2)$ for $\omega \ll \Delta$ (in units where $\hbar = 1$). This leads to characteristic temperature dependences for various low temperature properties with coefficients essentially determined by v_F and v_2. Specifically for $T \ll T_c$, the specific heat goes like $C(T) = c_1 T/(v_F v_2)$; the thermal conductivity goes like $\kappa(T)/T = c_2\,(v_F/v_2 + v_2/v_F)$ and the superfluid density goes like $\rho_s(T) = \rho_s(0) - c_3(v_F/v_2)T$. Here c_1 and c_2 are known constants while c_3 contains (a priori unknown) multiplicative Fermi liquid parameters [76,77].

Clearly ARPES has a unique ability to independently measure both v_F, using the dispersion of the nodal quasiparticle, and v_2, using the slope of the SC gap at the node. One can then make detailed comparison with various bulk measurements, and thereby test the validity of the description of the low temperature properties of the SC state in terms of weakly interacting quasiparticles [76–78]. This is particularly important given the lack of sharply defined quasiparticles in the normal state of the cuprates.

With these motivations in mind, a detailed measurement of the shape of the superconducting gap in Bi2212 as a function of doping was carried out by Mesot et al. [79]. Although these measurements were carried out at a time when the energy and momentum resolutions were lower than those currently available, they still give useful information because the gap over the full range of angles over the irreducible zone was measured. Using the simple BCS spectral function fits (described above) and taking into account the measured dispersion and the known energy and momentum resolutions, Mesot et al.found the results shown in Fig. 17.24. The simple d-wave gap $\Delta = \Delta_0 \cos(2\phi)$ (Fig. 17.23) is modified by the addition of the first harmonic $\Delta_{\text{k}} = \Delta_{\max}[B\cos(2\phi) + (1-B)\cos(6\phi)]$, with $0 \leq B \leq 1$. Note that the $\cos(6\phi)$ term in the Fermi surface harmonics can be shown to be closely related to the tight binding function $\cos(2k_x) - \cos(2k_y)$, which represents next nearest neighbors interaction, just as $\cos(2\phi)$ is closely related to the near neighbor pairing function $\cos(k_x) - \cos(k_y)$. From Fig. 17.24 we find that while the overdoped data are consistent with $B \simeq 1$, the parameter B decreases as a function of underdoping.

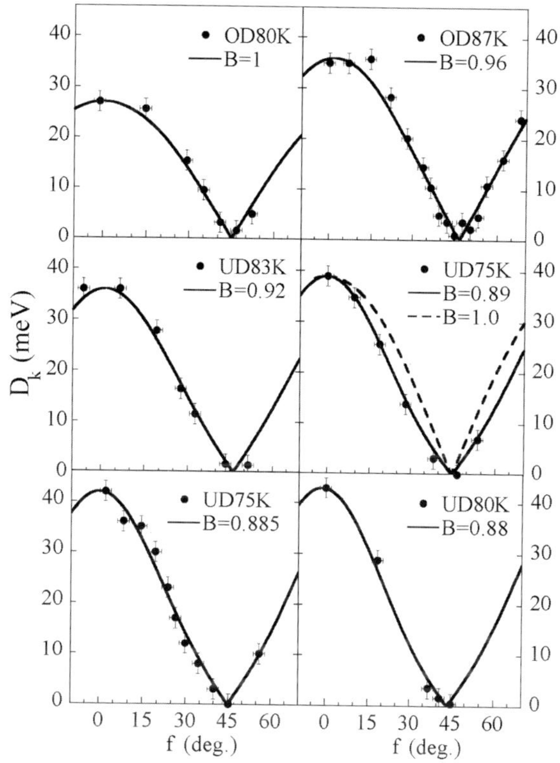

Fig. 17.24. Values of the superconducting gap as a function of the Fermi surface angle ϕ obtained for a series of Bi2212 samples with varying doping. Note two different UD75 K samples were measured, and the UD83 K sample has a larger doping due to sample aging [41]. The *solid lines* represent the best fit using the gap function: $\Delta_k = \Delta_{\max}[B\cos(2\phi) + (1-B)\cos(6\phi)]$ as explained in the text. The *dashed line in the panel* of an UD75 K sample represents the gap function with $B = 1$

Before discussing the doping dependence of the results in detail, let us first look at the comparisons between ARPES and other probes near optimality. Low temperature specific heat data on Bi2212 is not available, but the thermal conductivity has been measured by Taillefer and coworkers [78] on an optimal $T_c = 89$K sample. Now one can use the κ/T formula given above with the coefficient $c_2 = k_B^2 n/3\hbar d$, where n/d is the stacking density of CuO_2 planes. It is important to note that c_2 is *not* renormalized by either vertex corrections or Fermi liquid parameters [77]. Thus one infers $v_F/v_2 = 19$ from the thermal transport data [78], which is in remarkable agreement with the ARPES estimate [79] of $v_F/v_2 \simeq 20$ coming from the measured values of $v_F = 2.5 \times 10^7$ cm/s and $v_2 = 1.2 \times 10^6$ cm/s for the near optimal $T_c = 87$ K sample of Fig. 17.24. (We note that in [79] we used the notation v_Δ for one-half the gap slope at the node, which is related to v_2 defined above by $v_2 = 2v_\Delta/\hbar k_F$.)

The comparison of ARPES results with the slope of the superfluid density $\rho_s \sim 1/\lambda^2$ obtained from penetration depth measurements is more complicated, as discussed in more detail in [79], for two reasons. Experimentally, there seem to be discrepancies between the $d\lambda^{-2}/dT$ results of various groups, and theoretically the slope of ρ_s is renormalized by doping-dependent Fermi liquid parameters [76,77] which are not known a priori. These parameters characterize the residual interactions between the nodal quasiparticles in the superconducting state. Thus, e.g., even at optimality the slope $d\lambda^{-2}/dT$ obtained using the ARPES estimate of v_F/v_2 and ignoring Fermi liquid renormalization is almost three times as large as the experimentally measured value of $d\lambda^{-2}/dT$ in near optimal Bi2212. This indicates the importance of Fermi liquid renormalizations in order to make a quantitative comparison. For more details on the doping dependence of these renormalizations, see [76,79,81]. For penetration depth measurements on Bi2212, see [80].

Finally, let us return to the doping dependence of the results of Fig. 17.24. In contrast to the maximum gap (at the $(\pi, 0)-(\pi, \pi)$ Fermi surface crossing) increasing as a function of underdoping, noted earlier, we see the gap slope at the node, which determines v_2, decreasing with underdoping. This is a result of the doping dependence of the B parameter introduced above.

This surprising observation raises several questions. First, could the flattening at the node be, in fact, evidence for a "Fermi arc", a line of gapless excitations, in the underdoped materials, especially since such arcs are seen above T_c (see Sect. 17.6.3 for further discussion of Fermi arcs in underdoped materials). Given the error bars on gap estimates in Fig. 17.24, it is impossible to rule out arcs in all the samples. Nevertheless, it is clear that there are samples (especially OD87K, UD80K and UD75K) where there is clear evidence in favor of a point node rather than an arc at low temperatures. Furthermore, it is very important to note that a linear T dependence of $\rho_s(T)$ at low temperature, for all doping levels, in clean samples gives independent evidence for point nodes, at least in YBCO [82].

Second, is the change in gap anisotropy intrinsic, or related to impurity scattering [83]? We can eliminate the latter explanation on two grounds. The maximum gap *increases* as the doping is reduced, opposite to what would be expected from pair breaking due to impurities. Also, impurity scattering is expected to lead to a characteristic "tail" to the leading edge [84], for which there is no evidence in the observed spectra.

Thus the flattening near the nodes with underdoping does appear to be an intrinsic feature which may be related to the increased importance of longer range pairing interactions as one approaches the insulator. It would be of great interest to study the details of the doping dependence of the gap anisotropy with the new Scienta detectors which have greatly improved energy and momentum resolution.

17.6 Pseudogap

In this section we describe one of the most fascinating developments in the study of high T_c superconductors: the appearance of a pseudogap above T_c which is seen most prominently on the underdoped side of the cuprate phase diagram. Briefly the "pseudogap" phenomenon is the loss of low energy spectral weight in a window of temperatures $T_c < T < T^*$; see Fig. 17.25. The pseudogap regime has been probed by many techniques like NMR, optics, transport, tunneling, μSR and specific heat; for reviews and references, see [85,86]. ARPES, with its unique momentum-resolved capabilities, has played a central role in elucidating the pseudogap phenomenon [41,87–90].

We will discuss in this section ARPES results on the anisotropy of the pseudogap, its T-dependence, its doping dependence, and its effect on the normal state Fermi surface. We emphasize that for the most part we will focus on the "low energy" or leading edge pseudogap, and only mention ARPES evidence for the "high energy pseudogap" toward the end. We will conclude the Section with a summary of the constraints put by the ARPES data on various theoretical descriptions of the pseudogap.

17.6.1 Pseudogap near $(\pi, 0)$

In the underdoped materials, T_c is suppressed by lowering the carrier (hole) concentration as shown in Fig. 17.25. In the samples used by our group [41,89,90] underdoping was achieved by adjusting the oxygen partial pressure during annealing the float-zone grown crystals. These crystals also have structural coherence lengths of at least 1,250 Å as seen from x-ray diffraction, and optically flat surfaces upon cleaving, similar to the slightly overdoped T_c samples discussed above. We denote the underdoped (UD) samples by their onset T_c: the 83 K sample has a transition width of 2K and the highly underdoped 15 K and 10 K have transition widths > 5 K. Other groups have also studied samples where underdoping was achieved by cation substitution [87,88].

We now contrast the remarkable properties of the underdoped samples with the near-optimal Bi2212 samples which we have been mainly focusing on thus far. We will first focus on the behavior near the $(\pi, 0)$ point where the most dramatic effects occur, and come back to the very interesting **k**-dependence later. In Fig. 17.26 [74] we show the T-evolution of the ARPES spectrum at the $(\pi, 0) \to (\pi, \pi)$ Fermi crossing for an UD 83K sample. At sufficiently high temperature, the leading edge of the UD spectrum at $\mathbf{k}_F$ and the reference Pt spectrum coincide, but below a crossover temperature $T^* \simeq 180$K the leading edge midpoint of the spectrum shifts below the chemical potential. In Fig. 17.26 one can clearly see a loss of low energy spectral weight at 120K and 90K. It must be emphasized that this gap-like feature is seen in the normal (i.e., non-superconducting) state for $T_c = 83$ K $< T < T^* = 180$ K.

The doping dependence of the temperature T^*, below which a leading-edge pseudogap appears near $(\pi, 0)$, is shown in Fig. 17.25. Remarkably T^* increases with underdoping, in sharp contrast with T_c, but very similar to the low temperature SC gap, a point we will return to at the end of the Section. The region of the phase diagram between T_c and T^* is called the pseudogap region.

It is important to emphasize that our understanding of the lightly UD samples (e.g., the UD 83 K sample) is the best among the UD materials. In such samples all three regimes — the SC state below T_c, the pseudogap regime between T_c and T^* and the gapless "normal" regime above T^* — can be studied in detail. In contrast, in the heavily UD samples (e.g., the UD 10 K and UD 15 K samples), not only is the SC transition broad, one also has such low T_c's and such high T^*'s that only the pseudogap regime is experimentally accessible. Nevertheless, the results on the heavily underdoped samples appear to be a natural continuation of the weakly underdoped materials and the results (the trends of gap and T^*) on the low T_c samples are in qualitative agreement with those obtained from other probes (see [85,86]).

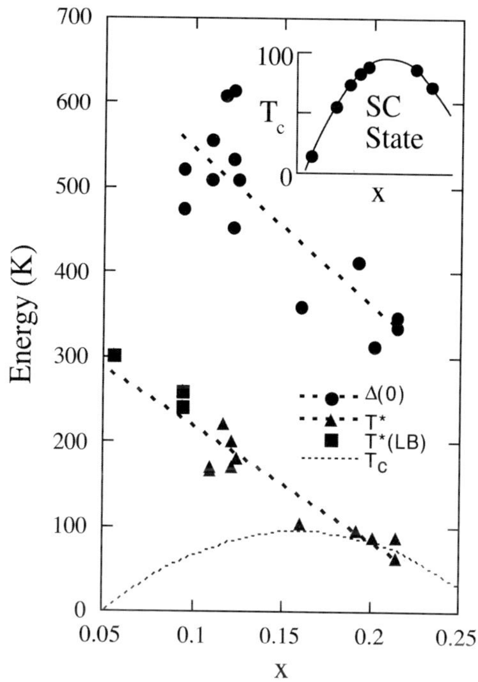

Fig. 17.25. T^* (*triangles* for determined values and *squares* for lower bounds) and T_c (*dashed line*) as a function of hole doping x. The x values for a measured T_c were obtained by using the empirical relation $T_c/T_c^{\max} = 1 - 82.6(x - 0.16)^2$ [91] with $T_c^{\max} = 95$ K. Also shown is the low temperature (maximum) superconducting gap $\Delta(0)$ (*circles*). Note the similar doping trends of $\Delta(0)$ and T^*

The T-dependence of the leading-edge midpoint shift appears to be completely smooth through the SC transition T_c. In other words, the normal state pseudogap evolves smoothly into the SC gap below T_c. Nevertheless, there *is* a characteristic change in the lineshape in passing through T_c associated with the appearance of a sharp feature below T_c in Fig. 17.26. This can be identified as the coherent quasiparticle peak for $T \ll T_c$. The existence of a SC state quasiparticle peak is quite remarkable given that the normal state spectra of UD materials are even broader than at optimality, and in fact become progressively broader with underdoping. In fact, the low tempera-

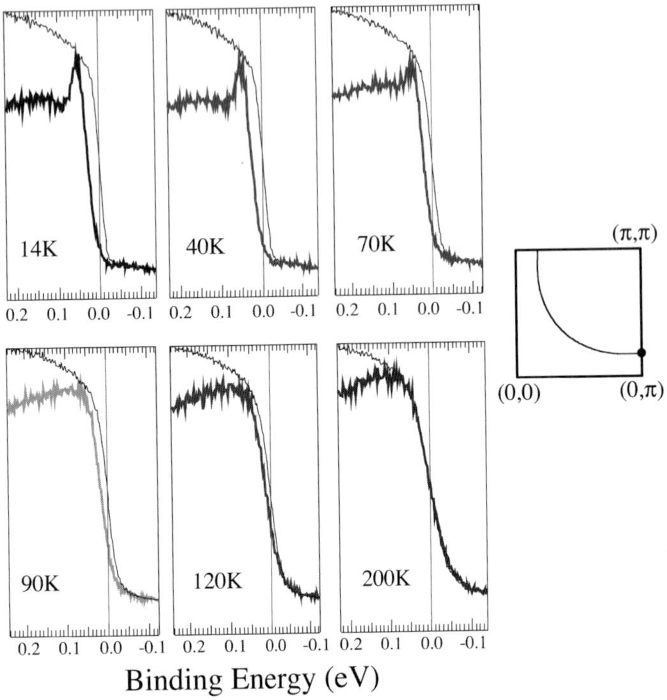

Fig. 17.26. ARPES spectra along the $(\pi, 0) \to (\pi, \pi)$ direction for an 83 K underdoped sample at various temperatures (*solid curves*). The *thin curves* in each *panel* are reference spectra from polycrystalline Pt used to accurately determine the zero of binding energy at each temperature

ture SC state spectra near $(\pi, 0)$ in the UD systems (see Fig. 17.50 of Sect. 17.7.6) are in many ways quite similar to those at optimal doping, with the one crucial difference that the spectral weight in the coherent quasiparticle peak diminishes rapidly with underdoping [92,93].

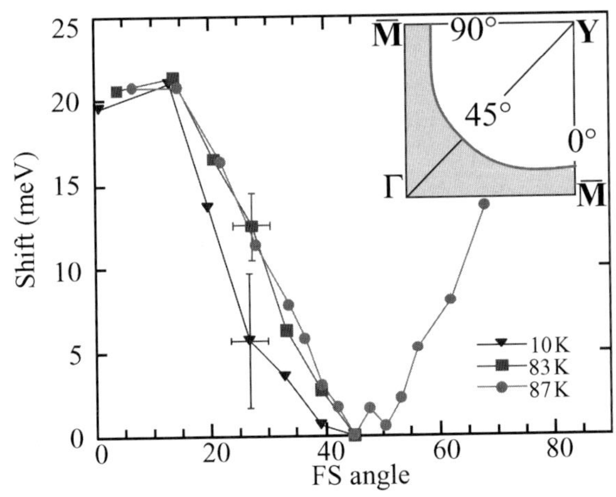

Fig. 17.27. Momentum dependence of the gap estimated from the leading-edge shift in samples with T_c's of 87 K (slightly overdoped), 83 K (UD) and 10 K (UD), measured at 14 K. For the sake of comparison between samples we made vertical offsets so that the shift at 45° is zero; the offsets are −3 meV for the 83 K and +2 meV for the 10 K sample. The *inset* shows the Brillouin zone with the large Fermi surface

17.6.2 Anisotropy of the Pseudogap

We have already indicated that the pseudogap above T_c near the $(\pi, 0)$ point of the zone evolves smoothly through T_c into the large SC gap below T_c, and thus the two also have the same magnitude. Since the SC gap has the d-wave anisotropy (discussed in detail in the preceding section), it is natural to ask: what is the $\mathbf{k}$-dependence of the pseudogap above T_c?

The first ARPES studies [87–89] showed that the pseudogap is also highly anisotropic and has a $\mathbf{k}$-dependence which is very similar to that of the SC gap below T_c. Later work [90] further clarified the situation by showing that the anisotropy has a very interesting temperature dependence. We now describe these developments in turn.

In Fig. 17.27 [89] we plot the leading edge shifts for three samples at 14 K: the slightly overdoped 87 K and UD 83 K samples are in their SC states while the UD 10 K sample is in the pseudogap regime. The gap estimate for each sample was made on the "minimum gap locus" (explained earlier in the context of the SC gap; see further below). The large error bars on the UD 10K sample come from the difficulty of accurately locating the midpoint of a broad spectrum. Also there is a flattening of the gap near the node, a feature that we discussed earlier for the SC gap in UD samples.

The remarkable conclusion is that the normal state pseudogap has a very similar **k**-dependence and magnitude as the SC gap below T_c.

17.6.3 Fermi Arcs

The T-dependence and anisotropy of the pseudogap was investigated in more detail in [90] motivated by the following question. Normal metallic systems are characterized by a Fermi surface, and optimally doped cuprates are no different despite the absence of sharp quasiparticles (see Sect. 17.4). On the underdoped side of the phase diagram, however, how does the opening of pseudogap affect the locus of low lying excitations in **k**-space?

In Fig. 17.28 we show ARPES spectra for an UD 83K sample at three **k** points on the Fermi surface for various temperatures. The superconducting gap, as estimated by the position of the sample leading edge midpoint at low T, is seen to decrease as one moves from point a near $(\pi, 0)$ to b to c, closer to the diagonal $(0, 0) \to (\pi, \pi)$ direction, consistent with a $d_{x^2-y^2}$ order parameter. At each **k** point the quasiparticle peak disappears above T_c as T increases, with the pseudogap persisting well above T_c, as noted earlier.

The striking feature which is apparent from Fig. 17.28 is that the pseudogap at different **k** points closes at different temperatures, with larger gaps per-

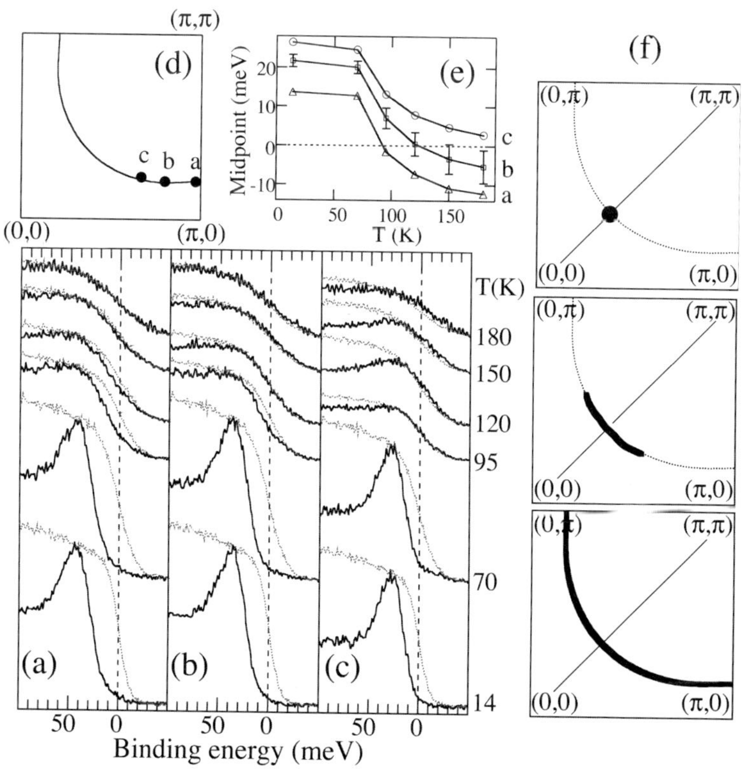

Fig. 17.28. (**a**), (**b**), (**c**): Spectra taken at three **k** points in the Y quadrant of the zone (shown in (**d**)) for an 83 K underdoped Bi2212 sample at various temperatures (*solid curves*). The *dotted curves* are reference spectra from polycrystalline Pt (in electrical contact with the sample) used to determine the chemical potential (zero binding energy). Note the closing of the spectral gap at different T for different **k**'s, which is also apparent in the plot (**e**) of the midpoint of the leading edge of the spectra as a function of T. (**f**) Schematic illustration of the temperature evolution of the Fermi surface in underdoped cuprates. The d-wave node below T_c (*top panel*) becomes a gapless arc above T_c (*middle panel*) which expands with increasing T to form the full Fermi surface at T^* (*bottom panel*)

sisting to higher T's. At point a, near $(\pi, 0)$, there is a pseudogap at all T's below 180 K, at which the Bi2212 leading edge matches that of Pt. As discussed above, this defines T^* above which the largest pseudogap has vanished within the resolution of our experiment, and a closed contour of gapless excitations — a Fermi surface — is obtained. The surprise is that if we move along this Fermi surface to point b the sample leading edge matches Pt at 120 K, which is smaller than T^*. Continuing to point c, about halfway to the diagonal direction, we find that the Bi2212 and Pt leading edges match at an even lower temperature of 95 K. In addition, spectra measured on the same sample along the Fermi contour near the $(0, 0) \rightarrow (\pi, \pi)$ line shows no gap at any T (even below T_c) consistent with $d_{x^2-y^2}$ anisotropy.

One simple way to quantify the behavior of the gap is to plot the midpoint of the leading edge of the spectrum; see Fig. 17.28(e). We note that a leading edge midpoint at a negative binding energy, particularly for $\mathbf{k}$ point c, indicates the formation of a peak in the spectral function at $\omega = 0$ at high T. Further, we will say that the pseudogap has closed at a $\mathbf{k}$ point when the midpoint equals zero energy, in accordance with the discussion above. A clearer way of determining this will be presented below when we discuss the symmetrization method, but the results will be the same. From Fig. 17.28, we find that the pseudogap closes at point a at a T above 1801K, at point b at 120 K, and at point c just below 95 K. If we now view these data as a function of decreasing T, the picture of Fig. 17.28(f) clearly emerges.

With decreasing T, the pseudogap first opens up near $(\pi, 0)$ and progressively gaps out larger portions of the Fermi contour. Thus one obtains gapless arcs which shrink as T is lowered, eventually leading to the four point nodes of the d-wave SC gap. The existence of such arcs is apparent from the first ARPES work on the pseudogap [87], where it was noted that the Fermi contours in the pseudogap phase did not extend all the way to the zone boundary (see Fig. 17.29). Whether the arcs shrink to a point precisely at T_c or below T_c is not clear from the existing data. As discussed in the preceding section, we do believe that arcs do not survive deep into the SC state where there is point node at $T \ll T_c$ in clean

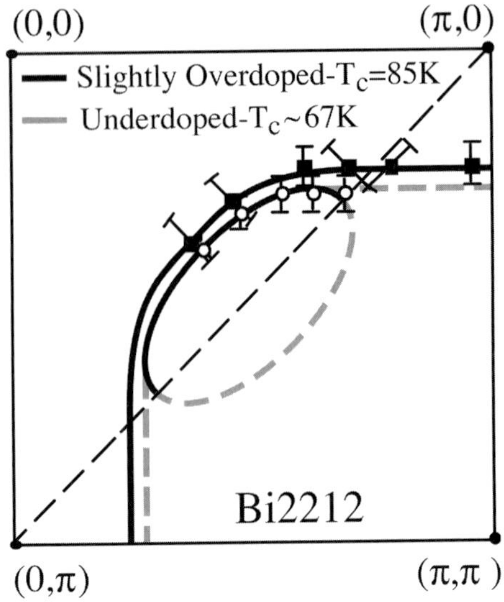

Fig. 17.29. Fermi level crossings from two Bi2212 samples of differing oxygen content. The entire zone can be reconstructed by fourfold rotation about (0,0) (from [87])

samples, as also evidenced by the linear T drop in the superfluid density at low T.

We next turn to a powerful visualization aid that makes these results very transparent. This is the symmetrization method introduced in [90], which effectively eliminates the Fermi function f from ARPES data and permits us to focus directly on the spectral function A. Given ARPES data described by [20] $I(\omega) = \sum_{\mathbf{k}} I_0 f(\omega) A(\mathbf{k}, \omega)$ with the sum over a small momentum window about the Fermi momentum $\mathbf{k}_F$, we can generate the symmetrized spectrum $I(\omega) + I(-\omega)$. Making the reasonable assumption of particle–hole (p–h) symmetry for a small range of ω and $\epsilon_{\mathbf{k}}$, we have $A(\epsilon_{\mathbf{k}}, \omega) = A(-\epsilon_{\mathbf{k}}, -\omega)$ for $|\omega|$, $|\epsilon|$ less than few tens of meV. It then follows, using the identity $f(-\omega) = 1 - f(\omega)$, that $I(\omega) + I(-\omega) = \sum_{\mathbf{k}} I_0 A(\mathbf{k}, \omega)$ which is true even after convolution with a (symmetric) energy resolution function; for details see the appendix of [8]. The symmetrized spectrum coincides with the raw data for $\omega \leq -2.2T_{\text{eff}}$, where $4.4T_{\text{eff}}$ is the 10%–90% width of the Pt leading edge, which includes the effects of both temperature and resolution. Non-trivial information is obtained for the range $|\omega| \leq 2.2T_{\text{eff}}$, which is then the scale on which p–h symmetry has

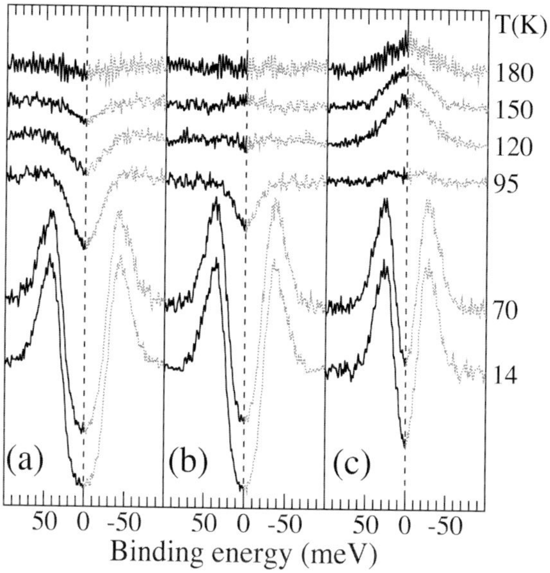

Fig. 17.30. Symmetrized spectra corresponding to the raw spectra (a), (b, (c) of Fig. 17.28. The gap closing in the raw spectrum of Fig. 17.28 corresponds to when the pseudogap depression disappears in the symmetrized spectrum. Note the appearance of a spectral peak at higher temperatures in (c)

to be valid. We have extensively checked this method, and studied in detail the errors introduced by incorrect determination of the chemical potential or of $\mathbf{k}_F$ (which lead to spurious narrow features in the symmetrized spectra), and the effect of the small (1° radius) **k**-window of the experiment (which was found to be small).

In Fig. 17.30 we show symmetrized data for the UD 83 K underdoped sample corresponding to the raw data of Fig. 17.28. To emphasize that the symmetry is put in by hand, we show the $\omega > 0$ curve as a dotted line. At **k** point a near $(\pi, 0)$ the sharp quasiparticle peak disappears above T_c but a strong pseudogap suppression, on the same scale as the superconducting gap, persists all the way up to 180K (T^*). Moving to panels (b) and (c) in Fig. 17.30 we again see pseudogap depressions on the scale of the superconducting gaps at those points, however the pseudogap fills up at lower temperatures: 120 K at b and 95 K at c. In panel c, moreover, a spectral peak at zero energy emerges as T is raised. All of the conclusions drawn from the raw data in Fig. 17.28 are immediately obvious from the simple symmetrization analysis of Fig. 17.30.

There are many important issues related to these results that will be taken up in Sect. 17.7.3 where we describe modeling the electron self-energy in the pseudogap state. We will discuss there the remarkable T-dependent lineshape changes and the T-dependence of the gap itself. Here we simply note that, without any detailed modeling, the data [90] clearly show qualitative differences in the T-dependence at different **k**-points. Near the $(\pi, 0)$ point the gap goes away with increasing temperature with the spectral weight filling-in, but no perceptible change in the gap scale with T. On the other hand, at $\mathbf{k}_F$ points halfway to the node, one sees a suppression of the gap scale with increasing temperature.

We conclude this discussion with a brief mention of the implications of our results. We believe that the unusual T-dependence of the pseudogap anisotropy will be a very important input in reconciling the different crossovers seen in the pseudogap regime by different probes. The point here is that each experiment is measuring a **k**-sum weighted with a different set of **k**-dependent matrix elements or kinematical factors (e.g., Fermi velocity). For instance, quantities which involve the Fermi velocity, like dc resistivity above T_c and the penetration depth below T_c (superfluid density), should be sensitive to the region near the zone diagonal, and would thus be affected by the behavior we see at **k** point c. Other types of measurements (e.g., specific heat and tunneling) are more "zone-averaged" and will have significant contributions from **k** points a and b as well, thus they should see a more pronounced pseudogap effect. Interestingly, other data we have indicate that the region in the Brillouin zone where behavior like **k** point c is seen shrinks as the doping is reduced, and thus appears to be correlated with the loss of superfluid density [94]. Further, we speculate that the disconnected Fermi arcs should have a profound influence on magnetotransport given the lack of a continuous Fermi contour in momentum space.

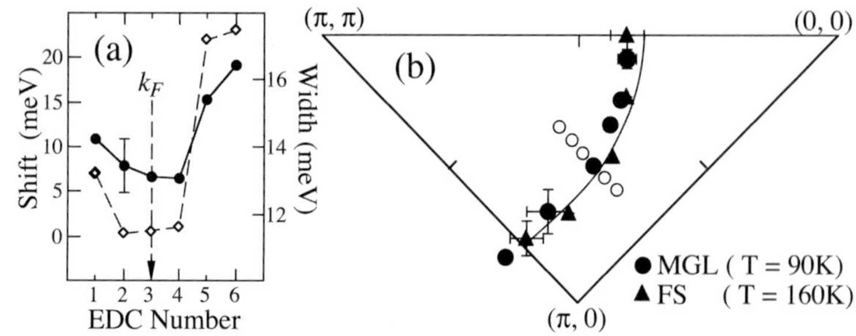

Fig. 17.31. (**a**) Midpoint shifts (*dots*) and widths (*diamonds*) for an UD 83 K sample at a photon energy of 22 eV at 90 K for a cut shown by the open dots in (**b**). (**b**) Fermi surface at 160 K (*solid triangles*) and minimum gap locus at 90 K (*solid dots*). Notice that the two surfaces coincide within error bars. The error bars represent uncertainties of Fermi crossings as well as possible sample misalignment. The *solid curve* is a rigid band estimate of the Fermi surface

17.6.4 Evolution of the Fermi Surface with Doping

We now discuss the doping dependence of the normal state Fermi surface on the underdoped side of the phase diagram. The first issue to face up to is: can the Fermi arcs described above be a manifestation of a Fermi surface with small closed contours centered about $(\pi/2, \pi/2)$? Such hole-pockets enclosing x holes (per planar Cu) are suggested by some theories of lightly doped Mott insulators (for a review, see [95]) as alternatives to the large Fermi surfaces containing $(1 + x)$ holes which would be consistent with the Luttinger counting.

The T-dependence of the arcs is by itself evidence against their being part of a pocket Fermi surface. Nevertheless, if there were such small hole pockets then one should observe two features in the ARPES data: a closure of the Fermi arc on the other side of $(\pi/2, \pi/2)$, which would be clear evidence for a "shadow band"-like dispersion ((π, π)-foldback of the main band) in the UD samples. In a variety of UD samples we have carefully searched for both these features and found no evidence for either [41]. However this is a tricky issue, given the very broad spectra and possible materials problems in the highly UD samples. Nevertheless, given the available evidence, the gapless arcs that we observe [90] are simply an intermediate state in the smooth evolution of d-wave nodes into a full Fermi surface. This smooth evolution was carefully checked on an UD 831 K sample where a dense mapping was done in **k** space at T = 90 K, revealing only a small Fermi arc just above T_c.

The other issue related to the Fermi surface is: what is its doping dependence above T^* where the pseudogap effects are absent. While one can easily compare the near optimal and lightly UD Fermi surfaces, the rapid rise of T^* with underdoping does not permit us to address this question. However, one can study the "minimum gap locus" in any gapped state, in close analogy with the manner in which this was defined in the SC state; see Sect. 17.5.

There is also a more fundamental reason to study the "minimum gap locus" in the pseudogap regime. One wants to know whether the pseudogap is "tied" to the Fermi surface, or if it has some other characteristic momentum **Q** (unrelated to $\mathbf{k}_f$). In Fig. 17.31(a) [41] we follow the dispersion of an UD 83K sample in the pseudogap regime. Moving perpendicular to the (expected) Fermi surface from occupied to unoccupied states, one finds that the dispersion first approaches the chemical potential and then recedes away from it. This locates a **k**-point on the minimum gap locus. For a lightly UD sample we find in Fig. 17.31(b) that this locus in the pseudogap regime coincides, within experimental error bars, with the Fermi surface determined above T^* where there is no pseudogap. The pseudogap is thus tied to the Fermi surface in the same way the SC gap is, and is in contrast with, say, charge or spin density waves, which are tied to other characteristic **Q** vectors.

In the more heavily underdoped samples, it is not possible to compare the minimum gap locus in the pseudogap state with the Fermi surface above T^*, or the minimum gap locus below T_c, since the latter

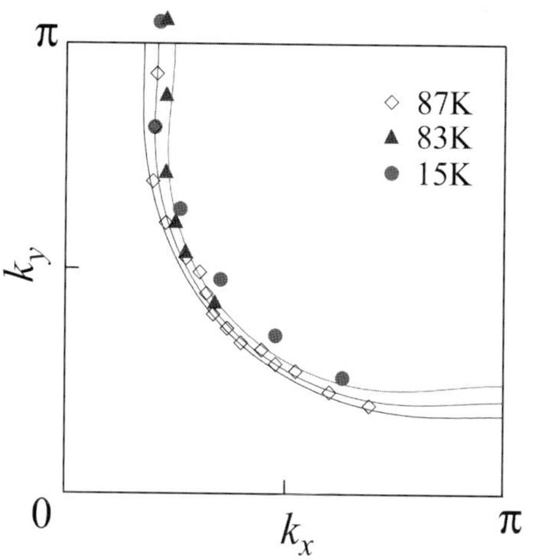

Fig. 17.32. Fermi surfaces of the 87 K, 83 K, and 15 K samples. All surfaces enclose a large area consistent with the Luttinger count (see text). The *solid lines* are tight binding estimates of the Fermi surface at 18%, 13%, and 6% doping assuming rigid band behavior

two are not measurable with T^* too high and T_c too low. Nevertheless, if one *assumes*, by continuity, that the minimum gap locus in the pseudogap state gives information about the Fermi surface that got gapped out, then even for an highly UD sample one finds a large underlying Fermi surface, satisfying the Luttinger count of $(1 + x)$ holes per planar Cu [41] as shown in Fig. 17.32.

The same conclusion has been recently reached by the Dresden group [42]. Figure 17.33 reproduces their Fermi energy intensity maps as a function of doping, where a large Fermi surface (plus its shadow band image) is always visible. They argue, though, that the volume is not quite $1 + x$, and they attribute this difference to the presence of bilayer splitting.

17.6.5 Low Energy Versus High Energy Pseudogaps

In all of the preceding discussion we have focussed on the "low energy" or leading-edge pseudogap. It is important to point out that the phrase pseudogap is (somewhat confusingly) also used to describe a higher energy feature, which we call the "high energy pseudogap".

The presence of a high energy pseudogap was evident in the first ARPES work on the pseudogap, reproduced in Fig. 17.34 [87]. As the doping is reduced from optimal doped, a gap opens up in a region around the $(\pi, 0)$ points of the zone. The energy of this gap is significantly higher than the leading edge gap emphasized in later work [88,89]. The resulting dispersion of this high energy feature looks reminiscent of what is expected for a spin density wave gap.

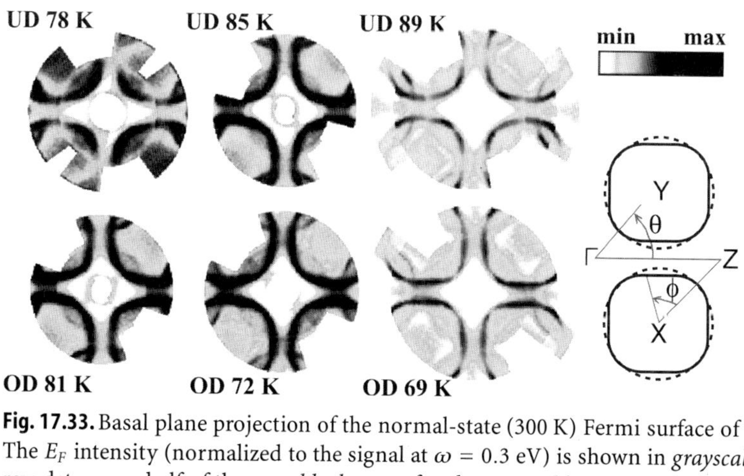

Fig. 17.33. Basal plane projection of the normal-state (300 K) Fermi surface of Bi(Pb)-2212 from high-resolution ARPES. The E_F intensity (normalized to the signal at $\omega = 0.3$ eV) is shown in *grayscale*. The T_c of each sample is indicated. The raw data cover half of the *gray-black* area of each map and have been rotated by 180° around the Γ point to give a better k-space overview. The line dividing raw and rotated data runs almost vertically for the UD76 K map and from top left to bottom right in all other maps. The *sketch* shows the Fermi surface for the OD69K data set as *black barrel-like* shapes defined by joining the maxima of fits to the normalized E_F MDCs (from [42])

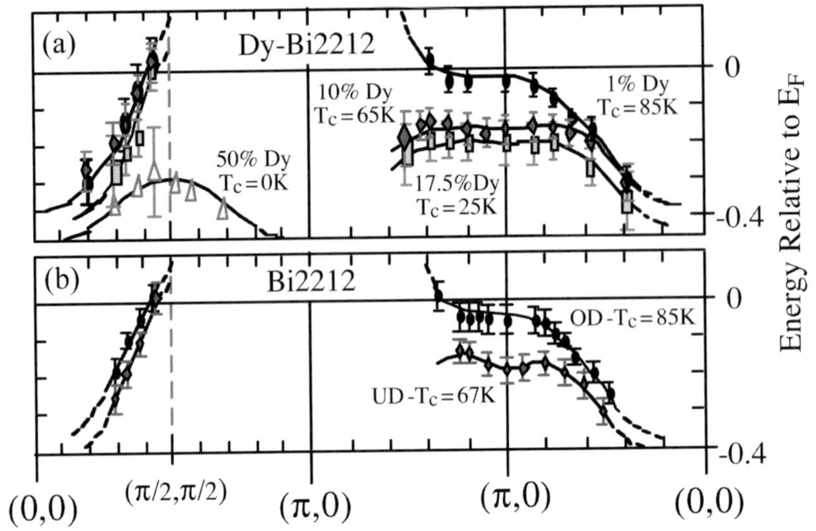

Fig. 17.34. (**a**) and (**b**) Map peak centroids vs k for $Bi_2Sr_2Ca_{1-x}Dy_xCu_2O_{8+\delta}$ thin films and deoxygenated $Bi_2Sr_2CaCu_2O_{8+\delta}$ bulk samples, respectively, with various hole doping levels. (**a**) *Filled oval*, 1% Dy near optimal doping with $T_c = 85$ K; *gray diamond*, 10% Dy underdoped with $T_c = 65$ K; *gray rectangle*, 17.5% Dy underdoped with $T_c = 25$ K; *triangle*, 50% Dy insulator. (**b**) *Filled oval*, 600 air annealed slightly overdoped with $T_c = 85$ K; *gray diamond*, 550 argon annealed underdoped with $T_c = 67$ K (from [87])

As the doping is further reduced, an energy gap then opens up along the (π, π) direction, and the material becomes truly insulating.

In Fig. 17.35 [59] we show the temperature dependence of the $(\pi, 0)$ spectrum for an UD 89 K sample. Note that there is no coherent quasiparticle peak until the system is cooled below T_c, with only a broad incoherent spectrum observed for all $T > T_c$. The leading edge pseudogap which develops below T^* is difficult to see on the energy scale of this figure; the midpoint shift at 135 K is 3 meV. However, a higher energy feature can easily be identified by a change in slope of the spectra as function of binding energy; this is also very clear in the data of Fig. 17.48 of Sect. 17.6). At the highest temperature $T = 247$ K this feature is just due to the Fermi function cutoff, but in the pseudogap regime, this feature actually represents the onset of loss of spectral weight on a high energy scale, and hence may be called the "high energy pseudogap". It can also be seen from Fig. 17.35 that the energy scale of this feature is very similar to that of the well-known $(\pi, 0)$-hump of the peak/dip/hump structure seen in the SC state. This connection will be discussed in detail in Sect. 17.6 where we also argue that the high energy pseudogap and the hump have similar dispersions [59].

17.6.6 Origin of the Pseudogap?

We conclude with a summary of ARPES results on the pseudogap and a brief discussion of its theoretical understanding.

As described above, the low-energy (leading edge) pseudogap has the following characteristics.

- The magnitude of the pseudogap near $(\pi, 0)$, i.e., the scale of which there is suppression of low energy spectral weight above T_c, is the same as the maximum SC gap at low temperatures. Further both have the same doping dependence.
- There is a crossover temperature scale T^* above which the full Fermi surface of gapless excitations is recovered. The pseudogap near $(\pi, 0)$ appears below T^*.
- The normal state pseudogap evolves smoothly through T_c into the SC gap as a function of decreasing temperature.
- The pseudogap is strongly anisotropic with **k**-dependence which resembles that of the d-wave

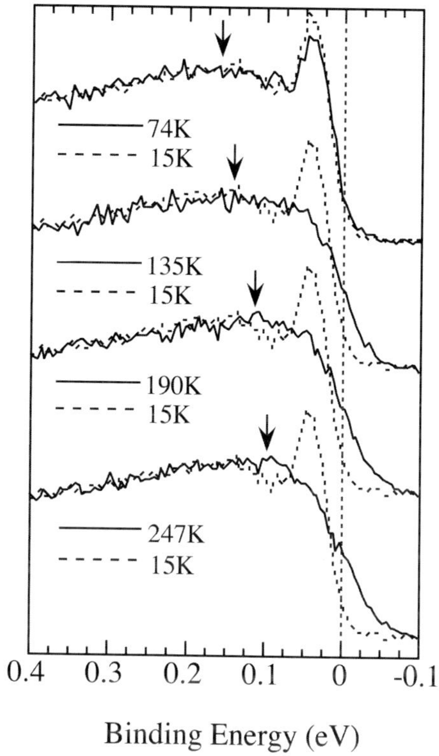

Fig. 17.35. Spectra at the $(\pi, 0)$ point for an UD 89 K sample at various temperatures compared with the low temperature (15 K) spectrum. The position of the high energy feature is marked by an *arrow*

SC gap. The anisotropy of the pseudogap seems to be T-dependent leading to the formation of disconnected Fermi arcs below T^*.

- The pseudogap is "tied" to the Fermi surface, i.e., the minimum gap locus in the pseudogap regime coincides with the Fermi surface above T^* and the minimum gap locus deep in the SC state, at least in those samples where all three loci can be measured.

The simplest theoretical explanation of the pseudogap, qualitatively consistent with the ARPES observations, is that it arises to due pairing fluctuations above T_c [86, 96]. The SC gap increases with underdoping while T_c decreases. Thus in the underdoped regime T_c is not controlled by the destruction of the pairing amplitude, as in conventional BCS theory, but rather by fluctuations of the phase [97] of the order parameter leading to the Uemura scaling $T_c \sim \rho_s$ [94]. Even though SC order is destroyed at T_c, the local pairing amplitude survives above T_c giving rise to the pseudogap features. A natural mechanism for such a pseudogap coming from spin pairing in a doped Mott insulator exists within the RVB framework [98], with possibility of additional chiral current fluctuations [99].

More recently the pairing origin of the pseudogap has been challenged. Some experiments [100] have been argued to suggest a non-pairing explanation with a competition between the pseudogap and the SC gap. A specific realization of this scenario is the staggered flux or d-density wave mechanism [101] in which T^* is actually a phase transition below which both time-reversal and translational invariance are broken. A more subtle phase transition with only broken time-reversal has also been proposed [102] as the origin of the pseudogap.

Although a qualitative understanding of some of the characteristics of the pseudogap within the non-pairing scenarios is not clear at this time, these theories make sharp predictions about broken symmetries below T^* which can be tested. A very recent ARPES study [103] of circular dichroism finds evidence in favor of broken time reversal, thus casting some doubt on the pairing fluctuation ideas. The last word has clearly not been said on this subject, either theoretically or experimentally, and the origin of the pseudogap remains one of the most important open questions in the field of high T_c superconductors.

17.7 Photoemission Lineshapes and the Electron Self-Energy

Under certain conditions, which were discussed in Sect. 17.2, ARPES measures the occupied part of the single particle spectral function, $A(\mathbf{k}, \omega)f(\omega)$, with $A = ImG/\pi$ where G is the Greens function. The latter can be expressed as $G^{-1} = \omega - \epsilon_{\mathbf{k}} - \Sigma(\mathbf{k}, \omega)$ where $\epsilon_{\mathbf{k}}$ is the single-particle energy (defined by the kinetic energy and single-particle potential energy terms of the Hamiltonian) and Σ is the Dyson self-energy (i.e., everything else). Often, this form is associated with a perturbative expansion used to estimate Σ, but of course the expression is itself tautological. The purpose of writing G in this form is that it isolates all many-body effects in the function Σ. An advantage of ARPES is that one has the possibility of extracting Σ directly from the data, allowing comparison to various microscopic predictions for Σ.

One of the more trivial examples of this is when one fits ARPES data to determine the superconducting gap, Δ. For instance, the work described in Sect. 17.5 [21] used a broadened form of BCS theory to fit the leading edge of the spectra. This is equivalent to $\Sigma = -i\Gamma + \Delta_{\mathbf{k}}^2/(\omega + \epsilon_{\mathbf{k}} + i\Gamma)$, $\Gamma = 0$ describing standard BCS theory. The advantage of this procedure is the actual gap function, Δ, is extracted from the data, rather than ill defined quantities, such as the often utilized leading edge shift (midpoint of the leading edge) which is not the same as Δ because of lifetime and resolution effects. When this is done, a Δ_k is obtained which has rather spectacular agreement with that expected for a d-wave order parameter. Although ARPES contains no phase information of the order parameter, the linear behavior of Δ_k along the Fermi surface near the gap zero (node) implies a sign change. Moreover, ARPES has the additional advantage of determining the shape of Δ_k in the Brillouin zone, which gives important information on the spatial range of the pairing interaction [79], as also discussed in Sect. 17.5.

Even when fitting data at low temperatures including energy and momentum resolution, a non-zero Γ is always needed. The origin of this residual Γ is still debated. It is larger than what is expected based on impurity scattering, and certainly larger than that implied by various conductivity probes (thermal, microwave, and infrared). Although the transport scattering rate is different from ImΣ (and in particular, only Umklapp processes contribute to electrical conductivity), the discrepancy is still large enough to be noticeable, even when taking into account the fact that in the simple approximation being employed here, Γ represents some average of ImΣ over a frequency range of order Δ.

Although it has been suggested that the residual Γ is due to surface inhomogeneity effects (in particular, a distribution of Δ due to local oxygen inhomogeneities [104]), a more likely possibility is that it is the same effect which is seen in normal metals like $TiTe_2$. In the latter case, it was convincingly argued that this was the expected final state lifetime contamination effect when attempting to extract Σ from ARPES spectra [105]. Although the latter is expected to vanish in the pure 2D limit, even small 3D effects can lead to a noticeable effect, since final state lifetimes are large. For instance, in simple models, its contribution to Γ is of order $(v_c^i/v_c^f)\Gamma_f$, where v_c^i is the c-axis velocity of the initial state, v_c^f that of the final state, and Γ_f is the final state lifetime [18]. Since Γ_f is typically of order 1 eV, then a velocity ratio of only 0.01 is sufficient to cause a residual Γ of 10 meV.

With this as an introduction, in this section, we desire to take a more serious look at the issue of extracting Σ from the data. The most commonly employed strategy is to come up with some model for Σ, and then see how well it fits the data, as illustrated by the simple example above. We will discuss this approach in more detail later. We start, though, with discussing an alternate approach which we have recently advocated.

17.7.1 Self-Energy Extraction

Let us first assume we know A. Given that, we can easily obtain Σ. A Kramers–Kronig transform of A will give us the real part of G

$$\mathrm{Re}G(\omega) = P \int_{-\infty}^{+\infty} d\omega' \frac{A(\omega')}{\omega' - \omega} , \tag{17.13}$$

where P denotes the principal part of the integral. Knowing now both ImG and ReG, then Σ can be directly read off from the definition of G.

$$\begin{aligned} \mathrm{Im}\Sigma &= \frac{\mathrm{Im}G}{(\mathrm{Re}G)^2 + (\mathrm{Im}G)^2} \\ \mathrm{Re}\Sigma &= \omega - \epsilon - \frac{\mathrm{Re}G}{(\mathrm{Re}G)^2 + (\mathrm{Im}G)^2} . \end{aligned} \tag{17.14}$$

To obtain ReG using Eq. (17.13), we need to know A for *all* energies. From ARPES, though, we only know the product of A and f. (While unoccupied states can be studied by inverse photoemission, its resolution at present is too poor to be useful for our purposes.) This is not a limitation if an occupied $\mathbf{k}$-state is being analyzed and one can either ignore the unoccupied weight or use a simple extrapolation for it (except that only $Re\Sigma + \epsilon$ is determined). On the other hand, one is usually interested in k vectors near the Fermi surface. Therefore a key assumption will have to be made. We can implement our procedure if we make the assumption of particle–hole symmetry, $A(\epsilon_{\mathbf{k}}, \omega) = A(-\epsilon_{\mathbf{k}}, -\omega)$, within the small $\mathbf{k}$-window centered at $\mathbf{k}_F$. Then, A is obtained by exploiting the identity $A(\epsilon_{\mathbf{k}}, \omega) f(\omega) + A(-\epsilon_{\mathbf{k}}, -\omega) f(-\omega) = A(\epsilon_{\mathbf{k}}, \omega)$, which holds even in the presence of the energy resolution convolution. Note, this can only be invoked at $\mathbf{k}_F$, and was used previously to remove the Fermi function from ARPES data [90], where it was denoted as the symmetrization procedure (note that the "symmetrized" data will correspond to the raw data for $\omega <\sim -2.2kT$). Although the particle–hole symmetry assumption is reasonable for small $|\omega|$ where it can be tested in the normal state by seeing whether the "symmetrized" spectrum has a maximum at the Fermi energy (E_F), it will almost certainly fail for sufficiently large $\omega > 0$. Nevertheless, since we only expect to derive Σ for $\omega < 0$, then the unoccupied spectral weight will affect the result only in two ways. The first is through the sum rule $\int d\omega A(\omega) = 1$ which must be used to eliminate the intensity prefactor of the ARPES photocurrent. From Eq. (17.14), we see that violation of the sum rule will simply rescale ImΣ, but not ReΣ due to the $\omega - \epsilon$ factor. Our normalization, though, is equivalent to assuming $n_{k_F} = 0.5$, and thus does not involve "symmetrized" data. The second influence comes from the Kramers–Kronig transformation in Eq. (17.13), which is a bigger problem. Fortunately, the contribution from large $\omega' > 0$, for which our assumption is least valid, is suppressed by $1/(\omega' - \omega)$. Further, for $\mathbf{k}_F$, $\epsilon_{\mathbf{k}} = 0$ and thus ReΣ is not plagued by an unknown constant.

When using real data, it is sometimes desirable to filter the noise out of the data, as well as to deconvolve the energy resolution, before employing the above procedure. These details can be found in [106]. Moreover, it is assumed that any "background" contribution (see Sect. 17.2) has been subtracted from the data as well.

In Fig. 17.36a, we show $T = 14$ K symmetrized data for a $T_c = 87$ K Bi2212 overdoped sample at the $(\pi, 0)$ point [106]. We note the important differences in this superconducting state spectrum, compared with the normal state one (which can be fit by a simple Lorentzian), due to the opening of the superconducting gap, with the appearance of a sharp quasiparticle peak displaced from E_F by the superconducting gap, followed by a spectral dip, then by a "hump" at higher binding energies. The resulting Σ is shown in Fig.s 17.36(b) and (c). At high binding energies, one obtains a constant ImΣ with a very large value (~ 300 meV). Near the spectral dip, ImΣ has a small peak followed by a sharp drop.

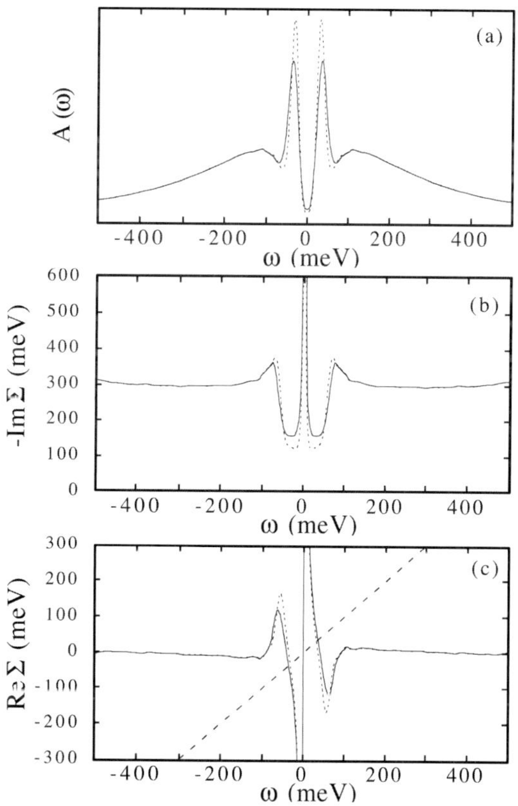

Fig. 17.36. (**a**) Symmetrized spectrum for overdoped Bi2212 ($T_c = 87$ K) at $T = 14$ K at $(\pi, 0)$ with (*dotted line*) and without (*solid line*) energy resolution deconvolution. The resulting ImΣ and ReΣ are shown in (**b**) and (**c**). The *dashed line* in (**c**) determines the condition Re$\Sigma = \omega$

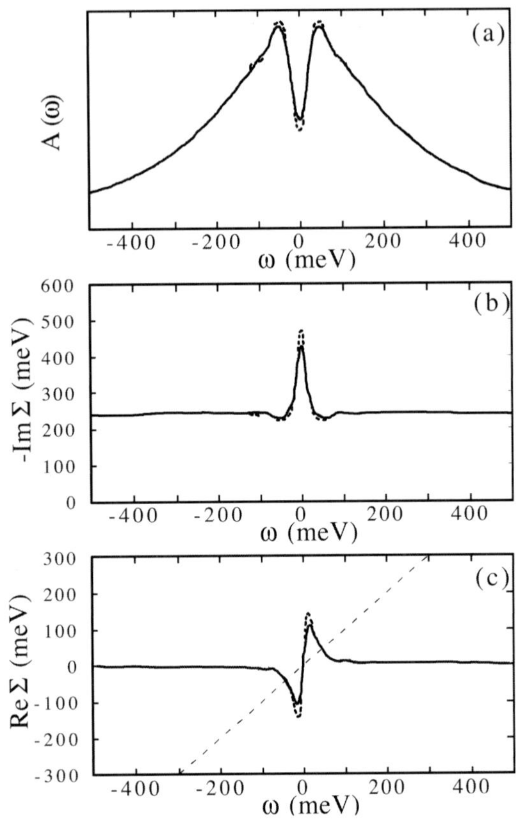

Fig. 17.37. (**a**) Symmetrized spectrum for underdoped Bi2212 (T_c = 85 K) at T = 95 K (pseudogap phase) at the $(\pi, 0) - (\pi, \pi)$ Fermi crossing with (*dotted line*) and without (*solid line*) energy resolution deconvolution. The resulting ImΣ and ReΣ are shown in (**b**) and (**c**). The *dashed line* in (**c**) determines the condition $Re\Sigma = \omega$

Despite this sharp drop below 70 meV, ImΣ remains quite large at low frequencies. Then, below 20 meV, there is a narrow spike in ImΣ. This is the imaginary part of the BCS self-energy, $\Delta^2/(\omega + i0^+)$, which kills the normal state pole at $\omega = 0$. The resulting $1/\omega$ divergence of the real part ReΣ, which creates new poles at $\pm\Delta$=32meV, is easily seen in Fig. 17.36(c). This is followed by a strong peak in ReΣ near the spectral dip energy, which follows from the Kramers–Kronig transformation of the sharp drop in ImΣ. The strong peak in *Re*Σ explains why the low energy peak in A is so narrow despite the large value of ImΣ. The halfwidth of the spectral peak is given by $\Gamma = z\mathrm{Im}\Sigma$ where $z^{-1} = 1 - \partial\mathrm{Re}\Sigma/\partial\omega$ (z is the quasiparticle residue). In the vicinity of the spectral peak, z^{-1} is large (~9), giving a Γ of ~14 meV. We note, though, that Γ is still quite sizeable, and thus the peak is not resolution limited, as discussed above.

We can contrast this result with that obtained in the pseudogap phase. In Fig. 17.37(a), we show T = 95 K symmetrized data from a T_c = 85 K underdoped Bi2212 sample at the $(\pi, 0) - (\pi, \pi)$ Fermi crossing. One again sees (Fig. 17.37b) a peak in ImΣ at ω=0, but it is broadened relative to that of the superconducting state, and the corresponding divergence of ReΣ (Fig. 17.37(c)) is smeared out. Such behavior would be consistent with replacing the BCS self-energy $\Delta^2/(\omega + i0^+)$ by $\Delta^2/(\omega + i\Gamma_0)$, and can be motivated by considering the presence of pair fluctuations above T_c, as will be discussed further below. Note from Fig. 17.37 that although the equation $\omega - \mathrm{Re}\Sigma(\omega) = 0$ is still satisfied at $|\omega| \sim \Delta$, zImΣ is so large that the spectral peak is strongly broadened in contrast to the sharp peak seen below T_c. Actually, to a good approximation, the spectral function is essentially the inverse of ImΣ in the range $|\omega| <\sim 2\Delta$. We can also contrast this case with data taken above T^*, the temperature at which the pseudogap "disappears". In that case, the spectrum is featureless, and the peak in ImΣ is strongly broadened. As the doping increases, this peak in ImΣ disappears. Further doping causes a depression in ImΣ to develop around $\omega = 0$, indicating a crossover to more Fermi liquid like behavior.

17.7.2 Temperature Dependence of Self-Energy

We now turn to the rather controversial issue of how the spectrum at $(\pi, 0)$ varies as a function of temperature, that is, how one interpolates between Figs. 17.36 and 17.37.

In Fig. 17.38, we show data taken for an optimal doped (T_c=90K) Bi2212 sample [107]. The leading edge of the spectral peak is determined by the superconducting gap, whose energy stays fairly fixed in temperature, and persists above T_c (the pseudogap). On the trailing edge, one sees a spectral dip, whose energy also remains fixed in temperature, and becomes filled in above T_c due to broadening of the trailing edge of the peak.

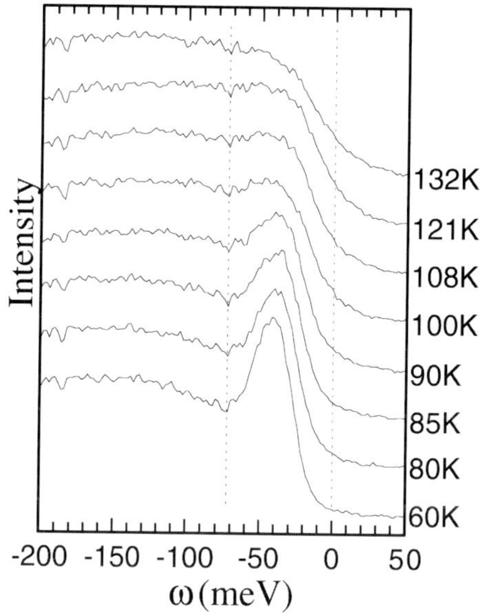

Fig. 17.38. Temperature dependence of ARPES data at $(\pi, 0)$ for a $T_c = 90$ K Bi2212 sample. The *vertical dotted lines* mark the spectral dip energy and the chemical potential

In Fig. 17.39(a), ImΣ is plotted for various temperatures. At low temperatures and energies, it is again characterized by a peak centered at zero energy due to the superconducting energy gap, and a "normal" part which can be treated as a constant plus an ω^2 term. A maximum in ImΣ occurs near the energy of the spectral dip. Beyond this, ImΣ has a large, nearly frequency independent, value. As the temperature is raised, the zero energy peak broadens, the constant term increases, and the ω^2 term goes away.

In Fig. 17.39(b), the quantity $\omega-\text{Re}\Sigma$ is plotted. At low temperatures and energies, there is a $1/\omega$ term due to the energy gap, and a "normal" part which is linear in ω. As expected, the zero crossing is near the location of the spectral peak. Beyond this, there is a minimum near the spectral dip energy, then the data are approximately linear again, but with a smaller slope than near the zero crossing. As the temperature is raised, the gap $(1/\omega)$ term broadens out and the low energy linear in ω term decreases, paralleling the behavior discussed above for ImΣ.

From Figs. 17.38 and 17.39, we see that rather than the spectral peak decreasing in weight with increasing temperature, it disappears by broadening strongly in energy. This can be seen directly by inspecting Fig. 17.39, in that as the temperature increases, ImΣ (Fig. 17.39(a)) in the vicinity of the peak increases in magnitude with T, and z^{-1} (roughly the slope in Fig. 17.39(b) near the zero crossing) decreases with T. In fact, it is the strong T variation of ImΣ and z^{-1}, and the fact that they operate in concert, which is responsible for the rapid variation in the effective width of the spectral peak with T.

The above analysis is important in that it shows how coherence is lost in the system. It is apparent from Figs. 17.38 and 17.39 that once a temperature

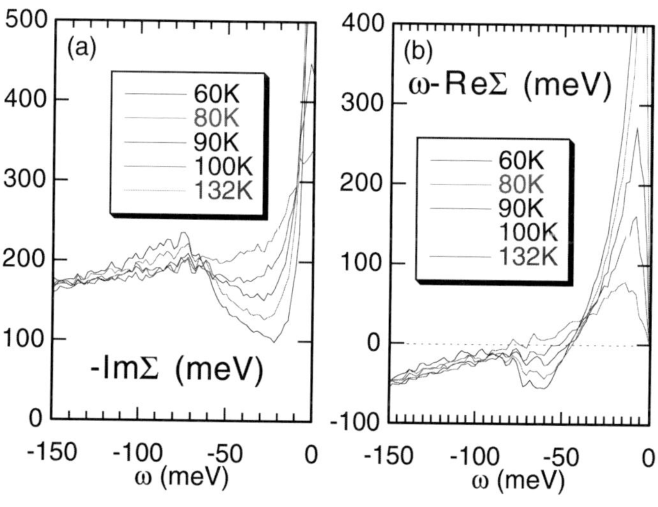

Fig. 17.39. Temperature dependence of (**a**) ImΣ and (**b**) ReΣ derived from the data of Fig. 17.38

is reached where the spectral peak is no longer discernable in the data, the difference in behavior of the self-energy between low energies and high energies is lost. That is, once the spectral dip is filled in, the low and high energy behaviors have merged, and the sharp peak and broad hump at low temperature is simply replaced by a single broad peak (with a leading edge gap due to the pseudogap). This is consistent with the spectral peak simply losing its integrity as the temperature is raised. The analysis does not support a picture of a well defined quasiparticle peak whose weight simply disappears upon heating, as has been suggested by other authors [92].

17.7.3 Modeling the Self-Energy

This behavior can be further quantified by fitting the self-energy for binding energies smaller than the dip energy to that expected for a superfluid Fermi liquid, and exploring the temperature dependence of the resulting parameters. The reader can find this analysis in [107]. Rather, we will discuss here a simpler analysis we performed where we contrasted the temperature dependence of overdoped and underdoped samples [108]. In this case, we chose to look at data at the $(\pi, 0)-(\pi, \pi)$ Fermi crossing (antinode). The dip/hump structure is considerably weaker here than at the $(\pi, 0)$ point, allowing us to concentrate on more general aspects of the spectra.

We begin with the overdoped sample, where there is no strong pseudogap effect. The simplest self-energy which can describe the low energy data at all T is

$$\Sigma(\mathbf{k}, \omega) = -i\Gamma_1 + \Delta^2/[(\omega + i0^+) + \epsilon(\mathbf{k})] \,. \quad (17.15)$$

Here Γ_1 is a single-particle scattering rate taken, for simplicity, to be an ω-independent constant. It is effectively an average of the (actual ω-dependent) Σ'' over the frequency range of the fit. (In [107], we generalized this to a constant plus an ω^2 term, and included the resulting linear ω contribution to Σ'). The second term is the BCS self-energy discussed previously.

In Fig. 17.40(a), we show symmetrized data for an overdoped $T_c = 82$ K sample at the antinode together with the fits obtained as follows [108]. Using Eq. (17.15), we calculate the spectral function

$$\pi A(\mathbf{k}, \omega) = \frac{\Sigma''(\mathbf{k}, \omega)}{(\omega - \epsilon_{\mathbf{k}} - \Sigma'(\mathbf{k}, \omega))^2 + \Sigma''(\mathbf{k}, \omega)^2} \,, \quad (17.16)$$

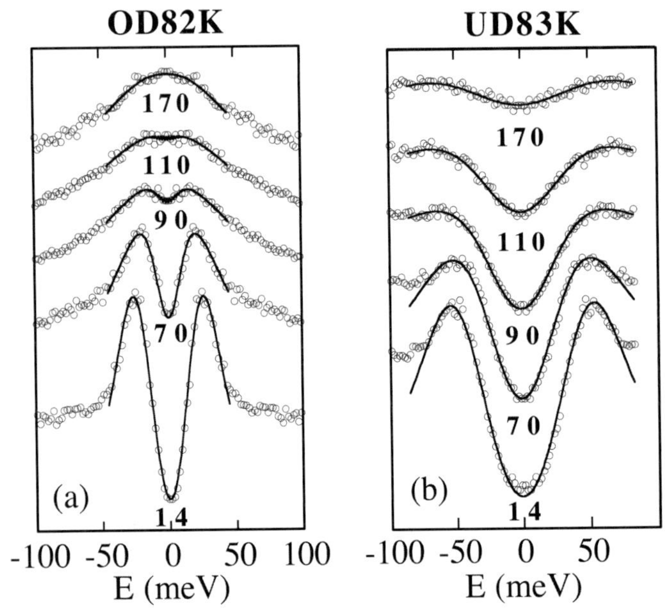

Fig. 17.40. Symmetrized data for (**a**) a $T_c = 82$ K overdoped sample and (**b**) a $T_c = 83$ K underdoped sample at the $(\pi, 0) - (\pi, \pi)$ Fermi crossing at five temperatures, compared to the model fits described in the text

convolve it with the experimental resolution, and fit to symmetrized data. The fit is restricted to a range of ±45 meV given the small gap in the overdoped case and the sharpness of the quasiparticle peaks below T_c. We find that Eq. (17.15) describes the low energy data quite well.

The T-variation of the fit parameters Δ and Γ_1 are shown in Fig. 17.41(a). $\Delta(T)$ decreases with T, and although small at T_c, it only vanishes above T_c, indicating the possibility of a weak pseudogap. This effect is sample dependent, in that several overdoped samples we have looked at, the gap vanishes closer to T_c. We caution that the error bars shown in Fig. 17.41(a) are based on the RMS error of the fits, but do not take into account experimental errors in μ and $\mathbf{k}_F$.

$\Gamma_1(T)$ is found to be relatively T-independent in the normal state. Below T_c, we see that Γ_1 decreases very rapidly, and can be perfectly fit to the form $a + bT^6$. This rapid drop in linewidth leading to sharp quasiparticle peaks at low T, which can be seen directly in the ARPES data, is consistent with microwave and thermal conductivity measurements, and implies that electron-electron interactions are responsible for Γ_1. Note the clear break in Γ_1 at T_c, despite the fact Δ has not quite vanished. We have seen similar behavior to that described above for a variety of overdoped samples at several $\mathbf{k}$ points.

We next turn to the more interesting underdoped case. We find that near $(\pi, 0)$ the self-energy equation (17.15) cannot give an adequate description of the data, in that it does not properly describe the pseudogap and its unusual "filling in" above T_c. Theoretically, we cannot have a divergence in $\Sigma(\mathbf{k}_F, \omega = 0)$ in a state without broken symmetry. A simple modification of the BCS self-energy rectifies both these problems:

$$\Sigma(\mathbf{k}, \omega) = -i\Gamma_1 + \Delta^2/[\omega + \epsilon(\mathbf{k}) + i\Gamma_0] \,. \qquad (17.17)$$

The new term $\Gamma_0(T)$ should be viewed as the inverse pair lifetime. The theoretical motivation for Eq. (17.17) is given in [108]. We stress that this three parameter form is again a minimal representation of the pseudogap self-energy. Since it is not obviously a unique representation, it is very important to see what one learns from the fits.

In Fig. 17.40(b), we show symmetrized data at the antinode for a T_c = 83 K underdoped sample. Below T_c we see quasiparticle peaks. Above T_c these peaks disappear but there is still a large suppression of spectral weight around ω=0. As T is raised further, the pseudogap fills in (rather than closing) leading to a flat spectrum at a temperature of T^* (200 K). The self-energy, Eq. (17.17), gives a good fit to the data. These fits were done below T_c over a larger energy

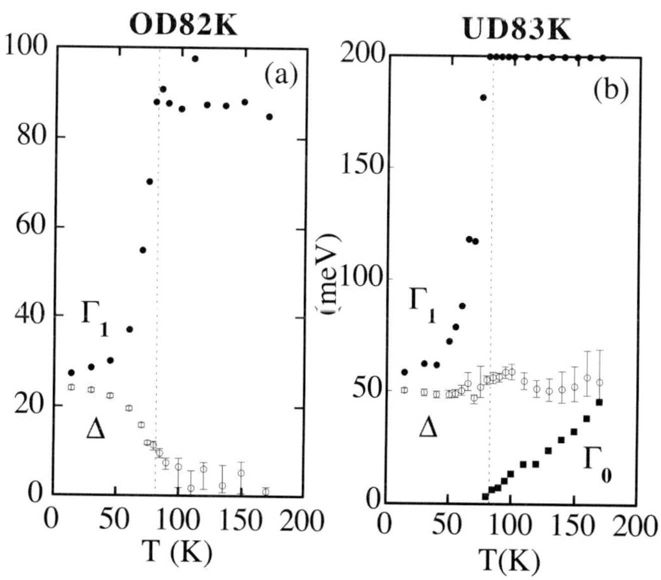

Fig. 17.41. Δ (*open circles*), Γ_1 (*solid circles*), and Γ_0 (*solid squares*) versus T at the $(\pi, 0)-(\pi, \pi)$ Fermi crossing for (**a**) a T_c = 82 K overdoped sample and (**b**) a T_c = 83 K underdoped sample. The *dashed line* marks T_c. The error bars for Δ are based on a 10% increase in the RMS error of the fits

range (±75 meV) than in the overdoped case because of the larger gap. The range above T_c was increased to ±85 meV so as to properly describe the pseudogap depression.

In Fig. 17.41(b), we show the T-dependence of the fit parameters. We find a number of surprises. First, Δ is independent of T within error bars. Similar behavior has been inferred from specific heat and tunneling data. This T-independence is in contrast to the behavior of the overdoped 82 K sample with almost identical T_c at the same **k** point. In addition, for the underdoped sample, the gap evolves smoothly through T_c.

The single-particle scattering rate $\Gamma_1(T)$ for the underdoped sample is found to be qualitatively similar to the overdoped case. It is consistent with being T-independent above T_c, but with a value over twice as large as the overdoped case (allowing Γ_1 to vary above T_c does not improve the RMS error of the fits). Second, we see the same rapid decrease in Γ_1 below T_c as in the overdoped case. Note again the clear break at T_c.

The most interesting result is $\Gamma_0(T)$. We find $\Gamma_0 = 0$ below T_c and proportional to $T - T_c$ above. This behavior is robust, and is seen in all the fits that we have tried. Moreover, a non-zero Γ_0 is needed above T_c to obtain a proper fit to the data (its effect cannot be reproduced by varying the other parameters). The fact that this T-dependence is exactly what one expects of an inverse pair lifetime is a non-trivial check on the validity of the physics underlying Eq. (17.17). Further, we observe from Fig. 17.41 that T^* corresponds to where $\Delta(T) \sim \Gamma_0(T)$. This condition can be understood from the small ω expansion of Eq. (17.17).

The next important question is whether the T-dependence at the antinode described above exists at other $\mathbf{k}_F$-points. To answer this, we have looked at T-dependent data for a number of underdoped samples at two different **k** vectors. All data at the antinode give results similar to those for the 83K sample. However at the second **k**-point, about halfway between the antinode and the node along the $(0,0)-(\pi,\pi)$ direction, we see quite different behavior. We demonstrate this in Fig. 17.42(a) where symmetrized data for a 77 K underdoped sample are shown. For the antinode, one clearly sees the gap fill in above T_c, with little evidence for any T-dependence of the position of the spectral feature defining the gap edge, just as for the 83K sample. In contrast, at the second **k** point, the gap is clearly closing, indicating a strong T-dependence of Δ. Similar behavior is seen in other underdoped samples with T_c between 75 K and 85 K.

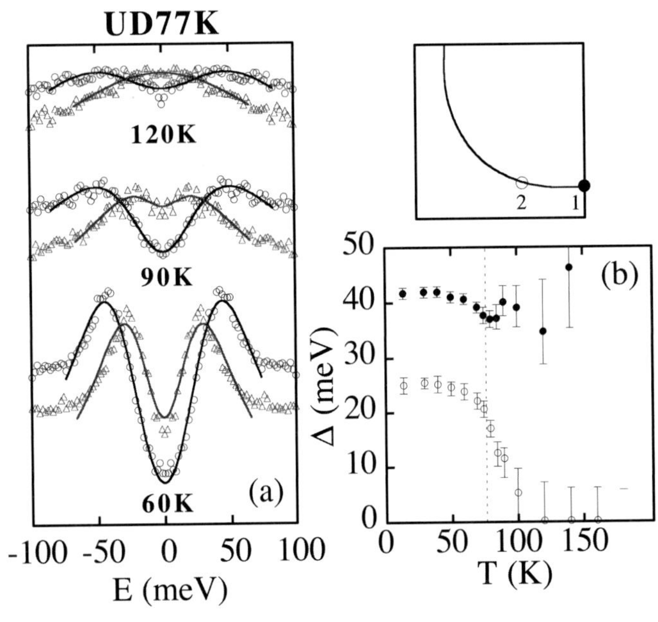

Fig. 17.42. (**a**) Symmetrized data for a $T_c = 77$ K underdoped sample for three temperatures at (*open circles*) $\mathbf{k}_F$ *point 1* in the *zone inset*, and at (*open triangles*) $\mathbf{k}_F$ *point 2*, compared to the model fits. (**b**) $\Delta(T)$ for these two k points (*filled and open circles*), with T_c marked by the *dashed line*

In Fig. 17.42(b), we show the T-dependence of Δ obtained from fits (over a range of ± 66 meV) at the second **k** point for the 77 K sample. Δ is found to be strongly T-dependent, being roughly constant below T_c, then dropping smoothly to zero above. The strong T-dependence of Δ makes it difficult to unambiguously determine Γ_0 from the fits at this **k**-point. On theoretical grounds, we expect that, here too, there is a non-zero Γ_0, and the closing of the pseudogap is again determined by $\Delta(T) \sim \Gamma_0(T)$, however this condition is satisfied by the rapid drop in $\Delta(T)$, rather than the rise in $\Gamma_0(T)$. For completeness, we also show $\Delta(T)$ for this sample at the antinode, which has a similar behavior to that of the 83 K sample.

We see that these results give further evidence for the unusual **k**-dependences first noted in [90]. Strong pairing correlations are seen over a very wide T-range near $(\pi, 0)$, but these effects are less pronounced and persist over a smaller T-range as one moves closer to the zone diagonal. This is clearly tied to the strong **k**-dependence of the effective interaction and the unusual (anomalously broad and non-dispersive) nature of electronic states near $(\pi, 0)$.

17.7.4 Peak/Dip/Hump – Experiment

We now turn to a detailed discussion of the peak/dip/hump lineshape. As mentioned above, a very broad normal state spectrum near the $(\pi, 0)$ point of the zone evolves quite rapidly for $T < T_c$ into a narrow quasiparticle peak, followed at higher binding energies by a dip then a hump, the latter corresponding to where the spectrum recovers to its normal state value [109]. Similar effects are observed in tunneling spectra [110].

In Fig. 17.43, we show spectra for a $T_c = 87$ K Bi2212 sample along $\Gamma - \bar{M} - Z$, i.e., $(0, 0) - (\pi, 0) - (2\pi, 0)$, in (a) the normal state (105 K) and (b) the superconducting state (13 K), from which we note two striking features [58]. First, we see that the low energy peak in the superconducting state persists over a large range in **k**-space, even when the normal state

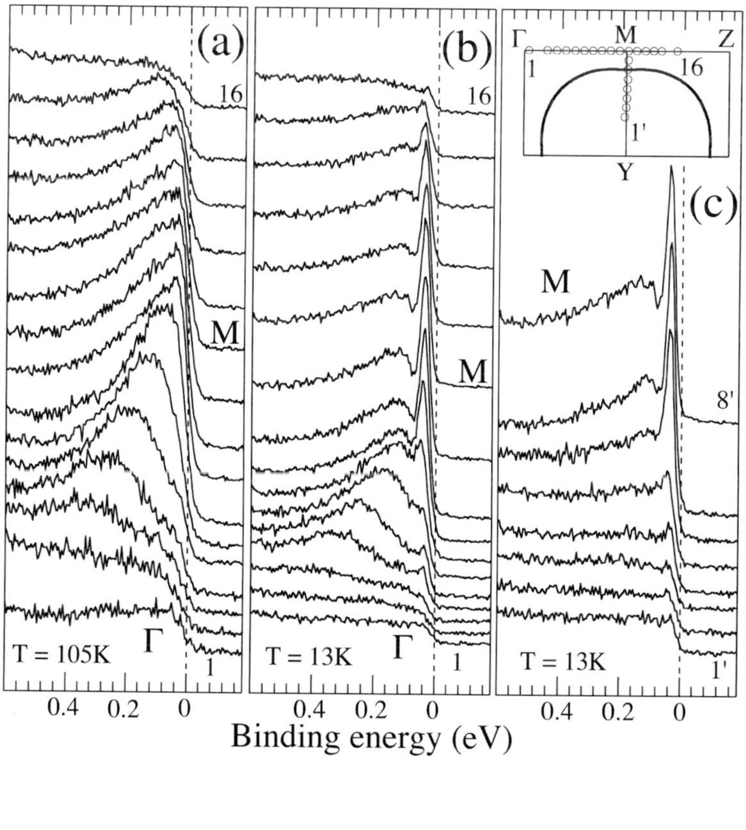

Fig. 17.43. Spectra in (**a**) the normal state (105 K) and (**b**) the superconducting state (13 K) along the line $\Gamma - \bar{M} - Z$, and (**c**) the superconducting state (13 K) along the line $\bar{M} - Y$ for an overdoped ($T_c = 87$ K) Bi2212 sample. The zone is shown as an *inset* in (**c**) with the *curved line* representing the observed Fermi surface

spectra have dispersed away from the Fermi energy. Second, when the hump in the superconducting state disperses, it essentially follows that of the normal state spectrum. This is accompanied by a transfer of weight to the hump from the low frequency peak, which is fairly fixed in energy. The same phenomena are also seen along $\bar{M}$ to Y (Fig. 17.43(c)). We will argue that the unusual dispersion seen in the superconducting state of Fig. 17.43 is closely tied to the lineshape change discussed earlier (Fig. 17.38).

The simplest explanation of the superconducting state spectra would be the presence of two bands (e.g., due to bilayer splitting), one responsible for the peak and the other for the hump. However, this explanation is untenable. (Bilayer splitting is discussed in Sect. 17.4.5.) First, if the sharp peak were associated with a second band, then this band should also appear above T_c. But there is no evidence for it in the normal state data. Second, if the peak and hump were from two different bands, then their intensities must be governed by different matrix elements. However, we found [36] that the intensities of both features scaled together as the photon polarization was varied from in to out of plane, as if they were governed by a common matrix element (Section 5.4.5). These arguments suggest that the unusual lineshape and dispersion represent a single electronic state governed by non-trivial many-body effects, as assumed in the previous discussion (Figs. 17.36–17.39). For more overdoped materials, though, bilayer splitting should be taken into account, as discussed in Sect. 17.4.5.

Under this assumption, the data are consistent with a strong reduction of the imaginary part of the self-energy (ImΣ) at low energies in the superconducting state (Fig. 17.36). If the scattering is electron–electron like in nature, then ImΣ at frequencies smaller than $\sim 3\Delta$ will be suppressed due to the opening of the superconducting gap [111]. On closer inspection, though, a more interesting story emerges. First, from Figs. 17.36 and 17.43, we see that the superconducting and normal state data match beyond 90 meV. From 90 meV, the dip is quickly reached at 70 meV, then one rises to the sharp peak. Notice that since the width of the peak is around 20 meV, then the change in behavior of the spectra (from hump, to dip, to the trailing edge of the peak) is occurring on the scale of the energy resolution. That means that the intrinsic dip must be quite sharp. This implies that the large ImΣ at high energies must drop to a small value over a narrow energy interval to be consistent with the data, i.e., there is essentially a step in ImΣ. In fact, the data are not only consistent with a step in ImΣ, but the depth of the dip is such that it is best fit by a peak in ImΣ at the dip energy, followed by a rapid drop to a small value. This behavior can again be seen from the independent analysis shown in Figs. 17.36 and 17.39.

What are the consequences of this behavior in ImΣ? If ImΣ has a sharp drop at $\tilde{\omega}$, then by Kramers–Kronig transformation, ReΣ will have a sharp peak at $\tilde{\omega}$ (Fig. 17.36). This peak can very simply explain the unusual dispersion shown in Fig. 17.43, as it will cause a low energy quasiparticle pole to appear even if the normal state binding energy is large. The most transparent way to appreciate this result is to note that a sharp step in ImΣ is equivalent to the problem of an electron interacting with a sharp (dispersionless) mode, since in that case, the mode makes no contribution to ImΣ for energies below the mode energy, and then makes a constant contribution for energies above. This problem has been treated by Engelsberg and Schrieffer [112], and extended to the superconducting state by Scalapino and coworkers [113]. The difference in our case is that since the effect only occurs *below* T_c, it is a consequence of the opening of the superconducting gap in the electronic energy spectrum, and thus of a collective origin, rather than a phonon.

To facilitate comparison to this classic work, in Fig. 17.44 we plot the position of the low energy peak and higher binding energy hump as a function of the energy of the single broad peak in the normal state. This plot has a striking resemblance to that predicted for electrons interacting with a sharp mode in the superconducting state, and one clearly sees the low energy pole which we associate with the peak in ReΣ. On general grounds, the flat dispersion of the low energy peak seen in Fig. 17.44 is a combination of two effects: (1) the peak in ReΣ, which provides an additional mass renormalization of the superconducting state relative to the normal state, and thus pushes spectral weight towards the Fermi energy, and (2) the

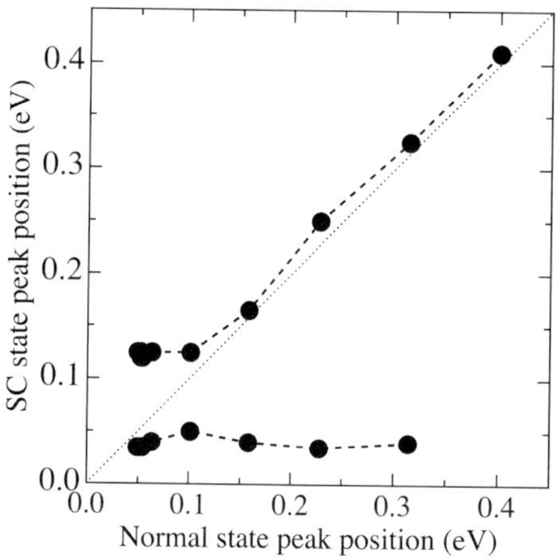

Fig. 17.44. Positions of the sharp peak and the broad hump in the superconducting state versus normal state peak position obtained from Figs. 17.43(a) and 17.43(b). *Solid points connected by a dashed line* are the data, the *dotted line* represents the normal state dispersion

superconducting gap, which pushes spectral weight away. This also explains the strong drop in intensity of the low energy peak as the higher binding energy hump disperses.

An important feature of the data is the dispersionless nature of the sharp peak. The mode picture discussed above would imply a dispersion of the peak from Δ_k to $\tilde{\omega} = \omega_0 + \Delta_k$ as the normal state binding energy increases (where ω_0 is the mode energy). However, this dispersion turns out to be weak. From the data of Fig. 17.36, we infer an $\omega_0 = 1.3\Delta_{\max}$, ω_0 being essentially the energy separation of the peak and dip. Since Δ_k is known to be of the $d_{x^2-y^2}$ form, then Δ_k should go to zero as we disperse towards the Γ point. Therefore, the predicted dispersion is only from $\Delta_{\max}$ to $1.3\Delta_{\max}$ (32 meV to 42 meV).

Since the dip/hump structure is most apparent at the $(\pi, 0)$ points, it is natural to assume that it has something to do with $Q = (\pi, \pi)$ scattering, as discussed by Shen and Schrieffer [114]. But here, we find a new effect. If one compares the data of Figs. 17.43(b) and 17.43(c), one sees that a low energy peak also exists along $(\pi, 0) - (\pi, \pi)$ for approximately the same momentum range as the one from $(\pi, 0) - (0, 0)$. That is, if there is a peak for momentum p, one also exists for momentum $p + Q$. This can be understood, since the self-energy equations for p and $p + Q$ will be strongly coupled if Q scattering is dominant.

17.7.5 Mode Model

For now, we ignore the complication of momentum dependence. The lowest order contribution to electron–electron scattering is represented by the Feynman diagram shown in the inset of Fig. 17.45. In the superconducting state, each internal line will be gapped by Δ. This implies that the scattering will be suppressed for $|\omega| < 3\Delta$. This explains the presence of a sharp quasiparticle peak at low temperatures. What is not so obvious is whether this in addition explains the strong spectral dip. Explicit calculations show only a weak dip-like feature [115]. To understand this in detail, we equate the bubble plus interaction lines (Fig. 17.45, inset) to an "α^2F" as in standard strong-coupling literature. In a marginal Fermi liquid (MFL) at T=0, $\alpha^2F(\Omega)$ is simply a constant in Ω. The effect of the gap is to force α^2F to zero for $\Omega < 2\Delta$. The question then arises where the gapped weight goes. It could be distributed to higher energies, but in light of the above discussion, we might expect it to appear as a collective mode inside of the 2Δ gap. For instance, if the bubble represents spin

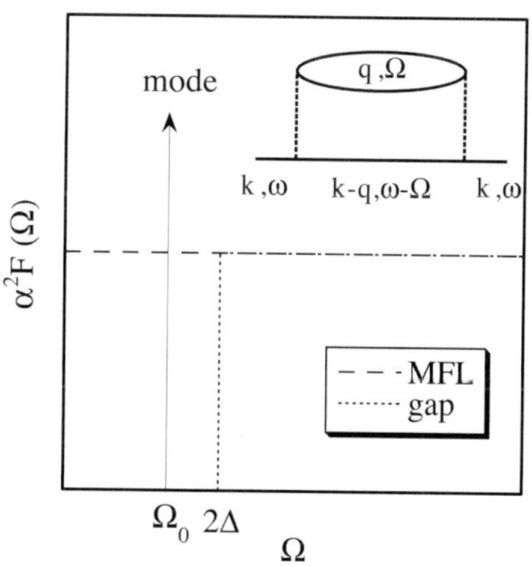

Fig. 17.45. α^2F for three models, MFL (*dashed line*), gapped MFL (*dotted line*), and gapped MFL plus mode (*dotted line* plus δ function). *Inset*: Feynman diagram for the lowest order contribution to Σ from electron–electron scattering

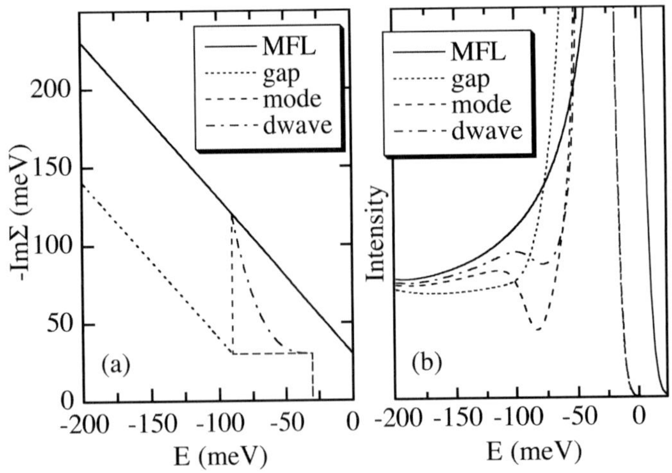

Fig. 17.46. (**a**) ImΣ for MFL (*solid line*), gapped MFL (*dotted line*), gapped MFL plus mode (*dashed line*), and simple *d*-wave model (*dashed-dotted* line). Parameters are $\alpha=1, \omega_c =$ 200 meV, Δ = 30 meV (0 for MFL), Ω_0 = *Delta*, and Γ_0 = 30 meV. (**b**) Spectral functions (times a Fermi function with T = 14 K) convolved with a resolution Gaussian of σ = 7.5 meV for these four cases (ϵ = -34 meV)

fluctuations, a sharp mode will appear if the condition $1 - U\chi_0(\mathbf{q}, \Omega) = 0$ is satisfied for $\Omega < 2\Delta$. These three cases (MFL, gapped MFL, gapped MFL plus mode) are illustrated in Fig. 17.45.

Σ is easy to obtain analytically if we ignore the complication of the superconducting density of states from the $\mathbf{k}-\mathbf{q}$ line of Fig. 17.45 and just replace this by a step function at Δ. The resulting ImΣ for the gapped MFL and gapped MFL plus mode models are shown in Fig. 17.46(a) [116] in comparison to the normal state MFL. Note that structure in α^2F at Ω appears in Σ at $|\omega| = \Omega + \Delta$ due to the gap in the $\mathbf{k} - \mathbf{q}$ line. Moreover, the MFL plus mode is simply the normal state MFL cutoff at 3Δ (this is obtained under the assumption that all the gapped weight in α^2F shows up in the mode). In contrast, the gapped MFL decays linearly to zero at 3Δ.

The Nambu spectral function is given by

$$A(\omega) = \frac{1}{\pi} Im \frac{Z\omega + \epsilon}{Z^2(\omega^2 - \Delta^2) - \epsilon^2} , \qquad (17.18)$$

with (a complex) $Z(\omega) = 1 - \Sigma(\omega)/\omega$. These are shown in Fig. 17.46(b) and were convolved with a gaussian of σ = 7.5 meV, typical of high resolution ARPES, with a constant ImΣ (Γ_0) added for $|\omega| > \Delta$ to reduce the size of the quasiparticle peak. We note that there is no dip as such for the gapped MFL model, whereas the addition of the mode causes a significant dip. The latter behavior is consistent with experiment. Moreover, the mode model has the additional advantage that ImΣ recovers back to the normal state value by 3Δ, which is also in agreement with experiment in that the normal and superconducting state spectra agree beyond 90 meV (Figs. 17.36 and 17.43).

We contrast this behavior with that expected for a simple *d*-wave model. To a first approximation, this can be obtained by replacing the step drop in ImΣ in the MFL plus mode model with $(|\omega| - \Delta)^3$ for $|\omega| < 3\Delta$ [117]. This is shown in Fig. 17.46(a) as well, with the resulting spectrum in Fig. 17.46(b). Only a weak dip appears. Moreover, we have analyzed models with the exponent 3 replaced by some n and have found that n must be large to obtain a dip as strong as seen in experiment. Therefore, the upshot is that at the least, something similar to a step is required in ImΣ to be consistent with experiment.

In principle, we could take the above MFL plus mode model and fit experiment with it. We consider a simpler model. There are several reasons for this. First, the MFL model has a number of adjustable parameters associated with it. There is the coupling constant (α), the cutoff frequency (ω_c), and the mode energy (which is not in general 2Δ). Moreover, the spectrum for **k** points near the $(\pi, 0)$ point does not appear to be MFL-like in nature. We have found that the normal state Bi2212 spectrum is fit very well by a Lorentzian plus a constant background in an energy range less than 0.5 eV. This is also true for Bi2201 spectrum where the normal state can be accessed to much lower temperatures.

In the resulting Lorentzian model, the normal state Σ is purely an imaginary constant, and α^2F is a mode at zero energy. In the superconducting state, this mode gets pushed back to some energy within 2Δ. This model is artificial in the sense that all the self-energy is being generated by the mode. That is why we went through the above discussion motivating the mode more properly as a rearrangement of α^2F due to the superconducting gap. In practice, though, the results are very similar to the MFL plus mode model, and has the further advantage of having the several parameters of that model collapse to just the mode strength (Γ_1) and mode position (Ω_0) of the Lorentzian model. Moreover, analytic results can still be obtained for Σ when the superconducting density of states for the $\mathbf{k}-\mathbf{q}$ line of Fig. 17.45 is taken into account. The result is [116]

$$\begin{aligned} -Im\Sigma(\omega) &= \Gamma_0 N(|\omega|) + \Gamma_1 N(|\omega| - \Omega_0), \\ &\quad |\omega| > \Omega_0 + \Delta \\ &= \Gamma_0 N(|\omega|),\ \Delta < |\omega| < \Omega_0 + \Delta \\ &= 0,\ |\omega| < \Delta\,, \end{aligned} \tag{17.19}$$

where $N(\omega) = \omega/\sqrt{\omega^2 - \Delta^2}$ is the BCS density of states, and

$$\begin{aligned} \pi \mathrm{Re}\Sigma(\omega) &= \Gamma_0 N(-\omega) \ln\left[|-\omega + \sqrt{\omega^2 - \Delta^2}|/\Delta\right] \\ &+ \Gamma_1 N(\Omega_0 - \omega) \ln\Big[|\Omega_0 - \omega \\ &\qquad + \sqrt{(\omega - \Omega_0)^2 - \Delta^2}|/\Delta\Big] \\ &- \{\omega \to -\omega\}\,. \end{aligned} \tag{17.20}$$

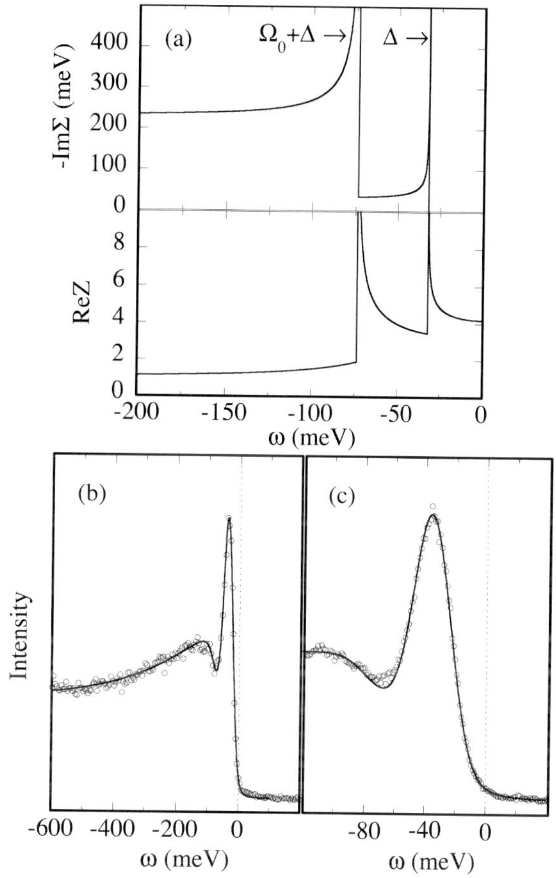

Fig. 17.47. (a) ImΣ and ReZ at $(\pi, 0)$ from Eqs. 17.19 and 17.20 ($\Gamma_1 = 200$ meV, $\Gamma_0 = 30$ meV, $\Delta = 32$ meV, $\Omega_0 = 1.3\Delta$). Comparison of the data at $(\pi, 0)$ for (**b**) wide and (**c**) narrow energy scans with calculations based on Eqs. 17.18–17.20, with an added step edge background contribution

Here, it has again been assumed that Δ is a real constant in frequency. An s-wave density of states has been used to obtain an analytic result. A d-wave density of states will not be that different. The advantage of an analytic result is that it is useful when having to take spectra and convolve with resolution to compare to experiment. Our results are not very sensitive to Γ_0, included again to damp the quasiparticle peak.

The resulting real (Eq. (17.20)) and imaginary (Eq. (17.19)) parts of Σ at $(\pi, 0)$ are shown in Fig. 17.47(a). Note the singular behaviors at Δ (peak energy) due to the Γ_0 term and at $\Omega_0 + \Delta$ (dip energy) due to the Γ_1 term. In both cases, step drops in ImΣ would also give singularities in $Re\Sigma$. The advantage of peaks in ImΣ (due to the SC density of states) is that it makes the dip deeper in better agreement with experiment. In Figs. 17.47(b) and 17.47(c), we show a comparison of the resulting spectral function (convolved with the experimental energy and momentum resolution) to experimental data at $(\pi, 0)$ for both wide and narrow energy scans, where a step edge background with a gap of Δ is added to the calculated spectrum. The resulting agreement is excellent.

It is interesting to note that the mode energy we infer from the data is 41 meV, equivalent to a

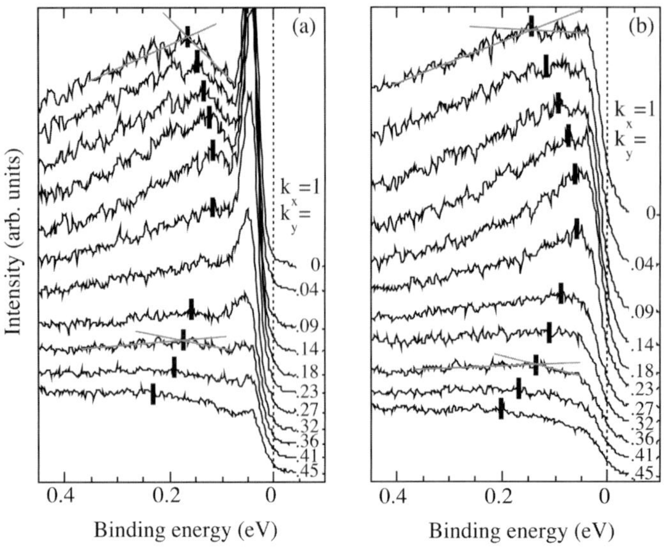

Fig. 17.48. Spectra along $(\pi, 0) \to (\pi, \pi)$ in (**a**) the superconducting state ($T = 60$ K), and (**b**) the pseudogap state ($T = 100$ K) for an underdoped 75 K sample (curves are labeled in units of π/a). The *thick vertical bar* indicates the position of the higher energy feature, at which the spectrum changes slope as highlighted by the *intersecting straight lines*

magnetic resonant mode energy observed in YBCO [118] and Bi2212 [119] by neutron scattering data at $\mathbf{Q} = (\pi, \pi)$. The Q dependence of this mode correlates well with the observations of Fig. 17.43. To explore this in greater detail, we now consider the doping dependence of the peak/dip/hump structure.

17.7.6 Doping Dependence

We show data along $(\pi, 0) \to (\pi, \pi)$ for an underdoped 75 K sample in the superconducting state (Fig. 17.48(a)) and in the pseudogap state (Fig. 17.48(b)) [59]. Below T_c, the sharp peak at low energy is essentially dispersionless, while the higher energy hump rapidly disperses from the $(\pi, 0)$ point towards the $(\pi, 0) \to (\pi, \pi)$ Fermi crossing seen above T^*. Beyond this, the intensity drops dramatically, but there is clear evidence that the hump disperses back to higher energy. In the pseudogap state, the high energy feature also shows strong dispersion, much like the hump below T_c, even though the leading edge is non-dispersive like the sharp peak in the superconducting state.

In Fig. 17.49 we show the dispersion of the sharp peak and hump (below T_c), for a variety of doping levels, in the vicinity of the $(\pi, 0)$ point along the two principal axes. The sharp peak at low energies is seen to be essentially non-dispersive along both directions for all doping levels, while the hump shows very interesting dispersion. Along $(\pi, 0) \to (0, 0)$ (Fig. 17.49(a)), the hump exhibits a maximum, with an eventual dispersion away from the Fermi energy, becoming rapidly equivalent to the binding energy of the broad peak in the normal state as one moves away from the region near $(\pi, 0)$. In the orthogonal direction (Fig. 17.49(b)), since the hump initially disperses towards the $(\pi, 0) \to (\pi, \pi)$ Fermi crossing, which is known to be a weak function of doping, one obtains the rather dramatic effect that the dispersion becomes stronger with underdoping. We also note that there is an energy separation between the peak and the hump due to the spectral dip. In essence, the hump disperses towards the spectral dip, but cannot cross it, with its weight dropping strongly as the dip energy is approached. Beyond this point, one sees evidence of the dispersion bending back to higher binding energy for more underdoped samples.

Fig. 17.50(a) shows the evolution of the low temperature spectra at the $(\pi, 0)$ point as a function of doping. The sharp quasiparticle peak moves to higher energy, indicating that the gap increases with underdoping (although this is difficult to see on the scale of Fig. 17.50(a)). We see that the hump moves rapidly to higher energy with underdoping. These trends can be seen very clearly in Fig. 17.50(b), where the energy of the peak and hump are shown as a func-

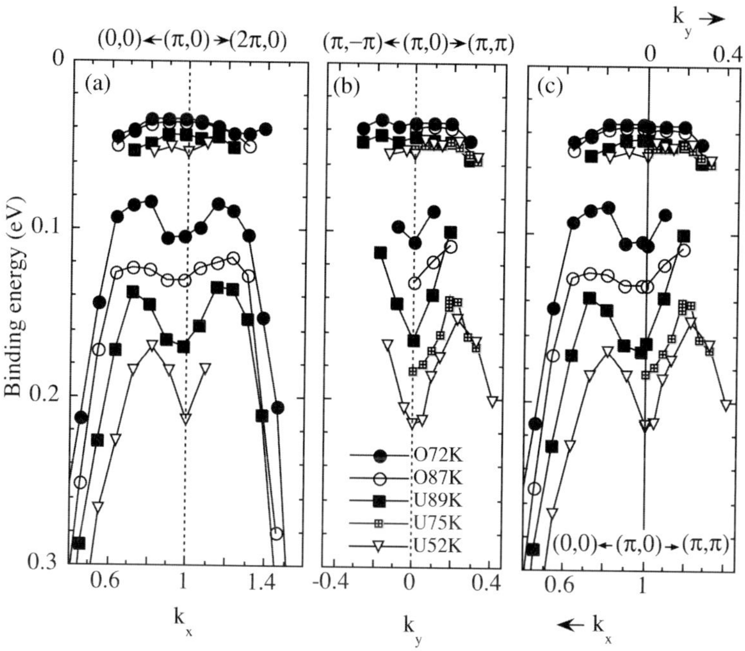

Fig. 17.49. Doping dependence of the dispersion from (**a**) $(\pi, 0) \rightarrow (\pi \pm \pi, 0)$. (**b**) $(\pi, 0) \rightarrow (\pi, \pm\pi)$, and (**c**) both directions, for the peak and hump in the superconducting state. U is underdoped and O is overdoped. Points were obtained by polynomial fits to the data and are consistent with the simpler criterion used in Fig. 17.48

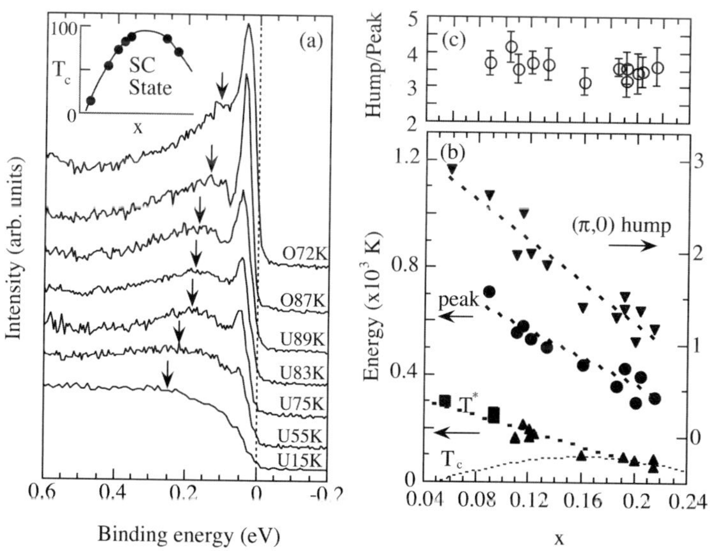

Fig. 17.50. Dependence of energy scale on carrier density. (**a**) Doping dependence of the spectra (T = 15 K) at the $(\pi, 0)$ point. The inset shows T_c vs. doping. (**b**) Doping dependence of T^*, and the peak and hump binding energies in the superconducting state along with their ratio (**c**), as a function of doping, x. The empirical relation between T_c and x is given by $T_c/T_c^{\max} = 1 - 82.6(x - 0.16)^2$ with $T_c^{\max}$ = 95 K. For T^* solid squares represent lower bounds

tion of doping for a large number of samples. Finally, we observe that the quasiparticle peak loses spectral weight with increasing underdoping, as expected for a doped Mott insulator; in addition the hump also loses spectral weight though less rapidly.

This effect has recently been quantified in greater detail, where it was found that the spectral weight of the peak varies linearly with doping, as reproduced in Fig. 17.51. [92,93]. We remark that Ding et al. [93] also found the unusual relation that the product of

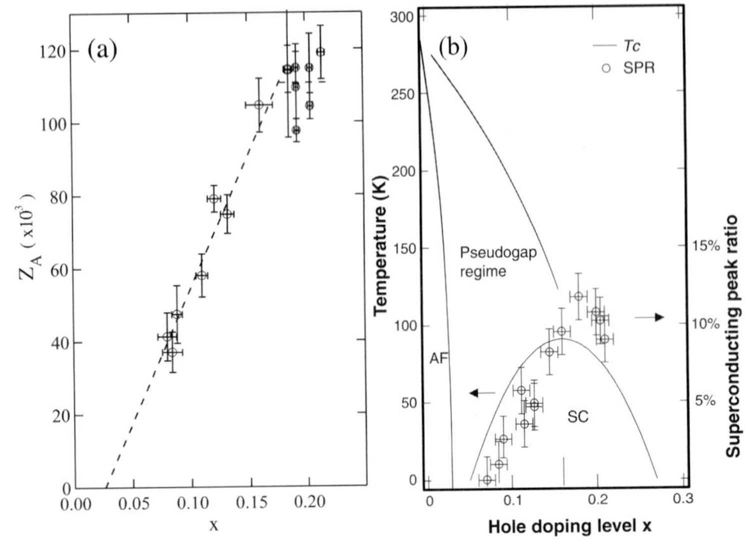

Fig. 17.51. (**a**) Doping dependence of the low-T (14 K) coherent weight (z_A). The *dashed line* is a guideline showing that z_A increases linearly on the underdoped side and tapers off on the overdoped side (from [93]). (**b**) The doping dependence of the superconducting peak ratio (SPR) is plotted over a typical Bi2212 phase diagram. The *solid line* is a guide to the eye. *Horizontal error bars* denote uncertainty in determining the doping level (± 0.01); *vertical error bars* denote uncertainty in determining the SPR ($\pm 1.5\%$). AF: antiferromagnetic regime. SC: superconducting regime (from [92])

the peak weight times the peak energy is constant with doping.

The hump below T_c is clearly related to the superconducting gap, given the weak doping dependence of the ratio between the hump and quasiparticle peak positions at $(\pi, 0)$, shown in Fig. 17.50(c). Tunneling data find this same correlation on a wide variety of high-T_c materials whose energy gaps vary by a factor of 30 [60].

To motivate the analysis below that firmly establishes the mode interpretation of the peak/dip/hump spectra and its connection with neutron data, we note that the spectral dip represents a pairing induced gap in the incoherent part of the spectral function at $(\pi, 0)$ occurring at an energy $\Delta + \Omega_0$, where Δ is the superconducting gap and Ω_0 is the mode energy. We can estimate the mode energy from ARPES data from the energy difference between the dip ($\Delta + \Omega_0$) and the quasiparticle peak (Δ).

In Fig. 17.52(b) we plot the mode energy as estimated from ARPES for various doping levels as a function of T_c and compare it with neutron measurements. We find striking agreement both in terms of the energy scale and its doping dependence. (Neutron results are reviewed by P. Bourges [120] and by Hayden, Chap. 18 in this volume.) The same agreement, in greater detail, has been recently found using tunneling data [121], as shown in Fig. 17.53. We note that the mode energy inferred from ARPES decreases with underdoping, just like the neutron data, unlike the gap energy (Fig. 17.50(b)), which increases. This can be seen directly in the raw data, shown in Fig. 17.52(a). This is also seen from the tunneling data, where they have found that the mode energy scales with doping as $5T_c$, just like the neutron resonance. An interesting point from the tunneling is that the ratio of the mode energy to the gap energy saturates to 2 in the overdoped limit, as would be expected for a collective mode sitting below a continuum with a gap of 2Δ. Moreover, there is strong correlation between the temperature dependences in the ARPES and neutron data. While neutrons see a sharp mode only below T_c, a smeared out remnant persists up to T^* [122]. As the sharpness of the mode is responsible for the sharp spectral dip, one then sees the correlation with ARPES where the dip disappears above T_c, but with a remnant of the hump persisting to T^*.

An important feature of the neutron data is that the mode only exists in a narrow momentum range about (π, π), and is magnetic in origin. To see a further connection with ARPES, we return to the results of Fig. 17.49. Note the dispersion along the two orthogonal directions are similar (Fig. 17.49(c)), unlike the dispersion inferred in the normal state. As

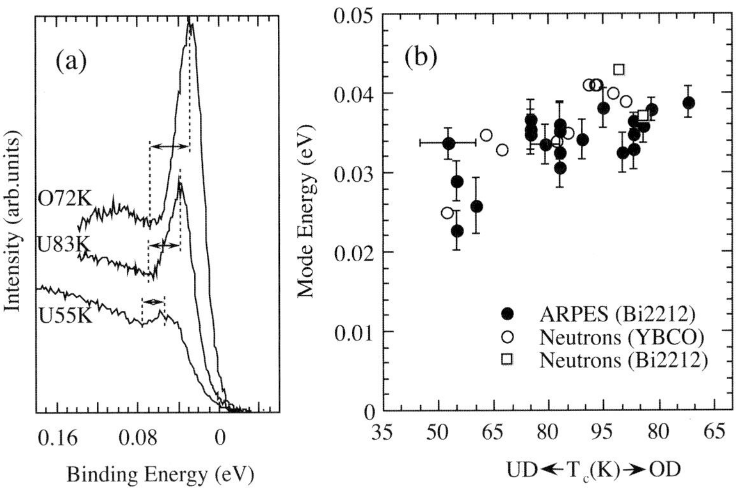

Fig. 17.52. Doping dependence of the mode energy. (**a**) Spectra at $(\pi, 0)$ showing the decrease in the energy separation of the peak and dip with underdoping. Peak and dip locations were obtained by independent polynomial fits and carefully checked for the effects of energy resolution. (**b**) Doping dependence of the collective mode energy inferred from ARPES together with that inferred from neutron data [120]

these two directions are related by a (π, π) translation $((x, 0) \equiv (0, -x); (0, -x) + (\pi, \pi) = (\pi, \pi - x))$, we see that the hump dispersion is clearly reflecting the (π, π) nature of the collective mode. This dispersion is also consistent with a number of models in the literature which identify the high energy feature in the pseudogap regime as a remnant of the insulating magnet [123]. We note, though, that the mode is due to quasiparticle pair creation and thus not just a continuation of the spin wave mode from the antiferromagnet [124].

This brings up a question that is at thc hcart of the high T_c problem: how can a feature which can be understood as a strong coupling effect of superconductivity, as discussed above, turn out to have a dispersion that resembles that of a magnetic insulator? The reason is that the collective mode has the same wavevector, (π, π), which characterizes the magnetic order of the insulator. It is easy to demonstrate that in the limit that the mode energy goes to zero (long range order), one actually reproduces a symmetric dispersion similar to that in Fig. 17.49(c), with the spectral gap determined by the strength of the mode [40]. This is in accord with the increase in the hump energy with underdoping (Fig. 17.50(b)) tracking the rise in the neutron mode intensity. Since the hump scales with the superconducting gap, the obvious implication is that the mode is intimately connected with pairing, a conclusion which can also be made by relating the mode to the superconducting condensation energy [125]. That is, high T_c superconductivity is likely due to the same magnetic correlations which characterize the insulator and give rise to the mode.

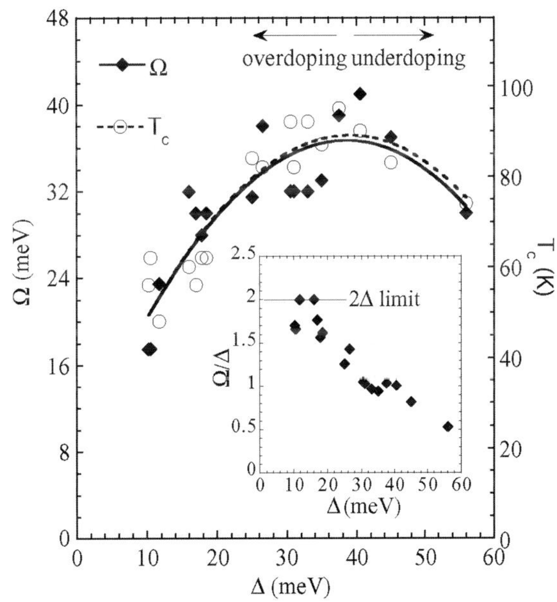

Fig. 17.53. Measured mode energy Ω and bulk T_c value vs. measured gap value Δ for 17 junctions over a wide doping range from UD74 K to OD48 K. *Solid* and *dashed lines* are quadratic fits of Ω and T_c vs. Δ. The *inset* shows Ω/Δ vs. Δ (from [121])

17.7.7 Dispersion Kink of Nodal Quasiparticles

So far, our discussion has largely been centered on behavior near the $(\pi, 0)$ point of the zone. We now turn to consideration of other **k** vectors.

Remarkably, we find that the effects discussed above are manifest even on the zone diagonal where the gap vanishes, with significant changes in both the spectral lineshape and dispersion below T_c, relative to the normal state where the nodal points exhibit quantum critical scaling [22]. Specifically, below T_c a kink in the dispersion develops along the diagonal at a finite energy (~70 meV) [23,24]. This is accompanied, as required by Kramers–Kronig relations, by a reduction in the linewidth leading to well-defined quasiparticles [32]. As one moves away from the node, the renormalization increases, and the kink in dispersion along the diagonal smoothly evolves into the spectral dip, with the same characteristic energy scale throughout the zone.

In Fig. 17.54(a), we plot the dispersion of the spectral peak above T_c obtained from constant **k** scans (energy distribution curves or EDCs), and the peak in momentum obtained from constant ω scans (momentum distribution curves or MDCs) [22] from data for a T_c = 90 K sample along the (π, π) direction [24]. We find that the EDC and MDC peak dispersions are very different, a consequence of the ω dependence of Σ.

To understand this, we start by noting that since $\epsilon_{\mathbf{k}} \simeq v_F^0(k - k_F)$, then from Eq. (17.8) the MDC at fixed ω is a Lorentzian centered at $k = k_F + [\omega - \Sigma'(\omega)] / v_F^0$, with a width (HWHM) $W_M = |\Sigma''(\omega)| / v_F^0$, *provided* (i) Σ is essentially independent of k normal to the Fermi surface, *and* (ii) the dipole matrix elements do not vary significantly with k over the range of interest. That these two conditions are fulfilled can be seen by the nearly Lorentzian MDC lineshape observed in ARPES [22].

On the other hand, in general, the EDC at fixed **k** has a non-Lorentzian lineshape reflecting the non-trivial ω-dependence of Σ, in addition to the Fermi cutoff at low energies. Thus the EDC peak is *not* given by $\omega - v_F^0(k - k_F) - \Sigma'(\omega) = 0$ but also involves Σ'', unlike the MDC peak. Further, if the EDC peak is sharp enough, making a Taylor expansion we find that its width (HWHM) is given by $W_E \simeq |\Sigma''(E_k)| / [1 - \partial\Sigma'/\partial\omega|_{E_k}]$, where E_k is the peak position.

We see that it is much simpler to interpret the MDC peak positions, and thus focus on the change in the MDC dispersion going from the normal (N) to the superconducting (SC) state shown in Fig. 17.54(b). The striking feature of Fig. 17.54(b) is the develop-

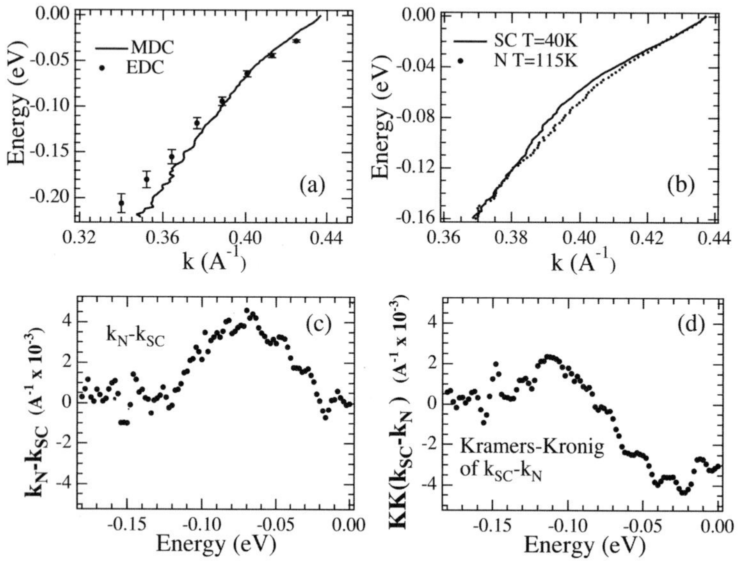

Fig. 17.54. ARPES data along the (π, π) direction at $h\nu$ = 28 eV. **(a)** EDC dispersion in the normal state compared to the MDC dispersion. The EDCs are shown in Fig. 17.55(d). **(b)** MDC dispersions in the superconducting state (T = 40 K) and normal state (T = 115 K). **(c)** Change in MDC dispersion from **(b)**. **(d)** Kramers–Kronig transform of **(c)**

ment of a kink in the dispersion below T_c. At fixed ω let the dispersion change from k_N to k_{SC}. Using $v_F^0(k_N - k_{SC}) = \Sigma'_{SC}(\omega) - \Sigma'_N(\omega)$, we directly obtain the change in real part of Σ plotted in Fig. 17.54(c). The Kramers–Kronig transformation of $\Sigma'_{SC} - \Sigma'_N$ then yields $\Sigma''_N - \Sigma''_{SC}$, plotted in Fig. 17.54(d), which shows that $|\Sigma''_{SC}|$ is smaller than $|\Sigma''_N|$ at low energies.

We compare these results in Fig. 17.55(a) with the $W_M = |\Sigma''|/v_F^0$ estimated directly from the MDC Lorentzian linewidths. The normal state curve was obtained from a linear fit to the corresponding MDC width data points in Fig. 17.55(a), and then the data from Fig. 17.54(d) was added to it to generate the low temperature curve. We are thus able to make a quantitative connection between the appearance of a kink in the (MDC) dispersion below T_c and a drop in the low energy scattering rate in the superconducting state relative to the normal state, which leads to the appearance of quasiparticles below T_c [32]. We emphasize that we have estimated these T-dependent changes in the complex self-energy without making fits to the EDC lineshape, thus avoiding the problem of modeling the ω dependence of Σ and the extrinsic background.

In Fig. 17.55(b), we plot the EDC width obtained as explained in [32] from Fig. 17.55(d). As an interesting exercise, we present in Fig. 17.55(c) the ratio of this EDC width to the MDC width of Fig. 17.55(a) (dotted lines), and compare it to the renormalized MDC velocities, $1/v \equiv dk/d\omega$, obtained directly by numerical differentiation of Fig. 17.54(b) (solid lines). We note that only for a sufficiently narrow EDC lineshape is the ratio $W_E/W_M \simeq v_F^0/[1 - \partial\Sigma'/\partial\omega] = v_F$. Interestingly, only in the superconducting state be-

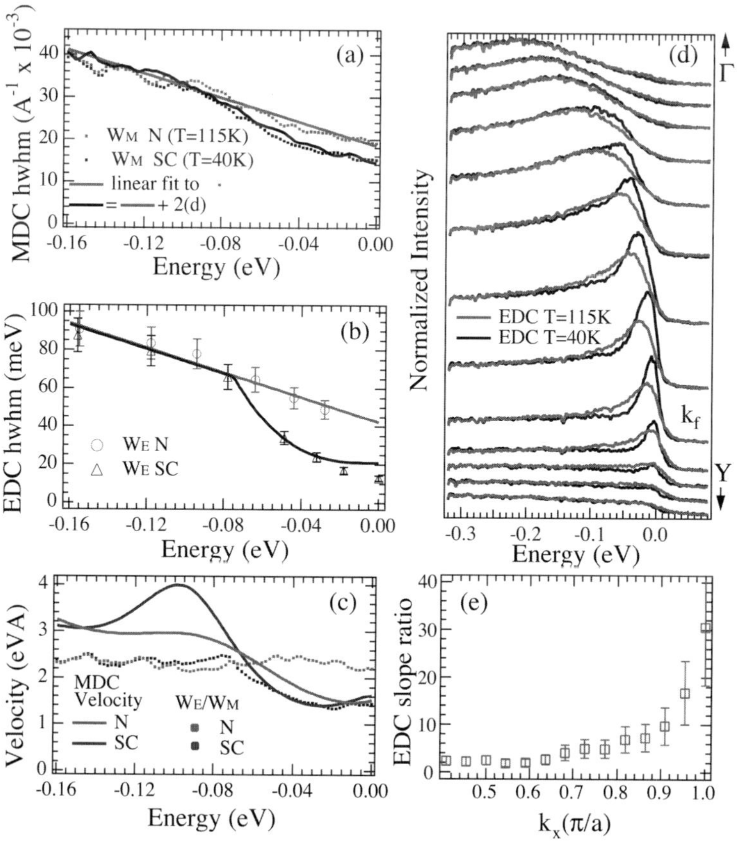

Fig. 17.55. (**a**) Comparison of change in Σ'' obtained directly from the MDC widths (HWHM) to the one obtained from the dispersion in Fig. 17.54d by using the Kramers–Kronig transform. (**b**) HWHM width obtained from EDCs shown in (**d**). Lines marked by fit are linear in normal state and linear/cubic in superconducting state. The data in (**b**) fall below the fits at low energies because of the Fermi cutoff of the EDCs. (**c**) Renormalized MDC velocity obtained from differentiating Fig. 17.54(b) (*solid lines*), compared to the ratio W_E/W_M from (**a**) and (**b**). (**e**) Ratio of EDC dispersion slopes above and below the kink energy at various points along the Fermi surface (from middle panels of Fig. 17.56)

low the kink energy do these two quantities agree, which implies that only in this case does one have a Fermi liquid.

Similar kinks in the dispersion have been seen by ARPES in normal metals due to the electron–phonon interaction [126]. Phonons cannot be the cause here, since the kink disappears above T_c. Rather, this effect is suggestive of coupling to an electronic collective excitation which only appears below T_c. Recently, this view has been challenged by Lanzara et al. [127], and we discuss this work at the end of this subsection.

We now study how the lineshape and dispersion evolve as we move along the Fermi surface. An analysis similar to the above is possible, but more complicated due to the presence of an energy gap [25]. We will thus confine ourselves here to a general description of the data. In Fig. 17.56, we plot raw intensities for a series of cuts parallel to the MY direction (normal state in left panels, superconducting state in middle panels). We start from the bottom row that corresponds to a cut close to the node and reveals the same kink described above. As we move towards $(\pi, 0)$, the dispersion kink (middle panels)

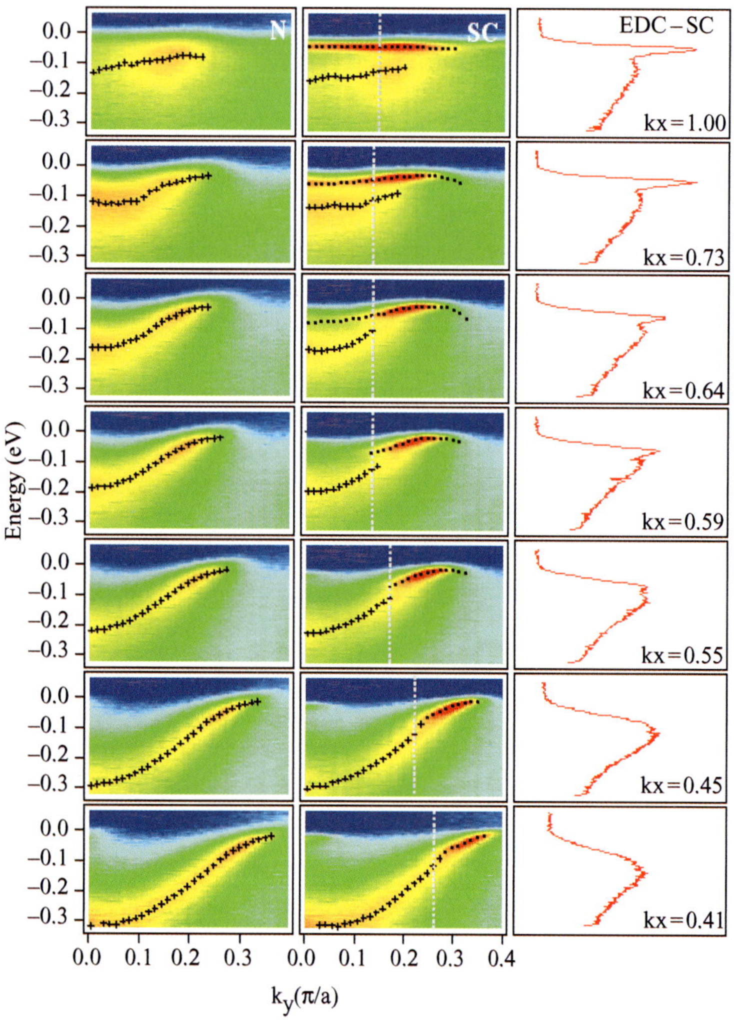

Fig. 17.56. *Left*: Log of normal state ($h\nu$ = 22 eV, T = 140 K) ARPES intensity along selected cuts parallel to MY. EDC peak positions are indicated by crosses. *Middle*: Log of superconducting state (T = 40 K) intensity at the same cuts as for left panels. *Crosses* indicate positions of broad high energy peaks, dots sharp low energy peaks. *Right*: EDCs at locations marked by the *vertical lines* in the *middle panels*

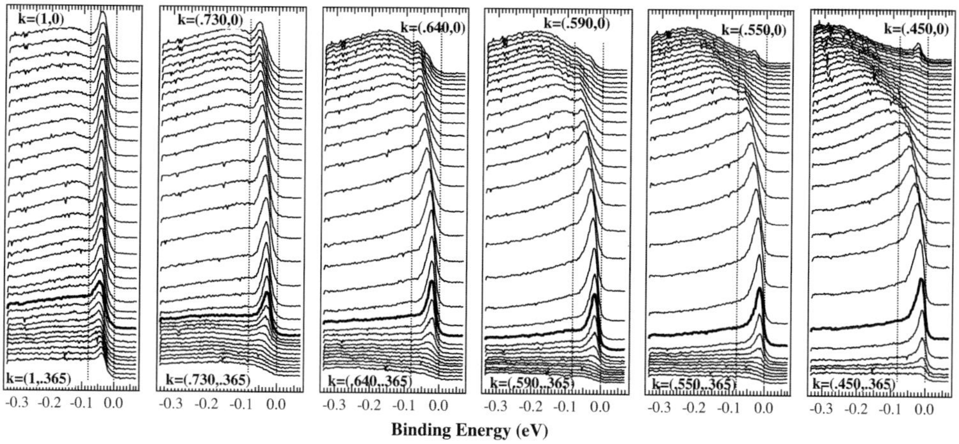

Fig. 17.57. ARPES intensity ($T = 40$ K) along selected cuts from Fig. 17.56. The *thick lined curves* correspond approximately to $\mathbf{k}_F$. *Vertical lines* are at 0 meV and -80 meV

becomes more pronounced and at around $k_x = 0.55$ develops into a break separating the faster dispersing high energy part of the spectrum from the slower dispersing low energy part. This break leads to the appearance of two features in the EDCs, shown in the right panels of Fig. 17.56. Further towards $(\pi, 0)$, the low energy feature, the quasiparticle peak, becomes almost dispersionless. At the $(\pi, 0)$ point, this break effect becomes the most pronounced, giving rise to the peak/dip/hump in the EDC. We note that there is a continuous evolution in the zone from kink to break, and these features all occur at exactly the same energy.

The above evolution is suggestive of the self-energy becoming stronger as the $(\pi, 0)$ point is approached. This can be quantified from the observed change in the dispersion. In Fig. 17.55(e) we plot the ratio of the EDC dispersion slope above and below the kink energy at various points along the Fermi surface obtained from middle panels of Fig. 17.56. Near the node, this ratio is around 2, but becomes large near the $(\pi, 0)$ point because of the nearly dispersionless quasiparticle peak. A different behavior was inferred in [127], but in their case, the cuts near $(\pi, 0)$ were perpendicular to ours, and thus not normal to the Fermi surface.

The lineshape also indicates that the self-energy is larger near $(\pi, 0)$, as is evident in Fig. 17.57. Along the diagonal, there is a gentle reduction in Σ'' at low energies, as shown in Figs. 17.55(a) and (b), with an onset at the dispersion kink energy scale. In contrast, near the $(\pi, 0)$ point there must be a very rapid change in Σ'' in order to produce a spectral dip, as discussed above. Despite these differences, it is important to note that these changes take place throughout the zone at the same characteristic energy scale (vertical line in Fig. 17.57).

As also discussed above, the $(\pi, 0)$ ARPES spectra can be naturally explained in terms of the interaction of the electron with a collective mode of electronic origin which only exists below T_c. It was further speculated that this mode was the neutron resonance. Here we have shown that dispersion and lineshape anomalies have a continuous evolution throughout the zone and are characterized by a single energy scale. This leads us to suggest that the same electron–mode interaction determines the superconducting lineshape and dispersion at all points in the zone, including the nodal direction. In essence, there is a suppression of the low energy scattering rate below the finite energy of the mode. Of course, since the neutron mode is characterized by a (π, π) wavevector, one would expect its effect on the lineshape to be much stronger at points in the zone which are spanned by (π, π), as observed here.

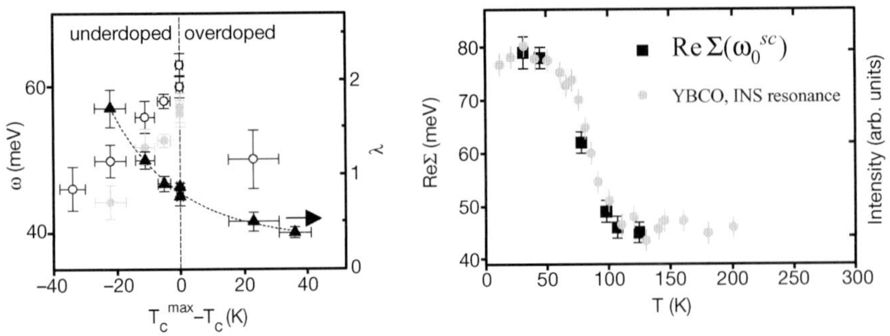

Fig. 17.58. *Left*: Plot of ω_0, the energy of the maximum value of ReΣ in the superconducting state (*open circles*), and ω_0^{SC} (*gray circles*), the energy of the maximum in difference between the superconducting and normal state values plotted as a function of T_c referenced to the maximum $T_c^{\max}$ (≡ 91 K). The coupling constant λ (*black triangles*) is referenced to the right-hand scale. *Right*: Temperature dependence of $\mathrm{Re}\Sigma(\omega_0^{SC})$ from the nodal line for the UD69 K sample (*black squares*) compared with the temperature dependence of the intensity of the resonance mode observed in inelastic neutron scattering studies of underdoped $YBa_2Cu_3O_{6+x}$, $T_c = 74$ K (*gray circles*) (adapted from [128])

A similar conclusion has been reached by Johnson et al. [128], where they find that the kink energy scales with doping like the neutron resonance (Fig. 17.58(a)), and that the temperature dependence of Σ' tracks that of the resonance intensity (Fig. 17.58(b)). Moreover, they find that Σ' increases with underdoping (Fig. 17.58(a)), much like that extracted from the peak/dip/hump lineshape at $(\pi, 0)$.

Detailed calculations which take into account the momentum dependence of the neutron resonance give an extremely good description of the experimental data [129]. These results can be understood by studying the Feynman diagram of Fig. 17.45. The key point is that the neutron resonance has a finite width in momentum space, corresponding to a short correlation length of order 2 lattice constants. Because of this, there is now an internal sum over momentum in the diagram, which will be dominated by the flat regions of the fermionic dispersion around the $(\pi, 0)$ points. This means that structure in the electron self-energy will occur at an energy of $\Delta_{\max}+\Omega_{\text{res}}$, independent of external momentum. This explains why the energy scale is invariant throughout the zone. On the other hand, as the external momentum is swept, the momentum dependence of the neutron form factor is probed. Since the latter peaks at $Q = (\pi, \pi)$, then the magnitude of the self-energy will be maximal at $(\pi, 0)$, since these points are connected by Q, and minimum at the node. This explains why the peak/dip/hump effect first weakens into a "break" effect and then into a dispersion kink as the node is approached. These calculations have been recently extended to incorporate bilayer splitting effects [130], and are able to explain a number of unusual lineshape and dispersion features present in data on heavily overdoped Bi2212 [62–64,131]

As mentioned above, this picture has been challenged by Lanzara et al. [127]. These authors claim that the kink is still present above T_c, except it is smeared in energy. Moreover, they find that a kink is present in a large variety of cuprates, including Bi2201 and LSCO, with an energy which is material and doping independent, as shown in Fig. 17.59. They argue that all of these observations are in support of a phonon interpretation of the kink.

Although initially attractive, there are some problems with this scenario. First, in regards to the kink above T_c, it has been claimed by Johnson et al. [128] that the "kink" above T_c is simply the curvature in the dispersion one expects based on marginal Fermi liquid theory. In support of this, they argue that the maximum in the real part of Σ is at a different energy in the normal state than in the superconducting state. Though this appears to be the case, there is indeed residual structure in the normal state self-energy at the kink energy in the optimal doped sample we have looked at. On the other hand, our normal state data actually corresponds to the pseudogap phase, and

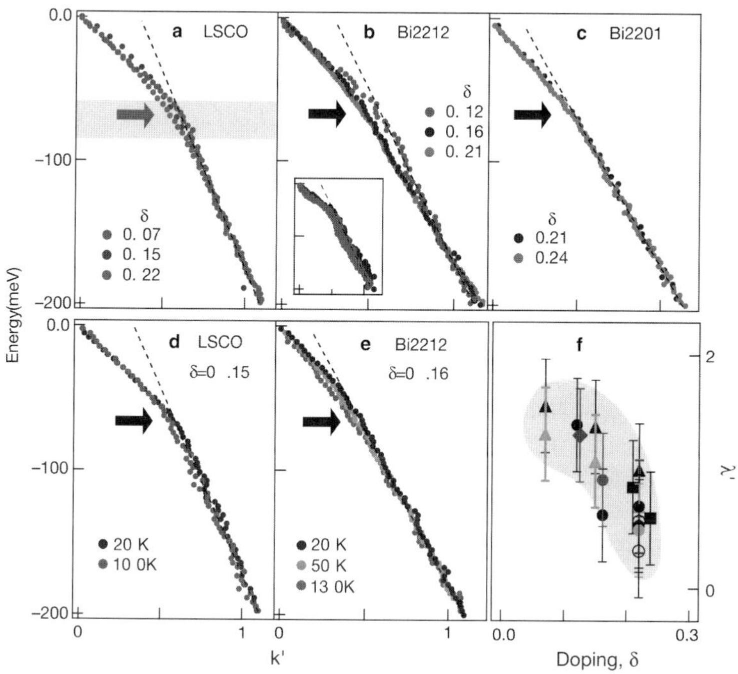

Fig. 17.59. Ubiquity of a sudden change ("kink") in the dispersion. *Top panels* are plots of the dispersion (derived from the momentum distribution curves) along $(0, 0) - (\pi, \pi)$ (except panel **b** inset, which is off this line) versus the rescaled momentum k' for different samples and at different doping levels. (**a**)–(**c**) Doping (δ) dependence of LSCO (at 20 K; (**a**)), Bi2212 (superconducting state, 20 K; (**b**)), and Bi2201 (normal state, 30 K; (**c**)). *Dotted lines* are guides to the eye. The kink position in (**a**) is compared with the phonon energy at $q = (\pi, 0)$ (*thick red arrow*) and the phonon width and dispersion (*shaded area*) from neutron data. The doping was determined from the T_c versus doping universal curve. The *inset* in (**b**), dispersions off the $(0, 0)$–(π, π) direction, showing also a sharpening of the kink on moving away from the nodal direction. The *black arrows* indicate the position of the kink in the dispersions. (**d**), (**e**) Temperature dependence of the dispersions for LSCO ((**d**), optimally doped) and Bi2212 ((**e**), optimally doped). (**f**) Doping dependence of λ along the (0, 0)–(π, π) direction as a function of doping. Data are shown for LSCO (*filled triangles*) and NdLSCO (1/8 doping; *filled diamonds*), Bi2201 (*filled squares*) and Bi2212 (*filled circles* in the first Brillouin zone, and *unfilled circles* in the second zone). The different *shadings* represent data obtained in different experimental runs. The *blue area* is a guide to the eye (adapted from [127])

as a residual of the neutron resonance is present in the pseudogap phase, the residual "kink" above T_c (if really there) does not rule out a magnetic interpretation. Moreover, as Johnson et al.convincingly showed (Fig. 17.58(b)), there is definitely a large component to the self-energy which follows the same order parameter like temperature variation that the neutron resonance intensity does. This observation is supported by recent work of Gromko et al. [131] concerning a dispersion kink in the bonding band of heavily overdoped Bi2212 near $(\pi, 0)$.

Second, in regards to constancy of the energy scale, this is indeed an interesting observation, though we note this statement contradicts that of Johnson et al.concerning the doping dependence of the kink energy mentioned above (Fig. 17.58). Also, even in a phonon model, the kink energy should occur at the sum of the maximum gap energy plus the phonon energy. Why the sum of these two numbers should be doping and material independent is a real puzzle (as it would be for a magnetic interpretation as well). And, why only one phonon would be relevant, despite the large number of phonons present in the cuprates, is another puzzle. Still, a phonon model for the kink has certain attractions, as discussed by Lanzara et al. [127]. Certainly, more work is needed to definitively resolve the controversies surrounding the origin of the dispersion kink.

17.7.8 Condensation Energy

We conclude this section by discussing the relation of ARPES data to the superconducting condensation energy.

We begin with the assumption that the condensation energy does not have a component due to phonons. To proceed, we assume an effective single-band Hamiltonian which involves only two particle interactions. Then, simply exploiting standard formulas for the internal energy $U = \langle H - \mu N \rangle$ (μ is the chemical potential, and N the number of particles) in terms of the one-particle Green's function, we obtain [132]

$$U_N - U_S = \sum_{\mathbf{k}} \int_{-\infty}^{+\infty} d\omega(\omega + \epsilon_k) f(\omega) \times [A_N(\mathbf{k},\omega) - A_S(\mathbf{k},\omega)] \,, \quad (17.21)$$

where the spin variable has been summed over. Here and below the subscript N stands for the normal state, S for the superconducting state. $A(\mathbf{k},\omega)$ is the single-particle spectral function, $f(\omega)$ the Fermi function, and ϵ_k the bare energy dispersion which defines the kinetic energy part of the Hamiltonian. Note that the μN term has been absorbed into ω and ϵ_k, that is, these quantities are defined relative to the appropriate chemical potential, μ_N or μ_S. In general, μ_N and μ_S will be different. This difference has to be taken into account, since the condensation energy is small.

The condensation energy is defined by the zero temperature limit of $U_N - U_S$ in the above expression. Note that this involves defining (or somehow extrapolating to) the normal state spectral function at $T = 0$. Such an extrapolation, which we return to below, is not specific to our approach, but required in all estimates of the condensation energy. We remark that Eq. (17.21) yields the correct condensation energy, $N(0)\Delta^2/2$, for the BCS theory of superconductivity.

We also note that Eq. (17.21) can also be broken up into two pieces to individually yield the thermal expectation value of the kinetic energy (using $2\epsilon_k$ in the parentheses in front of $f(\omega)$), and that of the potential energy (using $\omega - \epsilon_k$ instead).

The great advantage of Eq. (17.21) is that it involves just the occupied part of the single particle spectral function, which is measured by ARPES. Therefore, in principle, one should be able to derive the condensation energy from such data, if an appropriate extrapolation of the normal state spectral function to $T = 0$ can be made. On the other hand, a disadvantage is that the bare energies, ϵ_k, are a priori unknown. Note that these are not directly obtained from the measured ARPES dispersion, which already includes many-body renormalizations. Rather, they could be determined by projecting the kinetic energy operator onto the single-band subspace.

Some of the problems associated with an analysis based on experimental data can be appreciated. First, the condensation energy is obtained by subtracting two large numbers. Therefore, normalization of the data becomes a central concern. Problems in this regard when considering $n(\mathbf{k})$, which is the zeroth moment of the ARPES data, were discussed previously [133]. For the first moment, these problems are further amplified due to the ω weighting in the integrand. When analyzing real data, we have found that the high energy tail contribution to the first moment is very sensitive to how the data are normalized. Different choices of normalization can even lead to changes in sign of the first moment.

Another concern concerns the $\mathbf{k}$ sum in Eq. (17.21). ARPES has $\mathbf{k}$-dependent matrix elements, which lead to weighting factors not present in Eq. (17.21). These effects can in principle be factored out by either theoretical estimates of the matrix elements [129], or by comparing data at different photon energies to obtain information on them [8].

Another issue in connection with experimental data is an appropriate extrapolation of the normal state to zero temperature. Information on this can be obtained by analyzing the temperature dependence of the normal state data, remembering that the Fermi function will cause a temperature dependence of the data which should be factored out before attempting the $T = 0$ extrapolation. We finally note that the temperature dependence issue is strongly coupled to the normalization problem mentioned above. In ARPES, the absolute intensity can change due to temperature dependent changes in absorbed gasses, surface doping level, and sample location. Changes of background emission with temperature is another potential problem.

Despite these concerns, we believe that with careful experimentation, many of these difficulties can be overcome, and even if an exact determination of Eq. (17.21) is not possible, insights into the origin of the condensation energy will certainly be forthcoming from the data. This is particularly true for ARPES, which has the advantage of being **k** resolved and thus giving one information on the relative contribution of different **k** vectors to the condensation energy.

Insights into what real data might indicate have been offered by us [132] in the context of the "mode" model illustrated in Fig. 17.47. What we found was that for parameters characteristic of optimal doped ARPES data, the superconducting condensation was driven by kinetic energy lowering, as opposed to the potential energy lowering found in BCS theory. This occurs because n_k becomes sharper in the superconducting state than in the normal state. In essence, the normal state is a non Fermi liquid and the superconducting state is a Fermi liquid, so what occurs is that the effect of quasiparticle formation on sharpening n_k is greater than the effect of particle–hole mixing on smearing it. The net result is a sharpening, leading to a lowering in kinetic energy. In BCS theory, the normal state is a Fermi liquid, and thus only the particle–hole mixing effect is present, leading to a net smearing of n_k and thus an increase in the kinetic energy. The same model can be used to evaluate the optical sum rule [134], and what is found is a violation of the sum rule with a sign and magnitude consistent with recent optics experiments [135]. It will be of great interest to see whether these results can be confirmed directly from ARPES data, as speculated early on by Anderson [136].

17.8 Summary

We review angle resolved photoemission spectroscopy (ARPES) results on the high T_c superconductors, focusing primarily on results obtained on the quasi-two-dimensional cuprate $Bi_2Sr_2CaCu_2O_8$ and its single layer counterpart $Bi_2Sr_2CuO_6$. The topics treated include the basics of photoemission and methodologies for analyzing spectra, normal state electronic structure including the Fermi surface, the superconducting energy gap, the normal state pseudogap, and the electron self-energy as determined from photoemission lineshapes.

Acknowledgments

Much of the experimental work described in this article was done in collaboration with Hong Ding, Adam Kaminski, Helen Fretwell, Kazimierz Gofron, Joel Mesot, Stephan Rosenkranz, Tsunehiro Takeuchi, and the group of Takashi Takahashi, including Takafumi Sato and Takayoshi Yokoya. We were very fortunate to have available to us the samples from Kazuo Kadowaki, T. Mochiku, David Hinks, Prasenjit Guptasarma, Boyd Veal, Z.Z. Li, and Helene Raffy. We have also benefited from many interactions over the years with Phil Anderson, Alex Abrikosov, Jim Allen, Cliff Olson, Ole Andersen, Al Arko, Bertram Batlogg, Arun Bansil, Matthias Eschrig, Atsushi Fujimori, Peter Johnson, Bob Laughlin, Bob Schrieffer, Z.X. Shen, and Chandra Varma. This work was supported by the National Science Foundation, Grant No. DMR 9974401 (JCC) and the U.S. Department of Energy, Office of Science, under Contract No. W-31-109-ENG-38 (JCC and MRN). MR is grateful for partial support from the Indian DST through the Swarnajayanti scheme.

References

1. A. Chainani, T. Yokoya, T. Kiss, S. Shin, Phys. Rev. Lett. **85**, 1966 (2000)
2. A. Damascelli, Z.-X. Shen, Z. Hussain, Rev. Mod. Phys. 75, 473 (2003)
3. D.L. Lynch and C.G Olson, *Photoemission Studies of High-Temperature Superconductors* (Cambridge University Press, Cambridge, 1999)
4. S. Hüfner, *Photoelectron Spectroscopy* (Springer, Berlin, 1996) and references therein
5. C.N. Berglund and W.E. Spicer, Phys. Rev. **136**, A1030 and A1044 (1964)
6. J. Hermanson, Solid State Commun. **22**, 9 (1977)
7. For a review of the first five years of work on the cuprates, see Sec. 4 and 5 of Z. X. Shen and D. S. Dessau, Phys. Reps. **253**, 1 (1995)
8. J. Mesot, M. Randeria, M.R. Norman, A. Kaminski, H.M. Fretwell, J. C. Campuzano, H. Ding, T. Takeuchi, T. Sato, T. Yokoya, T. Takahashi, I. Chong, T. Terashima, M. Takano, T. Mochiku, and K. Kadowaki, Phys. Rev. B **63**, 224516 (2001)
9. H. Hertz, Ann. Physik **31**, 983 (1887)
10. W.L. Schaich and N.W. Ashcroft, Phys. Rev. B **3**, 2452 (1971)
11. C. Caroli, D. Lederer-Rozenblatt, B. Roulet, and D. Saint-James, Phys. Rev. B **8**, 4552 (1973)
12. J.B. Pendry, Surf. Sci. **57**, 679 (1976) For a review of applications of these methods to the cuprates, see: A. Bansil and M. Lindroos, J. Phys. Chem. Solids **56**, 1855 (1995)
13. C.G. Larsson, Surface Science **152/153**, 213 (1985); *ibid* **162**, 19 (1985)
14. P.A.P. Lindberg, L.I. Johansson and A.N. Christensen, Surf. Sci. **197**, 353 (1987)
15. F.L. Hopkinson, J.B. Pendry, and D.J. Titterington, Comput. Phys. Comm. **26**, 111 (1980)
16. A. Bansil, M. Lindroos, and J.C. Campuzano, Proc. Mater. Res. Soc. **253**, 519 (1992); M. Lindroos, A. Bansil, K. Gofron, H. Ding, J.C. Campuzano, R. Liu, and B.W. Veal, Physica C **212**, 347 (1993)
17. H.L. Edwards, J.T. Markert and A.L. de Lozanne, Phys. Rev. Lett. **69**, 2967 (1992)
18. N.V. Smith, P. Thiry, and Y. Petroff, Phys. Rev. B **47**, 15476 (1993)
19. See, e.g., L. Hedin and S. Lundquist, Solid State Physics **23**, 1 (Academic Press, New York, 1969)
20. M. Randeria, H. Ding, J-C. Campuzano, A. Bellman, G. Jennings, T. Yokoya, T. Takahashi, H. Katayama-Yoshida, T. Mochiku, and K. Kadowaki, Phys. Rev. Lett. **74**, 4951 (1995)
21. H. Ding, J. C. Campuzano, A. F. Bellman, T. Yokoya, M. R. Norman, M. Randeria, T. Takahashi, H. Katayama-Yoshida, T. Mochiku, K. Kadowaki, G. Jennings, Phys. Rev. Lett. **74**, 2784 (1995) and **75**, 1425 (E) (1995)
22. A.V. Federov, T. Valla, P.D. Johnson, Q. Li, G.D. Gu, N. Koshizuka, Phys. Rev. Lett. **82**, 2179 (1999); T. Valla, A.V. Fedorov, P.D. Johnson, B.O. Wells, S.L. Hulbert, Q. Li, G.D. Gu, and N. Koshizuka, Science **285**, 2110 (1999); T. Valla, A.V. Fedorov, P.D. Johnson, Q. Li, G. D. Gu, N. Koshizuka, Phys. Rev. Lett. **85**, 828 (2000)
23. P.V. Bogdanov, A. Lanzara, S.A. Kellar, X.J. Zhou, E.D. Lu, W.J. Zheng, G. Gu, J.-I. Shimoyama, K. Kishio, H. Ikeda, R. Yoshizaki, Z. Hussain, and Z. X. Shen, Phys. Rev. Lett. **85**, 2581 (2000)
24. A. Kaminski, M. Randeria, J.C. Campuzano, M.R. Norman, H. Fretwell, J. Mesot, T. Sato, T. Takahashi, and K. Kadowaki, Phys. Rev. Lett. **86**, 1070 (2001)
25. M.R. Norman, M. Eschrig, A. Kaminski, J. C. Campuzano, Phys. Rev. B **64**, 184508 (2001)
26. W.E. Pickett, Rev. Mod. Phys. **61**, 433 (1989)
27. P.W. Anderson, *The Theory of Superconductivity in the High-T_c Cuprates* (Princeton University Press, Princeton, 1997)
28. O.K. Andersen A.I. Liechtenstein, O. Jepsen, F. Paulsen, J. Phys. Chem. Solids **56**, 1573 (1995)
29. P.W. Anderson, Science **235**, 1196 (1987)
30. C.G. Olson, R. Liu, D.W. Lynch, R.S. List, A.J. Arko, B.W. Veal, Y.C. Chang, P.Z. Jiang, and A.P. Paulikas, Phys. Rev. B **42**, 381 (1990)
31. B.O. Wells, Z.-X. Shen, A. Matsuura, D.M. King, M.A. Kastner, M. Greven, and R.J. Birgeneau, Phys. Rev. Lett. **74**, 964 (1995)
32. A. Kaminski, J. Mesot, H. Fretwell, J.C. Campuzano, M.R. Norman, M. Randeria, H. Ding, T. Sato, T. Takahashi, T. Mochiku, K. Kadowaki, and H. Hoechst, Phys. Rev. Lett. **84**, 1788 (2000)
33. J.C. Campuzano, G. Jennings, M. Faiz, L. Beaulaigue, B.W. Veal, J.Z. Liu, A.P. Paulikas, K. Vandervoort, H. Claus, Phys. Rev. Lett. **64**, 2308 (1990)

34. D.M. King, Z.-X. Shen, D.S. Dessau, B.O. Wells, W.E. Spicer, A.J. Arko, D.S. Marshall, J. DiCarlo, A.G. Loeser, C.H. Park, E.R. Ratner, J.L. Peng, Z.Y. Li, R.L. Greene, Phys. Rev. Lett. **70**, 3159 (1993); R.O. Anderson, R. Claessen, J.W. Allen, C.G. Olson, C. Janowitz, L.Z. Liu, J.-H. Park, M.B. Maple, Y. Dalichaouch, M.C. de Andrade, R.F. Jardim, E.A. Early, S.-J. Oh, W.P. Ellis, *ibid* **70**, 3163 (1993); N.P. Armitage, D.H. Lu, C. Kim, A. Damascelli, K.M. Shen, F. Ronning, D.L. Feng, P. Bogdanov, Z.-X. Shen, Y. Onose, Y. Taguchi, Y. Tokura, P.K. Mang, N. Kaneko, M. Greven, *ibid* **87**, 147003 (2001); N.P. Armitage, F. Ronning, D.H. Lu, C. Kim, A. Damascelli, K.M. Shen, D.L. Feng, H. Eisaki, Z.-X. Shen, P.K. Mang, N. Kaneko, M. Greven, Y. Onose, Y. Taguchi, Y. Tokura, *ibid* **88**, 257001 (2002)
35. A. Ino, C. Kim, T. Mizokawa, Z.-X. Shen, A. Fujimori, M. Takaba, K. Tamasaku, H. Eisaki, S. Uchida, J. Phys. Soc. Japan **68**, 1496 (1999); A. Ino, C. Kim, M. Nakamura, T. Mizokawa, Z.-X. Shen, A. Fujimori, T. Kakeshita, H. Eisaki, S. Uchida, Phys. Rev. B **62**, 4137 (2000); A. Ino, C. Kim, M. Nakamura, T. Yoshida, T. Mizokawa, Z.-X. Shen, A. Fujimori, T. Kakeshita, H. Eisaki, S. Uchida, *ibid* **65**, 094504 (2002); T. Yoshida, X.J. Zhou, M. Nakamura, S.A. Kellar, P. V. Bogdanov, E.D. Lu, A. Lanzara, Z. Hussain, A. Ino, T. Mizokawa, A. Fujimori, H. Eisaki, C. Kim, Z.-X. Shen, T. Kakeshita, S. Uchida, *ibid*, **63**, 220501 (2001)
36. H. Ding, A.F. Bellman, J.C. Campuzano, M. Randeria, M.R. Norman, T. Yokoya, T. Takahashi, H. Katayama-Yoshida, T. Mochiku, K. Kadowaki, G. Jennings, and G.P. Brivio, Phys. Rev. Lett. **76**, 1533 (1996)
37. M.R. Norman, M. Randeria, H. Ding, and J.C. Campuzano, Phys. Rev. B **52**, 615 (1995)
38. R.L. Withers, J. G. Thompson, L. R. Wallenberg, J. D. Fitzgerald, J. S. Anderson, B.G. Hyde, J. Phys. **C 21**, 6067 (1988)
39. P. Aebi, J. Osterwalder, P. Schwaller, L. Schlapbach, M. Shimoda, T. Mochiku and K. Kadowaki, Phys. Rev. Lett. **72**, 2757 (1994); J. Osterwalder, P. Aebi, P. Schwaller, L. Schlapbach, M. Shimoda, T. Mochiku and K. Kadowaki, Appl. Phys. A **60**, 247 (1995)
40. A. Kampf and J.R. Schrieffer, Phys. Rev. **B 42**, 7967 (1990)
41. H. Ding, M.R. Norman, T. Yokoya, T. Takuechi, M. Randeria, J.C. Campuzano, T. Takahashi, T. Mochiku, and K. Kadowaki, Phys. Rev. Lett. **78**, 2628 (1997)
42. A.A. Kordyuk, S.V. Borisenko, M.S. Golden, S. Legner, K.A. Nenkov, M. Knupfer, J. Fink, H. Berger, L. Forro, R. Follath, Phys. Rev. B **66**, 014502 (2002)
43. D.S. Dessau, Z.X. Shen, D.M. King, D.S. Marshall, L.W. Lombardo, P.H. Dickinson, A.G. Loeser, J. DiCarlo, C.H. Park, A. Kapitulnik, and W.E. Spicer, Phys. Rev. Lett. **71**, 2781 (1993)
44. M.R. Norman, M. Randeria, H. Ding, J.C. Campuzano, and A.F. Bellman, Phys. Rev. B **52**, 15107 (1995)
45. T. Sato, T. Kamiyama, T. Takahashi, J. Mesot, A. Kaminski, J.C. Campuzano, H.M. Fretwell, T. Takeuchi, H. Ding, I. Chong, T. Terashima, M. Takano, Phys. Rev. B **64**, 054502 (2001)
46. N.L. Saini, J. Avila, A. Bianconi and A. Lanzara, M.C. Asensio, S. Tajima, G.D. Gu, and N. Koshizuka, Phys. Rev. Lett. **79**, 3467 (1997)
47. J. Mesot, M.R. Norman, H. Ding, J.C. Campuzano, Phys. Rev. Lett. **82**, 2618 (1999); N.L. Saini, A. Bianconi, A. Lanzara, J. Avila, M.C. Asensio, S. Tajima, G.D. Gu, N. Koshizuka, *ibid* 2619 (1999)
48. Y.-D. Chuang, A. D. Gromko, D. S. Dessau, Y. Aiura, Y. Yamaguchi, K. Oka, A. J. Arko, J. Joyce, H. Eisaki, S. I. Uchida, K. Nakamura, and Yoichi Ando, Phys. Rev. Lett. **83**, 3717 (1999)
49. H.M. Fretwell, A. Kaminski, J. Mesot, J.C. Campuzano, M.R. Norman, M. Randeria, T. Sato, R. Gatt, T. Takahashi, and K. Kadowaki, Phys. Rev. Lett. **84**, 4449 (2000)
50. S.V. Borisenko, M.S. Golden, S. Legner, T. Pichler, C. Durr, M. Knupfer, J. Fink, G. Yang, S. Abell, H. Berger, Phys. Rev. Lett. **84**, 4453 (2000); S.V. Borisenko, A.A. Kordyuk, S. Legner, C. Durr, M. Knupfer, M.S. Golden, J. Fink, K. Nenkov, D. Eckert, G. Yang, S. Abell, H. Berger, L. Forro, B. Liang, A. Mailouck, C.T. Lin, B. Keimer, Phys. Rev. B **64**, 094514 (2001)
51. D.L. Feng, W.J. Zheng, K.M. Shen, D.H. Lu, F. Ronning, J.-I. Shimoyama, K. Kishio, G. Gu, D. Van der Marel, Z.-X. Shen, cond-mat/9908056
52. J.C. Campuzano, H. Ding, M.R. Norman, M. Randeria, A.F. Bellman, T. Yokoya, T. Takahashi, H. Katayama-Yoshida, T. Mochiku, and K. Kadowaki, Phys. Rev. B **53**, R14737 (1996)
53. T. Straub, R. Claessen, P. Steiner, S. Hufner, V. Eyert, K. Friemelt and E. Bucher, Phys. Rev. B **55**, 13473 (1997)
54. M.C. Schabel, C.-H. Park, A. Matsuura, Z.-X. Shen, D.A. Bonn, Ruixing Liang, and W.N. Hardy, Phys. Rev. B **57**, 6107 (1998)
55. K. Gofron, J.C. Campuzano, H. Ding, C. Gu, R. Liu, B. Dabrowski, B. W. Veal, W. Cramer, G. Jennings, J. Phys. Chem. Sol. **54**, 1193 (1993); K. Gofron, J.C. Campuzano, A.A. Abrikosov, M. Lindroos, A. Bansil, H. Ding, D. Koelling, B. Dabrowski, Phys. Rev. Lett. **73**, 3302 (1994)

56. S. Massida, J. Yu, and A.J. Freeman, Physica **C 152**, 251 (1988); O.K. Andersen, O. Jepsen, A.I. Liechtenstein, I.I. Mazin, Phys. Rev. **B 49**, 4145 (1994)
57. S. Chakravarty, A. Sudbo, P.W. Anderson and S. Strong, Science **261**, 337 (1994)
58. M.R. Norman, H. Ding, J.C. Campuzano, T. Takeuchi, M. Randeria, T. Yokoya, T. Takahashi, T. Mochiku, and K. Kadowaki, Phys. Rev. Lett. **79**, 3506 (1997)
59. J.C. Campuzano, H. Ding, M.R. Norman, H.M. Fretwell, M. Randeria, A. Kaminski, J. Mesot, T. Takeuchi, T. Sato, T. Yokoya, T. Takahashi, K. Kadowaki, P. Guptasarma, D. G. Hinks, Z. Konstantinovic, Z.Z. Li, and H. Raffy, Phys. Rev. Lett. **83**, 3709 (1999)
60. J. Zasadzinski, L. Ozyuzer, Z. Yusof, J. Chen, K.E. Gray, R. Mogilevsky, D.G. Hinks, J.L. Cobb, J.T. Markert, SPIE **2696**, 338 (1996); See also J. Zasadzinski et al. (unpublished) in Fig. 1 of D. Coffey, J. Phys. Chem. Solids **54**, 1369 (1993)
61. P.W. Anderson, Science **256**, 1526 (1992)
62. D.L. Feng, N.P. Armitage, D.H. Lu, A. Damascelli, J.P Hu, P. Bogdanov, A. Lanzara, F. Ronning, K.M. Shen, H. Eisaki, C. Kim, and Z.-X. Shen, Phys. Rev. Lett. **86**, 5550 (2001); D.L. Feng, C. Kim, H. Eisaki, D.H. Lu, A. Damascelli, K.M. Shen, F. Ronning, N. P. Armitage, N. Kaneko, M. Greven, J.-I. Shimoyama, K. Kishio, R. Yoshizaki, G.D. Gu, Z.-X. Shen, Phys. Rev. B **65**, 220501 (2002)
63. Y.D. Chuang, A.D. Gromko, A. Fedorov, D.S. Dessau, Y. Aiura, K. Oka, Y. Ando, H. Eisaki, S.I. Uchida, Phys. Rev. Lett. **87**, 117002 (2001); Y.D. Chuang, A.D. Gromko, A. Fedorov, Y. Aiura, K. Oka, Y. Ando, D.S. Dessau, Phys. Rev. B**69**, 094515 (2004)
64. A.A. Kordyuk, S.V. Borisenko, T.K. Kim, K. Nenkov, M. Knupfer, M.S. Golden, J. Fink, H. Berger, R. Follath, Phys. Rev. Lett. **89**, 077003 (2002)
65. D.L. Feng, A. Damascelli, K.M. Shen, N. Motoyama, D.H. Lu, H. Eisaki, K. Shimizu, J.-I. Shimoyama, K. Kishio, N. Kaneko, M. Greven, G.D. Gu, X.J. Zhou, C. Kim, F. Ronning, N. P. Armitage, and Z.-X. Shen, Phys. Rev. Lett. **88**, 107001 (2002); D.L. Feng, H. Eisaki, K.M. Shen, A. Damascelli, C. Kim, D.H. Lu, Z.-X. Shen, K. Shimizu, J.-I. Shimoyama, K. Kishio, N. Motoyama, N. Kaneko, M. Greven, G.D. Gu, Int. J. Mod. Phys. **16**, 1691 (2002); T. Sato, H. Matsui, S. Nishina, T. Takahashi, T. Fujii, T. Watanabe, and A. Matsuda, Phys. Rev. Lett. **89**, 067005 (2002)
66. D. J. Van Harlingen, Rev. Mod. Phys **67**, 515 (1995); C. C. Tsuei and J.R. Kirtley, *ibid* **72**, 969 (2000)
67. Z.X. Shen, D.S. Dessau, B.O. Wells, D.M. King, W.E. Spicer, A.J. Arko, D. Marshall, L.W. Lombardo, A. Kapitulnik, P. Dickinson, S. Doniach, J. DiCarlo, A.G. Loeser, C.H. Park, Phys. Rev. Lett. **70**, 1553 (1993)
68. H. Ding, M.R. Norman, J.C. Campuzano, M. Randeria, A. Bellman, T. Yokoya, T. Takahashi, T. Mochiku, and K. Kadowaki, Phys. Rev. B **54**, R9678 (1996)
69. J.-M. Imer, F. Patthey, B. Dardel, W.-D. Schneider, Y. Baer, Y. Petroff, A. Zettl, Phys. Rev. Lett. **62**, 336 (1989)
70. C.G. Olson, R. Liu, A.-B. Yang, D.W. Lynch, A.J. Arko, R.S. List, B. W. Veal, Y.C. Chang, P.Z. Jiang, A.P. Paulikus, Science **245**, 731 (1989)
71. D. Pines and P. Monthoux, J. Phys. Chem. Solids **56**, 1651 (1995) and D.J. Scalapino, Phys. Rep. **250**, 329 (1995)
72. R.J. Kelley, J. Ma, C. Quitmann, G. Margaritondo, M. Onellion, Phys. Rev. B **50**, 590 (1994)
73. J.M. Harris, Z.-X. Shen, P. J. White, D.S. Marshall, M.C. Schabel, J.N. Eckstein, I. Bozovic, Phys. Rev. B **54**, R15665 (1996)
74. H. Ding, J.C. Campuzano, M.R. Norman, M. Randeria, T. Yokoya, T. Takahashi, T. Takeuchi, T. Mochiku, K. Kadowaki, P. Guptasarma, and D.G. Hinks, J. Phys. Chem. of Solids **59**, 1888 (1998)
75. C. Renner, B. Revaz, J.-Y. Genoud, K. Kadowaki, O. Fischer, Phys. Rev. Lett. **80**, 149 (1998); N. Miyakawa, P. Guptasarma, J.F. Zasadzinski, D.G. Hinks, K.E. Gray, *ibid* **80**, 157 (1998)
76. A.J. Millis, S.M. Girvin, L.B. Ioffe, and A.I. Larkin, J. Phys. Chem. Solids **59**, 1742 (1998)
77. A.C. Durst and P. A. Lee, Phys. Rev. **62**, 1270 (2000)
78. M. Chiao, P. Lambert, R.W. Hill, C. Lupien, R. Gagnon, L. Taillefer, P. Fournier, Phys. Rev. **62**, 3554 (2000)
79. J. Mesot, M.R. Norman, H. Ding, M. Randeria, J.C. Campuzano, A. Paramekanti, H.M. Fretwell, A. Kaminski, T. Takeuchi, T. Yokoya, T. Sato, T. Takahashi, T. Mochiku, and K. Kadowaki, Phys. Rev. Lett. **83**, 840 (1999)
80. S.-F. Lee, D.C. Morgan, R.J. Ormeno, D.M. Broun, R.A. Doyle, J.R. Waldram, K. Kadowaki, Phys. Rev. Lett. **77**, 735 (1996); T. Jacobs, S. Sridhar, Q. Li, G.D. Gu, N. Koshizuka, *ibid* **75**, 4516 (1995); O. Waldmann, F. Steinmeyer, P. Muller, J.J. Neumeier, F.X. Regi, H. Savary, J. Schneck, Phys. Rev. B **53**, 11825 (1996)
81. A. Paramekanti and M. Randeria, Physica C **341–348**, 827 (2000)
82. W.N. Hardy, D.A. Bonn, D.C. Morgan, R. Liang, K. Zhang, Phys. Rev. Lett. **70**, 3999 (1993); D.A. Bonn, S. Kamal, A. Bonakdarpour, R. Liang, W.N. Hardy, C.C. Homes, D.N. Basov, T. Timusk, Czech. J. Phys. **46**, 3195 (1996)

83. S. Haas, A.V. Balatsky, M. Sigrist and T.M. Rice, Phys. Rev. B **56**, 5108 (1997)
84. R. Fehrenbacher and M.R. Norman, Phys. Rev. B **50**, 3495 (1994)
85. T. Timusk and B.W. Statt, Rep. Prog. Phys. **62**, 61 (1999)
86. M. Randeria, In: *Proceedings of the International School of Physics "Enrico Fermi" Course CXXXVI on High Temperature Superconductors*, ed. by G. Iadonisi, J.R. Schrieffer, and M.L. Chiafalo (IOS Press, 1998) pp. 53–75; [Varenna Lectures, 1997] cond-mat/9710223
87. D.S. Marshall, D.S. Dessau, A.G. Loeser, C-H. Park, A.Y. Matsuura, J.N. Eckstein, I. Bozovic, P. Fournier, A. Kapitulnik, W.E. Spicer, and Z.-X. Shen, Phys. Rev. Lett. **76**, 4841 (1996)
88. A.G. Loeser, Z.-X. Shen, D.S. Dessau, D.S. Marshall, C.H. Park, P. Fournier, A. Kapitulnik, Science **273**, 325 (1996)
89. H. Ding, T. Yokoya, J.C. Campuzano, T. Takahashi, M. Randeria, M.R. Norman, T. Mochiku, K. Kadowaki, and J. Giapintzakis, Nature **382**, 51 (1996)
90. M.R. Norman, H. Ding, M. Randeria, J.C. Campuzano, T. Yokoya, T. Takeuchi, T. Takahashi, T. Mochiku, K. Kadowaki, P. Guptasarma, and D.G. Hinks, Nature **392**, 157 (1998)
91. M.R. Presland, J.L. Tallon, R.G. Buckley, R.S. Liu, and N.E. Flower, Physica C **176**, 95 (1991)
92. D.L. Feng, D.H. Lu, K.M. Shen, C. Kim, H. Eisaki, A. Damascelli, R. Yoshizaki, J.-I. Shimoyama, K. Kishio, G.D. Gu, S. Oh, A. Andrus, J. O'Donell, J.N. Eckstein, Z.-X. Shen, Science **289**, 277 (2000)
93. H. Ding, J.R. Engelbrecht, Z. Wang, J.C. Campuzano, S.-C. Wang, H.-B. Yang, R. Rogan, T. Takahashi, K. Kadowaki, and D. G. Hinks, Phys. Rev. Lett. **87**, 227001 (2001)
94. Y.J. Uemura, G.M. Luke, B.J. Sternlieb, J.H. Brewer, J.F. Carolan, W.N. Hardy, R. Kadono, J.R. Kempton, R.F. Kiefl, S.R. Kreitzman, P. Mulhern, T.M. Riseman, D.L. Williams, B.X. Yang, S. Uchida, H. Takagi, J. Gopalakrishnan, A.W. Sleight, M.A. Subramanian, C.L. Chien, M.Z. Cieplak, G. Xiao, V.Y. Lee, B.W. Statt, C.E. Stronach, W.J. Kossler, X.H. Yu, Phys. Rev. Lett. **62**, 2317 (1989)
95. P.A. Lee, Normal state properties of the oxide superconductors: a review, In: *High Temperature Superconductivity*, ed. by K.S. Bedell, D. Coffey, D.E. Meltzer, D. Pines, J.R. Schrieffer (Addison-Wesley, New York, 1990) pp. 96-116
96. M. Randeria, N. Trivedi, A. Moreo, and R.T. Scalettar, Phys. Rev. Lett. **69**, 2001 (1992); N. Trivedi and M. Randeria, Phys. Rev. Lett. **75**, 312 (1995)
97. V. Emery and S. Kivelson, Nature **374**, 434 (1995)
98. G. Baskaran, Z. Zou and P.W. Anderson, Solid State Commun. **63**, 973 (1987); G. Kotliar and J. Liu, Phys. Rev. B **38**, 5142 (1988); H. Fukuyama, Prog. Theor. Phys. Suppl. **108**, 287 (1992)
99. P.A. Lee and X.G. Wen, Phys. Rev. Lett. **76**, 503 (1996) and Phys. Rev. B **63**, 224517 (2001)
100. J. Tallon and J. Loram, Physica C **349**, 53 (2001)
101. S. Chakravarty, R.B. Laughlin, D.K. Morr and C. Nayak, Phys. Rev. B **63**, 094503 (2001)
102. C.M. Varma, Phys. Rev. Lett. **83**, 3538 (1999)
103. A. Kaminski, S. Rosenkranz, H. Fretwell, J.C. Campuzano, Z. Li, H. Raffy, W.G. Cullen, H. You, C.G. Olson, H. Hoechst, Nature **416**, 610 (2002)
104. S.H. Pan, J.P. O'Neal, R.L. Badzey, C. Chamon, H. Ding, J.R. Engelbrecht, Z. Wang, H. Eisaki, S. Uchida, A.K. Gupta, K.-W. Ng, E.W. Hudson, K.M. Lang, J.C. Davis, Nature **413**, 282 (2001)
105. L. Perfetti, C. Rojas, A. Reginelli, L. Gavioli, H. Berger, G. Margaritondo, M. Grioni, R. Gaal, L. Forro, F. Rullier Albenque, Phys. Rev. B **64**, 115102 (2001)
106. M.R. Norman, H. Ding, H. Fretwell, M. Randeria, and J.C. Campuzano, Phys. Rev. B **60**, 7585 (1999)
107. M.R. Norman, A. Kaminski, J. Mesot, J.C. Campuzano, Phys. Rev. B **63**, 140508 (2001)
108. M.R. Norman, M. Randeria, H. Ding, and J.C. Campuzano, Phys. Rev. B **57**, R11093 (1998)
109. D.S. Dessau, B.O. Wells, Z.-X. Shen, W.E. Spicer, A.J. Arko, R.S. List, D.B. Mitzi, and A. Kapitulnik, Phys. Rev. Lett. **66**, 2160 (1991); D.S. Dessau, Z.-X. Shen, B.O. Wells, D.M. King, W.E. Spicer, A.J. Arko, L.W. Lombardo, D.B. Mitzi, and A. Kapitulnik, Phys. Rev. B **45**, 5095 (1992)
110. Q. Huang, J.F. Zasadzinski, K.E. Gray, J.Z. Liu, and H. Claus, Phys. Rev. B **40**, 9366 (1989)
111. Y. Kuroda and C.M. Varma, Phys. Rev. B **42**, 8619 (1990)
112. S. Engelsberg and J.R. Schrieffer, Phys. Rev. **131**, 993 (1963)
113. D.J. Scalapino, In: *Superconductivity*, ed. by R.D. Parks (Marcel Dekker, New York, 1969) Vol 1, p. 449
114. Z.-X. Shen and J.R. Schrieffer, Phys. Rev. Lett. **78**, 1771 (1997)
115. P.B. Littlewood and C.M. Varma, Phys. Rev. B **46**, 405 (1992)

116. M.R. Norman and H. Ding, Phys. Rev. B **57**, R11089 (1998)
117. S.M. Quinlan, P.J. Hirschfeld, and D.J. Scalapino, Phys. Rev. B. **53**, 8575 (1996)
118. J. Rossat-Mignot, L.P. Regnault, C. Vettier, P. Bourges, P. Burlet, J. Bossy, J.Y. Henry, and G. Lapertot, Physica C **185-189**, 86 (1991); H.A. Mook, M. Yethiraj, G. Aeppli, T. E. Mason, T. Armstrong, Phys. Rev. Lett. **70**, 3490 (1993); Hung Fai Fong, B. Keimer, P.W. Anderson, D. Reznik, F. Dogan and I.A. Aksay, Phys. Rev. Lett. **75**, 316 (1995)
119. H.F. Fong, P. Bourges, Y. Sidis, L.P. Regnault, A. Ivanov, G.D. Gu, N. Koshizuka, and B. Keimer, Nature **398**, 588 (1999)
120. In: *The Gap Symmetry and Fluctuations in High T_c Superconductors*, ed. by J. Bok, G. Deutscher, D. Pavuna, and S.A. Wolf (Plenum, New York, 1998) p. 349
121. J.F. Zasadzinski, L. Ozyuzer, N. Miyakawa, K.E. Gray, D.G. Hinks, C. Kendziora, Phys. Rev. Lett. **87**, 067005 (2001)
122. P. Dai, M. Yethiraj, H. A. Mook, T.B. Lindemer, and F. Doğan, Phys. Rev. Lett. **77**, 5425 (1996)
123. R. Preuss, W. Hanke, C. Grober, and H.G. Evertz, Phys. Rev. Lett. **79**, 1122 (1997); J. Schmalian, D. Pines, and B. Stojkovic, *ibid.* **80**, 3839 (1998); X.-G. Wen and P.A. Lee, *ibid.* **76**, 503 (1996)
124. E. Demler and S.-C. Zhang, Phys. Rev. Lett. **75**, 4126 (1995)
125. P. Dai, H.A. Mook, S.M. Hayden, G. Aeppli, T.G. Perring, R.D. Hunt, and F. Doğan, Science **284**, 1344 (1999)
126. M. Hengsberger, D. Purdie, P. Segovia, M. Garnier, Y. Baer, Phys. Rev. Lett. **83**, 592 (1999); T. Valla, A.V. Fedorov, P.D. Johnson, S.L. Hulbert, *ibid.*, 2085 (1999)
127. A. Lanzara, P.V. Bogdanov, X.J. Zhou, S.A. Kellar, D.L. Feng, E.D. Lu, T. Yoshida, H. Eisaki, A. Fujimori, K. Kishio, J.-I. Shimoyama, T. Noda, S. Uchida, Z. Hussain, and Z.-X. Shen, Nature **412**, 510 (2001)
128. P.D. Johnson, T. Valla, A.V. Fedorov, Z. Yusof, B.O. Wells, Q. Li, A.R. Moodenbaugh, G.D. Gu, N. Koshizuka, C. Kendziora, Sha Jian, and D.G. Hinks, Phys. Rev. Lett. **87**, 177007 (2001)
129. M. Eschrig and M.R. Norman, Phys. Rev. Lett. **85**, 3261 (2000); Phys. Rev. B**67**, 144503 (2003)
130. M. Eschrig and M.R. Norman, Phys. Rev. Lett. **89**, 277005 (2002)
131. A.D. Gromko, A.V. Federov, Y.-D. Chuang, J.D. Koralek, Y. Aiura, Y. Yamaguchi, K. Oka, Y. Ando, D.S. Dessau, Phys. Rev. B**68**, 174520 (2003)
132. M.R. Norman, M. Randeria, B. Janko, J.C. Campuzano, Phys. Rev. B **61**, 14742 (2000)
133. J.C. Campuzano, H. Ding, H. Fretwell, J. Mesot, A. Kaminski, T. Sato, T. Takahashi, T. Mochiku, and K. Kadowaki, cond-mat/9811349
134. M.R. Norman and C. Pepin, Phys. Rev. B **66**, 100506 (2002)
135. H.J.A. Molegraaf, C. Presura, D. van der Marel, P.H. Kes, M. Li, Science **295**, 2239 (2002); A. F. Santander-Syro, R.P.S.M. Lobo, N. Bontemps, Z. Konstantinovic, Z.Z. Li, H. Raffy, Europhys. Lett. **62**, 568 (2003)
136. P.W. Anderson, Phys. Rev. B **42**, 2624 (1990)

18 Neutron Scattering and the Magnetic Response of Superconductors and Related Compounds

Stephen M.Hayden H.H. Wills Physics Laboratory, University of Bristol, Bristol, United Kingdom

18.1 Introduction

Neutron scattering has played a major role in understanding the properties of conventional and unconventional superconductors. The neutron is a versatile tool for studying condensed matter in general, and superconductors in particular, for a number of reasons. Firstly, the neutron's interaction with matter is weak and extremely well understood. The weak interaction between matter and neutrons means that the scattering cross-section can be interpreted using the Born approximation. Most elements have a small neutron absorption cross-section. Hence bulk material, with sample dimensions up to several cm, can be studied. Secondly, neutrons interact with matter through the strong and electromagnetic forces. They are able to investigate the position and motion of the atoms through the interaction of the neutron and the nuclei via the strong force. Their magnetic interaction with the electrons allows magnetic structures and excitations to be investigated. The magnetic interaction can be understood as an interaction between the neutron and the microscopic magnetic field distribution. Thus magnetic field structures, such as the Abrikosov flux lattice, and their excitations can be studied. A third reason which makes neutron scattering a powerful tool is that the neutron has a mass. This means that the energy and momentum of the neutron can be matched to the collective excitations of interest.

In view of the above, one might expect that neutron scattering would have made a significant impact on our understanding of superconductors. In this review we will concentrate on the study on spin correlations in superconductors and their related compounds. It is therefore pertinent to mention some areas *not* covered in detail by this review. These include the determination of the atomic structure of superconducting compounds, the study of lattice excitations (phonons) and the study of the flux line lattice by small angle neutron scattering (SANS).

18.2 The Neutron Scattering Technique

The neutron scattering technique and its interpretation has been described in detail in a number of classic textbooks [1–4]. Neutrons interact with condensed matter through the strong nuclear force and the electromagnetic force. Nuclear scattering is used to determine the atomic positions and motion in condensed matter systems. It is a powerful tool for determining phonon dispersion relations in solids. Magnetic neutron scattering arises because of the neutron's magnet moment. It is interesting to note that there are coherent cross terms between the nuclear and magnetic cross-sections which are also useful experimentally (see Sect. 18.3). In this section we summarize the basic features of magnetic neutron scattering. The results of magnetic neutron scattering experiments are most succinctly discussed in terms of the generalized magnetic susceptibility and the partial differential cross-section. We will introduce these quantities below.

18.2.1 Neutron Scattering Experiments

Neutrons for condensed matter research are produced either by a nuclear reactor or by a spallation source. It is well known that reactors produce neutrons through the fission of U-235 in a chain reaction. A spallation source produces neutrons when a high-energy (e.g. 800 MeV) proton beam hits a heavy metal (Ta, Hg, U) target. A spallation source can be easily pulsed, this allows the neutron's energy to be measured by time-of-flight methods. In this review, we are mainly concerned with the measurement of excitations using inelastic neutron scattering. Pulsed-spallation and reactor sources and their associated instrumentation each have their own advantages and disadvantages for such measurements. Reactors currently have more neutron flux and have better developed instrumentation for the use of polarized neutrons. Spectrometers with a low intrinsic background can be built at pulsed neutron sources. Pulsed spallation sources also lend themselves to experiments with epithermal neutrons. Figure 18.1(b) shows a schematic diagram of a triple-axis spec-

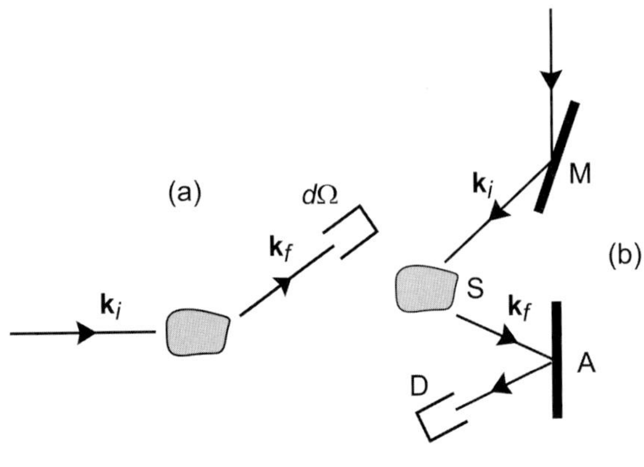

Fig. 18.1. (**a**) Illustration of the geometry of a neutron scattering experiment. Neutrons with wavevector $\mathbf{k}_i$ are incident on a sample (S). Scattered neutrons are detected in a solid angle dΩ over an energy range dE. (**b**) Schematic diagram of a triple-axis spectrometer, which is widely used at reactor sources for measuring excitations. The incident wavevector of the neutrons is selected by a monochromator (M) crystal (typically graphite). Neutrons scatter at the sample (S). The final wavevector (energy) of the scattered neutrons is selected by an analyzer (A) crystal. Neutrons are detected in a He-3 ionization detector

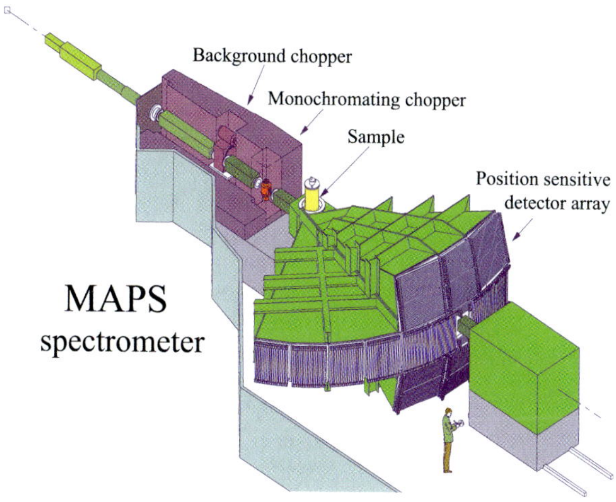

Fig. 18.2. The MAPS spectrometer at the ISIS pulsed spallation source of the Rutherford Appleton Laboratory. MAPS is an example of a direct-geometry time-of-time chopper spectrometer. Monochromated pulsed neutrons are produced using a Fermi chopper. Neutrons are detected by an array of position sensitive detectors (Figure: RAL.)

trometer, the workhorse for inelastic experiments on single crystals at reactor sources. The energies of the monochromated and scattered neutrons are determined by Bragg reflection from crystals of graphite, silicon, copper, or, in the case of polarized neutrons, a Heusler alloy. Neutrons are detected by a He-3 ionization chamber. Figure 18.2 shows a chopper spectrometer at a pulsed spallation source with position sensitive detectors (PSD) and time-of-flight energy analysis. This type of spectrometer is a relatively recent development which complements in many respects the three-axis spectrometer and is used for measuring excitations in single crystals.

18.2.2 The Partial Differential Cross-Section

Figure 18.1(a) shows the geometry of a typical neutron scattering experiment. Neutrons with (initial) wavevector $\mathbf{k}_i$ and energy E_i are incident on the sample. Some neutrons will be scattered such that they have final wavevector $\mathbf{k}_f$ and final energy E_f. We can determine the momentum and energy transfer between the neutron and an excitation,

$$\mathbf{Q} = \mathbf{k}_i - \mathbf{k}_f ,$$

$$\hbar\omega = E_i - E_f .$$

Neutron scattering experiments are best described in terms of a partial differential cross-section. This is a measure of the probability that a neutron scatters from an initial wavevector $\mathbf{k}_i$ to a final wavevector $\mathbf{k}_f$. Consider a beam of neutrons with the same energy E_i incident on a small sample. Let us set up a neutron counter and measure the number of neutrons scattered in a given direction as a function of their (final) energy E_f. Let the counter subtend a solid angle $\mathrm{d}\Omega$ at the sample (see Fig. 18.1(a)). The partial differential cross-section is defined by the equation

$$\frac{\mathrm{d}^2\sigma}{\mathrm{d}\Omega\,\mathrm{d}E} = \text{(number of neutrons scattered per second into } d\Omega \text{ with final energy between } E_f \text{ and } E_f + \mathrm{d}E)\,/\Phi\, d\Omega\,\mathrm{d}E ,$$

where Φ is the flux of incident neutrons, that is the number per unit area per second incident on the sample.

18.2.3 The Generalized Susceptibility

To most physicists, the magnetic susceptibility is the ratio of the average magnetisation $\mathbf{M}$ to the uniform field $\mathbf{H}$ which induces it. This susceptibility is typically measured using a SQUID or vibrating sample magnetometer. More generally, a system will respond to a magnetic field which varies in space and/or time, $H_{\mathbf{Q}\omega} \exp[i(\mathbf{Q}\cdot\mathbf{r} - \omega t)]$. If the spatial frequency of the magnetic field is comparable to the atomic spacing or the temporal frequency of the magnetic field is comparable to the magnetic excitations of the system, then the response will contain important information about the microscopic magnetic dynamics of the system. Such information about the magnetic dynamics can be obtained using a number of probes including nuclear magnetic resonance (NMR), muon spin resonance (μSR), Raman scattering and, of course, neutron scattering. The information obtained from these probes about the microscopic magnetic response is best described in terms of the generalized magnetic susceptibility, which we will now introduce.

In general, a susceptibility is used to characterize the response of a system to a probing perturbation or field. For example, the perturbation might be an electric field, a magnetic field or a force experienced by a particle. If the response is proportional to the perturbation, then we can define a susceptibility or response function as the ratio of a "generalized force" (the perturbation) to a "generalized displacement" (the response). In this review, we will be concerned with the magnetic susceptibility, thus, the relevant response is the magnetization $\mathbf{M}(\mathbf{r}, t)$ and the probing field is the magnetic field $\mathbf{H}(\mathbf{r}, t)$. Neutron scattering is usually most easily understood in the frequency domain. We therefore introduce the corresponding Fourier components $\mathbf{M}(\mathbf{Q}, \omega)$ and $\mathbf{H}(\mathbf{Q}, \omega)$. The generalized wavevector and frequency-dependent magnetic susceptibility $\chi(\mathbf{Q}, \omega)$ can now be defined through the equation

$$M_\nu(\mathbf{Q}, \omega) = \sum_\mu \chi_{\nu\mu}(\mathbf{Q}, \omega)\, H_\mu(\mathbf{Q}, \omega)\,, \qquad (18.1)$$

where we have only considered the response induced at the same $\mathbf{Q}$ as the perturbing field (i.e. $\chi_{\nu\mu}(\mathbf{Q}, \omega) \equiv \chi_{\nu\mu}(\mathbf{Q}, \mathbf{Q}, \omega)$) and the sum is over the different spatial directions $\nu, \mu = x, y$ and z.

The Reduced Susceptibility

The susceptibility defined in Eq. 18.1 is that associated with the *total* magnetization of the system. In model calculations (e.g. a calculation of spin waves in the Heisenberg model), we usually assume that the magnetic moment is either localized at atomic sites or completely itinerant. Such calculations yield a *reduced* susceptibility $\chi(\mathbf{Q}, \omega)_{\text{reduced}}$ [7]. The total susceptibility $\chi(\mathbf{Q}, \omega)_{\text{total}}$ of Eq. 18.1 includes the effect of spreading out the magnetization associated

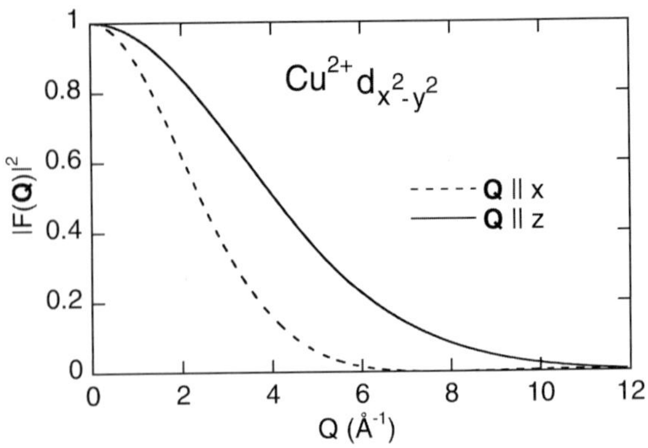

Fig. 18.3. The spin form factor of the Cu^{2+} $d_{x^2-y^2}$ orbital [5,6]. The *solid* and *dashed lines* correspond to wavevectors parallel to the x and z directions, respectively and demonstrate the anisotropy of the form factor in this case

with each atom. The electronic states (and hence the magnetisation distribution in space) are described by atomic orbitals or Wannier functions [7,8]. The total and reduced susceptibilities are related through a form factor $|F(\mathbf{Q})|^2$ which is the Fourier transform of the normalized magnetization associated with each site or orbital:

$$\chi(\mathbf{Q}, \omega)_{\text{total}} = |F(\mathbf{Q})|^2\, \chi(\mathbf{Q}, \omega)_{\text{reduced}}\,. \qquad (18.2)$$

The results of magnetic neutron scattering experiments are usually expressed in terms of the reduced susceptibility so that they can be directly compared with theory. Throughout the rest of this article we will use the symbol $\chi(\mathbf{Q}, \omega)$ to denote the *reduced* susceptibility unless noted explicitly. The measured response needs to be corrected for the form factor $|F(\mathbf{Q})|^2$ and this is done using an equation such as Eq. 18.9. The form factor is usually evaluated from published parameterizations based on atomic calculations [5]. Fig. 18.3 shows the form factor for the $d_{x^2-y^2}$ orbital of a Cu^{2+} ion. Note that the form factor is anisotropic in this case.

18.2.4 The Kramers–Kronig Relations, Wavevector-Dependent and Local Susceptibility

Information obtained from inelastic neutron scattering is most easily expressed in terms of the real and imaginary parts of the generalized susceptibility.

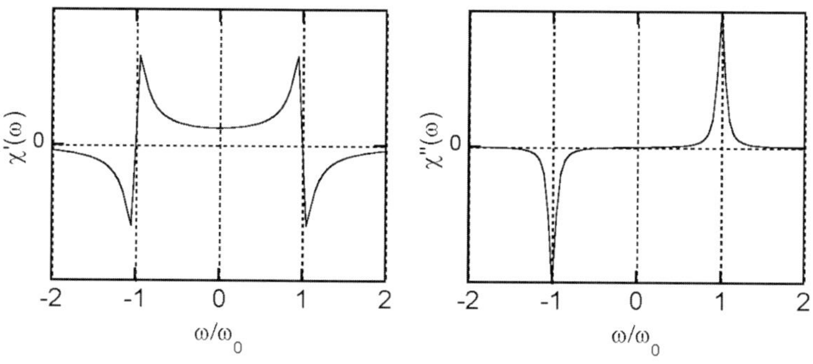

Fig. 18.4. The real and imaginary parts of the response function for a damped harmonic oscillator. The peak in the imaginary part of the response is used to estimate the frequency of the oscillator

Thus a few words about the physical interpretation of these qualities seems in order. The generalized susceptibility can be written in terms of its real part $\chi'(\mathbf{Q},\omega)$ and imaginary part $\chi''(\mathbf{Q},\omega)$,

$$\chi(\mathbf{Q},\omega) = \chi'(\mathbf{Q},\omega) + i\,\chi''(\mathbf{Q},\omega)\,. \tag{18.3}$$

To illustrate the interpretation of $\chi''(\mathbf{Q},\omega)$, we consider the case of a harmonic oscillator with a single pole (see Fig. 18.4). The imaginary part of the susceptibility is peaked near the frequency of the oscillator, while the real part changes sign at this frequency. Such a response function would apply to a well defined excitation such as a spin wave in a ferromagnet. Magnetic excitations can also be strongly damped, this is particularly the case in metals. The response function in this limit might be better described by an over-damped harmonic oscillator (Lorentzian) response. The real part of the susceptibility at zero frequency $\chi(\mathbf{Q}) = \chi'(\mathbf{Q},\omega=0)$ has a particular physical significance: it measures the response to a spatially varying *static* magnetic field.

It is well known that the real and imaginary parts of causal response functions are connected through the *Kramers–Kroning relations* [7],

$$\chi'(\mathbf{Q},\omega) - \chi'(\mathbf{Q},\infty) = \frac{1}{\pi}\int_{-\infty}^{\infty} \frac{\chi''(\mathbf{Q},\omega')}{\omega'-\omega}\,d\omega'\,, \tag{18.4}$$

$$\chi''(\mathbf{Q},\omega) = -\frac{1}{\pi}\int_{-\infty}^{\infty} \frac{\chi'(\mathbf{Q},\omega')-\chi'(\mathbf{Q},\infty)}{\omega'-\omega}\,d\omega'\,. \tag{18.5}$$

The Kramers–Kroning relations are useful because the absorption or scattering cross section is proportional to $\chi''(\mathbf{Q},\omega)$. By integrating over frequency we can obtain the real part $\chi'(\mathbf{Q},\omega)$.

Wavevector-Dependent Susceptibility

The wavevector-dependent susceptibility $\chi(\mathbf{Q}) = \chi'(\mathbf{Q},\omega=0)$ is the response to a static spatially varying magnetic field $\chi_{\nu\mu}(\mathbf{Q}) = M_\nu(\mathbf{Q})/H_\mu(\mathbf{Q})$. This may be obtained from the first Kramers–Kroning relation (Eq. 18.4),

$$\chi(\mathbf{Q}) = \frac{1}{\pi}\int_{-\infty}^{\infty} \frac{\chi''(\mathbf{Q},\omega=0)}{\omega}\,d\omega\,. \tag{18.6}$$

Local Susceptibility

Another useful quantity which can be derived from $\chi(\mathbf{Q},\omega)$ is the *local* (imaginary) susceptibility $\chi''(\omega)$. The local susceptibility is a "weighted density of states" obtained by integrating over the Brillouin zone for a given frequency,

$$\chi''(\omega) = \frac{\int_{\mathrm{BZ}} \chi''(\mathbf{Q},\omega)\,d^3\mathbf{Q}}{\int_{\mathrm{BZ}} d^3\mathbf{Q}}\,. \tag{18.7}$$

We shall see in Sect. 18.2.7 that $\langle M^2(\mathbf{Q},\omega)\rangle \propto \chi''(\mathbf{Q},\omega)$. Hence for a system with a single magnetic site in each unit cell, the moment fluctuating with a given frequency is proportional to $\chi''(\omega)$.

18.2.5 The Magnetic Cross-section

Neutron scattering directly measures the imaginary part of the generalized magnetic susceptibility $\chi''(\mathbf{Q},\omega)$. We will start by considering a system with a diagonal susceptibility tensor with two independent components i.e. $\chi_{xx}(\mathbf{Q},\omega)=\chi_{yy}(\mathbf{Q},\omega)=\chi_{\perp}(\mathbf{Q},\omega)$ and $\chi_{zz}(\mathbf{Q},\omega)$. The axial direction might represent the direction of an applied field, i.e. the quantization direction or the tetragonal axis of a crystal. The scattering cross-section for unpolarized neutrons is,

$$\begin{aligned}\frac{\mathrm{d}^2\sigma}{\mathrm{d}\Omega\,\mathrm{d}E_f} &= (\gamma r_e)^2\frac{k_f}{k_i}|F(\mathbf{Q})|^2 \\ &\times\left(\frac{1}{1-\exp(-\hbar\omega/kT)}\right)\frac{1}{\pi g^2\mu_B^2} \\ &\times\Big[(1+\hat{Q}^2)\chi''_{\perp}(\mathbf{Q},\omega) \\ &\quad+(1-\hat{Q}^2)\chi''_{zz}(\mathbf{Q},\omega)\Big]\,, \qquad (18.8)\end{aligned}$$

where $(\gamma r_e)^2$=0.2905 barn sr^{-1} μ_B^{-2}, $\mathbf{k}_i$ and $\mathbf{k}_f$ are the incident and final neutron wavevectors and $|F(\mathbf{Q})|^2$ is the magnetic form factor.

If the system is isotropic i.e. $\chi_{\perp}(\mathbf{Q},\omega)=\chi_{zz}(\mathbf{Q},\omega)$ then Eq. 18.8 becomes

$$\begin{aligned}\frac{\mathrm{d}^2\sigma}{\mathrm{d}\Omega\,\mathrm{d}E_f} &= (\gamma r_e)^2\frac{k_f}{k_i}|F(\mathbf{Q})|^2 \\ &\times\left(\frac{1}{1-\exp(-\hbar\omega/kT)}\right) \\ &\times\frac{2}{\pi g^2\mu_B^2}\chi''(\mathbf{Q},\omega)\,. \qquad (18.9)\end{aligned}$$

It is relatively easy to place neutron scattering measurements on an absolute scale unit either by measuring the elastic incoherent scattering from a vanadium standard or by measuring the low-frequency coherent phonon scattering from the sample [9]. Throughout this review, we label momentum transfers (Q_x, Q_y, Q_z) in units of Å^{-1} by their reciprocal space positions (h,k,l), where $\mathbf{Q}=h\mathbf{a}^*+k\mathbf{b}^*+l\mathbf{c}^*$.

18.2.6 Polarized Neutron Scattering

Magnetic neutron scattering has a number of unique features. Implicit in Eq. 18.8 is the fact that neutrons are only sensitive to moment or spin fluctuations perpendicular to the momentum change $\mathbf{Q}$. If the neutron spin is quantized along the z-axis, e.g. by a small magnetic field applied along the z-axis, then magnetic scattering can either involve a change of the neutron spin ($\Delta S=1$), or no change in the neutron spin ($\Delta S=0$). These are known as "spin-flip" and "non-spin-flip" scattering respectively. Spin-flip scattering is caused by moment fluctuations perpendicular to the quantization axis and non-spin-flip scattering by fluctuations parallel to the quantization axis. Polarized neutron scattering allows the separation of magnetic scattering from nuclear scattering and the separation of χ_{zz} from $\chi_{\perp}$. In systems without spin anisotropy (e.g. paramagnets and conventional superconductors), there will be no distinction between χ_{zz} and $\chi_{\perp}$ and scattering will, in general, be a mixture of spin-flip and non-spin-flip. However, even in this case, polarized neutron scattering can be a powerful tool in isolating the magnetic response. A number of geometries have been used for neutron scattering measurements (with polarization analysis) on high-T_c superconductors [10,11]. The cross sections for an isotropic system are summarized below,

$$\begin{aligned}\left(\frac{\mathrm{d}^2\sigma}{\mathrm{d}\Omega\,\mathrm{d}E}\right)_{\uparrow\to\uparrow,\mathbf{H}\|\mathbf{Q}} &= 0\,, \\ \left(\frac{\mathrm{d}^2\sigma}{\mathrm{d}\Omega\,\mathrm{d}E}\right)_{\uparrow\to\downarrow,\mathbf{H}\|\mathbf{Q}} &= (\gamma r_e)^2\frac{k_f}{k_i}|F(\mathbf{Q})|^2 \\ &\times\left(\frac{2/\pi g^2\mu_B^2}{1-\exp(-\hbar\omega/kT)}\right) \\ &\times\chi''(\mathbf{Q},\omega) \\ \left(\frac{\mathrm{d}^2\sigma}{\mathrm{d}\Omega\,\mathrm{d}E}\right)_{\uparrow\to\downarrow,\mathbf{H}\perp\mathbf{Q}} &= (\gamma r_e)^2\frac{k_f}{k_i}|F(\mathbf{Q})|^2 \\ &\times\left(\frac{1/\pi g^2\mu_B^2}{1-\exp(-\hbar\omega/kT)}\right) \\ &\times\chi''(\mathbf{Q},\omega)\,, \qquad (18.10)\end{aligned}$$

where the arrows ↑ and ↓ represent the spin polarization along the quantizing magnetic field $\mathbf{H}$ applied to the sample. In the first two configurations $\mathbf{H}$ is applied parallel to $\mathbf{Q}$ and all magnetic scattering is spin-flip.

18.2.7 The Moment Sum Rule and the Fluctuation-Dissipation Theorem

The best known example of the fluctuation-dissipation theorem (FDT) is probably Brownian motion. In a colloidal suspension, the particles move about randomly because they are bombarded by molecules in the liquid. If these particles were charged and we attempt to move them with an external field, then impacts of the molecules would cause the charged particles to experience a resistive force which is proportional to their velocity. The same mechanism (bombardment by molecules) is responsible for both the random fluctuations in position and the response to an external field. The relationship between the response of a system and (thermal or quantum) fluctuations within the system is governed by a *fluctuation-dissipation theorem* [12]. Here we consider only the magnetic case in which the fluctuations are in the magnetization $\mathbf{M}(\mathbf{r}, t)$ and the external field is the magnetic field $\mathbf{H}(\mathbf{r}, t)$. In this case, the *unsymmetrised* fluctuation-dissipation theorem (FDT) is [7],

$$\int_{-\infty}^{\infty} dt\, e^{-i\omega t} \left\langle M_\nu(\mathbf{Q}, t)\, M_\mu(-\mathbf{Q}, 0)\right\rangle = 2\hbar V \frac{\chi''_{\nu\mu}(\mathbf{Q}, \omega)}{\left(1 - e^{-\beta\hbar\omega}\right)}. \quad (18.11)$$

The unsymmetrised form of the FDT is most relevant to scattering experiments because the scattering cross-section is typically related to the left hand side of Eq. 18.11. If we integrate Eq. 18.11 with respect to frequency and wavevector and assume the system is isotropic, we obtain a moment sum rule for $\chi''(\mathbf{Q}, \omega)$:

$$\begin{aligned}\left\langle \mathbf{M}^2\right\rangle &= \left\langle M_x^2 + M_y^2 + M_z^2\right\rangle = \frac{\hbar}{\pi}\int_{BZ} d\mathbf{Q} \\ &\times \int_{-\infty}^{\infty} \frac{[\chi''_{xx}(\mathbf{Q}, \omega) + \chi''_{yy}(\mathbf{Q}, \omega) + \chi''_{zz}(\mathbf{Q}, \omega)]\, d\omega}{1 - \exp(-\hbar\omega/kT)} \\ &= \frac{3\hbar}{\pi}\int_{BZ} d\mathbf{Q} \int_{-\infty}^{\infty} \frac{\chi''(\mathbf{Q}, \omega)\, d\omega}{1 - \exp(-\hbar\omega/kT)}. \quad (18.12)\end{aligned}$$

We can compare the total moment obtained from the moment sum rule with the atomic moment of the constituent ions. For an ion with spin-only moment, we would expect the total moment to be,

$$\begin{aligned}\left\langle \mathbf{M}^2\right\rangle &= g^2 \mu_B^2 S(S+1) \\ &= 3\mu_B^2 \text{ for } S = \frac{1}{2},\ g = 2.\end{aligned}$$

18.3 The Static Spin Susceptibility of Superconductors

The static spin susceptibility $\chi_{\text{spin}} = \chi_{\text{spin}}(\mathbf{Q} = 0, \omega = 0)$ can provide very useful information about the nature of the superconducting state. In a conventional superconductor with spin-singlet pairing, the spin susceptibility is suppressed on entering the superconducting state because spins pair up. The singlet pairs have a total spin $S = 0$ and therefore do not respond to an external magnetic field. For $B \ll B_{c2}$, the temperature dependence of $\chi_{\text{spin}}(T)$ is described by the Yosida function [13], this function is plotted in Fig. 18.5(a). The spin susceptibility cannot be measured directly using a magnetometer in the superconducting state because diamagnetic supercurrents flow on the surface of the superconductor when a field is applied and screen out the response. However, magnetic neutron scattering and NMR (knight shift) do not suffer from this limitation. The measurement of χ_{spin} using neutrons is an interesting application of the interference between the magnetic and nuclear cross sections. Neutrons interact with condensed matter both through the strong nuclear interaction *and* through the electromagnetic interaction. If we satisfy the Bragg condition, then scattering occurs both because of the periodicity of the nuclear density and because of the microscopic periodicity of the magnetization density. The two scattered waves interfere. The sign of interference term depends on the relative orientation of the neutron spin and the sample magnetization. Thus, it may be isolated from other scattering by reversing the spin polarization of a spin-polarized neutron beam. Shull and Wedgwood [14] used this method to determine the susceptibility of the conventional superconductor V_3Si in the superconducting and normal states. Their results

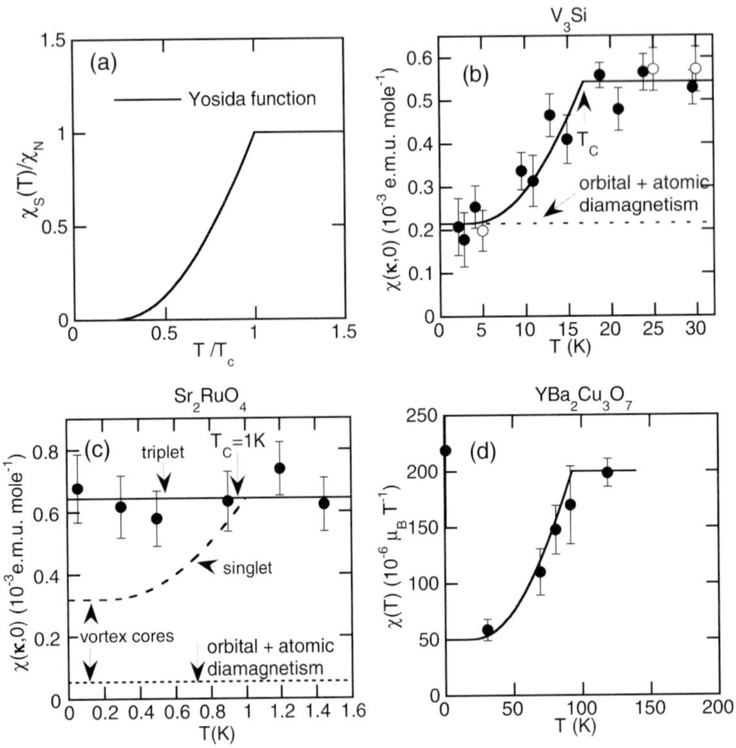

Fig. 18.5. (**a**) The spin susceptibility of a singlet superconductor (Yosida function) [13]. (**b**) The magnetic susceptibility of V_3Si, a conventional singlet superconductor measured in the superconducting state using polarized neutrons [14]. (**c**) Susceptibility of Sr_2RuO_4, believed to be a triplet-paired p-wave superconductor [15]. (**d**) Susceptibility of $YBa_2Cu_3O_7$, a singlet-paired d-wave superconductor [16]

are shown in Fig. 18.5(b). Shull and Wedgwood observed a drop of the susceptibility through T_c which was consistent with the Yosida function. However, their experiment also demonstrated that the orbital susceptibility remained finite as $T \to 0$. 3d transition metal compounds such as V_3Si have an orbital (van-Vleck) susceptibility which is comparable to their spin susceptibility.

Having discussed the case of a conventional spin-singlet superconductor, we now consider the case of triplet-paired (or odd-parity) superconductors. It is widely believed that Sr_2RuO_4 is such a system [17]. The pairing in triplet or odd-parity superconductors is somewhat more complicated than the singlet case because there are many possible triplet states. Rice and Sigrist [18] and Baskaran [19] proposed that Sr_2RuO_4 formed a state analogous to the "equal-spin pairing" (ESP) state of ^3He discussed by Anderson and Morel [20]. Within an EPS scenario the superconducting state is a superposition of the two possible ($S = 1$) parallel paired states. In an applied magnetic field, the Cooper pairs are simultaneously responsible for the superconductivity and the induced magnetization through the differing occupation of the $S_z = \pm 1$ states, leading to the same susceptibility as the normal state. Duffy and co-workers [15] used the polarized-beam measurements of the induced form factor to determine $\chi(T)$ for Sr_2RuO_4. Figure 18.5(c) shows $\chi(T)$ for Sr_2RuO_4 measured through the superconducting transition. On entering the superconducting state we find that there is no change in χ, supporting the notion that Sr_2RuO_4 has an EPS-type triplet paired state. Similar polarized-neutron experiments have been performed on the heavy-fermion superconductors UPt_3 and UBe_{13} [21] which also show no reduction in the spin susceptibility on entering the superconducting state suggesting that they too have odd-parity pairing. For completeness we note that complementary information is obtained from NMR Knight shift measurements which show no change in the spin susceptibility at T_c in UPt_3 or UBe_{13}.

Finally we discuss the case of the high-T_c superconductors which are generally believed to be d-wave with singlet spin pairing. In common with conventional s-wave singlet pairing, we expect $\chi(T)$ to follow a Yosida function. The complication with high-T_c materials is that the spin susceptibility can begin to drop at high temperatures because of the formation of the so-called "pseudo-gap". This is particularly the case for underdoped compositions. Nevertheless, the slightly overdoped $YBa_2Cu_3O_7$ sample studied by Boucherle and co-workers [16] shows a drop in $\chi(T)$ which is consistent with a Yosida function.

18.4 Magnetic Excitations in Metals and Weakly Coupled Superconductors

In this review we will discuss the magnetic excitations in conventional and more exotic superconductors such as the high-T_c materials. The purpose of this section is to introduce the theory of magnetic excitations in single-band weakly-correlated metals. We will consider the excitations of $S = \frac{1}{2}$ insulating antiferromagnets in Sect. 18.6.2. Both limits are relevant to the high-T_c materials where the parent compounds are antiferromagnets and overdoped materials show many of the properties of Fermi liquids. Of course, even in their normal state, optimally doped cuprates show anomalous properties and the electronic state is not well described by conventional one-electron theory. Nevertheless the next sections should provide some basis to qualitatively understand the cuprates and other superconductors.

18.4.1 Paramagnetic Metals

In metals with weak magnetic correlations such as sodium, the magnetic excitations are "electron-hole pairs". A magnetic field varying in space and time (such as that produced by a neutron moving through a sample) can excite electrons from below the Fermi energy to unoccupied states above it. This process creates an "electron-hole pair" or "Stoner" excitation. In the case of the sodium, the large band width and weak exchange interaction mean that the electron-hole excitations are spread out over several eV in energy. This makes them difficult to observe directly.

The response of a paramagnetic metal due to the electron-hole excitations is given by the so-called Lindhard functions [7,8,22],

$$\chi_0(\mathbf{q},\omega) = \frac{g^2\mu_B^2}{4N}\sum_k \frac{f(\epsilon_{\mathbf{k}+\mathbf{q}})-f(\epsilon_{\mathbf{k}})}{\hbar\omega-(\epsilon_{\mathbf{k}+\mathbf{q}}-\epsilon_{\mathbf{k}})+i\Gamma}, \quad (18.13)$$

$\epsilon(\mathbf{k})$ is the band energy, $f(E)$ the Fermi–Dirac function and there is no sum over spin states.

The Lindhard response function describes a continuum of electron-hole excitations. For small energy transfers the response can be related to electron states near the Fermi energy (cf. Eq. 18.13) and may therefore reflect the Fermi surface. The response can be strong in materials with highly nested Fermi surfaces or with an exchange interaction which favours (but does not result in) magnetic order. An example of the former is the $Cr_{1-x}V_x$ alloy system [23–25]. Chromium is an incommensurate antiferromagnet with a strongly nested Fermi surface. The antiferromagnetism may be destroyed by alloying with V, producing a highly-nested paramagnetic metal. Fawcett et al. [24] demonstrated the existence of strong overdamped excitations at incommensurate positions in the paramagnet $Cr_{0.95}V_{0.05}$, which have similarities to the low-frequency excitations in $La_{2-x}Sr_xCuO_4$. These "spin-density-wave paramagnons", as they were named, exist up to at least 400 meV [25].

18.4.2 The Superconducting Metal

One of the achievements of the BCS theory of superconductivity was to demonstrate that when a metal becomes superconducting, correlations are introduced between the motion of the electrons. In a "conventional" singlet s-wave superconductor, Cooper pairs are formed such that the states $|\mathbf{k}\uparrow\rangle$ and $|-\mathbf{k}\downarrow\rangle$ are simultaneously occupied. Since the spin susceptibility also measures correlations between electrons, it is not surprising that the formation of the superconducting state can dramatically effect $\chi''(\mathbf{q},\omega)$. In conventional superconductors, the small value of the superconducting gap has precluded direct measurements of changes in $\chi''(\mathbf{q},\omega)$ using inelastic neutron scattering. However, measurements of the spin susceptibility in the superconducting state have been made by other means. For example, nuclear magnetic resonance (NMR) measured the low-frequency local susceptibility and yielded the celebrated "Hebel–Slichter peak" [26]. BCS theory makes definite predictions for the magnetic response [27,28] (at least in a simple metal). We must effectively evaluate a Lindhard function for the superconductor. In the superconducting state, the quasiparticle excitation spectrum is given by,

$$E_{\mathbf{k}}^2 = \left[\epsilon_{\mathbf{k}} - \mu\right]^2 + |\Delta_{\mathbf{k}}|^2 , \qquad (18.14)$$

where $\epsilon(\mathbf{k})$ is the electron energy in the normal state and μ is the chemical potential. For a conventional s-wave superconductor the gap, $\Delta(\mathbf{k})$ is non-zero and $\mathbf{k}$-independent. The existence of the BCS pairing means that the $|\mathbf{k}\uparrow>$ and $|-\mathbf{k}\downarrow>$ states are coupled, this introduces "coherence factors" into the Lindhard expression. In the superconducting state, the magnetic response is given by [27,28],

$$\begin{aligned}\chi_0(\mathbf{q},\omega) &= \frac{g^2\mu_B^2}{4N} \\ &\times \left[\sum_k \frac{1}{2}\left(1+\frac{\xi_{\mathbf{k+q}}\xi_{\mathbf{k}}+\Delta_{\mathbf{k+q}}\Delta_{\mathbf{k}}}{E_{\mathbf{k+q}}E_{\mathbf{k}}}\right)\right. \\ &\times \frac{f(E_{\mathbf{k+q}})-f(E_{\mathbf{k}})}{\hbar\omega-(E_{\mathbf{k+q}}-E_{\mathbf{k}})+i\Gamma} \\ &+ \frac{1}{4}\left(1-\frac{\xi_{\mathbf{k+q}}\xi_{\mathbf{k}}+\Delta_{\mathbf{k+q}}\Delta_{\mathbf{k}}}{E_{\mathbf{k+q}}E_{\mathbf{k}}}\right) \\ &\times \frac{1-f(E_{\mathbf{k+q}})-f(E_{\mathbf{k}})}{\hbar\omega+(E_{\mathbf{k+q}}+E_{\mathbf{k}})+i\Gamma} \\ &+ \frac{1}{4}\left(1-\frac{\xi_{\mathbf{k+q}}\xi_{\mathbf{k}}+\Delta_{\mathbf{k+q}}\Delta_{\mathbf{k}}}{E_{\mathbf{k+q}}E_{\mathbf{k}}}\right) \\ &\times \left.\frac{f(E_{\mathbf{k+q}})+f(E_{\mathbf{k}})-1}{\hbar\omega-(E_{\mathbf{k+q}}+E_{\mathbf{k}})+i\Gamma}\right], \qquad (18.15)\end{aligned}$$

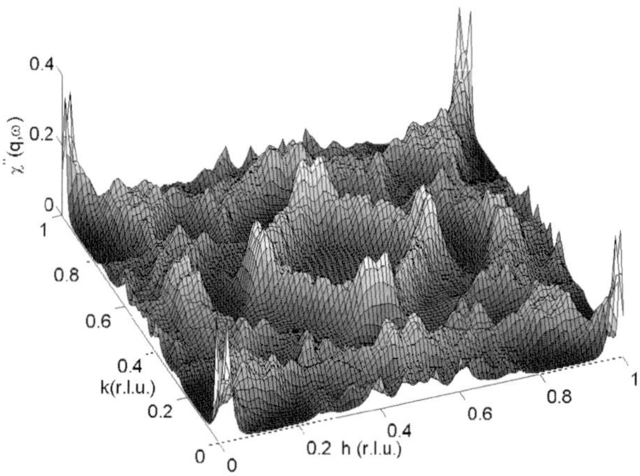

Fig. 18.6. Calculation of the bare Lindhard susceptibility $\chi_0''(\mathbf{q},\omega)$ with $\hbar\omega = 35$ meV for a 2D one-band tight-binding model with parameters used by Norman [35] and a d-wave superconducting gap $\Delta_0 = 35$ meV

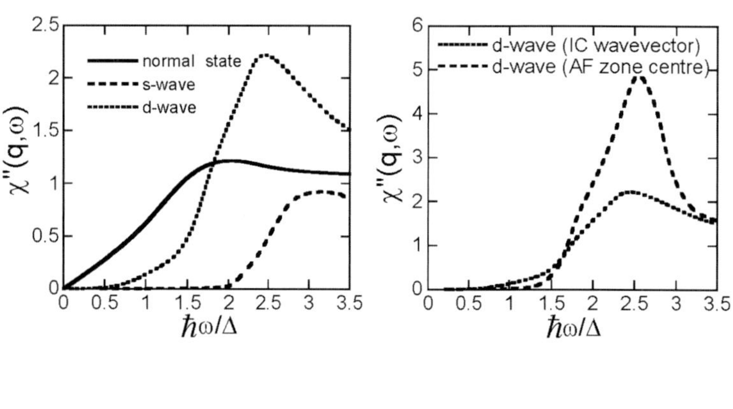

Fig. 18.7. The effect of superconductivity on the magnetic response $\chi_0''(\mathbf{q},\omega)$ of a simple metal illustrated through the calculation of the Lindhard functions with the BCS coherence factors (Eq. 18.15). Parameters as in Fig. 18.6. An s-wave gap leads to a suppression of the response below 2Δ, while a d-wave gap can lead to an enhancement of the response for energies near 2Δ. **(left panel)** $q = (1/2, 1/2)$, **(right panel)** $q = (1/2, 1/2)$ and $(1/2, 1/2+\delta)$

where $\xi_k = \epsilon_k - \mu$. The three terms correspond to quasiparticle scattering, quasiparticle pair annihilation and quasiparticle pair creation. In the limit of zero temperature, only the final term contributes to the response. Unfortunately, Eq. 18.15 neglects electron–electron interactions which are undoubtedly very important in the cuprates and many other types of superconductor. It implicity assumes that there are no exchange interactions. More complete models have been investigated in the literature [28–38]. Despite the approximations inherent in Eq. 18.15, it may still give some qualitative indications of how the magnetic response changes on entering the superconducting state. Figures 18.6 and 18.7 show a model calculation of $\chi''(\mathbf{q}, \omega)$ based on a 2D one-band tight-binding model with parameters [35] reputed to describe the cuprates. The model qualitatively reproduces the four incommensurate peaks located around the (1/2,1/2) position which are observed in $La_{2-x}Sr_xCuO_4$ and $YBa_2Cu_3O_{6+x}$. The effect of the superconductivity on the magnetic response is illustrated in Fig. 18.7. The panels shows the energy dependence of $\chi''(\mathbf{q}, \omega)$ at the peak positions in Fig. 18.6 and at (1/2,1/2). For an s-wave gap, there is a complete suppression of $\chi''(\mathbf{q}, \omega)$ for $\hbar\omega < 2\Delta$. In the d-wave singlet case, there is a $\mathbf{q}$-dependent suppression at lower energies and an increase in $\chi''(\mathbf{q}, \omega)$ for some wavevectors near $\hbar\omega \approx 2\Delta$.

18.5 Excitations and Superconductive Pairing

One of the remarkable successes of the BCS theory of superconductivity was the identification of the pairing mechanism. Within the BCS framework the exchange of virtual Bosonic excitations causes an attractive interaction between itinerant electrons which then pair up. In "conventional" superconductors such as Pb and Nb, the Bosonic excitations were identified as phonons and the pairing is in the s-wave channel. Neutron scattering and tunnelling measurements demonstrated the pairing boson in lead very directly. Superconductor-insulator-superconductor (SIS) tunnelling measurements allow the energy-dependent spectrum of the bosons which mediate the pairing to be extracted [41]. Figure 18.8(b) shows the experimentally determined $\alpha^2(\omega)F(\omega)$ for lead [41] ($\alpha^2(\omega)$ is a measure of the electron–boson coupling and $F(\omega)$ is the boson density of states). Figure 18.8(a) shows the phonon density of states determined from neutron scattering [40]. Comparing panels (a) and (b), one can immediately see that the total phonon density of states is reflected in $\alpha^2(\omega)F(\omega)$ identifying phonons as the mediating bosons. The neutron scattering measurements were even able to determine the contributions of the different phonon polarizations to the total DOS

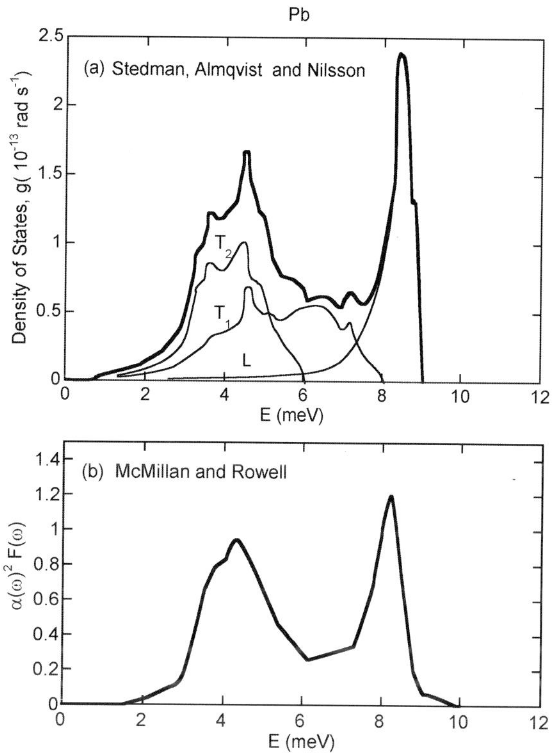

Fig. 18.8. (**a**) The density of phonon states determined from neutron scattering (after Stedman et al. [40]). In this experiment the phonon DOS was determined from the entire measured phonon dispersion for each phonon polarization. T1 and T2 are transverse modes, L is longitudinal. (**b**) The product of the electron–boson coupling and the boson DOS, $\alpha^2(\omega)F(\omega)$ determined from SIS tunnelling measurements (after McMillan and Rowell [41]). Note the similarity between the *total* phonon DOS and $\alpha^2(\omega)F(\omega)$, showing that both transverse and longitudinal phonons are involved in pairing

(see Fig. 18.8(a)). Since the peaks due to the longitudinal (L) and transverse (T) phonons occur in $\alpha^2(\omega)F(\omega)$, both polarizations must contribute to the superconductive pairing.

The antiferromagnetic ($T_N = 14.3$ K) heavy-fermion superconductor UPd_2Al_3 is an interesting compound because it shows a strong correspondence between its magnetic excitations and the tunnelling density of states. Dispersive crystal-field excitations (magnetic excitons) were observed by Petersen and co-workers [42] in INS experiments within the antiferromagnetic (AF) phase of UPd_2Al_3. Later higher-resolution experiments [43,44] revealed that a double-peak structure appears in the magnetic excitation spectrum below $T_c = 1.8$ K, with peak positions at $E \approx 0.35$ and 1.7 meV (see Fig. 18.9). The lower peak reduces with increasing temperature, and becomes a "quasielastic line" (that is, with spectral line-shape centred around $E = 0$). This dramatic redistribution of spectral weight as we pass through T_c suggests that the lower peak is associated with the itinerant heavy quasiparticles. The higher-energy peak shows considerable dispersion in wavevector on moving away from the from the antiferromagnetic zone centre, it has therefore been associated with a collective mode of the localized-$5f$ moment system due to the inter-site interactions. Thus the INS results demonstrate that a strong interaction exists between the localized and delocalized components of the $5f$-electron system. The tunnelling density of states of UPd_2Al_3 was measured by Jourdan and co-workers [45] using UPd_2Al_3-AlO_x-Pb tunnel junctions. Interestingly, strong-coupling features appear in the tunnelling DOS around 1.2 meV and a 2Δ peak appears near 0.3 meV. The neutron scattering and tunnelling together suggest that the bosons responsible for the superconductive pairing of the itinerant electrons in UPd_2Al_3 are the magnetic excitons. This scenario has been analyzed by Sato and co-workers [46] who were able to explain qualitative features of the magnetic excitation and the superconducting tunnelling spectra.

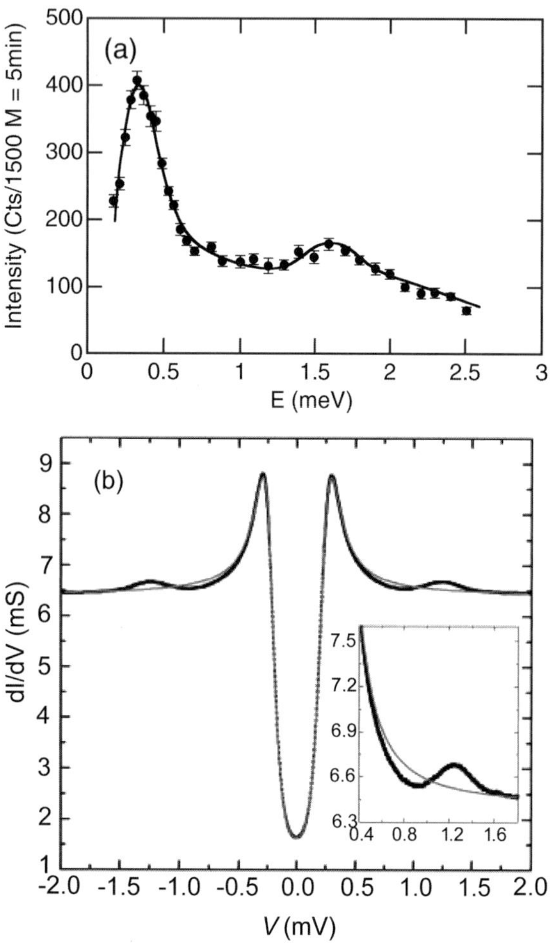

Fig. 18.9. (a) Magnetic excitations in UPd_2Al_3 measured by neutron scattering (after Bernhoeft et al. [44]). Data were collected near the antiferromagnetic ordering wavevector **Q** = (0,0,1/2) in the superconducting state (T=0.15 K). (b) Tunnelling DOS measured in a UPd_2Al_3-AlO_x-Pb tunnel junction (after Jourdan et al. [45])

18.6 High Temperature Superconductivity

High temperature superconductivity in CuO_2 layered cuprates was first discovered in $La_{2-x}Ba_xCuO_4$ [47]. Since this time, the phase diagram of this and other cuprates have since been explored in considerable detail (see, for example, the article by H.R. Ott in this series). A very schematic generic phase diagram is shown in Fig. 18.10. Soon after the original discovery of the high-T_c phenomenon, neutron scattering experiments performed by Vaknin et al. [48] demonstrated that the "undoped" parent compound La_2CuO_4 was antiferromagnetic. The

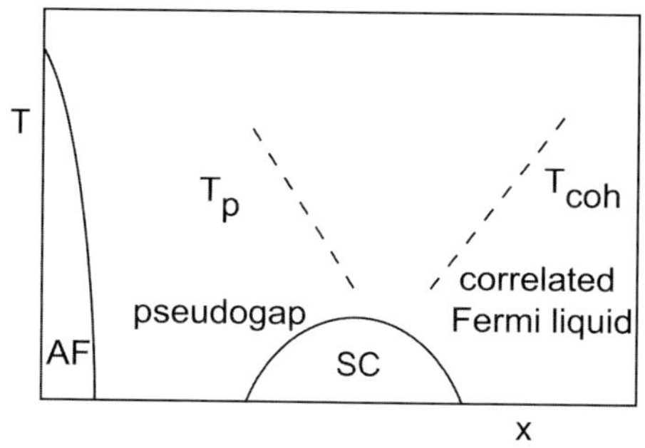

Fig. 18.10. The generic phase diagram of the cuprate superconductors

overdoped cuprates are strongly correlated (non-superconducting) metals which show a T^2 temperature dependence in their resistivity [49] and evidence of a Fermi surface [50]. The superconductivity occurs between these two regions and on phenological grounds should share some features with these other phases. In the next sections we review the magnetic excitations in the parent insulating antiferromagnets La_2CuO_4 and $YBa_2Cu_3O_6$. We will see in later sections that the magnetic excitations in the superconducting compositions show residual antiferromagnetic interactions. Thus, studies of the parent antiferromagnets afford an understanding of the underlying magnetic interactions.

For completeness we briefly mention the so-called pseudogap, a phenomenon which occurs in the underdoped region of the phase diagram. In the years following the discovery of high-T_c superconductors, it was found that the superconductivity had a large energy scale $2\Delta^*$ associated with it. Its importance was slowly recognized through the 1990s and it has become known as the "pseudogap" (see reviews by Timusk and Statt [51], and Tallon and Loram [52]). The pseudogap is often associated with a temperature scale T^* below which its effects are felt. The pseudogap was first observed in NMR experiments [53], although it is visible in many types of measurement (e.g. resistivity, infrared conductivity, Raman scattering and tunnelling data). The "pseudogap" and its associated temperature T^* are largest for underdoped compositions. Below T_c it merges with the superconducting gap, but it is visible far above T_c in many spectroscopic measurements.

18.6.1 Structures and Phase Diagrams

In this short review, we will concentrate on the two systems which have been studied most systematically by neutron scattering. Relatively large single crystals of $YBa_2Cu_3O_{6+x}$ and $La_{2-x}Sr_xCuO_4$ are available and this is the primary reason why more data is available for these systems.

Structure and Phase Diagram of $La_{2-x}Sr_xCuO_4$

The crystal structures observed in the $La_{2-x}Sr_xCuO_4$ system are shown in Fig. 18.11. At ambient temperature La_2CuO_4 is orthorhombic, due to a rotation of the octahedra centred on the copper atoms. La_2CuO_4 becomes tetragonal for temperatures above 530 K [48, 54] and with doping $La_{2-x}Sr_xCuO_4$ becomes tetragonal for $x > 0.21$ [55, 56]. For convenience and to allow comparison between different phases, we will generally use the tetragonal crys-

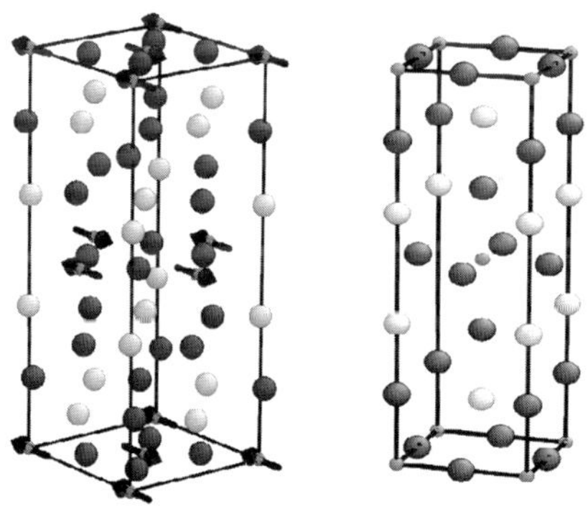

Fig. 18.11. *Left*: The low temperature orthorhombic (LTO) phase of La_2CuO_4. The LTO phase is stable at low temperature for Sr doping $x < 0.21$. The LTO phase is described by the *Cmca* (or isomorphic *Bmab*) space group [56]. In the antiferromagnetic state, the ordered moments are stacked in ferromagnetic sheets along [100] with the moments pointing along the [001] direction [48]. *Right*: At high temperatures and larger dopings the $La_{2-x}Sr_xCuO_4$ system forms the high temperature tetragonal phase (HTT). The HTT phase is described by the I4/*mmm* space group. To allow comparison between different HTC systems, we label reciprocal space using the tetragonal I4/*mmm* convention. In this convention $a = b \approx 3.8$ Å, $c \approx 13.2$ Å and [100] is parallel to the Cu-O bond within the CuO_2 planes

tallographic axes to describe the high-T_c cuprates and their parents. The tetragonal (I4/mmm) crystal structure of $La_{2-x}Sr_xCuO_4$ is shown in the right hand panel of Fig. 18.11. The electronic properties of the $La_{2-x}(Ba,Sr)_xCuO_4$ system are extremely anisotropic. Many properties are most easily understood by considering the structure as being made up of a series of square CuO_2 planes stacked along the **c** direction. There is weak overlap of the electronic orbitals in the **c** direction. The copper atoms are in the Cu^{2+} state with spin $S = \frac{1}{2}$. There is a strong exchange anisotropy, the coupling between copper spins in the same CuO_2 plane ($J_\parallel$) is several orders of magnitude stronger than the coupling between spins in neighbouring planes ($J_\perp$).

As mentioned above, experiments performed by Vaknin et al. [48] at Brookhaven National Laboratory showed that La_2CuO_4 is an antiferromagnet (see Fig. 18.11). The ordered moments point along the tetragonal [110] or orthorhombic [001] direction [48]. A number of effects are known to change the value from the 1 μ_B corresponding to $S = \frac{1}{2}$: (i) the g factor of Cu^{2+} [48]; (ii) covalency effects due to hybridisation of the copper and oxygen atoms in the planes [59,60]; (iii) variations due to oxygen stoichiometry [61]; (iv) reduction due to the quantum

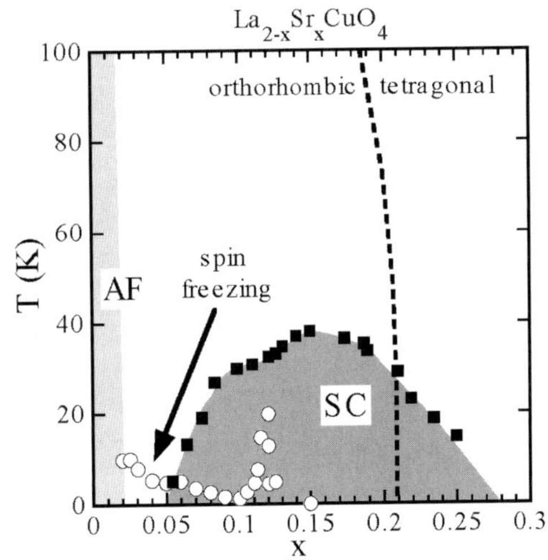

Fig. 18.12. The phase diagram of the $La_{2-x}Sr_xCuO_4$ system showing the antiferromagnetic phase, superconductivity, spin freezing and the orthorombic/tetragonal phase transition. Data collected from [57,58]

fluctuations expected in a low-dimensional antiferromagnet with small spin (see Sect. 18.6.2). Experiments [48,61,62] have yielded ordered moments in the range 0.3–0.6 μ_B.

LDA-bandstructure calculations [63] predict that La_2CuO_4 is a metal. In fact, correlation effects mean that it is an antiferromagnetic Mott insulator. The addition of barium or strontium creates a plethora of behaviour illustrated is Fig. 18.12. The effect of the Sr or Ba is to add holes, moving the system away from half filling and resulting in a transition to a metallic and superconducting state for $x \approx 0.05$. It is well known that both the normal metal and the superconductor are anomalous in many ways. Although the normal metal is characterized by an increasing resistivity with temperature, the resistivity is several orders of magnitude larger than for good metals. Further, it shows a linear variation with temperature signalling an extremely temperature dependent scattering mechanism.

Doping not only affects the electrical properties, but also has dramatic effects on the magnetic ground state. The long-range antiferromagnetic order is rapidly destroyed with increasing x. For $x > 0.02$, the antiferromagnetism is replaced by short range antiferromagnetic correlations at low temperatures. This state appears to be formed by "spin freezing" and has some analogies to the formation of a spin glass. This "antiferromagnetic spin glass" has been characterized both by local probes: μSR [64], NMR [65], Mössbauer [66] and by neutron scattering [67,68]. In the lightly doped regime, neutron scattering measurements on $La_{1.95}Ba_{0.05}CuO_4$ [67] have studied the dynamics of the spin-freezing process. Matsuda et al. [69] have made a systematic study of the doping dependence of the freezing wavevector in the $La_{2-x}Sr_xCuO_4$ system (see Fig. 18.25). At higher x, local probes indicate an enhanced island of static order in the $La_{2-x}Sr_xCuO_4$ phase diagram near $x = \frac{1}{8}$. Suzuki et al. [68] observed the corresponding incommensurate magnetic peaks. The elastic incommensurate peaks are strongly enhanced by an applied magnetic field [70–72]. Incommensurate magnetic superlattice peaks have also been observed in $La_{1.6-x}Nd_{0.4}Sr_xCuO_4$ [73], although in this case the material first undergoes a transition to a

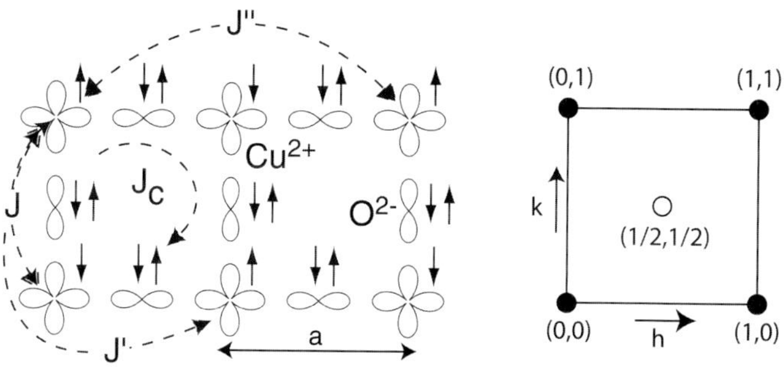

Fig. 18.13. *Left*: The CuO_2 plane showing the atomic orbitals (Cu $3d_{x^2-y^2}$ and O $2p_{x,y}$) involved in the magnetic interactions. J, J' and J'' are the first-, second- and third-nearest-neighbour exchanges and J_c is the cyclic interaction which couples spins at the corners of a square plaquette. *Arrows* indicate the spins of the valence electrons involved in the exchange. *Right*: Reciprocal space for a 2D square antiferromagnet such as a single CuO_2 plane in La_2CuO_4. *Solid circles* denote reciprocal lattice points. The *open circle* denotes the antiferromagnetic ordering wavevector

low-temperature tetragonal phase at higher temperatures.

Structure and Phase Diagram of $YBa_2Cu_3O_{6+x}$

The $YBa_2Cu_3O_{6+x}$ system is probably the most widely studied high-T_c superconductor, primarily because of its high-T_c and the availability of large single crystals. Figure 18.14 shows the unit cell of the antiferromagnetically ordered parent compound $YBa_2Cu_3O_6$. In common with many other high-T_c systems, the $YBa_2Cu_3O_{6+x}$ system has pairs of CuO_2 planes which have significant electronic coupling between the layers and therefore splitting in electronic bands [74]. The coupling between neighbouring bilayer blocks is usually neglected when modelling the magnetic excitations and the system treated as a two layer 2D system. Because there is a mirror (symmetry) plane between the layers, eigenstates, including magnetic excitations, will have either odd or even symmetry with respect to the mirror plane leading to two distinct types of magnetic excitation and therefore, for example, two dispersion branches. $YBa_2Cu_3O_{6+x}$ is hole doped by adding oxygen, the addition oxygen atoms

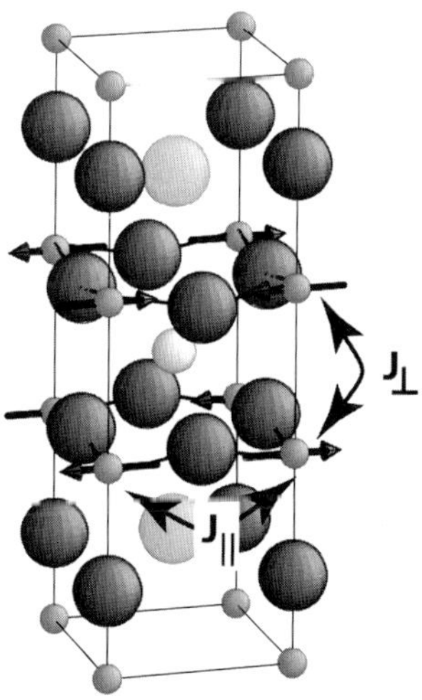

Fig. 18.14. The tetragonal unit cell of antiferromagnetic $YBa_2Cu_3O_6$. The two strongest exchange couplings are shown

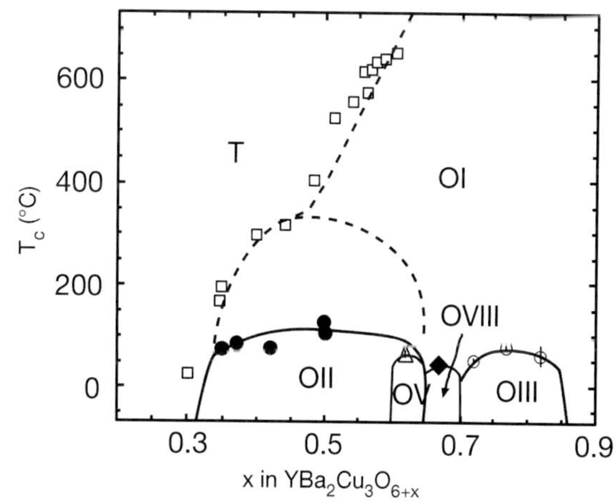

Fig. 18.15. Oxygen ordering phase diagram for $YBa_2Cu_3O_{6+x}$. Tetragonal phase (T) and various chain ordered phases (OI–OVIII) with different repeat distances along the a-axis. (After Zimmermann et al. [76])

occupy sites between the Cu atoms outside the CuO_2 planes. For light doping this occurs randomly and the structure remains tetragonal. For oxygen doping beyond $x \approx 0.3$, the additional oxygen atoms order into various chain structures (see Fig. 18.15). The ortho-II (underdoped) and ortho-I structure (optimally doped) structures are shown in Fig. 18.16. The chain ordered compositions have orthorhombic symmetry and crystals are typically highly twinned as grown. Crystals may be de-twinned into a single orthorhombic domain by annealing under uniaxial pressure. In common with $La_{2-x}Sr_xCuO_4$, $YBa_2Cu_3O_{6+x}$ also shows spin freezing behaviour [75] which persists into the superconducting state (see Fig. 18.17).

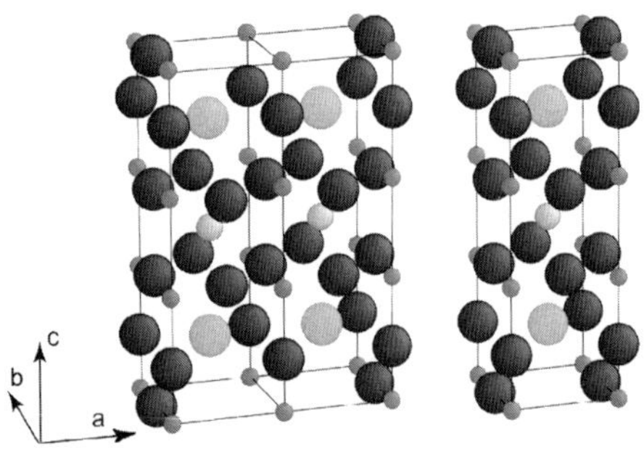

Fig. 18.16. The ortho-II (*left*) and ortho-I (*right*) structures of $YBa_2Cu_3O_{6+x}$. As oxygen is added to $YBa_2Cu_3O_6$, Cu-O chains are formed

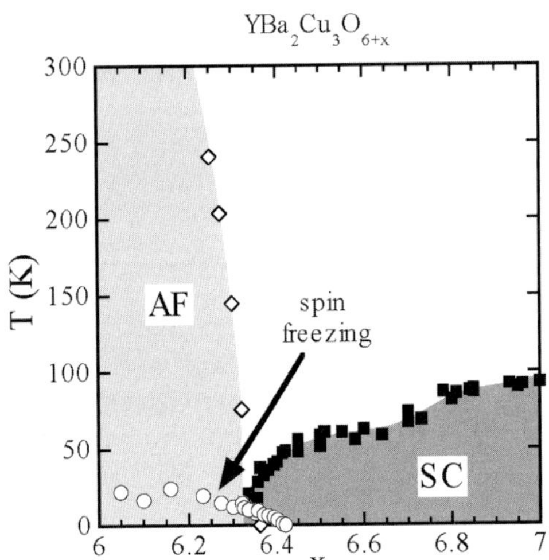

Fig. 18.17. The antiferromagnetic and superconducting phases of $YBa_2Cu_3O_{6+x}$. Data collected from [75,77–84]

18.6.2 Magnetic Excitations in the Parent Compounds of the High-T_c Superconductors

The crystal and electronic structure of La_2CuO_4 make it a good physical realization of a 2D $S = \frac{1}{2}$ Heisenberg antiferromagnet. We will start by reviewing the nature of the magnetic excitations for such a model system.

The 2D $S = \frac{1}{2}$ Square Lattice Antiferromagnet

The properties of Heisenberg antiferromagnets when quantum mechanics is taken into account have been investigated for much of this century. It is well known that in 1D, quantum fluctuations destabilize the Néel state completely. In higher dimensions, quantum fluctuations lead to significant corrections to the ground state energy [85,86] and spin-wave excitations [87]. However, the 2D $S = \frac{1}{2}$ square-lattice antiferromagnet is thought to be ordered at zero temperature. In practice, interplanar coupling leads to ordering at finite temperatures in physical realizations such as La_2CuO_4. The high anisotropy and large exchange constant make La_2CuO_4 a good realization of a 2D $S = \frac{1}{2}$ antiferromagnet, particularly when we consider the excitations on a wide energy scale comparable with $2J$. As we shall see below, there are deviations from ideal behaviour.

The magnetic scattering can be qualitatively separated into three contributions: (i) Bragg scattering;

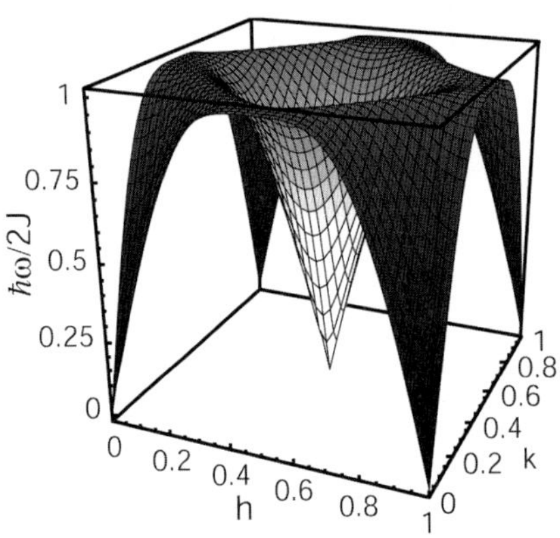

Fig. 18.18. Spin wave dispersion for a 2D $S = \frac{1}{2}$ antiferromagnet such as La_2CuO_4. The *shading* shows the intensity of the spin waves, *lighter shading* represents higher magnetic response

Table 18.1. Contributions to $\langle M^2 \rangle$ in a ferromagnet and $S = \frac{1}{2}$ 2D antiferromagnet. The table illustrates the total moment sum rule

	$\langle M^2 \rangle (\mu_B^2)$	
	Ferromagnet	2-D $S = \frac{1}{2}$ antiferromagnet
Bragg peaks	$g^2 S^2$	$(0.061gS)^2 = 0.37$
Spin waves	$g^2 S$	1.7
Multi-magnon	0	0.93
Total	$g^2 S(S+1)$	3

(ii) one-magnon excitations and (iii) multimagnon excitations. The sum of these contributions is governed by the moment sum rule (Eq. 18.12). Table 18.1 summarizes the spectral weights of the these contributions as determined from a $1/S$ expansion theory [88].

Linear Spin-Wave Theory for Nearest Neighbour Interactions

The coupling between neighbouring spins in a square-lattice antiferromagnet is described by the Heisenberg Hamiltonian,

$$H = \sum_{ij} J\, \mathbf{S}_i \cdot \mathbf{S}_j \,. \qquad (18.16)$$

Conventional spin-wave theory of the Holstein–Primakoff type, in the classical large-S limit, yields the transverse dynamic susceptibility

$$\chi''_{\perp}(\mathbf{Q}, \hbar\omega) = Z_d(\omega) \frac{\pi}{2} g^2 \mu_B^2 S \left(\frac{1-\gamma(\mathbf{Q})}{1+\gamma(\mathbf{Q})} \right)^{1/2} \times \delta\left(\hbar\omega \pm \hbar\omega(\mathbf{Q})\right) , \qquad (18.17)$$

where,

$$\hbar\omega(\mathbf{Q}) = 2Z_c J \left[1 - \gamma^2(\mathbf{Q})\right]^{1/2} , \qquad (18.18)$$

and $\gamma(\mathbf{Q}) = \frac{1}{2}[\cos(2\pi h) + \cos(2\pi k)]$. We have included a "quantum renormalisation" of the overall spin-wave amplitude $Z_d(\omega)$ (if Z_c is constant then $Z_d = Z_\chi / Z_c$). For the conventional linear spin-wave model applicable for large S, $Z_d(\omega) = 1$. In order to make comparisons with neutron scattering and nuclear resonance experiments, it is useful to calculate the "local" or "wavevector-integrated" energy-dependent susceptibility $\chi''(\omega)$ defined by Eq. 18.7. In the low frequency limit Eq. 18.17 yields,

$$\chi''_{\perp}(\omega) = \frac{Z_d(\omega) g^2 \mu_B^2 S}{2 Z_c J} \,. \qquad (18.19)$$

Magnetic Excitations in La_2CuO_4

Early measurements were performed on a conventional reactor source using an "energy integrating" technique [89,90] in which the two-dimensionality of the magnetic interactions was exploited and the magnetic response was integrated in energy up to about 30 meV. This method allowed the weak signal to be seen in relatively small single crystals. The spin fluctuations were shown to have a wide energy scale ($>$ 10 meV).

Subsequent experiments were performed using epithermal neutrons produced by the "hot" source at the Institute Laue–Langevin. The use of the hot source allowed the spin-wave dispersion to be followed up to 140 meV and the two counter-propagating branches to be resolved [91,92]. At T=5 K, the spin-wave velocity was found to be $v_s = 0.85 \pm 0.03$ eV Å. Using the low-frequency limit of Eq. 18.18, $\hbar v_s = Z_c \sqrt{8} S J a$ and assuming $Z_c = 1.18$ [88,93] yields a Heisenberg coupling constant $J = 136 \pm 5$ meV. In spite of the use of the hot source and 0.1 kg of aligned single crystal, the highest energy which could be achieved by reactor-based three-axis methods was about 140 meV. Spallation sources offer low background and high incident energies. The use of chopper spectrometers at the ISIS spallation source allowed the zone bound-

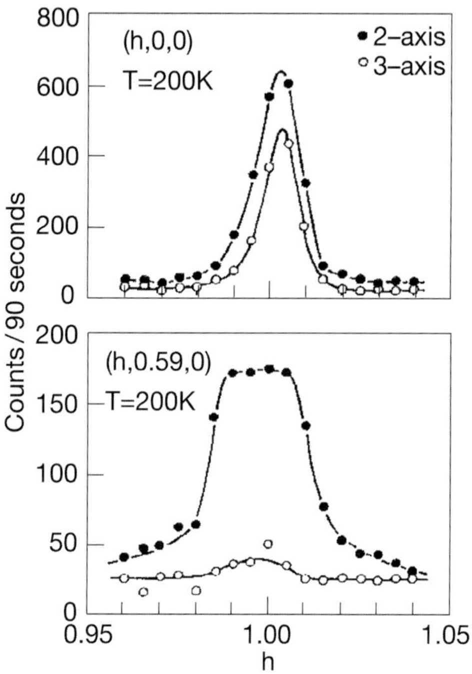

Fig. 18.19. Spin wave excitations in La_2CuO_4. *Open circles* are scans with an energy window of about 1 meV around zero. *Closed circles* correspond to an energy integration of about 10 meV. The *lower figure* demonstrates the existence of magnetic excitations with an energy scale greater than the integration window. After Shirane et al. [90]

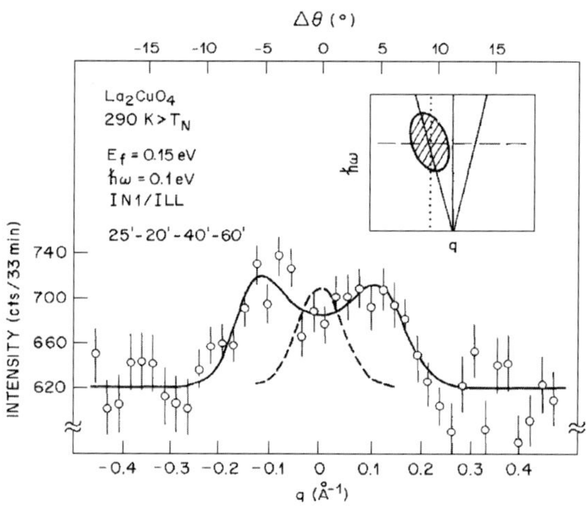

Fig. 18.20. Hot source measurements of spin waves in La_2CuO_4. The figure shows two resolved counter-propagating spin waves. After Aeppli et al. [91]

ary single-magnon excitation to be measured [94]. The value of J determined from this early ISIS experiment was $J = 132 \pm 4$ meV. Improvements in the ISIS instrumentation led to a considerable increase in the wavevector resolution of which allowed a subsequent measurement to resolve the detailed dispersion throughout the Brillouin zone.

The linear spin-wave theory for a 2D square lattice antiferromagnet (2DAF) with nearest neighbour interactions described in the previous section predicts that there is no dispersion in the magnon excitations around the antiferromagnetic Brillouin zone boundary, for example, along the line connecting $\mathbf{Q} = (1/2,0)$ and $(1/4,1/4)$ or $(3/4,1/4)$. This lack of dispersion is illustrated in Fig. 18.18. The measured magnon dispersion in La_2CuO_4 is shown in Fig. 18.21. In fact the magnon in La_2CuO_4 shows a 7.5 % *increase* in energy between $(3/4,1/4)$ [or $(1/4,1/4)$] and $(1/2,0)$ [95]. A deviation of 6 % in the opposite sense is observed in the another model $S = \frac{1}{2}$ 2D antiferromagnet, $Cu(DCOO)_2 \cdot 4H_20$ [96]. Two possible explanations of the observed deviations have been proposed. The first is that low-order spin wave theory fails to account for the high energy dynamics of the Heisenberg model because of the quantum fluctuations inherently present in the ground state of the 2D Heisenberg model [97–100]. The second is that a nearest-neighbour Heisenberg Hamiltonian does not describe La_2CuO_4 and higher order exchange terms such as cyclic exchange are required to describe La_2CuO_4. Monte Carlo simulations of a $S = \frac{1}{2}$ 2D Heisenberg model [98,99] suggest a 10 % *decrease* in the magnon energy between $(1/4,1/4)$ and $(1/2,0)$, the opposite to experiment. Thus the observed magnon dispersion must be largely due to longer range exchange interactions.

The exchange coupling constant J has been discussed widely in the literature (see, for example [101]). Our understanding is based on Anderson's theory of superexchange [102] in which $J \approx 4t^2/U$. The overlap of the copper d orbitals and the oxygen p orbitals is characterized by the hopping parameter t and U represents the on-site interaction (in the oxygen orbital). Fig. 18.13 illustrates the valence orbitals for a Cu_4O_4 plaquette in La_2CuO_4. The large orbital hybridization in the CuO_2 planes provides an

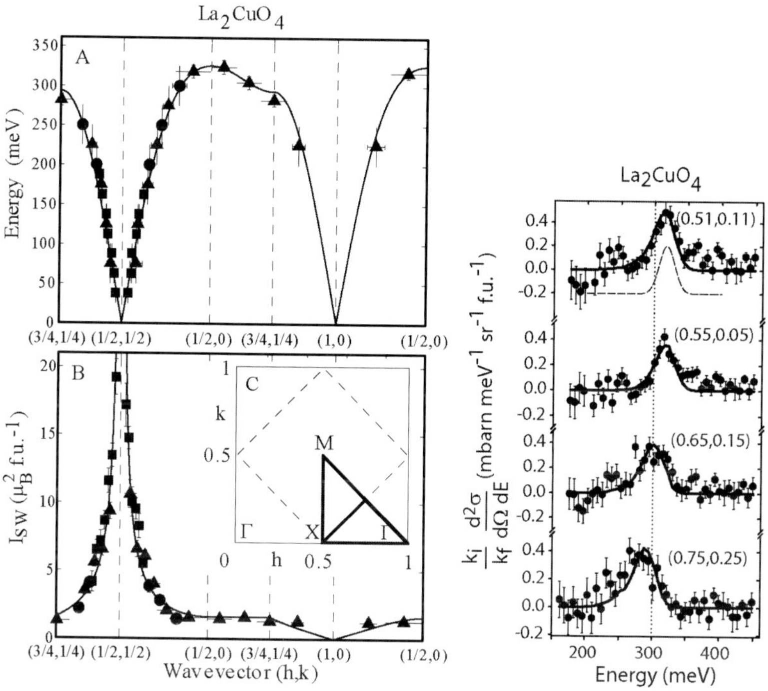

Fig. 18.21. *Left*: Spin wave dispersion of La_2CuO_4. *Right*: Zone boundary spin waves (after [95])

Table 18.2. Experimentally determined exchange constants for antiferromagnetic La_2CuO_4 and related compounds. J has been corrected for the effect of quantum fluctuations which are expected to renormalise the bare exchange constants $J^* = Z_c J$, with $Z_c \approx 1.18$ (see Sect. 18.6.2).

Composition	Ref.	T(K)	$J_\parallel$ (meV)	J_c (meV)
La_2CuO_4	Aeppli et al. [91]	290	136 ± 5	
La_2CuO_4	Hayden et al. [94]	296	132 ± 4	
La_2CuO_4	Coldea et al. [95]	295	138 ± 4	38 ± 8
La_2CuO_4	Coldea et al. [95]	10	146 ± 4	61 ± 8
Nd_2CuO_4	Bourges et al. [103]	17	155 ± 3	
Pr_2CuO_4	Bourges et al. [103]	18	121 ± 3	

exchange path to include all four spins at the corners of elementary Cu_4O_4 plaquettes. Second nearest neighbour Heisenberg exchange paths are also possible. Measurements of the magnon dispersion cannot distinguish between next-nearest neighbour two-spin exchange and cyclic exchange terms of the form [95]

$$\mathcal{H} = J_c \sum_{i,j,k,l} \{(\mathbf{S}_i \cdot \mathbf{S}_j)(\mathbf{S}_k \cdot \mathbf{S}_l) + (\mathbf{S}_i \cdot \mathbf{S}_l)(\mathbf{S}_k \cdot \mathbf{S}_j) - (\mathbf{S}_i \cdot \mathbf{S}_k)(\mathbf{S}_j \cdot \mathbf{S}_l)\} \,, \quad (18.20)$$

where the spins are labelled clockwise around a Cu_4O_4 square plaquette.

A one-band Hubbard model provides a natural understanding of the observed spin wave dispersion for La_2CuO_4 in terms of the Hubbard parameters t and U. If hopping is included up to order t^4 (i.e. four hops), a fit to the observed dispersion yields the values t =0.33 eV and U =2.9 eV [95]. The fact that higher order exchange terms are required to describe the dispersion relation in La_2CuO_4 suggests that longer range exchange interactions may also be important in doped and superconducting compositions.

Multi-Magnon Excitations

A Heisenberg ferromagnet has two interesting properties not shared by the Heisenberg antiferromagnet. The fully aligned ferromagnetic state is the ground state (and an eigenstate) of the Hamiltonian. In addition, spin wave states are also eigenstates of the Hamiltonian. In contrast, the Neel state (in which the moments on each site do not fluctuate with time) is not the ground state an antiferromagnetic Heisenberg Hamiltonian. For small S, quantum fluctuations [88] become important. Quantum fluctuations lead to a number of effects: (i) a reduction of the ordered moment; (ii) renormalisation of the spin-wave energies and intensities; (iii) the existence of multi-magnon excitations. Thus, in addition to the single-magnon modes, two-magnon excitations give rise to a continuum, as a function of energy, in the longitudinal response $\chi''_{zz}(\mathbf{q},\omega)$, and three-magnon excitations results in side bands (to the one-magnon excitations) in the transverse response $\chi''_{\perp}(\mathbf{q},\omega)$. Figure 18.22 shows a calculation of the two-magnon response for a 2D $S=\frac{1}{2}$ square lattice antiferromagnet [104,105].

Quantum effects lead to the renormalisation of the (one-magnon) spin-wave dispersion, the perpendicular susceptibility, the spin-stiffness constant, and the sub-lattice magnetization with respect to classical (large-S) spin-wave theory. These renormalisation factors have been estimated by a number of techniques. For the purpose of comparing theory and experiment in the present review, we include the renormalisation of the overall energy scale in the exchange constant, $J^* = Z_c J$, where J is the "bare" exchange constant occurring in Eq. 18.16. We have previously defined the renormalisation of the amplitude of the magnetic response in Eq. 18.17. We note that, in principle, these factors are energy dependent. The value of $Z_d(\omega)$ *can* be obtained from neutron scattering, if measurements are placed on an absolute intensity scale. However, Z_c *cannot* be measured directly from inelastic neutron scattering and must be estimated from theory. In the case of the $S=\frac{1}{2}$ square-lattice antiferromagnet, Singh [88] has estimated $Z_c = 1.18$ and $Z_d(\omega=0)=0.61$.

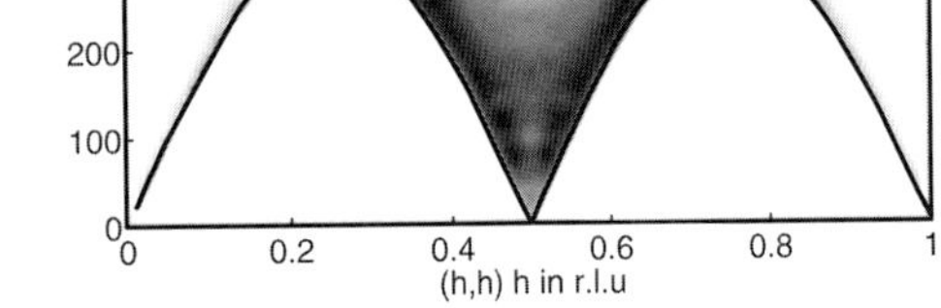

Fig. 18.22. Two-magnon excitations in a $S=\frac{1}{2}$ square lattice antiferromagnet. Calculation details are given in [104] and [105]

The 2D $S=\frac{1}{2}$ Antiferromagnet at Finite Temperatures

Since the 1950s, it has been widely believed that the 2D $S=\frac{1}{2}$ square-lattice antiferromagnet has long range order only at $T=0$. Anderson [85], argued that even through the staggered magnetization may be reduced, the Néel state is still stable against quantum fluctuations at $T=0$. Anderson's assertion means that this model system is an example of a "quantum critical" phase transition i.e. one which occurs at zero temperature and is therefore dominated by quantum, rather than classical, fluctuations in the order parameter. It was not surprising that the discovery of the antiferromagnetism in La_2CuO_4 and related compounds rejuvenated theoretical [93,106–108] and experimental [89] interest in this problem. The literature devoted to the $S=\frac{1}{2}$ square-lattice antiferromagnet is now vast, indeed there are several reviews [107] on this subject.

One of the novel features of low dimensional antiferromagnets, in particular 2D antiferromagnets, is the strong build up of "critical" magnetic fluctuations at temperatures considerably above the ordering temperature. This behaviour is not at all confined to $S=\frac{1}{2}$ (see, for example, the review by De Jongh and Miedema [109]). Figure 18.23 illustrates the magnetic excitation spectrum for a 2D $S=\frac{1}{2}$ antiferromagnet at finite temperature. For small frequencies and wavevectors, propagating excitations are believed not to exist, instead the response is

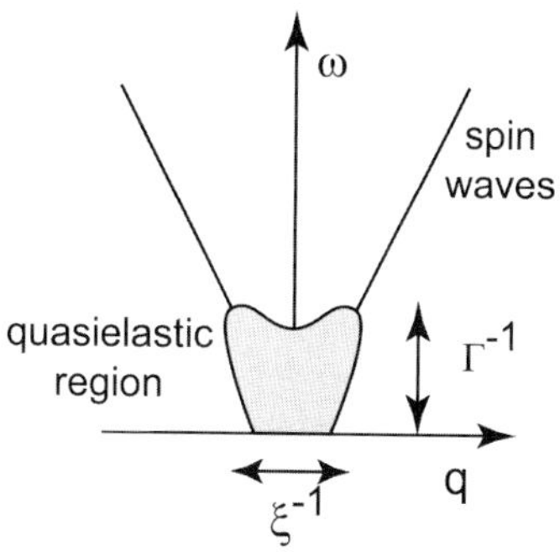

Fig. 18.23. Schematic representation of the magnetic response in a 2D antiferromagnet at finite temperature

dominated is a 'quasielastic peak' (i.e. $\chi''(\mathbf{q},\omega)/\omega$ is peaked at zero frequency). Since this peak dominates the response for wavevectors near the ordering wavevector, its width in wavevector is usually characterized by the inverse correlation length ξ^{-1} obtained from the width of $S(Q)$, the Fourier transform of the equal-time correlation function. In the regime $\hbar\omega \lesssim \hbar v_s \xi^{-1}$ (v_s is the spin-wave velocity) spin-wave interactions are the main mechanism for damping. Grempel [110] estimated the order-parameter relaxation rate for $q = 0$ using a coupled-mode calculation and found that $\hbar\Gamma_2 \simeq \hbar v_s \xi^{-1}(T/2\pi\rho_s)^{\frac{1}{2}}$. Chakravarty et al. [93, 106] have obtained definitive results for the low-frequency response. They showed that the long-wavelength, low-energy properties were well described by a mapping to a *classical* two-dimensional Heisenberg magnet. The effects of quantum fluctuations are absorbed into the coupling constants. Using this approach and Monte Carlo simulations they found that the temperature dependence of the correlation length ξ is,

$$\xi = C_\xi a \exp\left(2\pi\rho_s/k_B T\right) . \tag{18.21}$$

At energies larger than a few $\hbar v_s \xi^{-1}$, propagating spin waves should exist and the response is believed to be essentially as for $T = 0$, except that we must average the direction of the antiferromagnetic moment. The high frequency magnetic excitations see the moment in the "quasielastic peak" as frozen domains of size ξ.

Magnetic Excitations in Antiferromagnetic $YBa_2Cu_3O_6$

$YBa_2Cu_3O_6$ is made up of pairs of CuO_2 planes, these 'bilayers' are then stacked together. Band structure calculations [111, 112] suggest that there is a modest electronic coupling between neighbouring Cu atoms in different planes of the bilayer. This leads to a inter-planar hopping parameter $t_\perp \approx 0.3\ eV$. The resulting bilayer splitting in the band structure has been observed by angle-resolved photoemission spectroscopy (ARPES) [74] in the related bilayer high-T_c system $Bi_2Sr_2CaCu_2O_{8+\delta}$. The electron coupling between the two planes which make up the bilayer also results in an inter planar exchange coupling $J_\perp$. The structure of the tetragonal insulating phase of $YBa_2Cu_3O_6$ and the largest exchange couplings are illustrated in Fig. 18.14. In order to account for the exchange coupling between Cu spins in different layers we add a second term to Eq. 18.16,

$$H = \sum_{ij} J_\parallel \mathbf{S}_i \cdot \mathbf{S}_j + \sum_{ij'} J_\perp \mathbf{S}_i \cdot \mathbf{S}_{j'} . \tag{18.22}$$

The presence of the second term (i.e. $J_\perp \neq 0$) in Eq. 18.16 leads to the existence of two branches in the spin-wave dispersion which can be labelled according to whether neighbouring spins in different planes precess out-of-phase ("acoustic" or "odd" mode) or in-phase ("optical" or "even" mode) about their time-averaged (ordered) directions. In the conventional linear spin-wave approximation, the acoustic and optic modes have the response functions per formula unit [6]

$$\chi''_{\text{odd}}(\mathbf{Q},\omega) = Z_d\, \pi\, g^2\, \mu_B^2\, S \left(\frac{1-\gamma(\mathbf{Q})+J_\perp/2J_\parallel}{1+\gamma(\mathbf{Q})}\right)^{1/2} \times \sin^2\left(\frac{\pi\,\Delta z\, l}{c}\right) \delta\left(\hbar\omega \pm \hbar\omega_{\text{odd}}(\mathbf{Q})\right) \tag{18.23}$$

and

$$\chi''_{\text{even}}(\mathbf{Q},\omega) = Z_d\, \pi\, g^2\, \mu_B^2\, S \left(\frac{1-\gamma(\mathbf{Q})}{1+\gamma(\mathbf{Q})+J_\perp/2J_\parallel}\right)^{1/2} \times \cos^2\left(\frac{\pi\,\Delta z\, l}{c}\right) \delta\left(\hbar\omega \pm \hbar\omega_{\text{even}}(\mathbf{Q})\right) , \tag{18.24}$$

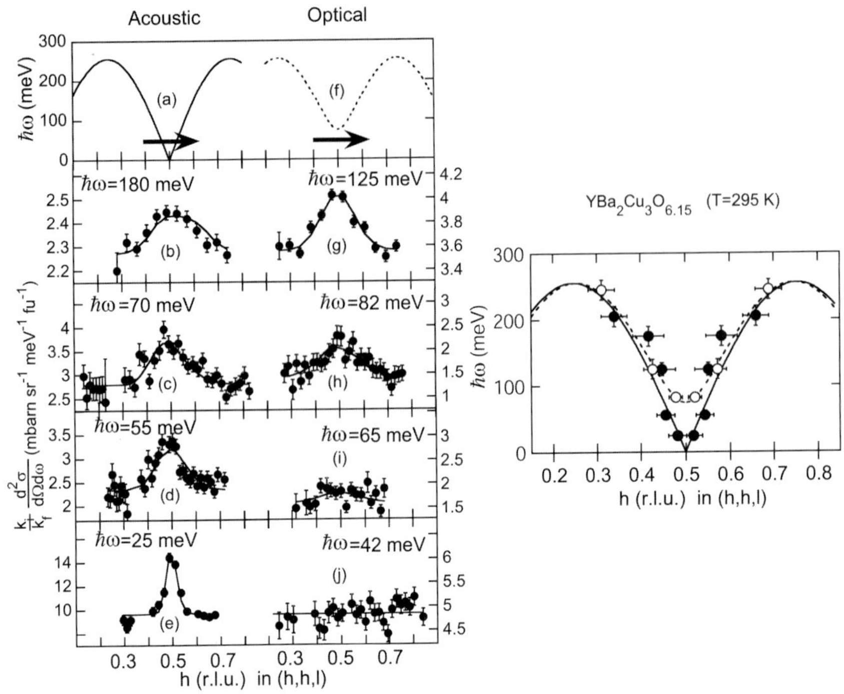

Fig. 18.24. *Left*: Constant energy scans showing the spin waves in $YBa_2Cu_3O_{6.15}$. Scans (b)–(e) are chosen to emphasize the acoustic mode and scans (g)–(i) to emphasize the optical mode. The onset of scattering for the optical (even) mode is at about 74 meV. *Right*: The spin wave dispersion in $YBa_2Cu_3O_{6.15}$: acoustic branch (*closed circles, solid line*), optic branch (*open circles, dashed line*). (After Hayden et al.) [116]

Table 18.3. Experimentally determined exchange constants for antiferromagnetic $YBa_2Cu_3O_{6+x}$. No corrections have been made for the effect of quantum fluctuations which are expected to renormalise the bare exchange constants $J^* = Z_c J$, with $Z_c \approx 1.18$ (see Sect. 18.6.2).

Composition	Ref.	$J_\parallel^*$ (meV)	$J_\perp^*$ (meV)	Z_d	$\hbar\omega_g$ (meV)
$YBa_2Cu_3O_{6.2}$	[114]	120 ± 20	> 8		
$YBa_2Cu_3O_{6.2}$	[115]		10		67 ± 5
$YBa_2Cu_3O_{6.15}$	[116]	125 ± 5	11 ± 2	0.4 ± 0.1	74 ± 5
$PrBa_2Cu_3O_{6.2}$	[117]	127 ± 10			

respectively. The dispersion relations are

$$\begin{aligned}\hbar\omega_{\text{odd}}(\mathbf{Q}) &= 2Z_c J_\parallel \\ &\quad \times \left\{1-\gamma^2(\mathbf{Q}) \right.\\ &\quad \left. + J_\perp/J_\parallel \left[1+\gamma(\mathbf{Q})\right]\right\}^{1/2} \\ \hbar\omega_{\text{even}}(\mathbf{Q}) &= 2Z_c J_\parallel \\ &\quad \times \left\{1-\gamma^2(\mathbf{Q}) \right.\\ &\quad \left. + J_\perp/J_\parallel \left[1-\gamma(\mathbf{Q})\right]\right\}^{1/2}, \end{aligned} \tag{18.25}$$

where $\gamma(\mathbf{Q}) = \frac{1}{2}[\cos(2\pi h)+\cos(2\pi k)]$ and $\Delta z = 3.2$ Å is the separation of the CuO_2 planes in a bilayer. The inter-planar coupling term in Eq. 18.22 leads to no additional dispersion along the z-direction, only a modulation in the amplitude of the response which nevertheless can be used to distinguish between the two modes. Thus the bilayer nature of this $YBa_2Cu_3O_{6+x}$ results in two collective modes with additional structure to $\chi''(\mathbf{q},\omega)$ as compared with the single layer material. The even or optic mode has a gap at the 2D magnetic zone centres such as $(\frac{1}{2}\frac{1}{2}l)$ of $\hbar\omega_g = 2\sqrt{J_\perp^* J_\parallel^*}$. Its intensity displays an overall modulation $\cos^2(\pi\,\Delta z\, l/c)$ which has maxima at positions $l = 0, 3.7, 7.3, \ldots$. Correspondingly, the even or acoustic mode has an overall modulation of $\sin^2(\pi\,\Delta z\, l/c)$ and therefore has maxima at l = 1.8, 5.5, ...

There have been several studies of the spin waves in strongly underdoped antiferromagnetic $YBa_2Cu_3O_{6+x}$ [6,113–116]. Measurements made with thermal neutrons at a reactor source by Shamoto and co-workers [114] were able to resolve the acoustic branches of the spin waves and estimate an exchange constant coupling of $J_{\parallel}^{*}=120\pm20$ meV. Higher energy measurements were carried out by Reznik et al. [115] and Hayden et al. [116] using hot neutrons at a reactor source and spallation neutrons, they were able to observe gaps in the optic (even) excitations of 67 ± 5 meV and 74 ± 5 respectively. Hayden et al. [116] also showed that the magnetic excitations persisted up to 275 meV and that the intensity of the spin wave excitations is strongly renormalised due to quantum fluctuations (see Sect. 18.6.2) with Z_d=0.4 ± 0.1.

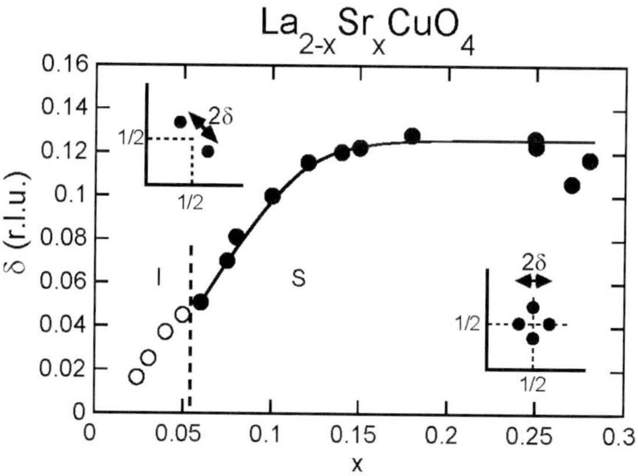

Fig. 18.25. The doping dependence of the incommensurability δ of the low frequency magnetic excitations in $La_{2-x}Sr_xCuO_4$. *Closed circles* refer to the superconducting or metallic state. *Open circles* refer to the incommensurate order observed for lightly doped insulating compositions. Data collected from [57,69,119,120]

18.6.3 Magnetic Excitations in $La_{2-x}Sr_xCuO_4$

Sr or Ba doping of La_2CuO_4 causes the long range antiferromagnetic Neel order to be replaced with a frozen spin state at low temperatures. As discussed above, this state has some of the characteristics of a spin glass and is suppressed by optimal doping. Thurston et al. [118] found that the low energy magnetic response of $La_{2-x}Sr_xCuO_4$ showed incommensurate peaks at finite frequencies near the (1/2,1/2) position. The location of these peaks was identified as $(1/2\pm\delta, 1/2)$ and $(1/2, 1/2\pm\delta)$ by Cheong et al. [119] for samples with x = 0.075 and x = 0.14.

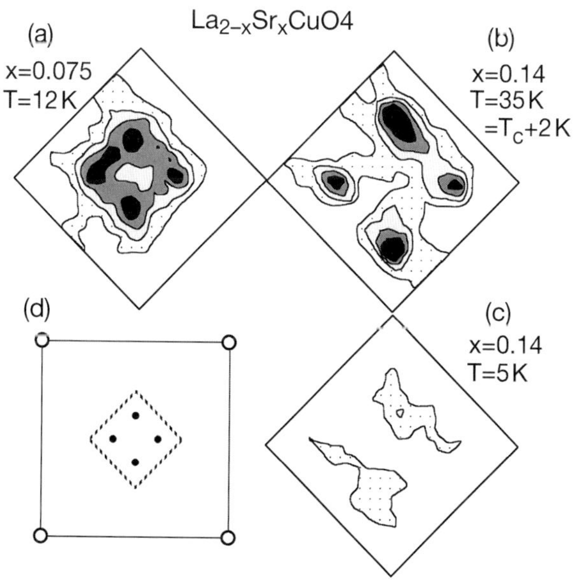

Fig. 18.26. Low energy magnetic excitations in $La_{2-x}Sr_xCuO_4$. (**a**) x = 0.075 and $\hbar\omega$ = 1 meV and T = 12 K (after Cheong et al. [119]). (**b**)–(**c**) x = 0.14 and $\hbar\omega$ = 2 meV for T above (**b**) and below T_c (**c**). Notice how superconductivity suppresses the low energy response

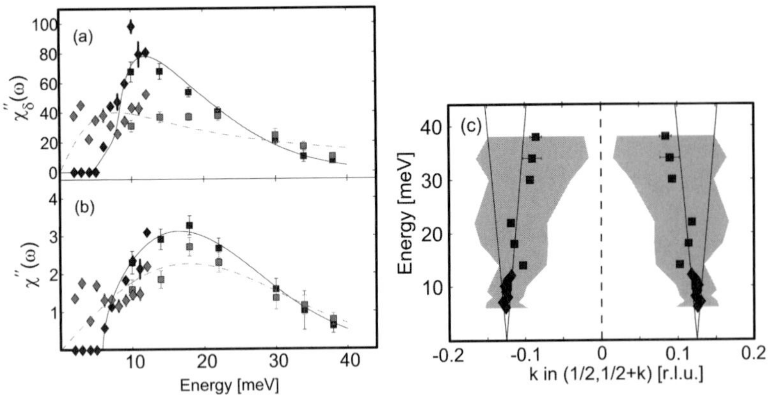

Fig. 18.27. Measurements of the low energy incommensurate magnetic excitations in $La_{1.84}Sr_{0.16}CuO_4$. (**a**) Energy dependence of $\chi''_\delta(\omega)$, the fitted peak intensity of $\chi''(\mathbf{q}, \omega)$ at the incommensurate peak position $\mathbf{Q}_\delta$. Superconducting phase, *filled symbols*; normal phase, *open symbols*. MAPS data, *squares*; triple-axis data, *diamonds*. (**b**) Local susceptibility $\chi''(\omega)$. (**c**) Dispersion of the excitations in LSCO. *Shaded regions* represent the fitted FWHM. *Solid lines* show the spin-wave velocity of La_2CuO_4 with $J^* = 156$ meV. *Open circle*: YBCO6.6. (After N.B. Christensen et al. [124])

The variation of the incommensurate wavevector has been studied systematically with doping by Yamada and co-workers [57,69,120]. The incommensurability increases up to about $\delta = 0.12$ and then saturates (see Fig. 18.25). The incommensurate fluctuations condense out near the $x = 1/8$ composition to give incommensurate static order. This coincides with a plateau in T_c versus doping in $La_{2-x}Sr_xCuO_4$ and a marked dip in $T_c(x)$ in $La_{2-x}Ba_xCuO_4$. Another interesting feature of the low-energy spin excitations is that they are strongly temperature dependent. As the temperature is lowered towards T_c the intensity of the low-energy incommensurate peaks increases and the width in wavevector decreases [121]. Thus it is clear that the incommensurate fluctuations are close to condensing and that they can be made to condense near the $x = 1/8$ composition. Tranquada et al. [73] have demonstrated that superconductivity can be completely destroyed and incommensurate static spin and charge density waves induced by substituting Nd for La in $La_{1.6-x}Nd_{0.4}Sr_xCuO_4$. However it should be noted that this composition has a different crystallographic structure to LTO structure of $La_{2-x}Sr_xCuO_4$. The low energy incommensurate excitations in $La_{2-x}(Ba,Sr)_xCuO_4$ broaden and disperse with increasing energy [122–124]. Figure 18.27 shows the recent data of Christensen et al. [124] which clearly demonstrate inward dispersion towards (1/2,1/2).

High Energy Response

The high-energy magnetic response in $La_{2-x}Sr_xCuO_4$ near optimal doping has been studied by spallation neutrons [125]: some of the results of this study are shown in Figs. 18.28 and 18.29. It is found that magnetic excitations persist up to 200–300 meV, although they are much weaker and broader in **Q**-space than for the parent insulator La_2CuO_4. Surprisingly the lower energy excitations are *stronger* in the superconductor than the insulator when the response is integrated over wavevector to yield the local susceptibility $\chi''(\omega)$. Thus the characteristic energy scale of the spin fluctuations is lowered by doping (at least up to $x \approx 0.14$).

Superconducting State

Clearly we would expect the low energy magnetic response to be suppressed on entering the superconducting state as other techniques such as NMR which probe the spin susceptibility in the superconducting state show a strong reduction. Early work on $La_{2-x}Sr_xCuO_4$ showed that the suppression of the

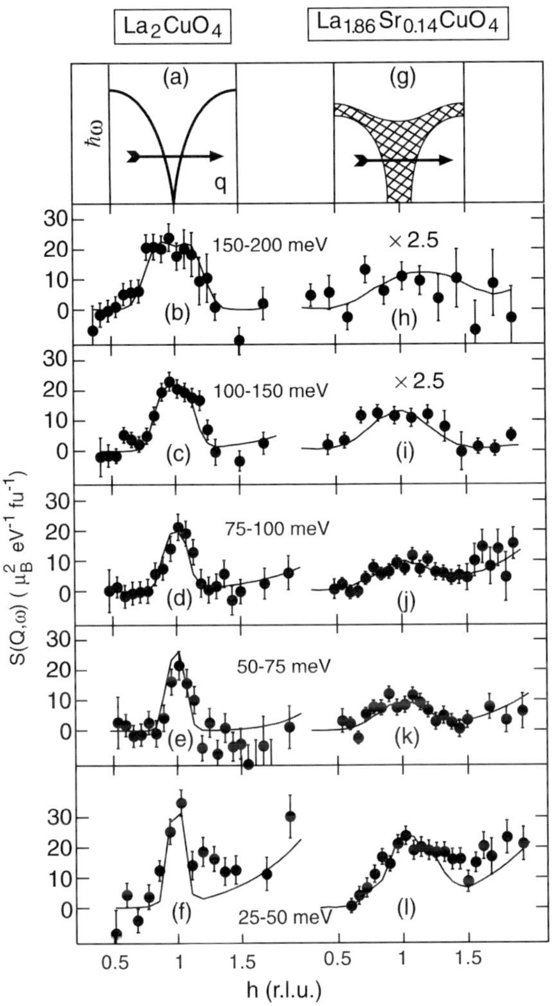

Fig. 18.28. A comparison of the magnetic excitations in insulating antiferromagnetic La_2CuO_4 and superconducting $La_{1.86}Sr_{0.14}CuO_4$. (After Hayden et al. [125])

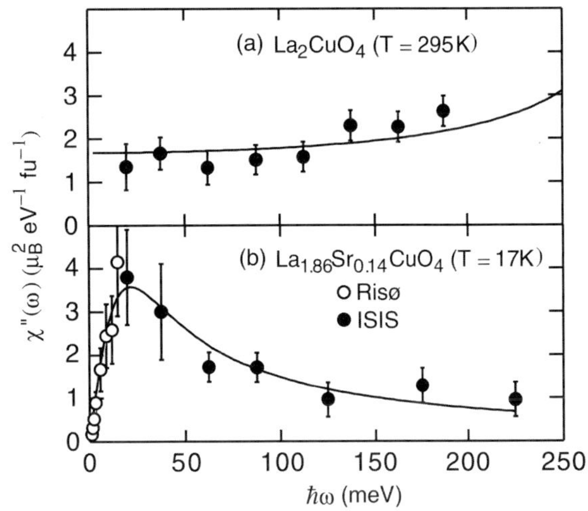

Fig. 18.29. The local susceptibility for insulating antiferromagnetic La_2CuO_4 and superconducting $La_{1.86}Sr_{0.14}CuO_4$. Doping leads to a suppression of the integrated spectral weight at higher energies. However, the integrated response at lower energies, $\hbar\omega \approx 15$ meV, is larger in the superconductor than the insulator. (After Hayden et al. [125])

magnetic response in the superconducting state is dependent on sample quality [119,126–130]. In clean samples, a suppression of the magnetic response is observed below about 7 meV for samples near optimal doping [124,127–130] with the lower energy incommensurate excitations being completely suppressed within the experimental error. At higher energies the incommensurate magnetic response is enhanced on entering the superconducting state [124, 129, 130]. Entering the superconducting state also causes the magnetic response to become sharper in wavevector [129, 130]. For optimally doped samples the greatest enhancement occurs at about 15 meV [124]. Fig. 18.27 shows data from Christensen et al. [124] illustrating these effects.

18.6.4 Magnetic Excitations in $YBa_2Cu_3O_{6+x}$

The bilayer structure of $YBa_2Cu_3O_{6+x}$ complicates both the measurement and interpretation of the spin excitations with respect to the $La_{2-x}Sr_xCuO_4$ system. As discussed in Sect. 18.6.2, the bilayer structure leads to two types of magnetic excitation (odd and even) in which corresponding spins in neighbouring bilayers precess in-phase or out-of-phase. Separating the two types of excitation requires the choice of particular components of momentum parallel to c^* (l values). This places kinematic constrains on the scattering experiment.

The most obvious feature of the magnetic response in $YBa_2Cu_3O_{6+x}$ is the so-called resonance peak [10,132,133] in $\chi''(\mathbf{q}, \omega)$ which occurs at $\mathbf{Q}_{2D}$= (1/2,1/2) in the odd channel and at an energy of 41 meV for optimal doping. The resonance develops near T_c at optimally doping [10, 133]. Figure 18.30

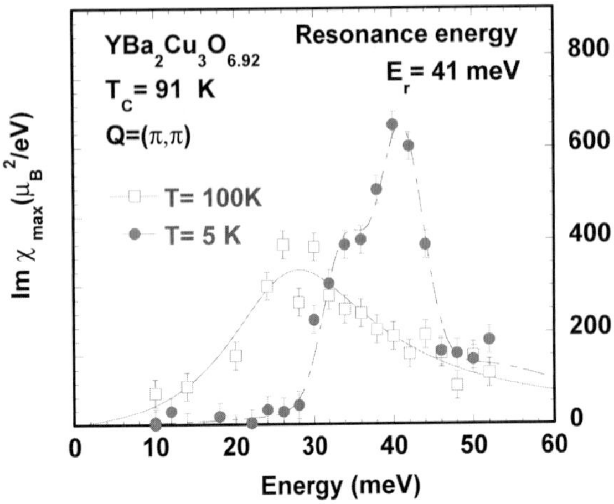

Fig. 18.30. The magnetic response $\chi''(\mathbf{Q}, \omega)$ in the normal (*open squares*) and superconducting (*closed circles*) state in $YBa_2Cu_3O_{6.92}$. (After Bourges et al. [131])

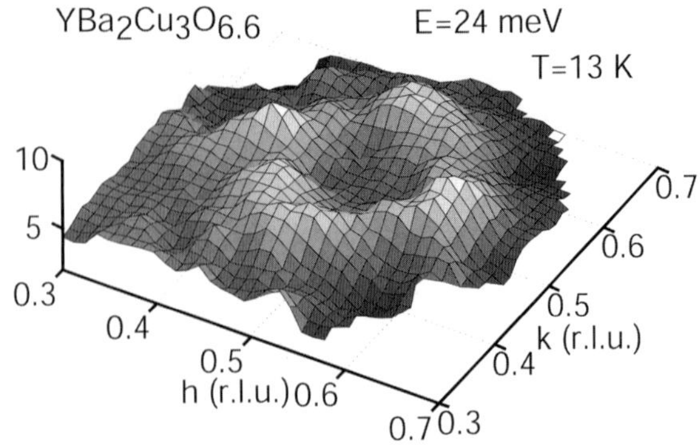

Fig. 18.32. The incommensurate response in $YBa_2Cu_3O_{6.6}$ for energies below the resonance (After Mook et al. [135])

shows the resonance for optimal doping. For underdoped compositions a peak in $\chi''(\mathbf{q}, \omega)$ as a function of ω develops at the resonance energy above T_c. This peak becomes sharper and more intense as the temperature is lowered through T_c [79,80]. Fig. 18.31 shows data on underdoped $YBa_2Cu_3O_{6.6}$ illustrating these points. The resonance has recently been studied in highly underdoped [82,84] and overdoped [134] samples. The energy of the (odd) resonance appears to approximately track T_c, with $E_{\text{res}} \approx 5.1kT_c$ as illustrated in Fig. 18.38.

It is now clear that the "resonance" is the most intense part of a mode which disperses downwards away from (1/2,1/2). The first evidence for the existence of such a mode was the observation of an incommensurate structure in wavevector for energies below the resonance energy [135,136]. Figure 18.32 shows incommensurate scattering measured at $\hbar\omega = 24$ meV for underdoped $YBa_2Cu_3O_{6.6}$. There have been several proposals [137–141] for the form of the dispersion of the resonance mode. Figures 18.33 and 18.34 show early proposals from Arai et al. [137] and Bourges et al. [138]. There is some debate about the existence of a weaker upward dispersing mode reported by some groups [139,140] but not others [141]. Figure 18.35 shows an image of the resonance mode in $YBa_2Cu_3O_{6.6}$ [141]. The incom-

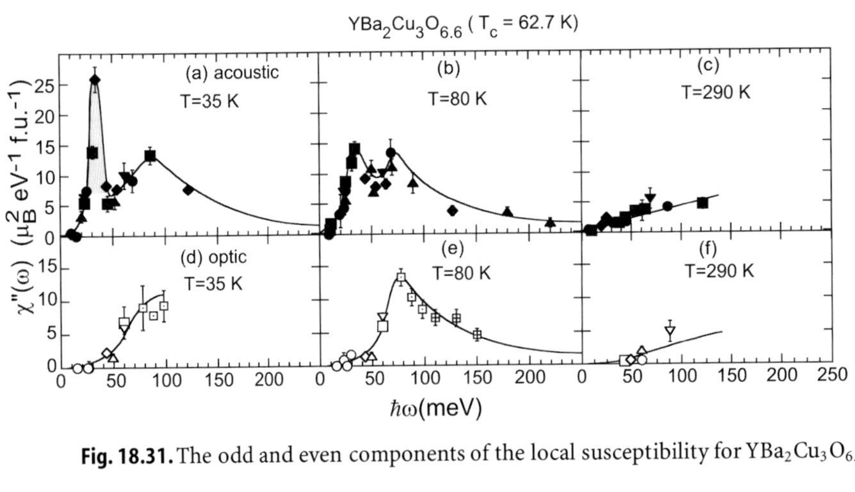

Fig. 18.31. The odd and even components of the local susceptibility for $YBa_2Cu_3O_{6.6}$ (After Dai et al. [80])

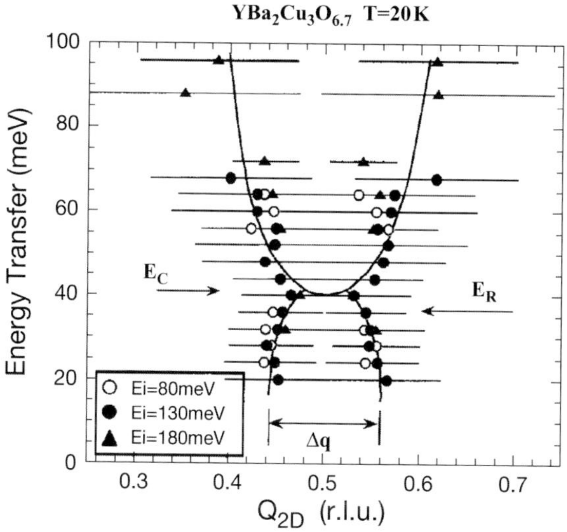

Fig. 18.33. The energy dependence of the incommensurate response in $YBa_2Cu_3O_{6.7}$. Peak positions are measured along $(h, h, 0)$ line. The *horizontal bars* show the widths of the peaks. (After M. Arai et al. [137])

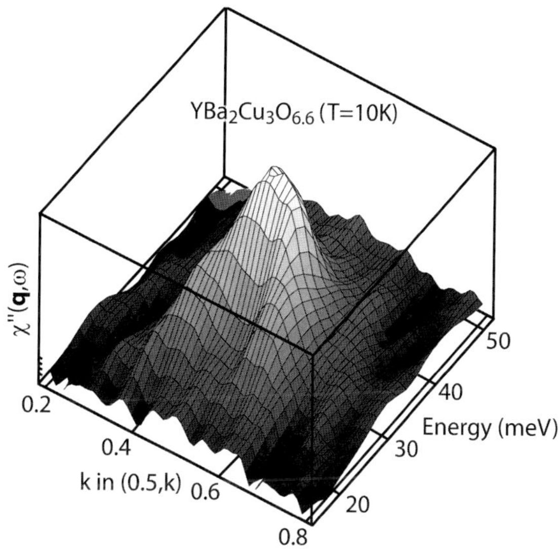

Fig. 18.35. The resonance "exciton" mode in the superconducting state of $YBa_2Cu_3O_{6.6}$ (T_c=63 K, T= 10 K). The plot is $\chi''(\mathbf{q}, \omega)$ for $\mathbf{Q}_{2D} = (1/2, k)$. (After Hayden et al. [141])

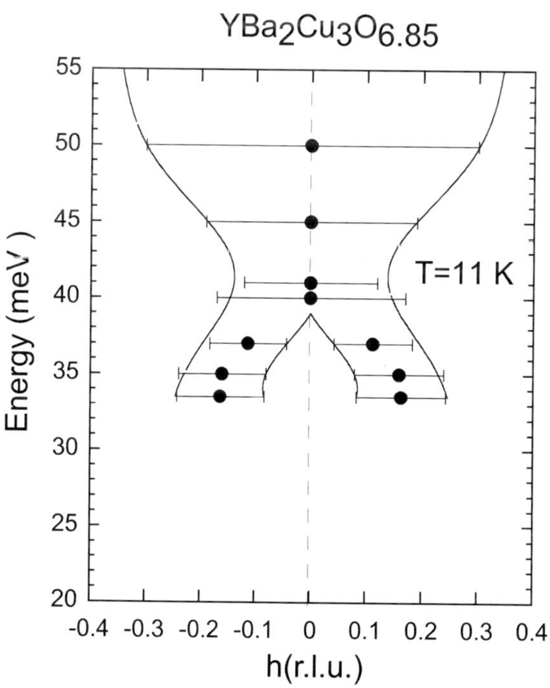

Fig. 18.34. The dispersion of the incommensurate peaks in $YBa_2Cu_3O_{6.85}$. Peak positions are measured along the (1/2+h,1/2,0) line. (After P. Bourges et al. [138])

mensurate scattering develops as the temperature is lowered through T_c. For underdoped compositions at T_c, there is evidence for magnetic scattering at energies where the incommensurate feature develops but a strong incommensurate structure is not seen [135].

As with $La_{2-x}Sr_xCuO_4$ there is a suppression of the magnetic scattering near (1/2,1/2) at the lowest energy transfers as the temperature is lowered through T_c [81,83,132,142,143]. This suppression is partly responsible for the formation of the incommensurate scattering described above and the resonance mode. We can define a spin gap (arbitrarily) as the energy below which the response decreases on entering the superconducting state. This spin gap energy is plotted in Fig. 18.39 and compared with measurements of the gap extracted from other probes.

The resonance mode is an extremely obvious feature in the magnetic response function of $YBa_2Cu_3O_{6+x}$. However when its integrated intensity is placed on an absolute scale we find that it only accounts for a small fraction of the total fluctuating moment per Cu atom expected for an ion with $S = 1/2$. Estimates for the fluctuating moment associated with the resonance from various groups are given in Table. 18.4. Since two of the Cu atoms in

Table 18.4. The spectral weight of the resonance obtain by integrating in wavevector and energy. The symmetry (Sym.) is denoted by o (odd) or e (even). †Hayden et al. [141] integrate over the whole resonance mode

Composition	Ref.	Sym.	$\langle \mathbf{M}^2 \rangle (\mu_B^2 \text{f.u.}^{-1})$
$YBa_2Cu_3O_{6.5}$	Stock et al. [144]	o	0.052
$YBa_2Cu_3O_{6.6}$	Dai et al. [80]	o	0.06 ± 0.04
$YBa_2Cu_3O_{6.6}$	Hayden et al. [141]	o	0.12 ± 0.02 †
$YBa_2Cu_3O_{6.7}$	Fong et al. [79]	o	0.056
$YBa_2Cu_3O_{6.93}$	Dai et al. [80]	o	0.06 ± 0.04
$YBa_2Cu_3O_7$	Fong et al. [79]	o	0.043
$Y_{0.9}Ca_{0.1}Ba_2Cu_3O_7$	Pailhès et al. [134]	o	0.042
		e	0.036

the formula unit are in the CuO_2 plane, we should compare $\langle \mathbf{M}^2 \rangle_{\text{res}}$ with

$$2 \times g^2 \mu_B^2 S(S+1) = 6\, \mu_B^2 .$$

Thus, the resonance only accounts for a small fraction of the total Cu^{2+} moment.

As discussed in Sect. 18.6.1 at higher doping levels $YBa_2Cu_3O_{6+x}$ is orthorhombic. The presence of Cu-O chains makes the structure anisotropic. Indeed this anisotropy is reflected in many physical properties of $YBa_2Cu_3O_{6+x}$, for example, the penetration depth [145] and the resistivity [146]. The first measurements of incommensurate structure for sub-resonance energies [135,136] were made on twinned samples. If $YBa_2Cu_3O_{6+x}$ is annealed without the application of uniaxial pressure then it forms a microstructure with two types of domains such that the a and b axes are interchanged when moving from one domain to the other. It is possible to prepare almost completely "detwinned" or single domain samples. Recent measurements on detwinned samples show that the low energy scattering reflects the anisotropy of the crystal lattice [147,148]. Specifically, the intensity is strongest near $\mathbf{Q}_{2D} = (1/2 \pm \delta, 1/2)$ rather than near $(1/2, 1/2 \pm \delta)$.

So far we have discussed the magnetic response for energies less than or equal to the resonance energy. The high-energy ($\hbar\omega > E_{\text{res}}$) response is of considerable interest for two reasons. Firstly, the cuprates are derived from parent compounds which are antiferromagnetic and have magnon excitations up to about 320 meV. We might expect that some residual interactions remain for superconducting compositions. Secondly, it is likely that high-energy excitations could be important for pairing. The high-energy response has been the subject of a number of studies [79,80,137,141,144,149,150]. Early work on the hot source IN1 spectrometer at the ILL [149] and on the HET and MARI chopper spectrometers at ISIS [80,150] showed that high-energy excitations persisted in underdoped $YBa_2Cu_3O_{6+x}$ for $x \approx 0.5$–0.6 both in the odd and even channels. With the availability of the MAPS spectrometer at ISIS, better high energy data has recently become available. Figure 18.36 shows data collected on underdoped $YBa_2Cu_3O_{6.6}$. For energies above the resonance energy (34 meV for this composition), the response exhibits a square-shaped diffuse pattern in wavevector, possibly with peaks near $\mathbf{Q}_{2D} = (1/2 \pm \epsilon, 1/2 \pm \epsilon)$ and $(1/2 \mp \epsilon, 1/2 \mp \epsilon)$. It is interesting to compare the wavevector and energy-integrated spectral weights of the high-energy (odd) response with the resonance and low-energy incommensurate structure. The two values obtained for $\langle \mathbf{M}^2 \rangle$ in a single experiment are $0.12 \pm 0.02 \mu_B^2$ per formula unit for the resonance and sub-resonance structure and $0.26 \pm 0.05 \mu_B^2$ for the high-energy scattering ($60 < \hbar\omega < 120$ meV). Thus, the higher-energy ($E > E_{\text{res}}$) response has a significantly greater contribution to the total fluctuating moment than the resonance structure [141].

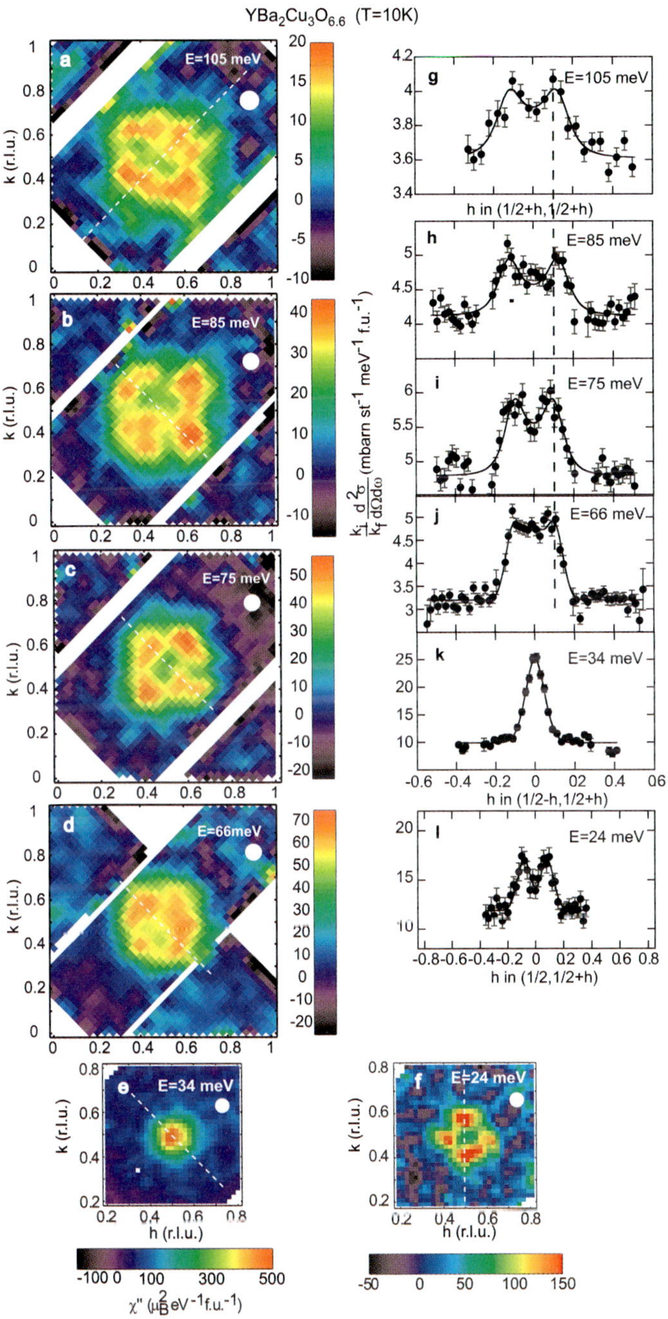

Fig. 18.36. The odd magnetic excitations in $YBa_2Cu_3O_{6.6}$. (After Hayden et al. [141])

The Effect of the Bilayer Structure

The effect of the bilayer structure is to create two types of excitation with odd and even symmetry. The even excitations are gapped (see Table 18.3) in the undoped parent compound with $\hbar\omega_{gap} \approx$ 70 meV. For underdoped superconductors, the onset frequency of the even scattering drops to about 60 meV [80,149,150]. The data of Dai et al. [80] illustrating the gap in the even excitations are shown in Fig. 18.31.

For higher doping levels the even excitations are weaker and more difficult to observe. However, there have been observations of an increase in scattering on entering the superconducting state for frequencies near or below the even onset energies. This "even resonance" is weaker and at a higher energy than its odd counterpart. Figure 18.37 shows measurements by Pailhès et al. [134] of the odd and even resonances in overdoped $Y_{0.9}Ca_{0.1}Ba_2Cu_3O_7$. The energy of the even resonance drops continuously as the doping increases. Figure 18.38 summarizes the doping dependence of the odd and even resonance energies.

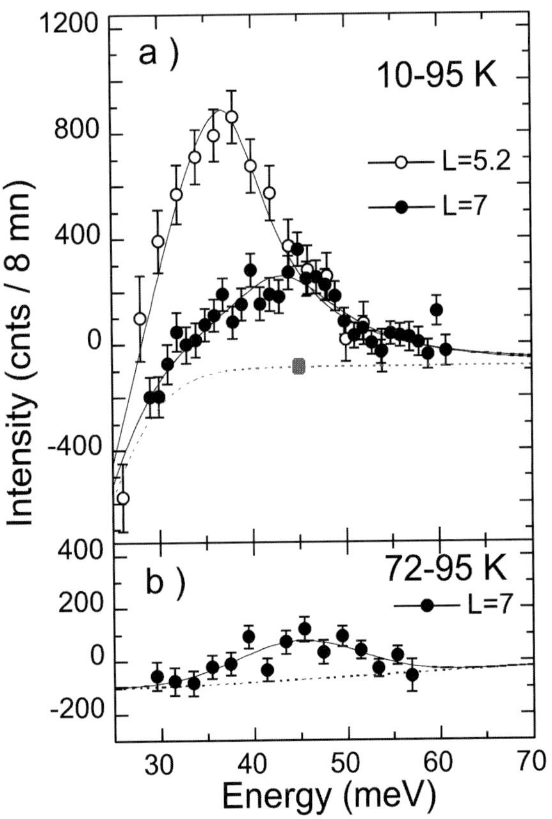

Fig. 18.37. The odd and even resonances in $Y_{0.9}Ca_{0.1}Ba_2Cu_3O_7$ as shown through the difference in the response above and below T_c. (**a**) Odd resonance (l = 5.2, *open circles*); Even resonance (l = 7, *closed circles*). (**b**) Even resonance with a smaller temperature difference. (After S. Pailhès et al. [134])

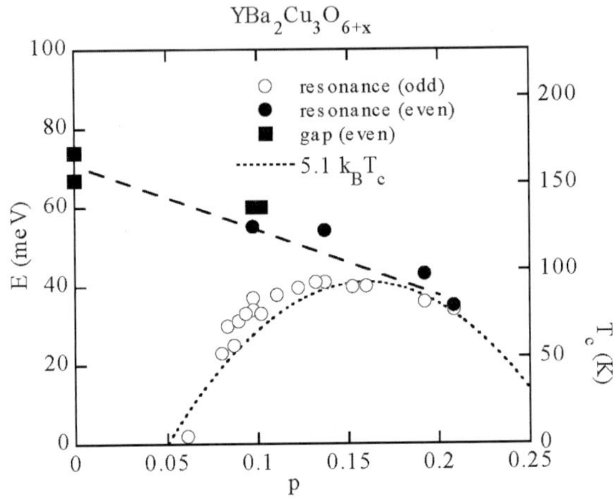

Fig. 18.38. The doping dependence of the odd (*open circles*) and even (*closed circles*) resonance energies in $YBa_2Cu_3O_{6+x}$. Also shown is the even gap in the excitations (*closed squares*). (Data collected from [79–84,115,116,134, 140,151])

18.6.5 Magnetic Excitations in Other High-T_c Systems

There is a large body of work on $YBa_2Cu_3O_{6+x}$ and $La_{2-x}(Sr,Ba)_xCuO_4$. In addition, a number of other high-T_c systems have been measured. However, less is known about the detailed $(\mathbf{Q}, \omega)$ dependence of their magnetic response. Fong et al. [152] report the resonance in optimally doped (T_c= 91 K) $Bi_2Sr_2CaCu_2O_{8+\delta}$ with E_{res} = 43 meV and $\langle m^2\rangle_{res}$=0.23 μ_B^2. He et al. [153] report that E_{res} drops in overdoped (T_c = 83 K) $Bi_2Sr_2CaCu_2O_{8+\delta}$. Significantly, a resonance peak has also been observed in the optimally doped (T_c = 92.5 K) single-layer compound $Tl_2Ba_2CuO_{6+\delta}$ [154] with E_{res} = 47 meV.

18.7 Discussion

The theoretical description of the electronic properties of cuprates still remains controversial. Since the magnetic response is intimately connected to the electron dynamics, it is inevitable that its interpretation will also remain controversial. Having said this, it should remembered that neutron scattering measures simple correlation functions which can be directly related to physical properties such as the magnetic susceptibility. Several excellent reviews of the theory relating to spin dynamics (in addition to those is this series!) have recently appeared [155–158].

The phase diagrams of the cuprates give a phenological indication of how the spin fluctuations might evolve with doping. The parent antiferromagnets are $S = \frac{1}{2}$ antiferromagnetic square-lattice insulators with fairly well understood magnon and multi-magnon excitations as discussed in Sect. 18.6.2. The overdoped cuprates show more metallic behaviour such as AMRO oscillations [50]. Thus we might expect that the excitations in overdoped cuprates could be described using a Lindhard picture (See Sect. 18.4.2). Throughout most of the phase diagram more complicated models are required, however the excitations may retain features from these respective limits.

Experimentally, we find that doping causes a collapse of the very-high ($\hbar\omega \approx 2J$) frequency response and leads to an increase in the lower energy response and a lowering of the characteristic energy of the magnetic excitations. This is illustrated in $La_{2-x}Sr_xCuO_4$, where ISIS measurements show that the response for $\hbar\omega \approx 15$ meV increases by a factor of two when x is changed from 0 to 0.14 (see Fig. 18.29). The doping also causes an incommensurate (IC) structure to develop in the $\chi''(\mathbf{q}, \omega)$ at low energies in both $La_{2-x}Sr_xCuO_4$ and $YBa_2Cu_3O_{6+x}$. However, the structure in LSCO is clearly developed at above T_c, whereas in YBCO the development of the IC structure appears to be associated with the resonance and pseudogap for underdoped compositions and the superconducting transition for optimally doped compositions. The incommensurate spin order and low-energy incommensurate spin fluctuations have been interpreted as stripe order and stripe fluctuations (see e.g. Kivelson et al. [155] for a review). A stripe picture can explain some features, such as the variation of the incommensurability at low doping (see Fig. 18.25), however, the incommensurability at low frequency persists to high doping levels both in LSCO [120] and YBCO [81,148]. Another model is that the incommensurate structure is due to underlying Fermi surface nesting [29–31]. The validity of this approach must be questioned as one moves towards the underdoped side of the phase diagram where the cuprates are anything but conventional metals. The nesting picture is also able to account for the incommensurate structure (see Fig. 18.36(b)) observed at higher frequencies ($E \approx$ 85 meV) in $YBa_2Cu_3O_{6+x}$ [35,141].

A universal feature of the cuprates seems to be that entering the superconducting state causes an increase in scattering near the energy $\hbar\omega \approx 5k_BT_c$. In LSCO this increase appears at the incommensurate positions and in YBCO at occurs at $\mathbf{Q}_{2D} = (\frac{1}{2}, \frac{1}{2})$. The effect occurs over a relatively small energy range in YBCO and the associated peak has become known as the "resonance" [10,132,133]. As pointed out in Sec. 18.4.2, the BCS coherence factors can give rise to an increase in scattering in a d-wave superconductor for energies near 2Δ. However, the sharpness of the resonance in YBCO suggests that it may be a spin-triplet exciton below $2\Delta_0$ [32–38,133]. This explanation also relies on the coherence factors [133]. It seems likely that the coherence factors are also responsible for the increase in the more gapless LSCO system (see Fig. 18.27). A second explanation proposed for the resonance is that it is a particle–particle antibound resonance which could only be observed below T_c [159]. This theory predicts an antibound resonance at higher energies which is not observed [156,160].

Another universal feature of the cuprates is that the magnetic response is suppressed at low energies on entering the superconducting state. Thus we may (arbitrarily) define a "spin-gap" $2\Delta_{\text{spin}}$ as the energy below which magnetic excitations are suppressed as we pass through T_c.

Within a BCS model (see Sect. 18.4.2), we would expect a suppression of the magnetic response at low energies due to the formation of a superconduct-

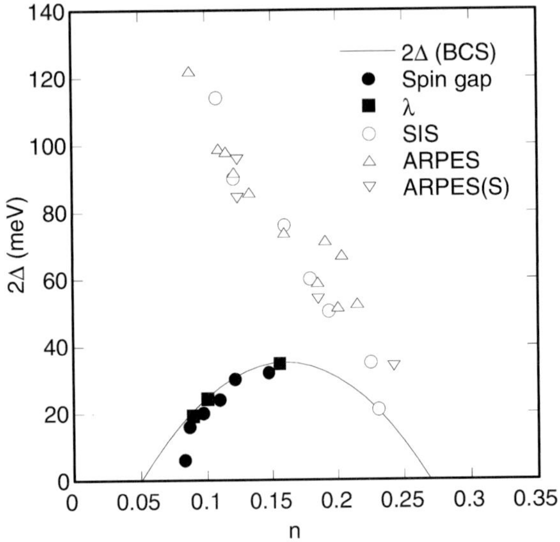

Fig. 18.39. Doping dependence of the spin gap $2\Delta_{\text{spin}}$ and the higher energy pseudogap $2\Delta^*$ obtained for various probes for YBCO and Bi-2212. Doping is estimated using the empirical relation $T_c/T_c^{\max} = 1 - 82.6(n - n_{\text{opt}})$ [161] with $T_c^{\max} = 94$ K. *The solid line* is the BCS weak coupling result $\Delta = 2.14 k_B T_c$ for a 2D d-wave superconductor [162]. The spin gap (YBCO) is determined from INS [81,134,141]. Penetration depth measurements (YBCO) [163]. In Bi-2212, the higher energy scale pseudogap is determined from SIS tunnelling [164] and by ARPES measurements in the superconducting [165] and the normal state [166–168]

ing gap. In the case of an s-wave superconductor, the suppression is complete at $T = 0$ for $\hbar\omega < 2\Delta$. For a d-wave superconductor the suppression is **Q**-dependent and may be incomplete. Model calculations suggest that the suppression occurs below approximately $2\Delta_0$ (see Sect. 18.4.2). Figure 18.39 show a comparison of the spin gap compared with the gap determined by other methods which yield significantly larger values for underdoped compositions. This suggests that the spin gap, measured by neutrons, is sensitive to different electronic correlations to other probes such as ARPES and tunnelling.

18.8 Final Remarks

Although magnetic excitations in cuprates have now been studied for some 20 years, progress continues to be made. The main reasons for this continuing progress are the ongoing development of neutron instrumentation and sources and the improvement of samples. Inelastic neutron scattering experiments are essentially flux limited because of the relatively small number of neutrons produced by neutron sources. The is no doubt that progress in instrumentation will continue unabated and many new results will be produced. It is likely that there is a significant part of the spectral weight that has not yet been observed, particulary in optimally doped cuprates.

Acknowledgements

It is a pleasure to thank my collaborators in this area G. Aeppli, S.-W. Cheong, N.B. Christensen, K.N. Clausen, R. Coldea, P. Dai, F. Dogan, C. Frost, J. Kulda, B. Lake, K. Lefmann, O. Lipscombe, Y. Maeno, T. E. Mason, D. F. McMorrow, H.A. Mook, T. G. Perring, H. N. Ronnow, H. Takagi, B. Vignolle, and Z. Fisk.

References

1. W. Marshall and S. W. Lovesey, *Theory of Thermal Neutron Scattering* (Oxford University Press, Oxford, 1971)
2. G. E. Bacon, *Neutron Diffraction*, 3rd edn. (Oxford University Press, Oxford, 1975)
3. S.W. Lovesey, *Theory of Neutron Scattering from Condensed Matter* (Clarendon Press, Oxford, 1984)
4. G.L. Squires, *Introduction to the Theory of Thermal Neutron Scattering* (Dover, New York, 1996)
5. P.J. Brown, In: *International Tables for Crystallography* (Kluwer, Dordrecht, 1992) Vol. C, p. 391
6. S. Shamato, M. Sato, J.M. Tranquada, B.J. Sternlieb, and G. Shirane, Phys. Rev. B **48**, 13817 (1993)
7. R.M. White, *Quantum Theory of Magnetism* (Springer, Berlin, 1983)
8. T. Izuymama, D.-J. Kim, and R. Kubo, J. Phys. Soc. Japan **18**, 1025 (1963)
9. O. Steinsvoll, C.F. Majkrzak, G. Shirane, and J. Wicksted, Phys. Rev. B **30**, 2377 (1984)
10. H.A. Mook, M. Yethiraj, G. Aeppli, T.E. Mason, and T. Armstrong, Phys. Rev. Lett. **70**, 3490 (1993)
11. H.F. Fong, B. Keimer, D. Reznik, D.L. Milius, and I.A. Aksay, Phys. Rev. B **54**, 6708 (1996)

12. H.B. Callen and T.A. Welton, Phys. Rev. **83**, 34 (1951)
13. K. Yosida, Phys. Rev. **110**, 769–770 (1958)
14. C.G. Shull and F.A. Wedgwood, Phys. Rev. Lett. **16**, 513–515 (1966)
15. J.A. Duffy, S.M. Hayden, Y. Maeno, Z. Mao, J. Kulda, and G.J. McIntyre, Phys. Rev. Lett. **85**, 5412 (2000)
16. J.X. Boucherle, J.Y. Henry, R.J. Papoular, J. Rossat-Mignod, J. Schweizer, F. Tasset, and G. Uimin, Physica B **192**, 25 (1993)
17. A.P. Mackenzie and Y. Maeno, Rev. Mod. Phys. **75**, 657 (2003)
18. T.M. Rice and M. Sigrist, J. Phys.-Condens. Mat. **7**, L643 (1995)
19. G. Baskaran, Physica B **224**, 490 (1996)
20. P.W. Anderson and P. Morel, Phys. Rev. **1**, 1911 (1961)
21. C. Stassis, J. Arthur, C.F. Majkrzak, J.D. Axe, B. Batlogg, J. Remeika, Z. Fisk, J.L. Smith, and A.S. Edelstein, Phys. Rev. B **34**, 4382 (1986)
22. J. Lindhard, 2005, Kgl. Danske Videnskab. Selskab, Mat.-fys. Medd. **28**. No. 8 (1954)
23. E. Fawcett, H.L. Alberts, V.Y. Galkin, D.R. Noakes, and J.V. Yakhmi, Rev. Mod. Phys. **66**, 25 (1994)
24. E. Fawcett, S.A. Werner, A. Goldman, and G. Shirane, Phys. Rev. Lett. **61**, 558 (1988)
25. S.M. Hayden, R. Doubble, G. Aeppli, T.G. Perring, and E. Fawcett, Phys. Rev. Lett. **84**, 999 (2000)
26. L.C. Hebel and C.P. Slichter, Phys. Rev. **113**, 1504 (1959)
27. J.R. Schrieffer, *Theory of Superconductivity* (Addison-Wesley, New York, 1983)
28. N. Bulut and D.J. Scalapino, Phys. Rev. B **47**, 3419 (1993)
29. P.B. Littlewood, J. Zaanen, G. Aeppli, and H. Monien, Phys. Rev. B **48**, 487 (1993)
30. Q.M. Si, Y.Y. Zha, K. Levin, and J.P. Lu, Phys. Rev. B **47**, 9055 (1993)
31. T. Tanamoto, H. Kohno, and H. Fukuyama, J. Phys. Soc. Jap. **62**, 1455 (1993)
32. A. Abanov and A.V. Chubukov, Phys. Rev. Lett. **83**, 1652 (1999)
33. J. Brinckmann and P.A. Lee, Phys. Rev. Lett. **82**, 2915 (1999)
34. Y.J. Kao, Q.M. Si, and K. Levin, Phys. Rev. B **61**, 11898 (2000)
35. M.R. Norman, Phys. Rev. B **61**, 14751 (2000)
36. A.V. Chubukov, B. Janko, and O. Tchernyshyov, Phys. Rev. B **63**, 180507 (2001)
37. D. Manske, I. Eremin, and K.H. Bennemann, Phys. Rev. B **6305**, 054517 (2001)
38. F. Onufrieva and P. Pfeuty, Phys. Rev. B **65**, 054515 (2002)
39. I. Eremin, D.K. Morr, A.V. Chubukov, K.H. Bennemann, and M.R. Norman, Phys. Rev. Lett. **94**, 147001 (2005)
40. R. Stedman, L. Almqvist, and G. Nilsson, Phys. Rev. **162**, 549 (1967)
41. W.L. McMillan and J.M. Rowell, Phys. Rev. Lett. **14**, 108 (1965)
42. T. Petersen, T.E. Mason, G. Aeppli, A.P. Ramirez, E. Bucher, and R.N. Kleiman, Physica B **199**, 151 (1994)
43. N. Metoki, Y. Haga, Y. Koike, and Y. Onuki, Phys. Rev. Lett. **80**, 5417 (1998)
44. N. Bernhoeft, N. Sato, B. Roessli, N. Aso, A. Hiess, G.H. Lander, Y. Endoh, and T. Komatsubara, Phys. Rev. Lett. **81**, 4244 (1998)
45. M. Jourdan, M. Huth, and H. Adrian, Nature **398**, 47 (1999)
46. N.K. Sato, N. Aso, K. Miyake, R. Shiina, P. Thalmeier, G. Varelogiannis, C. Geibel, F. Steglich, P. Fulde, and T. Komatsubara, Nature **410**, 340 (2001)
47. J.G. Bednorz and K.A. Müller, Z. Phys. B **64**, 189 (1986)
48. D. Vaknin, S.K. Sinha, D.E. Moncton, D.C. Johnston, J.M. Newsam, C.R. Safinya, and J.H.E. King, Phys. Rev. Lett. **58**, 2802 (1987)
49. S. Nakamae, K. Behnia, N. Mangkorntong, M. Nohara, H. Takagi, S.J.C. Yates, and N.E. Hussey, Phys. Rev. B **68**, 100502 (2003)
50. N. Hussey, M. Abdel-Jawad, A. Carrington, A. Mackenzie, and L. Balicas, Nature **425**, 814 (2003)
51. T. Timusk and B. Statt, Rep. Prog. Phys. **62**, 61 (1999)
52. J.L. Tallon and J.W. Loram, Physica C **349**, 53 (2001)
53. H. Alloul, T. Ohno, and P. Mendels, Phys. Rev. Lett. **63**, 1700 (1989)
54. P.G. Radaelli, J.D. Jorgensen, R. Kleb, B.A. Hunter, F.C. Chou, and D.C. Johnston, Phys. Rev. B **49**, 6239 (1994)
55. H. Takagi, R.J. Cava, M. Marezio, B. Battlog, J.J. Krajewski, and J.W.F. Peck, Phys. Rev. Lett. **68**, 3777 (1992)

56. P.G. Radaelli, D.G. Hinks, A.W. Mitchell, B.A. Hunter, J.L. Wagner, B. Dabrowski, K.G. Vandervoort, H.K. Viswanathan, and J.D. Jorgensen, Phys. Rev. B **49**, 4163 (1994)
57. K. Yamada, C.H. Lee, K. Kurahashi, J. Wada, S. Wakimoto, S. Ueki, H. Kimura, Y. Endoh, S. Hosoya, G. Shirane, R.J. Birgeneau, M. Greven, M.A. Kastner, and Y.J. Kim, Phys. Rev. B **57**, 6165 (1998)
58. M.-H. Julien, Physica B **329–333**, 693 (2003)
59. L.J.D. Jongh, Solid State Comm. **65**, 963 (1988)
60. T. Kaplan and S. Mahanti, J. App. Phys. **69**, 5382 (1991)
61. K. Yamada, E. Kudo, Y. Endoh, Y. Hidaka, M. Oda, M. Suzuki, and T. Murakami, Solid State Comm. **64**, 753 (1987)
62. S. Mitsuda, G. Shirane, S. Sinha, D. Johnson, M. Alvarez, D. Vaknin, and D. Moncton, Phys. Rev. B **41**, 1926 (1990)
63. J. Xu, T. Watson-Yang, J. Yu, and A. Freeman, Phys. Lett. **12A**, 489 (1987)
64. G. Luke, L. Le, B. Sternlieb, W. Wu, Y. Uemura, J. Brewer, T. Riseman, S. Ishbashi, and S. Uchida, Physica C **185–189**, 1175 (1991)
65. T. Goto, S. Kazama, K. Miyagawa, and T. Fukase, J. Phys. Soc. Jpn. **63**, 3494 (1994)
66. P. Imbert, G. Jéhanno, P. Debray, C. Garcin, and J. Hodges, J. Phys. (Paris) I **2**, 1405 (1992)
67. S.M. Hayden, G. Aeppli, H. Mook, D. Rytz, M.F. Hundley, and Z. Fisk, Phys. Rev. Lett. **66**, 821 (1991)
68. T. Suzuki, T. Goto, K. Chiba, T. Shinoda, T. Fukase, H. Kimura, K. Yamada, M. Ohashi, and Y. Yamaguchi, Phys. Rev. B **57**, R3229 (1998)
69. M. Matsuda, M. Fujita, K. Yamada, R.J. Birgeneau, M.A. Kastner, H. Hiraka, Y. Endoh, S. Wakimoto, and G. Shirane, Phys. Rev. B **62**, 9148 (2000)
70. S. Katano, M. Sato, K. Yamada, T. Suzuki, and T. Fukase, Phys. Rev. B **62**, R14677 (2000)
71. B. Lake, H. Ronnow, N. B. Christensen, G. Aeppli, K. Lefmann, D.F. McMorrow, P. Vorderwisch, P. Smeibidl, N. Mangkorntong, T. Sasagawa, M. Nohara, H. Takagi, and T.E. Mason, Nature **415**, 299 (2002)
72. B. Lake, K. Lefmann, N. B. Christensen, G. Aeppli, D.F. Mcmorrow, H.M. Ronnow, P. Vorderwisch, P. Smeibidl, N. Mangkorntong, T. Sasagawa, M. Nohara, and H. Takagi, Nat. Mater. **4**, 658 (2005)
73. J. Tranquada, B. Sternlieb, J. Axe, Y. Nakamura, and S. Uchida, Nature **375**, 561 (1995)
74. A. Damascelli, Z. Hussain, and Z.-X. Shen, Rev. Mod. Phys. **75**, 473 (2003)
75. S. Sanna, G. Allodi, G. Concas, A.D. Hillier, , and R.D. Renzi, Phys. Rev. B **93**, 207001 (2004)
76. M. v. Zimmermann, J.R. Schneider, T. Frello, N.H. Andersen, J. Madsen, M. Käll, H.F. Poulsen, R. Liang, P. Dosanjh, , and W.N. Hardy, Phys. Rev. B **68**, 104515 (2003)
77. R.J. Cava, A.W. Hewat, E.A. Hewat, B. Batlogg, M. Marezio, K.M. Rabe, J.J. Krajewski, W.F. Peck, Jr., and L.W. Rupp, Phys. Rev. B **165**, 419 (1990)
78. H. Casalta, P. Schleger, P. Harris, B. Lebech, N. H. Andersen, R. Liang, P. Dosanjh, and W.N. Hardy, Phys. Rev. B **258**, 321 (1996)
79. H.F. Fong, P. Bourges, Y. Sidis, L.P. Regnault, J. Bossy, A. Ivanov, D.L. Milius, I.A. Aksay, and B. Keimer, Phys. Rev. B **61**, 14773 (2000)
80. P.C. Dai, H.A. Mook, S.M. Hayden, G. Aeppli, T.G. Perring, R.D. Hunt, and F. Dogan, Science **284**, 1344 (1999)
81. P.C. Dai, H.A. Mook, R.D. Hunt, and F. Dogan, Phys. Rev. B **6305**, 054525 (2001)
82. H.A. Mook, P.C. Dai, and F. Dogan, Phys. Rev. Lett. **88**, 097004 (2002)
83. C. Stock, W.J.L. Buyers, R.A. Cowley, P.S. Clegg, R. Coldea, C.D. Frost, R. Liang, D. Peets, D. Bonn, W.N. Hardy, and R.J. Birgeneau, Phys. Rev. B **71**, 024522 (2005)
84. C. Stock, W.J.L. Buyers, Z. Yamani, C.L. Broholm, J.-H. Chung, Z. Tun, R. Liang, D. Bonn, W.N. Hardy, and R.J. Birgeneau, Phys. Rev. B **73**, 100504 (2006)
85. P. Anderson, Phys. Rev. **86**, 694 (1952)
86. R. Kubo, Phys. Rev. **87**, 568 (1952)
87. T. Oguchi, Phys. Rev. **117**, 117 (1960)
88. R. Singh, Phys. Rev. B **39**, 9760 (1989)
89. Y. Endoh, K. Yamada, R. Birgeneau, D. Gabbe, H. Jenssen, M. Kastner, C. Peters, P. Picone, T. Thurston, J. Tranquada, G. Shirane, Y. Hidaka, M. Oda, Y. Enomoto, M. Suzuki, and T. Murakami, Phys. Rev. B **37**, 7443 (1988)
90. G. Shirane, Y. Endoh, R.J. Birgeneau, M.A. Kastner, Y. Hidaka, M. Oda, M. Suzuki, and T. Murakami, Phys. Rev. Lett. **59**, 1613–1616 (1987)

91. G. Aeppli, S.M. Hayden, H.A. Mook, Z. Fisk, S.W. Cheong, D. Rytz, J.P. Remeika, G.P. Espinosa, and A.S. Cooper, Phys. Rev. Lett. **62**, 2052 (1989)
92. S.M. Hayden, G. Aeppli, H.A. Mook, S.W. Cheong, and Z. Fisk, Phys. Rev. B **42**, 10220 (1990)
93. S. Chakravarty, B. Halperin, and D. Nelson, Phys. Rev. B **38**, 2344 (1989)
94. S.M. Hayden, G. Aeppli, R. Osborn, A.D. Taylor, T.G. Perring, S.W. Cheong, and Z. Fisk, Phys. Rev. Lett. **67**, 3622 (1991)
95. R. Coldea, S.M. Hayden, G. Aeppli, T.G. Perring, C.D. Frost, T.E. Mason, S.W. Cheong, and Z. Fisk, Phys. Rev. Lett. **86**, 5377 (2001)
96. H.M. Ronnow, D.F. McMorrow, R. Coldea, A. Harrison, I.D. Youngson, T.G. Perring, G. Aeppli, O.S. sen, K. Lefmann, and C. Rischel, Phys. Rev. Lett. **87**, 037202 (2001)
97. R.R. Singh and M.P. Gelfand, Phys. Rev. B **52**, R15695 (1995)
98. O.F. Syljuåsen and H.M. Rønnow, J. Phys.-Condens. Mat. **12**, L405 (2000)
99. A.W. Sandvik and R.R.P. Singh, Phys. Rev. Lett. **86**, 528–531 (2001)
100. J. ichi Igarashi and T. Nagao, Phys. Rev. B 72, 014403 (2005)
101. H. Eskes and J.H. Jefferson, Phys. Rev. B **48**, 9788–9798 (1993)
102. P.W. Anderson, Phys. Rev. **79**, 350–356 (1950)
103. P. Bourges, H. Casalta, A.S. Ivanov, and D. Petitgrand, Phys. Rev. Lett. **79**, 4906 (1997)
104. I.U. Heilmann, J.K. Kjems, Y. Endoh, G.F. Reiter, G. Shirane, and R.J. Birgeneau, Phys. Rev. B **24**, 3939 (1981)
105. R. Coldea, D.A. Tennant, and Z. Tylczynski, Phys. Rev. B **68**, 134424 (2003)
106. S. Chakravarty, B. Halperin, and D. Nelson, Phys. Rev. Lett. **60**, 1057 (1988)
107. E. Manousakis, Rev. Mod. Phys. **63**, 1 (1991)
108. A. Chubukov, S. Sachdev, and J. Ye, Phys. Rev. **49**, 11919 (1994)
109. L. de Jongh and A. Miedema, Adv. Phys. **23**, 1 (1974)
110. D. Grempel, Phys. Rev. Lett. **61**, 1041 (1988)
111. O.K. Andersen, A.I. Liechtenstein, O. Jepsen, and F. Paulsen, J. Phys. Chem. Solids **56**, 1573 (1995)
112. A.I. Liechtenstein, O. Gunnarsson, O.K. Andersen, and R.M. Martin, Phys. Rev. B **54**, 12505–12508 (1996)
113. J. Rossat-Mignod, L.P. Regnault, C. Vettier, P. Burlet, J.Y. Henry, and G. Lapertot, Physica B **169**, 58 (1991)
114. S. Shamoto, M. Sato, J.M. Tranquada, B.J. Sternlieb, and G. Shirane, Phys. Rev. B **48**, 13817 (1993)
115. D. Reznik, P. Bourges, H.F. Fong, L.P. Regnault, J. Bossy, C. Vettier, D.L. Milius, I.A. Aksay, and B. Keimer, Phys. Rev. B **53**, R14741 (1996)
116. S.M. Hayden, G. Aeppli, T.G. Perring, H.A. Mook, and F. Dogan, Phys. Rev. B **54**, R6905 (1996)
117. A.T. Boothroyd, N.H. Andersen, B.H. Larsen, A.A. Zhokhov, C.D. Frost, and D.T. Adroja, Phys. Rev. B **71**, 094514 (2005)
118. T.R. Thurston, R.J. Birgeneau, M.A. Kastner, N.W. Preyer, G. Shirane, Y. Fujii, K. Yamada, Y. Endoh, K. Kakurai, M. Matsuda, and Y.H.T. Murakami, Phys. Rev. B **40**, 4585 (1989)
119. S.W. Cheong, G. Aeppli, T.E. Mason, H. Mook, S.M. Hayden, P.C. Canfield, Z. Fisk, K.N. Clausen, and J.L. Martinez, Phys. Rev. Lett. **67**, 1791 (1991)
120. S. Wakimoto, H. Zhang, K. Yamada, I. Swainson, H. Kim, and R.J. Birgeneau, Phys. Rev. Lett. **92**, 217004 (2004)
121. G. Aeppli, T.E. Mason, S.M. Hayden, H.A. Mook, and J. Kulda, Science **278**, 1432 (1997)
122. T.E. Mason, G. Aeppli, and H.A. Mook, Phys. Rev. Lett. **68**, 1414 (1992)
123. J.M. Tranquada, H. Woo, T.G. Perring, H. Goka, G.D. Gu, G. Xu, M. Fujita, and K. Yamada, Nature **429**, 534 (2004)
124. N.B. Christensen, D.F. McMorrow, H.M. Ronnow, B. Lake, S.M. Hayden, G. Aeppli, T.G. Perring, M. Mangkorntong, M. Nohara, and H. Takagi, Phys. Rev. Lett. **93**, 147002 (2004)
125. S.M. Hayden, G. Aeppli, H.A. Mook, T.G. Perring, T.E. Mason, S.W. Cheong, and Z. Fisk, Phys. Rev. Lett. **76**, 1344 (1996)
126. T.R. Thurston, P.M. Gehring, G. Shirane, R.J. Birgeneau, M.A. Kastner, Y. Endoh, M. Matsuda, K. Yamada, H. Kojima, and I. Tanaka, Phys. Rev. B **46**, 9128 (1992)
127. T.E. Mason, G. Aeppli, S.M. Hayden, A.P. Ramirez, and H.A. Mook, Phys. Rev. Lett. **71**, 919 (1993)
128. M. Matsuda, K. Yamada, Y. Endoh, T.R. Thurston, G. Shirane, R.J. Birgeneau, M.A. Kastner, I. Tanaka, and H. Kojima, Phys. Rev. B **49**, 6958 (1994)
129. T.E. Mason, A. Schroder, G. Aeppli, H.A. Mook, and S.M. Hayden, Phys. Rev. Lett. **77**, 1604 (1996)
130. B. Lake, G. Aeppli, T.E. Mason, A. Schroder, D.F. McMorrow, K. Lefmann, M. Isshiki, M. Nohara, H. Takagi, and S.M. Hayden, Nature **400**, 43 (1999)

131. P. Bourges, Y. Sidis, H. Fong, B. Keimer, L. Regnault, J. Bossy, A. Ivanov, D. Milius, and I. Aksay, 1999, In: *High Temperature Superconductivity*, ed. by S.E. Barnes et al. (CP483 American Institute of Physics, Amsterdam, 1999) pp. 207–212, cond-mat/9902067
132. J. Rossat-Mignod, L.P. Regnault, C. Vettier, P. Bourges, P. Burlet, J. Bossy, J.Y. Henry, and G. Lapertot, Physica C **185**, 86 (1991)
133. H.F. Fong, B. Keimer, P.W. Anderson, D. Reznik, F. Dogan, and I.A. Aksay, Phys. Rev. Lett. **75**, 316 (1995)
134. S. Pailhès, Y. Sidis, P. Bourges, C. Ulrich, V. Hinkov, L.P. Regnault, A. Ivanov, B. Liang, C.T. Lin, C. Bernhard, and B. Keimer, Phys. Rev. Lett. **91**, 237002 (2003)
135. H.A. Mook, P.C. Dai, S.M. Hayden, G. Aeppli, T.G. Perring, and F. Dogan, Nature **395**, 580 (1998)
136. P. Dai, H.A. Mook, and F. Doğan, Phys. Rev. Lett. **80**, 1738 (1998)
137. M. Arai, T. Nishijima, Y. Endoh, T. Egami, S. Tajima, K. Tomimoto, Y. Shiohara, M. Takahashi, A. Garrett, and S.M. Bennington, Phys. Rev. Lett. **83**, 608 (1999)
138. P. Bourges, Y. Sidis, H.F. Fong, L.P. Regnault, J. Bossy, A. Ivanov, and B. Keimer, Science **288**, 1234 (2000)
139. D. Reznik, P. Bourges, L. Pintschovius, Y. Endoh, Y. Sidis, T. Masui, and S. Tajima, Phys. Rev. Lett. **93**, 207003 (2004)
140. S. Pailhès, Y. Sidis, P. Bourges, V. Hinkov, A. Ivanov, C. Ulrich, L.P. Regnault, and B. Keimer, Phys. Rev. Lett. **93**, 167001 (2004)
141. S.M. Hayden, H.A. Mook, P.C. Dai, T.G. Perring, and F. Dogan, Nature **429**, 531 (2004)
142. P.M. Gehring, J.M. Tranquada, G. Shirane, J.R.D. Copley, R.W. Erwin, M. Sato, and S. Shamoto, Phys. Rev. B **44**, 2811 (1991)
143. P. Dai, M. Yethiraj, H.A. Mook, T.B. Lindemer, and F. Dogan, Phys. Rev. Lett. **77**, 5425 (1996)
144. C. Stock, W.J.L. Buyers, R. Liang, D. Peets, Z. Tun, D. Bonn, W.N. Hardy, and R.J. Birgeneau, Phys. Rev. B **69**, 014502 (2004)
145. K. Zhang, D.A. Bonn, S. Kamal, R. Liang, D.J. Baar, W.N. Hardy, D. Basov, and T. Timusk, Phys. Rev. Lett. **73**, 2484 (1994)
146. R. Gagnon, C. Lupien, and L. Taillefer, Phys. Rev. B **50**, 3458 (1994)
147. H.A. Mook, P. Dai, F. Doğan, and R.D. Hunt, Nature **404**, 729 (2000)
148. V. Hinkov, S. Pailhes, P. Bourges, Y. Sidis, A. Ivanov, A. Kulakov, C.T. Lin, D.P. Chen, C. Bernhard, and B. Keimer, Nature **430**, 650 (2004)
149. P. Bourges, H.F. Fong, L.P. Regnault, J. Bossy, C. Vettier, D.L. Milius, I.A. Aksay, and B. Keimer, Phys. Rev. B **56**, R11439 (1997)
150. S.M. Hayden, G. Aeppli, P. Dai, H.A. Mook, T.G. Perring, S.W. Cheong, Z. Fisk, F. Dogan, and T.E. Mason, Physica B **241**, 765 (1998)
151. S. Pailhès, C. Ulrich, B. Fauqué, V. Hinkov, Y. Sidis, A. Ivanov, C. Lin, B. Keimer, and P. Bourges, 2005, cond-mat/0512634.
152. H.F. Fong, P. Bourges, Y. Sidis, L.P. Regnault, A. Ivanov, G. D. Gu, N. Koshizuka, and B. Keimer, Nature **398**, 588 (1999)
153. H. He, Y. Sidis, P. Bourges, G.D. Gu, A. Ivanov, N. Koshizuka, B. Liang, C.T. Lin, L.P. Regnault, E. Schoenherr, and B. Keimer, Phys. Rev. Lett. **86**, 1610 (2001)
154. H. He, P. Bourges, Y. Sidis, C. Ulrich, L.P. Regnault, S. Pailhes, N.S. Berzigiarova, N.N. Kolesnikov, and B. Keimer, Science **295**, 1045 (2002)
155. S.A. Kivelson, I. Bindloss, E. Fradkin, V. Oganesyan, J. Tranquada, A. Kapitulnik, and C. Howard, Rev. Mod. Phys. **75**, 1201 (2003)
156. M.R. Norman and C. Pepin, Rep. Prog. Phys. **66**, 1547 (2003)
157. S. Sachdev, Rev. Mod. Phys. **75**, 913 (2003)
158. P.A. Lee, N. Nagaosa, and X.-G. Wen, Rev. Mod. Phys. **78**, 17 (2006)
159. E. Demler and S.C. Zhang, Phys. Rev. Lett. **75**, 4126 (1995)
160. O. Tchernyshyov, M. Norman, and A.V. Chubukov, Phys. Rev. B **63**, 144507 (2001)
161. J.L. Tallon, C. Bernhard, H. Shaked, R.L. Hitterman, and J.D. Jorgensen, Phys. Rev. B **51**, 12911 (1995)
162. H. Won and K. Maki, Phys. Rev. B **49**, 1397 (1994)
163. C. Panagopoulos and T. Xiang, Phys. Rev. Lett. **81**, 2336 (1998)
164. J.F. Zasadzinski, L. Ozyuzer, N. Miyakawa, K.E. Gray, D.G. Hinks, and C. Kendziora, Phys. Rev. Lett. **87**, 067005 (2001)
165. J.C. Campuzano, H. Ding, M.R. Norman, H.M. Fretwell, M. Randeria, A. Kaminski, J. Mesot, T. Takeuchi, T. Sato, T. Yokoya, T. Takahashi, T. Mochiku, K. Kadowaki, P. Guptasarma, D.G. Hinks, Z. Konstantinovic, Z.Z. Li, and H. Raffy, Phys. Rev. Lett. **83**, 3709 (1999)

166. M.R. Norman, H. Ding, M. Randeria, J.C. Campuzano, T. Yokoya, T. Takeuchi, T. Takahashi, T. Mochiku, K. Kadowaki, P. Guptasarma, and D.G. Hinks, Nature **392**, 157 (1998)
167. P.J. White, Z.X. Shen, C. Kim, J.M. Harris, A.G. Loeser, P. Fournier, and A. Kapitulnik, Phys. Rev. B **54**, 15669 (1996)
168. A.G. Loeser, Z.X. Shen, M.C. Schabel, C. Kim, M. Zhang, A. Kapitulnik, and P. Fournier, Phys. Rev. B **56**, 14185 (1997)

19 Heavy-Fermion Superconductivity

Peter S. Riseborough Dept. of Physics, Temple Univ., Philadelphia, Pennsylvania, USA
George M. Schmiedeshoff Dept. of Physics, Occidental College, Los Angeles, California, USA
James L. Smith Los Alamos National Laboratory, New Mexico, USA

19.1 Overview

When the BCS theory appeared in 1957, Bernd Matthias felt it did not do everything. He thought that it worked for the *p*-electron superconductors such as lead and tin, but the transition metal superconductors such as niobium and vanadium were still unexplained. Aside from his instinct, it was that here the isotope effect was all over the map. This led him to uranium, where his student Hunter Hill did the isotope effect for uranium and found that the orthorhombic α-phase had a backward mass squared dependence and that the cubic γ-phase looked BCS-like. This was Hunter's thesis work, and the experiment was difficult enough that those who did not like the result could ignore it. Bernd was also drawn to the U_6X superconducting compound with X = Mn, Co, Fe, and Ni. In these compounds, the T_c scaled with the moments, that is, they followed the Slater–Pauling curve. These were the first superconductors discovered that formed with 3*d* magnetic elements. These early interesting problems in superconductivity were simply hints of what was to come and what would be called highly correlated electron (heavy-fermion) behavior.

The heavy-fermion ground state arises at low temperature. Near room temperature, these metals have magnetic moments whose magnetic susceptibility suggest that they would order antiferromagnetically at low temperature. However somewhere below 50 K, unexpected behavior shows up. The electrical resistivity is high for a metal, and the magnetic ordering does not occur. The heat capacity below 10 K is quite large and seems consistent with the presence of magnetic moments. However, when some compounds became superconducting with the heat capacity indicating that the large value was due to the superconducting electrons instead of the magnetic moments, it seemed wrong. It was as if the rocks in a stream bed began to flow instead of the water. There is still no accepted explanation for such behavior.

In 1975 Bucher et al. noted that the compound UBe_{13} had become superconducting with a somewhat large critical field [1]. They said it was likely that pure uranium filaments were present in the sample, and the superconductivity was hence an impurity effect. In 1978, Franz et al. reported in a footnote that the compound $CeCu_2Si_2$ had become superconducting [2], but in this case there was not a simple explanation for an impurity effect. In both cases the authors were clearly wondering what had happened but knew it could not be correct. Steglich continued to pursue the superconductivity and in 1979 reported that is was genuine [3]. As a superconducting material, it seemed ordinary except for the heat capacity that was about a thousand times too large. Steglich coined the name of heavy-fermion superconductor by associating the huge heat capacity with a huge mass for the conduction electrons, which was the only parameter that could be adjusted. So there was a superconductor with a heat capacity for the superconducting electrons that was as large as that usually associated with fixed magnetic moments. People who were paying attention considered it an oddity, and little happened.

In 1983 Ott et al. rechecked Bucher's result and found UBe_{13} to have properties remarkably similar to those of $CeCu_2Si_2$ [4]. Polycrystalline and single-crystalline samples were made at Los Alamos and seen to be superconducting. Heat-capacity measurements at Zürich showed that indeed a huge heat capacity was in the superconducting electrons. The condensed-matter physics community began to take notice, and measurements of many other properties began in earnest on both compounds. It was noticed that the temperature dependence of the heat capacity in the superconducting state was not an exponential, as seen in known superconductors and as predicted by the Bardeen Cooper Schrieffer theory, but rather it was a power law. This suggested that a gap had not opened everywhere on the Fermi surface (Ott et al. 1984) [5]. This is similar to the pairing in superfluid ^{3}He, which is non-*s*-wave pairing. This was the first possible example of an exotic superconductor, one in which the Cooper pairs have a symmetry that is not isotropic or almost so. Some theorists sharpened their pencils, and the people in Los Alamos and Zürich were wondering about the analogy to ^{3}He because the smoking gun for low-symmetry superconductors would be more than one superconducting phase.

Then in 1984, Stewart et al. reported that UPt_3 was superconducting and was another candidate for a low-symmetry superconductor [6]. They had been studying it as a spin-fluctuating material when it went superconducting at low temperatures. While UBe_{13} and $CeCu_2Si_2$ were not good Fermi liquid metals (later to become its own field as non-Fermi liquid materials), UPt_3 fit the many-body models of Fermi liquids perfectly. The theorists began work and had much success at describing this ground state. However, how the state occurred was and is the mystery of heavy-fermion superconductors.

Also in 1984, Smith *et al.* began putting impurities into UBe_{13} and found a strange non-monotonic depression of the superconducting transition temperature with the addition of thorium [7]. In 1985, heat-capacity measurements in Zürich showed two huge superconducting transitions, and the analogy to ^{3}He and the proof of low-symmetry superconductivity was made [8]. The first material, $CeCu_2Si_2$, remained an s-wave superconductor. So by 1986, all of the measurements that anyone could think of were being performed around the world, and review articles were being printed such as Fisk et al. 1986 [9] and Fisk et al. 1988 [10]. These also cover non-superconducting heavy-fermion systems that became identifiable once the properties of the superconductors were well known. Many theses were being written, and the old men of superconductivity were thanking the younger experimentalists for breathing new life into superconductivity. Order seemed to be at hand.

Then late in 1986, Bednorz and Müller published their work on high-temperature superconductivity [11], and an avalanche cut a huge swath through condensed-matter physics leaving many subjects deserted. As it developed that high-temperature superconductors were also non-s-wave, theorists were well prepared. Only two example of transitions between superconducting states have been found in oxides (Movshovich et al. 1998 [12] and Mota et al. 1999 [13]), and these results are not widely known or accepted. However, all of the modern techniques that came from the original scanning tunneling microscope have, in the 1990s, permitted workers to directly measure the symmetry of the superconducting energy gap.

The compound UPt_3 had one more surprise left. In 1989, Fisher et al. discovered a second bump in the heat capacity of the superconducting transition [14]. Workers all over the world took the old samples out of their drawers to remeasure this because all had missed it, and it was confirmed to have two superconducting phases. It was soon seen that the magnetic-field-and-temperature phase diagram had three superconducting phases, and the theoretical description of them was under control quickly.

Now, we have heavy fermion superconductors and high-temperature superconductors both without an accepted model for how these ground states occur. New theory is still needed. One recent development is the possibility of quantum critical points at zero temperature that hold some promise of explaining the general phase diagrams for both of these superconductors, which usually have an antiferromagnetic state in them. There is a feeling that some larger theoretical picture may emerge and show us all new physics. It is also true that in Germany and Japan, there are now major experimental institutes engaged in the search for more heavy-fermion materials with the usual name now of highly-correlated electron systems. And of course, theoretical work is underway everywhere.

19.2 Introduction

The first heavy-fermion superconductor $CeCu_2Si_2$ was discovered by Steglich et al. [3] in 1979. Despite intense skepticism from members of the scientific community, Steglich demonstrated that stoichiometric $CeCu_2Si_2$ underwent a transition to a superconducting state at the critical temperature $T_c = 0.7$ K. This material was regarded as a unique anomaly until several other materials were discovered which had the same distinctive normal state properties. Some of these heavy-fermion materials [4,6] even have similar unusual superconducting phases. This class of heavy-fermion materials can be loosely categorized as systems that are in close proximity to a magnetic instability, and have the characteristic properties that at high temperatures, the system exhibits evidence of local moments; at low temperatures, the system resembles a Fermi liquid with very heavy quasi-particle masses. The heavy-fermion materials are based on elements from the lanthanide or actinide series which have incomplete f shells. The f derived electronic states retain a lot of their ionic character in that they are almost localized and experience large electronic interactions attributable to the smallness of the radius of the ionic f orbitals. The heavy-fermion materials are those in which there is a delicate balance between the strong ionic coulomb interactions that tend to localize the electrons and yield local magnetic moments, and the hybridization with extended band states that tends to delocalize the f electrons. The delicate balance is responsible for the temperature dependent cross-over from the high temperature local moment regime to the low temperature regime of itinerant f electron behavior. At high temperatures, the magnetic moments are manifested via the Curie–Weiss variation of the magnetic susceptibility with Curie constants almost equal the full ionic magnetic moments, and also through a Kondo-like logarithmic temperature variation of the electrical resistivity indicating resonant scattering of conduction electrons from independent local moments. In the low temperature regime, the strong electron–electron interactions show up as strong renormalizations of the properties of the itinerant electrons. The excitations of the interacting electron gas include quasi-particle excitations that closely resemble the excitations of a non-interacting electron gas. The quasi-particles have dispersion relations that form narrow correlated f bands close to the Fermi energy. The quasi-particle masses, as inferred from the coefficient, γ, of the low temperature linear T term of the specific heat, can be as large as 1000 free electron masses. The masses of the heavy quasi-particle are assumed to have evolved from the entropy released during the transformation of the high temperature magnetic moments. The specific heat coefficient γ determined from the normal state of $CeCu_2Si_2$ is roughly 1100 mJ/mole K^2, while that of the paramagnetic heavy-fermion $CeCu_6$ [15] system is 1300 mJ/mole K^2. The low temperature susceptibilities $\chi(T)$ are also enhanced above the values of the Pauli paramagnetic susceptibilities inferred from the density of states obtained via LDA calculations. The values of the low temperature susceptibilities can be as large as 28×10^{-3} e.m.u./mole for $CeCu_6$ and 16×10^{-3} e.m.u./mole for $CeRu_2Si_2$ [16]. A measure of the relative strengths of enhancements experienced by the susceptibility and the specific heat is given by the Wilson ratio,

$$R_W = \frac{\pi^2 k_B^2 \chi(0)}{3\mu_B^2 \gamma}, \tag{19.1}$$

which has the ideal value of unity for the non-interacting electron gas. However, the values of the Wilson ratio found for most heavy-fermion systems are also close to unity. Comparison of the γ values and the low temperature limit of the magnetic susceptibilities with the densities of states at the Fermi energy found in LDA electronic structure calculations [17] indicate that the quasi-particle masses are enhanced by factors as large as 25, presumably due to strong electron–electron interactions. This interpretation is supported by the observation of a $T^3 \ln T$ term in the specific heat of UPt_3, which is characteristic of spin-fluctuations driven by strong electron–electron interactions. In addition, the low temperature electrical resistivities of some materials, such as $CeCu_6$, show $\rho(T) = \rho(0) + A\,T^2$ variations where the coefficient A takes on large values. The large value of A is indicative of the formation of a Fermi liquid state in which the resistivity is dominated by scattering between the heavy quasi-particles. In this interpretation, the value of A is a measure of the inverse square of the renormalized Fermi energy. The most direct and definitive evidence proving the existence of the heavy Fermi liquid state is given by the measurement of de Haas–van Alphen oscillations [18,19]. In a number of materials, large quasi-particle mass enhancements have been found over significant portions of the Fermi surface. The large quasi-particle masses inferred for some or all portions of the Fermi surface correlate well with the measured value of γ. Evidence of the strong magnetic correlations is provided by the fact that, in many heavy-fermion systems, the addition of small amounts of impurities allows the strong electron interactions to drive the system magnetic. This suggests that the heavy-fermion systems may be in the vicinity of a quantum critical point and that the extremely large mass enhancements are produced by the nearly critical magnetic fluctuations. Unlike the U heavy-fermion systems, the Ce systems seem to show properties that are attributable to strongly localized fluctuating f magnetic moments, so that a large part of the mass enhancement for the Ce heavy-fermion compounds may result from the local moment fluctuations.

Large mass enhancements can also be inferred from the normal state specific heats of the uranium heavy-fermion superconductors UBe_{13} [4], UPt_3 [6] and URu_2Si_2 [20,21], which have magnitudes that are of the same order as in the heavy-fermion Ce compounds. The observed γ values are 1100 mJ/mole K^2 for UBe_{13}, 400 for UPt_3, and 60 mJ/mole K^2 for URu_2Si_2. The magnetic susceptibilities χ, in the low temperature normal state, are also enhanced. For UBe_{13}, which has a cubic structure, the low temperature normal state susceptibility has a value of 14×10^{-3} e.m.u./mole. In UPt_3, which has hexagonal symmetry, the susceptibility is anisotropic, having the value of 4×10^{-3} e.m.u./mole for fields along the

c-axis and 8.1×10^{-3} in the basal plane. In URu_2Si_2, which has tetragonal symmetry, the susceptibility is 4.9×10^{-3} for fields along the c-axis and 1.5×10^{-3} e.m.u./mole in the perpendicular directions. Fermi liquid analyses of the specific heat and susceptibility are complicated by the existence of unusual low temperature magnetic phases and correlations. In fact, while the large values of the χ and γ may be considered as indicative of the formation of a highly enhanced Fermi liquid, the resistivity of UBe_{13} is still large $\sim 100\ \mu\Omega$cm, and although rapidly varying, it shows no evidence of a T^2 variation setting in before the superconducting transition occurs.[1] The large magnitude of the specific heat jump that occurs at the superconducting transition temperature, T_c, shows that the electrons that take part in the formation of the heavy-fermion state form the Cooper pairs. That is, the normalized jump discontinuity $(C_s - C_n)/C_n$ is of the order of unity, similar to the B.C.S value of 1.43, which indicates that the superconducting electrons have heavy masses (see Fig. 19.1). A similar conclusion is arrived at by a thermodynamic analysis of the extremely large initial slopes of the upper critical fields $(\partial H_{c2}/\partial T)_{T_c}$ [22–24] which is about $-42\ T/K$ for UBe_{13}.

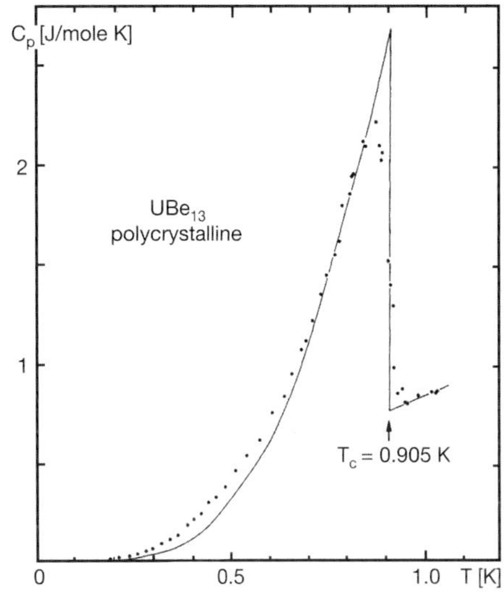

Fig. 19.1. Low temperature specific heat of UBe_{13} measured by Ott et al. [5]. The magnitude of the jump in the specific heat at T_c is comparable to the magnitude of the linear T term in the normal state. This suggests that the heavy quasi-particles of the normal state form the Cooper pairs. For comparison, the specific heat calculated for a strong coupling ABM p-wave superconductor is shown by the solid line. The temperature dependence is indicative that the quasi-particle density of states is gapless

The equilibrium superconducting state is that which minimizes the total energy, including the strong Coulomb interaction between the f electrons that gives rise to the enhanced masses in the Ce and U heavy-fermion materials. The superconducting electrons could form Cooper pairs with finite angular momentum, l, which could lower the Coulomb repulsion between the pairing electrons as the pair wave function vanishes at the origin. The decrease in the large Coulomb energy could offset the increase in kinetic energy due to the orbital motion. The finite angular momentum of the pairs could also lead to the superconducting gap at the Fermi energy falling to zero at either isolated points or lines, giving rise to a finite density of states for low-energy quasi-particle excitations. These point zeros and line zeros give rise to power law variations with T for various physical properties, which are in contrast to the exponentially activated behavior usually observed in s-wave superconductors. Such power law variations have been found in experiments [5,25–31] and show conclusive evidence of low-energy excitations. However, the observed power laws have not led to a consensus as to whether they are caused by either point zeros or line zeros in the order parameter. This lack of consensus could be due to the complications of either pair breaking effects of impurities in anisotropic superconductors or due to collective fluctuations. In either case, heavy-fermion superconductors are examples of exotic superconductors.

19.2.1 Multiple Superconducting Phases

Soon after the discovery of the uranium heavy-fermion superconductors [4], it was noticed that some heavy-fermion superconductors exhibited multiple superconducting phases. This discovery gave strong

[1] However, when magnetic fields large enough to suppress the superconductivity are applied, UBe_{13} does exhibit a T^2 term in the resistivity.

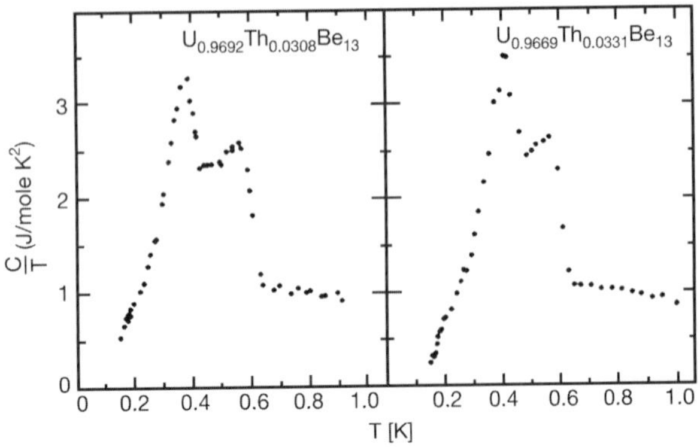

Fig. 19.2. The temperature dependence of the specific heat of $U_{1-x}Th_xBe_{13}$ compounds, for $x = 0.0308$ and $x = 0.0331$ showing the double transition measured by Ott et al. [8]

support to the hypothesis that the superconductivity involved exotic pairings. Motivated by the discovery of the power law temperature variations and the implications about the zeros in the gap, Smith et al. [7] doped the superconductors with impurities to find their effect on the superconducting transition temperature T_c, as it is well known that non-magnetic impurities suppress anisotropic pairing. It was found that substitution of just a few percent of Th atoms on the U sites in UBe_{13} produced a large non-monotonic change in T_c and also reduced the value of the resistivity at constant T [7]. Measurements of the specific heat [8] showed a spectacular result namely, that for Th concentrations in the range $0.01 < x < 0.06$, the specific heat shows two jumps that have comparable discontinuities (see Fig. 19.2). The upper transition is associated with a large Meissner effect and has a maximum T_{c1} near 0.6 K in this range of x, whereas the second transition occurs at T_{c2}, which is approximately 0.4 K. The existence of the second transition in the doped samples was confirmed by sound velocity and ultrasonic attenuation experiments [32]. The sound velocity showed a pronounced minimum at T_{c2}, while the attenuation showed a λ peak. The second superconducting phase was assigned as having some intrinsic weak magnetic character. The magnetic nature of the second transition was later confirmed via improved muon spin resonance measurements [33], which show that weak magnetic moments of the order 10^{-3} μ_B are formed below T_{c2} but only in the concentration range $0.01 < x < 0.06$. The phase diagram is shown in Fig. 19.3. The existence of two

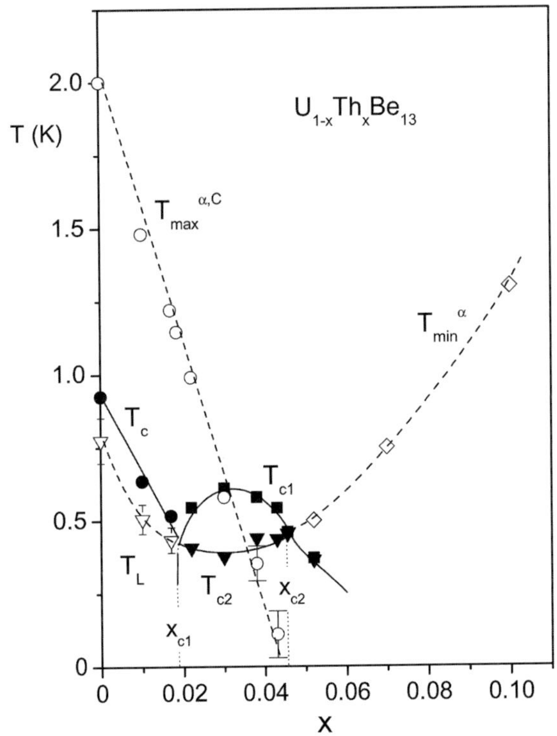

Fig. 19.3. A T–x phase diagram for $U_{1-x}Th_xBe_{13}$. *Filled symbols* denote phase transitions and *open symbols* indicate anomalies in the specific heat (C) and thermal expansion (α). The *solid vertical line* at $x = x_1$ represents a phase boundary established by the specific heat measurements under pressure of Zieve et al. [34]. [After Oeschler et al. [35]]

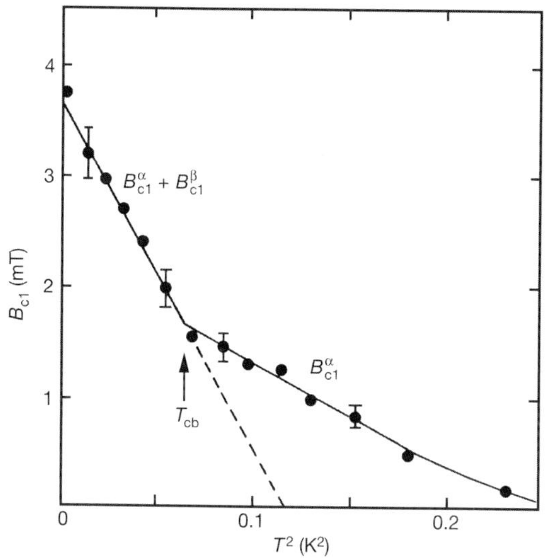

Fig. 19.4. The lower critical field H_{c1} of $U_{1-x}Th_xBe_{13}$ for $x = 0.03$, as measured by Rauchschwalbe et al. [38]. The temperature of the kink indicates the existence of a second transition

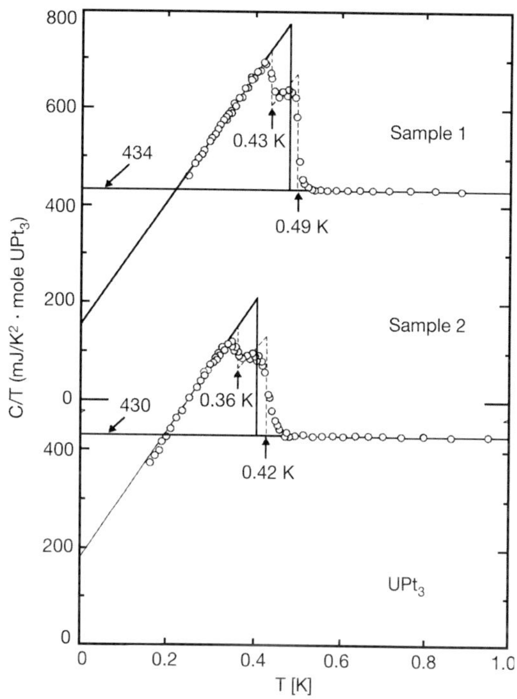

Fig. 19.5. The double transition in the specific heat found in two samples UPt_3 found by Fisher et al. [14]

superconducting phases, in analogy with the phase diagram of ^{3}He, is further evidence of the unconventional nature of the superconducting order parameter. If the system were an *s*-wave superconductor and the Fermi surface were isotropically gapped at T_{c1}, it seems very unlikely that the large jump in the specific heat at T_{c2} would be due to a weak magnetic transition that merely coexists with the superconducting transition. If the superconducting gap is anisotropic having nodes, the electronic states in the vicinity of nodes could take part in a magnetic transition and could result in the small size of the moments [36]. However, it still remains unlikely that magnetic ordering of these states could produce a large jump in the specific heat. An alternate possibility is that the second transition is between two distinct unconventional superconducting phases, one of which may involve broken time reversal symmetry [37]. This alternate possibility was given credence by the measurements performed by Rauchschwalbe et al. [38] where $H_{c1}(T)$ showed a quadratic T dependence with a coefficient that abruptly changes at T_{c2} (see Fig. 19.4). This suggests that the lower transition is between two superconducting phases that have different types of order parameters. The faster temperature variation of H_{c1} below T_{c2} is interpretable as due to an increase in the superconducting condensation energy.

Careful measurements on UPt_3 also revealed the presence of two jumps in the specific heat [14] with critical temperatures separated by 60 mK. The specific heat jumps for two different samples of UPt_3 are shown in Fig. 19.5. The application of a magnetic field in the basal plane reduces both transition temperatures [39], but the 60 mK splitting observed at H= 0 is diminished and the transitions merge at a critical point ($H_{c2}(T)$ = 5 kOe and T = 380 mK). The H–T phase diagram of UPt_3 is shown in Fig. 19.6. A definite kink in the temperature derivative of $H_{c2}(T)$ had been previously observed at this field [40, 41]. The splitting between the two jumps in the specific heat depends on the direction of the applied field. Application of a field parallel to the c direction does not reduce the splitting between the transitions, and for this field direction, the critical field does not show any discontinuity in the slope of $H_{c2}(T)$ up to the highest measured fields (7.5 kOe). The lower critical

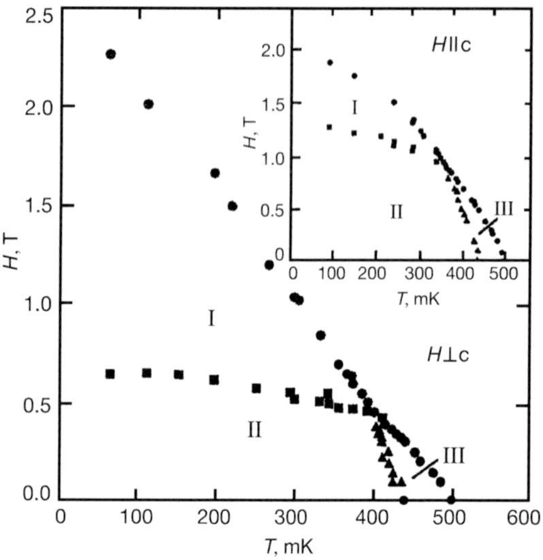

Fig. 19.6. The temperature-field phase diagram of UPt_3, as deduced from sound-velocity measurements of Adenwalla et al. [46]. The applied field is in the basal plane. The *inset* shows the phase diagram when the applied field is directed along the *c*-axis. The low field boundary separating the mixed and the Meissner phases is not shown

field, $H_{c1}(T)$, has a similar dependence on the orientation of the field [42]. A sharp kink in the temperature dependence of H_{c1} was observed for fields in the basal plane, and no kink was found for fields parallel to the *c*-axis. Evidence for further structure in the H–T phase diagram was provided by ultrasonic attenuation measurements. Early on, Qian et al. [43] and Müller et al. [44] observed a λ anomaly in the ultrasonic attenuation, just below T_c for longitudinal waves at H = 0. The position of the attenuation peak splits at higher fields, and the peak continues deep within the superconducting state as a cusp-like feature. Initially, the attenuation experiments were given scant attention by the scientific community, but given the preponderance of evidence for multiple superconducting phases, ultrasonic measurements are now recognized as providing excellent evidence for a phase boundary between different superconducting phases [45]. For H fields in the basal plane, this phase boundary appears to join up with the phase boundaries obtained from the two specific heat jumps at the very point where the splitting between the jumps disappears. The complete phase diagram of UPt_3 has

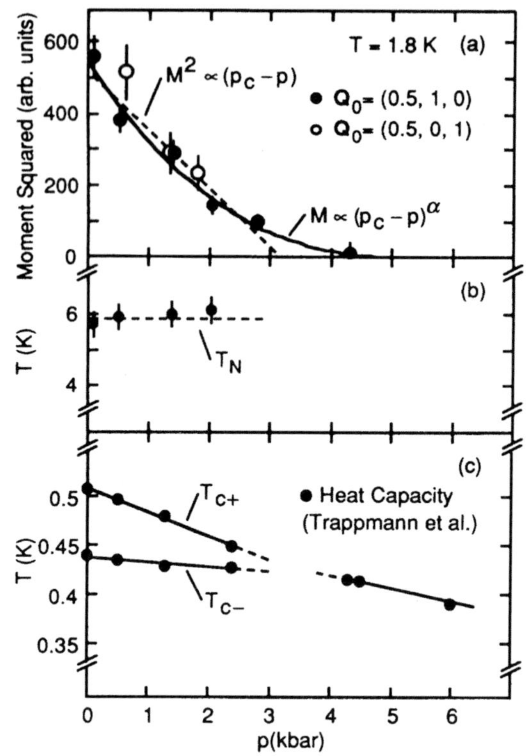

Fig. 19.7. The integrated intensity of the magnetic peaks $(\frac{1}{2}, 1, 0)$ and $(\frac{1}{2}, 0, 1)$ of UPt_3 at $T = 1.8$ K as a function of hydrostatic pressure [48] is shown in (**a**). The magnetic Bragg peak intensity should correspond to the square of the order parameter. The Néel temperature T_N, as determined from the integrated intensities, is shown as a function of pressure in (**b**). The critical pressure at which the magnetism disappears coincides with the pressure at which the specific heat jumps merge [49]

been obtained from anomalies in the measured velocity of acoustic phonons [46,47]. The phase diagram in the H–T plane shows the existence of five superconducting phases in contrast to the two usually observed in type II superconductors. The phase diagram contains two distinct Meissner phases in addition to three mixed phases seen in Fig 19.6. These phases of superconducting UPt_3 are discussed in more detail on page 1056. The transition to the superconducting phase of UPt_3, at zero field and ambient pressure, occurs at a temperature of 0.5 K, which is well below the Néel temperature of $T_N \sim 6$ K at which a small moment antiferromagnetic phase occurs. The application of uniaxial or hydro-static pressure shows that the splitting between the zero field superconducting transitions disappears above a critical pressure [48,50,51]. As shown in Fig. 19.7, the pressure at which the two jumps in the specific heat merge coincides with the pressure where the antiferromagnetism also disappears [48]. This suggests that the appearance of multiple superconducting phases in UPt_3 is intimately related to the occurrence of magnetic ordering.

19.2.2 Interplay of Superconductivity and Magnetism

A further notable characteristic of the heavy-fermion superconductors is that, with the exception of UBe_{13}, UGe_2, and URhGe, the superconducting phases coexist with antiferromagnetic correlations which have characteristic temperatures, usually T_N, that can be roughly an order of magnitude greater than the corresponding superconducting critical temperatures. The strengths of the antiferromagnetic correlations are weakest for the systems that show the largest mass enhancements, such as $CeCu_2Si_2$, UPt_3 and $U_{1-x}Th_xBe_{13}$, where the size of the moments is at most minute, of the order of $0.03\mu_B$. Muon spin resonance experiments on these three materials indicate that magnetic fluctuations have extremely long characteristic time scales. On the other hand, the more recently discovered compounds URu_2Si_2, UNi_2Al_3 [52] and UPd_2Al_3 [53] have much smaller γ coefficients which, if estimated above the respective Néel temperatures, are only as large as 150 mJ/mole K^2. The γ values below the Néel temperatures are reduced, indicating a partial gapping of the Fermi surface. These moderately enhanced materials have T_N's that can be as high as 14.5 K and have ordered magnetic moments that range up to 0.85 μ_B.

For a long time, $CeCu_2Si_2$ was the only known Ce based heavy-fermion superconductor at ambient pressure. However, this compound suffered from materials problems namely, a small change in stoichiometry could result in the ground state changing from superconducting to antiferromagnetic. Very recently it was found that $CeIrIn_5$ and $CeCoIn_5$ superconduct at $T_c = 0.4$ and $T_c = 2.3$ K, respectively [54,55]. Furthermore, these materials are almost always single crystals and apparently do not suffer from the same problems as exhibited by $CeCu_2Si_2$. These new heavy-fermion superconductors have a quasi-two-dimensional structure. They are quite anisotropic and exhibit well defined crystal field excitations at high temperatures [56] and at low temperatures show de Haas–van Alphen oscillations characteristic of anisotropic Fermi surfaces [57,58]. More important, they show that the superconductivity occurs in the vicinity of magnetism [59]. The quasi-two-dimensional nature of the materials and the anisotropy of the magnetically ordered states is favorable for the existence of large amplitude magnetic fluctuations in the superconducting state. In the superconducting state, the specific heat, thermal conductivities [60,61] and NMR $1T_1$ relaxation rates [62,63] show the power law temperature variations, which are consistent with the superconducting order parameter having lines of nodes. Since these materials share many common features with cubic $CeIn_3$, which is also antiferromagnetic and superconducts ($T_c \approx 200$ mK) at pressures greater than 25 kbar [64], together they form a family of materials in which the effect of structure (such as the role of dimensionality or magnetic anisotropy) on the interplay of superconductivity and magnetism can be investigated.

Since it was a commonly held belief that ferromagnetism is detrimental to superconductivity, it was a great surprise when superconductivity was discovered in the ferromagnetic phase of UGe_2 [66]. As shown in Fig. 19.8, the superconducting phase in UGe_2 occurs for pressures in the range of 1 to 1.5 GPa [67] where the material is ferromagnetically ordered. As the pressure is increased from 1 to 1.5 GPa, the critical temperature for ferromagnetic ordering shows indications of rapidly decreasing from about 30K to 0, while the maximum superconducting transition temperature is only 0.7 K and falls to zero at a pressure where the ferromagnetism disappears. URhGe which goes superconducting at ambient pressure for temperatures below 0.25 K, also has a ferromagnetic Curie temperature $T_c = 9.5$ K which is unusually small [68]. Since uniform internal magnetic fields are pair breaking for the singlet-pairs of a BCS superconductor, it has been suggested that in

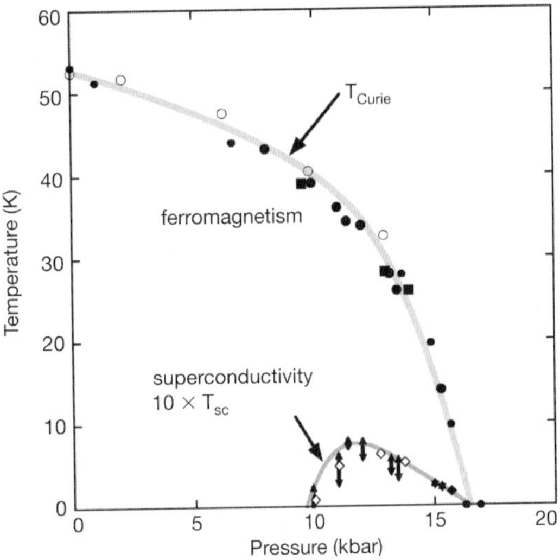

Fig. 19.8. The pressure-temperature phase diagram of the ferromagnetic superconductor UGe_2 [67]. The Curie temperatures were determined from the susceptibility (*filled circles*), resistivity (*open circles*) and neutron diffraction (*squares*). The onset and completion of the resistive superconducting transition are shown by the *filled triangles*. Note the change of scale for the superconducting transition temperature

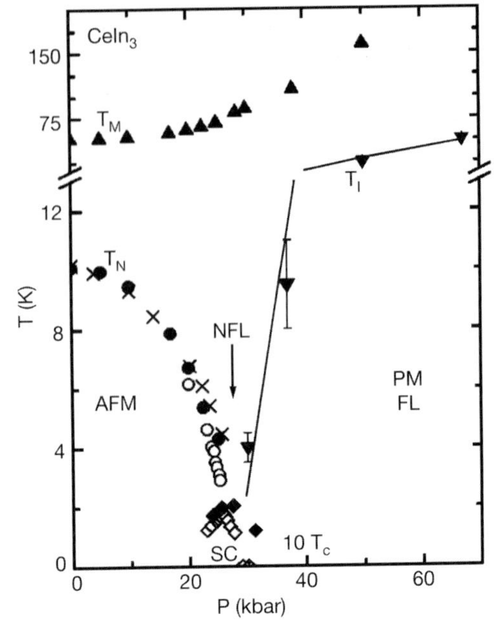

Fig. 19.9. The pressure-temperature phase diagram of the antiferromagnetic superconductor $CeIn_3$ [65]. The Neel temperature and superconducting transition temperatures are denoted by T_c and T_N, respectively. Note that T_c is scaled by a factor of 10. T_M denotes the temperature of the maximum in the resistivity, while T_I indicates a temperature at which the system crosses over into a Fermi liquid phase

these systems the Cooper pairs are in a triplet state. However, if this is the case, there seems to be no compelling reason as to why the superconducting phase should be restricted to only occur inside the ferromagnetic phase. On the other hand, if the ferromagnetic state can be described strictly as a local Fermi liquid [70], then it has been rigorously shown [71] that there is an attractive *s*-wave interaction and the triplet interaction is identically zero. Furthermore, a mean field analysis shows that a singlet superconducting state in a ferromagnet may survive if the pair breaking due to the uniform internal field is sufficiently strong [72].

The observation of the superconducting phase within ferromagnetic and antiferromagnetic phases with low transition temperatures shows that the heavy-fermion materials are often correlated with the proximity of a quantum critical point [73–75]. This point is illustrated by the phase diagrams shown in Figs. 19.8 through 19.10. At a quantum critical point, the large amplitude low frequency magnetic fluctuations could produce appreciable con-

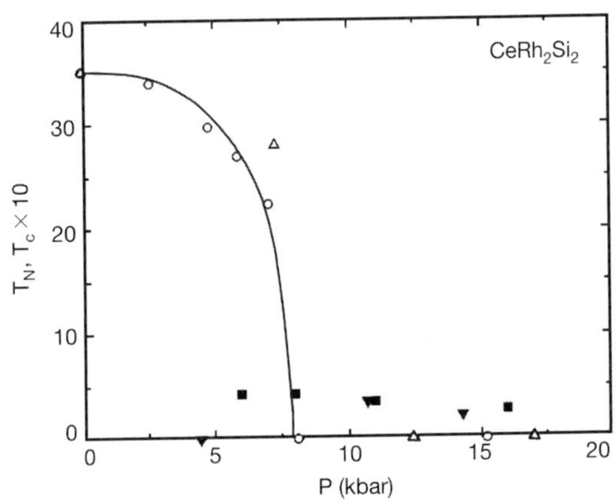

Fig. 19.10. The pressure-temperature phase diagram of the antiferromagnetic superconductor $CeRh_2Si_2$ [69]. The *solid symbols* represent the superconducting transition temperature T_c, while the *open symbols* denote the Neel temperature T_N

tributions to physical properties that are different from those expected of highly renormalized quasi-particles. The existence of large amplitude magnetic fluctuations associated with the quantum critical point leads to another exciting possibility namely, that the superconducting pairing mechanism for the quasi-particles is primarily mediated by low-energy spin-fluctuations [76]. Either the characteristic frequency of the low-energy spin-fluctuations or the mass renormalizations associated with the heavy quasi-particles presumably, could be responsible for setting the low values of the superconducting transition temperatures, T_c. Generally, heavy-fermion superconductors have superconducting transition temperatures in the range between 0.2 and 3 K. The compound $PuCoGa_5$ provides a notable exception to this statement, as it has a T_c of about 18.5 K, which is the highest reported T_c for a heavy-fermion superconductor [77]. The low values of T_c found in most heavy-fermion superconductors are in stark contrast with the very large critical temperatures found in the other well known examples of exotic superconductivity – the high temperature superconducting cuprates.

19.2.3 Quasi-Particles and Collective Excitations

At high temperatures, the properties of heavy-fermion systems can often be described in terms of a set of local moments coupled to a sea of conduction electrons. At temperatures below a characteristic temperature, sometimes known as the coherence temperature, the properties show evidence that the excitations have large spatial extents. Below the coherence temperature, the transport properties indicate that the scattering from the spin degrees of freedom start to freeze out and that the electronic excitations extend throughout the crystal. For sufficiently low temperatures, one may expect that the properties will be described by the quasi-particle excitations of Landau Fermi liquid theory. However, in a few of the heavy-fermion systems, the Fermi liquid state is never completely formed before superconductivity sets in. In these cases one expects that, below the coherence temperature, the properties may be determined by the low-energy excitations that include both the collective excitations, such as phonons and spin-waves, as well as the quasi-particle excitations.

Quasi-Particle Excitations

Elementary excitations can be categorized either as quasi-particles or as collective excitations. The quasi-particle excitations are in one-to-one correspondence with the excitations of the non-interacting system. The quasi-particle excitations have a close similarity to the single-electron excitations of a non-interacting electron gas. The properties of these quasi-particles are most readily seen through inspection of the one-electron thermal Green's function [78] defined by the expectation value of the time ordered product of the thermal average

$$G^{\alpha,\beta}_{\underline{k},\underline{k}'}(\tau) = -\frac{1}{\hbar} < |\hat{T} a_{\underline{k}',\beta}(\tau) a^{\dagger}_{\underline{k},\alpha}(0)| > . \qquad (19.2)$$

The Green's function represents the time evolution of the probability amplitude for a single electron to be added to the Bloch state with wave vector $\underline{k}$ and spin σ. The electron creation and annihilation operators are evaluated in the imaginary time representation where they evolve according to the prescription

$$a_{\underline{k},\alpha}(\tau) = \exp\left[+\frac{\hat{H}\tau}{\hbar}\right] a_{\underline{k},\alpha}(\tau) \exp\left[-\frac{\hat{H}\tau}{\hbar}\right] . \qquad (19.3)$$

Due to periodic translational invariance and spin rotational invariance of the normal state, the Green's function is diagonal in the wave vector and spin indices

$$G^{\alpha,\beta}_{\underline{k},\underline{k}'}(\tau) = \delta_{\alpha,\beta}\, \delta_{\underline{k},\underline{k}'}\, G(\underline{k};\tau) . \qquad (19.4)$$

The Fourier transform of the diagonal Green's function is defined via

$$G(\underline{k};\tau) = k_B T \sum_n G(\underline{k}; i\hbar\omega_n) \exp\left[-i\omega_n\tau\right] , \qquad (19.5)$$

where

$$\hbar\,\omega_n = k_B T \pi (2n+1) \qquad (19.6)$$

are the Matsubara frequencies. The interacting Green's function is expressed in terms of the non-interacting Green's function $G^0(\underline{k}, i\omega_n)$ and the self-energy $\Sigma(\underline{k}, i\omega_n)$ through Dyson's equation

$$\left[G(\underline{k}; i\hbar\omega_n) \right]^{-1} = \left[G^0(\underline{k}; i\hbar\omega_n) \right]^{-1} - \Sigma(\underline{k}; i\hbar\omega_n) , \tag{19.7}$$

where the non-interacting Green's function is evaluated as

$$\left[G^0(\underline{k}; i\hbar\omega_n) \right]^{-1} = i\hbar\, \omega_n - e(\underline{k}) + \mu . \tag{19.8}$$

The pole of the non-interacting Green's function is at the single-electron Bloch energy $e(\underline{k}) - \mu$, and measures the excitation energy relative to the Fermi energy. The self-energy represents the change in the Green's function due to the interactions. The Fermi energy of the interacting system, μ, is determined by the pole of the Green's function at $\omega_n = 0$, which leads to

$$\mu = e(\underline{k}_F) + \Sigma(\underline{k}_F; 0) , \tag{19.9}$$

where $\underline{k}_F$ is a Bloch vector on the Fermi surface. The thermal Green's function is related to the $T = 0$ Green's function via analytic continuation $i\hbar\omega_n \to E$. The correspondence between quasi-particle excitations and the single-electron excitations of the non-interacting system follows from the expansion of the self-energy near the Fermi energy. For energies close to the Fermi energy, the Green's function can be re-written as

$$\begin{aligned} \left[G(\underline{k}; E) \right]^{-1} &\approx E \left(1 - \frac{\partial \Sigma(\underline{k}; E)}{\partial E} \right) \Bigg|_{\omega=0} \\ &\quad - e(\underline{k}) - \Sigma(\underline{k}; 0) + \mu - i Im\Sigma(\underline{k}; E) \\ &= \left(1 - \frac{\partial \Sigma(\underline{k}; E)}{\partial E} \right) \Bigg|_{\omega=0} \\ &\quad \times \left[E - E_{\underline{k}} + \frac{i\hbar}{2\, \tau^*_{\underline{k}}(E)} \right] . \end{aligned} \tag{19.10}$$

The interaction produces a linear superposition of the electron in a Bloch state $\underline{k}$ with one-electron states surrounded by a cloud of electron-hole pairs. The quasi-particle weight, $Z_{\underline{k}}^{-1}$, represents the fraction of this superposition that corresponds to the bare Bloch electron. The fraction is less than unity $Z_{\underline{k}} > 1$, and $Z_{\underline{k}}$ is given in terms of the frequency derivative of the self-energy by

$$Z_{\underline{k}} = 1 - \frac{\partial \Sigma(\underline{k}; E)}{\partial E} \Bigg|_{E=0} . \tag{19.11}$$

The quasi-particle energy $E_{\underline{k}}$, measured from the Fermi energy μ, and the decay rate are given by

$$\begin{aligned} E_{\underline{k}} &= e(\underline{k}) + \Sigma(\underline{k}; E_{\underline{k}}) - \mu , \\ E_{\underline{k}} &\approx Z_{\underline{k}}^{-1} \bigg(e(\underline{k}) + \Sigma(\underline{k}; 0) - \mu \bigg) , \end{aligned} \tag{19.12}$$

and

$$\frac{\hbar}{2\, \tau^*_{\underline{k}}(E)} = -Z_{\underline{k}}^{-1} Im\Sigma(\underline{k}; E + i\delta) , \tag{19.13}$$

respectively. Since the imaginary part of the Green's function is proportional to the single-particle density of states, the self-energy can be viewed as a renormalization of the single-electron excitation energies $e(\underline{k})$, yielding the quasi-particle energy $E_{\underline{k}}$. The imaginary part of the self-energy can be viewed as providing the width or lifetime of the single-particle state. As shown by Luttinger [79], the imaginary part of the self-energy near the Fermi energy due to electron–electron interactions vanishes proportional to E^2, if perturbation theory converges. The small magnitude of the lifetime is due to the Pauli exclusion principle which reduces the phase space allowed for electron–electron scattering. The smallness of the lifetime of low-energy excitations has the effect that the spectrum resembles that of a non-interacting electron gas in which the quasi-particle masses are enhanced by a factor of $Z_{\underline{k}}$. The quasi-particle states are extremely long lived, not only because of the vanishing of the lifetime due to electron–electron interactions but also because the residual lifetime resulting from elastic scattering by impurities is enhanced by the factor of $Z_{\underline{k}}$. Due to the extremely small magnitude of the lifetime of quasi-particles at the Fermi energy, the Fermi energy is well defined and the effect of electron–electron interactions does not change the volume enclosed by the Fermi surface [79]. It is found that, in de Haas–van Alphen experiments on some of the heavy-fermion systems, the multi-sheeted Fermi surfaces enclose volumes that are consistent with Luttinger's theorem being valid. For heavy-fermion systems, the $\underline{k}$ dependence of the self-energy is considered to be small, and the frequency dependence is extremely rapid. Thus, the quasi-particles are expected to be extremely heavy and long-lived but have little spectral weight. Also, a large portion of the weight is expected to lie in the broad incoherent portion of the spectral density.

Since the quasi-particles are governed by Fermi–Dirac statistics, their contributions to thermodynamic quantities have asymptotic low temperature variations that are similar to those of the non-interacting electron gas. In particular, the entropy S of the gas of quasi-particles is given by

$$S = -k_B \sum_{\sigma,\underline{k}} \Big[(1 - f(E_{\underline{k}})) \ln[1 - f(E_{\underline{k}})] + f(E_{\underline{k}}) \ln[f(E_{\underline{k}})] \Big] . \quad (19.14)$$

Hence, the quasi-particles give rise to a linear T contribution in the low temperature electronic specific heat. However, the coefficient γ, instead of just reflecting the electronic density of states, is given by the density of quasi-particle energies at the Fermi surface

$$\rho_{qp}(E) = \sum_{\underline{k}} \delta(E - E_{\underline{k}}) ,$$
$$\rho_{qp}(0) \sim \sum_{\underline{k}} Z_{\underline{k}} \delta(\mu - e(\underline{k}) + \Sigma(\underline{k};0))) , \quad (19.15)$$

which is enhanced over the electronic density of states $\rho(\mu)$ by a factor Z similar to the Fermi surface average of $Z_{\underline{k}}$. Comparison of the low temperature electronic specific heat coefficient γ and electronic structure calculations, yields an estimate of the wave function renormalization Z of approximately 25 for highly enhanced systems such as UPt_3. The heavy quasi-particle masses in some or all parts of the Fermi surface can also be inferred from the amplitude of the de Haas–van Alphen oscillations in the magnetization [80]. The signatures of the gas of heavy quasi-particles may also be expected to show up in transport properties, albeit modified by the residual interactions between the quasi-particles.

The temperature dependence of the d.c. resistivity of the enhanced Fermi liquid state is dominated by the transport scattering rate. If the self-energy is roughly k independent, the transport scattering rate should coincide with the scattering rate $\frac{1}{\tau}$ found from the imaginary part of the self-energy since the conductivity vertex corrections are expected to be small. For low temperatures and samples of high purity, Matheissen's rule is expected to apply. In this case, the scattering rate is additive, and $\frac{1}{\tau}$ is expected to be composed of a sum of a temperature independent term $\frac{1}{\tau_0}$ due to the potential scattering from isolated impurities, a quadratic Baber term caused by the quasi-particles scattering off of each other, and a negligibly small T^5 term expected from electron–phonon scattering

$$\frac{1}{\tau} = \frac{1}{\tau_0} + AT^2 + BT^5 . \quad (19.16)$$

If the impurity scattering can be treated in the Born approximation, then, due to the approximate invariance of the density of states at the Fermi energy, the impurity scattering rate $\frac{1}{\tau_0}$ is not directly renormalized by the electron–electron interactions. The T^2 Baber term has its origin in the Pauli-exclusion principle limiting the phase space available for scattering of low-energy electrons, and is exactly the same physics behind the E^2 variation of the imaginary part of the self-energy. Since Baber scattering involves the scattering of two quasi-particles, the scattering rate is enhanced by a factor proportional to Z^2. The residual d.c. conductivity does not directly depend on the real part of the self-energy and is, therefore, un-renormalized. Alternately, the d.c. residual resistivity is un-renormalized due to the small magnitude of the velocity vertex correction, and due to the cancelation of the wave function renormalization in the ratio of the renormalized quantities

$$\frac{\tau^*}{m^*} \approx \frac{\tau}{m} \quad (19.17)$$

and also because the electron density n is unchanged by electron–electron interactions. This last fact is seen by noting that n is proportional to the Fermi surface volume which, according to Luttinger's theorem, is independent of the strength of electron–electron

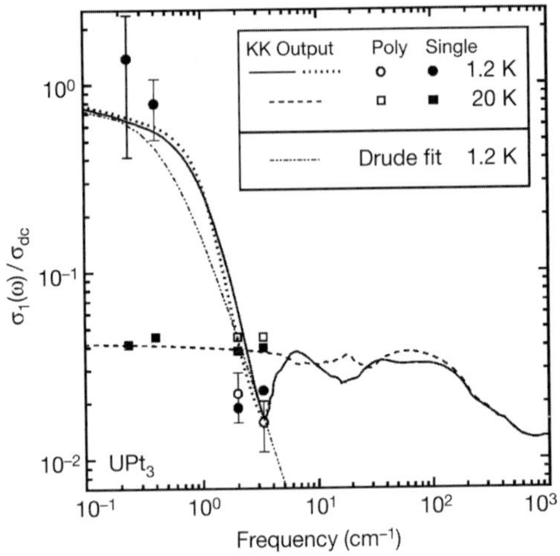

Fig. 19.11. The optical conductivity $\sigma(\omega)$ of single crystals and polycrystals of UPt_3 at $T = 1.2$ K and $T = 20$ K [81]. The data show the growth of the narrow quasi-particle Drude peak at low temperatures

interactions. Despite the absence of significant renormalization of the d.c. conductivity, the renormalized lifetimes do show up as extremely narrow widths of the Drude peak [81–83]

$$Re\Big[\sigma(\omega)\Big] = \frac{n\,e^2\tau^*}{m^*}\,\frac{1}{1+\omega^2\tau^{*\,2}} \tag{19.18}$$

observed in measurements of the dynamical conductivity $\sigma(\omega)$ at low temperatures. The frequency dependence of the measured conductivity $\sigma(\omega)$ for UPt_3 [81], is shown in Fig. 19.11 for low and high temperatures. Basically, if one has a fixed concentration of impurities, and hence a fixed mean free path, then the quasi-particle lifetime is just determined by the quasi-particle velocity $\hbar\, k_F/m^*$, which is reduced by the large quasi-particle mass. Therefore, the enhanced effective mass results in an enhancement of the lifetime due to impurity scattering. At higher frequencies, one expects that inelastic scattering processes should become important and the quasi-particle weight should acquire a frequency dependence. The frequency dependence of the scattering rate and the quasi-particle renormalization are expected to be related by causality and other requirements. In particular, the optical sum rule [84] relates the integral of the optical conductivity over all frequencies to the total number of electrons in the system

$$\int_0^\infty d\omega\; Re\Big[\sigma(\omega)\Big] = \frac{\pi}{2}\,\frac{n\,e^2}{m_e}\;, \tag{19.19}$$

where n is the density of electrons and m_e is the electron mass. Similarly, the integrated intensity of the low frequency Drude peak

$$\int_0^{\omega_0} d\omega Re\Big[\sigma(\omega)\Big] = \frac{\pi}{2}\,\frac{n(\omega_0)e^2}{m^*} \tag{19.20}$$

can be used to define the number of coherent quasi-particles $n(\omega_0)$ and their weight Z^{-1} [85], whereas the higher energy structure is due to the incoherent excitations. The existence of the quasi-particle Drude peak has been confirmed in UPt_3, $CeAl_3$ and $CeCoIn_5$ [81,82,86], however, it has not been observed in UBe_{13} [87] where it is doubtful that a Fermi liquid is formed at temperatures higher than the superconducting T_c. In the cases where the Fermi liquid is fully formed, it is not expected that good agreement will be found between the optical effective mass, whose definition involves the Fermi surface average of the inverse quasi-particle mass, and the Fermi surface average quasi-particle mass obtained from specific heat. The disagreement is expected to be marked specially if the de Haas–van Alphen experiments show both light and heavy quasi-particle excitations coexisting on the Fermi surface.

In systems like UBe_{13}, the Fermi liquid phase is not completely formed before superconductivity sets in, therefore, the thermodynamic and transport properties may be directly affected by the collective excitations of the electrons. In addition, the large mass renormalization Z of the quasi-particles might also be attributable to the existence of low frequency collective excitations, such as local spin-fluctuations or more extended magnetic excitations that are precursors of long-ranged magnetic ordering. The collective excitations are directly amenable to experimental observation and also may mediate residual interactions between the quasi-particles, and therefore, they could be responsible for the superconducting pairing.

Collective Excitations

Since the normal states of heavy-fermion materials are characterized by a large quasi-particle density of states near the Fermi-level, they are susceptible to entropy-driven instabilities, which reduce the density of states at the Fermi energy. This tendency is manifested by the sensitivity of the normal state to small amounts of added impurities that can lead to an instability towards states with spontaneously broken symmetries. If the interactions are short-ranged and the symmetry that is broken is continuous, Goldstone's theorem [88] is valid. Goldstone's theorem ensures that the system will support a branch of collective excitations with a zero threshold energy that dynamically restores the spontaneously broken symmetry. The order parameter acts as the collective coordinate for the zero energy collective excitations. The spin-waves with $\underline{q} \approx 0$ in a ferromagnet, the antiferromagnetic spin-waves near the critical wave vector(s) $\underline{Q}_c$, and the transverse sound waves in a periodic solid, form well known examples of these Goldstone collective modes. Similar boson-like collective excitations are expected to occur in the disordered or high temperature state as precursors to the instabilities. In the disordered state, these boson modes are expected to have extremely long lifetimes and have excitation spectra that form broad continua. A well known example of these precursor modes is provided by the paramagnon fluctuations in Pd, which occur as Pd is very close to an instability to a ferromagnetic state [89–91]. It is expected that, as the temperature is lowered through the instability, these pre-critical modes will merge together with the critical fluctuations and, eventually, the (Goldstone) spin-wave modes will emerge in the ordered state. There is a growing body of evidence that suggests that heavy-fermion systems are in the vicinity of a quantum critical point, implying that the system supports large amplitude critical fluctuations due to a nearby $T = 0$ phase transition. One expects that the properties of the material should show scaling behavior due to the quantum critical point. The critical fluctuations near a quantum critical point are expected to have a different nature than those associated with a finite temperature transition as they cannot be treated classically [73–75]. The zero-point quantum fluctuations replace the role of the thermally-driven fluctuations. Since the dynamics are inextricably linked to the statics at a quantum critical point, the phase space of the magnetic fluctuations is given by $(\omega, \underline{k})$. If the characteristic frequency scales as ξ^{-z}, where ξ is the magnetic correlation length and z is the dynamical exponent, then the effective dimensionality of the phase space for a $T = 0$ quantum critical point is given by $d_{\text{eff}} = d + z$. Hence, as the effective dimensionality d_{eff} differs from the dimensionality d of the classical critical point, one expects to find different types of scaling relations at a quantum critical point. The search for the ultimate description of quantum critical fluctuations is an actively ongoing field of research.

The simplest starting point for these theories of the collective spin-fluctuation modes lies in the random phase approximation (RPA) [92,93]. The RPA is the crudest approximation that captures the physics of the Gaussian fluctuations and is most certainly expected to fail near the quantum critical point. In most approaches, one assumes that the Coulomb interaction is highly screened, and this results in a Hubbard point contact interaction U between electrons of opposite spins. Within the quasi-particle treatment, the band energies $e(\underline{k})$ should be replaced by the quasi-particle energies $E_{\underline{k}}$, and the interaction should be expressed in terms of the Landau Fermi-liquid parameters. However, we shall, in the rest of this section, be consistent with the usual formulation of RPA as a one parameter Fermi liquid, in which the interaction between the quasi-particles is denoted by U. Since we shall neglect the vertex corrections to the susceptibility, we shall also suppress all the factors of Z^{-1}, while it is true that Fermi liquid corrections should renormalize U to UZ^{-2}. Although the results are derived on the basis of a one band Hubbard model, they can easily be extended to a two band or Anderson Lattice model [94]. The transverse susceptibilities are expressed in terms of multiple scattering processes involving an up-spin electron with a down-spin hole shown in Fig. 19.12 [92,93], yielding

$$\chi^{+-}(\underline{q};\omega) = \frac{\mu_B^2 \chi_0(\underline{q};\omega)}{1 - U\chi_0(\underline{q};\omega)} , \tag{19.21}$$

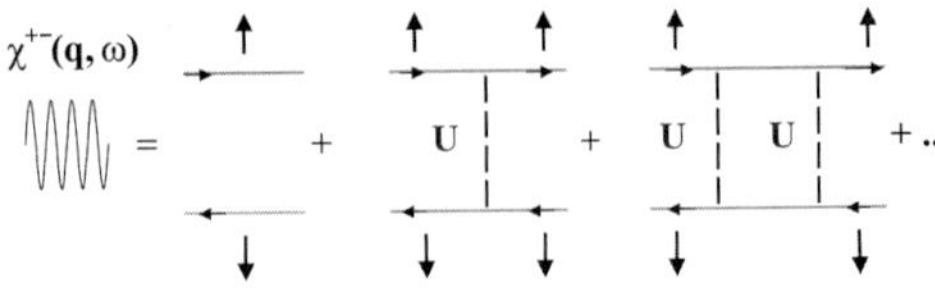

Fig. 19.12. Diagrammatic representation of the Feynmann diagrams leading to the RPA expression for the transverse spin-fluctuations. The *directed solid lines* denote the one-electron Green's functions. The *vertical arrows* represent the directions of the electronic spins. The *dashed vertical line* represents the on-site Coulomb interaction U

where $\chi_0(\underline{q};\omega)$ is the non-interacting Lindhard susceptibility, given by

$$\chi_0(\underline{q};\omega) = \frac{1}{N}\sum_k \times \left[\frac{f(e(\underline{k})) - f(e(\underline{k}+\underline{q}))}{\hbar\omega - e(\underline{k}) + e(\underline{k}+\underline{q}) + i\,\delta}\right]. \tag{19.22}$$

In the RPA, a magnetic instability of the paramagnetic state towards an ordered state with ordering wave vector $\underline{Q}$ is obtained when the static susceptibility $\chi(\underline{Q};0)$ diverges, which occurs due to the vanishing of a denominator. This happens when the generalized Stoner criterion is fulfilled,

$$1 = U\,\chi_0(\underline{Q};0)\,, \tag{19.23}$$

where $\chi_0(\underline{Q};0)$ is the reduced non-interacting susceptibility. The paramagnetic state is unstable for values of U greater than a critical value U_c where the equality of (19.23) first holds, at any value of $\underline{Q}$. The type of instability, determined by $\underline{Q}$ and the critical value of U at which it occurs, is governed by both the quasi-particle band structure and the state of occupation of the bands. If $\chi_0(\underline{Q};0)$ is largest at $\underline{Q} = 0$, the system is expected to become unstable to a ferromagnetic state at a critical value of U determined by the usual Stoner criterion for ferromagnetism, $1 = U_c\rho(\mu)$. For perfect nesting tight-binding bands at half filling, one finds that $\chi(\underline{Q};0)$ diverges for $\underline{Q} = \pi(1,1,1)$, leading to an instability towards an antiferromagnetic state for U greater than the critical value of $U_c = 0$ [95]. In general, for U values close to the critical value U_c, the static susceptibility evaluated at the relevant $\underline{Q}$ is enhanced, and

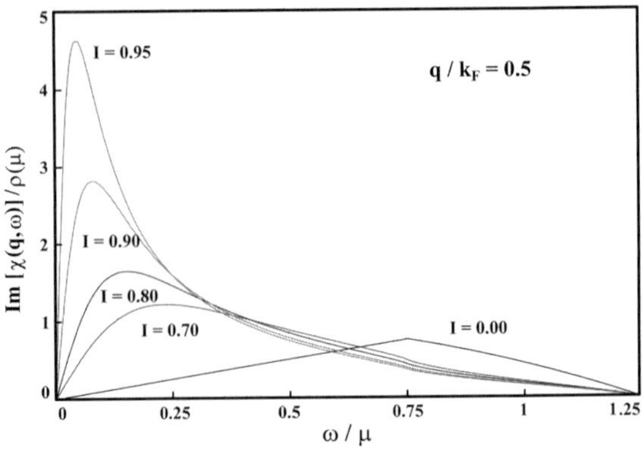

Fig. 19.13. The frequency variation of $Im\chi(\underline{q};\omega)$ for fixed $\underline{q}$. The parameter I is given by $I = U\rho(\mu)$. The damped spin-fluctuations soften and grow in amplitude as the quantum critical point is approached. [After Doniach [96]]

the imaginary part of the susceptibility undergoes a similar enhancement. Since the imaginary part of the susceptibility is a measure of the spectrum of magnetic excitations, the enhanced RPA expressions

$$Im\left[\chi^{+,-}(\underline{q};\omega)\right] = \frac{\mu_B^2 Im\,\chi_0^{+,-}(\underline{q};\omega)}{[1 - URe\,\chi_0^{+,-}(\underline{q};\omega)]^2 + [UIm\chi_0^{+,-}(\underline{q};\omega)]^2} \tag{19.24}$$

show the propensity for low frequency large amplitude spin-fluctuation excitations, as shown in Fig. 19.13. Near the instability, the magnetic excitation spectrum consists of a continuum of low-energy (quasi-elastic) and over-damped precritical fluctuations from which, on increasing U above U_c, a branch of sharp spin-wave excitations are expected to emerge in the magnetically ordered state.

The large amplitude spin-fluctuations are also expected to give rise to a renormalization of the quasi-particles. The change in the energies of the quasi-particles shows up in the RPA self-energy due to the emission and absorption of spin-waves, which flip the spin of the electron (see Fig. 19.14). The quasi-particle weight of the low frequency excitations is reduced as the scattering from the large amplitude spin-fluctuations reduces the probability that the electron remains in a spatially extended Bloch state. For a nearly ferromagnetic system, this leads

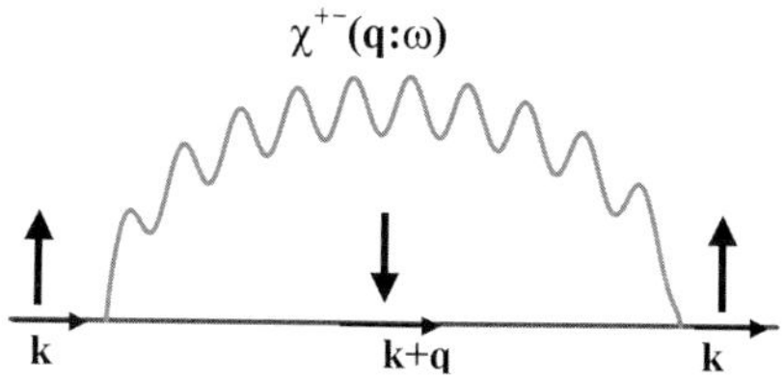

Fig. 19.14. The RPA expression for the up-spin electron self-energy $\Sigma(\underline{k};E)$. The one-electron Green's function is denoted by the *directed line*, and the spin-fluctuation by a *wavy line*. In this process, an up-spin electron of momentum $\underline{k}$ emits a spin-fluctuation of momentum $\underline{q}$, thereby flipping its spin

to a logarithmic enhancement of the linear term in specific heat [90,91] via

$$C \propto k_B T \ln \left| 1 - U\rho(\mu) \right| . \tag{19.25}$$

Using a Fermi liquid approach, Carneiro and Pethick [97] have shown that long wavelength collective fluctuations can also lead to a $T^3 \ln T$ term in the specific heat similar to that found in paramagnon theories [98]. Likewise, the collective fluctuations also can lead to an enhancement of the quasi-particle scattering rate which in turn, leads to an enhanced T^2 term in the electrical resistivity [89,99]. The precise form of the renormalization found in RPA does depend crucially on the type of magnetic instability (ferromagnetic, incommensurate spin density wave, antiferromagnetic) that is being approached. For example, in quantum phase transitions with finite ordering wave vectors $\underline{Q}$, electrons are resonantly scattered by magnetic fluctuations between portions of the Fermi surface that are connected by the vector $\underline{Q}$ (see Fig. 19.15). This leads to the occurrence of hot lines on the Fermi surface where the quasi-particles are extremely short lived and, therefore, are not well-defined. The size of these hot regions is proportional to $\sqrt{T}$. On the other hand, for a ferromagnetic quantum critical point, the entire Fermi surface is subject to critical scattering. The different nature of the critical scattering results in different power law temperature dependences of various physical quantities [100]. For example, the leading non-analytic part of the free energy ΔF is given by

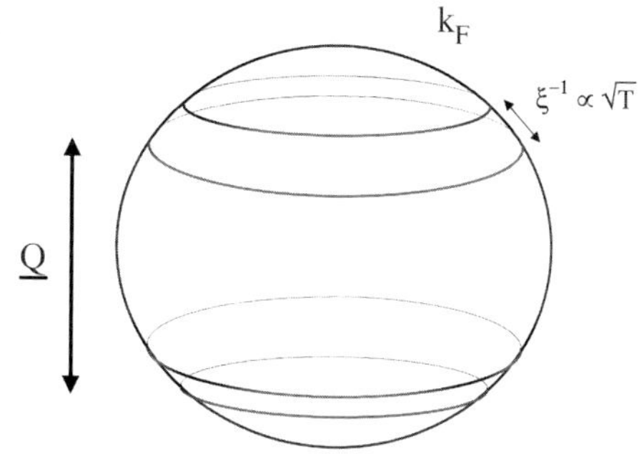

Fig. 19.15. Hot lines on a Fermi surface. The dynamic susceptibility becomes critical at wave vector $\underline{Q}$. The electrons on the two lines of the Fermi surface which are connected by wave vector $\underline{Q}$ are subject to resonant scattering. The width of the hot lines is given by ξ^{-1}, which is proportional to $\sqrt{T}$

$$\Delta F = \sum_{\underline{q}} \int_0^\infty \frac{d\omega}{\pi} \left[N(\omega) + \frac{1}{2} \right] Im \left[\ln U\chi(\underline{q};\omega) \right], \tag{19.26}$$

where $N(\omega)$ is the Bose–Einstein distribution function, and the spin-fluctuation propagator either has the form

$$\chi(\underline{q};\omega)^{-1} \sim (1-I) + a\left(\frac{q}{k_F}\right)^2 + ib\left(\frac{\hbar\omega k_F}{\mu q}\right) \tag{19.27}$$

near a ferromagnetic instability or has the form

$$\chi(\underline{q};\omega)^{-1} \sim (1-I) + a\,\frac{(\underline{q}-\underline{Q})^2}{k_F^2} + ib\left(\frac{\hbar\omega}{\mu}\right) \tag{19.28}$$

near an antiferromagnetic instability. For a d-dimensional system that is above the critical dimensionality, at a quantum critical point where $I = 1$, this produces a leading $T^{1+d/3}$ temperature dependence of ΔF for a ferromagnetic quantum critical point, but near an antiferromagnetic quantum critical point ΔF has a $T^{1+d/2}$ dependence. The temperature dependence of ΔF gives rise to the non-analytic temperature dependences of the C/T ratio.

Close to a quantum critical point, where both thermaland quantum critical fluctuations are important, the properties are expected to be significantly different from the properties calculated using simple

RPA. At finite temperatures and in the quantum critical region, the effect of coupling among the different modes of spin-fluctuations becomes important [101]. The theories of Moriya [93], Hertz [73] and Millis [75] predict that three-dimensional electronic systems at a quantum critical point have an effective dimensionality greater than the upper critical dimension and, hence, are dominated by Gaussian spin-fluctuations, albeit highly renormalized by mode–mode coupling. In this case, the hyperscaling relation is not expected to be obeyed. The strength of the mode–mode coupling is expected to vanish at zero temperature. It is generally believed that the region over which the Fermi liquid behavior is found, is smaller for systems which are close to exhibiting a magnetic instability. Furthermore, as the quantum critical point is approached, the Fermi liquid power laws are expected to be gradually replaced by other types of non-universal power laws. For example, the T^2 variation of the resistivity found in the Fermi liquid regime of a clean three-dimensional metal is expected to be replaced by a $T^{\frac{5}{3}}$ variation at a ferromagnetic quantum critical point or a $T^{\frac{3}{2}}$ variation at an antiferromagnetic quantum critical point. These power laws are intermediate between the low temperature Fermi liquid T^2 variation and the linear T variation found within RPA at higher temperatures and are consistent with expectations based on the shrinking temperature range over which Fermi liquid behavior is to be observed. The simple power laws obtained using self-consistent spin-wave theory are only expected to be recovered at low temperatures. Furthermore, the scaling behavior expected from the quantum critical point is expected to be severely modified by the effects of disorder [102,103]. In the case of an antiferromagnetic quantum critical point in a clean system, the hot lines are not expected to dominate drastically the low temperature physical properties, since the hot lines have a limited extent. For example, the contributions of the hot lines to the conductivity are expected to be shorted out by the normal regions of the Fermi surface [104].

Table 19.1. Quantum critical exponents for physical properties

$d = 3$	Ferro	Antiferro
C/T	$-\ln T$	$\gamma - \alpha\sqrt{T}$
$\Delta\chi_{\underline{Q}}^{-1}$	$T^{\frac{4}{3}}$	$T^{\frac{3}{2}}$
$\Delta\rho$	$T^{\frac{5}{3}}$	$T^{\frac{3}{2}}$
$d = 2$	Ferro	Antiferro
C/T	$T^{-\frac{1}{3}}$	$-\ln T$
$\Delta\chi_{\underline{Q}}^{-1}$	$-T\ln T$	$-T/\ln T$
$\Delta\rho$	$T^{\frac{4}{3}}$	T

However, the presence of impurities leads to $\underline{k}$ not being a good quantum number since electrons are inelastically scattered between different portions of the Fermi surface. Hence, the mixing between different $\underline{k}$ values results in all the electrons on the Fermi surface participating in the critical scattering [103].

In general, the power laws found in heavy-fermion systems do not coincide with those found in the above-mentioned type of two or three-dimensional theories (shown in Table 19.1). One possible cause for this discrepancy is perhaps due to the heaviness of the quasi-particle masses. That is, in the above theories, the electron dynamics are assumed to occur on a fast energy scale compared with the slow critical fluctuations, therefore the fast electron dynamics can be integrated out. For systems with high effective masses, such descriptions may no longer be appropriate [105]. Related ideas about the lack of scaling being caused by the breakdown of the quasi-particle concept have been expressed by Coleman [106]. An alternate possible cause for the discrepancy could be due to the nature of the assumed theoretical model.[2] Generally, it has been assumed that strong electron–electron interactions, which give rise to the formation of heavy quasi-particles and strong magnetic fluctuations, are unfavorable for the formation of superconducting pairs. However, the observation of superconducting phases only in the immediate vicinity of quantum critical points, as sketched in Fig. 19.16,

[2] The Moriya, Hertz, Millis theory assumes the validity of a non-degenerate one-band model, whereas multi-band models, with orbital degeneracy and strong spin-orbit coupling, appear to be more appropriate for describing heavy-fermion systems. These other models may be in a different universality class and, hence, have other critical exponents.

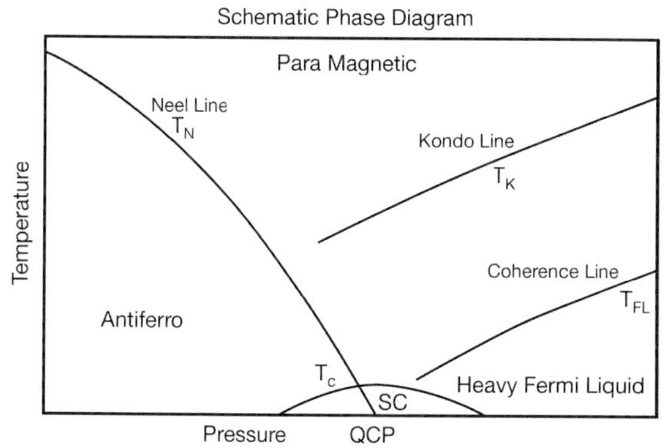

Fig. 19.16. A schematic pressure-temperature phase diagram near a quantum critical point (QCP). The *solid lines* T_N and T_c, respectively, denote the transition temperatures to the Néel and superconducting phases. The *dashed lines* represent the characteristic temperatures associated with the Kondo effect, T_K and the low temperature Fermi Liquid T_{FL}

challenges the assumption. These observations suggest that the large amplitude quantum critical fluctuations might even be responsible for the occurrence of the superconductivity. The two most outstanding questions about the superconductivity in heavy-fermion systems concern the nature of the pairing mechanism and the symmetry of the superconducting order parameter.

19.2.4 Possible Pairing Mechanisms

The interaction mechanism that is responsible for pairing electrons in common superconductors is mediated by phonons. Fröhlich [107] predicted that the superconducting transition temperature T_c should be proportional to a typical phonon frequency. Furthermore, as the phonon frequency squared is inversely proportional to the mass of the ions, M, Fröhlich predicted that the transition temperature should vary as

$$T_c \propto M^{-\frac{1}{2}} . \quad (19.29)$$

This dependence of T_c on the mass was confirmed by experiments by Maxwell [108] and Reynolds *et al.* [109] who measured T_c for samples composed of different isotopes. This isotope effect has been observed in a number of simple materials such as Hg, Pb, Mg, Sn, Tl. For these simple metals, the retarded electron–electron attraction, due to the charged ions over screening the Coulomb interaction [110], has a simple mass dependence. The isotope effect is much smaller or even almost absent in transition metals and compounds such as Ru and Os, where the electrons are more localized and the relative strength of the Coulomb repulsion is large [111]. In $\alpha - U$, a large isotope effect even occurs with a positive exponent [112], but detailed calculations show that the superconductivity is still phonon mediated [113]. The absence of an isotope effect does not necessarily imply non-phonon mediated electron–electron interactions but merely that simplifying circumstances that lead to Fröhlich's isotopic mass dependence are not present. As heavy-fermion systems have extremely heavy quasi-particle masses due to large electron–electron interactions, one does not expect isotope experiments to provide direct evidence of the nature of the pairing mechanism. Furthermore, for systems that appear to be on the verge of a magnetic instability [114,115], it is possible that the collective excitations of the spin system could provide an alternate or complementary mechanism to the phonon mediated interaction.

Many different pairing mechanisms have been proposed for heavy-fermion superconductors, ranging from electron-phonon coupling [116,117] to ferromagnetic and antiferromagnetic spin-fluctuations [118–120]. The main problem posed in developing a microscopic description of the superconductivity lies with the lack of knowledge of the normal state because of its strong electron correlations. A commonly used approach that describes the formation of the superconducting state starts from assuming the validity of a Fermi liquid description of the normal state. In what follows, we shall outline this approach to superconductivity. However, as some heavy-fermion superconductors show no evidence that a Fermi liquid state has formed before the superconducting transition has occurred, this approach is not on a firm basis. Second, as the Fermi liquid approach neglects the effect of the collective fluctuations, it does not address the role that the low-frequency spin-fluctuations play in suppressing the superconducting transition. The proper starting point for a

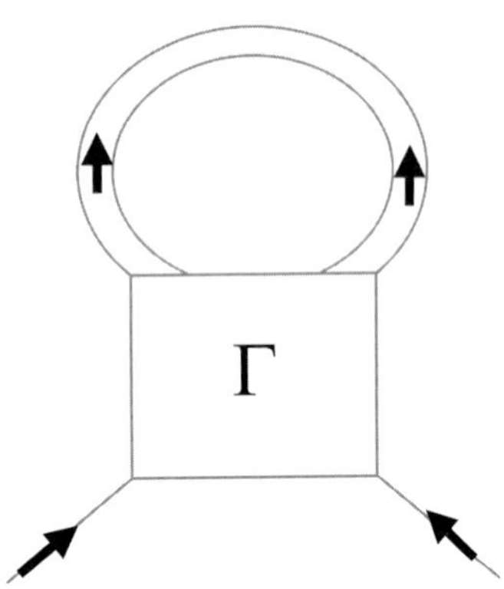

Fig. 19.17. The Feynmann diagram associated with the anomalous self-energy $\Sigma^{(a)}(\underline{k};\hbar\omega)$. The *solid square* represents the irreducible vertex interaction $\Gamma(\underline{q};\hbar\omega)$. The *directed double lines* represent the fully renormalized Green's function

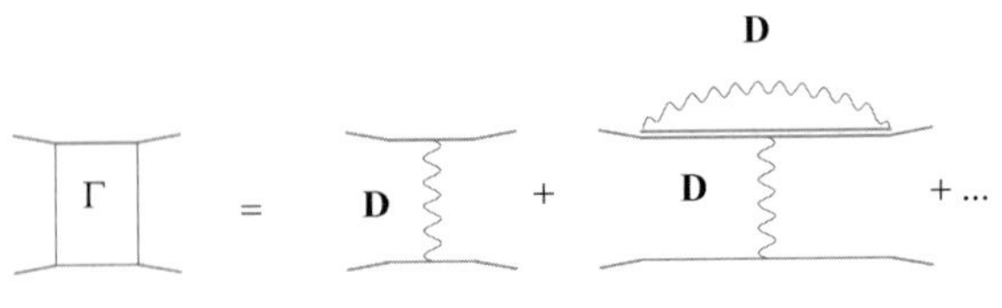

Fig. 19.18. The Feynmann diagram expansion of the irreducible vertex interaction $\Gamma(\underline{q};\hbar\omega)$ in terms of bosonic excitations. The boson propagators, $D(\underline{q}';\hbar\omega')$, are represented by the *wavy lines*

microscopic description of electron–phonon mediated superconductivity lies in the Eliashberg equations [121,122] in which the superconducting gap, or order parameter, is related to a time varying quantity $\Sigma^{(a)}_{\alpha,\beta}(\underline{k};\tau)$, which is the anomalous or pair self-energy in the superconducting state. The gap parameter in the quasi-particle spectra, $\Delta_{\alpha,\beta}(\underline{k};i\hbar\omega_n)$, is defined in terms of the anomalous self-energy and wave function renormalization via

$$\Delta(\underline{k};\omega_n) = \frac{\Sigma^{(a)}(\underline{k};i\hbar\omega_n)}{Z_{\underline{k}}(i\hbar\omega_n)} \,. \tag{19.30}$$

The self-energy $\Sigma^{(a)}_{\alpha,\beta}(\underline{k};i\hbar\omega_n)$ is depicted diagrammatically in Fig. 19.17. A rigorous derivation of the Eliashberg equations for the self energies relies on the validity of Migdal's theorem [123], which states that the vertex correction is tiny, of the order of the square root of the ratio of the electron to the ion mass $\sim 10^{-3}$. There are no analogous theorems known for couplings to other bosonic modes such as spin-waves [124], but nevertheless, it is still hoped that similar equations may describe superconducting pairing mediated by other bosonic mechanisms. In the Eliashberg equations, the gap can be expressed in terms of the spectral density associated with the bosonic pairing mechanism and the spectral density of the electrons forming the Cooper pairs. In the presence of strong interactions, the spectral densities include appropriate self energies, and the pairing interaction is manifested through an irreducible vertex interaction, $\Gamma(\underline{k},\underline{k}+\underline{q};i\hbar\omega_m,i\hbar\omega_n)_{\alpha,\beta;\mu,\nu}$. Terms corresponding to the irreducible vertex interaction due to the exchange of bosons are shown in Fig. 19.18. The self-energy, the gap, and the irreducible vertex interaction should all be calculated self consistently. The linearized gap equation, which determines T_c for the various pairings, is given by

$$\begin{aligned}\Delta_{\alpha,\beta}(\underline{k};i\hbar\omega_n) = -k_B T_c \frac{1}{N}\sum_m \sum_{\underline{q}} \sum_{\mu,\nu} \Gamma \\ \times (\underline{k},\underline{k}+\underline{q};i\hbar\omega_m,i\hbar\omega_n)_{\alpha,\beta;\mu,\nu} \\ \times G_\mu(\underline{k}+\underline{q};i\hbar\omega_m) \\ \times G_\nu(-\underline{k}-\underline{q};-i\hbar\omega_m) \\ \times \Delta_{\mu,\nu}(\underline{k}+\underline{q};i\hbar\omega_m) \,.\end{aligned} \tag{19.31}$$

The normal state becomes unstable to the superconducting state that has the highest T_c in the case of no degeneracy. The criterion for a further instability of superconducting phase to phases with other types of pairings must be determined on the basis of minimization of the Free energy. In the case of strong electron–phonon coupling, the vertex corrections are limited by Migdal's theorem [123] to be smaller than the bare vertex interaction by factors at least as small as 10^{-2}. These corrections are negligibly small and, therefore, have the effect that the self-energies can be calculated with extremely good accuracy. In such cases, the Eliashberg equations have been solved. Such calculations are reviewed in references [125,126]. For heavy-fermion superconductors, the physics is not so clear and, despite the absence of experimental confirmation, the validity of a Fermi liquid picture is often assumed. Under this assumption, the imaginary time Green's functions in the vicinity of the Fermi energy may be replaced by their quasi-particle contributions. In particular, the quasi-particle masses are changed from the band mass m_b by the wave function renormalization to $Z\, m_b$, the quasi-particle lifetimes τ_0 are increased to $Z\,\tau_0$ and the strength of the quasi-particle pole is reduced from unity by a factor Z^{-1}. The enhanced quasi-particle masses are simply absorbed into a re-definition of $e(\underline{k}) - \mu$ as the normal state quasi-particle energies $E_{\underline{k}}$. If the electron–phonon coupling was proven to be the mechanism responsible for heavy-fermion superconductivity, the ratio of the quasi-particle masses to the ionic masses are no longer negligible so, even in this case, Migdal's theorem and the Eliashberg equations may be of doubtful validity. Even though retardation plays an important role in superconductivity, an approximation that has been frequently used consists of replacing the vertex function by an appropriate interaction potential evaluated on the Fermi surface. In this approximation, the interaction potential contains the effect of the instantaneous Coulomb repulsion, and the linearized gap equation simplifies to take the weak-coupling BCS form [127],

$$\Delta_{\alpha,\beta}(\underline{k}) = Z^{-2}\frac{1}{N}\sum_{\underline{q}}\sum_{\mu,\nu}\Gamma(\underline{k},\underline{k}+\underline{q})_{\alpha,\beta;\mu,\nu} \times \Delta_{\mu,\nu}(\underline{k}+\underline{q}) \times \left[\frac{1-f(E_{\nu,\underline{k}+\underline{q}})-f(E_{\mu,-\underline{k}-\underline{q}})}{E_{\nu,\underline{k}+\underline{q}}+E_{\mu,-\underline{k}-\underline{q}}}\right]. \tag{19.32}$$

This differs from the usual BCS theory of simple superconductors in that the wave function renormalization Z^{-1} is now explicitly included, and also, there is an implicit difference in that the dispersion relations are those of the heavy quasi-particles. The spin dependence of the quasi-particle energies may be important for the occurrence of superconductivity within a magnetically ordered phase. The summation over $\underline{q}$ can be performed by introducing an integration over the density of states which is cut off at a frequency, ω_c, characteristic of the bosons responsible for the pairing

$$\begin{aligned}
\Delta_{\alpha,\beta}(\hat{k}) &= Z^{-2}\sum_{\mu,\nu}\int\frac{d\Omega}{4\pi}\Gamma(\hat{k},\hat{k}')_{\alpha,\beta;\mu,\nu}\Delta_{\mu,\nu}(\hat{k}') \\
&\quad\times\int_{-\hbar\omega_c}^{+\hbar\omega_c} dE\rho_{qp}(E)\left(\frac{1-2f(E)}{2E}\right) \qquad (19.33)\\
&= Z^{-1}\rho(\mu)\sum_{\mu,\nu}\int\frac{d\Omega}{4\pi}\Gamma(\hat{k},\hat{k}')_{\alpha,\beta;\mu,\nu} \\
&\quad\times\Delta_{\mu,\nu}(\hat{k}') \\
&\quad\times\left(\left[1-2f(\hbar\omega_c)\right]\ln\frac{\hbar\omega_c}{k_B T}\right. \\
&\quad\left.+2\int_0^{\frac{\hbar\omega_c}{k_B T}} dx\ln x\frac{\partial}{\partial x}\left(\frac{1}{\exp x+1}\right)\right) \\
&= Z^{-1}\rho(\mu)\sum_{\mu,\nu}\int\frac{d\Omega}{4\pi}\Gamma(\hat{k},\hat{k}')_{\alpha,\beta;\mu,\nu}\Delta_{\mu,\nu}(\hat{k}') \\
&\quad\times\left(\left[1-2f(\hbar\omega_c)\right]\ln\frac{\hbar\omega_c}{k_B T}+\ln\frac{2\gamma}{\pi}\right),
\end{aligned}$$

where we have ignored any spin polarization in the quasi-particle bands. Usually, the Fermi function $f(\hbar\omega_c)$ is neglected under the assumption that $\hbar\omega_c > k_B T_c$, which usually holds for phonon mediated pairing. However, this assumption may not be appropriate for spin-fluctuation mediated pairing near a quantum critical point. In such cases, a strong coupling approach may have to be used. In the above expression, the electron density of states $\rho(\mu)$ is un-enhanced by the interactions but is multiplied an explicit factor of Z^{-1}. As pointed out by Varma [128], this leads to an expression for T_c similar to the BCS weak-coupling form except that the exponent is increased by a factor of Z, i.e.,

$$k_B T_c = 1.14\hbar\omega_c \exp\left[-\left(\frac{Z}{\Gamma\rho(\mu)}\right)\right], \tag{19.34}$$

where Γ is the Fermi surface average of $\Gamma(\hat{k},\hat{k}')$. Thus, the mass renormalization Z could depress T_c. The value of T_c in heavy-fermion superconductors could also be low due to a small ω_c in the prefactor [129] as well as in the exponent due to the appearance of the Fermi functions.

The relationship between the nature of spin-fluctuations and the singlet or triplet character of the pairing is found by expanding both the on-Fermi surface vertex function and the order parameter in spherical harmonics. The expansions are

$$\Gamma(\hat{k}, \hat{k}')_{\alpha,\beta;\mu,\nu} = \sum_{l,m} \Gamma(l)_{\alpha,\beta;\mu,\nu} Y^l_m(\hat{k})^* Y^l_m(\hat{k}') \tag{19.35}$$

and

$$\Delta_{\mu,\nu}(\hat{k}) = \sum_{l,m} \Delta_{\mu,\nu}(l,m) Y^l_m(\hat{k}) \; . \tag{19.36}$$

The orthogonality of the spherical harmonics Y^l_m, leads to an independent linearized gap equation for each l value. Furthermore, if the spin dependence of the vertex function is of the form

$$\Gamma(l)_{\alpha,\beta;\mu,\nu} = \left(U_l \delta_{\alpha,\beta;\mu,\nu} - J_l \vec{\sigma}_{\alpha;\nu} . \vec{\sigma}_{\beta;\mu} \right) , \tag{19.37}$$

where U_l is an effective direct Coulomb repulsion and J_l is an effective exchange, then, with the aid of the identity

$$\left(S\,(S+1) - 2\frac{3}{4} \right) \delta_{\alpha,\beta;\mu,\nu} = \frac{1}{2} \vec{\sigma}_{\alpha;\nu} . \vec{\sigma}_{\beta;\mu} \; , \tag{19.38}$$

where S is the total spin of the Cooper pair, one can show that the equations for the different T_c's are

$$1 = \ln \frac{k_B T_c}{\hbar\omega_c} Z^{-1} \rho(\mu)(U_l + 3J_l) \tag{19.39}$$

for singlet T_c's with $S = 0$ and even l, while

$$1 = \ln \frac{k_B T_c}{\hbar\omega_c} Z^{-1} \rho(\mu)(U_l - J_l) \tag{19.40}$$

for triplet T_c's with $S = 1$ and odd l. The normal state becomes unstable to the angular momentum pairing state with the highest T_c. In the case where a set of order parameters have degenerate T_c's, the instability first occurs to the state with the linear combination of the order parameters that corresponds to the lowest free energy [130,131]. Large values of U_0, from the direct Coulomb repulsion, are unfavorable for s-wave pairing. However, U_0 represents the residual interactions between a pair of quasi-particles and is represented by a Fermi liquid parameter F_0 which is unknown, a priori, and it could have a quite small magnitude in which case s-wave pairing might still occur. It is also seen that the $\underline{q}$ dependence of the effective exchange interaction strongly influences the tendency for singlet versus triplet pairing [132,133].

In analogy to the paramagnetic spin-fluctuation pairing mechanisms [134–136] proposed for ^{3}He, a number of authors [118–120] have investigated the effect of incommensurate or antiferromagnetic spin-fluctuations within the RPA. The effective interaction between a pair of electrons with parallel spins is shown in Fig. 19.19(a). The dashed lines represent the local Coulomb interaction, and the directed lines represent the one-electron Green's functions. These diagrams only contain odd numbers of bubbles due to the spin-dependent nature of the Coulomb repulsion U and are related to the longitudinal or z–z components of the magnetic susceptibility. The resulting interaction between parallel spin electrons at the Fermi energy is given by

$$\Gamma(\underline{k}, \underline{k}')_{\sigma,\sigma:\sigma,\sigma} = - \frac{U^2 \chi_0(\underline{k} - \underline{k}'; 0)}{1 - U^2 \chi_0^2(\underline{k} - \underline{k}'; 0)} \; , \tag{19.41}$$

where $\chi_0(\underline{q}; 0)$ is the static limit of the appropriate transverse, non-interacting, reduced susceptibility, having the usual Lindhard form.

The effective interaction between electrons with anti-parallel spins is given by the sum of three terms, one being the on site Coulomb repulsion U, another term stemming from the transverse susceptibility, and the last term originates from the remaining part of the longitudinal susceptibility. These terms are depicted diagrammatically in Fig. 19.19(b) and are evaluated as

$$\Gamma(\underline{k}, \underline{k}')_{\sigma,-\sigma;-\sigma,\sigma} = U + \frac{U^3 \chi_0^2(\underline{k} - \underline{k}'; 0)}{1 - U^2 \chi_0^2(\underline{k} - \underline{k}'; 0)} + \frac{U^2 \chi_0(\underline{k} + \underline{k}'; 0)}{1 - U \chi_0(\underline{k} + \underline{k}'; 0)} \; . \tag{19.42}$$

For values of U in the vicinity of the critical value U_c, the susceptibilities for $\underline{q} \sim \underline{Q}$ are enhanced as are the effective quasi-particle interactions of Eqs. (19.41) and (19.42). Thus, one expects the effective interaction to be highly peaked at $\underline{q}$ values closely connected to the $\underline{Q}$ values of the quantum

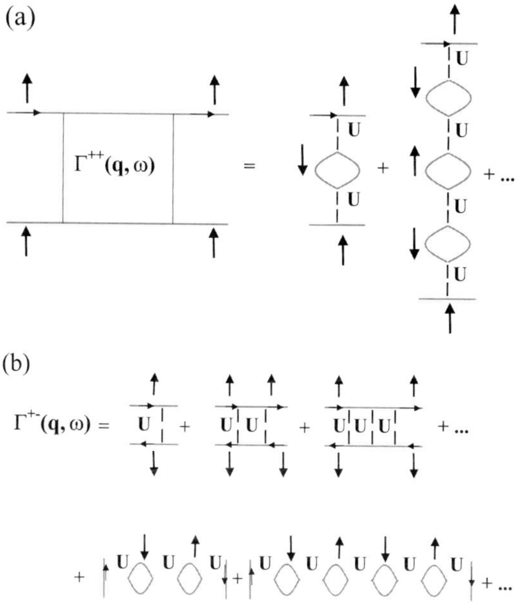

Fig. 19.19. The RPA expression for the irreducible interaction between a pair of electrons with parallel spins is depicted in terms of Feynmann diagrams in (**a**). The interaction between electrons with anti-parallel spins is shown in (**b**). The interaction not only involves the longitudinal spin-fluctuations but also transverse spin-fluctuations

critical point. Due to the increase in the amplitude of the spin-fluctuations as the quantum critical point is approached, the superconducting interaction in the paramagnetic phase is expected to be largest just at the quantum critical point. In the magnetically ordered phase, the transverse spin-fluctuations are expected to transform into a branch of undamped Goldstone modes. However, as shown by Schrieffer et al. [137], the Goldstone modes of the antiferromagnetically ordered state of an isotropic material do not produce superconducting pairing. Due to the loss of the low-energy transverse spin-fluctuation modes as a pairing mechanism in the ordered phase and as the amplitude (longitudinal) modes are expected to acquire a mass, one expects that the superconducting pairing will diminish deep within the magnetically ordered phase.[3] Hence, the superconductivity is expected to occur only in close proximity to a quantum critical point.

In comparing the relative tendencies of the nearly ferromagnetic and nearly antiferromagnetic spin-fluctuations in producing triplet or singlet pairings [132, 133], it is useful to re-write the momentum transfers for on-Fermi surface processes as $|\underline{k} \pm \underline{k}'|^2 = 2k_F^2(1 \pm \cos\theta)$, where $\cos\theta = \hat{k}.\hat{k}'$. Then, with the use of the addition theorem for spherical harmonics, one finds

$$J_l = \frac{2l+1}{4} \int_0^{\pi} d\theta \sin\theta \, P_l(\cos\theta) \times \frac{U^2 \chi_0(\underline{k}-\underline{k}';0)}{1 - U\chi_0(\underline{k}-\underline{k}';0)}, \tag{19.43}$$

where $P_l(x)$ are the Legendre polynomials. If the system is close to a ferromagnetic instability, the effective interaction due to the spin-fluctuations is enhanced and positive for momentum transfers of magnitude $q = 2k_F \sin\frac{\theta}{2} \sim 0$, and from the properties of the Legendre function, one expects that the strengths of J_l are positive ($P_l(1) = 1$) and decrease with increasing l due to the increasing number of nodes of $P_l(\cos\theta)$. This, combined with a large value of U_0, may forbid s-wave singlet pairing and leads to the conclusion that ferromagnetism may favor triplet spin pairing with $l = 1$. That is, a ferromagnetic exchange between distant electrons can stabilize triplet pairs [138]. On the other hand, if the proximity to a spin density wave phase enhances the susceptibility at non-zero $\underline{q}$ values in regions where the odd l values of $P_l(\cos\theta)$ are negative ($P_l(-1) = (-1)^l$), triplet pairing may be unfavorable. However, even l singlet pairing may still occur as it is stabilized by the antiferromagnetic exchange. Such a simple analysis of the strength of the various pairings is not expected to hold in real materials with multi-sheeted Fermi surfaces but is crucially controlled by the structure of the Fermi surface of the normal state [76].

[3] The combined effect of spin-orbit coupling and crystalline anisotropy in a heavy-fermion system may result in the magnetic order parameter losing its continuous symmetry. Hence, the nature of the soft-modes at the transition may change. In such cases, it might be expected that the soft-modes may remain effective in producing superconducting pairing within the magnetically ordered phase.

The usual weak-coupling description of superconducting states, developed by Bardeen Cooper and Schrieffer [127], has a starting point that assumes the normal state electronic system is in the Fermi liquid state. The superconducting pairing interaction is introduced, which then produces the instability to the superconducting state. To the extent that the pairing interactions can be treated in the mean-field approximation, the quasi-particle concept remains valid in the superconducting state. The absence of a significant lifetime for the quasi-particle states essentially follows from the same reasoning as in the normal state, namely, that the Pauli-exclusion principle significantly reduces the amount of phase space available for scattering processes. In this mean-field description, the quasi-particle dispersion relations $E_{\sigma,\underline{k}}$ have forms that depend on the nature of the superconducting order parameters. For the majority of conventional superconductors, the normal state is well described by Fermi liquid theory, and the transition to the superconducting state closely follows the mean-field predictions. However, measurements on some heavy-fermion systems show that the systems have not settled into the Fermi liquid phase before the superconducting transition occurs. Thus, the validity of the use of the quasi-particle concept is dubious. Nevertheless, it may be expected that some features of the heavy-fermion superconducting states are robust in that they only depend on underlying symmetries and thus, would be common to the mean-field description in terms of quasi-particles as well as a more rigorous treatment.

19.2.5 The Symmetry of the Order Parameter

The symmetry of the superconducting order parameter, or more precisely the vanishing of the gap at the Fermi energy, does lead to anomalous temperature dependences of specific heat and transport properties far below T_c. An early series of such experiments which were devised to elucidate the symmetry did lead to the conclusion that the superconducting gap does vanish somewhere on the Fermi surface [5,25–31] but was equivocal on the detailed nature of the order parameter. Nevertheless, in the following we shall examine the possible symmetries of the gap function and their experimental manifestations. These considerations shall then be applied to the various heavy-fermion compounds in subsequent sections.

The Cooper pairs are characterized by the pairing function, which is defined by the expectation value of the product of two electron annihilation operators, $a_{\sigma,\underline{k}}$, as

$$\Phi_{\alpha,\beta}(\underline{k}) = <|a_{\alpha,\underline{k}}a_{\beta,-\underline{k}}|> , \qquad (19.44)$$

where α, and β denote the spin values quantized along the z-axis. We shall consider the equilibrium state in which the center of mass of the pairs are at rest. For convenience we shall, as in the case of superfluid ^{3}He [139], assume spherical symmetry and also neglect spin-orbit scattering. Neither of these assumptions are valid for heavy-fermion systems [140]. For f electrons, spin-orbit coupling is known to be strong, and the anisotropy caused by the crystalline environment also needs to be accounted for. So, both these assumptions need to be abandoned in a proper description of heavy-fermion superconductors. As their abandonment only complicates the presentation [141,142] but does not invalidate the general approach being outlined, we shall make use of these assumptions for pedagogical purposes. This approach mirrors the historical development of the theoretical description of heavy-fermion superconductivity, which often proceeded by analogy with ^{3}He.

The singlet pairing function is written as the anti-symmetrized product

$$\Phi^s(\underline{k}) = <|(a_{\uparrow,\underline{k}}a_{\downarrow,-\underline{k}} - a_{\downarrow,\underline{k}}a_{\uparrow,-\underline{k}})|> . \qquad (19.45)$$

The triplet pairing functions can be expressed in terms of the components organized by the eigenvalues of S_z as

$$\Phi^t_{S_z=1}(\underline{k}) = <|a_{\uparrow,\underline{k}}\,a_{\uparrow,-\underline{k}}|> ,$$
$$\Phi^t_{S_z=0}(\underline{k}) = <|(a_{\uparrow,\underline{k}}\,a_{\downarrow,-\underline{k}} + a_{\downarrow,\underline{k}}\,a_{\uparrow,-\underline{k}})|> .$$

and

$$\Phi^t_{S_z=-1}(\underline{k}) = <|a_{\downarrow,\underline{k}}a_{\downarrow,-\underline{k}}|> . \qquad (19.46)$$

From consideration of the fermion anti-commutation relations, one finds that the singlet pairing function is even in $\underline{k}$ whereas the triplet wave function is odd in $\underline{k}$. By considering the effects of various symmetries on an arbitrary pairing wave function, one finds that in this case, one can parameterize an arbitrary pairing function in terms of a scalar singlet $\Phi^s(\underline{k})$ and a (vector) triplet $\vec{\Phi}^t(\underline{k})$ pairing function via

$$\Phi_{\alpha,\beta}(\underline{k}) = \frac{1}{2}\left[\left(\Phi^s(\underline{k}) + \vec{\sigma}.\vec{\Phi}^t(\underline{k})\right)i\sigma_y\right]_{\alpha,\beta} . \quad (19.47)$$

The unitary operator $i\sigma_y$ occurs as a consequence of the spin operator $\vec{\sigma}$ being odd under time reversal symmetry and also as the anti-symmetric Pauli matrix has the effect that

$$i\sigma_y\vec{\sigma}(-i\sigma_y) = -\vec{\sigma}^* . \quad (19.48)$$

The direction of the triplet ordering is given in terms of the matrix elements of the vector pairing function

$$\vec{\Phi}^t(\underline{k}) = \Big[- \Phi^t_{S_z=1}(\underline{k})(\hat{x} + i\hat{y}) \quad (19.49)$$

$$+ \Phi^t_{S_z=0}(\underline{k})\hat{z} + \Phi^t_{S_z=-1}(\underline{k})(\hat{x} - i\hat{y})\Big] .$$

If the pairing interaction is of the form

$$\hat{H}_{int} = \frac{1}{N}\sum_{\underline{k},\underline{q}}\sum_{\alpha,\beta,\mu,\nu} V(\underline{k},\underline{q})_{\beta,\alpha:\,\mu,\nu}$$

$$\times\, a^\dagger_{\beta,-\underline{k}}\, a^\dagger_{\alpha,\underline{k}}\, a_{\mu,\underline{q}}\, a_{\nu,-\underline{q}} , \quad (19.50)$$

then the pairing potential can be expressed as the average

$$\Delta_{\alpha,\beta}(\underline{k}) = \frac{1}{N}\sum_{\underline{q}}\sum_{\mu,\nu} V(\underline{k},\underline{q})_{\beta,\alpha:\,\mu,\nu}\Phi_{\mu,\nu}(\underline{q}) . \quad (19.51)$$

The above equation shows that the symmetry of the pairing potential is related to the pairing function. By analogy to Eq. (19.47), one can then write the pairing potential as

$$\Delta_{\alpha,\beta}(\underline{k}) = \frac{1}{2}\left[\left(D(\underline{k}) + \vec{\sigma}.\vec{d}(\underline{k})\right)i\sigma_y\right]_{\alpha,\beta} . \quad (19.52)$$

The three different S_z components of the spin triplet pairing potential are assigned the amplitudes which are related to the components of $\vec{d}^t$ via

(**a**) Line nodes of the d $x^2 - y^2$ order parameter

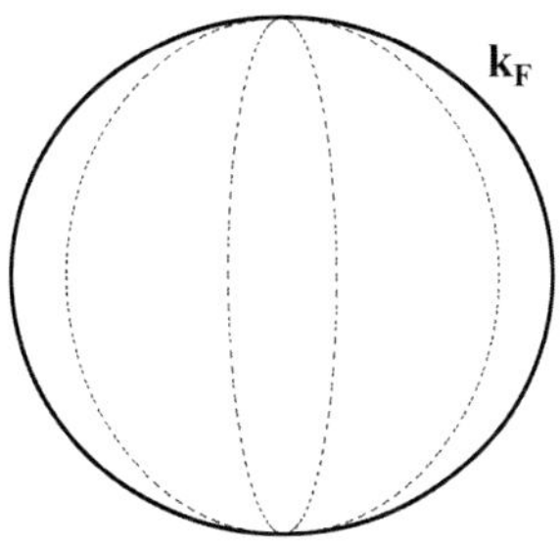

(**b**) Line nodes of the d $2z^2 - x^2 - y^2$ order parameter

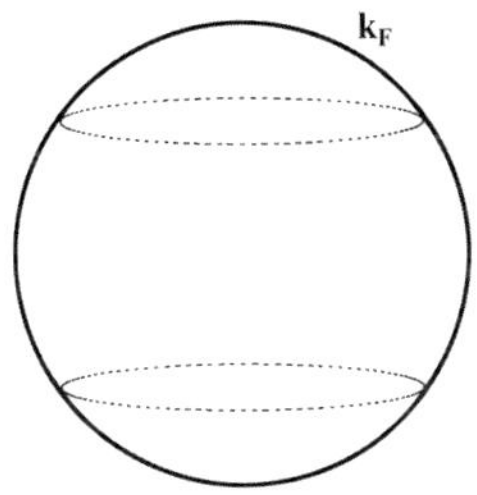

Fig. 19.20. Positions of the lines of nodes for d-wave superconductors, in momentum space. The dashed lines represent the lines of zeros of the superconducting order parameter on the spherical Fermi surface (indicated by the *solid line*). (**a**) The lines of nodes of the $d_{x^2-y^2}$ order parameter consists of a pair of great circles which intersect at the poles. (**b**) The lines of nodes for the $d_{2z^2-x^2-y^2}$ phase consists of two non-intersecting circles

$$\begin{aligned} -d_{S_z=1} &= d_x - i d_y , \\ d_{S_z=0} &= d_z , \\ d_{S_z=-1} &= d_x + i\, d_y . \end{aligned} \quad (19.53)$$

One can expand the singlet pairing potential in the spherical harmonics $Y^l_m(\hat{k})$ as

$$D(\underline{k}) = \sum_{l,m} D^l_m(k)\, Y^l_m(\hat{k}) , \quad (19.54)$$

where due to the symmetry imposed by fermion anti-commutation relations $D(\underline{k}) = D(-\underline{k})$, only even values of l contribute. Likewise, the components of the triplet pairing potential can also be expanded in the same way

$$d_{S_z}(\underline{k}) = \sum_{l,m} d^l_{m,S_z}(k) Y^l_m(\hat{k}) , \quad (19.55)$$

but as the pairing potential is anti-symmetric in $\underline{k}$ $d_{S_z}(\underline{k}) = -d_{S_z}(-\underline{k})$, only the odd values of l contribute. Thus, singlet pairing only occurs with even values of l and triplet pairing only occurs with odd values of l. It is noted that in anisotropic crystals, singlet and triplet pairing remain mutually exclusive if the symmetry group contains an inversion symmetry $\underline{k} \to -\underline{k}$.

The BCS mean-field Hamiltonian [127] can be expressed in terms of the pairing potentials as

$$\hat{H}_{MF} = \sum_{\alpha,\underline{k}} (e_\alpha(\underline{k}) - \mu) a^\dagger_{\alpha,\underline{k}} a_{\alpha,\underline{k}} + \sum_{\alpha,\beta,\underline{k}} \left(\Delta_{\alpha,\beta}(\underline{k}) a^\dagger_{\beta,-\underline{k}} a^\dagger_{\alpha,\underline{k}} + \Delta^*_{\alpha,\beta}(\underline{k}) a_{\alpha,\underline{k}} a_{\beta,-\underline{k}} \right) . \tag{19.56}$$

This apart from a constant term, can be re-expressed as

$$\hat{H}_{MF} = \frac{1}{2}\left[\sum_{\alpha,\underline{k}} (e_\alpha(-\underline{k}) - \mu) a^\dagger_{\alpha,-\underline{k}} a_{\alpha,-\underline{k}} - \sum_{\alpha,\underline{k}} (e_\alpha(\underline{k}) - \mu) \, a_{\alpha,\underline{k}} a^\dagger_{\alpha,\underline{k}} \right] + \sum_{\alpha,\beta,\underline{k}} \left(\Delta_{\alpha,\beta}(\underline{k}) a^\dagger_{\beta,-\underline{k}} a^\dagger_{\alpha,\underline{k}} + \Delta^*_{\alpha,\beta}(\underline{k}) a_{\alpha,\underline{k}} a_{\beta,-\underline{k}} \right) . \tag{19.57}$$

A four-component column operator, Ψ, is introduced

$$\Psi(\underline{k}) = \begin{pmatrix} a_{\uparrow,-\underline{k}} \\ a_{\downarrow,-\underline{k}} \\ a^\dagger_{\uparrow,\underline{k}} \\ a^\dagger_{\downarrow,\underline{k}} \end{pmatrix} ,$$

where the first two components (and the last two) are conjugate pairs, whereas the first and fourth (and second and third) are time reversal partners. This notation allows the BCS Hamiltonian to be written in matrix form

$$\hat{H}_{MF} = \frac{1}{2} \sum_{\underline{k}} \Psi^\dagger(\underline{k}) \tilde{H}_{MF}(\underline{k}) \Psi(\underline{k}) , \tag{19.58}$$

[4] The diagonal part of the self-energy has been neglected.

where the matrix is given by

$$\tilde{H}_{MF}(\underline{k}) = \begin{pmatrix} e_\uparrow(-\underline{k}) & 0 & -d_{S_z=1}(\underline{k}) & d_{S_z=0}(\underline{k}) + D(\underline{k}) \\ 0 & e_\downarrow(-\underline{k}) & d_{S_z=0}(\underline{k}) - D(\underline{k}) & d_{S_z=-1}(\underline{k}) \\ -d^*_{S_z=1}(\underline{k}) & d^*_{S_z=0}(\underline{k}) - D^*(\underline{k}) & -e_\uparrow(\underline{k}) & 0 \\ d^*_{S_z=0}(\underline{k}) + D^*(\underline{k}) & d^*_{S_z=-1}(\underline{k}) & 0 & -e_\downarrow(\underline{k}) \end{pmatrix} . \tag{19.59}$$

This mean-field Hamiltonian has quasi-particle excitation energies that can be expressed as

$$E_{\sigma,\underline{k}} = \pm\sqrt{(e(\underline{k}) - \mu)^2 + |D(\underline{k})|^2 + |\vec{d}(\underline{k})|^2 - \sigma |i\,\vec{d}(\underline{k}) \wedge \vec{d}^*(\underline{k})|} , \tag{19.60}$$

where $\sigma = \pm 1$. Consider the term inside the square root that is proportional to a vector product and σ, this term is real. When the vector product is zero the pairing is said to be unitary, and when it is non-zero the pairing is non-unitary. In the non-unitary state, the spin degeneracy is lifted and a staggered spin density will occur, which is a consequence of the spontaneous breaking of time reversal symmetry.

One important characteristic of unconventional superconductivity is that the gap at the Fermi surface may vanish for specific $\underline{k}$ values, which are the nodes of the order parameter. For singlet superconductivity, the gap vanishes when $|D(\underline{k})|^2 = 0$ at the Fermi surface $e(\underline{k}) = \mu$.[4] For triplet superconductivity, the gap vanishes when

$$|\vec{d}(\underline{k})|^2 \pm |i\,\vec{d}(\underline{k}) \wedge \vec{d}^*(\underline{k})| = 0 \tag{19.61}$$

at $e(\underline{k}) = \mu$. In the non-unitary state, it is possible that the gap may vanish for only one spin direction.

The vanishing of the superconducting gap at the nodes does produce a finite density of states for quasi-particle excitations that may be revealed in thermodynamic, transport, and spectroscopic measurements. Although several experiments were performed with the intention of establishing the existence of and characterizing the nodes, the results were often conflicting. In retrospect, the lack of consensus concerning the nature of the nodes is not surprising as even Nb based superconductors show various deviations from strict singlet s-wave BCS behavior.

For singlet s-wave superconductivity in an isotropic system, there are no nodes. Thus, for an isotropic singlet superconductor, the existence of nodes is related to the formation of pairs with non-zero angular momenta. The singlet pairing d-wave state, $d_{x^2-y^2}$, seems increasingly certain to be the symmetry of the high temperature superconducting phase of $YBa_2Cu_3O_7$, which has an extremely anisotropic perovskite structure. The order parameter for this phase has the long wavelength form $D(\underline{k}) \propto (k_x^2 - k_y^2)$, and so the gap vanishes at lines of nodes which form two great circles with a relative orientation of $\frac{\pi}{2}$ and the circles intersect at the poles as shown in Fig. 19.20(a). Another type of d-wave ordering is the $d_{2z^2-x^2-y^2}$ in which there are two separate planes where $D(\underline{k}) = 0$ that intersect with Fermi energy in circles. As shown in Fig. 19.20(b), one circle is in the $k_z = +k_f\sqrt{3}$ plane and the other in the $k_z = -k_f\sqrt{3}$ plane.

There are three triplet p-wave states which have been extensively studied in the context of ^{3}He. These are all unitary states, and the first one we shall consider is given by

$$\vec{d}\,(\underline{k}) \propto \hat{x}k_x + \hat{y}k_y + \hat{z}k_z\ , \tag{19.62}$$

which corresponds to the Balian–Werthamer (BW) phase [143]. This pairing corresponds to one with angular momentum $\vec{J} = \vec{L} + \vec{S} = 0$, as can be seen from re-writing the gap function as

$$\begin{aligned}\vec{d}\,(\underline{k}) &\propto k\Big(\sin\theta\cos\phi\,\hat{x} + \sin\theta\sin\phi\,\hat{y} + \cos\theta\,\hat{z}\Big)\\ &\propto \sqrt{\frac{2\pi}{3}}\,k\Big[\Big(\hat{x} - i\hat{y}\Big)Y_1^1 \\ &\quad - \Big(\hat{x} + i\hat{y}\Big)Y_{-1}^1 + \sqrt{2}\,\hat{z}\,Y_0^1\Big]\ ,\end{aligned} \tag{19.63}$$

which, on comparing with Eq. (19.49) and noting the normalization of $\Phi_{S_z=0}$, is easily identifiable as a state with $\vec{J} = 0$. The gap has a constant magnitude over the Fermi energy, and so the gap does not vanish for any $\underline{k}$ values in the BW state. The BW pairing is realized in the B phase of superfluid ^{3}He.

Another state is the Anderson–Brinkman–Morell (ABM) phase [144],

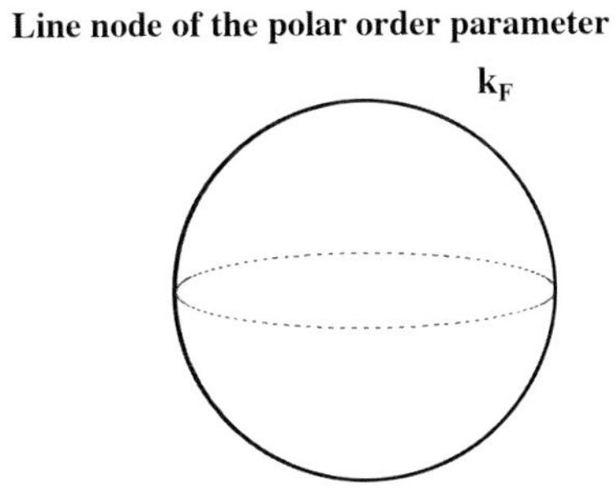

Fig. 19.21. The line of nodes of the order parameter for the polar phase of a p-wave superconductor is located on the equatorial circle on the Fermi surface

$$\vec{d}\,(\underline{k}) \propto \hat{z}(k_x + ik_y) = \hat{z}k\sin\theta\exp[i\varphi]\ , \tag{19.64}$$

which is identified as the state with $S_z = 0$ and $L_z = \hbar$. Thus, $\vec{S}$ and $\vec{L}$ are perpendicular to each other. For the ABM phase, the gap at the Fermi surface vanishes at the point nodes where $k_x = k_y = 0$. The ABM pairing is realized in the A phase of superfluid ^{3}He.

The third state is the polar phase, p_z, where

$$\vec{d}\,(\underline{k}) \propto \hat{z}\,k_z\ . \tag{19.65}$$

This state has $S_z = 0$ and $L_z = 0$. As shown in Fig. 19.21, the gap function vanishes at the line node in the equatorial plane where $k_z = 0$.

The above singlet and triplet states may be appropriate for isotropic systems but are not appropriate for crystals of lower symmetry. However, due to a symmetry breaking transition, the Cooper pairs may still have a point group symmetry lower than the full point group symmetry of the crystalline lattice. The various possible types of pairings in crystalline environments of different symmetries with spin-orbit coupling have been tabulated by Anderson [145], Volovik and Gor'kov [142], and also by Blount [141]. The possible pairings, neglecting spin-orbit coupling, were enumerated by Ozaki et al. [146]. We shall first discuss the effect of the periodic lattice and then the effect of spin-orbit coupling.

Crystal Symmetry

The periodic symmetry of the crystalline lattice breaks full rotational symmetry and, therefore, angular momentum is no longer a good quantum number. The basis functions of the rotational group can no longer be used to describe the pairing. However, the basis functions appropriate to the point group can be used. We shall focus the discussion on the O_h group, the tetragonal group D_{4h}, and the hexagonal group D_{6h}, which are relevant to UBe_{13}, $CeCu_2Si_2$ and URu_2Si_2, and UPt_3. For an irreducible representation with dimensionality d, the scalar or vector order parameter can be expressed as a linear combination of the degenerate partner basis functions $\phi_i(\hat{k})$ (scalar or vector), times functions of the invariants of the full point group,[5] $f_n(\hat{k})$. The degenerate partners are expressed in the form $\sum_n \phi_{i,n}(\hat{k}) f_n(\hat{k})$. Thus, for example, the vector order parameter is expressed as

$$\vec{d}(\hat{k}) = \sum_{i=1}^{i=d} C_i \left[\sum_n \vec{\phi}_{i,n}(\hat{k}) \, f_n(\hat{k}) \right] \tag{19.66}$$

and an analogous equation holds true for the singlet order parameter $D(\underline{k})$. The summation over n runs through d and $3d$ values respectively, for singlet and triple pairings [147]. A specific example found in Table(19.2) for singlet pairing in cubic symmetry is the doubly degenerate $\Gamma_3^+(E_g)$ representation in which the degenerate partners basis functions have the form

$$\phi_1(\hat{k}) = \sum_{n=2;4} (k_x^n + k_y^n - 2k_z^n) f_n(\hat{k}) ,$$
$$\phi_2(\hat{k}) = \sum_{n=2;4} (k_x^n - k_y^n) \, , f_n(\hat{k}) \tag{19.67}$$

where the $f_n(\hat{k})$ have the form $a_n + b_n k^2 + c_n(k_x^4 + k_y^4 + k_z^4) + \ldots$. Likewise, for cubic symmetry, the singly-degenerate triplet pairing $\Gamma_1^-(A_{1u})$ representation found in Table 19.3 has the basis function

$$\vec{\phi}_1(\hat{k}) = \sum_{n=1;3;5} (\hat{x}k_x^n + \hat{y}k_y^n + \hat{z}k_z^n) f_n(\hat{k}) \, . \tag{19.68}$$

The coefficients C_i multiplying the degenerate partner basis functions are arbitrary and are determined only by the spontaneous symmetry broken state. The invariant functions $f_n(\hat{k})$ corresponding to the admixtures of the different angular momentum components are well defined for each representation and should be determined by minimization of the free energy. The tetragonal symmetry singlet and triplet representation basis functions are given in Tables 19.4 and 19.5, respectively. The basis for the non-degenerate singlet pairings in the tetragonal $\Gamma_3^+(B_{1g})$ representation is written, with different notation $k_\pm = k_x \pm i k_y$, as

$$\phi_1(\hat{k}) = (k_x^2 - k_y^2) f(\hat{k}) = Re k_+^2 f(\hat{k}) \, , \tag{19.69}$$

where the expansion of the invariant function no longer has a quadratic term in the modulus of $\underline{k}$ but instead, has the different form $f(\hat{k}) = a + b k_z^2 + c(k_x^2 + k_y^2) + \ldots$. The triplet pairing in a D_{4h} structure belonging to the representation $\Gamma_1^-(A_{1u})$ has the form given by,

$$\vec{\phi}_1(\hat{k}) = \hat{z}k_z f_0(\hat{k}) + \sum_{n=1;3} (\hat{x}k_x^n + \hat{y}k_y^n) f_n(\hat{k}) \, . \tag{19.70}$$

Even though the crystal field mixes states of different angular momentum, one can usually identify the pairing states via the dominant components of the orbital angular momentum. For example, in the cubic group with spin singlet pairing shown in Table 19.2, we identify the nodeless one-dimensional

Table 19.2. Singlet pairing basis functions for the cubic group O_h. The forms of the degenerate partners are shown. As explained in the text, the components labelled by n must be multiplied by an invariant function $f_n(k)$ and summed over n to yield a degenerate partner. The degenerate partners are separated by commas

Representation	Basis Functions
$\Gamma_1^+(A_{1g})$	1
$\Gamma_2^+(A_{2g})$	$(k_x^2 - k_y^2)(k_y^2 - k_z^2)(k_z^2 - k_x^2)$
$\Gamma_3^+(E_g)$	$2k_z^n - k_x^n - k_y^n$, $k_x^n - k_y^n$ n = 2 ; 4
$\Gamma_4^+(T_{1g})$	$k_x k_y (k_x^n - k_y^n)$, $k_y k_z (k_y^n - k_z^n)$, $k_z k_x (k_z^n - k_x^n)$ n = 2 ; 4 ; 6
$\Gamma_5^+(T_{2g})$	$k_x k_y k_z^n$, $k_y k_z k_x^n$, $k_z k_x k_y^n$ n = 0 ; 2 ; 4

[5] It is possible that the functions $f_n(\hat{k})$ vanish at some $\underline{k}$ points, thereby producing nodes in the order parameter. However, the set of nodes that are produced in this way will still have the full point group symmetry of the lattice.

Table 19.3. Triplet pairing basis functions in the cubic O_h group. The vector $\underline{k}$ is just the radial vector given by $\hat{x}k_x+\hat{y}k_y+\hat{z}k_z$. The components of the symmetric partners are separated by colons, whereas the symmetric partners are separated by commas

Representation	Basis Functions
$\Gamma_1^-(A_{1u})$	$\hat{x}k_x^n + \hat{y}k_y^n + \hat{z}k_z^n$
	n = 1 ; 3 ; 5
$\Gamma_2^-(A_{2u})$	$\hat{x}k_x^n(k_y^2 - k_z^2) + \hat{y}k_y^n(k_z^2 - k_x^2) + \hat{z}k_z^n(k_x^2 - k_y^2)$
	n = 1 ; 3 ; 5
$\Gamma_3^-(E_u)$	$\hat{x}k_x^n - \hat{y}k_y^n$, $\hat{x}k_x^n + \hat{y}k_y^n - 2\hat{z}k_z^n$
	n = 1 ; 3 ; 5
	$(k_x^n - k_y^n)\underline{k}$, $(2k_z^n - k_x^n - k_y^n)\underline{k}$
	n = 2 ; 4
	$\hat{x}k_x^5k_z^2 - \hat{y}k_y^5k_z^2 + \hat{z}k_z^5(k_x^2 - k_y^2)$,
	$\hat{x}k_x^5(k_z^2 - 2k_y^2) + \hat{y}k_y^5(k_z^2 - 2k_x^2) + \hat{z}k_z^5(k_x^2 + k_y^2)$
$\Gamma_4^-(T_{1u})$	$k_x^m(\hat{y}k_z^n - \hat{z}k_y^n)$, $k_y^m(\hat{z}k_x^n - \hat{x}k_z^n)$, $k_z^m(\hat{x}k_y^n - \hat{y}k_x^n)$;
	n = 1 ; 3 ; 5 m = 0 ; 2
	$\hat{x}k_yk_z(k_y^2 - k_z^2)k_x^n$, $\hat{y}k_zk_x(k_z^2 - k_x^2)k_y^n$, $\hat{z}k_xk_y(k_x^2 - k_y^2)k_z^n$;
	n = 1 ; 3 ; 5
$\Gamma_5^-(T_{2u})$	$k_z^m(\hat{y}k_x^n + \hat{x}k_y^n)$, $k_x^m(\hat{z}k_y^n + \hat{y}k_z^n)$, $k_y^m(\hat{x}k_z^n + \hat{z}k_x^n)$;
	n = 1 ; 3 ; 5 m = 0 ; 2
	$\hat{z}k_xk_yk_z^n$, $\hat{x}k_yk_zk_x^n$, $\hat{y}k_xk_zk_y^n$;
	n = 1 ; 3 ; 5

$\Gamma_1^+(A_{1g})$ representation with s-wave pairing and the lowest components of the $\Gamma_3^+(E_g)$ $(n = 2)$ and $\Gamma_5^+(T_{2g})$ $(n = 0)$ representations as the crystal field split members of the five-dimensional d-wave pairing basis. Likewise, for triplet pairing in the cubic group shown in Table 19.3, we can identify the one-dimensional basis (formed by selecting the $n = 1$ component) of $\Gamma_1^-(A_{1u})$ and the two-dimensional basis (also formed by selecting $n = 1$) of $\Gamma_3^-(E_u)$ as the crystal field split three-dimensional p-wave pairing basis functions. The node structure for the non-degenerate representations can be found. For example, in a cubic structure with singlet pairing, the high symmetry reduces the number of non-degenerate representations to two. These are the $\Gamma_1^+(A_{1g})$ phase, which is nodeless and the $\Gamma_2^+(A_{2g})$ phase, where the order parameter vanishes on six distinct planes.

In the $\Gamma_2^+(A_{2g})$ phase, the six planes intersect the three-dimensional Fermi surface on six distinct arcs, giving six lines of nodes on the Fermi surface. In a structure with lower symmetry such as the tetragonal structure, the number of non-degenerate singlet representations is four. The possible singlet phases in D_{4h} are shown in Table 19.4. In addition to the nodeless $\Gamma_1^+(A_{1g})$ phase, one finds the two non degenerate $\Gamma_3^+(B_{1g})$, and $\Gamma_4^+(B_{2g})$ phases, which each produce two lines of nodes as expected for d-wave pairing, and finally the order parameter for the $\Gamma_2^+(A_{2g})$ pairing vanishes on four planes giving four lines of nodes.

For triplet pairing, the occurrence of nodes is generally rarer than for singlet pairings since the three components of $\vec{d}(k)$ must vanish simultaneously. The need to satisfy three conditions may be expected to reduce the dimensions of the nodes from a surface to a set of isolated points, which need not lie on the Fermi surface. Inspection of Table 19.3 shows that this is the case for the O_h $\Gamma_1^-(A_{1u})$ phase where the components vanish on orthogonal planes producing isolated zeros not on the Fermi surface, so the gap has no nodes. The components of the order parameter for the highly symmetric non-degenerate $\Gamma_2^-(A_{2u})$ phase each vanish on three planes. This leads

Table 19.4. Singlet pairing basis functions in the tetragonal D_{4h} group

Representation	Basis Functions
$\Gamma_1^+(A_{1g})$	1
$\Gamma_2^+(A_{2g})$	$k_y k_x (k_x^2 - k_y^2)$
$\Gamma_3^+(B_{1g})$	$k_x^2 - k_y^2$
$\Gamma_4^+(B_{2g})$	$k_x k_y$
$\Gamma_5^+(E_g)$	$k_z k_x \;;\; k_z k_y \;,\; k_z k_x^3 \;;\; k_z k_y^3$

Table 19.5. Triplet pairing basis functions in the tetragonal D_{4h} group. The value $k_x \pm i k_y$ is denoted by $k_\pm$ and the vector $\hat{x} \pm i\hat{y}$ is denoted by $\hat{r}_\pm$. The upper and lower signs $\pm$ in the doubly degenerate E_u representation correspond to the degenerate partners ($n = 0\,; 2$)

Representation	Basis Functions
$\Gamma_1^-(A_{1u})$	$\hat{z}k_z \;;\; Re\hat{r}_- k_+ \;;\; Re\hat{r}_+ k_+^3$
$\Gamma_2^-(A_{2u})$	$Im\hat{r}_- k_+ \;;\; Im\hat{r}_+ k_+^3 \;;\; \hat{z}k_z Im k_+^4$
$\Gamma_3^-(B_{1u})$	$\hat{z}k_z Re k_+^2 \;;\; Re\hat{r}_+ k_+ \;;\; Re\hat{r}_- k_+^3$
$\Gamma_4^-(B_{2u})$	$\hat{z}k_z Im k_+^2 \;;\; Im\hat{r}_+ k_+ \;;\; Im\hat{r}_- k_+^3$
$\Gamma_5^-(E_u)$	$\hat{z}k_\pm^{n+1} \;;\; k_z \hat{r}_\pm k_\pm^n \;;\; k_z \hat{r}_\pm k_\mp^{n+2}$

Table 19.6. Singlet pairing basis functions in the hexagonal D_{6h} group. The value $k_x \pm i k_y$ is denoted by $k_\pm$. The upper and lower signs $\pm$ in the doubly degenerate E_g representations correspond to the degenerate partners.

Representation	Basis Functions
$\Gamma_1^+(A_{1g})$	1
$\Gamma_2^+(A_{2g})$	$Im k_+^6$
$\Gamma_3^+(B_{1g})$	$k_z Im k_+^3$
$\Gamma_4^+(B_{2g})$	$k_z Re k_+^3$
$\Gamma_5^+(E_{1g})$	$k_z k_\pm \;;\; k_z k_\mp^5$
$\Gamma_6^+(E_{2g})$	$k_\pm^2 \;;\; k_\mp^4$

to the order parameter vanishing on the four lines $k_x = \pm k_y = \pm k_z$, which intersect the Fermi surface in eight points, resulting in eight point nodes.

The detailed node structure of degenerate representations depends on the spontaneous symmetry breaking between the degenerate partners. In the singlet case, the phase of the order parameter $D(\underline{k})$ can always be chosen to be zero and so $D(\underline{k})$, if it vanishes, does so on surfaces that intersect the Fermi surface in curves. Examples of this are found in both the triply degenerate O_h $\Gamma_5^+(T_{2g})$ and the doubly de-

Table 19.7. Triplet pairing basis functions in the hexagonal D_{6h} group. The value $k_x \pm i k_y$ is denoted by $k_\pm$ and the vector $\hat{x} \pm i\hat{y}$ is denoted by $\hat{r}_\pm$. The *Re* and *Im* terms in the doubly degenerate E_u representations correspond to the degenerate partners

Representation	Basis Functions
$\Gamma_1^-(A_{1u})$	$\hat{z}\,k_z \;;\; Re\,\hat{r}_+ k_- \;;\; Re\,\hat{r}_+ k_+^5$
$\Gamma_2^-(A_{2u})$	$Im\hat{r}_+ k_- \;;\; Im\,\hat{r}_+ k_+^5 \;;\; Im\hat{z}k_z k_-^6$
$\Gamma_3^-(B_{1u})$	$Im\hat{z}k_-^3 \;;\; Im\hat{r}_+ k_z k_+^2 \;;\; Im\hat{r}_+ k_z k_-^4$
$\Gamma_4^-(B_{2u})$	$Re\hat{z}k_-^3 \;;\; Re\hat{r}_+ k_z k_+^2 \;;\; Re\hat{r}_+ k_z k_-^4$
$\Gamma_5^-(E_{1u})$	$Re\hat{z}k_+ \;;\; Re\hat{r}_+ k_z \;;\; Re\hat{r}_- k_z k_+^2 \;;$
	$Re\,\hat{z}k_-^5 \;;\; Re\hat{r}_- k_z k_-^4 \;;\; Re\,\hat{r}_+ k_z k_-^6 \;,$
	$Im\hat{z}k_+ \;;\; Im\hat{r}_+ k_z \;;\; Im\hat{r}_- k_z k_+^2 \;;$
	$Im\,\hat{z}\,k_-^5 \;;\; Im\hat{r}_- k_z k_-^4 \;;\; Im\,\hat{r}_+ k_z k_-^6$
$\Gamma_6^-(E_{2u})$	$Re\hat{r}_+ k_+ \;;\; Re\hat{z}k_z k_+^2 \;;\; Re\hat{r}_- k_+^3 \;;$
	$Re\hat{r}_- k_-^3 \;;\; Re\hat{r}_+ k_-^5 \;;\; Re\hat{z}k_z k_-^4 \;,$
	$Im\hat{r}_+ k_+ \;;\; Im\hat{z}k_z k_+^2 \;;\; Im\hat{r}_- k_+^3 \;;$
	$Im\hat{r}_- k_-^3 \;;\; Im\hat{r}_+ k_-^5 \;;\; Im\hat{z}k_z k_-^4$

generate $\Gamma_3^+(E_g)$ singlet pairing representations. It is seen that if the system spontaneously selects any one of the degenerate partners, the order parameter will vanish on two distinct planes leading to two lines of nodes. Also, if the effects of the higher order angular momentum contributions are minimal, an arbitrary linear combination of degenerate partner basis functions also makes the order parameter vanish on two surfaces leading to two curved lines of nodes. The analogous case for the singlet D_{4h} symmetry is given by the doubly degenerate $\Gamma_5^+(E_g)$ representation, which also can yield lines of nodes. One of the line nodes is the intersection of the plane $k_z = 0$ with the Fermi surface.

With triplet pairing, the phase of the order parameter can also be chosen to be real in the unitary case. However, in this case, due to the stringent requirement that if $\vec{d}(\underline{k})$ is to vanish, three equations to be satisfied, the only nodes which occur usually are point nodes. These may be satisfied by symmetry, as in the tetragonal $\Gamma_2^-(A_{2u})$ pairing shown in Table 19.5 where, if $f_2 = f_3 = 0$, symmetry dictates that the order parameter vanishes on the line $k_x = k_y = 0$. In this case, the gap that develops on the Fermi surface has isolated nodes at the poles.

However, line nodes also can be obtained for triplet phases. As Blount has noted [141], the vanishing of the order parameter on a surface can be achieved since the rotational invariance in spin space can reduce the number of independent components of $\vec{d}$ to two and thus lead to a line of nodes on the Fermi surface. An example of this is given in the O_h group with the $\Gamma_3^-(E_u)$ irreducible representation where, if $\vec{d}$ spontaneously chooses to lie in the $\underline{k}$ direction, then $\vec{d}(\underline{k})$ vanishes on two surfaces and leads to two lines of nodes on the Fermi surface. This type of situation is expected to be forbidden when the effect of spin-orbit coupling is taken into consideration. The irreducible representations for singlet and triplet pairings of the hexagonal group D_{6h} are shown in Tables 19.6 and 19.7.

In addition to the point group symmetry, the superconducting order parameter must also have the same translational symmetry as the lattice. In particular, the order parameter must be periodic in the Brillouin zone and should also reflect any other translation symmetry element present in the space group. The functions given in the tables correspond to the long wavelength limit form of the order parameters. Some possible forms of the order parameters with the correct translational symmetry have been examined for the case of UPt_3 [148,149].

Spin-Orbit Coupling

In the presence of spin-orbit coupling, the electrons spin quantized about any fixed axis is no longer a good quantum number, however, the electron does have a well defined magnetic moment. The magnetic moment is an axial vector and can be expressed in terms of a $\underline{k}$ dependent gyromagnetic ratio $g_{i,j}$ which couples to the electron spins via

$$m_i(\underline{k}) = \frac{1}{2}\sum_j g_{i,j}(k)\sigma_j \,. \tag{19.71}$$

Then, in second quantized form, the magnetic moment operator is given by the expression

$$\vec{\hat{M}}(\underline{k}) = \frac{1}{2}\sum_{\alpha,\beta} a^\dagger_\alpha(\underline{k})\left(\vec{m}(\underline{k})\right)_{\alpha,\beta} a_\beta(\underline{k}) \tag{19.72}$$

The expectation value of this operator can be expressed in terms of the one-electron density matrix $\rho_{\alpha,\beta}(\underline{k}) =< |a^\dagger_\alpha(\underline{k})a_\beta(\underline{k})| >$, as

$$< |\vec{\hat{M}}(\underline{k})| >= Trace\left[\vec{m}(\underline{k})\rho(\underline{k})\right] . \tag{19.73}$$

Since the Zeeman splitting of the electron in a Bloch state $\underline{k}$ is given by the eigenvalues, $\pm\, E_z(\underline{k})$, of the Hamiltonian

$$H_Z = -\vec{M}(\underline{k}).\vec{H} \,. \tag{19.74}$$

Then the splitting of the eigenvalues is found from H_Z^2 independent of the choice of basis. As the Hamiltonian is a Hermitian operator, the eigenvalues are found in terms of the g factors as

$$E_Z(\underline{k})^2 = \frac{1}{4}\vec{H} \,.\, \tilde{g}(\underline{k})^t\, \tilde{g}(\underline{k}) \,.\, \vec{H} \,. \tag{19.75}$$

This shows that, even though the spin is not a good quantum number, the pairing states can be determined in terms of a quasi-spin, independent of the basis, as had been assumed by Anderson [145]. The order parameter still takes on the same form in that it is still composed of either a scalar or a scalar product of an axial vector with $\vec{m}(\underline{k})$

$$\Delta_{\alpha,\beta}(\underline{k}) = \left(\left[D(\underline{k}) + \vec{d}(\underline{k}) \,.\, \vec{m}(\underline{k})\right](i\,\sigma_y)\right)_{\alpha,\beta} . \tag{19.76}$$

The results derived in the presence of spin-orbit coupling go smoothly over into the zero spin-orbit case, as can be seen by expressing the order parameter in the form

$$\Delta^\dagger(\underline{k})\,\Delta(\underline{k}) = \left| D(\underline{k}) \right|^2 + \left(\vec{m}(\underline{k}) \,.\, \vec{d}(\underline{k})\right)\left(\vec{d}^*(\underline{k}) \,.\, \vec{m}(\underline{k})\right) . \tag{19.77}$$

With the use of the Pauli identity, this procedure yields the same expression for the quasi-particle eigenvalues as Eq. (19.60), but in which $\vec{d}(\underline{k})$ is merely replaced by $\vec{d}(\underline{k})\, g(\underline{k})$. Thus, most of the formal results for the superconducting quasi-particles found without spin-orbit coupling also hold when it is present. In particular, one expects that the characteristic energy dependence of the superconducting quasi-particle density of states will be unaffected by spin-orbit scattering.

Quasi-Particle Density of States

The density of states for single quasi-particle excitations for anisotropic superconducting states have distinctly different forms which are determined by whether the gap is nodeless, has isolated nodes, or lines of nodes and also the slope with which the gap falls to zero. The quasi-particle density of states $\rho_{qp}(E)$ is given by

$$\rho_{qp}(E) = \frac{1}{2} \sum_{\underline{k},\pm} \delta(E \pm E_{\sigma,\underline{k}}) . \qquad (19.78)$$

On assuming singlet s-wave pairing, one finds the BCS result

$$\rho_{\mathrm{BCS}}(E) = \rho(\mu) \frac{|E|}{\sqrt{E^2 - |\Delta_0|^2}} \qquad (19.79)$$

for $|E| > \Delta_0$, otherwise $\rho_{\mathrm{BCS}}(E) = 0$. Here $\rho(\mu)$ is the normal phase density of states that has been assumed to be constant near the Fermi energy and $D(k_F) = \Delta_0$. Since this s-wave phase shows a gap of width $2|\Delta_0|$ all the way around the position of the normal state Fermi surface, a number of physical quantities show simple exponentially activated behavior in the low temperature regime where the temperature dependence of the superconducting gap is small. For the anisotropic singlet $d_{x^2-y^2}$ pairing, the density of states is given by

$$\rho_{d_{x^2-y^2}}(E) = \frac{\rho(\mu)|E|}{4\pi} \int_0^{2\pi} \mathrm{d}\phi \int_0^{\pi} \mathrm{d}\theta \times \frac{\sin\theta}{\sqrt{E^2 - |\Delta_0|^2 \sin^4\theta \cos^2 2\phi}} . \qquad (19.80)$$

As the order parameter vanishes on two great circles, $\cos 2\phi = 0$, on the Fermi surface, the density of states is proportional to $E \ \ln E$ as $E/\Delta_0 \to 0$. The $d_{2z^2-x^2-y^2}$

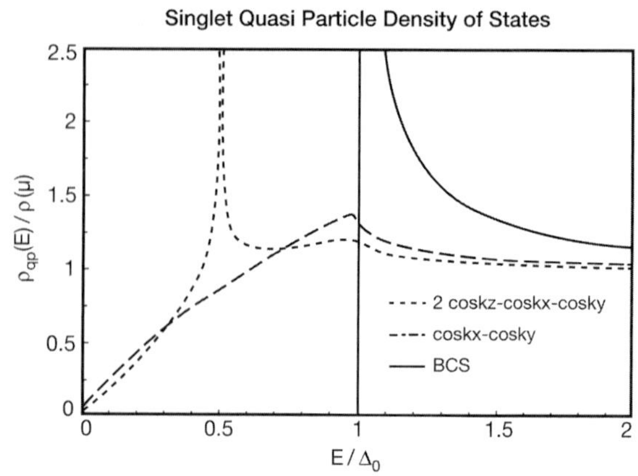

Fig. 19.22. The quasi-particle density of states $\rho_{qp}(E)$ for selected singlet superconducting phases, in the clean limit. The BCS s-wave state has a true gap in the quasi-particle density of states, and a singularity at the gap edge. The maximum superconducting gap is denoted by Δ_0. The d-wave phases have states with energies below Δ_0

order parameter of a cubic material also has two lines of nodes, which results in the density of states

$$\rho_{d_{2z^2-x^2-y^2}}(E) = \frac{\rho(\mu)|E|}{2} \int_0^{\pi} \mathrm{d}\theta \times \frac{\sin\theta}{\sqrt{E^2 - \frac{1}{4}|\Delta_0|^2(1 - 3\cos^2\theta)^2}} \qquad (19.81)$$

that also tends to zero linearly for small E. The quasi-particle density of states for these particular singlet phases are shown in Fig. 19.22. The quasi-particle density of states for the phases $d_{x,y}$, $d_{x,z}$ and $d_{y,z}$ should be identical with $\rho_{d_{x^2-y^2}}(E)$. For an isotropic system, the most stable d-wave phase [130] corresponds to a linear combination of the even m order parameters

$$D(\hat{k}) \propto \left[\frac{1}{\sqrt{2}} Y_0^2(\hat{k}) + \frac{1}{2}(Y_2^2(\hat{k}) - Y_{-2}^2(\hat{k})) \right] , \qquad (19.82)$$

which only has point nodes. Therefore, the quasi-particle density of states varies as E^2 for small E, as shown in Fig. 19.23.

For triplet pairing, the differing node structures yield different forms for the quasi-particle density of states. The density of states is given in terms of an integration over the direction of $\underline{k}$,

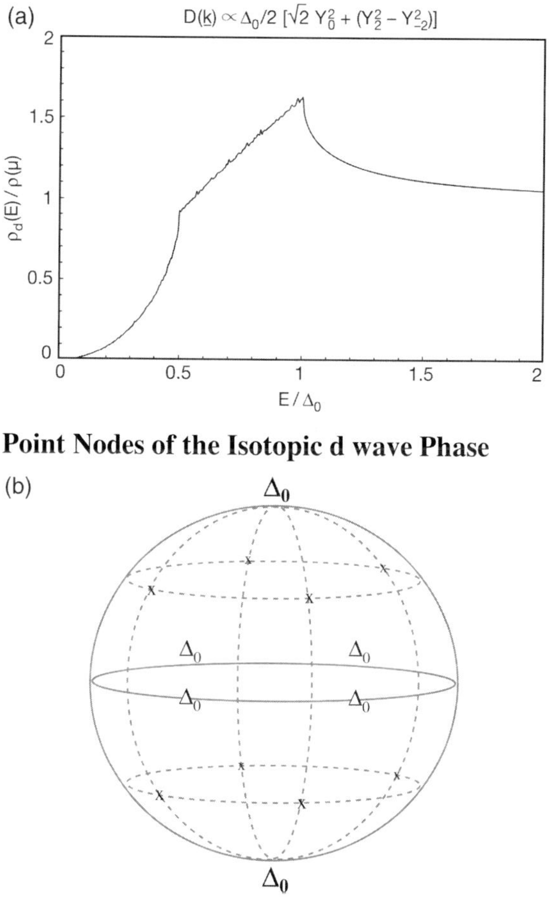

Fig. 19.23. (**a**) The quasi-particle density of states $\rho_d(E)$ for an isotropic d-wave phase, with the order parameter $|D(\hat{k})| = \frac{\Delta_0}{2}[(3\cos^2\theta - 1)^2 + 3\sin^2 2\varphi \sin^4\theta]^{\frac{1}{2}}$. Since the order parameter has eight point nodes as the quasi-particle density of states is proportional to E^2 for small E. Also, since the order parameter only attains its maximum value of Δ_0 at six isolated points, the density of states does not diverge, (**b**) Nodes of O.P.

$$\rho_{S=1}(E) = \frac{1}{8\pi}\rho(\mu)\int d\Omega \sum_{\sigma} Re \times \left[\frac{|E|}{\sqrt{E^2 + |\vec{d}(\hat{k})|^2 - \sigma|i\vec{d}(\hat{k}) \wedge \vec{d}^*(\hat{k})|}}\right]. \tag{19.83}$$

We shall describe the relation between the order parameter and the quasi-particle density of states for the well known p-wave states. The BW state, having an isotropic gap and no nodes, has the same density of states as the singlet s-wave phase given by the BCS expression, but where Δ_0 is given by the magnitude of $\vec{d}$ at the position of the normal state Fermi surface. The ABM state order parameter given in spherical polar coordinates is $\vec{d} = \hat{z}\Delta_0 \sin\theta e^{i\phi}$, and so the quasi-particle density of states in the ABM phase is found from

$$\begin{aligned}\rho_{\mathrm{ABM}}(E) &= \frac{\rho(\mu)\,|E|}{4\pi}\int_0^{2\pi} d\phi \int_0^{\pi} d\theta \frac{\sin\theta}{\sqrt{E^2 - |\Delta_0|^2\sin^2\theta}} \\ &= \frac{\rho(\mu)E}{2\Delta_0}\ln\left|\frac{E+\Delta_0}{E-\Delta_0}\right| .\end{aligned} \tag{19.84}$$

Thus, the density of states in the ABM phase has a weak logarithmic singularity at $E = \Delta_0$ and falls to zero as $E \to 0$ like E^2, due to the presence of isolated nodes. In the polar state, $\vec{d} = \hat{z}\Delta_0\cos\theta$, so the density of states is given by

$$\begin{aligned}\rho_{p_z}(E) &= \frac{\rho(\mu)|E|}{4\pi}\int_0^{2\pi} d\phi \int_0^{\pi} d\theta \frac{\sin\theta}{\sqrt{E^2 - |\Delta_0|^2\cos^2\theta}} \\ &= \frac{\pi\,\rho(\mu)|E|}{2\Delta_0} \qquad \Delta_0 > |E| \\ &= \frac{\rho(\mu)E}{\Delta_0}\arcsin\left[\frac{\Delta_0}{E}\right] \qquad \Delta_0 < |E| .\end{aligned} \tag{19.85}$$

Thus, due to the presence of a line of nodes in the polar phase, the quasi-particle density of states at lower energies only falls to zero linearly in E, as $E \to 0$. This should be contrasted with the ABM phase, which has isolated point nodes where the density of states goes to zero quadratically with E. The larger density of states at low energies for polar phase with its line nodes compared with the ABM phase with its point nodes, is compensated by the decrease in the peak of the quasi-particle density of states. That is, the polar state only shows a slight enhancement at Δ_0 where the density of states only has a cusp, but the ABM state shows a divergence. The quasi-particle density of states for the singlet BCS state is shown together with the $d_{2z^2-x^2-y^2}$ state and the $d_{x^2-y^2}$ in Fig. 19.22. The singularities and cusps are due to the gaps having extremal points at the position of the normal state

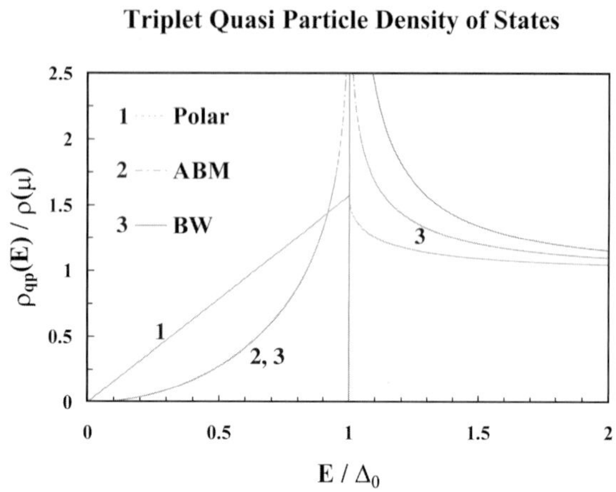

Fig. 19.24. The quasi-particle density of states $\rho_{qp}(E)$ for selected triplet p-wave superconducting states, in the clean limit. The density of states of the BW phase is identical to that of the BCS s-wave phase. The ABM phase and polar phases have states below Δ_0. The ABM or axial state shows a quadratic energy dependence for energies below Δ_0, while the polar state has a linear energy dependence

Fermi surface. The quasi-particle density of states for the triplet BW, ABM, and polar states are shown in Fig. 19.24. The low-energy behavior of these quasi-particle density of states are quite distinct, having different power law dependences in E. The different power laws are due to the differences in the node structure of the superconducting gap. These power law dependences may be expected to show up in thermodynamic properties far below T_c, albeit modified by the effects of impurity scattering and collective excitations.

Pair Breaking Impurities

In an s-wave superconductor, the effect of non-magnetic impurities is minimal[6] and results in a pair weakening effect, which can be absorbed into a reduction of the pairing potential or density of states. Anderson's theorem [151], which describes the effect of non-magnetic impurities in s-wave superconductors, can be stated as, "A static external perturbation, that neither breaks time reversal invariance nor produces a spatial variation of the order parameter, does not produce a change in the thermodynamic properties of a superconductor." Since the order parameters of anisotropic superconductors are inhomogeneous, Anderson's theorem no longer applies and the effect of impurities is analogous to that of magnetic impurities in s-wave superconductors [152,153]. The analogy is complete in that significant concentrations of impurities may produce significant changes in the low-energy density of states leading to the removal of any remaining gaps. The changes in the low-energy variation of the quasi-particle density of states lead to variances in the power law temperature dependences expected from analysis of the simple node structures of clean materials [154–156].

The types of effect caused by non-magnetic impurities in anisotropic superconductors can be seen by examining the density of states in the superconducting state in which the impurity potential, U_0, is short-ranged and is treated as a small perturbation. The density of states can be obtained from the trace of the Green's function in the superconducting state. To treat the pairing in the superconducting state, one introduces the four by four matrix Green's function $G(\underline{k};\tau)$ describing the four component fields, via

$$G(\underline{k};\tau) = -\frac{1}{\hbar} < |\hat{T}\Psi(\underline{k},\tau)\Psi^{\dagger}(\underline{k},0)| > , \quad (19.86)$$

where $\hat{T}$ is Wick's time ordering operator. The upper and lower two by two diagonal blocks have matrix elements that are related to the usual Green's functions

$$G_{\alpha,\beta}(\underline{k},\underline{k}';\tau) = -\frac{1}{\hbar} < |\hat{T}a_{\underline{k},\alpha}(\tau)a^{\dagger}_{\underline{k}',\beta}(0)| > . \quad (19.87)$$

The upper and lower two by two off-diagonal blocks represent the anomalous Green's functions, as introduced by Gor'kov [157]. The elements of the anomalous Green's functions are written as

[6] The variation of T_c for Sn with a concentration of In impurities was studied by Coles [150]. The T_c showed a sharp initial drop which saturated for In concentrations of about 2%. The initial drop of T_c is attributed to the impurities destroying the anisotropy of the Fermi surface. This is not a violation of Anderson's theorem since it strictly only applies to systems with isotropic Fermi surfaces. The almost constant value of T_c which is found for In concentrations greater than 2% is a verification of Anderson's theorem.

$$F_{\alpha,\beta}(\underline{k},\underline{k}';\tau) = \frac{1}{\hbar} < |\hat{T} a_{\underline{k},\alpha}(\tau) a_{-\underline{k}',\beta}(0)| > \quad (19.88)$$

and

$$F^{\dagger}_{\alpha,\beta}(\underline{k},\underline{k}';\tau) = \frac{1}{\hbar} < |\hat{T} a^{\dagger}_{-\underline{k},\alpha}(\tau) a^{\dagger}_{\underline{k}',\beta}(0)| > \ . \quad (19.89)$$

Within the mean-field approximation, the Fourier transform of the Green's functions are given by the solutions of the coupled equations

$$\begin{aligned}(i\hbar\,\omega_n - e(\underline{k}) + \mu) G_{\alpha,\beta}(\underline{k},\underline{k}';i\hbar\omega_n) \\ + \sum_{\gamma} \Delta_{\alpha,\gamma}(\underline{k}) F^{\dagger}_{\gamma,\beta}(\underline{k},\underline{k}';i\hbar\omega_n) \\ = \delta_{\alpha,\beta}\,\delta(\underline{k}-\underline{k}')\end{aligned} \quad (19.90)$$

and

$$\begin{aligned}(i\hbar\,\omega_n + e(\underline{k}) - \mu) F^{\dagger}_{\alpha,\beta}(\underline{k},\underline{k}';i\hbar\omega_n) \\ + \sum_{\gamma} \Delta^{*}_{\alpha,\gamma}(\underline{k}) G_{\gamma,\beta}(\underline{k},\underline{k}';i\hbar\omega_n) = 0\ .\end{aligned} \quad (19.91)$$

In the unitary states, these equations have the solution,

$$\begin{aligned}G_{\alpha,\beta}(\underline{k},\underline{k}';i\hbar\omega_n) = \\ - \frac{(i\hbar\,\omega_n + e(\underline{k}) - \mu)}{\hbar^2\omega_n^2 + (e(\underline{k}) - \mu)^2 + \Delta^{\dagger}(\underline{k})\Delta(\underline{k})} \delta_{\alpha\beta}\delta(\underline{k}-\underline{k}')\end{aligned} \quad (19.92)$$

and

$$\begin{aligned}F^{\dagger}_{\alpha,\beta}(\underline{k},\underline{k}';i\hbar\omega_n) = \\ + \frac{\Delta^{*}_{\alpha,\beta}(\underline{k})}{\hbar^2\omega_n^2 + (e(\underline{k}) - \mu)^2 + \Delta^{\dagger}(\underline{k})\Delta(\underline{k})} \delta(\underline{k}-\underline{k}')\ .\end{aligned} \quad (19.93)$$

The order parameter is determined from the non-linear equation

$$\Delta_{\alpha,\beta}(\underline{k}) = -k_B T \sum_{n,\underline{q}} \sum_{\mu,\nu} V(\underline{k},\underline{q})_{\beta,\alpha;\mu,\nu} F_{\mu,\nu}(\underline{q},\underline{q};i\hbar\omega_n), \quad (19.94)$$

which yields Δ as a function of temperature. We assume a low concentration of impurities, c_0, that are randomly distributed and act as point scatterers. The matrix self-energy can be calculated as

$$\Sigma(\underline{k};i\hbar\omega_n) = c_0 \sum_{\underline{p}} U_0 \rho_4 G(\underline{p};i\hbar\omega_n) \rho_4 U_0\ , \quad (19.95)$$

where ρ_4 is the 4×4 Dirac matrix that is written in block diagonal form as

$$\rho_4 = \begin{pmatrix} I & 0 \\ 0 & -I \end{pmatrix}. \quad (19.96)$$

Here, I is the unit two-by-two matrix. The Green's functions have to be calculated self-consistently. In the unitary case, the diagonal and non-diagonal elements in the self-energy can be evaluated and do give rise to a renormalization of the frequency and order parameter. The renormalizations are given by the solution of the coupled equations

$$\begin{aligned}i\hbar\,\tilde{\omega}_n &= i\hbar\,\omega_n + c_0 \sum_{\underline{p}} U_0^2 \\ &\times \frac{i\hbar\,\tilde{\omega}_n + (e(\underline{p}) - \mu)}{\hbar^2\tilde{\omega}_n^2 + (e(\underline{p}) - \mu)^2 + \tilde{\Delta}^{\dagger}(\underline{p})\tilde{\Delta}(\underline{p})}\ , \\ \tilde{\Delta}_{\alpha,\beta}(\underline{k}) &= \Delta_{\alpha,\beta}(\underline{k}) \\ &+ c_0 \sum_{\underline{p}} U_0^2 \frac{\tilde{\Delta}_{\alpha,\beta}(\underline{p})}{\hbar^2\tilde{\omega}_n^2 + (e(\underline{p}) - \mu)^2 + \tilde{\Delta}^{\dagger}(\underline{p})\tilde{\Delta}(\underline{p})}\ .\end{aligned} \quad (19.97)$$

The energy $e(\underline{p})$ is odd about the Fermi energy and so, for systems with electron–hole symmetry, the term proportional to $e(\underline{p}) - \mu$ drops out in the diagonal parts of the self-energy matrix. Thus, impurity scattering only results in a frequency dependent renormalization of ω_n. For s-wave scattering, subject to Anderson's theorem, the off-diagonal term produces an identical frequency dependent renormalization of the gap. Thus, for s-wave scattering in the Born approximation, one finds the explicit form for the renormalizations

$$\begin{aligned}i\tilde{\omega}_n &= i\omega_n + \pi c_0 U_0^2 \rho(\mu) \frac{i\tilde{\omega}_n}{\sqrt{\hbar^2\tilde{\omega}_n^2 + |\tilde{D}_s(k_F)|^2}}\ , \\ \tilde{D}_s(\underline{k}) &= D_s(\underline{k}) \\ &+ \pi c_0 U_0^2 \rho(\mu) \frac{\tilde{D}_s(k_F)}{\sqrt{\hbar^2\tilde{\omega}_n^2 + |\tilde{D}_s(k_F)|^2}}\ .\end{aligned} \quad (19.98)$$

Hence the ratio of $\tilde{\omega}_n/\tilde{D}_s$ is unchanged from the unrenormalized value ω_n/D_s. Furthermore, the quasiparticle density of states is unaltered since it only depends upon the ratio $\tilde{\omega}_n/\tilde{D}_s$. The equation that determines the order parameter remains unchanged. Thus, the critical temperature of an s-wave superconductor is not substantially reduced by the presence of the impurities. These conclusions are in accord with Anderson's theorem.

For triplet pairing, the gap is an odd function of $\underline{p}$ and so, due to the summation over the direction of $\underline{p}$, the off-diagonal self-energy vanishes identically. Hence, we have

$$i\tilde{\omega}_n = i\omega_n + \pi c_0 U_0^2 \rho(\mu) \int \frac{d\Omega_{\underline{p}}}{4\pi} \frac{i\tilde{\omega}_n}{\sqrt{\hbar^2\tilde{\omega}_n^2 + |\vec{d}(\underline{p})|^2}}$$

and

$$\tilde{\vec{d}}(\underline{k}) = \vec{d}(\underline{k}) \,. \qquad (19.99)$$

Therefore, Anderson's theorem does not apply to triplet superconductors. The effect of impurities reduce the critical temperature T_c, as can be seen from the linearized gap equation

$$1 = k_B T_c \pi \sum_{n=0}^{n_c} \frac{< V(k_F, k_F) > \rho(\mu)}{|\hbar\omega_n| + \pi c_0 |U_0|^2 \rho(\mu)} \,, \qquad (19.100)$$

where the summation is cut off at n_c, which is determined by a frequency range ω_c in which the pairing interaction is attractive $\hbar\,\omega_c = (2n_c + 1)\pi k_B T$, and $< V(k_F, k_F) >$ is the Fermi surface averaged value of the pairing interaction. The summation depends logarithmically on the cut off. The linearized gap equation can be re-written as

$$\frac{1}{k_B T_c \pi < V(k_F, k_F) > \rho(\mu)} = \sum_{n=0}^{n_c} \frac{1}{|\hbar\omega_n|} \qquad (19.101)$$
$$+ \sum_{n=0}^{\infty} \left(\frac{1}{|\hbar\omega_n| + \pi\, c_0 |U_0|^2 \rho(\mu)} - \frac{1}{|\hbar\omega_n|} \right) .$$

The critical temperature obtained for the dirty system should be compared with the expression for the critical temperature T_{c0} obtained for the clean system

$$1 = k_B T_{c0} \pi \sum_{n=0}^{n_c} \frac{< V(k_F, k_F) > \rho(\mu)}{|\hbar\omega_n|} \,. \qquad (19.102)$$

On identifying the $\frac{2\pi}{\hbar} c_0 |U_0|^2 \rho(\mu)$ as being equal to the impurity scattering rate $\frac{1}{\tau}$ asevaluated in the Born approximation, and using the expression for T_{c0} for the clean system, one finds that non-magnetic impurities depress T_c in an anisotropic superconductor according to

$$\ln\left(\frac{T_{c0}}{T_c}\right) = \psi\left(\frac{1}{2} + \frac{\hbar}{4\pi k_B T_c \tau}\right) - \psi\left(\frac{1}{2}\right) , \qquad (19.103)$$

where ψ is the digamma function. This formula is similar to the formula derived by Abrikosov and Gor'kov that describes the depression of T_c caused by spin-flip scattering by impurities in an s-wave superconductor [152]. The analogy to scattering by paramagnetic impurities in an s-wave superconductor is complete in that large enough impurity scattering can also lead to gapless superconductivity.

We shall explicitly examine the transition to the gapless phase for some triplet phases [155]. On analytically continuing from $i\hbar\omega_n \to E + i\delta$ one finds

$$\lim_{i\hbar\omega_n \to E} Im\left[i\hbar\tilde{\omega}_n \right] = \pi c_0 |U_0|^2 \rho_{qp}(E) \,, \qquad (19.104)$$

where the self-consistency equation for $s = i\hbar\tilde{\omega}_n$ is given by

$$s - E = i\frac{\hbar}{2\tau} \int \frac{d\Omega_{\underline{k}}}{4\pi} \frac{s}{\sqrt{s^2 - |\vec{d}(\hat{k})|^2}} \,. \qquad (19.105)$$

For the Balian–Werthamer phase, where $\vec{d}(\underline{k}) = \Delta_0 \hat{k}$, the above equation reduces to

$$s - E = i\frac{\hbar}{2\tau} \frac{s}{\sqrt{s^2 - |\Delta_0|^2}} \,. \qquad (19.106)$$

The gap energy is the threshold value of E_g below which s is real and is given by the critical value of s determined from the equation

$$\frac{\partial E}{\partial s} = 1 + i\frac{\hbar}{2\tau} \frac{|\Delta_0|^2}{(s^2 - |\Delta_0|^2)^{\frac{3}{2}}} = 0 \,. \qquad (19.107)$$

The solution for s results in the gap being given by

$$E_g = \Delta_0 \left[1 - \left(\frac{\hbar}{2\tau\Delta_0} \right)^{\frac{2}{3}} \right]^{\frac{3}{2}} . \qquad (19.108)$$

The gap is reduced by increasing the impurity scattering rate, and the gap first falls to zero when $\frac{\hbar}{2\tau} = \Delta_0$

in the regime where Δ_0 and T_c are still finite. For larger impurity scattering rates, the BW phase is gapless.

For the ABM phase, where $\vec{d}(\underline{k}) = \hat{z}\Delta_0 \sin\theta e^{i\,\phi}$, one finds

$$s - E = i\frac{\hbar}{4\tau\Delta_0} s \ln\left(\frac{s+\Delta_0}{s-\Delta_0}\right) . \qquad (19.109)$$

The density of states at $E = 0$ is found from the solution for $s(0)$

$$\begin{aligned} s(0) &= 0\,, \\ &\text{for} \quad \left(\frac{\hbar}{4\tau\Delta_0}\right) < \frac{1}{\pi} \\ s(0) &= i\,\Delta_0 \cot\left(\frac{2\tau\Delta_0}{\hbar}\right), \\ &\text{for} \quad \left(\frac{\hbar}{4\tau\Delta_0}\right) > \frac{1}{\pi}\,, \end{aligned} \qquad (19.110)$$

which shows the transition to a gapless phase for large impurity scattering rates. In the gapless phase, the zero energy quasi-particle density of states is given by

$$\begin{aligned} \rho_{\mathrm{ABM}}(0) &= -\frac{2\tau\Delta_0}{\hbar}\rho(\mu)\cot\left(\frac{2\tau\Delta_0}{\hbar}\right), \\ &\text{for} \quad \left(\frac{\hbar}{4\tau\Delta_0}\right) > \frac{1}{\pi}\,. \end{aligned} \qquad (19.111)$$

For finite E, the equation for $s(E)$ can be solved iteratively in powers of E. For low impurity scattering rates, one finds the solution

$$\begin{aligned} s &= \frac{E}{\left[1-\left(\frac{\pi\hbar}{4\tau\Delta_0}\right)\right]} \\ &\quad + i\frac{E^2\hbar}{2\tau\Delta_0^2}\frac{1}{\left[1-\left(\frac{\pi\hbar}{4\tau\Delta_0}\right)\right]^3} + \ldots \end{aligned} \qquad (19.112)$$

which yields a density of states which varies as E^2 for low energies

$$\rho_{\mathrm{ABM}}(E) = \rho(\mu)\left(\frac{E}{\Delta_0}\right)^2 \frac{1}{\left[1-\left(\frac{\pi\hbar}{4\tau\Delta_0}\right)\right]^3} + \ldots \qquad (19.113)$$

due to isolated point nodes. The coefficient of the E^2 term in the quasi-particle density of states diverges at a critical value of the scattering rate

$$1 = \left(\frac{\pi\hbar}{4\tau\Delta_0}\right) \qquad (19.114)$$

at which point the ABM phase becomes gapless as the isolated nodes grow into regions of finite area.

For the polar phase, a similar analysis yields

$$s - E = i\left(\frac{\hbar}{2\tau\Delta_0}\right) s \arcsin\left(\frac{\Delta_0}{s}\right) . \qquad (19.115)$$

This has the solution at $E = 0$ which is always given by

$$s(0) = i\frac{\Delta_0}{\sinh\left(\frac{2\tau\Delta_0}{\hbar}\right)} \qquad (19.116)$$

and hence, the quasi-particle density of states is

$$\rho_{p_z}(0) = \rho(\mu)\frac{\left(\frac{2\tau\Delta_0}{\hbar}\right)}{\sinh\left(\frac{2\tau\Delta_0}{\hbar}\right)} . \qquad (19.117)$$

This shows that the critical impurity scattering rate for producing a gapless phase is zero. Thus, the line nodes grow into regions of finite area, and the superconductivity becomes gapless, even for the slightest concentration of impurities. The calculated quasi-particle density of states for these three p-wave states are shown in Fig. 19.25, for various impurity scattering rates.

The above analysis were based on the assumption that the impurity scattering potential is sufficiently weak so that the Born approximation is adequate. However, in heavy-fermion systems, the large value of the quasi-particle density of states invalidates the above approach, as was first pointed out by Pethick and Pines [158]. The increased strength of the scattering mechanism can lead to marked modifications of the low-energy variation of the quasi-particle density of states. Within the limit of low impurity concentration, one can sum the series of multiple scattering processes involving the same impurity to yield the

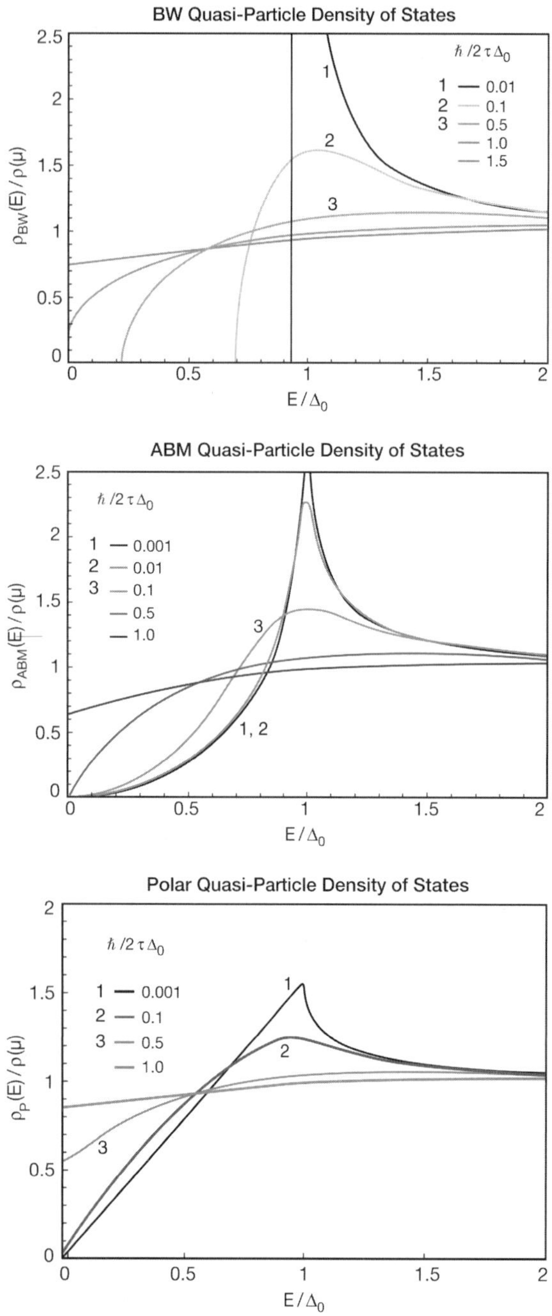

Fig. 19.25. The quasi-particle density of states for various phases of p-wave superconductors, and different strengths of the impurity scattering rate Γ. The impurity scattering rate is evaluated in the Born approximation [After Ueda and Rice [155]]

T-matrix result for the impurity self-energy

$$\Sigma(\underline{k}; i\hbar\omega_n) = c_0 \sum_{\underline{p}} U_0 \rho_4 G(\underline{p}; i\hbar\omega_n) \times \left[1 - \sum_{\underline{p}''} U_0 \rho_4 G(\underline{p}''; i\hbar\omega_n) \right]^{-1} \rho_4 U_0 \, . \tag{19.118}$$

A dimensionless measure of the strength of the scattering from a single impurity is given by the normal state phase shift δ_0, which is given by

$$\tan \delta_0 = -\pi U_0 \rho(\mu) \tag{19.119}$$

for the case of low-energy scattering and electron-hole symmetry. The limit of resonant scattering ($\delta_0 = \pm\frac{\pi}{2}$) can be treated similarly to the weak scattering results, and the quasi-particle density of states is given by

$$\rho_{qp}(E) = \rho(\mu) Re \left[\int \frac{d\Omega_{\underline{k}}}{4\pi} \frac{s}{\sqrt{s^2 - |\vec{d}(\hat{k})|^2}} \right] , \tag{19.120}$$

where $s(E)$ is found from

$$\frac{c_0}{\pi\, \rho(\mu)} \frac{i}{s - E} = \int \frac{d\Omega_{\underline{k}}}{4\pi} \frac{s}{\sqrt{s^2 - |\vec{d}(\hat{k})|^2}} \, . \tag{19.121}$$

In this limit, the low-energy density of states is different from the density of states found in the Born approximation. For example, the consistency relation for the ABM state can be solved at $E = 0$ to yield

$$s(0) = i \frac{1}{\pi} \sqrt{\frac{c_0}{2\rho(\mu)\Delta_0}} \, , \quad \text{for } 1 \gg |s(0)| \, . \tag{19.122}$$

Thus, the quasi-particle density of states is gapless for arbitrarily small concentrations of impurities and has the value

$$\rho_{ABM}(0) = \rho(\mu) \sqrt{\frac{c_0}{2\rho(\mu)\Delta_0}} \, , \tag{19.123}$$

which is a non-analytic function of the concentration, c_0. This result for resonant scattering is in contrast to the Born approximation result for the ABM

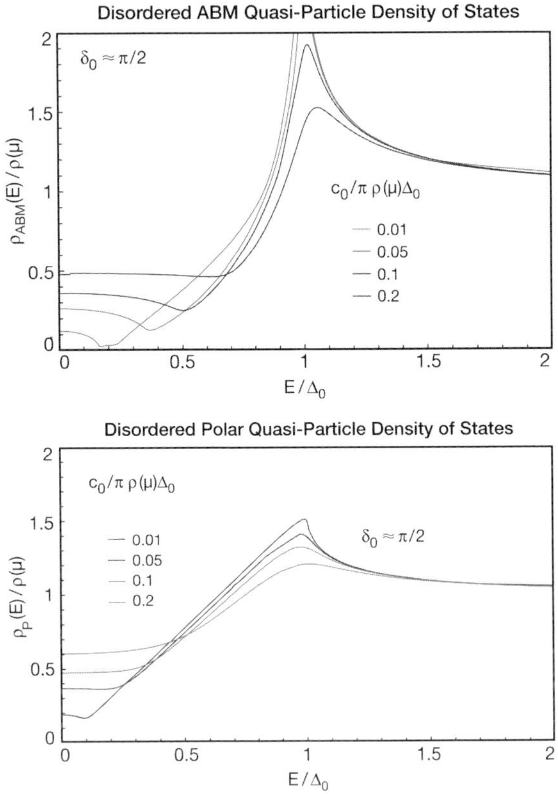

Fig. 19.26. The quasi-particle density of states for the ABM and polar p-wave superconducting phases, for different strengths of the impurity scattering. The scattering is treated in the T-matrix approximation. Resonant scattering occurs when the phase shift is given by $\delta_0 = \frac{\pi}{2}$, i.e., $\cot\delta_0 = 0.0$. The Born Approximation corresponds to the limit of large $\cot\delta_0$. [After Hirschfeld et al. [159]]

phase where there is a finite critical value of the impurity scattering rate. Furthermore, as shown in Fig. 19.26, for resonant impurity scattering, the ABM and polar states have low-energy maxima in the quasi-particle density of states [159]. As the order parameter is traceless, the balance between the diagonal and off-diagonal self-energy terms is also destroyed in the anisotropic singlet pairing phases. Generally, the different scaling behavior of the frequency and gap destroys the invariance of the quasi-particle density of states and leads to an increase in the number of low-energy states. Thereby, the simple power law dependences of the quasi-particle density of states and the concomitant power law dependences of the low temperature thermodynamic properties that are expected from the symmetry of the order parameter are spoiled. However, the power law dependences expected from the quasi-particle excitation spectrum could also be spoilt by the effect of collective excitations. The manner in which collective excitations affect the spectrum of excitations in the normal state and in the superconducting state is much less well understood.

19.3 Properties of the Normal State

Many properties of the normal state of heavy-fermion materials definitively show the existence of a characteristic temperature that marks the cross-over from the high temperature local moment phase to the low temperature non-magnetic phase. In the literature on Ce heavy-fermion materials, this characteristic temperature is often referred to as the Kondo temperature, T_K, since the high temperature phase shows properties characteristic of a site independent local moments that spin-flip scatter with the conduction electrons. In this high temperature regime, the local moments can be viewed in terms of the single-impurity Anderson model [160]. In the Anderson model, the local f states are broadened by the mixing with the conduction band density of states. The width of the virtual bound state Δ is approximately given by the Fermi–Golden rule expression

$$\Delta = \pi\rho(\mu)|V|^2 , \tag{19.124}$$

where $\rho(\mu)$ is the conduction band density of states at the Fermi energy, and V is the hybridization matrix element. An on-site local Coulomb interaction U tends to prevent multiple occupancy of the f level. When $U > \Delta$ and when the f level is almost singly occupied, a Hartree–Fock treatment [160] yields a solution which possesses a local magnetic moment. In the local moment or Kondo regime, the impurity f density of states is spin-split and broadened, as is shown in Fig. 19.27. However, this solution is only approximate since spin-rotational invariance has been broken locally, and the residual interactions between the local moment and the conduction electrons are expected to flip the local moment and hence, re-

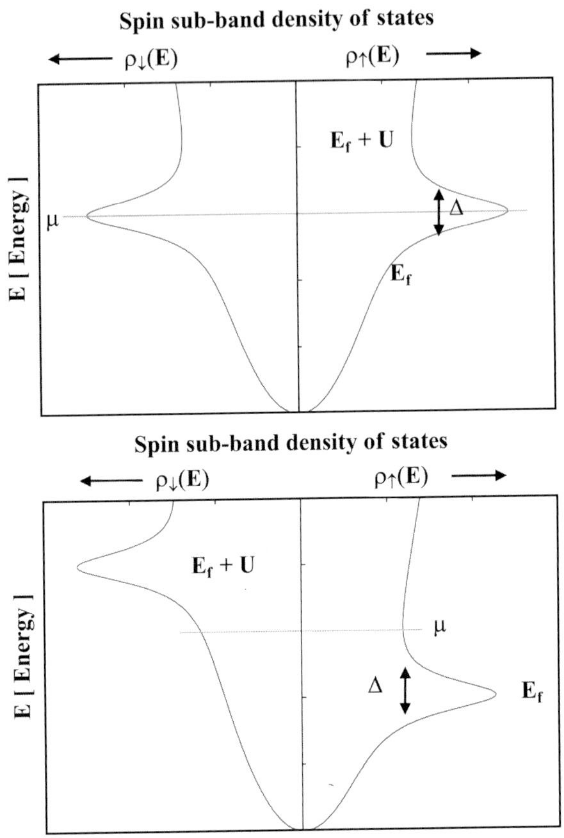

Fig. 19.27. The density of states of the single-impurity Anderson model, as calculated in the Hartree–Fock approximation. The density of states is composed of the sum of the conduction band contribution and a contribution from the f state on the impurity. In the mixed valent limit, shown in upper figure, the virtual bound state is unpolarized. The virtual bound state is located at the energy $E_f + Un_f/2$. In the Kondo limit where the local moment is present, the virtual bound state is spin split, as shown in lower figure. [After Anderson [160]]

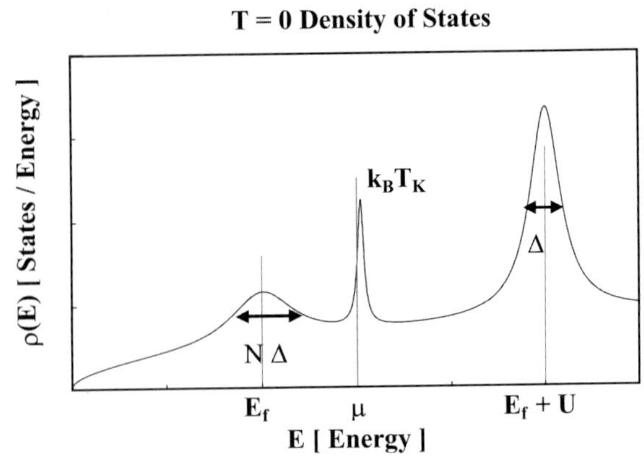

Fig. 19.28. A schematic plot of the $T = 0$ density of states for the N-fold degenerate single-impurity Anderson model, in the Kondo limit. The f contribution to the density of states has features at energies of E_f and $E_f + U$. In addition, at low temperatures, a narrow Kondo resonance of width $k_B T_K/N$ occurs at an energy $k_B T_K$ above the Fermi energy

store the broken symmetry. These residual interactions were investigated by Schrieffer and Wolf [161], who mapped the local moment limit of the single-impurity Anderson model onto the single-impurity Kondo model. In the single-impurity Kondo model [162], the repeated spin-flip scattering between the local moment and the conduction electrons results in the formation of a singlet bound state. The properties of the Kondo model show scaling with the energy of the bound state which is $k_B T_K$. The Kondo temperature for a spin one-half impurity is estimated to be of the order of

$$k_B T_K \sim \rho(\mu)^{-1} \exp\left[\frac{\pi(E_f - \mu)}{2\Delta}\right] , \qquad (19.125)$$

where E_f is the energy of a single electron in the f state of the isolated impurity. Due to the presence of the exponential factor, the Kondo temperature is expected to be much smaller than Δ. For temperatures above T_K, the bound state is thermally depopulated, and the high temperature properties may be adequately described by low order perturbation expansions [163]. However, as the temperature is decreased below the Kondo temperature, the resonant nature of the scattering increases and weak-coupling perturbation approximation fails, since the model scales to a strong-coupling fixed point [164,165]. Near this fixed point, the properties scale with T/T_K and are qualitatively described by a gas of non-interacting quasi-particles residing in a narrow virtual bound state of width $k_B T_K$ located near the Fermi energy (see Fig. 19.28). The exact Bethe–Ansatz solution [166–168] of the model shows that the low temperature properties are those of a local Fermi liquid. In the local Fermi liquid, the low temperature susceptibility is enhanced relative to the linear T coefficient of the specific heat by a factor that depends on the degeneracy, N, of the ground state magnetic configuration. For large degeneracies, the relative enhancement or Wilson ratio is simply $N/(N-1)$. Although the Wilson ratio and the scaling property of the Kondo impurity model are in reasonable agreement with experimental results on dilute Ce systems, Nozières has questioned whether the results of the simple Kondo model can be directly applied to concentrated Ce compounds [169]. In particular, in the Kondo model the screening of a local moment only involves electrons within $k_B T_K$ of the Fermi-level. For a concentrated compound in which the moments on each f ion are screened by a number of conduction electrons localized around each of them, the number of conduction electrons required for screening is far greater than the number $k_B T_K \rho(\mu)$ allowed for by the solution of the single impurity model. If instead, one assumes that the screening electrons remain itinerant and only form a resonance instead of a bound state, the spinflip interaction of the conduction electrons with the local moments may still produce a non-magnetic state in which the local moments are slowly fluctuating. That is, Nozières argument does not apply if the conduction electrons responsible for the screening remain itinerant [170]. In this case, one must abandon the purely local description at sufficiently low temperatures.

It is more likely that the Kondo paradigm contains the generic physics of local electronic correlations observed in the heavy-fermion materials. This has lent credence by the recent discovery, based on the infinite dimensional limit $d \to \infty$, that strongly correlated metals can have physical properties which qualitatively resemble those of localized magnetic impurities [171–173]. The limit $d \to \infty$ of most many-body models is usually exactly soluble via mean-field theory and can be scaled such that it can be exactly mapped onto a non-trivial effective local impurity model. In the resulting dynamical mean-field theory, the local fluctuations on the neighboring lattice sites are treated as being governed by an effective single-impurity Anderson model in which the hybridization to the conduction electron density of states has to be determined self-consistently [174, 175]. This theoretical approach is more promising than that of the single impurity Kondo model, as it decouples the temperature scale of the resonant magnetic scattering in the high temperature regime, and the temperature scale below which the low temperature Fermi liquid forms [176]. In addition to the questions posed by Nozières, it is quite unclear how the U compounds could be directly described by simple Kondo impurity models in which the $5f$ states are assigned as belonging to a unique magnetic configurations. It seems more likely that the U systems may be described by more general many-body models such as multi-band Hubbard models, which may also be described by dynamical mean-field theory. The application of dynamical mean-field theory to heavy-fermion systems is limited by its failure to describe non-local correlations. As dynamical mean-field theory omits the effect of non-local spin-fluctuations, and as quantum critical points occur when the non-local magnetic correlations dominate over the local spin-fluctuations [90], dynamical mean-field theory is expected to be inadequate close to quantum critical points.

For some heavy-fermion compounds, the characteristic temperatures T_K inferred from thermodynamic, transport, magnetic and other spectroscopic probes, are quite consistent. However, even for these systems, the low temperature thermodynamic properties do show the existence of other low-energy scales that are often attributed either to Fermi liquid formation or the onset of spatial coherence or spatial magnetic correlations. In the quantum critical point description, this coherence temperature is identified as the cross-over temperature at which the physics of the quantum critical point gives way to the physics of the Fermi liquid fixed point [177, 178]. In a phase space that is extended, as for instance when pressure is applied, the line of cross-over points is experimentally accessible since physical quantities should exhibit extrema on parallel lines. The phenomenon of coherence is most clearly manifested in transport phenomena, though it does also show up directly in high resolution spectroscopic measurements on single crystals [179]. The phenomenon of coherence falls beyond the scope of the exactly soluble single-impurity Kondo models or dynamical mean-field theory and, therefore, still lacks an adequate theoretical description.

19.3.1 Thermodynamic Properties

The Specific Heat

The low temperature specific heats of the normal state phase of heavy-fermion systems are dramatically different from ordinary metals. The extrapolated value of the ratio of the specific heat to the temperature

$$\lim_{T\to 0} \frac{C(T)}{T} = \gamma \tag{19.126}$$

can be of the order of 1 J/mole K^2, which is 1000 times larger than the corresponding ratio for Cu. In many heavy-fermion systems, notably UBe_{13}, $CeCu_2Si_2$, $CeAl_3$ and $CeCu_6$, the ratio $C(T)/T$ shows an increase with decreasing temperatures, at low temperatures, and shows signs of saturation. For UBe_{13} and $CeAl_3$ the upturn starts at a scale of the order of 10 K. The heat capacity of UBe_{13} is shown in Fig. 19.29. The large magnitudes of the extrapolated specific heat coefficients γ are attributed in part to the flat quasi-particle dispersion relation arising from the small hybridization between the f orbitals and the states on the ligand atoms and in part due to the average of the quasi-particle mass enhancement, $Z_{\underline{k}}$. The assignment of the large value of the γ coefficient as being due to enhanced quasi-particles is found to be consistent with de Haas–van Alphen measurements. For UPt_3 and UAl_2, the specific heat has the form

$$C(T) = \gamma T + \delta T^3 \ln T + \beta\, T^3 \tag{19.127}$$

for temperatures up to 20 K. The low temperature specific heat of UPt_3 and the fit to this form are shown in Fig. 19.30. The above form and the sign of δ are compatible with a Fermi liquid that exhibits large amplitude, long wavelength, collective fluctuations such as ferromagnetic spin-fluctuations [90,91,97,98]. The values of the coefficients β are also quite large compared with values expected from phonon contributions in normal metals and probably could also have a significant electronic component. The entropies calculated from the enhanced low temperature specific heats have values comparable

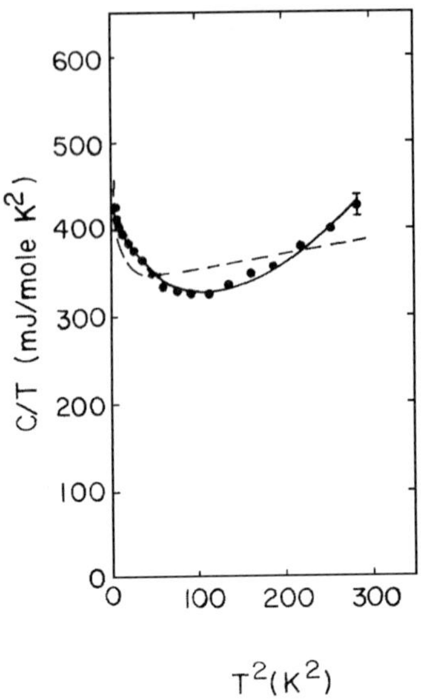

Fig. 19.30. The low temperature specific-heat ratio $C(T)/T$ versus T^2 for UPt_3. The *solid line* is a fit to $C(T) = \gamma T + \delta T^3 \ln T$. The *dashed line* is a fit to $\gamma T + \beta T^3 + \delta / T$. [After Stewart et al. [6]]

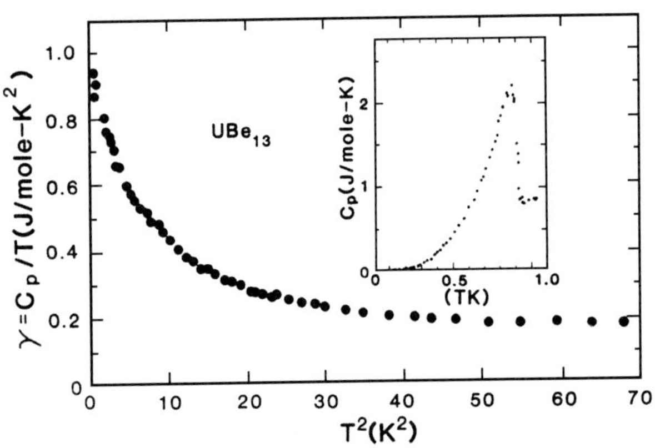

Fig. 19.29. The specific heat in the normal state of UBe_{13} divided by temperature as a function of T^2. [After Ott et al. [4]]. Note the upturn at low temperatures and the large extrapolated value at $T = 0$ as compared to the value of 0.07 mJ/mole/K^2 for Cu

to those of a disordered set of independent moments on the f ions, suggesting a magnetic origin of the enhancement. This conclusion is strengthened for UAl_2 by the finding [180, 181] that an applied magnetic field actually suppresses the low temperature upturn in C/T.

An unusual feature of the specific heat of some heavy-fermion systems is that they seem to indicate that the fluctuations are of a local character and reside mainly on the f ions. For example, on substituting La for Ce in $CeCu_6$, the temperature dependence of the specific heat of $Ce_xLa_{1-x}Cu_6$ per Ce ion is independent of x, for $1 > x > 0.5$, indicating that the large specific heat of this Ce compound is dominated by local processes [182]. A similar conclusion was previously drawn from dilution studies of $CePb_3$ [183], where it was found that the properties simply scale with the Ce concentration. The specific heats per mole Ce of diluted $CePb_3$ and $CeCu_6$ are shown in Fig. 19.31. This type of scaling behavior should be contrasted with the behavior found in dilution studies of $CeCu_2Si_2$, $CeAl_3$ and UBe_{13}. In stoichiometric $CeAl_3$, the C/T ratio shows a small maximum around $T = 0.3$ K [184], and on doping substitutionally for Ce with La, the peak moves up to higher temperatures. However, the Kondo temperature decreases with doping [185]. Calculations based on single impurity models show that the low temperature peak is not simply due to the Kondo effect in the presence of crystal fields [186, 187]. Therefore, it has been speculated that the peak is either due to the onset of magnetic correlations or is a manifestation of structure in the quasi-particle density of states caused by the coherence of the lattice.

There exists a similar low temperature peak in the specific heat of UBe_{13} near $T = 2$K which is suppressed and moves towards lower temperature on substituting 3 percent of U by Th [188]. Substitutional doping on the U sites of UBe_{13} by a few percent of impurities leads to a rapid decrease in the γ value. The introduction of roughly 3% concentration of impurities leads to a depression of γ by roughly 30 %. These studies suggest that the value of γ crucially depends on the U–Be separation. Stoichiometric $CeCu_2Si_2$ seems to be on the critical phase boundary between a phase that exhibits magnetic correla-

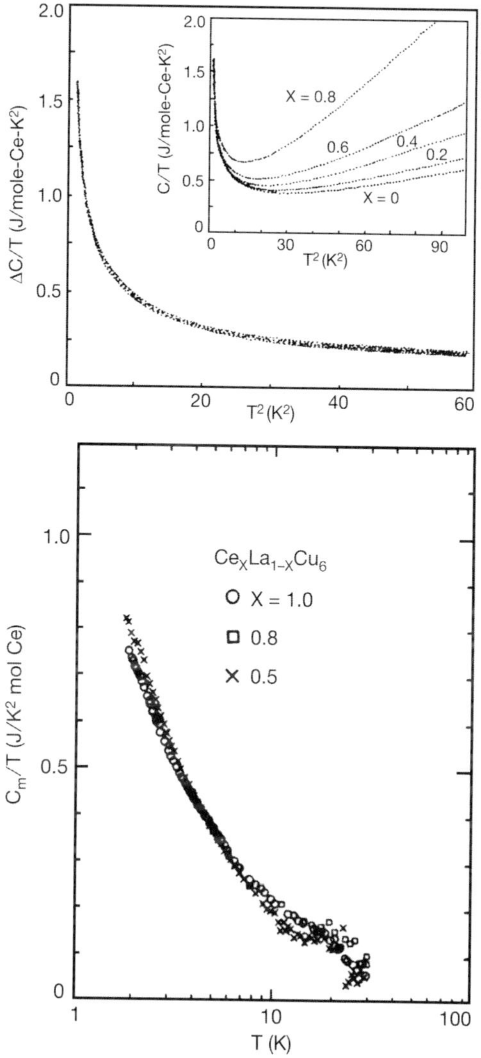

Fig. 19.31. The low temperature specific-heat ratio $C(T)/T$ versus T^2 for UPt_3. The *solid line* is a fit to $C(T) = \gamma T + \delta T^3 \ln T$. The *dashed line* is a fit to $\gamma T + \beta T^3 + \delta/T$. [After Stewart et al. [6]]

tions at low temperatures, the A phase and a phase that superconducts at low temperatures, the S phase. A generalized phase diagram of $CeCu_2Si_2$ is shown in Fig. 19.32. The A and S phases seem to coexist in stoichiometric single crystals [189]. In this case, altering the composition can result in the stabilization of the separate phases [190, 191]. The A phase shows a specific heat due to heavy quasi-particles perhaps

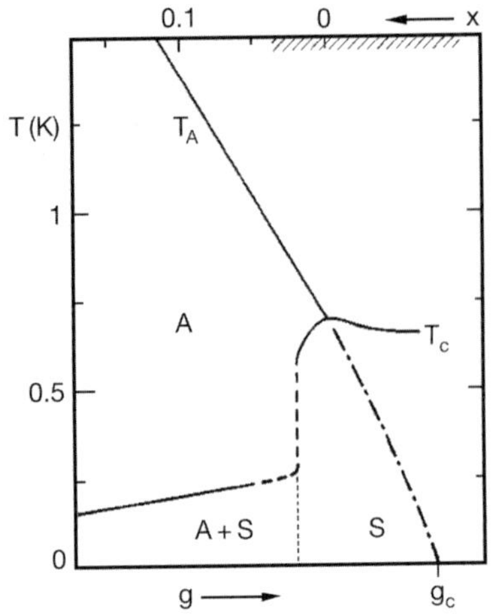

Fig. 19.32. A schematic phase diagram for $CeCu_2Si_2$ at zero field, indicating the existence ranges for a spin density wave-like phase (A), superconductivity (S), and the coexistence ranges ($A + S$). The abscissa is an effective coupling constant g which is proportional to x in $CeCu_2(Si_{1-x}Ge_x)_2$ or to the composition in homogeneous $CeCu_2Si_2$. [After Gegenwart et al. [193]]

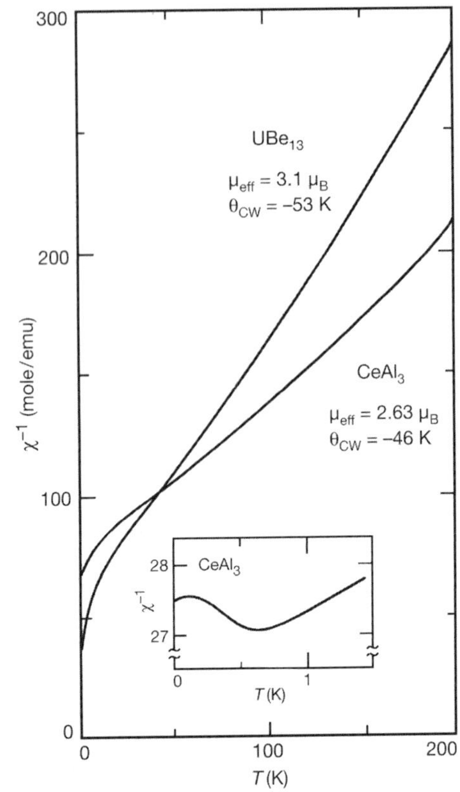

Fig. 19.33. The temperature dependence of the magnetic susceptibility $\chi(T)$ for UBe_{13} and $CeAl_3$ plotted as $\chi^{-1}(T)$. The effective moments and Curie–Weiss temperature obtained from high temperature fits are also shown. At low temperatures, the temperature variation of $\chi(T)$ for $CeAl_3$ weakens and becomes Pauli-paramagnetic, as shown in the *inset*

including a $T^3 \ln T$ contribution arising from ferromagnetic spin-fluctuations [192]. In the presence of applied magnetic fields sufficiently large to suppress the superconductivity, the S phase appears and has a non-Fermi-liquid like temperature dependence of the specific heat [193]

$$\frac{C}{T} = \gamma_0 - AT^{\frac{1}{2}} . \tag{19.128}$$

This non-analytic temperature variation indicates that the material is either close to a quantum critical point in which non-local correlations are developing, or that there is a significant amount of local disorder as has been inferred from neutron measurements of the nuclear pair distribution function [194].

The Magnetic Susceptibility

The magnetic susceptibilities of heavy-fermion systems show temperature driven cross-overs between forms reminiscent of local moments and low temperature Pauli-like susceptibilities. This is illustrated in Fig. 19.33, which shows the temperature dependence of the susceptibilities for UBe_{13} and $CeAl_3$. At high temperatures, the susceptibilities can be fit by the form

$$\chi(T) = n\frac{\mu_{\mathrm{eff}}^2}{3k_B(T-\Theta)} , \tag{19.129}$$

where μ_{eff} is the size of the effective magnetic moment, n is the density of local moments, and Θ is the Curie–Weiss temperature. The values of the effective moment μ_{eff} are nearly that of the free ionic moments, and the values of the Curie–Weiss temperature Θ are usually negative, suggesting antiferromagnetic interactions between the local moments.

At lower temperatures, the susceptibilities cross-over to Pauli-paramagnetic susceptibilities that show

Table 19.8. High Temperature properties of the normal state. Anisotropic properties of crystals are given along the principal directions in the form of "a/b/c".

Material	Θ (K)	μ_{eff} (μ_B)	$\rho(T = 300\,K)$ ($\mu\Omega$ cm)
UCu_5	284	3.52	286
UAl_2	245	3.1	190
UPt_3	50 / 170	2.5 / 2.37	230 / 135
U_2Zn_{17}	105	3.3	110
URu_2Si_2	65	3.51	324 / 169
UBe_{13}	53	3.1	107
UPd_2Al_3	47	3.2	140 / 200
UCd_{11}	23	3.45	80
UGe_2	-34	2.7	156
$CeCu_2Si_2$	140	2.68	90
$CeCu_6$	59 / 59 / 8	2.6 / 2.67 / 2.46	70
$CeAl_3$	46	2.63	65

large enhancements similar to those of the specific heat γ coefficients. The cross-over from Curie–Weiss like to Pauli-paramagnetic like susceptibilities seems to occur at higher temperatures than the temperatures at which the C/T ratios build up [195, 196]. The magnitudes of the measured γ and $\chi(0)$ can be compared indirectly by using the model of non-interacting electrons. For the gas of non-interacting electrons, the Pauli-susceptibility is given by

$$\chi(0) = 2\mu_B^2\rho(\mu)\,, \tag{19.130}$$

where $\rho(\mu)$ is the density of states, per spin, at the Fermi energy, and the specific heat coefficient is given by

$$\gamma = 2\frac{\pi^2}{3}k_B^2\,\rho(\mu)\,. \tag{19.131}$$

Therefore, a plot of γ versus $\chi(0)$ for hypothetical non-interacting electron systems should yield a straight line with slope given by

$$\frac{\pi^2}{3}\left(\frac{k_B}{\mu_B}\right)^2 . \tag{19.132}$$

As shown by Fisk et al. [197] (see Fig. 19.34), the heavy-fermion systems have γ and $\chi(0)$ values that lie close to the same line. Therefore, as the Wilson ratio is close to unity, this indicates that the enhancements of the susceptibilities have similar magnitudes to the specific heat enhancements. From a Fermi liquid point of view, one would expect that the uniform static susceptibility should show an enhancement, S, given by

$$S = \frac{m^*/m_b}{1 + F_0^a}\,, \tag{19.133}$$

where F_0^a is the spin antisymmetric interaction between a pair of quasi-particles with angular momentum $l = 0$. However, the specific heat is just enhanced by the factor

$$m^*/m_b\,. \tag{19.134}$$

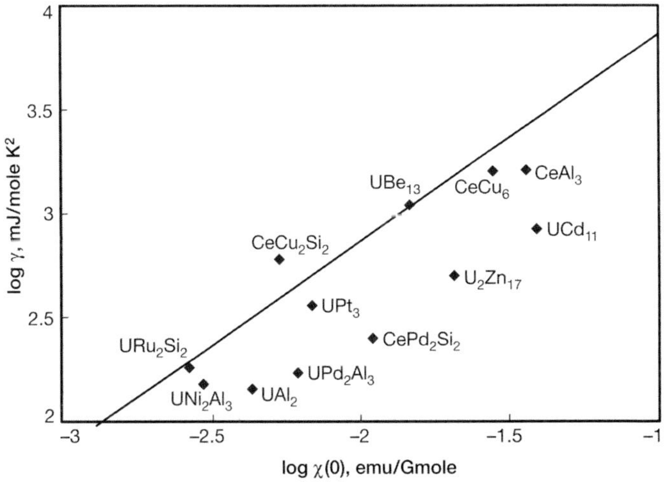

Fig. 19.34. A log–log plot of the limiting values, per mole f material, of C/T in mJ/mole K^2 and χ in emu/Gmole for selected heavy-fermion compounds. The *straight line* represents the relation for non-interacting electrons. [After Fisk et al. [197]]

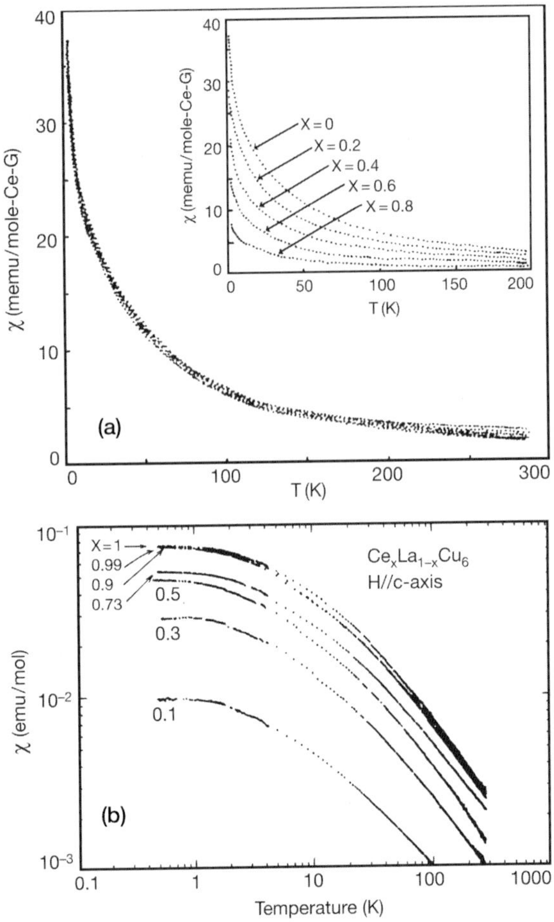

Fig. 19.35. (**a**) The magnetic susceptibility per mole Ce, $\chi(T)$ versus T for $Ce_{1-x}La_xPb_3$ samples with $x = 0.0, 0.2, 0.4, 0.6$ and 0.8. *Inset*: The total susceptibility per molar formula unit for the same samples. [After Lin et al. [183]]. (**b**) The magnetic susceptibility per mole formula unit for $Ce_xLa_{1-x}Cu_6$, plotted versus T. [After Ōnuki and Komatsubara [182]]

Thus, the χ to γ ratio yields information about the spin-dependent interaction between the quasi-particles. In most Ce compounds, both enhancements seem to be due to single-site processes. This can be seen by inspection of the susceptibility of $Ce_xLa_{1-x}Cu_6$ per Ce ion, which is independent of x in the range $1 > x > 0.1$. The susceptibilities of $Ce_xLa_{1-x}Cu_6$ and $Ce_xLa_{1-x}Pb_3$ are shown in Fig. 19.35. The similarity of the curves suggests that the large susceptibility, like the large specific heat, is dominated by local processes involving the f electrons. Since the spin-orbit split ground state of the Ce^{3+} ion corresponds to $J = 5/2$ and degeneracy of 6, the ratio of the susceptibility to γ term predicted by the single-site Kondo model should be very similar to the ratio for non-interacting electrons. On the other hand, in UBe_{13} doped substitutionally on the U sites [198], the susceptibility seems to scale linearly with the number of U atoms, while the γ coefficient per U atom is rapidly reduced by dilution. This suggests that an appreciable fraction of the mass enhancement in UBe_{13} is not due to a single-site magnetic Kondo process. A similar conclusion has been drawn from dilution studies of UPt_3 [199], which show that dilution can result in the suppression of the specific heat by up to 50% of the stoichiometric value while the uniform static susceptibility is barely affected. These conclusions are consistent with measurements of the de Haas–van Alphen oscillations on a number of U and Ce compounds, which show that the quasi-particles are itinerant and form part of the Fermi surface, albeit with a very small dispersion of the bands.

de Haas–van Alphen Oscillations

Measurements of the magnetization in high fields and at low temperatures show oscillations as a function of the inverse field. The presence of the oscillations has long been held as an indicator that the quasi-particles are itinerant, form part of the Fermi surface, and are long lived. The oscillations are due to the existence of Landau peaks in the density of states, which are a consequence of the quantized orbital motion of the quasi-particles in the presence of an applied field. As the applied field is increased, the energy separation between consecutive Landau peaks increases linearly with the field, and the oscillations occur as the peaks in the density of states move through the Fermi energy. The fundamental periods of the oscillations are related to the extremal cross-sectional areas of the Fermi surface perpendicular to the applied field. For non-interacting electrons, the fundamental oscillation stemming from each electron orbit around an extremal area is governed by the Lifschitz–Kosevich formula

$$M_z = A \cos\left(\frac{\hbar c S}{e H_z} + \phi\right) , \qquad (19.135)$$

where S is the extremal area of $\underline{k}$ space enclosed by the quasi-particle orbiting around the Fermi surface, and ϕ is a phase reflecting the Zeeman splitting between the bands of different spins. The amplitude of the oscillations, A, is given by

$$A \sim V \frac{k_B T S}{2\pi^2} \sqrt{\frac{e}{\hbar c\, H_z}} \times \exp\left[- \frac{2\pi^2\, m_b c}{\tau e H_z} \right] \operatorname{cosech}\left[\frac{2\pi^2 m_b c k_B T}{\hbar e H_z} \right] , \qquad (19.136)$$

where τ is the lifetime due to scattering processes and m_b is the effective mass deduced from the dispersion relation of the non-interacting electrons at the Fermi surface. The scattering rate, $1/\tau$, is usually introduced in an ad hoc manner. The contribution from each extremal area contributes to the signal with the weight given by

$$\sqrt{\frac{2\pi}{\left|\frac{\partial^2 S}{\partial k_z^2}\right|}} , \qquad (19.137)$$

where k_z is the component of the wave-vector parallel to the field. This factor expresses the weight of an extremal cross-section relative to that of an extremal section of a spherical Fermi surface of the same size, S. Since Fermi surfaces are often multi-sheeted, the measured areas are typically compared with the results of electronic structure calculations. For most simple metals, the experimentally determined areas are usually in excellent agreement with the LDA calculations. Since the oscillations originate from the Landau peaks in the density of states passing through the Fermi energy, broadening, either due to the Fermi function or from the effect of impurity scattering on the Landau peaks, is expected to wash out the oscillations. Therefore, one expects that the oscillations should only be measurable at extremely low temperatures ($T < 1K$), for high fields, and for very high quality samples having very low residual resistivities.

The Lifschitz–Kosevich formula must be modified when the effect of electron–electron interactions are introduced. First, the spin-splitting phase factor ϕ may be exchange enhanced and may become field-dependent [200]. Phenomenologically, one is also led to expect that the scattering rate should be identified with the imaginary part of the self-energy evaluated on the Fermi surface and that the real part of the self-energy may also be expected to appear as an addition to the imaginary part. In the thermal average which is performed by transforming to a sum over Matsubara frequencies ω_n, the real part of the self-energy contains a term proportional to $(1 - Z(0))\omega_n$, and therefore, produces a correction of order $\pi k_B T$. In the small H/T limit, the additional term originating from the real part of the self-energy can be combined with the $k_B T$ dependence in the argument of the cosech term in the non-interacting formulae, thereby replacing the band mass m_b by the quasi-particle mass m^*. Thus, the temperature dependence of the amplitude of the oscillations can be used to deduce the quasi-particle effective masses. As the linear T term in the specific heat-capacity is related to the average of the quasi-particle masses on the Fermi surface, one may be able to compare the average of the quasi-particle masses observed in de Haas–van Alphen experiments with the specific heat. However, in attempting to make this comparison, it is important to note that some portions of the Fermi surface may not have been observed in the de Haas–van Alphen experiments. As can seen from Eq. (19.136), the amplitudes of the heavier branches are expected to be quite small and may easily be missed. Also, in the case where the γ coefficient depends on the magnitude of the applied field, as it should if the quasi-particle mass enhancements are of magnetic origin, the average quasi-particle mass should be compared with the specific heat measured under comparable fields. It should be noted that the amplitude factor contains the ratio of the quasi-particle mass to the quasi-particle lifetime so, if the mass enhancements arise from a local process, the quasi-particle mass enhancement factor should cancel with the enhancement of the quasi-particle lifetime. This cancelation occurs much in the same way as the cancelation of the enhancements in the residual resistivity $\rho(0)$ that occurs when the self-energy is k independent. This implies that the impurity scattering should not be significantly more effective in reducing the amplitude of the oscillations than in metals with comparable density of states.

The de Haas–van Alphen oscillations were observed in measurements on $CeCu_6$, $CeCu_2Si_2$, UCd_{11}, UPt_3, UPd_2Al_3, URu_2Si_2 and $CeRu_2Si_2$ for temperatures between 200 and 100 mK and fields up to 18 T. Despite many valiant attempts, measurements on UBe_{13} have failed to yield de Haas van–Alphen oscillations [201]. In most of these materials, sheets of the Fermi surface with large effective masses have been found. The extremal areas inferred for UPt_3 are in good agreement with the Fermi surface geometry found from electronic structure calculations [202, 203]. The effective masses inferred from the amplitude can be factors of 10 to 30 times larger than those found in LDA band structure calculations [204, 205], and the heavy masses found range from $25m_e$ to $200m_e$. This suggests that the quasi-particle mass enhancements due to electron–electron interactions, are extremely large. The Fermi surface average of the observed effective masses produces a quasi-particle density of states that is in rough agreement with the γ value of UPt_3. Whereas in UPt_3 the quasi-particles at the Fermi surface have large effective masses, the effective masses found in $CeCu_6$ are smaller [208] and are in the range of 6 to 80 m_e. However, as the specific heat of $CeCu_6$ is appreciably larger than that of UPt_3, one suspects that the bands with heavier masses were not observed. This situation also seems to be definitely the case for $CeCu_2Si_2$ in which only small masses of 4 to 6 m_e have been observed [206]. The observation of the heavy mass sheets of the Fermi surface in UPt_3, but not in $CeCu_6$ and $CeCu_2Si_2$, is probably due to the higher quality of the UPt_3 samples. Also, part of the discrepancy between the de Haas–van Alphen effective masses and the specific heat of $CeCu_6$ measured in zero field [207, 208] is accounted for, if the field dependence of the specific heat is considered [209,210]. The decrease of the effective mass with applied magnetic field is taken as an indication that spinflip excitations are responsible for part of the mass renormalizations, and that application of the field quenches the spinflip scattering processes. The lack of agreement between the specific heat and de Haas–van Alphen effective masses in URu_2Si_2 is very notable in that the only effective masses that have been found, so far, are in the range of 15 to 25 m_e, and these only yield about 16 percent of the total γ value [211, 212]. Likewise for UCd_{11} [213], the only effective masses observed are only in the range of 2 to 11 m_e. As previously mentioned, the difficulty of observing heavy quasi-particle mass branches of the Fermi surface is due to the exponentially small amplitudes of their oscillations. However, both light and heavy quasi-particle masses can be found in the same compound such as in $CeRu_2Si_2$, where the observed masses are in the range of 1 to 120 m_e [214].

Some heavy-fermion materials undergo meta-magnetic like transitions at high fields in which the magnetizations undergo rapid change with respect to magnetic field at a critical value of the field H_m. The meta-magnetic transitions become sharper at lower temperatures. These highly non-linear susceptibilities indicate either that the systems are on the verge of undergoing an instability to a magnetically ordered phase[7] or that a spin-split branch of quasi-particles crosses the Fermi surface. For example, URu_2Si_2 may undergo as many as three tran-

[7] For a paramagnetic system close to a magnetic instability, one expects that the free-energy expressed as a function of the magnetization M could have a local minimum at $M \neq 0$, in addition to the global minimum at $M = 0$. The application of a magnetic field may cause the energy of the local minimum to be lowered, thereby stabilizing it at the critical field H_m associated with the meta-magnetic transition. The criterion for the occurrence of a meta-magnetic transition can be easily calculated for the Stoner model. In the Stoner model, the magnetization-field relation is given by

$$M = 2\mu_B^2 H S \rho(\mu) \left[1 + \frac{1}{6} \left[\left(\frac{\rho''}{\rho} \right) - 3 \left(\frac{\rho'}{\rho} \right)^2 \right] \mu_B^2 H^2 S^3 + \dots \right] , \tag{19.138}$$

where $S = (1 - U\,\rho(\mu))^{-1}$ is the Stoner enhancement factor. The terms non-linear in H can be large due to the appearance of extra powers of S. The cubic term may be negative either if the density of states at the Fermi energy is close to a maximum or if it is rapidly varying. If non-linear terms in M are negative, then meta-magnetic transitions may occur at critical fields. Thus, meta-magnetic transitions may be expected to occur in the vicinity of quantum critical points.

Table 19.9. Low temperature normal state properties (extrapolated to $T = 0$)

Material	γ (mJ mole f ion^{-1} K^{-2})	$\chi(0)$ (10^{-3} emu/mole)	A ($\mu\Omega$ cm K^{-2})	$\rho(0)$ ($\mu\Omega$ cm)
$CeAl_3$	1620	36	35	0.77
$CeCu_6$	1600	28	~ 120	15 / 9.2 / 8.3
UBe_{13}	1100	14.7	–	17
$CeCu_2Si_2$	600	4 / 8	10.7	4.8
UCd_{11}	840	39	–	
U_2Zn_{17}	500	19 / 24	–	
UPt_3	360	8.1 / 4.4	1.6 / 0.61	0.23 / 0.59
UPd_2Al_3	170	8.1 / 2.2	~ 0.3	3.5
URu_2Si_2	180	1.5 / 4.9	0.17 / 0.10	1.27
UNi_2Al_3	150	3.3 / 2.2		
UAl_2	142	4.3	10.7	4.8
UGe_2	100	–	~ 0.25	0.26
$CePd_2Si_2$	250	11	–	1.4
$CeRh_2Si_2$ (P=0)	21	2.9 – 3.2	0.8	0.75
$CeIn_3$	100	11	–	0.6
$CeRhIn_5$ (P=0)	60	4 – 22	–	0.4
$CeIrIn_5$	750	9 – 20	–	~ 0.1
$CeCoIn_5$	350	4 – 7	–	3

sitions, located at $H_{m1} = 35.9$ T, $H_{m2} = 36.1$ T and $H_{m3} = 39.7$ T. The three transitions in URu_2Si_2 have been observed in transport properties [215] as well as in the magnetization [216]. UPt_3 has a meta-magnetic transition at $H_m = 20$ T while $H_m = 18$ T for UPd_2Al_3; $H_m = 8$ T for $CeRu_2Si_2$ and a very much weaker meta-magnetic transition occurs in $CeCu_6$ at a critical field of $H_m = 2$ T. Above these fields, the effective masses rapidly decrease with increasing field [217,218]. For example, in UPd_2Al_3, the effective masses found below $H_m = 18$ T are as large as 45 m_e [219], but for fields above H_m, the effective masses found are 31 and 5.4 free electron masses [220].

De Haas van Alphen oscillations have also been observed in the mixed states of type II superconductors such as $NbSe_2$ [221], Nb_3Sn, V_3Si [222] and YNi_2B_2C [223]. This effect is probably due to the existence of unpaired quasi-particle particles which can be quantized into Landau levels similar to the manner in which the quasi-particles in the normal state are quantized into Landau levels. De Haas-van Alphen measurements have been performed in the superconducting states of UPd_2Al_3 and URu_2Si_2, and oscillations were observed to persist until the strength of the applied field was reduced substantially below H_{c2}. Since the estimated diameters of the cyclotron orbits are factors of 10 greater than the vortex lattice spacing, the unpaired quasi-particles apparently must traverse through the superconducting regions. The frequencies of the oscillations remain unchanged in the superconducting state, however, the quasi-particle masses and scattering rates inferred from a Fermi liquid like analysis are altered. For example, the branch with mass of 5.4 m_e found in the normal state of UPd_2Al_3 was also observed in the superconducting state, but the inferred effective mass was further reduced to 3.0 m_e. Phenomenologically, the amplitude of oscillations could be expected to be reduced due to the decrease in the number of unpaired quasi-particles and also due to the smearing of the Fermi energy in the superconducting state. The latter phenomenon could also appear as an increase in the quasi-particle scattering rate consistent with the observations in the superconducting state of URu_2Si_2 [212]. The appearance of de Haas–van Alphen oscillations in the superconducting state is surprising since it undermines a commonly held belief that observability of the oscillations is a signature of the existence of a sharply defined Fermi surface and the long lifetime of the quasi-particles. Furthermore, it is surprising as it definitively shows that such oscillations can also be observed in non-Fermi-liquid states.

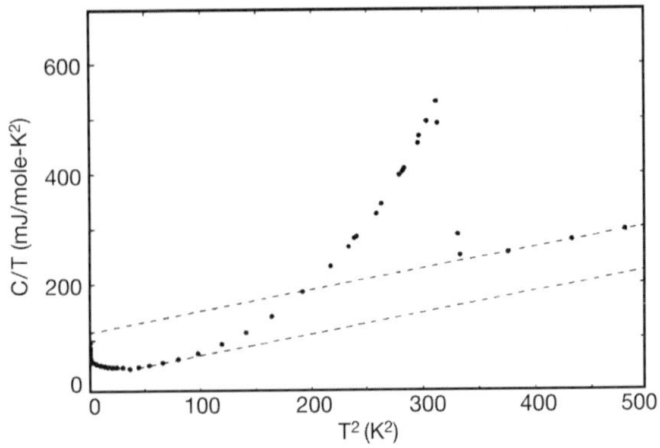

Fig. 19.36. The specific heat jump of the "hidden order" phase transition of URu_2Si_2. The specific heat ratio C/T is plotted as a function of T^2. A mean field-like jump occurs at $T_N \approx 17.5$ K which is associated with the "hidden order" phase transition. [After Maple et al. [21]]

Magnetic Instabilities

Several of the heavy-fermion materials undergo instabilities to magnetically ordered states at lower temperatures. The most frequently found type of ordering is antiferromagnetic, however, ferromagnetic phases have been identified in UGe_2, URhGe $CeRu_2Ge_2$, $CeGe_2$, and $CeSi_x$. The values of γ found from extrapolation above the ordering temperatures are generally about a factor of 1.2 to 3 times higher than the γ values found from extrapolating below the ordering temperatures. This indicates that the antiferromagnetic order may have produced a partial gapping of the Fermi surfaces. Exceptions to this are given by $CeAl_3$, UCu_5 and UPt_3, where the extrapolated γ values are larger in the magnetically ordered phases. These findings are consistent with the general observation that the entropy found by integrating below the ordering temperature is about 30% less than entropy expected from a simple Fermi liquid picture, i.e., γ may still be growing below T_N and the Fermi liquid may have not been completely formed when T has been decreased to T_N.

The values of the ordered moments are smaller than those associated with the free ion values (see Table 19.10). This is sometimes attributed to the screening of the f moments by the conduction electrons. However, in view of Nozières arguments for Ce compounds and the existence of reduced moment ordering in U compounds, it seems apparent that the itinerant nature of the f electrons is involved in the reduction of the moments. In some compounds such as $CeAl_3$ and URu_2Si_2, the magnetic ordering appears to be inhomogeneous, only occurring in a small fraction of the sample volume [224]. It seems likely that these samples are also inhomogeneous. The magnitude of the moments on the atoms in the magnetically ordered volume can be large, of the order of 0.5 μ_B per f atom but the moment averaged over the sample can be as small as 0.03 μ_B per f atom, as found for URu_2Si_2. It has been found [225] that in URu_2Si_2, application of pressures of up to 1.5 GPa produces an increase in the size of the ordered moment to a value of the order of $0.25\mu_B$ per U atom but only produces a slight increase in T_N. The magnetic ordering shows up in the specific heat of $CeAl_3$ as a faint and sample dependent anomaly at the ordering temperature that considerably increases in magnitude on dilution. As shown in Fig. 19.36, the specific heat jump observed in stoichiometric URu_2Si_2 is quite large [20, 21]. The small magnitude of the volume averaged ordered moment in URu_2Si_2 seems inconsistent with the large value of the entropy associated with the specific heat anomaly at T_N. The specific heat jump is about 5.82 J/mole-K, and the entropy ΔS associated with the anomaly has a magnitude of about $0.17\, Nk_B \ln 2$. On the other hand, Landau mean-field theory suggests that the magnitude of the entropy of the transition should be of the order of

$$\Delta S \sim Nk_B \left(\frac{\mu_{sat}}{\mu_{eff}} \right)^2 , \qquad (19.139)$$

Table 19.10. Properties of the magnetically ordered state

Material	T_N (K)	Ordered Moment M (μ_B)	γ (mJ K^{-2} mole f ion^{-1})
$CeAl_3$	1.6	–	1300
$CeCu_2Si_2$	0.8	~ 0.1	1000
UCd_{11}	5	–	250
U_2Zn_{17}	9.7	0.8	200
UPt_3	5.0	0.02	450
UPd_2Al_3	14.5	0.85	150
URu_2Si_2	17.5	–	65
UNi_2Al_3	4.6	0.24	120
UGe_2	$T_c(P = 1\ GPa) = 34$	$M(P = 0) = 1.4$	–
$CePd_2Si_2$	10.2	0.7	250
$CeRh_2Si_2$	36	1.8	21
$CeIn_3$	10.2	0.65	100
$CeRhIn_5$ (P=0)	3.8	0.37	60

where μ_{sat} is the $T = 0$ saturation value of the sublattice magnetization; and μ_{eff} is the paramagnetic moment ($\mu_{eff} \approx 3.51\mu_B$), as obtained from fitting the susceptibility above T_N to a Curie–Weiss law. The large discrepancy between the observed and inferred magnitudes of the anomalous entropy has been taken as indicating that another type of ordering occurs simultaneously with the antiferromagnetism. However, although the other type of ordering has been searched for, it has not yet been identified.

An alternate paradigm for heavy-fermion materials is provided by the model of a highly enhanced Fermi liquid close to a quantum critical point. In this picture, the large enhancements are caused by the slow, large amplitude, critical magnetic fluctuations. Due to the close proximity of the critical point, one may expect various physical quantities to obey scaling laws [74]. In this case, one expects that the properties may be expressed in terms of the $\underline{q}$ dependent magnetic susceptibility $\chi(\underline{q}, T)$ which expresses the response of the system to a staggered magnetic field $H(\underline{q})$. The absence of a large $\chi(0,0)/\gamma$ ratio indicates that the dominant magnetic fluctuations that occur are not localized around $q = 0$. This is consistent with the observation that most heavy-fermion systems undergo instabilities to antiferromagnetic phases. Likewise, since the transport scattering rate is dominated by processes involving large scattering angles θ due to the weighting factor of $(1 - \cos\theta)$, measurements of transport properties might also be expected to provide information on the magnitude of the wave-vectors of the magnetic fluctuations in the absence of competing instabilities and complicated multi-sheeted Fermi surfaces.

19.3.2 Transport Properties

Electrical Resistivity

The room temperature resistivities of heavy-fermion compounds are quite large, of the order of 100 $\mu\Omega cm^{-1}$, most probably due to spin disorder scattering. As seen in Fig. 19.37, the resistivity of some materials such as $CeCu_2Si_2$, $CeAl_3$, UBe_{13} and U_2Zn_{17}, shows an increase with decreasing temperature. This type of temperature dependence is unusual for simple metals but is characteristic of the Kondo effect often found in materials containing isolated and disordered magnetic impurities. In this subset of heavy-fermion materials, the upturn in the resistivity follows a logarithmic temperature dependence that results from the conduction electrons resonantly spin-flip scattering from isolated magnetic impurities, as calculated in third order perturbation theory [163].

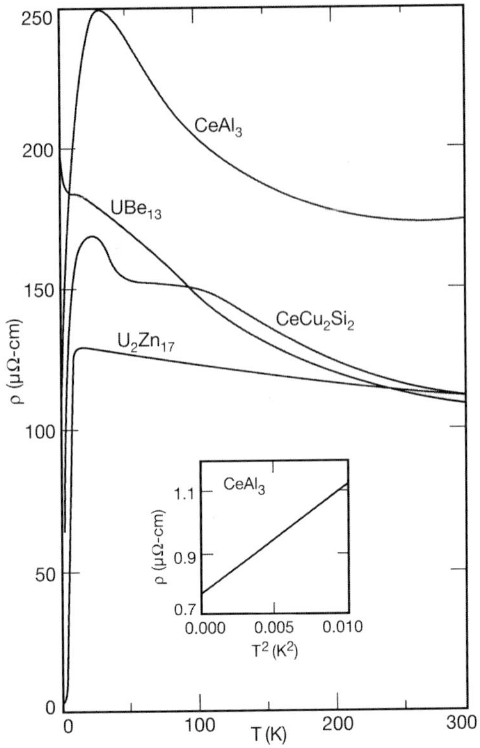

Fig. 19.37. The temperature dependence of the electrical resistivity of the heavy-fermion compounds U_2Zn_{17}, $CeAl_3$, $CeCu_2Si_2$ and UBe_{13}. The *inset* shows the low temperature T^2 variation of the resistivity found in $CeAl_3$

If, at high T, the conduction electrons are assumed to be scattered by localized moments, the conductivity is given by

$$\sigma(T) = \frac{e^2 k_F^3}{3\pi^2 m} I^{(0)} , \tag{19.140}$$

where

$$I^{(n)} = \int_{-\infty}^{+\infty} dE \left(- \frac{\partial f}{\partial E} \right) \frac{E^n}{Im\Sigma_c(E)} \tag{19.141}$$

and where $\Sigma_c(E)$ is the conduction electron self-energy. For independent local moments the imaginary part of the self-energy is given by

$$Im\Sigma_c(E) = \pi |V|^2 \rho_f(E, T) , \tag{19.142}$$

where $\rho_f(E, T)$ is the temperature dependent f density of states, and V is the hybridization matrix elements. For $T < T_K$, a narrow (possibly crystal field split) Kondo resonance forms at an energy T_K above μ. The temperature dependent crystal-field split $\rho_f(E, T)$ expected from the single-impurity Anderson model is shown in Fig. 19.38. Thus, the resistivity is expected to reflect any crystal-field splitting that may be present. As seen in Fig. 19.39, there are two peaks in the resistivity of $CeCu_2Si_2$ [226] both of which can be attributed to spinflip scattering from the $J = \frac{5}{2}$ crystal field split states of the Ce^{3+} ion. The position of the higher temperature peak ($\sim$ 160 K) is associated with the crystal field splitting between the multiplets of the lowest spin-orbit split $J = \frac{5}{2}$ level of the Ce^{3+} ions. The 160 K peak is produced by the freezing out of transitions into the higher crystal field split states [167,227]. For temperatures greater than the crystal field splitting, the conduction electrons participate in scattering events in which the Ce^{3+} ions are scattered between all the crystal field split states of the $J = \frac{5}{2}$. However, for temperatures below the crystal field splitting of the lowest multiplet, only the lowest crystal field multiplet contributes to the scattering rate. Hence, as the coefficient of the ln T term in the resistivity is determined by the effective degeneracy of the local moments, the drop in effective degeneracy produces the second peak in $\rho(T)$ for $CeCu_2Si_2$. In contrast to Ce systems [228–232], U based heavy-fermion systems show little evidence of crystal field splittings or, if they do, are ambiguous [233,235,236]. For single impurity spinflip scattering, the logarithmic temperature variation of the resistivity is expected to cease near T_K, below which the resistivity should show a Fermi liquid like $\rho(T) = \rho(0) - AT^2$ temperature variation with a resistivity maximum only occurring at zero temperature. In contrast to theoretical results for the single-impurity Kondo model, for heavy-fermion materials such as $CeAl_3$, $CeCu_2Si_2$, $CeCu_6$, URu_2Si_2 and UBe_{13}, the observed Kondoesque increase in the resistivity is followed by a rapid decrease with decreasing temperature. The decrease is often attributed to the onset of coherence in the set of f moments. Above the coherence temperature, the f moments are assumed to be independent and have uncorrelated fluctuations whereas at low temperatures, the moments are assumed to approximately align over large spatial

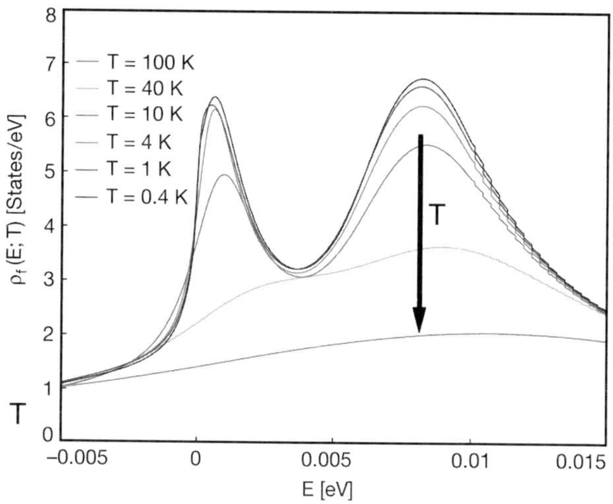

Fig. 19.38. The calculated energy dependence of the f electron density of states $\rho_f(E, T)$ for the single-impurity Anderson model, at various temperatures. The calculation is appropriate to describe a Ce ion in a cubic environment. The temperature dependent Kondo peak, above the Fermi energy, is split by the crystalline electric field. The spectrum was calculated using the non-crossing approximation

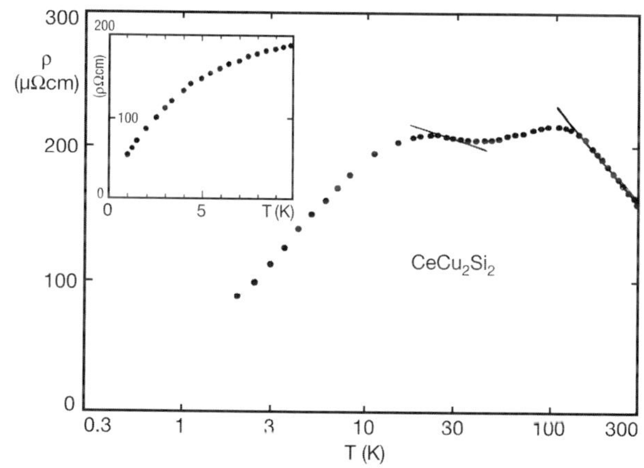

Fig. 19.39. The temperature dependence of the electrical resistivity of $CeCu_2Si_2$, plotted on a logarithmic temperature scale. The two regions where the resistivity shows an approximate ln T dependence are indicated by the *solid lines*. [After Franz et al. [226]]

regions. The fluctuations responsible for preventing perfect alignment are assumed to have a collective character, and it is these fluctuations that are assumed to provide the dominant inelastic scattering process that freezes out as T is reduced to zero. In URu_2Si_2, the broad peak in the resistivity occurs at 50 K where it attains a value of 3000 $\mu\Omega$cms and then shows a rapid decrease at lower temperatures. The decrease is briefly interrupted at $T_N = 17.5$ K, where the resistivity shows a slight, but abrupt, increase due to ordering [21]. The increase in the resistivity is understood in terms of a partial gapping of the Fermi surface and is consistent with the change in the extrapolated C/T ratio from 112 mJ/mole K^2 above T_N to 65.5 mJ/mole K^2 below T_N. In UBe_{13}, the peak in the resistivity is remarkably sharp; the peak occurs at 2.5 K and has a value of the order of 200 $\mu\Omega$cm. On the other hand, UPt_3 and UAl_2 do not show Kondoesque maxima in the resistivities but just decrease with decreasing temperature. The low temperature resistivities of $CeCu_6$, $CeAl_3$, UAl_2, UPt_3, UGe_2 under pressure and YbBiPt show $\rho(T) = \rho(0)+AT^2$ temperature variations characteristic of electron–electron scattering. The coefficients of the T^2 terms are enormous as they seem to scale with the square of the γ term in the specific heat [237], in agreement with the argument involving highly enhanced quasi-particles. The large A coefficients are, therefore, taken as indicating the mutual scattering of quasi-particles in a highly enhanced Fermi liquid. Although a large

T^2 coefficient was also observed in the resistivity of $CeCu_2Si_2$, it has been suggested that this is not a true manifestation of the existence of a Fermi liquid in that by suppressing the superconductivity through the application of a magnetic field, a $T^{3/2}$ temperature variation is observed in S type samples [193]. Since the specific heat of S type samples also shows non-Fermi-liquid like behavior, it has been suggested that the non-Fermi-liquid is caused by the existence of a quantum critical point, but could also be due to structural disorder [194].

The residual resistivities $\rho(0)$ of heavy-fermion materials are sensitive to the presence of disorder and impurities. Often the residual resistivity is found to have a magnitude comparable to that of common metals with similar levels of impurities. However, in cases where the f ion is substituted by a non-magnetic impurity, the resulting residual resistivities are quite large. This has led to the concept of the Kondo hole in which a substitutional non-magnetic impurity at an f site of a heavy-fermion system produces a scattering phase shift of $\frac{\pi}{2}$ relative to the coherent resonant scattering of the lattice of f ions. Since the difference of the phase shifts is close to $\frac{\pi}{2}$, the scattering of the substitutional non-magnetic impurity is similar to the scattering produced by a magnetic impurity in a simple metal [238]. Basically, the large residual resistance of a Kondo hole in a heavy-fermion metal can be attributed as being due to the persistence of the large mass renormalization of the current carrying quasi-particle in the presence of disorder. This is to be expected as current can only be carried by states that are spatially extended. On the other hand, the scattering time at the site of the non-magnetic impurity is not expected to be renormalized by the mass enhancement factor as the scattering is purely a local phenomenon taking place on the non-magnetic impurity ion. The identification of the wave function renormalization Z with the ratio $\Delta/k_B T_K$ then yields a residual resistivity that is governed by the characteristic Kondo scattering rate of the missing f ion.

Magneto-Resistance

Since large amplitude fluctuations of the magnetic moments are assumed to be responsible for the transport scattering rate, the field dependence of the resistivity or longitudinal magneto-resistance should be quite illuminating. Basically, the application of a sufficiently large magnetic field could result in a partial ordering of the moments thereby suppressing the magnetic fluctuations and producing a reduction of the resistivity. This expectation is borne out in experiments on $CeCu_6$ [239] and in the normal state of UBe_{13} [240,241] which show large negative magneto-resistances, $\Delta\rho(H)$,

$$\frac{\Delta\rho(H)}{\rho(0)} = \frac{\rho(H) - \rho(0)}{\rho(0)} . \tag{19.143}$$

On the other hand, for both $CeAl_3$ [242,243] and $CeCu_2Si_2$ [242], the magneto-resistance is negative at sufficiently high temperatures and changes sign at temperatures comparable to the cross-over temperature. The high temperature magneto-resistance shows a scaling with field which can be understood in terms of models of magnetic scattering from single impurities [244,245] in which Zeeman splitting suppresses the incoherent resonant scattering. However, the single impurity models fail at low temperatures where the incoherent magnetic scattering ceases [240,246]. In this low temperature limit, one expects that the resistivity will be dominated by impurity scattering and that the magneto-resistance should follow the scaling implied by Koehler's law. In UBe_{13}, the low temperature magneto-resistance is very large, negative, and anisotropic, which is indicative that the magnetic fluctuations responsible for the scattering of the conduction electrons are also anisotropic [247]. For UPt_3 and UAl_2, which have no maxima in the resistivity, the magneto-resistances are positive [248]. A positive transverse magneto-resistance is often found in common metals with multi-sheeted Fermi surfaces. In these common metals, the Lorentz force produces different shifts of the various sheets of Fermi surface that only produce a zero transverse current when their contributions are combined. The Lorentz force acting on the transverse current components produces a positive magneto-resistance. Therefore, the positive magneto-resistances found at low temperatures can be regarded as providing signatures that the quasi-particles are itinerant.

The Hall Effect

The application of a static magnetic field perpendicular to the direction current flow in a metal, produces a voltage drop in a direction that opposes the Lorentz force acting on the moving electrons. The Hall coefficient, R_H, describes the dependence of the ratio of the transverse electric field to the current on the strength of the applied field. In conventional metals, the Hall coefficient is almost temperature independent and provides information about the density and the sign of the charge carriers. However, in heavy-fermion systems, the Hall coefficient is about two orders of magnitude larger than in conventional metals and is also highly temperature dependent. At room temperature, the Hall coefficient is usually positive and initially increases with decreasing temperature and then goes through a maximum. The maximum has been observed in $CeCu_6$ [249], $CeCu_2Si_2$ [250], UBe_{13} [251] and UPt_3 [252]. The temperature dependence of the Hall coefficient of UPt_3 is shown in Fig. 19.40. For systems for which the resistivity has a maximum, the temperatures of the maximum in the Hall coefficients and the temperatures of the corresponding resistivity maxima are comparable [249]. The high temperature variation of the Hall coefficient is interpreted in terms of the combination of the usual Hall effect due to the Lorentz force and also from skew scattering. The skew scattering results from the scattering of conduction electrons from a set of independent magnetic impurities. The skew scattering involves the polarization of the moments by the applied magnetic field which also produces a spin-splitting in the impurity f density of states. The spin-orbit coupling on the impurity results in interference between the $l = 3$ and $l = 2$ angular momentum channels [253]. The extreme sensitivity of the narrow quasi-particle density of states at the Fermi energy has been identified [254] as the source of the very large magnitude of the Hall coefficient. The anomalous skew scattering from independent local moments does reproduce the high temperature behavior of the Hall coefficient but fails to predict the maximum in the Hall coefficient, which is followed by a rapid drop at lower temperatures. The drop in the Hall coefficient has been interpreted as

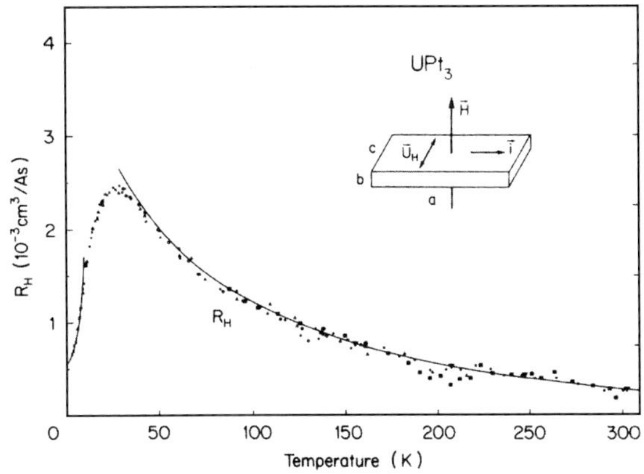

Fig. 19.40. The temperature dependence of the electrical resistivity of the heavy-fermion compounds U_2Zn_{17}, $CeAl_3$, $CeCu_2Si_2$ and UBe_{13}. The *inset* shows the low temperature T^2 variation of the resistivity found in $CeAl_3$

signifying the onset of coherence between the fluctuating parts of the local magnetic moments. In some heavy-fermion materials such as UAl_2, the Hall coefficient changes sign at low temperatures. Similar sign changes in R_H can be found in materials that have multi-sheeted Fermi surfaces.

Thermopower

It is expected that the low temperature limit of the thermopower S should be un-renormalized by the quasi-particle mass enhancement. Nevertheless, the measurement is expected to reflect the strong interactions as it should provide a measure of the logarithmic derivative of the imaginary part of the self-energy. This can be seen from the usual expression of the linear T contribution in terms of the ratio of two correlation functions

$$S = -\beta \frac{K^{11}}{K^{12}} , \tag{19.144}$$

where K^{11} involves the time ordered current–current correlation function, and K^{12} involves the heat flux–current correlation function

$$K^{11} = \frac{1}{3} \int_0^\beta d\tau \exp\left[i\,\omega\tau \right] < T_\tau \underline{j}(\tau) \,.\, \underline{j}(0) > ,$$

(19.145)

$$K^{12} = \frac{1}{3} \int_0^\beta d\tau \exp\left[i\omega\tau \right] < T_\tau \underline{j}_Q(\tau) \,.\, \underline{j}(0) > ,$$

where $\underline{j}$ is the current density and $\underline{j}_Q$ is the heat flux. The Seebeck coefficient can be approximately evaluated as

$$S = -\frac{1}{|e|T} \frac{I^{(1)}}{I^{(0)}} . \quad (19.146)$$

When the above expression is analyzed using the Sommerfeld expansion, one finds the relation

$$S = \frac{\pi^2}{3} \frac{k_B^2 T}{|e|} \frac{\left(\frac{\partial \, Im\Sigma(E)}{\partial E} \right)}{Im\,\Sigma(E)} \Bigg|_{E=0} , \quad (19.147)$$

which also shows that S should vanish linearly with T at low temperatures. The measured thermopowers show considerable variation both with compound and with temperature but do share one common characteristic, which is that the magnitude of S is about an order of magnitude larger than common metals. At 4 K, the thermopower of UBe_{13} is $-11\mu VK^{-1}$ [255] and has decreased to $-25\mu VK^{-1}$ just below 1 K where the superconductivity sets in. At high temperatures, the thermopowers of $CeCu_2Si_2$ [2] and UAl_2 [256] are both positive, and have magnitudes of 15 μVK^{-1} and 40 μVK^{-1}, respectively, while in the same temperature range S for UPt_3 is negative with a magnitude of about -12 μVK^{-1}. The compound $CeCu_2Si_2$ has a shallow maximum in S of about 20 μVK^{-1} at 170 K, which seems to be connected with the crystal field splitting observed in this compound, while the thermopower of UPt_3 appears to have an even shallower minimum of -12 μVK^{-1} near 200 K. The thermopower of UPd_2Al_3 is large and shows a monotonic decrease with decreasing temperature [257]. These slow variations of S in the high temperature region, are to be contrasted with the variations [258] found in $CeCu_6$ and $CeAl_3$, where S decreases from about 40 to 10 μVK^{-1} as T is increased from 50 and 300 K. The thermopower of UPt_3 changes sign at 24 K, and has a positive peak at about 8 K. The thermopower in $CeCu_2Si_2$ changes sign at 75 K from positive to negative with decreasing temperature, and there is a deep minimum ($-35\mu VK^{-1}$) at 20 K near the cross-over temperature. This is followed by a small positive maxima at a temperature of 0.2 K. The thermopower of $CeAl_3$ [259] is similar to $CeCu_2Si_2$ in that it goes through a minimum ($-4\mu VK^{-1}$ at 3.5 K), changes sign near the crossover temperature, and shows a small positive maximum at an even lower temperature ($\sim$ 0.3 K). However, $CeCu_6$ has three positive maxima [258], while UAl_2 only shows one negative minimum before tending to zero linearly with T [256]. This rich variety of behavior is not reconcilable with the behavior of a single impurity Kondo model in which the scattering is dominated by the position of the Kondo resonance with respect to the Fermi-level. For Ce impurities, the model predicts that S has a positive maximum at the Kondo or cross-over temperature and falls to zero as $T \to 0$. The richness of the low temperature variations of the thermo electric power, like the Hall effect and the de Haas van Alphen measurements, suggests that the electronic structure contains a number of quasi-particle bands with high effective masses near the Fermi energy.

Thermal Conductivity

The thermal conductivity of metals, unlike the electrical conductivity, has contributions from the non-equilibrium distribution of phonons as well as from the quasi-particles. Thus, one may write

$$\kappa = \kappa_L + \kappa_e , \quad (19.148)$$

where κ_L represents the lattice contribution and κ_e is the electronic contribution. There is no simple experimental way to decouple these two terms without recourse to theory. It is expected that, at sufficiently low temperatures, one will have

$$\kappa_L \propto T^2 , \quad (19.149)$$

when the phonon scattering is due to electron–phonon interactions, and if the mean free path is less than the size of the crystal. Otherwise, one expects

$$\kappa_L \propto T^3 \quad (19.150)$$

when the phonons mainly scatter from the surfaces of the crystal. At temperatures much greater than the Debye temperature, $T \gg \Theta_D$, one expects that the lattice contribution to the thermal conductivity will have the temperature dependence

$$\kappa_L \propto \frac{1}{T} . \tag{19.151}$$

These considerations lead one to conclude that, at sufficiently low temperatures, the quasi-particle contribution is the dominant term in the thermal conductivity. The quasi-particle contribution to the thermal conductivity is usually written as

$$\kappa_e = \frac{\pi^2}{9} k_B T \left(v^2 \tau \rho_{qp}(E) \right) \Big|_{E=0} , \tag{19.152}$$

where τ is the scattering rate, and v is the quasi-particle velocity ($\tau \sim Z\tau_0$ and $v \sim \frac{\hbar k_F}{Z m_b}$). The two factors of the mass renormalization Z, occurring in the quasi-particle velocities, are expected to cancel with a similar factor in the quasi-particle density of states and also with one factor of Z in the quasi-particle lifetime. Hence, the overall magnitude of the electronic contribution of the low temperature thermal conductivity should not differ significantly from that of a normal metal, for reasons similar to why the d.c. electrical conductivity is also unrenormalized. Thus, one expects that the Lorentz number, $L(T)$, defined by

$$L(T) = \frac{\kappa_e(T)}{T\,\sigma(T)} \tag{19.153}$$

should have a value similar to the value L_0 predicted for purely elastic scattering

$$L_0 = \frac{\pi^2}{3} \left(\frac{k_B}{e} \right)^2 . \tag{19.154}$$

For purely elastic scattering, the Lorentz number, involving the ratio of the appropriate conductivity tensors, is expected to be isotropic. Since inelastic scattering results in a reduction of the quasi-particles' energies, in addition to the reduction in the quasi-particle current, the Lorentz number is expected to be reduced below L_0 when inelastic scattering is also present. Since stoichiometric heavy-fermion materials often show sizeable T^2 terms in the low temperature electrical resistivity which are attributed to (inelastic) electron–electron scattering, one expects that when the T^2 term is larger than the residual resistivity, $L(T)$ should have a value close to that for purely elastic Baber scattering [260]

$$L_{\text{elas}} = L_0 \left(\frac{36}{\pi^2} - 3 \right) . \tag{19.155}$$

However, for lower temperatures such that the quasi-particle–quasi-particle scattering term becomes negligible compared with the residual resistivity, one should recover higher values of $L(T)$ close to the Lorentz number L_0. These expectation are very nearly borne out by experiments on the normal states of $CeAl_3$, $CeCu_6$, $CeCu_2Si_2$ and UPt_3. In $CeAl_3$ [259,261,262], the Lorentz number exhibits a minimum value of 0.75 L_0 at 0.5 K, which is larger than the theoretical limit of 0.65 L_0, and it overshoots the elastic limit yielding a weak maximum at 1.1 L_0 at 50 mK. The temperature of the minimum roughly agrees with the temperature where the T^2 term dominates the resistivity. For $CeCu_6$, the Lorentz number falls to a minimum value smaller than L_0 at a temperature of about 0.3 K, which is close to the temperature of 0.15 K where the T^2 term in the resistivity first becomes apparent as it cools [263]. The extrapolated data show that $L(T)$ approaches L_0 at 20 mK, in agreement with the arguments concerning the freezing out of the inelastic scattering processes. For $CeCu_2Si_2$, the low T value of $L(T)$ appears to be L_0 [2,264]. At low temperatures, the inferred phonon contribution to κ_L of $CeCu_2Si_2$ has a T^2 variation [264] as expected from electron–phonon scattering. The phonon contribution κ_L starts to become larger than the electronic component at the temperature of 1 K where on cooling the T^2 term in the resistivity first becomes apparent [264]. For UPt_3, the minimum value of 0.45 occurs at a temperature of the order of 2 K [216,265]. Since UBe_{13} does not form a Fermi liquid, it is not surprising that the Lorentz number is not constant but instead varies approximately linearly with temperature [28,266].

For temperatures greater than the coherence temperature, the Fermi liquid analysis of κ_e is expected to fail. However, in the temperature regime where $T > T_K$, one expects that Ce based heavy-fermion materials will resemble a metal in which the conduction electrons scatter off of a set of independent disordered magnetic moments. In this temperature regime, the thermal conductivity from the electrons may be calculated perturbatively [267]. The result for the Lorentz number $L(T)$ can be approximately expressed in the form

$$L(T) = \frac{1}{e^2 T}\left(\frac{I^{(2)}}{I^{(0)}} - \left(\frac{I^{(1)}}{I^{(0)}}\right)^2\right), \qquad (19.156)$$

where $I^{(n)}$ is defined by Eq. (19.141). The Lorentz number is also expected to depend strongly on the crystal field splittings [268]. The crystal field splitting in $CeCu_2Si_2$ is assumed to give rise to a minimum in the Lorentz number at the temperature around 80 K. In this high temperature regime, the Lorentz number can be appreciably greater than L_0. This result is caused by the large temperature induced shift of the f density of states away from the Fermi energy for $T > T_K$. The experiments are consistent with this trend: However, the lattice contribution to κ_L is no longer negligible and may be several times greater than the quasi-particle contribution. This hypothesis is supported by the comparison of measurements on $CeCu_2Si_2$ with the reference material $LaCu_2Si_2$. In fact, if the phonon contribution κ_L is completely ignored in the data analysis, the Lorentz number is about a factor of twenty times greater than L_0. The inferred presence of a large phonon contribution is also supported by the conclusion that the theoretical calculated values of $L(T)$, using the independent Kondo impurity model, are unable to account for electronic contributions to $L(T)$ of this magnitude. Likewise, for UPt_3, the phonon contribution has been identified as being important above 3.5 K [29] and may be responsible for up to a maximum of 4.5 L_0 of the Lorentz number at $T = 24$ K inferred from the raw data [216]. The Lorentz number for $CeCu_6$, in the high temperature regime [269] qualitatively resembles $L(T)$ for UPt_3, having a maximum at $T \sim 20$ K where it attains a value of about 2.5 L_0 but falls towards L_0 at higher temperatures.

Ultrasonic Attenuation

Ultrasonic experiments measure the attenuation and velocities of sound waves and, through the electron–phonon interaction, provide information about the electronic system. At temperatures much lower than the Debye temperature, the attenuation of the sound wave due to anharmonic phonon interactions should be negligible, and the electron–phonon interaction should provide the dominant contribution to the attenuation. In this case, one expects a strong similarity between the ultrasonic attenuation coefficient and the electrical conductivity. Just like the optical conductivity $\sigma(\omega)$ yields the lifetime of a photon, the ultrasonic attenuation $\alpha_{\underline{q},\hat{\epsilon}}(\omega)$ yields the lifetime of a phonon. This similarity is expected since, for many purposes, the effect of a phonon can be interpreted in terms of the effect of the accompanying electric field. In analogy with the conductivity, one expects that in the limit $q \to 0$, the renormalization of the scattering rate will cancel with the mass enhancement. Thus, in the long wavelength limit, the ultrasonic attenuation is expected to be unrenormalized. The most noticeable difference between ultrasonic attenuation and optical absorption occurs through the extremely different magnitudes of the velocities of the corresponding waves. If the conditions allow the heavy quasi-particles to co-move with the sound wave, it becomes possible for the quasi-particles to surf-ride and continuously absorb energy from the sound wave. For example, if a longitudinally polarized sound wave with phase velocity $\frac{\omega}{q}$ propagates through a gas of heavy quasi-particles, it strongly perturbs the quasi-particles with velocities almost parallel and almost equal to the phase velocity of the wave. In the frame of reference traveling with the wave, the quasi-particle is at rest and experiences an essentially time independent electric field. The electric field continuously transfers energy from the wave to the quasi-particles that have the same velocity. If there is a slight mismatch in the velocities, quasi-particles with lower velocities than the wave draw energy from the wave and accelerate, whereas quasi-particles that are moving faster lose energy and slow down. This results in the rate of energy loss of the wave being proportional to the derivative of the distribution of quasi-particle velocities, which is evaluated at the velocity of sound. At temperatures below those at which the heavy quasi-particles are being formed, the Fermi velocity is comparable to the sound velocity in contrast to normal metals. So the attenuation of sound waves may be expected to be larger than in ordinary metals. On the other hand, due to the approximate equality between the sound velocities [270], the Born–Oppenheimmer approximation no longer applies and the sound wave may have to adjust adiabatically to the quasi-particles' motions. The induced coupling of the phonon modes can also lead to damping. It is found that the ultrasonic attenuation for $CeCu_6$, UPt_3, and UBe_{13} have magnitudes comparable to that of ordinary metals [27]. This suggests that the magnitude of the coupling between the quasi-particle and sound wave is reduced from that observed in normal metals due to the appearance of vertex corrections. Nevertheless, the strong electron–phonon coupling shows up in other elastic properties. In ordinary metals, the attenuation is relatively featureless and is described by a Drude-like formula,

$$\alpha(\omega) = \frac{\omega^2\tau}{1+\omega^2\tau^2}\,\frac{\lambda^2\rho(\mu)}{\rho c^3}\,, \qquad (19.157)$$

where λ is the electron–phonon coupling, c is the velocity of sound, and ρ is the mass density. In heavy-fermion compounds at high temperatures, the magnitude and frequency dependence of the attenuation is reasonably similar to normal metals. However, anomalous temperature dependences are observed in the low temperature quasi-particle phases. The temperature dependence of the attenuation of longitudinal sound in UPt_3 shows an anomalous peak at 12 K [271], and has an amplitude that scales with the square of the frequency of sound. The temperature dependence of the attenuation coefficient is shown in Fig. 19.41. Since the anomaly is only present in the absorption of longitudinal sound but not of transverse sound, the origin of this peak has been ascribed by Schotte et al. [272] to the existence of a breathing mode. In the breathing mode, the lattice adjusts locally to the state of the f ion, and vice versa. Thus, the sound waves are coupled to the heavy quasi-particle bands. This type of coupling is also expected to modify the phonon dispersion relations due to hybridization with the quasi-particle bands. At temperatures above 12 K, the heavy quasi-particle masses have not fully formed, and so the coupling becomes incoherent. Thus, the effect of an increase in temperature results in a reduction of the attenuation coefficient. Since in the model of Schotte et al. the sound wave mainly couples to the thermally excited quasi-particles, the coupling is ineffective in attenuating sound at temperatures lower than 12 K. Therefore, the sound wave attenuation diminishes for temperatures lower than the peak temperature. At still lower temperatures, but still in the normal state, the attenuation coefficient of UPt_3 decreases proportional to T^2 with increasing temperatures [30]. This is indicative of the T^2 dependence of the quasi-particle scattering

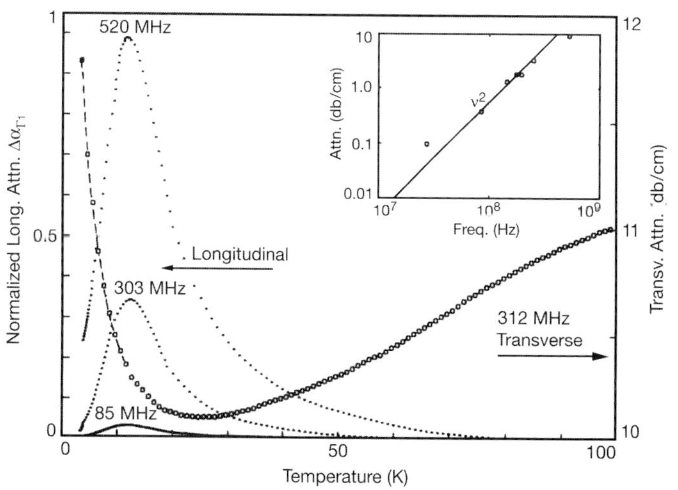

Fig. 19.41. The temperature dependence of the transverse (312MHz, q ∥ b, u∥c) and longitudinal ultrasonic attenuation coefficients $\alpha(\nu)$ of UPt_3 at various frequencies ν. The longitudinal data are normalized to the 520 MHz peak. The *inset* shows the ν^2 dependence. [After Müller et al. [271]]

rate. The observation of an anomalous peak in the phonon density of states [273] provides supporting evidence for the validity of the breathing mode picture of UPt_3. The attenuation of longitudinal sound waves in $CeCu_6$ [274] is somewhat similar to that of UPt_3 in that it shows a peak below a temperature of 10 K. However, unlike UPt_3, the intensity of the peak in $CeCu_6$ scales linearly with frequency. Furthermore, the peak broadens and shifts to higher temperatures with increasing phonon frequency.

19.3.3 Dynamic Magnetic Properties

Just as the effects of the collective magnetic fluctuations and the low temperature gas of heavy quasi-particles show up in the transport and thermodynamic properties of heavy-fermion materials, the effects of these excitations also show up in the magnetic properties.

Nuclear Magnetic Relaxation

In the presence of an applied static magnetic field, a nuclear spin relaxes to its ground state through interaction with the fluctuations of the electronic magnetic moments. On assuming that the coupling is weak so that the Born approximation is valid, the longitudinal relaxation rate $1/T_1$ can be expressed in terms of the hyperfine field coupling $A_{hf}(\underline{r} - \underline{r}')$ between the nuclear spin point $\underline{r}'$ to the magnetization density $\underline{M}(\underline{r})$ and the dynamic susceptibility $\chi^{+,-}(\underline{q}; \omega)$ via

$$\frac{\hbar}{T_1} = k_B T \frac{1}{N} \sum_{\underline{q}} \left| A_{hf}(\underline{q}) \right|^2 \left[\frac{Im \chi^{+,-}(\underline{q}; \omega_N)}{\hbar \omega_N} \right] . \tag{19.158}$$

In this, $A_{hf}(\underline{q})$ is the spatial Fourier transform of the hyperfine field, and ω_N is the nuclear Larmor frequency which, in conventional metals, can be neglected as it is much smaller than the lowest energy scale for the electrons. The dynamic magnetic susceptibility is the spatial and temporal Fourier transform of the quantity $\chi^{\alpha,\beta}(\underline{r}; t)$ defined by the causal correlation function involving the commutator of the components of the magnetization density

$$\chi^{\alpha,\beta}(\underline{r}; t) = -\frac{i}{\hbar} < | \left[M^{\alpha}(\underline{r}; t), M^{\beta}(0; 0) \right] | > \Theta(t) , \tag{19.159}$$

where $\Theta(t)$ is the Heaviside step function. In the paramagnetic state, the susceptibility tensor is spin rotationally invariant so one has

$$\chi^{x,x}(\underline{r}; t) = \chi^{y,y}(\underline{r}; t) = \chi^{z,z}(\underline{r}; t) = \frac{1}{2} \chi^{+,-}(\underline{r}; t) . \tag{19.160}$$

Due to the Shiba relation [275], one expects that, for a local paramagnetic Fermi liquid in the limit $\omega \to 0$, the imaginary part of the susceptibility should satisfy the equation

$$\lim_{\omega \to 0} (2J + 1) \frac{J(J + 1)}{3} \frac{Im \chi^{z,z}(0; \omega)}{\hbar \omega} = \pi \left(\frac{\chi^{z,z}(0; 0)}{g \mu_B} \right)^2 . \tag{19.161}$$

This suggests that, with a $\underline{q}$ independent hyperfine coupling, the longitudinal nuclear relaxation rate of heavy-fermion systems should vary as the square of the low temperature static susceptibility or the square of the γ coefficient in the specific heat, much the same as the correlation between the coefficient of the T^2 term of the resistivity and the linear T term in the specific heat found by Kadowaki and Woods [237]. For materials such as UBe_{13} [276], the scaling with the square of the quasi-particle density of states is not followed due to the $\underline{q}$ dependence of the coupling and vertex corrections. Nevertheless, magnetic relaxation experiments on heavy-fermion compounds at low temperatures do show the linear T Korringa relaxation rate with an enhanced magnitude, indicating the existence of heavy quasi-particles.

The Knight shift K is a shift in the nuclear resonance field which provides a measure of the induced field at the site of the nucleus due to the polarization of the electronic system. It is often found that the induced field is dominated by the bulk static susceptibility $\chi^{z,z}(0; 0)$ of the f electrons and so can be used as an estimate of the strength of the average hyperfine field

$$K = \frac{A_{hf}}{g_N \mu_N} \chi^{z,z}(0; 0) . \tag{19.162}$$

Assuming a local isotropic hyperfine field and isotropic $\underline{q}$ independent magnetic fluctuations, one finds the dimensionless Korringa product

$$S = k_B T K^2 \frac{T_1}{\hbar} \tag{19.163}$$

has the theoretical value

$$S_0 = \frac{(2J+1)J(J+1)}{6\pi}\left(\frac{g_e \mu_B}{g_N \mu_N}\right)^2 \tag{19.164}$$

for ions with magnetic moment J. Corrections must be made to this relation for heavy-fermions compounds with anisotropic susceptibilities. The magnitude of the Korringa ratio is defined as $S\,S_0$. Since the relaxation rate involves a sum over all $\underline{q}$, deviations from the ideal value of the Korringa ratio may be used to infer the relative magnitude of the $\underline{q}$ averaged susceptibility to the uniform ($\underline{q} = 0$) susceptibility.

The Knight shift and the longitudinal relaxation rate can be used to estimate the typical frequency of local moment fluctuations. Assuming a phenomenological $\underline{q}$ independent relaxational form for the dynamic susceptibility

$$\chi^{z,z}(\omega) = \chi^{z,z}(0)\frac{\Gamma}{\Gamma - i\hbar\omega}\,, \tag{19.165}$$

where Γ is the relaxation rate, one finds that, if $\hbar\omega_N \lll \Gamma$, the transverse relaxation rate can be expressed in terms of the limit

$$\lim_{\omega_N \to 0} \frac{Im\chi^{z,z}(\omega_N)}{\hbar\omega_N} = \frac{\chi^{z,z}(0)}{\Gamma}\,. \tag{19.166}$$

Thus, the spin-fluctuation energy is estimated as

$$\Gamma = 2k_B T \frac{T_1}{\hbar} |A_{hf}|^2 \chi^{z,z}(0)\,. \tag{19.167}$$

Above the Néel temperature, the NMR longitudinal relaxation rate of UPt_3 is linear in T [31], but below T_N, has a large anisotropic coefficient that varies from 1810 to 1050 s^{-1} K^{-1} for fields parallel and perpendicular to the crystalline c-axis [277]. The anisotropy of the relaxation rate (shown in Fig. 19.42) is accompanied by a large anisotropic Knight shift [278] which reflects the anisotropy of the bulk susceptibility. From estimates of the hyperfine field taken from the Knight shift, a $\underline{q}$ averaged fluctuation energy Γ of the order of 9 meV was deduced from the measurements. This energy is quite large when compared with the temperature scale over which the large specific heat γ value develops. The Korringa ratio was found to be of the order of 1.2, which indicates that the system supports low-energy paramagnon excitations, consistent with the finding of a $T^3 \ln T$ term in the low temperature specific heat.

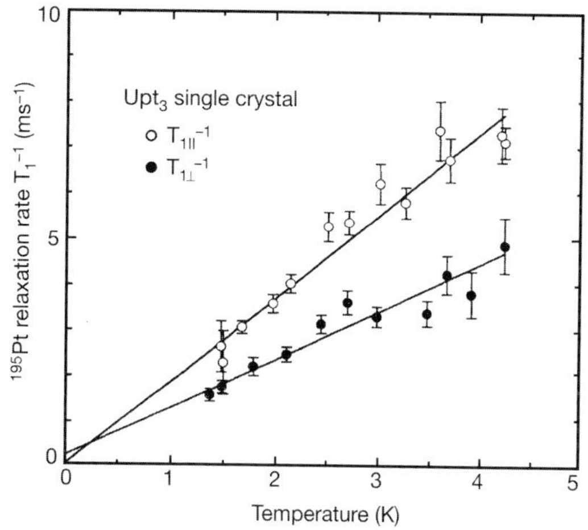

Fig. 19.42. The temperature dependence of the spin-lattice relaxation rate $1/T_1$ for ^{119}Pt in UPt_3. The relaxation rate is anisotropic and the results are measured with the static field applied parallel and perpendicular to the crystal c-axis. [After Vithayathil et al. [277]]

The linear T dependence of the longitudinal relaxation rate in UBe_{13} has been observed for temperatures less than $T = 1.5$ K, which is much lower than the temperature of 10 K, where the specific heat enhancement starts forming. The Korringa ratio is only 0.3, which is much reduced from the ideal value, suggesting the presence of strong antiferromagnetic correlations [279].

Below the temperature of about $T = 70$ K, the longitudinal relaxation rate of URu_2Si_2 follows the Korringa linear T dependence indicative of the formation of a Fermi liquid. This linear dependence is interrupted at the Néel temperature of $T_N = 17.5$ K, and the relaxation rate again follows a linear law a little below T_N, but with a reduced coefficient [280,281]. The temperature dependence of $1/T_1$ for URu_2Si_2 is

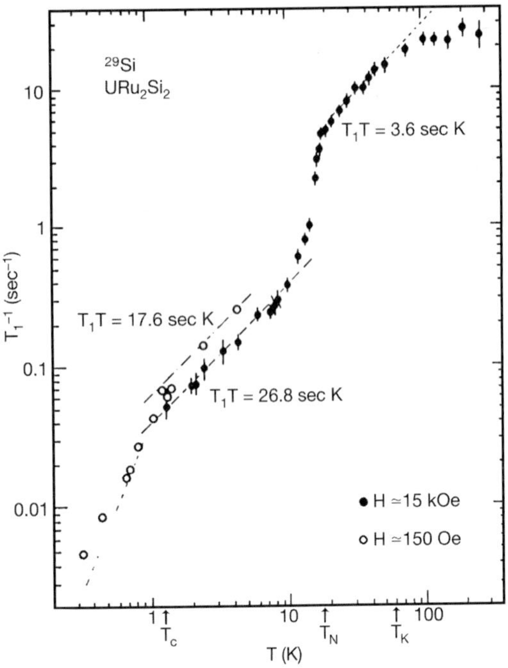

Fig. 19.43. The temperature dependence of the spin-lattice relaxation rate $1/T_1$ for ^{29}Si in external field (*open circles*) and ^{105}Ru in zero field (*solid circles*) in URu_2Si_2. The relaxation rate shows an abrupt change at the hidden ordering temperature T_N and at the superconducting transition temperature T_c. [After Kohara et al. [280]]

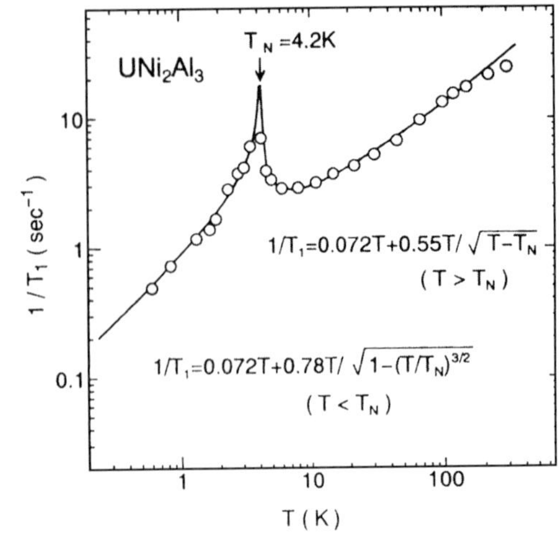

Fig. 19.44. The temperature dependence of the spin-lattice relaxation rate $1/T_1$ for ^{27}Al in UNi_2Al_3. [After Kyogaku et al. [282]]

shown in Fig. 19.43. The data does not show any sign of a spin-wave contribution to the relaxation process. For non-interacting quasi-particles, the coefficient of the T term in the Korringa rate is proportional to the square of the quasi-particle density of states. Thus, the drop in the coefficient of T found on entering the Néel state is indicative of a partial gapping of the Fermi surface. This is consistent with the drop in the specific heat γ from 180 mJ mol^{-1} K^{-2} above to 50 mJ mol^{-1} K^{-2} below the transition. Although no direct evidence of antiferromagnetic ordering was observed in NMR measurements at ambient pressure, experiments at $P = 8.3$ k bar [224] show a well defined splitting of the resonance line at T_N. The splitting is consistent with the staggered exchange splitting from a type I antiferromagnetic structure. In addition, the NMR spectra showed a substantial contribution from the un-split resonance line that continued, but diminished in intensity as the temperature was reduced below the Néel temperature. Since the magnitude of the staggered exchange field starts leveling off below 15 K, and as the ratio of intensities of the split and un-split lines continuously change with T, it has been suggested that regions of antiferromagnetism and paramagnetism coexist inside the single crystal sample but with temperature dependent volume fractions. The interpretation based on the temperature and pressure dependent volume fractions of the two phases is consistent with the spontaneous magnetic moments inferred from neutron diffraction experiments under pressure [225]. However, as the magnitude of the specific heat jump at T_N does not scale with the volume fraction, it has been suggested that another (as yet unknown) type of ordering also occurs at T_N.

The temperature dependence of $1/T_1$ of UPd_2Al_3 [282] is very similar to that found in URu_2Si_2, in that it shows a Korringa rate that shows a partial reduction of the density of states at the Fermi energy on entering the magnetically ordered state. The properties of UNi_2Al_3 are quite different from UPd_2Al_3, in that the Korringa law does not apply above $T_N = 4.2$ K, and any linear T coefficient is quite small. As seen in Fig. 19.44, the magnetic transition is marked by a peak in $1/T_1$, which is indicative of strongly antiferromagnetically coupled local moments. At lower

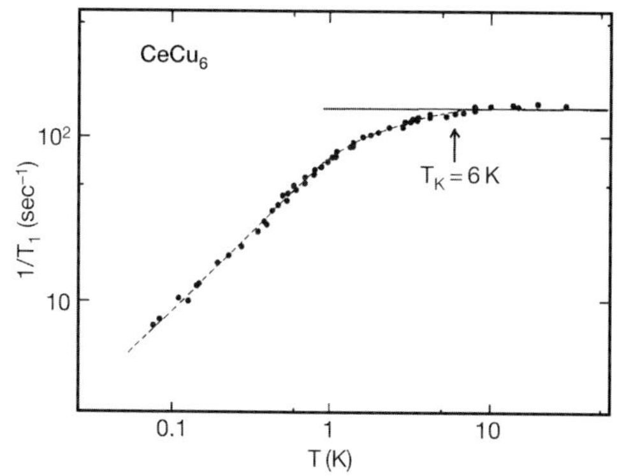

Fig. 19.45. The temperature dependence of the spin-lattice relaxation rate $1/T_1$ for ^{63}Cu in $CeCu_6$ and ^{29}Si in $CeRu_2Si_2$. [After Kitaoka et al. [283]]

temperatures, superconductivity occurs before any linear T region develops. When a sufficiently large magnetic field is applied to suppress the superconductivity, a linear T dependence can be resolved.

The longitudinal magnetic relaxation rates have been measured at the Cu site of $CeCu_6$ [283] and at the Cu and Si sites of $CeCu_2Si_2$ [276,284,285]. In $CeCu_2Si_2$, the low temperature relaxation rate varies linearly with T up to $T = 5$ K, but has a coefficient of proportionality of 5 s^{-1} K^{-1}. This large value of the slope should be contrasted with the corresponding slope of 0.025 s^{-1} K^{-1} found in the compound $LaCu_2Si_2$, which does not contain any f electrons. The enhanced value of the low temperature relaxation rate is even more apparent in $CeCu_6$, where $1/T_1$ follows a linear Korringa law below about $T = 0.2$ K with a slope of 88 s^{-1} K^{-1} [283]. The relaxation rates for $CeCu_6$ and $CeRu_2Si_2$ are shown in Fig. 19.45. At higher temperatures, the relaxation rate of $CeCu_6$ starts changing form and shows a plateau associated with the cross-over temperature T_K of 12 K, and above this, it shows a slow decrease with increasing temperature due to the Curie like variation of the susceptibility. From the temperature variation of the longitudinal relaxation rate, one finds that the low temperature Fermi liquid in $CeCu_6$ only starts forming at a temperature lower than the cross-over temperature by a factor of 60, whereas for UBe_{13} this ratio is estimated to be about 9. It is noteworthy that the result for $CeCu_6$ does not show the scaling expected from the single-impurity Anderson model, despite the fact that the non-local contributions to the magnetic response found from neutron scattering are estimated as being only 10%. However, the ratio found for UBe_{13} is more in line with the expectation based on the single-impurity Anderson model [286] despite the presence of strong antiferromagnetic correlations as indicated by the Korringa ratio.

The transverse magnetic relaxation rate $1/T_2$ provides information about processes whereby the nuclear spin precession is dephased and does not involve transitions where there is a change of Zeeman energy. On neglecting the effects of static random local fields, the transverse relaxation rate is given by perturbation theory as

$$\left(\frac{\hbar}{T_2}\right)^2 = \frac{3}{4} I(I+1)|A_{hf}|^4 \times \left[\frac{1}{N}\sum_{\underline{q}} \chi^2(\underline{q}) - \left(\frac{1}{N}\sum_{\underline{q}} \chi(\underline{q})\right)^2\right], \tag{19.168}$$

which measures the non-uniform component or $\underline{q}$ variation of the static susceptibility. For systems close to a magnetic instability in which $\chi(\underline{q})$ peaks up around a ordering vector $\underline{Q}$, one expects that the susceptibility may have an Ornstein–Zernicke form with a correlation length ξ,

$$\chi(\underline{q}) = \frac{\chi(\underline{Q})}{1 + |\underline{q} - \underline{Q}|^2 \xi^2} . \tag{19.169}$$

Thus, one finds that

$$\left(\frac{\hbar}{T_2}\right)^2 \propto \xi^{-3} \chi^2(\underline{Q}) \tag{19.170}$$

and as, from the simple scaling hypothesis $\chi(\underline{Q}) \propto \xi^2$, one has

$$\frac{1}{T_2} \propto \xi^{\frac{1}{2}} . \tag{19.171}$$

Thus, one expects $1/T_2$ to be enhanced in the vicinity of a magnetic phase transition. Furthermore, in the ordered state, the precession frequency is expected to be changed due to the presence of static magnetic moments or local fields. This information can be used to provide information about the size of the ordered moments.

In addition to $1/T_2$ relaxation rates, NMR linewidths are also sensitive to local static magnetic

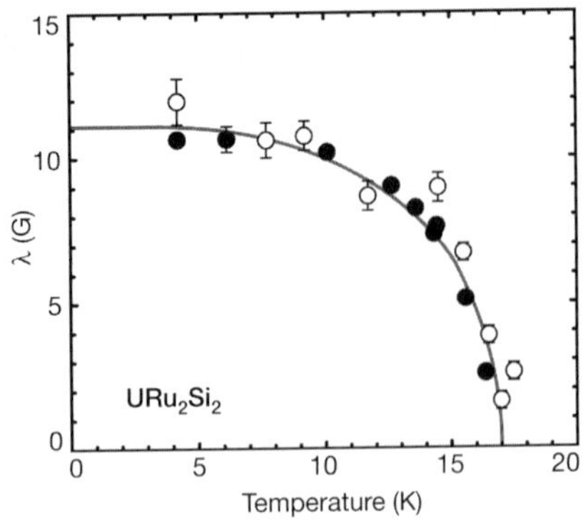

Fig. 19.46. The temperature dependence of the anomalous component of the ^{29}Si line width $\lambda(T)$ in URu_2Si_2. *Open circles* H ∥ to c, *solid circles* H ⊥ to c. [After Bernal et al. [287]]

fields [287]. This has provided some additional information on the $T = 17.5$ K transition in URu_2Si_2. In this compound it was found that, for temperatures above T_N, the linewidth is linearly-dependent on the field and has an anisotropy similar to that of the sample's magnetization. However, below T_N, the measurements also show an additional isotropic, field-independent, contribution to the linewidth which is shown in Fig. 19.46. The temperature dependence of the isotropic component is similar to that expected from a mean-field transition that occurs at T_N. Since at ambient pressures the signal is predominantly from the paramagnetic regions of the sample, and as (unlike the magnetic ordering) the field-independent contribution to the width is isotropic, the authors suggest that the width is due to the coupling to the unknown or "hidden" order parameter.

Muon spin relaxation works in much the same way as NMR, but the zero field precession rate is much more sensitive to ordered magnetic moments with small magnitudes. Also as the muon precession frequency is extremely slow, μSR is effective in discriminating between slowly varying magnetic order and static order. Therefore, muon spin relaxation measurements have played a particularly important role in identifying magnetic phase transitions, specially as the transition to magnetically ordered phases of heavy-fermions may or may not be accompanied by specific heat anomalies. In the case of UPt_3, Cooke et al. [288] discovered the existence of magnetic ordering at $T_N = 5$ K through the zero field muon spin relaxation line width $1/T_2$, even though no corresponding specific heat anomaly was observed at the transition. The transverse relaxation rate showed the growth of an additional contribution below T_N corresponding to an increase in the local fields due to a very small moment of order 0.01 μ_B. The small magnitude of the ordered moment is presumably responsible for the absence of a specific heat anomaly at T_N. The existence of magnetic ordering was rapidly confirmed by neutron scattering experiments [289]. However, later muon resonance experiments on better quality samples no longer showed this feature in the muon relaxation rate [290]. This discrepancy is presumably caused by the exact cancelation of the field due to the ordered moment at the muon sites which is disrupted by the presence of disorder and thereby revealing a finite staggered magnetization.

In URu_2Si_2, the magnetic transition is accompanied by a large change in the specific heat at $T_N = 17.5$ K, but the size of the average ordered magnetic moments found from neutron and x-ray scattering are only of the order of 0.03 μ_B. Although transverse relaxation rates and Knight shift measurements do indicate that magnetic ordering occurs, the static ordered magnetic moments inferred from the muon measurements [291,292] are an order of magnitude smaller than the average magnetic moments found from the neutron scattering experiments at ambient pressure. Due to the large magnitude of the specific heat jump and the small magnitude of the averaged ordered moments, it has been suggested [293–295] that the primary ordering is non-magnetic. However, polarized diffraction experiments and a symmetry analysis seem to rule out higher order multipolar transitions [296]. Also, measurements in applied fields do not yield evidence for exotic types of multiple spin correlations [297].

In $CeAl_3$, by contrast with URu_2Si_2 where a large specific heat anomaly is found at T_N, only a slight and sample dependent specific heat anomaly has been observed. However, the muon spin resonance spectrum shows evidence that magnetic order develops in $CeAl_3$ at $T = 1.5$ K [298,299]. Furthermore, near $T = 0.7$ K where the susceptibility shows a maximum, the ordered moment is quite sizeable and has been reported to be as large as 0.5 μ_B [300]. The lack of correlation between the size of the ordered moments and the size of the specific heat jump is not understood. However, the magnetic ordering does appear to show up in the resistivity of single crystals where a T^3 variation has been observed [301]. The T^3 variation is indicative of the existence of a low-energy branch of spin-wave excitations.

Evidence for static magnetic order in $CeCu_2Si_2$ in zero field and longitudinal field muon spin relaxation has been reported [302] just above the superconducting transition temperature $T_c = 0.7$ K. The inferred size of the ordered magnetic moments is of the order of $0.1\mu_B$. However, the intensity of the NQR line decreases monotonically with decreasing temperature and shows no anomalous broadening associated with ordered magnetic moments [303,304]. The magnetic ordering persists in the superconducting phase and competes with the superconductivity [305–307]. Investigations show evidence that the superconductivity and magnetic ordering may exist in separate thermodynamic phases, but these are phases for which T_c and T_N coincide in the best quality samples. Application of a magnetic field suppresses the superconductivity transition temperature but (initially) has no effect on T_N. The zero field muon spin relaxation experiments show two components that decay at rates that differ by about an order of magnitude [307]. The data can be interpreted in terms of the sample existing in two separate phases: the superconducting phase and a magnetic A phase. An estimate of the weight of these two phases can be made based on the amplitudes of the two components. For temperatures above the superconducting transition, the proportion of the A phase increases with decreasing T, reaching a maximum at T_c with a value of $\frac{3}{4}$. The magnitude of the specific heat jump at the superconducting transition seems to scale with the relative weights of the two phases. As the temperature is lowered, the volume fraction associated with the superconductor grows by expelling the A phase and saturates at about a volume fraction of $\frac{2}{3}$. The relative weights of the superconducting phase and the A phase are extremely sensitive to the deviations from stoichiometry of Ce [191]. Doping experiments have shown that substitutionally doping Th on the Ce sites can lead to the formation of an antiferromagnetic state with a significantly larger magnetic ordering temperature for Th concentrations of only about 7%.

Neutron Scattering Cross-Section

The most direct way of obtaining information about the magnetic excitations in a material is through inelastic neutron scattering experiments in which momentum $\hbar \underline{q}$ and energy $\hbar\omega$ are transferred between the neutron and the sample. The neutron interacts with the electronic spins via a dipolar interaction and yields information about the imaginary part of the dynamic susceptibility. The differential scattering cross-section for an unpolarized beam of neutrons due to purely magnetic scattering, is given by

$$\frac{d^2\sigma}{d\omega d\Omega} = N\left(\frac{g_n e^2}{g\mu_B m_e c^2}\right)^2 \frac{k}{k'}\left[1 + N(\omega)\right] \quad (19.172)$$
$$\times \left|F(q)\right|^2 \sum_{\alpha,\beta} Im\chi^{\alpha,\beta}(\underline{q};\omega)\left(\delta_{\alpha,\beta} - \hat{q}_\alpha \hat{q}_\beta\right) .$$

Here, $F(q)$ is the atomic form factor for the magnetic moments, k and k' are the magnitude of the incident and scattered neutron wave vectors, and $N(\omega)$ is the Bose–Einstein distribution function. The above expression should also be multiplied by the Debye–Waller factor due to zero point and thermally excited lattice vibrations.

Generally, the low-energy dynamic magnetic response of a heavy-fermion system is interpreted as the sum of a large quasi-elastic relaxation contribution that is $\underline{q}$ independent and $\underline{q}$ dependent terms associated with spatial magnetic correlations. The relaxational component is of the form

$$Im\left[\chi(q;\omega)\right] = A\frac{\hbar\omega\Gamma}{\hbar^2\omega^2 + \Gamma^2} , \quad (19.173)$$

which is indicative of local moment fluctuations. The width of the quasi-elastic peak seems to saturate at low temperatures but increases with increasing T at higher temperatures. The $\underline{q}$ independence and the temperature variation of the width Γ found in some Ce systems [229,230,308] are similar to the behaviors found in the single-impurity Anderson model [309], where

$$\Gamma(T) \sim \Gamma(0) + B\sqrt{T} . \quad (19.174)$$

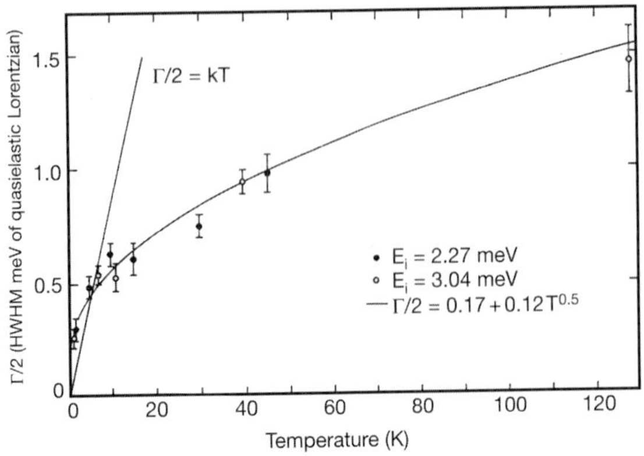

Fig. 19.47. The temperature dependence of the experimentally determined half-width Γ of the quasi-elastic peak in $CePb_3$. [After Balakrishnan et al. [312]]

The temperature dependence of the quasi-elastic linewidth of $CePb_3$ is shown in Fig. 19.47. The product of the zero temperature limit of the linewidth and the linear T term in specific heat, $\Gamma(0)\gamma$, should scale inversely with the Wilson ratio. For $CeAl_3$ [308], $CeCu_6$ and $CeRu_2Si_2$ [310, 311] respectively, the product has magnitudes of 0.77, 0.67 and 0.70 J meV/mole/K^2, even though the widths show considerable variation with $\underline{q}$. By contrast, in $CePb_3$ where no $\underline{q}$ dependence was observed [312], the product is only 0.034, whereas the Wilson ratio obtained from thermodynamic measurements at the same temperatures is of the order of unity [183]. Similar narrow quasi-elastic peaks are found in U heavy-fermion systems [313] and can have a wide range of values for the product $\Gamma(0)\gamma$. The narrow quasi-elastic response can also be interpreted as the response of a gas of heavy quasi-particles in a very narrow band. The $\underline{q}$ dependent contributions may signify fluctuations associated with long-ranged order or competing types of short-ranged order. In particular, the quasi-particle contribution to the susceptibility should show up at very low temperatures as a narrow quasi-elastic peak at small energy and momentum transfers. The intensity of the peak should be proportional to the quasi-particle weight but may be significantly reduced by spin-orbit scattering. For the cerium heavy-fermion compounds [228, 229, 231, 232, 234], the higher energy spectra also show inelastic peaks due to crystal field

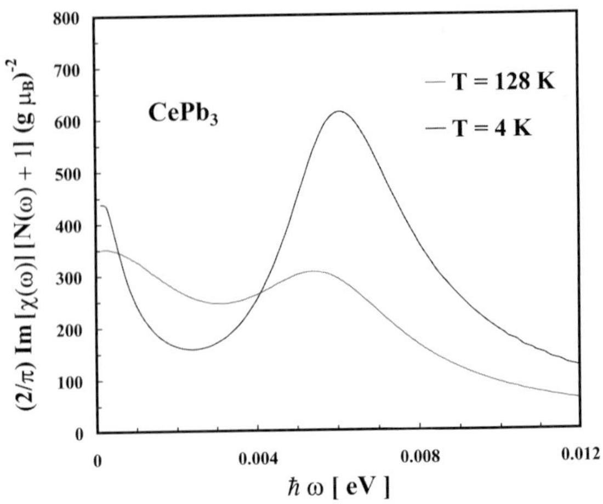

Fig. 19.48. The neutron inelastic scattering cross-section calculated in the NCA, for the single impurity Anderson model with parameters appropriate for $CePb_3$. The spectrum consists of a low energy quasi-elastic peak and an inelastic peak due to crystal field excitations in a cubic environment

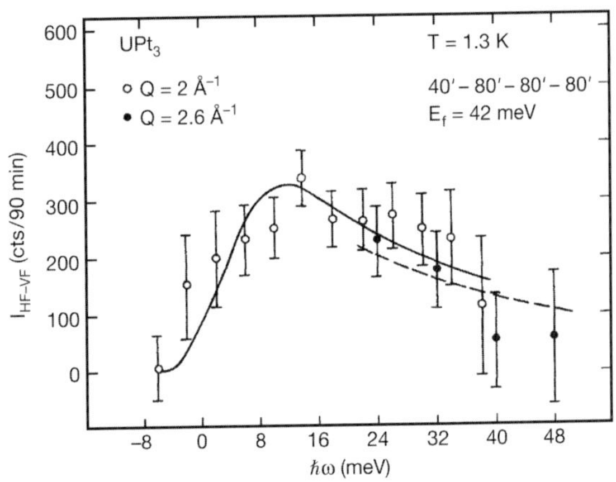

Fig. 19.49. The 9 meV quasi-elastic peak observed in polarized beam inelastic neutron scattering experiments on UPt_3. [After Aeppli et al. [314]]

excitations. As seen in Fig. 19.48, the localized crystal field excitations can be reasonably described by the single impurity Anderson model.

Inelastic neutron scattering experiments on UPt_3 show the existence of a local relaxational component to the response which has a width of $\Gamma = 9$ meV [314]. The experimental data are shown in Fig. 19.49. The $\underline{q}$ dependence is associated with three different types of magnetic correlations. Long wavelength ferromagnetic (0,0,0) fluctuations gradually start developing below 150 K [315] and show a $\underline{q}$ dependent energy width $\Gamma(q) \propto q$. These fluctuations are similar to paramagnons since, for $q \rightarrow 0$ fluctuations, one expects

$$1 - URe\chi_0(\underline{q};\omega) \sim 1 - U\rho(\mu) + O\left(\frac{q}{k_F}\right)^2 \quad (19.175)$$

and

$$Im\chi_0(\underline{q};\omega) \sim \frac{\pi}{4}\rho(\mu)\left(\frac{\hbar\omega k_F}{\mu q}\right). \quad (19.176)$$

Within RPA, this behavior of the quasi-particle susceptibility produces the pre-critical fluctuations appropriate to a damped, but conserved, order parameter. The paramagnon fluctuations could be responsible for the $T^3 \ln T$ term observed in the specific heat and the value of the large Korringa ratio [277,278]. In addition to the (0,0,0) fluctuations, there are two types of antiferromagnetic modes in the spectrum. A quasi-elastic peak starts developing below $T = 30$ K centered at momentum transfers of (0, 0, 1) and is characterized by a large energy width $\Gamma \sim 5$ meV [316,317]. This signifies the presence of a rapidly changing short-ranged magnetic order in which the two uranium ions in the unit cell are antiferromagnetically coupled. This produces an antiferromagnetic correlation between the spins along the c-axis and a ferromagnetic correlation between the spins in the basal plane. The intensity of the inelastic peak decreases as $\underline{q}$ is varied away from (0, 0, 1). The width of the peak $\Delta q \sim \xi^{-1}$ provides a measure of the correlation length ξ. The anisotropy of the correlation length in UPt_3 suggests that fluctuating in-plane ferromagnetic correlations are present for temperatures as large as $T = 100$ K. The second type of antiferromagnetic fluctuations is associated with a wave-vector of $(\frac{1}{2}, 0, 1)$ which develops at $T = 18$ K and peaks at a frequency of the order of 0.2 meV. Magnetic ordering occurs at this wave-vector below the critical temperature of $T_N = 5$ K, as evidenced by the development of magnetic Bragg peaks at $(\frac{1}{2}, 0, 1)$, since the magnetic Bragg peak intensities are (apart from the temperature dependent Debye–Waller factor) proportional to the square of the sub-lattice magnetization. The magnetic moments are in the basal plane of the hexagonal structure and are aligned parallel to the propagation vector $\underline{Q} \equiv (\frac{1}{2}, 0, 0)$. The magnitude of the ordered moments found through neutron diffraction experiments are anomalously small. The measured ordered moment only attain the value of 0.02 μ_B per U ion just above the superconducting transition temperature T_c [289]. The temperature dependence of the intensity of the magnetic Bragg peak of UPt_3 is shown in Fig. 19.50. The temperature dependence of the magnitude of the sub-lattice magnetization $\underline{M}_{\underline{Q}}(T)$ follows the scaling law

$$|\underline{M}_{\underline{Q}}(T)| \propto (T_N - T)^{\beta} \quad (19.177)$$

for $T_N > T$, where $\beta \sim 0.5$. The value of the critical exponent is not consistent with the values expected for three-dimensional ordering of localized magnetic moments. Similar values of β are also found

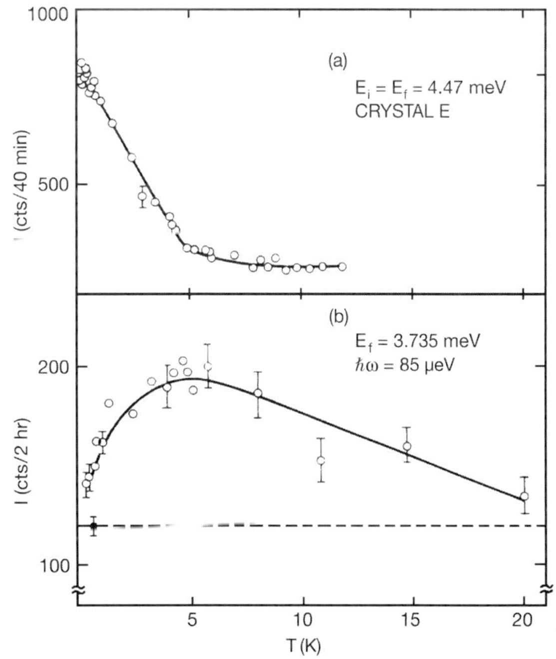

Fig. 19.50. (a) The temperature dependence of the elastic scattering intensity for UPt_3 at $q = (\frac{1}{2}, 0, 1)$. For purposes of comparison, (b) shows the temperature dependence of scattering with energy transfers of 85 μeV at $q = (0.52, 0, 0.99)$. [After Aeppli et al. [318]]

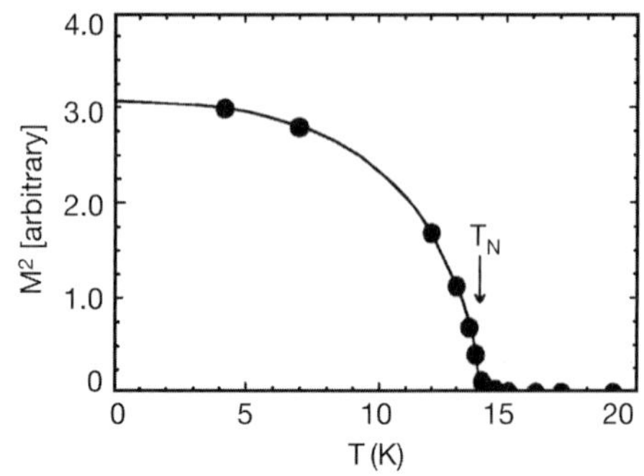

Fig. 19.51. The temperature dependence of the intensity of the $(0, 0, \frac{1}{2})$ antiferromagnetic Bragg peak in UPd_2Al_3. [After Petersen et al. [343]]

in other systems with anomalously low magnetizations such as URu_2Si_2 at ambient pressure. The critical exponents β found for other magnetically ordered heavy-fermion systems with larger moments such as UNi_2Al_3, UPd_2Al_3 and U_2Zn_{17}, have values close to 0.35 which are within the range expected for three-dimensional ordering of localized moments. For comparison, we show the temperature dependence of the magnetic Bragg peak intensity of UPd_2Al_3 in Fig. 19.51. The correlation lengths associated with the Bragg peaks of UPt_3 remain finite above and below T_N, and are of the order of 250–500 Å [318]. As the cross-section for elastic Bragg scattering is proportional to the energy conserving Dirac delta function $\delta(\omega)$, measurements of the elastic Bragg peaks involve windows of energy transfers which includes the point $\omega = 0$. Thus, the Bragg intensity also contains contributions from low-energy critical scattering, which have finite correlation lengths. Since the Bragg peaks were observed to have finite correlation lengths, questions are raised as to the nature of the ordering whether it is quasi-static short-ranged ordering, or whether it is long-ranged ordering that is interrupted by the presence of defects [319]. Muon spin resonance experiments confirmed the presence of magnetic ordering at $T_N \sim 5$ K [288,320]. As the temperature is reduced from just above the superconducting transition temperature, the intensity of the antiferromagnetic Bragg peaks shows a sudden change of slope as it starts to decrease. This indicates that the antiferromagnetic ordering coexists and competes with the superconductivity [321].

Inelastic neutron scattering experiments on UBe_{13} [322,323] at $T \sim 10$ K show an approximately $\underline{q}$ independent relaxational paramagnetic response with a large quasi-elastic width $\Gamma = 14$ meV. The approximate $\underline{q}$ independence of the inelastic scattering cross-section is indicative of the localized nature of the magnetic fluctuations. At energy transfers less than 2 meV [313], there is another roughly $\underline{q}$ independent quasi-elastic component to the spectrum, of width $\Gamma \sim 1.6$ meV at $T = 1.0$ K, which decreases to 1.0 meV at $T = 0.6$ K. If these smaller values of $\Gamma(0)$ are combined with the specific heat coefficient γ, the product $\gamma\,\Gamma(0)$ has a value of 1.7 which is closer to the values found for Ce heavy-fermion compounds. Antiferromagnetic like correlations are apparent in the quasi-elastic spectrum at momentum transfers $(\frac{1}{2}, \frac{1}{2}, 0)$ below a temperature of 30 K [324], but no long-ranged magnetic order was found.

Neutron scattering experiments on URu_2Si_2 show an instability to an antiferromagnetically ordered state below $T_N = 17.5$ K with average ordered magnetic moments of only $\mu \sim 0.03\mu_B$ [325] or $0.023\mu_B$ per U [326]. The value of the ordered moment increases when pressure is applied, becoming as large as $0.25\,\mu_B$ per U atom at $P = 1.3$ GPa [225]. The measured value of the critical exponent for magnetic order parameter is $\beta \sim 0.5$ at ambient pressure. However, for URu_2Si_2, the comparison of NMR and neutron diffraction experiments under pressure [224] indicate that magnetism only occurs in a small temperature dependent fraction of the sample volume and that, if the temperature dependence of the volume is accounted for, the critical exponent β falls within the normal range. Initial reports suggested that the correlation length ξ remained finite and was of the order 200 to 400 Å. Later, it was found that the correlation length is sample dependent and that the Bragg peaks of the highest quality samples were resolution limited [327]. The ordering found is consistent with a Type I antiferromagnetic structure in which the spins align parallel in planes perpendicular to the tetragonal c-axis and are anti-parallel between planes. Since "hidden ordering" has been suggested as the cause of the large specific heat jump at T_N, neutron diffraction measurements were performed using polarized beams and applied fields. However,

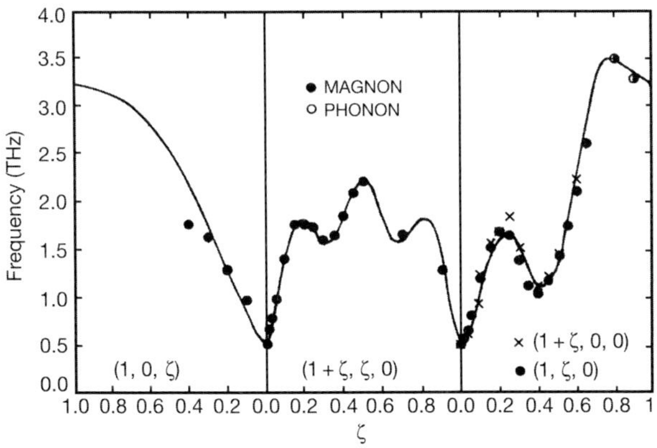

Fig. 19.52. The dispersion relation of a sharp magnetic excitation observed in URu_2Si_2. The dispersion relation is shown in the $(1, 0, \zeta)$, $(1 + \zeta, \zeta, 0)$ and $(1, 0, \zeta)$ directions. It should be noted that the dispersion relation does not go soft, as is expected for isotropic antiferromagnetic spin waves, but instead shows a gap. The existence of a gap is indicative of magnetic anisotropy. [After Broholm et al. [329]]

these experiments show no evidence of the "hidden ordering" [296,297].

At temperatures of the order of $T = 50$ K, the inelastic scattering spectrum of URu_2Si_2 [328] shows the presence of a broad quasi-elastic response of width $\Gamma \sim 6$ meV, suggesting that the spins have relaxational dynamics. Below the Néel temperature, the inelastic scattering spectra show that both sharp spin-wave-like excitations and broad relaxational spin-fluctuations coexist in the antiferromagnetic state. At $T = 1$ K, there are well defined dispersive spin-wave peaks along the $(1, q, 0)$ direction that have a minimum excitation energy of 1.8 meV at the Antiferromagnetic zone boundary [329] which is located at $q = 0$. The dispersion relation for these magnetic excitations is shown in Fig. 19.52. The longitudinal character of the spin-wave excitations, the Ising-like nature of the magnetic ordering, and the existence of a gap, all indicate that there is considerable anisotropy in the magnetic interactions at low temperatures. The spin-wave excitations are rapidly damped out above the Néel temperature T_N. The spectrum of magnetic fluctuations in the energy range between 4 and 12 meV appears to have a short correlation length, of the order of the lattice spacing, in that the intensity decreases slowly as $\underline{q}$ is varied away from $\underline{Q} = (1, 0, 0)$. The response in this energy range seems to be associated with Stoner excitations of the gas of heavy quasi-particles. Since URu_2Si_2 has a large magnetic anisotropy, and as it may play an important role in the formation of superconductivity, inelastic scattering measurements were made with the view of identifying the source of the anisotropy [330]. Although crystal-field-like features were observed at the high energies of 49, 99 and 158 meV, these features had widths of 64, 36 and 89 meV, which are comparable with the splittings. This indicates that the f states in URu_2Si_2 are much more strongly hybridized with the conduction band states than in the Ce heavy-fermion compounds, and perhaps, have a large mixed valent character. Like UPt_3, the intensity of the antiferromagnetic Bragg peak in URu_2Si_2 is diminished at the onset of the superconducting transition [331].

The inelastic neutron scattering spectra of cerium based heavy-fermion compounds are different from the uranium based heavy-fermion compounds in that they show clear evidence of crystal field splittings. An example is given by the inelastic neutron scattering spectra of $CeAl_3$ [234], which is shown in Fig. 19.53. Like the uranium compounds, the cerium compounds also show the existence of an almost $\underline{q}$ independent quasi-elastic peak which develops at low temperatures, but in the low temperature limit, the widths attain smaller values. For example, the width found in $CeCu_2Si_2$ [229,332] is of the order of 1 meV, and is less than 0.5 meV for $CeAl_3$ [308].

In $CeCu_6$, the magnetic fluctuations exhibit an Ising-like anisotropy along the b-axis. Two different components of quasi-elastic magnetic scattering were identified [333,334]. The first component which becomes apparent at temperatures below 10 K, has an almost $\underline{q}$ independent quasi-elastic response with a width that follows a $\sqrt{T}$ temperature variation and

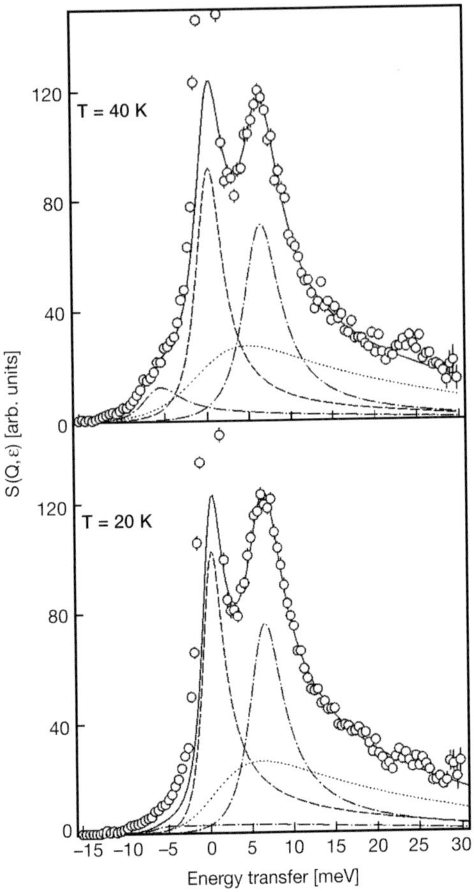

Fig. 19.53. The inelastic neutron scattering cross-section of $CeAl_3$. The spectrum shows a quasi-elastic contribution and a crystal field excitation. [After Goremychkin et al. [234]]

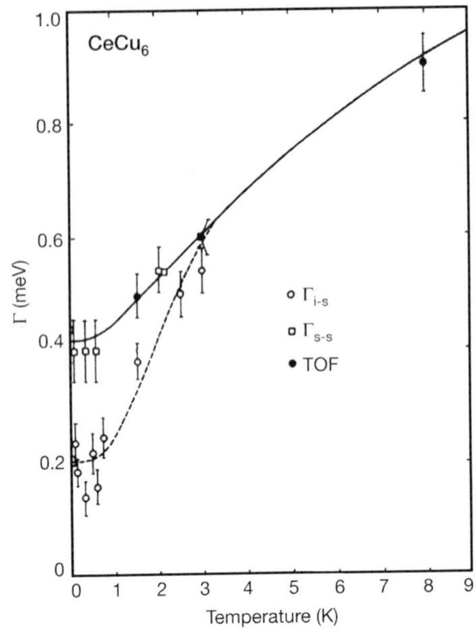

Fig. 19.54. The temperature variation of the single-site and inter-site contributions to the line-width $\Gamma(T)$, for $CeCu_6$. [After Rossat-Mignod et al. [333]]

has a residual width, $\Gamma(0)$, of the order of 0.4 meV. The second component consists of a narrower $\underline{q}$ dependent quasi-elastic peak that starts developing below $T = 3$ K. For the Ce compounds, unlike the U compounds, the magnitude of the residual $\underline{q}$ independent quasi-elastic width seems to correlate well with the large low temperature specific heat γ values. The magnetic spectrum of $CeCu_6$ has a $\underline{q}$ dependent component which shows a peak at a frequency of the order of 0.2 meV, and the $\underline{q}$ dependence is indicative of incommensurate antiferromagnetic fluctuations with the pair of propagation vectors $(1 \pm \delta, 0, 0)$, where $\delta = 0.15$. The correlation length, ξ, in $CeCu_6$

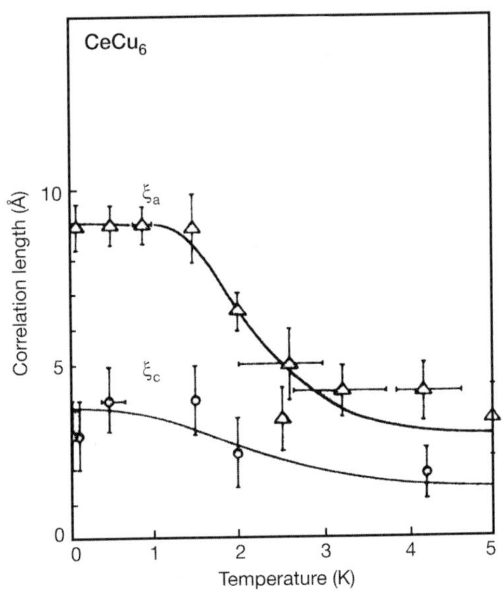

Fig. 19.55. The temperature variation of the correlation lengths $\xi(T)$ for inter-site correlations along the a and c-axes, for $CeCu_6$. [After Rossat-Mignod et al. [333]]

was found to be anisotropic, and saturates below $T = 1.5$ K to values of the order of 4 Åalong the c-axis and 10 Åalong the a-axis [333], indicating that the order is definitely short-ranged. The temperature dependence of the various widths are shown in Fig. 19.54, and the temperature dependence of the correlation lengths are shown in Fig. 19.55. The $\underline{q}$ dependent component of the total spectrum is estimated to be much smaller than the local component by a factor of about 10.

The introduction of Au impurities substitutionally for Cu can drive paramagnetic $CeCu_{6-x}Au_x$ to an antiferromagnetic quantum critical point. The critical concentration is $x_c \approx 0.1$. The thermodynamic and transport properties of the disordered compound have anomalous temperature dependences in the vicinity of the quantum critical point. For example, the specific heat contains a $T \ln T$ term and the low temperature resistivity is linear in T [335]. For $x > x_c$, the system has incommensurate order with three pairs of Bragg peaks located very near to the values $(1 \pm 0.15, 0, 0)$, identified in the stoichiometric compound [336]. At the quantum critical point, the Bragg peaks broaden into ridges which connect into a "butterfly" shaped critical line [337]. The extended region of q space were this type of critical scattering occurs is shown in Fig. 19.56. The anomalous critical exponents at the quantum critical point led Rosch et al. [338] to speculate that the magnetic fluctuations were two-dimensional, despite the underlying electronic system being three-dimensional. At the quantum critical point, the width of the magnetic response may be expected to vanish at $T = 0$, for an isotropic magnetic system, as the magnetic excitations soften. This has led Schröder et al. [339] to examine the energy dependence for $\underline{q}$ on the critical "butterfly" and look for a $\hbar\omega/k_B T$ scaling, with the functional form

$$Im\chi(\underline{Q}; \omega, T) \sim T^{-\alpha} g(\omega/T) , \qquad (19.178)$$

where $g(\omega/T)$ is a universal scaling function. The scaling behavior is shown on a double logarithmic plot, for the best fit value of the exponent $\alpha = \frac{3}{4}$. The scaled data are shown in Fig. 19.57. This critical scaling behavior seems not to be just confined to the region of critical scattering since the static suscepti-

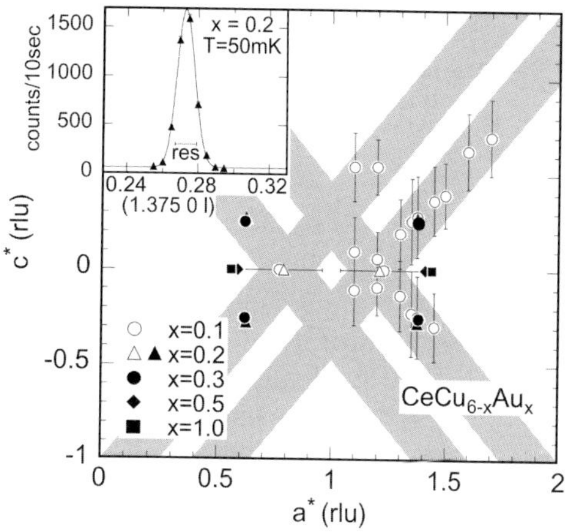

Fig. 19.56. The region of q space where critical scattering occurs for $Ce(Au_xCu_{6-x})$. The region of critical scattering forms a "dog's bone" or "butterfly". The *inset* shows a resolution limited Bragg peak for $x = 0.2$ and $T = 50$ mK. [After Stockert et al. [337]]

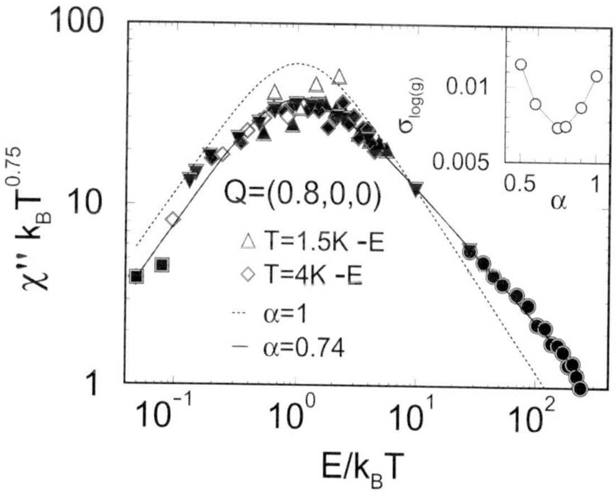

Fig. 19.57. The ω/T scaling of $Im[\chi(\omega, T)]$ on the critical "butterfly" for $Ce(Au_xCu_{6-x})$. The scaling is shown on a log–log plot. The *inset* shows that the optimal value of α is close to 0.74. [After Schroeder et al. [339]]

bility also follows a modified Curie–Weiss form

$$\chi(T)^{-1} = \chi(0)^{-1} + aT^{\alpha} \qquad (19.179)$$

with the exponent $\alpha = \frac{3}{4}$, as opposed to the Curie–Weiss value, $\alpha = 1$. Similar scaling has also been observed in the scattering cross-section of non-Fermi-liquid disordered compound $UCu_{5-x}Pd_x$ [340], with $x \sim 1$ and $x \sim 1.5$. However, in this case, the best critical exponent is $\alpha = \frac{1}{3}$ which indicates that the form of the scaling is non-universal. The magnetic response for $x \sim 1$ shows a $\underline{q}$ dependence which is similar to that of the stoichiometric, antiferromagnetic, compound with $x = 0$. However, for $x = 1.5$, this $\underline{q}$ dependence was not observed [341]. This was interpreted as signalling the cross-over from a quantum critical region to a quantum disorder region. The implications of these non-universal scaling results are not currently understood.

Neutron diffraction experiments on UPd_2Al_3 and UNi_2Al_3 show that these materials both undergo transitions to antiferromagnetic ordered phases [342]. UPd_2Al_3 orders magnetically at $T_N = 14.3$ K, and has a magnetic structure comprised of spins ferromagnetically aligned in the basal plane parallel to the a direction. The sheets of ferromagnetically aligned spins are coupled antiferromagnetically with the sheets of spins in the neighboring planes. This arrangement corresponds to the wave-vector $\underline{Q} = (0, 0, \frac{1}{2})$. The quasi-elastic peaks centered on the magnetic ordering vectors become resolution limited Bragg peaks in the antiferromagnetic phase. The Bragg peak intensities are consistent with the saturated magnetic moment having the magnitude of 0.85 μ_B per U. The critical exponent β is close to the value 0.326 expected for a three-dimensional Ising system. The inelastic spectrum clearly shows the existence of a spin-wave mode with a peak position that disperses almost linearly from an energy transfer close to zero at $\underline{q} = (0, 0, \frac{1}{2})$ to 3 meV at $\underline{q} = (0, 0, 0.6)$. Thus, to within the experimental resolution, the spectrum shows the existence of dispersive spin-wave like excitations [343]. The line shape resembles an inelastic peak and, unlike conventional spin-waves, is subjected to significant broadening which is of the order of 1 meV. Higher resolution measurements show that as the temperature is lowered below T_N, in addition to the peak due to the propagating spin-waves, a quasi-elastic peak with a $\underline{q}$ dependent width of about 0.5 meV can be resolved [344,345].

At the superconducting transition of UPd_2Al_3, the intensity of the $(0, 0, \frac{1}{2})$ Bragg peak shows that the sub-lattice magnetizationundergoes a sharp but small (1%) decrease with decreasing temperature. This indicates that superconductivity and magnetism coexist and compete [342,346]. A superconducting gap was observed in the spin-wave spectrum below T_c [347]. The gap is marked by an inelastic peak at the magnetic Bragg vector which shows a sharp increase in intensity below T_c. The gap switches on more rapidly than expected from the BCS theory and saturates at a value of $\Delta(0) \geq 0.36$ meV. This results in the ratio $2\Delta(0)/k_B T_c \geq 4.6$ instead of the value 3.5 expected from BCS theory [348]. The observation of the superconducting gap in the inelastic neutron scattering spectra makes UPd_2Al_3 unique, and the observation of any $\underline{q}$ dependence should enable the symmetry of the order parameter to be obtained by direct methods.

Magnetic ordering in UNi_2Al_3 is found to occur below a temperature $T_N = 4.6$ K. Neutron diffraction measurements [349] show incommensurate magnetic order with ordering wave vectors $\underline{Q} = (\frac{1}{2} \pm \delta, 0, \frac{1}{2})$ where $\delta = 0.11$ and the ordered moments are only 0.24 μ_B per U. The critical exponent was found to have the value $\beta \sim 0.345$ as expected for a three-dimensional $x - y$ system. The widths of the Bragg peaks correspond to long coherence lengths of the order of 400 Å. The field dependence of the magnetic structure suggests that the magnetic moments align along the a direction in the basal plane and that the magnitude and sense of the moments are modulated [350]. Inelastic neutron scattering experiments [351] were unable to identify any spin-wave like excitations but did show a low-energy quasi-elastic peak which is strongly localized around the incommensurate wave vector. The energy width of the quasi-elastic peak decreases with decreasing temperatures, saturating to a value of the order of 0.6 meV. However, a more recent study [352] has investigated the spectrum at energy transfers in the range of 2 to 10 meV. It was found that there was a ridge of high intensity on a line segment near the incommensurate ordering vectors $(\frac{1}{2} \pm \delta, 0, \frac{1}{2})$ and there was also a well localized maximum around the wave-vector $(0, 0, \frac{1}{2})$. These peaks have widths of the order of 6 meV and start developing at temperatures below 80 K. In fact, for energy transfers greater than

0.5 meV ($\approx k_B T_N$), the intensity of the commensurate peak dominates over the incommensurate peak. That is, above this energy, the magnetic fluctuations of incommensurate UNi_2Al_3 strongly resemble those of commensurate UPd_2Al_3. In contrast to UPd_2Al_3, no superconducting gap was observable in the magnetic excitation spectrum of UNi_2Al_3 below the transition temperature.

19.4 Properties of the Superconducting State

19.4.1 Thermodynamic Properties

The Specific Heat

The entropy of the gas of quasi-particles is given by the formula

$$S = -k_B \sum_{\sigma,\underline{k}} \left[(1 - f(E_{\sigma,\underline{k}})) \ln[1 - f(E_{\sigma,\underline{k}})] + f(E_{\sigma,\underline{k}}) \ln[f(E_{\sigma,\underline{k}})] \right] . \quad (19.180)$$

Therefore, for unitary phases, the quasi-particle contribution to the specific heat is given by

$$C_{qp}(T) = \frac{4}{T} \int_0^\infty \mathrm{d}E \rho_{qp}(E) \left(E^2 - \frac{T}{2} \frac{\partial \Delta_0^2}{\partial T} \right) \left(- \frac{\partial f}{\partial E} \right) , \quad (19.181)$$

which involves the average of the temperature derivative of the square of the quasi-particle energy, evaluated at the position of the normal state Fermi surface. Since, in the mean-field approximation, the square of the gap has a finite slope for T just below T_c and is zero above, the specific heat has a discontinuity at T_c. In BCS theory, the magnitude of the jump has the value given by $3.03\Delta_0^2(0)\rho(\mu)/T_c$. Thus, the value of the specific heat jump found in weak-coupling BCS theory when normalized by the normal state specific heat, is given by

$$\frac{\Delta C(T_c)}{C(T_c)} = \frac{C_s - C_n}{C_n} = \frac{12}{7\xi(3)} = 1.426 . \quad (19.182)$$

This ratio is a measure of the quantity

$$\frac{1}{2k_B^2 T_c} \left(\frac{\partial \Delta_0^2(T)}{\partial T} \right) \Bigg|_{T_c} \sim \left(\frac{\Delta_0(0)}{k_B T_c} \right)^2 . \quad (19.183)$$

The values of the specific heat jumps for strong coupling materials tend to be higher than the BCS value. For example the normalized jump for Pb is as large as 2.71. This trend is understood as being due to inelastic scattering processes [353, 354] which tend to suppress T_c more than $\Delta_0(0)$. With the exception of UBe_{13}, the heavy-fermion superconductors have normalized specific heat discontinuities which either are of the same order as or are smaller than the BCS ratio. The values of the specific heat jumps show large variations which are mainly due to differences in sample quality. In UPt_3, the largest specific heat jumps are found in the samples with the sharpest transitions and highest T_c's. In UPt_3 and URu_2Si_2, the transitions found are usually broad, while UBe_{13} shows the sharpest transition. The broadness of the transition initially found in UPt_3 was responsible for masking the double peak structure of the specific heat observed in higher quality samples. The deviation of the jump from the BCS value is indicative of anisotropic superconductivity as weak-coupling calculations indicate that the magnitude of the jump is diminished as the density of low-energy quasi-particle states increases.[8] Thus, for example, the normalized specific heat jumps for the pristine BW, ABM and polar triplet states are, respectively, calculated as $\frac{12}{7\xi(3)} \sim 1.426$, $\frac{10}{7\xi(3)} \sim 1.188$ and $\frac{20}{21\xi(3)} \sim 0.792$, and decrease along with T_c as the concentration of non-magnetic impurities is increased. The temperature dependence of the specific heats for various p-wave superconducting phases are shown in Fig. 19.58. In heavy-fermion superconductors, which generally are type II superconductors, the jump in the specific heat occurs at lower temperatures when magnetic fields, H, are applied. The jump marks the instability of the normal state to a vortex state. The relation between the temperature and field provides the temperature

[8] More precisely, assuming a separable pairing interaction, the normalized specific heat jump yields a measure of the angular variation of the gap over the Fermi surface. It is equal to the BCS value of the discontinuity times the ratio of the square of the second moment of the gap to the fourth moment.

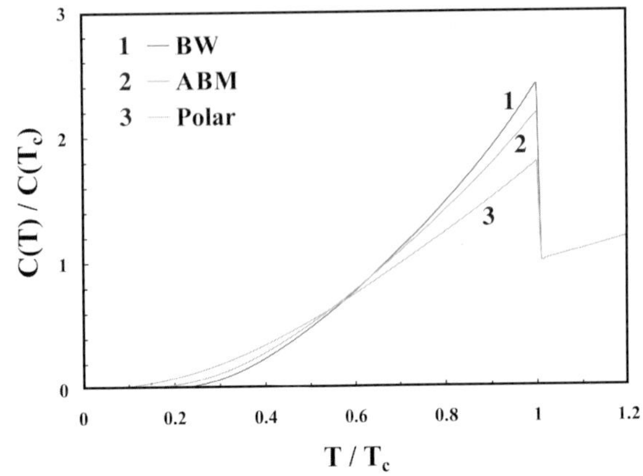

Fig. 19.58. The temperature dependent specific heat normalized to the normal state specific heat at T_c calculated for various p-wave superconducting phases. The magnitude of the normalized discontinuity at T_c decreases from ≈ 1.4, to 1.2 to 0.8 along the series BW, ABM and polar. The low temperature specific heat shows an exponential variation for the BW state, whereas the low temperature specific heats of the ABM and polar phases show power law temperature variations which are consequences of the types of nodes in the order parameters

dependence of the upper critical field $H_{c2}(T)$. A λ-like anomaly is expected to occur in the specific heat at a lower temperature where the applied field H is equal to the lower critical field $H_{c1}(T)$. At this temperature, the usual Abrikosov magnetization-field relation has an infinite slope which implies a similar temperature variation of the entropy, and produces the λ anomaly. Good examples of this second anomaly are found in V and Nb superconductors [355,356].

Low Temperatures

The low-energy power law energy dependence of the quasi-particle density of states could be expected to show up in power law temperature dependence of the low temperature electronic specific heat, for $T \ll T_c$. For these temperatures, the order parameter is expected to have saturated, and so then, if one considers the Fermi liquid as being well formed, the quasi-particle contribution is given by

$$C_{qp}(T) = \frac{4}{T} \int_0^\infty \mathrm{d}E \rho_{qp}(E) E^2 \left(- \frac{\partial f}{\partial E} \right) . \qquad (19.184)$$

This yield an exponentially activated behavior of the specific heat, $C_{qp}(T) \sim 9.17 \gamma T_c \exp[-\Delta_0 / k_B T\ ]$, for states like the BCS or BW states which have no nodes; a T^3 variation for the low temperature specific heat in the ABM phase with isolated nodes and a T^2 variation for the polar (p_z) or $d_{2z^2-x^2-y^2}$ phase with lines of nodes. However, the experimentally determined low temperature specific heats of the superconducting heavy-fermion uranium compounds are described by

$$C(T) = \gamma_s T + \epsilon_s T^n , \qquad (19.185)$$

where γ_s is sample dependent. The residual γ_s has often been attributed to either the existence of a normal component, or due to a gapless phase caused by pair breaking impurities. For example, in UPt_3 the coefficient of the residual linear T term takes values between 265 and 56 mJ mole^{-1} K^{-2} depending on sample quality [14]. This suggests that between 50 % and 16 % of the normal state Fermi surface is ungapped for these samples. The value of the exponent n is 2 for UPt_3, whereas $n = 3$ for UBe_{13} and UPd_2Al_3. The specific heat of URu_2Si_2 shows a quadratic temperature variation [357], and the smallest γ_s value reported is 4 mJ mole^{-1} K^{-2} and the largest is 13 mJ mole^{-1} K^{-2}. The smaller magnitude of the ratio γ_s/γ for URu_2Si_2 is perhaps not surprising since it has been estimated that about 30% of the Fermi surface remains ungapped below the 17.5 K transition. In $CeCu_2Si_2$, the temperature dependence of the specific heat below T_c is non-exponential, and can be fitted via a T^2 variation close to the critical temperature but shows a T^3 variation at low temperatures down to 50 mK. A residual value of γ_s of the order of 24 mJ mole^{-1} K^{-2} has been reported [358]. This number suggests that only 4% of the original Fermi surface is ungapped in the superconducting state of $CeCu_2Si_2$. The values of the exponents, n, suggests that there are nodes in the superconducting gap, however, the pair breaking effects of impurities may alter the temperature dependence [154–156] and may also lead to a gapless phase. This is consistent with the observation that materials with the broadest transitions tend to have the largest values of γ_s. As the heavy-fermion superconductors may be in the vicinity of quantum critical point, the Fermi liquid may not have been fully formed before the superconducting transition occurs as is true for UBe_{13}, and as the specific heat provides a measure of the density of all low-energy excitations and not just quasi-particles, it may prove difficult to disentangle the various factors which affect the temperature dependence of the specific heat.

The Critical Fields

The thermodynamic critical field H_c of a superconductor is determined by equating the energy of the magnetic field to the difference in free energies between the superconducting and normal states

$$V_o \frac{H_c^2(T)}{8\pi} = F_n - F_s \,, \tag{19.186}$$

where V_o is the volume. At $T = 0$ this reduces to the condensation energy of the superconducting state

$$V_o \frac{H_c^2(T)}{8\pi} = U_n - U_s = \frac{1}{2}\rho(\mu)\int \frac{d\Omega}{4\pi} \,\text{trace}\, \Delta(\underline{k})\Delta^\dagger(\underline{k}) \,. \tag{19.187}$$

Here, $\rho(\mu)$ is the single-particle density of states at the Fermi energy and $\Delta(\underline{k})$ is the gap function. For an s-wave superconductor at finite temperatures $F = U - TS$ with $U_n - U_s = \frac{1}{2}\rho(\mu)\Delta_0^2(T)$ and the temperature dependent gap is determined from

$$\frac{1}{\rho(\mu)V} = \int_0^{\hbar\omega_c} d\xi \, \frac{\tanh(E/2k_BT)}{E} \,, \tag{19.188}$$

where $-V$ is the potential for scattering a pair of electrons (assumed constant in BCS theory), ω_c is a cutoff frequency (usually the Debye frequency), $E = \sqrt{\xi^2 + \Delta_0^2}$, and the entropy of the superconducting state is

$$S_s = -2k_B\rho(\mu)\int_0^{\hbar\omega_c} d\xi \, \big[f(E)\ln f(E) + (1 - f(E))\ln(1 - f(E))\big] \,, \tag{19.189}$$

where $f(E) = 1/(e^{E/k_BT} + 1)$.

When solved numerically, in either the strong or weak coupling limit, the solution of the above equations differs from

$$H_c(T) = H_c(0)\left[1 - \left(\frac{T}{T_c}\right)^2\right] \tag{19.190}$$

by no more than a few percent. Meissner flux expulsion and zero resistivity disappear at H_c and the normal state is recovered. (It is interesting to note that Eq. (19.190), a good approximation to the BCS result, follows from a two-fluid model of superconductivity proposed by Gorter and Casimir in 1934. In this model the free energy of the normal state is taken to be $-\frac{1}{2}\gamma T^2$ and the free energy of the superconducting state is a constant [359].)

In a type-I superconductor the only critical field is H_c discussed above. In a type-II superconductor, quantized flux vortices appear in fields greater than the lower critical field H_{c1}, and superconductivity persists to fields as high as the upper critical field H_{c2} where the normal cores of the vortex lattice overlap and the zero resistance state is destroyed. The creation of the vortex state becomes energetically favorable when the Ginzburg–Landau parameter $\kappa > 1/\sqrt{2}$ ($\kappa = \lambda/\xi$ where λ is the penetration depth and ξ is hereafter the coherence length. When $\kappa \gg 1$ (an appropriate limit for the heavy-fermion superconductors) the Ginzburg–Landau equations can be solved to give

$$H_c = \frac{\Phi_0}{2\sqrt{2\pi}\mu_0\lambda\xi} \,, \tag{19.191}$$

$$H_{c1} = H_c\frac{\ln\kappa}{\sqrt{2}\kappa} = \frac{\Phi_0}{4\pi\mu_0\lambda^2}\ln\left(\frac{\lambda}{\xi}\right) , \tag{19.192}$$

and

$$H_{c2} = \sqrt{2}\,\kappa\, H_c = \frac{\Phi_0}{2\pi\mu_0\xi^2} \,. \tag{19.193}$$

Here, Φ_0 is the flux quantum (similar results for anisotropic systems are summarized elsewhere [360]). In the high κ limit generally appropriate for the heavy-fermion superconductors, H_c is very nearly the geometric mean of H_{c1} and H_{c2}. In fact, there are three Ginzburg–Landau parameters which depend on T, l/ξ (where l is the mean free path), and the degree of anisotropy of the impurity scattering. However, the various κ_i differ from κ by no more than 20% [361], and we will not pursue the matter here.

Many early theories of the upper critical field focused on the maximum attainable field associated with a single limiting factor. An example of such a limiting factor, frequently encountered in the heavy-fermion superconductors, is the field dependence of the normal state free energy caused by the coupling to the paramagnetic spins. Since the spin susceptibility of spin-singlet superconductors vanishes at $T = 0$, the Pauli limiting field $H_p(0)$ is defined as the field at which the free energy of the normal state with spin susceptibility χ_n equals the condensation energy of the superconducting state. At $T = 0$, the definition takes the form

$$\frac{1}{2}\chi_n H_p^2(0) = \frac{1}{2}\rho(\mu)\Delta_0^2(0) . \tag{19.194}$$

H_p, sometimes known as the Chandrasekhar [362] or Clogston [363] limit, can be significantly enhanced when spin-orbit coupling is strong. At $T = 0$

$$H_{p,\mathrm{so}}(0) = \left(\frac{\hbar}{6\lambda_{\mathrm{so}}\Delta(0)}\right)^{1/2} H_p(0) , \tag{19.195}$$

where λ_{so} is the spin-orbit coupling parameter. Since the zero temperature limit of the spin susceptibilities of spin-triplet superconductors are generally expected to be either non-zero or anisotropic, the upper critical field may be expected to exceed $H_p(0)$ in spin-triplet phases. In particular, for a spin-triplet superconductor, a field applied parallel to the direction of the Cooper pairs spin will not be subject to the Pauli limit but fields with components in the perpendicular directions will produce pair breaking [364].

Perhaps the most commonly used and successful framework to interpret $H_{c2}(T)$ measurements on "classic" type II superconductors is that developed by Werthamer, Helfand, Hohenberg, and Maki (WHHM) [365, 366]. WHHM solved the Ginzburg–Landau equations incorporating both orbital and spin paramagnetic effects in addition to both non-magnetic and spinorbit scattering. The resulting H_{c2} curves are qualitatively similar to the shape of Eq. (19.190) featuring a linear temperature dependence below T_c that gently rolls over to a constant as $T \to 0$ (in other words, $H_{c2}(T)$ exhibits negative curvature or none at all). Since these predictions, which work so well with classic type II superconductors, generally fail (except in the broadest terms) with heavy-fermion superconductors, we will not pursue this approach here but refer the reader to the overview and compilation of formulas presented by Orlando et al. [367]. Strong coupling can also cause significant deviations from the behaviors discussed above and we refer the reader to the overview of Carbotte [126]. Critical field anisotropy can originate from an anisotropic Fermi surface (sometimes characterized by an effective mass tensor [360]), an anisotropic order parameter, or magnetism for example.

Specific predictions for the upper critical fields resulting from anisotropic order parameters (or combinations thereof) have been made by a number of authors. In many cases, such as the p-wave ABM state for example [368, 369], these order parameters lead to a characteristic anisotropy of H_{c2} even though the temperature dependence remains similar to that of the WHHM model. In other cases the temperature dependence is also affected, and can result in the positive curvature which is so inconsistent with the WHHM model [370, 371]. A good overview of these issues can be found in the review article by Sigrist and Ueda [156].

The appearance of a non-uniform superconducting state, independently predicted in the mid 1960s by Fulde and Ferrell [372] and Larkin and Ovchinnikov [373], below $T/T_c \approx 0.55$ and in fields near H_{c2} has sometimes been invoked to explain the unusual shapes of the upper critical fields exhibited by the heavy-fermion superconductors. The so-called FFLO state is one in which the internal magnetic field breaks a number of the spin-singlet Cooper pairs, and produces a uniform flow of the condensate. This state is stabilized when the current of the Cooper pair condensate is compensated by a counter flow of the quasi-particle gas. This FFLO state is suppressed by impurities and results in an additional first order phase transition to the (relatively more uniform) vortex lattice state [374].

Many features of the models discussed above are consistent with the heavy electrons themselves entering the superconducting state. For example, standard thermodynamics shows that [375]

$$\left(\frac{\Delta C}{T}\right)\bigg|_{H_{c2}(T)} = \left(\frac{\partial H_{c2}}{\partial T}\right)^2 \left[\left(\frac{\partial M}{\partial H}\right)_s - \left(\frac{\partial M}{\partial H}\right)_n\right] . \tag{19.196}$$

Table 19.11. The superconducting parameters of some heavy-fermion superconductors. Values separated by slashes characterize *a*-axis/*c*-axis anisotropy (* under pressure)

Material	T_c (K)	$\frac{\Delta C(T_c)}{C(T_c)}$	$H_{c2}(0)$ (T)	$-\frac{dH_{c2}}{dT}\|_{T_c}$ (T/K)	$H_{c1}(0)$ (mT)	Refs.
UPt_3	0.55	1.0	2.5 / 2.0	4.5 / 7.7	3.0	[376,477,489]
$CeCu_2Si_2$	0.69	1.27	2 / 2.4	23	2.3	[386,477,489]
UBe_{13}	0.9	2.5	12	42	4.5	[386,477,489]
URu_2Si_2	1.5	0.82	13 / 3	4 / 14	1.4	[477,489]
UPd_2Al_3	2.0	1.2	3.0 / 3.6	4.3	1.0	[477,489]
UNi_2Al_3	1.0	0.5	1.5	1.4	1.4	[477,489]
$CeRhIn_5$	2.1*	0.36	~ 16	14		[59,516]
$CeIrIn_5$	0.4	0.76	9.3 / 5.3	4.8 / 2.5		[517]
$CeCoIn_5$	2.3	4.5	11.7 / 5.0	24.0 / 11.4	2.5 / 9.5	[517,518,520]
$CeIn_3$	0.18*		0.45	3.2		[65,524]
$CePd_2Si_2$	0.52*		2	13		[524,526]
$CeRh_2Si_2$	0.35*		.35	1.4		[69]
$CeCu_2Ge_2$	0.7*		2	11		[529]
$PrOs_4Sb_{12}$	1.85	3	2.2	1.9	23	[531,535,536]
U_2PtC_2	1.47			9		[537]

Thus, the large initial slope of the upper critical field and the low values of T_c are related to the large effective masses of the electrons forming the Cooper pairs. Parameters characterizing the critical field curves of the heavy-fermion superconductors are collected in Table 19.11.

The upper critical field of UPt_3 reflects the three underlying superconducting states meeting a tetracritical point (at 0.39 K and 0.44 T) which manifests as a sharp kink in $H_{c2}(T)$ when fields are applied along the *c*-axis. The anisotropy of $H_{c2}(T)$ reverses sign near 0.2 K: above (below) this temperature H_{c2} is larger (smaller) with fields applied along the *c*-axis than with fields applied within the basal plane. There have been fewer measurements of the lower critical field of UPt_3 owing, in part perhaps, to the difficulty of these measurements. All of the measurements are consistent with a sharp increase in slope near the transition between the lower temperature B phase and higher temperature A phase though the existence and size of the anisotropy is somewhat less certain. An increase in slope is consistent with the appearance of a second order parameter which would increase the condensation energy. A deeper discussion of the critical fields of UPt_3 can be found in the recent review article by Joynt and Taillefer [376].

The upper critical field of the best single crystal samples of UBe_{13} shows a broad region of positive curvature setting in near $T_c/2$ which has been interpreted as evidence for the FFLO state discussed above [377,378]. Sharp features in $H_{c2}(T)$ (or, more correctly, the irreversibility curve) of UBe_{13}, suggesting additional phases in either the superconducting or normal state have been reported [379] but as yet there is no consensus on the existence of such features. An anisotropy consistent with lines of zeros for the gap function has been reported [380,381] but has not been confirmed [382]. Despite the unusual shape of the upper critical field, the lower critical field of a polycrystal sample of UBe_{13} exhibits the usual quadratic temperature dependence [383], however, we are unaware of any measurements on single crystals at this time.

In $CeCu_2Si_2$, the upper critical field exhibits a broad, shallow maximum centered near 0.2 K the origin of which is uncertain but suggestions have included "(1) exchange-enhanced 'polarization fields' between (residual) Ce moments, (2) 'Kondo-type' pair breaking from these residual moments, and (3) competition between the (phonon-mediated) attraction and Coulomb repulsion between those slowly moving heavy-fermions whose velocity is already comparable to the velocity of sound" [22]. At low temperatures, the upper critical field with fields applied along the c-axis is about 20% larger than when they are applied along the a-axis. The size of this anisotropy decreases monotonically as T_c is approached. In earlier studies, the anisotropy vanished near T_c near which the slopes of the critical field curves were equal [22]. More recently, however, a small anisotropy has been observed to exist very close to T_c where the slopes vary from 23 T/K with fields along the a-axis to 18.5 T/K with fields along the c-axis [384]. The few measurements of the lower critical field of $CeCu_2Si_2$ show a short region of positive curvature just below T_c [383,385].

19.4.2 Transport Properties

At sufficiently low temperatures, transport properties are determined by the properties of the quasi-particle excitations and, unlike thermodynamic properties, are essentially affected by scattering. If the temperature is so low that inelastic scattering from low-energy collective fluctuations is frozen out, then elastic impurity scattering may be expected to provide the dominant scattering mechanism. However, it is not appropriate to treat the impurity scattering in the Born approximation since it leads to a scattering time that has the same energy dependence as the quasi-particle density of states and which would lead to similar temperature dependences as in the normal state. This is in contradiction to experimental observation. Hence, as pointed out by Pethick and Pines [158], it appears as if the impurity scattering is resonant.

Thermal Conductivity

The electronic contribution to the thermal conductivity can be calculated via linear response theory. The thermal conductivity can be expressed in terms of the heat flux correlation function. The thermal conductivity tensor $\kappa_{i,j}(T)$ can be written as the limit $\omega \to 0$ of the expression [387]

$$\kappa_{i,j}(T) = -\frac{1}{\hbar T}\left(\frac{\partial Im\, K^R_{i,j}(\omega)}{\partial \omega}\right)\Bigg|_{\omega=0}, \qquad (19.197)$$

where

$$\begin{aligned} K^R_{i,j}(\omega) = \frac{\hbar^2}{8m^2}\int\frac{dE_1}{2\pi}\int\frac{dE_2}{2\pi}\frac{(E_1+E_2)^2}{E_1-\hbar\omega-E_2-i\delta} \\ \times\Big(f(E_2)-f(E_1)\Big)\sum_{\underline{k}} k_i k_j \,\text{trace} \\ \times\Big(ImG(\underline{k};E_1)\rho_4 ImG(\underline{k};E_2)\rho_4\Big)_{av} \end{aligned} \qquad (19.198)$$

and where the Trace is over the matrix Green's function $G(\underline{k};\omega)$ defined by Eq. (19.86). This expression is to be averaged over the distribution of impurities. This type of analysis produces the standard result[9]

$$\kappa_{i,j} = \frac{1}{T}\sum_{\sigma,\underline{k}} v_i(\underline{k})\, v_j(\underline{k})\tau(E_{\underline{k}})E^2_{\underline{k}}\left(-\frac{\partial f}{\partial E_{\underline{k}}}\right) \qquad (19.199)$$

in which $\underline{v}(\underline{k})$, $E_{\underline{k}}$ and $\tau(E_{\underline{k}})$, respectively, are the quasi-particle velocity, the quasi-particle energy and the quasi-particle lifetime.

Singlet Pairing

For the case of isotropic singlet pairing and in the Born approximation, the rate of scattering of a quasi-particle from state $\underline{k}$ to state $\underline{k}'$ by non-magnetic impurities is given in terms of the quasi-particle density of states by

$$\frac{1}{\tau(E_{\underline{k}})} = \frac{1}{\tau_0}\frac{\rho_{BCS}(E_{\underline{k}})}{\rho(\mu)}\left(1-\frac{|\Delta_0|^2}{E^2_{\underline{k}}}\right), \qquad (19.200)$$

where the normal state scattering rate due to a concentration of randomly distributed non-magnetic impurities, c_0, is given by

[9] This expression for the thermal conductivity of a superconductor was derived by O. Betbeder-Matibet and P. Nozières, using a Boltzmann equation approach [388]). This result is obtained when the self-energy due to impurity scattering is small compared with $k_B T$.

$$\frac{1}{\tau_0} = \frac{2\,\pi}{\hbar} c_0\,\rho(\mu)|U_0|^2 \,. \tag{19.201}$$

The quasi-particle velocity is defined by

$$\underline{v}(\underline{k}) = \frac{1}{\hbar}\frac{\partial E_{\underline{k}}}{\partial \underline{k}} = \frac{\hbar \underline{k}}{m}\left(\frac{e(\underline{k}) - \mu}{E_{\underline{k}}}\right) , \tag{19.202}$$

where the order parameter has been assumed to be approximately independent of the magnitude of $\underline{k}$. Hence, the coherence factor in the expression for the quasi-particle lifetime $\tau(E_{\underline{k}})$ will cancel with a factor of

$$\left(\frac{e(\underline{k}) - \mu}{E_{\underline{k}}}\right)^2 \tag{19.203}$$

in the expression for $v_i(\underline{k})\, v_j(\underline{k})$. On combining these expressions, and on replacing the sum over $\underline{k}$ by integrals over the solid angle and the magnitude of k, one finds that the conductivity tensor is given by

$$\kappa_{i,j} = \frac{2\tau_0}{T}\left(\frac{\hbar k_F}{m}\right)^2 \int \frac{d\Omega}{4\pi}\hat{k}_i\hat{k}_j \int_{-\infty}^{+\infty} de \frac{\rho^2(e)}{\rho_{BCS}(E)} E^2 \left(- \frac{\partial f}{\partial E}\right) . \tag{19.204}$$

The integral over the solid angle $d\Omega$ causes the off-diagonal components of the tensor to vanish, and yields equal diagonal components. The tensor is given by

$$\kappa_{i,j} \approx \delta_{i,j}\frac{4\tau_0}{3T}\left(\frac{\hbar k_F}{m}\right)^2 \rho(\mu) \int_{\Delta_0}^{\infty} dE\, E^2 \left(- \frac{\partial f}{\partial E}\right) . \tag{19.205}$$

In this expression, the factor of the quasi-particle density of states is absent as we have used the Born approximation for the quasi-particle lifetime. On evaluating the integral at low temperatures, where $k_B\, T \ll \Delta_0$, one finds

$$\kappa_{i,j}(T) = \delta_{i,j}\,\frac{4\,\tau_0}{3\,T}\left(\frac{\hbar k_F}{m}\right)^2 \rho(\mu)\,\Delta_0^2 \exp\left[- \beta\,\Delta_0 \right] . \tag{19.206}$$

Therefore, the quasi-particle component to the thermal conductivity tensor of a BCS superconductor is thermally activated and is isotropic.

For the case of anisotropic pairing, the Born approximation for the scattering rate is given in terms of the quasi-particle density of states by

$$\frac{1}{\tau(E_{\underline{k}})} = \frac{1}{\tau_0}\frac{\rho_{qp}(E_{\underline{k}})}{\rho(\mu)} , \tag{19.207}$$

since the angular integration over the final state momentum $\underline{k}'$ has caused the term in the coherence factor which is bi-linear in the order parameter to vanish. Since the cancelation between the coherence factor in the lifetime and the factors in the quasi-particle velocity does not occur for anisotropic pairing, the quasi-particle contribution to the thermal conductivity tensor is given by

$$\begin{aligned}\kappa_{i,j} &= \frac{2\,\tau_0}{T}\left(\frac{\hbar k_F}{m}\right)^2 \int \frac{d\Omega}{4\,\pi}\,\hat{k}_i\,\hat{k}_j \\ &\times \int_{-\infty}^{+\infty} de (e-\mu)^2 \frac{\rho^2(e)}{\rho_{qp}(E)}\left(- \frac{\partial f}{\partial E}\right) \\ &= \frac{4\,\tau_0}{T}\left(\frac{\hbar k_F}{m}\right)^2 \rho^2(\mu) \int \frac{d\Omega}{4\pi}\hat{k}_i\hat{k}_j \\ &\times \int_{|D(\hat{k})|}^{\infty} dE \frac{E\sqrt{E^2 - |D(\hat{k})|^2}}{\rho_{qp}(E)}\left(- \frac{\partial f}{\partial E}\right) .\end{aligned} \tag{19.208}$$

On interchanging the order of the integrations, one obtains the result

$$\kappa_{i,j} = \frac{4\tau_0}{T}\left(\frac{\hbar k_F}{m}\right)^2 \rho^2(\mu) \int_0^{\infty} dE \frac{E}{\rho_{qp}(E)}\left(- \frac{\partial f}{\partial E}\right) I_{i,j}(E) , \tag{19.209}$$

where the factor $I_{i,j}(E)$ describes the anisotropy, and is given by

$$I_{i,j}(E) = \int \frac{d\Omega}{4\pi}\hat{k}_i\hat{k}_j\sqrt{E^2 - |D(\hat{k})|^2} \,. \tag{19.210}$$

The integration only runs over the solid angles for which the argument of the square root is positive. Assuming the validity of this expression, one observes that there exists a possibility that the anisotropy of the thermal conductivity tensor may provide information about the locations of the nodes of the order parameter.

For illustrative purposes, we shall consider the d-wave order parameter with $m = 1$,

$$D(\hat{k}) = \Delta_0 \sin 2\theta \exp[i\varphi] \,. \tag{19.211}$$

This order parameter has a line of nodes in the equatorial plane and two isolated point nodes, one at each of the poles. Hence, the quasi-particle density of states is linearly proportional to E at low energies and is logarithmically divergent at $E = \Delta_0$. The thermal conductivity tensor only has diagonal components and is isotropic in the x–y plane. The anisotropy of the tensor is governed by the angular integrals

$$I_{z,z}(E) = \frac{\Delta_0}{2} \int d\theta \sin\theta \cos^2\theta \sqrt{\left(\frac{E}{\Delta_0}\right)^2 - \sin^2 2\theta} \,,$$
$$I_{x,x}(E) = \frac{\Delta_0}{4} \int d\theta \sin\theta \sin^2\theta \sqrt{\left(\frac{E}{\Delta_0}\right)^2 - \sin^2 2\theta} \,, \tag{19.212}$$

where the integration only runs over the ranges of θ where the argument of the square root is positive. For small E, these integrals are proportional to E^4/Δ_0^4 and E^2/Δ_0^2 respectively. The differences in the behavior occurs because, at low energies, the line of nodes has more weight than the point nodes and the factor of $\cos^2\theta$ in the z component of the tensor is limited to be smaller than $E^2/4\Delta_0^2$ while $\sin^2\theta \approx 1$. Hence, for temperatures such that $k_B T \ll 0.2\Delta_0$, one expects that $\kappa_{z,z}(T) \propto T^3$, while $\kappa_{x,x}(T) = \kappa_{y,y}(T) \propto T$. The different temperature dependence of the components of the thermal conductivity tensor reflect the presence of the line of nodes in the equatorial plane.

Any anisotropic behavior of $\kappa(T)$ is expected to be extremely difficult to observe experimentally, since a slight misalignment of the crystal, the presence of domains or grain boundaries, etc., may cause the measured thermal conductivity to be dominated by the largest component of the tensor.

Triplet Pairing

In the case of triplet pairing, the quasi-particle contribution to the thermal conductivity can be written in a form analogous to the thermal conductivity for anisotropic singlet pairing. The tensor can be expressed as

$$\kappa_{ij}(T) = \frac{\hbar^2 k_F^2}{m^2 k_B T^2}\rho(\mu) \int_0^\infty dE E \operatorname{sech}^2\frac{\beta E}{2}\tau(E) I_{i,j}(E) \,, \tag{19.213}$$

where $\tau(E)$ is the scattering time and $I_{i,j}(E)$ is given by

$$I_{i,j}(E) = \int \frac{d\Omega}{4\pi} \hat{k}_i \hat{k}_j \sqrt{E^2 - |\vec{d}(\hat{k})|^2} \,. \tag{19.214}$$

In the BW phase, the order parameter is nodeless and so, the thermal conductivity is diagonal, isotropic and exponentially activated. Thus, the thermal conductivity in the BW phase is similar to the thermal conductivity of the BCS state.

For the ABM state, the low temperature thermal conductivity is anisotropic. As the magnitude of the order parameter varies as $|\vec{d}(\hat{k})| = \Delta_0 \sin\theta$, the factor of k_i^2 in the diagonal components of the thermal conductivity tensor introduces an extra factor of E^2/Δ_0^2 in the low-temperature limit of the transverse components compared with the longitudinal component. The longitudinal thermal conductivity is estimated as

$$\kappa_\parallel(T) = \frac{\hbar^2 k_F^2}{k_B T^2 m^2}\rho(\mu) \int_0^\infty dE E \operatorname{sech}^2\frac{\beta E}{2}\tau(E)\frac{E^3}{|\Delta_0|^2} \tag{19.215}$$

and the transverse components are estimated as

$$\kappa_\perp(T) = \frac{\hbar^2 k_F^2}{k_B T^2 m^2}\rho(\mu) \int_0^\infty dE E \operatorname{sech}^2\frac{\beta E}{2}\tau(E)\frac{E^5}{|\Delta_0|^4} \,. \tag{19.216}$$

In the Born approximation, since $\rho_{ABM}(E)$ varies as E^2 and thus $\tau(E) \propto E^{-2}$, the longitudinal component of the thermal varies as $\kappa_\parallel(T) \propto T$ whereas the transverse components vary as $\kappa_\perp(T) \propto T^3$.

For the polar state, where $|\vec{d}(\hat{k})| = \Delta_0 \cos\theta$ which vanishes along the equatorial plane, the transverse and longitudinal components of the thermal conductivity tensor are given by

$$\kappa_\perp(T) = \frac{\hbar^2 k_F^2}{k_B T^2 m^2}\rho(\mu) \int_0^\infty dE E \operatorname{sech}^2\frac{\beta E}{2}\tau(E)\frac{E^2}{|\Delta_0|} \tag{19.217}$$

and

$$\kappa_{\|}(T) = \frac{\hbar^2 k_F^2}{k_B T^2 m^2} \rho(\mu) \int_0^\infty \mathrm{d}E E \operatorname{sech}^2 \frac{\beta E}{2} \tau(E) \frac{E^4}{|\Delta_0|^3} . \tag{19.218}$$

In the polar state, the quasi-particle density of states is given by $\rho_{p_z}(E) \propto E$, hence $\tau(E) \propto E^{-1}$. Thus, one has $\kappa_{\|}(T) \propto T^3$ and $\kappa_{\perp}(T) \propto T$. It should be noted that the anisotropy in the T dependence of $\kappa(T)$ is switched between the polar and the ABM states [389].

Since $\Delta_0 \gg k_B T$, the above results imply that the linear T term will dominate the experimentally determined temperature variation of the thermal conductivity of either a polar or an ABM state. In general, a linear T dependence is expected for states with either point or line nodes. Also, if the Lorentz number found in the normal state is constant, as is expected if impurities provide the dominant scattering mechanism, the thermal conductivity in the superconducting state should have a magnitude which is very similar to the normal state value. The above results were obtained using the Born approximation. However, it has been argued that, even in the normal state, it is not valid to treat impurity scattering within the Born approximation since the impurity scattering may be resonant, i.e., close to the unitarity limit.

If the impurity scattering is resonant and $k_B T \gg \hbar/2\tau$, then in the ABM phase one has $\tau(E) \sim \tau_0$, while $\tau(E) \propto E \ln^2 E$ in the polar phase. Thus, for resonant scattering, the leading T dependences for the components of the thermal conductivity for the ABM phase are expected to be $\kappa_{\|}(T) \propto T^3$ and $\kappa_{\perp}(T) \propto T^5$. The polar phase has the opposite anisotropy to the ABM phase, however, additional factors of $\ln^2 T$ multiply the leading low-temperature powers of T. That is, for the polar phase, $\kappa_{\|}(T) \propto T^5 \ln^2 T$ and $\kappa_{\perp}(T) \propto T^3 \ln^2 T$. In general, it is expected that for strong scattering by dilute impurities such that the T-matrix approximation is valid [159,390], the orientationally averaged thermal conductivity $\kappa(T)$ should follow a T^3 law if the order parameter has point nodes or should follow a T^2 law if the order parameter has line nodes.

However, the resonant impurity scattering may be sufficiently large to cause the superconducting quasi-particle density of states to be finite at zero energy. In this case, it has been shown [391] that the low-energy superconducting quasi-particles may be strongly localized, even when the normal state quasi-particles are not. This results in a universal expression for the d.c. limit of the electrical conductivity of the superconducting state which is finite and independent of the disorder. As a consequence, for $k_B T \ll \hbar/2\tau$, a Wiedemann–Franz like relation holds [392]. This relation shows that, in the very low temperature limit, the superconductor's thermal conductivity $\kappa(T)$ should also be linear in T, irrespective of the node structure. Furthermore, the largest component of $\kappa_{i,j}(T)$ can also be independent of impurity scattering for certain types of order parameter [392]. These results can be understood by recognizing that the effect of strong impurity scattering is to randomize the Fermi surface, thereby acting to restore its approximate spherical symmetry.

The electronic contribution to the thermal conductivity of UBe_{13} does not follow the exponentially activated law expected from BCS theory but shows a T^2 variation in the superconducting state at relatively high temperatures (between 0.8 and 0.1 K) which could be indicative of lines of nodes in the order parameter [28]. However, this assignment is contrary to the assignment of point nodes made on the basis of the T^3 variation observed in the specific heat. Experiments on some samples show that, below 100 mK, $\kappa(T)$ varies linearly with T

$$\kappa(T) = \alpha_s T , \tag{19.219}$$

where $\alpha_s = 0.03$ mW K^{-2}cm^{-1} [266]. The temperature dependence of κ for the two different temperature regimes found in this sample of UBe_{13} are shown in Fig. 19.59. However, other samples do not show any noticeable linear temperature dependences [393]. This seems to indicate that the thermal conductivity is extremely sensitive to the quality of the sample.

The temperature dependence of the thermal conductivity of UPt_3 also does not follow an exponential activation law but has a temperature variation which can be fit by the sum of a T and a T^2 term [29,394,395]

$$\kappa(T) = \alpha_s T + \beta_s T^2 . \tag{19.220}$$

The value of the coefficient of the linear T term, α_s, was found to vary from sample to sample, varying

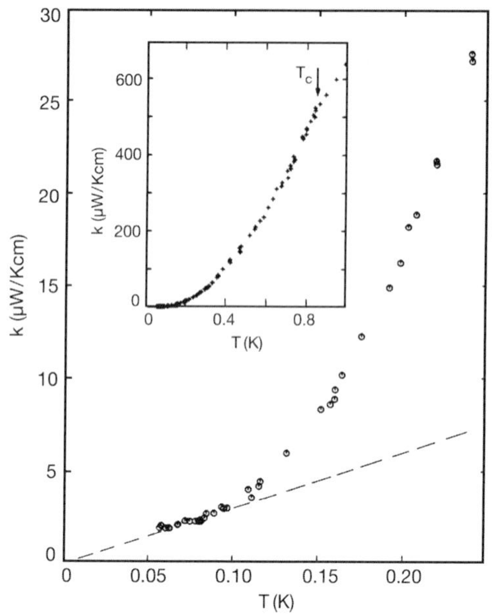

Fig. 19.59. The temperature dependence of the thermal conductivity κ of UBe_{13}, at low temperatures. The *dashed line* represents a possible linear term $\kappa(T) = \alpha_s T$. The *inset* shows the temperature dependence of κ in a wider temperature range which extends up to T_c. [After Ravex et al. [266]]

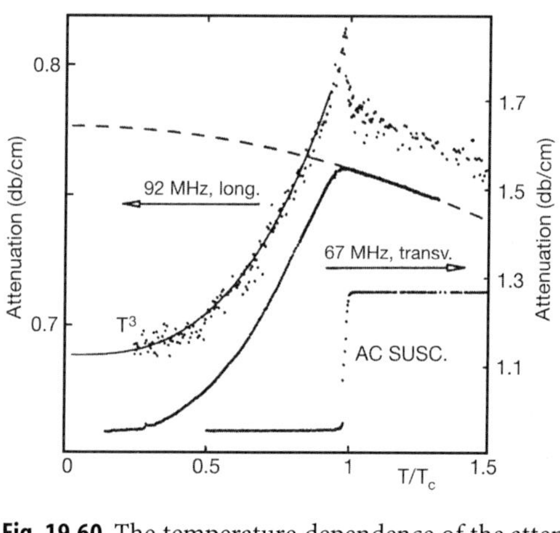

Fig. 19.60. The temperature dependence of the attenuation rate for longitudinal and transverse sound in UPt_3. [After Müller et. al [402]]

from 0.55 to 12 mW K^{-2} cm^{-1}. The magnitude of the coefficient β_s of the T^2 term in UPt_3 can be as large as 60 mW K^{-3} cm^{-1} [395] which is two orders of magnitude greater than the corresponding coefficient in UBe_{13}. The strength of the T^2 term in the $\kappa(T)$ of UPt_3 is also much larger than the corresponding term found from the $\kappa(T)$ of $CeCu_2Si_2$ [396], for which $\beta_s \sim 1.8$ mW K^{-3} cm^{-1} and $\alpha_s \sim 0.28$ mW K^{-2} cm^{-1}. Although an anisotropy of $\kappa(T)$ was detected in UPt_3 [397], the anisotropy is fully developed in the normal state and $\kappa(T)$ becomes more isotropic for temperatures below the second superconducting transition. An analysis of the measurements favors an order parameter with a form similar to $Y_1^2(\theta, \varphi)$ but did not conclusively identify the symmetry of the gap. Figure 19.60 gives results for the attenuation in UPt_3.

The lack of agreement with expectations based on simple quasi-particle theories and elastic impurity scattering is perhaps not surprising, since the validity of the quasi-particle description of the normal state of both UBe_{13} and $CeCu_2Si_2$ is questionable.

Ultrasonic Attenuation

Measurements of the ultrasonic attenuation should allow the temperature dependence and the anisotropy of the gap to be determined. Within BCS theory, and even with *s*-wave pairing, the attenuation coefficients for longitudinal and transverse sound are different. This difference occurs since a longitudinal sound wave merely produces a time-dependent scalar potential which acts on the electrons, whereas a transverse sound wave produces a transverse electromagnetic field and hence, is subject to the Meissner effect. The Meissner effect has the effect of reducing the coupling strength between the electrons and the transverse phonons and, therefore, reduces the transverse attenuation rate just below the superconducting transition [398,399]. This mechanism is held to be responsible for an almost discontinuous drop in the attenuation rate for shear waves in Al [400] and In [401] at T_c. Below the transition temperature, the evolution of the superconducting gap leads to a further reduction of the attenuation rate. This behavior is in direct contrast with the ultrasonic attenuation of the heavy-fermion superconductors UPt_3 [44, 402–404], UBe_{13} [26] and URu_2Si_2 [405], where peaks have been found in the ultrasonic attenuation just below T_c.

A sharp peak in the longitudinal attenuation coefficient was observed in UBe_{13} in the form of a λ-like anomaly at a temperature just below the superconducting transition temperature of 0.8 K [26]. A similar sharp λ-like peak for the attenuation of longitudinal sound waves in UPt_3 was found at a temperature of about 500 mK [402]. A less pronounced and more rounded peak was also seen in the attenuation of transverse sound waves in UPt_3 [403,404]. The peak in URu_2Si_2 [405] is of the form of a rounded λ-like anomaly and has a maximum at a temperature of $\sim$ 1.25 K, which is below the superconducting transition T_c of 1.35 K. Initially, it had been speculated that these peaks have their origin in the collective modes of the superconducting order parameter, similar to the collective modes observed in ^{3}He. However, in this case, the analogy with ^{3}He is flawed. Since the crystals do not posses continuous rotational symmetries, but only have the discrete rotational symmetries contained in their point groups, the collective excitations are not Goldstone modes. If spin-orbit scattering is strong in these U compounds, the frequencies of the collective modes of the order parameter could have magnitudes similar to the superconducting gaps. Furthermore, these collective modes are only expected to be well defined [132,406] at very low temperatures in very clean materials. Since the peaks in the attenuation coefficient occur at temperatures just below the critical temperatures, and since there appears to be a residual zero energy density of states due to impurity scattering, the collective modes are expected to be overdamped, therefore, this explanation seems unlikely. An alternate viewpoint was provided by Coffey [407], who related the peak in the attenuation to the enhanced quasi-particle masses which makes the velocity of sound comparable to the Fermi velocity. In this picture, the conductivity enters the expression for the longitudinal attenuation rate, thereby producing a screening of the electron–phonon coupling. As in the normal state conductivity, the mass enhancement factor produces a sharp low-frequency Drude peak in the conductivity. Although this effect is also predicted for a BCS state, it is stronger in anisotropic superconducting states where the low-energy quasi-particle density of states is finite, and so, the screening is more effective.

Low Temperatures

As a precursor to the discussion of the anisotropy in $\alpha_{\underline{q}}(T)$, we shall first examine the ultrasonic attenuation in an ordinary singlet superconductor. For a singlet superconductor, the attenuation rate is approximated by the Fermi Golden Rule expression

$$\alpha_{\underline{q},\hat{\epsilon}} = 2\pi|\lambda_{\underline{q},\hat{\epsilon}}|^2 \sum_{\underline{k},\sigma} \left(1 + \frac{(e(\underline{k})-\mu)(e(\underline{k}+\underline{q})-\mu) - D(\underline{k})\, D(\underline{k}+\underline{q})}{E_{\underline{k}} E_{\underline{k}+\underline{q}}}\right) \times \left(f(E_{\underline{k}}) - f(E_{\underline{k}+\underline{q}})\right)\delta\left(E_{\underline{k}+\underline{q}} - E_{\underline{k}} - \hbar\omega_{\underline{q},\hat{\epsilon}}\right) , \quad (19.221)$$

where $\lambda_{\underline{q},\hat{\epsilon}}$ is the electron–phonon coupling strength, and $\omega_{\underline{q},\hat{\epsilon}}$ is the frequency of a phonon with wave-vector $\underline{q}$ and polarization $\hat{\epsilon}$. The summation over $\underline{k}$ can be separated into an integral over the magnitude of k (or $e(\underline{k})$) and an integral over the polar angle, θ, defined with respect to $\underline{q}$. For an s-wave superconductor, the angular integral can be replaced by an integral over the magnitude of $e(\underline{k}+\underline{q})$ where the range of integration runs between $e(k-q)$ and $e(k+q)$. The range of this transformed integral is shown in Fig. 19.61.

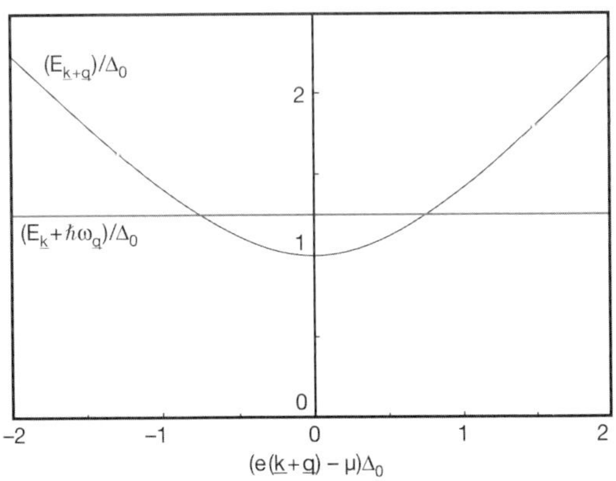

Fig. 19.61. The range of the integration over energy in Eq. (19.222), representing the integration over the polar angle θ

Conservation of energy $E_{\underline{k}+\underline{q}} = E_{\underline{k}} + \hbar\omega_{\underline{q}}$ demands that the only $\underline{k}$ values that contribute to the attenuation are those which are close to the equatorial plane of the Fermi surface, if $v_F > c$ and $k_F > q$, as shown in Fig. 19.62. Since the term in the coherence factor

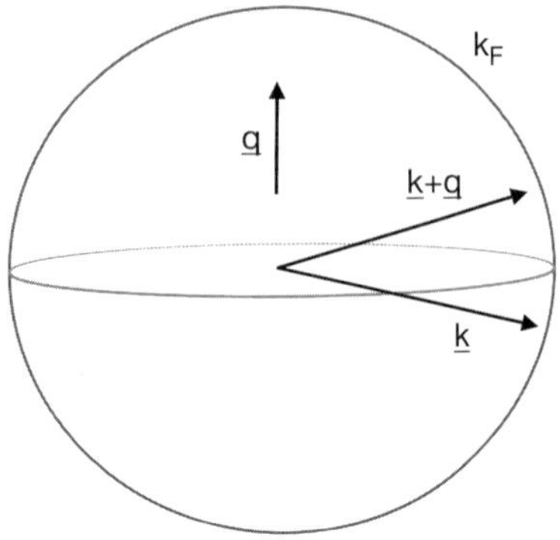

Fig. 19.62. Typical reciprocal space geometry of electrons involved in the attenuation of sound with wave vector $\underline{q}$, in the clean limit. The electrons which are primarily involved in the attenuation process are located near the equator of the Fermi surface

proportional to $e(\underline{k}+\underline{q})-\mu$ is anti-symmetric under the integration, it vanishes, thereby yielding

$$\alpha_{\underline{q},\hat{\epsilon}} = \frac{m^2}{\pi q \hbar^4}|\Lambda_{\underline{q},\hat{\epsilon}}|^2 \int_0^\infty \mathrm{d}e(\underline{k}) \int_{e(k-q)}^{e(k+q)} \mathrm{d}e(\underline{k}+\underline{q}) \times \left(1 - \frac{D(\underline{k})D(\underline{k}+\underline{q})}{E_{\underline{k}}E_{\underline{k}+\underline{q}}}\right) \times (f(E_{\underline{k}}) - f(E_{\underline{k}+\underline{q}}))\delta(E_{\underline{k}+\underline{q}} - E_{\underline{k}} - \hbar\omega_{\underline{q},\hat{\epsilon}}) , \tag{19.222}$$

which, since $\omega_{\underline{q}}$ is small, simplifies to

$$\begin{aligned}\alpha_{\underline{q},\hat{\epsilon}} &= \frac{m^2\,\omega_{\underline{q}}}{\pi\, q\, \hbar^4} \mid \Lambda_{\underline{q},\hat{\epsilon}} \mid^2 \int_0^\infty \mathrm{d}e(\underline{k}) \int_{e(k-q)}^{e(k+q)} \mathrm{d}e(\underline{k}+\underline{q}) \\ &\times \left(\frac{e(\underline{k}) - \mu}{E_{\underline{k}}}\right)^2 \left(-\frac{\partial f(E_{\underline{k}})}{\partial E_{\underline{k}}}\right) \\ &\times \delta(E_{\underline{k}+\underline{q}} - E_{\underline{k}} - \hbar\omega_{\underline{q},\hat{\epsilon}}) \\ &= \frac{m^2\,\omega_{\underline{q}}}{\pi\, q\, \hbar^4} \mid \Lambda_{\underline{q},\hat{\epsilon}}\mid^2 \int_{\Delta_0}^\infty \mathrm{d}E_{\underline{k}} \left(-\frac{\partial f(E_{\underline{k}})}{\partial E_{\underline{k}}}\right) .\end{aligned} \tag{19.223}$$

In this expression, the coherence factors have canceled with the quasi-particle density of states. Thus, in a BCS superconductor, the attenuation coefficient $\alpha(T)$ should follow the law

$$\alpha(T) \propto f(\Delta_0(T)) , \tag{19.224}$$

where $f(x)$ is the Fermi function and $\Delta_0(T)$ is the superconducting gap. Thus, for temperature far below T_c, the attenuation coefficient should be thermally activated

$$\alpha(T) \sim \exp\left[-\beta\Delta_0\right] . \tag{19.225}$$

The above derivation emphasizes that, if $\underline{q}$ is defined as the polar axis, it is the electrons which have $\underline{k}$ vectors near the equatorial plane of the spherical Fermi surface that provide the dominant contribution to the attenuation of sound waves. Furthermore, since the contribution from this region of $\underline{k}$ space is weighted by a function depending on $E_{\underline{k}}/k_B T$, the presence of nodes of the order parameter within this equatorial plane may result in a power law temperature dependence of $\alpha_{\underline{q}}(T)$.

To illustrate the origin of the anisotropy, consider the clean limit of a polar-like superconductor with a line of nodes in the equatorial plane of the Fermi surface [408]. When the propagation vector $\underline{q}$ is aligned precisely along the polar axis $\vec{d}$, the attenuation process primarily involves excitations on the line of nodes and, therefore, the attenuation coefficient resembles that of the normal state. However, if the direction of $\underline{q}$ is slightly rotated, the attenuation will only involve the low energy excitations near two isolated points on the line of nodes. Hence, one expects that the attenuation coefficient $\alpha_{\underline{q}}(T)$ will vary linearly with T. The coefficient of the linear T term will depend sensitively on the angle $(\vec{d}\,.\,\hat{q})$. For an ABM-like state, one may expect that if the $\underline{q}$ vector is oriented in a plane exactly perpendicular to the direction of $\vec{d}$, the low-energy excitations will predominately be near the isolated point nodes. In this case, one expects that in this case, the attenuation coefficient should vary linearly with T. For other orientations the density of low energy excitations is reduced, so $\alpha_{\underline{q}}(T)$ is expected to be proportional to T^2.

Most of the heavy-fermion superconducting materials are not in the clean limit, with the exceptions of UPt_3 and $CeCoIn_5$. However, similar types of anisotropic results and pseudo-selection rules have been obtained for systems with resonant impurity scattering and high concentrations [159,409,410]. For example, in the dirty limit, the attenuation of transverse sound waves [411] is proportional to

$$\alpha_{\underline{q},\hat{\epsilon}}(\omega) \propto \omega^2 \sum_{\underline{k}} \left(- \frac{\partial f}{\partial E_{\underline{k}}} \right) \tau_{\underline{k}}(E) \left(\frac{\xi(\underline{k})}{E_{\underline{k}}} \right)^2 \left[\hat{k}.\hat{q} \right]^2 \left[\hat{k}.\hat{\epsilon} \right]^2 , \tag{19.226}$$

where $\xi(\underline{k}) \equiv e(\underline{k}) - \mu$. For a superconductor with a line of nodes and at low temperatures, the largest contribution comes from $\underline{k}$ values close to the nodes. Due to the geometric factors, the attenuation is small if either the propagation vector $\underline{q}$ or the polarization $\hat{\epsilon}$ is directed towards the nodes and is larger for other orientations. Hence, even in the dirty limit, the anisotropy can be used to determine the positions of the nodes in the superconductors order parameter.

In the heavy-fermion superconductors UBe_{13} and UPt_3, the low temperature attenuation coefficients have power law temperature dependences, suggesting that the gaps are anisotropic and have nodes. The exponent of the power law temperature-dependence not only varies with the direction of propagation but also depends upon the polarization of the sound waves.

For the hexagonal compound UPt_3, Müller et al. [402] measured the attenuation coefficient for both longitudinal and transverse sound propagating along the c-axis. At temperatures sufficiently below T_c, the transverse [402] and longitudinal waves [27,30] seemed to show a T^2 dependence of the attenuation rate. However, a fit to the attenuation of longitudinal sound over a wider temperature range, yields a T^3 dependence [402]. The attenuation of transverse sound waves propagating in the basal plane is dependent upon the direction of the polarization. Transverse sound waves propagating in the basal plane but with polarizations also in the basal plane, showed an attenuation which is linear in T [403]. This is in direct contrast with the attenuation of transverse waves propagating in the basal plane, but with the polarization along the c-axis, which have an attenuation coefficient that shows a T^2 or T^3 [404] temperature dependence. Since the polarization dependence vanishes in the normal state, this is strong evidence that the anisotropy is a feature of the superconducting state. However, the nature of the order parameter is not easily discerned from these experiments. For a polar-like state with lines of nodes, the longitudinal attenuation should have a quasi-linear T dependence or, for an ABM-like state with point nodes, the attenuation should vary as T^2 [159,409]. The experimental observations of the longitudinal attenuation rate, therefore, favor an interpretation based on ABM-like states [408,412] and rule out polar-like states. However, the T dependences found in the transverse sound wave experiments of Shivaram et al. [403] do not fit the predictions for ABM-like states with point nodes but instead fit the theoretical results for polar-like states with lines of nodes.

For UBe_{13}, measurements below 0.9 K [26] showed that the attenuation of longitudinal sound waves followed a T^2 variation which implies the existence of an ABM-like state with point nodes.

Electromagnetic Response

The electromagnetic absorption spectra of an s-wave superconductor has been calculated by Mattis and Bardeen [413], using the Kubo formulae. On assuming non-conservation of momentum, either caused by the presence of impurities or by the samples surface, the absorption spectrum is given by the integral

$$\frac{\sigma_s}{\sigma_n} = \int_{-\infty}^{\infty} \mathrm{d}E \rho_{BCS}(E) \rho_{BCS}(E - \hbar\omega) \times \left(\frac{f(E - \hbar\omega) - f(E)}{\hbar\omega} \right) \left(1 + \frac{\Delta_0^2}{E(E - \hbar\omega)} \right) \tag{19.227}$$

for $\hbar\omega > 2\Delta_0$. At $T = 0$, the absorption is evaluated in terms of the complete elliptical integral functions $E(x)$ and $K(x)$ as

$$\frac{\sigma_s}{\sigma_n} = \left(1 + \frac{2\,\Delta_0}{\hbar\omega} \right) E\left(\frac{\hbar\omega - 2\Delta_0}{\hbar\omega + 2\Delta_0} \right) - \left(\frac{4\,\Delta_0}{\hbar\omega} \right) K\left(\frac{\hbar\omega - 2\Delta_0}{\hbar\omega + 2\Delta_0} \right) \tag{19.228}$$

for $\hbar\omega > 2\Delta_0$. At $T = 0$ where no thermally excited quasi-particles are present, the optical conductivity is zero for frequencies in the range $2\Delta_0 > \hbar\omega > 0$. At higher frequencies, the incident photon can be absorbed through a process in which a Cooper pair is broken and two quasi-particles are created. Since the

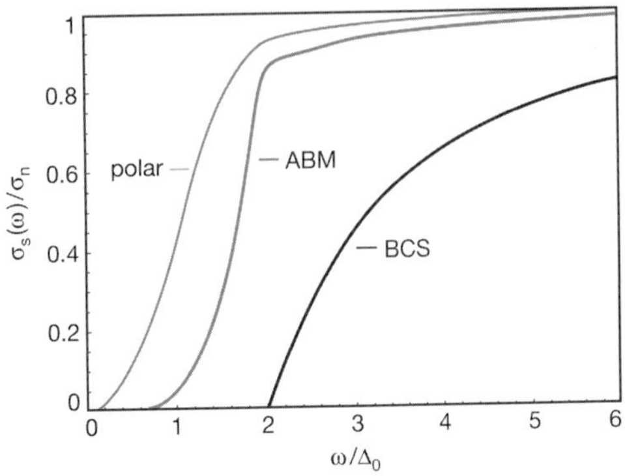

Fig. 19.63. The frequency dependence of the real part of the $T = 0$ optical conductivity calculated for various superconducting phases. The BCS phase conductivity exhibits a threshold at $\hbar\omega = 2\Delta_0$. The ABM and polar phases have nodes in the order parameter, therefore, the respective conductivities should show power law variations at low frequencies

coherence factors are reduced near the gap edge, the a.c. conductivity shows a continuous increase above the threshold frequency and approaches the normal state conductivity at frequencies much greater than the gap frequency. Thus, as seen in Fig. 19.63, the area under the theoretically predicted curve $\sigma_s(\omega)$ is smaller than found in the normal state. The ω variation of $\sigma(\omega)$ calculated by Mattis and Bardeen is in agreement with experiments on In [414]. There is also good agreement between the theoretical and experimentally observed temperature dependence of $\sigma(\omega)$ for example, as found from measurements of the microwave surface impedance of Zn [415].

As shown by Tinkham and Ferrell [416], the inequality $\sigma_s(\omega) < \sigma_n(\omega)$, which is valid for all finite frequencies, has the consequence that the optical sum rule

$$\int_0^\infty \mathrm{d}\omega\sigma(\omega) = \frac{\pi\, n\, e^2}{2\, m} \tag{19.229}$$

is not satisfied, if the integration is performed down to a finite cut off ϵ at the lower limit. Hence, the conductivity must have a zero frequency delta function spike in order to satisfy the optical sum rule. This spike represents the absorption of energy from d.c. electric fields which produce accelerated supercurrents. This spike is of special significance as causality connects the $\delta(\omega)$ contribution to the real part of the conductivity to a $1/\omega$ variation in the imaginary part of the conductivity. The existence of the $1/\omega$ term shows that supercurrent will flow in response to a uniform d.c. vector potential $\vec{A}$, thereby screening the external magnetic field. The ensuing non-local relation is similar to the (local) London relation

$$\vec{J}(\underline{r}) = -\frac{c}{4\pi\lambda_L^2}\vec{A}(\underline{r}) \,, \tag{19.230}$$

which involves the London penetration depth λ_L. Within the BCS theory, the penetration depth $\lambda(T)$ should have a leading exponential low temperature variation [417]. The calculated temperature dependence of the penetration depth is given by

$$\frac{\lambda(T)}{\lambda(0)} - 1 \propto T^{-\frac{1}{2}} \exp\left[-\frac{\Delta_0}{k_B T}\right] , \tag{19.231}$$

since the Fermi surface is completely gapped in s-wave superconductors.

For anisotropic superconductors which have nodes in the order parameter, one expects very different results. First, since the angular average of the order parameter vanishes, and since the tensorial nature of the conductivity has been ignored, the coherence factors are reduced to unity. Hence, the expression analogous to that of Mattis and Bardeen for the dirty limit of the absorption spectrum is just given by the convolution of the quasi-particle density of states

$$\frac{\sigma_s}{\sigma_n} = \int_0^{\hbar\omega} \mathrm{d}E\rho_{qp}(E)\rho_{qp}(E-\hbar\omega)\left(\frac{f(E-\hbar\omega)-f(E)}{\hbar\omega}\right) . \tag{19.232}$$

Second, due to the existence of nodes, the quasi-particle density of states is finite and, therefore, the conductivity is expected to be finite down to zero frequency. For an order parameter with point nodes, one expects that the above expression for the conductivity should vary as ω^4 at low frequencies while for an order parameter with line nodes, one expects an ω^2 variation. The electromagnetic absorption spectra for the BCS, ABM and polar states are shown in Fig. 19.63. For a more general description of electromagnetic absorption, incorporating the tensorial nature of the conductivity and the effects of impurity scattering, the interested reader is referred to the article of Hirschfeld et al. [418]. From an analysis of this type, Lee [391] found the surprising result that the $\omega \rightarrow 0$ limit of the conductivity has a value which is universal and independent of the impurity scattering rate. However, if the impurity scattering is sufficiently strong, Lee [391] has shown that the electronic states in the superconducting phase may be localized even though the electronic states in the normal state are not. The increased tendency for localization of the superconducting electrons is caused by the reduction of the effective dimension of the space in which the quasi-particles close to the nodes are free to move. The localization leads to the a.c. conductivity being dominated by the thermal activation of quasi-particle excitations across the mobility gap, despite any power law energy variation of the quasi-particle density of states near the Fermi energy.

The temperature dependence of the surface-impedance of UBe_{13} has been measured [419], and compared with theoretical predictions based on the BCS state. The value of Δ_0/k_BT_c was treated as an adjustable parameter. The experimentally determined surface-reactance shows a coherence peak just below T_c, which is well reproduced by Mattis–Bardeen theory [413], and less well so by the theory of Skalski et al. [420] which incorporates the effect of a finite spin-flip scattering rate caused by magnetic impurities. However, a single value of Δ_0/k_BT_c does not fit the data at all frequencies, which indicates that the order parameter corresponds to $l \neq 0$. In fact, subsequently anomalous structure (shown in Fig. 19.64) was identified [421] in the surface-resistance of UBe_{13} below 0.8 T_c, which was attributed to collective excitations of an $l \neq 0$ superconducting order parameter.

Since, as seen in Fig. 19.63, the spectral weight below twice the maximal gap $2\Delta_0$ is diminished below the normal state conductivity, the zero-frequency delta function present in $\sigma_s(\omega)$ is also expected to exist in the anisotropic superconducting phases, albeit with reduced weight. On employing the Kramers–Kronig relations, one can infer that the penetration depth should have a power law temperature dependence of the form

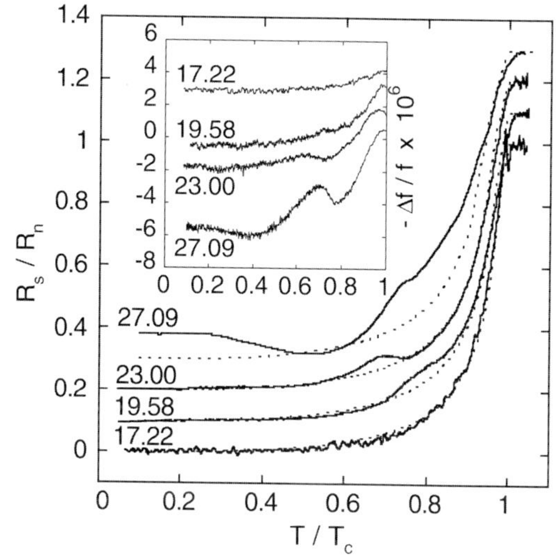

Fig. 19.64. The temperature dependence of the surface resistance of UBe_{13}, for various frequencies. The surface resistance was normalized to unity at T_c and zero at $T = 0$. Each curve is labelled by the measurement frequency in GHz. Anomalous structures are seen for temperatures in the interval $T_c > T > 0.8T_c$. The *dotted lines* are fits to the Mattis–Bardeen theory for an *s*-wave superconductor. The inset shows the resonant frequency shift. [After Feller et al. [421]]

$$\frac{\lambda(T)}{\lambda(0)} - 1 \propto \left(\frac{T}{T_c}\right)^n , \qquad (19.233)$$

where the exponent n depends on the node structure of the superconducting order parameter. Power law temperature variations of the penetration depth have been observed in heavy-fermion superconductors for example, a T^2 power law dependence has been observed in UBe_{13} [422]. By contrast, in UPt_3 the superfluid density measured by μSR experiments [423, 424] was also found to have a power law temperature dependence but one in which the exponent is anisotropic. Since the main contribution to the superfluid density is governed by the nodes in the superconducting order parameter, the magnetic field dependence of the penetration depth can be significant. Yip and Sauls [425] have suggested that this non-linear Meissner effect might be used to probe the momentum space positions of the nodes in the order parameter. However, Li et al. [426] find that in general, the non-local nature of the electromagnetic response does mask any non-linear Meissner effect, except under special circumstances.

Tunneling Measurements

Measurements of quasi-particle tunneling at sufficiently low temperatures are expected to yield the quasi-particle density of states $\rho_{qp}(E)$. We assume that the tunneling occurs between a normal metal with a relatively featureless density of state $\rho_N(E) \sim \rho(\mu)$ and a superconductor, and that the tunneling matrix element T is independent of $\underline{k}$. If a bias voltage V is applied across the junction, the tunneling current I is given by

$$I = \frac{2\pi}{\hbar} e |T|^2 \int_{-\infty}^{\infty} \mathrm{d}E \rho_N(E - eV) \rho_{qp}(E) \times \Big[f(E - eV) - f(E) \Big] . \quad (19.234)$$

The coherence factors do not enter the above expression [427], since the quasi-particle wave functions are reduced to the normal state wave functions in the barrier region. Hence, if the differential conductance σ is defined by

$$\sigma = \left(\frac{\mathrm{d}I}{\mathrm{d}V} \right) , \quad (19.235)$$

the differential tunneling conductance is evaluated as

$$\sigma(V) = \frac{2\pi}{\hbar} e^2 |T|^2 \rho(\mu) \int_{-\infty}^{\infty} \mathrm{d}E \rho_{qp}(E) \left(- \frac{\partial f}{\partial E} \right) \Bigg|_{E-eV} . \quad (19.236)$$

At sufficiently low temperatures, the expression for the tunneling conductance reduces to

$$\sigma(V) \propto \rho_{qp}(eV) . \quad (19.237)$$

Hence, a measurement of the voltage dependence of the quasi-particle tunneling conductance should yield the quasi-particle density of states [428]. Likewise, a similar analysis of quasi-particle tunneling between two superconductors should yield the joint density of states with features at the sum and difference of the characteristic energies of the superconducting density of states for the two materials. For two s-wave superconductors at finite temperatures, a discontinuity occurs in the tunneling current for a voltage V_+ corresponding to the sum of the two superconducting gaps, while a weak temperature dependent logarithmic singularity occurs in the current at the voltage V_- corresponding to the difference of the gaps [428,429]

$$eV_+ = \Delta_1 + \Delta_2 ,$$
$$eV_- = |\Delta_1 - \Delta_2| . \quad (19.238)$$

The observation of the quasi-particle density of states and, therefore, the characteristic van–Hove singularities and the low-energy variation, should provide information about the nature of the order parameter. However, it should be noted that the presence of the surface may locally distort the superconducting phase [430]. In this case, the tunneling matrix elements may become orientational dependent and the local density of states may differ from the bulk quasi-particle density of states.

The best tunneling measurements obtained to date are those on UPd_2Al_3. The junctions consisted of high quality UPd_2Al_3 films which were separated from the strong coupling superconductor Pb by a thin layer of Al [431]. A small magnetic field was applied in order to suppress the superconductivity of Pb. For temperatures below the superconducting transition of UPd_2Al_3, the tunneling conductance developed a minimum at zero bias voltages and a large peak structure at a voltage of the order of 0.24 mV. The large peak structure resembles the van Hove singularities often found in superconducting quasi-particle density of states at Δ_0 (see Figs. 19.23 and 19.24). On identifying this voltage with the gap energy, one finds that the universal ratio $\Delta_0(0)/k_B T$ is of the order of 3.71 [432]. The differential conductance also showed a fainter peak at a voltage of 1.25 mV, which was attributed to strong coupling corrections. These features in the density of states are assumed to originate from coupling to unusual antiferromagnetic spin excitations seen in inelastic neutron scattering experiments [348,433]. By contrast, the interpretation of tunneling experiments on $CeCu_2Si_2$

is not so clear cut. The tunnel junctions were constructed from $CeCu_2Si_2$ which were separated by a thin oxide layer from an overlayer of Pb. The measurements showed that, for temperatures below the T_c of $CeCu_2Si_2$, the tunneling conductance resembles the superconducting density of states of Pb but is superimposed with very small features that were attributed to a gapless superconducting quasi-particle density of states of $CeCu_2Si_2$ [434].

Point Contact Spectroscopy

The difficulty of fabricating good tunnel junctions and the small tunneling probabilities associated with oxide layers has hampered tunneling experiments. On the other hand, measurements of the differential conductance of point-contacts between a superconductor and metal are expected to produce larger currents. In this case, the Cooper pair wave function of the superconductor extends into the metal and Andreev reflection may occur [435] along with the usual single-particle transfer process. That is, an electron of momentum $\underline{k}$ incident on the boundary between the metal and superconductor may combine with a normal state electron of momentum $-\underline{k}$ to form a Cooper pair that subsequently resides within the superconducting condensate. This process can be viewed as an incident electron, with energy $E < \Delta_0$, inside the metal being scattered from the surface of the superconductor and emerging in the metal as a reflected hole, thereby transferring a net charge of $2e$ across the junction. The role of Andreev reflection in point contact spectroscopy on conventional s-wave superconductors was pointed out by Blonder et al. [436]. Since the conductance depends on an interface potential which in part models the decay of the pairing potential in the normal metal, the conductance smoothly evolves between the result for classic tunneling expressed by Eq. (19.236) and the extreme Andreev limit. In the extreme Andreev limit, the conductance of an s-wave superconductor–metal contact shows a zero bias peak which is twice as great as the normal state conductance of the junction. The magnitude of the peak height of the normalized conductance is two since a charge of $2e$ is transferred across the junction for each normal-metal electron with energy $E < \Delta_0$ that falls incident on the surface. The conductance recovers to the normal state value at higher voltages ($eV > \Delta_0$) as other electron transfer processes gradually take over. Bruder extended the analysis of Andreev reflection to the case of unconventional superconductors [437] and found that the energy of the peak is extremely dependent on the type of superconducting order parameter and the direction of the incident electron's momentum. In the usual geometry of point contact measurement, the electrons have varying angles of incidence with the boundary and, therefore, conductance measurements yield weighted orientational averages of functions of the order parameter.

Early Point contact tunnelling measurements were made on UPt_3 and UBe_{13} and various metals [438,439] but did not lead to reliable estimates of the gaps for the heavy-fermion superconductors. Later, measurements on UPt_3 [440–442] showed that distinct minima in the differential conductance–voltage relation occur for current flow along the c-axis but are generally absent for current flow in the basal plane. The differential conductance shows significant variation with the surface treatment. These results indicate that the superconducting order parameter has an anisotropic dependence on $\underline{k}$. The amplitude of the zero bias peak, when normalized to the conductance at large voltages, is an order of magnitude smaller than expected from theory. A similar discrepancy with the theoretical magnitude of the peak was found in URu_2Si_2 [443]. On defining the strength of the zero bias conductance peak as the area between $\sigma(V)$ in the superconductor and the normal state background, the strength vanished close to T_{c2}. The vanishing of the strength between T_{c1} and T_{c2} is very surface sensitive but indicates that Andreev reflection is suppressed in the A phase. A similar reduction of the strength was found when a magnetic field was applied in the basal plane.

Point contact measurements on URu_2Si_2 [443] are of particular interest since they showed a gap with magnitude of approximately 10 meV opened up in the conductance spectrum near the transition temperature of 17.5 K. As also observed in point contact measurements in UPt_3 [441], the shape of the zero bias conduction peak for URu_2Si_2 is consistent with an order parameter that has lines of nodes [443].

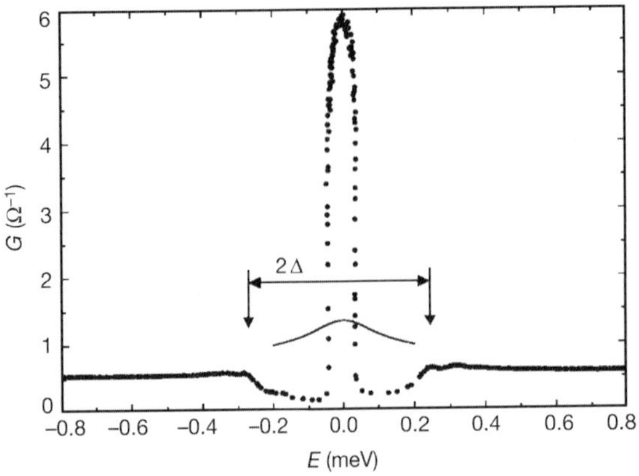

Fig. 19.65. The differential conductivity of a UBe_{13}-Au point contact at $T = 330$ mK, as a function of voltage. The height of the zero-bias anomaly is roughly five times greater than is expected for an s-wave superconductor. The *solid line* shows an estimate of the Kondo contribution to $G(V)$. [After Wälti et al. [444]]

In contrast to the very small magnitude of the zero bias conductance peak found in UPt_3 and URu_2Si_2, the magnitude of the conductance peak in UBe_{13} [444] is roughly a factor of 5 times larger than the theoretical value for an s-wave superconductor. The enhancement of the zero bias anomaly was attributed to the formation of an Andreev surface state, which can occur if there is a non-trivial dependence on the angle subtended by the nodes of the order parameter to the normal to the interface [445]. As the bias voltage was increased, the zero bias conductance peak of UBe_{13} was followed by a minimum which was reduced from the normal state value and then showed a sharp increase as it recovered to the normal state value. The variation of the differential conductivity with voltage is shown in Fig. 19.65. The sharp increase occurred at a temperature-dependent voltage which was of the order of magnitude 0.25 mV at 330 mK. Such sharp changes of the conductance have not been observed for s-wave superconductors but have been predicted to occur at the gap voltage $eV = \Delta_0$ for junctions with anisotropic superconductors [445]. If the temperature dependent characteristic voltage is interpreted as being the magnitude of the order parameter $\Delta_0(T)$, then UBe_{13} would correspond to a universal ratio $2\Delta_0(0)/k_B T_c$ of about 6.7 which is far greater than the BCS value of 3.5, and is also greater than the ratio predicted for most p-wave phases (ABM 4.6, polar 4.9) but is more comparable with the value of 6.2 predicted for the $m = 0$ d-wave phase.

Josephson Tunneling

Josephson tunneling involves the tunneling of Cooper pairs between two superconductors [446]. A d.c. tunnel current may appear even in the absence of an applied voltage but also appears as an a.c. current if a voltage is applied. The a.c. current is expressed in terms of the phase difference between the wave functions of the two superconductors

$$I(t) = I_0 \sin\left(\frac{2eVt}{\hbar} + \Delta\varphi\right) . \qquad (19.239)$$

The phase difference not only has contributions from the intrinsic phases of the superconductors but also has contributions from any magnetic vector potential that may be present at the junction. When integrated over the finite size of the contact, this term can give rise to a Frauenhofer diffraction pattern [447]. The leading order contribution to the Josephson coupling can be calculated using second order perturbation theory [446]. To lowest order there should be no Josephson current between a singlet and triplet superconductor [448], since spin is conserved in the tunneling process. However, even in this case, spin-orbit interaction and higher order processes may produce a non-zero Josephson current. If the tunneling is due to a fourth order process, the Josephson frequency is expected to be given by $4eV/h$. Since the presence of the surface locally violates certain rotational symmetries contained in the point group, it is possible for relatively large Josephson currents to flow between superconductors with different orbital symmetries. In fact, the magnitude of the order parameter at the surface may even vary with the relative orientation of the boundary and the node directions. Due to this, it is possible to extract the symmetry of the order parameter from Josephson tunneling measurements [449]. However, the Josephson effect is a very weak effect and sensitive to impurities and inhomogeneities, since the total coupling energy for a conventional Josephson junction is estimated to be only of the order of 1 eV. For the heavy-fermion systems, the preparation and characterization of good surfaces and junctions represent formidable experimental problems. Therefore, there has been only limited progress in the investigation of the Josephson effect in heavy-fermion superconductors.

A d.c. Josephson current was found to exist between $CeCu_2Si_2$ and Al [438]. The Josephson current vanished at the T_c of $CeCu_2Si_2$ with applied magnetic fields greater than the upper critical field of Al but smaller than the critical field of $CeCu_2Si_2$. Hence, the tunneling current involved the superconducting phases of both materials. The application of magnetic fields produced irregular Frauenhofer diffraction patterns which were smeared out. The smearing was attributed to the irregular geometry of the junctions. As the observed value of the critical current was quite large, having a maximum magnitude that is 80% of the BCS value, it appears as if the Josephson coupling is of second order [450] and not of fourth order. Furthermore, since Al is known to be a singlet superconductor, the large coupling strength indicates that $CeCu_2Si_2$ is also a singlet superconductor.

No Josephson currents were found to flow across weak links between UPt_3 surfaces [438]. However, a Josephson current, with an irregular Frauenhofer pattern and well defined Shapiro steps was observed between UPt_3 and Nb in a superconductor–normal metal–superconductor junction [451]. The critical current measured for junctions where the current flow is primarily along the c-axis of the UPt_3 single crystal is significantly larger than for junctions where the current flows along the basal plane. The critical current temperature relations show clear kinks at the lower critical temperature T_{c2}. For temperatures below T_{c2}, the slope of the critical current-temperature relation is large for current flow along the c direction, and smaller for current flow in the basal plane. For temperatures above T_{c2}, the slope of the critical current temperature relation for flow in the basal plane changes to a larger value but the curve representing flow along the c-axis is rather flat. The change in the anisotropy at T_{c2} seems to indicate that the anisotropy in the current flow is related to the unconventional nature of the superconductivity in UPt_3. It should be noted that, since the anisotropy measurements were made on junctions with different surfaces, the apparent anisotropy possibly could reflect a difference in the properties of the interfaces. A much larger anisotropy was inferred from measurements of Josephson currents through point contacts between URu_2Si_2 and Nb [452] where no current was observed for contacts aligned parallel to the c-axis but finite currents were found for contacts aligned parallel to the a–b directions. The absence of Josephson currents flowing along the c-axis could, however, have many other possible causes.

No Josephson tunneling currents were observed to flow across a weak link between two surfaces of UBe_{13}. In a junction between Al and UBe_{13}, superconductivity was introduced in UBe_{13} by the proximity effect for temperatures below the superconducting transition of Al but above the superconducting transition for UBe_{13}. In this temperature regime, a weak Josephson current was observed between these materials. A proximity induced Josephson current was also found to occur across a point contact between UBe_{13} and Ta at temperatures below T_c for Ta and above T_c for UBe_{13}. This Josephson current was destroyed as the temperature was lowered below the T_c of UBe_{13} [453, 454]. This was taken as evidence that the superconducting order parameter of UBe_{13} competes with the superconductivity of Ta and hence, are of different symmetry. The analysis presented in [454] suggested that the Cooper pairs were spin-triplet, however, spin singlet d-wave pairing would be more consistent with the experimental observations. In particular, when illuminated with microwave radiation of frequency ω, the current d.c. voltage relation exhibited Shapiro steps of magnitude $\Delta V = \hbar\omega/2e$, instead of $\Delta V = \hbar\omega/4e$ as would be expected for a fourth order Josephson coupling between singlet and triplet superconductors [448,455]. The interpretation of these experiments is enigmatic since a conventional Josephson tunneling current, with an irregular Frauenhofer pattern (see Figs. 19.66 and 19.67) and conventional Shapiro steps (shown in Fig. 19.68), was observed at temperatures below 0.94 K between UBe_{13} and Nb in a fairly well defined superconductor–normal-metal–superconductor geometry [456]. The Shapiro steps were observed to have a magnitude of $\Delta V = \hbar\omega/2e$, as expected for a second order Josephson coupling between two singlet superconductors. On the other hand, the magni-

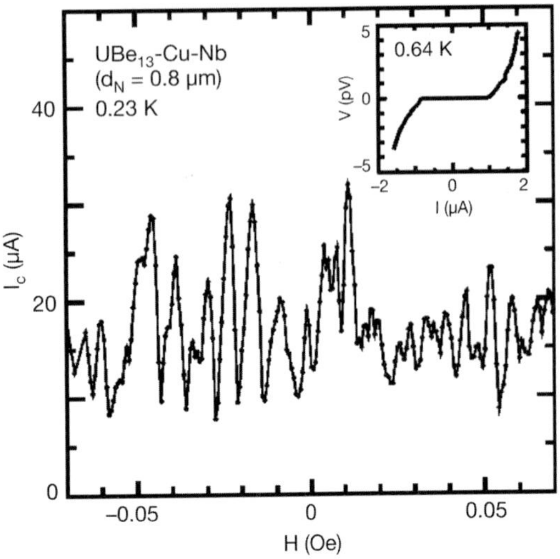

Fig. 19.66. The field dependence of the Josephson current through a UBe_{13}-Cu-Nb junction. The irregular Frauenhoffer pattern suggests that the junction is not uniform. The *inset shows* the $I - V$ curve showing the critical Josephson current. [After Shibata et al. [456]]

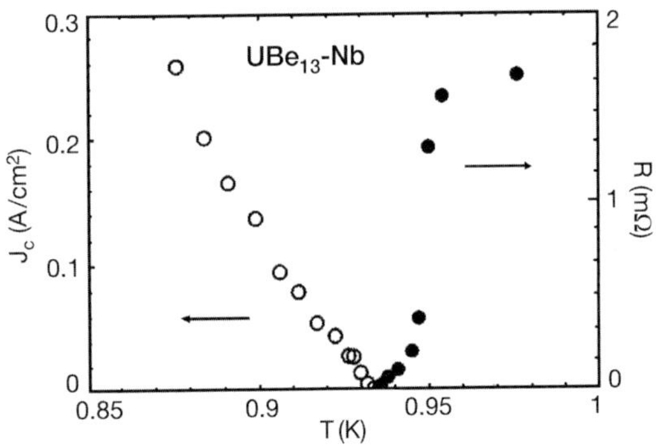

Fig. 19.67. The temperature dependence of the critical Josephson current J_c and the junction resistance R, near T_c for a UBe_{13}-Nb junction. After Shibata et al. [456]

tudes of the critical currents were remarkably small, being two orders of magnitude smaller than the critical currents observed in UPt_3. However, as the critical currents were limited by heating, it was not possible to determine whether the magnitude of the Josephson coupling in UBe_{13} was consistent with the tunneling being due to a fourth order process or a second order process.

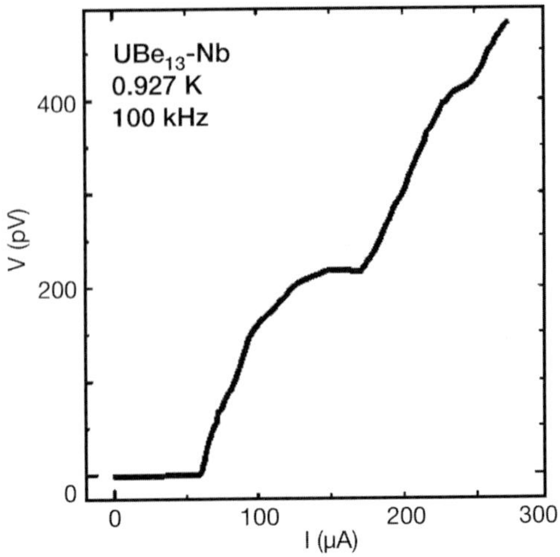

Fig. 19.68. The $I(V)$ relation for a UBe_{13}-Nb junction, showing conventional Shapiro steps when an ac current of frequency 100 kHz is superimposed on the dc current. [After Shibata et al. [456]]

19.4.3 Dynamic Magnetic Properties

Nuclear Magnetic Resonance

As previously mentioned, nuclear magnetic resonance experiments yield several important quantities: the Knight shift and the nuclear spin relaxation rates which are $1/T_1$ the longitudinal relaxation rate, and the transverse relaxation rate $1/T_2$. The changes in behavior of these quantities on entering the superconducting phase does give information about the nature of the density of magnetic excitations in the superconductor.

The Knight Shift

The Knight shift has many contributions, one important contribution provides a measure of the local part of the static susceptibility of the itinerant electrons which produces a magnetic polarization at the nuclear site. For heavy-fermion superconductors, the normal state Knight shift is expected to be dominated by the enhanced Pauli spin susceptibility, though spin-orbit coupling will introduce an orbital part. However, in the superconducting state, the Knight shift is expected to be dominated by the diamagnetic susceptibility produced by the supercurrent shielding the external field. The Pauli spin susceptibility will be modified by the superconductivity, and provide information about the pairing. The zero field susceptibility is defined as a derivative of the magnetization $\chi(T) = (\frac{\partial M}{\partial H})$. The magnetization that is produced by the electronic spins aligning with a magnetic field applied along the z-axis is given by

$$M_z = \left(\frac{g\mu_B}{2}\right)\sum_{\underline{k}}\left[f(E_{\uparrow,\underline{k}}) - f(E_{\downarrow,\underline{k}})\right], \quad (19.240)$$

which is expressed in terms of the Fermi-distribution for quasi-particles with spin σ and quasi-particle energy $E_{\sigma,\underline{k}}$. The field dependence of the quasi-particle energies depends on the type of spin pairing, so we shall discuss the different types of spin pairings separately.

Singlet Pairing

For singlet pairing, the magnetic field couples to the spins of the quasi-particles via the Zeeman energies. As can be seen from inspection of the matrix in Eq. (19.58), only the time reversal partners pair when $\vec{d}(\underline{k}) \equiv 0$. The quasi-particles consist of broken pairs, i.e., electrons of spin σ and holes of spin $-\sigma$. Since a down-spin hole has the same Zeeman energy as an up-spin electron, the quasi-particle energies depend on the magnetic field through

$$E_{\sigma,\underline{k}} = E_{H=0,\underline{k}} - \left(\frac{g\mu_B\sigma H}{2}\right) \quad (19.241)$$

and so the spin susceptibility takes the usual form

$$\chi_{S=0}(T) = 4\left(\frac{g\mu_B}{2}\right)^2\int_0^\infty \mathrm{d}E\ \rho_{S=0}(E)\left(-\frac{\partial f}{\partial E}\right). \quad (19.242)$$

The BCS density of states should be used in the above expression, for an s-wave superconductor. In this case, the susceptibility tends to zero as $T \to 0$ in an exponentially activated way $\chi_{\mathrm{BCS}}(T) \sim \exp[-\Delta_0/k_BT]$. The exponential vanishing of the spin susceptibility occurs as the electrons form singlet pairs in the ground state, and the finite spin moment is caused by thermal population of quasi-particles [457]. For a singlet superconductor that has point nodes, the density of states varies as E^2 at low energies and, therefore, one expects that $\chi_{S=0}(T) \propto T^2$ for $T \ll T_c$. Whereas for a singlet superconductor which has line nodes, the density of states is given by $\rho_{S=0}(E) \propto E$ at low energies, hence one obtains $\chi_{S=0}(T) \propto T$. The temperature dependence of the calculated spin susceptibility expected for various singlet spin superconducting phases is shown in Fig. 19.69.

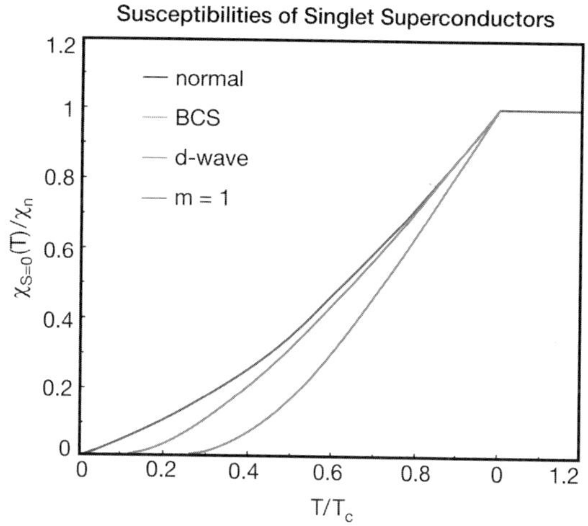

Fig. 19.69. The calculated spin susceptibilities of singlet superconducting phases. The susceptibilities are normalized to the normals state susceptibility. The susceptibilities in the singlet superconducting phases all vanish as $T \to 0$. The susceptibility of the BCS phase is given by the Yosida function, which vanishes exponentially at low temperatures. The susceptibilities of the isotropic and $m = 1$ d-wave phases vary as powers of the temperature, at low temperatures

Thus, in the spin singlet phases, the spin susceptibility could be expected to vanish as $T \to 0$. However, spin-orbit coupling will produce a residual susceptibility that depends on the ratio of the superconducting coherence length, ξ_0, to the mean free path due to spinorbit scattering, l_{so}. In the presence of spin-orbit coupling, the spin is no longer a good quantum number for the single-particle eigenstates and the spin-up and spin-down states are mixed. In the limit that the strength of the spin-orbit coupling

$\lambda \vec{L}.\vec{S}$ is so large that $\lambda \gg \Delta_0$, the average value of σ_z for a single particle state tends to zero. The spin susceptibility is, therefore, reduced. From the Kramers–Kronig relations, one observes that a significant contribution to the normal state $\chi(T)$ comes from single-particle states with excitation energies of the order of the spin-orbit scattering rate $\hbar/\tau_{so}$ which is, by assumption, greater than Δ_0. As an opening up of a superconducting gap at the Fermi energy is not expected to change the contribution of these higher energy states, one finds that the susceptibility in the superconducting state can remain comparable in magnitude to the normal state value. According to Anderson [458, 459], the normalized susceptibility should have the two limits. The limit of strong spin-orbit scattering is defined by $l_{so} \ll \pi\xi_0$, where $\xi_0 = \hbar v_F/\pi\Delta_0$ and $l_{so} = v_F\tau_{so}$, and the limit of weak spin-orbit scattering is defined by $l_{so} \gg \pi\xi_0$. In the limit of strong spinorbit scattering, the normalized susceptibility is given by [460]

$$\frac{\chi_{BCS}(0)}{\chi_n} \approx 1 - \frac{8\, l_{so}}{3\pi\xi_0} \,. \tag{19.243}$$

For weak spin-orbit scattering, the susceptibility ratio is given by

$$\frac{\chi_{BCS}(0)}{\chi_n} \approx \frac{3\pi\xi_0}{8 l_{so}} \,. \tag{19.244}$$

Hence, the existence of a partial Meissner effect at $T = 0$ does not necessarily imply the presence of either triplet or gapless superconducting phases.

Triplet Pairing

For triplet pairing, the above argument indicates that a finite susceptibility could be expected to arise due to the $S = 1$ spin of the Cooper pairs. The way the susceptibility depends on the orientation of the field $\vec{H}$ relative to the direction of $\vec{d}$, for a direction of $\vec{d}$ independent of $\underline{k}$, is described below.

First, if $\vec{d}$ is parallel to $\vec{H}$ for all $\underline{k}$, then, as $\vec{H}$ is the direction of quantization of σ, $\vec{d}$ is parallel to $\hat{z}$ and so $S_z = 0$. Since $d_{S_z=0}$ is the only finite component of the order parameter, the off-diagonal blocks in the mean-field Hamiltonian Eq. (19.58) are proportional to σ_x. And so, as in the singlet case, only the same time reversal partner states are coupled. Thus, one obtains the same Zeeman type of coupling as the singlet case, $E_{\sigma,\underline{k}} = E_{H=0,\underline{k}} - \frac{1}{2}(g\mu_B\sigma H)$, and the susceptibility is given by the same expression as Eq. (19.242) but with the appropriate triplet density of states.

$$\begin{aligned}\chi_{S=1,\parallel}(T) &= 2\left(\frac{g\mu_B}{2}\right)^2 \sum_{\underline{k}} \left(-\frac{\partial f}{\partial E_{\underline{k}}}\right) \\ &= 4\left(\frac{g\mu_B}{2}\right)^2 \int_0^\infty \mathrm{d}E \rho_{S=1}(E)\left(-\frac{\partial f}{\partial E}\right) .\end{aligned} \tag{19.245}$$

Hence, the spin susceptibility has a linear T variation for a density of states with lines of nodes or a T^2 variation with point nodes.

Second, if $\vec{H}$ is perpendicular to $\vec{d}$ for all $\underline{k}$, then, as $\vec{d}$ is perpendicular to $\hat{z}$, the pairs are in a linear combination of states with $S_z = \pm 1$. The off-diagonal blocks of the mean-field Hamiltonian are of diagonal form and only couple electrons and holes with the same spin directions. Then, the quasi-particle energies are given by

$$E^2_{\sigma,\underline{k}} = \left[e(\underline{k}) - \mu - \left(\frac{g\mu_B\sigma H}{2}\right)\right]^2 + \left| d_x(\underline{k}) + i\sigma d_y(\underline{k})\right|^2 . \tag{19.246}$$

Furthermore, since

$$\left(\frac{\partial E_{\sigma,\underline{k}}}{\partial H}\right) = -\left(\frac{g\mu_B\sigma}{2}\right)\left(\frac{\partial E_{\sigma,\underline{k}}}{\partial e_\sigma(\underline{k})}\right) , \tag{19.247}$$

the expression for $\chi_{S=1,\perp}(T)$ has the form

$$\begin{aligned}\chi_{S=1,\perp}(T) &= \sum_\sigma \left(\frac{g\mu_B\sigma}{2}\right)^2 \rho(\mu) \int_{-\infty}^{+\infty} \mathrm{d}e \\ &\times \left(\frac{\partial E_\sigma}{\partial e}\right)\left(-\frac{\partial f(E_\sigma)}{\partial E_\sigma}\right) \\ &= 2\left(\frac{g\mu_B}{2}\right)^2 \rho(\mu) .\end{aligned} \tag{19.248}$$

Hence, in this case, the uniform static spin susceptibility retains the normal state value. In general, for triplet pairing, the orientational average of $\chi_{S=1}(T)$ will have a magnitude of 2/3 the normal state susceptibility.

The sum of the quasi-particle energies are lowered when $\vec{H}$ is perpendicular to $\vec{d}(\underline{k})$. For an isotropic system like ^{3}He, and a pairing vector which has a direction independent of $\underline{k}$ such as in the ABM phase, the vector $\vec{d}$ will rotate to remain perpendicular to the field and the susceptibility will remain constant. On the other hand, heavy-fermion systems are expected to have strong spin-orbit coupling and crystalline anisotropy, therefore, the direction of $\vec{d}$ is expected to remain constant when $\vec{H}$ is applied. If the material has an anisotropy such that spins have an easy axis or plane, then $\chi_{S=1}$ for that phase will be reduced if $\vec{H}$ tilts away from the axis or plane. In such cases of strong anisotropy, the susceptibility tensor is given by the directional average

$$\chi^{i,j}_{S=1}(T) = 2\left(\frac{g\mu_B}{2}\right)^2 \sum_{\underline{k}} \left(-\frac{\partial f}{\partial E_{\underline{k}}}\right) \times \left[\frac{d^*_i(\underline{k})d_j(\underline{k})}{\underline{d}^*(\underline{k}).\underline{d}(\underline{k})} + \left(\frac{\partial E_{\underline{k}}}{\partial e_{\underline{k}}}\right)\left(\delta_{i,j} - \frac{d^*_i(\underline{k})d_j(\underline{k})}{\underline{d}^*(\underline{k}).\underline{d}(\underline{k})}\right)\right] . \quad (19.249)$$

In the BW phase the zero temperature susceptibility is 2/3 of the normal state value, this occurs as the $\underline{k}$ averaged value of $\vec{d}^{\,2}$ has one component parallel to $\vec{H}$ which is suppressed, the other two components retain their normal values. The calculated spin susceptibilities expected for various triplet superconducting phases is shown in Fig. 19.70. The above discussion indicates that, if the behavior of the hyperfine field is known (perhaps extrapolated from the normal state), the temperature dependence of the Knight shift in the superconducting state can yield information about the nature of the pairing.

In pure UBe_{13}, the muon Knight shift [320] exhibits a large reduction below T_c with a magnitude similar to that found in weak coupling s-wave superconductors. The Knight shifts found for Th [320] and B doped samples [461] show no appreciable changes on entering the superconducting states which is consistent with the formation of spin-singlet Cooper pairs but with strong spin-orbit scattering or gapless phases caused by impurity scattering. The anisotropic Knight shift observed in $CeCu_2Si_2$ [462,463] undergoes a significant reduction in the superconducting state, similar to the

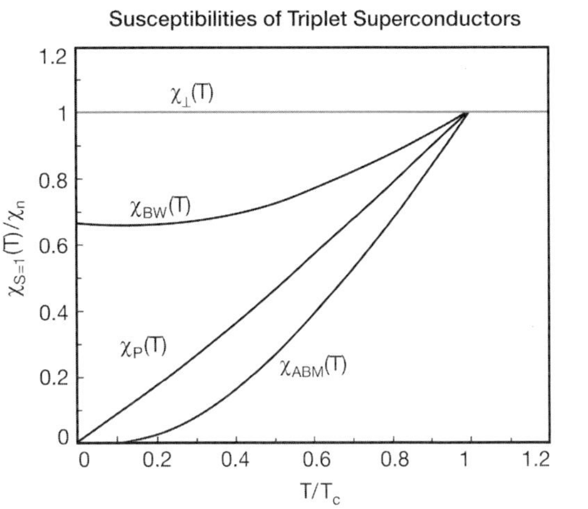

Fig. 19.70. The calculated spin susceptibilities of triplet superconducting phases. The susceptibility for the BW phase is isotropic. The susceptibilities of the ABM and polar phases are anisotropic, if the order parameter is pinned by the lattice. The susceptibilities vanish at low temperatures for applied fields which are parallel to the Cooper pair spin but remain unaffected for fields oriented perpendicular to the spin

behavior observed in UBe_{13}. The reduction found in $CeCu_2Si_2$ is consistent with the vanishing of the spin susceptibility at zero temperature, as expected for a spin-singlet superconducting phase. In $CeCoIn_5$ [62], the decrease of the ^{59}Co Knight shift in the superconducting state is large for fields along the c-axis and is hardly noticeable for fields in the perpendicular directions. However, the lack of any decrease in the perpendicular directions was attributed to a large temperature independent contribution to the susceptibility from the Co 3d orbitals. The observation of a smaller but isotropic decrease in the Knight shift of ^{115}In below T_c is supportive of a superconducting state which involves spin-singlet Cooper pairs.

The Knight shift has been measured in the superconducting state of single crystals of UPt_3 [464,465]. For large fields [464], no appreciable change was observed on entering the superconducting phase for all directions of the applied field. However, a later study [465] showed a very small decrease occurred for fields aligned along either the b-axis or c-axis but only for fields such that $H_a < 5$ kOe and $H_c < 2$ kOe. This was interpreted as showing evidence for spin-

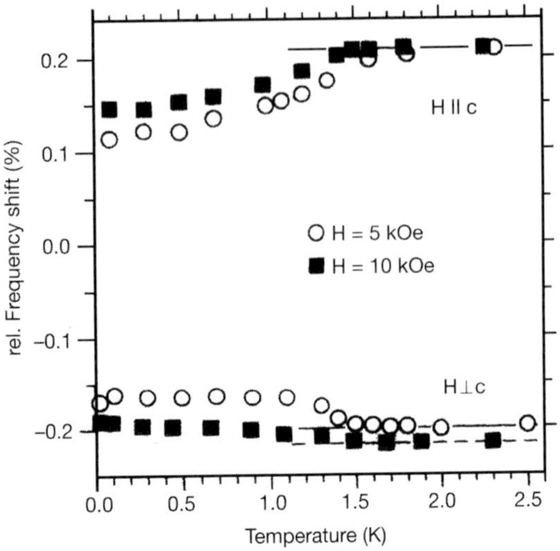

Fig. 19.71. Temperature dependence of the muon Knight shifts in UPd_2Al_3, for fields applied parallel and perpendicular to the c-axis. [After Feyerherm et al. [466]]

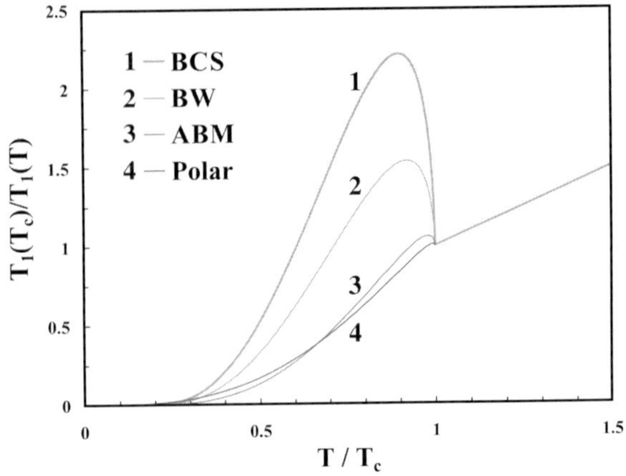

Fig. 19.72. The calculated temperature dependence of the spin-lattice relaxation rate $1/T_1$ for the BCS, BW, ABM and polar phases. The rates are normalized to the normal state relaxation rate at T_c. A Hebel–Slichter peak occurs below T_c for the BCS and BW phases, but is absent for the ABM and polar phases

triplet pairing, in which the superconducting vector order parameter is pinned to the lattice for small applied fields but for sufficiently large fields is free to rotate so as to follow the direction of the applied field. This explanation rests on the unjustified assumption of very weak spin-orbit coupling in UPt_3 and also does not explain the extremely small magnitude of the change of Knight shift that occurs when the order parameter is pinned. However, since the crystals were of very high quality, the absence of a reduction of the spin susceptibility at low temperatures is not attributable to spin-orbit or to impurity scattering in a spin-singlet superconducting phase.

Although the normal state susceptibility and the μSR Knight shift of UPd_2Al_3 are anisotropic (see Fig. 19.71), the small and anisotropic change in the Knight shift which occurs below T_c has been taken as evidence that the electrons form spin-singlet Cooper pairs. An analysis of the normal state data indicates that part of the anisotropy originates in the coupling. The dipole contribution is estimated to be three times stronger than the hyperfine field coupling. Although a small anisotropic change in the Knight shift is observed as the temperature is decreased below T_c, the analysis attributes the change to an isotropic reduction of the f spin susceptibility [466]. The reduction has a magnitude of 11% which is only compatible with spin-singlet pairing of the quasi-particles if one partitions the f susceptibility into separate quasi-particle and local moment contributions. This is in direct contrast with conclusions from Al NMR Knight shift studies of the isostructural compound UNi_2Al_3. The Knight shift measurements of UNi_2Al_3 have been interpreted as indicating that the quasi-particles form triplet-spin Cooper pairs [467]. Since the crystals are not axially symmetric, the crystals contain two inequivalent Al sites. As a consequence, the experiments on UNi_2Al_3 show two distinct NMR peaks for magnetic fields applied parallel to the a-axis. In UNi_2Al_3, the Knight shift predominantly comes from the hyperfine coupling which has a similar magnitude to that found in UPd_2Al_3. However, on reducing T below T_c, the Knight shift observed for one peak did not show a change comparable to that found in UPd_2Al_3 and hence, it is argued that these measurements provide evidence for triplet-spin superconductivity. Due to the much larger $1/T_1$ contribution to the peak width for fields along the c-axis, these experiments were unable to determine if the Knight shift exhibits the full anisotropic Meissner behavior expected from a spin-

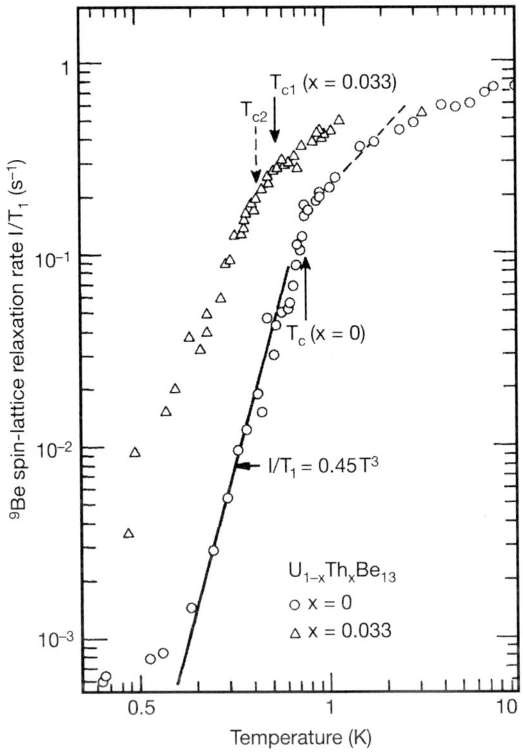

Fig. 19.73. The temperature dependence of the ^{9}Be nuclear spin relaxation rate $1/T_1$ in $U_{1-x}Th_xBe_{13}$, for $x = 0$ and $x = 0.033$. The solid line is a fit to a T^3 power law below T_c, and the dashed line is the normal state Korringa law. [After MacLaughlin et al [25]]

triplet superconductor. Further results are shown in Figs. 19.72 and 19.73.

Longitudinal Nuclear Spin Relaxation

The temperature variation of the Spin Lattice relaxation rate $1/T_1$ can be drastically changed at the onset of superconductivity. This occurs as a result of the change in the quasi-particle excitation spectrum but also is due to a modification of the matrix elements for the coupling of the quasi-particle excitations to the nuclear spins. As the modification of the matrix elements of the coupling interaction, known as the coherence factor, involves the gap function, $1/T_1$ could provide additional information about the symmetry of the superconducting phase. The relaxation rate for a spin distribution aligned along the z-axis is driven by spin-flip fluctuations, and is given by the expression

$$1/T_1 = k_B T (g\mu_B)^2 \frac{1}{N} \sum_{\underline{q}} A(\underline{q})^2 \left[\frac{Im\chi^{+,-}(\underline{q};\omega_N)}{\hbar\omega_N} \right] , \qquad (19.250)$$

where ω_N is the nuclear Larmor frequency, and $A(\underline{q})$ is the averaged strength of the local hyperfine field. The factor $Im\chi^{+,-}(\underline{q};\omega)$ represents the density of states for magnetic spin-flip fluctuations. A typical value of the nuclear Larmor frequency is $\hbar\omega_N \sim 10^{-7}$ meV and so, for all but in systems with exceptionally slow spin-fluctuations, one may approximate ω_N by $\omega_N = 0$. This approximation could be expected to fail near a quantum critical point. Just to illustrate the nature of the coupling to the superconductivity, we shall replace the full susceptibility by the lowest order polarization part or quasi-particle susceptibility.

The first step in evaluating the imaginary part of the susceptibility is to express the spin-flip operators in terms of the four component fields $\Psi(\underline{k})$, so the spin raising operator $\hat{\sigma}^+(\underline{q})$ is given by

$$\hat{\sigma}^+(\underline{q}) = \Psi^\dagger(-\underline{k}-\underline{q}) \begin{pmatrix} \sigma^+ & 0 \\ 0 & -\sigma^- \end{pmatrix} \Psi(-\underline{k}) , \qquad (19.251)$$

where σ^+ and σ^- are the 2×2 Pauli spin matrices. The spin lowering operator $\hat{\sigma}^-(\underline{q})$ is given by

$$\hat{\sigma}^-(\underline{q}) = \Psi^\dagger(-\underline{k}-\underline{q}) \begin{pmatrix} \sigma^- & 0 \\ 0 & -\sigma^+ \end{pmatrix} \Psi(-\underline{k}) . \qquad (19.252)$$

In general, the matrix $\tilde{\sigma}^i$ corresponding to a spin operator $\hat{\sigma}^i$ is a 2×2 block diagonal matrix with the upper and lower diagonal elements given, respectively, by σ^i and $(i\sigma_y)\sigma_i(\ i\sigma_y)$, where the lower diagonal element reflects the effect of time reversal on the spins. The lowest order contribution to the susceptibility tensor $\chi^{\alpha,\beta}(\underline{q};\omega)$ is expressed in terms of the Fourier transform of the imaginary time single-particle Green's functions. These Green's functions are defined in terms of the Wick's time ordered product of two four-component fields

$$G(\underline{k};\tau) = - < |\hat{T}\ \Psi(\underline{k},\tau)\Psi^\dagger(\underline{k},0)| > . \qquad (19.253)$$

The susceptibility tensor can be written as the analytic continuation onto the real frequency axis of the trace

$$\chi^{\alpha,\beta}(\underline{q};\nu_m) = k_B T \left(\frac{g\mu_B}{2}\right)^2 \frac{1}{N} \sum_{n,\underline{k}} \frac{1}{2} \text{trace}$$

$$\times\, G(\underline{k}+\underline{q}; i\hbar(\omega_n+\nu_m))\tilde{\sigma}^{\beta} G(\underline{k}; i\hbar\omega_n)\tilde{\sigma}^{\alpha} \,, \tag{19.254}$$

where ω_n are the fermion Matsubara frequencies, $\hbar\omega_n = \pi k_B T(2n+1)$ and ν_m is a boson Matsubara frequency. For unitary states where $\tilde{H}^2_{MF}$ is diagonal, the quasi-particle contribution to the Fourier transformed Green's function can be written as

$$G(\underline{k}; i\hbar\omega_n) = -Z^{-1} \frac{(i\hbar\omega_n + \tilde{H}_{MF})}{\hbar^2\omega_n^2 + E^2_{\underline{k}}} \,. \tag{19.255}$$

Hence, on evaluating Eq. (19.254) and continuing the imaginary frequency $i\nu_m$ onto the real ω axis, the energy difference components of the imaginary part of the susceptibility tensor are given by the expression,

$$Im\left[\chi^{\alpha,\beta}(\underline{q};\omega)\right] = Z^{-2}\pi \left(\frac{g\mu_B}{2}\right)^2 \frac{1}{N} \sum_{\underline{k}} \int_{-\infty}^{+\infty} \mathrm{d}E$$

$$\times \left[f(E) - f(E+\hbar\omega)\right] \frac{1}{2}\,\text{trace}$$

$$\times \left(\frac{(E+\tilde{H}_{MF})\tilde{\sigma}^{\beta}(E+\hbar\omega+\tilde{H}_{MF})\tilde{\sigma}^{\alpha}}{4E(E+\hbar\omega)}\right)$$

$$\times \Big[\delta(E-E_{\underline{k}})\delta(E+\hbar\omega-E_{\underline{k}+\underline{q}})$$

$$+\, \delta(E+E_{\underline{k}})\delta(E+\hbar\omega+E_{\underline{k}+\underline{q}})\Big] \,. \tag{19.256}$$

This represents a quasi-elastic process involving scattering from the thermal population of quasi-particles. The factors in front of the spin operators, as modified by the matrix nature of $\tilde{\sigma}$, produce the coherence factors. The coherence factors differ between the singlet and triplet pairing cases.

Singlet Pairing

In this case, the spin-flip excitation spectrum is obtained as

$$Im\left[\chi^{+,-}(\underline{q};\omega)\right] = \pi Z^{-2} \left(\frac{g\mu_B}{2}\right)^2 \frac{1}{N} \sum_{\underline{k}} \int_{-\infty}^{+\infty} \mathrm{d}E$$

$$\times \left[f(E) - f(E+\hbar\omega)\right]$$

$$\times \left(1 + \frac{\xi(\underline{k})\xi(\underline{k}+\underline{q}) + D(\underline{k})D(\underline{k}+\underline{q})}{E_{\underline{k}}E_{\underline{k}+\underline{q}}}\right)$$

$$\times \Big[\delta(E-E_{\underline{k}})\delta(E+\hbar\omega-E_{\underline{k}+\underline{q}})$$

$$+\, \delta(E+E_{\underline{k}})\delta(E+\hbar\omega+E_{\underline{k}+\underline{q}})\Big] \,. \tag{19.257}$$

The two summations over $\underline{k}$ and $\underline{k}+\underline{q}$ in Eq. (19.250) when combined with Eqn. (19.257) have the effect that the bi-linear term in $\xi(\underline{k}) = e(\underline{k}) - \mu$, which is anti-symmetric about the Fermi energy, vanishes. Also, since for singlet superconductivity, the order parameter is an even function of $\underline{k}$ these terms are retained. Then, in the s-wave phase, the relaxation rate is given by

$$\hbar/T_1 = \pi k_B T Z^{-2}(g\mu_B)^2 A^2 \int_0^{\infty} \mathrm{d}E\left(-\frac{\partial f}{\partial E}\right)$$

$$\times \left[1 + \left(\frac{\Delta_0}{E}\right)^2\right] \rho_{\text{BCS}}(E)^2 \,, \tag{19.258}$$

where we have set $\omega_N = 0$, and A is the local hyperfine field. The above expression indicates that a peak in $1/T_1$ should occur just at the superconducting transition. The peak occurs since the BCS density of states diverges at $E = \Delta_0$, which yields a divergence in the integrand in the superconducting state. The logarithmic divergence of $1/T_1$ in the superconducting state is actually suppressed by small residual anisotropy in the gap and also by the quasi-particle lifetimes. As the derivative of the Fermi function

$$\left(-\frac{\partial f}{\partial E}\right)\bigg|_{E=\Delta_0} \tag{19.259}$$

has its largest value in the superconducting state at $T = T_c$, the above argument indicate that a peak in $1/T_1$ occurs just at T_c. The resulting experimentally observed peak is known as the Hebel–Slichter peak [468], and is often seen in weak-coupling s-wave superconductors. The peak actually occurs at roughly 0.8 T_c. This reduced temperature and the suppression of the logarithmic divergence is well described by interactions between thermally excited phonons and the quasi-particles [469]. Fibich found that the temperature dependence of the rate could be expressed as

$$\frac{\hbar}{T_1} \propto 2k_B T f(\Delta_0)\left[1 + \frac{\Delta_0}{k_B T}(1 - f(\Delta_0)) \times \ln\left(2\left(\frac{2\Delta_0}{\Gamma(T)}\right)^{4/3}\right)\right], \quad (19.260)$$

where $\Gamma(T)$ represents the imaginary part of the gap which is a characteristic energy for quasi-particle scattering from a thermal distribution of phonons, $\Gamma(T) \sim (k_B T)^2/\hbar\omega_D$. However, in strong-coupling s-wave systems, quasi-particle interactions and lifetime effects can substantially suppress the peak. At low temperatures, $T \ll T_c$, the relaxation rate is expected to show a thermally activated temperature variation, $1/T_1 \propto \exp[-\Delta_0/k_B T]$.

Triplet Pairing

The coherence factors in the triplet or p-wave pairing phases should be replaced by

$$\left(1 + \frac{Re\left[\xi(\underline{k})\xi(\underline{k}+\underline{q}) - (\hat{z}\,.\,\vec{d}\,(\underline{k}))(\vec{d}^{\,*}(\underline{k}+\underline{q})\,.\,\hat{z})\right]}{E_{\underline{k}}E_{\underline{k}+\underline{q}}}\right). \quad (19.261)$$

Since $\xi(\underline{k}) = e(\underline{k}) - \mu$ is an odd function about the Fermi energy and $\vec{d}\,(\underline{k})$ is an odd function of $\underline{k}$, the bi-linear terms vanish, leaving a coherence factor of unity. Hence, the rate is given in terms of an integration over an appropriate quasi-particle density of states squared for the p-wave phase,

$$\hbar/T_1 = \pi\, k_B T\, Z^{-2}\,(g\mu_B)^2\, A^2 \int_0^\infty \mathrm{d}E \left(-\frac{\partial f}{\partial E}\right) \rho_{S=1}(E)^2\,. \quad (19.262)$$

We note that the weak logarithmic divergence of the density of states in the ABM phase is integrable, and the Hebel–Slichter peak is also completely absent in the polar phase.

At low temperatures in the BW phase, the same type of exponentially activated behavior is recovered as in the BCS phase. The low temperature form of $1/T_1$ in the ABM phase is given by

$$\hbar/T_1 = 4!\pi\,\frac{(g\mu_B)^2 A^2 Z^{-2}\rho(\mu)^2}{\Delta_0^4}(k_B T)^5 \quad (19.263)$$

and in the polar phase

$$\hbar/T_1 = \pi\,\frac{\pi^2 (g\mu_B)^2\, A^2\, Z^{-2}\rho(\mu)^2}{2\Delta_0^2}(k_B T)^3\,, \quad (19.264)$$

where the inequality $T \ll T_c$ should hold for both expressions. Thus, the characteristic energy dependence associated with the distribution of nodes could govern the low temperature quasi-particle contribution to the spin-lattice relaxation rate. Also, the quasi-particle contribution to the relaxation rate should remain unrenormalized. However, it should be noted that only the lowest order polarization part of the susceptibility has been considered but the higher order interactions between the quasi-particles and vertex corrections are also important. The vertex corrections may be expected to partially cancel the effect of the self-energy which is responsible for Z. The interactions are responsible for adding the low-energy collective spin-fluctuations to the continuum of single quasi-particle Stoner excitations.

The spin-lattice relaxation rates of the heavy-fermion superconductors UBe_{13} [25], UPt_3 [470], $CeCu_2Si_2$ [472], UPd_2Al_3 [471] and $CeIrIn_5$ [62,63] do not show any evidence of a Hebel–Slichter peak at T_c, but instead show a cross-over between a linear T-like Korringa variation above T_c to an approximate T^3 variation in the superconducting state. In UPt_3, the T^3 law is only followed closely in the restricted temperature range between 0.1 and 0.3 K, which is far below the superconducting transition temperature ($T_c \sim 0.5$ K). However, in UBe_{13}, UPd_2Al_3 and to a lesser extent $CeCu_2Si_2$, the T^3 dependence holds right up to T_c. Since the gap $\Delta_0(T)$ is expected to be rapidly varying with T in this temperature region, the T^3 variation should not be interpreted in terms of the power law energy dependence of the quasi-particle density of states. The power law T dependence due to the energy dependence of the quasi-particle density of states is expected to be restricted to occur in a temperature regime far below T_c. In UBe_{13}, the approximate T^3 variation crosses over to a linear T variation at the lowest measured temperatures [473]. The coefficient of the linear T term is observed to increase with doping [461], which is consistent with the picture of a gapless superconducting phase caused by resonant impurity scattering of Cooper pairs with non-zero angular momentum, $l \neq 0$. Although no low temperature linear T regimes were identified in the relaxation rates of UPd_2Al_3 [471], $CeCu_2Si_2$ [472] and $CeCoIn_5$ [62], at the lowest measured temperatures, the data do show indications of departures from the T^3 power law variation which are consistent with linear T variations.

Neutron Scattering Cross-Section

Perhaps one of the potentially most powerful probes of the nature of the order parameter would be inelastic neutron scattering measurements. The scattering cross-section not only provides a measure of the spectrum of the magnetic fluctuations of the quasi-particles but also, via the dependence on polarization and the direction of the momentum transfer $\hbar\underline{q}$, the anisotropy in the coherence factors. The main limitations of this method are due to the strong residual interactions between the quasi-particles that result in the large amplitude spin-fluctuations, and the loss of intensity needed for polarized scattering. The gaps that could be observed in the inelastic neutron scattering cross-section may be expected to be of the order of $2\Delta_0$ which, for the heavy-fermion superconductors, could be expected to range between 0.3 meV and 1 meV.

The magnetic scattering cross-section of an incident beam of unpolarized neutrons is related the imaginary part of the dynamic susceptibility via,

$$\frac{d^2\sigma}{d\omega d\Omega}(\underline{q};\omega) = \gamma_0^2 \frac{k_f}{k_i}\frac{|F(\underline{q})|^2}{g^2\mu_B^2}[1+N(\omega)]\sum_{\alpha,\beta} Im \times [\chi^{\alpha,\beta}(\underline{q};\omega)](\delta_{\alpha,\beta}-\hat{q}_\alpha\hat{q}_\beta) , \qquad (19.265)$$

where $\hbar\underline{q}$ and $\hbar\omega$ refer to the momentum and energy transfer of the neutron, and $F(q)$ is the atomic form factor. The summation over β corresponds to experiments in which the polarization of the scattered beam is not observed. It should be noted that the dipole nature of the interaction between the neutron and the electron spins results in the neutron only interacting with the spin components transverse to the momentum transfer $\underline{q}$. The imaginary part of the dynamic susceptibility contains the Raman scattering type of process expressed in Eq. (19.256), and a pair-breaking scattering process. The quasi-elastic or Raman scattering process involves the scattering of the neutrons from thermally excited quasi-particles. These low-energy fluctuations formed the basis for the discussion of NMR experiments, and are expected to vanish in the limit of zero temperature. The spectrum originating from processes in which pairs in the condensate are broken and where the resulting quasi-particles are then scattered across the gap is given by the energy sum terms in Eq. (19.254). These terms remain finite at zero temperature. The energy loss contribution to the pair-breaking process is evaluated as

$$\begin{aligned} Im[\chi^{\alpha,\beta}(\underline{q};\omega)] &= Z^{-2}\pi\left(\frac{g\mu_B}{2}\right)^2\frac{1}{N} \\ &\times \sum_{\underline{k}}\left[1-f(E_{\underline{k}})-f(E_{\underline{k}+\underline{q}})\right]\frac{1}{2}\text{trace} \\ &\times \left(\frac{(E_{\underline{k}}-\tilde{H}_{MF})\tilde{\sigma}^\beta(E_{\underline{k}+\underline{q}}+\tilde{H}_{MF})\tilde{\sigma}^\alpha}{4E_{\underline{k}}E_{\underline{k}+\underline{q}}}\right) \\ &\times \delta(\hbar\omega - E_{\underline{k}} - E_{\underline{k}+\underline{q}}) . \end{aligned} \qquad (19.266)$$

Here, we have only considered the quasi-particle contribution. The value of the trace depends upon the character of the superconductivity. Again, we consider the singlet pairing case separately from triplet pairing.

Singlet Pairing

Conventional singlet pairing superconductors are paramagnetic, so one retains spin rotational invariance

$$\frac{1}{2} Im\left[\chi^{+,-}(\underline{q};\omega)\right] = Im\left[\chi^{z,z}(\underline{q};\omega)\right] = Im\left[\chi^{x,x}(\underline{q};\omega)\right] = Im\left[\chi^{y,y}(\underline{q};\omega)\right] . \quad (19.267)$$

The imaginary part of the quasi-particle susceptibility is found as

$$Im\left[\chi^{+,-}(\underline{q};\omega)\right] = \pi(g\mu_B)^2 Z^{-2} \times \frac{1}{N}\sum_{\underline{k}}\left[1 - f(E_{\underline{k}}) - f(E_{\underline{k}+\underline{q}})\right] \times \left(\frac{(E_{\underline{k}} - \xi(\underline{k}))(E_{\underline{k}+\underline{q}} + \xi(\underline{k}+\underline{q})) - D(\underline{k})D(\underline{k}+\underline{q})}{4E_{\underline{k}}E_{\underline{k}+\underline{q}}}\right) \times \delta(\hbar\omega - E_{\underline{k}} - E_{\underline{k}+\underline{q}}) . \quad (19.268)$$

The scattering cross-section for pair-breaking scattering processes vanishes identically in the limit $\underline{q} \to 0$ so, for this $\underline{q}$ value, neutron scattering only occurs through processes involving the Raman scattering from thermally activated quasi-particles. In this limit, the non-zero part of the quasi-particle contribution involves coherence factors of unity and would manifest itself by a narrow zero energy delta function like peak with strength proportional to Z^{-1}, in the absence of spinorbit scattering.

For finite $\underline{q}$, the cancelation in the coherence factors for the pair breaking processes is incomplete. For s-wave pairing, the pair breaking process has a threshold at the maximum value of the superconducting gap $\hbar\omega = 2\Delta_0$, for $0 < q < 2k_F$. For non s-wave pairing phases where there are nodes in the order parameter, interesting possibilities occur when the system is close to an antiferromagnetic instability. In this case, one expects that in the normal state, there should be a quasi-elastic peak from magnetic fluctuations that should show softening at the $\underline{q}$ value $\underline{Q}_c$ which corresponds to the magnetic Bragg peak of the (hypothetical) ordered state [474]. The magnetic fluctuations are expected to decay into quasi-particle excitations. Therefore, the height and width of the quasi-elastic peak is governed by the quasi-particle density of states. As the temperature is lowered and the normal state becomes unstable to a superconducting state, the low-energy quasi-particle density of state is suppressed since the superconducting gap opens up on portions of the Fermi surface. This may result in the quasi-elastic neutron scattering peak at $\underline{Q}_c$ transforming into a narrow low-energy resonance for $\hbar\omega < 2\Delta_0$. The height of the peak is expected to have an inverse correlation with the maximum value of the superconducting gap. Furthermore, the ω dependence of the low-energy edge of the resonance should either show a ω or ω^2 variation depending upon the existence of either line nodes or point nodes in the gap. The sharp peak is expected to rapid broaden and disperse to higher energies as $\underline{q}$ is varied away from $\underline{Q}_c$. Excitations of a similar type have been observed in the superconducting state of magnetically ordered UPd_2Al_3 [345,347,433,475]. The spectra of the transverse excitations found in the polarized inelastic neutron scattering spectra, at $\underline{q} = \underline{Q}_c$, are shown in Fig. 19.74.

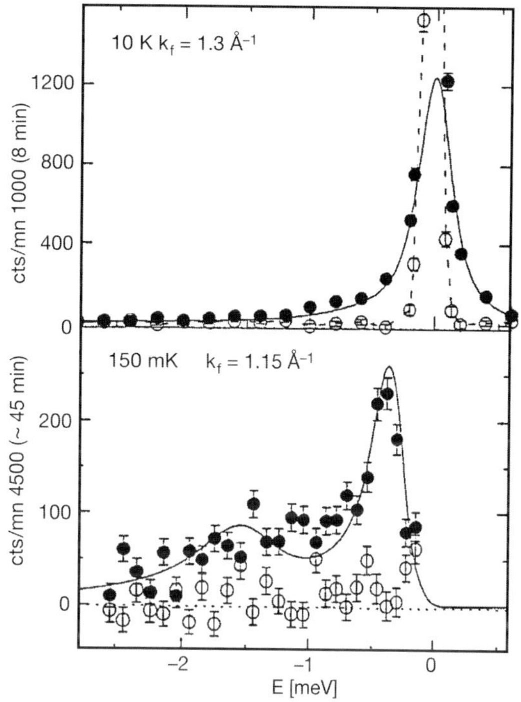

Fig. 19.74. The inelastic polarized neutron scattering cross-section for UPd_2Al_3 at the antiferromagnetic Bragg point $\underline{Q}$. The longitudinal and transverse response is shown by *open and closed circles*, respectively. At this $\underline{q}$ value, the spectra shows that the quasi-elastic peak due to an overdamped spin-wave evolves into a resonant mode as T is decreased below the superconducting transition temperature. [After Bernhoeft et al. [433]]

Triplet Pairing

The pair breaking scattering cross-section for triplet superconductors is governed by

$$Im\left[\chi^{i,i}(\underline{q};\omega)\right] = \pi(g\mu_B)^2 Z^{-2} \qquad (19.269)$$
$$\times \frac{1}{N}\sum_{\underline{k}}\left[1 - f(E_{\underline{k}}) - f(E_{\underline{k}+\underline{q}})\right]\delta(\hbar\omega - E_{\underline{k}} - E_{\underline{k}+\underline{q}})$$
$$\times Re\left(\frac{(E_{\underline{k}} - \xi(\underline{k}))(E_{\underline{k}+\underline{q}} + \xi(\underline{k}+\underline{q})) + \vec{d}(\underline{k})\vec{d}^*(\underline{k}+\underline{q}) - 2d_i(\underline{k})d_i^*(\underline{k}+\underline{q})}{4E_{\underline{k}}E_{\underline{k}+\underline{q}}}\right).$$

In the limit $\underline{q} \to 0$, the coherence factors relevant to unpolarized experiments remain finite, in contrast with the singlet phase. This is because the vector $\vec{d}$ takes over the role of the momentum transfer $\underline{q}$, in defining the coupling of the neutron's spin to the electronic magnetic moments. The energy dependence of the various limiting forms of the unpolarized neutron cross-section are given by

$$\propto \pi\rho(\mu)\frac{\Delta_0^2}{\hbar\omega}\frac{2}{\sqrt{\hbar^2\omega^2 - 4\Delta_0^2}}, \qquad \hbar\omega \geq 2\Delta_0\ , \quad (19.270)$$

for the BW phase,

$$\propto \frac{\pi\,\rho(\mu)}{4\,\hbar\omega\,\Delta_0}\left[\left(\hbar^2\omega^2 + 4\,\Delta_0^2\right)\ln\left[\frac{\sqrt{\hbar\omega + 2\Delta_0}}{\sqrt{\hbar\omega - 2\Delta_0}}\right] - \frac{\Delta_0\hbar\omega}{2}\right] \qquad (19.271)$$

in the ABM phase, while the polar phase cross-section is proportional to

$$\propto \frac{\pi^2\rho(\mu)\hbar\omega}{8\Delta_0}\ , \quad 2\Delta_0 \geq \hbar\omega$$
$$\propto \pi\rho(\mu)\left(\sin^{-1}\frac{2\Delta_0}{\hbar\omega} - \frac{2\Delta_0}{\hbar^2\omega^2}\sqrt{\hbar^2\omega^2 - 4\Delta_0^2}\right),$$
$$2\Delta_0 \leq \hbar\omega\ . \qquad (19.272)$$

These are shown in Fig. 19.75. The non-analyticities stem from the characteristic energy dependences of the quasi-particle density of states. The $q = 0$ spectra could lead to a unique identification of triplet superconducting phases, however, neutron scattering experiments at low momentum transfers are difficult to perform. Therefore, the observation of the energy and momentum dependences of low-energy resonances corresponding to the quasi-elastic paramagnon or anti-paramagnon excitations of the normal state could be used to determine the nature of superconducting states. In particular, in triplet superconductors, there exists a possibility for observing a double-peaked structure in the low-energy magnetic response, corresponding to the different behavior of magnetic fluctuations perpendicular and parallel to the vector superconducting order parameter.

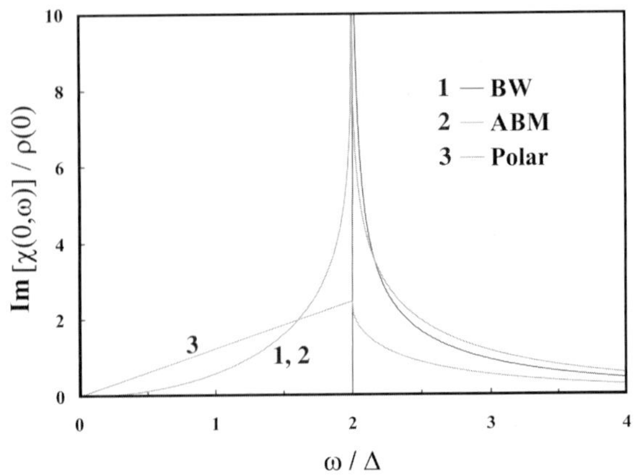

Fig. 19.75. The calculated $T \to 0$ limit of the imaginary part of the uniform ($q = 0$) frequency dependent susceptibility of p-wave superconducting phases, normalized to the density of states at the Fermi energy. The BW phase exhibits a square root singularity at $\hbar\omega = 2\Delta_0$. The imaginary part of the $q = 0$ dynamic susceptibility for the ABM phase exhibits a square root singularity at the maximum gap and a ω^2 variation at low frequencies. The polar phase shows a linear ω variation at low frequencies

19.5 Heavy Fermion Superconducting Compounds

Elsewhere in this chapter, the underlying theoretical and experimental approaches to the problem of heavy fermion superconductivity are discussed in detail. In this section we consider the heavy fermion superconductors on a compound-by-compound basis. For each compound we provide a brief introduction and an overall "snapshot" of the physical picture that is developing for each compound at the time this review was written (mid 2003). Our focus is on the nature and origin of the superconducting states themselves. One cannot, of course, completely understand the superconducting states without a clear understanding of the normal state that they arise from, so the fact that the heavy fermion problem has not yet been solved represents an intrinsic limitation in our endeavor. We assume that the reader is familiar with the general characteristics of the heavy fermion state either from reading other sections in this chapter or from any of the excellent review articles that exist such as those by Stewart [476], Hess, Riseborough, and Smith [477], or Grewe and Steglich [300] for example.

In a recent review article focused on the heavy fermion superconductors, Brison et al. [478] (citing an earlier review by Ott [479]) identify the central questions in the field, questions that date from the earliest heavy fermion superconductors studied:

- Does the pairing mechanism involve magnetic interactions?
- What is the symmetry of the superconducting order parameter and the topology of the gap function?
- What is the nature of the interplay between magnetic order (when it exists) and superconductivity?

In another recent review of unconventional pairing in superconductors Annett [480] identifies five "classes" of experimental evidence for unconventional superconductivity:

- Class 1: Experiments consistent with a multi-component order parameter, such as those experiments showing multiple superconducting phases in UPt_3 ([14] and [46], for example). These experiments constitute "definitive proof" of an unconventional superconducting state.
- Class 2: Experiments that measure the macroscopic order parameter symmetry, such as Josephson interference effects. Such experiments have been very useful in studies of the cuprates. Several heavy fermion compounds show Josephson critical currents. Measurements consistent with anisotropic order parameters exist but overall these difficult experiments have not yet been accomplished in the heavy fermion compounds as successfully as in the cuprates.
- Class 3: Measurements of the electronic spin susceptibility in the superconducting state which tends to zero as $T \rightarrow 0$ for singlet pairing and remains finite for triplet or higher order pairing, measurements such as the Knight shift of thoriated UBe_{13} [481,482], for example.
- Class 4: Experiments demonstrating that the Fermi surface average of the gap function is zero (usually through the failure of Anderson's theorem), experiments such as resistivity measurements probing the sensitivity of T_c to non-magnetic impurities. If the average is non-zero, Anderson's theorem can be applied and the superconducting state is not easily destroyed by non-magnetic impurities. Results consistent with this effect have been reported for UPt_3 [483,484].
- Class 5: Experiments consistent with the gap function vanishing at line and/or point nodes on the Fermi surface. Such a structure results in power-law temperature dependences for a host of measurable quantities such as the specific heat, ultrasonic attenuation, or thermal conductivity [485]. Such power-law dependences (as opposed to the exponential temperature dependence associated with a uniform *s*-wave gap) are ubiquitous in the heavy-fermion superconductors.

19.5.1 Uranium Compounds

UPt_3

Since the discovery of superconductivity by Stewart et al. [6], UPt_3 has arguably become the most extensively studied of all the heavy fermion superconductors. It now seems safe to assert that unconventional superconductivity in UPt_3 has been unambiguously established. A comprehensive review on UPt_3 by Joynt and Taillefer [376] has recently appeared, it contains an extensive survey of the literature and an in-depth analysis of the state of contemporary research.

Superconductivity in hexagonal UPt_3 (T_c^+ = 0.53 K) condenses out of a "heavy" normal state characterized by a resistivity that decreases monotonically with falling temperature and a spin fluctuation-like term in the specific heat, both of which are unusual in heavy fermion superconductors. Experiments show that UPt_3 is an extreme Type II superconductor (with a Ginzburg–Landau parameter of about 44 in the A-phase) in the clean limit [376].

A tendency toward magnetism manifests as a peak in the magnetic susceptibility near 18 K which evolves into a metamagnetic transition near 20 T at low temperatures. A subtle, small-moment ($\sim 0.02\mu_B$/U-atom oriented along a* in the basal plane with domain sizes of about 150 Å), antiferromagnetic order appears at $T_N \sim 5$ K which seems to be intimately related to the superconducting phase diagram as discussed below. The inability to detect a feature associated with T_N using techniques other than neutron scattering (such as μSR, NMR, or virtually any thermodynamic measurement) suggests that the moments are not static, but fluctuate at a very high rate [486].

UPt_3 exhibits three distinct superconducting phases: the A phase occurs at the onset of superconductivity in ambient fields, the B phase appears at $T_c^- = T_c^+ - 50$ mK, below the onset of the A phase and extends to $T = 0$ (the B-phase is the ground state). The proximity of T_c^+ and T_c^- suggests that they may arise from degenerate states split by some perturbation, the most likely candidate being the antiferromagnetic order. The C phase exists in magnetic fields above about 0.4 T. (One could argue that there are, in fact, five phases since the A and B phases each have a Meissner state below H_{c1} and a vortex state above H_{c1}.) A schematic of the high field superconducting phase diagram is shown in Fig. 19.76, the four phase boundaries shown meet at a true tetracritical point near (0.38 K, 0.8 T) with $H \parallel c$ and (0.43 K, 0.8 T) with $H \perp c$. Also shown in Fig. 19.76 are figures characterizing the gap structures of the E_{2u} model [487], a solid candidate for the multi-component order parameter characterizing the superconducting states of UPt_3 in terms of spin-triplet, f-wave pairing. This approach describes the unusual shape and anisotropy of the upper critical field very well with paramagnetic limiting for fields along the c-axis [488].

The interplay between superconductivity and magnetism is demonstrated most clearly by neutron-scattering and specific heat experiments under pressure: the antiferromagnetism is suppressed while the two superconducting transitions at T_c^+ and T_c^- appear to merge into one. Within experimental resolution, the superconducting transitions merge at the same pressure (about 3 kbar) that destroys antiferromagnetism [48], a most compelling result.

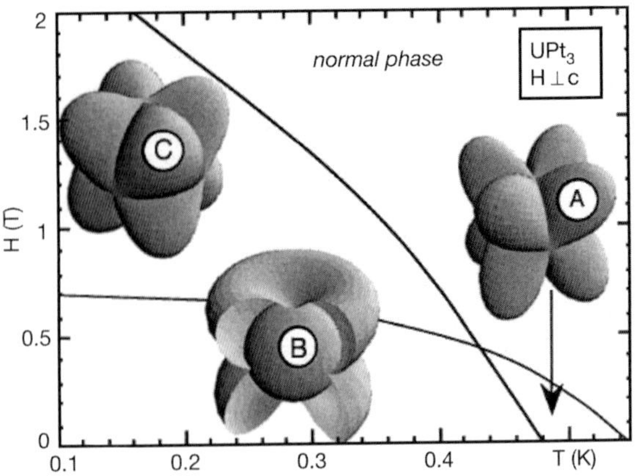

Fig. 19.76. A schematic phase diagram of the three high field superconducting states of UPt_3, together with the hypothetical corresponding gap structures of the E_{2u} model: a rotation of the azimuthal line nodes distinguishes the A and C phases, the B phase has only a second-order point node on the c-axis. [After Brison et al. [478]]

Candidate theories for the order parameter of UPt_3 include two-dimensional representations where a small coupling to the antiferromagnetism splits the energies of two degenerate states, mixed representations where the two zero field superconducting transitions involve gap functions that are not related by symmetry (so the splitting of T_c^+ and T_c^- is "accidental", unrelated to the antiferromagnetism), and others. Virtually all of the candidate theories result in an odd-parity order parameter; the reader is referred to Joynt and Taillefer [376] for an in-depth discussion.

The identification of the pairing mechanism remains one of the most important and least understood aspects of the work on the heavy fermion superconductors in general and the situation for UPt_3 is no exception. It is generally assumed that mechanism involving the exchange of spin fluctuations, similar to that of superfluid ^{3}He [139], and that the Cooper pairs are in a triplet state.

UBe_{13}

UBe_{13} (along with $CeCu_2Si_2$ and other heavy fermion superconductors that also show a magnetic phase transition) is frequently characterized as a non-Fermi Liquid superconductor [35] because the superconducting state appears (near $T_c = 0.9$ K) before Fermi liquid behavior has a chance to set in (though the application of modest pressures and/or magnetic fields causes the Fermi liquid state to emerge above T_c [490]). Of all the known heavy fermion superconductors UBe_{13} has the largest effective mass and one of the few compounds with cubic symmetry. UBe_{13} does not appear to order magnetically, though there is some evidence for field-induced small-moment magnetism coexisting with the superconducting state above about 5 T it has not been confirmed [491]. The resistivity of UBe_{13} is enormous, near 100 μΩ–cm just above T_c, leading to an electronic mean-free-path of only a few lattice spacings in length and (not surprisingly) dirty-limit superconductivity.

Many of the initial measurements on UBe_{13}, such as the specific heat, NMR spin-lattice relaxation rate, and London penetration depth exhibit power-law temperature dependences suggesting the presence of line-nodes in the gap function. Tunneling measurements have also resulted in evidence for Andreev surface bound-states [444]. More recent measurements of the specific heat and thermal expansion on higher quality single crystals show an unusual temperature dependence suggestive of a line of "anomalies" in the H–T plane that has been interpreted in terms of antiferromagnetic fluctuations associated with a quantum critical point [35]. This behavior is especially intriguing in samples doped with thorium, a subject we will return to in another section.

Measurements of the lower critical field show the quadratic temperature dependence characteristic of a conventional superconducting state, but the shape of the upper critical field is quite unusual, exhibiting significant positive curvature in the best samples. Recent measurements of the upper critical field under pressure yield a series of H_{c2}(T,P) curves that are well fit by a strong coupling theory incorporating an FFLO state in high magnetic fields. These results are also consistent with a non-phonon-mediated coupling mechanism [378].

Evidence for unconventional superconductivity in UBe_{13} is mounting but the nature of the anomalies described above have yet to be incorporated into a coherent picture of the superconducting state.

Thoriated UBe_{13}

The phase diagram of $U_{1-x}Th_xBe_{13}$ is shown in Fig. 19.3. Doping non-magnetic thorium onto the uranium sites depresses T_c non-monotonically and a second phase transition appears at a lower temperature (T_{c2}) in the thorium concentration range of about 2% to 4%. Though not as well established experimentally as the multiple superconducting phase diagram of UPt_3, pressure studies, lower critical field measurements (showing a sharp upward kink at T_{c2} [38]) and the size of the specific heat jumps suggest that the colder phase transition separates one kind of superconducting state from another. On the other hand, the observation of small moment magnetism below T_{c2} ($\mu \sim 10^{-3}\mu_B$/U-atom [33]) suggests that this lower state may be magnetic, perhaps a spin-density-wave state coexisting with the superconductivity.

Recent phase diagram studies utilizing both specific heat and thermal expansion measurements are shown in Fig. 19.3 where the open symbols denote the "anomalies" discussed in the section on pure UBe_{13} and the solid symbols represent phase transitions [35]. There is an intriguing correlation suggesting that the anomalies in the pure compound evolve into the colder phase of the thoriated compounds. Also shown in Fig. 19.3 is a vertical solid line near a thorium concentration $x_{c1} \approx 0.02$ representing the pressure studies of Zieve et al. [34] who find clear evidence for a vertical phase boundary separating the two regions.

Theoretical efforts to understand this unusual phase diagram include multiple superconducting states and the onset of antiferromagnetism [492], broken time-reversal symmetry at T_{c2} [493], a "freezing" of the antiferromagnetic fluctuations associated with proximity to a quantum critical point [35], and a "ferrisuperconductivity" theory involving a multi-component order parameter and the appearance of a charge density wave state at T_{c2} [494].

URu_2Si_2

Tetragonal URu_2Si_2 enters the superconducting state near 1.5 K, a temperature about 10 times colder than the temperature where it enters into a state of "hidden order". The term "hidden order" denotes an unknown phase with which the superconductivity coexists and which is presently the subject of much current study and debate. Long thought to be a transition to a small-moment ($\mu = 0.03\mu_B$/U-atom) antiferromagnetic state, it has also been recognized for a long time that the size of the specific heat jump is much larger than one would expect for such a transition. Recent neutron diffraction [225] and NMR measurements under pressure [224] have led to the suggestion that the magnetism is from a minority phase, though this matter is still under active discussion [495]. A resolution of this issue will impact our understanding of the superconducting state which, according to recent neutron scattering measurements, is coupled to the magnetic order [496].

Evidence for unconventional superconductivity manifests in specific heat measurements consistent with a line of nodes in the gap function [497] and by anisotropic point contact measurements of the Josephson current between URu_2Si_2 and Nb, a result which suggests that the order parameter must have even parity [452] (e.g. *d*-wave). The upper critical field is strongly anisotropic, strongly Pauli limited along the *c*-axis, and exhibits a positive curvature consistent with *d*-wave symmetry [498]. It has been pointed out that this interpretation violates the symmetry required of a simple antiferromagnet [499], but with recent measurements suggesting that the magnetism may not be intrinsic the original argument remains an important part of the picture of unconventional superconductivity developing for this compound [500].

UPd_2Al_3

The superconducting state of UPd_2Al_3 condenses near $T_c = 2.0$ K out of a coexisting antiferromagnetic state ($T_N = 14.5$ K) with atomic-like moments ($\mu = 0.85\mu_B$/U-atom) aligned in the hexagonal basal plane [53,342,501]. NMR measurements of the nuclear relaxation time [502] and the thermal conductivity [503] are consistent with a line of nodes in the gap function, while Knight shift measurements and a Pauli limited upper critical field indicate even parity pairing [504]. But the most important development in UPd_2Al_3 derives from the fact that relatively high quality thin films, with a T_c near 1.6 K only slightly reduced from the bulk T_c, have been fabricated for tunneling experiments. These experiments reveal a strong coupling to an excitation whose energy nearly matches a magnetic excitation clearly visible in neutron scattering measurements [431,475]. Taken together, these exciting experimental results constitute the strongest evidence for a magnetic mechanism (in this case magnetic excitons) for superconductivity yet discovered.

UNi_2Al_3

Though similar in many respects to isostructural UPd_2Al_3, there are clear differences in the unconventional superconducting state of UNi_2Al_3. Superconductivity appears near 1 K and coexists with an incommensurate spin density wave state with a (relatively) small magnetic moment (about 0.2 μ_B/U-atom) that appears near 4.5 K [505,506]. As with the Pd compound, NMR measurements of the temperature dependence of T_1 suggest the presence of line nodes in the gap [507]. The upper critical field of UNi_2Al_3 is well fit by the conventional WHHM theory without paramagnetic limiting [508,509] while measurements of the Knight shift show that the spin susceptibility does not change on crossing T_c and down to 50 mK [467], both of these results are consistent with a spin-triplet superconducting state.

U_2PtC_2

The lowest effective mass of all the heavy fermion compounds we discuss belongs to U_2PtC_2 which may be considered as intermediate between the heavy fermion superconductors and less-anomalous uranium based superconductors such as U_6Fe. Though the critical temperature is appreciable ($T_c = 1.47$ K) we are unaware of any clear evidence for unconventional superconductivity in this compound owing to a lack of experimental studies.

19.5.2 Cerium Compounds

$CeCu_2Si_2$

Tetragonal $CeCu_2Si_2$ was the first heavy fermion superconductor discovered [3]. The close proximity of superconductivity and magnetism in $CeCu_2Si_2$ is highlighted by the sensitivity of the ground state to the details of sample preparation and stoichiometry. Quite a bit of work was required to sort out the ground state which can be either bulk superconducting (S-type), magnetic (A-type), both (A/S-type), or "a related phase" (X-type) which is probably magnetic; there is also a high field magnetic phase (the B-phase) and recent high field magnetization studies are consistent with the A and B phases being spin density wave phases [511]. The best single crystals are of the A/S type [512] which becomes superconducting at 0.68 K just below the onset of the magnetic A-phase at 0.70 K. μSR measurements show that the superconducting state competes with the magnetic state (consistent with an SDW state which will compete with superconductivity for the Fermi surface) and that the two states do not coexist microscopically [307]. Given this proximity of low temperature states it is not surprising that $CeCu_2Si_2$ exhibits non Fermi-liquid behavior in its specific heat and electrical resistivity [193]. It is also not surprising that the superconducting properties of $CeCu_2Si_2$ show significant sample dependences.

In studies of several off-stoichiometry single crystal samples of $CeCu_2Si_2$, the anisotropic upper critical fields can sometimes be fit by a standard s-wave model with significant Pauli limiting and sometimes not [386] though the overall shape is qualitatively consistent with conventional superconductivity. The superconductivity of $CeCu_2Si_2$ is in the clean limit and exhibits a strong type II character with a typical Ginzburg-Landau parameter in the 50–56 range.

Early studies of the superconducting states of a wide range of samples do not show significant evidence for unconventional superconductivity [386]. A more recent study, however, shows evidence for two different kinds of superconducting states from samples with only slightly different stoichiometry. The two sample groups show a T^2 dependence for the superconducting specific heat but opposite pressure dependences for T_c, d-wave superconductivity has been proposed [513].

$CeRhIn_5$

The tetragonal crystal structure of $CeRhIn_5$ (and that of $CeIrIn_5$ and $CeCoIn_5$ discussed below) can be thought of as layers of $CeIn_3$ separated by layers of $RhIn_2$ leading to a quasi-two-dimensional electronic structure. Some features of this sub-family of heavy fermion superconductors, such as a proximity to antiferromagnetism, may therefore be similar to the cuprate superconductors [59]. However, unlike the isostructural compounds $CeIrIn_5$ and $CeCoIn_5$ (both of which are heavy fermion superconductors at ambient pressure), $CeRhIn_5$ is a heavy fermion antiferromagnet at ambient pressure with a Néel temperature $T_N = 3.8$ K. In the antiferromagnetic state the Ce spins (0.8 μ_B/Ce-atom [510]) form a helical structure along the c-axis, a structure characterized by an incommensurate wave vector [514]. The magnitude of the internal field created by the Ce spins decreases linearly with applied pressure while T_N remains nearly constant [515]. At a critical pressure near $P_c \sim 15$ kbar specific heat measurements suggest that the antiferromagnetic state is abruptly replaced by a superconducting state whose critical temperature (about 2.1 K) is surprisingly pressure independent [516]. More recent NQR measurements show that the two states coexist homogeneously at a pressure of 17.5 kbar. At 21 kbar the spin-lattice relaxation rate exhibits a T^3 temperature dependence and no coherence peak, suggesting a line node in the gap function. A T^2 term in the superconducting specific heat is also consistent with line nodes [516].

$CeIrIn_5$

Nearly antiferromagnetic $CeIrIn_5$ shows quadratic and linear temperature dependences of the superconducting specific heat and thermal conductivity (in the "universal limit") respectively [60] along with a T^3 temperature dependence of the spin-lattice relaxation rate with no Hebel–Slichter coherence peak [62, 63]. All of these experimental results are consistent with line nodes in the gap function. The upper critical field is anisotropic but of a conventional shape that does not yield strong experimental evidence either for or against Pauli limiting [517].

$CeCoIn_5$

Just at the edge of an antiferromagnetic quantum critical point, tetragonal $CeCoIn_5$ exhibits the highest critical temperature ($T_c = 2.3$ K) of the heavy fermion superconductors and an enormous specific heat jump at T_c. Several experimental results are consistent with line nodes in the energy gap: T^2 and T^3 temperature dependences of the superconducting specific heat and thermal conductivity respectively [60] and a T^3 temperature dependence of the spin-lattice relaxation rate with no Hebel–Slichter coherence peak [62]. The Knight shift decreases for both parallel and perpendicular directions to the tetragonal c-axis in the superconducting state, which shows that the spin susceptibility decreases with decreasing temperature consistent with spin singlet pairing [62]. A $d_{x^2+y^2}$ state has been proposed for $CeCoIn_5$ based on thermal conductivity measurements which reveal a fourfold symmetry in the a–b plane, characteristic of a superconducting gap with nodes along the ($\pm\pi$, $\pm\pi$) directions [61]. Also consistent with such a state, several groups report that the phase transition at H_{c2} is first order in high fields [61, 519].

The lower critical field shows a linear temperature dependence [520], while the upper critical field exhibits an unusual low temperature shape and hysteresis [521]. Both critical fields are anisotropic. Strong indications of an FFLO state in $CeCoIn_5$ [519] have been confirmed with the observation of a second order phase boundary just below H_{c2} at low temperature [522].

$CeIn_3$

It is not unreasonable to think of cubic $CeIn_3$ as the parent compound of the previous three heavy fermion superconductors. Under ambient pressure $CeIn_3$ is a heavy fermion antiferromagnet with an ordering temperature $T_N = 10.2$ K, an ordered moment of 0.65 μ_B/Ce-atom, and a wave vector in the [111] direction [523]. As pressure is applied T_N is depressed and eventually vanishes near a temperature of about 0.18 K at a critical pressure of about 26 kbar, below this temperature antiferromagnetism is replaced by a superconducting state [524]. This P–T phase diagram, similar to that observed in several high-T_c compounds, has been shown to be characteristic of magnetically mediated superconductivity [524]. Measurements of the spin-lattice relaxation time at a pressure of 26.5 kbar do not show a coherence peak [525] though the upper critical field (at the same pressure) can be described by a conventional strong coupling model in the clean limit [65]. Existing experimental results are not inconsistent with unconventional superconductivity in this important compound, especially given its P–T phase diagram, but, as with most of the heavy fermion superconductors (especially those below), more work is required to identify the symmetry of the order parameter.

$CePd_2Si_2$

Evolving from a heavy fermion antiferromagnet to a heavy fermion superconductor under applied pressure, the P–T phase diagram of $CePd_2Si_2$ is similar to that of $CeIn_3$ and suggests that the superconductivity of $CePd_2Si_2$ is also magnetically mediated [524]. The disappearance of antiferromagnetism does not quite coincide with the maximum superconducting critical temperature (as it does in $CeIn_3$) of about 0.52 K which occurs near an applied pressure of 5 GPa. The shape of the upper critical field at this pressure is not inconsistent with conventional superconductivity though there appear to be some deviations at the lowest temperatures [526]. Another study shows that the upper critical field is well described by a weak-coupling, clean-limit model with a slightly anisotropic orbital limit and a strongly anisotropic paramagnetic one [527].

$CeRh_2Si_2$

The P–T phase diagram of $CeRh_2Si_2$ is qualitatively similar to that of $CePd_2Si_2$ and $CeIn_3$ (discussed above) where antiferromagnetism (ambient pressure $T_N = 36$ K) is replaced by superconductivity near a critical pressure of 9 kbar and a critical temperature of 0.35 K [69]. The upper critical field is not inconsistent with that of a conventional Type II superconductor though the transition widths are wide and the data are sparse.

$CeCu_2Ge_2$

Superconductivity at very high pressures has been reported for $CeCu_2Ge_2$. Similar to other Ce compounds discussed above, the antiferromagnetic order ($T_N = 4.7$ K at ambient pressure) is suppressed between 7 and 10 GPa and is replaced by a superconducting state which appears below a temperature of about 0.6 K. The critical temperature is nearly pressure independent up to about 13 GPa after which it increases rapidly to a maximum value of almost 2 K near 16 GPa before vanishing near 20 GPa [528]. The upper critical field shows significant Pauli limiting consistent with singlet pairing [529].

19.5.3 Praseodymium Compounds

$PrOs_4Sb_{12}$

Discovered just last year, cubic $PrOs_4Sb_{12}$ is the only known Pr based heavy fermion superconductor so far. Thermal conductivity [530] and specific heat [531] measurements suggest the presence of two distinct superconducting states featuring point-nodes in the gap function (an unusual feature in heavy fermion superconductors). No Hebel–Slichter peak is observed in the temperature dependence of the spin-lattice relaxation time, although an exponential temperature dependence at low temperatures suggests that the gap function is isotropic [532]. An exponential temperature dependence is also found for the magnetic penetration depth as deduced from μSR measurements [533]. A small internal magnetic field, appearing just below T_c is clearly visible in the μSR measurements of Aoki et al. [534], consistent with a multi-component order parameter which, in turn, is consistent with multiple superconducting states. The upper critical field is of a conventional shape and does not show evidence of Pauli limiting [535]. The lower critical field, on the other hand, exhibits a pronounced positive curvature even at the lowest temperatures [536]. Lastly, we note that the presence of a field-induced antiferro-quadrupolar ordered phase just above the upper critical field suggests that quantum quadrupole fluctuations of the Pr ions may play a role in the superconducting mechanism.

19.5.4 Related Materials

Lastly, we discuss several superconductors that are not usually considered to be heavy fermion superconductors but which show a "family resemblance" in the sense that all are examples of superconductivity in strongly correlated electron systems.

UGe_2, URhGe, and $ZrZn_2$

The topic of ferromagnetic superconductivity might be said to date from 1957 and the seminal work of Ginzburg [538]; a number of excellent reviews covering the work of the intervening decades exist (see, for example, the articles collected in [539]) most recently by Flouquet et al. [540]. A most intriguing possibility for the following compounds is that superconductivity and ferromagnetism involve the *same* electrons.

Superconductivity in orthorhombic UGe_2 appears near a pressure of 12 kbar just below the pressure at which ferromagnetism is suppressed [66]. The maximum critical temperature $T_c = 0.7$ K occurs at 12 kbar where the Curie temperature is still about 30 K and the size of the ordered moment is about 1 μ_B/U-atom. Estimates of the internal field created by the ferromagnetic order are of order 100 T so triplet pairing seems likely. The shape of the upper critical field is anisotropic and strongly pressure dependent, exhibiting re entrant behavior near a pressure of 13.2 kbar [541]. A p-wave superconducting state has been proposed that coexists with the itinerant ferromagnetism [542].

At ambient pressure, ferromagnetism appears at 9.5 K in orthorhombic URhGe followed by superconductivity near 0.3 K [68]. The upper critical field of a polycrystal exhibits a conventional shape. Triplet superconductivity is expected given the large internal fields associated with the 0.5 μ_B/U-atom moments.

Cubic $ZrZn_2$ has been reported to be an ambient pressure superconductor ($T_c = 0.3$ K) where the superconductivity appears out of an itinerant ferromagnetic state ($T_C = 28.5$ K with an ordered moment of 0.17 μ_B/f.u.) [543]. The upper critical field shows a qualitatively conventional shape. Recent theoretical work suggests a multi-component order parameter with nodes in the gap function [544], but the reader is cautioned that very recent work suggests that the superconductivity may be due to surface effects and is not a property of bulk $ZrZn_2$ [545]. Clearly a great deal of work remains to be done to characterize the superconducting states of these intriguing compounds.

$PuCoGa_5$

The final compound we discuss in this section is the tetragonal plutonium based compound $PuCoGa_5$ which exhibits the highest critical temperature ($T_c = 18.5$ K) ever observed for an f-electron compound [77]. $PuCoGa_5$ (and $PuRhGa_5$ which is reported to superconduct at 8.7 K [546]) exhibits the same crystal structure as the Ce 1–1–5 compounds discussed above, aspects of the Fermi surface structure are also quite similar [547]. Recent electronic structure calculations suggest that the superconductivity is caused by the pairing of plutonium 5f electrons [548]. Measurements of the upper critical field near T_c, fit to a conventional WHHM form, suggest that the critical field reaches 74 T at $T = 0$ which would exceed the Pauli limit. The lower critical field at 5 K is estimated to be about 350 Oe which, in turn, permits an estimate of 32 for the Ginzburg–Landau parameter.

We close this section by referring to Table 19.11 containing various superconducting parameters of the heavy fermion superconductors we have discussed above. The significant number of empty spaces is just one piece of evidence suggesting the opportunities still awaiting experimentalists in this mature yet steadily growing field of research.

19.6 The Conclusion

Superconductivity is one of nature's most ubiquitous collective phenomena in many-particle systems. Superconductivity in metals was explained by the theory of Bardeen, Cooper, and Schrieffer (BCS) in 1957. The basic picture was the notion of Cooper pairing where two fermions (electrons in metals) can form a bound pair in a relative s-wave state (conventional superconductivity) in the neighborhood of the Fermi surface. The interaction that causes the binding is the exchange of a boson (a phonon in a metal) between the fermions, which is strong enough to overcome the repulsive interactions between the fermions. With the great success of the BCS theory, it was not long before the notion of Cooper pairing "tunneled its way" into other fields of physics. This included the role of Cooper pairs in nuclear structure, the pairing of nucleons in nuclear matter via the exchange of pions and other bosonic particles, which is important in nuclear astrophysics, and the proposal for color superconductivity in a quark gluon plasma, to name a few. What binds the Cooper phenomena together in these different fields is that the dominant mechanism for pairing of the fermions is the exchange of some bosonic particle that coexists along side the fermions, like the electrons in an elastic medium or the gluons in the quark-gluon soup. It is here that heavy-fermion superconductivity parts company from the standard model for Cooper pairing.

Heavy-fermion superconductors were the first electronic example of Cooper pairs being formed predominantly from the exchange of collective excitations of the many-body medium; liquid ^{3}He is the charge neutral superfluid example. In a many-body system, bosonic excitations are formed from collective excitations of the fermions, like density and spin-density fluctuations. Two fermions can interact by the exchange of these bosons. This effective interaction between the fermions can give rise to either repulsive interactions or attractive interactions that form Cooper pairs. This, in general, is an intractable many-body problem because the Cooper pairs will be formed by excitations made up from the same fermions that would like to form other kinds of collective pairs in addition to the Cooper pairs. What made the heavy-fermion problem, as well as liquid ^{3}He, special was the separation of the high-energy electronic scales from the low-energy collective scales, which made it possible to develop microscopic and phenomenological theories of the interacting system. This large separation of energy scales is also responsible for the unconventional nature of the superconductivity. The underlying force between the electrons is the strong short-range repulsive coulomb interaction between the electrons in a relative s-wave state. Thus, it is no surprise that s-wave superconductivity is not seen in heavy-fermion superconductors. The unconventional nature of the superconducting state in heavy-fermion superconductors is the Cooper pair trying to find a state that maximizes the attractive interactions while minimizing the strong coulomb repulsion.

There is no other example in condensed-matter physics where the interaction between theory and experiment played such a crucial role in uncovering the unconventional nature of the pairing and the many-body origins of the pairing interaction. Even today, with all of the attention paid to it, there is far less theoretical insight into the physics of the high-temperature superconductor problem than we have in the heavy-fermion problem. Heavy-fermion superconductivity still remains as the model problem for how to do physics in strongly interacting many-body systems.

The field remains healthy, and young scientists still have the hope of exciting new superconductors. The evidence is in last section on the first plutonium-containing superconductor that was discovered recently, and it had the highest transition temperature of any heavy-fermion superconductor. This was absolutely unexpected. There is still much experimental and theoretical work to be done before this subfield matures.

Acknowledgements

We thank K.S. Bedell and J.C. Lashley for discussions. The work at Temple University was supported by the US Department of Energy, Office of Basic Energy Sciences, through award no. DEFG02-01ER45872. The work at Occidental College was supported by the National Science Foundation under DMR-0305397. One of us (GMS) gratefully acknowledges the hospitality of the National High Magnetic Field Laboratory at Los Alamos National Laboratory (supported by the National Science Foundation, the State of Florida and the U.S. Department of Energy respectively.) Work at Los Alamos was performed under the auspices of the U.S. Dept. of Energy.

References

1. E. Bucher, J. P. Maita, G.W. Hull, R.C. Fulton, and A.S. Cooper, Phys. Rev. B **11**, 440 (1975)
2. W. Franz, A. Grießel, F. Steglich, and D. Wohlleben, Z. für Physik B **31**, 7 (1978)
3. F. Steglich, J. Aarts, C. Bredl, W. Lieke, D. Meschede, W. Franz, and H. Schäfer, Phys. Rev. Lett. **43**, 1892, (1979)
4. H.R. Ott, H. Rudigier, Z. Fisk, and J.L. Smith, Phys. Rev. Lett. **50**, 1595 (1983)
5. H.R. Ott, H. Rudigier, T.M. Rice, K. Ueda, Z. Fisk, and J.L. Smith, Phys. Rev. Lett. **52**, 1915 (1984)
6. G.R. Stewart, Z. Fisk, J.O. Willis, and J.L. Smith, Phys. Rev. Lett. **52**, 679 (1984)
7. J.L. Smith, Z. Fisk, J.O. Willis, B. Batlogg, and H.R. Ott, J. Appl. Phys. **55**, 1996 (1984)
8. H.R. Ott, H. Rudigier, Z. Fisk, and J.L. Smith, Phys. Rev. B **31**, 1651 (1985)
9. Z. Fisk, H.R. Ott, T.M. Rice, and J.L. Smith, Nature **320**, 124 (1986)
10. Z. Fisk, D.W. Hess, C.J. Pethick, D. Pines, J.L. Smith, J.D. Thompson, and J.O. Willis, Science **239**, 33 (1988)
11. J.G. Bednorz, and K.A. Müller, Z. für Physik B, **64**, 189 (1986)
12. R. Movshovich, M.A. Hubbard, M.B. Salamon, A.V. Balatsky, R. Yoshizaki, J.L. Sarrao, and M. Jaime, Phys. Rev. Lett. **80**, 1968 (1988)
13. A.C. Mota, E. Dumont, A. Amann, and Y. Maeno, Physica B **261**, 934 (1999)
14. R.A. Fisher, S. Kim, B.W. Woodfield, N.E. Phillips, L. Taillefer, K. Hasselbach, J. Flouquet, A.L. Giorgi, and J.L. Smith, Phys. Rev. Lett. **62**, 1411 (1989)
15. G.R. Stewart, Z. Fisk, and M.S. Wire, Phys. Rev. B **30**, 482 (1984)
16. J.-M. Mignot, J. Flouquet, P. Haen, F. Lapierre, L. Puech, and J. Voiron, J. Mag. Mag. Mat., **76–77**, 97 (1988)
17. T. Oguchi, A.J. Freeman, and G. Crabtree, Phys. Lett. A **117**, 428 (1986)

18. L. Taillefer, R. Newbury, G.G. Lonzarich, Z. Fisk, and J.L. Smith, J. Mag. Mag. Mat. **63-64**, 372 (1987)
19. L. Taillefer, and G.G. Lonzarich, Phys. Rev. Lett. **60**, 1570 (1988)
20. T.T.M. Palstra, A.M. Menovsky, A.J. Dirkmaat, P.H. Kes, G.J. Nieuwenhuys, and J.A. Mydosh, Phys. Rev. Lett. **55**, 2727 (1985)
21. M.B. Maple, J.W. Chen, Y. Dalichaouch, T. Kohara, C. Rossel, M.S. Torikachvili, M.W. McElfresh, and J.D. Thompson, Phys. Rev. Lett. **56**, 185 (1986)
22. W. Assmus, M. Herrmann, U. Rauchschwalbe, S. Riegel, W. Lieke, H. Spille, S. Horn, G. Weber, F. Steglich, and G. Cordier, Phys. Rev. Lett. **52**, 469 (1984)
23. M.B. Maple, J.W. Chen, S.E. Lambert, Z. Fisk, J.L. Smith, H.R. Ott, J.S. Brooks, and M.J. Naughton, Phys. Rev. Lett. **54**, 477 (1985)
24. J.W. Chen, S.E. Lambert, M.B. Maple, Z. Fisk, J.L. Smith, G.R. Stewart, and J.O. Willis, Phys. Rev. B, **30**, 1583 (1984)
25. D.E. MacLaughlin, Cheng Tien, W.G. Clark, M.D. Lan, Z. Fisk, J.L. Smith, and H.R. Ott, Phys. Rev. Lett. **53**, 1833 (1984)
26. B. Golding, D.J. Bishop, B. Batlogg, W.H. Haemmerle, Z. Fisk, J.L. Smith, and H.R. Ott, Phys. Rev. Lett. **55**, 2479 (1985)
27. B. Batlogg, D.J. Bishop, B. Golding, E. Bucher, J. Hufnagl, Z. Fisk, J.L. Smith, and H.R. Ott, Phys. Rev. B **33**, 5906 (1986)
28. D. Jaccard, J. Flouquet, Z. Fisk, J.L. Smith, and H.R. Ott, J. Phys. (Paris) **46**, L 811 (1985)
29. J.J.M. Franse, A. Menovsky, A. de Visser, C.D. Bredl, U. Gottwick, W. Lieke, H.M. Mayer, U. Rauchschwalbe, G. Sparn, and F. Steglich, Z. für Physik B **59**, 15 (1985)
30. D.J. Bishop, C.M. Varma, B. Batlogg, E. Bucher, Z. Fisk, and J.L. Smith, Phys. Rev. Lett. **53**, 1009 (1984)
31. Y. Kohori, T. Kohara, H. Shibai, Y. Oda, T. Kaneko, Y. Kitaoka, and K. Asayama, J. Phys. Soc. Jpn. **56**, 2263 (1987)
32. B. Batlogg, D. Bishop, B. Golding, C.M. Varma, Z. Fisk, J.L. Smith, and H.R. Ott, Phys. Rev. Lett. **55**, 1319 (1985)
33. R.H. Heffner, J.L. Smith, J.O. Willis, P. Birrer, C. Baines, F.N. Gygax, B. Hitti, E. Lippelt, H.R. Ott, A. Schenck, E.A. Knetsch, J.A. Mydosh, and D.E. MacLaughlin, Phys. Rev. Lett. **65**, 2816 (1990)
34. R.J. Zieve, D.S. Jin, T.F. Rosenbaum, J. S. Kim and G.R. Stewart, Phys. Rev. Lett. **72**, 756 (1994); R.J. Zieve, R. Duke, and J.L. Smith, Phys. Rev. B, **69**, 144503 (2003)
35. N. Oeschler, F. Kromer, T. Tayama, K. Tenya, P. Gegenwart, G. Sparn, F. Steglich, M. Lang, G.R. Stewart, Acta Phys. Pol. B **34**, 255 (2003)
36. K. Machida, and M. Kato, Phys. Rev. Lett. **58** (1986 (1987)
37. M. Gulácsi, and Zs. Gulácsi, Phys. Rev. B **39** 12352 (1989)
38. U. Rauchschwalbe, F. Steglich, G.R. Stewart, A.L. Giorgi, P. Fulde, and K. Maki, Europhys. Lett. **3**, 751 (1987)
39. K. Hasselbach, L. Taillefer, and J. Flouquet, Phys. Rev. Lett. **63**, 93 (1989)
40. U. Rauchschwalbe, U. Ahlen, F. Steglich, D. Rainer, and J.J.M. Franse, Z. für Physik B, **66**, 379 (1985)
41. B.S. Shivaram, T.F. Rosenbaum, and D. Hinks, Phys. Rev. Lett. **57**, 1259 (1986)
42. B.S. Shivaram J.J. Ganon Jr., and D. Hnks, Phys. Rev. Lett. **63**, 1723 (1989)
43. Y.J. Qian, M.F. Xu, A. Schenstrom, H.P. Braun, J.B. Ketterson, D. Hinks, M. Levy, and B.K. Sarma, Solid State Commun. **63**, 599 (1987)
44. V. Müller, C, Roth, D. Maurer, E.W. Scheidt, K. Lüders, E. Bucher, and H.E. Bömmel, Phys. Rev. Lett. **58**, 1224 (1987)
45. A.M. Shenstrom, M.F. Xu, Y. Hong, D. Bein, M. Levy, B.K. Sarma, S. Adenwalla, Z. Zhao, T. Tokayasu, D.W. Hess, J.B. Ketterson, J.A. Sauls, and D.G. Hinks, Phys. Rev. Lett. **62**, 332 (1989)
46. S. Adenwalla, S.W. Lin, Q.Z. Ran, Z. Zhao, J.B. Ketterson, J.A. Sauls, D. Hinks, M. Levy, and B.K. Sarma, Phys. Rev. Lett. **65**, 2298 (1990)
47. G. Bruls, D. Weber, B. Wolf, P. Thalmeier, B. Lüthi, A. de Visser, A.A. Menovsky, Phys. Rev. Lett. **65**, 2294 (1990)
48. S.M. Hayden, L. Taillefer, C. Vettier, and J. Flouquet, Phys. Rev. B **46**, 8675 (1992)
49. T. Trappmann, H. von Löhneysen, and L. Taillefer, Phys. Rev. B **43**, 13714 (1991)
50. D.S. Jin, S.A. Carter, B. Ellman, T.F. Rosenbaum, and D.G. Hinks, Phys. Rev. Lett. **68**, 1579 (1992)
51. M. Bouhkny, G.L. Bullock, B.S. Shivaram, and D.G. Hinks, Phys. Rev. B **50**, 8985 (1994)
52. C. Geibel, S. Theis, D. Kaczorowski, A. Mehner, A. Grauel, B. Seidel, U. Ahlheim, R. Helfrich, K. Petersen, C.D. Bredl, and F. Steglich, Z. für Physik B **83**, 305 (1991)
53. C. Geibel, C. Shank, S. Theis, H. Kitazawa, C.D. Bredl, A. Bohm, M. Rau, A. Grauel, R. Caspary, R. Helfrich, U. Ahlheim, G. Weber, and F. Steglich, Z. für Physik B **84**, 1 (1991)
54. C. Petrovic, R. Movshovich, M. Jaime, P.G. Pagliuso, M.F. Hundley, J.L. Sarrao Z. Fisk, and J.D. Thompson, Europhys. Lett. **53**, 354 (2001)

55. C. Petrovic, P. Pagliuso, M.F. Hundley, R. Movshovich, J.D. Thompson, Z. Fisk, and P. Monthoux, J. Phys. C.M. **13**, L 337 (2001)
56. T. Takeuchi, T. Inoure, K. Sugiyama, D. Aoki, Y. Haga, K. Kindo, and Y. Õnuki, J. Phys. Soc. Jpn. **70**, 877 (2001)
57. Y. Haga, Y. Inada, H. Harima, K. Oikawa, M. Murakawa, H. Nakawaki, Y. Tokiwa, D. Aoki, H. Shishido, S. Ikeda, N. Watanabe, and Y. Õnuki, Phys. Rev. B, **63**, 060503(R) (2001)
58. H. Shishido, R. Settai, D. Aoki, S. Ikeda, H. Nakawaki, N. Nakamaura, T. Iizuka, Y. Inada, K. Sugiyama, T. Takeuchi, K. Kindo, T.C. Kobayashi, Y. Haga, H. Harima, Y. Aoki, T. Namiki, H. Sato, and Y. Õnuki, J. Phys. Soc. Jpn., **71**, 162 (2002)
59. H. Hegger, C. Petrovic, E.G. Moshpoulo, M.F. Hundley, J.L. Sarrao, Z. Fisk, and J.D. Thompson, Phys. Rev. Lett. **84**, 4986 (2000)
60. R. Movshovich, M. Jaime, J.D. Thompson, C. Petrovic, P.G. Pagliuso, Z. Fisk, and J.L. Sarrao, Phys. Rev. Lett. **86**, 5152 (2001)
61. K. Izawa, H. Yamaguchi, Y. Matsuda, H. Shishido, R. Settai, and Y. Õnuki, Phys. Rev. Lett. , **87**, 057002 (2001)
62. Y. Kohori, Y. Yamato, Y. Iwamoto, T. Kohara, E.D. Bauer, M.B. Maple, and J.L. Sarrao, Phys. Rev. B, **64**, 134526 (2001)
63. G.-Q. Zheng, K. Tanabe, T. Mito, S. Kawasaki, Y. Kitaoka, D. Aoki, Y. Haga, and Y. Õnuki, Phys. Rev. Lett. **85**, 4664 (2001)
64. I.R. Walker, F.M. Grosche, D.M. Freye, and G.G. Lonzarich, Physica C **282**, 303 (1997)
65. G. Knebel, D. Braithwaite, P.C. Canfield, G. Lapertot, and J. Flouquet, Phys. Rev. B **65**, 024425 (2001)
66. S.S. Saxena, P. Agarwal, K. Ahilan, F.M. Grosche, R.K.W. Haselwimmer, M.J. Steiner, E. Pugh, I.R. Walker, S.R. Julian, P. Monthoux, G.G. Lonzarich, A. Huxley, I. Sheikin, D. Braithwaite, and J. Flouquet, Nature **406**, 587 (2000)
67. A. Huxley, I. Sheikin, E. Ressouche, N. Kernavanois, D. Braithwaite, R. Calemczuk, and J. Flouquet, Phys. Rev. B **63**, 144519 (2001)
68. D. Aoki, A. Huxley, E. Ressouche, D. Braithwaite, J. Flouquet, J.-P. Brison, E. Lhotel, and C. Paulsen, Nature **413**, 613 (2001)
69. R. Movshovich, T. Graf, D. Mandrus, J.D. Thompson, J.L. Smith, and Z. Fisk, Phys. Rev. B **53**, 8241 (1996)
70. I.E. Dzyaloshinskii, and P.S. Kondratenko, Sov. Phys. J.E.T.P. **43**, 1036 (1976)
71. K.B. Blagoev, J.R. Engelbrecht, and K.S. Bedell, Phil. Mag. Lett. **78**, 169 (1998); K.B. Blagoev, J.R. Engelbrecht, and K.S. Bedell, Phys. Rev. Lett. **82**, 133 (1999)
72. N.I. Karchev, K.B. Blagoev, K.S. Bedell and P.B. Littlewood, Phys. Rev. Lett. **86**, 846 (2001)
73. J.A. Hertz, Phys. Rev. B **14**, 1165 (1976)
74. M.A. Continentino, Phys. Rep. **239**, 180 (1994)
75. A.J. Millis, Phys. Rev. B **48**, 7183 (1993)
76. D. Fay, and J. Appel, Phys. Rev. B **32**, 6071 (1985)
77. J.L. Sarrao, L.A. Morales, J.D. Thompson, B.L. Scott, G.R. Stewart, F. Wastin, J. Rebizant, P. Boulet, E. Colineau, and G.H. Lander, Nature **420**, 297 (2002)
78. A.A. Abrikosov, L.P. Gor'kov, and I.E. Dzyaloshinki, *Methods of Quantum Field Theory in Statistical Physics* (Dover Publications, New York, 1975)
79. J.M. Luttinger, and J.C. Ward, Phys. Rev., 118, 1417 (1960); J.M. Luttinger, Phys. Rev. **119**, 1153 (1960); J.M. Luttinger, Phys. Rev. **121**, 942 (1961)
80. M. Springford, and A. Wasserman, Adv. Phys. **45**, 471 (1996)
81. S. Donovan, A. Schwartz, and G. Grüner, Phys. Rev. Lett. **79**, 1401 (1997)
82. A.M. Awasthi, L. DeGiorgi, G. Grüner, Y. Dalichaouch, and M.B. Maple, Phys. Rev. B **48**, 10 692 (1993)
83. L. DeGiorgi, M. Dressel, G. Grüner, P. Wachter, N. Sato, and T. Komatsubara, Europhys. Lett. **25**, 3111 (1994)
84. W. Kohn, Phys. Rev. **A 171**, 133 (1964)
85. L. DeGiorgi, Rev. Mod. Phys. **71**, 687 (1999)
86. E.J. Singley, D.N. Basov, E.D. Bauer, and M.B. Maple, Phys. Rev. B **65**, 161101(R) (2002)
87. F. Bommeli, L. DeGiorgi, P. Wachter, F.B. Anders, and A.V. Mitin, Phys. Rev. B **56**, R1001 (1997)
88. J. Goldstone, Nuovo Cimento **19**, 154 (1961)
89. A.I. Schindler, and B.R. Coles, J. Appl. Phys. **39**, 956 (1968)
90. S. Doniach, and S. Engelsberg, Phys. Rev. Lett. **17**, 750 (1966)
91. N.F. Berk, and J.R. Schrieffer, Phys. Rev. Lett. **17**, 433 (1966)
92. T. Izuyama, D.J. Kim, and R. Kubo, J. Phys. Soc. Jpn. **18**, 1025 (1963)
93. T. Moriya, *Spin Fluctuations in Itinerant Magnets* (Springer, Berlin, 1985)

94. P.S. Riseborough, and D.L. Mills, Phys. Rev. B **21**, 5338 (1980)
95. D.R. Penn, Phys. Rev. **142**, 350 (1966)
96. S. Doniach, Physica B **91**, 231 (1967)
97. G.M. Carneiro, and C.J. Pethick, Phys. Rev. B, **11**, 1106 (1975)
98. W.F. Brinkman, and S. Engelsberg, Phys. Rev., **169**, 417 (1968)
99. P. Lederer, and D.L. Mills, Phys. Rev. **165**, 837 (1968)
100. T. Moriya, Phys. Rev. Lett. **24**, 1433 (1970)
101. T. Moriya, and T. Takimoto, J. Phys. Soc. Jpn., **64**, 960 (1995)
102. S. Sachdev, and J. Ye, Phys. Rev. Lett. **69**, 2411 (1999)
103. A. Rosch, Phys. Rev. Lett. **82**, 4280 (1999)
104. R. Hlubina, and T.M. Rice, Phys. Rev. B **51**, 9534 (1995)
105. B.L. Altshuler, L.B. Ioffe, and A.J. Millis, Phys. Rev. B **52**, 5563 (1995)
106. P. Coleman, Physica B **259–261**, 353 (1999)
107. H. Fröhlich, Phys. Rev. **79**, 845 (1950)
108. E. Maxwell, Phys. Rev. **78**, 477 (1950)
109. C.A. Reynolds, B. Serin, W.H. Wright, and L.B. Nesbitt, Phys. Rev. **78**, 487 (1950)
110. J. Bardeen, and D. Pines, Phys. Rev. **99**, 1140 (1955)
111. J.W. Garland, Phys. Rev. **153**, 460 (1967)
112. R.D. Fowler, J.D.G. Lindsay, R.W. White, H.H. Hill, B.T. Mathias, Phys. Rev. Lett. **19**, 892 (1967)
113. H. Capellmann, and J.R. Schrieffer, Phys. Rev. Lett. **21**, 1060 (1968)
114. J.O. Willis, J.D. Thompson, Z. Fisk, A. de Visser, J.J.M. Franse, and A.A. Menovsky, Phys. Rev. B **31**, 1654 (1985)
115. N. Tateiwa, T.C. Kobayashi, K. Hanazono, K. Amaya, Y. Haga, R. Settai, and Y. Ōnuki, J. Phys. C.M. **13**, L17 (2001)
116. H. Razafimandimby, P. Fulde, and J. Keller, Z. für Physik B **54**, 111 (1984)
117. H. Jichu, A.D.S. Nagi, B. Jin, T. Matsuura, and Y. Kuroda, Phys. Rev. B **35**, 1692 (1987)
118. M.T. Beal Monod, C. Bourbonnais, and V.J. Emery, Phys. Rev. B **34**, 7716 (1986)
119. D.J. Scalapino, E. Loh, and J.E. Hirsch, Phys. Rev. B **34**, 8190 (1986); also D.J. Scalapino, E. Loh, and J.E. Hirsch, Phys. Rev. B **35**, 6694 (1987)
120. K. Miyake, S. Schmitt-Rink, and C.M. Varma, Phys. Rev. B **34**, 6554 (1986)
121. G.M. Eliashberg, Sov. Phys. J.E.T.P. Lett. **11**, 696 (1960)
122. D.J. Scalapino, J.R. Schrieffer, and J.W. Wilkins, Phys. Rev. **148**, 263 (1966); D.J. Scalapino, In: R.D. Parks, *Superconductivity* (Marcel Dekker, New York, 1969)
123. A.B. Migdal, Sov. Phys. J.E.T.P. **7**, 996 (1958)
124. J.A. Hertz, K. Levin, and M.T. Beal-Monod, Solid State Commun. **18**, 803 (1976)
125. P.B. Allen, and B. Mitrović, *Solid State Physics*, ed. by H. Ehrenreich, F. Seitz, and D. Turnbull, (Academic Press, New York, 1982) Vol. 37
126. J.P. Carbotte, Rev. Mod. Phys. **62**, 1027 (1990)
127. J. Bardeen, L.N. Cooper, and J.R. Schrieffer, Phys. Rev. **108**, 1175 (1957)
128. C.M. Varma, unpublished
129. G.A. Gehring, and L.E. Major, J. Phys. C.M. **6**, 495 (1994)
130. D.J. Thouless, Ann. Phys. **10**, 553 (1960)
131. N.D. Mermin, Phys. Rev. A **9**, 868 (1974)
132. P. Wölfle, Rep. Prog. Phys. **42**, 269 (1979)
133. J.W. Serene, and D. Rainer, Phys. Reports **101**, 221 (1983)
134. P.W. Anderson, and W.F. Brinkman, Phys. Rev. Lett. **30**, 1108 (1973)
135. W.F. Brinkman, J.W. Serene, and P.W. Anderson, Phys. Rev. A **10**, 2386 (1974)
136. D. Fay, Phys. Lett. A **51 A**, 356 (1975)
137. J.R. Schrieffer, X.G. Wen, and S.C. Zhang, Phys. Rev. B **39**, 11663 (1989)
138. S. Nakajima, Prog. Theor. Phys. **60**, 1101 (1973)
139. A.J. Leggett, Rev. Mod. Phys. **47**, 331 (1975)
140. P.W. Anderson, Phys. Rev. B **30**, 1549 (1984)
141. E.I. Blount, Phys. Rev. B **32**, 2935 (1985)

142. G.E. Volovik, and L.P. Gor'kov, Sov. Phys. J.E.T.P. **61**, 843 (1985)
143. R. Balian, and N.R. Werthamer, Phys. Rev. **131**, 1553 (1963)
144. P.W. Anderson, and P. Morel, Phys. Rev. **123**, 1911 (1961)
145. P.W. Anderson, Phys. Rev. B **30**, 4000 (1984)
146. M. Ozaki, K. Machida, and T. Ohmi, Prog. Theor. Phys. **74**, 221 (1985); M. Ozaki, K. Machida, and T. Ohmi, Prog. Theor. Phys. **745**, 442 (1986)
147. S. Yip, and A. Garg, Phys. Rev. B **48**, 3304 (1993)
148. M.R. Norman, Phys. Rev. B **41**, 170 (1990)
149. M.R. Norman, Phys. Rev. B **50**, 6904 (1994)
150. B.R. Coles, IBM. J. Res. Dev. **6**, 68 (1962)
151. P.W. Anderson, J. Phys. Chem. Solids **11**, 26 (1959)
152. A.A. Abrikosov, and L.P. Gor'kov, Sov. Phys. J.E.T.P. Lett. **12**, 143 (1963)
153. K. Maki, In: *Superconductivity*, ed. by R.D. Parks (Marcel Dekker, New York, 1969)
154. L.P. Gor'kov, and P.A. Kalugin, Sov. Phys. J.E.T.P. **41**, 253 (1985)
155. K. Ueda, and T.M. Rice, *Theory of Heavy Fermions and Valence Fluctutaions* (Springer, Berlin, 1985)
156. M. Sigrist, and K. Ueda, Rev. Mod. Phys. **63**, 239 (1991)
157. L.P. Gor'kov , Sov. Phys. J.E.T.P. **7**, 505 (1958)
158. C.J. Pethick, and D. Pines, Phys. Rev. Lett. **57**, 118 (1986)
159. P. Hirschfeld, D. Vollhardt, and P. Wölfle, Solid State Commun. **59**, 111 (1986)
160. P.W. Anderson, Phys. Rev. **124**, 41 (1961)
161. J.R. Schrieffer, and P.A. Wolf, Phys. Rev. **149**, 491 (1966)
162. A.C. Hewson, *The Kondo Problem to Heavy Fermions* (Cambridge University Press, Cambridge, 1993)
163. J. Kondo, Prog. Theor. Phys. **32**, 37 (1964)
164. P.W. Anderson, J. Phys. C **3**, 2439 (1970)
165. K.G. Wilson, Rev. Mod. Phys. **47**, 773 (1975)
166. N. Andrei, K. Furuya, and J.H. Lowenstein, Rev. Mod. Phys. **55**, 331 (1983)
167. A. Tsvelik, and P. Wiegmann, Adv. Phys. **32**, 453 (1983)
168. P. Schlottmann, Phys. Reps. **181**, 1 (1989)
169. P. Nozières, Ann. Phys. **10**, 19 (1985)
170. P. Nozières, Eur. J. Phys. B **6**, 447 (1998)
171. J. Friedel, and C.M. Sayers, J. Phys. (Paris), **38**, 697 (1977)
172. E. Muller-Hartmannn, Z. für Phys. B **74**, 507 (1989)
173. D. Vollhardt, and W. Metzner, Phys. Rev. Lett. **62**, 324 (1989)
174. A. Georges, G. Kotliar, W. Krauth, and J. Marcello, Rev. Mod. Phys. **68**, 13 (1996)
175. Th. Pruschke, M. Jarrell, and J.K. Freericks, Adv. Phys. **44**, 187 (1995)
176. S. Burdin, A. Georges, and D. Grempel, Phys. Rev. Lett. **85**, 1048 (2000)
177. M. Continentino, G. Japiassu, and A. Troper, Phys. Rev. B **39**, 9734 (1989)
178. M. Continentino, Phys. Rev. B **47**, 11587 (1993)
179. A.J. Arko, P.S. Riseborough, A.B. Andrews, J.J. Joyce, A.N. Tahvildar-Zadeh, and M. Jarrell, *Handbook of the Physics and Chemistry of Rare Earths*, ed. by K.A. Gschneidner and L. Eyring (Elsevier, Amsterdam, 1999) Vol. 26
180. R.J. Trainor, M.B. Brodsky, and H.V. Culbert, Phys. Rev. Lett. **34**, 1019 (1975)
181. G.R. Stewart, A.L. Giorgi, B.L. Brandt, S. Foner, and A.J. Arko, Phys. Rev. B **28**, 1524 (1983)
182. Y. Ōnuki, and T. Komatsubara, J. Mag. Mag. Mat. **63&64**, 281 (1987)
183. C.L. Lin, A. Wallash, J.E. Crow, T. Mihalisin, and P. Schlottmann, Phys. Rev. Lett. **58**, 1232 (1987)
184. G. Lapertot, R. Calemczuk, C. Marcenant, J.Y. Henry, J.X. Boucherle, J. Flouquet, J. Hammann, R. Cibin, J. Cors, D. Jaccard, and J. Sierro, Physica B **186–188**, 454 (1993)
185. B. Andraka, C.S. Jee, and G.R. Stewart, Phys. Rev. B **52**, 9462 (1995)
186. R. Pietri, K. Ingersent, and B. Andraka, Phys. Rev. Lett. **86**, 1090 (2001)
187. P.S. Riseborough, Phys. Rev. B **67**, 045102 (2003)
188. J.S. Kim, and G.R. Stewart, Phys. Rev. B **51**, 16 190 (1995)

189. G. Bruls, B. Wolf, D. Finsterbusch, P. Thalmeier, I. Kouroudis, W. Sun, W. Assmus, B. Lüthi, M. Lang, K. Gloos, F. Steglich, and R. Modler, Phys. Rev. Lett. **72**, 1754 (1994)
190. F. Steglich, P. Buschinger, P. Gegenwart, M. Lohmann, R. Helfrich, C. Langhammer, P. Hellmann, L. Donnevert, S. Thomas, A. Link, C. Geibel, M. Lang, G. Sparn, and W. Assmus, J. Phys. C.M. **8**, 9909 (1996)
191. R. Modler, M. Lang, C. Geibel, C. Schank, R. Muller-Reisener, P. Hellmann, A. Link, G. Sparn, W. Assmus, F. Steglich, Physica B **206& 207**, 586 (1995)
192. M. Ishikawa, N. Takeda, P. Ahmet, V. Karaki, and H. Ishimoto, J. Phys. C.M. **13**, L25 (2001)
193. P. Gegenwart, C. Langhammer, C. Geibel, R. Helfrich, M. Lang, G. Sparn, F. Steglich, R. Horn, L. Donnevert, A. Link, W. Assmus, Phys. Rev. Lett. **81**, 1501 (1998); see also P. Gegenwart, M. Lang, A. Link, G. Sparn, C. Geibel, F. Steglich, and W. Assmus, Physica B **259&261**, 403 (1999); see also P. Gegenwart, M. Lohmann, M. Lang, R. Helfrich, C. Langhammer, M. Koppen, C. Geibel, F. Steglich, and W. Assmus, Physica B **230&232**, 527 (1997)
194. J.D. Thompson, H. Hegger, D. Louca, G.H. Kwei, R. Movshovich, C. Petrovic, and J.L. Sarrao, J. Alloys Compd. **303&304**, 239 (2000)
195. M. McElfresh, M.B. Maple, J.O. Willis, D. Schiferl, J.L. Smith, Z. Fisk, and D.L. Cox, Phys. Rev. B **48**, 10395 (1993)
196. B. Andraka, Phys. Rev. B **52**, 16031 (1995)
197. Z. Fisk, H.R. Ott, and J.L. Smith, J. Less-Common Metals **133**, 99 (1987)
198. J.S. Kim, B. Andraka, C.S. Jee, S.B. Roy, and G.R. Stewart, Phys. Rev. B **41**, 11 073 (1990)
199. G. Frauenberger, M. Baldus, and G.R. Stewart, Phys. Rev. B **47**, 3204 (1993)
200. P.S. Riseborough, Phys. Stat. Solidi B **122**, 161 (1984)
201. N. Harrison, A.L. Cornelius, H. Harima, K. Takegahara, J.A. Detwiler, G.M. Schmiedeshoff, J.C. Cooley, and J.L. Smith, Phys. Rev. B **61**, 1779 (2000)
202. L. Taillefer, R. Newbury, G.G. Lonzarich, Z. Fisk, and J.L. Smith, J. Mag. Mag. Mat. **63&64**, 372 (1987)
203. N. Kimura, T. Komatsubara, D. Aoki, Y. Õnuki, Y. Haga, E. Yamamaoto, H. Aoki, and H. Harima, J. Phys. Soc. Jpn. **67**, 2185 (1998)
204. C.S. Wang, M.R. Norman, R.C. Albers, A.M. Boring, W.E. Pickett, H. Krakauer, N.E. Christiensen, Phys. Rev. B **35**, 7260 (1987)
205. M.R. Norman, R.C. Albers, A.M. Boring, and N.E. Christiensen, Solid State Commun. **68**, 245 (1988)
206. M. Hunt, P. Meeson, P.-A. Probst, P. Reinders, M. Springford, W. Assmus, and W. Sun, J. Phys. C.M., **2**, 6859 (1990)
207. P.H.P. Reinders, M. Springford, P.T. Coleridge, R. Boulet, and D. Ravot, Phys. Rev. Lett. **57**, 1631 (1986)
208. S. Chapman, M. Hunt, P. Meeson, P.H.P. Reinders, M. Springford, and M. Norman, J. Phys. C.M. **2**, 8123 (1990)
209. G.R. Stewart, B. Andraka, C. Quitmann, B. Treadway, Y. Shapira, and E.J. McNiff, Jr., Phys. Rev. B, **37**, 3344 (1988)
210. K. Satoh, T. Fujita, Y. Maeno, Y. Õnuki, and T. Komatsubara, J. Mag. Mag. Mat. **76&77**, 128 (1988)
211. C. Bergemann, S.R. Julian, G.J. McMullan, B.K. Howard, G.G. Lonzarich, P. Lejay, J.P. Brison, and J. Flouquet, Physica B **230**, 348 (1997)
212. H. Ohkuni, Y. Inada, Y. Tokiwa, K. Sakurai, R. Settai, T. Honma, Y. Haga, E. Yamamoto, Y. Õnuki, H. Yamagami, S. Takahashi, and T. Yanagisawa, Phil. Mag. B, **79**, 1045 (1999)
213. A.L. Cornelius, A.J. Arko, J.L. Sarrao, and N. Harrison, Phys. Rev. B **59**, 13542 (1999)
214. G.G. Lonzarich, J. Mag. Mag. Mat. **76&77**, 1 (1988)
215. K. Bakker, A. de Visser, A.A. Menovsky, and J.J.M. Franse, Physica B **186–188** 720 (1993)
216. A. de Visser, A. Menovsky, and J.J.M. Franse, Physica B **147**, 81 (1987)
217. H. Aoki, S. Uji, A.K. Albesaard, and Y. Õnuki, Phys. Rev. Lett. **71**, 2110 (1993)
218. S.R. Julian, F.S. Tautz, G.J. McMullan, and G.G. Lonzarich, Physica B **199&200**, 63 (1994)
219. Y. Inada, A. Ishiguro, J. Kimura, N. Sato, A. Sawada, T. Komatsubara, and H. Yamagami, Physica B, **206&207**, 33 (1995)
220. T. Terashima, C. Haworth, M. Takashita, H. Aoki, N. Sato, and T. Komatsubara, Phys. Rev. B, **55**, 13369 (1997)
221. J.E. Graebner, and M. Robbins, Phys. Rev. Lett. **36**, 422 (1976)
222. F.M. Mueller, D.H. Lowndes, Y.K. Chang, A.J. Arko, and R.S. List, Phys. Rev. Lett. **68**, 3928 (1992)
223. M. Heinecke, and K. Winzer, Z. für Physik B **98**, 147 (1999)
224. K. Matsuda, Y. Kohori, T. Kohara, K. Kuwahara, and H. Amitsuka, Phys. Rev. Lett. **87**, 2031 (2001)
225. H. Amitsuka, M. Sato, N. Metoki, M. Yokohama. K. Kuwahara, T. Sakakibara, H. Morimoto, S. Kawarazaki, Y. Miyako, and J.A. Mydosh, Phys. Rev. Lett. **83**, 5114 (1999)
226. W. Franz, F. Steglich, and D. Wohlleben, J. Phys. (Paris) Colloque **C5**, 342 (1970)

227. B. Cornut, and B. Coqblin, Phys. Rev. B **5**, 4541 (1972)
228. A.P. Murani, K. Knorr, K.H.J. Buschow, *Crystal Field Effects in Metals and Alloys*, ed. by A. Furrer (Plenum Press, New York, 1977) p. 268
229. S. Horn, E. Holland-Moritz, M. Lowenhaupt, F. Steglich, H. Scheuer, A. Benoit, and J. Flouquet, Phys. Rev. B **23**, 3171 (1981)
230. A. Loidl, G. Knopp, H. Spille, F. Steglich, and A.P. Murani, Physica B **156&157**, 794 (1989)
231. E.A. Goremychkin, and R. Osborn, Phys. Rev. B **47**, 14280 (1993)
232. E.A. Goremychkin, and R. Osborn, Phys. Rev. B, **47**, 14580 (1993)
233. G.J. Nieuwenhuys, Phys. Rev. B **35**, 5260 (1987)
234. E.A. Goremychkin, R. Osborn and I.L. Sashin, J. Appl. Phys., **85**, 6046 (1999)
235. P. Santini, and G. Amoretti, Phys. Rev. Lett. **73**, 1027 (1994)
236. H. Amitsuka, and T. Sakakibara, J. Phys. Soc. Jpn. **63**, 736 (1994)
237. K. Kadowaki, and S.B. Woods, Solid State Commun. **58**, 507 (1986)
238. J.M. Lawrence, J.D. Thompson, and Y.Y. Chen, Phys. Rev. Lett. **54**, 2537 (1985)
239. Y. Õnuki, Y. Shimizu, and T. Komatsubara, J. Phys. Soc. Jpn. **54**, 304 (1985)
240. B. Batlogg, D.J. Bishop, E. Bucher, B. Golding Jr., A.P. Ramierez, Z. Fisk, and J.L. Smith, J. Mag. Mag. Mat. **63&64**, 441 (1987)
241. B. Andraka, G.R. Stewart, Phys. Rev. B **49**, 1259 (1994)
242. U. Rauchschwalbe, F. Steglich, A. de Visser, and J.J.M. Franse, J. Mag. Mag. Mat., **63&64**, 347 (1987)
243. G. Remenyi, A. Briggs, J. Flouquet, O. Laborde, and F. Lapierre, J. Mag. Mag. Mat. **31–34**, 407 (1983)
244. B. Horvatic, and V. Zlatic, Phys. Rev. B **30**, 6717 (1984)
245. P. Schlottmann, Phys. Rev. B **35**, 5279 (1987)
246. B. Andraka, Phys. Rev. B **52**, 16031 (1995)
247. Y. Haga, E. Yamamoto, T. Honma, A. Nakamura, M. Hedo, S. Araki, H. Ohkuni, and Y. Õnuki, Physica B **259-261**, 627 (1999)
248. G.R. Stewart, Z. Fisk, J.O. Willis, and J.L. Smith, Physica B **127**, 448 (1984)
249. F.P. Milliken, T. Penney, F. Holtzberg, and Z. Fisk, J. Mag. Mag. Mat. **76&77**, 201 (1988)
250. F.G. Aliev, N.B. Brandt, V.V. Moschalkov, and S.M. Chudinkov, Solid State Commun. **47**, 693 (1983)
251. T. Penney, J. Stankiewicz, S. von Molnar, Z. Fisk, J.L. Smith, and H.R. Ott, J. Mag. Mag. Mat., **54-57**, 370 (1986)
252. J. Schoenes, and J.J.M. Franse, Phys. Rev. B **33**, 5138 (1986)
253. A. Fert, and O. Jaoul, Phys. Rev. Lett. **28**, 303 (1972)
254. P. Coleman, P.W. Anderson, and T.V. Ramakrishnan, Phys. Rev. Lett. **55**, 414 (1985)
255. D. Jaccard, J. Flouquet, Z. Fisk, and J.L. Smith, J. Phys. Lett. **46**, L811 (1985)
256. H. Armbruster, W. Franz, W. Schlabitz, and F. Steglich, J. Phys. Colloq. **C4**, 40, 150 (1979)
257. A. Grauel, D. Fromm, C. Geibel, F. Steglich, N. Sato, and T. Komatsubara, Int. J. Mod. Phys. B, **7**, 50 (1993)
258. F. Steglich, U. Rauchschwalbe, U. Gottwick, H.M. Mayer, G. Sparn, N. Grewe, U. Poppe, and J.J.M. Franse, J. Appl. Phys. **57**, 3054 (1985)
259. D. Jaccard, and J. Flouquet, J. Mag. Mag. Mat. **47&48**, 45 (1985)
260. L. Colquitt Jr., H.R. Frankhauser, and F.J. Blatt, Phys. Rev. B **4**, 292 (1971)
261. H.R. Ott, O. Marti, and F. Hulliger, Solid State Commun. **49**, 1129 (1984)
262. J. Flouquet, P. Haen, C. Marcenat, P. Lejay, A. Amato, D. Jacaard, and E. Walker, J. Mag. Mag. Mat., **52**, 85 (1985)
263. H. Sato, J. Zhao, W.P. Pratt Jr., Y. Õnuki, and T. Komatsubara, Phys. Rev. B **36**, 8841 (1987)
264. G. Sparn, W. Leike, U. Gottwick, F. Steglich, and N. Grewe, J. Mag. Mag. Mat. **47&48**, 521 (1985)
265. A. Sulpice, P. Candit, J. Chaussy, J. Flouquet, D. Jacaard, P. LeJay, and J.L. Tholence, J. Low Temp. Phys. **62**, 39 (1986)
266. A. Ravex, J. Flouquet, J.L. Tholence, D. Jaccard, and A. Mayer, J. Mag. Mag. Mat. **63&64**, 400 (1987)
267. A.K. Bhattacharjee, and B. Coqblin, Phys. Rev. B **13**, 3441 (1976)
268. A.K. Bhattacharjee, and B. Coqblin, Phys. Rev. B **38**, 338 (1988)
269. E. Brück, A. Nowack, N. Hohn, H. Paulus, and A. Freimuth, Z. für Physik B **63**, 155 (1986)
270. B. Lüthi, J. Mag. Mag. Mat. **52**, 70 (1985)
271. V. Müller, D. Maurer, K. de Groot, E. Bucher, and H.E. Bömmel, Phys. Rev. Lett. **56**, 248 (1986)
272. K.D. Schotte, D. Föster, and U. Schotte, Z. für Physik B **64**, 165 (1986)

273. R. Mock, and G. Güntherodt, Z. für Physik B **74**, 315 (1989)
274. S.-W. Lin, S. Adenwalla, J.B. Ketterson, M. Levy, and B.K. Sarma, J. Low. Temp. Phys. **89**, 217 (1992)
275. H. Shiba, Prog. Theor. Phys. **54**, 967 (1975)
276. D.E. MacLaughlin, J. Mag. Mag. Mat. **47&48**, 121 (1985)
277. J.P. Vithayathil, D.E. MacLaughlin, E. Koster, D.Ll. Williams, and E. Bucher, Phys. Rev. B, **44**, 4705 (1991)
278. Y. Kohori, H. Shibai, T. Kohara, Y. Oda, Y. Kitaoka, and K. Asayama, J. Mag. Mag. Mat., **76&77**, 478 (1988)
279. H.R. Ott, H. Rudigier, E. Felner, Z. Fisk, and J.L. Smith, Phys. Rev. B **33**, 126 (1986)
280. T. Kohara, Y. Kohori, K. Assayama, Y. Kitaoka, M.B. Maple, and M.S. Torikachvilli, Jpn. J. Appl. Phys., **26**, 1247 (1987)
281. Y. Kohori, J. Phys. Soc. Jpn. **65**, 679 (1996)
282. M. Kyogaku, Y. Kitaoka, K. Assayama, C. Geibel, C. Schank, and F. Steglich, J. Phys. Soc. Jpn., **62**, 4016 (1993)
283. Y. Kitaoka, Y. Fujiwara, Y. Kohori, K. Asayama, Y. Õnuki, and T. Komatsubara, J. Phys. Soc. Jpn. **54**, 3686 (1985)
284. J. Aarts, F.R. de Boer, and D.E. MacLaughlin, Physica B **121**, 162 (1983)
285. Y. Kitaoka, K. Ueda, T. Kohara, K. Asayama, Y. Õnuki, and T. Komatsubara, J. Mag. Mag. Mat., **52**, 341 (1985)
286. M. Jarrell, J. Gubernatis, and R.N. Silver, Phys. Rev. B **44**, 5347 (1991)
287. O.O. Bernal, C. Rodrigues, A. Martinez, H.G. Lukefahr, A.A. Menovsky, and J.A. Mydosh, Phys. Rev. Lett. **87**, 196402/3 (2001)
288. D.W. Cooke, R.H. Heffner, R.L. Hutson, M.E. Schillaci, J.L. Smith, J.O. Willis. D.E. MacLaughlin, C. Boekema, R.L. Lichti, A.B. Denison, and J. Oostens, Hyperfine Interact. **31**, 425 (1986)
289. G. Aeppli, E. Bucher, A.I. Goldman, G. Shirane C. Broholm, and J.K. Kjems, J. Mag. Mag. Mat., **76&77**, 385 (1988)
290. P. Dalmas de Réotier, A. Yaouanc, R.H. Heffner, P.C.M. Gubbens, and C.T. Kaiser, Phys. Rev. B, **61**, 6377 (2000)
291. D.E. MacLaughlin, D.W. Cooke, R.H. Heffner, R.L. Hutson, M.W. McElfresh, M.E. Schillaci, H.D. Rempp, J.L. Smith, J.O. Willis, E. Zirngiebl, C. Boekema, R.L. Lichti, and J. Oostens, Phys. Rev. B **37**, 3153 (1988)
292. E.A. Knetsch, A.A. Menovsky, G.J. Nieuwenhuys, J.A. Mydosh, A. Amato, R. Feyerherm, F.N. Gygax, A. Schenck, R.H. Heffner, and D.E. MacLaughlin, Physica B, **186-188**, 300 (1993)
293. L.P. Gor'kov, Europhys. Lett. **16**, 301 (1991)
294. L.P. Gor'kov, and R. Sokol, Phys. Rev. Lett. **69**, 2586 (1992)
295. V. Barzykin, and L.P. Gor'kov, Phys. Rev. Lett. , **70**, 2479 (1993)
296. M.B. Walker, W.J.L. Buyers, Z. Tun, W. Que, A.A. Menovsky, and J.D. Garrett, Phys. Rev. Lett. **71**, 2630 (1993)
297. T.E. Mason, W.J.L. Buyers, T. Petersen, and J.D. Garrett, J. Phys. C.M. **7**, 5089 (1995)
298. S. Barth, H.R. Ott, F.N. Gygax, B. Hitti, E. Lippelt A. Schenck, C. Baines, B. van der Brandt, T. Konter, and S. Mango, Phys. Rev. Lett. **59**, 2991 (1987)
299. S. Barth, H.R. Ott, F.N. Gygax, B. Hitti, E. Lippelt, and A. Schenck, J. Mag. Mag. Mat. **76&77**, 455 (1988)
300. F. Steglich, and N. Grewe, *Handbook on the Physics and Chemistry of Rare Earths*, ed. by K.A. Gschneidner Jr. and L. Eyring, (North Holland, Amsterdam, 1991) Vol. 14
301. D. Jaccard, R. Cibin, J.L. Jorda, and J. Flouquet, Jpn. J. Appl. Phys. **26**, 517 (1987)
302. Y.J. Uemura, W.J. Kossler, X.H. Yu, H.E. Schone, J.R. Kempton, C.E. Stronach, S. Barth, F.N. Gygax, B. Hitti, A. Schenck, C. Baines, W.F. Lankford, Y. Õnuki, and T. Komatsubara, Phys. Rev. B **39**, 4726 (1989)
303. H. Nakamura, Y. Kitaoka, T. Iwai, H. Yamada, and K. Asayama, J. Phys. C.M. **4**, 473 (1992)
304. Y. Kitaoka, H. Nakamura, T. Iwai, K. Asayama, U. Ahlheim, C. Geibel, C. Schank, and F. Steglich, J. Phys. Soc. Jpn. **60**, 2122 (1992)
305. H. Nakamura, Y. Kitaoka, H. Yamada, and K. Asayama, J. Mag. Mag. Mat. **76&77**, 517 (1988)
306. G.M. Luke, A. Keren, K. Kojima, L.P. Le, B.J. Sternlieb, W.D. Wu, Y.J. Uemura, Y. Õnuki, and T. Komatsubara, Phys. Rev. Lett. **73**, 1853 (1994)
307. R. Feyerherm, A. Amato, C. Geibel, F.N. Gygax, P. Hellmann, R.H. Heffner, D.E. MacLaughlin, R. Mueller-Reisener, G.J. Nieuwenhuys, A. Schenck, and F. Steglich, Physica B, **206&207**, 596 (1995); see also R. Feyerherm, A. Amato, C. Geibel, F.N. Gygax, P. Hellmann, R.H. Heffner, D.E. MacLaughlin, R. Mueller-Reisener, G.J. Nieuwenhuys, A. Schenck, and F. Steglich, Phys. Rev. B **56**, 699 (1997)
308. A.P. Murani, K. Knorr, K.H.J. Buschow, A. Benoit, and J. Flouquet, Solid State Commun. **36**, 523 (1980)
309. N.E. Bickers, D.L. Cox and J.W. Wilkins, Phys. Rev. Lett. **54**, 230 (1987); see also N.E. Bickers, D.L. Cox, and J.W. Wilkins, Phys. Rev. B **36**, 2306 (1987)
310. L.P. Regnault, W.A.C. Erkelens, J Rossat-Mignod, P. Lejay, and J. Flouquet, Phys. Rev. B **38**, 4481 (1988)

311. J.L. Jacoud, L.P. Regnault, J. Rossay-Mignod, C. Vettier, P. LeJay, and J. Flouquet, Physica B, **156&157**, 818 (1989)
312. G. Balakrishnan, D. McK Paul, and N.R. Bernhoeft, Physica B **156&157**, 817 (1988)
313. G.H. Lander, S.M. Shapiro, C. Vettier, and A.J. Dianoux, Phys. Rev. B **46**, 5387 (1992)
314. G. Aeppli, E. Bucher, and G. Shirane, Phys. Rev. B **32**, 7579 (1985)
315. N.R. Bernhoeft, and G.G. Lonzarich, J. Phys. C.M. **7**, 7325 (1995)
316. G. Aeppli, A.I. Goldman, G. Shirane, G. Aeppli, E. Bucher, and M.C. Lux-Steiner, Phys. Rev. Lett. **58**, 808 (1987)
317. A.I. Goldman, G. Shirane, G. Aeppli, E. Bucher, and J. Hufnagl, Phys. Rev. B **36**, 8523 (1987)
318. G. Aeppli, E. Bucher, C. Broholm, J.K. Kjems, J. Baumann, and J. Hufnagl, Phys. Rev. Lett. **60**, 615 (1988)
319. P.A. Midgely, S.M. Hayden, L. Taillefer, B. Bogenberger, and H. von Löhneysen, Phys. Rev. Lett. **70**, 678 (1993)
320. R.H. Heffner, D.W. Cooke, A.L. Giorgi, R.L. Hutson, M.E. Schillaci, H.D. Rempp, J.L. Smith, J.O. Willis, D.E. MacLaughlin, C. Boekema, R.L. Lichti, J. Oostens, and A.B. Dension, Phys. Rev. B **39**, 11345 (1989)
321. G. Aeppli, D. Bishop, C. Broholm, E. Bucher, K. Siemensmeyer, M. Steiner, and N. Stuesser, Phys. Rev. Lett. **63**, 676 (1989)
322. A.I. Goldman, S.M. Shapiro, D.E. Cox, J.L. Smith, and Z. Fisk, Phys. Rev. B **32**, 6042 (1985)
323. A.I. Goldman, S.M. Shapiro, G. Shirane, J.L. Smith, and Z. Fisk, Phys. Rev. B **33**, 1627 (1986)
324. S. Coad, A. Heiss, D.F. McMorrow, G.H. Lander, G. Aeppli, Z. Fisk and G.R. Stewart, S.M. Hayden, and H.A. Mook, Physica B **276-278**, 764 (2000)
325. C. Broholm, J.K. Kjems, W.J.L. Buyers, P. Mathews, T.T.M. Palstra, A.A. Menovsky, and J.A. Mydosh, Phys. Rev. Lett. **59**, 1467 (1987)
326. B. Fak, C. Vettier, J. Flouquet, P. Lejay, and J.M. Mignot, Physica B **206&207**, 415 (1995)
327. T.E. Mason, B.D. Gaulin, J.D. Garrett, Z. Tun, W.J.L. Buyers, and E.D. Isaacs, Phys. Rev. Lett. **65**, 3189 (1990)
328. U. Walter, M. Lowenhaupt, W. Schlabitz, and C.K. Loong, Phys. Rev. B **33**, 7875 (1986)
329. C. Broholm, H. Lin, P.T. Mathews, T.E. Mason, W.J.L. Buyers, M.F. Collins, A.A. Menovsky, J.A. Mydosh, and J.K. Kjems, Phys. Rev. B **43**, 12809 (1991)
330. J.-G. Park, K.A. McEwen, and M.J. Bull, Phys. Rev. B, **66**, 094502 (2002)
331. T. Honma, Y. Haga, E. Yamamoto, N. Metoki, Y. Koike, H. Ohkuni, and Y. Õnuki, Physica B, **259**, 646 (1999)
332. C. Stassis, B. Batlogg, J.P. Reimka, J.D. Axe, G. Shirane, and Y.J. Uemura, Phys. Rev. B **33**, 1680 (1986)
333. J. Rossat-Mignod, L.P. Regnault, J.L. Jacoud, C. Vettier, P. Lejay, J. Floquet, E. Walker, D. Jaccard, and A. Amato, J. Mag. Mag. Mat. **76&77**, 376 (1988)
334. G. Aeppli, H. Yoshizawa, Y. Endoh, E. Bucher, J. Hufnagl, Y. Õnuki, and T. Komatsubara, Phys. Rev. Lett. **57**, 122 (1986)
335. H. von Löhneysen, J. Phys. C.M. **8**, 9689 (1996)
336. A. Schröder, J.W. Lynn, R.W. Erwin, M. Lowenhaupt, and H. von Löhneysen, Physica B **199-200**, 47 (1994)
337. O. Stockert, H. von Löhneysen, A. Rosch, N. Pyka, and M. Lowenhaupt, Phys. Rev. Lett. **80**, 5627 (1998)
338. A. Rosch, A. Schröder, O. Stockert, and H. von Löhneysen, Phys. Rev. Lett. **79**, 159 (1997)
339. A. Schröder, G. Aeppli, E. Bucher, R. Ramazashvili, and P. Coleman, Phys. Rev. Lett. **80**, 5623 (1998)
340. M.C. Aronson, R. Osborne, R.A. Robinson, J.W. Lynn, R. Chau, C.L. Seaman, and M.B. Maple, Phys. Rev. Lett. **75**, 725 (1995)
341. R. Osborn, M.C. Aronson, B.D. Rainford, M.B. Maple, R. Chau, and K.H. Andersen, Physica B **241**, 859 (1997)
342. A. Krimmel, P. Fischer, B. Roessli, H. Maletta, C. Geibel, C. Schank, A. Grauel, A. Loidl, and F. Steglich, Z. für Physik B **86**, 161 (1992)
343. T. Petersen, T.E. Mason, G. Aeppli, A.P. Ramierez, E. Bucher, and R.R. Kleimann, Physica B **199&200**, 151 (1994)
344. N. Sato, N. Aso, G.H. Lander, B. Roessli, T. Komatsubara, and Y. Endoh, J. Phys. Soc. Jpn., 66, 1884 (1997)
345. N. Bernhoeft, B. Roessli, N. Sato, N. Aso, A. Heiss, G.H. Lander, Y. Endoh, and T. Komatsubara, Physica B **259-261**, 614 (1999)
346. N. Metoki, Y. Haga, Y. Koike, N. Aso, and Y. Õnuki, J. Phys. Soc. Jpn. **66**, 2560 (1997)
347. N. Metoki, Y. Haga, Y. Koike, and Y. Õnuki, Phys. Rev. Lett. **80**, 5417 (1998)
348. N. Metoki, Y. Koike, Y. Haga, T. Honma, M. Sato, Y. Õnuki, N. Kimura, K. Maesawa, H. Amitsuka, M. Yokoyama, K. Kuwahara, and Y. Miyako, Physica B **280**, 362 (2000)
349. A. Schroder, J.G. Lussier, B.D. Gaulin, J.D. Garrett, W.J.L. Buyers, L. Rebelsky, and S.M. Shapiro, Phys. Rev. Lett. **72**, 136 (1994)

350. J.G. Lussier, M. Mao, A. Schroder, J.D. Garrett, B.D. Gaulin, S.M. Shapiro, and W.J.L. Buyers, Phys. Rev. B **56**, 11749 (1997)
351. N. Aso, B. Roessli, N. Bernhoeft, R. Calemczuk, N.K. Sato, Y. Endoh, T. Komatsubara, A. Heiss, G.H. Lander, and H. Kadowaki, Phys. Rev. B **61**, 11867 (2000)
352. B.D. Gaulin, M. Mao, C.R. Weibe, Y. Qiu, S.M. Shapiro, S.H. Lee, and J.D. Garrett, Phys. Rev. B, **66**, 174520 (2002)
353. Y. Wada, Rev. Mod. Phys. **36**, 253 (1964)
354. J.C. Swihart, D.J. Scalapino, and Y. Wada, Phys. Rev. Lett. **14**, 106 (1965)
355. D.K. Finnemore, T.F. Stromberg, and C.A. Swenson, Phys. Rev. **149**, 231 (1966)
356. R. Radebaugh, and P.H. Keesom, Phys. Rev. **149**, 217 (1966)
357. W. Schlabitz, J. Baumann, B. Pollit, U. Rauchschwalbe, H.M. Mayer, U. Ahlheim, and C.D. Bredl, Z. für Physik B **62**, 171 (1986)
358. F. Steglich, C.D. Bredl, W. Lieke, U. Rauchschwalbe, and G. Sparn, Physica B **126**, 82 (1984)
359. J.R. Schrieffer, *Theory of Superconductivity* (W.A. Benjamin, Reading, MA, 1964)
360. Terry P. Orlando, and Kevin A. Delin, *Foundations of Applied Superconductivity* (Addison-Wesley, Reading, MA, 1991)
361. G. Eilenberger, Phys. Rev. **153**, 584 (1967)
362. B.S. Chandrasekhar, Appl. Phys. Lett. **1**, 7 (1962)
363. A.M. Clogston, Phys. Rev. Lett. **9**, 261 (1962)
364. I.A. Luk'yanchuk, and V.P. Minnev, Sov. Phys. J.E.T.P. **66**, 1168 (1987)
365. N.R. Werthamer, E. Helfand, and P.C. Hohenberg, Phys. Rev. **147**, 295 (1966)
366. K. Maki, Phys. Rev. **148**, 362 (1966)
367. T.P. Orlando, E.J. McNiff Jr., S. Foner, and M.R. Beasley, Phys. Rev. B **19**, 4545 (1979)
368. K. Scharnberg, and R.A. Klemm, Phys. Rev. B **22**, 5233 (1980)
369. K. Scharnberg, and R.A. Klemm, Phys. Rev. Lett. **54**, 2445 (1985)
370. Robert Joynt, Phys. Rev. B **41**, 4271 (1990)
371. M. Prohammer, and J.P. Carbotte, Phys. Rev. B **42**, 2032 (1990)
372. P. Fulde, and R.A. Ferrell, Phys. Rev. **135**, 550 (1964)
373. A.I. Larkin, and Y.N. Ovchinnikov, Sov. Phys. J.E.T.P. **20**, 762 (1965)
374. L.W. Gruenberg, and L. Gunther, Phys. Rev. Lett. **16**, 996 (1966)
375. B.B. Goodman, Phys. Lett. **1**, 215 (1962)
376. R. Joynt, and L. Taillefer, Rev. Mod. Phys. **74**, 235 (2002)
377. F. Thomas, B. Wand, T. Luhmann, P. Gegenwart, G.R. Stewart, F. Steglich, J.P. Brison, A. Buzdin, L. Glemot, and J. Flouquet, J. Low Temp. Phys. **102**, 117 (1996)
378. L. Glemot, J.P. Brison, J. Flouquet, A.I. Buzdin, I. Sheikin, D. Jaccard, C. Thessieu, and F. Thomas, Phys. Rev. Lett. **82**, 169 (1999)
379. G.M. Schmiedeshoff, Z. Fisk, and J.L. Smith, Phys. Rev. B **45**, 10544 (1992)
380. N.E. Alekseevskii, V.I. Nizhankovskii, V.N. Narozhnyi, E.P. Khlybov, and A.V. Mitin, J. Low Temp. Phys. **64**, 87 (1986)
381. F.G. Aliev, V. Kovachik, V.V. Moshchalkov, V.V. Pryadun, N.E. Alekseevskii, A.V. Mitin, N. Agrait, S. Vieira, and R. Villar, J. Low Temp. Phys. **85**, 359 (1991)
382. P.J.C. Signore, B. Andraka, G.R. Stewart, and M.W. Meisel, Phys. Rev. B **52**, 10315 (1995)
383. U. Rauchschwalbe, W. Lieke, C.D. Bredl, F. Steglich , J. Aarts, K.M. Martini, and A.C. Mota, Phys. Rev. Lett. **49**, 1448 (1982)
384. I. Sheikin, D. Braithwaite, J.-P. Brison, W. Assmus, and J. Flouquet, J. Low Temp. Phys. **118**, 113 (2000)
385. A.Pollini, A.C. Mota, P. Visani, R. Pittini, G. Juri, and T. Teruzzi, J. Low Temp. Phys. **9**, 15 (1993)
386. U. Rauchschwalbe, Physica B **147**, 1 (1987)
387. V. Ambegokar, and A. Griffin, Phys. Rev. **137**, A 1151 (1965)
388. O. Betbeder-Matibet, and P. Nozières, Ann. Phys. (NY) **51**, 392 (1969)
389. L. Coffey, T.M. Rice, and K. Ueda, J. Phys. C **18**, L 813 (1985)
390. S. Schmitt-Rink, K. Miyake, and C.M. Varma, Phys. Rev. Lett. **57**, 2575 (1986)
391. P.A. Lee, Phys. Rev. Lett. **71**, 1887 (1993)
392. M.J. Graf, S.-K. Yip, J.A. Sauls, and D. Rainer, Phys. Rev. B **53**, 15147 (1996)

393. H.R. Ott, E. Felder, A. Bernasconi, Z. Fisk, J.L. Smith, L. Taillefer, and G. Lonzarich, Jpn. J. Appl. Phys. **26**, Suppl. **26-3**, 1217 (1987)
394. D. Jaccard, J. Flouquet, P. Lejay, and J.L. Tholence, J. Appl. Phys. **57**, 3082 (1985)
395. K. Behnia, L. Taillefer, J. Flouquet, D. Jaccard, K. Maki, and Z. Fisk, J. Low. Temp. Phys. **84**, 261 (1991)
396. F. Steglich, U. Ahlheim, J.J.M. Franse, N. Grewe, D. Rainer, and U. Rauchschwalbe, J. Mag. Mag. Mat. **52**, 54 (1985)
397. B. Lussier, B. Ellman, and L. Taillefer, Phys. Rev. Lett. **73**, 3295 (1994)
398. T. Tsuneto, Phys. Rev. **121**, 402 (1961)
399. L.P. Kadanoff, and I.I. Falko, Phys. Rev. **136**, A 1170 (1964)
400. L.T. Claiborne Jr., and R.W. Morse, Phys. Rev. **136**, A 893 (1964)
401. K. Fossheim, Phys. Rev. Lett. **19**, 81 (1967)
402. V. Müller, D. Maurer, E.W. Scheidt, Ch. Roth, K. Lüders, E. Bucher, and H.E. Bömmel, Solid State Commun. **57**, 319 (1986)
403. B.S. Shivaram, J.H. Jeong, T.F. Rosenbaum, and D.G. Hinks, Phys. Rev. Lett. **56**, 1078 (1986)
404. P. Thalmeier, B. Wolf, D. Weber, R. Blick, G. Bruls, and B. Lüthi, J. Mag. Mag. Mat. **108**, 109 (1992)
405. K.J. Sun, A. Schenstrom, B.K. Sarma, M. Levy, and D.G. Hinks, Phys. Rev. B **40**, 11284 (1989)
406. L.D. Landau, Sov. Phys. J.E.T.P. **5**, 101 (1957)
407. L. Coffey, Phys. Rev. B **40**, 715 (1989); L. Coffey, Phys. Rev. B **35**, 8440 (1987)
408. S.N. Coppersmith, and R.A. Klemm, Phys. Rev. Lett. **56**, 1870 (1986)
409. K. Scharnberg, D. Walker, H. Monien, L. Tewordt, and R.A. Klemm, Solid State Commun. **60**, 535 (1986)
410. B. Arfi, and C.J. Pethick, Phys. Rev. B **38**, 2312 (1988)
411. J. Moreno, and P. Coleman, Phys. Rev. Lett. **53**, R2995 (1996)
412. J.P. Rodriguez, Phys. Rev. Lett. **55**, 250 (1985)
413. D.C. Mattis, and J. Bardeen, Phys. Rev. **111**, 412 (1958)
414. D.M. Ginsberg, and M. Tinkham, Phys. Rev. **118**, 990 (1960)
415. S.A. Zemon, and H.A. Boorse, Phys. Rev. **146**, 309 (1966)
416. M. Tinkham, and R.A. Ferrell, Phys. Rev. Lett. **2**, 331 (1959)
417. J. Halbritter, Z. für Physik **243**, 201 (1971)
418. P.J. Hirschfeld, P. Wölfle, J.A. Sauls, D. Einzel, and W.O. Putikka, Phys. Rev. B **40**, 6695 (1989)
419. S.-W. Lin, C.T. Lin, B.K. Sarma, G.R. Stewart, and J.B. Ketterson, Phys. Lett. A **217**, 161 (1996)
420. S. Skalski, O. Betbeder-Matibet, and P.R. Weiss, Phys. Rev. **136**, A 1500 (1964)
421. J.R. Feller, C.-C. Tsai, J.B. Ketterson, J.L. Smith, and B.K. Sarma, Phys. Rev. Lett. **88**, 247005 (2002)
422. D. Einzel, P.J. Hirschfeld, F. Gross, B.S. Chandrasekhar, K. Andres, H.R. Ott, J. Beuers, Z. Fisk, and J.L. Smith, Phys. Rev. Lett. **56**, 2513 (1986)
423. C. Broholm, G. Aeppli, R.N. Kleinman, D.R. Harshman, D.J. Bishop, E. Bucher, D.Ll. Williams, E.J. Ansaldo, and R.H. Heffner, Phys. Rev. Lett. **65**, 2062 (1990)
424. A. Yaouanc, P. Dalmas de Réotier, A. Huxley, J. Flouquet, P. Bonville, P.C.M. Gubbens, and A.M. Mulders, J. Phys. C.M. **10**, 9791 (1998)
425. S.K. Yip, and J.A. Sauls, Phys. Rev. Lett. **69**, 2264 (1992)
426. M.R. Li, P.J. Hirschfeld, and P. Wölfle, Phys. Rev. Lett. **81**, 5640 (1998)
427. J. Bardeen, Phys. Rev. Lett. **9**, 147 (1962); J. Bardeen, Phys. Rev. Lett. **6**, 57 (1960)
428. I. Giaever, Phys. Rev. Lett. **5**, 147 (1960); I. Giaever, and K. Megerle, Phys, Rev. **122**, 1101 (1961)
429. J. Nicol, S. Shapiro, and P.H. Smith, Phys. Rev. Lett. **5**, 461 (1960)
430. V. Ambegaokar, P.G. deGennes, and D. Rainer, Phys. Rev. A **9**, 2676 (1974)
431. M. Jourdan, M. Huth, and H. Adrian, Nature **398**, 47 (1999)
432. M. Huth, M. Jourdan, and H. Adrian, Physica B, **281&282**, 882 (2000)
433. N. Bernhoeft, N. Sato, B. Roessli, N. Aso, A. Heiss, G.H. Lander, Y. Endoh, and T. Komatsubara, Phys. Rev. Lett. **81**, 4244 (1998)
434. I. Iguchi, T. Yasuda, Y. Ōnuki, and T. Komatsubara, Phys. Rev. B **35**, 8873 (1987)
435. A.F. Andreev, Sov. Phys. J.E.T.P. **19**, 1228 (1964)
436. G.E. Blonder, M. Tinkham, and T.M. Klapwijk, Phys. Rev. B **25**, 4515 (1981)
437. Ch. Bruder, Phys. Rev. B **41**, 4017 (1990)

438. U. Poppe, J. Mag. Mag. Mat. **52**, 157 (1985)
439. A. Nowack, A. Heinz, F. Oster, D. Wohlleben, G. Güntherodt, Z. Fisk, and A. Menovsky, Phys. Rev. B **36**, 2436 (1987)
440. G. Goll, H. von Löhneysen, I.K. Yanson, and L. Taillefer, Phys. Rev. Lett. **70**, 2008 (1993)
441. G. Goll, C. Bruder, and H. von Löhneysen, Phys. Rev. B **52**, 6801 (1995)
442. C. Obermair, G. Goll, H. von Löhneysen, I.K. Yanson, and L. Taillefer, Phys. Rev. B, **57**, 7506 (1998)
443. K. Hasselbach, J.R. Kirtley, and P. Lejay, Phys. Rev. B **46**, 5826 (1992)
444. Ch. Wälti, H.R. Ott, Z. Fisk, and J.L. Smith, Phys. Rev. Lett. **84**, 5616 (2000)
445. Y. Tanaka, and S.S. Kashiwaya, Phys. Rev. Lett. **74**, 3451 (1995)
446. B.D. Josephson, Phys. Lett. **1**, 251 (1962)
447. R.C. Jaklevic, J. Lambe, J.E. Mercereau, and A.H. Silver, Phys. Rev. A **140**, 1628 (1965)
448. J.A. Pals, W. van Haeringen, and M.H. van Maaren, Phys. Rev. B **15**, 2952 (1977)
449. J.R. Kirtley, C.C. Tsuei, J.Z. Sun, C.C. Chi, L.S. Yu-Jahes, A. Gupta, M. Rupp, and M.B. Ketchan, Nature **373**, 225 (1995)
450. V. Ambegaokar, and A. Baratoff, Phys. Rev. Lett. **10**, 486 (1963)
451. A. Sumiyama, S. Shibata, Y. Oda, N. Kimura, Y. Haga, and Y. Ōnuki, Phys. Rev. Lett. **81**, 5213 (1998)
452. S. Wasser, A. Nowack, W. Schlabitz, A. Freimuth, O.E. Kvitnitskaya, A.A. Menovsky, and C. Bruder, Phys. Rev. Lett. **81**, 898 (1998)
453. S. Han, K.W. Ng, E.L. Wolf, H.F. Braun, L. Tanner, Z. Fisk, J.L. Smith, and M.R. Beasley, Phys. Rev. B **32**, 7567 (1981)
454. S. Han, K.W. Ng, E.L. Wolf, A.J. Millis, J.L. Smith, and Z. Fisk, Phys. Rev. Lett. **57**, 238 (1986)
455. E.V. Thuneberg, and V. Ambegaokar, Phys. Rev. Lett. **60**, 365 (1988)
456. S. Shibata, A. Sumiyama, Y. Oda, Y. Haga, and Y. Ōnuki, Phys. Rev. B **60**, 3076 (1999)
457. K. Yosida, Phys. Rev. **110**, 769 (1958)
458. P.W. Anderson, Phys. Rev. Lett. **3**, 325 (1959)
459. A.A. Abrikosov, and L.P. Gor'kov, Sov. Phys. J.E.T.P. **15**, 752 (1962)
460. K. Miyake, and K. Asayama, Phys. Rev. B **62**, 11363 (2000)
461. E.T. Ahrens, R.H. Heffner, P.C. Hammel, A.P. Reyes, J.D. Thompson, J.L. Smith, and W.G. Clark, Phys. Rev. B **59**, 1432 (1999)
462. Y. Kitaoka, H. Tou, K. Ishida, N. Kimura, Y. Ōnuki, E. Yamamoto, Y. Haga, and K. Maezawa, Physica B **281&282**, 878 (2000)
463. K. Ueda, Y. Kitaoka, H. Yamada, Y. Kohori, T. Kohara, and K. Asayama, J. Phys. Soc. Jpn. **56**, 867 (1987)
464. H. Tou, Y. Kitaoka, K. Asayama, N. Kimura, Y. Ōnuki, E. Yamamoto, and K. Maezawa, Phys. Rev. Lett. **77**, 1374 (1996)
465. H. Tou, Y. Kitaoka, K. Ishida, K. Asayama, N. Kimura, Y. Ōnuki, E. Yamamoto, Y. Haga, and K. Maezawa, Phys. Rev. Lett. **80**, 3129 (1998)
466. R. Feyerherm, A. Amato, F.N. Gygax, A. Schenck, C. Geibel, F. Steglich, N. Sato, and T. Komatsubara, Phys. Rev. Lett. **73**, 1849 (1994)
467. K. Ishida, D. Ozaki, T. Kamatsuka, H. Tou, M. Kyogaku, Y. Kitaoka, N. Tateiawa, N.K. Sato, N. Aso, C. Geibel, and F. Steglich, Phys. Rev. Lett. **89**, 37002 (2002)
468. L.C. Hebel, and C.P. Slichter, Phys. Rev. **107**, 901 (1957); L.C. Hebel, and C.P. Slichter, Phys. Rev. **113**, 1504 (1959)
469. M. Fibich, Phys. Rev. Lett. **14**, 561 (1965)
470. Y. Kohori, T. Kohara, H. Shibai, Y. Oda, Y. Kitaoka, and K. Asayama, J. Phys. Soc. Jpn., **57**, 395 (1988)
471. K. Matsuda, Y. Kohori, and T. Kohara, Phys. Rev. B **55**, 15223 (1997)
472. K. Ishida, Y. Kawasaki, K. Tabuchi, K. Kashima, Y. Kitaoka, K. Asayama, C. Geibel, and F. Steglich, Phys. Rev. Lett. **82**, 5353 (1999)
473. C. Tien, and I.M. Jiang, Phys. Rev. B **40**, 229 (1989)
474. P.S. Riseborough, J. Jpn. Phys. Soc. Suppl. **71**, 344 (2002); see also P.S. Riseborough, Physica B **312-313**, 19 (2002)
475. N. Sato, N. Aso, K. Miyake, R. Shiina, P. Thalmeier, G. Varlogiannis, C. Geibel, F. Steglich, P. Fulde, and T. Komatsubara, Nature **410**, 340 (2001)
476. G.R. Stewart, Rev. Mod. Phys. **56**, 755 (1984)
477. D.W. Hess, P. Riseborough, and J.L. Smith, In: *Encyclopedia of Applied Physics* (VCH, Weinheim, 1993) Vol. 7, p. 435
478. J.P. Brison, L. Glémot, H. Suderow, A. Huxley, S. Kambe, and J. Flouquet, Physica B **280**, 165 (2000)
479. H.R. Ott, *Progress in Low Temperature Physics* (Elsevier, Amsterdam, 1987) Vol. 11, p. 215
480. J.F. Annett, Physica C **317-318**, 1 (1999)

481. R.H. Heffner, D.W. Cooke, Z. Fisk, R.L. Hutson, M.E. Schillaci, J.L. Smith, J.O. Willis, D.E. MacLaughlin, C. Boekema, R.L. Lichti, A.B. Denison, and J. Oostens, Phys. Rev. Lett. **57**, 1255 (1986)
482. J.E. Sonier, R.H. Heffner, D.E. MacLaughlin, G.J. Nieuwenhuys, O. Bernal, R. Movshovich, P.G. Pagliuso, J. Cooley, J.L. Smith and J.D. Thompson, Phys. Rev. Lett. **85**, 2821 (2000)
483. M.C. Aronson, T. Vorenkamp, Z. Koziol, A. de Visser, K. Bakker, J.J.M. Franse, and J.L. Smith, J. Appl. Phys. **69** 5487 (1991)
484. H.G.M. Duijn, N.H. van Dijk, A. de Visser, and J.J.M. Franse, Physica B **224**, 44 (1996)
485. C.M. Varma, Comments Solid State Phys. **11**, 221 (1985)
486. R.J. Keizer, A. de Visser, A.A. Menovsky, J.J.M. Franse, B.Fak, and J.-M. Mignot, Phys. Rev. B **60**, 6668 (1999)
487. J.A. Sauls, Adv. Phys. **43**, 113 (1994)
488. C.H. Choi and J.A. Sauls, Phys. Rev. Lett. **66**, 484 (1991)
489. R.H. Heffner and M.R. Norman, Comments Cond. Mat. Phys. **17**, 361 (1996)
490. J.D. Thompson, M.W. McElfresh, J.O. Willis, Z. Fisk, J.L. Smith, and M. B. Maple, Phys. Rev. B **35**, 48 (1987)
491. G.M. Schmiedeshoff, Z. Fisk, and J.L. Smith, Phys. Rev. B **48**, 16417 (1992)
492. P. Kumar, and P. Wölfle, Phys. Rev. Lett. **59**, 1954 (1987)
493. M. Sigrist, and T.M. Rice, Phys. Rev. B **39**, 2200 (1989)
494. V. Martisovits, G. Zaránd, and D.L. Cox, Phys. Rev. Lett. **84** 5872 (2000)
495. N. Bernhoeft, G.H. Lander, M.J. Longfield, S. Langridge, D. Mannix, E. Lidström, E. Colineau, A. Hiess, C. Vettier, F. Wastin, J. Rebizant, and P. Lejay, Acta Phys. Pol. **34**, 1367 (2003)
496. Tetsuo Honma, Yoshinori Haga, Etsuji Yamamoto, Naoto Metoki, Yoshihiro Koike, Hitoshi Ohkuni, Norihito Suzuki and Yoshichika Ōnuki, J. Phys. Soc. Jpn. **68**, 338 (1999)
497. K. Hasselbach, J.R. Kirtley, and J. Flouquet, Phys. Rev. B **47**, 509 (1993)
498. L.E. de Long, D. Togliatti, G.W. Lehman, W.-K. Kwok, G.W. Crabtree, K.G. Vandervoort, D.G. Hinks, Physica B **163**, 499 (1990)
499. P. Thalmeier, Phys. Rev. B **44** 7120 (1991)
500. R. Joynt, W.K. Kwok, G.W. Crabtree, D.G. Hinks, and L. DeLong, Phys. Rev. B **44**, 7122 (1991)
501. L. Paolasini, J.A. Paixao, G.H. Lander, A. Delapalme, N. Sato, and T. Komatsubara, J. Phys. Cond. Mat. **5**, 8905 (1993)
502. K. Gloos, R. Modler, H. Schimanski, C.D. Bredl, C. Geibel, F. Steglich, A.I. Buzdin, N. Sato, and T. Komatsubara, Phys. Rev. Lett. **70**, 501 (1993)
503. M. Hiroi, M. Sera, N. Kobayashi, Y. Haga, E. Yamamoto, and Y. Ōnuki, J. Phys. Soc. Jpn. **66**, 1595 (1997)
504. M. Jourdan, M. Huth, S. Mouloud, and H. Adrian, J. Mag. Mag. Mat. **177-181**, 294 (1998)
505. A. Schröder, J.G. Lussier, B.D. Gaulin, J.D. Garrett, W.J.L. Buyers, L. Rebelsky and S.M. Shapiro, Phys. Rev. Lett. **72**, 136 (1994)
506. Y. J. Uemura and G. M. Luke, Physica B **186-188**, 223 (1993)
507. H. Tou, Y. Kitaoka, T. Kamatsuka, K. Asayama, C. Geibel, F. Steglich, S. Süllow and J.A. Mydosh, Physica B **230-232**, 361 (1997)
508. Y. Dalichaouch, M.C. de Andrade, and M.B. Maple, Phys. Rev. B **46**, 8761 (1992)
509. N. Sato, N. Koga, and T. Komatsubara, J. Phys. Soc. Jpn. **65**, 1555 (1996)
510. A. Llobet, J.S. Gardner, E.G. Moshopoulou, J.-M. Mignot, M. Nicklas, W. Bao, N.O. Moreno, P.G. Pagliuso, I.N. Goncharenko, J.L. Sarrao, and J.D. Thompson, Phys. Rev. B **69**, 024403 (2004)
511. T. Tayama, M. Lang, T. Lühmann, F. Steglich, and W. Assmus, Phys. Rev. B **67**, 214504 (2003)
512. F. Steglich, P. Gegenwart, C. Geibel, R. Helfrich, P. Hellmann, M. Lang, A. Link, R. Modler, G. Sparn, N. Buttgen, A. Loidl, Physica B **223 & 224**, 1 (1996)
513. M. Ishikawa, N. Takeda, M. Koeda, M. Hedo, and Y. Uwatoko, Phys. Rev. B **68**, 024522 (2003)
514. Wei Bao, P.G. Pagliuso, J.L. Sarrao, J.D. Thompson, Z. Fisk , J.W. Lynn and R.W. Erwin, Phys. Rev. B **62**, R14621 (2000); Wei Bao, P.G. Pagliuso, J.L. Sarrao, J.D. Thompson, Z. Fisk, J.W. Lynn, and R.W. Erwin, Phys. Rev. B **63**, 219901 (2001)
515. T. Mito, S. Kawasaki, G.-Q. Zheng, Y. Kawasaki, K. Ishida, Y. Kitaoka, D. Aoki, Y. Haga, and Y. Ōnuki, Phys. Rev. B **63**, 220507 (2001)
516. R.A. Fisher, F. Bouquet, N.E. Phillips, M.F. Hundley, P.G. Pagliuso, J.L. Sarrao, Z. Fisk, and J.D. Thompson, Phys. Rev. B **65**, 224509 (2002)

517. R. Movshovich, A. Bianchi, M. Jaime, M.F. Hundley, J.D. Thompson, N. Curro, P.C. Hammel, Z. Fisk, P.G. Pagliuso, and J.L. Sarrao, Physica B **312-313**, 7 (2002)
518. S. Ikeda, H. Shishido, M. Nakashima, R. Settai, D. Aoki, Y. Haga, H. Harima, Y. Aoki, T. Namiki, H. Sato, Y. Õnuki, J. Phys. Soc. Jpn. **70**, 2248 (2001)
519. A. Bianchi, R. Movshovich, N. Oeschler, P. Gegenwart, F. Steglich, J.D. Thompson, P.G. Pagliuso, and J.L. Sarrao, Phys. Rev. Lett. **87**, 057002 (2002)
520. S. Majumdar, M.R. Lees, G. Balakrishnan, and D.McK. Paul, Phys. Rev. B **68**, 012504 (2003)
521. T.P. Murphy, Donavan Hall, E.C. Palm, S.W. Tozer, C. Petrovic, Z. Fisk, R.G. Goodrich, P.G. Pagliuso, J.L. Sarrao, and J.D. Thompson, Phys. Rev. B **65**, 100514 (2002)
522. H.A. Radovan, N.A. Fortune, T.P. Murphy, S.T. Hannahs, E.C. Palm, S.W. Tozer, and D. Hall, Nature **425**, 51 (2003)
523. J.M. Lawrence and S.M. Shapiro, Phys. Rev. B **22**, 4379 (1980)
524. N.D. Mathur, F.M. Grosche, S.R. Julian, I.R. Walker, D.M. Freye, R.K.W. Haselwimmer, and G.G. Lonzarich, Nature **394**, 39 (1998)
525. S. Kawasaki, T. Mito, Y. Kawasaki, G.-q. Zheng, Y. Kitaoka, H. Shishido, S. Araki, R. Settai, and Y. Õnuki, Phys. Rev. B **66**, 054521 (2002)
526. S. Raymond, and D. Jaccard, Phys. Rev. B **61**, 8679 (2000)
527. I. Sheikin, E. Steep, D. Braithwaite, J.-P. Brison, S. Raymond, D. Jaccard, and J. Flouquet, J. Low Temp. Phys. **122**, 591 (2001)
528. E. Vargoz and D. Jaccard, J. Mag. Mag. Mat. **177-181**, 294 (1998)
529. T.C. Kobayashi, T. Miyazu, N. Takeshita, K. Shimizu, K. Amaya, Y. Kitaoka, and Y. Õnuki, J. Phys. Soc. Jpn. **67**, 996 (1998)
530. K. Izawa, Y. Nakajima, J. Goryo, Y. Matsuda, S. Osaki, H. Sugawara, H. Sato, P. Thalmeier, and K. Maki, Phys. Rev. Lett. **90**, 117001/1 (2003)
531. R. Vollmer, A. Faißt, C. Pfleiderer, H.v. Löhneysen, E.D. Bauer, P.-C. Ho, V. Zapf, and M.B. Maple, Phys. Rev. Lett. **90**, 057001/1 (2003)
532. H. Kotegawa, M. Yogi, Y. Imamura , Y. Kawasaki, G.-q. Zheng, Y. Kitaoka, S. Ohsaki, H. Sugawara, Y. Aoki, and H. Sato, Phys. Rev. Lett. **90**, 027001/1 (2003)
533. D.E. MacLaughlin, J.E. Sonier, R.H. Heffner, O.O. Bernal, Ben-Li Young, M.S. Rose, G.D. Morris, E.D. Bauer, T.D. Do, and M.B. Maple, Phys. Rev. Lett. **89**, 157001/1 (2002)
534. Y. Aoki, A. Tsuchiya, T. Kanayama, S.R. Saha, H. Sugawara, H. Sato, W. Higemoto, A. Koda, K. Ohishi, K. Nishiyama, and R. Kadono, Phys. Rev. Lett. **91**, 067003/1 (2003)
535. E.D. Bauer, N.A. Frederick, P.-C. Ho, V.S. Zapf, and M.B. Maple, Phys. Rev. B **65**, 100506 (2002)
536. P.-C. Ho, N.A. Frederick, V.S. Zapf, E.D. Bauer, T.D. Do, M.B. Maple, A. D. Christianson and A.H. Lacerda, Phys. Rev. B **67** 180508 (2003)
537. G.P. Meisner, A.L. Giorgi, A.C. Lawson, G.R. Stewart, J.O. Willis, M.S. Wire, and J.L. Smith, Phys. Rev. Lett. **53**, 1829 (1984)
538. V. Ginzburg, Sov. Phys. JETP **4**, 153 (1957)
539. *Superconductivity in Ternary Compounds*, ed. by M.B. Maple and Ø. Fischer (Springer, Berlin, 1982) Vol. 2
540. J. Flouquet, A. Huxley, D. Braithwaite, F. Hardy, G. Knebel, V. Mineev, E. Ressouche, D. Aoki, and J.P. Brison, Acta Phys. Pol. B **34**, 275 (2003)
541. I. Sheikin, A. Huxley, D. Braithwaite, J.P. Brison, S. Watanabe, K. Miyake, and J. Flouquet, Phys. Rev. B **64**, 220503 (2001)
542. T.R. Kirkpatrick, and D. Belitz, Phys. Rev. B **67**, 024515 (2003)
543. C. Pfleiderer, M. Uhlarz, S.M. Hayden, R. Vollmer, H.v. Löhneysen, N.R. Bernhoeft, and G. G. Lonzarich, Nature **412**, 58 (2001)
544. K.V. Samokhin and M.B. Walker, Phys. Rev. B **66**, 174501 (2002)
545. E.A. Yelland, S.M. Hayden, S.J.C. Yates, C.Pfleiderer, M.Uhlarz, R.Vollmer, H.v. Lohneysen, N.R. Bernhoeft, R.P. Smith, S.S. Saxena, N.Kimura, preprint, http://xxx.lanl.gov/abs/cond-mat/0502341 (2005)
546. F. Wastin et al., private comunication
547. Takahiro Maehira, Takashi Hotta, Kazuo Ueda, and Akira Hasegawa, Phys. Rev. Lett. **90**, 207007/1 (2003)
548. I. Opahle and P.M. Oppeneer, Phys. Rev. Lett. **90**, 157001/1 (2003)

20 Organic Superconductors

Michael Lang Phys.-Institut der J. W. Goethe Univ. Frankfurt, Germany
Jens Müller Max–Planck–Institut für Chemische Physik fester Stoffe, Dresden, Germany

20.1 Introduction

Over the past 20 years the research on organic conductors has developed into a most active branch of modern condensed-matter physics. The main difference of molecular conductors compared to conventional metals is that the former are made up of building blocks constructed from carbon atoms and their combinations with other elements such as sulfur, selenium or oxygen. As a result, when forming a crystal, these molecular units preserve to a large extent their specific features such as the molecular orbitals, the ionization energy and the intramolecular vibrational modes. Molecular materials thus have the potential of providing a flexible building-block system where the physical properties can be tuned by small modifications in the arrangements and bridging of these functional units. In contrast to the molecular crystals formed by weakly van der Waals-bond neutral entities, organic conductors consist of open-shell molecular units which are the result of a partial oxidation and reduction of the donor and acceptor molecules in the crystal-growth process. It is the unpaired electron residing in the π-molecular orbital (π-hole) of the donor unit which is responsible for the electronic properties of these charge-transfer salts. Due to a π-orbital overlap between adjacent molecules, the π-holes can delocalize throughout the crystal giving rise to metallic conductivity.

The unique chemistry of carbon, which, on the one hand, provides a rich basis of potential organic donor molecules, and the manifold possibilities of combining them with charge-compensating acceptor units, on the other hand, have enabled the synthesis of an enormous number of conducting organic charge-transfer salts. The geometry of the building blocks and the way they are packed together in the crystal determine the effective dimensionality of the electronic structure of the compound. The planar shape of the TMTSF (tetramethyltetraselenafulvalene) molecule, a derivative of the prototype TTF (tetrathiafulvalene), permits an infinite stacking of these units in the crystal structure. As a consequence, a significant intermolecular overlap of the π-orbitals occurs only along the stacking axis giving rise to a quasi-one-dimensional conduction band. In contrast, packing of the larger BEDT-TTF (bisethylenedithio-tetrathiafulvalene) molecules often results in a quasi-two-dimensional electronic structure.

These low-dimensional organic conductors constitute a peculiar class of materials which has gained strong interest among scientists due to the wealth of interesting phenomena that have been observed over the last two decades or so. Above all it was the discovery of superconductivity in pressurized $(TMTSF)_2PF_6$ by Jérome et al. in 1979 [1] which created much excitement among both solid-state chemists and physicists. Indeed, this finding was a result of interdisciplinary efforts in synthesizing and characterizing organic conductors; activities which started up in the early 1950s by the pioneering work of Akamatu et al. [2] and which received a fresh impetus in 1964 thanks to the work of Little [3]. He proposed that in a suitably designed one-dimensional conductor embedded in a polarizable medium, superconductivity at high temperatures should be possible. Although it soon became clear that the nature of the superconducting state in the $(TMTSF)_2PF_6$ salt is of a fundamentally different type from what had been discussed in Little's model, this finding set the stage for the discovery of a large number of conducting charge-transfer salts including up to the latest count, more than 80 superconductors. Many of these systems superconduct even at ambient pressure with the highest T_c value of about 12 K received in the quasi-two-dimensional BEDT-TTF-based systems. Besides superconductivity, the organic charge-transfer salts reveal a variety of other interesting collective phenomena including metal-to-insulator transitions, spin-Peierls and density-wave states as well as phases with localized charges and commensurate-type antiferromagnetic order. These states have been found to depend most sensitively on factors such as the acceptor ions, the magnetic field or external pressure.

One of the key features underlying the above phenomena is the low dimensionality of the materials. The confinement of the carrier motion to one or two spatial dimensions together with the low charge-carrier concentration enhance the effect of the interactions between the electrons. Another important feature specific to these molecular systems is a considerably strong coupling of the charge carriers to the lattice degrees of freedom. For molecular crystals, this coupling includes both the interactions with high-frequency intramolecular modes and the low-lying intermolecular vibrations. The present organic metals thus represent ideal model systems for exploring the interplay of strong electron–electron and electron–phonon interactions in reduced dimensions. In particular, the close proximity of superconductivity to a magnetically-ordered state encountered not only in the organic materials but also in other strongly-correlated electron systems such as the heavy-fermion metals and high-temperature superconductors continues to challenge our understanding of superconductivity. While for the latter systems evidence is growing that, indeed, the magnetism and superconductivity are intimately related to each other, the nature of the superconducting state for the present materials is far from being understood.

The discussion of the superconducting-state properties for the present organic materials is complicated by the existence of contradictory experimental evidences. In some cases, this controversy even encompasses results on the same quantity when the measurements have been carried out by different groups. While about one half of the data seem to support an anisotropic superconducting state, with a

d-wave order parameter being the most favored one, the other half are consistent with an order parameter which is finite everywhere on the Fermi surface.

Closely related to the issue of superconductivity in the proximity to antiferromagnetic order is the nature of the state above T_c. For the quasi-two-dimensional organic conductors, for example, unusual metallic properties have been observed posing the question whether this phenomenon is related to the equally unusual behavior seen in the other class of quasi-two-dimensional superconductors, the high-T_c cuprates [4].

This article will give an overview on the normal-state and superconducting-state properties of organic superconductors. There are a number of review articles on this subject [5–10], mostly focussing on either the quasi-one-dimensional or two-dimensional materials. The intention of the present review is therefore to provide a discussion which covers aspects common to both families on the same footing. Instead of reviewing the whole diversity of behaviors found among the various compounds with all their structural and chemical modifications, we will mainly focus on selected compounds of both families. These are the most extensively studied and best characterized $(TMTSF)_2X$ and $(BEDT\text{-}TTF)_2X$ slats whose properties are representative for a wide class of materials. Special attention is paid to the more recent developments including the controversial discussions of some aspects, in particular the discussion on the nature and symmetry of the superconducting state.

20.2 Characteristics of Organic Charge-Transfer Conductors

20.2.1 Molecular Building Blocks

The prerequisites of forming conducting molecular solids are essentially (i) the creation of unpaired electrons and (ii) their delocalization throughout the crystal. For the organic superconductors discussed in this article, condition (i) is satisfied by a partial transfer of charge between the two constituent parts of a charge-transfer complex: an organic electron-donor molecule D is combined with an, in most cases inorganic, electron-acceptor complex X according to the reaction $[D_m] + [X_n] \rightarrow [D_m]^{+\delta} + [X_n]^{-\delta}$, where m and n are integers. Since the π-electrons of the D molecule with orbits extending perpendicularly to the planes of the molecules have low binding energies, much lower than those of the σ-electrons, can easily be excited which can be seen in the charge-transfer process. Accordingly, the so created charge carriers have a π-electron (hole) character. In most cases, the negatively charged anions $[X_n]^{-\delta}$ adopt a closed-shell configuration and thus do not contribute to the electrical conductivity. When the crystal is formed, a delocalization of the charge carriers (ii) may be obtained by a dense packing of the donor molecules. As a result, the π-orbitals of the partially filled outer molecular shells overlap and an electronic band structure is formed.

The conductivity of solids depends on both the number of free carriers and their mobility in the crystal. In organic charge-transfer salts, the carrier concentration is determined by the electronegativity of the donor, the electron affinity of the acceptor molecule and the chemical bonds. A high mobility is obtained by a relatively large bandwidth which is the result of a considerably strong overlap of the π-orbitals from adjacent molecules.

Organic superconductors have been derived from a variety of different organic electron-donor molecules, where most of them are derivatives of the archetype TTF molecule. Its combination with the electron acceptor TCNQ (tetracyanoquinodimethane) in 1973 led to the synthesis of the first quasi-one-dimensional (quasi-1D) organic conductor TTF-TCNQ [11,12]. This material and its derivatives served for a long time as model systems for exploring the physical properties of quasi-1D conductors.

In the vast majority of D_mX_n salts, the donor–acceptor molecular ratio $m{:}\,n$ is fixed to 2: 1, i.e., two donor molecules transfer one electron to the acceptor X. Organic and organometallic chemistry have provided an enormous number of donor molecules which may serve as building blocks for organic con-

[1] TMTTF stands for tetramethyltetrathiafulvalene.

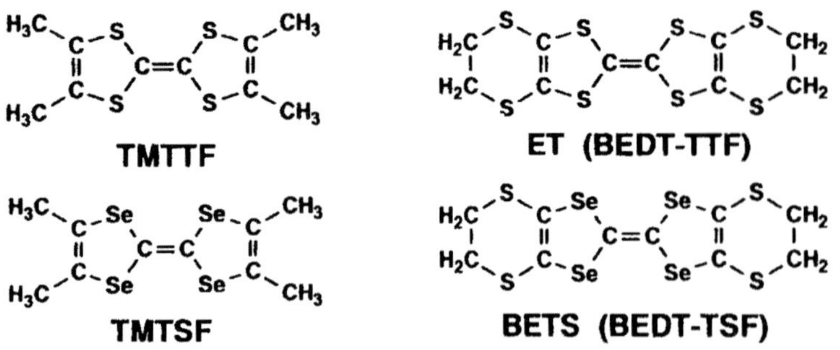

Fig. 20.1. Principal structures of donor molecules that furnish organic superconductors

ductors and superconductors. The main examples are the TTF derivatives TMTSF, TMTTF[1], BEDT-TTF (commonly abbreviated as ET) and BEDT-TSF[2] (or simply BETS), see Fig. 20.1. The TMTSF differs from TMTTF in that the four sulfur atoms of the latter have been replaced by selenium. In the same way, BEDT-TSF is derived from BEDT-TTF. It is interesting to note that organic superconductors have been synthesized not only by using these symmetric molecules, but also on the basis of asymmetric complexes. These are the DMET (dimethylethylenedithio-diselenedithiafulvalene) [13,14], which is a hybrid of TMTSF and BEDT-TTF, MDT-TTF (methylenedithio-TTF) [13, 14] and BEDO-TTF (bisethylenedithioxy-TTF) [15], see [5] for an overview.

The TMTSF molecule provides the basis for the so-called Bechgaard salts $(TMTSF)_2X$ which form with a variety of inorganic monovalent acceptor molecules X. Indeed, it was $(TMTSF)_2PF_6$ where in 1979 superconductivity had been observed for the first time in an organic compound [1]. At ambient-pressure conditions, the material was found to undergo a metal-to-insulator transition around 12 K which had been identified as a spin-density-wave (SDW) ordering [6]. By the application of hydrostatic pressure of 12 kbar the SDW instability can be suppressed and superconductivity forms below $T_c = 0.9$ K [1, 16]. A replacement of PF_6 by ClO_4 has resulted in the first and – till now – only member of the quasi-1D salts which becomes superconducting at ambient pressure [17].

Apparently, the strong tendency of Bechgaard's salts to undergo a metal–insulator transition, inherent to quasi-1D electron systems, counteracts with their ability to become superconducting. In order to achieve ambient-pressure superconductivity in these organic complexes with possibly even higher transition temperatures it is necessary to increase their dimensionality. According to this strategy, in 1982 the first metallic compound based on the new electron donor molecule BEDT-TTF [18], cf. Fig. 20.1, was synthesized by Saito et al. [19]. The underlying idea was to enhance the overlap between π-orbitals of adjacent molecules by enlarging the π-electron system on each molecule. This has been accomplished by adding rings of carbon and sulfur atoms at the outer ends of the TTF skeleton. In contrast to the Bechgaard salts where the donor molecules form infinite stacks, steric effects specific to the BEDT-TTF molecules prevent such an infinite face-to-face stacking in the $(BEDT\text{-}TTF)_2X$ salts. As a consequence, the side-by-side overlap between π-orbitals of adjacent molecules becomes stronger and, in some cases, comparable to the face-to-face interaction resulting in a quasi-2D electronic structure of the BEDT-TTF-based salts. The combination of BEDT-TTF with the monovalent anion X = $Cu(NCS)_2$ achieved in 1988 led to the discovery of the first ambient-pressure superconductor in this class of materials, the second generation of organic superconductors, with transition temperatures in the range of 10 K [20].

The quasi-1D and quasi-2D systems tend to grow in needle-like and plate-like shapes, with the latter usually showing well developed smooth and shiny metallic surfaces. The synthesis of organic molecular conductors can be divided into two steps: (1) the synthesis of the neutral donor and acceptor molecules and (2) the oxidation and reduction of the donors

[2] BEDT-TSF stands for bisethylenedithio-tetraselenafulvalene.

and acceptors to radical cations and radical anions. For step (1) several independent methods have been discussed and used even for selected donor and acceptor molecules, see e.g. [18]. Conversely, to perform step (2), the redox process, the electrocrystallization technique has proved to be the method of choice for synthesizing high-quality crystals. For details on the chemical synthesis and crystal growth techniques, the reader is referred to [5,18,21] and references cited therein.

20.2.2 Structural Aspects

Figure 20.2 shows exemplarily the crystal structure of $(TMTSF)_2PF_6$ viewed almost perpendicular to the stacking axis (a-axis); the direction of highest conductivity. All members of the $(TM)_2X$ family, where TM stands for TMTSF and TMTTF, are isostructural with triclinic symmetry. The conducting stacks are separated by the anions which are located at inversion centers of the lattice. The donor molecules which are nearly planar and almost perpendicular to the chain axis are arranged in a zigzag-type manner along the chains with two slightly different intrachain distances corresponding to a weak dimerization. The fairly close Se · · · Se distance of 3.87 Å along the b-axis being smaller than the sum of the van der Waals radii of 4 Å results in a weak interchain overlap [6] and thus a weakly 2D electronic character. An important structural aspect of the $(TM)_2X$ compounds is related to the symmetry of the anion X. Where these anions are centrosymmetric such as the octahedral X = PF_6, AsF_6, SbF_6 or TaF_6 complexes, their orientation is fixed in the structure. In contrast, $(TM)_2X$ salts formed with non-centrosymmetric tetrahedral anions such as ClO_4 or FeO_4 undergo a structural

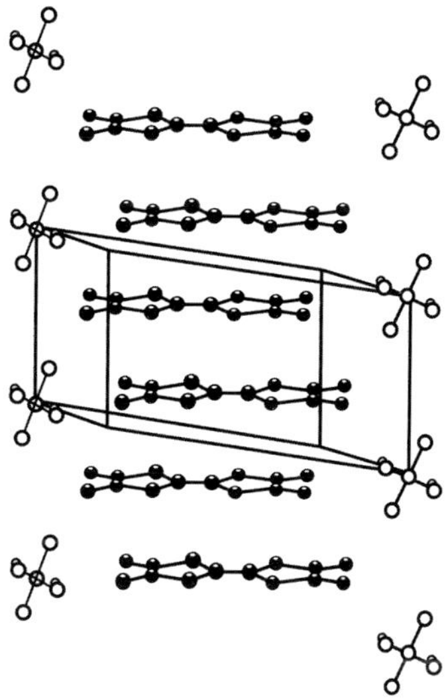

Fig. 20.2. Crystal structure of $(TMTSF)_2PF_6$ (side-viewed tilted). The axis with the highest conductivity is the vertical a'-axis, the one with the lowest conductivity the horizontal c-axis. a' is the projection of the a-axis onto the direction perpendicular to the bc-plane

transition from a disordered high-temperature state to an ordered low-temperature phase [22].

While all members of the $(TM)_2X$ family share the same crystal structure, the rather loose intra-stack coupling of the $(BEDT\text{-}TTF)_2X$ salts gives rise to a variety of polymorphic phases (packing motifs) which are distinguished by Greek characters; the most important amongst them are the α, β, and κ-phases, see e.g. [23] for a comprehensive review of the structural properties of these salts. In some cases, as for instance realized in the compound with the linear anion X = I_3, various structural modifications exist even for the same anion, cf. Fig. 20.3. The α-type

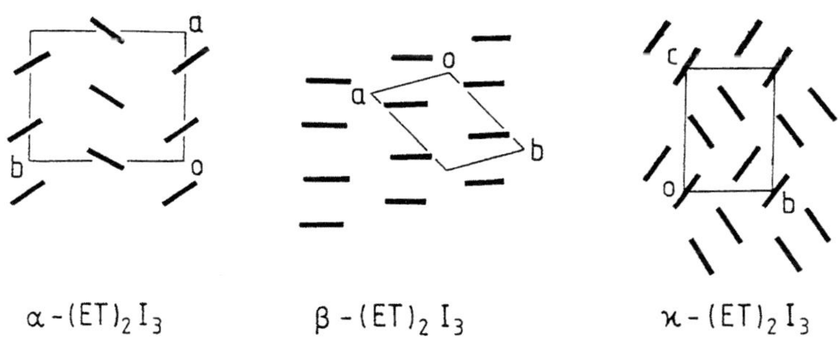

Fig. 20.3. Schematic packing motifs of the ET molecules in the α, β and κ-phases of $(ET)_2I_3$ viewed along the long axis of the ET molecules

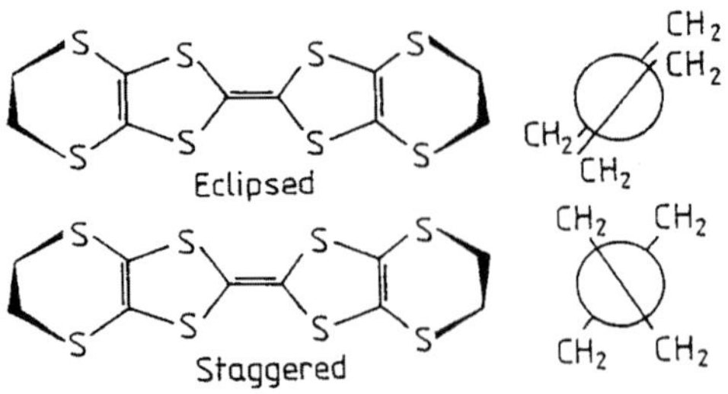

Fig. 20.4. Schematic view of the relative orientations of the ethylene endgroups $[(CH_2)_2]$ of the ET molecule. The *right side* shows the view along the long axis of the molecule

structure consists of stacks arranged in a herring-bone pattern. The ET molecules are connected via S···S contacts being shorter than the sum of the van-der-Waals radii of 3.6 Å. The β-type packing is reminiscent of the stacking arrangement found in the Bechgaard salts. However, the smaller inter-stack distances in β-$(ET)_2I_3$ lead to a more two-dimensional electronic structure. The κ-phase is unique in that it does not consist of interacting stacks but rather of interacting dimers formed by two face-to-face aligned ET molecules. Adjacent dimers are arranged almost orthogonal to each other so that the *intra-* and *inter-*dimer interactions are of the same size. This results in a quasi-2D electronic structure with a small in-plane anisotropy. The κ-type compounds with polymer-like anions are of particular interest with respect to their superconducting properties as they exhibit the highest transition temperatures.

In forming the crystal, apart from the ethylene $[(CH_2)_2]$ groups at the outer ends of the molecules, the charged ET molecules $C_6S_8[(CH_2)_2]_2$ untwist at their centre and become planar. As shown schematically in Fig. 20.4, the relative orientation of the outer C–C bonds can either be parallel (eclipsed) or canted (staggered). At high temperatures, the ethylene end-groups become disordered due to the strong thermal vibrations. Upon cooling to low temperatures, the endgroups adopt one of the two possible conformations, depending on the anion and the crystal structure. As will be discussed in Sect. 20.3.4 for the κ-$(ET)_2X$ salts, disorder in the conformation of the $[(CH_2)_2]$ groups can have a severe influence on the electronic properties in these compounds, in particular the superconductivity.

The planar C_6S_8 skeleton of the ET molecules permits a rather dense packing with a variety of possible packing arrangements. As a result, the interdimer interaction becomes comparable to that within the dimers giving rise to a quasi-2D electronic structure. Besides the intermolecular S···S contacts, i.e. the donor–donor interaction, the donor–acceptor couplings also play an important role for the physical properties of these multilayer systems. The latter interaction is provided by electrostatic forces as a consequence of the charged molecules and the hydrogen bonds joining between the carbon atoms at the donor site and the sulfur, carbon or nitrogen atoms being located at the acceptor site. The relative strength of

Table 20.1. Room-temperature crystallographic data of some $(ET)_2X$ superconductors including the space group SG, lattice parameters *a,b* and *c*, unit-cell volume *V*, number of formular units per unit cell *z* as well as T_c values. In the case of quasi-2D $(ET)_2X$ salts the lattice parameter perpendicular to the conducting planes is underlined

	SG	*a* (Å)	*b* (Å)	*c* (Å)	*V* (Å^3)	z	T_c (K)
$(TMTSF)_2PF_6$	$P\bar{1}$	7.297	7.711	13.522	713.14	1	1.1 (6.5 kbar)
κ-$(ET)_2Cu(NCS)_2$	$P2_1$	<u>16.248</u>	8.440	13.124	1688	2	10.4
κ-$(ET)_2Cu[N(CN)_2]Br$	*Pnma*	12.949	<u>30.016</u>	8.539	3317	4	11.2
κ-$(ET)_2Cu[N(CN)_2]Cl$	*Pnma*	12.977	<u>29.977</u>	8.480	3299	4	12.8 (300 bar)
β''-$(ET)_2SF_5CH_2CF_2SO_3$	$P\bar{1}$	9.260	11.635	<u>17.572</u>	1836	2	5.3
α-$(ET)_2NH_4Hg(SCN)_4$	$P\bar{1}$	10.091	<u>20.595</u>	9.963	2008	2	1.1
κ-$(ET)_2I_3$	$P2_1/c$	<u>16.387</u>	8.466	12.832	1688	2	3.5
λ-$(BETS)_2GaCl_4$	$P\bar{1}$	16.141	18.58	6.594	1774	2	6

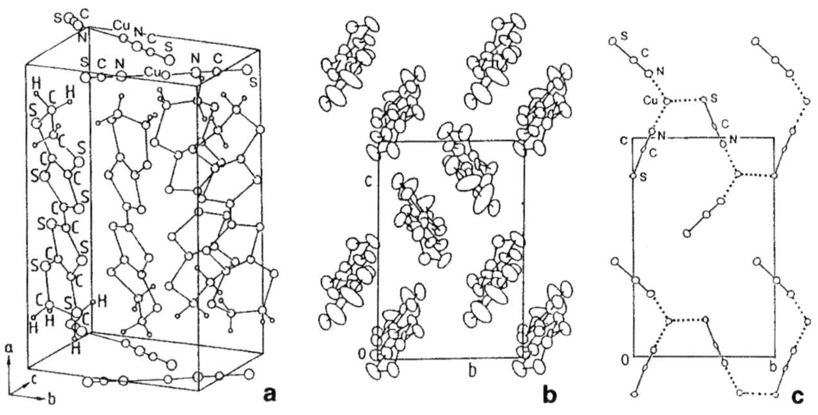

Fig. 20.5. (**a**) Crystal structure of κ-(BEDT-TTF)$_2$Cu(NCS)$_2$. The arrangement of the ET molecules (**b**) and the Cu(NCS)$_2$ anions (**c**) when viewed along the a^* direction, i.e. perpendicular to the conducting planes. The a-axis is slightly tilted from the a^*-axis which is normal to the conducting bc-plane

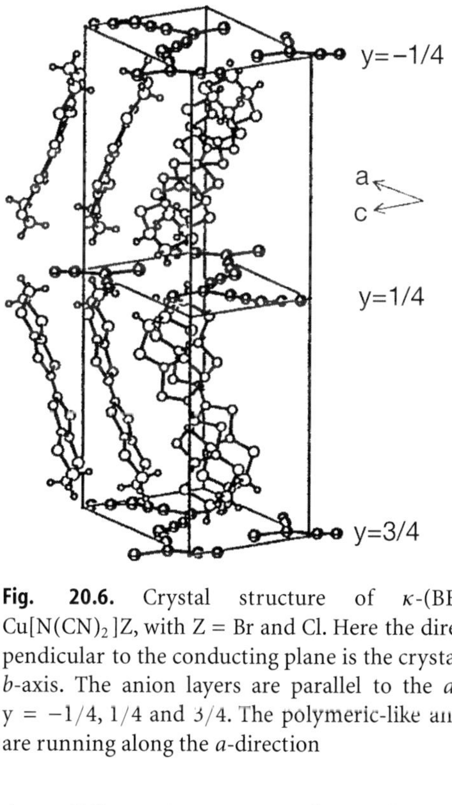

Fig. 20.6. Crystal structure of κ-(BEDT-TTF)$_2$ Cu[N(CN)$_2$]Z, with Z = Br and Cl. Here the direction perpendicular to the conducting plane is the crystallographic b-axis. The anion layers are parallel to the ac-plane at y = −1/4, 1/4 and 3/4. The polymeric-like anion chains are running along the a-direction

these different interactions, the conformational degrees of freedom of the ethylene groups along with the flexibility of the molecular framework give rise to a variety of different ET complexes [5,23].

Despite their complex crystal structure with rather low symmetry (cf. Table 20.1) it is convenient to think of the (ET)$_2$X compounds as layered systems consisting of conducting sheets formed by the ET molecules which are intersected by more or less thick insulating anion layers. Prime examples are the κ-phase (ET)$_2$X salts with X=Cu(NCS)$_2$, Cu[N(CN)$_2$]Br and Cu[N(CN)$_2$]Cl which are the most intensively studied and best characterized members of this class of materials. These compounds are of particular interest not only because of their relatively high superconducting transition temperatures but also owing to certain similarities in their normal-state and superconducting-state properties with those of the high-temperature cuprate superconductors [4,9]. Figures 20.5 and 20.6 display the crystal structures of κ-(ET)$_2$Cu(NCS)$_2$ and κ-(ET)$_2$Cu[N(CN)$_2$]Z. In both cases the layered structure consists of conducting planes with the characteristic κ-type arrangement of the ET molecules separated by insulating anion layers. While the crystal structure of κ-(ET)$_2$Cu(NCS)$_2$ has monoclinic symmetry with two dimers, i.e. two formula units per unit cell, the κ-(ET)$_2$Cu[N(CN)$_2$]Z salts are orthorhombic with a unit cell containing four dimers, see Table 20.1. Due to the particular arrangement of their polymeric anions, these crystals lack a center of inversion symmetry.

Subtle changes in the intermolecular spacing or relative orientation of the ET molecules as e.g. induced by either external pressure or anion substitution may significantly alter the π-electron overlap between adjacent molecules. This can have a severe influence on the electronic properties as demonstrated for the κ-(ET)$_2$Cu[N(CN)$_2$]Z system for various Z: while the compound with Z = Br is a superconductor with $T_c = 11.2$ K [24], replacement of Br by the slightly smaller Cl results in an antiferromagnetic insulating ground state. On the other hand, the application of hydrostatic pressure of only about 300 bar drives the latter system to a superconductor with a T_c of 12.8 K [25–27], the highest transition temperature found among this class of materials so far.

A new class of materials which has recently gained considerable interest is based on the donor molecule BETS and its combination with the discrete anions MX_4 (M = Fe, Ga, In; X = Cl, Br). Two structural modifications have been found. These are the orthorhombic κ-type structure (Pnma) which results in plate-like crystals and the triclinic ($P\bar{1}$) λ-type variant which grow in a needle-like manner [28–30]. The λ-$(BETS)_2GaCl_4$ salt is a superconductor with $T_c = 6$ K [31]. Upon substituting Ga by Fe in λ-$(BETS)_2Fe_xGa_{1-x}Cl_4$ superconductivity becomes continuously suppressed with increasing x [32] and, for $x \geq 0.5$, replaced by an antiferromagnetic insulating ground state.

Some structural data for a selection of organic superconductors are summarized in Table 20.1.

20.3 Normal-State Properties

20.3.1 Electronic Structure

As for ordinary metals, the electronic properties of organic charge-transfer salts are determined by the quasiparticles at the Fermi surface (FS).[3] The energy-band structures for both the quasi-1D $(TM)_2X$ and the quasi-2D $(ET)_2X$ salts have been calculated employing a tight-binding scheme with a few simplifications. For a comprehensive overview on the band-structure calculations and Fermi-surface studies see [5] and references cited therein. The calculations are based on the assumption that the *intra*molecular interactions are much stronger than the interactions *between* adjacent molecules reducing the complexity of the problem enormously. In a first step, σ- and π-molecular orbitals are constructed using linear combinations of atomic s and p-orbitals of the constituent atoms. In the molecular-orbital (MO) approximation, the electrons (holes) are considered to be spread over the whole molecule and only those electrons (holes) near the Fermi surface in the highest occupied (HOMO) and lowest unoccupied molecular orbitals (LUMO) are taken into account. Due to the overlap between molecular orbitals of adjacent molecules, the corresponding π-electrons (holes) are delocalized. Using available structural data, the overlap integrals and transfer energies can be obtained from quantum chemistry. These are input parameters for a standard tight-binding calculation based on molecular orbitals obtained by the extended Hückel approximation (EHA) from which the band structure and Fermi surfaces are derived [23,34–36].

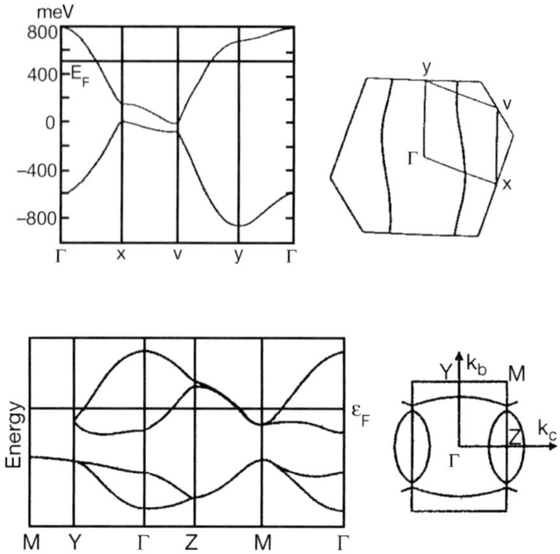

Fig. 20.7. Calculated energy dispersion and Fermi surface of $(TMTSF)_2X$ [37] (*upper panel*) and κ-$(ET)_2Cu(NCS)_2$ [41,42] (*lower panel*)

Based on the above approximations, Grant et al. have calculated a model band structure for the quasi-1D materials $(TM)_2X$ [37], see upper panel of Fig. 20.7. The FS consists of two open sheets which are slightly corrugated due to weak interactions perpendicular to the stacking axis. While the standard magnetic-quantum-oscillation studies cannot be used for these quasi-1D metals, some important information on the FS can still be derived

[3] This implies the applicability of the Fermi-liquid concept which is questionable for the most anisotropic $(TM)_2X$ salts, see e.g. [33].

from angular-dependent magnetoresistance measurements, see e.g. [38]. Of crucial importance are the topological aspects of the FS, i.e. the nesting properties, and the band filling. The conduction band can accommodate four electrons per $(TM)_2$ unit. Due to the weak structural dimerization which is more pronounced in the TMTTF compared to the TMTSF salts, a dimerization gap splits the conduction band into two parts. Therefore, removing one electron per unit cell in the charge-transfer process results in a half-filled conduction band.

The FS topology of the quasi-2D materials has been studied in great detail employing measurements of the de Haas–van Alphen (dHvA) and Shubnikov–de Haas (SdH) effect, the angular-dependent magnetorestistance (AMRO) and the cyclotron resonance, see [39,40]. These results clearly demonstrate the presence of a well-defined Fermi surface and quasiparticle excitations in accordance with the Fermi-liquid theory.

The lower panel of Fig. 20.7 shows the results of EHA band-structure calculations for the superconductor κ-$(ET)_2Cu(NCS)_2$. Since any *inter*layer electron transfer has been neglected in these calculations, the resulting FS is strictly two dimensional. Despite the various simplifications employed, the main features of the so-derived FS are generally found to be in remarkable agreement with the experimental results [39], although a more elaborated analysis reveals certain details which are not adequately described [10,43].

The four bands correspond to the four ET molecules in the unit cell, each represented by its HOMO. Due to the lack of a center-of-inversion symmetry, an energy gap opens at the Z-M zone boundary. As a consequence, the FS consists of closed hole-like quasi-2D orbits (α-pockets) and a pair of open quasi-1D corrugated sheets. According to a charge transfer of one electron per pair of ET molecules, the conduction band is three quarters filled. Due to the strong dimerization of the ET molecules in the κ-type structure, the conduction bands split up so that the upper band becomes half filled. Band-structure calculations based on high-temperature crystallographic data reveal FS topologies which are very similar among the various κ-$(ET)_2X$ systems [35,36], except for the degeneracy of the upper two bands along the Z-M zone boundary for the linear anion X $= I_3$. While the α-phase and κ-phase $(ET)_2X$ salts still combine quasi-1D and quasi-2D bands, the FS of β-type salts is even more simple. It is of almost cylindrical shape and closed within the first Brillouin zone, reflecting the isotropic in-plane interactions between adjacent ET molecules.

Fermi-Surface Studies

In a magnetic field B, the transverse motion of electrons becomes quantized and their allowed states in k-space are confined to so-called Landau levels. Periodic oscillations in magnetization and resistivity as a function of $1/B$ arise from the oscillatory behavior of the density of states at the Fermi level E_F as the Landau levels pass through the Fermi surface. From the oscillation period $\Delta(1/B)$ the extremal area of the FS cross section, S_F, perpendicular to the magnetic-field direction can be derived [44]:

$$S_F = \frac{2\pi e}{\hbar} \frac{1}{\Delta(1/B)} , \tag{20.1}$$

where $-e$ is the electron charge and $\hbar$ the Planck constant. A quantitative description of the oscillatory magnetization was given by Lifshitz and Kosevich [45]. According to their work, the amplitude of the oscillations is given by:

$$A \propto \frac{T}{\sqrt{B}} \frac{\exp\left(-\frac{\lambda m_c^* T_D}{m_e B}\right)}{\sinh\left(\frac{\lambda m_c^* T}{m_e B}\right)} , \tag{20.2}$$

where the effect on the electron spin has been neglected. m_e denotes the free electron mass and m_c^* the cyclotron effective mass, $\lambda = 2\pi^2 m_e k_B/(e\hbar)$ with k_B being the Boltzmann constant. $T_D = \hbar/(2\pi k_B \tau)$ is the Dingle temperature which accounts for the broadening of the Landau levels due to scattering of the electrons where τ is the relaxation time averaged over a cyclotron orbit. For state-of-the-art crystals of the $(ET)_2X$ salts, the Dingle temperatures are usually far below about 1 K, as e.g. $T_D \sim 0.5$ K as reported for κ-$(ET)_2Cu(NCS)_2$ [46], which reflect the high quality of these materials.

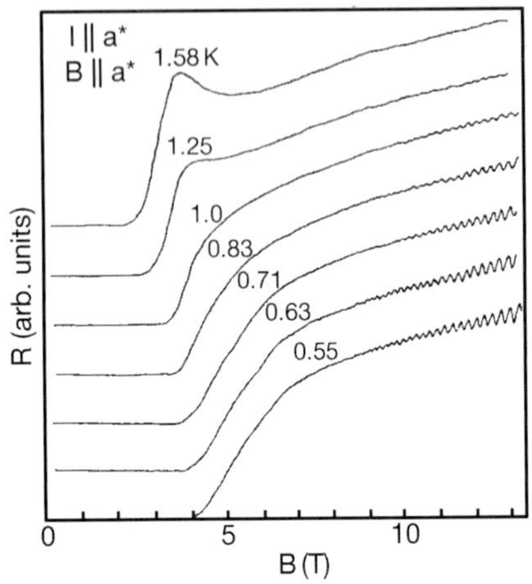

Fig. 20.8. Magnetorestistance of κ-$(ET)_2Cu(NCS)_2$ at different temperatures. Electrical current and magnetic field were applied along the a^*-axis, i.e. perpendicular to the conducting planes. Shubnikov–de Haas oscillations starts to become visible below 1 K, taken from [41]

Figure 20.8 shows early magnetoresistance data on κ-$(ET)_2Cu(NCS)_2$ [41]; see also [40] for more recent data. At the low-field side of the data sets, the transition from the superconducting to the normal state is visible. With increasing the field and at temperatures below 1 K, Subnikov–de Haas (SdH) oscillations caused by the closed α-orbits of the FS are superimposed. As expected from the simple FS topology (lower panel of Fig. 20.7), a single frequency according to only one extremal orbit (α-orbit) dominates the oscillatory behavior at lower fields. At higher magnetic fields, however, a second high-frequency component becomes superimposed [47,48]. The latter corresponds to the so-called magnetic breakdown effect which is due to tunneling of charge carriers across the energy gaps at the FS. It is common to refer to the magnetic-breakdown orbit which encompasses the whole FS as the β-orbit. For a detailed description of the FS studies on quasi-1D and quasi-2D charge-transfer salts, see [39,40] and references therein.

Effective Masses and Renormalization Effects

The experimentally derived effective cyclotron masses m_c^* for the various $(ET)_2X$ salts are significantly larger than the band masses m_b predicted by the above band-structure calculations, which are of the order of the free-electron mass m_e (see below). For the κ-$(ET)_2Cu(NCS)_2$ salt, for example, experiments reveal $m_c^* = (3.5 \pm 0.1)\, m_e$ for the α-orbit and (6.9 ± 0.8) for the magnetic-breakdown β-orbit [47]. A mass enhancement of comparable size is observed also for the thermodynamic effective mass m_{th}^* as determined by measurements of the specific heat $C(T)$. For a quasi-2D material consisting of stacks of metallic planes with interlayer spacing s, the Sommerfeld coefficient $\gamma^{2D} = C/T$ is given by:

$$\gamma^{2D} = \frac{\pi k_B^2}{3} \frac{m_{th}^*}{\hbar^2} \frac{1}{s} \,. \tag{20.3}$$

For κ-$(ET)_2Cu(NCS)_2$ one finds $\gamma = (23 \pm 1)\,\mathrm{mJ/mol\,K^2}$ [49,50] which corresponds to $m_{th}^* = (4.7 \pm 0.2)\, m_e$. Both m_c^* and m_{th}^* are renormalized compared to the band mass m_b. The latter takes into account the fact that the electrons are moving in a periodic potential associated with the crystal lattice. The band masses estimated from tight-binding calculations and interband optical measurements are of the order of the free electron mass: Caulfield et al. applied the effective dimer model to κ-$(ET)_2Cu(NCS)_2$ and found $m_b^\alpha = 0.64\, m_e$ and $m_b^\beta = 1.27\, m_e$ [46] corresponding to a width of the conduction band of $W = 0.5 \sim 0.7$ eV. These values have to be compared with $m_b^\alpha = (1.72 \pm 0.05)\, m_e$ and $m_b^\beta = (3.05 \pm 0.1)\, m_e$ as derived from first-principles self-consistent local-density calculations [51]. For a discussion on the band masses derived from band-structure calculations and their relation to the cyclotron effective masses, see e.g. [52]. The substantial enhancement of the cyclotron masses compared to the band masses suggest an appreciable quasiparticle renormalization due to many-body effects, i.e. electron–electron and electron–phonon interactions. It has been proposed that a direct tool to determine the relative role of electron–electron correlations in the mass enhancement is provided by cyclotron resonance measurements [53]. According to the Kohn theorem the effective mass determined by cyclotron resonance experiments, m_{cr}^*, is independent of the electron–electron interactions. As a consequence, the experimental finding of $m_{cr}^* \approx m_b$ has been attributed to a dominant role of the Coulomb interaction for the

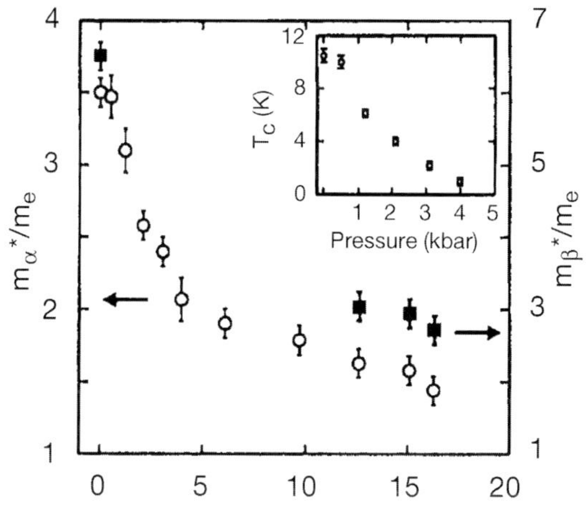

Fig. 20.9. Cyclotron effective masses of the α-orbits (*open circles, left scale*) and β-panels (*full squares, right scale*) as a function of hydrostatic pressure for κ-$(ET)_2Cu(NCS)_2$. The *inset* shows T_c against pressure, taken from [46]

mass renormalization [46]. However, recent studies on different $(ET)_2X$ systems along with theoretical calculations showed that the general applicability of the Kohn theorem for the quasi-2D organic superconductors is questionable, see [10,40].

On the other hand, various experiments such as optical studies [54–59], thermal conductivity [50,60,61] as well as inelastic neutron scattering experiments [62] indicate a substantial coupling of the charge carriers to the lattice vibrations. Taken together, it is likely that for the present molecular conductors both electron–electron as well as electron–phonon interactions are responsible for the mass-renormalization.

By means of pressure-dependent Shubnikov–de Haas experiments a striking interrelation between the suppression of superconductivity and changes in the effective masses have been found [46]. Figure 20.9 shows the hydrostatic-pressure dependence of the effective masses for κ-$(ET)_2Cu(NCS)_2$ [46]. As the pressure increases, m_c^* rapidly decreases, an effect which has been observed also for other $(ET)_2X$ compounds [63]. Above some critical pressure of about $4 \sim 5$ kbar the rate of suppression of m_c^* becomes much weaker. As this is about the same pressure value above which superconductivity becomes completely suppressed (see inset of Fig. 20.9), an intimate interrelation between mass enhancement and superconductivity has been suggested [46]. This is consistent with recent results of reflectivity measurements which showed that the pressure dependence of the "optical masses", which are closely related to the bare band masses, do not show such a crossover behaviour [64]. Consequently, the pressure-induced reduction of the effective cyclotron masses has to be associated with a decrease in the strength of the electron–electron and/or electron–phonon interactions.

20.3.2 Transport and Optical Properties

Electrical Resistivity

The organic superconductors discussed in this article are fairly good metals at room temperature with resistivities that vary over wide ranges depending on the particular compound and the current direction in respect to the crystal axes. The pronounced anisotropies found in the electrical properties are direct manifestations of the strongly directional-dependent overlap integrals.

For the $(TM)_2X$ series (cf. Fig. 20.2) one typically finds $\rho_a : \rho_b : \rho_c$ of the order of 1 : 200 : 30.000, where a is along the stacking axis. These numbers correspond to a ratio of the overlap integrals $t_a : t_b : t_c$ of about 10: 1: 0.1 with $t_a \approx 0.1 \sim 0.24$ eV and 0.36 eV for the TMTTF and TMTSF compounds, respectively [6,33]. Figure 20.10 compiles temperature profiles of the resistivity for various $(TM)_2X$ compounds. Below a temperature T_ρ, depending on the anion, the resistivity of the sulfur-containing $(TMTTF)_2X$ compounds changes from a metallic-like high-temperature into a thermally-activated low-T behavior. Upon further cooling through $T_{SP} \simeq 20K < T_\rho$ (not shown), the $(TMTTF)_2PF_6$ salt undergoes a phase transition into a spin-Peierls (SP)-distorted nonmagnetic ground state, see e.g. [65] and references cited therein. With the application of moderate pressure both T_ρ and T_{SP} were found to decrease. By increasing the pressure to $p \geq 10$ kbar, the spin-Peierls ground state becomes replaced by an antiferromagnetic Néel state similar to the one found at normal conditions in $(TMTTF)_2Br$.

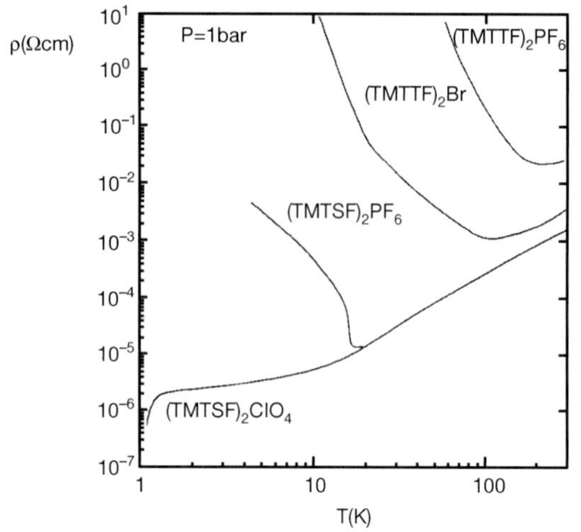

Fig. 20.10. Resistivity vs temperature for various $(TMTTF)_2X$ and $(TMTSF)_2X$ salts at ambient-pressure conditions in a double-logarithmic plot [65]

In the selenium-containing $(TMTSF)_2PF_6$ salt the metallic range extends down to lower temperatures until the sudden increase in the resistivity indicates the transition into an insulating SDW ground state. Above a critical pressure of about 6 kbar, the SDW state of $(TMTSF)_2PF_6$ becomes unstable giving way to superconductivity at lower temperatures, cf. the phase diagrams Figs. 20.22 and 20.23 in Sect. 20.3.5.

Interestingly enough, when the spin-Peierls salt $(TMTTF)_2PF_6$ is exposed to sufficiently high pressure in excess of 43.5 kbar, a superconducting state can be stabilized [66,67], which completes, for a single compound, the sequence of ground states indicated in the generic phase diagram in Fig. 20.22.

In their metallic regime the resistivity of the $(TM)_2X$ salts along the most conducting direction decreases monotonically with a power-law temperature dependence $\rho \propto T^{\alpha}$, where the exponent α, depending on the temperature interval, varies between 1 and 2, see e.g. [33,68]. For $(TMTSF)_2PF_6$ for example, $\alpha \approx 1.8$ between 300 and 100 K and approaches approximately 2 at lower temperatures down to the metal-SDW transition [33]. A T^2 dependence in the resistivity has been frequently observed not only in the quasi-1D [68] but also for the various $(ET)_2X$ salts, see below.

A question of high current interest for the present quasi-1D conductors concerns the nature of their low-energy excitations. Is a Fermi-liquid approach still adequate or do we have to treat these materials within the framework of a Tomonaga–Luttinger liquid (a concept proposed for dimensionality $D = 1$) [69–71]? Arguments in favor of a Luttinger-liquid behavior have been derived from various observations, not all of which have been generally accepted. Undisputed are, however, the non-Fermi-liquid features in the sulfur compounds $(TMTTF)_2X$: these materials undergo a charge localization at elevated temperatures $T_{\rho} = 250$ K for X = PF_6 and 100 K for X = Br (cf. Fig. 20.10) which leaves the static magnetic susceptibility unaffected [6,33]. This apparent separation of spin and charge degrees of freedom is one of the hallmarks of a Luttinger liquid, see e.g. [72]. Indications for a spin and charge separation have been reported also for other $(TM)_2X$ salts from optical-conductivity and thermal-conductivity experiments [73,74]. The other signatures of a Luttinger liquid are (ii) a power-law decay at long distances of the spin or charge correlation functions which suppresses long-range order in 1D systems and (iii) the absence of any discontinuity in the distribution function for electron states at the Fermi energy. Indications for (ii) and (iii) have been reported from NMR [33], photoemission [75,76] as well as transverse (c-axis) dc-resistivity [77] measurements.

The resistivity for the various $(ET)_2X$ and related compounds can be roughly classified into two distinct types of temperature dependences. While some of the materials show a more or less normal metallic-like T behavior, i.e. a monotonic decrease of $\rho(T)$ upon cooling, a pronounced $\rho(T)$ maximum above about 80 K has been found for a number of $(ET)_2X$ compounds. Among them are the κ-$(ET)_2X$ salts with polymer-like anions such as X = $Cu(NCS)_2$ or $Cu[N(CN)_2]Br$ [24,78–80], the α-$(ET)_2NH_4Hg(SCN)_4$ [81] as well as the κ-type and λ-type BETS salts [29]. The occurrence of the same kind of $\rho(T)$ anomaly in $(DMET)_2AuBr_2$ [82] demonstrates that (i) this feature is not a property specific to ET-based or BETS-based salts and (ii) does not rely on the presence of Cu ions. Figure 20.11 shows the in-plane resistivity of κ-$(ET)_2Cu(NCS)_2$ as a func-

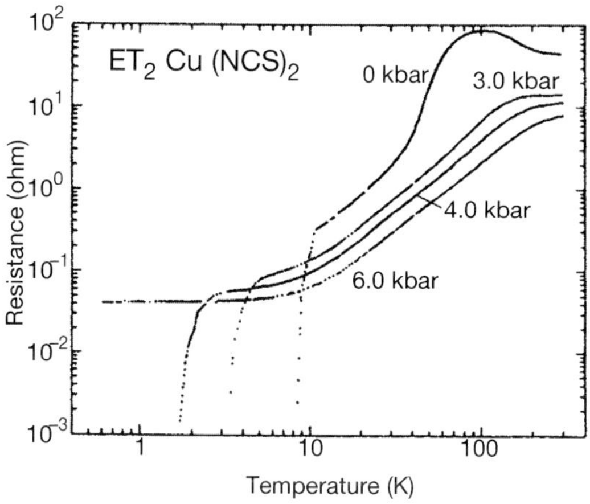

Fig. 20.11. Temperature dependence of the resistivity measured along the in-plane b-axis of κ-$(ET)_2Cu(NCS)_2$ at various pressures, taken from [78]

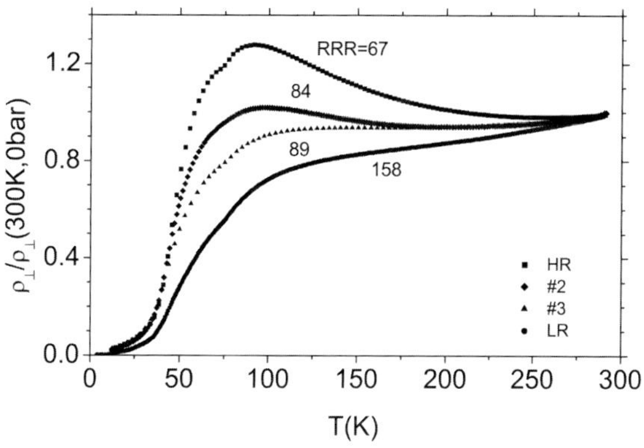

Fig. 20.12. Temperature dependence of the normalized interlayer resistivity of various single crystalline samples of κ-$(ET)_2Cu[N(CN)_2]Br$. The RRR values denote the residual resistivity ratio $\rho_\perp(300K)/\rho_\perp(0)$ with the residual resistivity $\rho_\perp(0)$ determined by extrapolating the normal-state resistivity to $T = 0$, taken from [92]

tion of temperature at various pressures [78]. With decreasing temperatures, $\rho(T)$ first increases to a maximum at around 100 K before a metallic behavior sets in at lower temperatures. Under hydrostatic pressure, the maximum shifts to higher temperatures and becomes progressively suppressed. This is accompanied by a significant reduction of T_c (see also inset of Fig. 20.9). The origin of the anomalous $\rho(T)$ hump has been discussed by many authors and various explanations have been suggested including the formation of small polarons [83], a metal–metal phase transition [84], a valence instability of Cu [85], an order–disorder transition of the terminal ethylene groups of the ET molecules [86–88] as well as a crossover from localized small-polaron to coherent large-polaron behavior [89]. In this context it is interesting to note that for the κ-$(ET)_2Cu[N(CN)_2]Br$ system this maximum has been found to be sample dependent: using a different synthesis route, Thoma et al. [90] and Montgomery et al. [91] succeeded in preparing superconducting crystals which lack the anomalous resistance hump. Those differently prepared κ-$(ET)_2Cu[N(CN)_2]Br$ crystals have been the subject of a recent comparative resistivity study [92], see Fig. 20.12. It has been found that irrespective of the markedly different resistivity profiles, especially at intermediate temperatures and in the low-T range where the resistance decreases quadratically with temperature $\rho \propto AT^2$, these crystals reveal almost the same, high superconducting transition temperatures. In the absence of significant differences in the crystals' structural parameters and chemical compositions, as probed by X-ray and electron-probe-microanalysis, respectively, these results indicate that real-structure phenomena such as disorder and/or defects may strongly affect the inelastic scattering in these compounds [92].

A closer look on the resistivity of κ-$(ET)_2Cu(NCS)_2$ below the maximum discloses an abrupt change in the slope at temperatures around 45 ∼ 50 K [93]. A similar behavior is found for κ-$(ET)_2Cu[N(CN)_2]Br$ [94] and has been interpreted as a crossover from a regime of antiferromagnetic fluctuations of localized spins at high temperatures to a low-T Fermi-liquid regime [95,96]. Within a dynamical mean field approach, Merino et al. [97] have attributed this change in the charge response to a smooth crossover from coherent Fermi liquid excitations at low temperatures to incoherent (*bad metal*)-excitations at higher temperatures. Using such DMFT calculations for a simple Hubbard model, Limelette et al. [98] recently attempted to provide even a quantitative description of the whole anomalous $\rho(T)$ behavior for pressurized κ-(BEDT-TTF)$_2$Cu[N(CN)$_2$]Cl covering also the resistivity maximum and the semiconducting regime at higher temperatures. Alternatively, it has been proposed that the rather drastic change in the resistivity, which coincides with sharp features in thermodynamic quantities [92], marks a cooperative phenomenon such as a density-wave-type *phase transition* [99–101], see Sect. 20.3.3 for a detailed discussion.

At temperatures below the inflection point, the resistivity turns into an approximate $\rho(T) = \rho_0 + AT^2$ behavior until superconductivity sets in around 10 K. As mentioned above, a resistivity roughly following a T^2 law, even at elevated temperatures, is not an exception in the present molecular conductors, see e.g. [5]. In some high-quality crystals of β-type and κ-type $(ET)_2I_3$, it has been observed over an extraordinarily wide temperature range up to temperatures as high as 100 K [102]. It has been argued that the T^2 dependence of the resistivity indicates a dominant role of electron–electron scattering in these materials [7]. On the other hand, for such a mechanism to predominate the resistivity at temperatures as high as 45 K for the X = $Cu(NCS)_2$ and even 100 K for the X = I_3 salt implies that there is only a minor contribution from electron–phonon scattering. In light of the considerable electron–phonon coupling in these materials as proved by various experiments, such a scenario appears questionable. Alternatively, the T^2 law has been attributed to the scattering of electrons by phonons via electron–libron [103] or a novel electron–phonon scattering mechanism proposed for the high-T_c cuprates [102] which invokes electron–electron interactions [104, 105]. For the discussion of the temperature dependence of the resistivity, it is important to bear in mind, however, that due to the large pressure coefficients of the resistivity of about $\partial \ln \rho / \partial p \simeq -20$ %/kbar at room temperature together with the extraordinary strong thermal contraction, it is difficult to make a comparison with theoretical predictions. Since the theory usually describes the temperature dependence at constant volume, a detailed comparison is meaningful only after transforming the constant-pressure into constant-volume profiles by taking into account the thermal expansion of the material.

Similar to the quasi-1D $(TM)_2X$ salts, the room-temperature resistivities of the quasi-2D $(BEDT\text{-}TTF)_2X$ materials are generally rather high. For the κ-$(ET)_2Cu(NCS)_2$ salt for example, one finds $\rho_b \approx 6 \cdot 10^4\,\mu\Omega$cm and $\rho_c \approx 3 \cdot 10^4\,\mu\Omega$cm [106], which exceed the values for Cu by several orders of magnitude. This is partly due to the relatively low charge-carrier concentration of only about 10^{21} cm^{-3}.

In accordance with their quasi-2D electronic structure, a pronounced in-plane vs out-of-plane anisotropy has been observed which amounts to $10^{-3} \sim 10^{-5}$ [40]. In this respect it is interesting to ask whether under these conditions the interlayer transport is coherent or not, i.e. whether there is a coherent motion of band states associated with well-defined wave vectors or if the motion from layer to layer is diffusive and a Fermi velocity perpendicular to the layers cannot be defined [107]. This question has been addressed in recent magnetoresistance studies on the κ-$(ET)_2Cu(NCS)_2$ salt [108]. Here the interlayer-transfer integral has been estimated to be $t_\perp \approx 0.04$ meV [108] as compared with $t_\parallel \sim 150$ meV for the intralayer transfer [46]. According to this work, the Fermi surface is extended along the interlayer direction corresponding to a coherent transport.

Optical Conductivity

Optical investigations by means of infrared and Raman measurements provide important information on the electronic parameters such as the plasma frequency, the optical masses and also the bandwidths and collision times for the carriers. In addition, they permit an investigation of vibrational properties and their coupling to the charge carriers. Using polarized light it is also possible to look for anisotropies in these quantities, as e.g. in the effective masses. The optical properties of quasi-1D and quasi-2D organic conductors have been reviewed by several authors [6, 33, 109–112], see also [5, 10]. For a detailed discussion on the normal-state and superconducting-state optical properties of the $(ET)_2X$ salts see [113]. A summary of Raman results on $(ET)_2X$ salts is given in [114, 115].

First extensive optical studies of the electronic properties of $(TM)_2X$ by Jacobsen et al. [116] provided information on the energy of charge-transfer processes and on the electron–phonon coupling: the

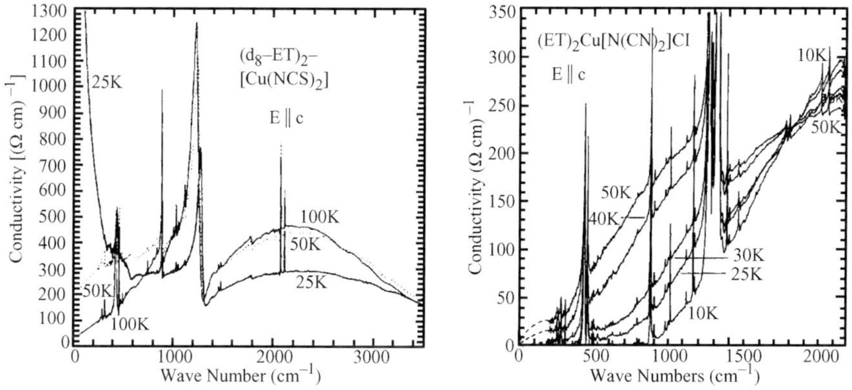

Fig. 20.13. Optical conductivity of κ-$(D_8\text{-ET})_2Cu(NCS)_2$ (*left panel*) and κ-$(ET)_2Cu[N(CN)_2]Cl$ (*right panel*) at various temperatures for $E \parallel c$. Taken from [120] and [121], respectively

large absorption features observed in the optical conductivity of the TMTTF salts have been assigned to intramolecular and intermolecular vibrations [117]. These studies have been supplemented by a series of more detailed investigations covering also the low frequency range, see [112] and references therein. In accordance with the expectations for a strongly anisotropic material with open Fermi surface, the optical response of the Bechgaards salts $(TMTSF)_2X$ was found to deviate strongly from that of a simple metal. The main features are a gap-like structure around 25 meV for $X = PF_6$ and a zero-frequency mode which grows upon decreasing the temperature. The latter contribution, having only a small spectral weight corresponding to 1 % of the carriers, is responsible for the metallic conductivity of the compound [112]. At high frequencies, i.e. at energies in excess of the interchain transfer integral t_b but below the intraband width $4t_a$, the data for the optical conductivity follow a $\sigma_1(\omega) \propto \omega^{-\gamma}$ dependence with $\gamma = 1.3$ [112]. Here $\gamma = 4n^2K_\rho - 5$, where K_ρ is the Luttinger-liquid-correlation parameter and n the degree of commensurability. From the experimentally derived γ value and assuming $n = 2$, i.e. a dominant quarter-filled band Umklapp scattering, $K_\rho \simeq 0.23$ has been determined which agrees reasonably well with photoemission [75, 118, 119] and transport data [77]. These observations are consistent with a dimensional crossover in $(TMTSF)_2PF_6$ from a high-temperature Luttinger-liquid phase to a low-temperature (anisotropic) 3D Fermi liquid induced by interchain coupling.

For the various ET salts, the reflectance spectra are generally characterized by intensive sharp features due to molecular vibrations superimposed on a broad electronic background with the plasma edge at a frequency of about $4500 \sim 5000\,\mathrm{cm}^{-1}$. Figure 20.13 shows optical conductivity data of superconducting deuterated κ-$(D_8\text{-ET})_2Cu(NCS)_2$ (left panel) and insulating κ-$(ET)_2Cu[N(CN)_2]Cl$ (right panel) at various temperatures obtained from reflectivity measurements after a Kramers–Kronig analysis [120, 121]. The far-infrared conductivities have been found to agree reasonably well with the dc-conductivities in showing a rapid increase below 50 K for κ-$(ET)_2Cu(NCS)_2$. The low-frequency feature has been interpreted as a Drude peak which increases with decreasing temperature or increasing pressure, i.e. when the metallic character of the material increases. The data are in good agreement with results of other studies, see [10, 89, 113] and references cited therein. However, the interpretation of the spectra may vary from author to author: Kornelsen et al. attributed the mid-infrared peak $(1000 \sim 4000\,\mathrm{cm}^{-1})$ to interband transitions superimposed on the free-carrier tail [120]. Wang et al. [89] argued that the sharp Drude peak in the conductivity that develops at low temperature, together with the large mid-infrared spectral weight indicate polaronic effects. According to this interpretation, the change from non-metallic to metallic behavior around $90 \sim 100$ K as observed in the resistivity is due to a crossover from localized small-polaron to coherent large-polaron behavior [89]. In contrast to the metallic/superconducting κ-$(ET)_2Cu(NCS)_2$ salt, the optical conductivity in the low-energy region for insulating κ-$(ET)_2Cu[N(CN)_2]Cl$ rapidly decreases below 50 K, indicating a large temperature-dependent semiconducting energy gap with a value of about 900 cm^{-1} at 10 K [121].

Electron–Phonon Coupling

Infrared reflectivity measurements can also be used to obtain information on the electron–phonon interaction. In molecular crystals, the coupling between the conduction electrons and the phonons is twofold. One kind of interaction, the so-called electron-molecular-vibration (EMV) coupling, involves the *intra*molecular vibrations which are characteristic of the molecular structure. This has to be distinguished from the electron-*inter*molecular-vibration coupling which refers to the interaction of the charge carriers with motions of almost rigid molecules around their equilibrium positions and orientations (translational and librational modes).

Electron-Molecular-Vibration Coupling

It is well known [122] that electrons in the HOMO's of the TTF molecule and its derivatives couple strongly to the totally symmetric (A_g) molecular vibrations via the modulation of the HOMO energy, E_{HOMO}, by the atomic displacements. The linear EMV coupling constant g_i for mode i is defined as:

$$g_i = \frac{1}{h\nu_i}\frac{\partial E_{\mathrm{HOMO}}}{\partial Q_i}\,, \tag{20.4}$$

where Q_i is an intramolecular normal coordinate and ν_i the mode frequency. The effective electron-intramolecular-phonon coupling constant λ_i can then be calculated using $\lambda_i = 2g_i^2 h\nu_i N(E_F)$ where $N(E_F)$ is the density of states at the Fermi level.

The sharp features superimposed on the electronic background in Fig. 20.13 can be attributed to molecular vibrations. An assignment of these features to the various vibrational modes[4] is feasible by comparing spectra of different isotopically labeled salts with calculations based on a valence-force-field model by Kozlov et al. [125,126]. For the ET molecule, the modes with the strongest coupling constants are those involving the central carbon and sulfur atoms [125–127] at which the HOMO's have the largest amplitudes [34]. The frequencies of these C=C stretching and ring-breathing modes are ν_2 = 1465 cm^{-1} (g_2 = 0.165), ν_3 = 1427 cm^{-1} (g_3 = 0.746) and ν_9 = 508 cm^{-1} (g_9 = 0.476), where the calculated coupling constants are given in the brackets. Despite these sizable coupling constants, several studies, especially those of the mass isotope shifts on T_c for the ET salts [128], indicate that the EMV coupling seems to play only a minor role in mediating the attractive electron–electron interaction, cf. Sect. 20.4.5.

Electron-Intermolecular-Vibration Coupling

While much experimental data are available on the EMV coupling, relatively little is known about the coupling of the charge carriers to the low-lying intermolecular phonons. This interaction is provided by the modulation of the charge-transfer integrals t_{eff} between neighboring molecules during their translational or librational motions. Within the Eliashberg theory, the dimensionless electron-intermolecular-phonon coupling constant λ is given by

$$\lambda = 2\int \frac{\alpha^2(\omega)}{\omega}F(\omega)\mathrm{d}\omega\,, \tag{20.5}$$

where $\alpha(\omega)$ is the electron–phonon coupling constant, ω the phonon frequency and $F(\omega)$ the phonon density of states. The Eliashberg function $\alpha^2(\omega)F(\omega)$ can, in principle, be derived from tunneling characteristics of strong-coupling superconductors or via point-contact measurements. The latter experiments have been carried out by Nowack et al. on the β-type $(ET)_2X$ salts with X = I_3 and AuI_2 [129, 130] yielding $\lambda \simeq 1$. Some of the frequencies of the intermolecular modes have been determined by employing Raman and far-infrared measurements [59, 131–133]. More recent studies including inelastic-neutron [62] and Raman-scattering [56–59] have focused on investigation

[4] Important references are [54,120,123,124], see [9,113] for both earlier and more recent citations.

of the role of intermolecular phonons for superconductivity. These experiments yielded quite sizable superconductivity-induced phonon-renormalization effects which clearly indicate a significant coupling of the superconducting charge carriers to the intermolecular phonons and suggest an important role of these modes in the pairing interaction, cf. Sect. 20.4.5.

20.3.3 Thermal and Magnetic Properties

Common to both the quasi-1D and quasi-2D charge-transfer salts is the variety of ground states the systems can adopt depending on parameters such as the chemical composition or external pressure. Most remarkable is the fact that for both families the superconducting phase shares a common phase boundary with a long-range magnetically ordered state.

For the $(TMTSF)_2PF_6$ salt, cooling at ambient pressure leads to a metal–insulator transition at $T_{MI} \sim$ 12 K. The insulating ground state in the $(TMTSF)_2X$ series has been identified via NMR [134, 135] and susceptibility [136] measurements as a spin-density-wave ordering whose wave vector $\boldsymbol{q} = 2\,\boldsymbol{k}_F$ is determined by the optimum nesting condition of the quasi-1D Fermi surface [6]. The application of hydrostatic pressure to the X = PF_6 salt suppresses the magnetic state by destroying the nesting properties and instead stabilizes superconductivity below about $T_c = 1.1$ K at 6.5 kbar [137]. As has been shown again by NMR experiments, short-range spin-fluctuations with the same wave vector $\boldsymbol{q}$ remain active up to $\sim$ 100 K, i.e. far above the SDW ordering in this salt. Moreover, these fluctuations are present also in the normal state of the ClO_4 compound despite its superconducting ground state [138].

For the quasi-2D κ-phase $(ET)_2X$ compounds, the nesting properties are expected to be less strongly pronounced, cf. the Fermi surface in Fig. 20.7. Nevertheless, as in the case of the quasi-1D salts, superconductivity lies next to an antiferromagnetic insulating state in the pressure-temperature phase diagram, see Sect. 20.3.5. While the compounds with the complex anions X=$Cu(NCS)_2$ and $Cu[N(CN)_2]Br$ are superconductors with T_c values of 10.4 K and 11.2 K, respectively, κ-$(ET)_2Cu[N(CN)_2]Cl$ is an antiferromagnetic insulator with $T_N = 27$ K which can be transformed into a superconductor with $T_c = 12.8$ K by the application of a small hydrostatic pressure of only 300 bar [25]. Likewise seen in the quasi-1D salts, the metallic state above T_c in these quasi-2D systems reveals indications for magnetic fluctuations. This has been demonstrated by NMR measurements on the various κ-$(ET)_2X$ salts [139–143]. As shown in Fig. 20.14 the spin-lattice relaxation rate divided by temperature, $(T_1T)^{-1}$, for the superconductors κ-$(ET)_2Cu(NCS)_2$ and κ-$(ET)_2Cu[N(CN)_2]Br$ behaves quite differently from what would be expected for a simple metal and realized to a good approximation in some other organic superconductors as, e.g. α-$(ET)_2NH_4Hg(SCN)_4$ [96, 141]. For both κ-phase compounds, the $(T_1T)^{-1}$ values at higher temperatures are enhanced by a factor 5 $\sim$ 10 compared to a conventional Korringa-type behavior. Upon cooling, $(T_1T)^{-1}$ gradually increases down to a temperature $T^* \simeq 50$ K, below which a steep decrease sets in. Both the overall enhancement of $(T_1T)^{-1}$ along with its anomalous peak around 50 K have been assigned to the effect of strong antiferromagnetic spin fluctuations with a finite wave vector [96, 139, 141, 144]. The latter might be related to the

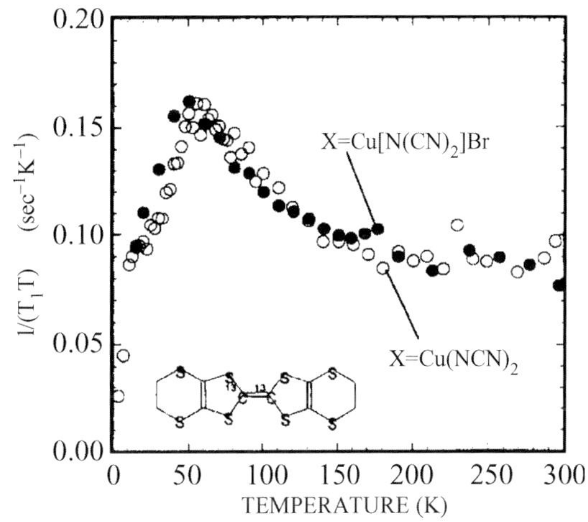

Fig. 20.14. Spin-lattice relaxation rate divided by temperature, $(T_1T)^{-1}$, for κ-$(ET)_2Cu(NCS)_2$ and κ-$(ET)_2Cu[N(CN)_2]Br$ as a function of temperature, taken from [141]. For the ^{13}C-NMR measurements, the ^{12}C atoms of the central C = C double bond in the ET molecule have been replaced by ^{13}C

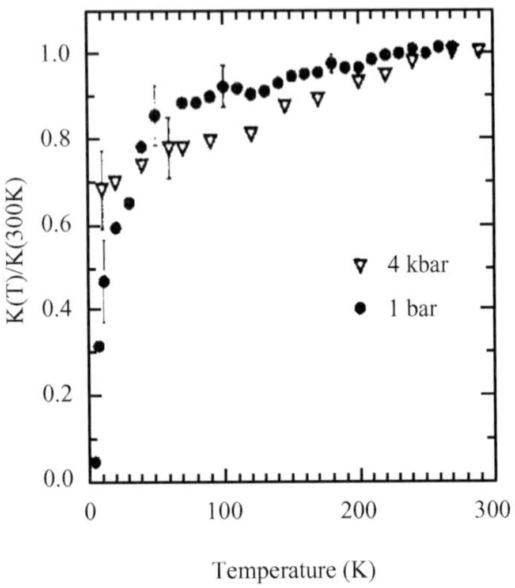

Fig. 20.15. Knight-shift, K_S, for κ-(ET)$_2$Cu[N(CN)$_2$]Br as a function of temperature at ambient pressure and p = 4 kbar, taken from [139]

ordering wave vector that characterizes the AF phase of κ-(ET)$_2$Cu[N(CN)$_2$]Cl [96, 139, 144]. It has been emphasized by these authors that despite the rapid drop below $T^* \simeq 50$ K the overall enhancement of $(T_1T)^{-1}$ persists until the onset of superconductivity indicative of the highly-correlated nature of the metallic state in these compounds [96,141]. NMR investigations performed at various pressures revealed that with increasing the pressure, the maximum at T^* shifts to higher temperatures while its size becomes progressively reduced. At pressures above 4 kbar the peak is replaced by a normal Korringa-type behavior, i.e. $(T_1T)^{-1}$ = const. [139,140]. In [95,96] the abrupt reduction of $(T_1T)^{-1}$ below $T^* \simeq 50$ has been linked phenomenologically to a spin-gap behavior as discussed also in connection with the high-T_c cuprates. Indeed, the formation of a pseudogap was first proposed by Kataev et al. [145] on the basis of their ESR measurements. These authors observed a reduction of the spin susceptibility around 50 K, indicating a decrease of the electronic density of states at the Fermi level $N(E_F)$ near T^*. The same conclusion has been drawn from results of the static magnetic susceptibility [141] and Knight-shift (K_S) measurements [139]. Figure 20.15 shows K_S as a function of temperature for the κ-(ET)$_2$Cu[N(CN)$_2$]Br salt. While the data at ambient pressure reveal a clear drop below about 50K, a rather smooth behavior with a gradual reduction upon cooling was found at 4 kbar [139]. The variation of the spin susceptibility in the high-temperature region has been attributed to the lattice contraction [144]. Although these experiments show clear evidence for a reduction of the density of states at E_F below $T^* \simeq 50$, the nature of this phenomenon and its interrelation to superconductivity are still unclear.

Cooling through $T^* \simeq 50$ does not only cause anomalies in the above magnetic properties but also leads to clear signatures in transport, acoustic, optical and thermodynamic quantities. As mentioned above, for both superconducting compounds a distinct peak shows up in the temperature derivative of the electrical resistivity $d\rho/dT$ [93, 94] indicating a change in the density of states at E_F. A pronounced softening of ultrasound modes for κ-(ET)$_2$Cu[N(CN)$_2$]Br and κ-(ET)$_2$Cu(NCS)$_2$ with distinct minima at $T^* \simeq 38$ K and 46 K, respectively, have been attributed to a coupling between acoustic phonons and antiferromagnetic fluctuations [146, 147]. An interaction between the phonon system and magnetism has also been suggested by Lin et al. based on their Raman scattering experiments [115, 148]. Recent theoretical studies have attempted to explain both the acoustic and Raman experiments by a correlation-induced crossover from a coherent Fermi liquid at low temperatures to an incoherent bad metal at high temperatures [97, 149]. According to this work, pronounced phonon anomalies as well as anomalous transport and thermodynamic properties are expected to occur at the crossover temperature T^*. Based on their NMR results, Kawamoto et al. [142] and Kanoda [95, 96] argued that T^* marks the crossover temperature from a region of antiferromagnetic fluctuations of localized spins at high T to a Fermi-liquid regime at low temperatures. This differs from the interpretation given by the Orsay group [139, 140, 144] who analyzed their NMR results in terms of strong antiferromagnetic fluctuations enhanced by Coulomb repulsion and the nesting properties of the Fermi surface.

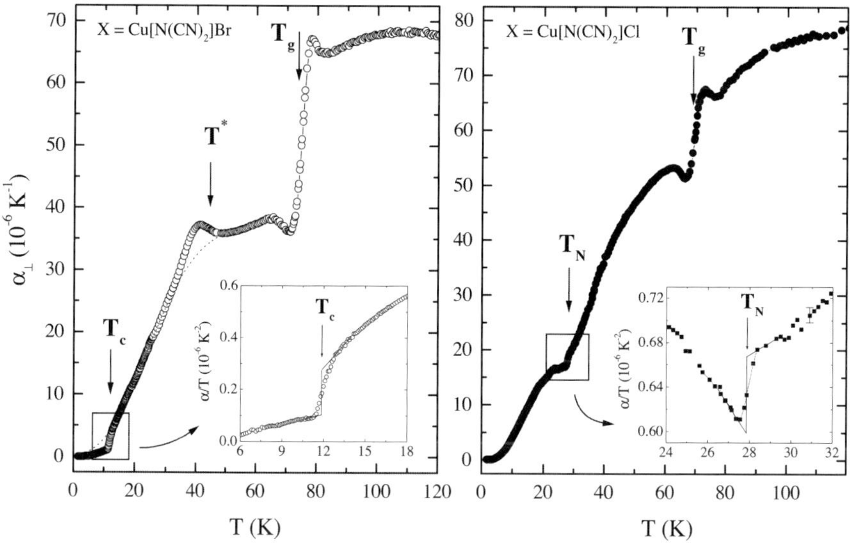

Fig. 20.16. Linear thermal expansion coefficient perpendicular to the planes of highest conductivity, $\alpha_\perp$, as a function of temperature for superconducting κ-$(ET)_2Cu[N(CN)_2]Br$ (*left panel*) and insulating κ-$(ET)_2Cu[N(CN)_2]Cl$ (*right panel*). The *inset* shows details of $\alpha_\perp$ for both salts as $\alpha_\perp/T$ vs T at the superconducting (X=$Cu[N(CN)_2]Br$) and antiferromagnetic (X = $Cu[N(CN)_2]Cl$) phase transition. *Arrows* indicate different kinds of anomalies as explained in the text, taken from [99]

More insight into the nature of the anomaly at T^* and its interrelation with superconductivity can be obtained by studying the coupling to the lattice degrees of freedom. This has been done by employing high-resolution thermal expansion measurements which also allow for studying directional-dependent effects [87,99,150]. Figure 20.16 compares the linear coefficient of thermal expansion, $\alpha(T) = \partial \ln l(T)/\partial T$, where $l(T)$ is the sample length, perpendicular to the planes for the superconducting κ-$(ET)_2Cu[N(CN)_2]Br$ (left panel) with that of the non-metallic κ-$(ET)_2Cu[N(CN)_2]Cl$ salt (right panel). For both compounds, various anomalies have been observed as indicated by the arrows [99, 151]. These are (i) large step-like anomalies at $T_g = 70 \sim 80$ K which are due to a kinetic, glass-like transition associated with the ethylene endgroups, cf. Sect. 20.3.4 and (ii) a distinct peak in $\alpha(T)$ at T^*. The latter feature, also observed for the superconductor κ-$(ET)_2Cu(NCS)_2$ (not shown) is absent in the non-metallic κ-$(ET)_2Cu[N(CN)_2]Cl$ salt, cf. right panel of Fig. 20.16 [99]. As demonstrated in the insets of Fig. 20.16, the transitions into the superconducting ($T_c = 11.8$ K) and antiferromagnetic ($T_N = 27.8$ K) ground states for the X = $Cu[N(CN)_2]Br$ and X = $Cu[N(CN)_2]Cl$ salts, respectively, are accompanied by distinct second-order phase transition anomalies in the coefficient of thermal expansion.

Figure 20.17 shows the anomalous contribution, $\delta\alpha_i(T) = \alpha_i(T) - \alpha_{ib}(T)$, to the uniaxial thermal expansion coefficients along the principal axes, $\alpha_i(T)$, at T_c and T^* for the superconducting salts X=$Cu[N(CN)_2]Br$ (left panel) and X=$Cu(NCS)_2$ (right panel) obtained after subtracting a smooth background α_{ib} (dotted line in the left panel of Fig. 20.16) [100]. Judging from the shape of the anomalies at T^*, i.e. their sharpness and magnitude, it has been suggested that this feature be assigned to a second-order phase transition [100]. Figure 20.17 uncovers an intimate interrelation between the phase-transition anomalies at T_c and T^*: while both features are correlated in size, i.e. a large (small) anomaly at T_c complies with a large (small) one at T^*, they are anticorrelated in sign. A positive peak at T_c goes along

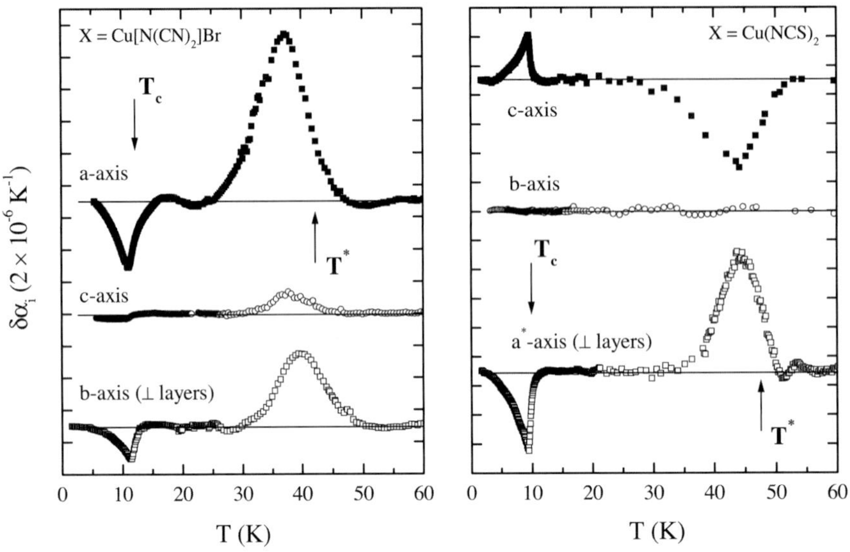

Fig. 20.17. Anomalous contributions, $\delta\alpha_i(T) = \alpha_i(T) - \alpha_{ib}(T)$, to the uniaxial thermal expansion coefficients, $\alpha_i(T)$, along the three principal axes for the superconducting salts κ-$(ET)_2Cu[N(CN)_2]Br$ (*left panel*) and κ-$(ET)_2Cu(NCS)_2$ (*right panel*), taken from [100]

with a negative anomaly at T^* and vice versa. According to the Ehrenfest relation

$$\left(\frac{\partial T^\star}{\partial p_i}\right)_{p_i\to 0} = V_{\text{mol}} \cdot T^\star \cdot \frac{\Delta\alpha_i}{\Delta C}\,, \tag{20.6}$$

which relates the uniaxial-pressure dependence of a second-order phase-transition temperature $T^\star$ to the discontinuities in $\alpha_i, \Delta\alpha_i$, and that of the specific heat, ΔC, the above findings imply that the uniaxial-pressure coefficients of T_c and T^* are strictly anticorrelated. In [99,100] it has been argued that the transition at T^* is not of structural but of electronic origin and related to the Fermi-surface topology. Based on the above uniaxial-pressure results, it has been proposed that T_c and T^* mark competing instabilities on disjunct parts of the Fermi surface [99,100]: while the instability at T^* most likely involves only the minor quasi-1D fractions (see Fig. 20.7 and [139]), the major quasi-2D parts are subject to the superconducting instability at lower temperatures. As a result, these studies hint at the opening of a real gap associated with T^* on the small 1D-parts of the FS as opposed to a pseudogap on the major quasi-2D fractions. The condensation of parts of the FS into a density-wave below T^* would imply the onset of anisotropies in magnetic and transport properties. In fact, this has been found in recent orientational-dependent stud-

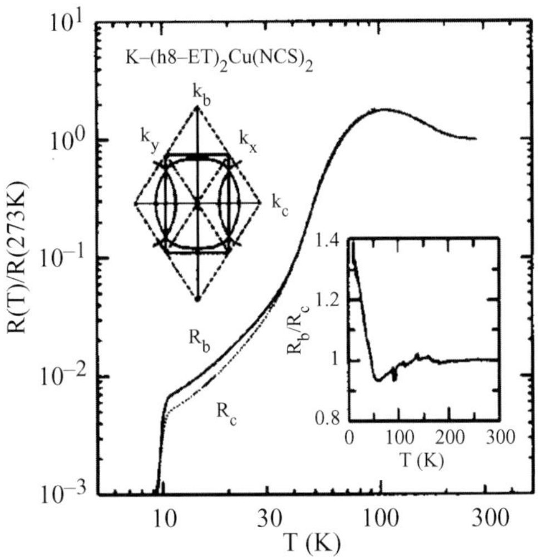

Fig. 20.18. Temperature dependence of the normalized resistance along the in-plane b and c-axis of κ-$(ET)_2Cu(NCS)_2$. The *top left inset* shows the FS and the Brillouin zone, taken from [101]

ies on both compounds [101]: cooling through T^* is accompanied by the onset of a small but distinct anisotropy in the magnetic susceptibility. As can be seen in Fig. 20.18, T^* affects also the charge degrees of freedom where below T^* the b-axis transport becomes more resistive compared to that along the c-axis [101]. These authors proposed that below T^* a static or fluctuating charge-density-wave (CDW) on minor parts of the FS coexists with the metallic phase on the remaining quasi-2D fractions.

The above discussion on the nature of the anomalies at T^* for the superconducting salts κ-$(ET)_2Cu[N(CN)_2]Br$ and κ-$(ET)_2Cu(NCS)_2$ poses the question whether this phenomenon is related to the magnetic signatures found in the non-metallic κ-$(ET)_2Cu[N(CN)_2]Cl$. Earlier magnetization measurements on κ-$(ET)_2Cu[N(CN)_2]Cl$ revealed a shallow decrease below 45 K which was interpreted as the onset of an antiferromagnetic order [152]. In addition, indications were found for a weak ferromagnetic state at 22 K with a small saturation moment of $8 \cdot 10^{-4}\,\mu_B$/dimer. However, according to more recent NMR experiments, the spins order in a commensurate antiferromagnetic structure below $T_N \approx 27$ K with a sizable magnetic moment of $(0.4 - 1.0)\,\mu_B$/dimer [153]. From these measurements, along with magnetization studies, it has been inferred that the easy magnetic axis is aligned perpendicular to the planes and that a small canting of the spins causes a weak ferromagnetic moment parallel to the planes below about $22 \sim 23$ K, see also [154]. Recent ^{13}C-NMR experiments confirmed the commensurate character of the magnetic structure yielding a moment of $0.45\,\mu_B/(ET)_2$ [155].

Three different proposals have been put forward on the origin of the magnetic moments and the nature of the antiferromagnetic insulating state in κ-$(ET)_2Cu[N(CN)_2]Cl$: (i) electron localization due to lattice disorder accompanied by an incomplete compensation of their spins, i.e. an inhomogeneous frozen-in magnetic state [156], (ii) an itinerant SDW-type magnetism associated with the good nesting properties of the quasi-1D parts of the Fermi surface [121, 144, 157] and (iii) a correlation-induced Mott–Hubbard type metal–insulator transition leading to a magnetic state characterized by localized spins [153].

Although proposal (i) has been ruled out by a recent thermal expansion study providing clear thermodynamic evidence for a phase transition at T_N [99] (see inset of the right panel of Fig. 20.16), the nature of the ordered state is still unclear. So far the results of optical, thermal and magnetic properties seem to indicate that certain elements of the models (ii) and (iii) would be applicable to κ-$(ET)_2Cu[N(CN)_2]Cl$.

20.3.4 Anion Ordering and Glassy Phenomena

In discussing molecular conductors and superconductors, an important issue which should not be overlooked is disorder and its possible implications on the electronic properties. In this respect we have to distinguish between different kinds of imperfections. The *extrinsic* disorder, i.e. impurity concentrations, contaminations or crystal defects can be vastly controlled in the preparation process although some aspects remain puzzling, see e.g. the discussion on the resistivity maximum in Sect. 20.3.2. In a study of the alloy series β-$(ET)_2X$ with $X = (I_3)_{1-x}(IBr_2)_x$ ($0 \leq x \leq 1$), where the salts with the two limiting compositions with $x = 0$ and $x = 1$ are superconductors, a clear correlation between the residual-resistivity ratio (RRR) and T_c was found [158]. These experiments show that superconductivity is very sensitive to the induced random potentials which lead to electron localization. The effect of random potentials created by radiation damage effects, resulting in a suppression of superconductivity, has been studied for the Bechgaard salts as well as for β-$(ET)_2I_3$. For more details, the reader is referred to [5] and references therein.

$(TM)_2X$ Salts

However, certain kinds of *intrinsic* disorder are unavoidable and can be of particular importance for experiments attempting to explore superconducting-state properties. The latter type of imperfections concerns materials where, by symmetry, certain structural elements can adopt one of two possible orientations which are almost degenerate in energy [159, 160].

This can be seen in the $(TM)_2X$ salts with non-centrosymmetric anions such as tetrahedral ClO_4. As a result, these anions are disordered at room temperature with an equal occupation for both orientations. Upon cooling, entropy is gained by a more or less perfect ordering of the anions, depending on how fast the

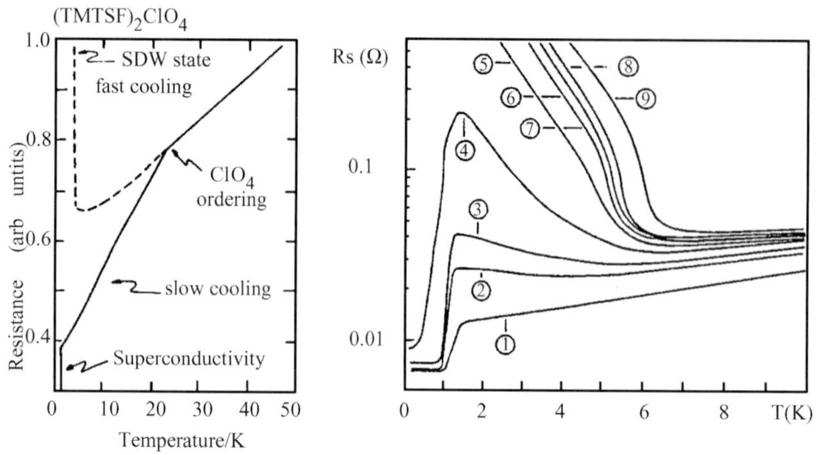

Fig. 20.19. *Left panel*: Effects of anion ordering in $(TMTSF)_2ClO_4$ at $T_{AO} = 24\,K$, on the resistivity, from [161]. *Right panel*: Semilogarithmic plot of the resistance vs temperature in states with various degrees of disorder characterized by T_Q numbered from 1 to 9 (see text), from [162]

system is cooled through the ordering temperature T_{AO}. A perfect long-range anion ordering, realized to a good approximation when cooled sufficiently slowly, then introduces a new periodicity of the lattice. Depending on the anion this can have quite different implications on the electronic properties: for the $(TMTSF)_2ClO_4$ salt, for example, the anion ordering below 24 K is accompanied by a doubling of the periodicity along the *b*-axis, i.e. perpendicular to the stacking axis which leaves the conducting properties almost unaffected. In contrast, the anion ordering for $(TMTSF)_2ReO_4$ opens up a large gap at the Fermi level leading to an insulating ground state. If the compound had been cooled quickly through T_{AO}, a disordered "quenched" state is adopted at low temperatures whose properties are quite different from the "relaxed" state obtained after slow cooling.

The left panel of Fig. 20.19 shows the effect of anion ordering on $(TMTSF)_2ClO_4$. For a slowly cooled system, the anion-ordering phase transition occurs at $T_{AO} = 24\,K$. This is accompanied by a decrease of the resistivity due to the reduction of scattering by randomly distributed anion potentials. Upon further cooling, superconductivity sets in with $T_c = (1.2 \pm 0.2)\,K$. In contrast, for a crystal cooled rapidly, the high-temperature disordered state becomes quenched and the system undergoes a metal-to-insulator transition at $T_{MI} = 6.05\,K$. The insulating quenched state has been identified via NMR and ESR measurements [163,164] as a SDW state with an energy gap $2\,\Delta_0/k_B T_{SDW} = 3.64$ close to the mean-field value of 3.52 [162]. The phase transition at T_{SDW} has been explored recently by specific heat measurements [165]. The right panel of Fig. 20.19 shows that the low-temperature resistivity may change over several orders of magnitude depending on the thermal history of the sample. Intermediate states with various degrees of frozen anion disorder have been produced by rapidly cooling the crystal from different temperatures $T_Q \geq T_{AO}$. Recent specific heat measurements showed that there is no difference between the specific heat anomalies at T_{AO} of slowly cooled and those of quenched-cooled samples, i.e. the structural transition is independent of the kinetic conditions. From these results, the authors concluded that the reordering transition of the anions is of second order and occurs within the experimental timescales [165]. This is contrary to what would be expected for a glass-like transition. Such a glassy behavior, however, has been observed at the ordering temperatures of the compounds with X = ReO_4 ($T_{AO} = 180\,K$) and FSO_3 ($T_{AO} = 86\,K$). For more details see [5,33] and references therein.

κ-(BEDT-TTF)$_2$X Salts

Indications for frozen-in disorder have been also reported for the quasi-2D salts of the κ-$(ET)_2X$ family. In an ac-calorimetry study, a glass-like transition has been found for κ-$(ET)_2Cu[N(CN)_2]Br$

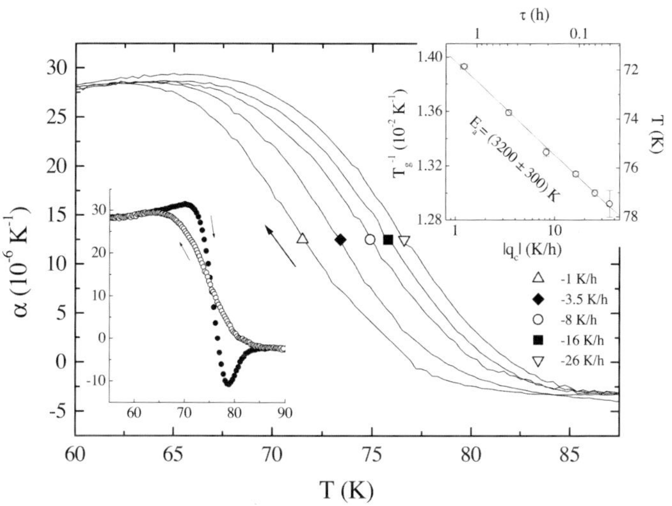

Fig. 20.20. Linear thermal expansion coefficient, α, vs T measured parallel to the conducting planes of κ-$(ET)_2Cu[N(CN)_2]Br$ in the vicinity of the glass transition defined as the midpoint of the step-like change in α for varying cooling rates q_c. *Insets*: hysteresis between heating and cooling curves around T_g (*left side*) and Arrhenius plot of T_g^{-1} vs $|q_c|$ and τ (right side), where $|q_c|$ is the cooling rate and τ the relaxation time (taken from [99])

and κ-$(ET)_2Cu[N(CN)_2]Cl$ [166–168]. The authors observed step-like anomalies in the heat capacity around 100 K, which have been attributed to a freezing out of the intramolecular motions of the ethylene endgroups at the ET molecules. Clear evidence for a glass-like transition have also been derived from thermal expansion measurements [99] which are extremely sensitive to structural rearrangements such as those involved in the glass-like freezing process. The outcome of this study confirmed, on the one hand, the above-mentioned specific heat results and clarified on the other, the nature of the thermal expansion anomalies previously reported by Kund et al. [87,150]. In addition, this study showed that a glass-like transition also exists for the κ-$(ET)_2Cu(NCS)_2$ salt. A glass transition is due to a relaxation process where, below a characteristic temperature T_g the relaxation time $\tau(T)$ of certain structural elements or molecules becomes so large that they can no longer reach thermodynamic equilibrium. As a result, a short range order, which is characteristic for this temperature T_g, becomes frozen in. Figure 20.20 shows exemplarily the linear coefficient of thermal expansion for κ-$(ET)_2Cu[N(CN)_2]Br$ measured parallel to the conducting planes at temperatures near T_g. As discussed in [99] the anomaly at $T_g \simeq 75$ K shows all the characteristics expected for a glassy transition [169], i.e. a step in the thermal expansion coefficient, a pronounced hysteresis between heating and cooling (left inset) and a cooling-rate dependent characteristic temperature T_g. The inset on the right side of Fig. 20.20 shows in an Arrhenius plot the inverse of the glass-transition temperatures, T_g^{-1}, vs the cooling rate $|q_c|$. The data nicely follow a linear behavior as expected for a thermally activated relaxation time [170,171]:

$$\tau(T) = \nu_0^{-1} \cdot \exp\left(\frac{E_a}{k_B T}\right) , \qquad (20.7)$$

where E_a denotes the activation energy barrier. The prefactor represents an attempt frequency ν_0. A linear fit to the data of Fig. 20.20 yields $E_a = (3200 \pm 300)$ K.

The characteristic activation energy of the $[(CH_2)_2]$ conformational motion (cf. Fig. 20.4) was determined to $E_a = 2650$ K by ^{1}H-NMR measurements [144]. The similar size of the activation energy derived from Fig. 20.20 along with the observation of a mass-isotope shift when replacing the hydrogen atoms in $[(CH_2)_2]$ by deuterium provide clear evidence that the ethylene endgroups are the relevant entities involved in the relaxation process [99].

An influence of the thermal history of the samples around $70 \sim 80$ K on the electronic properties had been realized by various authors and interpreted in different ways. Based on resistance measurements of structural relaxation kinetics on κ-$(ET)_2Cu[N(CN)_2]Br$, Tanatar et al. [88] claimed that the ethylene-endgroup ordering is associated

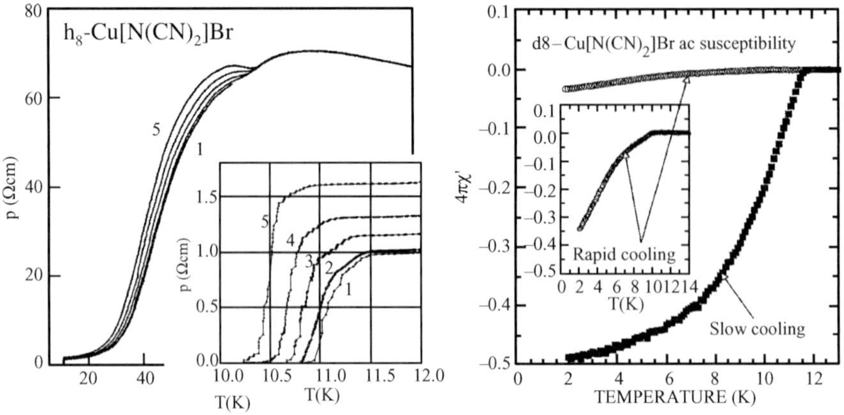

Fig. 20.21. *Left panel*: resistivity as a function of temperature for a sample of κ-(H_8-ET)$_2$Cu[N(CN)$_2$]Br cooled at different rates ranging from about 0.5 K/min, lower curve (1), to 60 K/min upper curve (5). The *inset* shows an expansion of the data near the superconducting transition. Reproduced from [172]. Right panel: ac-susceptibility of κ-(D_8-ET)$_2$Cu[N(CN)$_2$]Br after slow and rapid cooling (taken from [173])

with a sequence of first-order phase transitions around 75 K. A pronounced kink in the resistivity at 75 K accompanied by hysteresis between heating and cooling had been reported early on for κ-(ET)$_2$Cu[N(CN)$_2$]Br [79, 174]. Similar behavior in the vicinity of T_g was observed also for κ-(ET)$_2$Cu(NCS)$_2$ [86]. Besides that, more recent measurements on κ-(ET)$_2$Cu[N(CN)$_2$]Br yielded interesting time dependences affecting not only the electronic properties around 70 $\sim$ 80 K but also the properties at lower temperatures: Su et al. reported relaxation effects in $R(T)$ and a separation of the curves below about 80 K as a function of the cooling rate q_c. As shown in the inset of Fig. 20.21 (left panel) the residual resistivity increases with increasing $|q_c|$ [172,175]. The inset also shows that the way of cooling through 70 $\sim$ 80 K may influence the superconducting properties such that T_c decreases on increasing $|q_c|$. In addition, magnetization measurements revealed that with increasing $|q_c|$, a growing amount of disorder is induced causing an enlarged penetration depth [176]. In reference [177] an ac-susceptibility investigation of the magnetic penetration depths and their dependence on the cooling-rate-dependent intrinsic disorder have been performed. The authors found that the superconducting-state properties are critically determined by the time scale of the experiment around T_g.

For the deuterated κ-(ET)$_2$Cu[N(CN)$_2$]Br salt, it has been reported that rapid cooling through 80 K drives the superconducting ground state into an insulating antiferromagnetic state [173, 178].[5] As shown in the right panel of Fig. 20.21 the ac-susceptibility data reveal a strong suppression of the superconducting volume fraction with increasing cooling rate [173]. It is tempting to assign the apparent deterioration of superconductivity to the frozen disorder at T_g: via the C-H$\cdots$ donor and C-H$\cdots$ anion contact interactions, disorder in the ethylene groups introduces a random potential that may alter the effective transfer integrals t_{eff} and, by this, may destroy superconductivity, see also [36].

[5] The ground state of deuterated κ-(D_8-ET)$_2$Cu[N(CN)$_2$]Br is strongly sample dependent; there are both superconducting as well as non-superconducting samples. In [173] it is claimed that the crystals always contain superconducting and non-superconducting components, whereas the latter have a magnetic character, possibly similar to that of κ-(ET)$_2$Cu[N(CN)$_2$]Cl. It is thus believed that the system is situated in the critical region of the phase diagram just between the superconducting and antiferromagnetic phases, see Sect. 20.3.5.

20.3.5 Phase Diagrams

$(TM)_2X$ Salts

Figure 20.22 comprises in a pressure-temperature plane results of various experiments on the quasi-1D charge-transfer salts $(TMTTF)_2X$ and their selenium analogues $(TMTSF)_2X$. The arrows indicate the position of the various salts at ambient pressure (cf. also Fig. 20.10). The generic character of the phase diagram, which was first proposed by Jérome et al. [138], has been demonstrated recently by pressure studies on the $(TMTTF)_2PF_6$ salt [66,67,179] for which the ambient-pressure ground state is a dimerized spin-Peierls state. With increasing pressure, the system was found to pass through the whole sequence of ground states as shown in Fig. 20.23 and eventually becoming superconducting at high pressures above 43.5 kbar [67,179]. On the low-pressure (left) side of the phase diagram in Figs. 20.22 and 20.23 the molecular stacks can be considered as only weakly-coupled chains, i.e. the system is close to be truly 1D. In fact upon cooling, the $(TMTTF)_2PF_6$ compound behaves very much like canonical 1D conductors where spin and charge degrees of freedom are decoupled: below $T_\rho = 250\,\text{K}$ the resistivity increases by several orders of magnitude due to charge localization while the spin susceptibility remains unaffected [6,33,77]. The phase below T_ρ has been interpreted as a Mott insulating state. Upon further cooling to $T_{SP} = 19\,\text{K}$ a spin gap opens and the system enters a distorted spin-Peierls ground state. In $(TMTTF)_2Br$, position (b) Figs. 20.22 and 20.23, a long-range magnetic order is established below the spin-density-wave transition at $T_{SDW} \approx 13\,\text{K}$. Toward the right side of the phase diagram, which can be carried out either by varying the anion or by the application of hydrostatic pressure, inter-stack interactions become more important. In this region of the phase diagram the electron–phonon interaction is less dominant and electron–electron interactions along with the good nesting properties of the Fermi surface (cf. Fig. 20.7) lead to

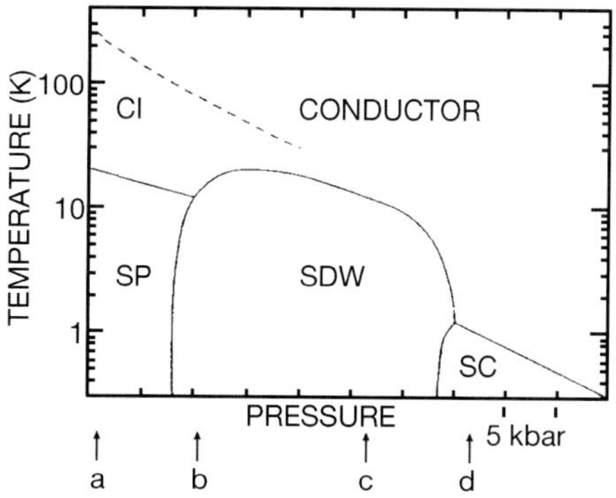

Fig. 20.22. Generalized phase diagram for the quasi-1D $(TM)_2X$ salts as proposed first by Jérome [138]. *Arrows* indicate the positions of the salts with different anions X at ambient pressure: (**a**) $(TMTTF)_2PF_6$, (**b**) $(TMTTF)_2Br$, (**c**) $(TMTSF)_2PF_6$, (**d**) $(TMTSF)_2ClO_4$. The following abbreviations are used: charge-localised insulator (CL), spin Peierls (SP), incommensurate spin-density-wave (SDW) and superconductivity (SC) (taken from [180])

a spin-density-wave ground state as observed, e.g. in the Bechgaard salt $(TMTSF)_2PF_6$ at ambient pressure. After suppression of the SP phase in $(TMTTF)_2PF_6$ with increasing pressure, a commensurate antiferromagnetic state is adopted before an incommensurate SDW phase is stabilized. With increasing pressure, T_{SDW} becomes progressively reduced until, above some critical pressure, the systems remain metallic and superconductivity replaces the SDW ground state. The effect of pressure is to increase the π-orbital overlap also in the transverse direction, i.e. perpendicular to the stacking axis. As a result the almost perfect nesting properties are destroyed and the systems become more 3D in character.[6] According to NMR experiments [33,65] and recent transport measurements under hydrostatic pressure [67,179] strong SDW correlations are still active in the metallic state even when the SDW instability is replaced by superconductivity in $(TMTSF)_2ClO_4$. The range of strong SDW correlations for $(TMTTF)_2PF_6$ derived from these experiments is indicated in Fig. 20.23

6 Recent measurements under uniaxial strain revealed that the SDW transition is most strongly suppressed and superconductivity can be induced by the strain along the stacking a-axis [181,182]. Considerably smaller effects were observed for strain along the b' and c'-axes, although the former direction is directly associated with the denesting of the quasi-1D Fermi surface [181]. This has been interpreted in terms of a pressure-induced decrease in the density of states at the FS [183].

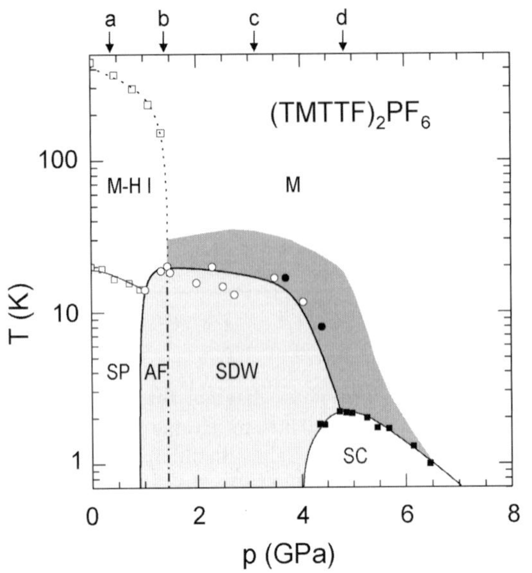

Fig. 20.23. Temperature-hydrostatic-pressure phase diagram for $(TMTTF)_2PF_6$. The abbreviations are: Mott–Hubbard insulating state (M-H I), metallic (M) and superconducting (SC) state, spin-Peierls (SP), commensurate (AF) and incommensurate (SDW) antiferromagnetic spin-density-wave state. *Arrows*: ambient-pressure ground-state location of other salts; (**a**) $(TMTTF)_2BF_4$, (**b**) $(TMTTF)_2Br$, (**c**) $(TMTSF)_2PF_6$, (**d**) $(TMTSF)_2ClO_4$ (taken from [67])

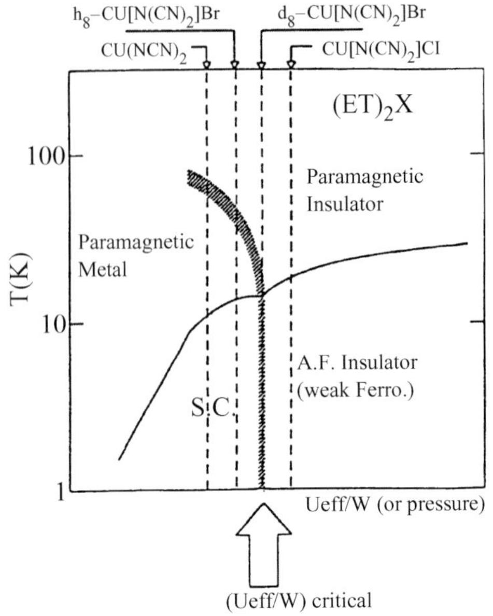

Fig. 20.24. Conceptual phase diagram for κ-phase $(ET)_2X$ as proposed by Kanoda [96]. Note that hydrostatic pressure decreases the ratio $U_{\rm eff}/W$, i.e. the low pressure side is on the right end of the phase diagram. The arrows indicate the ambient-pressure position of the various systems and AF and SC denote an antiferromagnetic insulator and superconductor, respectively

by the shaded region above the SDW and SC phase boundaries.

κ-(BEDT-TTF)$_2$X Salts

Figure 20.24 shows a conceptual phase diagram proposed by Kanoda for the dimerized κ-type BEDT-TTF salts. Here it has been assumed that it is the effective on-site (dimer) Coulomb interaction $U_{\rm eff}$ normalized to the bandwidth W which is the key factor associated with the various phases and phase transitions [95,96]. The positions of the various salts are determined by their ambient-pressure ground-state properties. The deuterated κ-$(D_8$-$ET)_2Cu[N(CN)_2]Br$ salt is situated right at the AFI/SC border. The system lies in between the antiferromagnetic insulating X = $Cu[N(CN)_2]Cl$ and the superconducting hydrogenated κ-$(H_8$-$ET)_2Cu[N(CN)_2]Br$ salts. It has been proposed that a *partial* substitution of the 2 × 4 H-atoms by D-atoms allows for fine-tuning the κ-$(ET)_2Cu[N(CN)_2]Br$ system across the AFI/SC border [184–186]. The close proximity of an antiferromagnetic insulating to a superconducting phase has been considered, in analogy to the high-T_c cuprates, as a strong indication that both phenomena are closely connected to each other [4].

Figure 20.25 summarizes experimental data of a detailed thermodynamic study on the various κ-$(ET)_2X$ compounds in a pressure-temperature phase diagram [99,187]. The positions of the various salts at ambient pressure are indicated by the arrows.[7] The solid lines representing the phase boundaries

[7] Note that hydrostatic pressure has been used as an abscissa for the purpose of compatibility with the conceptual phase diagram in Fig. 20.24. It has been found, however, that the uniaxial-pressure dependences for the various phase boundaries are strongly anisotropic [99,100,151] with a non-uniform behavior for the uniaxial-pressure coefficients of both the density-wave instability at T^* and those of T_c, cf. Figs. 20.17 and 20.32.

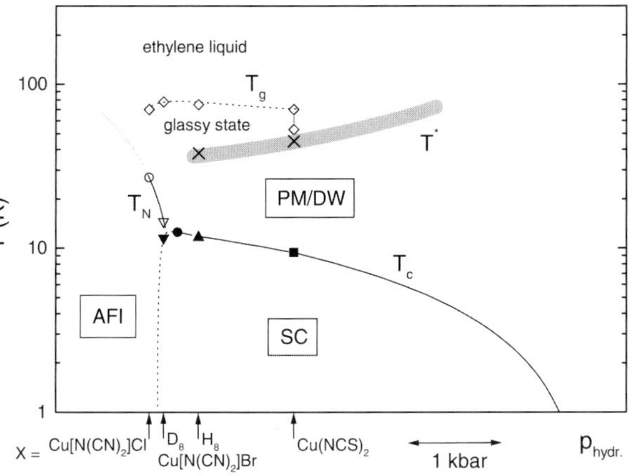

Fig. 20.25. Temperature-hydrostatic-pressure phase diagram for the κ-$(ET)_2X$ compounds. *Arrows* indicate the positions of the various compounds at ambient pressure. *Circles* correspond to results on κ-$(ET)_2Cu[N(CN)_2]Cl$ while *down and up triangles* indicate phase-transition temperatures on deuterated and hydrogenated κ-$(ET)_2Cu[N(CN)_2]Br$, respectively, and *squares* stand for results on κ-$(ET)_2Cu(NCS)_2$. The transitions into the superconducting and antiferromagnetic states are represented by *closed and open symbols*, respectively. *Diamonds* denote the glass-like freezing of ethylene disorder and crosses the density-wave (DW) transition on minor parts of the Fermi surface. These anomalies coincide with various features observed in magnetic, transport and acoustic properties (shaded area, see Sect. 20.3.3) (taken from [99,187])

between the paramagnetic (PM) and the superconducting (SC) or antiferromagnetic insulating (AFI) states refer to the results of hydrostatic-pressure studies of T_c and T_N [80, 188, 189]. The metal–insulator transition and the coexistence range of metallic/superconducting and insulating phases as deduced from hydrostatic pressure experiments by Lefebvre et al. [191] and Limelette et al. [98], see also Fig. 20.26 below, is not addressed in this phase diagram.

At elevated temperatures, a glass-like transition at a temperature T_g (dotted line) has been identified. It marks the boundary between an ethylene-liquid at $T > T_g$ and a glassy state at $T < T_g$. While at temperatures above T_g, the motional degrees of freedom of the ethylene endgroups are excited with an equal occupancy for the two possible ethylene conformations, a certain disorder becomes frozen in at temperatures below T_g. The glass-like transition which is structural in nature has been shown to cause time dependences in electronic properties and may have severe implications on the ground-state properties of the κ-$(ET)_2Cu[N(CN)_2]Br$ salt depending on the cooling rate employed at T_g (see Sect. 20.3.4).

At intermediate temperatures T^*, anomalies in the coefficient of thermal expansion have been found and assigned to a density-wave transition involving only the quasi-1D parts of the Fermi surface. These anomalies coincide with various features observed in magnetic, transport and acoustic properties (thick shaded line in Fig. 20.25, see also Sect. 20.3.3). In [99,100] it has been proposed that instead of a pseudogap on the quasi-2D parts of the Fermi surface, a real gap associated with a density wave opens on the minor quasi-1D parts below T^*, see also [101]. This scenario implies that the density wave and superconductivity involve disjunct parts on the Fermi surface and compete for stability.

Details of the pressure-temperature phase diagram of the antiferromagnetic insulating salt κ-$(ET)_2Cu[N(CN)_2]Cl$ have been reported by Ito et al. [190], Lefebvre et al. [191], and, more recently, by some other authors [98, 192, 193]. In [191], it has been shown that the superconducting and antiferromagnetic phases overlap through a first-order boundary that separates two regions of an inhomogeneous phase coexistence [191]. It has been argued that this boundary curve merges with a first-order

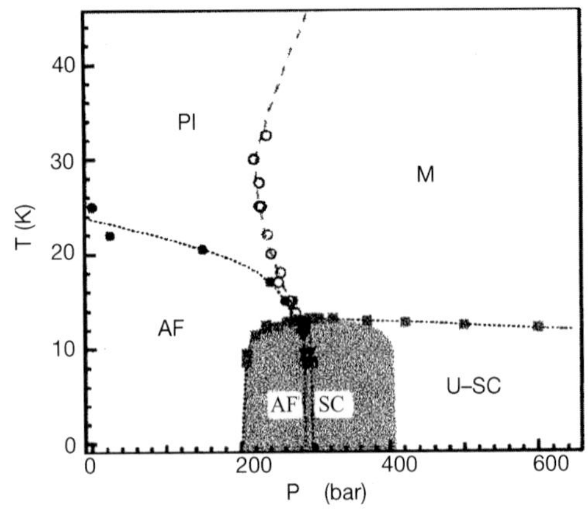

Fig. 20.26. Pressure-temperature phase diagram for κ-$(ET)_2Cu[N(CN)_2]Cl$. The antiferromagnetic (AF) transition temperature $T_N(p)$ (*closed circles*) were determined from the NMR relaxation rate while $T_c(p)$ (*closed squares*) and $T_{MI}(p)$ (*open circles*) were obtained from ac-susceptibility measurements. U-SC denotes uniform superconductivity; the AF-SC boundary line separates two regions of inhomogeneous phase coexistence (*shaded area*) (taken from [191])

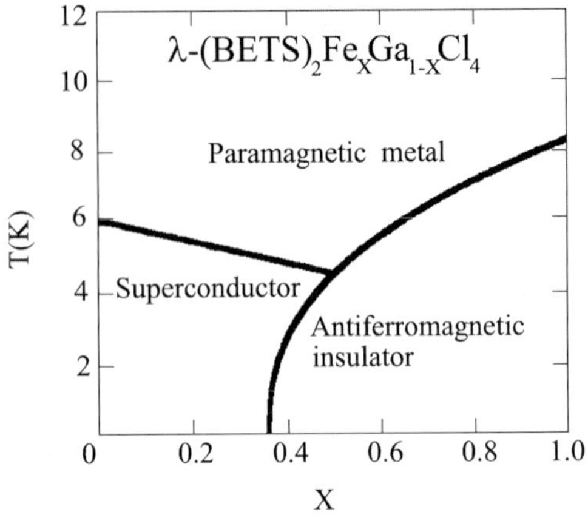

Fig. 20.27. Phase diagram for λ-$(BETS)_2Fe_xGa_{1-x}Cl_4$ taken from [32]

line of the metal–insulator transition and that this line ends at a critical point at higher temperature, see Fig. 20.26. The figure also suggests the existence of a point-like region where the metallic, insulating, antiferromagnetic as well as superconducting phases all meet. This would imply the absence of a boundary between metallic and complete antiferromagnetic phases which would be incompatible with an itinerant type of magnetism [191]. By measurements of the ultrasonic velocity on pressurized κ-(BEDT-TTF)$_2Cu[N(CN)_2]Cl$ [192] a dramatic anomaly has been observed near 34 K and at $P \simeq 210$ bar which defines the critical point in the pressure-temperature plane.

$(BEDT\text{-}TSF)_2X$ Salts

Figure 20.27 shows the phase diagram for the quasi-2D alloy system λ-$(BETS)_2Fe_xGa_{1-x}Cl_4$ which has recently gained strong interest due to its interesting magnetic and superconducting properties [194]. The system is based on the donor molecule BEDT-TSF or simply BETS which represents the Se analogue to BEDT-TTF, see Sect. 20.2. λ-$(BETS)_2$ $GaCl_4$ on the left side is a nonmagnetic salt which becomes superconducting at $T_c = 6$ K [31]. A magnetic field of 13 T aligned parallel to the highly conducting planes destroys superconductivity and stabilizes a paramagnetic metallic state. Conversely, λ-$(BETS)_2FeCl_4$ shows a metal–insulator (M–I) transition around 8 K which is accompanied by an antiferromagnetic order of the Fe^{3+} moments in the anion layers [28]. Applying a magnetic field in excess of about 10 T destabilizes the insulating phase and the paramagnetic phase recovers [195, 196]. In the mixed series λ-$(BETS)_2Fe_xGa_{1-x}Cl_4$, the M–I transition becomes suppressed as the concentration x of magnetic Fe ions decreases and a superconducting ground state is formed for $x \leq 0.35$. A striking feature is the metal–superconductor–insulator transition for $0.35 \leq x \leq 0.5$, see Fig. 20.27. Apparently, the various phases contained in the above phase diagram originate from an intimate coupling between the magnetic moments of the Fe^{3+} 3d electrons and the π-conduction-electron spins of the BETS molecule, see e.g. [32, 196, 197].

20.4 Superconducting-State Properties

Since the discovery of superconductivity in pressurized $(TMTSF)_2PF_6$ in 1979 [1], continuing efforts to design new potential donor and acceptor molecules have led to more than about 80 organic superconductors.[8] In the vast majority of cases, the donor molecules are derivatives of the prototype TMTSF molecule including BEDT-TTF as well as the selenium and oxygen substituted variants BEDT-TSF and BEDO-TTF, respectively. In addition, superconductors have been derived using asymmetric hybrids such as DMET and MDT-TTF. For a comprehensive list of organic superconductors the reader is referred to [5].

In the following discussion of superconducting properties we will confine ourselves to a few selected examples. These are the $(TMTSF)_2X$ and the quasi-2D $(BEDT\text{-}TTF)_2X$ and $(BEDT\text{-}TSF)_2X$ salts which are representative for a wide class of materials.

20.4.1 The Superconducting Phase Transition

Organic superconductors are characterized by a highly anisotropic electronic structure, a low charge carrier concentration and unusual lattice properties. As will be discussed below, the combination of these unique material parameters lead to a variety of remarkable phenomena of the superconducting state such as pronounced thermal fluctuations, an extraordinarily high sensitivity to external pressure and anomalous mixed-state properties.

Superconducting Anisotropy

The abrupt disappearance of the electrical resistance is one of the hallmarks that manifests the transition from the normal into the superconducting state for usual 3D superconductors. For the present low-dimensional organic superconductors, as in the layered high-T_c cuprates, however, strong fluctuations of both the amplitude and phase of the superconducting order parameter may cause a substantial broadening of the superconducting transition. This becomes particularly clear when a strong magnetic field is applied. As a consequence, for these materials zero-resistance is no longer a good measure of the superconducting transition temperature in a finite magnetic field. Therefore, for a precise determination of the upper critical fields, thermodynamic investigations such as magnetization, specific heat or thermal expansion measurements are necessary.

For materials with strongly directional-dependent electronic properties, a highly anisotropic superconducting state is expected as well. In an attempt to account for these anisotropies, the phenomenological Ginzburg–Landau and London models have been extended by employing an effective-mass tensor [199]. In the extreme case of a quasi-2D superconductor characterized by a superconducting coherence length perpendicular to the planes, $\xi_\perp$, being even smaller than the spacing between the conducting layers, s, these anisotropic 3D models are no longer valid. Instead, the superconductor has to be described by a model that takes the discreteness of the structure into account. Such a description is provided by the phenomenological Lawrence–Doniach model [200] which encloses the above anisotropic Ginzburg–Landau and London theories as limiting cases for $\xi_\perp > s$. This model considers a set of superconducting layers separated by thin insulating sheets implying that the 3D phase coherence is maintained by Josephson currents running across the insulating layers. In fact, the presence of an intrinsic Josephson effect has been demonstrated for several layered superconductors including some of the high-T_c cuprates and the present $\kappa\text{-}(ET)_2Cu(NCS)_2$ salt [201,202].

To quantify the degree of anisotropy, it is convenient to compare the results of orientational-dependent measurements with the above anisotropic models. For layered systems such as the present $(BEDT\text{-}TTF)_2X$ compounds, it is customary to use the effective-mass ratio $\Gamma = m_\perp^*/m_\parallel^*$, where $m_\perp^*$ and $m_\parallel^*$ denote the effective masses for the superconducting carriers moving perpendicular and parallel to the conducting planes, respectively. In the London and Ginzburg–Landau theories, Γ is directly related to the anisotropies in the magnetic penetration depths λ and coherence lengths ξ by

$$\sqrt{\Gamma} = \frac{\lambda_\perp}{\lambda_\parallel} = \frac{\xi_\parallel}{\xi_\perp} . \tag{20.8}$$

[8] The materials discussed in this article have to be distinguished from another class of molecular superconductors; the alkali-metal-doped fullerenes discovered in 1991 [198], which are usually not referred to as organic materials as they contain only carbon atoms.

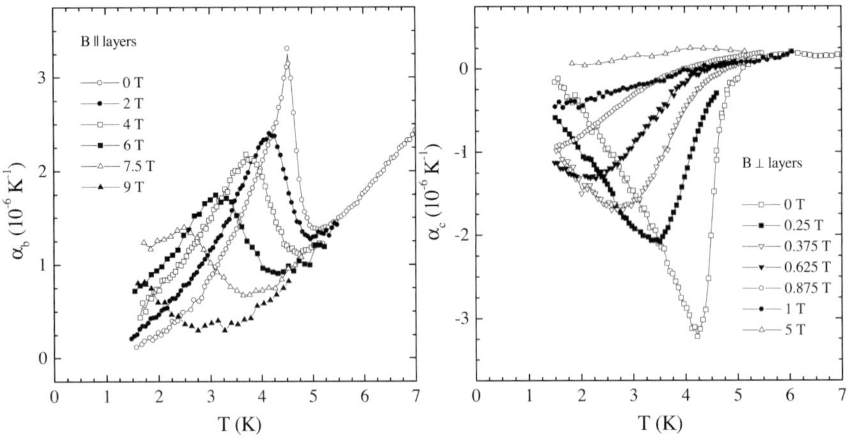

Fig. 20.28. Coefficient of thermal expansion of β''-$(ET)_2SF_5CH_2CF_2SO_3$ measured parallel (*left panel*) and perpendicular (*right panel*) to the conducting planes in varying fields applied along the measuring directions (taken from [151])

As an example for the highly anisotropic response of the superconducting transition to a magnetic field, we show in Fig. 20.28 results of the coefficient of thermal expansion $\alpha(T)$ near T_c for β''-$(ET)_2SF_5CH_2CF_2SO_3$.[9] While for fields parallel to the planes (left panel) the phase transition in $\alpha(T)$ is still visible even in $B = 9$ T, a field of 1 T applied perpendicular to the conducting planes is sufficient to suppress almost completely superconductivity (right panel of Fig. 20.28). In addition, for this field orientation, a pronounced rounding of the phase-transition anomaly even for very small fields can be observed. From these measurements the upper critical fields, B_{c_2}, can be determined permitting an estimate of the anisotropy parameter Γ:

$$B_{c_2}^{\perp'} = \left|\frac{\mathrm{d}B_{c_2}^{\perp}}{\mathrm{d}T}\right| = \frac{\phi_0}{2\pi\xi_{\parallel}^2 T_c} \quad \text{and}$$
$$\frac{B_{c_2}^{\perp'}}{B_{c_2}^{\parallel'}} = \frac{\xi_{\perp}}{\xi_{\parallel}} = \frac{1}{\sqrt{\Gamma}}, \tag{20.9}$$

where $B_{c_2}^{\perp'}$ and $B_{c_2}^{\parallel'}$ are the initial slopes of the upper critical fields for B perpendicular and parallel to the conducting planes, respectively [205,206] and ϕ_0 is the flux quantum. For β''-$(ET)_2SF_5CH_2CF_2SO_3$ one finds $\xi_{\parallel} = (144\pm9)$ Å, $\xi_{\perp} = (7.9\pm1.5)$ Å and $\Gamma \approx 330$ [151] (cf. Table 20.1 in Sect. 20.2.2 and Table 20.2 in Sect. 20.4.2), which underlines the quasi-2D character of the superconducting state in this material. The large anisotropy parameter $\Gamma \approx 330$ exceeds the value of $\Gamma \approx 100$ found for the κ-$(ET)_2Cu(NCS)_2$ salt in dc-magnetization experiments [207].

The so-derived Γ values, however, may serve only as a rough estimate of the actual anisotropy parameters. The latter can be probed most sensitively by employing torque-magnetometry. For κ-$(ET)_2Cu(NCS)_2$ for example, Γ values ranging from 200 to 350 have been reported [208,209] which place this material in the same class of quasi-2D superconductors as the most anisotropic high-T_c cuprates with $\Gamma = 150 \sim 420$ for $Bi_2Sr_2CaCu_2O_{8+x}$ [210,211].

Fluctuation Effects

The highly anisotropic response of the present quasi-2D superconductors to a magnetic field is also demonstrated in Fig. 20.29 where the temperature dependence of the magnetization around the superconducting transition is shown for κ-$(ET)_2Cu(NCS)_2$. While for fields aligned perpendicular to the planes (left panel) the transition considerably broadens with increasing field strength, there is only a little effect on the transition for fields parallel to the layers (right panel) [207], cf. also Fig. 20.28. This behavior is quite different from that which is found in a usual 3D superconductor and indicates the presence of strong superconducting fluctuations which

[9] This salt of the β''-type structure contains large discrete anions and is unique in being the first superconductor free of any metal atoms [203,204].

are strongly enhanced in systems with reduced dimensionality [212].

A measure of the strength of thermal order-parameter fluctuations is provided by the so-called Ginzburg number

$$G = \frac{|T - T_c|}{T_c} = \frac{1}{2}\left(\frac{k_B T}{B_{c_{th}}^2(0)\xi_{\|}^2 \xi_{\perp}}\right)^2 , \qquad (20.10)$$

where $B_{c_{th}}(0)$ is the thermodynamic critical field. G measures the ratio of thermal energy to the condensation energy per coherence volume. For classical 3D superconductors like niobium G amounts to about $G \sim 10^{-11}$. In contrast, for the present compounds and some of the high-T_c cuprates one finds $G \sim 10^{-2} - 10^{-3}$ [207,213,214]. Here, the relatively high transition temperatures together with a low charge-carrier concentration (the latter results in a small Fermi velocity, v_F, and thus a short coherence length $\xi \propto \hbar v_F/(k_B T_c)$) enhance the effect of superconducting fluctuations. The strong rounding of the phase-transition anomaly with increasing magnetic fields aligned perpendicular to the conducting planes is then understood to be a result of a field-induced dimensional crossover: while the electronic state in small fields is quasi-2D, the confinement of the quasiparticles to their lower Landau levels in high fields leads to a quasi-zero-dimensional situation [215,216]. As a result, the relatively sharp phase-transition anomaly in zero field becomes progressively rounded and smeared out with increasing field. Instead of a well defined phase boundary between normal and superconducting states, the high-field range of the B–T phase diagram is characterized by a crossover behavior with extended critical fluctuations. Here, the assertion of a mean-field transition temperature, $T_c^{\mathrm{mf}}(B)$, from the raw data is difficult and a fluctuation analysis has to be invoked.

The effect of fluctuations on transport and thermodynamic properties has been studied by several authors [217,218]. Assuming the lowest-Landau-level approximation and taking into account only non-interacting Gaussian fluctuations, Ullah and Dorsey

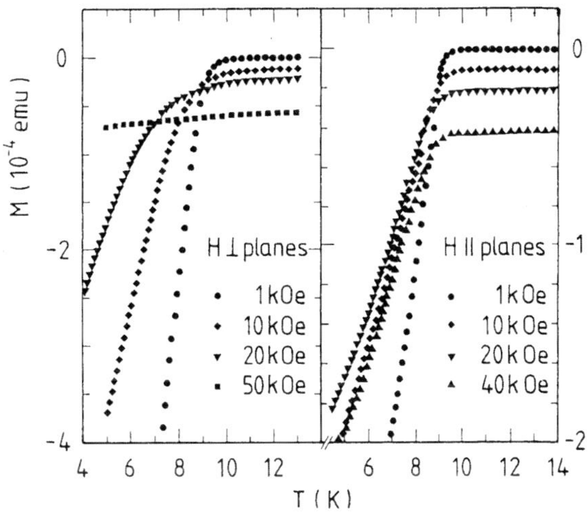

Fig. 20.29. Raw data of the dc-magnetization of single crystalline κ-$(ET)_2Cu(NCS)_2$ in various fields perpendicular (*left panel*) and parallel (*right panel*) to the conducting planes. The offset of each curve is due to contributions from the core electrons, the spin susceptibility as well as a small background signal (taken from [207])

obtained an expression for a scaling function of various thermodynamic quantities as the magnetization M or the specific heat C:

$$\Xi_i = F_i\left(A\frac{T - T_c^{\mathrm{mf}}(B)}{(TB)^n}\right) , \qquad (20.11)$$

with $\Xi_i = M/(TB)^n$ or C/T [213]. F_i is an unknown scaling function, A a temperature-independent and field-independent coefficient characterizing the transition width and $n = 2/3$ for anisotropic 3D materials and $n = 1/2$ for a 2D system. Thus from a scaling analysis both the actual dimensionality as well as the mean-field-transition temperature $T_c^{\mathrm{mf}}(B)$ can be determined.

Figure 20.30 shows the data of the dc-magnetization of κ-$(ET)_2Cu(NCS)_2$ (Fig. 20.29) and thermal expansion of β''-$(ET)_2SF_5CH_2CF_2SO_3$ (Fig. 20.28) taken in varying fields in the proper scaling forms $M/(TH)^{(n)}$ vs $(T - T_c^{\mathrm{mf}}(B))/(TB)^n$ and α_c^{sc}/T vs $(T - T_c^{\mathrm{mf}}(B))/(TB)^{1/2}$, respectively, where α_c^{sc} denotes the superconducting contribution to the coefficient of

[10] Since the volume coefficient of thermal expansion, $\beta(T)$, is related to the specific heat via the Grüneisen relation $\beta(T) = \gamma \cdot \frac{\kappa_T}{V_{mol}} \cdot C_V(T)$, where κ_T denotes the isothermal compressibility, V_{mol} the molar volume and γ a field-independent and temperature-independent Grüneisen parameter, the scaling form holds also for α/T.

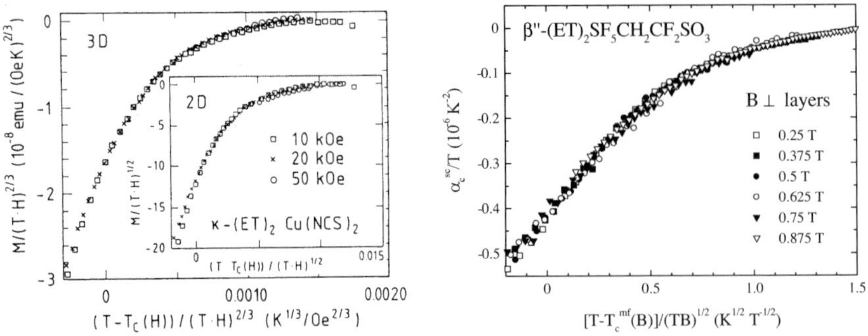

Fig. 20.30. Scaling behavior of the superconducting contribution to the magnetization of κ-$(ET)_2Cu(NCS)_2$ (*left panel*) [207] and the linear coefficient of thermal expansion of β''-$(ET)_2SF_5CH_2CF_2SO_3$ (*right panel*) [151] in magnetic fields applied perpendicular to the planes

thermal expansion.[10] As shown in Fig. 20.30 the various field curves $\alpha_c^{sc}(T, B)$ show the 2D scaling over a rather wide temperature and field range, see also [213,219]. According to the scaling analysis of the high-field magnetization in Fig. 20.30 as well as the high-field conductivity in [220], κ-$(ET)_2Cu(NCS)_2$ is at the threshold from being a strongly anisotropic 3D to a 2D superconductor. On the other hand, a distinct 2D behavior has been claimed from a scaling analysis of low-field magnetization data by Ito et al. [221].

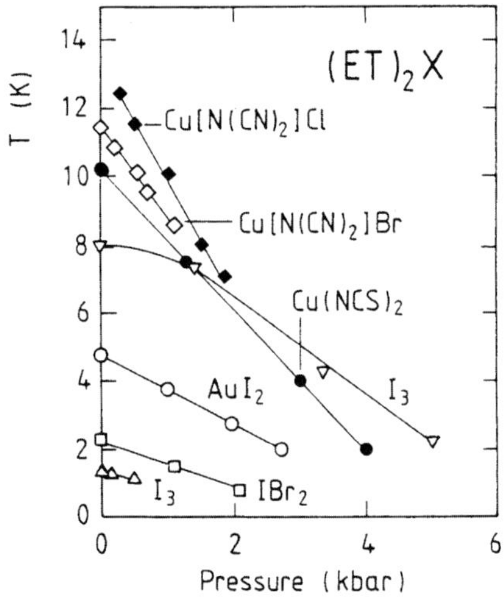

Fig. 20.31. Hydrostatic-pressure dependence of T_c for various β-type and κ-type $(ET)_2X$ superconductors, reproduced from [5]

Pressure Dependence of T_c

By applying pressure to a superconductor, one can study the volume dependence of the pairing interaction through changes of T_c. For the $(TMTSF)_2X$ and $(ET)_2X$ superconductors, one generally finds an extraordinarily high sensitivity to external pressure and, in the vast majority of cases, a rapid decrease of T_c with pressure. Figure 20.31 shows the variation of T_c for a selection of β-type and κ-type $(ET)_2X$ salts under hydrostatic-pressure conditions. The initial slope of the pressure dependence of T_c, $(\partial T_c/\partial p)_{p\to 0}$, determined via resistivity measurements ranges from -0.25 K/kbar for α-$(ET)_2NH_4Hg(SCN)_4$ ($T_c = 1$ K) [222] to -3.2 K/kbar for κ-$(ET)_2Cu[N(CN)_2]Cl$ ($T_c = 12.8$ K at 0.3 kbar) [80]. For $(TMTSF)_2PF_6$, one finds $-(8 \pm 1) \cdot 10^{-2}$ K/kbar ($T_c = 1.1$ K at 6.5 kbar) [137]. At first glance a strong pressure dependence of T_c appears not surprising in view of the weak van der Waals bonds between the organic molecules, giving rise to a highly compressible crystal lattice. In fact, the isothermal compressibility $\kappa_T = -\partial \ln V/\partial p$ for κ-$(ET)_2Cu(NCS)_2$ of $\kappa_T = (122\,\text{kbar})^{-1}$ [223,224] exceeds the values found for ordinary metals by about a factor of five. To account for this "lattice effect" one should, therefore, consider the physically more meaningful volume dependence of T_c:

$$\frac{\partial \ln T_c}{\partial \ln V} = \frac{V}{T_c} \cdot \frac{\partial T_c}{\partial V} = -\frac{1}{\kappa_T \cdot T_c} \cdot \frac{\partial T_c}{\partial p} . \tag{20.12}$$

Using the above isothermal compressibility, one finds $\partial \ln T_c/\partial \ln V \approx 40$ for κ-$(ET)_2Cu(NCS)_2$ [151] which exceeds the values found for ordinary metallic superconductors, as e.g. for Pb with $\partial \ln T_c/\partial \ln V = 2.4$ [225], or the layered copper-oxides with $-(0.36 \sim 0.6)$ reported for $YBa_2Cu_3O_7$ [226] by $1 \sim 2$ orders of magnitude.

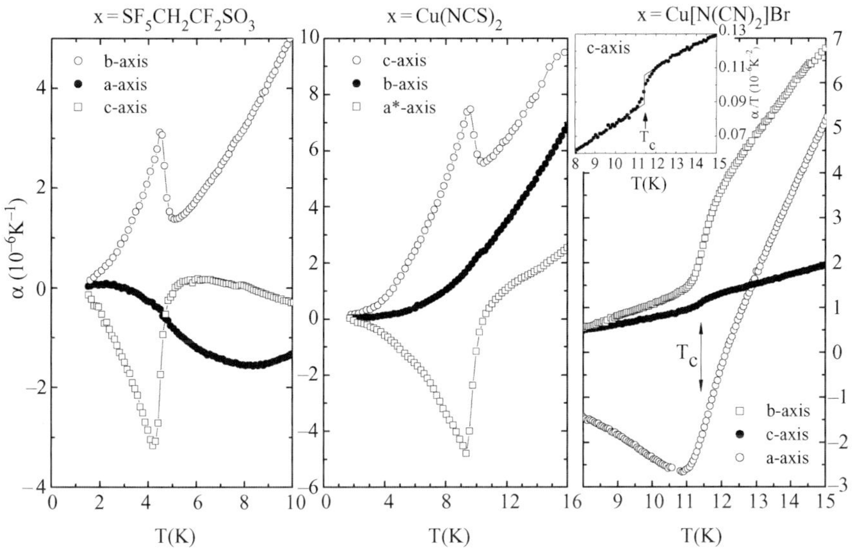

Fig. 20.32. Uniaxial coefficients of thermal expansion, α_i, vs temperature around the superconducting transition for $(ET)_2X$ with X = $SF_5CH_2CF_2SO_3$ (*left panel*), $Cu(NCS)_2$ (middle) and $Cu[N(CN)_2]Br$ (*right panel*), taken from [99, 151, 187]. *Open squares* indicate α data perpendicular to the planes; *open and closed circles* correspond to the in-plane expansion coefficients

For strongly anisotropic superconductors like those discussed here, even more information on the relevant microscopic couplings can be obtained by studying the effect of uniaxial pressure on T_c. Different techniques have been employed to determine the uniaxial-pressure coefficients of T_c for the $(TMTSF)_2X$ and $(ET)_2X$ salts including measurements under uniaxial strain or stress [227–230] or by using a thermodynamic analysis of ambient-pressure thermal expansion and specific heat data [99, 151, 231, 232]. The latter approach is based on the Ehrenfest relation which connects the pressure coefficients of T_c for uniaxial pressure along the i-axis (in the limit of vanishing pressure) to the phase-transition anomalies at T_c in the coefficient of thermal expansion, $\Delta\alpha_i$, and specific heat, ΔC:

$$\left(\frac{\partial T_c}{\partial p_i}\right)_{p_i \to 0} = V_{\text{mol}} \cdot T_c \cdot \frac{\Delta\alpha_i}{\Delta C} , \qquad (20.13)$$

with V_{mol} being the molar volume. Figure 20.32 shows results of the linear thermal expansion coefficients along the three principal crystal axes of the superconductors β''-$(ET)_2SF_5CH_2CF_2SO_3$, κ-$(D_8\text{-}ET)_2Cu(NCS)_2$ and κ-$(ET)_2Cu[N(CN)_2]Br$. In all three cases, the uniaxial expansion coefficients are strongly anisotropic with striking similarities in the α_i's for the β'' and κ-$(ET)_2Cu(NCS)_2$ salts.

Using (20.13), the uniaxial-pressure coefficients can be derived [99, 151]. For κ-$(ET)_2Cu[N(CN)_2]Br$ one finds $\partial T_c/\partial p_b = -(1.26 \pm 0.25)$ K/kbar for the out-of-plane coefficient and $\partial T_c/\partial p_a = -(1.16 \pm 0.2)$ K/kbar and $\partial T_c/\partial p_c = -(0.12 \pm 0.05)$ K/kbar for the in-plane coefficients, employing a jump height in the specific heat ΔC as reported by [233]. In the same way, one obtains for β''-$(ET)_2SF_5CH_2CF_2SO_3$ $-(5.9 \pm 0.25)$ K/kbar along the out-of-plane c-axis and $+(3.9 \pm 0.15)$ K/kbar and $+(0.39 \pm 0.1)$ K/kbar for the in-plane coefficients along the b- and a-axis, respectively, using the ΔC value given in [234]. To check for consistency, the hydrostatic-pressure dependences can be calculated by summing over the uniaxial-pressure coefficients yielding $\partial T_c/\partial p_{\text{hydr}} = \sum_i(\partial T_c/\partial p_i) = -(2.54 \pm 0.5)$ K/kbar and $-(1.6 \pm 0.5)$ K/kbar for the κ and the β'' salt, respectively. These values are found to be in good agreement with the results obtained by hydrostatic-pressure experiments, i.e. $-(2.4 \sim 2.8)$ K/kbar for κ-

[11] Figure 20.32 shows that there is a strikingly similar behavior for the uniaxial thermal expansion coefficients and thus the uniaxial-pressure coefficients of T_c of β''-$(ET)_2SF_5CH_2CF_2SO_3$ and κ-$(D_8\text{-}ET)_2Cu(NCS)_2$. The same observation was made also for the hydrogenated κ-$(H_8\text{-}ET)_2Cu(NCS)_2$ compound [187, 232]. Due to the lack of specific heat data

$(ET)_2Cu[N(CN)_2]Br$ [94,189] and −1.43 K/ kbar for β''-$(ET)_2SF_5CH_2CF_2SO_3$ [234].[11]

An obvious step towards a microscopic understanding of this class of superconductors is to trace out those uniaxial-pressure effects which are common to all systems and thus reflect a generic property. Such information would provide a most useful check of theoretical models attempting to explain superconductivity and its interrelation with the various other instabilities in the pressure-temperature plane. As Fig. 20.32 demonstrates, an extraordinarily large negative uniaxial-pressure coefficient of T_c for uniaxial pressure perpendicular to the conducting planes is common to all three superconductors shown there. Apparently, it is this huge component which predominates the large response of T_c under hydrostatic pressure. On the other hand, and in contrast to what has been frequently assumed, the systems behave quite non-uniformly concerning the in-plane pressure effects. While for κ-$(ET)_2Cu(NCS)_2$ the in-plane pressure coefficients of T_c are either vanishingly small or positive, they are both negative for the related κ-$(ET)_2Cu[N(CN)_2]Br$ system [99].

The above finding of a large negative uniaxial-pressure coefficient of T_c for pressure perpendicular to the planes as the only universal feature common to the κ-$(ET)_2X$ family is supported by results on the related κ-$(ET)_2I_3$ salt, see [236].[12]

As has been discussed in [9], a large cross-plane pressure effect on T_c may arise from several factors: (i) Pressure-induced changes in the interlayer interaction. This effect includes changes of both the interlayer coupling, i.e. the degree of two-dimensionality, as well as changes in the electron–electron and electron–phonon coupling constants and (ii) changes in the phonon frequencies. Likewise, changes in the vibrational properties could be of relevance for the intraplane-pressure effects on T_c. In addition, in-plane stress effectively modifies the electronic degrees of freedom by changing the transfer integrals between the HOMO's of the nearest-neighbor ET molecules. Most remarkably, for some compounds like κ-$(ET)_2Cu(NCS)_2$ the in-plane-stress effect is either positive or zero. This makes a purely density-of-states effect account for the pressure-induced T_c shifts very unlikely: pressure-induced changes in the density-of-states should be strongest for in-plane stress owing to the quasi-2D electronic band structure. According to the simple BCS relation [237]

$$T_c = 1.13\,\Theta_D \exp\left(-\frac{1}{\lambda}\right) \quad \text{with } \lambda = \frac{N(E_F)\langle I^2\rangle}{M\bar{\omega}^2}\,, \tag{20.14}$$

where Θ_D denotes the Debye temperature, $\langle I^2\rangle$ the electron–phonon matrix element averaged over the Fermi surface, M the ionic mass and $\bar{\omega}$ an average phonon energy, an in-plane-stress-induced increase in the π-orbital overlap, i.e. a reduced density-of-states at the Fermi level $N(E_F)$ is expected to cause a reduction of T_c. This is in contrast to the experimental observations.

An important piece of information contained in the above uniaxial-pressure results is that there is *no* uniform behavior in the intralayer-pressure effects on T_c for the various $(ET)_2X$ superconductors. It is especially the results on κ-$(ET)_2Cu(NCS)_2$ which show that in-plane pressure can even cause an increase of T_c [151,230]. This is in contrast to what has been assumed in the 2D electronic models discussed in [238,239]. In addition, the studies revealed a predominant effect of uniaxial pressure perpendicular to the planes clearly demonstrating that attempts to model the pressure-temperature phase diagrams by solely considering *in-plane* electronic degrees of freedom are inappropriate, see also [240].

for the deuterated salt, the uniaxial-pressure coefficients of κ-$(ET)_2Cu(NCS)_2$ can only be discussed qualitatively. According to a recent comparative study on the pressure dependences of the normal-state and superconducting-state properties of hydrogenated and deuterated κ-$(ET)_2Cu(NCS)_2$, the latter compound reveals an even stronger pressure dependence of T_c [235].

12 A different situation is encountered for the α-$(ET)_2MHg(SCN)_4$ salt, where uniaxial pressure perpendicular to the planes is found to either induce superconductivity by suppressing an ambient-pressure density-wave ground state for $M = K$, or enhance T_c for $M = NH_4$ [187,228]. This behavior is most likely related to the exceptionally thick anion layers specific to this compound resulting in a strong decoupling of the conducting layers.

Isotope Substitution

Studying the effect of isotope substitutions on the superconducting transition temperature is one of the key experiments to illuminate the role of phonons in the pairing mechanism. For elementary superconductors, the observation of a $M^{-1/2}$ dependence of T_c where M is the isotopic mass, provided convincing evidence that the attractive interaction between the electrons of a Cooper pair is mediated by the exchange of lattice deformations, i.e. by phonons.

For the κ-phase $(ET)_2X$ compounds, the mass-isotope effect on T_c has been intensively studied, see [5], including isotope substitutions in both the ET donor molecule as well as the charge compensating anions. A most comprehensive study has been performed by the Argonne group on κ-$(ET)_2Cu(NCS)_2$ where overall seven isotopically labeled BEDT-TTF derivatives, with partial substitutions of ^{13}S, ^{34}C and ^{2}D, as well as isotopically labeled anions $[Cu(^{15}N^{13}CS)_2]^-$ have been used [128]. As will be discussed below in Sect. 20.4.5, these studies revealed a genuine mass-isotope effect on T_c.

An "inverse" isotope effect on T_c has been observed for κ-$(ET)_2Cu(NCS)_2$ where T_c of deuterated samples κ-$(D_8\text{-}ET)_2Cu(NCS)_2$ was found to be higher than that of hydrogenated salts, see [5]. This effect has been confirmed and quantified by the above mentioned study where particular care has been taken to guarantee otherwise comparable quality of both the labeled and unlabeled crystals [128]. The physical reason for the inverse isotope effect is still unclear. A geometric H-D isotope effect has also been found for two other $(ET)_2X$ compounds κ_L-$(ET)_2Ag(CF_3)_4$(solvent) and β''-$(ET)_2SF_5CH_2CF_2SO_3$ having different packing motifs and anion structures. Although the T_c values vary considerably among these salts ranging from 2.9 K to 9.2 K the investigations reveal an almost identical "universal" shift of T_c of $\Delta T_c = +(0.26 \pm 0.06)$ K [241,242]. Taking into account the results of thermal expansion and X-ray studies of the lattice parameters [151,243], it has been proposed that the inverse isotope effect is not directly related to the pairing mechanism. Instead it has been attributed to a geometrical isotope effect, i.e. changes in the internal chemical pressure: provided that the interlayer lattice parameter is identical for both compounds, the effectively shorter C–D bond of the deuterated salt [244] corresponds to a higher chemical pressure perpendicular to the planes for the hydrogenated salt. The negative values of $\partial T_c/\partial p_\perp$ then result in a higher T_c for the deuterated compound [241,242], see also [187].

An alternative explanation has been proposed recently by Biggs et al. based on their measurements of the Shubnikov–de Haas effect focusing on pressure-induced changes of the Fermi-surface topology of deuterated and protonated κ-$(ET)_2Cu(NCS)_2$ [235]. It has been suggested that the superconducting mechanism is most sensitively influenced by the exact topology of the Fermi surface. Since the latter has been found to change more drastically with pressure in the deuterated salt, this effect might also be responsible for the inverse isotope effect [235]. In addition from recent millimeter-wave magnetoconductivity experiments it has been inferred that the quasi-one-dimensional FS sheets (see Fig. 20.7) are more corrugated in the deuterated salt (higher T_c) suggesting that the "nestability" of the FS may be important for T_c [245].

20.4.2 Superconducting Parameters

$(TMTSF)_2X$ Salts

As a consequence of the highly anisotropic electronic structure, strong directional dependences are also expected for the superconducting-state properties such as the lower and upper critical fields. Among the $(TMTSF)_2X$ salts, the latter have been extensively studied for the ambient-pressure superconductor X = ClO_4, see [5,33] and, more recently, also for pressurized $(TMTSF)_2PF_6$, see also Sect. 20.4.5 below. For $(TMTSF)_2ClO_4$ the Meissner and diamagnetic shielding effects have been examined for magnetic fields aligned along the three principal axes [246]. From these experiments the lower critical field values B_{c_1} (at 50 mK) have been determined to 0.2, 1 and 10 (in units of 10^{-4} T) along the a-axis, b-axis and c-axis, respectively. The thermodynamic critical field, as estimated from the condensation energy, amounts to $B_{c_{th}} = (44 \pm 2) \cdot 10^{-4}$ T [247]. Figure 20.33

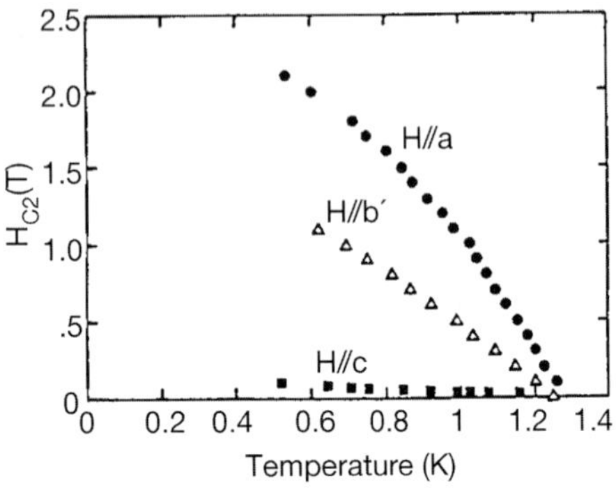

Fig. 20.33. Upper critical fields B_{c_2} determined from resistivity measurements as a function of temperature for $(TMTSF)_2ClO_4$ for the three principal crystal axes [248]

shows the temperature dependence of the upper critical fields, $B_{c_2}(T)$, for $(TMTSF)_2ClO_4$ as determined by early resistivity measurements [248]. The $B_{c_2}(0)$ values of 2.8 T, 2.1 T and 0.16 T for the a-axis, b-axis and c-axis, respectively, have been obtained from the data of Fig. 20.33 by an extrapolation to zero temperature. The so-derived value for fields aligned parallel to the a-axis, where B_{c_2} is the largest, is close to the Pauli limiting field B_P which is the critical value required to break spin-singlet (S = 0) Cooper pairs. For a non-interacting electron gas this is the case when the Zeeman energy just equals the condensation energy, i.e.

$$B_P = \Delta_0 \frac{1}{\sqrt{2}\mu_B} , \qquad (20.15)$$

where μ_B denotes the Bohr magneton [249,250]. Assuming a BCS ratio for the energy gap $\Delta_0 = 1.76\, k_B T_c$ yields B_P (in Tesla) $= 1.84 \times T_c$ (in K) $= 2.3$ T for $(TMTSF)_2ClO_4$. The fairly good coincidence with the experimentally derived critical field has been taken as an indication for a spin-singlet pairing state [248,251,252]. As will be discussed in more detail in Sect. 20.4.5, more recent resistivity measurements on the pressurized X = PF_6 salt down to lower temperatures revealed upper critical field curves which show an upward curvature for $T \rightarrow 0$ with no sign of saturation down to 0.1 K [253]. It has been argued in [253] that this unusual enhancement of B_{c_2} which exceeds the Pauli paramagnetic limit by a factor of 4 is strongly suggestive of a spin-triplet (S=1) pairing state.

Yet from the initial slopes of the upper critical field curves in Fig. 20.33, $B^{i'}_{c_2}$, the Ginzburg–Landau coherence lengths $\xi_i(0)$ can be derived using the following relation: $B^{i'}_{c_2} = \phi_0/(2\pi\xi_j\xi_k T_c)$, where i, j and k can be a, b and c [205,206]. The so-derived ξ_c value of about 20 Å being much smaller than the numbers for the a-axis and b-axis coherence lengths of 700 Å and 335 Å respectively, but comparable to the lattice parameter c = 13.5 Å indicates that superconductivity has, in fact, a more quasi-2D character. For the London penetration depth for B parallel to the a-axis, the axis of highest conductivity, a value of $\bar{\lambda} = 40\ \mu$m has been reported [254]. This number exceeds the GL coherence lengths by orders of magnitude indicating that the present system is an extreme type-II superconductor.

$(BEDT\text{-}TTF)_2X$ and $(BEDT\text{-}TSF)_2X$ Salts

Due to the strong effects of fluctuations in these superconductors of reduced dimensionality, an accurate determination of the upper critical fields is difficult and in many cases not free of ambiguities. This holds true in particular for resistivity measurements in finite fields aligned perpendicular to the highly conducting planes as phase fluctuations of the order parameter give rise to a resistive state which tends to descend far below the mean-field transition temperature. A more reliable way to determine the upper critical fields is provided by measuring thermodynamic properties and employing a fluctuation analysis as described in Sect. 20.4.1. The left panel of Fig. 20.34 shows B_{c_2} curves for κ-$(ET)_2Cu[N(CN)_2]Br$ as determined from dc-magnetization measurements [9,207] using (20.11), cf. Figs. 20.29 and 20.30. The $B_{c_2}(T)$ curve for B aligned parallel to the highly conducting planes as determined from resistivity measurements are shown in the right panel of Fig. 20.34 over an extended field range. For the layered superconductors with negligible in-plane anisotropy, the expression (20.9) can be used to determine the GL coherence lengths perpendicular and parallel to the conducting planes.

Table 20.2 compiles the B_{c_2} values together with other superconducting parameters for the above BEDT-TTF compounds. For comparison, the table

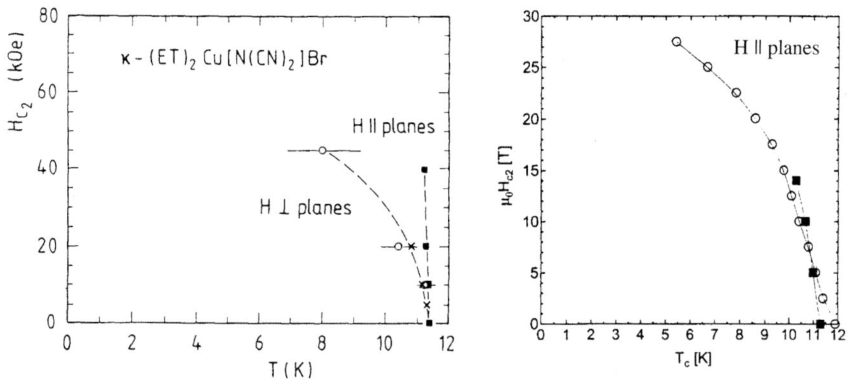

Fig. 20.34. Upper critical fields of κ-(ET)$_2$Cu[N(CN)$_2$]Br. *Left panel*: anisotropy of B_{c_2} as a function of temperature as determined from dc-magnetization measurements, taken from [207]. *Right panel*: $B_{c_2}(T)$ for fields aligned parallel to the conducting planes as determined from resistivity measurements using different criteria (*closed and open symbols*), reproduced from [255]

Table 20.2. Superconducting-state parameters of two representative κ-phase (ET)$_2$X salts X = Cu(NCS)$_2$ and Cu[N(CN)$_2$]Br as well as λ-(BETS)GaCl$_4$

	κ-(ET)$_2$Cu(NCS)$_2$	κ-(ET)$_2$Cu[N(CN)$_2$]Br	λ-(ET)$_2$GaCl$_4$
T_c	8.7 ~ 10.4	11.0 ~ 11.8	5 ~ 6
$B_{c_2}^{\perp}(0)$ (T) [13]	6	8 ~ 10	3
$B_{c_2}^{\parallel}(0)$ (T) [14]	30 ~ 35	> 30	12
$B_{c_1}^{\perp}(0)$ (mT) [15]	6.5	3	
$B_{c_1}^{\parallel}(0)$ (mT) [16]	0.2		
$B_{c_{th}}(0)$ (mT) [17]	54	65	
$\xi_\perp(0)$ (Å) [18]	5 ~ 9	5 ~ 12	9 ~ 14
$\xi_\parallel(0)$ (Å) [19]	53 ~ 74	28 ~ 64	143
$\xi_\parallel(0)$ (Å) [20]	74	60	105
$\lambda_\perp(0)$ (μm) [21,22]	40 ~ 200	38 ~ 133	
$\lambda_\parallel(0)$ (Å) [23]	5100 ~ 20000	6500 ~ 15000	1500
$\kappa_\parallel$ [24]	100 ~ 200	200 ~ 300	107

13 [9,256–258]
14 [255,258–261]
15 [39,262]
16 [39]
17 $B_{c_{th}}(0) = T_c \cdot \sqrt{\frac{\mu_0 \gamma}{2 V_{\mathrm{mol}}}}$
18 [9,39,187,207,256,258,263]
19 [9,39,207,258,263]
20 $\xi_\parallel(0) = \sqrt{\frac{\phi_0}{2\pi B_{c_2}^{\perp}}}$
21 $\lambda_\perp$ and $\lambda_\parallel$ denote the screening of supercurrents flowing perpendicular and parallel to the conducting planes, respectively, and *not* the direction of the magnetic field.
22 [263–265]
23 [256,263,265–269]
24 [9,256]

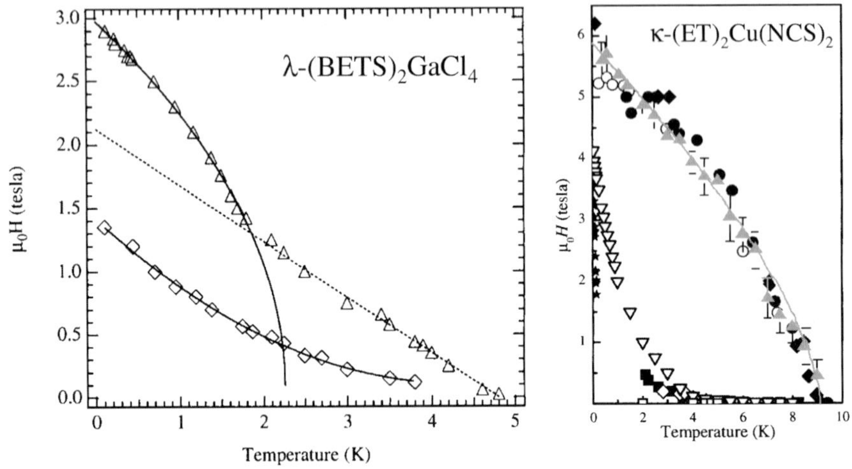

Fig. 20.35. *Left panel*: Upper critical field for $B \perp$ to the planes (*triangles*) of λ-(BETS)$GaCl_4$. *Upper solid curve* is $B_{c_2} \propto (T^* - T)^{1/2}$; the *dashed curve* $B_{c_2} \propto (T_c - T)$ (see text). *Diamonds* and the *lower solid curve* indicate the flux-lattice melting line (see also Sect. 20.4.3). *Right panel*: $B_{c_2}(T)$ for B $\perp$ to the planes of κ-$(ET)_2Cu(NCS)_2$ as determined from different experimental techniques, see [256]. The *solid curve* is $B_{c_2} \propto (T_c - T)^{2/3}$ (see text). Also shown are transition and crossover lines in the mixed state, cf. Figs. 20.36 and 20.37 in Sect. 20.4.3 below. Reproduced from [256]

also contains data for the BETS-based system λ-(BETS)$GaCl_4$. As indicated in the table, the transition temperatures reported in the literature show a considerably large variation depending on the method and criterion used to determine T_c. This may partly be related to the fact that the superconducting transition is usually found to be relatively broad. Even for high quality single crystals with an in-plane mean free path of typically $\sim$ 2000 Å [108], the transition can be broadened due to internal strain fields as a consequence of the extraordinarily large pressure dependence of T_c. In addition, the quasi-2D nature of the electronic structure gives rise to pronounced fluctuations which cause a rounding of the transition [10].

According to recent magnetoresistivity studies, the in-plane Fermi surface of λ-(BETS)$GaCl_4$ strongly resembles that of κ-$(ET)_2Cu(NCS)_2$ with the effective masses being almost identical for both compounds [256]. However, the interplane transfer integral of $t_\perp \approx 0.21$ meV for the BETS salt is about a factor of 5 larger [256], indicating that λ-(BETS)$GaCl_4$ is more three-dimensional. Figure 20.35 compares the magnetic field-temperature phase diagrams of the above compounds for fields aligned perpendicular to the planes. A dimensional crossover has been suggested to account for the unusual temperature dependence of $B_{c_2}^{\perp}$(T) observed for the BETS-based compound (left panel): B_{c_2}(T) shows a 3D-like linear behavior close to T_c which turns into a power-law dependence characteristic for a 2D superconductor with weakly coupled layers below some crossover temperature labeled as T^* [10,256]. In contrast, B_{c_2}(T) for the κ-$(ET)_2Cu(NCS)_2$ salt follows a $B_{c_2}(T) \propto (T - T_c)^{2/3}$ over the whole temperature range. The additional crossover and transition lines in the mixed state below the $B_{c_2}(T)$ curves indicated in Fig. 20.35 will be discussed in Sect. 20.4.3.

Important information on the spin state of the Cooper pairs can be gained by comparing the experimentally determined $B_{c_2}(T)$ for $T \to 0$ with the Pauli-limiting field, B_P, as defined in (20.15). Using this formula, which neglects any orbital effects, and assuming a weak-coupling BCS ratio for the gap, i.e. $\Delta_0 = 1.76\, k_B T_c$, B_P amounts to $\sim$ 18 T and $\sim$ 21 T for κ-$(ET)_2Cu(NCS)_2$ and κ-$(ET)_2Cu[N(CN)_2]Br$, respectively. Apparently, these numbers are significantly smaller than the $B_{c_2}^{\parallel}(0)$ values found experimentally and listed in Table 20.2. On the other hand, clear evidence for a spin-singlet pairing state has been inferred from Knight shift measurements on κ-$(ET)_2Cu[N(CN)_2]Br$ yielding a vanishingly small spin susceptibility at low temperatures [257]. These deviations might find a natural explanation by recalling that (20.15) is valid only in the weak-coupling limit. For a strong-coupling superconductor, on the other hand, the density of states is renormalized leading to a Pauli field which is enhanced by a factor of $(1 + \lambda)^{1/2}$ [205, 249], where λ denotes the interaction parameter, see (20.14). Clear evidence for a strong-coupling type of superconductivity has been found in specific heat experiments on the κ-$(ET)_2Cu[N(CN)_2]Br$ and κ-$(ET)_2Cu(NCS)_2$ salts [49,50,233,270], see Sect. 20.4.5. As an alternative mechanism to account for a $B_{c_2}^{\|}(0)$ value in excess of the Pauli field, a transition into a Fulde–Ferrell–Larkin–Ovchinnikov (FFLO) state [271,272] has been proposed [260,273] and discussed controversially, see e.g. [10,274]. In such a scenario, a superconductor with suitable materials parameter adopts a new state at sufficiently high fields where the order parameter is spatially modulated. A more detailed discussion on this issue will be given in Sect. 20.4.3 below.

Besides T_c and the upper critical fields, Table 20.2 contains further superconducting-state parameters such as the lower and the thermodynamic critical fields B_{c_1} and $B_{c_{th}}$, respectively, as well as the GL coherence lengths $\xi_{\|,\perp}$ and the London penetration depths $\lambda_{\|,\perp}$. B_{c_1} is usually determined by measuring the magnetization as a function of field under isothermal conditions where B_{c_1} corresponds to the field above which flux starts to penetrate the sample. Due to the smallness of B_{c_1}, the plate-like shapes of the crystals and the peculiar pinning properties of these layered superconductors, an accurate determination of B_{c_1} is difficult. A more reliable way to determine $B_{c_1}^{\perp}(0)$ has been proposed by Hagel et al. based on a model for thermally activated flux creep yielding $B_{c_1}^{\perp}(0) = (3 \pm 0.5)$ mT for κ-$(ET)_2Cu[N(CN)_2]Br$ [262]. The values for the thermodynamic critical fields, $B_{c_{th}}$, in Table 20.2 are estimated from specific heat results using $B_{c_{th}}(0) = T_c \cdot \sqrt{\mu_0 \gamma/(2 V_{mol})}$ where γ is the Sommerfeld coefficient. These values roughly agree with those calculated from $B_{c_{th}} = B_{c_2}/(\kappa\sqrt{2}) = B_{c_1}\sqrt{2}\kappa/\ln\kappa$, where $\kappa = \lambda/\xi$ is the GL parameter. The large numbers of κ reflect the extreme type-II character of these superconductors.

Also listed in Table 20.2 are values of the magnetic penetration depth λ, the characteristic length over which magnetic fields are attenuated in the superconductor. While the absolute values for λ derived from various experimental techniques are in fair agreement within a factor $4 \sim 5$, no consensus has yet been achieved concerning its temperature dependence, see Sect. 20.4.5 below.

20.4.3 Mixed State

The peculiar material parameters of the present organic superconductors such as the pronounced anisotropy of the electronic states, the small coherence lengths and large magnetic penetration depths give rise to highly anomalous mixed-state properties and a rich B-T phase diagram in the superconducting state. Exploring the unusual features of extreme type-II layered superconductors continues to be a subject of considerable interest owing to the variety of exciting phenomena that has been found in these materials, see e.g. [275]. Among them is a first-order melting transition of the Abrikosov vortex lattice into a vortex-liquid phase [276,277] not known for usual 3D superconductors. One of the striking early observations related to the anomalous mixed-state properties was the appearance of a so-called irreversibility line, B_{irr}, which separates the B-T plane into an extended range $B_{irr} < B < B_{c_2}$ where the magnetization is entirely reversible from a magnetically irreversible state at $B < B_{irr}$. When a layered superconductor is exposed to a magnetic field $B > B_{c_1}$ aligned perpendicular to the planes, the confinement of the screening currents to the superconducting layers results in a segmentation of the flux lines into two-dimensional objects, the so-called vortex pancakes [278]. The coupling between vortex segments of adjacent layers is provided by their magnetic interaction and the Josephson coupling. The latter interaction drives tunneling currents when two vortex segments are displaced relative to each other. As a result of both effects, the vortex pancakes tend to align thereby forming extended stacks.

A quite different situation arises for fields aligned parallel to the layers. In the limiting case of a quasi-2D superconductor characterized by a cross-plane coherence length $\xi_\perp$ being smaller than the interlayer distance s, the vortex cores slip into the insulating layers where the superconducting order parameter is small. Such a Josephson vortex has an elliptically deformed cross section and lacks a normal core. Since the Josephson screening currents across the insulating layers are very weak, the material is almost transparent for fields parallel to the planes corresponding to a large value for the upper critical field $B_{c_2}^{\parallel}$, cf. Table 20.2. Below we will discuss some of the anomalous mixed-state properties such as the vortex-lattice melting transition, the irreversibility line, the lock-in transition as well as the possible realization of an anomalous high-field state.

Muon spin rotation measurements on the κ-$(ET)_2Cu(NCS)_2$ salt have shown that a 3D flux-line lattice exists only at very low fields $B < 7$ mT [279].[25] Using a decoration technique, Vinnikov et al. succeeded in imaging the vortex lattice in the low-field range [280]. Upon increasing the field to above some dimensional crossover field, B_{2D}, the vortex lattices of adjacent layers become effectively decoupled. Theory predicts that the crossover field is related to the anisotropy parameter Γ and the interlayer distance s by $B_{2D} = \phi_0/(\Gamma^2 s^2)$ [275], which for the κ-$(ET)_2Cu(NCS)_2$ salt results in $B_{2D} = 7.3 \sim 30$ mT [279,281]. An anomalous second peak in magnetization curves indicating a redistribution of pancake vortices at more suitable pinning centers, has been associated with B_{2D} [281,282]. According to measurements of the interlayer Josephson-plasma resonance [283], a long-range quasi-2D order among vortices within the individual layers characterizes the state above B_{2D} and persists up to the irreversibility line. In this region of the B-T plane the pancake vortices of adjacent layers become effectively decoupled leading to a pinned quasi-2D vortex solid in each layer with no correlations between the locations of vortices among the layers [284]. A somewhat different point of view is taken in [282,286], where B_{2D} marks the crossover from a 3D flux line lattice below B_{2D} to a state with less strong interlayer coupling on a long range scale above, where a coupling between the layers is, to some extent, still present.

Another striking property common to the present quasi-2D superconductors is an extended vortex-liquid phase above B_{irr}. Here the magnetization behaves entirely reversible upon increasing and decreasing the magnetic field, indicating that in this range flux pinning is ineffective. The abrupt onset of magnetic hysteresis at $B \leq B_{irr}$ indicates a drastic increase in the pinning capability. The temperature dependence of the irreversibility field has been studied in detail for κ-$(ET)_2Cu(NCS)_2$ and κ-$(ET)_2Cu[N(CN)_2]Br$ using a variety of techniques including ac-susceptibility [262], dc-magnetization [207], magnetic torque [281, 287] or Josephson-plasma-resonance experiments [283]. Figure 20.36 shows on a linear scale the irreversibility line in the B-T phase diagram of κ-$(ET)_2Cu(NCS)_2$ deduced from torque-magnetometry measurements in fields perpendicular to the planes [287]. As demonstrated in the left panel of Fig. 20.36, the irreversibility field at the lowest temperatures lies well below the upper critical field. Thus quantum fluctuations of the vortices as opposed to thermally driven motions are responsible for the vortex liquid state in this region of the phase diagram [287]. The crossover from quantum to thermal fluctuations manifests itself in the temperature dependence of $B_{irr}(T)$. Below ~ 1 K where quantum fluctuations are predominant, $B_{irr}(T)$ varies linearly with temperature [287], whereas in the thermal fluctuation regime an exponential behavior $B_{irr} = B_0 \exp(-AT/T_c)$ has been observed above the dimensional crossover field B_{2D} in contrast to a power-law behavior in the 3D vortex line lattice region below B_{2D} [207,281]. A similar behavior has been observed also for κ-$(ET)_2Cu[N(CN)_2]Br$ where the crossover in the temperature dependence of the irreversibility line has been interpreted as a crossover from 2D to 3D pinning [262].

Indications for a first-order phase transition associated with a melting and/or a decoupling of the quasi-2D vortex lattice near the irreversibility line, similar to what has been found in some

[25] Here the sample was arranged so that the superconducting planes enclose an angle of 45° in respect to the magnetic field.

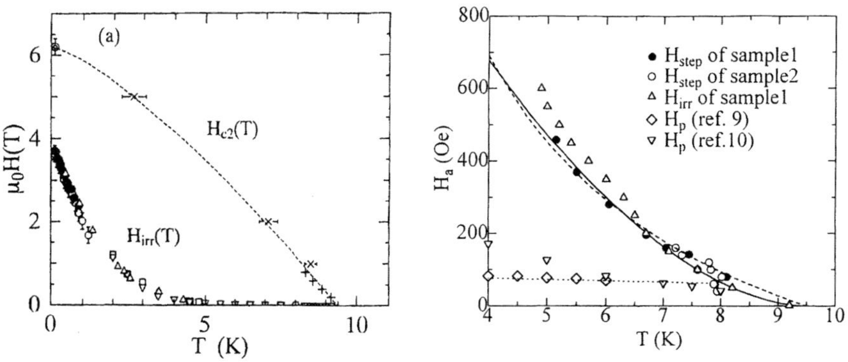

Fig. 20.36. Details of the B–T phase diagram for fields perpendicular to the conducting planes of κ-(ET)$_2$Cu(NCS)$_2$. *Left panel*: temperature dependence of the irreversibility field and the upper critical field taken from [287]. *Right panel*: first-order transition line (*open and closed circles*) determined by a step in the local induction. *Solid and dashed curves* represent theoretical predictions for the melting and decoupling transitions, respectively, see text. *Open diamonds and down triangles* indicate an anomalous second peak in magnetization associated with the dimensional-crossover field B_{2D} of the vortex system, taken from [282]

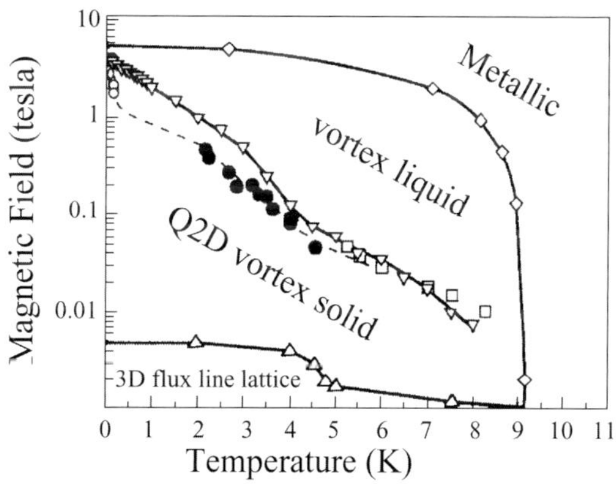

Fig. 20.37. Mixed-state B-T phase diagram for κ-(ET)$_2$Cu(NCS)$_2$ for B $\perp$ planes in a semi-logarithmic representation. B_{irr} is indicated by *down triangles*, the dimensional-crossover field by *up triangles*. *Open squares* denote a first order decoupling and/or melting transition. *Closed and open circles* refer to thermal melting or depinning and quantum melting transition from a quasi-2D vortex lattice to a liquid phase, respectively, taken from [284]

high-T_c cuprates [276,277], have been reported for κ-(ET)$_2$Cu(NCS)$_2$ [282,288]. This refers to results from micro Hall-probe experiments where the local induction as a function of temperature at constant fields shows step-like changes. As shown in the right panel of Fig. 20.36 the first-order transition line can be fitted equally well by a melting or a decoupling transition. Steps in the equilibrium and local magnetization have been observed also in SQUID and micro-hall-probe measurements, respectively, for κ-(ET)$_2$Cu[N(CN)$_2$]Br [286,289]. In addition, Josephson-plasma-resonance experiments revealed evidence that at the first-order transition, the interlayer coherence becomes lost indicating that the melting and the decoupling occur simultaneously [286].

Figure 20.37 shows in a semi-logarithmic plot the B-T phase diagram for κ-(ET)$_2$Cu(NCS)$_2$ as proposed by [284]. Note that here the crossover field B_{2D} separating the 3D flux-line lattice from the quasi-2D vortex solid is temperature dependent in contrast to the results shown in Fig. 20.36. A first-order melting transition driven by quantum fluctuations has been inferred from torque-magnetization measurements at very low temperatures [284].

An interesting situation arises when the magnetic field is aligned parallel to the planes enabling the vortices to slip in between the superconducting layers where the order parameter is small. For small tilt angles of the field in respect to the exact alignment, the vortex lattice can gain energy by remaining in this parallel "lock-in" configuration. This new state remains stable until the perpendicular field component, $B_\perp$, exceeds a threshold field at which the energy required to expel $B_\perp$ exceeds that associated with the creation of normal cores in the layers. Evidence for coreless Josephson vortices parallel to the superconducting layers and a lock-in state has been reported from ac-susceptibility measurements for κ-(ET)$_2$Cu(NCS)$_2$ [264] and from torque magnetometry on κ-(ET)$_2$Cu(NCS)$_2$ and κ-(ET)$_2$Cu[N(CN)$_2$]Br [209,290], see also [9].

Further interest in the behavior of the present quasi-2D organic superconductors in fields precisely aligned parallel to the planes arose from the proposal that these systems are possible candidates for a Fulde–Ferrell–Larkin–Ovchinnikov (FFLO) state [291]. Under suitable conditions, a spin-singlet superconductor can reduce the pair-breaking effect of a magnetic field by adopting a spatially modulated order-parameter along the field direction [271,272]. The wavelength of the modulation is of the order of the coherence length which results in a periodic array of nodal planes perpendicular to the vortices [271,272]. In the case of an anisotropic superconductor, calculations show that the FFLO state might lead to an enhancement of the upper critical field to between 1.5 and 2.5 times the Pauli paramagnetic limit [291,292].

By studying the magnetic behavior and the resistivity of κ-(ET)$_2$Cu(NCS)$_2$ in high magnetic fields employing a tuned circuit differential susceptometer, changes in the rigidity of the vortex arrangement at certain fields B_L *within* the superconducting state have been found for fields precisely aligned parallel to the planes [260,273]. These effects have been interpreted as the manifestation of a phase transition from the superconducting mixed state into an FFLO state.

According to [271,272], see also [294,295], the stabilization of a FFLO state requires: (i) a large electronic mean free path $l > \xi_0$, i.e. clean limit, (ii) a Pauli limiting that predominates the orbital pair-breaking effect, (iii) a Zeeman energy that overcompensates the loss of superconducting condensation energy and (iv) a short coherence length (or a large GL parameter $\kappa = \lambda/\xi$). As already discussed in Sect. 20.4.2, the criteria (i) and (iv) are met for the present κ-(ET)$_2$X salts. Condition (ii) is fulfilled since $H_P = \frac{1}{2}H_c(\pi\chi_{\rm Spin})^{-1/2} \approx 17$ T is considerably smaller than the orbital critical field as derived from (20.9) taking the values from Table 20.2. To check for condition (iii) one has to compare the Zeemann energy with the condensation energy. Using $\chi_{\rm Spin} = 4 \cdot 10^{-4}$ emu/mol for κ-(ET)$_2$Cu(NCS)$_2$ [85] and the upper critical field of 35 T yields a Zeeman-energy density of $E_Z = \frac{1}{2}\chi_{\rm Spin}H_{c_2}^2 = 5\,\mathrm{mJ/cm^3}$ (= $5 \cdot 10^4\,\mathrm{erg/cm^3}$), which exceeds the condensation-energy density of $E_c = H_c^2/(8\pi) = \frac{1}{4}\cdot\gamma/V_{\rm mol}\cdot T_c^2 = 1\,\mathrm{mJ/cm^3}$ (= $10^4\,\mathrm{erg/cm^3}$) calculated by employing the experimentally determined Sommerfeld coefficient $\gamma = (23 \pm 1)\,\mathrm{mJ/mol\,K^2}$ for κ-(ET)$_2$Cu(NCS)$_2$ [49,50].

Figure 20.38 shows the temperature dependence of B_L and B_P, the latter serves as a rough estimate of B_{c_2} [260,273], see inset of Fig. 20.38. Comparing the results with theoretical calculations derived for a generic quasi-2D metal [293] (solid and dotted

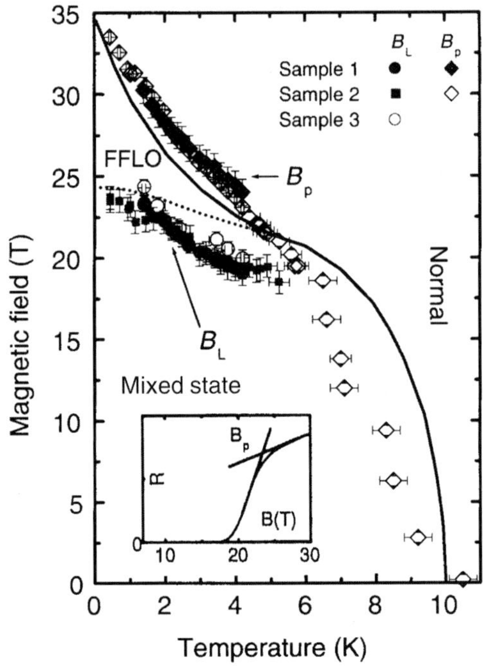

Fig. 20.38. Magnetic field-temperature phase diagram of κ-(ET)$_2$Cu(NCS)$_2$ for B aligned parallel to the planes compared with the theoretical FFLO phase diagram discussed in [293]. B_P as defined in the inset serves as a rough estimate for B_{c_2}. B_L indicates changes in the rigidity of the vortex arrangement within the superconducting state. Reproduced from [10,260]

lines in Fig. 20.38) using the parameter $B_{c_2}(0) = 35\,\mathrm{T}$ yielded a fairly good agreement with the predictions of the FFLO model. In particular, the temperature T^* below which the new state is stabilized was found to meet the theoretical predictions of $T^* = 0.56\,T_c$. However, recent magnetic torque measurements failed to detect any indication for such a transition [284]. Also, in studies of the critical field under pressure using the tunnel diode oscillator technique, no evidence for a FFLO phase has been seen and the data suggest that κ-$(ET)_2Cu(NCS)_2$ is always Pauli limited [285].

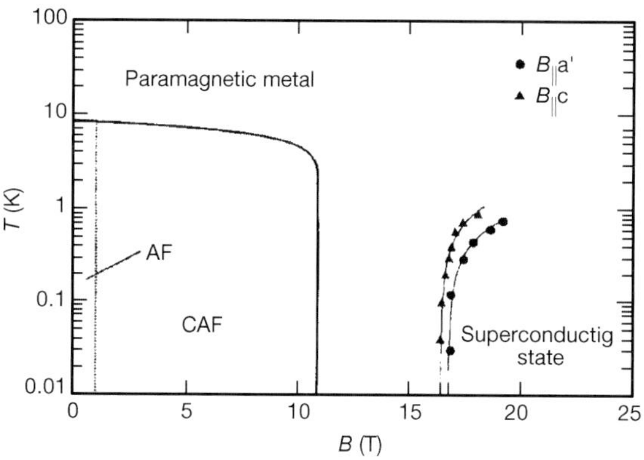

Fig. 20.39. Temperature vs magnetic field phase diagram for λ-$(BETS)_2FeCl_4$ with the magnetic field aligned parallel to the conducting planes. AF and CAF indicate the antiferromagnetic and canted antiferromagnetic phase, respectively (taken from [296])

20.4.4 Magnetic-Field-Induced Superconductivity

The λ-$(BETS)_2FeCl_4$ systems has attracted strong interest recently owing to the intriguing behavior found when the system is exposed to a magnetic field [296]. With increasing field aligned parallel to the planes, the low-temperature antiferromagnetic insulating state becomes suppressed and vanishes above about 10 T. At higher fields the paramagnetic metallic state which governs the $B = 0\,\mathrm{T}$ behavior above the Néel temperature of about 10 K is restored. For fields aligned exactly parallel to the highly conducting planes, a further increase to above 17 T has been found to induce a superconducting state at low temperatures, see Fig. 20.39. For this field configuration, the orbital effect is strongly suppressed so that superconductivity is limited only by the Zeeman effect, i.e. the Pauli limit, resulting in a high upper critical field. The field-induced superconducting state which is stable up to 41 T has been attributed to the Jaccarino–Peter effect, where the external magnetic field compensates the exchange field of the aligned Fe^{3+} moments [297]. The field-induced superconductivity is suppressed if the magnetic field is tilted from the conducting plane.

In the concentration range $0.35 \leq x \leq 0.5$ of the series λ-$(BETS)_2Fe_xGa_{1-x}Cl_4$ (cf. Fig. 20.27 in Sect. 20.3.5), where superconductivity shares a common phase boundary with an antiferromagnetically ordered insulating state, a field-induced afm insulator-to-superconductor transition has been observed [298]. It has been suggested that the pairing interaction arises from the magnetic fluctuations through the paramagnetic Fe moments [296].

20.4.5 The Superconducting State: Pairing Mechanism and Order-Parameter Symmetry

Understanding the nature of superconductivity in all its variants continues to pose a challenging problem of enormous complexity. Elementary superconductors like Al or Zn are well described by the BCS model [299] which envisages pairing between quasiparticles in a relative zero angular momentum (L = 0) and spin-singlet (S = 0) state, with an attractive pairing interaction mediated by the exchange of virtual phonons. In the ground state, all electron pairs (Cooper pairs) are in the same quantum-mechanical state described by a single macroscopic wavefunction. Excitations, i.e. the creation of unpaired quasiparticles, require an energy in excess of $2\,\Delta$, where Δ is the energy gap, known as the order parameter. In the BCS model the gap amplitude $\Delta(\boldsymbol{k})$ is uniform over the whole Fermi surface. In addition, the following quantitative relations are implied in the BCS model: a universal ratio $\Delta_0/k_BT_c = 1.76$, where Δ_0 is the energy gap at zero temperature, and $\Delta C/\gamma T_c = 1.43$ with ΔC being the jump height in the specific heat at T_c and γ the Sommerfeld coefficient.

In recent decades, novel classes of superconductors have been discovered in complex materials such as heavy-fermion metals [300], cuprates [301], ruthenates [302] and the present organic materials [1]. These systems have properties which are markedly different from those of the simple elements Al or Zn. The differences manifest themselves particularly clearly in the role of electron–electron interactions, which are strong in the new superconductors but of no relevance in the simple metals. The question at hand is whether superconductivity in the new materials is of a fundamentally different type to what is described in the BCS model and proved valid to a good approximation in so many superconductors. As for the present organic materials this question covers the following aspects

(i) Are the superconducting carriers "BCS-type" pairs of electrons that form below T_c (Cooper-pairs)? Or do tightly bound electron pairs preexist already at higher temperatures $T > T_c$ which undergo a Bose–Einstein condensation?
(ii) Is the relevant pairing mechanism different from a conventional phononic mechanism? The exchange of antiferromagnetic spin fluctuations as one of the models discussed for the high-T_c cuprates and also for the present organic materials is an example for an unconventional pairing interaction.
(iii) Is the symmetry of the order parameter lower than that of the Fermi surface? It is customary to refer to an unconventional superconducting state as one where the order parameter breaks at least one of the crystal symmetries.
(iv) Is the spin part of the superconducting pair wavefunction an antisymmetric singlet or a symmetric triplet state?

Although a local pairing scheme as distinguished from the BCS-type pairing has been discussed also for the present organic systems [303,304], the bulk of experimental observations do not support such a scenario. Rather the formation of Cooper-pairs below T_c implying an energy gap Δ with a maximum at $T = 0$ that vanishes for $T \geq T_c$ has been widely established, see e.g. [9].

The discussion of the above issues (ii)–(iv) in the context of the organic superconductors is complicated by the fact that even for the most fundamental properties contradictory experimental evidences exist. Since some of these discrepancies might simply reflect extrinsic effects as a consequence of an incomplete sample characterization, we shall therefore concentrate the discussion on those materials which are best characterized and where most data are available. These are the κ-phase $(ET)_2X$ and the $(TMTSF)_2X$ salts.

On the Nature of the Superconducting State

The central question for the present organic superconductors concerns the nature of the pairing interaction. The excitonic mechanism proposed by Little [3, 305] for certain quasi-1D organic polymers was of great importance historically in giving the initial impetus to the development of the field. From today's perspective, however, it is fair to say that the search for materials with suitable chemical and physical properties for such a mechanism to work has been unsuccessful so far.

In the discussion of the transport and optical properties in Sect. 20.3.2 it became clear that there is a substantial coupling of the charge carriers to both intramolecular as well as intermolecular phonons. Consequently, some researchers in the field believe in a conventional electron–phonon coupling mechanism. On the other hand, the fact that for the present materials, as for the cuprates and heavy-fermion systems, the superconducting region in the phase diagram lies next to a magnetically ordered state suggests that magnetic interactions are involved in the pairing mechanism.

Models Considering the Role of Phonons in the Pairing Mechanism

One example for a conventional electron–phonon pairing scenario has been discussed by Yamaji [306]. In this model he considers an attractive interaction mediated by several high-frequency internal molecular vibrations in addition to one low-frequency intermolecular phonon.

A more general account for the complex role of phonons for superconductivity has been given by Girlando et al. [59,307,308]. By calculating the lattice phonons for the κ-$(ET)_2I_3$ and β-$(ET)_2I_3$ superconductors using the quasi-harmonic-lattice-dynamics method and evaluating the coupling to the electrons,

λ, they succeeded in reproducing all available experimental data related to the phonon dynamics, for example the lattice specific heat. They showed that a lattice mode that couples particularly strongly to the electrons is one in which the relative displacement of the ET molecule is along the long axis of the molecule, i.e. perpendicular to the conducting planes. As an interesting side result of their study on κ-$(ET)_2I_3$, it is mentioned, that the coupling of acoustic phonons is very anisotropic and likely to cause gap anisotropies, though of a conventional type. In addition, it has been shown that the lattice phonons alone cannot account for the T_c values in these compounds. High-frequency intramolecular phonons modulating the on-site electronic energies have to be taken into account to reproduce the critical temperatures [307].

Recently, a distinct kind of phonon-mediated pairing has been suggested for the $(ET)_2X$ salts [309]. It is based on the idea that in a system in which Coulomb correlations are screened to be short range, i.e. Hubbard type, the electron–phonon scattering is dominated by forward processes. This results in an effective small-q pairing potential. Subsequent self-consistent solutions of the BCS gap equation using a band structure based on the effective-dimer approximation, lead to a gap structure where d-wave and anisotropic s-wave states are nearly degenerate. Furthermore it has been argued that the conflicting experimental results about the gap symmetry may originate in the decorrelation of superconductivity on various parts of the Fermi surface (a consequence of small-q dominated pairing) and the near degeneracy of s-wave and d-wave superconducting gap states [309].

Models That Consider a Magnetic Interaction

On the other hand, the rich phase diagram and the anomalous properties of the metallic state of these materials may suggest that the key elements dominating the physical behavior are the layered structure and the strong interactions between the electrons [4]. As a consequence, some researchers even resign from considering any coupling to the phonons and instead consider mechanisms which are solely based on two-dimensional electronic interactions. Most of these proposals deal with a spin-fluctuation-mediated pairing mechanism. The latter is motivated by the close proximity of the superconducting region in the phase diagram to an antiferromagnetic insulating state which, in analogy to the high-T_c cuprates, suggests that both phenomena are closely connected to each other [4,238,310].

An approach in this direction has been proposed by Kino and Fukuyama [238] who studied the effects of on-site Coulomb interaction and dimer structure in a strictly two-dimensional system within the Hartree–Fock approximation, see also [311]. In their picture, the antiferromagnetic insulating state of κ-$(ET)_2X$ is a Mott insulator. The Mott–Hubbard scenario for the present organic superconductors implies a half-filled conduction band, together with strong electron correlations. Because of the approximate square-lattice configuration of the dimers the authors expect a similar spin-fluctuation mediated superconductivity with probably $d_{x^2-y^2}$ symmetry as in the cuprates [238].

Such a possibility has been studied in detail by Schmalian [310]. Using a two-band Hubbard model to describe the antibonding orbitals on the ET dimer he succeeded in creating a superconducting state with $T_c \simeq 10$ K mediated by short-ranged antiferromagnetic spin fluctuations. It has been argued that despite the frustrating interactions and in-plane anisotropies which distinguish the organic materials from the high-T_c cuprates, the origin of superconductivity is very similar for both material classes [310].

A spin-fluctuation-based superconductivity similar to that of the cuprates has been claimed also by Kondo and Moriya [312–314]. They investigated the properties of a half-filled Hubbard model in a fluctuation exchange approximation (FLEX) with a right-angle isosceles triangular lattice with transfer matrices $-\tau'$ and $-\tau$. They revealed an energy gap of $d_{x^2-y^2}$ symmetry which upon cooling grows much faster compared to that expected in the BCS model. In addition they showed, that the appearance of d-wave superconductivity near an antiferromagnetic instability requires a suitable electronic structure, i.e. $\tau'/\tau > 0.3$ [239,312,313].

A spin-fluctuation mediated d-wave superconducting state has been found also by several other approaches including FLEX, perturbation theory or quantum Monte Carlo simulation applied to κ-$(ET)_2X$ [315–318] or the quasi-one-dimensional $(TM)_2X$ salts [317,319]. In [320] and [319] an explanation for the pseudogap behavior at elevated temperatures T^* has been proposed in terms of strong antiferromagnetic spin-fluctuations.

While the starting point of the above models is in the limit of strong correlations, i.e. near the Mott-insulating state, a somewhat different viewpoint is taken in the work by Louati et al. [321]. These authors studied the effect of spin fluctuations in a two-dimensional model in the weak correlation regime by varying the bandwidth and the nesting properties of the Fermi surface. They argued that spin fluctuations are enhanced by the good nesting properties which may account for the anomalous NMR relaxation rate observed at temperatures T^* above T_c in the κ-$(ET)_2Cu[N(CN)_2]Br$ salt. Furthermore they found that spin fluctuations can induce a superconducting coupling with d-wave symmetry that lies next to a spin-density-wave instability [321].

The above models address systems at half filling, which is realized in the dimerized κ-phase $(ET)_2X$ and $(TMTSF)_2X$ salts, suggesting spin-fluctuation-mediated superconductivity with a $d_{x^2-y^2}$ symmetry. A different scenario has been proposed for the θ and β'' structures [322]. Here the ET molecules are not dimerized which results in a quarter-filled hole band. In this case, a nearby insulating phase is believed to be due to a charge ordering, driven by strong inter-site Coulomb correlations. Applying slave-boson theory to an extended Hubbard model at quarter filling, superconductivity mediated by charge fluctuations has been found. This results in a d_{xy} symmetry of the superconducting state [322] as opposed to the $d_{x^2-y^2}$ symmetry for the above spin-fluctuation mechanism.

Experiments Probing the Superconducting State

On the experimental side, the determination of the actual pairing mechanism is a most difficult task as there is no decisive probe to pin down the relevant pairing interaction. There are, however, some crucial experiments which may help to delineate the various possibilities. Investigating the mass-isotope effect on T_c is such a key experiment. Another one is to study the phonon system to probe the role of electron–phonon interactions. If phonons are involved in the pairing interaction this would result in renormalization effects in the temperature dependences of the phonon frequencies and linewidths upon cooling through T_c. Likewise, if a non-phononic mechanism is at work leading to an anisotropic gap with nodes along certain directions on the Fermi surface, a determination of the orientation of the gap zeroes by angular-dependent measurements can provide important information on the pairing mechanism.

For classical phonon-mediated superconductors the gap amplitude $\Delta(\boldsymbol{k})$ is assumed to be isotropic or at least to have an isotropic component combined with a $\boldsymbol{k}$-dependent part which obeys all symmetries of the crystal lattice. In contrast to such a conventional "finite-gap" state, the above mentioned electronic coupling schemes lead to a pairing state with higher angular momentum where $L = 2$ (d-wave) being the most favored one. In this case the amplitude of the Cooper-pair wave function vanishes at the origin of the relative coordinate which keeps the constituent quasiparticles of the Cooper pair apart. Therefore, $L \neq 0$ pairing states are good candidates for materials with strong on-site Coulomb repulsion. The gap-function of such an $L \neq 0$ state has a $\boldsymbol{k}$-dependence which is given by the spherical harmonics of the same angular momentum. For those states where the $\Delta(\boldsymbol{k})$ functions vanish at certain $\boldsymbol{k}$-vectors at the Fermi surface, the quasiparticle excitation spectrum at low energies is markedly different from that of an isotropic finite gap state. For the above d-wave order parameter $\Delta(\boldsymbol{k}) = \Delta_0 \left(\cos(k_x a) - \cos(k_y b)\right)$ the zero crossings along the diagonals correspond to line nodes at the Fermi surface. This should be reflected in all quantities that depend on the number of thermally excited quasiparticles such as specific heat, NMR relaxation rate, magnetic penetration depth, etc. in the form of simple power-law dependences at sufficiently low temperatures. In contrast, an exponentially weak T-dependence in these quantities characterizes an isotropic non-vanishing order parameter. Thus, careful measurements of the above thermal properties should, in principle, provide a handle on the order-parameter symmetry, or at least permit to discriminate whether gap zeroes exist or not. In this context, it should be noted, however, that the observation of a non-exponential or power-law T-dependence in one of these quantities does not necessarily imply a gap structure with nodes. As an example, we mention gapless superconductivity via pair breaking which destroys an exponential behavior [323, 324]. Likewise, a power-law T-dependence could be of conventional origin, as e.g. a T^3 dependence in the specific heat for $0.2\,T_c < T < T_c$ which has been attributed to strong-coupling effects [325,326].

To our knowledge, the only example where a non-phononic mechanism has been clearly identified is the heavy-fermion superconductor UPd_2Al_3. Here, the combination of tunnel spectroscopy [327] and neutron scattering experiments [328] has provided sound evidence for a magnetic pairing interaction, i.e. the exchange of magnetic excitons.

Although numerous experiments have been devoted to the issue of the order parameter symmetry for the organic materials, no consensus has yet been achieved. It is fair to say that for the present systems direct evidence for a non-phononic mechanism such as the one mentioned above does not exist. Also phase-sensitive experiments as those applied successfully to the high-T_c cuprates [329, 330] have not been performed so far for the organic materials.

In what follows we shall give a discussion of a selection of experimental results on the pairing mechanism and the symmetry of the order parameter. To begin with we shall focus on the $(ET)_2X$ salts.

Experiments on the Pairing Mechanism: Isotope Effect and Electron–Phonon Coupling

The effect of isotope substitution has been studied for various members of the $(ET)_2X$ superconductors. Isotopes have been substituted in the ET molecule by replacing ^1H by ^2D in the ethylene endgroups, by a partial exchange of ^{12}C by ^{13}C or ^{32}S by ^{34}S atoms in the inner skeleton of the molecule. In addition, systems have been studied where the acceptor molecule has been isotopically labeled. First experiments focused on the role of the electron-molecular-vibration (EMV) coupling by substituting ^{13}C for ^{12}C in the central double bond of the ET molecule. The large decrease of T_c of $\Delta T_c/T_c = -2.5\,\%$ found for the high-temperature variant (β_H) of β-$(ET)_2I_3$ by the Orsay group [331] could not be reproduced by a subsequent study where no systematic decrease of T_c could be detected [332]. Most intensive studies on the isotope effect have been carried out on the κ-$(ET)_2Cu(NCS)_2$ salt by the Argonne group [128]. Their investigations include isotope substitutions on both the BEDT-TTF molecules (all together seven differently labeled BEDT-TTF derivatives) as well as the anions. In each case, a batch of unlabeled samples has been synthesized under strictly parallel experimental conditions. These crystals serve as a reference for comparison. By sampling a large number of crystals, typically eight or more of both labeled and unlabeled material, a genuine mass-isotope effect on T_c has been found: upon replacing all eight sulfur atoms by ^{34}S and the outer-ring-carbon atoms of the $[(CH_2)_2]$ endgroups by ^{13}C, which corresponds to a 5 % increase of the ET molecule's mass, a shift of $\Delta T_c = -(0.12 \pm 0.08)$ K has been observed. Assuming a BCS-like mass-isotope effect $T_c \propto M^{-\alpha}$ with the whole ET molecule as the relevant mass entity M, this shift corresponds to $\alpha = 0.26 \pm 0.11$. This experiment provides clear evidence that the *inter*molecular (lattice) phonon modes are strongly involved in the pairing mechanism. On the other hand, the same group demonstrated the absence of a comparable isotope effect on T_c for κ-$(ET)_2Cu(NCS)_2$ and κ-$(ET)_2Cu[N(CN)_2]Br$ upon a partial substitution of the central C=C and also a simultaneous replacement of both the central and ring C=C atoms. The same holds true for a substitution of the eight sulfur atoms, see [128] and references therein. These results indicate that the *intra*molecular C=C and C–S bond-stretching vibrational modes of the ET molecule do not provide a substantial contribution to the Cooper pairing.

Further indications for a strong electron–phonon coupling have been inferred from measurements of the thermal conductivity on κ-$(ET)_2Cu(NCS)_2$

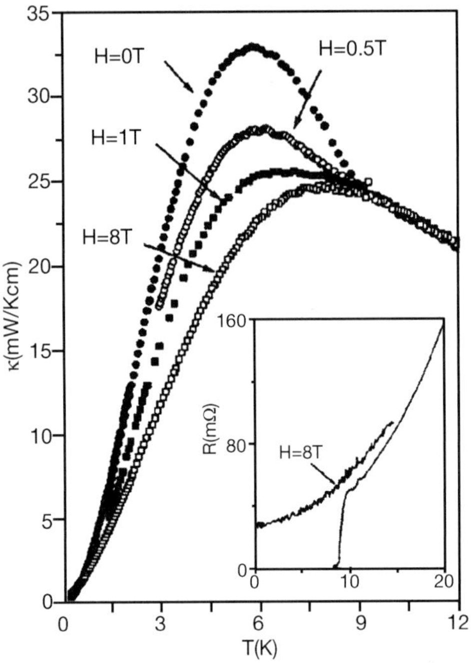

Fig. 20.40. Temperature dependence of the thermal conductivity of κ-$(ET)_2Cu(NCS)_2$ for B applied perpendicular to the planes. Inset: temperature dependence of the resistance, taken from [60]

and κ-$(ET)_2Cu[N(CN)_2]Br$ [50, 60, 61]. As shown in Fig. 20.40, the thermal conductivity $\kappa(T)$ of κ-$(ET)_2Cu(NCS)_2$ exhibits an upturn at the onset of superconductivity followed by a pronounced maximum just below T_c. It has been convincingly argued by the authors that the enhancement of $\kappa(T)$ in the superconducting state is a consequence of the condensation of electrons into Cooper pairs which strengthens the heat transport by freezing out the scattering of the main heat carriers, the phonons. By employing the Wiedemann–Franz law it has been found that just above T_c the electronic contribution amounts to only 5 % of the total thermal conductivity [60]. Figure 20.40 also shows $\kappa(T)$ data taken at varying fields applied perpendicular to the conducting planes. In the normal state, a magnetic field of 8 T does not affect the thermal conductivity within the resolution of the experiment but induces a sizeable decrease in the charge conductivity (see inset) which is an additional indication of lattice-dominated thermal conductivity in the vicinity of T_c.

Temperature-dependent Raman scattering studies of the phonon dynamics of κ-$(ET)_2Cu(NCS)_2$ and κ-$(ET)_2Cu[N(CN)_2]Br$ substantiate the strong coupling of the superconducting charge carriers to intermolecular phonons [54–59]. The observed anomalous temperature dependence of the low-frequency phonons around and below T_c were found to be consistent with an isotropic gap $2\Delta_0$ close to 2.8 meV [58]. From the reported frequency shifts the electron–phonon coupling constants λ_i have been calculated yielding a total coupling constant of $\lambda_{tot} = 0.97 \pm 0.11$ [58].

Superconductivity-Induced Phonon Renormalization

As a consequence of the interaction of charge carriers with the phonon system, the opening of a gap in the electronic density of states below T_c induces changes in the phonon frequencies and linewidths. These effects were first observed in the classical superconductors Nb_3Sn and Nb [333,334]. The results of these studies support the generally accepted picture that superconductivity in these materials is phonon-mediated. Inelastic neutron-scattering experiments have been performed on single crystals of κ-$(ET)_2Cu(NCS)_2$ on both hydrogenated and deuterated crystals [62,335]. Due to the extraordinarily large incoherent cross section of the protons, the study of κ-$(H_8$-$ET)_2Cu(NCS)_2$ allows for a selective investigation of those vibrational modes that involve the hydrogen atoms at the terminal ethylene groups. The analysis of measurements above and below T_c suggest a substantial coupling of these modes to the superconducting charge carriers [335]. Figure 20.41 shows the results of inelastic neutron-scattering experiments on deuterated κ-$(ET)_2Cu(NCS)_2$ carried out by Pintschovius et al. [62]. The data reveal a sudden increase of the frequencies of transverse acoustic phonons upon cooling through T_c. This phonon hardening was found to be most pronounced for the wave vector $\boldsymbol{q} = (-0.225, 0, 0.45)$ and a phonon energy 2.4 meV. As discussed by Zeyher and Zwicknagl [336], significant changes are expected only for those phonons whose energy $\hbar\omega$ is close to

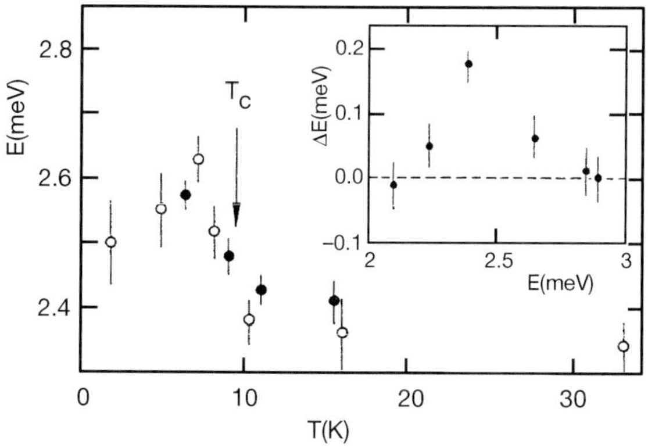

Fig. 20.41. Temperature dependence of the energy of the transverse acoustic phonon with wave vector q = (−0.225, 0, 0.45) derived from inelastic neutron scattering on two different single crystals (*open and closed circles*) of deuterated κ-$(ET)_2Cu(NCS)_2$. *Inset*: observed frequency shifts [$\Delta E = E(T < T_c) - E(T > T_c)$] of transverse acoustic phonons in the [$-\zeta$, 0, 2ζ] direction, reproduced from [62]

the gap value 2Δ with a softening (hardening) for $\hbar\omega < 2\Delta$ ($\hbar\omega > 2\Delta$) [336]. The above results thus imply $2\Delta \simeq 2.4$ meV, i.e. $2\Delta/k_BT_c \simeq 3.1$, which is close to the BCS ratio of 3.52. The salient feature of this study is that intermolecular modes strongly couple to the superconducting charge carriers and may thus provide a substantial contribution to the pairing interaction [62].

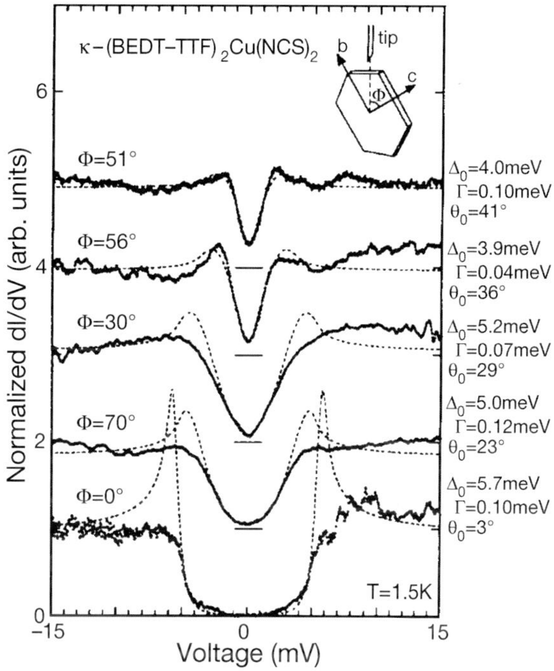

Fig. 20.42. dI/dV-V curves taken at 1.5 K on the lateral surfaces of κ-$(ET)_2Cu(NCS)_2$ single crystals. Data have been taken along various tunneling directions at different angles ϕ as defined in the *inset*. The *dashed line* represents the calculated curve based on the d-wave gap model with a $\mathbf{k}$ dependent tunneling, taken from [338]

On the Order-Parameter Symmetry in $(BEDT\text{-}TTF)_2X$: Measurements of the Gap Anisotropy

A new development in the investigation of the order-parameter symmetry is to probe the gap anisotropy directly by using orientational-dependent measurements. For the present organic superconductors, these techniques include mm-wave transmission [337], STM spectroscopy [338] and thermal conductivity [61] studies.

An *mm-wave magneto-optical technique* was used to determine the angle dependence of the high-frequency conductivity of κ-$(ET)_2Cu(NCS)_2$ [337, 339]. The results have been interpreted to support an anisotropic gap with "X shape", i.e. with nodes along the b-direction and c-direction [337], consistent with a $d_{x^2-y^2}$ symmetry of the order parameter as theoretically suggested by Schmalian [310]. However, these results have been critically commented upon by other groups [340,341].

The superconducting gap structure of the same compound has been investigated using *STM spectroscopy* by Arai et al. [338]. The tunneling curves observed on the bc-plane (parallel to the conducting layers) in the low-energy region could be well fitted by a d-wave gap model. The corresponding $2\Delta_0/k_BT_c$ ratio was found to be 6.7 which is smaller than a previously reported value of 9 [342] but substantially larger than the BCS value of 3.52. In addition, the in-plane gap anisotropy was investigated, see Fig. 20.42. The dI/dV-V curves observed on the lateral surfaces were found to be also consistent with a d-wave gap. For this configuration a very large $2\Delta_0/k_BT_c$ ratio of 8.7 ~ 12.9 has been obtained. The analysis of the angular dependence revealed that the direction of the line nodes of the gap is $\pi/4$ from the k_b and k_c-

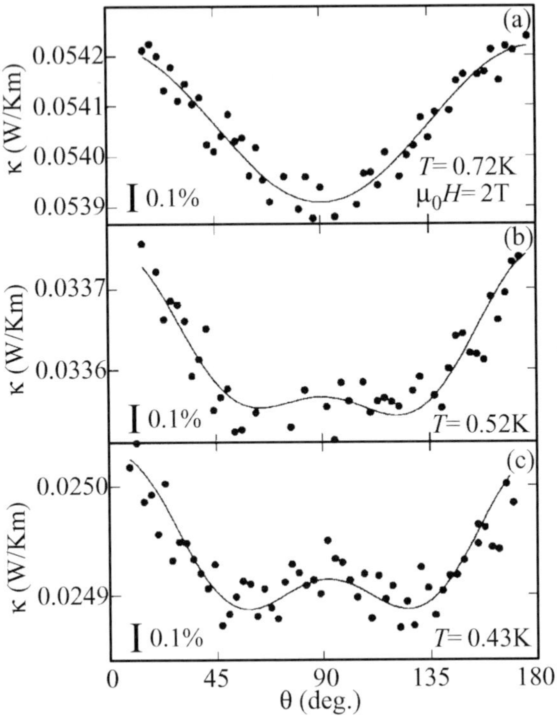

Fig. 20.43. Angular variation of $\kappa(B, \Theta)$ at 2 T for different temperatures where Θ denotes the angle between a rotating magnetic field in respect to the heat current flowing along the b-axis of the κ-$(ET)_2Cu(NCS)_2$ crystal. The *solid lines* represent the results of a fit using the function $\kappa(B, \Theta) = C_0 + C_{2\Theta} \cos 2\Theta + C_{4\Theta} \cos 4\Theta$, where C_0, $C_{2\Theta}$ and $C_{4\Theta}$ are constants, taken from [61]

axes, i.e. the gap has $d_{x^2-y^2}$ symmetry [338]. It has been noted that these orientations of the gap nodes are at variance with those inferred from the above mm-wave-transmission experiments.

The *thermal conductivity* has been used as another directional-dependent probe. When compared to STM measurements, for example, this quantity has the advantage that it is free of surface effects. The implications of the symmetry of gap zeroes on the thermal conductivity in the vortex state have been theoretically investigated by various authors, see e.g. Won and Maki [343]. Measurements have been performed for the κ-$(ET)_2Cu(NCS)_2$ salt in a magnetic field rotating within the 2D superconducting plane. Figure 20.43 shows the angular variation of κ at a fixed field of $B = 2$ T. Θ denotes the angle between the heat current flowing along the crystallographic b direction and the magnetic field, i.e. $\Theta = 0°$ for $B \parallel b$. The salient result of this study is the occurrence of a $\kappa(\theta)$ contribution with a fourfold symmetry, $\kappa_{4\theta}$, at low temperatures $T \leq 0.52\,K$ that adds to a predominant term with twofold symmetry. While the latter has been interpreted as being mainly phononic in origin, it is argued that the former is of a purely electronic nature and reflects the nodal gap structure [61]. Their analysis revealed that the gap zeroes are oriented along the directions rotated by 45° relative to the b and c-axes. It has been pointed out in [61] that this nodal structure is inconsistent with the theories based on antiferromagnetic spin fluctuation where the nodes are expected to be along the b and c-directions. Based on this observation Izawa et al. proposed a d_{xy} symmetry (referring to the magnetic Brillouin zone, see inset of Fig. 1 in [61]) which has been theoretically suggested for a charge-fluctuation scenario [322,344].

NMR Measurements

A more indirect information on the symmetry of the superconducting order parameter is provided by temperature-dependent measurements of quantities which depend on the quasiparticle excitation spectrum. In this context NMR experiments, i.e. measurements of the Knight shift K_S and the spin-lattice relaxation rate $(T_1)^{-1}$ are of particular interest. The ^{13}C spin-lattice relaxation rate and Knight shift of κ-$(ET)_2Cu[N(CN)_2]Br$ have been investigated by various groups [143, 257, 345] with similar results. In these experiments single crystalline material was used where both ^{12}C atoms in the central carbon double bond of the ET molecule had been replaced by ^{13}C. For the investigation of electronic properties, these nuclei are superior since their coupling to the π-electron system is much stronger than that of the protons in the ethylene endgroups of the ET molecules. The salient results of these studies are: (i) Knight-shift measurements performed in fields aligned parallel to the conducting planes reveal a spin susceptibility that tends to zero at low temperatures. Since any contributions from the pancake vortices have been excluded for this field configuration, the above re-

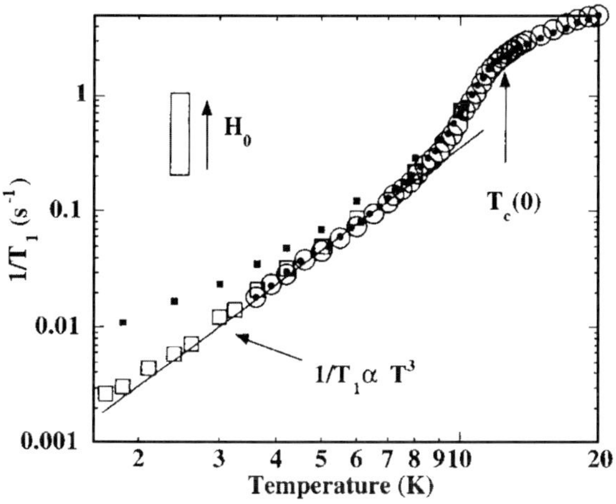

Fig. 20.44. Spin-lattice-relaxation rate $(T_1)^{-1}$ at fields of 5.6 T (*open circles*), 7.8 T (*black circles*) and 7 T (*open squares*) applied parallel to the conducting planes of κ-$(ET)_2Cu[N(CN)_2]Br$. *Black squares* correspond to a field of 7 T with a small misalignment, taken from [257]

sults have been taken as evidence for the spin-singlet character of the pairing state. (ii) The spin-lattice relaxation rate, $(T_1)^{-1}$, measured in the same parallel field configuration lacks a Hebel–Slichter peak and shows a power-law T^n behavior at low T, with n being close to 3, see Fig. 20.44. For the experimental conditions chosen, the authors ascribed the dominant source of relaxation to the quasiparticle excitations in the superconducting state. Consequently, the power-law temperature dependence in $(T_1)^{-1}$ has been interpreted as indicating an anisotropic pairing with nodes in the gap function [143,257,345].

Thermal Conductivity

Investigations of the thermal conductivity on quasi-2D organic superconductors at low temperatures have been performed first on κ-$(ET)_2Cu(NCS)_2$ [60] and more recently also on κ-$(ET)_2Cu[N(CN)_2]Br$ [50]. These studies reveal that the onset of superconductivity is associated with a sudden increase of $\kappa(T)$ which can be suppressed by a moderate magnetic field. The enhancement of $\kappa(T)$ at the onset of superconductivity has been attributed to a strengthening of the phonon heat transport by reducing the scattering due to the gap formation. Their argument is based on a quantitative analysis of the data employing the Wiedemann–Franz law. It showed that just above T_c the electronic contribution amounts to only 5 % of the total thermal conductivity. A lattice-dominated thermal conductivity around T_c is also consistent with the absence of a magnetic field dependence of κ in this temperature range [60]. As for the question of the gap symmetry, the data at low temperature have been interpreted as indicating an excitation spectrum with gap zeroes: an extrapolation of the data for κ-$(ET)_2Cu(NCS)_2$ to $T \to 0$ revealed a finite in-T linear term which has been attributed to a residual electronic contribution [60]. The latter is expected for an unconventional superconductor due to impurity scattering of residual quasiparticles [346,347].

On the other hand, recent thermal conductivity measurements on the κ-$(ET)_2Cu[N(CN)_2]Br$ salt showed that down to the lowest temperatures the phonon scattering length is strongly influenced by quasiparticle scattering [50] which renders the analysis of the data on κ-$(ET)_2Cu(NCS)_2$ [60] as being questionable.

Magnetic Penetration Depth

The quantity which has been studied most intensively for the κ-$(ET)_2X$ superconductors in connection with the question on the order-parameter symmetry is the magnetic penetration depth.

According to the London theory, the penetration depth λ_L in the limit $T \to 0$ is directly related to the density of superconducting electrons n_s via

$$\lambda_L(0) = \sqrt{\frac{m^* c^2}{4\pi n_s e^2}} , \qquad (20.16)$$

where m^* is the effective mass of the superconducting carriers [206]. Employing a two-fluid model with $n_e = n_s(T) + n_n(T)$ and n_e the density of conduction electrons, the temperature dependence of $\lambda_L(T)$ provides information on the normal-conducting component $n_n(T)$, i.e. the quasiparticle excitation spectrum. Since $n_s(T \to 0) = n_e$, the low-temperature value $\lambda(T \to 0)$ is a measure of the pair condensate, i.e. $\lambda_L^2(0) \propto m^*/n_s(0)$. For a conventional weak-coupling superconductor, the BCS theory predicts a mean-field temperature dependence of λ_L around T_c and an exponentially small variation at low temperatures $T \ll T_c$ [348]:

$$\lambda_L(T) \simeq \lambda(0)\left[1+\left(\frac{2\pi\Delta}{k_B T}\right)^{\frac{1}{2}} \exp\left(-\frac{\Delta}{k_B T}\right)\right]. \tag{20.17}$$

This holds true also for an anisotropic gap function without nodes, where for $k_B T \ll \Delta_{\min}$ the exponential low-temperature behavior is governed by the minimum value of the gap $\Delta_{\min}$. In contrast, an energy gap which vanishes along lines or at points at the Fermi surface will result in a power-law dependence of $\lambda_L(T)$ for $T \ll T_c$.

For the present materials the magnetic penetration depth has been determined by a variety of different techniques including ac-susceptibility [265,349], muon-spin relaxation [266, 350], dc-magnetization [267,268], surface impedance [263,269] and a related high-frequency technique using a tunnel diode oscillator [351]. The results of these studies, however, are quite inconsistent regarding both the temperature dependence as well as the absolute values of $\lambda_L(T \to 0)$ (see Table 20.2 in Sect. 20.4.2) and have led to quite different conclusions as to the symmetry of the superconducting order parameter.

Interestingly enough, these inconsistencies do not only involve results from different experimental techniques. Contradictory conclusions have been drawn also on the basis of seemingly identical experiments performed by different groups. This is the case for surface impedance studies where the penetration depth can be extracted from the complex conductivity. The latter is derived from the frequency shifts and variations of the quality factor of the resonator caused by the sample. While the surface impedance studies using a microwave perturbation technique on κ-$(ET)_2Cu(NCS)_2$ and κ-$(ET)_2Cu[N(CN)_2]Br$ by Klein et al. [352] and Dressel et al. [263] were found to be in good agreement with the BCS predictions, other studies by Achkir et al. [269] on κ-$(ET)_2Cu(NCS)_2$ revealed an in-T linear behavior at low temperatures indicative of an order parameter with zeroes on the Fermi surface. Deviations from an exponential temperature dependence of $\lambda_L(T)$ for the above two κ-$(ET)_2X$ compounds have been observed also in

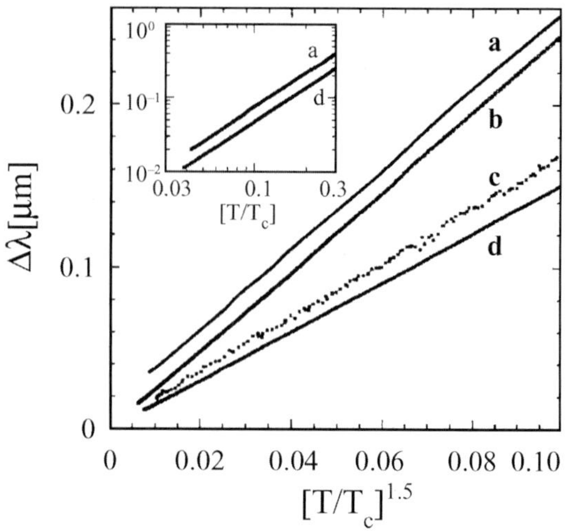

Fig. 20.45. Changes of the in-plane penetration depth $\Delta\lambda_{\parallel}(T)$ of κ-$(ET)_2Cu[N(CN)_2]Br$ [two samples (**a**), (**b**)] and κ-$(ET)_2Cu(NCS)_2$ [(c),(d)] plotted versus $(T/T_c)^{\frac{3}{2}}$. The data have been offset, taken from [351]

a more recent experiment using an rf tunnel-diode oscillator [351]. In contrast to the above measurements by Achkir et al., however, their data of the in-plane penetration depth rather follow a $T^{\frac{3}{2}}$ power law (see Fig. 20.45). As has been argued by the authors, the data would still be consistent with a quasi-linear variation of the superfluid density as expected for a d-wave superconductor with impurities or a small residual gap [351]. Alternatively, the authors point out that the exponent 3/2 may arise naturally in a model proposed for short-coherence-length superconductors exhibiting a pseudogap [353].

An inconsistency exists also for μSR experiments performed by different groups. Here λ_L can be determined by measuring the field inhomogeneities in the mixed state, i.e. the spatial variation of the local induction of the vortex lattice. This technique was first applied to κ-$(ET)_2Cu(NCS)_2$ by Harshman et al. [266] who could fit their data by a BCS temperature dependence. Subsequently, Le et al. [350] carried out similar experiments on the same system as well as on the κ-$(ET)_2Cu[N(CN)_2]Br$ salt and found an in-T linear variation for the in-plane penetration depth at low temperatures $\lambda_{\parallel}(T) \approx 1 + \alpha \cdot (T/T_c)$.

A power law temperature dependence of $\lambda_{\parallel}(T)$ consistent with d-wave superconductivity has been observed also by ac-susceptibility measurements performed by different groups [177,265,349]. The latter experiments as well as the surface impedance studies are operating at very small external magnetic fields ($B < B_{c_1}$) attempting to probe the Meissner state. Possible difficulties in these experiments that may arise from flux-pinning-related phenomena near the surface of the superconductor, i.e. an inhomogeneous superconducting state, have been discussed in [9].

An alternative way to determine the penetration depth is to make use of the reversible mixed-state magnetization, a peculiarity of these strongly anisotropic superconductors with short coherence lengths.

According to the London model, the field dependence of the magnetization is given by

$$\frac{\mathrm{d}M}{\mathrm{d}(\ln H)} = \frac{\phi_0}{32\pi^2\lambda_{\mathrm{eff}}^2}, \qquad (20.18)$$

where $\lambda_{\mathrm{eff}}^2 = \lambda_{\|}^2$ for $B \perp$ planes and $\lambda_{\mathrm{eff}}^2 = \lambda_{\|}\lambda_{\perp}$ for $B \parallel$ planes. In 3D superconductors, vortex pinning usually gives rise to an inhomogeneous distribution of the vortices in the mixed state and thus to an irreversible behavior of the magnetization upon increasing and decreasing the field. This may cause substantial uncertainties in determining the penetration depth from magnetization data. On the contrary, for quasi-2D superconductors with short coherence length the magnetization is entirely reversible over an extended field range, i.e. $B_{c_1} < B_{\mathrm{irr}} < B < B_{c_2}$ with B_{irr} being the temperature-dependent irreversibility line (see Sect. 20.4.3). For κ-$(ET)_2Cu(NCS)_2$ and κ-$(ET)_2Cu[N(CN)_2]Br$ a reversible magnetization has been observed over an extended range in the B–T plane which thus allows for a precise determination of the in-plane penetration depth [267,268]. The in-plane penetration depths $\lambda_{\|}(T)$ were determined, see Fig. 20.46, from the slopes $\mathrm{d}M/\mathrm{d}(\ln H)$ of the isotherms taken at different temperatures and using (20.18). The solid line represents a BCS fit [348] to the data. For both systems, the data reveal only a weak variation with temperature at low T consistent with an exponential temperature dependence as expected for a finite gap.

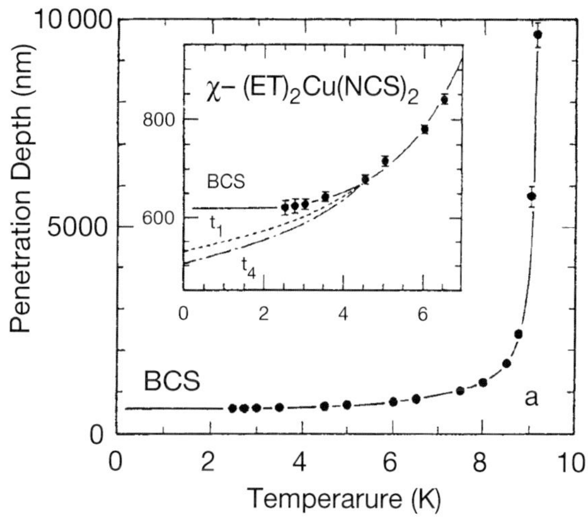

Fig. 20.46. In-plane penetration depth for single crystalline κ-$(ET)_2Cu(NCS)_2$. The *solid line* represents a BCS fit. The model calculations labelled t_1 and t_4 represent those anisotropic states proposed by Hasegawa et al. [354] and used by Le et al. to explain their μSR data [350] which have the weakest and strongest T dependences, respectively, taken from [267]

Specific Heat

The above variety of contradictory results on the magnetic penetration depth indicate an extraordinarily high sensitivity of this quantity to extrinsic effects such as disorder or pinning-related phenomena. As a possible source, we mention the disorder associated with the glass transition of the ethylene endgroups, see Sect. 20.3.4 and [177].

A quantity which is less sensitive to the above problems but can still provide fundamental information on the gap structure is the specific heat. In case this integral thermodynamic quantity were to find a low temperature electronic quasiparticle contribution, C_{es}, that varies exponentially weakly with the temperature, the existence of gap zeroes on the Fermi surface could be definitely ruled out. On the other hand, the observation of a non-exponential temperature dependence does not necessarily prove the existence of gap zeroes as this result might originate in extraneous contributions such as impurity phases, normal-conducting regions or pair-breaking

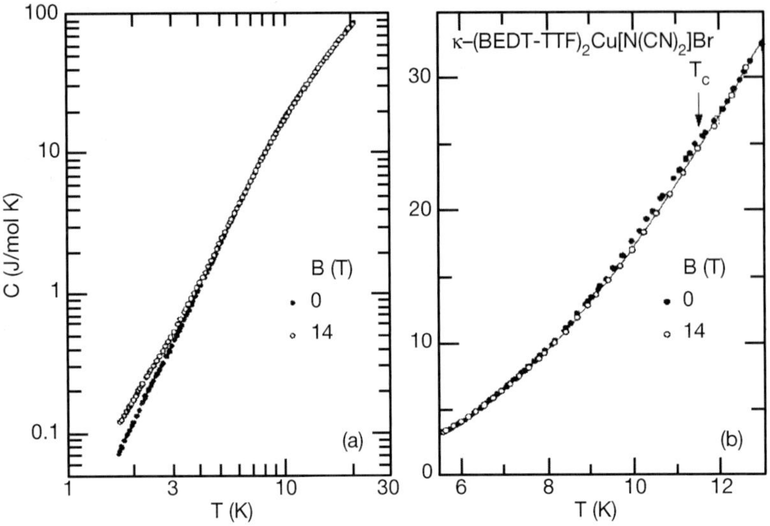

Fig. 20.47. Temperature dependence of the specific heat of κ-$(ET)_2Cu[N(CN)_2]Br$ in the superconducting ($B = 0$) and normal state ($B = 14\,T$) over an extended temperature range (*left panel*) and in the vicinity of $T_c = 11.5\,K$ (*right panel*). The *solid line* is a polynomial fit to the 14 T data, taken from [233]

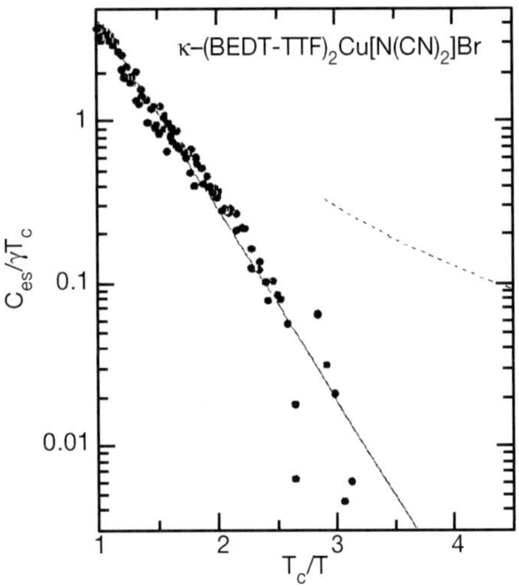

Fig. 20.48. A semi-logarithmic plot of $C_{es}/(\gamma T_c)$ vs T_c/T as determined from the data shown in Fig. 20.47. The *solid line* indicates the exponential variation of C_{es}. The *dashed line* corresponds to C_{es} as determined by Nakazawa et al. [356] (see text and footnote 27), taken from [233]

effects. First specific heat measurements were focussing on the determination of the discontinuity at T_c which provides information on the coupling strength. From the results of Andraka et al. [355] and Graebner et al. [270] on κ-$(ET)_2Cu(NCS)_2$ yielding a ratio of $\Delta C/\gamma T_c > 2$ a strong-coupling behavior has been inferred for this salt. In a series of subsequent experiments, the temperature dependence of the electronic contribution, C_{es}, at lower temperatures was at the focus of the investigations. From experiments on κ-$(ET)_2Cu[N(CN)_2]Br$ Nakazawa et al. [356] reported a quadratic temperature dependence of C_{es} at low temperatures, which was taken as an indication for line nodes in the gap. However, recent high-resolution specific heat measurements on the same compound revealed an exponentially weak low-T electronic contribution to the specific heat implying a finite energy gap [233]. Moreover it has been shown in the latter study that the T^2 dependence in the C_{es} data by Nakazawa et al. [356] most likely originates in their incorrect determination of the phonon background[26]. Figures 20.47 and

[26] In [356] the lattice specific heat of κ-$(ET)_2Cu[N(CN)_2]Br$ was estimated by measuring a second, deuterated sample after quench cooling to suppress superconductivity. The data analysis in [356] is based on the assumption that by this procedure the lattice specific heat remains unaffected and is identical to that of the hydrogenated superconducting compound. In [233] it has been shown, however, that the so-derived phonon background differs substantially from that determined in an overcritical field. The reason for this might be related to the glass-like transition at T_g observed in this system [99], see Sect. 20.3.4.

20.48 show the results of specific heat measurements performed by Elsinger et al. [233]. The phonon contribution, which predominates the specific heat near T_c has been determined from measurements in an overcritical field. This standard procedure is valid as long as there are no magnetic contributions to the specific heat which might change with the field. From the absence of any measurable field dependence in the data above T_c this assumption appears justified. Figure 20.48 shows the exponential decrease of C_{es} with decreasing temperatures. The lack of a finite C_{es} for $T \rightarrow 0$ rules out the existence of zeroes in the energy gap.

A similar behavior has been observed also for κ-$(ET)_2Cu(NCS)_2$ [49,50]. Figure 20.49 shows the difference $\Delta C(T) = C(T, B = 0) - C(T, B = 8\,T > B_{c_2})$ used to analyze the specific heat data. The advantage of using this quantity means that the unknown phonon contribution drops out. As Fig. 20.49 demonstrates, $\Delta C(T)$ deviates markedly from the weak-coupling BCS behavior in both the jump height at T_c as well as the overall temperature dependence. However, as was found also for κ-$(ET)_2Cu[N(CN)_2]Br$ [233], a much better description of the data is obtained by using the semi-empirical extension of the BCS formalism to strong-coupling superconductors; the so-called α-model [357]. It contains a single free parameter $\alpha \equiv \Delta_0/k_B T_c$ which scales the BCS energy gap $\Delta(T) = (\alpha/\alpha_{BCS}) \cdot \Delta_{BCS}(T)$ with $\alpha_{BCS} = 1.764$. As Fig. 20.49 clearly demonstrates, the strong-coupling BCS model with $\alpha = 2.8 \pm 0.1$ provides an excellent description of the data over the entire temperature range investigated [49]. Similar to what has been found for κ-$(ET)_2Cu[N(CN)_2]Br$ (Fig. 20.48), the data for κ-$(ET)_2Cu(NCS)_2$ (inset of Fig. 20.49) are fully consistent with an exponentially small C_{es} at low temperatures, i.e. an energy gap without zeroes at the Fermi surface. The same behavior has been previously observed for other $(ET)_2X$ superconductors [50,359,360]. The above findings of an exponentially weak specific heat at low temperatures are clearly incompatible with the existence of gap zeroes as claimed by various of the above mentioned experiments. It has to be shown by future studies whether or not this controversy can be removed by taking properly into account the influence of magnetic fields even when applied parallel to the planes and other extraneous effects such as the cooling-rate-dependent disorder associated with the ethylene groups.

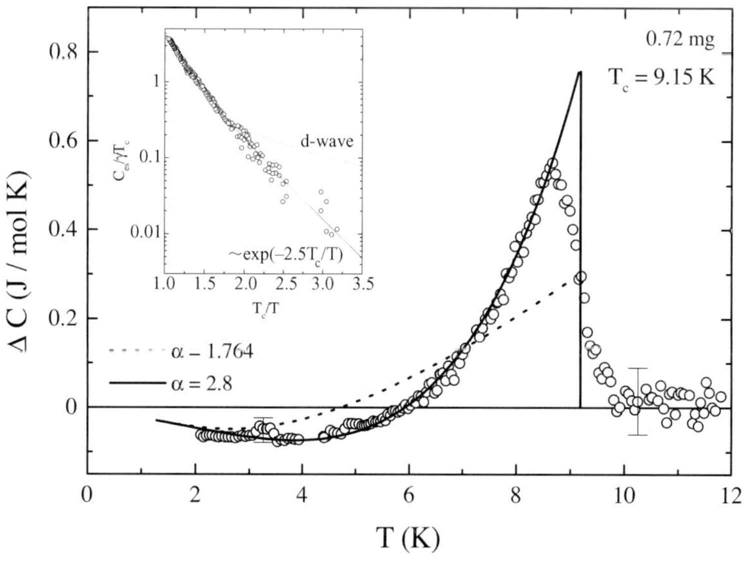

Fig. 20.49. Specific heat difference $\Delta C = C(0\,T) - C(8\,T)$ of a κ-$(ET)_2Cu(NCS)_2$ single crystal of m = 0.72 mg (main panel). The *dotted and solid thick lines* represent the BCS curves for weak and strong coupling, respectively. The *inset* shows the quasiparticle contribution to the specific heat in the superconducting state as $C_{es}/\gamma T_c$ vs T_c/T in a semi-logarithmic representation. Here the *solid line* represents the strong-coupling BCS behavior while the *dotted line* indicates a T^2 behavior as expected for a *d*-wave order parameter [358], taken from [49,187]

On the Pairing State in $(TMTSF)_2X$

As for the BEDT-TTF systems, the nature of superconductivity in the Bechgaard salts $(TMTSF)_2X$ is still far from being understood and continues to be a controversial issue. Early suggestions of a spin-triplet ($S = 1$) state were based on the proximity of superconductivity to a SDW state in the pressure-temperature phase diagram (cf. Sect. 20.3.5). This situation resembles the theoretical expectation for an interacting 1D electron gas [361] where a SDW phase lies next to triplet superconductivity. Further arguments for a spin-triplet p-wave superconductivity were derived from the observation that T_c is extremely sensitive to the introduction of nonmagnetic defects [362, 363] and substitutional impurities [364, 365], see also [366, 367]. As discussed in Sect. 20.3.5, depending on external pressure and magnetic field, the Bechgaard salts can be tuned to either a superconducting or a SDW ground state which has led to the proposal that the order parameters of both phases are not independent of each other. The superconducting properties of the Bechgaard salts have been reviewed by several authors [5, 33, 368]. Since these articles are comprehensive up to 1998, we shall give only a brief overview on the early results and concentrate the discussion on the more recent developments including the possibility of spin-triplet superconductivity.

Among the $(TMTSF)_2X$ superconductors, the X = PF_6 and ClO_4 salts are the most extensively studied materials although the number of experimental investigations of the superconducting properties is much less compared to that of the BEDT-TTF salts and their derivatives. The reason for this is most likely related to the low T_c values of the former systems which require extensive low-temperature equipment and, in the case of X = PF_6, external pressure of $p \geq 5.8$ kbar [369] to stabilize the superconducting state. For the ambient-pressure superconductor $(TMTSF)ClO_4$ it is the anion ordering which renders the experimental situation difficult (see Sect. 20.3.4). By slowly cooling through the anion-ordering temperature T_{AO} at around 24 K an ambient pressure superconducting state below $T_c = (1.2 \pm 0.2)$ K can be stabilized. In this case, the anions are believed to be well ordered. On the other hand, for samples that have been cooled rapidly across T_{AO}, an insulating state forms below $T_{MI} \simeq 6.1$ K.

Early specific heat measurements on $(TMTSF)_2 ClO_4$ focusing on the temperature range close to T_c revealed a discontinuity at T_c, $\Delta C/\gamma T_c = 1.67$ which is in fair agreement with the BCS value of 1.43 [247]. On the other hand, deviations from a BCS-type of superconductivity have been observed in NMR measurements by Takigawa et al. [370]. These authors reported the absence of a Hebel–Schlichter peak and a T^3 dependence in the spin-lattice relaxation rate. This has led to the proposal of a d-wave order parameter with a gap that vanishes along lines on the Fermi surface [370]. These results are at variance with more recent thermal conductivity data on the same salt, showing a rapid decrease of the electronic contribution to the heat transport below T_c which indicates the absence of low-lying excitations [371]. Their results provide strong evidence for a nodeless gap function. However, as pointed out by the authors, this result is not necessarily associated with an s-wave order parameter.

By enumerating possible gap functions for quasi-1D systems, Hasegawa and Fukuyama [354] showed that besides an anisotropic spin-singlet d-wave also a spin-triplet p-wave state (in both cases the order parameters vanish along lines on the FS) is possible. The authors suggested the possibility of an antiferromagnetic spin-fluctuation pairing mechanism for the Bechgaard salts.

Arguments in favor of such a spin-fluctuation-mediated superconductivity with d-wave symmetry have been derived from a recent resistivity study under pressure [67, 179]. According to these results, a minimum in the resistivity $\rho_a(T)$ at $T_{\min}$ marks the onset of AF fluctuations before, at lower temperatures, an itinerant antiferromagnetic state (SDW) is stabilized. The width of the region of critical AF fluctuations in the T–p phase diagram (see Fig. 20.23 in Sect. 20.3.5) is enhanced when the SDW ground state is approached from the high-pressure side, where the system is close to the SDW/SC border and largest where $T_c(p)$ reaches its optimum value. The correlation between the spin-fluctuation regime above the onset of superconductivity and the T_c value is taken

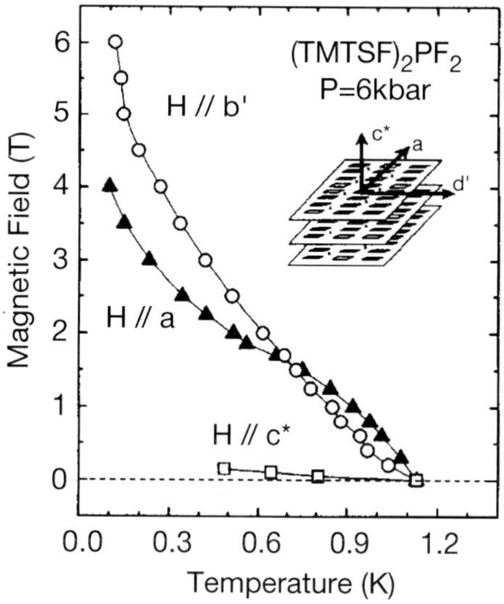

Fig. 20.50. *B-T* phase diagram of $(TMTSF)_2PF_6$ at a pressure of 6 kbar for magnetic fields aligned along three perpendicular axes as defined in the *inset*, taken from [253]

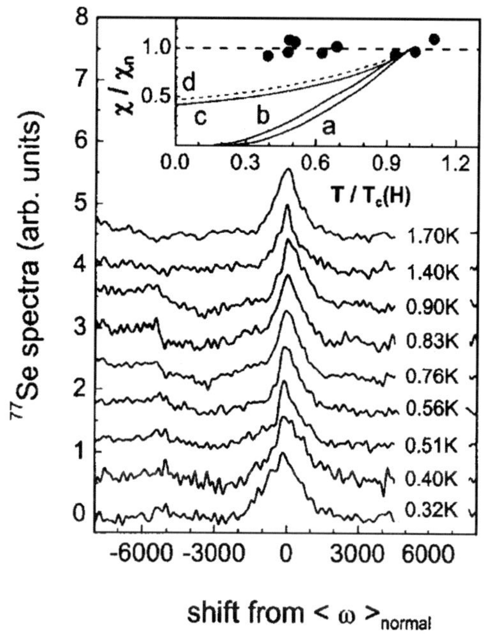

Fig. 20.51. ^{77}Se NMR spectra of $(TMTSF)_2PF_6$ under 7 kbar at various temperatures above and below T_c which is 0.81 K at the applied field of 1.43 T aligned along the most conductive *a*-axis. The inset shows normalized χ_{Spin} data compared with theoretical calculations. Curves *a* and *b* are for a singlet state in applied fields B/B_{c_2} near zero and 0.63, respectively, as calculated by Fulde and Maki [373] while curves *c* and *d* represent a scenario where the vortex cores induce normal regions of a fraction equal to B/B_{c_2}, taken from [374]

as a strong argument for a pairing mechanism mediated by the exchange of these fluctuations [67,179].

In a recent series of papers on the upper critical fields, the discussion of a possible realization of a triplet-pairing state in the Bechgaard salts has again been raised [253,369,372]. Lee et al. examined the upper critical field $B_{c_2}(T)$ and its directional dependence in $(TMTSF)_2PF_6$ under pressure via resistivity measurements. The resulting magnetic field-temperature phase diagram is depicted in Fig. 20.50 for fields aligned along the three crystal axes [253]. While the upper-critical-field curves near T_c were found to be consistent with earlier results on the same and related compounds, which indicated a more conventional behavior (see also Fig. 20.33 in Sect. 20.4.2) [248,251,252], the extension of the measurements to lower temperatures uncovered important new features: (i) the upper critical fields for *B* aligned along the *a* and *b*-directions display strong positive curvatures *without* saturation. This behavior was found to be independent of the criterion used to extract B_{c_2} from the resistance profiles. (ii) $B^b_{c_2}$ becomes larger than $B^a_{c_2}$, i.e. the anisotropy inverts upon cooling, and (iii), the critical fields in both directions exceed the Pauli limiting field of $B_P = 1.84 \times T_c \simeq 2.2\,\text{T}$ [249,250]. In a subsequent resistivity study under fields along the *b*-axis and optimum pressure settings [369], the onset of superconductivity was found to persist even up to 9 T, which is more than four times B_P. The authors discussed various proposals attempting to explain an upturn in the upper critical field. Among them are strong spin-orbit scattering [375,376], a magnetic-field-induced dimensional crossover from 3D to 2D [377,378] and the formation of a spatially inhomogeneous Fulde–Ferrell–Larkin–Ovchinnikov (FFLO) state [271,272]. It has been argued by Lee et al. that, even with a field-induced dimensional crossover which greatly enhances the orbital critical field, an additional effect is required to exceed the paramagnetic limit. This could be either the formation of the FFLO state or triplet superconductivity. The fact that no indications for a first-order phase transition into the FFLO state have been observed thus points to the possibility of a spin-triplet superconducting state.

This interpretation has gained further support from recent NMR measurements of the Knight shift K_S in the superconducting state of pressurized $(TMTSF)_2PF_6$ [374]. Since K_S is proportional to the spin-susceptibility χ_{Spin}, which is different for spin singlet ($\chi_{Spin} \to 0$ for $T \to 0$) and triplet (χ_{Spin} remains unchanged on cooling through T_c) superconductors, measurements of K_S provide a direct probe of the spin parity of the superconductor. Figure 20.51 shows that in contrast to the expectations for the various scenarios involving singlet superconductivity there is no significant change in K_S for $B \parallel a$ on cooling into the superconducting state. These observations together with similar results of a previous study for $B \parallel b$ [379] thus argue for a spin-triplet pairing. Besides K_S, the above studies included also measurements on the spin-lattice relaxation rate $1/T_1$ which exhibits a small peak near T_c. In light of the absence of such a peak in previous reports of zero field proton-NMR measurements [370] its identification with a Hebel–Schlichter peak is unclear.

20.5 Epilogue

Studies on superconductivity in organic charge-transfer salts have been carried out for more than 20 years. The joint efforts of researchers from the fields of synthetic chemistry and solid-state physics have created this new class of molecular conductors which exhibits fascinating properties. Key features are a pronounced anisotropy of the electronic states spanning the whole range from quasi-1D to anisotropic 3D, a low charge carrier concentration and a strong coupling of the charge carriers to the lattice degrees of freedom. The combination of these extraordinary material parameters make the organic conductors ideal model systems for exploring the interplay of strong electron–electron and electron–phonon interactions in reduced dimensions. In fact, this material class has enriched the solid-state physics by a wealth of interesting collective phenomena such as spin-Peierls or density-wave instabilities, Mott-type metal–insulator transitions, and, above all, superconductivity. Most remarkably, these systems offer a unique opportunity to scan through the various ground states either in discrete steps by chemical means, or even in a continuous way through changes of the applied pressure.

Thanks to the intense interdisciplinary efforts from research groups in chemistry and physics, our level of understanding of these materials has progressed substantially. The basis for this development was accomplished by advances in organic and organometallic chemistry which have provided a rich and varied supply of molecules serving as building blocks for conductors and superconductors. By employing state-of-the-art preparation techniques, it is possible nowadays to synthesize crystals of unprecedented high quality which permit highly accurate measurements of their electronic parameters. In particular, these clean materials enable direct experimental access to the electronic band structure with a high level of accuracy. For the 2D materials, the various experiments agree in that the normal-state electronic properties are well described by quasiparticles within a Fermi-liquid approach. These quasiparticles reveal a considerable mass renormalization due to both electron–electron as well as electron–phonon interactions. This contrasts with the behavior found in the quasi-1D salts, where non-Fermi-liquid properties have been reported for transport and optical properties. Indeed, these results indicate the separation of spin and charge degrees of freedom; one of the hallmarks of a Luttinger liquid.

The phase diagrams of the quasi-1D and 2D materials have been mapped out in great detail utilizing a wide spectrum of experimental techniques; some of which have been applied under extreme conditions such as high magnetic fields and high pressure. These studies confirmed, on the one hand, the universal character of the phase diagrams for both families and revealed, on the other hand, intriguing new details of the interplay of the various phases. Despite such detailed information, in particular on the phase boundary between superconductivity and magnetism, there is still no generally accepted picture on the nature of the superconducting state. The fact that both phases share a common phase boundary have led some researchers to believe that both phenomena have the same origin. In such a scenario, the attractive pairing interaction would have to be provided by the exchange of antiferromagnetic spin fluctuations. In fact, some of the experimental data seem to indicate an anisotropic pairing state compatible with such a magnetic pairing interaction. While for the quasi-2D salts a spin-singlet state with d-wave symmetry is the most favored one among the anisotropic states proposed, a number of experiments on the quasi-1D materials even suggest triplet superconductivity. However, for the quasi-2D salts, the notion of an anisotropic superconducting state is in clear conflict with results of other experiments, in particular specific heat measurements, which demonstrate the existence of a finite superconducting energy gap on the whole Fermi surface.

It is remarkable that despite the continuous experimental efforts to unravel the fundamental features of the superconducting state, the above controversy still persists. These efforts include the application of standard experimental techniques, some of which have been driven to an extraordinarily high resolution, as well as new, more sophisticated methods such as angular-dependent investigations of the gap functions.

An important lesson to be learnt from this controversy concerns the role of disorder, in particular the *intrinsic*-type of disorder which is inherent to many of the materials even when prepared under ideal conditions. As was recognized early on for the quasi-1D materials but has now proved to be true also for the quasi-2D systems, *intrinsic* disorder can be of crucial importance and should not be overlooked in exploring and discussing superconducting-state properties. This is of concern for all those materials where, by the symmetry, certain structural elements can adopt more than one orientation. Since these states are almost degenerate in energy, a cooling-rate dependent frozen *intrinsic* disorder will result at low temperatures which may have a severe influence on both the electronic and also the elastic properties of the sample. It is to be hoped that part of the above controversy can be removed by taking these disorder effects into account properly.

In viewing the field of organic conductors and superconductors as a whole, it is clear that these materials have provided a considerable contribution to our understanding of strongly interacting electrons in low dimensions, though important questions still remain open. This concerns, in particular, the nature of the metallic state above T_c and the role of electron–electron and electron–phonon interactions in the pairing mechanism. In light of the progress we have made in recent years, one can be optimistic that more systematic investigations on the relationship between molecular properties and crystal structure on the one hand, and the collective features of the bulk, on the other, will not only provide an answer to these questions but also will be a guidance for the synthesis of new materials with hopefully even more fascinating properties.

Acknowledgements

We are indebted to many colleagues and friends for collaborations, discussions and the support over the last years, especially M. Dressel, A.M. Kini, M. Köppen, F. Kromer, C. Langhammer, K. Maki, T. Sasaki, J.A. Schlueter, D. Schweitzer, F. Steglich, N. Toyota and J. Wosnitza. The technical assistance of C. Agnew is greatly acknowledged. Part of the work is supported by the Deutsche Forschungsgemeinschaft under the auspieces of the FOR 412.

References

1. D. Jérome, A. Mazaud, M. Ribault, K. Bechgaard, J. Physique Lett. **41**, L95 (1980)
2. H. Akamatu, H. Inokuchi, Y. Matsunaga, Nature **173**, 168 (1954)
3. W.A. Little, Phys. Rev. A **134**, 1416 (1964)
4. R.H. McKenzie, Science **278**, 820 (1997); Comments Cond. Mat. Phys. **18**, 309 (1998)
5. T. Ishiguro, K. Yamaji, G. Saito, *Organic Superconductors*, 2nd edn. (Springer, Berlin, 1998)
6. The review Organic conductors and superconductors by D. Jérome and H.J. Schulz, Adv. Phys. **31**, 229 (1982) on the quasi-1D organic conductors has been republished in: Adv. Phys. **51**, 293 (2002)
7. L.N. Bulaevskii, Adv. Phys. **37**, 443 (1988)
8. J.-P. Farges (editor), *Organic Conductors — Fundamentals and Applications*, (Marcel Dekker, New York, 1994)
9. M. Lang, Quasi-twodimensional organic superconductors, Superconduct. Rev. **2**, 1 (1996)
10. J. Singleton, C. Mielke, Quasi-two-dimensional organic superconductors: a review, Contemp. Phys. **43**, 63 (2002)
11. L.B. Coleman, M.J. Cohen, D.J. Sandman, F.G. Yamagishi, A.F. Garito, A.J. Heeger, Solid State Commun. **12**, 1125 (1973)
12. J. Ferraris, D.O. Cowan, V. Walatka, Jr., J.H. Perlstein, J. Am. Chem. Soc. **95**, 948 (1973)
13. G.C. Papavassiliou, G.A. Mousdis, J.S. Zambounis, A. Terzis, A. Hountas, B. Hilti, C.W. Mayer, J. Pfeiffer, Synth. Met. **27**, B379 (1988)
14. A.M. Kini, M.A. Beno, D. Son, H.H. Wang, K.D. Carlson, L.C. Porter, U. Welp, B.A. Vogt, J.M. Williams, D. Jung, M. Evain, M.-H. Whangbo, D.L. Overmyer, J.E. Schirber, Solid State Commun. **69**, 503 (1989)
15. M.A. Beno, H.H. Wang, A.M. Kini, K.D. Carlson, U. Geiser, W.K. Kwok, J.E. Thompson, J.M. Williams, J. Ren, M.-H. Whangbo, Inorg. Chem. **29**, 1599 (1990)
16. K. Bechgaard, C.S. Jacobsen, K. Mortensen, H.J. Pedersen, N. Thorup, Solid State Commun. **33**, 1119 (1980)
17. K. Bechgaard, K. Carneiro, M. Olsen, F.B. Rasmussen, C.S. Jacobsen, Phys. Rev. Lett. **46**, 852 (1981)
18. L.K. Montgomery, Chemical synthesis and crystal growth techniques. In: *Organic Conductors — Fundamentals and Applications*, ed. by J.-P. Farges (Marcel Dekker, New York, 1994) pp. 115–145
19. G. Saito, T. Enoki, K. Toriumi, H. Inokuchi, Solid State Commun. **42**, 557 (1982)
20. K. Oshima, H. Urayama, H. Yamochi, G. Saito, Physica C **153–155**, 1148 (1988)
21. J.M. Williams, J.R. Ferraro, R.J. Thorn, K.D. Carlson, U. Geiser, H.H. Wang, A.M. Kini, M.-H. Whangbo, *Organic Superconductors (Including Fullerenes)* (Prentice Hall, Englewood Cliffs, NJ, 1992) Chap. 2
22. R. Moret, J.P. Pouget, In: *Crystal Chemistry and Properties of Materials with Quasi-One-Dimensional Structures*, ed. by J. Rouxel (D. Reidel, Dordrecht, 1986) pp. 87–143
23. T. Mori, Bull. Chem. Soc. Japan **71**, 2509 (1998); T. Mori, H. Mori, S. Tanaka, *ibid.* **72**, 179 (1999); T. Mori, *ibid.* **72**, 2011 (1999)
24. A.M. Kini, U. Geiser, H.H. Wang, K.D. Carlson, J.M. Williams, W.K. Kwok, K.D. Vandervoort, J.E. Thompson, D.L. Supka, D. Jung, M.-H. Whangbo, Inorg. Chem. **29**, 2555 (1990)
25. J.M. Williams, A.M. Kini, H.H. Wang, K.D. Carlson, U. Geiser, L.K. Montgomery, G.J. Pyrka, D.M. Watkins, J.M. Kommers, S.J. Boryschuk, A.V. Strieby Crouch, W.K. Kwok, J.E. Schirber, D.L. Overmyer, D. Jung, M.-H. Whangbo, Inorg. Chem. **29**, 3272 (1990)
26. H.H. Wang, K.D. Carlson, U. Geiser, A.M. Kini, A.J. Schultz, J.M. Williams, L.K. Montgomery, W.K. Kwok, U. Welp, K.G. Vandervoort, S.J. Boryshuk, A.V. Strieby Crouch, J.M. Kommers, D.M. Watkins, J.E. Schirber, D.L. Overmyer, D. Jung, J.J. Novoa, M.-H. Whangbo, Synth. Met. **41-43**, 1983 (1991)
27. Yu.V. Sushko, K. Andres, Phys. Rev. B **47**, 330 (1993)
28. A. Kobayashi, R. Kato, T. Naito, H. Kobayashi, Synth. Met. **56**, 2078 (1993)
29. A. Kobayashi, T. Udagawa, H. Tomita, T. Naito, H. Kobayashi, Chem. Lett. **1993**, 2179 (1993)
30. L.K. Montgomery, T. Burgin, C. Husting, L. Tilley, J.C. Huffmann, K.D. Carlson, J.D. Dudek, G.A. Yaconi, U. Geiser, J.M. Williams, Mol. Cryst. Liq. Cryst. **211**, 283 (1992)
31. H. Kobayashi, H. Akutsu, E. Arai, H. Tanaka, A. Kobayashi, Phys. Rev. B **56**, R8526 (1997)
32. S. Uji, C. Terakura, T. Terashima, T. Yakabe, Y. Terai, M. Tokumoto, A. Kobayashi, F. Sakai, H. Tanaka, H. Kobayashi, Phys. Rev. B **65**, 113101 (2002)
33. D. Jérome, Organic superconductors: From $(TMTSF)_2PF_6$ to fullerenes'. In: *Organic Conductors — Fundamentals and Applications*, ed. by J.-P. Farges (Marcel Dekker, New York, 1994) pp. 405–494

34. T. Mori, A. Kobayashi, Y. Sasaki, H. Kobayashi, G. Saito, H. Inokuchi, Bull. Chem. Soc. Jpn. **57**, 627 (1984)
35. M.-H. Whangbo, J.J. Novoa, D. Jung, J.M. Williams, A.M. Kini, H.H. Wang, U. Geiser, M.A. Beno, K.D. Carlson, In: *Organic Superconductivity*, ed. by V.Z. Kresin and W.A. Little (Plenum Press, New York, 1990) p. 243
36. U. Geiser, A.J. Schultz, H.H. Wang, D.M. Watkins, D.L. Stupka, J.M. Williams, J.E. Schirber, D.L. Overmyer, D. Jung, J.J. Novoa, M.-H. Whangbo, Physica C **174**, 475 (1991)
37. P.M. Grant, J. Phys. C3 **44**, 847 (1983)
38. G.M. Danner, W. Kang, P.M. Chaikin, Phys. Rev. Lett. **72**, 3714 (1994)
39. J. Wosnitza, *Fermi Surfaces of Low-Dimensional Organic Metals and Superconductors* (Springer, Berlin, 1996)
40. J. Singleton, Studies of quasi-two-dimensional organic conductors based on BEDT-TTF using high magnetic fields, Rep. Prog. Phys. **63**, 1111 (2000)
41. K. Oshima, T. Mori, H. Inokuchi, H. Urayama, H. Yamochi, G. Saito, Phys. Rev. B **38**, 938 (1988)
42. H. Mori, S. Tanaka, M. Oshima, G. Saito, T. Mori, Y. Maruyama, H. Inokuchi, Bull. Chem. Soc. Jpn. **63**, 2183 (1990)
43. N. Harrison, E. Rzepniewski, J. Singleton, J.P. Gee, M.M. Honold, P. Day, M. Kurmoo, J. Phys.: Cond. Mat. **11**, 7227 (1999)
44. L. Onsager, Phil. Mag. **43**, 1006 (1952)
45. L.F. Lifshitz, A.M. Kosevich, Zh. Eksp. Teor. Fiz. **29**, 730 (1955); Sov. Phys. JETP **2**, 636 (1956)
46. J. Caulfield, W. Lubczynski, F.L. Pratt, J. Singleton, D.Y.K. Ko, W. Hayes, M. Kurmoo, P. Day, J. Phys., Cond. Mat. **6**, 2911 (1994)
47. T. Sasaki, H. Sato, N. Toyota, Solid State Commun. **76**, 507 (1990)
48. T. Sasaki, H. Sato, N. Toyota, Physica C **185–189**, 2687 (1991)
49. J. Müller, M. Lang, R. Helfrich, F. Steglich, T. Sasaki, Phys. Rev. B **65**, R140509 (2002)
50. J. Wosnitza, S. Wanka, J. Hagel, M. Reibelt, D. Schweitzer, J.A. Schlueter, Synth. Met. **133-134**, 201 (2003)
51. Y.-N. Xu, W.Y. Ching, Y.C. Jean, Y. Lou, Phys. Rev. B **52**, 12946 (1995)
52. J. Merino, R.H. McKenzie, Phys. Rev. B **62**, 2416 (2000)
53. J. Singleton, F.L. Pratt, M. Doporto, T.J.B.M. Janssen, M. Kurmoo, J.A.A.J. Perenboom, W. Hayes, P. Day, Phys. Rev. Lett. **68**, 2500 (1992)
54. J.E. Eldridge, Y. Xie, H.H. Wang, J.M. Williams, A.M.Kini, J.A. Schlueter, Spectrochim. Acta A **52**, 45 (1996)
55. J.E. Eldridge, Y. Xie, Y. Lin, H.H. Wang, J.M. Williams, J.A. Schlueter, Synth. Met. **86**, 2067 (1997)
56. D. Pedron, G. Visentini, R. Bozio, J.M. Williams, J.A. Schlueter, Physica C **276**, 1 (1997)
57. D. Pedron, R. Bozio, J.A. Schlueter, M.E. Kelly, A.M. Kini, J.M. Williams, Synth. Met. **103**, 2220 (1999)
58. E. Faulques, V.G. Ivanov, C. Mézière, P. Batail, Phys. Rev. B **62**, R9291 (2000)
59. A. Girlando, M. Masino, G. Visentini, A. Brillante, R.G. Della Valle, E. Venuti, Synth. Met. **109**, 13 (2000)
60. S. Belin, K. Behnia, A. Deluzet, Phys. Rev. Lett. **81**, 4728 (1998)
61. K. Izawa, H. Yamaguchi, T. Sasaki, Y. Matsuda, Phys. Rev. Lett. **88**, 027002 (2002)
62. L. Pintschovius, H. Rietschel, T. Sasaki, H. Mori, S. Tanaka, N. Toyota, M. Lang, F. Steglich, Europhys. Lett. **37**, 627 (1997)
63. H. Weiss, M.V. Kartsovnik, W. Biberacher, E. Steep, E. Balthes, A.G.M. Janssen, K. Andres, N.D. Kushch, Phys. Rev. B **59**, 12370 (1999)
64. A.-K. Klehe, R.D. McDonald, A.F. Goncharov, V.V. Struzhkin, Ho-kwang Mao, R.J. Hemley, T. Sasaki, W. Hayes, J. Singleton, J. Phys.: Cond. Mat. **12**, L247 (2000)
65. P. Wzietek, F. Creuzet, C. Bourbonnais, D. Jérome, K. Bechgaard, P. Batail, J. Phys. I France **3**, 171 (1993)
66. T. Adachi, E. Ojima, K. Kato, H. Kobayashi, J. Am. Chem. Soc. **122**, 3238 (2000)
67. H. Wilhelm, D. Jaccard, R. Duprat, C. Bourbonnais, D. Jérome, J. Moser, C. Carcel, J.M. Fabre, Eur. Phys. J. B **21**, 175 (2001); H. Wilhelm, D. Jaccard, D. Jérome, J. Moser, R. Duprat, C. Bourbonnais, In: *Frontiers of High Pressure Research II: Application of High Pressure to Low-Dimensional Novel Electronic Materials*, ed. by H.D. Hochheimer et al. (Kluwer Academic Publishers, 2001) p. 423
68. J.R. Cooper, B. Korin-Hamzić, Organic metals, In: *Organic Conductors — Fundamentals and Applications*. ed. by J.-P. Farges (Marcel Dekker, New York, 1994) pp. 359–403
69. S. Tomonaga, Prog. Theor. Phys. **5**, 554 (1950)
70. J. Luttinger, J. Math. Phys. **4**, 1154 (1963)
71. H.J. Schulz, Int. J. Mod. Phys. B **5**, 57 (1991)

72. C. Bourbonnais, D. Jérome, Physics World (September 1998) p. 41
73. V. Vescoli, L. Degiorgi, W. Henderson, G. Grüner, K.P. Starkey, L.K. Montgomery, Nature **281**, 1181 (1998)
74. T. Lorenz, M. Hofmann, M. Grüninger, A. Freimuth, G.S. Uhrig, M. Dumm, M. Dressel, Nature **418**, 614 (2002)
75. B. Dardel, D. Malterre, M. Grioni, P. Weibel, Y. Baer, J. Voit, D. Jérome, Europhys. Lett. **24**, 687 (1993)
76. V. Vescoli, F. Zwick, W. Henderson, L. Degiorgi, M. Grioni, G. Gruner, L.K. Montgomery, Eur. Phys. J. B **13**, 503 (2000)
77. J. Moser, M. Gabay, P. Auban-Senzier, D. Jérome, K. Bechgaard, J.M. Fabre, Eur. Phys. J. B **1**, 39 (1998)
78. K. Murata, M. Tokumoto, H. Anzai, Y. Honda, N. Kinoshita, T. Ishiguro, N. Toyota, T. Sasaki, Y. Muto, Synth. Met **27**, A263 (1988)
79. H. Sato, T. Sasaki, N. Toyota, Physica C **185-189**, 2679 (1991)
80. J.E. Schirber, D.L. Overmyer, K.D. Carlson, J.M. Williams, A.M. Kini, H.H. Wang, H.A. Charlier, B.J. Love, D.M. Watkins, G.A. Yaconi, Phys. Rev. B **44**, 4666 (1991)
81. M. Oshima, H. Mori, G. Saito, K. Oshima, In: *The Physics and Chemistry of Organic Superconductors*, ed. by G. Saito, S. Kagoshima (Springer, Berlin, 1990) p. 257.
82. K. Kikuchi, Y. Honda, Y. Ishikawa, K. Saito, I. Ikemoto, K. Murata, H. Anzai, T. Ishiguro, Solid State Commun. **66**, 405 (1988)
83. K. Yamaji, Synth. Met. A **27**, 115 (1988)
84. S. Gärtner, E. Gogu, I. Heinen, H.J. Keller, T. Klutz, D. Schweitzer, Solid State Commun. **65**, 1531 (1988)
85. N. Toyota, T. Sasaki, H. Sato, Y. Watanabe, Physica C **178**, 339 (1991)
86. I.D. Parker, R.H. Friend, M. Kurmoo, P. Day, C. Lenoir, P. Batail, J. Phys. Cond. Mat. **1**, 4479 (1989)
87. M. Kund, H. Müller, W. Biberacher, K. Andres, Physica C **191**, 274 (1993)
88. M.A. Tanatar, T. Ishiguro, T. Kondo, G. Saito, Phys. Rev. B **59**, 3841 (1999)
89. N.L. Wang, B.P. Clayman, H. Mori, S. Tanaka, J. Phys.: Cond. Mat. **12**, 2867 (2000)
90. B. Thoma, Diploma Thesis, Stuttgart University (1996) unpublished.
91. L.K. Montgomery, R.M. Vestal, K.P. Starkey, B.W. Fravel, M.J. Samide, D.G. Peters, C.H. Mielke, J.D. Thompson, Synth. Met. **103**, 1878 (1999)
92. Ch. Strack, C. Akinci, V. Pashchenko, B. Wolf, E. Uhrig, W. Assmus, M. Lang, J. Schreuer, L. Wiehl, J.A. Schlueter, J. Wosnitza, D. Schweitzer, J. Müller, J. Wykhoff, Phys. Rev. B **72**, 054511 (2005)
93. K. Murata, M. Ishibashi, Y. Honda, N.A. Fortune, M. Tokumoto, N. Kinoshita, H. Anzai, Solid State Commun. **76**, 377 (1990)
94. Y.V. Sushko, V.A. Bondarenko, R.A. Petrosov, N.D. Kushch, E.B. Yagubskii, J. Phys. I France **1**, 1375 (1991)
95. K. Kanoda, Hyperfine Int. **104**, 235 (1997)
96. K. Kanoda, Physica C **282–287**, 299 (1997)
97. J. Merino, R.H. McKenzie, Phys. Rev. B **61**, 7996 (2000)
98. P. Limelette, P. Wzietek, S. Florens, A. Georges, T.A. Costi, C. Pasquier, D. Jérome, C. Mézière, and P. Batail, Phys. Rev. Lett. **91**, 016401 (2003)
99. J. Müller, M. Lang, F. Steglich, J.A. Schlueter, A.M. Kini, T. Sasaki, Phys. Rev. B **65**, 144521 (2002)
100. M. Lang, J. Müller, F. Steglich, J.A. Schlueter, A.M. Kini, T. Sasaki, Synth. Met. (2002) in press.
101. T. Sasaki, N. Yoneyama, A. Matsuyama, N. Kobayashi, Phys. Rev. B **65**, 060505 (2002)
102. M. Weger, M. Tittelbach, E. Balthes, D. Schweitzer, H.J. Keller, J. Phys.: Cond. Mat. **5**, 8569 (1993)
103. M. Weger, M. Kaveh, H. Gutfreund, Solid State Commun. **37**, 421 (1981)
104. M. Weger, D. Schweitzer, Synth. Met. **70**, 889 (1995)
105. M. Weger, Acta Physica Polonica A **87**, 723 (1995)
106. L.I. Buravov, A.V. Zvarykina, N.D. Kushch, V.N. Laukhin, V.A. Merzhanov, A.G. Khomenko, E.B. Yagubskii, Sov. Phys. JETP **68**, 182 (1989)
107. R.H. McKenzie, P. Moses, Phys. Rev. Lett. **81**, 4492 (1998)
108. J. Singleton, P.A. Goddard, A. Ardavan, N. Harrison, S.J. Blundell, J.A. Schlueter, A.M. Kini, Phys. Rev. Lett. **88**, 037001 (2002)
109. C.S. Jacobsen, In: *Low-Dimensional Conductors and Superconductors*, ed. by D. Jérome and L.G. Caron (Plenum Press, New York 1987) p. 253
110. A. Graja, Optical properties, In: *Organic Conductors, Fundamentals and Applications*, ed. by J.-P. Farges (Marcel Dekker, New York, 1994) pp. 229–267.

111. M. Dressel, Synth. Met. **85**, 1503 (1997)
112. L. Degiorgi, V. Vescoli, W. Henderson, G. Gruner, L.K. Montgomery, J. Phys. IV France **10**, Pr3-103 (2000)
113. M. Dressel, Optical properties of layered organic superconductors. In: *Studies of High Temperature Superconductors* (Nova Science, New York, 2000) Vol. 34.
114. J.E. Eldridge, Y. Lin, H.H. Wang, J.M. Williams, A.M. Kini, Phys. Rev. B **57**, 597 (1998)
115. Y. Lin, J.E. Eldridge, J.A. Schlueter, H.H. Wang, A.M. Kini, Phys. Rev. B **64**, 024506 (2001)
116. C.S. Jacobsen, D.B. Danner, K. Bechgaard, Phys. Rev. Lett. **46**, 1142 (1981); Phys. Rev B **28**, 7019 (1983); C.S. Jacobsen, K. Mortensen, M. Weger, K. Bechgaard, Solid State Commun. **38**, 423 (1981)
117. D. Pedron, R. Bozio, M. Meneghetti, C. Pecile, Phys. Rev B **49**, 10893 (1994)
118. F. Zwick, S. Brown, G. Margaritondo, C. Merlic, M. Onellion, J. Voit, M. Grioni, Phys. Rev. Lett. **79**, 3982 (1997)
119. F. Zwick, D. Jérome, G. Margaritondo, M. Onellion, J. Voit, M. Grioni, Phys. Rev. Lett. **81**, 2974 (1998)
120. K. Kornelsen, J.E. Eldridge, H.H. Wang, J.M. Williams, Phys. Rev. B **44**, 5235 (1991)
121. K. Kornelsen, J.E. Eldridge, H.H. Wang, H.A. Charlier, J.M. Williams, Solid State Commun. **81**, 343 (1992)
122. N.O. Lipari, M.J. Rice, C.B. Duke, R. Bozio, A. Girlando, C. Pecile, Int. Quantum Chem. Symp. **11**, 583 (1977)
123. J.E. Eldridge, C.C. Homes, J.M. Williams, A.M. Kini, H.H. Wang, Spectrochim. Acta A **51**, 947 (1995)
124. G. Visentini, M. Masino, C. Bellitto, A. Girlando, Phys. Rev. B **58**, 9460 (1998)
125. M.E. Kozlov, K.I. Pokhodnia, A.A. Yurchenko, Spectrochim. Acta A **43**, 323 (1987)
126. M.E. Kozlov, K.I. Pokhodnia, A.A. Yurchenko, Spectrochim. Acta A **45**, 437 (1989)
127. M. Meneghetti, R. Bozio, C. Pecile, J. Phys. **47**, 1377 (1986)
128. A.M.Kini, K.D. Carlson, H.H. Wang, J.A. Schlueter, J.D. Dudek, S.A. Sirchio, U. Geiser, K.R. Lykke, J.M. Williams, Physica C **264**, 81 (1996); A.M.Kini, K.D. Carlson, J.D. Dudek, U. Geiser, H.H. Wang, J.M. Williams, Synth. Met. **85**, 1617 (1997)
129. A. Nowack, M. Weger, D. Schweitzer, H. Keller, Solid State Commun. **60**, 199 (1986)
130. A. Nowack, U. Poppe, M. Weger, D. Schweitzer, H. Schenk, Z. Phys. B **68**, 41 (1987)
131. S. Sugai, H. Mori, H. Yamochi, G. Saito, Phys. Rev. B **47**, 14374 (1993)
132. K.I. Pokhodnia, A. Graja, M. Weger, D. Schweitzer, Z. Phys. B **90**, 127 (1993)
133. M. Dressel, J.E. Eldridge, J.M. Williams, H.H. Wang, Physica C **203**, 247 (1992)
134. A. Andrieux, D. Jérome, K. Bechgaard, J. Phys. (Paris) Lett. **42**, L87 (1981)
135. E. Barthel, G. Quirion, P. Wzietek, D. Jérome, J.B. Christensen, M. Jorgensen, K. Bechgaard, Europhys. Lett. **21**, 87 (1993)
136. K. Mortensen, Y. Tomkiewicz, K. Bechgaard, Phys. Rev. B **25**, 3319 (1982)
137. R.L. Greene, E.M. Engler, Phys. Rev. Lett. **45**, 1587 (1980)
138. D. Jérome, Science **252**, 1509 (1991)
139. H. Mayaffre, P. Wzietek, C. Lenoir, D. Jérome, P. Batail, Europhys. Lett. **28**, 205 (1994)
140. H. Mayaffre, P. Wzietek, S. Charfi-Kaddour, C. Lenoir, D. Jérome, P. Batail, Physica B **206 & 207**, 767 (1995)
141. A. Kawamoto, K. Miyagawa, Y. Nakazawa, K. Kanoda, Phys. Rev. Lett. **74**, 3455 (1995)
142. A. Kawamoto, K. Miyagawa, Y. Nakazawa, K. Kanoda, Phys. Rev. B **52**, 15522 (1995)
143. S.M. De Soto, C.P. Slichter, A.M. Kini, H.H. Wang, U. Geiser, J.M. Williams, Phys. Rev. B **52**, 10364 (1995)
144. P. Wzietek, H. Mayaffre, D. Jérome, S. Brazovskii, J. Phys. I France **6**, 2011 (1996)
145. V. Kataev, G. Winkel, D. Khomskii, D. Wohlleben, W. Crump, K.F. Tebbe, J. Hahn, Solid State Commun. **83**, 435 (1992)
146. K. Frikach, M. Poirier, M. Castonguay, K.D. Truong, Phys. Rev. B **61**, R6491 (2000)
147. T. Shimizu, N. Yoshimoto, M. Nakamura, M. Yoshizawa, Physica B **281 & 282**, 896 (2000)
148. Y. Lin, J.E. Eldridge, H.H. Wang, A.M. Kini, J.M. Williams, J.A. Schlueter, Phys. Rev. B **58**, R599 (1998)
149. J. Merino, R.H. McKenzie, Phys. Rev. B **62**, 16442 (2000)
150. M. Kund, K. Andres, H. Müller, G. Saito, Physica B **203**, 129 (1994)
151. J. Müller, M. Lang, F. Steglich, J.A. Schlueter, A.M. Kini, U. Geiser, J. Mohtasham, R.W. Winter, G.L. Gard, T. Sasaki, N. Toyota, Phys. Rev. B **61**, 11739 (2000)
152. U. Welp, S. Fleshler, W.K. Kwok, G.W. Crabtree, K.D. Carlson, H.H. Wang, U. Geiser, J.M. Williams, V.M. Hitsman, Phys. Rev. Lett. **69**, 840 (1992)
153. K. Miyagawa, A. Kawamoto, Y. Nakazawa, K. Kanoda, Phys. Rev. Lett. **75**, 1174 (1995)
154. M. Pinterić, M. Miljak, N. Biškup, O. Milat, I. Aviani, S. Tomić, D. Schweitzer, W. Strunz, I. Heinen, Eur. Phys. J B **11**, 217 (1999)

155. K. Miyagawa, A. Kawamoto, K. Uchida, K. Kanoda, Physica B **284–288**, 1589 (2000)
156. H. Posselt, H. Müller, K. Andres, G. Saito, Phys. Rev. B **49**, 15849 (1994)
157. M.A. Tanatar, T. Ishiguro, H. Ito, M. Kubota, G. Saito, Phys. Rev. B **55**, 12529 (1997)
158. M. Tokumoto, H. Anzai, K. Murata, K. Kajimura, T. Ishiguro, Japan J. Appl. Phys. **26** Suppl. 26–3, 1977 (1987)
159. J.P. Pouget, S. Ravy, J. Phys. I France **6**, 1501 (1996)
160. S. Ravy, R. Moret, J.P. Pouget, Phys. Rev. B **38**, 4469 (1988)
161. S. Tomic et al., J. Phys. C3 **44**, 1083 (1983)
162. H. Schwenk, K. Andres, F. Wudl, Phys. Rev. B **29**, 500 (1984)
163. T. Takahashi, D. Jérome, K. Bechgaard, J. Physique Lett. **43**, L565 (1982)
164. K. Mortensen, C.S. Jacobsen, A. Lindegaard-Andersen, K. Bechgaard, J. Physique **44**, C3-963 (1983)
165. H. Yang, J.C. Lasjaunias, P. Monceau, J. Phys.: Cond. Mat. **12**, 7183 (2000)
166. K. Saito, H. Akutsu, M. Sorai, Solid State Commun. **111**, 471 (1999)
167. H. Akutsu, K. Saito, M. Sorai, Phys. Rev. B **61**, 4346 (2000)
168. A. Sato, H. Akutsu, K. Saito, M. Sorai, Synth. Met. **120**, 1035 (2001)
169. For a review, see, e.g., R.W. Cahn (editor): Glasses and amorphous materials. In: *Materials Science and Technology* (Verlag Chemie, Weinheim, 1991) Vol. 9, p. 137; W. Kauzmann, The nature of the glassy state and the behavior of liquids at low temperatures, Chem. Rev. **43**, 219 (1948)
170. M.A. deBolt, A.J. Easteal, P.B. Macedo, and C.T. Moynihan, J. Am. Ceram. Soc. **59**, 16 (1976)
171. P. Nagel, V. Pasler, C. Meingast, A.I. Rykov, S. Tajima, Phys. Rev. Lett. **85**, 2376 (2000)
172. X. Su, F. Zuo, J.A. Schlueter, M.E. Kelly, J.M. Williams, Phys. Rev. B **57**, R14056 (1998)
173. A. Kawamoto, K. Miyagawa, K. Kanoda, Phys. Rev. B **55**, 14140 (1997)
174. Y. Watanabe, H. Sato, T. Sasaki, N. Toyota, J. Phys. Soc. Japan **60**, 3608 (1991)
175. X. Su, F. Zuo, J.A. Schlueter, M.E. Kelly, and J.M. Williams, Solid State Commun. **107**, 731 (1998)
176. A. Aburto, L. Fruchter, C. Pasquier, Physica C **303**, 185 (1998)
177. M. Pinterić, S. Tomić, M. Prester, D. Drobac, K. Maki, Phys. Rev. B **66**, 174521 (2002)
178. H. Ito, T. Ishiguro, T. Kondo, G. Saito, Phys. Rev. B **61**, 3243 (2000)
179. D. Jaccard, H. Wilhelm, D. Jérome, J. Moser, C. Carcel, J.M. Fabre, J. Phys.: Cond. Mat. **13**, L1 (2001)
180. L. Balicas, K. Behnia, W. Kang, E. Canadell, P. Auban-Senzier, D. Jérome, M. Ribault, J.M. Fabre, J. Phys. I France **4**, 1539 (1994)
181. F. Guo, K. Murata, A. Oda, Y. Mizuno, H. Yoshino, J. Phy. Soc. Japan, **69**, 2164 (2000)
182. H. Hirai, R. Kondo, M. Maesato, T. Shibata, S. Kagoshima, Synth. Met. **120**, 963 (2001)
183. M. Miyazaki, K. Kishigi, Y. Hasegawa, J. Phy. Soc. Japan, **69**, 997 (2000)
184. A. Kawamoto, H. Taniguchi, K. Kanoda, J. Am. Chem. Soc. **120**, 10984 (1998)
185. H. Taniguchi, A. Kawamoto, K. Kanoda, Phys. Rev. B **59**, 8424 (1999); Physica B **284–288**, 519 (2000)
186. Y. Nakazawa, H. Taniguchi, A. Kawamoto, and K. Kanoda, Phys. Rev. B **61**, R16295 (2000)
187. J. Müller, Thermodynamische Untersuchungen an quasi-zweidimensionalen organischen Supraleitern. PhD Thesis, Dresden University, Dresden (2001) (Shaker Verlag, Aachen, 2002)
188. J.E. Schirber, E.L. Venturini, A.M. Kini, H.H. Wang, J.R. Witworth, J.M. Williams, Physica C **152**, 157 (1988)
189. J.E. Schirber, D.L. Overmyer, J.M. Williams, A.M. Kini, H.H. Wang, Physica C **170**, 231 (1990)
190. H. Ito, T. Ishiguro, M. Kubota, G. Saito, J. Phys. Soc. Japan **65**, 2987 (1996)
191. S. Lefebvre, P. Wzietek, S. Brown, C. Bourbonnais, D. Jérome, C. Mézière, M. Fourmigué, P. Batail, Phys. Rev. Lett. **85**, 5420 (2000)
192. D. Fournier, M. Poirier, M. Castonguay, and K.D. Truong, Phys. Rev. Lett. **90**, 127002 (2003)
193. F. Kagawa, T. Itou, K. Miyagawa, and K. Kanoda, Phys. Rev. B. **69**, 064511 (2004)
194. A. Sato, E. Ojima, H. Akutsu, H. Kobayashi, A. Kobayashi, P. Cassoux, Chem. Lett. **1998**, 673 (1998)
195. F. Goze, V.N. Laukhin, L. Brossard, A. Audouard, J.P. Ulmet, S. Azkenazy, T. Naito, H. Kobayashi, A. Kobayashi, M. Tokumoto, P. Cassoux, Physica B **211**, 290 (1995)
196. L. Brossard, R. Clerac, C. Coulon, M. Tokumoto, T. Ziman, D.K. Petrov, V.N. Laukhin, M.J. Naughton, A. Audouard, F. Goze, A. Kobayashi, H. Kobayashi, P. Cassoux, Eur. Phys. J. B **1**, 439 (1998)
197. C. Hotta, H. Fukuyama, J. Phys. Soc. Japan **69**, 2577 (2000)

198. A.F. Hebard, M.J. Rosseinsky, R.C. Haddon, D.W. Murphy, S.H. Glarum, T.T.M. Palstra, A.P. Ramirez, A.R. Kortan, Nature (London) **350**, 600 (1991)
199. V.G. Kogan, Phys. Rev. B **24**, 1572 (1981)
200. W.E. Lawrence, S. Doniach, In: *Proceedings of the 12th International Conference on Low-Temperature Physics*, ed. by E. Kanada (Keigaku, Tokyo, 1971) p. 361
201. R. Kleiner, F. Steinmeyer, G. Kunkel, P. Müller, Phys. Rev. Lett. **68**, 2394 (1992)
202. P. Müller, Advances in Solid State Physics **34** (Vieweg-Verlag, Braunschweig 1994)
203. U. Geiser, J.A. Schlueter, H.H. Wang, A.M. Kini, J.M. Williams, P.P. Sche, H.I. Zakowicz, M.L. VanZile, J.D. Dudek, P.G. Nixon, R.W. Winter, G.L. Gard, J. Ren, M.-H. Whangbo, J. Am. Chem. Soc. **118**, 9996 (1996)
204. J.A. Schlueter, U. Geiser, J.M. Williams, J.D. Dudek, M.E. Kelly, J.P. Flynn, R.R. Wilson, H.I. Zakowicz, P.P. Sche, D. Naumann, T. Roy, P.G. Nixon, R.W. Winter, G.L. Gard, Synth. Met. **85**, 1453 (1997)
205. T.P. Orlando, E.J. McNiff, Jr., S. Foner, M.R. Beasly, *Phys. Rev. B* **19**, 4545 (1979)
206. M. Tinkham, *Introduction to Superconductivity*, 2nd edn. (McGraw–Hill, New York, 1996)
207. M. Lang, F. Steglich, N. Toyota, T. Sasaki, Phys. Rev. B **49**, 15227 (1994)
208. D.E. Farrell, C.J. Allen, R.C. Haddon, S.V. Chichester, Phys. Rev. B **42**, 8694 (1990)
209. S. Kawamata, K. Okuda, T. Sasaki, N. Toyota, Solid State Commun. **89**, 955 (1994)
210. J.C. Martinez, S.H. Brongersma, A. Koshelev, B. Ivlev, P.H. Kes, R.P. Griessen, D.G. de Groot, Z. Tarnavski, A.A. Menovsky, Phys. Rev. Lett. **69**, 2276 (1992)
211. Y. Matsuda, Marat B. Gaifullin, K. Kumagai, M. Kosugi, K. Hirata, Phys. Rev. Lett. **78**, 1972 (1997)
212. W.J. Skocpol, M. Tinkham, Rep. Prog. Phys. **38**, 1049 (1975)
213. U. Welp, S. Fleshler, W.K. Kwok, R.A. Klemm, V.M. Vinokur, J. Downey, B. Veal, G.W. Crabtree, Phys. Rev. Lett. **67**, 3180 (1991)
214. V. Pasler, PhD Thesis, Karlsruhe University, Karlsruhe (2000)
215. G. Bergmann, Z. Phys. **225**, 430 (1969)
216. P.A. Lee, S.R. Shenoy, Phys. Rev. Lett. **28**, 1025 (1972)
217. R. Ikeda, T. Ohmi, T. Tsuneto, J. Phys. Soc. Japan **59**, 1397 (1990)
218. S. Ullah, A.T. Dorsey, Phys. Rev. B **44**, 262 (1991)
219. Q. Li, In: *Physical Properties of High-Temperature Superconductors V*, ed. by D.M. Ginsberg (World Scientific, Singapore, 1996)
220. S. Friemel, C. Pasquier, Y. Loirat, D. Jérome, Physica C **259**, 181 (1996)
221. H. Ito, M. Watanabe, Y. Nogami, T. Ishiguro, T. Komatsu, G. Saito, N. Hosoito, Jpn. J. Appl. Phys. Series **7**, 419 (1992)
222. J.S. Brooks, X. Chen, S.J. Klepper, S. Valfells, G.J. Athas, Y. Tanaka, T. Kinoshita, N. Kinoshita, M. Tokumoto, H. Anzai, C.C. Agosta, Phys. Rev. B **52**, 14457 (1995)
223. D. Chasseau, J. Gaultier, M. Rahal, L. Ducasse, M. Kurmoo, P. Day, Synth. Met. **41–43**, 2039 (1991)
224. M. Rahal, D. Chasseau, J. Gaultier, L. Ducasse, M. Kurmoo, P. Day, Acta Cryst. B **53**, 159 (1997)
225. G. Gladstone, M.A. Jensen, J.R. Schrieffer, In: *Superconductivity*, ed. by R.D. Parks (Marcel Dekker, New York, 1969) Vol. 2, p. 655
226. C. Meingast, B. Blank, H.Bürkle, B. Obst, T. Wolf, H. Wühl, V. Selvamanickam, K. Salama, Phys. Rev. B **41**, 11299 (1990)
227. H. Kusuhara, Y. Sakata, Y. Ueba, K. Tada, M. Kaji, T. Ishiguro, Solid State Commun. **74**, 251 (1990)
228. C.E. Campos, J.S. Brooks, P.J.M. van Bentum, J.A.A.J. Perenboom, S.J. Klepper, P.S. Sandhu, S. Valfells, Y. Tanaka, T. Kinoshita, N. Kinoshita, M. Tokumoto, H. Anzai, Phys. Rev. B **52**, R7014 (1995)
229. S. Kagoshima, M. Maesato, R. Kondo, T. Shibata, H. Hirai, Synth. Met. **120**, 683 (2001); M. Maesato, Y. Kaga, R. Kondo, H. Hirai, S. Kagoshima, Synth. Met. **120**, 941 (2001)
230. E.S. Choi, J.S. Brooks, S.Y. Han, L. Balicas. J.S. Qualls, Phil. Mag. B **81**, 399 (2001)
231. M. Kund, H. Veith, H. Müller, K. Andres, G. Saito, Physica C **221**, 119 (1994)
232. M. Kund, J. Lehrke, W. Biberacher, A. Lerf, K. Andres, Synth. Met. **70**, 949 (1995)
233. H. Elsinger, J. Wosnitza, S. Wanka, J. Hagel, D. Schweitzer, W. Strunz, Phys. Rev. Lett. **84**, 6098 (2000)
234. S. Sadewasser, C. Looney, J.S. Schilling, J.A. Schlueter, J.M. Williams, P.G. Nixon, R.W. Winter, G.L. Gard, Solid State Commun. **102**, 571 (1997)
235. T. Biggs, A.-K. Klehe, J. Singleton, D. Bakker, J. Symington, P. Goddard, A. Ardavan, W. Hayes, J.A. Schlueter, T. Sasaki, M. Kurmoo, J. Phys.: Cond. Mat. **14**, L495 (2002)

236. J. Müller, M. Lang, J.A. Schlueter, U. Geiser, D. Schweitzer, Synth. Met. **120**, 855 (2001)
237. P.B. Allen, R.C. Dynes, Phys. Rev. B **12**, 905 (1975)
238. H. Kino, H. Fukuyama, J. Phys. Soc. Japan **65**, 2158 (1996)
239. H. Kondo, T. Moriya, Physica B **281 & 282**, 940 (2000)
240. A. Painelli, A. Girlando, A. Fortunelli, Phys. Rev. B **64**, 054509 (2001)
241. A.M. Kini, J.A. Schlueter, B.H. Ward, U. Geiser, H.H. Wang, Synth. Met. **120**, 713 (2001)
242. J.A. Schlueter, A.M. Kini, B.H. Ward, U. Geiser, H.H. Wang, J. Mohtasham, R.W. Winter, G.L. Gard, Physica C **351**, 261 (2001)
243. Y. Watanabe, T. Shimazu, T. Sasaki, N. Toyota, Synth. Met. **86**, 1917 (1997)
244. J.M. Williams, A.J. Schultz, U. Geiser, K.D. Carlson, A.M. Kini, H.H. Wang, W.K. Kwok, M.-H. Whangbo, J.E. Schirber, Science **252**, 1501 (1991)
245. R.S. Edwards, A. Narduzzo, J. Singleton, A. Ardavan, J.A. Schlueter, J. Phys. Condens. Mat. **15** L483 (2003)
246. D. Mailly, M. Ribault, K. Bechgaard, J. Physique **44**, C3-1037 (1983)
247. P. Garoche, R. Brusetti, D. Jérome, K. Bechgaard, J. Physique Lett. **43**, L147 (1982)
248. K. Murata, M. Tokumoto, H. Anzai, K. Kajimura, T. Ishiguro, Jpn. J. Appl. Phys. **26**, Suppl. 26–3, 1367 (1987)
249. A.M. Clogston, Phys. Rev. Lett. **9**, 266 (1962)
250. B.S. Chandrasekhar, Appl. Phys. Lett. **1**, 7 (1962)
251. R.L. Greene, P. Haen, S.Z. Huang, E.M. Engler, M.Y. Choi, P.M Chaikin, Mol. Cryst. Liq. Cryst. **79**, 539 (1982)
252. P.M. Chaikin, M.Y. Choi, R.L. Greene, J. Magn. Magn. Mater. **31**, 1268 (1983)
253. I.J. Lee, M.J. Naughton, G.M. Danner, P.M. Chaikin, Phys. Rev. Lett. **78**, 3555 (1997)
254. H. Schwenk, K. Andres, F. Wudl, E. Aharon-Shalom, Solid State Commun. **45**, 767 (1983)
255. Y. Shimojo, A.E. Kovalev, S. Kamiya, E. Ohmichi, T. Ishiguro, H. Yamochi, G. Saito, A. Ayari, P. Monceau, Physica B **294–295**, 427 (2001)
256. C. Mielke, J. Singleton, M.-S. Nam, N. Harrison, C.C. Agosta, B. Fravel, L.K. Montgomery, J. Phys.: Cond. Mat. **13**, 8325 (2001)
257. H. Mayaffre, P. Wzietek, D. Jérome, C. Lenoir, P. Batail, Phys. Rev. Lett. **75**, 4122 (1995)
258. T. Kawasaki, M.A. Tanatar, T. Ishiguro, H. Tanaka, A. Kobayashi, H. Kobayashi, Synth. Met. **120**, 771 (2001)
259. F. Zuo, J.S. Brooks, R.H. McKenzie, J.A. Schlueter, J.M. Williams, Phys. Rev. B **61**, 750 (2000)
260. J. Singleton, J.A. Symington, M.S. Nam, A. Ardavan, M. Kurmoo, P. Day, J. Phys.: Cond. Mat. **12**, L641 (2001)
261. M.A. Tanatar, T. Ishiguro, H. Tanaka, A. Kobayashi, H. Kobayashi, J. Supercond. **12**, 511 (1999)
262. J. Hagel, S. Wanka, D. Beckmann, J. Wosnitza, D. Schweitzer, W. Strunz, M. Thumfart, Physica C **291**, 213 (1997)
263. M. Dressel, O. Klein, G. Grüner, K.D. Carlson, H.H. Wang, J.M. Williams, Phys. Rev. B **50**, 13603 (1994)
264. P.A. Mansky, P.M. Chaikin, R.C. Haddon, Phys. Rev. B **50**, 15929 (1994)
265. M. Pinterić, S. Tomić, M. Prester, D. Drobac, O. Milat, K. Maki, D. Schweitzer, I. Heinen, W. Strunz, Phys. Rev. B **61**, 7033 (2000)
266. D.R. Harshman, R.N. Kleiman, R.C. Haddon, S.V. Chichester-Hicks, M.L. Kaplan, L.W. Rupp Jr., T. Pfiz, D.L. Williams, D.B. Mitzi, Phys. Rev. Lett. **64**, 1293 (1990)
267. M. Lang, N. Toyota, T. Sasaki, H. Sato, Phys. Rev. Lett. **69**, 1443 (1992)
268. M. Lang, N. Toyota, T. Sasaki, H. Sato, Phys. Rev. B **46**, 5822 (1992)
269. D. Achkir, M. Poirier, C. Bourbonnais, G. Quirion, C. Lenoir, P. Batail, D. Jérome, Phys. Rev. B **47**, 11595 (1993)
270. J.E. Graebner, R.C. Haddon, S.V. Chichester, S.H. Glarum, Phys. Rev. B **41**, 4808 (1990)
271. P. Fulde, R.A. Ferrell, Phys. Rev. **135**, A550 (1964)
272. A.I. Larkin, Y.N. Ovchinnikov, Sov. Phys. JETP **20**, 762 (1965)
273. J.A. Symington, J. Singleton, M.-S. Nam, A. Ardavan, M. Kurmoo, P. Day, Physica B **294–295**, 418 (2001)
274. T. Ishiguro, J. Phys. IV France **10**, Pr3–139 (2000)
275. G. Blatter, M.V. Feigel'man, V.B. Geshkenbein, A.I. Larkin, V.M. Vinokur, Rev. Mod. Phys. **66**, 1125 (1994)
276. E. Zeldov, D. Majer, M. Konczykowski, V.B. Geshkenbein, V.M. Vinokur, H. Shtrikman, Nature **375**, 373 (1995)
277. S. Schilling, R.A. Fisher, N.E. Phillips, U. Welp, D. Dasgupta, W.K. Kwok, G.W. Crabtree, Nature **382**, 791 (1996)
278. J.R. Clem, Phys. Rev. B **43**, 7837 (1991)
279. S.L. Lee, F.L. Pratt, S.J. Blundell, C.M. Aegerter, P.A. Pattenden, K.H. Chow, E.M. Forgan, T. Sasaki, W. Hayes, H. Keller, Phys. Rev. Lett. **79**, 1563 (1997)

280. L.Ya. Vinnikov, T.L. Barkov, M.V. Kartsovnik, N.D. Kushch, Phys. Rev. B **61**, 14358 (2000)
281. T. Nishizaki, T. Sasaki, T. Fukase, N. Kobayashi, Phys. Rev. B **54**, R3760 (1996)
282. M. Inada, T. Sasaki, T. Nishizaki, N. Kobayashi, S. Yamada, T. Fukase, J. Low Temp. Phys. **117**, 1423 (1999)
283. M.M. Mola, J.T. King, C.P. McRaven, S. Hill, J.S. Qualls, J.S. Brooks, Phys. Rev. B **62**, 5965 (2000)
284. M.M. Mola, S. Hill, J.S. Brooks, J.S. Qualls, Phys. Rev. Lett. **86**, 2130 (2001)
285. C. Martin, C.C. Agosta, S.W. Tozer, H.A. Radovan, T. Kinoshita, and M. Tokumoto, J. Low Temp. Phys. **138**, 1025 (2005)
286. T. Shibauchi, M. Sato, S. Ooi, T. Tamegai, Phys. Rev. B **57**, R5622 (1998)
287. T. Sasaki, W. Biberacher, K. Neumaier, W. Hehn, K. Andres, T. Fukase, Phys. Rev. B **56**, 10889 (1998)
288. N. Yoneyama, A. Higashihara, T. Sasaki, N. Kobayashi, S. Yamada, Synth. Met. **133**, 223 (2003)
289. L. Fruchter, A. Aburto, C. Pham-Phu, Phys. Rev. B **56**, R2936 (1997)
290. F. Steinmeyer, R. Kleiner, P. Müller, H. Müller, K. Winzer, Europhys. Lett. **25**, 459 (1994)
291. H. Shimahara, J. Phys. Soc. Japan **66**, 541 (1997)
292. H. Shimahara, J. Phys. Soc. Japan **67**, 1872 (1998)
293. H. Shimahara, Phys. Rev. B **50**, 12760 (1994)
294. M. Tachiki, S. Takahashi, P. Gegenwart, M. Weiden, M. Lang, C. Geibel, F. Steglich, R. Modler, C. Paulsen, Y. Ōnuki, Z. Phys. B **100**, 369 (1996)
295. P. Gegenwart, M. Deppe, M. Köppen, F. Kromer, M. Lang, R. Modler, M. Weiden, C. Geibel, F. Steglich, T. Fukase, N. Toyota, Ann. Physik **5**, 307 (1996)
296. S. Uji, H. Shinagawa, T. Terashima, T. Yakabe, Y. Terai, M. Tokumoto, A. Kobayashi, H. Tanaka, H. Kobayashi, Nature, **410**, 908 (2001)
297. L. Balicas, J.S. Brooks, K. Storr, S. Uji, M. Tokumoto, H. Tanaka, H. Kobayashi, A. Kobayashi, V. Barzykin, L.P. Gor'kov, Phys. Rev. Lett. **87**, 0670021 (2001)
298. S. Uji, C. Terakura, T. Terashima, T. Yakabe, Y. Imanaka, Y. Terai, S. Yasuzuka, M. Tokumoto, F. Sakai, A. Kobayashi, H. Tanaka, H. Kobayashi, L. Balicas, J.S. Brooks, Synth. Met. **133**, 481 (2003)
299. J. Bardeen, L.N. Cooper, J.R. Schrieffer, Phys. Rev. **108**, 1175 (1957)
300. F. Steglich, J. Aarts, C.D. Bredl, W. Lieke, D. Meschede, W. Franz, H. Schäfer, Phys. Rev. Lett. **43**, 1892 (1979)
301. J.G. Bednorz, K.A. Müller, Z. Phys. B **64**, 189 (1986)
302. Y. Maeno, H. Hashimoto, K. Yoshida, D. Nishizaki, T. Fujita, J.G. Bednorz, F. Lichtenberg, Nature **372**, 532 (1994)
303. S. Mazumdar, Solid State Commun. **66**, 427 (1988)
304. Y.J. Uemura, L.P. Le, G.M. Luke, B.J. Sternlieb, W.D. Wu, J.H. Brewer, T.M. Riseman, C.L. Seaman, M.B. Maple, M. Ishikawa, D.G. Hinks, J.D. Jorgensen, G. Saito, H. Yamochi, Phys. Rev. Lett. **66**, 2665 (1991)
305. W.A. Little, Organic conductors: An overview, In: *Organic Conductors, Fundamentals and Applications*, ed. by J.-P. Farges (Marcel Dekker, New York, 1994) pp. 1–24.
306. K. Yamaji, Solid State Commun. **61**, 413 (1987)
307. A. Girlando, M. Masino, A. Brillante, R.G. Della Valle, E. Venuti, Phys. Rev. B **66**, 100507(R) (2002)
308. A. Girlando, M. Masino, G. Visentini, R.G. Della Valle, A. Brillante, E. Venuti, Phys. Rev. B **62**, 14476 (2000)
309. G. Varelogiannis, Phys. Rev. Lett. **88**, 117005 (2002)
310. J. Schmalian, Phys. Rev. Lett. **81**, 4232 (1998)
311. E. Demiralp, W.A. Goddard III, Phys. Rev. B **56**, 11907 (1997)
312. H. Kondo, T. Moriya, J. Phys. Soc. Japan **67**, 3695 (1998)
313. H. Kondo, T. Moriya, J. Phys. Cond. Mat. **11**, L363 (1999)
314. H. Kondo, T. Moriya, J. Phys. Soc. Japan **68**, 3170 (1999)
315. H. Kino, H. Kontani, J. Phys. Soc. Japan **67**, 3691 (1998)
316. M. Vojta, E. Dagotto, Phys. Rev. B **59**, R713 (1999)
317. K. Kuroki, H. Aoki, Phys. Rev. B **60**, 3060 (1999)
318. T. Jujo, K. Yamada, J. Phys. Soc. Japan **68**, 1331 (1999)
319. H. Kino, H. Kontani, J. Phys. Soc. Japan **68**, 1481 (1999)
320. T. Jujo, K. Yamada, J. Phys. Soc. Japan **68**, 2198 (1999)
321. R. Louati, S. Charfi-Kaddour, A. Ben Ali, R. Bennaceur, M. Héritier, Phys. Rev. B **62**, 5957 (2000)
322. J. Marino, R.H. McKenzie, Phys. Rev. Lett. **87**, 237002 (2001)
323. K. Maki, In: *Superconductivity* Vol. **2**. ed. by R.D. Parks (M. Dekker, New York 1969) p. 1035.

324. A.A. Abrikosov, L.P. Gor'kov, Sov. Phys. JETP **12**, 1243 (1969)
325. H. Keiber, C. Geibel, B. Renker, H. Rietschel, H. Schmidt, H. Wühl, G.R. Stewart, Phys. Rev. B **30**, 2542 (1984)
326. B. Lüthi, M. Herrmann, W. Assmus, H. Schmidt, H. Rietschel, H. Wühl, U. Gottwick, G. Sparn, F. Steglich, Z. Phys. B. **60**, 387 (1985)
327. M. Jourdan, M. Huth, H. Adrian, Nature **398**, 47 (1999)
328. N.K. Sato, N. Aso, K. Miyake, S. Shiina, P. Thalmeier, G. Varelogiannis, C. Geibel, F. Steglich, P. Fulde, T. Komatsubara, Nature 410, 340 (2001)
329. C.C. Tsuei, J.R. Kirtley, C.C. Chi, Lock See Yu-Jahnes, A. Gupta, T. Shaw, J.Z. Sun, M.B. Ketchen, Phys. Rev. Lett. **73**, 593 (1994)
330. D.J. van Harlingen, Phase-sensitive tests of the symmetry of the pairing state in the high-temperature superconductors — Evidence for $d_{x^2-y^2}$ symmetry, Rev. Mod. Phys. **67**, 515 (1995)
331. P. Auban-Senzier, C. Bourbonnais, D. Jérome, C. Lenoir, P. Batail, E. Canadell, J.P. Buisson, S. Lefrant, J. Phys. I France **3**, 871 (1993)
332. K.D. Carlson, J.N. Williams, U. Geiser, A.M. Kini, H.H. Wang, R.A, Klemm, S.K. Kumar, J.A. Schlueter, J.R. Ferraro, K.R. Lykke, P. Wurz, D.H. Parker, J.D.M. Sutin, Inorg. Chem. **114**, 10069 (1992)
333. J.D. Axe, G. Shirane, Phys. Rev. B **8**, 1965 (1973)
334. S.M. Shapiro, G. Shirane, Phys. Rev. B **12**, 4899 (1975)
335. N. Toyota, M. Lang, S. Ikeda, T. Kajitani, T. Shimazu, T. Sasaki, K. Shibata, Synth. Met. **86**, 2009 (1997)
336. R. Zeyher, G. Zwicknagl, Z. Phys. B **78**, 175 (1990)
337. J.M. Schrama, E. Rzepniewski, R.S. Edwards, J. Singleton, A. Ardavan, M. Kurmoo, P. Day, Phys. Rev. Lett. **83**, 3041 (1999)
338. T. Arai, K. Ichimura, K. Nomura, S. Takasaki, J. Yamada, S. Nakatsuji, H. Anzai, Phys. Rev. B **63**, 104518 (2001); Solid State Commun. **116**, 679 (2000)
339. J.M. Schrama, J. Singleton, Phys. Rev. Lett. **86**, 3453 (2001)
340. S. Hill, N. Harrison, M. Mola, J. Wosnitza, Phys. Rev. Lett. **86**, 3451 (2001)
341. T. Shibauchi, Y. Matsuda, M.B. Gaifullin, T. Tamegai, Phys. Rev. Lett. **86**, 3452 (2001)
342. H. Bando, S. Kashiwaya, H. Tokumoto, H. Anzai, N. Kinoshita, K. Kajimura, J. Vac. Sci. Technol. A **8**, 479 (1990)
343. H. Won, K. Maki, Physica B **312–313**, 44 (2002)
344. D.J. Scalapino, E. Loh, Jr., J.E. Hirsch, Phys. Rev. B **35**, 6694 (1987)
345. K. Kanoda, K. Miyagawa, A. Kawamoto, Y. Nakazawa, Phys. Rev. B **54**, 76 (1996)
346. M.J. Graf, S.-K. Yip, J.A. Sauls, D. Rainer , Phys. Rev. B **53**, 15147 (1996)
347. M.R. Norman, P.J. Hirschfeld, Phys. Rev. B **53**, 5706 (1996)
348. B. Mühlschlegel, Z. Phys. **155**, 313 (1959)
349. K. Kanoda, K. Akiba, K. Suzuki, T. Takahashi, G. Saito, Phys. Rev. Lett. **65**, 1271 (1990)
350. L.P. Le, G.M. Luke, B.J. Sternlieb, W.D. Wu, Y.J. Uemura, J.H. Brewer, T.M. Riseman, C.E. Stronach, G. Saito, H. Yamochi, H.H. Wang, A.M. Kini, K.D. Carlson, J.M. Williams, Phys. Rev. Lett. **68**, 1923 (1992)
351. A. Carrington, I.J. Bonalde, R. Prozorov, R.W. Giannetta, A.M. Kini, J.A. Schlueter, H.H. Wang, U. Geiser, J.M. Williams, Phys. Rev. Lett. **83**, 4172 (1999)
352. O. Klein, K. Holczer, G. Grüner, J.J. Chang, F. Wudl, Phys. Rev. Lett. **66**, 655 (1991)
353. I. Kosztin, Q.J. Chen, B. Janko, K. Levin, Phys. Rev. B **58**, R5936 (1998); Q.J. Chen, I. Kosztin, B. Janko, K. Levin, Phys. Rev. Lett. **81**, 4708 (1998); Q. Chen, I. Kosztin, K. Levin, Phys. Rev. Lett. **85**, 2801 (2000)
354. Y. Hasegawa, H. Fukuyama, J. Phys. Soc. Japan **56**, 877 (1987)
355. B. Andraka, J.S. Kim, G.R. Stewart, K.D. Carlson, H.H. Wang, J.M. Williams, Phys. Rev. B **40**, 11345 (1989)
356. Y. Nakazawa, K. Kanoda, Phys. Rev. B **55**, R8670 (1997)
357. H. Padamsee, J.E. Neighbor, C.A. Shiffman, J. Low Temp. Phys. **12**, 387 (1973)
358. N. Momono, M. Ido, Physica C **264**, 311 (1996)
359. J. Wosnitza, X. Liu, D. Schweitzer, H.J. Keller, Phys. Rev. B **50**, 12747 (1994)
360. S. Wanka, J. Hagel, D. Beckmann, J. Wosnitza, J.A. Schlueter, J.M. Williams, P.G. Nixon, R.W. Winter, G.L. Gard, Phys. Rev. B **57**, 3084 (1998)
361. J. Solyom, Adv. Phys. **28**, 201 (1979)
362. M.Y. Choi, P.M. Chaikin, S.Z. Huang, P. Haen, E.M. Engler, R.L. Greene, Phys. Rev. B **25**, 6208 (1982)

363. S. Bouffard, M. Ribault, R. Brusetti, D. Jérome, K. Bechgaard, J. Phys. C **15**, 2951 (1982)
364. C. Coulon, P. Delhaes, J. Amiell, J.P. Manceau, J.M. Fabre, L. Giral, J. Phys. (Paris) **43**, 1721 (1982)
365. S. Tomić, D. Jérome, D. Mailly, M. Ribault, K. Bechgaard, J. Phys. (Paris) Colloq. **44**, C3-1075 (1983)
366. A.A. Abrikosov, J. Low Temp. Phys. **53**, 359 (1983)
367. L.P. Gor'kov, D. Jérome, J. Phys. (Paris) Lett. **46**, L643 (1985)
368. M. Dressel, Physica C **317–318**, 89 (1999)
369. I.J. Lee, P.M. Chaikin, M.J. Naughton, Phys. Rev. B **62**, R14669 (2000)
370. M. Takigawa, H. Yasuoka, G. Saito, J. Phys. Soc. Jpn. **56**, 873 (1987)
371. S. Belin, K. Behnia, Phys. Rev. Lett. **79**, 2125 (1997)
372. I.J. Lee, P.M. Chaikin, M.J. Naughton, Phys. Rev. B **65**, 180502 (2002)
373. P. Fulde, K. Maki, Phys. Rev. **139**, A788 (1965)
374. I.J. Lee, S.E. Brown, W.G. Clark, M.J. Strouse, M.J. Naughton, W. Kang, P.M. Chaikin, Phys. Rev. Lett. **88**, 017004 (2002)
375. R.A. Klemm, A. Luther, M.R. Beasley, Phys. Rev. B **12**, 877 (1975)
376. X. Huang, K. Maki, Phys. Rev. B **39**, 6459 (1989)
377. A.G. Lebed', JETP Lett. **44**, 114 (1986); L.I. Burlachkov, L.P. Gor'kov, A.G. Lebed', Europhys. Lett. **4**, 941 (1987)
378. N. Dupuis, G. Montambaux, C.A.R. Sá de Melo, Phys. Rev. Lett. **70**, 2613 (1993)
379. I.J. Lee, D.S. Chow, W.G. Clark, J. Strouse, M.J. Naughton, P.M. Chaikin, S.E. Brown, Phys. Rev. B **68**, 092510 (2003)

21 Concepts in High Temperature Superconductivity

E. W. Carlson Purdue University, Dept. of Physics, IN, USA
V. J. Emery Brookhaven National Laboratory Dept. of Physics, New York, USA
S. A. Kivelson Stanford University, Dept. of Physics, Ca, USA
D. Orgad The Hebrew Univ. of Jerusalem, Racah Institute of Physics, Israel

Preface

In this paper, we discuss the core theoretical issues associated with the mechanism of high temperature superconductivity. Our primary focus is on physics at intermediate temperature scales of order T_c (as well as the somewhat larger "pseudogap" temperature) and energies of order the gap maximum, Δ_0, in the context of rigorous theoretical analysis. We also briefly discuss these ideas in the experimental context of the cuprates, and close with a set of open issues. Note, high temperature superconductivity is difficult to achieve, even in principle. It requires that the energy scales of both pairing and phase coherence be simultaneously elevated, yet model calculations and empirics reveal an inherent competition between the two. Furthermore, in a highly correlated system, the electronic structure is prone to local inhomogeneity (and therefore also to competing orders). However, such mesoscale structure can mitigate the deleterious effects of the Coulomb repulsion, while providing a novel route to superconductivity through kinetic-energy driven pairing.

21.1 Introduction

The virtues of BCS theory are extolled.

Conventional superconductors are good metals in their normal states, and are well described by Fermi liquid theory. They also exhibit a hierarchy of energy scales, $E_F \gg \hbar\omega_D \gg k_B T_c$, where E_F and $\hbar\omega_D$ are the Fermi and Debye energies, respectively, and T_c is the superconducting transition temperature. E_FFermi energy ω_DDebye frequency k_BBoltzmann's constant Moreover, one typically does not have to think about the interplay between superconductivity and any other sort of collective ordering, since in most cases the only weak coupling instability of a Fermi liquid is to superconductivity. These reasons underlie the success of the BCS-Eliashberg–Migdal theory in describing metallic superconductors [1].

The assumptions of BCS theory are violated by the high temperature superconductors.
By contrast, the cuprate high temperature superconductors [2] (and various other newly discovered materials with high superconducting transition temperatures) are highly correlated "bad metals," [3,4] with normal state properties that are not at all those of a Fermi liquid. There is compelling evidence that they are better thought of as doped Mott insulators, rather than as strongly interacting versions of conventional metals [5–7]. The cuprates also exhibit numerous types of low temperature order which interact strongly with the superconductivity, the most prominent being antiferromagnetism and the unidirectional charge and spin density wave "stripe" order. These orders can compete or coexist with superconductivity. Furthermore, whereas phase fluctuations of the superconducting order parameter are negligibly small in conventional superconductors, fluctuation effects are of order one in the high temperature superconductors because of their much smaller superfluid stiffness.

Apparently, none of this complicates the fundamental character of the superconducting order parameter: it is still a charge 2e scalar field, although it transforms according to a nontrivial representation of the point group symmetry of the crystal—it is a "*d*-wave superconductor." At asymptotically low temperatures and energies, there is every reason to expect that the physics is dominated by nodal quasiparticles that are similar to those that one might find in a BCS superconductor of the same symmetry. Indeed, there is considerable direct experimental evidence that this expectation is realized [8–11]. However, the failure of Fermi liquid theory to describe the normal state and the presence of competing orders necessitates an entirely different approach to understanding much of the physics, especially at intermediate scales of order $k_B T_c$, which is the relevant scale for the mechanism of high temperature superconductivity.

The purpose of this paper.
It is the purpose of this paper to address the physics of high temperature superconductivity at these intermediate scales. We pay particular attention to the problem of charge dynamics in doped Mott insulators. We also stress the physics of quasi-one-dimensional superconductors, in part because that is the one *theoretically* well understood limit in which superconductivity emerges from a non-Fermi liquid normal state. To the extent that the physics evolves adiabatically from the quasi-one to the quasi-two-dimensional limit, this case provides considerable insight into the actual problem of interest. The soundness of this approach can be argued from the observation that $YBa_2Cu_3O_{7-\delta}$ (YBCO) (which is strongly orthorhombic) exhibits very similar physics to that of the more tetragonal cuprates. Since the conductivity and the superfluid density in YBCO exhibit a factor of 2 or greater anisotropy within the plane [12,13], this material is already part way toward the quasi-one-dimensional limit without substantial changes in the physics! In the second place, because of the delicate interplay between stripe and superconducting orders observed in the cuprates, it is reasonable to speculate that the electronic structure may be literally quasi-one-dimensional at the local level, even when little of this anisotropy is apparent at the macroscopic scale.

Mesoscale electronic structure is emphasized.
A prominent theme of this article is the role of *mesoscale structure* [14]. Because the kinetic energy is strongly dominant in good metals, their wavefunctions are very rigid and hence the electron density is highly homogeneous in real space, even in the presence of a spatially varying external potential (e.g. disorder). In a highly correlated system, the electronic structure is much more prone to inhomo-

geneity [15–17], and intermediate scale structures (stripes are an example) are likely an integral piece of the physics. Indeed, based on the systematics of local superconducting correlations in exact solutions of various limiting models and in numerical "experiments" on $t - J$ and Hubbard models, we have come to the conclusion that mesoscale structure may be *essential* to a mechanism of high temperature superconducting pairing. (See Sects. 21.10 and 21.11.) This is a potentially important guiding principle in the search for new high temperature superconductors.

A kinetic energy driven mechanism is called for.
This is related to a concept that we believe is central to the mechanism of high temperature superconductivity: the condensation is driven by a lowering of kinetic energy. A Fermi liquid normal state is essentially the ground state of the electron kinetic energy, so any superconducting state which emerges from it must have *higher* kinetic energy. The energy gain which powers the superconducting transition from a Fermi liquid must therefore be energy of interaction—this underlies any BCS-like approach to the problem. In the opposite limit of strong repulsive interactions between electrons, the normal state has high kinetic energy. It is thus possible to conceive of a *kinetic energy driven mechanism of superconductivity*, in which the strong frustration of the kinetic energy is partially relieved upon entering the superconducting state [18–24]. Such a mechanism does not require subtle induced attractions, but derives directly from the strong repulsion between electrons. As will be discussed in Sect. 21.10, the proximity effect in the conventional theory of superconductivity is a prototypical example of such a kinetic energy driven mechanism: when a superconductor and a normal metal are placed in contact with each other, the electrons in the metal pair (even if the interactions between them are repulsive) in order to lower their zero point kinetic energy by delocalizing across the interface. A related phenomenon, which we have called the "spin gap proximity effect" [20,25] (see Sect. 21.10.4), produces strong superconducting correlations in $t - J$ and Hubbard ladders [26], where the reduction of kinetic energy transverse to the ladder direction drives pairing. It is unclear to us whether experiments can unambiguously distinguish between a potential energy and a kinetic energy driven mechanism.[1] But since the interaction between electrons is strongly repulsive for the systems in question, we feel that the a priori case for a kinetic energy driven mechanism is very strong.

The plan of the article is discussed.
Our approach in this article is first to analyze various aspects of high temperature superconductivity as abstract problems in theoretical physics, and then to discuss their specific application to the cuprate high temperature superconductors.[2] We have also attempted to make each section self contained. Although many readers no doubt will be drawn to read this compelling article in its entirety, we have also tried to make it useful for those readers who are interested in learning about one or another more specific issue. The first eleven sections focus on theoretical issues, except for Sect. 21.4, where we briefly sketch the mechanism in light of *our* view of the phase diagram of the cuprate superconductors. In the final section, we focus more directly on the physics of high temperature superconductivity in the cuprates, and summarize some of the experimental issues that remain, in our opinion, unsettled.

$\hbar = 1$
$k_B = 1.$
Except where dimensional arguments are important, we will henceforth work with units in which $\hbar = k_B = 1$.

[1] Recent papers by Molegraaf et al. [27] and Santadner-Syro et al. [28] present very plausible experimental evidence of a kinetic energy driven mechanism of superconductivity in at least certain high temperature superconductors.

[2] While examples of similar behavior can be found in other materials, for ease of exposition we have focused on this single example.

21.2 High Temperature Superconductivity is Hard to Attain

Catch 22
Superconductivity in metals is the result of two distinct quantum phenomena: pairing and long range phase coherence. In conventional homogeneous superconductors, the phase stiffness is so great that these two phenomena occur simultaneously. On the other hand, in granular superconductors and Josephson junction arrays, pairing occurs at the bulk transition temperature of the constituent metal, while long range phase coherence, if it occurs at all, obtains at a much lower temperature characteristic of the Josephson coupling between superconducting grains. To achieve high temperature superconductivity requires that both scales be elevated simultaneously. However, given that the bare interactions between electrons are strongly repulsive, it is somewhat miraculous that electron pairing occurs at all. Strong interactions, which might enable pairing at high scales, typically also have the effect of strongly suppressing the phase stiffness, and moreover typically induce other orders[3] in the system which compete with superconductivity.

BCS is not for high T_c superconductivity.
It is important in any discussion of the theory of high temperature superconductivity to have clearly in mind why conventional metallic superconductors, which are so completely understood in the context of the Fermi liquid based BCS-Eliashberg theory, rarely have T_c's above 15K, and never above 30 K. In this section, we briefly discuss the principal reasons why a straightforward extension of the BCS-Eliashberg theory *does not* provide a framework for understanding high temperature superconductivity, whether in the cuprate superconductors, or in C_{60}, or *possibly* even BaKBiO.

21.2.1 Effects of the Coulomb Repulsion and Retardation on Pairing

In conventional BCS superconductors, the instantaneous interactions between electrons are typically repulsive (or at best very weakly attractive)—it is only because the phonon induced attraction is retarded that it (barely) dominates at low frequencies. Even if new types of intermediate bosons are invoked to replace phonons in a straightforward variant of the BCS mechanism, the instantaneous interactions will still be repulsive, so any induced attraction is typically weak, and only operative at low frequencies.

Never forget the Coulomb interaction.
Strangely enough, the deleterious effects of the Coulomb interaction on high temperature superconductivity has been largely ignored in the theoretical literature. The suggestion has been made that high pairing scales can be achieved by replacing the relatively low frequency phonons which mediate the pairing in conventional metals by higher frequency bosonic modes, such as the spin waves in the high temperature superconductors [29–32] or the shape modes [33,34] of C_{60} molecules. However, in most theoretical treatments of this idea, the Coulomb pseudopotential is either neglected or treated in a cavalier manner. That is, models are considered in which the instantaneous interactions between electrons are strongly attractive. This is almost certainly [14,20,35–37] an unphysical assumption!

In Sect. 21.9, we use modern renormalization group (RG) methods [38,39] to derive the conventional expression for the Coulomb pseudopotential, and how it enters the effective pairing interaction at frequencies lower than the Debye frequency, ω_D. This theory is well controlled so long as $\omega_D \ll E_F$ and the interaction strengths are not too large. It is worth reflecting on a well known, but remarkably profound result that emerges from this analysis: As electronic states are integrated out between the microscopic scale E_F and the intermediate scale, ω_D, the electron-phonon interaction is unrenormalized (and so can be well estimated from microscopic considerations), but the Coulomb repulsion is reduced from a bare value, μ, to a renormalized value,

$$\mu^* = \mu/[1 + \mu \log(E_F/\omega_D)]. \tag{21.1}$$

Here, as is traditional, μ and μ^* are the dimensionless measures of the interaction strength obtained

[3] That is, magnetic, structural, etc.

by multiplying the interaction strength by the density of states. We define λ in an analogous manner for the electron-phonon interaction. Thus, even if the instantaneous interaction is repulsive (i.e. $\lambda - \mu < 0$), the effective interaction at the scale ω_D will nonetheless be attractive ($\lambda - \mu^* > 0$) for $\omega_D \ll E_F$. Below this scale, the standard RG analysis yields the familiar weak coupling estimate of the pairing scale T_p:

$$T_p \sim \omega_D \exp[-1/(\lambda - \mu^*)]. \tag{21.2}$$

Retardation is an essential feature of the BCS mechanism.
The essential role of retardation is made clear if one considers the dependence of T_p on ω_D:

$$\frac{d\log[T_p]}{d\log[\omega_D]} = 1 - \left[\mu^* \log\left(\frac{T_p}{\omega_D}\right)\right]^2. \tag{21.3}$$

So long as $\omega_D \ll E_F \exp[-(1-\lambda)/\lambda\mu]$, we have $\frac{d\log[T_p]}{d\log[\omega_D]} \approx 1$, and T_p is a linearly rising function of ω_D, giving rise to the conventional isotope effect.[4] However, when $\omega_D > T_p \exp[1/\mu^*]$, we have $\frac{d\log[T_p]}{d\log[\omega_D]} < 0$, and T_p becomes a decreasing function of ω_D! Clearly, unless ω_D is exponentially smaller than E_F, superconducting pairing is impossible by the conventional mechanism.[5]

This problem is particularly vexing in the cuprate high temperature superconductors and similar materials, which have low electron densities, and incipient or apparent Mott insulating behavior. This means that screening of the Coulomb interaction is typically poor, and μ is thus expected to be large. Specifically, from the inverse Fourier transform of the $\boldsymbol{k}$ dependent gap function measured [40] in angle resolved photoemission spectroscopy (ARPES) on $Bi_2Sr_2CaCu_2O_{8+\delta}$, it is possible to conclude (at least at the level of the BCS gap equation) that the dominant pairing interactions have a range equal to the nearest neighbor copper distance.

Pairing's Bane
Since this distance is less than the distance between doped holes, it is difficult to believe that metallic screening is very effective at these distances. From cluster calculations and an analysis of various local spectroscopies, a crude estimate [20] of the Coulomb repulsion at this distance is of order 0.5 eV or more. To obtain pairing from a conventional mechanism with relatively little retardation, it is necessary that the effective attraction be considerably larger than this!

We are therefore led to the conclusion that the only way a BCS mechanism can produce a high pairing scale is if the effective attraction, λ, is very large indeed. This, however, brings other problems with it.

21.2.2 Pairing Versus Phase Ordering

In most cases, it is unphysical to assume the existence of strong attractive interactions between electrons. However, even supposing we ignore this, strong attractive interactions bring about other problems for high temperature superconductivity: 1) There is a concomitant strong reduction of the phase ordering temperature and thus of T_c. 2) There is the possibility of competing orders. We discuss the first problem here, and the second in Sect. 21.2.3.

Strong attractive interactions typically result in a large increase in the effective mass, and a corresponding reduction of the phase ordering temperature. Consider, for example, the strong coupling limit of the negative U Hubbard model [41] or the Holstein model [42], discussed in Sect. 21.10. In both cases, pairs have a large binding energy, but they typically Bose condense at a very low temperature because of the large effective mass of a tightly bound pair—the effective mass is proportional to $|U|$ in the Hubbard model and is exponentially large in the Holstein model (see Sect. 21.10.)

[4] Recall, for phonons, $d\log[\omega_D]/d\log[M] = -1/2$.
[5] In the present discussion we have imagined varying ω_D while keeping fixed the electron-phonon coupling constant, $\lambda = \frac{C}{M\omega_D^2} = \frac{C}{K}$, where C is proportional to the (squared) gradient of the electron-ion potential and K is the "spring constant" between the ions. If we consider instead the effect of increasing ω_D at fixed C/M, it leads to a decrease in λ and hence a very rapid suppression of the pairing scale.

Phase ordering is a serious business in the cuprates. Whereas in conventional superconductors, the bare superfluid stiffness is so great that even a substantial renormalization of the effective mass would hardly matter, in the cuprate high temperature superconductors, the superfluid stiffness is small, and a substantial mass renormalization would be catastrophic. The point can be made most simply by considering the result of simple dimensional analysis. The density of doped holes per plane in an optimally doped high temperature superconductor is approximately $n_{2d} = 10^{14}\text{cm}^{-2}$. Assuming a density of hole pairs that is half this, and taking the rough estimate for the pair effective mass, $m^* = 2m_e$, we find a phase ordering scale,

$$T_\theta = \hbar^2 n_{2d}/2m^* \approx 10^{-2} eV \approx 100\,\text{K}\,. \quad (21.4)$$

Since this is in the neighborhood of the actual T_c, it clearly implies that any large mass renormalization would be incompatible with a high transition temperature. What about conventional superconductors? A similar estimate in a $W = 10$ Å thick Pb film gives $T_\theta = \hbar^2 n_{3d} W/2m^* \approx 1 eV \approx 10,000$ K! Clearly, phase fluctuations are unimportant in Pb. This issue is addressed in detail in Sect. 21.8.

A general principle is proposed: "optimal" T_c occurs as a crossover.
We have seen how T_p and T_θ have opposite dependence on coupling strength. If this is a general trend, then it is likely that any material in which T_c has been optimized has effectively been tuned to a crossover point between pairing and condensation. A modification of the material which produces stronger effective interactions will increase phase fluctuations and thereby reduce T_c, while weaker interactions will lower T_c because of pair breaking. In Sect. 21.8 it will be shown that optimal doping in the cuprate superconductors corresponds to precisely this sort of crossover from a regime in which T_c is determined by phase ordering to a pairing dominated regime.

21.2.3 Competing Orders

A Fermi liquid is a remarkably robust state of matter. In the absence of nesting, it is stable for a range[6] of repulsive interactions; the Cooper instability is its only weak coupling instability. The phase diagram of simple metals consists of a high temperature metallic phase and a low temperature superconducting state. When the superconductivity is suppressed by either a magnetic field or appropriate disorder (e.g. paramagnetic impurities), the system remains metallic down to the lowest temperatures.

The situation becomes considerably more complex for sufficiently strong interactions between electrons. In this case, the Fermi liquid description of the normal or high temperature phase breaks down[7] and many possible phases compete. In addition to metallic and superconducting phases, one would generally expect various sorts of electronic "crystalline" phases, including charge ordered phases (i.e. a charge density wave (CDW) of which the Wigner crystal is the simplest example) and spin ordered phases (i.e. a spin density wave (SDW) of which the Néel state is the simplest example).

Typically, one thinks of such phases as insulating, but it is certainly possible for charge and spin order to coexist with metallic or even superconducting electron transport. For example, this can occur in a conventional weak coupling theory if the density wave order opens a gap on only part of the Fermi surface, leaving other parts gapless [43]. It can also occur in a multicomponent system, in which the density wave order involves one set of electronic orbitals, and the conduction occurs through others—this is the traditional understanding of the coexisting superconducting and magnetic order in the Chevrel compounds [44].

"Stripe" order
Such coexistence is also possible for less conventional orders. One particular class of competing orders is known loosely as "stripe" order. Stripe order refers to unidirectional density wave order, i.e. order which spontaneously breaks translational symmetry

[6] As long as the interactions are not too strong.
[7] Whether it breaks down for fundamental or practical reasons is unimportant.

in one direction but not in others. We will refer to charge stripe order, if the broken symmetry leads to charge density modulations and spin stripe order if the broken symmetry leads to spin density modulations, as well. Charge stripe order can occur without spin order, but spin order (in a sense that will be made precise, below) implies charge order [45]. Both are known on theoretical and experimental grounds to be a prominent feature of doped Mott insulators in general, and the high temperature superconductors in particular [6,46–51]. Each of these orders can occur in an insulating, metallic, or superconducting state.

In recent years there has been considerable theoretical interest in other types of order that could be induced by strong interactions. From the perspective of stripe phases, it is natural to consider various partially melted "stripe liquid" phases, and to classify such phases, in analogy with the classification of phases of classical liquid crystals, according to their broken symmetries [52]. For instance, one can imagine a phase that breaks rotational symmetry (or, in a crystal, the point group symmetry) but not translational symmetry, i.e. quantum (ground state) analogues of nematic or hexatic liquid crystalline phases. Still more exotic phases, such as those with ground state orbital currents [53–58] or topological order [59], have also been suggested as the explanation for various observed features of the phenomenology of the high temperature superconductors.

Competition matters...
Given the complex character of the phase diagram of highly correlated electrons, it is clear that the conventional approach to superconductivity, which focuses solely on the properties of the normal metal and the pure superconducting phase, is suspect. A more global approach, which takes into account some (or all) of the competing phases is called for. Moreover, even the term "competing" carries with it a prejudice that must not be accepted without thought. In a weakly correlated system, in which any low temperature ordered state occurs as a Fermi surface instability, different orders generally *do* compete: if one order produces a gap on part of the Fermi surface, there are fewer remaining low energy degrees of freedom to participate in the formation of another type of order.

... and so does symbiosis.
For highly correlated electrons, however, the sign of the interaction between different types of order is less clear. It can happen [60] that under one set of circumstances, a given order tends to enhance superconductivity and under others, to suppress it.

The issue of competing orders, of course, is not new. In a Fermi liquid, strong effective attractions typically lead to lattice instabilities, charge or spin density wave order, etc. Here the problem is that the system either becomes an insulator or, if it remains metallic, the residual attraction is typically weak. For instance, lattice instability has been seen to limit the superconducting transition temperature of the A15 compounds, the high temperature superconductors of a previous generation. Indeed, the previous generation of BCS based theories which addressed the issue always concluded that competing orders suppress superconductivity [44].

More recently it has been argued that near an instability to an ordered state there is a low lying collective mode (the incipient Goldstone mode of the ordered phase) which can play the role of the phonon in a BCS-like mechanism of superconductivity [29,61,62]. In an interesting variant of this idea, it has been argued that in the neighborhood of a zero temperature transition to an ordered phase, quantum critical fluctuations can mediate superconducting pairing in a more or less traditional way [63–65]. There are reasons to expect this type of fluctuation mediated pair binding to lead to a depression of T_c. If the collective modes are nearly Goldstone modes (as opposed to relaxational "critical modes"), general considerations governing the couplings of such modes in the ordered phase imply that the superconducting transition temperature is depressed substantially from any naive estimate by large vertex corrections [66]. Moreover, in a regime of large fluctuations to a nearby ordered phase, one generally expects a density of states reduction due to the development of a pseudogap; feeding this pseudogapped density of states back into the BCS-Eliashberg theory will again result in a significant reduction of T_c.

21.3 Superconductivity in the Cuprates: General Considerations

While the principal focus of the present article is theoretical, the choice of topics and models and the approaches are very much motivated by our interest in the experimentally observed properties of the cuprate high temperature superconductors. In this section, we discuss briefly some of the most dramatic (and least controversial) aspects of the phenomenology of these materials, and what sorts of constraints those observations imply for theory. As we are primarily interested in the origin of high temperature superconductivity, we will deal here almost exclusively with experiments in the temperature and energy ranges between about $T_c/2$ and a few times T_c.

Before starting, there are a number of descriptive terms that warrant definition. The parent state of each family of the high temperature superconductors is an antiferromagnetic "Mott" insulator with one hole (and spin $1/2$) per planar copper.[8] These insulators are transformed into superconductors by introducing a concentration, x, of "doped holes" into the copper oxide planes. As a function of increasing x, the antiferromagnetic transition temperature is rapidly suppressed to zero, then the superconducting transition temperature rises from zero to a maximum and then drops down again (see Fig. 21.1). Where T_c is an increasing function of x, the materials are said to be "underdoped." They are "optimally doped" where T_c reaches its maximum at $x \approx 0.15$, and they are "overdoped" for larger x. In the underdoped regime there are a variety of crossover phenomena observed [81,82] at temperatures above T_c in which various forms of spectral weight at low energies are apparently suppressed—these phenomena are associated with the opening of a "psuedogap." There are various families of high temperature superconductors, all of which have the same nearly square copper oxide planes, but different structures in the regions between the planes. One characteristic that seems to have a fairly direct connection with T_c is the number of copper-oxide planes that are close enough to each other that interplane coupling may be significant; T_c seems generally to increase with number of planes within a homologous series, at least as one progresses from "single layer" to "bilayer," to "trilayer" materials [4,83].

21.3.1 A Fermi Surface Instability Requires a Fermi Surface

As has been stressed, for instance, by Schrieffer [1], BCS theory relies heavily on the accuracy with which the normal state is described by Fermi liquid theory. BCS superconductivity is a Fermi surface instability, which is only a reasonable concept if there is a well defined Fermi surface. BCS-Eliashberg theory relies on the dominance of a certain class of diagrams, summed to all orders in perturbation theory.

[8] The term "Mott insulator" means many things to many people. One definition is that a Mott insulator is insulating because of interactions between electrons, rather than because a noninteracting band is filled. This is not a precise definition. For example, a Mott insulating state can arise due to a spontaneously broken symmetry which increases the size of the unit cell. However, this is adiabatically connected to the weak coupling limit, and can be qualitatively understood via generalized Hartree–Fock theory. There is still a *quantitative* distinction between a weak coupling "simple" insulator on the one hand, which has an insulating gap that is directly related to the order parameter which characterizes the broken symmetry, and the "Mott" insulator on the other hand, which has an insulating gap which is large due to the strong repulsion between electrons. In the latter case, the resistivity begins to grow very large compared to the quantum of resistance well above the temperature at which the broken symmetry occurs. The undoped cuprate superconductors are clearly Mott insulators in the quantitative sense that the insulating gap is of order 2eV, while the antiferromagnetic ordering temperatures are around 30 meV.
However, for those who prefer [67] a sharp, *qualitative* distinction, the term "Mott insulator" is reserved for "spin liquid" states which are distinct zero temperature phases of matter, do not break symmetries, and cannot be understood in terms of any straightforward Hartree–Fock description. Many such exotic states have been theoretically envisaged, including the long [5,68] and short ranged [69–71] RVB liquids, the chiral spin liquid [72–74], the nodal spin liquid [75,76] and various other fractionalized states with topological order [77,78]. Very recently, in the first "proof of principle," a concrete model with a well defined short ranged RVB *phase* has been discovered [79,80].

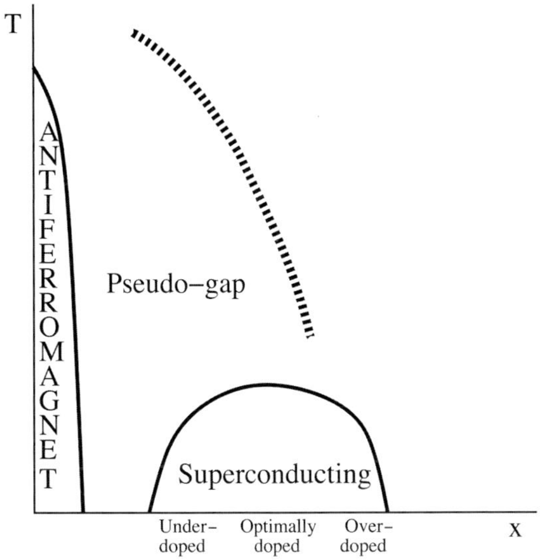

Fig. 21.1. Schematic phase diagram of a cuprate high temperature superconductor as a function of temperature and x—the density of doped holes per planar Cu. The *solid lines* represent phase transitions into the antiferromagnetic (AF) and superconducting (SC) states. The *dashed line* marks the openning of a pseudogap (PG). The latter crossover is not sharply defined and there is still debate on its position; see [81,82]

This can be justified from phase space considerations for a Fermi liquid, but need not be valid more generally. To put it most physically, BCS theory pairs well defined quasiparticles, and therefore requires well defined quasiparticles in the normal state.

We belabor the need for a non-Fermi liquid based approach.
There is ample evidence that in optimally and underdoped cuprates, at least, there are no well defined quasiparticles in the normal state. This can be deduced directly from ARPES studies of the single particle spectral function [84–91], or indirectly from an analysis of various spin, current, and density response functions of the system [3,4]. (Many, though not all, of these response functions have been successfully described [92–94] by the "marginal Fermi liquid" phenomenology.) Because we understand the nature of a Fermi liquid so well, it is relatively straightforward to establish that a system is a non-Fermi liquid, at least in extreme cases. It is much harder to establish the cause of this behavior—it could be due to the proximity of a fundamentally new non-Fermi liquid ground state phase of matter, or it could be because the characteristic coherence temperature, below which well defined quasiparticles dominate the physics, is lower than the temperatures of interest. Regardless of the reason for the breakdown of Fermi liquid theory, a description of the physics at scales of temperatures comparable to T_c can clearly *not* be based on a quasiparticle description, and thus cannot rely on BCS theory.

21.3.2 There is No Room for Retardation

As stressed in Sect. 21.2.1, retardation plays a pivotal role in the BCS mechanism. In the typical metallic superconductor, the Fermi energy is of order 10eV, while phonon frequencies are of order 10^{-2}eV, so $E_F/\omega_D \sim 10^3$! Since the renormalization of the Coulomb pseudopotential is logarithmic, this large value of the retardation is needed. In the cuprate superconductors, the bandwidth measured in ARPES is roughly $E_F \approx 0.3$eV—this is a renormalized bandwidth of sorts, but this is presumably what determines the quasiparticle dynamics. Independent of anything else, the induced interaction must clearly be fast compared to the gap scale, $\omega_D > 2\Delta_0$, where Δ_0 is the magnitude of the superconducting gap. From either ARPES [95,96] or tunneling [97] experiments, we can estimate $2\Delta_0 \approx 0.06$eV. Thus, a rough upper bound $E_F/\omega_D < E_F/2\Delta_0 \sim 5$ can be established on how retarded an interaction in the cuprates can possibly be. That is almost not retarded at all!

21.3.3 Pairing is Collective!

For the most part, the superconducting coherence length, ξ_0, cannot be directly measured in the high temperature superconductors because, for $T \lll T_c$, the upper critical field, H_{c2}, is too high to access readily. However, it can be inferred indirectly [98–102] in various ways, and for the most part people have concluded that ξ_0 is approximately 2 or 3 lattice constants in typical optimally doped materials. This has lead many people to conclude that these materials are nearly in a "real space pairing" limit [103–107], in which pairs of holes form actual two particle bound states, and then Bose condense at T_c. This notion is based on the observation that if x is the density of "doped holes" per site, then the number of pairs per coherence area, $N_p = (1/2)x\pi\xi_0^2/a^2$, is a number which is approximately equal to 1 for "optimal doping," $x \approx 0.15 - 0.20$.

Real space pairs are dismissed.
However, there are strong a priori and empirical reasons to discard this viewpoint.

On theoretical grounds: In a system dominated by strong repulsive interactions between electrons, it is clear that pairing must be a collective phenomenon. The Coulomb interaction between an isolated pair of doped holes would seem to be prohibitively large, and it seems unlikely that a strong enough effective attraction can emerge to make such a strong binding possible. (Some numerical studies of this have been carried out, in the context of ladder systems, by Dagotto and collaborators [108].) Moreover, it is far from clear that the dimensional argument used above makes any sense: Why should we only count doped holes in making this estimate? What are the rest of the holes doing all this time? If we use the density of holes per site $(1 + x)$, which is consistent with the area enclosed by the Fermi surface seen in ARPES [109], the resulting N_p is an order of magnitude larger than the above estimate. A theory of real space pairs which includes all the electrons and the repulsive interactions between them can be caricatured as a hard core quantum dimer model [70]. Here the pairing is collective, due to the high density of pairs. Indeed, N_p involves all of the electrons (the doped holes are not paired at all), but the superfluid density is small, involving only the density of doped holes. This contrasts markedly with the case of clean metallic superconductors where the density of pairs (that is, the density of electrons whose state is significantly altered by pairing) is small, $\sim N(E_F)\Delta_0$, while the superfluid density is large and involves all the electrons. There is some evidence that the former situation in fact pertains to the high temperature superconductors [110].

On experimental grounds: The essential defining feature of real space pairing is that the chemical potential moves below the bottom of the band. Incipient real space pairing must thus be associated with significant motion of the chemical potential toward the band bottom with pairing [103, 104, 111, 112]. However, experimentally, the chemical potential is found to lie in the middle of the band, where the enclosed area of the Brillouin zone satisfies Luttinger's theorem, and no significant motion at T_c (or at any pseudogap temperature in underdoped materials) has been observed [113–116]. This fact, alone, establishes that the physics is nowhere near the real space pairing limit. A rather larger value of the chemical potential shift at T_c, but still very small compared to the bandwidth, was obtained from high precision measurements of the work function [117].

21.3.4 What Determines the Symmetry of the Pair Wavefunction?

Theory has had its triumphs.
Independent of but contemporary with the discovery of high temperature superconductivity in the cuprates, Scalapino, Loh, and Hirsch [118], in a prescient work suggested the possibility of superconductivity in the two-dimensional Hubbard model in the neighborhood of the antiferromagnetic state at half filling. This work, which was in spirit a realization of the ideas of Kohn and Luttinger [119], concluded that the dominant superconducting instability should have $d_{(x^2-y^2)}$ symmetry, as opposed to s symmetry. Immediately after the discovery of high temperature superconductivity, a large number of other theorists [29, 120–125] came to the same conclusion, based on a variety of purely theoretical analyses, although at the time the experimental evidence of such pairing was ambiguous, at best. By now it seems very clear that this idea was correct, *at least* for a majority of the cuprate superconductors, based on a variety of phase sensitive measurements [126–128]. This represents one of the great triumphs of theory in this field. (There are still some experiments which appear to contradict this symmetry assignment [129], so the subject cannot be said to be completely closed, but it seems very unlikely that the basic conclusion will be overturned.)

d-wave pairing is defined.
While the names "*s*" and "*d*" relate to the rotational symmetries of free space, it is important to understand what is meant by *s*-wave and *d*-wave in a lattice system which, in place of continuous rotational symmetry, has the discrete point group symmetry of the crystal. Consequently, the possible pairing symmetries correspond to the irreducible representations of the point group: singlet orders are even under inversion and triplet orders are odd. In the case of a square crystal,[9] the possible singlet orders (all corresponding to one-dimensional representations) are colloquially called s, $d_{(x^2-y^2)}$, $d_{(xy)}$, and g, and transform like 1, (x^2-y^2), (xy), and $(x^2-y^2)(xy)$, respectively. As a function of angle, the gap parameter in an *s*-wave order always has a unique sign, the *d*-wave gap changes sign four times, and the *g*-wave changes sign 8 times. A fifth type of order is sometimes discussed, called extended-*s*, in which the gap function changes sign as a function of the magnitude of $\boldsymbol{k}$, rather than as a function of its direction—this is not a true symmetry classification, and in any generic model there is always finite mixing between *s* and extended *s*.

"d-wave-like" pairing is defined.
In crystals with lower symmetry, there are fewer truly distinct irreducible representations. For instance, if the square lattice is replaced with a rectangular one, the distinction between s and $d_{(x^2-y^2)}$ is lost (they mix), as is that between $d_{(xy)}$ and g. On the other hand, if the elementary squares are sheared to form rhombuses, then the s and $d_{(xy)}$ symmetries are mixed, as are $d_{(x^2-y^2)}$, and g. Both of these lower symmetries correspond to a form of orthorhombic distortion observed in the cuprates—the former is the correct symmetry group for $YBa_2Cu_3O_{7-\delta}$ and the latter for $La_{2-x}Sr_xCuO_4$. However, so long as the physics does not change fundamentally as the lattice symmetry is reduced, it is reasonable to classify order parameters as "*d*-wave-like" or "*s*-wave-like." We define an order parameter as being *d*-wave-like if it changes sign under 90° rotation, although it is only a true *d*-wave if its magnitude is invariant under this transformation. Conversely, it is *s*-wave-like if its sign does not change under this rotation, or when reflected through any approximate symmetry plane. In almost all cases what is really being seen in phase sensitive measurements on the cuprates is that the order parameter is *d*-wave-like. (It is worth noting that in $t-J$ and Hubbard ladders, *d*-wave-like pairing is the dominant form of pairing observed in both analytic and numerical studies, as discussed below.)

Strong repulsion does not necessarily lead to d-wave pairing.
There is a widespread belief that *d*-wave symmetry follows directly from the presence of strong short range interactions between electrons, irrespective of details such as band structure. The essential idea here follows from the observation that the pair wavefunction, at the level of BCS mean field theory, is expressed in terms of the gap parameter, Δ_k, and the quasiparticle spectrum, E_k, as

$$\phi_{\text{pair}}(\boldsymbol{r}) = \sum_{k} \frac{1}{L^{d/2}} \, e^{i\boldsymbol{k}\cdot\boldsymbol{r}} \, \frac{\Delta_k}{2E_k} \,. \tag{21.5}$$

In the presence of strong short range repulsion (and weaker longer range attraction) between electrons, it is favorable for ϕ_{pair} to vanish at $\boldsymbol{r} = \boldsymbol{0}$, which it does automatically if the pairing is not *s*-wave. While this argument makes some physical sense, it is ultimately *wrong*. In the limit of dilute electrons, where the coherence length is much smaller than the inter-electron distance, the pairing problem reduces to a two particle problem. It is well known that in the continuum the lowest energy two particle spin singlet bound state is nodeless. Given certain mild conditions on the band structure one can also prove it on the lattice.[10] Therefore, in this limit, the order parameter is necessarily *s*-wave-like!

The above discrepancy teaches us that it is the presence of the kinematical constraints imposed by

[9] The pairing symmetries should really be classified according to the point group of a tetragonal crystal, but since the cuprates are quasi-two-dimensional, it is conventional, and probably reasonable, to classify them according to the symmetries of a square lattice.

[10] This is true under conditions that the hopping matrix, i.e. the band structure, satisfies a Peron–Frobenius condition.

the Fermi sea that allows for non s-wave pairing. The ultimate pairing symmetry is a reflection of the distribution in momentum space of the low energy single particle spectral weight. The reason for this is clear within BCS theory where the energy gain, which drives the transition, comes from the interaction term

$$\text{potential energy} = \sum_{k,k'} V_{k,k'} \frac{\Delta_k}{2E_k} \frac{\Delta_{k'}}{2E_{k'}} , \qquad (21.6)$$

which is maximized by a gap function that peaks in regions of high density of states unless the pairing potential that connects these regions is particularly small. (Although we do not know of an explicit justification of this argument for a non-BCS theory, for example one which is driven by gain in kinetic energy, we feel that the physical consideration behind it is robust.)

Nodal quasiparticles do not a d-wave mean.
Finally, there is another issue which is related to order parameter symmetry in a manner that is more complex than is usually thought—this is the issue of the existence of nodal quasiparticles. While nodal quasiparticles are natural in a d-wave superconductor, d-wave superconductors can be nodeless, and s-wave superconductors can be nodal. To see this, it is possible to work entirely in the weak coupling limit where BCS theory is reliable. The quasiparticle excitation spectrum can thus be expressed as

$$E_k = \sqrt{\varepsilon_k^2 + \Delta_k^2} , \qquad (21.7)$$

where ε_k is the quasiparticle dispersion in the normal state (measured from the Fermi energy). Nodal quasiparticles occur wherever the Fermi surface, that is the locus of points where $\varepsilon_k = 0$, crosses a line of gap nodes, the locus of points where $\Delta_k = 0$. If the Fermi surface is closed around the origin, $\boldsymbol{k} = 0$, or about the Brillouin zone center, $\boldsymbol{k} = (\pi, \pi)$ (as it is most likely in optimally doped $Bi_2Sr_2CaCu_2O_{8+\delta}$ [130]), then the d-wave symmetry of $\Delta_k = 0$ implies the existence of nodes. However, if the Fermi surface were closed about $\boldsymbol{k} = (0, \pi)$ (and symmetry related points), there would be no nodal quasiparticles [131]. Indeed, it is relatively easy to characterize [132,133] the quantum phase transition between a nodal and nodeless d-wave superconductor which occurs as a parameter that alters the underlying band structure is varied. Conversely, it is possible to have lines of gap nodes for an extended s-wave superconductor, and if these cross the Fermi surface, the superconductor will posses nodal quasiparticles.

21.3.5 What Does the Pseudogap Mean?

What Experiments Define the Pseudogap?

The pseudogap.
One of the most prominent, and most discussed features of the cuprate superconductors is a set of crossover phenomena [54,81,82] which are widely observed in underdoped cuprates and, to various extents, in optimally and even slightly overdoped materials. Among the experimental probes which are used to locate the pseudogap temperature in different materials are:

1) ARPES and c-axis Tunnelling:
There is a suppression of the low energy single particle spectral weight, shown in Figs. 21.2 and 21.3 at temperatures above T_c as detected, primarily, in c-axis tunneling [134] and ARPES [95,96] experiments. The scale of energies and the momentum dependence of this suppression are very reminiscent of the d-wave superconducting gap observed in the same materials at temperatures well below T_c. This is highly suggestive of an identification between the pseudogap and some form of local superconducting pairing. Although a pseudogap energy scale is easily deduced from these experiments, it is not so clear to us that an unambiguous temperature scale can be cleanly obtained from them. (The c-axis here, and henceforth, refers to the direction perpendicular to the copper-oxide planes, which are also referred to, crystallographically, as the ab plane.)

2) Cu NMR:
There is a suppression of low energy spin fluctuations as detected [135] primarily in Cu NMR. In some cases, two rather different temperature scales are deduced from these experiments: an upper crossover temperature, at which a peak occurs in χ', the real part of the uniform spin susceptibility (i.e. the

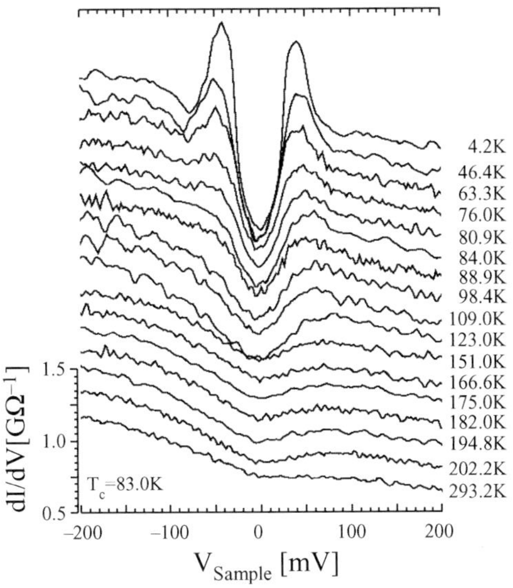

Fig. 21.2. Tunneling density of states in a sample of underdoped $Bi_2Sr_2CaCu_2O_{8+\delta}$ ($T_c = 83$ K) as a function of temperature. Note that there is no tendency for the gap to close as T_c is approached from below, but that the sharp "coherence peaks" in the spectrum do vanish at T_c. From [97]

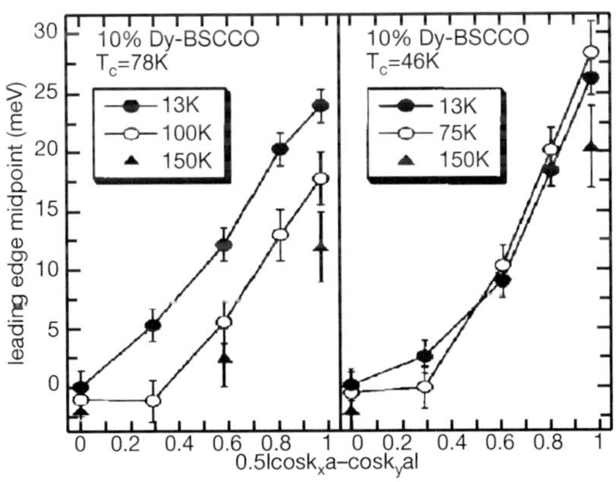

Fig. 21.3. The angular dependence of the gap in the normal and superconducting states of underdoped $Bi_2Sr_2Ca_{1-x}Dy_xCu_2O_{8+\delta}$ as deduced from the leading edge energy of the single hole spectral function $A^<(\boldsymbol{k}, \omega)$ measured by ARPES. A straight line in this plot would correspond to the simplest $d_{x^2-y^2}$ gap, $|\Delta_k| = \Delta_0|\cos(k_x) - \cos(k_y)|$. From [95]

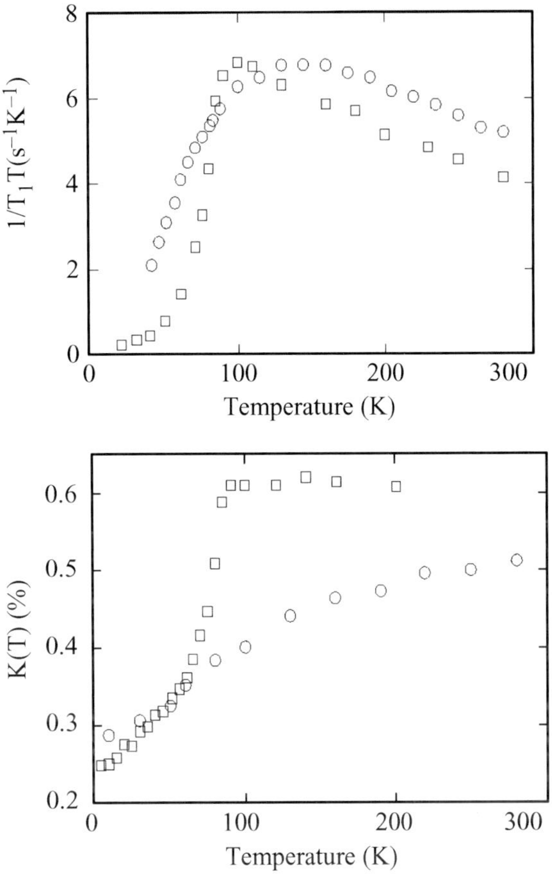

Fig. 21.4. Temperature dependence of the planar ^{63}Cu relaxation rate $1/T_1T$ and Knight shift K in optimally doped $YBa_2Cu_3O_{6.95}$ (*squares*) and underdoped $YBa_2Cu_3O_{6.64}$ (*circles*). From [81]

Knight shift), and a lower crossover temperature, below which $1/T_1T$ drops precipitously (see Fig. 21.4). Note that $1/T_1T \propto \lim_{\omega\to 0} \int \mathrm{d}\boldsymbol{k} f(\boldsymbol{k})\chi''(\boldsymbol{k}, \omega)/\omega$, the $\boldsymbol{k}$ averaged density of states for magnetic excitations, where $f(\boldsymbol{k})$ is an appropriate form factor which reflects the local hyperfine coupling. Although the temperature scale deduced from χ' is more or less in accordance with the pseudogap scale deduced from a number of other spectroscopies, it is actually a measure of the reactive response of the spin system. The notion of a gap can be more directly identified with a feature in χ''. (*A note of warning*: while the structure in $1/T_1T$ can be fairly sharp at times, the observed maxima in χ' are always very broad and do not yield a sharply defined temperature scale without further analysis.)

3) Resistivity:

There is a significant deviation [136, 137] of the resistivity in the *ab* plane from the *T* linear temperature dependence which is universally observed at high temperatures. A pseudogap temperature is then identified as the point below which $d\rho_{xx}/dT$ deviates (increases) significantly from its high temperature value (see Fig. 21.5). In some cases, a similar temperature scale can be inferred from a scaling analysis of the Hall resistance, as well.

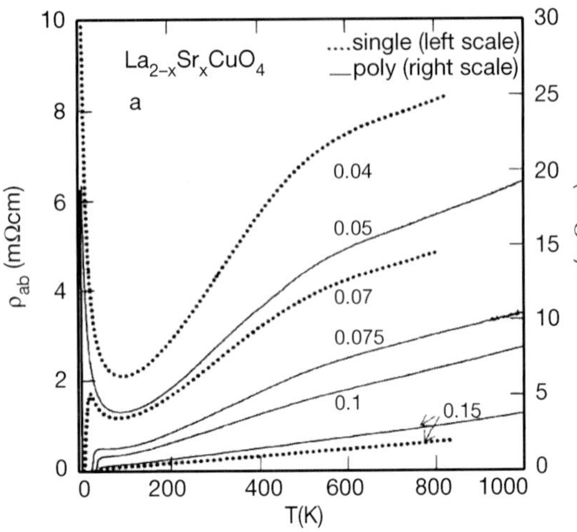

Fig. 21.5. The temperature dependence of the longitudinal resistivity in underdoped and optimally doped $La_{2-x}Sr_xCuO_4$. The *dotted lines* correspond to the in-plane resistivity (ρ_{ab}) of single crystal films while the *solid lines* depict the resistivity (ρ) of polycrystalline samples. The doping levels are indicated next to the *curves*. From [136]

The pseudogap also appears in the *c*-axis resitivity, although in a somewhat different manner [138, 139]. In this direction, the pseudogap results in a strong increase in the resistivity, reminiscent of the behavior of a narrow gap semiconductor, as shown in Fig. 21.6. If we imagine that the *c*-axis transport is dominated by tunneling events between neighboring planes, it is reasonable that a bulk measurement of ρ_c will reflect the pseudogap in much the same way as the *c*-axis tunneling does.

4) Specific Heat:

There is a suppression of the expected electronic specific heat [82]. Above the pseudogap scale, the specific heat is generally found to be linear in temperature, $C_V \approx \gamma T$, but below the pseudogap temperature, C_V/T begins to decrease with decreasing temperature (see Fig. 21.7). Interestingly, since the value of γ above the pseudogap temperature appears to be roughly doping independent, the drop in the specific at lower temperatures can be interpreted as a doping dependent loss of entropy,

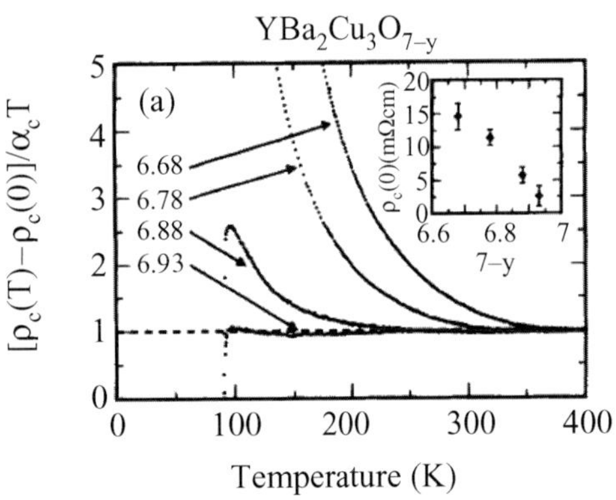

Fig. 21.6. The temperature dependence of the *c*-axis resistivity in underdoped and optimally doped $YBa_2Cu_3O_{7-\delta}$. Here α_c and $\rho_c(0)$ are the slope and the intercept, respectively, when the metalic part of ρ_c is approximated by a linear-*T* behavior. The *inset* shows how $\rho_c(0)$ varies with oxygen content. From [138]

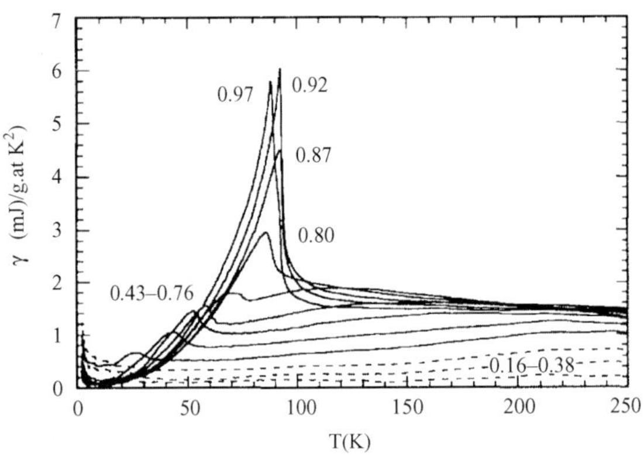

Fig. 21.7. Thermal density of "electronic" states, $\gamma \equiv C_V/T$ as a function of temperature for various oxygen concentrations in underdoped $YBa_2Cu_3O_{6+x}$. From [140]. As discussed in [140], a complicated proceedure has been used to subtract the large nonelectronic component of the measured specific heat

$\Delta S(x) \equiv S(x, T) - S(x_{\text{optimal}}, T)$, with a magnitude which is independent of temperature for any $T > T^*$. This is the origin of the famous (and still not understood) observation of Loram and collaborators [140] that there is a large entropy, $k_B/2$, which is somehow associated with each doped hole. *A word of warning*: except at the lowest temperatures, the electronic specific heat is always a small fraction of the total specific heat, and complicated empirical subtraction procedures, for which the theoretical justification is not always clear to us, are necessary to extract the electronic contribution.

5) Infrared Conductivity:

There is an anomalous motion of infrared spectral weight to low energies [141, 142]. The pseudogap is most clearly identified by plotting [142] the frequency dependent scattering rate, defined either as $1/\tau^*(\omega) \equiv \omega\sigma'_{ab}(\omega)/\sigma''_{ab}(\omega)$, or as $1/\tau(\omega) = [\omega_P^2/4\pi]\text{Re}[1/\sigma(\omega)]$ where ω_P is the plasma frequency; the pseudogap is rather harder to pick out from the in-plane conductivity, σ'_{ab}, itself. At large ω, one generally sees $1/\tau(\omega) \approx A\omega$, and it then drops to much smaller values, $1/\tau \ll \omega$, below a characteristic pseudogap frequency, see Fig. 21.8. (A is generally a bit larger than 1 in underdoped materials and roughly equal to 1 in optimally doped ones.)

While in optimally doped materials, this manifestation of a pseudogap is only observed at temperatures less than T_c, in underdoped materials, it is seen to persist well above T_c, and indeed to be not strongly temperature dependent near T_c. A characteristic pseudogap energy is easily identified from this data, but, again, it is not clear to us to what extent it is possible to identify a clear pseudogap temperature from this data. A pseudogap can also be deduced directly [143, 144] from an analysis of $\sigma'_c(\omega)$, where it manifests itself as a suppressed response at low frequencies, as shown in Fig. 21.9.

6) Inelastic Neutron Scattering:

There are temperature dependent changes in the dynamic spin structure factor as measured by inelastic neutron scattering. Here, both features associated with low energy incommensurate magnetic correlations (possibly associated with stripes) [145] and the so-called "resonant peak" are found to emerge below a temperature which is very close to T_c in optimally doped materials, but which rises considerably above T_c in underdoped materials [146] (see Fig. 21.10).

What Does the Pseudogap Imply for Theory?

It is generally accepted that the pseudogap, in one way or another, reflects the collective physics associated with the growth of electronic correlations. This, more than any other aspect of the data, has focused attention on theories of the collective variables representing the order parameters of various possible broken symmetry states [20, 51–54, 62, 77, 147–158]. Among these theories, there are two rather different

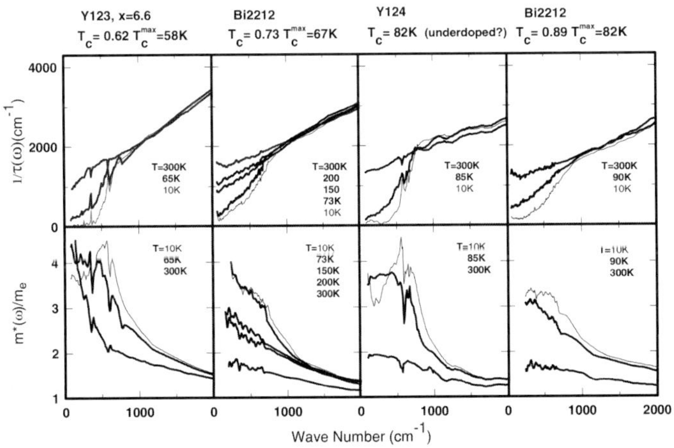

Fig. 21.8. *Upper panels*: Frequency dependent scattering rate for a series of underdoped cuprate superconductors above, near and below the superconducting transition temperature. *Lower panels*: The effective mass enhancement $m^*/m_e = 1 + \lambda(\omega)$. Both are deduced from fitting infrared conductivity data to an extended Drude model $\sigma = (\omega_P^2/4\pi)/[1/\tau(\omega) - i\omega(1 + \lambda(\omega))]$. From [142]

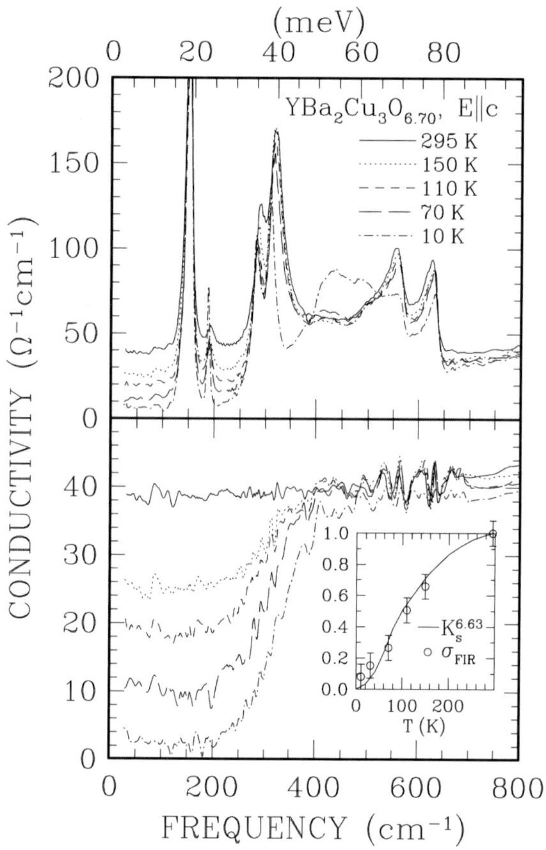

Fig. 21.9. The *c*-axis optical conductivity of underdoped $YBa_2Cu_3O_{7-\delta}$ (T_c = 63 K) as a function of temperature (*top panel*). The optical conductivity after the substraction of the phonon features is presented in the *lower panel*. The *inset* compares the low frequency conductivity with the Knight shift. From [143]

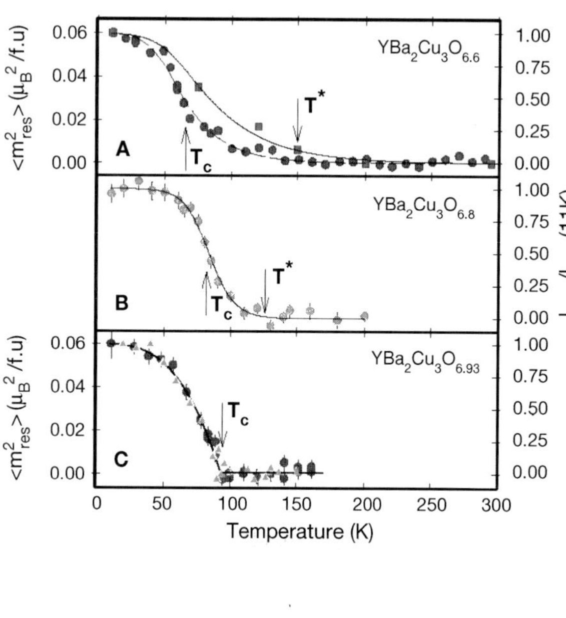

Fig. 21.10. The temperature dependence of the intensity of the so called resonant peak observed in neutron scattering in underdoped $YBa_2Cu_3O_{7-\delta}$. From [146]

classes of ways to interpret the pseudogap phenomena.

(1)
It is well known that fluctuation effects can produce local order which, under appropriate circumstances, can extend well into the disordered phase. Such fluctuations produce in the disordered phase some of the local characteristics of the ordered phase, and if there is a gap in the ordered phase, a pseudogap as a fluctuation effect is eminently reasonable—see Fig. 21.1. As is discussed in Sect. 21.8, the small superfluid density of the cuprates leads to the unavoidable conclusion that superconducting fluctuations are an order 1 effect in these materials, so it is quite reasonable to associate some pseudogap phenomena with these fluctuations. However, as the system is progressively underdoped, it gets closer and closer to the antiferromagnetic insulating state, and indeed there is fairly direct NMR evidence of increasingly strong local antiferromagnetic correlations [159]. It is thus plausible that there are significant effects of antiferromagnetic fluctuations, and since the antiferromagnetic state also has a gap, one might expect these fluctuations to contribute to the pseudogap phenomena as well. There are significant incommensurate charge and spin density (stripe) fluctuations observed directly in scattering experiments on a variety of underdoped materials [47,145,160–162], as well as the occasional stripe ordered phase [163–167]. These fluctuations, too, certainly contribute to the observed pseudogap phenomena. Finally, fluctuations associated with more exotic phases, especially the "staggered flux phase" (which we will discuss momentarily) have been proposed [148,168] as contributing to the pseudogap as well.

Crossovers can be murky.
There has been a tremendous amount of controversy in the literature concerning which of these various fluctuation effects best account for the observed pseudogap phenomena. Critical phenomena, which are clearly associated with the phase fluctuations of the superconducting order parameter, have been observed [169–172] in regions that extend between 10% to 40% above and below the superconducting T_c in optimally and underdoped samples of $YBa_2Cu_3O_{7-\delta}$ and $Bi_2Sr_2CaCu_2O_{8+\delta}$; in our opinion, the dominance of superconducting fluctuations in this substantial range of temperatures is now beyond question. However, pseudogap phenomena are clearly observed in a much larger range of temperatures. Even if fluctuation effects are ultimately the correct explanation for all the pseudogap phenomena, there may not truly be one type of fluctuation which dominates the physics over the entire range of temperatures.

One cannot always tell a fluctuating superconductor from a fluctuating insulator!
To illustrate this point explicitly, consider a one-dimensional electron gas (at an incommensurate density) with weak attractive backscattering interactions. (See Sect. 21.5.) If the backscattering interactions are attractive ($g_1 < 0$), they produce a spin gap Δ_s. This gap persists as a pseudogap in the spectrum up to temperatures of order $\Delta_s/2$. Now, because of the nature of fluctuations in one dimension, the system can never actually order at any finite temperature. However, there is a very real sense in which one can view the pseudogap as an effect of superconducting fluctuations, since at low temperatures, the superconducting susceptibility is proportional to Δ_s. The problem is that one can equally well view the pseudogap as an effect of CDW fluctuations. One could arbitrarily declare that where the CDW susceptibility is the most divergent, the pseudogap should be viewed as an effect of local CDW order, while when the superconducting susceptibility is more divergent, it is an effect of local pairing. However, this position is untenable; by varying the strength of the forward scattering (g_2), it is possible to pass smoothly from one regime to the other without changing Δ_s in any way!

(2)
There are several theoretical proposals [52–54] on the table which suggest that there is a heretofore undetected electronic phase transition in underdoped materials with a transition temperature well above the superconducting T_c. As a function of doping, this transition temperature is pictured as decreasing, and tending to zero at a quantum critical point some-

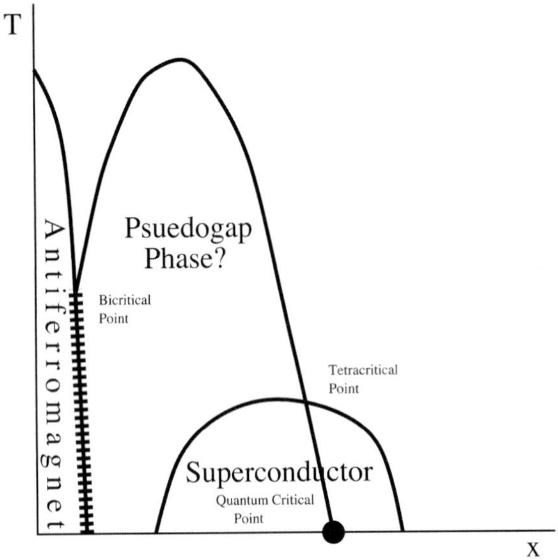

Fig. 21.11. There are many ideas concerning the meaning of the pseudogap. Defined purely phenomenologically, as shown in Fig. 21.1, it is a region in which there is a general reduction in the density of low energy excitations, and hence is bounded by an ill-defined crossover line. It is also possible that, to some extent, the pseudogap reflects the presence of a broken symmetry, in which case it must be bounded by a precise phase boundary, as shown in the present figure. There are many ways such a pseudogap phase could interact with the other well established phases. For purposes of illustration, we have shown a tetracritical and a bicritical point where the pseudogap meets, respectively, the superconducting and antiferromagnetic phases. One consequence of the assumption that the transition into the pseudogap phase is continuous is the exisence of a quantum critical point (indicated by the *heavy circle*) somewhere under the superconducting dome. See, for example, [20,52,54,62,173]

where in the neighborhood of optimal doping, as shown schematically in Fig. 21.11.

Covert phase transitions are considered.
If such a transition occurs, it would be natural to associate at least some of the observed pseudogap phenomena with it. Since these scenarios involve a new broken symmetry, they make predictions which are, in principle, sharply defined and falsifiable by experiment. However, there is an important piece of phenomenology which these theories must address: if there is a phase transition underlying pseudogap formation, why hasn't direct thermodynamic evidence (i.e. nonanalytic behavior of the specific heat, the susceptibility, or some other correlation function of the system) been seen in existing experiments? Possible answers to this question typically invoke disorder broadening of the proposed phase transition [54], rounding of the transition by a symmetry breaking field [52], or possibly the intrinsic weakness of the thermodynamic signatures of the transition under discussion [53,174].

21.4 Preview: Our View of the Phase Diagram

Clearly, the pseudogap phenomena described above are just the tip of the iceberg, and any understanding of the physics of the cuprate high temperature superconductors will necessarily be complicated. For this reason, we have arranged this article to focus primarily on high temperature superconductivity as an abstract theoretical issue, and only really discuss how these ideas apply to the cuprates in Sect. 21.13. However, to orient the reader, we will take a moment here to briefly sketch *our* understanding of how these abstract issues determine the behavior, especially the high temperature superconductivity of the cuprates.

Figure 21.12 is a schematic representation of the temperature vs. doping phase diagram of a representative cuprate. There are four energy scales relevant to the mechanism of superconductivity, marked as T^*_{stripe}, T^*_{pair}, T^*_{3D} and T_c. Away from the peak of the superconducting dome, these energy scales are often well separated. At least some of the pseudogap phenomena are, presumably, associated with the two crossover scales, T^*_{pair} and T^*_{stripe}.

Stripe Formation T^*_{stripe}:
The kinetic energy of doped holes is frustrated in an antiferromagnet. As the temperature is lowered through T^*_{stripe}, the doped holes are effectively ejected from the antiferromagnet to form metallic regions, thus relieving some of this frustration. Being charged objects, the holes can only phase separate on short length scales, since the Coulomb repulsion enforces charge homogeneity at long length scales. As a result, at T^*_{stripe}, the material develops significant one-

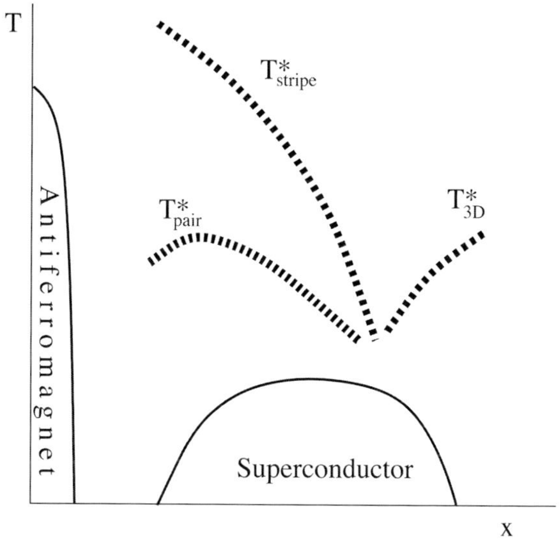

Fig. 21.12. Phase diagram as a function of temperature and doping within the stripes scenario discussed here

dimensional charge modulations, which we refer to as charge stripes. This can be an actual phase transition (e.g. to a "nematic phase"), or a crossover scale at which significant local charge stripe correlations develop.

Pair Formation T^*_{pair}:
While stripe formation permits hole delocalization in one direction, hole motion transverse to the stripe is still restricted. It is thus favorable, under appropriate circumstances, for the holes to pair so that the pairs can spread out somewhat into the antiferromagnetic neighborhood of the stripe. This "spin gap proximity effect" [20] (see Sect. 21.10.4), which is much like the proximity effect at the interface between a normal metal and a conventional superconductor, results in the opening of a spin gap and an enhancement of the superconducting susceptibility on a single stripe. In other words, T^*_{pair} marks a crossover below which the superconducting order parameter amplitude (and therefore a superconducting pseudo gap) has developed, but without global phase coherence.

Superconductivity T_c:
Superconducting long ranged order onsets as the phase of the superconducting order parameter on each charge stripe becomes correlated across the sample. Since it is triggered by Josephson tunneling between stripes, this is a kinetic energy driven phase ordering transition.

Dimensional Crossover T^*_{3D}:
Superconducting long range order implies coherence in all three dimensions, and hence the existence of well defined electron-like quasiparticles [21, 149, 175]. Where the stripe order is sufficiently strong (in the underdoped regime), the dimensional crossover to 3D physics is directly associated with the onset of superconducting order. However, in overdoped materials, where the electron dynamics is less strongly influenced by stripe formation, we expect the dimensional crossover to occur well above T_c. (See Sect. 21.5.)

21.5 Quasi-1D Superconductors

In this section we address the physics of the one-dimensional electron gas and quasi-one-dimensional systems consisting of higher dimensional arrays of weakly coupled chains. Our motivation is twofold. Firstly, these systems offer a concrete realization of various non-Fermi liquid phenomena and are amenable to *controlled* theoretical treatments. As such they constitute a unique theoretical laboratory for studying strong correlations. In particular, for whatever reason, much of the experimentally observed behavior of the cuprate superconductors is strongly reminiscent [84,86,149] of a quasi-1D superconductor. Secondly, we are motivated by a growing body of experimental evidence for the existence of electron smectic and nematic phases in the high temperature superconductors, manganites and quantum Hall systems [6,176–179]. It is possible that these materials actually are quasi 1D on a local scale. Preliminary evidence of the existence of nematic order in $La_{2-x}Sr_xCuO_4$ and $YBa_2Cu_3O_{7-\delta}$ can be found in [98].

Our emphasis will be on quasi-one-dimensional superconductors, the different unconventional signatures they exhibit as a function of temperature, and the conditions for their expression and stability. We will, however, include some discussion of other quasi-one-dimensional phases which typically tend to suppress superconductivity. It is also worth noting that, for the most part, the discussion is simply generalized to quasi-1D systems with different types of order, including quasi-1D CDW insulators.

21.5.1 Elementary Excitations of the 1DEG

We begin by considering the continuum model of an interacting one-dimensional electron gas (1DEG). It consists of approximating the 1DEG by a pair of linearly dispersing branches of left ($\eta = -1$) and right ($\eta = 1$) moving spin 1/2 ($\sigma = \pm 1$ denotes the z spin component) fermions constructed around the left and right Fermi points of the 1DEG. This approximation correctly describes the physics in the limit of low energy and long wavelength where the only important processes are those involving electrons in the vicinity of the Fermi points. The Hamiltonian density of the model is

$$\begin{aligned}\mathcal{H} = &- iv_F \sum_{\eta,\sigma=\pm 1} \eta \psi^\dagger_{\eta,\sigma} \partial_x \psi_{\eta,\sigma} \\ &+ \frac{g_4}{2} \sum_{\eta,\sigma=\pm 1} \psi^\dagger_{\eta,\sigma} \psi^\dagger_{\eta,-\sigma} \psi_{\eta,-\sigma} \psi_{\eta,\sigma} \\ &+ g_2 \sum_{\sigma,\sigma'=\pm 1} \psi^\dagger_{1,\sigma} \psi^\dagger_{-1,\sigma'} \psi_{-1,\sigma'} \psi_{1,\sigma} \\ &+ g_{1\parallel} \sum_{\sigma=\pm 1} \psi^\dagger_{1,\sigma} \psi^\dagger_{-1,\sigma} \psi_{1,\sigma} \psi_{-1,\sigma} \\ &+ g_{1\perp} \sum_{\sigma=\pm 1} \psi^\dagger_{1,\sigma} \psi^\dagger_{-1,-\sigma} \psi_{1,-\sigma} \psi_{-1,\sigma} \,, \end{aligned} \tag{21.8}$$

where, e.g. $\psi_{1,1}$ destroys a right moving electron of spin 1/2. The g_4 term describes forward scattering events of electrons in a single branch. The g_2 term corresponds to similar events but involving electrons on both branches. Finally, the $g_{1\parallel}$ and $g_{1\perp}$ terms allow for backscattering from one branch to the other. The system is invariant under $SU(2)$ spin rotations provided $g_{1\parallel} = g_{1\perp} = g_1$. In the following we consider mostly this case.

Umklapp processes of the form

$$g_3 \psi^\dagger_{-1,\uparrow} \psi^\dagger_{-1,\downarrow} \psi_{1,\downarrow} \psi_{1,\uparrow} e^{i(4k_F - G)x} + \text{H.c.}\,,$$

are important only when $4k_F$ equals a reciprocal lattice vector G. When the 1DEG is incommensurate ($4k_F \neq G$), the rapid phase oscillations in this term render it irrelevant in the renormalization group sense. We will assume such incommensurability and correspondingly ignore this term. We will also neglect single particle scattering between branches (for example due to disorder) and terms that do not conserve the z component of the spin.

It is important to stress [180] that in considering this model we are focusing on the long distance physics that can be precisely derived from an effective field theory. However, all the coupling constants that appear in Eq. (21.8) are effective parameters which implicitly include much of the high energy physics. For instance, the bare velocity which enters the model, v_F, is not necessarily simply related to the dispersion of the band electrons in a zeroth order, noninteracting model, but instead includes all sorts of finite renormalizations due to the interactions. The weak coupling perturbative renormalization group treatment of this model is discussed in Sect. 21.9, below; the most important result from this analysis is that the Fermi liquid fixed point is always unstable, so that an entirely new, nonperturbative method must be employed to reveal the low energy physics.

Bosonization

Fortunately, such a solution is possible; the Hamiltonian in Eq. (21.8) is equivalent to a model of two independent bosonic fields, one representing the charge and the other the spin degrees of freedom in the system. (For reviews and recent perspectives see [38, 180–186].) The two representations are related via the bosonization identity

$$\psi_{\eta,\sigma} = \frac{1}{\sqrt{2\pi a}} F_{\eta,\sigma} \exp[-i\Phi_{\eta,\sigma}(x)]\,, \tag{21.9}$$

which expresses the fermionic fields in terms of self dual fields $\Phi_{\eta,\sigma}(x)$ obeying $[\Phi_{\eta,\sigma}(x), \Phi_{\eta',\sigma'}(x')] = -i\pi\delta_{\eta,\eta'}\delta_{\sigma,\sigma'}\text{sign}(x - x')$. They in turn are combinations of the bosonic fields ϕ_c and ϕ_s and their conjugate momenta $\partial_x \theta_c$ and $\partial_x \theta_s$

$$\Phi_{\eta,\sigma} = \sqrt{\pi/2}\,[(\theta_c - \eta\phi_c) + \sigma(\theta_s - \eta\phi_s)]\,. \tag{21.10}$$

Physically, ϕ_c and ϕ_s are, respectively, the phases of the $2k_F$ charge density wave (CDW) and spin density wave (SDW) fluctuations, and θ_c is the superconducting phase. In terms of them the long wavelength component of the charge and spin densities are given by

$$\rho(x) = \sum_{\eta,\sigma} \psi^\dagger_{\eta,\sigma}\psi_{\eta,\sigma} - \frac{2k_F}{\pi} = \sqrt{\frac{2}{\pi}}\partial_x\phi_c\,, \quad (21.11)$$

$$S_z(x) = \frac{1}{2}\sum_{\eta,\sigma} \sigma\psi^\dagger_{\eta,\sigma}\psi_{\eta,\sigma} = \sqrt{\frac{1}{2\pi}}\partial_x\phi_s\,. \quad (21.12)$$

The Klein factors $F_{\eta,\sigma}$ in Eq. (21.9) are responsible for reproducing the correct anticommutation relations between different fermionic species and a is a short distance cutoff that is taken to zero at the end of the calculation.

In 1D spin and charge separate.
The widely discussed separation of charge and spin in this problem is formally a statement that the Hamiltonian density can be expressed as a sum of two pieces, each of the sine-Gordon variety, involving only charge or spin fields

$$\mathcal{H} = \sum_{\alpha=c,s}\left\{\frac{v_\alpha}{2}\left[K_\alpha(\partial_x\theta_\alpha)^2 + \frac{(\partial_x\phi_\alpha)^2}{K_\alpha}\right] + V_\alpha\cos(\sqrt{8\pi}\phi_\alpha)\right\}\,. \quad (21.13)$$

When the Hamiltonian is separable, wavefunctions, and therefore correlation functions, factor. (See Eqs. (21.24) and (21.25).) In terms of the parameters of the fermionic formulation Eq. (21.8) the charge and spin velocities are given by

$$v_c = \frac{1}{2\pi}\sqrt{(2\pi v_F + g_4)^2 - (g_{1\parallel} - 2g_2)^2}\,, \quad (21.14)$$

$$v_s = \frac{1}{2\pi}\sqrt{(2\pi v_F - g_4)^2 - g_{1\parallel}^2}\,, \quad (21.15)$$

while the Luttinger parameters K_α, which determine the power law behavior of the correlation functions, are

$$K_c = \sqrt{\frac{2\pi v_F + g_4 - 2g_2 + g_{1\parallel}}{2\pi v_F + g_4 + 2g_2 - g_{1\parallel}}}\,, \quad (21.16)$$

$$K_s = \sqrt{\frac{2\pi v_F - g_4 + g_{1\parallel}}{2\pi v_F - g_4 - g_{1\parallel}}}\,. \quad (21.17)$$

The cosine term in the spin sector of the bosonized version of the Hamiltonian (Eq. (21.13)) originates from the back scattering term in Eq. (21.8) where the amplitudes are related according to

$$V_s = \frac{g_{1\perp}}{2(\pi a)^2}\,. \quad (21.18)$$

The corresponding term in the charge sector describes umklapp processes and in view of our assumption will be set to zero $V_c = 0$. Equations (21.14)–(21.18) complete the exact mapping between the fermionic and bosonic field theories.

In the absence of back scattering ($g_1 = 0$) this model is usually called the Tomonaga-Luttinger model. Since $\partial_x\theta_{c,s}$ and $\phi_{c,s}$ are canonically conjugate, it is clear from the form of the bosonized Hamiltonian (Eq. (21.13)) that it describes a collection of independent charge and spin density waves with linear dispersion $\omega_{c,s} = v_{c,s}k$. The quadratic nature of the theory and the coherent representation (Eq. (21.9)) of the electronic operators in terms of the bosonic fields allow for a straightforward evaluation of various electronic correlation functions.

For $g_1 \neq 0$ the spin sector of the theory turns into a sine-Gordon theory whose renormalization group flow is well known [187]. In particular, for repulsive interactions ($g_1 > 0$) the backscattering amplitude is renormalized to zero in the long wavelength low energy limit and consequently at the fixed point $K_s = 1$. On the other hand, in the presence of attractive interactions ($g_1 < 0$) the model flows to strong (negative) coupling where the cosine term in Eq. (21.13) is relevant. As a result ϕ_s is pinned in the sense that in the ground state, it executes only small amplitude fluctuations about its classical ground state value (i.e. one of the minima of the cosine). There is a spin gap to both extended phonon-like small amplitude oscillations about this minimum and large amplitude soliton excitations that are domain walls at which ϕ_s changes between two adjacent minima.

The susceptibility of the interacting one-dimensional electron gas to various instabilities can be investigated by calculating the correlation functions of the operators that describe its possible orders. They include, among others, the $2k_F$ CDW and SDW operators

$$O_{\mathrm{CDW}}(x) = e^{-i2k_F x} \sum_{\tau} \psi^{\dagger}_{1,\tau}(x)\psi_{-1,\tau}(x) \,, \qquad (21.19)$$

$$O_{\mathrm{SDW}_\alpha}(x) = e^{-i2k_F x} \sum_{\tau,\tau'} \psi^{\dagger}_{1,\tau}(x)\sigma^{\alpha}_{\tau,\tau'}\psi_{-1,\tau'}(x) \,, \qquad (21.20)$$

where $\boldsymbol{\sigma}$ are the Pauli matrices, the $4k_F$ CDW (or Wigner crystal) order

$$O_{4k_F}(x) = e^{-i4k_F x} \sum_{\tau} \psi^{\dagger}_{1,\tau}(x)\psi^{\dagger}_{1,-\tau}(x) \times \psi_{-1,-\tau}(x)\psi_{-1,\tau}(x) \,, \qquad (21.21)$$

and the singlet (SS) and triplet (TS) pair annihilation operators

$$O_{\mathrm{SS}}(x) = \sum_{\tau} \tau\psi_{1,\tau}(x)\psi_{-1,-\tau}(x) \,, \qquad (21.22)$$

$$O_{\mathrm{TS}_\alpha}(x) = \sum_{\tau,\tau'} \tau\psi_{1,\tau}(x)\sigma^{\alpha}_{\tau,\tau'}\psi_{-1,-\tau'}(x) \,. \qquad (21.23)$$

They can also be written in a suggestive bosonized form. For example the CDW and the singlet pairing operators are expressed as[11]

$$O_{\mathrm{CDW}}(x) = \frac{e^{-2ik_F x}}{\pi a} \cos[\sqrt{2\pi}\phi_s(x)] e^{-i\sqrt{2\pi}\phi_c(x)} \,, \qquad (21.24)$$

$$O_{\mathrm{SS}}(x) = \frac{1}{\pi a} \cos[\sqrt{2\pi}\phi_s(x)] e^{-i\sqrt{2\pi}\theta_c(x)} \,. \qquad (21.25)$$

1D order parameters have "spin" amplitudes and "charge" phases.
The distinct roles of spin and charge are vividly apparent in these expressions: the amplitude of the order parameters is a function of the spin fields while their phase is determined by the charge degrees of freedom. Similar relations are found for the SDW and triplet pairing operators. However, the $4k_F$ CDW order is independent of the spin fields.

If in the bare Hamiltonian, $g_1 > 0$ and V_s is not too large, the system flows to the Gaussian fixed point with $K_s = 1$ and no spin gap. The gapless fluctuations of the amplitude (spin) and phase (charge) of the various orders lead then to an algebraic decay of their zero temperature space-time correlation functions (with logarithmic corrections which reflect the slow renormalization of marginally irrelevant operators near the fixed point [188]):

$$\begin{aligned}
\langle O^{\dagger}_{\mathrm{CDW}}(x) O_{\mathrm{CDW}}(0)\rangle &\propto e^{2ik_F x} x^{-(1+K_c)} \ln^{-3/2}(x) \,, \\
\langle O^{\dagger}_{\mathrm{SDW}_\alpha}(x) O_{\mathrm{SDW}_\alpha}(0)\rangle &\propto e^{2ik_F x} x^{-(1+K_c)} \ln^{1/2}(x) \,, \\
\langle O^{\dagger}_{4k_F}(x) O_{4k_F}(0)\rangle &\propto e^{4ik_F x} x^{-4K_c} \,, \\
\langle O^{\dagger}_{\mathrm{SS}}(x) O_{\mathrm{SS}}(0)\rangle &\propto x^{-(1+1/K_c)} \ln^{-3/2}(x) \,, \\
\langle O^{\dagger}_{\mathrm{TS}_\alpha}(x) O_{\mathrm{TS}_\alpha}(0)\rangle &\propto x^{-(1+1/K_c)} \ln^{1/2}(x) \,, \qquad (21.26)
\end{aligned}$$

where the proportionality involves model dependent constants and where sub-leading terms have been omitted. In the presence of interactions that break spin rotation symmetry ($g_{1\parallel} \neq g_{1\perp}$) the model flows, for moderately repulsive bare $g_{1\parallel}$, to a point on a fixed line with $V_s = 0$ and $K_s > 1$. Correspondingly, the spin contribution to the decay exponent of the correlation functions (see Eq. (21.26)) changes from 1 to K_s for the CDW, SS, and the z component of the SDW order, and from 1 to $1/K_s$ for TS and the x and y components of the SDW order. (For $K_s \neq 1$, there are no logarithmic corrections and the leading behavior is that of a pure power law [188].)

The temporal dependence of the above correlation functions is easily obtained owing to the Lorentz invariance of the model (Eq. (21.13)). By Fourier transforming them one obtains the related susceptibilities whose low temperature behavior for the spin rotationally invariant case is given according to

$$\begin{aligned}
\chi_{\mathrm{CDW}} &\propto T^{K_c-1} |\ln(T)|^{-3/2} \,, \\
\chi_{\mathrm{SDW}} &\propto T^{K_c-1} |\ln(T)|^{1/2} \,, \\
\chi_{4k_F} &\propto T^{4K_c-2} \,, \\
\chi_{\mathrm{SS}} &\propto T^{1/K_c-1} |\ln(T)|^{-3/2} \,, \\
\chi_{\mathrm{TS}} &\propto T^{1/K_c-1} |\ln(T)|^{1/2} \,. \qquad (21.27)
\end{aligned}$$

Without a spin gap, SDW and triplet pairing fluctuations are most relevant.
Therefore in the absence of a spin gap and for $1/3 < K_c < 1$, the $2k_F$ fluctuations are the most divergent, and the SDW is slightly more divergent than the CDW. In the presence of strong repulsive interactions when $K_c < 1/3$, the $4k_F$ correlations dom-

[11] For a discussion of some delicate points involving Klein factors in such expressions see [183] and [185].

inate. If $K_c > 1$, the pairing susceptibilities diverge at low temperatures and triplet pairing is the dominant channel.

When $g_1 < 0$, a spin gap opens of magnitude

$$\Delta_s \sim \frac{v_s}{a} \left(\frac{|g_1|}{2\pi v_s} \right)^{1/(2-2K_s)} . \tag{21.28}$$

This can be explicitly demonstrated at the special Luther-Emery point [189] $K_s = 1/2$, where the spin sector is equivalent to a massive free Dirac theory. At this point, a new set of spinless fermions can be defined

$$\Psi_\eta \equiv \frac{1}{\sqrt{2\pi a}} F_\eta \exp[i\sqrt{\pi/2}(\theta_s - 2\eta\phi_s)] , \tag{21.29}$$

in terms of which the spin part of the Hamiltonian can be refermionized

$$\mathcal{H}_s = -iv_s \sum_\eta \eta \Psi_\eta^\dagger \partial_x \Psi_\eta + \Delta_s (\Psi_1^\dagger \Psi_{-1} + H.c.) , \tag{21.30}$$

and readily diagonalized to obtain the spin excitation spectrum

$$E_s = \sqrt{v_s^2 k^2 + \Delta_s^2} . \tag{21.31}$$

With a spin gap, CDW or singlet pairing fluctuations are the most relevant.
In the spin gapped phase, correlations involving spin 1 order parameters, such as SDW and triplet pairing, decay exponentially with correlation length $\xi_s = v_s/\Delta_s$. On the other hand the amplitude of the CDW and SS order parameters acquire a vacuum expectation value. Actual long range order, however, does not occur due to the phase fluctuations associated with the still gapless charge modes. Nevertheless, the CDW and SS susceptibilities are enhanced compared to the case with no spin gap and in a spin rotationally invariant system are given by

$$\chi_{\text{CDW}} \propto \Delta_s T^{K_c - 2} ,$$
$$\chi_{\text{SS}} \propto \Delta_s T^{1/K_c - 2} . \tag{21.32}$$

As long as $K_c > 1/2$ the singlet pairing susceptibility is divergent but it becomes more divergent

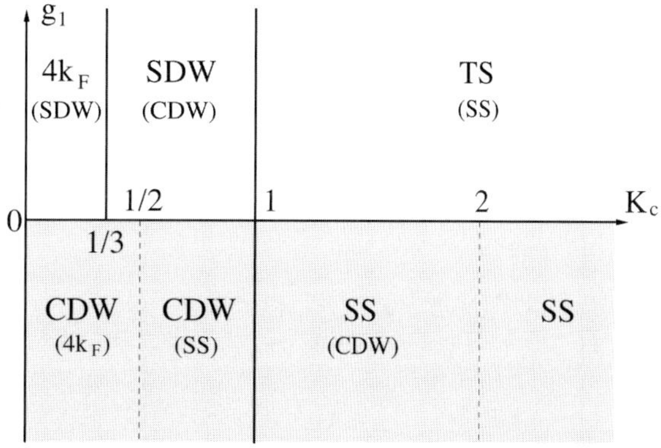

Fig. 21.13. Phase diagram for the one-dimensional spin rotationally invariant electron gas showing where various zero temperature correlations diverge. Parentheses indicate subdivergent correlations and the shaded region contains the spin gapped phases. The order parameters that appear in the figure are: singlet superconductivity (SS); triplet superconductivity (TS); $2k_F$ spin density wave (SDW); $2k_F$ charge density wave (CDW); and $4k_F$ charge density wave ($4k_F$)

than the CDW susceptibility only when $K_c > 1$. The latter diverges for $K_c < 2$ and is the predominant channel provided $K_c < 1$. Figure 21.13 summarizes the situation for low temperatures showing where in parameter space each type of correlation diverges.

Concerning the sign of the effective interactions.
We see that the low energy behavior of a system with a spin gap is basically determined by a single parameter K_c. For a Hubbard chain with repulsive interactions, it is well known [190] that $K_c < 1$, but this is not a general physical bound. For instance, numerical experiments on two leg Hubbard ladders (which are spin gapped systems as we discuss in Sects. 21.10 and 21.11) have found a power law decay $r^{-\theta}$ of the singlet d-wave pairing correlations along the ladder. Figure 21.14 presents the minimal value of the decay exponent θ obtained for ladders with varying ratio of inter- to intra-leg hopping $t_\perp/t$ as a function of the relative interaction strength U/t [191]. By comparing it with the corresponding exponent $\theta = 1/K_c$ calculated for a spin gapped one-dimensional system, one can see that $K_c > 1/2$ over the entire range of parameters and that for some ranges $K_c > 1$. Our point is that in a multicomponent 1DEG, it is possible

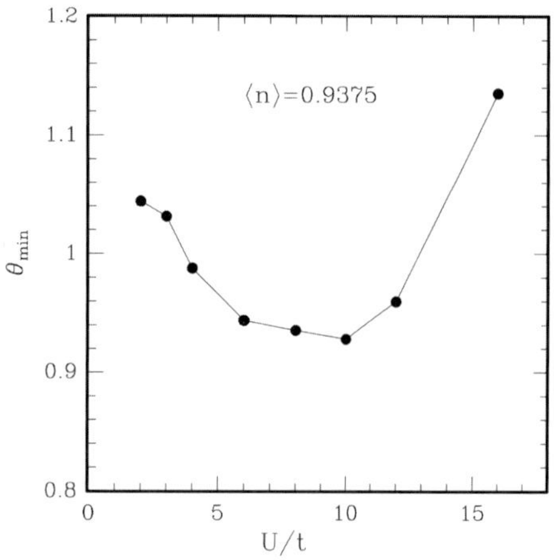

Fig. 21.14. Minimal value of the decay exponent, $\theta = 1/K_c$, of the d-wave singlet pairing correlations in a two leg ladder with varying hopping ratio $t_\perp/t$ as a function of U/t. The electron filling is $\langle n\rangle = 0.9375$. (From Noack et al. [191])

to have $K_c > 1$ (and thus singlet superconductivity as the most divergent susceptibility) even for repulsive interactions.

21.5.2 Spectral Functions of the 1DEG–Signatures of Fractionalization

The fact that one can obtain a strong (power law) divergence of the superconducting susceptibility from repulsive interactions between electrons is certainly reason enough to look to the 1DEG for clues concerning the origins of high temperature superconductivity—we will further pursue this in Sects. 21.10 and 21.11, below. What we will do now is to continue to study the 1DEG as a solved model of a non-Fermi liquid.

In a Fermi liquid the elementary excitations have the quantum numbers of an electron and a nonvanishing overlap with the state created by the electronic creation operator acting on the ground state. As a result the single particle spectral function, $A(\boldsymbol{k},\omega)$, is peaked at $\omega = \epsilon(\boldsymbol{k}) = \boldsymbol{v}_F(\boldsymbol{k}_F)\cdot(\boldsymbol{k}-\boldsymbol{k}_F)$, where $\epsilon(\boldsymbol{k})$ is the quasiparticle dispersion relation. This peak can be and has been [192] directly observed using angle resolved photoemission spectroscopy (ARPES) which measures the single hole piece of the spectral function

$$A^<(\boldsymbol{k},\omega) = \int_{-\infty}^{\infty} \mathrm{d}\boldsymbol{r}\,\mathrm{d}t\, e^{i(\boldsymbol{k}\cdot\boldsymbol{r}+\omega t)}\langle\psi_\sigma^\dagger(\boldsymbol{r},t)\psi_\sigma(0,0)\rangle \,. \tag{21.33}$$

The lifetime of the quasiparticle, $\tau(\boldsymbol{k})$, can be determined from the width of the peak in the "energy distribution curve" (EDC) defined by considering $A^<(\boldsymbol{k},\omega)$ at fixed $\boldsymbol{k}$ as a function of ω:

$$1/\tau = \Delta\omega \,. \tag{21.34}$$

In a Fermi liquid, so long as the quasiparticle excitation is well defined (i.e. the decay rate is small compared to the binding energy) this width is related via the Fermi velocity to the peak width Δk in the "momentum distribution curve" (MDC). This curve is defined as a cross section of $A^<(\boldsymbol{k},\omega)$ taken at constant binding energy, ω. Explicitly

$$\Delta\omega = v_F\Delta k \,. \tag{21.35}$$

There are no stable excitations of the 1DEG with quantum numbers of an electron.
A very different situation occurs in the theory of the 1DEG where the elementary excitations, charge and spin density waves, do not have the quantum numbers of a hole. Despite the fact that the elementary excitations are bosons, they give rise to a linear in T specific heat that is not qualitatively different from that of a Fermi liquid. However, because of the separation of charge and spin, the creation of a hole (or an electron) necessarily implies the creation of two or more elementary excitations, of which one or more carries its spin and one or more carries its charge. Consequently, $A^<(k,\omega)$ does not have a pole contribution, but rather consists of a multiparticle continuum which is distributed over a wide region in the (k,ω) plane. The shape of this region is determined predominantly by the kinematics. The energy and momentum of an added electron are distributed between the constituent charge and spin pieces. In the case where both of them are gapless [see Figs. 21.15(a) and 21.15(b)] this means

$$\begin{aligned} E &= v_c|k_c| + v_s|k_s| \,, \\ k &= k_c + k_s \,, \end{aligned} \tag{21.36}$$

where energy and momentum are measured relative to E_F and k_F respectively. Consequently any point above the dispersion curve of the slower excitation (taken here to be the spin) may be reached by placing an appropriate amount of energy and momentum into the spin degrees of freedom, and the remaining energy and momentum into the charge degrees of freedom, as shown in Fig. 21.15(c). The addition of a hole is similarly constrained kinematically, and the corresponding zero temperature ARPES response has weight only within the shaded regions of Fig. 21.15(d).

Further constraints on the distribution of spectral weight may arise from symmetries. In the spin rotationally invariant case, at the fixed point $K_s = 1$, the spin correlators do not mix left and right moving pieces. As a consequence, $A^<(T = 0)$ for a right moving hole vanishes when ω is in the range $v_s k \leq |\omega| \leq v_c k$ (assuming $v_s < v_c$ and $k > 0$), even if the kinematic conditions are satisfied; see Fig. 21.15(d).[12] If in addition $K_c = 1$, so that the charge piece also does not mix left and right movers, $A^<(T = 0)$ vanishes unless $k < 0$ and $v_s|k| \leq |\omega| \leq v_c|k|$, (the darkest region in Fig. 21.15(d)). While $K_s = 1$ is fixed by symmetry, there is no reason why K_c should be precisely equal to 1. However, if the interactions are weak, (i.e. if K_c is near 1) most of the spectral weight is still concentrated in this region. It spreads throughout the rest of the triangle with increasing interaction strength.

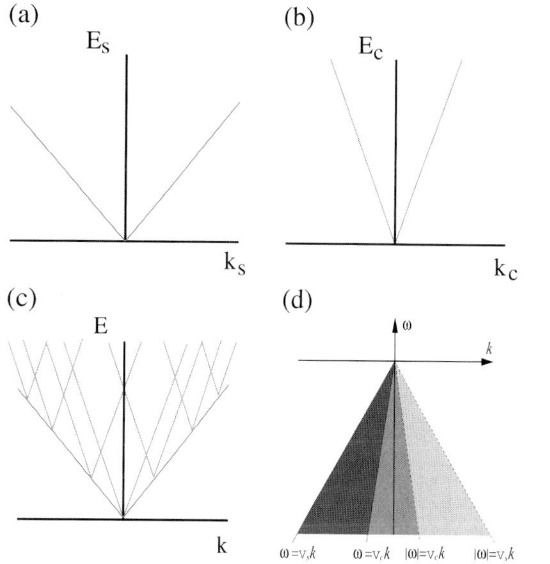

Fig. 21.15. Kinematics of the 1DEG: (**a**) Dispersion of the spin excitations. (**b**) Dispersion of the charge excitations. (**c**) The available electronic states. (**d**) Kinematic constraints on the spectral function: $A^<(k, \omega)$ for the 1DEG is nonzero at $T = 0$ only in the shaded region of the (k, ω) plane. In the spin rotationally invariant case, $K_s = 1$, $A^<(k, \omega) = 0$ in the lightly shaded region, as well. If in addition, $K_c = 1$, $A^<(k, \omega) = 0$ outside of the darkest region. We have assumed $v_c > v_s$, which is usually the case in realistic systems

A dichotomy between sharp MDC's and broad EDC's is a telltale sign of electron fractionalization.
Clearly, the total width of the MDC is bounded by kinematics and is at most $\Delta k_{\max} = 2|\omega|/\min(v_c, v_s)$. Any peak in the MDC will have a width which equals a fraction of this, depending on the interactions and symmetries of the problem, but in any case will vanish as the Fermi energy is approached. By contrast, at $k = 0$, the shape of the EDC is not given by the kinematics at all, but is rather determined by the details of the matrix elements linking the one hole state to the various multi particle-hole states which form the continuum. In this case, the spectrum has a nonuniversal power law behavior with exponents determined by the interactions in the 1DEG. Whenever such a dichotomy between the MDC and EDC is present, it can be taken as evidence of electron fractionalization [86].

These general considerations can be substantiated by examining the explicit expression for the spectral function of the Tomonaga-Luttinger model [193–196]. The quantum criticality and the spin-charge separation of the model imply a scaling form for its correlation functions

$$A^<(k, \omega) \propto T^{2(\gamma_c+\gamma_s)+1} \int \mathrm{d}q\, \mathrm{d}\nu\, G_c(q, \nu) \times G_s(\tilde{k} - rq, \tilde{\omega} - \nu)\,, \qquad (21.37)$$

where we introduce the velocity ratio $r = v_s/v_c$ and define the scaling variables

[12] While the kinematic constraints are symmetric under $k \to -k$, the dynamical considerations are not, since although we have shifted the origin of k, we are in fact considering a right moving electron, i.e. one with momentum near $+k_F$.

$$\tilde{k} = \frac{v_s k}{\pi T} \quad , \quad \tilde{\omega} = \frac{\omega}{\pi T} \; . \tag{21.38}$$

Since the spin and charge sectors are formally invariant under separate Lorentz transformations, the functions G_α, ($\alpha = c, s$) also split into right and left moving parts

$$G_\alpha(k, \omega) = \frac{1}{2} h_{\gamma_\alpha + \frac{1}{2}} \left(\frac{\omega + k}{2} \right) h_{\gamma_\alpha} \left(\frac{\omega - k}{2} \right) , \tag{21.39}$$

where h_γ is expressed via the beta function

$$h_\gamma(k) = \mathrm{Re}\left[(2i)^\gamma B\left(\frac{\gamma - ik}{2}, 1 - \gamma \right) \right] , \tag{21.40}$$

and the exponents

$$\gamma_\alpha = \frac{1}{8}(K_\alpha + K_\alpha^{-1} - 2) , \tag{21.41}$$

are defined so that $\gamma_\alpha = 0$ for noninteracting fermions.

Figure 21.16 depicts MDC's at the Fermi energy ($\omega = 0$) and EDC's at the Fermi wavevector ($k = 0$) for a spin rotationally invariant ($\gamma_s = 0$) Tomonaga-Luttinger model for various values of the parameter γ_c. While the MDC's broaden somewhat with increasing interaction strength they remain relatively sharp with a well defined peak structure. On the other hand any corresponding structure in the EDC's is completely wiped out in the presence of strong interactions. Such behavior has been observed in ARPES studies of quasi-one-dimensional compounds as depicted in Fig. 21.17 as well as in the cuprate high temperature superconductors [86].

Away from the Fermi energy and Fermi wavevector and for not too strong interactions the peaks in the MDC and EDC split into a double peak structure, one dispersing with v_s and the other with v_c. If observed this can be taken as further evidence for spin-charge separation.

We now turn to the interesting case in which the superconducting susceptibility is enhanced due to the opening of a spin gap, Δ_s. At temperatures large compared to Δ_s, the spin gap can be ignored, and the spectral function is well approximated by that of the Tomonaga–Luttinger liquid. However, even

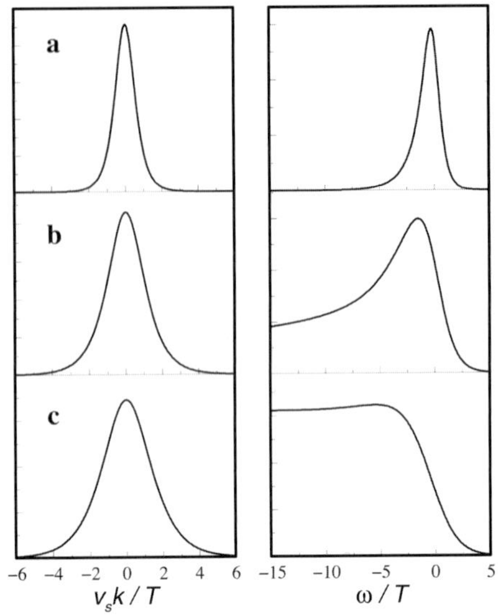

Fig. 21.16. MDC's at $\omega = 0$ (left) and EDC's at $k = 0$ (*right*), for a spin rotationally invariant Tomonaga-Luttinger liquid, with $v_c/v_s = 3$ and a) $\gamma_c = 0$, b) $\gamma_c = 0.25$, and c) $\gamma_c = 0.5$

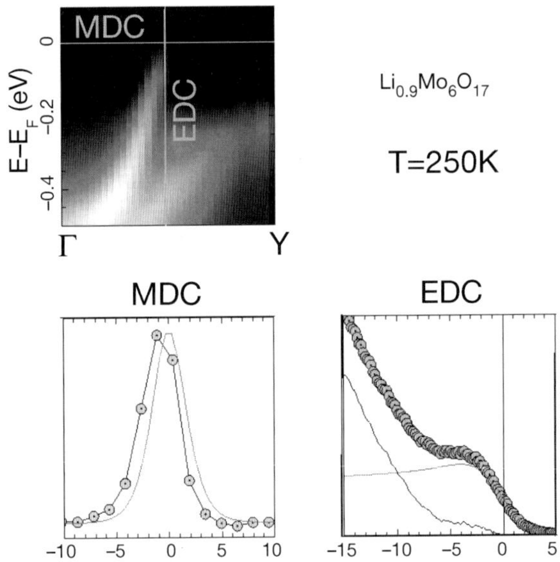

Fig. 21.17. ARPES intensity map for the purple bronze $Li_{0.9}Mo_9O_{17}$. The *lower left panel* depicts the MDC at the Fermi energy together with a Tomonaga–Luttinger theoretical curve. The *lower right panel* contains the EDC at the Fermi wavevector. The *red line* corresponds to the Tomonaga–Luttinger result and the *black curve* is its deviation from the experimental data. (From [197])

below the spin gap scale, many of the characteristics of the Tomonaga–Luttinger spectral function are retained. Spin-charge separation still holds in the spin gapped Luther-Emery liquids and there are no stable "electron-like" excitations. The charge excitations are still the gapless charge density waves of the Tomonaga–Luttinger liquid but the spin excitations now consist of massive spin solitons with dispersion $E_s(k) = \sqrt{v_s^2 k^2 + \Delta_s^2}$. As a result the spin piece of the spectral function is modified and from kinematics it follows that it consists of a coherent one spin soliton piece and an incoherent multisoliton part

$$G_s(k, \omega) = Z_s(k)\delta[\omega + E_s(k)] + G_s^{(\mathrm{multi})}(k, \omega) , \tag{21.42}$$

where the multisoliton piece is proportional (at $T = 0$) to $\Theta[-\omega - 3E_s(k/3)]$. (For $K_s < 1/2$ formation of spin soliton-antisoliton bound states, "breathers", may shift the threshold energy for multisoliton excitations somewhat). The form of $Z_s(k)$ has been calculated explicitly [198], but a simple scaling argument gleans the essential physics [149]. It follows from the fact that the Luther-Emery liquid is asymptotically free that at high energies and short distances compared to the spin gap, the physics looks the same as in the gapless state. Therefore the dependence of the correlation functions on high energy physics, such as the short distance cutoff a, cannot change with the opening of the gap. Since in the gapless system G_s is proportional to $a^{2\gamma_s - 1/2}$, it is a matter of dimensional analysis to see that

$$Z_s(k) = (\xi_s/a)^{\frac{1}{2} - 2\gamma_s} f_s(k\xi_s) , \tag{21.43}$$

where f_s is a scaling function and $\xi_s = v_s/\Delta_s$ is the spin correlation length.

The Luther-Emery liquid is a pseudogap state. Despite the appearance of a coherent piece in the spin sector, the spectral function (Eq. 21.37) still exhibits an overall incoherent response owing to the convolution with the incoherent charge part. The result is grossly similar to the gapless case, aside from the fact that the Fermi edge (the tip of the triangular support of $A^<$ in Fig. 21.15(d)) is pushed back from the Fermi energy by the magnitude of the spin gap (thus rounding the tip of the triangle). If, as suggested in Sect. 21.3, the Luther-Emery liquid is the paradigmatic example of a pseudogap state, clearly the above spectral function gives us an impression of what to expect the signature of the pseudogap to be in the one electron properties.

21.5.3 Dimensional Crossover in a Quasi-1D Superconductor

Continuous global symmetries cannot be spontaneously broken in one dimension, even at $T = 0$. Since the one-dimensional Hamiltonian (Eq. (21.8)) is invariant under translations and spin $SU(2)$ and charge $U(1)$ transformations, no CDW, SDW, or superconducting long range order can exist in its ground state. Therefore, in a quasi-one-dimensional system made out of an array of coupled 1DEG's, a transition into an ordered state necessarily signifies a dimensional crossover at which, owing to relevant interchain couplings, phases of individual chains lock together [23,149]. The ultimate low temperature fate of the system is fixed by the identity of the first phase to do so. This, in turn, is determined by the relative strength of the various couplings and the nature of the low energy correlations in the 1DEG.

In the spin gapped phase, which we consider in the rest of this section, both the CDW and the superconducting susceptibilities are enhanced. To begin with, we will analyze the simplest model of a quasi-one-dimensional superconductor. We defer until the following section any serious discussion of the competition between CDW and superconducting order. We will also defer until then any discussion of the richer possibilities which arise when the quasi-one-dimensional physics arises from a self-organized structure, i.e. stripes, with their own additional degrees of freedom.

Interchain Coupling and the Onset of Order

The simplest and most widely studied model of a quasi-one-dimensional spin gapped fluid is

$$\begin{aligned}\mathcal{H} = \sum_j \mathcal{H}_j + \mathcal{J} \sum_{<i,j>} [O_{\mathrm{SS}}^\dagger(i, x) O_{\mathrm{SS}}(j, x) + \mathrm{H.C.}] \\ + \mathcal{V} \sum_{<i,j>} [O_{\mathrm{CDW}}^\dagger(i, x) O_{\mathrm{CDW}}(j, x) + \mathrm{H.C.}] ,\end{aligned} \tag{21.44}$$

where $\mathcal{H}_j$ describes the Luther-Emery liquid on chain, pairs of nearest neighbor chains are denoted $< i, j >$, and $O_\alpha(j, x)$ is the appropriate order parameter field on chain j. The bosonized form of these operators is given in Eqs. (21.24) and (21.25), above. It is assumed that the interchain couplings, $\mathcal{J}$ and $\mathcal{V}$, are small compared to all intrachain energies. There are two more or less complementary ways of approaching this problem:

(1)
The first is to perform a perturbative renormalization group (RG) analysis about the decoupled fixed point, i.e. compute the beta function perturbatively in powers of the interchain couplings. To lowest order, the beta function is simply determined by the scaling dimension, D_α, of each operator; if $D_\alpha < 2$, it means that O_α is perturbatively relevant, and otherwise it is irrelevant. It turns out that the CDW and SC orders are dual to each other, so that

$$D_{SS} = 1/K_c \quad , \quad D_{CDW} = K_c \,. \tag{21.45}$$

This has the implication that one, or the other, or both of the interchain couplings is always relevant. From this, we conclude with a high level of confidence that at low temperature, even if the interchain couplings are arbitrarily weak, the system eventually undergoes a phase transition to a higher dimensional ordered state. An estimate of T_c can be derived from these equations in the standard way, by identifying the transition temperature with the scale at which an initially weak interchain coupling grows to be of order 1. In this way, for $D_{SS} < 2$, one obtains an estimate of the superconducting transition temperature

$$T_c \sim E_F[\mathcal{J}/E_F]^{1/(2-D_{SS})} = \mathcal{J}[\mathcal{J}/E_F]^{(D_{SS}-1)/(2-D_{SS})}, \tag{21.46}$$

and similarly for the CDW ordering temperature. Note that as $D_{SS} \to 2^-$, $T_c \to 0$, and that $T_c \gg \mathcal{J}$ for $D_{SS} < 1$. Clearly, the power law dependence of T_c on coupling constant offers the promise of a high T_c when compared with the exponential dependence in BCS theory.

(2)
The other way is to use interchain mean field theory [199]. Here, one treats the one-dimensional fluctuations exactly, but the interchain couplings in mean field theory. For instance, in the case of interchain SS ordering, one considers each chain in the presence of an external field

$$\mathcal{H}^{\text{eff}} = \mathcal{H}_j + [\Delta^*_{SS} O_{SS}(j, x) + \text{H.C.}] \,, \tag{21.47}$$

where Δ_{SS} is determined self-consistently:

$$\Delta_{SS} = z\mathcal{J}\langle O_{SS}(j, x)\rangle \,, \tag{21.48}$$

where z is the number of nearest neighbor chains and the expectation value is taken with respect to the effective Hamiltonian. This mean field theory is exact [149,200] in the limit of large z, and is expected to be reliable so long as the interchain coupling is weak. It can be shown to give exact results in the limit of extreme anisotropy for the Ising model, even in two dimensions (where $z = 2$) [199]. More generally, it is a well controlled approximation at least for temperatures $T \gg \mathcal{J}$ (which includes temperatures in the neighborhood of T_c as long as $D_{SS} < 1$). This approach gives an estimate of T_c which is related to the susceptibility of the single chain,

$$1 = z\mathcal{J}\chi_{SS}(T_c) \,, \tag{21.49}$$

which, from the expression in Eq. (21.32), can be seen to produce qualitatively the same estimate for T_c as the perturbative RG treatment. The advantage of the mean field treatment is not only that it gives an explicit, and very physical expression for T_c, but that it permits us to compute explicitly the effect of ordering on various response functions, including the one particle spectral function. The case of CDW ordering is a straightforward extension.

Emergence of the Quasiparticle in the Ordered State

Superconducting order binds fractionalized excitations into "ordinary" quasiparticles.
The excitation spectrum changes dramatically below T_c when the interchain "Josephson" coupling $\mathcal{J}$ triggers long range order [149]. The fractionalized excitations of the Tomonaga-Luttinger and the Luther-Emery liquids are replaced by new excitations with familiar "BCS" quantum numbers. Formally, superconducting order leads to a confinement phenomenon. While the spin gap in the Luther-Emery state already implies suppressed fluctuations of ϕ_s

on each chain, and correspondingly a finite amplitude $\cos(\sqrt{2\pi}\phi_s)$ of the superconducting order parameter, it is the interchain Josephson coupling that tends to lock its phase θ_c from one chain to the next.

Operating with the hole operator, Eq. (21.9), on the ground state at the position of the jth chain creates a pair of kinks (solitons) of magnitude $\sqrt{\pi/2}$ in both the charge and spin fields θ_c and ϕ_s of this chain. As a result the phase of the order parameter changes by π upon passing either the spin or the charge soliton. This introduces a negative Josephson coupling between the affected chain and its neighbors along the entire distance between the charge and spin solitons. The energy penalty due to this frustration grows linearly with the separation between the solitons and causes a bound pair to form. In fact, all solitonic excitations are confined into pairs, including charge-charge and spin-spin pairs. The bound state between the charge and the spin pieces restores the electron, or more precisely the Bogoliubov quasiparticle, as an elementary excitation, causing a coherent (delta function) peak to appear in the single particle spectral function.

An explicit expression for the spectral function in the superconducting state can be obtained in the context of the effective Hamiltonian in Eq. (21.47):

$$A^<(k,\omega) = Z(k)\delta[\omega - E(k)] + A^{(\text{incoherent})}(k,\omega)\,, \tag{21.50}$$

where $E(k) = \sqrt{v_s^2 k^2 + \Delta_0^2}$. Here $\Delta_0 = \Delta_s + \Delta_c/2$ is the creation energy of the bound state where $\Delta_c \propto \Delta_{SS}$ is the mean field gap ($\Delta_c \ll \Delta_s$) that opens in the charge sector below T_c [149]. The multiparticle incoherent piece has a threshold slightly above the single hole threshold at $\omega = E(k) + 2\Delta_c$. The shape of $A^<(k,\omega)$ at $T = 0$ is presented schematically in Fig. 21.18.

Once again, we may employ the asymptotic freedom of the system to construct a scaling argument. In this case, high energy physics dependent upon either the cutoff or the spin gap (which is by assumption much larger than T_c) cannot change upon entering the superconducting state. Comparing the form of the spectral response in the normal spin gapped state with that of the superconductor reveals the weight of the coherent peak

$$Z(k) = Z_s(0)(\xi_c/a)^{-\frac{1}{2}-2\gamma_c} f(k\xi_c)\,, \tag{21.51}$$

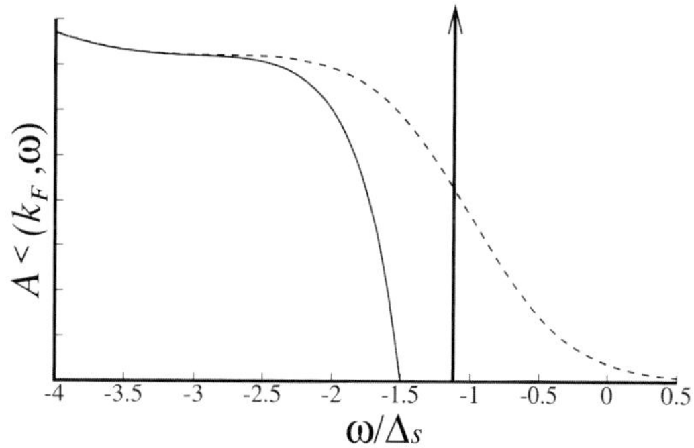

Fig. 21.18. The temperature evolution of the spectral function. The *dashed line* depicts $A^<$ at intermediate temperatures below the spin gap Δ_s but above T_c. The *solid line* represents the spectral function at zero temperature. A coherent delta function peak onsets near T_c at energy $\Delta_0 = \Delta_s + \Delta_c(0)/2$. The multiparticle piece starts at a threshold $2\Delta_c(0)$ away from the coherent peak

where f is a scaling function and $\xi_c = v_c/\Delta_c$ is the charge correlation length. Physically, the dependence of the weight on Δ_c, which also equals the (local) superfluid density [149], reflects the fact that the superfluid stiffness between chains controls the strength of the bound state forming the quasiparticle.

Since the superfluid density is a rapid function of temperature upon entering the superconducting state, the weight of the coherent peak will also rapidly increase as the temperature is lowered. Because the Josephson coupling is weak, the energy of the bound state is largely set by the spin gap, so that the energy of the coherent peak will not be a strong function of temperature in the neighborhood of T_c. Likewise, since the gap is not changing rapidly, the scattering rate and therefore the lifetime of the new quasiparticle will not have strong temperature dependence either. All of the above signatures have been observed in ARPES measurements of the coherent peak in $Bi_2Sr_2CaCu_2O_{8+\delta}$ [89,91,201,202] and $YBa_2Cu_3O_{7-\delta}$ [12].

The temperature evolution of the spectral function is in marked contrast with that in a BCS superconductor

The behavior we have just described is in sharp contrast to that of a conventional superconductor, where the gap opens precisely at T_c. Since in that case the gap is a rapid function of temperature, so is the energy of the conventional quasiparticle peak. Moreover, scattering processes are rapidly gapped out upon entering the BCS superconducting state, so that the quasiparticle often sharpens substantially as the temperature is lowered below T_c. Most importantly, in the conventional case, quasiparticles exist above the transition temperature, so the intensity (Z factor) of the peak does not change much upon entering the superconducting state. By contrast, in a quasi-one-dimensional superconductor, there are no quasiparticle excitations in the normal state. The existence of the quasiparticle is due to the dimensional crossover to the three-dimensional state, and is an entirely collective effect!

21.5.4 Alternative Routes to Dimensional Crossover

Until now, we have assumed that the spin gap is large compared to the interchain couplings, and this assumption leads inevitably to the existence of a quasi-1D pseudogap regime above T_c and a dimensional crossover associated with the phase ordering at T_c. Since under some circumstances, the spin gap in 1D can be zero or exponentially small compared to E_F, it is possible for a system to be quasi-1D, in the sense that the interchain couplings are small compared to the intrachain interactions, and yet have the dimensional crossover occur above any putative spin gap scale. In this case, most likely the dimensional crossover is triggered by the relevance of the interchain, single particle hopping operator—since any spin gap is negligible, the previous argument for its irrelevance is invalidated. What this means is that there is a dimensional crossover, T^*_{3D} (see Fig. 21.19), at which the system transforms from a Luttinger liquid at high temperatures to a Fermi liquid at lower temperatures (see Fig. 21.12). If there are residual effective attractive interactions, the system will ultimately become a superconductor at still lower temperatures. However, in this case, the transition will be more or less of the BCS type—a Fermi surface instability (albeit on a highly anisotropic Fermi surface) with well defined quasiparticles existing both above and below T_c.

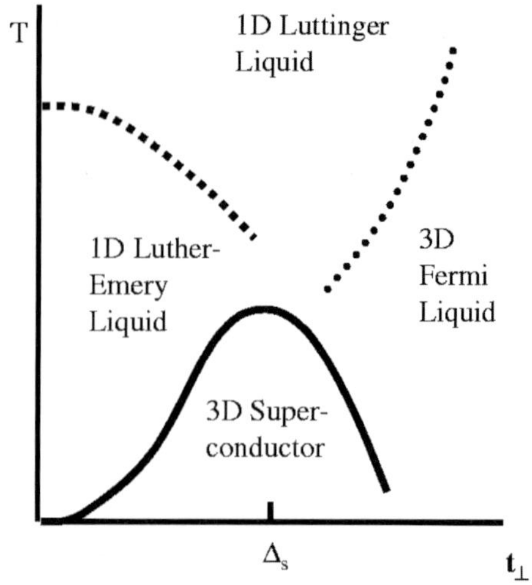

Fig. 21.19. *Two routes to dimensional crossover.* In an array of multicomponent 1DEG's, for temperatures large compared to the transverse single particle tunneling, $t_\perp$, the system behaves as a collection of independent (1D) Luttinger liquids. For weak $t_\perp$, the dimensional crossover may proceed as described in Sect. 21.5.3, with a crossover first to a (1D) Luther–Emery liquid, and a lower temperature dimensional crossover to a (3D) superconductor. For large $t_\perp$, there may be a dimensional crossover into a (3D) Fermi liquid, before the system becomes a (3D) superconductor

The case where dimensional crossover to a Fermi liquid occurs well above T_c may serve as a model for the overdoped cuprates.

The crossover from a Luttinger liquid to a Fermi liquid is not as well characterized, theoretically, as the crossover to a superconductor. The reason is that no simple form of interchain mean field theory can be employed to study it. Various energy scales associated with the crossover can be readily obtained from a scaling analysis. A recent interesting advance [200, 203, 204] has been made on this problem using "dynamical mean field theory," again based on the idea of using $1/z$ (where z is the number of neighboring chains) as a small parameter, which gives some justification for a widely used RPA-like approximation for the spectral function [184]. However, there are still serious shortcomings with this approximation [200, 205]. Clearly more interesting work remains to be done to sort out the physics in this limit, which may be a caricature of the physics of the overdoped cuprates. More complicated routes to dimensional crossover can also be studied [132], relevant to systems with more than one flavor of chain. For instance, it has recently been found that it is possible for a two component quasi-1D system to produce a superconducting state which supports gapless "nodal quasiparticles," even in the limit of extreme anisotropy [132].

21.6 Quasi-1D Physics in a Dynamical Stripe Array

Competition between CDW and SS is key in quasi-1D systems.
While in strictly 1D systems CDW and superconducting fluctuations happily coexist, interchain coupling typically leads to a low temperature state where one is ordered and the other is suppressed. As mentioned before, in the simplest microscopic realizations of the 1DEG with repulsive interactions, $0 < K_c < 1$ and hence the CDW susceptibility is the most divergent as $T \to 0$ (see Eq. (21.32)). This seemingly implies that the typical fate of a quasi-one-dimensional system with a spin gap is to wind up a CDW insulator in which CDW modulations on neighboring chains phase lock to each other. And, indeed, many quasi-one-dimensional metals in nature suffer precisely this fate—the competition between CDW and SS order is a real feature of quasi-1D systems. Recall, however, that in more complicated realizations of the 1DEG, K_c can be greater than 1 even for repulsive interactions, as shown in Fig. 21.14 above.

What we will examine in this section is another way in which the balance between CDW and SS ordering can be affected [52, 206, 207]. Specifically, we will show below that transverse fluctuations of the backbone on which the quasi-1D system lives may significantly enhance the tendency to SS while suppressing CDW ordering. Such fluctuations are unimportant in conventional quasi-one-dimensional solids, where the constituent molecules, upon which the electrons move, have a large mass and a rigid structure. But when the 1DEG's live along highly quantum electronic textures, or "stripes," transverse stripe fluctuations are probably always large.

21.6.1 Ordering in the Presence of Quasi-static Stripe Fluctuations

Consider a two-dimensional array of stripes that run along the x direction, and imagine that there is a 1DEG which lives on each stripe. To begin with, we will consider the case in which the stripe fluctuations are sufficiently slow that they can be treated as static—in other words, we consider an array of imperfectly ordered stripes, over whose meanderings we will eventually take an equilibrium (annealed) average. We will use a coordinate system in which points on the stripes are labeled by the coordinate x, the stripe number j, and in which transverse displacements of the stripe in the y direction are labeled by $h_j(x)$, as shown in Fig. 21.20. We therefore ignore the possibility of overhangs which is a safe assumption in the ordered state.

We now consider the effect that stripe geometry fluctuations have on the inter-stripe couplings. Because the CDW order (and any other $2k_F$ or $4k_F$ orders) occurs at a large wave vector, the geometric fluctuations profoundly affect its phase. As shown in the cartoon in Fig. 21.20, the transverse stripe fluctuations lead to a dephasing of the CDW order on neighboring stripes. To study this effect formally, we note that the local CDW order parameter has the form:

$$O_{\mathrm{CDW}}(j,x) = \frac{e^{-2ik_F L_j(x)}}{\pi a} \times \cos[\sqrt{2\pi}\phi_s(j,x)]\, e^{-i\sqrt{2\pi}\phi_c(j,x)}\,, \tag{21.52}$$

where

$$L_j(x) = \int_0^x \mathrm{d}x' \sqrt{1 + (\partial_{x'} h_j)^2}\,, \tag{21.53}$$

is the arc length, i.e. the distance measured along stripe j to point x, and $k_F = (\pi/2)n$, with n the average linear electronic density along the stripe. In contrast, the superconducting order parameter O_{SS} and any other $k = 0$ order are unaffected by the geometric fluctuations. This results in a fundamental difference in the way CDW and Josephson inter-stripe couplings evolve with growing stripe fluctuations.

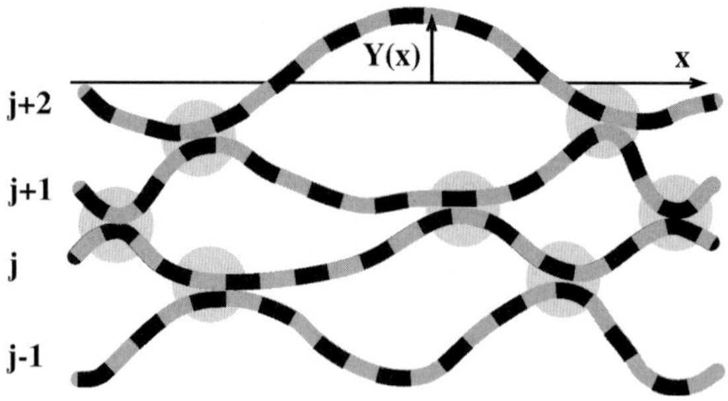

Fig. 21.20. Schematic representation of a smectic stripe phase. Stripe collisions are denoted by the *light grey circles*, the *top left* of which shows a favorable relative phase for a CDW along neighboring stripes. However, once this favorable phase is set, the *top right* collision is out of phase by half of a wavelength, and maximally unfavorable. As described in the text, any ordering with a finite wavevector along the stripe direction experiences this kind of destructive interference due to transverse stripe fluctuations. With the CDW ground state thus avoided, superconductivity (a $k = 0$ order) can flourish. Stripe collisions also enhance the Josephson coupling, further encouraging superconductivity

The CDW and Josephson couplings between neighboring stripes are of the form

$$H_{\mathcal{V}} = \frac{1}{(2\pi a)^2} \sum_j \int dx \mathcal{V}(\Delta_j h) \times \cos[\sqrt{2\pi}\phi_s(j,x)] \cos[\sqrt{2\pi}\phi_s(j+1,x)] \times \cos[\sqrt{2\pi}\Delta_j \phi_c + 2k_f \Delta_j L] , \quad (21.54)$$

$$H_{\mathcal{J}} = -\frac{1}{(2\pi a)^2} \sum_j \int dx \mathcal{J}(\Delta_j h) \times \cos[\sqrt{2\pi}\phi_s(j,x)] \cos[\sqrt{2\pi}\phi_s(j+1,x)] \times \cos[\sqrt{2\pi}\Delta_j \theta_c] , \quad (21.55)$$

where $\Delta_j h \equiv h(j+1,x) - h(j,x)$, etc. Here we have assumed that the average electronic density, and therefore k_F, is the same on neighboring stripes. The coupling constants $\mathcal{V}(\Delta_j h)$ and $\mathcal{J}(\Delta_j h)$, depend on the local spacing between adjacent stripes, since they are more strongly coupled when they are close together than when they are far apart. This is particularly important for the Josephson coupling which depends on the pair tunneling amplitude and therefore roughly exponentially on the local spacing between the stripes

$$\mathcal{J}(\Delta_j h) \approx \mathcal{J}_0 e^{-\alpha \Delta_j h} . \quad (21.56)$$

Stripe fluctuations dephase CDW order...
By integrating out the stripe fluctuations h one obtains the effective Hamiltonian of an equivalent rigid system of stripes. To first order in $\mathcal{V}$ the CDW coupling is similar to Eq. (21.55) but with $\Delta_j L$ set equal to 0 in the last term and $\mathcal{V}(\Delta_j h)$ replaced by

$$\langle \mathcal{V}(\Delta_j h) \rangle \exp[-2k_F^2 \langle (\Delta_j L)^2 \rangle] , \quad (21.57)$$

where $\langle \ \rangle$ signifies averaging over transverse stripe fluctuations. Since $\Delta_j L = L_{j+1}(x) - L_j(x)$ is a sum of contributions with random signs, which are more or less independently distributed along the distance $|x|$, we expect it to grow roughly as in a random walk, i.e. $\langle (\Delta_j L)^2 \rangle \sim D|x|$, where D is a constant. Indeed one can show [52] that at finite temperature $\langle (\Delta_j L)^2 \rangle \sim T|x|$ while at $T = 0$ $\langle (\Delta_j L)^2 \rangle \sim \hbar\bar{\omega} \log |x|$, where $\hbar\bar{\omega}$ is a suitable measure of the transverse stripe zero point energy. As a result of this dephasing effect[13], coupling between CDW's vanishes rapidly except in a narrow region near the ends of the stripes and hence can be ignored in the thermodynamic limit. In short, transverse stripe fluctuations cause destructive interference of $k \neq 0$ order on neighboring chains, strongly suppressing those orders.

... but they enhance SS order.
The effects of stripe fluctuations on the Josephson coupling can be analyzed in the same way. To first order in the inter-stripe coupling, $\mathcal{J}(\Delta_j h)$ is simply replaced by its average value, $\bar{\mathcal{J}} \equiv < \mathcal{J}(\Delta_j h) >$. In other words, once quasi-static stripe fluctuations are integrated out, the result is once again the Hamiltonian we studied in Eq. (21.44), above, but with $\mathcal{V} = 0$ and $\mathcal{J} = \bar{\mathcal{J}}$. Moreover, due to the exponential dependence of $\mathcal{J}(\Delta_j h)$ on $(\Delta_j h)$, it is clear that $\bar{\mathcal{J}} > \mathcal{J}(0)$,

[13] The effect is absent, at least to lowest order in the strength of the stripe fluctuations, in case the total number of electrons, rather than the average density, is taken to be the same on different stripes [208].

i.e. transverse stripe fluctuations strongly enhance the Josephson coupling between stripes. (There is a similar enhancement of the CDW coupling but it is overwhelmed by the dephasing effect.) Physically, this enhancement reflects the fact that the mean value of $\mathcal{J}$ is dominated by regions where neighboring stripes come close together. In the case of small amplitude fluctuations, this enhancement can be viewed as an inverse Debye-Waller factor,

$$\langle \mathcal{J} \rangle \approx \mathcal{J}_0 e^{\frac{\alpha^2}{2}\langle (\Delta_j h)^2 \rangle} . \tag{21.58}$$

Where the transverse stripe fluctuations are comparable in magnitude to the inter-stripe spacing, the mean Josephson coupling is geometrically determined by the mean density of points at which neighboring stripes actually "bump" (i.e. are separated by about one lattice constant a, see Fig. 21.20). In this limit, treating the stripe fluctuations as a random walk yields the estimate

$$\mathcal{J} \sim \left(\frac{a}{R}\right)^2 \mathcal{J}_0 , \tag{21.59}$$

where R is the mean distance between stripes.

21.6.2 The General Smectic Fixed Point

The quasi-static limit discussed above is presumably inadequate at low enough temperatures, where the quantum dynamics of stripe fluctuations must always be relevant. Progress towards understanding of the complete problem, in which both the stripe dynamics and the dynamics of the 1DEG's are treated on an equal footing, has been made in [208], where a single weakly fluctuating 1DEG has been considered. The problem of the strongly fluctuating smectic remains unsolved. However, since in a crystalline background, the stripe fluctuations are typically not gapless, we expect that at low enough temperatures, the stripe fluctuations can be treated as fast, and be integrated out to produce new effective interactions. So long as the stripes are reasonably smooth, these induced interactions will consist of long wavelength (around $k = 0$) density-density and current-current interactions between the neighboring Luttinger liquids—interactions that we have ignored until now. These interactions should undoubtedly be present in the bare model, as well, even in the absence of stripe fluctuations. They are marginal operators and should be included in the fixed point action.

The effect of these marginal interactions has been treated in several recent papers and the interested reader is referred to [206,209]. The results bear some similarity to the quasi-static results described in the previous section. Specifically, repulsive interchain interactions suppress the tendency to CDW order, thereby enhancing the possibility of superconductivity from repulsive interactions. An exotic "smectic metal" phase with neither superconducting nor CDW order, but infinite anisotropy at low energies, has also been found on the basis of this analysis.

Extensions of this model to a three-dimensional array of chains [210] and the inclusion of a magnetic field [211] have been considered as well. In particular, it is found that the magnetic field suppresses the region of superconducting order in the phase diagram, thus expanding the regime in which the smectic metal is stable. Similar considerations lead one naturally to consider other states obtained when the stripe fluctuations become still more violent. Assuming that the long range stripe order is destroyed by such fluctuations, while the short distance physics remains that of quasi-1DEG's living along the locally defined stripes, one is led to investigate the physics of electron nematic and stripe liquid phases. We shall return to this point in the final section.

21.7 Electron Fractionalization in $D > 1$ as a Mechanism of High Temperature Superconductivity

We briefly discuss here a remarkable set of ideas for a novel mechanism of high temperature superconductivity based on higher dimensional generalizations of the 1D notion of spin-charge separation. Boasting a high pairing scale as well as crisp experimental predictions, these theories have many attractive features. They also bear a strong family resemblance to the "spin gap proximity effect mechanism," which we develop in some detail in Sect. 21.10.4. These appealing ideas, while valid, require the proximity of a spin liquid phase which in turn appears to be a fragile state of matter; for this reason, and others which will be made clear below, it is our opinion that these ideas are probably not applicable to the cuprate superconductors. The discussion in this section is therefore somewhat disconnected from the development in the rest of the paper. We merely sketch the central ideas, without providing any derivations. There are a number of recent papers dealing with this subject to which the interested reader can refer; see [77,80,182,212,213].

21.7.1 RVB and Spin-Charge Separation in Two Dimensions

Immediately following the discovery of high temperature superconductivity [2], Anderson proposed [5] that the key to the problem lay in the occurrence of a never before documented state of matter (in $D > 1$), a spin liquid or "resonating valence bond" (RVB) state, related to a state he originally proposed [214] for quantum antiferromagnets on a triangular (or similarly frustrating) lattice. In this context [80], a spin liquid is defined to be an insulating state with an odd number of electrons per unit cell (and a charge gap) which breaks neither spin rotational nor translational symmetry. Building on this proposal, Kivelson, Rokhsar, and Sethna [69] showed that a consequence of the existence of such a spin liquid state is that there exist quasiparticles with reversed charge spin relations, just like the solitons in the 1DEG discussed in Sect. 21.5, above. Specifically, there exist charge 0 spin 1/2 "spinons" and charge e spin 0 "holons." Indeed, these quasiparticles were recognized as having a topological character [69, 215] analogous to that of the Laughlin quasiparticles in the quantum Hall effect.

There was a debate at the time concerning the proper exchange statistics, with proposals presented identifying the holon as a boson [68,69], a fermion [216], and a semion [217]. It is now clear that all sides of this debate were correct, in the sense that there is no universal answer to the question. The statistics of the fractionalized quasiparticles is dynamically determined, and is sensitive to a form of "topological order" [59,78,212,216,218] which differentiates various spin liquids. There are even transitions between states in which the holon has different statistics [218,219].

Two features of this proposal are particularly attractive:

(1).
It is possible to envisage a high pairing scale in the Mott insulating parent state, since the strong repulsive interactions between electrons, which result in the insulating behavior, are insensitive to any subtler correlations between electrons. Thus, the "μ^* issue" does not arise: the spin liquid can be viewed as an insulating liquid of preformed cooper pairs [5,69,70], or equivalently a superconductor with zero superfluid density.[14] If this pairing scale is somehow preserved upon doping, then the transition temperature of the doped system is determined by superfluid stiffness and is not limited by a low pairing scale, as it would be in a BCS superconductor. Indeed, as in the case of the 1D Luther-Emery liquid discussed in Sect. 21.5, pairing becomes primarily a property of the spin degrees of freedom, and involves little or no pairing of actual charge.

(2).
When the holons are bosonic, their density directly determines the superfluid density. Thus the superconducting T_c can be crudely viewed as the Bose condensation temperature of the holons. The result is that for small concentration of doped holes x [5], the transition temperature is proportional to a positive power of x (presumably [69] $T_c \sim x$ in 2D), in contrast to the exponential dependence on parameters in a BCS superconductor.

In short, many of the same features that would make a quasi-1D system attractive from the point of view of high temperature superconductivity (see Sects. 21.5 and 21.10) would make a doped spin liquid even more attractive. However, there are both theoretical and phenomenological reasons for discounting this idea in the context of the cuprates.

[14] An oxymoron since in this case $T_\theta = T_c = 0$, but the intuitive notion is clear: we refer to a state which is derived from a superconductor by taking the limit of zero superfluid density while holding the pairing scale fixed.

21.7.2 Is an Insulating Spin Liquid Ground State Possible in $D > 1$?

Is this simply angels dancing on the head of a pin?
The most basic theoretical issue concerning the applicability of the fractionalization idea is whether a spin liquid state occurs at all in $D > 1$. The typical consequence of the Mott physics is an antiferromagnetically ordered ("spin crystalline") state, especially the Néel state, which indeed occurs at $x = 0$ in the cuprates. Moreover, the most straightforward quantum disordering of an antiferromagnet will lead to a spin Peierls state, rather than a spin liquid, as was elegantly demonstrated by Haldane [220] and Read and Sachdev [71]. Indeed, despite many heroic efforts, the theoretical "proof of principle," i.e. a theoretically tractable microscopic model with plausible short range interactions which exhibits a spin liquid ground state phase, was difficult to achieve. A liquid is an intermediate phase, between solid and gas, and so cannot readily be understood in a strong or weak coupling limit [80].

Very recently, Moessner and Sondhi [79] have managed to demonstrate just this point of principle! They have considered a model [70] on a triangular lattice (thus returning very closely to the original proposal of Anderson) which is a bit of a caricature in the sense that the constituents are not single electrons, but rather valence bonds (hard core dimers), much in the spirit pioneered by Pauling.[15] The model is sufficiently well motivated microscopically, and the spin liquid character robust enough, that it is reasonable to declare the spin liquid a theoretical possibility. The spin liquid state of Moessner and Sondhi does not break any obvious symmetry.[16]

Spin liquids are fragile.
That said, the difficulty in finding such a spin liquid ground state in model calculations is still a telling point. A time reversal invariant insulating state cannot be adiabatically connected to a problem of noninteracting quasiparticles with an effective band structure[17]—band insulators always have an even number of electrons per unit cell. Thus, an insulating spin liquid is actually quite an exotic state of matter. Presumably, it only occurs when all more obvious types of ordered states are frustrated, i.e. those which break spin rotational symmetry, translational symmetry, or both. The best indications at present are that this occurs in an exceedingly small corner of model space, and that consequently spin liquids are likely to be rather delicate phenomena, if they occur at all in nature. This, in our opinion, is the basic theoretical reason for discarding this appealing idea in the cuprates, where high temperature superconductivity is an amazingly robust phenomenon.

The cuprates appear to be doped spin crystals, not doped spin liquids.
One could still imagine that the insulating state is magnetically ordered, as indeed it is in the cuprates, but that upon doping, once the magnetic order is suppressed, the system looks more like a doped spin liquid than a doped antiferromagnet. In this context, there are a number of phenomenological points about the cuprates that strongly discourage this viewpoint. In the first place, the undoped system is not only an ordered antiferromagnet, it is a nearly classical one: its ground state and elementary excitation spectrum [221–224] are quantitatively understood using lowest order spin wave theory.[18] This state is as far from a spin liquid as can be imagined! Moreover, even in the doped system, spin glass and other types of magnetic order are seen to persist up to (and even into) the superconducting state, often with frozen moments with magnitude comparable to the ordered moments in the undoped system [224,227–229]. These and other indications show that the doped system "remembers" that it is a doped antiferromagnet, rather a doped spin liquid.

15 Indeed, it is tempting to interpret the dimer model as the strong coupling, high density limit of a fluid of Cooper pairs [70].

16 This work was, to some extent, anticipated in studies of large N generalizations of the Heisenberg antiferromagnet [71].

17 In a time reversal symmetry broken state, the band structure need not exhibit the Kramer's degeneracy, so that a weak coupling state with an odd number of electrons per unit cell is possible.

18 Recent experiments on $La_{2-x}Ba_xCuO_4$ at $x = 1/8$ [225] are in agreement with spin wave calculations [226] throughout the entire measured energy range.

Where to look for spin liquids
Regardless of applicability to the cuprates, it would be worthwhile to search for materials that do exhibit spin liquid states, and even more so to look for superconductivity when they are doped. Numerical studies [230–233] indicate that good candidates for this are electrons on a triangular lattice with substantial longer range ring exchange interactions, such as may occur in a 2D Wigner crystal near to its quantum melting point [234], and the Kagomé lattice. It is also possible, as discussed in Sect. 21.11, to look for superconductivity in systems that exhibit some form of spin-charge separation at intermediate length scales (see also [14]).

21.7.3 Topological Order and Electron Fractionalization

Finally, we address the problem of classifying phases in which true electron fractionalization occurs, e.g. in which spinons are deconfined. It is now clear from the work of Wen [78] and Senthil and Fisher [59] that the best macroscopic characterization of fractionalized phases in two or more dimensions is topological, since they frequently possess no local order parameter. Specifically, a fractionalized phase exhibits certain predictable ground state degeneracies on various closed surfaces—degeneracies which Senthil and Fisher have given a physical interpretation in terms of "vision expulsion." Unlike the degeneracies associated with conventional broken symmetries, these degeneracies are not lifted by small external fields which break either translational or spin rotational symmetry. It has even been shown [59,218,235] (as funny as this may sound) that topological order is amenable to experimental detection. Once topological classification is accepted, the one to one relation between spin liquids and electron fractionalization, implied in our previous discussion, is eliminated. Indeed, it is possible to imagine [59,76] ordered (broken symmetry) states, proximate to a spin liquid phase, which will preserve the ground state degeneracies of the nearby spin liquid, and hence will exhibit spin-charge separation.

21.8 Superconductors with Small Superfluid Density

A hallmark of BCS theory is that pairing precipitates order. But it is possible for the two phenomena to happen separately: pairing can occur at a higher temperature than superconductivity. In this case, there is an intermediate temperature range described by electron pairs which have not condensed. In the order parameter language, this corresponds to a well developed amplitude of the order parameter, but with a phase which varies throughout the sample. Superconductivity then occurs with the onset of long range phase coherence. (This is how ordering occurs in a quasi-1D superconductor, as discussed in Sect. 21.5, above.) Such superconductors, while they may have a large pairing scale, have a small stiffness to phase fluctuations, or equivalently a small superfluid density.

21.8.1 What Ground State Properties Predict T_c?

When the normal state is understood, it is reasonable to describe superconductivity as an instability of the normal state as temperature is lowered, which BCS theory does quite successfully in simple metals. Another approach, useful especially when the normal state is not well understood, is to consider which thermal fluctuations degrade the superconducting order as the temperature is raised. Put another way, we address the question, "What measurable ground state ($T = 0$) properties permit us to predict T_c?"

Two classes of thermal excitations are responsible for disordering the ground state of a superconductor: amplitude fluctuations of the complex order parameter (associated with pair breaking), and fluctuations of the phase (associated with pair currents).

Pairing is one energy scale...
The strength of the pairing at $T = 0$ is quantifiable as a typical gap value, Δ_0, where

$$T_p \equiv \Delta_0/2 , \qquad (21.60)$$

is the characteristic temperature at which the pairs fall apart. In a BCS superconductor, it is possible to estimate that $T_c \approx T_p$. (The factor 1/2 in this definition approximates the weak coupling BCS expression, $T_c = \Delta_0/1.78$.) Certainly, more generally, T_p marks a loose upper bound to T_c, since if there is no pairing, there is probably no superconductivity.

... the superfluid phase stiffness sets another.
We can construct another ground state energy scale as follows: Divide the sample into blocks of linear dimension, L, and ask how much energy it costs to flip the sign of the superconducting order parameter at the center of one such region. So long as L is larger than the coherence length, ξ_0, the cheapest way to do this is by winding the phase of the order parameter, so the energy is determined by the superfluid phase stiffness

$$T_\theta = \frac{1}{2} A \gamma L^{d-2} , \qquad (21.61)$$

where d is the number of spatial dimensions, A is a geometry dependent dimensionless number of order 1 and the "helicity modulus", γ, is traditionally expressed in terms of the ratio of the superfluid density, n_s, to the pair effective mass, m^*:

$$\gamma \equiv \frac{\hbar^2 n_s}{m^*} . \qquad (21.62)$$

(We will discuss the quantitative aspects of this relation in Subsection 21.8.3.) Note that for $d = 2$, this energy is independent of L, while for $d = 3$ it is minimized for the smallest allowable value of $L \sim \xi_0$. Clearly, when the temperature is comparable to T_θ, thermal agitation will produce random phase changes from block to block, and hence destroy any long range order. Again, a rough upper bound to T_c is obtained in this way.

In short, it is possible to conclude on very general grounds that

$$T_c \leq \min[T_p, T_\theta] . \qquad (21.63)$$

When $T_p \ll T_\theta$, phase fluctuations can be completely neglected except in the immediate neighborhood of T_c—this is the case in BCS superconductors. If $T_p \gg T_\theta$, quasiparticle excitations, i.e. the broken Cooper pairs, play no significant thermodynamic role up to T_c. In this case a considerable amount of local pairing, and consequently a pseudogap, must persist to temperatures well above T_c. When both T_p and T_θ are comparable to T_c, as is the case in most optimally doped high temperature superconductors, neither class of thermal excitation can be safely neglected.

Of this there is no possible doubt whatever.
In Table 21.1, following [279], we tabulate T_θ, T_p, and T_c for various superconducting materials. Clearly, in bulk Pb, phase fluctuations are not terribly important, while in the cuprate superconductors (and the ET superconductors), phase fluctuations are an order 1 effect. Of this there is no possible doubt! Looking more closely at the table, one sees that the ratio of T_θ/T_c is generally smaller for the underdoped materials, and larger for overdoped, which implies that phase fluctuations are progressively less dominant with increasing doping. The ratio of T_p/T_c varies in the opposite manner with doping.

The obvious implication of the trends exhibited in Table 21.1 is that optimal doping marks a gradual crossover from an underdoped regime, where T_c is predominantly a phase ordering transition, to an overdoped regime in which it is predominantly a pairing transition. This also implies that both pairing and phase fluctuation physics play a nonnegligible role, except in the regimes of extreme underdoping or overdoping where $T_c \to 0$.

21.8.2 An Illustrative Example: Granular Superconductors

We now turn to a beautiful set of experiments carried out by Merchant et al. [280] on granular Pb films with a thin coating of Ag. This is a system in which the microscopic physics is well understood. The T_c of bulk Pb is 7.2K while Ag remains normal down to the lowest accessible temperatures, so that T_θ can be varied with respect to T_p by changing the thickness of Ag. In this way, the system can be tuned from an "underdoped" regime, where T_c is a phase ordering transition and pairing persists to much higher temperatures, to an "overdoped" regime, where the transition is very BCS-like.

Table 21.1. *Zero temperature properties of the superconducting state as predictors of T_c.* Here, T_p is computed from Eq. (21.60) using values of Δ_0 obtained from either tunneling or ARPES, except for overdoped Tl-2201, for which we have used Raman data. In computing T_θ from Eq. (21.61) for nearly isotropic materials (those above the *double line*), we have taken $d = 3, A = 2.2, L = \sqrt{\pi}\xi_0$, and $n_s/2m^* = (8\pi)^{-1}(c/e)^2\lambda_L^{-2}$ where λ_L and ξ_0 are the zero temperature London penetration depth and coherence length, respectively. For layered materials, we have taken $d = 2, A = 0.9$, and the areal superfluid density $n_s/2m^* = (8\pi)^{-1}(c/e)^2 L\lambda_L^{-2}$ where L is now the mean spacing between layers and λ_L is the in-plane London penetration depth. The precise numerical values of A and the factor of $\sqrt{\pi}$ should not be taken seriously—they depend on microscopic details, which can vary from material to material as discussed in Sect. 21.8.3. Penetration depth measurements on Y-123 refer to polycrystalline $Y_{0.8}Ca_{0.2}Ba_2Cu_3O_{7-\delta}$, and report λ_{ab}. The two entries for Hg-1223 assume that the superfluid density resides in all three planes (L = 5.3Å), or the outer two planes only (L = 7.9Å). In the case of the high temperature superconductors, the notations "ud", "op", and "od" refer to under, optimally, and overdoped materials, respectively

Material	(L) [Å]	(λ_L)[Å]	(T_p)[K]	(T_c)[K]	(T_θ)[K]	Ref.
Pb	830	390	7.9	7.2	6×10^5	[236,237]
Nb_3Sn	60	640	18.7	17.8	2×10^4	[238]
UBe_{13}	140	10,000	0.8	0.9	10^2	[239–241]
$Ba_{0.6}K_{0.4}BiO_3$	40	3000	17.4	26	5×10^2	[242,243]
K_3C_{60}	30	4800	26	20	10^2	[244–246]
MgB_2	50	1400	41	39	1.4×10^3	[247–250]
ET	15.2	8000	17.4	10.4	15	[251,252]
PCCO	6.2	2800	23	23	86	[253–255]
Tl-2201 (op)	11.6	1650	122	91	150	[256,257]
Tl-2201 (od)	11.6	2000		80	160	[252,258]
Tl-2201 (od)	11.6	2200		48	130	[252,258]
Tl-2201 (od)	11.6		26	25		[259]
Tl-2201 (od)	11.6	4000		13	40	[252,258]
Bi-2212 (ud)	7.5	3220		65	40	[260]
Bi-2212 (ud) x=.11	7.5		275	83		[97,261]
Bi-2212 (ud) x=.15	7.5		190	90		[262]
Bi-2212 (op)	7.5		220	95		[261]
Bi-2212 (op)	7.5	2700		90-93	60	[253,263]
Bi-2212 (op)	7.5	1800		84	130	[264,265]
Bi-2212 (op)	7.5		242.53	93		[266]
Bi-2212 (op)	7.5	2300		87.3	80	[260]
Bi-2212 (od) x=.19	7.5		143	82		[261]
Bi-2212 (od) x=.21	7.5		145	90		[262]
Bi-2212 (od)	7.5	2220		82	90	[267]
Bi-2212 (od)	7.5	1920		77	120	[260]
Bi-2212 (od) x=.225	7.5		104	62		[261]
Y-123 (ud) x=.075	5.9	2800		38	42	[268]
Y-123 (ud) x= .1	5.9	1900		64	90	[268]
Y-123 (op) x=.16	5.9	1500		85.5	140	[268,269]
Y-123 (op)	5.9	1100	116	91-92	260	[99,270]
Y-123 (op)	5.9	1120		93	250	[271]
Y-123 (od) x=.19	5.9	1300		79	180	[268]
Y-123 (od) x=.23	5.9	1500		55	140	[268]
Y-248	6.8	1600		82	150	[272]
Hg-1201 (op)	9.5	1700	192	95-97	180	[269,273]
Hg-1212 (op)	6.4	1700	290	108	130	[273,274]
Hg-1223 (op)	5.3	1500	435	132-135	130	[269,273,274]
Hg-1223 (op)	7.9	1500		135	190	[269,274]
LSCO (ud) x=.1	6.6	2800	75	30	47	[275–277]
LSCO (op) x=.15	6.6	2600	58	38	54	[275,276]
LSCO (od) x=.16	6.6		46.42	39		[278]
LSCO (od) x=.20	6.6	1950	30	34	96	[262,276]
LSCO (od) x=.22	6.6	1900		27	100	[276]
LSCO (od) x=.24	6.6	1900		20	100	[276]

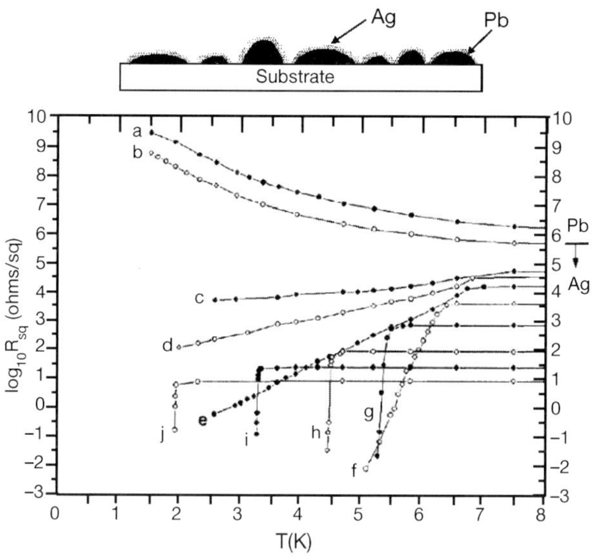

Fig. 21.21. The logarithm of the resistance vs. temperature for a sequence of films, starting with a granular Pb film (a) to which is added successively larger coverage of Ag. From Fig. 5 of Merchant et al. [280]

Figure 21.21 shows the log of the resistance vs. temperature for a sequence of films (a–j) obtained by adding successive layers of Ag to a granular Pb substrate. Films a and b are seen to be globally insulating, despite being locally superconducting below $7.2K$. Films g–j are clearly superconductors. Films c-f are anomalous metals of some still not understood variety. It is important to note that Fig. 21.21 is plotted on a log-linear scale, so that although it is unclear whether films c–f will ever become truly superconducting, films e and f, for example, have low temperature resistances which are 5 or 6 orders of magnitude lower than their normal state values, due to significant superconducting fluctuations; see Fig. 21.22.

Figure 21.23 shows I–V curves obtained from planar tunneling in the direction perpendicular to the

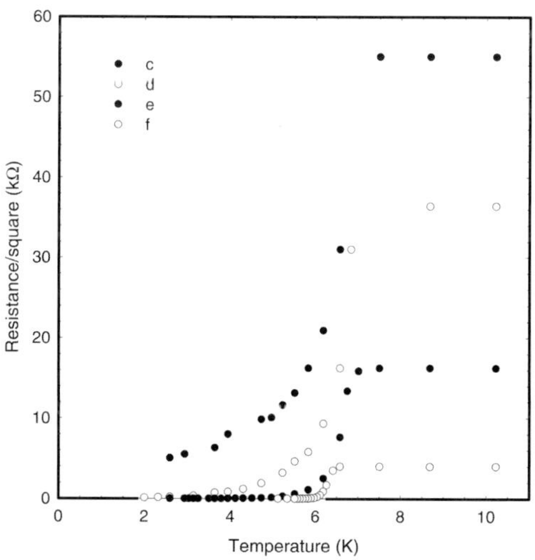

Fig. 21.22. The same data as in Fig. (21.21), but on a linear, as opposed to a logarithmic, scale of resistivity

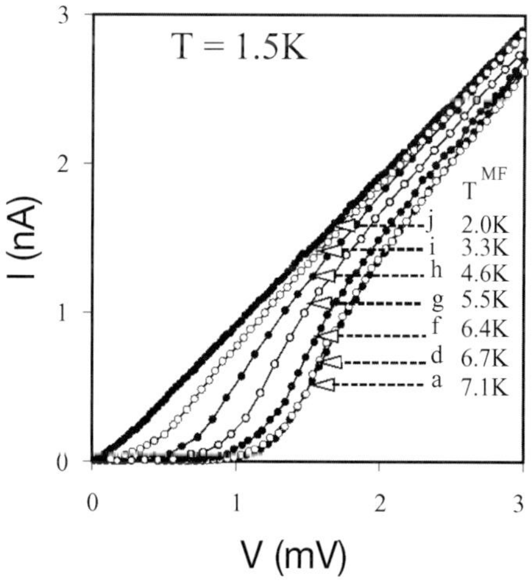

Fig. 21.23. I–V curves from planar tunneling into the same sequence of films shown in Fig. (21.21). From Fig. 6 of Merchant et al [280]

same set of films. As $\mathrm{d}I/\mathrm{d}V$ is proportional to the single particle density of states at energy V, this can be interpreted as the analogue of an ARPES or tunneling experiment in the high temperature superconductors. Among other things, the gap seen in films a–d is roughly independent of Ag coverage, and looks precisely like the gap that is seen upon tunneling into thick Pb films. In these films, the gap seen in tunneling is clearly a superconducting pseudogap.

The analogy between the behavior of these films as a function of Ag coverage, and the cuprate high temperature superconductors as a function of hole concentration is immediately apparent:

"T_c" increases with increasing Ag…
With little or no Ag, the typical Josephson coupling, J, between far separated grains of Pb is small; thermal phase fluctuations preclude any possibility of long range phase order for $T > J$. Clearly, increasing Ag coverage increases the coupling between grains, or more correctly, since the granular character of the films is gradually obscured with increasing Ag coverage, it increases the phase stiffness or superfluid density. This causes the phase ordering temperature to rise, much like the underdoped regime of the cuprates.

…and then T_c decreases.
However, the pairing scale, or equivalently the mean field T_c, is a decreasing function of Ag coverage due to the proximity effect. Since the Pb grains are small compared to the bulk coherence length, ξ_0, the granularity of the films has little effect on the BCS gap equation. The pairing scale is equivalent to that of a homogeneous system with an effective pairing interaction,

$$\lambda^{\mathrm{eff}} = \lambda_{\mathrm{Pb}} \times f_{\mathrm{Pb}} + 0 \times f_{\mathrm{Ag}} , \qquad (21.64)$$

where f_{Pb} and f_{Ag} are, respectively, the volume fraction of Pb and Ag. Consequently, the pairing gap,

$$\Delta_0 \sim \exp[-1/(\lambda^{\mathrm{eff}} - \mu^*)] \qquad (21.65)$$

is a decreasing function of Ag coverage. So long as $f_{\mathrm{Ag}} \ll 1$ (films a-f) this effect is rather slight, as can be seen directly from the figures, but then the gap value can be seen to plummet with increasing Ag coverage. In films g–j, this leads to a decrease of T_c, reminiscent of overdoped cuprates.

Of course, it is clear that there is more going on in the experiment than this simple theoretical discussion implies:

Things we swept under the rug.
(1) *Disorder:* The effects of disorder are neglected in this discussion. A priori these should be strong, especially at low Ag coverage.

(2) *Coulomb Blockade:* As best one can tell from the existing data, films a–f are not superconductors with a reduced T_c—in fact films a and b appear to be headed toward an insulating ground state, presumably due to quantum phase fluctuations induced by the charging energy of the grains. The energy to transfer a Cooper pair (charge 2e) between grains is

$$V_C = 4\alpha e^2/L , \qquad (21.66)$$

where L is the grain size and α is a dimensionless constant which takes into account the grain shape and screening. When $V_C > J$, the number of pairs per grain becomes fixed at low temperature and the ground state is a type of paired Mott insulator. Since the number of pairs and the phase are quantum mechanically conjugate on each grain, when number fluctuations are suppressed by the charging energy, quantum phase fluctuations flourish, and prevent superconducting order. The screening of the Coulomb interaction can mitigate this effect. Screening clearly improves with increasing Ag coverage, so coverage dependent effects of quantum phase fluctuations contribute to the evolution observed in the experiments, as well.

(3) *Dissipation:* There is even more to this story than the $\omega = 0$ charging energies. In contrast with classical statistical mechanics, the dynamics and the thermodynamics are inexorably linked in quantum statistical mechanics, and finite frequency physics becomes important. This issue has been addressed experimentally by Rimberg et al. [281]. While there has been considerable progress in understanding the theory of quantum phase fluctuations (See, for example, [282] for a recent review), there are still many

basic issues that are unresolved. For instance, films c-f show no sign of becoming truly superconducting or insulating as $T \to 0$!

A mysterious ground state
What is the nature of this intermediate state? This is a widely observed phenomenon in systems which are expected to be undergoing a superconductor to insulator transition [282,283]. The physics of this anomalous metallic state is not understood *at all*, even in systems, such as the present one, where the microscopic physics is believed to be understood. (See Sect. 21.8.4 for a taste of the theoretical subtleties involved.)

21.8.3 Classical Phase Fluctuations

We now undertake a critical analysis of thermal phase fluctuations. We will for now ignore the effects of thermal quasiparticle excitations, as well as the quantum dynamics which certainly dominate the phase mode physics at temperatures low compared to its effective Debye temperature. These important omissions will be addressed in Sect. 21.8.4.

Superconductors and Classical XY Models

The superfluid density sets the phase stiffness.
When $T_\theta \ll T_p$, the superconducting transition temperature $T_c \approx T_\theta$, and the transition can be well described by a phase only model. On general symmetry grounds, the free energy associated with time independent deformations of the phase must be of the form

$$V_{\text{phase}} = (\gamma/2) \int \mathrm{d}\boldsymbol{r} (\nabla\theta)^2 \ , \tag{21.67}$$

where the helicity modulus, γ, is given by the superfluid density, n_s, and the effective pair mass, m^*, according to Eq. (21.62). Since $v_s = \frac{\hbar}{m^*}\nabla\theta$ is the superfluid velocity, V_{phase} is easily seen to have an interpretation as the kinetic energy of the superfluid, $V_{\text{phase}} = \int \mathrm{d}\boldsymbol{r} n_s m^* v_s^2/2$, so that classical phase fluctuations correspond to thermally induced pair currents. Equations (21.67) and (21.62) establish the sense in which the superfluid density controls the stiffness to phase fluctuations.

Equation (21.67) is the continuum form of the classical XY model. Both in a superconductor and in the XY model, θ is a periodic variable (defined modulo 2π). Thus, we must handle the short distance physics with some care to permit the vortex excitations which are the expression of that periodicity. When this is done, typically by defining the model on a lattice, it captures the essential physics of the transition between a low temperature ordered and a high temperature disordered state.

To be concrete, let us consider an XY model on a d dimensional hypercubic lattice

$$H_{XY} = -\sum_{<i,j>} \mathcal{V}(\theta_i - \theta_j) \ , \tag{21.68}$$

where $< i,j >$ are nearest neighbor sites and $\mathcal{V}$ is an even, periodic function $\mathcal{V}(\theta) = \mathcal{V}(\theta + 2\pi) = \mathcal{V}(-\theta)$, with a maximum at $\theta = 0$ such that the Hamiltonian is minimized by the uniform state. The lattice constant, a, in this model has a physical interpretation—it defines the size of the vortex core. To generalize this model to the case of an anisotropic (e.g. layered) superconductor, we let both the lattice constant, a_μ, and the potential, $\mathcal{V}_\mu(\theta)$, depend on the direction, μ.

At zero temperature, the helicity modulus can be simply computed:

$$\gamma_\mu(T = 0) = 2[a_\mu^2/v]\mathcal{V}''_\mu(0) \ , \tag{21.69}$$

where $v = (\prod_\nu a_\nu)$ is the unit cell volume. Thus, the relation between $\gamma(0)$ and T_θ, the ordering temperature of the model, depends both on the detailed form of $\mathcal{V}$ and on the lattice cutoff. In constructing Table 21.1 above, we have taken $\mathcal{V} = V\cos(\theta)$, and identified the area of the vortex core, $\pi\xi_0^2$, with the plaquette area, a^2 - this is the origin of the somewhat arbitrary $\sqrt{\pi}$ which appears in the three-dimensional expression for T_θ. Fortuitously, for layered materials, $\gamma_x = \gamma_y \equiv \gamma_{xy}$ depends only on the spacing between planes, a_z, and not on the in-plane lattice constant.

One can, in principle, handle the short distance physics in a more systematic way by solving the microscopic problem (probably numerically) on large systems (large compared to ξ_0), and then matching the results with the short distance behavior of the XY model. In this way, one could, in principle, derive explicit expressions for $\mathcal{V}$ and a_μ in terms of the microscopic properties of a given material. However, no one (to the best of our knowledge) has carried through such an analysis for any relevant microscopic model.

How much does the detailed shape of $\mathcal{V}$ matter?
What we [284] have done, instead, is to keep at most the first 2 terms in a Fourier cosine series of Eq. (21.68). With the cuprates in mind, we have studied planar systems:

$$H = -J_\parallel \sum_{<ij>_\parallel} \left\{\cos(\theta_{ij}) + \delta \cos(2\theta_{ij})\right\} - J_\perp \sum_{<ij>_\perp} \left\{\cos(\theta_{ij})\right\} , \qquad (21.70)$$

where $< ij >_\parallel$ denotes nearest neighbors within a plane, and $< ij >_\perp$ denotes nearest neighbors between planes. It is assumed that $J_\parallel$, $J_\perp$, and δ are positive, since there is no reason to expect any frustration in the problem, and that $\delta \le 0.25$, since for $\delta > 0.25$ there is a secondary minimum in the potential for $\theta_{ij} = \pi$, which is probably unphysical. Since dimensional analysis arguments of the sort made above are essentially independent of δ, varying δ permits us to obtain some feeling for how quantitatively robust the results are with regard to "microscopic details." Electron correlations could change this assumption [285].

Properties of Classical XY Models

The XY model is one of the most studied models in physics [286]. We [284] have recently carried out a series of quantitative analytic and numerical studies of XY models (using Eq. (21.70)). In particular, we have focused on the thermal evolution of the superfluid density and the relation between the superfluid density and the ordering temperature.

As long as $J_\perp$ is nonzero, this model is in the universality class of the 3D XY model, and near enough to T_c, $\gamma(T) \sim |T_c - T|^\nu$, where ν is the correlation length exponent of the 3DXY model, $\nu \approx .67$. For sufficient anisotropy, there may be a crossover from 2D critical behavior close (but not too close) to T_c, to 3D critical behavior very near T_c. In practice, this crossover is very hard to see due to the special character of the critical phenomena of the 2D XY model; even a very weak $J_\perp$ significantly increases the transition temperature.

To see this, consider the case in which $J_\parallel \gg J_\perp$; in this limit, one can study the physics of the system using an asymptotically exact interplane mean field theory [199]. We define the order parameter, $m(T) \equiv \langle\cos[\theta_j]\rangle$, and consider the behavior of a single decoupled planar XY model in the presence of an external field, $h(T) = 2J_\perp m(T)$ due to the mean field of the neighboring two planes. The self-consistency condition thus reads

$$m(T) = m_{2D}(T, h) , \qquad (21.71)$$

where $m_{2D}(T, h)$ is computed for the 2D model. A simple estimate for T_c can be obtained by linearizing this equation:

$$1 = 2J_\perp \chi_{2D}(T_c) . \qquad (21.72)$$

2D critical behavior may be hard to see.
Here the 2D susceptibility is

$$\chi_{2D} \sim T_{2D}^{-1} \exp\left\{A_\chi \sqrt{T_{2D}/(T_c - T_{2D})}\right\} , \qquad (21.73)$$

where T_{2D} is the Kosterlitz–Thouless transition temperature and A_χ is a nonuniversal number of order 1. A consequence of this is that even a very small interlayer coupling leads to a very large fractional increase in T_c

$$T_c - T_{2D} \sim T_{2D} A_\chi^2 / \log^2[J_\parallel/J_\perp] . \qquad (21.74)$$

Only if $(T_c - T_{2D})/T_{2D} \ll 1$ will there be clear 2D critical behavior observed in the thermodynamics.

To make contact with a range of experiments it is necessary that we focus attention not only on universal critical properties, but also on other properties which are at least relatively robust to changes in microscopic details. One such property is the width of the critical region, but we are not aware of any systematic studies of the factors that influence this. For the simple ($\delta = 0$) isotropic 3D XY model, the critical region certainly does not extend further than 10% away from T_c.

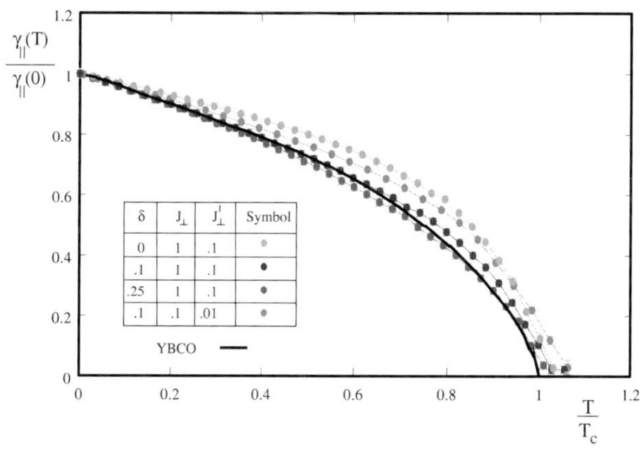

Fig. 21.24. Superfluid density vs. temperature, scaled by the zero temperature superfluid density and by T_c, respectively, from [284]. Experimental data on *YBCO* is depicted by the black line, and is taken from Kamal et al. [289] (The data are essentially the same for a range of doping concentration.) Our Monte Carlo results for system size 16 × 16 × 16 are the filled symbols. Calculations are for two planes per unit cell, with coupling $J_\parallel = 1$ within each plane, and $J_\perp$ and $J'_\perp$ between alternate planes. Monte Carlo points above T_c are nonzero due to finite size effects. Except where explicitly shown, error bars are smaller than symbol size

The superfluid density is linear at low T.
Another such property is the low temperature slope of superfluid density curves as a function of temperature. Using linear spin wave theory [287, 288], one can obtain a low temperature expansion of the in-plane helicity modulus,

$$\frac{\gamma_\parallel(T)}{a_\perp} = J_\parallel(1+4\delta) - \frac{\alpha(1-16\delta)}{4(1+\delta)} T + \mathcal{O}(T^2) , \quad (21.75)$$

where we have used $a_x = a_y \equiv a_\perp$ and $\gamma_x = \gamma_y \equiv \gamma_\parallel$ for a planar system and α is a nonuniversal number which depends on $J_\perp/J_\parallel$. It is easy to show [284] that $\alpha = 1$ in the two-dimensional limit ($J_\perp/J_\parallel = 0$), and that $\alpha = 2/3$ in the three-dimensional limit ($J_\perp = J_\parallel(1+4\delta)$). The T-linear term is independent of $J_\parallel$, so that we expect the slope of scaled superfluid density curves, $\gamma_\parallel(T)/\gamma_\parallel(0)$ vs. T/T_c, to be much less sensitive to microscopic parameters (i.e. material dependent properties such as doping in the cuprates) than is $\gamma_\parallel(0)$. That this expectation is realized can be seen from our Monte Carlo simulation results presented in Fig. 21.24.

In addition, we find that there is a *characteristic shape* to the superfluid density vs. temperature curves in XY models. We have used Monte Carlo simulations to focus on two other dimensionless nonuniversal parameters: $A_1 = T_c/\gamma_\parallel(0)$ and $A_2 = T_c\gamma'_\parallel(0)/\gamma_\parallel(0)$, where $\gamma'_\parallel(0) = d\gamma_\parallel(0)/dT$. A_1 is a measure of how well the ground state property $\gamma_\parallel(0)$ (measurable through the superfluid density) predicts T_c, which is equivalent to T_θ in this model. A_2 can be expressed in the more intuitive form $A_2 = T_c/T_{ex}$, where $T_{ex} \equiv \gamma_\parallel(0)/\gamma'_\parallel(0)$, is the estimate of T_c one would obtain by extrapolating from the low temperature slope of $\gamma_\parallel(T)$ to the point at which the superfluid stiffness would vanish.

The shape of $\gamma(T)$ is robust!
Over *orders of magnitude* of couplings ($0 \le J_\perp/J_\parallel \le .1$), and throughout the range $0 \le \delta \le .25$, A_1 and A_2 are remarkably robust: $A_1 \sim .6 - 1.7$, and $A_2 \sim .2 - .5$.

21.8.4 Quantum Considerations

In quantum systems, the dynamics affects the thermodynamics. However, the role of quantum effects on the phase dynamics is a large topic, and one in which many uncertainties remain. We will briefly discuss the simplest case here, mostly to illustrate the complexity of the problem.

Let us consider a simple two fluid model [3] in which a phase fluctuating superconductor is capacitively coupled to a normal fluid. The continuum limit of the effective action obtained upon integrating out the normal fluid can be derived from simple hydrodynamic considerations. From the Josephson relation, it follows that the electric field

$$\boldsymbol{E} = -(\hbar/2e)\nabla\dot{\theta} . \quad (21.76)$$

The Euclidean effective action is obtained by augmenting the classical action, Eq. (21.67), with the Maxwell term, and analytically continuing to imaginary time:

$$S[\theta] = \int_0^\beta d\tau \left\{ \int dr \mathcal{L}_{\text{quantum}} + V_{\text{phase}} \right\} ,$$
$$\mathcal{L}_{\text{quantum}} = \boldsymbol{E} \cdot \boldsymbol{D}/8\pi , \qquad (21.77)$$

where $\beta = 1/T$, $\boldsymbol{D}(\boldsymbol{k}, \omega) = \epsilon_0(\boldsymbol{k}, \omega)\boldsymbol{E}(\boldsymbol{k}, \omega)$, and ϵ_0 is the normal fluid dielectric function (analytically continued to imaginary time). Again, this effective action must be cutoff at short distances in such a way as to preserve the periodicity of θ by allowing vortex excitations.

The order of limits matters.
An analysis of the Maxwell term, S_{quantum}, allows us to illustrate some of the complexity of this problem. At $\boldsymbol{k} = \boldsymbol{0}$ and small ω, $\epsilon_0 \approx 4\pi\sigma_0/i\omega$, where σ_0 is the D.C. conductivity of the normal fluid. Thus, if we first consider the spatial continuum limit before going to low frequencies, $S_{\text{quantum}} \sim \sum_{\omega_n} \int dr \sigma_0 |\omega_n| |\nabla\theta|^2$, where $\omega_n = 2\pi n T$ are the Matsubara frequencies. We recognize the resulting action as the continuum limit of an array of resistively shunted Josephson junctions [290, 291] (RSJ). Here, the normal fluid plays the role of an "Ohmic heat bath."

On the other hand, if we first take $\omega = 0$, and then k small, $\epsilon \approx (k_{TF}/k)^2$ where k_{TF} is the Thomas-Fermi screening length. In this limit, the Maxwell term has the form of a phase kinetic energy, $\mathcal{L}_{\text{quantum}} \sim (M_\theta/2)|\dot{\theta}|^2$, with an effective mass, $M_\theta \propto [e^2/k_{TF}^2]^{-1}$ inversely proportional to an appropriately defined local charging energy. The resulting effective action is the continuum limit of the "lattice quantum rotor" (QR) model, also a widely studied problem [292]. The RSJ and QR models have quite different behavior at low temperatures. Without a rather complete understanding of the physics of the normal fluid, it is impossible in general to determine which, if either, of these limits captures the essential quantum physics.

The classical to quantum crossover temperature is estimated.
There is nonetheless one important issue which can be addressed in a theoretically straightforward fashion: the temperature scale below which quantum effects dominate. The classical physics we studied in the previous section is readily obtained from the quantum model by suppressing all fluctuations with nonzero Matsubara frequency. We thus estimate a classical to quantum crossover temperature, T_{cl}, by comparing the classical ($\omega = 0$) and first finite frequency ($\omega = \omega_1 = 2\pi T$) contributions to $S[\theta]$. This leads to the implicit equation for T_{cl}:

$$T_{cl} = \sqrt{e^2 n_s/\epsilon_0 m^*} , \qquad (21.78)$$

where ϵ_0 is evaluated at temperature $T = T_{cl}$, frequency $\omega \sim 2\pi T_{cl}$, and a typical momentum, $k \sim 1/a$. So long as $T \gg T_{cl}$, the imaginary time independent (classical) field configurations dominate the thermodynamics. Clearly, depending on how good the screening is, T_{cl} can be much smaller or much larger than T_c. If we approximate ϵ_0 by its finite frequency, $k \to 0$ form, this estimate can be recast in an intuitively appealing form [284]:

$$T_{cl} \sim \left(\frac{\sigma_Q}{\sigma_0} \right) T_\theta , \qquad (21.79)$$

where $\sigma_Q = e^2/(ha)$ is the quantum of conductance in which the vortex core radius enters as the quantum of length.

Recent theoretical developments have uncovered yet more subtleties. Although the low energy physics involves only phase fluctuations, phase slips (short imaginary time events where the phase spontaneously "slips" by 2π) involve amplitude fluctuations. In the presence of an ohmic heat bath, there are subtle, long time consequences of these amplitude fluctuations [293–295]. Another interesting possibility is electron fractionalization. Under some circumstances, it has been proposed [59] that hc/e vortices may be energetically preferred to the usual $hc/2e$ vortices, leading to a fractionalized state.

This is an important unsolved problem!
Combine this exciting but incomplete jumble of theoretical ideas with the remarkably simple but entirely unexplained behavior observed experimentally in granular superconducting films as they crossover from superconducting to insulating behavior, and one is forced to concede that the theory of quantum phase fluctuations is seriously incomplete.

21.8.5 Applicability to the Cuprates

Both phase and pair breaking fluctuations are more prevalent at low T in the cuprate superconductors than in conventional BCS superconductors. The low superfluid density provides only a weak stiffness to thermal phase fluctuations of the order parameter. In addition, the nodes in the gap mean that there are low energy quasiparticle excitations down to arbitrarily low temperature. However, it is important to remember that nodal quasiparticles occupy only a small fraction of the Brillouin zone so long as $\Delta_o \gg T$.

T_c Is Unrelated to the Gap in Underdoped Cuprates

As mentioned in Sect. 21.3, in underdoped cuprates, many probes detect a pseudogap in the normal state, such as NMR, STM, junction tunneling, and ARPES. Whereas BCS theory would predict $T_c \sim \Delta_o/2$, where Δ_o is the superconducting gap maximum at zero temperature, the low temperature magnitude of the single particle gap as measured by ARPES or tunneling experiments does not follow this relation, qualitatively or quantitatively. On the underdoped side, T_c increases with increasing doping, whereas Δ_o moves in the opposite direction in all cases studied to date. Even at optimal doping, T_c is always considerably smaller than the BCS value of $\Delta_o/2$. In optimally doped BSCCO, for example, $T_c \sim \Delta_o/5$, where Δ_o is the peak energy observed in low temperature tunneling experiments [150,296,297] (see also Table 21.1).

There is no signature of the transition in the single particle gap.
The ARPES experiments provide k-space information demonstrating that the gap, above and below T_c, has an anisotropy consistent with a d-wave order parameter. Furthermore, $\Delta_o(T)$ is largely undiminished in going from $T = 0$ to $T = T_c$ in underdoped samples, and the size and shape of the gap are basically unchanged through the transition. Add to this the contravariance of T_c with the low temperature magnitude of the gap as the doping is changed, and it appears the gap and T_c are simply independent energies [134,298]. The gap decreases with overdoping, which may be responsible for the depression of T_c in that region, so that the transition may be more conventional on the overdoped side.

T_c Is Set by the Superfluid Density in Underdoped Cuprates

As emphasized above, the superfluid density in cuprates is much smaller than in conventional superconductors [279]. In addition, when the superfluid density is converted to an energy scale, it is comparable to T_c, whereas in conventional superconductors this phase stiffness energy scale is far above the transition temperature. In those conventional cases, BCS theory works quite well, but in the cuprates, the phase stiffness energy scale should also be considered.

This is further emphasized by the Uemura plot [107], which compares the transition temperature to the superfluid density. For underdoped systems, the relationship is an approximately linear over some range of doping within experimental errors. This is strong evidence that T_c is determined by the superfluid density, and therefore set by phase ordering.

In the limit that $T_c \to 0$, the relation between T_c and the superfluid density is necessarily more subtle. In particular, the nature of the transition is likely to be different depending on potentially material specific factors, such as whether quenched disorder is a controlling consideration or not, whether the $T = 0$ quantum transition is a superconductor to insulator or a superconductor to metal transition, and what the role is of states with other types of order, as for instance stripe states. Many apparently consistent theoretical scenarios can be imagined; even the question of whether the transition (in the limit of vanishing disorder strength) is first order or second order is debatable. The experimental situation is rich, interesting, but still unsettled. There is certainly evidence [299–305] that the character of the transition is different in $La_{2-x}Sr_xCuO_4$, $Bi_2Sr_2CaCu_2O_{8+\delta}$ and $YBa_2Cu_3O_{7-\delta}$. The existence in all currently known cases of a non-zero critical value of doping for superconductivity certainly implies that any linear relation between doping level and T_c must break-down for small enough T_c - in simple models without quenched disorder, this observation by itself implies that quantum effects must cause a breakdown of the

Uemura relation as $T_c \to 0$, and preliminary data supports this expectation [305]. This is clearly an area which warrants further focused theoretical and experimental investigation.

Experimental Signatures of Phase Fluctuations

In YBCO, $3DXY$ critical fluctuations have been observed in the superfluid density within 10% of T_c [172,306], implying that the temperature dependence of the superfluid density below and near T_c is governed by phase fluctuations. It needs to be stressed that in conventional superconductors, such fluctuations that are seen are Gaussian in character—that is they involve fluctuations of both the amplitude and the phase of the order parameter. (An interesting way to identify separate Gaussian and phase fluctuation regimes in YBCO is presented in [307] (f. a recent analysis of the doping dependence of ξ_0 see [307]). See also [79].) The purely critical phase fluctuations observed in YBCO are entirely different.

Over a broader range of temperatures, in order to detect signatures of superconducting fluctuations, one needs to find the most sensitive possible probe of local superconductivity. Fluctuation conductivity has been the method of choice in conventional superconductors. However, this is generally a small correction to the normal-state conductivity, and so can only be clearly identified when the normal contribution, which is a large "background" signal, is extremely well understood. No unambiguous division into normal and fluctuation contributions to the conductivity has been made for the cuprates. Fluctuation diamagnetism is a more promising method, as the normal state magnetic response is miniscule. Recently, Li et al. [308] and Wang et al. [309] have measured fluctuation diamagnetism in a broad range of temperatures above T_c (up to several times T_c) in both LSCO an BSCCO. Indeed, many of the striking features of the B and T dependences of the magnetization (especially in BSCCO) are well reproduced by the theory of 2D superconducting phase fluctuations [310]. On this basis, Wang et al. [311] were able to infer a crossover scale associated with the onset of detectable superconducting correlations, something like $T^\star_{\text{pair}}$ in Fig. 21.12, which rises to a maximum value several times T_c at $x \sim 1/8$, and then decreases, probably extrapolating to 0 only at values of x smaller than the minimum for superconductivity. Moreover, the fluctuation diamagnetism corresponds closely to an anomalous contribution to the Nernst effect, the "vortex Nernst effect", which had earlier been identified [312,313] as a signature of phase fluctuations.

At low temperature (as low as $T = 1$K [314]), the superfluid density is a linearly decreasing function of temperature [9]. While this linear behavior is generally believed to be the result of quasi-particle fluctuations of an order parameter with nodes, it is difficult [148,151,315,316] from this perspective to understand why the slope is nearly independent of x and of Δ_0/T_c. This feature of the data is naturally explained if it is assumed that the linear temperature dependence, too, arises from classical phase fluctuations, but then it is hard to understand [284,317] why quantum effects would not quench these fluctuations at such low temperatures.

21.9 Lessons from Weak Coupling

21.9.1 Perturbative RG Approach in $D > 1$

In recent years, Fermi liquid theory, and with it the characterization of the BCS instability, has been recast in the language of a perturbative renormalization group (RG) treatment. We will adopt this approach as we reconsider the conventional BCS-Eliashberg theory of the phonon mediated mechanism of superconductivity in simple metals. In particular, we are interested in exploring the interplay between a short ranged instantaneous electron-electron repulsion of strength μ and a retarded attraction (which we can think of as being mediated by the exchange of phonons) of strength λ, which operates only below a frequency scale ω_D. Although we will make use of a perturbative expression for the beta function which is valid only for μ and λ small compared to 1, the results are nonperturbative in the sense that we will recover the nonanalytic behavior of the pairing scale, T_p, expected from BCS mean field theory. The results are valid for any relative strength of μ/λ and, moreover, the corrections due to higher order terms in the beta function are generally smooth, and so are not expected to have large qualitative effects on the results so long as μ and λ are not large compared to 1.

All the results obtained in this section have been well understood by experts since the golden age of many-body theory, along with some of the most important higher order corrections which occur for λ of order 1 (which will be entirely neglected here). Our principal purpose in including this section is to provide a simple derivation of these results in a language that may be more accessible to the modern reader. A most insightful exposition of this approach is available in the articles by Polchinski [39], and Shankar [38], which can be consulted wherever the reader is curious about parts of the analysis we have skipped over. The one technical modification we adopt here is to employ an energy shell RG transformation, rather than the momentum shell approach adopted in [38]; this method allows us to handle the retarded and instantaneous interactions on an equal footing. It can also be viewed as an extension of the analogous treatment of the 1D problem adopted in [318], as discussed in the next subsection.

We start by defining a scale invariant (fixed point) Euclidean action for a noninteracting Fermi gas

$$S_{fp}[\Psi_\uparrow, \Psi_\downarrow] = (2\pi)^{-(d+1)} k_F^{d-1} \sum_\sigma \int d\omega d\hat{k} dk \mathcal{L}_0[\Psi_\sigma] ,$$

$$\mathcal{L}_0[\Psi_\sigma] = \bar{\Psi}_\sigma[i\omega + v_F(\hat{k})k]\Psi_\sigma , \qquad (21.80)$$

where $d\boldsymbol{k} = k_F^{d-1} d\hat{k} dk$, the unit vector $\hat{k}$ is the direction of $\boldsymbol{k}$ and k is the displacement from the Fermi surface; we have assumed a simple spherical Fermi surface. The treatment that we present here breaks down when the Fermi surface is nested or contains Van Hove singularities. To regularize the theory, it is necessary to cut off the integrals; whereas Shankar confines k to a narrow shell about the Fermi surface, $|k| < \Lambda \ll k_F$, we allow k to vary from $-\infty$ to $+\infty$, but confine the ω integral to a narrow shell $|\omega| < \Omega \ll E_F$.

We now introduce electron-electron interactions. Naive power counting leads to the conclusion that the four fermion terms are marginal, and all higher order terms are irrelevant, so we take

$$\begin{aligned} S_{int} = \sum_{\sigma,\sigma'} \int \prod_{j=1}^{3} \frac{d\boldsymbol{k}_j d\omega_j}{(2\pi)^{d+1}} & \bar{\Psi}_\sigma(\boldsymbol{k}_1, \omega_1)\bar{\Psi}_{\sigma'}(\boldsymbol{k}_2, \omega_2) \\ & \times [g(\boldsymbol{k}_2 - \boldsymbol{k}_3) \\ & + \Theta(\omega_D - |\omega_2 - \omega_3|)\tilde{g}(\boldsymbol{k}_2 - \boldsymbol{k}_3)] \\ & \times \Psi_{\sigma'}(\boldsymbol{k}_3, \omega_3) \\ & \times \Psi_\sigma(\boldsymbol{k}_1 + \boldsymbol{k}_2 - \boldsymbol{k}_3, \omega_1 + \omega_2 - \omega_3) , \end{aligned} \qquad (21.81)$$

where Θ is the Heavyside function, and g and $\tilde{g}$ are, respectively, the instantaneous and retarded interactions. Signs are such that positive g corresponds to repulsive interactions. The distinction between retarded and instantaneous interactions is important so long as $\Omega \gg \omega_D$. We have invoked spin rotation invariance in order to ignore the dependence of g and $\tilde{g}$ on the spin indices.

It should be stressed, as already mentioned in Sect. 21.5, that this should already be interpreted as an effective field theory, in which the microscopic properties that depend on the band structure away from the Fermi surface such as mixing with other bands, more complicated three and four-body interactions, etc. have already fed into the parameters that appear in the model. What we do now is to address the question of what further changes in the effective interactions are produced when we integrate out electronic modes in a narrow shell between Ω and $\Omega e^{-\ell}$, ($\ell > 0$ and small), and then rescale all frequencies according to

$$\omega \to e^\ell \omega, \quad k \to e^\ell k \text{ and } \Psi \to e^{-(3/2+\eta_F)\ell}\Psi , \qquad (21.82)$$

to restore the cutoff to its original form and where, as usual, η_F is a critical exponent that is determined by the properties of the interacting fixed point. We will carry this procedure out perturbatively in powers of g and $\tilde{g}$ to the one loop order we (and everyone else) analyzes, $\eta_F = 0$.

To first order in perturbation theory, simple power counting insures that the entire effective action is invariant under the RG transformation, other than the parameter ω_D which changes according to

$$d\omega_D/d\ell = \omega_D . \qquad (21.83)$$

To second (one loop) order, the forward scattering interactions are still unchanged; they produce the Fermi liquid parameters, and should actually be included as part of the fixed point action and treated nonpeturbatively. This can be done straightforwardly, but for simplicity will be ignored here. The one loop diagrams which potentially produce contributions to the beta function are shown in Fig. 21.25. All internal legs of the diagrams refer to electron propagators at arbitrary momenta but with their frequencies constrained to lie in the shell which is being integrated out, $\Omega > |\omega| \geq \Omega e^{-\ell}$. The dashed lines represent interactions. All external legs are taken to lie on or near the Fermi surface. Clearly, the energy transfer along the interaction lines in the Cooper channel, Figs. 21.25(a) and 21.25(b), is of order Ω, and so for $\Omega \gg \omega_D$, $\tilde{g}$ does not contribute, while in Figs. 21.25(c) and 21.25(d) there is zero frequency transfer along the interaction lines, and so g and $\tilde{g}$ contribute equally.

Since $\Omega \ll E_F$, we can classify the magnitude of each diagram in powers of Ω; any term of order $|\Omega|^{-1}$ makes a logarithmically divergent contribution to the effective interaction upon integration over frequency, while any terms that are proportional to E_F^{-1} are much smaller and make only finite contributions which can be ignored for the present purposes. When the Cooper diagrams, shown in Figs. 21.25(a) and 21.25(b), are evaluated for zero center of mass momentum, (i.e. if the momenta on the external legs are $\boldsymbol{k}_F$ and $-\boldsymbol{k}_F$), the bubble is easily seen to be proportional to Ω^{-1}. However, if the center of mass momentum is nonzero (i.e. if the external momenta are $\boldsymbol{k}_F + \boldsymbol{q}$ and $-\boldsymbol{k}_F$), the same bubble is proportional to $1/v_F|\boldsymbol{q}|$, and hence is negligible. The particle-hole diagrams in Figs. 21.25(c) and 21.25(d) are a bit more complicated. The bubble is zero for total momentum 0, and proportional to $1/v_F k_F$ for momentum transfer near $2k_F$. Thus, in more than one dimension, the particle hole bubbles can be neglected entirely. (We will treat the 1d case separately, below.) Putting all this together in the usual manner, we are left with the one-loop RG equations for the interactions between electrons on opposing sides of the Fermi surface,

$$\frac{\mathrm{d}g_l}{\mathrm{d}\ell} = -\frac{1}{\pi v_F} g_l^2 \, , \quad \frac{\mathrm{d}\tilde{g}_l}{\mathrm{d}\ell} = 0 \, , \qquad (21.84)$$

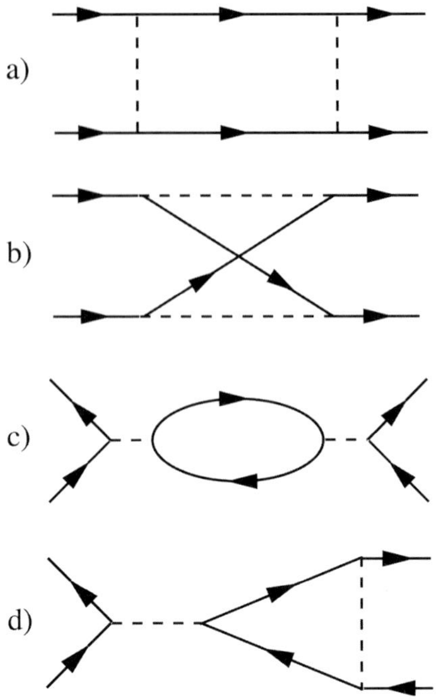

Fig. 21.25. The one loop diagrams that are invoked in the discussion of the renormalization of the effective interactions. (**a**) and (**b**) are referred to as the "Cooper channel" and (**c**) and (**d**) as "particle-hole channels". The loop is made out of electronic propagators with frequencies in the shell which is being integrated. The *dashed lines* represent interactions

where l refers to the appropriate Fermi surface harmonic; for the case of a circular Fermi surface in two dimensions, l is simply angular momentum. (Implicit in this is the fact that odd l are associated with interactions in the triplet channel while even l are in the singlet channel.)

These equations describe the changes in the effective interactions upon an infinitesimal RG transformation. They can be easily integrated to obtain expressions for the scale dependent interactions. However, these equations are only valid so long as all the interactions are weak (to justify perturbation theory) and so long as $\Omega \gg \omega_D$.

Note the nonrenormalization of λ for $\Omega > \omega_D$. Assuming that it is the second condition that is violated first, we can obtain expression for the effective interactions at this scale by integrating to the point at which $\Omega = \omega_D$; the result is

$$\mu(\omega_D) = \frac{\mu_0}{1 + \mu_0 \log(\Omega_0/\omega_D)} , \quad \lambda(\omega_D) = \lambda_0 , \tag{21.85}$$

where $\mu = g/\pi v_F$, $\lambda = \tilde{g}/\pi v_F$, the symmetry labels on g and $\tilde{g}$ are left implicit, and the subscript "0" refers to the initial values of the couplings at a microscopic scale, $\Omega_0 \sim E_F$.

The fact that the retarded interactions do not renormalize is certainly as noteworthy as the famous renormalization of μ. This means that it is possible to estimate λ from microscopic calculations or from high temperature measurements, such as resistivity measurements in the quasi-classical regime where $\rho \propto \lambda T$.

Once the scale $\Omega = \omega_D$ is reached, a new RG procedure must be adopted. At this point, the retarded and instantaneous interactions are not distinguishable, so we must simply add them to obtain a new, effective interaction, $g^{\text{eff}}(\omega_D) = g(\omega_D) + \tilde{g}$, which upon further reduction of Ω renormalizes as a non-retarded interaction. If $g^{\text{eff}}(\omega_D)$ is repulsive, it will be further reduced with decreasing Ω.

Fermi liquid behavior breaks down at the pairing scale.
However, if it is attractive in any channel, the RG flows carry the system to stronger couplings, and eventually the perturbation theory breaks down. We can estimate the characteristic energy scale at which this breakdown occurs by integrating the one loop equations until the running coupling constant reaches a certain finite value $-1/\alpha$:

$$\Omega_1 = \omega_D e^{\alpha} \exp[-1/|g^{\text{eff}}(\omega_D)|] . \tag{21.86}$$

Of course, the RG approach does not tell us how to interpret this energy scale, other than that it is the scale at which Fermi liquid behavior breaks down. However, we know on other grounds that this scale is the pairing scale, and that the breakdown of Fermi liquid behavior is associated with the onset of superconducting behavior.

21.9.2 Perturbative RG Approach in $D = 1$

The One Loop Beta Function

In one dimension, the structure of the perturbative beta function is very different from in higher dimensions. In addition to the familiar logarithmic divergences in the particle-particle (or Cooper) channel, there appear similar logarithms in the particle-hole channel. That these lead to a serious breakdown of Fermi liquid theory can be deduced directly from the perturbation theory, although it is only through the magic of bosonization (discussed in Sect. 21.5) that it is possible to understand what these divergences lead to.

To highlight the differences with the higher dimensional case, we will treat the 1d case using the perturbative RG approach, but now taking into account the dimension specific interference between the Cooper and particle-hole channels. However, having belabored the derivation of the perturbative beta function for the higher dimensional case, we will simply write down the result for the 1d case; the reader interested in the details of the derivation is referred to [318] and [319].

In 1d, there are only two potentially important momentum transfers which scatter electrons at the Fermi surface, as contrasted with the continuum of possibilities in high dimension. It is conventional to indicate by g_1 the interaction with momentum transfer $2k_F$, and by g_2 that with zero momentum transfer. If we are interested in the case of a nearly half filled band, we also need to keep track of the umklapp scattering, g_3, which involves a momentum transfer 2π to the lattice (see Sect. 21.5). Consequently, we must introduce a chemical potential, μ, defined such that $\mu = 0$ corresponds to the half filled band. Finally, we consider the retarded interactions, $\tilde{g}_1$, $\tilde{g}_2$, and $\tilde{g}_3$ which operate at frequencies less than ω_D. For simplicity, we consider only the case of spin rotationally invariant interactions.

The one loop RG equations (obtained by evaluating precisely the diagrams in Fig. 21.25), under conditions $\Omega \gg \omega_D, \mu$, are

$$\frac{dg_1}{d\ell} = -\frac{g_1^2}{\pi v_F}\,, \quad \frac{dg_c}{d\ell} = -\frac{g_3^2}{\pi v_F}\,, \quad \frac{dg_3}{d\ell} = -\frac{g_3 g_c}{\pi v_F}\,,$$
$$\frac{d\tilde{g}_\pm}{d\ell} = -\frac{g_\pm}{\pi v_F}[\frac{3}{2}g_1 \pm g_3 + \frac{1}{2}g_c + \tilde{g}_\pm]\,, \quad \frac{d\tilde{g}_2}{d\ell} = 0\,,$$
$$\frac{d\mu}{d\ell} = \mu\,, \quad \frac{d\omega_D}{d\ell} = [1 + \frac{\tilde{g}_+}{\pi v_F}]\,\omega_D\,, \qquad (21.87)$$

where $g_c \equiv g_1 - 2g_2$ and $\tilde{g}_\pm = \tilde{g}_1 \pm \tilde{g}_3$. For $\mu \gg \Omega \gg \omega_D$, the same equations apply, except now we must set $g_3 = \tilde{g}_3 = 0$. And, of course, if $\omega_D > \Omega$, we simply drop the notion of retarded interactions, altogether.

The electron-phonon interaction in a non-Fermi liquid can be strongly renormalized.
There are many remarkable qualitative aspects to these equations, many of which differ markedly from the analogous equations in higher dimensions. The most obvious feature is that the retarded interactions are strongly renormalized, even when the states being eliminated have energies large compared to ω_D. What this means is that in one dimension, the effective electron-phonon interaction at low energies is *not* simply related to the microscopic interaction strength. Some of the effects of this strong coupling on the spectral properties of quasi-one-dimensional systems can be found in [319–321].

Away From Half Filling

To see how this works out, let us consider the typical case in which the nonretarded interactions are repulsive (g_1, and $g_2 > 0$) and the retarded interactions are attractive ($\tilde{g}_\pm < 0$) and strongly retarded, $\omega_D/E_F \ll 1$. Far from half filling, we can also set $g_3 = \tilde{g}_3 = 0$. The presence or absence of a spin gap is determined by the sign of g_1. Thus, just as in the 3d case, in order to derive the effective theory with nonretarded interactions which is appropriate to study the low energy physics at scales small compared to ω_D, we integrate out the fermionic degrees of freedom at scales between E_F and ω_D, and then compute the effective backscattering interaction,

$$g_1^{\text{eff}} = g_1(\omega_D) + \tilde{g}_1(\omega_D)\,. \qquad (21.88)$$

If $g_1^{\text{eff}} > 0$ (i.e. if $g_1(\omega_D) > |\tilde{g}_1(\omega_D)|$), then the Luttinger liquid is a stable fixed point, and in particular no spin gap develops. If $g_1^{\text{eff}} < 0$, however, the Luttinger liquid fixed point is unstable; now, the system flows to a Luther-Emery fixed point with a spin gap which can be determined in the familiar way to be

$$\Delta_s \sim \omega_D \exp[-\pi v_F / g_1^{\text{eff}}]\,. \qquad (21.89)$$

This looks very much like the BCS result from high dimensions. The parallel with BCS theory goes even a bit further, since under the RG transformation, a repulsive g_1 scales to weaker values in just the same way as the Coulomb pseudopotential in higher dimensions:

$$g_1(\omega_D) = \frac{g_1^0}{1 + (g_1^0/\pi v_F)\log(E_F/\omega_D)}\,, \qquad (21.90)$$

where $g_1^0 \equiv g_1(E_F)$. However, in contrast to the higher dimensional case, $\tilde{g}_1$ is strongly renormalized; integrating the one-loop equations, it is easy to show that

$$\tilde{g}_1(\omega_D) = \left(\frac{\tilde{g}_1^0}{1 + \tilde{g}_1^0 L}\right)\left(\frac{g_1(\omega_D)}{g_1^0}\right)^{3/2}\left(\frac{E_F}{\omega_D}\right)^{-g_c/2\pi v_F}\,, \qquad (21.91)$$

$$L = \int_0^{\log(E_F/\omega_D)} \frac{dx}{\pi v_F}\,\frac{\exp[-g_c x/2\pi v_F]}{[1 + (g_1^0/\pi v_F)x]^{3/2}}\,. \qquad (21.92)$$

Various limits of this expression can easily be analyzed—we will not give an exhaustive analysis here. For $g_1 = g_c = 0$, Eq. (21.92) reduces to the same logarithmic expression, Eq. (21.90), as for g_1, although because $\tilde{g}_1$ has the opposite sign, the result is a logarithmic *increase* of the effective interaction; this is simply the familiar Peierls renormalization of the electron-phonon interaction. For $g_c < 0$, this renormalization is substantially amplified. Thus, in marked contrast to the higher dimensional case, strong repulsive interactions actually enhance the effects of weak retarded attractions!

Repulsive interactions enhance the effects of weak retarded attractions.
Finally, there is bad news as well as good news. As discussed in Sect. 21.5, the behavior of the charge modes is largely determined by the "charge Luttinger exponent, K_c, which is in turn determined by the effective interaction

$$g_c^{\text{eff}} = g_c + \tilde{g}_1^{\text{eff}} - 2\tilde{g}_2 \,, \tag{21.93}$$

according to the relation (See Eq. (21.16).)

$$K_c = \sqrt{\frac{1 + (g_c^{\text{eff}}/\pi v_F)}{1 - (g_c^{\text{eff}}/\pi v_F)}} \,. \tag{21.94}$$

In particular, the relative strength of the superconducting and CDW fluctuations are determined by K_c; the smaller K_c, i.e. the more negative g_c^{eff}, the more dominant are the CDW fluctuations. It therefore follows from Eq. (21.93) that a large negative value of $\tilde{g}_1^{\text{eff}}$ due to the renormalization of the electron-phonon interaction only throws the balance more strongly in favor of the CDW order. For this reason, most quasi 1D systems with a spin gap are CDW insulators, rather than superconductors.

Half Filling

Near half filling, the interference between the retarded and instantaneous interactions becomes even stronger. In the presence of Umklapp scattering, an initially negative g_c renormalizes to stronger coupling, as does g_3 itself. Without loss of generality, we can take $g_3 > 0$ since its sign can be reversed by a change of basis. Then we can see that both g_3 and g_c contribute to an inflationary growth of $\tilde{g}_-$. The RG equations have been integrated in [322], and we will not repeat the analysis here. The point is that all the effects discussed above apply still more strongly near half filling. In addition, we now encounter an entirely novel phenomenon—we find that the effective electron-phonon interaction strength at energy scale ω_D is strongly doping dependent, as well. It is possible [322], as indeed seems to be the case in the model conducting polymer polyacetylene, for the electron-phonon coupling to be sufficiently strong to open a Peierls gap of magnitude 2eV (roughly, 1/5 of the π-band width) at half filling, and yet be so weak at a microscopic scale that for doping concentrations greater than 5%, no sign of a Peierls gap is seen down to temperatures of order 1K!

The effective electron-phonon coupling can even be strongly doping dependent.

How many of the features seen from this study of the 1DEG are specific to one-dimensional systems is not presently clear. Conversely, these results prove by example that familiar properties of Fermi liquids cannot be taken as generic. In particular, strongly energy and doping dependent electron phonon interactions are certainly possibilities that should be taken seriously in systems that are not Fermi liquids.

21.10 Lessons from Strong Coupling

In certain special cases, well controlled analytic results can be obtained in the limit in which the bare electron-electron interactions are nonperturbative. We discuss several such models.

21.10.1 The Holstein Model of Interacting Electrons and Phonons

The simplest model of strong electron-phonon coupling is the Holstein model of an optic phonon, treated as an Einstein oscillator, coupled to a single tight binding electron band,

$$H_{\text{Hol}} = -t \sum_{<i,j>,\sigma} [c_{i,\sigma}^\dagger c_{j,\sigma} + \text{H.C.}] + \alpha \sum_j x_j \hat{n}_j + \sum_j \left[\frac{P_j^2}{2M} + \frac{K x_j^2}{2} \right] , \tag{21.95}$$

where $\hat{n}_j = \sum_\sigma c_{j,\sigma}^\dagger c_{j,\sigma}$ is the electron density operator and P_j is the momentum conjugate to x_j.

In treating the interesting strong coupling physics of this problem, it is sometimes useful to transform this model so that the phonon displacements are defined relative to their instantaneous ground state configuration. This is done by means of the unitary transformation,

$$U = \prod_j \exp[i(\alpha/K)P_j \hat{n}_j] \,, \tag{21.96}$$

which shifts the origin of oscillation as $U^\dagger x_j U = x_j - (\alpha/K)\hat{n}_j$. Consequently, the transformed Hamiltonian has the form

$$U^\dagger H_{\text{Hol}} U = -t \sum_{<i,j>,\sigma} [\hat{S}_{ij} c_{i,\sigma}^\dagger c_{j,\sigma} + \text{H.C.}] \tag{21.97}$$

$$- \frac{U_{\text{eff}}}{2} \sum_j [\hat{n}_j]^2 + \sum_j \left[\frac{P_j^2}{2M} + \frac{K x_j^2}{2} \right] ,$$

where $\hat{S}_{i,j} = \exp[-i(\alpha/K)(P_i - P_j)]$ and $U_{\text{eff}} = \alpha^2/K$.

There are several limits in which this model can be readily analyzed:

Adiabatic Limit: $E_F \gg \omega_D$

In the limit $t \gg \omega_D$, where $\omega_D = \sqrt{K/M}$ is the phonon frequency and for α not too large, this is just the sort of model considered in the weak coupling section, or any other conventional treatment of the electron-phonon problem. Here, Migdal's theorem provides us with guidance, and at least for not too strong coupling, the BCS-Eliashberg treatment discussed in Sect. 21.9 can be applied. While U_{eff} is, indeed, the effective interaction which enters the BCS expression for the superconducting T_c, because the fluctuations of P_i are large if M is large, it is not useful to work with the transformed version of the Hamiltonian.

Inverse Adiabatic Limit; Negative U Hubbard model

In the inverse adiabatic limit, $M \to 0$, fluctuations of P_j are negligible, so that $\hat{S}_{ij} \to 1$. Hence, in this limit, the Holstein model is precisely equivalent to the Hubbard model, but with an effective negative U. If $U_{\text{eff}} \ll t$, this is again a weak coupling model, and will yield a superconducting T_c given by the usual BCS expression, although in this case with a prefactor proportional to t rather than ω_D.

In contrast, if $U_{\text{eff}} \gg t$, a strong coupling expansion is required. Here, we first find the (degenerate) ground states of the unperturbed model with $t = 0$, and then perform perturbation theory in small t/U_{eff}. In the zeroth order ground states, each site is either unoccupied, or is occupied by a singlet pair of electrons. The energy of this state is $-U_{\text{eff}}N^{el}$, where N^{el} is the number of electrons. These states can be thought of as the states of infinite mass, hard core charge 2e bosons on the lattice. There is a gap to the first excited state of magnitude U_{eff}. Second order perturbation theory in the ground state manifold straightforwardly yields an effective Hamiltonian which is equivalent to a model of hard core bosons ($[b_i^\dagger, b_j] = \delta_{i,j}$)

$$H_{\text{boson}} = -t_{\text{eff}} \sum_{<i,j>} [b_i^\dagger b_j + \text{H.C.}] + V_{\text{eff}} \sum_{<i,j>} b_i^\dagger b_i b_j^\dagger b_j + [\infty] \sum_j b_j^\dagger b_j [b_j^\dagger b_j - 1] \,, \tag{21.98}$$

with nearest neighbor hopping $t_{\text{eff}} = 2t^2/U_{\text{eff}}$ and nearest neighbor repulsion $V_{\text{eff}} = 2t_{\text{eff}}$. This effective model is applicable for energies and temperatures small compared to U_{eff}. (Clearly, $\tilde{b}_j \equiv c_{j\uparrow}c_{j\downarrow}$ does not satisfy the same-site piece of the bosonic commutation relation, but the hard core constraint on the b_j bosons corrects any errors introduced by neglecting this.)

Strong attractions impede coherent motion, and enhance charge ordering.
The properties of this bosonic Hamiltonian, and closely related models where additional interactions between bosons are included, have been widely studied [323,324] (see also [547]). It has a large number of possible phases, including superconducting, crystalline, and striped or liquid crystalline phases. The equivalence between hard core bosons and spin-1/2 operators can be used to relate this model to various spin models that have been studied in their own right. However, for the present purposes, there are two clear lessons we wish to draw from this exercise. The first is that there are ordered states, in particular insulating charge ordered states, which can compete very successfully with the superconducting state in strong coupling. The second is that, even if the system does manage to achieve a superconducting ground state, the characteristic superconducting T_c will be proportional to t_{eff}, and hence to the small parameter, t/U_{eff}.

Large U_{eff}: Bipolarons

More generally, in the strong coupling limit, $U_{\text{eff}} \gg t$, a perturbative approach in powers of t/U_{eff} can be undertaken, regardless of the value of M. Once again, the zeroth order ground states are those of charge $2e$ hard core bosons, as in Eq. (21.98). However, now the phonons make a contribution to the ground state—the ground state energy is $-U_{\text{eff}}N^{el} + (1/2)\omega_D N$

where N is the number of sites, and the gap to the first excited state is the smaller of U_{eff} and ω_D. Still, we can study the properties of the model at energies and temperatures small compared to the gap in terms of the hard core bosonic model. Now, however,

$$t_{\text{eff}} = 2\frac{t^2}{U_{\text{eff}}}F_+(X) \ ,$$

$$V_{\text{eff}} = 4\frac{t^2}{U_{\text{eff}}}F_-(X) \ , \qquad (21.99)$$

where $X \equiv \frac{U_{\text{eff}}}{\omega_D}$ and

$$F_\pm(X) = \int_0^\infty \mathrm{d}t \exp\{-t - X[1 \pm \exp(-t/X)]\} \ . \qquad (21.100)$$

This is often referred to as a model of bipolarons. In the inverse adiabatic limit, $F_\pm(X) \to 1$ as $X \to 0$, and hence these expressions reduce to those of the previous subsection. However, in the adiabatic limit, $X \gg 1, F_+(X) \sim e^{-2X}$, so t_{eff} is exponentially reduced by a Frank–Condon factor! However, $F_-(X) \to 1$ as $X \to \infty$, so V_{eff} remains substantial. Clearly, the lessons concerning the difficulty of obtaining high temperature superconductivity from strong coupling drawn from the negative U Hubbard model apply even more strongly to the case in which the phonon frequency is small. A bipolaron mechanism of superconductivity is simply impossible unless the phonon frequency is greater than or comparable to U_{eff}; in the opposite limit, the exponential suppression of t_{eff} relative to the effective interactions, V_{eff}, strongly suppresses the coherent Bose-condensed state, and favors various types of insulating, charge ordered states.

21.10.2 Insulating Quantum Antiferromagnets

We now turn to models with repulsive interactions. To begin with, we discuss the "Mott limit" of the antiferromagnetic insulating state. Here, we imagine that there is one electron per site, and such strong interactions between them that charge fluctuations can be treated perturbatively. In this limit, as is well known, the only low energy degrees of freedom involve the electron spins, and hence the problem reduces to that of an effective quantum Heisenberg antiferromagnet.

Quantum Antiferromagnets in More Than One Dimension

In more than one dimension, it is a solved problem. In recent years, there has been considerable interest [76,79,80,212,231–233,325,326] in the many remarkable quantum states that can occur in quantum spin models with sufficiently strong frustration—these studies are beyond the scope of the present review. On a hypercubic lattice (probably on any simple, bipartite lattice) and in dimension 2 or greater, there is by now no doubt that even the spin 1/2 model (in which quantum fluctuations are the most severe) has a Nèel ordered ground state [221]. Consequently, the properties of such systems at temperatures and energies low compared to the antiferromagnetic exchange energy, J, are determined by the properties of interacting spin waves. This physics, in turn, is well described in terms of a simple field theory, known as the $O(3)$ nonlinear sigma model. While interesting work is still ongoing on this problem, it is in essence a solved problem, and excellent modern reviews exist [327].

Antiferromagnetic order is bad for superconductivity. In its ordered phase, the antiferromagnet has: (i) gapless spin wave excitations, and (ii) reduced tendency to phase ordering due to the frustration of charge motion. Since the superconducting state possesses a spin gap (or, for d-wave, a partial gap) and is characterized by the extreme coherence of charge motion, it is clear that both these features of the antiferromagnet are disadvantageous for superconductivity.[19]

[19] There is a very interesting line of reasoning [154] which takes the opposite viewpoint: it is argued that the important point to focus on is that both the superconductor and the antiferromagnet have gapless Goldstone modes, not whether those modes are spinless or spinful. In this line of thought there is a near symmetry, which turns out to be SO(5), between the d-wave superconducting and the Néel ordered antiferromagnetic states. This is an attractive notion, but it is not clear to us precisely how this line of reasoning relates to the more microscopic considerations discussed here.

There is a body of thought [29–32] that holds that it is possible, at sufficiently strong doping of an antiferromagnet to reach a state in which the antiferromagnetic order and the consequent low energy spin fluctuations are eliminated and electron itineracy is restored, which yet has vestiges of the high energy spin wave excitations of the parent ordered state that can serve to induce a sufficiently strong effective attraction between electrons for high temperature superconductivity. Various strong critiques of this approach have also been articulated [18].

We feel that the theoretical viability of this "spin fluctuation exchange" idea has yet to be firmly established. As an example of how this could be done, one could imagine studying a two component system consisting of a planar, Heisenberg antiferromagnet coupled to a planar Fermi liquid. One would like to see that, as some well articulated measure of the strength of the antiferromagnetism is increased, the superconducting pairing scale likewise increases. If such a system could be shown to be a high temperature superconductor, it would establish the point of principal. However, it has been shown by Schrieffer [66] that Ward identities, which are ultimately related to Goldstone's theorem, imply that long wavelength spin waves cannot produce any pairing interaction at all. A model of this sort that has been analyzed in detail is the one-dimensional Kondo–Heisenberg model, which is the 1D analogue of this system [328–330]. This system does not exhibit significant superconducting fluctuations of any conventional kind. While there certainly does not exist a "no-go" theorem, it does not seem likely to us that an exchange of spin waves in a nearly anitferromagnetic system can ever give rise to high temperature superconductivity.[20]

Spin Gap in Even Leg Heisenberg Ladders

The physics of quantum antiferromagnets in one dimension is quite different from that in higher dimension, since the ground state is not magnetically ordered. However, its general features have been well understood for many years. In particular, for spin-1/2 Heisenberg ladders or cylinders with an even number of sites on a rung, quantum fluctuations result in a state with a spin gap. This is a special case of a general result [332], known as "Haldane's conjecture," that any 1D spin system with an even integer number of electrons per unit cell has a spin rotationally invariant ground state and a finite spin gap in the excitation spectrum. This conjecture has not been proven, but has been validated in many limits and there are no known exceptions[21].

The physics of interacting electrons on ladders—i.e. "fat" 1D systems, will be discussed at length below. We believe this is an important, paradigmatic system for understanding the physics of high temperature superconductivity. The fact that even the undoped (insulating) ladder has a spin gap can be interpreted as a form of incipient superconducting pairing. Where that gap is large, i.e. a substantial fraction of the exchange energy, J, it is reasonable to hope that doping it will lead to a conducting state which inherits from the parent insulating state this large gap, now directly interpretable as a pairing gap.

Let us start by considering an N leg spin-1/2 Heisenberg model

$$H = \sum_{<i,j>} J_{ij} \mathbf{S}_i \cdot \mathbf{S}_j \,, \tag{21.101}$$

where $\mathbf{S}_i$ is the spin operator on site i, so for a, b, $c = \{x, y, z\}$, $[S_i^a, S_j^b] = i\delta_{ij}\epsilon^{abc}S_i^c$ and $\mathbf{S}_i \cdot \mathbf{S}_i = 3/4$. Here, we still take the lattice to be infinite in one ("parallel") direction but of width N sites in the other.

[20] Under circumstances in which antiferromagnetic correlations are very short ranged, it may still be possible to think of an effective attraction between electrons mediated by the exchange of very local spin excitations [31]. This escapes most of the critiques discussed above—neither Ward identities nor the general incompatibility between antiferromagnetism and easy electron itineracy have any crisp meaning at short distances. By the same token, however, it is not easy to unambiguously show that such short range magnetic correlations are the origin of strong superconducting correlations in any system, despite some recent progress along these lines [331].

[21] One can hardly fail to notice that the Haldane conjecture is closely related to the conventional band structure view that insulators are systems with a gap to both charge and spin excitations due to the fact that there are an even number of electrons per unit cell and all bands are either full or empty.

At times, we will distinguish between a ladder, with open boundary conditions in the "perpendicular" direction, and a cylinder, with periodic boundary conditions in this direction. We will typically consider isotropic antiferromagnetic couplings, $J_{ij} = J > 0$.

Ladders with Many Legs:
In the limit of large N, it is clear that the model can be viewed as a two-dimensional antiferromagnet up to a crossover scale, beyond which the asymptotic one-dimensional behavior is manifest. This viewpoint was exploited by Chakravarty [333] to obtain a remarkably accurate analytic estimate of the crossover scale. His approach was to first employ the equivalence between the Heisenberg model and the quantum nonlinear sigma model.

The spin gap falls exponentially with N.
One feature of this mapping is that the thermodynamic properties of the d dimensional Heisenberg model are related to a $d+1$ dimensional sigma model, with an imaginary time direction which, by suitable rescaling, is *precisely* equivalent to any of the spatial directions. The properties of the Heisenberg model at finite temperatures are then related to the sigma model on a generalized cylinder, which is periodic in the imaginary time direction with circumference $\hbar v_s/T$ where v_s is the spin wave velocity. What Chakravarty pointed out is that, through this mapping, there is an equivalence between the Heisenberg cylinder with circumference $L = Na$ at zero temperature and the infinite planar Heisenberg magnet at temperature, $T = v_s/L$. From the well known exponential divergence of the correlation length with decreasing temperature in the 2d system, he obtained the asymptotic expression for the dimensional crossover length in the cylinder,

$$\xi_{\text{dim}} \sim a \exp[0.682N] . \qquad (21.102)$$

As this estimate is obtained from the continuum theory, it is only well justified in the large N limit. However, comparison with numerical experiments described in Sect. 21.11 (some of which predated the analytic theory [334]) reveal that it is amazingly accurate, even for $N = 2$, and that the distinction between ladders and cylinders is not very significant, either.

This result is worth contemplating. It implies that the special physics of one-dimensional magnets is only manifest at exponentially long distances in fat systems. Correspondingly, it means that these effects are confined to energies (or temperatures) smaller than the characteristic scale

$$\Delta_{\text{dim}} = v_s/\xi_{\text{dim}} . \qquad (21.103)$$

As a practical matter, it means that only the very narrowest systems, with N no bigger than 3 or 4, will exhibit the peculiarities of one-dimensional magnetism at any reasonable temperature.

To understand more physically what these crossover scales mean, one needs to know something about the behavior of one-dimensional magnets. Since even leg ladders and cylinders have a spin gap, it is intuitively clear (and correct) that Δ_{dim} is nothing but the spin gap and ξ_{dim} the correlation length associated with the exponential fall of magnetic correlations at $T = 0$. For odd leg ladders, ξ_{dim} is analogous to a Josephson length, where correlations crossover from the two-dimensional power law behavior associated with the existence of Goldstone modes, to the peculiar quantum critical behavior of the one-dimensional spin 1/2 Heisenberg chain.

The Two Leg Ladder:
It is often useful in developing intuition to consider limiting cases in which the mathematics becomes trivial, although one must always be sensitive to the danger of being overly influenced by the naive intuitions that result.

In the case of the two leg ladder, there exists such a limit, $J_\perp \gg J_\parallel$, where $J_\perp$ and $J_\parallel$ are, respectively, the exchange couplings across the rungs, and along the sides of the ladder. Here the zeroth order ground state is a direct product of singlet pairs (valence bonds) on the rungs of the ladder. Perturbative corrections to the ground state cause these valence bonds to resonate, locally, but do not fundamentally affect the character of the ground state. The ground state energy per site is

$$E_0 = -(3/8)J_\perp[1 + (J_\parallel/J_\perp)^2 + \ldots] . \qquad (21.104)$$

Since each valence bond is nothing but a singlet pair of electrons, this makes it clear that there is a very direct sense in which the two leg ladder can be thought of as a paired insulator. The lowest lying spin-1 excited states are a superposition of bond triplets on different rungs, and have a dispersion relation which can easily be derived in perturbation theory:

$$E_{\text{triplet}} = J_\perp + J_\parallel \cos(k) + \mathcal{O}(J_\parallel^2/J_\parallel) . \qquad (21.105)$$

This, too, reveals some features that are more general, such as a minimal spin gap of magnitude $\Delta_s = J_\perp[1 - (J_\parallel/J_\perp) + \mathcal{O}(J_\parallel^2/J_\parallel^2)]$ at what would be the antiferromagnetic ordering wavevector $k = \pi$.

21.10.3 The Isolated Square

While we are considering mathematically trivial problems, it is worth taking a minute to discuss the solution of the $t - J$ model (defined in Eq. (21.119), below) on an isolated 4-site square. The pedagogic value of this problem, which is exactly diagonalizable, was first stressed by Trugman and Scalapino [335]. This idea was recently carried further by Auerbach and collaborators [336,337], who have attempted to build a theory of the 2D $t - J$ model by linking together fundamental squares. The main properties of the lowest energy states of this system are given in Table 21.2 for any number of doped holes.

The "undoped" state of this system (i.e. with four electrons) is a singlet with ground state energy $E_0 = -3J$. However, interestingly, it is not in the identity representation of the symmetry group of the problem—it is odd under 90° rotation. If we number the sites of the square sequentially from 1 to 4, then the ground state wavefunction is

$$|4 - \text{electron}\rangle = [\hat{P}^\dagger_{1,2}\hat{P}^\dagger_{3,4} - \hat{P}^\dagger_{1,4}\hat{P}^\dagger_{2,3}]|0\rangle \qquad (21.106)$$

where $\hat{P}^\dagger_{i,j} = \hat{P}^\dagger_{j,i} = [c^\dagger_{i,\uparrow}c^\dagger_{j,\downarrow} + c^\dagger_{j,\uparrow}c^\dagger_{i,\downarrow}]/\sqrt{2}$ creates a singlet pair of electrons on the bond between sites i and j. Manifestly, $|4\text{–electron}\rangle$ has the form of an odd superposition of nearest neighbor valence bond states—in this sense, it is the quintessential resonating valence bond state. The lowest lying excitation is a spin-1 state with energy $-2J$, so the spin gap is J.

There are level crossings as a function of J/t in the "one hole" (3 electron) spectrum. For $0 < J/t < (8 - \sqrt{52})/3 \approx 0.263$ the ground state is a spin 3/2 multiplet with energy $E_1 = -2t$. It is orbitally nondegenerate with zero momentum (we consider the square as a 4-site chain with periodic boundary conditions and refer to the momentum along the chain.) For $(8 - \sqrt{52})/3 < J/t < 2$ the ground state has spin 1/2, is twofold degenerate with crystal momentum $\pm\pi/2$, and has energy $E_1 = -[2J + \sqrt{J^2 + 12t^2}]/2$. For $2 < J/t$, the ground state has spin 1/2, zero momentum, and energy $E_1 = -3J/2 - t$.

The two hole (2 electron) ground state has energy $E_2 = -[J + \sqrt{J^2 + 32t^2}]/2$, and spin 0. It lies in the identity representation of the symmetry group. The lowest excitation is a spin 1 state. For $0 < J/t < 2$ it has crystal momentum $k = \pm\pi/2$ (i.e. it has a twofold orbital and threefold spin degeneracy) and has energy $E_2(S = 1) = -2t$. For $2 < J/t$ it is orbitally nondegenerate with energy $E_2(S = 1) = -J$.

Pair field correlations have $d_{x^2-y^2}$ symmetry. One important consequence of this, which follows directly from the Wigner–Eckhart theorem, is that the pair annihilation operator that connects the zero hole and the two hole ground states must transform as $d_{x^2-y^2}$. This is, perhaps, the most important result of this exercise. It shows the robustness of the d wave character of the pairing in a broad class of highly correlated systems. The dominant component of this operator is of the form

$$\phi_1 = \hat{P}_{12} - \hat{P}_{23} + \hat{P}_{34} - \hat{P}_{41} . \qquad (21.107)$$

It also includes terms that create holes on next nearest neighbor diagonal sites [338,339].

There are a few other interesting aspects of this solution. In the single hole sector, the ground state is maximally polarized, in agreement with Nagaoka's theorem, for sufficiently large t/J, but there is a level crossing to a state with smaller spin when t/J is still moderately large. Moreover, even when the single hole state is maximally polarized, the two hole state, like the zero hole state, is always a spin singlet. Both of these features have been observed in numerical studies on larger $t - J$ clusters [340].

If we look still more closely at the $J/t \to 0$ limit, there is another interesting aspect of the physics: It is intuitively clear that in this limit, the holes should

Table 21.2. The low energy spectrum of the four-site $t-J$ square for 0 holes (four electrons), one hole (three electrons), and two holes (two electrons). The three-hole and four-hole problems are left as an exercise for the reader

	Energy	Spin	Momentum
0 holes			
g.s.	$E=-3J$	$S=0$	$P=\pi$
1st e.s.	$E=-2J$	$S=1$	$P=0$
1 hole			
$0<J/t<0.263$			
g.s.	$E=-2t$	$S=3/2$	$P=0$
1st e.s.	$E=-J-\sqrt{J^2/4+3t^2}$	$S=1/2$	$P=\pm\pi/2$
$0.263<J/t<2/3$			
g.s.	$E=-J-\sqrt{J^2/4+3t^2}$	$S=1/2$	$P=\pm\pi/2$
1st e.s.	$E=-2t$	$S=3/2$	$P=0$
$2/3<J/t<2$			
g.s.	$E=-J-\sqrt{J^2/4+3t^2}$	$S=1/2$	$P=\pm\pi/2$
1st e.s.	$E=-3J/2-t$	$S=1/2$	$P=\pi$
$2<J/t$			
g.s.	$E=-3J/2-t$	$S=1/2$	$P=\pi$
1st e.s.	$E=-J-\sqrt{J^2/4+3t^2}$	$S=1/2$	$P=\pm\pi/2$
2 holes			
$0<J/t<2$			
g.s.	$E=-J/2-\sqrt{J^2/4+8t^2}$	$S=0$	$P=0$
1st e.s.	$E=-2t$	$S=1$	$P=\pm\pi/2$
$2<J/t$			
g.s.	$E=-J/2-\sqrt{J^2/4+8t^2}$	$S=0$	$P=0$
1st e.s.	$E=-J$	$S=0$	$P=\pi,\pm\pi/2$

behave as spinless fermions. This statement requires no apology in the maximum spin state. Thus, the lowest energy spin-1 state with two holes has energy $E_2(S=1)=-2t$ in this limit. It corresponds to a state in which one spinless fermion has crystal momentum $k=0$ and energy $-2t$, and the other has crystal momentum $\pm\pi/2$ and energy 0. However, what is more interesting is that there is also a simple interpretation of the two hole ground state in the same representation. The antisymmetry of the spins in their singlet state means that they affect the hole dynamics through a Berry's phase as if half a magnetic flux quantum were threaded through the square. This Berry's phase implies that the spinless fermions satisfy antiperiodic boundary conditions. The ground state is thus formed by occupying the single particle states with $k=\pm\pi/4$ for a total ground state energy of $E_2=-2\sqrt{2}t$, precisely the $J/t\to 0$ limit of the expression obtained above. The interesting thing is that, in this case, it is the hole kinetic energy, and *not* the exchange energy, which favors the singlet over the triplet state. This simple exercise provides an intuitive motivation for the existence of various forms of "flux phase" in strongly interacting systems [341].

Finally, it is worth noting that pair binding occurs, in the sense that $2E_1 - E_0 - E_2 > 0$, so long as $J/t > \sqrt{(39-\sqrt{491})/\sqrt{3}} \approx 0.2068$. We will return to the issue of pair binding in Sect. 21.11 where we will show a similar behavior in Hubbard and $t-J$ ladders.

21.10.4 The Spin Gap Proximity Effect Mechanism

The final strong coupling model we will consider consists of two inequivalent 1DEG's weakly coupled together—a generalization of a two leg ladder. Each 1DEG is represented by an appropriate bosonized field theory—either a Luttinger liquid or a Luther-Emery liquid. Most importantly, the two systems are assumed to have substantially different values of the Fermi momentum, k_F and $\tilde{k}_F$. We consider the case in which the interactions between the two systems are weak, but the interactions within each 1DEG may be arbitrarily large. The issue we address is what changes in the properties of the coupled system are induced by these interactions. (For all technical details, see [20] and [25].)

Intuitive description of the spin gap proximity effect...
There is an important intuitive reason to expect this system to exhibit a novel form of kinetic energy driven superconducting pairing. Because $k_F \neq \tilde{k}_F$, single particle tunneling between the two 1DEG's is not a low energy process—it is irrelevant in the renormalization group sense, and can be ignored as anything but a high energy virtual fluctuation. The same conclusion holds for any weak coupling between the $2k_F$ or $4k_F$ density wave fluctuations. There are only two types of coupling that are potentially important at low energies: pair tunneling, since the relevant pairs have 0 momentum, and coupling between long wavelength spin fluctuations.

The magnetic interactions are marginal to leading order in a perturbative RG analysis—they turn out to be marginally relevant if the interactions are antiferromagnetic and marginally irrelevant if ferromagnetic [328,330]. The effect of purely magnetic interactions has been widely studied in the context of Kondo-Heisenberg chains, but will not be discussed here. The effect of triplet pair tunneling has only been superficially analyzed in the literature [25,342,343]—it would be worthwhile extending this analysis, as it may provide some insight into the origin of the triplet superconductivity that has been observed recently in certain highly correlated materials. However, in the interest of brevity, we will ignore these interactions.

... as a kinetic energy driven mechanism of pairing.
Singlet pair tunneling interactions between the two 1DEG's have a scaling dimension which depends on the nature of the correlations in the decoupled system. Under appropriate circumstances, they can be relevant. When this is the case, the coupled system scales to a new strong coupling fixed point which exhibits a total spin gap and strong global superconducting fluctuations. This is what we refer to as the spin gap proximity effect, because the underlying physics is analogous to the proximity effect in conventional superconductors. The point is that even if it is energetically costly to form pairs in one or both of the 1DEGs, once the pairs are formed they can coherently tunnel between the two systems, thereby lowering their zero point kinetic energy. Under appropriate circumstances, the kinetic energy gain outweighs the cost of pairing. This mechanism is quite distinct from any relative of the BCS mechanism—it does not involve an induced attraction.

The explicit model which is analyzed here is expressed in terms of four bosonic fields: ϕ_c and ϕ_s represent the charge and spin degrees of freedom of the first 1DEG, and $\tilde{\phi}_c$ and $\tilde{\phi}_s$ of the other, as is discussed in Sect. 21.5, above. The Hamiltonian of the decoupled system is the general bosonized Hamiltonian described in that section, with appropriate velocities and charge Luttinger exponents, v_s, v_c, $\tilde{v}_s$, $\tilde{v}_c$, K_c, and $\tilde{K}_c$ if both are Luttinger liquids, and values of the spin gap, Δ_s and $\tilde{\Delta}_s$ in the case of Luther-Emery liquids (i.e. if the cosine potential in the sine-Gordon theory for the spin degrees of freedom is relevant). If we ignore the long wavelength magnetic couplings and triplet pair tunneling between the two systems, the remaining possibly important interactions at low energy,

$$H_{\text{inter}} = \int dx[\mathcal{H}_{for} + \mathcal{H}_{\text{pair}}] , \quad (21.108)$$

are the forward scattering (density-density and current-current) interactions in the charge sector

$$\mathcal{H}_{for} = V_1 \partial_x \phi_c \partial_x \tilde{\phi}_c + V_2 \partial_x \theta_c \partial_x \tilde{\theta}_c , \quad (21.109)$$

where θ designates the field dual to ϕ (see Sect. 21.5), and the singlet pair tunneling

$$\mathcal{H}_{\text{pair}} = \mathcal{J} \cos[\sqrt{2\pi}\phi_s] \cos[\sqrt{2\pi}\tilde{\phi}_s] \times \cos[\sqrt{2\pi}(\theta_c - \tilde{\theta}_c)] . \quad (21.110)$$

As discussed previously, the singlet pair creation operator involves both the spin and the charge fields.

The forward scattering interactions are precisely marginal, and should properly be incorporated in the definition of the fixed point Hamiltonian. $\mathcal{H}_{\text{pair}}$ is a nonlinear interaction; the coupled problem with nonzero $\mathcal{J}$ has not been exactly solved. However, it is relatively straightforward to asses the perturbative relevance of this interaction, and to deduce the properties of the most likely strong coupling fixed point (large $\mathcal{J}$) which governs the low energy physics when it is relevant.

The scaling dimension of the pair tunneling interaction is introduced.
The general expression for the scaling dimension of $\mathcal{H}_{\text{pair}}$ is a complicated analytic combination of the parameters of the decoupled problem

$$\delta_{\text{pair}} = \frac{1}{2}\left[\frac{A}{K_c} + \frac{B}{\tilde{K}_c} + K_s + \tilde{K}_s\right] , \quad (21.111)$$

where $A = 1$ and $B = 1$ in the absence of intersystem forward scattering interactions, but more generally A and B are complicated functions of the coupling constants. For illustrative purposes, one can consider the explicit expression for these functions under the special circumstances $V_2 = -(\tilde{v}_c/v_c)(K_c\tilde{K}_c)V_1$; then $A = \sqrt{1 - (V_1^2 K_c \tilde{K}_c / v_c \tilde{v}_c)}$ and $B = (v_c - V_1 K_c)^2 / \sqrt{v_c^4 - V_1^2 v_c \tilde{v}_c K_c \tilde{K}_c}$. Here, if both 1DEG's are Luttinger liquids, spin rotational invariance implies that $K_s = \tilde{K}_s = 1$. If one or the other 1DEG is a Luther-Emery liquid, one should substitute $K_s = 0$ or $\tilde{K}_s = 0$ in the above expression.

Pair tunneling is perturbatively relevant if $\delta_{\text{pair}} < 2$, and irrelevant otherwise. Clearly, having a preexisting spin gap in either of the 1DEG's dramatically decreases δ_{pair}—if there is already pairing in one subsystem, then it stands to reason that pair tunneling will more easily produce pairing in the other.

The physical effects which make pair tunneling relevant are described.
However, even if neither system has a preexisting spin gap, there are a wide set of physical circumstances for which $\delta_{\text{pair}} < 2$. Notice, in particular, that repulsive intersystem interactions, $V_1 > 0$, produce a reduction of δ_{pair}. Again, the physics of this is intuitive—an induced anticorrelation between regions of higher than average electron density in the two 1DEG's means that where there is a pair in one system, there tends to be a low density region on the other which is just waiting for a pair to tunnel into it. (See, also, Sect. 21.6.)

The implications of strong pair tunneling are discussed.
In the limit that $\mathcal{J}$ is large, the spin fields in both 1DEG's are locked, which implies a total spin gap, and the out-of-phase fluctuations of the dual charge phases are gapped as well. This means that the only possible gapless modes of the system involve the total charge phase, $\phi \equiv [\phi_c + \tilde{\phi}_c]/\sqrt{2}$, and its dual, $\theta \equiv [\theta_c + \tilde{\theta}_c]/\sqrt{2}$. θ is simply the total superconducting phase of the coupled system, and ϕ the total CDW phase. At the end of the day, this strong coupling fixed point of the coupled system is a Luther-Emery liquid, and consequently has a strong tendency to superconductivity. In general, there will be substantial renormalization of the effective parameters as the system scales from the weak to the strong coupling fixed point. Thus, it is difficult to estimate the effective Luttinger parameters which govern the charge modes of the resulting Luther-Emery liquid. A naive estimate, which may well be unreliable, can be made by simply setting $\mathcal{J} \to \infty$. In this case, all the induced gaps are

infinite, and the velocity and Luttinger exponent that govern the dynamics of the remaining mode are

$$K_c^{\text{total}} = \sqrt{\frac{v_c K_c + \tilde{v}_c \tilde{K}_c + 2V_2}{v_c / K_c + \tilde{v}_c / \tilde{K}_c + 2V_1}} , \tag{21.112}$$

$$v_c^{\text{total}} = \frac{1}{4} \sqrt{[v_c K_c + \tilde{v}_c \tilde{K}_c + 2V_2][v_c / K_c + \tilde{v}_c / \tilde{K}_c + 2V_1]} .$$

21.10.5 Optimal Inhomogeneity for Superconductivity

In this subsection (and to some extent, in the next), we have discussed the solution of the strong coupling problem in certain special restricted geometries. It is notable that, even with repulsive interactions, these systems can show a strong tendency toward pairing. However, zero-dimensional systems, such as an isolated Hubbard square, or one-dimensional systems, such as a ladder, cannot have a finite temperature superconducting transition. So, it is somewhat uncertain what lessons one should take away from this exercise. Specifically, the issue arises whether studying these special geometries is just a calculational trick which permits a controlled theoretical study of the same physics that occurs in an extended two or three-dimensional system, or if there is some physical significance to these smaller structures. It is certainly true that most finite Hubbard clusters *do not* exhibit pair-binding for repulsive U - for instance, of the 3, 4, and 6 member Hubbard rings (which could be viewed as representative of a plaquette for the triangular, square, or honeycomb lattice), only the 4 membered ring (the square) exhibits pair binding. Thus, it is not clear how one should imagine extrapolating from the results of small Hubbard clusters to the behavior of the two-dimensional Hubbard model.

The key unsettled issue is whether the 2D Hubbard model (or any other uniform 2D model with strong repulsive interactions) is a high temperature superconductor or not [344]. One way to think about this problem is to imagine the phase diagram of a 2D Hubbard model on a checkerboard version of the square lattice, with 4 atoms per unit cell [345]. Here, the hopping matrix between nearest-neighbor sites within a unit cell is t and between nearest-neighbor sites in adjacent unit cells is t'. In the limit $t' = 0$, this model consists of an array of disconnected Hubbard squares, while for $t' = t$ it is the usual square-lattice Hubbard model. It is possible to show [345] that, for U in an appropriate optimal range, and for $t' \ll t$, this model has a d-wave superconducting ground-state and a superconducting transition temperature, $T_c \sim xt'$, which rises from 0 with increasing t'. If the uniform model is, by itself, superconducting, then extrapolating the small t' result to the limit $t' = t$ might be a reasonable way to understand this behavior. On the other hand, if the uniform model does not support high temperature superconductivity, then there must be an intermediate value of t' in the range $0 < t' < t$ at which T_c is maximal. This would imply that there is an optimal degree of inhomogeneity for high temperature superconductivity. In the years since the first version of this article was published, there has been a number of papers which have pursued this idea [346–349]. This subject has recently been critically reviewed in [349].

21.11 Lessons from Numerical Studies of Hubbard and Related Models

Numerical studies are motivated...
High temperature superconductivity is a result of strong electronic correlations. Couple this prevailing thesis with the lack of controlled analytic methods for most relevant models, and the strong motivation for numerical approaches becomes evident. Such numerical studies are limited to relatively small systems, due to a rapid growth in complexity with system size. However, many of the interesting aspects of the high temperature superconductors, especially those which relate to the "mechanism" of pairing, are moderately local, involving physics on the length scale of the superconducting coherence length ξ. Since ξ is typically a few lattice spacings in the high T_c compounds, one expects that numerical solutions of model problems on clusters with as few as 50-100 sites should be able to reveal the salient features of high temperature superconductivity, if it exists in these models. Moreover, numerical studies can guide our mesoscale intuition, and serve as important tests of analytic predictions.

Notwithstanding these merits, a few words of caution are in order. Even the largest systems that have been studied so far[22] are still relatively small. Therefore, the results are manifestly sensitive to the shape and size of the cluster and other finite size effects. Some features, especially with regard to stripes, appear particularly sensitive to small changes in the model such as the presence of second neighbor hopping [353, 354], the type of boundary conditions [355], etc. Less subtle modifications seem to have important consequences, too [358], most notably the inclusion of long range Coulomb forces (although this has been much less studied). This sensitivity has resulted in considerable controversy in the field concerning the true ground state phase diagrams of the stated models in the thermodynamic limit; see [355–357] and [359], among others.

The best numerical data, especially in terms of system size, exists for narrow Hubbard and $t-J$ ladders. We therefore begin by considering them. Apart from their intrinsic appeal, these systems also offer several lessons which we believe are pertinent to the two-dimensional models. The second part of this section provides a brief review of the conflicting results and views which have emerged from attempts to extrapolate from fat ladders and small periodic clusters to the entire plane.

What do we learn from numerical studies?

We feel that numerical studies are *essential* in order to explore the important mesoscale physics of highly correlated systems, but except in the few cases where a careful finite size scaling analysis has been possible over a wide range of system sizes, conclusions concerning the long distance physics should be viewed as speculative. Even where the extrapolation to the thermodynamic limit has been convincingly established for a given model, the established fact that there are so many closely competing phases in the strong correlation limit carries with it the corollary that small changes in the Hamiltonian can sometimes tip the balance one way or the other. Thus, there are significant limitations concerning the conclusions that can be drawn from numerical studies. In the present section we focus on the reproducible features of the local correlations that follow robustly from the physics of strong, short ranged repulsions between electrons, paying somewhat less attention to the various controversies concerning the actual phase diagram of this or that model.

We entirely omit any discussion of the technical details of the numerical calculations. Methods that have been used include exact diagonalization by Lanczos techniques, Monte Carlo simulations of various sorts, numerical renormalization group approaches, and variational ansatz. The reader who is interested in such aspects is invited to consult [360–366].

21.11.1 Properties of Doped Ladders

Ladder systems, that is, quasi-one-dimensional systems obtained by assembling chains one next to the other, constitute a bridge between the essentially understood behavior of strictly one-dimensional models and the incompletely understood behavior in two dimensions. Such systems are not merely a theoretical creation but are realized in nature [367, 368]. For example, two leg $S = 1/2$ ladders (two coupled spin-1/2 chains) are found in vanadyl pyrophosphate $(VO)_2P_2O_7$. Similarly, the cuprate compounds $SrCu_2O_3$ and $Sr_2Cu_3O_5$ consist of weakly coupled arrays of 2-leg and 3-leg ladders, respectively. It is likely that ladder physics is also relevant to the high temperature superconductors, at least in the underdoped regime, where ample experimental evidence exists for the formation of self-organized stripes.

Synopsis of findings

In this section we review some of the most prominent features of Hubbard and especially $t-J$ ladders. As we shall see the data offers extensive support in favor of the contention that a purely electronic mechanism of superconductivity requires mesoscale structure [14]. Specifically, we will find that spin gap formation and pairing correlations, with robust d-wave-like character, are intimately connected. Both of these signatures of local superconductivity appear

[22] The largest are about 250 sites [350,351] using the density matrix renormalization group method (DMRG) and up to approximately 800 sites in Green function Monte Carlo simulations [352].

as distinct and universal features in the physics of doped ladders. Nevertheless, they tend to diminish, in some cases very rapidly, with the lateral extent of the ladder, thus strongly suggesting that such structures are *essential* for the attainment of high temperature superconductivity. In addition we shall demonstrate the tendency of these systems to develop charge density wave correlations upon doping; it is natural to imagine that as the transverse width of the ladder tends to infinity, these density wave correlations will evolve into true two-dimensional stripe order.

Spin Gap and Pairing Correlations

Hubbard Chains:
The purely one-dimensional Hubbard model can be solved exactly using Bethe ansatz [369,370] and thus may seem out of place in this section. However, like other models in this section, it is a lattice fermion model. In analyzing it, we will encounter many of the concepts that will figure prominently in our discussion of the other models treated here, most notably the importance of intermediate scales. Anyway, in many cases, the Bethe ansatz equations themselves must be solved numerically, so we can view this as simply a more efficient numerical algorithm which permits us to study larger systems (up to 1000 sites [14] or more).

The Hubbard Hamiltonian is

$$H_U = -t \sum_{\langle i,j \rangle, s} (c^\dagger_{i,s} c_{j,s} + h.c.) + U \sum_i n_{i,\uparrow} n_{i,\downarrow} \ , \tag{21.113}$$

where $\langle\ \rangle$ denotes nearest neighbors on a ring with an even number of sites N and $N + Q$ electrons. We define $E(Q, S)$ to be the lowest lying energy eigenvalue with total spin S and "charge" Q. Whenever the ground state is a spin singlet we can define the spin gap Δ_s as the energy gap to the lowest $S = 1$ excitation

$$\Delta_s(Q) = E(Q, 1) - E(Q, 0) \ . \tag{21.114}$$

The pair binding energy is defined as

$$E_{pb}(Q) = 2E(Q + 1) - E(Q + 2) - E(Q) \ , \tag{21.115}$$

where $E(Q)$ has been minimized with respect to S. A positive pair binding energy means that given $2(N + Q + 1)$ electrons and two clusters, it is energetically more favorable to place $N + Q + 2$ electrons on one cluster and $N + Q$ on the other than it is to put $N + Q + 1$ electrons on each cluster. In this sense, a positive E_{pb} signifies an effective *attraction* between electrons. The exact particle-hole symmetry of the Hubbard model on a bipartite lattice implies that electron doping $Q > 0$ is equivalent to hole doping $Q < 0$.

Intermediate scales play an important role.
Figure 21.26 displays the pair binding energy for electrons added to $Q = 0$ rings. The role of intermediate scales is apparent: E_{pb} vanishes for large N and is maximal at an intermediate value of N. (The fact that pair binding occurs for $N = 4n$ rings but not when $N = 4n + 2$ is readily understood from low order perturbation theory in U/t [14]). Moreover, the spin gap Δ_s reaches a maximum at intermediate interaction strength, and then decreases for large values of U, as expected from its proportionality to the exchange constant $J = 4t^2/U$ in this limit. The pair binding energy E_{pb} follows suit with a similar dependence, as seen from Fig. 21.27.

We have already seen the intimate relation between the spin gap and the superconducting susceptibility in the context of quasi-one-dimensional superconductors (see Sect. 21.5). Further understanding of the relation between pair binding and the spin gap can be gained by using bosonization to study the Hubbard model in the large N limit [14, 370]. The result for $N = 4n \gg 1$ is

$$\Delta_s = \frac{v_s}{N} \left[B_1 \ln^{1/2}(N) + B_2 \right] + \ldots \tag{21.116}$$

$$E_{pb} = \Delta_s + B_3 \frac{v_s}{N} - \frac{B_4}{N^2} \left[\frac{{v_c}^2}{\Delta_c} \right] + \ldots \tag{21.117}$$

The spin gap and pairing are related.
Here, v_s and v_c are the spin and charge velocities, respectively (in units in which the lattice constant is unity), and Δ_c is the charge gap in the $N \to \infty$ limit. The constants, B_j, are numbers of order unity. The important lesson of this analysis is that pair binding is closely related to the phenomenon of spin gap formation. Indeed, for large N, $E_{pb} \approx \Delta_s$.

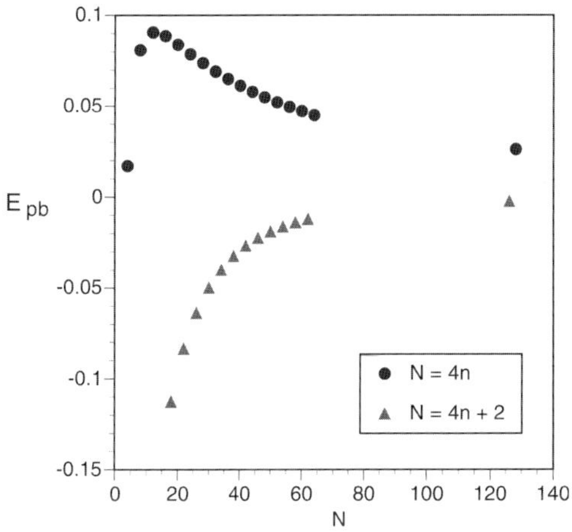

Fig. 21.26. Pair binding energy, E_{pb}, of $N = 4n$ and $N = 4n + 2$ site Hubbard rings with $t = 1$ and $U = 4$. (From Chakravarty and Kivelson [14])

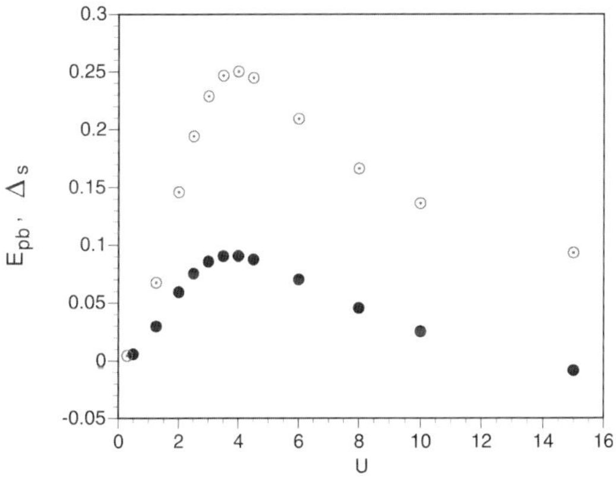

Fig. 21.27. Pair binding energy, E_{pb} (*solid symbols*), and spin gap, Δ_s (*open symbols*), of a 12 site $Q = 0$ Hubbard ring as a function of U in units of $t = 1$. (From Chakravarty and Kivelson [14])

Hubbard and $t - J$ Ladders:

In the thermodynamic limit, where the number of sites $N \to \infty$, Hubbard chains, and their strong coupling descendants the $t - J$ chains, have no spin gap and a small superconducting susceptibility, irrespective of the doping level. In contrast, ladder systems can exhibit both a spin gap and a strong tendency towards superconducting order even in the thermodynamic limit. While these systems are infinite in extent, the mesoscopic physics comes in through the finiteness of the transverse dimension.

In the large U limit and at half filling (one electron per site) the Hubbard ladder is equivalent to the spin-1/2 Heisenberg ladder

$$H_J = J \sum_{\langle i,j \rangle} \mathbf{S}_i \cdot \mathbf{S}_j \,, \qquad (21.118)$$

where $\mathbf{S}_i$ is a spin 1/2 operator, $J = 4t^2/U \ll t$ is the antiferromagnetic exchange interaction, and $\langle i, j \rangle$ now signifies nearest neighbor sites of spacing a on the ladder. As discussed in Sect. 21.10, there is a marked difference between the behavior of ladders with even and odd numbers of chains or "legs".

The number of legs matters!

While even leg ladders are spin gapped with exponentially decaying spin-spin correlations, odd leg ladders are gapless and exhibit power law falloff of these correlations (up to logarithmic corrections). This difference is clearly demonstrated in Fig. 21.28. The spin gaps for the first few even leg ladders are known numerically [334,371].

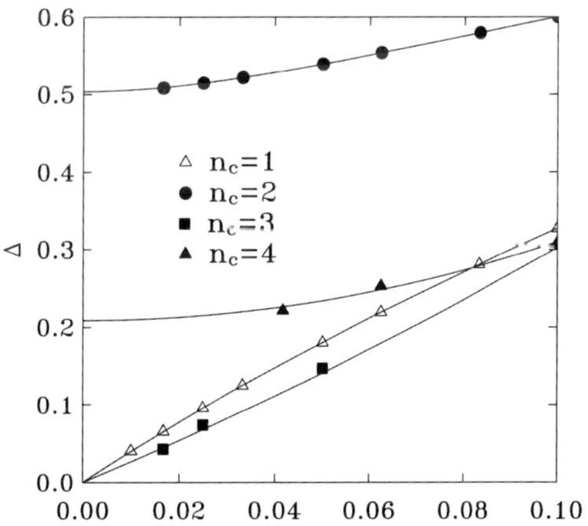

Fig. 21.28. Spin gaps as a function of system size L for open $L \times n_c$ Heisenberg ladders. (From White et al. [334])

For the two, four, and six leg ladders, $\Delta_s = 0.51(1)J$, $\Delta_s = 0.17(1)J$, and $\Delta_s = 0.05(1)J$, respectively. This gap appears to vanish exponentially with the width W of the system, in accordance with the theoretical estimate [333] $\Delta_s \sim 3.35J \exp[-0.682(W/a)]$, as discussed in Sect. 21.10.

Widening the ladder closes the gap.
Although odd leg Heisenberg ladders are gapless, they are characterized by an energy scale which has the same functional dependence on W as Δ_s. Below this energy, the excitations are gapless spinons analogous to those in the Heisenberg chain [333], while above it they are weakly interacting spin waves. Based on our experience with the Hubbard rings we expect that spin gap formation is related to superconductivity. As we shall see below this is indeed the case once the ladders are doped with holes. On the face of it, this implies that only rather narrow ladders are good candidates for the mesoscopic building blocks of a high temperature superconductors.

When the Hubbard ladder is doped with holes away from half filling, its strong coupling description is modified from the Heisenberg model (Eq. (21.118)) to the $t-J$ model

$$H_{t-J} = -t \sum_{\langle i,j\rangle,s} (c^\dagger_{i,s} c_{j,s} + h.c.) + J \sum_{\langle i,j\rangle} \left[\mathbf{S}_i \cdot \mathbf{S}_j - \frac{1}{4} n_i n_j \right] , \quad (21.119)$$

which is defined with the supplementary constraint of no doubly occupied sites. This is the version which has been most extensively studied numerically. Unless otherwise stated, we will quote results for representative values of J/t in the range $J/t = 0.35$ to 0.5.

Numerical studies of the two leg Hubbard model [372, 373] have demonstrated that doping tends to decrease the spin gap *continuously* from its value in the undoped system but it persists down to at least an average filling of $\langle n\rangle = 0.75$, as can be seen from the inset in Fig. 21.29. A similar behavior is observed in the $t-J$ ladder although the precise evolution of the spin gap upon doping depends on details of the model [353].

Holes like to d-pair.
Pairs of holes in two leg Hubbard or $t-J$ ladders form bound pairs as can be seen both from the fact that the pair binding energy is positive, and from the fact that positional correlations between holes are indicative of a bound state. The pairs have a predominant $d_{x^2-y^2}$ symmetry as is revealed by the relative minus sign between the ground state to ground state amplitudes for adding a singlet pair on neighboring sites along and across the legs [191, 372]. It seems that the dom-

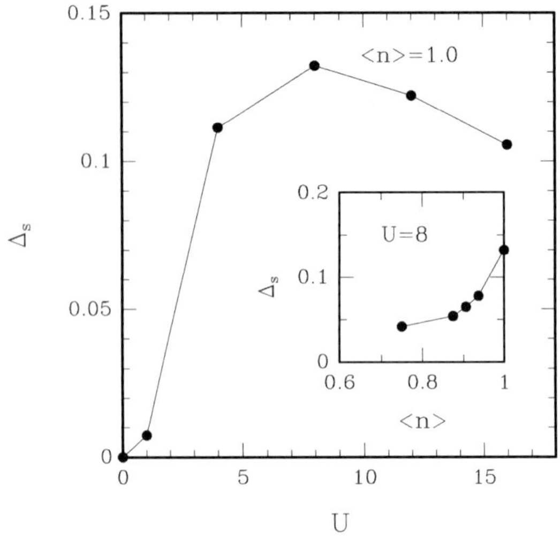

Fig. 21.29. The spin gap as a function of U for a half filled 2×32 Hubbard ladder. The *inset* shows Δ_s as a function of filling $\langle n\rangle$ for $U = 8$. Energies are measured in units of $t = 1$. (From Noack et al. [373])

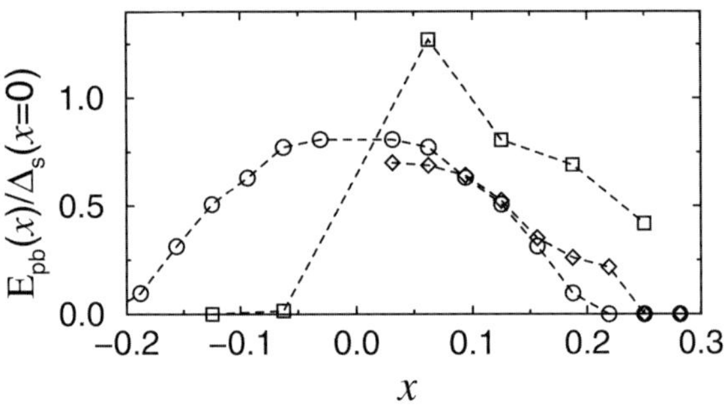

Fig. 21.30. The ratio of the pair binding energy to the undoped spin gap as a function of hole doping $x = 1 - \langle n\rangle$. The *diamonds* are for a 32×2 $t-J$ ladder with $J/t = 0.3$. The *circles* are for a one-band 32×2 Hubbard ladder with $U/t = 12$. The squares are for a three-band Hubbard model of a two leg Cu-O ladder, i.e. a ladder made of Cu sites where nearest neighbor sites are connected by a link containing an O atom. Here $U_d/t_{pd} = 8$, where U_d is the on-site Cu Coulomb interaction and t_{pd} is the hopping matrix element between the O and Cu sites. The energy difference between the O and Cu sites is $(\epsilon_p - \epsilon_d)/t_{pd} = 2$, and the calculation is done on a 16×2 ladder. (From Jeckelmann et al. [377])

inance of the $d_{x^2-y^2}$ channel is universally shared by all models over the entire range of doping that has been studied. (See Sect. 21.10.3 for a discussion of this phenomenon in the 2×2 plaquette.)

The doping dependence of the pair binding energy roughly follows the spin gap in various versions of the two leg ladder as shown in Fig. 21.30. The correlation function $D(l)$ of the pair field

$$\Delta_i^\dagger = (c_{i1\uparrow}^\dagger c_{i2\downarrow}^\dagger - c_{i1\downarrow}^\dagger c_{i2\uparrow}^\dagger) , \qquad (21.120)$$

exhibits behavior consistent with a power law decay [191,372,374–376]

$$D(l) = \langle \Delta_{i+l} \Delta_i^\dagger \rangle \sim l^{-\theta} . \qquad (21.121)$$

There exists less data concerning its doping dependence, but from the relevant studies [191,372] we can conclude that the pair correlations increase from the undoped system to a maximum at $x \sim 0.0625$ and then decrease when more holes are added to the system.

Details and their importance

Both the spin gap and the pairing correlations in doped Hubbard and $t-J$ ladders can be appreciably enhanced by slight generalizations of the models. For example, the exponent θ in Eq. (21.121), which depends on the coupling strengths U/t or J/t and the doping level x, is also sensitive to the ratio of the hopping amplitudes between neighboring sites on a rung and within a chain $t_\perp/t$. By varying this parameter, the exponent θ can be tuned over the range $0.9 \le \theta \le 2.1$. In particular, for $x = 0.0625$ and intermediate values of the (repulsive) interaction $5 \le U/t \le 15$, it can be made smaller than 1 [191]; see Fig. 21.14. This is significant since, as we saw in Sect. 21.5.1, whenever $\theta < 1$ the superconducting susceptibility is the most divergent among the various susceptibilities of the ladder. Adding a nearest neighbor exchange coupling, J, to H_U also leads to stronger superconducting signatures owing to an increase in the pair mobility and binding energy [378].

Another lesson in humility

The moral here is that details are important as far as they reveal the nonuniversal properties of the Hamiltonians that we study, and indicate relevant directions in model space. It should also imprint on us a sense of humility when attempting to fit real world data with such theoretical results.

Odd and want a gap? –Dope!

We already noted that, in contrast to the two leg ladder, the three leg system does not possess a spin gap at half filling. This situation persists up to hole doping of about $x = 1 - \langle n \rangle = 0.05$, as can be seen in Fig. 21.31.[23] However, with moderate doping a spin gap is formed which reaches a maximum value at a doping level of $x = 0.125$. For the system shown here, with $J/t = 0.35$, the gap is only 20 percent smaller than that of the undoped two leg Heisenberg ladder. Upon further doping, the spin gap decreases and possibly vanishes as x gets to be 0.2 or larger.

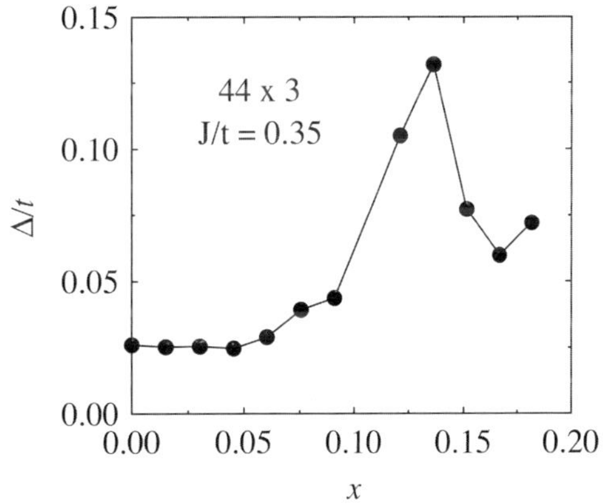

Fig. 21.31. Spin gap for a 44×3 ladder with open boundary conditions and $J/t = 0.35$ as a function of doping. (From White and Scalapino [379])

The same goes for pairing.

The establishment of a spin gap is concurrent with the onset of pairing correlations in the system. While two holes introduced into a long, half filled three chain ladder do not bind [338], indications of pairing emerge as soon as the spin gap builds up [379,380]. As an example, Fig. 21.32 plots the pair field–pair

[23] The nonvanishing spin gap in this region is presumably a finite size effect; see Fig. 21.28.

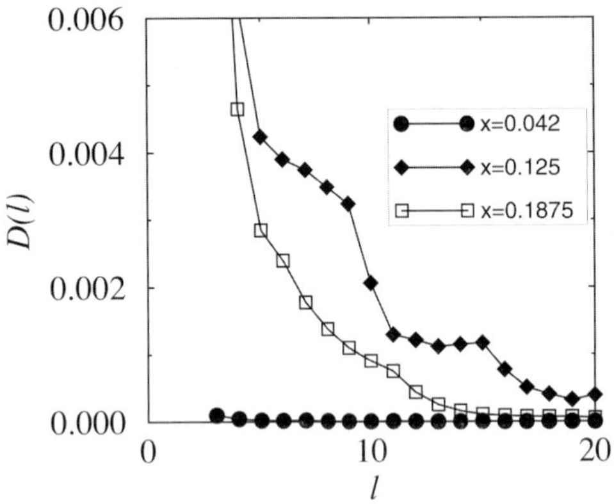

Fig. 21.32. The $d_{x^2-y^2}$ pair field correlations $D(l)$ for three different densities, calculated on 32×3 ($x = 0.1875$) and 48×3 ($x = 0.042$, $x = 0.125$) open $t-J$ ladders with $J/t = 0.35$. (From White and Scalapino [379])

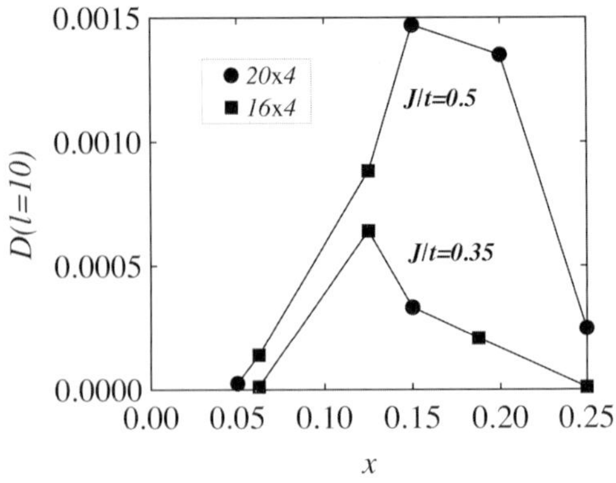

Fig. 21.33. The $d_{x^2-y^2}$ pair field correlation $D(l)$ at a separation of $l = 10$ rungs as a function of doping x, for 20×4 and 16×4 open ladders with $J/t = 0.35$ and 0.5. (From White and Scalapino [381])

field correlation function of Eq. (21.121) for various values of the hole doping, defined with

$$\Delta_i^\dagger = c_{i,2\uparrow}^\dagger(c_{i+1,2\downarrow}^\dagger + c_{i-1,2\downarrow}^\dagger - c_{i,1\downarrow}^\dagger - c_{i,3\downarrow}^\dagger) - (\uparrow\leftrightarrow\downarrow) \tag{21.122}$$

which creates a $d_{x^2-y^2}$ pair around the ith site of the middle leg (the leg index runs from 1 to 3).[24] In the regime of low doping $x \le 0.05$, the pair field correlations are negligible. However, clear pair field correlations are present at $x = 0.125$, where they are comparable to those in a two leg ladder under similar conditions. The pair field correlations are less strong at $x = 0.1875$; they follow an approximate power law decay as a function of the distance [375,379]. (The oscillations in $D(l)$ are produced by the open boundary conditions used in this calculation.) This behavior can be understood from strong coupling bosonization considerations [20] in which the two even modes (with respect to reflection about the center leg) form a spin gapped two leg ladder and for small doping the holes enter the odd mode giving rise to a gapless one-dimensional electron gas. As the doping increases, pair hopping between the two subsystems may induce a gap in the gapless channel via the spin gap proximity effect [20].

Increasing the number of legs from three to four leads to behavior similar to that exhibited by the two leg ladder. The system is spin gapped and two holes in a half filled four leg ladder tend to bind. The pair exhibits features common to all pairs in an antiferromagnetic environment, including a d-wave-like symmetry [338]. Further similarity with the two leg ladder is seen in the d-wave pair field correlations $D(l)$. Figure 21.33 shows $D(l = 10)$ for a $t-J$ four leg ladder as a function of doping (extended s-wave correlations are much smaller in magnitude). The pairing correlations for $J/t = 0.5$ increase with doping, reaching a maximum between $x = 0.15$ and $x = 0.2$, and then decrease.

Four legs are good; two legs are better.
The magnitude of the correlations near the maximum is similar to that of a two leg Hubbard ladder with $U = 8t$ (corresponding to $J \sim 4t^2/U = 0.5$) with the same doping, but smaller than the maximum in the two leg ladder which occurs at smaller doping [191,372]. For $J/t = 0.35$ the peak is reduced in magnitude and occurs at lower doping. The behavior of $D(l)$ near the maximum is consistent with power law decay for short to moderate distances but seems to fall more rapidly at long distances (perhaps even exponentially [382]).

Lastly, we present in Fig. 21.34 the response of a few ladder systems to a proximity pairing field

$$H_1 = d\sum_i(c_{i,\uparrow}^\dagger c_{i+\hat{y},\downarrow}^\dagger - c_{i,\downarrow}^\dagger c_{i+\hat{y},\uparrow}^\dagger + h.c.)\,, \tag{21.123}$$

[24] There also exists a small s-wave component in the pair field due to the one-dimensional nature of the cluster.

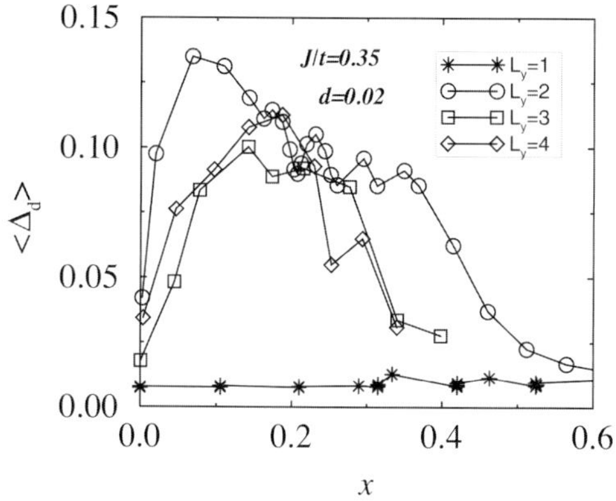

Fig. 21.34. The $d_{x^2-y^2}$ pairing response to a proximity pair field operator as a function of doping for a single chain and two, three, and four leg ladders. For the single chain, near neighbor pairing is measured. (From White and Scalapino [379])

which adds and destroys a singlet electron pair along the ladder. The response is given by the average $d_{x^2-y^2}$ pair field

$$\langle \Delta_d \rangle = \frac{1}{N} \sum_i \langle \Delta_i \rangle \,, \tag{21.124}$$

with Δ_i defined in Eq. (21.120). We see that the pair field response tends to decrease somewhat with the width of the system but is overall similar for the two, three and four leg ladders. We suspect it gets rapidly smaller for wider ladders.

Phase Separation and Stripe Formation in Ladders

We now address the issue of whether there is any apparent tendency to form charge density and/or spin density wave order in ladder systems, and whether there is a tendency of the doped holes to phase separate. Since incommensurate density-wave long-range order, like superconducting order, is destroyed by quantum fluctuations in one dimension, we will again be looking primarily at local correlations, rather than actual ordered states. Of course, we have in mind that local correlations and enhanced susceptibilities in a one-dimensional context can be interpreted as indications that in two dimensions true superconductivity, stripe order, or phase separation may occur.

Phase Separation:
Phase separation was first found in the one-dimensional chain [383, 384] and subsequently in the two leg ladder [385–387]. As a rule, the phase separation line has been determined by calculating the coupling J at which the compressibility diverges. (See, however, [352].) This is in principle an incorrect criterion. The compressibility only diverges at the consolute point. Thermodynamically appropriate criteria for identifying regimes of phase separation from finite size studies include the Maxwell construction (discussed explicitly in Sect. 21.12, below), and measurements of the surface tension in the presence of boundary conditions that force phase coexistence. The divergent compressibility is most directly related to the spinodal line, which is not even strictly well defined beyond mean field theory. Thus, while in many cases the phase diagrams obtained in this way may be qualitatively correct, they are always subject to some uncertainty.

More recently Rommer, White, and Scalapino [388] have used DMRG methods to extend the study to ladders of up to six legs. Since these calculations are carried out with open boundary conditions, which break the translational symmetry of the system, they have used as their criterion the appearance of an inhomogeneous state with a hole rich region at one edge of the ladder and hole free regions near the other, which is a thermodynamically correct criterion for phase separation. However, where the hole rich phase has relatively low hole density, and in all cases for the six leg ladder, they were forced to use a different criterion which is not thermodynamic in character, but is at least intuitively appealing. From earlier studies (which we discuss below) it appears that the "uniform density" phase, which replaces the phase separated state for J/t less than the critical value for phase separation, is a "striped" state, in which the holes congregate into puddles (identified as stripes) with fixed number of holes, but with the density of stripes determined by the mean hole density on the ladder. With this in mind, Rommer et al. computed the interaction energy between two stripes, and estimated the phase separation boundary as the point at which this interaction turns from repulsive to attractive. The results, summarized in

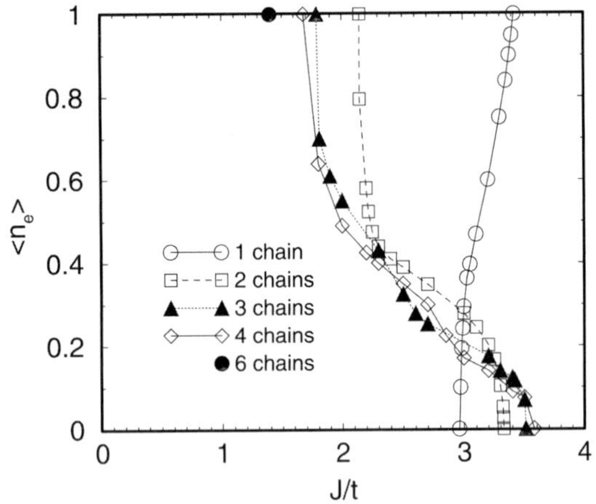

Fig. 21.35. Boundary to phase separated region in $t-J$ ladders. Open boundary conditions were used in both the leg and rung directions except for the six leg ladder where periodic boundary conditions were imposed along the rung. Phase separation is realized to the right of the *curves*. $\langle n_e \rangle$ is the total electron density in the system. (From Rommer et al. [388])

Fig. 21.35, agree with the thermodynamically determined phase boundary where they can be compared.

Ladders phase separate for large enough J/t.
For large enough values of J/t, both the single chain and the ladders are fully phase separated into a Heisenberg phase ($\langle n_e \rangle = 1$) and an empty phase ($\langle n_e \rangle = 0$). However, the evolution of this state as J/t is reduced is apparently different for the two cases. For the chain, the Heisenberg phase is destroyed first by holes that diffuse into it; this presumably reflects the fact that hole motion is not significantly frustrated in the single chain system. In the ladders, on the other hand, the empty phase is the one that becomes unstable due to the sublimation of electron pairs from the Heisenberg region. This difference is evident in Fig. 21.35 where the phase separation boundary occurs first at high electron density in the chain and high hole density in the ladders. It is also clear from looking at this figure that the value of J/t at which phase separation first occurs for small electron densities is hardly sensitive to the width of the ladder. However, as more electrons are added to the system (removing holes), phase separation is realized for smaller values of J/t in wider ladders. Whether this is an indication that phase separation takes place at arbitrarily small J/t for small enough hole densities in the two-dimensional system is currently under debate, as we discuss in Sect. 21.11.2.

"Stripes" in Ladders: Stripes appear at smaller J/t.
At intermediate values of J/t, and not too close to half filling, the doped holes tend to segregate into puddles which straddle the ladders, as is apparent from the spatial modulation of the mean charge density along the ladder. Intuitively, we can think of this state as consisting of an array of stripes with a spacing which is determined by the doped hole density. From this perspective, the total number of doped holes associated with each puddle, $N_{\text{puddle}} = \varrho L$, is interpreted as arising from a stripe with a mean linear density of holes, ϱ, times the length of the stripe, L.[25] (L is also the width of the ladder.) In the thermodynamic limit, long wavelength quantum fluctuations of the stripe array would presumably result in a uniform charge density, but the ladder ends, even in the longest systems studied to date, are a sufficiently strong perturbation that they pin the stripe array [389]. In two and three leg ladders, the observed stripes apparently always have $\varrho = 1$. For the four leg ladder, typically $\varrho = 1$, but under appropriate circumstances (especially for $x = 1/8$), $\varrho = 1/2$ stripes are observed. In six and eight leg ladders, the charge density oscillations are particularly strong, and correspond to stripes with $\varrho = 2/3$ and $1/2$, respectively. Various arguments have been presented to identify certain of these stripe arrays as being "vertical" (i.e. preferentially oriented along the rungs of the ladder) or "diagonal" (i.e. preferentially oriented at 45° to the rung), but these arguments, while intuitively appealing, do not have a rigorous basis.

We will return to the results on the wider ladders, below, where we discuss attempts to extrapolate these results to two dimensions.

[25] For instance, on a long, N site, 4 leg ladder with $4n$ holes, where $n \ll N$, one typically observes n or $2n$ distinct peaks in the rung-averaged charge density, which is then interpreted as indicating a stripe array with $\varrho = 1$ or $\varrho = 1/2$, respectively.

21.11.2 Properties of the Two-Dimensional $t-J$ and Hubbard Models

It is a subtle affair to draw conclusions about the properties of the two-dimensional Hubbard and $t-J$ models from numerical studies of finite systems. The present numerical capabilities do not generally permit a systematic finite size scaling analysis. As a result, extrapolating results from small clusters with periodic boundary conditions, typically used when utilizing Monte Carlo or Lanczos techniques, or from strips with open boundary conditions as used in DMRG studies, is susceptible to criticism [355,359]. It comes as no surprise then that several key issues concerning the ground state properties of the two-dimensional models are under dispute. In the following we present a brief account of some of the conflicting results and views. However, at least two things do not seem to be in dispute: 1) there is a strong tendency for doped holes in an antiferromagnet to clump in order relieve the frustration of hole motion [390], and 2) where it occurs, hole pairing has a $d_{x^2-y^2}$ character. Thus, in one way or another, the local correlations that lead to stripe formation and d-wave superconductivity are clearly present in $t-J$-like models!

Phase Separation and Stripe Formation

Everybody agrees on the phase separation boundary for $x \sim 1$.
The question of phase separation in the $t-J$ model has been addressed by a number of studies. Most of them agree on the behavior in the regime of very low electron density $n_e = 1 - x \ll 1$. The critical J/t value for phase separation at vanishingly small n_e was calculated very accurately by Hellberg and Manousakis [391] and was found to be $J/t = 3.4367$. However, there are conflicting results for systems close to half filling ($n_e \sim 1$) and with small $t-J$. This is the most delicate region where high numerical accuracy is hard to obtain. Consequently, there is no agreement on whether the two-dimensional $t-J$ model phase separates for all values of J/t at sufficiently low hole doping x.

The situation for $x \sim 0$ is murkier, but...
Emery et al. [392,393] presented a variational argument (recently extended and substantially improved by Eisenberg et al. [394]) that for $J/t \ll 1$ and for x less than a critical concentration, $x_c \sim \sqrt{J/t}$, phase separation occurs between a hole free antiferromagnetic and a metallic ferromagnetic state. Since for large J/t there is clearly phase separation for all x, they proposed that for sufficiently small x, phase separation is likely to occur for all J/t. To test this, they computed the ground state energy by exact diagonalization of 4×4 doped $t-J$ clusters. If taken at face value and interpreted via a Maxwell construction, these results imply that for any $x < 1/8$, phase separation occurs at least for all $J/t > 0.2$. Hellberg and Manousakis [352,361] calculated the ground state energy on larger clusters of up to 28×28 sites using Green function Monte Carlo methods. By implementing a Maxwell construction, they reached the similar conclusion that the $t-J$ model phase separates for all values of J/t in the low hole doping regime.

On the other hand, Putikka et al. [395] studied this problem using a high temperature series expansion extrapolated to $T = 0$ and concluded that phase separation only occurs above a line extending from $J/t = 3.8$ at zero filling to $J/t = 1.2$ at half filling. In other words, they concluded that there is no phase separation for any x so long as $J/t < 1.2$. Exact diagonalization results for the compressibility and the binding energy of n-hole clusters in systems of up to 26 sites by Poilblanc [396] were interpreted as suggesting that the ground state is phase separated close to half filling only if $J/t > 1$. Quantum Monte Carlo simulations of up to 242 sites using stochastic reconfiguration by Calandra et al. [397] have found a phase separation instability for $J/t \sim 0.5$ at similar doping levels, but no phase separation for $J/t < 0.5$, while earlier variational Monte Carlo calculations [398] reported a critical value of $J/t = 1.5$. Using Lanczos techniques to calculate the ground state energy on lattices of up to 122 sites, Shih et al. [399,400] estimate the lower critical value for phase separation as $J/t = 0.3 - 0.5$, a somewhat lower bound than previously found using similar numerical methods [401]. Finally, DMRG calculations on wide ladders with open boundary conditions in one direction by White and Scalapino [350,351] found striped ground states for $J/t = 0.35$ and $0 < x < 0.3$, but no indication of phase separation.

... it seems that the model is either phase separated, or very close to it.
For comparison, we have gathered a few of the results mentioned above in Fig. 21.36. The scatter of the data at the upper left corner of the $n_e - J/t$ plane is a reflection of the near linearity of the ground state energy as a function of doping in this region [359]. High numerical accuracy is needed in order to establish a true linear behavior which would be indicative of phase separation. While there is currently no definitive answer concerning phase separation at small doping, it seems clear that in this region the two-dimensional $t-J$ model is in delicate balance, either in or close to a phase separation instability.

The nature of the ground state for moderately small J/t beyond any phase separated regime is also

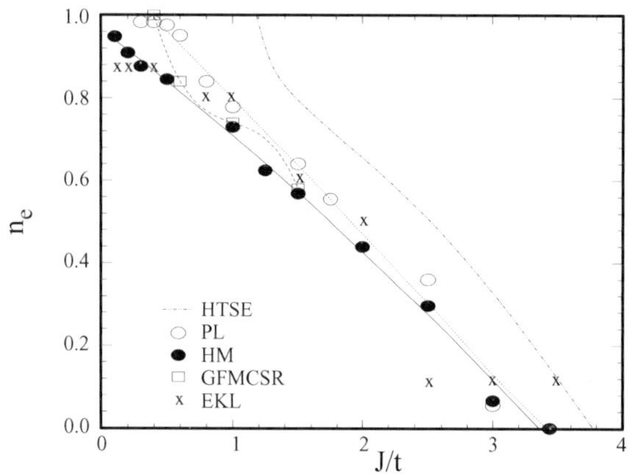

Fig. 21.36. Phase separation boundary of the two-dimensional *t-J* model according to various numerical studies. The *dashed-dotted line* represents the high temperature series expansion results by Puttika et al. [395]. Also shown are results from calculations using the Power–Lanczos method by Shin et al. [400] (*open circles*), Greens function Monte Carlo simulations by Hellberg and Manousakis [361] (*closed circles*) and by Calandra et al. [397] (*open squares*), and exact diagonalization of 4 × 4 clusters by Emery et al. [392] (x's). (Adapted from Shin et al. [400])

in dispute. While DMRG calculations on fat ladders [350,351] find striped ground states for $J/t = 0.35$ and $x = 1/8$, Monte Carlo simulations on a torus [355] exhibit stripes only as excited states. Whether this discrepancy is due to finite size effects or the type of boundary conditions used is still not settled. (The fixed node Monte Carlo studies of Becca et al. [402] likewise conclude that stripes do not occur in the ground state, although they can be induced by the addition of rather modest anisotropy into the $t-J$ model, suggesting that they are at least energetically competitive.)

Stripes are important low energy configurations of the $t-J$ model.
While these conflicting conclusions may be difficult to resolve, it seems inescapable to us that stripes are important low energy configurations of the two-dimensional $t-J$ model for small doping and moderately small J/t.

Typically stripes are quarter-filled antiphase domain walls.
The most reliable results concerning the internal structure of the stripes themselves come from studies of fat $t-J$ ladders, where stripes are certainly a prominent part of the electronic structure. In all studies of ladders, the doped holes aggregate into "stripes" which are oriented either perpendicular or parallel to the extended direction of the ladder, depending on boundary conditions. In many cases the spin correlations in the hole poor regions between stripes locally resemble those in the undoped antiferromagnet but suffer a π-phase shift across the hole rich stripe. This magnetic structure is vividly apparent in studies for which the low energy orientational fluctuations of the spins are suppressed by the application of staggered magnetic fields on certain boundary sites of the ladders—then, these magnetic correlations are directly seen in the expectation values of the spins [403]. However, such findings are not universal: in the case of the four leg ladder, with stripes along the ladder rungs, Arrigoni et al. [358] recently showed that in long systems (up to 4 × 27), these antiphase magnetic correlations are weak or nonexistent, despite strong evidence of charge stripe

correlations. Ladder studies have also demonstrated that stripes tend to favor a linear charge density of $\rho = 1/2$ along each stripe.[26] Specifically, by applying boundary conditions which force a single stripe to lie along the long axis of the ladder, White et al. [351] were able to study the energy of a stripe as a function of ρ. They found an energy which is apparently a smooth function of ρ (i.e. with no evidence of a nonanalyticity which would lock ρ to a specific value), but with a pronounced minimum at $\rho = 1/2$. Moreover, with boundary conditions favoring stripes perpendicular to the ladder axis, they found that for $x \leq 1/8$ stripes tend to form with $\rho = 1/2$ so that the spacing between neighboring stripes is approximately $1/2x$, while at larger x, a first order transition occurs to "empty domain walls" with $\rho = 1$ and an inter-stripe spacing of $1/x$. In the region $0.125 < x < 0.17$ the two types of stripes can coexist.

It is worth noting that the original indications of stripe order came from Hartree–Fock treatments [405–408]. Hartree–Fock stripes are primarily spin textures. In comparison to the DMRG results on ladders, they correspond to "empty" ($\rho = 1$) antiphase (π-phase shifted) domain walls, and so are insulating and overemphasize the spin component of the stripe order, but otherwise capture much of the physics of stripe formation remarkably accurately.

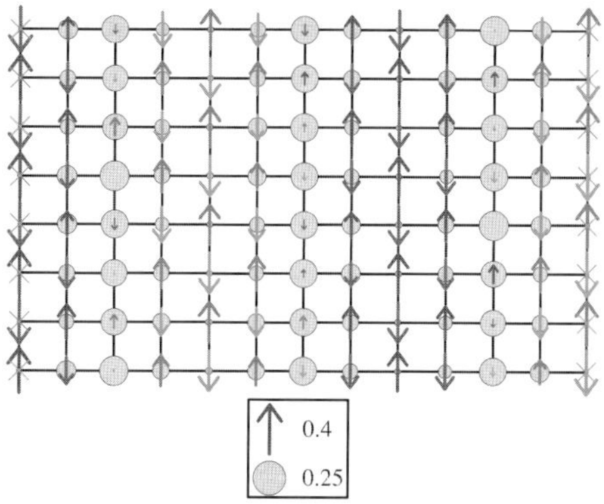

Fig. 21.37. Hole density and spin moments on a 13×8 cylinder with 12 holes, $J/t = 0.35$, periodic boundary conditions along the y direction and π-shifted staggered magnetic field of magnitude $0.1t$ on the open edges. The diameter of the *circles* is proportional to the hole density $1 - \langle n_i \rangle$ and the length of the *arrows* is proportional to $\langle S_i^z \rangle$. (From White and Scalapino [403])

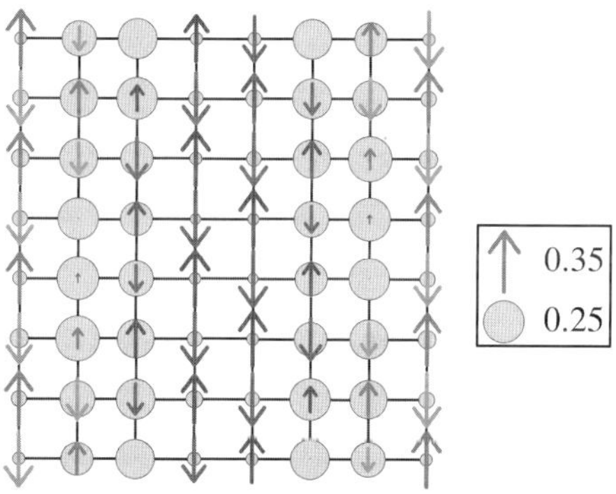

Fig. 21.38. Hole density and spin moments on a central section of a 16×8 cylinder with 16 holes, $J/t = 0.35$, with periodic boundary conditions along the y direction and staggered magnetic field of magnitude $0.1t$ on the open edges. The notation is similar to Fig. 21.37. (From White and Scalapino [403])

Stripes can be site- or bond-centered.
Further insight into the physics that generates the domain walls can be gained by looking more closely at their hole density and spin structures. Both site-centered and bond-centered stripes are observed. They are close in energy and each type can be stabilized by adjusting the boundary conditions [350]. Figure 21.37 depicts three site-centered stripes in a 13×8 system with 12 holes, periodic boundary conditions along the y direction and a π-shifted staggered magnetic field on the open ends of magnitude $0.1t$. These stripes are quarter-filled antiphase domain walls. Figure 21.38 shows a central section of a 16×8 cluster containing two bond-centered domain walls. This system is similar to the one considered above except that the magnetic field on the open ends is not π-shifted. Like their site-centered counterparts, the bond-centered stripes are antiphase domain walls, but with one hole per two domain wall unit cells.

[26] At about the same time, Nayak and Wilczek [404] presented an interesting analytic argument which leads to the same bottom line.

The topological character of spin stripes can be inferred from local considerations.
The π-phase shift in the exchange field across the stripe can probably be traced, in both the bond- and site-centered cases, to a gain in the transverse kinetic energy of the holes. To demonstrate this point consider a pair of holes in a 2×2 $t-J$ plaquette, as was done in Sect. 21.10.3. One can simulate the effect of the exchange field running on both sides of the plaquette through a mean field h which couples to the spins on the square [403]. For the in-phase domain wall such a coupling introduces a perturbation $h(S_1^z - S_2^z - S_3^z + S_4^z)$ which, to lowest order in h, lowers the ground state energy by $-h^2/\sqrt{J^2+32t^2}$. For the π-shifted stripe the perturbation is $h(S_1^z + S_2^z - S_3^z - S_4^z)$ with a gain of $-4h^2/\sqrt{J^2+32t^2}$ in energy, thereby being more advantageous for the pair. Indeed, this physics has been confirmed by several serious studies, which combine analytic and numerical work, by Zachar [409], Liu and Fradkin [410], and Chernyshev et al. [411] (see also [445]). These studies indicate that there is a transition from a tendency for in-phase magnetic order across a stripe for small ρ, when the direct magnetic interactions are dominant, to antiphase magnetic order for $\rho > 0.3$, when the transverse hole kinetic energy is dominant.

In contrast to the $t-J$ model there have been relatively few numerical studies of large two-dimensional Hubbard model clusters. Monte Carlo simulations [360,412,413] of systems of up to 16×16 sites have reached somewhat conflicting conclusions. While variational "fixed node" calculations by Cosentini et al. [412] are suggestive of phase separation at small x, the work of Becca et al. [413] claims a spatially homogeneous ground state up to values of $U/t = 10$. Such findings are in conflict with the very latest DMRG studies of 6-leg Hubbard ladders.

Stripes appear to be a robust feature of fat Hubbard ladders at strong coupling
White and Scalapino [414] have shown that a narrow stripe appears in the ground state of a 7×6-site cluster with average hole density $x \approx 0.095$ for $U/t \geq 6$. For weaker couplings the hole and spin densities show structures which were interpreted as a broad stripe. In a recent tour de force DMRG study by Hager et al. [415] 6-leg Hubbard ladders with $x \approx 0.095$ and lengths of up to 28 sites were studied with careful analysis of numerical errors and finite size scaling. This work gives strong evidence that stripes exist and are robust in the ground state of the 6-leg Hubbard ladder for strong coupling ($U/t = 12$), while the structures found at weaker coupling ($U/t = 3$) are probably an artifact of the DMRG approach. The properties found in the studies of stripes in the $t-J$ model, such as them being anti-phase domain walls in the antiferromagnetic background and the near-degeneracy in energy between site-centered and bond-centered stripes, have been demonstrated in the context of the Hubbard ladder as well.

Superconductivity and Stripes

There is no evidence for superconductivity in the Hubbard model.
There is no unambiguous evidence for superconductivity in the Hubbard model. The original finite temperature Monte Carlo simulations on small periodic clusters with $U/t = 4$ and $x = 0.15$ [360,416] found only short range pair-pair correlations. The same conclusion was reached by a later zero temperature constrained path Monte Carlo calculation [417]. The above mentioned DMRG studies of the 6-leg Hubbard ladder did not include a calculation of the pairing correlations in this system. However, if the robust static charge-density-wave correlations found in the ladder persist in the two-dimensional model it is likely that they work against the establishment of long-range superconducting order in the plane.

There are conflicting results concerning the question of superconductivity in the $t-J$ model.

In the unphysical region of large J/t, solid conclusions can be reached: Emery et al. [392] have shown that proximate to the phase separation boundary at $J/t \leq 3.8$, the hole rich phase (which is actually a dilute electron phase with $x \sim 1$) has an s-wave superconducting ground state. This result was confirmed and extended by Hellberg and Manousakis [352], who further argued that in the dilute electron limit, $x \to 1^-$, there is a transition from an s-wave state for $2 < J/t < 3.5$ to a p-wave superconducting state for $J/t < 2$, possibly with a d-wave state at intermediate J/t.

There is conflicting evidence for superconductivity in the $t-J$ model.
Early Lanczos calculations were carried out by Dagotto and Riera [360,418,419] in which various quantities, such as the pair field correlation function and the superfluid density, were computed to search for signs of superconductivity in 4×4 $t-J$ clusters. In agreement with the analytic results, these studies gave strong evidence of superconductivity for large J/t. Interestingly, the strongest signatures of superconductivity were found for $J/t = 3$ and $x = 0.5$ and decayed rapidly for larger J/t. This was interpreted as due to a transition into the phase separation region. (Note, however, that *all* the studies summarized in Fig. 21.36 suggest that $x = 0.5$ is already inside the region that, in the thermodynamic limit, would be unstable to phase separation.)

More recent Green's function Monte Carlo simulations by Sorella et al. [420,421] showed evidence for long range superconducting order in $J/t = 0.4$ clusters of up to 242 sites with periodic boundary conditions and for a range of $x > 0.1$, as shown in Fig. 21.39. No signs of static stripes have been found in the parameter region that was investigated in these studies. A slight tendency towards incommensurability appears in the spin structure factor at (and sometimes above) optimal doping, suggesting perhaps very weak dynamical stripe correlations. Both the evidence for superconductivity and the absence of stripes in this study are in conflict with the results of other studies, and it was suggested [422,423] that the choice of the guiding trial wave function for the calculation may bias the system towards a superconducting state. For example, Shih et al. [400] have shown that the pair binding energy on small clusters decreases with cluster size, and extrapolates to positive values in the thermodynamic limit for $J/t < 0.8$. (A similar conclusion was reached earlier by Boninsegni and Manousakis [424].) Stripe ground states were found for the same values of parameters in DMRG [350,351] and other calculations [425].

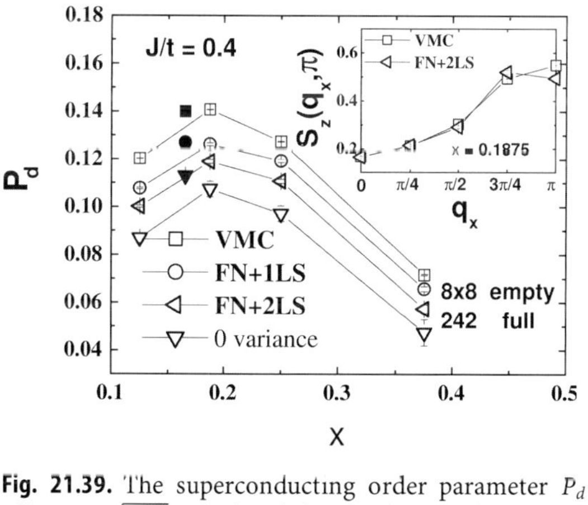

Fig. 21.39. The superconducting order parameter $P_d = 2\lim_{l\to\infty}\sqrt{D(l)}$ calculated for the largest distance on a 8×8, $J/t = 0.4$ cluster as function of hole doping x. Results for $x = 0.17$ on a 242 site cluster are also shown. The different sets of data correspond to various Monte Carlo techniques. The *inset* shows the spin structure factor at $x = 0.1875$. (From Sorella et al. [421])

Static stripes hamper superconductivity, but dynamic stripes may enhance it.
Notwithstanding this controversy, these results seem to add to the general consensus that *static* stripe order and superconductivity compete. This is not to say that stripes and superconductivity cannot coexist. As we saw, evidence for both stripes and pairing have been found in three and four leg $t-J$ ladders [379,381]. In fact pairing is enhanced in both of these systems when stripes are formed compared to the unstriped states found at small doping levels. Because of the open boundary conditions that were used in these studies the stripes were open ended and more dynamic. Imposing periodic boundary conditions in wider ladders (and also the four leg ladder) results in stripes that wrap around the periodic direction. These stripes appear to be more static, and pairing correlations are suppressed. A similar behavior is observed when the stripes are pinned by external potentials.

Further evidence for the delicate interplay between stripes and pairing comes from studies of the $t-t'-J$ model in which a diagonal, single particle, next nearest neighbor hopping t' is added to the basic $t-J$ model [354,425]. Stripes destabilize for either sign of t'. This is probably due to the enhanced mobility of the holes that can now hop on the same sublattice without interfering with the antiferromagnetic background. Pairing is suppressed for $t' < 0$

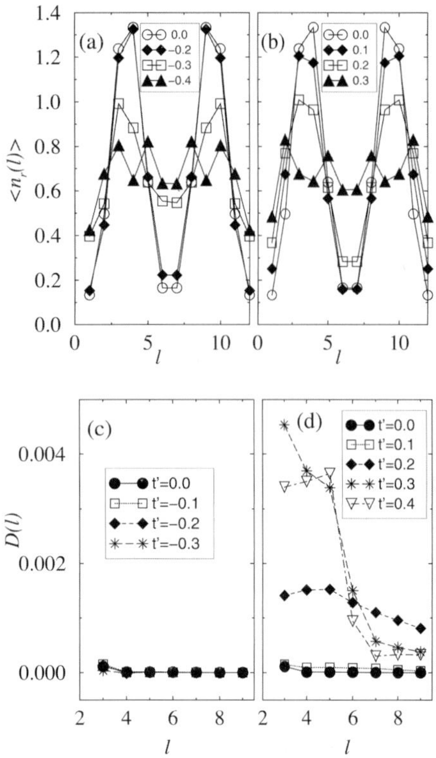

Fig. 21.40. Hole density per rung for a 12 × 6 ladder with periodic boundary conditions along the rungs, 8 holes, $J/t = 0.35$ and (**a**) $t' \leq 0$ and (**b**) $t' \geq 0$. (**c**) and (**d**) depict the d-wave pairing correlations for the same systems. (From White and Scalapino [354])

and enhanced for $t' > 0$.[27] It is not clear whether the complete elimination of stripes or only a slight destabilization is more favorable to pairing correlations. Figure 21.40 suggests that optimal pairing occurs in between the strongly modulated ladder and the homogeneous system.

Finally, allowing for extra hopping terms in the Hamiltonian is not the only way tip the balance between static charge order and superconductivity. So far we have not mentioned the effects of long range Coulomb interactions on the properties of Hubbard related systems. This is not a coincidence since the treatment of such interactions in any standard numerical method is difficult. Nevertheless, a recent DMRG study of four leg ladders with open and periodic boundary conditions which takes into account the Coulomb potential in a self-consistent Hartree way [358], gives interesting results. It suggests that the inclusion of Coulomb interactions suppresses the charge modulations associated with stripes while enhancing the long range superconducting pairing correlations. At the same time the local superconducting pairing is not suppressed. Taken together, these facts support the notion that enhanced correlations come from long range phase ordering between stripes with well-established pairing. This enhanced phase stiffness is presumably due to pair tunneling between stripes produced by increased stripe fluctuations.

21.12 Doped Antiferromagnets

The undoped state of the cuprate superconductors is a strongly insulating antiferromagnet. It is now widely believed that the existence of such a parent correlated insulator is an essential feature of high temperature superconductivity, as was emphasized in some of the earliest studies of this problem [5,120]. However, the doped antiferromagnet is a complicated theoretical problem—to even cursorily review what is known about it would more than double the size of this document. In this section we very briefly discuss the aspects of this problem which we consider most germane to the cuprates, and in particular to the physics of stripes. More extensive reviews of the subject can be found in [6,15,390,426].

21.12.1 Frustration of the Motion of Dilute Holes in an Antiferromagnet

The most important local interactions in a doped antiferromagnet are well represented by the large U Hubbard model, the $t-J$ model, and their various relatives. To be concrete, we will focus on the $t-J-V$ model [393] (a slight generalization of the $t-J$ model, Eq. (21.119), to which it reduces for $V = -J/4$.)

[27] This is surprising since T_c is generally higher for hole doped cuprates (believed to have $t' < 0$) than it is for electron doped cuprates (which have $t' > 0$).

$$H = -t \sum_{<i,j>,\sigma} \left\{ c^\dagger_{i,\sigma} c_{j,\sigma} + \text{h.c.} \right\} + \sum_{<i,j>} \left\{ J \boldsymbol{S}_i \cdot \boldsymbol{S}_j + V n_i n_j \right\} , \quad (21.125)$$

where $\boldsymbol{S}_i = \sum_{\sigma,\sigma'} c^\dagger_{i,\sigma} \boldsymbol{\sigma}_{\sigma,\sigma'} c_{i,\sigma'}$ is the spin of an electron on site i. Here $\boldsymbol{\sigma}$ are the Pauli matrices and $< i,j >$ signifies nearest neighbor sites on a hypercubic lattice in d dimensions. There is a constraint of no double occupancy on any site,

$$n_i = \Sigma_\sigma c^\dagger_{i,\sigma} c_{i,\sigma} = 0, 1 . \quad (21.126)$$

The concentration of doped holes, x, is taken to be much smaller than 1, and is defined as

$$x = N^{-1} \sum_j n_j , \quad (21.127)$$

where N is the number of sites.

The essential feature of this model is that it embodies a strong, short range repulsion between electrons, manifest in the constraint of no double occupancy. The exchange integral J arises through virtual processes wherein the intermediate state has a doubly occupied site, producing an antiferromagnetic coupling. Doping is assumed to remove electrons thereby producing a "hole" or missing spin which is mobile because neighboring electrons can hop into its place with amplitude t.

Like a good game, the rules are simple: antialign adjacent spins, and let holes hop. And like any good game, the winning strategy is complex. The ground state of this model must simultaneously minimize the zero point kinetic energy of the doped holes and the exchange energy, but the two terms compete.

For $t > J > xt$, the problem is highly frustrated.
The spatially confined wavefunction of a localized hole has a high kinetic energy; the t term accounts for the tendency of a doped hole to delocalize by hopping from site to site. However, as holes move through an antiferromagnet they scramble the spins: each time a hole hops from one site to its nearest neighbor, a spin is also moved one register in the lattice, *onto the wrong sublattice*. So it is impossible to minimize both energies simultaneously in $d > 1$. Moreover, in the physically relevant range of parameters, $t > J > tx$, neither energy is dominant. On the one hand, because $t > J$, one cannot simply perturb about the $t = 0$ state which minimizes the exchange energy. On the other hand, because $J > tx$ one cannot simply perturb about the ground state of the kinetic energy.

A number of strategies, usually involving further generalizations of the model, have been applied to the study of this problem, including: large n [427], large S [428,429], large d [430], small t/J [392], large t/J [392,431,432], and various numerical studies of finite size clusters. (Some of the latter are reviewed in Sect. 21.11.) For pedagogic purposes, we will frame aspects of the ensuing discussion in terms of the large d behavior of the model since it is tractable, and involves no additional theoretical technology, but similar conclusions can be drawn from a study of any of the analytically tractable limits listed above.[28] One common feature[29] of these solutions is a tendency of the doped holes to phase separate at small x. The reason for this is intuitive: in a phase separated state, the holes are expelled from the pure antiferromagnetic fraction of the system, where the exchange energy is minimized and the hole kinetic energy is not an issue, while in the hole rich regions, the kinetic energy of the holes is minimized, and the exchange energy can be neglected to zeroth order since $J < tx_{\text{rich}}$, where x_{rich} is the concentration of doped holes in the hole rich regions.

A large dimension expansion
We employ the following large dimension strategy. We take as the unperturbed Hamiltonian the Ising piece of the interaction:

$$H_o = \sum_{<i,j>} \left\{ J_z S^z_i S^z_j + V n_i n_j \right\} , \quad (21.128)$$

and treat as perturbations the XY piece of the interaction and the hopping:

[28] In some ways, the large S limit is the most physically transparent of all these approaches; see [429] for further discussion.
[29] It is still controversial whether or not phase separation is universal in $d = 2$ and 3 at small enough x; see [352,359,390,399,433–435].

$$H_1 = \frac{J_\perp}{2} \sum_{<i,j>} \left\{ S_i^+ S_j^- + \text{h.c.} \right\} , \quad (21.129)$$

$$H_2 = -t \sum_{<i,j>,\sigma} \left\{ c_{i,\sigma}^\dagger c_{j,\sigma} + \text{h.c.} \right\} . \quad (21.130)$$

Expansions derived in powers of $J_\perp/J_z$ and t/J_z can be reorganized in powers of $1/d$ [430], at which point we will again set $J_\perp = J_z \equiv J$ as in the original model (Eq. (21.125)), and allow the ratio t/J to assume physical values.

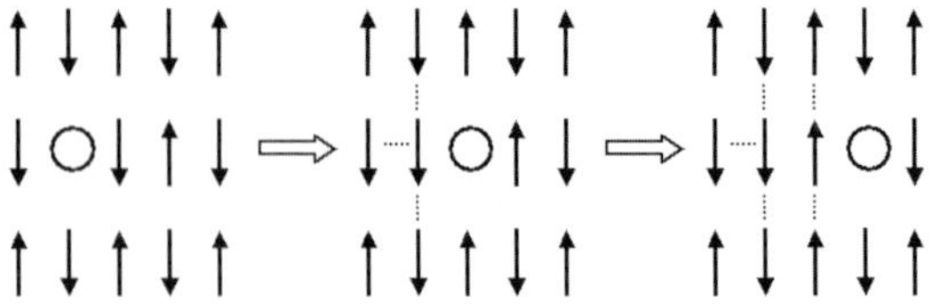

Fig. 21.41. *Frustration of one hole's motion in an antiferromagnet.* As the hole hops, it leaves behind a string of frustrated bonds designated here by *dashed lines*

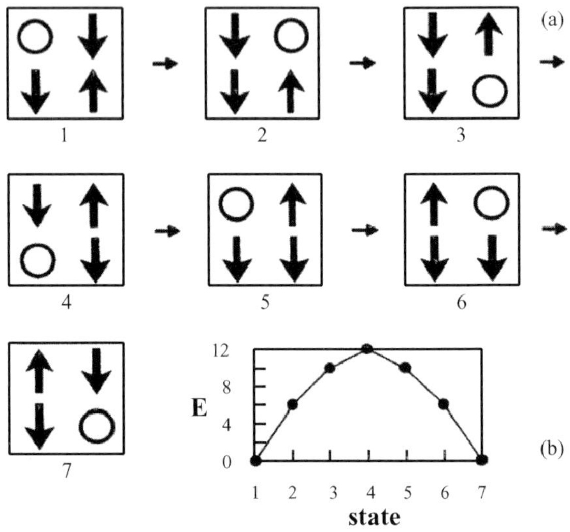

Fig. 21.42. *Trugman terms.* (**a**) A hole moving one and a half times around a plaquette translates a degenerate ground state without leaving a frustrated string of spins behind. (**b**) The energy of the intermediate states in units of J. The hole has to tunnel through this barrier as it moves. From [436]

One Hole in an Antiferromagnet

It is universally recognized that a key principle governing the physics of doped antiferromagnets is that the motion of a single hole is highly frustrated.

The motion of one hole in an antiferromagnet is frustrated.
To illustrate this point, it is convenient to examine it from the perspective of a large dimension treatment in which the motion of one hole in an antiferromagnet is seen to be frustrated by a "string" left in its wake (see Fig. 21.41), which costs an energy of order $(d-1)J$ times the length of the string. The unperturbed ground state of one hole on, say, the "black" sublattice, is $N/2$-fold degenerate (equal to the number of black sublattice sites), once a direction for the Néel order is chosen (the other $N/2$ degenerate ground states describing a hole on the "red" sublattice form a disjoint Hilbert space under the operation of H_1 and H_2). These ground states are only connected in degenerate perturbation theory of third or higher order, via, e.g. two operations of H_2 and one of H_1. They are connected in perturbation theory of sixth or higher order by operations solely of the hopping term H_1 via the Trugman [436] terms, in which a hole traces any closed, nonintersecting path two steps less than two full circuits; see Fig. 21.42 for an example (such paths become important when $J \ll t$). In this manner a hole can "eat its own string". Owing to such processes a hole can propagate through an antiferromagnet. However, the high order in the perturbation series and the energetic barriers involved render the effective hopping matrix elements significantly smaller than their unperturbed values.

Two Holes in an Antiferromagnet

In early work on high temperature superconductivity, it was often claimed that, whereas the motion of a single hole is inhibited by antiferromagnetic order, pair motion appears to be entirely unfrustrated. It was even suggested [19] that this might be the basis of a novel, kinetic energy driven mechanism of pairing—perhaps the first such suggestion. However, a flaw with this argument was revealed in the work of

Trugman [436], who showed that this mode of propagation of the hole pair is frustrated by a quantum effect which originates from the fermionic character of the background spins. While Trugman's original argument was based on a careful analysis of numerical studies in $d = 2$, the same essential effect can be seen analytically in the context of a large d expansion. The effective Hamiltonian of two holes can be written as follows [430]:

$$H_2^{\text{eff}} = U^{\text{eff}} \sum_{<i,j>} c_i^\dagger c_j^\dagger c_j c_i - T^{\text{eff}} \sum_{<i,j,k>} c_j^\dagger c_i^\dagger c_j c_k + \mathcal{O}(1/d^2) \,, \tag{21.131}$$

where $< i, j, k >$ signifies a set of sites such that i and k are both nearest neighbors of j, and the $c_i^\dagger$ creates a hole at site i. To lowest order in $(1/d)$, $U^{\text{eff}} = V - J/4$ and $T^{\text{eff}} = t^2/Jd$. For states with the two holes as nearest neighbors, H_2^{eff} can be block diagonalized by Fourier transform, yielding d bands of eigenstates labeled by a band index and a Bloch wavevector $\boldsymbol{k}$. The result is that $d - 1$ of these bands have energy U^{eff} and do not disperse. The remaining band has energy $U^{\text{eff}} + 4T^{\text{eff}} \sum_{a=1}^{d} \sin^2(k_a/2)$, where k_a is the component of $\boldsymbol{k}$ along a. This final band, which feels the effects of pair propagation, has the largest energy.

Two holes are no less frustrated.
This counterintuitive result follows from the fermionic nature of the background spins. A similar calculation for bosons would differ by a minus sign: in that case, the final band has energy $U^{\text{eff}} - 4T^{\text{eff}} \sum_{a=1}^{d} \sin^2(k_a/2)$, which is much closer to what one might have expected.[30] The interference effect for the fermionic problem is illustrated in Fig. 21.43. Different paths that carry the system from one hole pair configuration to another generally interfere with each other, and when two such paths differ by the exchange of two electrons, they interfere destructively in the fermionic case and constructively in the bosonic. It follows from this argument that pair motion, too, is frustrated—it actually results in an effective kinetic repulsion between holes, rather than in pair binding.[31]

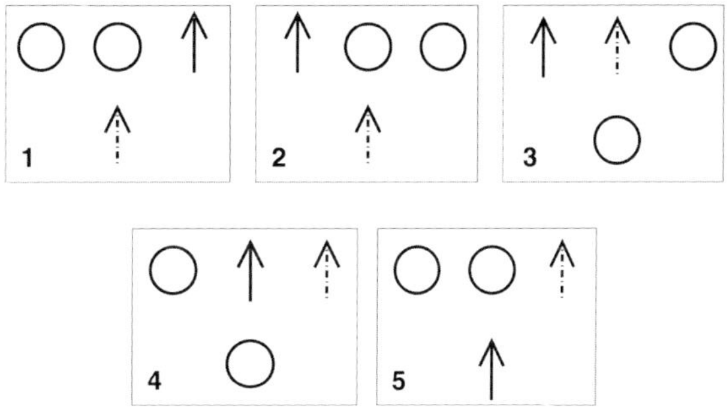

Fig. 21.43. *Frustration of a hole pair's motion in an antiferromagnet.* The figure shows a sequence of snapshots in a process that takes a pair of holes back to their original position, but with a pair of spins switched. The sequence is as follows: (**1**) Initial two hole state. (**2**) A spin has moved two sites to the left. (**3**) The other spin has moved one site up. (**4**) A hole has moved two sites to the left. (**5**) A hole has moved up. Due to the fermionic nature of the spins, the above process leads to an increase in the pair energy, so that pair propagation is not an effective mechanism of pair binding

Many Holes: Phase Separation

In large d, the frustration of the kinetic energy of doped holes in an antiferromagnet leads to a miscibility gap [430]. Perhaps this should not be surprising, since phase separation is the generic fate of mixtures at low temperatures. At any finite temperature, two-phase coexistence occurs whenever the chemical potentials of the two phases are equal. In the present case, one of the phases, the undoped antiferromagnet, is incompressible, which means that at $T = 0$ its chemical potential lies at an indeterminate point within the Mott gap. Under these circumstances, phase coexistence is instead established by considering the total energy of the system:

[30] This corrects similar expressions in [430].

[31] It is apparent that second neighbor hopping terms, t', produce less frustration of the single particle motion, and "pair hopping" terms, which arise naturally in the t/U expansion of the Hubbard model, lead to unfrustrated pair motion [156]. However, t' is generally substantially smaller than t, and if pair hopping is derived from the Hubbard model, it is of order J, and hence relatively small.

$$E_{\text{tot}} = N_{\text{AF}}e_{\text{AF}} + N_h e_h = Ne_{\text{AF}} + N_h(e_h - e_{\text{AF}}) \,, \qquad (21.132)$$

where N_{AF} and N_h are the number of sites occupied by the undoped antiferromagnet and by the hole rich phase, respectively; $N = N_{\text{AF}} + N_h$; e_{AF} is the energy per site of the antiferromagnet and e_h is the energy per site of the hole rich phase, in which the concentration of doped holes is $x_{\text{rich}} = x(N/N_h) \geq x$. If E_{tot} has a minimum with respect to N_h at a value $N_h < N$, there is phase coexistence. This minimization leads to the equation

$$\mu = \frac{e_{\text{AF}} - e_h(\mu)}{1 - n(\mu)} \,, \qquad (21.133)$$

where μ is the chemical potential of the hole rich phase, and $n = 1 - x_c$ is the electron density in the hole rich phase.

As we shall see, in the limit of large dimension, $n(\mu)$ (and hence e_h as well) is either 0 or exponentially small, so Eq. (21.133) reduces to

$$\mu \approx e_{\text{AF}} \,. \qquad (21.134)$$

Phase separation occurs below a critical concentration of doped holes.

We can see already how phase separation can transpire. As the electron density is raised from zero (i.e. starting from $x = 1$ and lowering x), the chemical potential of the electron gas increases. Once μ reaches e_{AF}, the added electrons must go into the antiferromagnetic phase, and the density of the electron gas stops increasing. We can employ a small k expansion of the electronic dispersion, $\epsilon(\boldsymbol{k}) = -2td + tk^2 + \ldots$, to determine that $\mu \approx -2td + tk_F^2$. Thus if $e_{\text{AF}} < -2td$, the electron gas is completely unstable, and there is phase separation into the pure antiferromagnet, and an insulating hole rich phase with $n = 0$. In this case, $x_c = 1$. Otherwise, the density of the electron gas is

$$n = \frac{2A_d}{d}\left(\frac{k_F}{2\pi}\right)^d = \frac{2A_d}{d}\left(\frac{\sqrt{(\mu + 2td)/t}}{2\pi}\right)^d . \qquad (21.135)$$

Here A_d is the hypersurface area of a d dimensional unit sphere. In large d, the energy per site of the pure antiferromagnet approaches that of the classical Néel state:

$$e_{\text{AF}} = -d\left(\frac{J}{4} - V\right)[1 + \mathcal{O}(1/d^2)] \,. \qquad (21.136)$$

From this, it follows that the hole rich phase is insulating (i.e. it has no electrons) if $J - 4V > 8t$ and it is metallic ($x_c < 1$) if $J - 4V < 8t$. However, even when the hole rich phase is metallic, its electron density is exponentially small (as promised):

$$n = 1 - x_c = \frac{2}{\sqrt{\pi d}}\left[\frac{e}{\pi}\left(1 - \left[\frac{J - 4V}{8t}\right]\right)\right]^{d/2} \times [1 + \mathcal{O}(1/d)] \,, \qquad (21.137)$$

where we have used the asymptotic large d expression [430] $A_d \approx \sqrt{\frac{d}{\pi}}(\frac{2\pi e}{d})^{d/2}$. As illustrated in Fig. 21.44 in large d, so long as $0 < x < x_c$, the ground state of the $t - J - V$ model is phase separated, with an undoped antiferromagnetic region and a hole rich region which, if $8t > J - 4V$, is a Fermi liquid of dilute electrons, or if $8t < J - 4V$, is an insulator. (Under these same circumstances, if $x_c < x < 1$, the ground state is a uniform, Fermi liquid metal.[32])

In the low dimensions of physical interest, such as $d = 2$ and $d = 3$, the quantitative accuracy of a large dimension expansion is certainly suspect. Nonetheless, we expect the qualitative physics of $d = 2$ and $d = 3$ to be captured in a large dimension treatment, since the lower critical dimension of most long range $T = 0$ ordered states is $d = 1$. For comparison, in Fig. 21.45 we reproduce the phase diagram of the 2D $t - J$ model which was proposed by Hellberg and Manousakis [352] on the basis of Monte Carlo studies of systems with up to 60 electrons. There is clearly substantial similarity between this and the large D result in Fig. 21.44.

In one sense phase separation certainly can be thought of as a strong attractive interaction between holes, although in reality the mechanism is more properly regarded as the ejection of holes from the

[32] This statement neglects a possible subtlety due to the Kohn–Luttinger theorem.

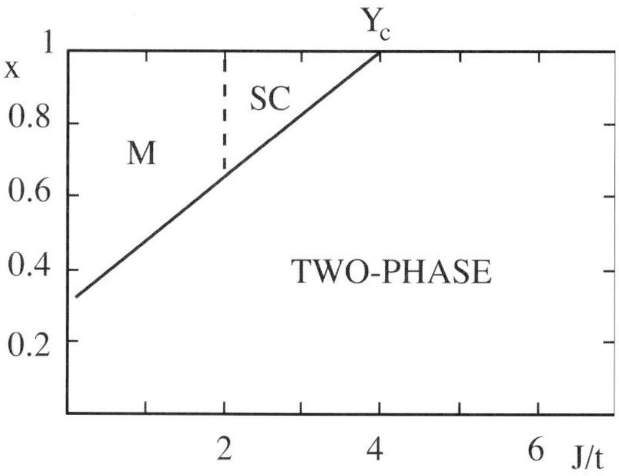

Fig. 21.44. Phase diagram of the $t-J$ model deduced from large the d expansion. In the figure, we have set $d = 2$. "Two-phase" labels the region in which phase separation occurs between the pure antiferromagnet and a hole rich phase, "SC" labels a region of s-wave superconductivity, and "M" labels a region of metallic behavior. At parametrically small $J/t \propto 1/\sqrt{d}$, a ferromagnetic phase intervenes at small doping. From [430]

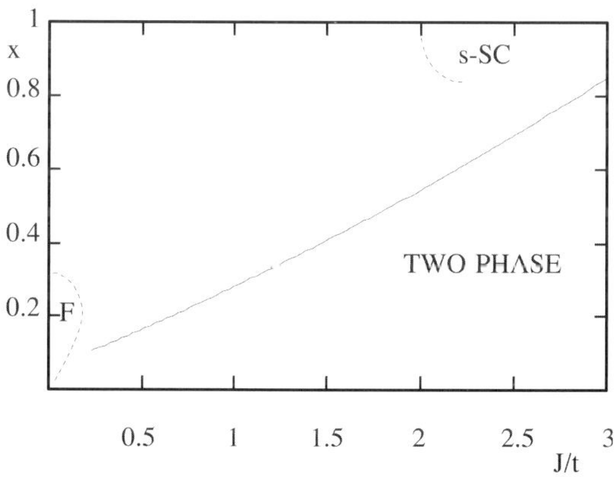

Fig. 21.45. Phase diagram of the $t-J$ model in two dimensions at zero temperature, deduced from numerical studies with up to 60 electrons. *Two-phase* labels the region of phase separation, *s-SC* labels a region of s-wave superconductivity, and *F* labels a region of ferromagnetism. This figure is abstracted from Hellberg and Manousakis [352]

antiferromagnet.[33] The characteristic energy scale of this interaction is set by magnetic energies, so one expects to see phase separation only at temperatures that are small compared to the antiferromagnetic exchange energy J.

21.12.2 Coulomb Frustrated Phase Separation and Stripes

Were holes neutral, phase separation would be a physically reasonable solution to the problem of frustrated hole motion in an antiferromagnet. But there is another competition if the holes carry charge. In this case, full phase separation is impossible because of the infinite Coulomb energy density it would entail. Thus, there is a second competition between the short range tendency to phase separation embodied in the $t-J$ model, and the long range piece of the Coulomb interaction. The compromise solution to this second level of frustration results in an emergent length scale [437]—a crossover between phase separation on short length scales, and the required homogeneity on long length scales.

Stripes are a unidirectional density wave.
Depending upon microscopic details, many solutions are possible [438] which are inhomogeneous on intermediate length scales, such as checkerboard patterns, stripes, bubbles, or others.

Of these, the stripe solution is remarkably stable in simple models [392,430,439], and moreover is widely observed in the cuprates [6]. A stripe state is a unidirectional density wave state—we think of such a state, at an intuitive level, as consisting of alternating strips of hole rich and hole poor phase. A fully ordered stripe phase has charge density wave and spin density wave order interleaved.

Certain aspects of stripe states can be made precise on the basis of long distance considerations. If we consider the Landau theory [45] of coupled order parameters for a spin density wave $\boldsymbol{S}$ with ordering vector $\boldsymbol{k}$ and a charge density wave ρ with ordering vector $\boldsymbol{q}$, then if $2\boldsymbol{k} \equiv \boldsymbol{q}$ (where $\equiv$, in this case, means equal modulo a reciprocal lattice vector), then there is a cubic term in the Landau free energy allowed by symmetry,

$$F_{\text{coupling}} = \gamma_{\text{stripe}} \left[\rho_{-q} S_k \cdot S_k + \text{C.C.} \right] . \qquad (21.138)$$

[33] Like salt crystallizing from a solution of salt water, the spin crystal is pure.

There are two important consequences of this term. Firstly, the system can lower its energy by locking the ordering vectors of the spin and charge density wave components of the order, such that the period of the spin order is twice that of the charge order. At order parameter level, is the origin of the antiphase character of the stripe order.[34] Secondly, because this term is linear in ρ, it means that if there is spin order, $< S_k > \neq 0$, there must *necessarily*[35] be charge order, $< \rho_{2k} > \neq 0$, although the converse is not true.

The Landau theory also allows us to distinguish three macroscopically distinct scenarios for the onset of stripe order. If charge order onsets at a higher critical temperature, and spin order either does not occur, or onsets at a lower critical temperature, the stripe order can be called "charge driven." If spin and charge order onset at the same critical temperature, but the charge order is parasitic, in the sense that $< \rho_{2k} > \sim < S_k >^2$, the stripe order is "spin driven." Finally, if charge and spin order onset simultaneously by a first order transition, the stripe order is driven by the symbiosis between charge and spin order. This is discussed in more detail in [45].

The antiphase nature of the stripes was first predicted by the Hartree–Fock theory and has been confirmed as being the most probable outcome in various later, more detailed studies of the problem [350,409,410,441]. In this case, the spin texture undergoes a π phase shift across every charge stripe, so that every other spin stripe has the opposite Néel vector, canceling out any magnetic intensity at the commensurate wavevector, $< \pi, \pi >$. This situation [442,443] has been called "topological doping." And, indeed, the predicted factor of two ratio between the spin and charge periodicities has been observed in all well established experimental realizations of stripe order in doped antiferromagnets [47]. Still, it is important to remember that nontopological stripes are also a logical possibility [409,410,433,440,444,445], and we should keep our eyes open for this form of order, as well.[36]

The Coulomb interaction sets the stripe spacing.
In the context of frustrated phase separation, the formation of inhomogeneous structures is predominantly a statement about the charge density, and its scale is set by the Coulomb interaction. This has several implications. Firstly, this means that charge stripes may begin to self-organize (at least locally) at relatively high temperatures, i.e. they are charge driven in the sense described above.[37] Secondly, charge density wave order always couples linearly to lattice distortions, so we should expect dramatic signatures of stripe formation to show up in the phonon spectrum. Indeed, phonons may significantly affect the energetics of stripe formation [446]. Thirdly, although we are used to thinking of density wave states as insulating, or at least as having a dramatically reduced density of states at the Fermi energy, this is not necessarily true.

Competition sets the hole concentration on a stripe.
If the average hole concentration on each stripe is determined primarily by the competition between the Coulomb interaction and the local tendency to phase separation, the linear hole density per site along each stripe can vary as a function of x and consequently there is no reason to expect the Fermi energy to lie in a gap or pseudogap. In essence, stripes may be intrinsically metallic, or even superconducting. Moreover, such compressible stripes are highly prone to lattice commensurability effects which tend to pin the interstripe spacing at commensurate values. Conversely, if the stripes are a consequence of some sort of Fermi surface nesting, as is the case in the Hartree-Fock studies [405,408,447] of stripe formation, the stripe

[34] In the context of Landau–Ginzberg theory, the situation is somewhat more complex, and whether the spin and charge order have this relation, or have the same period turns out to depend on short distance physics, see footnote 36 and [440].

[35] Here, we exclude the possibility of perfectly circular spiral spin order, in which $Re\{< S >\} \cdot Im\{< S >\} = 0$ and $[Re\{< S >\}]^2 = [Im\{< S >\}]^2 \neq 0$.

[36] For example, an analogous Landau theory of stripes near the Néel state must include the order parameter S_π, which favors in-phase domain walls [440].

[37] In Hartree-Fock theory, stripes are spin driven.

period always adjusts precisely so as to maintain a gap or pseudogap at the Fermi surface: there is always one doped hole per site along each charge stripe. This insulating behavior is likely a generic feature of all local models of stripe formation [442], although more sophisticated treatments can lead to other preferred linear hole densities along a stripe [350,404].

In short, stripe order is theoretically expected to be a common form of self-organized charge ordering in doped antiferromagnets. In a d-dimensional striped state, the doped holes are concentrated in an ordered array of parallel $(d-1)$ dimensional hypersurfaces: solitons in $d=1$, "rivers of charge" in $d=2$, and sheets of charge in $d=3$. This "charge stripe order" can either coexist with antiferromagnetism with twice the period (topological doping) or with the same period as the charge order, or the magnetic order can be destroyed by quantum or thermal fluctuations of the spins.

"Stripe glasses" and "stripe liquids" are also possible. Moreover, the stripes can be insulating, conducting, or even superconducting. It is important to recall that for $d<4$ quenched disorder is always a relevant perturbation for charge density waves [448], so rather than stripe ordered states, real experiments may often require interpretation in terms of a "stripe glass" [449–452]. Finally, for many purposes, it is useful to think of systems that are not quite ordered, but have substantial short range stripe order as low frequency fluctuations, as a "fluctuating stripe liquid". We will present an example of such a state in the next subsection.

21.12.3 Avoided Critical Phenomena

Let us examine a simple model of Coulomb frustrated phase separation. We seek to embody a system with two coexisting phases, which are forced to interleave due to the charged nature of one of the phases. To account for the short range tendency to phase separation, we include a short range "ferromagnetic" interaction which encourages nearest neighbor regions to be of the same phase, and also a long range "antiferromagnetic" interaction which prevents any domain from growing too large:

$$H=-L\sum_{<i,j>}\mathbf{S}_i\cdot\mathbf{S}_j+\frac{Qa^{d-2}}{2}\sum_{i\neq j}\frac{\mathbf{S}_i\cdot\mathbf{S}_j}{|\mathbf{R}_i-\mathbf{R}_j|^{d-2}}\,. \tag{21.139}$$

Here $\mathbf{S}_j$ is an N component unit vector, $\mathbf{S}_i\cdot\mathbf{S}_i=1$, L is a nearest neighbor ferromagnetic interaction, Q is an antiferromagnetic "Coulomb" term which represents the frustration (and is always assumed small, $Q\ll L$), d is the spatial dimension, $<i,j>$ signifies nearest neighbor sites, a is the lattice constant, and $\mathbf{R}_j$ is the location of lattice site j. The Ising ($N=1$) version of this model is the simplest coarse grained model [390,453] of Coulomb frustrated phase separation, in which $S_j=1$ represents a hole rich, and $S_j=-1$ a hole poor region. In this case, $L>0$ is the surface tension of an interface between the two phases, and Q is the strength of the Coulomb frustration. While the phase diagram of this model has been analyzed [453] at $T=0$, it is fairly complicated, and its extension to finite temperature has only been attempted numerically [454]. However, all the thermodynamic properties of this model can be obtained [455,456] exactly in the large N limit.

Figure 21.46 shows the phase diagram for this model. Both for $Q=0$ and $Q\neq 0$, there is a low temperature ordered state, but the ordered state is fundamentally different for the two cases. For the unfrustrated case, the ordered state is homogeneous,

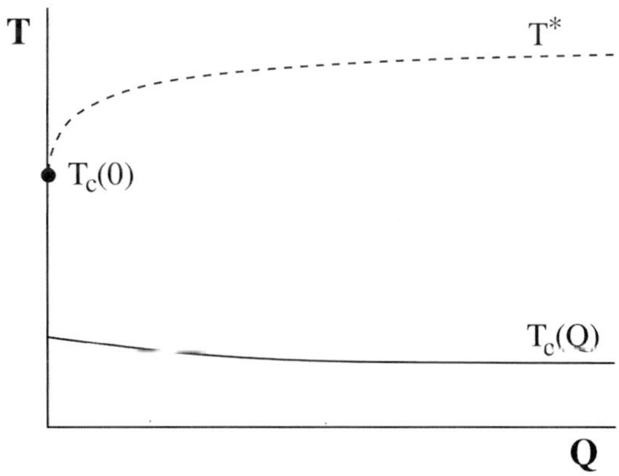

Fig. 21.46. Schematic phase diagram of the model in Eq. (21.139) of avoided critical phenomena. The thick black dot marks $T_c(Q=0)$, the ordering temperature in the absence of frustration; this is "the avoided critical point". Notice that $T_c(Q\to 0)<T_c(Q=0)$. From [456]

whereas with frustration, there is an emergent length scale in the ordered state which governs the modulation of the order parameter. To be specific, in dimensions $d > 2$ and for $N > 2$, there is a low temperature ordered unidirectional spiral phase, which one can think of as a sort of stripe ordered phase [456]. Clearly, as $Q \to 0$, the modulation length scale must diverge, so that the homogeneous ordered state is recovered. However, like an antiferromagnet doped with neutral holes, there is a discontinuous change in the physics from $Q = 0$ to any finite Q: for $d \leq 3$, $\lim_{Q \to 0} T_c(Q) \equiv T_c(0^+)$ is strictly less than $T_c(0)$. In other words, an infinitesimal amount of frustration depresses the ordering temperature discontinuously.

This model exhibits a "fluctuating stripe" phase.
Although for any finite Q the system does not experience a phase transition as the temperature is lowered through $T_c(0)$, the avoided critical point heavily influences the short range physics. For temperatures in the range $T_c(0) > T > T_c(0^+)$, substantial local order develops. An explicit expression for the spin-spin correlator can be obtained in this temperature range: At distances less than the correlation length $\xi_0(T)$ of the unfrustrated magnet, $R_{ij} < \xi_0(T)$, the correlator is critical,

$$\langle \boldsymbol{S}_i \cdot \boldsymbol{S}_j \rangle \sim \left(a/R_{ij}\right)^{d-2-\eta} , \qquad (21.140)$$

but for longer distances, $R_{ij} > \xi_0(T)$, it exhibits a damped version of the Goldstone behavior of a fluctuating stripe phase,

$$\langle \boldsymbol{S}_i \cdot \boldsymbol{S}_j \rangle \sim \left(a/R_{ij}\right)^{\frac{d-1}{2}} \cos[KR_{ij}] \exp[-\kappa R_{ij}] . \qquad (21.141)$$

At $T_c(Q)$, the wavevector K is equal to the stripe ordering wavevector of the low temperature ordered state, $K(T_c) = (Q/L)^{1/4}$. As the temperature is raised, K decreases until it vanishes at a disorder line marked T^* in the figure. The inverse domain size is given by

$$\kappa(T) = \sqrt{(Q/L)^{1/4} - K^2(T)} . \qquad (21.142)$$

For a broad range of temperatures (which does not narrow as $Q \to 0$), this model is in a fluctuating stripe phase in a sense that can be made arbitrarily precise for small enough Q.

21.12.4 The Cuprates as Doped Antiferromagnets

General Considerations

Our theoretical understanding of the undoped antiferromagnets is extolled.
There is no question that the undoped parents of the high temperature superconductors are Mott insulators, in which the strong short range repulsion between electrons is responsible for the insulating behavior, and the residual effects of the electron kinetic energy (superexchange) lead to the observed antiferromagnetism. Indeed, one of the great theoretical triumphs of the field is the complete description, based on interacting spin waves and the resulting nonlinear sigma model, of the magnetism in these materials [222,223,327].

However, it is certainly less clear that one should inevitably view the superconducting materials as doped antiferromagnets, especially given that we have presented strong reasons to expect a first order phase transition between $x = 0$ and $x > 0$. Nonetheless, many experiments on the cuprates are suggestive of a doped antiferromagnetic character. In the first place, various measurements of the density of mobile charge, including the superfluid density [107,243], the "Drude weight" measured in optical conductivity [457], and the Hall number [458,459], are all consistent with a density proportional to the *doped hole* density, x, rather than the total hole density, $1 + x$, expected from a band structure approach. Moreover, over a broad range of doping, the cuprates retain a clear memory of the antiferromagnetism of the parent correlated insulator. Local magnetism abounds. NMR, μSR, and neutron scattering find evidence (some of which is summarized in Sect. 21.13.1) of static, or slowly fluctuating, spin patterns, including stripes, spin glasses, and perhaps staggered orbital currents.

Why the cuprates should be viewed as doped antiferromagnets
Static magnetic moments, or slowly fluctuating ones, are hard to reconcile with a Fermi liquid picture. There is also some evidence from STM of local electronic inhomogeneity [100, 101, 460] in BSCCO, indicative of the short range tendency to phase sepa-

rate. The Fermi liquid state in a simple metal is highly structured in k-space, and so is highly homogeneous (rigid) in real space. This is certainly in contrast with experiments on the cuprates which indicate significant real space structure.

Stripes

There is increasingly strong evidence that stripe correlations, as a specific feature of doped antiferromagnets, occur in *at least some* high temperature superconducting materials.

Another triumph of theory!
The occurrence of stripe phases in the high temperature superconductors in particular, and in doped antiferromagnets more generally, was successfully predicted[38] by theory [405,408,447]. Indeed, it is clear that a fair fraction of the theoretical inferences discussed in Sect. 21.12.2 are, at least in broad outline, applicable to a large number of materials, including at least some high temperature superconductors [6]. In particular, the seminal discovery [463] that in $La_{1.6-x}Nd_{0.4}Sr_xCuO_4$, first charge stripe order, then spin stripe order, and then superconductivity onset at successively lower critical temperatures is consistent with Coulomb frustrated phase separation (see Fig. 21.47 in Sect. 21.13.1). Somewhat earlier work on the closely related nickelates [464] established that the charge stripes are, indeed, antiphase domain walls in the spin order.

Controversy remains as to how universal stripe phases are in the cuprate superconducting materials, and even how the observed phases should be precisely characterized. This is also an exciting topic, on which there is considerable ongoing theoretical and experimental study. We will defer further discussion of this topic to Sect. 21.13.

21.12.5 Additional Considerations and Alternative Perspectives

There are a number of additional aspects of this problem which we have not discussed here, but which we feel warrant a mention. In each case, clear discussions exist in the literature to which the interested reader is directed for a fuller exposition.

Phonons

There is no doubt that strong electron-phonon coupling can drive a system to phase separate. Strong correlation effects necessarily enhance such tendencies, since they reduce the rigidity of the electron wavefunction to spatial modulation. (See, e.g. the 1D example in Sect. 21.9.2.) In particular, when there is already a tendency to some form of charge ordering, on very general grounds we expect it to be strongly enhanced by electron-phonon interactions.

This observation makes us very leery of any attempt at a *quantitative* comparison between results on phase separation or stripe formation in the $t-J$ or Hubbard models with experiments in the cuprates, where the electron-phonon interaction is manifestly strong [465]. Conversely, there should generally be substantial signatures of various stripe-related phenomena in the phonon dynamics, and this can be used to obtain an experimental handle on these behaviors [160]. Indeed, there exists a parallel development of stripe-related theories of high temperature superconductivity based on Coulomb frustration of a phase separation instability which is driven by strong electron-phonon interactions [16,61,466]. The similarity between many of the notions that have emerged from these studies, and those that have grown out of studies of doped antiferromagnets illustrates both how robust the consequences of frustrated phase separation are in highly correlated systems, and how difficult it is to unambiguously identify a "mechanism" for it. For a recent discussion of many of the same phenomena discussed here from this alternative viewpoint, see [62].

Spin-Peierls Order

Another approach to this problem, which emerges naturally from an analysis of the large N limit [71], is

[38] The theoretical predictions predated any clear body of well accepted experimental facts, although in all fairness it must be admitted that there was some empirical evidence of stripe-like structures which predated all of the theoretical inquiry: Even at the time of the first Hartree–Fock studies, there was already dramatic experimental evidence [461,462] of incommensurate magnetic structure in $La_{2-x}Sr_xCuO_4$.

to view the doped system as a "spin-Peierls" insulator, by which we mean a quantum disordered magnet in which the unit cell size is doubled but spin rotational invariance is preserved.[39] While the undoped system is certainly antiferromagnetically ordered, it is argued that when the doping exceeds the critical value at which spin rotational symmetry is restored, the doped Mott insulating features of the resulting state are better viewed as if they arose from a doped spin-Peierls state. Moreover, since the spin-Peierls state has a spin gap, it can profitably be treated as a crystal of Cooper pairs, which makes the connection to superconductivity very natural. Finally, as mentioned in Sect. 21.7, this approach has a natural connection with various spin liquid ideas.

Interestingly, it turns out that the doped spin-Peierls state also generically phase separates [427, 469–471]. When the effect of long range Coulomb interactions are included, the result is a staircase of commensurate stripe phases [471]. Again, the convergence of the pictures emerging from diverse starting points convinces us of the generality of stripey physics in correlated systems. For a recent discussion of the physics of stripe phases, and their connection to the cuprate high temperature superconductors approached from the large N/spin-Peierls perspective, see [472].

Stripes in Other Systems

It is not only the robustness of stripes in various theories that warrants mention, but also the fact that they are observed, in one way or another, in diverse physical realizations of correlated electrons. Stripes, and even a tendency to electronic phase separation, are by now well documented in the manganites—the colossal magnetoresistance materials. (For recent discussions, which review some of the literature, see [17,473] and [108].) This system, like the nickelates and cuprates, is a doped antiferromagnet, so the analogy is quite precise.

Although the microscopic physics of quantum Hall systems is quite different from that of doped antiferromagnets, it has been realized for some time [474, 475] that in higher Landau levels, a similar drama occurs due to the interplay between a short ranged attraction and a long range repulsion between electrons which gives rise to stripe and bubble phases. Evidence of these, as well as quantum Hall nematic phases [177,476], has become increasingly compelling in recent years. (For a recent review, see [178].) On a more speculative note, it has been noticed that such behavior may be expected in the neighborhood of many first order transitions in electronic systems, and it has been suggested that various charge inhomogeneous states may play a role in the apparent metal-insulator transition observed in the two-dimensional electron gas [477].

21.13 Stripes and High Temperature Superconductivity

In this article, we have analyzed the problem of high temperature superconductivity in a highly correlated electron liquid, with particular emphasis on doped antiferromagnets. We have identified theoretical issues, and even some solutions. We have also discussed aspects of the physics that elude a BCS description. This is progress.

However, we have not presented a single, unified solution to the problem. Contrast this with BCS, a theory so elegant it may captured in haiku:

Instability
Of a tranquil Fermi sea –
Broken symmetry.

Of course, to obtain a more quantitative understanding of particular materials would require a few more verses—we might need to study the Eliashberg equations to treat the phonon dynamics in a more realistic fashion, and we may need to include Fermi liquid corrections, and we may also have to wave our hands a bit about μ^*, etc. But basically, in the context of a single approximate solution of a very simple model problem, we obtain a remarkably detailed and satisfactory understanding of the physics. And while we may not be able to compute T_c very accurately—it does, after all, depend exponentially on parameters—we can understand what sort of metals will tend to be

[39] Alternatively, this state can be viewed as a bond-centered charge density wave [467,468].

good superconductors: metals with strong electron-phonon coupling, and consequently high room temperature resistances, are good candidates, as are metals with large density of states at the Fermi energy. We can also compute various dimensionless ratios of physical quantities, predict dramatic coherence effects (which *do* depend on microscopic details), and understand the qualitative effects of disorder.

The theory of high temperature superconductivity presented here reads more like a Russian novel, with exciting chapters and fascinating characters, but there are many intricate subplots, and the pages are awash in familiars, diminutives, and patronymics. To some extent, this is probably unavoidable. Fluctuation effects matter in the superconducting state: the phase ordering temperature, T_θ, is approximately equal to T_c, and the zero temperature coherence length, ξ_0, is a couple of lattice constants. In addition, the existence of one or more physical pseudogap scales (the T^*'s) in addition to T_c means that there are multiple distinct qualitative changes in the physics in going from high temperature to $T = 0$. Moreover, various other types of ordered states are seen in close proximity to or in coexistence with the superconducting state. Thus, it is more plausible that we will weave together a qualitative understanding of the basic physics in terms of a number of effective field theories, each capturing the important physics in some range of energy and length scales. Ideally, these different theories will be nested, with each effective Hamiltonian derived as the low energy limit of the preceding one.

While not as satisfying as the unified description of BCS-Eliashberg–Migdal theory, there is certainly ample precedent for the validity of this kind of multiscale approach. The number of quantitative predictions may be limited, but we should expect the approach to provide a simple understanding of a large number of qualitative observations. In fact, we may never be able to predict T_c reliably, or even whether a particular material, if made, will be a good superconductor, but a successful theory should certainly give us some guidance concerning what *types* of new materials are good candidate high temperature superconductors [478,479].

Before we continue, we wish to state a major change of emphasis. Up until this point, we have presented only results that we consider to be on secure theoretical footing. That is, we have presented a valid theory.[40] We now allow ourselves free rein to discuss the *applicability* of these ideas to the real world. In particular, we discuss the cuprate high temperature superconductors, and whether the salient physics therein finds a natural explanation in terms of stripes in doped antiferromagnets. Various open issues are laid out, as well as some general strategies for addressing them.

21.13.1 Experimental Signatures of Stripes

At the simplest level, stripes refer to a broken symmetry state in which the discrete translational symmetry of the crystal is broken in one direction: stripes is a term for a *unidirectional density wave*. "Charge stripes" refer to a unidirectional charge density wave (CDW). "Spin stripes" are unidirectional colinear spin density waves (SDW).[41] More subtle local forms of stripes, such as stripe liquids, nematics and glasses are addressed in Sect. 21.13.2.

Where Do Stripes Occur in the Phase Diagram?

As discussed in Sects. 21.11 and 21.12, holes doped into an antiferromagnet have a tendency to self-organize into rivers of charge, and these charge stripes tend to associate with antiphase domain walls in the spin texture. As shown in Sect. 21.12.2, stripe order is typically either "charge driven," in which case

[40] High temperature superconductivity being a contentious field, it will not surprise the reader to learn that there is controversy over how important each of the issues discussed above is to the physics of the cuprates. As the field progresses, and especially as new data are brought to light, it may be that in a future version of this article we, too, might change matters of *emphasis*, but we are confident that no new understanding will challenge the *validity* of the theoretical constructs discussed until now.

[41] Spiral SDW order has somewhat different character, even when unidirectional, and is not generally included in the class of striped states.

spin order onsets (if at all) at a temperature less than the charge ordering temperature, or "spin driven," if the charge order onsets as a weak parasitic order at the same temperature as the spin order. To the extent that stripes are indeed a consequence of Coulomb frustrated phase separation, we expect them to be charge driven, in this sense.

Experimental evidence of stripes has been detected in: Neutron scattering has proven the most useful probe for unambiguously detecting stripe order. Neutrons can scatter directly from the electron spins. However, neutrons (and, for practical reasons, X-rays as well) can only detect charge stripes indirectly by imaging the induced lattice distortions. Alternatively, (as discussed in Sect. 21.12.2) since spin stripe order implies charge order, the magnetic neutron scattering itself can be viewed as an indirect measure of charge order. Since stripe order is unidirectional, it should ideally show up in a diffraction experiment as pairs of new Bragg peaks at positions $\boldsymbol{k}_{\pm} = \boldsymbol{Q} \pm 2\pi\hat{e}/\lambda$ where $\hat{e}$ is the unit vector perpendicular to the stripe direction, λ is the stripe period, and $\boldsymbol{Q}$ is an appropriate fiduciary point. For charge stripes, $\boldsymbol{Q}$ is any reciprocal lattice vector of the underlying crystal, while for spin stripes, $\boldsymbol{Q}$ is offset from this by the Néel ordering vector, $< \pi, \pi >$. Where both spin and charge order are present, the fact that the charge stripes are associated with magnetic antiphase domain walls is reflected in the fact that $\lambda_{\text{spin}} = 2\lambda_{\text{charge}}$, or equivalently $\boldsymbol{k}_{\text{charge}} = 2\boldsymbol{k}_{\text{spin}}$.

LNSCO

$La_{1.6-x}Nd_{0.4}Sr_xCuO_4$ (LNSCO) is stripe ordered, and the onset of stripe ordering with temperature is clear. Furthermore, the charge ordering has been confirmed with X-ray scattering [480, 481]. Figure 21.47 shows data from neutron scattering, NQR, and susceptibility measurements [450]. In this material, charge stripes form at a higher temperature than spin stripes. Note also that static charge and spin stripes coexist with superconductivity throughout the superconducting dome. In fact experiments reveal quartets of new Bragg peaks, at $\boldsymbol{Q} \pm 2\pi\hat{x}/\lambda$ and

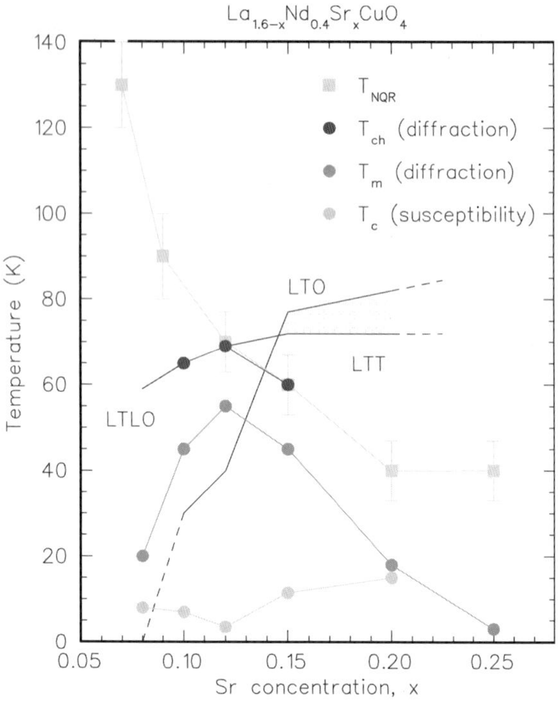

Fig. 21.47. *Blue data points* refer to the onset of charge inhomogeneity. *Red data points* denote the onset of incommensurate magnetic peaks. *Green data points* are the superconducting T_c. From Ichikawa et al. [450]

$\boldsymbol{Q} \pm 2\pi\hat{y}/\lambda$. In this material, the reason for this is understood to be a bilayer effect—there is a crystallographically imposed tendency for the stripes on neighboring planes to be oriented at right angles to each other, giving rise to two equivalent pairs of peaks.

LSCO

Spin stripe order has also been observed from elastic neutron scattering in $La_{2-x}Sr_xCuO_4$ (LSCO) for dopings between $x = .02$ and $x = .05$ where the material is not superconducting at any T; these stripes are called diagonal, because they lie along a direction rotated 45° to the Cu-O bond direction [164]. Above $x = .05$ [482], the stripes are vertical[42], i.e. along the Cu-O bond direction, and the samples are super-

[42] We should say *mostly* vertical. Careful neutron scattering work [165, 483] on LSCO and LCO has shown that the incommensurate peaks are slightly rotated from the Cu-O bond direction, corresponding to the orthorhombicity of the crystal.

conducting at low temperature. For dopings between $x = .05$ and $x = .13$, the stripes have an ordered (static) component. In the region $x = .13$ to $x = .25$, incommensurate magnetic peaks have been detected with inelastic neutron scattering. Because of the close resemblance between these peaks and the static order observed at lower doping, this can be unambiguously interpreted as being due to slowly fluctuating stripes.

Neutron scattering has also detected spin stripes in $La_2CuO_{4+\delta}$ (LCO) with $\delta = .12$ [483]. In this material, static stripes coexist with superconductivity even at optimal doping. In the $T_c = 42K$ samples (the highest T_c for this family thus far), superconductivity and spin stripe order onset simultaneously [166,483]. Application of a magnetic field suppresses the superconducting transition temperature, but has little effect on the ordering temperature of the spins [484].

In very underdoped nonsuperconducting LSCO, because the stripes lie along one of the orthorhombic axes, it has been possible to confirm [485,486] that stripe order leads, as expected, to pairs of equivalent Bragg spots, indicating unidirectional density wave order. In both superconducting LSCO and LCO, quartets of equivalent Bragg peaks are observed whenever stripe order occurs. This could be due to a bilayer effect, as in LNSCO, or due to a large distance domain structure of the stripes within a given plane, such that different domains contribute weight to one or the other of the two pairs of peaks. However, because the stripe character in these materials so closely resembles that in LNSCO, there is no real doubt that the observed ordering peaks are associated with stripe order, as opposed to some form of checkerboard order.

In $YBaCu_2O_{6+y}$ (YBCO), incommensurate spin fluctuations have been identified throughout the superconducting doping range [145, 160, 163, 488]. By themselves, these peaks (which are only observed at frequencies above a rather substantial spin gap) are subject to more than one possible interpretation [489], although their similarity [490] to the stripe signals seen in LSCO is strong circumstantial evidence that they are associated with stripe fluctuations. Recently, this interpretation has been strongly reinforced by several additional observations. Neutron scattering evidence [163] has been found of static charge stripe order in underdoped YBCO with $y = .35$ and $T_c = 39$K. The charge peaks persist to at least $300K$. The presence of a static stripe phase in YBCO means that inelastic peaks seen at higher doping are very likely fluctuations of this ordered phase. In addition, phonon anomalies have been linked to the static charge stripes at $y = .35$, and used to detect charge fluctuations at $y = .6$ [160]. By studying a partially detwinned sample with $y = .6$, with a 2: 1 ratio of domains of crystallographic orientation, Mook and collaborators were able to show that the quartet of incommensurate magnetic peaks consists of two inequivalent pairs, also with a 2:1 ratio of intensities in the two directions [491]. This confirms that in YBCO, as well, the signal arises from unidirectional spin and charge modulations (stripes), and not from a checkerboard-like pattern.

LBCO: Towards a unification

Recent high energy neutron scattering experiments have found a striking universality in the dispersion of magnetic excitations in YBCO, LBCO, and LSCO. The magnetic excitations in LBCO at $x = 1/8$ (where T_c is suppressed [492] due to spin and charge ordering [493–495]) have been measured up to 200meV [225]. (The charge ordering has been confirmed with resonant soft X-ray scattering [496]). The response of this stripe-ordered compound shows striking similarity at high energy to the excitations in $YBa_2Cu_3O_{6.6}$, which have been measured up to 105meV [497]. In particular, both display a resonance peak, with incommensurate branches emanating from the resonance at high and low energy. The dispersions (for twinned samples) are shown in Fig. 21.48, with each material's excitations normalized to the value of the superexchange energy J in the undoped parent antiferromagnet [487]. Also plotted is data for $La_{1.84}Sr_{0.16}CuO_4$, the low energy response of which follows the same universal dispersion.

Empirically, charge stripe formation precedes spin stripe formation as the temperature is lowered, and charge stripes also form at higher temperatures than T_c. Both types of stripe formation may be a phase transition, or may simply be a crossover of local stripe ordering, depending upon the material and doping. Where it can be detected, charge stripe for-

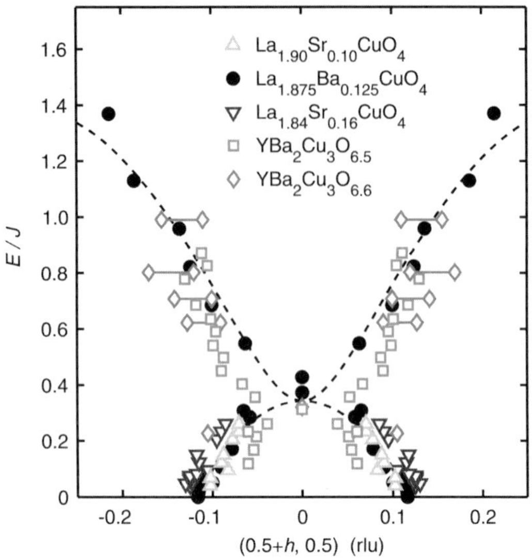

Fig. 21.48. Universal magnetic excitations of YBCO, LSCO, and LBCO, measured with neutron scattering. From Tranquada [487]

mation occurs at a higher temperature than the formation of the pairing gap,[43] consistent with the spin gap proximity effect (see Sect. 21.10.4).

BSCCO

Although some neutron scattering has been done on $Bi_2Sr_2CaCu_2O_{8+\delta}$ (BSCCO), the probe has only produced weak evidence of significant incommensurate structure [498]. The weak coupling of planes in BSCCO makes neutron scattering difficult, as it is difficult to grow the requisite large crystals. However, BSCCO is very well suited to surface probes such as ARPES and STM. Recent STM data, both with [499] and without [500, 501] an external magnetic field have revealed a *static* modulation in the local density of states that is very reminiscent of the incommensurate peaks observed with neutron scattering. Indeed, in both cases, the Fourier transform of the STM image exhibits a clear quartet of incommensurate peaks, just like those seen in neutron scattering in LSCO, LBCO, and YBCO. Here, however, unlike in the neutron scattering data, phase information is available in that Fourier transform. Using standard image enhancement methods, this phase information can be exploited [500] to directly confirm that the quartet of intensity peaks is a consequence of a domain structure, in which the observed density of states modulation is locally one-dimensional, but with an orientation that switches from domain to domain. However, disorder makes it difficult to distinguish stripes from a checkerboard modulation solely from STM measurements [502]. The use of STM as a probe of charge order is new, and there is much about the method that needs to be better understood [503] before definitive conclusions can be reached, but the results to date certainly look very promising.

Na-CCOC

Similar evidence of charge modulations have been detected via STM in $Ca_{2-x}Na_xCuO_2Cl_2$ [504], although X-ray scattering studies indicate that the charge order is either very weak, or only on the surface [505].

Preliminary evidence of nematic order has been detected, as well.

Finally, striking evidence of electronic anisotropy has been seen in untwinned crystals of $La_{2-x}Sr_xCuO_4$ ($x = 0.02 - 0.04$) and YBa_2CuO_{6+y} ($y = 0.35 - 1.0$) by Ando and collaborators [98]. The resistivity differs in the two in-plane directions in a way that cannot be readily accounted for by crystalline anisotropy alone. It is notable that in YBCO, the anisotropy increases as y is decreased. That is, the electrical anisotropy *increases* as the orthorhombicity is *reduced*. In some cases, substantial anisotropy persists up to temperatures as high as 300K. Furthermore, for $y < 0.6$, the anisotropy increases with decreasing temperature, much as would be expected [506] for an electron nematic. These observations from transport correlate well with the evidence from neutron scattering [491], discussed above, of substantial orientational order of the stripe correlations in YBCO, and with the substantial, and largely temperature independent anisotropy of the superfluid density observed in the same material [13]. Together, these observations constitute important, but still preliminary evidence of a nematic stripe phase in the cuprates.

[43] See our discussion of the pseudogap(s) in Sect. 21.3.5.

21.13.2 Stripe Crystals, Fluids, and Electronic Liquid Crystals

Stripe ordered phases are precisely defined in terms of broken symmetry. A charge stripe phase spontaneously breaks the discrete translational symmetry and typically also the point group symmetry (e.g. fourfold rotational symmetry) of the host crystal. A spin stripe phase breaks spin rotational symmetry as well. While experiments to detect these orders in one or another specific material may be difficult to implement for practical reasons and because of the complicating effects of quenched disorder, the issues are unambiguous. Where these broken symmetries occur, it is certainly reasonable to conclude that the existence of stripe order is an established fact. That this can be said to be the case in a number of superconducting cuprates is responsible for the upsurge of interest in stripe physics.

It is much more complicated to define precisely the intuitive notion of a "stripe fluid".[44] Operationally, it means there is sufficient short ranged stripe order that, for the purposes of understanding the mesoscale physics, it is possible to treat the system *as if* it were stripe ordered, even though translational symmetry is not actually broken. It is possible to imagine intermediate stripe liquid phases which are translationally invariant, but which still break some symmetries which directly reflect the existence of local stripe order. The simplest example of this is an "electron nematic" phase.

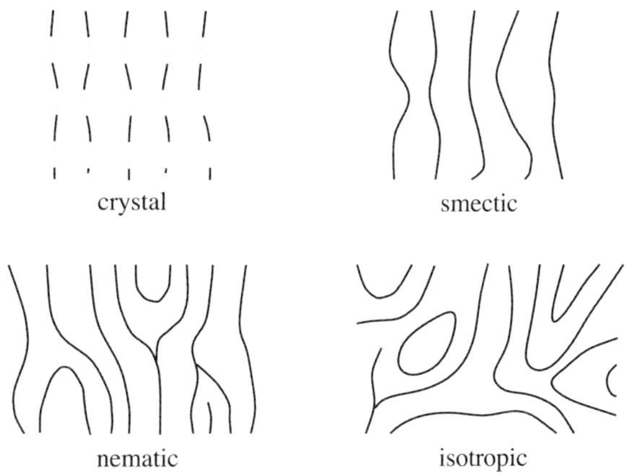

Fig. 21.49. Schematic representation of various stripe phases in two dimensions. The *broken lines* represent density modulations along the stripes. In the electronic crystal, density waves on neighboring stripes are locked in phase and pinned. The resulting state is insulating and breaks translation symmetry in all directions. *Solid lines* represent metallic stripes along which electrons can flow. They execute increasingly violent transverse fluctuations as the system is driven towards the transition into the nematic phase. The transition itself is associated with unbinding of dislocations that are seen in the snapshot of the nematic state. The isotropic stripe fluid breaks no spatial symmetries of the host crystal, but retains a local vestige of stripe order

Some stripe liquids break rotational symmetry.
In classical liquid crystals, the nematic phase occurs when the constituent molecules are more or less cigar shaped. It can be thought of as a phase in which the cigars are preferentially aligned in one direction, so that the rotational symmetry of free space is broken (leaving only rotation by π intact) but translational symmetry is unbroken. In a very direct sense, this pattern of macroscopic symmetry breaking is thus encoding information about the microscopic constituents of the liquid. In a similar fashion, we can envisage an electron nematic as consisting of a melted stripe ordered phase in which the stripes meander, and even break into finite segments, but maintain some degree of orientational order—for instance, the stripes are more likely to lie in the x rather than the y direction; see Fig. 21.49.

Melting stripes
One way to think about different types of stripe order is to imagine starting with an initial "classical" ordered state, with coexisting unidirectional SDW and CDW order. As quantum fluctuations are increased (metaphorically, by increasing $\hbar$), one can envisage that the soft orientational fluctuations of the spins will first cause the spin order to quantum melt, while the charge order remains. If the charge order, too, is to quantum melt in a continuous phase transition, the resulting state will still have the stripes gener-

[44] For the present purposes, the term "fluctuating stripes" is taken to be synonymous with a stripe fluid. See, for example, [442] and [507].

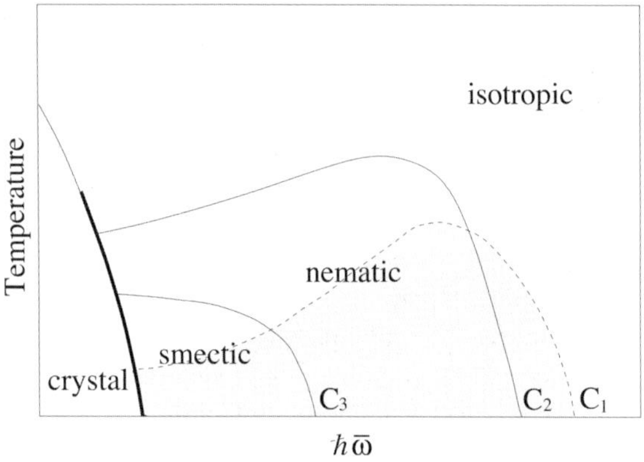

Fig. 21.50. Schematic phase diagram of a fluctuating stripe array in a (tetragonal) system with fourfold rotational symmetry in $D = 2$. Here $\hbar\bar{\omega}$ is a measure of the magnitude of the transverse zero point stripe fluctuations. Thin lines represent continuous transitions and the thick line a first order transition. We have assumed that the superconducting susceptibility on an isolated stripe diverges as $T \to 0$, so that at finite stripe density, there will be a transition to a globally superconducting state below a finite transition temperature. On the basis of qualitative arguments, discussed in the text, we have sketched a boundary of the superconducting phase, indicated by the *shaded region*. Depending on microscopic details the positions of the quantum critical points C_1 and C_2 could be interchanged. Distinctions between various possible commensurate and incommensurate stripe crystalline and smectic phases are not indicated in the figure. Similarly, all forms of spin order are neglected in the interest of simplicity

ally oriented in the same direction as in the ordered state, but with unbound dislocations which restore translational symmetry.[45] If the underlying crystal is tetragonal [512], this state still spontaneously breaks the crystal point group symmetry. In analogy with the corresponding classical state, it has been called an electron nematic, but it could also be viewed as an electronically driven orthorhombicity. This is still a state with broken symmetry, so in principle its existence should be unambiguously identifiable from experiment.[46] The order parameter can be identified with the matrix elements of any traceless symmetric tensor quantity, for instance the traceless piece of the dielectric or conductivity tensors.

With this physics in mind, we have sketched a qualitative phase diagram, shown in Fig. 21.50, which provides a physical picture of the consequences of melting a stripe ordered phase. As a function of increasing quantum and thermal transverse stripe fluctuations one expects the insulating electronic crystal, which exists at low temperatures and small $\hbar$, to evolve eventually into an isotropic disordered phase. At zero temperature this melting occurs in a sequence of quantum transitions [52]. The first is a first order transition into a smectic phase, then by dislocation unbinding a continuous transition leads to a nematic phase that eventually evolves (by a transition that can be continuous in $D = 2$, but is first order in a cubic system) into the isotropic phase. Similar transitions exist at finite temperature as indicated in Fig. 21.50.

Superconducting electronic liquid crystals

We have also sketched a superconducting phase boundary in the same figure. Provided that there is a spin gap on each stripe, and that the charge Luttinger exponent $K_c > 1/2$, then (as discussed in Sect. 21.5) there is a divergent superconducting susceptibility on an isolated stripe. In this case, the superconducting T_c is determined by the Josephson coupling between stripes. Since, as discussed in Sect. 21.6, the mean Josephson coupling increases with increasing stripe fluctuations, T_c also rises with increasing $\hbar$

[45] It is also possible to view the electron nematic from a weak coupling perspective, where it occurs as a Fermi surface instability [508], sometimes referred to as a Pomeranchuk instability [509,510]. This instability is "natural" when the Fermi surface lies near a Van Hove singularity. The relation between the weak coupling and the stripe fluid pictures is currently a subject of ongoing investigation [511].

[46] It is probable that when nematic order is lost, the resulting stripe liquid phase is not thermodynamically distinct from a conventional metallic phase, although the local order is sufficiently different that one might expect them to be separated by a first order transition. However, it is also possible that some more subtle form of order could distinguish a stripe liquid from other electron liquid phases—for instance, it has been proposed by Zaanen and collaborators [513] that a stripe liquid might posses an interesting, discrete topological order which is a vestige of the antiphase character of the magnetic correlations across a stripe.

throughout the smectic phase. While there is currently no well developed theory of the superconducting properties of the nematic phase,[47] to the extent that we can think of the nematic as being locally smectic, it is reasonable to expect a continued increase in T_c across much, or all of the nematic phase, as shown in the figure. However, as the stripes lose their local integrity toward the transition to the isotropic phase we expect, assuming that stripes are essential to the mechanism of pairing, that T_c will decrease, as shown.

The study of electronic liquid crystalline phases is in its infancy. Increasingly unambiguous experimental evidence of the existence of nematic phases has been recently reported in quantum Hall systems [177,179,476,506,515] in addition to the preliminary evidence of such phases in highly underdoped cuprates discussed above. Other more exotic electronic liquid crystalline phases are being studied theoretically. This is a very promising area for obtaining precise answers to well posed questions that may yield critical information concerning the important mesoscale physics of the high temperature superconductors.

21.13.3 Our View of the Phase Diagram—Reprise

It's all about kinetic energy.

Since the motion of dilute holes in a doped antiferromagnet is frustrated, the minimization of their kinetic energy is a complicated, multistage process. We have argued that this is accomplished in three stages: (a) the formation of static or dynamical charge inhomogeneity (stripes) at T^*_{stripe}, (b) the creation of local spin pairs at T^*_{pair}, which creates a spin gap, and (c) the establishment of a phase-coherent superconducting state at T_c. The zero point kinetic energy is lowered along a stripe in the first stage, and perpendicular to the stripe in the second and third stages. Steps (a), (b), and (c) above are clearcut only if the energy scales are well separated, that is, if $T^*_{\text{stripe}} \gg T^*_{\text{pair}} \gg T_c$. On the underdoped side at least, if we identify T^*_{stripe} and T^*_{pair} with the appropriate observed pseudogap phenomena (see Sect. 21.3.5) there is a substantial (if not enormous) separation of these temperature scales.

Pseudogap Scales

At high temperatures, the system must be disordered. As temperature is lowered, the antiferromagnet ejects holes, and charge stripe correlations develop. This may be either a phase transition or a crossover. We have called this temperature T^*_{stripe} in Fig. 21.12. Even if it is a phase transition, for instance a transition to a stripe nematic state, local order may develop above the ordering temperature, and probes on various time scales may yield different answers for T^*_{stripe}. As the antiferromagnet ejects holes, local antiferromagnet correlations are allowed to develop. Probes bearing on this temperature include the Knight shift, NQR, and diffraction. At a lower temperature, through communication with the locally antiferromagnetic environment, a spin gap develops on stripes. We identify this spin gap with the pairing gap, and have labeled this temperature (which is always a crossover) T^*_{pair}. Probes bearing on this temperature measure the single particle gap, and include ARPES, tunneling, and NMR.

Dimensional Crossovers

Dimensional crossovers are a necessary consequence of stripe physics.

Looking at this evolution from a broader perspective, there are many consequences that can be understood based entirely on the notion that the effective dimensionality of the coherent electronic motion is temperature dependent. At high temperatures, before local stripe order occurs, the electronic motion is largely incoherent— i.e the physics is entirely local. Below T^*_{stripe}, the motion crosses over from quasi

[47] Some very promising recent progress toward developing a microscopic theory of the electron nematic phase has been reported in [508] and [514].

0D to quasi 1D behavior.[48] Here, significant $\boldsymbol{k}$ space structure of various response functions is expected, and there may well emerge a degree of coherence and possibly pseudogaps, but the electron is not an elementary excitation, so broad spectral functions and non-Fermi liquid behavior should be the rule. Then, at a still lower temperature, a 1D to 3D crossover occurs as coherent electronic motion between stripes becomes possible. At this point coherent quasiparticles come to dominate the single particle spectrum, and more familiar metallic and/or superconducting physics will emerge. If the spin gap is larger than this crossover temperature (as it presumably is in underdoped materials), then this crossover occurs in the neighborhood of T_c. However, if the spin gap is small, then the dimensional crossover will likely occur at temperatures well above T_c, and T_c itself will have a more nearly BCS character, as discussed in Sect. 21.5.3—this seems to be crudely what happens in the overdoped materials [519]. Since once there are well developed quasiparticles, there is every reason to expect them to be able to move coherently between planes, there is actually no substantial region of quasi 2D behavior expected. Although it may be hard, without a macroscopically oriented stripe array, to study the dimensional crossover by measuring in-plane response functions, the dimensional crossover can be studied by comparing in-plane to out-of plane behavior.[49]

The Cuprates as Quasi-1D Superconductors

When $T^*_{\text{stripe}} \gg T^*_{\text{pair}} \gg T_c$, the model of a quasi-one-dimensional superconductor introduced in Sect. 21.5.3 is applicable in the entire temperature range below T^*_{stripe}. The application of these results to the overdoped side is suspect, since that is where all of these energy scales appear to crash into each other.

ARPES and stripes

The temperature dependence of the spectral response of a quasi-one-dimensional superconductor may be described as follows: At temperatures high compared to both the Josephson coupling and the spin gap, the system behaves as a collection of independent (gapless) Luttinger liquids. Spin-charge separation holds, so that an added hole dissolves into a spin part and a charge part. Consequently the spectral response exhibits broad EDC's and sharp MDC's.[50] In the intermediate temperature regime (below the spin gap), spin-charge separation still holds, and the ARPES response still exhibits fractionalized spectra, but with a pseudogap. In the low temperature phase, Josephson coupling between stripes confines spin and charge excitations, restoring the electron as an elementary excitation, and a sharp coherent peak emerges from the incoherent background, with weight proportional to the coupling between stripes.

There is a wealth of ARPES data on BSCCO, a material which lends itself more to surface probes than to diffraction. However, as mentioned previously, the presently available evidence of stripes in this material is compelling, but not definitive, so it requires a leap of faith to interpret the ARPES data in terms of stripes. The best evidence of stripes comes from STM data which is suggestive of local stripe correlations [499,500]. Since STM observes a static modulation, any stripes observed in STM can certainly be considered static as far as ARPES is concerned.[51] As long as the stripes have integrity over a length scale at least as large as $\xi_s = v_s/\Delta_s$, it is possible for the

[48] It is intuitively clear that kinetic energy driven stripe formation should lead to increased hole mobility, as is observed, but how the famous T-linear resistivity can emerges from local quasi-0D physics is not yet clear. See, however, [390,516–518].

[49] Much of the successful phenomenology of dimensional crossover developed in conjunction with the interlayer pairing mechanism of superconductivity [21] is explained in this way in the context of a stripe theory.

[50] See Sect. 21.5 for a description of EDC's and MDC's.

[51] Unfortunately,there is currently little direct experimental information concerning the temperature dependence of the stripe order in BSCCO, although what neutron scattering evidence does exist [498], suggests that substantial stripe correlations survive to temperatures well above the superconducting T_c.

stripes to support superconducting pairing through the spin gap proximity effect.

ARPES spectra from the antinodal region resemble a quasi-1D superconductor.
At any rate, many features of the ARPES spectra, especially those for $\boldsymbol{k}$ in the antinodal region of the Brillouin zone (near $(\pi, 0)$) in BSCCO are unlike anything seen in a conventional metal, and highly reminiscent of a quasi-1D superconductor. Above T_c, ARPES spectra reveal sharp MDC's and broad EDC's. We take this [86] as evidence of electron fractionalization. Below T_c, a well defined quasiparticle peak emerges [89], whose features are strikingly similar to those derived in this model. The quasiparticle peak is nearly dispersionless along the $(0, 0)$ to $(\pi, 0)$ direction, and within experimental bounds its energy and lifetime are temperature independent. The only strongly temperature dependent part of the spectrum is the intensity associated with the superconducting peak. The temperature dependence of the intensity is consistent with its being proportional to a fractional power of the local condensate fraction or the superfluid density. Similar behavior has been measured now in an untwinned single crystal of YBCO [12] as well.

Stripes and superconductivity involve the same regions of k-space
The most dramatic signatures of superconducting phenomena in ARPES experiments, both the development of the gap *and* the striking onset of the superconducting peak with phase coherence, occur in the *same* regions of k-space most associated with stripes: Specifically, an array of "horizontal" charge stripes embedded in a locally antiferromagnetic environment [520–523] has most of its low energy spectral weight concentrated near the $(\pi, 0)$ regions of k-space. Similarly, the strongest gap develops in the $(\pi, 0)$ regions, and in both BSCCO and YBCO, the only dramatic change in the ARPES response upon entering the superconducting state is the coherent peak seen in these same regions.

The ARPES spectrum from the nodal region ($\boldsymbol{k}$ near $(\pi/2, \pi/2)$) is less obviously one-dimensional in character, although the nodal spectrum is certainly *consistent* with the existence of stripes, as has been demonstrated in several model calculations [133, 520–522, 524]. However, to a large extent, the spectrum in the nodal region is insensitive to stripe correlations [133]. Nodal quasiparticles are certainly important for the low temperature properties of the superconducting state. Moreover, there is indirect evidence that they dominate the in-plane transport above T_c. But the fact that the ARPES spectrum in the nodal direction does not change [525] in any dramatic fashion from above to below T_c, as one would have deduced even from the simplest BCS considerations, suggests that they do not play a direct role in the mechanism of superconductivity. This observation, however, must not be accepted unconditionally. There is an apparent contradiction between the smooth evolution of the spectral function observed in ARPES and the evolution inferred from macroscopic transport experiments [526, 527] (see also [549]); the latter suggest that a catastrophic change in the nodal quasiparticle lifetime occurs in the immediate neighborhood of T_c.

Inherent Competition

Finally, it should be made clear that a stripes based mechanism of high temperature superconductivity *predicts* competition between stripes and superconductivity:

We, too, think stripes compete with superconductivity. static stripes may be good for pairing, but are certainly bad for the Josephson coupling (superfluid stiffness) between stripes. On the other hand, fluctuating stripes produce better Josephson coupling, but weaker pairing. The dependence of the gap on stripe fluctuations finds its origin in the spin gap proximity effect, where the development of the spin gap hinges on the one dimensionality of the electronic degrees of freedom [20], whereas stripe fluctuations cause the system to be more two-dimensional. In addition, as described in Sect. 21.6, stripe fluctuations work against $2k_F$ CDW order along a stripe, but strengthen the Josephson coupling.

This is consistent with the empirical phase diagram: on the underdoped side there is a large gap, small superfluid stiffness, small transition temperature, and static stripes have been observed. With increasing doping, stripes fluctuate more, reducing the pairing gap, but increasing the Josephson coupling between stripes. This is a specific example of the doping dependent crossover scenario proposed in [110, 279], in which underdoped cuprates have a strong pairing scale but weak phase stiffness and T_c is determined more or less by T_θ, whereas the overdoped cuprates have a strong phase stiffness but weak pairing scale and T_c is more closely associated with T^*_{pair}. Optimal doping is a crossover between a dominantly phase ordering transition and a dominantly pairing transition. Recently, direct evidence that optimal doping is a crossover of two energy scales was reported in an ARPES study of homologous compounds [528].

21.13.4 Some Open Questions

Absence of evidence is not evidence of absence.

As has been stressed by many authors, the cuprate superconductors are exceedingly complex systems. Crisp theoretical statements can be made concerning the behavior of simplified models of these systems, but it is probably ultimately impossible to make clean predictions about whether the results will actually be found in any given material. We are therefore reliant on experiment to establish certain basic empirical *facts*. In this subsection, we will discuss some of the fundamental issues of fact that are pertinent to the stripe scenario presented above, and make a few comments about the present state of knowledge concerning them. A word of caution is in order before we begin: positive results have clearer implications than negative results. Especially in these complicated materials, there can be many reasons to fail to see an effect.

Are Stripes Universal in the Cuprate Superconductors?

If stripes are not, in some sense, universal in the high temperature superconductors, then they cannot be, in any sense, essential to the mechanism of high temperature superconductivity. So an important experimental issue is whether stripes are universal in the cuprate superconductors.

The evidence from neutron scattering is discussed in Sect. 21.13.1: Incommensurate (IC) spin peaks (whether elastic or inelastic) have been detected throughout the doping range of superconductivity in the lanthanum compounds. In YBCO, IC spin peaks are seen with inelastic scattering, but it is presently unclear how much of that scattering intensity should be associated with stripe fluctuations, and how much should be associated with the "resonance peak". Neutron scattering has produced some evidence [498] of IC spin peaks in BSCCO, but this result is controversial [529]. No such peaks have been reported in TlBaCaCuO or HgBaCaCuO, although little or no neutron scattering has yet been done on crystals of these materials.

CDW order turns out to be much harder to observe, even when we know it is there. Charge stripe order has only been observed directly in $La_{1.6-x}Nd_{0.4}Sr_xCuO_4$ [450] and very underdoped $YBa_2Cu_3O_{7-\delta}$ [163], although the general argument presented in Sect. 21.12.2 implies that it must occur wherever spin stripe order exists. Given the difficulty in observing the charge order where we know it exists, we consider an important open question to be: Where does charge stripe order exist in the general phase diagram of the cuprate superconductors?

As mentioned before, STM experiments point to local charge stripes in Na-CCOC [504] and in BSCCO, both with [499] and without [500, 501] a magnetic field. But there is nowhere near enough systematic data to know whether charge stripes are ubiquitous as a function of doping and in all the superconducting cuprates, how pronounced it is, and over what range of temperatures significant stripe correlations exist, even where we know they exist at low temperatures. Perhaps, in the future, this issue can be addressed further with STM, or even with ARPES or new and improved X-ray scattering experiments.

Are Stripes an Unimportant Low Temperature Complication?

There is a general tendency for increasingly subtle forms of order to appear as systems are cooled—involving residual low energy degrees of freedom that remain after the correlations that are the central features of the physics have developed. (A classic example of this is transitions involving ordering of the nuclear moments at ultra-low temperatures in a metal.) While such forms of order are fascinating in their own right, one would not, typically, view them as important aspects of the basic materials physics of the studied system. There is a body of thought that holds that the various forms of stripe order that have been observed are in this class of phenomena—interesting side shows, but not the main event. It is also true that actual, static stripe order has only been observed under rather restrictive conditions—mostly in highly underdoped materials or materials with significantly depressed superconducting T_c's, and at temperatures less than or of order the optimal superconducting T_c.

To be central to the physics of high temperature superconductivity, charge stripes must occur at high enough energies and temperatures that they are relevant to zeroth order. Specifically, we want to look for evidence that local stripes persist up to temperatures which are greater than or equal to T_c. If stripes are universal, then there must be a characteristic crossover scale below which significant stripe correlations emerge—clearly, at high enough temperature, no significant self-organization is possible. Undoubtedly, there is a high energy scale associated with one or more pseudogap crossovers in many underdoped materials—can we associate some of this crossover with the scale at which local stripe correlations become significant? If so, then manifestly stripes are a central player in the drama. If not, and if no still higher energy scale can be identified at which stripe physics begins, it would become increasingly difficult to envisage a starring role for stripes in the physics of the cuprates.

This issue has not been unambiguously resolved. There is substantial (yet not definitive) evidence that local stripe order persists to rather high temperatures. Evidence of local stripe order from observed [530] infrared active phonon modes has been seen to persist to at least 300K in highly underdoped $La_{2-x}Sr_xCuO_4$. Phonon anomalies, which have been tentatively associated with stripes, have been observed in neutron scattering experiments in slightly underdoped $YBa_2Cu_3O_{7-\delta}$ up to comparable temperatures [160]. Still more indirect evidence also abounds. This is a key question, and much more work is necessary to resolve it.

Are the Length and Time Scales Reasonable?

As emphasized above, to understand the mechanism of high temperature superconductivity, we are primarily concerned with mesoscopic physics, on length scales a few times the superconducting coherence length and time scales a few times $\hbar/\Delta_0$. So the real question we want to address is: Does stripe order exist on these length and time scales? Given that it is so difficult to determine where long range charge stripe order occurs, it is clearly still more complicated to determine where substantial stripey short range order occurs, or even precisely how much short range order is sufficient.

Are Stripes Conducting or Insulating?

The earliest theoretical studies which predicted stripes as a general feature of doped antiferromagnets envisaged insulating stripes [405,407,408]. These stripes are conceptually close relatives of conventional CDW's in that they are obtained as a Fermi surface instability due to near perfect nesting of the Fermi surface. Such stripes have no low lying fermionic excitations. This perspective has led to an interesting theory of superconductivity which relies on stripe defects for charge transport [531].

The strongest evidence that charge stripes are incompressible, and therefore insulating, comes from plotting the magnetic incommensurability against the doping concentration. If this relationship is strictly linear, it implies that the concentration of holes on a stripe does not change, but rather the only effect of further doping is to change the concentration of stripes in a plane, bringing the stripes closer together. The data for LSCO are close to linear in the range $.024 \leq x \leq .12$, despite the change in orientation from diagonal to vertical at $x = .05$ [167,532], but the small deviation from linearity below $x = .06$ does exceed the error bars. At present, the data leave open the possibility that the relationship is not strictly linear, and is also consistent with compressible (metallic) stripes throughout the doping range where they are observed (see, e.g. Fig. 7 of [482]).

Most of the other experiments we have mentioned support the notion that the stripes are intrinsically metallic. Of course, the observed coexistence of static stripe order and superconductivity is a strong indicator of this, as presumably it would be hard to attribute long distance charge transport to stripe motion.[52] The situation is most dramatic in nonsuperconducting LSCO with $0.02 < x < 0.05$, where the stripes are ordered [167,532], and far enough separated that the intrinsic properties of an individual stripe must surely determine the electronic structure—the mean stripe spacing [532] grows to be as large as 350 Å or so for $x = 0.02$.[53] These materials exhibit [98, 533] a metallic (linear in T) temperature dependence of the resistivity down to moderate temperatures. More remarkably, as shown [533] by Ando et al., although the magnitude of the resistance is large compared to the quantum of resistance at all temperatures, when interpreted in terms of a model in which the conduction occurs along dilute, metallic stripes, the inferred electron mobility within a stripe is nearly the same as that observed in optimally doped LSCO!

Are Stripes Good or Bad for Superconductivity?

The Uemura plot and the Yamada plot may be about the same physics.

Striking empirical evidence which suggests that stripes and superconductivity are related comes from the Yamada plot [482], which reports T_c vs. the incommensurability seen in neutron scattering, i.e. the inverse spacing between stripes. First noted in LSCO, the relationship is remarkably linear for the underdoped region of the lanthanum compounds [482]. For far separated stripes, the transition temperature is depressed. As the stripes move closer together, and the Josephson coupling between them increases, T_c increases. In addition, the similarities between the Yamada plot and the Uemura plot [107], which shows a linear relationship between T_c and the superfluid density, indirectly imply that the Josephson coupling between stripes plays an important role in determining the macroscopic superfluid density.

It has been argued that since stripes compete with superconductivity, they cannot be involved in the mechanism of superconductivity [51]. (We would point out that, at the very least, such competition must imply that stripes and superconductivity are strongly connected.) The empirics are presently unclear on the issue. There is some evidence that static stripes compete with superconductivity, whereas fluctuating stripes enhance it. In instances where stripes are pinned, T_c is generally suppressed, such as with Nd doping, Zn doping, or at the 1/8 anomaly. An exception to this trend occurs in the LCO family, which exhibits its highest T_c for a static stripe ordered material. Recently, Ichikawa et al. [450] have argued that it is spin stripe order, rather than charge stripe order, which competes with superconductivity. Whatever the details, the gross trend in materials other than LCO seems to be that the highest transition temperatures are achieved for dopings that presumably do not support actual (static) stripe order. It is also worth noting that in LSCO [534] and YBCO [146], neutron scattering shows a gap developing in the incommensurate magnetic fluctuations at T_c, perhaps indicating that superconductivity favors fluctuating stripes.

On the other hand, T_c is a nonmonotonic function of x, and pretty clearly determined by the lesser of two distinct energy scales. But the superconducting gap, as deduced from low temperature tunneling or ARPES experiments deep in the superconducting state, is a monotonically decreasing function of x. It is generally believed that stripe correlations are similarly strongest when x is small and vanish with sufficient overdoping, although in truth the direct experimental evidence for this intuitively obvious statement is not strong. Thus, there is at least a generally positive correlation between the degree of local stripe order and the most obvious scale characterizing pairing. This leads us to our next question:

[52] One could envisage stripe defect motion which transports charge perpendicular to the stripes [531], but certainly the effective number of carriers due to this effect must be small.

[53] This is equivalent to 64 (orthorhombic) lattice constants, $b^*_{\mathrm{ortho}} = 5.41$ Å [164].

Do Stripes Produce Pairing?

It is well known that the physics of an antiferromagnet is kinetic energy driven, and phase coherence must be kinetic energy driven when $T_{\mathrm{pair}} \gg T_c$, since spatial fluctuations of the phase drive pair currents. But can pair formation be kinetic energy driven? In particular, do stripes produce pairing? As reviewed in Sect. 21.11, numerical studies do find pairing in "fat" 1D systems.

However, there is no experiment we can point to that proves the pairing is either kinetic energy driven[54] or due to stripes. Nor is it clear what such an experiment would be. There are ways to falsify the conjecture that stripes produce pairing, such as a demonstration that stripes are not in some sense ubiquitous in the cuprates, or a demonstration that pairing generally precedes local charge stripe formation as the temperature is lowered. We have discussed many predictions which find some support in experiments, such as the fact that static stripes are good for pairing but bad for phase coherence, and vice versa, and the systematics of the superconducting coherence peak. But these interpretations are not necessarily unique. Much of the phenomenology is *consistent* with a spin gap proximity effect mechanism of pairing, but we see no smoking gun.

Do Stripes Really Make the Electronic Structure Quasi-1D?

Does the existence of stripes provide a sufficient excuse to treat the cuprates as self-organized quasi-1D conductors? If so, then we can apply many of the insights we have obtained directly, and without apology to the interpretation of experiment. As has been discussed in previous sections, and in considerably more detail in other places [6,20,86,149,520,521,523], there are many striking experiments in the cuprates that can be simply and naturally understood in this way. But do they actually affect the electronic structure so profoundly as to render it quasi-1D?

The most direct evidence comes from the ARPES results of Shen and collaborators [87] on the stripe ordered material, $La_{1.6-x}Nd_{0.4}Sr_xCuO_4$. These experiments reveal a remarkable confinement of the majority of the electronic spectral weight inside a dramatically 1D Fermi surface. This experiment probes fairly high energy excitations, and so demonstrates a profound effect of an ordered stripe array on all aspects of the electronic structure. More generally, studies have shown [86,149,520,523] that many of the most striking features of the ARPES spectra of the cuprates are readily rationalized on the basis of an assumed, locally quasi-1D electronic structure.

Transport measurements are macroscopic, so even if locally the electronic structure is strongly quasi-1D, the effects of stripe meandering, domain formation, and disorder will always produce a substantially reduced effective anisotropy at long distances.

Macroscopic anisotropy
From this perspective, the order 1, strongly temperature dependent transport anisotropies observed by Ando and collaborators [98] in LSCO and YBCO provide tangible evidence of a strong susceptibility of the electron liquid in the copper oxide planes to develop anisotropies in tensor response functions. Less direct, but even more dramatic evidence that stripes make the electron dynamics quasi-1D has been adduced from Hall effect measurements on the stripe ordered material, $La_{1.6-x}Nd_{0.4}Sr_xCuO_4$, by Noda et al. [535] They have observed that the Hall coefficient, R_H, which is relatively weakly temperature depen-

[54] The brilliantly conceived high precision measurements of the optical conductivity of van der Marel and collaborators [27], and more recently by Bontemps and collaborators [28], are highly suggestive in this regard. In optimally doped BSCCO, they observe a strongly temperature dependent change in the optical spectral weight integrated up to frequencies two orders of magnitude greater than T_c—if interpreted in terms of the single band sum rule, this observation implies a decrease in the kinetic energy upon entering the superconducting state of a magnitude comparable to reasonable estimates of the condensation energy. This is very striking, since in a BCS superconductor, the kinetic energy would *increase* by precisely this amount. However, neither the single band sum rule, nor the notion of a condensation energy are unambiguously applicable in the present problem. This is *the best* existing evidence that the mechanism of superconductivity is kinetic energy driven, but it is not yet evidence that would stand up in court.

dent above the stripe ordering transition temperature, T_{co}, drops dramatically for $T < T_{co}$, such that $R_H \to 0$ as $T \to 0$ for doped hole concentration, $x \leq 1/8$, and R_H tends to a reduced but finite value for $x > 1/8$. This observation was initially interpreted [535] as evidence that ordered stripes prevent coherent transverse motion of electrons within the copper oxide plane; this interpretation was later shown to be not entirely correct [206], although the basic conclusion that the stripes render the electron dynamics quasi-one-dimensional is probably sound. Further evidence that stripe formation inhibits transverse electronic motion is strongly suggested by the observed suppression of c-axis coherent charge motion in the stripe ordered state of the same materials [536].

However, it would be very desirable to develop new strategies to directly address this issue. For instance, a defect, such as a twin boundary, could purposely be introduced to locally aline the stripe orientation, and the induced electronic anisotropy then be detected with STM.

What About Overdoping?

On the underdoped side of the phase diagram of the cuprates, the energy scales of T^*_{stripe}, T^*_{pair}, and T_c are generally sufficiently separated to make the application of many of these ideas plausible. Yet on the overdoped side, the energy scales seem to come crashing into each other, depressing T_c. Furthermore, on the overdoped side, we have $T_\theta > T_{\text{pair}}$, in violation of a common assumption we have made throughout this article. The very existence of stripes on the overdoped side is questionable. The Uemura and Yamada plot are not satisfied there. If there are no stripes, and yet there is superconductivity, this does not bode well for a stripes based mechanism. Indeed, it is easier to believe that a mean field like solution is crudely applicable on the overdoped side, where T_c is closer to T_{pair} than it is to T_θ.

One possibility is that the superconducting state far on the overdoped side (especially, where T_c is low and the normal state ARPES spectrum begins to look more Fermi liquid-like) is best approached in terms of a Fermi surface instability and a BCS-Eliashberg mechanism, while on the underdoped side it is best viewed from a stripes perspective. In keeping with the multiscale approach advocated above, it may be no simple matter to unify these approaches in a smooth way.

However, there is an attractive possibility that is worth mentioning here. As we have mentioned, in a stripe liquid, so long as the characteristic stripe fluctuations frequency, $\hbar\bar{\omega}$, is small compared to the superconducting gap scale, the stripes can be treated as quasi-static for the purposes of understanding the mechanism of pairing. Conversely, when $\hbar\bar{\omega} \gg \Delta_0$, the stripe fluctuations can be integrated out to yield an effectively homogeneous system with an induced interaction between electrons. Indeed, it has previously been proposed [125] that stripe fluctuations themselves are a candidate for the "glue" that mediates an effective attraction between electrons. It is easy to imagine that $\hbar\bar{\omega}/\Delta_0$ is a strongly increasing function of x. A sort of unification of the two limits could be achieved if stripe fluctuations play the role of the intermediate boson which mediates the pairing in highly overdoped materials, while in underdoped and optimally doped materials the system can be broken up into quasi-1D ladders, which exhibit the spin gap proximity effect.

How Large is the Regime of Substantial Fluctuation Superconductivity?

This important question is fundamentally ill-defined. It is important, because its answer determines the point of view we take with regard to a number of key experiments. But it is ill-defined in the following sense: in the neighborhood of any phase transition, there is a region above T_c where substantial local order exists, but how broad the fluctuation region is said to be depends on exactly how "substantial" is defined, or measured. There has been an enormous amount written on this subject already, so we will just make a few brief observations.

Because in one dimension, phase fluctuations always reduce the superconducting T_c to zero, in a quasi-one-dimensional superconductor (i.e. in the limit of large anisotropy), there is necessarily a parametrically large fluctuation regime between the mean field transition temperature and the actual ordering temperature.

The finite frequency superfluid density measured in BSCCO [170] with $T_c = 74K$ shows a local superfluid density persists up to at least $90K$, indicative of fluctuation superconductivity in that regime. Both microwave absorption [171] and thermal expansivity measurements [169] on optimally doped YBCO detect significant critical superconducting fluctuations within $\pm 10\%$ of T_c. All of these experiments are well accounted for in terms of the critical properties of a phase-only (XY) model, and are not well described as Gaussian fluctuations of a Landau-Ginzberg theory. Thus, there is no question that there is a well defined magnitude of the order parameter, and substantial local superconducting order for at least 10K to 20K above T_c, and a correspondingly broad range of substantial phase fluctuations below T_c.

There are, however, some experiments that suggest that substantial local pairing persists in a much broader range of temperatures. [537] Nernst measurements [312,313] have detected vortex-like signals up to $100K$ above T_c in LSCO and YBCO, i.e. to temperatures up to 5 times T_c! As discussed in Sect. 21.5.3, magnetization experiments have recently confirmed that this Nernst signal is, in fact, a superconducting precursor [308–310, 312, 313]. However, the final word has yet to be spoken concerning the proper interpretation of these intriguing experiments [538]. ARPES [96,539] and tunneling [97,540] studies find that the gap in BSCCO persists up to $100K$ above T_c, i.e. to temperatures of order two or more times T_c.

Finally, there are preliminary indications that there may be substantial local superconducting order in severely underdoped materials in which no macroscopic indications of superconductivity appear at any temperature. Presumably, if this is the case, long range phase coherence has been suppressed in these materials by quantum phase fluctuations [297] which proliferate due to the small bare superfluid stiffness and the poor screening of the Coulomb potential. In particular, experiments on films of severely underdoped nonsuperconducting YBCO have revealed that a metastable superconducting state can be induced by photodoping. This has permitted the patterning of small scale superconducting structures, in which it has been shown [541] that substantial Josephson coupling between two superconducting regions can persist even when they are separated by as much as 1000 Å. This "anomalous proximity effect" implies that there is a substantial pair field susceptibility in this nonsuperconducting material.

What About Phonons?

Phonons are clearly strongly coupled to the electron gas in the cuprates. Certainly, when there is charge order of any sort, it is unavoidable that it induces (or is induced by) lattice distortions. Manifestly, phonons will enhance any electronic tendency to phase separation or stripe formation [124]. They will also tend to make any stripes "heavy," and so suppress quantum fluctuations—likely, this leads to a depression of superconductivity. There is a dramatic isotope effect anomaly seen [542] in some materials when the doped hole density, $x = 1/8$; presumably, this is related to just such a phonon-induced pinning of the stripe order [463]. Recently, there has been considerable controversy generated by the suggestion [465] that certain features of the ARPES spectrum of a wide class of cuprates reflect the effects of strong electron-phonon coupling. This is clearly an area in which much work remains to be done. In our opinion, other than in 1D, the effects of electron-phonon coupling in a strongly correlated electron gas is an entirely unsolved problem.

What Are the Effects of Quenched Disorder?

We have said essentially nothing about the effects of quenched disorder on the materials of interest, although the materials are complicated, and disorder is always present. There are even some theories which consider the disorder to be essential to the mechanism of high temperature superconductivity [543]. A strong case against this proposition is made by the observation that as increasingly well ordered materials are produced, including some which are stoichiometric and so do not have any of the intrinsic disorder associated with a random alloy, the superconducting properties are not fundamentally altered, and that if anything T_c and the superfluid density both seem to rise very slightly as disorder is decreased.

However, other properties of the system are manifestly sensitive to disorder. Since disorder couples to spatial symmetry breaking order parameters in the same way that a random field couples to a magnetic order parameter, it is generally a relevant perturbation. Among other things, this means that none of the stripe orders discussed above will ever occur as true long range order, and the putative transitions are rounded and rendered glassy [451,544–546]. So even the supposedly sharp statements discussed above are only sharp, in practice, if we can study such highly perfect crystals that they approximate the disorder→0 limit. This is a general problem. Progress has been made in recent years in growing more and more perfect single crystals of particular stoichiometric superconductors. Clearly, advances in this area are an essential component of the ongoing effort to unravel the physics of these materials.

Acknowledgements

We would especially like to acknowledge the profound influence on our understanding of every aspect of the physics discussed in this paper of discussions with our colleagues and collaborators, John Tranquada, Vadim Oganesyan, and Eduardo Fradkin. We also want to explicitly aknowledge the extremely helpful suggestions and critiques we obtained from J.W. Allen, Y. Ando, N.P. Armitage, A. Auerbach, D.A. Bonn, R.J. Birgeneau, E. Dagotto, A.H. Castro Neto, E. Fradkin, S. Sachdev, D.J. Scalapino, S.L. Sondhi, J.M. Tranquada, and O. Zachar. Finally, we have benefited greatly from discussion of the ideas presented herein with more colleagues than we can hope to acknowledge, but we feel we must at least acknowledge our intellectual debts to G. Aeppli, J.W. Allen, P.B. Allen, Y. Ando, D.N. Basov, M.R. Beasley, A.H. Castro Neto, S. Chakravarty, E. Daggatto, P.C. Dai, J.C.S. Davis, C. Di Castro, R.C. Dynes, H. Eisaki, A. Finkelstein, M.P.A. Fisher, T.H. Geballe, M. Granath, P.D. Johnson, C. Kallin, A. Kapitulnik, H.-Y. Kee, Y.-B. Kim, R.B. Laughlin, D-H. Lee, Y.S. Lee, K.A. Moller, H.A. Mook, C. Nayak, Z. Nussinov, N.P. Ong, J. Orenstein, S. Sachdev, D.J. Scalapino, J.R. Schrieffer, Z-X. Shen, S.L. Sondhi, B.I. Spivak, T. Timusk, T. Valla, J. Zaanen, O. Zachar, S-C. Zhang, and X.J. Zhou. We are grateful to B. Phillabaum for assistance in updating Table 21.1. This work was supported, in part, by NSF grant DMR-0110329 at UCLA, DOE grant DE-FG03-00ER45798 at UCLA and BNL, NSF grant DMR-97-12765, BSF 2004-162 at the Hebrew University of Jerusalem, the Office of the Provost at Boston University, and the Purdue Research Foundation at Purdue University. EWC is a Cottrell Scholar of Research Corporation.

List of Symbols

Symbol	Definition
$A^{<}(\boldsymbol{k}, \omega)$	Single hole spectral function
CDW	Charge density wave
d	Dimension
$D(l)$	Correlation function of the pair field
D_α	Scaling dimension of O_α
E_F	Fermi energy
E_s	Spin spectrum
$F_{\eta,\sigma}$	Klein factor
G	Reciprocal lattice vector
g_1	Backscattering
$g_{1\parallel}$	Backscattering of same spin particles
$g_{1\perp}$	Backscattering of opposite spin particles
g_2	Forward scattering on both branches
g_3	Umklapp scattering
g_4	Forward scattering within same branch
G_c	Charge piece of the one hole spectral function
G_s	Spin piece of the one hole spectral function
$h_j(x)$	Transverse stripe displacement
J	Nearest neighbor exchange coupling
k_B	Boltzmann's constant
K_c	Charge Luttinger parameter
k_F	Fermi wavevector
K_s	Spin Luttinger parameter
$L_j(x)$	Arc length
m^*	Effective pair mass
n_{2d}	Density of doped holes per plane
n_e	Electron density

$N(E_F)$	Density of states at E_F
n_s	Superfluid density
SDW	Spin density wave
SS	Singlet superconductivity
$S_z(x)$	Spin density
T	Temperature
$1/T_1T$	NMR relaxation rate
t	Nearest neighbor hopping
T_c	Superconducting transition temperature
T_p	Pairing scale
$t_\perp$	Interchain single particle tunneling
TS	Triplet superconductivity
T^*_{pair}	Crossover temperature at which pairs form
T^*_{stripe}	Crossover temperature at which charge stripes form
T_θ	Phase ordering scale
U	On-site Hubbard interaction
V	Nearest neighbor interaction
v_c	Charge velocity
v_F	Fermi velocity
$V_{k,k'}$	BCS pair potential
v_s	Spin velocity
x	Hole doping
Z	Coherent quasiparticle weight
Z_s	Coherent spin soliton weight
Δ_k	BCS gap parameter
Δ_0	Superconducting gap maximum
Δ_s	Spin gap
$\partial_x\theta_c$	Conjugate momentum of ϕ_c
$\partial_x\theta_c$	Conjugate momentum of ϕ_s
γ	Helicity modulus
γ_c	Charge Luttinger exponent
γ_s	Spin Luttinger exponent
λ	Bare electron-phonon coupling
μ	Bare Coulomb repulsion
μ^*	Renormalized Coulomb repulsion
ω_D	Debye frequency
ω_P	Plasma frequency
ϕ_c	Bosonic charge field
ϕ_s	Bosonic spin field
π	3.141592653589793238462643...
$\psi^\dagger_{\eta,\sigma}$	Fermion creation operator in the 1DEG
$\rho(x)$	Charge density
ϱ	Hole density on a stripe
τ	Quasiparticle lifetime
ξ_0	Superconducting coherence length
ξ_c	Charge correlation length
ξ_s	Spin correlation length
χ	Susceptibility

References

1. J.R. Schrieffer, *Theory of Superconductivity*, Frontiers in Physics (Addison-Wesley, Reading, MA, 1988)
2. J.G. Bednorz and K.A. Muller, Possible high T_c superconductivity in the Ba-La-Cu-O system, Z. Phys. B **64**, 189–193 (1986)
3. V.J. Emery and S.A. Kivelson, Superconductivity in bad metals, Phys. Rev. Lett. **74**, 3253–3256 (1995)
4. P.W. Anderson, Experimental constraints on the theory of high-T_c superconductivity, Science **256**, 1526–1531 (1992)
5. P.W. Anderson, The resonating valence bond state in La_2CuO_4 and superconductivity, Science **235**, 1169–1198 (1987)
6. V.J. Emery, S.A. Kivelson, and J.M. Tranquada, Stripe phases in high-temperature superconductors, Proc. Natl. Acad. Sci. **96**, 8814–8817 (1999)
7. P.A. Lee, Pseudogaps in underdoped cuprates, Physica C **317-318**, 194–204 (1999)
8. L. Taillefer, B. Lussier, R. Gagnon, K. Behnia, and H. Aubin, Universal heat conduction in $YBa_2Cu_3O_{6.9}$, Phys. Rev. Lett. **79**, 483–486 (1997)
9. P.J. Turner, R. Harris, S. Kamal, M.E. Hayden, D.M. Broun, D.C. Morgan, A. Hosseini, P. Dosanjh, J.S. Preston, R. Liang, D.A. Bonn, and W.N. Hardy, Broadband microwave spectroscopy of d-wave quasiparticles in oxygen-ordered $YBa_2Cu_3O_{6.50}$, Phys. Rev. Lett. **90**, 237005 (2003)
10. K.A. Moller, D.J. Baar, J.S. Urbach, R. Liang, W.N. Hardy, and A. Kapitulnik, Magnetic field dependence of the density of states of $YBa_2Cu_3O_{6.95}$ as determined from the specific heat, Phys. Rev. Lett. **73**, 2744–2747 (1994)
11. B. Revaz, J.-Y. Genoud, J.K. Neumaier, A. Erb, and E. Walker, d-wave scaling relations in the mixed-state specific heat of $YBa_2Cu_3O_7$, Phys. Rev. Lett. **80**, 3364–3368 (1998)

12. D.H. Lu, D.L. Feng, N.P. Armitage, K.M. Shen, A. Damascelli, C. Kim, F. Ronning, Z.X. Shen, D.A. Bonn, R. Liang, W.N. Hardy, A.I. Rykov, and S. Tajima, Superconducting gap and strong in-plane anisotropy in untwinned $YBa_2Cu_3O_{7-\delta}$, Phys. Rev. Lett. **86**(19), 4370–4373 (2001)
13. D. Basov, R. Liang, D.A. Bonn, W.N. Hardy, B. Dabrowski, D.B.T.M. Quijada, J.P. Rice, D.M. Ginsberg, and T.Timusk, In-plane anisotropy of the penetration depth in $YBa_2Cu_3O_{7-x}$ and $YBa_2Cu_4O_8$, Phys. Rev. Lett. **74**, 598–601 (1995)
14. S. Chakravarty and S. Kivelson, Electronic mechanism of superconductivity in the cuprates, C_{60}, and polyacenes, Phys. Rev. B **64**, 064511 (2001)
15. V.J. Emery and S.A. Kivelson, Frustrated electronic phase separation and high-temperature superconductors, Physica C **209**, 597–621 (1993)
16. S. Caprara, C. Castellani, C. Di Castro, and M. Grilli, Phase separation and superconductivity in strongly interacting electron systems, Physica C **235-240**, 2155–2156 (1994)
17. A. Moreo, S. Yunoki, and E. Dagotto, Phase separation scenario for manganese oxides and related materials, Science **283**, 2034–2040 (1999)
18. P.W. Anderson, A re-examination of concepts in magnetic metals: the 'nearly antiferromagnetic Fermi liquid', Adv. Phys. **46**, 3–11 (1997)
19. J.E. Hirsch, Antiferromagnetism, localization, and pairing in a two-dimensional model for CuO2, Phys. Rev. Lett. **59**, 228–231 (1987)
20. V.J. Emery, S.A. Kivelson, and O. Zachar, Spin-gap proximity effect mechanism of high-temperature superconductivity, Phys. Rev. B **56**, 6120–6147 (1997)
21. S. Chakravarty, A. Sudbo, P.W. Anderson, and S. Strong, Interlayer tunneling and gap anisotropy in high-temperature superconductors, Science **261**, 337–340 (1993)
22. E. Demler and S.-C. Zhang, Quantitative test of a microscopic mechanism of high-temperature superconductivity, Nature **396**, 733–737 (1998)
23. A.H. Castro Neto and F. Guinea, Superconductivity, Josephson coupling, and order parameter symmetry in striped cuprates, Phys. Rev. Lett. **80**, 4040–4043 (1998)
24. J.E. Hirsch, The true colors of cuprates, Science **295**, 2226–2227 (2002)
25. V.J. Emery, S.A. Kivelson, and O. Zachar, Classification and stability of phases of the multicomponent one-dimensional electron gas, Phys. Rev. B **59**, 15641–15653 (1999)
26. D.J. Scalapino, The 2-leg Hubbard ladder: Computational studies of new materials, Physica B **318**, 92–96 (2002)
27. H.J.A. Molegraaf, C. Presura, D. van der Marel, P.H. Kes, and M. Li, Superconductivity-induced transfer of in-plane spectral weight in $Bi_2Sr_2CaCu_2O_{8+\delta}$, Science **295**, 2239–2241 (2002)
28. A.F. Santander-Syro, R.P.S.M. Lobo, N. Bontemps, Z. Konstantinovic, Z.Z. Li, and H. Raffy, Pairing in cuprates from high energy electronic states, Europhys. Lett. **62**, 568–574 (2001)
29. P. Monthoux, A.V. Balatsky, and D. Pines, Toward a theory of high-temperature superconductivity in the antiferromagnetically correlated cuprate oxides, Phys. Rev. Lett. **67**, 3448–3451 (1991)
30. N. Bulut and D.J. Scalapino, $d(x^2 - y^2)$ symmetry and the pairing mechanism, Phys. Rev. B **54**, 14971–14973 (1996)
31. D.J. Scalapino, Superconductivity and spin fluctuations, J. Low Temp. Phys. **117**, 179–188 (1999), International Conference on Physics and Chemistry of Molecular and Oxide Superconductors, MOS'99, Stockholm, 28 July–2 Aug, 1999 (Kluwer Academic/Plenum)
32. R.J. Radtke, S. Ullah, K. Levin, and N.R. Norman, Constraints on superconducting transition temperatures in the cuprates: antiferromagnetic spin fluctuations, Phys. Rev. B **46**, 11975–11985 (1992)
33. C.M. Varma, J. Zaanen, and K. Raghavachari, Superconductivity in the fullerenes, Science **254**, 989–992 (1991)
34. M. Schluter, M. Lannoo, M. Needels, G.A. Baraff, and D. Tomanek, Electron–phonon coupling and superconductivity in alkali-intercalated C_{60} solid, Phys. Rev. Lett. **68**, 526–529 (1992)
35. P. Morel and P.W. Anderson, Calculation of the superconducting state parameters with retarded electron-phonon interaction, Phys. Rev. **125**, 1263–1271 (1962)
36. J.R. Schrieffer, D.J. Scalapino, and J.W. Wilkins, Effective tunneling density of states in superconductors, Phys. Rev. Lett. **10** (1963)
37. S. Chakravarty, S. Khlebnikov, and S. Kivelson, Comment on "Electron-phonon coupling and superconductivity in alkali-intercalated C_{60} solid", Phys. Rev. Lett. **69**, 212 (1992)
38. R. Shankar, Renormalization-group approach to interacting fermions, Rev. Mod. Phys. **66**, 129–192 (1994)

39. J. Polchinski, Renormalization and effective lagrangians, Nucl. Phys. B **231**, 269–295 (1984)
40. Z.-X. Shen, D.S. Dessau, B.O. Wells, D.M. King, W.E. Spicer, A.J. Arko, D. Marshal, L.W. Lombardo, A. Kapitulnik, P. Dickinson, S. Doniach, J. DiCarlo, A.G. Losser, and C.H. Park, Anomalously large gap anisotropy in the $a-b$ plane of bi2212, Phys. Rev. Lett. **70**, 1553–1556 (1993)
41. R. Micnas, J. Ranninger, and S. Robaszkiewicz, Superconductivity in narrow-band systems with local nonretarded attractive interactions, Rev. Mod. Phys. **62**, 113–234 (1990)
42. T. Holstein, Studies of polaron motion: I, Ann. Phys. **8**, 325–342 (1959)
43. G. Gruner, *Density Waves in Solids* (Perseus Books Group, Jackson, TN, 2000)
44. G. Bilbro and W.L. McMillan, Theoretical model of superconductivity and the martensitic transformation in A15 compounds, Phys. Rev. B, **14**, 1887–1892 (1976)
45. O. Zachar, S.A. Kivelson, and V.J. Emery, Landau theory of stripe phases in cuprates and nickelates, Phys. Rev. B **57**, 1422–1426 (1998)
46. J.M. Tranquada, Phase separation, charge segregation and superconductivity in layered cuprates,, In: *Neutron Scattering in Layered Copper-Oxide Superconductorsr*, ed. by A. Furrer (Kluwer, Dordrecht, 1998) Vol. 83, pp. 225–260
47. J.M. Tranquada, Experimental evidence for topological doping in the cuprates, *AIP Conf. Proc.*, Vol. 483, pp. 336–340 (American Institute of Physics, 1999)
48. J. Zaanen, High-temperature superconductivity: stripes defeat the Fermi liquid, Nature **404**, 714 (2000)
49. S. Sachdev, Quantum criticality: competing ground states in low dimensions, Science **288**, 475–80 (2000)
50. J. Orenstein and A.J. Millis, Advances in the physics of high-temperature superconductivity, Science **288**, 468–474 (2000)
51. G. Baskaran, Competition between superconductivity and charge stripe order in high-T_c cuprates, Mod. Phys. Lett. B **14**, 377–384 (2000)
52. S.A. Kivelson, E. Fradkin, and V.J. Emery, Electronic liquid-crystal phases of a doped Mott insulator, Nature **393**, 550–553 (1998)
53. C.M. Varma, Non Fermi-liquid states and pairing of a general model of copper-oxide metals, Phys. Rev. B **55**, 14554–14580 (1997)
54. S. Chakravarty, R.B. Laughlin, D.K. Morr, and C. Nayak, Hidden order in the cuprates, Phys. Rev. B **63**, 094503 (2001)
55. I. Affleck and J.B. Marston, Large-n limit of the Heisenberg–Hubbard model: Implications for high-T_c superconductors, Phys. Rev. B, **37**, 3774–3777 (1988)
56. G. Kotliar, Resonating valence bonds and d-wave superconductivity, Phys. Rev. B **37**, 3664–3666 (1998)
57. D.A. Ivanov, P.A. Lee, and X.-G. Wen, Staggered-vorticity correlations in a lightly doped t–J model: A variational approach, Phys. Rev. Lett. **34**, 3958–3961 (2000)
58. Q.H. Wang, J.H. Han, and D.H. Lee, Staggered currents in the mixed state, Phys. Rev. Lett. **87**, 7004–7007 (2001)
59. T. Senthil and M.P.A. Fisher, Fractionalization in the cuprates: Detecting the topological order, Phys. Rev. Lett. **86**, 292–295 (2000)
60. S.A. Kivelson, G. Aeppli, and V.J. Emery, Thermodynamics of the interplay between magnetism and high-temperature superconductivity, Proc. Nat. Acad. Sci. **98**(21), 11903–11907 (2001)
61. C. Castellani, C. Di Castro, and M. Grilli, Singular quasiparticle scattering in the proximity of charge instabilities, Phys. Rev. Lett. **75**, 4650–4653 (1995)
62. S. Andergassen, S. Caprara, C. Di Castro, and M. Grilli, Anomalous isotopic effect near the charge-ordering quantum criticality, Phys. Rev. Lett. **87**, 56401–56403 (2001)
63. A.V. Chubukov, S. Sachdev, and J. Ye, Theory of two-dimensional quantum Heisenberg antiferromagnets with a nearly critical ground state, Phys. Rev. B **49**, 11919–11961 (1994)
64. N.D. Mathur, F.M. Grosche, S.R. Julian, I.R. Walker, D.M. Freye, R.K.W. Haselwimmer, and G.G. Lonzarich, Magnetically mediated superconductivity in heavy fermion compounds, Nature **394**, 39–43 (1998)
65. S.A. Grigera, R.S. Perry, A.J. Schofield, M. Chiao, S.R. Julian, G.G. Lonzarich, S.I. Ikeda, Y. Maeno, A.J. Millis, and A.P. Mackenzie, Magnetic field-tuned quantum criticality in the metallic ruthenate $Sr_3Ru_2O_7$, Science **294**, 329–332 (2001)
66. J.R. Schrieffer, Ward's identity and the suppression of spin fluctuation superconductivity, J. Low Temp. Phys. **99**, 397–402 (1995)
67. R.B. Laughlin, A critique of two metals, Adv. Phys. **47**, 943–958 (1998)

68. G. Baskaran, Z. Zou, and R.B. Laughlin, The resonating valence bond state and high T_c superconductivity — a mean field theory, Solid State Comm. **63**, 973–976 (1987)
69. S.A. Kivelson, D.S. Rokhsar, and J.P. Sethna, Topology of the resonating valence-bond state: Solitons and high-T_c superconductivity, Phys. Rev. B **35**, 8865–8868 (1987)
70. D.S. Rokhsar and S.A. Kivelson, Superconductivity and the quantum hard-core dimer gas, Phys. Rev. Lett. **61**, 2376–2379 (1988)
71. N. Read and S. Sachdev, Large-N expansion for frustrated quantum antiferromagnets, Phys. Rev. Lett. **66**, 1773–1776 (1991)
72. V. Kalmeyer and R.B. Laughlin, Theory of the spin liquid state of the Heisenberg antiferromagnet, Phys. Rev. B **39**, 11879–11899 (1989)
73. X.G. Wen, F. Wilcek, and A. Zee, Chiral spin states and superconductivity, Phys. Rev. B **39**, 11413–11423 (1989)
74. P.B. Wiegmann, Superconductivity in strongly correlated electronic systems and confinement versus deconfinement phenomenon, Phys. Rev. Lett. **60**, 821–824 (1988)
75. L. Balents, M.P.A. Fisher, and C. Nayak, Nodal liquid theory of the pseudo-gap phase of high-T_c superconductors, Int. J. Mod. Phys. B **12**, 1033–1068 (1998)
76. L. Balents, M.P.A. Fisher, and C. Nayak, Dual order parameter for the nodal liquid, Phys. Rev. B **60**, 1654–1667 (1999)
77. T. Senthil and M.P.A. Fisher, Z2 gauge theory of electron fractionalization in strongly correlated systems, Phys. Rev. B **62**, 7850–7881 (2000)
78. X.G. Wen, Topological orders in rigid states, Int. J. Mod. Phys. B **4**, 239–271 (1990)
79. R. Moesner and S.L. Sondhi, Resonating valence bond phase in the triangular lattice quantum dimer model, Phys. Rev. Lett. **86**, 1881–1884 (2001)
80. R. Moessner, S.L. Sondhi, and E. Fradkin, Short-ranged resonating valance bond physics, quantum dimer models, and Ising gauge theories, Phys. Rev. B **65**, 024504 (2002)
81. T. Timusk and B. Statt, The pseudogap in high-temperature superconductors: an experimental survey, Rep. Prog. Phys. **62**, 61–122 (1999)
82. J.L. Tallon and J.W. Loram, The doping dependence of T^* — What is the real high T_c phase diagram? Physica C **349**, 53–68 (2001)
83. A. Leggett, Cuprate superconductivity: Dependence of T_c on the c-axis layering structure, Phys. Rev. Lett. **83**, 392–395 (1999)
84. P.W. Anderson, *The Theory of Superconductivity in the Cuprates* (Princeton University Press, Princeton, NJ, 1997)
85. R.B. Laughlin, Evidence for quasiparticle decay in photoemission from underdoped cuprates, Phys. Rev. Lett. **79**, 1726–1729 (1997)
86. D. Orgad, S.A. Kivelson, E.W. Carlson, V.J. Emery, X.J. Zhou, and Z.X. Shen, Evidence of electron fractionalization from photoemission spectra in the high temperature superconductors, Phys. Rev. Lett. **86**, 4362–4365 (2001)
87. X.J. Zhou, P. Bogdanov, S.A. Kellar, T. Noda, H. Eisaki, S. Uchida, Z. Hussain, and Z.-X. Shen, One-dimensional electronic structure and suppression of d-wave node state in $La_{1.28}Nd_{0.6}Sr_{0.12}CuO_4$, Science **286**, 268–272 (1999)
88. T. Valla, A.V. Fedorov, P.D. Johnson, Q. Li, G.D. Gu, and N. Koshizuka, Temperature dependent scattering rates at the Fermi surface of optimally doped $Bi_2Sr_2CaCu_2O_{8+\delta}$, Phys. Rev. Lett. **85**, 828–831 (2000)
89. A.V. Fedorov, T. Valla, P.D. Johnson, Q. Li, G.D. Gu, and N. Koshizuka, Temperature dependent photoemission studies of optimally doped $Bi_2Sr_2CaCu_2O_8$, Phys. Rev. Lett. **82**, 2179–2182 (1999)
90. D.L. Feng, D.H. Lu, K.M. Shen, S. Oh, A. Andrus, J. O'Donnell, J.N. Eckstein, J. Shimoyama, K. Kishio, and Z.X. Shen, On the similarity of the spectral weight pattern of $Bi_2Sr_2CaCuO_{8+\delta}$ and $La_{1.48}Nd_{0.4}Sr_{0.12}CuO_4$, Physica C **341**, 2097–2098 (2000)
91. A.G. Loeser, Z.-X. Shen, M.C. Schabel, C. Kim, M. Zhang, A. Kapitulnik, and P. Fournier, Temperature and doping dependence of the Bi-Sr-Ca-Cu-O electronic structure and fluctuation effects, Phys. Rev. B **56**, 14185–14189 (1997)
92. C.M. Varma, P.B. Littlewood, S. Schmittrink, E. Abrahams, and A.E. Ruckenstein, Phenomenology of the normal state of the Cu-O high-temperature superconductors, Phys. Rev. Lett. **63**, 1996–1999 (1989)
93. C.M. Varma, Z. Nussinov, and W. van Saarloos, Singular Fermi liquids, Phys. Rep. **361**, 267–417 (2002)
94. E. Abrahams and C.M. Varma, What angle-resolved photoemission experiments tell us about the microscopic theory for high-temperature superconductors, Proc. Natl. Acad. Sci. (2000)

95. J.M. Harris, Z.-X. Shen, P.J. White, D.S. Marshall, M.C. Schabel, J.N. Eckstein, and I. Bozovic, Anomalous superconducting state gap size versus Tc behavior in underdoped $Bi_2Sr_2Ca_{1-x}Dy_xCu_2O_8$, Phys. Rev. B **54**, R15665–R15668 (1996)
96. H. Ding, T. Yokoya, J.C. Campuzano, T. Takahashi, M. Randeria, M.R. Norman, and T. Mochiku, K. Kadowaki, and J. Giapintzakis, Spectroscopic evidence for a pseudogap in the normal state of underdoped high-T_c superconductors, Nature **382**, 51–54 (1996)
97. C. Renner, B. Revaz, J.-Y. Genoud, K. Kadowaki, and O. Fischer, Pseudogap precursor of the superconducting gap in under- and overdoped $Bi_2Sr_2CaCu_2O_{8+\delta}$, Phys. Rev. Lett. **80**, 149–152 (1998)
98. Y. Ando, K. Segawa, S. Komiya, and A.N. Lavrov, Electrical resistivity anisotropy from self-organized one-dimensionality in high-temperature superconductors, Phys. Rev. Lett. **88**, 137005 (2002)
99. I. Maggio-Aprile, C. Renner, A. Erb, E. Walker, and O. Fischer, Direct vortex lattice imaging and tunneling spectroscopy of flux lines on $YBa_2Cu_3O_{7-\delta}$, Phys. Rev. Lett. **75**, 2754–2757 (1995)
100. C. Howald, P. Fournier, and A. Kapitulnik, Inherent inhomogeneities in tunneling spectra of $Bi_2Sr_2CaCu_2O_{8-x}$ crystals in the superconducting state, Phys. Rev. B **64**, 100504–100507 (2001)
101. S.H. Pan, J.P. O'Neal, R.L. Badzey, C. Chamon, H. Ding, J.R. Engelbrecht, Z. Wang, H. Eisaki, S. Uchida, A.K. Guptak, K.W. Ng, E.W. Hudson, K.M. Lang, and J.C. Davis, Microscopic electronic inhomogeneity in the high T_c superconductor $Bi_2Sr_2CaCu_2O_{8+x}$, Nature **413**, 282–285 (2001)
102. J.E. Sonier, J.H. Brewer, R.F. Kiefl, D.A. Bonn, S.R. Dunsiger, W.N. Hardy, R. Liang, W.A. MacFarlane, R.I. Miller, and T.M. Riseman, Measurements of the fundamental length scales in the vortex state of $YBa_2Cu_3O_{6.60}$, Phys. Rev. Lett. **79**, 2875–2878 (1997)
103. C.A.R.S. de Melo, M. Randeria, and J.R. Engelbrecht, Crossover from BCS to Bose superconductivity: Transition temperature and time-dependent Ginzburg-Landau theory, Phys. Rev. Lett. **71**, 3202–3205 (1993)
104. Q. Chen, I. Kosztin, B. Janko, and K. Levin, Pairing fluctuation theory of superconducting properties in underdoped to overdoped cuprates, Phys. Rev. Lett. **81**, 4708–4711 (1998)
105. A.S. Alexandrov and N.F. Mott, Thermal transport in a charged Bose gas and in high-T_c oxides, Phys. Rev. Lett. **71**, 1075–1078 (1993)
106. A.S. Alexandrov and N.F. Mott, Bipolarons, Rep. Prog. Phys. **57**, 1197–1288 (1994)
107. Y.J. Uemura, G.M. Luke, B.J. Sternlieb, J.H. Brewer, J.F. Carolan, W.N. Hardy, R. Kadono, J.R. Kempton, R.F. Kiefl, S.R. Kreitzman, P. Mulhern, T.M. Riseman, D.L. Williams, B.X. Yang, S. Uchida, H. Takagi, J. Gopalakrishnan, A.W. Sleight, M.A. Subramanian, C.L. Chien, M.Z. Cieplak, G. Xiao, V.Y. Lee, B.W. Statt, C.E. Stronach, W.J. Kossler, and X.H. Yu, Universal correlations between Tc and $ns/m*$ (carrier density over effective mass) in high-T_c cuprate superconductors, Phys. Rev. Lett. **62**, 2317–2320 (1989)
108. E. Dagotto, T. Hotta, and A. Moreo, Colossal magnetoresistant materials: the key role of phase separation, Phys. Rep. **344**, 1–153 (2001)
109. J.M. Luttinger, Fermi surface and some simple equilibrium properties of a system of interacting fermions, Phys. Rev. **119**, 1153–1163 (1960)
110. V.J. Emery and S.A. Kivelson, Crossovers and phase coherence in cuprate superconductors, J. Phys. Chem. Solids **59**, 1705–1710 (1998)
111. A.N. Kocharian, C. Yang, and Y.L. Chiang, Self-consistent and exact studies of pairing correlations and crossover in the one-dimensional attractive Hubbard model, Phys. Rev. B **59**, 7458–7472 (1999)
112. A.N. Kocharian, C. Yang, and Y.L. Chiang, Phase diagram and BCS-Bose condensation crossover in 1D and 2D Hubbard models, Physica C **364-365**, 131–133 (2001)
113. J.W. Allen, C.G. Olson, M.B. Maple, J.-S. Kang, L.Z. Liu, J.-H. Park, R.O. Anderson, W.P. Ellis, J.T. Market, Y. Dalichaouch, and R. Liu, Resonant photoemission study of $Nd_{2-x}Ce_xCuO_{4-y}$: Nature of electronic states near the Fermi level, Phys. Rev. Lett. **64**, 595–598 (1990)
114. R.O. Anderson, R. Cleassen, J.W. Allen, C.G. Olson, C. Janowitz, L.Z. Liu, J.-H. Park, M.B. Maple, Y. Dalichaouch, M.C. de Andrade, R.F. Jardim, E.A. Early, S.-J. Ho, and W.P. Ellis, Luttinger Fermi surface of metallic gap spectral weight in $Nd_{1.85}Ce_{0.15}CuO_{4-y}$, Phys. Rev. Lett. **70**, 3163–3166 (1993)
115. T. Watanabe, T. Takahashi, S. Suzuki, S. Sato, and H. Katayama-Yoshida, Inverse-photoemission study of hole-concentration dependence of the electronic structure in $Bi_2Sr_2Ca_{1-x}Y_xCu_2O_8$ (x=0.0-0.05), Phys. Rev. B **44**, 5316–5317 (1991)

116. A. Ino, C. Kim, M. Nakamura, T. Yoshida, T. Mizokawa, Z.-X. Shen, A. Fujimori, T. Kakeshita, H. Eisaki, and S. Uchida, Electronic structure of $La_{2-x}Sr_xCuO_4$ in the vicinity of the superconductor-insulator transition, Phys. Rev. B **62**, 4137–4141 (2000)
117. G. Rietveld, N.Y. Chen, and D. van der Marel, Anomalous temperature dependence of the work function in $YBa_2Cu_3O_{7-\delta}$, Phys. Rev. B **69**, 2578–2581 (1992)
118. D.J. Scalapino, J.E. Loh, and J.E. Hirsch, d-wave pairing near a spin density wave instability, Phys. Rev. B **34**, 8190–8192 (1986)
119. W. Kohn and J.M. Luttinger, New mechanism for superconductivity, Phys. Rev. Lett. **15**, 524–526 (1965)
120. V.J. Emery, Theory of high-T_c superconductivity in oxides, Phys. Rev. Lett. **58**, 2794–2797 (1987)
121. G. Kotliar and J. Liu, Superexchange mechanism and d-wave superconductivity, Phys. Rev. B **38**, 5142–5145 (1988)
122. C. Gros, R. Joynt, and T.M. Rice, Superconductivity instability in the large-U limit of the two-dimensional Hubbard model, Z. Phys. B **68**, 425–432 (1987)
123. D.J. Scalapino, E. Loh, and J.E. Hirsch, Fermi-surface instabilities and superconducting d-wave pairing, Phys. Rev. B **35**, 6694–6698 (1987)
124. M. Grilli, R. Raimondi, C. Castellani, C. Di Castro, and G. Kotliar, Superconductivity, phase separation, and charge-transfer instability in the $U = \infty$ limit of the three-band model of the CuO_2 planes, Phys. Rev. Lett. **67**, 259–262 (1991)
125. A. Perali, C. Castellani, C. Di Castro, and M. Grilli, d-wave superconductivity near charge instabilities, Phys. Rev. B **54**, 16216–16225 (1996)
126. D.A. Wollman, D.J.V. Harlingen, W.C. Lee, D.M. Ginsberg, and A.J. Leggett, Experimental determination of the superconducting pairing state in YBCO from phase coherence in YBCO-Pb SQUIDs, Phys. Rev. Lett. **71**, 2134–2147 (1993)
127. C.C. Tsuei, J.R. Kirtley, C.C. Chi, L.S. Yu-Jahnes, A. Gupta, T. Shaw, J.Z. Sun, and M.B. Ketchen, Pairing symmetry and flux quantization in a tricrystal superconducting ring of $YBa_2Cu_3O_{7-\delta}$, Phys. Rev. Lett. **73**, 593–596 (1994)
128. K.A. Kouznetsov, A.G. Sun, B. Chen, A.S. Katz, S.R. Bahcall, J. Clarke, R.C. Dynes, D.A. Gajewski, S.H. Han, M.B. Maple, J. Giapintzakis, J.-T. Kim, and D.M. Ginsberg, c-axis Josephson tunneling between $YBa_2Cu_3O_{7-\delta}$ and Pb: Direct evidence for mixed order parameter symmetry in a high-T_c superconductor, Phys. Rev. Lett. **79**, 3050–3053 (1997)
129. R.A. Klemm, G. Arnold, A. Bille, and K. Scharnberg, Theory of c-axis twist Bi2212 Josephson junctions: Strong evidence for incoherent tunneling and s-wave superconductivity, Physica **C 341**, 1663–1664 (2000)
130. A. Damascelli, D.H. Lu, and Z.-X. Shen, From Mott insulator to overdoped superconductor: Evolution of the electronic structure of cuprates studied by ARPES, J. Electron Spectr. Relat. Phenom. **165**, 117–118 (2001)
131. J.R. Schrieffer, S.C. Zhang, and X.G. Wen, Spin-bag mechanism of high-temperature superconductivity, Phys. Rev. Lett. **60**, 944–947 (1988)
132. M. Granath, V. Oganesyan, S.A. Kivelson, E. Fradkin, and V.J. Emery, Nodal quasiparticles in stripe ordered superconductors, Phys. Rev. Lett. **87**, 167011–167014 (2001)
133. M. Vojta, Y. Zhang, and S. Sachdev, Quantum phase transitions in d-wave superconductors, Phys. Rev. Lett. **85**, 4940–4943 (2000)
134. M. Kugler, O. Fischer, C. Renner, S. Ono, and Y. Ando, Scanning tunneling spectroscopy of $Bi_2Sr_2CuO6 + \delta$: New evidence for the common origin of the pseudogap and superconductivity, Phys. Rev. Lett. **86**, 4911 (2001)
135. N.J. Curro, P.C. Hammel, B.J. Suh, M. Hucker, B. Buchner, U. Ammerahl, and A. Revcolevschi, Inhomogeneous low frequency spin dynamics in $La_{1.65}Eu_{0.2}Sr_{0.15}CuO_4$, Phys. Rev. Lett. **85**, 642–645 (2000)
136. H. Takagi, B. Batlogg, H.L. Kao, J. Kwo, R.J. Cava, J.J. Krajewski, and W.F.P. Jr., Systematic evolution of temperature-dependent resistivity in $La_{2-x}Sr_xCuO_4$, Phys. Rev. Lett. **69**, 2975–2978 (1992)
137. B. Bucher, P. Steiner, J. Karpinski, E. Kaldis, and P. Wachter, Influence of the spin gap on the normal state transport in $YBa_2Cu_4O_8$, Phys. Rev. Lett. **70**, 2012–2015 (1993)
138. K. Takenaka, K. Mizuhashi, H. Takagi, and S. Uchida, Interplane charge transport in $YBa_2Cu_3O_{7-y}$: Spin-gap effect on in-plane and out-of-plane resistivity, Phys. Rev. B **50**, 6534–6537 (1994)
139. A.N. Lavrov, Y. Ando, and S. Ono, Two mechanisms of pseudogap formation in Bi-2201: evidence from the c-axis magnetoresistance, Euro. Phys. Lett **57**, 267–273 (2002)
140. J.W. Loram, K.A. Mirza, J.R. Cooper, and W.Y. Liang, Electronic specific heat of $YBa_2Cu_3O_{6+x}$ from 1.8 to 300 K, Phys. Rev. Lett. **71**, 1740–1743 (1993)

141. J. Orenstein, G.A. Thomas, A.J. Millis, S.L. Cooper, D.H. Rapkine, T. Timusk, L.F. Schneemeyer, and J.V. Waszczak, Frequency- and temperature-dependent conductivity in $YBa_2Cu_3O_{6+x}$ crystals, Phys. Rev. B **42**, 6342–6362 (1990)
142. A. Puchkov, D.N. Basov, and T. Timusk, Pseudogap state in high-T_c superconductors: An infrared study, J. Phys. Cond. Matt. **8**, 10049 (1996)
143. C.C. Homes, T. Timusk, R. Liang, D.A. Bonn, and W.N. Hardy, Optical conductivity of c axis oriented $YBa_2Cu_3O_{6.70}$: Evidence for a pseudogap, Phys. Rev. Lett. **71**, 1645–1648 (1993)
144. D. Basov, H.A. Mook, B. Dabrowski, and T. Timusk, c-axis response of single- and double-layered cuprates, Phys. Rev. B **52**, R13141–R13144 (1995)
145. M. Arai, T. Nishijima, Y. Endoh, T. Egami, S. Tajima, K. Tomimoto, Y. Shiohara, M. Takahashi, A. Garrett, and S.M. Bennington, Incommensurate spin dynamics of underdoped superconductor $YBa_2Cu_3O_{6.7}$, Phys. Rev. Lett. **83**, 608–611 (1999)
146. P.C. Dai, H.A. Mook, S.M. Hayden, G. Aeppli, T.G. Perring, R.D. Hunt, and F. Doğan, The magnetic excitation spectrum and thermodynamics of high T_c superconductors, Science **284**, 1344–1347 (1999)
147. B. Batlogg and V.J. Emery, Crossovers in cuprates, Nature **382**, 20 (1996)
148. X.G. Wen and P.A. Lee, Theory of underdoped cuprates, Phys. Rev. Lett. **76**, 503–506 (1996)
149. E.W. Carlson, D. Orgad, S.A. Kivelson, and V.J. Emery, Dimensional crossover in quasi-one-dimensional and high T_c superconductors, Phys. Rev. B **62**(5), 3422–3437 (2000)
150. B.G. Levi, Physics Today **49**, 17 (1996)
151. D.-H. Lee, Superconductivity in a doped Mott insulator, Phys. Rev. Lett. **84**, 2694–2697 (2000)
152. S. Sachdev, Kagome-acute- and triangular-lattice Heisenberg antiferromagnets: Ordering from quantum fluctuations and quantum-disordered ground states with unconfined bosonic spinons, Phys. Rev. B **45**, 12377–12396 (1992)
153. M.U. Ubbens and P.A. Lee, Superconductivity phase diagram in the gauge-field description of the t-J model, Phys. Rev. B **49**, 6853–6863 (1994)
154. S.-C. Zhang, SO(5) quantum nonlinear sigma model theory of the high T_c superconductivity, Science **275**, 1089–1096 (1997)
155. S.-C. Zhang, J.-P. Hu, E. Arrigoni, W. Hanke, and A. Auerbach, Projected SO(5) models, Phys. Rev. B **60**, 13070–13084 (1999)
156. A. Auerbach and E. Altman, Projected SO(5) Hamiltonian for cuprates and its applications, Int. J. Mod. Phys. B **15**, 2509–2518 (2001)
157. J. Ye, Thermally generated vortices, gauge invariance, and electron spectral function in the pseudogap regime, Phys. Rev. Lett. **87**, 227003–227006 (2001)
158. A. Abanaov, A.V. Chubukov, and J. Schmalian, Fingerprints of spin mediated pairing in cuprates, J. Electron. Spectrosc. **117-118**, 129–151 (2001)
159. M. Takigawa, A.P. Reyes, P.C. Hammel, J.D. Thompson, R.H. Heffner, Z. Fisk, and K.C. Ott, Cu and O NMR studies of the magnetic properties of $YBa_2Cu_3O_{6.63}$ (T_c=62 K), Phys. Rev. B **43**, 247–257 (1991)
160. H.A. Mook and F. Dogan, Charge fluctuations in $YBa2Cu3O_{7-x}$ high-temperature superconductors, Nature **401**, 145–148 (1999)
161. P. Dai, H.A. Mook, and F. Dogan, Incommensurate magnetic fluctuations in $YBa_2Cu3O_{6.6}$, Phys. Rev. B **80**, 1738–1741 (1998)
162. J.M. Tranquada, P.M. Gehring, G. Shirane, S. Shamoto, and M. Sato, Neutron-scattering study of the dynamical spin susceptibility in $YBa_2Cu_3O_{6.6}$, Phys. Rev. B **46**, 5561–5575 (1992)
163. H.A. Mook, P. Dai, and F. Dogan, Charge and spin structure in $YBa_2Cu_3O_{6.35}$, Phys. Rev. Lett. **88**, 097004 (2002)
164. S. Wakimoto, R.J. Birgeneau, M.A. Kastner, Y.S. Lee, R. Erwin, P.M. Gehring, S.H. Lee, M. Fujita, K. Yamada, Y. Endoh, K. Hirota, and G. Shirane, Direct observation of a one-dimensional static spin modulation in insulating $La_{1.95}Sr_{0.05}CuO_4$, Phys. Rev. B **61**, 3699–3706 (2000)
165. H. Kimura, H. Matsushita, K. Hirota, Y. Endoh, K. Yamada, G. Shirane, Y.S. Lee, M.A. Kastner, and R.J. Birgeneau, Incommensurate geometry of the elastic magnetic peaks in superconducting $La_{1.88}Sr_{0.12}CuO_4$, Phys. Rev. B **61**(21), 14366–14369 (2000)
166. B.O. Wells, Y.S. Lee, M.A. Kastner, R.J. Christianson, R.J. Birgeneau, K. Yamada, Y. Endoh, and G. Shirane, Incommensurate spin fluctuations in high-transition temperature superconductors, Science **277**, 1067–1071 (1997)

167. M. Matsuda, M. Fujita, K. Yamada, R.J. Birgeneau, Y. Endoh, and G. Shirane, Electronic phase separation in lightly doped $La_{2-x}Sr_xCuO_4$, Phys. Rev. B **65**, 134515 (2002)
168. P.A. Lee and N. Nagaosa, Gauge theory of the normal state of high-T_c superconductors, Phys. Rev. B **46**, 5621–5639 (1992)
169. C. Meingast, V. Pasler, P. Nagel, A. Rykov, S. Tajima, and P. Olsson, Phase fluctuations and the pseudogap in $YBa_2Cu_3O_{7-\delta}$, Phys. Rev. Lett. **86**, 1606–1609 (2001)
170. J. Corson, R. Mallozzi, J. Orenstein, J.N. Eckstein, and I. Bozovic, Vanishing of phase coherence in underdoped $Bi_2Sr_2CaCu_2O_{8+\delta}$, Nature **398**, 221–223 (1999)
171. D.A. Bonn, S. Kamal, A. Bonakdarpour, L. Ruixing, W.N. Hardy, C.C. Homes, and D.N.B.T. Timusk, Surface impedance studies of YBCO, Czech. J. Phys. **46**, 3195–3202 (1996)
172. V. Pasler, P. Schweiss, C. Meingast, B. Obst, H. Wuhl, A.I. Rykov, and S. Tajima, 3D-XY critical fluctuations of the thermal expansivity in detwinned $YBa_2Cu_3O_{7-\delta}$ single crystals near optimal doping, Phys. Rev. Lett. **81**, 1094–1097 (1998)
173. C. Varma, Pseudogap phase and the quantum-critical point in copper-oxide metals, Phys. Rev. Lett. **83**, 3538–3541 (1999)
174. S. Chakravarty, private comunication
175. T. Valla, P.D. Johnson, Z. Yusof, B. Wells, Q. Li, S.M. Loureiro, R.J. Cava, M. Mikami, Y. Mori, M. Yoshimura, and T. Sasaki, Coherence-incoherence and dimensional crossover in layered strongly correlated metals, Nature **417**, 627 (2002)
176. J. Zaanen, Superconductivity: Self-organized one dimensionality, Science **286**, 251–252 (1999)
177. J.P. Eisenstein, M.P. Lilly, K.B. Cooper, L.N. Pfeiffer, and K.W. West, New collective states of 2D electrons in high Landau levels, Physica E **9**, 1–8 (2001)
178. M.M. Fogler, Stripe and bubble phases in quantum Hall systems, cond-mat/0111001, 98–138 (2002)
179. M.M. Fogler, Quantum Hall liquid crystals, Int. J. Mod. Phys. B, **16**, 2924–2929 (2002)
180. V.J. Emery, Theory of the one-dimensional electron gas, In: *Highly Conducting One-Dimensional Solids,* ed. by J.T. Devreese, R.P. Evrard, and V.E. van Doren, eds. (Plenum, New York, 1979) Vol. 327
181. J. Solyom, The Fermi gas model of one-dimensional conductors, Adv. Phys. **28**, 201–303 (1979)
182. E. Fradkin, *Field Theories of Condensed Matter Systems*, (Addison-Wesley, Reading, MA, 1991)
183. J. von Delft and H. Schoeller, Bosonization for beginners - Refermionization for experts, Annalen Phys. **7**, 225–305 (1998)
184. A.O. Gogolin, A.A. Nersesyan, and A.M. Tsvelik, *Bosonization and Strongly Correlated Systems* (Cambridge University Press, Cambridge, 1998)
185. H.J. Schulz, G. Guniberti, and P. Pieri, Fermi liquids and Luttinger liquids, cond-mat/9807366 (1998)
186. J. Voit, One-dimensional Fermi liquids, Rep. Prog. Phys. **58**, 977–1116 (1995)
187. J.M. Kosterlitz and D.J. Thouless, Ordering, metastability and phase transitions in two-dimensional systems, J. Phys. C **6**, 1181–1203 (1973)
188. T. Giamarchi and H.J. Schulz, Correlation functions of one-dimensional quantum systems, Phys. Rev. B **39**, 4620–4629 (1989)
189. A. Luther and V.J. Emery, Backward scattering in the one-dimensional electron gas, Phys. Rev. Lett. **33**, 589–592 (1974)
190. N. Kawakami and S.K. Yang, Luttinger anomaly exponent of momentum distribution in the Hubbard chain, Phys. Lett. A **148**, 359–362 (1990)
191. R.M. Noack, N. Bulut, D.J. Scalapino, and M.G. Zacher, Enhanced $d_{x^2-y^2}$ pairing correlations in the two-leg Hubbard ladder, Phys. Rev. B **56**, 7162–7166 (1997)
192. T. Valla, A.V. Fedorov, P.D. Johnson, and S.L. Hulber, Many-body effects in angle-resolved photoemission: Quasiparticle energy and lifetime of a Mo(110) surface state, Phys. Rev. Lett. **83**, 2085–2088 (1999)
193. A. Luther and I. Peschel, Single-particle states, Kohn anomaly and pairing fluctuations in one dimension, Phys. Rev. B **9**, 2911–2919 (1974)
194. V. Meden and K. Schönhammer, Spectral functions for the Tomonaga-Luttinger model, Phys. Rev. B **46**, 15753–15760 (1992)
195. J. Voit, Charge-spin separation and the spectral properties of Luttinger liquids, Phys. Rev. B **47**, 6740–6743 (1993)
196. D. Orgad, Spectral functions for the Tomonga-Luttinger and Luther-Emery liquids, Phil. Mag. B **81**, 377–398 (2001)

197. G.-H. Gweon, J.W. Allen, and J.D. Denlinger, Generalized spectral signatures of electron fractionalization in quasi-one- and two-dimensional molybdenum bronzes and superconducting cuprates, Phys. Rev. B **68**, 195117 (2003)
198. F.H.L. Essler and A.M. Tsvelik, Weakly coupled one-dimensional Mott insulators, Phys. Rev. B **65**, 115117 (2002)
199. D.J. Scalapino, Y. Imry, and P. Pincus, Generalized Ginzburg–Landau theory of pseudo-one-dimensional systems, Phys. Rev. B **11**, 2042–2048 (1975)
200. E. Arrigoni, Crossover to Fermi-liquid behavior for weakly coupled Luttinger liquids in the anisotropic large-dimension limit, Phys. Rev. B **61**, 7909–7029 (2000)
201. D.L. Feng, D.H. Lu, K.M. Shen, C. Kim, H. Eisaki, A. Damascelli, R. Yoshizaki, J. Shimoyama, K. Kishio, G.D. Gu, S. Oh, A. Andrus, J. O'Donnell, J.N. Eckstein, and Z.X. Shen, Signature of superfluid density in the single-particle excitation spectrum of $Bi_2Sr_2CaCu_2O_{8+\delta}$, Science **289**, 277–281 (2000)
202. H. Ding, J.R. Engelbrecht, Z. Wang, J.C. Campuzano, S.-C. Wang, H.-B. Yang, R. Rogan, T. Takahashi, K. Kadowaki, and D.G. Hinks, Coherent quasiparticle weight and its connection to high-T_c superconductivity from angle-resolved photoemission, Phys. Rev. Lett. **87**, 227001–227004 (2000)
203. S. Biermann, A. Georges, A. Lichtenstein, and T. Giamarchi, Deconfinement transition and Luttinger to Fermi liquid crossover in quasi-one-dimensional systems, Phys. Rev. Lett. **87**, 276405 (2001)
204. S. Biermann, A. Georges, T. Giamarchi, and A. Lichtenstein, Quasi-one-dimensional organic conductors: dimensional crossover and some puzzles, cond-mat/0201542 (2002)
205. L. Yin and S. Chakravarty, Spectral anomaly and high temperature superconductors, Int. J. Mod. Phys. B **7**, 805–845 (1996)
206. V.J. Emery, E. Fradkin, S.A. Kivelson, and T.C. Lubensky, Quantum theory of the smectic metal state in stripe phases, Phys. Rev. Lett. **85**, 2160–2163 (2000)
207. A.H. Castro Neto, Stripes, vibrations, and superconductivity, Phys. Rev. B **64**, 104509–104535 (2001)
208. D. Orgad, Interacting electrons on a fluctuating string, Phys. Rev. B **69**, 205104 (2004)
209. A. Vishwanath and D. Carpentier, Two-dimensional anisotropic non-Fermi-liquid phase of coupled Luttinger liquids, Phys. Rev. Lett. **86**, 676–679 (2001)
210. R. Mukhopadhyay, C.L. Kane, and T.C. Lubensky, Sliding Luttinger liquid phases, Phys. Rev. B **64**, 045120–045137 (2001)
211. S.L. Sondhi and K. Yang, Sliding phases via magnetic fields, Phys. Rev. B **63**, 054430–054436 (2001)
212. S.A. Kivelson, Electron fractionalization, Synth. Met. **125** (2001)
213. T. Senthil and O. Motrunich, Microscopic models for fractionalized phases in strongly correlated systems, Phys. Rev. B **66**, 205104 (2002)
214. P. Fazekas and P.W. Anderson, On the ground state properties of the anisotropic triangular antiferromagnet. (Application of anisotropic Heisenberg model), Phil. Mag. **30**, 423–440 (1974)
215. R.B. Laughlin, The relationship between high-temperature superconductivity and the fractional quantum Hall effect, Science, **242**, 525–533 (1988)
216. N. Read and B. Chakraborty, Statistics of the excitations of the resonating-valence-bond state, Phys. Rev. B **40**, 7133–7140 (1989)
217. V. Kalmeyer and R.B. Laughlin, Equivalence of the resonating-valence-bond and fractional quantum Hall states, Phys. Rev. Lett. **59**, 2095–2098 (1987)
218. S.A. Kivelson, Statistics of holons in the quantum hard-core dimer gas, Phys. Rev. B **39**, 259–264 (1989)
219. E. Demler, C. Nayak, H.-Y. Kee, Y.B. Kim, and T. Senthil, Fractionalization patterns in strongly correlated electron systems: Spin-charge separation and beyond, Phys. Rev. B **65**, 155103 (2002)
220. F.D.M. Haldane, O(3) nonlinear sigma model and the topological distinction between integer- and half-integer-spin antiferromagnets in two dimensions, Phys. Rev. Lett. **61**, 1029–1032 (1988)
221. S. Chakravarty, B.I. Halperin, and D. Nelson, Two-dimensional quantum Heisenberg antiferromagnet at low temperatures, Phys. Rev. B **39**, 2344–2371 (1989)
222. S. Chakravarty, B.I. Halperin, and D. Nelson, Low-temperature behavior of two-dimensional quantum antiferromagnets, Phys. Rev. Lett. **60**, 1057–1060 (1988)
223. R. Coldea, S.M. Hayden, G. Aeppli, T.G. Perring, C.D. Frost, T.E. Mason, S.-W. Cheong, and Z. Fisk, Spin waves and electronic interactions in La_2CuO_4, Phys. Rev. Lett. **86** (2001)

224. R.J. Birgeneau, A. Aharony, N.R. Belk, F.C. Chou, Y. Endoh, M. Greven, S. Hosoya, M.A. Kastner, C.H. Lee, Y.S. Lee, G. Shirane, S. Wakimoto, B.O. Wells, and K. Yamada, Magnetism and magnetic fluctuations in $La_{2-x}Sr_xCuO_4$ for $x = 0$ (2D antiferromagnet), 0.04 (3D spin glass) and $x = 0.15$ (superconductor), J. Phys. Chem. Solids **56**, 1912–1919 (1995)
225. J.M. Tranquada, H. Woo, T.G. Perring, H. Goka, G.D. Gu, G. Xu, M. Fujita, and K. Yamada, Quantum magnetic excitations from stripes in copper oxide superconductors, Nature **429**, 534 (2004)
226. D.X. Yao, E.W. Carlson, and D.K. Campbell, Magnetic excitations of stripes near a quantum critical point, Phys. Rev. Lett. **97**, 017004 (2006)
227. A. Weidinger, C. Niedermayer, A. Golnik, R. Simon, E. Recknagel, J.I. Budnick, B. Chamberland, and C. Baines, Observation of magnetic ordering in superconducting $La_{2-x}Sr_xCuO_4$ by muon spin rotation, Phys. Rev. Lett. **62** (1989)
228. C. Niedermayer, C. Bernhard, T. Blasius, A. Golnik, A. Moodenbaugh, and J.I. Budnick, Common phase diagram for antiferromagnetism in $La_{2-x}Sr_xCuO_4$ and $Y_{1-x}Ca_xBa_2Cu_3O_6$ as seen by muon spin rotation, Phys. Rev. Lett. **80** (1999)
229. C. Panagopoulos, J.L. Tallon, B.D. Rainford, T. Xiang, J.R. Cooper, and C.A. Scott, Evidence for a generic novel quantum transition in high-Tc cuprates, Phys. Rev. B **66**, 064501 (2002)
230. P. Azaria, C. Hooley, P. Lecheminant, C. Lhuillier, and A.M. Tsvelik, Kagome lattice antiferromagnet stripped to its basics, Phys. Rev. Lett. **81**, 1694–1697 (1998)
231. W. LiMing, G. Misguich, P. Sindzingre, and C. Lhuillier, From Neel long-range order to spin liquids in the multiple-spin exchange model, Phys. Rev. B **62**, 6372–6377 (2000)
232. G. Misguich, B. Bernu, C. Lhuillier, and C. Waldtmann, Spin liquid in the multiple-spin exchange model on the triangular lattice: ^{3}He on graphite, Phys. Rev. Lett. **81**, 1098–1101 (1998)
233. C. Lhuillier and G. Misguich, Frustrated quantum magnets, cond-mat/0109146 (2001)
234. S. Chakravarty, S.A. Kivelson, C. Nayak, , and K. Volker, Wigner glass, spin liquids and the metal-insulator transition, Phil. Mag. B **79** (1999)
235. A.V. Chubukov, S. Sachdev, and T. Senthil, Quantum phase transitions in frustrated quantumantiferromagnets, Nucl. Phys. B **426**, 601–643 (1994)
236. R. Meservey and B.B. Schwartz, Equilibrium properties: Comparison of experimental results with predictions of the BCS theory, In: Superconductivity, ed. by R.D. Parks (Marcel Deckker, New York, 1969) Vol. 1
237. E.A. Lynton, *Superconductivity* (Methuen, London, 1962)
238. T.P. Orlando, E.J. McNiff, Jr., S. Foner, and M.R. Beasley, Critical fields, Pauli paramagnetic limiting, and material parameters of Nb3Sn and V3Si, Phys. Rev. B **19**, 4545–4561 (1979)
239. Y.G. Naidyuk and I.K. Yanson, Point-contact spectroscopy of heavy-fermion systems, J. Phys. Cond. Matt. **10**, 8905–8938 (1998)
240. M.B. Maple, J.W. Chen, S.E. Lambert, Z. Fisk, J.L. Smith, H.R. Ott, J.S. Brooks, and M.J. Naughton, Upper critical magnetic field of the heavy-fermion superconductor UBe13, Phys. Rev. Lett. **54**, 477–480 (1985)
241. F. Gross, K. Andres, and S. Chandrasekhar, Experimental determination of the absolute value of the London penetration depth in the heavy fermion superconductors UBe_{13} and UPt_3, Physica C **162-164**, 419–420 (1989)
242. F. Sharifi, A. Pargellis, R.C. Dynes, B. Miller, E.S. Hellman, J. Rosamilia, and E.H.Hartford, Jr., Electron tunneling in the high-T_c bismuthate superconductors, Phys. Rev. B **44**, 12521–12524 (1991)
243. Y.J. Uemura, L.P. Le, G.M. Luke, B.J. Sternlieb, W.D. Wu, J.H. Brewer, T.M. Riseman, C.L. Seaman, M.B. Maple, M. Ishikawa, D.G. Hinks, J.D. Jorgensen, G. Saito, and H. Yamochi, Basic similarities among cuprate, bismuthate, organic, chevrel-phase, and heavy-fermion superconductors shown by penetration-depth measurements, Phys. Rev. Lett. **66**, 2665–2668 (1991)
244. O. Gunnarsson, Superconductivity in fullerides, Rev. Mod. Phys. **69**, 575–606 (1997)
245. Y.J. Uemura, A. Keren, L.P. Le, G.M. Luke, B.J. Sternlieb, W.D. Wu, J.H. Brewer, R.L. Whetten, S.M. Huang, S. Lin, R.B. Kaner, F. Diederich, S. Donovan, G. Gruner, and K. Holczer, Magnetic-field penetration depth in K_3C_{60} measured by muon spin relaxation, Nature **352**, 605–607 (1991)
246. A.P. Ramirez, Superconduct. Rev. **1**, 1–101 (1994)
247. A.S. Alexandrov, Nonadiabatic polaronic superconductivity in MgB_2 and cuprates, Physica C **363**, 231–236 (2001)
248. H. Schmidt, J.F. Zasadzinski, K.E. Gray, and D.G. Hinks, Evidence for two-band superconductivity from break junction tunneling on MgB_2, Phys. Rev. Lett. **88**, 127002 (2002)

249. A.V. Sologubenko, J. Jun, S.M. Kazakov, J. Karpinski, and H.R. Ott, Temperature dependence and anisotropy of the bulk upper critical field H_{c2} of MgB_2, Phys. Rev. B **65**, R180505–R180508 (2002)
250. S. Lee, Z.G. Khim, Y. Chong, S.H. Moon, H.N. Lee, H.G. Kim, B. Oh, and E.J. Choi, Measurements of the superconducting gap of MgB_2 by point contact of spectroscopy, Physica C **377**, 202–207 (2002)
251. T. Arai, K. Ichimura, K. Nomur, S. Takasaki, J. Yamada, S. Nakatsuji, and H.Anzai, Superconducting and normal-state gaps in κ-(BEDT-TTF)$_2$Cu(NCS)$_2$ studied by STM spectroscopy, Solid State Commun. **116**, 679–682 (2000)
252. Y.J. Uemura, A. Keren, L.P. Le, G.M. Luke, W.D. Wu, Y. Kubo, T. Manako, Y. Shimakawa, M. Subramanian, J.L. Cobb, and J.T. Markert, Magnetic-field penetration depth in $Tl_2Ba_2CuO_{6+\delta}$ in the overdoped regime, Nature **364**, 605–607 (1993)
253. R. Prozorov, R.W. Giannetta, A. Carrington, P. Fournier, R.L. Greene, P. Guptasarma, D.G. Hinks, and A.R. Banks, Measurements of the absolute value of the penetration depth in high-T_c superconductors using a low-T_c superconductive coating, Appl. Phys. Lett. **77**, 4202 (2000)
254. A. Biswas, P. Fournier, V.N. Smolyaninova, R.C. Budhani, J.S. Higgins, and R.L. Greene, Gapped tunneling spectra in the normal state of $Pr_{2-x}Ce_xCuO_4$, Phys. Rev. B **64**, 104519–104526 (2001)
255. T. Kajitani, K. Hiraga, S. Hosoya, T. Fukuda, K. Oh-ishi, and Y. Syono, Structural study of oxygen-saturated or quenched $Pr_{2-x}Ce_xCuO_4$ with $x \leqq 0.15$, Physica C **178**, 397–404 (1991)
256. L. Ozyuzer, Z. Yusof, J.F. Zasadzinski, R. Mogilevsky, D.G. Hinks, and K.E. Gray, Evidence of $d_{x^2-y^2}$ symmetry in the tunneling conductance density of states of $Tl_2Ba_2CuO_6$, Phys. Rev. B **57**, R3245–R3248 (1998)
257. Y.T. Wang and A.M. Hermann, Pairing symmetry of Tl-2201 from measurements of temperature and magnetic field dependencies of the anisotropic penetration depth, Physica C **335**, 134–138 (2000)
258. C. Niedermayer, C. Bernhard, U. Binninger, H. Glückler, J.L. Tallon, E.J. Ansaldo, and J.I. Budnick, Muon spin rotation study of the correlation between Tc and $ns/m*$ in overdoped $Tl_2Ba_2CuO_{6+\delta}$, Phys. Rev. Lett. **71**, 1764 (1993)
259. M. Kang, G. Blumberg, M.V. Klein, and N.N. Kolesnikov, Resonance Raman study of the superconducting gap and low energy excitations in $Tl_2Ba_2CuO_{6+\delta}$ superconductors, Phys. Rev. Lett. **77**, 4434–4437 (1996)
260. M. Li, C.J. van der Beek, M. Konczykowski, A.A. Menovsky, and P.H. Kes, Superconducting properties of strongle underdoped $Bi_2Sr_2CaCu_2O_{8+x}$ single crystals, Phys. Rev. B **66**, 024502 (2002)
261. N. Miyakawa, P. Guptasarma, J.F. Zasadzinski, D.G. Hinks, and K.E. Gray, Strong dependence of the superconducting gap on oxygen doping from tunneling measurements on $Bi_2Sr_2CaCu_2O_{8-\delta}$, Phys. Rev. Lett. **80**, 157–160 (1998)
262. N. Momono, R. Dipasupil, H. Ishiguro, S. Saigo, T. Nakano, M. Oda, and M. Ido, Crossover behavior of in-plane and out-of-plane resistivity in $La_{2-x}Sr_xCuO_4$, Physica C **317-318**, 603–606 (1999)
263. M. Niderost, R. Frassanito, M. Saalfrank, A.C. Mota, G. Blatter, V.N. Zavaritsky, T.W. Li, and P.H. Kes, Lower critical field H_{c1} and barriers for vortex entry in $Bi_2Sr_2CaCu_2O_{8+\delta}$ crystals, Phys. Rev. Lett. **81**, 3231 (1998)
264. S.L. Lee, P. Zimmermann, H. Keller, M. Warden, I.M. Savic, R. Schauwecker, D. Zech, R. Cubitt, E.M. Forgan, P.H. Kes, T.W. Li, A.A. Menovsky, and Z. Tarnawski, Evidence for flux-lattice melting and a dimensional crossover in single-crystal $Bi_{2.15}Sr_{1.85}CaCu_2O_{8+\delta}$ from muon spin rotation studies, Phys. Rev. Lett. **71**, 3862–3865 (1993)
265. M. Weber, P. Birrer, F.N. Gygax, B. Hitti, E. Lippelt, H. Maletta, and A. Schenck, Measurements of the London penetration depth in Bi-based high-T_c compounds, Hyp. Int. **63**, 93 (1993)
266. A. Sugimoto, S. Kahiwaya, H. Eisaki, H. Yamaguchi, K. Oka, H. Kahiwaya, H. Tsuchiura, and Y. Tanaka, Correlation between modulation structure and electronic inhomogeneity on Pb-doped Bi-2212 single crystals, Physica C **426-431**, 390–395 (2005)
267. G. Villard, D. Pelloquin, and A. Maignan, Doping dependence of the in-plane penetration depth and fishtail in $Bi_2Sr_2Ca_{1-x}Y_xCu_2O_{8+\delta}$, Phys. Rev. B **58**, 15231–15237 (1998)
268. C. Bernhard, J.L. Tallon, T. Blasius, A. Golnik, and C. Neidermayer, Anomalous peak in the superconducting condensate density of cuprate high-Tc superconductors at a unique doping state, Phys. Rev. Lett. **86**, 1614–1617 (2001)
269. C. Panagopoulos, J.R. Cooper, and T. Xiang, Systematic behavior of the in-plane penetration depth in d-wave cuprates, Phys. Rev. B **57**, 13422–13425 (1998)
270. J.E. Sonier, J.H. Brewer, R.F. Kiefl, G.D. Morris, R.I. Miller, D.A. Bonn, J. Chakhalian, R.H. Heffner, W.N. Hardy, and R. Liang, Field induced reduction of the low-temperature superfluid density in $YBa_2Cu_3O_{6.95}$, Phys. Rev. Lett. **83**, 4156–4159 (1999)
271. T. Pereg-Barnea, P.J. Turner, R. Harris, G.K. Mullins, J.S. Bobowski, M. Raudsepp, R. Liang, D.A. Bonn, and W.N. Hardy, Absolute values of the London penetration depth in $YBa_2Cu_3O_{6+y}$ measured by zero field ESR spectroscopy on Gd doped single crystals, Phys. Rev. B **69**, 184513 (2004)

272. P. Zimmermann, H. Keller, S.I. Lee, I.M. Savic, M. Warden, D. Zech, R. Cubitt, E.M. Forgan, E. Kaldis, J. Karpinski, and C. Kruger, Muon-spin-rotation studies of the temperature dependence of the magnetic penetration depth in the $YBa_2Cu_3O_x$ family and related compounds, Phys. Rev. B **52**, 541–552 (1995)
273. J.Y.T. Wei, C.C. Tsuei, P.J.M. van Bentum, Z. Xiong, C.W. Chu, and M.K. Wu, Quasiparticle tunneling spectra of the high-Tc mercury cuprates: Implications of the d-wave two-dimensional van Hove scenario, Phys. Rev. B **57**, 3650–3662 (1998)
274. L. Fàbrega, A. Calleja, A. Sin, S.P. nol, X. Obradors, J. Fontcuberta, and P.J.C. King, Muon spin relaxation in Re-substituted $HgA_2Ca_{n-1}Cu_nO_{2n+2+x}$ (A = Sr,Ba; n = 2,3) superconductors, Phys. Rev. B **60**, 7579–7584 (1999)
275. A. Fujimori, A. Ino, T. Yoshida, T. Mizokawa, M. Nakamura, C. Kim, Z.X. Shen, T. Kakeshita, H. Eisaki, and S. Uchida, Fermi surface, pseudogap and superconducting gap in $La_{2-x}Sr_xCuO_4$, Physica C **341-348**, 2067 (2000)
276. C. Panagopoulos, B.D. Rainford, J.R. Cooper, W. Lo, J.L. Tallon, J.W. Loram, J. Betouras, Y.S. Wang, and C.W. Chu, Effects of carrier concentration on the superfluid density of high-T_c cuprates, Phys. Rev. B **60**, 14617–14620 (1999)
277. P.G. Radaelli, D.G. Hinks, A.W. Mitchell, B.A. Hunter, J.L. Wagner, B. Dabrowski, K.G. Vandervoort, H.K. Viswanathan, and J.D. Jorgensen, Structural and superconducting properties of $La_{2-x}Sr_xCuO_4$ as a function of Sr content, Phys. Rev. B **49**, 4163–4175 (1994)
278. T. Kato, H. Morimoto, A. Katagiri, S. Okitsu, and H. Sakata, Scanning tunneling microscopy and spectroscopy on $La_{2-x}Sr_xCuO_4$, Physica C **392-396**, 221–223 (2003)
279. V.J. Emery and S.A. Kivelson, Importance of phase fluctuations in superconductors with small superfluid density, Nature **374**, 434–437 (1995)
280. L.M. Merchant, J. Ostrick, R.P. Barber, Jr., and R.C. Dynes, Crossover from phase fluctuation to amplitude-dominated superconductivity: A model system, Phys. Rev. B **63**, 134508–134514 (2001)
281. A.J. Rimberg, T.R. Ho, C. Kurdak, J. Clarke, K.L. Campman, and A.C. Gossard, Dissipation-driven superconductor-insulator transition in a two-dimensional Josephson-junction array, Phys. Rev. Lett. **78**, 2632–2635 (1997)
282. A. Kapitulnik, N. Mason, S.A. Kivelson, and S. Chakravarty, Effects of dissipation on quantum phase transitions, Phys. Rev. B **63**, 125322–125333 (2001)
283. N. Mason and A. Kapitulnik, True superconductivity in a two-dimensional superconducting-insulating system, Phys. Rev. B **64**, 60504–60508 (2001)
284. E.W. Carlson, S.A. Kivelson, V.J. Emery, and E. Manousakis, Classical phase fluctuations in high temperature superconductors, Phys. Rev. Lett. **83**, 612–615 (2000)
285. B.I. Spivak and S.A. Kivelson, Aharonov–Bohm oscillations with period hc/4e and negative magnetoresistance in dirty superconductors, Phys. Rev. B **45**, 10490–10495 (1992)
286. P.M. Chaikin and T.C. Lubensky, *Principles of Condensed Matter Physics* (Cambridge University Press, Cambridge, 1995)
287. E. Roddick and D. Stroud, Effect of phase fluctuations on the low-temperature penetration depth of high-T_c superconductors, Phys. Rev. Lett. **74**, 1430–1433 (1995)
288. M.W. Coffey, Effect of superconductor phase fluctuations upon penetration depth, Phys. Lett. A **200**, 195–200 (1995)
289. S. Kamal, D.A. Bonn, N. Goldenfeld, P.J. Hirschfeld, R. Liang, and W.N. Hardy, Penetration depth measurements of 3D XY critical behavior in $YBa_2Cu_3O_{6.95}$ crystals, Phys. Rev. Lett. **73**, 1845–1848 (1994)
290. S. Chakravarty, G. Ingold, S.A. Kivelson, and A. Luther, Onset of global phase coherence in Josephson-junction arrays: a dissipative phase transition, Phys. Rev. Lett. **56** (1986)
291. K.-H. Wagenblast, A.V. Otterlo, G. Schon, and G.T. Zimanyi, Superconductor-insulator transition in a tunable dissipative environment, Phys. Rev. Lett. **79** (1997)
292. S. Sachdev, *Quantum Phase Transitions* (Cambridge University Press, Cambridge, 1999)
293. M.V. Feigel'man, A.I. Larkin, and A.I. and M.A. Skvortsov, Quantum superconductor-metal transition in a proximity array, Phys. Rev. Lett. **86**, 1869–1872 (2001)
294. B.I. Spivak, A. Zyuzin, and M. Hruska, Quantum superconductor-metal transition, Phys. Rev. B **64**, 132502–132505 (2001)
295. Y. Oreg and E. Demler, Fermions and bosons in superconducting amorphous wires, cond-mat/0106645 (2001)
296. L. Ozyuzer, J.F. Zasadzinski, and N. Miyakawa, Tunneling spectra and superconducting gap in $Bi_2Sr_2CaCu_2O_{8+\delta}$ and $Tl_2Ba_2CuO_{6+\delta}$, Int. J. Mod. Phys. B **13**, 3721–3724 (1999)

297. S. Doniach and M. Inui, Long-range Coulomb interactions and the onset of superconductivity in the high-T_c materials, Phys. Rev. B **41**, 6668–6678 (1990)
298. J.M. Harris, P.J. White, Z.-X. Shen, H. Ikeda, R. Yoshizaki, H. Eisaki, S. Uchida, W.D. Si, J.W. Xiong, Z.-X. Zhao, and D.S. Dessau, Measurement of an anisotropic energy gap in single plane $Bi_2Sr_{2-x}La_xCuO_{6+\delta}$, Phys. Rev. Lett. **79**, 143–146 (1997)
299. X.F. Sun, S. Komiya, J. Takeya, and Y. Ando, Magnetic-field-induced localization of quasiparticles in underdoped $La_{2-x}Sr_xCuO_4$ single crystals, Phys. Rev. Lett. **90**, 117004 (2003)
300. D.G. Hawthorn, R.W. Hill, C. Proust, F. Ronning, M. Sutherland, E. Boaknin, C. Lupien, M.A. Tanatar, J. Paglione, S. Wakimoto, H. Zhang, L.Taillefer, T.Kimura, M.Nohara, H.Takagi, and N.E. Hussey, Field-induced thermal metal-to-insulator transition in underdoped $La_{2-x}Sr_xCuO_{4+\delta}$, Phys. Rev. Lett. **90**, 197004 (2003)
301. X.F. Sun, K. Segawa, and Y. Ando, Low-temperature nodal-quasiparticle transport in lightly doped $YBa_2Cu_3O_y$ near the edge of the superconducting doping regime, Phys. Rev. B **72**, 100502 (2005)
302. M. Sutherland, S.Y. Li, D.G. Hawthorn, R.W. Hill, F. Ronning, M.A. Tanatar, J. Paglione, H. Zhang, L. Taillefer, J. DeBenedictis, R. Liang, D.A. Bonn, and W.N. Hardy, Delocalized fermions in underdoped cuprate superconductors, Phys. Rev. Lett. **94**(14), 147004 (2005)
303. C. Proust, K. Behnia, R. Bel, D. Maude, and S.I. Vedeneev, Heat transport in $Bi_{2+x}Sr_{2-x}CuO_{6+\delta}$: Departure from the Wiedemann-Franz law in the vicinity of the metal-insulator transition, Phys. Rev. B **72**, 214511 (2005)
304. X.F. Sun, S. Ono, Y. Abe, S. Komiya, K. Segawa, and Y. Ando, Electronic inhomogeneity and breakdown of the universal thermal conductivity of cuprate superconductors, Phys. Rev. Lett. **96**(1), 017008 (2006)
305. D.M. Broun, P.J. Turner, W.A. Huttema, S. Ozcan, B. Morgan, R. Liang, W.N. Hardy, and D.A. Bonn, The superfluid density reveals a quantum critical point between d-wave superconductivity and a Mott insulator
306. T. Schneider and H. Keller, Extreme type II superconductors. Universal properties and trends, Physica **207**, 366–380 (1993)
307. Y. Ando and K. Segawa, Magnetoresistance of untwinned $YBa_2Cu_3O_y$ single crystals in a wide range of doping: anomalous hole-doping dependence of the coherence length, Phys. Rev. Lett. **88**, 167005 (2002)
308. L. Li, Y.Y. Wang, M.J. Naughton, S. Ono, Y. Ando, and N.P. Ong, Strongly nonlinear magnetization above T_c in $Bi_2Sr_2CaCu_2O_{8+\delta}$, Europhys. Lett. 72, 451–457 (2005)
309. Y. Wang, L. Li, M.J. Naughton, G.D. Gu, S. Uchida, and N.P. Ong, Field-enhanced diamagnetism in the pseudogap state of the cuprate $Bi_2Sr_2CaCu_2O_{8+\delta}$ superconductor in an intense magnetic field, Phys. Rev. Lett. **95**(24), 247002 (2005)
310. V. Oganesyan, D.A. Huse, and S.L. Sondhi, Theory of diamagnetic response of the vortex liquid phase of two-dimensional superconductors, Phys. Rev. B 73(9), 094503 (2006)
311. Y. Wang, L. Li, and N.P. Ong, Nernst effect in high-T_c superconductors, Phys. Rev. B 73(2), 024510 (2006)
312. Y. Wang, Z.A. Xu, T. Kakeshita, S. Uchida, S. Ono, Y. Ando, and N.P. Ong, Onset of the vortexlike Nernst signal above T_c in $La_{2-x}Sr_xCuO_4$ and $Bi_2Sr_{2-y}LaCuO_6$, Phys. Rev. B **64**, 224519 (2001)
313. Z.A. Xu, N. Ong, Y. Wang, T. Kakeshita, and S. Uchida, Vortex-like excitations and the onset of superconducting phase fluctuation in underdoped $La_{2-x}Sr_xCuO_4$, Nature **406**, 486–488 (2000)
314. C.P. Bidinosti, W.N. Hardy, D.A. Bonn, and R.Liang, Magnetic field dependence of lambda in $YBa_2Cu_3O_{6.95}$, Phys. Rev. Lett. **83**, 3277–3280 (1999)
315. X.-G. Wen and P.A. Lee, Theory of quasiparticles in the underdoped high-T_c superconducting state, Phys. Rev. Lett. **80** 2193–2196 (1998)
316. A.J. Millis, S.M. Girvin, L.B. Ioffe, and A.I. Larkin, Anomalous charge dynamics in the superconducting state of underdoped cuprates, Jour. Phys. Chem. Solids **59**, 1742–1744 (1998)
317. L. Benfatto, S. Caprara, C. Castellani, A. Paramekanti, and M. Randeria, Phase fluctuations, dissipation, and superfluid stiffness in d-wave superconductors, Phys. Rev. B **63**, 174513–21 (2001)
318. G.T. Zimanyi, S.A. Kivelson, and A. Luther, Superconductivity from predominantly repulsive interactions in quasi one-dimensional systems, Phys. Rev. Lett. **60**, 2089–2092 (1988)
319. S.A. Kivelson and G.T. Zimanyi, High temperature superconductors, RVB, and conducting polymers, Mol. Cryst. Liq. Cryst **160**, 457–481 (1988)
320. A.M. Finkel'stein and S.A. Brazovsky, Interchain coupling in linear conductors, J. Phys. C **14**, 847–857 (1981)
321. S.A. Brazovsky and A.M. Finkel'stein, The influence of phonons on the optical properties and conductivity of quasi-one-dimensional metals, Solid State Comm. **38**, 745 (1981)

322. S.A. Kivelson and M.I. Salkola, Metal-non-metal transition in polyacetylene, Synth. Met. **44**, 281–291 (1991)
323. G.G. Batrouni, R.T. Scalettar, G.T. Zimanyi, and A.P. Kampf, Supersolids in the Bose-Hubbard Hamiltonian, Phys. Rev. Lett. **74**, 2527–2530 (1995)
324. M.P.A. Fisher, P.B. Weichman, G. Grinstein, and D.S. Fisher, Boson localization and the superfluid-insulator transition, Phys. Rev. B **40**, 546–570 (1989)
325. C. Nayak and K. Shtengel, Microscopic models of two-dimensional magnets with fractionalized excitations, Phys. Rev. B **64**, 064422 (2001)
326. R. Coldea, D.A. Tennant, A.M. Tsvelik, and Z. Tylczynski, Experimental realization of a 2D fractional quantum spin liquid, Phys. Rev. Lett. **86**, 1335–1338 (2001)
327. S. Chakravarty, Magnetic Properties of La_2CuO_4, (Addison-Wesley, Redwood City, CA, 1990)
328. A.E. Sikkema, I. Affleck, and S.R. White, Spin gap in a doped Kondo chain, Phys. Rev. Lett. **79**, 929–932 (1997)
329. O. Zachar, V.J. Emery, and S.A. Kivelson, Exact results for a 1D Kondo lattice from bosonization, Phys. Rev. Lett. **77**, 1342–1345 (1996)
330. O. Zachar, Staggered liquid phases of the one-dimensional Kondo–Heisenberg lattice model, Phys. Rev. B **63**, 205104–205113 (2001)
331. D.J. Scalapino and S.R. White, Numerical results for the Hubbard model: implications for the high T_c pairing mechanism, Found. Phys. **31**, 27–39 (2001)
332. F.D.M. Haldane, Continuum dynamics of the 1-D Heisenberg antiferromagnet: Identification with the O(3) nonlinear sigma model, Phys. Lett. A **93**, 464–468 (1983)
333. S. Chakravarty, Dimensional crossover in quantum antiferromagnets, Phys. Rev. Lett. **77**, 4446–4449 (1996)
334. S.R. White, R.M. Noack, and D.J. Scalapino, Resonating valence bond theory of coupled Heisenberg chains, Phys. Rev. Lett. **73**, 886–889 (1994)
335. D.J. Scalapino and S.A. Trugman, Local antiferromagnetic correlations and $d_{x^2-y^2}$ pairing, Philos. Mag. B **74**, 607–610 (1996)
336. M. Havilio and A. Auerbach, Superconductivity and quantum spin disorder in cuprates, Phys. Rev. Lett. **83**, 4848–4851 (1999)
337. M. Havilio and A. Auerbach, Correlations in doped antiferromagnets, Phys. Rev. B **62**, 324–336 (2000)
338. S.R. White and D.J. Scalapino, Superconductivity in ladders and coupled planes, Phys. Rev. B **45**, 5744–5747 (1992)
339. D. Poilblanc, Internal structure of the singlet $d_{x^2-y^2}$ hole pair in an antiferromagnet, Phys. Rev. B **495**, 1477–1479 (1993)
340. Y. Fang, A.E. Ruckenstein, E. Dagotto, and S. Schmitt-Rink, Holes in the infinite-U Hubbard model: Instability of the Nagaoka state, Phys. Rev. B **40**, 7406–7409 (1989); see also [548].
341. D. Rokhsar, Quadratic quantum antiferromagnets in the fermionic large-N limit, Phys. Rev. B **42**, 2526–2531 (1990)
342. C.M. Varma and A. Zawadowskii, Scaling in an interacting two-component (valence-fluctuation) electron gas, Phys. Rev. B **32**, 7399–7407 (1985)
343. H.-H. Lin, L. Balents, and M.P.A. Fisher, N-chain Hubbard model in weak coupling, Phys. Rev. B **56**, 6569–6593 (1997)
344. L.P. Pryadko, S.A. Kivelson, and O. Zachar, Incipient order in the t–J model at high temperatures, Phys. Rev. Lett. **92**(6), 067002 (2004)
345. W.-F. Tsai and S.A. Kivelson, Superconductivity in inhomogeneous Hubbard models, Phys. Rev. B **73**(21), 214510 (2006)
346. E. Arrigoni, E. Fradkin, and S.A. Kivelson, Mechanism of high-temperature superconductivity in a striped Hubbard model, Phys. Rev. B **69**(21), 214519 (2004)
347. I. Martin, D. Podolsky, and S.A. Kivelson, Enhancement of superconductivity by local inhomogeneities, Phys. Rev. B **72**(6), 060502 (2005)
348. E. Arrigoni and S.A. Kivelson, Optimal inhomogeneity for superconductivity, Phys. Rev. B **68**(18), 180503 (2003)
349. E.F. Steven A. Kivelson, How optimal inhomogeneity produces high temperature superconductivity, To appear as a chapter in "Treatise of High Temperature Superconductivity" ed. by J.R. Schrieffer (Springer, 2006)
350. S.R. White and D.J. Scalapino, Density matrix renormalization group study of the striped phase in the 2D $t-J$ model, Phys. Rev. Lett. **80**, 1272–1275 (1998)
351. S.R. White and D.J. Scalapino, Energetics of domain walls in the 2D $t-J$ model, Phys. Rev. Lett. **81**, 3227–3230 (1998)

352. S. Hellberg and E.Manousakis, Phase separation at all interaction strengths in the $t-J$ model, Phys. Rev. Lett. **78**, 4609–4612 (1997)
353. D. Poilblanc, O. Chiappa, J. Riera, S.R. White, and D.J. Scalapino, Evolution of the spin gap upon doping a 2-leg ladder, Phys. Rev. B **62**, R14633–R14636 (2000)
354. S.R. White and D.J. Scalapino, Competition between stripes and pairing in a $t-t'-J$ model, Phys. Rev. B **60**, R753–R756 (1999)
355. S.C. Hellberg and E. Manousakis, Stripes and the $t-J$ model, Phys. Rev. Lett. **83**, 132–135 (1999), see also [356,357]
356. S.R. White and D.J. Scalapino, Comment on "Stripes and the $t-J$ Model", Phys. Rev. Lett. **84**, 3021 (2000)
357. S.C. Hellberg and E. Manousakis, Reply: Hellberg and Manousakis, Phys. Rev. Lett. **84**, 3022 (2000)
358. E. Arrigoni, A.P. Harju, W. Hanke, B. Brendel, and S.A. Kivelson, Stripes and superconducting pairing in the $t-J$ model with Coulomb interactions, Phys. Rev. B **65**, 134503 (2002)
359. S.R. White and D.J. Scalapino, Phase separation and stripe formation in the two-dimensional $t-J$ model: A comparison of numerical results, Phys. Rev. B **61**, 6320–6326 (2000)
360. E. Daggoto, Correlated electrons in high-temperature superconductors, Rev. Mod. Phys. **66**, 763–840 (1994)
361. S.C. Hellberg and E. Manousakis, Green's-function Monte Carlo for lattice fermions: Application to the $t-J$ model, Phys. Rev. B **61**, 11787–11806 (2000)
362. S. Sorella, Green function Monte Carlo with stochastic reconfiguration, Phys. Rev. Lett. **80**, 4558–4561 (1998)
363. Y.C. Chen and T.K. Lee, $t-J$ model studied by the power Lanćzos method, Phys. Rev. B **51**, 6723–6726 (1995)
364. S.R. White, Density-matrix algorithms for quantum renormalization groups, Phys. Rev. B **48**, 10345–10356 (1993)
365. S.R. White, Spin gaps in a frustrated Heisenberg model for CaV_4O_9, Phys. Rev. Lett. **77**, 3633–3636 (1996)
366. U. SchollwRck, The density-matrix renormalization group, Rev. Mod. Phys. **77**, 259–314 (2005)
367. E. Dagotto and T.M. Rice, Surprises on the way from one- to two-dimensional quantum magnets: the ladder materials, Science **271**, 618–623 (1996)
368. E. Dagotto, Experiments on ladders reveal a complex interplay between a spin-gapped normal state and superconductivity, Rep. Prog. Phys. **62**, 1525–1571 (1999)
369. E. Lieb and F.Y. Wu, Absence of Mott transition in an exact solution of the short-range, one-band model in one dimension, Phys. Rev. Lett. **20**, 1445–1448 (1968)
370. H.J. Schulz, In: Proceedings of the 9th Jerusalem Winter School for Theoretical Physics,, ed. by V.J. Emery (World Scientific, New York, 1993)
371. B. Frischmuth, B. Ammon, and M. Troyer, Susceptibility and low-temperature thermodynamics of spin-1/2 Heisenberg ladders, Phys. Rev. B **54**, R3714–R3717 (1996)
372. R.M. Noack, S.R. White, and D.J. Scalapino, Correlations in a two-chain Hubbard model, Phys. Rev. Lett. **73**, 882–885 (1994)
373. R.M. Noack, D.J. Scalapino, and S.R. White, The ground state of the two-leg Hubbard ladder. A density-matrix renormalization group study, Physica C **270**, 281–296 (1996)
374. E. Dagotto, J. Riera, and D. Scalapino, Superconductivity in ladders and coupled planes, Phys. Rev. B **45**, 5744–5747 (1992)
375. T. Kimura, K. Kuroki, and H. Aoki, Pairing correlation in the three-leg Hubbard ladder-renormalization group and quantum Monte Carlo studies, J. Phys. Soc. Jpn. **67**, 1377–1390 (1998)
376. S. Nishimoto, E. Jeckelmann, and D.J. Scalapino, Differences between hole and electron doping of a two-leg CuO ladder, Phys. Rev. B, **66**, 245109 (2002)
377. E. Jeckelmann, D.J. Scalapino, and S.R. White, Comparison of different ladder models, Phys. Rev. B **58**, 9492–9497 (1998)
378. S. Daul, D.J. Scalapino, and S.R. White, Pairing correlations on $t-U-J$ ladders, Phys. Rev. Lett. **84**, 4188–4191 (2000)
379. S.R. White and D.J. Scalapino, Ground-state properties of the doped three-leg $t-J$ ladder, Phys. Rev. B **57**, 3031–3037 (1998)
380. T.M. Rice, S. Haas, M. Sigrist, and F.-C. Zhang, Lightly doped $t-J$ three-leg ladders: An analog for the underdoped cuprates, Phys. Rev. B **56**, 14655–14667 (1997)
381. W.R. White and D.J. Scalapino, Ground states of the doped four-leg $t-J$ ladder, Phys. Rev. B **55**, R14701–R14704 (1997)
382. D.J. Scalapino, private communication

383. M. Ogata, M.U. Luchini, S. Sorella, and F.F. Assaad, Phase diagram of the one-dimensional $t-J$ model, Phys. Rev. Lett. **66**, 2388–2391 (1991)
384. C.S. Hellberg and E.J. Mele, Luttinger-liquid instability in the one-dimensional $t-J$ model, Phys. Rev. B **48**, 646–649 (1993)
385. M. Troyer, H. Tsunetsugu, and T.M. Rice, Properties of lightly doped $t-J$ two-leg ladders, Phys. Rev. B **53**, 251–267 (1996)
386. C.A. Hayward and D. Poilblanc, Luttinger-liquid behavior and superconducting correlations in $t-J$ ladders, Phys. Rev. B **53**, 11721–11728 (1996)
387. G. Sierra, M.A. Martin-Delgado, J. Dukelsky, S.R. White, and D.J. Scalapino, Dimer-hole-RVB state of the two-leg $t-J$ ladder: A recurrent variational ansatz, Phys. Rev. B **57**, 11666–11673 (1998)
388. S. Rommer, S.R. White, and D.J. Scalapino, Phase separation in $t-J$ ladders, Phys. Rev. B **61**, 13424–13430 (2000)
389. S.R. White, I. Affleck, and D.J. Scalapino, Friedel oscillations and charge density waves in chains and ladders, Phys. Rev. B **65**, 165122 (2002)
390. S.A. Kivelson and V.J. Emery, Strongly Correlated Electronic Materials: The Los Alamos Symposium 1993 (Addison-Wesley, Redwood City, CA, 1994)
391. S. Hellberg and E. Manousakis, 2-Dimensional $t-J$ model at low electron density, Phys. Rev. B **52**, 4639–4642 (1995)
392. V.J. Emery, S.A. Kivelson, and H.Q. Lin, Phase separation in the t-J model, Phys. Rev. Lett. **64**, 475–478 (1990)
393. S.A. Kivelson, V.J. Emery, and H.Q. Lin, Doped antiferromagnets in the weak-hopping limit, Phys. Rev. B **42**, 6523–6530 (1990)
394. E. Eisenberg, R. Berkovits, D.A. Huse, and B.L. Altshuler, The breakdown of the Nagaoka phase in the 2D $t-J$ model, Phys. Rev. B **65**, 134437 (2002)
395. W.O. Putikka, M.U. Luchini, and T.M. Rice, Aspects of the phase diagram of the two-dimensional $t-J$ model, Phys. Rev. Lett. **68**, 538–541 (1992)
396. D. Poilblanc, Phase diagram of the two-dimensional $t-J$ model at low doping, Phys. Rev. B **52**, 9201–9204 (1995)
397. M. Calandra, F. Becca, and S. Sorella, Charge fluctuations close to phase separation in the two-dimensional $t-J$ model, Phys. Rev. Lett. **81**, 5185–5188 (1998)
398. H. Yokoyama and M. Ogata, Phase diagram and pairing symmetry of the two-dimensional t-J model by a variation theory, J. Phys. Soc. Jpn. **65**, 3615–3629 (1996)
399. C.T. Shih, Y.C. Chen, and T.K. Lee, Phase separation of the two-dimensional t-J model, Phys. Rev. B **57**, 627–631 (1998)
400. C.S. Shin, Y.C. Chen, and T.K. Lee, Revisit phase separation of the two-dimensional $t-J$ model by the Power–Lanczos method, J. Phys. Chem. Solids **62**, 1797–1811 (2001)
401. M. Khono, Ground-state properties of the two-dimensional $t-J$ model, Phys. Rev. B **55**, 1435–1441 (1997)
402. F. Becca, L. Capriotti, and S. Sorella, Stripes and spin incommensurabilities are favored by lattice anisotropies, Phys. Rev. Lett. **87**, 167005 (2001)
403. S.R. White and D.J. Scalapino, Why do stripes form in doped antiferromagnets and what is their relationship to superconductivity? cond-mat/0006071 (2000)
404. C. Nayak and F. Wilczek, Possible electronic structure of domain walls in Mott insulators, Int. J. Mod. Phys. B **10**, 2125–2136 (1996)
405. J. Zaanen and O. Gunnarsson, Charged magnetic domain lines and the magnetism of high-T_c oxides, Phys. Rev. B **40**, 7391–7394 (1989)
406. D. Poilblanc and T.M. Rice, Charged solitons in the Hartree–Fock approximation to the large-U Hubbard model, Phys. Rev. B **39**, 9749–9752 (1989)
407. H.J. Schulz, Domain walls in a doped antiferromagnet, J. de Physique **50**, 2833–2849 (1989)
408. K. Machida, Magnetism in La_2CuO_4 based compounds, Physica C **158**, 192–196 (1989)
409. O. Zachar, Stripes: Why hole rich lines are antiphase domain walls Phys. Rev. B **65**, 174411 (2002)
410. W.V. Liu and E. Fradkin, Antiferromagnetic spin ladders effectively coupled by one-dimensional electron liquids, Phys. Rev. Lett. **86**, 1865–1868 (2001)
411. A.L. Chernyshev, S.R. White, and A.H. Castro Neto, Charge stripe in an antiferromagnet: 1d band of composite excitations, Physica B **312-313**, 566–568 (2002)
412. A.C. Cosentini, M. Capone, L. Guidoni, and G.B. Bachelet, Phase separation in the two-dimensional Hubbard model: A fixed-node quantum Monte Carlo study, Phys. Rev. B **58**, R14685–R14688 (1998)

413. F. Becca, M. Capone, and S. Sorella, Spatially homogeneous ground state of the two-dimensional Hubbard model, Phys. Rev. B **62**, 12700 (2001)
414. S.R. White and D.J. Scalapino, Stripes on a 6-leg Hubbard ladder, Phys. Rev. Lett. **91**, 136403 (2003)
415. G. Hager, G. Wellein, E. Jeckelmann, and H. Fehske, Stripe formation in doped Hubbard ladders, Phys. Rev. B **71**, 075108 (2005)
416. A. Moreo, Pairing correlations in the two-dimensional Hubbard model, Phys. Rev. B **45**, 5059–5061 (1992)
417. S.-C. Zhang, J. Carlson, and J.E. Gubernatis, Constrained path Monte Carlo method for fermion ground states, Phys. Rev. B **55**, 7464–7477 (1997)
418. E. Dagotto and J. Riera, Indications of $d_{2^x-y^2}$ superconductivity in the two-dimensional $t-J$ model, Phys. Rev. Lett. **70**, 682–685 (1993)
419. E. Dagotto, A. Moreo, F. Ortolani, D. Poilblanc, and J. Riera, Static and dynamical properties of doped Hubbard clusters, Phys. Rev. B, **45**, 10741–10760 (1992)
420. M. Calandra and S. Sorella, From antiferromagnetism to d-wave superconductivity in the two-dimensional $t-J$ model, Phys. Rev. B **61**, R11894–R11897 (2000)
421. S. Sorella, G.B. Martins, F. Becca, C. Gazza, L. Capriotti, A. Parola, and E. Dagotto, Superconductivity in the two-dimensional $t-J$ model, Phys. Rev. Lett. **88**, 117002 (2002); see also [422,423]
422. T.K. Lee, C.T. Shih, Y.C. Chen, and H.Q. Lin, Comment on "Superconductivity in the two-dimensional $t-J$ model", Phys. Rev. Lett. **89**, 279702 (2002)
423. S. Sorella, A. Parola, F. Becca, L. Capriotti, C. Gazza, E. Dagotto, and G. Martins, Sorella et al. reply, Phys. Rev. Lett. **89**, 279703 (2002)
424. M. Boninsegni and E. Manousakis, Two-hole d-wave binding in the physical region of the $t-J$ model: A Green's-function Monte Carlo study, Phys. Rev. B **47**, 11897–11904 (1993)
425. T. Tohyama, C.G. C.T. Shih, Y.C. Chen, T.K. Lee, S. Maekawa, and E. Dagotto, Stripe stability in the extended t-J model on planes and four-leg ladders, Phys. Rev. B **59**, R11649–R11652 (1999)
426. E.L. Nagaev, *Physics of Magnetic Semiconductors* (Mir, Moscow, 1983)
427. A. Auerbach and B.E. Larson, Doped antiferromagnet: The instability of homogeneous magnetic phases, Phys. Rev. B **43**, 7800–7809 (1991)
428. A. Auerbach and B.E. Larson, Small-polaron theory of doped antiferromagnets, Phys. Rev. Lett. **66**, 2262–2265 (1990)
429. A. Auerbach, Spin tunneling, Berry phases, and doped antiferromagnets, Phys. Rev. B **48**, 3287–3289 (1993)
430. E.W. Carlson, S.A. Kivelson, Z. Nussinov, and V.J. Emery, Doped antiferromagnets in high dimension, Phys. Rev. B **57**, 14704–14721 (1998)
431. P.B. Visscher, Phase separation instability in the Hubbard model, Phys. Rev. B **10**, 943–945 (1974)
432. L.B. Ioffe and A.I. Larkin, Two-dimensional Hubbard model with strong electron repulsion, Phys. Rev. B **37**, 5730–5737 (1988)
433. L. Pryadko, D. Hone, and S.A. Kivelson, Instability of charge ordered states in doped antiferromagnets, Phys. Rev. Lett. **80**, 5651–5654 (1998)
434. J.K. Freericks, E.H. Lieb, and D. Ueltschi, Segregation in the Falicov-Kimball model, Commun. Math. Phys. **227**, 243–279 (2002)
435. E.L. Nagaev, Lanthanum manganites and other giant-magnetoresistance magnetic conductors, Usp. Fiz. Nauk. **166**, 833 (1996)
436. S. Trugman, Interaction of holes in a Hubbard antiferromagnet and high-temperature superconductivity, Phys. Rev. B **37**, 1597–1603 (1988)
437. A.H. Castro Neto, Landau theory of phase separation in cuprates, Phys. Rev. B **51**, 3254–3256 (1995)
438. M. Seul and D. Andelman, Domain shapes and patterns — the phenomenology of modulated phases, Science **267**, 476–483 (1995)
439. B.P. Stojkovic, Z.G. Yu, A.R. Bishop, A.H. Castro Neto, and N. Gronbech-Jensen, Charge ordering and long-range interactions in layered transition metal oxides, Phys. Rev. Lett. **82**, 4679–4682 (1999)
440. L.P. Pryadko, S.A. Kivelson, V.J. Emery, Y.B. Bazaliy, and E.A. Demler, Topological doping and the stability of stripe phases, Phys. Rev. B **60**, 7541–7557 (1999)
441. A.L. Chernyshev, A.H. Castro Neto, and A.R. Bishop, Metallic stripe in two dimensions: stability and spin-charge separation, Phys. Rev. Lett. **84**, 4922–4925 (2000)

442. S.A. Kivelson and V.J. Emery, Topological doping of correlated insulators, Synth. Metals **80**, 151–158 (1996)
443. J. Zaanen and Z. Nussinov, Stripes and nodal fermions as two sides of the same coin, cond-mat/0006193 (2000)
444. J. Han, Q.H. Wang, and D.H. Lee, Antiferromagnetism, stripes, and superconductivity in the $t-J$ model with Coulomb interaction, Int. J. Mod. Phys. B **15**, 1117–1126 (2001)
445. A.H. Castro Neto, Luttinger stripes in antiferromagnets, Z. Phys. B **103**, 185–192 (1997)
446. J. Zaanen and P.B. Littlewood, Freezing electronic correlations by polaronic instabilities in doped La_2NiO, Phys. Rev. B **50**, 7222–7225 (1994)
447. H.J. Schulz, Incommensurate antiferromagnetism in the two-dimensional Hubbard model, Phys. Rev. Lett. **64**, 1445–1448 (1990)
448. A.I. Larkin, Effect of inhomogeneities on the structure of the mixed state of superconductors, Sov. Phys. JETP **31**, 784–786 (1970)
449. V.J. Emery and S.A. Kivelson, Charge inhomogeneity and high temperature superconductivity, J. Phys. Chem. Sol. **61**, 467–471 (2000)
450. N. Ichikawa, S. Uchida, J.M. Tranquada, T. Niemoller, P.M. Gehring, S.H. Lee, and J.R. Schneider, Local magnetic order vs superconductivity in a layered cuprate, Phys. Rev. Lett. **85**, 1738–1741 (2000)
451. J. Schmalian and P.G. Wolynes, Stripe glasses: Self-generated randomness in a uniformly frustrated system, Phys. Rev. Lett. **85**, 836–839 (2000)
452. J. Burgy, M. Mayr, V. Martin-Mayor, A. Moreo, and E. Dagotto, Colossal effects in transition metal oxides caused by intrinsic inhomogeneities, Phys. Rev. Lett. **87**, 277202 (2001)
453. U. Löw, V.J. Emery, K. Fabricius, and S.A. Kivelson, Study of an Ising model with competing long- and short-range interactions, Phys. Rev. Lett. **72**, 1918–1921 (1994)
454. M. Grousson, G. Tarjus, and P. Viot, Monte Carlo study of the three-dimensional Coulomb frustrated Ising ferromagnet, Phys. Rev. E **64**, 036109 (2001)
455. L. Chayes, V.J. Emery, S.A. Kivelson, Z. Nussinov, and G. Tarjus, Avoided critical behavior in a uniformly frustrated system, Physica A **225**, 129–153 (1996)
456. Z. Nussinov, J. Rudnick, S.A. Kivelson, and L.N. Chayes, Avoided critical behavior in O(n) systems, Phys. Rev. Lett. **83**, 472–475 (1999)
457. D.B. Tanner and T. Timusk, Optical properties of high-temperature superconductors, In: Physical Properties of High Temperature Superconductors, ed. by D.M. Ginsberg (World Scientific, Singapore, 1992) Vol. III
458. N.P. Ong, Transport properties of high T_c cuprates, In: Physical Properties of High Temperature Superconductors, ed. by D.M. Ginsberg (World Scientific, Singapore, 1992) Vol. II
459. K. Krishana, J.M. Harris, and N.P. Ong, Quasiparticle mean free path in $YBa_2Cu_3O_7$ measured by the thermal Hall conductivity, Phys. Rev. Lett. **75**, 3529–3532 (1995)
460. K.M. Lang, V. Madhavan, J.E. Hoffman, E.W. Hudson, H. Eisaki, S. Uchida, and J.C. Davis, Imaging the granular structure of high-T_c superconductivity in underdoped $Bi_2Sr_2CaCu_2O_{8+delta}$, Nature **415**, 412–416 (2002)
461. T.R. Thurston, R.J. Birgeneau, M.A. Kastner, N.W. Preyer, G. Shirane, Y.Fujii, K.Yamada, Y.Endoh, K.Kakurai, M.Matsuda, Y.Hidaka, and T. Murakami, Neutron scattering study of the magnetic excitations in metallic and superconducting $La_{2-x}Sr_xCuO_{4-y}$, Phys. Rev. B **40**, 4585–4595 (1989)
462. S.-W. Cheong, G.Aeppli, T.E. Mason, H. Mook, S.M. Hayden, P.C. Canfield, Z. Fisk, K.N. Clausen, and J.L. Martinez, Incommensurate magnetic fluctuations in $La_{2-x}Sr_xCuO_4$, Phys. Rev. Lett. **67**, 1791–1794 (1991)
463. J.M. Tranquada, B.J. Sternlieb, J.D. Axe, Y. Nakamura, and S. Uchida, Evidence for stripe correlations of spins and holes in copper oxide superconductors, Nature **375**, 561–563 (1995)
464. J.M. Tranquada, D.J. Buttrey, V. Sachan, and J.E. Lorenzo, Simultaneous ordering of holes and spins in $La_2NiO_{4.125}$, Phys. Rev. Lett. **73**, 1003–1006 (1994)
465. A. Lanzara, P.V. Bogdanov, X.J. Zhou, S.A. Kellar, D.L. Feng, E.D. Lu, T. Yoshida, H. Eisaki, A. Fujimori, K. Kishio, J.-I. Shimoyama, T. Noda, S. Uchida, Z. Hussain, and Z.-X. Shen, Evidence for ubiquitous strong electron-phonon coupling in high-temperature superconductors, Nature **412**, 510–514 (2001)
466. L.P. Gor'kov and A.V. Sokol, On the problem of the phase diagram of new superconductors, J. Physique **50**, 2823–32 (1989)
467. A.J. Heeger, S.A. Kivelson, J.R. Schrieffer, and W. Su, Solitons in conducting polymers, Rev. Mod. Phys. **60**, 781–850 (1988)

468. D.K. Campbell, A.R. Bishop, and K. Fesser, Polarons in quasi-one-dimensional systems, Phys. Rev. B **26**, 6862–6874 (1982)
469. L.B. Ioffe and A. Larkin, Superconductivity in the liquid-dimer valence-bond state, Phys. Rev. B **40**, 6941–6947 (1989)
470. M. Marder, N. Papanicolaou, and G.C. Psaltakis, Phase separation in a $t - J$ model, Phys. Rev. B **41**, 6920–32 (1990)
471. M. Vojta and S. Sachdev, Charge order, superconductivity and a global phase diagram of doped antiferromagnets, Phys. Rev. Lett. **83**, 3916–3919 (1999)
472. S. Sachdev and S.C. Zhang, Tuning order in cuprate superconductors, Science **295**, 452 (2002)
473. D.I. Khomskii and K.I. Kugel, Why stripes? Spontaneous formation of inhomogeneous structures due to elastic interactions, Europhys. Lett. **55**, 208–213 (2001)
474. A.A. Koulakov, M.M. Fogler, and B.I. Shklovskii, Charge density wave in two-dimensional electron liquid in weak magnetic field, Phys. Rev. Lett. **76**, 499–502 (1996)
475. R. Moessner and T.J. Chalker, Exact results for interacting electrons in high Landau levels, Phys. Rev. B **54**, 5006–5015 (1996)
476. M.P. Lilly, K.B. Cooper, J.P. Eisenstein, L.N. Pfeiffer, and K.W. West, Evidence for an anisotropic state of two-dimensional electrons in high Landau levels, Phys. Rev. Lett. **82**, 394–397 (1999)
477. B. Spivak, Phase separation in the two-dimensional electron liquid in MOSFETs, cond-mat/0205127 (2002)
478. S.A. Kivelson, Making high T_c higher: A theoretical proposal, Physica B **61-67**, 61–67 (2002)
479. T.H. Geballe and B.Y. Moyzhes, Qualitative understanding of the highest T_c cuprates, Physica C **342**, 1821–1824 (2000)
480. T. Niemöller, N. Ichikawa, T. Frello, H. Hünnefeld, N.H. Andersen, S. Uchida, J.R. Schneider, and J. Tranquada, Charge stripes seen with X-rays in $La_{1.45}Nd_{0.4}Sr_{0.15}CuO_4$, Eur. Phys. J. B **12**, 509–513 (1999)
481. M. v. Zimmermann, A. Vigliante, T. Niemöller, N. Ichikawa, T. Frello, J. Madsen, P. Wochner, S. Uchida, N.H. Andersen, J.M. Tranquada, D. Gibbs, and J.R. Schneider, Hard-X-ray diffraction study of charge stripe order in $La_{1.48} Nd_{0.4} Sr_{0.12} CuO_4$, Europhys. Lett **41**, 629–634 (1998)
482. K. Yamada, C.H. Lee, K. Kurahashi, J. Wada, S. Wakimoto, S. Ueki, H. Kimura, Y. Endoh, S. Hosoya, G. Shirane, R.J. Birgeneau, M. Greven, M.A. Kastner, and Y.J. Kim, Doping dependence of the spatially modulated dynamical spin correlations and the superconducting-transition temperature in $La_{2-x}Sr_xCuO_4$, Phys. Rev. B **57**, 6165–6172 (1998)
483. Y.S. Lee, R.J. Birgeneau, M.A. Kastner, Y. Endoh, S. Wakimoto, K. Yamada, R.W. Erwin, S.-H. Lee, and G. Shirane, Neutron-scattering study of spin-density wave order in the superconducting state of excess-oxygen-doped La_2CuO_{4+y}, Phys. Rev. B **60**, 3643–3654 (1999)
484. B. Khaykovich, Y.S. Lee, R. Erwin, S.-H. Lee, S. Wakimoto, K.J. Thomas, M.A. Kastner, and R.J. Birgeneau, Enhancement of long-range magnetic order by magnetic field in superconducting La_2CuO_{4+y}, Phys. Rev. B **66**, 014528 (2002)
485. H. Kimura, K. Hirota, H. Matsushita, K. Yamada, Y. Endoh, S.-H. Lee, C.F. Majkrzak, R. Erwin, G. Shirane, M. Greven, Y.S. Lee, M.A. Kastner, and R.J. Birgeneau, Neutron-scattering study of static antiferromagnetic correlations in $La_{2-x}Sr_xCu_{1-y}Zn_yO_4$, Phys. Rev. B **59**, 6517–6523 (1999)
486. M. Fujita, K. Yamada, H. Hiraka, P.M. Gehring, S.H. Lee, S. Wakimoto, and G. Shirane, Static magnetic correlations near the insulating-superconducting phase boundary in $La_{2-x}Sr_xCuO_4$, Phys. Rev. B **65**, 064505–064511 (2002)
487. J.M. Tranquada, Neutron scattering studies of antiferromagnetic correlations in cuprates (2005)
488. P. Dai, H.A. Mook, R.D. Hunt, and F. Dogan, Evolution of the resonance and incommensurate spin fluctuations in superconducting $YBa_2Cu_3O_{6+x}$, Phys. Rev. B **63**, 054525–054544 (2001)
489. P. Bourges, Y. Sidis, H.F. Fong, L.P. Regnault, J. Bossy, A. Ivanov, and B. Keimer, The spin excitation spectrum in superconducting $YBa_2Cu_3O_{6.85}$, Science **288**, 1234–1237 (2000)
490. G. Aeppli, S.M. Hayden, P. Dai, H.A. Mook, R.D. Hunt, T.G. Perring, and F. Dogan, The weights of various features in the magnetic spectra of cuprates, Phys. Stat. Sol. **215**, 519–522 (1999)
491. H.A. Mook, P.C. Dai, F. Dogan, and R.D. Hunt, One-dimensional nature of the magnetic fluctuations in $YBa_2Cu_3O_6$, Nature, **404**, 729–731 (2000)
492. A. Moodenbaugh, Y. Xu, M. Suenaga, T.J. Folkerts, and R.N. Shelton, Superconducting properties of $La_{2-x}Ba_xCuO_4$, Phys. Rev. B **38**, 4596–4600 (1988)
493. M. Fujita, H. Goka, K. Yamada, and M. Matsuda, Competition between charge/spin-density-wave orders and superconductivity in $La_{1.875}Ba_{0.125-x}Sr_xCuO_4$, Phys. Rev. Lett. **88**, 167008 (2002)
494. M. Fujita, H. Goka, K. Yamada, J.M. Tranquada, and L.P. Regnault, Stripe order, depinning, and fluctuations in $La_{1.875}Ba_{0.125}CuO_4$ and $La_{1.875}Ba_{0.075}Sr_{0.050}CuO_4$, Phys. Rev. B **70**, 104517 (2004)

495. S.A. Kivelson, E. Fradkin, V. Oganesyan, I.P. Bindloss, J.M. Tranquada, A. Kapitulnik, and C. Howald, How to detect fluctuating order in the high-temperature superconductors, Rev. Mod. Phys. **71**, 1201–1241 (2003)
496. P. Abbamonte, A. Rusydi, S. Smadici, G.D. Gu, G.A. Sawatzky, and D.L. Feng, Spatially modulated 'Mottness' in La2-xBaxCuO4, Nature Physics **1**, 155–158 (2005)
497. S.M. Hayden, H.A. Mook, P. Dai, T.G. Perring, and F. Dogan, The structure of the high-energy spin excitations in a high-transition-temperature superconductor, Nature **429**, 531 (2004)
498. H.A. Mook and B.C. Chakoumakos, Incommensurate fluctuations in $Bi_2Sr_2CaCu_2O_8$, J. Superconduct. **10**, 389–392 (1997)
499. J.E. Hoffman, E.W. Hudson, K.M. Lang, V.Madhavan, H.Eisaki, S.Uchida, and J.C. Davis, A four unit cell periodic pattern of quasi-particle states surrounding vortex cores in $Bi_2Sr_2CaCu_2O_{8+x}$, Science **295**, 466–469 (2002)
500. C. Howald, H. Eisaki, N. Kaneko, and A. Kapitulnik, Coexistence of charged stripes and superconductivity in $Bi_2Sr_2CaCu_2O_{8+\delta}$, Proc. Nat. Acad. Sci. **100**, 9705–9709 (2003)
501. C. Howald, H. Eisaki, N. Kaneko, M. Greven, and A. Kapitulnik, Periodic density-of-states modulations in superconducting $Bi_2Sr_2CaCu_2O_{8+\delta}$, Phys. Rev. B **67**, 014533 (2003)
502. J.A. Robertson, S.A. Kivelson, E. Fradkin, A.C. Fang, and A. Kapitulnik, Distinguishing Patterns of Charge Order: Stripes or Checkerboards
503. J.E. Hoffman, K.McElroy, D.-H. Lee, K.M. Lang, H. Eisaki, S. Uchida, and J.C. Davis, Imaging quasiparticle quantum interference in $Bi_2Sr_2CaCu_2O_{8+\delta}$, Science **297**, 1148–1151 (2002)
504. T. Hanguri, C. Lupien, Y. Kohsaka, D.H. Lee, M. Azuma, M. Takano, H. Takagi, and J.C. Davis, A 'checkerboard' electronic crystal state in lightly hole-doped $Ca_{2-x}Na_xCuO_2Cl_2$, Nature **430**, 1001–1005 (2004)
505. S. Smadici, P. Abbamonte, M. Taguchi, T. Sasagawa, M. Takano, and H. Takagi, Soft x-ray absorption and scattering from $Na_xCa_{2-x}CuO_2Cl_2$ at $x = 0.08$
506. E. Fradkin, S.A. Kivelson, E. Manousakis, and K. Nho, Nematic phase of the two-dimensional electron gas in a magnetic field, Phys. Rev. Lett. **84**, 1982–1985 (2000)
507. J.Zaanen, M.L. Horbach, and W.V. Saarloos, Charged domain-wall dynamics in doped antiferromagnets and spin fluctuations in cuprate superconductors, Phys. Rev. B **53**, 8671–8680 (1996)
508. V. Oganesyan, S.A. Kivelson, and E.Fradkin, Quantum theory of a nematic Fermi fluid, Phys. Rev. B **64**, 195109–195114 (2001)
509. C. Halboth and W. Metzner, d-wave superconductivity and Pomeranchuk instability in the two-dimensional Hubbard model, Phys. Rev. Lett. **85**, 5162–5165 (2000)
510. V. Hankevych, I. Grote, and F. Wegner, Pomeranchuk and other instabilities in the $t - t'$ Hubbard model at the Van Hove filling, Phys. Rev. B **66**, 094516 (2002)
511. V. Oganesyan, E. Fradkin, and S.A. Kivelson, work in progress (2002)
512. A. Abanov, V. Kalatsky, and V.L. Pokrovsky, Phase diagram of ultrathin ferromagnetic films with perpendicular anisotropy, Phys. Rev. B **51**, 1023–1038 (1995)
513. J. Zaanen, O.Y. Osman, H.V. Kruis, Z. Nussinov, and J. Tworzydlo, The geometric order of stripes and Luttinger liquids, Phil. Mag. B **81**, 1485–1531 (2002)
514. Y.-B. Kim and H.-Y. Kee, Pairing instability in a nematic Fermi liquid, J. Phys. Cond. Matt. **16**, 3139 (2004)
515. K.B. Cooper, M. Lilly, J.P. Eisenstein, L.N. Pfeiffer, and K.W. West, Onset of anisotropic transport of two-dimensional electrons in high Landau levels: Possible isotropic-to-nematic liquid crystal phase transition, Phys. Rev. B **65**, 241313 (2002)
516. V.J. Emery and S. Kivelson, Mapping of the two-channel Kondo problem to a resonant-level model, Phys. Rev. B **46**, 10812–10817 (1992)
517. V.J. Emery and S.A. Kivelson, Solution of an orbital Kondo array, Phys. Rev. Lett. **71**, 3701–3704 (1993)
518. D. Chang and D.H. Lee, Transport in inhomogeneous strongly correlated systems, cond-mat/0205057 (2002)
519. Z.M. Yusof, B.O. Wells, T. Valla, A.V. Fedorov, P.D. Johnson, Q. Li, C. Kendziora, S. Jian, and D.G. Hinks, Quasiparticle liquid in the highly overdoped $Bi_2Sr_2CaCu_2O_{8+\delta}$, Phys. Rev. Lett. **88**, 167006–167009 (2002)
520. M. Granath, V. Oganesyan, D. Orgad, and S.A. Kivelson, Distribution of spectral weight in a system with disordered stripes, Phys. Rev. B **65**, 184501–184510 (2002)
521. M.I. Salkola, V.J. Emery, and S.A. Kivelson, Implications of charge ordering for single-particle properties of high T_c superconductors, Phys. Rev. Lett. **77**(1), 155–158 (1996)

522. M.G. Zacher, R. Eder, E. Arrigoni, and W. Hanke, Stripes in doped antiferromagnets: single-particle spectral weight, Phys. Rev. Lett. **85**, 2585–2588 (2000)
523. M.G. Zacher, R. Eder, E. Arrigoni, and W. Hanke, Evolution of the stripe phase as a function of doping from a theoretical analysis of angle-resolved photoemission data, Phys. Rev. B **65**, 045109 (2002)
524. M. Vojta, Y. Zhang, and S. Sachdev, Renormalization group analysis of quantum critical points in d-wave superconductors, Int. J. Mod. Phys.B **14**, 3719–3734 (2000)
525. T. Valla, A.V. Fedorov, P.D. Johnson, B.O. Wells, S.L. Hulbert, Q.Li, G.D. Gu, and N. Koshizuka, Evidence for quantum critical behavior in the optimally doped cuprate $Bi_2Sr_2CaCu_2O_8$, Science **285**, 2110–2113 (1999)
526. Y. Zhang, N.P. Ong, P.W. Anderson, D.A. Bonn, R. Liang, and W.N. Hardy, Giant enhancement of the thermal Hall conductivity kappa(xy) in the superconductor $YBa_2Cu_3O_7$, Phys. Rev. Lett. **86**, 890–893 (2001)
527. D.A. Bonn, P. Dosanjh, R. Liang, and W.N. Hardy, Evidence for rapid suppression of quasi-particle scattering below T_c in $YBa_2Cu_3O_7$, Phys. Rev. Lett. **68**, 2390–2393 (1992)
528. D.L. Feng, A. Damascelli, K.M. Shen, N. Motoyama, D.H. Lu, H. Eisaki, K.Shimizu, J.i.Shimoyama, K.Kishio, N.Kaneko, M.Greven, G.D. Gu, X.J. Zhou, C. Kim, F. Ronning, N.P. Armitage, and Z.-X. Shen, Electronic structure of the trilayer cuprate superconductor $Bi_2Sr_2Ca_2Cu_3O_{10+\delta}$, Phys. Rev. Lett. **88**, 107001 (2001)
529. H.F. Fong, P. Bourges, Y. Sidis, L.P. Regnault, A. Ivanov, G.D. Gul, N. Koshizuka, and B. Keimer, Neutron scattering from magnetic excitation in $Bi_2Sr_2CaCu_2O_{8+\delta}$, Nature **398**, 588–591 (1999)
530. M. Dumm, D.N. Basov, S. Komiya, and Y. Ando, Anistropic electromagnetic in-plane response of $La_{1.97}Sr_{0.03}CuO_4$, Physica C **388**, 335–336 (2003)
531. H.V. Kruis, Z. Nussinov, and J. Zaanen, Emergent Z_2 gauge symmetry and spin-charge separation in one-dimensional physics, cond-mat/0110055 (2001)
532. M. Matsuda, M. Fujita, K. Yamada, R.J. Birgeneau, M.A. Kastner, H. Hiraka, Y. Endoh, S. Wakimoto, and G. Shirane, Static and dynamic spin correlations in the spin-glass phase of slightly doped $La_{2-x}Sr_xCuO_4$, Phys. Rev. B **62**, 9148–9154 (2000)
533. A.N. Lavrov, Y. Ando, S. Komiya, and I. Tsukada, Unusual magnetic susceptibility anisotropy in untwinned $La_{2-x}Sr_xCuO_4$ single crystals in the lightly doped region, Phys. Rev. Lett. **87**, 017007 (2001)
534. B. Lake, G. Aeppli, T.E. Mason, A. Schroder, D.F. McMorrow, K. Lefmann, M. Isshiki, M. Nohara, H. Takagi, and S.M. Hayden, Spin gap and magnetic coherence in a clean high-temperature superconductor, Nature **400**, 43–46 (1999)
535. T. Noda, H. Eisaki, and S.I. Uchida, Evidence for one-dimensional charge transport in $La_{2-x-y}Nd_ySr_xCuO_4$, Science **286**, 265–268 (1999)
536. S. Tajima, T. Noda, H. Eisaki, and S. Uchida, c-axis optical response in the static stripe ordered phase of the cuprates, Phys. Rev. Lett. **86**, 500–503 (2001)
537. I. Iguchi, T. Yamaguchi, and A. Sugimoto, Diamagnetic activity above T_c as a precursor to superconductivity in $La_{2-x}Sr_xCuO_4$, Nature **412**, 420–23 (2001)
538. I. Ussishkin, S.L. Sondhi, , and D.A. Huse, Gaussian superconducting fluctuations, thermal transport, and the Nernst effect, Phys. Rev. Lett. **89**, 287001 (2002)
539. A.G. Loeser, Z.-X. Shen, D.S. Dessau, D.S. Marshall, C.H. Park, P. Fournier, and A. Kapitulnik, Excitation Gap in the Normal State of Underdoped $Bi_2Sr_2CaCu_2O_{8+\delta}$, Science **273**, 325–329 (1996)
540. N. Momono, M. Oda, M. Ido, et al., Energy gap evolution over wide temperature and doping ranges in $Bi_2Sr_2CaCu_2O_8+\delta$, Mod. Phys. Lett. B **17**, 401–414 (2003)
541. R.S. Decca, H.D. Drew, E. Osquiguil, B. Maiorov, and J. Guimpel, Anomalous proximity effect in underdoped $YBa_2Cu_3O_{6+x}$ josephson junctions, Phys. Rev. Lett. **85**, 3708–3711 (2000)
542. M.K. Crawford, M.N. Kunchur, W.E. Farneth, and E.M. McCarron, Anomalous oxygen isotope effect in $La_{2-x}Sr_xCuO_4$, Phys. Rev. B **41**, 282–287 (1990)
543. J.C. Phillips and J. Jung, Nanodomain structure and function of high temperature superconductors, Phil. Mag. B **81**, 745–756 (2001)
544. S.A. Kivelson and V.J. Emery, Stripes and Related Phenomena (Kluwer Academic/Plenum, New York, 2000)
545. O. Zachar, Stripes disorder and correlation lengths in doped antiferromagnets, Phys. Rev. B **62**, 13836–13839 (2000)
546. N. Hasselmann, A.H. Castro Neto, C. Morais Smith, and Y. Dimashko, Striped phase in the presence of disorder and lattice potentials, Phys. Rev. Lett. **82**, 2135 (1999)

547. R.T. Scalettar, G.G. Batrouni, A.P. Kampf, and G.T. Zimanyi, Simultaneous diagonal and off-diagonal order in the Bose–Hubbard Hamiltonian, Phys. Rev. B **51**, 8467–8480 (1995)
548. S.R. White and I. Affleck, Density matrix renormalization group analysis of the Nagaoka polaron in the two-dimensional $t - J$ model, Phys. Rev. B **64**, 024411–024416 (2001)
549. D.A. Bonn, K. Zhang, S. Kamal, R. Liang, P. Dosanjh, W.N. Hardy, C. Kallin, and A.J. Berlinsky, Evidence for rapid suppression of quasi-particle scattering below T_c in $YBa_2CuO_{7-\delta}$, Phys. Rev. Lett. **72**, 1391–1391 (1994)

22 A Spin Fluctuation Model for *d*-Wave Superconductivity

A. V. Chubukov Univ. of Wisconsin, Dept. of Physics, Madison, USA
D. Pines Los Alamos National Laboratory CNLS, USA
J. Schmalian Ames Laboratory, Dept. of Physics and Astronomy, USA

22.1 Introduction and Overview

The identification of the microscopic mechanisms responsible for superconductivity and the nature of the superconducting pairing state continues to represent one of the most exciting challenges in theoretical physics [1]. In the so-called conventional superconductors, at frequencies less than or comparable to the Debye frequency, the attractive phonon-induced interaction between electrons wins out over the repulsive screened Coulomb interaction [2] and brings about superconductivity [3]. The pairing of electrons in the superconducting state is in an *s*-wave channel. The primacy of phonon-induced interaction in conventional superconductors has been demonstrated with great clarity. The phonon density of states, obtained by inelastic neutron scattering experiments, and the spectrum of the bosons which mediate pairing, as deduced from tunneling experiments, agree very well in systems like Pb [4,5]. In addition to the isotope effect [6,7], this comparison of two independent experiments is generally considered to be a very reliable proof of a phonon-mediated pairing state.

The analysis of the tunneling data relies heavily on the existence of strong coupling effects in the quasiparticle density of states and assumes the validity of the Eliashberg approach to superconductivity [8]. Eliashberg theory for conventional superconductors is extremely robust due to the decoupling of typical electron and phonon time scales caused by the small ratio of the velocity of sound and the Fermi velocity. This smallness also implies that the Debye frequency is much smaller than the Fermi energy, and hence the quasiparticles that participate in the pairing are low-energy quasiparticles, located in the near vicinity of the Fermi surface. Landau [9] showed that the low energy properties of a normal Fermi liquid are characterized by a small number of parameters and are independent of the details of the underlying lattice Hamiltonian. The success of the BCS-Eliashberg theory for conventional phonon mediated superconductors is almost unique for interacting many body systems.

It is well known however that the pairing state in the Bardeen–Cooper–Schrieffer theory does not need to be caused by the interaction between electrons and lattice vibrations. Generally there are two distinct classes of theories of non-phonon pairing. The first, and more conservative approach is to replace phonons by another collective bosonic excitation of the solid. This approach successfully describes the physics of superfluid ^{3}He [10], where the intermediate bosons are failed ferromagnetic spin fluctuations (ferromagnetic paramagnons); a large value of the ferromagnetic correlation length is not even required in view of the Kohn–Luttinger effect [11]. This magnetically-mediated interaction causes pairing in a state with angular momentum $l = 1$ (p-wave pairing) and leads to a rich phase diagram and a large class of new physical phenomena [10, 12, 13]. Magnetically-mediated superconductivity has been proposed for various organic and heavy fermion superconductors by a number of authors [14–20]. A second approach to unconventional pairing is more phenomenological, and is based on the assumption that the superconducting condensation energy is not determined by the attractive interaction mediated by some boson but rather by the energy gain due to feedback effects associated with pairing. The latter may, in principle, occur even for a purely repulsive pairing interaction. In general, this approach assumes a non-Fermi-liquid behavior in the normal state. Two examples are the inter-layer tunneling model of Anderson et al. [21,22] and the mid-infrared model of Legett [23]. In both models, the expectation value of the Hamiltonian is drastically different in the superconducting and the normal states, and the energy gain due to pairing apparently cannot be traced back to some boson-mediated attraction.

The two approaches to unconventional pairing are not necessarily in contradiction with each other. First, at strong coupling, the pairing interaction mediated by a low-energy bosonic mode is highly retarded and is a complex function of frequency, so it is not straightforward to determine whether it is repulsive or attractive. Second the pairing obviously changes the form of the fermionic self-energy and hence affects the kinetic energy. And third, if the bosonic mode is itself made out of fermions, then the propagator of this mode also changes when the system becomes superconducting. This change affects the potential energy of the system. Which of the two effects is larger depends on the details of the system behavior, but in any case, there are clear similarities between a strong-coupling theory which involves a pairing boson, and a scenario based entirely on energy gain due to feedback from pairing.

In what follows we adopt the first approach and investigate the role of antiferromagnetic spin fluctuations (antiferromagnetic paramagnons) as a possible cause for both anomalous normal state behavior and unconventional i.e. non-s-wave superconductivity. This approach is chiefly motivated by the physics of the high temperature cuprate superconductors which have been shown to exhibit both highly anomalous normal state behavior and an unconventional pairing state with angular momentum $l = 2$ (d-wave pairing) [24–26]. As may be seen in

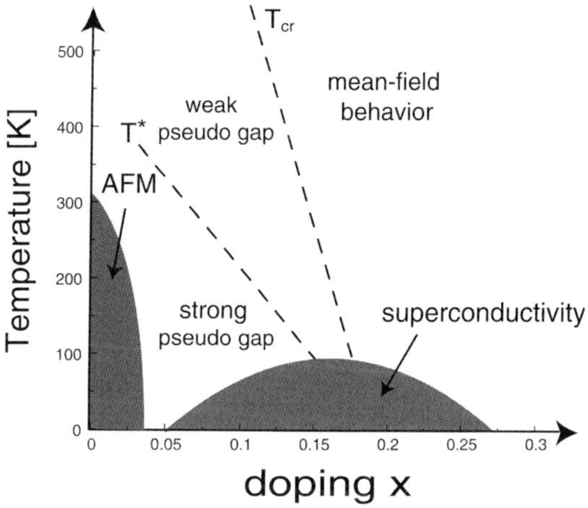

Fig. 22.1. Generic phase diagram of high temperature superconducting cuprates. The thermodynamic phases (antiferromagnetic at low doping and superconducting at higher doping) are depicted by the *shaded regions*. The remaining *lines* are either phase transitions or crossovers, visible in a variety of experiments

Fig. 22.1, the materials with the highest T_c are located reasonably close to an antiferromagnetic state and have been shown in nuclear magnetic resonance and inelastic neutron scattering experiments to exhibit significant antiferromagnetic correlations in the paramagnetic state. [27] We will show that in a quasi two-dimensional material where those correlations are significant (e.g. a spin correlation length larger than a lattice constant) the normal state behavior is anomalous while for Fermi surface parameters appropriate for the cuprates, one always gets a $d_{x^2-y^2}$ superconducting pairing state. We discuss other materials below, following a brief historical overview of the developments in the spin-fluctuation approach over the last decade. References to earlier works can be found in the papers cited below.

A $d_{x^2-y^2}$ pairing state in two dimensions due to the exchange of near-antiferromagnetic spin fluctuations was found in the detailed Hubbard model calculations of Bickers et al. [18]. For parameters believed to be relevant for cuprates in 1987, the superconducting transition temperature was comparatively low (< 40 K) under what seemed to be optimal conditions. Furthermore, T_c decreased as one increased the planar hole concentration from a low level, in contrast to experiment. These results, when taken together with the early penetration-depth experiments that supported an s-wave pairing state, were responsible for the fact that the magnetic mechanism and $d_{x^2-y^2}$ pairing had been abandoned by most of the high temperature superconductivity community by the end of 1989 (Bedell et al. [28]).

At about this time, theoretical groups in Tokyo [29,30] and Urbana [31,32] independently began developing a semi-phenomenological, theory of spin-mediated pairing. Both groups assumed that the magnetic interaction between the planar quasiparticles was responsible for the anomalous normal state properties and found a superconducting transition to a $d_{x^2-y^2}$ pairing state at a significantly higher temperature than those achieved using the Hubbard model. Moriya et al. [29, 30] used a self-consistent renormalization group approach to characterize the dynamic spin susceptibility. The resulting effective magnetic interaction between the planar quasiparticles was then used to calculate T_c and the normal state resistivity. Monthoux et al. [32] did not attempt a first-principles calculation of the planar quasiparticle interaction. Rather, they turned to experiment and used quasiparticles whose spectra was determined by fits to band structure calculations and angular resolved photoemission spectroscopy (ARPES) experiments. The effective magnetic interaction between these quasiparticles was assumed to be proportional to a mean field dynamic spin susceptibility of the form developed by Millis et al. [33] that had been shown to provide a good description of NMR experiments on the $YBa_2Cu_3O_{7-y}$ system [27,33].

Both groups followed up their initial weak coupling calculations with strong coupling (Eliashberg) calculations [34–37] that enabled them to take into account lifetime effects brought about by the strong magnetic interaction. These calculations showed that $d_{x^2-y^2}$ superconductivity at high-T_c is a robust phenomenon. Monthoux and Pines [36] also found in a strong coupling calculation that they could obtain an approximately correct magnitude and temperature dependence of the planar resistivity of optimally doped $YBa_2Cu_3O_{7-y}$ using the same coupling constant (and the same parameters to characterize the quasiparticle and spin spectrum) that had yielded a T_c of approximately 90 K. They concluded that they had established a "proof of concept" for a nearly antiferromagnetic Fermi liquid (NAFL) description of the anomalous normal state behavior and a spin fluctuation mechanism for high temperature superconductivity. Referring back to Fig. 22.1, these calculations should apply to the right of the T_{cr}-line, where the normal state is an unconventional Fermi liquid in which the characteristic energy above which quasiparticles lose their Fermi liquid behavior is low compared to the fermionic bandwidth.

Since their calculations unambiguously predicted a $d_{x^2-y^2}$ pairing state, Monthoux and Pines challenged the experimental community to find unambiguously the symmetry of the pairing state. At that time (1991–1992), only NMR Knight shift and Cu spin-lattice relaxation rate results supported $d_{x^2-y^2}$ pairing [38–42]. However within the next year or so, the tide turned dramatically away from s-wave pairing, with ARPES [43], penetration depth [44] and new NMR experiments [45–47] on the oxygen spin-lattice relaxation time and the anisotropy of the copper spin-lattice relaxation time all supporting a $d_{x^2-y^2}$ state. The decisive experiments were the direct phase-sensitive tests of pairing symmetry carried out by van Harlingen and his group in Urbana [24] as well as by Kirtley, Tsuei, and their collaborators [25].

In subsequent work on the spin-fluctuation mechanism a microscopic, Hamiltonian approach to the problem was developed [48, 49]. It was shown that the low-energy physics of spin-mediated pairing is fully captured by a model that describes the interaction of low energy fermionic quasiparticles with their own collective spin excitations (the spin-fermion model) [48, 49]. In particular, it was demonstrated that the phenomenological interaction between quasiparticles could be derived in a controllable way, even at strong coupling, by expanding either in the inverse number of hot spots in the Brillouin zone (= 8 for the physical case), or in the inverse number of fermionic flavors. We discuss this theory in detail in Sect. 22.4. As will be seen there, the spin-fermion model contains only a small number of parameters. These uniquely determine system behavior that is fully universal in the sense that it does not depend on the behavior of the underlying electronic system at energies comparable to the fermionic bandwidth. It is therefore possible to verify its applicability by first using a few experimental results to determine these parameters, and then comparing the predictions of the resulting parameter-free theory with the larger subset of experimental results obtained at temperatures and frequencies which are much smaller than the fermionic bandwidth. A major prediction of the spin fermion model is that the upper energy scale for the Fermi liquid behavior progressively shifts down as the system approaches a quantum-critical point at $T = 0$, and there emerges a large intermediate range of frequencies where, on the one hand, the system behavior is still a low-energy one and universal, and on the other hand, it is quantum-critical and not a Fermi liquid.

Cuprate superconductors are not the only candidate materials for spin fluctuation mediated pairing and non-Fermi liquid behavior. A number of organic superconductors are anisotropic quasi-two-dimensional materials that exhibit many of the anomalies typical of a system with an unconventional (i.e. non s-wave) pairing state. The phase diagram of a quasi two-dimensional organic compound κ-BEDT-TTF is shown in Fig. 22.2. One can see that, as in the case of the cuprates, the superconducting phase is found in the vicinity of an antiferromagnetic phase. Several groups [51–54] have used spin-fluctuation theory to predict the position of nodes of the superconducting order parameter of these materials. An unconventional order parameter with nodes of the gap is indeed supported by NMR [55–57], thermal conductivity [58], millimeter transmission [59] and STM [60] experiments. However, the last two experiments seem to come to different conclusions as far as the position of the nodes is concerned. [61] also finds nodes, but at a position that is not consistent with the prediction of a spin fluctuation induced pairing state. Finally, penetration depth experiments [62] and recent specific heat data [63] appear to support a conventional s-wave gap. Given those contradictory experiments, whether quasi-two dimensional organic superconductors exhibit an unconventional pairing state is, as of this writing, an open question.

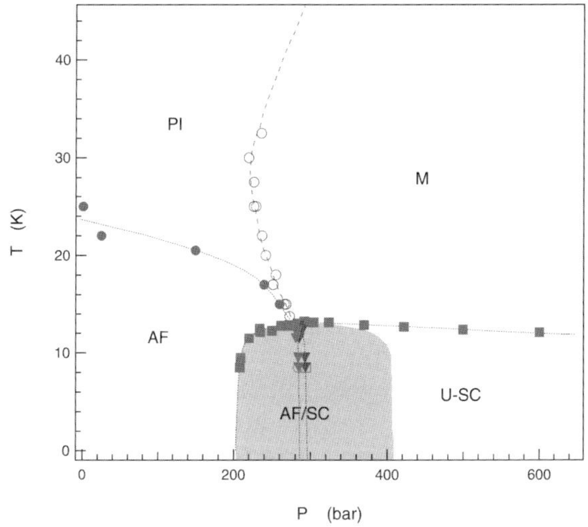

Fig. 22.2. The phase diagram of the layered organic superconductor κ-$(ET)_2Cu[N(CN)_2]Cl$ in the units of temperature and pressure (from [50]). PI refers to a paramagnetic insulating regime, M to a metallic regime, AF to an antiferromagnetic regime, and SC to a superconducting regime. In the region AF-SC, superconductivity and antiferromagnetism co-exist. In the region U-SC, the system is an unconventional superconductor

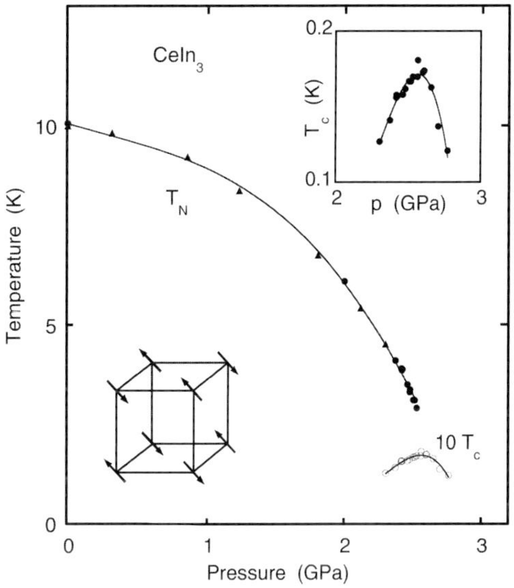

Fig. 22.3. The phase diagram of $CeIn_3$ in the units of temperature and pressure (from [65])

Cerium-based heavy electron superconductors represent another class of strongly correlated electron superconductors for which a spin-fluctuation induced interaction between quasiparticles is a strong candidate for the superconducting mechanism. Examples include $CeCu_2Si_2$ [64], $CePd_2Si_2$, $CeIn_3$ [65] and the newly discovered 1-1-5 materials $CeXIn_5$ with X = Co, Rh and Ir [66] or mixtures thereof. As may be seen in the phase diagrams of Figs. 22.3 and 22.4, all these materials are close to antiferromagnetism, with superconductivity occurring close to the critical pressure or alloy concentration at which the magnetic ordering disappears. Moreover, thermal conductivity measurements of $CeCoIn_5$, which becomes superconducting at 2.4 K at ambient pressures, the highest known value of T_c for a heavy fermion based system, strongly support a superconducting gap with nodes along the $(\pm\pi, \pm\pi)$ directions, as found in a $d_{x^2-y^2}$ pairing state [67]. Another exciting aspect of these systems is that by changing the relative compositions of Ir and Rh in 1-1-5 materials $CeRh_{1-x}Ir_xIn_5$, one can move the system from an antiferromagnetic to a superconducting state at ambient pressure.

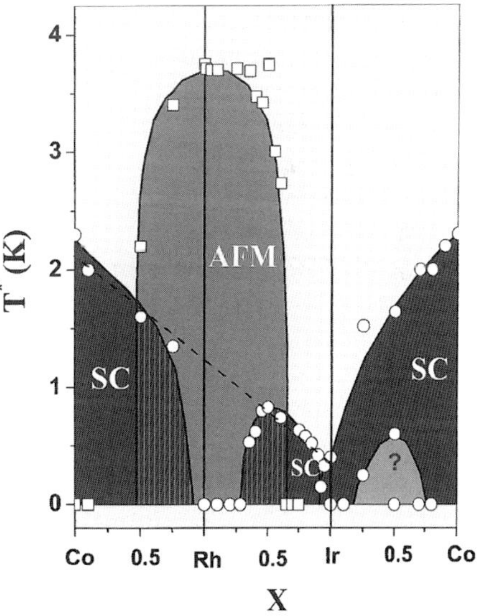

Fig. 22.4. The phase diagram of $Ce_2X\ In_5$ with X = Co, Rh, and Ir in the units of temperature and doping (from Pagliuso et al. [68])

Another widely studied material in which pairing is possibly due to spin fluctuation exchange is

Sr_2RuO_4 [69–72], where NMR Knight shift experiments [73] and spin-polarized neutron scattering measurements [74] reveal that the spin susceptibility is unchanged upon entering the superconducting state, consistent with spin-triplet superconductivity.

In summary, the cuprates, the 1-1-5 heavy fermion materials and the layered organic superconductors are strongly correlated materials that exhibit unconventional normal state and superconducting behavior, while the superconducting phases are located in the vicinity of magnetic instabilities in their corresponding phase diagrams. It is then quite natural to assume that in all three cases magnetic interactions may be responsible for the pairing.

The presence of antiferromagnetic and superconducting regions in the phase diagram raises the question of whether antiferromagnetism and superconductivity should be treated on equal footing in a spin fluctuation approach. If they should, the theoretical analysis would be complex. Fortunately, this is not the case, at least as long as the characteristic energy scales for the magnetic interactions are smaller than the fermionic bandwidth. The point is that superconductivity is generally a low-energy phenomenon associated with fermions in the near vicinity of the Fermi surface. On the other hand, antiferromagnetism originates in fermions with energies comparable to the bandwidth. Perhaps the easiest way to see this is to formally compute the static spin susceptibility in the random phase approximation (RPA). An RPA analysis yields $\chi^{-1}(\mathbf{q}) \propto 1 - g_{\mathrm{eff}}(\mathbf{q})\Pi(\mathbf{q})$ where $g_{\mathrm{eff}}(\mathbf{q})$ is some effective interaction, and $\Pi(\mathbf{q})$ is the static spin polarization operator (a particle-hole bubble with Pauli matrices in the vertices). For an antiferromagnetic instability we need $g_{\mathrm{eff}}(\mathbf{Q})\Pi(\mathbf{Q}) = 1$. One can easily make sure, by evaluating $\Pi(\mathbf{Q})$ for free fermions, that the momentum/frequency integration in the particle-hole bubble is dominated by the upper energy limit that is the fermionic bandwidth. This implies that whether or not a system orders antiferromagnetically is primarily determined by high-energy fermions that are located far away from the Fermi surface, and hence the antiferromagnetic correlation length, that measures the proximity of a material to a nearby antiferromagnetic region in the phase diagram, should not be calculated but rather be taken as an input for any low-energy analysis. We discuss the practical meaning of this separation of energies in Sect. 22.4.

A more subtle but important issue is whether the dynamical part of the spin susceptibility should be considered simply as an input for a low-energy model (as in the case for phonons), or whether the spin dynamics is produced by the same electrons that are responsible for the superconductivity and hence needs to be determined consistently within the low-energy theory. The first issue one has to consider here is whether a one-band description is valid, i.e. whether localized electrons remain quenched near the antiferromagnetic instability and form a single large Fermi surface together with the conduction electrons to which they are strongly coupled [75,76], or whether the magnetic instability is accompanied by the unquenching of local moments. In the latter case, the volume of the Fermi surface may change discontinuously at the magnetic transition and could, e.g. cause a jump in the Hall coefficient [77]. The quenching versus unquenching issue is currently a subject of intensive debate in heavy fermion materials [77,78]. In cuprates, however, this issue does not seem to play a role; it is widely accepted that the formation of Zhang–Rice singlets [79] gives rise to a single electronic degree of freedom. Similarly, in organic materials, the charge transport in the metallic and superconducting parts of the phase diagram is due to the same missing electrons in otherwise closed filled molecular orbital states. Whether or not the spin dynamics originates in low-energy fermions then reduces to the geometry of a single, large Fermi surface. For a Fermi surface with hot spots, points connected by the wave vector at which the spin fluctuation spectrum peaks, the low-energy spin dynamics is dominated by a process in which a collective spin excitation decays into a particle–hole pair. By virtue of energy conservation, this process involves fermions with frequencies comparable to the frequency of a spin excitation. Consequently, the spin dynamics is not an input. If, however, the Fermi surface does not contain hot spots, spin damping is forbidden at low-energies and spin fluctuations are magnon-like propagating excitations. It is easy to show that in the latter situation, the full dynamic spin propagator comes from particle–hole excitations at energies comparable to the bandwidth and therefore should be considered as an input for the low-energy theory.

In this chapter we consider in detail the scenario in which the Fermi surface contains hot spots and the spin damping by quasiparticles is allowed. Our approach to the cuprates is largely justified by the results of extensive ARPES and NMR and neutron measurements that indicate that the Fermi surface possesses hot spots, and that spin excitations are overdamped in the normal state.

Whether or not spin fluctuations are overdamped is also of significant conceptual importance for spin mediated pairing, since this mechanism requires that quasiparticles be strongly coupled to the collective spin excitation mode. At first glance, the undamped (magnon) form of the spin propagator appears more favorable for spin-mediated pairing than the overdamped form. Indeed, if one assumes that the spin-mediated interaction is just proportional to the spin susceptibility, the magnon-like form is preferable. In the antiferromagnetically ordered state, the transverse spin susceptibility $\chi(\mathbf{q})$ (that yields an attraction in the $d_{x^2-y^2}$ channel) even diverges as $\mathbf{q}$ approaches the antiferromagnetic momentum $\mathbf{Q}$, hence the d-wave attraction appears to be the strongest. This reasoning, however, is incorrect. Schrieffer and his collaborators [80, 81] and others [82, 83] have shown that the Goldstone modes of an ordered antiferromagnet cannot give rise to a strong d-wave pairing because the full spin mediated interaction is the product of the spin susceptibility and the square of the fully renormalized coupling constant between fermions and magnons. The latter vanishes in the ordered SDW state at $\mathbf{q} = \mathbf{Q}$ and this effect exactly compensates the divergence of the static susceptibility. The vanishing of the effective coupling is a consequence of the Adler principle which states that true Goldstone modes always decouple from other excitations in a system [84].

Schrieffer later argued [81] that the near cancelation between the enhancement of the spin susceptibility and the reduction of the effective magnon-fermion interaction persists in the paramagnetic state as long as spin fluctuations remain propagating excitations. This would substantially reduce (although not eliminate [85]) the spin-mediated d-wave attraction. This argument is however inapplicable to overdamped spin fluctuations. These are *not* Goldstone modes although they become gapless at the magnetic instability. Goldstone modes appear only in the ordered state at the smallest $\mathbf{q} - \mathbf{Q}$ values [83]. For near-gapless, but overdamped spin excitations, the Adler principle does not work. Consequently, the spin–fermion vertex does not vanish at the magnetic transition and hence cannot cancel out the enhancement of the d-wave interaction due to the increase of the spin susceptibility near $\mathbf{Q}$. Thus, overdamped spin fluctuations are better for spin-mediated pairing than magnon-like excitations.

Another aspect of the fact that spin dynamics is made out of low-energy fermions is that the retarded interaction which causes the pairing changes when fermions acquire a superconducting gap. This feedback from quasiparticle pairing on the form of the pairing interaction distinguishes pairing mediated by overdamped spin fluctuations from conventional phonon induced pairing. In the latter the bosonic propagator is an input and is only very weakly affected by the opening of the gap in the quasiparticle spectrum. We will discuss in detail how feedback forces one to go beyond an approach in which one solely replaces a phonon by a spin fluctuation, and requires that one consistently calculates the spin dynamics at low energies. While doing this is a theoretical challenge, the approach is appealing since it reduces the number of unknown parameters in the problem. In particular, we will see that in the superconducting state, the propagator of spin fluctuations acquires the same form as for optical phonons, but the collective mode that is the analog of the phonon frequency is fully determined by the superconducting gap and the normal state spin damping. This gives rise to new, unique "fingerprints" of spin mediated pairing, whose presence can be checked experimentally.

What is the role of dimensionality? As noted above, many of the candidates for spin-mediated pairing are strongly anisotropic, quasi-two-dimensional systems. This not only holds for the cuprates, but also for a large class of organic superconductors. Also, heavy fermion superconductors such as $CeCoIn_5$

display a considerable spatial anisotropy. On the other hand, $CeIn_3$ and to a lesser extent $CeCu_2Si_2$ do not display appreciable quasi two-dimensionality in their electronic properties. The dimensionality of the electronic system is important to the spin fluctuation model for both normal state and superconducting behavior. We will see that the dynamics of the fermions in the normal state is very differently affected by antiferromagnetic spin fluctuations in two-dimensional and in three-dimensional systems. While in the latter case only small (logarithmic) corrections to the ideal Fermi gas behavior occur in the vicinity of hot spots, we shall see that in 2D systems, the strong interaction between fermions and spins gives rise to non-Fermi liquid, diffusive behavior of low energy fermions as the quantum critical point is approached. The importance of dimensionality for superconductivity has been emphasized by Monthoux and Lonzarich [72] who have shown that it exerts a considerable influence on the superconducting transition temperature. They pointed out that in three dimensions one cannot avoid repulsive contributions to the pairing interaction in choosing a pairing state with nodes, so that the same spin-mediated quasiparticle interaction is far less effective in bringing about superconductivity in three dimensions than in two.

Since the non-Fermi-liquid behavior of fermionic quasiparticles extends down to progressively lower frequencies as one approaches the magnetic transition at $T = 0$, one can inquire whether pairing near this quantum-critical point is caused by the fermions at the lowest energies that are still coherent, or comes from those at higher energies (that are still smaller than the bandwidth) that display non-Fermi-liquid behavior. If only coherent fermions are involved in the pairing, then, according to McMillan's extension of the BCS theory [86], the resulting superconducting transition temperature, T_c^{FL}, is comparable to the upper energy cutoff of the Fermi liquid regime, and thus will be of the order of the spin fluctuation energy. This energy vanishes at the critical point, and therefore magnetic criticality is unaffected by pairing (see the left panel in Fig. 22.5). If, however, "non-Fermi-liquid" fermions can give rise to a pairing instability, then the onset temperature of this instability in the particle–particle channel, that we will identify with T_{cr} in the phase diagram of Fig. 22.1, generally scales

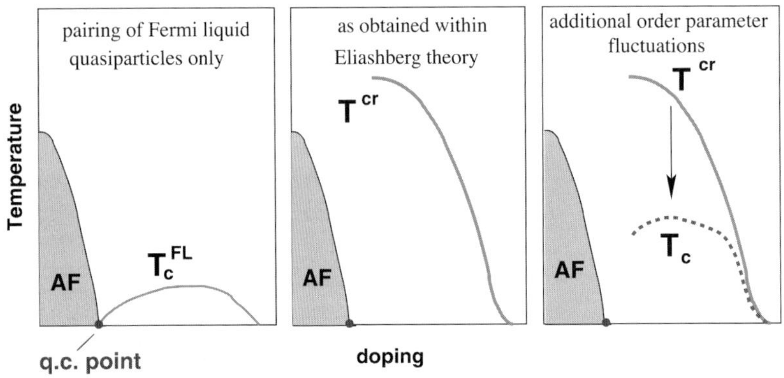

Fig. 22.5. The candidate phase diagrams in the units of temperature and doping of a one-band electronic system near an antiferromagnetic quantum critical point. *Left panel*: the phase diagram for the hypothetical situation when only coherent, Fermi liquid quasiparticles contribute to the pairing (the McMillan theory applied to spin-fluctuations). The antiferromagnetic and superconducting regions are completely decoupled. *Central panel*: the solution of the coupled set of the Eliashberg equations for the onset of spin-mediated pairing instability [87]. The solution shows that at strong coupling, the pairing instability is predominantly produced by incoherent fermions, and the instability temperature remains finite at $\xi = \infty$. *Right panel*: the proposed phase diagram based on the solution of the Eliashberg equations below the pairing instability and general arguments about superconducting fluctuations [89]

with the upper cutoff energy for the quantum-critical behavior and remains finite at criticality [87–89]. In this situation, the quantum critical point is necessarily surrounded by a dome beneath which pairing correlations cannot be neglected as shown in the central panel of Fig. 22.5. The critical behavior inside and outside the dome is different, and the "primary" critical behavior (which gives rise to pairing) can only be detected outside the dome. We will demonstrate below that T_{cr} saturates at a finite value when the magnetic correlation length diverges. Furthermore, for parameters relevant to cuprates at low doping, this temperature is of order of the magnetic exchange interaction J, i.e., it is not small.

A related issue is whether the pairing instability at T_{cr} implies the onset of true superconductivity (i.e. $T_{cr} \equiv T_c$), or whether it marks the onset of pseudogap behavior. In the latter scenario, for which we will see there is considerable experimental support, fluctuations prevent a superfluid stiffness ρ_s from developing a nonzero value until one reaches a much smaller T_c. Abanov et al. [89] conjectured that a pseudogap regime is a universal feature of the spin fluctuation scenario, as at strong spin–fermion coupling, incoherent fermions are paired into singlets below T_{cr}, but still remain incoherent and cannot carry a supercurrent. True superconductivity is reached only at much smaller T_c where the systems recovers coherent, Fermi-liquid behavior (see the right panel in Fig. 22.5).

The term pseudogap was introduced by Friedel [90] to describe the fact that in the underdoped regime of the cuprates, the planar quasiparticles begin to develop a gap-like structure well above T_c. This behavior was first seen in Knight shift measurements of the uniform spin susceptibility [91], and later detected in almost all measured properties of underdoped cuprates. At present, the physics of the pseudogap phase in the underdoped cuprate superconductors is not yet fully understood and its origin continues to be an open question. We believe that the "magnetic scenario" for the pseudogap provides a reasonable explanation, but many details still need to be worked out.

A distinguished feature of the spin-fluctuation scenario for the pseudogap is that the pseudogap is determined by fermions with energies well below the bandwidth. Other proposals associate the pseudogap with the Mott physics. These scenarios include Anderson's RVB theory [92], the phase fluctuation theory [93–97], and the theories associated with bond currents [98–100]. A somewhat more general phenomenological possibility discussed by several researchers is that there exists an additional quantum critical point of yet unknown origin slightly above optimal doping [101–104] (a number of experiments suggest that this point is at doping concentration $x \approx 0.19$). The pseudogap and Fermi liquid phases are assumed to be to the left and to the right of this new quantum critical point, respectively.

Our main goal is to discuss in detail the "primary" quantum-critical behavior within the magnetic scenario and how it gives rise to pairing at T_{cr}. A detailed theory of the pseudogap state of high temperature superconductors is beyond the scope of this Chapter. However, in the interest of providing a base line against which to compare both experiment and future theoretical developments, we summarize the predictions of the spin–fermion model for the pseudogap in Sect. 22.6 and discuss other alternatives.

To spell out the expected regions of applicability to the superconducting cuprates (in doping and temperature) of the spin-fluctuation theory without pseudogap physics involved, we return to the candidate generic phase diagram in Fig. 22.1. The two lines, T_{cr} and T^* determine distinct regimes of physical behavior. Above T_{cr}, pseudogap physics plays no role; the theory of a nearly antiferromagnetic Fermi liquid (NAFL) presented in this Chapter should be applicable for both the normal state and the superconducting state. Since T_{cr} crosses T_c near the optimal doping concentration, the theory with no pseudogap involved is roughly applicable at and above optimal doping (from an experimental perspective, optimally doped materials do show some pseudogap behavior, but only over a very limited temperature regime). For the overdoped and nearly optimally doped cuprates the transition is then from a nearly antiferromagnetic Fermi liquid to a superconductor with $d_{x^2-y^2}$-pairing symmetry. We will argue in Sect. 22.6 that there is a great deal of experimental evidence that at and above optimal doping the normal state is indeed a NAFL, and that the pairing is of magnetic origin. It is also likely that the theory can also be extended into a so-called "weak pseudogap" regime between T_{cr} and T^* [89,105], but we will not discuss this issue here.

In Sect. 22.2 we introduce and motivate the spin fermion model that we use to study spin fluctuation induced pairing. We discuss the weak coupling approach to the pairing problem and the symmetry of the magnetically mediated pairing state. In Sect. 22.3 we review the main results and arguments used to justify Eliashberg theory for conventional phonon superconductors. In particular, we discuss the physical origin of the Migdal theorem that allows a controlled approach to phonon-induced pairing. In Sect. 22.4 we then analyze in detail the strong coupling theory for the spin–fermion model. We first discuss the normal state properties of this model and calculate the low frequency dynamics of quasiparticles and spin fluctuations. We next consider spin-fluctuation induced superconductivity. We show that for magnetically-mediated superconductivity one can again analyze the pairing problem in controlled calculations that on the level of the equations involved resemble the Eliashberg equations for electron–phonon superconductivity [49,106]. We demonstrate that the actual physical origin of the applicability of a generalized Eliashberg approach for spin mediated pairing is qualitatively different from the phonon case, and is associated with the overdamped nature of the spin excitations. We solve the resulting equations in certain limits and investigate the role of quantum critical pairing. In Sect. 22.5 we present a general discussion of some of the observable fingerprints of spin fluctuation induced superconductivity, and in Sect. 22.6 we compare our results with experiments and discuss to what extent the fingerprints of spin mediated pairing have already been seen in optimally doped cuprate superconductors. In our concluding section 22.7 we summarize our results and comment on several topics that are of interest for a further understanding of spin mediated pairing, including the extent to which our theory can be extended to address the physics of the pseudogap state in underdoped cuprates.

22.2 Spin–Fermion Model

22.2.1 Physical Motivation of the Spin Fermion Model

We first discuss the formal strategy one has to follow to *derive* an effective low-energy model from a microscopic Hubbard-type Hamiltonian

$$\mathcal{H} = \sum_{\mathbf{k},\alpha} \varepsilon_{\mathbf{k}} \psi^{\dagger}_{\mathbf{k},\alpha} \psi_{\mathbf{k},\alpha} + \frac{1}{2} \sum_{\mathbf{k}_i,\alpha_i} U^{\alpha_1,\alpha_2,\alpha_3,\alpha_4}_{\mathbf{k}_1,\mathbf{k}_2,\mathbf{k}_3,\mathbf{k}_4} \times \psi^{\dagger}_{\mathbf{k}_1,\alpha_1} \psi^{\dagger}_{\mathbf{k}_2,\alpha_2} \psi_{\mathbf{k}_3,\alpha_3} \psi_{\mathbf{k}_4,\alpha_4} \,.$$

Here $U^{\alpha_1,\alpha_2,\alpha_3,\alpha_4}_{\mathbf{k}_1,\mathbf{k}_2,\mathbf{k}_3,\mathbf{k}_4}$ is the four-fermion interaction, $\psi^{\dagger}_{\mathbf{k},\alpha}$ is the creation operator for fermions with spin α and momentum $\mathbf{k}$, and $\varepsilon_{\mathbf{k}}$ is the band-structure dispersion. For a one-band Hubbard model with local Coulomb interaction,

$$U^{\alpha_1,\alpha_2,\alpha_3,\alpha_4}_{\mathbf{k}_1,\mathbf{k}_2,\mathbf{k}_3,\mathbf{k}_4} = U \delta_{\mathbf{k}_1+\mathbf{k}_2-\mathbf{k}_3-\mathbf{k}_4} \times \left(\delta_{\alpha_1\alpha_4}\delta_{\alpha_2\alpha_3} - \delta_{\alpha_1\alpha_3}\delta_{\alpha_2\alpha_4}\right) \,.$$

In a perturbation theory for (22.1) involving U and the fermion band width, the contributions from large and small fermionic energies are mixed. However, near a magnetic instability much of the non-trivial physics is associated (at any U) with the system behavior at low energies. To single out this low-energy sector, one can borrow a strategy from field theory: introduce a characteristic energy cut off, Λ, and generate an effective low energy model by eliminating all degrees of freedom above Λ. By itself, this does not guarantee that there exists a universal physics confined to low energies and independent on the actual choice of Λ. This we will have to prove. This also does not mean that the system is in the weak coupling regime, as near the antiferromagnetic transition we will find a strong, near-divergent contribution to the fermionic self-energy that comes from low frequencies. What the separation of scales actually implies (to the extent that we find universal, low-energy physics) is that Mott physics does not play a major role. In particular, in our analysis the Fermi surface in the normal state remains large, and its volume satisfies Luttinger theorem. How well this approximation is satisfied depends on doping for a given material and also varies from one material to another. Most of our experimental comparisons will be made with the cuprates. In cuprates, the Hubbard U in the effective one-band model for CuO_2 unit (a charge transfer gap) is estimated to be between 1 and 2 eV. The bandwidth, measured by ARPES and resonant Raman experiments, roughly has the same value. This suggests that lattice effects do, indeed, play some role. At half-filling, lattice effects are crucial as evidenced by the fact that half-filled materials are both Mott insulators and antiferromagnets with local (nearest-neighbor) spin correlations. Doping a Mott insulator almost certainly initially produces a small Fermi surface (hole or electron pockets). This small Fermi surface evolves as doping increases and eventually transforms into a large, "Luttinger" Fermi surface. How this evolution actually occurs is still a subject of debate. From our perspective, it is essential that at and above optimal doping, all ARPES data indicate that the Fermi surface is large. Correspondingly, magneto-oscillation experiments in BEDT-TTF based organic superconductors also show that the Fermi surface of these materials is large. We believe that in this situation, lattice effects influence the system behavior quantitatively but not qualitatively, and the neglect of lattice effects is justified. We emphasize however that our analysis certainly needs to be modified to incorporate Mott physics close to half-filling.

Several aspects of our approach have a close similarity to the fluctuation exchange approximation (FLEX), which in case of a single band Hubbard model corresponds to a self-consistent summation of bubble and ladder diagrams [107, 108]. Specifically, the emergence of a sharp resonance mode in the spin excitation spectrum of a d-wave superconductor, the feedback of this mode on the fermions and the anomalous normal state behavior of low energy fermions close to an antiferromagnetic instability are very similar in both approaches [109–112]. On the other hand, the FLEX approach attempts to determine the static spin response and thus the actual position of the magnetic quantum critical point in terms of the bare parameters of the model such as the local or additional nonlocal Coulomb repulsions as well as the band structure $\varepsilon_{\mathbf{k}}$. As discussed above, the static spin response, characterized by the correlation length ξ, strongly depends on the behavior of fermions with large energy. Details of the underlying microscopic model, which are hard to specify uniquely, as well as uncontrolled approximations in the treatment of the high energy behavior, strongly affect the static spin response within the FLEX approach, making it hard to discriminate model dependent aspects from universal behavior. It is this latter aspect which is resolved in our approach which concentrates exclusively on the universal low energy physics for a given ξ.

What should be the form of the low-energy action? Clearly, it should involve fermions which live near the Fermi surface. It also should involve collective spin bosonic degrees of freedom with momenta near $\mathbf{Q}$, as these excitations become gapless at the magnetic transition. The most straightforward way to obtain this action is to introduce a spin-1 Bose field $\mathbf{S}$ and decouple the four-fermion interaction using the Hubbard–Stratonovich procedure [113, 114]. This yields

$$\begin{aligned}\mathcal{H} = &\sum_{\mathbf{k},\alpha} \varepsilon_{\mathbf{k}} \psi^{\dagger}_{\mathbf{k},\alpha} \psi_{\mathbf{k},\alpha} + \sum_{\mathbf{q}} U(\mathbf{q})\, \mathbf{S}_{\mathbf{q}} \cdot \mathbf{S}_{-\mathbf{q}} \\ &+ \sum_{\mathbf{k},\mathbf{q},\alpha,\beta} U(\mathbf{q})\, \psi^{\dagger}_{\mathbf{k}+\mathbf{q},\alpha} \sigma_{\alpha\beta} \psi_{\mathbf{k},\beta} \cdot \mathbf{S}_{-\mathbf{q}}\,, \end{aligned} \tag{22.1}$$

where the $\sigma_{\alpha\beta}$ are Pauli matrices and we assumed that the four fermion interaction only makes a contribution in the spin channel with momentum transfer $\mathbf{q}$. Integrating formally over energies larger than Λ we obtain the effective action in the form (see, e.g. [115])

$$\begin{aligned}\mathcal{S} = &- \int^{\Lambda} \mathrm{d}k G_0^{-1}(k)\, \psi^{\dagger}_{k,\alpha} \psi_{k,\alpha} \\ &+ \frac{1}{2} \int^{\Lambda} \mathrm{d}q \chi_0^{-1}(q)\, \mathbf{S}_{\mathbf{q}} \cdot \mathbf{S}_{-\mathbf{q}} \\ &+ g \int^{\Lambda} \mathrm{d}k\mathrm{d}q \psi^{\dagger}_{k+q,\alpha} \sigma_{\alpha\beta} \psi_{k,\beta} \cdot \mathbf{S}_{-\mathbf{q}} + O(S^4)\,. \end{aligned} \tag{22.2}$$

The last term is a symbolic notation for all terms with higher powers of S. In dimensions $d \geq 2$ these higher order terms are irrelevant (marginal for $d = 2$) and can therefore be neglected [75, 76].

The integration over k and q in (22.3) is over 2 + 1 dimensional vectors $q = (\mathbf{q}, i\omega_m)$ with Matsubara frequency ω_m. In explicit form, the integrals read

$$\int^{\Lambda} dq \ldots = \int\limits_{|\mathbf{q}-\mathbf{Q}|<\Lambda} \frac{d^d\mathbf{q}}{(2\pi)^d} T \sum_m \ldots \quad (22.3)$$

in the boson case, and

$$\int^{\Lambda} dk \ldots = \int\limits_{|\mathbf{k}-\mathbf{k}_F|<\Lambda} \frac{d^d\mathbf{k}}{(2\pi)^d} T \sum_m \ldots \quad (22.4)$$

in the fermion case. Further, g is the effective coupling constant, which differs from U because of high-energy renormalization, $G_0(k)$ is the bare low-energy fermion propagator, and $\chi_0(q)$ is the bare low-energy collective spin boson propagator. As we have emphasized, a controlled derivation of g, $G_0(k)$, and $\chi_0(q)$ is impossible. We therefore will not try to calculate g, etc. Rather, we use the fact that antiferromagnetism predominantly comes from high-energy fermions and further assume that the integration over high energies does not produce singularities in both bosonic and fermionic propagators. Then, quite generally, $G_0(k)$, and $\chi_0(q)$ should have Fermi-liquid and Ornstein–Zernike forms, respectively

$$G_0(k) = \frac{z_0}{i\omega_m - \epsilon_\mathbf{k}}, \quad (22.5)$$

$$\chi_0(q) = \frac{\alpha}{\xi_0^{-2} + (\mathbf{q}-\mathbf{Q})^2 + \omega_m^2/c^2}. \quad (22.6)$$

Here, ξ is the spin correlation length, and the other notations are self-explanatory. It is essential that the bare spin propagator does not contain a term linear in ω. The latter will only appear when we consider the interaction *within* the low-energy model.

We also assume that (i) the momentum dependence of the effective coupling g is non-singular and can be neglected (recall that we are only interested in a narrow range of bosonic momenta near $\mathbf{Q}$), (ii) the low-energy fermionic dispersion can be linearized in $\mathbf{k}-\mathbf{k}_F$: $\epsilon_\mathbf{k} = \mathbf{v}_F \cdot (\mathbf{k}-\mathbf{k}_F)$, and (iii) that the Fermi velocity is non-singular near hot spots and to first approximation its magnitude can be approximated by a constant. In doing this we neglect effects due to a van Hove singularity in the density of states. Finally, the exact values of z_0 and α are not relevant, as both can be absorbed into an effective coupling with dimension of an energy

$$\bar{g} = g^2 z_0^2 \alpha, \quad (22.7)$$

while v_F and ξ will always appear only in the combination $v_F\xi^{-1}$.

We see therefore that the input parameters in (22.3) are the effective coupling energy $\bar{g}$, the typical quasiparticle energy $v_F\xi^{-1}$, and the upper cutoff Λ. An additional parameter is the angle ϕ_0 between the Fermi velocities at the two hot spots separated by $\mathbf{Q}$, but this angle does not enter the theory in any significant manner as long as it is different from 0 or π. When hot spots are located near $(0,\pi)$ and $(\pi,0)$ points, as in optimally doped cuprates, ϕ_0 is close to $\pi/2$, the value we use in what follows. One more initial parameter is the spin velocity c, but, as we will see, the ω^2 term in (22.6) is in fact irrelevant as it will be completely overshadowed by bosonic self-energy. As we have emphasized, we will demonstrate that the low-energy properties of the model are universal and do not depend on Λ, which then can be set to infinity. Out of the two parameters that are left, one can construct a doping dependent dimensionless ratio

$$\lambda = \frac{3}{4\pi} \frac{\bar{g}}{v_F\xi^{-1}}, \quad (22.8)$$

which will turn out to be the effective dimensionless coupling constant of the problem (the factor $3/4\pi$ is introduced for further convenience). The fact that λ scales with ξ immediately implies that close enough to a magnetic transition $\lambda > 1$, i.e. the system will necessarily be in a strong coupling limit. Besides λ the only other free parameter of the theory is an overall energy scale, i.e. $\bar{g}$. All physical quantities that we discuss will be expressed in terms of these two parameters only.

Equations (22.3)–(22.6) determine the structure of the perturbation theory of the model. The interaction between fermions and collective spin excitations yields self-energy corrections to both bosonic and fermionic propagators. We will show below that at strong coupling, the fermionic self-energy strongly depends on frequency and also displays some dependence on the momentum along the Fermi surface. However, its dependence on the momentum transverse to the Fermi surface can be neglected together with vertex corrections. The fermionic and bosonic propagators are then given by the Gor'kov expressions, which for generality we present in the superconducting state,

$$G_{\mathbf{k}}(i\omega) = \frac{i\omega + \Sigma_{\mathbf{k}}(i\omega) + \varepsilon_{\mathbf{k}}}{((i\omega + \Sigma_{\mathbf{k}}(i\omega))^2 - \Phi^2_{\mathbf{k}}(i\omega) - \varepsilon^2_{\mathbf{k}}}, \quad (22.9)$$

$$F_{\mathbf{k}}(i\omega) = -\frac{\Phi_{\mathbf{k}}(i\omega)}{(i\omega + \Sigma_{\mathbf{k}}(i\omega))^2 - \Phi^2_{\mathbf{k}}(i\omega) - \varepsilon^2_{\mathbf{k}}}, \quad (22.10)$$

$$\chi_{\mathbf{q}}(i\omega) = \frac{\alpha\xi^2}{1 + \xi^2(\mathbf{q} - \mathbf{Q})^2 - \Pi_{\mathbf{Q}}(i\omega)}. \quad (22.11)$$

Here $\Sigma_{\mathbf{k}}(i\omega)$ and $\Pi_{\mathbf{Q}}(i\omega)$ are fermionic and bosonic self-energies (**k** stands for the component along the Fermi surface), and $F_{\mathbf{k}}(i\omega)$ and $\Phi_{\mathbf{k}}(i\omega)$ are the anomalous Green's function and the anomalous self-energy, respectively. In the next sections we compute $\Sigma_{\mathbf{k}}(i\omega)$, $\Pi_{\mathbf{Q}}(i\omega)$ and $\Phi_{\mathbf{k}}(i\omega)$.

One can also motivate the spin fermion model by following the approach developed by Landau in his theory of Fermi liquids [9], i.e. by assuming that the influence of the other fermionic quasiparticles on a given quasiparticle can be described in terms of a set of molecular fields [10,36]. In the present case the dominant molecular field is an exchange field produced by the Coulomb interaction $U(\mathbf{q})$. However, one has to consider this field as dynamic, not static. The corresponding part of the action contains

$$\begin{aligned}\mathcal{S} = &-\int^{\Lambda} \mathrm{d}k G_0^{-1}(k)\, \psi^\dagger_{k,\alpha}\psi_{k,\alpha} \quad (22.12)\\ &+ \frac{1}{2}\int^{\Lambda}\mathrm{d}q\int^{\Lambda}\mathrm{d}k\int^{\Lambda}\mathrm{d}k'\, \psi^\dagger_{k+q,\alpha}\psi_{k,\beta}\psi^\dagger_{k'+q,\alpha}\\ &\times \psi_{k',\beta}\,\Gamma^0_{\alpha\beta,\gamma\delta}(k,k',q)\,,\end{aligned}$$

with bare four fermion interaction at low energies:

$$\Gamma^0_{\alpha\beta,\gamma\delta}(k,k',q) = -g^2\chi_0(q)\,\sigma_{\alpha\beta}\sigma_{\gamma\delta}\,. \quad (22.12a)$$

Formally, this can be obtained from (22.3) by integrating out the collective degrees of freedom **S**. A relation between this purely fermionic approach and the bosonic spin susceptibility $\chi(\mathbf{q})$ of (22.11) can be established by evaluating the four point vertex in the spin channel within perturbation theory. We find

$$\Gamma_{\alpha\beta,\gamma\delta}(k,k',q) = -V_{\mathrm{eff}}(q)\,\sigma_{\alpha\beta}\sigma_{\gamma\delta}\,, \quad (22.13)$$

where the renormalized quasiparticle interaction is proportional to the renormalized spin propagator

$$V_{\mathrm{eff}}(q) = g^2\chi(q) = \quad (22.14)$$
$$\times \frac{g^2\alpha\xi^2}{1 + \xi^2(\mathbf{q} - \mathbf{Q})^2 - \Pi_{\mathbf{Q}}(i\omega)}.$$

22.2.2 Weak-Coupling Approach to the Pairing Instability

One of the most appealing aspects of the spin-fluctuation theory is that it inevitably yields an attraction in the $d_{x^2-y^2}$ channel. As with any unconventional pairing, this attraction is the result of a specific momentum dependence of the interaction, not of its overall sign which for spin fluctuation exchange is positive, i.e. repulsive in the s-wave channel. This differentiates spin induced pairing from the pairing mediated by phonons. In the latter case, the effective interaction between fermions is negative up to a Debye frequency. In configuration space the spin-fluctuation interaction, (22.15), can easily seen to be repulsive at the origin and alternate between attraction and repulsion as one goes to nearest neighbors. It is always repulsive along the diagonals, and this is why a $d_{x^2-y^2}$ state, in which the nodes of the gap are along the diagonals, is the energetically preferred pairing state due to spin-fluctuation exchange [116].

Suppose first that the spin–fermion coupling is small enough such that conventional perturbation theory is valid. To second order in the spin–fermion coupling, the spin mediated interaction has the form of (22.12a).

The total antisymmetry of the interaction implies that the orbital part is symmetric when the spin part is antisymmetric and vice versa. The orbital part of the interaction at low frequencies has the same sign as the propagator of optical phonons at $\omega < \omega_D$. However, the spin part involves a convolution of the Pauli matrices, and is different from phononic $\delta_{\alpha\beta}\delta_{\gamma\delta} - \delta_{\alpha\delta}\delta_{\beta\gamma}$. Using $\sigma_{\alpha\beta}\sigma_{\gamma\delta} = 2\delta_{\alpha\delta}\delta_{\beta\gamma} - \delta_{\alpha\beta}\delta_{\gamma\delta}$ we find

$$\Gamma^0_{\alpha\gamma,\beta\delta}(q) = -g^2\chi_0(q)\,(T_{\alpha\gamma,\beta\delta} - 3S_{\alpha\gamma,\beta\delta})\,, \quad (22.15)$$

where $T_{\alpha\gamma,\beta\delta} = (\delta_{\alpha\beta}\delta_{\gamma\delta} + \delta_{\alpha\delta}\delta_{\beta\gamma})/2$ and $S_{\alpha\gamma,\beta\delta} = (\delta_{\alpha\beta}\delta_{\gamma\delta} - \delta_{\alpha\delta}\delta_{\beta\gamma})/2$ are triplet and singlet spin configurations, respectively. We see that there is an extra minus sign in the singlet channel. This obviously implies that in distinction to phonons, the isotropic s-wave component of the interaction is repulsive, i.e. isotropic s-wave pairing due to spin fluctuation exchange is impossible. There are two other possibilities: unconventional singlet pairing for which we will need a partial component of $\chi_0(q)$ to be negative, or triplet pairing, for which the corresponding partial component of $\chi_0(q)$ should be positive.

It turns out that for a spin susceptibility peaked at or near the antiferromagnetic momentum $\mathbf{Q}$, the triplet components are all negative, and triplet pairing is impossible. We therefore focus on singlet pairing. The linearized equation for the pairing vertex $F_{\mathbf{k}}(i\omega_m)$ in the singlet channel is

$$F(k) = -3g \int_{k'}^{\Lambda} F(k')\chi_0(k-k')G_0(k')G_0(-k') \,. \quad (22.16)$$

If $\chi(k-k')$ was independent of momentum, a solution of this equation would not exist because of the overall minus sign. In our situation, however, the interaction predominantly occurs between fermions that are separated in momentum space by $\mathbf{Q} = (\pi, \pi)$ (see Fig. 22.6). Since both $\mathbf{k}$ and $\mathbf{k} + \mathbf{Q}$ should be near the Fermi surface, the pairing predominantly involves fermions near hot spots $\mathbf{k}_{\mathrm{hs}}$ and $\mathbf{k}_{\mathrm{hs}} + \mathbf{Q}$. As the hot spots are well separated in momentum space, we can change the sign in the r.h.s. of (22.16) by requiring that the pairing gap changes sign between $\mathbf{k}_{\mathrm{hs}}$ and $\mathbf{k}_{\mathrm{hs}} + \mathbf{Q}$. Indeed, substituting $F_{\mathbf{k}+\mathbf{Q}}(i\omega_m) = -F_{\mathbf{k}}(i\omega_m)$ into (22.16), we find that both sides of this equation have the same sign. The pairing problem then becomes almost identical to that for phonons, and the gap equation has a solution at some T_c.

Altogether, there are $N = 8$ hot spots in the Brillouin zone; these form four pairs separated by $\mathbf{Q}$, between which the gap should change sign. This still does not specify the relative sign of the gap between adjacent hot spots. However, if the hot spots are close to $(0, \pi)$ and symmetry related points, as in optimally doped cuprates, the gap is unlikely to change sign between them. This leaves as the only possibility a gap that vanishes along diagonal directions $k_x = \pm k_y$, i.e. it obeys $d_{x^2-y^2}$ symmetry.

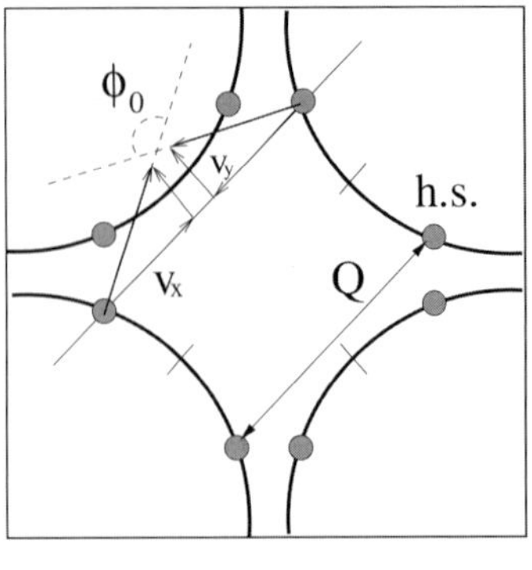

Fig. 22.6. A representative Brillouin zone and Fermi surface for a near-optimally doped cuprate superconductor. The hot spots are connected by the antiferromagnetic wave vector $\mathbf{Q} = (\pi, \pi)$. For a weakly incommensurate magnetic response, the number of hot spots doubles but simultaneously the intensity of the magnetic response at momenta connecting pairs of hot spots drops by a factor of two. The net effect then remains the same as for the commensurate response. The figure is taken from [49]

We emphasize that although we found that the pairing gap necessarily should have $d_{x^2-y^2}$ symmetry, its momentum dependence is generally more complex than simply $\cos k_x - \cos k_y$. To understand this, we consider how one can generally analyze (22.16) [87]. A standard strategy is to expand both the pairing vertex and the interaction in the eigenfunctions of the representations of the D_{4h} symmetry group of the square lattice. There are four singlet representations of D_{4h} labeled A_{1g}, B_{1g}, B_{2g} and A_{2g}. The basic eigenfunctions in each representation are 1 in A_{1g}, $\cos k_x - \cos k_y$ in B_{1g}, $\sin k_x \sin k_y$ in B_{2g} and $(\cos k_x - \cos k_y)\sin k_x \sin k_y$ in A_{2g}. Other eigenfunctions in each representation are obtained by combining the basic eigenfunction with the full set of eigenfunctions with full D_{4h} symmetry. Obviously, the eigenfunctions with $d_{x^2-y^2}$ symmetry belong to B_{1g} channel. The orthogonal functions in the set can be chosen as $d_n(\mathbf{k}) = \cos nk_x - \cos nk_y$.

One can easily make sure that the pairing problem close to T_c decouples between different representations. Numerical calculations show that B_{1g} components are the largest. We therefore neglect other channels and focus only on B_{1g} pairing. Still, there are an infinite number of B_{1g} eigenfunctions, and the pairing problem does not decouple between them. Indeed, expanding $F_{\mathbf{k}}(\omega_m)$ and $\chi_0(\mathbf{k})$ in $d_n(\mathbf{k})$ as $F_{\mathbf{k}} = \sum_p F_p d_p(\mathbf{k})$, $\chi_0(\mathbf{k}-\mathbf{k}') = \sum_p \chi_p d_p(\mathbf{k}) d_p(\mathbf{k}')$ and substituting the result into (22.16) we obtain [87]

$$F_n(i\omega_m) = -3g^2 T \sum_{p,m'} F_p\, \chi_n \Gamma_{n,p}\,, \qquad (22.17)$$

where

$$\Gamma_{n,p} = \int \frac{d^2k'}{4\pi^2} \frac{d_n(\mathbf{k}') d_p(\mathbf{k}')}{(\omega'_m)^2 + (\epsilon'_{\mathbf{k}})^2}\,. \qquad (22.18)$$

We see that all $\Gamma_{n,p}$ are nonzero for arbitrary n and p as the products $d_n(k)d_p(k)$ and $G_0(k)G_0(-k) = 1/((\omega_m)^2 + (\epsilon_{\mathbf{k}})^2)$ are symmetric under D_{4h}. As a result, (22.17) couples together an infinite number of partial components F_n. As long as the partial components of the susceptibility are dominated by momenta relatively far from $\mathbf{Q}$, all χ_n and hence F_n are of the same order. In some particular lattice models an RPA analysis shows that at $\xi \sim 1$, χ_1 is numerically significantly larger than all other χ_n [36]. In this situation, $F_{\mathbf{k}} \approx F_1 d_1(\mathbf{k})$, i.e. the momentum dependence of the gap should closely resemble the $\cos k_x - \cos k_y$ form. In real space, this implies that the pairing is local and involves fermions from the nearest neighbor sites on the lattice. In models with different high-energy physics, other χ_n may dominate. Clearly, the momentum dependence of the gap is not universal as long as the pairing is local in real space. Universality is recovered when the susceptibility becomes strongly peaked at $\mathbf{Q}$. In this limit (in which we can also rigorously justify restricting our attention to the B_{1g} channel [87]), all partial components χ_n scale as $\log \xi$ and differ only in sub-leading non-logarithmic terms. A straightforward trigonometric exercise shows that for near-equal χ_n, the pairing gap is very different from $\cos k_x - \cos k_y$: it is reduced near the nodes and enhanced near the maxima at hot spots with the ratio of the gap maximum and the slope near the nodes increasing as ξ^2. In this situation the pairing problem is confined to hot spots.

22.3 Summary of Strong-Coupling Theory for Electron–Phonon Pairing

For phonon-mediated superconductivity, Eliashberg theory offers a successful approach to study the system behavior at strong coupling. It is justified by Migdal's theorem which states that the vertex corrections, $\delta g/g$ and $(1/v_F)d\Sigma(\mathbf{k}, i\omega)/dk$, are small due to the smallness of the ratio of sound velocity and Fermi velocity stemming from the smallness of the ratio of the electronic and ionic masses. Eliashberg has demonstrated that for $\Sigma(\mathbf{k}, i\omega) \approx \Sigma(i\omega)$ and $g_{\text{tot}} = g + \delta g \simeq g$, the phonon-mediated pairing problem can be solved exactly. A review of Eliashberg theory for the electron–phonon problem is given in [117]. As we already pointed out, for spin fluctuation induced pairing there is no Migdal theorem simply because the spin propagator is made out of electrons and hence a typical spin velocity is of the same order as the Fermi velocity. It is therefore natural to inquire whether this implies that an Eliashberg-type treatment is inapplicable to the spin problem. To properly address this issue, we review the case of electron–phonon interactions and examine why the smallness of the mass ratio affects $d\Sigma(\mathbf{k}, i\omega)/d\mathbf{k}$ but not $d\Sigma(\mathbf{k}, i\omega)/d\omega$.

The electron–phonon interaction is generally rather involved and has been discussed in detail in the literature [5]. For our purposes, it is sufficient to consider the simplest Fröhlich-type model of the electron–phonon interaction in which low-energy electrons are coupled to optical phonons by a momentum, frequency and polarization independent interaction g_{ep}. Despite its simplicity, this model captures the key physics of phonon-mediated pairing.

The propagator of an optical phonon has the form

$$D(\mathbf{q}, i\omega_m) = \frac{\omega_0}{\omega_0^2 + \omega_m^2 + (v_s \mathbf{q})^2}\,. \qquad (22.19)$$

Here ω_0 is a typical phonon frequency, which is of the order of the Debye frequency, and v_s is the sound velocity. Both ω_0 and v_s are input parameters, unrelated to fermions. The ratio v_s/v_F scales as $(m/M)^{1/2}$ where M is the ionic mass, and m is the electron mass. In practice, $v_s/v_F \sim 10^{-2}$ [5]. This is a real physically motivated small parameter for the electron–phonon problem.

To make the analogy with spin fluctuations more transparent, we will consider below the case when $\omega_0 \ll v_s/a$, i.e. the correlation length for optical phonons is large. This last assumption allows us to simplify the calculations but will not change their conclusions.

In the presence of two different velocities, it is not *a priori* obvious what is the best choice of a dimensionless expansion parameter in the theory. Two candidate dimensionless parameters are

$$\lambda_{\rm ep} = g_{\rm ep}^2/(4\pi(v_s v_F)) \qquad (22.20)$$

(the factor 4π is chosen for further convenience), and

$$\tilde{\lambda}_{\rm ep} = \lambda_{\rm ep} v_s/v_F \sim (g_{\rm ep}/v_F)^2 . \qquad (22.21)$$

For simplicity, we set the lattice constant $a = 1$. Obviously, $\lambda_{\rm ep} \gg \tilde{\lambda}_{\rm ep}$. In practice, $\tilde{\lambda}_{\rm ep} \ll 1$ for all reasonable $g_{\rm ep}$ while $\lambda_{\rm ep}$ can be either small or large.

Which of the two dimensionless couplings appears in perturbation theory? Consider first the lowest-order diagram for the spin–fermion vertex at zero external frequency and zero bosonic momentum with all fermionic momenta at the Fermi surface. Choosing $T = 0$ and the x axis along the direction of $\mathbf{v}_F$ of an external fermion, we obtain the vertex correction in the form:

$$\frac{\delta g_{\rm ep}}{g_{\rm ep}} = \frac{g_{\rm ep}^2}{8\pi^3} \int \frac{d\omega_m d^2q}{(i\omega_m - v_F q_x)^2} \frac{\omega_0}{\omega_0^2 + \omega_m^2 + (v_s \mathbf{q})^2} . \qquad (22.22)$$

The evaluation of the momentum and frequency integrals is straightforward. On performing the integration we obtain for $\omega_0 \ll v_s$,

$$\frac{\delta g_{\rm ep}}{g_{\rm ep}} = \tilde{\lambda}_{\rm ep} = \lambda_{\rm ep} \frac{v_s}{v_F} . \qquad (22.23)$$

We see that the vertex correction scales as $\tilde{\lambda}_{\rm ep}$ and is small. For $\omega_0 \sim v_s$, the expression is more complex but still is of order $\tilde{\lambda}$. One can easily check that higher-order diagrams yield higher powers of $\tilde{\lambda}_{\rm ep}$, i.e. are also small. This result, first obtained by Migdal [118], is Migdal's theorem.

The physics behind this result is transparent and can be directly deduced from an analysis of the integrand in (22.22). We see that the product of the two fermionic Green's functions yields a double pole at $q_x = i\omega_m/v_F \sim \omega_0/v_F$. The integral over q_x is then finite only due to the presence of the pole in the bosonic propagator which is located at a much larger momentum $q_x \sim i\omega_0/v_s$. This implies that the electrons that contribute to the vertex correction are oscillating not at their own fast frequencies but at much slower phonon frequencies. At vanishing v_s/v_F, fermions can be considered as static objects with a typical $q_x \sim q_y \sim \omega_0/v_s$. The two dimensional momentum integration over the product of two fermionic propagators then yields $O(1/v_F^2)$. Simultaneously, the frequency integral over the bosonic propagator yields $D(q \sim \omega_0/v_s, 0) = O(1)$. Combining the two results, we obtain $\delta g_{\rm ep}/g_{\rm ep} = O(\tilde{\lambda})$ as in (22.23).

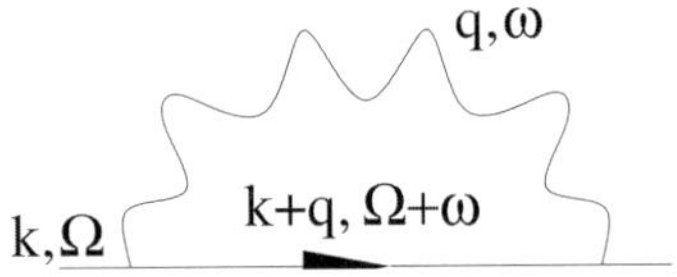

We next consider the fermionic self-energy, see figure above. Evaluating the Feynman diagram and subtracting the contribution at $\omega = 0$, $\mathbf{k} = \mathbf{k}_F$, we obtain

$$\Sigma(\mathbf{k}, i\omega_m) = (i\omega_m - \epsilon_{\mathbf{k}}) I(\mathbf{k}, i\omega_m) , \qquad (22.24)$$

where

$$I(\mathbf{k}, i\omega_m) = g_{\rm ep}^2 \int \frac{d\Omega_m d^2q}{8\pi^3} \frac{\omega_0}{\omega_0^2 + \Omega_m^2 + (v_s \mathbf{q})^2} \times \frac{1}{i\Omega_m - v_F q_x} \frac{1}{i(\omega_m + \Omega_m) - \epsilon_k - v_F q_x} . \qquad (22.25)$$

We again have chosen the q_x axis along the Fermi velocity $\mathbf{v}_F$.

Suppose for a moment that $I(\mathbf{k}, i\omega_m)$ is nonsingular in the limit $\mathbf{k} \to 0$ and $\omega_m \to 0$. Then the wave function renormalization and the velocity renormalization are both expressed via $I(0, 0)$. This quantity can be evaluated in exactly the same way as the vertex correction. Performing the integration, we find

$$I(0,0) = -\tilde{\lambda}_{\mathrm{ep}} \,. \qquad (22.26)$$

Substituting this into (22.24) we find $v_F^{-1} d\Sigma/dk = -d\Sigma/d(i\omega) = \tilde{\lambda}_{\mathrm{ep}}$. As $\tilde{\lambda}_{\mathrm{ep}} \ll 1$, this result, if true, would imply that both derivatives of the self-energy are small, i.e. the system is fully in the weak coupling regime. The physical interpretation of this result would parallel that for the vertex correction: fast electrons vibrate at slow phonon velocities, and this well-out-of-resonance vibration cannot substantially affect electronic properties.

However, this result is incorrect; $\mathrm{d}\Sigma/\mathrm{d}\omega$ is of order λ_{ep}, not $\tilde{\lambda}_{\mathrm{ep}}$. The origin of this apparent contradiction lies in the singular behavior of $I(\mathbf{k}, i\omega_m)$ in the limit of vanishing $\mathbf{k}$ and ω_m. To see this, we need to compute $I(\mathbf{k}, i\omega)$ in (22.25) at small but finite ω and $\epsilon_{\mathbf{k}}$. An explicit computation yields

$$I(\mathbf{k}, i\omega_m) = -\tilde{\lambda}_{\mathrm{ep}} + \lambda_{\mathrm{ep}} \frac{i\omega_m}{i\omega_m - \epsilon_{\mathbf{k}}} \,, \qquad (22.27)$$

demonstrating that $I(0,0)$ and the limit of $I(\mathbf{k}, i\omega)$ at $\omega, \epsilon_{\mathbf{k}} \to 0$ do not coincide. The additional contribution in (22.27) comes from the regularization of the double pole in the integrand in (22.25) (mathematically this is similar to the chiral anomaly in quantum chromodynamics [119]).

It is clear from (22.27) that $I(\mathbf{k}, i\omega_m)$ has singular low frequency and small momentum limits. Substituting (22.27) into (22.24), we find that the actual self-energy is

$$\Sigma(\mathbf{k}, i\omega_m) = i\lambda_{\mathrm{ep}}\, \omega_m - \tilde{\lambda}_{\mathrm{ep}}(i\omega_m - \epsilon_{\mathbf{k}}) \,. \qquad (22.28)$$

We see that $d\Sigma/d\omega$ scales with λ_{ep} whereas $v_F^{-1}\mathrm{d}\Sigma/\mathrm{d}k$ is proportional to $\tilde{\lambda}_{\mathrm{ep}}$. At strong coupling, $\lambda_{\mathrm{ep}} \geq 1$, this self-energy gives rise to a strong renormalization of both the quasiparticle mass and the quasiparticle wave function. Still, vertex corrections and the renormalization of the Fermi velocity scale with $\tilde{\lambda}_{\mathrm{ep}}$ and are small.

To understand the physical origin of the distinction between the frequency and momentum dependence of the fermionic self-energy, it is essential to realize that the second, $O(\lambda_{\mathrm{ep}})$, term in (22.27) is not caused by real electron–phonon scattering. Rather, a careful examination of the structure of the expression for $I(\mathbf{k}, \omega_m)$ shows that this term accounts for the pole in the fermionic particle hole polarization bubble at small momentum and frequency [120,121]. This implies that the fermionic self-energy is actually caused by coupling of fermions to their own bosonic collective modes, while phonons just mediate this coupling.

The neglect of vertex corrections and $v_F^{-1}\mathrm{d}\Sigma/\mathrm{d}k$ leads to the well known Eliashberg equations for $\Sigma_m = \Sigma(i\omega_m)$ and (in the superconducting state) the pairing self-energy $\Phi_m = \Phi(i\omega_m)$ that also depends only on frequency. For phonon-mediated superconductors, these equations have the form

$$\Sigma_m = i\,\frac{\lambda_{\mathrm{ep}} T}{2} \sum_n \frac{(\omega_n - i\Sigma_n)\, D(i\omega_{m-n})}{\sqrt{\Phi_n^2 + (\omega_n - i\Sigma_n)^2}} \,, \qquad (22.29)$$

$$\Phi_m = \frac{\lambda_{\mathrm{ep}} T}{2} \sum_n \frac{\Phi_n D(i\omega_{m-n})}{\sqrt{\Phi_n^2 + (\omega_n - i\Sigma_n)^2}} \,. \qquad (22.30)$$

This set of equations is solved for a given $D(i\omega_m)$ which by itself is only very weakly affected by fermions. In the normal state and at $\omega \ll \omega_0$, the self-energy equation (22.29) reduces to the first term in (22.28).

We have reviewed the derivation of these well known equations to underline the physical origin of the applicability of the Eliashberg approach. We see that the electron–phonon interaction gives rise to two physically distinct classes of interaction processes that contribute very differently to the fermionic self-energy and vertex corrections. In the first class, fast electrons are forced to vibrate at slow phonon frequencies (i.e. phonons are in resonance, but electrons are far from resonance). This gives rise to both vertex and self-energy corrections, but these are small ($\propto v_s/v_F$) and are neglected in the Eliashberg theory. In the second class, phonons mediate an effective coupling between fermions and their bosonic collective modes. This process involves fermions with energies near resonance and phonons far away from resonance, and yields a strong renormalization of the fermionic propagator. The applicability of the Eliashberg approach is then based on the fact that one has to include the phonon-mediated scattering on collective modes, but can neglect the direct scattering of electrons by phonons.

These insights into the origin of the applicability of the Eliashberg theory will now be used to justify a generalized Eliashberg approach for spin mediated pairing.

22.4 Strong-Coupling Approach to Spin–Fermion Interaction

As shown in the previous section, for electron phonon interactions the smallness of vertex and velocity renormalizations is caused by the small ratio of the Bose velocity and the Fermi velocity. The spin problem is qualitatively different. The bare spin-fluctuation propagator, (22.6), describes propagating magnons whose velocity c is expected to be of the same order as the Fermi velocity, i.e. $c \sim v_F$. There is then no a priori reason to neglect vertex and velocity renormalizations. Fortunately, this argument is not correct for the following reasons: First, as we just found in case of electron–phonon interaction, the fermionic self-energy in the Eliashberg theory is insensitive to the ratio of sound and Fermi velocities. The small ratio of v_s/v_F is only necessary to eliminate regular terms in the self-energy, which are due to real scattering by phonons. For these terms to be small, bosons must be slow modes compared to fermions. Second, phonons are propagating modes, i.e. they are weakly damped. Spin fluctuations are not propagating modes, as a strong spin–fermion interaction at low-energies changes the spin dynamics from propagating to diffusive. Diffusive modes have a different relationship between typical momenta and energies compared to ballistic ones, making them slow modes compared to electrons simply because typical spin fluctuation energies ω scale as q^2 rather than q. Consequently spin fluctuations are slow compared to fermions for small q despite $c \sim V_F$, and the regular terms in the fermionic self-energy again become smaller than singular ones.

We will show that in dimensions between $d = 2$ and $d = 3$, $\mathrm{d}\Sigma(\omega)/\mathrm{d}\omega$ scales as λ^{3-d} where, we recall, $\lambda \propto \xi$ is the dimensionless spin–fermion coupling, while $v_F^{-1} d\Sigma/dk$ remains non-singular at $\xi = \infty$ in $d > 2$ and only logarithmically increases for $d = 2$. This implies that an Eliashberg-type approach near a magnetic instability is fully justified for $d > 2$ and is "almost" justified for $d = 2$. In the latter case (which is the most interesting because of the cuprates), we will have to invoke an extra approximation (an extension to large N) to be able to perform calculations in a controllable way.

Our strategy is the following. We first establish that one can indeed develop a controlled approach to the spin fluctuation problem in the normal state, see also [106]. Next we apply this theory to the pairing problem and show that there is indeed d-wave superconductivity caused by antiferromagnetic spin fluctuations. We discuss the value of T_c near optimal doping and the properties of the superconducting state, particularly the new effects associated with the feedback from superconductivity on the bosonic propagator, since these distinguish between spin-mediated and phonon-mediated pairings.

22.4.1 Normal State Behavior

For our normal state analysis we follow [49] and perform computations assuming that the Eliashberg theory is valid, analyze the strong coupling results, and then show that vertex corrections and $v_F^{-1} d\Sigma_{\mathbf{k}}(\omega = 0)/d\mathbf{k}$ are relatively small at strong coupling.

We begin by obtaining the full form of the dynamical spin susceptibility as it should undergo qualitative changes due to interactions with fermions. The self-energy for the spin susceptibility (that is the spin polarization operator) is given by the convolution of the two fermionic propagators with the momentum difference near $\mathbf{Q}$ and the Pauli matrices in the vertices. Collecting all combinatorial factors, we obtain:

$$\Pi_{\mathbf{q}}(i\omega_m) = -16\bar{g}\xi^2 \int \frac{\mathrm{d}^2 k \mathrm{d}\Omega_m}{(2\pi)^3} G_{\mathbf{k}}(i\Omega_m) \times G_{\mathbf{k}+\mathbf{q}}(i\omega_m + i\Omega_m) \,. \tag{22.31}$$

Here $G_{\mathbf{k}}(i\omega_m)$ is a full fermionic propagator

$$G_{\mathbf{k}}(i\omega_m) = \frac{1}{i\omega_m + \Sigma(i\omega_m) - \varepsilon_{\mathbf{k}}} \tag{22.32}$$

and $\Sigma(i\omega_m)$ is the self-energy which remains to be determined. In principle, even in Eliashberg theory, this self-energy depends on the momentum component along the Fermi surface. However, for computations of $\Pi_Q(i\omega_m)$ this dependence can be safely neglected; both fermions in the fermion polarization bubble should be close to the Fermi surface, and hence the momentum integration is necessarily confined to a narrow region around hot spots.

The 2 + 1-dimensional integral over momentum and frequency in (22.31) is ultraviolet divergent; this implies that its dominant piece comes from highest energies which we have chosen to be $O(\Lambda)$. This contribution, however, should not be counted as it is already incorporated into the bare susceptibility. In addition, $\Pi_Q(i\omega_m)$ contains a universal piece which remains finite even if we extend the momentum and frequency integration to infinity. The most straightforward way to obtain this contribution is to first integrate over momentum and then over frequency. One can easily make sure that for this order of limits, the high-energy contribution is absent, and the full $\Pi_q(i\omega_m)$ coincides with the universal part. The integration is straightforward and on performing it we obtain

$$\Pi_Q(i\omega_m) = -\frac{|\omega_m|}{\omega_{sf}}, \tag{22.33}$$

where the characteristic energy scale is the spin fluctuation frequency,

$$\omega_{sf} = \frac{\pi}{4}\frac{(v_F\xi^{-1})^2}{\bar{g}\sin\phi_0}. \tag{22.34}$$

We recall that ϕ_0 is the angle between the Fermi velocities at the two hot spots separated by $\mathbf{q} \sim \mathbf{Q}$. In optimally doped cuprates, $\phi_0 \approx \pi/2$ and weakly depends on $\mathbf{q}$ as long as one is far from momenta connecting Fermi points along the zone diagonal. Setting $\phi_0 = \pi/2$ and making use of (22.8), we obtain

$$\omega_{sf} = \frac{\pi}{4}\frac{v_F^2\xi^{-2}}{\bar{g}} = \frac{3}{16}\frac{v_F\xi^{-1}}{\lambda}. \tag{22.35}$$

An analytical continuation of (22.33) to the real frequency axis yields $\Pi_Q(\omega) = i|\omega|/\omega_{sf}$. As we anticipated, the coupling to low-energy fermions gives rise to a finite decay rate for a spin fluctuation. As typical spin frequencies are of order $c\xi^{-1} \sim v_F\xi^{-1}$, at strong coupling, $\lambda \geq 1$, the induced damping term $\omega/\omega_{sf} \sim \lambda\,\omega/(v_F\xi^{-1})$ is large compared to the frequency dependent $\omega^2/(c\xi^{-1})^2$ term in the bare susceptibility, i.e. the interaction with low energy fermions changes the form of the spin dynamics from a propagating one to a relaxational one. Neglecting the bare frequency term, we obtain

$$\chi_q(\omega + i0^+) = \frac{\alpha\xi^2}{1 + \xi^2(\mathbf{q}-\mathbf{Q})^2 - i\omega/\omega_{sf}}, \tag{22.36}$$

The same purely relaxational dynamic spin susceptibility was introduced phenomenologically by Millis, Monien and Pines [33] to describe the spin dynamics observed in NMR experiments on the cuprates.

We emphasize that the polarization operator, (22.33), is independent of the actual form of the fermionic self-energy $\Sigma_{\mathbf{k}_{hs}}(\omega)$. The explanation of why a $\mathbf{k}$–independent fermionic self-energy does not affect the polarization bubble at a finite momentum transfer was given by Kadanoff [122] who analyzed a similar problem for phonons. He pointed out that the linearization of the fermionic energy around k_F is equivalent to imposing an approximate Migdal sum rule on the spectral function $A(\mathbf{k},\omega) = -(1/\pi)\,\mathrm{Im}G_R(\mathbf{k},\omega)$

$$\int d\epsilon_{\mathbf{k}} A(\mathbf{k},\omega) = 1, \tag{22.37}$$

where $\epsilon_{\mathbf{k}} = \mathbf{v}_{\mathbf{k}}\cdot(\mathbf{k}-\mathbf{k}_F)$ in our case. Expressing $G(\mathbf{k},\omega_m)$ in terms of $A(\mathbf{k},\omega)$ via the Kramers–Kronig relation and making use of (22.37), one finds

$$\Pi(\mathbf{q},\omega) = -\frac{i\omega}{\omega_{sf}}\int_{-\infty}^{\infty} d\omega' \frac{df(\omega')}{d\omega'} + \mathcal{O}(\omega^2), \tag{22.38}$$

where $f(\omega')$ is the Fermi function. Equation (22.33) then follows from the fact that $f(\omega')$ is 1 at $\omega' = -\infty$ and 0 at $\omega' = +\infty$.

We next determine the fermionic self-energy $\Sigma_{\mathbf{k}}(i\omega)$. In Eliashberg theory it is given by

$$\Sigma_{\mathbf{k}}(i\omega_m) = -3g^2 \int \frac{d^2\mathbf{q}d\omega_{m'}}{(2\pi)^3} \chi_{\mathbf{q}}(i\omega_{m'}) \times G_{\mathbf{k+q}}(i\omega_m + i\omega_{m'}) . \quad (22.39)$$

Consider first the self-energy at a hot spot. Introducing $i\widetilde{\Sigma}_n = i\omega_n + \Sigma_{\mathbf{k}_{\text{hs}}}(i\omega_n)$ and $\mathbf{q} - \mathbf{Q} = \tilde{\mathbf{q}}$, we obtain from (22.39)

$$\Sigma_{\mathbf{k}_{\text{hs}}}(i\omega_m) = \frac{-3\bar{g}}{(2\pi)^3} \int \frac{d\omega_{m'} d^2\tilde{q}}{\tilde{\mathbf{q}}^2 + \xi^{-2}\left(1 + \frac{|\omega_{m'}|}{\omega_{\text{sf}}}\right)} \times \frac{1}{\left(i\widetilde{\Sigma}_{m+m'} - v_F \tilde{q}_\perp\right)} , \quad (22.40)$$

with $\tilde{\mathbf{q}}^2 = \tilde{q}_\parallel^2 + \tilde{q}_\perp^2$. The integration over the component $\tilde{q}_\parallel$, parallel to the velocity $\mathbf{v}_{\mathbf{k}_{\text{hs}}}$ at $\mathbf{k}_{\text{hs}}$, is elementary and yields

$$\Sigma_{\mathbf{k}_{\text{hs}}}(i\omega_m) = \frac{3\bar{g}}{8\pi^2 v_F^2} \int d\omega_{m'}\, d\tilde{q}_\perp \times \frac{1}{\sqrt{\tilde{q}_\perp^2 + q_0^2}\,(\tilde{q}_\perp - iq_E)} , \quad (22.41)$$

where $q_0 = \xi^{-1}\left(1 + \frac{|\omega_{m'}|}{\omega_{\text{sf}}}\right)^{1/2}$ and $q_E = \widetilde{\Sigma}_{m+m'}/v_F$.

We next assume, and then verify, that a typical $\tilde{q}_\perp \ll q_0$. $q_\perp^2$ in the first term can then be ignored, and the momentum integration yields

$$\int d\tilde{q}_\perp \frac{1}{\sqrt{\tilde{q}_\perp^2 + q_0^2}} \frac{1}{\tilde{q}_\perp - iq_E} \approx \frac{1}{q_0} \int_{-\infty}^{\infty} d\tilde{q}_\perp \frac{1}{(\tilde{q}_\perp - iq_E)} = i\pi \operatorname{sign} q_E = i\pi \operatorname{sign}(\omega_m + \omega_{m'}) . \quad (22.42)$$

On substituting this result into (22.42) and splitting the frequency integral into several pieces, depending on the sign of $\omega_m + \omega'_m$, we obtain

$$\Sigma_{\mathbf{k}_{\text{hs}}}(i\omega_m) = i\lambda \int_{-\omega_m}^{0} d\omega_{m'} \frac{1}{\sqrt{1 + |\omega_{m'}|/\omega_{\text{sf}}}} , \quad (22.43)$$

where λ is given by (22.8). The remaining frequency integration is elementary and yields (see Fig. 22.7)

$$\Sigma_{\mathbf{k}}(i\omega_m) = i\lambda \frac{2\omega_m}{1 + \sqrt{1 + \frac{|\omega_m|}{\omega_{\text{sf}}}}} . \quad (22.44)$$

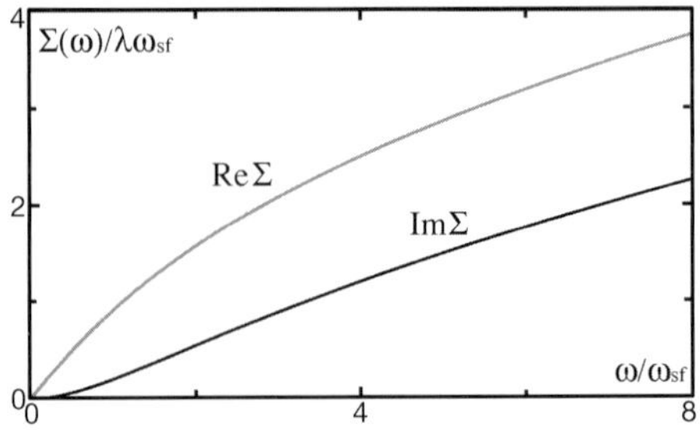

Fig. 22.7. The frequency dependence of the normal state electron self-energy due to spin-fluctuation–fermion interaction for a quasi-particle at a hot spot (from [49])

We now return to the approximation we made above, that a typical $\tilde{q}_\perp \ll q_0$. Since the typical values for $\tilde{q}_\perp$, i.e. those which dominate the integral in (22.42), are of order q_E, the approximation is valid if $q_0 \gg q_E$. Further, a typical internal frequency $\omega_{m'}$ is of order ω_m, hence $q_E \sim |\omega_m + \Sigma(\omega_m)|/v_F$ and $q_0 \sim (\omega_{sf}\xi^2)^{-1}(|\omega_m| + \omega_{\text{sf}})^{1/2} \sim \bar{g}^{1/2}(|\omega_m| + \omega_{\text{sf}})^{1/2}/v_F$. At weak coupling, $\lambda \ll 1$, the condition $q_0 \gg q_E$ implies that $\omega \ll \bar{g}/\lambda$, i.e. at energies smaller that $\bar{g}$, the approximation is well justified. At strong coupling and $\omega \ll \omega_{\text{sf}}$, $q_E \approx \lambda\omega/v_F$, $q_0 \sim \lambda\omega_{\text{sf}}/v_F$, and the criterion $q_0 \gg q_E$ is again satisfied. Finally, at strong coupling and $\omega > \omega_{sf}$, $q_E \sim q_0$ (and both ore of order $(\bar{g}\omega_m)^{1/2}/v_F$. In this limit, the approximation we made is qualitatively, but not quantitatively correct. In order to develop a well controlled theoretical framework for this limit, Abanov et al. [87,123] developed a controllable $1/N$ approach by extending the model to a large number of hot spots, N, in the Brillouin zone ($N = 8$ in the physical case), or, equivalently, to a large number of fermionic flavors. This allows an expansion in terms of $1/N$. For large N, $q_0/q_E \sim N^2$, i.e. the approximation $q_0 \gg q_E$ is justified. Another appealing feature of this $\frac{1}{N}$ expansion is that within it, vertex corrections and the dependence of the self-energy of the momentum component transverse to the Fermi surface are also small in $\frac{1}{N}$ and can be computed systematically together with the corrections to the frequency dependent part of the self-energy [49]. To keep our discussion focused on the key results, we will not discuss further the details of the $1/N$ approach. Rather, we just emphasize that (i) (22.44) is quantitatively correct even if

$q_0 \sim q_E$ and (ii) numerically the difference between approximate and more involved "exact" results for the fermionic $\Sigma_{\mathbf{k}_{hs}}(\omega)$ is only few percent already for a physical N [49].

We next analyze the functional form of (22.44). We see that $\Sigma_{\mathbf{k}_{hs}}(i\omega)$ scales with λ and at strong coupling exceeds the bare $i\omega$ term in the inverse fermion propagator $G^{-1}_{\mathbf{k}_{hs}}(i\omega) = i\omega - \Sigma_{\mathbf{k}_{hs}}(i\omega)$. As was the case with phonons, it originates in the scattering on zero-sound-like vibrations of the electronic subsystem, while spin-fluctuations just mediate this interaction. Further, $\Sigma_{\mathbf{k}_{hs}}(i\omega)$ (see Fig. 22.7) evolves at the same typical energy ω_{sf} as the bosonic self-energy. This interconnection between bosonic and fermionic propagators is one of the key "fingerprints" of the spin–fermion model.

In Fig. 22.8 we show the behavior of the quasiparticle spectral function $A(\epsilon_{\mathbf{k}}, \omega)$ at various $\epsilon_{\mathbf{k}}$. We see that the spectral weight of the quasiparticle peak rapidly decreases as $\epsilon_{\mathbf{k}}$ becomes larger than ω_{sf}.

For small frequencies, $\omega \ll \omega_{sf}$, the spin susceptibility can be approximated by its static form. For the fermionic self-energy we find after analytical continuation to real frequencies

$$\Sigma_{\mathbf{k}_{hs}}(\omega) = \lambda\left(\omega + \frac{i\omega\,|\omega|}{4\omega_{sf}}\right); \quad (\omega \ll \omega_{sf})\,. \quad (22.45)$$

We see that the quasiparticle damping term, although quadratic in ω as it should be in a Fermi liquid, scales inversely with ω_{sf}, not with the Fermi energy as in conventional metals. As ω_{sf} vanishes at the critical point, the width of the Fermi liquid region, where damping is small compared to ω, progressively shrinks as ξ increases. The quasiparticle renormalization factor $d\Sigma_{\mathbf{k}_{hs}}(\omega)/d\omega|_{\omega=0} = \lambda \propto \xi$ increases as the system approaches the magnetic quantum critical point. The quasiparticle z factor simultaneously decreases as

$$z_{\mathbf{k}_{hs}} = \frac{1}{1 + \left.\frac{\partial \Sigma_{\mathbf{k}_{hs}}(\omega)}{\partial \omega}\right|_{\omega=0}} = \frac{1}{1+\lambda} \quad (22.46)$$

and vanishes at criticality.

At frequencies above ω_{sf}, the imaginary part of the fermionic self-energy resembles a linear function of ω over a substantial frequency range up to

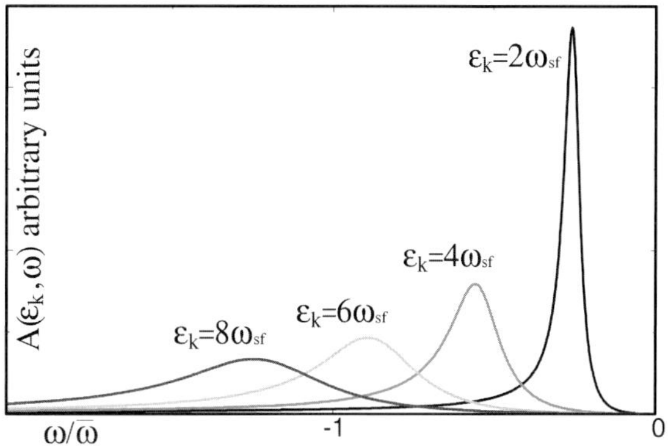

Fig. 22.8. The normal state spectral function. Note the absence of a quasiparticle peak. This is the consequence of the proximity to an antiferromagnetic quantum critical point. The figure is taken from [49]

about $8\omega_{sf}$. At larger ω, the self-energy eventually crosses over to

$$\Sigma_{\mathbf{k}_{hs}}(\omega) = \text{sign}\omega\ (i\bar{\omega}|\omega|)^{1/2}\,, \quad (22.47)$$

where

$$\bar{\omega} = 4\lambda^2\omega_{sf} = \frac{9}{16\pi}\bar{g}\,. \quad (22.48)$$

Observe that $\bar{\omega}$ remains finite as $\xi \to \infty$. At the critical point $\omega_{sf} \to 0$, the self-energy displays the non-Fermi-liquid behavior of (22.47) down to $\omega = 0$. A plot of the fermionic self-energy is presented in Fig. 22.7. The intermediate quasi-linear regime is clearly visible. Note also that the deviations from Fermi liquid behavior starts already at small $\omega \sim \omega_{sf}/2$.

We now consider how well the Eliashberg approximation is satisfied, i.e. whether vertex corrections and the momentum dependent piece in the fermionic self-energy are relatively small. To do this one needs to evaluate $\Sigma_{\mathbf{k}}(\omega = 0)$ at $\mathbf{k} \neq \mathbf{k}_{hs}$ and the vertex correction $\delta g/g$. The details of the derivation can be found in [49]. We have

$$\Sigma_{\mathbf{k}}(\omega = 0) = \left[\frac{3}{4\pi}\log\lambda\right]\varepsilon_{\mathbf{k}+\mathbf{Q}} \quad (22.49)$$

and

$$\frac{\delta g}{g} = \frac{Q(\phi_0)}{8}\log\lambda\,, \quad (22.50)$$

where $Q(\phi_0) = 2\phi_0/\pi$. For $\phi_0 \approx \pi/2, Q \approx 1$. We see that these two corrections depend only logarithmically on the coupling and at large λ are parametrically small compared to the frequency dependence of the self-energy. Furthermore, at large N, both $\delta g/g$ and $\Sigma_{\mathbf{k}}$ contain an extra factor of $1/N$, i.e., scale as $(1/N)\log\lambda$. We see therefore that $(1/N)\log\lambda$ is the analog of the second coupling constant $\tilde{\lambda}_{ep}$ for the phonon case. Just as for phonons, the applicability of the Eliashberg theory is related to the fact that this second coupling constant is much smaller than the primary coupling λ. In our case, this requires that $\lambda \gg (\log\lambda)/N$. We however emphasize that the smallness of the two "couplings" is not the result of the smallness of the velocity ratio but the consequence of the proximity to a critical point, when relevant frequencies are small, and the $O(\omega)$ damping term in the spin propagator dominates over the ω^2 term. Also, in distinction to phonons, the second coupling still diverges at the critical point and therefore corrections to Eliashberg theory cannot be neglected close to the antiferromagnetic transition. These corrections have been analyzed within a renormalization group approach in [49,124] and found to be of minor relevance at intermediate coupling discussed here.

It is also instructive to explicitly compute the fermionic self-energy in the same way as we did in Sect. 22.3 for systems with electron–phonon interaction, and verify that the reason for the dominance of the frequency dependence of the self energy is common to that for the phonon case. To see this, we introduce, by analogy with (22.24)

$$\Sigma(\mathbf{k}, i\omega_m) = (i\omega_m - \epsilon_{\mathbf{k+Q}}) I(\mathbf{k}, i\omega_m)\,. \qquad (22.51)$$

Evaluating $I(\mathbf{k}_{\text{hs}}, i\omega_m)$ in the same way as in the phonon case [49], we find at $\omega \ll \omega_{sf}$

$$I(\mathbf{k}_{\text{hs}}, i\omega_m) = I_{\text{reg}} + \lambda \frac{i\omega_m}{i\omega_m - \epsilon_{\mathbf{k+Q}}}\,, \qquad (22.52)$$

where $I_{\text{reg}} = O(\log\lambda)$. This form is the same as that found for phonons (see (22.27)). The analogy implies that the dominant, $O(\lambda)$, contribution to the fermionic self-energy comes from magnetically mediated interactions between fermions and their collective particle-hole excitations, whereas the actual spin–fermion scattering process in which fermions at forced to vibrate at typical spin frequencies yields a smaller $O(\log\lambda)$ contribution.

Finally, away from hot spots but still at the Fermi surface, the fermionic self-energy is given by the same expression, (22.44), as at a hot spot, but with a momentum dependent coupling constant $\lambda_{\mathbf{k}}$ and energy scale $\omega_{\text{sf}}(\mathbf{k})$ which obey

$$\lambda_{\mathbf{k}} = \lambda/(1 + \delta k\xi),$$
$$\omega_{\text{sf}}(\mathbf{k}) = \omega_{\text{sf}}(1 + \delta k\xi)^2\,, \qquad (22.53)$$

where $\delta k = |\mathbf{k}_F - \mathbf{k}_{\text{hs}}|$ is the momentum deviation from a hot spot along the Fermi surface. We see that the effective coupling decreases upon deviation from a hot spot, while the upper energy scale for the Fermi liquid behavior increases. The increase of ω_{sf}, however, is counterbalanced by the fact that $\omega_{\text{sf}} \propto \sin\phi_0$, and ϕ_0, which we had set to be $\approx \pi/2$, increases away from a hot spot [125].

We see from (22.53) that the width of the region where $\Sigma_{\mathbf{k}}(\omega)$ is independent of $\mathbf{k}$ (i.e. the "size" of a hot spot) depends on frequency. At the lowest frequencies, $\omega < \omega_{\text{sf}}(\mathbf{k})$, $\Sigma_{\mathbf{k}}(\omega) = i\lambda_{\mathbf{k}}\omega$, and the hot spot physics is confined to a region of width ξ^{-1} which progressively shrinks as ξ increases. However, at frequencies above $\omega_{\text{sf}}(\mathbf{k})$, $\Sigma_{\mathbf{k}}(\omega) \sim \lambda_{\mathbf{k}}(i\omega\omega_{\text{sf}}(\mathbf{k}))^{1/2} = \lambda(i\omega\omega_{\text{sf}})^{1/2}$ is independent of $\mathbf{k}$. Accordingly, physical processes that happen on these scales are isotropic (apart from the dependence on ϕ_0). In this sense the whole Fermi surface acts as one big "hot spot".

One can easily perform the above analysis in dimensions larger than $d = 2$. One finds that the quasiparticle spectral weight behaves for large λ and the lowest energies $\omega < \omega_{\text{sf}}$ as

$$z_{\mathbf{k}_{\text{hs}}} \propto \lambda^{d-3} \qquad (22.54)$$

for $2 < d < 3$, and vanishes logarithmically for $d = 3$. Correspondingly, in the quantum critical regime we find for the self-energy along "hot lines"

$$\Sigma_{\mathbf{k}_{\text{hs}}}(i\omega) \propto i\omega\,|\omega|^{\frac{d-3}{2}}\,. \qquad (22.55)$$

This expression transforms into $\Sigma_{\mathbf{k}_{\text{hs}}}(i\omega) \propto i\omega\log(|\omega|)$ for $d = 3$. Above $d = 3$ no non-Fermi liquid effects result from the proximity to the quantum critical point. This demonstrates that many of the effects caused by the incoherent nature of the fermions are peculiar to 2d and are considerably less pronounced in three-dimensional systems. In 22.8 results for the spectral function are given.

22.4.2 The $d_{x^2-y^2}$ Pairing Instability Temperature

We now consider the development of the pairing instability in the spin–fermion model. We follow [87]. It is customary in an analysis of the pairing problem to introduce an infinitesimally small particle–particle vertex $\Phi^{(0)}_{\mathbf{k},-\mathbf{k}}(\omega,-\omega) \equiv \Phi^{(0)}_{\mathbf{k}}(\omega)$ and study its renormalization by the pairing interaction. The corresponding diagrams are presented in Fig. 22.9.

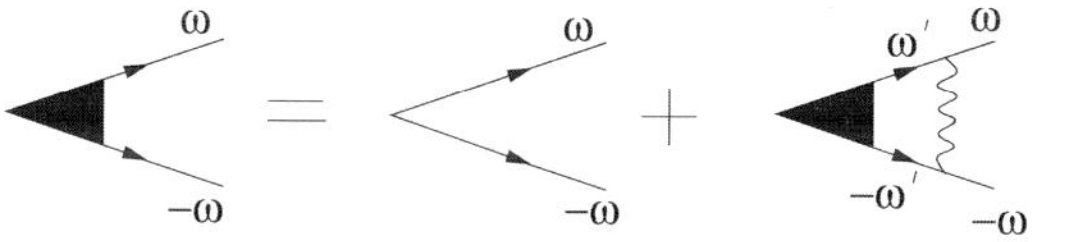

Fig. 22.9. Diagrammatic representation for the pairing vertex [87]. The *solid and wavy lines* are fermionic and spin-fluctuation propagators, respectively

The temperature at which the renormalized vertex diverges, i.e. when the equation for the full $\Phi_{\mathbf{k}}(\omega)$ has a nontrivial solution at vanishing $\Phi^{(0)}_{\mathbf{k}}(\omega)$, marks the onset of pairing. As noted above, the spin-mediated pairing interaction gives rise to $d_{x^2-y^2}$ superconductivity. We argued above that near the magnetic instability, the gap is maximum near the hot spots. One can check (see [87]) that the pairing problem is confined to the hot regions in the sense that the momentum integration never extends to $|\mathbf{k}-\mathbf{k}_{\rm hs}| \sim k_F$, where the momentum dependence of the self-energy or of the pairing vertex becomes relevant. We can then assume that the pairing vertex is flat near the hot spots. The underlying d-wave symmetry then implies that the gap changes sign between two hot regions separated by the antiferromagnetic $\mathbf{Q}$.

The momentum independence of the pairing problem is due to the fact that, as we will see, the pairing predominantly comes from frequencies of order $\bar{g}$. For these frequencies, the width of the hot region is constrained by the requirement that $\omega_{\rm sf}(\mathbf{k}) < \bar{g}$. The boundaries are set by $\omega_{\rm sf}(\mathbf{k}) \sim \bar{g}$, i.e. by $|\mathbf{k}-\mathbf{k}_{\rm hs}| \sim k_F(\bar{g}/W)$. Here, W is the bandwidth. The condition $|\mathbf{k}-\mathbf{k}_{\rm hs}| < k_F$ then implies that the effective coupling is smaller than the fermionic bandwidth. As discussed in the Introduction, the latter is a necessary condition for the separation between high and low energies, and we assume it to hold. We comment below on what happens if typical $|\mathbf{k}-\mathbf{k}_{\rm hs}| > k_F$, i.e. hot and cold regions cannot be separated.

The value of the transition temperature depends sensitively on the behavior of fermions that are paired by the spin-mediated interaction. Our analysis of the normal state has shown that the character of the fermionic degrees of freedom changes at energies of order $\omega_{\rm sf}$. For energies smaller than $\omega_{\rm sf}$, fermions display Fermi liquid behavior, while at higher energies they display behavior that is different from that in a Fermi liquid. In the BCS theory of superconductivity only Fermi liquid degrees of freedom contribute to the pairing. Let us suppose that this also holds in the present case. Then the pairing problem would be qualitatively similar to that of BCS, since for frequencies smaller than $\omega_{\rm sf}$, the spin susceptibility that mediates pairing can be approximated by its static value. The linearized equation for the pairing vertex then has the form

$$\Phi(\omega) = \frac{\lambda}{1+\lambda} \int_{T_c^{\rm FL}}^{\omega_{\rm sf}} d\omega' \frac{\Phi(\omega')}{\omega'}, \qquad (22.56)$$

where the $1+\lambda$ factor in the denominator is the result of mass renormalization in the Fermi liquid regime ($\Sigma(\omega) \approx \lambda\omega$). The solution of this equation [86] yields

$$T_c^{\rm FL} \sim \omega_{\rm sf} \exp\left(-\frac{1+\lambda}{\lambda}\right). \qquad (22.57)$$

At weak coupling, this is just the BCS result. At strong coupling, the mass renormalization compensates the coupling constant, and $T_c^{\rm FL}$ saturates at $T_c^{\rm FL} \sim \omega_{\rm sf} \propto \xi^{-2}$. This result, if correct, would imply that the pairing fluctuations become progressively less relevant as one approaches the quantum critical point $\xi^{-1} \to 0$ (see the left panel in Fig. 22.5). At a first glance, this is what happens, because pairing of non Fermi-liquid degrees of freedom seems hard to accomplish. Indeed, at frequencies larger than $\omega_{\rm sf}$, the pairing interaction decreases as $(1+|\omega_m|/\omega_{\rm sf})^{-1/2}$, and this apparently makes the frequency integral ultraviolet convergent, i.e. the "logarithmic" pairing problem does not appear to extend above $\omega_{\rm sf}$. The flaw in this argument is that when the interaction decreases, the mass renormalization produced by the *same* interaction also decreases, and the large overall λ is no longer compensated by $1 + \Sigma(i\omega_m)/i\omega_m$. Indeed, for $\omega_m \gg \omega_{\rm sf}$, $\Sigma(i\omega_m) = i(\bar{\omega}\omega_m)^{1/2}$, where, we recall, $\bar{\omega} = (9/16\pi)\bar{g}$, and the mass renormalization is $1 + (\bar{\omega}/\omega_m)^{1/2} = 1 + 2\lambda(\omega_{\rm sf}/\omega_m)^{1/2} \ll \lambda$. Furthermore, we see that at frequencies between $\omega_{\rm sf}$ and $\bar{\omega}$, the effective mass and effective interaction both scale as $(\omega)^{-1/2}$. The product of the two then scales as $1/\omega$, i.e. the $1/\omega_m$ behavior of the pairing kernel extends to frequencies of the order of $\bar{\omega}$ which, we recall, remains finite at $\xi = \infty$.

By itself, this effect does not guarantee that the pairing instability temperature is of order $\bar{\omega}$ as the pairing interaction depends on the transferred frequency $\omega - \omega'$, and the linearized equation for the pairing vertex becomes an integral equation in frequency. In particular, for $\xi = \infty$, and hence $\omega_{\rm sf} = 0$, we need to solve

$$\Phi(\omega) = \frac{1}{4}\int_{T_{\rm cr}}^{\bar{\omega}} d\omega' \Phi(\omega') \frac{1}{\sqrt{\omega'}} \frac{1}{\sqrt{|\omega - \omega'|}} . \quad (22.58)$$

Observe that this equation does not have any adjustable parameter and is therefore fully universal when T is expressed in units of $\bar{\omega}$. Equation (22.58) has been analyzed in detail by Finkel'stein, Abanov and one of us [87]. They found that it does have a nontrivial solution at

$$T_{\rm cr} \sim 0.2\bar{\omega} . \quad (22.59)$$

They also analyzed the pairing problem at a finite λ and found that incoherent fermions dominate pairing down to a surprisingly small $\lambda \sim 0.5$. The McMillan like formula, (22.57), becomes valid only at smaller λ. A plot of $T_{\rm cr}$ versus λ is presented in Fig. 22.10.

Several comments are in order here. First, at these values of the coupling constant (and doping) $T_{\rm cr}$ does not coincide with the onset temperature for superconductivity, T_c, but rather represents the onset temperature for pseudogap behavior; the actual T_c is lower, as discussed below. Second, we have neglected fluctuation effects due to the quasi-two-dimensionality. The latter are expected to yield Kosterlitz–Thouless physics [126]. Third, (22.59), is only valid when $\bar{g} < W$. In the opposite limit $\bar{g} > W$, lattice effects are important and controlled analytical calculations are difficult to perform. One can, however, easily estimate that in this limit $T_{\rm cr} \propto v_F^2/\bar{g} \propto \omega_{sf}\xi^2$ (we set interatomic spacing $a = 1$). This estimate coincides with the result of Monthoux and Pines who extracted $T_{\rm cr} \propto \omega_{sf}\xi^2$ from their numerical analysis [37]. Since v_F is proportional to the hopping matrix element, and $\bar{g}$ scales with the Hubbard U, it follows that for $U >> W$, the spin-mediated pairing yields $T_{\rm cr} \propto J$, where J is the magnetic exchange integral of the corresponding Heisenberg model which describes antiferromagnetism at half-filling.

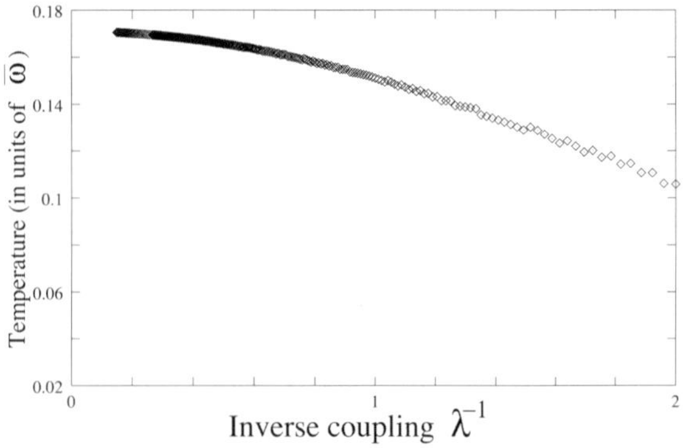

Fig. 22.10. The results for the instability temperature $T_{\rm cr}$ obtained from the solution of the linearized Eliashberg equations for different values of the coupling constant λ. The figure is taken from [87]

Equation (22.59) demonstrates that the d-wave pairing instability temperature of a two dimensional system at an antiferromagnetic quantum critical point is finite. The behavior near a ferromagnetic quantum-critical point is more complex, see [88].

22.4.3 Superconducting State

We next extend the Eliashberg theory to the spin-fluctuation induced superconducting state. The discussion in this section follows [89]. We derive a generalized set of Eliashberg equations for the fermionic self-energy and the gap function that include an additional coupled equation for the spin polarization operator. The latter, as discussed in the Introduction, is produced by low-energy fermions and has to be determined self consistently.

We will see that in the superconducting state, the momentum integration is also confined to hot regions $|\mathbf{k} - \mathbf{k}_{hs}| \sim k_F(\bar{g}/W) \leq k_F$. We can then safely neglect the weak momentum dependence of both $\Sigma(i\omega)$ and $\Phi(i\omega)$, as we did above in calculating T_{cr}. Subtle effects due to this weak momentum dependence will be considered in the next section. We will not attempt to discuss the behavior of the gap near the nodes. The nodal behavior is central for the interpretation of the experimental data at the lowest temperatures and frequencies, but not at energies comparable or larger than the maximum pairing gap which we consider below.

Generalized Eliashberg Equations

The derivation of the Eliashberg equations is straightforward. In the superconducting state, the normal and anomalous fermionic Green's functions $G_\mathbf{k}(i\omega_n)$ and $F_\mathbf{k}(i\omega_n)$ and the dynamical spin susceptibility are given by (22.9)–(22.11). It is convenient to rewrite $G_\mathbf{k}(i\omega_n)$ and $F_\mathbf{k}(i\omega_n)$ as

$$\begin{aligned} G_\mathbf{k}(i\omega_n) &= -\frac{\epsilon_\mathbf{k} + i\widetilde{\Sigma}_n}{\epsilon_\mathbf{k}^2 + \widetilde{\Sigma}_n^2 + \Phi^2(i\omega_n)}, \\ F_\mathbf{k}(i\omega_n) &= i\frac{\Phi(i\omega_n)}{\epsilon_\mathbf{k}^2 + \widetilde{\Sigma}_n^2 + \Phi^2(i\omega_n)}, \\ F_{\mathbf{k}+\mathbf{Q}}(i\omega_n) &= -i\frac{\Phi(i\omega_n)}{\epsilon_{\mathbf{k}+\mathbf{Q}}^2 + \widetilde{\Sigma}_n^2 + \Phi^2(i\omega_n)}, \end{aligned} \tag{22.60}$$

where $i\widetilde{\Sigma}_n = i\omega_n + \Sigma_{\mathbf{k}_{hs}}(i\omega_n)$ (in real frequencies, $\widetilde{\Sigma}(\omega) = \omega + \Sigma_{\mathbf{k}_{hs}}(\omega)$). Without losing generality we can set $\epsilon_\mathbf{k} = v_x\tilde{k}_x + v_y\tilde{k}_y$ and $\epsilon_{\mathbf{k}+\mathbf{Q}} = -v_x\tilde{k}_x + v_y\tilde{k}_y$ where $\tilde{k} = k - k_{hs}$. The sign change between $F_\mathbf{k}$ and $F_{\mathbf{k}+\mathbf{Q}}$ is the result of $d_{x^2-y^2}$ symmetry. The spin susceptibility, we recall, is given by

$$\chi_\mathbf{q}(i\omega_m) = \frac{\alpha\xi^2}{1 + \xi^2(\mathbf{q}-\mathbf{Q})^2 - \Pi_\mathbf{Q}(i\omega_m)}. \tag{22.61}$$

We need to obtain the equations for the fermionic self-energy $\Sigma_{\mathbf{k}_{hs}}(i\omega_m)$, the anomalous vertex $\Phi_{\mathbf{k}_{hs}}(i\omega_m)$, and the spin polarization operator $\Pi_\mathbf{Q}(i\omega_m)$. The spin polarization operator is obtained in the same way as in the normal state, but now there are two particle–hole bubbles: one is the convolution of $G_\mathbf{k}G_{\mathbf{k}+\mathbf{Q}}$ and the other is the convolution of $F_\mathbf{k}F_{\mathbf{k}+\mathbf{Q}}$. We have

$$\begin{aligned} \Pi_\mathbf{Q}(i\omega_n) = -16\bar{g}\xi^2 T \sum_m \int \frac{\mathrm{d}^2k}{(2\pi)^2} \\ \times \left[G_{\mathbf{k}+\mathbf{Q}}(i\omega_{n+m})\,G_\mathbf{k}(i\omega_m) \right. \\ \left. - F_{\mathbf{k}+\mathbf{Q}}(i\omega_{n+m})\,F_\mathbf{k}(i\omega_m)\right] \end{aligned} \tag{22.62}$$

(the negative sign between the two terms originates in the summation over the spin components). The momentum integration can be performed explicitly and yields

$$\begin{aligned} \Pi_\mathbf{Q}(i\omega_n) = -\frac{4\bar{g}\xi^2}{v_F^2} T \sum_m \left[1 - g(i\omega_m)\,g(i\omega_{n+m}) \right. \\ \left. - f(i\omega_m) f(i\omega_{n+m})\right], \end{aligned} \tag{22.63}$$

where

$$g(i\omega_m) = \frac{\widetilde{\Sigma}_m}{\sqrt{\widetilde{\Sigma}_m^2 + \Phi^2(i\omega_m)}}, \tag{22.64}$$

and

$$f(i\omega_m) = \frac{\Phi(i\omega_m)}{\sqrt{\widetilde{\Sigma}_m^2 + \Phi^2(i\omega_m)}}. \tag{22.65}$$

The first term in (22.63) is the result of the regularization of the ultraviolet singularity. The minus sign factor in $f(i\omega_m)f(i\omega_{n+m})$ term in (22.63) is due to the d-wave form of $F_\mathbf{k}$.

Equation (22.63) takes into account the change of the low energy spin dynamics in the pairing state. In the case of electron–phonon interaction a corresponding change of the phonon dynamics exists as well, causing a shift of the phonon frequency and line width below T_c. While this effect is only a minor correction to the phononic dynamics and is often neglected [117], it eads to a dramatic change of the spin dynamics in our case.

The other two equations are formally the same as for phonon-mediated superconductors. The fermionic self-energy $\Sigma(\omega_n)$ is given by

$$\Sigma(i\omega_n) = -3g^2 T \sum_m \int \frac{d^2q}{(2\pi)^2} \chi(\mathbf{q}, i\omega_m) \times G_{k+q}(\omega_{n+m}) \,. \tag{22.66}$$

Performing the momentum integration along the same lines as in the normal state calculations we find

$$\Sigma(i\omega_n) = \frac{3g^2}{2v_F} T \sum_m D(i\omega_m)\, g(i\omega_n + i\omega_m) \,, \tag{22.67}$$

where D is the effective bosonic propagator that is obtained by integrating the dynamical spin susceptibility over the momentum component along the Fermi surface and setting other momentum components to $\mathbf{Q}$ (the last step is equivalent to the approximation we discussed below (22.42)). We have

$$D(i\omega_m) = \int \frac{dq_\parallel}{2\pi} \chi(\mathbf{q}, i\omega_m)\bigg|_{q_\perp = Q} = \frac{\alpha\xi}{2\sqrt{1 - \Pi_Q(i\omega_m)}} \,. \tag{22.68}$$

An analogous equation is obtained for the anomalous vertex

$$\Phi(i\omega_n) = \frac{3g^2}{2v_F} T \sum_m D(i\omega_m) f(i\omega_n + i\omega_m) \,. \tag{22.69}$$

Equations (22.63), (22.67), and (22.69) constitute the full set of Eliashberg equations for the spin-mediated superconductivity. Alternatively to $\Phi(\omega)$ and $\Sigma(\omega)$ we can also introduce

$$Z(\omega) = 1 + \frac{\Sigma(\omega)}{\omega}\,, \quad \Delta(\omega) = \frac{\Phi(\omega)}{Z(\omega)} \,. \tag{22.70}$$

The complex function $\Delta(\omega)$ reduces to the superconducting gap Δ in BCS theory where we also have $Z(\omega) = 1$. In Eliashberg theory, the superconducting gap, defined as a frequency where the density of states has a peak, is the solution of $\Delta(\omega = \Delta) = \omega$.

We again emphasize that the Eliashberg equations are valid for fermionic momenta which deviate from hot spots by less than $\bar{g}/W$. For these momenta, the pairing vertex can be approximated by a $\mathbf{k}$-independent function which however changes sign between two hot regions separated by $\mathbf{Q}$. For larger deviations, the anomalous vertex rapidly goes down and eventually vanishes along zone diagonals.

Solution of the Eliashberg Equations

We discuss the general structure of the solutions of the set of Eliashberg equations, and then present the results of their numerical solution. First, we see from (22.63) that, as in the normal state, $\Pi(\omega = 0) = 0$ for any $\Sigma(\omega)$ and $\Phi(\omega)$. This physically implies that the development of the gap does not change the magnetic correlation length. This result becomes evident if one notices that d-wave pairing involves fermions from opposite sub-lattices. Second, the opening of the superconducting gap changes the low frequency spin dynamics. Now quasiparticles near hot spots are gapped, and a spin fluctuation can decay into a particle–hole pair only when it can pull two particles out of the condensate of Cooper pairs. This implies that the decay into particle–hole excitations is only possible if the external frequency is larger than 2Δ. At smaller frequencies, we should have $\Pi''(\omega) = 0$ at $T = 0$ [123,128]. This result indeed readily follows from (22.63). The Kramers–Kronig relation $\Pi'(\omega) = (2/\pi)\int_0^\infty \Pi''(x)/(x^2 - \omega^2)$ then implies that because of a drop in $\Pi''(\omega)$, the spin polarization operator in a superconductor acquires a real part, which at low ω is quadratic in frequency and is of order $\omega^2/(\Delta\omega_{\rm sf})$. Substituting this result into (22.11), we find that at low energies, spin excitations in a d-wave superconductor are propagating, gapped magnon-like quasiparticles

$$\chi(\mathbf{q}, \omega) \propto \frac{\Delta_s^2}{\Delta_s^2 + c_s^2(\mathbf{q} - \mathbf{Q})^2 - \omega^2} \,, \tag{22.71}$$

where

$$\Delta_s \sim (\Delta\omega_{\rm sf})^{1/2} \tag{22.72}$$

and $c_s^2 \sim v_{\rm F}^2 \Delta/\bar{g}$. The re-emergence of propagating spin dynamics implies that the dynamical spin susceptibility acquires a resonance peak which at $\mathbf{q} = \mathbf{Q}$ is located at $\omega = \Delta_s$.

Equation (22.71) is indeed meaningful only if $\Delta_s \le \Delta$, i.e. $\omega_{\rm sf} \le \Delta$. Otherwise the use of the quadratic form for $\Pi(\omega)$ is not justified. To find out how Δ depends on the coupling constant, one needs to carefully analyze the full set of equations (22.63)–(22.69). This analysis is rather involved [87,89], and is not directly related to the goal of this chapter. We skip the details and quote the result. It turns out that at strong coupling, $\lambda \geq 1$, i.e. for optimally and underdoped cuprates, the condition $\Delta > \omega_{sf}$ is satisfied: the gap scales with $\bar{\omega}$ and saturates at $\Delta \approx 0.35\bar{\omega} = 0.06\bar{g}$ at $\lambda \to \infty$, while $\omega_{sf} \propto \lambda^{-2} \to 0$. In this situation, the spin excitations in a superconductor are propagating, particle-like modes with a gap Δ_s. However, in distinction to phonons, these propagating magnons get their identity from a strong coupling feedback effect in the superconducting state.

At weak coupling, the superconducting problem is of the BCS type, and $\Delta \ll \omega_{sf}$. This result is intuitively obvious as ω_{sf} plays the role of the Debye frequency in the sense that the bosonic mode that mediates pairing decreases at frequencies above ω_{sf}. For $\Delta \ll \omega_{sf}$, $\chi(\mathbf{Q}, \omega)$ does not have a pole at frequencies where $\Pi(\omega) \propto \omega^2$. Still, a pole in $\chi''(\mathbf{Q}, \omega)$ does exist even at weak coupling [123, 128–132]. To see this, note that at $\omega \approx 2\Delta$ one can simultaneously set both fermionic frequencies in the bubble to be close to Δ, and make both propagators singular due to the vanishing of $\sqrt{\tilde{\Sigma}^2 - \Phi^2}$ where, we recall, $\tilde{\Sigma} = \omega + \Sigma(\omega)$. Substituting $\tilde{\Sigma}^2(\omega) - \Phi^2(\omega) \propto \omega - \Delta$ into (22.63) and using the spectral representation, we obtain for $\omega = 2\Delta + \epsilon$

$$\Pi''(\omega) \propto \int_0^{\epsilon} \frac{dx}{(x(\epsilon - x))^{1/2}} . \tag{22.73}$$

Evaluating the integral, we find that Π'' undergoes a finite jump at $\omega = 2\Delta$. By the Kramers–Kronig relation, this jump gives rise to a logarithmic singularity in $\Pi'(\omega)$ at $\omega = 2\Delta$:

$$\Pi'(\omega) = \frac{2}{\pi} \int_{2\Delta}^{\infty} dx \frac{\Pi''(x)}{x^2 - \omega^2} \propto \Delta \log \frac{2\Delta}{|\omega - 2\Delta|} . \tag{22.74}$$

The behavior of $\Pi'(\omega)$ and $\Pi''(\omega)$ is schematically shown in Fig. 22.11.

The fact that $\Pi'(\omega)$ diverges logarithmically at 2Δ implies that no matter how small Δ/ω_{sf} is, $\chi(\mathbf{Q}, \omega)$ has a pole at $\Delta_s < 2\Delta$, when $\Pi''(\omega)$ is still zero. Simple estimates show that for weak coupling, where $\omega_{sf} \gg \Delta$, the singularity occurs at $\Delta_s = 2\Delta(1 - Z_s)$ where $Z_s \propto e^{-\omega_{sf}/(2\Delta)}$ is also the spectral weight of the resonance peak in this limit.

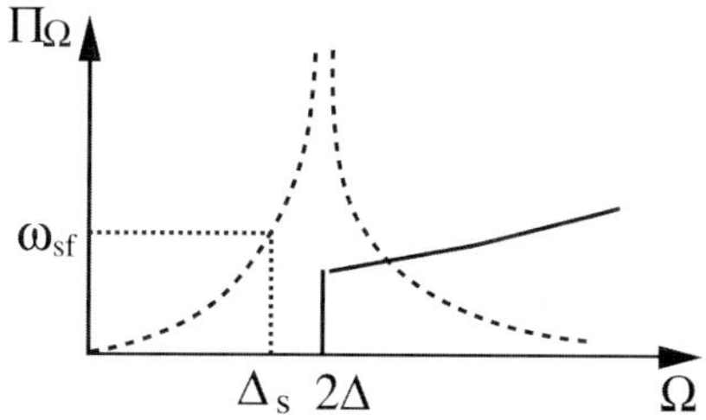

Fig. 22.11. Schematic behavior of the real (*dashed line*) and imaginary (*solid line*) parts of the particle hole bubble in the superconducting state. Due to the discontinuous behavior of $\Pi''(\omega)$ at $\omega = 2\Delta$, the real part $\Pi'(\omega)$ is logarithmically divergent at 2Δ. For small ω, the real part behaves like ω^2/Δ. The figure is taken from [133]

We see therefore that the resonance in the spin susceptibility exists both at weak and at strong coupling. At strong coupling, the resonance frequency is $\Delta_s \sim \Delta/\lambda \ll \Delta$, i.e. the resonance occurs in the frequency range where spin excitations behave as propagating magnon-like excitations. At weak coupling, the resonance occurs very near 2Δ due to the logarithmic singularity in $\Pi'(\omega)$. In practice, however, the resonance at weak coupling can hardly be observed because the residue of the peak in the spin susceptibility Z_s is exponentially small.

Figure 22.12 shows the results for $\chi(\mathbf{Q}, \omega)$ obtained from the full solution of the set of three cou-

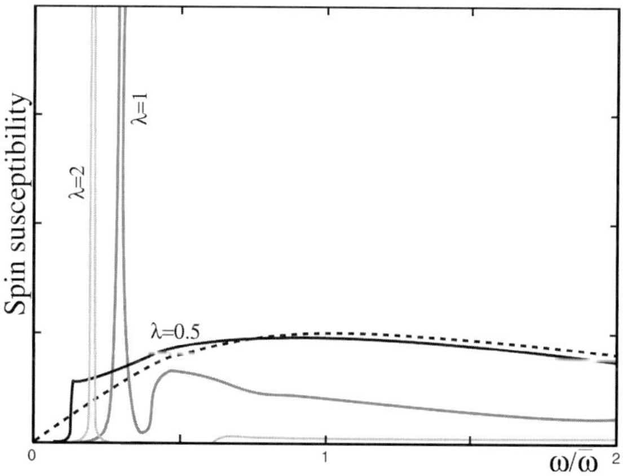

Fig. 22.12. Imaginary part of the dynamical spin susceptibility in the superconducting state at $T \ll T_c$ obtained from the solution of the set of three Eliashberg equations for coupling constants $\lambda = 0.5$, $\lambda = 1$, and $\lambda = 2$. The figure is taken from [133]

pled equations at $T \approx 0$ and three different coupling constants [89]. For $\lambda \geq 1$, the spin susceptibility has a sharp peak at $\omega = \Delta_s$. The peak gets sharper when it moves away from 2Δ. At the same time, for $\lambda = 0.5$, corresponding to weak coupling, the peak is very weak and is washed out by a small thermal damping. In this case, χ'' only displays a discontinuity at 2Δ.

We next show that the resonance peak does not exist for s-wave superconductors [134]. In the latter case, the spin polarization operator is given by almost the same expression as in (22.63), but with a different sign of the ff-term; recall that the original sign in (22.63) originated from the fact that the two fermions in the spin polarization bubble differ in momentum by $\mathbf{Q}$, and the d-wave gap changes sign under $\mathbf{k} \to \mathbf{k} + \mathbf{Q}$. One can immediately check that for a different sign of the anomalous term, Π'' is continuous at 2Δ. Accordingly, $\Pi'(\omega)$ does not diverge at 2Δ, and hence there is no resonance at weak coupling. Still, however, one could expect the resonance at strong coupling as at small frequencies $\Pi'(\omega)$ is quadratic in ω by virtue of the existence of the threshold for Π''. It turns out, however, that for s-wave pairing the resonance is precluded by the fact that $\Pi(\omega = 0)$ becomes finite and negative in the superconducting state. This negative term overshadows the positive ω^2 term in $\Pi(\omega)$ in such a way that for all frequencies below 2Δ, $\Pi(\omega) < 0$ and hence the resonance simply does not exist. That $\Pi(\omega = 0) < 0$ in s-wave superconductors can be easily explained: a negative $\Pi(0)$ implies that the spin correlation length decreases as the system becomes superconducting. This is exactly what one should expect as s-wave pairing involves fermions both from different magnetic sub-lattices as well as from the same sub-lattice. The pairing of fermions from the same sub-lattice into a spin-singlet state obviously reduces the antiferromagnetic correlation length. The d-wave pairing, on the contrary, only involves fermions from different sub-lattices, and $\Pi(\omega = 0) = 0$.

We also comment on the dispersion of the resonance peak. In (22.71) we assumed that Δ_s is a constant. In fact, Δ_s depends on $\mathbf{q}$ since for any given $\mathbf{q}$, $\Delta_s^2 \propto \Delta(\mathbf{q})$ where $\Delta(\mathbf{q})$ is a d-wave gap at the points at the Fermi surface which are connected by $\mathbf{q}$. In particular, Δ_s should vanish at $\mathbf{q} = \mathbf{Q}_{\min}$ which connects the nodal points. This effect accounts for the "negative" dispersion of the resonance peak [129,134]. The latter certainly overshadows the positive dispersion due to $(\mathbf{q} - \mathbf{Q})^2$ term for $\mathbf{q}$ close to $\mathbf{Q}_{\min}$ and may do so even for $\mathbf{q}$ near $\mathbf{Q}$ if the correlation length is not large enough. This effect is, however, not a part of the quantum-critical description (it should become progressively less relevant for $\mathbf{q} \neq \mathbf{Q}_{min}$ when ξ increases), and we ignore it in the subsequent analysis. Note, however, that the negative dispersion of the peak implies that the peak exists only for a small range of momenta between $\mathbf{Q}$ and $\mathbf{Q}_{\min}$. In optimally doped cuprates, $\mathbf{Q}_{\min} \approx (0.8\pi, 0.8\pi)$ [136,137], and the momentum range for the peak does not exceed 4% of the Brillouin zone. The actual $\mathbf{q}$ region where the peak is observable is even smaller as the intensity of the peak also decreases when $\mathbf{q}$ approaches $\mathbf{Q}_{\min}$. The smallness of the $\mathbf{q}$-range for the peak accounts for small overall spectral intensity $I_0 = \int S(\mathbf{q}, \omega) d^2q d\omega/(2\pi)^3$ that turns out to be substantially smaller than $S(S+1)/3 = 1/4$. Still, at $\mathbf{Q}$, the intensity of the peak is not small (experimentally, $\int S(\mathbf{Q}, \omega) d\omega \sim 1.5$ in optimally doped YBCO [138,139]), and we have verified that for the frequencies that we consider below, the typical $\mathbf{q} - \mathbf{Q}$ that account for the feedback on the fermions are well within the $\mathbf{q}$ range between $\mathbf{Q}$ and $\mathbf{Q}_{\min}$. In other words, the small overall intensity of the resonance peak does not preclude strong feedback effects from the resonance peak on fermionic variables.

For completeness, in Figs. 22.13 and 22.14 we present results for the fermionic self-energy and the pairing vertex for the smallest T. We see that the real parts of $\Phi(\omega)$ and $\Delta(\omega)$ are finite at $\omega = 0$ as should be the case in the superconducting state. The imaginary parts of $\Phi(\omega)$ and $\Sigma(\omega)$ (and of $\Delta(\omega)$ and $Z(\omega)$) vanish at small frequencies and appear only above the threshold frequency that is precisely $\Delta + \Delta_s$. Furthermore, all variables have a complex internal structure at large frequencies. In the next section we discuss the physical origin of the threshold at $\Delta + \Delta_s$ and also show that one can extract 3Δ from the derivative of $\Sigma''(\omega)$.

Few words about the numbers. For $\lambda = 2$, $\Delta \approx 0.3\bar{\omega}$ and $\Delta_s \approx 0.2\bar{\omega}$, i.e. Δ and Δ_s are comparable to each other. For $\lambda \gg 1$ a numerical solution of the

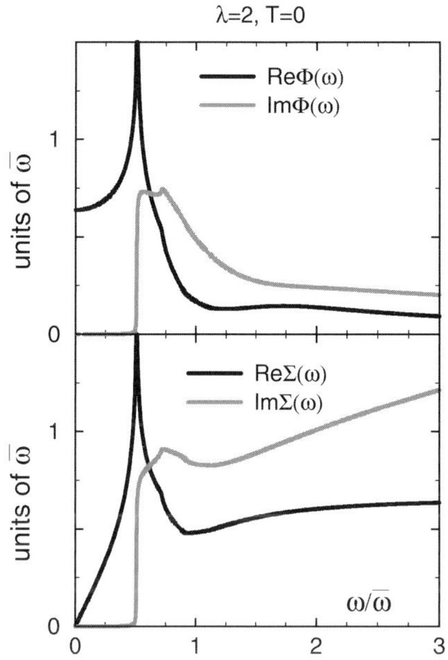

Fig. 22.13. The real and imaginary parts of the fermionic self-energy $\Sigma(\omega)$ and the pairing vertex $\Phi(\omega)$ for $\lambda = 2$ and the lowest T. The results are from [135]

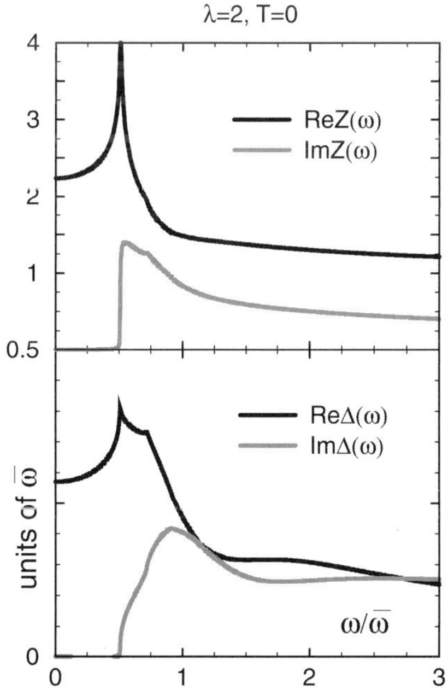

Fig. 22.14. The real and imaginary parts of the effective gap $\Delta(\omega)$ and the quasiparticle renormalization factor $Z(\omega)$ for $\lambda = 2$ and the lowest T. The results are from [135]

Eliashberg equations leads to $\Delta \sim 2T_{\mathrm{cr}} \sim 0.35\bar{\omega}$, and $\Delta_s \sim 0.25\bar{\omega}/\lambda \ll \Delta$.

22.5 Fingerprints of Spin Fermion Pairing

In this Section, we discuss the extent to which the "fingerprints" of spin-mediated pairing can be extracted from experiments on materials that are candidates for a magnetically-mediated superconductivity. Due to strong spin–fermion coupling, there is unusually strong feedback from spin excitations on fermions, specific to d-wave superconductors. The origin of this feedback is the emergence of a propagating collective spin bosonic mode below T_{cr}. This mode is present for any coupling strength, and its gap Δ_s is smaller than the minimum energy $\sim 2\Delta$ that is necessary to break a Cooper pair. In the vicinity of the antiferromagnetic phase, $\Delta_s \propto \xi^{-1}$ where ξ is the magnetic correlation length. We show that this propagating spin mode changes the onset frequency for single particle scattering, gives rise to the "peak-dip-hump" features in the quasiparticle spectral function, the "dip-peak" features in tunneling SIS and SIN conductances, and to singularities and fine structures in the optical conductivity. In Sect. 22.6, we apply these results to cuprate superconductors and argue that (i) these features have been observed [136,137,140–145], (ii) ARPES [136,137,140,141], tunneling [142,143], and conductivity data [144,145] are consistent with each other, and (iii) the value of Δ_s extracted from these various experiments coincides with the resonance frequency measured directly in neutron scattering experiments [138,139,146].

22.5.1 The Physical Origin of the Effect

The physical effect that accounts for dips and humps in the density of states and spectral function of cuprates is not new and is known for conventional s-wave superconductors as the Holstein effect [5,147,148]. Consider a clean s-wave superconductor, and suppose that the residual interaction between fermions occurs via the exchange of an Einstein phonon. Assume for simplicity that the fully

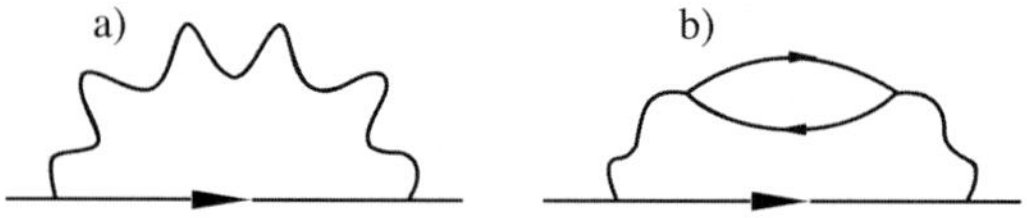

Fig. 22.15. (**a**) The exchange diagram for a boson mediated interaction. The *solid line* stands for a propagating fermion. The *wiggly line* is a phonon propagator in the case of electron–phonon interaction, and a magnon line in the case of a spin-fluctuation mediated interaction. (**b**) The lowest order diagram for the fermionic self energy due to a direct four fermion interaction, also represented by a *wiggly line*. The figure is taken from [133]

renormalized electron phonon coupling is some constant g_{ep}, and that the phonon propagator $D(q,\omega)$ is independent of momentum q and has a single pole at a phonon frequency ω_0 (the Holstein model [147–149]). Phonon exchange gives rise to a fermionic self-energy (see Fig. 22.15(a))

$$\Sigma(i\omega_m) = -g_{ep}^2 T \sum_n \int \frac{d^d k}{(2\pi)^d} G_k(i\omega_n) D(i\omega_m - i\omega_n)\,, \tag{22.75}$$

which is a convolution of $D(\omega) = 1/(\omega_0^2 - (\omega + i0^+)^2)$ with the full fermionic propagator $G_k(\omega)$, which in a superconductor is given by (22.9):

$$G_k(\omega) = \frac{\tilde{\Sigma}(\omega) + \varepsilon_k}{\tilde{\Sigma}(\omega)^2 - \Phi^2(\omega) - \varepsilon_k^2}\,. \tag{22.76}$$

As before, $\Phi(\omega)$ is the pairing vertex, and ε_k is the band dispersion of the fermions. At $T = 0$ both $\Sigma''(\omega)$ and $\Phi''(\omega)$ obviously vanish for $\omega \le \Delta$. This implies that the fermionic spectral function $A_k(\omega) = \left|G''_k(\omega)\right|/\pi$ for particles at the Fermi surface ($\mathbf{k} = \mathbf{k}_F$) has a δ-function peak at $\omega = \Delta$, i.e. Δ is a sharp gap at zero temperature. The fermionic density of states in a superconductor

$$N(\omega) = N_0\,\mathrm{Im}\left[\frac{\tilde{\Sigma}(\omega)}{(\Phi^2(\omega) - \tilde{\Sigma}^2(\omega))^{1/2}}\right] \tag{22.77}$$

vanishes for $\omega < \Delta$ and has a square root singularity $N(\omega) \propto (\omega - \Delta)^{-1/2}$ for frequencies above the gap (N_0 is the normal state density of states).

The onset of the imaginary part of the self-energy, (22.75), can be easily obtained by using the spectral representation for fermionic and bosonic propagators in (22.75) and re-expressing the momentum integration in terms of an integration over ε_k. At $T = 0$ we obtain

$$\Sigma''(\omega > 0) \propto \int_0^\omega d\omega' N(\omega') D''(\omega - \omega')\,. \tag{22.78}$$

Since for positive frequencies, $D''(\omega) = (\pi D_0/2\omega_0) \times \delta(\omega - \omega_0)$, the frequency integration is elementary and yields

$$\Sigma''(\omega > 0) \propto N(\omega - \omega_0)\,. \tag{22.79}$$

We see that the single particle scattering rate is directly proportional to the density of states shifted by the phonon frequency. Clearly, the imaginary part of the fermionic self-energy emerges only when ω exceeds the threshold

$$\Omega_t \equiv \Delta + \omega_0\,, \tag{22.80}$$

which is the sum of the superconducting gap and the phonon frequency. Right above this threshold, $\tilde{\Sigma}''(\omega) \propto (\omega - \Omega_t)^{-1/2}$. By the Kramers–Kronig relation, this non-analyticity causes a square root divergence of $\tilde{\Sigma}'(\omega)$ at $\omega < \Omega_t$. Combining the two results, we find that near the threshold, $\tilde{\Sigma}(\omega) = A + C/\sqrt{\Omega_t - \omega}$ where A and C are real numbers. By the same reasoning, the pairing vertex $\Phi(\omega)$ also possesses a square root singularity at Ω_t. Near $\omega = \Omega_t$, $\Phi(\omega) = B + C/\sqrt{\Omega_t - \omega}$ with real B. Since $\Omega_t > \Delta$, we have $A > B$.

The singularity in the fermionic self-energy gives rise to an extra dip-hump structure of the fermionic spectral function at $\mathbf{k} = \mathbf{k}_F$. Below Ω_t, the spectral function is zero except for $\omega = \Delta$, where it has a δ-functional peak. Immediately above Ω_t, $A(\omega) \propto \mathrm{Im}(\tilde{\Sigma}(\omega)/(\tilde{\Sigma}^2(\omega) - \Phi^2(\omega)))$ takes the form $A(\omega) \propto (\omega - \Omega_t)^{1/2}$. At larger frequencies, $A(\omega)$ passes through a maximum, and eventually vanishes. Adding a small damping introduced by either impurities or finite temperatures, one obtains the spectral function with a peak at $\omega = \Delta$, a dip at $\omega \approx \Omega_t$, and a hump at a somewhat larger frequency. This behavior is shown schematically in Fig. 22.16.

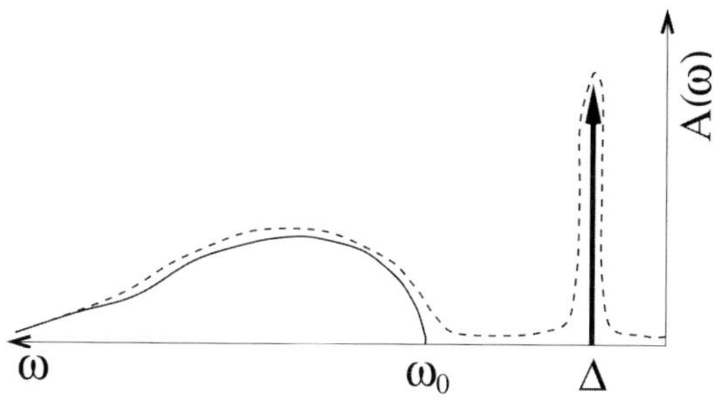

Fig. 22.16. The schematic form of the quasiparticle spectral function in an s-wave superconductor. *Solid line*: $T = 0$. *Dashed line*: at finite T. $\Omega_t = \Delta + \omega_0$ (from [133])

The singularities in $\tilde{\Sigma}(\omega)$ and $\Phi(\omega)$ affect other observables such as the fermionic DOS, optical conductivity, Raman response, and the SIS tunneling dynamical conductance [148,150].

For a more complex phonon propagator, which depends on both frequency and momentum, the singularities in the fermionic self-energy and other observables are weaker and may only show up in the derivatives over frequency [4,151]. Still, the opening of the new relaxational channel at Ω_t gives rise to singularities in the electronic properties of a phonon-mediated s-wave superconductor.

22.5.2 Similarities and Discrepancies Between d-Wave and s-Wave-Superconductors

As we already discussed, for magnetically mediated d-wave superconductivity, spin fluctuations play the role of phonons. Below T_c, spin excitations are propagating, magnon-like modes with the gap Δ_s. This Δ_s obviously plays the same role as ω_0 for phonons, and hence we expect that for spin-mediated pairing, the spectral function should display a peak-dip-hump structure as well. Furthermore, we will demonstrate below that for observables such as the DOS, Raman intensity and the optical conductivity, which measure the response averaged over the Fermi surface, the angular dependence of the d-wave gap $\Delta(\theta) \propto \cos(2\theta)$ softens the singularities, but does not wash them out over a finite frequency range. Indeed, we will find that the positions of the singularities are not determined by some averaged gap amplitude but by the maximum value of the d-wave gap, $\Delta_{max} = \Delta$, i.e. the Holstein effect is still present for a d-wave superconductor.

Despite many similarities, the feedback effects for s-wave and d-wave superconductors are not equivalent as we now demonstrate. The point is that for s-wave superconductors, the exchange process shown in Fig. 22.15(a) is not the only possible source for the fermionic decay; there exists another process, shown in Fig. 22.15(b), in which a fermion decays into three other fermions. This process is due to a residual four-fermion interaction [148,150]. One can easily make sure that this second process also gives rise to the fermionic decay when the external ω exceeds a minimum energy of 3Δ, necessary to pull all three intermediate particles out of the condensate of Cooper pairs. At the threshold, the fermionic spectral function is non-analytic, much like that found at $\Delta + \omega_0$. This implies that in s-wave superconductors, there are two physically distinct singularities, at $\Delta + \omega_0$ and at 3Δ, which come from *different* processes and therefore are independent of each other. Which of the two threshold frequencies is larger depends on the strength of the coupling and on the shape of the phonon density of states. At weak coupling, ω_0 is exponentially larger than Δ, hence the 3Δ threshold comes first. At strong coupling, ω_0 and Δ are comparable, but calculations within the Eliashberg formalism show that for real materials (e.g. for lead or niobium), still $3\Delta < \Delta + \omega_0$ [152]. This result is fully consistent with the photoemission data for these materials [153].

For magnetically mediated d-wave superconductors the situation is different. As we discussed in Sect. 22.2, in the one-band model for cuprates, which we adopt, the underlying interaction is a Hubbard-type four-fermion interaction. The introduction of a spin fluctuation as an extra degree of freedom is just a way to account for the fact that there exists a particular interaction channel, where the effective interaction between fermions is the strongest due to the proximity to a magnetic instability. This implies that the spin fluctuation propagator is made out of particle–hole bubbles like those in Fig. 22.15(b). Then, to the lowest order in the interaction, the fermionic self-energy is given by the diagram in Fig. 22.15(b). Higher-order terms convert a particle–

hole bubble in Fig. 22.15(b). into a wiggly line, and transform this diagram into the one in Fig. 22.15(a). Clearly then, inclusion of both diagrams would be double counting, i.e. there is only a *single* process which gives rise to the threshold in the fermionic self-energy. We will show that this process generates two singularities: at 3Δ, and at $\Delta + \Delta_s < 3\Delta$. The fact that this is an internal effect implies that Δ_s *depends* on Δ. The experimental verification of this dependence can then be considered as a "fingerprint" of the spin-fluctuation mechanism. Furthermore, as the singularities at 3Δ and $\Delta + \Delta_s$ are due to the same interaction, their relative intensity is another gauge of the magnetic mechanism for the pairing. We will argue below that some experiments on cuprates, particularly measurements of the optical conductivity [145], allow one to detect both singularities, and that their calculated relative intensity is consistent with the data.

We now discuss separately the behavior of the electronic spectral function, the density of states, SIS tunneling, the Raman intensity and the optical conductivity. To account for all features associated with d-wave pairing, we will keep the momentum dependence of the fermionic self-energy and the pairing vertex on momenta along the Fermi surface, although this dependence is indeed weak near hot spots. For simplicity, we assume a circular Fermi surface. In this situation, the k-dependence of the self-energy and the pairing vertex reduces to the angular dependence, i.e. $\Sigma = \Sigma(\theta, \omega)$ and $\Phi = \Phi(\theta, \omega)$.

22.5.3 The Spectral Function

We first consider the spectral function $A_{\mathbf{k}}(\omega) = (1/\pi)|G''_{\mathbf{k}}(\omega)|$. In the superconducting state, for quasiparticles near the Fermi surface

$$A_{\mathbf{k}}(\omega > 0) = \frac{1}{\pi}\,\mathrm{Im} \times \left[\frac{\omega + \Sigma(\theta, \omega) + \varepsilon_{\mathbf{k}}}{(\omega + \Sigma(\theta, \omega))^2 - \Phi^2(\theta, \omega) - \varepsilon_{\mathbf{k}}^2}\right]. \tag{22.81}$$

By definition, $A_{\mathbf{k}}(-\omega) = A_{\mathbf{k}}(\omega)$.

In a Fermi gas with d-wave pairing, $\Sigma(\theta, \omega) = 0$, and $\Phi(\theta, \omega) = \Delta(\theta) \propto \cos(2\theta)$. The spectral function then has a δ−function peak at $\omega = (\Delta^2(\theta) + \varepsilon_{\mathbf{k}}^2)^{1/2}$. It is obvious, but essential for comparison with the strong coupling case, that the peak disperses with **k** and that far away from the Fermi surface one recovers normal state dispersion.

For strong coupling we consider the spectral function for fermions located near hot spots, $\theta = \theta_{hs}$, where the gap $\Delta(\theta)$ (defined as a solution of $\tilde{\Sigma}'(\omega = \Delta, \theta_{hs}) = \Phi'(\omega = \Delta, \theta_{hs})$) is maximum. As discussed above, we expect the spectral function to possess a peak at $\omega = \Delta$ and a singularity at

$$\omega = \Omega_{\mathrm{t}} = \Delta + \Delta_s\,. \tag{22.82}$$

The behavior of $A(\omega)$ near the singularity is robust and can be obtained without a precise knowledge of the frequency dependence of $\tilde{\Sigma}(\omega)$ and $\Phi(\omega)$. All we need to know is that near $\omega = \Delta$, $\tilde{\Sigma}^2(\omega, \theta_{hs}) - \Phi^2(\omega, \theta_{hs}) \propto \omega - \Delta$. Substituting this form into (22.67) and converting to the real axis using the spectral representation, we obtain for $\omega = \Omega_{\mathrm{t}} + \epsilon$

$$\tilde{\Sigma}''(\omega) \propto \int_0^{\epsilon} \frac{dx}{(x(\epsilon - x))^{1/2}}\,. \tag{22.83}$$

This integral is the same as in (22.73), hence $\tilde{\Sigma}''$ undergoes a finite jump at $\omega = \Omega_{\mathrm{t}}$, just as the spin polarization operator does at $\omega = 2\Delta$. By the Kramers–Kronig relation, this jump gives rise to a logarithmic divergence of $\tilde{\Sigma}'$. The same singular behavior holds for the pairing vertex $\Phi(\omega)$, with exactly the same prefactor in front of the logarithm. The last result implies that $\tilde{\Sigma}(\omega) - \Phi(\omega)$ is non-singular at $\omega = \Omega_{\mathrm{t}}$. Substituting these results into (22.82), we find that the spectral function $A(\omega)$ behaves at $\omega > \Omega_{\mathrm{t}}$ as $1/\log^2(\omega - \Omega_{\mathrm{t}})$, i.e. almost discontinuously. Obviously, at a small but finite T, the spectral function should have a dip very near $\omega = \Omega_{\mathrm{t}}$, and a hump at a somewhat higher frequency. Note that the logarithmic divergence of Σ and Φ is the consequence of the momentum dependence of the dynamic spin susceptibility. Without it, i.e. for a flat susceptibility, we would obtain the square root divergencies, like for interaction with an Einstein phonon.

In Fig. 22.17 we present the result for $A(\omega)$ obtained from a solution of the set of three coupled Eliashberg equations at $T \ll T_c$ [133]. This solu-

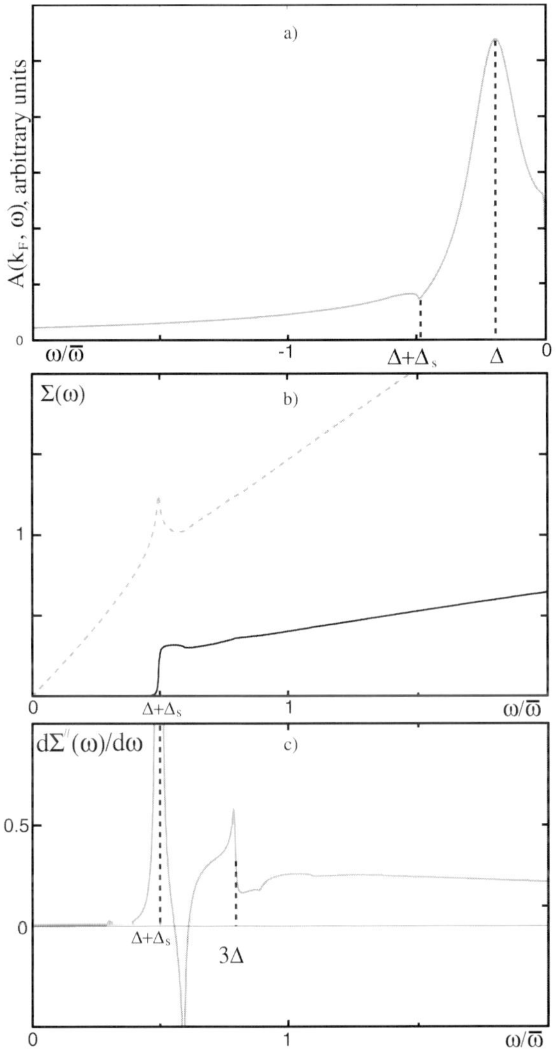

Fig. 22.17. *Upper panel*: the quasiparticle spectral function determined by solving the coupled Eliashberg equations for $\lambda = 1$. The peak-dip-hump structure of $A(\omega)$ is clearly visible but not dramatic. *Middle panel*: real and imaginary parts of the fermionic self-energy (*dashed and solid lines*, respectively). *Lower panel*: the frequency derivative of $\Sigma''(\omega)$. The extra structure at 3Δ is clearly visible. The figure is taken from [133]

tion is consistent with our analytical estimate. We clearly see that the fermionic spectral function has a peak-dip-hump structure, and the peak-dip distance equals Δ_s. We also see in Fig. 22.17 that the fermionic self-energy is non-analytic at $\omega = 3\Delta$. As we discussed above, this last non-analyticity originates in the non-analyticity of the dynamical spin susceptibility at $\omega = 2\Delta$.

Another "fingerprint" of the spin-fluctuation scattering can be found by studying the evolution of the spectral function as one moves away from the Fermi surface. The argument here goes as follows: at strong coupling, where $\Delta \geq \omega_{sf}$, probing the fermionic spectral function at frequencies progressively larger than Δ, one eventually probes the normal state fermionic self-energy at $\omega \gg \omega_{sf}$. Substituting the self-energy (22.44) into the fermionic propagator, we find that up to $\omega \sim \bar{\omega}$, the spectral function in the normal state does not have a quasiparticle peak at $\omega = \varepsilon_{\mathbf{k}}$. Instead, it only displays a broad maximum at $\omega = \varepsilon_{\mathbf{k}}^2/\bar{\omega}$ (see Fig. 22.8). The absence of a quasiparticle peak in the normal state implies that the sharp quasiparticle peak that we found at $\omega = \Delta$ for momenta at the Fermi surface cannot simply disperse with $\mathbf{k}$, as it does for noninteracting fermions with a d-wave gap. Specifically, the quasiparticle peak cannot move further in energy than $\Delta + \Delta_s$ since at larger frequencies, spin scattering rapidly increases, and the fermionic spectral function should display roughly the same non-Fermi-liquid behavior as in the normal state.

In Fig. 22.18(a) we present plots for the spectral function as the momentum moves away from the Fermi surface. We see the behavior we just described: the quasiparticle peak does not move further than $\Delta + \Delta_s$. Instead, when $\mathbf{k} - \mathbf{k}_F$ increases, it gets pinned at $\Delta + \Delta_s$ and gradually looses its spectral weight. At the same time, the hump disperses with $\mathbf{k}$ and for frequencies larger than $\Delta + \Delta_s$ gradually transforms into a broad maximum at $\omega = \varepsilon_{\mathbf{k}}^2/\bar{\omega}$. The positions of the peak and the dip versus $\mathbf{k} - \mathbf{k}_F$ are presented in Fig. 22.18(b).

22.5.4 The Density of States

The quasiparticle density of states, $N(\omega)$, is the momentum integral of the spectral function:

$$N(\omega) = \int \frac{d^2\mathbf{k}}{4\pi^2} A_{\mathbf{k}}(\omega)\,. \qquad (22.84)$$

Substituting $A_{\mathbf{k}}(\omega)$ from (22.82) and integrating over $\varepsilon_{\mathbf{k}}$, one obtains

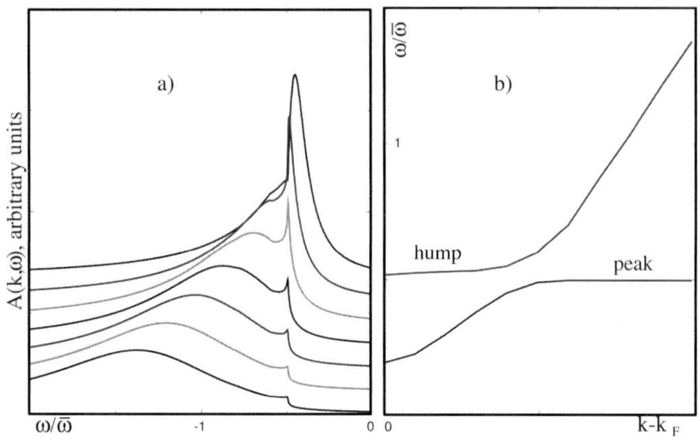

Fig. 22.18. (**a**) Frequency dependence of the spectral function in the superconducting state for different ϵ_k. The curve at the bottom has a highest ϵ_k. No coherent quasiparticle peak occurs for energies larger than $\Delta+\Delta_s$. Instead, the spectral function displays a broad maximum, similar to that in the normal state. (**b**) (From [133])

$$N(\omega) \propto \mathrm{Im} \int_0^{2\pi} \mathrm{d}\theta \, \frac{\omega + \Sigma(\theta,\omega)}{(\Phi^2(\theta,\omega) - (\omega+\Sigma(\theta,\omega))^2)^{1/2}} . \quad (22.85)$$

We first consider $N(\omega)$ in a d-wave gas, and then discuss strong coupling effects. In a d-wave gas, $\Sigma = 0$ and $\Delta_{\mathbf{k}} = \Delta\cos(2\theta)$. Integrating in (22.85) over θ we obtain [154,155]

$$N(\omega) = N_0 \, \mathrm{Re}\left[\frac{\omega}{2\pi}\int_0^{2\pi} \frac{\mathrm{d}\theta}{\sqrt{\omega^2 - \Delta^2\cos^2(2\theta)}}\right] = \frac{2N_0}{\pi}\begin{cases} K(\Delta/\omega) & , \text{for } \omega > \Delta \\ (\omega/\Delta)K(\omega/\Delta) & , \text{for } \omega < \Delta \end{cases} , \quad (22.86)$$

where $K(x)$ is the elliptic integral of first kind. We see that $N(\omega) \sim \omega$ for $\omega \ll \Delta$ and diverges logarithmically as $(1/\pi)\ln(8\Delta/|\Delta-\omega|)$ for $\omega \approx \Delta$. At larger frequencies, $N(\omega)$ gradually decreases towards the frequency independent, normal state value of the DOS, which we have normalized to unity. The plot of $N(\omega)$ in a d-wave BCS superconductor is presented in Fig. 22.19.

For comparison, in an s-wave superconductor, the DOS vanishes at $\omega < \Delta$ and diverges as $N(\omega) \propto (\omega-\Delta)^{-1/2}$ at $\omega \geq \Delta$. We see that a d-wave superconductor is different in that (i) the DOS is finite down to the smallest frequencies, and (ii) the singularity at $\omega = \Delta$ is weaker (logarithmic). Still, however, $N(\omega)$ is singular only at a frequency which equals to the largest value of the d-wave gap. This illustrates a point made earlier: the angular dependence of the d-wave gap reduces the strength of the singularity at $\omega = \Delta_{\max}(\theta)$, but does not wash it out over a finite frequency range.

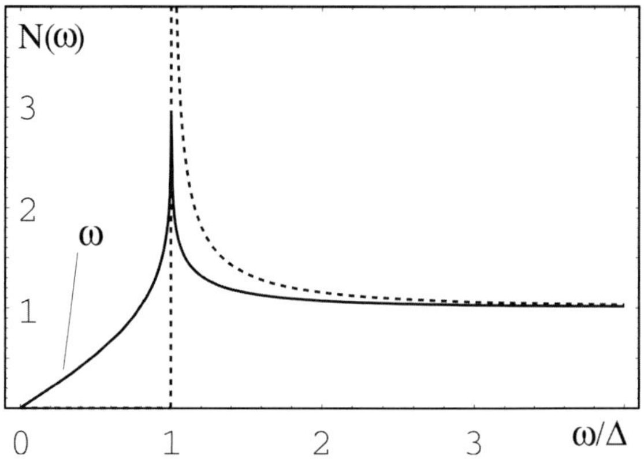

Fig. 22.19. Density of states of a noninteracting Fermi gas with a d-wave gap (*solid line*) and with an s-wave gap (*dashed line*). (From [133])

We now turn to strong coupling. We first demonstrate that the DOS possesses extra peak-dip features, associated with the singularities in $\tilde{\Sigma}(\omega)$ and $\Phi(\omega)$ at $\omega = \Omega_t = \Delta+\Delta_s$. An analytical approach proceeds as follows [156]. Consider first a case when the gap is totally flat near a hot spot, i.e. $\Delta(\theta) = \Delta$. At $\omega = \Omega_t$, both $\tilde{\Sigma}(\omega)$ and $\Phi(\omega)$ diverge logarithmically. On substituting these forms into (22.85), we find that $N(\omega)$ has a logarithmic singularity:

$$N_{\mathrm{sing}}(\omega) \propto \left(\log\frac{1}{|\omega-\Omega_t|}\right)^{1/2} . \quad (22.87)$$

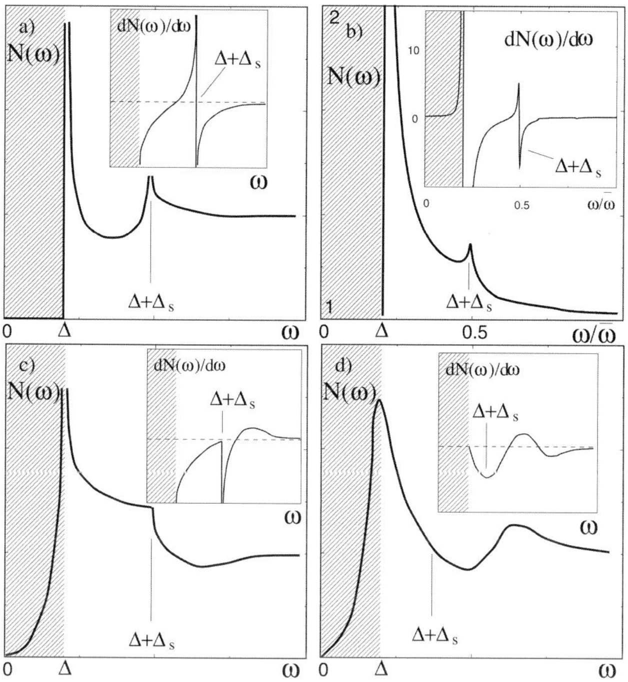

Fig. 22.20. (**a**) The behavior of the SIN tunneling conductance (i.e., DOS) in a strongly coupled d-wave superconductor. *Main pictures*: $N(\omega)$, *insets*: $dN(\omega)/d\omega$. (**a**) The schematic behavior of the DOS for a flat gap. (**b**) The solution of the Eliashberg-type equations for a flat gap. The *shaded regions* are the ones in which the flat gap approximation is incorrect as the physics is dominated by nodal quasiparticles. (**c**) The schematic behavior of $N(\omega)$ for the quadratic variation of the gap near its maxima. (**d**) The expected behavior of the DOS in a real situation when singularities are softened out by finite T or impurity scattering. The position of $\Delta + \Delta_s$ roughly corresponds to a minimum of $dN(\omega)/d\omega$. The figure is taken from [133]

This singularity gives rise to a strong divergence of $dN(\omega)/d\omega$ at $\omega = \Omega_t$, a behavior schematically shown in Fig. 22.20(a). In part (b) of this figure we present the result for $N(\omega)$ obtained by the solution of the Eliashberg-type equations (22.63)–(22.69). A small but finite temperature was used to smear out divergences. We recall that the Eliashberg set does not include the angular dependence of the gap near hot spots, and hence the numerical result for the DOS in Fig. 22.20(b) should be compared with Fig. 22.20(a). We clearly see that $N(\omega)$ has a second peak at $\omega = \Omega_t$. This peak strongly affects the frequency derivative of $N(\omega)$ which become singular near Ω_t.

The relatively small magnitude of the singularity in $N(\omega)$ is a consequence of the linearization of the fermionic dispersion near the Fermi surface. For an actual ϵ_k chosen to fit ARPES data [129], nonlinearities in the fermionic dispersion occur at energies comparable to Ω_t. This is due to the fact that hot spots are located close to $(0, \pi)$ and related points at which the Fermi velocity vanishes. As a consequence, the momentum integration in the spectral function should have a less pronounced smearing effect than found in our calculations, and the frequency dependence of $N(\omega)$ should more resemble that of $A(\omega)$ for momenta where the gap is at maximum.

For a momentum dependent gap $\Delta(\theta)$, the behavior of fermions near hot spots is the same as when the gap is flat, but now Ω_t depends on θ as both Δ and Δ_s vary as one moves away from a hot spot. The variation of Δ is obvious, the variation of Δ_s is due to the fact that this frequency scales as $\Delta^{1/2}$. Since both Δ and Δ_s are maximal at a hot spot, we can model the momentum dependence by replacing

$$\Omega_t \to \Omega_t - a\tilde{\theta}^2 , \qquad (22.88)$$

where $\tilde{\theta} = \theta - \theta_{hs}$, and $a > 0$. The singular pieces of the self-energy and the pairing vertex then behave as $\log(\Omega_t - \omega - a\tilde{\theta}^2)^{-1}$. Substituting these forms into (22.85) and using the fact that $\tilde{\Sigma}(\omega) - \Phi(\omega) \approx$ const at $\omega \approx \Omega_t$, we obtain

$$N_{\text{sing}}(\omega) \propto \text{Re} \int d\tilde{\theta} \left[\log(\Omega_t - \omega - a\tilde{\theta}^2)^{-1}\right]^{-1/2} . \qquad (22.89)$$

A straightforward analysis of the integral shows that now $N(\omega)$ displays a one-sided non-analyticity at $\omega = \Omega_t$:

$$N_{\rm sing}(\omega) = -B\Theta(\omega - \Omega_t)\left(\frac{\omega - \Omega_t}{|\log(\omega - \Omega_t)|}\right)^{1/2}, \tag{22.90}$$

where $B > 0$, and $\Theta(x) = 1$ for $x > 0$, and $\Theta(x) = 0$ for $x < 0$. This non-analyticity gives rise to a cusp in $N(\omega)$ right above Ω_t, and one-sided square root divergence of the frequency derivative of the DOS. This behavior is shown schematically in Fig. 22.20(c). Comparing this behavior with that shown in Fig. 22.20(a) for a flat gap, we observe that the angular dependence of the gap predominantly affects the form of $N(\omega)$ at $\omega \leq \Omega_t$. At these frequencies, the angular variation of the gap completely eliminates the singularity in $N(\omega)$. At the same time, above Ω_t, the angular dependence of the gap softens the singularity, but, still, the DOS sharply drops above Ω_t in such a way that the derivative of the DOS diverges on approaching Ω_t from above. We see again that in a d-wave superconductor, the singularity in the DOS is softened by the angular dependence of the gap, but still persists at a particular frequency related to the maximum value of the gap. This point is essential as it enables us to read off the maximum gap value directly from the experimental data without any "deconvolution" of momentum averages.

For real materials, in which singularities are removed by e.g., impurity scattering, the location of Ω_t is best described as a point where the frequency derivative of the DOS passes through a minimum (Fig. 22.20(d)). The singularity in $dN(\omega)/d\omega$ at Ω_t, and the dip-hump structure of $N(\omega)$ at $\omega \geq \Omega_t$ are additional "fingerprints" of the spin-fluctuation mechanism in the single particle response.

22.5.5 SIS Tunneling

Measurements of the dynamical conductance dI/dV through a superconductor-insulator-superconductor (SIS) junction offer another tool to search for the fingerprints of the spin-fluctuation mechanism. The conductance through this junction is the derivative over voltage of the convolution of the two DOS [152]: $dI/dV \propto S(\omega)$ where

$$S(\omega) = N_0^{-2} \int_0^{\omega} d\Omega N(\omega - \Omega)\, \partial_\Omega N(\Omega)\,. \tag{22.91}$$

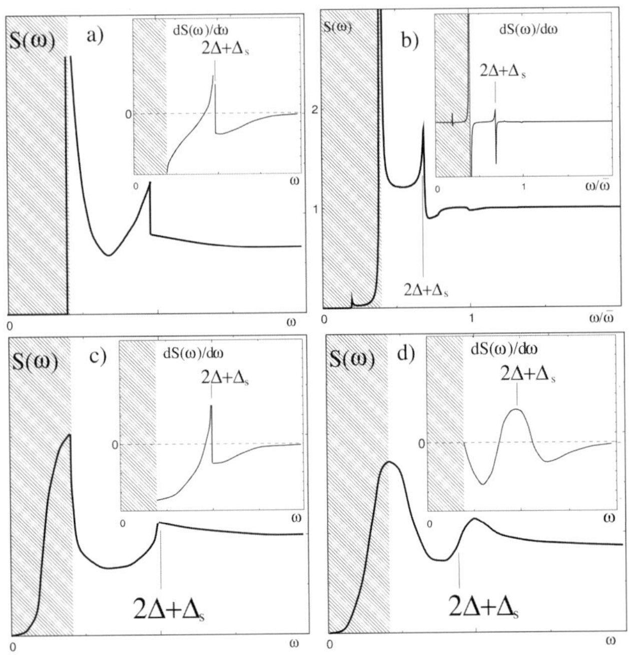

Fig. 22.21. (a) The schematic behavior of the SIS tunneling conductance, $S(\omega)$, in a strongly coupled d-wave superconductor. *Main pictures*: $S(\omega)$, *insets*: $dS(\omega)/d\omega$. (a) The schematic behavior of $S(\omega)$ for a flat gap. (b) The solution of the Eliashberg-type equations for a flat gap using the DOS from the previous subsection. The shaded regions are the ones in which the flat gap approximation is incorrect as the physics is dominated by nodal quasiparticles. (c) The schematic behavior of $S(\omega)$ for a quadratic variation of the gap near its maxima. (d) The expected behavior of the SIS conductance in a real situation when singularities are softened out by finite T or by impurity scattering. $2\Delta + \Delta_s$ roughly corresponds to the maximum of $dS(\omega)/d\omega$. (From [133])

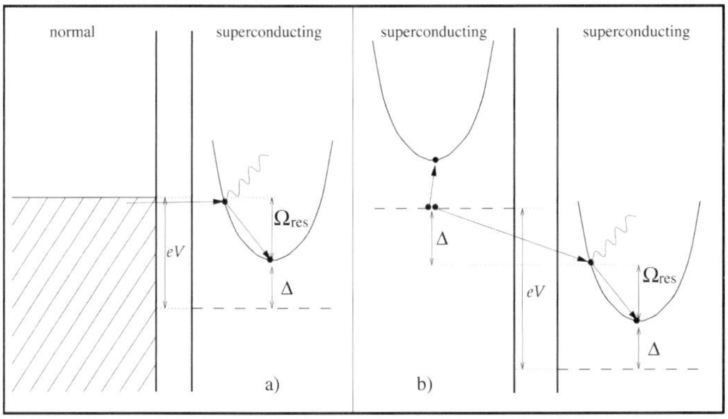

Fig. 22.22. The schematic diagram for the dip features in SIN and SIS tunneling conductances ((**a**) and (**b**), respectively). For SIN tunneling, which measures the fermionic DOS, the electron which tunnels from a normal metal can emit a propagating magnon if the voltage $eV = \Delta + \Delta_s$. After emitting a magnon, the electron falls to the bottom of the band. This leads to a sharp reduction of the current and produces a drop in dI/dV. For SIS tunneling, the physics is similar, but one first has to break an electron pair, which costs energy 2Δ (taken from [156])

The DOS in a d-wave gas is given in (22.86). Substituting this form into (22.91) and integrating over frequency we obtain the result presented in Fig. 22.21. At small ω, $S(\omega)$ is quadratic in ω [154]. This is an obvious consequence of the fact that the DOS is linear in ω. At $\omega = 2\Delta$, $S(\omega)$ undergoes a finite jump. This jump is related to the fact that near 2Δ, the integral over the two DOS includes the region around $\Omega = \Delta$ where both $N(\Omega)$ and $N(\omega - \Omega)$ are logarithmically singular, and $\partial_\Omega N(\Omega)$ diverges as $1/(\Omega - \Delta)$. The singular contribution to $S(\omega)$ from this region can be evaluated analytically and yields

$$\begin{aligned} S(\omega) &= -\frac{1}{\pi^2} P \int_{-\infty}^{\infty} \frac{dx \, \log|x|}{x + \omega - 2\Delta} \\ &= -\frac{1}{2} \mathrm{sign}(\omega - 2\Delta) \, . \end{aligned} \tag{22.92}$$

Observe that the amplitude of the jump in the SIS conductance is a universal number which does not depend on the value of Δ. At larger frequencies, $S(\omega)$ continuously goes down and eventually approaches a value of $S(\omega \to \infty) = 1$.

In the case of strong coupling one finds again that the quadratic behavior at low frequencies and the discontinuity at 2Δ survive at arbitrary coupling. Indeed, the quadratic behavior at low ω is just a consequence of the linearity of $N(\omega)$ at low frequencies. Therefore, just as we did for the density of states we concentrate on behavior above 2Δ that is sensitive to strong coupling effects.

Consider first how the singularity in $\tilde{\Sigma}(\omega)$ at Ω_t affects $S(\omega)$. From a physical perspective, we would expect a singularity in $S(\omega)$ at $\omega = \Delta + \Omega_t = 2\Delta + \Delta_s$. Indeed, to get a nonzero SIS conductance, one has to first break a Cooper pair, which costs an energy of 2Δ. After a pair is broken, one of the electrons becomes a quasiparticle in a superconductor and takes the energy Δ, while the other tunnels. If the tunneling voltage equals $\Delta + \Omega_t$, the electron which tunnels through a barrier has energy Ω_t, and can emit a spin excitation and fall to the bottom of the band (see Fig. 22.22). This behavior is responsible for the drop in dI/dV and is schematically shown in Fig. 22.21.

Consider this effect in more detail [133, 156]. We first note that $\omega = \Delta + \Omega_t$ is special for (22.91) because both $dN(\Omega)/d\Omega$ and $N(\omega - \Omega)$ diverge at the same energy, $\Omega = \Omega_t$. Substituting the general forms of $N(\omega)$ near $\omega = \Omega_t$ and $\omega = \Delta$, we obtain after simple manipulations that for a flat gap, $S(\omega)$ has a one-sided divergence at $\omega = \Omega_t + \Delta = 2\Delta + \Delta_s$.

$$S_{\text{sing}}(\epsilon) \propto \frac{\Theta(-\epsilon)}{\sqrt{-\epsilon}} \, , \tag{22.93}$$

where $\epsilon = \omega - (\Omega_t + \Delta)$. This obviously causes a divergence of the frequency derivative of $S(\omega)$ (i.e. of

d^2I/dV^2). This behavior is schematically shown in Fig. 22.21(a). In Fig. 22.21(b) we present the results for $S(\omega)$ obtained by integrating theoretical $N(\omega)$ from Fig. 22.20(b). We clearly see that $S(\omega)$ and its frequency derivative are singular at $\omega = 2\Delta + \Delta_s$, in agreement with the analytical prediction.

For a quadratic variation of the gap near the maxima, calculations similar to those for the SIN tunneling yield the result that $S(\omega)$ is continuous through $2\Delta + \Delta_s$, but its frequency derivative diverges as

$$\frac{\mathrm{d}S(\omega)}{\mathrm{d}\omega} \propto P\int_0^{\Delta} \frac{\mathrm{d}x}{(x|\log x|)^{1/2}(x-\epsilon)} \sim \frac{\Theta(-\epsilon)}{|\epsilon \log|\epsilon||^{1/2}} . \quad (22.94)$$

The singularity in the derivative implies that near $\epsilon = 0$

$$S(\epsilon) = S(0) - C\,\Theta(-\epsilon)\left(\frac{-\epsilon}{|\log(-\epsilon)|}\right)^{1/2} , \quad (22.95)$$

where $C > 0$. This behavior is schematically presented in Fig. 22.21(d). We again see that the angular dependence of the gap softens the strength of the singularity, but the singularity remains confined to a single frequency $\omega = 2\Delta + \Delta_s$.

In real materials, the singularity in $S(\omega)$ is softened and transforms into a dip slightly below $2\Delta+\Delta_s$, and a hump at a frequency larger than $2\Delta + \Delta_s$. The frequency $2\Delta + \Delta_s$ roughly corresponds to a maximum of the frequency derivative of the SIS conductance.

22.5.6 Optical Conductivity and Raman Response

Other phenomena sensitive to Ω_t are the optical conductivity, $\sigma(\omega)$ and the Raman response, $R(\omega)$. Both are proportional to the fully renormalized particle–hole polarization bubble, but with different signs attributed to the bubble composed of anomalous propagators. Specifically, after integrating in the particle–hole bubble over $\varepsilon_{\mathbf{k}}$, one obtains

$$R(\omega) = \mathrm{Im}\int \mathrm{d}\omega' \mathrm{d}\theta V^2(\theta)\Pi_r(\theta,\omega,\omega') , \quad (22.96)$$

$$\sigma(\omega) = \mathrm{Re}\left(\frac{i}{\omega + i\delta}\int \mathrm{d}\omega' \mathrm{d}\theta \Pi_\sigma(\theta,\omega,\omega')\right) ,$$

where $V(\theta)$ is a Raman vertex which depends on the scattering geometry [157], and

$$\Pi_{r,\sigma}(\theta,\omega,\omega') = \frac{\tilde{\Sigma}_+\tilde{\Sigma}_- + \alpha\Phi_+\Phi_- + D_+D_-}{D_+D_-(D_+ + D_-)} . \quad (22.97)$$

Here $\alpha = -1$ for Π_r, and $\alpha = 1$ for Π_σ. Also, $\tilde{\Sigma}_\pm = \tilde{\Sigma}(\omega_\pm)$ and $\Phi_\pm = \Phi(\omega_\pm)$, where $\omega_\pm = \omega' \pm \omega/2$. We also introduced $D_\pm = (\Phi_\pm^2 - \tilde{\Sigma}_\pm^2)^{1/2}$. Note that $\tilde{\Sigma}$ and Φ depend on ω and θ.

In a superconducting gas, the optical conductivity vanishes identically for any nonzero frequency due to the absence of a physical scattering between quasiparticles in a gas. The presence of a superconducting condensate, however, gives rise to a δ functional term in σ at $\omega = 0$: $\sigma(\omega) = \pi\delta(\omega)\int \mathrm{d}\theta \mathrm{d}\omega' \Pi_\sigma(\theta, 0, \omega')$. This behavior is typical for any BCS superconductor [158]. The behavior of $\sigma(\omega)$ for a d-wave gas with additional impurities, causing inelastic scattering, is more complex and has been discussed in [159].

The form of the Raman intensity depends on the scattering geometry. For the scattering in the B_{1g} channel, the Raman vertex has the same angular dependence as the d-wave gap, i.e. $V(\theta) \propto \cos(2\theta)$ [157, 160]. Straightforward computations then show that at low frequencies, $R(\omega) \propto \omega^3$ [160]. For a constant $V(\theta)$, we would have $R(\omega) \propto \omega$.

Near $\omega = 2\Delta$, the B_{1g} Raman intensity is singular. For this frequency, both D_+ and D_- vanish at $\omega' = 0$ and $\theta = 0$. This causes the integral for $R(\omega)$ to be divergent. The singular contribution to $R(\omega)$ can be obtained analytically by expanding in the integrand to leading order in ω' and in θ. Using the spectral representation, we then obtain, for $\omega = 2\Delta + \delta$ [157]

$$R(\omega) = \int_0^{\epsilon} \mathrm{d}\Omega \int \mathrm{d}\tilde{\theta}\, \frac{1}{\sqrt{\Omega + a\tilde{\theta}^2}\sqrt{\delta - \Omega + a\tilde{\theta}^2}} \times \frac{1}{(\sqrt{\Omega + a\tilde{\theta}^2} + \sqrt{\delta - \Omega + a\tilde{\theta}^2})} , \quad (22.98)$$

where, as before, $\tilde{\theta} = \theta - \theta_{hs}$ For a flat band ($a = 0$), $R(\omega) \propto Re[(\omega - 2\Delta)^{-1/2}]$. For $a \neq 0$, i.e. for a quadratic variation of the gap near its maximum, the 2D integration in (22.98) is elementary, and yields

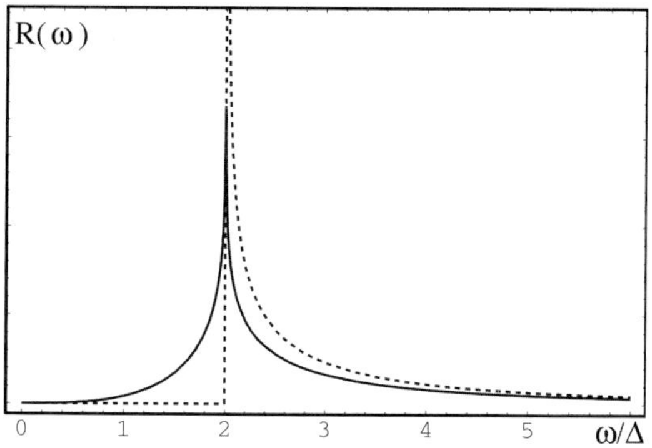

Fig. 22.23. The behavior of the Raman response in a BCS superconductor with a flat gap (*dashed line*), and for a quadratic variation of the gap near its maximum (*solid line*)

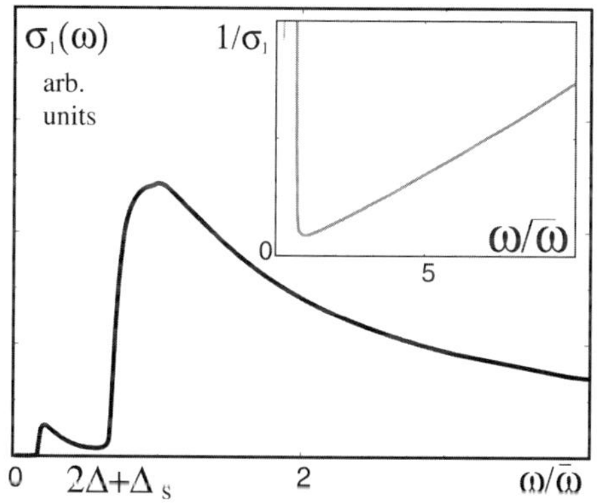

Fig. 22.24. The real part of the optical conductivity $\sigma_1(\omega)$ at the lowest T obtained using the self energy and the pairing vertex from the solution of the Eliashberg equations for $\lambda = 1$. The onset of the optical response is $\omega = 2\Delta + \Delta_s$. The contributions from nodal regions (not included in calculations) yield a nonzero conductivity at all ω. *Inset*: the behavior of the inverse conductivity vs frequency. (From [161])

$R(\omega) \propto \log|\omega - 2\Delta|$. At larger frequencies $R(\omega)$ gradually decreases.

The behavior of $R(\omega)$ in a d-wave gas is shown in Fig. 22.23. Observe that due to the interplay of numerical factors, the logarithmic singularity shows up only in the near vicinity of 2Δ, while at somewhat larger ω, the angular dependence of the gap becomes irrelevant, and $R(\omega)$ behaves as $(\omega - 2\Delta)^{1/2}$, i.e. the same as for a flat gap [155].

We now consider strong coupling effects. A nonzero fermionic self-energy mostly affects the optical conductivity for the simple reason that it becomes finite in the presence of spin scattering which can relax fermionic momenta. For a momentum-independent gap, a finite conductivity emerges above a sharp threshold. This threshold stems from the fact that at least one of the two fermions in the conductivity bubble should have a finite $\tilde{\Sigma}''$, i.e. its energy should be larger than Ω_t. Another fermion should be able to propagate, i.e. its energy should be larger than Δ. The combination of the two requirements yields the threshold for $\sigma(\omega > 0)$ at $2\Delta + \Omega_t$, i.e. at the same frequency where the SIS tunneling conductance is singular. One can easily demonstrate that for a flat gap, the conductivity behaves above the threshold as $\epsilon^{1/2}/\log^2\epsilon$, where $\epsilon = \omega - (\Delta + \Omega_t) = \omega - (2\Delta + \Delta_s)$. This singularity obviously causes a divergence of the first derivative of the conductivity at $\epsilon = +0$.

In Fig. 22.24 we show the result for the conductivity obtained by solving the set of coupled Eliashberg-type equations, (22.63)–(22.69) [161,162]. We see the expected singularity at $2\Delta + \Delta_s$. The insert shows the behavior of the inverse conductivity $1/\sigma(\omega)$ Observe that $1/\sigma(\omega)$ is linear in ω over a rather wide frequency range [162].

For a d-wave gap $\Delta = \Delta(\theta)$, the conductivity is finite for all frequencies simply because the angular integration in (22.97) involves the region near the nodes, where $\tilde{\Sigma}''$ is nonzero down to the lowest frequencies. Still, the conductivity is singular at $\Omega_t + \Delta$ as we now demonstrate. Indeed, as we already discussed, at deviations from $\theta = \theta_{hs}$, where the gap is at maximum, both Δ and Δ_s decrease, hence $\Omega_t(\theta) = \Omega_t - a\tilde{\theta}^2$, where $\tilde{\theta} = \theta - \theta_{hs}$ and $a > 0$. The singular pieces in $\tilde{\Sigma}(\omega)$ and $\Phi(\omega)$ then behave as $|\log(\Omega_t - \omega - a\theta^2)|$. Substituting these forms into the particle–hole bubble and integrating over θ, we find that the conductivity and its first derivative are continuous at $\omega = 2\Delta + \Delta_s$, but the second derivative of the conductivity diverges as $d^2\sigma/d\omega^2 \propto 1/(|\epsilon|\log^2\epsilon)$. We see that the singularity is weakened by the angular dependence of the gap, but is still located exactly at $\Omega_t + \Delta = 2\Delta + \Delta_s$.

The same reasoning can be applied to a region near 4Δ. The singularity at 4Δ is also weakened by the angular dependence of the gap, but is not shifted and still should show up in the second derivative of the conductivity.

For the Raman intensity, strong coupling effects are less relevant. First, one can prove along the same lines as in previous subsections that the cubic behavior at low frequencies for B_{1g} scattering (and the linear behavior for angular independent vertices), and the logarithmic singularity at 2Δ, are general properties of a d-wave superconductor, which survive for all couplings. Thus, analogous to the density of states and the SIS-tunneling spectrum, the Raman response below 2Δ is not sensitive to strong coupling effects. Second, near $\omega_0 + \Delta$, singular contributions which come from $\tilde{\Sigma}_+\tilde{\Sigma}_-$ and $\Phi_+\Phi_-$ terms in Π_r in (22.97) cancel each other. As a result, for a flat gap, only the second derivative of $R(\omega)$ diverges at $\Delta + \Omega_t$. For a quadratic variation of a gap near its maximum, the singularity is even weaker and shows up only in the third derivative of $R(\omega)$. Obviously, this is a very weak effect, and its experimental determination is difficult.

We now argue that measurements of the optical conductivity allow one not only to verify the magnetic scenario, but also to determine both Δ_s and Δ independently in the same experiment. In the magnetic scenario, the fermionic self-energy is singular at two frequencies: at $\Omega_t = \Delta + \Delta_s$, which is the onset frequency for spin-fluctuation scattering near hot spots, and at $\omega = 3\Delta$, where fermionic damping near hot spots first emerges due to a direct four-fermion interaction. Since in the spin-fluctuation mechanism both singularities are due to the same underlying interaction, their relative intensity can be obtained within the model.

In general, the singularity at 3Δ is much weaker at strong coupling, and can be detected only in the analysis of the derivatives of the fermionic self-energy. We recall that the singularity in $\tilde{\Sigma}(\omega)$ at Ω_t gives rise to singularity in the conductivity at $\Delta + \Omega_t$, while the 3Δ singularity in $\tilde{\Sigma}(\omega)$ obviously causes a singularity in conductivity at $\omega = 4\Delta$. In addition, we expect a singularity in $\sigma(\omega)$ at $2\Omega_t$, as at this frequency both fermions in the bubble have a singular $\tilde{\Sigma}(\Omega_t)$.

For superconductors with pairing due to electron–phonon interaction the fine structure of the optical conductivity has been analyzed by studying the second derivative of conductivity via $W(\omega) = \frac{d^2}{d^2\omega}(\omega \mathrm{Re}\sigma^{-1}(\omega))$ which is proportional to $\alpha^2(\omega)F(\omega)$ where $\alpha(\omega)$ is an effective electron–phonon coupling, and $F(\omega)$ is a phonon DOS [117]. In Fig. 22.25 we present the result of the strong coupling calculations of $W(\omega)$ [161]. There is a sharp maximum in $W(\omega)$ near $2\Delta + \Delta_s$, which is followed by a deep minimum. This form is consistent with our analytical observation that for a flat gap (which we used in our numerical analysis), the first derivative of conductivity diverges at $\omega = 2\Delta + \Delta_s$. At a finite T (a necessary attribute of a numerical solution), the singularity is smoothed, and the divergence is transformed into a maximum. Accordingly, the second derivative of the conductivity should have a maximum and a minimum near $2\Delta + \Delta_s$. The numerical analysis shows that the maximum shifts to lower frequencies with increasing T, but the minimum moves very little from $2\Delta + \Delta_s$, and is therefore a good measure of a magnetic "fingerprint".

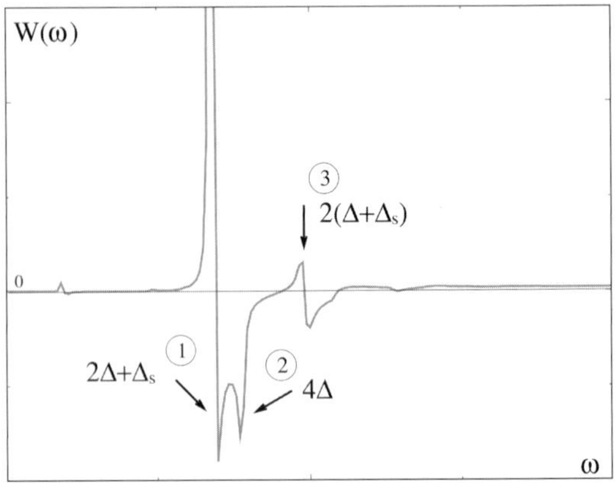

Fig. 22.25. The calculated frequency dependence of $W(\omega) = \frac{d^2}{d^2\omega}[\omega \mathrm{Re}[1/\sigma(\omega)]]$ at $T \to 0$. This quantity is a sensitive measure of the fine structure in the optical response. The locations of the extrema are: (1) $2\Delta + \Delta_s$, (2) 4Δ, (3) $2\Delta + 2\Delta_s$. Observe that the maximum shifts to a lower temperature, but the minimum remains at $2\Delta + \Delta_s$. (From [161])

Second, we note from Fig. 22.25 that in addition to the maximum and the minimum near $2\Delta + \Delta_s$, $W(\omega)$ has extra extrema at 4Δ and $2\Omega_t = 2\Delta + 2\Delta_s$. These are precisely the extra features that we expect: they are a primary effect due to a singularity in $\tilde{\Sigma}(\omega)$ at $\omega = 3\Delta$ and a secondary effect due to a singularity in $\tilde{\Sigma}(\omega)$ at $\omega = \Omega_t$. The experimental discovery of these features will be a further argument in favor of spin-mediated pairing and the applicability of the spin–fermion model.

22.6 Comparison with the Experiments on Cuprates

In this section we compare the theoretical results for the spin–fermion model of the nearly antiferromagnetic Fermi liquid with the experimental data for optimally doped members of the $Bi_2Sr_2CaCu_2O_8$ and $YBa_2Cu_3O_{7-y}$ families of cuprate superconductors. We make the assumption that at this doping level the normal state behavior of $Bi_2Sr_2CaCu_2O_8$ will resemble closely that of $YBa_2Cu_3O_{7-y}$. This enables us to take the two input parameters of the model from fits to NMR in the latter material. We then can compare theory and experiment in the normal state and as $T \to 0$ in the superconducting state. Finally, we discuss the general phase diagram of the cuprates and the pseudogap physics of these materials.

22.6.1 Parameters of the Model

The two input parameters of the theory are the coupling constant λ and the overall energy scale $\bar{\omega} = 4\lambda^2\omega_{sf}$. Alternatively, we can re-express λ as $\lambda = 3v_F\xi^{-1}/(16\omega_{sf})$ and use $v_F\xi^{-1}$ and ω_{sf} as inputs. The values of ω_{sf} and ξ can be extracted from the NMR measurements of the longitudinal and transverse spin-lattice relaxation rates, and from neutron scattering data, which measure $S(\mathbf{q},\omega) \propto \omega/((1 + (\mathbf{q}-\mathbf{Q})^2\xi^2)^2 + (\omega/\omega_{sf})^2)$. We will primarily rely on NMR data for near optimally doped $YBa_2Cu_3O_{6+\delta}$. The NMR analysis [163,164] yields a moderately temperature dependent ω_{sf} and ξ which take the values $\omega_{sf} \sim 15-20$meV and $\xi \sim 2a$ in the vicinity of T_{cr}, which for slightly overdoped materials will be close to T_c. The neutron data from inelastic scattering (INS) experiments on the normal state are more difficult to analyze because of the background which increases the measured width of the neutron peak and because of the possible influence of weak intrinsic inhomogeneities on a global probe such as INS. The data show [165] that the dynamical structure factor in the normal state is indeed peaked at $\mathbf{q} = \mathbf{Q} = (\pi/a, \pi/a)$, and that the width of the peak increases with frequency and at $\omega = 50$meV reaches 1.5 of its value at $\omega = 0$. A straightforward fit to the theory yields $\omega_{sf} \sim 35$–40 meV and a weakly temperature dependent $\xi \sim a$ which are, as expected, larger than the ω_{sf} and smaller than the ξ values extracted from NMR. We will be using $\omega_{sf} \sim 20$ meV and $\xi = 2a$ for further estimates.

The value of the Fermi velocity can be obtained from the photoemission data on $Bi_2Sr_2CaCu_2O_8$ at high frequencies, where the self-energy corrections to the fermionic dispersion become relatively minor. We note that because of problems related to the surface reconstruction in $YBa_2Cu_3O_{6+\delta}$ the vast majority of high quality angular resolved photoemission spectroscopy (ARPES) experiments are performed on $Bi_2Sr_2CaCu_2O_8$, the material where there are much less reliable NMR experiments in part because of superstructure induced line broadening. The three groups that report MDC (momentum distribution curve) data for $Bi_2Sr_2CaCu_2O_8$ and momenta along the zone diagonal [137, 166, 167] all agree that the value of the bare Fermi velocity along the diagonal (determined at higher energies where mass renormalization is assumed to be small) is rather high: $2.5-3$eVÅ, or $0.7-0.8$eVa where $a \simeq 3.8$Å is the Cu – Cu distance. We can use the $t-t'$ tight binding model for the electronic dispersion to relate this velocity with that at hot spots. Using the experimental facts that the Fermi surface is located at $\mathbf{k} \approx (0.4\pi/a, 0.4\pi/a)$ for momenta along the zone diagonal and at $\mathbf{k} \approx (\pi/a, 0.2\pi/a)$ for k_x along the zone boundary, we find $t \sim 0.2-0.25$eV, $t' \approx -0.35t$ and $\mu \approx -1.1t$. These numbers agree with those used in other numerical studies [168]. The hot spots are located at $\mathbf{k}_{hs} = (0.16\pi, 0.84\pi)$ and symmetry related points, and the velocity at a hot spot is then approximately half of that along zone diagonal. This yields $v_F \approx 0.35-0.4$ eVa.

Combining the results for v_F, ξ and ω_{sf}, we obtain $\lambda \sim 1.5-2$. This in turn yields $\bar{\omega} \sim 0.2-0.3$eV. As an independent check of the internal consistency of these estimates, we compare theoretical and experimental values of the resonance spin frequency Δ_s. As we said at the end of Sect. 22.4, $\Delta_s \simeq 0.2\bar{\omega}$ for $\lambda = 2$. Substituting the value of $\bar{\omega}$, we obtain Δ_s close to the experimental value of 40meV. A smaller $\omega_{sf} = 15$ meV would require a slightly larger λ, but variations of this magnitude are certainly beyond the quantitative accuracy of our theory.

Away from hot spots, the effective coupling decreases as $\lambda(\mathbf{k}) = \lambda/(1 + (\delta k\xi)^2)^{1/2}$ where δk is the momentum deviation from a hot spot along the Fermi surface. The largest $\delta k\xi$ is for $\mathbf{k}$ vectors along the zone diagonals. At optimal doping, ARPES data yield $\delta k^{\max} \sim 0.2\pi/a \approx 0.6/a$ [169]. We see that λ is reduced by at most 1.7 as one moves from hot spots to the zone diagonal. A prediction of the model is that $\omega_{sf}(k)$ increases at deviations from hot spots. This increase, however, should be at least partly compensated by the fact that $\omega_{sf} \propto \sin\phi_0$, where ϕ_0 is the angle between Fermi velocities at $\mathbf{k}$ and $\mathbf{k} + \mathbf{Q}$, with $\phi_0 \simeq \frac{\pi}{2}$ in the vicinity of hot spots. ϕ_0 tends to π as $\mathbf{k}$ approaches the zone diagonal, and this *reduces* ω_{sf}. In view of this competing effect which we cannot fully control, we believe that the effective $\omega_{sf}(\mathbf{k})$ can best be obtained from the fit to the photoemission data, particularly from the MDC measurements of the electronic dispersion $\omega + \Sigma'(\omega) = \epsilon_{\mathbf{k}}$. In Fig. 22.26 we compare our $(1 + \partial\Sigma'(\omega)/\partial\omega)$ with the measured variation of the effective velocity $v_F(\omega)$ of the electronic dispersion along zone diagonal [137]. We see that the theoretical dispersion has a bump (a "soft kink") at $\omega \sim 3\omega_{sf}(\mathbf{k}_{diag})$. The experimental curves look quite similar and show a "kink" at $\sim 70 - 80$meV [137, 166, 167]. This yields $\omega_{sf}(\mathbf{k}_{diag}) \sim 25$meV, a value only slightly larger than that near hot spots.

Note in passing that although $\delta k\xi$ does not vary much when k moves along the Fermi surface, the fact that the Fermi velocity is fairly large implies that along the zone diagonal, $\epsilon_{\mathbf{k}_F+\mathbf{Q}}$ is roughly $\sqrt{2}v_F 0.2\pi/a \approx 0.8eV$, i.e. it is comparable to the bandwidth. This implies that the Fermi-surface is very different from the near-perfect square that one would obtain for only nearest neighbor hopping. Furthermore, the fact that the Fermi velocity is large implies the physics at energies up to few hundred meV is confined to the near vicinity of the Fermi surface, when one can safely expand $\epsilon_{\mathbf{k}}$ to linear order in $\mathbf{k} - \mathbf{k}_F$. Finally, van-Hove singularities (which we neglected) do play some role [170, 171] but as $\epsilon_{(0,\pi/a)} \approx 0.34t \sim 85$meV $\gg \omega_{sf}$, we expect that the van-Hove singularity softens due to fermionic incoherence and should not substantially affect the physics. The value of $\epsilon_{(0,\pi/a)}$ might however be affected by an additional bilayer splitting which moves one of the bands closer to $\mathbf{k} = (0, \pi/a)$.

Finally, in the analysis of the spin–fermion model we have neglected the temperature dependence of the correlation length, and thus of ω_{sf}. Fits to NMR experiments on the near optimally doped member of the $YBa_2Cu_3O_{6+\delta}$ family show that at $T_{cr} \simeq T_c$ both ω_{sf} and ξ display mean field behavior with $\xi^{-2} \simeq \xi_0^{-2}(1 + T/T_0)$ and $\omega_{sf}\xi^2 \simeq 70$meV almost independent on T.

From a theoretical perspective, the leading temperature dependence of ξ arises from an interaction between spin-fluctuations and near the critical point in two dimensions has the form $\xi^{-2}(T) =$

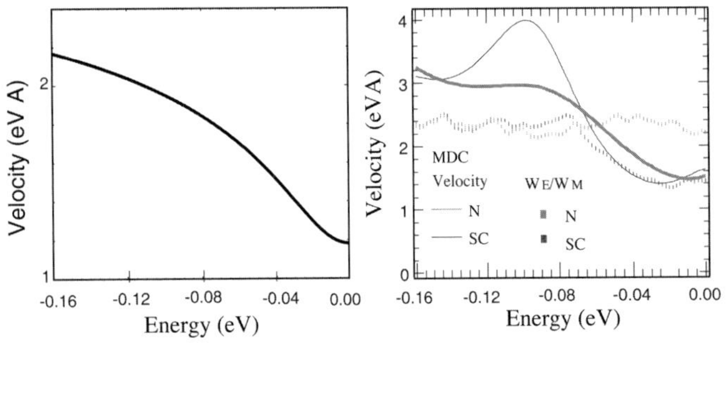

Fig. 22.26. (**a**) The theoretical result for the effective velocity of the quasiparticles $v_F^* = v_F/(1 + \partial\Sigma'(\omega)/\partial\omega)$. For definiteness we used $\omega_{sf} = 20$ meV, $\lambda = 1.7$ and bare velocity $v_F = 3$ eVA along the diagonal. (**b**) Experimental result for the effective velocity, extracted from the MDC dispersion [137] along the zone diagonal. Observe the bump in the frequency dependence of the velocity at 70 − 80 meV in the data and at about $3 - 4\omega_{sf}$ in the theory

$\xi^{-2}(T = 0) + B\Gamma_4 T \log T$ where $B = O(1)$, and Γ_4 is the effective four boson interaction that is made out of fermions [75,76,124]. The $T \log T$ factor is the universal contribution from the bosonic loop, confined to momenta near $\mathbf{Q}$. The four-boson interaction has two contributions: one comes from low-energy fermions and is universal; the other comes from high-energy fermions and depends on the fermionic bandwidth W. One can show [124] that the temperature correction to ξ involves only the high-energy part of the interaction (this is what we labeled as Γ_4). The magnitude of Γ_4 can be easily estimated to be $\bar{g}^2/W^3$. Accordingly, the temperature correction to ξ scales as $\frac{T}{\bar{g}}(\bar{g}/W)^3$. As we have remarked, the theory is universal as long as $\bar{g} \ll W$. In this limit, the temperature dependence of ξ is obviously small and can be neglected. This is what we will do. Notice however that in the opposite limit, when $\bar{g} \gg W$, the full four-boson interaction differs from the lowest order term in $\bar{g}$ and might be estimated within an RPA-type summation. Estimates show that in this limit, the full Γ_4 is fully determined within the low-energy sector and scales as $O(1/J)$ where $J \sim W^2/\bar{g}$ is the magnetic exchange integral. This in turn yields a much stronger temperature dependence of ξ: $\xi^{-2}(T) - \xi^{-2}(T = 0) \sim (T/J) \log T$. This result is similar to that obtained using a nonlinear σ-model approach to near antiferromagnetism [127,172,173]. The agreement becomes obvious in the limit of a large spin–fermion interaction (which, we recall, is the Hubbard U if we derive the spin–fermion model within the RPA); double occupancy is energetically unfavorable and the spin susceptibility obeys the constraint $\int d^2q d\omega \chi(q, i\omega) \propto 1 - x$. This is equivalent to imposing a constraint on the length of the spin field in the nonlinear σ-model [127,172,173].

22.6.2 The Normal State

In this section we compare the experimental and theoretical forms of the fermionic spectral function and optical conductivity in the normal state. No free parameters remain, since those which are needed to specify the model completely have been taken from NMR and ARPES experiments. The discussion will follow [49].

The Spectral Function

The quasiparticle spectral function at various momenta is measured in angle resolved photoemission experiments. In a sudden approximation (an electron, hit by light, leaves the crystal without further interactions with other electrons and without paying attention to selection rules for the optical transition to its final state), the photoemission intensity is given by $I_{\mathbf{k}}(\omega) = A_{\mathbf{k}}(\omega) n_F(\omega)$ where n_F is the Fermi function and $A_{\mathbf{k}}(\Omega) = (1/\pi)\, |\mathrm{Im} G(\mathbf{k}, \Omega)|$ is the spectral function.

We first use our form of the fermionic self-energy to fit MDC data which measure the width of the photoemission peak as a function of $\mathbf{k}$ at a given frequency. In Fig. 22.27 we compare the theoretical results for $\Delta k = \Sigma''(\mathbf{k}, \Omega)/v_F$ with the measured Δk versus frequency at $T \sim 100$K [137] and versus temperature at $\Omega \to 0$ [166]. We used $\lambda = 1.7$ and $\omega_{sf} = 20$ meV. The slope of Δk is chiefly controlled by λ. We obtain rather good agreement with the data, both for the frequency and temperature dependence of the self-energy. On the other hand, the magnitude of our Σ'' is smaller than that found in the experimental data [174]. To account for the values of Δk, we had to *add* a constant of about 70meV to Σ''. The origin of this constant is unclear and explaining it is presently a challenge to the theory. It may be the effect of elastic scattering by impurities [175], but the large value of this constant implies that it is more likely the contribution from scattering channels that we ignored. It is essential, however, that the functional dependence of $\Sigma''(\Omega, T)$ can be captured in the spin-fluctuation approach.

In Fig. 22.28 we present the results for the half width at half maximum of the EDC (energy distribution curve) which measures fermionic $I_{\mathbf{k}}(\Omega) = A_{\mathbf{k}}(\Omega) n_F(\Omega)$ as a function of frequency at a given $\mathbf{k}$. For a Lorentzian line-shape, the EDC hwhm is given by $\Sigma''(\Omega)/(1 + \Sigma'(\Omega)/\Omega)$. The data are taken at $T = 115$K [137]. We see that the theoretical slope agrees reasonably well with the experimental one. The visible discrepancy is most likely associated with the fact that the experimental line-shape is not a perfect Lorentzian.

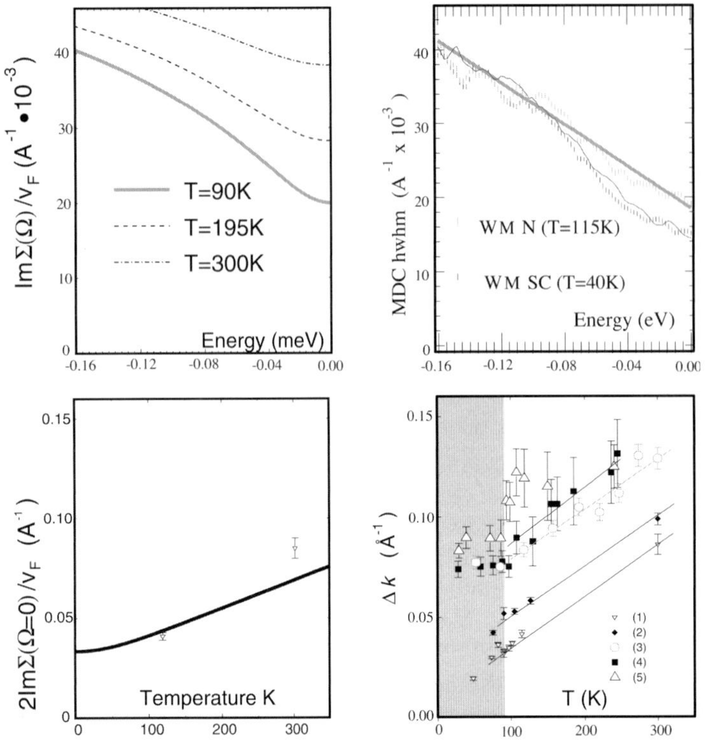

Fig. 22.27. A comparison between the theoretical results of the spin–fermion model and the photoemission MDC data. For the Lorentzian line-shape of the MDC signal, observed in experiments, the MDC hwhm equals to Σ''/v_F. *Upper panel*: the results for the MDC hwhm vs frequency at a given T. The experimental results are taken from [137]. *Lower panel*: the MDC fwhm vs temperature at $\Omega = 0$. The experimental results (*right figure* and the *points on the left figure*) are taken from [166]. The figure is taken from [49]

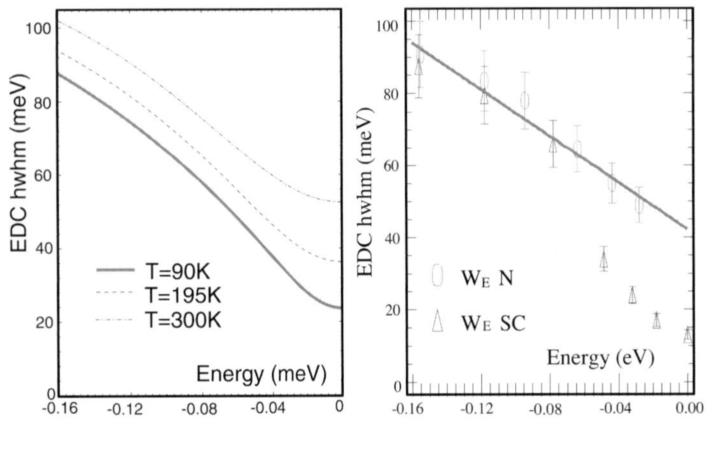

Fig. 22.28. A comparison of the theoretical result for $\Sigma''(\Omega)/(1 + \Sigma'(\Omega)/\Omega)$ with the experimental hwhm of the EDC dispersion from [137]. The figure is taken from [49]

The Optical Conductivity

In Fig. 22.29 we compare the theoretical results for the conductivity [161] with the experimental data for σ_1 and σ_2 at different temperatures [176]. The theoretical results are obtained using the same $\lambda = 1.7$ and $\omega_{sf} = 20$meV as in the fit to the photoemission data along zone diagonals. Changing λ affects the ratio σ_2/σ_1 at high frequencies, but does not change the functional forms of the conductivities [161]. The theoretical value of the plasma frequency was adjusted [161] to match the d.c. conductivity [144,176,177]. We see that theoretical calculations of $\sigma_1(\omega)$ and $\sigma_2(\omega)$ capture the essential fea-

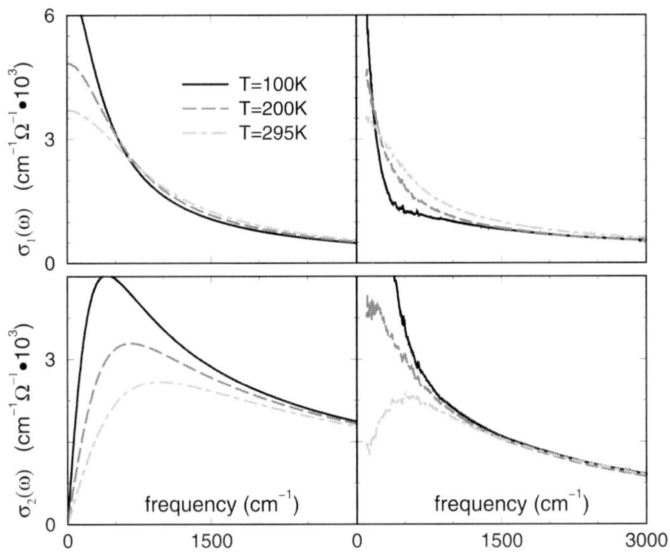

Fig. 22.29. The theoretical and experimental results for the real and imaginary parts of optical conductivity. The data are from [176]. The figure is taken from [49]

tures of the measured forms of the conductivities. In particular, the curves of σ_1 at different temperatures cross such that at the lowest frequencies, the conductivity decreases with T while at larger frequencies it increases with T, a behavior seen in [177,179]. The imaginary part of conductivity decreases with T at any frequency, and the peak in $\sigma_2(\omega)$ increases in magnitude and shifts to lower T with decreasing T [176–178]. At frequencies above 1500cm^{-1} both σ_1 and σ_2 depend weakly on T and are comparable in amplitude.

To make the comparison more quantitative, in Fig. 22.30 we present experimental and theoretical results for the imaginary part of the full particle–hole polarization bubble $\Pi''_\sigma(\omega) = 4\pi\sigma_1\omega/\omega^2_{\rm pl}$. Theoretically, at $T = 0$, $\Pi''_\sigma(\omega)$ saturates at a value of about 0.2 *independently* of λ and remains almost independent of frequency over a very wide frequency range [161]. The experimental data also clearly show a near saturation of Π''_σ at a value close to 0.2.

The agreement between theory and experiment is, however, not a perfect one. In Fig. 22.31 we show theoretical and experimental results for $1/\tau^* = \omega\sigma_1/\sigma_2$. The advantage of comparing $1/\tau^*$ is that this quantity does not depend on the unknown plasma frequency. We see that while both experimental and theoretical curves are linear in frequency, the slopes are off roughly by a factor of 3. This discrepancy is possibly related to the fact that in the spin-fermion model, $\Pi'_\sigma(\omega)$ at high enough frequencies is roughly 3 times

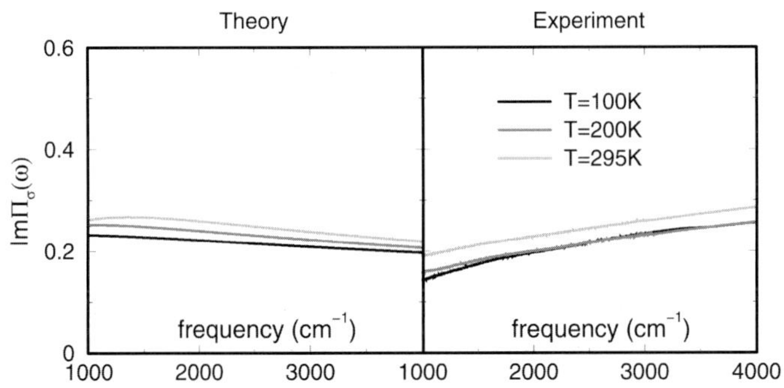

Fig. 22.30. The theoretical and experimental results for $\Pi''_\sigma(\omega) = 4\pi\sigma_1\omega/\omega^2_{pl}$ (from [49]). The data are from [176]

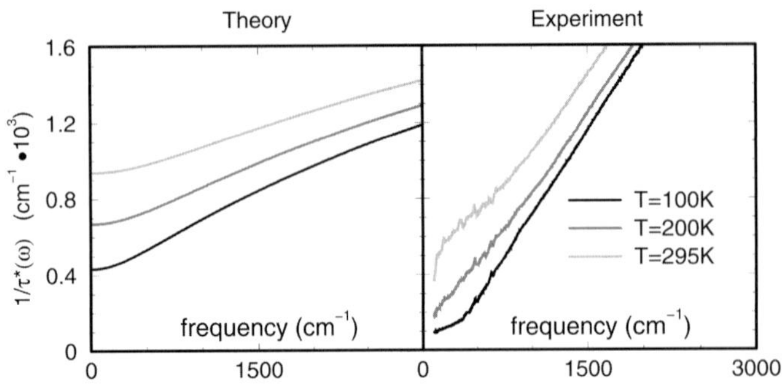

Fig. 22.31. The theoretical and experimental results for $1/\tau^* = \omega \, \mathrm{Re}\, \sigma / \mathrm{Im}\, \sigma$ (from [49]). The data are from [176]

larger than $\Pi''_\sigma(\omega)$ [161], and hence $\sigma_2/\sigma_1 \sim 3$, whereas experimentally σ_1 and σ_2 are comparable in magnitude at high frequencies. The discrepancy in $1/\tau^*$ indicates that either the averaging over the Fermi surface, vertex corrections inside a particle–hole bubble, or RPA-type corrections to the conductivity [180] play some role. Still, Figs. 22.29 and 22.30 indicate that the general trends of the behavior of the conductivities near optimal doping are reasonably well captured within the spin-fluctuation approach.

22.6.3 The Superconducting State

In this section, we apply our results from Sect. 22.5 to cuprates and examine to what extent the "fingerprints" of spin-fluctuation pairing have been detected in experiments on optimally doped high-T_c materials.

The Spin Susceptibility

The major prediction of the spin fermion model for the spin susceptibility in the superconducting state is the emergence of the resonance peak in $\chi''(\mathbf{Q}, \omega)$ at $\omega = \Delta_s$ if $\Delta_s < 2\Delta$. The magnitude Δ_s is fully determined within the theory and is chiefly set by the magnitude of the superconducting gap as well as the energy scale of magnetic fluctuations in the normal state, ω_{sf}. For small doping concentration $\Delta_s \propto \xi^{-1}$ must decrease as one approaches the antiferromagnetic state. The resonance mode is confined to a small region in momentum space (where it is of high intensity). For momenta away from $\mathbf{Q}$ and its close vicinity, magnetic excitations couple to gapless, nodal quasiparticles and become overdamped, eliminating the resonance mode.

In Fig. 22.32 we show representative experimental data for $\chi''(\mathbf{Q}, \omega)$ showing the resonance peak at $\omega \approx 41$ meV for optimally doped $YBa_2Cu_3O_{6.9}$ [181]. As noted earlier, the position of the peak is consistent with the prediction of the spin fermion model. Similar behavior is found in $Bi_2Sr_2CaCu_2O_8$ [146]; here the peak is at 43 meV. With underdoping, the measured resonance energy goes down [138, 139]. In strongly underdoped $YBa_2Cu_3O_{6.6}$, it is approximately 25 meV [138]. The existence of the peak and the downturn with underdoping agree with the predictions of the spin-fluctuation theory. Further, the measured amplitude of $\chi''(\mathbf{Q}, \omega)$ decreases above the peak, but increases again for 60–80 meV [139, 181]. This might indeed be a 2Δ effect, which appears naturally within the model.

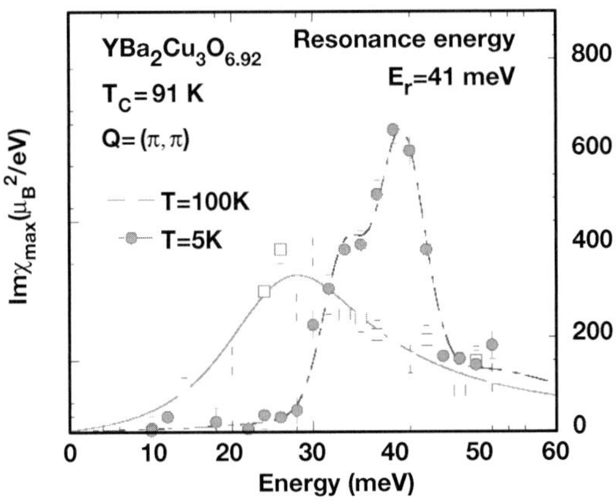

Fig. 22.32. Inelastic neutron scattering intensity for momentum $\mathbf{Q} = (\pi, \pi)$ as function of frequency for $YBa_2Cu_3O_{6.92}$. Data from [181]

The full analysis of the resonance peak requires more care as (i) the peak is only observed in two-layer materials, and only in the odd channel, (ii) the momentum dispersion of the peak is more complex than that for magnons [165], (iii) the peak broadens with underdoping [138, 139], and (iv) in underdoped materials, the peak emerges at the onset of the pseudogap and only sharpens up at T_c [139,181]. All these features, already present on the level of a weak coupling approach [128–130], have been explained within the spin–fermion model [89,134]. The broadening of the peak was recently studied in detail in [132]. The explanation of these effects, however, requires careful analysis of the details of the electronic structure and is beyond the scope of this chapter.

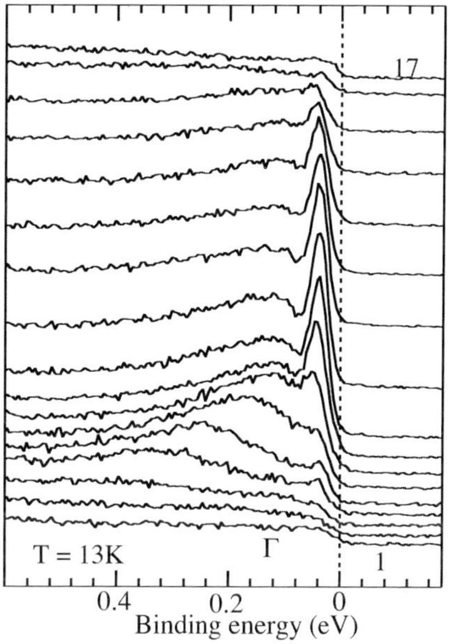

Fig. 22.33. ARPES spectrum for near optimally doped Bi2212 for momenta close to the hot spots. Data from [140]

The Spectral Function

The predictions of our approach are a peak-dip structure of the spectral function, with a weakly dispersing peak at $\omega \approx \Delta$ and a peak–dip distance $\approx \Delta_s$. On the other hand we expect a broad incoherent peak which disperses like $\epsilon_{\mathbf{k}}^2/\bar{\omega}$. In Fig. 22.33 we present ARPES data for near optimally doped Bi2212 with $T_c = 87$K for momenta near a hot spot [140]. The intensity displays the predicted peak/dip/hump structure. A sharp peak is located at ~ 40meV, and the dip is at 80meV such that the peak–dip distance is 42meV [140]. In the spin-fluctuation theory, the peak-dip distance is the energy of the INS resonance peak frequency [123, 128]. The neutron scattering data on Bi2212 with nearly the same $T_c = 91$ K yield [146] $\Delta_s = 43$ meV, in excellent agreement with this prediction.

Furthermore, with underdoping, the peak-dip energy difference decreases and, within error bars, remains equal to Δ_s. This behavior is illustrated in Fig. 22.34.

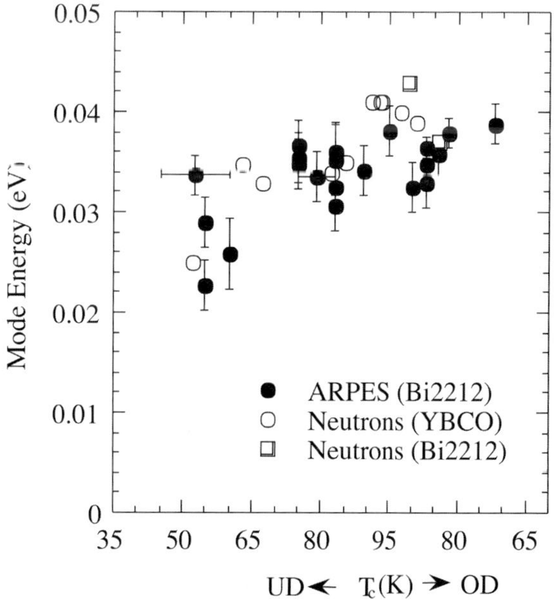

Fig. 22.34. The experimental peak–dip distance at various doping concentrations compared with Δ_s extracted from neutron measurements. Data from [30]. The theoretical result is presented in Fig. 22.18

In Fig. 22.35 we present experimental results for the variation of the peak and hump positions with the deviation from the Fermi surface. These show that the hump disperses with $\mathbf{k} - \mathbf{k}_F$ and eventually recovers the position of the broad maximum in the normal state. At the same time, the peak shows little dispersion, and does not move further in energy than $\Delta + \Delta_s$. Instead, the amplitude of the peak dies off as

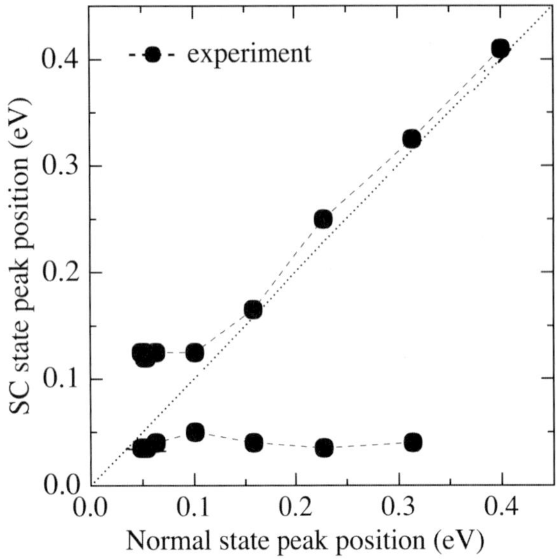

Fig. 22.35. The dependence of the experimental peak (flat curve) and hump (dispersing curve) positions on the deviation from the Fermi surface. The hump disperses with $\mathbf{k} - \mathbf{k}_F$ (*dotted line*) and eventually recovers the position of the broad maximum in the normal state, while the peak position changes little with the deviation from $\mathbf{k}_F$. Data from [140]. The theoretical result is presented in Fig. 22.18

$\mathbf{k}$ moves away from $\mathbf{k}_F$. This behavior is again fully consistent with the theoretical predictions [49,156].

We regard the presence of the dip at $\Delta+\Delta_s$, and the absence of the dispersion of the quasiparticle peak as two major "fingerprints" of strong spin-fluctuation scattering in the spectral density of cuprate superconductors.

The Density of States

The fermionic DOS $N(\omega)$ is proportional to the dynamical conductance dI/dV through a superconductor-insulator-normal metal (SIN) measured at $\omega = eV$, where V is the applied voltage [152]. The key prediction of our approach is the occurrence of a dip in the DOS at an energy $\approx \Delta_s$ away from the peak at $\omega = \Delta$. The drop in the DOS at $\Omega_t = \Delta + \Delta_s$ can be understood in terms of SIN conductance as follows: when the applied voltage, V, equals Ω_t/e an electron that tunnels from a normal metal can emit a spin excitation and fall to the bottom of the band, losing its group velocity. This loss leads to a sharp reduction of the current and produces a drop in dI/dV. This process is shown schematically in Fig. 22.22.

SIN tunneling experiments have been performed on YBCO and Bi2212 materials [142]. We reproduce these data in Fig. 22.36. Similar results have been recently obtained by Davis et al. [182]. At low and moderate frequencies, the SIN conductance displays a behavior which is generally expected in a d-wave superconductor, i.e. it is linear in voltage for small voltages, and has a peak at $eV = \Delta$ where Δ is the maximum value of the d-wave gap [142,182] The value of Δ extracted from tunneling agrees well with the maximum value of the gap extracted from ARPES measurements [136,137]. At frequencies larger than Δ, the measured SIN conductance clearly displays an extra dip-hump feature which become visible at around optimal doping, and grows in amplitude with underdoping [142]. At optimal doping, the distance between the peak at Δ and the dip is around 40meV. This is consistent with Δ_s extracted from neutron measurements.

SIS Tunneling

The major prediction of the spin–fermion model for the SIS tunneling conductance, $S(\omega)$, is the emergence of a singularity at $\omega = 2\Delta + \Delta_s$. As mentioned above, this singularity is likely softened due to thermal excitations or non-magnetic scattering processes and transforms into a dip slightly below $2\Delta+\Delta_s$, and a hump at a frequency larger than $2\Delta+\Delta_s$. Recently, Zasadzinski et al. obtained both new data and carefully examined their previous SIS tunneling data for a set of Bi2212 materials ranging from overdoped to underdoped [143]. Their data, presented in Fig. 22.37 show that in addition to the peak at 2Δ, the SIS conductance displays the dip and the hump at larger frequencies. The distance between the peak and the dip (which approximately equals Δ_s in the spin fluctuation model [123,133]) is close to 2Δ in overdoped Bi2212 materials, but goes down with underdoping. Near optimal doping, this distance is around 40meV. For an underdoped, $T_c = 74$K, material, the peak-dip distance is reduced to about 30meV. These results are in qualitative and quanti-

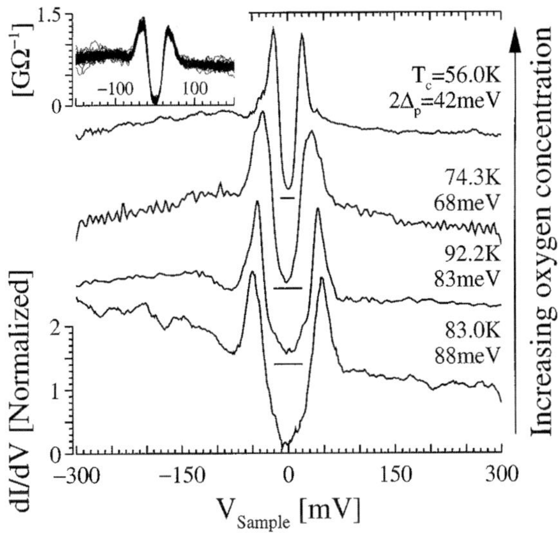

Fig. 22.36. The experimental result for the differential conductance through the SIN tunneling junction. This conductance is proportional to the quasiparticle DOS. The data are for Bi2212 and are taken from Ref. [142]. The theoretical result is presented in Fig. 22.20(d)

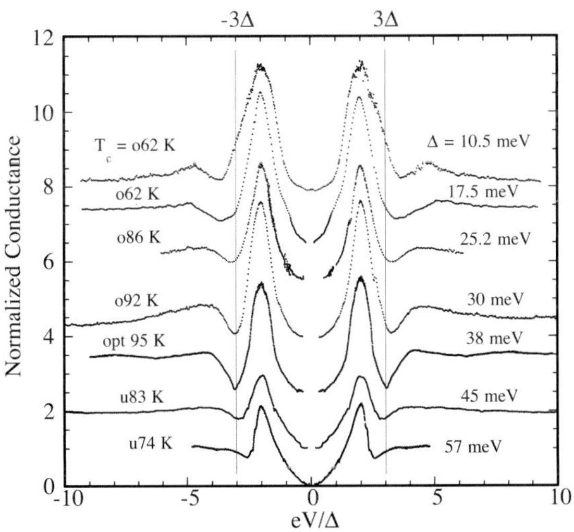

Fig. 22.37. SIS tunneling conductance normalized by Δ for Bi2212 materials ranging from overdoped (*top curves*) to underdoped (*bottom curves*) from [143]. The peak dip distance increases for increasing doping and saturates at around 3Δ as expected in our theory. The corresponding theoretical result is presented in Fig. 22.21

tative agreement with ARPES and neutron scattering data, as well as with the theoretical estimates. The most important aspect is that with underdoping, the experimentally measured peak-dip distance progressively shifts down from 2Δ. This downturn deviation from 2Δ is a key feature of the spin-fluctuation mechanism. We regard the experimental verification of this feature in the SIS tunneling data as an additional strong argument in favor of the magnetic scenario for superconductivity.

Optical and Raman Response

Theoretical considerations show that optical measurements are much better suited than Raman measurements to search for the "fingerprints" of a magnetic scenario [133]. For the optical conductivity we predict a singular behavior at energies $2\Delta + \Delta_s$, 4Δ, $2\Delta + 2\Delta_s$, which can be amplified if one considers the second derivative of conductivity via $W(\omega) = \frac{d^2}{d^2\omega}(\omega \mathrm{Re}\sigma^{-1}(\omega))$. Evidence for strong coupling effects in the optical conductivity in superconducting cuprates has been reported in [144,145,183,184].

We first discuss the form of $\sigma_1(\omega)$. In Fig. 22.38 we compare the theoretical result for $\sigma_1(\omega)$ [133,161] with the experimental data by Puchkov et al. [144] for optimally doped $YBa_2Cu_3O_{6+\delta}$ in the superconducting state. The parameters are the same as in the normal state fits. As the theoretical formula does

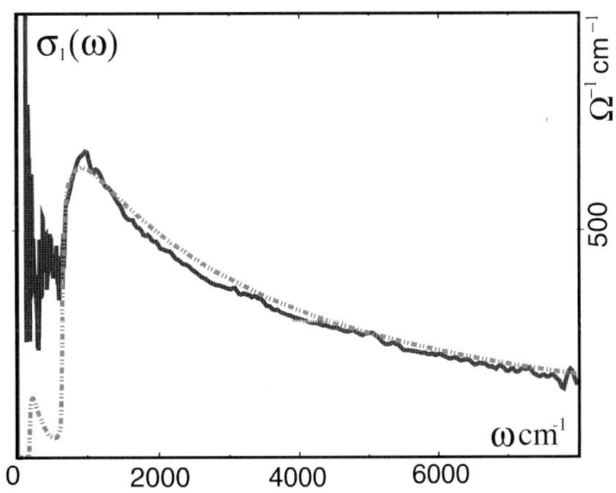

Fig. 22.38. A comparison between theoretical and experimental results for the optical conductivity (from [161]). The experimental data are from [144]

not include the contributions from the nodes, the comparison is meaningful only for $\omega > 2\Delta$. We see that the frequency dependence of the conductivity at high frequencies agrees well with the data. The measured conductivity drops at about 100meV in rough agreement with $2\Delta + \Delta_s$ which for $\Delta \approx 30$ meV and $\Delta_s \approx 40$ meV is also around 100 meV. The good agreement between theory and experiment is also supportive of our argument that the momentum dependence of the fermionic dynamics becomes irrelevant at high frequencies, and fermions from all over the Fermi surface behave as if they were at hot spots.

We next consider the singularities in the frequency dependence of the conductivity in more detail and compare the theoretical and experimental results for $W(\omega) = \frac{d^2}{d^2\omega}(\omega \mathrm{Re}\sigma^{-1}(\omega))$. The theoretical result for $W(\omega)$ is presented in Fig. 22.25 The experimental result for $W(\omega)$ in YBCO is shown in Fig. 22.39. We see that the theoretical and experimental plots of $W(\omega)$ look rather similar, and the relative intensities of the peaks are at least qualitatively consistent with the theory. We identify (see explanations below) $2\Delta + \Delta_s$ with the deep minimum in $W(\omega)$. This identification, that is consistent with the analysis of $\sigma_1(\omega)$, yields $2\Delta + \Delta_s \approx 100$ meV. Identifying the extra extrema in the experimental $W(\omega)$ with 4Δ and $2\Delta + 2\Delta_s$, respectively, we obtain $4\Delta \sim 130$ meV, and $2\Delta + 2\Delta_s \sim 150$ meV. We see that three sets of data are consistent with each other and yield $\Delta \sim 30$ meV and $\Delta_s \sim 40$–45 meV. The value of Δ is in good agreement with tunneling measurements [185], and Δ_s agrees well with the resonance frequency extracted from neutron measurements [138]. Indeed, the analysis of a second derivative of a measured quantity is a very subtle procedure. The good agreement between the theory and experiment is promising but has to be verified in further experimental studies. Still, theoretical calculations [133,161] clearly demonstrate the presence and observability of these "higher harmonics" of the optical response at 4Δ and $2\Delta + 2\Delta_s$.

Finally, we comment on the position of the $2\Delta + \Delta_s$ peak and compare the results of Abanov et al [133, 161] with those by Carbotte et al. [145]. Theoretically, at $T = 0$ and in clean limit, the maximum and minimum in $W(\omega)$ are located at the same frequency. At a finite T, however, they quickly move apart (see

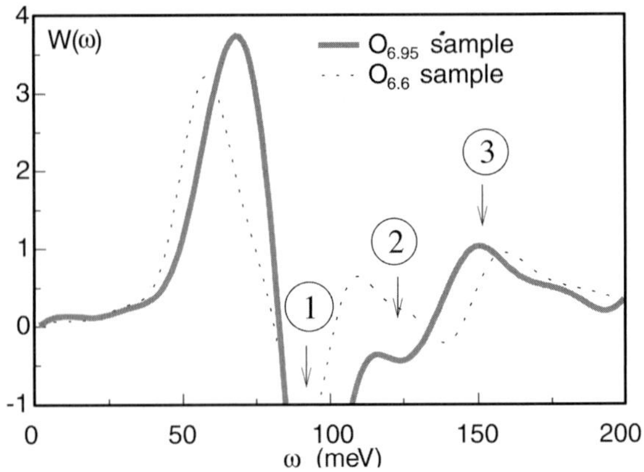

Fig. 22.39. Experimental results for $W(\omega) = \frac{d^2}{d^2\omega} \times (\omega \mathrm{Re}\sigma^{-1}(\omega))$ from [145]. The theoretical result is presented in Fig. 22.25. The position of the deep minimum agrees well with $2\Delta + \Delta_s$. The extrema at higher frequencies are consistent with 4Δ and $2(\Delta + \Delta_s)$ predicted by the theory

Fig. 22.25). Carbotte et al. [145] focused on the maximum in $W(\omega)$ and argued that it is located at $\Delta + \Delta_s$ instead of $2\Delta + \Delta_s$. We see from Fig. 22.25 that the maximum in $W(\omega)$ shifts to a lower frequency with increasing temperature and over some T range is located close to $\Delta + \Delta_s$. On the other hand, the minimum in $W(\omega)$ moves very little with increasing T and virtually remains at the same frequency as at $T = 0$. This result suggests that the minimum in $W(\omega)$ is a more reliable feature for comparisons with experiments. This conclusion is in agreement with recent conductivity data on optimally doped Bi2212 [176]. $W(\omega)$ extracted from these data shows a strong downturn variation of the maximum in $W(\omega)$ with increasing temperature, but the minimum in $W(\omega)$ is located at around 110meV for all temperatures.

22.6.4 Experimental Facts That We Cannot Yet Describe

There are several experimental results that we do not understand. First are the results by Ando, Boebinger and collaborators [186–188] on the behavior of the Lanthanum and Bismuthate based superconductors in magnetic fields sufficiently strong to (almost) destroy superconductivity. For doping levels close to the optimal one, they found that the resistivity at low temperatures continues to be linear in T with the same slope seen at higher temperatures. If the assumption that the magnetic field destroys superconductivity but otherwise does not affect the system properties is correct, this result poses a problem for the spin-fluctuation model as the latter yields a linear in T resistivity over a wide range of temperatures, but only for T larger than a fraction of ω_{sf}. To account for these data one might have to invoke some kind of quantum-critical physics associated with the opening of the pseudogap (see below).

Another experiment that is not yet understood is the measurement of the Hall angle, $\theta_H \equiv \rho_{xy}/\rho_{xx}$, which shows an incredibly simple behavior, $\cot\theta_H \propto T^2$ [189] and also displays a particular frequency behavior [190–192]. The orbital magnetoresistance $\Delta\rho/\rho$ also behaves in quite an unusual way, violating Kohler's rule, according to which $\Delta\rho/\rho$ is a function of H^2/ρ^2, independent of T, where H is the applied magnetic field. Some of this physics is already captured in the semi-phenomenological calculations by Stojkovich and Pines [193]; however problems remain. In the description based on the spin–fermion model the technical problem not yet solved is how to include in a controlled way vertex corrections which are not small; in one of the vertices for the Hall conductivity the momentum transfer is small. Some progress with these calculations have been recently made by Katami and collaborators [194]. Another explanation of the Hall data has recently been proposed by Abrahams and Varma [190].

Yet another unanswered question, already noted above, is the origin of a large (almost 100 meV), frequency and temperature independent contribution to the self-energy that one has to invoke in order to fit conductivity and ARPES data (see, however, [174]).

Electronic Raman scattering reveals further puzzling behavior: in all geometries one observes a frequency independent behavior over a very large energy scale, frequently referred to as the positive background. Moreover, the overall size of the background is very different in different geometries [157, 160]. There are also uncertainties associated with reconciling the incommensurability of the magnetic response in the normal state of 214 materials [195] with the commensurate peaks required to obtain a consistent explanation of ^{17}O and ^{63}Cu NMR experiments, but these are not likely to pose fundamental problems to the spin-fluctuation approach [163,196].

Finally, the claim of universality of the low-energy behavior relies heavily on the existence of a quantum critical point at which the antiferromagnetic correlation length diverges. In real materials there are indications that the transition to antiferromagnetism is actually of first order. In this situation, the theory we described is valid only if there still exists a substantial region in parameter space where the system is critical before it changes its behavior discontinuously. NMR and neutron scattering experiments on optimally doped cuprates seem to support such behavior. Another reason for concern is the role of disorder and inhomogeneities. Despite enormous progress in sample fabrication, cuprates often tend to be very heterogeneous materials. It has been established in several cases that these aspects are actually intrinsic, forcing one to include effects due to inhomogeneities and disorder into the theoretical description [197].

22.6.5 Phase Diagram

In this section we discuss in detail the experimental phase diagram of cuprate superconductors and comment on the origin of the pseudogap behavior found for small charge carrier concentrations.

From a general perspective, the key to understanding of cuprate superconductors is identifying the nature of the protected behavior of the novel states of matter encountered in the insulating, conducting, and superconducting states as one varies doping and temperature, including the possible existence of one or more quantum critical points. Consider first the $YBa_2Cu_3O_{7-\delta}$ system on which the generic phase diagram of Fig. 22.1 was based [163]. A somewhat similar diagram based on transport measurements was independently proposed by Hwang et al. [198], while one based on specific heat and susceptibility measurements has been proposed recently by Tallon et al. [199]. As discussed in the Introduction, in addition to the T_c line, there are two crossover or phase transition lines in Fig. 22.1. The upper line $T = T_{cr}$ is defined experimentally by a maximum in the temperature dependent uniform magnetic susceptibility, χ_0. It has been further characterized [163] as the temperature at which the antiferromagnetic correlation length ξ is of the order Cu-Cu lattice spacing (Barzykin and Pines used a criterion $\xi(T_{\rm cr}) = 2a$). The lower line $T = T^*$ may be defined experimentally as the temperature at which the product of the copper spin-lattice relaxation time,$^{63}T_1$ and the temperature,T, reaches its minimum value. In the $Bi_2Sr_2CaCu_4O_8$ counterparts of the $YBa_2Cu_3O_{7-\delta}$ system, it corresponds to the temperature at which the leading edge gap found in ARPES experiments for quasiparticles near $(\pi, 0)$ becomes fully open, effectively gapping that portion of the quasiparticle Fermi surface. To a first approximation, on making use of the experimental results for optimally and underdoped $YBa_2Cu_3O_{7-\delta}$ materials one finds that $T^* \approx T_{\rm cr}/3$.

The superconducting T_c in Fig. 22.1 is obtained using the empirical relation [199]

$$x = 0.16 \pm 0.11\sqrt{1 - \frac{Tc}{T_c^{\max}}}, \tag{22.99}$$

where x is the doping level, and $T_c^{\max}$ is the maximal transition temperature for a given class of materials. The location of $T_{\rm cr} > T_c$ can well be fitted by another empirical relation.

$$T_{\rm cr} \approx 1250\,\mathrm{K}\left(1 - \frac{x}{x_{\rm cr}}\right), \tag{22.100}$$

where $x_{\rm cr} \approx 0.19$. Similar expressions are found for the $La_{2-x}Sr_xCuO_4$ and $Bi_2Sr_2CaCu_4O_8$ materials. This expression for $T_{\rm cr}$ is, in both its magnitude and doping dependence, close to the pseudogap temperature obtained by Loram and his collaborators [200] from an analysis of specific heat experiments. A remarkable result of this purely phenomenological analysis is that the crossover temperature $T_{\rm cr}$ extrapolates at zero doping to the known value of the antiferromagnetic super-exchange interaction J.

The fit to $T_{\rm cr}$ by (22.100) raises the issue of whether T^* and $T_{\rm cr}$ are independent of T_c and would extrapolate to the origin at a doping level $x = x_{\rm cr}$ if superconductivity was absent. The system then would have an additional quantum critical point at $x = x_{\rm cr}$ with a new kind of ordered state for $x < x_{\rm cr}$. This issue is currently open and is a subject of active research. Support for a phase diagram with an additional quantum critical point at $x = x_{\rm cr}$ comes from the work of Loram , Tallon, and their collaborators [199,200], who have proposed such behavior based on a detailed analysis of their specific heat experiments on underdoped and overdoped systems. Moreover, as Loram, Tanner, Panagopoulos and others have emphasized [199–201], in the superconducting state of the low-doping side of $T_{\rm cr}$ one has "weak" superconducting behavior, with a superfluid density ρ_s decreasing with decreasing doping, while on the high doping side one has a "conventional" superconductivity, and a value of ρ_s that is nearly independent of the doping concentration. Further support for the idea of an additional quantum-critical point comes from the well established fact that optimally doped cuprates are the ones for which the extension of the linear resistivity to $T = 0$ yields very small residual resistivity, and from the experiments of [186,187] which, we recall, show that in the absence of superconductivity the linear temperature dependence of the resistivity extends to lower T indicating that at some doping the resistivity can be linear down to $T = 0$. As Laughlin et al. [202] have emphasized, the presence of a quantum critical point with a large domain of influence, together with superconductivity, serves to conceal the nature of the non-superconducting ground states on either side of the quantum critical point. One might hope that ARPES experiments near optimal doping would distinguish between a quantum critical behavior with quantum critical point at around optimal doping and a spin fermion scenario with antiferromagnetic quantum critical point at considerably smaller doping concentration. However, a recent analysis of Haslinger et al. [125] showed that fits to current experiments with either model is possible and requires in both cases the introduction of a large temperature and frequency independent scattering rate, as noted earlier.

The variety of experimental results for the pseudogap allows one to understand it phenomenologically, without invoking a particular microscopic mechanism. First, as T^* and $T_{\rm cr}$ scale with each other, it is natural to attribute both T^* and $T_{\rm cr}$ to different aspects of the same physical phenomenon which begins at T_{cr} and gains full strength at T^*. This idea is fully consistent with NMR data which show the onset of changes in T_1T at T_{cr}, which eventually give rise to a sign change of the temperature derivative of $^{63}T_1T$ at T^*. Second, ARPES data on the leading edge gap clearly demonstrate that the pseudogap physics below T^* is associated with the redistribution of the spectral weight for hot quasiparticles; quasiparticles near the nodes are almost unaffected by the development of the pseudogap. In the ARPES literature, this effect is described as a progressive development of the arcs of the Fermi surface centered around nodal points. The evolution of the full Fermi surface into the arcs begins at around T^*, and at T_c the whole Fermi surface becomes gapped. The "gapping" of hot fermions obviously affects NMR relaxation rates dominated by momenta near **Q** (such as $^{63}Cu\ T_1T$) as a spectral weight transfer would lead to a reduction of a decay rate of a spin fluctuation into a particle–hole pair [48, 105]. NMR experiments by Curro et al. [203] and Haase et al. [204] show that this is indeed the case. The gapping of hot quasiparticles should also lead to a temperature-dependent reduction in the uniform magnetic susceptibility [91].

The phenomenological description is of course not enough as it leaves open the key question, namely what causes the spectral weight transfer for hot quasiparticles. We now discuss how the experimental phase diagram fits into the spin-fluctuation scenario.

First of all a general phase diagram based on spin fluctuation approach should distinguish between weakly and strongly antiferromagnetic materials. Weakly antiferromagnetic materials are those far to the right of T_{cr} for which the dimensionless coupling constant λ is smaller than unity, which corresponds to a correlation length smaller than a few lattice constants. For these materials, the normal state is a renormalized Fermi liquid, the nearly antiferromagnetic Fermi liquid, and T_c signals a transition to a BCS-like superconducting state with a $d_{x^2-y^2}$ order parameter. For strongly antiferromagnetic materials, on the other hand, the dimensionless coupling constant $\lambda \geq 1$. In this situation, the normal state behavior deviates from a Fermi liquid already at comparatively small ω and T although at the lowest frequencies the system still would display a Fermi liquid behavior if indeed one could extend the normal state down to $T = 0$. For these systems, we also know that the pairing instability temperature T_{cr} increases with decreasing doping and for large enough coupling saturates at a value comparable to the magnetic J. Applying this to the experimental phase diagram, we see that optimally doped materials are at the borderline between being weakly and strongly antiferromagnetic: on the one hand, the dimensionless coupling is already not small, on the other hand, the pseudogap phase extends at best over a T range which is only a fraction of T_{cr}.

As we have noted above, these results of the spin–fermion model make it a strong candidate for the microscopic description of the pseudogap phase: T_{cr} saturates at a finite value at the magnetic transition; for $\lambda \geq 1$, the pairing involves non-Fermi liquid fermions; at $T = 0$ there are two distinct energy scales in the problem, a fermionic gap $\Delta \propto T_{cr}$ and a bosonic gap $\Delta_s \propto T_{cr}\lambda^{-1} \ll T_{cr}$. The central issue is whether the pairing of incoherent fermions only gradually changes the fermionic self-energy, or whether it creates a feedback on fermions which immediately gives rise to a coherent quasiparticle behavior at the lowest frequencies, as happens in dirty superconductors where $\Sigma(\omega) = i\gamma$ in the normal state transforms below T_c into a mass renormalization at the smallest ω

$$\Sigma(\omega) = i\gamma \frac{\omega}{(\omega^2 - \Delta^2)^{1/2}} \simeq \gamma \frac{\omega}{\Delta} + O\left(\omega^3\right) . \quad (22.101)$$

If the feedback is gradual, then the pairing creates bound states of incoherent fermions with $S = 0$. In this situation, there is a reduction in the density of states below T_{cr}, but a full superconducting gap does not develop until a smaller temperature, T_c. A simple toy model which describes this physics would be one [48] in which fermions in the normal state display a quantum critical behavior with $\tilde{\Sigma}(\omega) \approx \Sigma(\omega) = (i\omega\bar{\omega})^{1/2}$, and pairing creates a nonzero flat pairing vertex Φ but does not affect $\Sigma(\omega)$. In this situation, the fermionic propagator acquires a gap at a finite but imaginary frequency:

$$G_{\mathbf{k}}(\omega) \propto \frac{\sqrt{i\omega\bar{\omega}} + \epsilon_{\mathbf{k}}}{i\omega - E_{\mathbf{k}}} , \quad (22.102)$$

where

$$E_{\mathbf{k}} = (\Phi^2 + \epsilon_{\mathbf{k}}^2)/\bar{\omega}\,. \qquad (22.103)$$

The spectral function at $\mathbf{k} = \mathbf{k}_F$ and the fermionic DOS both have broad maxima at $\omega = \Delta = \Phi^2/\bar{\omega}$, but the spectral weight is finite at any finite ω, although reduced at low frequencies. Within this model, the transition to the true superconducting state can be understood as a rotation of the pole in (22.102) from the imaginary to the real frequency axis such that at the lowest frequencies the $i\omega$ term becomes purely real. The frequency up to which $N(\omega) = 0$ then give an estimate for the actual T_c.

It is unclear to what extent the results of this toy model reflect the physics of the spin–fermion model below T_{cr}. Without elaborating on this subject of current research we mention that as long as an Eliashberg approach is justified, phase fluctuations of the superconducting order parameter cannot substantially reduce T_c compared to T_{cr}. Behavior different from that in dirty superconductors could emerge only if longitudinal fluctuations of the superconducting order parameter are soft and able to destroy the superconducting coherence at $T_c \ll T_{\text{cr}}$. We have both numerical [89] and analytical [205] evidence that longitudinal gap fluctuations are indeed strong in the limit $\lambda = \infty$. Still, this subject is far from being fully understood and clearly requires further study.

Another subtle issue is whether the spin–fermion model displays quantum critical behavior at $x = x_{\text{cr}}$. Physically, this would imply that the pairing of incoherent fermions at T_{cr} and the pairing of coherent fermions at T_c are uncorrelated phenomena; the first gives rise only to the pseudogap, while the latter yields BCS-type superconductivity. Since the pairing of incoherent fermions is not a perturbative phenomenon and requires the interaction to exceed a threshold value [87], the pairing of incoherent fermions involves only quasiparticles in some *finite* region around a hot spot and would form a dome on top of a magnetic quantum critical point and vanish at a finite x. It is not clear how well one can separate coherent and incoherent pairings.

Another issue related to the possible explanation of the pseudogap within the spin–fermion model is whether one can smoothly interpolate between the limit $\bar{g} \ll W$ (W is the fermionic bandwidth), where one can perform calculations in a controlled fashion, and $\bar{g} \gg W$ where Mott physics become relevant. In essence the issue is whether or not there is a qualitative difference between limits in which the effective interaction is either much larger or much smaller than the fermionic bandwidth. A sign that these two limits may describe some aspects of the basic physics similarly is that, as shown earlier, T_{cr} scales with $\bar{g}$ for $\bar{g} < W$, but crosses over to $W^2/\bar{g} \propto J$ for $\bar{g} > W$. At the same time, the limit $\bar{g} > W$ is probably more rich than the small $\bar{g}$ limit as the physics associated with the localization in the Mott insulator is not included into our analysis. We speculate that due to this Mott physics, bound singlet pairs of fermions that emerge below T_{cr} could order for example in columnar fashion as suggested by Sachdev and collaborators [206] who arrived at a columnar phase by studying weakly doped Mott antiferromagnets. This ordering in turn would imply that T_{cr} is a true phase transition line below which Z_4 symmetry is broken. Columnar ordering also opens a link between our approach and the approaches which depart from Mott insulator at half-filling. In particular, columnar ordering of bound electron pairs naturally leads to stripe physics [206]. An alternative possibility is that singlet pairs remain spatially disordered [92,207]. In any event, the role of localization effects certainly increases as the system approaches half-filling. Whether they remain strong near optimal doping in the normal state is a subject of debate, but still, strong localization effects should reduce the number of low energy carriers and therefore change the volume of the Fermi surface or increase their mass. ARPES experiments on the other hand indicate that in the normal phase, the Fermi surface is large and obeys Luttinger's theorem without dramatic mass renormalizations. We therefore believe that near optimal doping localization effects are at best moderate.

We conclude this discussion of the phase diagram by mentioning two alternative scenarios for the pseudogap and anomalous normal state properties. The first scenario, pioneered by P.W. Anderson [92], X.-G. Wen and P. A. Lee [208] and others [207,210–216] and later modified chiefly by M.P.A. Fisher and T. Senthil [209], assumes spin-charge separation at half-filling and explains the whole phase diagram as a result of weak doping of a Mott insulator. A second scenario, on the contrary, assumes that one can understand the phase diagram within mean-field theory; as Chakravarty et al. [100] have proposed the pseudogap might then be a result of the breaking of a hidden symmetry.

22.7 Conclusions

In this chapter we have demonstrated that superconducting pairing mediated by the exchange of spin fluctuations is a viable alternative to conventional phonon-mediated pairing. We discussed in detail the normal state properties, the pairing instability and the superconducting behavior of a material near an antiferromagnetic instability, when the dominant interaction between quasiparticles is of electronic origin and, at energies much smaller than the fermionic bandwidth, can be viewed as being due to the emission and absorption of a collective, soft spin degree of freedom. We argued that the spin-fluctuation exchange yields an attraction in the $d_{x^2-y^2}$ channel in agreement with what nearly all researchers now believe is the pairing symmetry in the cuprates.

We demonstrated that the physics is qualitatively different depending on whether or not Fermi surface geometry allows a process in which a collective mode decays into a particle and a hole. For this to be possible, the Fermi surface should contain hot spots. We focused on the case in which the Fermi surface does contain hot spots (as the photoemission experiments in cuprates indicate) and showed that spin fluctuations are then overdamped and that their diffusive dynamics should be analyzed in a consistent manner with the low energy dynamics of the fermions. We further argued that contrary to naive expectations, this case is better for d-wave pairing than one in which spin fluctuations are propagating, magnon-like quasiparticles.

We showed that the low-energy theory for fermions interacting with overdamped collective spin excitations is universal, independent of the details of the underlying lattice Hamiltonian and is characterized by only two input parameters: the dimensionless coupling constant, λ, and an overall energy scale, $\bar{\omega}$, proportional to the effective spin–fermion interaction $\bar{g}$. In so doing, we have developed the microscopic justification for the NAFL. The coupling constant, λ, scales with the magnetic correlation length, so that close enough to a magnetic transition, the system falls into a strong coupling regime. A universal description is valid if $\bar{g}$ is smaller than the fermionic bandwidth W (that would correspond to a weak coupling limit if the system was far away from a magnetic instability). In the opposite limit which we did not discuss in detail, lattice effects become important, and the universality is lost.

At sufficiently low temperatures and energies, the nearly antiferromagnetic Fermi liquid (NAFL) is a Fermi liquid, in the sense that one can obtain a one-to-one correspondence between the low-lying states of a Fermi gas and the Fermi liquid, as though the particle interaction was turned on adiabatically [121]. However, because of the closeness to an antiferromagnetic instability, it is an unconventional Fermi liquid, in that the characteristic energy above which this description is no longer valid is not the Fermi energy, but is the much lower spin-fluctuation energy $\omega_{sf} \propto \bar{\omega}\lambda^2$ that is typically two orders of magnitude smaller than the Fermi energy. For energies (or temperatures) between ω_{sf} and the Fermi energy, we have seen using the spin–fermion model that the system behavior is again universal and depends only on a very limited number of phenomenological parameters.

We compared in detail the spin fluctuation approach with the Eliashberg approach to phonon superconductors and showed that despite the absence of the small electron to ionic mass ratio that justified Eliashberg theory for phonons, an Eliashberg-type approach to the spin–fermion model is still justified, but only at strong coupling $\lambda \geq 1$.

We showed that at large λ, there are two distinct energy scales for the normal state problem: $\bar{\omega} \sim \bar{g}$ which at $\bar{g} \sim W$ is of the order of the Fermi energy (even though numerically smaller by about an order of magnitude), and a much smaller $\omega_{sf} = \bar{\omega}/(4\lambda^2) \ll \bar{\omega}$. Conventional Fermi liquid behavior with $\Sigma'' \propto \omega^2 + (\pi T)^2$ and an almost temperature independent static spin susceptibility exists only at frequencies and temperatures smaller that ω_{sf}. At frequencies between ω_{sf} and $\bar{\omega}$, the system crosses over into a regime in which spin susceptibility is diffusive and $\Sigma'' \propto \omega^{\alpha}$. We found that $\alpha = \frac{1}{2}$, while the behavior in the crossover region from Fermi liquid to non Fermi liquid behavior resembles a linear in frequency dependence. In the diffusive regime, both spin and fermionic propagators become independent of λ, i.e. the system displays magnetic quantum-critical behavior. Finally, at $\omega > \bar{\omega}$, the self-energy gets smaller than the bare ω, although $\Sigma'' \propto \omega^{1/2}$ still holds.

We next argued that the existence of the two distinct energy scales has a strong impact on the pairing problem: the pairing instability temperature T_{cr} is predominantly determined by incoherent, non-Fermi liquid fermions with energies between ω_{sf} and $\overline{\omega}$. We then considered the superconducting state, found the pairing gap $\Delta \propto T_{cr}$ ($2\Delta/T_{cr} \sim 4$ for $\lambda \gg 1$), and argued that one of the signatures of the spin-fluctuation mechanism is the change of the low-frequency spin dynamics from relaxational to propagating due to the feedback from the pairing. This give rise to the emergence of the resonance peak in the spin propagator at a frequency Δ_s which at large λ scales as $\Delta_s \sim 1.5T_{cr}/\lambda$. For smaller λ, the resonance frequency goes up but always remain smaller than 2Δ. This restriction (not found for phonon superconductors) is a consequence of feedback between fermionic and bosonic dynamics.

We discussed how the emergence of the resonance mode affects fermionic properties and identified several "fingerprints" of the strong coupling spin-fluctuation scenario. These include, but are not limited to, the peak/dip/hump in the spectral function, the near-absence of the dispersion of the quasiparticle peak, the peak/dip features in the SIN and SIS tunneling conductances, and in the second derivative of the optical conductivity.

Finally, we compared the theoretical results with the normal state data for near optimally-doped cuprates, and found that a large number of experimental results can be understood within the spin-fluctuation scenario. In particular, we argued that the tunneling, photoemission and optical data for the superconducting state display the expected "fingerprints" of the spin-fluctuation mechanism for superconductivity in the cuprates.

The spin-fluctuation theory should be equally applicable to quasi two dimensional organic superconductors or compounds of the family $CeXIn_5$ with X=Co, Rh, Ir, if the Fermi surface of these systems possesses hot spots. A detailed experimental investigation of spectroscopic properties of these systems in the superconducting state will provide important clues on whether they are indeed magnetically mediated superconductors.

Despite the successful identification of the "fingerprints" of spin mediated pairing in the cuprates, we listed in Sect. 22.6.2 some properties of these materials which we have not explained within the spin-fluctuation scenario. Perhaps the most important remaining question is to what extent the spin-fluctuation theory explains the pseudogap physics in underdoped cuprates. The existence of the two distinct energy scales in the spin–fermion model and the presence of fermionic incoherence in the normal state make this model a good candidate model for the pseudogap. The plausible argument here is that while the pairing of the incoherent fermions will create spin singlets, fermions will still behave incoherently and hence not superconduct until a smaller T_c is reached when feedback from the pairing restores fermionic coherence. Particularly relevant here is whether or not there exists an additional critical point on the phase diagram at around $x = 0.19$. This critical point emerges in the spin-fluctuation theory if the pairing of incoherent fermions from frequencies above ω_{sf} and the pairing of coherent fermions from frequencies below ω_{sf} are separate phenomena (the first then definitely gives rise to a pseudogap physics while the second yields a true superconductivity). The pairing of incoherent fermions is a threshold phenomenon and thus occurs at $T_{cr}(x)$. We still however need to understand whether incoherent pairing and coherent pairing can be totally separated from each other.

Another unresolved issue is to what extent the fact that the parent compounds of the cuprates (La_2CuO_4 or $YBa_2Cu_3O_6$) are Mott insulators and nearest-neighbor Heisenberg antiferromagnets affects the behavior of doped materials. For the spin–fermion model, this question could be reformulated as whether the limit $\bar{g} \ll W$, in which we can separate low and high energies and perform controlled calculations, and the limit $\bar{g} \gg W$, are only quantitatively different or are qualitatively different. An encouraging sign that the two limits may not be very different is the result that T_{cr} smoothly interpolates between $O(\bar{g})$ for $\bar{g} \ll W$ to $O(W^2/\bar{g})$ for $\bar{g} \gg W$, in the latter case T_{cr} becomes of order of a magnetic exchange integral J. On the other hand, the localization effects which accompany a transition to a Mott insulator are not included in the theory presented here. In particular, we cannot predict what would happen with singlet pairs below T_{cr}, if it indeed is the onset of the pseudogap, whether they remain disordered or form columnar stripes as Sachdev and collaborators [206] suggested.

The enduring presence and richness of unsolved problems in the field of unconventional superconductivity makes us optimistic that we will continue to see unexpected experimental observations and new, creative theoretical concepts in this field of research. We have enjoyed over a number of years our respective collaborations on the spin fluctuation approach because of its clarity and ability to make falsifiable predictions for experiment. We hope that the readers of this Chapter will share our excitement for this approach to magnetically mediated superconductivity.

Acknowledgements

It is our pleasure to thank our immediate collaborators Artem Abanov, Victor Barzykin, Alexander Finkelstein, Rob Haslinger, Philippe Monthoux and Branko Stojkovic for stimulating discussions on essentially all aspects of the spin fluctuation approach presented here. We are also thankful to E. Abrahams, A.A. Abrikosov, G. Aeppli, B.L. Altshuler, N. Andrei, A. Balatsky, D. Basov, K.H. Bennemann, G. Blumberg, J. Brinckmann, J.C. Campuzano, P. Chandra, P. Coleman, E. Dagotto, J.C. Davis, E. Demler, M.P.A. Fisher, A. Georges, L.P. Gor'kov, M. Grilli, L. Ioffe, P. Johnson, R. Joynt, A. Kaminski, B. Keimer, D. Khveschenko, S. Kivelson, G. Kotliar, R. Laughlin, M. Lavagna, P.A. Lee, G. Lonzarich, H. von Löhneysen, A. Millis, H. Monien, D.K. Morr, M. Norman, C. Pépin, A. Rosch, S. Sachdev, D. Scalapino, J.R. Schrieffer, T. Senthil, Q. Si, C.P. Slichter, S. Sondhi, Z. Tesanovic, O. Tchernyshyov, A. Tsvelik, J. Tu, C. Varma , P.G. Wolynes, A. Yazdani, J. Zasadzinski and S.-C. Zhang for useful conversations. We further wish to thank A. Abanov, D. Basov, J.C. Davis and J. Tu for sharing their unpublished results with us.

The research which led to this review was supported in part by NSF DMR-9979749 (A. Ch.), by the Institute for Complex Adaptive Matter, an independent unit of the University of California, by Project DR 200153 on *Emergent Properties of Correlated Electron Superconductors* and general DOE support to Los Alamos National Laboratory (D. P.), and by the Ames Laboratory, operated for the U.S. Department of Energy by Iowa State University under contract No. W-7405-Eng-82 (J.S).

22.8 Note Added

The first edition of the "Physics of Superconductors" was released in 2003. Since that time, there have been many new exciting developments in the field of high temperature superconductivity. Below we list the several developments that are most closely related to the issues that we considered in the review. This list is by no means complete.

1. There has been a strong interest in the dispersion anomalies in the cuprates. Of particular interest is the understanding of the kink in the fermionic dispersion [169, 217]. The kink was first observed along the nodal direction, but later it was found that the effect is the strongest along antinodal direction, where the dispersion has an S-shape below T_c Two different explanations for the kink were proposed. Within a magnetic scenario [218], the S–shape dispersion in the superconducting and pseudogap states is the result of the interaction with the resonance mode. The absence of the S-shape dispersion along the nodal direction is a natural consequence of the fact that a spin response does not have a peak structure at a bosonic momentum connecting a nodal fermion with any other point at the Fermi surface. As a result, the fermionic dispersion only has a kink (a discontinuity in the derivative) near $\omega = 2\Delta$. An alternative explanation identifies the

dispersion anomalies with the electron-phonon interaction [219]. The difference between nodal and antinodal dispersion was attributed to interactions with two different phonon modes: breathing mode for nodal fermions, and buckling mode for antinodal fermions. We believe that magnetic explanation is more natural, particularly as the magnetic scenario explains the data rather well for the same value of the coupling that yields a right T_c.

2. There have been spectacular improvements of the accuracy of the optics data, mostly associated with ellipsometry technique [220,221]. As a result, it has become possible to measure the optical integral $K(\omega) = \int_{0+}^{\omega} \sigma_1(\omega)d\omega$ to an accuracy of 10^{-3} and analyze in detail the redistribution of the spectral weight between the normal and superconducting state. In clean BCS superconductors, the spectral weight in the δ-functional peak below T_c comes from frequencies $\sim 2\Delta$, and $\Delta K(\omega) = K_{sc}(\omega) - K_n(\omega)$ is negative at low frequencies, changes the sign at $\omega \leq 2\Delta$ and remains positive at higher frequencies. This behavior has been observed in the overdoped regime [222]. The data for optimally-doped and underdoped regimes are still somewhat controversial and at least partly depend on how normal state data are extrapolated to $T = 0$ [220–222]. Still, all groups agree that the scale, at which differential optical integral $\Delta K(\omega)$ changes sign, increases with underdoping and possibly becomes larger than the fermionic bandwidth. Spin-fluctuation theory predicts this kind of behavior [223]. The scale at which $\Delta K = 0$ saturates at strong coupling is of order $\bar{\omega}$.

   In another development of optics studies, both real and imaginary parts of the optical conductivity in the normal state have been carefully measured at intermediate $\omega \sim 300 - 600 meV$, using the advanced technique to estimate ϵ_∞ [224]. The results show that the complex conductivity $\sigma(\omega)$ in this frequency range is well described by $\sigma \propto (i\omega)^{-\gamma}$, where $\gamma \sim 0.6 - 0.7$. These data were first interpreted as an evidence for quantum-critical scaling [224], but later were re-interpreted as a generic result of the interaction between fermions and a bosonic continuum [225]. The constraints on the bosonic continuum are rather loose, and spin-fluctuation is one the candidate bosons.

3. The accuracy of the ARPES experiments also improved substantially in recent years. In particular, the Dresden group managed to separate an intrinsic peak-dip-hump behavior of the spectral function from the effects of the bilayer splitting. [226] They identified peak-dip distance as 40 meV, which is the resonance mode frequency. In a different development, the improved accuracy of the MDC measurements and the measurements using a novel laser ARPES technique [227] both demonstrated that the intrinsic width of the ARPES spectral function is smaller than originally thought. This partly eliminates the issue of the large frequency and temperature independent contribution to the self-energy that one earlier has to invoke in order to fit ARPES data. Furthermore, the new data show that at above optimal doping, $Im\Sigma \propto \omega^2$, as in a Fermi liquid.

   In neutron scattering, the developments related to our theory are the detection of the resonance peak in even channel, and the measurement of the second resonance peak at diagonal deviations from (π, π) larger than to nodal–nodal distance $(0.8\pi, 0.8\pi)$ [228]. Both of these effects are naturally explained within spin-fluctuation scenario [229]. The novel resonance is also a spin exciton.

4. Recently there has been a surge of interest in *electron-doped* cuprates, $Nd_{2-x}Ce_xCuO_4$ and $Pr_{2-x}Ce_xCuO_4$. Their phase diagram [230] has a wider region of antiferromagnetism, superconducting $T_c(x)$ forms a dome above the antiferromagnetic quantum-critical point [231], and pseudogap T^* [232] tracks the Neel temperature [232], and so is likely just a property of a quasi-2D antiferromagnet [233, 234]. Optical data show [235] the sharp charge-transfer gap near 2eV at small x melting away near the optimal doping $x \sim 0.15$, where the superconductivity appears. This implies that electron-doped materials are even better for the application of

spin–fermion model than hole-doped materials. Recent theoretical results [236] are optimistic.

5. In theory, the development most closely related to our review is the discussion on the value of the spin–fermion coupling constant g. We argued above that $g \sim 0.7$ eV. Kee et al. [237] argued that the effective spin–fermion coupling is much weaker, or order $g \sim 0.01 eV$. The difference between the two results is related to different choices for the fermionic density of states N_0: a large value of g is obtained by assuming a large, Luttinger Fermi surface, with $N_0 \sim 1\ \mathrm{eV}^{-1}$, while a smaller g results by assuming that the density of states $N_0 \sim J^{-1} \sim 10\ \mathrm{eV}^{-1}$ is the same as in weakly doped quantum AF. We believe that experimental measurements of $p_F \sim 0.6\pi/a$ and $v_F \sim 1$ eV$\times a$ prove that $N_0 \sim 1\ \mathrm{eV}^{-1}$ Kee et al. also argued that the small overall spectral weight of the resonance peak precludes strong feedback effects on the fermions of the type we considered in the review. We believe [238] that the overall weight of the peak is small because the momentum range for the peak is small (roughly between (π, π) and $(0.8\iota, 0.8\pi)$ what constitutes only 4% of the Brillouin zone). At the same time, for energies $\omega \sim 100$ meV, typical fermionic momenta $|\pi - q| \sim \omega/v_F \sim 0.1/a$ are well within the range where the resonance has been observed.
6. There are continuous theoretical efforts to understand the connection between spin–fermion model (applicable in the case where the Fermi surface is large and when Mott physics is not essential) and the physics very near half-filling, where the spin excitations are governed by the two dimensional Heisenberg model and strong coupling lattice effects become important [239–242].

We also refer the reader to recent detailed reviews [243–246] on the interplay between fermions and low-energy bosons that mediate fermion–fermion interaction.

References

1. See for example: *Proceedings of Materials and Mechanisms of Superconductivity: High Temperature Superconductors VI* in Physica C (Amsterdam), **341–348** (2000)
2. J. Bardeen and D. Pines, Phys. Rev. **99**, 1140–1150 (1955)
3. J. Bardeen, L. Cooper, J.R. Schrieffer, Phys. Rev. **108**, 1175–1204 (1957)
4. W.L. McMillan and J.M. Rowell, In: *Superconductivity*, ed. by R.D. Parks (Marcell-Dekker, New York, 1969) Vol. 1, p. 561
5. D.J. Scalapino, In: *Superconductivity*, ed. by R.D. Parks (Marcell-Dekker, New York, 1969) Vol. 1, p. 449
6. E. Maxwell, Phys. Rev. **78**, 477 (1950)
7. A. Reynolds, B. Serin, W.H. Wright, and L.B. Nesbit, Phys. Rev. **78**, 487 (1950)
8. G.M. Eliashberg, Sov. Phys. JETP **11**, 696 (1960)
9. L.D. Landau, Sov. Phys. JETP **3**, 920-925 (1957)
10. A.J. Leggett, Rev. Mod. Phys. **47**, 331 (1975)
11. W. Kohn and J.M. Luttinger, Phys. Rev. Lett. **15**, 524 (1965)
12. D. Vollhardt, and P. Wölfle, In: *The Superfluid Phases of Helium 3* (Taylor and Francis, London, 1990)
13. G.E. Volovik, In: *Exotic Properties of* ^{3}*He* (World Scientific, Singapore, 1991)
14. V.J. Emery, J. Phys. (Paris) Colloq. **44**, C3-977 (1983); Synth. Met. **13**, 21 (1986)
15. J.E. Hirsch, Phys. Rev. Lett. **54**, 1317 (1985)
16. K. Miyake, S. Schmitt-Rink, and C.M. Varma, Phys. Rev. B **34**, 6554 (1986)
17. D.J. Scalapino, E. Loh, Jr. and J.E. Hirsch, Phys. Rev. B **34**, 8190 (1986); *ibid* **35** 6694 (1987)
18. N.E. Bickers, D.J. Scalapino, and R.T. Scalettar, Int. J. Mod. Phys B **1**, 687 (1987)
19. A.J. Millis, S. Sachdev, and C.M. Varma, Phys. Rev. B **37**, 4975 (1988)
20. G. Kotliar, Phys. Rev. B **37**, 3664 (1988)
21. P.W. Anderson, S. Chakravarty, S. Strong and A. Sudb, Science **261**, 337 (1993)

22. P.W. Anderson, In: *The Theory of Superconductivity in the High-T_c Cuprate Superconductors* (Princeton University Press, Princeton, NJ, 1997)
23. A.J. Leggett, Proc. Nat. Acad. Sci. **96**, 8365-8372 (2000)
24. D.A. Wollmann, D.J. Van Harlingen, W.C. Lee, D.M. Ginsberg, and A.J. Leggett, Phys. Rev. Lett. **71**, 2134 (1993)
25. C.C. Tsuei, J.R. Kirtley, C.C. Chi, Lock See Yu–Jahnes, A. Gupta, T. Shaw, J.Z. Sun, and M.B. Ketchen, Phys. Rev. Lett. **73**, 593 (1994)
26. D.A. Brawner and H.R. Ott, Phys. Rev. B **50**, 6530 (1994)
27. C.P. Slichter, In: *Strongly Correlated Electron Systems*, ed. by K.S. Bedell et al. (Addison-Wesley, Reading, MA, 1994)
28. See, e.g., In: *High Temperature Superconductivity*, ed. by K. S. Bedell, D. Coffey, D. Melzer, D. Pines, and J.R. Schrieffer (Addison-Wesley, Reading, MA, 1990)
29. T. Moriya, Y. Takahashi and K. Ueda: J. Phys. Soc. Jpn. **59**, 2905 (1990)
30. T. Moriya, Y. Takahashi and K. Ueda, Physica C **185-189**, 114 (1991)
31. D. Pines, Physica B **163**, 78 (1990)
32. P. Monthoux, A.V. Balatsky and D. Pines, Phys. Rev. Lett. **67**, 3448 (1991)
33. A.J. Millis, H. Monien, and D. Pines, Phys. Rev. B **42**, 167 (1990)
34. K. Ueda, T. Moriya and Y. Takahashi, In: *Electronic Properties and Mechanisms of High-T_c Superconductors* (Tsukuba Symposium, 1991), ed. by T. Oguchi, et al. (North Holland, Amsterdam, 1992) p. 145; J. Phys, Chem. Solids **53** 1515 (1992)
35. P. Monthoux and D. Pines: Phys. Rev. Lett. **69**, 961 (1992)
36. P. Monthoux and D. Pines, Phys. Rev. B **48**, 6069 (1993)
37. P. Monthoux and D. Pines, Phys. Rev. B **50**, 16015 (1994)
38. S.E. Barrett, D.J. Durand, C.H. Pennington, C.P. Slichter, T.A. Friedmann, J.P. Rice, and D.M. Ginsberg, Phys. Rev. B **41**, 6283–6296 (1990)
39. M. Takigawa, P.C. Hammel, R.H. Heffner, and Z. Fisk, Phys. Rev. B **39**, 7371–7374 (1989)
40. H. Monien and D. Pines, Phys. Rev. B **41**, 6297 (1990)
41. R.E. Walstedt, W.W. Warren, Jr., R.F. Bell, G.F. Brennert, G.P. Espinosa, J.P. Remeika, R.J. Cava, and E.A. Rietman, Phys. Rev. B **36**, 5727–5730 (1987)
42. W.W. Warren, Jr., R.E. Walstedt, G.F. Brennert, G.P. Espinosa, and J.P. Remeika, Phys. Rev. Lett. **59**, 1860–1863 (1987)
43. Z.X. Shen, D.S. Dessau, B.O. Wells, D.M. King, W.E. Spicer, A.J. Arko, D. Marshall, L.W. Lombardo, A. Kapitulnik, P. Dickinson, S. Doniach, J. DiCarlo, T. Loeser, and C.H. Park, Phys. Rev. Lett. **70** 1553 (1993)
44. W.N. Hardy, D.A. Bonn, D.C. Morgan, Ruixing Liang, and Kuan Zhang, Phys. Rev. Lett. **70**, 3999–4002 (1993)
45. J.A. Martindale, S.E. Barrett, K.E. O'Hara, C.P. Slichter, W.C. Lee, and D.M. Ginsberg, Phys. Rev. B **47**, 9155–9157 (1993)
46. N. Bulut and D.J. Scalapino, Phys. Rev. Lett. **68**, 706–709 (1992)
47. D. Thelen, D. Pines and Jian Ping Lu, Phys. Rev. B **47**, 9151–9154 (1993)
48. A.V. Chubukov, Phys. Rev. B **52**, R3840 (1995); Europhys. Lett. **44**, 655 (1998)
49. A. Abanov, A.V. Chubukov, J. Schmalian, Adv. Phys. **52**, 119 (2003)
50. S. Lefebvre, P. Wzietek, S. Brown, C. Bourbonnais, D. Jerome, C. Meziere, M. Fourmigue, P. Batail, Phys. Rev. Lett. **85**, 5420 (2000)
51. J. Schmalian, Phys. Rev. Lett. **81**, 4232 (1998)
52. H. Kino and H. Kontani, J. Phys. Soc. Jpn. **67**, 3691 (1998)
53. H. Kondo and T. Moriya, J. Phys. Soc. Jpn. **67**, 3695 (1998)
54. K. Kuroki and H. Aoki, Phys. Rev. B 60, 3060 (1999)
55. S.M. De Soto, C.P. Slichter, A.M. Kini, H.H. Wang, U. Geiser, and J.M. Williams, Phys. Rev. B **52**, 10364 (1995)
56. H. Mayaffre, P. Wzietek, D. Jerome, C. Lenoir, and P. Batail , Phys. Rev. Lett. **75**, 4122 (1995)
57. K. Kanoda, K. Miyagawa, A. Kawamoto, Y. Nakazawa, Phys. Rev. B **54**, 76 (1996)
58. S. Belin, K. Behnia, and A. Deluzet, Phys. Rev. Lett. **81**, 4728 (1998)
59. J.M. Schrama, E. Rzepniewski, R.S. Edwards, J. Singleton, A. Ardavan, M. Kurmoo, and P. Day, Phys. Rev. Lett. **83**, 3041 (1999)
60. T. Arai, K. Ichimura, K. Nomura, Takasaki, J. Yamada, S. Nakatsuji, and H. Anzai, Phys. Rev. B **63**, 104518 (2001)
61. K. Izawa et al., Phys. Rev. Lett. **86**, 2653 (2001)
62. M. Lang, N. Toyota, T. Sasaki, and H. Sato, Phys. Rev. Lett. **69**, 1443 (1992)

63. J. Müller, M. Lang, R. Helfrich, F. Steglich, T. Sasaki, cond-mat/0109030
64. F. Steglich, J. Aarts, C.D. Bredl, W. Lieke, D. Meschede, W. Franz, and H. Schäfer, Phys. Rev. Lett. **43**, 1892 (1979)
65. N.D. Mathur, F.M. Grosche, S.R. Julian, I.R. Walker, D.M. Freye, R.K.W. Haselwimmer, G.G. Lonzarich, Nature (London) **394**, 39–43 (1998)
66. C. Petrovic, P.G. Pagliuso, M.F. Hundley, R. Movshovich, J.L. Sarrao, J.D. Thompson, Z. Fisk, and P. Monthoux, J. Phys. Condens. Mat. **13**, L337 (2001); C. Petrovic, R. Movshovich, M. Jaime, P.G. Pagliuso, M.F. Hundley, J.L. Sarrao, Z. Fisk, and J.D. Thompson, Europhys. Lett. **53** 354 (2001); H. Hegger, C. Petrovic, E.G. Moshopoulou, M.F. Hundley, J.L. Sarrao, Z. Fisk, and J.D. Thompson , Phys. Rev. Lett. **84** 4986 (2000)
67. K. Izawa, H. Yamaguchi, Yuji Matsuda, H. Shishido, R. Settai, and Y. Onuki, Phys. Rev. Lett. **87**, 057002 (2001)
68. P.G. Pagliuso, R. Movshovich, A.D. Bianchi, M. Nicklas, N.O. Moreno, J.D. Thompson, M.F. Hundley, J.L. Sarrao, Z. Fisk, cond-mat/0107266
69. Y. Maeno, H. Hashimoto, Y. Yoshida. S. Nishizaki, T. Fujita, J.G. Bednorz, and F. Lichtenberg, Nature (London) 372, 532 (1994)
70. Y. Maeno, T.M. Rice, and M. Sigrist, Phys. Today **54**, 42 (2001)
71. T.M. Rice and M. Sigrist, J. Phys.: Condens. Matter **7**, L643 (1995)
72. P. Monthoux and G.G. Lonzarich, Phys. Rev. B **63**, 054529 (2001)
73. K. Ishida, H. Mukuda, Y. Kitaoka, K. Asayama, Z.Q. Mao, Y. Mori, and Y. Maeno, Nature (London) **96**, 658 (1998)
74. J.A. Duffy, S.M. Hayden, Y. Maeno, Z. Mao, J. Kulda, and G.J. McIntyre, Phys. Rev. Lett. **85**, 5412 (2000)
75. J.A. Hertz, Phys. Rev. B **14**, 1165 (1976)
76. A.J. Millis, Phys. Rev. B **48**, 7183 (1993)
77. P. Coleman, C. Pepin, Q. Si, R. Ramazashvili, J. Phys. Cond. Matt. **13**, 723 (2001)
78. Q. Si, JMMM **226-230**, 30 (2001)
79. F.C. Zhang, T.M. Rice, Phys. Rev. B **37**, 3759 (1988)
80. J.R. Schrieffer, X.G. Wen, and S.C. Zhang, Phys. Rev. B **39**, 11663 (1989)
81. J.R. Schrieffer, J. Low Temp. Phys. **99**, 397 (1995)
82. A.V. Chubukov and D. Frenkel, Phys. Rev. B **46**, 11884 (1992)
83. A.V. Chubukov and D.K. Morr, Phys. Rep. **288**, 355 (1997)
84. S.L. Adler, Phys. Rev. **137**, B1022 (1965)
85. M.Yu. Kuchiev, O.P. Sushkov, Phys. Rev. B. **53** 443 (1996)
86. W.L. McMillan, Phys. Rev. **167** 331 (1968)
87. A. Abanov, A.V. Chubukov and A.M. Finkel'stein, Europhys. Lett. **54**, 488 (1999)
88. R. Roussev and A.J. Millis, Phys. Rev. B **63**, 140504 (2001)
89. A. Abanov, A.V. Chubukov, J. Schmalian, Europhys. Lett. **55** 369 (2001)
90. J. Friedel, J. Phys. Condens. Matter **1**, 7757 (1989)
91. H. Alloul, P. Mendels, G. Collin, and Ph. Monod, Physica C **156**, 355 (1988)
92. P.W. Anderson, Science, **235**, 4793 (1987); Mater. Res. Bull. **8**, 153 (1973)
93. V.J. Emery and S.A. Kivelson, Nature, London **374**, 434 (1995); Phys. Rev. Lett. **74**, 3253 (1995)
94. M. Franz and A.J. Millis, Phys. Rev. B **58**, 14572 (1998)
95. H.-J. Kwon and A.T. Dorsey, Phys. Rev. B**59**, 6438 (1999)
96. L.B. Ioffe and A.J. Millis, Science **285**, 1241 (1999)
97. M. Franz and Z. Tesanovic, Phys. Rev. Lett. **87**, 257003 (2001)
98. H.J. Schulz, Phys. Rev. B **39**, 2940 (1989)
99. J.B. Marston and I. Affleck, Phys. Rev. B **43**, 11538 (1989); B **61**, 14773 (2000)
100. S. Chakravarty, R. Laughlin, D.K. Morr, C. Nayak, Phys. Rev. B **63**, 094503 (2001)
101. C.M. Varma, P.B. Littlewood, S. Schmitt-Rink, E. Abrahams, A.E. Ruckenstein, Phys. Rev. B **63**, 1996 (1989); Phys. Rev. Lett. **64**, 497 (1990)
102. C.M. Varma, Phys. Rev. Lett. **83**, 3538 (1999); Phys. Rev. B **55**, 14554 (1997)
103. C. Castellani, C. DiCastro, and M. Grilli, Z. Phys. B **103**, 137 (1997)
104. M. Vojta, Y. Zhang, and S. Sachdev, Phys. Rev. Lett. **85**, 4940 (2000)
105. J. Schmalian, D. Pines, B. Stojkovich, Phys. Rev. Lett. **80**, 3839 (1998)
106. L.B. Ioffe and A.J. Millis, Uso. Fiz. Nauk. **41**, 595 (1998)

107. P. Monthoux and D.J. Scalapino, Phys. Rev. Lett. **72**, 1874 (1994)
108. C.-H. Pao and N.E. Bickers, Phys. Rev. Lett. **72**, 1870 (1994)
109. T. Dahm, L. Tewordt, Phys. Rev. Lett. **74**, 793 (1997)
110. T. Dahm, D. Manske, L. Tewordt, Europhys. Lett. **39**, 201 (1997); D. Manske, I. Eremin, and K. H. Bennemann, Phys. Rev. B **63**, 054517 (2001)
111. S. Grabowski, M. Langer, J. Schmalian, and K.H. Bennemann, Europhys. Lett. **34**, (1996)
112. M. Langer, J. Schmalian, S. Grabowski, and K.H. Bennemann, Phys. Rev. Lett. **75**, 4508 (1995)
113. R.L. Stratonovich, Dokl. Akad. Nauk SSSR **115**, 1097 (1957) [Soviet Phys. Doklady 2, **416** (1958)]
114. J. Hubbard, Phys. Rev. Lett. **3**, 77 (1959)
115. A.H. Castro Neto and B. Jones, Phys. Rev. B **62**, 14975 (2000)
116. P. Monthoux and D. Pines, J. Phys. Chem. Solids, **56**, 1651 (1995)
117. J.P. Carbotte, Rev. Mod. Phys. **62**, 1027 (1990); see also F. Marsiglio and J.P. Carbotte, Chap. 3 in volume one of this book
118. A. Migdal, Sov. Phys. JETP **7**, 996–1001 (1958)
119. S.B. Treiman, R. Jackiw, and D.J. Gross, In: *Lectures on Current Algebra and its Applications* (Princeton University Press, Princeton, 1972)
120. L.D. Landau, E.M. Lifshits and L.P. Pitaevskii, In: *Statistical Physics, Part 2* (Pergamon, New York, 1969)
121. D. Pines and P. Nozieres, In: *The Theory of Quantum Liquids* (W.A. Benjamin, New York, 1966)
122. L. Kadanoff, Phys. Rev. **132**, 2073 (1963)
123. A. Abanov and A. Chubukov, Phys. Rev. Lett. **83**, 1652 (1999)
124. A. Abanov and A. Chubukov, Phys. Rev. Lett. **84**, 5608 (2000)
125. R. Haslinger, A. Abanov, and A. Chubukov, Europhys. Lett. **58**, 271 (2002)
126. J.M. Kosterlitz and D.J. Thouless, J. Phys. C **6**, 1181 (1973)
127. S. Sachdev, A.V. Chubukov, A. Sokol, Phys. Rev. B **51**, 14874 (1995)
128. M.R. Norman and H. Ding, Phys. Rev. B **57** R11089 (1998)
129. M. R. Norman, Phys. Rev. B **61**, 14751 (2000); M. Eschrig and M.R. Norman, Phys. Rev. Lett. **85**, 3261 (2000)
130. C. Stemmann, C. Pépin, and M. Lavagna, Phys. Rev. B **50**, 4075 (1994); D.Z. Liu, Y. Zha, and K. Levin, Phys. Rev. B **75**, 4130 (1995); I. Mazin and V. Yakovenko, *ibid* **75**, 4134 (1995); A. Millis and H. Monien, Phys. Rev. B**54**, 16172 (1996); N. Bulut and D. Scalapino, Phys. Rev. B **53**, 5149 (1996); D.K. Morr and D. Pines, Phys. Rev. Lett. **81**, 1086 (1998); S. Sachdev and M. Vojta, Physica B **280**, 333 (2000); Y.-J. Kao et al., Phys. Rev. B **61**, R11898 (2000); F. Onufrieva and P. Pfeuty, cond-mat/9903097
131. O. Tchernyshyov, M.R. Norman, A.V. Chubukov, Phys. Rev. B **63**, 144507 (2001)
132. J. Brinckmann and P.A. Lee, Phys. Rev. Lett. **82**, 2915 (1999)
133. A. Abanov, A.V. Chubukov, J. Schmalian, J. Elect. Spectrosc. Relat. Phenom. **117**, 129 (2001)
134. A.V. Chubukov, B. Janko, O. Tchernyshyov, Phys. Rev. B **63**, 180507 (2001)
135. A. Abanov, unpublished
136. A.V. Fedorov, T. Valla, and P.D. Johnson, Q. Li, G.D. Gu, N. Koshizuka, Phys. Rev. Lett. **82**, 2179 (1999)
137. A. Kaminski, J. Mesot, H. Fretwell, J.C. Campuzano, M.R. Norman, M. Randeria, H. Ding, T. Sato, T. Takahashi, T. Mochiku, K. Kadowaki, and H. Hoechst, Phys. Rev. Lett. **84**, 1788 (2000)
138. H.F. Fong, B. Keimer, D. Reznik, D.L. Milius, and I.A. Aksay, Phys. Rev. B **54**, 6708 (1996); H.F. Fong, P. Bourges, Y. Sidis, L.P. Regnault, J. Bossy, A. Ivanov, L. Milius, I.A. Aksay, and B. Keimer, Phys. Rev. B **61**, 14773 (2000)
139. P. Dai, H.A. Mook, S.M. Hayden, G. Aeppli, T.G. Perring, D. Hunt, and F. Dogan, Science **284**, 1344 (1999) P. Dai, H.A. Mook, R.D. Hunt, F. Dogan, Phys. Rev. Lett. **63**, 054525 (2001)
140. M.R. Norman, H. Ding, J.C. Campuzano, T. Takeuchi, M. Randeria, T. Yokoya, T. Takahashi, T. Mochiku, and K. Kadowaki Phys. Rev. Lett. **79**, 3506 (1997)
141. Z-X. Shen et al., Science **280**, 259 (1998)
142. Ch. Renner, B. Revaz, J.-Y. Genoud, K. Kadowaki, and O. Fischer, Phys. Rev. Lett. **80**, 149 (1998)
143. J.F. Zasadzinski, L. Ozyuzer, N. Miyakawa, K.E. Gray, D.G. Hinks and C. Kendzora, Phys. Rev. Lett. **87**, 067005 (2001)
144. A.V. Puchkov et al. J. Phys. Chem. Solids **59**, 1907 (1998); D.N. Basov, R. Liang, B. Dabrovski, D.A. Bonn, W.N. Hardy, and T. Timusk, Phys. Rev. Lett. **77**, 4090 (1996)
145. J.P. Carbotte, E. Schachinger, D.N. Basov, Nature (London) **401**, 354 (1999)

146. H.F. Fong, P. Bourges, Y. Sidis, L.P. Regnault, A. Ivanov, G.D. Gu, N. Koshizuka, and B. Keimer, Nature **398**, 588 (1999)
147. T. Holstein, Phys. Rev. **96**, 535 (1954); P.B. Allen, Phys. Rev. B **3**, 305 (1971)
148. P. Littlewood and C.M. Varma, Phys. Rev. B **46**, 405 (1992)
149. D.J. Scalapino, J.R. Schrieffer and J.W. Wilkins, Phys. Rev. **148**, 263 (1966)
150. D. Coffey, Phys. Rev. B **42**, 6040 (1990)
151. W. Shaw and J.C. Swihart, Phys. Rev. Lett. **20**, 1000 (1968)
152. G.D. Mahan, In: *Many-Particle Physics* (Plenum Press, New York, 1990)
153. A. Chainani, T. Yokoya, T. Kiss, and S. Shin, Phys. Rev. Lett. **85**, 1966 (2000), and references therein.
154. H. Won and K. Maki, Phys. Rev. B **49**, 1397 (1994); Ye Sun and K. Maki, Phys. Rev. B **51**, 6059 (1995)
155. A. Abanov, A.V. Chubukov, N. Gemelke, Phys. Rev. B **61**, R787 (2000)
156. A. Abanov and A. Chubukov, Phys. Rev. B **61** ,R9241 (2000)
157. A. Chubukov, G. Blumberg, and D. Morr, Solid State Comm. **112**, 183 (1999)
158. J.R. Schrieffer, In: *Theory of Superconductivity*(W.A. Benjamin, Reading, MA, 1966)
159. S.M. Quinlan, P.J. Hirschfeld, D.J. Scalapino, Phys. Rev. B **53**, 8575 (1996)
160. T.P. Devereaux, D. Einzel, B. Stadlober, R. Hackl, D.H. Leach, and J.J. Neumeier Phys. Rev. Lett. **72**, 396 (1994); *ibid* Phys. Rev. B **54**, 12 523 (1996)
161. A. Abanov, A.V. Chubukov, J. Schmalian, Phys. Rev. B **63**, 180510 (2001)
162. R. Haslinger, A. Chubukov, and A. Abanov, Phys. Rev. B **63**, 020503 (2001)
163. V. Barzykin and D. Pines, Phys. Rev. B **54**, 7567 (1995)
164. V. Barzykin, D. Pines, A. Sokol, and D. Thelen, Phys. Rev. B **49**, 1544 (1994)
165. P. Bourges, Y. Sidis, H.F. Fong, L.P. Regnault, J. Bossy, A. Ivanov, and B. Keimer, Science **288**, 1234 (2000)
166. P.D. Johnson, T. Valla, A. Fedorov, Z. Yusof, B.O. Wells, Q. Li, A.R. Moodenbaugh, G.D. Gu, N. Koshizuka, C. Kendziora, Sha Jian, D.G. Hinks, Phys. Rev. Lett. **87**, 177007 (2001)
167. P.V. Bogdanov, A. Lanzara, S. A. Kellar, X.J. Zhou, E.D. Lu, W. Zheng, G. Gu, K. Kishio, J.-I. Shimoyama, Z. Hussain, Z.X. Shen, Phys. Rev. Lett. **85**, 2581 (2000)
168. E. Dagotto, A. Nazarenko, and M. Boninsegni, Phys. Rev. Lett. **73**, 728 (1994); E. Dagotto, A. Nazarenko, and A. Moreo, Phys. Rev. Lett. **74**, 310 (1995); A. Nazarenko, K.J.E. Vos, S. Haas, E. Dagotto, and R.J. Gooding, Phys. Rev. B, **51**, 8676 (1995)
169. T. Valla, A.V. Fedorov, P.D. Johnson, B.O. Wells, S.L. Hulbert, Q. Li, G.D. Gu, and N. Koshizuka, Science **285**, 2210 (1999)
170. F. Bouis, M.N. Kiselev, F. Onufrieva, and P. Pfeuty, cond-mat/9906369
171. V. Yu Irkhin, A.A. Katanin and M.I. Katsnelson, cond-mat/0102381 and references therein
172. S. Sachdev, In: *Quantum Phase Transitions* (Cambridge Univ Press., Cambridge, 1999)
173. M. Lavagna, Phil. Mag. B **81**, 1469 (2001)
174. More recent ARPES data [226,227] show that the magnitude of Σ' is actually smaller than as thought before
175. E. Abrahams and C. Varma, Proc. Nat. Acad. Sci. **97**, 5714 (2000)
176. J.J. Tu, C.C. Homes, G.D. Gu, D.N. Basov, S.M. Loureiro, R.J. Cava, M. Strongin, cond-mat/0104208
177. M.A. Quijada, D.B. Tanner, R.J. Kelley, M. Onellion, H. Berger and G. Margaritondo, Phys. Rev. B **60**, 14917 (1999)
178. S. Spielman, D. Parks, J. Orenstein, D.T. Nemeth, F. Ludwig, J. Clarke, P. Merchant, and D.J. Lew, Phys. Rev. Lett. **73**, 1537 (1994)
179. A. Puchkov, D. Basov, and T. Timusk, J. Phys.: Cond. Matter **8**, 10049 (1996)
180. L. Ioffe and A.I. Larkin, Phys. Rev. B **39**, 8988 (1989)
181. P. Bourges, Y. Sidis, H.F. Fong, B. Keimer, L.P. Regnault, J. Bossy, A.S. Ivanov, D.L. Milius, I.A. Aksay, In: *High Temperature Superconductivity*, ed. by S.E. Barnes et al. (American Institute of Physics, Amsterdam, 1999) pp. 207–212
182. S.H. Pan, E.W. Hudson, K.M. Lang, H. Eisaki, S. Uchida, J.C. Davis, Nature, **403**, 746 (2000)
183. D.A. Bonn, P. Dosanjh, R. Liang, and W.N. Hardy, Phys. Rev. Lett. **68**, 2390 (1992)
184. M.C. Nuss, P.M. Mankiewich, M.L. O'Malley and E.H. Westerwick, Phys. Rev. Lett. **66**, 3305 (1991)
185. N. Miyakawa, J.F. Zasadzinski, L. Ozyuzer, P. Guptasarma, D.G. Hinks, C. Kendziora, and K.E. Gray, Phys. Rev. Lett. **83**, 1018 (1999)
186. Y. Ando, G.S. Boebinger, A. Passner, T. Kimura, and K. Kishio, Phys. Rev. Lett. **75**, 4662 (1995)
187. G.S. Boebinger, Y. Ando, A. Passner, T. Kimura, M. Okuya, J. Shimoyama, K. Kishio, K. Tamasaku, N. Ichikawa, and S. Uchida, Phys. Rev. Lett. **77**, 5417 (1996)

188. S. Ono, Y. Ando, T. Murayama, F.F. Balakirev, J.B. Betts, and G.S. Boebinger, Phys. Rev. Lett. **85**, 638 (2000)
189. P.W. Anderson, Phys. Rev. Lett. **67**, 2092 (1991)
190. C.M. Varma and E. Abrahams, Phys. Rev. Lett. **86**, 4652 (2001)
191. J. Cerne, M. Grayson, D.C. Schmadel, G.S. Jenkins, H.D. Drew, R. Hughes, A. Dabkowski, J.S. Preston and P.-J. Kung, Phys. Rev. Lett. **84**, 3418 (2000)
192. N.P. Ong, In: *Physical Properties of High Temperature Superconductors*, ed. by D.M. Ginsberg (World Scientific, Singapore, 1990) Vol. 2
193. B. Stojkovic and D. Pines, Phys. Rev. Lett. **70**, 811 (1996); Phys. Rev. B **55**, 8576 (1997)
194. H. Kontani, K. Kanki, K. Ueda, Phys. Rev. B **59**, 14723 (1999)
195. G. Aeppli, T.E. Mason, S.M. Hayden, H.A. Mook, J. Kulda, Science **278**, 1432 (1997)
196. Y. Zha, V. Barzykin and D. Pines, Phys. Rev. B **54**, 7561 (1996)
197. J. Schmalian and P.G. Wolynes, Phys. Rev. Lett. **85**, 836 (2000)
198. H.Y. Hwang, B. Batlogg, H. Takagi, H.L. Kao, J. Kwo, R.J. Cava, J.J. Krajewski, and W.F. Peck, Jr., Phys. Rev. Lett. **72**, 2636 (1994)
199. J.L. Tallon, J.W. Loram, G.V.M. Williams, J.R. Cooper, I.R. Fisher, J.D. Johnson, M.P. Staines, C. Bernhard, Phys. Stat. Sol. (b)**215**, 531 (1999)
200. J. Loram, K.A. Mirza, J.R. Cooper, and W.Y. Liang, Phys. Rev. Lett. **71**, 1740 (1993); J.W. Loram, K.A. Mirza, J.R. Cooper, W.Y. Liang, J.M. Wade, J. Supercond. **7**, 243 (1994)
201. C. Panagopoulos, B.D. Rainford, J.R. Cooper, and C.A. Scott, Physica C **341–348**, 843 (2000)
202. R.B. Laughlin, G. Lonzarich, P. Monthoux, and D. Pines, Adv. Phys. **50**, 361 (2001)
203. N.J. Curro, P.C. Hammel, B.J. Suh, M. Hücker, B.B. üchner, U. Ammerahl, and A. Revcolevschi, Phys. Rev. Lett. **85**, 642 (2000)
204. J. Haase, C.P. Slichter, R. Stern, C.T. Milling, D.G. Hinks, J. Supercond. **13**, 723 (2000)
205. A. Abanov, B. Altshuler, E. Yuzbashyan and A.V. Chubukov, unpublished
206. M. Vojta and S. Sachdev, Phys. Rev. Lett. **83**, 3916 (1999); N. Read and S. Sachdev, Phys. Rev. Lett. **62**, 1694 (1989); Phys. Rev. B **42**, 4568 (1990)
207. S. Kivelson, D.S. Rokhsar, and J. Sethna, Phys. Rev. B **35**, 8865 (1987)
208. X.-G. Wen and P.A. Lee, Phys. Rev. Lett. **76** 503 (1996)
209. T. Senthil and M.P.A. Fisher, Phys. Rev. B **62**, 7850 (2000)
210. I. Affleck and J.B. Marston, Phys. Rev. B **37**, 3774 (1988)
211. G. Kotliar and J. Liu, Phys. Rev. B **38**, 5142 (1988)
212. Y. Suzumura, Y. Hasegawa, and H. Fukuyama, J. Phys. Soc. Jpn. **57**, 2768 (1988)
213. G. Baskaran and P.W. Anderson, Phys. Rev. B **37**, 580 (1988)
214. P.A. Lee and N. Nagaosa, Phys. Rev. B **46**, 5621 (1992)
215. I. Affleck, Z. Zou, T. Hsu, and P.W. Anderson, Phys. Rev. B. **38**, 745 (1988)
216. D. Poilblanc and Y. Hasegawa, Phys. Rev. B. **41**, 6989 (1990)
217. A. Kaminski, M. Randeria, J.C. Campuzano, M.R. Norman, H. Fretwell, J. Mesot, T. Sato, T. Takahashi, and K. Kadowaki, Phys. Rev. Lett. **86**, 1070 (2001); A. Lanzara, P.V. Bogdanov, X.J. Zhou, S.A. Kellar, D.L. Feng, E.D. Lu, T. Yoshida, H. Eisaki, A. Fujimori, K. Kishio, J.-I. Shimoyama, T. Noda, S. Uchida, Z. Hussain, and Z.-X. Shen, Nature (London) **412**, 510 (2001); P.D. Johnson, T. Valla, A.V. Fedorov, Z. Yusof, B.O. Wells, Q. Li, A.R. Moodenbaugh, G.D. Gu, N. Koshizuka, C. Kendziora, Sha Jian, and D.G. Hinks, Phys. Rev. Lett. **87**, 177007 (2001); F. Ronning, T. Sasagawa, Y. Kohsaka, K.M. Shen, A. Damascelli, C. Kim, T. Yoshida, N.P. Armitage, D.H. Lu, D.L. Feng, L.L. Miller, H. Takagi and Z.-X. Shen, Phys. Rev. B **67**, 165101 (2003); X.J. Zhou, T. Yoshida, A. Lanzara, P.V. Bogdanov, S.A. Kellar, K.M. Shen, W.L. Yang, F. Ronning, T. Sasagawa, T. Kakeshita, T. Noda, H. Eisaki, S. Uchida, C.T. Lin, F. Zhou, J.W. Xiong, W.X. Ti, Z.X. Zhao, A. Fujimori, Z. Hussain and Z.-X. Shen, Nature (London) **423**, 398 (2003); T. Sato, H. Matsui, T. Takahashi, H. Ding, H.-B. Yang, S.-C. Wang, T. Fujii, T. Watanabe, A. Matsuda, T. Terashima and K. Kadowaki, Phys. Rev. Lett. **91**, 157003 (2003); T.K. Kim, A.A. Kordyuk, S.V. Borisenko, A. Koitzsch, M. Knupfer, H. Berger, and J. Fink, Phys. Rev. Lett. **91**, 167002 (2003); A.D. Gromko, A.V. Federov, Y.-D. Chuang, J.D. Koralek, Y. Aiura, Y. Yamaguchi, K. Oka, Y. Ando, and D.S. Dessau, Phys. Rev. B **68** 174520 (2003)
218. A.V. Chubukov and M.R. Norman, Phys. Rev. B, **70**, 174505 (2004)
219. T.P. Devereaux et al., Phys. Rev. Lett. 93, 117004 (2004); T. Cuk et al., Phys. Stat. Sol. (b) **242**, 11 (2005)

220. H.J.A. Molegraaf, C. Presura, D. van der Marel, P.H. Kes, and M. Li, Science **295** 2239 (2002)
221. A.V. Boris et al., Science **304**, 708 (2004)
222. A.F. Santander-Syro, R.P.S.M. Lobo, N. Bontemps, Z. Konstantinovic, Z.Z. Li, and H. Raffy, Europhys. Lett. **62**, 568 (2003); N. Bontempts, private communication
223. A. Abanov and A. Chubukov, Phys. Rev. B **70**, 100504 (2004)
224. D. van der Marel, H.J.A. Molegraaf, J. Zaanen, Z. Nussinov, F. Carbone, A. Damascelli, H. Eisaki, M. Greven, P.H. Kes and M. Li, Nature **425**, 271 (2003)
225. M.R. Norman and A.V. Chubukov, Phys. Rev. B **73**, 140501 (2006)
226. A.A. Kordyuk et al., Pys. Rev. B **67**, 064504 (2003)
227. J.D. Koralek et al., cond-mat/0508404)
228. S. Pailhes et al., cond-mat/0403609; S.M. Hayden et al., Nature (London) **429**, 531 (2004)
229. I. Eremin et al., Phys. Rev. Lett., **94**, 147001 (2005)
230. Y. Tokura, H. Takagi, S. Uchida , Nature **337**, 345 (1989)
231. Y. Dagan et al., Phys. Rev. Lett. **92**, 167001 (2004)
232. A. Koitzsch et al., Phys. Rev. B, **67**, 184522 (2003)
233. A. Zimmers et al., Europhys. Lett. **70**, 225 (2005)
234. L. Alff et al., Nature **422**, 698 (2003)
235. Y. Onose et al., Phys. Rev. B **69**, 024504 (2004)
236. P. Krotkov and A. Chubukov, Phys. Rev. Lett. **97**, 106002 (2006)
237. H.Y. Kee, S.A. Kivelson, and G. Aeppli, Phys. Rev. Lett. **88**, 257002 (2002)
238. A. Abanov et al., Phys. Rev. Lett. **89**, 177002 (2002)
239. C. Timm, D. Manske, and K.H. Bennemann, Phys. Rev. B **66**, 094515 (2002)
240. A. V. Chubukov and J. Schmalian, Phys. Rev. B 72, 174520 (2005)
241. T.A. Maier, M.S. Jarrell, and D.J. Scalapino, Phys. Rev. Lett. **96**, 047005 (2006)
242. V. Barzykin and D. Pines, preprint cond-mat/0601396
243. D.N. Basov and T. Timusk, Rev. Mod. Phys. **77**, 721 (2005)
244. A. Damascelli, Z. Hussain, and Z.-X. Shen, Rev. Mod. Phys. **75**, 473 (2003)
245. M.R. Norman and C. Pepin, Rep. Prog. Phys. **66**, 1574 (2003)
246. M.R. Norman, D. Pines and C. Kallin, Adv. Phys. 54, 715 (2005)

23 Electronic Theory for Superconductivity in High-T_c Cuprates and Sr_2RuO_4

D. Manske Max Planck Institute for Solid State Research, Stuttgart, Germany
I. Eremin Institut für Mathematische Physik, TU-Braunschweig, Germany
K. H. Bennemann Institut für Theoretische Physik, Freie Universität Berlin, Germany

23.1 Introduction

Superconductivity remains one of the most interesting problems in solid state physics. During the last years in particular high-transition temperature superconductivity in the cuprates with non-s-wave pairing symmetry and unconventional superconductivity in the isostructural strontiumruthenate (Sr_2RuO_4) also with a non-s-wave symmetry of the superconducting order parameter and triplet Cooper-pairing have been studied. Cuprates and ruthenates have an oxide layered structure with CuO_2 and RuO_2-planes, respectively. These seem to play an important role regarding superconductivity and magnetic activity. For an illustration in Fig. 23.1 we show the crystal structures (in Sr_2RuO_4 the basic element is a RuO_2-plane and the superconducting transition temperature is T_c=1.5 K). Undoped La_2CuO_4 is an antiferromagnetic insulator. Superconductivity occurs only upon doping with Ba and reaches a maximum for hole doping $x = 0.15$ with a transition temperature $T_c \approx 40$ K [1]. The different behavior of Sr_2RuO_4 and $La_{2-x}Ba_xCuO_4$ is definitely related to the electronic structure of RuO_2 and CuO_2-planes. Let us emphasize also some differences in the crystal structures of Sr_2RuO_4 and $La_{2-x}Ba_xCuO_4$. Sr_2RuO_4 has a pure tetragonal crystal structure. The RuO_6-octahedra are not distorted. On the other hand, in $La_{2-x}Ba_xCuO_4$ the corresponding CuO_6-octahedra are sufficiently distorted. In general, it is important to understand the dependence of T_c based on a crystal structure. In Fig. 23.2 we show the crystallographic structures of Bi-based cuprates with various CuO_2-layers per unit cell. T_c increases almost linearly from

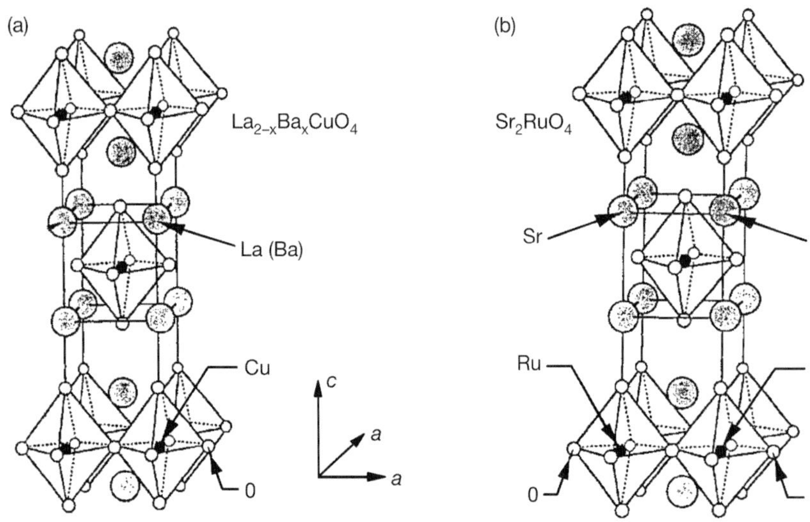

Fig. 23.1. (**a**) Cuprate singlet superconductor $La_{2-x}Ba_xCuO_4$. (**b**) Layered perovskite structure of the triplet superconductor Sr_2RuO_4

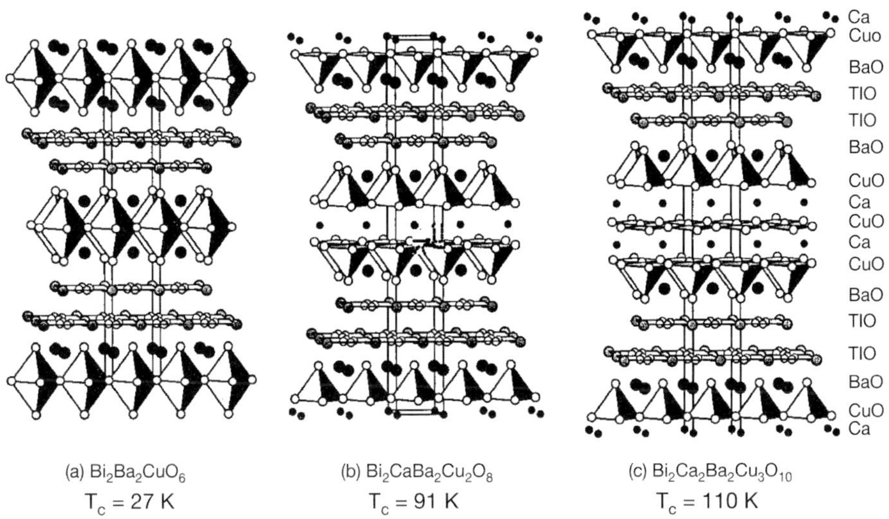

Fig. 23.2. T_c of cuprate crystal structures with various CuO_2-layers: (a) one-layer, (b) two-layer, and (c) three-layer

$Bi_2Ba_2CuO_6$ to the $Bi_2Ca_2Ba_2Cu_3O_{10}$. Note that not only the number of CuO_2-layers changes, but also other properties like the strength of the orthorhombic distortions in these materials.

Cuprates

The parent compounds of the cuprate superconductors such as La_2CuO_4 or $YBa_2Cu_3O_6$ are quasi-two-dimensional antiferromagnetic insulators. Doping these systems with holes, for example, La_2CuO_4 with strontium or $YBa_2Cu_3O_6$ with oxygen, causes a transition from a Mott insulator to a metallic phase. The metallic phase exhibits unconventional superconductivity with non s-wave symmetry of the order parameter and the transition temperature T_c increases with increasing doping towards a maximum value. Further doping decreases T_c. Finally, superconductivity disappears in the overdoped region. Remarkably, despite differences between the phase diagrams of various hole-doped cuprates one can map all of them into one so-called generic phase diagram as shown in Fig. 23.3. This assumes that Cooper-pairing and normal state dynamics already occur essentially in the CuO_2-plane. Then, the physical properties of each cuprate compound depend mainly on the number of carriers within a CuO_2 plane and the correlation between them.

Besides crystal structures, the doping dependence $T_c(x)$, and the phase diagram, many important properties have been observed. For example, various experiments indicate for the cuprates $d_{x^2-y^2}$-symmetry of the superconducting order parameter and singlet Cooper-pairing:

$$\Delta_{\mathrm{k}} = \frac{\Delta_0}{2} \left(\cos k_x - \cos k_y\right) . \qquad (23.1)$$

The superconducting transition temperatures may be as high as 150 K. Considerable electronic correlations yield a non-Fermi liquid behavior of the elementary excitations in the normal state (below T^*). Generally, the behavior of the cuprates depends significantly on the doping concentration x.

Note, regarding the symmetry of the superconducting order parameter in the cuprates, evidence for $d_{x^2-y^2}$-wave pairing in hole-doped cuprates comes from several sources, in particular from nuclear magnetic resonance (NMR) [2], phase sensitive measurements [3–6], penetration depth measurements [7, 8], ARPES [9], and polarization-dependent Raman scattering experiments [10–13]. In contrast to the conventional superconductors, the behavior of the $d_{x^2-y^2}$-wave symmetry of the superconducting order parameter in the first Brillouin zone (BZ) is very peculiar. In particular, the presence of the line nodes along the diagonal of the BZ and the corresponding

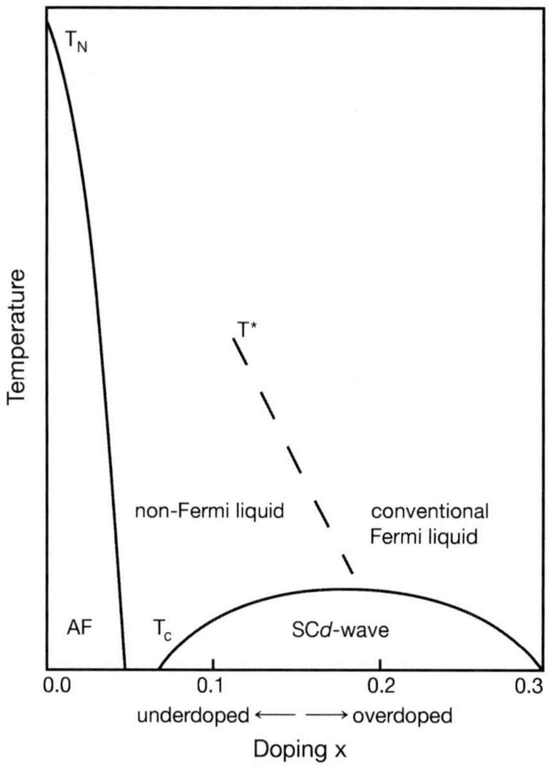

Fig. 23.3. Schematic phase diagram of hole-doped cuprates. High-T_c superconductivity always occurs in the vicinity of an antiferromagnetic phase transition (AF). The superconducting transition temperature as a function of the hole concentration $T_c(x)$ has a characteristic parabola-like shape. The superconducting order parameter is of d-wave symmetry. The normal state can be divided into two parts: In the overdoped region the cuprates behave like a conventional a Fermi-liquid, whereas in the underdoped regime below a temperature T^* one finds strong antiferromagnetic correlations and non-Fermi-liquid behavior. T^* is a crossover temperature and is called the pseudogap temperature, because a normal-state gap opens up in the spin and charge degrees of freedom. T^* enters the T_c curve roughly at optimum doping concentration or at overdoped values dependent on the number of CuO_2-planes per unit cell

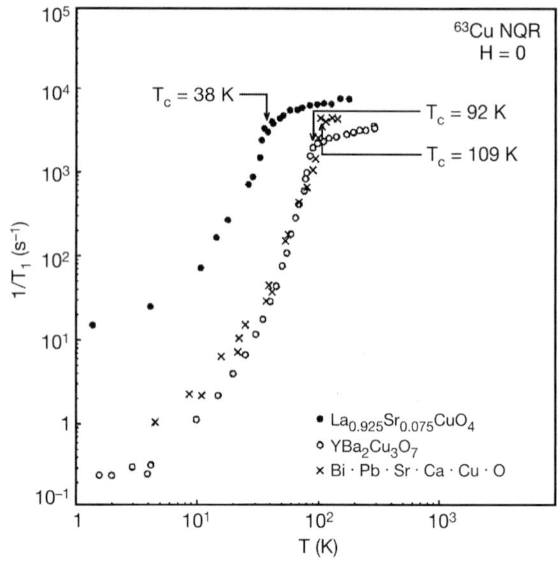

Fig. 23.4. Temperature dependence of the ^{63}Cu NQR spin-lattice relaxation rate for various cuprate superconductors revealing singlet d-wave Cooper-pairing (taken from [2])

changes of sign in the gap function allow different experimental techniques to distinguish a $d_{x^2-y^2}$-wave symmetry from other symmetries, for example from isotropic s-wave symmetry.

In Fig. 23.4 we show NMR/NQR results of [2] for the temperature dependence of the spin-lattice relaxation rate $1/T_1$ of ^{63}Cu for three different high-T_c cuprate superconductors below T_c. A signature of unconventional superconductivity is the absence of the so-called Hebel–Slichter peak at $T \simeq T_c$ and also the temperature dependence of $1/T_1$. In contrast to a low temperature exponential decay in the case of s-wave pairing, the relaxation rate $1/T_1$ has a rather different dependence on T, namely a power law that one would expect for d-wave symmetry of the superconducting order parameter. Note that for a d-wave gap with line nodes along the diagonals the density of states is not fully gaped and thus allows the quasiparticles to occupy states for energies $|\omega| < 2\Delta_0$ (Δ_0 is the maximal gap at $T = 0$). In an NMR experiments one averages over the first BZ and thus one takes also into account the states at the line nodes. As a result, the temperature dependence of the spin-lattice relaxation follows a power law ($\propto T^3$). As can be seen in Fig. 23.4, this is indeed the case for high-T_c cuprate superconductors.

There is another important signature of unconventional superconductivity in the cuprates. Usually, one observes $T_c \propto \Delta_0$ for conventional superconductors ($\Delta_0 = \Delta(T = 0)$). However, in hole-doped cuprates this relation is not generally satisfied and is only observed for $x > x_0$. For $x < x_0$ (x_0 = optimal

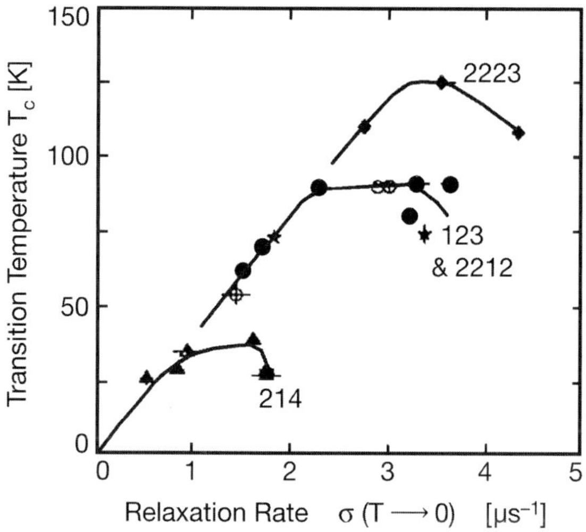

Fig. 23.5. The superconducting transition temperature T_c plotted vs low-temperature muon-spin-relaxation rate σ measured for different high-T_c superconductors, see [14] (214: $La_{2-x}Sr_xCuO_4$, 123: $YBa_2Cu_3O_7$, 2212: $Bi_2Sr_2Ca_2CuO_8$, 2223: $Bi_2Sr_2Ca_2Cu_3O_{10}$). Note that $\sigma(0)$ is proportional to the superfluid density, n_s/m^*

doping), i.e. in underdoped cuprates, T_c decreases with decreasing doping, while the measured Δ_0 becomes larger. As noted by Uemura et al. [14] in underdoped cuprates one finds $T_c \propto n_s$, where n_s is a superfluid density. This is shown in Fig. 23.5.

In general, the normal state of the hole-doped cuprates depends sensitively on doping, see Fig. 23.6. The overdoped cuprates behave like a conventional Fermi-liquid, whereas the underdoped ones, in particular below the pseudogap temperature T^*, exhibit some unusual properties. For example, a gap is present in the elementary excitation spectrum, strong anisotropies are observed (mainly caused by the quasi-two-dimensional nature of the system), and short-range antiferromagnetic fluctuations exist. This different behavior of underdoped and overdoped crystals can be seen, for example, from the ^{63}Cu spin-lattice relaxation rate and the inelastic neutron scattering (INS) intensity in hole-doped cuprates. In the overdoped cuprates $1/T_1T$ increases monotonously as T decreases down to T_c, while in the underdoped ones $1/T_1T$ passes through a maximum at the spin-gap temperature T^* for decreasing T [16]. At the same time similar behavior is observed by INS where in the underdoped cuprates the imaginary part of the spin susceptibility $\mathrm{Im}\,\chi(\mathbf{Q},\omega)$ passes at fixed small frequency ω ($\simeq$ 10–15 meV) through a maximum at T^* for decreasing T [17].

In addition, angle-resolved photoemission spectroscopy (ARPES) experiments on underdoped $Bi_2Sr_2CaCu_2O_{8+\delta}$ show the presence of a "pseudo"-gap with $d_{x^2-y^2}$-wave symmetry well above T_c also in the charge excitation spectrum [18, 19]. Recent tunneling experiments also indicate a presence of a pseudogap in the tunneling spectral density well above T_c [20]. This gap opens below the temperature T^*. Several other experiments including heat capacity [21], transport [22], Raman scattering [23], and, in particular, scanning tunneling spectroscopy [24] have confirmed the existence of a pseudogap in the elementary excitations below T^*. To clarify the origin of the pseudogap a few phenomenological models like the marginal-Fermi-liquid (MFL) [25], the nested-Fermi-liquid [26, 27], and nearly-antiferromagnetic-liquid (NAFL) [28] have been developed. These describe the unusual non-Fermi-liquid properties in the normal state of hole-doped cuprates.

In the cuprate superconductors the CuO_2-plane seems to be most important. The surrounding environment of the CuO_2-plane, usually called charge reservoirs, somehow play a passive role supplying charge carriers for the planes. For some cuprates it is possible that also electrons can be doped into the CuO_2-plane. It is remarkable that these so-called electron-doped cuprates differ drastically from the hole-doped ones. Note the large asymmetry in their phase diagrams, see Fig. 23.6:

1. The antiferromagnetism persists in the electron-doped compounds up to $x = 0.12$ or 0.14 [29], while in hole-doped compound $La_{2-x}Sr_xCuO_4$ (LSCO) antiferromagnetism disappears already at $x = 0.05$.
2. Superconductivity in electron-doped cuprate systems occurs in a very narrow doping range ($0.14 < x < 0.17$ for $Nd_{2-x}Ce_xCuO_4$(NCCO) [30, 31] and has a smaller transition temperature T_c), in the hole-doped LSCO system the doping range for which superconductivity occurs is broader ($0.05 < x < 0.3$). In addition,

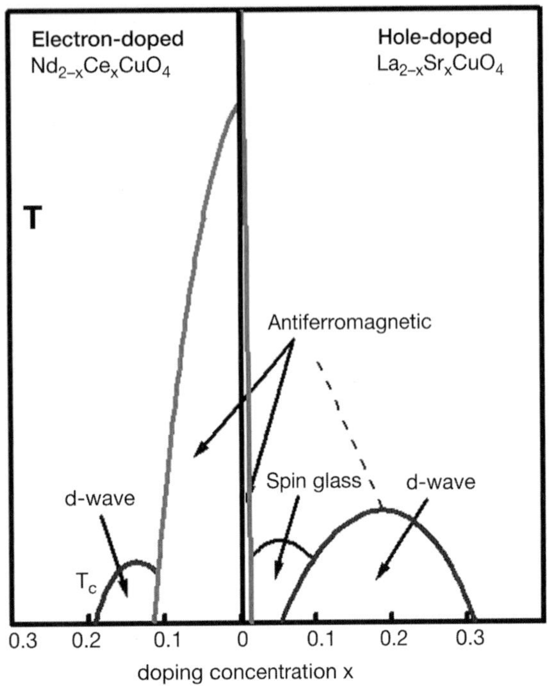

Fig. 23.6. Illustration of the phase diagram of hole (LSCO) and electron (NCCO)-doped cuprates, temperature T versus doping concentration x. Note that the asymmetry between electron-doped and hole-doped cuprates. In the case of hole doping the carriers mainly occupy the oxygen p-band, while for electron doping the carriers occupy the upper Hubbard band of the copper. The antiferromagnetism is for rather different doping concentration destroyed in electron-doped and hole-doped cuprates. Note that it would be interesting to compare LSCO directly with its electron-doped counterpart $La_{2-x}Ce_xCuO_4$. However, a structural phase transition at $x = 0.05$ changes the electronic properties of the latter significantly [15]

the behavior of the electron-doped cuprates in the normal and superconducting state is very different in comparison to the hole-doped ones. For example, in the normal state a pseudogap in the excitation spectrum in hole-underdoped cuprates is seen in many experiments [32]. In electron-doped ones the existence of the pseudogap is unclear. In optimally-doped $YBa_2Cu_3O_{6+y}$ (YBCO) and LSCO the in-plane resistivity increases linearly with temperature over a wide range [33] indicating a non-Fermi-liquid behavior. In NCCO ($x = 0.15$) the in-plane resistivity is quadratic in temperature [34]. Photoemission spectroscopy studies of a CuO_2-plane show flat energy bands near the Fermi surface (FS) of the high-T_c hole doped cuprates [35]. However, in NCCO one finds a flat band only 300 meV below the FS [36].

Therefore, the question arises: Are the pairing symmetries of the electron and hole-doped cuprates different? It is widely believed that hole-doped are $d_{x^2-y^2}$-wave superconductors. In electron-doped cuprates the symmetry of the order parameter was for a long time under debate. Early measurements of in-plane penetration depth in NCCO have been fitted with an exponential temperature dependence suggesting isotropic BCS type s-wave gap [37] rather than the power law decay expected for unconventional superconductors with nodes in the gap function. Recent experiments indicate non-s-wave symmetry and rather $d_{x^2-y^2}$-wave symmetry of the superconducting order parameter. Most recent measurements of the magnetic penetration depth [38,39] have shown that the best fit occurs for dominant the $d_{x^2-y^2}$-wave plus small admixture of the s-wave contribution for the underlying gap symmetry. The most convincing evidence for dominant $\mathrm{d}_{x^2-y^2}$-wave symmetry in electron-doped NCCO and PCCO at optimal doping came very recently from phase sensitive experiments [31]. Note, however, that with increasing doping the symmetry of the superconducting order parameter seems to change towards anisotropic s-wave type [40].

In general, for increasing doping itinerancy increases and antiferromagnetic activity disappears. The strong interplay of antiferromagnetic excitations and superconductivity is in particular seen by inelastic neutron scattering (INS) and elementary excitations measured by angle-resolved photoemission spectroscopy (ARPES) experiments. The feedback of superconductivity on the spin susceptibility $\chi(\mathbf{q}, \omega)$ and the recently observed "kink" in the dispersions is illustrated in Figs. 23.7 and 23.8, respectively.

In Fig. 23.7 we show the INS results for optimally-doped $YBa_2Cu_3O_{6.92}$ [41]. Employing linear response theory, the intensity in INS is proportional to the magnetic structure factor that yields the imaginary

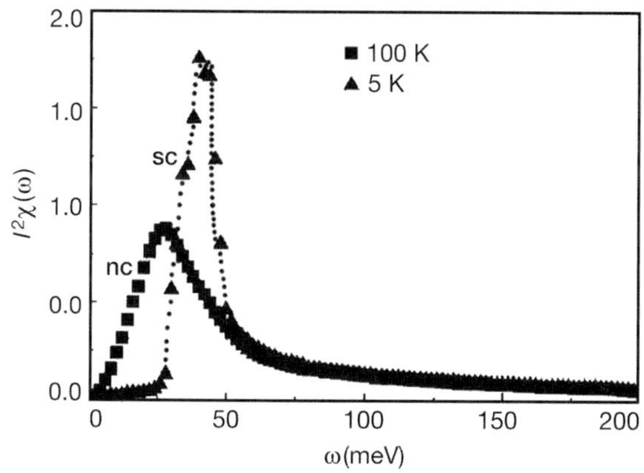

Fig. 23.7. Spin excitation spectrum $I^2\chi(\mathbf{q} = \mathbf{Q}, \omega)$ vs frequency for optimally-doped YBCO as observed by INS experiments in cuprates [41]. In the normal state an Ornstein–Zernike behavior is observed. Note the feedback of superconductivity ($T = 5\text{K} < T_c$) on the spin susceptibility and the formation of the so-called "resonance" peak at $\omega_{\text{res}} = 41$ meV

part of the dynamical spin susceptibility. Thus, INS provides a direct probe of the spin susceptibility, $\chi(\mathbf{q}, \omega)$. The normal state spin dynamics can be described approximately at the important antiferromagnetic wave vector $\mathbf{Q} = (\pi, \pi)$ by the Ornstein–Zernicke response function

$$\text{Im}\chi \propto \frac{\omega\,\omega_{sf}}{\omega^2 + \omega_{sf}^2} . \tag{23.2}$$

Note, the rapid increase of $\text{Im}\chi$ and the maximum at $\omega = \omega_{sf}$. In the superconducting state the behavior of $\text{Im}\chi$ is very different. The spin excitation spectrum renormalizes strongly and becomes gaped at lower frequencies and then exhibits a sharp so-called "resonance" peak at an energy ω_{res}. In the case of the optimally-doped YBCO one gets $\omega_{\text{res}} = 41$ meV.

The appearance of the resonance peak firstly reported in [42] suggests a strong renormalization of the spin fluctuation spectrum in the superconducting state that cannot be accounted for by the simple phenomenological form of the spin susceptibility, $\chi(\mathbf{q}, \omega)$ (see (23.2)). Note that phonon-mediated superconductivity is difficult to reconcile with the appearance of a resonance peak. Thus, the important resonance peak is a fingerprint of Cooper-pairing due to spin excitations. In a recent INS measurements [43] a resonant spin excitation in the overdoped and optimally-doped $Bi_2Sr_2CaCu_2O_{8+y}$ (BSCCO) is also observed with mode energies $E_{\text{res}} = 38$ meV and 43 meV, respectively.[1] This observation together with the previous measurements in underdoped phase of YBCO [41, 44] indicates a constant ratio (for two CuO_2-planes per unit cell) for the whole doping range for the odd mode

$$\frac{E_{res}}{k_B T_c} = 5.4 . \tag{23.3}$$

The resonance peak emerged as a key property of the copper oxide superconductors. The observation of a resonance like feature in the superconducting state of single CuO_2-layer cuprate superconductor $Tl_2Ba_2CuO_{6+x}$ near optimal doping seems very important [45]. Note that no resonance peak is observed in electron-doped cuprates.

In view of this one may conclude that the analysis of the spin fluctuation spectrum above and below T_c and its doping dependence is significant for a microscopic theory of cuprate superconductivity.

Similarly important is the behavior of the elementary excitations above and below T_c. Regarding the "kink" structure shown in Fig. 23.8 we note the following. With certain assumptions ARPES measures the spectral density, $A(\mathbf{k}, \omega)$, of the electrons in the CuO_2-planes. Due to recent developments in ARPES the spectral density can be studied with high accuracy versus frequency at fixed momentum (so-called energy distribution curves (EDC)) as well as a function of momentum at fixed frequency (so-called momentum distribution curves (MDC)). One of the most important information that one gets by analyzing MDC and EDC is the renormalized energy dispersion $\omega_{\mathbf{k}}$ for the elementary excitations. Note that MDC curves are better suited for determining the energy dispersion due to a weaker momentum than frequency dependence of the self-energy, $\Sigma(\mathbf{k}, \omega)$.

In Fig. 23.8 we show the results of Lanzara et al. [46] for the $(0, 0) \to (\pi, \pi)$ direction in the first

[1] Due to the bilayer structure of YBCO and BSCCO a resonating mode is expected in both the even and odd channels of the spin excitation spectrum, respectively. Here, we refer only to the odd channel which exhibits a much larger intensity.

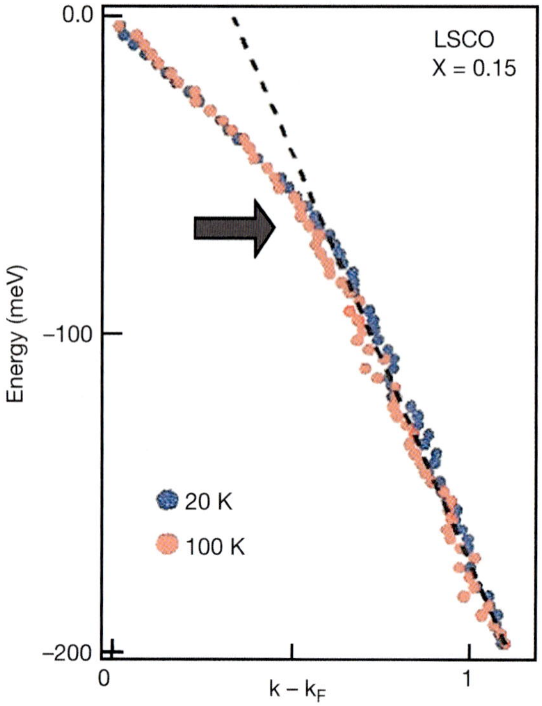

Fig. 23.8. Energy dispersion $\omega(\mathbf{k}-\mathbf{k}_F)$ in the nodal direction with respect to the Fermi energy in the first BZ as observed by ARPES for the hole-doped cuprate compound BISSCO at different temperatures [46] vs the normalized momentum relative to the Fermi level. Note that the slope of the curve gives the renormalized group velocities

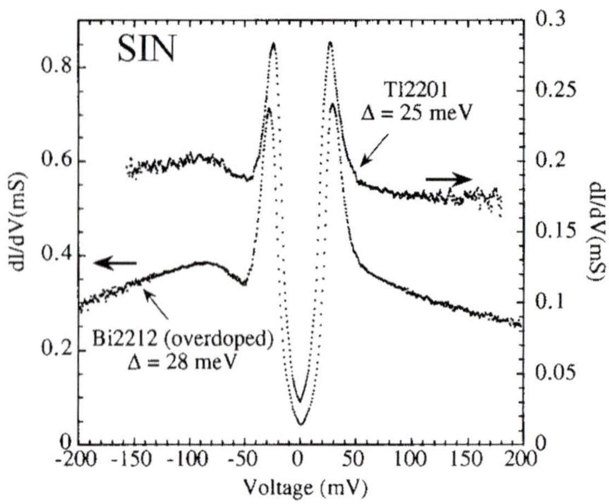

Fig. 23.9. Tunneling Spectra: Comparison of SIN conductances of Tl2201 and slightly overdoped Bi2212 at T = 4.2 K. The results are taken from [49]

Brillouin zone (BZ). One clearly sees that the curve shows a change in its slope around 50 ± 15 meV. This so-called "kink" feature was observed by several groups.[2] Note, however, that bilayer effects and matrix elements for various incident photon energies make the spectra more complicated to analyze. Changes in the elementary excitation spectrum due to coupling of the carriers to spin fluctuations or other (collective) modes are generally expected. The coupling to spin fluctuations should result in rather different frequency and momentum dependencies of the self-energy than in the *phonon case of coupling*. To understand this "kink" feature in detail is one of the important problems in the theory of high-T_c cuprates.

Note, tunneling spectroscopy allows also to determine the structure of the elementary excitations in cuprates. Originally, the successful analysis of tunneling spectra in conventional superconductors like Pb led to the common acceptance of the electron–phonon mechanism and Eliashberg theory of superconductivity [50]. Similarly, an understanding of the structure in the tunneling spectra in high-T_c cuprates is a central problem.

In Fig. 23.9 we show results for the superconductor-insulator-normal metal (SIN) tunneling conductance $(dI/dV)_s$ normalized to its normal state value for two different cuprate superconductors, $Bi_2Sr_2CaCu_2O_{8+\delta}$ (BISCCO) and $Tl_2Ba_2CuO_6$ (Tl2201) having two and one CuO_2-plane per unit cell, respectively [49]. In both cases the supercon-

[2] One attempted to relate the kink feature to a coupling of the itinerant carriers to phonons, in particular, to a longitudinal optical phonon mode at 60 meV, which was observed to behave anomalously in several experiments [47]. However, this interpretation has several disadvantages. The first one relates to the fact that the inelastic scattering rate in hole-doped cuprates is linear with frequency or temperature (whichever is larger), which can be hardly explained using conventional electron–phonon coupling. Also in electron-doped cuprates there is no kink present [48] and the electronic part of the resistivity follows a quadratic behavior at low temperatures indicating conventional Fermi-liquid behavior. It is not clear how both features can be explained assuming the same electron–phonon coupling.

ducting gaps and T_c are demonstrating the essential independence of the observed features on the number of CuO_2-layers. Notice again the non-s-wave character of the superconducting gap, since the tunneling spectra have a characteristic V-shape indicating a presence of states inside the superconducting gap. In addition, further important details are related to the pairing interaction. Immediately after the coherence peak a "dip-hump" structure is present with a "dip" approximately at energy $3\Delta_0$, where Δ_0 is the superconducting gap at $T = 0$. This structure is asymmetric with respect to the zero bias, since it occurs only below T_c. It is natural to suggest that the observed structure results from the pairing interaction and reflects the important frequency range of the spin fluctuations relevant for Cooper-pairing. One has to understand the relation of the "dip-hump" feature to the "kink" structure seen by ARPES in order to connect both experiments with each other.

This summarizes some key facts about superconductivity in hole-doped and electron-doped cuprates. We will present an electronic theory for the central problems by studying the interaction between carriers and spin excitations. Our theory is able to explain basic and characteristic features in the normal and superconducting state of cuprates including the kink seen in ARPES and the resonance peak observed in INS experiments.

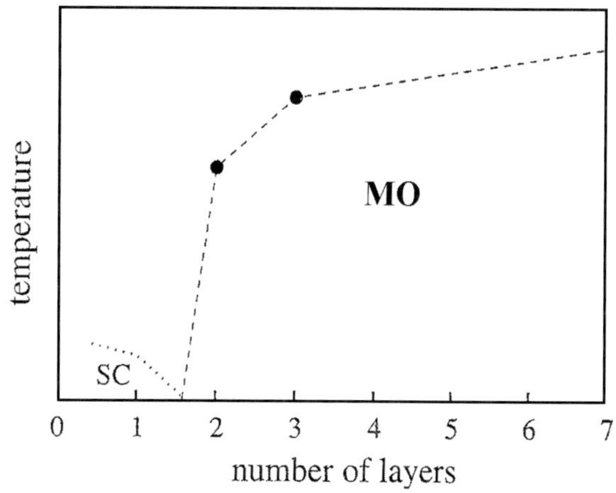

Fig. 23.10. Schematic phase diagram of the ferromagnetic and superconducting systems (Rodllesen–Popper crystals) $Sr_{n+1}Ru_nO_{3n+1}$, taken from [51]. The number of layers is the parameter that determines the transition between the two phases: MO = magnetically ordered and SC = superconducting

Ruthenates

Let us describe now essential properties of Cooper-pairing in Sr_2RuO_4. Superconductivity and magnetic activity in the ruthenates seem intimately related. The character of the magnetic activity is still being discussed. In view of triplet Cooper-pairing one expects that ferromagnetic spin fluctuations are present in Sr_2RuO_4. This is indirectly confirmed by the results shown in Fig. 23.10 where we plot a schematic phase diagram of the ferromagnetic and superconducting members of the Ruddlesen–Popper series ($Sr_{n+1}Ru_nO_{3n+1}$) as a function of the number of RuO_2-layers per unit cell, n. The infinite layer ($SrRuO_3$) is a ferromagnet with $T_{Curie} \approx 165$ K. For $n = 3$ one finds $T_{Curie} \approx 148$ K and for $n = 2$ $T_{Curie} \approx 102$ K. This demonstrates the tendency that T_{Curie} is reduced with decreasing layer number n. This suggests that even for $n = 1$, when superconductivity occurs, one expects significant ferromagnetic fluctuations. These may play an important role for superconductivity in Sr_2RuO_4.

Similar to the case of cuprates, inelastic neutron scattering (INS) studies have revealed the formation of strong incommensurate antiferromagnetic fluctuations (IAF) for the wave vector $\mathbf{Q}= (2\pi/3, 2\pi/3)$ at energy $\omega_{sf} = 6$ meV [52] and no signature of ferromagnetic fluctuations. In addition, results for the spin susceptibility $\chi(\mathbf{q}, \omega)$ obtained by means of NMR [53] indicate a remarkable temperature-dependent magnetic anisotropy; for an illustration see Fig. 23.11.

In general, the spin-lattice relaxation as measured by NMR allows one to study different components of the magnetic susceptibility like their transverse and longitudinal parts. One finds that at room temperature all components are the same. Then, for decreasing temperature, an isotropic response in the ruthenates is observed. Upon cooling the spin-lattice relaxation starts to differ and the anisotropy reaches a maximum at $T = T_c$. Therefore, at low tempera-

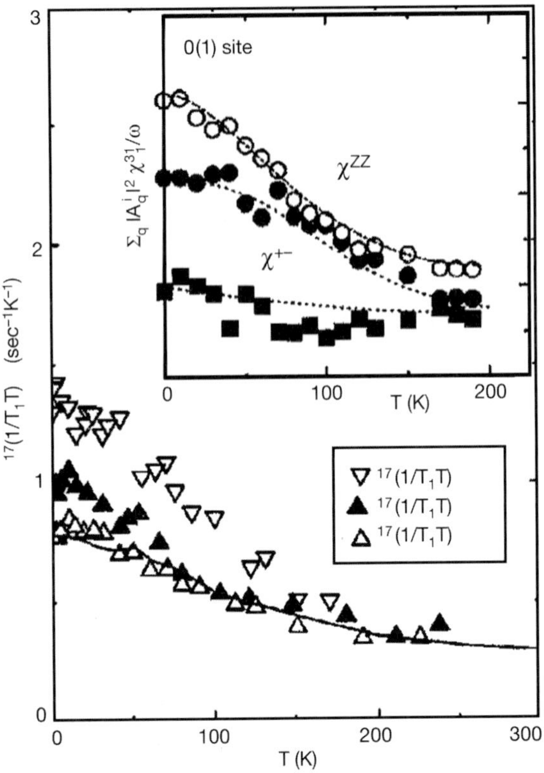

Fig. 23.11. Anisotropic magnetic response of a single RuO_2-plane in the triplet superconductor Sr_2RuO_4 vs temperature as observed by NMR experiments [53]. Note that the data are displayed only for the normal state, i.e. $T \geq T_c = 1.5$ K. Results are shown for the spin-lattice relaxation on the oxygen nuclei ^{17}O in the RuO_2-plane in terms of $(T_1T)^{-1}$. In the *inset* the results for the corresponding components of the spin susceptibility (transverse and longitudinal ones) are shown indicating the increase of the magnetic anisotropy upon cooling. χ^{zz} and χ^{+-} refer to the response of the system to the external magnetic field parallel and perpendicular to the RuO_2-plane, respectively

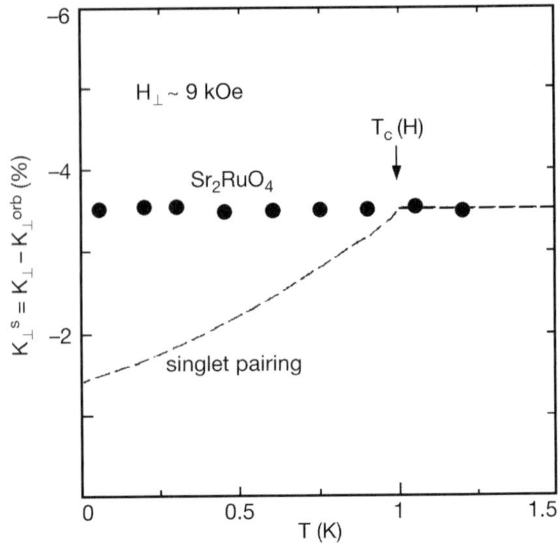

Fig. 23.12. Results for the uniform spin susceptibility, $\chi(\mathbf{q} = 0, \omega = 0)$ in the superconducting state of cuprates ($YBa_2Cu_3O_7$, *dashed curve*) and ruthenates measured by NMR ^{99}Ru Knight shift [53]. One can clearly see that in contrast to the singlet Cooper-pairing where the spin susceptibility decreases upon cooling, the Knight shift in Sr_2RuO_4 is unchanged by lowering T below T_c. In the case of triplet Cooper-pairing the Knight shift does note decrease below T_c, since the polarization induced by the external magnetic field does not change in the superconducting state

tures one expects that magnetic anisotropy plays an important role for Cooper-pairing.

What happens in the superconducting state? In Fig. 23.12 we show experimental results suggesting triplet Cooper-pairing in Sr_2RuO_4. In conventional superconductors the spin part of the Knight shift measured by NMR decreases rapidly below T_c due to formation of singlet Cooper-pairs. On the other hand, in a triplet superconductor with $S = 1$ the spin part of the Knight shift should not change below T_c, since the polarization induced by the weak external magnetic field and probed by NMR does not change. This behavior was observed in Sr_2RuO_4 by Ishida et al. [53] and provides evidence for triplet Cooper-pairing.[3]

Note that shortly after this discovery p-wave symmetry of the superconducting order parameter and triplet pairing was suggested by Sigrist and Rice [54]

[3] Although one has to wait for further decisive and unambiguous evidence for spin-triplet Cooper-pairing from in-plane ^{17}O and ^{99}Ru Knight shift data, it is important to note that $H_{c2}^{||}$ (parallel to the c-axis) is too small for NMR measurements. Therefore, strictly speaking the present Knight shift data do not fully reinforce Sr_2RuO_4 to be a triplet superconductor. In order to complete the picture, further studies, for example tunneling experiments along the c-axis, are necessary.

using a phenomenological theory and some analogies between the normal state of ^{3}He and Sr_2RuO_4. In the case of p-wave pairing one has

$$\Delta_p = \Delta_0\,(k_x + ik_y)\,, \tag{23.4}$$

and thus an order parameter where $|\Delta|$ has no line nodes. Yet it must be remarked that the microscopic theory of triplet Cooper-pairing in Sr_2RuO_4 is still under discussion [55].

In general, assuming an exchange of magnetic fluctuations as responsible for Cooper-pairing like in cuprates, the structure of the spin susceptibility $\chi(\mathbf{q}, \omega)$ and the Fermi surface topology should control the superconductivity in the ruthenates. Note that ferromagnetic spin fluctuations dominating over antiferromagnetic ones will cause then triplet paring. This is analogous to the case of triplet pairing in superfluid ^{3}He in which p-wave symmetry is present.

The behavior mentioned above, i.e. the magnetic anisotropy, temperature dependence of the Knight shift as well as of the specific heat and other properties, and of the electronic structure (Fermi surface topology) are important characteristics of Sr_2RuO_4 and essential for an electronic theory.

Electronic Theory

Obviously, it is of great interest to have an electronic theory explaining the important normal state behavior and the pairing mechanism for superconductivity in cuprates and ruthenates. In this article we present results obtained by using a Hubbard-like model Hamiltonian and by taking into account the coupling of the elementary excitations to spin fluctuations (which are partially generated by the quasiparticles itself). This coupling may play the most important role for Cooper-pairing. Using an effective second-order perturbation theory, the interaction of the electrons (or holes) and spin fluctuations leads to the effective electron–electron interaction shown in Fig. 23.13. Applying a generalized Eliashberg-type theory in the strong coupling limit we describe important behavior of hole-doped and electron-doped cuprates and Sr_2RuO_4. We compare our results critically with experiments. Note that in contrast to phonon-mediated superconductors no Migdal theorem can be used [56]. In general, Tremblay et al. [57], and Chubukov et al. [58] have studied the role if vertex corrections of quasiparticle couple to spin fluctuations. They have pointed out that due to $\omega_{sf}/E_F \sim 10^{-2}$ and a flat dispersion around $(\pi, 0)$ they do only change the magnitude of the effective Coulomb interaction U, but the dynamics of vertex corrections may be neglected ($U \rightarrow U_{\mathrm{eff}}$). Note that in the optimally and overdoped cuprates the vertex corrections are small and have positive sign of the Eliashberg coupling. This even enhances the coupling to spin fluctuations [59], but in underdoped cuprates the vertex corrections may become important. However, simply speaking, as long as $U\chi(\mathbf{q}, \omega) < 1$, perturbation theory for constructing a pairing potential in the paramagnetic phase should work, unless poles of the bosonic mode (i.e. in χ) enhance the pair-breaking contribution of the vertex function drastically.[4] Concerning the dynamics, note that the pairing potential is mediated by the same quasiparticles that form the Copper-pairs. Therefore, one has to solve this many-body problem self-consistently. As we will discuss below, the spin susceptibility χ, resulting from d-electrons of Cu, and the gap function that is dominated by a hybridization of p-electrons and d-electrons, strongly interact with each other. Thus, in general one has coupled equations for $\chi\{G, \Delta\}$ and $\Delta\{G, \chi\}$ where G is the Green's function of the corresponding electron. Note that in a simple spin–fermion model the feedback of the elementary excitations on the pairing potential is neglected [58]. Simply speaking, electrons (holes) do not only condense into Cooper-pairs, but are also involved in the magnetic activity. Thus, a self-consistent treatment is indeed required.

By employing the trial-and-error principle we hope to find the correct route for a theory that may play the same role as the BCS theory for the case of electron–phonon interaction with s-wave symmetry singlet Cooper-pairing. Of course, not only the dop-

[4] If so, then the usual perturbation theory breaks down [60] and a non-perturbative approach has to be used as suggested for example in [61].

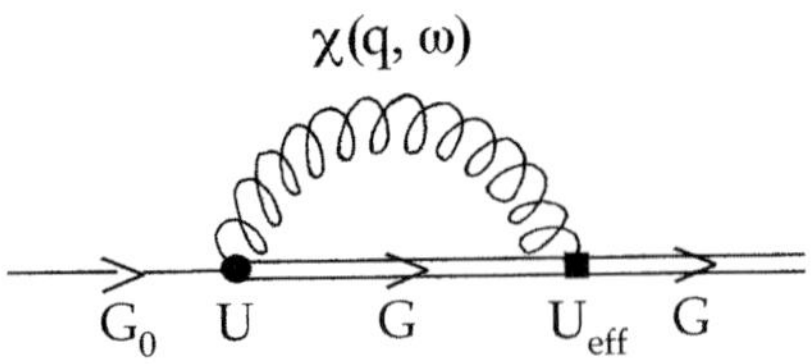

Fig. 23.13. Feynman diagram representing for the cuprates and ruthenates the coupling of the carriers described by Green's function G (G_0 = bare propagator) to magnetic spin fluctuations described by the spin susceptibility $\chi(\mathbf{q}, \omega)$. U (U_{eff}) is an effective (renormalized) coupling constant that includes vertex corrections

ing dependence of superconductivity in the cuprates, but also that of the elementary excitations and of the spin susceptibility $\chi(\mathbf{q}, \omega)$ are important problems to investigate. Cuprates and ruthenates, behave differently due to the differences in their electronic properties. Of course, the electronic structure, the energy dispersions and resultant Fermi surface will play an important role as input for the calculations. We will see later that in an effective one-band description of the itinerant quasiparticles the coupling constant U (see Fig. 23.13) and a tight-binding energy dispersion will be the only parameters that enter the theory.

Basic Electronic Structure

For all cuprate compounds (both hole-doped and electron-doped) the degeneracy between $3d$-orbitals is removed by the lattice structure. After some straightforward calculations it can be shown (see Fig. 23.14(a)) that the hybridized copper and oxygen orbitals separate. The state with highest (lowest) energy has mainly $d_{x^2-y^2}$-wave character. The missing electron (i.e. a hole) gives the Cu-ion a spin $\frac{1}{2}$. Thus, in the absence of doping, the cuprate material is well described by a model of mostly localized spin-$\frac{1}{2}$ states. The other orbitals are occupied and therefore can be neglected. However, the strong Coulomb repulsion between holes in the same orbital has so far not been taken into account and must be included.[5] Due to large Coulomb repulsion $d_{x^2-y^2}$-orbitals split into two so-called lower and upper Hubbard bands (LHB and UHB), respectively, as shown in Fig. 23.14(b). Then, at half-filling the system becomes an insulator. Furthermore, due to a very large U_d the splitting of the $d_{x^2-y^2}$-band is so large that the oxygen p-band lies between the UHB and the LHB. The charge transfer gap in cuprates δ ($\delta = \epsilon_p - \epsilon_d$) is smaller than U_d [62,63].

This has a very interesting consequence for the doping of the CuO_2-plane with holes or electrons. Due to the large Coulomb repulsion the doped hole is mainly in the oxygen p-band. Therefore, in hole-doped cuprates the carriers (holes) occupy mainly hybridized oxygen p-states. On the contrary, electron doping means effective doping of the $d_{x^2-y^2}$-orbital in UHB. Thus, the carriers in the electron-doped cuprates sit largely at Cu $3d$-sites.

Figure 23.14(b) suggests that hole-doped and electron-doped cuprates may under certain assumptions be mapped on to an effective one-band Hubbard model. The difference between hole and electron doping will be reflected by the different parameters of the model and differences of the character of the involved orbitals of the hybridization between p-states and d-states. For hole doping the Fermi level lies in the p-band and for electron doping in UHB. The spin at Cu-sites will induce a spin polarization at the neighboring oxygen sites upon hole doping (see the Zhang–Rice singlet formation). This is not the case for electron doping (at Cu-sites) in which a dilution of the spin system occurs. On the other hand, in hole-doped cuprates one decreases effectively only the number of Cu-spins and the system becomes more itinerant. Consequently, one gets a different behavior of $\chi(\mathbf{q}, \omega)$ and different doping dependence of antiferromagnetism in the hole-doped and electron-doped cuprates.

Similarly as for the cuprates we use also a Hubbard-like Hamiltonian for the ruthenates (Sr_2RuO_4). However, we include three bands (α, β, γ) crossing the Fermi-level into the calculations. In Sr_2RuO_4 the formal valence of the ruthenium ion is Ru^{4+}. This leaves four electrons remaining in the $4d$-shell. Furthermore, the Ru ion sits at the center of

[5] Actually neglecting interactions, one would have expected La_2CuO_4 to be metallic with a half-filled conduction band. As we know this is not true and this material is an antiferromagnetic insulator. Therefore, double occupancy should be energetically not favorable due to Coulomb repulsion.

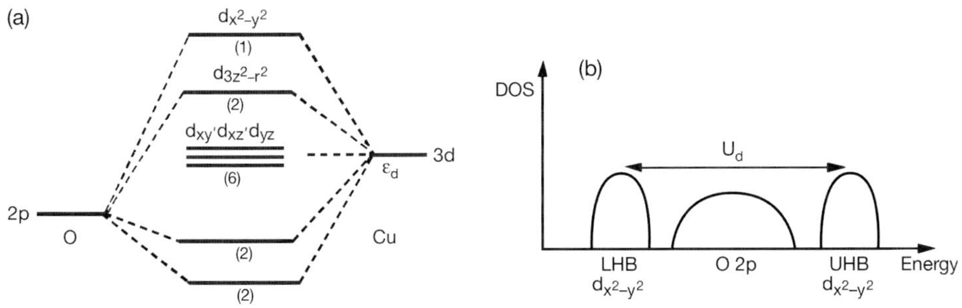

Fig. 23.14. Schematic electronic structure of CuO_2-planes reflecting bonding between a Cu^{2+} and two O^{2-} ions. (**a**) Initial splitting of the bands taking into account crystal field interaction only. We consider only d-electrons (holes) of Cu and p_x and p_y orbitals of the oxygen. The numbers indicate the occupations of the different levels in the undoped compound. (**b**) Role of the on-site Coulomb repulsion and the splitting of the $3d_{x^2-y^2}$-orbital into a lower (LHB) and upper Hubbard band (UHB). U_d is the Coulomb repulsion between electrons at Copper. Due to $\delta = \epsilon_p - \epsilon_d < U_d$ the cuprates are Mott–Hubbard insulators without doping

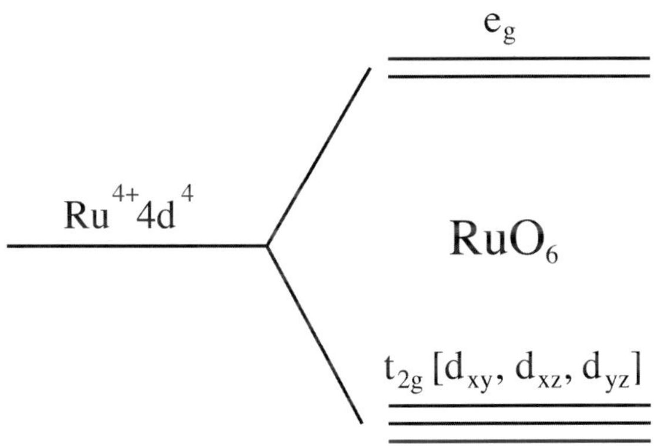

Fig. 23.15. Electronic structure of Sr_2RuO_4. The Ru^{4+} ion corresponds to a $4d^4$ level. The splitting of the e_g and t_{2g} subshells is due to the RuO_6 crystal field. The orbitals d_{xy}, d_{xz}, and d_{yz} cross the Fermi level. Magnetic activity results from the t_{2g} subshell. The RuO_2-plane structure plays an important role

a RuO_6-octahedron and the crystal field of the O^{2-} ions splits the five $4d$-states into threefold (t_{2g}) and twofold (e_g) subshells as illustrated in Fig. 23.15. The negative charge of the O^{2-} ions causes the t_{2g} subshell to lie lower in energy. Note that these orbitals (xz, yz, and xy) have lobes that point between the O^{2-} ions lying along the x, y, and z-axes. Electrons of these orbitals form the Fermi surface and as we have already mentioned the α, β, and γ-bands. We again assume that the most important interaction of the carriers is with spin excitations described by the spin susceptibility $\chi(\mathbf{q}, \omega)$ and resulting from Ru α, β, and γ-states.

To explain the experimentally observed large anisotropy of the magnetic activity in the RuO_2-plane, see $\chi^{+-}(\mathbf{q}, \omega)$ and $\chi^{zz}(\mathbf{q}, \omega)$ along the z-direction perpendicular to the planes, we include also spin–orbit coupling for the significant α, β, and γ-bands. Again, we hope that our trial-and-error type philosophy and critical comparison with experiments will help us to find the correct theory and will explain triplet superconductivity.

Summary

In summary, by using a Hubbard-like Hamiltonian we attempt to describe the electronic properties of the layered superconductors Cu-oxides and Ru-oxides. The theory for spin excitation-mediated superconductivity in cuprates and ruthenates may also help us to find a theory for the other novel superconductors such as the heavy-fermion ones, $ZrZn_2$, UPt_3, UGe_2, etc.

In the next sections we describe the electronic theory for calculating (a) the elementary excitations, the self-energy and spectral density; (b) the spin susceptibility $\chi(\mathbf{q}, \omega)$; and (c) superconductivity and the symmetry of the superconducting order parameter. We present results for various important properties. Finally, we discuss the limitation of our model for the cuprates and ruthenates and give an outlook.

23.2 Electronic Theory for Hole-Doped and Electron-Doped Cuprates

We present now the theory for the interaction between quasiparticles and spin fluctuations in the cuprates and in Sr_2RuO_4. For simplicity we assume that the electronic dynamics is controlled by the CuO_2 and RuO_2-planes, respectively.

For the description of singlet superconductivity in the cuprates we use a one-band model Hamiltonian. The interaction between the carriers (quasiparticles) is described by an effective Coulomb coupling U and the itinerancy of the carriers by a hopping integral t. *We assume that these can be described by the effective one-band with* U. In the case of hole doping the states of the effective band are mainly the p-states originating from O-atoms. The spin fluctuations arise from the magnetic activity of the Cu-spins and the induced spin polarization of the p-states. In the case of electron doping the effective one-band Hamiltonian describes the hybridized d-states of the UHB (see Fig. 23.14(b)) and U is an effective coupling between the electrons in these states. The spin fluctuations result from the Cu-spins. Upon doping the Cu-spins are quenched and the spin polarization of the band is reduced. We assume itinerant magnetism for the UHB. We will see later from a comparison of the theoretical results with experiments in the optimally and overdoped case that this is a useful model.

For describing Cooper-pairing in Sr_2RuO_4 we also employ a Hubbard Hamiltonian. In contrast to cuprates no doping occurs. In order to derive an electronic theory it is important to take the electronic band structure of the t_{2g}-subshell into account and thus to include the states of the d_{xy}, d_{yz}, and d_{xz}-bands that cross the Fermi level. Magnetic activity arises from itinerant electrons in the Ru d-orbitals. Due to a different topology of the corresponding Fermi surfaces a strong magnetic anisotropy occurs ($\chi^{+-} > \chi^{zz}$) in the normal state. This mainly results from different values of the g-factor for the transverse and longitudinal components of the spin susceptibility (i.e. the matrix elements differ) and from a change of the bare energy dispersion of the d-electrons. Obviously, this will have consequences for the symmetry of the superconducting order parameter. In order to describe this magnetic anisotropy, one must also take into account spin–orbit coupling.

23.2.1 Electronic Structure and the Hamiltonian

In the case of the cuprates the calculation of the electronic properties is complicated due to strong electronic correlations.[6] In order to account for these we use a simplifying model that includes the states originating from the CuO_2-planes and play the most significant role. First, we neglect interlayer coupling of CuO_2-planes.

In Fig. 23.16 we show the electronic structure of a CuO_2-plane. The undoped case corresponds to a half-filled band, i.e. one hole per copper site. The holes mainly occupy the $d_{x^2-y^2}$-orbitals and interact with each other via superexchange yielding an antiferromagnetic ground state. Due to a large Coulomb repulsion $U = U_d$ further holes that are doped into the CuO_2-plane are mainly placed on the four oxygen p-states surrounding the copper sites. Thus, the center of mass of these new quasiparticles is still located on the copper sites. Obviously, the spin polarization of the doped holes tends to suppress the antiferromagnetism. In addition, spin frustration takes place for small hole-doping.

In the case of electron-doped cuprates the doped carriers directly occupy copper $d_{x^2-y^2}$-states. Therefore, the antiferromagnetism is weakened by diluting the copper spins. No spin frustration occurs.

Finally, after doping a reasonable number of carriers into a CuO_2-plane, in both cases, electron or hole doping, one ends up with a non-ordered paramagnetic (metallic) phase that still reveals (short-range) antiferromagnetic correlations. The resulting spin excitations are responsible for singlet Cooper-

[6] As we know from the experiment the parent cuprate compounds are antiferromagnetic insulators. At the same time by applying a local-density-approximation calculation on undoped cuprates one finds that the system should be metallic with one hole per unit cell that satisfies the formal valencies of La^{3+}, O^{2-}, and Cu^{2+} [64] that is inconsistent with experiment. The neglection of the electronic correlations due to strong local Coulomb repulsion is responsible for this.

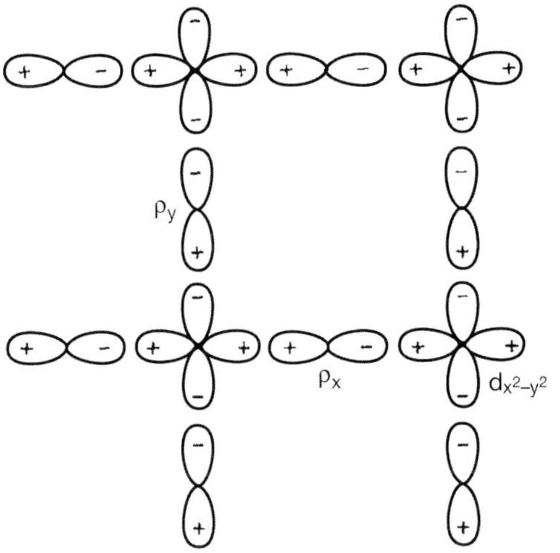

Fig. 23.16. Schematic representation of the electronic structure of a single CuO_2-plane consisting of Cu $d_{x^2-y^2}$-orbitals and the oxygen p_σ-orbitals. Without doping one hole is present in the Cu $d_{x^2-y^2}$-orbital. Via hybridization with oxygen (i.e. superexchange) p-orbitals of two neighboring copper holes interact antiferromagnetically. This leads to antiferromagnetic ordering of the Cu-spins. The doped holes are mainly in the oxygen p-orbitals due to a large Coulomb repulsion $U = U_d$ and destroy antiferromagnetism of the copper spins and spin frustration occurs. In contrast to this, doping the CuO_2-plane with electrons means effectively occupation of the UHB band and dilution of the Cu-spins and thus weakening of the antiferromagnetism

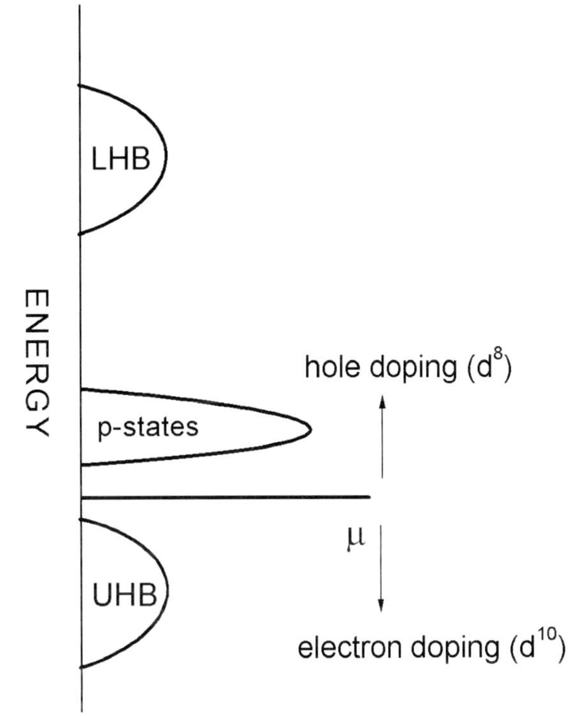

Fig. 23.17. Illustration of the hole and electron doping in cuprates. Without doping due to strong electronic correlations the chemical potential μ lies between the upper Hubbard band (UHB) of the copper $d_{x^2-y^2}$-orbital and oxygen p-states. Upon hole doping the chemical potential shifts towards the oxygen p-band leading to an insulator–metal transition. In the case of electron doping the chemical potential lies in the copper UHB. Note that the lower Hubbard band (LHB) of $d_{x^2-y^2}$-states is separated largely in energy from the oxygen p-states due to $U_d \gg \delta,\ \delta = \epsilon_p - \epsilon_d$

pairing in the cuprates. In our electronic theory we assume that these can be described mainly by itinerant carriers.[7]

We start by using the three-band Hubbard-like model [65]

$$H = \epsilon_d \sum_{i,\sigma} d^+_{i\sigma} d_{i\sigma} + \epsilon_p \sum_{i,\sigma} p^+_{i\sigma} p_{i\sigma} + \sum_{i,j,\sigma} t^{ij}_{pd} \left(d^+_{i\sigma} p_{j\sigma} + \text{h.c}\right) + \sum_{i,j,\sigma} t^{ij}_{pp} \left(p^+_{i\sigma} p_{j\sigma} + \text{h.c}\right) + U_d \sum_i n^d_{i\uparrow} n^d_{i\downarrow} + U_p \sum_i n^p_{i\uparrow} n^p_{i\downarrow} + U_{pd} \sum_{i,j} n^p_{i\uparrow} n^d_{j\downarrow} \tag{23.5}$$

for describing the hole-doped and electron-doped cuprates. Here, the sums are performed over the copper and oxygen lattice positions both labeled by i and σ refers to the spin. $t^{ij}_{\alpha\beta}$ describes the hopping between the sites i and j and states α and β. $n^d_{i\sigma} = d^+_{i\sigma} d_{i\sigma}$ and $n^p_{i\sigma} = p^+_{i\sigma} p_{i\sigma}$ are the Cu $3d$ and O $2p$ hole densities for site i, respectively. U_d and U_p refer to on-site copper and oxygen Coulomb repulsion, respectively, and U_{pd} accounts for the copper–oxygen interaction.

As we have discussed above, due to strong electronic correlations the undoped cuprates are antiferromagnetic insulators. The chemical potential lies between the upper Hubbard band of the $d_{x^2-y^2}$-orbital and oxygen p-band (see Fig. 23.17 for an illus-

[7] We assume that the Cu-spins are part of the $p-d$ hybridized states.

tration). Upon hole doping the chemical potential lies in the oxygen p-band. The latter hybridizes strongly with copper d-states. As a result the bandwidth of the oxygen band reduces and Zhang–Rice singlets are formed [66]. In the case of electron doping, the chemical potential lies in the copper upper Hubbard band (UHB). Furthermore, due to large values of U_d and of the charge-transfer gap δ, the copper UHB and LHB are well separated in energy. This suggests the use of an effective one-band Hubbard-like Hamiltonian[8] for doped cuprates:

$$H = -\sum_{i,j} t_{ij} \left(c^+_{i\sigma} c_{j\sigma} + \text{h.c.}\right) + U \sum_i n_{i\uparrow} n_{i\downarrow} + \dots \tag{23.6}$$

describing the dynamics of holes in hybridized p-states and of electrons in effective d-states of the UHB. In (23.6) the $c^+_{i\sigma}$ are fermion creation operators for spin σ on sites i of a two-dimensional square lattice, and $n_{i\sigma} = c^+_{i\sigma} c_{i\sigma}$ denotes the density for spin σ. The Hubbard U describes the effective Coulomb interaction between the quasiparticles in the conduction band and thus should be a good description of the paramagnetic state with strong antiferrromagnetic correlations.

Note that the effective interaction U might be different in hole-doped and electron-doped cuprates. In the latter case both itinerant behavior and magnetism result from the electrons in the copper d-states of the UHB. In the hole-doped case the hole carriers are located mainly on the oxygen p-states, and the magnetic activity arises mainly from Cu-spins, the localized copper d-holes. These interact with the spins of the doped holes that are present at the oxygen sites. In (23.6) the interaction of the itinerant electrons with the Cu-spins results from U and the remaining not explicitly given terms take care of the local spin character. Note that the interaction of local spins $\mathbf{S}_i$ with itinerant carriers having spin σ should be of the form $(S_i \cdot \boldsymbol{\sigma})$. For simplicity one is tempted to simulate this in the case of hole doping by an effective interaction between the band electrons (holes). Consequently the dots in (23.6) refer to residual interactions, for example resulting from the local character of the Cu-spins.

Using this physical picture we develop the Eliashberg theory for Cooper-pairing in hole-doped and electron-doped cuprates. First, we rewrite the effective one-band Hubbard Hamiltonian for the itinerant carriers in momentum space:

$$H = \sum_{\mathbf{k}\sigma} \epsilon_{\mathbf{k}} c^\dagger_{\mathbf{k},\sigma} c_{\mathbf{k},\sigma} + \frac{U}{2} \sum_{\mathbf{k}\mathbf{k}'\mathbf{q}\sigma} c^\dagger_{\mathbf{k},\sigma} c^\dagger_{\mathbf{k}',-\sigma} c_{\mathbf{k}'+\mathbf{q},-\sigma} c_{\mathbf{k}-\mathbf{q},\sigma} , \tag{23.7}$$

with the one-band electron (or hole) energy

$$\epsilon_{\mathbf{k}} = -2t \left[\cos(k_x) + \cos(k_y) - 2t' \cos(k_x)\cos(k_y) + \frac{\mu}{2}\right] . \tag{23.8}$$

From now on the coupling constant U denotes an effective interaction for the carriers in the $p-d$-states including largely the effects due to $(S_i \cdot \boldsymbol{\sigma})$. For a description of the itinerant quasiparticles we assume nearest neighbor and next nearest neighbor hopping described by t and t', respectively. Here and in the following we set the lattice constant equal to one.

Note that the general form of the unrenormalized tight-binding energy dispersion is the same for holes and electrons, but with different parameters used for t and t'. These have to be chosen in agreement with the experimentally observed Fermi surface in hole-doped and electron-doped cuprates.

23.2.2 Eliashberg Theory for the Interaction Between Quasiparticles and Antiferromagnetic Spin Fluctuations

In general, the interaction between the carriers and spin fluctuations in the cuprates is given by

$$H_{\text{int}} = \sum_{\mathbf{q}} \tilde{U}_{\text{eff}}(\mathbf{q}) \vec{s}(\mathbf{q}) \vec{s}(-\mathbf{q}) . \tag{23.9}$$

Here, $\tilde{U}_{\text{eff}}$ denotes the effective coupling and $s(\mathbf{q})$ is the spin operator of the itinerant carriers,

$$\vec{s}(\mathbf{q}) = \frac{1}{2} \sum_{\mathbf{k}} \sum_{\sigma\sigma'} c^+_{\mathbf{k}+\mathbf{q},\sigma} \tau_{\sigma\sigma'} c_{\mathbf{k},\sigma'} , \tag{23.10}$$

[8] Note, the validity of the effective one-band Hamiltonian for the hole-doped cuprates has been also justified from a microscopic consideration of Baumgärtel, Schmalian and Bennemann [67] on the basis of $p-d$ Hamiltonian using a slave-boson technique [68].

where τ represents the Pauli matrices. $S(\mathbf{Q})$ is the spin operator referring to the Cu-spins. The corresponding spin–spin correlation function gives the dynamical spin susceptibility, $\chi(\mathbf{q},\omega)$. Due to itinerancy of the Cu spin electrons the susceptibility χ may be determined by the random phase approximation (RPA). Then,

$$\chi_{\mathrm{RPA}}(\mathbf{q},\omega) = \frac{\chi_0(\mathbf{q},\omega)}{1-\tilde{U}\chi_0(\mathbf{q},\omega)} . \quad (23.11)$$

Here, χ_0 is the bare spin susceptibility which is defined via retarded Green's functions for the quasiparticles creating the spin excitation. Then, in second-order perturbation theory the effective interaction between carriers has the form (see Fig. 23.13):

$$V_{\mathrm{eff}} = \tilde{U}_{\mathrm{eff}}\, \chi_{\mathrm{RPA}}(\mathbf{q},\omega)\, \tilde{U} , \quad (23.12)$$

where $\tilde{U}_{\mathrm{eff}}$ and $\tilde{U}$ refer to the coupling between the carriers and the spin excitations. The Hubbard-like interaction $\tilde{U}_{\mathrm{eff}}$ includes vertex corrections whereas $\tilde{U}$ denotes the effective coupling. Then, the Eliashberg-like system of equations for a superconductor are given by the Green's function ($G^{-1} = G_0^{-1} - \Sigma$) with self-energy:

$$\Sigma(\mathbf{k},\omega) = \tilde{U}_{\mathrm{eff}}\, \tilde{U} \sum_{\mathbf{k}'}\sum_{\omega'} \chi_{\mathrm{RPA}}(\mathbf{k}-\mathbf{k}',\omega-\omega') \times G(\mathbf{k}',\omega') \quad (23.13)$$

and anomalous self-energy

$$\Phi(\mathbf{k},\omega) = -\tilde{U}_{\mathrm{eff}}\, \tilde{U} \sum_{\mathbf{k}'}\sum_{\omega'} \chi_{\mathrm{RPA}}(\mathbf{k}-\mathbf{k}',\omega-\omega') \times G(\mathbf{k}',\omega')\, G(-\mathbf{k}',-\omega')\Phi(\mathbf{k}',\omega') . \quad (23.14)$$

Here, $\Sigma(\mathbf{k},\omega)$ is the self-energy of the one-particle Green's function $G(\mathbf{k},\omega)$ in the normal state and $\Phi(\mathbf{k},\omega) = \Delta(\mathbf{k},\omega)Z(\mathbf{k},\omega)$ is the corresponding anomalous part. The superconducting order parameter is $\Delta(\mathbf{k},\omega)$. $Z(\mathbf{k},\omega)$ denotes the renormalization function describing the effective mass of the quasiparticles. We employ the Dyson equation yielding

$$G(\mathbf{k},\omega) = \frac{1}{\omega - \epsilon_{\mathbf{k}} + \mu - \Sigma(\mathbf{k},\omega)} \equiv G(k) . \quad (23.15)$$

This system of equations allows one to find the superconducting gap and the superconducting transition temperature T_c (defined through a finite off-diagonal self-energy, i.e. $\Delta(T < T_c) \neq 0$). The spin fluctuations and $\chi(\mathbf{q},\omega)$ would act as an input.

Note that on the RPA level also the charge fluctuations are included. This may be of interest if a charge-density-wave (CDW) or stripe formation occurs. Both may compete for small doping. The superposition of spin and charge fluctuations leads to $\chi = \chi^c + \chi^s$ with

$$\chi^c_{\mathrm{RPA}}(\mathbf{q},\omega) = \frac{\chi_0(\mathbf{q},\omega)}{1+\tilde{U}\chi_0(\mathbf{q},\omega)} . \quad (23.16)$$

Then, in the Nambu representation [69] the 2×2 matrix $\Sigma(\mathbf{k})$ is given by the expression

$$\Sigma(k) = \sum_{k'} V_s(k-k')\,\tau_0\, G(k')\,\tau_0 + \sum_{k'} V_c(k-k')\,\tau_3\, G(k')\,\tau_3 , \quad (23.17)$$

where τ_α ($\alpha = 0, 1, 2, 3$) are the Pauli matrices and V_s and V_c are the matrix elements of the electron–electron interaction due to spin and charge fluctuations shown in Fig. 23.18. Omitting the first diagram, since it yields a shift in the chemical potential μ (non-retarded) and has no d-wave contribution, one finds

$$V_s(q) = \frac{3}{2}\tilde{U}_{\mathrm{eff}}\, \tilde{U}\, \chi_s(q), \qquad V_c(q) = \frac{1}{2}\tilde{U}_{\mathrm{eff}}\, \tilde{U}\, \chi_c(q) , \quad (23.18)$$

where χ_s and χ_c are the dynamical spin and charge (dielectric) susceptibilities which within RPA are given by

$$\chi_s(q) = \frac{\chi_{s0}(q)}{1-\tilde{U}\chi_{s0}(q)} , \qquad \chi_c(q) = \frac{\chi_{c0}(q)}{1+\tilde{U}\chi_{c0}(q)} . \quad (23.19)$$

Here, $\chi_{s0}(q)$ and $\chi_{c0}(q)$ are the corresponding irreducible parts calculated from unrenormalized Green's functions[9], i.e. $\chi_0 \sim \sum_k G_0(k)G_0(k+q)$.

[9] Note that including charge fluctuations effects might be of interest in connection with lattice distortions and stripe formation. Superconductivity and charge density wave formation may coexist to some extent in the cuprates.

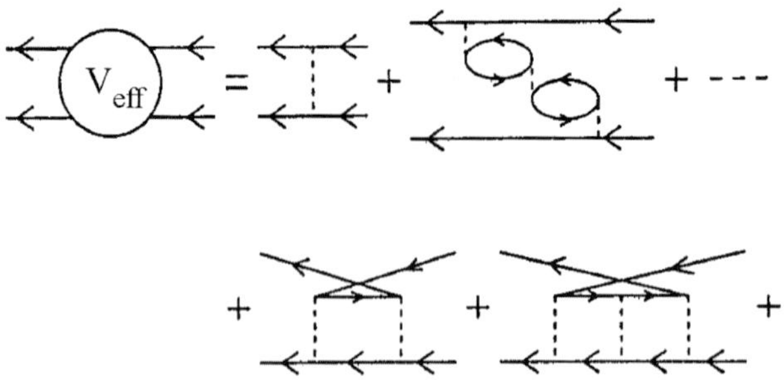

Fig. 23.18. RPA diagrams for the effective pairing interaction V_{eff} for singlet pairing in cuprates resulting from the exchange of longitudinal and transverse spin and charge fluctuations. The *solid lines* refer to the one-particle Green's function that are taken in accordance with (23.15) and (23.17) and the *dashed lines* denote an effective Coulomb interaction $\tilde{U}$. The first diagram leads to a renormalized chemical potential. The diagrams refer to Cooper-pairing due to spin fluctuations. Note that for singlet pairing only an even number of bubble diagrams occur due to Pauli's principle describing both spin and charge fluctuations. The charge and spin fluctuations can be separated from each other and give rise to the first and second terms in (23.21)–(23.22), respectively. Note that the selection of diagrams requires an approximation: we solve the particle–particle (i.e. the Cooper) channel assuming that V_{eff} is *only* particle-hole-like. A comprehensive study would require solving in addition the particle-hole channel for the magnetic instability using *all* particle–particle contributions for V_{eff}. This would yield the set of Parquet-equations (if further mode–mode coupling is taken into account)

Obviously, the effective pairing interaction defined in Fig. 23.18 is related to the self-energy Σ shown in Fig. 23.13 and thus to (23.17) by $V_{\text{eff}} = \delta\Sigma/\delta G$. A similar relation holds for the vertex corrections leading to $\tilde{U} \to \tilde{U}_{\text{eff}}$. To a good approximation they have a weak ω-dependence, and therefore do not change the dynamics of the elementary excitations and the spectrum of the superconducting gap too much. Note also that in general the Green's functions used for V_{eff} may be different from those of the carriers that participate in Cooper-pairing. Furthermore, also the form and value of the bare susceptibilities may differ for localized and itinerant magnetism.

As mentioned above, the matrix Green's function $\hat{G}(k)$ and the self-energy are related by the Dyson equation (23.15):

$$\hat{G}^{-1} = \begin{pmatrix} G_0^{-1} - \Sigma^G & -\Sigma^F \\ -\bar{\Sigma}^F & \bar{G}_0^{-1} - \bar{\Sigma}^G \end{pmatrix}, \qquad (23.20)$$

where Σ^G and Σ^F denote the normal and anomalous part of the self-energy, respectively. Using the diagrams displayed in Fig. 23.18 and employing Matsubara frequencies the self-energies are given by

$$\begin{aligned}\Sigma^G(\mathbf{k}, \imath\omega_n) = \frac{1}{\beta N} \sum_{\mathbf{k}' i\omega'_n} \Big[\frac{1}{2} \tilde{U}_{\text{eff}} \tilde{U} \\ \times \frac{\chi_{c0}(\mathbf{q}, i\nu_m)}{1 + \tilde{U}\chi_{c0}(\mathbf{q}, i\nu_m)} + \frac{3}{2} \tilde{U}_{\text{eff}} \tilde{U} \frac{\chi_{s0}(\mathbf{q}, i\nu_m)}{1 - \tilde{U}\chi_{s0}(\mathbf{q}, i\nu_m)} \\ + \tilde{U}_{\text{eff}} \tilde{U} \chi_G(\mathbf{q}, i\nu_m) \Big] G(\mathbf{k}', i\omega'_n) \end{aligned} \qquad (23.21)$$

and

$$\begin{aligned}\Sigma^F(\mathbf{k}, \imath\omega_n) = -\frac{1}{\beta N} \sum_{\mathbf{k}' i\omega'_n} \Big[\frac{1}{2} \tilde{U}_{\text{eff}} \tilde{U} \\ \times \frac{\chi_{c0}(\mathbf{q}, i\nu_m)}{1 + \tilde{U}\chi_{c0}(\mathbf{q}, i\nu_m)} - \frac{3}{2} \tilde{U}_{\text{eff}} \tilde{U} \frac{\chi_{s0}(\mathbf{q}, i\nu_m)}{1 - \tilde{U}\chi_{s0}(\mathbf{q}, i\nu_m)} \\ - \tilde{U}_{\text{eff}} \tilde{U} \chi_F(\mathbf{q}, i\nu_m) \Big] G(\mathbf{k}', i\omega'_n), \end{aligned} \qquad (23.22)$$

where $i\nu_m = i\omega_n - i\omega'_n$. The first and second terms refer to charge and spin fluctuations, respectively. Note, the bare susceptibilities can be defined in terms of $\chi_{G,F}$, i.e. $\chi_{c0} = \chi_G - \chi_F$ and $\chi_{s0} = \chi_G + \chi_F$. Both are defined in Appendix A. The term $\tilde{U}_{\text{eff}}\tilde{U}\chi_{G,F}$ on the right-hand side compensates for double counting that occurs in the second order.

So far, the Eliashberg-like equations have been derived on the imaginary axis and for finite temperatures $\beta^{-1} = k_B T$. This is convenient because of the spectral properties of the corresponding Green's function. However, in order to compare directly with experiment we need the solution of the self-energy equations on the real ω-axis. One possibility would be to solve the generalized Eliashberg equations slightly above the real axis [70]. However, we further want to avoid continuation to the real ω-axis, $z = i\omega_n \to \omega + i0^+$, after a solution has been obtained. This can lead to numerical uncertainties and instabilities. Therefore, we formulate the generalized Eliashberg equations from the beginning directly on the real frequency axis. To do this we use spectral representation of the one-particle Green's functions [71,72]. Although this requires a large numerical effort, it is not at all a problem with today's computer power. Details of the calculation are given in Appendix A. After performing the sum over the Matsubara frequencies [73] we arrive at the following set of equations for the quasiparticle self-energy components Σ_ν ($\nu = 0, 3, 1$) with respect to the Pauli matrices τ_ν in the Nambu representation:

$$\Sigma_\nu(\mathbf{k}, \omega) = N^{-1} \sum_{\mathbf{k}'} \int_0^\infty d\Omega \left[P_s(\mathbf{k} - \mathbf{k}', \Omega) + (\delta_{\nu 0} + \delta_{\nu 3} - \delta_{\nu 1}) P_c(\mathbf{k} - \mathbf{k}', \Omega)\right] \times \int_{-\infty}^{\infty} d\omega' \, I(\omega, \Omega, \omega') A_\nu(\mathbf{k}', \omega') . \quad (23.23)$$

P_s and P_c refer to spin and charge fluctuations, respectively. The kernel I and the spectral functions A_ν are given by

$$I(\omega, \Omega, \omega') = \frac{f(-\omega') + b(\Omega)}{\omega + i\delta - \Omega - \omega'} + \frac{f(\omega') + b(\Omega)}{\omega + i\delta + \Omega - \omega'} \quad (23.24)$$

and

$$A_\nu(\mathbf{k}, \omega) = -\pi^{-1} \, \mathrm{Im} \, [a_\nu(\mathbf{k}, \omega)/D(\mathbf{k}, \omega)] . \quad (23.25)$$

The denominator is given by

$$D = [\omega Z]^2 - \left[\epsilon(\mathbf{k}) + \xi\right]^2 - \phi^2 ,$$

Also we use $a_0 = \omega Z$, $a_3 = \epsilon(\mathbf{k}) + \xi$, and $a_1 = \Delta Z$. In (23.24), f and b denote the Fermi and Bose distribution function, respectively.

The band filling

$$n = 1/N \sum_{\mathbf{k}} n_{\mathbf{k}}$$

is determined with the help of the **k**-dependent occupation number

$$n_{\mathbf{k}} = 2 \int_{-\infty}^{\infty} d\omega f(\omega) [A_0(\mathbf{k}, \omega) + A_3(\mathbf{k}, \omega)] ,$$

which is calculated self-consistently. Note that this introduces the doping dependence of the superconducting gap, the elementary excitations and spin excitations. The case $n = 1$ corresponds to the half-filling. The interaction between quasiparticles due to spin and charge fluctuation is given by

$$P_s = (2\pi)^{-1} \, \tilde{U}_{\mathrm{eff}} \, \tilde{U} \, \mathrm{Im}(3\chi_s - \chi_{s0}) ,$$
$$P_c = (2\pi)^{-1} \, \tilde{U}_{\mathrm{eff}} \, \tilde{U} \, \mathrm{Im}(3\chi_c - \chi_{c0}) . \quad (23.26)$$

So far our formulation has been quite general and applies similarly to any system with strong interaction of the carriers with spin fluctuations. In case of strong hybridization between the Cu d-states and the oxygen p-states one may approximately use the same Green's functions for determining $G(\mathbf{k}, \omega)$, $\Sigma(\mathbf{q}, \omega)$, and $\chi(\mathbf{q}, \omega)$. In this effective one-band picture one uses $\tilde{U} = U$.[10] Further neglection of vertex corrections yields $\tilde{U}_{\mathrm{eff}} = \tilde{U} = U$.

From now on the local character of the Cu spins (characterized by $\mathbf{S}_i \cdot \boldsymbol{\sigma}$) and the hybridized states resulting from d-electrons and p-electrons are incorporated into the one-band states with the effective coupling U. It remains the *only* parameter (in addition to a bare tight-binding band structure $\epsilon_{\mathbf{k}}$), and we will assume $U \simeq 4t$ in most of the cases; χ_0 now refers to the susceptibility of this band described by (23.7) and (23.8). Then, the above expressions constitute a system of coupled equations for the quasiparticles Green's function and Cooper-pairs. These equations must be solved numerically by iterations for a temperature given and fixed parameters t, U,

[10] Note that $\tilde{U} = U_{\mathrm{eff}}$ neglects differences between the itinerant carriers and the (local) Cu-spins.

as well as doping x.[11] The spin susceptibility is the important input for determining Σ and G. In general, one expects that $\chi(\mathbf{q}, \omega)$ is approximately given by RPA theory, with

$$\chi_{s0} = -\frac{1}{2} T \sum_k \text{Tr} \left[G(k) \tau_0 G(k+q) \tau_0 \right] \quad (23.27)$$

and

$$\chi_{c0} = -\frac{1}{2} T \sum_k \text{Tr} \left[G(k) \tau_3 G(k+q) \tau_3 \right] . \quad (23.28)$$

Here, Tr stands for the trace of the corresponding 2×2 matrix.

In the case of electron-doped cuprates we may safely assume that $\chi(\mathbf{q}, \omega)$ results from the itinerant quasiparticle states in the upper Hubbard band. Thus, we have $\chi = \chi\{G\}$ with χ_0 given by (23.27) and (23.28). This yields a Berk–Schrieffer-type theory for the coupling between electrons described by G and the spin fluctuations described by

$$\chi = \chi_0 + U_{\text{eff}} \chi_0 \chi . \quad (23.29)$$

On the other hand, in the case of hole-doped cuprates the hybridization between the p and d-states suggests for simplicity that the spin excitations may be determined (at least approximately) by effective quasiparticle states described by $G(\mathbf{k}, \omega)$. Hence, we use $\chi = \chi\{G\}$ and for χ the RPA-form and for χ_0 again (23.27) and (23.28). This model is supported by the observation of a strong feedback effect of G on χ [41].

In the case of ruthenates the spin excitations described by $\chi(\mathbf{q}, \omega)$ result from the quasiparticles of the α, β, and γ-states. Hence, the Green's function G is also used for calculating the spin susceptibility, i.e. $\chi = \chi\{G\}$.

Summary

In summary, we use for the cuprates and ruthenates a Berk–Schrieffer-type theory for the behavior of the quasiparticles coupling to spin excitations:

$$G = G\{\chi\} , \qquad \chi = \chi\{G\} , \quad (23.30)$$

and determine χ from an electronic theory within RPA. Obviously, the calculation demands a high degree of self-consistency for determining G, the resultant elementary excitations, the spin susceptibility, and for superconductivity. Details of this procedure are given in Appendix B.

The dynamical spin susceptibility χ^s determines the properties of the electronic system. Its peak structure at $\mathbf{Q}_{AF} = (\pi, \pi)$ is important. The latter arises due to the nesting properties of the Fermi surface. The RPA denominator enhances further this peak structure of the spin susceptibility. The charge fluctuations arise also from the nesting properties, but are not that much enhanced by the RPA denominator. Therefore, the charge fluctuations play a smaller role for Cooper-pairing. However, some other interactions (like electron–phonon interaction or long-range Coulomb repulsion) may change the situation completely and may lead to the enhancement of the dynamical charge susceptibility. This may happen at low doping where the screening is smaller and the lattice distortions can be more important. Although we do not consider these effects explicitly in our electronic theory one can incorporate them relatively easily in our Eliashberg-like theory by just changing mainly the form of χ^c.

23.2.3 Elementary Excitations

For understanding the high-T_c cuprates their elementary excitations are of central significance. First we discuss those in hole-doped cuprates. Using the Green's functions for the elementary excitations (i.e. the local density of states) one gets after continuation to the real ω-axis and for a fixed temperature for the spectral density

[11] Note that one has to use the solution of the generalized Eliashberg equations in order to solve the corresponding two-particle vertex equations. This yields vertex corrections and thus a renormalized Coulomb interaction $U \to U_{\text{eff}}$. In [57] it is argued that vertex corrections are, to a good approximation, frequency independent, and therefore the dynamics of the coupling of elementary excitations and spin excitations remain unchanged. In view of our calculations this seems to be in good agreement with experiments.

$$A(\mathbf{k},\omega) = -\frac{1}{\pi}\frac{\Sigma''(\mathbf{k},\omega)}{[\omega-\epsilon_{\mathbf{k}}-\Sigma'(\mathbf{k},\omega)]^2+[\Sigma''(\mathbf{k},\omega)]^2}\,. \tag{23.31}$$

Here, $\epsilon_{\mathbf{k}}$ is the bare energy dispersion given by (23.8) and $\Sigma'(\mathbf{k},\omega)$ and $\Sigma''(\mathbf{k},\omega)$ are the real and imaginary parts of the self-energy, respectively. The renormalized energy dispersion $\omega_{\mathbf{k}}$ of the quasiparticles is given by

$$\omega_{\mathbf{k}} = \epsilon_{\mathbf{k}} + \mathrm{Re}\Sigma(\mathbf{k},\omega=\omega_{\mathbf{k}})\,. \tag{23.32}$$

Here, the doping dependence occurs due to a change in the chemical potential of $\epsilon_{\mathbf{k}}$ and in particular from the frequency dependence and also from the doping dependence of the dynamical spin susceptibility $\chi(\mathbf{q},\omega)$.

In Fig. 23.19 we demonstrate the importance of many-body effects due to the self-energy. The renormalization of the dispersion by Σ causes structure in the dispersion, the so-called kink feature. We show the unrenormalized tight-binding energy dispersion $\epsilon_{\mathbf{k}}$ in the normal state along the route $(0,0)\to(\pi,0)\to(\pi,\pi)\to(0,0)$ of the first BZ. Below the Fermi level the measured energy dispersion $\omega_{\mathbf{k}}$ along $(0,0)\to(\pi,\pi)$ changes compared to the bare tight-binding case due to the self-energy corrections. We define the kink at the inflection point of the dashed curve where the renormalized dispersion tends to approach again the bare dispersion $\epsilon_{\mathbf{k}}$. Note that the renormalization of $\epsilon_{\mathbf{k}}$ is expected to be anisotropic. Recent studies by Dessau and co-workers [74] revealed no kink structure along $(\pi,0)\to(\pi,\pi)$ direction in the normal state, but only in the superconducting state.

The self-energy $\Sigma(\mathbf{k},\omega)$ for cuprates is calculated self-consistently using the 2D one-band Hubbard Hamiltonian for the CuO_2-planes and assuming coupling of the quasiparticles to spin fluctuations. The superconducting gap function $\Delta(\mathbf{k},\omega)$ is also calculated self-consistently. As described earlier the full momentum and frequency dependence of the quantities is kept. No further parameter is introduced. It is important to realize that due to the combined effects of Fermi surface topology and $\chi(\mathbf{q}=\mathbf{Q},\omega)$ at the antiferromagnetic wave-vector $\mathbf{Q}_{AF}=(\pi,\pi)$, the $\mathbf{k}$ and ω dependence of $\Sigma(\mathbf{k},\omega)$ become very pronounced and change the dispersion $\omega(\mathbf{k})$.

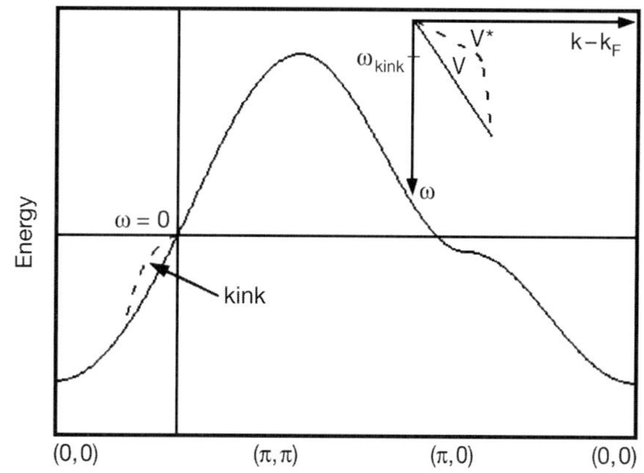

Fig. 23.19. Calculated bare tight-binding energy dispersion for hole-doped cuprates in the normal state using (23.8). The *dashed curve* illustrates the changes due to renormalization, $\omega_{\mathbf{k}}=\epsilon_{\mathbf{k}}+\mathrm{Re}\Sigma(\mathbf{k},\omega=\omega_{\mathbf{k}})$. Due to the structure in $\mathrm{Re}\Sigma$ the energy dispersion shows a kink feature along the $(0,0)\to(\pi,\pi)$ and also other directions in the first Brillouin zone. The *inset* shows an enlargement of the kink structure. The maximum deviation from the bare dispersion and also the renormalized velocity of the quasiparticles v^* reflect the behavior of the self-energy and thus the coupling to spin fluctuations (or in general also possibly to phonons)

Most importantly, the changes in the electronic dispersion are anisotropic in the first BZ. This is illustrated in Fig. 23.20 where we show the Fermi surface of optimally hole-doped cuprates. Due to the fact that the spin susceptibility $\chi(\mathbf{q},\omega)$ has structure mainly at ω_{sf} and for the transferred wave vector $\mathbf{q}=\mathbf{Q}$ the influence of $\chi(\mathbf{q},\omega)$ on the elementary excitations is expected to be very anisotropic with respect to the different parts of the Brillouin Zone. First, the antiferromagnetic wave vector $\mathbf{Q}=(\pi,\pi)$ connects exactly quasiparticles at the Fermi level close to the $(0,\pi)$ points of the first Brillouin Zone. These quasiparticles experience the strongest coupling to antiferromagnetic spin fluctuations. Quasiparticles at the diagonals of the BZ are *not* connected by the $\mathbf{Q}$ and thus have smaller scattering by spin fluctuations at the Fermi level. The corresponding point on the Fermi surface is called a *cold spot* or a *nodal point.* Furthermore, the quasiparticle states at the Fermi level close to the $M=(0,\pi)$ point in the first BZ are

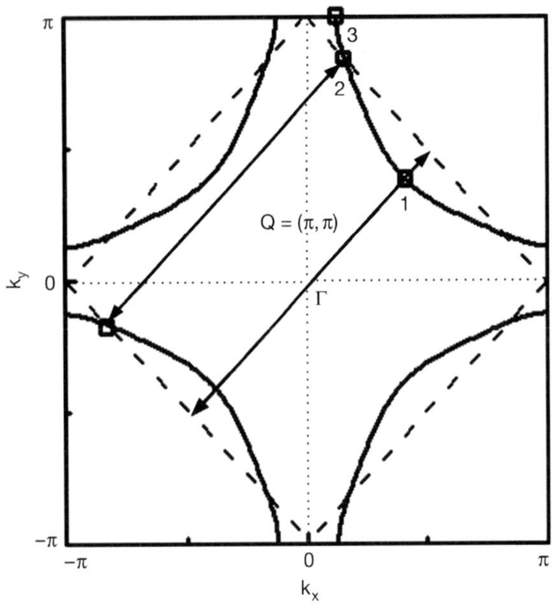

Fig. 23.20. Illustration of the anisotropy of the elementary excitations using the calculated Fermi surface for hole-doped cuprates in the first Brillouin (BZ) zone for a single CuO_2-layer. The anisotropy can be characterized by three different points on the Fermi surface. *Point number 1* refers to the diagonal part of the BZ (the so-called nodal point or cold spot) and *point number 3* refers to the so-called antinodal point close to $(0, \pi)$. The *dashed line* denotes the magnetic BZ crossing the electronic Fermi surface exactly at the "hot spots" (labeled *number 2*). The antiferromagnetic wave vector **Q** connects via scattering by spin fluctuations the two pieces of the Fermi surface mainly at the hot spots. At the $(0, \pi)$ points and along the diagonals the wave vector **Q** connects quasiparticle states below the Fermi level only. Note that the characteristic anisotropic behavior of the elementary excitations may help to distinguish the resultant kink structure from the one which may be caused by electron–phonon coupling

usually called *antinodal* points. Note that the quasiparticle states at cold and hot spots connected by **Q** lie *close to* the Fermi level. This will be important later when discussing the kink feature. Therefore, we may define three characteristic regions at the Fermi surface regarding their sensitivity to coupling to antiferromagnetic spin fluctuations (see Fig. 23.20).

Furthermore, this feature in $\Sigma(\mathbf{k}, \omega)$ changes the spectral density and the energy dispersion. Using Fig. 23.13 the self-energy on the real axis within the weak-coupling limit is approximately given by (see Appendix C)

$$\Sigma(\mathbf{k}, \omega) \approx -\frac{UU_{\text{eff}}}{4} \times \sum_{\mathbf{k}'} \int_0^\infty d\omega' \frac{\text{Im}\chi_{\text{RPA}}(\mathbf{k} - \mathbf{k}', \omega')}{\omega - \omega' - \epsilon_{\mathbf{k}'}} \times \left[\coth\frac{\omega'}{2T} - \tanh\frac{\omega' - \omega}{2T}\right] . \quad (23.33)$$

Due to the momentum and energy conservation one expects a characteristic structure in Σ. In the following we set $U_{\text{eff}} = U$.

It is important to mention that the self-energy has a strong frequency dependence, while the bare energy dispersion of the carriers $\epsilon_{\mathbf{k}}$ has mostly no frequency dependence. Thus, already in the normal state, $\Sigma(\mathbf{k}, \omega)$ has a maximum reflecting a corresponding maximum of $\text{Im}\,\chi$ at $\mathbf{q} \approx \mathbf{Q}$ and $\omega \approx \omega_{sf}$. Then, the kink position follows from the pole of the denominator of (23.34). This leads to the "kink condition"

$$\omega_{\text{kink}} \approx E_{\mathbf{k}-\mathbf{Q}} + \omega_{sf}(x) . \quad (23.34)$$

This gives an estimate of the position of the kink. We will discuss later how ω_{sf} depends on the doping concentration. Note that in the underdoped regime of hole-doped cuprates a normal-state pseudogap with $d_{x^2-y^2}$-wave symmetry opens up below T^* around the antinodal point $(0, \pi)$. This leads to a smaller density of states and therefore to a smaller coupling of holes to spin fluctuations. Furthermore, if the pseudogap is larger than the superconducting gap, no kink structure below T_c should be present along the antinodal direction.

What happens in the superconducting state? Below T_c the superconducting gap $\phi(\mathbf{k}, \omega)$ opens rapidly for decreasing temperature T and becomes maximal in momentum space around the M point reflecting the momentum dependence of the effective pairing interaction. Therefore, the occurrence of a kink structure *only* below T_c in the antinodal direction is a direct fingerprint of the spin excitation spectrum. Furthermore, as we will discuss below $\text{Im}\,\chi(\mathbf{Q}, \omega)$ in (23.34) peaks at the resonance frequency ω_{res} (roughly at $\omega_{sf} + \Delta$). Therefore, the kink condition in the superconducting states is given by

$$\omega_{\text{kink}} \approx E_{\mathbf{k}-\mathbf{Q}} + \omega_{\text{res}}(x)\,. \qquad (23.35)$$

Obviously, ω_{sf}, $\Delta(\mathbf{k}, \omega)$, and thus ω_{res} depend on the doping concentration x. Their behavior is intimately connected to the doping dependence of the dynamical spin susceptibility $\chi(\mathbf{q}, \omega)$ as will be discussed below.

Finally, in order to compare the spectral density of the elementary excitations directly with results for the local density of states measured by angle-resolved photoemission spectroscopy one also needs information about the corresponding matrix elements that are anisotropic and are dependent on the energy of the incident photon. In addition, R. Manzke et al. pointed out that polarization-dependent photoemission also reveals a fine structure in the low energy excitation spectrum that complicates the line-shape analysis of the data [75]. In particular, for a single-layer high-T_c superconductor at optimum hole-doping two maxima dependent on the orientation of the polarization vector are found. Note that if the tetragonal symmetry of the crystal is broken by the occurrence of a charge-density wave, see related formation of stripes for example, polarization-dependent effects are to be expected.

In underdoped cuprates we study the influence of the pseudogap formation on the spin fluctuations by assuming an additional (to a superconducting) d-wave gap in the electronic excitations

$$E_g(\mathbf{k}) = E_g(0)(\cos k_x - \cos k_y)/2\,. \qquad (23.36)$$

Below the pseudogap temperature T* the electronic density of states at the Fermi level is reduced due to the gap in the electronic dispersion. Since $E_g(\mathbf{k})$ is strongly momentum-dependent, the "kink" structure in the dispersion will reveal some changes along (0,0)→ $(\pi, 0)$ in the pseudogap regime. While the origin of the pseudogap has not yet been clarified experimentally, it is instructive to see how the phenomenological form used in (23.36) affects the elementary excitations and the spin excitations.

23.2.4 Dynamical Spin Susceptibility

One of the most important quantity of our electronic theory is the dynamical spin susceptibility. Note that χ controls the corresponding Cooper-pairing, for example. We have already discussed that we assume the magnetic activity of the Cu-spins and of the O-atoms may be described by the states of the UHB in the case of electron doping and by the hybridized p and d-states of the one-band Hamiltonian in the case of hole doping. (This leads to an effective coupling constant U (see the discussion in connection with (23.26)–(23.30).) Then, from the elementary excitations of the quasiparticles we calculate the irreducible parts of the lowest order (Lindhard) spin and charge susceptibility in the normal state

$$\chi_{s0,c0}(\mathbf{q}, \omega) = -\sum_{\mathbf{k}} \frac{f(\epsilon_{\mathbf{k}}) - f(\epsilon_{\mathbf{k}+\mathbf{q}})}{\epsilon_{\mathbf{k}} - \epsilon_{\mathbf{k}+\mathbf{q}} + i\delta}\,. \qquad (23.37)$$

Again, f denotes the Fermi distribution function and $\epsilon_{\mathbf{k}}$ is the energy dispersion. Due to the nesting properties of the Fermi surface χ_{s0} and χ_{c0} (in the following abbreviated as χ_0) are mainly peaked at the antiferromagnetic wave vector $\mathbf{Q} = (\pi, \pi)$. (Note that, in general, χ_{s0} and χ_{c0} can be peaked at different wave vectors depending on the additional interaction included in a model.) Using the random phase approximation (RPA) this will be further enhanced, since

$$\chi_{\text{RPA}}(\mathbf{q}, \omega) = \frac{\chi_0(\mathbf{q}, \omega)}{1 - U\,\chi_0(\mathbf{q}, \omega)}\,. \qquad (23.38)$$

Like already the Lindhard susceptibility also χ_{RPA} may yield low-energy antiferromagnetic spin fluctuations at a frequency $\omega = \omega_{sf}$. (Note that for holes in the p-band U refers to their effective interaction. Note that the doping dependence of χ is controlled by the chemical potential via band filling. On general grounds we expect that for doping $x \to 0$ in particular spin density excitations and charge density excitations compete with each other.)

In order to compare our results with inelastic neutron scattering (INS) experiments, we calculate the imaginary part of $\chi_{\text{RPA}}(\mathbf{Q}, \omega)$:

$$\text{Im}\,\chi_{\text{RPA}}(\mathbf{Q}, \omega) = \frac{\text{Im}\,\chi_0(\mathbf{Q}, \omega)}{(1 - U\,\text{Re}\,\chi_0(\mathbf{Q}, \omega))^2 + U^2\,(\text{Im}\,\chi_0(\mathbf{Q}, \omega))^2}\,. \qquad (23.39)$$

Note that due to the vicinity to an antiferromagnetic instability, Im χ_{RPA} is characterized approximately by an Ornstein–Zernicke behavior [28]

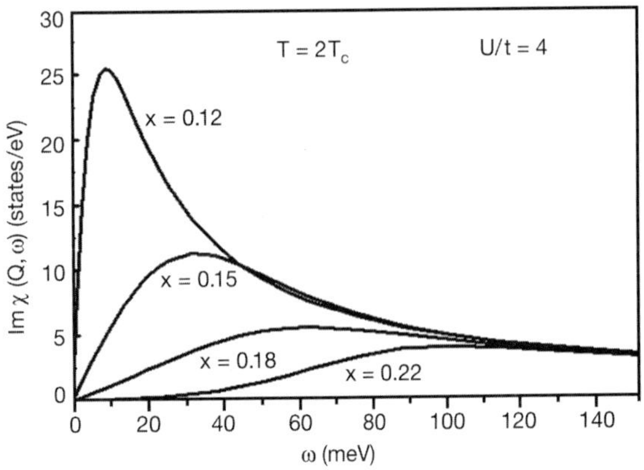

Fig. 23.21. Dynamical spin susceptibility $\mathrm{Im}\,\chi_{\mathrm{RPA}}(\mathbf{Q},\omega)$ calculated using RPA and the generalized Eliashberg equations for hole-doped cuprates in the normal state at a temperature $T = 2T_c$ for various doping concentrations, $x = 0.12$ (underdoped), $x = 0.15$ (optimal doping), and $x = 0.18, x = 0.22$ (overdoped)

$$\mathrm{Im}\,\chi_{\mathrm{RPA}}(\mathbf{Q},\omega) \propto \frac{\omega\,\omega_{sf}}{\omega^2+\omega_{sf}^2}\,, \tag{23.40}$$

with a pronounced peak at the wave vector $\mathbf{q} = \mathbf{Q}$ and $\omega = \omega_{sf}$. In order to illustrate this in Fig. 23.21 we show the doping dependence of $\mathrm{Im}\,\chi_{\mathrm{RPA}}$ at the antiferromagnetic wave vector $\mathbf{Q}$ versus frequency in the normal state. Using $t = 200$ meV we find a typical value of $\omega_{sf} \simeq 30$ meV for the case of optimal doping ($x = 0.15$).

In the superconducting state the situation is different due to the presence of the superconducting gap $\phi(\mathbf{k},\omega) = Z(\mathbf{k},\omega)\Delta(\mathbf{k},\omega)$. The essential behavior follows already from the Lindhard susceptibility χ_0. In the weak-coupling limit ($Z \equiv 1$) χ_0 is of BCS-form [76]:

$$\begin{aligned}\chi_0(\mathbf{q},\omega) = \sum_{\mathbf{k}} \Bigg\{ &\frac{1}{2}\left[1+\frac{\epsilon_{\mathbf{k}}\epsilon_{\mathbf{k+q}}+\Delta_{\mathbf{k}}\Delta_{k+q}}{E_{\mathbf{k}}E_{\mathbf{k+q}}}\right] \\ \times &\frac{f(E_{\mathbf{k+q}})-f(E_{\mathbf{k}})}{\omega-E_{\mathbf{k+q}}+E_{\mathbf{k}}+i\delta} + \frac{1}{4}\left[1-\frac{\epsilon_{\mathbf{k}}\epsilon_{\mathbf{k+q}}+\Delta_{\mathbf{k}}\Delta_{k+q}}{E_{\mathbf{k}}E_{\mathbf{k+q}}}\right] \\ \times &\frac{1-f(E_{\mathbf{k+q}})-f(E_{\mathbf{k}})}{\omega+E_{\mathbf{k+q}}-E_{\mathbf{k}}+i\delta} + \frac{1}{4}\left[1-\frac{\epsilon_{\mathbf{k}}\epsilon_{\mathbf{k+q}}+\Delta_{\mathbf{k}}\Delta_{k+q}}{E_{\mathbf{k}}E_{\mathbf{k+q}}}\right] \\ \times &\frac{f(E_{\mathbf{k+q}})+f(E_{\mathbf{k}})-1}{\omega-E_{\mathbf{k+q}}+E_{\mathbf{k}}+i\delta}\Bigg\}\,. \end{aligned}\tag{23.41}$$

Thus, assuming d-wave symmetry ($\Delta_{\mathbf{k+Q}} = -\Delta_{\mathbf{k}}$) and perfect nesting ($\epsilon_{\mathbf{k+Q}} = -\epsilon_{\mathbf{k}}$), the imaginary part at the wave vector $\mathbf{Q} = (\pi,\pi)$ is given by

$$\begin{aligned}\mathrm{Im}\chi_0(\mathbf{Q},\omega) = \frac{1}{2}\sum_{\mathbf{k}}\Big\{&\left[1-f(E_{\mathbf{k}})\right]\delta(\omega+2E_{\mathbf{k}}) \\ &+\left[f(E_{\mathbf{k}})-1\right]\delta(\omega-2E_{\mathbf{k}})\Big\}\,.\end{aligned}\tag{23.42}$$

Here, the energy $E_{\mathbf{k}} = \sqrt{\epsilon_{\mathbf{k}}^2+\Delta_{\mathbf{k}}^2}$ is the dispersion relation of the Cooper-pairs in the superconducting state. Thus, for $T = 0$ and $\omega > 0$, the spin susceptibility reveals a threshold at $\omega_{DOS} \approx 2\Delta_0(x)$.

A closer inspection of (23.40) shows an interesting resonant-like behavior of $\mathrm{Im}\,\chi(\mathbf{q},\omega)$. Generally, one finds that the structure of $\mathrm{Im}\,\chi$ is determined by $\mathrm{Im}\,\chi_0$, if $(U\mathrm{Re}\,\chi_0) \neq 1$, and by $(U\mathrm{Re}\,\chi_0) = 1$, if this can be fulfilled. Furthermore, one finds for increasing U that the peak in $\mathrm{Im}\,\chi$ shifts to lower energies. Most importantly, χ_0 becomes resonant if the condition

$$\frac{1}{U_{cr}} = \mathrm{Re}\chi_0(\mathbf{q}=\mathbf{Q},\omega=\omega_{\mathrm{res}}) \tag{23.43}$$

is satisfied. Note that this would signal the occurrence of a spin density wave collective mode with energy $\hbar\omega_{\mathrm{res}}$. This resonance energy is slightly lower than ω_{DOS} and the collective mode can be viewed as a bound state inside the gap region $2\Delta(x,T)$. The real part is given (at $T = 0$) by:

$$\begin{aligned}&\mathrm{Re}\,\chi_0(\mathbf{Q},\omega_{\mathrm{res}}) \\ &= \sum_{\mathbf{k}} \frac{E_{\mathbf{k}}E_{\mathbf{k+Q}}-\epsilon_{\mathbf{k}}\epsilon_{\mathbf{k+Q}}-\Delta_{\mathbf{k}}\Delta_{\mathbf{k+Q}}}{\left(E_k+E_{k+Q}\right)^2-\omega_{\mathrm{res}}^2}\,\frac{E_{\mathbf{k}}+E_{\mathbf{k+Q}}}{2E_{\mathbf{k}}E_{\mathbf{k+Q}}}\,.\end{aligned}\tag{23.44}$$

Therefore, due to the feedback effect of superconductivity on $\mathrm{Im}\,\chi$ its behavior differs drastically from the normal state and yields the formation of the resonance feature at $\mathbf{q} = \mathbf{Q}$ and $\omega = \omega_{\mathrm{res}}$ as it is observed in INS experiments.

Physically speaking, the resonance peak appearing in INS experiments only below T_c is mainly determined by the maximum of the superconducting gap, but renormalized by normal state spin excitations. Obviously, Δ_0 and ω_{sf} are doping dependent quantities. While in the overdoped case the renormalization due to ω_{sf} can be neglected, we find for a too small Δ_0 (i.e. the strongly underdoped case) that the resonance condition of (23.43) cannot be fulfilled and the nominator of (23.40) determines the peak position. Thus, on general grounds one expects that

$$\omega_{\text{res}}(x) \propto \begin{cases} \omega_{sf}(x), & \text{underdoped cuprates} \\ 2\Delta_0(x), & \text{overdoped cuprates} \end{cases} . \tag{23.45}$$

Note that the pseudogap will also affect the properties of the dynamical spin susceptibility below T*. For example, the uniform part of the spin susceptibility $\chi(\mathbf{q} = 0, \omega = 0)$ which is proportional to the density of states will strongly decrease upon decreasing temperature below T* due to the gap in the electronic spectrum. Later, we will discuss in detail the influence of the pseudogap by using a new energy dispersion $\epsilon_{\mathbf{k}} = \sqrt{(\epsilon^0_{\mathbf{k}})^2 + E_g(\mathbf{k})^2}$. Similarly the spectrum of the antiferromagnetic spin fluctuations $\text{Im}\chi(\mathbf{Q}, \omega)$ will show some structure at energies about $2E_g(0)$ as follows from (23.42). Furthermore, the resonant condition, (23.43), can be fulfilled also below T*, since the pseudogap influences the spin fluctuation spectrum as the superconducting gap below T_c.

In the superconducting state the influence of the pseudogap on the dynamical spin susceptibility is expected to be small, since it is believed that both are of d-wave type and that the superconducting gap and pseudogap are similar in magnitude. In electron-doped cuprates there seems to be no clear evidence for a pseudogap behavior in the elementary excitations and in the spin susceptibility below a certain temperature T^*. Only recent tunneling experiments on thin films show a pseudogap after a strong external magnetic field ($\simeq H_{c2}$) has been applied [77].

23.2.5 Gap Equation: Singlet and Triplet Cooper-Pairing

We want to investigate now in more detail how unconventional superconductivity can arise from the exchange of spin fluctuations, from a purely electronic mechanism.

Singlet Cooper-Pairing

For this purpose let us first consider the simplified weak-coupling gap equation for singlet pairing ($T = 0$)

$$\Delta(\mathbf{k}) = -\sum_{\mathbf{k}'} \frac{V_s^{\text{eff}}(\mathbf{k}-\mathbf{k}')}{2E(\mathbf{k}')} \Delta(\mathbf{k}') , \tag{23.46}$$

which is a self-consistency equation for the superconducting order parameter $\Delta(\mathbf{k})$ in momentum space where $\mathbf{k}$ is defined in the first Brillouin zone. $V_s^{\text{eff}}(\mathbf{k}-\mathbf{k}')$ represents the effective two-particle pairing interaction in the singlet pairing channel. This is, to a good approximation, proportional to χ_{RPA} if Cooper-pairing due to spin fluctuations is present. Again, the energy $E(\mathbf{k}) = \sqrt{\Delta^2(\mathbf{k}) + \epsilon^2(\mathbf{k})}$ is the dispersion relation of the Bogolubov quasiparticles where $\epsilon(\mathbf{k})$ denotes the bare electronic dispersion of the electrons in the normal state. In contrast to the pairing via electron–phonon coupling the momentum dependence cannot be neglected due to the strong anisotropy of the pairing interaction. Note that in the BCS theory $V_s^{\text{eff}} < 0$ (and V_t^{eff}) is taken as a constant. Therefore, one obtains a solution for $\Delta(\mathbf{k})$ of (23.46) which is structureless in momentum space [50,76].

Let us now investigate how it is possible to solve (23.46) with a *repulsive* pairing potential assuming singlet pairing. Due to the fact that $V_s^{\text{eff}} > 0$, one would naively think that there exists no solution. However, this is not so. If one takes into account that $V_s^{\text{eff}}(\mathbf{k}-\mathbf{k}')$ might have a strong momentum dependence, it is easily seen that indeed the gap equation has a solution. This is illustrated in Fig. 23.22 where we show the first Brillouin zone (BZ) and a Fermi surface that corresponds to the half-filled case of the two-dimensional one-band Hubbard model. This Fermi surface resembles the one measured in the high-T_c superconductor $La_{2-x}Sr_xCuO_4$. For simplicity, we also assume an underlying tetragonal symmetry of the crystal. The behavior of a superconducting order parameter with $d_{x^2-y^2}$-wave symmetry and resultant $\Delta > 0$ and $\Delta < 0$ areas is indicated. If one now assumes that the effective pairing interaction $V_s^{\text{eff}}(\mathbf{k}-\mathbf{k}')$ has a strong momentum dependence, for example, a large peak at the antiferromagnetic wave

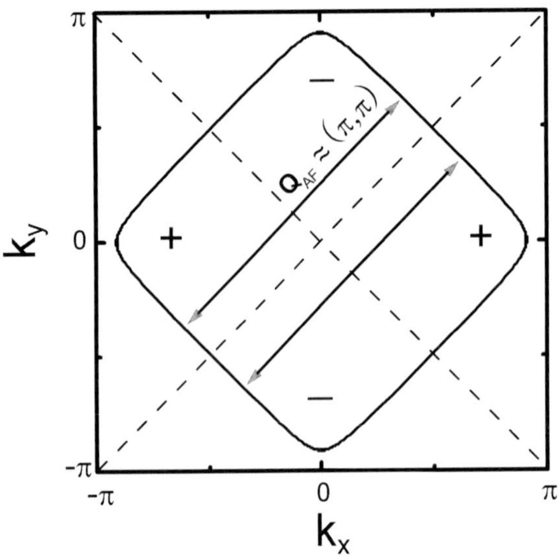

Fig. 23.22. Illustration of the $d_{x^2-y^2}$-symmetry of the superconducting order parameter in the first BZ of hole-doped and electron-doped cuprates. The +(−) sign corresponds to the sign of the order parameter. Two electrons (holes) at the Fermi surface (solid curve) are coupled by the exchange of antiferromagnetic spin fluctuations peaked at (π, π) yielding the formation of Cooper-pairs only in the case if the order parameter changes sign. The *dashed lines* refer to the nodes in the *d*-wave order parameter $\Delta(\mathbf{k}) = \Delta_0[\cos k_x - \cos k_y]/2$

vector $\mathbf{q}_{AF} = \mathbf{Q} = (\pi, \pi)$ connecting different parts of the Fermi surface, then the gap equation has a solution for $V_s^{\text{eff}} > 0$. Note that the wave vector $\mathbf{Q}$ connects areas where Δ is positive with those where $\Delta < 0$. Thus, one indeed finds a solution of (23.46).

The simplest solution is the $d_{x^2-y^2}$-symmetry gap

$$\Delta(\mathbf{k}) = \Delta_0 \left[\cos k_x - \cos k_y\right]/2\,, \qquad (23.47)$$

which has nodes along the diagonals. Note, for an effective pairing interaction which is structureless in momentum space such a solution of the gap equation for a *d*-wave order parameter would not be possible. In fact an order parameter belonging to an anisotropic *s*-wave representation (possibly with nodes) would not satisfy the pairing condition mentioned above. In order to solve (23.46) for a pairing interaction that is peaked at (π, π) one definitely needs an order parameter that changes sign. Therefore, we see that the superconducting gap has less symmetry than the underlying Fermi surface. In addition to a broken gauge invariance ($U(1)$) due to the occurrence of superconductivity a further symmetry is broken (in our case the invariance under a rotation of 90°) we call this "unconventional". Notice that this neither implies an electronic pairing mechanism nor a corresponding large value of T_c, but a distinction from electron–phonon pairing.

Triplet Cooper-Pairing

In the case of triplet Cooper-pairing one has to solve the gap equation

$$\Delta(\mathbf{k}) = +\sum_{\mathbf{k}'} \frac{V_t^{\text{eff}}(\mathbf{k}-\mathbf{k}')}{2E(\mathbf{k}')}\Delta(\mathbf{k}')\,, \qquad (23.48)$$

where $V_t^{\text{eff}}(\mathbf{k}-\mathbf{k}')$ denotes the effective two-particle pairing interaction in the triplet pairing channel. Most importantly, Pauli's principle yields the plus sign on the right-hand side of (23.48), which then requires $V_t^{\text{eff}} < 0$, i.e. an *attractive* pairing interaction in momentum space. Note that this is similar to pairing via electron–phonon interaction where the minus sign in front of the right-hand side of (23.48) is also canceled. A solution of the gap equation is then relatively easy. In particular, if the transferred momentum $\mathbf{q} = \mathbf{k} - \mathbf{k}'$ is small, one obtains *p*-wave symmetry of the superconducting order parameter:

$$\Delta_p(\mathbf{k}) = \Delta_0(\sin k_x + i \sin k_y)\,. \qquad (23.49)$$

Note that the condition $\mathbf{q} \approx 0$ is indeed fulfilled in the case of superfluid ^{3}He that is close to a ferromagnetic transition. A similar situation is present in Sr_2RuO_4 (see its phase diagram in Fig. 23.10 in which the γ-band forms a circle and has a high density of states). This already suggests triplet *p*-wave symmetry for the order parameter of Sr_2RuO_4. Obviously, the results for $\chi(\mathbf{q}, \omega)$ in cuprates exclude triplet Cooper-pairing.

General Situation

Finally, let us critically mention that the arguments given above for cuprates and ruthenates are weak-coupling arguments (i.e. $\omega = 0$). However, these may be used also in the strong-coupling limit of the pairing process. For example, the linearized version of the gap equation for high-T_c cuprates in the strong-coupling limit is given by

$$\begin{aligned}
&\lambda \operatorname{Im} \Delta(\mathbf{k}, \omega) Z(\mathbf{k}, \omega) \\
&= -\frac{1}{N} \sum_{\mathbf{k}} \int_{-\infty}^{\infty} d\omega' \left[b(\Omega) + f(\Omega - \omega) \right] \\
&\times \left[P_s(\mathbf{q}, \omega) - P_c(\mathbf{q}, \omega) \right] \\
&\times \operatorname{Im} \left[\frac{\Delta(\mathbf{k}', \omega') Z(\mathbf{k}', \omega')}{(\omega Z)^2 - \left(\epsilon_{\mathbf{k}'} + \xi \right)^2} \right], \qquad (23.50)
\end{aligned}$$

where P_s and P_c have been defined in (23.23). However, although lifetime effects of the electrons will lead to a renormalization of the quasiparticles, the weak-coupling arguments given above still remain valid at $\omega = \omega_{sf}$. As in the weak-coupling case, for decreasing temperature T the eigenvalue $\lambda(T)$ increases and passes through unity at $T = T_c$. The derivation of (23.50) is given in Appendix A.

In general, the superconducting order parameter $\Delta(\mathbf{k}, \omega)$ does not only yield a gap in the single-particle density of states, it also sets the scale for the condensation energy due to the formation of Cooper-pairs. Therefore, one might expect

$$T_c \propto \Delta_0 ,$$

as in conventional superconductors. This is so in overdoped cuprates where the superconducting transition is of mean-field type with a corresponding coherence length of Cooper-pairs ($\xi \sim 100$Å) and thus global uniform phase of the Cooper-pairing. In contrast to this we will see below that in the underdoped case the number of carriers and density of Cooper-pairs is small and thus the wave functions of Cooper-pairs do not overlap significantly. Under such conditions, also because of the layered structure of cuprates, it is well known that classical (thermal) fluctuations of the phases of the Cooper-pairs play an important role in determining T_c. Their corresponding energy scale is given by the superfluid density (or phase stiffness) n_s [70,78,79]. In such a situation, the Kosterlitz–Thouless theory or the XY model should be applicable. Therefore, in the underdoped regime one already expects from general arguments

$$T_c \propto n_s ,$$

as observed experimentally by Uemura et al. [14]. The calculation of n_s, which controls many properties, is described in the following section.

To summarize, we note on general grounds that for Cooper-pairing via spin fluctuations the underlying Fermi surface topology plays an important role. In particular, for singlet pairing one expects a $d_{x^2-y^2}$-wave order parameter if nesting properties are present. Without nesting one expects no solution for a repulsive pairing interaction. Furthermore, as we will show later the nesting properties of the electron-doped cuprates are weak. Therefore, a competition between repulsive spin-fluctuation-mediated interaction and attractive electron–phonon interaction may result. A a consequence a transition from $d_{x^2-y^2}$-wave order parameter towards anisotropic s-wave symmetry may occur for $T_c \to 0$ and as a function of doping. In the case of (attractive) triplet pairing no nesting properties are needed and p-wave symmetry for the superconducting order parameter would naturally occur if the pairing is dominated by nearly ferromagnetic spin fluctuations. However, if strong nesting were present, an order parameter with f-wave symmetry could win over p-wave symmetry. We will see later that the symmetry of the superconducting order parameter calculated from a microscopic electronic theory will indeed support these general arguments.

23.2.6 Superfluid Density n_s

We discuss now the behavior of the superfluid density $n_s(\omega, T, x)$. The frequency ω reflects the dynamics of the corresponding quasiparticles and thus lifetime effects. Note that n_s controls the doping dependence of the phase coherence, thermodynamic behavior, the penetration depth, the Nernst effect, Cooper-pair phase fluctuations, etc.

Obviously, the superfluid density n_s is a central property for understanding the superconductivity and dynamics of the cuprates, in particular of underdoped cuprates. The superfluid density $n_s(\omega, T, x)$ can be calculated using the current–current correlation function [73] or equivalently from

$$\frac{n_s}{m} = \frac{2t}{\hbar^2} \left(S_N - S_S \right) , \qquad (23.51)$$

where we have introduced for convenience the oscillator strength

$$S_N = \frac{\hbar^2 c}{2\pi e^2 t} \int_{0^+}^{\infty} \sigma_1(\omega) \mathrm{d}\omega \,. \qquad (23.52)$$

S_S is the corresponding value of (23.52) in the superconducting state. In (23.52) we have used the f-sum rule for the real part of the conductivity $\sigma_1(\omega)$

$$\int_0^{\infty} \sigma_1(\omega)\, \mathrm{d}\omega = \pi e^2 n / 2m^* \,,$$

where n is the 3D hole density and m^* denotes the effective band mass (in our case for the tight-binding band considered). The conductivity $\sigma(\omega)$ that enters the equations above is calculated in the normal and superconducting state using the Kubo formula [73,80]

$$\sigma(\omega) = \frac{2e^2}{\hbar c} \frac{\pi}{\omega} \int_{-\infty}^{\infty} \mathrm{d}\omega' \left[f(\omega') - f(\omega' + \omega)\right] \times \frac{1}{N} \sum_{\mathbf{k}} \left[v_{k,x}^2 + v_{k,y}^2\right] \qquad (23.53)$$
$$\times \left[N(\mathbf{k}, \omega' + \omega)\, N(\mathbf{k}, \omega') A_1(\mathbf{k}, \omega' + \omega)\, A_1(\mathbf{k}, \omega')\right] .$$

Here, $v_{k,i} = \partial\epsilon_k / \partial k_i$ are the calculated band velocities within the CuO_2-plane and c is the c-axis lattice constant. The spectral functions N and A_1 are defined in (23.168). Vertex corrections have been neglected.

Physically speaking, we are looking for the loss of spectral weight of the Drude peak at $\omega = 0$ that corresponds to excited quasiparticles above the superconducting condensate for temperatures $T < T_c^*$. Then, with the help of above equations we calculate the dynamical conductivity $\sigma(\omega)$ via (e = elementary charge) [73]

$$\frac{n_s(\omega)}{m} = \frac{1}{e^2} \omega \operatorname{Im} \sigma(\omega) \,, \qquad (23.54)$$

where $\operatorname{Im} \sigma(\omega)$ has been obtained using (23.54) from the current–current correlation function and the Kubo formula. In a next step the penetration depth $\lambda(x, T)$ is calculated within the London theory [81] by

$$\lambda^{-2} \propto n_s$$

This will be discussed later.

Of course, in particular for underdoped cuprates phase fluctuations of the Cooper-pairs must be included into the calculation of n_s. A generalization of (23.54) leads to

$$n_s(\mathbf{q}, \omega) = \omega \operatorname{Im} \sigma(\mathbf{q}, \omega) \,. \qquad (23.55)$$

(Note that the derivation requires Fermi-liquid-like arguments assuming a thermodynamical phase transition which might not be valid if the limit $T \to 0$ in the vicinity of a quantum critical point (QCP) is considered. This will be discussed in Sect. 23.4.7.) Since the conductivity σ can be expressed in terms of a current–current correlation function, n_s is related to $\langle \mathbf{J}(\mathbf{r}, t) \cdot \mathbf{J}(0, 0) \rangle$, where $J \propto (\psi^* \nabla \psi - \psi \nabla \psi^*)$. Using $\psi(\mathbf{r}, t) = |\psi| e^{i\phi(\mathbf{r}, t)}$ one arrives at ($\nabla\phi(0, 0) = \frac{2e}{c\hbar} \vec{A}(0, 0)$ for frozen amplitude fluctuations, see Appendix D.)

$$n_s \sim \langle \nabla\phi(\mathbf{r}, t) \cdot \nabla\phi(0, 0) \rangle \,. \qquad (23.56)$$

On general grounds one expects that the phase correlation function $C = \langle \nabla\phi \cdot \nabla\phi \rangle_\phi$ behaves as $C \to 1$ for large doping x and that a Meissner effect occurs only for $C(T) \to 1$ at T_c.[12] (Here, $\langle \nabla\phi \cdot \nabla\phi \rangle_\phi$ means phase averaging).

From this we can calculate $n_s(\mathbf{q}, \omega)$ and its doping dependence. For overdoped cuprates and for $T \to 0$ we expect that phase fluctuations play a minor role, the see detailed discussion in the next section, while for underdoped cuprates ($x \to 0$) Cooper-pair phase fluctuations should become more and more important. However, for frequencies $\omega \sim 1/time > \omega_{phase}$ the behavior of $n_s(\mathbf{q}, \omega)$ should correspond to the

[12] In analogy to ferromagnetism like in Ni we expect $n_s^0/n_s = \langle \nabla\Phi(\mathbf{r}, t) \nabla\Phi(0) \rangle$. Note that it is straightforward to map our electronic theory on a lattice (Wannier type representation) and then to derive from the product of the anomalous Green's functions, $\{\mathcal{F} \cdot \mathcal{F}^*\}$, a contribution to the free energy of the form $n_s \cos\Theta_{ij}$ as used by Chakraverty et al. [78]. Here, Θ_{ij} is the angle between the phase of neighboring Cooper-pairs. Approximately, one has $n_s = n_s^0 \bar{\Theta}_{ij}$, with $\bar{\Theta}_{ij} = 1$ for $T < T_c$ and $\bar{\Theta}_{ij} = 0$ for $T > T_c$.

case where no phase fluctuations are seen. In this connection $\omega_{phase}^{-1} = \tau_1$ refers to the lifetime of the phase fluctuations. In underdoped cuprates it describes the dynamics of n_s. We estimate the lifetime from $\tau_1 \propto \Delta^{-1} \propto T_c^{-1}$, since the energy change involved in the excitations is of the order of $\Delta e^{i\phi}$. Note that above T_c one has $\langle e^{i\phi} \rangle = 0$ due to phase-incoherent Cooper-pairs.

In summary, our mean field results for $n_s(x, T)$ are expected to be approximately valid for overdoped cuprates and also for underdoped cuprates if $T \to 0$ for times shorter than the lifetime of the phase fluctuations. If Cooper-pair phase fluctuations become important the superfluid density is given by (23.56). Note that for $\mathbf{q} \to 0, \omega \to 0$ after phase averaging of $\langle \nabla\phi(\mathbf{r}, t) \cdot \nabla\phi(0, 0) \rangle \to 0$ one arrives at $T = T_c$. Thus, the Meissner effect occurs only for phase-coherent Cooper-pairs. In the following we safely approximate the phase correlation function $\langle \nabla\phi \cdot \nabla\phi \rangle \approx 0$ above T_c. Below T_c one takes $\langle \nabla\phi\nabla\phi \rangle \approx 1$ for simplicity. This is approximately true for Gaussian fluctuations.

23.2.7 Cooper-Pair Phase Fluctuations

As mentioned above the cuprates consist of weakly coupled two-dimensional CuO_2 planes. For doping $x \to 0$ the Cooper-pairs do not overlap much. Therefore, one expects that Cooper-pair phase fluctuations are enhanced. These fluctuations of phase incoherent Cooper-pairs destroy the long-range superconducting order and the Meissner effect. In conventional bulk superconductors Cooper-pair phase fluctuations are not relevant, since the large superfluid density leads to a strong overlap of Cooper-pairs and the phase fluctuations cost much energy. Thus, in conventional superconductors the superconducting transition is due to the destruction of the Cooper-pairs and T_c is proportional to Δ_0 [76]. On the other hand, the observation by Uemura et al. that $T_c \propto n_s$ in underdoped hole-doped cuprates [14] indicates that the phase fluctuations drive the transition in these regime. Then, at $T \to T_c$ where $\Delta F_{phase} \simeq T_c$ phase coherent Cooper-pairs occur. ΔF_{phase} denotes the free energy contribution due to phase fluctuations. Cooper-pairs might exist above T_c and break only up at a crossover temperature around $T_c^* > T_c$. Here, T_c^*

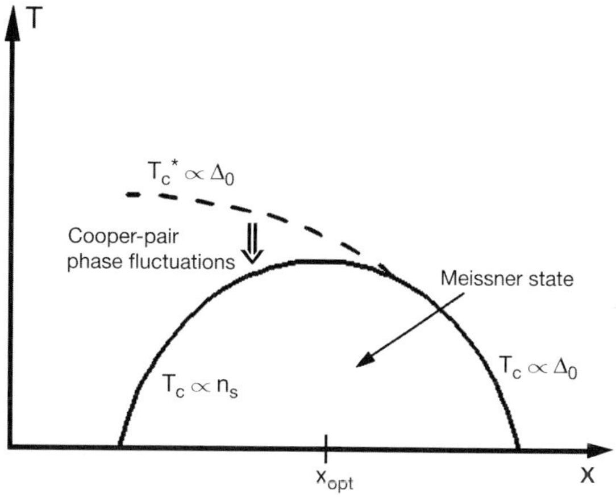

Fig. 23.23. Illustration of the significance of Cooper-pair phase fluctuations in underdoped cuprates with low superfluid density n_s. With the help of the generalized Eliashberg theory below the crossover temperature T_c^* one finds local Cooper-pairs without long-range phase coherence ("preformed pairs"). The temperature T_c denotes the superconducting phase transition temperature below which phase coherent Cooper-pairs and the Meissner effect occur. For $x < x_{opt}$ one gets $T_c^* \to T_c$ due to phase fluctuations

is approximately given by the transition temperature that one would obtain without phase fluctuations. Note that the Meissner effect takes place only below T_c. This situation is illustrated in Fig. 23.23.

The bulk transition temperature T_c at which phase coherence of the Cooper-pairs occurs is determined by the Ginzburg–Landau free energy functional $\Delta F\{n_s, \Delta\}$ where the superfluid density divided by the effective mass m, $n_s(x, T)/m$, is calculated self-consistently from the current–current correlation function. First, we consider the static case, i.e. $n_s \neq n_s(\omega)$. The doping dependence of the Ginzburg–Landau-like free-energy change (see $n_s(x, T)$) is given by

$$\Delta F = F_S - F_N = -\frac{1}{\beta} \ln \left[\int \mathcal{D}\Delta^\dagger \mathcal{D}\Delta \, e^{-\frac{1}{\hbar} S_{\text{eff}}[\Delta^\dagger, \Delta]} \right] . \tag{23.57}$$

The functional integral is taken over the fields $\mathcal{D}\Delta^\dagger = \prod_{\mathbf{r}_i, t_i} d\Delta^\dagger(\mathbf{r}_i, t_i)$. The effective action S_{eff} is defined in Appendix D where we also derive the Dyson equation for the Cooper-pair phase fluctuations on a microscopic level. On phenomenological grounds the

Gor'kov equations including the phase fluctuations were derived by Berthod and Giovannini [82]. Equation (23.57) is the basis for determining T_c in the presence of Cooper-pair phase fluctuations. For simplicity, we neglect first the interaction between the amplitude and the phase of Cooper-pairs (see footnote on page 1442 and Appendix D for details of the interplay between amplitude and phase fluctuations). Then, we write

$$\Delta F = \Delta F_{\text{cond}} + \Delta F_{\text{phase}} , \tag{23.58}$$

where $\Delta F_{\text{cond}} \approx \alpha\{n_s/m^*\}\Delta_0(x)$ is the condensation energy due to Cooper-pairing and $\Delta F_{\text{phase}} \approx \frac{\hbar^2 n_s}{2m^*}$ is the loss in energy due to phase incoherent Cooper-pairs. Here, α denotes the available phase space for Cooper-pairs normalized per unit volume and can be estimated in strongly overdoped regime. In contrast to conventional superconductors in cuprates α is smaller mainly due to large size of the unit cell [83–85]. This will be discussed later in connection with Fig. 23.52.

From the interplay between condensation energy and phase fluctuation energy we can estimate the role of phase fluctuations for determining T_c at different doping. One expects that for underdoped cuprates $0.15 > x \to 0$, phase fluctuations get more important and cost less energy. Thus $\Delta F_{\text{phase}} < \Delta F_{\text{cond}}$ and consequently $T_c \sim \Delta F_{\text{phase}} < \Delta F_{\text{cond}} \sim T_c^*$. Note that Cooper-pairs break up thermally at T_c^*. Since $\Delta F_{\text{phase}} \propto n_s$ one gets due to Cooper-pair phase fluctuations $T_c(x) \propto n_s(x)$ in the underdoped regime.

In the overdoped case the situation is described by the fact that Cooper-pair phase fluctuations cost too much energy. Consequently, the system undergoes a mean-field transition according to a gain in condensation energy ΔF_{cond} and then $T_c(x) \propto \Delta_0(x)$. Note that the explicit derivation of the free-energy functional is given in Appendix D. The Meissner effect only occurs if after phase averaging $n_s(\omega \to 0) \neq 0$ yielding a London penetration depth

$$\frac{1}{\lambda^2(T)} \propto n_s(T, \omega = 0) .$$

Here, n_s refers to the phase averaged superfluid density. Within standard (time-dependent) Ginzburg–Landau theory (see footnote on page 1442), the superfluid density n_s can be calculated through (23.56).

To summarize, using this physical picture one already obtains on general grounds a superconducting phase diagram and in particular an optimum T_c and doping concentration as it is illustrated in Fig. 23.23. In particular, calculating ΔF_{cond} and ΔF_{phase} from the generalized Eliashberg equations one finds an optimal doping concentration at $x_{\text{opt}} \simeq 0.15$. We get $T_c \propto \Delta F_{\text{cond}} \sim \Delta_0$ for overdoped cuprates and $T_c \propto \Delta F_{\text{phase}} \sim n_s$ for underdoped cuprates. Clearly, the Meissner effect occurs only for phase coherent Cooper-pairing and $\lambda^{-2} \propto n_s$.

BKT Theory for Cooper-Pair Phase Fluctuations

Due to the layered structure of the cuprates and weak interlayer coupling the underdoped cuprates should behave in accordance with the 2D-XY model (except in a narrow critical range around T_c where 3D-XY is more appropriate) [86,87]. Thus, the standard theory for the 2D-XY model, the Berezinskii–Kosterlitz–Thouless (BKT) renormalization group theory, should be a reasonable starting point for a study of Cooper-pair phase fluctuations [88–91]. In particular we discuss the determination of T_c with the help of the BKT theory similar to the work by Timm et al. [92].

The superconducting transition predicted by BKT theory is due to unbinding of fluctuating vortex–antivortex pairs in the superconducting order parameter. Gaussian phase fluctuations are not important, since they do not shift T_c. (This is true in three dimensions. Of course, in the 2D case Gaussian fluctuations destroy the long-range order yielding $T_c = 0$, but the mean-field transition is still unchanged.) Since we are mainly interested in the determination of T_c, we first consider the static case where the phase ϕ does not depend on frequency. Then, the phase action is given by

$$S_{\text{phase}} = \frac{\tilde{K}}{2} \int d^2 r \, (\nabla\phi)^2 , \tag{23.59}$$

where the phase stiffness is related to the superfluid density through $\tilde{K} = n_s(T)\hbar^2/(4m^*T)$. For the cuprates it turns out that the relevant parameters for the unbinding of thermally created vortex–antivortex pairs are the dimensionless stiffness K and the vortex core energy E_{core}. The vortex core energy refers to the energy amount that is needed for the suppression of the superconducting order parameter in the vortex core, i.e. $\Delta(\mathbf{r} \to \mathbf{r}_{\text{core}}) = 0$. This, in general will renormalize the phase stiffness yielding $K \to K_r$. The stiffness is related to n_s by

$$K(T) = \beta\hbar^2 \frac{n_s(T)}{m^*} \frac{d}{4}, \qquad (23.60)$$

where β denotes again the inverse temperature and m is again the effective mass. d is the average spacing between CuO_2 layers. In our calculations we set d to half the height of the unit cell of $YBa_2Cu_3O_{6+x}$, see, for example [92]. Note that the renormalized stiffness K_r is related to the superfluid velocity through

$$\frac{1}{K_r} = \frac{4m^2}{\hbar^2} \int \mathrm{d}^2r \, \langle \mathbf{v}_s(\mathbf{r}) \cdot \mathbf{v}_s(0) \rangle . \qquad (23.61)$$

For determining T_c one has to solve the Kosterlitz recursion relations ("flow equations") [88,92]

$$\frac{\mathrm{d}y}{\mathrm{d}l} = (2 - \pi K)\, y \qquad (23.62)$$

and

$$\frac{\mathrm{d}K}{\mathrm{d}l} = -4\pi^3 y^2 K^2 . \qquad (23.63)$$

Here, $y = e^{-\beta E_c}$ denotes the vortex fugacity. For the vortex core energy we use the approximate result by Blatter et al., i.e. [87]

$$E_{\text{core}} = \pi k_B T K \ln \kappa , \qquad (23.64)$$

where κ is the Ginzburg parameter. $l = (r/r_0)$ is a logarithmic length scale which relates K to the strength of the vortex–antivortex interaction. Note that for $T > T_c$, K tends to zero for $l \to \infty$. The interaction at large distances is screened and the largest vortex–antivortex pair unbind. This destroys the Meissner effect and leads to dissipation. On the other hand, bound Cooper-pairs reduce K and thus n_s, but do not destroy superconductivity. After solving (23.62)-(23.64) it turns out that the renormalization of K is very small [92]. Thus, within a good approximation, Timm et al. obtain $T_c(x)$ from the simple criterion [88,92–94][13]

$$K(T_c) = \frac{2}{\pi}, \qquad (23.65)$$

or

$$\frac{n_s(T_c, x)}{m^*} = \frac{2}{\pi} \frac{4k_B T_c}{\hbar^2 d} . \qquad (23.66)$$

The superfluid density $n_s(T, x)$ entering (23.66) is calculated from the generalized Eliashberg equations using (23.56)–(23.54).

Finally, in order to compare our results also with time-dependent measurements we calculate $n_s(\omega, T)$. Regarding the frequency dependence of n_s we mention the following: A dynamical generalization of the BKT theory has been developed by Ambegaokar et al. [90,95]. It turns out that the critical size for a vortex–antivortex pair is given by

$$r_\omega = \sqrt{\frac{D_v}{2\pi\omega}}, \qquad (23.67)$$

where D_v denotes the corresponding diffusion constant. Only pairs with size $r \le r_\omega$ contribute to the screening. Unfortunately, D_v is not easy to calculate. In the absence of pinning, the theory by Bardeen and Stephen yields [96]

$$D_v^0 = \frac{2\pi c^2 \xi_{ab}^2 \rho_n T}{\phi_0^2 \tilde{d}}, \qquad (23.68)$$

where c is the speed of light, $\xi_{ab} \sim r_0/2$ corresponds to the coherence length, ρ_n denotes the normal-state resistivity, $\phi_0 = hc/2e$ is the elementary superconducting flux quantum, and $\tilde{d}$ corresponds to an effective layer thickness. However, in cuprate superconductors, pinning becomes important. Thus, we assume a simple Arrhenius law

$$D_v = D_v^0 \exp\left(\frac{-E_p}{T}\right), \qquad (23.69)$$

[13] In FLEX theory the fluctuations of the ordered antiferromagnetic state in the paramagnetic metallic regime are treated *beyond* the mean-field level. However, the fluctuations of the superconducting condensate were neglected in earlier treatments. A detailed comparison between $n_s(\omega)$ within the XY model and the FLEX approximation is given in [92,94].

where E_p denotes the corresponding pinning energy barrier. Equations (23.69) and (23.68) are inserted into (23.67) yielding a new length scale l in (23.62) and (23.63), namely a new upper limit $l = \ln(r_\omega/r_0)$. Thus, we get $n_s(\omega, T)/m^* = \omega \operatorname{Im} \sigma(\omega)/e^2$ where the conductivity $\sigma(\omega)$ is obtained from the current–current correlation function. Since $1/\omega \sim$ *time*, and we refer to the time of an observation, we expect phase coherence already for $\omega\tau \gg 1$, where τ refers to the lifetime of phase fluctuations.

Resumé

In summary, we have demonstrated on general grounds that the interplay between Cooper-pair condensation energy and phase fluctuation energy already yields an optimal doping concentration. Thus, the phase diagram can be described in the overdoped part by $T_c \propto \Delta_0$ and in the underdoped case by $T_c \propto n_s$. The superconducting transition temperature T_c in underdoped cuprates might be estimated by the BKT theory. It is interesting that $n_s = (m/e^2)\omega\sigma''(\omega)$ gives $n_s \propto \sigma_{DC}T_c$ if $(\omega\sigma''/\sigma_{DC}T_c)_{\omega=0} = \text{const}$. Such a scaling was noted by C. Homes et al. [97]. For underdoped cuprates one observes $(\sigma_{DC}T_c)_{T\to T_c}$ to be independent of T.

The *dynamical stiffness* $\propto n_s(\omega)$ provides information about the time-scale of Cooper-pair phase fluctuations. In our theory these are treated classically and the coupling between amplitude and phase has been neglected. The results for $n_s(\omega)$ can be compared with time-resolved measurements. Note that for $\omega > \omega_{\text{phase}}$, where ω_{phase} refer to the phase fluctuations, one expects a behavior as in the absence of the phase fluctuations. Thus $n_s(\omega) \to n_s(T, x)$ in the absence of the Copper-pair phase fluctuations. We further would like to mention that phase fluctuations are also related to the formation of vortices. For details see Appendix D.

23.2.8 Tunneling Spectra

A most important verification of the approximate validity of the conventional Eliashberg theory came from electron tunneling experiments. They probe the **k**-averaged quasiparticle excitation spectra in the normal and superconducting state. Therefore, in the case of singlet pairing in high-T_c cuprates and triplet pairing in ruthenates also important fingerprints of the spin-fluctuation-mediated pairing interaction can be seen. Hence, we now analyze tunneling in superconductor–insulator–normal metal (SIN) and superconductor–insulator–superconductor (SIS) junctions.

Using standard tunneling theory [98] we have

$$H = H_L + H_R + H_T \,, \tag{23.70}$$

where $H_L(H_R)$ is the Hamiltonian for particles on the left-hand side (right-hand side) of the tunnel junction. Note, H contains all many-body effects. Here, in general, $[H_L, H_R] = 0$. H_T denotes the tunnel operator with a matrix element T_{pk}. The total current through the tunnel junction can be expressed as the average rate of change per time t of the number of particles. For example, on the left-hand-side of the junction ($\hbar = 1$),

$$\begin{aligned} I_{\text{tot}}(t) &= -e\,\langle \dot{N}_L(t)\rangle \qquad (23.71)\\ &= -e\,i\int_{-\infty}^{t} \mathrm{d}t'\,\langle\,[N_L(t), H_T(t')]\,\rangle\,. \end{aligned}$$

After some algebra [73] one arrives at

$$I_{\text{tot}}(t) = I_s(t) + I_J(t)\,,$$

where

$$\begin{aligned} I_s(t) = e &\int_{-\infty}^{\infty} \mathrm{d}t'\,\theta(t-t') \qquad (23.72)\\ &\times \Big(e^{ieV(t'-t)}\langle\,[A(t), A^+(t')]\,\rangle \\ &\quad - e^{ieV(t-t')}\langle\,[A^+(t), A(t')]\,\rangle\Big)\,, \end{aligned}$$

and

$$\begin{aligned} I_J(t) = e &\int_{-\infty}^{\infty} \mathrm{d}t'\,\theta(t-t') \qquad (23.73)\\ &\times \Big(e^{-ieV(t'+t)}\langle\,[A(t), A(t')]\,\rangle \\ &\quad - e^{ieV(t+t')}\langle\,[A^+(t), A^+(t')]\,\rangle\Big)\,. \end{aligned}$$

Here, I_s refers to the single-particle tunneling and I_J describes tunneling of Cooper-pairs, the Josephson effect. The operator A is given by $A(t) =$

$\sum_{\mathbf{k},\mathbf{p}} T_{\mathbf{kp}} c^+_{\mathbf{k}}(t) c_{\mathbf{p}}(t)$ where $c^+_{\mathbf{k}}$ ($c_{\mathbf{p}}$) refers to the single-particle creation (annihilation) operator entering the Hubbard Hamiltonian, and eV is the applied voltage.

In order to study the single-particle excitations we focus in the following on I_s. A closer inspection of (23.73) shows that the current can be expressed as $I \propto (G_{\text{red}} - G_{\text{adv}})$ where G_{red} (G_{adv}) is the retarded (advanced) Green's function that correspond to the operator A. This would lead to $I_s = -2e \,\text{Im}\,(G_{\text{red}})$ and is important for SIS junctions. In the case of SIN tunneling one obtains [73,76]

$$I_s = 2e \sum_{\mathbf{k},\mathbf{p}} |T_{\mathbf{k},\mathbf{p}}|^2 \int_{-\infty}^{\infty} \frac{d\omega}{2\pi} A_R(\mathbf{k}, \omega) \times A_L(\mathbf{p}, \omega + \text{eV}) \left[f(\omega) - f(\omega + \text{eV}) \right] , \qquad (23.74)$$

where $f(\omega)$ again denotes the Fermi distribution function and A_L and A_R is the spectral density of the left-hand-side (sc-state) and right-hand-side (n-state) material, respectively. The factor 2 arises from the spin summation. The matrix element $T_{\mathbf{k},\mathbf{p}}$ (i.e. the transfer rate) varies on the energy scale E/E_F which should be negligible if $E \simeq \Delta_0$. Thus, an adequate approximation should be to take $T_{\mathbf{k},\mathbf{p}} = T_0 = \text{const.}$

Note the elementary excitations are entering in (23.75) via $A(\mathbf{k}, \omega)$. In particular, the frequency dependence of the gap, $\phi(\omega)$, reflecting the dynamics of the Cooper-pairing is of importance. Performing the sum over momenta we obtain the important result that the single-particle density of states $N(\omega)$ is proportional to the dynamical conductivity dI_s/dV (normalized by its normal-state value) for a SIN tunneling junction system measured at $\omega = \text{eV}$ with V being the applied voltage:

$$\sigma_{\text{tun}}(\text{eV}) = \frac{dI_s(\text{eV})}{dV} \propto \sum_{\mathbf{k}} A(\mathbf{k}, \omega) = N(\omega) . \quad (23.75)$$

Obviously, below T_c the tunneling density of states σ_{tun} (eV) in cuprates reveals interesting structures due to the interaction of quasiparticles with antiferromagnetic spin fluctuations and the opening of the superconducting gap. What is the energy position of these structures? In the superconducting state the spin fluctuation spectrum is peaked at the resonance energy ω_{res}. Thus, the structure in the NIS tunneling occurs at

$$\omega_{\text{res}} + \Delta_0 \text{ (NIS)} . \qquad (23.76)$$

This follows immediately from (23.75), since the spectral density $A(\mathbf{k}, \omega)$ in the superconductor reflects ω_{res} and the shift of the quasiparticle density by Δ_0.

In the case of SIS tunneling the characteristic feature in the tunneling spectrum is shifted to the higher energy

$$\omega_{\text{res}} + 2\Delta_0 \text{ (SIS)} . \qquad (23.77)$$

Again, in addition to ω_{res} each spectral density shifts the energy by Δ_0.

The maximum gap value Δ_0 and the resonance frequency ω_{res} are both doping dependent quantities. As mentioned earlier to a good approximation one finds $\omega_{\text{res}}(x) \propto 2\Delta_0(x)$ and $T_c \simeq T_c^* \propto \Delta_0$ in the overdoped case and $\omega_{\text{res}}(x) \propto \omega_{sf}(x)$ in the underdoped cuprates. Thus, in overdoped cuprates a reduction of spectral weight ("dip") is expected at around $\omega_{\text{res}} + \Delta_0 \simeq 3\Delta_0$ (SIN), and in underdoped cuprates at $\omega_{\text{res}} + \Delta_0 \simeq \omega_{sf} + \Delta_0$ (SIN). Similarly we get for overdoped cuprates structure at $\omega_{\text{res}} + 2\Delta_0 \simeq 4\Delta_0$ (SIS), and for underdoped ones at $\omega_{\text{res}} + 2\Delta_0 \simeq \omega_{sf} + 2\Delta_0$ (SIS). We will see later in Sect. 23.4.8 that this agrees well with available experimental data [20,99]. Therefore, we may safely conclude that as in the BCS and conventional Eliashberg theory the single-particle tunneling spectrum reflects the density of states and thus provides insight into the strength and dynamics of the Cooper-pairing.

Finally, we would like to mention that for electron–phonon mediated superconductivity the spectrum of the pairing potential is mainly unchanged in the superconducting state with respect to the normal state. Thus, structure in the tunneling spectrum is expected at an energy $\hbar\omega_D$ where ω_D is the Debye frequency. In contrast to this, in the case of spin-fluctuation-mediated Cooper-pairing strong feedback effects are present yielding anisotropic elementary excitations and a renormalized spin spectrum ($\omega_{sf} \rightarrow \omega_{\text{res}}$). This leads to the structure described above and is clearly a direct fingerprint of Cooper-pairing by spin fluctuations and is expected for both hole-dope and electron-doped cuprates. In addition, also the momentum dependence of the superconducting order parameter is reflected in the tunneling spectrum. In particular, for an order parameter with s-wave symmetry resulting from electron–phonon interaction, one finds an isotropic density of states with a gap for frequencies $|\omega| < 2\Delta_0$. In contrast, in the case of unconventional superconductivity where the order parameter has nodes, thermal excitations with energies smaller than $2\Delta_0$ are possible (around the nodes). Using thermodynamic arguments one can easily show that the density of states has a linear slope for $\omega \to 0$ [27]. Note that additional structure might occur due to Andreev bound-states in SIN and SIS tunneling, which cannot be described by our theory.

Note, the spectral density $A(\mathbf{k}, \omega)$ exhibits particle-hole asymmetry reflecting the correlations amongst the quasiparticles. (see an early discussion of this by Sawatzky et al. [100] and Bennemann et al., and others.)

23.2.9 Bilayer Effects

Many high-T_c cuprates contain more than one CuO_2-plane per unit cell like for example YBCO or BISSCO compounds (see Fig. 23.2). So far, our theory has focused mainly on the physical properties of a single CuO_2-plane. Of course, it is desirable to understand how the properties change if several CuO_2-planes per unit cell are present and how the number of CuO_2-layers per unit cell affects $T_c(x)$, for example. Note that the bilayer coupling $t_\perp$ causes changes of the Fermi surface topology and the dispersion $\epsilon_\mathbf{k}$. There is also an interlayer coupling of Cooper-pairs. For several CuO_2-planes per unit cell also effects due to varying hole doping are expected. The simplest situation is illustrated in Fig. 23.24 where we show schematically the coupling of two CuO_2-planes via the coupling parameter $t_\perp$.

In the following we will present an extension of the generalized Eliashberg-like theory for the case of two coupled CuO_2-planes [101,102]. We start from a Hamiltonian that contains nearest and next nearest neighbor hopping t and tt', respectively, as well as an intra-layer Coulomb repulsion U, an inter-layer interaction V and an inter-layer hopping $t_\perp$. The corresponding Hamiltonian is given by

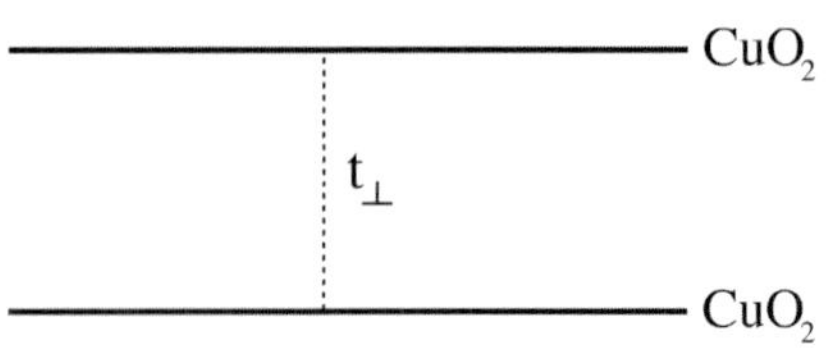

Fig. 23.24. Illustration of the interlayer coupling $t_\perp$ affecting the energy dispersion $\epsilon_\mathbf{k}$, Fermi surface and Cooper-pairing across CuO_2-layers. The magnetic coupling between the planes, $J_\perp$, is not displayed. It is one order of magnitude smaller than the in-plane (Heisenberg-like) superexchange $J_\parallel$ and thus is not considered in our electronic theory. However, $J_\perp$ will split the spin-susceptibility into an even and odd part, see neutron scattering experiments

$$H = H_{\text{int}} + H_{\text{hop}}\,, \tag{23.78}$$

and with $U_{\text{eff}} = U$

$$H_{\text{int}} = U \sum_i n_{il\uparrow} n_{il\downarrow} + \frac{V}{2} \sum_{i \neq j \sigma \sigma' l} n_{il\sigma} n_{jl\sigma'}\,, \tag{23.79}$$

and

$$H_{\text{hop}} = -\sum_{\langle i,j \rangle l \sigma} t_l\, c^\dagger_{il\sigma} c_{jl\sigma} - t_\perp \sum_{i\sigma} c^\dagger_{i1\sigma} c_{i2\sigma}\,. \tag{23.80}$$

Here, $c^\dagger_{il\sigma}$ ($c_{il\sigma}$) creates (annihilates) an electron on site i in the layer $l = 1$ or $l = 2$ with spin σ and $n_{il\sigma} = c^\dagger_{il\sigma} c_{il\sigma}$. We diagonalize the Hamiltonian with respect to the hopping term. Thus, we perform a transformation in term of the operators $d_{\mathbf{k}\nu\sigma}$ with the help of

$$c_{\mathbf{k}l\sigma} = \sum_\nu S_{l\nu}\, d_{\mathbf{k}\nu\sigma}\,,$$
$$S_{l\nu} = \frac{1}{\sqrt{2}} \begin{pmatrix} e^{-ik_z c} & e^{-ik_z c} \\ e^{-ik_z c} & -e^{-ik_z c} \end{pmatrix}, \tag{23.81}$$

where $\nu = +$ and $\nu = -$ denotes the bonding and antibonding band, respectively, $2c$ is the distance between adjacent planes, and $\mathbf{k}$ is the in-plane momentum. Thus, (23.80) can be rewritten as

$$H_{\text{hop}} = \sum_{\mathbf{k}\sigma} (d_{\mathbf{k}1\sigma}, d_{\mathbf{k}2\sigma}) \begin{pmatrix} \epsilon_+(\mathbf{k}) & 0 \\ 0 & \epsilon_-(\mathbf{k}) \end{pmatrix} \begin{pmatrix} d_{\mathbf{k}1\sigma} \\ d_{\mathbf{k}2\sigma} \end{pmatrix}, \tag{23.82}$$

with

$$\epsilon_{\pm} = -2t\left[\cos(k_x) + \cos(k_y) - 2t'\cos(k_x)\cos(k_y) + \frac{\mu}{2}\right] \mp t_{\perp} . \quad (23.83)$$

One is now able to extend the generalized Eliashberg equations as described at the beginning of this chapter. The corresponding Bethe–Salpeter equations separate with respect to band indices into two sets of coupled equations [101,103]. If, for simplicity the inter-plane Coulomb repulsion V is neglected we find six coupled equations for the self-energy components:

$$\Sigma_{i\nu}(\mathbf{k},\omega) = N^{-1}\sum_{\mathbf{k}'}\sum_{\mu=+,-}\int_0^\infty \mathrm{d}\Omega\,\left[P_s^{\mu\nu}(\mathbf{k}-\mathbf{k}',\Omega) + (\delta_{\nu 0}+\delta_{\nu 3}-\delta_{\nu 1})\,P_c^{\mu\nu}(\mathbf{k}-\mathbf{k}',\Omega)\right] \times \int_{-\infty}^{\infty}\mathrm{d}\omega'\, I(\omega,\Omega,\omega')A_{i\nu}(\mathbf{k}',\omega') , \quad (23.84)$$

(for a comparison, see (23.23)). In a shorthand notation we get

$$\Sigma_{ll'}(k) = \sum_{k'} V_{ll'}(k-k')\,G_{ll'}(k') , \quad (23.85)$$

where $V_{ll'}$ is the effective pairing interaction due to spin (and also charge) fluctuations.

Obviously, V and thus the dynamical spin susceptibilities $\chi_{s,c}$ are 2×2 matrices in the layer indices, having elements diagonal (χ^{aa},χ^{bb}) and off-diagonal (χ^{ab},χ^{ba}) with respect to the layer indices. The components of the spin susceptibility transforming as even and odd with respect to the layer indices are given by $\chi^e = \chi^{aa}+\chi^{bb}$ and $\chi^o = \chi^{ab}+\chi^{ba}$. For identical planes $\chi^{aa} = \chi^{bb}$ and $\chi^{ab} = \chi^{ba}$. The measured susceptibility is then given

$$\chi(\mathbf{q},q_z,\omega) = \chi^e(\mathbf{q},\omega)\cos^2\left(\frac{q_z c}{2}\right) + \chi^o(\mathbf{q},\omega)\sin^2\left(\frac{q_z c}{2}\right) , \quad (23.86)$$

where c is the separation of the layers within a unit cell, $\mathbf{q}$ is the in-plane wave vector, and q_z denotes the component perpendicular to the planes. Note, if in cuprates or ruthenates the spin response in INS experiments would dominate at the wave vector $q_z = \pi$, for example, one would expect that $\chi^o \gg \chi^e$ in (23.86).

23.2.10 Electron-Doped Cuprates

Although we discussed in the previous chapters already the differences and similarities of electron-doped and hole-doped cuprates (in particular the evolution of antiferromagnetism as a function of the doping concentration) it is of interest to summarize the significant properties of the electron-doped compounds. The understanding of the phase diagram of the electron-doped cuprates should play a role in resolving the physics of cuprates in general.

In Fig. 23.25 we show the phase diagram for the metallic phase of electron-doped cuprates. The behavior of T_c upon doping follows a parabolic-like curve with a narrow doping range for which superconductivity occurs. Therefore, despite the rather different behavior of the spin fluctuations in hole-doped (localized) and electron-doped (itinerant) cuprates the doping dependence of the T_c curve looks similar in hole-doped and electron-doped cuprates. This suggests the same mechanism for Cooper-pairing and, most importantly, indicates the similar role played by the localized and itinerant spin fluctuations. The appearance of the "pseudogap" in the electronic spectrum of electron-doped cuprates has been debated for a long time. Recently its existence in the presence of a high magnetic field has been confirmed by careful tunneling spectroscopy by Alff et al. [77]. In particular, it has been found that the pseudogap shows a very similar doping dependence as in hole-doped cuprates, but its temperature range is lower. Moreover, as seen from Fig. 23.25 as in hole-cuprates the energy scale of the pseudogap and superconductivity crosses each other around the optimal doping and the pseudogap disappears in slightly overdoped cuprates. This confirms the close similarity between electron-doped and hole-doped cuprates regarding most of the properties. It also suggests that the mechanism of superconductivity, symmetry of the superconducting order parameter, and the anomalous nor-

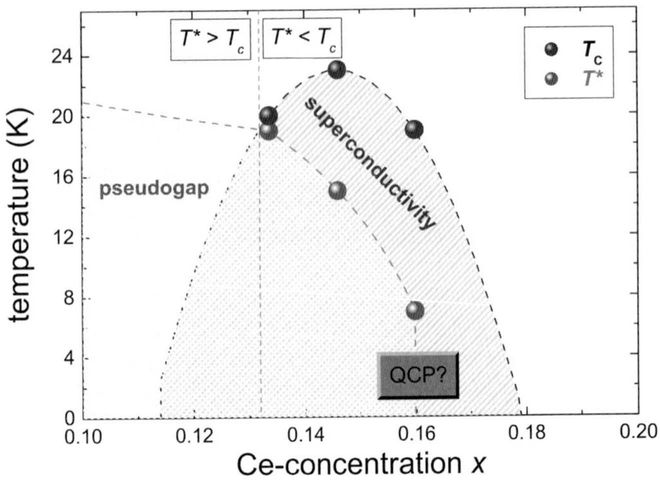

Fig. 23.25. Phase diagram of the electron-doped cuprates measured by tunneling experiments on thin films (taken from [77]). Note the narrow range of superconductivity in electron-doped cuprates, and the smaller energy scale of the pseudogap. Both hole-doped and electron-doped cuprates show a similar phase diagram indicating the same mechanism for superconductivity below T_c and pseudogap below formation T^*. The doping concentration where the pseudogap disappears may be related to a quantum critical point (QCP)

mal state properties should have the same origin in both compounds.

In order to discuss the main difference in Fig. 23.26 we show our results for the Fermi surface of the hole-doped and electron-doped cuprates. These agree well with available experimental data [104]. Remarkably, the differences of the Fermi surfaces in the first BZ in hole and electron-doped cuprates have important consequences for their superconducting and normal state properties. In particular, the antiferromagnetic spin fluctuations which are peaked at $\mathbf{Q} = (\pi, \pi)$ connect different portions of the Fermi surface in hole-doped and electron-doped cuprates as one can see in Fig. 23.26. In the hole-doped cuprates the antiferromagnetic wave vector connects the carriers that lie close to the $M = (\pi, 0)$ point of the first BZ. This means that the main scattering by the spin fluctuations occurs in hole-doped cuprates around these points (so-called "hot spots"). On the other hand, in the electron-doped cuprates the antiferromagnetic wave vector $\mathbf{Q} = (\pi, \pi)$ connects carriers that are close to the diagonal of the BZ. Therefore, the main effect of the interaction will occur there. This will be reflected in the asymmetrical behavior of the hole-doped and electron-doped cuprates.

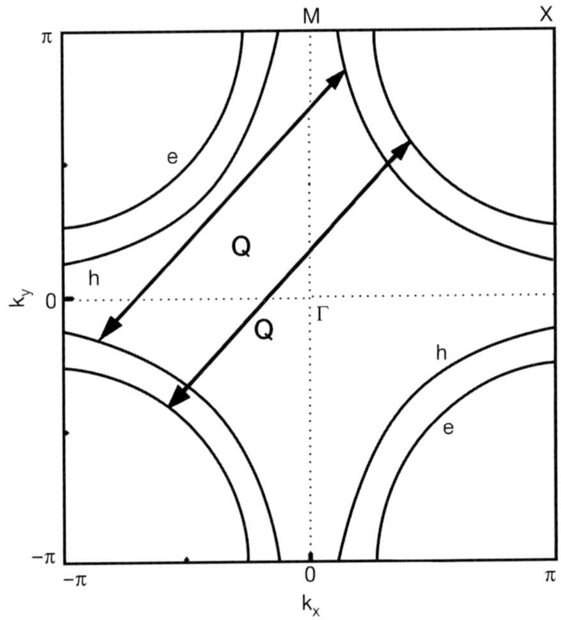

Fig. 23.26. Tight-binding results for the Fermi surface for hole-doped (h) and electron-doped (e) cuprates at optimal doping. $\Gamma = (0, 0)$, $M = (\pi, 0)$, and $X = (\pi, \pi)$ denote the symmetry points of the first Brillouin zone. Remarkably the topology of the Fermi surface is hole-like in both cases. $Q = (\pi, \pi)$ refers to the antiferromagnetic wave vector

What are the consequences for the superconducting state? In particular, we expect the appearance of the higher harmonics in the $d_{x^2-y^2}$-wave superconducting order parameter in the electron-doped cuprates, since the Cooper-pairing occurs far from the antinodal M points and close to the diagonals of the Brillouin zone. Furthermore, due to a weaker (weaker than in hole-doped cuprates) nesting of the

Fermi surface the non-Fermi-liquid behavior and antiferromagnetic fluctuations should also be less pronounced in the electron-doped cuprates. Lattice vibrations may play a role for superconductivity in the overdoped part of the electron-doped phase diagram. For $T_c \to 0$ a competition between quasiparticles coupling to spin fluctuations and phonons may for energetic reasons in principle cause a change from $d_{x^2-y^2}$-symmetry to s-symmetry of the superconducting order parameter. Such a scenario could be formulated using Ginzburg–Landau theory.

To summarize, we show that the differences between hole-doped and electron-doped cuprates are expected due to differences in their Fermi surfaces and density of states. As a result of this, the scattering by spin fluctuations plays a more important role around $M = (\pi, 0)$ points of the BZ in hole-doped cuprates. On the contrary, the scattering by spin fluctuations occurs close to the diagonals of the BZ in the electron-doped cuprates. This will be reflected in the asymmetric behavior between hole-doped and electron-doped cuprates in the normal and superconducting states.

23.3 Electronic Theory for Ruthenates (Sr_2RuO_4)

We discuss Sr_2RuO_4 in detail, since triplet Cooper-pairing seems present as maybe also realized in some heavy fermion systems. What may cause triplet pairing, since both ferromagnetic and antiferromagnetic spin fluctuations are present in the ruthenates? What is the effect of the strong magnetic anisotropy on superconductivity and its thermodynamic properties?

23.3.1 Electronic Structure and the Hamiltonian

We present now the electronic theory for Sr_2RuO_4 which exhibits a superconducting transition temperature $T_c = 1.5$ K. Effects involving all bands crossing the Fermi level are expected to play an important role in determining the triplet superconductivity in this compound. We will start the analysis by calculating the electronic structure of Sr_2RuO_4. In Fig. 23.27 we illustrate the basic electronic structure of Sr_2RuO_4. The highly planar structure of Sr_2RuO_4 prevents substantial energy dispersion along the z-direction. Note the large interplanar separation of the RuO_6 octahedra. At the same time in the ab-plane neighboring RuO_6-octahedra share O ions which in turn are π-bonded to the Ru ions. Thus, the xy-orbital will form a two-dimensional band, while the xz and yz-bands have only a restricted one-dimensional character. For the metallic properties one has to take into account intermediate electronic correlations.

Then an effective Hamiltonian for describing the physics in Sr_2RuO_4 is a two-dimensional three-band Hubbard Hamiltonian

$$H = \sum_{\mathbf{k},\sigma} \sum_{\alpha} \epsilon_{\mathbf{k}\alpha} a^{+}_{\mathbf{k},\alpha\sigma} a_{\mathbf{k},\alpha\sigma} + \sum_{i,\alpha} U_{\alpha}\, n_{i\alpha\uparrow} n_{i\alpha\downarrow}\,, \tag{23.87}$$

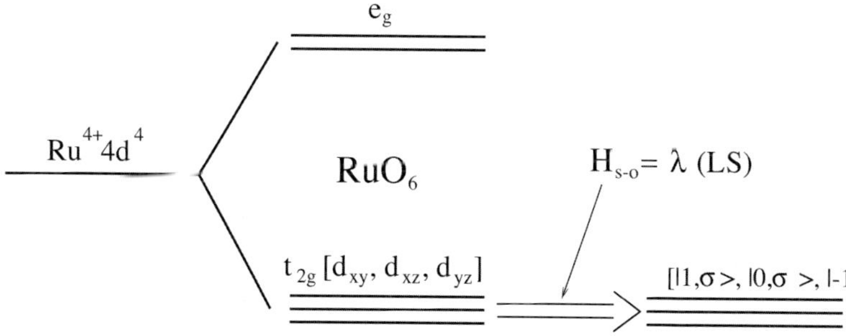

Fig. 23.27. Electronic structure of Sr_2RuO_4 including an effective spin-orbit coupling between d_{xy}, d_{yz}, and d_{xz} states. Due to a mixing of spin and orbital degrees of freedom the new states are characterized by pseudo-spin and pseudo-orbital quantum numbers. Since the spin-orbit coupling H_{s-o} does not break the time-reversal symmetry the Kramers degeneracy between the spin up and down is not removed

where $a_{\mathbf{k},\alpha\sigma}$ is the Fourier transform of the annihilation operator for the d_α orbital electrons ($\alpha = xy, yz, zx$) and U_α is an effective on-site Coulomb repulsion. The hopping integrals $t_{\mathbf{k}\alpha}$ denote the energy dispersions of the tight-binding bands

$$\epsilon_{\mathbf{k}\alpha} = -\epsilon_0 - 2t_x \cos k_x - 2t_y \cos k_y + 4t' \cos k_x \cos k_y \,. \tag{23.88}$$

In accordance with LDA calculations [105] and experimental measurements of the Fermi surface and energy dispersions [106] one chooses for d_{xy}, d_{zx}, and d_{yz}-orbitals the values for the parameter set $(\epsilon_0, t_x, t_y, t')$ as $(0.5, 0.42, 0.44, 0.14)$, $(0.23, 0.31, 0.045, 0.01)$, and $(0.23, 0.045, 0.31, 0.01)$ eV. This model was recently applied to Sr_2RuO_4 [107–109]. Its electronic properties can explain some features of the spin excitation spectrum in the compound Sr_2RuO_4. However, this model neglects spin–orbit coupling and fails to explain the magnetic anisotropy at low temperatures observed in NMR experiments [110].

It is known that the spin–orbit coupling plays an important role in the superconducting state of Sr_2RuO_4. In particular it was shown by Ng and Sigrist that the spin–orbit coupling lowers the free-energy and effects superconductivity and the symmetry of the superconducting order parameter [108]. The superconducting state is characterized by the orbital momentum of the Cooper-pair wave function aligning along the z-direction. This so-called chiral state of the superconducting order parameter is given by

$$\Delta(\mathbf{k}) = \Delta_0 \tilde{\mathbf{z}}(\sin k_x \pm \sin k_y) \,. \tag{23.89}$$

Another indication of the importance of the spin–orbit coupling is the recent observation of the large spin–orbit coupling in the related insulating compound Ca_2RuO_4 [111].

Therefore, we extend the theory by adding to the Hamiltonian given in (23.87) the spin–orbit coupling:

$$H_{s-o} = \lambda \sum_i L_i \cdot \sigma_i \,. \tag{23.90}$$

Here, the effective angular momentum L_i operates on the three t_{2g}-orbitals on the site i and $\mathbf{s}_i$ are the electron spins. As in an earlier approach [108] we restrict ourselves to the three orbitals, ignoring e_{2g}-orbitals and choose the coupling constant λ such that the t_{2g}-states behave like an $l = 1$ angular momentum representation. Moreover, it is known that the quasi-two-dimensional xy-band is separated from the quasi-one-dimensional xz and yz-bands. Then, one expects that the effect of spin–orbit coupling is small for the xy-band and can be neglected for simplicity. Therefore, we consider the effect of the spin–orbit coupling on xz and yz-bands only. Then, the kinetic part of the Hamiltonian $H_t + H_{so}$ can be diagonalized and the new energy dispersions are

$$\begin{aligned} \epsilon^\sigma_{\mathbf{k},yz} &= (t_{\mathbf{k},yz} + t_{\mathbf{k},xz} + A_\mathbf{k})/2 \,, \\ \epsilon^\sigma_{\mathbf{k},xz} &= (t_{\mathbf{k},yz} + t_{\mathbf{k},xz} - A_\mathbf{k})/2 \,, \end{aligned} \tag{23.91}$$

where $A_\mathbf{k} = \sqrt{(t_{\mathbf{k},yz} - t_{\mathbf{k},xz})^2 + \lambda^2}$, and σ refers to a pseudo-spin quantum number. The spin–orbit coupling does not break the time-reversal symmetry and therefore the Kramers degeneracy between the spin *up* and *down* is not removed. The resultant Fermi surface consists of three sheets as observed in experiment [106]. Most importantly spin–orbit coupling together with (23.87) leads to a new quasiparticle which we label by pseudo-spin and pseudo-orbital indices. The unitary transformation $\tilde{U}_\mathbf{k}$ connecting old and new quasiparticle states is defined for each wave vector and leads to the following relation between them:

$$\begin{aligned} c^+_{\mathbf{k},yz+} &= u_{1\mathbf{k}} a^+_{\mathbf{k},yz+} - iv_{1\mathbf{k}} a^+_{\mathbf{k},xz+} \,, \\ c^+_{\mathbf{k},xz+} &= u_{2\mathbf{k}} a^+_{\mathbf{k},yz+} - iv_{2\mathbf{k}} a^+_{\mathbf{k},xz+} \,, \\ c^+_{\mathbf{k},yz-} &= u_{1\mathbf{k}} a^+_{\mathbf{k},yz-} + iv_{1\mathbf{k}} a^+_{\mathbf{k},xz-} \,, \\ c^+_{\mathbf{k},xz-} &= u_{2\mathbf{k}} a^+_{\mathbf{k},yz-} + iv_{2\mathbf{k}} a^+_{\mathbf{k},xz-} \,, \end{aligned} \tag{23.92}$$

where

$$u_{m\mathbf{k}} = \frac{\lambda}{\sqrt{(t_{\mathbf{k},yz} - t_{\mathbf{k},xz} \mp A_\mathbf{k})^2 + \lambda^2}} \,, \tag{23.93}$$

and

$$v_{m\mathbf{k}} = \frac{t_{\mathbf{k},yz} - t_{\mathbf{k},xz} \mp A_\mathbf{k}}{\sqrt{(t_{\mathbf{k},yz} - t_{\mathbf{k},xz} \mp A_\mathbf{k})^2 + \lambda^2}} \,. \tag{23.94}$$

The "-" and "+" signs refer to $m = 1$ and $m = 2$, respectively.

Note that despite the spin–orbit coupling causing the spin and orbit quantum numbers not to be good ones we can still identify the Cooper-pairing to be triplet or singlet one. This refers then to the pseudospin quantum numbers. At the same time, the magnetic behavior of Sr_2RuO_4 becomes very anisotropic due to the fact that both one-particle Green's functions and Lande's g-factors will be different if the magnetic field is applied along the c-axis or in the ab-RuO_2-plane. In particular, the anisotropy arises mainly from the calculations of the Lande's g-factors and in particular their orbital parts. The factors $g_z = \tilde{l}_z + 2s_z$ and $g_+ = \tilde{l}_+ + 2s_+$ are calculated using the new quasiparticle states. The latter consist, for example, of xz and yz-orbitals which in turn are the combinations of the initial orbital states $|2, +1\rangle$ and $|2, -1\rangle$ mixed due to the crystal field. Then, the matrix elements $\langle|l_+|\rangle$ ($\langle|l_-|\rangle$) are zero for the xz- and yz-orbitals while $\langle|l_z|\rangle$ matrix element is not. Therefore, the longitudinal components of the spin susceptibility of the xz and yz-band get enhanced in comparison to the transverse one. An interesting question that we will analyze later is the effect of spin–orbit coupling on the antiferromagnetic and ferromagnetic fluctuations. This provides insight into a microscopic explanation of the pairing mechanism and allows to calculate the spatial structure of the superconducting order parameter.

23.3.2 Eliashberg-Like Theory for the Coupling of Quasiparticles to Ferromagnetic and Incommensurate Antiferromagnetic Spin Fluctuations

For the analysis of the interaction between quasiparticles and spin fluctuations (ferromagnetic and incommensurate antiferromagnetic) we proceed as follows. The form of the electronic theory remains the same as for the case of cuprates and we extend the electronic theory towards a three-band theory as observed in Sr_2RuO_4 [112,113]. In the normal state both the self-energy and the thermal Green's functions become a matrix of 3×3 form, *i.e.* $G_{i,j,m}(\mathbf{k}, \omega_n)$ and $\Sigma_{i,j,m}(\mathbf{k}, \omega_n)$, where i, j, m refer to the band indexes of the xy, yz, and xz-orbitals. The corresponding Dyson equation is given by

$$\left[\hat{G}(\mathbf{k}, \omega_n)\right]^{-1} = \left[\hat{G}^0(\mathbf{k}, \omega_n)\right]^{-1} - \hat{\Sigma}(\mathbf{k}, \omega_n) , \quad (23.95)$$

where $\hat{G}^0_{i,j,m}(\mathbf{k}, \omega_n)$ is the matrix of the bare Green's function determined via the tight-binding energy dispersions for the xy, yz, and xz-bands. The self-energy is given by [113]

$$\Sigma_{i,j,m}(\mathbf{k}, \omega_n) = T \sum_{q,l} V^1_{i,j,m}(\mathbf{q}, \omega_l) \times G_{i,j,m}(\mathbf{k} - \mathbf{q}, \omega_n - \omega_l) , \quad (23.96)$$

where $V^1_{i,j,m}(\mathbf{q}, \omega_l)$ is an effective interaction between quasiparticles and spin fluctuations. Similarly as for cuprates it consists of an infinite series of diagram including charge and spin fluctuations. However, some important differences are included in the random phase approximation.

Most importantly, we consider now the diagrams shown in Fig. 23.28 with an odd number of bubbles that contribute to the triplet pairing. This is in contrast to singlet pairing in cuprates where an even number of bubbles occur (see Fig. 23.18). Furthermore, due to the inclusion of the spin–orbit coupling we expect that the transverse (+-) and the longitudinal (zz) parts of the spin susceptibility are different. Thus, in the RPA series they have to be summed separately. Then the effective pairing interaction in the 3×3 form including transverse and longitudinal spin fluctuations and also charge fluctuations is given by

$$V^1_{i,j,m}(\mathbf{q}, \omega_l) = \frac{1}{2} V^{sp,zz}_{i,j,m}(\mathbf{q}, \omega_l) + V^{+-}_{i,j,m}(\mathbf{q}, \omega_l) - \frac{1}{2} V^{\text{ch}}_{i,j,m}(\mathbf{q}, \omega_l) , \quad (23.97)$$

where

$$V^{sp,zz} = U^2 \frac{\chi_0^{sp,zz}}{1 - U\chi_0^{sp,zz}} , \qquad V^{sp,+-} = U^2 \frac{\chi_0^{sp,+-}}{1 - U\chi_0^{sp,+-}} \quad (23.98)$$

describe coupling to spin density fluctuations and

$$V^{\text{ch}} = U^2 \frac{\chi_0^{\text{ch}}}{1 + U\chi_0^{\text{ch}}} \quad (23.99)$$

to charge density fluctuations. Here, χ_0^{ch}, $\chi_0^{sp,+-}$ are the irreducible parts of the spin and charge susceptibilities [114]. Note that in the Bethe–Salpeter equation shown diagrammatically in Fig. 23.18 for singlet

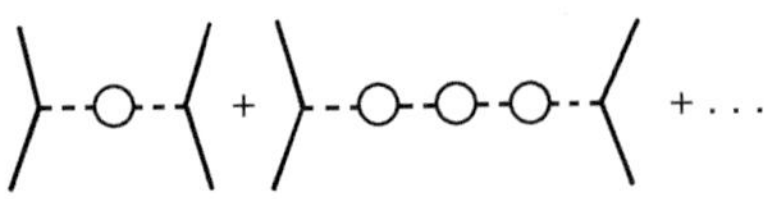

Fig. 23.28. RPA diagrams for triplet Cooper-pairing with an odd number of bubbles that refer to longitudinal charge and spin fluctuations

Cooper-pairing an even number of bubbles and also ladder diagrams occur. In the case of triplet Cooper-pairing the contribution of the ladder diagrams is zero and an odd number bubbles diagrams occur due to Pauli's principle shown in Fig. 23.28. Since the Feynman rules require a (-1) for each loop an extra minus sign enters the gap equation via V_t^{eff} in (23.48). As described earlier this makes a solution of the gap equation relatively easy.

The magnetism in the ruthenates resulting from the quasiparticles in the t_{2g}-orbital is itinerant and thus the magnetic response is created by the same electrons that also form the Cooper-pairs. Then, for example, the irreducible part of the charge susceptibility χ_0^{ch} is defined in terms of the electronic Green's functions (on a square lattice with size N)

$$\chi_0^{\text{ch}}(\mathbf{q}) = \frac{1}{N} \sum_{\mathbf{k}} G(\mathbf{k}+\mathbf{q}) G(\mathbf{k}) , \qquad (23.100)$$

where $G(\mathbf{k})$ is the one-electron Green's function (here we omit the band indices for simplicity). The longitudinal and transverse components of the spin susceptibilities are also calculated in terms of the electronic Green's functions. In contrast to the cuprates, $\chi_0^{sp,zz}$ and $\chi_0^{sp,+-}$ in Sr_2RuO_4 are different due to the magnetic anisotropy resulting from spin–orbit coupling.

To determine the superconducting transition temperature T_c one must solve the following set of linearized gap equations with eigenvalue $\lambda_\mu(T)$, where μ refers to the band index ($\mu = \alpha, \beta, \gamma$):

$$\begin{aligned} &\lambda_\mu \Delta_{\mu,l,m}(\mathbf{k}, \omega_n) \\ &= -\frac{T}{N} \sum_{k',\omega_j,\mu'} V^{(2)}_{\mu,l,m}(\mathbf{k}-\mathbf{k}', \omega_n - \omega_j)\, G_\mu(\mathbf{k}', \omega_j) \\ &\times G_{\mu'}(-\mathbf{k}', -\omega_j)\, \Delta_{\mu',l,m}(\mathbf{k}', \omega_j) . \qquad (23.101) \end{aligned}$$

Here, the quantum numbers l and m refer to the orbital and spin state of a Cooper-pair, respectively. Note that T_c is determined from $\lambda_\mu(T_c) = 1$ and the interband coupling will provide a single T_c for all three bands (of course the largest energy gain will be due to the band having the larger density of states).

In order to simplify the discussion we now focus mainly on the γ-band due to its large density of states. In the case of singlet pairing one has $l = 0, 2, \ldots$ and $m = 0$, and for triplet pairing one gets $l = 1, 3, \ldots$ and $m = \pm 1$ or $m = 0$. The pairing potential $V^{(2)}$ (taken for singlet or triplet pairing) controls which state gives the lowest energy and determines singlet or triplet Cooper-pairing. Three possibilities may occur:

(a) singlet pairing,
(b) triplet pairing with the total spin $S_z = 0$ of the Cooper-pair wave function,
(c) triplet pairing with the total spin $S_z = \pm 1$.

Note that a priori we cannot judge which pairing state is realized in the ruthenates due to the presence of both antiferromagnetic and ferromagnetic fluctuations. Therefore, one has to solve the gap equations for all three possibilities.

Thus, for singlet d-wave pairing we would use for the pairing interaction

$$V_s^{(2)} = \frac{1}{2} V_{sp}^{zz} + V_{sp}^{+-} - \frac{1}{2} V_{\text{ch}} . \qquad (23.102)$$

leading to the linearized gap equation ($\mu = \gamma, l = 2, m = 0$)

$$\lambda \Delta = -\frac{T}{N} \sum V_s^{(2)} G G \Delta . \qquad (23.103)$$

Note that due to magnetic anisotropy we separate the longitudinal and transverse parts of spin fluctuations.

As we will discuss below the magnetic anisotropy that is due to spin–orbit coupling is in particular very important for triplet pairing [108]. We will see in the following that it will lift the degeneracy of the three possible triplet states. In order to demonstrate this we separate the gap equation into two parts for $|m| = 1$ and $m = 0$. For triplet pairing with $S_z = \pm 1$ we have

$$V_{\text{tr1}}^{(2)} = -\frac{1}{2} V_{sp}^{zz} - \frac{1}{2} V_{\text{ch}} \qquad (23.104)$$

resulting from the set of diagrams with odd number of bubbles (see Fig. 23.28, all other diagrams are identical to zero) yielding

$$\lambda\Delta = -\frac{T}{N}\sum V^{(2)}_{\text{tr1}} GG\Delta\,. \qquad (23.105)$$

Finally, for triplet pairing with $S_z = 0$ we get

$$V^{(2)}_{\text{tr0}} = \frac{1}{2}\left(V^{zz}_{sp} - V^{+-}_{sp} - V_{\text{ch}}\right)\,, \qquad (23.106)$$

and the gap equation

$$\lambda\Delta = -\frac{T}{N}\sum V^{(2)}_{\text{tr0}} GG\Delta\,. \qquad (23.107)$$

This clearly demonstrates the importance of spin-orbit coupling for the magnetic anisotropy and thus for the pairing interaction. Obviously, if no magnetic anisotropy would be present, i.e. $\chi^{+-} = 2\chi^{zz}$, one obtains the *same* gap equation for $|m| = 1$ and $m = 0$. Therefore, we safely conclude that the important magnetic anisotropy selects one of the triplet state ($\chi^{zz} > \chi^{+-}$) lifting the degeneracy. As we will discuss in Sect. 23.5.4, ferromagnetic fluctuations favoring triplet pairing are mainly present in χ^{+-} and thus in V^{+-}_{sp}. This will lower the energy for the $m = 0$ triplet state.

Obviously, the magnetic anisotropy described above that enters the Eliashberg-like theory will lead to interesting effects if an external magnetic field h_{ext} is applied. For example, the anisotropy of the ratio $h^{\perp}_{c2}/h^{||}_{c2}$ and its connection to the internal structure of the superconducting order parameter will be discussed in Sects. 23.3.4 and 23.5.4.

23.3.3 Elementary Excitations in Sr_2RuO_4

The behavior of the elementary excitations in ruthenates is controlled by the spin fluctuations:

$$\omega_{\mathbf{k}} = \epsilon_{\mathbf{k}} + \Sigma(\mathbf{k}, \omega = \omega_{\mathbf{k}})\,. \qquad (23.108)$$

Here, the self-energy $\Sigma(\mathbf{k}, \omega)$ results mainly from the coupling to spin fluctuations. Clearly, since the anisotropic spin susceptibility $\chi(\mathbf{k}, \omega)$ is the important input for Σ, the self-energy and thus the elementary excitations are affected by the magnetic anisotropy. Note that the spin fluctuations in ruthenates are highly anisotropic. Strong *incommensurate*

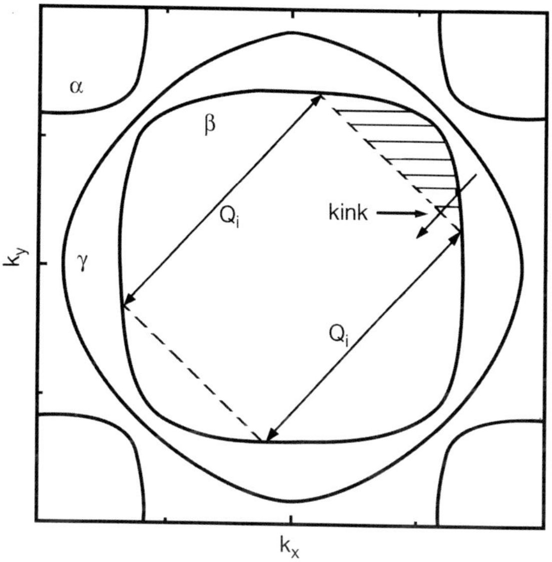

Fig. 23.29. Expected "kink" structure in Sr_2RuO_4. The Fermi surface in Sr_2RuO_4 consists of three bands. The nesting properties of the $yz(\beta)$-band reveal the formation of two-dimensional incommensurate spin fluctuations at $\mathbf{Q}_i = (2\pi/3, 2\pi/3)$ and $\omega_{sf} \approx 6$ meV. The quasiparticles at the β-band reveal a strong renormalization due to scattering by spin fluctuations. Thus, a corresponding "kink" formation for occupied states below E_F (*shaded area*) along certain direction in the first BZ as indicated by the arrow may occur

antiferromagnetic fluctuations at the wave vector $\mathbf{Q}_i = (2\pi/3, 2\pi/3)$ and $\omega_{sf} = 6$meV are present in Sr_2RuO_4 only in the quasi-one-dimensional $xz(\alpha)$ and $yz(\beta)$-bands. Similar to the cuprates, we expect also the formation of the "kink" feature due to spin fluctuations as shown in Fig. 23.29. On general grounds a "kink" structure occurs in the energy dispersion of the β-band below E_F for the quasiparticles that are connected by the corresponding wave vector $\mathbf{Q}_i$ and ω_{sf}. Note that according to the relation

$$\omega_{\text{kink}} = \omega_{sf} + \epsilon^{yz}_{\mathbf{k}+\mathbf{Q}_i} \qquad (23.109)$$

the "kink" energy will be smaller than in cuprates due to a lower value of ω_{sf} in the ruthenates.

So far, a "kink" has only been observed in the γ-band [115]. This is interesting because the quasiparticles in the $xy(\gamma)$-band do not experience the scattering by antiferromagnetic spin fluctuations and interact mainly with weak ferromagnetic fluctuations. Thus, one expects mainly a Fermi-liquid-type of behavior for the quasiparticles in the xy-band. Furthermore, due to a small value of the superconducting gap ($\Delta_0 \ll \omega_{sf}$) there will be no noticeable feedback of superconductivity on the "kink" feature.

23.3.4 Dynamical Spin Susceptibility: Magnetic Anisotropy

Let us analyze the dynamical spin susceptibility in Sr_2RuO_4. Due to the mixing of the spin and orbital degrees of freedom, the magnetic susceptibility also involves the orbital magnetism which is very anisotropic.

For the calculation of the transverse, χ_l^{+-}, and longitudinal, χ_l^{zz}, components of the spin susceptibility of each band l we use the diagrammatic representation shown in Fig. 23.30. Note that the Kramers degeneracy is not removed by the spin–orbit coupling, since the latter does not remove the time-reversal symmetry and $G^l_{k+} = G^l_{kz}$. Thus, the anisotropy arises mainly from the calculations of the Lande's g-factors and in particular their orbital parts. The factors $g_z = \tilde{l}_z + 2s_z$ and $g_+ = \tilde{l}_+ + 2s_+$ are calculated using the new quasiparticle states. The latter consist, for example, of xz and yz-orbitals, which in turn are the combinations of the initial orbital states $|2,+1\rangle$ and $|2,-1\rangle$ mixed due to the crystal field. Then, the matrix elements $\langle |l_+|\rangle$ ($\langle |l_-|\rangle$) are zero for the xz and yz-orbitals while $\langle |l_z|\rangle$ matrix element is not. Therefore, the longitudinal components of the spin susceptibility of the xz and yz-band are enhanced in comparison to the transverse one. We obtain, for example, for the $|xz\rangle$-states for the transverse susceptibility

$$\chi^{+-}_{0,xz}(\mathbf{q},\omega) = -\frac{4}{N}\sum_{\mathbf{k}}(u_{2\mathbf{k}}u_{2\mathbf{k+q}} - v_{2\mathbf{k}}v_{2\mathbf{k+q}})^2 \times \frac{f(\epsilon^+_{\mathbf{k}xz}) - f(\epsilon^-_{\mathbf{k+q}xz})}{\epsilon^+_{\mathbf{k}xz} - \epsilon^-_{\mathbf{k+q}xz} + \omega + iO^+}, \quad (23.110)$$

and for the longitudinal susceptibility

$$\begin{aligned}\chi^{zz}_{0,xz}(\mathbf{q},\omega) &= \chi^{\uparrow}_{xz}(\mathbf{q},\omega) + \chi^{\downarrow}_{xz}(\mathbf{q},\omega) \\ &= -\frac{2}{N}\sum_{\mathbf{k}}\Big[u_{2\mathbf{k}}u_{2\mathbf{k+q}} + v_{2\mathbf{k}}v_{2\mathbf{k+q}} \\ &\quad + \sqrt{2}(u_{2\mathbf{k}}v_{2\mathbf{k+q}} + v_{2\mathbf{k}}u_{2\mathbf{k+q}})\Big]^2 \\ &\quad \times \frac{f(\epsilon^+_{\mathbf{k}xz}) - f(\epsilon^+_{\mathbf{k+q}xz})}{\epsilon^+_{\mathbf{k}xz} - \epsilon^+_{\mathbf{k+q}xz} + \omega + iO^+}. \quad (23.111)\end{aligned}$$

Here, $f(x)$ is again the Fermi function and $u^2_{\mathbf{k}}$ and $v^2_{\mathbf{k}}$ are the corresponding coherence factors defined in (23.93)–(23.94).

For all other orbitals the calculations are similar and straightforward. Note that the magnetic response of the xy-band remains mainly isotropic. This is due to the absence of the nesting features in the xy-band.

Then, one gets within RPA the following expressions for the transverse susceptibility ($\tilde{l} = -l$):

$$\chi^{+-}_{\mathrm{RPA},l}(\mathbf{q},\omega) = \frac{\chi^{+-}_{0,\tilde{l}}(\mathbf{q},\omega)}{1 - U\chi^{+-}_{0,\tilde{l}}(\mathbf{q},\omega)}, \quad (23.112)$$

and for the longitudinal susceptibility

$$\chi^{zz}_{\mathrm{RPA},l}(\mathbf{q},\omega) = \frac{\chi^{zz}_{0,l}(\mathbf{q},\omega)}{1 - U\chi^{zz}_{0,l}(\mathbf{q},\omega)}, \quad (23.113)$$

where $\chi^{zz}_{0,l} = \chi^{++}_{0,l} + \chi^{--}_{0,l}$.

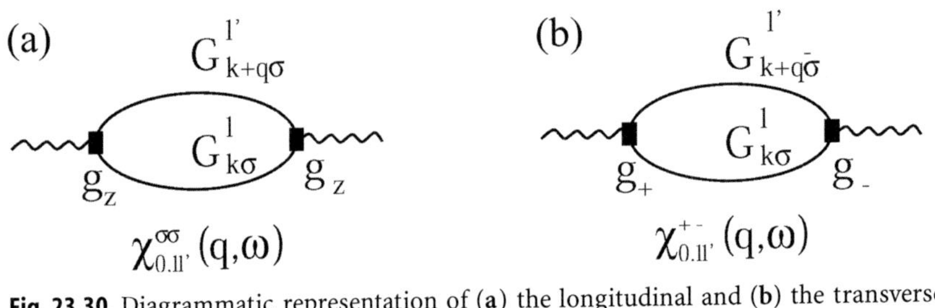

Fig. 23.30. Diagrammatic representation of (**a**) the longitudinal and (**b**) the transverse magnetic susceptibility. The *full lines* represent the electron Green's function with pseudo-spin σ and pseudo-orbital $\tilde{l}$ quantum numbers. The Lande's g-factors are denoted by $g_+ = \tilde{l}_+ + 2s_+$ ($g_- = \tilde{l}_- + 2s_-$) and $g_z = \tilde{l}_z + 2s_z$

These susceptibilities are used in the corresponding pairing interaction for triplet pairing. In order to compare our results with NMR and INS experiments we take

$$\chi^{zz}_{\text{tot}} = \sum_l \chi^{zz}_{\text{RPA},l} \tag{23.114}$$

and

$$\chi^{+-}_{\text{tot}} = \sum_l \chi^{+-}_{\text{RPA},l} \,. \tag{23.115}$$

23.3.5 Gap Equation: Triplet Pairing and Symmetry of the Superconducting Order Parameter

The interesting question in the ruthenates is how can triplet Cooper-pairing with the possible pair states $\psi^\pm$

$$|\uparrow\uparrow\rangle \,, \; m_z = +1$$
$$|\downarrow\downarrow\rangle \,, \; m_z = -1$$

and ψ^0

$$\frac{1}{\sqrt{2}}\left(|\uparrow\downarrow+|\downarrow\uparrow\rangle\right) \,, \; m_z = 0 \tag{23.116}$$

occur in the presence of strong incommensurate antiferromagnetic spin fluctuations at $\mathbf{Q}_i = (2\pi/3, 2\pi/3)$ originating from the xz and yz-bands and weaker ferromagnetic fluctuations arising from the xy-band. For the interplay of the antiferromagnetic and ferromagnetic-like spin excitations the magnetic anisotropy, $\chi^{zz}(\mathbf{Q}_i) > \chi^{\pm}(\mathbf{Q}_i)$, is very significant. This anisotropy will also lift the degeneracy of the three possible triplet states and favor for the Cooper-pairs the state $\psi^0 \sim \frac{1}{\sqrt{2}}(|\uparrow\downarrow+|\downarrow\uparrow\rangle)$. Note that the magnetic anisotropy is caused by spin-orbit coupling. The above three triplet Cooper-pair states with magnetic quantum numbers $m_z = \pm 1, 0$ result from the transverse and longitudinal spin excitations. Therefore, the important normal-state result $\chi^{zz}(\mathbf{Q}_i) > \chi^{+-}(\mathbf{Q}_i)$ already suggests lifting of the triplet states degeneracy and the realization of the $m_z = 0$ state $\frac{1}{\sqrt{2}}(|\uparrow\downarrow+|\downarrow\uparrow\rangle)$.

In the case of a dominating γ-band, with a large density of states, it follows almost directly from the Fermi surface topology illustrated in Fig. 23.31 that triplet Cooper-pairing mediated by ferromagnetic excitations occurs.

Of course, the gap equation determines the occurrence of the triplet state with $m_z = 0$. Note that the order parameter for triplet Cooper-pairing can be written in the form

$$\Delta_{\mathbf{k}} = \sum_\mu d_\mu(\mathbf{k})(\sigma_\mu i\sigma_2) = \begin{pmatrix} -d_x + id_y & d_z \\ d_z & d_x + id_y \end{pmatrix} . \tag{23.117}$$

Again as in the cuprates all chosen symmetries must be irreducible representations of the D_{4h} crystallographic point group symmetry, since the crystal field potential is very strong in ruthenates as well as in cuprates. The full set of them was found by Sigrist and Rice [116]. Experimentally it is confirmed that in Sr_2RuO_4 (close to T_c) only the $d_z = \hat{\mathbf{z}}$ component of the superconducting order parameter is present while the $d_x = \hat{\mathbf{x}}$ and $d_y = \hat{\mathbf{y}}$ component is strongly suppressed [53].

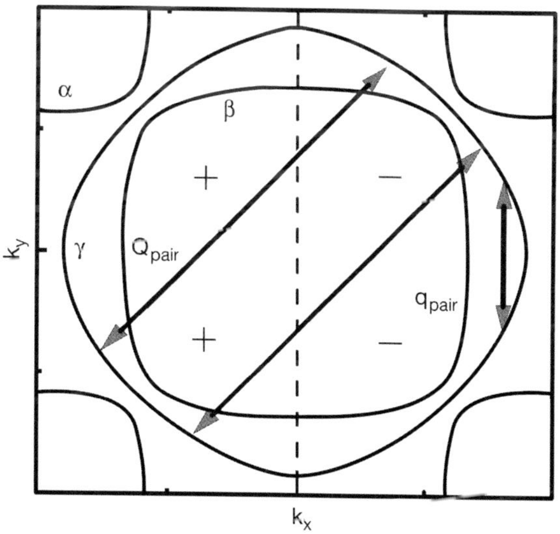

Fig. 23.31. Calculated Fermi surface (FS) topology for Sr_2RuO_4 and symmetry analysis of the superconducting order parameter Δ in the first Brillouin zone. The real part of a p-wave order parameter has the node along k_x=0. +(-) and the *dashed lines* refer to the signs of the momentum dependent order parameter of Δ. α, β, and γ denote the FS of the corresponding (hybridized) bands. The Cooper-pairing wave vectors $\mathbf{Q}_i$ and $\mathbf{q}_i$ refer to maxima in the spin susceptibility and provide the main contribution to the pairing instability

For a discussion of the symmetry of the order parameter it is instructive to consider the Fermi surface topology and the structure of $\chi(\mathbf{q})$. Thus, in order to investigate triplet pairing in Sr_2RuO_4 in more detail, in Fig. 23.31 we show its corresponding Fermi surface topology obtained from the three-band Hubbard Hamiltonian discussed earlier in this chapter. However, for simplicity we discuss here only the γ-band, which has a high density of states. The effects due to the other bands (α and β) will be analyzed in detail later. For the moment and for simplicity let us discuss only the γ-band in order to study the differences between Sr_2RuO_4 and cuprates.

Sr_2RuO_4 shows a two-dimensional electronic structure, which indicates that the Cooper-pairing mainly occurs in the RuO_2-plane. Thus, we discuss first the superconductivity in the RuO_2-plane and then analyze what happens along the c-direction.

A closer inspection of (23.49) shows that $|\Delta_p|^2$ has no nodes. However, Re Δ_p (also Im Δ_p) indeed has a nodal line also displayed in Fig. 23.31. This has important consequences if strong nesting is present. Then, f-wave symmetry of the superconducting order parameter wins over p-symmetry. This can be seen as follows (for simplicity we restrict the discussion to the case of equal spin pairing, i.e. $m_z = \pm 1$): The summation over $\mathbf{k}'$ in the first BZ is dominated by the contributions due to $\mathbf{Q}_{\text{pair}}$ and the one due to a smaller wave vector $\mathbf{q}_{\text{pair}}$ (see $\chi(\mathbf{q}, \omega)$). Thus, we obtain approximately for the γ-band contribution (angular quantum number $l = f$ or p)

$$\Delta_l(\mathbf{k}) \approx \sum_i \frac{V^{\text{eff}}_{\text{tr1}}(\mathbf{Q}_i)}{2\epsilon^{\gamma}_{\mathbf{k}+\mathbf{Q}_i}} \Delta_l(\mathbf{k} + \mathbf{Q}_i) + \sum_i \frac{V^{\text{eff}}_{\text{tr1}}(\mathbf{q}_i)}{2\epsilon^{\gamma}_{\mathbf{k}+\mathbf{q}_i}} \Delta_l(\mathbf{k} + \mathbf{q}_i) , \qquad (23.118)$$

where the sum is over all contributions due to $\mathbf{Q}_i$ and $\mathbf{q}_{pair}$. The wave vectors $\mathbf{Q}_{\text{pair}}$ bridge portions of the FS where Re Δ_p has the opposite sign. Since the smaller wave vector $\mathbf{q}_{\text{pair}}$ bridges areas on the Fermi surface with same sign. Contributions with opposite signs occur and thus

$$\sum_i \frac{V^{\text{eff}}_{\text{tr1}}(\mathbf{q}_i)}{2\epsilon^{\gamma}_{\mathbf{k}+\mathbf{q}_i}} \Delta_l(\mathbf{k} + \mathbf{q}_i) \approx 0 . \qquad (23.119)$$

Hence, we find a gap equation where Δ_l is expected to change its sign for an attractive interaction and dominating $\mathbf{Q}_i$ transitions. However, this is not possible! In other words, if the corresponding pairing interaction involves strong nesting similar to cuprates, i.e. a peak of χ_{RPA} at $q \approx \mathbf{Q}_{\text{pair}}$, Cooper-pairing and a p-symmetry solution of (23.49) would not be possible due to $V_t^{\text{eff}} < 0$. Nesting properties would suppress the p-wave and favor the f-wave, and then one would get (for f-symmetry)

$$\Delta_f(\mathbf{k}) = \Delta_0 \hat{\mathbf{z}}(\cos k_x - \cos k_y)(\sin k_x + i \sin k_y) . \qquad (23.120)$$

Like the $d_{x^2-y^2}$-wave order parameter in cuprates the f-wave symmetry order parameter has also nodes along the diagonals. Note that $\tilde{\mathbf{z}}$ indicates that only the d_z component of the superconducting order parameter is present.

However, note that for weaker nesting and for ferromagnetic spin fluctuations one expects that p-wave symmetry wins over f-wave symmetry triplet Cooper-pairing. Obviously, the magnetic anisotropy observed in Sr_2RuO_4 is important for this.

The odd parity of the superconducting order parameter and strong reduction of T_c by non-magnetic impurities suggests that electron–phonon interaction does not cause superconductivity in Sr_2RuO_4. Note that the recently observed negative isotope effect ($\alpha = -0.15$) indicates the complicated dependence of the isotope mass from interaction parameters [117].

One expects on general grounds that the singlet Cooper-pairing will not be realized in the ruthenates. The simple reason behind this is that the antiferromagnetic fluctuations favoring singlet pairing with $d_{x^2-y^2}$-wave symmetry of the superconducting order parameter originate from the quasi-one-dimensional xz and yz-bands. However, the xy-band with only weak ferromagnetic fluctuations has a much larger density of states at the Fermi level. This indicates its large contribution to superconductivity. Therefore, the xy-band will not be unstable with respect to the singlet Cooper-pairing. This is illustrated in Fig. 23.31. Note that for $d_{x^2-y^2}$-wave symmetry we get a change of sign of the order parameter upon crossing the diagonals of the BZ. For the singlet Cooper-pairing the wave vectors around $\mathbf{Q}_i$ connecting areas (+) and (-) contribute constructively to the pairing. Contributions due to $\mathbf{q}_i$ and the background connecting areas with the same sign subtract from the pairing (see Fig. 23.31 with nodes at the diagonals). Therefore, we get that the four contributions due to $\mathbf{q}_i$ in the xy-band do not allow to have $d_{x^2-y^2}$-wave symmetry in the xy-band. Despite the pair-building contribution due to $\mathbf{Q}_i$ one gets that the eigenvalue of the $d_{x^2-y^2}$-wave symmetry in the xy-band is smaller than for the $f_{x^2-y^2}$-wave symmetry. This is due to the large contribution from $\mathbf{Q}_i$ to the cross-terms for the triplet pairing, which are absent for the singlet pairing. For d_{xy}-symmetry where the nodes are along (π,0) and (0,π) directions we argue similarly. Thus, we may exclude this symmetry.

What is the situation with the RuO_2-planes? First, due to a weak dispersion the electronic bands along c-direction one may expect a weak dispersion of the superconducting p-wave gap:

$$\Delta_{\mathrm{k}} = \Delta_0 \tilde{\mathrm{z}} \left(\sin \frac{k_x a}{2} \cos \frac{k_y a}{2} + i \sin \frac{k_y a}{2} \cos \frac{k_x a}{2} \right) \times \cos \frac{k_z c}{2} . \tag{23.121}$$

Note that due to the cosine function the superconducting gap will have a node for the $k_z = \pi/c$. On the other hand, the formation of the node in the superconducting order parameter usually reflects the underlying structure of the pairing potential. In our case this results from the spin susceptibility, which becomes anisotropic due to spin-orbit coupling. In particular, the antiferromagnetic spin fluctuations are polarized along the c-direction, while ferromagnetic fluctuations are mainly present in the RuO_2-plane. Then, the polarized antiferromagnetic spin fluctuations could be one of the reasons for the node formation away from the RuO_2-plane in Sr_2RuO_4. This we will discuss later.

What happens if an external magnetic field is applied? As in superfluid ^{3}He there exists an interesting effect concerning the possible change of the orientation of the unit vector $d_z = \hat{\mathbf{z}}$ of the superconducting order parameter. In general, the spin susceptibility is a tensor in spin space with principal axes along the preferred directions. Therefore the magnetic energy density

$$f_h = -\frac{1}{2} \sum_{\mu\nu} \chi_{\mu\nu} h_\mu h_\nu \tag{23.122}$$

depends on the orientation of the susceptibility tensor and hence on the order parameter itself. As we have discussed in the Introduction the spin susceptibility of Sr_2RuO_4 in the superconducting state behaves differently for different orientation of the magnetic field. For example for $\mathbf{h} \perp \mathbf{d}$ (i.e an external magnetic field is applied along the a or b-direction) the susceptibility is unchanged with respect to its normal state value χ_N [118]. On the contrary, when $\mathbf{h} || \mathbf{d}$ the susceptibility becomes gapped, the temperature dependence of which is described by the so-called Yoshida function [118]. Therefore, the spin susceptibility in Sr_2RuO_4 depends strongly on the relative orientation of $\mathbf{d}$ and $\mathbf{h}$. In its general form the susceptibility tensor reads

$$\chi_{\mu\nu} = \chi_N^0 \{ \delta_{\mu\nu} - \hat{d}_\mu \hat{d}_\nu \left[1 - Y_0(T) \right] \} , \tag{23.123}$$

where $Y_0(T) = -\int_{-\infty}^{\infty} \mathrm{d}E_{\mathrm{k}} \frac{\mathrm{d}f_{\mathrm{k}}}{\mathrm{d}E_{\mathrm{k}}}$ is the Yoshida function.

One finds that for the magnetic field applied along the c-direction, the susceptibility and consequently the magnetic energy gets a contribution which tends to orient $\mathbf{d}$ perpendicular to the magnetic field

$$\Delta f_h = \frac{1}{2} \Delta\chi (\mathbf{d} \cdot \mathbf{h})^2 , \tag{23.124}$$

where $\Delta\chi = \chi_N^0 (1 - Y_0(T))$. This means that for a relatively large magnetic field applied along the c-direction the significant d_x and d_y components in the superconducting order parameter in Sr_2RuO_4 can be induced. This possibility is already known from superfluid ^{3}He. Of course, as in ^{3}He, this effect is relatively small but observable. This is a subject of further experimental studies.

In the following we present results for important properties of the superconducting state. First we discuss superconductivity of cuprates, then that of ruthenates.

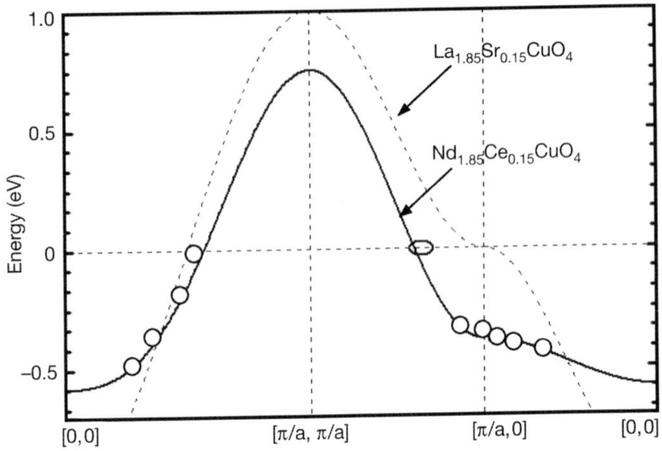

Fig. 23.32. Calculated energy dispersion for the hole-doped and electron-doped cuprates at optimal doping concentration. We use $t = 250$ meV and $t' = 0$ for the hole-doped cuprates (*dashed curve*) and $t = 128$ meV and $t' = -0.3$ for the electron-doped cuprates (*solid curve*). The experimental data (*open symbols*) taken from [104]. The position of the flat bands in the electron-doped cuprates is 300 meV below E_F. On the other hand, in most of the hole-doped cuprates the van-Hove singularity lies close to the Fermi level

23.4 Results for Hole-Doped and Electron-Doped Cuprates

We present results for key properties of cuprate superconductivity using as a model the Hubbard Hamiltonian and Eliashberg-like theory and assuming Cooper-pairing due to spin fluctuations. Important input for the calculations is the electronic band structure. A comparison of the calculations with experiments should further justify the validity of our theory. We find $d_{x^2-y^2}$-wave superconductivity for both hole-doped and electron-doped cuprates. The elementary excitations and the doping dependent phase diagram are central results.

23.4.1 Bare Electronic Structure

In Fig. 23.32 we present results for the energy dispersion along the route $(0,0) \to (\pi,0) \to (\pi\pi) \to (0,0)$ of the first Brillouin zone for hole-doped and electron-doped cuprates at optimal doping. As a typical example we refer to $La_{2-x}Sr_xCuO_4$ and $Nd_{2-x}Ce_xCuO_4$. It is known that the position of the flat bands with respect to the Fermi level is described by the ratio of t'/t. Using the fact that in hole-doped cuprates the flat bands lie close to the Fermi energy while in the electron-doped ones they lie much lower, this ratio is expected to be small in hole-doped cuprates, while in electron-doped cuprates we approximate $t'/t = -0.3$ in order to describe the experimental results.

The corresponding densities of states are shown in Fig. 23.33. The flat bands reflecting the so-called van-Hove singularity are a common feature of cuprates having a two-dimensional square lattice. Note again that the bands are at different energies in hole-doped and electron-doped cuprates. Due to the fact that in hole-doped cuprates the van-Hove singularity lies near the Fermi energy, these are closer to an antiferromagnetic instability than the electron-doped ones. Therefore, antiferromagnetic spin fluctuations

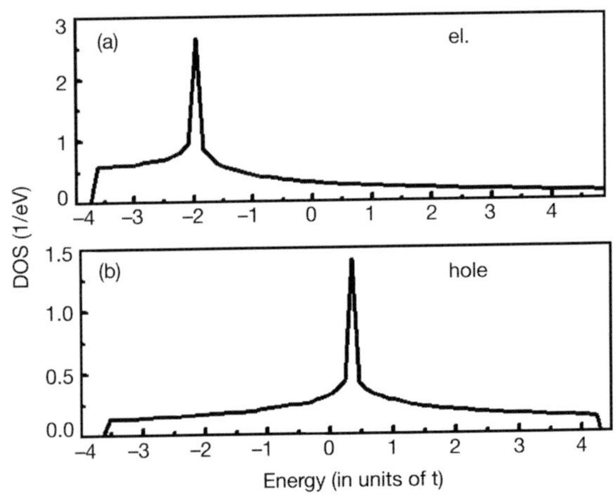

Fig. 23.33. Results for the density of states in (**a**) electron-doped and (**b**) hole-doped cuprates using the energy dispersion shown in Fig. 23.32. The so-called van-Hove singularity that originates from the flat part of the conduction bands in hole-doped cuprates lies much closer to the Fermi energy ($\epsilon_F = 0$) than in electron-doped cuprates

in the hole-doped cuprates are expected to be much stronger due to the nesting. As a consequence more phase space is available for Cooper-pairing.

Note that this does not contradict the experimental data, which indicate a large doping range of antiferromagnetism in electron-doped cuprates. As has been shown earlier [63,119] the small doping range of antiferromagnetism in hole-doped cuprates is due to spin frustration that arises from holes at the oxygen sites. In contrast in the electron-doped cuprates only a dilution of the Cu-spins takes place and thus as a consequence antiferromagnetism occurs over larger doping range.

The input (material-dependent) parameters of the theory will play an important role in our calculations and will explain the observed asymmetry between electron-doped and hole-doped cuprates. Moreover, since the topology of the Fermi surface and the corresponding position of the flat bands change with doping, we also expect the characteristic changes in the behavior of hole-doped and electron-doped cuprates with doping.

We would like also to note that many of the hole-doped cuprates contain two CuO_2-planes per unit cell and that this requires the inclusion of bilayer effects into our theory. We will study this later, but would like to stress that the main features that will be obtained within a one-band calculation will already describe the main experimental facts of the cuprates.

23.4.2 Elementary Excitations

Let us first discuss the elementary excitations. The results for the spectral density due to the scattering of the carriers by antiferromagnetic spin fluctuations are expected to exhibit fundamental behavior. As we have shown the dispersion of the carriers will be modified by the self-energy, Σ, arising from the coupling of the quasiparticles to antiferromagnetic spin fluctuations:

$$\omega_{\mathbf{k}} = \epsilon_{\mathbf{k}} + \mathrm{Re}\Sigma(\mathbf{k}, \omega = \omega_{\mathbf{k}}) .$$

Note that the renormalization of $\epsilon_{\mathbf{k}}$ by $\Sigma(\mathbf{k}, \omega = \omega_{\mathbf{k}})$ will be characteristically anisotropic.

1. Nodal Direction

We start the discussion analyzing the spectral density of hole-doped superconductors in the normal state. The spectral density reveals the elementary excitations and in particular the renormalized energy dispersion. First, we present our results for the spectral density along the nodal $(0,0) \rightarrow (\pi,\pi)$-direction in the first BZ.

In Fig. 23.34 we show the calculated spectral density $N(\mathbf{k}, \omega)$, i.e. the local density of states, as a function of frequency and momentum $\mathbf{k} - \mathbf{k}_F$. The peak positions correspond to the renormalized energy dispersion. Due to coupling of holes to antiferromagnetic spin fluctuations the quasiparticle dispersion changes its slope and shows a pronounced kink feature at the energy $\omega_{\text{kink}} \approx 75 \pm 15$ meV.

How can one understand the kink feature in a simple way? At a first glance the occurrence of a kink in the nodal direction seems to be surprising, since the main interaction of the carriers with spin fluctuations occurs at the hot spots. Note that the kink feature is present along the diagonal of the BZ close to the cold spots. However, away from the Fermi level, but close to it (along $(0,0) \rightarrow (\pi,\pi)$), the quasi-

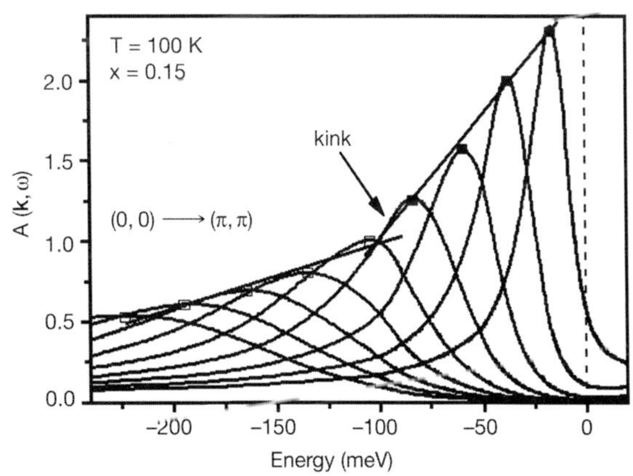

Fig. 23.34. Calculated spectral density $A(\mathbf{k}, \omega)$ in the normal state along the nodal $(0,0) \rightarrow (\pi,\pi)$ direction (from left to right) as a function of frequency in the first Brillouin zone. The peak positions (connected by the *solid line* to guide the eye) refer to the renormalized energy dispersion $\omega_{\mathbf{k}}$. One clearly sees the kink structure at an energy approximately $\omega_{\text{kink}} = 75 \pm 15$ meV that results from coupling of the quasiparticles to spin fluctuations

particles couple strongly to spin fluctuations. Most importantly, as follows from Fig. 23.20, the largest scattering will occur at values of $\mathbf{k} - \mathbf{k}_F = \mathbf{Q}$ and $\omega = \omega_{sf}$. From (23.170) it follows that

$$\Sigma(\mathbf{k}, i\omega_n) = -T^2 \sum_{\omega_m, \nu_m} \sum_{\mathbf{k}', \mathbf{q}} \tilde{\tau}_0 \times G(\mathbf{k} - \mathbf{k}', i\omega_n - i\nu_m)\tilde{\tau}_0 U^2 \times \frac{1}{2} \mathrm{Tr}\,[\tilde{\tau}_0 G(\mathbf{k} + \mathbf{q}, i\omega_m + i\nu_m)\tilde{\tau}_0 \times G(\mathbf{q}, i\omega_m)] \,. \quad (23.125)$$

Then approximating the Green's function

$$G(\mathbf{k}, i\omega_n) \approx G_0(\mathbf{k}, i\omega_m) = \frac{i\omega_n \tilde{\tau}_0 + \epsilon_{\mathbf{k}}\tilde{\tau}_3 - \phi_{\mathbf{k}}\tilde{\tau}_1}{(i\omega_n)^2 - E_{\mathbf{k}}^2} \,, \quad (23.126)$$

after some algebra one obtains with $E_{\mathbf{k}}^2 = \epsilon_{\mathbf{k}}^2 + \phi_{\mathbf{k}}^2$ on the real axis

$$\Sigma(\mathbf{k}, \omega) \approx -\frac{U^2}{4} \sum_{\mathbf{k}'} \int_0^\infty \mathrm{d}\omega' \frac{\mathrm{Im}\chi_{\mathrm{RPA}}(\mathbf{k} - \mathbf{k}', \omega')}{\omega - \omega' - E_{\mathbf{k}'}} \times \left[\coth\left(\frac{\omega'}{2T}\right) - \tanh\left(\frac{\omega' - \omega}{2T}\right)\right] . \quad (23.127)$$

In a simple view, in which the feedback effect of superconductivity on χ is neglected, the imaginary part of the spin susceptibility is approximately given by the Ornstein–Zernicke expression, which has a peak structure at the wave vector $\mathbf{q}$ and frequency $\omega = \omega_{sf}$. Furthermore, the self-energy is mainly frequency-dependent, while the bare dispersion of the carriers is not. Then, already in the normal state the self-energy $\Sigma(\mathbf{k}, \omega)$ has a maximum reflecting a corresponding maximum of Im χ at $\mathbf{q} \approx \mathbf{Q}$ and $\omega' \approx \omega_{sf}$. Then, the kink position follows from the pole of the denominator of (23.127). This leads to the "kink condition"

$$\omega_{\mathrm{kink}} \approx E_{\mathbf{k}-\mathbf{Q}} + \omega_{sf}(x) \,. \quad (23.128)$$

Note that this gives an estimate of the position of the kink and explains the behavior of the spectral density $A(\mathbf{k}, \omega)$ shown in Fig. 23.34. Furthermore, since the superconducting gap is zero for $\omega = 0$, but not for $\omega = \omega_{sf}$, the kink feature along the nodal direction $(0, 0) \rightarrow (\pi, \pi)$ will change only slightly below T_c.

2. (0, 0) ⟶ (0, π) Direction

In order to see whether the kink feature is present in other directions of the Brillouin Zone, in Fig. 23.35 we show the evolution of the spectral density along the $(0, 0) \rightarrow (0, \pi)$ direction. Despite the fact that along this direction we do not cross the Fermi level, the kink feature is still present and is found at an energy similar to the one for the nodal $(0, 0) \rightarrow (\pi, \pi)$-direction. This indicates that the kink feature occurs not only along the $(0, 0) \rightarrow (\pi, \pi)$-direction. Instead, the kink is characteristic for all direction where $\mathbf{k} - \mathbf{k}_F \simeq \mathbf{Q}$ and $\omega \simeq \omega_{sf}$. Also below T_c we find that the kink feature is present in the $(0, 0) \rightarrow (0, \pi)$-direction (not shown). Note that our results are in fair agreement with experimental data [46].

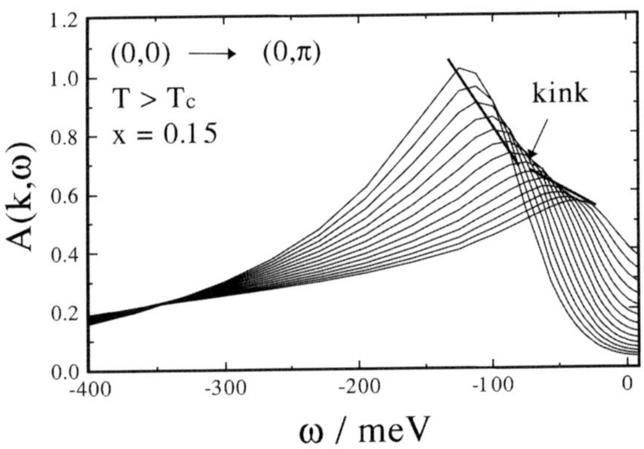

Fig. 23.35. Spectral density $A(\mathbf{k}, \omega)$ as a function of frequency along the $(0, 0) \rightarrow (0, \pi)$ direction of the first BZ in the normal state calculated from the generalized Eliashberg equations. Again, the peak positions reveal the renormalized energy dispersion $\omega_{\mathbf{k}}$. A kink occurs at similar energy as in the nodal direction. Due to inelastic scattering of holes by spin fluctuations close to $(0, \pi)$, $A(\mathbf{k}, \omega)$ also becomes broader. Note that, in contrast to the nodal direction, one does not cross the Fermi level in the $(0, 0) \rightarrow (0, \pi)$-direction. Instead, one reaches the flat part of the tight-binding band

3. Antinodal Direction

In Fig. 23.36(a) we show our results for $N(\mathbf{k}, \omega)$ along the $(\pi, 0) \rightarrow (\pi, \pi)$ route, i.e. the antinodal direction, of the first BZ in the normal state. Note, that the

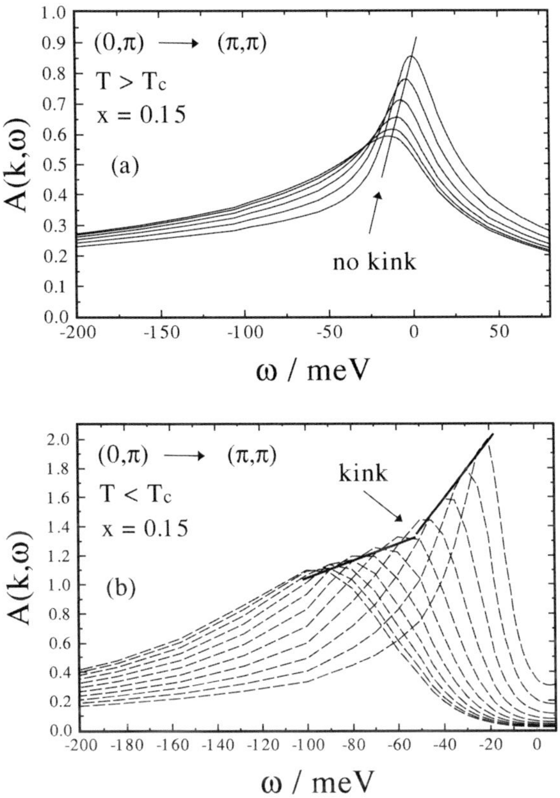

Fig. 23.36. Calculated spectral density $A(\mathbf{k}, \omega)$ along the antinodal $(\pi, 0) \to (\pi, \pi)$ direction in the first BZ as a function of frequency in the normal (**a**) and superconducting (**b**) state. Due to the flat band close to the Fermi level the spectral density shows no kink structure in the normal state. Below T_c the superconducting gap $\phi(\omega)$ opens yielding a kink structure in the spectral density that occurs at the energies $\omega_{\text{kink}} \approx 50 \pm 10$ meV for optimal doping

spectral density at the $(0, \pi)$ point is broader than at the antinodal point due to stronger coupling to spin excitations peaked at $\mathbf{q} = \mathbf{Q} = (\pi, \pi)$ as discussed in Fig. 23.20. Clearly, no kink is present.

The absence of a kink structure can be explained with the flat structure of the CuO_2-plane around the M point (see Fig. 23.32) and with the fact that the wavevector is not large enough to bridge antinodal points in the BZ. Simply speaking, for a flat band the frequency dependence of Σ in (23.32) does not play a significant role and therefore no change of the velocity and no kink structure is present.

What happens in the superconducting state? Below T_c the superconducting gap $\phi(\mathbf{k}, \omega)$ opens rapidly for decreasing temperature T and becomes maximal in momentum space around the M point reflecting the momentum dependence of the effective pairing interaction (see (23.170)). In addition, due to the frequency dependence of the gap the flat band around M disappears.

In Fig. 23.36(b) we show results for $A(\mathbf{k}, \omega)$ at a temperature $T = 0.5T_c$ where the superconducting gap has opened. A kink structure around $\omega_{\text{kink}} \approx 50 \pm 10$ meV is present reflecting the magnitude of ϕ. Hence, in the $(\pi, 0) \to (\pi, \pi)$ direction this kink feature is only present below T_c and connected to the feedback effect of ϕ on the elementary excitations. We will show later that this feedback is also important for the resonance peak seen in INS. Note that the superconducting gap $\phi(\mathbf{k}, \omega)$ is calculated self-consistently in our theory and reflecting the underlying spin fluctuations which dominate the pairing potential V_{eff}. Therefore, the occurrence of a kink structure *only* below T_c in the antinodal direction is a *direct* fingerprint of the spin excitation spectrum. Furthermore, as we will discuss below, Im $\chi(\mathbf{Q}, \omega)$ entering (23.127) is peaked at the resonance frequency ω_{res}. Therefore, the kink condition is given by

$$\omega_{\text{kink}} \approx E_{\mathbf{k}-\mathbf{Q}} + \omega_{\text{res}}(x) \,. \tag{23.129}$$

In Fig. 23.37(a) the frequency dependence of Re $\Sigma(\mathbf{k}_a, \omega)$ in the normal and superconducting state at the antinodal point $\mathbf{k} = \mathbf{k}_a$ is shown. Due to the occurrence of the resonance feature in Im $\chi(\mathbf{Q}, \omega)$ and the related feedback of the superconducting gap $\phi(\omega)$, Re Σ shows a pronounced structure below T_c at energies of about $\omega_{\text{res}} + \Delta_0$. Also the corresponding imaginary part Im $\Sigma(\mathbf{k} = \mathbf{k}_a, \omega)$ shows a peak below T_c (see Fig. 23.37(b)). This pronounced behavior is responsible for the kink formation along $(\pi, 0) \to (\pi, \pi)$ direction in the BZ.

Therefore, while the kink features are present along $(0, 0) \to (\pi, \pi)$ and $(\pi, 0) \to (\pi, \pi)$-directions in the superconducting state of hole-doped cuprates, their nature is qualitatively different. Along the nodal direction the superconducting gap is zero (for $\omega = 0$) and thus the feedback effect of superconductivity on the elementary and spin excitations

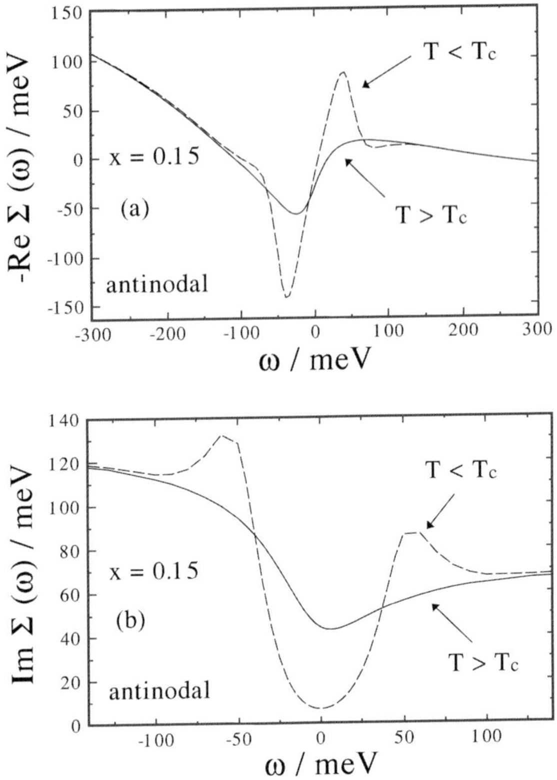

Fig. 23.37. (**a**) Calculated frequency dependence of the self-energy Re $\Sigma(\mathbf{k}_a, \omega)$ at the antinodal point $\mathbf{k}_a$ of the first BZ in the normal (*solid curve*) and superconducting state (*dashed curve*). Due to the feedback effect of the superconducting gap $\phi(\omega)$, a peak (dip) occurs for $\omega > 0$ ($\omega < 0$), which roughly defines the position of the kink structure. (**b**) The corresponding imaginary part at the antinodal point Im $\Sigma(\mathbf{k} = \mathbf{k}_a, \omega)$ is shown. Again, due to the feedback effect of $\phi(\omega)$, a maximum occurs below T_c. Note that both Re Σ and Im Σ are not fully antisymmetric (symmetric) with respect to ω at optimum doping $x = 0.15$

is small. Therefore, ω_{sf} determines mainly the kink feature. On the other hand, along the antinodal direction the gap is maximal and yields a strong feedback of superconductivity on χ. Thus, in the superconducting state ω_{res} and Δ_0 yield the kink structure along $(\pi, 0) \rightarrow (\pi, \pi)$ direction that is not present in the normal state.

Doping Dependence of the Elementary Excitations and Their Energy Renormalization

The different reasons for the kink structures in hole-doped cuprates along different directions in the first BZ will be also reflected in their doping dependence. The results we have shown so far were for optimal doping concentration $x = 0.15$. This refers to a band filling of $n = 0.85$. Note that the superconducting transition temperature T_c behaves differently in the overdoped (OD) and underdoped (UD) regime:

$$T_c \propto \Delta(T \rightarrow 0), \ \text{OD-regime}$$
$$T_c \propto n_s(T \rightarrow 0), \ \text{UD-regime},$$

where n_s is the superfluid density calculated self-consistently from the generalized Eliashberg equations.

In the antinodal $(0, \pi) \rightarrow (\pi, \pi)$-direction the kink is only present below T_c due to the feedback of $\phi(\omega)$. In the OD case the gap $\phi(\omega)$ decreases reflecting a mean-field-like behavior. Thus, the energy where the kink occurs must decrease with overdoping:

$$\omega_{\text{kink}}(x) \propto \Delta_0(x)\,. \tag{23.130}$$

This behavior was indeed observed by Dessau and co-workers [74]. Note that the above argument remains true also in the strongly OD case where no resonance peak in Im $\chi(\mathbf{Q}, \omega)$ occurs because the feedback effect of $\phi(\omega)$ should always be present.

Regarding the kink along the nodal $(0, 0) \rightarrow (\pi, \pi)$-direction we note the following: ω_{sf} increases with increasing doping from underdoped to overdoped cuprates. Since ω_{sf} determines the kink position along $(0, 0) \rightarrow (\pi, \pi)$ direction we expect

$$\omega_{\text{kink}}(x) \propto \omega_{sf}(x)\,. \tag{23.131}$$

This is in qualitative agreement with experimental data [120] (for underdoped regime and optimally doped superconductors). On the other hand, the spectral weight of Im $\chi(\mathbf{Q}, \omega)$ decreases drastically with overdoping. Therefore, the coupling of the quasiparticles to spin fluctuations becomes much weaker in the OD case. These two competing effects seem responsible for the non-monotonic and weak doping dependence of the kink position in the nodal direction [74].

Elementary Excitations in Electron-Doped Cuprate

Another interesting observation is the asymmetric behavior of hole and electron-doped cuprates. Note that no kink feature has been reported in the electron-doped cuprates [48]. It is believed that the electron–phonon coupling is much more pronounced in electron-doped cuprates than in hole-doped ones. This is reflected, for example, by the behavior of the resistivity $\rho \propto T^2$ in the normal state at optimum doping and by the transition between $d_{x^2-y^2}$-wave symmetry of the superconducting gap towards anisotropic s-wave, as has been observed in several experiments [121]. Simply speaking, the spin fluctuations in electron-doped cuprates are weaker than in the hole-doped ones, yielding a smaller T_c and a smaller superconducting gap [122]. Thus, no kink is present in the nodal direction and also no kink occurs in the $(0, \pi) \rightarrow (\pi, \pi)$-direction below T_c. This is related to the fact that the flat band around $(0, \pi)$ lies in electron-doped cuprates well below the Fermi level and, therefore, it cannot be softened due to $\phi(\omega)$.

Summary

Regarding the superconducting state note that in the hole-doped cuprates a strong renormalization of the spin fluctuation spectra occurs due to the feedback effect of superconductivity. (Note that this leads also to a resonance peak at $\omega = \omega_{\text{res}}$.) In electron-doped cuprates, only a rearrangement of spectral weight occurs below T_c.Note that the kink feature is intimately connected with the resonance peak resulting for Im χ, see the discussion below. There is only a small feedback of superconductivity below T_c on Im χ in the electron-doped cuprates due to $\omega_{sf} \gg \Delta_0$. Thus, we find no kink feature in the superconducting state of electron-doped cuprates in the antinodal direction. Due to correlations the hole-doped and electron spectral density is different.

Anisotropic Scattering Rates

The behavior of the self-energy $\Sigma(\mathbf{k}, \omega)$ implies of course corresponding anisotropy for the scattering rate of the quasiparticles and for the conductivity. We discuss now the anisotropy of the scattering rate $\tau^{-1}(\omega)$ of hole-doped cuprates at different points on the Fermi surface.

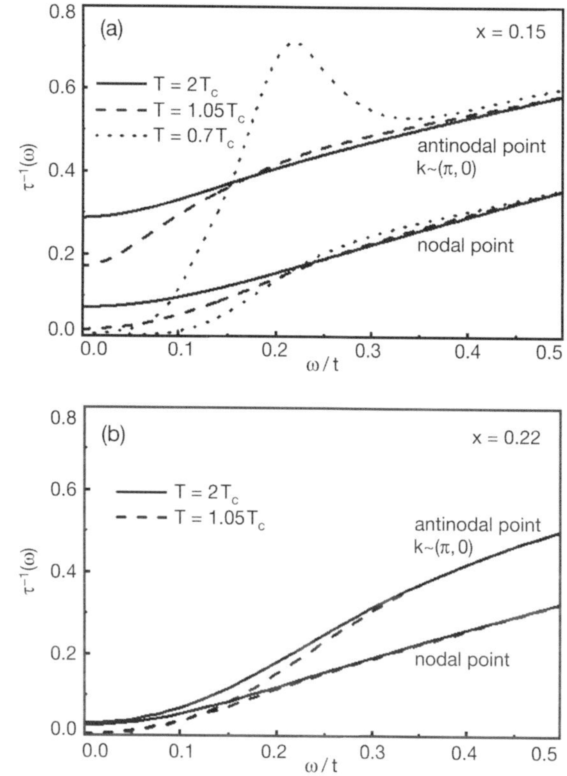

Fig. 23.38. Scattering rate $\tau^{-1}(\omega)$ of optimally (**a**) and overdoped (**b**) hole-doped cuprates versus frequency at the nodal and antinodal point of the BZ and calculated for various temperatures. The anisotropy results from coupling to spin fluctuations and disappears in the overdoped case. Thus, a crossover from a non-Fermi-liquid to a Fermi-liquid behavior occurs. Note also the feedback effect of superconductivity for different parts of the BZ at optimal doping

In Fig. 23.38 we show our results for $\tau^{-1}(\omega)$ at the antinodal point and the nodal point, respectively, for optimal doping (a) and for the overdoped case, and (b) for various temperatures. In Fig. 23.38(a) one clearly sees that the scattering rate is very anisotropic on the Fermi surface reflecting the anisotropy of the coupling of elementary excitations to spin fluctuations. In particular, $\tau^{-1}(\omega)$ in the normal state is almost three times larger at the antinodal point than at the nodal point. This agrees with recent ARPES experiments [123]. Furthermore, we find that for $\omega \to 0$,

$$\mathrm{Im}\,\Sigma \propto \omega\,.$$

This demonstrates non-Fermi-liquid behavior in the underdoped and optimally hole-doped cuprates.

In the overdoped cuprates the anisotropy between nodal and antinodal points is strongly reduced and disappears almost for $\omega \to 0$. Most importantly the system then behaves more Fermi-liquid-like. The latter is seen from Fig. 23.38(b), where one observes a crossover from the $\mathrm{Im}\,\Sigma \propto \omega$ to the $\mathrm{Im}\,\Sigma \propto \omega^2$ behavior. This is also in agreement with experimental observations [124]. Note, results shown in Fig. 23.38 suggest that for increasing doping the spectral density gets sharper around hot spots, while broader in modal direction. This results from the F.G.-topology and Q.

Below T_c at the antinodal point $\tau^{-1}(\omega)$ for optimal doped reveals a strong feedback at energies

$$\omega \sim \omega_{\mathrm{res}} + \Delta_0\,,$$

where ω_{res} refers to the resonant frequency of the spin fluctuations (see $\mathrm{Im}\,\chi$). This relation is due to the simple fact that the self-energy Σ is a convolution of the spin susceptibility χ_s (peaked at ω_{sf}) and the Green's function of the dressed quasiparticles G (peaked at Δ_0). Furthermore, note that the power law for $\omega \to 0$ changes to $\Sigma \propto \omega^3$ and that at the nodal point the effect of superconductivity is rather weak due to the node of the $d_{x^2-y^2}$-wave symmetry of the superconducting order parameter.

In the overdoped cuprates below T_c both at the nodal and antinodal points no structure in the frequency dependence of $\tau^{-1}(\omega)$ occurs. The latter is also clear, since spin fluctuations spectrum weakens and does not reveal resonant condition below T_c. This is consistent with the behavior of $\tau^{-1}(\omega)$ above T_c where crossover to Fermi-liquid behavior takes place.

To summarize the doping dependence of the quasiparticle scattering rate we find pronounced non-Fermi-liquid behavior in underdoped and optimally doped cuprates ($\mathrm{Im}\,\Sigma \propto \omega$) due to strong scattering of quasiparticles by spin fluctuations. For all

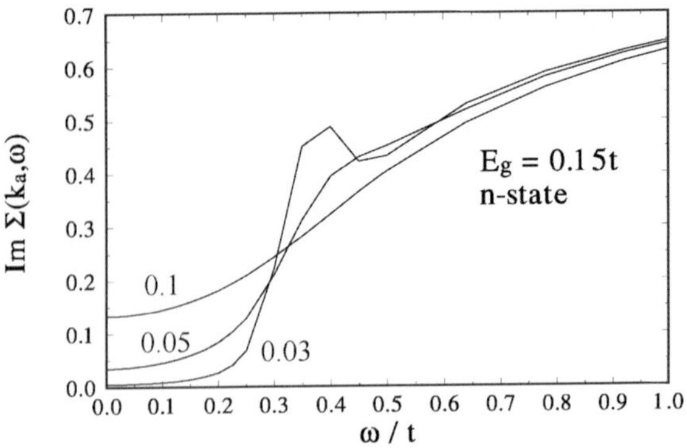

Fig. 23.39. Calculated imaginary part of the self-energy $\mathrm{Im}\Sigma(\mathbf{k}_a, \omega)$ at the antinodal wave vector $\mathbf{k}_a$ and assuming a pseudogap $E_g = 0.15t$ for various temperatures $T = 0.1$, 0.05, and 0.03t in units of t

doping regions we find qualitatively agreement with transport experiments on hole-doped cuprates (see, for example, the discussion of Figs. 29–31 in [32]). In particular, in the overdoped case the crossover to Fermi-liquid-like behavior occurs ($\mathrm{Im}\Sigma \propto \omega^2$) that reflects the weakening of the spin fluctuations with increasing doping. In the superconducting state the pronounced spin fluctuations reveal structure in $\tau^{-1}(\omega)$ at the antinodal point of the Brillouin zone for underdoped and optimally doped cuprates. Due to weakening of spin fluctuations this structure disappears with increase of doping towards overdoped cuprates. Our results for Fermi-liquid versus non-Fermi-liquid seem quite general and could be used as a basis for systematic studies.

In order to illustrate the influence of the pseudogap E_g on the elementary excitations we have calculated the self-energy including in the energetic dispersion the pseudogap $E_g(\mathbf{k}) = E_g(\cos k_x - \cos k_y)$ ($T^* = 0.1t = 250$ K).

In Fig. 23.39 we show the evolution of the imaginary part of the self-energy $\mathrm{Im}\Sigma(\mathbf{k}_a, \omega)$ for decreasing temperature at the antinodal point of the BZ where the pseudogap is maximal. In the pseudogap regime the self-energy shows a gap-like feature up to $2E_g$. The renormalization of the quasiparticle scattering rate leads to the peak at $3E_g$ and it becomes more pronounced for decreasing temperature. Correspondingly this peak in the imaginary part of the self energy at $3E_g$ reveals an enhancement of the $\mathrm{Re}\Sigma$.

As a result the quasiparticle effective mass m^* enhances in comparison to its value at $\omega = 0$. Our results are in qualitative agreement with complex optical conductivity results on underdoped Bi2212, YBCO, and LSCO compounds [125].

23.4.3 Dynamical Spin Susceptibility

An important quantity in our model is the dynamical spin susceptibility, which is calculated in the random-phase-approximation (RPA) for renormalized Green's functions, as discussed in the previous section. Concerning the validity of the results obtained by the model hamiltonian (see (23.6) and (23.7)) one should remember that physically the magnetic activity in the CuO_2-planes results from the Cu-spins (and the spin polarizations induced by them into the p-band). The behaviour of the Cu-magnetic moments in the T–x plane (phase diagram) need to be studied directly. One of the most important questions is how the spin excitation spectrum changes upon doping in high-T_c cuprates. Due to the nesting properties of the Fermi surface at the wave vector **Q** both electron and hole-doped cuprates show an enhancement of the spin response at the antiferromagnetic wave vector, which is further augmented by the RPA denominator. These antiferromagnetic fluctuations induced by the itinerant carriers can be characterized by the frequency ω_{sf} of the spin fluctuations where ω_{sf} refers roughly to the peak position in the dynamical spin susceptibility.

In Fig. 23.21 we show the results of our calculations for $\mathrm{Im}\chi(\mathbf{Q}, \omega)$ versus frequency at different doping concentration. As one sees in the underdoped cuprates $\mathrm{Im}\chi$ shows a sharp peak and follows an Ornstein–Zernicke behavior. The position of the peak defines the spin fluctuation frequency ω_{sf}. With increasing doping ω_{sf} shifts towards larger frequencies and the peak becomes less pronounced. This agrees well with experimental data of INS. This behavior is expected, since for increasing doping the system goes away from the antiferromagnetic instability and therefore ω_{sf} becomes larger. This, for example, indicates that in the overdoped regime the renormalization of the elementary excitations will be isotropic at different parts of the Fermi surface, while

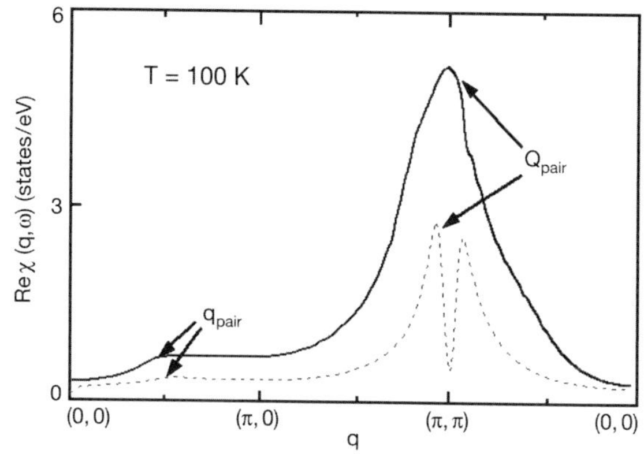

Fig. 23.40. Momentum dependence of the real part of the spin susceptibility for the optimally electron-doped cuprates $\mathrm{Re}\chi(\mathbf{q}, \omega)$ along the Brillouin zone route $(0, 0) \to (\pi, 0) \to (\pi, \pi) \to (0, 0)$ at $U/t = 4$, $T = 100$ K and $\omega = 0$ (*solid curve*) and $\omega = \omega_{sf} = 0.47t$ (*dashed curve*). The main contributions to the Cooper-pairing interaction come from the wave vectors $\mathbf{q}_{\mathrm{pair}}$ and $\mathbf{Q}_{\mathrm{pair}}$

in optimally and underdoped materials the strongest scattering takes place at the parts of the Fermi surface that are connected by the wave vector (π, π) (so-called "hot spots"). These results are consistent with the behavior of the elementary excitations that we have discussed previously.

What is the spin spectrum of the electron-doped cuprates? As expected the electron-doped cuprates demonstrate also the presence of the antiferromagnetic fluctuations at **Q**. This is seen in Fig. 23.40 where we plot the behavior of the $\mathrm{Re}\chi(\mathbf{q}, \omega)$ along the route $(0, 0) \to (\pi, 0) \to (\pi, \pi) \to (0, 0)$ of the first BZ at $\omega = 0$ and $\omega = \omega_{sf}$. In both cases $\mathrm{Re}\chi$ yields a peak at wave vector **Q**. However, this peak is much weaker than in hole-doped cuprates due to the weaker nesting properties of the corresponding Fermi surface. The latter can be seen by analyzing the position of the ω_{sf} in the electron-doped cuprates. While in hole-doped cuprates at optimal doping it is of the order of

$$\omega_{sf}^h \approx 20\,\mathrm{meV},$$

in the electron-doped cuprates it is approximately

$$\omega_{sf}^e \approx 60\,\mathrm{meV}.$$

Therefore, the effect of the spin fluctuations on the elementary excitations will be weak in the electron-doped cuprates. For example, since the energy scale of the spin fluctuations is large, the low-energy properties in the electron-doped case will behave more like in the standard Fermi-liquid theory (i.e. $\Sigma'' \sim \omega^2$). This is suggested experimentally by the quadratic temperature dependence of the electronic part of the resistivity in electron-doped cuprates.

Another important feature is that electron-doped cuprates will behave more like overdoped cuprates, since their spin dynamics is very similar. A consequence is the absence of the "kink" feature and "hot spots" in the electron-doped cuprates.

23.4.4 Order Parameter: Doping Dependence

Let us now discuss the superconducting properties of the cuprates. The most important question concerns the symmetry of the superconducting order parameter and its doping dependence in hole-doped and electron-doped cuprates. Due to the repulsive nature of the spin fluctuation mediated pairing interaction a conventional isotropic s-wave pairing is not possible. Therefore, for singlet pairing the orbital state of the Cooper-pair wave function should be the $l = 2$ representation that corresponds to a d-symmetry of the order parameter. The particular symmetries that might occur in the CuO_2 plane are the ones belonging to the corresponding tetragonal crystal field symmetry D_{4h}. The corresponding representations can be substituted into the gap equation and the symmetry that yields the eigenvalue $\lambda = 1$ at highest temperature is the realized symmetry of the superconducting order parameter in layered cuprates.

The solutions of the generalized Eliashberg equations within the FLEX approximation give the $d_{x^2-y^2}$-wave symmetry of the superconducting order parameter in the CuO_2-plane for both hole and electron doping as shown for demonstration in Fig. 23.41(a) for the electron-doped cuprates at optimal doping. This is easy to understand if one remembers that the main contribution to the pairing comes from the spin susceptibility that is peaked at $\mathbf{Q} \approx (\pi, \pi)$. The latter connects in the hole-doped cuprates the parts of the Fermi surface close to the M points of the BZ. Keeping in mind the repulsive nature of the pairing interaction one sees that in order to have a Cooper-pairing the order parameter has to have different sign at the corresponding parts of the Fermi surface connected by the antiferromagnetic wave vector **Q**. Then, $d_{x^2-y^2}$-wave symmetry is the most natural solution.

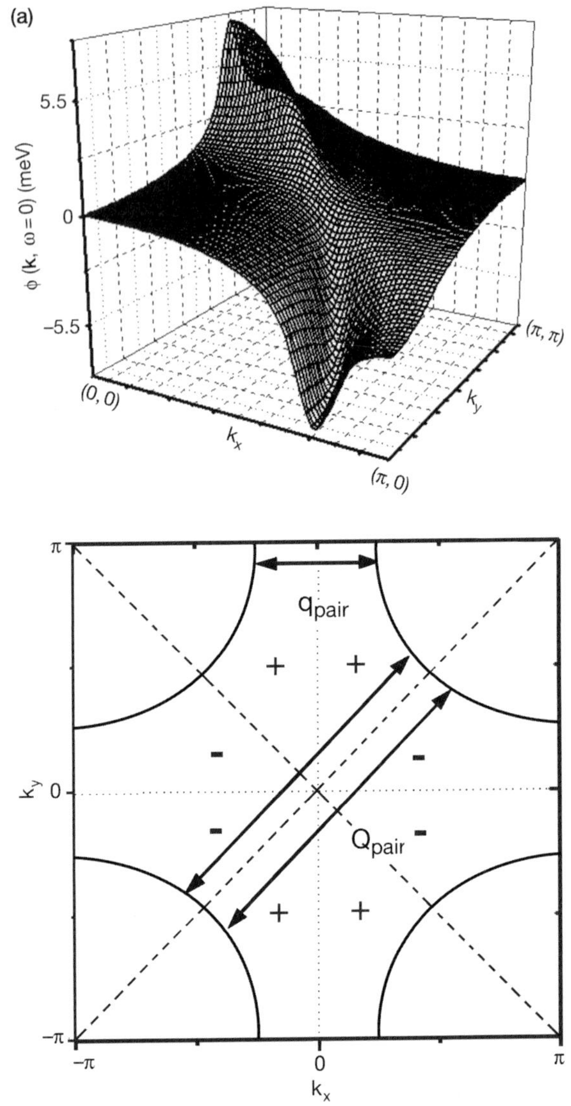

Fig. 23.41. (a) Calculated $d_{x^2-y^2}$-wave superconducting gap at $T = 0.8\ T_c$ and $x = 0.15$ in the first square of the BZ. (b) Calculated Fermi surface for the optimally electron-doped cuprates. The +(-) signs and *dashed lines* refer to the signs of the momentum dependence $d_{x^2-y^2}$-wave superconducting gap $\Delta(\mathbf{k}, \omega = 0)$ and its nodes, respectively

Since the topology of the Fermi surface does not change significantly in the hole-doped cuprates at

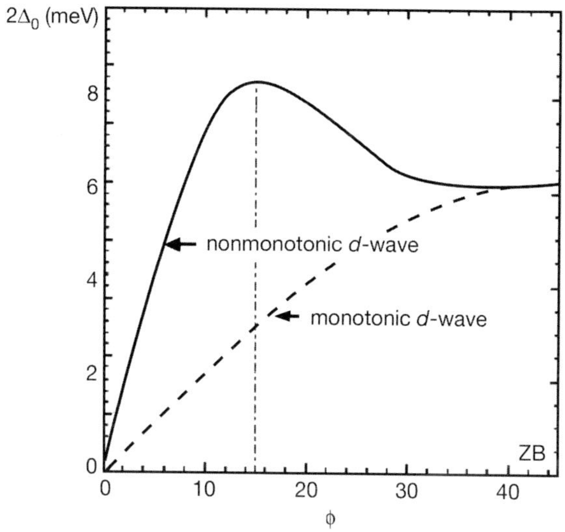

Fig. 23.42. Illustration of the non-monotonic d-wave behavior of the superconducting gap below T_c at $T = 5$ K in optimally electron-doped cuprates ($x = 0.15$) along the Fermi surface (the superconducting gap Δ_0 changes from the diagonal of the BZ towards the zone boundary(ZB) as a function of angle ϕ [126]

least in optimally and overdoped cuprates the $d_{x^2-y^2}$-wave symmetry is always the most prominent solution of the gap equation. The situation is different in the electron-doped cuprates where phonons are more important and weaker nesting is present. Assuming an exchange of antiferromagnetic spin fluctuations as responsible for Cooper-pairing in electron-doped cuprates we also obtain a $d_{x^2-y^2}$-wave symmetry of the underlying order parameter for the electron-doped cuprates at optimal doping. However, as we pointed out earlier, the topology of the electron-doped cuprates is such that the parts of the Fermi surface connected by the $\mathbf{Q}$ are closer to the diagonals of the BZ rather than to M points as shown in Fig. 23.41(b). Therefore, even though the symmetry of the superconducting order parameter is still $d_{x^2-y^2}$, the gap structure has some additional features (higher harmonics, see Fig. 23.42) that reflect the structure of the susceptibility χ and the corresponding topology of the Fermi surface.

Hence, it is possible that the symmetry of the superconducting order parameter can change with doping. This in particular may be the case for the overdoped compounds where the nesting becomes weaker. In addition, if the electron–phonon coupling plays a role it may overcome the repulsive interaction due to spin fluctuations and favor s-wave pairing. In principle, it could happen that the gap symmetry changes from $d_{x^2-y^2}$-wave to the s-wave for $T_c \to 0$, since more Cooper-pair condensation energy may result in nodeless s-wave symmetry.

It is interesting to mention that phonons may also assist in principle the d-wave pairing if one assumes that the important phonon mode participating in the superconductivity is peaked at $\mathbf{q} \approx (0.4\pi, 0)$, as shown in Fig. 23.41(b). Indeed, the parts of the Fermi surface connected by the wave vector $\mathbf{q}$ have the same sign even for the $d_{x^2-y^2}$-wave pairing. Therefore, an attractive electron–phonon interaction could contribute to d-wave pairing [122,127].

23.4.5 Resonance Peak and Magnetic Coherence

An important consequence of the spin-fluctuation-mediated Cooper-pairing and $d_{x^2-y^2}$-wave symmetry of the superconducting order parameter is the peculiar behavior of the spin dynamics in the superconducting state. The feedback of superconductivity on χ occurs due to the fact that the spin fluctuation frequency ω_{sf} is of the same order as the superconducting gap Δ_0, at least in the hole-doped cuprates. One expects that the feedback effect of superconductivity on the spin excitations will be important if the quasiparticles condensing into Cooper-pairs are also partly involved in the spin fluctuation dynamics. Thus, it is important to see whether the experimentally observed "resonance" peak and "magnetic coherence" effect for $\chi(\mathbf{q}, \omega)$ can be explained within our self-consistent electronic theory.

Let us start our study with the analysis of the imaginary part of the bare BCS-like expression for the spin susceptibility (see (23.41)) for the $d_{x^2-y^2}$-wave case at the wave vector $\mathbf{Q}=(\pi, \pi)$; see Eq. (23.42):

$$\mathrm{Im}\chi_0(\mathbf{Q}, \omega) = \frac{1}{2}\sum_{\mathbf{k}} \Big\{ \left[1 - f(E_{\mathbf{k}})\right] \delta(\omega + 2E_{\mathbf{k}}) + \left[f(E_{\mathbf{k}}) - 1\right] \delta(\omega - 2E_{\mathbf{k}}) \Big\} . \quad (23.132)$$

Here, $f(E_{\mathbf{k}})$ denotes the Fermi function and $E_{\mathbf{k}} = \sqrt{\epsilon_{\mathbf{k}}^2 + \Delta_{\mathbf{k}}^2}$ is the dispersion of the Cooper-pairs in the superconducting state. As one can see from (23.132) the imaginary part of the BCS spin susceptibility has an important characteristic frequency ω_{DOS} that arises from the density of states of the quasiparticles in the superconducting state which have a gap in their spectrum due to superconductivity. Note that $\omega_{\mathrm{DOS}} \approx 2\Delta(x, T)$. In order to describe the effect of spin fluctuations one uses in the simplest case the RPA spin susceptibility. Then, in addition to ω_{DOS} a second characteristic frequency occurs in the RPA spin susceptibility. In particular, one finds that the structure in $\mathrm{Im}\chi(\mathbf{Q}, \omega)$ can also be determined by the condition

$$1 = U\mathrm{Re}\chi_0(\mathbf{Q}, \omega = \omega_{\mathrm{res}}) \text{ and}$$
$$\mathrm{Im}\chi_0(\omega = \omega_{\mathrm{res}}) = 0\,,$$

see $\chi = \chi_0(1 - U\chi_0)^{-1}$ and $\chi'' = \chi_0''([1 - U\chi_0']^2 + [U\chi_0'']^2)^{-1}$. Therefore, with increasing U one sees that the peak resulting from ω_{DOS} will shift to lower energies and most importantly will become resonant for $U = U_{\mathrm{cr}}$. Note that this signals the occurrence of a spin density wave collective mode.

In Fig. 23.43(a) we present our weak coupling results for different values of U for $\mathrm{Im}\chi(\mathbf{Q}, \omega)$ (i.e. without taking into account the lifetime effects). In the superconducting state $\mathrm{Im}\chi$ becomes gaped approximately at $2\Delta_0$. With increasing U the peak in $\mathrm{Im}\chi$ shifts to lower frequencies and at $U_{\mathrm{cr}} = 4t$ it becomes resonant. Therefore, the resonance peak observed in INS results in our model due to occurrence of superconductivity, but is renormalized by the normal state spin excitations.

Another interesting feature at optimal doping relates to the momentum dependence of $\mathrm{Im}\chi(\mathbf{q}, \omega)$ as observed in LSCO. In contrast to some other cuprates the susceptibility is peaked at the incommensurate wave vector $\mathbf{Q}_i = (\pi \pm \delta, \pi \pm \delta)$. In the superconducting state these incommensurate peaks behave rather unusually. With decreasing frequency they first become much more pronounced in comparison to the normal state and at low enough frequency they disappear [128, 129]. This phenomenon was originally called "magnetic coherence" due to the similarity of the observed effect with the quasiparticle coherence peak that occurs below T_c in conventional superconductors. Using the parameters of the tight-binding dispersion for LSCO we found the same behavior of $\mathrm{Im}\chi$ in our calculations [130], as shown in Fig. 23.43(b).

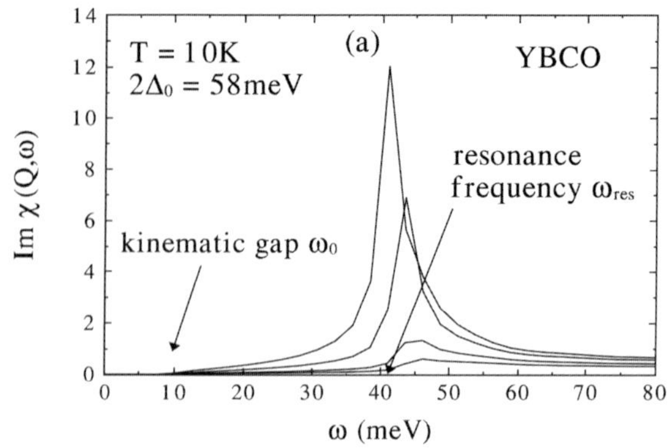

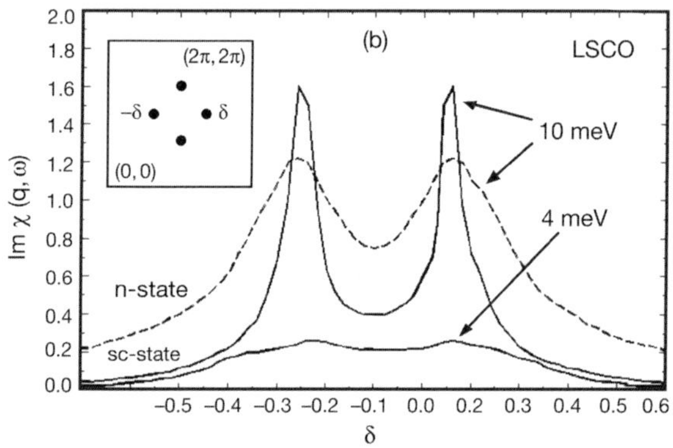

Fig. 23.43. Numerical results for the resonance peak and magnetic coherence in the weak-coupling limit. (**a**) Imaginary part of the RPA spin susceptibility in units of states/eV versus ω in the superconducting state at wave vector $\mathbf{Q}=(\pi, \pi)$ for U/t=1,2,3, and 4 from *bottom to top*. We find ω_{res}=41 meV. Below the kinematic gap ω_0, $\mathrm{Im}\chi(\mathbf{Q}, \omega)$ is zero. (**b**) Calculated magnetic coherence: the *solid curves* correspond to the superconducting state whereas the *dotted curve* is calculated in the normal state. The observed four peaks occur at $\mathbf{Q}_i = (1 \pm \delta, 1 \pm \delta)\pi$ (see *inset*) and in the figure we show only the peaks at $\mathbf{Q}_i = (1, 1 \pm \delta)\pi$. In our calculations we find for the parameter of incommensurability $\delta = 0.18$

The same is the case for the resonance peak. One sees that in LSCO ω_{sf} is comparable to the superconducting gap and therefore, there is a strong renormal-

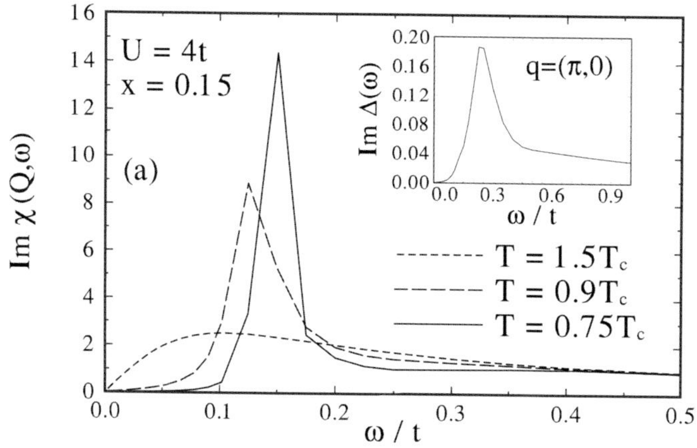

Fig. 23.44. Resonance peak: Imaginary part of the RPA spin susceptibility at $\mathbf{Q} = (\pi, \pi)$ calculated using the FLEX approximation. For the normal state we get $\omega_{sf} = 0.1t$ and in the superconducting state we obtain $\omega_{\text{res}} = 0.15t$. Assuming $t = 250$ meV we find that $0.16t = 40$ meV. *Inset*: Imaginary part of the gap function at $T = 0.75T_c$ for wave vector $=(\pi, 0)$

ization of the spin spectrum in the superconducting gap. Due to the reduced scattering below T_c the incommensurate peaks at $\mathbf{Q}_i$ become sharper than in the normal state. Furthermore, for decreasing energy the intensity of the peaks will decrease due to a superconducting gap. At energies of about 2 meV we do not obtain any well-resolved peaks. This is shown in Fig. 23.43(b).

Using the self-consistent FLEX approach we see that our weak-coupling results do not change drastically. Many experimental facts can be already understood on the basis of the $d_{x^2-y^2}$-wave symmetry of the superconducting order parameter. However, in the strong-coupling Eliashberg approach the spin excitation spectrum can be studied self-consistently. For example, in Fig. 23.44 we show the FLEX results for $\text{Im}\chi(\mathbf{Q}, \omega)$ in the normal and superconducting state. In the normal state the spin fluctuation spectrum shows a peak at $\omega_{sf} = 0.1t$. In the superconducting state the resonance peak occurs at ω_{res} =41 meV and remarkably already at $T = 0.7_c$ the resonance peak is fully developed and does further not change its position in frequency with lowering temperature. This reflects that in the strong-coupling Eliashberg approach the superconducting gap evolves rapidly below T_c and reaches almost its maximum already at $T = 0.7T_c$.

Physically speaking, the resonance peak results from the frequency dependence of the superconducting gap function, which is calculated self-consistently, and from the vicinity to a spin-density-wave collective mode that satisfies (23.43). Since the quasiparticle scattering rate is reduced in the superconducting state the strong and weak-coupling results do not differ drastically.

In Fig. 23.45 we show results (see I. Eremin et al. [131]) for the resonance at incommensurate momentum. Note that in particular the structure near $\mathbf{q} \simeq \mu(\pi, \pi)$ for $\mu \simeq 0.8$ and 1.2. RPA results give that this structure has maximal intensity along the BZ diagonal directions.

The spin-susceptibility in bilayered cuprates will also exhibit resonance peaks for the odd and even parts of χ at $T < T_c$ ($\chi(\mathbf{q}, \omega) = \chi^e(\mathbf{q}, \omega) \cos^2 (q_z d/2) + \chi^0(\mathbf{q}, \omega) \sin^2(q_z d/2)$). The resonances of χ^e and χ^0 occur at different frequencies due to $J_\perp$ and hopping $t_\perp$ (Note that $t_\perp$ causes two Fermi-surfaces representing bonding and antibonding states). For underdoped cuprates the resonance frequency ω^e_{res} increases, while ω^0_{res} decreases for decreasing doping.

It is of general interest to study the doping dependence (in particular for monolayered cuprates) of the resonance peak in the overdoped and underdoped cuprates. In Fig. 23.46 we show our results for the doping dependence of ω_{res} from underdoped to the overdoped regimes. Note that ω_{res} decreases away from optimal doping. In the overdoped regime the resonance peak is mainly determined by the superconducting gap, since the spin fluctuations get weaker and play a less significant role. The superconducting gap decreases in the overdoped cuprates and

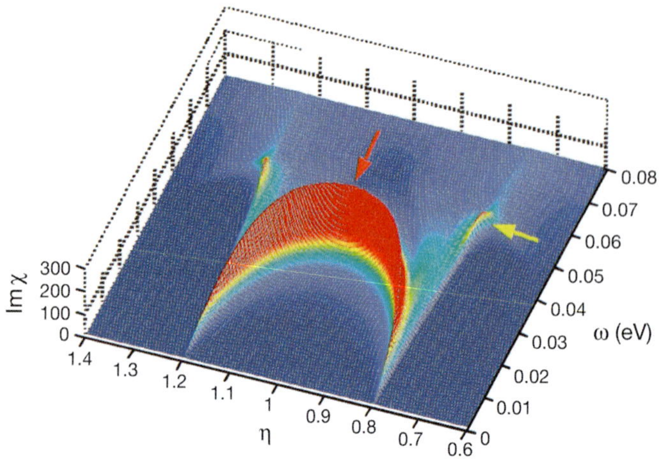

Fig. 23.45. RPA results for magnetic excitations in a $d_{x^2-y^2}$ superconductor. Imχ obtained from Eq. (23.38) with $U = g(\mathbf{q})$ as a function of momentum [along $\mathbf{q}=\eta(\pi,\pi)$] and frequency in the SC state. We use $\Delta_{\mathbf{k}} = \Delta_0(\cos k_x - \cos k_y)/2$ with $\Delta_0 = 42$ meV, and $g(\mathbf{q})$ weakly peaked at (π,π): $g(\mathbf{q})=q_0[1 - 0.1(\cos q_x + \cos q_y)]$ with $q_0 = 0.572$ eV, in order to reproduce the correct energy position of the Q^* mode near $0.8(\pi,\pi$ and the Q mode at (π,π))

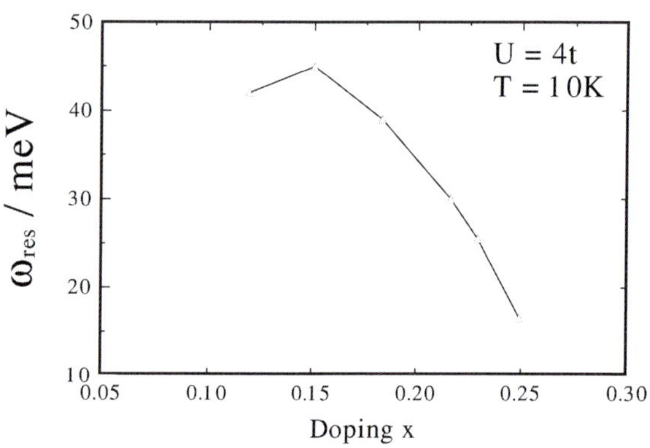

Fig. 23.46. FLEX results for ω_{res} versus doping in monolayered cuprates. In the overdoped regime with $T_c \propto \Delta_0$ we find a constant ratio of ω_{res}/T_c. Note that $t = 275$ meV has been used

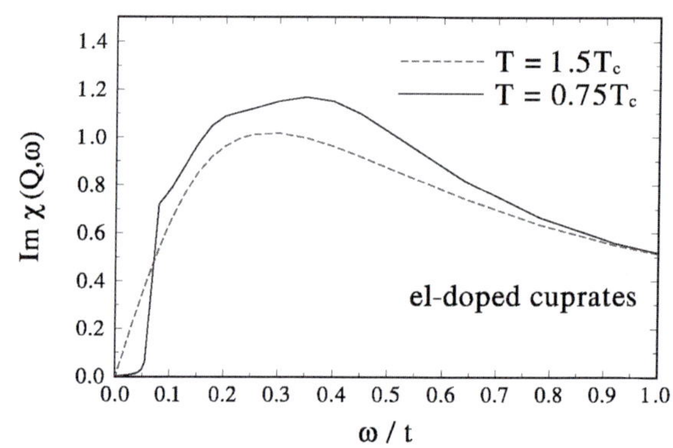

Fig. 23.47. The imaginary part of the spin susceptibility for the electron-doped cuprates above and below T_c at $\mathbf{q} = \mathbf{Q} = (\pi,\pi)$ calculated using the FLEX theory. The calculations are performed for $U = 4t$ and for an optimal doping concentration $x = 0.15$. No resonance peak is found. Only a rearrangement of spectral weight due to $\Delta(\omega)$ occurs

thus the energy position of the resonance peak will also decrease. In the underdoped cuprates the situation is different. In this case, the system is very close to an antiferromagnetic instability and thus the effect of superconductivity on the formation of the resonance peak consists only of reducing the damping. This helps to satisfy the resonant condition. The frequency position of the resonance peak will be rather set by the normal state frequency ω_{sf}. Since ω_{sf} gets smaller with decreasing doping the resonance peak position also decreases in the underdoped cuprates.

To summarize, we obtain for the position of the resonance:

$$\omega_{\text{res}} \approx \begin{cases} \omega_{sf}, & \text{underdoped} \\ 2\Delta_0, & \text{overdoped}\,. \end{cases} \qquad (23.133)$$

In the optimal doped case (which corresponds approximately to $x = 0.15$ holes per copper site) the resonance energy is slightly lower than $2\Delta_0$. Physically, this corresponds to a bound state inside the gap region. Let us also comment on the spin dynamics of the electron-doped cuprates. As we have already mentioned the nesting properties in the electron-doped cuprates are much weaker and fewer density of states are available, and thus they are far from the antiferromagnetic instability. As a result, calculating $\mathrm{Im}\chi(\mathbf{Q}, \omega)$ we find that ω_{sf} becomes larger than in hole-doped cuprates and less pronounced as shown in Fig. 23.47. This has an important consequence also for the feedback of superconductivity. In particular, we find that the resonance condition, i.e. (23.43), in the electron-doped cuprates cannot be fulfilled and the spin excitation spectrum is only slightly renormalized due to the superconducting gap. (Note that one might argue that due to the singularity in $\mathrm{Re}\chi(\mathbf{Q}, \omega = 2\Delta_0)$, a resonance condition could *always* be fulfilled and thus a resonance peak must occur for a $d_{x^2-y^2}$-wave gap. However, this singularity is of logarithmic-type and thus less spectral weight is available for an increasing U. Numerical calculation show that already for $U \simeq 4t$ there exits no resonance peak at $T = 0$.)

We would like to stress that despite the absence of the resonance peak in electron-doped cuprates the superconducting instability is nevertheless driven by the spin fluctuations.

To summarize, we have shown that the spin dynamics in cuprates above and below T_c can be successfully explained on the basis of the one-band Hubbard Hamiltonian and RPA spin susceptibility employing the FLEX approximation. Important features like resonance peak and magnetic coherence occur in the superconducting state of hole-doped cuprates due to the feedback of superconductivity and rearrangement of the normal state spin excitation spectrum calculated within the self-consistent Eliashberg approach for various doping concentrations. Our results are also qualitatively similar to the those of the spin-fermion model [58]. Improved calculations of $\chi(q, \omega)$ should include the local character of the Cu-spins.

23.4.6 Phase Diagram for Electron-Doped and Hole-Doped Cuprates

We demonstrate that our electronic theory assuming the exchange of antiferromagnetic spin fluctuations as a relevant pairing mechanism for singlet pairing in cuprates can account for the main features in the phase diagram of hole-doped cuprates. In particular, we determine the doping dependence of the characteristic temperatures of the phase diagram, namely $T_c^*(x)$, $T_c(x)$, and also T^* at which a gap appears in the spectral density. Below T_c^* we indeed find incoherent Cooper-pairs ("pre-formed pairs"), which become phase coherent only below the critical temperature T_c. We will show that phase fluctuations contributing $\Delta F_{\mathrm{phase}}$ to the free energy lead to a decreasing critical temperature in the underdoped regime, $T_c^* \rightarrow T_c$, and thus to the appearance of an optimal doping x_{opt}. It is shown that this result is due to the small superfluid density $n_s(T)$ in the system. Most importantly, for $x > x_{\mathrm{opt}}$ we find that Cooper-pair phase fluctuations cost too much energy ($\Delta F_{\mathrm{phase}} \propto n_s$) and thus $T_c \propto \Delta_0$, while for $x < x_{\mathrm{opt}}$ one gets $T_c \propto \Delta F_{\mathrm{phase}}$. Note that at T_c the Meissner effect occurs. ΔF_{cond} denotes the contribution to the free energy due to Cooper-pair formation and $\Delta F_{\mathrm{phase}}$ refers to the contribution due to phase fluctuations of the Cooper-pairs. We compare our results with those of Berezinskii–Kosterlitz–Thouless (BKT) theory as well as with the *XY* model.

In Fig. 23.48 we show our results for the phase diagram of hole-doped cuprates together with the experimental curve. T_c^* was obtained from the linearized Eliashberg equations. In optimally and overdoped cuprates one clearly sees that the results of FLEX calculations and experiments agree well with each other, $T_c^* \approx T_c$ and $T_c \propto \Delta_0(x)$. Due to a weakening of the nesting properties of the Fermi surface upon increasing doping and corresponding suppression of antiferromagnetic spin fluctuations the superconducting transition temperature goes down and we do not find any superconductivity above $x = 0.3$. On the other hand, in the underdoped regime the T_c curve in the FLEX approach still slightly increases while in experiments it decreases and superconductivity does not exist below $x = 0.05$.

The slight increase of T_c upon lowering the doping below optimal one is the result of the shape of the Fermi surface and of the large scattering rate that prevent more and more Cooper-pairing. In our model the Fermi surface becomes closer to nesting and thus the corresponding spin fluctuations responsible for superconductivity get enhanced. This indi-

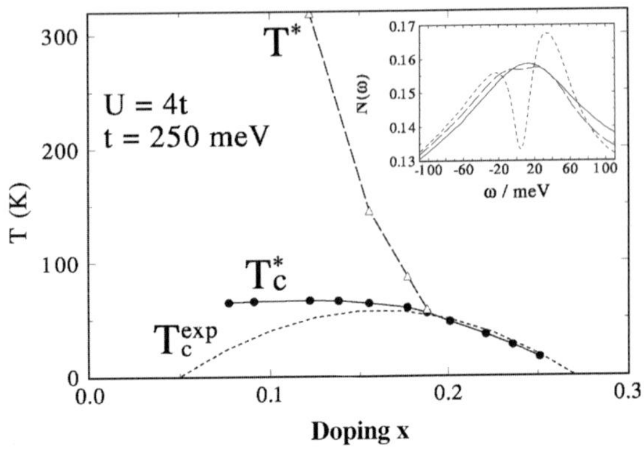

Fig. 23.48. Results for the phase diagram of hole-doped cuprates using the one-band Hubbard model. Cooper-pair phase fluctuations are neglected. Note, $x = 1 - n$ and on-site Coulomb repulsion is $U = 4t$, where t is the nearest neighbor hopping integral. Below the pseudogap temperature T^* (*dashed line*) one finds a reduction of spectral weight in the density of states at $\omega = 0$ (see *inset*). T_c^* (*solid line*) denotes the superconducting transition temperature neglecting Cooper-pair phase fluctuations. The *dashed-dotted line* corresponds to a phenomenological fit to experimental data T_C^{exp} [132]. *Inset*: Density of states $N(\omega)$ in units $1/t$ for doping $x = 0.15$ at temperature $T = 4.5T_c^*$ (*solid line*), $T = 2.3T_c^* \equiv T^*$ (*dashed line*), and $T = 1.01T_c^*$ (*dotted line*)

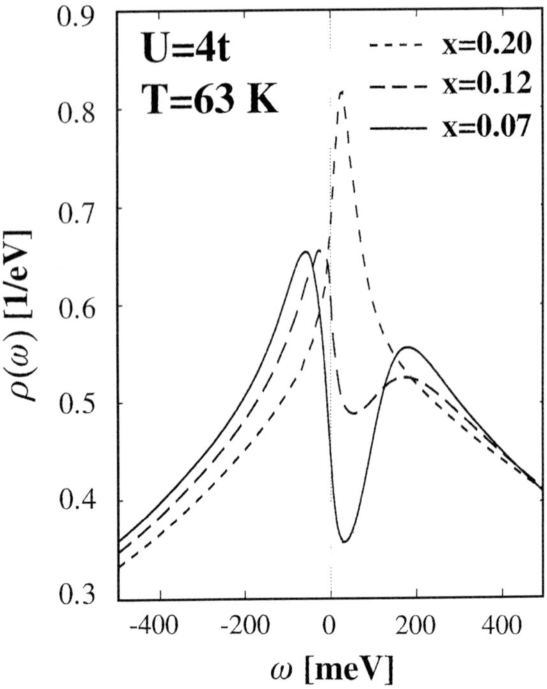

Fig. 23.49. Calculated doping dependence of the pseudogap that occurs below T^* in the electronic density of states in the normal state of hole-doped cuprates for a fixed temperature $T = 63$ K. For small doping this pseudogap appears due to the scattering of quasiparticles by antiferromagnetic spin fluctuations. With decreasing doping the pseudogap increases. This agrees well with experimental data [133]. Moreover, for underdoped cuprates the value of the calculated pseudogap (taken as a distance between the edges) $\Delta_p \approx 80$ meV is similar to that measured in underdoped cuprates$\Delta_p^{\mathrm{exp}} \approx 88$ meV [133]

cates that the model we have used in the mean-field is not applicable for the underdoped cuprates. In the vicinity of antiferromagnetism the effects of the electronic correlations play an important role. The perturbative approach we are using will not properly account of the correlations. For example, it is known from ARPES experiments [134] that the shape of the Fermi surface changes in hole underdoped cuprates and forms pockets around $(\pi/2, \pi/2)$ points of BZ, which is believed to be a result of the electronic correlations. Also, note that a pseudogap opens up in the elementary excitation spectrum, see Fig. 23.49. Furthermore, Cooper-pair phase fluctuations matter and suppress T_c. We will present results for this case later, see Fig. 23.51.

In the Eliashberg formulation of the Cooper-pairing the carriers not only pair, but also scatter due to spin fluctuations. If the scattering is strong some of the quasiparticles will be away from the Fermi level already in the normal state. This forms a dip, i.e. the reduction of the spectral weight in the density of states at the Fermi level as one can see from the inset of Fig. 23.48. In order to investigate the doping dependence of such a pseudogap in the density of states more in detail, in Fig. 23.49 we show the calculated results for $N(\omega)$ for various doping. One clearly sees that in the overdoped region the density of states is unrenormalized due to weakness of the antiferromagnetic correlation. For decreasing doping a pseudogap at the Fermi level occurs and it increases upon decreasing doping. The pseudogap opens below $T = T^*$. Results for T^* using $U = 4t$ as a function of hole-doping are shown in Fig. 23.48.

Table 23.1. Results for the calculated pseudogap formation temperature T^* resulting from the scattering of carriers by spin fluctuations for different doping in comparison with experimental data on T^*_{exp} found in YBaCuO-compounds [135]

x	T* (K)	T^*_{exp} (K)
0.16	100	90
0.15	150	160
0.14	200	190
0.13	310	240

Most importantly, our results show a good agreement for the doping dependence of the pseudogap. Similar to the NIS tunneling experimental data [133] the pseudogap increases with decreasing doping from overdoped to underdoped cuprates. In Table 23.1 we show our calculated results for T^* using $U = 4t$ for different doping in comparison with experimental data on YBaCuO-compounds [135]. One notices that our results agree also well on the quantitative level. Moreover, for the underdoped cuprates the calculated magnitude of the pseudogap taken as a distance between edges of two peaks ($\Delta_p \approx 80$ meV at $x = 0.07$) agrees well with the experimental value $\Delta_p^{exp} = 88$ meV found for underdoped cuprates [133]. For completeness it remains to be seen whether the same theory can explain also the phase diagram of electron-doped cuprates. In Fig. 23.50 we present the results of FLEX calculations for electron-doped cuprates [122]. Most importantly our mean-field results for the phase diagram are in good agreement with experimental data again for optimally and overdoped region. For example, one sees that the maximal T_c is smaller than in the hole-doped case. The reasons for this are poorer nesting properties of the Fermi surface in electron-doped case and also the fact that a flat bands lies in NCCO is approximately 300 meV below the Fermi level. Another important feature of the phase diagram of the electron-doped cuprates is the narrower range of superconductivity than in the hole-doped case ($0.12 < x < 0.18$ in the electron-doped case and $0.05 < x < 0.3$ for hole-doped cuprates). The reason for this is also clear: in the overdoped regime the nesting properties of electron-doped NCCO rapidly become very poor and cannot drive superconductivity. In the underdoped regime one should keep in mind that antiferromagnetism persists in the electron-doped case for larger doping. Thus, the superconductivity region will be further destroyed by the antiferromagnetic instability that competes with the superconducting one.

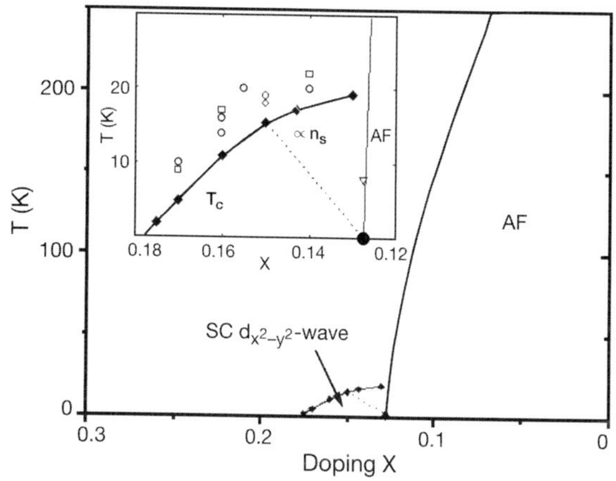

Fig. 23.50. Results for the phase diagram of electron-doped cuprates using the one-band Hubbard model. We neglect Cooper-pair phase and amplitude fluctuations, thus $T_c^* = T_c$. The hole doping concentration is $x = 1 - n$ and the on-site Coulomb repulsion $U = 4t$, where t is the nearest neighbor hopping integral. The *inset* shows the region of superconductivity.The antiferromagnetic (AF) transition curve is taken from [136]. The *solid curve* corresponds to our calculated T_c values [122]. The experimental data are *squares* from [137], *triangles* from [138], *circles* from [139]. The *dashed curve* corresponds to $T_c \propto n_s$ as in hole-doped cuprates

Similarly to hole-doped cuprates the important question is how T_c is lowered in the underdoped regime. Our calculations have shown in agreement with experiment that there is no weak pseudogap above T_c in the underdoped regime of the electron-doped cuprates in contrast to the hole-doped ones. The reason is clear: The spin fluctuations are less pronounced in the electron-doped cuprates and thus do not produce a dip in the spectral density at the Fermi level.

We now present results for $T_c(x)$ in the underdoped cuprates ($x < x_{opt}$) including Cooper-pair phase fluctuations. The region $x < x_{opt}$ can be experimentally characterized by a T_c that decreases with decreasing hole density x. Furthermore, the so-

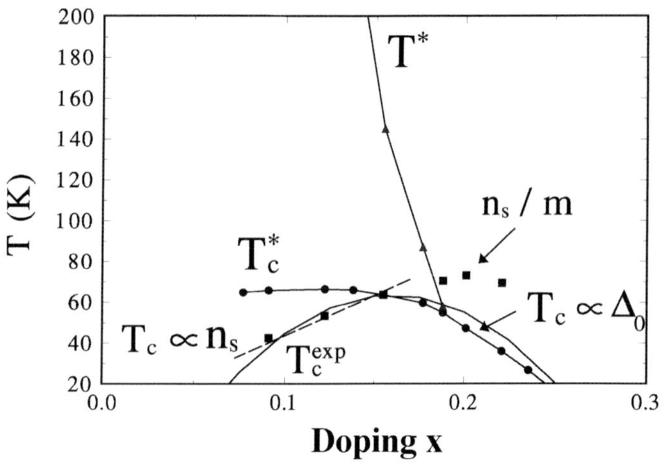

Fig. 23.51. Phase diagram for high-T_c superconductors resulting from a spin fluctuation induced Cooper-pairing including their phase fluctuations. The calculated values for $n_s(0)/m$ (*squares*) are in good agreement with muon-spin rotation experiments [141]. They determine the energy costs for the Cooper-pair phase fluctuations. T_c^* denotes the temperature below which Cooper-pairs are formed. The *dashed curve* refers to our calculated results using $n_s(x)$ and gives the observed Uemura scaling $T_c \propto n_s(T = 0, x)$ [14]. As shown in Figs. 23.50 and 23.49 below T^* we get a gap structure in the spectral density as observed in tunneling spectroscopy [20,142]. The *solid curve* T_c^{exp} refers to experiments describing many hole-doped superconductors [132]

called Uemura scaling holds [14]. It was recognized early on that a small n_s leads to a reduced stiffness against fluctuations of the phase of the superconducting order parameter [78,79,140]. Furthermore, cuprate superconductors consist of weakly-coupled CuO_2 planes so that Cooper-pair phase fluctuations are enhanced also by the reduced dimensionality. Note that in conventional superconductors this is not relevant, since the large superfluid density leads to a typical energy scale of Cooper-pair phase fluctuations much larger than the superconducting energy gap Δ that governs the thermal breaking of a Cooper-pair. Thus, in conventional superconductors $T_c \propto \Delta(T = 0)$ [76]. In contrast to this, the observation $T_c \propto n_s$ in underdoped (hole-doped) cuprates indicates that Cooper-pair phase fluctuations drive the superconducting instability. The Cooper-pairs only break up at a crossover temperature $T_c^* > T_c$ and between T_c and T_c^* local phase-incoherent Cooper-pairs exist [70,78,79,93,140]. Then a transition to superconductivity with phase-coherent Cooper-pairs occurs at $T_c \propto n_s$ [70,78,79].

Using results for $n_s(x, T)$ calculated from Kubo formula for the underdoped cuprates we obtain the phase diagram shown in Fig. 23.51. In the overdoped regime T_c^* is identical with the bulk transition temperature T_c below which a Meissner effect is found. In agreement with experiments from our electronic theory we get $T_c \propto \Delta(T \to 0)$ for overdoped cuprates. In the overdoped cuprates the energy costs due to Cooper-pair phase fluctuations ($\propto n_s/m$) are higher than the condensation energy. Then, Cooper-pair phase fluctuations play a minor role in determining T_c in the overdoped cuprates. In contrast to this, in the underdoped cuprates we find $T_c \propto n_s(x) \propto T_c^*$. This results from the low superfluid density n_s. At T_c^* phase-incoherent Cooper-pairs are formed and only at $T_c < T_c^*$ does one get phase coherence (quantum phase fluctuations at $T_c = 0$ have not been included).

Our theory shows that for optimal doping occurs as a result of replacing, determining T_c, the free energy change $\Delta F_{\text{cond}}(x)$ due to Cooper-pair condensation for overdoped cuprates by the free energy change ΔF_{phase} gained from phase coherence of the Cooper-pairs. Thus, T_c^{opt} occurs when $\Delta F_{\text{cond}} = \Delta F_{\text{phase}} \sim n_s$.

Results for $\Delta F(x)$ are shown in n Fig. 23.52. We find that ΔF_{cond} mainly follows the doping dependence of the mean-field transition temperature T_c^*. The doping dependence of $\Delta F_{\text{phase}}(x)$ is given by $n_s(0)/m$. Note that the energy cost due to phase fluctuations has the opposite behavior to the energy gain due to Cooper-pair condensation with respect to the doping concentration x. It is remarkable that from our electronic theory we get a crossing of the two energy contributions ΔF_{cond} and ΔF_{phase} at $x \simeq 0.15$ for which the largest T_c is observed. The consequence of this is that we find a non-monotonic doping de-

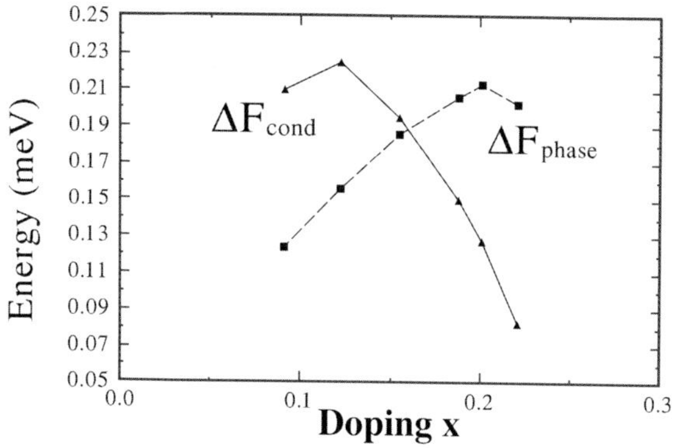

Fig. 23.52. Calculated crossover of the phase-stiffness energy and $\Delta F_{\rm cond}$. $\Delta F_{\rm phase} \propto n_s/m$ and the condensation energy is $\Delta F_{\rm cond} \simeq \alpha\{n_s/m\}\Delta_0(x)$. Here, we estimate $\alpha \simeq 2.5 \times 10^{-3}$. (For comparison in conventional superconductor Al $\alpha \approx 0.4$ [143]. Note that in cuprates α is small for two reasons, namely, the number of holes condensed into Cooper-pairs is maximum 50% of the whole number of carriers available and, secondly, due to the large size of the unit cell of cuprates). Our results for α agree well with other theoretical and experimental estimates [83–85]. $\Delta F_{\rm phase} < \Delta F_{\rm cond}$ implies the two characteristic temperatures T_c^*, where Cooper-pairs are formed at $T_c \sim \Delta_0$, and $T_c \approx \Delta F_{\rm phase} \propto n_s$, where Cooper-pairs become phase coherent

pendence of $T_c(x)$ and optimal doping at $x \simeq 0.15$ and $T_c \propto n_s$ for underdoped cuprates (and thus the Uemura scaling).

Physically speaking in the overdoped regime we find a large $\Delta F_{\rm phase}$, which means that Cooper-pair phase fluctuations cost much energy. Thus, the system will undergo a mean-field transition involving phase coherent Cooper-pairs and a small condensation energy $\Delta F_{\rm cond}$ ($\Delta F_{\rm cond} \sim T_c \propto \Delta_0 < \Delta F_{\rm phase}$). In the underdoped cuprate superconductors the energy gain due to the formation of Cooper-pairs is such that they are broken up to T_c^* and $T_c^* > T_c$. Only at $T \leq T_c$ does the phase coherence of the Cooper-pairs occur.

Note that the coherence length in electron-doped cuprates estimated from H_{c2} measurements is $\xi_{ab} \simeq 30$Å [77], much larger than a typical value for the hole-doped ones. This means that all Cooper-pairs overlap well with each other and that almost all electrons are involved in superconductivity. Therefore, one expects that the Cooper-pair phase fluctuations play a much less important role in the electron-doped cuprates than in hole-doped ones.

To summarize, we have solved the generalized Eliashberg equations self-consistently and extended them by including Cooper-pair phase fluctuations to calculate some basic properties of the hole-doped cuprate superconductors. The doping dependence of $T_c(x)$ and an improved phase diagram is obtained. In particular we show also results for the condensation energy and for the weak pseudogap temperature T^* below which a small reduction in the spectral density at the Fermi level appears.

23.4.7 Doping Dependence of the Superfluid Density $n_s(x, T)$

In Fig. 23.53 results for the temperature dependence of the superfluid density $n_s(\omega = 0)/n$ (n denotes the normal-state band filling) are given below T_c^* for various doping concentrations. We calculate n_s using the current–current correlation function and standard many-body theory [73], see Eq. (23.54):

$$\frac{n_s(\omega)}{m} = \frac{1}{e^2}\,\omega\,\mathrm{Im}\,\sigma(\omega)\,. \qquad (23.134)$$

Note that according to London's theory, $\lambda_L \propto n_s$, the ratio n_s/n is related to measurements of the (in-plane) penetration depth; for an example see microwave experiments. Note that our calculations show that roughly only 1/3 of the holes become superconducting, even for $T \to 0$. This is typical for a strongly interacting system. For decreasing doping ($x \to 0$) the curves $n_s(T)$ get flatter and smaller, as experimentally observed.

For underdoped cuprates Cooper-pair phase fluctuations become important ($T_c^* < T_c$). Then, after averaging over the phases of the Cooper-pairs one expects of course $n_s \to 0$ for T$\to T_c$. We illustrate this in Fig. 23.53 where we show experimental results and FLEX results for underdoped cuprates. Then, a qualitative agreement with the data of Bonn, Hardy and co-workers on $\lambda^2(T = 0)/\lambda^2(T)$ concerning the slope of the curves in the vicinity of T_c^* and the linear behavior for $T \to 0$ is found [8].

In particular the FLEX approximation for the generalized Eliashberg equations yields close to T_c^* a

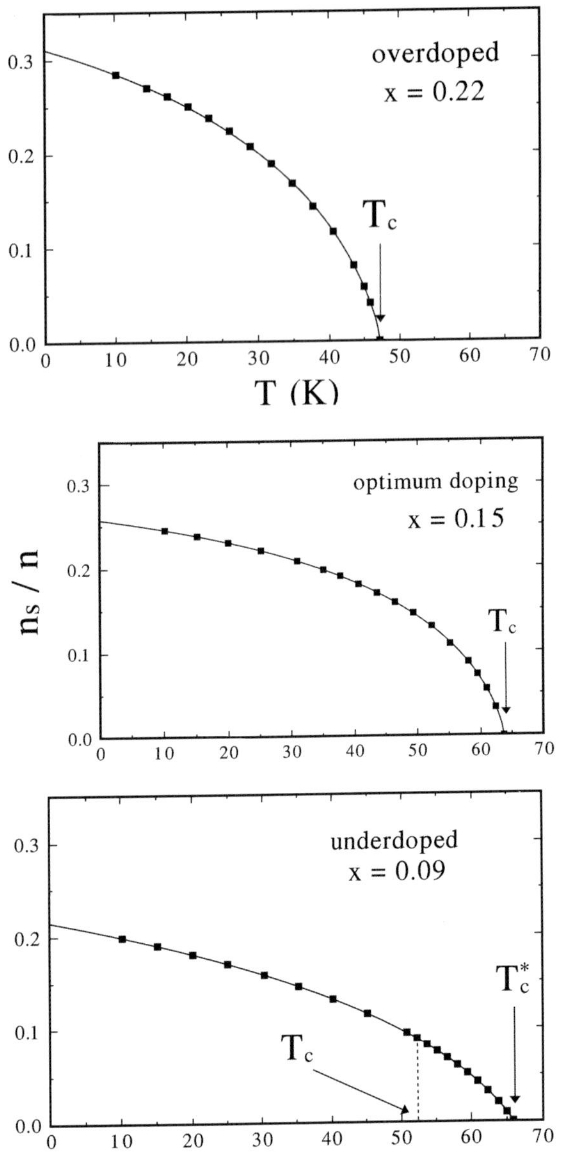

Fig. 23.53. Temperature dependence of the superfluid density $n_s(x, T)$ calculated with the help of (23.51)–(23.54) for various hole doping concentrations x. We extrapolate the results to $T \rightarrow 0$. At temperature T such that $T_c < T < T_c^*$ Cooper-pairs become phase incoherent, $n_s^0 \rightarrow 0$. Our results are in fair agreement with measurements of the in-plane penetration depth by Bonn, Hardy and co-workers [8]. The *dashed curve* for the underdoped case illustrates the effect of Cooper-pair phase fluctuations according to the (static) Kosterlitz–Thouless theory

relation $\lambda^3(T = 0)/\lambda^3(T) \propto (T_c^* - T)$. The same power law was found by Kamal et al. and has been attributed to critical fluctuations starting about 10 K below T_c, since the slope coincides with the critical exponent for the 3D-*XY* model [8]. Here, we obtain this power law from the generalized Eliashberg equations using the FLEX approximation, which is purely 2D and does not contain critical fluctuations. Instead, the rapid increase of n_s below T_c^* is due to the self-consistent treatment of the superconducting gap function $\Delta(\omega)$. Thus, we conclude that while 3D critical fluctuations are expected in a very narrow temperature range close to T_c, these are not the origin of the observed power law on the scale of 10 K.

In Fig. 23.54 we present the FLEX results for Cooper-pairing phase averaged superfluid density

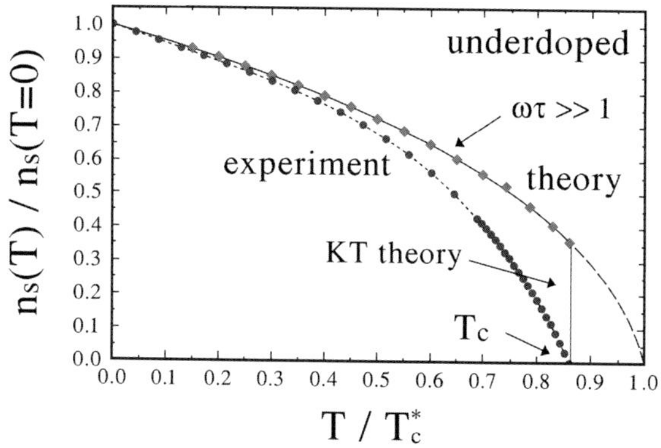

Fig. 23.54. Phase averaged superfluid density for underdoped cuprates. The deviation of experimental results (*circles*) from FLEX results is due to Cooper-pair phase fluctuations with lifetime τ. Taking these into account within BKT theory one gets results indicated by the *vertical line*. For the dynamical superfluid density ($t \sim \omega^{-1}$) one expects $n_s(t) \to n_s(\text{FLEX})$ if $\omega\tau \gg 1$

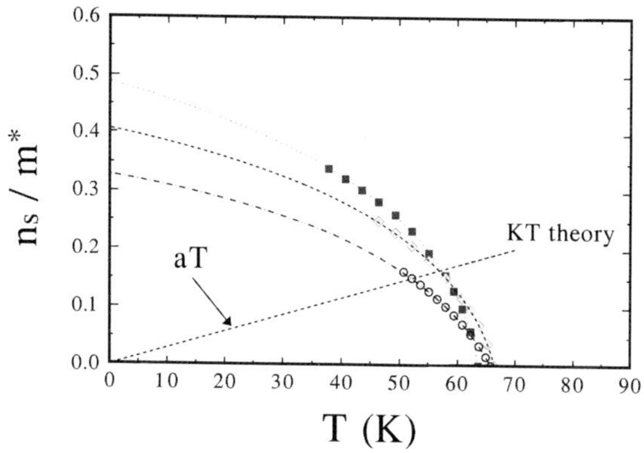

Fig. 23.55. Results for the superfluid density divided by m in units $10^{-2} m_e^{-1} \text{Å}^{-2}$ as a function of temperature T for various doping concentrations: $x = 0.155$ (*squares*), $x = 0.122$ (*diamonds*), $x = 0.091$ (*circles*). T_c is given as indicated by the intersection of $n_s(x, T)$ with aT, where $1/a = \pi/2$ (BKT theory)

in underdoped superconductors. The effect of phase fluctuations can be seen from the discrepancy between experiment [8] and theory. One clearly sees that no Meissner effect results for $T > T_c$, since the phase correlation function $< \nabla\phi(\mathbf{r}) \cdot \nabla\phi(0) >$ becomes zero. The results obtained using static Kosterlitz–Thouless theory (straight line) and the FLEX results (diamonds) for $n_s(x, T)$ are shown. The FLEX results neglect phase fluctuations, which is appropriate for $\omega\tau \to \infty$. Of course, in Kosterlitz–Thouless theory we get $n_s \to 0$ at T_c. Note that in our full FLEX calculations we have to use for the phase averaged superfluid density $n_s = n_s^0 \langle \nabla\phi(\mathbf{r}) \cdot \nabla\phi(0)\rangle$ and then also $n_s \to 0$ at T_c. Note that for $T < T_c$ the discrepancy between experimental results and our FLEX calculations indicates the importance of Cooper-pair phase fluctuations below T_c.

In Fig. 23.55 we show results for n_s renormalized due to phase (vortex) fluctuations (for details see Appendix D and [92]). The Berezinski–Kosterlitz–Thouless (BKT) theory describes the unbinding of thermally created pancake vortex–antivortex pairs. The relevant parameters are the dimensionless stiffness K and the core energy E_c of vortices. The stiffness is related to n_s

$$K(T) = \frac{\hbar^2}{k_B T} \frac{n_s(T)}{m^*} \frac{d}{4}, \tag{23.135}$$

where d is the average spacing between neighboring CuO_2-planes. Hence, $T_c \sim n_s(T_c)K(T_c)$ and $K(T_c)$ is given by BKT. Starting from the smallest vortex–antivortex pairs of size r_0, the pairs are integrated out and their effect is incorporated by an approximate renormalization of K and the fugacity $y = e^{-E_c/k_B T}$. As mentioned earlier this leads to the Kosterlitz recursion relation [92]

$$\frac{dy}{dl} = (2 - \pi K)y, \tag{23.136}$$

$$\frac{dK}{dl} = -4\pi^3 y^2 K^2, \tag{23.137}$$

where $l = \ln(r/r_0)$ is a logarithmic length scale. Bound pairs reduce K and thus n_s, but do not destroy superconductivity. At T_c, K jumps from a universal value of $2/\pi$ to zero. The values of T_c shown in Fig. 23.56 are obtained by numerically integrating (23.137) with n_s taken from FLEX results.

In Fig. 23.56 we plot results by Timm et al. for the transition temperature T_c obtained with the help of BKT theory and the temperature T_c^* where Cooper-pairs form. For decreasing doping x, T_c^* becomes nearly constant and decreases slightly for the lowest doping level, consistent with the strong decrease of the onset temperature of vortex effects at even lower doping. On the other hand, T_c decreases again in the

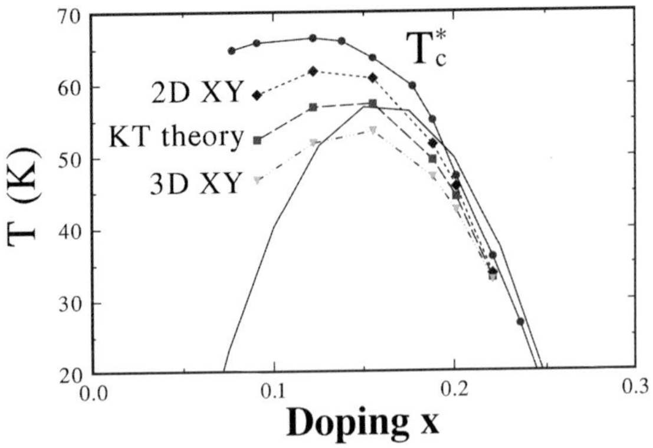

Fig. 23.56. Calculated doping dependence for T_c calculated by C. Timm et al. with the help of the Berezinski–Kosterlitz–Thouless (BKT) theory. For comparison the T_c curves obtained within 2D and 3D-*XY* models, and also the mean-field result T_c^* are displayed

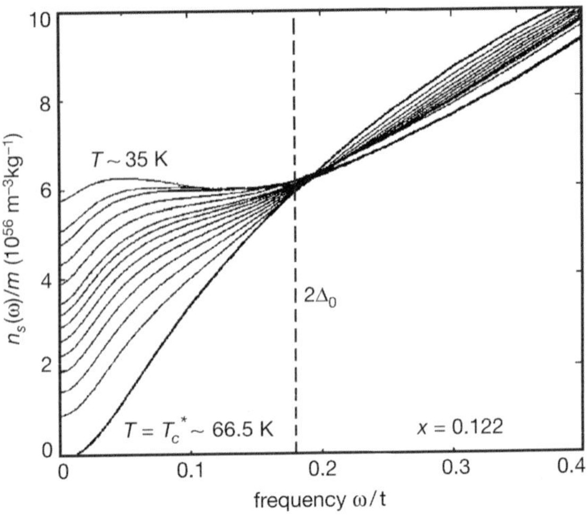

Fig. 23.57. Frequency dependent phase stiffness $n_s(\omega)/m^*$ for doping $x = 0.12$ (underdoped) and various temperatures k_BT/t =0.012, 0.015, 0.016, 0.017, 0.018, 0.019, 0.0195, 0.02, 0.0205, 0.021, 0.0215, 0.022, 0.0225, 0.023 (with decreasing $n_s(0)/m^*$). Here, $t = 250$ meV is the hopping integral. The frequency is given in units of t ($\hbar = 1$). At $T_c^* \approx 0.023t/k_B = 66.5$ K, Cooper-pairs start to form. Below T_c^* there is a marked transfer of weight from energies above $2\Delta_0$ to energies below, where Δ_0 is the maximum gap at low temperatures as obtained from the FLEX calculations

underdoped regime. We have also calculated the superconducting gap Δ_0 extrapolated to $T \to 0$ (not shown). Δ_0 is here defined as half the peak-to-peak separation in the density of states. We find approximately $\Delta_0 \approx T_c^*$. For a comparison the T_c curves obtained within 2D and 3D *XY* models are also shown. We again find that Cooper-pair phase fluctuations are unimportant in the overdoped regime. Note that $n_s(T, x) \to 0$ for $T \to T_c^*$, since Cooper-pairs disappear at T_c^*.

The transition temperature T_c has to be determined by averaging over the Cooper-pair phase fluctuations. In a Ginzburg–Landau (GL) treatment the phase is given by the GL wave function $\psi(\mathbf{r}, t) = |\psi|e^{i\phi(\mathbf{r},t)}$, where n_s is related to the correlation function $\langle \nabla\phi(\mathbf{r}, t) \cdot \nabla\phi(0, 0)\rangle$. In the presence of spatial phase fluctuations of $\psi(\mathbf{r})$ the averaged superfluid density $\bar{n}_s = |\bar{\psi}(\mathbf{r})|^2\langle\nabla\phi(\mathbf{r}) \cdot \nabla\phi(0)\rangle$ will vanish at T_c so that no Meißner effect occurs above T_c. BKT theory as well as the 2D and 3D-*XY* models [147] predict that T_c is proportional to $n_s(T_c)/m$. Note that T_c as obtained within the 3D-*XY* model is larger than the 2D values, since fluctuations are less important in 3D.

Finally, we calculate the dynamical phase stiffness. Note that the superfluid density can be obtained from the imaginary part of the conductivity

$$\frac{n_s}{m^*} = \frac{1}{e^2} \lim_{\omega \to 0} \omega\sigma_2^S(\omega) \,. \tag{23.138}$$

This can be shown with the help of Kramers–Kronig relations. We have recalculated n_s/m^* in this way and find results identical to those obtained from (23.51). Note that for the small wave vector $\mathbf{q} \to 0$ the (23.138) can be written as [92]

$$\frac{n_s(\omega)}{m^*} = \frac{1}{e^2} \omega\sigma_2^S(\omega)\bigg|_{\omega\to 0} \,. \tag{23.139}$$

The imaginary part $\sigma_2^S(\omega)$ of the dynamical conductivity is obtained from the FLEX approximation for the dynamical current–current correlation function using the Kubo formula. This is discussed in Appendix D.

Note that it follows immediately from (23.139) that $n_s = A\sigma_{DC} T_c$ with $A = \frac{m^*}{e^2}(\omega\sigma_2^s/\sigma_{DC} T_c)_{\omega\to 0, T\to T_c}$.

The resulting phase stiffness $n_s(\omega)/m^*$ is shown in Fig. 23.57 for $x = 0.12$ (underdoped) at various temperatures. At higher doping the results (not shown) are similar, only the typical frequency scale, which turns out to be the low-temperature superconducting gap Δ_0, is reduced. For temperatures $T < T_c^*$ the superfluid density is non-zero at $\omega = 0$, indicating the presence of the Cooper-pair phase fluctuations in the system. On the contrary, for $T = T_c^*$ $n_s(\omega = 0)/m^* = 0$ and thus no dynamical Cooper-pair phase fluctuations are present. Note that for $\omega > 0$ the superfluid density is finite even for $T > T_c^*$. At at a first glance this is surprising, since the phase is not well-defined for $\Delta = 0$. However, the corresponding action of the Cooper-pair phases also contains a contribution from the time derivative of the phase besides the stiffness term. While the total action vanishes, each term on its own does not. Another interesting observation is that all curves merge together at energies of about $2\Delta_0$. This is expected, since the superfluid density is a current–current correlation function, which has a characteristic energy scale at about $2\Delta_0$ in the superconducting state.

Even slightly below T_c^*, $n_s(\omega = 0)/m^*$ obtains a significant finite value, leading to the Meissner effect, and there is a considerable redistribution of weight from energies roughly above twice the low-temperature gap, $2\Delta_0$, to energies below $2\Delta_0$. This redistribution increases with decreasing temperature. Also, a peak develops slightly below Δ_0 followed by a dip around $2\Delta_0$, this structure being most pronounced in the underdoped case. Since Δ_0 is smaller in the overdoped regime, $n_s(\omega)/m^*$ changes more rapidly for small ω in this case. Of course it is not surprising that $2\Delta_0$ is the characteristic frequency of changes in $n_s(\omega)/m^*$ related to the formation of Cooper-pairs.

To summarize, we have analyzed the behavior of the superfluid density $n_s(\omega, T, x)$, which controls the doping dependence of the phase coherence, the thermodynamic behavior, the penetration depth, the Nernst effect, Cooper-pair phase fluctuations, etc. Note that we find that the superfluid density plays the most significant role in the underdoped cuprates and determines the superconductivity. We find our results to be consistent with the BKT theory. Also the recently noted universal scaling relation by Homes

$$n_s \propto \sigma_{DC} T_c \tag{23.140}$$

in hole-doped and electron-doped superconductors, in which the dc conductivity σ_{DC} is measured approximately at $T = T_c$ [97], is contained in the general expression for n_s. We immediately obtain in the underdoped high-T_c cuprates

$$n_s(T = 0) \propto \sigma_{DC}(T = T_c)\, T_c\,, \tag{23.141}$$

as observed in experiment. Note that (23.139) is taken in the limit $T \to 0$ assuming a thermodynamical phase transition and Fermi liquid-like behavior. However, in the case that quantum fluctuations are important close to $T = 0$, the simple form of (23.139) is no longer valid because of logarithmic correction terms [148–150].

Another interesting question in cuprates is whether there exists a quantum critical point (QCP) at around $x = 0.19$ doping concentration. The QCP should be reflected characteristically by a universal behavior of some physical properties such as the optical conductivity (independent of time and length scales). A possible interplay of QCP and thermodynamical points needs further study.

23.4.8 SIS and NIS Tunneling Spectra

Tunneling spectra are a good measure of the superconducting state and the elementary excitations in general. They shed light on the validity of our electronic theory. Here, we calculate the tunneling spectra for SIS (superconductor/insulator/superconductor) and NIS (normal-state metal/insulator/superconductor). Let us see to which extent the spin fluctuation model for superconductivity can explain the fine structures seen by tunneling spectroscopy below T_c in the cuprates (remember the significant success of the BCS and Eliashberg theory of superconductivity due to electron–phonon coupling resulting from the explanation of the fine structures seen by tunneling spectroscopy in conventional superconductors, which exhibits the single-particle excitations spectrum [50]). It is important to relate these features to the spin fluctuation frequency ω_{sf} and resonance frequency ω_{res}. One expects that the remarkable dip structure seen in the single-particle excitation spectrum of various cuprates might be an intrinsic *fingerprint* of a

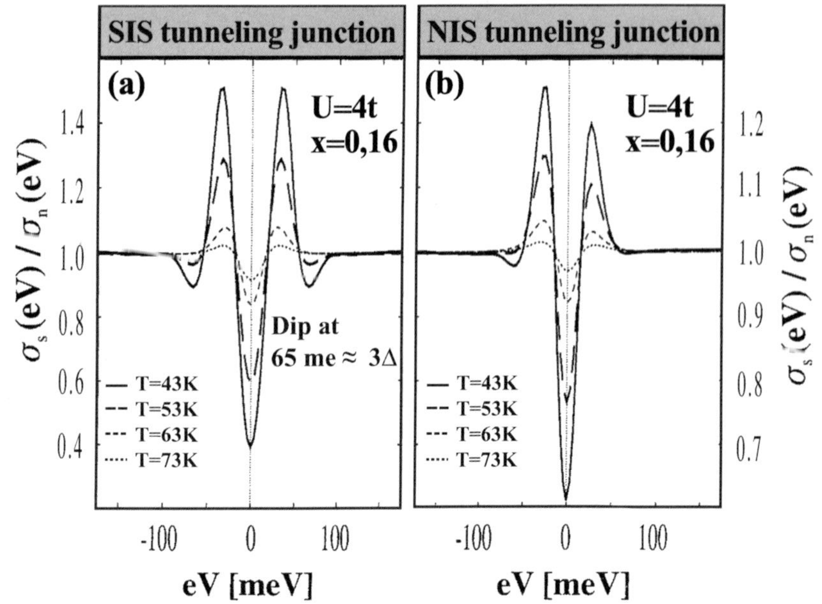

Fig. 23.58. Results for (**a**) SIS and (**b**) NIS tunneling conductances $\sigma_s(eV)$ using Cooper-pairing theory due to spin-fluctuation exchange. Both quantities are normalized by the normal state σ_n (eV) taken at $T = T_c$. This is in fair agreement with experiments [20,99,151]

spin-fluctuation induced Cooper-pairing in high-T_c cuprates [99]. It is also interesting to understand why these dip structures occur at different energies in the in-plane NIS and SIS measurements. Furthermore, while the SIS spectrum is symmetric with respect to the sign of applied voltage the NIS spectra are asymmetric [151–153].

In Fig. 23.58 we present results [154] for the SIS and NIS spectra at various temperatures obtained using the generalized Eliashberg equations. Remarkably, both NIS and SIS tunneling curves show the dip features at $\omega = \omega_{\text{res}} + \Delta_0$ and $\omega = \omega_{\text{res}} + 2\Delta_0$, respectively. This agrees with experimental data [20,99,151]. In order to better understand the origin of these peaks let us remember that the experimental tunneling conductance in NIS is determined by the quasiparticle density of states as discussed in the section on theory. As we already know, the density of states in cuprates is affected by the scattering of the carriers by antiferromagnetic spin fluctuations, which are peaked in the superconducting state at ω_{res}. Taking into account that the whole spectrum will be shifted by the value of the superconducting gap Δ_0, one concludes that the dip seen in the experiments results from the peak in the spin excitations spectra at (π, π) shifted by the superconducting gap. Therefore, the position of the dip in NIS is given by:

$$\text{NIS–dip: } \omega \approx \omega_{\text{res}} + \Delta_0 \, .$$

In the SIS tunneling both convoluted density of states have to be taken in the superconducting state and thus the dip in SIS occurs at:

$$\text{SIS–dip: } \omega \approx \omega_{\text{res}} + 2\Delta_0 \, .$$

This agrees with experiments [99]. Furthermore, in NIS tunneling the curves are asymmetric due to the fact that the density of states of the normal state is asymmetric itself $(t', t'' \neq 0)$. Thus, the effect will be visible only for the unoccupied states and also depends on the shape of the Fermi surface for the different materials. Therefore, depending on the particular doping level and material the effect may show up for the positive bias voltage or on its negative side.

Our results for SIN and SIS tunneling are similar to those obtained by the spin–fermion model [156]. Regarding the doping dependence of the "dip" feature we note that it correlates with the doping dependence of the "resonance peak" seen in INS. Therefore, the position of the "dip" feature should have qualitatively similar doping dependence as expected from Fig. 23.46. Finally, we would also like to note that recently it was argued [157] that in many cases the "dip

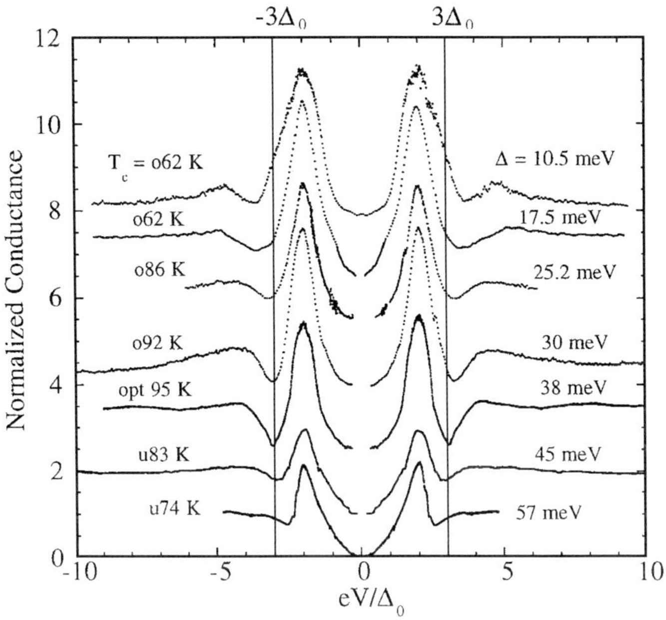

Fig. 23.59. SIS tunneling conductance at various doping levels showing the shift of the dip feature with T_c. The voltage axis is rescaled in units of Δ_0/e, taken from [155]. Note that SIS spectra reveal no pseudogap

feature" seen in bilayer cuprates can be produced by the effects of bilayer splitting. However, recent studies of the one-layer Tl-based cuprates [158] confirm the presence of the "dip" feature indicating its intrinsic relation to the 'resonance' peak. In order to identify the "dip" position seen in bilayer cuprates and its relation to the "resonance" peak further studies are necessary.

In Fig. 23.59 we show the experimental results on SIS tunneling [155]. One can clearly see that the position of the dip does not occur exactly at $3\Delta_0$ but, as expected, at $2\Delta_0 + \omega_{res}$. Furthermore as was argued in [155] the position of the dip scales similarly to the scaling of the "resonance" peak seen by INS. Therefore, it appears that the dip structure can be considered a strong coupling effect analogous to phonon structures, but the excitations that are responsible for this feature are the spin fluctuations. Note that the antiferromagnetic spin fluctuations are much weaker in the electron-doped cuprates. As a result the "dip-hump" feature has not been observed in these compounds so far.

23.4.9 Bilayer Effects

Another interesting effect is the bilayer splitting observed in high-T_c cuprates. For a long time it was believed that the enhancement of T_c in two-layer cuprates resulted from interlayer coupling in the elementary unit cell. The bilayer splitting was observed only recently by means of ARPES experiments [74,160,161]. As expected the bilayer splitting can be described by the interlayer hopping term

$$2t_\perp \cos k_z \left(\cos k_x - \cos k_y\right)^2 ,$$

where $t_\perp = 70$ meV. The matrix element $t_\perp$ describes hopping term between the CuO_2 planes and $k_z = 0$ or π corresponds to the bonding and antibonding bands, respectively. As a result there are two bands crossing the Fermi level as shown in Fig. 23.60.

The change in the Fermi surface topology is important for our theory, since the spin fluctuations are dependent on the amount of nesting. In the bilayered cuprates, due to the introduced hopping between neighboring CuO_2-planes, the topology of the Fermi surface changes slightly and most importantly the nesting shifts to the three-dimensional antiferro-

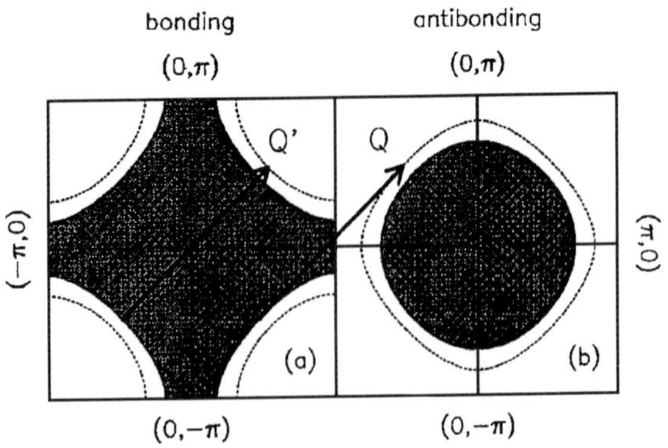

Fig. 23.60. (**a**) Bonding and (**b**) antibonding equal energy contours taken from [159] at optimal doping. The *solid lines* bounding the shaded regions are the bonding and antibonding Fermi surfaces, respectively. The *arrow* labeled **Q** corresponds to a (π, π, π) momentum transfer, which dynamically nests the bonding Fermi surface with an excited antibonding state shown by the *dotted curve*. A momentum transfer **Q'**=$(\pi, \pi, 0)$ now fails to provide nesting. This yields a smaller T_c for increasing $t_\perp$ [101]

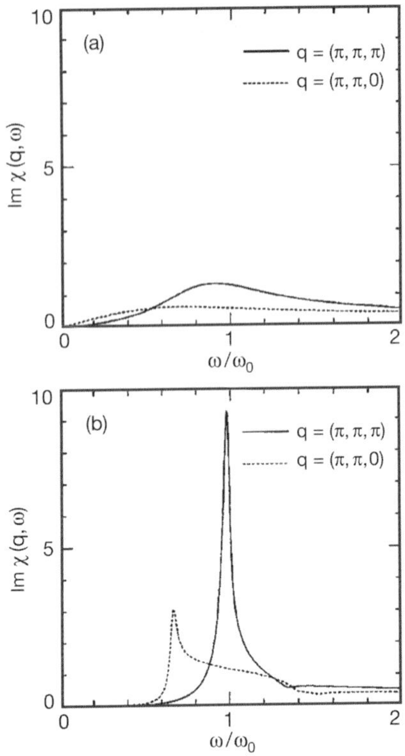

Fig. 23.61. Change of nesting: Results for spin susceptibility Im$\chi(\mathbf{Q}, \omega)$ versus ω for **Q'**=(π, π, π) (*solid curve*) and **Q**= $(\pi, \pi, 0)$ (*dotted curve*) for (**a**) the normal state at T=T_c and (**b**) the superconducting state at T=$T_c/2$ taken from [159]. The scales in (**a**) and (**b**) are the same for comparison

magnetic wave vector $\mathbf{Q}' = (\pi, \pi, \pi)$. This is shown in Fig. 23.61.

The change of nesting is also seen in the calculation of the spin susceptibility Im$\chi(\mathbf{q}, \omega)$. This is illustrated in Fig. 23.61(a). The peak in Imχ in the normal state is more pronounced at **Q'** rather than for **Q**, indicating that in the bilayer cuprates the antiferromagnetic fluctuations are essentially three-dimensional. Another interesting feature relates to the resonance peak formation below T_c. Similar to the normal state the "resonance" peak will be more pronounced for the wave vector **Q'** and stronger in its intensity (see Fig. 23.61(b)) as compared to the one-layer materials. This probably explains why in most cases the resonance peak is visible only in the bilayer cuprates.

What happens with further increase of the number of CuO_2-planes per unit cell? Due to crystallographic position of the inner CuO_2-plane, already for the three-layer materials its doping concentration is sufficiently lower than for the two outer ones, as illustrated in Fig. 23.62. A further increase of the number of CuO_2 layers per unit cell does not result in the enhancement of T_c for the same reason. Namely, the inner planes will be less and less metallic and already for five-layered compounds the inner CuO_2

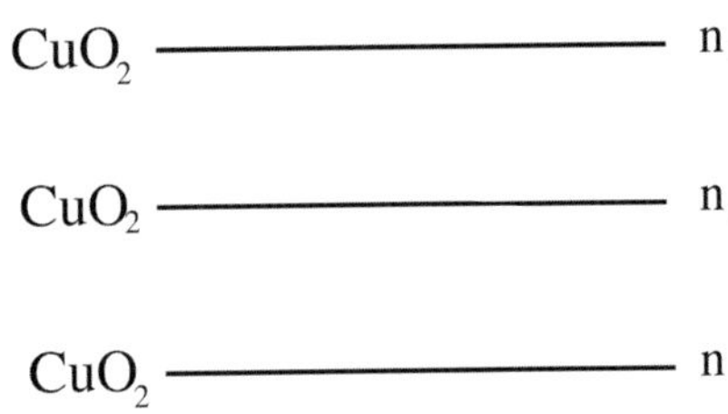

Fig. 23.62. Illustration of the change of the doping concentration of CuO_2-planes in trilayer cuprates. Due to its crystallographic position the inner CuO_2-plane has a smaller doping than the outer ones. Therefore, the inner CuO_2-plane will be underdoped. This may effectively reduce the T_c of the system if a further increase in the number of layers takes place. For example, it is known that already in five-layered cuprate superconductors the inner CuO_2-plane is undoped and remains insulating

is non-superconducting. This explains the reduction of T_c with an increasing number of CuO_2-planes per unit cell, n ($n > 3$).

A final remark we would like to make is on the value of calculated T_c in bilayered cuprates. In the spin fluctuation scenario and using the FLEX approximation, T_c decreases in the bilayer cuprates due to a smaller nesting and the corresponding spin fluctuations at **Q'**. However, this depends on the momentum dependence of the hopping term. Furthermore, the spin susceptibility χ_s in the odd channel becomes larger than its counterpart in the even channel (roughly factor 5 for realistic parameter values). This has consequences for the resulting magnitude of the gap function Δ_0, which also turns out to be larger in the odd channel. But, all in all, in bilayer cuprates the essential physics is unchanged. Therefore, the corresponding "fingerprints" of the spin fluctuation Cooper-pairing like the "resonance" peak and "dip" feature seen in INS, and the kink structure in the elementary excitations seen in ARPES, and SIS/NIS tunneling spectroscopy should be present. Further experimental and theoretical studies are expected to support this.

23.5 Results for Sr_2RuO_4

23.5.1 Electronic Structure

We begin the discussion of our results for Sr_2RuO_4 by analyzing their electronic structure. In contrast to the cuprates the electronic structure of the ruthenates is more complicated. For example, according to the LDA calculations there are three bands that cross the Fermi level [107]. Moreover, while one band is quasi-two-dimensional (xy), the other two bands (xz, yz) are quasi-one-dimensional. This is shown in Fig. 23.63 where we present our results for the energy dispersion of the Sr_2RuO_4 in the RuO_2-plane ($k_z = 0$) using the tight-binding parameters and spin–orbit coupling as described before.

The main effect of the spin–orbit coupling on the dispersion is the removal of the degeneracy between xz and yz-bands as present in the LDA calculation. Therefore, one may say that the spin–orbit coupling acts like a hybridization. On the other hand,

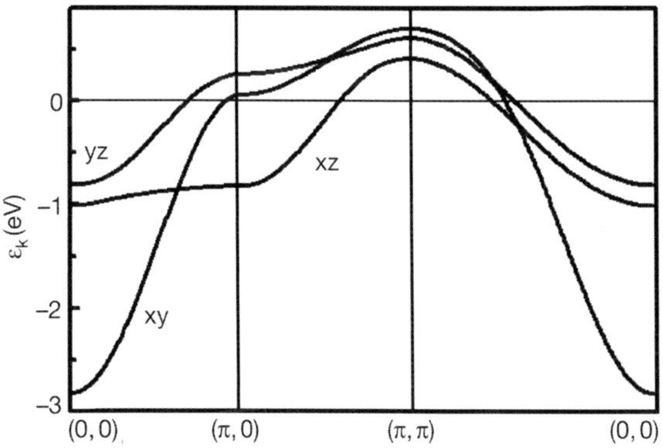

Fig. 23.63. Calculated energy dispersion of the xy, yz, and xz-bands in Sr_2RuO_4 along the route $(0,0) \rightarrow (\pi,0) \rightarrow (\pi,\pi) \rightarrow (0,0)$ in the first BZ. The tight-binding parameters are used. The spin–orbit coupling is taken into account. Note that the energy dispersion of the $xy(\gamma)$-band is two-dimensional, while $xz(\alpha)$ and $yz(\beta)$-bands are sufficiently one-dimensional as seen from their dispersion

the character of the xz and yz-bands still remains one-dimensional. This can also be recognized in Fig. 23.63. For example, one clearly sees that the xz and yz-bands are weakly momentum dependent along the $(0,0) \rightarrow (\pi,0)$ and $(\pi,0) \rightarrow (\pi,\pi)$-directions, respectively. At the same time the xy-band is two-dimensional. The reason for this is the strong overlap between the d_{xy} Ru orbital with the $2p_x$ and $2p_y$ oxygen orbitals. Since Sr_2RuO_4 shows a quasi-two-dimensional behavior, it was proposed originally [54] that only the xy-band plays an important role in determining the physical properties of Sr_2RuO_4 in the normal and superconducting states. Actually, this is further supported by the fact that the flat region of the xy-band leading to a logarithmic singularity in the density of states (the so-called Van-Hove singularity) lies very close to the Fermi level.

Yet the question about the role of the xz and yz-bands for determining the physical properties of Sr_2RuO_4 remains unclear. In Fig. 23.64 we show the results for the two-dimensional Fermi surface in ruthenates, which consists of two electron-like Fermi surfaces of yz nd xy-bands and a hole-like Fermi surface of the xz-band. The filling of each band is approximately 2/3. The important feature that we would like to stress is a strong nesting of the $yz(\beta)$-band

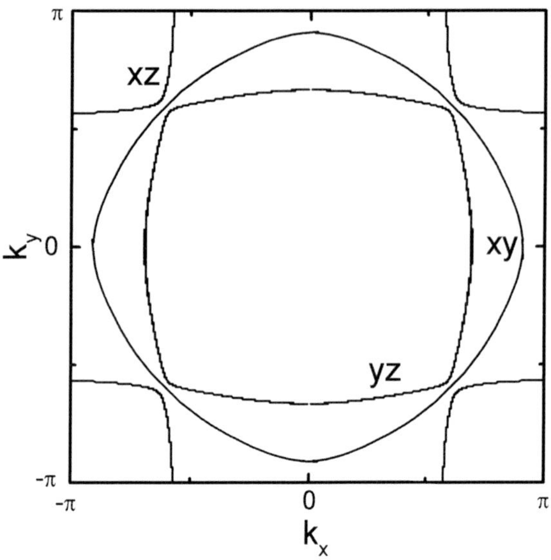

Fig. 23.64. Calculated two-dimensional Fermi surface of Sr_2RuO_4 by taking into account spin–orbit coupling. The Fermi surface consists of three sheets: two with electron-like topology and one with hole-like topology. Spin–orbit coupling acts like a hybridization and removes the degeneracy of the $\alpha(xz)$ and $\beta(yz)$-bands

at the incommensurate antiferromagnetic wave vector $\mathbf{Q}_i = (2\pi/3, 2\pi/3)$. This is extremely interesting, since it first indicates an importance of the band structure for determining the normal and superconducting properties of Sr_2RuO_4 and second it shows the necessity to include all three bands into the theoretical analysis. This is in contrast to the cuprates where the one-band approximation seems to be in order mostly.

23.5.2 Elementary Excitations: Effects of Spin Fluctuations

Unfortunately, the complete analysis of the elementary excitations including all three bands and spin–orbit coupling has not been performed so far in Sr_2RuO_4. However, there have been several attempts to analyze the behavior of the elementary excitations in the "most important" γ(xy)-band using a perturbation theory up to third order with respect to the on-site Coulomb interaction U. A three-band Hubbard model has been used [162,163].

In Fig. 23.65 we show the results for the real and imaginary parts of the self-energy versus frequency for different values of Coulomb repulsion (intraband U and interband U') and Hund's coupling (J and J') calculated at $\mathbf{k}_F$ for the γ-band. As one can see this band demonstrates the usual Fermi-liquid behavior for different values of the inter-orbit coupling, $\mathrm{Re}\Sigma(\omega) \sim \omega$ and $\mathrm{Im}\Sigma(\omega) \sim \omega^2$ at low energies. Therefore, it is likely that the nesting in α and β-bands does not strongly influence the low-energy excitations of the γ-band that arise due to a presence of the ferromagnetic-like fluctuations also originating from the γ-band.

On the other hand, at higher energies the self-energy is sensitive to the inter-orbital interaction, which can be further observed in the optical conductivity, for example. Another important question is the behavior of the self-energy in the α and β-bands. For example, one naively expects that due to a presence of the strong incommensurate antiferromagnetic fluctuations (IAF) at $(2\pi/3, 2\pi/3)$ and the quasi-one-dimensional character of these bands their self-energy may show a pronounced non-Fermi-liquid behavior like in cuprates. Note that experiments do not show any significant deviation from the Fermi-liquid behavior. Further studies are needed.

23.5.3 Dynamical Spin Susceptibility, NMR, and INS

Perhaps one of the most interesting property of the ruthenates is the dynamical spin susceptibility in the normal state. As we discussed in the Introduction, the understanding of the temperature dependence of $\chi(\mathbf{q}, \omega)$ is of key significance for determining the superconducting properties in Sr_2RuO_4. Let us study how the two-dimensional three-band Hubbard Hamiltonian together with the spin–orbit coupling can explain the existing experimental data.

In Fig. 23.66 we show the results for the Lindhard response functions calculated for the different bands using their tight-binding dispersions taken from LDA calculations [105]. Note that the spin–orbit coupling is *not* taken into account. As expected due to the pronounced nesting of the *xz* and *yz*-bands, their susceptibilities display peaks at $\mathbf{Q}_i = (2\pi/3, 2\pi/3)$,

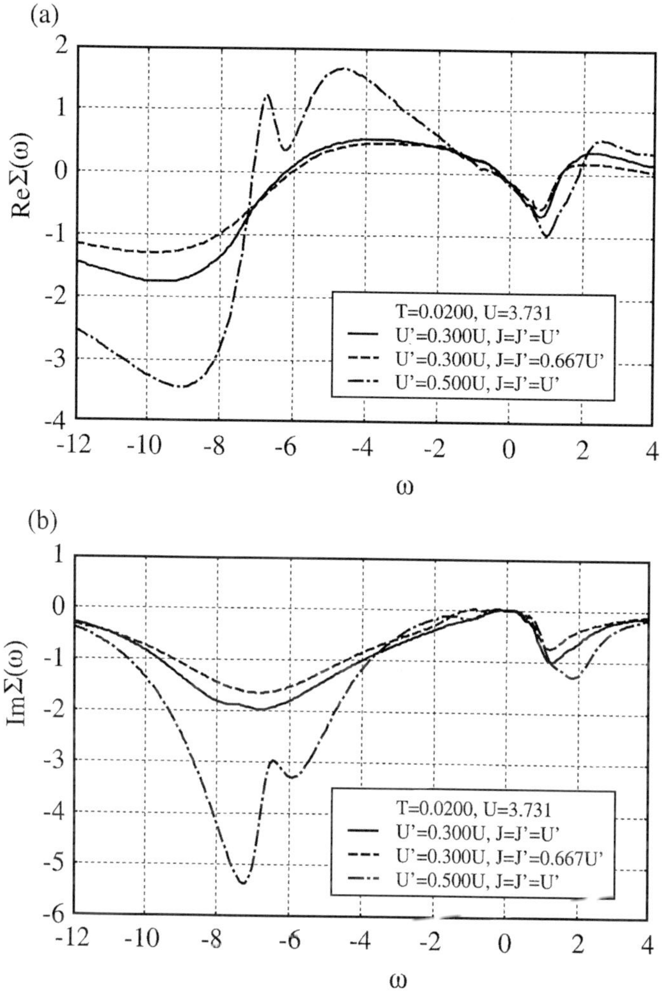

Fig. 23.65. The normal state (**a**) real and (**b**) imaginary part of the self-energy for the γ-band taken from [162]. U, U', and J, J' refer to the intraband and interband Coulomb and Hund's coupling, respectively. The self-energy $\Sigma(\omega)$ and ω are given in terms of the hopping integral, t

while the xy-band does not show any significant feature. The response of the xy-band is enhanced due to the presence of the van-Hove singularity close to the Fermi level. It becomes clear that the features observed by INS relate mainly to the magnetic response of the xz and yz-bands. However, the present results cannot account for the observed magnetic anisotropy, since both longitudinal and transverse components of the total spin susceptibility are the same if the term transferring the anisotropy from the orbital subspace into the spin subspace is neglected. In order to do this we include spin–orbit coupling.

Using the random phase approximation (RPA) for every particular band we calculate the longitudinal and transverse components of the total susceptibility in the RuO_2-plane using (23.114) and (23.115) in Sect. 23.3.4. The real part at $\omega = 0$ is shown in Fig. 23.67. As a result of the spin–orbit coupling the magnetic response becomes very anisotropic along the whole Brillouin zone. Since the spin and orbital degrees of freedom are now mixed, the orbital anisotropy will be reflected in the magnetic susceptibility. As one can see in Fig. 23.67 the longitudinal component has a pronounced peak at $\mathbf{Q}_i$, while the

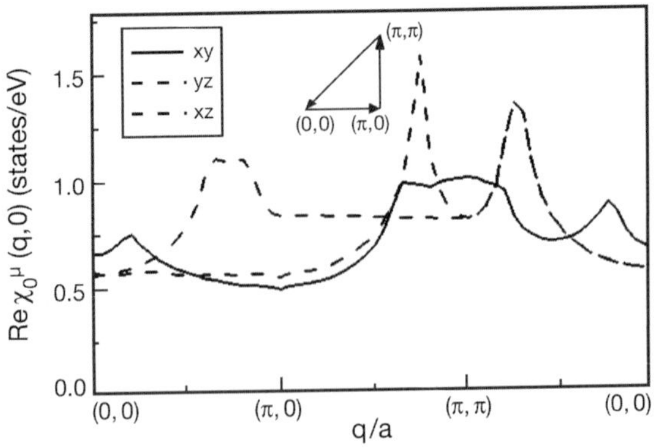

Fig. 23.66. Results without spin–orbit coupling: Calculated real part of the Lindhard spin susceptibility χ_0^μ (μ refers to the band indices) in the normal state of Sr_2RuO_4 along the route $(0,0) \to (\pi,0) \to (\pi,\pi) \to (0,0)$ in the first BZ for the three different orbitals (*xz*,*yz*, and *xy*) crossing the Fermi level. Due to the nesting of the *xz* and *yz*-bands their susceptibilities show an enhancement at the incommensurate antiferromagnetic wave vector $\mathbf{Q}_i = (2\pi/3, 2\pi/3)$. The response of the *xy*-band is more isotropic, but significantly larger than in the normal metal due to the vicinity of the Van-Hove singularity

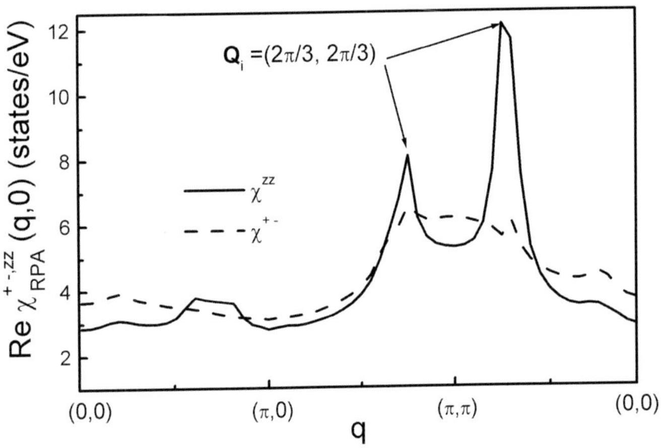

Fig. 23.67. Calculated real part of the longitudinal and transverse components of the total RPA spin susceptibility $\chi_{\mathrm{RPA}}^{+-,zz} = \sum_\mu \chi_{\mu,\,\mathrm{RPA}}^{+-,zz}$ (μ refers to the band indices) in the normal state of Sr_2RuO_4 along the route $(0,0) \to (\pi,0) \to (\pi,\pi) \to (0,0)$ in the first BZ. Due to spin–orbit interaction the transverse component (χ^{+-}) contains not incommensurate antiferromagnetic fluctuations (IAF) from *xz* and *yz*-orbitals, while in the longitudinal component (χ^{zz}) they are enhanced

transverse one does not show these features at all and is almost isotropic (it mostly reproduces the response of the *xy*-band). In order to understand why the longitudinal susceptibility mostly shows the nesting features of the *xz* and *yz*-bands one has to remember that due to spin–orbit coupling the orbital component of the magnetic susceptibility cannot be excluded. Therefore, matrix elements such as $\langle i|l_z|j\rangle$ and $\langle i|l_{+(-)}|j\rangle$ have to be taken into account. At the same time the *xz* and *yz*-bands consist of the $|2,+1\rangle$ and $|2,-1\rangle$ orbital states. One can see that while the longitudinal component gets an extra term due to l_z, the transverse component does not, since the transitions between states $|2,+1\rangle$ and $|2,-1\rangle$ are not allowed. Thus, the contribution of the nesting of the *xz* and *yz* orbitals to the longitudinal component of the susceptibility is larger than to the transverse one and incommensurate antiferromagnetic fluctuations are almost absent.

In order to see this in the inset of Fig. 23.68 we show the frequency dependence of the longitudinal and transverse components of Imχ calculated for the wave vector $\mathbf{Q}_i = (2\pi/3, 2\pi/3)$. Despite some structure seen in the real part of χ_{RPA}^{+-} at $(2\pi/3, 2\pi/3)$ (see Fig. 23.67) there are no real incommensurate antiferromagnetic fluctuations (IAF) at this wave vector. On the other hand, the structure in χ_{RPA}^{zz} at the same wave vector refers to the IAF fluctuations. The longitudinal component of Imχ reveals a peak at approximately $\omega_{sf} = 6$ meV in quantitative agreement with INS experimental data [52]. The transverse component is featureless showing the absence of incommensurate antiferromagnetic spin fluctuations. Thus, the fluctuations in the transverse susceptibil-

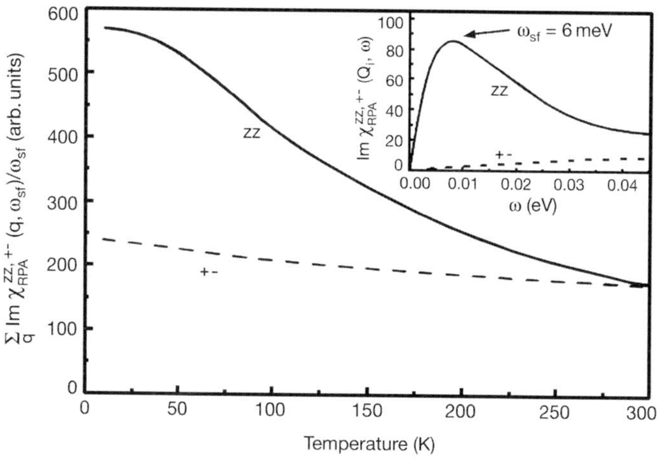

Fig. 23.68. Magnetic anisotropy: Temperature dependence of the imaginary part of the spin susceptibility divided by ω_{sf} and summed over **q**. *zz* and +- refer to the out-of-plane (*solid curve*) and in-plane (*dashed curve*) components of the RPA spin susceptibility. In the *inset* we show the corresponding frequency dependence of $\mathrm{Im}\chi_{\mathrm{RPA}}(\mathbf{Q}_i, \omega)$ at the IAF wave vector $\mathbf{Q}_i = (2\pi/3, 2\pi/3)$ and for $\omega_{sf} = 6$ meV

ity are isotropic and ferromagnetic-like. Therefore, antiferromagnetic fluctuations are present only perpendicular to the RuO_2-plane.

In order to demonstrate the temperature dependence of the magnetic anisotropy induced by the spin–orbit coupling in Fig. 23.68 we display the temperature dependence of the quantity

$$\sum_{\mathbf{q}} \frac{Im\chi_{\mathrm{RPA}}(\mathbf{q}, \omega_{sf})}{\omega_{sf}}$$

for both components. At room temperature both longitudinal and transverse susceptibilities are almost identical, since thermal effects wash out the influence of the spin–orbit interaction. With decreasing temperature the magnetic anisotropy increases and at low temperatures we find the important result that the out-of-plane component χ^{zz} is about twice as large as the in-plane one:

$$\chi^{zz} > \chi^{+-}/2 . \tag{23.142}$$

Of course, this relation is only valid close to the wavevector $\mathbf{q} = \frac{2}{3}(\pi, \pi)$. We also note that our results are in accordance with earlier estimations made by Ng and Sigrist [108]. However, there is one important difference. In their work it was found that the IAF are slightly enhanced in the longitudinal components of the *xz* and *yz*-bands in comparison to the transverse one. In our case we have found that the longitudinal component of the magnetic susceptibility is strongly enhanced due to orbital contributions. Moreover, by taking into account the correlation effects within RPA we show that the IAF are even further enhanced in the *z*-direction.

Finally, in order to compare our results with experimental data we calculate the nuclear spin-lattice relaxation rate T_1^{-1} for the ^{17}O ion in the RuO_2-plane for different external magnetic field orientation (i = *a*, *b*, and *c*):

$$\left[\frac{1}{T_1 T}\right]_i = \frac{2k_B\gamma_n^2}{(\gamma_e\hbar)^2} \sum_{\mathbf{q}} |A_{\mathbf{q}}^p|^2 \frac{\chi_p''(\mathbf{q}, \omega_{sf})}{\omega_{sf}} , \tag{23.143}$$

where $A_{\mathbf{q}}^p$ is the q-dependent hyperfine-coupling constant perpendicular to the i-direction.

In Fig. 23.69 we show the calculated temperature dependence of the spin-lattice relaxation for an external magnetic field within and perpendicular to the RuO_2-plane together with experimental data. At $T = 250$ K the spin-lattice relaxation rate is almost isotropic. Due to the anisotropy in the spin susceptibilities arising from spin-orbit coupling the relaxation rates become increasingly different for decreasing temperature. The largest anisotropy occurs close to the superconducting transition temperature, in good agreement with experimental data [110].

It is still important to see whether our theory yields the weak ferromagnetic fluctuations observed by means of the NMR Knight shift [164]. The latter

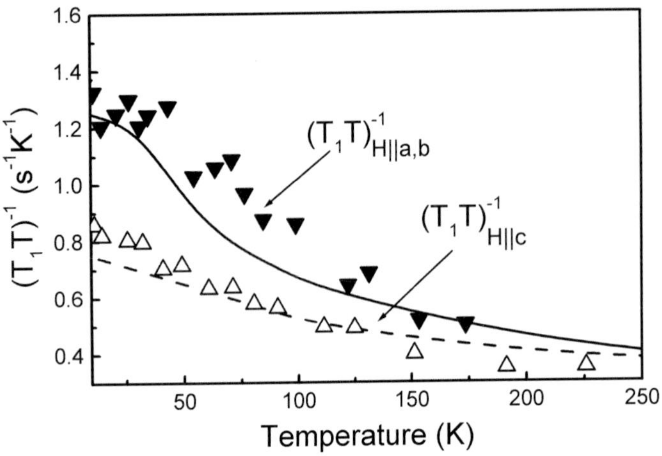

Fig. 23.69. Calculated normal state temperature dependence of the nuclear spin-lattice relaxation rate T_1^{-1} of ^{17}O in the RuO_2-plane for the external magnetic field applied along the *c*-axis (*dashed curve*) and along the *ab*-plane (*solid curve*). *Down and up-triangles* are experimental results taken from [110]

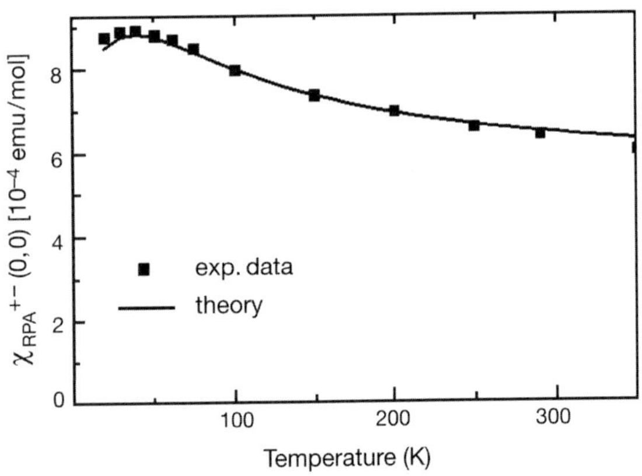

Fig. 23.70. The temperature dependence of the uniform spin susceptibility $\chi^{+-}_{RPA}(0, 0)$ calculated using our RPA theory in comparison with the ^{17}O Knight shift measurements in the RuO_2-plane. The peak is due to thermal activation involving *xy* and *yz*, *xz*-bands

probes the magnetic fluctuations around $\mathbf{q} = 0$. In Fig. 23.70 we show the results for the temperature dependence of the uniform susceptibility $\chi^{+-}_{RPA}(0, 0)$. One can see that a weak temperature dependence in fair agreement with experimental data [164] occurs and arises due to ferromagnetic-like fluctuations mainly of the *xy*-band. Note that very recently ferromagnetic-like excitations were also observed by means of INS experiments [165]. This also agrees with our results.

23.5.4 Superconducting Order Parameter

Triplet superconductivity in Sr_2RuO_4 is of particular interest. It is important to understand in detail how the triplet Cooper-pairing is realized in this system with ferromagnetic and incommensurate antiferromagnetic spin fluctuations at $\mathbf{Q}_i = (2\pi/3, 2\pi/3)$. We note that the total spin susceptibility $\chi(\mathbf{q}, \omega)$ controls the symmetry of the superconducting order parameter. Therefore, the spin–orbit coupling plays an important role in the superconducting state and in determining the symmetry of the superconducting state. First, we present the analysis of the superconducting gap without taking the spin–orbit coupling into account. We assume that exchange of spin fluctuations is responsible for Cooper-pairing. Why is triplet pairing realized despite the strong IAF at $\mathbf{Q}_i \approx (2\pi/3, 2\pi/3)$?

For the analysis of superconductivity in Sr_2RuO_4 we take into account that experiment observes non-*s*-wave symmetry of the order parameter, which strongly suggests spin-fluctuation-mediated Cooper pairing. Then assuming spin-fluctuation-induced pairing it is possible to analyze the symmetry of the superconducting state from the gap equation and our calculated results for $\chi(\mathbf{q}, \omega)$ with pronounced wave vectors at $\mathbf{Q}_i$ and $\mathbf{q}_i$. The gap equation is for the band *i* given by:

$$\Delta^i_{\mathbf{k}} = -\sum_{\mathbf{k}',j} [V^{\text{eff}}_\sigma(\mathbf{k},\mathbf{k}')]^{ij} \frac{\Delta^j_{\mathbf{k}'}}{2E^j_{\mathbf{k}'}} \tanh\left(\frac{E^j_{\mathbf{k}'}}{2k_BT}\right), \quad (23.144)$$

where $E^i_{\mathbf{k}} = \sqrt{\epsilon^{i\,2}_{\mathbf{k}} + [\Delta^i_{\mathbf{k}}]^2}$ is the energy dispersion of the band i. Note that the pairing potential $V^{\text{eff}}_\sigma(\mathbf{k},\mathbf{k}')$ is different for singlet ($\sigma = s$) and triplet ($\sigma = tr1$) Cooper-pairing. In order to keep the discussion of the triplet state relatively simple we restrict ourself to the case of equal spin pairing (i.e. V^{eff}_{tr1} defined in (23.104)). The eigenvalue analysis of (23.144) will yield the symmetry with the lowest energy.

For the determination of the pairing potential we use the calculated FS, the corresponding density of states, and the spin susceptibility for Sr_2RuO_4 and also follow the analysis by Anderson and Brinkmann for ^{3}He [114] and by Scalapino for the cuprates [166]. We compare the results obtained by assuming singlet or triplet pairing.

For the equal-spin triplet pairing the effective pairing interaction is given by

$$\left[V^{\text{eff}}_{\text{tr1}}(\mathbf{k},\mathbf{k}')\right]^{ij} = -\frac{U^2}{2}\left(\frac{\chi^{ii}_0(\mathbf{k}-\mathbf{k}',0)}{1-U\chi^{ii}_0(\mathbf{k}-\mathbf{k}',0)} + \frac{\chi^{ii}_0(\mathbf{k}-\mathbf{k}',0)}{1+U\chi^{ii}_0(\mathbf{k}-\mathbf{k}',0)}\right) - 2U^2\chi^{ij}_0(\mathbf{k}-\mathbf{k}',0)\,, \quad (23.145)$$

where the last term occurs only for α and β-bands. Note that the first two terms refer to the intraband spin and charge fluctuations, respectively.

For singlet pairing the effective pairing interaction has the form

$$\left[V^{\text{eff}}_s(\mathbf{k},\mathbf{k}')\right]^{ij} = \frac{U^2\chi^{ii}_0(\mathbf{k}-\mathbf{k}',0)}{1-U\chi^{ii}_0(\mathbf{k}-\mathbf{k}',0)} + \frac{U^3\left[\chi^{ii}_0(\mathbf{k}-\mathbf{k}',0)\right]^2}{1-U^2\left[\chi^{ii}_0(\mathbf{k}-\mathbf{k}',0)\right]^2} - U^2\chi^{ii}(\mathbf{k}-\mathbf{k}',0)\,. \quad (23.146)$$

Again, the first two terms refer to the spin and charge fluctuations, respectively, and the last term avoids double counting. Note that the cross-terms including χ^{ij} are small due to the Pauli principle and thus we neglect them for singlet pairing. Note, the second term in V_0 comes from ladder diagrams which due to the Pauli principle do not contribute to the triplet pairing.

Using appropriate symmetry representations [167] we discuss the solutions of (23.144) assuming p, d, or f-wave symmetry of the order parameter in the RuO_2-plane:

$$\Delta_p(\mathbf{k}) = \Delta_0\hat{\mathbf{z}}(\sin k_xa + i\sin k_ya), \quad (23.147)$$

$$\Delta_d(\mathbf{k}) = \Delta_0(\cos k_xa - \cos k_ya), \quad (23.148)$$

$$\Delta_{f_{x^2-y^2}}(\mathbf{k}) = \Delta_0\hat{\mathbf{z}}(\cos k_xa - \cos k_ya) \times (\sin k_xa + i\sin k_ya)\,. \quad (23.149)$$

Here the $\hat{\mathbf{z}}$-unit vector refers to the d_z component of the Cooper-pairs as observed in the experiments [168]. We present the possible explanation for this later. These symmetries of the superconducting order parameter must be substituted into (23.144). For simplicity we first consider the solution of (23.144) in the RuO_2-planes and then discuss what is happening along the c-axis. Note that the largest eigenvalue of (23.144) will yield the symmetry of the superconducting order parameter Δ_l in Sr_2RuO_4.

Solving (23.144) in the first BZ down to 5 K we found that p-wave symmetry yields the largest energy gain for the xy-band, while for the xz and yz-bands the situation is more complicated.

One can see that the gap equations for the xy and yz, xz-bands can be separated due to their weak interaction. For the xy band the p-wave is the most stable solution, while for the xz and yz-bands $f_{x^2-y^2}$-wave symmetry yields the largest eigenvalue due to stronger nesting.

Figure 23.71 characterizes the solutions of (23.144). Note that a good approximation is to linearize (23.144) with respect to Δ^i_l, i.e. $E^i_{\mathbf{k}'} \to \epsilon^i_{\mathbf{k}'}$. We also use $\tanh(\epsilon^i_{\mathbf{k}'}/2k_BT) = 1$. Thus, the main contribution to the pairing comes from the Fermi levels. We present our results for the Fermi surface in the RuO_2 plane. The wave vectors $\mathbf{Q}_i$ and $\mathbf{q}_i$ refer to the peaks in $\chi(\mathbf{q},\omega)$. The areas with $\Delta_{f_{x^2-y^2}} > 0$ and $\Delta_{f_{x^2-y^2}} < 0$ are denoted by (+) and (-), respectively. Note that the minus sign in (23.144) is canceled for triplet pairing. The summation over $\mathbf{k}'$ in the first BZ is dominated by the contributions due to $\mathbf{Q}_i$ for the α and β-bands and the one due to $\mathbf{q}_i$ for the γ-band.

As can be seen from Fig. 23.71 in the case of $f_{x^2-y^2}$-wave symmetry for the xy-band the wave vector $\mathbf{q}_i$ bridges the same number of portions of the FS with

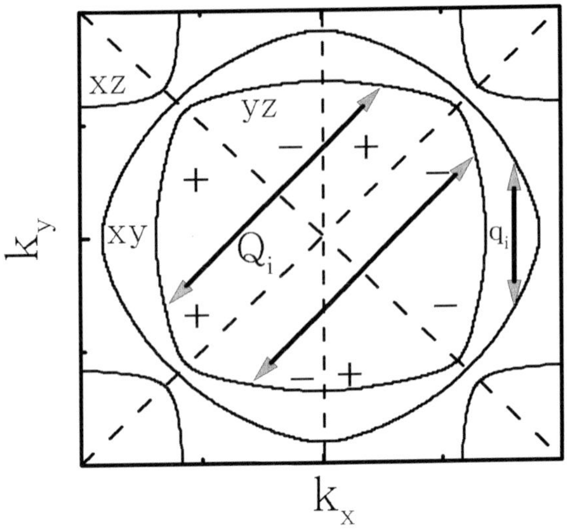

Fig. 23.71. Symmetry analysis of the order parameter for triplet pairing with $| m |= 1$ in the first BZ for $k_z = 0$. The wave vectors $\mathbf{Q}_i$ and $\mathbf{q}_i$ are the pronounced wave vectors resulting from the susceptibility (without spin–orbit coupling). For $f_{x^2-y^2}$-wave symmetry the nodes of the real part of the order parameter are shown (*dashed lines*) and the regions +(−) where the $f_{x^2-y^2}$-wave superconducting gap is positive (negative). Note that for the real part of a p-wave order parameter the node would occur only along $k_x = 0$

opposite and equal sign. Therefore, for the xy-band we get no solution with $f_{x^2-y^2}$-wave symmetry. On the other hand, for the xz and yz-bands the wave vector $\mathbf{Q}_i$ bridges portions of the FS with *equal* signs of the $f_{x^2-y^2}$-superconducting order parameter. The eigenvalue of this order parameter is also enhanced due the interband nesting for xz and yz-bands. Thus, superconductivity in the xz and yz-bands in the RuO_2 plane would be possible with the $f_{x^2-y^2}$-wave order parameter. Therefore, solving (23.144) for the three-band picture we found a competition between triplet p-wave and $f_{x^2-y^2}$-wave superconductivity in the RuO_2-plane.

Also using (23.147) we can rule out singlet pairing. In particular, assuming $d_{x^2-y^2}$-symmetry for Sr_2RuO_4 (23.144) yields an eigenvalue that is lower than in the case of triplet pairing. As can be seen using Fig. 23.71 this is plausible. Note that for $d_{x^2-y^2}$-wave symmetry we get a change of sign of the order parameter upon crossing the diagonals of the BZ. According to (23.144) wave vectors around $\mathbf{Q}_i$ connecting areas (+) and (−) contribute constructively to the pairing. Contributions due to $\mathbf{q}_i$ and the background connecting areas with the same sign would subtract from the pairing (see Fig. 23.71 with nodes at the diagonals). Therefore, we get that the four contributions due to $\mathbf{q}_i$ in the xy-band do not allow to have $d_{x^2-y^2}$-wave symmetry in the xy-band. Despite the pair-building contribution due to $\mathbf{Q}_i$ one gets that the eigenvalue of the $d_{x^2-y^2}$-wave symmetry in the xy-band is smaller than for the $f_{x^2-y^2}$-wave symmetry. This is due to the large contribution from $\mathbf{Q}_i$ to the cross-terms for the triplet pairing, which are absent for the singlet pairing. For d_{xy}-symmetry where the nodes are along the $(\pi,0)$ and $(0,\pi)$-directions we argue similarly. Thus, we may exclude this symmetry. We safely conclude that as a result of the topology of the FS and the spin susceptibility for p and f-wave symmetry we seem to get the strongest pairing. We can definitely exclude d-wave pairing in the RuO_2-plane.

Since we have found a strong competition between p-wave and $f_{x^2-y^2}$-wave superconductivity in the RuO_2-plane the question arises as to what is the role of the c-axis in determining the superconducting properties in Sr_2RuO_4. Does this question require an understanding of the role of spin–orbit coupling for the determining the symmetry of the superconducting order parameter?

First, the spin–orbit coupling affects the spin dynamics and as we have shown induces the anisotropy in the spin subspace. In particular, the two-dimensional IAF at $\mathbf{Q}_i = (2\pi/3, 2\pi/3)$ align along the c-direction. This simply means that the antiferromagnetic moments associated with these fluctuations align along the z-direction. At the same time the ferromagnetic fluctuations are in the ab-plane. Since the interaction in the RuO_2-plane (k_z=0) is mainly ferromagnetic nodeless p-wave pairing is indeed possible and the superconducting order parameter in all three bands will not have nodes and not only for xy-band as before. Note also that the γ-band reveals a high density of states. Therefore, the triplet p-wave pairing *without* line-nodes will be realized in the RuO_2-plane.

On the other hand, the state between the RuO_2-plane will be determined mainly by the IAF that are polarized along the c-direction. This implies that the magnetic interaction between the planes has to be rather antiferromagnetic than ferromagnetic. This antiferromagnetic interaction suggests a line of nodes for $\Delta(\mathbf{k})$ between the neighboring RuO_2-planes. In a phenomenological approach this was shown by two groups [169–171]. For example, it was proposed that the symmetry of the superconducting order parameter in Sr_2RuO_4 will have the form

$$\Delta_{\mathrm{k}} = \Delta_0 \tilde{\mathbf{z}} \left(\sin \frac{k_x a}{2} \cos \frac{k_y a}{2} + i \sin \frac{k_y a}{2} \cos \frac{k_x a}{2} \right) \times \cos \frac{k_z c}{2} \,. \tag{23.150}$$

Then, due to the magnetic anisotropy induced by spin–orbit coupling a nodeless p-wave pairing is possible in the RuO_2-plane, as observed experimentally [168]. Thus, a node would lie between the RuO_2-planes. This is illustrated in Fig. 23.72. Indeed as observed by thermal conductivity measurements [168, 169], the nodes in the superconducting order parameter are between the RuO_2-planes.

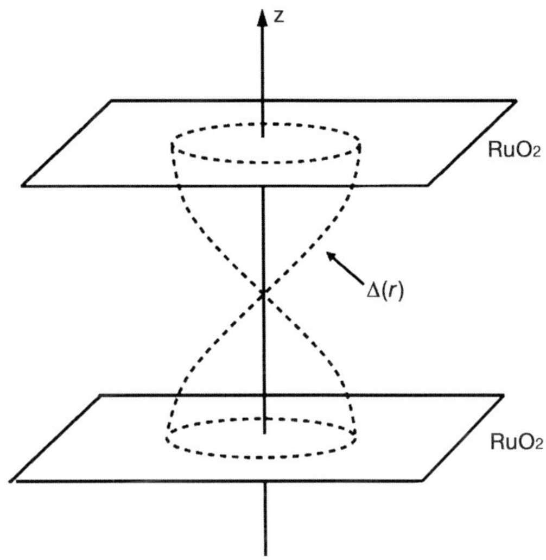

Fig. 23.72. Schematic representation of the possible node formation in the order parameter between the RuO_2-planes (in real coordinate representation) as resulting from the magnetic anisotropy. Here, the amplitude of the order parameter along the z-direction has been drawn in cylindrical coordinates between RuO_2-planes. This seems in agreement with thermal conductivity measurements below T_c [168, 169]

Note that the spin–orbit coupling also determines the orientation of the orbital moment of the Cooper-pair. In particular, it is convenient to combine the three spin component into a vector $\mathbf{d}(\mathbf{k})$ in spin space. As mentioned earlier this is achieved by the expanding Δ_{k} in the three symmetric 2×2 matrices $i\sigma_\mu \sigma_2$, $\mu = 1, 2, 3$, where σ_μ are the Pauli matrices as

$$\Delta_{\mathrm{k}} = \sum_\mu d_\mu(\mathbf{k})(\sigma_\mu i \sigma_2) = \begin{pmatrix} -d_x + i d_y & d_z \\ d_z & d_x + i d_y \end{pmatrix} . \tag{23.151}$$

Then, in a triplet superconductor, the pairing state can be represented by the three-dimensional vector $\mathbf{d}(\mathbf{k})$, whose magnitude and direction may vary over the Fermi surface in $\mathbf{k}$-space. Like in ^3He several pairing states related to the different orientation of the total spin moment of the Cooper-pair d_z, d_y, d_z may have the same condensation energy under weak-coupling conditions. This degeneracy is related to the spin rotation symmetry being present in ^3He (which leads to the presence of a so-called B-phase), but this is lifted by the spin–orbit coupling in Sr_2RuO_4 [108]. Thus, the analogy of the A-phase in ^3He occurs in Sr_2RuO_4 ($(\mathbf{d}(\mathbf{k})_A = (0,0,d_z(k_x \pm ik_y)))$, while the B-phase might be suppressed ($\mathbf{d}(\mathbf{k})_B = (d_x k_x, d_y k_y, 0)$). This B-phase corresponds to $S_z = 0$. As one can see from (23.106), we get that the transverse susceptibility and thus V_{sp}^{+-} is larger than V_{sp}^{zz} for $\mathbf{q} \simeq 0$ (see Fig. 23.67).[14] This tends to suppress the B-phase.

The different orientation of the spin and orbital angular momenta of the Cooper-pair is illustrated in Fig. 23.73. The important consequence of this order parameter is the uncompensated orbital angular momentum of the Cooper-pair (the so-called "chiral" state). Why is the chiral state $\Delta(\mathbf{k}) = d_z(\sin k_x \pm i \sin k_y)$ realized? To understand this, one has to consider the additional contribution of the spin–orbit coupling $\lambda(\mathbf{l}_{\mathrm{CP}} \cdot \mathbf{s})$ to the free energy in the supercon-

[14] We note that for the sake of simplicity the momentum dependence of the gap function was illustrated in the previous sections by the use of (23.104), which corresponds to $| m | = 1$ rather than (23.106), which requires a more sophisticated analysis.

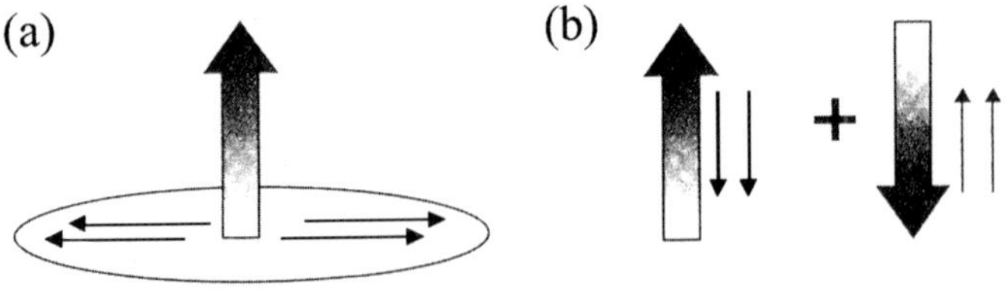

Fig. 23.73. Illustration of the possible superconducting states in Sr_2RuO_4 taken from [172]. Spin and total angular momentum for (**a**) $\Delta(\mathbf{k}) = d_z(\sin k_x + i \sin k_y)$ and (**b**) $\Delta(\mathbf{k}) = d_x \sin k_x + d_y \sin k_y$. The $\Delta(\mathbf{k}) = d_z(\sin k_x + i \sin k_y)$ superconducting state has an angular momentum along the c-axis (*thick arrow*) and the spin lying in the RuO_2-plane. Note that the chiral state displayed in (**a**) is obtained from our eigenvalue analysis

ducting state. Here, $\mathbf{l}_{CP}$ refers to the orbital moment of a Cooper-pair and $\mathbf{s}$ denotes the electron spin. In a simple mean-field analysis the components of $\mathbf{s}$ are replaced by $\langle s_i \rangle \sim \chi_{ij} \cdot H_j$ where χ_{ij} refers to the transversal (χ_{+-}) and longitudinal (χ_{zz}) component of the magnetic susceptibility. Obviously, due to $\chi_{zz} > \chi_{+-}$ (see Figs. 23.67 and 23.68) the degeneracy of the states $\Delta(\mathbf{k})$ is lifted, since one gains more energy if the orbital moment of a Cooper-pairs is aligned along the z-direction. Thus, the chiral state has the lowest energy [108,171].

In analogy to ^{3}He, interesting effects can be observed by applying an external magnetic field or external pressure along ab-plane to Sr_2RuO_4. For example, in ^{3}He under pressure the transition from B to A phase has been observed. In Sr_2RuO_4 the situation is the opposite. Superconductivity in Sr_2RuO_4 is already in the A-like phase and therefore, by applying pressure or magnetic field one could cause the transition into B-like phase. However, if the γ-band becomes more nested due to external pressure, an f-wave order parameter is possible. Another interesting effect concerns the possible change of the orientation of the unit vector $d_z = \hat{\mathbf{z}}$ of the superconducting order parameter induced by the external magnetic field discussed in Sect. 23.3.5. This means that for a relatively large magnetic field applied along the c-direction the significant d_x and d_y components in the superconducting order parameter in Sr_2RuO_4 can be induced. Of course, this effect is relatively small (as also in ^{3}He) but can be observed.

In summary, Sr_2RuO_4 is a remarkable example of a system where the interplay of ferromagnetic and incommensurate antiferromagnetic fluctuations may result in a non-trivial symmetry of the superconducting order parameter. Note that the nesting of α and β-bands may, as we have shown, lead to a node formation in the superconducting gap. Although its existence has not yet been definitely concluded experimentally it can be verified by applying a magnetic field or pressure. Since the antiferromagnetic fluctuations are not helpful for the triplet pairing, one may expect an enhancement of the superconducting transition temperature and the nodeless superconducting gap if the nesting is suppressed. Interestingly, this phenomenon was proposed to be responsible for the occurrence of superconductivity in the ferromagnetic phase [173] of unconventional superconductors like $ZrZn_2$ and UGe_2. In particular, we have shown that spin–orbit coupling plays an essential role in stabilization of the A-like phase in Sr_2RuO_4 and might be responsible for the possible node formation between neighboring RuO_2-planes.

23.5.5 Doping Dependence of Ruthenates

Comparing ruthenates and cuprates one can clearly see that their behavior differs drastically in the normal state. The ruthenates are good Fermi-liquid systems despite of the presence of IAF at $\mathbf{Q}_i = (2\pi/3, 2\pi/3)$. On the other hand, cuprates can be characterized as systems that may behave mainly non-Fermi-liquid-like. The most remarkable is the doping concentration dependence of the cuprates properties. The electronic correlations are more important in cuprates. The on-site Coulomb repulsion U (which can be roughly estimated as e^2/r where r refers to the radius of the corresponding electronic shell) is much larger than the bandwidth. In the

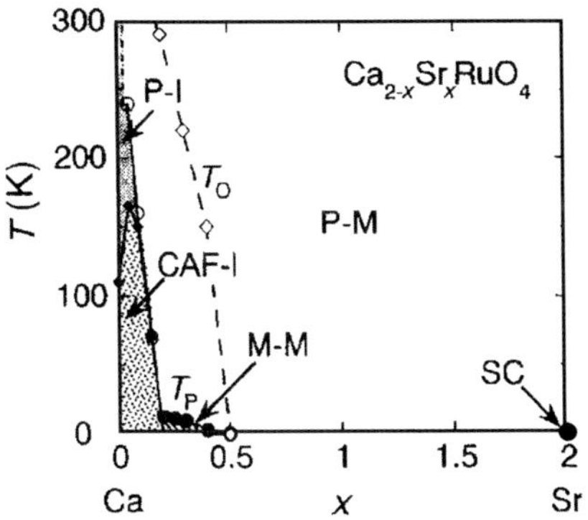

Fig. 23.74. Phase diagram of $Ca_{2-x}Sr_xRuO_4$. P refers to paramagnetic, CAF to canted antiferromagnetic, M to magnetic, SC to the superconducting phase, –M to the metallic phase, and –I to the insulating phase

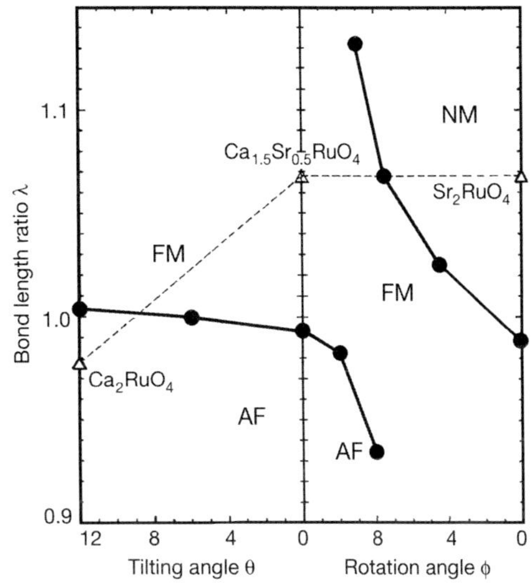

Fig. 23.75. The calculated magnetic phase diagram of Sr_2RuO_4 including structural distortions, taken from [175]. The *solid lines* are calculated phase boundaries, while the *triangles linked by dashed lines* correspond to the experimental data

ruthenates that are $4d$-electron systems the effect of correlations is weaker than in cuprates and U is of order of the electronic bandwidth, W. Therefore, one concludes that in ruthenates we are in the intermediate correlation regime. As a result, in the ruthenates the correlations are not strong enough to violate Fermi-liquid theory. Nevertheless, correlations in the ruthenates also occur. This is obvious from Fig. 23.74.

In Fig. 23.74 we show results for the magnetic phase diagram obtained experimentally by Nakatsuji and Maeno [174] for $Ca_{2-x}Sr_xRuO_4$. Despite the fact that introducing Ca does not change the valence in the system the magnetic phase diagram changes completely. First, superconductivity disappears and systems remain paramagnetic metals up to $x = 0.5$. Then, the system becomes a ferromagnetic metal. Most remarkably, upon further doping an insulating antiferromagnetic transition occurs.

Note that introducing a Ca instead of an Sr ion allows to have structural distortion of the pure tetragonal crystal lattice of Sr_2RuO_4 due to the smaller radius of the Ca ionic sphere. This was discussed by Fang and Terakura [175] by means of LDA calculations. Their main results are shown in Fig. 23.75. First, between $0 < x < 0.5$ the RuO_6 octahedra gets rotated around the c-axis. This basically changes the hopping term of the xy-band and most importantly brings the van Hove singularity closer to the Fermi level enhancing the ferromagnetic fluctuations and leading to a ferromagnetic state at $x = 0.5$. Further doping with Ca allows one to have not only the rotation of the RuO_6 octahedra, but also its tilting. Thus, hopping in the xz and yz-bands reduces. As a consequence, the Coulomb repulsion U will be larger than the bandwidth of the conduction bands. The resulting increase of correlations cause a Mott insulator transition driven by the IAF fluctuations. Further studies of the resistivity in Ca doped samples will justify the proposed scenario.

Summary

To summarize in this section we have studied triplet superconductivity. Sr_2RuO_4 is a solid state system for which unconventional triplet superconductivity may occur. We have shown that spin fluctuations may explain qualitatively the occurrence of p-wave superconductivity with line nodes between the RuO_2-planes. Furthermore, due to the presence of ferromagnetic and incommensurate antiferromagnetic fluctuations Sr_2RuO_4 seems to be an ideal system where the role and competitions of various spin fluctuations for the formation of superconductivity can be studied in detail. In particular, we have seen that incommensurate antiferromagnetic spin fluctuations are not helpful for triplet pairing and may result in the formation of line nodes between neighboring RuO_2-planes. The relative simplicity of this compound in comparison to heavy-fermion superconductors could make Sr_2RuO_4 the model system where further interesting and exciting physics (in particular regarding thermodynamical properties, effects due to magnetic and non-magnetic impurities, etc.) will be discovered.

23.6 Summary and Outlook

In this review we present a microscopic theory using a model Hamiltonian for the interaction between quasiparticles and spin fluctuations. This is applied to the layered cuprates and Sr_2RuO_4. For simplicity we assume that the electronic dynamics is essentially controlled by the CuO_2 and RuO_2-planes, respectively.

For the description of singlet superconductivity in the cuprates we employ a one-band model Hamiltonian. The interaction between the carriers (quasiparticles) is described by an effective Coulomb coupling U and the itinerancy of the carriers by a hopping integral t. In the case of hole doping the states of the effective band are mainly the hybridized p-states originating from the O-atoms. The spin fluctuations arise from the magnetic activity of the Cu-spins and the induced spin polarization of the p-states. In the case of electron doping the effective one-band Hamiltonian describes the hybridized d-states of the UHB and U is an effective coupling between the electrons in these states. The spin fluctuations result from the Cu-spins. Upon doping the Cu-spins are quenched and the spin polarization of the band is reduced. We assume itinerant magnetism for the carriers in the UHB.

Using this model we have found that the phase diagram, the symmetry of the superconducting order parameter, and the behavior of the elementary and spin excitations in hole-doped and electron-doped cuprates can be largely understood. We have found a particularly interesting behavior for the underdoped cuprates. (Note, however, that for smaller doping our theory needs to be extended.) Due to strong scattering of quasiparticles by antiferromagnetic spin fluctuations a "weak" pseudogap appears at the Fermi level in the density of states. It increases with decreasing doping concentration. The spectral density has different weights for holes and electrons. Cooper-pair phase fluctuations in underdoped cuprates are enhanced by the reduced dimensionality and also since phase-incoherent Cooper-pairs do not overlap well, resulting in a small n_s. These fluctuations of phase-incoherent Cooper-pairs destroy the long-range superconducting order and the Meissner effect. In conventional bulk superconductors this is not relevant, since the large superfluid density leads to a typical energy scale of phase fluctuations much larger than the superconducting energy gap Δ that governs the thermal breaking of Cooper-pairs. Thus, in conventional superconductors the superconducting transition is due to the destruction of the Cooper-pairs and T_c is proportional to Δ_0. We have found $T_c \propto n_s$ in underdoped hole-doped cuprates indicating that the phase fluctuations drive the transition. The Cooper-pairs only break up at a crossover temperature around T_c^*, $T_c^* > T_c$. T_c^* is approximately given by the transition temperature one would obtain without phase fluctuations. Between T_c and T_c^* local Cooper-pairs without long-range phase coherence may be present. Therefore, we conclude that we may have a qualitative understanding of the phase diagram in the cuprates. Furthermore, the important elementary excitations and their interdependence with spin excitations (resonance peak) can be well understood within our theory. Our results also shed light on the asymmetry of hole-doped and electron-doped cuprates.

For describing triplet Cooper-pairing in Sr_2RuO_4 we also employ a Hubbard Hamiltonian. In order to derive an electronic theory we take the electronic band structure of Sr_2RuO_4 into account. Magnetic activity arises from the itinerant electrons in the Ru d-orbitals. Due to spin–orbit coupling a strong magnetic anisotropy occurs ($\chi^{+-} < \chi^{zz}$) in the normal state. This mainly results from different values of the g-factor for the transverse and longitudinal components of the spin susceptibility (i.e. the matrix elements differ) and from a change of the bare energy dispersion of the d-electrons. Obviously, this and the presence of incommensurate antiferromagnetic and ferromagnetic fluctuations have consequences for the symmetry of the superconducting order parameter. In particular, using spin-fluctuation-induced Cooper-pairing one can understand how p-wave Cooper-pairing with line nodes between neighboring RuO_2-planes may occur.

Despite the relative success of the Eliashberg-like treatment of superconductivity via spin fluctuations exchange in cuprates and ruthenates there remain critical questions. In particular, in the theory of cuprates it is necessary to clarify the role of the vertex corrections and to which extent an effective second-order perturbation theory can be applied. It might be useful to take the original Hamiltonian referring to Cu-spins and polarized p-states at oxygen sites and derive the magnetic activity (spin-susceptibility, stability of local Cu-spins, etc.) and the itinerancy of the carriers. Also canonical transformation theory and slave-boson type theory should be used as alternative treatments. In the underdoped cuprates the origin of the pseudogap and the coexistence of short-range weak antiferromagnetism and superconductivity has to be understood. Also the symmetry of the superconducting order parameter in overdoped electron-doped cuprates must be clarified. In the ruthenates the thermodynamical properties remain to be studied in detail. It is of interest to apply our theory also to superconductivity in heavy-fermion systems exhibiting spin fluctuations as well.

Appendix A: Derivation of the Generalized Eliashberg Equations for the Interaction of Quasiparticles and Spin Fluctuations

Here, we present the derivation of the generalized Eliashberg equations for the interaction of quasiparticles and spin fluctuations in a detailed form, however, without discussing the relevancy of the approximations we used. This is discussed in the main text [176–179].

We start with Green's function for itinerant electrons (holes). Using the Nambu notation for the superconducting state [69], the one-particle Green's function is a 2×2-matrix

$$\hat{G}(\mathbf{k}, i\omega_n) = \int_0^\beta \mathrm{d}\tau \, e^{i\omega_n\tau} \hat{G}(\mathbf{k}, \tau) , \qquad (23.152)$$

where

$$\begin{aligned} \hat{G}(\mathbf{k}, \tau) &= -\langle T_\tau \, \psi_\mathbf{k}(\tau) \psi_\mathbf{k}^\dagger(0) \rangle , \\ \psi_\mathbf{k}(\tau) &= \begin{pmatrix} c_{\mathbf{k}\uparrow}(\tau) \\ c_{-\mathbf{k}\downarrow}^\dagger(\tau) \end{pmatrix} . \end{aligned} \qquad (23.153)$$

We employ the Heisenberg picture:

$$c_{\mathbf{k}\sigma}(\tau) \equiv e^{\tau H} \, c_{\mathbf{k}\sigma} \, e^{-\tau H} , \qquad (23.154)$$

and

$$\bar{c}_{\mathbf{k}\sigma}(\tau) \equiv e^{\tau H} \, c_{\mathbf{k}\sigma}^\dagger \, e^{-\tau H} , \qquad (23.155)$$

where τ corresponds to an imaginary time. Thus, we define the matrix elements of the one-particle Green's functions as

$$\begin{aligned} G_{\alpha\beta}(\mathbf{k}, \tau) &= -\langle T_\tau \, c_{\mathbf{k}\alpha}(\tau) c_{\mathbf{k}\beta}^\dagger \rangle , \\ \bar{G}_{\alpha\beta}(\mathbf{k}, \tau) &= -\langle T_\tau \, \bar{c}_{-\mathbf{k}\alpha}(\tau) c_{-\mathbf{k}\beta} \rangle , \\ F_{\alpha\beta}(\mathbf{k}, \tau) &= -\langle T_\tau \, c_{\mathbf{k}\alpha}(\tau) c_{-\mathbf{k}\beta} \rangle , \\ \bar{F}_{\alpha\beta}(\mathbf{k}, \tau) &= -\langle T_\tau \, \bar{c}_{-\mathbf{k}\alpha}(\tau) c_{\mathbf{k}\beta}^\dagger \rangle . \end{aligned} \qquad (23.156)$$

In order to obtain the renormalized single-particle Green's function we solve the Dyson equation

$$\hat{G}^{-1} = \hat{G}_0^{-1} - \hat{\Sigma} . \qquad (23.157)$$

This relates the matrix representing the bare propagator $\hat{G}_0$ and its renormalization due to the sum of all irreducible self-energy diagrams $\hat{\Sigma}$ to the dressed propagator $\hat{G}$. The Dyson equation reads in Nambu space

$$\hat{G}^{-1} = \begin{pmatrix} G_0^{-1} - \Sigma^G & -\Sigma^F \\ -\bar{\Sigma}^F & \bar{G}_0^{-1} - \bar{\Sigma}^G \end{pmatrix} , \qquad (23.158)$$

where the inverse bare propagator is given by $\hat{G}_0^{-1} = i\omega_n - \epsilon_\mathbf{k}$. In the case where the effective pairing interaction V_{eff} is given by the diagrams shown in Fig. 23.18 the self-energy is (for simplicity we put

$U_{\mathrm{eff}} = U$)

$$\Sigma^G(\mathbf{k}, \imath\omega_n) = \frac{1}{\beta N} \sum_{\mathbf{k}' i\omega'_n} \left[\frac{1}{2} U^2 \frac{\chi_{c0}(\mathbf{q}, i\nu_m)}{1 + U\chi_{c0}(\mathbf{q}, i\nu_m)} + \frac{3}{2} U^2 \frac{\chi_{s0}(\mathbf{q}, i\nu_m)}{1 - U\chi_{s0}(\mathbf{q}, i\nu_m)} + U^2 \chi_G(\mathbf{q}, i\nu_m) \right] \times G(\mathbf{k}', i\omega'_n) \tag{23.159}$$

and

$$\Sigma^F(\mathbf{k}, \imath\omega_n) = -\frac{1}{\beta N} \sum_{\mathbf{k}' i\omega'_n} \left[\frac{1}{2} U^2 \frac{\chi_{c0}(\mathbf{q}, i\nu_m)}{1 + U\chi_{c0}(\mathbf{q}, i\nu_m)} - \frac{3}{2} U^2 \frac{\chi_{s0}(\mathbf{q}, i\nu_m)}{1 - U\chi_{s0}(\mathbf{q}, i\nu_m)} - U^2 \chi_F(\mathbf{q}, i\nu_m) \right] G(\mathbf{k}', i\omega'_n) , \tag{23.160}$$

Here, $i\nu_m = i\omega_n - i\omega'_n$ and N denotes size of a square lattice. The term $U^2\chi_{G,F}$ on the right-hand side compensates for double counting that occurs in second order. Σ^G and Σ^F denote the normal and superconducting part of the self-energy, respectively [176]. χ_{c0} and χ_{s0} correspond to the irreducible charge and spin susceptibility, respectively, and are given by

$$\chi_{c0}(\mathbf{q}, i\nu_m) = -\frac{1}{\beta N} \sum_{\mathbf{k} i\omega_n} \times [G(\mathbf{k}+\mathbf{q}, i\omega_n + i\nu_m) G(\mathbf{k}, i\omega_n) - F(\mathbf{k}+\mathbf{q}, i\omega_n + i\nu_m) F^\dagger(-\mathbf{k}, i\omega_n)] \equiv \chi_G(\mathbf{q}, i\nu_m) - \chi_F(\mathbf{q}, i\nu_m) \tag{23.161}$$

and

$$\chi_{s0}(\mathbf{q}, i\nu_m) \equiv \chi_G(\mathbf{q}, i\nu_m) + \chi_F(\mathbf{q}, i\nu_m) . \tag{23.162}$$

Note that on the RPA level also the charge fluctuations are included. This may be of interest if a charge-density-wave (CDW) or stripe formation occurs. Both may compete for small doping. For example, it was pointed out that polarization-dependent photoemission also reveals a fine-structure in the low energy excitation spectrum [75]. In particular, for a single-layer high-T_c superconductor at optimum hole-doping two different behaviors of the spectral density at $(\pi, 0)$ and $(0, \pi)$ were found indicating their difference. This may indicate the importance of charge fluctuations in cuprates.

Because Re Σ is symmetric and Im Σ is antisymmetric with respect to $i\omega_n$, it is possible to put:

$$\Sigma^G(\mathbf{k}, i\omega_n) = i\omega_n [1 - Z(\mathbf{k}, i\omega_n)] + \xi(\mathbf{k}, i\omega_n) , \tag{23.163}$$

$$\Sigma^F(\mathbf{k}, i\omega_n) = \phi(\mathbf{k}, i\omega_n) , \tag{23.164}$$

where Z and ξ are real and also symmetric with respect to $i\omega_n$. Z describes the renormalization of the quasiparticle mass and ξ corresponds to the renormalization of the bare band energy $\epsilon_\mathbf{k}$. This will be also discussed in Appendix C. $\phi(\mathbf{k}, i\omega_n) = Z(\mathbf{k}, i\omega_n)\Delta(\mathbf{k}, i\omega_n)$ denotes the strong-coupling superconducting gap function. In general, the interdependence of elementary excitations with spin excitations leads to strong self-energy effects. If lifetime effects of the elementary excitations are included the superconducting gap will also have a strong frequency dependence.

Note that in the case where Cooper-pair phase fluctuations are unimportant one may choose

$$F(\mathbf{k}, i\omega_n) = e^{2i\alpha} F^*(\mathbf{k}, -i\omega_n) \tag{23.165}$$

and put the phase equal to zero. Thus, no extra equation occurs for $\bar{\phi}(\mathbf{k}, i\omega_n)$.

Then, with the help of (23.163) and (23.164) we are able to write down the diagonal and off-diagonal parts of the one-particle Green's function [76]:

$$G(\mathbf{k}, i\omega_n) = \frac{i\omega_n Z(\mathbf{k}, i\omega_n) + \epsilon_\mathbf{k} + \xi(\mathbf{k}, i\omega_n)}{(i\omega_n Z(\mathbf{k}, i\omega_n))^2 - (\epsilon_\mathbf{k} + \xi(\mathbf{k}, i\omega_n))^2 - \phi^2(\mathbf{k}, i\omega_n)} , \tag{23.166}$$

$$F(\mathbf{k}, i\omega_n) = \frac{\phi(\mathbf{k}, i\omega_n)}{(i\omega_n Z(\mathbf{k}, i\omega_n))^2 - (\epsilon_\mathbf{k} + \xi(\mathbf{k}, i\omega_n))^2 - \phi^2(\mathbf{k}, i\omega_n)} . \tag{23.167}$$

Since we want to solve the generalized Eliashberg equations on the real ω-axis rather than on the Matsubara points on the imaginary axis, we now use the spectral representation of the one-particle Green's function

$$G(\mathbf{k}, i\omega_n) = \int_{-\infty}^{\infty} d\omega \frac{N(\mathbf{k}, \omega)}{i\omega_n - \omega},$$

$$F(\mathbf{k}, i\omega_n) = \int_{-\infty}^{\infty} d\omega \frac{A_1(\mathbf{k}, \omega)}{i\omega_n - \omega}, \quad (23.168)$$

and inserting (23.168) into (23.161) and (23.162). After performing the sum over the Matsubara frequencies [73] we arrive at

$$\begin{aligned}\mathrm{Im}\,\chi_{s0,c0}(\mathbf{q}, \omega) &= \frac{\pi}{N} \int_{-\infty}^{\infty} d\omega' \left[f(\omega') - f(\omega' + \omega)\right] \\ &\times \sum_{\mathbf{k}} \left[N(\mathbf{k} + \mathbf{q}, \omega' + \omega) N(\mathbf{k}, \omega')\right. \\ &\times \left. A_1(\mathbf{k} + \mathbf{q}, \omega' + \omega) A_1(\mathbf{k}, \omega')\right].\end{aligned} \quad (23.169)$$

f denotes the Fermi distribution function.

Thus we arrive at the following set of equations for the quasiparticle self-energy Σ_ν $\nu = 0, 3, 1$ in the Nambu representation:

$$\begin{aligned}\Sigma_\nu(\mathbf{k}, \omega) = N^{-1} &\sum_{\mathbf{k}'} \int_0^{\infty} d\Omega \left[P_s(\mathbf{k} - \mathbf{k}', \Omega)\right. \\ &+ \left. (\delta_{\nu 0} + \delta_{\nu 3} - \delta_{\nu 1}) P_c(\mathbf{k} - \mathbf{k}', \Omega)\right] \\ &\times \int_{-\infty}^{\infty} d\omega' I(\omega, \Omega, \omega') A_\nu(\mathbf{k}', \omega').\end{aligned} \quad (23.170)$$

These equations have been discussed in the main text. Here, the spin and charge fluctuation interactions are given by

$$\begin{aligned}P_s &= (2\pi)^{-1} U^2 \,\mathrm{Im}\left(3\chi_s - \chi_{s0}\right), \\ \chi_s &= \chi_{s0}\left(1 - U\chi_{s0}\right)^{-1}\end{aligned} \quad (23.171)$$

and

$$\begin{aligned}P_c &= (2\pi)^{-1} U^2 \,\mathrm{Im}\left(3\chi_c - \chi_{c0}\right), \\ \chi_s &= \chi_{c0}\left(1 + U\chi_{c0}\right)^{-1}.\end{aligned} \quad (23.172)$$

A_ν was defined in (23.25). Then,

$$\begin{aligned}\Delta(\mathbf{k}, \omega)\, Z(\mathbf{k}, \omega) = \\ &- \frac{1}{N\pi} \sum_{\mathbf{k}'} \int_0^{\infty} d\Omega \left[P_s(\mathbf{k} - \mathbf{k}', \Omega) - P_c(\mathbf{k} - \mathbf{k}', \Omega)\right] \\ &\times \int_{-\infty}^{\infty} d\omega' \left[\frac{f(-\omega') + b(\Omega)}{\omega + i\delta - \Omega - \omega'} + \frac{f(\omega') + b(\Omega)}{\omega + i\delta + \Omega - \omega'}\right] \\ &\times \mathrm{Im}\left[\frac{Z(\mathbf{k}', \omega')\Delta(\mathbf{k}', \omega')}{D(\mathbf{k}', \omega')}\right],\end{aligned} \quad (23.173)$$

$$\begin{aligned}\omega\, Z(\mathbf{k}, \omega) = \\ &- \frac{1}{N\pi} \sum_{\mathbf{k}'} \int_0^{\infty} d\Omega \left[P_s(\mathbf{k} - \mathbf{k}', \Omega) + P_c(\mathbf{k} - \mathbf{k}', \Omega)\right] \\ &\times \int_{-\infty}^{\infty} d\omega' \left[\frac{f(-\omega') + b(\Omega)}{\omega + i\delta - \Omega - \omega'} + \frac{f(\omega') + b(\Omega)}{\omega + i\delta + \Omega - \omega'}\right] \\ &\times \mathrm{Im}\left[\frac{Z(\mathbf{k}', \omega')\omega'}{D(\mathbf{k}', \omega')}\right]\end{aligned} \quad (23.174)$$

and

$$\begin{aligned}\epsilon_{\mathbf{k}} + \xi(\mathbf{k}, \omega) = \\ &- (N\pi)^{-1} \sum_{\mathbf{k}'} \int_0^{\infty} d\Omega \left[P_s(\mathbf{k} - \mathbf{k}', \Omega) + P_c(\mathbf{k} - \mathbf{k}', \Omega)\right] \\ &\times \int_{-\infty}^{\infty} d\omega' \left[\frac{f(-\omega') + b(\Omega)}{\omega + i\delta - \Omega - \omega'} + \frac{f(\omega') + b(\Omega)}{\omega + i\delta + \Omega - \omega'}\right] \\ &\times \mathrm{Im}\left[\frac{\epsilon_{\mathbf{k}'} + \xi(\mathbf{k}', \omega')}{D(\mathbf{k}', \omega')}\right].\end{aligned} \quad (23.175)$$

Here, $D(\mathbf{k}, \omega) = [\omega Z(\mathbf{k}, \omega)]^2 + [\Delta(\mathbf{k}, \omega) Z(\mathbf{k}, \omega)]^2 - \left[\epsilon_{\mathbf{k}} + \xi(\mathbf{k}, \omega)\right]^2$ and $b(\omega)$, $f(\omega)$ are the Bose and Fermi distribution functions, respectively.

These are the so-called generalized Eliashberg equations on the real frequency axis. Their detailed numerical solution will be discussed in Appendix B. Note that in contrast to the electron–phonon case the momentum dependence cannot be restricted to the Fermi surface due to the strong anisotropy of the elementary excitations and the pairing interaction.

Appendix B: Numerical Solution Method for the Generalized Eliashberg Equations

We discuss the details of the algorithm for the numerical solution of the Eliashberg equations. This is helpful for the programming of the equations. The expressions are presented line by line as they occur in the main program. To achieve the highest accuracy we avoid continuation methods like the Pade-approximation and solve the Eliashberg equation directly on the real-frequency axis. First, one starts with the calculation of the imaginary part of the spin susceptibilities using the tight-binding unrenormalized energy dispersion $\epsilon_{\mathbf{k}}$:

$$\begin{aligned}\mathrm{Im}\chi_G(\mathbf{q},\omega) &= \frac{\pi}{N}\sum_{\mathbf{k}}\int_{-\infty}^{\infty}\mathrm{d}\omega \\ &\times\left[f(\omega)-f(\omega+\Omega)\right]N(\mathbf{k},\omega) \\ &\times N(\mathbf{k}+\mathbf{q},\omega+\Omega)\,,\end{aligned} \tag{23.176}$$

$$\mathrm{Im}\chi_F(\mathbf{q},\omega) = \frac{\pi}{N}\sum_{\mathbf{k}}\int_{-\infty}^{\infty}\left[f(\omega)-f(\omega+\Omega)\right] A_1(\mathbf{k},\omega)\,A_1\,(\mathbf{k}+\mathbf{q},\omega+\Omega). \tag{23.177}$$

Then, we obtain their real parts via Kramers–Kronig relation:

$$\mathrm{Re}\,\chi_{G,F}(\mathbf{q},\omega) = \frac{1}{\pi}P\int_{-\infty}^{\infty}\mathrm{d}\omega\,\frac{\mathrm{Im}\,\chi_{G,F}(\mathbf{q},\omega)}{\omega-\Omega}\,, \tag{23.178}$$

and define the irreducible charge and spin susceptibilities

$$\chi_{c0}=\chi_G-\chi_F,\ \chi_{s0}=\chi_G+\chi_F\,. \tag{23.179}$$

In the next step we calculate the kernels of the Eliashberg equations:

$$\begin{aligned}P_s &\equiv \frac{1}{\pi}\,\mathrm{Im}\left[\frac{3}{2}\frac{U^2\chi_{s0}}{1-U\chi_{s0}}\right],\\ P_c &\equiv \frac{1}{\pi}\,\mathrm{Im}\left[\frac{3}{2}\frac{U^2\chi_{c0}}{1+U\chi_{c0}}\right],\end{aligned} \tag{23.180}$$

$$P_G \equiv P_s+P_c-\frac{1}{\pi}\,\mathrm{Im}\left[\frac{1}{2}U^2\left(\chi_{s0}+\chi_{c0}\right)\right], \tag{23.181}$$

$$P_F \equiv -P_s+P_c+\frac{1}{\pi}\,\mathrm{Im}\left[\frac{1}{2}U^2\left(\chi_{s0}+\chi_{c0}\right)\right], \tag{23.182}$$

and plug them into the equations for determining the imaginary parts of the self-energies

$$\mathrm{Im}\phi(\mathbf{k},\omega) = \frac{\pi}{N}\sum_{\mathbf{q}}\int_{-\infty}^{\infty}\mathrm{d}\Omega \left[b(\Omega)+f(\Omega-\omega)\right]P_F(\mathbf{q},\omega)A_1(\mathbf{k}-\mathbf{q},\omega-\Omega) \tag{23.183}$$

and

$$\mathrm{Im}\Sigma(\mathbf{k},\omega) = -\frac{\pi}{N}\sum_{\mathbf{q}}\int_{-\infty}^{\infty}\mathrm{d}\Omega \left[b(\Omega)+f(\Omega-\omega)\right]P_G(\mathbf{q},\omega)N(\mathbf{k}-\mathbf{q},\omega-\Omega)\,. \tag{23.184}$$

The real parts are obtained again via Kramers–Kronig relations:

$$\mathrm{Re}\,\phi(\mathbf{k},\omega) = \frac{1}{\pi}P\int_{-\infty}^{\infty}\mathrm{d}\omega'\,\frac{\mathrm{Im}\,\phi(\mathbf{k},\omega')}{\omega'-\omega}\,, \tag{23.185}$$

$$\mathrm{Re}\,\Sigma^G(\mathbf{k},\omega) = \frac{1}{\pi}P\int_{-\infty}^{\infty}\mathrm{d}\omega'\,\frac{\mathrm{Im}\,\Sigma^G(\mathbf{k},\omega')}{\omega'-\omega}\,. \tag{23.186}$$

The next step calculates the energy shift

$$\xi \equiv \frac{1}{2}\left[\Sigma^G(\mathbf{k},\omega)+\Sigma^{G*}(\mathbf{k},-\omega)\right], \tag{23.187}$$

and the mass renormalization function

$$\omega Z \equiv \omega-\frac{1}{2}\left[\Sigma^G(\mathbf{k},\omega)-\Sigma^{G*}(\mathbf{k},-\omega)\right]. \tag{23.188}$$

This completes the first iteration.

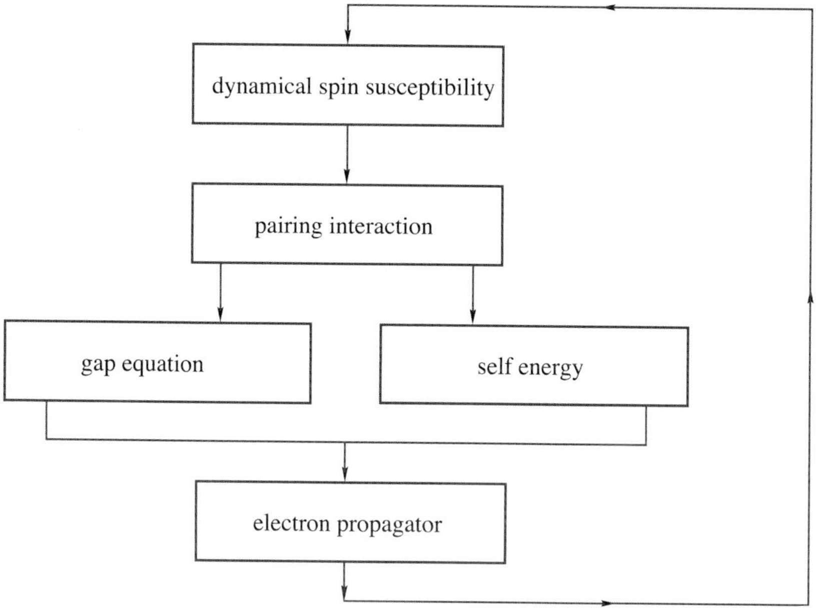

Fig. 23.76. Illustration of the procedure in order to solve (23.170). The full momentum and frequency dependence of the quantities are kept. Thus, our calculations include pair breaking effects for the Cooper-pairs resulting from lifetime effects of the elementary excitations

With the obtained results for the energy shift and the renormalization function we calculate the new spectral densities

$$N(\mathbf{k},\omega) = -\frac{1}{\pi}\text{Im}\,\frac{\omega Z + \epsilon_{\mathbf{k}} + \xi}{(\omega Z)^2 - \left(\epsilon_{\mathbf{k}} + \xi\right)^2 - \phi^2} = A_0(\mathbf{k},\omega) + A_3(\mathbf{k},\omega)\,, \qquad (23.189)$$

$$A_1(\mathbf{k},\omega) = -\frac{1}{\pi}\text{Im}\,\frac{\phi}{(\omega Z)^2 - \left(\epsilon_{\mathbf{k}} + \xi\right)^2 - \phi^2}\,, \qquad (23.190)$$

and then return to the (23.176). The procedure is repeated until the convergence is reached. The numerical calculations are performed on a square lattice with 256 × 256 points in the Brillouin zone and with 200 points on the real ω axis up to $16t$ with an almost logarithmic mesh. The full momentum and frequency dependence of the quantities is kept. The convolutions in $\mathbf{k}$ space are carried out with fast Fourier transforms. As have we mentioned, the real parts of the susceptibilities, gap, and self-energy are calculated with the help of the Kramers–Kronig relations (P denotes the corresponding principal value integration). The spectral functions A_1, $P_{G,F}$, Im ϕ, Im ξ, and Im $\chi_{G,F}$ are antisymmetric and the corresponding real parts are symmetric with respect to ω. Thus the spectral function for the interacting electrons (or holes) $N(\mathbf{k},\omega)$ is separated into a symmetric part A_0 and an antisymmetric part A_3 (see (23.25)). The corresponding equations for the normal state can be recovered by setting the off-diagonal terms of the self-energy, ϕ, A_1, and χ_F, identically to zero.

For an illustration in Fig. 23.76 we show how (23.170)–(23.172) are solved: One starts with a dynamical spin susceptibility $\chi(\mathbf{q},\omega)$ and constructs the effective pairing interaction using (23.18). Then, the strong-coupling gap equation for the superconducting order parameter $\phi(\mathbf{k},\omega)$ and the corresponding Dyson equation $G^{-1}(\mathbf{k},\omega) = G_0^{-1}(\mathbf{k},\omega) - \Sigma(\mathbf{k},\omega)$ have to be solved, respectively. Having solved these two equations one has new appropriate starting input values for an electron propagator G. This is again used to calculate χ. This procedure is repeated until all equations are solved.

In order to determine the superconducting transition temperature T_c we solve the linearized gap equation, see (23.185):

$$\lambda(T)\,\mathrm{Im}\,\phi(\mathbf{k},\omega) = -\frac{1}{N}\sum_{\mathbf{k}}\int_{-\infty}^{\infty} \mathrm{d}\omega' \left[b(\Omega) + f(\Omega-\omega)\right]$$
$$\times P_F(\mathbf{q},\omega)\mathrm{Im}\left[\frac{\phi(\mathbf{k}',\omega')}{(\omega Z)^2 - \left(\epsilon_{\mathbf{k}'}+\xi\right)^2}\right],$$
$$\lambda(T_c) = 1\,. \quad (23.191)$$

Here, χ_F in P_F is set to zero identically. For decreasing temperature T the eigenvalue $\lambda(T)$ increases and reaches unity at $T = T_c$. The spectral functions calculated during the solution of (23.191) are then appropriate starting values for the solution of the equations below T_c. Solving the generalized Eliashberg equations for cuprates we find that below T_c the superconducting gap function has $d_{x^2-y^2}$-wave symmetry for both hole-doped and electron-doped superconductors. Vertex corrections for the two-particle correlation function are not included. The calculations for Sr_2RuO_4 are performed similarly.

As mentioned earlier there exists an important feedback effect of the self-energy on the spectral functions of the corresponding quasi-particles (dressed holes or electrons). This is in particular the case in the superconducting state. Below T_c a gap appears in the spectral density of the quasi-particles. These not only condensed into Cooper-pairs, but also provide the effective pairing interaction $V^{\mathrm{eff}}(\mathbf{q},\omega)$. The superconducting gap function $\phi(\mathbf{k},\omega) = Z(\mathbf{k},\omega)\Delta(\mathbf{k},\omega)$ will be calculated self-consistently including all scattering and damping effects. This strong feedback effect has important consequences: for example, the dynamical spin susceptibility $\mathrm{Im}\,\chi_s(\mathbf{Q},\omega)$shows a rearrangement of its spectral weight for small frequencies $\omega < 2\phi(\mathbf{Q},\omega)$ followed by a peak at approximately $\omega = 2\phi$. Note that for simplicity we restrict our discussion to the antiferromagnetic nesting vector $\mathbf{q} = \mathbf{Q} = (\pi,\pi)$. As discussed in the Introduction this behavior provides an explanation for the observed "resonance peak" in inelastic neutron scattering in many cuprate superconductors. Therefore, the strong feedback effect of superconductivity on the elementary excitations and thus on $\mathrm{Im}\,\chi_s(\mathbf{Q},\omega)$ can be viewed as a general phenomenon of a strongly interacting system and in particular as an important fingerprint of spin-fluctuation-mediated pairing in cuprate high-T_c superconductors. The opening of the superconducting gap also helps to obtain a kink in the antinodal direction. To demonstrate this interdependence the self-consistent procedure shown in Fig. 23.76 is required.

Appendix C: Derivation of the Quasiparticle Self-Energy (Weak-Coupling Case)

In order to discuss the anisotropic $\mathbf{k}$-dependence of the elementary excitations, $\epsilon_{\mathbf{k}} = \epsilon^0_{\mathbf{k}} + \Sigma(\mathbf{k},\omega)$, it is instructive to consider the weak-coupling limit ($Z = 1$) of the generalized Eliashberg equations. Using a one-loop approximation the self-energy is given by:

$$\Sigma(\mathbf{k},\omega) = -T^2 \sum_{\omega_n,\nu_m}\sum_{\mathbf{p},\mathbf{q}} \hat{\tau}_i\, G(\mathbf{k}-\mathbf{p}, i\omega_n - i\nu_m)$$
$$\times \hat{\tau}_i\, U_{\mathrm{eff}}(\mathbf{p})U(\mathbf{p}) \quad (23.192)$$
$$\times \frac{1}{2}\mathrm{Tr}\big[\hat{\tau}_i\, G(\mathbf{p}+\mathbf{q}, i\omega_n + i\nu_m)$$
$$\times \hat{\tau}_i\, G(\mathbf{k}, i\omega_n)\big]\,.$$

For spin fluctuations we have $\tau_i = \tau_0$. For simplicity we use $U_{\mathrm{eff}}U = U^2$ and assume that the Coulomb interaction $U(\mathbf{p})$ is constant. The full Green's function is approximated by its non-interacting part

$$G(\mathbf{k}, i\omega_n) \simeq G_0(\mathbf{k}, i\omega_n) = \frac{i\omega_n\hat{\tau}_0 + \epsilon_k\hat{\tau}_3 - \Delta_k\hat{\tau}_1}{(i\omega_n)^2 - \epsilon_k^2 - \Delta_k^2}\,. \quad (23.193)$$

Using (23.193) and introducing the quasiparticle energy dispersion in the superconducting state $E_p^2 = \epsilon_p^2 + \Delta_p^2$ and the coherence factors

$$c_{\pm}(\mathbf{p}+\mathbf{q},\mathbf{q}) = 1 \pm \frac{\epsilon_{p+q}\epsilon_q + \Delta_{p+q}\Delta_q}{E_{p+q}E_q}\,, \quad (23.194)$$

we can perform the sum over ω_n in (23.193). Noting that

$$T\sum_{\omega_n}\frac{1}{(i\omega_n + i\nu_m - y)(i\omega_n - x)} = \frac{f(x) - f(y)}{i\nu_m + x - y}\,, \quad (23.195)$$

one obtains

$$T\sum_{\omega_n}\frac{1}{2}\,\mathrm{Tr}\,[G(\mathbf{p}+\mathbf{q},i\omega_n+i\nu_m)\,G(\mathbf{q},i\omega_n)] =$$

$$\frac{1}{4}\left[\left[\frac{f(E_q)-f(E_{p+q})}{i\nu_m+E_q-E_{p+q}}+\frac{f(-E_q)-f(-E_{p+q})}{i\nu_m-E_q+E_{p+q}}\right]c_+(\mathbf{p}+\mathbf{q},\mathbf{q})\right.$$
$$\left.+\left[\frac{f(-E_q)-f(E_{p+q})}{i\nu_m-E_q-E_{p+q}}+\frac{f(E_q)-f(-E_{p+q})}{i\nu_m+E_q+E_{p+q}}\right]c_-(\mathbf{p}+\mathbf{q},\mathbf{q})\right]. \tag{23.196}$$

Equation (23.196) has to be multiplied by $\frac{i\omega_n-i\nu_m}{(i\omega_n-i\nu_m)^2-E_{k-p}^2}$ for the $\hat{\tau}_0$-component and by $\frac{1}{(i\omega_n-i\nu_m)^2-E_{k-p}^2}$ for the $\hat{\tau}_3$ and $\hat{\tau}_1$-components, respectively, and summed over ν_m. Using (n = Bose function) $n(-x) = -(1+n(x))$, $f(-x) = 1-f(x)$ and $-T\sum_{\nu_m}\frac{1}{(i\nu_m-x)(i\nu_m-y)} = \frac{n(x)-n(y)}{x-y}$ in the form

$$-T\sum_{\nu_m}\frac{1}{(i\nu_m-i\omega_n-x)(i\nu_m-y)} = \frac{n(i\omega_n+x)-n(y)}{i\omega_n+x-y} = -\frac{f(x)+n(y)}{i\omega_n+x-y},$$

after the analytic continuation $i\omega_n \to \omega+i\eta$ we obtain the self-energy. For example, the imaginary part of the self-energy is

$$\mathrm{Im}\,\Sigma_0(\mathbf{k},\omega) = \frac{\pi U^2}{8} \tag{23.197}$$
$$\times\left[\sum_{\mathbf{p}}\left[f(E_{k-p})+n(E_{p+q}+\omega)\right]\right.$$
$$\times\sum_{\mathbf{q}}\Big[c_+(\mathbf{p}+\mathbf{q},\mathbf{q})\left[f(E_q)-f(E_{p+q})\right]$$
$$\times\left[\delta(\omega+E_{k-p}-E_q+E_{p+q})\right.$$
$$\left.-\,\delta(\omega+E_{k-p}+E_q-E_{p+q})\right]$$
$$+\,c_-(\mathbf{p}+\mathbf{q},\mathbf{q})\left[1-f(E_q)-f(E_{p+q})\right]$$
$$\times\left[\delta(\omega+E_{k-p}+E_q+E_{p+q})\right.$$
$$\left.\left.-\,\delta(\omega+E_{k-p}-E_q-E_{p+q})\right]\Big]\right]$$
$$+\sum_{\mathbf{p}}\left[f(E_{k-p})+n(E_{k-p}-\omega)\right]$$
$$\times\sum_{\mathbf{q}}\Big[c_+(\mathbf{p}+\mathbf{q},\mathbf{q})\left[f(E_q)-f(E_{p+q})\right]$$
$$\times\left[\delta(\omega-E_{k-p}+E_q-E_{p+q})\right.$$
$$\left.-\,\delta(\omega-E_{k-p}-E_q+E_{p+q})\right]$$
$$+\,c_-(\mathbf{p}+\mathbf{q},\mathbf{q})\left[1-f(E_q)-f(E_{p+q})\right]$$
$$\times\left[\delta(\omega-E_{k-p}-E_q-E_{p+q})\right.$$
$$\left.-\,\delta(\omega-E_{k-p}+E_q+E_{p+q})\right]\Big]\Big].$$

We introduce the susceptibility in the superconducting state

$$\chi(\mathbf{p},i\nu_m) = -T\sum_{\omega_n}\sum_{\mathbf{q}}\frac{1}{2}\,\mathrm{Tr} \tag{23.198}$$
$$\times\,[G(\mathbf{p}+\mathbf{q},i\omega_n+i\nu_m)G(\mathbf{q},i\omega_n)]$$
$$= -\frac{1}{4}\sum_{\mathbf{q}}\Big[c_+(\mathbf{p}+\mathbf{q},\mathbf{q})\left[f(E_q)-f(E_{p+q})\right]$$
$$\times\left[\frac{1}{i\nu_m+E_q-E_{p+q}}\,\frac{1}{i\nu_m-E_q+E_{p+q}}\right]$$
$$+\,c_-(\mathbf{p}+\mathbf{q},\mathbf{q})\left[1-f(E_q)-f(E_{p+q})\right]$$
$$\times\left[\frac{1}{i\nu_m-E_q-E_{p+q}}\,\frac{1}{i\nu_m+E_q+E_{p+q}}\right]\Big].$$

Then,

$$\mathrm{Im}\Sigma_0(\mathbf{k},\omega) = \frac{U^2}{2}\sum_{\mathbf{p}}\left[-\left[f(E_{k-p})\right.\right. \tag{23.199}$$
$$\left.+\,n(E_{k-p}+\omega)\right]\mathrm{Im}\,\chi(\mathbf{p},\omega+E_{k-p})$$
$$\times\left[f(E_{k-p})+n(E_{k-p}-\omega)\right]$$
$$\left.\times\,\mathrm{Im}\,\chi(\mathbf{p},\omega-E_{k-p})\right].$$

With the help of $f(x)+n(y) = \frac{1}{2}\left[\coth\left(\frac{y}{2T}\right)-\tanh\left(\frac{x}{2T}\right)\right]$ one finds the important result

$$\mathrm{Im}\Sigma_0(\mathbf{k},\omega) = -\frac{U^2}{4}\sum_{\mathbf{p}}\int_{-\infty}^{\infty}\mathrm{d}\omega' \tag{23.200}$$
$$\times\left[\coth\left(\frac{\omega'}{2T}\right)-\tanh\left(\frac{\omega'-\omega}{2T}\right)\right]$$
$$\times\,\mathrm{Im}\,\chi(\mathbf{k}-\mathbf{p},\omega')\,\delta(|\omega-\omega'|-E_p).$$

Therefore, the anisotropy of the spin excitations (weighted by the density of states) is intimately related to the anisotropy of the self-energy of the elementary excitations. The strength of this anisotropy is controlled by U^2. Thus, it is clear that spin dynamics will control the quasiparticle behavior in cuprates. As a result one expects an important feedback of spin fluctuations on quasiparticle dynamics. In particular the characteristic spin fluctuation frequencies in the normal (ω_{sf}) and superconducting (ω_{res}) will determine the characteristic features in the energy dispersion.

For the $\hat{\tau}_1$ and $\hat{\tau}_3$-components of Σ one obtains similar results. It is instructive to write them in a unified way using the nesting approximation. In this approximation one assumes that the main contribution to the momentum sum comes from Im χ at the nesting vector $\mathbf{Q} = (\pi, \pi)$. Then, the self-energy no longer depends on $\mathbf{k}$ and one can write ($\nu = 0, 1, 3$):

$$\text{Im}\,\Sigma_\nu(\omega) = -\frac{U^2}{4}\int_{-\infty}^{\infty} d\omega' \times \left[\coth\left(\frac{\omega'}{2T}\right) - \tanh\left(\frac{\omega'-\omega}{2T}\right)\right] \times \text{Im}\,\chi(\mathbf{Q},\omega)\,N_\nu(|\omega-\omega'|)\,, \tag{23.201}$$

where

$$N_0(x) = \sum_{\mathbf{p}} \delta(|x| - E_p)\,,$$

$$N_1(x) = -\text{signx}\sum_{\mathbf{p}} \frac{\Delta_p}{E_p}\delta(|x| - E_p)\,,$$

$$N_3(x) = \text{signx}\sum_{\mathbf{p}} \frac{\epsilon_p}{E_p}\,\delta(|x| - E_p)\,. \tag{23.202}$$

This completes the derivation of the self-energy in the weak-coupling case. In a strong-coupling case the procedure described in Appendix A has to be used. Then, it is possible to obtain the self-energy only numerically.

Appendix D: Cooper-Pair Phase Fluctuations

As suggested by the observation of $T_c \propto n_s$ for the underdoped cuprates [14] and by the small superfluid density, $n_s << n$ for $x \to 0$, and also by the weakly coupled two-dimensional CuO_2 planes, one expects that Cooper-pair phase fluctuations are important. These fluctuations of phase-incoherent Cooper-pairs will destroy the long-range superconducting order and the Meissner effect. In conventional, bulk superconductors these phase fluctuations are not relevant, since for large superfluid density the phase fluctuations are too costly energetically. Thus, in conventional superconductors the superconducting transition is due to the destruction of the Cooper pairs and T_c is proportional to Δ_0 [76]. Note that the superconducting energy gap Δ governs the thermal breaking of Cooper-pairs. Clearly, the observation by Uemura et al. that $T_c \propto n_s$ in underdoped hole-doped cuprates [14] indicates that Cooper-pair phase fluctuations are important [78,79]. The Cooper-pairs break up only at a crossover temperature around $T_c^* > T_c$ where T_c^* is approximately given by the transition temperature one would obtain without phase fluctuations and where at T_c Cooper-pairs become phase coherent. Between T_c and T_c^* local Cooper-pairs without long-range phase coherence are present.

In the following, we discuss phase fluctuations and their influence on elementary excitations, Meissner effect, interplay of phase and amplitude fluctuations, and superconducting transition temperature T_c.

Role of Phase and Amplitude Fluctuations for Determining T_c

We start from the expression for the free energy of a superconductor:

$$\Delta F \equiv F_S - F_N = -\frac{1}{\beta}\ln\left\{\int \mathcal{D}\Delta^\dagger \mathcal{D}\Delta e^{-\frac{1}{\hbar}S_{\text{eff}}[\Delta^\dagger,\Delta]}\right\}\,, \tag{23.203}$$

where $S_{\text{eff}}[\Delta^\dagger, \Delta]$ is the effective action obtained after integrating out the fermion variables. This is a functional of the Cooper-pair field $\Delta(t, \mathbf{r})$. Note that (23.203) is *exact*. The BCS theory is obtained from the saddle-point of the effective action. The simplest case corresponds to taking $\Delta = \Delta_0 = \text{const}$. Then the free energy becomes

$$\Delta F = -\frac{1}{\beta}\operatorname{Tr}\ln[\mathbf{G}^{-1}(\Delta_0^*, \Delta_0)] + \frac{\Omega}{V_0}|\Delta_0|^2\,, \quad (23.204)$$

where $\mathbf{G}$ is the Nambu–Gorkov matrix Green's function, $V_0 > 0$ is the coupling constant, and Ω is the volume of the system. Minimization of ΔF gives the BCS gap equation.

To start the study the effect of phase fluctuations in superconductors we make the change of variables

$$\Delta(\tau, \mathbf{r}) = \rho(\tau, \mathbf{r})e^{i\theta(\tau,\mathbf{r})}\,. \quad (23.205)$$

Let us write $\rho(\tau, \mathbf{r}) = \rho_0 + \delta\rho(\tau, \mathbf{r})$ and for the phase $\theta(\tau, \mathbf{r}) = \theta_0 + \delta\theta(\tau, \mathbf{r})$ and expand the effective action up to quadratic fluctuations in $\delta\rho$ and $\delta\theta$. The effective action then becomes

$$S_{\text{eff}} \approx \frac{\beta}{2}\sum_{\mathbf{q},n}[\Pi_{\rho\rho}(\omega_n, \mathbf{q})\delta\rho(\omega_n, \mathbf{q})\delta\rho(-\omega_n, -\mathbf{q}) \\ +\Pi_{\theta\theta}(\omega_n, \mathbf{q})\delta\theta(\omega_n, \mathbf{q})\delta\theta(-\omega_n, -\mathbf{q})]\,. \quad (23.206)$$

Note that we do not consider terms involving both the phase and amplitude fluctuations. These terms are often very small at low frequencies. Thus, to lowest order we can write

$$\Delta F = \Delta F_{\text{cond}} + \Delta F_{\text{phase}}\,, \quad (23.207)$$

where

$$\Delta F_{\text{cond}} = -\frac{1}{\beta}\ln\int\mathcal{D}\delta\rho \\ \exp\left[-\frac{\beta}{2}\sum_{\mathbf{q},n}\Pi_{\rho\rho}(\omega_n, \mathbf{q})\delta\rho(\omega_n, \mathbf{q})\delta\rho(-\omega_n, -\mathbf{q})\right] \quad (23.208)$$

gives the condensation energy, and

$$\Delta F_{\text{phase}} = -\frac{1}{\beta}\ln\int\mathcal{D}\delta\theta \\ \exp\left[-\frac{\beta}{2}\sum_{\mathbf{q},n}\Pi_{\theta\theta}(\omega_n, \mathbf{q})\delta\theta(\omega_n, \mathbf{q})\delta\theta(-\omega_n, -\mathbf{q})\right], \quad (23.209)$$

is the Cooper-pair energy gained due to phase coherence. In the low frequency and momentum limit we have

$$\Pi_{\theta\theta}(\omega_n, \mathbf{q}) = -\chi(\omega_n, \mathbf{q})\omega_n^2 + \Lambda_{\mu\nu}(\omega_n, \mathbf{q})q_\mu q_\nu\,, \quad (23.210)$$

where $\chi(\omega_n, \mathbf{q})$ is the density–density correlation function and

$$\Lambda_{\mu\nu} = \frac{n}{m}\delta_{\mu\nu} - \beta\langle j_\mu(\omega_n, \mathbf{q})j_\nu(-\omega_n, \mathbf{q})\rangle \quad (23.211)$$

is the phase stiffness. In the limit $\omega \to 0$ and $\mathbf{q} \to 0$ the expression $\Lambda_{\mu\nu}$ gives the superfluid density divided by the effective mass, n_s/m. Thus, one gets approximately from $\Delta F_{\text{phase}}(0) \sim k_B T_c$ and $\Delta F_{\text{phase}} \propto n_s$, which results for $\Delta F_{\text{cond}} > \Delta F_{\text{phase}}$, that $T_c \propto n_s$.

Quantum mechanically the role of the phase fluctuations is expressed by phase and particle number which are conjugate operators:

$$[\hat{\theta}, \hat{N}] = i\hbar\,. \quad (23.212)$$

Here, $\hat{N}$ denotes the Cooper-pair particle number operator. Thus, an immediate consequence of (23.212) is the uncertainty relation

$$\Delta\theta\Delta N \sim \hbar\,. \quad (23.213)$$

A rough estimate gives $\Delta N \sim n_s(0)$ and thus

$$\Delta\theta \sim \frac{1}{n_s(0)}\,. \quad (23.214)$$

Therefore, phase fluctuations are very important in systems having a small value of the zero-temperature superfluid density. This is precisely the case for the cuprate superconductors in the underdoped regime [79]. Indeed, since $n_s(0) \propto x$, where x is the hole concentration, we see that $\Delta\theta \sim 1/x$. As a result of Cooper-pair phase fluctuations and vortex condensation we have that

$$\Delta\theta \propto \frac{1}{T_c}\,. \quad (23.215)$$

Therefore, we obtain from (23.214) that the reduction of T_c in underdoped cuprates due to Cooper-pair phase fluctuations is given by

$$T_c \sim n_s(0)\,. \quad (23.216)$$

Therefore, for small doping phase fluctuations are very important.

Meissner Effect

Let us now estimate the role played by the coherent phase fluctuations for the Meissner effect. Writing $\psi = \sqrt{n_s}e^{i\phi}$ and taking into account the phase fluctuations only, one finds that the free energy is given by

$$F[n_s, \phi] = \int_0^\beta \mathrm{d}\tau \int \mathrm{d}^2 r n_s \times \left[\frac{1}{4}\left(\nabla_\mu \phi + \frac{2e}{c\hbar}A_\mu\right)^2 + \eta(\partial_\tau \phi)^2\right]. \quad (23.217)$$

Then using

$$\sigma_{\nu\mu} = \frac{\hbar}{\omega_\nu}\int \mathrm{d}^2 r \int \mathrm{d}\tau \frac{\delta^2 \ln Z}{\delta A_\mu(\tau, \mathbf{r})\delta A_\nu(0)} e^{i\omega_\nu \tau + i\mathbf{qr}}, \quad (23.218)$$

where Z is the partition function ($Z = \int \mathcal{D}\psi^* \mathcal{D}\psi \times \exp(-F[\psi^*, \psi])$), we find for the conductivity after phase averaging

$$\sigma_{\mu\nu}(\mathbf{q}, i\omega_\nu) = \frac{\pi n_s}{R_Q \omega_\nu}\left(-\delta_{\mu\nu} + \frac{q_\nu q_\mu}{4\eta\omega_\nu^2 + q^2}\right). \quad (23.219)$$

The analytic continuation to the real frequencies yields for $q \to 0$

$$\sigma_{\mu\nu}(\mathbf{q} = 0, \omega_\nu) = -\frac{\pi n_s}{R_Q}\delta_{\mu\nu}\left(\frac{i}{\omega} + \pi\delta(\omega)\right), \quad (23.220)$$

which implies perfect conductivity and for $\omega \to 0$

$$< j_\mu > = -\frac{\pi n_s}{R_Q}\left(\delta_{\mu\nu} - \frac{q_\mu q_\nu}{q^2}\right)A_\nu. \quad (23.221)$$

This demonstrates the occurrence of a Meissner effect below T_c in the superconducting phase in presence of the phase fluctuations and shows that phase fluctuations are indeed important.

Effect of the Phase Fluctuations on the Elementary Excitations: Dyson's Equation

In order to calculate the density of states for a d-wave superconductor in the presence of Cooper-pair phase fluctuations, one has to derive the corresponding Dyson equation for the elementary excitations. If one considers the Fourier transformation of the spectral density of the pairing interaction P_s, it turns out that the effective pairing potential V_{eff} is *attractive* on nearest neighbor sites in real space [166]. Thus we start with

$$H = -t\sum_{\langle i,j\rangle}\sum_{\sigma} c_{i\sigma}^\dagger c_{j\sigma} + \text{h.c.} - \mu\sum_i n_i + V_{\text{eff}}\sum_{\langle i,j\rangle} c_{i\uparrow}^\dagger c_{j\downarrow}^\dagger c_{j\downarrow} c_{i\uparrow}, \quad (23.222)$$

where $V_{\text{eff}} < 0$. In order to evaluate the free energy and correlation functions of the system one has to evaluate the partition function (Φ, Φ^* = Grassmann variables):

$$\mathcal{Z} = \int \mathcal{D}(\Phi, \Phi^*)\, \mathrm{e}^{-\mathcal{S}[\Phi^*, \Phi]},$$

where the action is given by

$$\mathcal{S}[\Phi^*, \Phi] = \int_0^\beta d\tau \times \left[\sum_{i\sigma} \Phi_{i\sigma}^*(\partial_\tau - \mu)\Phi_{i\sigma} + t\sum_{\langle ij\rangle\sigma}\Phi_{i\sigma}^*\Phi_{j\sigma} + V_{\text{eff}}\sum_{\langle ij\rangle}\Phi_{i\uparrow}^*\Phi_{j\downarrow}^*\Phi_{j\downarrow}\Phi_{i\uparrow}\right].$$

We now come to the derivation of the propagator for Cooper-pair phase fluctuations and the corresponding self-energy equations. For this purpose one has to integrate out the electronic degrees of freedom. This leads to

$$\mathcal{S}_{\text{eff}}[\varphi] = \mathcal{S}_{\text{eff}}^{BCS}(\Delta^0) + \frac{1}{\beta}\sum_{qn}\varphi(q, i\nu_n) \times \left[\mathcal{D}_\varphi(q, i\nu_n)\right]^{-1}\varphi(-q, -i\nu_n), \quad (23.223)$$

or diagrammatically

$$\mathcal{S}_{\text{eff}}[\varphi] - \mathcal{S}_{\text{eff}}^{BCS}(\Delta^0) = \tag{23.224}$$

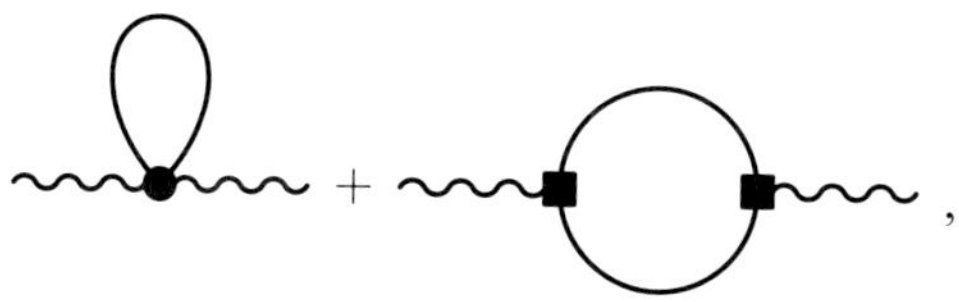

where the vertices, which couple the quasiparticles (holes) and the phases, are given by

$$\hat{\Gamma}_0^1(k, k') = \frac{i}{2}\left[(ik - ik')\,\hat{\tau}_3 - (\epsilon_{\mathrm{k}} - \epsilon_{\mathrm{k}'})\,\hat{\tau}_0\right]$$

$$= \tag{23.225}$$

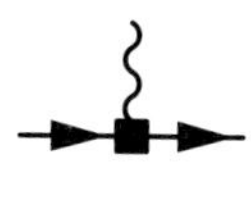

$$\hat{\Gamma}_0^2(k, k', k'') = \frac{1}{8}\left[\epsilon_{\mathrm{k}} - \epsilon_{\mathrm{k}-\mathrm{k}''} - \epsilon_{\mathrm{k}''+\mathrm{k}'} + \epsilon_{k'}\right]\hat{\tau}_3$$

$$= \quad . \tag{23.226}$$

and $\hat{\tau}_i$ are the corresponding Pauli matrices. The BCS part of the effective action is given by

$$\mathcal{S}_{\text{eff}}^{BCS}(\Delta^0) = -\mathrm{Tr}\ln\left[-\hat{\mathcal{G}}_0^{-1}\right] + \beta N\,\frac{|\Delta^0|^2}{|V_{\text{eff}}|}. \tag{23.227}$$

In (23.223) $\mathcal{D}_\varphi(q, i\nu_n) = \langle\langle\varphi(q, i\nu_n)\;\varphi(-q, -i\nu_n)\rangle\rangle$ is the propagator of the corresponding phase of a Cooper-pair given by [93]

$$\mathcal{D}_\varphi(q, i\nu_n) = 8\Bigg[-\chi_0(q, i\nu_n)\,(i\nu_n)^2 \tag{23.228}$$
$$-(i\nu_n)\sum_\alpha\left[\Pi^{\alpha\rho}(q, i\nu_n) + \Pi^{\rho\alpha}(q, i\nu_n)\right]2\sin\frac{q_\alpha}{2}$$
$$+\sum_{\alpha\beta}\rho^{\alpha\beta}(q, i\nu_n)\,(2\,\sin\frac{q_\alpha}{2})\,(2\,\sin\frac{q_\beta}{2})\Bigg]^{-1},$$

with the phase stiffness $\rho^{\alpha\beta}(q, i\nu_n) = \mathrm{T}^{\alpha\beta} + \Pi^{\alpha\beta}$. Π and T are the current–current correlation function and mass operator, respectively [73]. It is

$$\Pi^{\alpha\beta}(q, i\nu_n) = \frac{1}{\beta N}\sum_{kn'} v^\alpha_{k-\frac{q}{2}}\,v^\beta_{k-\frac{q}{2}}\,\mathrm{Tr} \tag{23.229}$$
$$\times\left[\hat{\mathcal{G}}_0(k, i\omega_{n'})\,\hat{\mathcal{G}}_0(k-q, i\omega_{n'} - i\nu_n)\right],$$

$$\mathrm{T}^{\alpha\beta} = \frac{1}{N}\sum_k t_k^{\alpha\beta}\,(n_{k\uparrow} + n_{k\downarrow})\,. \tag{23.230}$$

In analogy to Kadanoff and Baym we are now able to calculate the dressed Green's functions and their corresponding self-energies [177,178]. If we consider Cooper-pair phase fluctuations and treat them up to second order, it is [93]

$$\hat{\mathcal{G}}(k, i\omega_n) = \hat{\mathcal{G}}_0(k, i\omega_n) \tag{23.231}$$
$$-\,\hat{\mathcal{G}}_0(k, i\omega_n)\,\hat{\Sigma}^{\varphi\varphi}(k, i\omega_n)\,\hat{\mathcal{G}}(k, i\omega_n)\,,$$

or diagrammatically

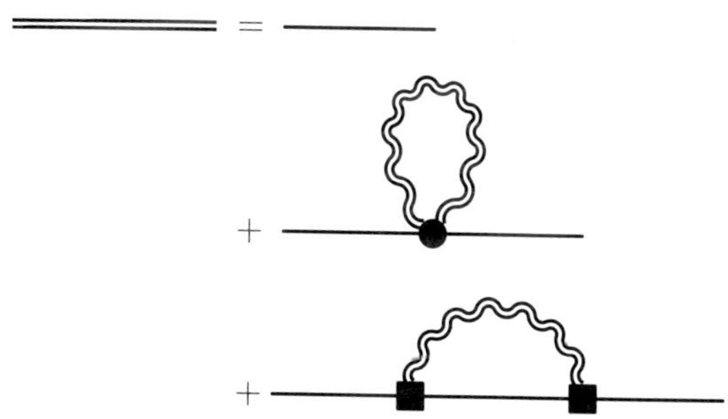

(23.232)

where the normal and anomalous part of the self-energy due to Cooper-pair phase fluctuations are given by [179]

$$\Sigma^{\mathcal{G}}(k, i\omega_n) = \frac{1}{4\beta N}\sum_{q,n'}[(i\nu_{n'})^2 - 2\,i\nu_{n'} \tag{23.233}$$
$$\times\,(\epsilon_k - \epsilon_{k-q}) + (\epsilon_k - \epsilon_{k-q})^2]$$
$$\times\;\mathcal{G}_0(k-q, i\omega_n - i\nu_{n'})\,\Pi^{\varphi\varphi}(q, i\nu_n)\,,$$

$$\Sigma^{\mathcal{F}}(k, i\omega_n) = \frac{1}{4\beta N}\sum_{q,n'}[(i\nu_{n'})^2 + (\epsilon_k - \epsilon_{k-q})^2]$$
$$\times\;\mathcal{F}_0(k-q, i\omega_n - i\nu_{n'})\,\Pi^{\varphi\varphi}(q, i\nu_n)\,. \tag{23.234}$$

The vertices are given above. Furthermore, in the diagrammatic representation $\hat{\mathcal{G}}_0$ is the dressed Green's function (given in Appendix A) within the FLEX approximation that includes, in particular, the scattering by spin fluctuations.

Finally, these self-energy corrections must be added to their counterparts Σ^G and Σ^F resulting from the exchange of spin fluctuations alone in (23.158). Then, for $\hat{\mathcal{G}}$ one can perform an analytical continuation to the real ω-axis and solve the corresponding equations. Of course, the current–current correlation function Π has to be calculated self-consistently using (23.229) describing the propagation of Cooper-pair phase fluctuations.

Interplay of Phase and Amplitude Fluctuations

So far we assume that the phase and amplitude fluctuations are decoupled. The next step is to study the interplay between them. At low frequencies and momenta this is equivalent to consider an incompressible charged fluid with $\nabla \cdot \mathbf{v}_s = 0$, where $\mathbf{v}_s$ is the superfluid velocity. Therefore, in the hydrodynamic limit the phase-decoupled free energy leads to a constant charge density, i.e. the amplitude fluctuations are essentially frozen in the low energy limit. To see this in a simple way, note that the hydrodynamics of a charged fluid is described by the non-linear Schrödinger-like equation of the Cooper-pair wave function:

$$\frac{\partial \rho^2}{\partial t} + \nabla \cdot (\rho^2 \mathbf{v}_s) = 0 \tag{23.235}$$

and

$$\hbar \frac{\partial \theta}{\partial t} = -\frac{m}{2}\mathbf{v}_s^2 + 2eV - \frac{\hbar^2}{2m}\left(\frac{1}{\rho}\nabla^2 \rho\right) . \tag{23.236}$$

Here, V is the electric potential and generally $\mathbf{v}_s = \hbar[\nabla\theta - (2e/c\hbar)\mathbf{A}]/m$, with $\mathbf{A}$ being the vector potential minimally coupled to the phase gradient. Equation (23.235) is just the continuity equation. In the above hydrodynamic equations the phase and amplitude of the Cooper pair are coupled. When the amplitude is frozen, the continuity equation gives us the incompressibility condition $\nabla \cdot \mathbf{v}_s = 0$, which implies other gauge-equivalent solution $\nabla\theta = 2e\mathbf{A}/(c\hbar)$. In this case (23.235) yields

$$\hbar \frac{\partial \theta}{\partial t} = 2\,\mathrm{eV} , \tag{23.237}$$

which is just Josephson's voltage-phase relation.

What happens to the interplay if the doping concentration is changed? At optimal doping the phase and amplitude fluctuations are equally important. Note that the critical temperature T_c is then at its highest value. However, with further increasing doping the critical temperature decreases, now not due to phase fluctuations which become negligible in the overdoping regime, but due to increasing itinerancy with $\chi \to 0$, i.e. the pairing field disappears.

Note that, in general, like in magnetism (ferromagnetism) phase and amplitude fluctuations of the order parameter are coupled and one would get for $\Delta = |\Delta| e^{i\phi}$ that $|\Delta| \to 0$ affects $\langle \nabla\phi \nabla\phi \rangle$ and vica versa.

Using Ginzburg–Landau theory it is easy to see that amplitude fluctuations decrease T_c. Consider the inverse susceptibility $\chi^{-1}(T)$, with $T > T_c$, in a Ginzburg–Landau (GL) description. We have

$$\chi^{-1}(T) = a_0(T - T_0) + u\langle|\psi|^2\rangle , \tag{23.238}$$

where T_0 is the mean-field critical temperature and u is the GL interaction. In the overdoped regime the CuO_2 planes are more strongly coupled and the Cooper-pairs tunnel coherently between the planes. Thus, we calculate $\langle|\psi|^2\rangle$ in three dimensions. Considering an isotropic system for simplicity we have the self-consistent equation

$$\begin{aligned}\chi^{-1}(T) &= a_0(T - T_0) + u \int \frac{\mathrm{d}^3 p}{(2\pi)^3} \frac{1}{p^2 + \chi^{-1}(T)} \\ &= a_0(T - T_0) + \frac{u\Lambda}{2\pi^2} - \frac{u}{4\pi}\sqrt{\chi^{-1}(T)} ,\end{aligned} \tag{23.239}$$

where Λ is a cutoff. We see that the critical temperature is

$$T_c = T_0 - \frac{u\Lambda}{2\pi^2} , \tag{23.240}$$

which is indeed lower than T_0.

This T_c reduction and the one due to phase fluctuations are, however, fundamentally different regarding the way by which the superconducting state is destroyed. In the underdoped cuprates the phase fluctuations are expected to relate to the formation of vortices and the superconducting phase is destroyed due to the proliferation of thermally and quantum excited vortices. Thus, the non-superconducting phase can be thought of as a phase where the vortices are condensed. Note that the vortex condensation in underdoped cuprates may in part be responsible for the suppression of spectral weight and non-Fermi-liquid behavior.

In a simple Eliashberg-like treatment of T_c in which fluctuations of the superconducting condensate are neglected (see Fig. 23.48), the temperature $T_c^* \propto \Delta(T \to 0)$ is equal to the mean-field temperature T_0. Therefore, fluctuations are reducing $T_c < T_c^*$.

The Role of Transverse Phase Fluctuations in the BKT Theory and in Determining T_c

How can one connect the Berezinskii–Kosterlitz–Thouless (BKT) theory to the Eliashberg-like theory in the presence of vortices? For this let us write the following simple effective Hamiltonian density containing both the phase and Cooper-pair particle number:

$$\mathcal{H} = \frac{1}{2g}\hat{N}^2 + \frac{n_s(T)\hbar^2}{4m}(\nabla\hat{\theta})^2 . \qquad (23.241)$$

Note that the above Hamiltonian describing a continuum limit is also an array of Josephson weak-link junctions. If we identify $\hat{N}$ as the "momentum" operator conjugated to the phase operator $\hat{\theta}$ (see (23.212)), we see that in (23.241) the first term on the RHS corresponds to the kinetic energy, while the second one is the potential energy. The partition function of the system can be written as the functional integral

$$Z = \int [\prod_{\tau,\mathbf{r}} \mathrm{d}\theta(\tau,\mathbf{r})\mathrm{d}N(\tau,\mathbf{r})] \exp\left[-\frac{1}{\hbar}S(\theta,N)\right] \qquad (23.242)$$

and the action is given by

$$S(\theta,N) = \int_0^\beta \mathrm{d}\tau \int \mathrm{d}^2r(-iN\,\partial_\tau\theta + \mathcal{H}) . \qquad (23.243)$$

After performing the Gaussian integral over $N(\tau,\mathbf{r})$ we obtain the effective action containing only the phase $\theta(\tau,\mathbf{r})$:

$$S_{\text{phase}} = \int_0^\beta \mathrm{d}\tau \int \mathrm{d}^2r\mathcal{L}(\theta) , \qquad (23.244)$$

where the Lagrangian is given by

$$\mathcal{L} = \frac{g}{2}(\partial_\tau\theta)^2 + \frac{n_s(T)\hbar^2}{4m}(\nabla\theta)^2 . \qquad (23.245)$$

Let us now turn to a determination of T_c with the help of the BKT theory. We make the approximation that θ is independent of the Matsubara time. Then we can write the phase action as

$$S_{\text{phase}} = \frac{\tilde{K}}{2}\int \mathrm{d}^2r(\nabla\theta)^2 , \qquad (23.246)$$

where the phase stiffness $\tilde{K}$ is related to the superfluid density through $\tilde{K} = n_s(T)\hbar^2/(4mT)$. The superfluid density can be computed from a Hamiltonian like the Hubbard model employing a many-body treatment as the FLEX approximation and the Kubo formula. In the case of the cuprates it is useful to introduce an additional parameter d giving the average distance between the CuO_2 planes such that the new phase stiffness is given by

$$K(T) = \beta\hbar^2\frac{n_s(T)}{m}\frac{d}{4} , \qquad (23.247)$$

In our calculations we set d to half the height of the unit cell of $YBa_2Cu_3O_{6+x}$. The renormalized phase stiffness can be expressed as

$$\begin{aligned}\frac{1}{K_R} &= \frac{4m^2T}{\hbar^2 d n_s(T)} \\ &= \frac{4m^2}{\hbar^2}\int \mathrm{d}^2r\langle\mathbf{v}_s(\mathbf{r})\cdot\mathbf{v}_s(0)\rangle . \end{aligned} \qquad (23.248)$$

In order to compute the correlation function involving the superfluid velocity, we have to split the gradient of the phase between the longitudinal (regular) and transverse (singular) contributions. The transverse fluctuations account for the vortices. We then have

$$v_{s,\mu}(\mathbf{r}) = \partial_\mu\varphi(\mathbf{r}) + 2\pi\varepsilon_{\mu\nu}\partial_\nu\int G(\mathbf{r}-\mathbf{r}')n(\mathbf{r}') , \qquad (23.249)$$

where $G(\mathbf{r}-\mathbf{r}')$ satisfies the equation $-\nabla^2G(\mathbf{r}-\mathbf{r}') = \delta^2(\mathbf{r}-\mathbf{r}')$. The *vortex density* can be written as

$n(\mathbf{r}) = \sum_i q_i \delta^2(\mathbf{r} - \mathbf{r}_i)$, where $q_i = \pm 1$. As is well known, the statistical mechanics of two-dimensional vortices can be mapped on a Coulomb gas [180]. Using the Coulomb gas scenario the superfluid density can be related to the vortex density correlation function in momentum space. One finds [181]

$$\frac{n_s}{n_s^0} = 1 - \frac{2\pi}{T}\Lambda^2 \lim_{k\to 0}\langle n(k)n(-k)\rangle\,, \tag{23.250}$$

where n_s^0 is the bare superfluid density and Λ is a cutoff.

Starting from the smallest vortex–antivortex pairs of size r_0, the pairs are integrated out and their effect is incorporated by an approximate renormalization of K and the fugacity $y = e^{-\beta E_c}$. Then, from (23.248) and (23.249) we obtain the BKT recursion relations [180–182]:

$$\frac{\mathrm{d}y}{\mathrm{d}l} = (2 - \pi K)\,y\,, \qquad \frac{\mathrm{d}K}{\mathrm{d}l} = -4\pi^3 y^2 K^2\,. \tag{23.251}$$

For the vortex core energy we use the approximate result of Blatter et al., i.e.

$$E_{\text{core}} = \pi k_B T K \ln\kappa\,, \tag{23.252}$$

where κ is the Ginzburg parameter. $l = \ln(r/r_0)$ is a logarithmic length scale which relates K to the strength of the vortex–antivortex interaction. For $T > T_c$, K tends to zero for $l \to \infty$, so that the interaction at large distances is screened and the largest vortex–antivortex pair unbind. This destroys the Meissner effect and leads to dissipation. On the other hand, bound Cooper-pairs reduce K and thus n_s, but do not destroy superconductivity. After solving (23.251) and (23.252) it turns out that the renormalization of K is very small. Thus, within a good approximation, one can obtain $T_c(x)$ from the simple criterion

$$K(T_c) = \frac{2}{\pi}\,, \quad \text{or} \quad \frac{n_s(T_c, x)}{m} = \frac{2}{\pi}\frac{4k_B T_c}{\hbar^2 d}\,, \tag{23.253}$$

which we have used in the main text.

Let us turn now to the ω-dependence of the Cooper-pair phase fluctuations. A dynamical generalization of the BKT theory was developed by Ambegaokar et al. [95, 181]. It turns out that the critical size for a vortex–antivortex pair is given by

$$r_\omega = \sqrt{\frac{D_v}{2\pi\omega}}\,,$$

where D_v denotes the corresponding diffusion constant. In the absence of pinning, the theory by Bardeen and Stephen yields [96]

$$D_v^0 = \frac{2\pi c^2 \xi_{ab}^2 \rho_n T}{\phi_0^2 \tilde{d}}\,,$$

where c is the speed of light, $\xi_{ab} \sim r_0/2$ corresponds to the coherence length, ρ_n denotes the normal-state resistivity, $\phi_0 = hc/2e$ is the elementary superconducting flux quantum, and $\tilde{d}$ corresponds to an effective layer thickness. For the pinning we assume an Arrhenius law. Then, one finds

$$\frac{n_s(\omega)}{m} = \frac{1}{e^2}\,\omega\,\mathrm{Im}\,\sigma_S(\omega)\,, \tag{23.254}$$

where $\sigma_S(\omega)$ is the difference between the total dynamical conductivity in the superconducting state and its value in the normal state. It can be obtained from the current–current correlation function and the Kubo formula using the generalized Eliashberg equations described in Appendix A.

Acknowledgements

We thank L. Tewordt, C. Timm, J. Schmalian, F. Nogueira, D. Morr, J. Fink and M. Peter for many useful discussions.

References

1. J.G. Bednorz and K.A. Müller, Z. Phys. B **64**, 189 (1986)
2. Y. Kitaoka, K. Ishida, F. Fujiwara, T. Kondo, K. Asayama, M. Horvatic, Y. Berthier, P. Butaud, P. Segransan, C. Berthier, H. Katayana-Yoshida, Y. Okabe, and T. Takahashi, In: *Strong Correlations and Superconductivity*, Springer Series in Solid-State Sciences 89, ed. by H. Fukuyama, S. Maekawa, and A.P. Malozemoff (Springer, Berlin, 1989), p. 262
3. D.A. Wollmann, D.J. van Harlingen, W.C. Lee, D.M. Ginsberg, and A.J. Leggett, Phys. Rev. Lett. **71**, 2134 (1993)
4. J.R. Kirtley, C.C. Tsuei, J.Z. Sun, C.C. Chi, L.S. Yu-Jahnes, A. Gupta, M. Rupp, and M.B. Ketchen,, Nature (London) **373**, 225 (1995)
5. C.C. Tsuei, J.R. Kirtley, M. Rupp, J.Z. Sun, A. Gupta, and M.B. Ketchen, Science **272**, 329 (1996)
6. C.C. Tsuei, J.R. Kirtley, Rev. Mod. Phys. **72**, 969 (2000)
7. W.N. Hardy, D.A. Bonn, D.C. Morgan, R. Liang, and K. Zhang, Phys. Rev. Lett. **70**, 3999 (1994)
8. S. Kamal, Ruixing Liang, A. Hosseini, D.A. Bonn, and W.N. Hardy, Phys. Rev. B **58**, R8933 (1998); S, Kamal, D.A. Bonn, Nigel Goldenfeld, P.J. Hirschfeld, Ruixing Liang, and W.N. Hardy, Phys. Rev. Lett. **73**, 1845 (1994)
9. Z.X. Shen et al., Phys. Rev. Lett. **70**, 1553 (1993)
10. M.C. Krantz and M. Cardona, J. Low Temp. Phys. **99** 205 (1995)
11. D. Einzel and R. Hackl, J. Raman Spectrosc. **27**, 307 (1996)
12. T.P. Devereaux and D. Einzel, Phys. Rev. B **51**, 16336 (1995); **54** 15547 (1996)
13. D. Manske, C. T. Rieck, R. Das Sharma, A. Bock, and D. Fay, Phys. Rev. B **56**, R2940 (1997)
14. Y.J. Uemura et al., Phys. Rev. Lett. **62**, 2317 (1989)
15. A. Tsukada, Y. Krockenberger, M. Noda, H. Yamamoto, D. Manske, L. Alff and M. Naito, Solid State Comm. **133**, 427 (2005)
16. For a review, see C.P. Slichter, In: *Strongly Correlated Electronic Systems*, ed. by K. S. Bedell *et al.* (Addison-Wesley, Reading, MA, 1994)
17. L.P. Regnault, P. Bourges, P. Burlet, J.Y. Henry, J. Rossat-Mignod, Y. Sidis, C. Vettier, Physica B **213–214**, 48 (1995)
18. D.S. Marshall, D.S. Dessau, A.G. Loeser, C-H. Park, A.Y. Matsuura, J.N. Eckstein, I. Bozovic, P. Fournier, A. Kapitulnik, W.E. Spicer, and Z.-X. Shen, Phys. Rev. Lett. **76**, 4841 (1996)
19. J.M. Harris, Z.-X. Shen, P.J. White, D.S. Marshall, M.C. Schabel, J.N. Eckstein, and I. Bozovic, Phys. Rev. B **54**, R15655 (1996)
20. Ch. Renner, B. Revaz, J.-Y. Genoud, K. Kadowaki, and Ø. Fischer, Phys. Rev. Lett. **80**, 149 (1998)
21. J.W. Loram, K.A. Mirza, J.R. Cooper, and W.Y. Liang , Phys. Rev. Lett. **71**, 1740 (1993); J.W. Loram et al., J. Superconductivity **7**, 234 (1994)
22. B. Batlogg, H.Y. Hwang, H. Takagi, R.J. Cava, H.L. Kao, and J. Kwo, Physica C **235–240**, 130 (1994)
23. R. Nemetschek, M. Opel, C. Hoffmann, P.F. Mueller, R. Hackl, H. Berger, L. Forro, A. Erb, and E. Walker, Phys. Rev. Lett. **78**, 4837 (1997); G. Blumberg, M. Kang, M.V. Klein, K. Kadowaki, and C. Kendziora, Science **278**, 1427 (1997)
24. A.K. Gupta and K.-W. Ng, Phys. Rev. B **58**, R8901 (1998)
25. C.M. Varma, Phys. Rev. Lett. **63**, 1996 (1989)
26. A. Virosztek and J. Ruvalds, Phys. Rev. B **42**, 4064 (1990)
27. J. Ruvalds, C.T. Rieck, J. Zhang and A. Virosztek, Science **256**, 1664 (1992)
28. A.J. Millis, H. Monien, and D. Pines, Phys. Rev. B **42**, 167 (1990)
29. G.M. Luke, L.P. Le, B.J. Sternlieb, Y.J. Uemura, J.H. Brewer, R. Kadono, R.F. Kiefl, S.R. Kreitzman, T.M. Riseman, C.E. Stronach, M.R. Davis, S. Uchida, H. Takagi, Y. Tokura, Y. Hidaka, T. Murakami, J. Gopalakrishnan, A.W. Sleight, M.A. Subramanian, E.A. Early, J.T. Markert, M.B. Maple, and C.L. Seaman, Phys. Rev. B **42**, 7981 (1990)
30. Y. Tokura, H. Takagi, and S. Uchida, Nature **337**, 345 (1989)
31. C.C. Tsuei and J.R. Kirtley, Phys. Rev. Lett. **85**, 182 (2000)
32. T. Timusk and B. Statt, Rep. Prog. Phys. **62**, 61 (1999)
33. H. Takagi, B. Batlogg, H.L. Kao, J. Kwo, R.J. Cava, J.J. Krajewski, and W.F. Peck, Phys. Rev. Lett. **69**, 2975 (1992)
34. C.C. Tsuei, A. Gupta, and G. Koren, Physica C **161**, 415 (1989)
35. D.S. Dessau, Z.-X. Shen, D.M. King, D.S. Marshall, L.W. Lombardo, P.H. Dickinson, A.G. Loeser, J. DiCarlo, C.-H Park, A. Kapitulnik, and W.E. Spicer, Phys. Rev. Lett. **71**, 2781 (1993); A.A. Abrikosov, J.C. Campuzano, and K. Gofron, Physica C **214**, 73 (1993)

36. D.M. King, Z.-X. Shen, D.S. Dessau, B.O. Wells, W.E. Spicer, A.J. Arko, D.S. Marshall, J. DiCarlo, A.G. Loeser, C.H. Park, E.R. Ratner, J.L. Peng, Z.Y. Li, and R.L. Greene, Phys. Rev. Lett. **70**, 3159 (1993)
37. S.M. Anlage, D.-H. Wu, J. Mao, S.N. Mao, X.X. Xi, T. Venkatesan, J.L. Peng, and R.L. Greene, Phys. Rev. B **50**, 523 (1994); D.H. Wu, J. Mao, S.N. Mao, J.L. Peng, X.X. Xi, T. Venkatesan, R.L. Greene, and S.M. Anlage, Phys. Rev. Lett. **70**, 85 (1993)
38. R. Prozorov, R.W. Gianetta, P. Furnier, and R.L. Greene, Phys. Rev. Lett. **85**, 3700 (2000)
39. J.D. Kokales, P. Furnier, L.V. Mercaldo, V.V. Talanov, R.L. Greene, and S.M. Anlage, Phys. Rev. Lett. **85**, 3696 (2000)
40. J.A. Skinta, M.-S. Kim, T.R. Lemberger, T. Greibe, and M. Naito, Phys. Rev. Lett. **88**, 207005 (2002)
41. P. Bourges, In: *Gap Symmetry and Fluctuatios in High-Tc Superconductors*, ed. by J. Bok, G. Deutscher, D. Pavuna and S. Wolf (Plenum, New York, 1998)
42. J. Rossat-Mignod, L.P. Regnault, C. Vettier, P. Bourges, P. Burlet, J. Bossy, J.Y. Henry, L. Lapertot, Physica C **185–189**, 86 (1991)
43. H. He, Y. Sidis, P. Bourges, G.D. Gu, A. Ivanov, N. Koshizuka, B. Liang, C.T. Lin, L.P. Regnault, E. Schoenherr, and B. Keimer, Phys. Rev. Lett. **86**, 1610 (2001)
44. H.F. Fong, P. Bourges, Y. Sildis, L.P. Regnault, A. Ivanov, G.D. Gu, N. Koshizuka, and B. Keimer, Nature **398**, 588 (1999); P. Dai, H.A. Mook, S.M. Hayden, G. Aeppli, T.G. Perring, R.D. Hunt, and F. Dogan, Science **284**, 1344 (1999)
45. H. He, P. Bourges, Y. Sidis, C. Ulrich, L.P. Regnault, S. Pailhes, N. . Berzigiarova, N.N. Kolesnikov, and B. Keimer, Science **295**, 1045 (2002)
46. A. Lanzara, P.V. Bogdanov, X.J. Zhou, S.A. Kellar, D.L. Feng, E.D. Lu, T. Yoshida, H. Eisaki, A. Fujimori, K. Kishio, J.-I. Shimoyama, T. Nodall, S. Uchida, Z. Hussain, Z.-X. Shen, Nature (London) **412**, 510 (2001)
47. R. J. McQueeney, Y. Petrov, T. Egami, M. Yethiraj, G. Shirane, and Y. Endoh, Phys. Rev. Lett. **82**, 628 (1999)
48. Z.-X. Shen, A. Lanzara, S. Ishihara, N. Nagaosa, Philosophical Magazine B **82**, 1349 (2002)
49. J.F. Zasadzinski, L. Ozyuzer, M. Miyakawa, D.G. Hinks, and K.E. Gray, Physica C **341–348**, 867 (2000)
50. J.R. Schrieffer, D.J. Scalapino, and J.W. Wilkins, Phys. Rev. Lett. **10**, 336 (1963)
51. M. Sigrist, D. Agterberg, A. Furusaki, C. Honerkamp, K.K. Ng, T.M. Rice, and M.E. Zhitomirsky, Physica C **317–318**, 134 (1999)
52. Y. Sidis, M. Braden, P. Bourges, B. Hennion, S. NishiZaki, Y. Maeno, and Y. Mori, Phys. Rev. Lett. **83**, 3320 (1999)
53. K. Ishida, H. Mukuda, Y. Kitaoka, Z.Q. Mao, H. Fukuzawa, and Y. Maeno, Phys. Rev. B **63**, 060507 (2001)
54. T.M. Rice and T.M. Sigrist, J. Phys.: Condens. Matter **7**, L643 (1995)
55. I. Eremin, D. Manske, and K.H. Bennemann, Europhys. Lett. **58**, 871 (2002)
56. A.B. Migdal, Zh. Eksp. Teor. Fiz **34**, 1438 (Soviet Phys. JETP **34**, 996 (1958))
57. Y.M. Vilk and A.-M.S. Tremblay, J. Phys. I France **7**, 1309 (1997)
58. A.V. Chubukov, D. Pines, and J. Schmalian, In: *The Physics of Superconductors, Vol.1*, ed. by K.H. Bennemann, J.B. Ketterson (Springer, Berlin, 2002)
59. P. Monthoux, Phys. Rev. B **55**, 15261 (1997); A. Chubukov, P. Monthoux, and D.K. Morr, Phys. Rev. B **56**, 7789 (1997)
60. J.R. Schrieffer, J. Low Temp. Phys. **99**, 397 (1995)
61. N.E. Bickers and D.J. Scalapino, cond-mat/0010480
62. J. Zaanen, G.A. Sawatzky, and J.W. Allen, Phys. Rev. Lett. **55**, 418 (1985)
63. G. Baumgärtel, J. Schmalian, and K.H. Bennemann, Phys. Rev. B **48**, 3983 (1993)
64. W.E. Pickett, Rev. Mod. Phys. **61**, 433 (1989)
65. V.J. Emery, Phys. Rev. Lett. **58**, 2794 (1987)
66. F.C. Zhang and T.M. Rice, Phys. Rev. B **37**, 3759 (1988)
67. G. Baumgärtel, J. Schmalian, and K.H. Bennemann, Phys. Rev. B **48**, 3983 (1993)
68. G. Kotliar, and A.E. Ruckenstein, Phys. Rev. Lett. **57**, 1362 (1986)
69. Y. Nambu, Phys. Rev. **117**, 648 (1960)
70. J. Schmalian, S. Grabowski, and K.H. Bennemann, Phys. Rev. B **56**, R509 (1997)
71. T. Dahm and L. Tewordt, Phys. Rev. Lett. **74**, 793 (1995)
72. M. Langer, J. Schmalian, S. Grabowski, and K.H. Bennemann, Phys. Rev. Lett. **75**, 4508 (1995)
73. G.D. Mahan *Many-Particle Physics* (Plenum, New York, 1981)
74. Y.-D. Chuang, A.D. Gromko, A. Fedorov, Y. Aiura, K. Oka, Y. Ando, H. Eisaki, S.I. Uchida, and D.S. Dessau, Phys. Rev. Lett. **87**, 117002 (2001)
75. R. Manzke, R. Müller, C. Janowitz, M. Schneider, A. Krapf, and H. Dwelk, Phys. Rev. B **63**, 100504 (2001)

76. J.R Schrieffer, *Theory of Superconductivity* (Addison-Wesley, Redwood City, MA, 1964)
77. L. Alff, Y. Krockenberger, B. Welter, R. Gross, D. Manske, and M. Naito, Nature **422**, 698 (2003)
78. B.K. Chakraverty, A. Taraphder, and M. Avignon, Physica C **235-240**, 2323 (1994); B.K. Chakraverty, and T.V. Ramakrishnan, *ibid*, **282-287**, 290 (1997)
79. V.J. Emery and S.A. Kivelson, Nature (London) **374**, 434 (1995)
80. S. Wermbter and L. Tewordt, Physica C **211**, 132 (1993)
81. F. London and H. London, Proc. Roy. Soc. (London) **A149**, 71 (1935)
82. B. Giovannini, and C. Berthod, Phys. Rev. B **63**, 144516 (2001)
83. M.R. Norman, M. Randeria, B. Janko, and J.C. Campuzano, Phys. Rev. B **61**, 14742 (2000)
84. R. Haslinger and A.V. Chubukov, Phys. Rev. B **67**, 140504 (2003) and Phys. Rev. B **68**, 214508 (2003)
85. J.L. Tallon and J.W. Loram, Physica C **349**, 53 (2001)
86. M.V. Feigel'mann, Zh. Eksp. Teo. Fiz. **76**, 784 [ov. Phys. JETP **49**, 395 (1989)]
87. G. Blatter, M.V. Feigel'mann, V.B. Geshkenbein, A.I. Larkin, and V.M. Vinokur, Rev. Mod. Phys. **66**, 1125 (1994)
88. V.L. Berezinskii, Zh. Eksp. Teo. Fiz. **61**, 1144 (1971) [Sov. Phys. JETP **34**, 610 (1972)]; J.M. Kosterlitz and D.J. Thouless, J. Phys. C **6**, 1181 (1973); J.M. Kosterlitz, *ibid* **7**, 1046 (1974)
89. M.R. Beasley, J.E. Mooij, and T.P. Orlando, Phys. Rev. Lett. **42**, 1165 (1979); S. Doniach and B.A. Hubermann, Phys. Rev. Lett. **42**, 1169 (1979); B.I. Halperin and D. R. Nelson, J. Low Temp. Phys. **36**, 599 (1979); L.A. Turkevich, J. Phys. C **12**, L385 (1979)
90. P. Minnhagen, Rev. Mod. Phys. **59**, 1001 (1987)
91. A. Kovner and B. Rosenstein, Phys. Rev. B **42**, 4748 (1990)
92. C. Timm, D. Manske, and K.H. Bennemann, Phys. Rev. B **66**, 094515 (2002)
93. F. Schäfer, C. Timm, D. Manske, and K.H. Bennemann, J. Low Temp. Phys. **117**, 223 (1999)
94. D. Manske and K.H. Bennemann, Physica C **341-348**, 83-86 (2000); D. Manske, T. Dahm, and K.H. Bennemann, Phys. Rev. B **64**, 144520 (2001)
95. V. Ambegaokar, B.I. Halperin, D.R. Nelson, and E.D. Siggia, Phys. Rev. Lett. **40**, 783 (1978); V. Ambegaokar and D. Seitel, Phys. Rev. B **19**, 1667 (1979); V. Ambegaokar, Halperin, D.R. Nelson, and E.D. Siggia, Phys. Rev. Lett. **21**, 1806 (1980)
96. J. Bardeen and M.J. Stephen, Phys. Rev. **140**, 1197A (1965)
97. C.C. Homes, S.V. Dordevic, M. Stongin, D.A. Bonn, R. Liang, W.N. Hardy, S. Komiya, Y. Ando, G. Yu, N. Kaneko, X. Zhao, M. Greven, D.N. Basov, and T. Timusk, Nature **430**, 539 (2004)
98. M.H. Cohen, L.M. Falicov, and J.C. Phillips, Phys. Rev. Lett. **8**, 316 (1962)
99. J. Zasadzinski, Chap. 15 of this volume
100. M.B.J. Meinders, H. Eshes, and G.A. Sawatzky, Phys. Rev. **48**, 3916 (1993)
101. T. Dahm, D. Manske, and L. Tewordt, Phys. Rev. **B** 54, 6640 (1996)
102. G. Hildebrand, E. Arrigoni, J. Schmalian, and W. Hanke, Eur. Phys. J. B **8**, 195 (1999)
103. S. Grabowski, J. Schmalian, M. Langer, and K.H. Bennemann, Phys. Rev. B **55**, 2784 (1997)
104. D.M. King, Z.-X. Shen, D.S. Dessau, B.O. Wells, W.E. Spicer, A.J. Arko, D.S. Marshall, J. DiCarlo, A.G. Loeser, C.H. Park, E.R. Ratner, J.L. Peng, Z.Y. Li, and R.L. Greene, Phys. Rev. Lett. **70**, 3159 (1993)
105. A. Liebsch, and A. Lichtenstein, Phys. Rev. Lett. **84**, 001591 (2000)
106. A. Damascelli et al., Phys. Rev. Lett. **85**, 5194 (2000)
107. I.I. Mazin and D.J. Singh, Phys. Rev. Lett. **82**, 4324 (1999)
108. K.K. Ng, and M. Sigrist, Europhys. Lett. **49**, 473 (2000)
109. D.K. Morr, P.F. Trautmann, and M.J. Graf, Phys. Rev. Lett. **86**, 5978 (2001)
110. K. Ishida, H. Mukuda, Y. Minami, Y. Kitaoka, Z.Q. Mao, H. Fukazawa, and Y. Maeno, Phys. Rev. B**64**, 100501(R) (2001)
111. T. Mizokawa, L.H. Tjeng, G.A. Sawatzky, G. Ghiringhelli, O. Tjernberg, N.B. Brookes, H. Fukazawa, S. Nakatsuji, and Y. Maeno, Phys. Rev. Lett. **87**, 077202 (2001)
112. H. Kontani, and K. Ueda, Phys. Rev. Lett. **80**, 5619 (1998)
113. K. Kuroki, M. Ogata, R. Arita, and H. Aoki, Phys. Rev. B **63**, 060506(R) (2001)
114. A.J. Layzer and D. Fay, *Proc. Int. Low Temp. Conf.*, St. Andrews (LT-11) (Academic Press, London, 1968); P.W. Anderson and W.F. Brinkmann, In: *The Physics of Liquid and Solid Helium*, ed. by K.H. Bennemann and J.B. Ketterson (Wiley-Interscience, New York, 1978), Vol. 2

115. Y. Aiura, Y. Yoshida, I. Hase, S. I. Ikeda, M. Higashiguchi, X. Y. Cui, K. Shimada, H. Namatame, M. Taniguchi, and H. Bando, Phys. Rev. Lett. **93**, 117005 (2004)
116. M. Sigrist and T.M. Rice, Z. Phys.B: Condensed Matter **68**, 9 (1987)
117. Z.Q. Mao, Y. Maeno, Y. Mori, S. Sakita, S. Nimori, and M. Udagawa, Phys. Rev. B **63**, 144514 (2001)
118. For a more detailed discussion of the behavior of the spin susceptibility see D, Vollhardt and P. Wölfle, In: *The Superfluid Phases of Helium*^{3} (Taylor and Francis, London, 1990)
119. C. Timm and K.H. Bennemann, Phys. Rev. Lett. **84**, 4994 (2000)
120. T. Valla, A.V. Fedorov, P.D. Johnson, and S.L. Hulbert, Phys. Rev. Lett. **82**, 2179 (1999)
121. J.A. Skinta, T.R. Lemberger, T. Greibe, and M. Naito, Phys. Rev. Lett. **88**, 207003 (2002); J.A. Skinta, M.-S. Kim, T. Greibe, and M. Naito, *ibid*, 207005 (2002)
122. D. Manske, I. Eremin, and K.H. Bennemann, Phys. Rev. B **62**, 13922 (2000)
123. S.V. Borisenko, A.A. Kordyuk, T.K. Kim, A. Koitzsch, M. Knupfer, M.S. Golden, J. Fink, M. Eschrig, H. Berger, and R. Follath, Phys. Rev. Lett. **90**, 207001 (2003)
124. Z.M. Yusof, B.O. Wells, T. Valla, A.V. Fedorov, P.D. Johnson, Q. Li, C. Kendziora, S. Jian, and D. G. Hinks, Phys. Rev. Lett. **88**, 167006 (2002)
125. A.V. Puchkov, D.N. Basov, and T. Timusk, J. Phys. Condens. Matter **8**, 10049 (1996)
126. G. Blumberg, A. Koitzsch, A. Gozar, B.S. Dennis, C.A. Kendziora, P. Fournier, and R.L. Greene, Phys. Rev. Lett. **88**, 107002 (2002)
127. T.S. Nunner, J. Schmalian, and K.H. Bennemann, Phys. Rev. B **59**, 8859 (1999)
128. T.E. Mason, A. Schröder, G. Aeppli, H.A. Mook, and S.M. Hayden, Phys. Rev. Lett. **77**, 1604 (1996)
129. B. Lake, G. Aeppli, T.E. Mason, A. Schroder, D.F McMorrow, K. Lefmann, M. Isshiki, M. Nohara, H. Takagi, and S.M. Haydon, Nature (London) **400**, 43 (1999)
130. D. Manske, I. Eremin, and K.H. Bennemann, Phys. Rev. B **63**, 054517 (2001)
131. I. Eremin, D.K. Morr, A.V. Chubukov, K. Bennemann, and M.R. Norman Phys. Rev. Lett. **94**, 147001 (2005)
132. J.L. Tallon, C. Bernhard, H. Shaked, R.L. Hittermann, and J.D. Jorgensen, Phys. Rev. B **51**, 12911 (1995)
133. Ch. Renner, B. Revaz, J.-Y. Genoud, K. Kadowaki, and Ø. Fischer, Phys. Rev. Lett. **80**, 149 (1998)
134. A. Ino, C. Kim, M. Nakamura, T. Yoshida, T. Mizokawa, A. Fujimori, Z.-X. Shen, T. Kakeshita, H. Eisaki, and S. Uchida, Phys. Rev. B **65**, 094504 (2002)
135. G.V.M. Williams, J.L. Tallon, and J.W. Loram, Phys. Rev. B **58**, 15053 (1998)
136. G. Baumgärtel, J. Schmalian, and K.H. Bennemann, Phys. Rev. B **48**, 3983 (1993)
137. E.F. Paulus et al., Solid State Commun. **73**, 791 (1990)
138. G. Liang, J. Chen, M. Croft, K.V. Ramanujachary, M. Greenblatt, and M. Hegde, Phys. Rev. B **40**, 2646 (1989)
139. H. Takagi, S. Uchida, and Y. Tokura, Phys. Rev. Lett. **62**, 1197 (1989)
140. S. Doniach and M. Inui, Phys. Rev. B **41**, 6668 (1990)
141. C. Bernhard, J.L. Tallon, Th. Blasisus, A. Golnik, and Ch. Niedermeyer, Phys. Rev. Lett. **86**, 1614 (2001)
142. N. Miyakawa, J.F. Zasadzinski, L. Ozyuzer, P. Guptasarma, D.G. Hinks, C. Kendziora, and K.E. Gray, Phys. Rev. Lett. **83**, 1018 (1999)
143. J.C. Swihart, D.J. Scalapino, and Y. Wada, Phys. Rev. Lett. **14**, 106 (1965)
144. Z.A. Xu, N.P. Ong, Y. Wang, T. Kakeshita, S. Uchida, Nature **406**, 486 (2000)
145. Y. Wang, Z.A. Xu, T. Kekeshita, S. Uchida, S. Ono, Y. Ando, and N.P. Ong, Phys. Rev. B **64**, 224519 (2001)
146. P. Olsson and P. Minnhagen, Physica Scripta **43**, 203 (1991)
147. J. Adler, C. Holm, and W. Janke, Physica A **201**, 581 (1993)
148. A.V. Chubukov, C. Pépin , and J. Rech, Phys. Rev. Lett. **92**, 147003 (2004)
149. A.V. Chubukov, D.L. Maslov, Phys. Rev. B **69**, 121102 (2004)
150. A.V. Chubukov, D.L. Maslov, Phys. Rev. B **68**, 155113 (2003)
151. M. Kugler, Ø. Fischer, Ch. Renner, S. Ono, and Y. Ando, Phys. Rev. Lett. **86**, 4911 (2001)
152. D. Mandrus, L. Forro, and D. Koller, and L. Mihaly, Nature **351**, 460 (1991)
153. J.F. Zasadzinski, Bull. Am. Phys. Soc. **37**, 668 (1992)
154. S. Grabowski, Ph.D. thesis, Berlin (1996)
155. J.F. Zasadzinski, L. Qzyuzer, N. Miyakawa, K.E. Gray, D.G. Hinks, and C. Kendziora, Phys. Rev. Lett. **87**, 067005 (2001)
156. A.V. Chubukov, D. Pines, and J. Schmalian, this book, Vol. 1, p. 495.

157. A.A. Kordyuk, S.V. Borisenko, T.K. Kim, K. Nenkov, M. Knupfer, M.S. Golden, J. Fink, H. Berger, R. Follath, Phys. Rev. Lett. **89**, 077003 (2002)
158. J.F. Zasadzinski, L. Ozyuzer, N. Miyakawa, D.G. Hinks, and K.E. Gray, Physica C **341-348**, 867 (2000)
159. N. Bulut and D.J. Scalapino, Phys. Rev. B **53**, 5149 (1996)
160. A.A. Kordyuk, S.V. Borisenko, M.S. Golden, S. Legner, K.A. Nenkov, M. Knupfer, J. Fink, H. Berger, L. Forro, and R. Follath, Phys. Rev. B **66**, 014502 (2002)
161. D.L. Feng et al., Phys. Rev. Lett. **86**, 5550 (2001)
162. T. Nomura and K. Yamada, J. Phys. Soc. Jpn. **69**, 1856 (2000); ibid, cond-mat/0203453
163. K. Kuroki, M. Ogata, R. Arita, and H. Aoki, Phys. Rev. B **63**, 060506 (2001)
164. T. Imai et al., Phys. Rev. Lett. **81**, 3006 (1998)
165. M. Braden, Y. Sidis, P. Bourges, P. Pfeuty, J. Kulda, Z. Mao, and Y. Maeno, Phys. Rev. Lett. **92**, 097402 (2004)
166. D.J. Scalapino, Phys. Rep. **250**, 329 (1995)
167. Y. Hasegawa, K. Machida, and M. Ozaki, J. Phys. Soc. Jpn. **69**, 336 (2000)
168. K. Izawa, H. Takahashi, H. Yamaguchi, Y. Matsuda, M. Suzuki, T. Sasaki, T. Fukase, Y. Yoshida, R. Settai, and Y. Onuki, Phys. Rev. Lett. **86**, 2653 (2001)
169. M.A. Tanatar, M. Suzuki, S. Nagai, Z.Q. Mao, Y. Maeno, and T. Ishiguro, Phys. Rev. Lett. **86**, 2649 (2001)
170. M.E. Zhitomirsky, and T.M. Rice, Phys. Rev. Lett. **87**, 057001 (2001)
171. J.F. Annett, G. Litak, B.L. Gyorffy, and K.I. Wysokinski, Phys. Rev. B **66**, 134514 (2002)
172. Y. Maeno, T.M. Rice, and M. Sigrist, Phys. Today **54**, 42 (2001)
173. K.G. Sandeman, G.G. Lonzarich, A.J. Schofield, Phys. Rev. Lett. **90**, 167005 (2003)
174. S. Nakatsuji, and Y. Maeno, Phys. Rev. Lett. **84**, 2666 (2000)
175. Z. Fang and K. Terakura, Phys. Rev. B **64**, 020509(R) (2001)
176. L. Tewordt, J. Low Temp. Phys. **15**, 349 (1974)
177. G. Baym and L. P. Kadanoff, Phys. Rev. **124**, 287 (1961)
178. G. Baym, Phys. Rev. **127**, 1391 (1962)
179. F. Schäfer, D. Manske, and K.H. Bennemann, FU Report (2000)
180. J.M. Kosterlitz and D.J. Thouless, J. Phys. C **6**, 1181 (1973)
181. P. Minnhagen, Phys. Rev. B **23**, 5745 (1981)
182. D. R. Nelson and J. M. Kosterlitz, Phys. Rev. Lett. **39**, 1201 (1977)

24 Superfluid ^{3}He and the Cuprate Superconductors

A. J. Leggett Department of Physics, University of Illinois, USA

24.1 Introduction: Bose Condensation and Cooper Pairing

This chapter is devoted to a comparison of two apparently very different physical systems, superfluid liquid ^{3}He and the cuprate superconductors. What these two systems have in common, at least according to almost universal belief, is that they are systems of Fermi particles, already very degenerate, in which not only are the "quasi-dielectronic molecules" which we call Cooper pairs formed, but once formed are automatically forced to occupy a *single* two-particle state, as regards both their center-of-mass motion and their internal structure. This phenomenon is very reminiscent of the phenomenon of Bose condensation which is almost universally believed to occur in the boson system liquid ^{4}He below the λ-point, and has in the last few years been spectacularly demonstrated in a second Bose system, trapped ultracold monatomic alkali gases. It is therefore appropriate to start this chapter with a brief discussion of the more easily understood phenomenon of simple Bose condensation and its relationship to the apparently more complex behavior associated with Cooper pairing in degenerate Fermi systems.

Let us consider, first, a simple system of N bosons for which the internal structure and any internal degrees of freedom can be neglected. Such a system is realized, at least in the context of the low-energy phenomena of interest to us, in liquid ^{4}He and also in the boson alkali gases to the extent that one and only one hyperfine species is present (and we can safely neglect effects such as the bending of the quantization axis in space, etc. (see [1], Sect. IV.E). The total particle number N is assumed to be large compared to unity. While various definitions of Bose or Bose–Einstein condensation (BEC) have been proposed in the literature, the conceptually simplest (and the one most appropriate to the alkali gases) seems to me to be that given many years ago by Penrose and Onsager [2], which goes roughly as follows. Consider an arbitrary many-body state of the system (which need not be in or even close to thermal equilibrium) and define the single-particle density matrix $\hat{\rho}_1(\mathbf{r}, \mathbf{r}', t)$ in the standard way. Since this quantity, regarded as a matrix in its indices, is from its definition Hermitian, it may be diagonalized by choosing an appropriate basis of single-particle eigenfunctions (which need not be eigenfunctions of the Hamiltonian, and will in the general case be time-dependent). Let's now examine the eigenvalues of $\hat{\rho}_1$. There are two cases of prime interest: (A) All eigenvalues are of order unity (not of order of the total particle number N). We call such a state of the system "normal" (i.e. not Bose-condensed). (B) One eigenvalue is of order N, the rest of order 1. In this case we say we have ("simple") Bose–Einstein condensation (BEC). The large eigenvalue is called the "number of condensed particles" and denoted N_o so that the ratio N_o/N is the "condensate fraction", and the unique single-particle eigenfunction (in general time-dependent) associated with this special eigenvalue is called the "condensate wave function", $\chi_o(\mathbf{r}, t)$ or when multiplied by a factor, the "order parameter". The above discussion (or considerations involving the so-called "spontaneous breaking of U(1) symmetry", which are in effect equivalent though to my mind conceptually less satisfactory) is the standard basis for discussion of BEC systems in the literature. It should be noted that by itself it fails to answer two obvious questions, namely: (1) why should *any* eigenvalue of the single-particle density matrix ever be of order N?, (2) why should not *more than one* eigenvalue be of order N? The most obvious answer to question (1) is that (as shown by Einstein in his original 1925 paper) for a free Bose gas in thermal equilibrium the result follows from an explicit calculation, and that (contrary to the view expressed by some of his contemporaries) it is reasonable to expect this result to be robust against small departures from equilibrium and/or weak interatomic interactions. (In fact, it was shown many years ago by Gavoret and Nozières [3] that so long as perturbation theory in the interactions converges, the system indeed shows BEC in equilibrium at $T = 0$). The second question is more tricky. For a Bose system with weak *repulsive* interatomic interactions, it turns out that macroscopic occupation of more than one single-particle state is energetically very costly, because of the "Fock" term in the mean-field energy (which has a sign opposite to that in the more familiar Fermi case). Consequently, the effect of a weak interatomic interaction is actually to reinforce and stabilize "simple" BEC. By contrast, a weak attraction between atoms favors macroscopic occupation of several different single-particle states (a phenomenon sometimes called "fragmentation"), or what turns out to be effectively equivalent, "collapse" of the system in real (coordinate) space. The general belief, which seems consistent with existing experiment, is that in an alkali Bose gas with weakly attractive interatomic interactions no stable BEC state is possible.

If we confine ourselves from now on to the repulsive case, the existence of BEC has many spectacular consequences. The most direct one is a characteristic phenomenon seen in the alkali gases when they are confined, as is the usual experimental situation, called a harmonic trap: Above the temperature of transition to the BEC state the atoms have a thermal distribution which is little different from the classical Maxwell one and the atomic cloud therefore has a width determined by the condition that the potential at its edge should be of order kT. When BEC sets in, however, a finite fraction of the atoms (which at low temperatures is nearly equal to one) by definition occupies a *single* one-particle state. In thermal equilibrium, for a noninteracting gas, this would be the harmonic-oscillator ground state, and since the width of the latter is much smaller than that of the thermal cloud, a very narrow and high peak in the density should rise out of the thermal background. In real life the repulsive interactions broaden the (self-consistently determined) condensate wave function so that it is appreciably wider than the harmonic groundstate, but the effect is still spectacular; it was in fact the most direct evidence for the onset of BEC.

A second spectacular manifestation of the effects of BEC is the complex of phenomena which are usually lumped together under the name of "superfluidity", and which generally speaking take a cleaner form in He-II (the phase of liquid ^{4}He below the λ-point) than in the alkali gases. The reason for this has to do with the different potential "geometries" in the two cases. The question of exactly what are the conditions for each of the characteristic phenomena of superfluidity to occur in a given system and geometry is a quite subtle one (see, e.g. [1]), Sect. VI), but it seems clear that a necessary condition for the most striking manifestation, that of persistent flow in a ring geometry, is a sufficiently strong interatomic repulsion. It is easily shown that a noninteracting Bose gas, even when Bose-condensed, will not show this effect.

Finally, it should be remarked that while the ^{4}He atom has no "internal" degrees of freedom (at least to the extent that it remains in its electronic groundstate), in the case of the BEC alkali gases there exists such an internal degree of freedom, namely that associated with the electron and nuclear spin, neither of which is zero. As a result, in the general case the description of BEC in these systems is more complicated than that given above, and in particular, even when all interactions are repulsive, the occurrence of "fragmented" condensates cannot be excluded.

I now turn to the Fermi case. As pointed out long ago by Yang [4], in this case there are rigorous theorems which forbid the single-particle density matrix to have a macroscopic eigenvalue or indeed any eigenvalue greater than 1, as one would intuitively expect. However, there is nothing to prevent the *two-particle* density matrix defined explicitly below from having one or more macroscopic eigenvalues. We can look at this question intuitively by considering our N-fermion system at zero temperature and make (at present for no obviously good reason!) the following ansatz for the groundstate

$$\Psi_N(\mathbf{r}_1\mathbf{r}_2\ldots\mathbf{r}_N:\sigma_1\sigma_2\ldots,\sigma_N) = \mathcal{A}\,\phi_i(\mathbf{r}_1,\mathbf{r}_2\sigma_1\sigma_2)\phi_j(\mathbf{r}_3\mathbf{r}_4\sigma_3\sigma_4)\ldots \phi_w(\mathbf{r}_{N-1}\mathbf{r}_N\sigma_{N-1}\sigma_N)\,, \tag{24.1}$$

where $\mathcal{A}$ is the operator which antisymmetrizes the expression with respect to simultaneous interchange of the coordinates and spins of any two fermions. We do not yet specify whether any two or more of the subscripts $i, j \ldots w$ are identical. Provided the antisymmetrization is properly carried out, there are no a priori constraints in this respect. Intuitively, an ansatz of the form (24.1) (with an appropriately "bound" form of the ϕ's) could correspond to pairing of the fermions (say for definiteness electrons) into a sort of "di-electronic molecules" in as many different states as there are different values of the subscript. We can now ask for the number of "molecules" in each different state, and in particular whether one or several states are macroscopically occupied. The analogy with BEC of simple bosonic systems is then obvious.

Actually, the above approach, while perhaps helpful to one's intuition, is of limited practical usefulness. The reason is that the literal interpretation of a many-body wave function of the form (24.1) in terms of "molecules" which are in the states ϕ_i is useful, if at all, only in the limit that the effective "radius" of the pair states described by the functions $\phi_i(\mathbf{r}_1\mathbf{r}_2\sigma_1\sigma_2)$. That is the range of the variable $|\,\mathbf{r}_1 - \mathbf{r}_2\,|$ over which the function is of appreciable weight is small compared to the mean interparticle distance. A condition which may be fulfilled in for example liquid deuterium but is certainly not in most of the systems of interest in the present context which are predominantly in precisely the opposite limit (range of ϕ much *greater* than mean interparticle distance). In this limit (sometimes called the "BCS" as opposed to the "Bose" limit), the "molecular" wave functions ϕ_i, while certainly definable, have little physical significance. A much more useful quantity is the two-particle density matrix $\hat{\rho}_2$ defined (at $T = 0$) by

$$\hat{\rho}_2(\mathbf{r}_1\sigma_1\mathbf{r}_2\sigma_2:\mathbf{r}'_1\sigma'_1\mathbf{r}'_2\sigma'_2:t) \equiv \langle\, \psi^\dagger_{\sigma_1}(\mathbf{r}_1)\psi^\dagger_{\sigma_2}(\mathbf{r}_2)\psi_{\sigma_3}(\mathbf{r}_3)\psi_{\sigma_4}(\mathbf{r}_4)\,\rangle\,. \tag{24.2}$$

This in some sense automatically takes into account the complications introduced by the need for antisymmetrization. For example, the expressions for expectation values of two-particle quantities such as the potential energy, or more precisely for the contributions to them from the formation of "pairs", look horribly complicated when expressed in terms of the ϕ_i. However, when written in terms of $\hat{\rho}_2$ have a very simple form, in fact one exactly analogous to the familiar expressions for the two particle case (see, e.g. [5]). Thus, for any quantitative purpose it is highly desirable to work in terms of $\hat{\rho}_2$.

The quantity $\hat{\rho}_2$ is defined for an arbitrary state of the system as a function of the four coordinate-plus-spin variables $\mathbf{r}_1\sigma_1$, $\mathbf{r}_2\sigma_2$ and the time t. Let us for pedagogical convenience temporarily ignore the spin coordinates, fix t and regard $\hat{\rho}_2$ as a matrix with respect to the *pair* of indices $\mathbf{r}_1\mathbf{r}_2$, which replaces the single index $\mathbf{r}$ in the Bose case. (We could of course equally well have chosen to consider the pair $\mathbf{r}_1$, $\mathbf{r}_3$, but there is a reason for the present choice). As in the Bose case, the matrix is Hermitian and can thus be diagonalized. Let the eigenvalues be denoted n_i and the corresponding eigenfunctions be χ_i. The a priori constraints on the n_i are quite weak (see Yang's original paper [4]), and in particular there is nothing to forbid one or more of them being of order N. Thus, just as in the Bose case, it is possible to define (simple) "quasi-BEC" by the condition that one *and only one* of the n_i's is of order N and also, if we wish, "fragmented quasi-BEC" by the statement that more than one is of order N. However, it turns out that, at least for the systems of interest in the present context, fragmented quasi-BEC is likely to be relevant only under conditions so extreme that they have probably not been realized experimentally [6].

As is well-known, the state realized in conventional superconductors, and also in superfluid ^{3}He and the cuprate superconductors, is almost universally believed to correspond in the above language to "simple BEC", and the entities which are condensed are called Cooper pairs. The concept of a Cooper pair in the abstract perhaps not totally well defined, but one can as a matter of definition think of it as the object whose wave function $\chi_o(\mathbf{r}_1\mathbf{r}_2) \equiv F(\mathbf{r}_1\mathbf{r}_2)$ is just the eigenfunction of $\hat{\rho}_2$ corresponding to the single macroscopic eigenvalue. It is worth noting that the actual eigenvalue N_0 (which one might regard as the "number of condensed Cooper pairs"), while it is "of order of" the total particle number N, is quite a small fraction of order kT_c/ϵ_F of N, at least in the cases of interest to us here. It may also be noted that in contrast to the Bose case an attractive interparticle interaction is usually conducive to (simple) quasi-BEC (as in the original BCS theory) rather than the reverse.

The "pair wave function" or "order parameter" $F(\mathbf{r}_1\mathbf{r}_2)$ (often written as $\langle\psi(\mathbf{r}_1)\psi(\mathbf{r}_2)\rangle$) may be expressed as a function of the center-of-mass coordinate $\mathbf{R} = (\mathbf{r}_1 + \mathbf{r}_2)/2$ and the relative coordinate $\mathbf{r} \equiv \mathbf{r}_1 - \mathbf{r}_2$ (and the spins, which we can now safely put back in). In the case of the conventional superconductors the general belief is that the dependence on the "internal" degrees of freedom (relative coordinate and spin) is uniquely fixed, in the groundstate and more generally in thermal equilibrium, by the energetics, so that the Cooper pairs resemble the ^{4}He atom in having effectively no internal variable. On the other hand, there is nothing which excludes a priori the possibility that one or more "internal" symmetries are spontaneously broken in the groundstate, so that the pairs have one or more "orientational" degrees of freedom, analogously to the hyperfine degree of freedom in the BEC alkali gases. We will see below that this situation is in fact realized both in superfluid ^{3}He and in a sense in the cuprates.

Obvious Differences and Some Similarities

At first sight the physical systems formed by the superfluid phases of liquid ^{3}He on the one hand, and the cuprate superconductors on the other could hardly be more different. In fact, as already remarked, at the level of superficial inspection just about the only thing they have in common is that both are Fermi systems in which the formation of Cooper pairs leads to the characteristic phenomena of superfluidity (superconductivity). There are many obvious differences. I now list some of them.

The first and most obvious difference is of course that in the case of the cuprates, as in all metallic superconductors, the relevant fermions (electrons) are charged, whereas in liquid ^{3}He they are neutral. This has the consequence that the "macroscopic quantum phenomena" characteristic of the paired superfluid state look at first sight very different in the two systems. However, this is something we are already used to in the context of the analogy between superfluidity in ^{4}He and "old-fashioned" superconductivity (see, e.g. [7]), so it is not necessary to discuss it further here.

A second glaring difference is that the electrons in the cuprates as in all metals move in the background of a periodic crystalline lattice, whereas in liquid ^{3}He the atoms move in free space. Actually, the difference is even more pronounced; whereas in most traditional superconductors it is at least not a bad zeroth-order approximation to treat the conduction electrons as moving freely in three-dimensional space, the cuprates have a strongly layered structure, so that the approximation of free motion, to the extent that it is valid at all, can be applied only for the motion parallel to the CuO_2 planes; for motion along the third axis (c-axis) it is certainly essential to take the lattice into account from the start.

A third difference lies in the degree of sensitivity to parameters which can be varied. Liquid ^{3}He at temperatures of the order of the superfluid transition temperature T_c is about as pure a system as one can find in nature: any chemical or even isotopic (^{4}He) impurities rapidly condense on the walls of the experimental cell, so that the only parameters which can be meaningfully varied are the macroscopic thermodynamic ones, pressure and magnetic field and of course temperature. As we shall see, the onset temperature and nature of the superfluid state does in fact depend in an interesting way on those variables. By contrast, superconductivity in the cuprates, while to some extent sensitive to pressure and field, is much more dramatically influenced by other variables, in particular by the details of the chemical stoichiometry, which in most cases is variable over a wide range and by the presence of various trace elements such as Zn.

Why, then, is there any hope at all that any of the extensive research done on superfluid ^{3}He could have any relevance to the cuprates? One can think immediately of three arguments in this direction, although to be sure none of them is a priori overwhelming. First, consider the ratio of the transition temperature T_c to the Fermi temperature T_F (the fact that there is no completely unambiguous definition of the latter quantity for a strongly interacting system need not worry us, since we are interested only in orders of magnitude). For the classic pre-1975 superconductors T_c always lies below 25 K, while T_F is typically $\sim 3\times10^4$ K. Thus, the ratio T_c/T_F is never greater than $\sim10^{-3}$ and often considerably smaller. For the cuprates T_c can be up to 160 K, and T_F can be plausibly argued to be as small as a few thousand K. Thus, T_c/T_F is typically of order 3×10^{-2}, Finally, for liquid ^{3}He T_c is of the order 3 mK and T_F of order 0.5 K, so that the ratio is $\sim10^{-2}$. We see that in this respect the cuprates are considerably closer to ^{3}He than they are to the classic superconductors. The heavy-fermion systems have an even larger ratio ($\sim$0.2) of T_c/T_F, if T_F is defined in terms of the low-temperature effective mass.

A second and perhaps more fundamental reason has to do with possible mechanisms of Cooper pairing. As is well known, in the BCS theory which is believed to describe the classic superconductors the mechanism is an effective attraction resulting from the exchange of virtual phonons. It is an important feature of the BCS theory that the mediating excitation (the phonon) is primarily associated with the ionic cores, i.e. with a system different from the conduction electrons which form the Cooper pairs. On the other hand, it seems almost certain for the heavy-fermion superconductors, and at least probable for the cuprates, that phonons do not play the dominant role and that the origin of Cooper pairing must therefore be sought in the conduction-electron system itself. In this respect the situation is similar to that in liquid ^{3}He, where there simply is nothing outside the ^{3}He atoms. In this system, trivially, the mechanism of pairing must reside within the system of paired fermions.

Finally, there is the question of the symmetry of the order parameter (pair wave function). In the classic superconductors this is believed to be isotropic, or more accurately to have the symmetry of the lattice. Formally, the order parameter transforms according to the identity representation of the symmetry group of the Hamiltonian. By contrast, in superfluid ^{3}He it is by now almost universally believed that the order parameter is of "exotic" form, i.e. does not have the (orbital) symmetry (in this case SO(3)) of the underlying Hamiltonian, and more specifically that is of the p-wave form, though different in detail for the different superfluid phases (see Sect. 24.5). It is by now generally though not universally believed that in the cuprates, where the symmetry group of the CuO_2 planes is exactly or approximately that of the square, the order parameter has the so-called $d_{x^2-y^2}$ form. So again it transforms according to a nontrivial representation of the symmetry group of the Hamiltonian. Once more, some heavy-fermion systems may be even closer to superfluid ^{3}He systems in this respect. The pairing state in UPt_3 is widely believed to be of p-wave type. Actually, this resemblance between ^{3}He and the cuprates or the heavy-fermion systems may not be quite as significant as it looks at first sight, see Sect. 24.5. However, it has undoubtedly played a role in the historical development of the analysis of cuprate superconductivity.

24.2 The Normal State: Is the Fermi-Liquid Picture Valid?

It is convenient for the purpose of the present discussion to focus on a single-plane cuprate such as Tl-2201 or Hg-1201, thus setting aside some questions which are specific to the multilayer systems such as YBCO. Furthermore, I shall make two assumptions which, while not universally agreed, would probably win the consent of a large majority of the relevant community: (a) In the normal or in the superconducting state, most of the interesting phenomena are primarily associated with the behavior of the conduction electrons in the CuO_2 planes. Moreover, the process of electron transfer from one plane to the next is sufficiently weak that it may be neglected to a first approximation. (b) The kinetic energy of the ions (i.e. phonons) does not play an important role. Then, at least to a first approximation, the system of fermions we need to consider in the cuprates consists of the conduction electrons which move within the single CuO_2 planes in the "external" potential provided by the static ionic lattice and interact via the standard Coulomb interaction. This is screened by the core electrons associated with the ions. To a first approximation it is reasonable to approximate the effects of the screening by a single number the real high frequency dielectric constant ϵ_∞. At a similar level of approximation liquid ^{3}He may be viewed as a system of neutral atoms with total (nuclear) spin 1/2 obeying Fermi statistics and interacting via a short-range potential of the van der Waals type. (The small nuclear dipole–dipole interaction plays an important role in the superfluid phase, see Sect. 24.5, but can be neglected to a first approximation in the normal phase). In both cases the inter-fermion interactions are in some sense "strong". The repulsive part of the van der Waals interaction in ^{3}He approximates a hard core at interatomic distances $\lesssim$ 2.5 Å, while the interatomic spacing in the liquid at melting pressure is $\sim$ 3.4 Å. In the case of the cuprates the interaction of two conduction electrons on the same lattice site is believed to be of the order of 10 eV. This is more than an order of magnitude greater than the Fermi energy of a noninteracting system of the same density.

It was precisely in the context of a three-dimensional neutral Fermi system, such as (normal) liquid ^{3}He, that Landau developed his celebrated theory of a Fermi liquid. [8] While it is conventional to define what one means by a Fermi liquid in terms of the behavior of the microscopic Green's functions of the system, for our purposes a less technical description similar to that used in Landau's first paper [8] will suffice. Crudely speaking, a Fermi liquid is a Fermi system such that if we start from the noninteracting gas, perturbation theory in the interaction strength converges. As a result the character of the groundstate and low-lying excited states (or more accurately the vast majority of the latter) is similar to that of the corresponding states of the noninteracting gas. To be a little more concrete, the excited energy eigenstates of the latter can be specified quite generally by enumerating the *difference* $\delta n(\mathbf{p}\sigma)$ between the number $n(\mathbf{p}, \sigma)$ of particles in momentum state $\mathbf{p}$ and with spin[1] σ and its value in the groundstate: evidently $\delta n(\mathbf{p}\sigma)$ is 0 or +1 for $|\mathbf{p}| > p_F$ and 0 or -1 for $|\mathbf{p}| < p_F$, where the Fermi momentum p_F is uniquely fixed as a function of density $n(p_F^3 = 3\pi 2n)$. The quantity $n(\mathbf{p}\sigma)$ defined in this way (taking $n(\mathbf{p}\sigma) \equiv \theta(p_F - |\mathbf{p}|)$ by definition in the groundstate) is usually called the "number of quasiparticles in state $(\mathbf{p}\sigma)$". It is clear that generally speaking the lowest-lying energy eigenstates will be those where only a few $\delta n(\mathbf{p}\sigma)$ are different from zero. The interacting system is now said to be a Fermi liquid if the structure of the (interacting) groundstate and *low-lying* excited states is such that to a good approximation the latter can be labeled by a set of quantities $\{\delta n(\mathbf{p}\sigma)\}$ having the above property ($\delta n(\mathbf{p}\sigma) = 0$ or +1 for $|\mathbf{p}| > p_F$, $\delta n(\mathbf{p}\sigma) = 0$ or -1 for $|\mathbf{p}| < p_F$). Furthermore the energy can be expanded in the $\delta n(\mathbf{p}\sigma)$ as follows

[1] For notational simplicity I ignore here the fact that the spin should strictly be treated as a vector, in particular as regards its occurrence in f: see, e.g. [9].

$$E\{\delta n(\mathbf{p}\sigma)\} = E_o + \sum_{\mathbf{p}\sigma} \epsilon(\mathbf{p})\delta n(\mathbf{p}\sigma) + \frac{1}{2} \sum_{pp'\sigma\sigma'} f(\mathbf{pp}'\sigma\sigma')\delta n(\mathbf{p}\sigma)\delta n(\mathbf{p}'\sigma'), \tag{24.3}$$

where the groundstate energy E_o is a constant. The coefficients ϵ and f have as $|\mathbf{p}| \to p_F$ the following behavior

$$\epsilon(\mathbf{p}) = \text{const.}(|\mathbf{p}| - p_F), \tag{24.4}$$

$$f(\mathbf{pp}'\sigma\sigma') = \text{independent of } |\mathbf{p}|, |\mathbf{p}'|,$$

$$f \equiv (\mathrm{d}n/\mathrm{d}\epsilon)^{-1} \sum_i \left(F_l^s + F_l^a \sigma\sigma'\right) P_l(\hat{\mathbf{p}} \cdot \hat{\mathbf{p}}'),$$

where $\mathrm{d}n/\mathrm{d}\epsilon$ is the single-particle density of states at the Fermi surface and P_l the Legendre polynomial. (Clearly, the noninteracting system is the special case $\epsilon = (p_F/m)(|\mathbf{p}| - p_F)$, $f = 0$). The above expression for the energy, which is strictly diagonal in the quantities $\delta\mathrm{n}(\mathbf{p}\sigma)$, is never exactly valid in an interacting system, because of terms in the Hamiltonian which lead to mutual scattering of quasiparticles between states with different $\mathbf{p}$,σ; however, arguments based on the Pauli principle show that it is self-consistent to take the probability of scattering to tend to zero as $(|\mathbf{p}| - p_F)^2$, that is, faster than the excitation energy of the quasiparticle in question (see e.g., [9]).

It is worth noting that in the above discussion it is *not* assumed that the interaction terms in the Hamiltonian conserve the total momentum $\mathbf{P}$ (or spin $\mathbf{S}$) of the system. The question of whether or not there is a one–one correspondence, as above, between the groundstate and low-lying states of the interacting and noninteracting systems is independent of that consideration. If the total momentum is in fact conserved, as e.g. in the case of liquid ^{3}He, then the quasiparticle states (low-lying approximate energy eigenstates) are exact eigenstates of total momentum with eigenvalue

$$\mathbf{P} = \sum_p \mathbf{p}\delta n(\mathbf{p}\sigma) \tag{24.5}$$

(and one can make a similar statement for the total spin). In the opposite case, the labeling is still valid, but the total momentum operator is in general not diagonal in the basis formed by the quantity $\delta n(\mathbf{p}\sigma)$.

It should be strongly emphasized that for any given real-life system of fermions the validity of Landau Fermi-liquid theory is an *assumption*, which can almost never be proved from first principles, but at best can be substantiated by comparison of the predictions made on that hypothesis with experiment. It should be stressed that some of these predictions are nontrivial, in the sense that the number of predicted relations between different experimental quantities is larger than the number of a priori undetermined parameters needed to characterize the fundamental quantities $\epsilon(\mathbf{p})$ and $f(\mathbf{pp}'\sigma\sigma')$.

Is liquid ^{3}He in fact a Fermi liquid? If in the definitions one were to treat "low-lying" as meaning "with (single-quasiparticle) energies much less than 1 K but much greater than 3 mK", then all the evidence is that the answer is yes. Whenever the properties of liquid ^{3}He in the normal phase well below 100 mK have been measured, it has always proved possible to fit the results into the framework of Fermi-liquid theory. However, the very existence of the superfluid phases shows that by a technically accurate definition liquid ^{3}He is ironically *not* a Fermi liquid. Indeed, the excitation energies of the *very* low-lying excited states relative to the groundstate are of the generalized BCS form (see Sect. 24.4) and certainly do not correspond to expressions (24.3) and (24.4). This situation is usually summed up by saying that liquid ^{3}He in the *normal* state behaves as a Fermi liquid. It then follows that the phenomenon of Cooper pairing can be discussed on the basis of this picture, see Sect. 24.4.

As already remarked, the Landau Fermi-liquid theory was originally developed with specific reference to liquid ^{3}He, which is both neutral and translation-invariant. The generalization of the the ory to a charged but still translationally invariant system was made soon thereafter by Silin [10]. In effect, one simply generalizes the philosophy of the standard random phase approximation by using the original Landau theory to calculate the "local" responses and then, when necessary, replacing the external electric field by the sum of the external field and that produced self-consistently by the charge response of the system itself. The generalization to the case of a finite crystalline potential is a little more tricky and is discussed, for example, in [11]. The simplest formulation is probably to start by ignoring the interconduction electron interactions and solving for the eigenstates of the single-particle terms in the Hamiltonian (kinetic plus lattice potential energy); in this way one generates the usual Bloch-wave states, which are eigenstates not of total momentum **P** but rather of total quasimomentum **K**, and are labeled correspondingly by a quasimomentum index **k** and spin σ. The resulting system is the analog of the noninteracting gas in the translation-invariant case. Then one turns on the interaction and *assumes* that the resulting groundstate and low-lying excited states can be put into one–one correspondence with these of the starting ("noninteracting") state. It is important to note that, because of the possibility of umklapp processes, the interaction does not conserve the total quasimomentum **K**; hence while one can still *label* the resulting states by $\delta n(\mathbf{k}, \sigma)$, **k** no longer has the physical interpretation of true quasimomentum (and in fact the operator $\hat{\mathbf{K}}$ is nondiagonal in the $\delta \mathrm{n}(\mathbf{k}, \sigma)$ basis). This observation has some importance in the context of the discussion of the "mid-infrared" peak in the cuprates: see next section.

Is the normal state of the cuprates a Fermi liquid (in the same sense as that in which the normal state of ^{3}He can be said to be a Fermi liquid)? This is one of the most hotly debated questions in the theory of cuprate superconductivity, and to discuss it we need to define it carefully. Let us first ask the question: Suppose we could somehow cool the system down to zero temperature while forbidding the superconducting transition to occur. Would the resulting "groundstate" and low-lying excited states conform to the definition of a Fermi liquid? In the case of liquid ^{3}He (where there is apparently no practical way of suppressing the superfluid transition, so that this really is only a thought-experiment) the answer is almost certainly yes. In the case of the cuprates, there are real-life possibilities of suppressing the superconducting transition, in particular (a) by the application of a sufficiently strong magnetic field and (b) by doping the CuO_2 planes with certain types of impurity such as Zn. Although for practical reasons rather few systematic studies have been done of the resultant state an answer seems to be emerging which is rather intriguing [12]: There appears to be a critical concentration $p_c \sim 0.19$ of the hole-doping parameter p (number of holes per CuO_2 units) such that for $p > p_c$ the lowest-lying states indeed have Fermi-liquid like character. However, to the left of this point in the phase diagram (i.e. for $p < p_c$) the behavior appears to be qualitatively different, with the system tending continuously towards an *insulating* state as the temperature falls towards zero. If this is really the correct picture, then it is not entirely clear that the question "does the normal state of the cuprate superconductors behave as a Fermi liquid?" has any clearly defined meaning. Of course one could in any case argue that the system placed in a strong ($\sim$ 30–60 T)magnetic field, or doped substantially with Zn impurities, is not "the same" system as a true (superconducting) cuprate, and that therefore no conclusions can be drawn from experiments of the above type. In that case it is conceivable that the true situation is the following: If one, by theoretical fiat rather than by any realistically available experimental means, could forbid the formation of Cooper pairs in the (pure, zero-field) system while cooling it to zero, then the "groundstate" and low-lying states would indeed be of Fermi-liquid type, as in the corresponding thought-experiment with ^{3}He. However, in both cases we should expect the Fermi-liquid description to break down badly when the temperature becomes some (small) fraction of T_F, let us say for example $0.1 T_F$. Now, in the case of ^{3}He, ($T_c/T_F \sim 10^{-2}$) this "breakdown" temperature is still some way above T_c, so that there is a fairly wide temperature regime (3mK $\lesssim$ T $\lesssim$ 50 mK) where a well-defined Fermi-liquid behavior is seen. On the other hand, in the cuprates the breakdown temperature could be so little above the observed T_c that no clear Fermi-liquid regime is seen. It has to be said, however, that this scenario, while perhaps plausible for the higher-temperature compounds such as the Tℓ and Hg series, is much less so for compounds such as (optimally doped) Bi-2201 where T_c is only in the range 10–30 K, and to defend it one has to argue that such compounds are in some way untypical, even in the normal state, of the cuprates as a whole.

A point which has been widely noticed is that while at optimal doping the resistivity of the cuprates is rather accurately linear in T, it appears to cross over to the T^2 form characteristic of Fermi-liquid theory as one increases the doping p ("overdopes"). Other experimental properties, where they have been measured, also appear to approximate better the Fermi-liquid predictions with overdoping. This observation appears entirely consistent with the hypothesis that the crossover observed as a function of p in samples where T_c has been suppressed to zero is indeed characteristic of the behavior of a (hypothetical) pure, zero-field sample in the absence of pair formation. Exactly what are the implications of this scenario for the interpretation of the finite-T behavior as a function of p is at present a hotly debated issue.

24.3 Response Functions: The MIR Peak in the Cuprates

One of the most universal properties of the superconducting cuprates (though it is also found in nonsuperconducting ones) is the so-called midinfrared peak in the optical spectrum. The most dramatic way of bringing out this feature is to plot as a function of frequency not, as is conventionally done, the raw data (reflectance) or the (inferred) real part of the conductivity, but rather the so-called loss function $L(\omega) \equiv -\mathrm{Im}\,\epsilon^{-1}(\omega)$ (where $\epsilon(\omega)$ is the $\mathbf{q} \to 0$ limit of the complex transverse dielectric constant $\epsilon_{\perp}(\mathbf{q}\omega)$). For a conventional metal described by the Drude theory this quantity would have a narrow peak at the plasma frequency ω_p, which is typically of the order of a few eV, and for ω small compared to ω_p (but large compared to the scattering rate Γ) would be very small. For a more realistic (band) model $L(\omega)$ should of course have features corresponding to inter-band transitions, but unless the band gap ΔE is anomalously small there should still be a regime $\Gamma << \omega << \Delta E$ when it is small. The experimental behavior of $L(\omega)$ in the cuprates is very characteristic and very different from these expectations. $L(\omega)$ is small below $\sim$0.1 eV, but there is a wide and strong peak extending from about that point to $\sim$ 1–1.5 eV (the exact upper limit depends on the system), at which point L drops precipitously almost to zero, thereafter slowly increasing again (see, for example [13]).

Because this "midinfrared" (MIR) peak occurs in nonsuperconducting cuprates as well as the superconducting ones, it has often been regarded as irrelevant to the mechanism of superconductivity. However, already several years ago it was established [14] that the transition to the superconducting state is accompanied by a change in the reflectance in the MIR regime of the order of 1%, about two orders of magnitude larger than expected in a naive BCS-type theory, and recent ellipsometric measurements [15,16] have strongly confirmed this result. Thus, while it is at present unclear exactly what relation these phenomena have to the mechanism of superconductivity, they can certainly not be dismissed as totally irrelevant to it. Thus it becomes a more urgent question to understand the origin of the MIR peak.

It is tempting to regard this peak, along with the linear-in-T resistivity, the anomalous T-dependence of the Hall effect, etc., as evidence that the normal state of the cuprates cannot be a Fermi liquid. I shall now demonstrate, using the analogy with liquid ^{3}He, that this argument is not convincing and that it is perfectly consistent to interpret the MIR peak as simply the "incoherent background" on a Fermi-liquid picture. Since the quantity $\epsilon(\omega)$ is, apart from a factor of ω^{-2}, just the (transverse) current–current correlation function in the limit $\mathbf{q} \to 0$, it is necessary to analyze the behavior of such functions in the Fermi-liquid scenario.

Consider then some quantity $A(\mathbf{r}t)$ which may or may not be locally conserved. Define the corresponding response function $\chi_{AA}(\mathbf{r}t,\mathbf{r}'t')$ in the standard way (see [17]), take its Fourier transform $\chi_{AA}(\mathbf{q}\omega)$ with respect to the variable $\mathbf{r} - \mathbf{r}'$ and $t - t'$, and consider the limit $\mathbf{q} \to 0$. It now turns out that in Fermi-liquid theory one has the following result. If $A(\mathbf{r}t)$ is a locally conserved quantity, then in the static limit the correlation function $\chi_{AA}(\mathbf{q}\omega)$ can be expressed entirely in terms of the properties of the low-lying (quasiparticle) states, and generally speaking has a rather simple generic form (which however involves the values of relevant Landau parameters F_s^l, F_a^l). In particular, the region of ω for which there is finite spectral weight ($Im\chi_{AA}(\mathbf{q}\omega) \neq 0$) tends to zero as $\mathbf{q} \to 0$. Thus, in the case of liquid ^{3}He, we expect that the correlation functions of the three conserved quantities N (particle number), **S** and **P** will have no spectral weight at finite frequencies in the limit $\mathbf{q} \to 0$. Experiments on ultrasound properties appear consistent with this expectation in the case of N. The correlation functions of **S** and **P** are more difficult to measure. On the other hand, consider some quantity such as the spin current Λ which is not locally conserved (in the case of spin current it is easily verified that it can be changed in a collision, even though the total spin and total current are each locally conserved). In this case there will be some contribution to $\chi_{AA}(\mathbf{q}\omega)$ from the quasiparticle states, and this contribution will have zero weight at finite frequencies as $\mathbf{q} \to 0$. However, there is also a contribution from nonquasiparticle states, sometimes called the "background" contribution, even in the limit $\mathbf{q} \to 0$. Physically, the process involved may be regarded as the "undressing" of a quasiparticle by turning off part of its screening cloud. In the case of the spin current it turns out that one can actually calculate the relative contribution Z_b of this background to the sum rule

$$\int Im \frac{\chi_{\Lambda\Lambda}(\mathbf{q}\omega)}{\omega} = \frac{N}{m} .$$

In fact we have

$$Z_b = 1 - \frac{m}{m^*} \frac{1}{1 + F_1^a/3} . \qquad (24.6)$$

Since for ^{3}He at melting pressure the ratio m^*/m is about 6, while F_1^a is not very different from zero, the contribution of the background is about 80% of the whole!

Now let us turn to the cuprates. In this case, while total particle number and spin are still locally conserved, the current **J** is not, because of the effects of the crystalline potential. One's first instinct in this situation is to re-express **J** as a sum of terms respectively diagonal and nondiagonal in the band index, and then to argue that the nondiagonal terms involve interband transitions. Thus these will have zero weight below the energy ΔE corresponding to the relevant band gap, while the diagonal terms can be treated just like the (true) current in the translation-invariant case and thus will have zero weight at finite ω in the limit of $\mathbf{q} \to 0$. However, this is not correct, because in the presence of umklapp processes, although the total quasimomentum **K** commutes with the single-particle terms in the Hamiltonian, it does not commute with the interaction. Hence the quantity $\chi_{JJ}(\mathbf{q}\omega)$ can have finite spectral weight in the region $\omega \ll \Delta E$, even in the limit $\mathbf{q} \to 0$. (It should perhaps be cautioned that "ΔE" here should refer to the band gap of the *noninteracting* system: in the presence of strong interaction it is not obvious that ΔE even retains any clear meaning). Thus, there is no a priori objection to fitting the MIR peak into a Fermi-liquid picture.

24.4 The Cooper-Paired States of Superfluid ^{3}He and the Cuprates: General Considerations

As noted in Sect. 24.2, in the case of liquid ^{3}He there is a regime of more than a decade in temperature where the system appears to behave as a Fermi liquid in the sense of Landau. Indeed, normal ^{3}He is often regarded as the paradigm of such a system. This region is bounded below by the onset of superfluidity, which is almost universally believed to be a consequence of the onset of Cooper pairing. It is therefore natural to try to construct a theory of the paired (superfluid) state which as it were takes off from the Fermi liquid description of the normal state. A system so described is often referred to as a "superfluid Fermi liquid". Before doing so I very briefly review the most salient properties of the superfluid states. For details see, e.g. [19], or Chap. 4 of [9].

Actually, what is most striking about the phase diagram of liquid ^{3}He below 3 mK is that the superfluid "state" actually corresponds not to one but to three different phases, which are conventionally known as the A, B, and A_1 phases. The B phase is stable over most of the P–T plane in magnetic fields of less than a few kG; the A phase is stable, in zero field, at high pressures and temperatures $\gtrsim 0.7T_c$, and also stable in fields $\gtrsim 6kG$ for arbitrary pressure, while the A_1 phase occupies, in finite magnetic field, a thin sliver of the phase diagram between the A (or B) and normal phases. The B phase appears to be isotropic as regards most of its properties and its thermodynamics appear to be qualitatively similar to those of a classic (BCS) superconductor. In particular, the data on the specific heat are consistent with its vanishing exponentially as $T \to 0$. By contrast, the A phase shows considerable anisotropy, e.g. in the ultrasonic attenuation, and its thermodynamics is different from the standard BCS-like behavior. In particular, the specific heat appears to tend to zero as $T \to 0$ as a power law rather than exponentially. As regards the paramagnetic susceptibility (which, remember, in the case of ^{3}He is due entirely to the nuclear spins), this falls rapidly in the B phase to a value as $T \to 0$ of about 1/3 of the normal-state value, while in the A phase it is independent of temperature and very nearly (but not quite!)equal to the normal-state value. The properties of the A_1 phase are generally similar to those of the A phase. It should be mentioned that although these three new phases have been referred to from the very earliest days as the "superfluid" phases of ^{3}He, the evidence for superfluidity as such is very much more circumstantial than in the case of ^{4}He. In fact, had one not had theoretical ideas developed in the context of ^{4}He and metallic superconductors to guide one, it seems likely that the phenomenon would have remained hidden for even longer than the 15 years or so for which it went undetected in ^{4}He(He-II).

Let us now discuss how to formulate the theory of Cooper pairing in a system such as ^{3}He which above T_c appears to be behaving as a normal Fermi liquid. A simple semi-phenomenological way to do this is as follows. We simply define operators $\alpha^+_{\mathbf{p}\sigma}$, $\alpha_{\mathbf{p}\sigma}$ which create and destroy Landau quasiparticles rather than real atoms (e.g. when acting on the interacting groundstate, the operator $\alpha^+_{\mathbf{p}\sigma}\alpha_{\mathbf{p}'\sigma'}$ ($|\mathbf{p}| > p_F$, $|\mathbf{p}'| < p_F$) creates a state with $\delta n(\mathbf{p}\sigma) = +1$ and $\delta n(\mathbf{p}'\sigma') = -1$). These operators rather obviously satisfy Fermi statistics, and in general have the same status in the context of the groundstate and low-lying states of the interacting system as do the original fermion operators $a^+_{\mathbf{p}\sigma}$, $a_{\mathbf{p}\sigma}$ with respect to the corresponding states of the noninteracting system. Then, in addition to the "Fermi-liquid" terms (24.3) in the Hamiltonian, which are diagonal in the δn-representation, one introduces a BCS-like "pairing" term of the usual general form. Formulated in terms of the quasiparticle operators $\alpha^+_{p\sigma}$, $\alpha_{p\sigma}$ one has

$$\hat{V}_{pair} = \sum_{\substack{\mathbf{pp}' \\ \alpha\beta\gamma\delta}} V(\mathbf{p}, \mathbf{p}' : \alpha, \beta, \gamma, \delta)\alpha^+_{\mathbf{p}\alpha} \times \alpha^+_{-\mathbf{p}\beta}\alpha_{-\mathbf{p}'\gamma}\alpha_{\mathbf{p}'\delta} \,. \tag{24.7}$$

This term is actually a special case of the "scattering" terms which as already pointed out are left out in (24.3). It is intuitively clear that the effect of such a term is likely to be qualitatively similar to that of the BCS pairing term in a Fermi gas. It will induce a collective bound state of the Cooper type, but with the individual fermion components of the pair being Landau quasiparticles rather than real fermions. However, the internal configuration of the pair need not have s-wave symmetry as in the classic superconductors. Rather, the effective angular spin and angular momentum of the pair will correspond to that "channel" in which the projection of the pairing interaction V is most attractive (negative). A system which is well described by this procedure is often called a "superfluid Fermi liquid" (SFL).

It is clear that the role of the single-particle energy in all this is just as in the BCS case, the only difference being that the free mass m is replaced by the effective mass m^*. What about the last term in (24.3), involving the Landau interaction function $f(\mathbf{pp}'\sigma\sigma')$, which has no obvious analog in the BCS theory? The crucial point is that, as discussed in [18], this term only comes into play when there is some macroscopic "polarization" of the system, and the formation of Cooper pairs does not lead to such polarization (at least in the approximation of "particle-hole symmetry". This is thought to be very good for ^{3}He). Thus, at first sight at least, the value of the transition temperature T_c should be a function only of m^* and the coefficients V in (24.7), and should be totally insensitive to the Landau f-function. This should also be true of these properties of the paired state (for example the specific heat) which do not refer, either explicitly or implicitly, to any macroscopic polarization; however, these properties (e.g. the spin susceptibility and normal density) which do involve such polarization should be sensitive to the Landau function, i.e. to the relevant F^s_l's or F^a_l's. As an example, the ratio of the spin susceptibility $\chi(T)$ in an (s-wave-paired) superfluid state to its normal-state value χ_n which in the BCS theory is just the Yoshida function Y(T), should be given for a superfluid Fermi liquid by the expression

$$\frac{\chi(T)}{\chi_n} = \left(1 + F_0^a\right) \frac{Y(T)}{1 + F_0^a Y(T)} . \qquad (24.8)$$

Although this expression tends to 1 for $T \to T_c$ and to zero for $T \to 0$, its shape as a function of temperature may be quite different from that of the function Y(T). Similar remarks apply to the normal density $\rho_n(T)$ as a function of T. For a much more detailed discussion, see, e.g. [18], which uses the notation $Z_0 \equiv 4F_0^a$.

At first sight, then the theory of a superfluid Fermi liquid is nothing but the theory of a superfluid Fermi "gas" as described by BCS plus the set of molecular fields associated with the Landau parameters F_l, F_l^a. And indeed, it turns out that once the nature of the pairing states is currently chosen and T_c itself fitted from experiment, such a description gives predictions for the properties of the superfluid phase in question as a function of T/T_c which are quantitatively accurate typically at the level of a few percent and often better. As we shall see, this observation is actually quite puzzling, but let us first briefly pause to identify the relevant pairing states.

It is by now almost universally believed, on the basis of NMR and other experimental signatures, that the pairing states corresponding to the observed A, B and A_1 phases of ^{3}He all correspond to Cooper pairs formed in a state with total spin 1 and total (apparent) orbital angular momentum 1 (i.e. a *p*-state), but with different combinations of spins and orbital angular momenta within this class. More specifically, the Fourier-transformed order parameter (pair wave function) $F_{\mathbf{k},\alpha\beta} \equiv \langle a_{\mathbf{k}\alpha} a_{-\mathbf{k}\beta} \rangle$ is in the case of the A phase a product of a spin function which by an appropriate choice of axis can be written $2^{-1/2}$ $(\uparrow_1\downarrow_2 + \downarrow_1\uparrow_2)$, i.e. corresponds to $S = 1$, $S_z = 0$, an orbital function which, by an appropriate choice of the orbital axis can be written as proportional to $k_x + ik_y$, i.e. corresponds to $L = L_z = 1$. The A_1 phase has an identical orbital wave function, but the spin function can now be written as simply $\uparrow_1\uparrow_2$ ($S = S_z = 1$: only one spin species undergoes pairing!). The B phase is a bit more subtle; it is what in atomic physics would be called a $^3P_0$ state, but with the spins rotated through a finite angle relative to the orbital angular momenta. For more details see Sect. 24.5.

The rather good agreement of the predictions of the simple SFL picture with the experimentally observed properties of superfluid ^{3}He is actually rather surprising at first sight, for the following reason: Unlike in the classic metallic superconductors where there is a natural energy cutoff of the order of the Debye energy for the pairing interaction $\hat{V}_{\text{pair}}$, in liquid ^{3}He not only is there no obvious such cutoff, but we do not know a priori where the pairing terms came from in the first place. Recall, this is a pairing interaction between *quasiparticles*, not between real atoms, and we have very little quantitative knowledge of the "structure" of a quasiparticle. In any event, there seems no good reason why the pairing interaction should not extend up to energies where the BCS assumption of a constant density of states breaks down badly, or why it should not itself depend substantially on the magnitude of the momenta $\mathbf{p}$ and $\mathbf{p}'$. Why do these complications not affect the temperature-dependence of the measured quantities?

The answer to this dilemma (or at least a large part of it) was given in a classic 1961 paper [20] by Anderson and Morel, which long predates the experimental discovery of pairing in ^{3}He. Imagine that the original inter-quasiparticle pairing interaction indeed has the above properties, but that the characteristic scale ϵ_0 over which the strength of V, the density of states, etc., varies is much larger than the experimentally measured T_c. Then one can choose an intermediate energy cutoff ϵ_c such that $T_c \ll \epsilon_c \ll \epsilon_0$, and take into account the "scattering" into pair states of energy $|\epsilon| > \epsilon_c$ by a standard multiple-scattering (t-matrix) procedure. As a result we get a problem where the pairing interaction operates only within a "shell" of width $2\epsilon_c$ around the (quasiparticle) Fermi energy, with a (single-spin) density of states $N(0) \equiv \frac{1}{2}\left(\frac{dn}{d\epsilon}\right)$ and pairing interaction V which is constant as a function of energy (i.e. of $|\mathbf{p}|$ within this shell). This is just the original BCS assumption generalized to include possible angular dependence of V. Thus, all the unknowns are buried in the quantities N(0) and $V(\hat{\mathbf{p}}, \hat{\mathbf{p}}')$. The transition temperature T_c is determined by N(0) and the magnitude of the most attractive component V_ℓ of V. The angular dependence of the order parameter reflects the identity of this "most attractive" component. Thus once T_c and the symmetry is determined everything else goes exactly according to the BCS prescriptions and there are no further unknowns.

The above discussion leaves one major question unanswered: Why should there be more than one stable superfluid phase, even in zero external magnetic field? As mentioned above, all the evidence is that both the A and B phases are spin triplet p-wave states but with different forms of the order parameter. Such a situation cannot occur in a simple BCS-type theory with a single fixed value V_1 of the interaction in the p-wave channel. In such a theory the B phase is the unique free energy minimum. The generally accepted explanation, originally given by Anderson and Brinkman [21], is that because the effective interaction between quasiparticles is strongly renormalized and the renormalization process is itself sensitive to the pairing state, the "effective" value of V_1 is slightly different for the A and B phases, and this small difference is enough to stabilize the A phase, even at zero field, in a small region of the P–T plane. This effect is formally of order $(T_c/T_F)^2$ and is a correction to the simple superfluid Fermi-liquid picture. A comprehensive discussion of all corrections which appear to this order has been given by Serene and Rainer [22].

Now let us turn to the superconducting cuprates, and suppose that we take the transition temperature T_c of a given compound from experiment and consider the superconducting-state properties as functions of the reduced temperature T/T_c. Consider first those bulk properties which within the BCS theory are only indirectly sensitive to the energy gap Δ(T). Examples are the specific heat, spin susceptibility and superfluid density which is measured in the standard way by the penetration depth. Then the most salient qualitative observation is that the general pattern of the temperature-dependence of these quantities, at least in the optimally doped and overdoped regimes, is very similar to that expected in a superfluid Fermi liquid. Indeed, if we restrict ourselves to "not too low" temperatures (say $T/T_c \gtrsim 0.2$), it is actually quite difficult to distinguish the system from a classic BCS superconductor with s-wave pairing! The characteristic effects associated with the "exotic" nature of the order parameter mostly come into play only at the lowest temperatures, see next section. For a reason to be discussed in the next section, it does not make a lot of sense to ask whether we can fit the temperature-dependences at say the 1% level by using a SFL model. In the "underdoped" region of the phase diagram the situation is less clear. For example, the specific heat and the spin susceptibility show considerable "precursor" effects at temperatures above T_c. From this alone it is clear that they cannot follow the SFL predictions.

If one considers phenomena such as angle-resolved photoemission or tunnelling which, at least in a BCS superconductor, are direct measures of the energy gap Δ(T), then even at optimal doping (or in the overdoped regime) the agreement with the SFL model is more problematic. The ratio $\Delta(0)/T_c$ inferred from this data is often much larger than the BCS value 1.76, and in addition it is not always even clear that the "gap" Δ(T) inferred from these data actually tends to zero in the limit $T \to T_c$. This pattern is amplified in the underdoped regime, where a feature which looks suspiciously reminiscent of an energy gap (the "pseudogap") persists in the data up to several times T_c. To summarize, the data seem to be qualitatively consistent with an SFL picture at optimal doping and in the overdoped regime, but not in the underdoped regime. This is not particularly surprising, since we saw in Sects. 24.2 and 24.3 that a similar statement could be made about the normal phase.

It is interesting to ask, does there exist a quantitative test of SFL theory in the cuprates? Recall that in the case of normal ^{3}He, the normal state does permit such a quantitative test, see Sec. 24.2. Rather serendipitously, it turns out that the answer is yes, as pointed out by Chiao et al. [23]. They note that according to recent theory [24], in a superconductor with the order parameter symmetry believed to be characteristic of the cuprates and described by the SFL model the low-temperature limit of the thermal conductivity divided by T is independent of the impurity concentration and a function only of two parameters which are measured independently in ARPES experiments. Thus, if the SFL model applies, from the ARPES data it should be possible to predict the thermal conductivity with no free parameters. It turns out that such a prediction works essentially perfectly for optimally doped BSCCO, so that one can say that the SFL model is at least not excluded for this system. One cannot of course say that this proves it to be right, since it is possible that the SFL model is a special case of a more general class of models, which makes the same prediction. In principle, further tests of this quantitative type are possible by using data on e.g. the low-temperature specific heat, but the quality of the data at present is inadequate for this. It should further be noted that while the prediction of the SFL model for the $T \to 0$ temperature-dependence of the penetration depth involves a Landau parameter which cannot be independently measured, the value required for this parameter in order to fit the data is quite "reasonable" (i.e. ~ 1, not ~ 10 or 0.1!).

A final question of interest concerns the "size" of the Cooper pairs, and the related issue of their possible "pre-formation". Although the definition of the "size" (or "radius") ξ of the pairs is to some extent a matter of convention, a natural choice is that it is the mean square radius associated with the so-called "anomalous average" $F(r) \equiv \langle \psi_\alpha(0)\psi_\beta(\mathbf{r})\rangle$ which characterizes the paired state (the real-space Fourier transform of the quantity entering (24.10) below, i.e. the single macroscopically occupied eigenfunction of $\hat{\rho}_2$). For a simple BCS model ξ is then only weakly temperature-dependent and of order of magnitude $\hbar v_F/\pi k_B T_c$ (i.e. equal, up to a numerical factor of order unity, to the "Pippard coherence length" ξ_0).

As is well known, for the classic superconductors such as Al the quantity ξ is of order 10^3–10^4 Å. This is typically several orders of magnitude greater than the interparticle spacing a. It is this state of affairs which is the key to the remarkable success of the BCS (mean-field) theory of superconductivity. An equivalent way of looking at it is to note that the ration a/ξ_0 is up to a factor of order unity equal to $\Delta/\epsilon_F(\sim T_c/T_F)$. The pair binding energy Δ is very small compared to the Fermi energy ϵ_F, so that the pair formation process affects only a fairly thin shell of states around the Fermi surface. If on the other hand it were to turn out that ξ is comparable to a (or Δ or kT_c to ϵ_F) then we should expect fluctuations around the mean-field solution to play a major role. In particular we might get a substantial range of temperature where Cooper pairs are formed locally, but do not undergo the "quasi-Bose–Einstein condensation" characteristic of the BCS state. The scenario is sometimes referred to as that of "pre-formed pairs." In the extreme limit $\xi \ll a$, the picture would be one of tightly bound di-electronic (bosonic) "molecules" which form at some high temperature T_0 and then undergo Bose condensation at a much lower, unrelated, temperature $T_{BEC} \ll T_0$.

As remarked in Sect. 24.1, the ratio T_c/T_F is considerably larger for both ^{3}He and the cuprate superconductors than for classic superconductors such as Al. If we work in terms[2] of the Cooper-pair radius ξ, then it turns out that for ^{3}He ξ varies from about 200 Å at melting pressure to about 1200 Å at saturated vapor pressure. This is large compared to the interparticle spacing (~ 3 Å). It is therefore not surprising that superfluid ^{3}He appears to be very well described by the mean-field BCS theory. In the case of the cuprates, it is not entirely clear that we should continue to use the estimate $\xi \sim \hbar v_F/m\pi k T_c$. If however we do so, and take the value of v_F (for BSCCO) from ARPES measurements, see, e.g. [23], we find that at optimal doping ξ is typically ~ 50 Å, while if we take the relevant "interparticle spacing" to be that of the extra holes doped into the CuO_2 planes it is about 8 Å at optimal doping. Thus, the ratio ξ/a is only ~ 6. Away from optimal doping it increases. If we take this result seriously, it implies that while we are by no means in the limit of "di-electronic" molecules, we should expect fluctuations to give substantial corrections to the mean-field theory; such effects are indeed seen in the cuprates, most spectacularly in the macroscopic electromagnetic properties. A particular prediction, as noted above, would be that there should be a fairly substantial regime of temperature where pairs are formed but are not Bose condensed. In this regime the "local" properties of the system

[2] Working in terms of a/ξ rather than T_c/T_F has the advantage that the quantity a ($\equiv n^{-1/3}$ where n is the relevant particle density) is better defined than T_F.

should be similar to those of the superconducting phase, but there should be no long-range order and hence no superconductivity (in the sense of zero resistance or a Meissner effect). A substantial school of thought [25] claims that the so-called "strange metal" regime of the cuprates which appears on the underdoped side of the phase diagram is exactly the instantiation of this prediction, and in particular that the approximately defined line of anomalies in the experimental properties usually denoted by $T^*(p)$ corresponds to the onset of pair formation (which, just as in the case of the formation of ordinary diatomic molecules, is expected to be gradual rather than by a phase transition). Whether such a scenario can explain the properties of the "strange metal" regime quantitatively or even qualitatively is at present the subject of vigorous controversy. Parenthetically, it is amusing that following early suggestions along these or similar lines for the cuprates, it has been proposed [26] that normal ^{3}He, usually taken to be our paradigmatic Fermi liquid, see Sect. 24.2, may itself be described by a scenario of this general type.

24.5 Symmetry of the Order Parameter

Perhaps the most striking difference between superfluid ^{3}He and the cuprates on the one hand, and the classic superconductors on the other, is that while in the latter case the order parameter is believed to have the same symmetry as the lattice, in both ^{3}He and the cuprates the general belief is that it is "exotic", that is, transforms according to a nontrivial representation of the symmetry group G of the Hamiltonian. This may well also be true in the case of certain heavy-fermion and organic superconductors. For the moment, I will assume that in each case the order parameter corresponds to a single irreducible representation of G. This point will be briefly re-examined below.

In the case of liquid ^{3}He, where there is no lattice, the symmetry group is (to the extent that we ignore the small nuclear dipole-dipole interaction) a product of the group $SO(3)_{orb}$ of rotations in orbital coordinate space and the group $SU(2)_{sp}$ corresponding to rotations in spin space; we need SU(2) rather then SO(3) because of the double-valued character of the spin functions. For a complex (Cooper pair) of two particles of spin 1/2 the relevant irreducible representations of SU(2) correspond to a singlet and a triplet, and on the basis of the behavior of the spin susceptibility it is almost universally believed that both the A and the B phases correspond to the triplet. Under all but very special circumstances, the Pauli principle then implies that the orbital wave function is odd-parity, i.e. corresponds to "angular momentum" $\ell = 1, 3, 5 \ldots$ All the evidence is that in both A and B phases ℓ is equal to 1. Thus, in both the A and B phases of ^{3}He the Cooper pairs have $\ell = S = 1$. Of course, this statement does not characterize the pairing state uniquely, since we have three possible Zeeman substates corresponding to spin and three to orbital angular momentum, thus nine possible states in all. The most compact characterization of the order parameter turns out to be through the quantity

$$\mathbf{d}(\mathbf{k}) \equiv (i\sigma_y\boldsymbol{\sigma})_{\alpha\beta}\langle a^+_{\mathbf{k}\beta}a^+_{-\mathbf{k}\alpha}\rangle\,, \tag{24.9}$$

where the components of $\boldsymbol{\sigma}$ are the Pauli matrices. The quantity $\mathbf{d}(\mathbf{k})$ automatically behaves as a vector in spin space, and because of the $\ell = 1$ requirement must also transform as a vector in orbital space. The forms of $\mathbf{d}(\mathbf{k})$ inferred from NMR and other evidence to correspond to the A and B phases of ^{3}He are as follows: For the A phase $\mathbf{d}(\mathbf{k})$ is given up to a multiplying constant by

$$\mathbf{d}(\mathbf{k}) = \hat{\mathbf{d}}(\mathbf{k}\cdot(\hat{\mathbf{n}}_1 + i\hat{\mathbf{n}}_2))\,, \tag{24.10}$$

where $\hat{\mathbf{d}}$ is a real vector in spin space and $\hat{\mathbf{n}}_1$ and $\hat{\mathbf{n}}_2$ are a pair of mutually orthogonal (real) unit vectors in orbital space. It is conventional to define a third unit vector $\hat{\ell}$ by $\hat{\ell} \equiv \hat{\mathbf{n}}_1 \times \hat{\mathbf{n}}_2$. The physical interpretation is that in the A phase the pairs have total spin 1 and spin projection 0 on the (spin-space) axis $\hat{\mathbf{d}}$, and in orbital space have "apparent" angular momentum 1 along the(orbital-space) axis ℓ. To discuss the B phase, it is convenient to start from the state $\mathbf{d}(\mathbf{k}) = \mathbf{k}$. In this state the apparent orbital angular momentum is always directed oppositely to the spin, so this is what in the language of atomic physics would be denoted as a $^3P_0$ state. The actual state of ^{3}He-B is believed to be obtained by rotating the spin coordinate system relative to the orbital one through an angle θ around an axis $\hat{\omega}$. To minimize the dipole–dipole energy θ has to be $\cos^{-1}(-1/4) \approx 104^\circ$, but the axis $\hat{\omega}$ is fixed only by higher-order effects. The resulting state, while it has $\langle \mathbf{L} \rangle = \langle \mathbf{S} \rangle = 0$, has a finite value of the quantity $\langle \mathbf{L} \times \mathbf{S} \rangle$ directed along $\hat{\omega}$. For further details see, e.g., [9].

In the case of the single-plane cuprates, it is most natural to regard the pair wave function as defined within a single CuO_2 plane, with transfer between neighboring planes effected by a Josephson-type mechanism. Then to the extent that we neglect spin–orbit interaction, the relevant symmetry group is the product of the orbital symmetry group, which is exactly or approximately that of the square, C_{4v}, times the group $SU(2)_{spin}$. Again, the possible spin representations are singlet or triplet, and the evidence from the NMR behavior strongly favors the singlet. This then suggests that the orbital state has even parity. The even-parity irreducible representations of C_{4v} are four in number, and are denoted in the formal notation of group theory A_{1g}, A_{2g}, B_{1g} and B_{2g}, or informally s (or s^+), g(or s^-), $d_{x^2-y^2}$ and d_{xy}, respectively. There is by now rather convincing evidence (see, e.g. [27]) that, at least in these cuprates where phase interference experiments have been done (YBCO, Tℓ-2201, Bi-2212...) the symmetry of the order parameter is $d_{x^2-y^2}$ (B_{1g}). There is a strong suspicion that this is so for all the superconducting cuprates.

Here it is necessary to emphasize an important distinction: The orbital symmetry group SO(3) of ^{3}He (as well as the spin group $SU(2)_{sp}$) is continuous. Thus, once one has specified not just the irreducible representation involved (in this case $\ell = S = 1$) but also the "nature" of the state (e.g. A-like or B-like), the only freedom left over is the "orientation" in orbital and/or spin space, e.g. for an A-like state the vectors $\hat{\ell}$ and $\hat{\mathbf{d}}$. It is clear that the free energy must be independent of those in the neglect of the small nuclear dipole–dipole interaction. This has the consequence that if one makes the hypothesis that the SFL model is exact for ^{3}He and chooses a particular state, e.g. A or B, one can calculate the free energy and most of the relevant experimental quantities exactly as a function of T/T_c and of the Landau parameters. It thus makes sense to try to quantify the small discrepancies from these predictions, cf., e.g. [22]. By contrast, the orbital group C_{4v} of the cuprates consists of discrete operations ($\pi/2$ rotations and reflections), so the statement that the symmetry of the order parameter is $d_{x^2-y^2}$ tells us only that it changes sign under $\pi/2$ rotations and also under reflections in a 45° crystal axis. This still allows a wide variety of dependencies on $\mathbf{k}$, from a simple form to a complicated form with several nodes other than the 4 in the 45° direction which are imposed by symmetry alone. Consequently, the statement that the system behaves as a SFL does not uniquely determine the free energy and other quantities as functions of T/T_c and the Landau parameters: to obtain such a result one needs a specific microscopic model which fixes quantities such as the single-quasiparticle energy and the effective pairing interactions as detailed functions of $\mathbf{k}$ as the latter varies around the Fermi surface.

A second difference associated in part with the different structure of the groups SO(3) and C_{4v} is that in the former case "continuous" mixing-in of irreducible representations other than the principal one allowed, whereas in the latter this requires a second phase transition (on this point, see Sect. 3 of [28]). Thus the absence of any observed second phase transition in the essentially pure tetragonal[3] crystal Tℓ-2201 is strong evidence that the bulk order parameter is pure $d_{x^2-y^2}$, while the pairing state of ^{3}He-A (though not ^{3}He-B) may have a small admixture of other odd-parity orbital states ($\ell = 3, 5 \ldots$). In practice this seems very unlikely to produce quantitatively significant corrections.

In general, the fact that the order parameter of a given system is "exotic" (i.e. has symmetry lower than that of the Hamiltonian) has a number of interesting consequences. First, in most such cases the order parameter, and hence the energy gap which is crudely speaking proportional to it, will have nodes at one or more points on the Fermi surface. This means that the number of normal quasiparticles excited at low temperatures will be proportional to a power of T rather then exponentially small. Consequently, ex-

[3] For cuprates such as YBCO with appreciable orthorhombic anisotropy the analysis becomes rather messier, see e.g., [28].

perimental quantities such as the specific heat which are crudely proportional to the amount of normal component will themselves have a power-law behavior. Such power-law behavior is indeed observed in the cuprates, in particular in the NMR properties and the deviation of the penetration depth from its zero-temperature limit. This is strong evidence for the existence of nodes of the gap which are also observed more directly in ARPES experiments. To be sure, the mere existence of nodes is not peculiar to exotic pairing states, an s-wave (A_{1g}) state, provided it is of the so-called "extended" type, can also have them, but the salient point is that a $d_{x^2-y^2}$ state, because it must (by definition!) change sign under reflection in any 45° axis, necessarily has nodes at these points, which is exactly where they are observed in ARPES.

In the above argument, it was taken for granted that the energy gap $\Delta(\mathbf{k})$ is simply proportional to the orbital order parameter. This is indeed the case for spin singlet pairing, where the order parameter can be expressed as a simple product of a trivial spin factor (the singlet state) and an orbital factor. For spin triplet pairing the situation is in general a little more complicated. In this case it turns out (at least for the so-called "unitary" class of states which includes the A and B states) that the gap $\Delta(k)$ is proportional to the quantity $|\mathbf{d}(\mathbf{k})|^{1/2}$. We see that the B-phase gap is *uniform* over the Fermi surface and thus has no nodes, while the A-phase gap is proportional to $(\mathbf{k} \cdot \ell)2$ and so has two nodes in the directions $\pm\hat{\ell}$. Consequently, we expect that the properties of ^{3}He-B should behave as a function of temperature, similarly to those of a classic (s-wave) superconductor, (e.g. the specific heat should decrease exponentially in the limit $T \rightarrow 0$), while for ^{3}He-A we should get power-law behavior qualitatively similar to that seen in the cuprates. But at the very lowest temperatures (lower than currently accessible) the thermodynamics should be dominated by bosonic collective excitations such as spin waves. This prediction appears consistent with existing experiments, although lower-temperature measurements would be needed to confirm it definitively [9].

A second consequence of exotic pairing in general is that various experimentally measured quantities may show an anisotropy over and above that prescribed by the Hamiltonian. Examples of such behavior in superfluid ^{3}He include the A-phase ultrasonic attenuation, which depends on the angle of the sound propagation vector relative to the characteristic orbital vector $\hat{\ell}$, and the magnetic susceptibility, which depends on the angle of the field relative to the characteristic spin vector $\hat{\mathbf{d}}$. In ^{3}He-B the orbital and spin properties are individually isotropic, but the effect of the rather subtle anisotropy in the correlation of the spin and orbital coordinates comes out in the NMR behavior, which is sensitive to just such correlations and indeed depends on the direction of the static field relative to the "spin-orbit rotation vector" $\hat{\omega}$. In the cuprates the most obvious signature of anisotropy is in the ARPES data. While one would expect (and finds) the spectra to be somewhat different along the crystal axes and at 45° even in the normal phase, this difference is strongly accentuated in the superconducting phase, since as already described a gap develops which is finite along the axes but zero in the 45° directions.

A third class of phenomena related to exotic pairing has to do with possible variations of the characteristic "orientation" in space and/or in time. Here there is a very substantial difference between ^{3}He and the cuprates. In the former case, if for definiteness we consider the A phase, the characteristic orbital vector $\hat{\ell}$ and spin vector $\ddot{\mathbf{d}}$ are pinned only very weakly in bulk (by effects such as the nuclear dipole–dipole interaction, residual heat flows, etc., see [29], Sect. X), and as a result such spatial and/or temporal variation of the orientation is quite commonplace and of great interest both experimentally and theoretically, see, e.g. [9]. By contrast, in the cuprates there is no question of spin "orientation" since the state is a singlet, and the "orientation" of the orbital wave function is firmly pinned by the crystal axes. The only remaining freedom is to exchange the + or – signs on the four "lobes" of the pair wave function. Since this merely multiplies the total pair wave function by a – sign, one might think it would have no physical significance. Indeed for a spatially uniform situation this is true. However, it is precisely this subtle degree of freedom which allows, in a spatially nonuniform setup, the rather spectacular Josephson (quantum phase interference) experiments which are the most convincing evidence in favor of the $d_{x^2-y^2}$ symmetry assignment, see, e.g. [27].

24.6 Conclusion

From the discussion in this chapter the reader might draw the conclusion that the phenomenon of superfluidity in liquid ^{3}He has been no help in understanding cuprate superconductivity. Note, the vast differences between the two systems and the few similarities. However, such a conclusion would, I think, miss a crucial point: In the history of research on Cooper pairing in Fermi systems, superfluid ^{3}He was the very first case of a system where the pairing was neither of the conventional (s-wave) symmetry nor due to exchange of entities (phonons) different from those being paired. These features had been anticipated theoretically. It is difficult to overestimate the role of the experimental discovery of superfluidity in ^{3}He in opening the minds of the community to the idea that superconducting (or superfluid) long-range order might be a very generic phenomenon and that the detailed properties of systems sharing it might be very different from those of the classic superconductors. Certainly, such an expectation has been spectacularly realized in the cuprates; and to my mind it is entirely possible that it will in the future have other equally spectacular realizations elsewhere.

24.7 Summary

This chapter compared the phenomenon of superfluidity in liquid ^{3}He with that of superconductivity in the cuprates. It discussed, inter alia, the similarities and differences of the two systems in the normal state with particular attention to the ^{3}He analog of the "mid-infrared peak" seen in the cuprates, the extent to which the superfluid state in either or both systems can be regarded as a "superfluid Fermi liquid", and the differences in the consequences of the "exotic" order-parameter symmetry believed to be common to the two superfluid phases.

Acknowledgements

This work was supported by the National Science Foundation under grant NSF–DMR–99–86199.

References

1. A.J. Leggett, Rev. Mod. Phys. **73**, 307 (2001)
2. O. Penrose and L. Onsager, Phys. Rev. **104**, 576 (1956)
3. P. Gavoret and P. Nozières, Ann. Phys. **28**, 349 (1964)
4. C.N. Yang, Rev. Mod. Phys. **34**, 694 (1952)
5. A.J. Leggett, In: *Electron*, ed. by M. Springford (Cambridge University Press, Cambridge, 1997)
6. A.J. Leggett, In: *Bose–Einstein Condensation*, ed. by A. Griffin, D.W. Snoke, and S. Stringain (Cambridge University Press, Cambridge, 1995)
7. W.F. Vinen, In: *Superconductivity*, ed. by R.D. Parks (Marcel-Dekker, New York, 1969) Vol. II, p. 1167
8. L.D. Landau, Zh. Eksp. Teor. Fiz. **30**, 1058 (1956): Translation, Soviet Physics JETP **3**, 920 (1957)
9. D. Vollhardt and P. Wölfle, *The Superfluid Phases of ^{3}He* (Taylor and Francis, London, 1990)
10. V.P. Silin, Zh. Eksp. Teor. Fiz. **33**, 495 (1957): Translation, Soviet Physics JETP **6**, 387 (1958)
11. B.L. Jones and J.W. McClure, Phys. Rev. **143**, 133 (1966)
12. G.S. Boebinger et al., Phys. Rev. Lett. **77**, 5417 (1996)
13. T. Timusk and D.B. Tanner, In: *Physical Properties of High-Temperature Superconductors*, ed. by D.M. Ginsberg (World Scientific, Singapore 1989) Vol. 1, p. 339
14. M.J. Holcomb, C.L. Perry, J.P. Collman and W.A. Little, Phys. Rev. B **53**, 6739 (1996)
15. M. Rübhausen, A. Gozar, M.V. Klein, P. Guptasarma and D.G. Hinks, Phys. Rev. B **63**, 224514 (2001)
16. H.J. Molegraaf, C. Presura, D. van der Marel, P.H. Kes and M. Li, Science **295** 2239 (2002)
17. D. Pines and P. Nozières, *The Theory of Quantum Liquids* (W.A. Benjamin, New York, 1966)
18. A.J. Leggett, Phys. Rev. **140** A, 1869 (1965)
19. J.C. Wheatley, Revs. Mod. Phys. **47**, 415 (1975)
20. P.W. Anderson and P. Morel, Phys. Rev. **123**, 1911 (1961)
21. P.W. Anderson and W.F. Brinkman, Phys. Rev. **30**, 1108 (1973)
22. J.W. Serene and D. Rainer, Phys. Rep. **101**, 221 (1983)
23. M. Chiao, R.W. Hill, C. Lupien, L. Taillaer, P. Lambert, R. Gagnon and P. Fournier, Phys. Rev. B **62**, 3554 (2000)
24. A.C. Durst and P.A. Lee, Phys. Rev. B **62**, 1270 (2000)
25. M. Randeria and N. Trivedi, J. Phys. Chem. Solids **59**, 1754 (1998)
26. J.P. Bouchand and C. Lhuillier, Z. Phys. B **75**, 283 (1989)
27. C.C. Tsuei and J.R. Kirtley, Rev. Mod. Phys. **72**, 969 (2000)
28. J.F. Annett, N. D. Goldenfeld and A.J. Leggett, *Physical Properties of High Temperature Superconductors*, ed. by D.M. Ginsberg (World Scientific, Singapore, 1996) Vol. V
29. A.J. Leggett, Rev. Mod. Phys. **47**, 331 (1975)

Author Index

Subject Index